Advanced Organic Chemistry

FOURTH EDITION

Part B: Reactions and Synthesis

Nam Lee

Advanced Organic Chemistry

PART A: Structure and Mechanisms
PART B: Reactions and Synthesis

Advanced Organic Chemistry

FOURTH EDITION

Part B: Reactions and Synthesis

FRANCIS A. CAREY
and RICHARD J. SUNDBERG
University of Virginia
Charlottesville, Virginia

Kluwer Academic / Plenum Publishers
New York, Boston, Dordrecht, London, Moscow

Library of Congress Cataloging-in-Publication Data

Carey, Francis A., 1937–
Advanced organic chemistry/Francis A. Carey and Richard J. Sundberg.—4th ed.
p. cm.
Includes bibliographical references and index.
Contents: pt. B. Reactions and synthesis
ISBN 00-306-46244-3 (Hardcover)—ISBN 0-306-46245-1 (pbk.)
I. Sundberg, Richard J., 1938– II. Title.

QD251.2 .C36 2000
547—dc21

00-059652

The front cover shows the key orbital interactions in the Diels-Alder reaction. The highest-occupied molecular orbital (HOMO) of 1,3-butadiene and the lowest-unoccupied molecular orbital (LUMO) of ethylene overlap in phase with one another allowing the reaction to occur in a single step.

ISBN: 0-306-46244-3 (Hardbound)
ISBN: 0-306-46245-1 (Paperback)

©2001, 1990, 1984, 1977 Kluwer Academic / Plenum Publishers, New York
233 Spring Street, New York, New York 10013

http://www.wkap.nl

10 9 8 7 6 5 4

A C.I.P. record for this book is available from the Library of Congress

All rights reserved

No part of this book may be reproduced, stored in a retrieval system, or transmitted in any form or by any means, electronic, mechanical, photocopying, microfilming, recording, or otherwise, without written permission from the Publisher.

Printed in the United States of America

Preface to the Fourth Edition

Part B emphasizes the most important reactions used in organic synthesis. The material is organized by reaction type. Chapters 1 and 2 discuss the alkylation, conjugate addition and carbonyl addition/condensation reactions of enolates and other carbon nucleophiles. Chapter 3 covers the use of nucleophilic substitution, both at saturated carbon and at carbonyl groups, in functional group of interconversions. Chapter 4 discusses electrophilic additions to alkenes and alkynes, including hydroboration. Chapter 5 discusses reduction reactions, emphasizing alkene and carbonyl-group reductions. Concerted reactions, especially Diels–Alder and other cycloadditions and sigmatropic rearrangements, are considered in Chapter 6. Chapters 7, 8, and 9 cover organometallic reagents and intermediates in synthesis. The main-group elements lithium and magnesium as well as zinc are covered in Chapter 7. Chapter 8 deals with the transition metals, especially copper, palladium, and nickel. Chapter 9 discusses synthetic reactions involving boranes, silanes, and stannanes. Synthetic reactions which involve highly reactive intermediates—carbocations, carbenes, and radicals—are discussed in Chapter 10. Aromatic substitution by both electrophilic and nucleophilic reagents is the topic of Chapter 11. Chapter 12 discusses the most important synthetic procedures for oxidizing organic compounds. In each of these chapters, the most widely used reactions are illustrated by a number of specific examples of typical procedures. Chapter 13 introduces the concept of synthetic planning, including the use of protective groups and synthetic equivalents. Multistep syntheses are illustrated with several syntheses of juvabione, longifolene, Prelog–Djerassi lactone, Taxol, and epothilone. The chapter concludes with a discussion of solid-phase synthesis and its application in the synthesis of polypeptides and oligonucleotides, as well as to combinatorial synthesis.

The control of reactivity to achieve specific syntheses is one of the overarching goals of organic chemistry. In the decade since the publication of the third edition, major advances have been made in the development of efficient new methods, particularly catalytic processes, and in means for control of reaction stereochemistry. For example, the scope and efficiency of palladium- catalyzed cross coupling have been greatly improved by optimization of catalysts by ligand modification. Among the developments in stereocontrol are catalysts for enantioselective reduction of ketones, improved methods for control of the stereoselectivity of Diels–Alder reactions, and improved catalysts for enantioselective hydroxylation and epoxidation of alkenes.

This volume assumes a level of familiarity with structural and mechanistic concepts comparable to that in the companion volume, *Part A, Structure and Mechanisms*. Together, the two volumes are intended to provide the advanced undergraduate or beginning graduate student in chemistry a sufficient foundation to comprehend and use the research literature in organic chemistry.

Contents of Part B

Advanced Organic Chemistry

FOURTH EDITION

Part B: Reactions and Synthesis

1

Alkylation of Nucleophilic Carbon Intermediates

Introduction

Carbon–carbon bond formation is the basis for the construction of the molecular framework of organic molecules by synthesis. One of the fundamental processes for carbon–carbon bond formation is a reaction between a nucleophilic carbon and an electrophilic one. The focus in this chapter is on *enolate ions*, *imine anions*, and *enamines*, which are the most useful kinds of carbon nucleophiles, and on their reactions with *alkylating agents*. Mechanistically, these are usually S_N2 reactions in which the carbon nucleophile displaces a halide or other leaving group. Successful carbon–carbon bond formation requires that the S_N2 alkylation be the dominant reaction. The crucial factors which must be considered include (1) the conditions for generation of the carbon nucleophile; (2) the effect of the reaction conditions on the structure and reactivity of the nucleophile; (3) the regio- and stereoselectivity of the alkylation reaction; and (4) the role of solvents, counterions, and other components of the reaction media that can influence the rate of competing reactions.

1.1. Generation of Carbanions by Deprotonation

A very important means of generating carbon nucleophiles involves removal of a proton from a carbon by a Brønsted base. The anions produced are *carbanions*. Both the rate of deprotonation and the stability of the resulting carbanion are enhanced by the presence of substituent groups that can stabilize negative charge. A carbonyl group bonded directly to the anionic carbon can delocalize the negative charge by resonance, and carbonyl compounds are especially important in carbanion chemistry. The anions formed by deprotonation of the carbon *alpha* to a carbonyl group bear most of their negative charge on oxygen and are referred to as *enolates*. Several typical examples of proton-abstraction equilibria are listed in Scheme 1.1. Electron delocalization in the corresponding carbanions is represented by the resonance structures presented in Scheme 1.2.

Scheme 1.1. Generation of Carbon Nucleophiles by Deprotonation

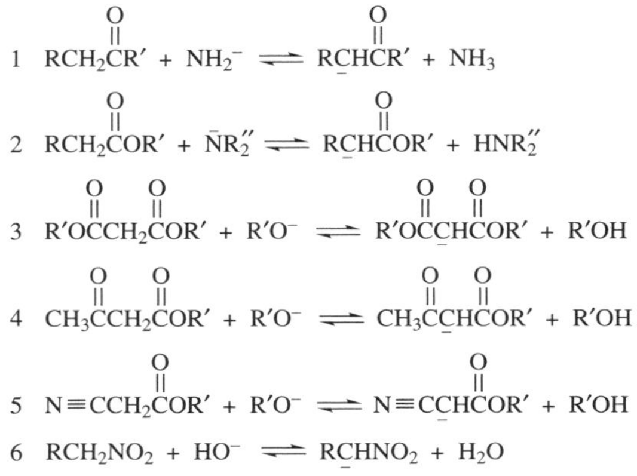

1 $RCH_2\overset{O}{\overset{\|}{C}}R' + NH_2^- \rightleftharpoons R\underline{C}H\overset{O}{\overset{\|}{C}}R' + NH_3$

2 $RCH_2\overset{O}{\overset{\|}{C}}OR' + \bar{N}R''_2 \rightleftharpoons R\underline{C}H\overset{O}{\overset{\|}{C}}OR' + HNR''_2$

3 $R'O\overset{O}{\overset{\|}{C}}CH_2\overset{O}{\overset{\|}{C}}OR' + R'O^- \rightleftharpoons R'O\overset{O}{\overset{\|}{C}}\underline{C}H\overset{O}{\overset{\|}{C}}OR' + R'OH$

4 $CH_3\overset{O}{\overset{\|}{C}}CH_2\overset{O}{\overset{\|}{C}}OR' + R'O^- \rightleftharpoons CH_3\overset{O}{\overset{\|}{C}}\underline{C}H\overset{O}{\overset{\|}{C}}OR' + R'OH$

5 $N{\equiv}CCH_2\overset{O}{\overset{\|}{C}}OR' + R'O^- \rightleftharpoons N{\equiv}C\underline{C}H\overset{O}{\overset{\|}{C}}OR' + R'OH$

6 $RCH_2NO_2 + HO^- \rightleftharpoons R\underline{C}HNO_2 + H_2O$

Scheme 1.2. Resonance in Some Carbanions

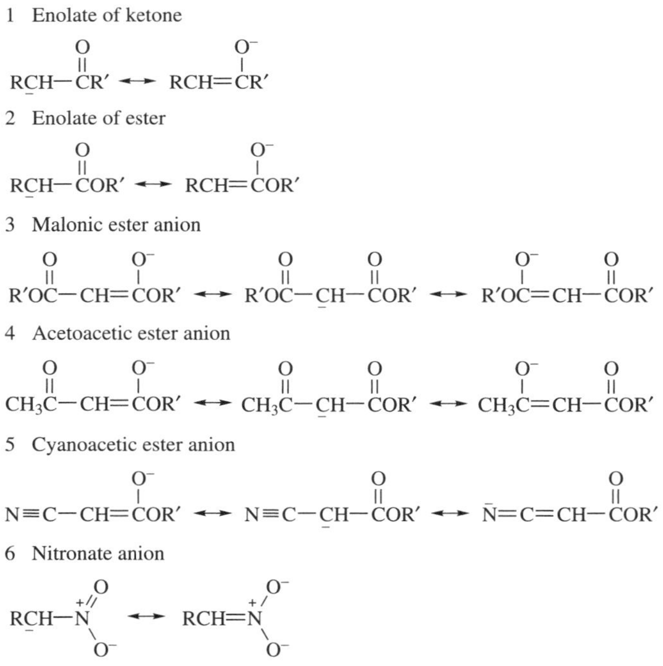

1 Enolate of ketone

$R\underline{C}H{-}\overset{O}{\overset{\|}{C}}R' \leftrightarrow RCH{=}\overset{O^-}{\overset{|}{C}}R'$

2 Enolate of ester

$R\underline{C}H{-}\overset{O}{\overset{\|}{C}}OR' \leftrightarrow RCH{=}\overset{O^-}{\overset{|}{C}}OR'$

3 Malonic ester anion

$R'O\overset{O}{\overset{\|}{C}}{-}CH{=}\overset{O^-}{\overset{|}{C}}OR' \leftrightarrow R'O\overset{O}{\overset{\|}{C}}{-}\underline{C}H{-}\overset{O}{\overset{\|}{C}}OR' \leftrightarrow R'O\overset{O^-}{\overset{|}{C}}{=}CH{-}\overset{O}{\overset{\|}{C}}OR'$

4 Acetoacetic ester anion

$CH_3\overset{O}{\overset{\|}{C}}{-}CH{=}\overset{O^-}{\overset{|}{C}}OR' \leftrightarrow CH_3\overset{O}{\overset{\|}{C}}{-}\underline{C}H{-}\overset{O}{\overset{\|}{C}}OR' \leftrightarrow CH_3\overset{O^-}{\overset{|}{C}}{=}CH{-}\overset{O}{\overset{\|}{C}}OR'$

5 Cyanoacetic ester anion

$N{\equiv}C{-}CH{=}\overset{O^-}{\overset{|}{C}}OR' \leftrightarrow N{\equiv}C{-}\underline{C}H{-}\overset{O}{\overset{\|}{C}}OR' \leftrightarrow \bar{N}{=}C{=}CH{-}\overset{O}{\overset{\|}{C}}OR'$

6 Nitronate anion

$R\underline{C}H{-}\overset{+}{N}({=}O)O^- \leftrightarrow RCH{=}\overset{+}{N}(O^-)O^-$

The efficient generation of a significant equilibrium concentration of a carbanion requires choice of a proper Brønsted base. The equilibrium will favor carbanion formation only when the acidity of the carbon acid is greater than that of the conjugate acid corresponding to the base used for deprotonation. Acidity is quantitatively expressed as pK_a, which is equal to $-\log K_a$ and applies, by definition, to dilute *aqueous* solution. Because most important carbon acids are quite weak acids ($pK_a > 15$), accurate measurement of their acidity in aqueous solutions is impossible, and acidities are determined in organic solvents and referenced to the pK_a in an approximate way. The data produced are not true pK_a's, and their approximate nature is indicated by referring to them as simply pK values. Table 1.1 presents a list of pK data for some typical carbon acids. The table also includes examples of the bases which are often used for deprotonation. The strongest acids appear at the top of the table, and the strongest bases at the bottom. A favorable equilibrium between a carbon acid and its carbanion will be established if the base which is used appears below the acid in the table. Also included in the table are pK values determined in dimethyl sulfoxide (pK_{DMSO}). The range of acidities that can be directly measured in dimethyl sulfoxide (DMSO) is much greater than in aqueous media, thereby allowing direct comparisons between compounds to be made more confidently. The pK values in DMSO are normally greater than in water because water stabilizes anions more effectively, by hydrogen bonding, than does DMSO. Stated another way, many anions are more strongly basic in DMSO than in water. At the present time, the pK_{DMSO} scale includes the widest variety of structural types of synthetic interest.[1] From the pK values collected in Table 1.1, an ordering of some important substituents with respect to their ability to stabilize carbanions can be established. The order suggested is $NO_2 > COR > CN \sim CO_2R > SO_2R > SOR > Ph \sim SR > H > R$.

By comparing the approximate pK values of the conjugate acids of the bases with those of the carbon acid of interest, it is possible to estimate the position of the acid–base equilibrium for a given reactant–base combination. If we consider the case of a simple alkyl ketone in a protic solvent, for example, it can be seen that hydroxide ion and primary alkoxide ions will convert only a small fraction of such a ketone to its anion.

$$\mathrm{R\overset{\overset{\displaystyle O}{\|}}{C}CH_3 + RCH_2O^- \rightleftharpoons R\overset{\overset{\displaystyle O^-}{|}}{C}{=}CH_2 + RCH_2OH} \qquad K < 1$$

The slightly more basic tertiary alkoxides are comparable to the enolates in basicity, and a somewhat more favorable equilibrium will be established with such bases:

$$\mathrm{R\overset{\overset{\displaystyle O}{\|}}{C}CH_3 + R_3CO^- \rightleftharpoons R\overset{\overset{\displaystyle O^-}{|}}{C}{=}CH_2 + R_3COH} \qquad K \approx 1$$

To obtain complete conversion of ketones to enolates, it is necessary to use aprotic solvents so that solvent deprotonation does not compete with enolate formation. Stronger bases, such as amide anion ($^-NH_2$), the conjugate base of DMSO (sometimes referred to as the "dimsyl" anion),[2] and triphenylmethyl anion, are capable of effecting essentially complete conversion of a ketone to its enolate. Lithium diisopropylamide (LDA), which is generated by addition of *n*-butyllithium to diisopropylamine, is widely used as a strong

1. F. G. Bordwell, *Acc. Chem. Res.* **21**:456 (1988).
2. E. J. Corey and M. Chaykovsky, *J. Am. Chem. Soc.* **87**:1345 (1965).

Table 1.1. Approximate pK Values for Some Carbon Acids and Some Common Bases[a]

Carbon acid	pK	pK_{DMSO}	Common bases	pK	pK_{DMSO}
$O_2NCH_2NO_2$	3.6		$CH_3CO_2^-$	4.2	11.6
$CH_3COCH_2NO_2$	5.1				
$PhCH_2NO_2$		12.3			
$CH_3CH_2NO_2$	8.6	16.7			
$CH_3COCH_2COCH_3$	9				
$PhCOCH_2COCH_3$	9.6		PhO^-	9.9	16.4
CH_3NO_2	10.2	17.2			
$CH_3COCH_2CO_2CH_2CH_3$	10.7	14.2	$(CH_3CH_2)_3N$	10.7	
$CH_3COCH(CH_3)COCH_3$	11		$(CH_3CH_2)_2NH$	11	
$NCCH_2CN$	11.2	11.0			
$CH_2(SO_2CH_2CH_3)_2$	12.2	14.4			
$PhCH_2NO_2$	12.3				
$CH_2(CO_2CH_2CH_3)_2$	12.7	16.4			
Cyclopentadiene	15		CH_3O^-	15.5	29.0
$PhSCH_2COCH_3$		18.7	HO^-	15.7	31.4
$PhCH_2COCH_3$		19.9	$CH_3CH_2O^-$	15.9	29.8
$CH_3CH_2CH(CO_2CH_2CH_3)_2$	15		$(CH_3)_2CHO^-$		30.3
$PhSCH_2CN$		20.8	$(CH_3)_3CO^-$	19	32.2
$PhCH_2CN$		21.9			
$(PhCH_2)_2SO_2$		23.9			
$PhCOCH_3$	15.8	24.7			
CH_3COCH_3	20	26.5			
$CH_3CH_2COCH_2CH_3$		27.1			
Fluorene	20.5	22.6			
$PhSO_2CH_3$		29.0			
$PhCH_2SOCH_3$		29.0			
CH_3CN	25	31.3			
Ph_2CH_2		32.2			
Ph_3CH	33	30.6	NH_2^-	35	41
			$CH_3SOCH_2^-$	35	35.1
			$(CH_3CH_2)_2N^-$	36	
$PhCH_3$		43			
CH_4		56			

a. F. G. Bordwell, *Acc. Chem. Res.* **21**:456 (1988).

base in synthetic procedures.[3] It is a very strong base, yet it is sufficiently bulky so as to be relatively nonnucleophilic, a feature that is important in minimizing side reactions. The lithium, sodium and potassium salts of hexamethyldisilazane, $[(CH_3)_3Si]_2NH$, are easily prepared and handled compounds with properties similar to those of lithium diisopropylamide and also find extensive use in synthesis.[4] These bases must be used in aprotic solvents such as ether, tetrahydrofuran (THF), or dimethoxyethane (DME).

$$\underset{}{RC(=O)CH_3} + \underset{\text{LDA}}{[(CH_3)_2CH]_2NLi} \rightleftharpoons RC(OLi){=}CH_2 + [(CH_3)_2CH]_2NH \qquad K > 1$$

3. H. O. House, W. V. Phillips, T. S. B. Sayer, and C.-C. Yau, *J. Org. Chem.* **43**:700 (1978).
4. E. H. Amonoco-Neizer, R. A. Shaw, D. O. Skovlin, and B. C. Smith, *J. Chem. Soc.* **1965**:2997; C. R. Kruger and E. G. Rochow, *J. Organmet. Chem.* **1**:476 (1964).

Sodium hydride and potassium hydride can also be used to prepare enolates from ketones. The reactivity of the metal hydrides is somewhat dependent on the means of preparation and purification of the hydride.[5]

The data in Table 1.1 allow one to estimate the position of the equilibrium for any of the other carbon acids with a given base. It is important to keep in mind the position of such equilibria as other aspects of reactions of carbanions are considered. The base and solvent used will determine the extent of deprotonation. There is another important physical characteristic which needs to be kept in mind, and that is the degree of aggregation of the carbanion. Both the solvent and the cation will influence the state of aggregation, as will be discussed further in Section 1.6.

1.2. Regioselectivity and Stereoselectivity in Enolate Formation

An unsymmetrical dialkyl ketone can form two *regioisomeric* enolates on deprotonation:

$$R_2CHC(=O)CH_2R' \xrightarrow{B^-} R_2C{=}C(O^-)CH_2R' \quad \text{or} \quad R_2CHC(O^-){=}CHR'$$

In order to exploit fully the synthetic potential of enolate ions, control over the regioselectivity of their formation is required. Although it may not be possible to direct deprotonation so as to form one enolate to the exclusion of the other, experimental conditions can often be chosen to provide a substantial preference for the desired regioisomer. To understand why a particular set of experimental conditions leads to the preferential formation of one enolate while other conditions lead to the regioisomer, we need to examine the process of enolate generation in more detail.

The composition of an enolate mixture may be governed by kinetic or thermodynamic factors. The enolate ratio is governed by *kinetic control* when the product composition is determined by the relative *rates* of the two or more competing proton-abstraction reactions.

$$R_2CHC(=O)CH_2R' + B^- \xrightarrow{k_a} R_2C{=}C(O^-)CH_2R' \ (\mathbf{A})$$

$$R_2CHC(=O)CH_2R' + B^- \xrightarrow{k_b} R_2CHC(O^-){=}CHR' \ (\mathbf{B})$$

$$\frac{[\mathbf{A}]}{[\mathbf{B}]} = \frac{k_a}{k_b}$$

Kinetic control of isomeric enolate composition

On the other hand, if enolates **A** and **B** can be interconverted readily, equilibrium is established and the product composition reflects the relative thermodynamic stability of the

5. C. A. Brown, *J. Org. Chem.* **39:**1324 (1974); R. Pi. T. Friedl, P. v. R. Schleyer, P. Klusener, and L. Brandsma, *J. Org. Chem.* **52:**4299 (1987); T. L. Macdonald, K. J. Natalie, Jr., G. Prasad, and J. S. Sawyer, *J. Org. Chem.* **51:**1124 (1986).

Scheme 1.3. Composition of Enolate Mixtures

1[a]

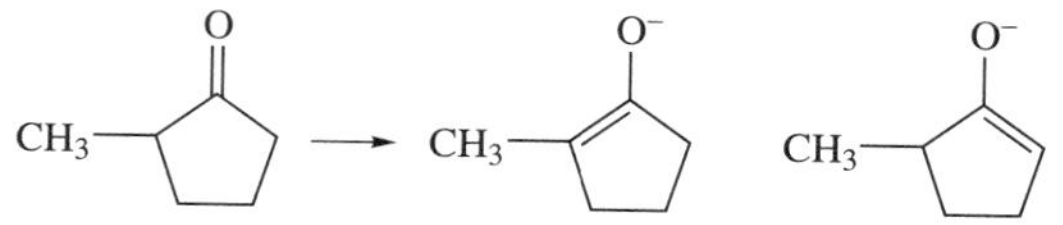

	CH_3 (cyclopentanone, O) →	CH_3, O^-	CH_3, O^-
Kinetic control (Ph_3CLi/dimethoxyethane)		28	72
Thermodynamic control (Ph_3CLi/equilibration in the presence of excess ketone)		94	6

2[b,c]

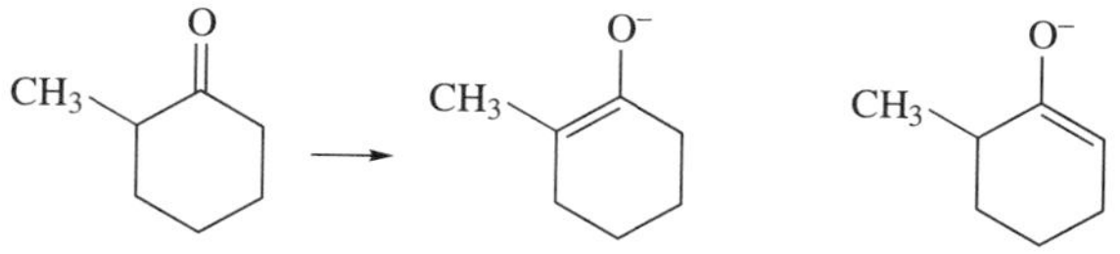

	CH_3 (cyclohexanone, O) →	CH_3, O^-	CH_3, O^-
Kinetic control (LDA/dimethoxyethane)		1	99
Thermodynamic control (Et_3N/DMF)		78	22

3[d]

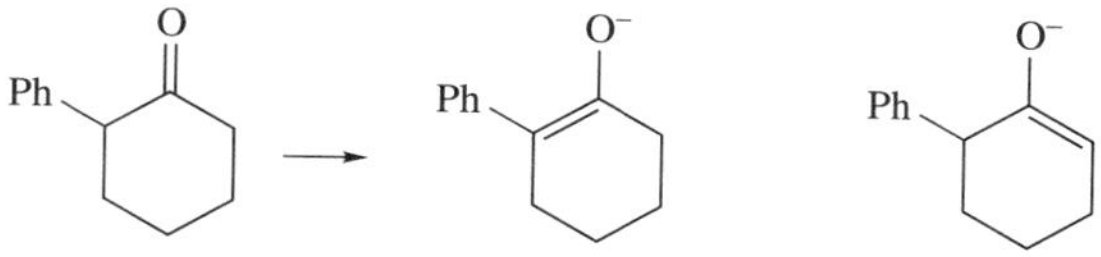

	Ph (cyclohexanone, O) →	Ph, O^-	Ph, O^-
Kinetic control (LDA/tetrahydrofuran, –70°C)[d]			Only enolate
Thermodynamic control (KH, tetrahydrofuran)[c]		Only enolate	

4[a]

	H, H, O →	H, H, O^-	H, H, O^-
Kinetic control (Ph_3CLi/dimethoxyethane)		13	87
Thermodynamic control (equilibration in the presence of excess ketone)		53	47

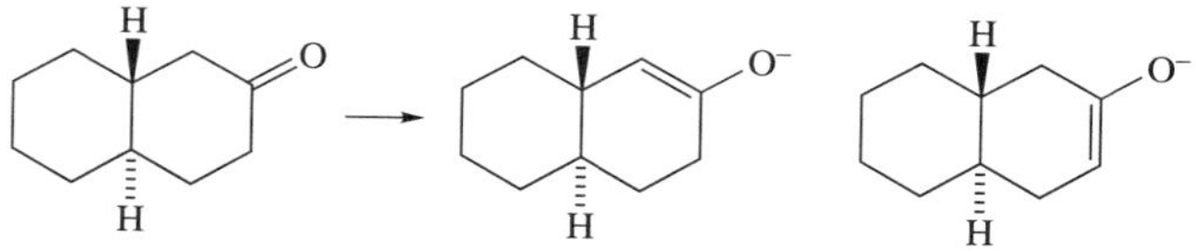

5[e]

$CH_3CH_2CH_2\overset{O}{\overset{\|}{C}}CH_3 \longrightarrow CH_3CH_2CH_2\overset{O^-}{\overset{|}{C}}{=}CH_2$

Kinetic control (LDA/tetrahydrofuran, –78°C) — Only enolate

6[f]

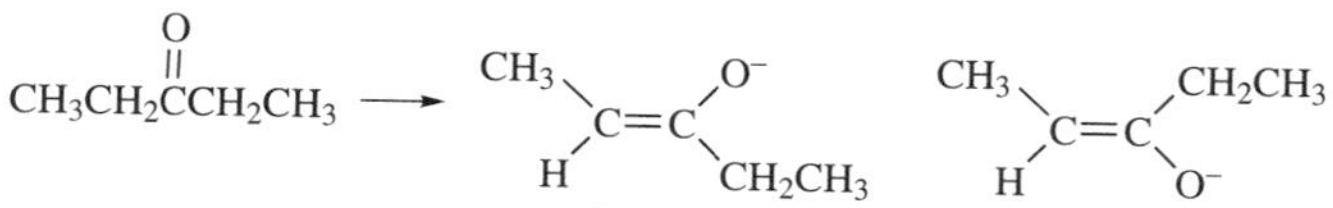

$CH_3CH_2\overset{O}{\overset{\|}{C}}CH_2CH_3 \longrightarrow$

	Z-enolate (CH_3, H, C=C, O^-, CH_2CH_3)	*E*-enolate (CH_3, H, C=C, CH_2CH_3, O^-)
Kinetic control (lithium 2,2,6,6-tetramethylpiperidide/tetrahydrofuran)	13	87
Thermodynamic control (equilibration in the presence of excess ketone)	84	16

7[g] $CH_3CH_2C(=O)C(CH_3)_3 \longrightarrow$ Z-$CH_3CH=C(O^-)C(CH_3)_3$ + E-$CH_3CH=C(O^-)C(CH_3)_3$

	Z	E
Kinetic control (LDA/ tetrahydrofuran)	>98	<2

8[g] $CH_3CH_2C(=O)Ph \longrightarrow$ Z-$CH_3CH=C(O^-)Ph$ + E-$CH_3CH=C(O^-)Ph$

	Z	E
Kinetic control (LDA/ tetrahydrofuran)	>98	<2

9[h] $CH_3CH_2CH_2C(=O)CH_3 \longrightarrow CH_3(CH_2)_4C(O^-)=CH_2$ + $CH_3(CH_2)_3CH=C(O^-)CH_3$

	$CH_3(CH_2)_4C(O^-)=CH_2$	$CH_3(CH_2)_3CH=C(O^-)CH_3$
Kinetic control (LDA, –78°C)	74	26
Kinetic control (LDA TMSCl)	95	5
Kinetic control (NDA/tetramethylenediamine)	91	9
Thermodynamic control (KH, tetrahydrofuran, 20°C)	46	54

a. H. O. House and B. M. Trost, *J. Org. Chem.* **30:**1341 (1965).
b. H. O. House, M. Gall, and H. D. Olmstead, *J. Org. Chem.* **36:**2361 (1971).
c. H. O. House, L. J. Czuba, M. Gall, and H. D. Olmstead, *J. Org. Chem.***34:**2324 (1969).
d. E. Vedejs, *J. Am. Chem. Soc.* **96:**5944 (1974); H. J. Reich, J. M. Renga, and I. L. Reich, *J. Am. Chem. Soc.* **97:**5434 (1975).
e. G. Stork, G. A. Kraus, and G. A. Garcia, *J. Org. Chem.* **39:**3459 (1974).
f. Z. A. Fataftah, I. E. Kopka, and M. W. Rathke, *J. Am. Chem. Soc.* **102:**3959 (1980); Y. Balamraju, C. D. Sharp, W. Gammill, N. Manue, and L. M. Pratt, *Tetrahedron* **54:**7357 (1998).
g. C. H. Heathcock, C. T. Buse, W. A. Kleschick, M. C. Pirrung, J. E. Sohn, and J. Lampe, *J. Org. Chem.* **45:**1066 (1980).
h. R. D. Clark and C. H. Heathcock, *J. Org. Chem.* **41:**1396 (1976); C. A. Brown, *J. Org. Chem.* **39:**3913 (1974); E. J. Corey and A. W. Gross, *Tetrahedron Lett.* **25:**495 (1984); P. C. Andrews, N. D. R. Barnett, R. E. Malvey, W. Clegg, P. A. D. Neil, D. Barr, L. Couton, A. J. Dawson, and B. J. Wakefield, *J. Organomet. Chem.* **518:**85 (1996).

enolates. The enolate ratio is then governed by *thermodynamic control.*

$$R_2CHC(=O)CH_2R' + B^- \underset{k_{-a}}{\overset{k_a}{\rightleftharpoons}} \underset{\mathbf{A}}{R_2C=C(O^-)CH_2R'}$$

$$R_2CHC(=O)CH_2R' + B^- \underset{k_{-b}}{\overset{k_b}{\rightleftharpoons}} \underset{\mathbf{B}}{R_2CHC(O^-)=CHR'}$$

$$\mathbf{A} \overset{K}{\rightleftharpoons} \mathbf{B} \qquad \frac{[\mathbf{A}]}{[\mathbf{B}]} = K$$

Thermodynamic control of isomeric enolate composition

By adjusting the conditions under which an enolate mixture is formed from a ketone, it is possible to establish either kinetic or thermodynamic control. *Ideal conditions for kinetic control of enolate formation are those in which deprotonation is rapid, quantitative, and irreversible.*[6] This ideal is approached experimentally by using a very strong base such as LDA or hexamethyldisilyamide (HMDS) in an aprotic solvent in the absence of excess ketone. Lithium is a better counterion than sodium or potassium for regioselective generation of the kinetic enolate. Lithium maintains a tighter coordination at oxygen and reduces the rate of proton exchange. Aprotic solvents are essential because protic solvents permit enolate equilibration by reversible protonation–deprotonation, which gives rise to the thermodynamically controlled enolate composition. Excess ketone also catalyzes the equilibration by proton exchange. Scheme 1.3 shows data for the regioselectivity of enolate formation for several ketones under various reaction conditions.

A quite consistent relationship is found in these and related data. *Conditions of kinetic control usually favor the less substituted enolate.* The principal reason for this result is that removal of the less hindered hydrogen is faster, for steric reasons, than removal of more hindered protons. Removal of the less hindered proton leads to the less substituted enolate. Steric factors in ketone deprotonation can be accentuated by using more highly hindered bases. The most widely used base is the hexamethyldisilylamide ion, as a lithium or sodium salt. Even more hindered disilylamides such as hexaethyldisilylamide[7] and bis(dimethylphenylsilyl)amide[8] may be useful for specific cases. *On the other hand, at equilibrium the more substituted enolate is usually the dominant species.* The stability of carbon–carbon double bonds increases with increasing substitution, and this effect leads to the greater stability of the more substituted enolate.

The terms *kinetic control* and *thermodynamic control* are applicable to other reactions besides enolate formation; the general concept was covered in Part A, Section 4.4. In discussions of other reactions in this chapter, it may be stated that a given reagent or set of conditions favors the "thermodynamic product." This statement means that the mechanism operating is such that the various possible products are equilibrated after initial formation. When this is true, the dominant product can be predicted by considering the relative stabilities of the various possible products. On the other hand, if a given reaction is under "kinetic control," prediction or interpretation of the relative amounts of products must be made by analyzing the competing rates of product formation.

For many ketones, *stereoisomeric* as well as regioisomeric enolates can be formed, as is illustrated by entries 6, 7, and 8 of Scheme 1.3. The *stereoselectivity* of enolate formation, under conditions of either kinetic or thermodynamic control, can also be controlled to some extent. We will return to this topic in more detail in Chapter 2.

It is also possible to achieve *enantioselective* enolate formation by using chiral bases. Enantioselective deprotonation requires discrimination between two enantiotopic hydrogens, such as in *cis*-2,6-dimethylcyclohexanone or 4-(*t*-butyl)cyclohexanone.

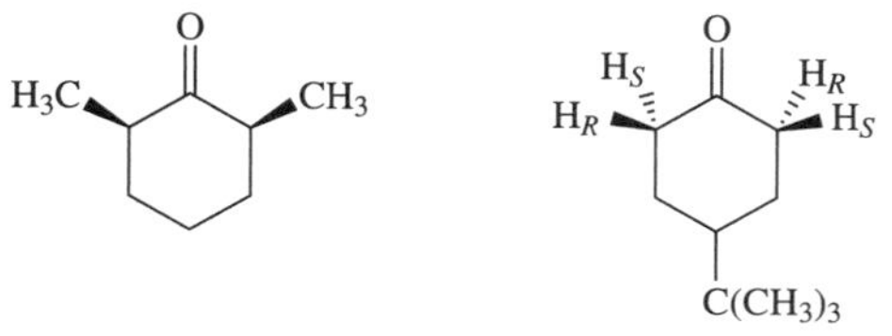

O
H_3C
CH_3

O
H_S
H_R
H_R
H_S
$C(CH_3)_3$

6. For a review, see J. d'Angelo, *Tetrahedron* **32:**2979 (1976).
7. S. Masamune, J. W. Ellingboe, and W. Choy, *J. Am. Chem. Soc.* **104:**5526 (1982).
8. S. R. Angle, J. M. Fevig, S. D. Knight, R. W. Marquis, Jr., and L. E. Overman, *J. Am. Chem. Soc.* **115:**3966 (1993).

The most studied bases are chiral amides such as **C–F**.[9]

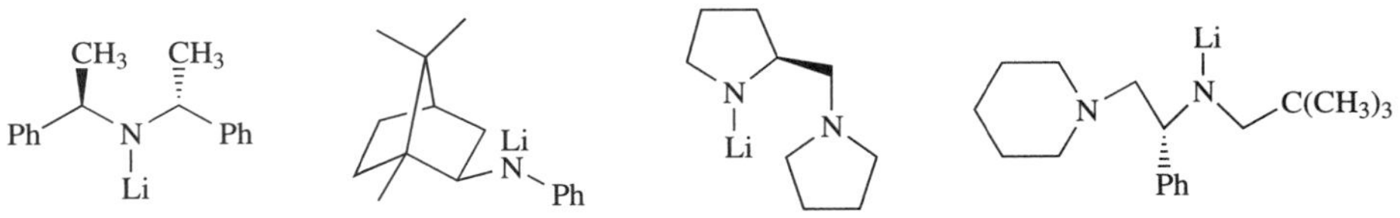

C[10] **D**[11] **E**[12] **F**[13]

Enantioselective enolate formation can also be achieved by *kinetic resolution* by preferential reaction of one of the enantiomers of a racemic chiral ketone such as 2-(*t*-butyl)cyclohexanone (see Part A, Section 2.2 to review the principles of kinetic resolution).

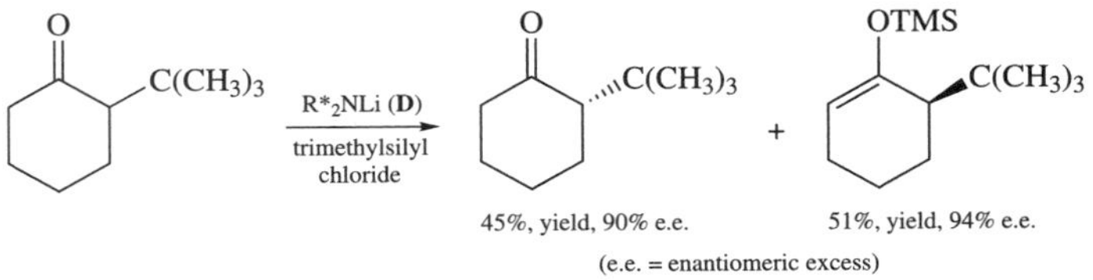

R*$_2$NLi (**D**), trimethylsilyl chloride — Ref. 14

45%, yield, 90% e.e. + 51%, yield, 94% e.e.

(e.e. = enantiomeric excess)

Such enantioselective deprotonations depend upon kinetic selection between prochiral or enantiomeric protons and the chiral base resulting from differences in diastereomeric transition states.[15] For example, transition state **G** has been proposed for deprotonation of 4-substituted cyclohexanones by base **F**.[16]

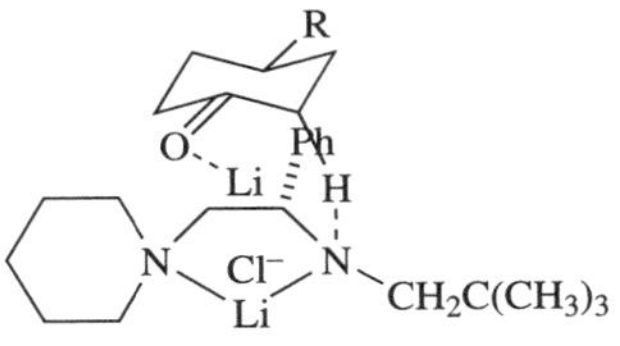

G

Kinetically controlled deprotonation of α,β-unsaturated ketones usually occurs preferentially at the α' carbon adjacent to the carbonyl group. The polar effect of the

9. P. O'Brien, *J. Chem. Soc., Perkin Trans 1* **1998:**1439; H. J. Geis, *Methods of Organic Chemistry (Houben-Weyl)*, Vol. E21a, G. Thiemer, Stuttgart, 1996, p. 589.
10. P. J. Cox and N. S. Simpkins, *Tetrahedron Asymmetry*, **2:**1 (1991); N. S. Simpkins, *Pure Appl. Chem.* **68:**691 (1996); B. J. Bunn and N. S. Simpkins, *J. Org. Chem.* **58:**533 (1993).
11. C. M. Cain, R. P. C. Cousins, G. Coumbarides, and N. S. Simpkins, *Tetrahedron* **46:**523 (1990).
12. D. Sato, H. Kawasaki, T. Shimada, Y. Arata, K. Okamura, T. Date, and K. Koga, *J. Am. Chem. Soc.* **114:**761 (1992); T. Yamashita, D. Sato, T. Kiyoto, A. Kumar, and K. Koga, *Tetrahedron Lett.* **37:**8195 (1996); H. Chatani, M. Nakajima, H. Kawasaki, and K. Koga, *Heterocycles* **46:**53 (1997); R. Shirai, D. Sato, K. Aoki, M. Tanaka, H. Kawasaki, and K. Koga, *Tetrahedron* **53:**5963 (1997).
13. M. Asami, *Bull. Chem. Soc. Jpn.* **63:**721 (1996).
14. H. Kim, H. Kawasaki, M. Nakajima, and K. Koga, *Tetrahedron Lett.* **30:**6537 (1989); D. Sato, H. Kawasaki, T. Shimada, Y. Arata, K. Okamura, T. Date, and K. Koga, *J. Am. Chem. Soc.* **114:**761 (1992).
15. A. Corruble, J.-Y. Valnot, J. Maddaluno, Y. Prigent, D. Davoust, and P. Duhamel, *J. Am. Chem. Soc.* **119:**10042 (1997); D. Sato, H. Kawasaki, and K. Koga, *Chem. Pharm. Bull.* **45:**1399 (1997); K. Sugasawa, M. Shindo, H. Noguchi, and K. Koga, *Tetrahedron Lett.* **37:**7377 (1996).
16. M. Toriyama, K. Sugasawa, M. Shindo, N. Tokutake, and K. Koga, *Tetrahedron Lett.* **38:**567 (1997).

carbonyl group is probably responsible for the faster deprotonation at this position.

$$\text{3,3-dimethylcyclohexenone} \xrightarrow[\text{THF, 0°C}]{[\text{C}_6\text{H}_{11}\text{NCH(CH}_3)_2]\text{Li}^+} \text{O}^-\text{Li}^+ \text{ dienolate (only enolate)}$$

Ref. 17

Under conditions of thermodynamic control, however, it is the enolate corresponding to deprotonation of the γ carbon that is present in the greater amount.

$$(\overset{\gamma}{\text{CH}_3})_2\overset{\beta}{\text{C}}{=}\overset{\alpha}{\text{C}}\text{H}\overset{\text{O}}{\overset{\|}{\text{C}}}\overset{\alpha'}{\text{C}}\text{H}_3 \xrightarrow[\text{NH}_3]{\text{NaNH}_2} \overset{\gamma}{\text{C}}\text{H}_2{=}\underset{\text{CH}_3}{\text{C}}{-}\overset{\alpha}{\text{C}}\text{H}{=}\overset{\text{O}^-}{\text{C}}\overset{\alpha'}{\text{C}}\text{H}_3 > (\overset{\gamma}{\text{CH}_3})_2\text{C}{=}\overset{\alpha}{\text{C}}\text{H}{-}\overset{\text{O}^-}{\text{C}}{=}\overset{\alpha'}{\text{C}}\text{H}_2$$

major enolate (more stable) **H**; (less stable) **I**

Ref. 18

These isomeric enolates differ in stability in that **H** is fully conjugated, whereas the π system in **I** is cross-conjugated. In isomer **I**, the delocalization of the negative charge is restricted to the oxygen and the α' carbon, whereas in the conjugated system of **H**, the negative charge is delocalized on oxygen and both the α and the γ carbon.

1.3. Other Means of Generating Enolates

The recognition of conditions under which lithium enolates are stable and do not equilibrate with regioisomers allows the use of other reactions in addition to proton abstraction to generate specific enolates. Several methods are shown in Scheme 1.4. Cleavage of trimethylsilyl enol ethers or enol acetates by methyllithium (entries 1 and 3, Scheme 1.4) is a route to specific enolate formation that depends on the availability of these starting materials in high purity. The composition of the trimethylsilyl enol ethers prepared from an enolate mixture will reflect the enolate composition. If the enolate formation can be done with high regioselection, the corresponding trimethylsilyl enol ether can be obtained in high purity. If not, the silyl enol ether mixture must be separated. Trimethylsilyl enol ethers can be cleaved by tetraalkylammonium fluoride salts (entry 2, Scheme 1.4). The driving force for this reaction is the formation of the very strong Si–F bond, which has a bond energy of 142 kcal/mol.[19]

Trimethylsilyl enol ethers can be prepared directly from ketones. One procedure involves reaction with trimethylsilyl chloride and a tertiary amine.[20] This procedure gives the regioisomers in a ratio favoring the thermodynamically more stable enol ether. Use of

17. R. A. Lee, C. McAndrews, K. M. Patel, and W. Reusch, *Tetrahedron Lett.* **1973:**965.
18. G. Büchi and H. Wuest, *J. Am. Chem. Soc.* **96:**7573 (1974).
19. For reviews of the chemistry of *O*-silyl enol ethers, see J. K. Rasmussen, *Synthesis* **1977:**91; P. Brownbridge, *Synthesis* **1:**85 (1983); I. Kuwajima and E. Nakamura, *Acc. Chem. Res.* **18:**181 (1985).
20. H. O. House, L. J. Czuba, M. Gall, and H. D. Olmstead, *J. Org. Chem.* **34:**2324 (1969); R. D. Miller and D. R. McKean, *Synthesis* **1979:**730.

t-butyldimethylsilyl chloride with potassium hydride as the base also seems to favor the thermodynamic product.[21] Trimethylsilyl trifluoromethanesulfonate (TMS triflate), which is more reactive, gives primarily the less substituted trimethylsilyl enol ether.[22] Higher ratios of less substituted to more substituted enol ether are obtained by treating a mixture of ketone and trimethylsilyl chloride with LDA at $-78°C$.[23] Under these conditions, the kinetically preferred enolate is immediately trapped by reaction with trimethylsilyl chloride. Even greater preferences for the less substituted silyl enol ether can be obtained by using the more hindered amide from *t*-octyl-*t*-butylamine.

Trimethylsilyl enol ethers can also be prepared by 1,4-reduction of enones using silanes as reductants. Several effective catalysts have been found.[24] The most versatile of these catalysts appears to be a Pt complex of divinyltetramethyldisiloxane.[25] This catalyst gives good yields of substituted silyl enol ethers.

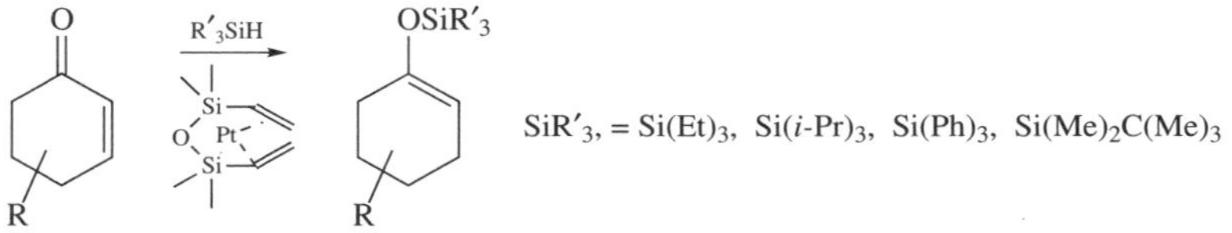

R'_3SiH ; $OSiR'_3$; SiR'_3, = $Si(Et)_3$, $Si(i\text{-}Pr)_3$, $Si(Ph)_3$, $Si(Me)_2C(Me)_3$

Lithium–ammonia reduction of α,β-unsaturated ketones (entry 6, Scheme 1.4) provides a very useful method for generating specific enolates.[26] The desired starting materials are often readily available, and the position of the double bond in the enone determines the structure of the resulting enolate. This and other reductive methods for generating enolates from enones will be discussed more fully in Chapter 5. Another very important method for specific enolate generation, the addition of organometallic reagents to enones, will be discussed in Chapter 8.

1.4. Alkylation of Enolates

Alkylation of enolate is an important synthetic method.[27] The alkylation of relatively acidic compounds such as β-diketones, β-ketoesters, and esters of malonic acid can be carried out in alcohols as solvents using metal alkoxides as bases. The presence of two electron-withdrawing substituents facilitates formation of the enolate resulting from removal of a proton from the carbon situated between them. Alkylation then occurs by an S_N2 process. Some examples of alkylation reactions involving relatively acidic carbon acids are shown in Scheme 1.5. These reactions are all mechanistically similar in that a

21. J. Orban. J. V. Turner, and B. Twitchin, *Tetrahedron Lett.* **25:**5099 (1984).
22. H. Emde, A. Götz, K. Hofmann, and G. Simchen, *Justus Liebigs Ann. Chem.* **1981:**1643; see also E. J. Corey, H. Cho, C. Rücker, and D. Hua *Tetrahedron Lett.* **1981:**3455.
23. E. J. Corey and A. W. Gross, *Tetrahedron Lett.* **25:**495 (1984).
24. I. Ojima and T. Kogure, *Organometallics* **1:**1390 (1982); T. H. Chan and G. Z. Zheng, *Tetrahedron Lett.* **34:**3095 (1993); D. E. Cane and M. Tandon, *Tetrahedron Lett.* **35:**5351 (1994).
25. C. R. Johnson and R. K. Raheja, *J. Org. Chem.* **59:**2287 (1994).
26. For a review of α,β-enone reduction, see D. Caine, *Org. React.* **23:**1 (1976).
27. D. Caine, in *Carbon–Carbon Bond Formation*, Vol. 1, R. L. Augustine, ed., Marcel Dekker, New York, 1979, Chapter 2.

Scheme 1.4. Generation of Specific Enolates

A. Cleavage of trimethylsilyl enol esters

1[a] OSiMe$_3$ / CH(CH$_3$)$_2$ / CH$_3$ / CH$_3$ —(CH_3Li, dimethoxyethane)→ O$^-$Li$^+$ / CH(CH$_3$)$_2$ / CH$_3$ / CH$_3$ + $(CH_3)_4Si$

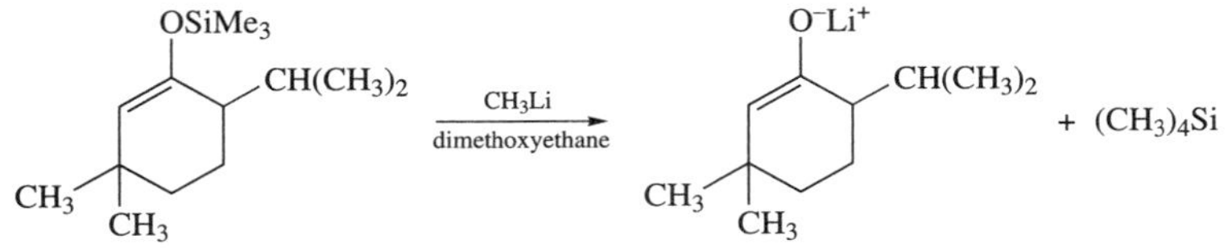

2[b] OSi(CH$_3$)$_3$ / H$_3$C —($PhCH_2\overset{+}{N}(CH_3)_3F$, THF)→ O$^-$ $PhCH_2\overset{+}{N}(CH_3)_3$ / CH$_3$ + $(CH_3)_3SiF$

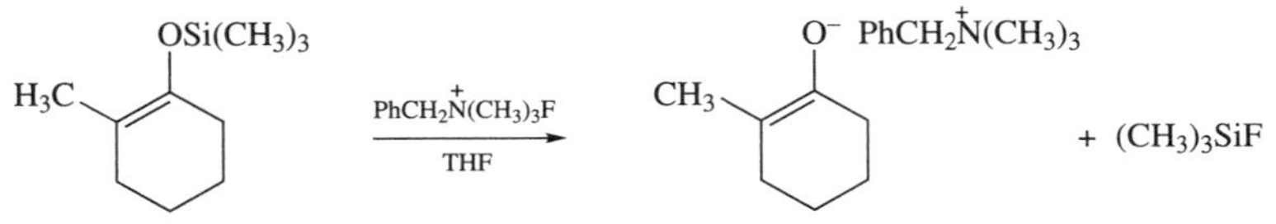

B. Cleavage of enol acetates

3[c] $PhCH{=}C(CH_3)OC(=O)CH_3$ —(2 equiv CH_3Li, dimethoxyethane)→ $PhCH{=}C(CH_3)O^-Li^+$ + $(CH_3)_3COLi$

C. Regioselective silylation of ketones by *in situ* enolate trapping

4[d] $C_6H_{13}C(=O)CH_3$ —($(CH_3)_3SiCl$, add LDA at −78°C)→ $C_6H_{13}C(OSi(CH_3)_3){=}CH_2$ (95%) + $C_5H_{11}CH{=}C(OSi(CH_3)_3)CH_3$ (5%)

5[e] $(CH_3)_2CHC(=O)CH_3$ —($(CH_3)_3SiO_3SCF_3$, 20°C, $(C_2H_5)_3N$)→ $(CH_3)_2CHC(OSi(CH_3)_3){=}CH_2$ (84%) + $(CH_3)_2C{=}C(OSi(CH_3)_3)CH_3$ (16%)

D. Reduction of a,b-unsaturated ketones

6[f] O + Li —(NH_3)→ $^-$O … $^-$ —(NH_3)→ $^+$Li$^-$O

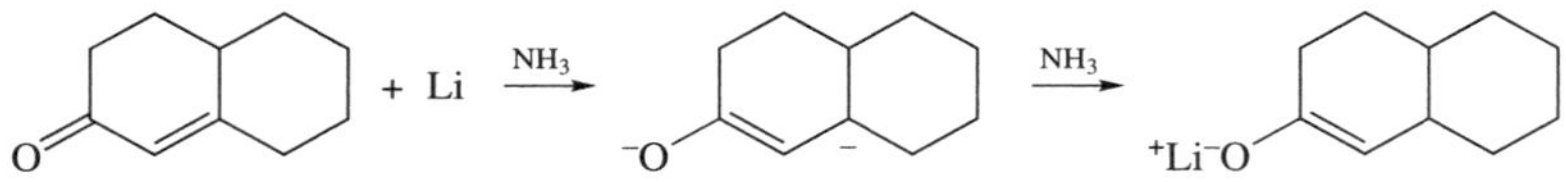

7[g] O / O / O —($(i\text{-}Pr)_3SiH$, $Pt[CH_2{=}CHSi(CH_3)_2]_2O$)→ OSi($i$-Pr)$_3$ / O / O

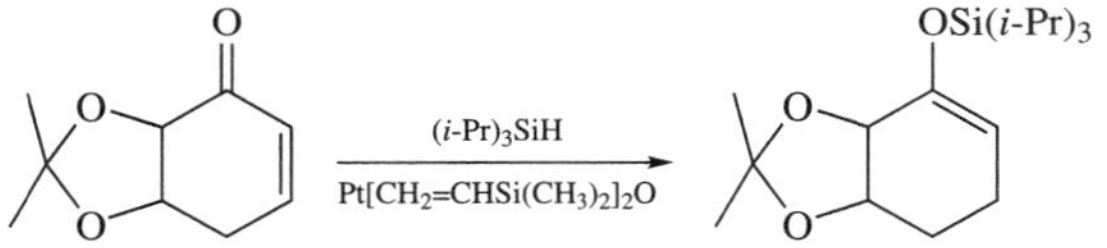

a. G. Stork and P. F. Hudrlik, *J. Am. Chem. Soc.* **90:**4464 (1968); see also H. O. House, L. J. Czuba, M. Gall, and H. D. Olmstead, *J. Org. Chem.* **34:**2324 (1969).
b. I. Kuwajima and E. Nakamura, *J. Am. Chem. Soc.* **97:**3258 (1975).
c. G. Stork and S. R. Dowd, *Org. Synth.* **55:**46 (1976); see also H. O. House and B. M. Trost, *J. Org. Chem.* **30:**2502 (1965).
d. E. J. Corey and A. W. Gross, *Tetrahedron Lett.* **25:**495 (1984).
e. H. Emde, A. Götz, K. Hofmann, and G. Simchen, *Justus Liebigs Ann. Chem.* **1981:**1643.
f. G. Stork, P. Rosen, N. Goldman, R. V. Coombs, and J. Tsuji, *J. Am. Chem. Soc.* **87:**275 (1965).
g. C. R. Johnson and R. K. Raheja, *J. Org. Chem.* **59:**2287 (1994).

carbanion, formed by deprotonation using a suitable base, reacts with an electrophile by an S_N2 mechanism. The alkylating agent must be reactive toward nucleophilic displacement. Primary halides and sulfonates, especially allylic and benzylic ones, are the most reactive alkylating agents. Secondary systems react more slowly and often give only moderate yields because of competing elimination. Tertiary halides give only elimination products.

Methylene groups can be dialkylated if sufficient base and alkylating agent are used. Dialkylation can be an undesirable side reaction if the monoalkyl derivative is the desired product. Use of dihaloalkanes as the alkylating reagent leads to ring formation, as illustrated by the diethyl cyclobutanedicarboxylate synthesis (entry 7) shown in Scheme 1.5. This example illustrates the synthesis of cyclic compounds by *intramolecular* alkylation reactions. The relative rates of cyclization for ω-haloalkyl malonate esters are $650{,}000:1:6500:5$ for formation of three-, four-, five-, and six-membered rings, respectively.[28] (See Section 3.9 of Part A to review the effect of ring size on S_N2 reactions.)

Relatively acidic carbon acids such as malonic esters and β-keto esters were the first class of carbanions for which reliable conditions for alkylation were developed. The reason being that these carbanions are formed using easily accessible alkoxide ions. The preparation of 2-substiuted β-keto esters (entries 1, 4, and 8) and 2-substituted derivatives of malonic ester (entries 2 and 7) by the methods illustrated in Scheme 1.5 are useful for the synthesis of ketones and carboxylic acids, since both β-ketoacids and malonic acids undergo facile decarboxylation:

$$\mathrm{X{-}C(=O){-}C(R)(R'){-}C(=O)OH} \xrightarrow{-CO_2} \mathrm{X{-}C(OH){=}C(R)R'} \longrightarrow \mathrm{X{-}C(=O){-}CH(R)R'}$$

β = keto acid: X = alkyl or aryl = ketone

substituted malonic acid: X = OH = substituted acetic acid

Examples of this approach to the synthesis of ketones and carboxylic acids are presented in Scheme 1.6. In these procedures, an ester group is removed by hydrolysis and decarboxylation after the alkylation step. The malonate and acetoacetate carbanions are the *synthetic equivalents* of the simpler carbanions lacking the ester substituents. In the preparation of 2-heptanone (entries 1, Schemes 1.5 and 1.6), for example, ethyl acetoacetate functions as the synthetic equivalent of acetone. It is also possible to use the dilithium derivative of acetoacetic acid as the synthetic equivalent of acetone enolate.[29] In this case, the hydrolysis step is unnecessary, and decarboxylation can be done directly on the alkylation product.

$$\mathrm{CH_3\overset{O}{\overset{\|}{C}}CH_2CO_2H} \xrightarrow{2\ n\text{-BuLi}} \mathrm{CH_3\overset{O^-\,Li^+}{\overset{|}{C}}{=}CH\overset{O}{\overset{\|}{C}}O^-Li^+} \xrightarrow[\substack{2)\ H^+ \\ (-CO_2)}]{1)\ R{-}X} \mathrm{CH_3\overset{O}{\overset{\|}{C}}CH_2R}$$

28. A. C. Knipe and C. J. Stirling, *J. Chem. Soc., B* **1968:**67; for a discussion of factors which affect intramolecular alkylation of enolates, see J. Janjatovic and Z. Majerski, *J. Org. Chem.* **45:**4892 (1980).
29. R. A. Kjonaas and D. D. Patel, *Tetrahedron Lett.* **25:**5467 (1984).

Scheme 1.5. Alkylations of Relatively Acidic Carbon Acids

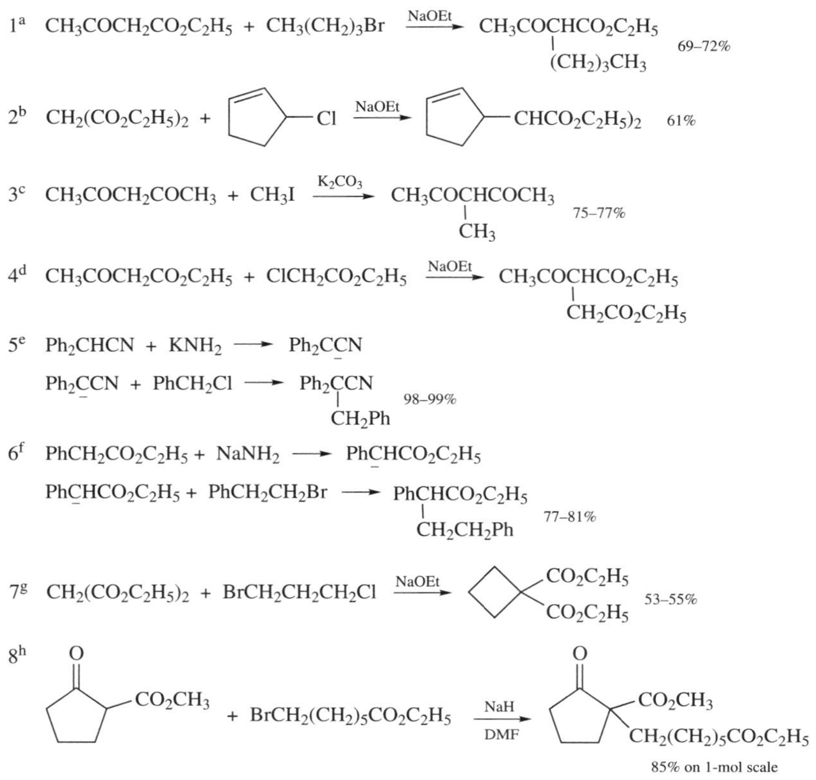

1[a] $CH_3COCH_2CO_2C_2H_5 + CH_3(CH_2)_3Br \xrightarrow{NaOEt} CH_3COCH(CO_2C_2H_5)(CH_2)_3CH_3$ 69–72%

2[b] $CH_2(CO_2C_2H_5)_2$ + 3-chlorocyclopentene $\xrightarrow{NaOEt}$ 2-cyclopentenyl-$CH(CO_2C_2H_5)_2$ 61%

3[c] $CH_3COCH_2COCH_3 + CH_3I \xrightarrow{K_2CO_3} CH_3COCH(CH_3)COCH_3$ 75–77%

4[d] $CH_3COCH_2CO_2C_2H_5 + ClCH_2CO_2C_2H_5 \xrightarrow{NaOEt} CH_3COCH(CO_2C_2H_5)CH_2CO_2C_2H_5$

5[e] $Ph_2CHCN + KNH_2 \longrightarrow Ph_2\bar{C}CN$

$Ph_2\bar{C}CN + PhCH_2Cl \longrightarrow Ph_2C(CN)CH_2Ph$ 98–99%

6[f] $PhCH_2CO_2C_2H_5 + NaNH_2 \longrightarrow Ph\bar{C}HCO_2C_2H_5$

$Ph\bar{C}HCO_2C_2H_5 + PhCH_2CH_2Br \longrightarrow PhCH(CO_2C_2H_5)CH_2CH_2Ph$ 77–81%

7[g] $CH_2(CO_2C_2H_5)_2 + BrCH_2CH_2CH_2Cl \xrightarrow{NaOEt}$ cyclobutane-1,1-$(CO_2C_2H_5)_2$ 53–55%

8[h] 2-(CO_2CH_3)cyclopentanone $+ BrCH_2(CH_2)_5CO_2C_2H_5 \xrightarrow[DMF]{NaH}$ 2-(CO_2CH_3)-2-($CH_2(CH_2)_5CO_2C_2H_5$)cyclopentanone 85% on 1-mol scale

a. C. S. Marvel and F. D. Hager, *Org. Synth.* **I:**248 (1941).
b. R. B. Moffett, *Org. Synth.* **IV:**291 (1963).
c. A. W. Johnson, E. Markham, and R. Price, *Org. Synth.* **42:**75 (1962).
d. H. Adkins, N. Isbell, and B. Wojcik, *Org. Synth.* **II:**262 (1943).
e. C. R. Hauser and W. R. Dunnavant, *Org. Synth.* **IV:**962 (1963).
f. E. M. Kaiser, W. G. Kenyon, and C. R. Hauser, *Org. Synth.* **47:**72 (1967).
g. R. P. Mariella and R. Raube, *Org. Synth.* **IV:**288 (1963).
h. K. F. Bernardy. J. F. Poletto, J. Nocera, P. Miranda, R. E. Schaub, and M. J. Weiss, *J. Org. Chem.* **45:**4702 (1980).

Similarly, the dilithium salt of monoethyl malonic dianion is easily alkylated and the product decarboxylates on acidification.[30]

The use of β-ketoesters and malonic ester enolates has largely been supplanted by the development of the newer procedures based on selective enolate formation that permit direct alkylation of ketone and ester enolates and avoid the hydrolysis and decarboxylation of ketoesters intermediates. Most enolate alkylations are carried out by deprotonating the ketone under conditions that are appropriate for kinetic or thermodynamic control. Enolates can also be prepared from silyl enol ethers and by reduction of enones (see Section 1.3). Alkylation also can be carried out using silyl enol ethers by reaction with fluoride ion.[31] Tetraalkylammonium fluoride salts in anhydrous solvents are normally the

30. J. E. McMurry and J. H. Musser, *J. Org. Chem.* **40:**2556 (1975).
31. I. Kuwajima, E. Nakamura, and M. Shimizu, *J. Am. Chem. Soc.* **104:**1025 (1982).

Scheme 1.6. Synthesis of Ketones and Carboxylic Acid Derivatives via Alkylation Followed by Decarboxylation

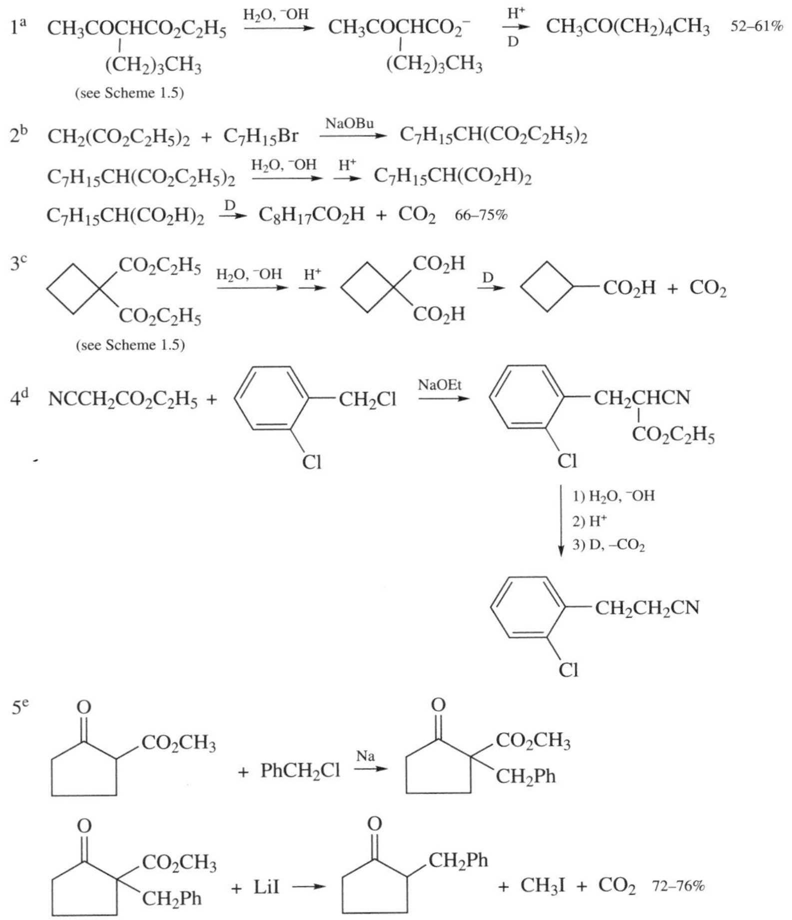

1[a] $CH_3COCH(CO_2C_2H_5)(CH_2)_3CH_3 \xrightarrow{H_2O,\ ^-OH} CH_3COCH(CO_2^-)(CH_2)_3CH_3 \xrightarrow[D]{H^+} CH_3CO(CH_2)_4CH_3$ 52–61%

(see Scheme 1.5)

2[b] $CH_2(CO_2C_2H_5)_2 + C_7H_{15}Br \xrightarrow{NaOBu} C_7H_{15}CH(CO_2C_2H_5)_2$

$C_7H_{15}CH(CO_2C_2H_5)_2 \xrightarrow{H_2O,\ ^-OH} \xrightarrow{H^+} C_7H_{15}CH(CO_2H)_2$

$C_7H_{15}CH(CO_2H)_2 \xrightarrow{D} C_8H_{17}CO_2H + CO_2$ 66–75%

3[c] cyclobutane-1,1-$(CO_2C_2H_5)_2$ $\xrightarrow{H_2O,\ ^-OH} \xrightarrow{H^+}$ cyclobutane-1,1-$(CO_2H)_2$ $\xrightarrow{D}$ cyclobutane–$CO_2H + CO_2$

(see Scheme 1.5)

4[d] $NCCH_2CO_2C_2H_5$ + 2-Cl-C_6H_4–CH_2Cl $\xrightarrow{NaOEt}$ 2-Cl-C_6H_4–$CH_2CH(CN)CO_2C_2H_5$

1) H_2O, ^-OH
2) H^+
3) D, $-CO_2$

2-Cl-C_6H_4–CH_2CH_2CN

5[e] 2-(CO_2CH_3)cyclopentanone + $PhCH_2Cl$ $\xrightarrow{Na}$ 2-(CO_2CH_3)-2-(CH_2Ph)cyclopentanone

2-(CO_2CH_3)-2-(CH_2Ph)cyclopentanone + LiI ⟶ 2-(CH_2Ph)cyclopentanone + $CH_3I + CO_2$ 72–76%

a. J. R. Johnson and F. D. Hager, *Org. Synth.* **I:**351 (1941).
b. E. E. Reid and J. R. Ruhoff, *Org. Synth.* **II:**474 (1943).
c. G. B. Heisig and F. H. Stodola, *Org. Synth.* **III:**213 (1955).
d. J. A. Skorcz and F. E. Kaminski, *Org. Synth.* **48:**53 (1968).
e. F. Elsinger, *Org. Synth.* **V:**76 (1973).

fluoride ion source.

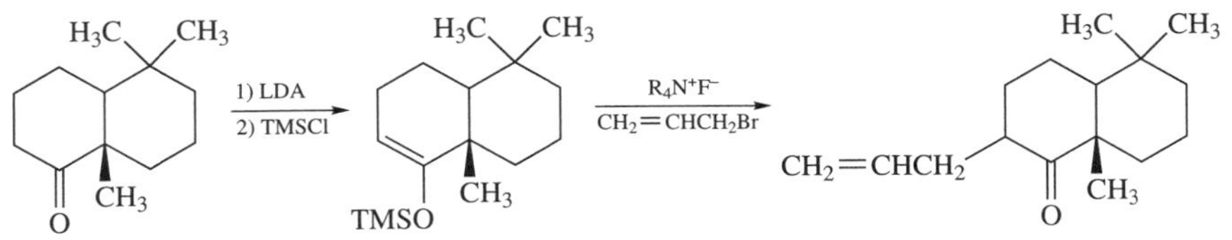

$\xrightarrow[\text{2) TMSCl}]{\text{1) LDA}}$ (TMSO enol ether) $\xrightarrow[CH_2=CHCH_2Br]{R_4N^+F^-}$ ($CH_2=CHCH_2$ ketone) Ref. 32

Several examples of alkylation of ketone enolates are given in Scheme 1.7.

32. A. B. Smith III and R. Mewshaw, *J. Org. Chem.* **49:**3685 (1984).

Scheme 1.7. Regioselective Enolate Alkylation

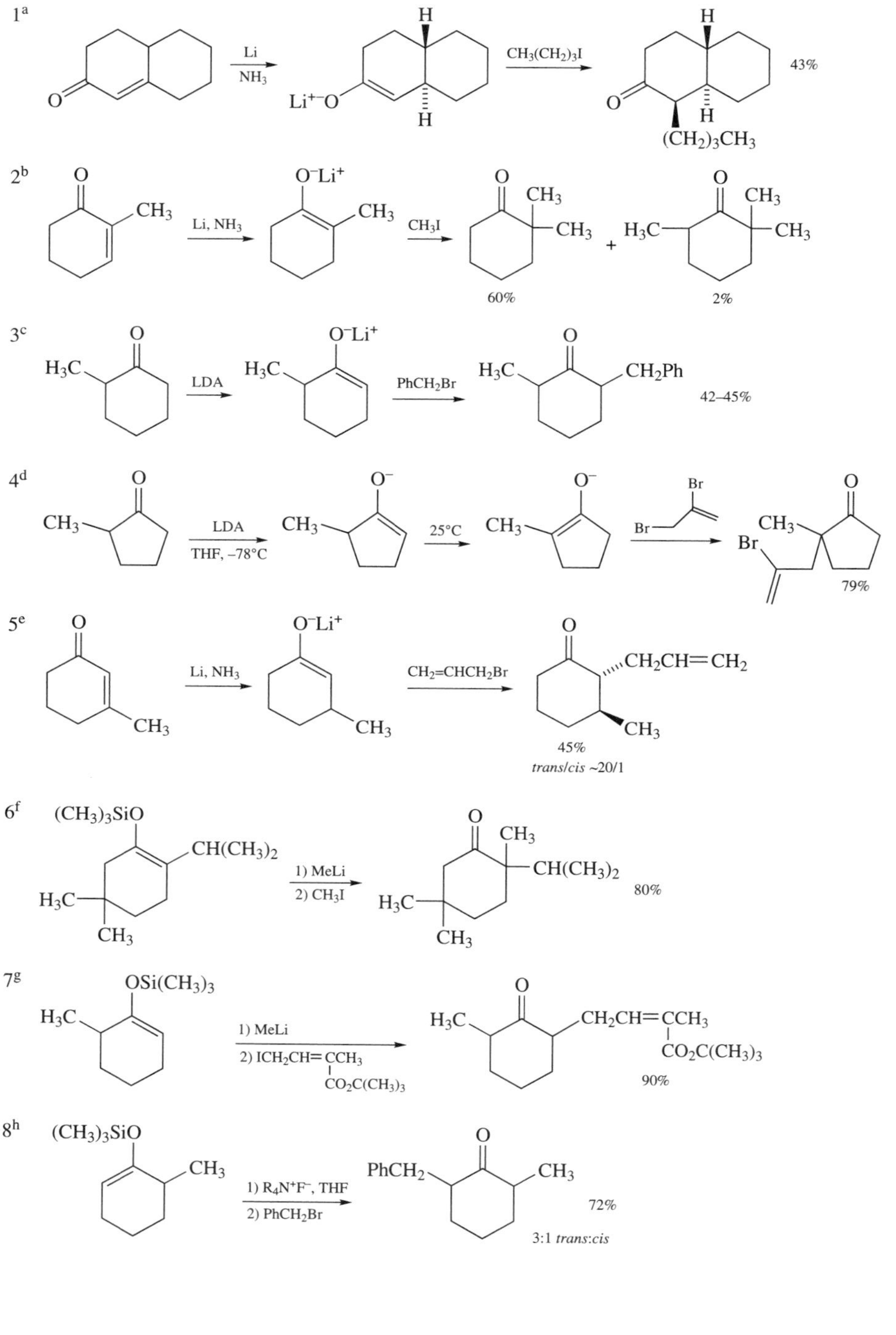

1[a]
Li, NH_3; $CH_3(CH_2)_3I$ — 43%

2[b]
Li, NH_3; CH_3I — 60% + 2%

3[c]
LDA; $PhCH_2Br$ — 42–45%

4[d]
LDA, THF, –78°C; 25°C; $BrCH_2C(Br)=CH_2$ — 79%

5[e]
Li, NH_3; $CH_2=CHCH_2Br$ — 45%, *trans/cis* ~20/1

6[f]
1) MeLi 2) CH_3I — 80%

7[g]
1) MeLi 2) $ICH_2CH=C(CH_3)CO_2C(CH_3)_3$ — 90%

8[h]
1) $R_4N^+F^-$, THF 2) $PhCH_2Br$ — 72%, 3:1 *trans:cis*

9[i]

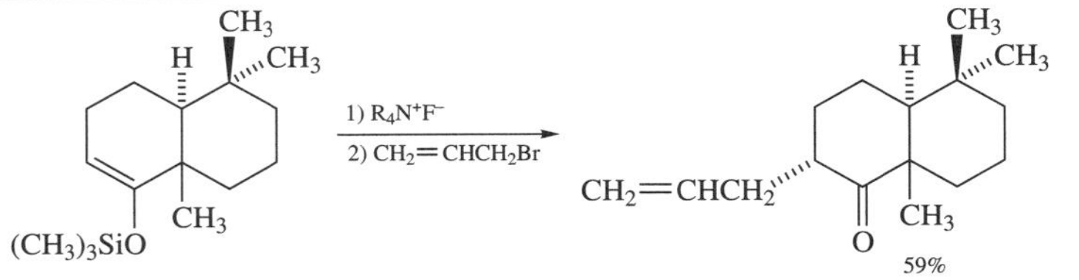

1) $R_4N^+F^-$
2) $CH_2{=}CHCH_2Br$

59%

a. G. Stork, P. Rosen, N. Goldman, R. V. Coombs, and J. Tsujii, *J. Am. Chem. Soc.* **87**:275 (1965).
b. H. A. Smith, B. J. L. Huff, W. J. Powers III, and D. Caine, *J. Org. Chem.* **32**:2851 (1967).
c. M. Gall and H. O. House, *Org. Synth.* **52**:39 (1972).
d. S. C. Welch and S. Chayabunjonglerd, *J. Am. Chem. Soc.* **101**:6768 (1979).
e. D. Caine, S. T. Chao, and H. A. Smith, *Org. Synth.* **56**:52 (1977).
f. G. Stork and P. F. Hudrlik, *J. Am. Chem. Soc.* **90**:4464 (1968).
g. P. L. Stotter and K. A. Hill, *J. Am. Chem. Soc.* **96**:6524 (1974).
h. I. Kuwajima, E. Nakamura, and M. Shimizu, *J. Am. Chem. Soc.* **104**:1025 (1982).
i. A. B. Smith III and R. Mewshaw, *J. Org. Chem.* **49**:3685 (1984).

The development of conditions for stoichiometric formation of both kinetically and thermodynamically controlled enolates has permitted the extensive use of enolate alkylation reactions in multistep synthesis of complex molecules. One aspect of the reaction which is crucial in many cases is the stereoselectivity. The alkylation step has a stereoelectronic preference for approach of the electrophile perpendicular to the plane of the enolate, since the electrons which are involved in bond formation are the π electrons. A major factor in determining the stereoselectivity of ketone enolate alkylations is the difference in steric hindrance on the two faces of the enolate. The electrophile will approach from the less hindered of the two faces, and the degree of stereoselectivity depends upon the steric differentiation. For simple, conformationally based cyclohexanone enolates such as that from 4-*t*-butylcyclohexanone, there is little steric differentiation. The alkylation product is a nearly 1 : 1 mixture of the *cis* and *trans* isomers.

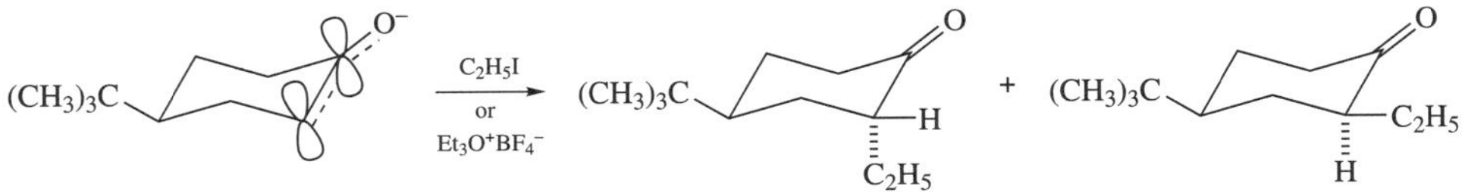

C_2H_5I or $Et_3O^+BF_4^-$

Ref. 33

The *cis* product must be formed through a transition state with a twistlike conformation to adhere to the requirements of stereoelectronic control. The fact that this pathway is not disfavored is consistent with other evidence that the transition state in enolate alkylations occurs *early* and reflects primarily the structural features of the reactant, not the product. A late transition state should disfavor the formation of the *cis* isomer because of the strain energy associated with the nonchair conformation of the product.

The introduction of an alkyl substituents at the α carbon in the enolate enhances stereoselectivity somewhat. This is attributed to a steric effect in the enolate. To minimize steric interaction with the solvated oxygen, the alkyl group is distorted somewhat from coplanarity. This biases the enolate toward attack from the axial direction. The alternative approach from the upper face would enhance the steric interaction by forcing the alkyl

33. H. O. House, B. A. Terfertiller, and H. D. Olmstead, *J. Org. Chem.* **33**:935 (1968).

group to become eclipsed with the enolate oxygen.[34]

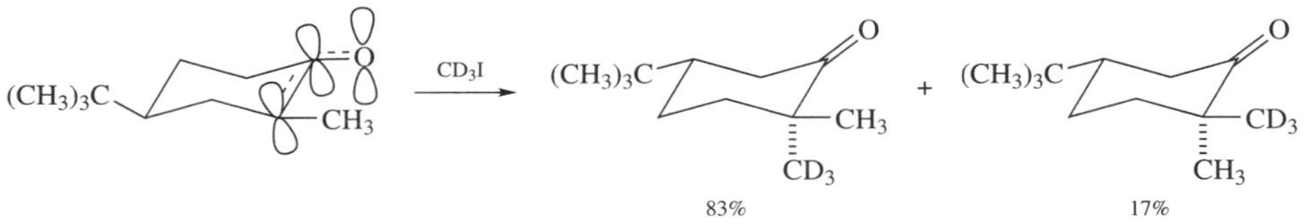

When an additional methyl substituent is placed at C-3, there is a strong preference for alkylation *anti* to the 3-methyl group. This can be attributed to the conformation of the enolate, which places the methyl in a pseudoaxial conformation because of allylic strain (see Part A, Section 3.3.) The C-3 methyl then shields the lower face of the enolate.[35]

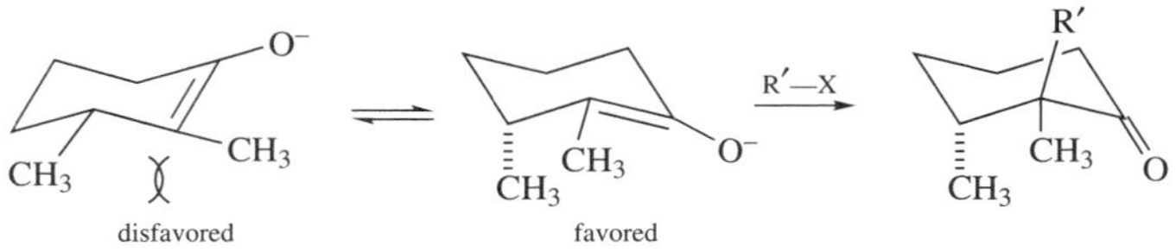

The enolates of 1- and 2-decalone derivatives provide further insights into the factors governing stereoselectivity in enolate alkylations. The 1(9)-enolate of 1-decalone shows a preference for alkylation to give the *cis* ring juncture. This is believed to be due primarily to a steric effect. The upper face of the enolate presents three hydrogens in a 1,3-diaxial relationship to the approaching electrophile. The corresponding hydrogens on the lower face are equatorial.[36]

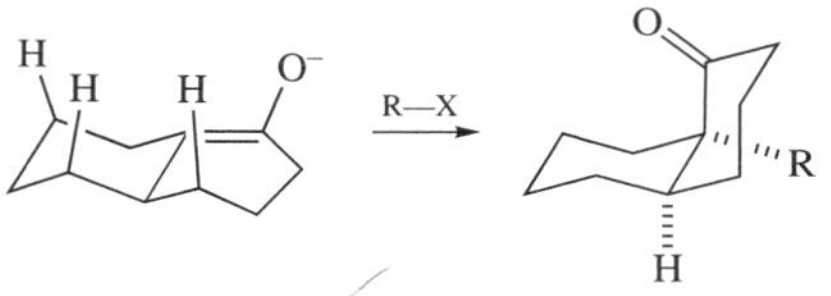

The 2(1)-enolate of *trans*-2-decalone is preferentially alkylated by an axial approach of the electrophile.

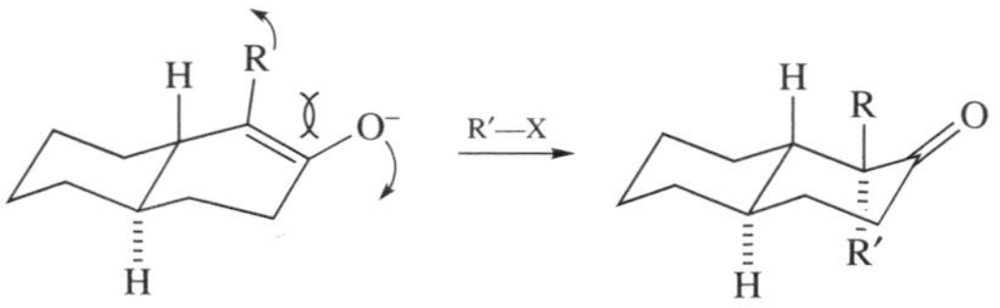

The stereoselectivity is enhanced if there is an alkyl substituent at C-1. The factors operating in this case are similar to those described for 4-*t*-butylcyclohexanone. The *trans*-decalone framework is conformationally rigid. Axial attack from the lower face leads directly to the chair conformation of the product. The 1-alkyl group enhances this

34. H. O. House and M. J. Umen, *J. Org. Chem.* **38:**1000 (1973).
35. R. K. Boeckman, Jr., *J. Org. Chem.* **38:**4450 (1973).
36. H. O. House and B. M. Trost, *J. Org. Chem.* **30:**2502 (1965).

stereoselectivity because a steric interaction with the solvated enolate oxygen distorts the enolate in such a way as to favor the axial attack.[37] The placement of an axial methyl group at C-10 in a 2(1)-decalone enolate introduces a 1,3-diaxial interaction with the approaching electrophile. The preferred alkylation product results from approach on the upper face of the enolate.

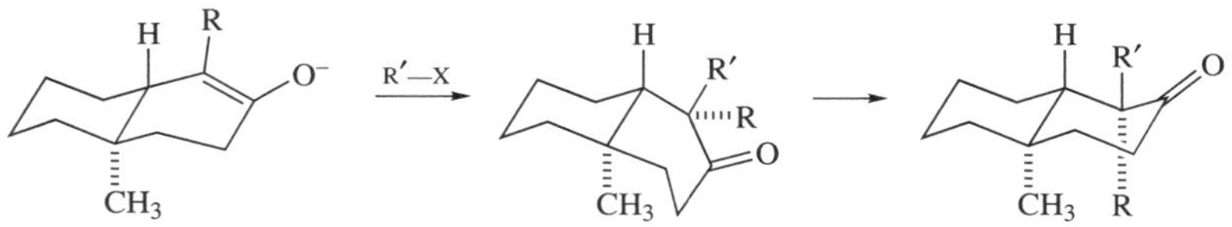

The prediction and interpretation of alkylation stereochemistry also depends on consideration of conformational effects in the enolate. The decalone enolate **1** was found to have a strong preference for alkylation to give the *cis* ring junction, with alkylation occurring *syn* to the *t*-butyl substituent.[38]

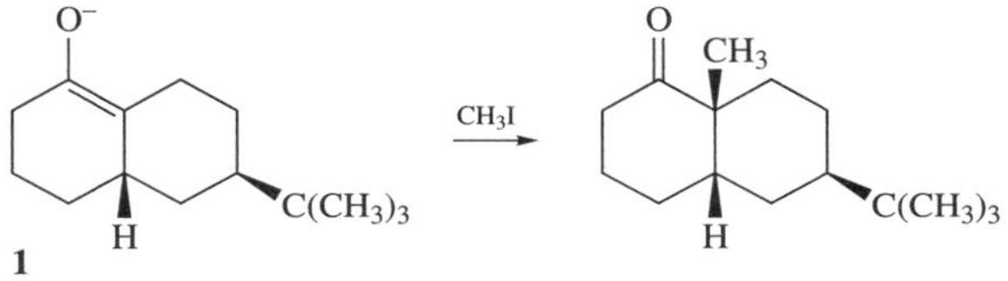

According to molecular mechanics calculations, the minimum-energy conformation of the enolate is a twist-boat conformation (because the chair leads to an axial orientation of the *t*-butyl group). The enolate is convex in shape, with the second ring shielding the lower face of the enolate, and alkylation therefore occurs from the top.

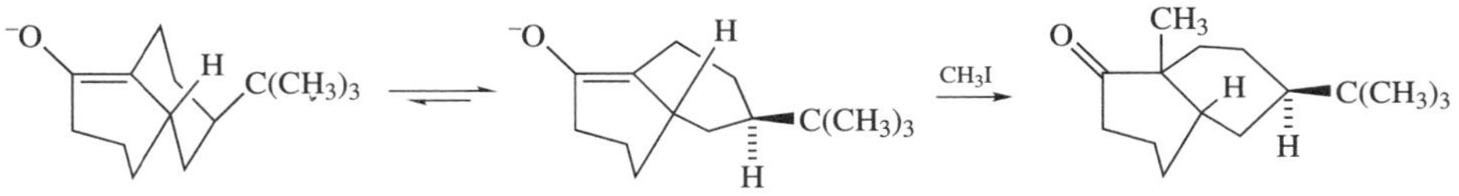

If the alkylation is intramolecular, additional conformational restrictions on the direction of approach of the electrophile to the enolate become important. Baldwin *et al.* have summarized the general principles that govern the energetics of intramolecular ring-closure reactions.[39] (See Part A, Section 3.9). The intramolecular alkylation reaction of **2** gives exclusively **3**.[40] The transition state must achieve a geometry that permits interaction of the π orbital of the enolate to achieve an approximately collinear alignment with the sulfonate leaving group. The alkylation probably occurs through a transition state like **J**. The transition state **K** for formation of the *trans* ring junction would be more

37. R. S. Mathews, S. S. Grigenti, and E. A. Folkers, *J. Chem. Soc., Chem. Commun.* **1970:**708; P. Lansbury and G. E. DuBois, *Tetrahedron Lett.* **1972:**3305.
38. H. O. House, W. V. Phillips, and D. Van Derveer, *J. Org. Chem.* **44:**2400 (1979).
39. J. E. Baldwin, R. C. Thomas, L. I. Kruse, and L. Silberman, *J. Org. Chem.* **42:**3846 (1977).
40. J. M. Conia and F. Rouessac, *Tetrahedron* **16:**45 (1961).

strained because of the necessity to span the opposite face of the enolate π system.

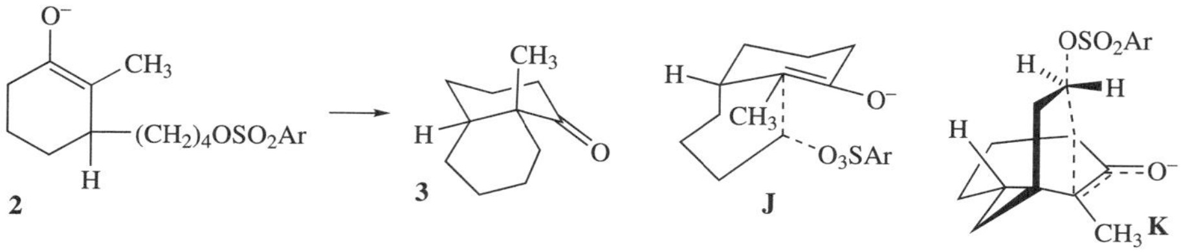

These examples illustrate the issues which must be considered in analyzing the stereoselectivity of enolate alkylation. The major factors are the conformation of the enolate, the stereoelectronic requirement for an approximately perpendicular trajectory, and the steric preference for the least hindered path of approach.

1.5. Generation and Alkylation of Dianions

In the presence of a sufficiently strong base, such as an alkyllithium, sodium or potassium hydride, sodium or potassium amide, or LDA, 1,3-dicarbonyl compounds can be converted to their *dianions* by two sequential deprotonations.[41] For example, reaction of benzoylacetone with sodium amide leads first to the enolate generated by deprotonation at the methylene group between the two carbonyl groups. A second equivalent of base deprotonates the benzyl methylene group to give a diendiolate.

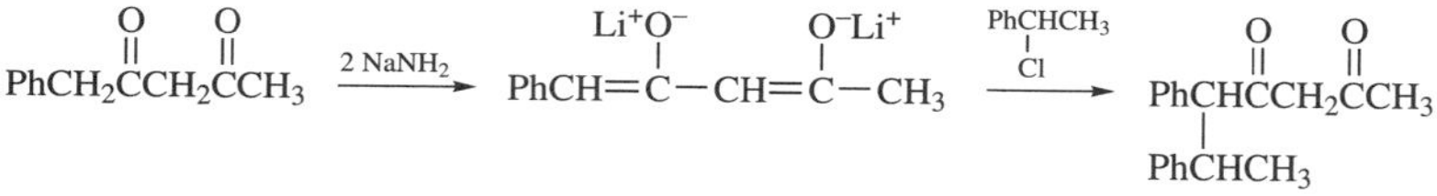

$$PhCH_2\overset{O}{\overset{\|}{C}}CH_2\overset{O}{\overset{\|}{C}}CH_3 \xrightarrow{2\ NaNH_2} PhCH{=}\overset{Li^+O^-}{\overset{|}{C}}{-}CH{=}\overset{O^-Li^+}{\overset{|}{C}}{-}CH_3 \xrightarrow{PhCH(Cl)CH_3} PhCH(CH(CH_3)Ph)\overset{O}{\overset{\|}{C}}CH_2\overset{O}{\overset{\|}{C}}CH_3 \qquad \text{Ref. 42}$$

Alkylation reactions of dianions occur at the *more basic* carbon. This technique allows alkylation of 1,3-dicarbonyl compounds to be carried out cleanly at the less acidic position. Because, as discussed earlier, alkylation of the monoanion occurs at the carbon between the two carbonyl groups, the site of monoalkylation can be controlled by choice of the amount and nature of the base. A few examples of the formation and alkylation of dianions are collected in Scheme 1.8.

1.6. Medium Effects in the Alkylation of Enolates

The rate of alkylation of enolate ions is strongly dependent on the solvent in which the reaction is carried out.[43] The relative rates of reaction of the sodium enolate of diethyl *n*-butylmalonate with *n*-butyl bromide are shown in Table 1.2.

41. For reviews, see T. M. Harris and C. M. Harris, *Org. React.* **17:**155 (1969); E. M. Kaiser, J. D. Petty, and P. L. A. Knutson, *Synthesis* **1977:**509; C. M. Thompson and D. L. C. Green, *Tetrahedron* **47:**4223 (1991); C. M. Thompson, *Dianion Chemistry in Organic Synthesis*, CRC Press, Boca Raton, Florida, 1994.
42. D. M. von Schriltz, K. G. Hamton, and C. R. Hauser, *J. Org. Chem.* **34:**2509 (1969).
43. For reviews, see (a) A. J. Parker, *Chem. Rev.* **69:**1 (1969); (b) L. M. Jackmamn and B. C. Lange, *Tetrahedron* **33:**2737 (1977).

Scheme 1.8. Generation and Alkylation of Dianions

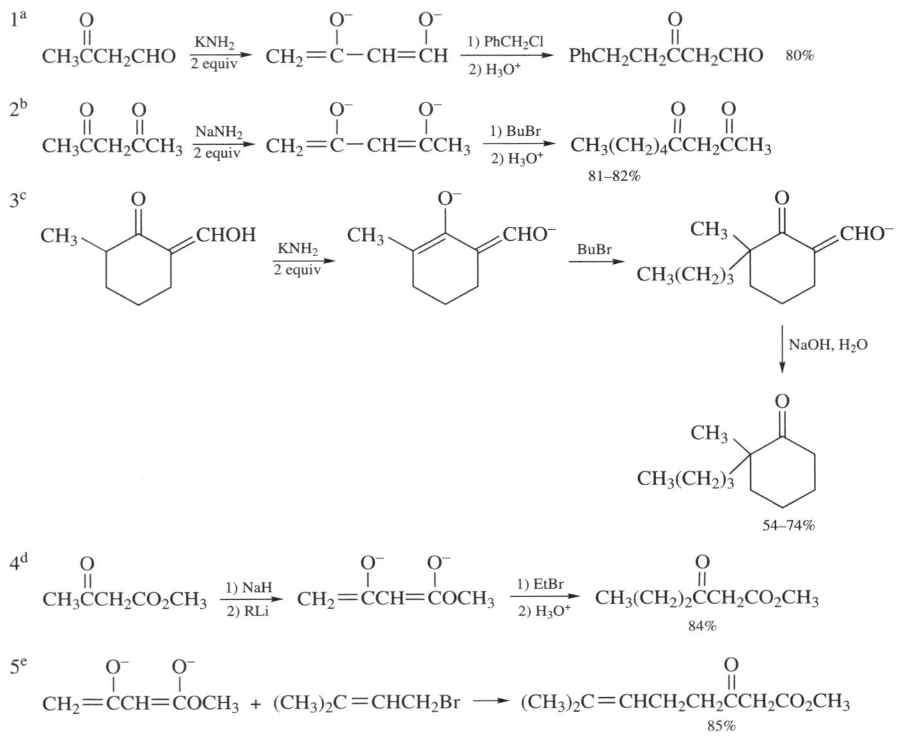

1[a] $CH_3COCH_2CHO \xrightarrow[\text{2 equiv}]{KNH_2} CH_2{=}C(O^-){-}CH{=}CH(O^-) \xrightarrow[\text{2) } H_3O^+]{\text{1) } PhCH_2Cl} PhCH_2CH_2COCH_2CHO$ 80%

2[b] $CH_3COCH_2COCH_3 \xrightarrow[\text{2 equiv}]{NaNH_2} CH_2{=}C(O^-){-}CH{=}C(O^-)CH_3 \xrightarrow[\text{2) } H_3O^+]{\text{1) BuBr}} CH_3(CH_2)_4COCH_2COCH_3$ 81–82%

3[c] 2-(hydroxymethylene)-6-methylcyclohexanone (CH3, CHOH) $\xrightarrow[\text{2 equiv}]{KNH_2}$ dianion (CH3, O⁻, CHO⁻) $\xrightarrow{BuBr}$ 6-butyl-6-methyl-2-(oxidomethylene)cyclohexanone (CH_3, $CH_3(CH_2)_3$, CHO^-) $\xrightarrow{NaOH,\ H_2O}$ 2-butyl-2-methylcyclohexanone (CH_3, $CH_3(CH_2)_3$) 54–74%

4[d] $CH_3COCH_2CO_2CH_3 \xrightarrow[\text{2) RLi}]{\text{1) NaH}} CH_2{=}C(O^-)CH{=}C(O^-)OCH_3 \xrightarrow[\text{2) } H_3O^+]{\text{1) EtBr}} CH_3(CH_2)_2COCH_2CO_2CH_3$ 84%

5[e] $CH_2{=}C(O^-)CH{=}C(O^-)OCH_3 + (CH_3)_2C{=}CHCH_2Br \longrightarrow (CH_3)_2C{=}CHCH_2CH_2COCH_2CO_2CH_3$ 85%

a. T. M. Harris, S. Boatman, and C. R. Hauser, *J. Am. Chem. Soc.* **85**:3273 (1963); S. Boatman, T. M. Harris, and C. R. Hauser, *J. Am. Chem. Soc.* **87**:82 (1965); K. G. Hampton, T. M. Harris, and C. R. Hauser, *J. Org. Chem.* **28**:1946 (1963).
b. K. G. Hampton, T. M. Harris, and C. R. Hauser, *Org. Synth.* **47**:92 (1967).
c. S. Boatman, T. M. Harris, and C. R. Hauser, *Org. Synth.* **48**:40 (1968).
d. S. N. Huckin and L. Weiler, *J. Am. Chem. Soc.* **96**:1082 (1974).
e. F. W. Sum and L. Weiler, *J. Am. Chem. Soc.* **101**:4401 (1979).

DMSO and *N,N*-dimethylformamide (DMF) are particularly effective in enhancing the reactivity of enolate ions, as Table 1.2 shows. Both of these compounds belong to the *polar aprotic* class of solvents. Other members of this class that are used as solvents in reactions between carbanions and alkyl halides include *N*-methylpyrrolidone (NMP) and hexamethylphosphoric triamide (HMPA). Polar aprotic solvents, as their name implies, are materials which have high dielectric constants but which lack hydroxyl groups or other

Table 1.2 Relative Alkylation Rates of Sodium Diethyl *n*-Butylmalonate in Various Solvents[a]

Solvent	Dielectric constant, ε	Relative rate
Benzene	2.3	1
Tetrahydrofuran	7.3	14
Dimethoxyethane	6.8	80
Dimethylformamide	37	970
Dimethyl sulfoxide	47	1420

a. From H. E. Zaugg, *J. Am. Chem. Soc.* **83**:837 (1961).

hydrogen-bonding groups. Polar aprotic solvents possess excellent metal-cation coordination ability, so they can solvate and dissociate enolates and other carbanions from ion pairs and clusters.

$H_3C-\overset{O^-}{\underset{+}{S}}-CH_3$
dimethyl sulfoxide (DMSO)
e = 47

$H-\overset{O}{\overset{||}{C}}-N(CH_3)_2$
N,N-dimethylformamide (DMF)
e = 37

N-methylpyrrolidone (NMP)
e = 32

$O{=}P[N(CH_3)_2]_3$
hexamethylphosphoric triamide (HMPA)
e = 30

The reactivity of alkali-metal (Li^+, Na^+, K^+) enolates is very sensitive to the state of aggregation, which is, in turn, influenced by the reaction medium. The highest level of reactivity, which can be approached but not achieved in solution, is that of the "bare" unsolvated enolate anion. For an enolate–metal ion pair in solution, the maximum reactivity would be expected in a medium in which the cation was strongly solvated and the enolate was very weakly solvated. Polar aprotic solvents are good cation solvators and poor anion solvators. Each one (DMSO, DMF, HMPA, and NMP) has a negatively polarized oxygen available for coordination to the alkali-metal cation. Coordination to the enolate ion is much less effective because the positively polarized atom of these molecules is not nearly as exposed as the oxygen. Thus, these solvents provide a medium in which enolate–metal ion pairs are dissociated to give a less encumbered, more reactive enolate.

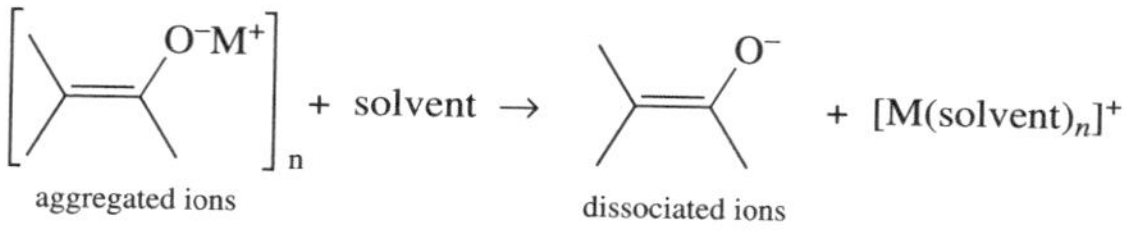

$[\text{enolate } O^-M^+]_n$ + solvent → enolate O^- + $[M(\text{solvent})_n]^+$

aggregated ions — dissociated ions

Polar protic solvents also possess a pronounced ability to separate ion pairs but are less favorable as solvents for enolate alkylation reactions because they coordinate to both the metal cation and the enolate ion. Solvation of the enolate anion occurs through hydrogen bonding. The solvated enolate is relatively less reactive because the hydrogen-bonded enolate must be disrupted during alkylation. Enolates generated in polar protic solvents such as water, alcohols, or ammonia are therefore less reactive than the same enolate in a polar aprotic solvent such as DMSO.

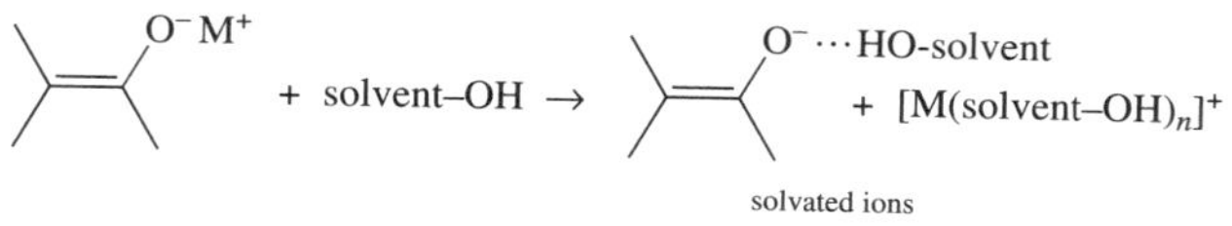

enolate O^-M^+ + solvent–OH → enolate $O^-\cdots$HO-solvent + $[M(\text{solvent–OH})_n]^+$

solvated ions

THF and DME are slightly polar solvents which are moderately good cation solvators. Coordination to the metal cation involves the oxygen lone pairs. These solvents, because of their lower dielectric constants, are less effective at separating ion pairs and higher aggregates than are the polar aprotic solvents. The crystal structures of the lithium and potassium enolates of methyl *t*-butyl ketone have been determined by X-ray crystallography.[44] The structures are shown in Figs. 1.1 and 1.2. While these represent the solid-state structural situation, the hexameric clusters are a good indication of the nature of the enolates in relatively weakly coordinating solvents. Despite the somewhat reduced reactivity of aggregated enolates, THF and DME are the most commonly used solvents for synthetic reactions involving enolate alkylation. They are the most suitable solvents for kinetic enolate generation and also have advantages in terms of product workup and purification over the polar aprotic solvents. Enolate reactivity in these solvents can often be enhanced by adding a reagent that can bind alkali-metal cations more strongly. Popular choices are HMPA, tetramethylethylenediamine (TMEDA), and the crown ethers.[45] TMEDA can chelate metal ions through the electron pairs on nitrogen. The crown ethers can coordinate metal ions in structures in which the metal ion is encapsulated by the ether oxygens. The 18-crown-6 structure is of such a size as to allow sodium or potassium ions to fit comfortably in the cavity. The smaller 12-crown-4 binds Li^+ preferentially. The cation complexing agents lower the degree of aggregation of the enolate–metal-cation ion pairs and result in enhanced reactivity.

The reactivity of enolates is also affected by the metal counterion. Among the most commonly used ions, the order of reactivity is $Mg^{2+} < Li^+ < Na^+ < K^+$. The factors that are responsible for this order are closely related to those described for solvents. The smaller, harder Mg^{2+} and Li^+ cations are more tightly associated with the enolate than are the Na^+ and K^+ ions. The tighter coordination decreases the reactivity of the enolate and gives rise to more highly associated species.

1.7. Oxygen versus Carbon as the Site of Alkylation

Enolate anions are *ambident nucleophiles*. Alkylation of an enolate can occur at either carbon or oxygen. Because most of the negative charge of an enolate is on the oxygen atom, it might be supposed that O-alkylation would dominate. A number of factors other than charge density affect the C/O-alkylation ratio, and it is normally possible to establish reaction conditions that favor alkylation on carbon.

C-alkylation $$RC(O^-)\!=\!CH_2 + R'X \longrightarrow RC(=O)CH_2R'$$

O-alkylation $$RC(O^-)\!=\!CH_2 + R'X \longrightarrow RC(OR')\!=\!CH_2$$

O-Alkylation is most pronounced when the enolate is dissociated. When the potassium salt of ethyl acetoacetate is treated with diethyl sulfate in the polar aprotic solvent HMPA, the major product (83%) is the O-alkylated one. In THF, where ion clustering occurs, all of the product is C-alkylated. In *t*-butanol, where the acetoacetate

44. P. G. Williard and G. B. Carpenter, *J. Am. Chem. Soc.* **108:**462 (1986).
45. C. L. Liotta and T. C. Caruso, *Tetrahedron Lett.* **26:**1599 (1985).

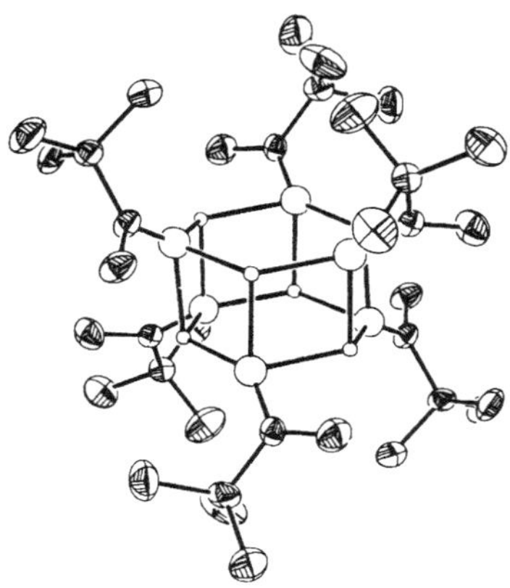

Fig. 1.1. Unsolvated hexameric aggregate of lithium enolate of methyl *t*-butyl ketone; large circles = oxygen, small circles = lithium. (Reproduced with permission from Ref. 44. Copyright 1986 American Chemical Society.)

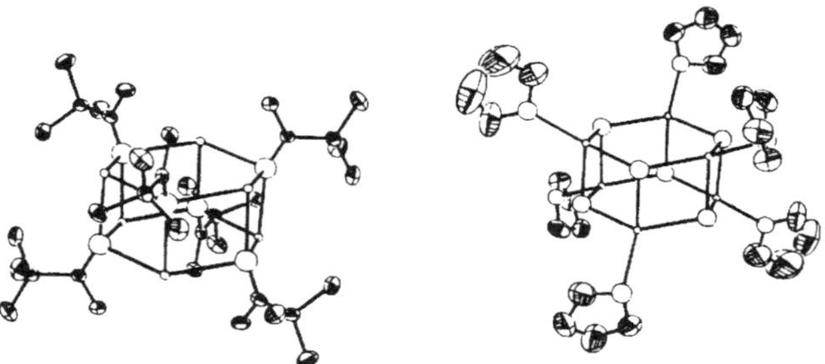

Fig. 1.2. Potassium enolate of methyl *t*-butyl ketone; large cirlces = oxygen, small circles = potassium. (a) Left-hand plot shows only methyl *t*-butyl ketone residues. (b) Right-hand plot shows only the solvating THF molecules. The crystal is a composite of these two structures. (Reproduced with permission from Ref. 44. Copyright 1986 American Chemical Society.)

anion is hydrogen-bonded by solvent, again only C-alkylation is observed.[46]

$$CH_3C(O^-K^+){=}CHCO_2CH_2CH_3 + (CH_3CH_2O)_2SO_2 \longrightarrow CH_3C(OCH_2CH_3){=}CHCO_2CH_2CH_3 + CH_3C(=O)CH(CH_2CH_3)CO_2CH_2CH_3$$

in HMPA	83%	15% (2% dialkyl)
in *t*-butanol	0%	94% (6% dialkyl)
in THF	0%	94% (6% dialkyl)

46. A. L. Kurts, A. Masias, N. K. Genkina, I. P. Beletskaya, and O. A. Reutov, *Dokl. Akad. Nauk. SSR (Engl. Transl.)* **187:**595 (1969).

Higher C/O-alkylation ratios are observed with alkyl halides than with alkyl sulfonates and sulfates. The highest C/O-alkylation ratios are obtained with alkyl iodides. For ethylation of the potassium salt of ethyl acetoacetate in HMPA, the product compositions shown below were obtained.[47]

$$CH_3C(O^-K^+){=}CHCO_2CH_2CH_3 + CH_3CH_2X \xrightarrow{HMPA} CH_3C(OCH_2CH_3){=}CHCO_2CH_2CH_3 + CH_3C(=O)CH(CH_2CH_3)CO_2CH_2CH_3$$

X = OTs	88%	11% (1% dialkyl)
X = Cl	60%	32% (8% dialkyl)
X = Br	39%	38% (23% dialkyl)
X = I	13%	71% (16% dialkyl)

Leaving-group effects on the ratio of C- to O-alkylation can be correlated by reference to the "hard–soft-acid–base" (HSAB) rationale.[48] Of the two nucleophilic sites in an enolate ion, oxygen is harder than carbon. Nucleophilic substitution reactions of the S_N2 type proceed best when the nucleophile and leaving group are either both hard or both soft.[49] Consequently, ethyl iodide, with the very soft leaving group iodide, reacts preferentially with the softer carbon site rather than the harder oxygen. Oxygen leaving groups, such as sulfonate and sulfate, are harder, and alkyl sulfonates and sulfates react preferentially at the hard oxygen site of the enolate. The hard–hard combination is favored by an early transition state, where the charge distribution is the most important factor. The soft–soft combination is favored by a later transition state, where partial bond formation is the dominant factor. The C-alkylation product is more stable than the O-alkylation product (because the bond energy of $C{=}O + C{-}C$ is greater than that of $C{=}C + C{-}O$). Therefore, conditions that favor a dissociated, more reactive enolate favor O-alkylation.

Similar effects are also seen with enolates of simple ketones. For isopropyl phenyl ketone, the inclusion of one equivalent of 12-crown-4 in a DME solution of the lithium enolate changes the C/O-alkylation ratio from 1.2 : 1 to 1 : 3, with methyl sulfate as the alkylating agent.[50] With methyl iodide as the alkylating agent, C-alkylation is strongly favored with or without 12-crown-4.

$$PhC(O^-Li^+){=}C(CH_3)_2 \xrightarrow{(CH_3O)_2SO_2} PhC(=O)C(CH_3)_3 + PhC(OCH_3){=}C(CH_3)_2$$

favored by added crown ether

To summarize, the amount of *O-alkylation* is maximized by use of an alkyl sulfate or alkyl sulfonate in a polar aprotic solvent. The amount of *C-alkylation* is maximized by use of an alkyl halide in a less polar or protic solvent. The majority of synthetic operations involving ketone enolates are carried out in THF or DME using an alkyl bromide or alkyl iodide, and C-alkylation is favored.

47. A. L. Kurts, N. K. Genkina, A. Masias, I. P. Beletskaya, and O. A. Reutov, *Tetrahedron* **27**:4777 (1971).
48. T.-L. Ho, *Hard and Soft Acids and Bases Principle in Organic Chemistry*, Academic Press, New York, 1977.
49. R. G. Pearson and J. Songstad, *J. Am. Chem. Soc.* **89**:1827 (1967).
50. L. M. Jackman and B. C. Lange, *J. Am. Chem. Soc.* **103**:4494 (1981).

Intramolecular alkylation of enolates leads to formation of cyclic products. In addition to the other factors that govern C/O-alkylation ratios, the element of stereoelectronic control comes into play in such cases. The following reactions illustrate this point.[51]

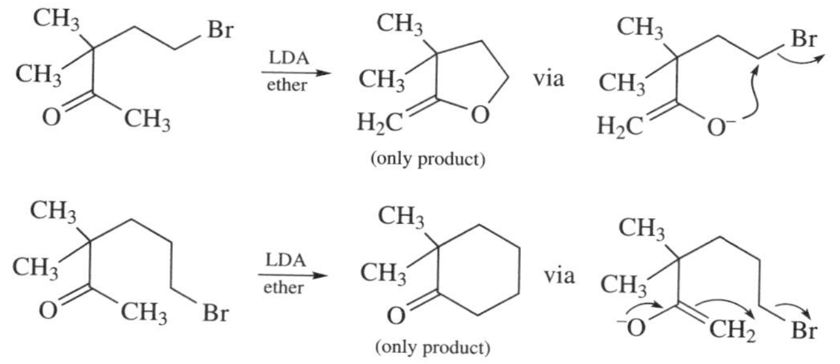

LDA ether (only product) via

LDA ether (only product) via

In order for C-alkylation to occur, the p orbital at the α carbon must be aligned with the C−Br bond in the linear geometry associated with the S_N2 transition state. When the ring to be closed is six-membered, this geometry is accessible, and cyclization to the cyclohexanone occurs. With five-membered rings, colinearity cannot be achieved easily. Cyclization at oxygen then occurs faster than does cyclopentanone formation. The transition state for O-alkylation involves an oxygen lone-pair orbital and is less strained than the transition state for C-alkylation.

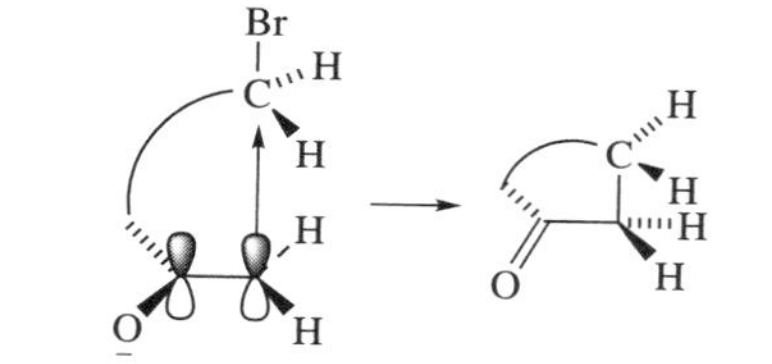

geometry required for intramolecular C-alklation of enolate

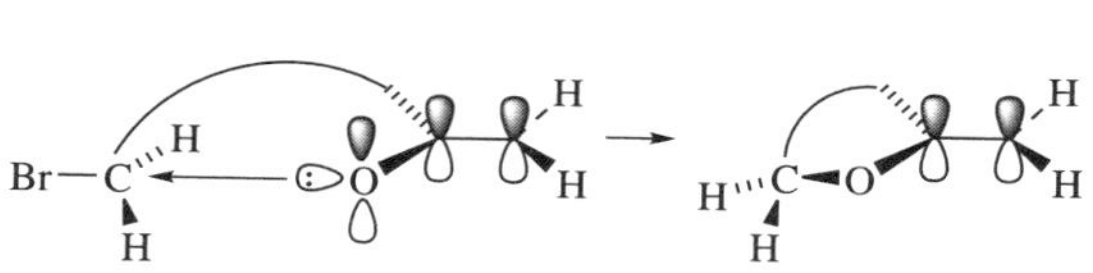

geometry required for intramolecular O-alklation of enolate

In enolates formed by proton abstraction from α,β-unsaturated ketones, there are three potential sites for attack by electrophiles: the oxygen, the α carbon, and the γ carbon. The kinetically preferred site for both protonation and alkylation is the α carbon.

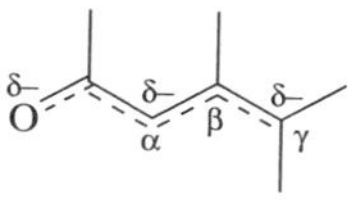

51. J. E. Baldwin and L. I. Kruse, *J. Chem. Soc., Chem. Commun.* **1977**:233.

Protonation of the enolate provides a method for converting α,β-unsaturated ketones and esters to the less stable β,γ-unsaturated isomers.

$$\xrightarrow[H_2O]{AcOH}$$

(major) (minor)

Ref. 52

$$CH_3CH{=}CHCO_2C_2H_5 \xrightarrow{LiNR_2} \xrightarrow{H_2O} \underset{87\%}{CH_2{=}CHCH_2CO_2C_2H_5} + \underset{13\%}{CH_3CH{=}CHCO_2C_2H_5}$$ Ref. 53

Alkylation also takes place selectively at the α carbon.[17] The selectivity for electrophilic attack at the α carbon presumably reflects a greater negative charge, as compared with the γ carbon.

$$(CH_3)_2C{=}CHCOCH_3 + CH_2{=}CHC(CH_3){=}CHCH_2Br \xrightarrow[NH_3]{NaNH_2} \underset{88\%}{CH_2{=}CHC(CH_3){=}CHCH_2{-}CH(C(CH_3){=}CH_2)COCH_3}$$

Phenoxide ions are a special case related to enolate anions but with a strong preference for O-alkylation because C-alkylation disrupts aromatic conjugation.

Phenoxides undergo O-alkylation in solvents such as DMSO, DMF, ethers, and alcohols. In water and trifluoroethanol, however, extensive C-alkylation occurs.[54] These latter solvents form particularly strong hydrogen bonds with the oxygen atom of the phenolate

52. J. H. Ringold and S. K. Malhotra, *Tetrahedron Lett.* **1962:**669; S. K. Malhotra and H. J. Ringold, *J. Am. Chem. Soc.* **85:**1538 (1963).
53. M. W. Rathke and D. Sullivan, *Tetrahedron Lett.* **1972:**4249.
54. N. Kornblum, P. J. Berrigan, and W. J. LeNoble, *J. Am. Chem. Soc.* **85:**1141 (1963); N. Kornblum, R. Seltzer, and P. Haberfield, *J. Am. Chem. Soc.* **85:**1148 (1963).

anion. This strong solvation decreases the reactivity at oxygen and favors C-alkylation.

$$\text{2-naphthoxide } (O^-) + PhCH_2Br \xrightarrow{DMF} \text{2-}OCH_2Ph\text{-naphthalene } (97\%)$$

$$\text{2-naphthoxide } (O^-) + PhCH_2Br \xrightarrow{CF_3CH_2OH} \text{1-}CH_2Ph\text{-2-}OH\text{-naphthalene } (85\%) + \text{2-}OCH_2Ph\text{-naphthalene } (7\%)$$

1.8. Alkylation of Aldehydes, Esters, Amides, and Nitriles

Among the compounds capable of forming enolates, the alkylation of ketones has been most widely studied and used synthetically. Similar reactions of esters, amides, and nitriles have also been developed. Alkylation of aldehyde enolates is not very common. One limitation is the fact that aldehydes are rapidly converted to aldol condensation products by base (see Chapter 2 for more discussion of this reaction). Only when the enolate can be rapidly and quantitatively formed is aldol condensation avoided. Success has been reported using potassium amide in liquid ammonia[55] and potassium hydride in THF. Alkylation via enamines or enamine anions provides a more general method for alkylation of aldehydes. These reactions will be discussed in Section 1.9.

$$(CH_3)_2CHCHO \xrightarrow[\text{2) } BrCH_2CH{=}C(CH_3)_2]{\text{1) KH, THF}} (CH_3)_2C(CHO)CH_2CH{=}(CH_3)_2 \quad 88\% \qquad \text{Ref. 56}$$

Alkylations of simple esters require a strong base because relatively weak bases such as alkoxides promote condensation reactions (see Chapter 2). The successful formation of ester enolates typically involves an amide base, usually LDA or potassium hexamethyldisilylamide (KHMDS) at low temperature.[57] The resulting enolates can be successfully alkylated with alkyl bromides or iodides. HMPA is sometimes added to accelerate the reaction. Some examples are given in Scheme 1.9.

Carboxylic acids can be directly alkylated by conversion to dianions by two equivalents of LDA. The dianions are alkylated at the α carbon as would be expected.[58]

$$(CH_3)_2CHCO_2H \xrightarrow{2LDA} (CH_3)_2C{=}C(O^-Li^+)_2 \xrightarrow[H^+]{CH_3(CH_2)_3Br} CH_3(CH_2)_3C(CH_3)_2{-}CO_2H \quad (80\%)$$

55. S. A. G. De Graaf, P. E. R. Oosterhof, and A. van der Gen, *Tetrahedron Lett.* **1974:**1653.
56. P. Groenewegen, H. Kallenberg, and A. van der Gen, *Tetrahedron Lett.* **1978:**491.
57. (a) M. W. Rathke and A. Lindert, *J. Am. Chem. Soc.* **93:**2318 (1971); (b) R. J. Cregge, J. L. Herrmann, C. S. Lee, J. E. Richman, and R. H. Schlessinger, *Tetrahedron Lett.* **1973:**2425; (c) J. L. Herrmann and R. H. Schlessinger, *J. Chem. Soc., Chem. Commun.* **1973:**711.
58. P. L. Creger, *J. Am. Chem. Soc.* **89:**2500 (1967); P. L. Creger, *Org. Synth.* **50:**58 (1970); P. L. Creger, *J. Org. Chem.* **37:**1907 (1972).

1[a] cyclohexyl–CO_2CH_3 → 1) LDA, THF, −70°C; 2) $CH_3(CH_2)_6I$, HMPA, 25°C → 1-CO_2CH_3-1-$(CH_2)_6CH_3$-cyclohexane ~90%

2[b] $CH_3(CH_2)_4CO_2C_2H_5$ → 1) cyclohexyl–$NCH(CH_3)_2$ Li^+, −78°C; 2) $CH_3CH_2CH_2CH_2Br$ → $CH_3(CH_2)_3CHCO_2C_2H_5$ with $CH_2CH_2CH_2CH_3$ 75%

3[c] H, CH_3, O, O, CH_3 → 1) LDA, DME; 2) $CH_2{=}CH(CH_2)_3Br$; 3) LDA, DME; 4) CH_3I → H, CH_3, O, O, H_3C, $(CH_2)_3CH{=}CH_2$, CH_3 86%

4[d] O, H, O, H → 1) LDA; 2) CH_3I, HMPA → CH_3, O, H, O, H 82%

5[e] HO, O, O → 1) 2 LDA, THF, −78°C; 2) 2 CH_3I, HMPA, −45°C → CH_3O, CH_3, O, O 65%

6[f] O, $PhCH_2$—N → 1) LDA; 2) $(CH_3)_2C{=}CH(CH_2)_4O_3SAr$ → O, $(CH_2)_4CH{=}C(CH_3)_2$, $PhCH_2$—N 83%

7[g] O, O, O, N, Ph, CH_3 → 1) LDA; 2) $PhCH_2Br$ → O, O, O, N, CH_2Ph, Ph, CH_3 78%

8[h] O, O, O, N → 1) NaHMDS; 2) CH_3I → O, O, O, N, CH_3 77%

Scheme 1.9. (*continued*)

9[i] 1) LDA, THF, HMPA 2) $Br(CH_2)_4OTMS$ — 83%

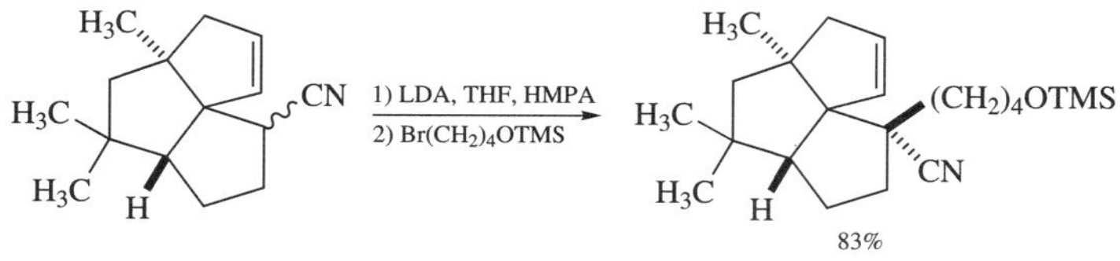

10[j] NaHMDS — 83%

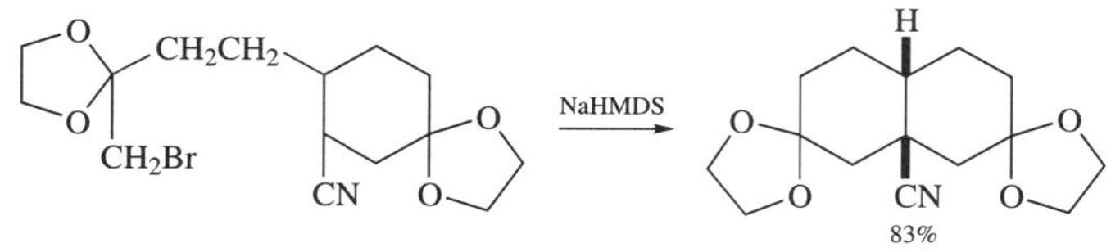

a. T. R. Williams and L. M. Sirvio, *J. Org. Chem.* **45:**5082 (1980).
b. M. W. Rathke and A. Lindert, *J. Am. Chem. Soc.* **93:**2320 (1971).
c. S. C. Welch, A. S. C. Prakasa Rao, C. G. Gibbs, and R. Y. Wong, *J. Org. Chem.* **45:**4077 (1980).
d. W. H. Pirkle and P. E. Adams, *J. Org. Chem.* **45:**4111 (1980).
e. H.-M. Shieh and G. D. Prestwich, *J. Org. Chem.* **46:**4319 (1981).
f. D. Kim, H. S. Kim, and J. Y. Yoo, *Tetrahedron Lett.* **32:**1577 (1991).
g. D. A. Evans, M. D. Ennis, and D. J. Mathre, *J. Am. Chem. Soc.* **104:**73 (1982).
h. A. Fadel, *Synlett.* **1992:**48.
i. L. A. Paquette, M. E. Okazaki, and J.-C. Caille, *J. Org. Chem.* **53:**477 (1988).

A method for enantioselective synthesis of carboxylic acid derivatives is based on alkylation of the enolates of *N*-acyl oxazolidinones.[59] The lithium enolates have the structures shown because of the tendency for the metal cation to form a chelate.

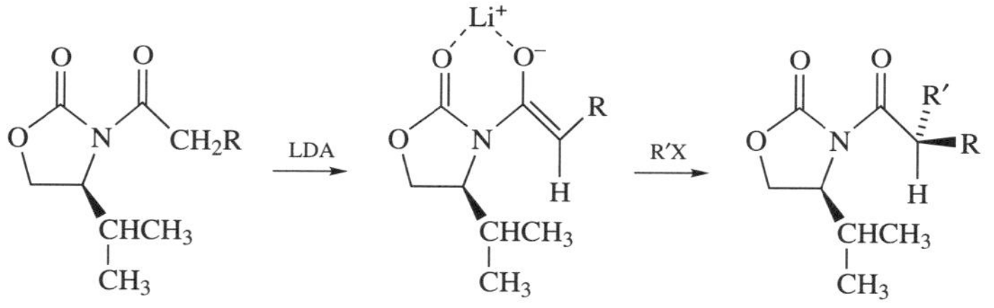

LDA → **4** → R′X

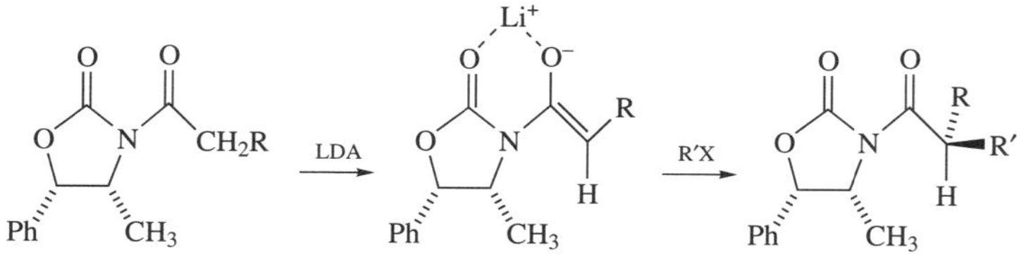

5

In **4** the upper face is shielded by the isopropyl group whereas in **5** the lower face is shielded by the methyl and phenyl groups. As a result, alkylation of the two derivatives gives products of the opposite configuration. Subsequent hydrolysis or alcoholysis provides acids or esters in enantiomerically enriched form. The initial alkylation product

59. D. A. Evans, M. D. Ennis, and D. J. Mathre, *J. Am. Chem. Soc.* **104:**1737 (1982); D. J. Ager, I. Prakash, and D. R. Schaad, *Chem. Rev.* **96:**835 (1996); D. J. Ager, I. Prakash, and D. R. Schaad, *Aldrichimica Acta* **30:**3 (1997).

ratios are typically 95 : 5 in favor of the major isomer. Because the intermediates are diastereomeric mixtures, they can be separated. The final products can then be obtained in >99% enantiomeric purity. Several other oxazolidinones have been developed for use as chiral auxiliaries. 5,5-Diaryl derivatives are quite promising.[60]

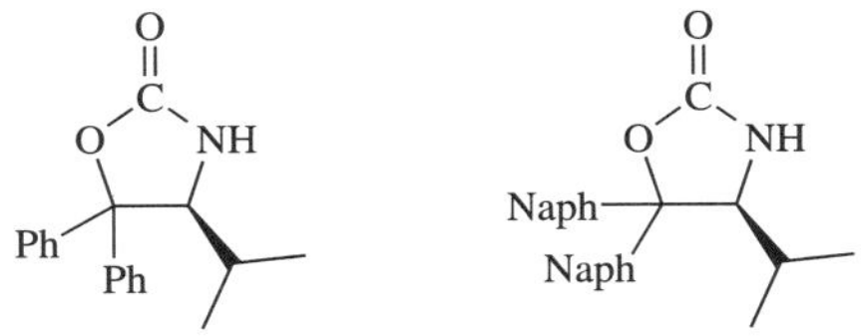

Acetonitrile ($pK_{DMSO} = 31.3$) can be deprotonated, provided a strong nonnucleophilic base such as LDA is used.

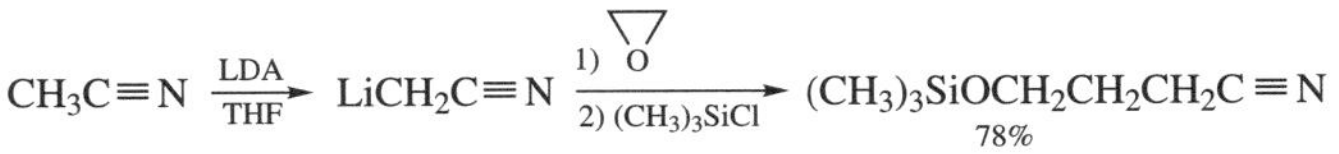

$$CH_3C{\equiv}N \xrightarrow[THF]{LDA} LiCH_2C{\equiv}N \xrightarrow[2)\ (CH_3)_3SiCl]{1)\ \text{ethylene oxide}} (CH_3)_3SiOCH_2CH_2CH_2C{\equiv}N \quad 78\%$$

Ref. 61

Phenylacetonitrile ($pK_{DMSO} = 21.9$) is considerably more acidic than acetonitrile. Deprotonation can be done with sodium amide. Dialkylation has been used in the synthesis of meperidine, an analgesic substance.[62]

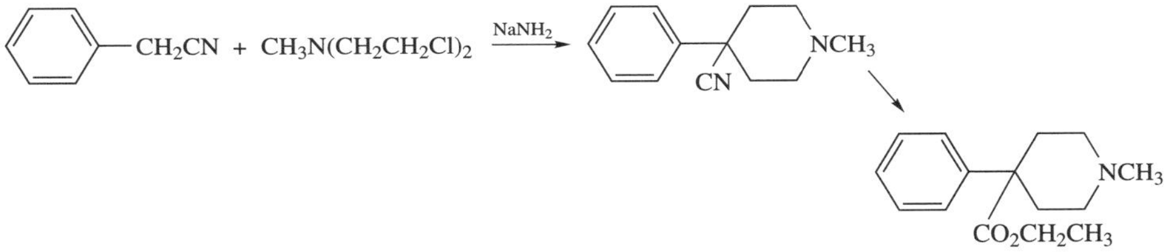

$$C_6H_5CH_2CN + CH_3N(CH_2CH_2Cl)_2 \xrightarrow{NaNH_2}$$

meperidene

1.9. The Nitrogen Analogs of Enols and Enolates—Enamines and Imine Anions

The nitrogen analogs of ketones and aldehydes are called *imines*, *azomethines*, or *Schiff bases*. *Imine* is the preferred name and will be used here. These compounds can be prepared by condensation of primary amines with ketones and aldehydes.[63]

$$R{-}\overset{O}{\overset{\|}{C}}{-}R + RNH_2 \rightarrow R{-}\overset{N-R}{\overset{\|}{C}}{-}R + H_2O$$

60. T. Hintermann and D. Seebach, *Helv. Chim. Acta* **81:**2093 (1998); C. L. Gibson, K. Gillon, and S. Cook, *Tetrahedron Lett.* **39:**6733 (1998).
61. S. Murata and I. Matsuda, *Synthesis* **1978:**221.
62. O. Eisleb, *Berichte* **74:**1433 (1941); cited in H. Kagi and K. Miescher, *Helv. Chim. Acta* **32:**2489 (1949).
63. For general reviews of imines and enamines see P. Y. Sollenberger and R. B. Martin, in *The Chemistry of the Amino Group*, S. Patai, ed., John Wiley & Sons, 1968, Chapter 7; G. Pitacco and E. Valentin, in *The Chemistry of Amino, Nitroso and Nitro Groups and Their Derivatives*, Part 1, S. Patai, ed., John Wiley & Sons, New York, 1982, Chapter 15; P. W. Hickmott, *Tetrahedron* **38:**3363 (1982); A. G. Cook, ed., *Enamines: Synthesis, Structure and Reactions*, Marcel Dekker, New York, 1988.

When secondary amines are heated with ketones or aldehydes in the presence of an acidic catalyst, a related condensation reaction occurs and can be driven to completion by removal of water by azetropic distillation or use of molecular sieves. The condensation product is a substituted vinylamine or *enamine*.

$$\mathrm{R\overset{O}{\overset{\|}{C}}CHR_2 + R'_2NH \rightleftharpoons R'_2N{-}\underset{R}{\underset{|}{\overset{OH}{\overset{|}{C}}}}{-}CHR_2 \overset{H^+}{\rightleftharpoons} R'_2\overset{+}{N}{=}\underset{R}{\underset{|}{C}}{-}CHR_2 \overset{-H^+}{\rightleftharpoons} R_2N{-}\underset{R}{\underset{|}{C}}{=}CR_2}$$

There are other methods for preparing enamines from ketones that utilize strong dehydrating reagents to drive the reaction to completion. For example, mixing carbonyl compounds and secondary amines followed by addition of titanium tetrachloride rapidly gives enamines. This method is especially applicable to hindered amines.[64] Triethoxysilane can also be used.[65] Another procedure involves converting the secondary amine to its *N*-trimethylsilyl derivative. Because of the higher affinity of silicon for oxygen than nitrogen, enamine formation is favored and takes place under mild conditions.[66]

$$\mathrm{(CH_3)_2CHCH_2CH{=}O + (CH_3)_3SiN(CH_3)_2 \longrightarrow (CH_3)_2CHCH{=}CHN(CH_3)_2}\ \ 88\%$$

The β-carbon atom of an enamine is a nucleophilic site because of conjugation with the nitrogen atom. Protonation of enamines takes place at the β carbon, giving an iminium ion.

$$\mathrm{R'_2N{-}\underset{R}{\underset{|}{C}}{=}CR_2 \longleftrightarrow R'_2\overset{+}{N}{=}\underset{R}{\underset{|}{C}}{-}\overset{-}{C}R_2 \qquad R'_2N{-}\underset{R}{\underset{|}{C}}{=}CR_2 \xrightarrow{H^+} R'_2\overset{+}{N}{=}\underset{R}{\underset{|}{C}}{-}CHR_2}$$

The nucleophilicity of the β-carbon atoms permits enamines to be used in synthetically useful alkylation reactions:

$$\mathrm{R'_2N{-}\underset{R}{\underset{|}{C}}{=}CR_2\ \ \underset{R''}{\underset{|}{CH_2}}{-}X \longrightarrow R'_2\overset{+}{N}{=}\underset{R}{\underset{|}{C}}{-}\underset{R}{\underset{|}{\overset{R}{\overset{|}{C}}}}{-}CH_2R'' \xrightarrow{H_2O} R\overset{O}{\overset{\|}{C}}{-}\underset{R}{\underset{|}{\overset{R}{\overset{|}{C}}}}{-}CH_2R''}$$

The enamines derived from cyclohexanones have been of particular interest. The pyrrolidine enamine is most frequently used for synthetic applications. In the enamine mixture formed from pyrrolidine and 2-methylcyclohexanone, structure **6** is predominant.[67] The tendency for the less substituted enamine to predominate is quite general. A steric effect is responsible for this preference. Conjugation between the nitrogen atom and the π orbitals of the double bond favors coplanarity of the bonds that are darkened in the structures. A serious nonbonded repulsion ($A^{1,3}$ strain) destabilizes isomer **7**. Furthermore, in isomer **6** the methyl group adopts a quasi-axial conformation to avoid steric interaction

64. W. A. White and H. Weingarten, *J. Org. Chem.* **32:**213 (1967); R. Carlson, R. Phan-Tan-Luu, D. Mathieu, F. S. Ahounde, A. Babadjamian, and J. Metzger, *Acta Chem. Scand.* **B32:**335 (1978); R. Carlson, A. Nilsson, and M. Stromqvist, *Acta Chem. Scand.* **B37:**7 (1983); R. Carlson and A. Nilsson, *Acta Chem. Scand.* **B38:**49 (1984); S. Schubert, P. Renaud, P.-A. Carrupt, and K. Schenk, *Helv. Chim. Acta* **76:**2473 (1993).
65. B. E. Love and J. Ren, *J. Org. Chem.* **58:**5556 (1993).
66. R. Comi, R. W. Franck, M. Reitano, and S. M. Weinreb, *Tetrahedron Lett.* **1973:**3107.
67. W. D. Gurowitz and M. A. Joseph, *J. Org. Chem.* **32:**3289 (1967).

with the amine substituents.[68]

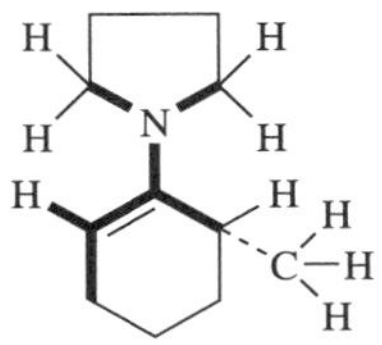

6 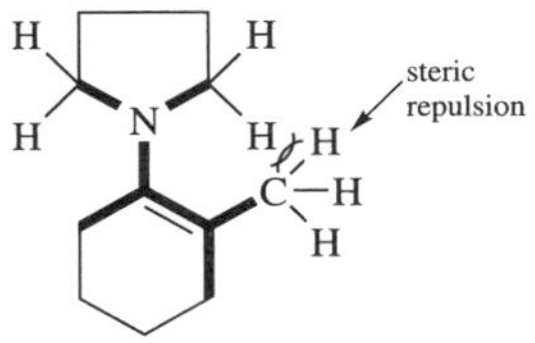 **7**

steric repulsion

Because of the predominance of the less substituted enamine, alkylations occur primarily at the less substituted α carbon. Synthetic advantage can be taken of this selectivity to prepare 2,6-disubstituted cyclohexanones. The iminium ions resulting from C-alkylation are hydrolyzed in the workup procedure.

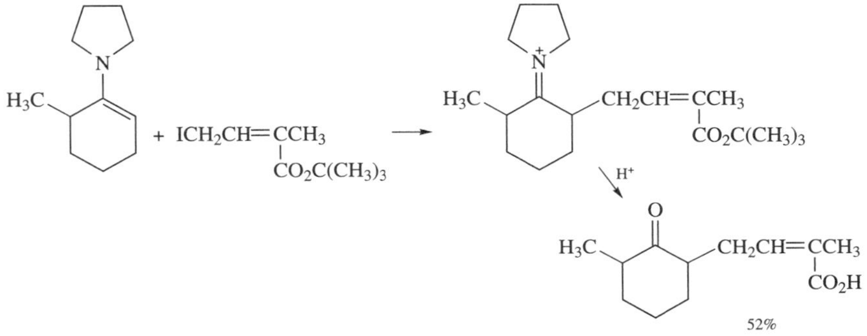

$+ \ ICH_2CH{=}C(CO_2C(CH_3)_3)CH_3 \longrightarrow$ iminium salt with $CH_2CH{=}C(CO_2C(CH_3)_3)CH_3$ $\xrightarrow{H^+}$ 2-methyl-6-($CH_2CH{=}C(CO_2H)CH_3$)cyclohexanone, 52% Ref. 69

Alkylation of enamines requires relatively reactive alkylating agents for good results. Methyl iodide, allylic and benzylic halides, α-haloesters, α-haloethers, and α-haloketones are the most successful alkylating agents. Some typical examples of enamine alkylation reactions are shown in Scheme 1.10.

Enamines also react with electrophilic alkenes. This aspect of their chemistry will be described in Section 1.10.

Imines can be deprotonated at the α carbon by strong bases to give the nitrogen analogs of enolates. Originally, Grignard reagents were used for deprotonation, but LDA is now commonly used. These anions are usually referred to as *imine anions* or *metalloenamines*.[70] Imine anions are isoelectronic and structurally analogous to both enolates and allyl anions and can also be called *azaallyl anions*.

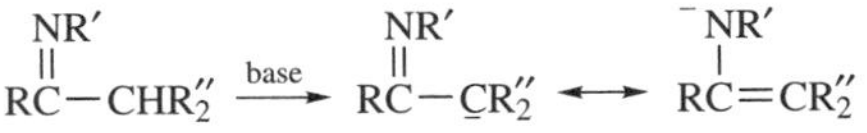

$$RC(=NR')-CHR''_2 \xrightarrow{\text{base}} RC(=NR')-\bar{C}R''_2 \longleftrightarrow RC(-\bar{N}R')=CR''_2$$

Spectroscopic investigations of the lithium derivatives of cyclohexanone *N*-phenylimine indicate that it exists as a dimer in toluene and that as a better donor solvent, THF, is added, equilibrium with a monomeric structure is established. The monomer is favored at high THF concentrations.[71] A crystal structure determination has been done on the

68. F. Johnson, L. G. Duquette, A. Whitehead, and L. C. Dorman, *Tetrahedron* **30:**3241 (1974); K. Muller, F. Previdoli, and H. Desilvestro, *Helv. Chim. Acta* **64:**2497 (1981); J. E. Anderson, D. Casarini, and L. Lunazzi, *Tetrahedron Lett.* **25:**3141 (1988).
69. P. L. Stotter and K. A. Hill, *J. Am. Chem. Soc.* **96:**6524 (1974).
70. For a general review of imine anions, see J. K. Whitesell and M. A. Whitesell, *Synthesis* **1983:**517.

Scheme 1.10. Enamine Alkylation

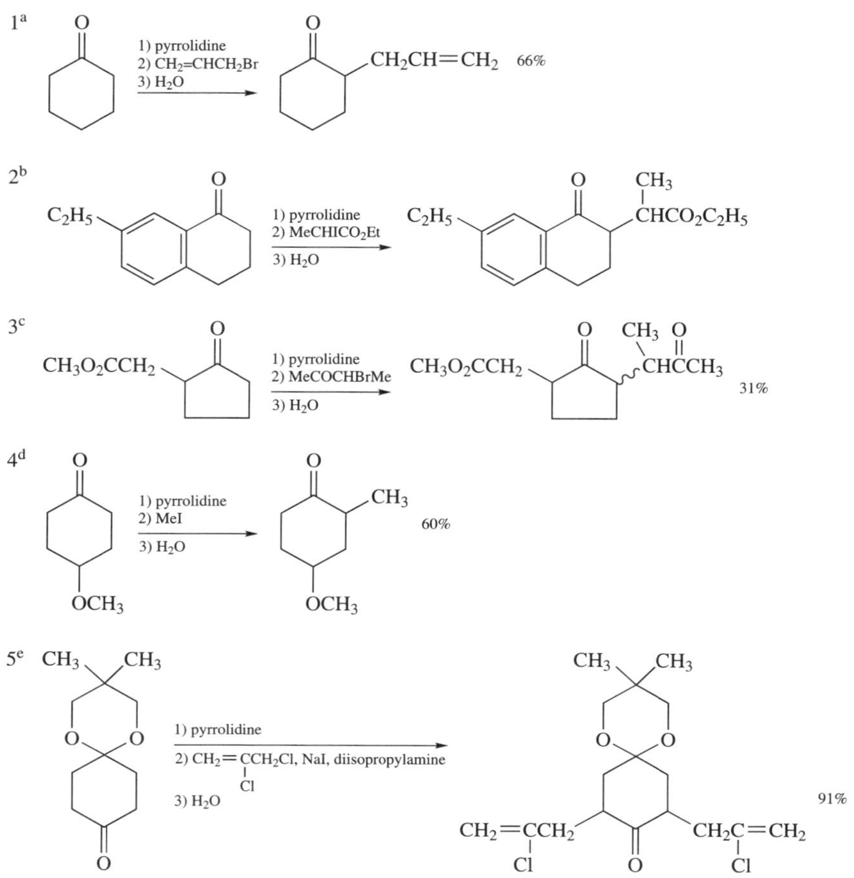

1[a] Cyclohexanone → 1) pyrrolidine 2) $CH_2{=}CHCH_2Br$ 3) H_2O → 2-($CH_2CH{=}CH_2$)cyclohexanone 66%

2[b] → 1) pyrrolidine 2) $MeCHICO_2Et$ 3) H_2O → product with C_2H_5, O, CH_3, $CHCO_2C_2H_5$

3[c] $CH_3O_2CCH_2$-cyclopentanone → 1) pyrrolidine 2) MeCOCHBrMe 3) H_2O → $CH_3O_2CCH_2$, O, CH_3 O, $CHCCH_3$ 31%

4[d] 4-OCH_3-cyclohexanone → 1) pyrrolidine 2) MeI 3) H_2O → 2-CH_3-4-OCH_3-cyclohexanone 60%

5[e] CH_3 CH_3 spiroketal → 1) pyrrolidine 2) $CH_2{=}C(Cl)CH_2Cl$, NaI, diisopropylamine 3) H_2O → CH_3 CH_3, O O, $CH_2{=}C(Cl)CH_2$, O, $CH_2C(Cl){=}CH_2$ 91%

a. G. Stork, A. Brizzolara, H. Landesman, J. Szmuszkovicz, and R. Terrell, *J. Am. Chem. Soc.* **85:**207 (1963).
b. D. M. Locke and S. W. Pelletier, *J. Am. Chem. Soc.* **80:**2588 (1958).
c. K. Sisido. S. Kurozumi, and K. Utimoto, *J. Am. Chem. Soc.* **34:**2661 (1969).
d. G. Stork and S. D. Darling, *J. Am. Chem. Soc.* **86:**1761 (1964).
e. J. A. Marshall and D. A. Flynn, *J. Org. Chem.* **44:**1391 (1979).

lithiated *N*-phenylimine of methyl *t*-butyl ketone. It is a dimeric structure with the lithium cation positioned above the nitrogen and closer to the phenyl ring than to the β carbon of the imine anion.[72] The structure is shown in Fig. 1.3.

Just as enamines are more nucleophilic than enols, imine anions are more nucleophilic than enolates and react efficiently with alkyl halides. One application of imine

71. N. Kallman and D. B. Collum, *J. Am. Chem. Soc.* **109:**7466 (1987).
72. H. Dietrich, W. Mahdi, and R. Knorr, *J. Am. Chem. Soc.* **108:**2462 (1986).

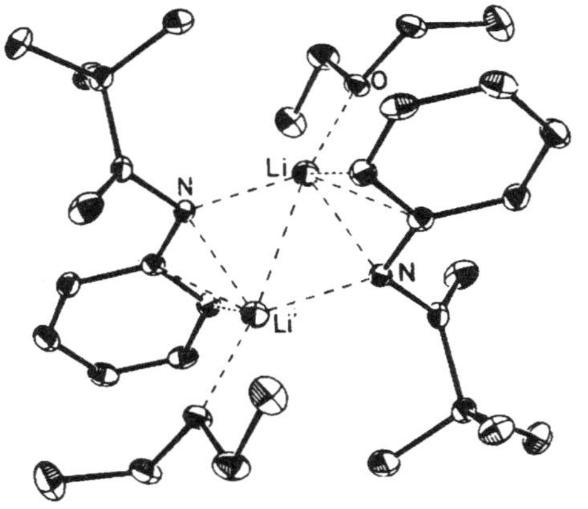

Fig. 1.3. Crystal structure of dimer of lithium derivative of *N*-phenyl imine of methyl *t*-butyl ketone. (Reproduced with permission from Ref. 72. Copyright 1986 American Chemical Society.)

anions is for the alkylation of aldehydes.

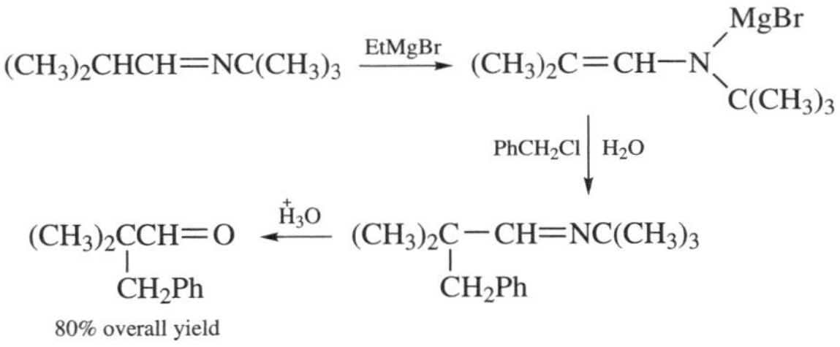

$$(CH_3)_2CHCH{=}NC(CH_3)_3 \xrightarrow{EtMgBr} (CH_3)_2C{=}CH{-}N(MgBr)C(CH_3)_3$$

$$\xrightarrow{PhCH_2Cl;\ H_2O} (CH_3)_2C(CH_2Ph){-}CH{=}NC(CH_3)_3 \xrightarrow{H_3\overset{+}{O}} (CH_3)_2C(CH_2Ph)CH{=}O$$

80% overall yield

Ref. 73

CH₃CH=CH—CH=N—(cyclohexyl) → 1) LDA; 2) 2-(2-iodoethyl)-2,5,5-trimethyl-1,3-dioxane (H₃C, ICH₂CH₂, CH₃, CH₃); 3) H₂O → CH₃CH=C(CH₂CH₂–dioxanyl)—CH=O

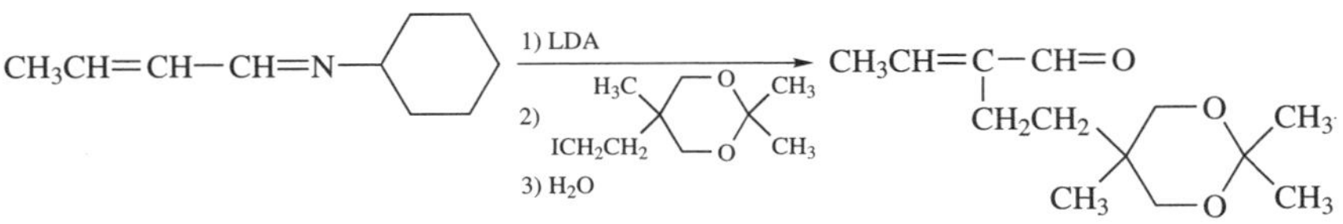

Ref. 74

Ketone imine anions can also be alkylated. The prediction of the regioselectivity of lithioenamine formation is somewhat more complex than for the case of kinetic ketone enolate formation. One of the complicating factors is that there are two imine stereoisomers, each of which can give rise to two regioisomeric imine anions. The isomers in

73. G. Stork and S. R. Dowd, *J. Am. Chem. Soc.* **85**:2178 (1963).
74. T. Kametani, Y. Suzuki, H. Furuyama, and T. Honda, *J. Org. Chem.* **48**:31 (1983).

which the nitrogen substituent R′ is *syn* to the double bond are the more stable.[75]

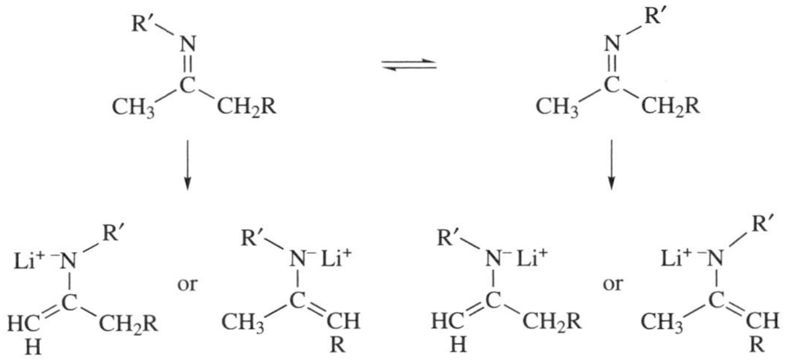

For methyl ketimines, good regiochemical control in favor of methyl deprotonation, regardless of imine stereochemistry, is observed using LDA at $-78°C$. With larger substituents, deprotonation at 25°C occurs *anti* to the nitrogen substituent.[76]

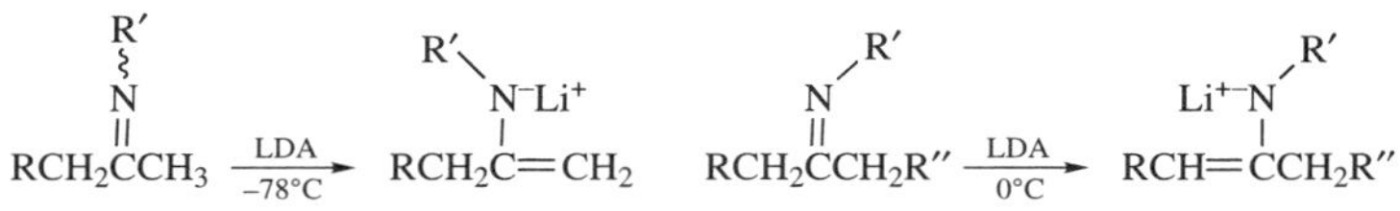

However, the *syn* and *anti* isomers of imines are easily thermally equilibrated. They cannot be prepared as single stereoisomers directly from ketones and amines so this method cannot be used to control regiochemistry of deprotonation. By allowing lithiated ketimines to come to room temperature, the thermodynamic composition is established. The most stable structures are those shown below, which in each case represent the less substituted isomer.

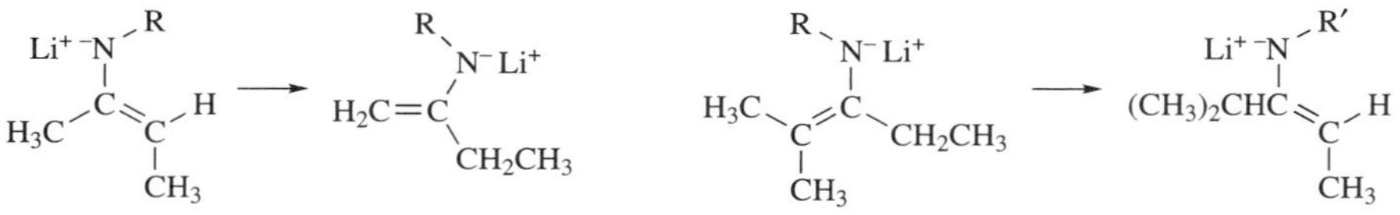

The complete interpretation of regiochemistry and stereochemistry of imine deprotonation also requires consideration of the state of aggregation and solvation of the base.[77]

One of the most useful aspects of the imine anions is that they can be readily prepared from enantiomerically pure amines. When imines derived from these amines are alkylated, the new carbon–carbon bond is formed with a bias for one of the two possible stereochemical configurations. Hydrolysis of the imine then leads to enantiomerically enriched ketone. Table 1.3 lists some reported examples.[78]

75. K. N. Houk, R. W. Stozier, N. G. Rondan, R. R. Frazier, and N. Chauqui-Ottermans, *J. Am. Chem. Soc.* **102:**1426 (1980).
76. J. K. Smith, M. Newcomb, D. E. Bergbreiter, D. R. Williams, and A. I. Meyer, *Tetrahedron Lett.* **24:**3559 (1983); J. K. Smith, D. E. Bergbreiter, and M. Newcomb, *J. Am. Chem. Soc.* **105:**4396 (1983); A. Hosomi, Y. Araki, and H. Sakurai, *J. Am. Chem. Soc.* **104:**2081 (1982).
77. M. P. Bernstein and D. B. Collum, *J. Am. Chem. Soc.* **115:**8008 (1993).
78. For a review, see D. E. Bergbreiter and M. Newcomb, in *Asymmetric Synthesis*, Vol. 2, J. D. Morrison, ed., Academic Press, New York, 1983, Chapter 9.

The interpretation and prediction of the relationship between the configuration of the newly formed chiral center and the configuration of the amine are usually based on steric differentiation of the two faces of the imine anion. Most imine anions that show high stereoselectivity incorporate a substituent which can hold the metal cation in a compact transition state by chelation. In the case of entry 2 in Table 1.3, for example, the observed enantioselectivity is rationalized on the basis of transition state **L**.

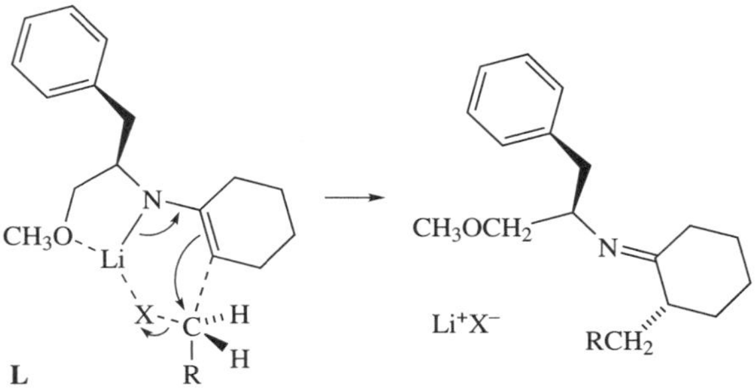

CH_3O Li N X C H H R **L** → CH_3OCH_2 N Li^+X^- RCH_2

The fundamental features of this transition state are (1) the chelation of the methoxy group with the lithium ion, which establishes a rigid transition state; (2) the interaction of the

Table 1.3. Enantioselective Alkylation of Ketimines

Amine	Ketone	Alkyl group	Yield	% E.E.	Reference
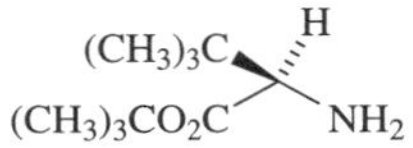 $(CH_3)_3C$, H, $(CH_3)_3CO_2C$, NH_2	Cyclohexanone	$CH_2{=}CHCH_2Br$	75	84	a
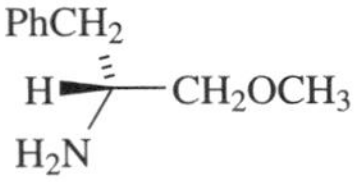 $PhCH_2$, H, CH_2OCH_3, H_2N	Cyclohexanone	$CH_2{=}CHCH_2Br$	80	>99	b
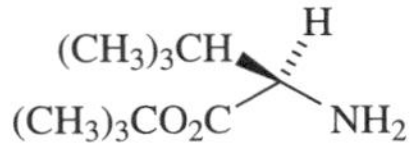 $(CH_3)_3CH$, H, $(CH_3)_3CO_2C$, NH_2	2-Carbomethoxy-cyclohexanone	CH_3I	57	>99	c
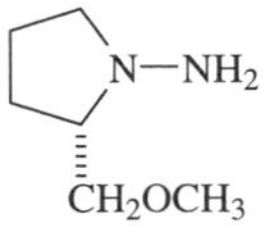 $N{-}NH_2$, CH_2OCH_3	3-pentanone	$CH_3CH_2CH_2I$	57	97	d
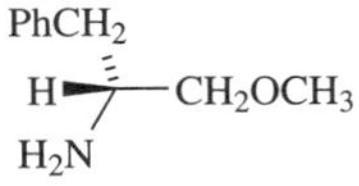 $PhCH_2$, H, CH_2OCH_3, H_2N	5-Nonanone	$CH_2{=}CHCH_2Br$	80	94	e

a. S. Hashimoto and K. Koga, *Tetrahedron Lett.* **1978**:573.
b. A. I. Meyers, D. R. Williams, G. W. Erickson, S. White, and M. Druelinger, *J. Am. Chem. Soc.* **103**:3081 (1981).
c. K. Tomioka, K. Ando, Y. Takemasa, and K. Koga, *J. Am. Chem. Soc.* **106**:2718 (1984).
d. D. Enders, H. Kipphardt, and P. Fey, *Org. Synth.* **65**:183 (1987).
e. A. I. Meyers, D. R. Williams, S. White, and G. W. Erickson, *J. Am. Chem. Soc.* **103**:3088 (1981).

lithium ion with the bromide leaving group; and (3) the steric effect of the benzyl group, which makes the underside the preferred direction of approach for the alkylating agent.

Hydrazones can also be deprotonated to give lithium salts which are reactive toward alkylation at the β carbon. Hydrazones are more stable than alkylimines and therefore have some advantages in synthesis.[79] The *N,N*-dimethylhydrazones of methyl ketones are kinetically deprotonated at the methyl group. This regioselectivity is independent of the stereochemistry of the hydrazone.[80] Two successive alkylations of the *N,N*-dimethylhydrazone of acetone can provide unsymmetrical ketones.

$CH_3C(=NN(CH_3)_2)CH_3$ —1) *n*-BuLi, 0°C; 2) $C_5H_{11}I$→ $CH_3(CH_2)_5C(=NN(CH_3)_2)CH_3$ —1) *n*-BuLi, −5°C; 2) $BrCH_2CH{=}CH_2$; 3) H^+, H_2O→ $CH_3(CH_2)_5C(=O)CH_2CH_2CH{=}CH_2$ Ref. 81

The anion of cyclohexanone *N,N*-dimethylhydrazone shows a strong preference for axial alkylation.[82] 2-Methylcyclohexanone *N,N*-dimethylhydrazone is alkylated by methyl iodide to give *cis*-2,6-dimethylcyclohexanone. The methyl group in the hydrazone occupies a pseudoaxial orientation. Alkylation apparently is preferred *anti* to the lithium cation, which is on the face opposite the 2-methyl substituent.

N, $N(CH_3)_2$, CH_3 —LDA→ N, Li, $N(CH_3)_2$, CH_3 —CH_3I→ N, $N(CH_3)_2$, H_3C, CH_3 —H_2O→ CH_3, H_3C, O

Chiral hydrazones have also been developed for enantioselective alkylation of ketones. The hydrazones can be converted to the lithium salt, alkylated, and then hydrolyzed to give alkylated ketone in good chemical yield and with high enantioselectivity[83] (see entry 4 in Table 1.3).

Hydrazones are substantially more stable toward hydrolysis than imines or enamines. Several procedures have been developed for conversion of the hydrazones back to ketones.[81–83] Mild conditions are particularly important when stereochemical configuration must be maintained at the enolizable position adjacent to the carbonyl group.

A procedure for enantioselective synthesis of carboxylic acids is based on sequential alkylation of the oxazoline **8** via its lithium salt. Chelation by the methoxy group leads preferentially to the transition state in which the lithium is located as shown. The lithium acts as a Lewis acid in directing the approach of the alkyl halide. This is reinforced by a steric effect from the phenyl substituent. As a result, alkylation occurs predominantly from the lower face of the anion. The sequence in which the groups R and R′ are introduced

79. E. J. Corey and D. Enders, *Tetrahedron Lett.* **1976:**3.
80. D. E. Bergbreiter and M. Newcomb, *Tetrahedron Lett.* **1979:**4145; M. E. Jung and T. J. Shaw, *Tetrahedron Lett.* **1979:**4149.
81. M. Yamashita, K. Matsumiya, M. Tanabe, and R. Suetmitsu, *Bull. Chem. Soc. Jpn.* **58:**407 (1985).
82. D. B. Collum, D. Kahne, S. A. Gut, R. T. DePue, F. Mohamadi, R. A. Wanat, J. Clardy, and G. VanDuyne, *J. Am. Chem. Soc.* **106:**4865 (1984).
83. D. Enders, H. Eichenauer, U. Bas, H. Schubert, and K. A. M. Kremer, *Tetrahedron* **40:**1345 (1984); D. Enders, H. Kipphardt, and P. Fey, *Org. Synth.* **65:**183 (1987); D. Enders and M. Klatt, *Synthesis* **1996:**1403.

determines the chirality of the product. The enantiomeric purity of disubstituted acetic acids obtained after hydrolysis is in the range of 70–90%.[84]

H_3C–(oxazoline, Ph, CH_2OCH_3) **8** $\xrightarrow{\text{1) } LiNR_2 \text{; 2) } R–X}$ RCH_2–(oxazoline, Ph, CH_2OCH_3) $\xrightarrow{LiNR_2}$ (R, H, Li, CH_2, O, CH_3 chelate) $\xrightarrow{R'\text{-}X}$ H, R, R'–(oxazoline, Ph, CH_2OCH_3)

1.10. Alkylation of Carbon Nucleophiles by Conjugate Addition

The previous sections have dealt primarily with reactions in which the new carbon–carbon bond is formed by an S_N2 reaction between the nucleophilic carbanions and the alkylating reagent. Another important method for alkylation of carbon involves the addition of a nucleophilic carbon species to an electrophilic multiple bond. The electrophilic reaction partner is typically an α,β-unsaturated ketone, aldehyde, or ester, but other electron-withdrawing substituents such as nitro, cyano, or sulfonyl also activate carbon–carbon double and triple bonds to nucleophilic attack. The reaction is called *conjugate addition* or the *Michael reaction*. Other kinds of nucleophiles such as amines, alkoxides, and sulfide anions also react similarly, but we will focus on the carbon–carbon bond-forming reactions.

In contrast to the reaction of an enolate anion with an alkyl halide, which requires one equivalent of base, conjugate addition of enolates can be carried out with a catalytic amount of base. All the steps are reversible.

$$RC(=O)CHR_2 + B^- \rightleftharpoons RC(O^-)=CR_2 + BH$$

$$RC(O^-)=CR_2 + \;>C=C(X)< \;\rightleftharpoons RC(=O)-CR_2-C-C^-(X)$$

$$RC(=O)-CR_2-C-C^-(X) + BH \rightleftharpoons RC(=O)-CR_2-C-C(X)-H + B^-$$

When a catalytic amount of base is used, the reaction proceeds with thermodynamic control of enolate formation. The most effective nucleophiles under these conditions are carbanions derived from relatively acidic compounds such as β-ketoesters or malonate esters. The adduct anions are more basic and are protonated under the reaction conditions. Scheme 1.11 provides some examples.

84. A. I. Meyers, G. Knaus, K. Kamata, and M. E. Ford, *J. Am. Chem. Soc.* **98**:567 (1976).

Scheme 1.11. Alkylation of Carbon by Conjugate Addition

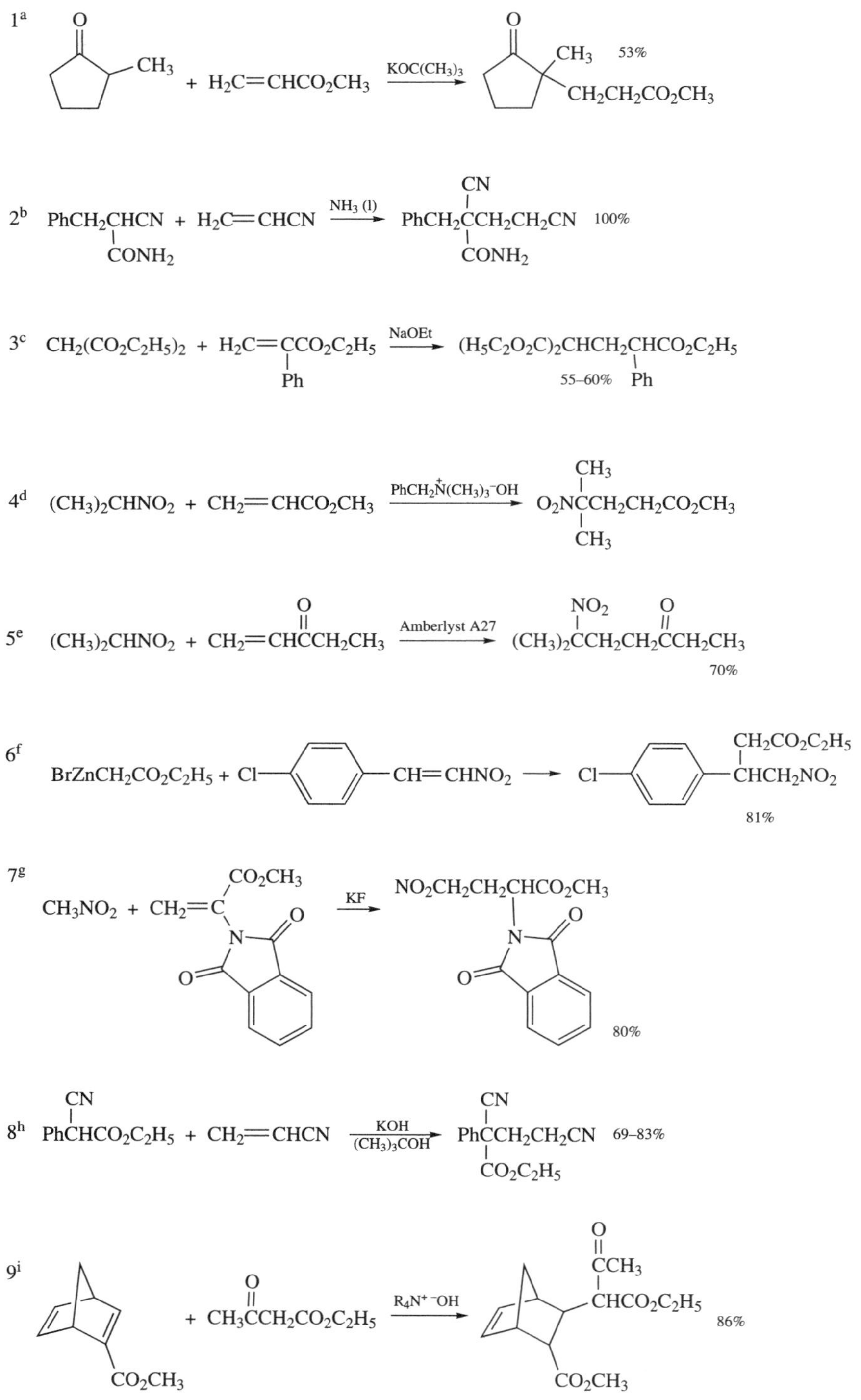

1[a] 2-methylcyclopentanone + $H_2C{=}CHCO_2CH_3$ —$KOC(CH_3)_3$→ 2-methyl-2-($CH_2CH_2CO_2CH_3$)cyclopentanone 53%

2[b] $PhCH_2CH(CONH_2)CN$ + $H_2C{=}CHCN$ —NH_3 (l)→ $PhCH_2C(CN)(CONH_2)CH_2CH_2CN$ 100%

3[c] $CH_2(CO_2C_2H_5)_2$ + $H_2C{=}C(Ph)CO_2C_2H_5$ —NaOEt→ $(H_5C_2O_2C)_2CHCH_2CH(Ph)CO_2C_2H_5$ 55–60%

4[d] $(CH_3)_2CHNO_2$ + $CH_2{=}CHCO_2CH_3$ —$PhCH_2\overset{+}{N}(CH_3)_3\,^{-}OH$→ $O_2NC(CH_3)_2CH_2CH_2CO_2CH_3$

5[e] $(CH_3)_2CHNO_2$ + $CH_2{=}CHC(O)CH_2CH_3$ —Amberlyst A27→ $(CH_3)_2C(NO_2)CH_2CH_2C(O)CH_2CH_3$ 70%

6[f] $BrZnCH_2CO_2C_2H_5$ + $Cl{-}C_6H_4{-}CH{=}CHNO_2$ → $Cl{-}C_6H_4{-}CH(CH_2CO_2C_2H_5)CH_2NO_2$ 81%

7[g] CH_3NO_2 + $CH_2{=}C(CO_2CH_3)$(N-phthalimido) —KF→ $NO_2CH_2CH_2CH(CO_2CH_3)$(N-phthalimido) 80%

8[h] $PhCH(CN)CO_2C_2H_5$ + $CH_2{=}CHCN$ —KOH, $(CH_3)_3COH$→ $PhC(CN)(CO_2C_2H_5)CH_2CH_2CN$ 69–83%

9[i] methyl norbornadiene-2-carboxylate (CO_2CH_3) + $CH_3C(O)CH_2CO_2C_2H_5$ —$R_4N^+\,^{-}OH$→ norbornene bearing $CH(C(O)CH_3)CO_2C_2H_5$ and CO_2CH_3 86%

10[j] 2-methyl-5-isopropyl-1-pyrrolidinylcyclohexene + $CH_2{=}CHCCH_3$ (C=O) $\xrightarrow[\text{2) NaOAc, HOAc, } H_2O \text{ reflux}]{\text{1) dioxane, 16 h}}$ 2-methyl-5-isopropyl-6-($CH_2CH_2CCH_3$, C=O)cyclohexanone, 66%

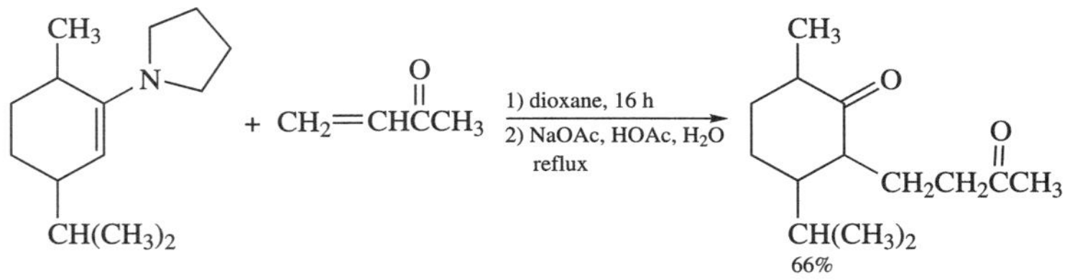

a. H. O. House, W. L. Roelofs, and B. M. Trost, *J. Org. Chem.* **31:**646 (1966).
b. S. Wakamatsu, *J. Org. Chem.* **27:**1285 (1962).
c. E. M. Kaiser, C. L. Mao, C. F. Hauser, and C. R. Hauser, *J. Org. Chem.* **35:**410 (1970).
d. R. B. Moffett, *Org. Synth.* **IV:**652 (1963).
e. R. Ballini, P. Marziali, and A. Mozziacafreddo, *J. Org. Chem.* **61:**3209 (1996).
f. R. Menicagli and S. Samaritani, *Tetrahedron* **52:**1425 (1996).
g. M. J. Crossley, Y. M. Fung, J. J. Potter, and A. W. Stamford, *J. Chem. Soc., Perkin Trans 1* **1998:**1113.
h. E. C. Horning and A. F. Finelli, *Org. Synth.* **IV:**776 (1963).
i. K. Alder, H. Wirtz, and H. Koppelberg, *Justus Liebigs Ann. Chem.* **601:**138 (1956).

The fluoride ion is an effective catalyst for Michael additions involving relatively acidic carbon compounds.[85] The reactions can be done in the presence of excess fluoride, where the formation of the $[F{-}H{-}F^-]$ ion occurs, or by use of a tetraalkylammonium fluoride in an aprotic solvent.

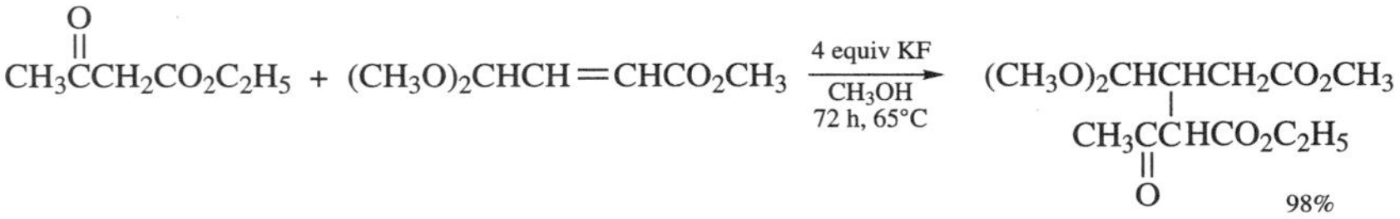

$$CH_3\overset{O}{\overset{\|}{C}}CH_2CO_2C_2H_5 + (CH_3O)_2CHCH{=}CHCO_2CH_3 \xrightarrow[\text{72 h, 65°C}]{\text{4 equiv KF, } CH_3OH} (CH_3O)_2CHCHCH_2CO_2CH_3 \text{ with } CH_3\overset{O}{\overset{\|}{C}}CHCO_2C_2H_5 \quad 98\%$$ Ref. 86

$$(CH_3)_2CHNO_2 + CH_2{=}CHCOCH_3 \xrightarrow[\text{2 h, 25°C}]{\text{0.5 equiv } R_4N^+F^-} O_2NC(CH_3)_2CH_2CH_2\overset{O}{\overset{\|}{C}}CH_3 \quad 95\%$$ Ref. 87

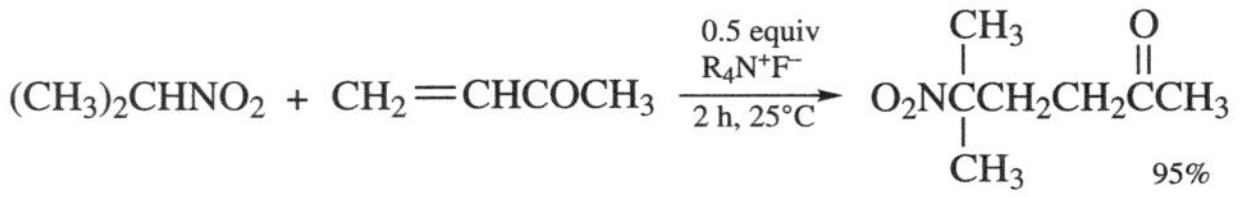

Fluoride ion can also induce reaction of silyl enol ethers with electrophilic alkenes.

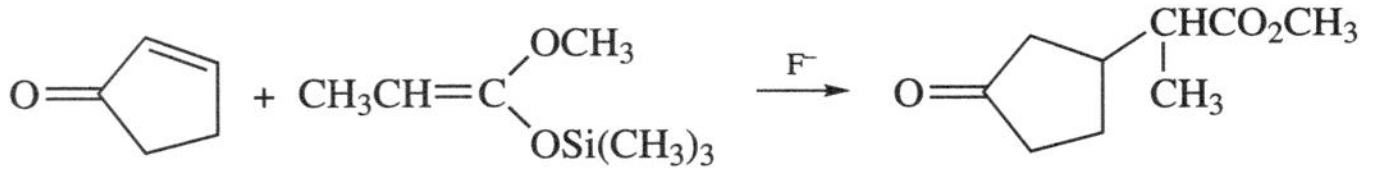

$$\text{cyclopent-2-enone} + CH_3CH{=}C(OCH_3)(OSi(CH_3)_3) \xrightarrow{F^-} \text{3-}(CH(CH_3)CO_2CH_3)\text{cyclopentanone}$$ Ref. 88

The hindered aluminum tris(2,6-diphenylphenoxide) is an effective promoter of Michael additions of enolates to enones.[89]

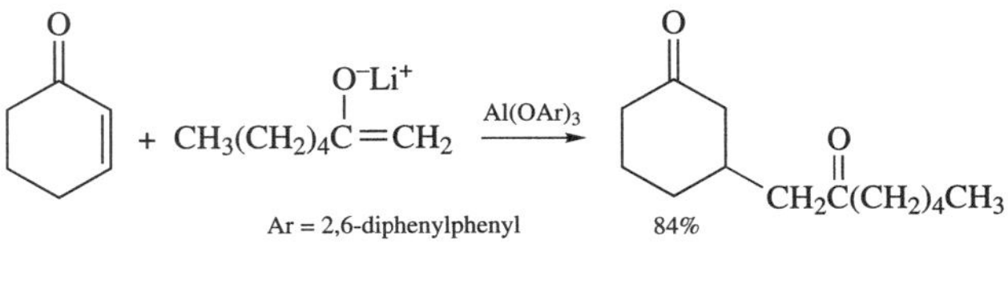

$$\text{cyclohex-2-enone} + CH_3(CH_2)_4C(O^-Li^+){=}CH_2 \xrightarrow{Al(OAr)_3} \text{3-}(CH_2\overset{O}{\overset{\|}{C}}(CH_2)_4CH_3)\text{cyclohexanone} \quad 84\%$$

Ar = 2,6-diphenylphenyl

85. J. H. Clark, *Chem. Rev.* **80:**429 (1980).
86. S. Tori, H. Tanaka, and Y. Kobayashi, *J. Org. Chem.* **42:**3473 (1977).
87. J. H. Clark, J. M. Miller, and K.-H. So, *J. Chem. Soc., Perkin Trans. 1* **1978:**941.
88. T. V. Rajan Babu, *J. Org. Chem.* **49:**2083 (1984).
89. S. Saito, I. Shimada, Y. Takamori, M. Tanaka, K. Maruoka, and H. Yamamoto, *Bull. Chem. Soc. Jpn.* **70:**1671 (1997).

Conjugate addition can also be carried out by completely forming the nucleophilic enolate under kinetic conditions. Ketone enolates formed by reaction with LDA in THF react with enones to give 1,5-diketones (entries 1 and 2, Scheme 1.12). Esters of 1,5-dicarboxylic acids are obtained by addition of ester enolates to α,β-unsaturated esters (entry 5, Scheme 1.12).

Among Michael acceptors that have been demonstrated to react with ketone and ester enolates under kinetic conditions are methyl α-trimethylsilylvinyl ketone[90] methyl α-methylthioacrylate,[91] methyl methylthiovinyl sulfoxide,[92] and ethyl α-cyanoacrylate.[93] The latter class of acceptors has been shown to be capable of generating contiguous quanternary carbon centers.

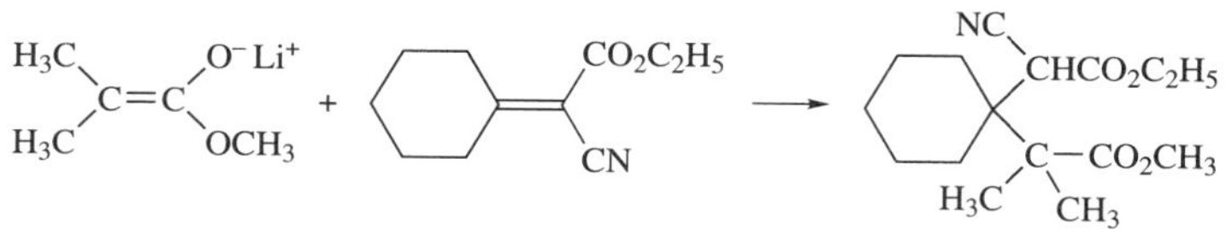

Ref. 93

Several other examples of conjugate addition of carbanions carried out under kinetically controlled conditions are given in Scheme 1.12.

There have been several studies of the stereochemistry of conjugate addition reactions. If there are substituents on both the nucleophilic enolate and the acceptor, either *syn* or *anti* adducts can be formed.

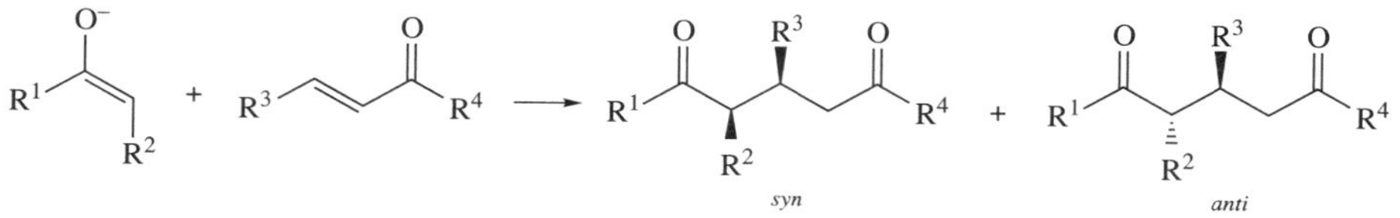

syn　　*anti*

The reaction shows a dependence on the *E*- or *Z*-stereochemistry of the enolate. *Z*-Enolates favor *anti* adducts and *E*-enolates favor *syn* adducts. These tendencies can be understood in terms of a chelated transition state.[94]

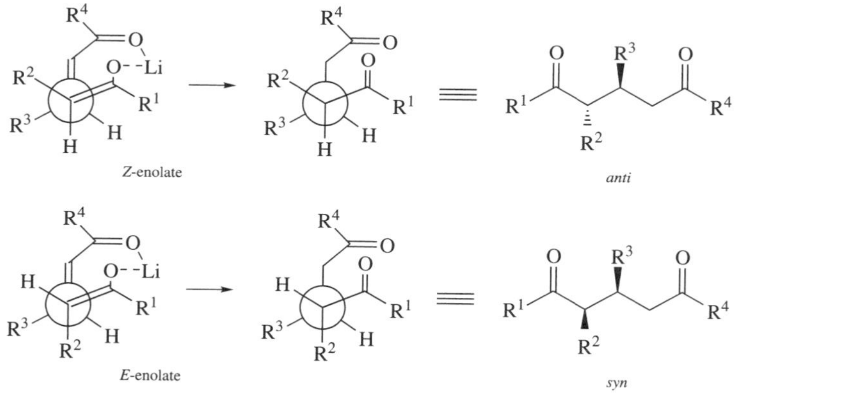

Z-enolate　　*anti*

E-enolate　　*syn*

90. G. Stork and B. Ganem, *J. Am. Chem. Soc.* **95:**6152 (1973).
91. R. J. Cregge, J. L. Herrmann, and R. H. Schlessinger, *Tetrahedron Lett.* **1973:**2603.
92. J. L. Hermann, G. R. Kieczykowski, R. F. Romanet, P. J. Wepple, and R. H. Schlessinger, *Tetrahedron Lett.* **1973:**4711.
93. R. A. Holton, A. D. Williams, and R. M. Kennedy, *J. Org. Chem.* **51:**5480 (1986).
94. D. Oare and C. H. Heathcock, *J. Org. Chem.* **55:**157 (1990); D. A. Oare and C. H. Heathcock, *Top. Stereochem.* **19:**227 (1989).

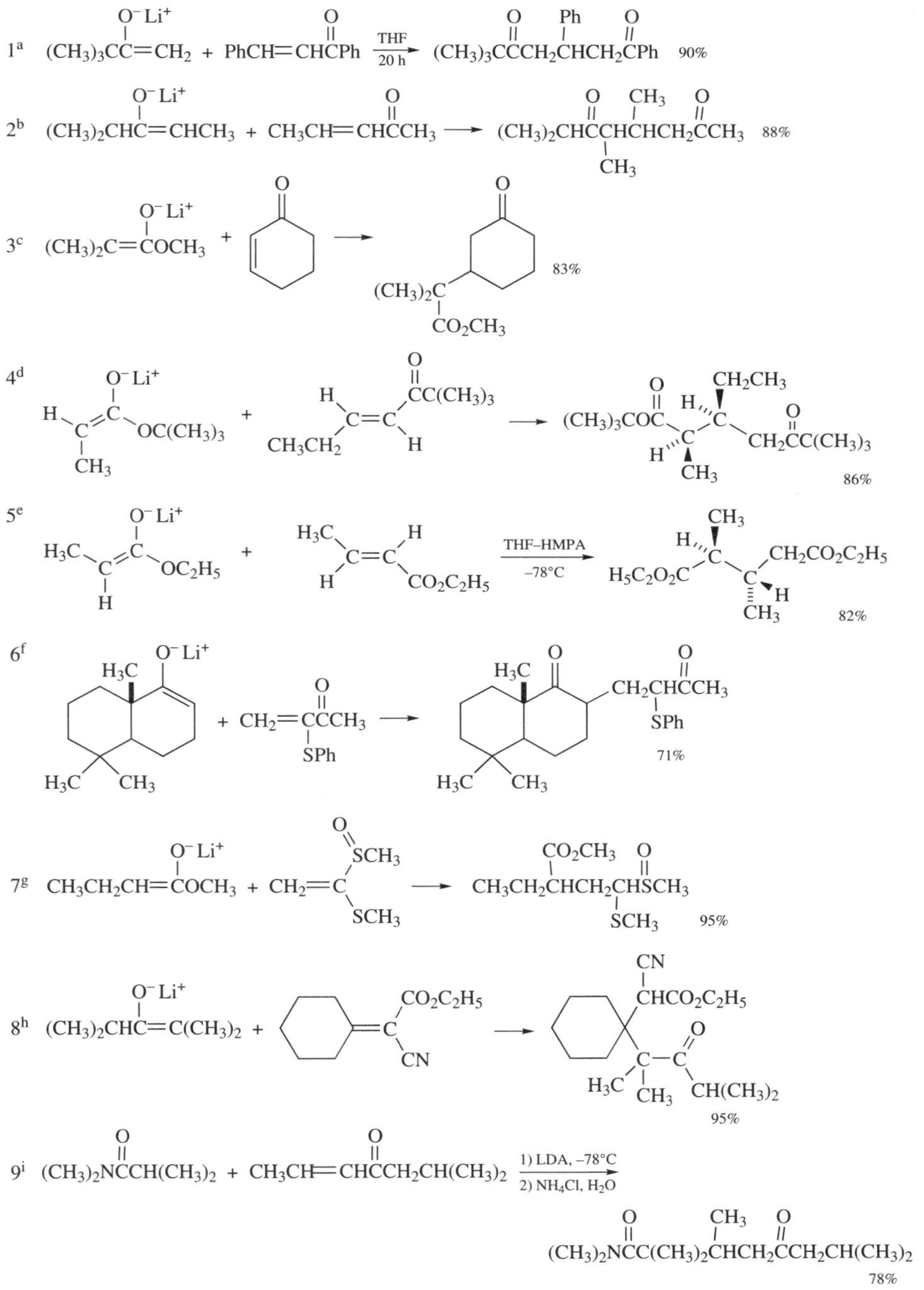

Scheme 1.12. Michael Additions under Kinetic Conditions

1[a] $(CH_3)_3C(O^-Li^+)C{=}CH_2$ + $PhCH{=}CHC(O)Ph$ $\xrightarrow[20\ h]{THF}$ $(CH_3)_3CC(O)CH_2CH(Ph)CH_2C(O)Ph$ 90%

2[b] $(CH_3)_2CHC(O^-Li^+){=}CHCH_3$ + $CH_3CH{=}CHC(O)CH_3$ $\longrightarrow$ $(CH_3)_2CHC(O)CH(CH_3)CH(CH_3)CH_2C(O)CH_3$ 88%

3[c] $(CH_3)_2C{=}C(O^-Li^+)OCH_3$ + cyclohexenone $\longrightarrow$ 3-[$(CH_3)_2C(CO_2CH_3)$]cyclohexanone 83%

4[d] $H(CH_3)C{=}C(O^-Li^+)OC(CH_3)_3$ + $CH_3CH_2CH{=}CHC(O)C(CH_3)_3$ $\longrightarrow$ $(CH_3)_3COC(O)CH(CH_3)CH(CH_2CH_3)CH_2C(O)C(CH_3)_3$ 86%

5[e] $H_3C(H)C{=}C(O^-Li^+)OC_2H_5$ + $H_3CCH{=}CHCO_2C_2H_5$ $\xrightarrow[-78°C]{THF-HMPA}$ $H_5C_2O_2CCH(CH_3)CH(CH_3)CH_2CO_2C_2H_5$ 82%

6[f] lithium enolate of H_3C, H_3C, CH_3-substituted octalone + $CH_2{=}C(SPh)C(O)CH_3$ $\longrightarrow$ ketone with $CH_2CH(SPh)C(O)CH_3$ side chain 71%

7[g] $CH_3CH_2CH{=}C(O^-Li^+)OCH_3$ + $CH_2{=}C(S(O)CH_3)SCH_3$ $\longrightarrow$ $CH_3CH_2CH(CO_2CH_3)CH_2CH(SCH_3)S(O)CH_3$ 95%

8[h] $(CH_3)_2CHC(O^-Li^+){=}C(CH_3)_2$ + cyclohexylidene$C(CO_2C_2H_5)CN$ $\longrightarrow$ 1-[$CH(CN)CO_2C_2H_5$]-1-[$C(CH_3)_2C(O)CH(CH_3)_2$]cyclohexane 95%

9[i] $(CH_3)_2NC(O)CH(CH_3)_2$ + $CH_3CH{=}CHC(O)CH_2CH(CH_3)_2$ $\xrightarrow[2)\ NH_4Cl,\ H_2O]{1)\ LDA,\ -78°C}$ $(CH_3)_2NC(O)C(CH_3)_2CH(CH_3)CH_2C(O)CH_2CH(CH_3)_2$ 78%

a. J. Bertrand, L. Gorrichon, and P. Maroni, *Tetrahedron* **40:**4127 (1984).
b. D. A. Oare and C. H. Heathcock, *Tetrahedron Lett.* **27:**6169 (1986).
c. A. G. Schultz and Y. K. Yee, *J. Org. Chem.* **41:**4044 (1976).
d. C. H. Heathcock and D. A. Oare, *J. Org. Chem.* **50:**3022 (1985).
e. M. Yamaguchi, M. Tsukamoto, S. Tanaka, and I. Hirao, *Tetrahedron Lett.* **25:**5661 (1984).
f. K. Takaki, M. Ohsugi, M. Okada, M. Yasumura, and K. Negoro, *J. Chem. Soc., Perkin Trans. 1* **1984:**741.
g. J. L. Herrmann, G. R. Kieczykowski, R. F. Romanet, P. J. Wepplo, R. H. Schlessinger, *Tetrahedron Lett.* **1973:**4711.
h. R. A. Holton, A. D. Williams, and R. M. Kennedy, *J. Org. Chem.* **51:**5480 (1986).
i. D. A. Oare, M. A. Henderson, M. A. Sanner, and C. H. Heathcock, *J. Org. Chem.* **55:**132 (1990).

The stereoselectivity can be enhanced by addition of $Ti(O\text{-}i\text{-}Pr)_4$. The active nucleophile under these conditions is expected to be an "ate" complex in which the much larger $Ti(O\text{-}i\text{-}Pr)_3$ group replaces Li^+.[95] Here too, the *syn* : *anti* ratio depends on the stereochemistry of the enolate.

O^-Li^+ / R, CH_3 —$Ti(O\text{-}i\text{-}Pr)_4$→ $OTi(O\text{-}i\text{-}Pr)_3$ / R, CH_3 —$R^4CH{=}CHCR^1$ (C=O)→ *anti* + *syn*

R	enolate	R^1	R^4	*anti* : *syn*	yield (%)
Et	*Z*	*t*-Bu	Ph	95 : 5	69
Ph	*Z*	Me	Ph	> 97 : 3	70
Ph	*Z*	*t*-Bu	Ph	> 92 : 8	85
i-Pr	*Z*	*t*-Bu	Ph	> 97 : 3	65
i-Pr	*E*	*t*-Bu	Ph	17 : 83	91

When the conjugate addition is carried out under kinetic conditions with stoichiometric formation of the enolate, the adduct is also an enolate until the reaction mixture is quenched with a proton source. It should therefore be possible to effect a second reaction of the enolate if an electrophile is added prior to protonation of the enolate. This can be done by adding an alkyl halide to the solution of the adduct enolate, which results in an alkylation. Two or more successive reactions conducted in this way are referred to as *tandem reactions*.

$H_3C{=}C(O^-Li^+)OC(CH_3)_3$ + $CH_3CH{=}CHCO_2C_2H_5$ —−78°C→ $(CH_3)_3CO_2CCH_2CH(CH_3)CH{=}C(O^-Li^+)OC_2H_5$ —CH_3I, HMPA→ $(CH_3)_3CO_2CCH_2CH(CH_3)CH(CH_3)CO_2C_2H_5$ 60% Ref. 96

$H_3C{=}C(CH_3){-}CH{=}C(O^-Li^+)OC_2H_5$ + cyclopentenone → O^- enolate with $CH(CO_2C_2H_5){-}C(CH_3){=}CH_2$ —$H_2C{=}CHCH_2Br$→ 2-($CH_2CH{=}CH_2$)-3-[$CH(CO_2C_2H_5){-}C(CH_3){=}CH_2$]cyclopentanone Ref. 97

95. A. Bernardi, P. Dotti, G. Poli, and C. Scolastico, *Tetrahedron* **48:**5597 (1992); A. Bernardi, *Gazz. Chim. Ital.* **125:**539 (1995).
96. M. Yamaguchi, M. Tsukamoto, and I. Hirao, *Tetrahedron Lett.* *26:*1723 (1985).
97. W. Oppolzer, R. P. Heloud, G. Bernardinelli, and K. Baettig, *Tetrahedron Lett.* **24:**4975 (1983).

78% Ref. 98

Tandem conjugate addition–alkylation has proven to be an efficient means of introducing both α and β substituents at enones.[99]

Conditions for effecting conjugate addition in the presence of Lewis acids have also been developed. Trimethylsilyl enol ethers can be caused to react with electrophilic alkenes by use of $TiCl_4$. These reactions proceed rapidly even at $-78°C$.[100]

$$PhCCH{=}C(CH_3)_2 + CH_2{=}C(OSi(CH_3)_3)Ph \xrightarrow{TiCl_4} PhCCH_2C(CH_3)_2CH_2CPh$$ 72–78% Ref. 101

Similarly, titanium tetrachloride or stannic tetrachloride induces addition of silyl enol ethers to nitroalkenes. The initial adduct is trapped in cyclic form by trimethylsilylation.[102] Hydrolysis of this intermediate regenerates the carbonyl group.[103]

Other Lewis acids can also effect conjugate addition of silyl enol ethers to electrophilic alkenes. For example, $Mg(ClO_4)_2$ catalyzes addition of ketene silyl acetals:

$$(TMSO)(CH_3O)C{=}C(CH_3)_2 + H_2C{=}CHCCH_3 \xrightarrow{Mg(ClO_4)_2} CH_3O_2CC(CH_3)_2CH_2CH_2CCH_3$$ Ref. 104

Lanthanide salts have been found to catalyze addition of α-nitroesters, even in aqueous solution.[105]

$$O_2NCH(CH_3)CO_2CH_3 + H_2C{=}CHCCH_3 \xrightarrow[10\ mol\ \%]{Yb(O_3SCF_3)_3} CH_3O_2CC(CH_3)(NO_2)CH_2CH_2CCH_3$$ 99% Ref. 106

98. C. H. Heathcock, M. M. Hansen, R. B. Ruggeri, and J. C. Kath, *J. Org. Chem.* **57**:2545 (1992).
99. For additional examples, see M. C. Chapdelaine and M. Hulce, *Org. React.* **38**:225 (1990).
100. K. Narasaka, K. Soai, Y. Aikawa, and T. Mukaiyama, *Bull. Chem. Soc. Jpn.* **49**:779 (1976).
101. K. Narasaka, *Org. Synth.* **65**:12 (1987).
102. A. F. Mateos and J. A. de la Fuento Blanco, *J. Org. Chem.* **55**:1349 (1990).
103. M. Miyashita, T. Yanami, T. Kumazawa, and A. Yoshikoshi, *J. Am. Chem. Soc.* **106**:2149 (1984).
104. S. Fukuzumi, T. Okamoto, K. Yasui, T. Suenobu, S. Itoh, and J. Otera, *Chem. Lett.* **1997**:667.
105. J. B. N. F. Engberts, B. L. Feringa, E. Keller, and S. Otto, *Rec. Trav. Chim. Pays-Bas* **115**:457 (1996).
106. E. Keller and B. L. Feringa *Synlett.* **1997**:842.

Cyanide ion acts as a carbon nucleophile in the conjugate addition reaction. An alcoholic solution of potassium or sodium cyanide is suitable for simple enones.

KCN, NH_4Cl, EtOH—H_2O; 12%; 42% Ref. 107

Triethylaluminum–hydrogen cyanide and diethylaluminum cyanide are also useful reagents for conjugate addition of cyanide. The latter is the more reactive of the two reagents. These reactions presumably involve the coordination of the aluminum reagent as a Lewis acid at the carbonyl oxygen.

Et_3Al—HCN; 92–93% Ref. 108

$(C_2H_5)_2AlCN$ Ref. 109

Diethylaluminum cyanide mediates conjugate addition of cyanide to α,β-unsaturated oxazolines. With a chiral oxazoline, 30–50% diastereomeric excess (d.e.) can be achieved. Hydrolysis gives partially resolved α-substituted succinic acids.

Et_2AlCN, ^-CN; HCl, H_2O Ref. 110

R = CH_3, Ph

R = CH_3, d.e. = 50–56%; e.e. = 45–50%
R = Ph, d.e. = 45–52%; e.e. = 57%

Enamines also react with electrophilic alkenes to give conjugate addition products. The addition reactions of enamines of cyclohexanones show a strong preference for attack from the axial direction.[111] This is anticipated on stereoelectronic grounds because the π

107. O. R. Rodig and N. J. Johnston, *J. Org. Chem.* **34:**1942 (1969).
108. W. Nagata and M. Yoshioka, *Org. Synth.* **52:**100 (1972).
109. W. Nagata, M. Yoshioka, and S. Hirai, *J. Am. Chem. Soc.* **94:**4635 (1972).
110. M. Dahuron and N. Langlois, *Synlett.* **1996:**51.

orbital of the enamine is the site of nucleophilicity.

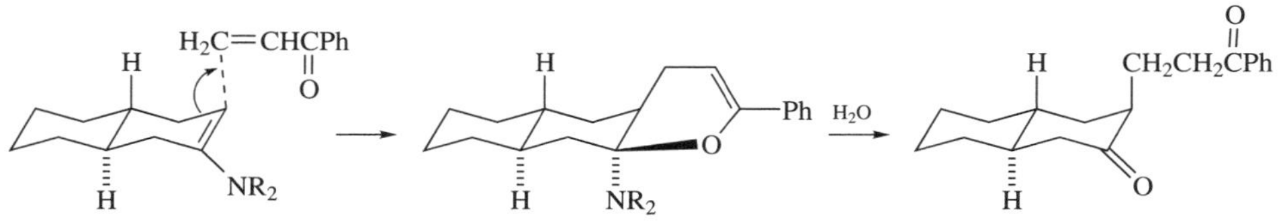

Another very important method for adding a carbon chain at the β-carbon of α,β-unsaturated carbonyl system involves organometallic reagents, particularly organocopper intermediates. This reaction will be discussed in Chapter 8.

General References

D. E. Bergbreiter and M. Newcomb, in *Asymmetric Synthesis*, J. D. Morrison, ed., Academic Press, New York, 1983, Chapter 9.
D. Caine, in *Carbon–Carbon Bond Formation*, Vol. 1, R. L. Augustine, ed., Marcel Dekker, New York, 1979, Chapter 2.
A. G. Cook, ed., *Enamines: Synthesis, Structure and Reactions*, 2nd ed., Marcel Dekker, New York, 1988.
H. O. House, *Modern Synthetic Reactions*, 2nd ed., W. A. Benjamin, Menlo Park, California, 1972, Chapter 9.
P. Perlmutter, *Conjugate Addition Reactions in Organic Synthesis*, Pergamon Press, New York, 1992.
V. Snieckus, ed., *Advances in Carbanion Chemistry*, Vol. 1, JAI Press, Greenwich, Connecticut, 1992.
J. C. Stowell, *Carbanions in Organic Synthesis*, Wiley-Interscience, New York, 1979.

Problems

(*References for these problems will be found on page 923.*)

1. Arrange each series of compounds in order of decreasing acidity:

(a) $CH_3CH_2NO_2$, $(CH_3)_2CH\overset{O}{C}Ph$, CH_3CH_2CN, $CH_2(CN)_2$

(b) $[(CH_3)_2CH]_2NH$, $(CH_3)_2CHOH$, $(CH_3)_2CH_2$, $(CH_3)_2CHPh$

(c) $CH_3\overset{O}{C}CH_2CO_2CH_3$, $CH_3\overset{O}{C}CH_2\overset{O}{C}CH_3$, $CH_3O\overset{O}{C}CH_2Ph$, $CH_3\overset{O}{C}OCH_2Ph$

(d) $Ph\overset{O}{C}CH_2Ph$, $(CH_3)_3C\overset{O}{C}CH_3$, $(CH_3)_3C\overset{O}{C}CH(CH_3)_2$, $Ph\overset{O}{C}CH_2CH_2CH_3$

111. E. Valentin, G. Pitacco, F. P. Colonna, and A. Risalti, *Tetrahedron* **30:**2741 (1974); M. Forchiassin, A. Risalti, C. Russo, M. Calligaris, and G. Pitacco, *J. Chem. Soc.* **1974:**660.

2. Write the structures of all possible enolates for each ketone. Indicate which you would expect to be favored in a kinetically controlled deprotonation. Which would you expect to be the most stable enolate in each case?

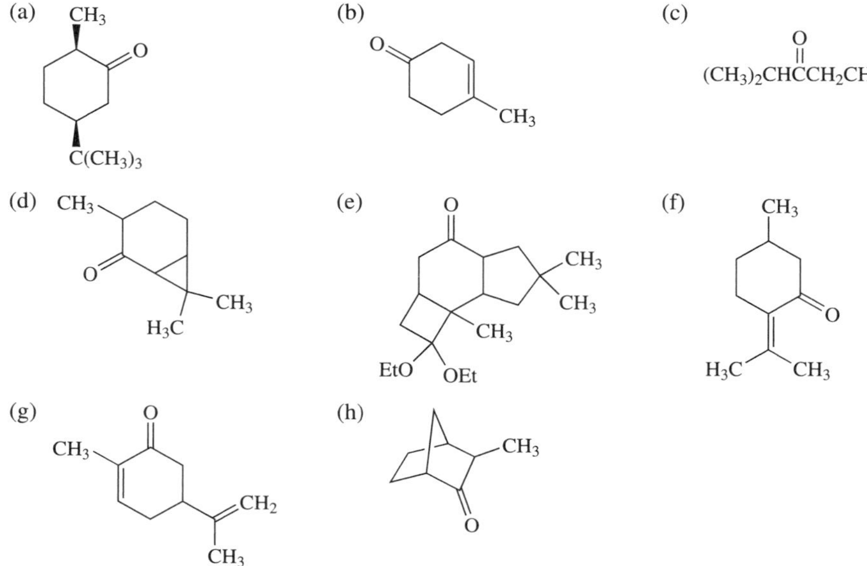

3. Suggest reagents and reaction conditions suitable for effecting each of the following conversions:

(a) 2-methylcyclohexanone to 2-benzyl-6-methylcyclohexanone.

(b) to

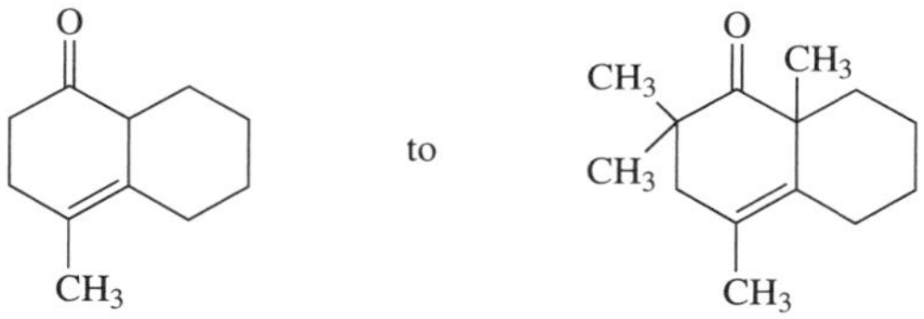

(c) to

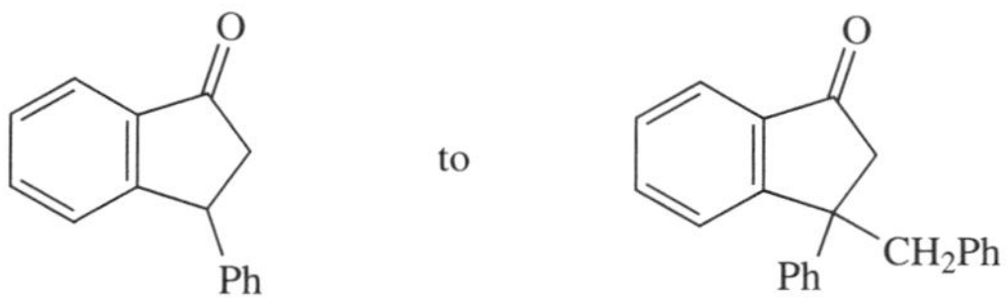

(d) to

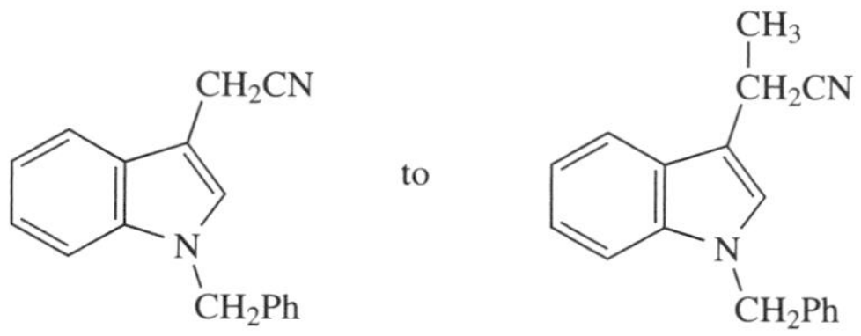

(e) 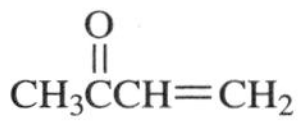 $CH_3C(=O)CH{=}CH_2$ to $CH_2{=}C(OSi(CH_3)_3){-}CH{=}CH_2$

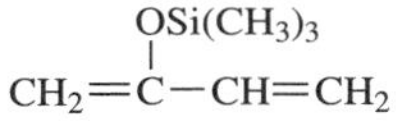

(f) 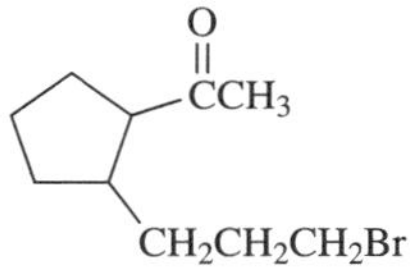 CCH_3 (O) / $CH_2CH_2CH_2Br$ to O

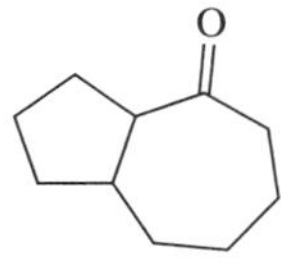

(g) 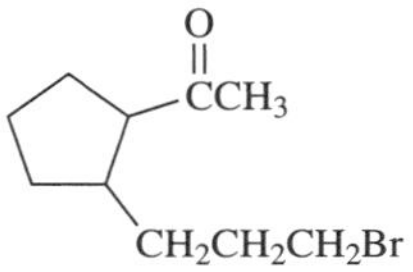 CCH_3 (O) / $CH_2CH_2CH_2Br$ to O=C–CH_3

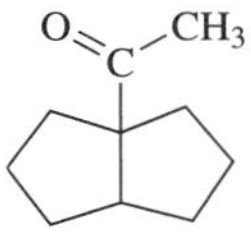

4. Intramolecular alkylation of enolates has been used to advantage in synthesis of bi- and tricyclic compounds. Indicate how such a procedure could be used to synthesize each of the following molecules by drawing the structure of a suitable precursor.

(a)  O

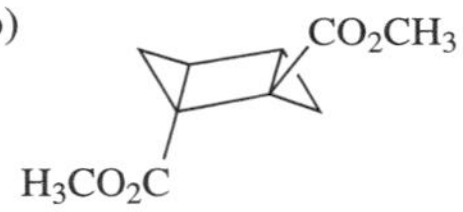

(c) 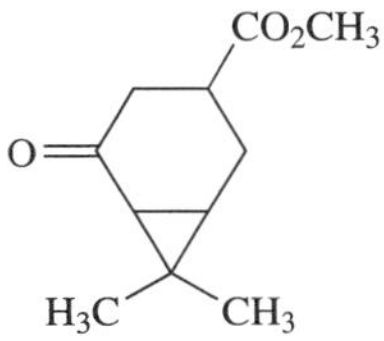 CO_2CH_3, O=, H_3C, CH_3

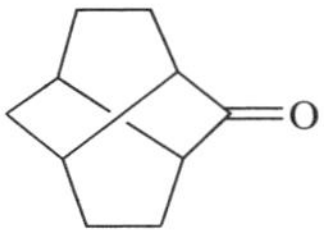

(e) 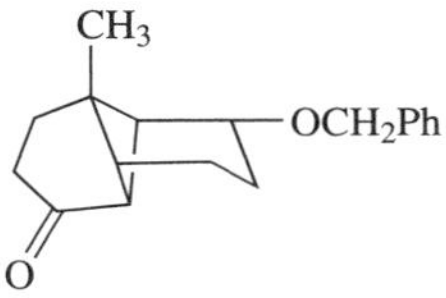 CH_3, OCH_2Ph, O

(b) CO_2CH_3, H_3CO_2C

(d) =O

(f) 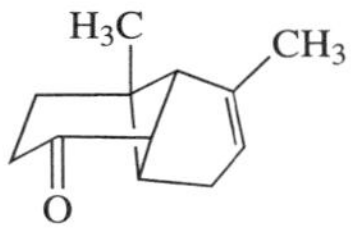 H_3C, CH_3, O

5. Predict the major product of each of the following reactions:

(a) 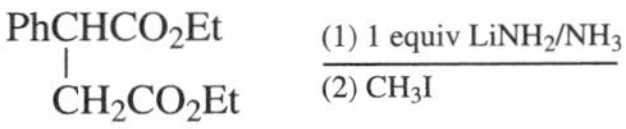 $PhCHCO_2Et$ / CH_2CO_2Et $\xrightarrow[\text{(2) } CH_3I]{\text{(1) 1 equiv } LiNH_2/NH_3}$

(b) 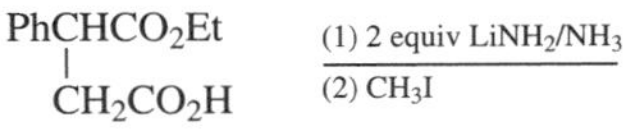 $PhCHCO_2Et$ / CH_2CO_2H $\xrightarrow[\text{(2) } CH_3I]{\text{(1) 2 equiv } LiNH_2/NH_3}$

(c) 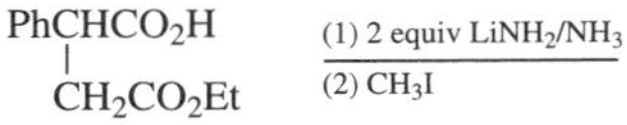 $PhCHCO_2H$ / CH_2CO_2Et $\xrightarrow[\text{(2) } CH_3I]{\text{(1) 2 equiv } LiNH_2/NH_3}$

6. Treatment of 2,3,3-triphenylpropionitrile with 1 equiv of potassium amide in liquid ammonia followed by addition of benzyl chloride affords 2-benzyl-2,3,3-triphenylpropionitrile in 97% yield. Use of 2 equiv of potassium amide gives an 80% yield of 2,3,3,4-tetraphenylbutyronitrile under the same reaction conditions. Explain.

7. Suggest readily available starting materials and reaction conditions suitable for obtaining each of the following compounds by a procedure involving alkylation of

nucleophilic carbon.

(a) $PhCH_2CH_2CH(CN)Ph$

(b) $(CH_3)_2C{=}CHCH_2CH_2C(=O)CH_2CO_2CH_3$

(c) 5,5-dimethyl-1,3-dioxocyclohexane-2-CH_2CO_2H (ring with two CH_3 groups, two C=O, and CH_2CO_2H)

(d) $CH_2{=}CHCH{=}CHCH_2CH_2CO_2H$

(e) 2,3-diphenylpropanoic acid

(f) 2,6-diallylcyclohexanone

(g) 7-CH_3O tetralone bearing CH_3CH_2 and $CH_2CH{=}CH_2$ on the carbon α to C=O

(h) $H_2C{=}CHCH_2C(CN)(Ph)C(=O)NH_2$

(i) 2-(CH_3CO)-2-($CH_3C(=O)CH_2CH_2$)cyclohexane-1,3-dione

(j) $CH_2{=}CHCH(CO_2CH_2CH_3)CH_2C{\equiv}CH$

8. Suggest starting materials and reaction conditions suitable for obtaining each of the following compounds by a procedure involving a Michael reaction.

(a) 4,4-dimethyl-5-nitropentan-2-one
(b) diethyl 2,3-diphenylglutarate
(c) ethyl 2-benzoyl-4-(2-pyridyl)butyrate
(d) 2-phenyl-3-oxocyclohexaneacetic acid

(e) cyclohexanone with $NCCH_2$ and CH_2CH_2CN at the 2- and 6-positions

(f) cycloheptanone with $CH_2C(=O)CH_3$ at the 2-position

(g) $CH_3CH_2CH(NO_2)CH_2CH_2C(=O)CH_3$

(h) $(CH_3)_2CHCH(CH{=}O)CH_2CH_2CO_2CH_2CH_3$

(i) cyclohexanone with $CH(Ph)CH_2NO_2$ at the 2-position

(k) Ph-benzylidene acetal pyranoside bearing OCH_3, NO_2, and $CH(NO_2)CH_3$ substituents

(j) 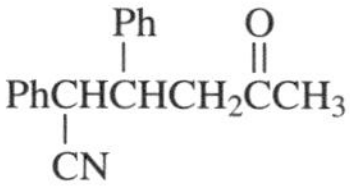 Ph, O; $PhCHCHCH_2CCH_3$ (with Ph on C-2, O on C=O), CN

(l) 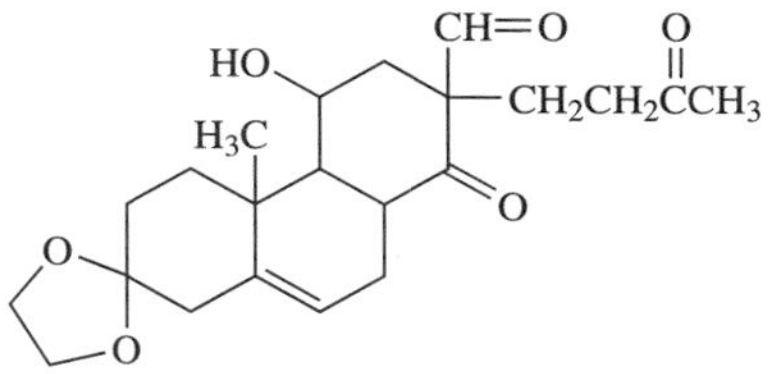 HO, H_3C, CH=O, O, $CH_2CH_2CCH_3$, O, O, O

9. In planning a synthesis, the most effective approach is to reason backwards from the target molecule to some readily available starting material. This is called *retrosynthetic analysis* and is indicated by an open arrow of the type shown below. In each of the following problems, the target molecule is shown on the left and the starting material on the right. Determine how you could prepare the target molecule from the indicated starting material using any necessary organic or inorganic reagents. In some cases, more than one step is necessary.

(a) 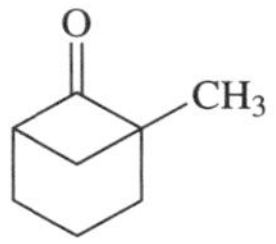 O, CH_3 $\Rightarrow$ O, $CO_2C_2H_5$

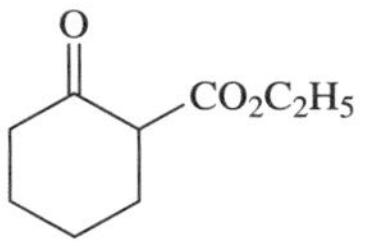

(b) 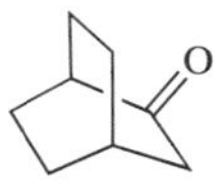 O $\Rightarrow$ O

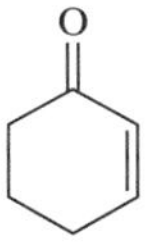

(c) 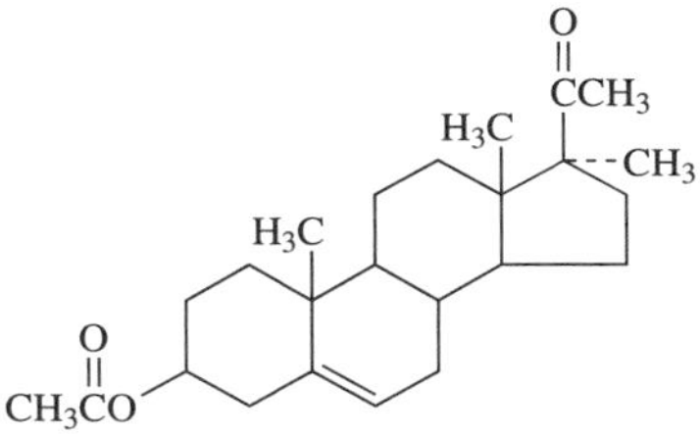 O, CCH_3, H_3C, CH_3, H_3C, O, CH_3CO $\Rightarrow$ O, CCH_3, H_3C, H_3C, O, CH_3CO

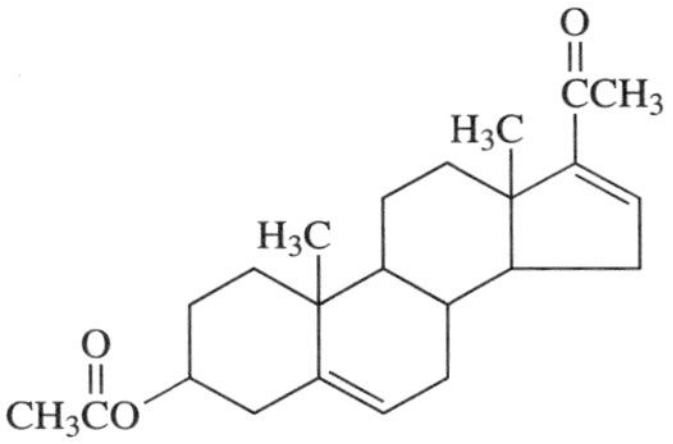

(d) $(CH_3O)_2\overset{O}{\overset{\|}{P}}CH_2\overset{O}{\overset{\|}{C}}(CH_2)_4CH_3$ $\Rightarrow$ $(CH_3O)_2\overset{O}{\overset{\|}{P}}CH_2\overset{O}{\overset{\|}{C}}CH_3$

(e) $PhCH_2CH_2\underset{Ph}{\underset{|}{C}}HCO_2C_2H_5$ $\Rightarrow$ $PhCH_2CO_2C_2H_5$

(f) 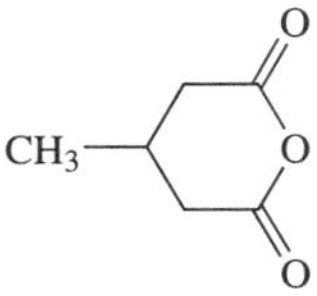 O, CH_3, O, O $\Rightarrow$ $CH_3CH{=}CHCO_2CH_3$

(g) 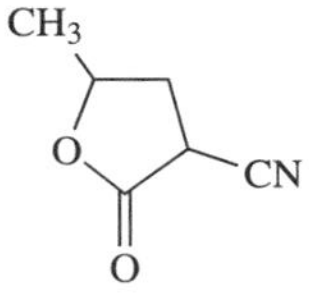 CH_3, O, CN, O $\Rightarrow$ $NCCH_2CO_2C_2H_5$

(h) $OCH_2CH{=}CH_2$ ⇒ HO, HO, OH

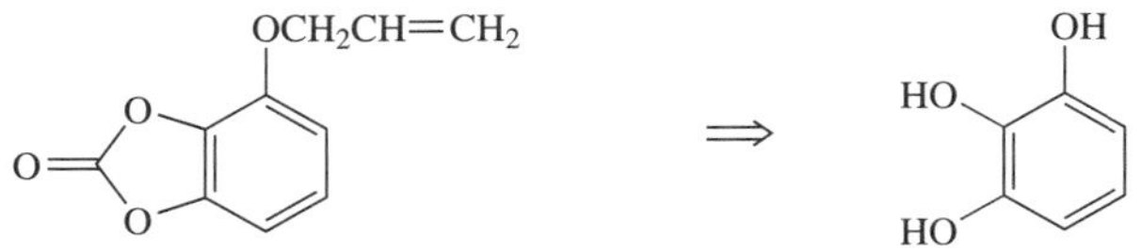

(i) CH_3, $\underset{\|}{C}CH_2CH_2C{=}CH_2$ O, CH_3 ⇒ CH_3, $\underset{\|}{C}CH_2CO_2CH_2CH_3$ O

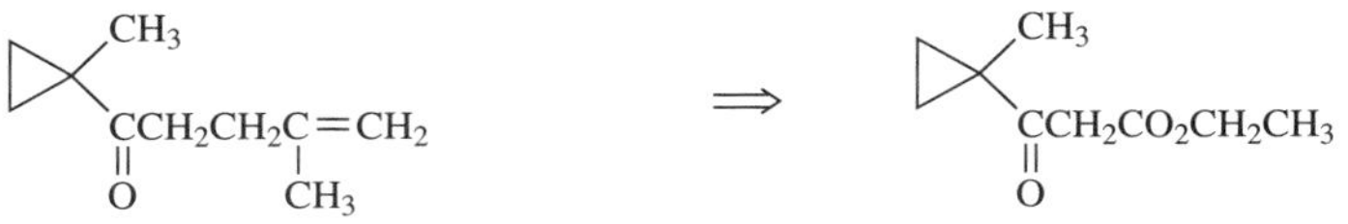

10. In a synthesis of diterpenes via compound **C**, a key intermediate **B** was obtained from carboxylic acid **A**. Suggest a series of reactions for obtaining **B** from **A**.

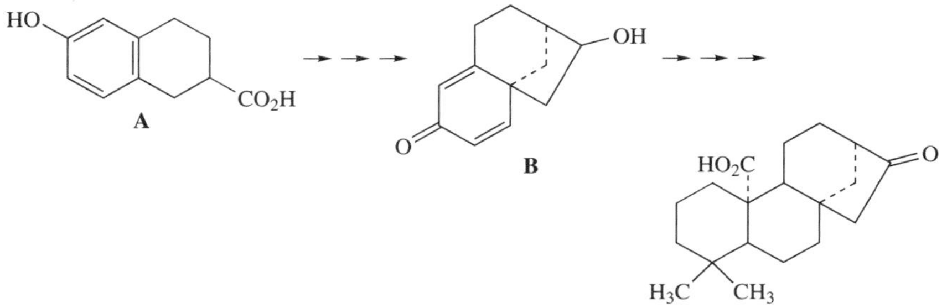

HO, CO_2H **A** → → → O, OH **B** → → → HO_2C, O, H_3C, CH_3 **C**

11. In a synthesis of the terpene longifolene, the tricyclic intermediate **D** was obtained from a bicyclic intermediate by an intermolecular Michael addition. Deduce the possible structure(s) of the bicyclic precursor.

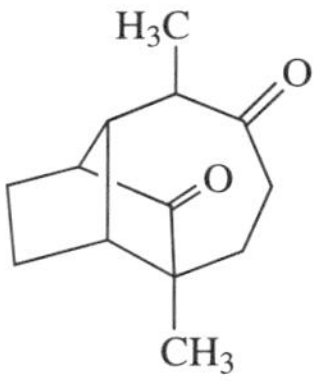

H_3C, O, O, CH_3 **D**

12. Substituted acetophenones react with ethyl phenylpropiolate under the conditions of the Michael reaction to give pyrones. Formulate a mechanism.

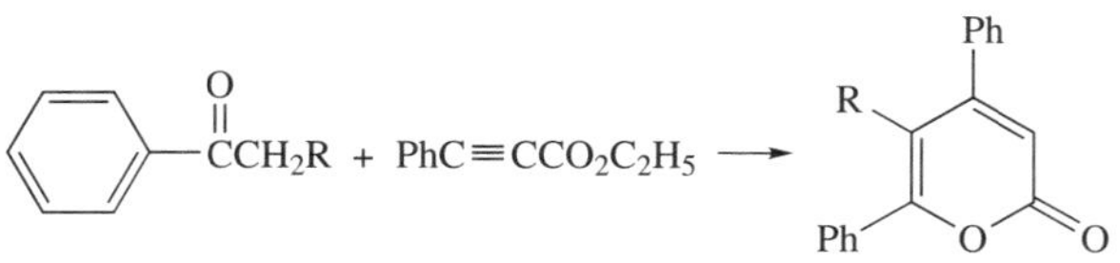

$$C_6H_5\overset{O}{\overset{\|}{C}}CH_2R + PhC{\equiv}CCO_2C_2H_5 \longrightarrow$$ (pyrone with R, Ph, Ph, O, O)

13. The reaction of simple ketones such as 2-butanone or phenylacetone with α,β-unsaturated ketones gives cyclohexenones when the reaction is effected by heating in methanol with potassium methoxide. Explain how the cyclohexenones are formed. What structures are possible for the cyclohexenones? Can you suggest means for distinguishing between possible isomeric cyclohexenones?

14. Analyze the factors that would be expected to control the stereochemistry of the following reactions, and predict the stereochemistry of the product(s).

(a) $\xrightarrow{\text{EtAlCN}}$

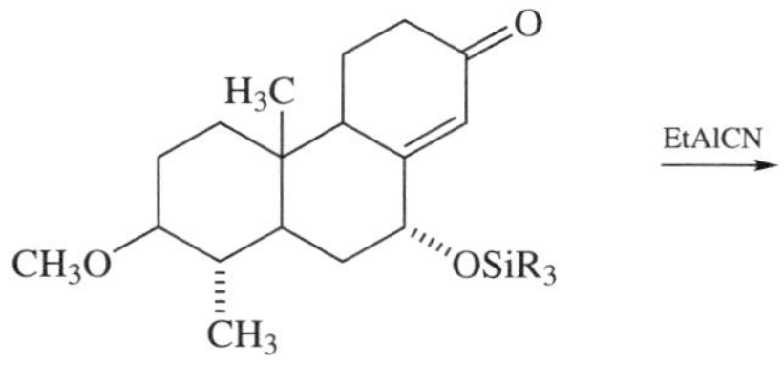

(b) $\xrightarrow[\text{2) } CH_3I]{\text{1) } K^{+}\,{}^{-}N(SiMe_3)_2,\ 25°C}$

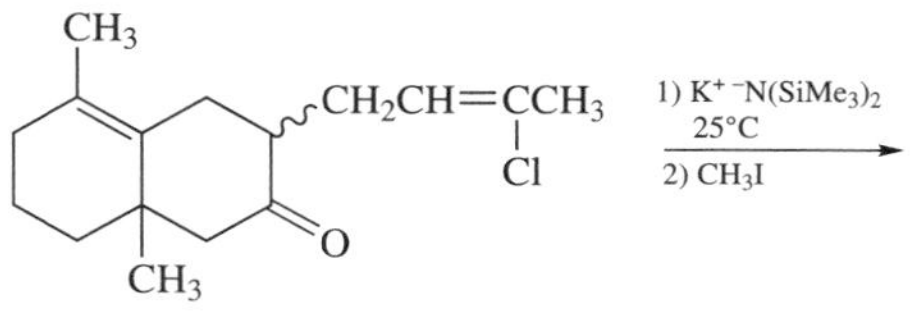

(c) $\xrightarrow[\text{2) } CH_3I]{\text{1) NaH}}$ $R = CH_3CH_2OCH(CH_3)$—

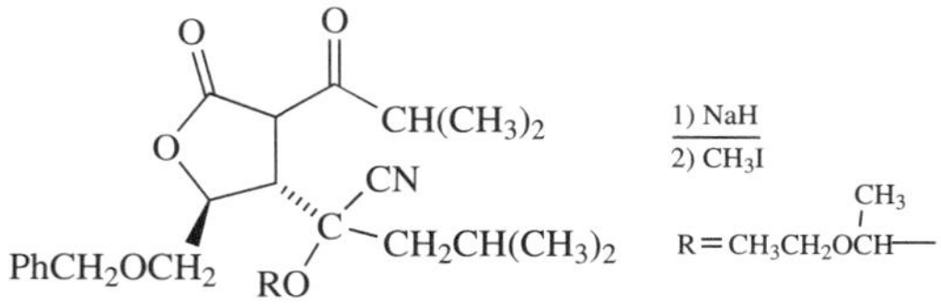

(d) $\xrightarrow[\text{2) } BrCH_2CH{=}CH_2]{\text{1) LDA}}$

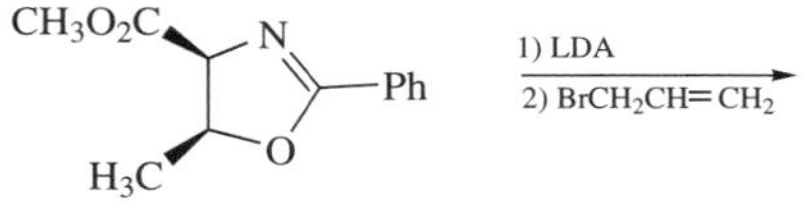

(e) $\xrightarrow[\text{2) } BrCH_2C(CH_3){=}CH_2]{\text{1) NaH}}$

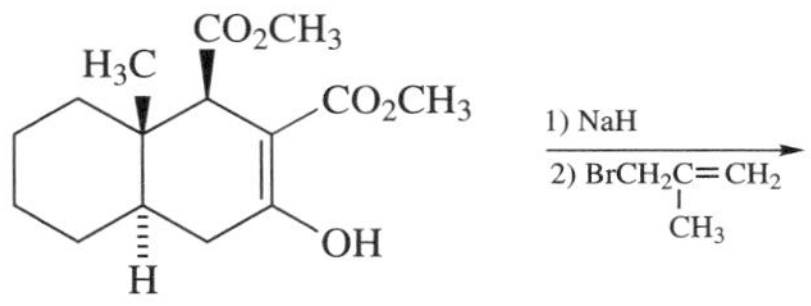

(f) $\xrightarrow[\text{LiNH}_2]{CH_3I}$

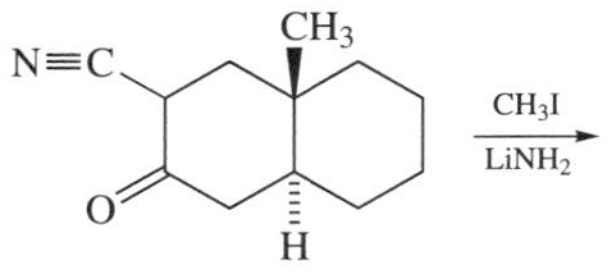

(g) $\xrightarrow[\text{2) } CH_2{=}CHCH_2I]{\text{1) } NaN[Si(CH_3)_3]_2}$

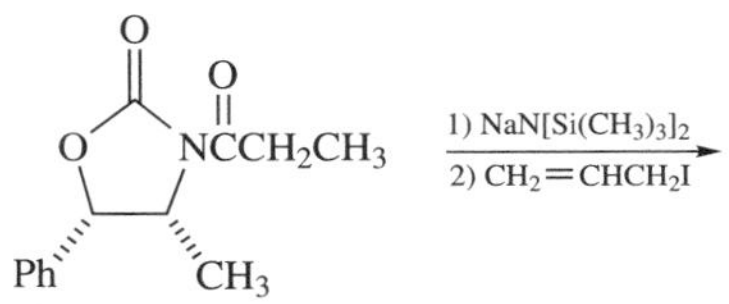

(h) $\xrightarrow[\text{2) LDA}/CH_2{=}CHCH_2Br]{\text{1) LDA}/CH_3I}$

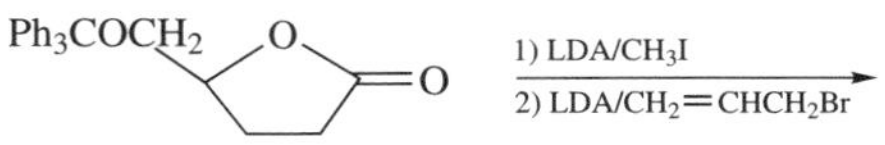

(i) $\xrightarrow[\text{2) } C_2H_5I]{\text{1) LDA/HMPA}}$

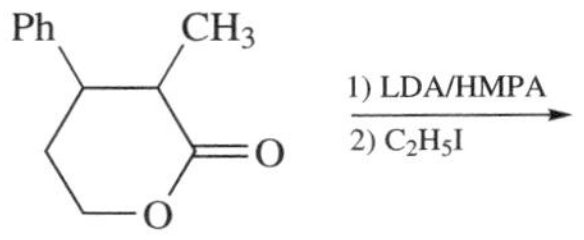

15. Indicate reaction sequences and approximate conditions that could be used to effect the following transformations. More than one step may be required.

(a) IH_2C, CH_3, O, O, H_3C, CH_3 ⟶ $CH_3CH{=}CCH{=}O$, H_2C, CH_3, O, O, H_3C, CH_3

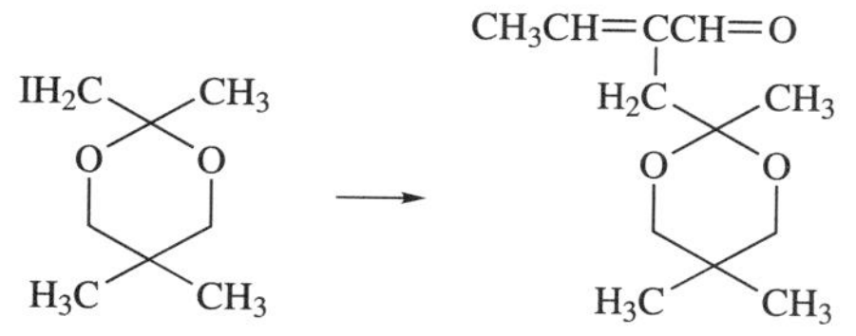

(b) $CH_3\overset{O}{\overset{\|}{C}}CH_2CO_2H$ ⟶ $CH_3\overset{O}{\overset{\|}{C}}CH_2CH_2CH{=}CH_2$

(c) $(CH_3)_2CHCH_2CH_2\overset{O}{\overset{\|}{C}}CH_2CO_2CH_3$ ⟶ $(CH_3)_2CHCH_2CH\overset{O}{\overset{\|}{C}}CH_2CO_2CH_3$ with $CH_3CH_2CH_2$

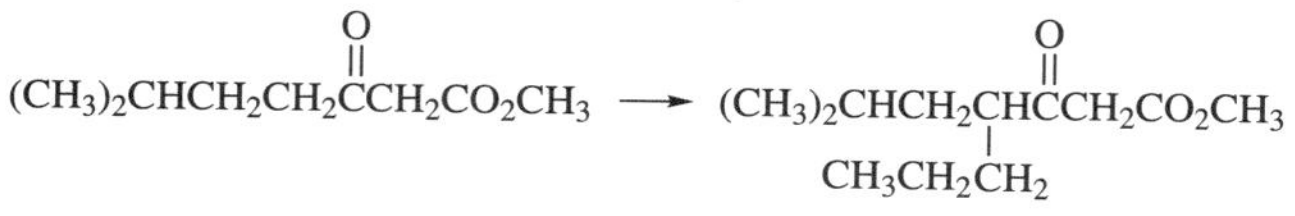

(d) H_3C, H, C=C, H, CO_2CH_3 ⟶ H_3C, H, CH_3O_2C, C, C, $CH_2CO_2C(CH_3)_3$, H, CH_3

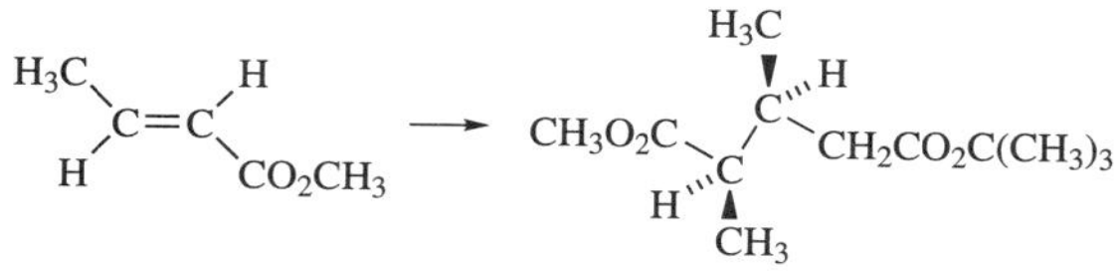

(e) H_3C, O, O, O, CH_3 ⟶ H_3C, O, O, O, H_3C, CH_3

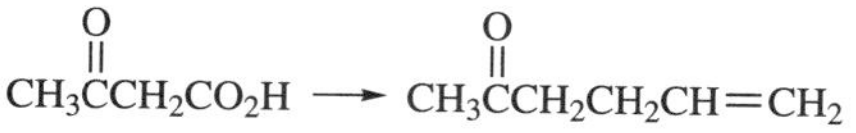

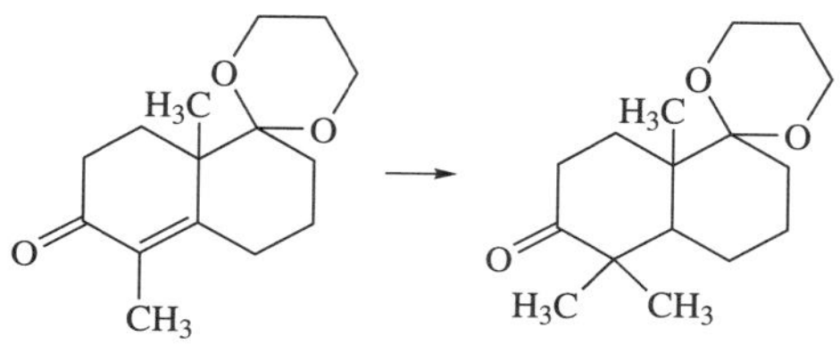

(f) O, CH_3, $(CH_2)_3Cl$ ⟶ O, $CH_2{=}CCH_2$, H_3C, CH_3

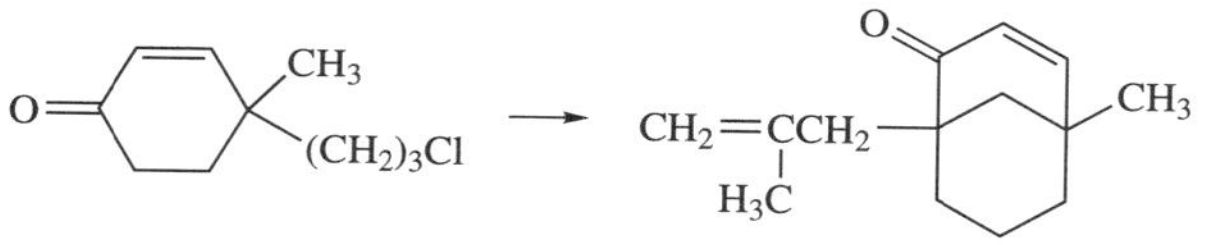

16. Offer an explanation for the stereoselectivity observed in the following reactions.

(a) $(CH_3)_3CO_2C$, CO_2CH_2Ph, CO_2CH_2Ph —1) NaH, DME; 2) O, CH_2Br, N, CH_3, O⟶ $(CH_3)_3CO_2C$, CO_2CH_2Ph, CO_2CH_2Ph, O, O, N, CH_3

syn:anti = 1:4

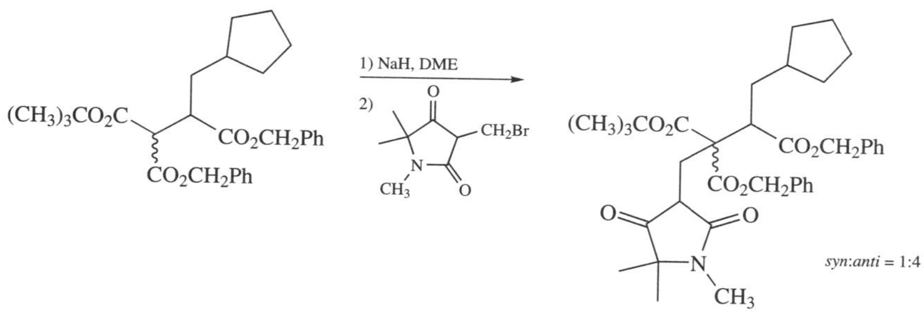

(b) CH_3O ... 1) Ph, Li⁺ N⁻, Ph, CH_3 CH_3 2) $(C_2H_5)_3SiCl$ → CH_3O ... $OSi(C_2H_5)_3$ H high e.e.

(c) 1) LiCl, THF 2) Ph, Li^+ N^-, Ph, CH_3 CH_3 3) $(C_2H_5)_3SiCl$ → $OSi(C_2H_5)_3$

(d) 1) LiHMDS 2) n-C_4H_9I → $(CH_2)_3CH_3$ 56% + $(CH_2)_3CH_3$ 5%

17. One of the compounds shown below undergoes intramolecular cyclization to give a tricyclic ketone on being treated with $[(CH_3)_3Si]_2NNa$. The other does not. Suggest a structure for the product. Explain the difference in reactivity.

$CH_2CH_2CH_2OTs$ $CH_2CH_2CH_2OTs$

18. The alkylation of 3-methyl-2-cyclohexenone with several dibromides led to the products shown below. Discuss the course of each reaction and suggest an explanation for the dependence of the product structure on the structure of the dihalide.

CH_3 1) $NaNH_2$ 2) $Br(CH_2)_nBr$ → (n = 2) CH_3 31% + CH_2 25% + starting material 42%

n = 3 CH_2 55%

n = 4 CH_2 42%

19. Treatment of ethyl 2-azidobutanoate with catalytic quantities of lithium ethoxide in tetrahydrofuran leads to the evolution of nitrogen. On quenching the resulting solution with 3 *N* hydrochloride acid, ethyl 2-oxobutanonate is isolated in 86% yield. Suggest a mechanism for this process.

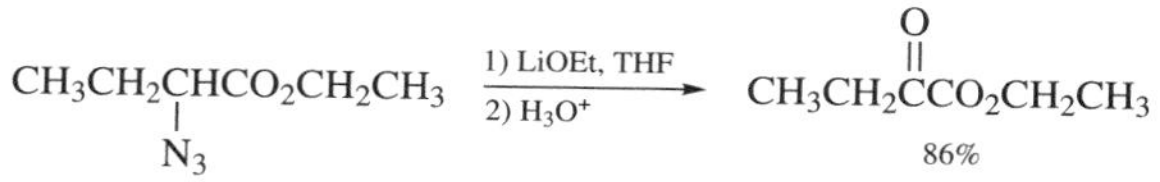

$$CH_3CH_2CH(N_3)CO_2CH_2CH_3 \xrightarrow[2)\ H_3O^+]{1)\ LiOEt,\ THF} CH_3CH_2C(=O)CO_2CH_2CH_3 \quad 86\%$$

20. Suggest a mechanism for the reaction

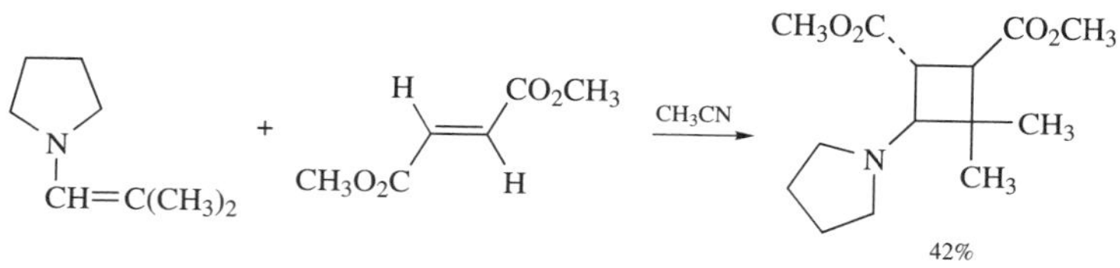

N-pyrrolidinyl $CH{=}C(CH_3)_2$ + CH_3O_2C–CH=CH–CO_2CH_3 (H, H trans) $\xrightarrow{CH_3CN}$ cyclobutane bearing CH_3O_2C, CO_2CH_3, CH_3, CH_3 and pyrrolidin-1-yl N; 42%

21. Suggest a route for the *enantioselective* synthesis of the following substance.

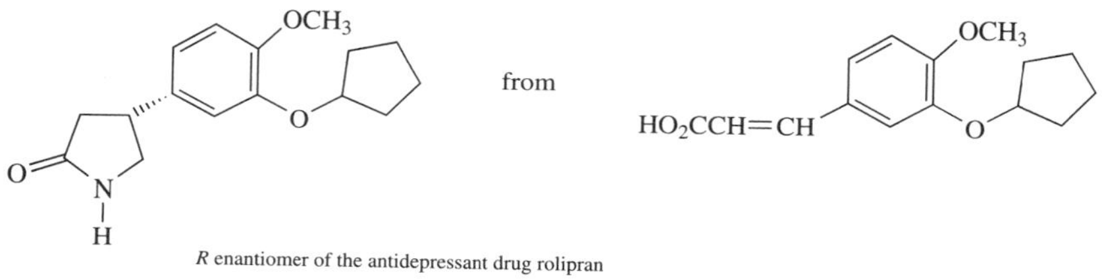

[4-(3-cyclopentyloxy-4-methoxyphenyl)pyrrolidin-2-one, OCH_3, O, N–H] from $HO_2CCH{=}CH$–(3-cyclopentyloxy-4-methoxyphenyl) (OCH_3, O)

R enantiomer of the antidepressant drug rolipran

2

Reaction of Carbon Nucleophiles with Carbonyl Groups

Introduction

The reactions described in this chapter include some of the most useful synthetic methods for carbon–carbon bond formation: *the aldol and Claisen condensations, the Robinson annulation, and the Wittig reaction and related olefination methods*. All of these reactions begin by the addition of a carbon nucleophile to a carbonyl group. The product which is isolated depends on the nature of the substituent (X) on the carbon nucleophile, the substituents (A and B) on the carbonyl group, and the ways in which A, B, and X interact to control the reaction pathways available to the addition intermediate.

$$\text{X}{-}\overset{|}{\underset{|}{\text{C}^-}} + \text{A}{-}\overset{\text{O}}{\overset{\|}{\text{C}}}{-}\text{B} \longrightarrow {-}\overset{\text{X}}{\overset{|}{\underset{|}{\text{C}}}}{-}\overset{\text{O}^-}{\overset{|}{\underset{\underset{\text{A}}{|}}{\text{C}}}}{-}\text{B} \longrightarrow \text{product}$$

The fundamental mechanistic concepts underlying these reactions were introduced in Chapter 8 of Part A. Here we will explore the scope and synthetic utility of these reactions.

2.1. Aldol Addition and Condensation Reactions

2.1.1. The General Mechanism

The prototypical aldol addition reaction is the acid- or base-catalyzed dimerization of a ketone or aldehyde.[1] Under certain conditions, the reaction product may undergo

1. A. T. Nielsen and W. J. Houlihan, *Org. Rect.* **16**: 1 (1968); R. L. Reeves in *Chemistry of the Carbonyl Group*, S. Patai, ed., Interscience, New York, 166, pp. 580–593; H. O. House, *Modern Synthetic Reactions* 2nd ed., W. A. Benjamin, Menlo Park, California, 1972, pp. 629–682.

dehydration leading to an α,β-unsaturated aldehyde or ketone.

$$2\,RCH_2\overset{O}{\overset{\|}{C}}R' \longrightarrow RCH_2\underset{R'}{\overset{OH}{C}}-\underset{R}{C}H\overset{O}{\overset{\|}{C}}R' \xrightarrow{-H_2O} RCH_2\underset{R'}{C}=\underset{R}{C}\overset{O}{\overset{\|}{C}}R'$$

R′ = H or alkyl or aryl

The mechanism of the base-catalyzed reaction involves equilibrium formation of the enolate ion, followed by addition of the enolate to a carbonyl group of the aldehyde or ketone.

Base catalyzed mechanism

1. Addition phase
   a. Enolate formation:

$$RCH_2\underset{O}{\underset{\|}{C}}R' + B^- \rightleftharpoons RCH=\underset{O^-}{C}HR' + BH$$

   b. Nucleophilic addition:

$$RCH_2\underset{O}{\underset{\|}{C}}R' + RCH=\overset{O^-}{C}R' \rightleftharpoons RCH_2\underset{{}^-O}{\overset{R'}{C}}-\underset{R}{C}H\overset{O}{\overset{\|}{C}}R'$$

   c. Proton transfer:

$$RCH_2\underset{{}^-O}{\overset{R'}{C}}-\underset{R}{C}H\overset{O}{\overset{\|}{C}}R' + BH \rightleftharpoons RCH_2\underset{HO}{\overset{R'}{C}}-\underset{R}{C}H\overset{O}{\overset{\|}{C}}R' + B^-$$

2. Dehydration phase

$$RCH_2\underset{HO}{\overset{R'}{C}}-\underset{R}{C}H\overset{O}{\overset{\|}{C}}R' + B^- \longrightarrow (R')(RCH_2)C=C(R)(\overset{O}{\overset{\|}{C}}R') + BH + HO^-$$

Entries 1 and 2 in Scheme 2.1 illustrate the preparation of aldol reaction products by the base-catalyzed mechanism. In entry 1, the product is a β-hydroxyaldehyde, whereas in entry 2 dehydration has occurred and the product is an α,β-unsaturated aldehyde.

Under conditions of acid catalysis, it is the enol form of the aldehyde or ketone which functions as the nucleophile. The carbonyl group is activated toward nucleophilic attack by

Scheme 2.1. Aldol Condensation of Simple Aldehydes and Ketones

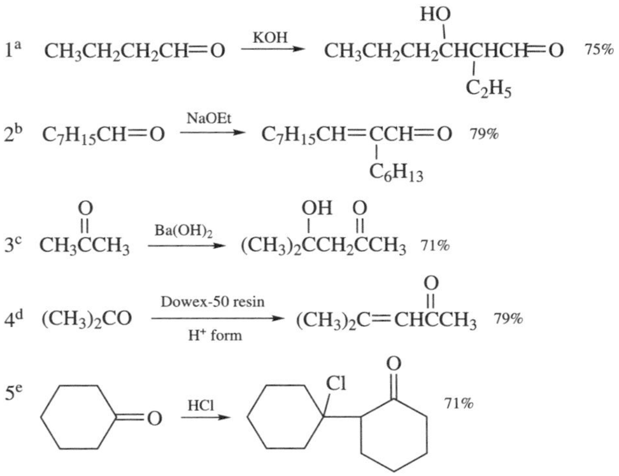

1[a] $CH_3CH_2CH_2CH{=}O \xrightarrow{KOH} CH_3CH_2CH_2CH(OH)CH(C_2H_5)CH{=}O$ 75%

2[b] $C_7H_{15}CH{=}O \xrightarrow{NaOEt} C_7H_{15}CH{=}C(C_6H_{13})CH{=}O$ 79%

3[c] $CH_3C(O)CH_3 \xrightarrow{Ba(OH)_2} (CH_3)_2C(OH)CH_2C(O)CH_3$ 71%

4[d] $(CH_3)_2CO \xrightarrow[H^+ \text{ form}]{\text{Dowex-50 resin}} (CH_3)_2C{=}CHC(O)CH_3$ 79%

5[e] cyclohexanone $\xrightarrow{HCl}$ 2-(1-chlorocyclohexyl)cyclohexanone 71%

a. V. Grignard and A. Vesterman, *Bull. Chim. Soc. Fr.* **37:**425 (1925).
b. F. J. Villani and F. F. Nord, *J. Am. Chem. Soc.* **69:**2605 (1947).
c. J. B. Conant and N. Tuttle, *Org. Synth.* **1:**199 (1941).
d. N. B. Lorette, *J. Org. Chem.* **22:**346 (1957).
e. O. Wallach, *Berichte* **40:**70 (1907); E. Wenkert, S. K. Bhattacharya, and E. M. Wilson, *J. Chem. Soc.* **1964:**5617.

oxygen protonation.

Acid catalyzed mechanism

1. Addition phase

a. Enolization:

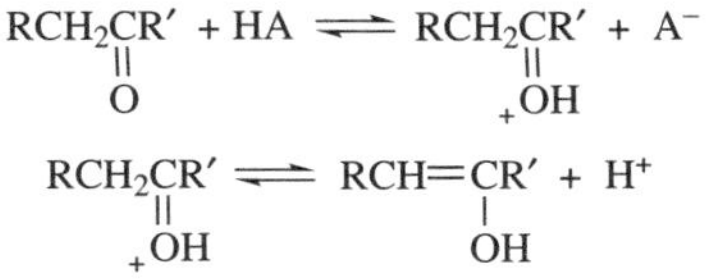

$$RCH_2C(=O)R' + HA \rightleftharpoons RCH_2C(={}^+OH)R' + A^-$$

$$RCH_2C(={}^+OH)R' \rightleftharpoons RCH{=}C(OH)R' + H^+$$

b. Nucleophilic addition:

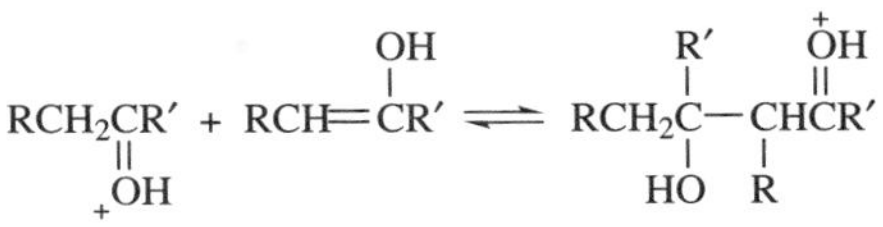

$$RCH_2C(={}^+OH)R' + RCH{=}C(OH)R' \rightleftharpoons RCH_2C(R')(HO){-}CH(R)C(={}^+OH)R'$$

c. Proton transfer:

$$RCH_2C(R')(HO){-}CH(R)C(={}^+OH)R' \rightleftharpoons RCH_2C(R')(HO){-}CH(R)C(=O)R' + H^+$$

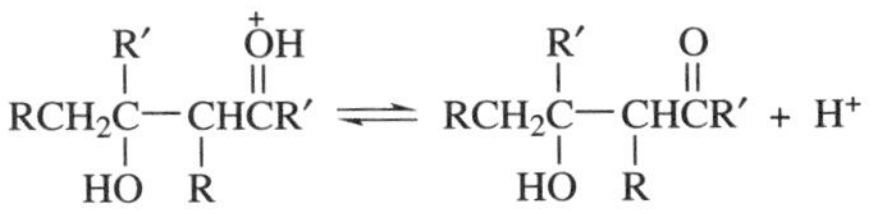

2. Dehydration phase

$$RCH_2C(R')(HO){-}CH(R)C(=O)R' + H^+ \longrightarrow (R')(RCH_2)C{=}C(R)C(=O)R' + H_2O$$

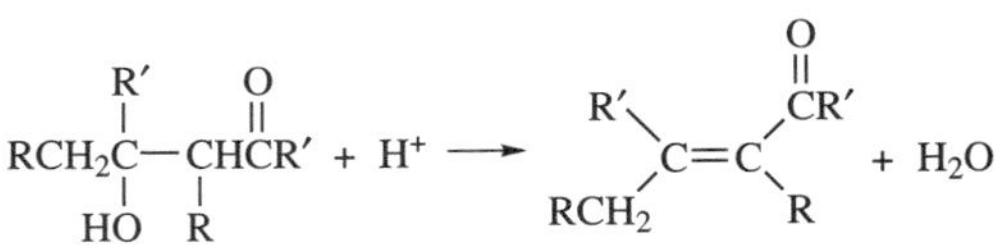

Entries 4 and 5 in Scheme 2.1 depict acid-catalyzed aldol reactions. In entry 4, condensation is accompanied by dehydration. In entry 5, a β-chloroketone is formed by addition of hydrogen chloride to the enone.

In general, the reactions in the addition phase of both the base- and acid-catalyzed mechanisms are reversible. The equilibrium constant for addition is usually unfavorable for acyclic ketones. The equilibrium constant for the dehydration phase is usually favorable, because of the conjugated α,β-unsaturated carbonyl system that is formed. When the reaction conditions are sufficiently vigorous to cause dehydration, the overall reaction will go to completion, even if the equilibrium constant for the addition step is unfavorable. Entry 3 in Scheme 2.1 illustrates a clever way of overcoming the unfavorable equilibrium of the addition step. The basic catalyst is contained in a separate compartment of a Soxhlet extractor. Acetone is repeatedly passed over the basic catalyst by distillation and then returns to the reaction flask. The concentration of the addition product builds up in the reaction flask as the more volatile acetone distills preferentially. Because there is no catalyst in the reaction flask, the adduct remains stable.

2.1.2. Mixed Aldol Condensations with Aromatic Aldehydes

Aldol addition and condensation reactions involving two different carbonyl compounds are called *mixed aldol reactions*. For these reactions to be useful as a method for synthesis, there must be some basis for controlling which carbonyl component serves as the electrophile and which acts as the enolate precursor. One of the most general mixed aldol condensations involves the use of aromatic aldehydes with alkyl ketones or aldehydes. Aromatic aldehydes are incapable of enolization and cannot function as the nucleophilic component. Furthermore, dehydration is especially favorable because the resulting enone is conjugated with the aromatic ring.

$$\mathrm{ArCH{=}O} + \mathrm{RCH_2\overset{O}{\overset{\|}{C}}R'} \rightleftharpoons \mathrm{Ar\overset{OH}{\overset{|}{C}}H\underset{R}{\underset{|}{C}}H\overset{O}{\overset{\|}{C}}R'} \xrightarrow{-H_2O} \mathrm{ArCH{=}C(R)\overset{O}{\overset{\|}{C}}R'}$$

There are numerous examples of both acid- and base-catalyzed mixed aldol condensations involving aromatic aldehydes. The reaction is sometimes referred to as the *Claisen–Schmidt condensation*. Scheme 2.2 presents some representative examples.

There is a pronounced preference for the formation of a *trans* double bond in the Claisen–Schmidt condensation of methyl ketones. This stereoselectivity arises in the dehydration step. In the transition state for elimination to a *cis* double bond, an unfavorable steric interaction between the ketone substituent (R) and the phenyl group occurs. This interaction is absent in the transition state for elimination to the *trans* double bond.

less favorable

more favorable

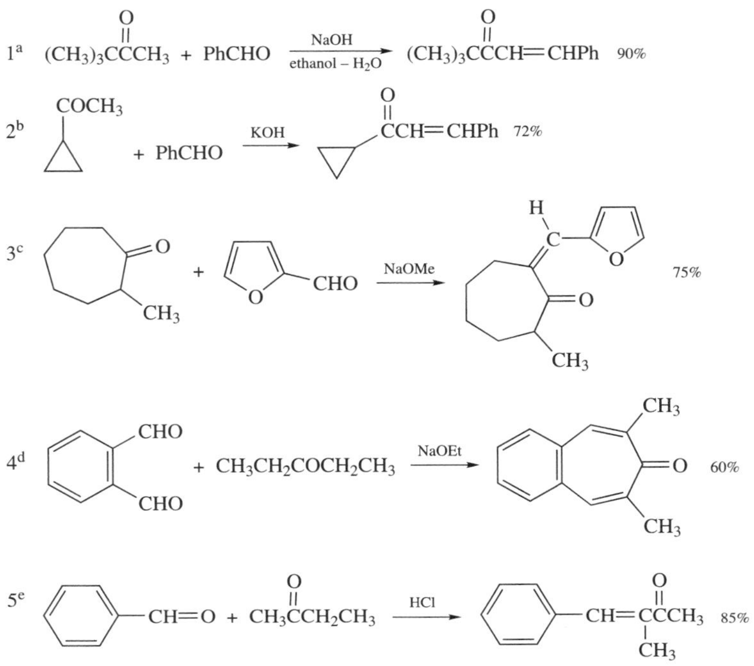

Scheme 2.2. Mixed Condensation of Aromatic Aldehydes with Ketones

1[a] $(CH_3)_3CCCH_3$ (C=O) + PhCHO $\xrightarrow[\text{ethanol} - H_2O]{\text{NaOH}}$ $(CH_3)_3CCCH{=}CHPh$ (C=O) 90%

2[b] cyclopropyl–$COCH_3$ + PhCHO $\xrightarrow{\text{KOH}}$ cyclopropyl–$CCH{=}CHPh$ (C=O) 72%

3[c] 2-methylcycloheptanone + furan-2-CHO $\xrightarrow{\text{NaOMe}}$ 2-(furylmethylene)-7-methylcycloheptanone 75%

4[d] o-phthalaldehyde (CHO, CHO) + $CH_3CH_2COCH_2CH_3$ $\xrightarrow{\text{NaOEt}}$ 6,8-dimethylbenzocyclohepten-7-one (CH_3, =O, CH_3) 60%

5[e] $C_6H_5CH{=}O$ + $CH_3CCH_2CH_3$ (C=O) $\xrightarrow{\text{HCl}}$ $C_6H_5CH{=}C(CH_3)CCH_3$ (C=O) 85%

a. G. A. Hill and G. Bramann, *Org. Synth.* **I:**81 (1941).
b. S. C. Bunce. H. J. Dorsman, and F. D. Popp, *J. Chem. Soc.* **1963:**303.
c. A. M. Islam and M. T. Zemaity, *J. Am. Chem. Soc.* **79:**6023 (1957).
d. D. Meuche, H. Strauss, and E. Heilbronner, *Helv. Chim. Acta* **41:**2220 (1958).
e. M. E. Kronenberg and E. Havinga, *Rec. Trav. Chim.* **84:**17, 979 (1965).

Additional insight into the factors affecting product structure was obtained by study of the condensation of 2-butanone with benzaldehyde.[2]

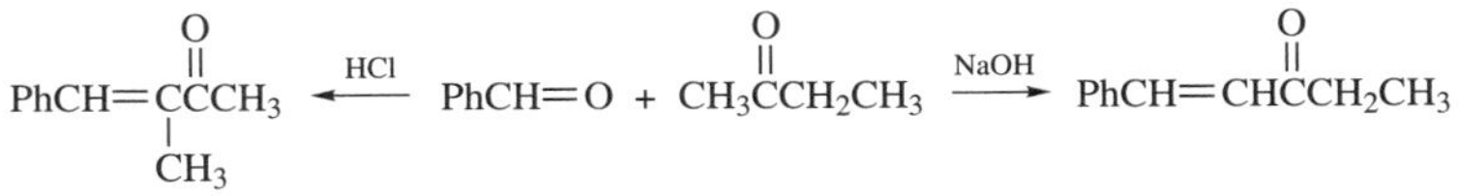

$$PhCH{=}\underset{CH_3}{C}\overset{O}{\overset{\|}{C}}CH_3 \xleftarrow{\text{HCl}} PhCH{=}O + CH_3\overset{O}{\overset{\|}{C}}CH_2CH_3 \xrightarrow{\text{NaOH}} PhCH{=}CH\overset{O}{\overset{\|}{C}}CH_2CH_3$$

The results indicate that the product ratio is determined by the competition between the various reaction steps. Under base-catalyzed conditions, 2-butanone reacts with benzaldehyde at the methyl group to give 1-phenylpent-1-en-3-one. Under acid-catalyzed conditions, the product is the result of condensation at the methylene group, namely, 3-methyl-4-phenylbut-3-en-2-one. Under the reaction conditions used, it is not possible to isolate the intermediate ketols, because the addition step is rate-limiting. These intermediates can be

2. M. Stiles, D. Wolf, and G. V. Hudson, *J. Am. Chem. Soc.* **81**: 628 (1959); D. S. Noyce and W. L. Reed, *J. Am. Chem. Soc.* **81**: 618, 620, 624 (1959).

prepared by alternative methods, and they behave as shown in the following equations:

$$\mathrm{PhCH(OH)CH_2C(=O)CH_2CH_3} \xrightarrow{\mathrm{NaOH}} \mathrm{PhCH{=}CHC(=O)CH_2CH_3} + \mathrm{PhCH{=}O} + \mathrm{CH_3CH_2C(=O)CH_3}$$

$$\mathrm{PhCH(OH)CH(CH_3)C(=O)CH_3} \xrightarrow{\mathrm{NaOH}} \mathrm{PhCH{=}O} + \mathrm{CH_3CH_2C(=O)CH_3}$$

$$\mathrm{PhCH(OH)CH_2C(=O)CH_2CH_3} \xrightarrow{\mathrm{HCl}} \mathrm{PhCH{=}CHC(=O)CH_2CH_3} + \mathrm{PhCH{=}O} + \mathrm{CH_3C(=O)CH_2CH_3}$$

$$\mathrm{PhCH(OH)CH(CH_3)C(=O)CH_3} \xrightarrow{\mathrm{HCl}} \mathrm{PhCH{=}C(CH_3)C(=O)CH_3} + \mathrm{PhCH{=}O} + \mathrm{CH_3C(=O)CH_2CH_3}$$

These results establish that the base-catalyzed dehydration is slow relative to the reverse of the addition phase for the branched-chain isomer. The reason for selective formation of the straight-chain product under conditions of base catalysis is then apparent. In base, the straight-chain ketol is the only intermediate which is dehydrated. The branched-chain ketol reverts to starting material. Under acid conditions, both intermediates are dehydrated; however, the branched-chain ketol is formed most rapidly, because of the preference for acid-catalyzed enolization to give the more substituted enol (see Section 7.3 of Part A).

$$\mathrm{CH_3C(=O)CH_2CH_3} \xrightarrow{\mathrm{H^+}} \underset{\text{major}}{\mathrm{CH_3C(OH){=}CHCH_3}} + \underset{\text{minor}}{\mathrm{CH_2{=}C(OH)CH_2CH_3}}$$

$$\mathrm{CH_3C(OH){=}CHCH_3} + \mathrm{PhCH{=}O} \xrightarrow{\text{slow}} \mathrm{PhCH(OH)CH(CH_3)C(=O)CH_3} \xrightarrow{\text{fast}} \mathrm{PhCH{=}C(CH_3)C(=O)CH_3}$$

In general, the product ratio of a mixed aldol condensation will depend upon the individual reaction rates. Most ketones show a pattern similar to butanone in reactions with aromatic aldehydes. Base catalysis favors reaction at a methyl position over a methylene group, whereas acid catalysis gives the opposite preference.

2.1.3. Control of Regiochemistry and Stereochemistry of Mixed Aldol Reactions of Aliphatic Aldehydes and Ketones

2.1.3.1. Lithium Enolates. The control of mixed aldol additions between aldehydes and ketones that present several possible sites for enolization is a challenging problem. Such reactions are normally carried out by complete conversion of the carbonyl compound that is to serve as the nucleophile to an enolate, silyl enol ether, or imine anion. The reactive nucleophile is then allowed to react with the second reaction component. As long as the addition step is faster than proton transfer, or other mechanisms of interconversion of the nucleophilic and electrophilic components, the adduct will have the desired structure. The term *directed aldol reaction* is given to these reactions.[3] Directed aldol reactions must be carried out under conditions designed to ensure that the desired product is obtained. In general, this requires that the product structure be controlled by *kinetic factors*, both in the formation of the enolate and in the addition step, and that equilibration by reversibility of either step be avoided. Scheme 2.3 illustrates some of the procedures which have been developed to achieve this goal.

Scheme 2.3. Directed Aldol Additions

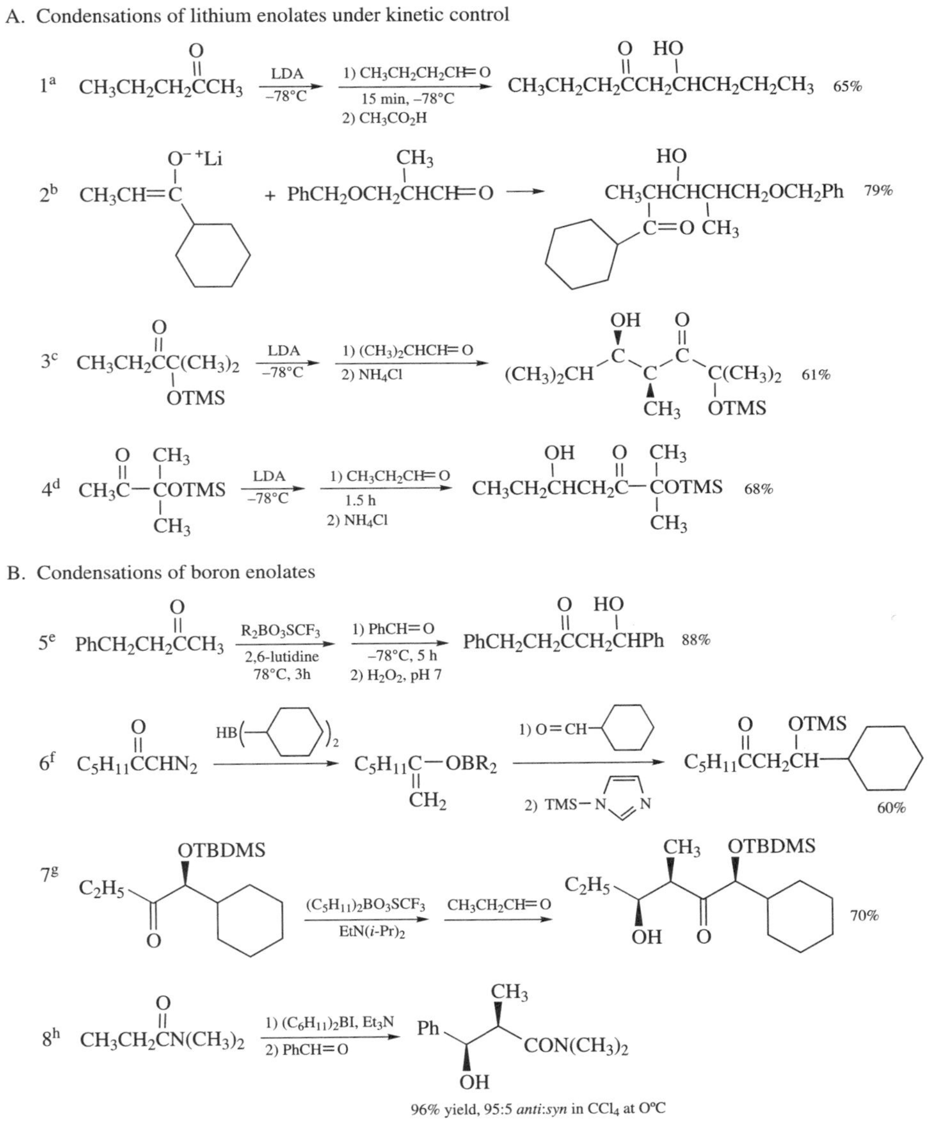

A. Condensations of lithium enolates under kinetic control

1[a] $CH_3CH_2CH_2C(=O)CH_3$ —LDA, −78°C→ 1) $CH_3CH_2CH_2CH{=}O$, 15 min, −78°C; 2) CH_3CO_2H → $CH_3CH_2CH_2C(=O)CH_2CH(OH)CH_2CH_2CH_3$ 65%

2[b] $CH_3CH{=}C(O^-\,{}^+Li)$(cyclohexyl) + $PhCH_2OCH_2CH(CH_3)CH{=}O$ → $CH_3CH(C(=O)\text{-cyclohexyl})CH(OH)CH(CH_3)CH_2OCH_2Ph$ 79%

3[c] $CH_3CH_2C(=O)C(CH_3)_2(OTMS)$ —LDA, −78°C→ 1) $(CH_3)_2CHCH{=}O$; 2) NH_4Cl → $(CH_3)_2CH\text{–}CH(OH)\text{–}CH(CH_3)\text{–}C(=O)\text{–}C(CH_3)_2(OTMS)$ 61%

4[d] $CH_3C(=O)\text{–}C(CH_3)_2OTMS$ —LDA, −78°C→ 1) $CH_3CH_2CH{=}O$, 1.5 h; 2) NH_4Cl → $CH_3CH_2CH(OH)CH_2C(=O)\text{–}C(CH_3)_2OTMS$ 68%

B. Condensations of boron enolates

5[e] $PhCH_2CH_2C(=O)CH_3$ —$R_2BO_3SCF_3$, 2,6-lutidine, 78°C, 3h→ 1) PhCH=O, −78°C, 5 h; 2) H_2O_2, pH 7 → $PhCH_2CH_2C(=O)CH_2CH(OH)Ph$ 88%

6[f] $C_5H_{11}C(=O)CHN_2$ —$HB(\text{cyclohexyl})_2$→ $C_5H_{11}C(=CH_2)\text{–}OBR_2$ —1) O=CH–cyclohexyl; 2) TMS–imidazole→ $C_5H_{11}C(=O)CH_2CH(OTMS)\text{–}$cyclohexyl 60%

7[g] $C_2H_5C(=O)CH(CH_3)$–CH(OTBDMS)–cyclohexyl —$(C_5H_{11})_2BO_3SCF_3$, $EtN(i\text{-}Pr)_2$→ $CH_3CH_2CH{=}O$ → $C_2H_5CH(OH)CH(CH_3)C(=O)CH(CH_3)CH(OTBDMS)$–cyclohexyl 70%

8[h] $CH_3CH_2C(=O)N(CH_3)_2$ —1) $(C_6H_{11})_2BI$, Et_3N; 2) PhCH=O→ PhCH(OH)CH(CH_3)CON(CH_3)_2

96% yield, 95:5 *anti:syn* in CCl_4 at 0°C

3. T. Mukaiyama, *Org. React.* **28**: 203 (1982).

Scheme 2.3. *(continued)*

9[i] 1) $Bu_2BO_3SCF_3$, EtN; 2) furan-CH=O → 92%

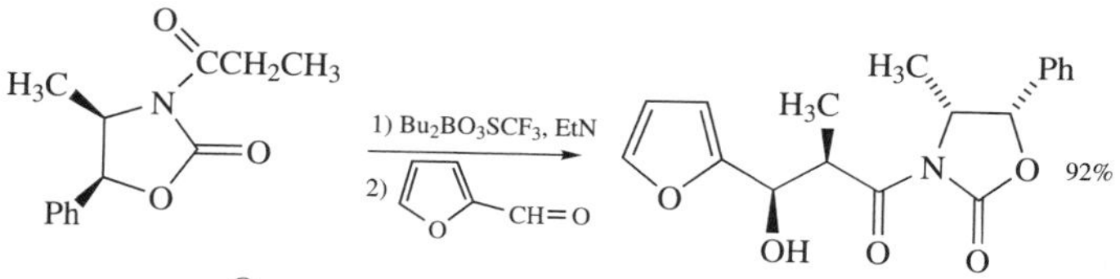

10[j] $R_2BO_3SCF_3$, $EtN(i\text{-}Pr)_2$; O=CH, H_3C, CH_2OCH_2Ph, H; $NaOCH_3$ | CH_3OH → OH, CH_3O_2C, CH_2OCH_2Ph, CH_3, CH_3

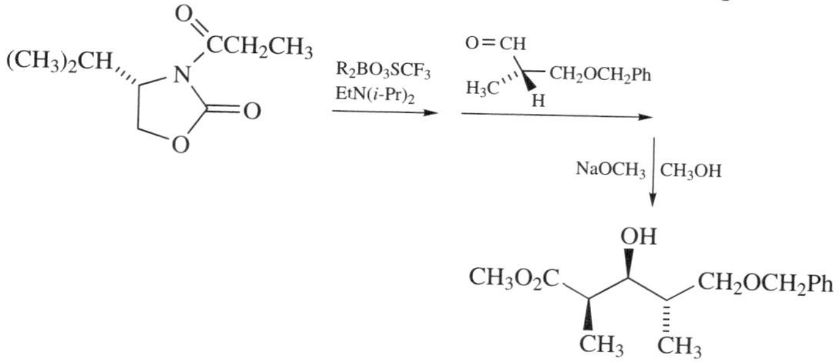

C. Tin, titanium, and zirconium enolates

11[k] 1) $TiCl_4(i\text{-}Pr)_2NEt$; 2) PhCH=O → 81% yields, 96:4 *syn:anti*

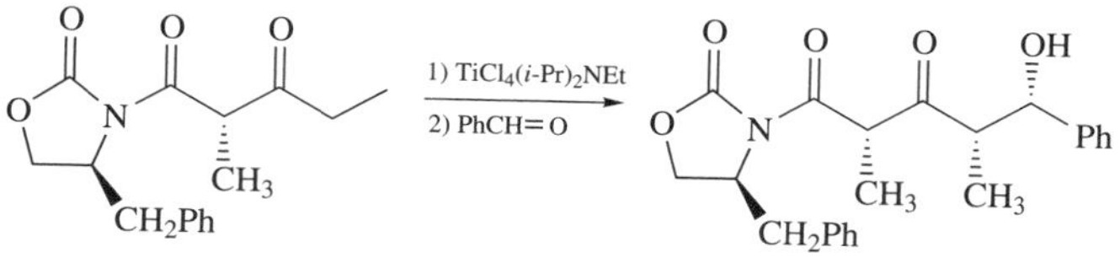

a. G. Stork, G. A Kraus, and G. A. Garcia, *J. Org. Chem.* **39:**3459 (1974).
b. S. Masamune, J. W. Ellingboe, and W. Choy, *J. Am. Chem. Soc.* **104:**5526 (1982).
c. R. Bal, C. T. Buse, K. Smith, and C. Heathcock, *Org. Synth.* **63:**89 (1984).
d. P. J. Jerris and A. P. Smith III, *J. Org. Chem.* **46:**577 (1981).
e. T. Inoue, T. Uchimaru, and T. Mukaiyama, *Chem. Lett.* **1977:**153.
f. J. Hooz, J. Oudenes, J. L. Roberts, and A. Benderly, *J. Org. Chem.* **52:**1347 (1987).
g. S. Masamune, W. Choy, F. A. J. Kerdesky, and B. Imperiali, *J. Am. Chem. Soc.* **103:**1566 (1981).
h. K. Ganesan and H. C. Brown, *J. Org. Chem.* **59:**7346 (1994).
i. S. F. Martin and D. E. Guinn, *J. Org. Chem.* **52:**5588 (1987).
j. D. Seebach, H.-F. Chow, R. F. W. Jackson, K. Lawson, M. A. Sutter, S. Thaisrivongs, and J. Zimmermann, *J. Am. Chem. Soc.* **107:**5292 (1985).

Entries 1–4 in Scheme 2.3 represent cases in which the nucleophilic component is converted to the enolate under kinetically controlled conditions by the methods discussed in Section 1.2. Such enolates are usually highly reactive toward aldehydes so that addition occurs rapidly when the aldehyde is added, even at low temperature. When the addition step is complete, the reaction is stopped by neutralization and the product is isolated. The guiding mechanistic concept for reactions carried out under these conditions is that they occur through a cyclic transition state in which lithium or another metal cation is coordinated to both the enolate oxygen and the carbonyl oxygen.[4]

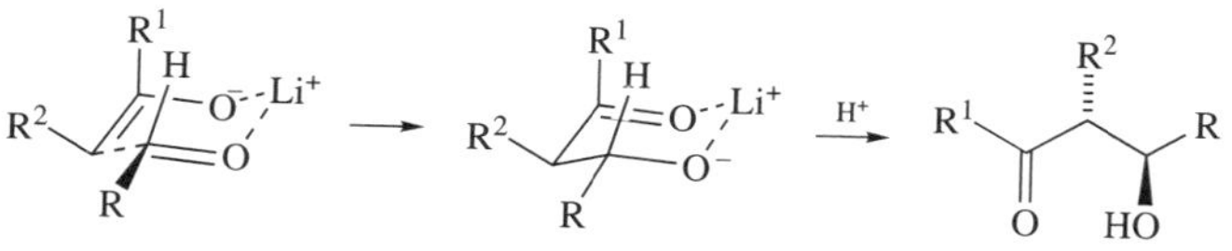

R^1, H, R^2, O^-, Li^+, O, R → R^1, H, R^2, O, Li^+, O^-, R → H^+ → R^1, R^2, R, O, HO

4. H. E. Zimmerman and M. D. Traxler, *J. Am. Chem. Soc.* **79:**1920 (1957); C. H. Heathcock, C. T. Buse, W. A. Kleschick, M. C. Pirrung, J. E. Sohn, and J. Lampe, *J. Org. Chem.* **45:**1066 (1980).

This transition-state model has been the basis both for development of other reaction conditions and for the interpretation of the stereochemistry of the reaction.

Most enolates can exist as two stereoisomers. Also, most aldol condensation products formed from a ketone enolate and an aldehyde can have two diastereomeric structures. These are designated as *syn* and *anti*. The cyclic-transition-state model provides a basis for understanding the relationship between enolate geometry and the stereochemistry of the aldol product.

Z-enolate *E*-enolate *syn*-ketol *anti*-ketol

The enolate formed from 2,2-dimethyl-3-pentanone under kinetically controlled conditions is the *Z*-isomer. When it reacts with benzaldehyde, only the *syn* aldol is formed.[4] This stereochemical relationship is accounted for by a cyclic transition state with a chair-like conformation. The product stereochemistry is correctly predicted if the aldehyde is in a conformation such that the phenyl substituent occupies an equatorial position in the cyclic transition state.

PhCHO, −72°C

78% yield, 100% *syn*

A similar preference for formation of the *syn* aldol is found for other *Z*-enolates derived from ketones in which one of the carbonyl substituents is bulky.[5] Ketone enolates in which the other carbonyl substituent is less bulky show a decreasing stereoselectivity in the order *t*-butyl > *i*-propyl > ethyl.[4] This trend reflects a decreasing preference for formation of the *Z*-enolate.

CH_3CH_2COR —LDA→ *E* enolate + *Z* enolate —PhCH=O→ *anti* + *syn*

R =	*E*	*Z*	*anti*	*syn*
C_2H_5	70	30	36	64
$CH(CH_3)_2$	40	60	18	82
$C(CH_3)_3$	2	98	2	98

The *E*:*Z* ratio can be modified by the precise conditions for formation of the enolate. For example, the *E*:*Z* ratio can be increased for 3-pentanone and 2-methyl-3-pentanone by use of a 1 : 1 lithium tetramethylpiperidide (LiTMP)–LiBr mixture for kinetic enolization.[6] The precise mechanism of this effect is not clear, but it probably is due to an aggregate

5. P. Fellman and J. E. Dubois, *Tetrahedron* **34:**1349 (1978).
6. P. L. Hall, J. H. Gilchrist, and D. B. Collum, *J. Am. Chem. Soc.* **113:**9571 (1991).

species containing bromide acting as the base.[7]

	E:*Z* Stereoselectivity		
	LDA	LiTMP	LiTMP + LiBr
$CH_3CH_2C(=O)CH_2CH_3$	3.3 : 1	5 : 1	50 : 1
$(CH_3)_2CHC(=O)CH_2CH_3$	1.7 : 1	2 : 1	21 : 1
$(CH_3)_3CC(=O)CH_2CH_3$	1 : >50	1 : >20	1 : >20

The enolates derived from cyclic ketones are necessarily *E*-isomers. The enolate of cyclohexanone reacts with benzaldehyde to give both possible stereoisomeric products under kinetically controlled conditions. The stereochemistry can be raised to about 6 : 1 in favor of the *anti* isomer under optimum conditions.[8]

O⁻ Li⁺ + PhCH=O ⟶ 16% + 84%

From these and related examples, the following generalizations have been drawn about kinetic stereoselection in aldol additions.[9] (1) The chair transition-state model provides a basis for explaining the stereoselectivity observed in aldol reactions of ketones having one bulky substituent. The preference is *Z*-enolate → *syn* aldol; *E*-enolate → *anti* aldol. (2) When the enolate has no bulky substituents, stereoselectivity is low. (3) *Z*-Enolates are more stereoselective than *E*-enolates. Table 2.1 gives some illustrative data.

Because the aldol reaction is reversible, it is possible to adjust reaction conditions so that the two stereoisomeric aldol products equilibrate. This can be done in the case of lithium enolates by keeping the reaction mixture at room temperature until the product composition reaches equilibrium. This has been done, for example, for the product from the reaction of the enolate of ethyl *t*-butyl ketone and benzaldehyde.

PhCH=O, fast ⟶ *syn* ⟶ 25°C, slow ⟶ *anti*

syn ⇌ *anti*

7. F. S. Mair, W. Clegg, and P. A. O'Neil, *J. Am. Chem. Soc.* **115**:3388 (1993).
8. M. Majewski and D. M. Gleave, *Tetrahedron Lett.* **30**:5681 (1989).
9. D. A. Evans, J. V. Nelson, and T. R. Taber, *Top. Stereochem.* **13**:1 (1982); C. H. Heathcock, in *Comprehensive Carbanion Chemistry*, Part B, E. Buncel and T. Durst, eds., Elsevier, Amsterdam, 1984, pp. 177–237; C. H. Heathcock, in *Asymmetric Synthesis*, Vol. 3, J. D. Morrison, ed., Academic Press, New York, 1984, pp. 111–212.

Table 2.1. Stereoselectivity of Lithium Enolates toward Benzaldehyde[a]

R^1	Enolate geometry, Z/E	Aldol stereostructure, *syn* : *anti*
H	100 : 0	50 : 50
H	0 : 100	65 : 35
Et	30 : 70	64 : 36
Et	66 : 34	77 : 23
i-Pr	> 98 : 2	90 : 10
i-Pr	32 : 68	58 : 42
*i*Pr	0 : 100	45 : 55
t-Bu	> 98 : 2	> 98 : 2
1-Adamantyl	> 98 : 2	> 98 : 2
Ph	> 98 : 2	88 : 12
Mesityl	8 : 92	8 : 92
Mesityl	87 : 13	88 : 12

a. From C. H. Heathcock, in *Asymmetric Synthesis*, Vol. 3, J. D. Morrison, ed., Academic Press, New York, 1984, Chapter 2.

For synthetic efficiency, it is useful to add $MgBr_2$.

1) LDA
2) $(CH_3)_2CHCH{=}O$
3) $MgBr_2$

syn + *anti* — Ref. 10

kinetic 31:69 *syn* : *anti*
thermodynamic ($MgBr_2$) 9:91 *syn* : *anti*

The greater stability of the *anti* isomer is attributed to the pseudoequatorial position of the methyl group in the chair-like chelate. With larger substituent groups, the thermodynamic preference for the *anti* isomer is still greater.[11] Thermodynamic equilibration can be used to control product composition if one of the desired stereoisomers is significantly more stable than the other.

The requirement that an enolate have at least one bulky substituent restricts the types of compounds that can be expected to give highly stereoselective aldol additions. Furthermore, only the enolate formed by kinetic deprotonation is directly available. Ketones with one tertiary alkyl substituent give mainly the *Z*-enolate. However, less highly substituted ketones usually give mixtures of *E*- and *Z*-enolates.[12] Therefore, efforts aimed at expanding the scope of stereoselective aldol condensations have been directed at

10. K. A. Swiss, W.-B. Choi, D. C. Liotta, A. F. Abdel-Magid, and C. A. Maryanoff, *J. Org. Chem.* **56:**5978 (1991).
11. C. H. Heathcock and J. Lampe, *J. Org. Chem.* **48:**4330 (1983).
12. R. E. Ireland, R. H. Mueller, and A. K. Willard, *J. Am. Chem. Soc.* **98:**2868 (1976); W. A. Kleschick, C. T. Buse, and C. H. Heathcock, *J. Am. Chem. Soc.* **99:**247 (1977); Z. A. Fataftah, I. E. Kopka, and M. W. Rathke, *J. Am. Chem. Soc.* **102:**3959 (1980).

two facets of the problem: (1) control of enolate stereochemistry and (2) enhancement of the stereoselectivity in the addition step. We will return to this topic in Section 2.1.3.5.

The enolates of other carbonyl compounds can be used in mixed aldol condensations. Extensive use has been made of the enolates of esters, thioesters, amides, nitriles, and nitroalkanes. Scheme 2.4 gives a selection of such reactions.

Because of their usefulness in aldol additions and other synthetic methods (see especially Section 6.5.2), there has been a good deal of interest in the factors that control the stereoselectivity of enolate formation from esters. For simple esters such as ethyl propanoate, the *E*-enolate is preferred under kinetic conditions using a strong base such as LDA in THF solution. Inclusion of a strong cation solvating co-solvent, such as HMPA or tetrahydro-1,3-dimethyl-2(1*H*)pyrimidone (DMPU) favors the *Z*-enolate.[13]

$CH_3CH_2CO_2CH_2CH_3$ $\xrightarrow[\text{THF}]{\text{LDA}}$ $\xrightarrow{\text{TMSCl}}$ $CH_3CH{=}C(OSi(CH_3)_3)(OCH_2CH_3)$

E-ketene silyl acetal

$CH_3CH_2CO_2CH_2CH_3$ $\xrightarrow[\text{THF, HMPA}]{\text{LDA}}$ $\xrightarrow{\text{TMSCl}}$ $CH_3CH{=}C(OCH_2CH_3)(OSi(CH_3)_3)$

Z-ketene silyl acetal

These observations are explained in terms of a cyclic transition state for the LDA/THF conditions and an open transition state in the presence of an aprotic dipolar solvent.

E-enolate

Z-enolate

Simple alkyl esters show rather low stereoselectivity. However, highly hindered esters derived from 2,6-dimethylphenol or 2,6-di-*t*-butyl-4-methylphenol provide the *anti* stereoisomers.

$R^2CH_2CO_2R^1$ $\xrightarrow{\text{LDA}}$ $R^2CH{=}C(O^-\,{}^+Li)(OR^1)$ $\xrightarrow{\text{RCH=O}}$ $RCH(OH)CH(R^2)CO_2R^1$ + $RCH(OH)CH(R^2)CO_2R^1$

Some illustrative data are given in Table 2.2.

The lithium enolates of α-alkoxy esters have been extensively explored, and several cases in which high stereoselectivity is observed have been documented.[14] This stereoselectivity can be explained in terms of a chelated ester enolate which is approached by the

13. R. E. Ireland, P. Wipf, and J. D. Armstrong III, *J. Org. Chem.* **56**:650 (1991).
14. A. I. Meyers and P. J. Reider, *J. Am. Chem. Soc.* **101**:2501 (1979); C. H. Heathcock, M. C. Pirrung, S. D. Young, J. P. Hagen, E. T. Jarvi, U. Badertscher, H.-P. Märki, and S. H. Montgomery, *J. Am. Chem. Soc.* **106**:8161 (1984).

Scheme 2.4. Addition Reactions of Carbanions Derived from Esters, Carboxylic Acids, Amides, and Nitriles

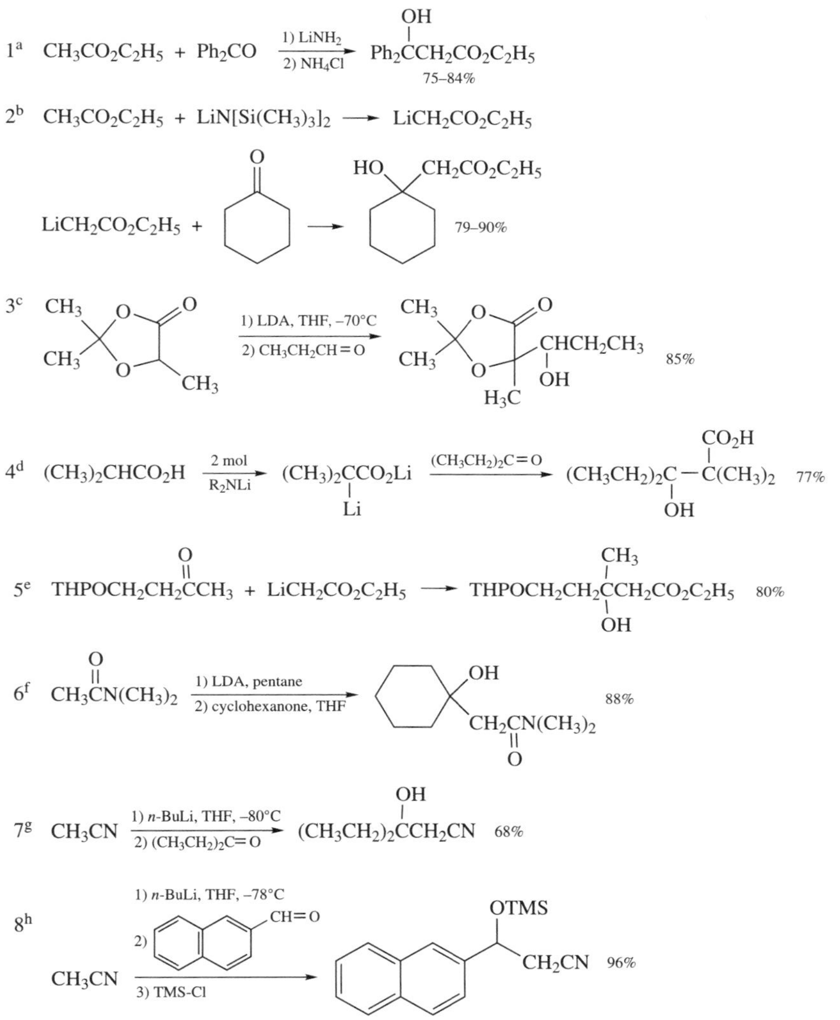

1[a] $CH_3CO_2C_2H_5$ + Ph_2CO $\xrightarrow[\text{2) } NH_4Cl]{\text{1) } LiNH_2}$ $Ph_2C(OH)CH_2CO_2C_2H_5$ 75–84%

2[b] $CH_3CO_2C_2H_5$ + $LiN[Si(CH_3)_3]_2$ ⟶ $LiCH_2CO_2C_2H_5$

$LiCH_2CO_2C_2H_5$ + cyclohexanone ⟶ 1-hydroxy-1-($CH_2CO_2C_2H_5$)cyclohexane (HO, $CH_2CO_2C_2H_5$) 79–90%

3[c] 2,2-dimethyl-5-methyl-1,3-dioxolan-4-one (CH_3, CH_3, O, O, CH_3) $\xrightarrow[\text{2) } CH_3CH_2CH=O]{\text{1) LDA, THF, –70°C}}$ product with $CHCH_2CH_3$, OH, H_3C 85%

4[d] $(CH_3)_2CHCO_2H$ $\xrightarrow[R_2NLi]{\text{2 mol}}$ $(CH_3)_2C(Li)CO_2Li$ $\xrightarrow{(CH_3CH_2)_2C=O}$ $(CH_3CH_2)_2C(OH)-C(CO_2H)(CH_3)_2$ 77%

5[e] $THPOCH_2CH_2C(=O)CH_3$ + $LiCH_2CO_2C_2H_5$ ⟶ $THPOCH_2CH_2C(CH_3)(OH)CH_2CO_2C_2H_5$ 80%

6[f] $CH_3C(=O)N(CH_3)_2$ $\xrightarrow[\text{2) cyclohexanone, THF}]{\text{1) LDA, pentane}}$ 1-hydroxy-1-($CH_2C(=O)N(CH_3)_2$)cyclohexane 88%

7[g] CH_3CN $\xrightarrow[\text{2) } (CH_3CH_2)_2C=O]{\text{1) } n\text{-BuLi, THF, –80°C}}$ $(CH_3CH_2)_2C(OH)CH_2CN$ 68%

8[h] CH_3CN $\xrightarrow{\text{1) } n\text{-BuLi, THF, –78°C; 2) 2-naphthyl-}CH=O\text{; 3) TMS-Cl}}$ 2-naphthyl-CH(OTMS)CH_2CN 96%

a. W. R. Dunnavant and C. R. Hauser, *Org. Synth.* **V**:564 (1973).
b. M. W. Rathke, *Org. Synth.* **53**:66 (1973).
c. C. H. Heathcock, S. D. Young, J. P. Hagen, M. C. Pirrung, C. T. White, and D. VanDerveer, *J. Org. Chem.* **45**:3846 (1980).
d. G. W. Moersch and A. R. Burkett, *J. Org. Chem.* **36**:1149 (1971).
e. J. D. White, M. A. Avery, and J. P. Carter, *J. Am. Chem. Soc.* **104**:5486 (1982).
f. R. P. Woodbury and M. W. Rathke, *J. Org. Chem.* **42**:1688 (1977).
g. E. M. Kaiser and C. R. Hauser, *J. Org. Chem.* **33**:3402 (1968).
h. J. J. P. Zhou, B. Zhong, and R. B. Silverman, *J. Org. Chem.* **60**:2261 (1995).

Table 2.2. Stereoselectivity of Ester Enolates toward Aldehydes[a]

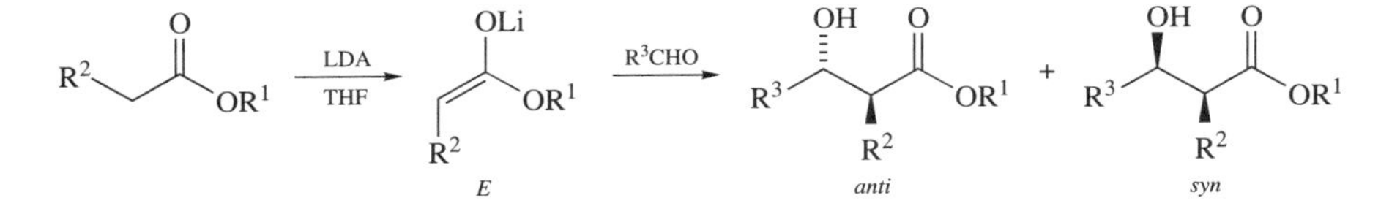

R^1	R^2	R^3	*anti/syn*	Reference
Me	Me	Ph	55 : 45	c
Me	Me	*i*-Pr	55 : 45	c
Me	Me	Me	57 : 43	c
$MeOCH_2$	Me	*i*-Pr	90 : 10	c
$MeOCH_2$	Me	Me	67 : 33	c
DMP[b]	Me	Ph	88 : 12	d
DMP	$H_2C{=}CHCH_2$	Ph	91 : 9	d
DMP	Me	n-C_5H_{11}	86 : 14	d
DMP	$H_2C{=}CHCH_2$	Et	84 : 16	d
DMP	Me	*i*-Pr	> 98 : 2	d
DMP	Et	*i*-Pr	> 98 : 2	d
DMP	$H_2C{=}CHCH_2$	*i*-Pr	> 98 : 2	d
DMP	Me	*t*-Bu	> 98 : 2	d
BHT[b]	Me	Ph	> 98 : 2	d
BHT	$H_2C{=}CHCH_2$	Ph	> 94 : 6	d
BHT	$H_2C{=}CHCH_2$	Et	> 98 : 2	d
BHT	Me	*i*-Pr	> 98 : 2	d
BHT	$H_2C{=}CHCH_2$	*i*-Pr	> 98 : 2	d

a. From a more extensive compilation by C. H. Heathcock, in *Asymmetric Synthesis*, Vol. 3, J. D. Morrison, ed., Academic Press, New York, 1984, Chapter 2.
b. DMP = 2,6-dimethylphenyl; BHT = 2,6-di-*t*-butyl-4-methylphenyl.
c. A. I. Meyers and P. Reider, *J. Am. Chem. Soc.* **101**:2501 (1979).
d. C. H. Heathcock, M. C. Pirrung, S. H. Montgomery, and J. Lampe, *Tetrahedron* **37**:4087 (1981).

aldehyde in such a manner that the aldehyde R group avoids being between the α-alkoxy and the methyl group in the ester enolate. When the ester alkyl group R becomes very bulky, the stereoselectivity is reversed.

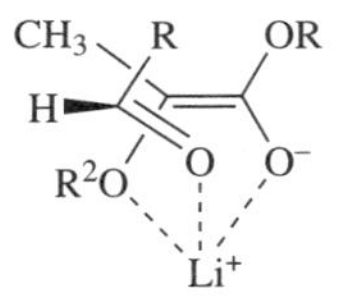

favored 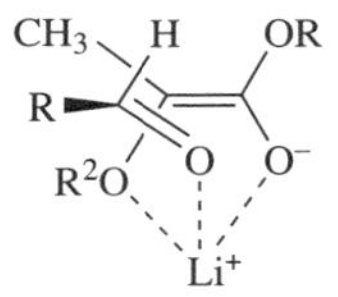 disfavored

Regioselective aldol addition of α,β-unsaturated aldehydes has been achieved using a method in which the enal and the carbonyl acceptor are treated first with a bulky Lewis acid, aluminum *tris*-(2,6-diphenoxide), and then LDA is added.

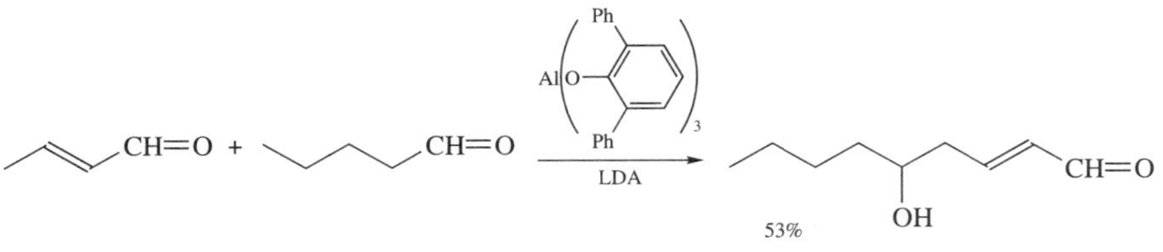

Ref. 15

53%

15. S. Saito, M. Shiozawa, M. Ito, and H. Yamamoto, *J. Am. Chem. Soc.* **120**:813 (1998).

This selectivity presumably reflects several circumstances. Both carbonyl oxygens are presumably complexed by aluminum. The allylic stabilization of the γ-deprotonation product can then lead to kinetic selectivity in the deprotonation. Selectivity for γ-attack by the dienolate is accentuated by the steric bulk near the α position.

2.1.3.2. Boron Enolates. Another important version of the aldol reaction involves the use of boron enolates. A cyclic transition state is believed to be involved, and, in general, the stereoselectivity is higher than for lithium enolates. The O–B bond distances are shorter than the O–Li bond in the lithium enolates, and this leads to a more compact transition state, which magnifies the steric interactions that control stereoselectivity.

E-enolate → *anti*; *Z*-enolate → *syn*

Boron enolates can be prepared by reaction of the ketone with a dialkylboron trifluoromethanesulfonate (triflate) and a tertiary amine.[16] The *Z*-stereoisomer is formed preferentially for ethyl ketones with various R^1 substituents. The resulting aldol products are predominantly the *syn* stereoisomers.

$$R^1COCH_2CH_3 \xrightarrow[(i\text{-Pr})_2NEt]{L_2BOSO_2CF_3} (L_2BO)(R^1)C{=}C(CH_3)(H) \xrightarrow{RCH=O} R^1COCH(CH_3)CH(OH)R$$

The *E*-boron enolate from cyclohexanone shows a preference for the *anti* ketol product.

$$\text{OBL}_2\text{-cyclohexene} + RCH{=}O \longrightarrow \text{major} + \text{minor}$$

The exact ratio of stereoisomeric ketols is a function of the substituents on boron and the solvent.

The *E*-boron enolates of some ketones can be preferentially obtained with the use of dialkylboron chlorides.[17] The data in Table 2.3 pertaining to 3-pentanone and 2-methyl-3-pentanone illustrate this method. Use of boron triflates with a more hindered amine favors the *Z*-enolate. The contrasting stereoselectivity of the boron triflates and chlorides has been discussed in terms of reactant conformation and the stereoelectronic requirement for perpendicular alignment of the hydrogen being removed with the carbonyl group.[18] The

16. D. A. Evans, E. Vogel, and J. V. Nelson, *J. Am. Chem. Soc.* **101:**6120 (1979); D. A. Evans, J. V. Nelson, E. Vogel, and T. R. Taber, *J. Am. Chem. Soc.* **103:**3099 (1981).
17. H. C. Brown, R. K. Dhar, R. K. Bakshi, P. K. Pandiarajan, and B. Singaram, *J. Am. Chem. Soc.* **111:**3441 (1989); H. C. Brown, R. K. Dhar, K. Ganesan, and B. Singaram, *J. Org. Chem.* **57:**499 (1992); H. C. Brown, R. K. Dhar, K.Ganesan, and B. Singaram, *J. Org. Chem.* **57:**2716 (1992); H. C. Brown, K. Ganesan, and R. K. Dhar, *J. Org. Chem.* **58:**147 (1993); K. Ganesan and H. C. Brown, *J. Org. Chem.* **58:**7162 (1993).
18. J. M. Goodman and I. Paterson, *Tetrahedron Lett.* **33:**7223 (1992).

two preferred transitions states are shown below.

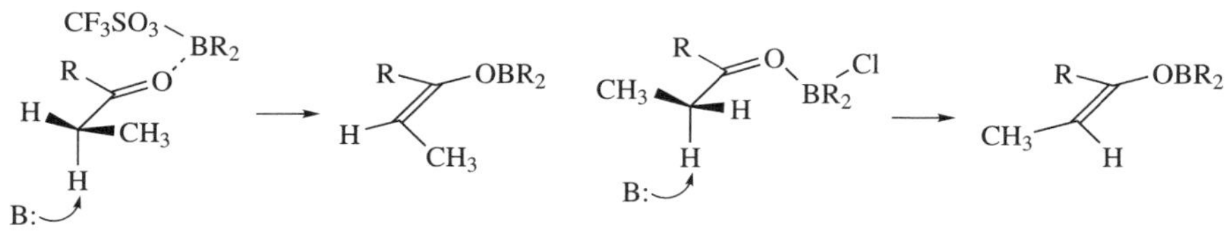

Other methods are also available for generation of boron enolates. Dialkylboranes react with acyclic enones to give *Z*-enolates by a 1,4-reduction.[19] The preferred *Z*-stereochemistry is attributed to a cyclic mechanism for hydride transfer:

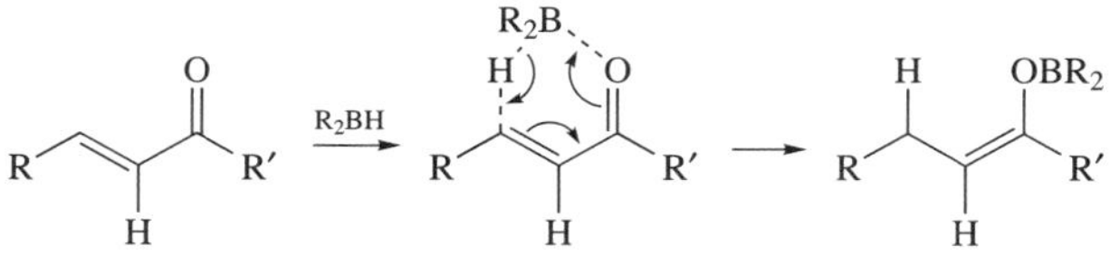

Z-Boron enolates can also be obtained from silyl enol ethers. This method is necessary for ketones such as ethyl *t*-butyl ketone, which gives *E*-boron enolates by other methods. The *Z*-stereoisomer is formed from either the *Z*- or *E*-silyl enol ether.[20]

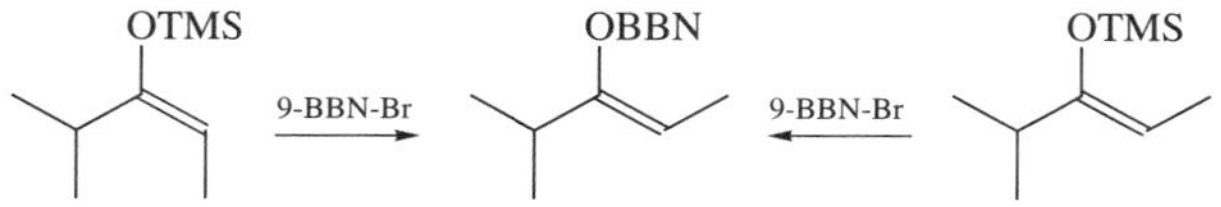

The *E*-boron enolates show a modest preference for formation of the *anti* aldol product.

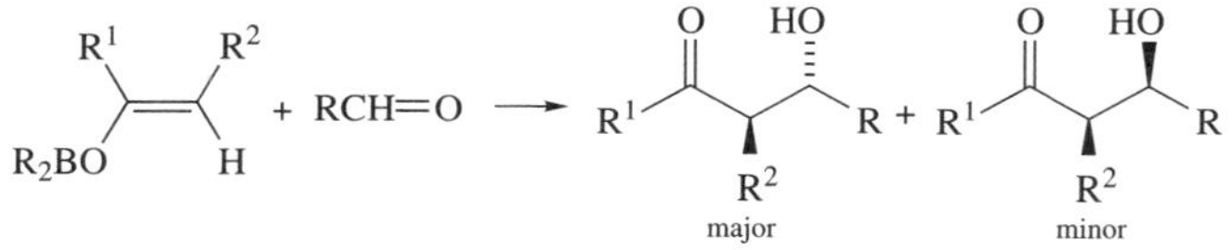

The general trend then is that boron enolates *parallel* lithium enolates in their stereoselectivity but show *enhanced stereoselectivity*. They also have the advantage of providing access to both stereoisomeric enol derivatives. Table 2.3 gives a compilation of some of the data on stereoselectivity of aldol reactions with boron enolates.

Boron enolates can also be obtained from esters[21] and amides,[22] and these too undergo aldol addition reactions. Various combinations of boronating reagents and amines have been used, and the $E:Z$ ratios are dependent on the reagents and conditions. In most

19. D. A. Evans and G. C. Fu, *J. Org. Chem.* **55:**5678 (1990); G. P. Boldrini, M. Bortolotti, F. Mancini, E. Tagliavini, C. Trombini, and A. Umani-Ronchi, *J. Org. Chem.* **56:**5820 (1991).
20. J. L. Duffy, T. P. Yoon, and D. A. Evans, *Tetrahedron Lett.* **36:**9245 (1993).
21. K. Ganesan and H. C. Brown, *J. Org. Chem.* **59:**2336 (1994).
22. K. Ganesan and H. C. Brown, *J. Org. Chem.* **59:**7346 (1994).

Table 2.3. Stereoselectivity of Boron Enolates toward Aldehydes[a]

O=C(R^1)Et —L_2BOTf→ Z-enolate (OBL_2, R^1) + E-enolate (OBL_2, R^1) —R^2CHO→ syn aldol (OH, O; R^2, R^1) + anti aldol (OH, O; R^2, R^1)

Z   *E*   *syn*   *anti*

R^1	L[b]	R^2	*Z/E*	*syn* : *anti*	Reference
Et	n-C_4H_9	Ph	> 97 : 3	> 97 : 3	c
Et	c-C_5H_9	Ph	82 : 18	84 : 16	c
Et	n-C_4H_9	Ph	69 : 31	72 : 28	c
Et	n-C_4H_9	n-Pr	> 97 : 3	> 97 : 3	c
Et	n-C_4H_9	t-Bu	> 97 : 3	> 97 : 3	c
Et	n-C_4H_9	$H_2C{=}C(CH_3)$	> 97 : 3	92 : 8	c
Et	n-C_4H_9	(E)-C_4H_7	> 97 : 3	93 : 7	c
i-Bu	n-C_4H_9	Ph	> 99 : 1	> 97 : 3	c
i-Bu	c-C_5H_9	Ph	–	84 : 16	c
i-Pr	n-C_4H_9	Ph	45 : 55	44 : 56	c
i-Pr	c-C_5H_9	Ph	19 : 81	18 : 82	c
t-Bu	n-C_4H_9	Ph	> 99 : 1	> 97 : 3	c
c-C_6H_{11}	c-C_5H_9	Ph	12 : 88	14 : 86	d
c-C_6H_{11}	9-BBN	Ph	> 99 : 1	> 97 : 3	d
Ph	n-C_4H_9	Ph	99 : 1	> 97 : 3	c
Et	9-BBN	Ph		> 97 : 3	c
i-Pr	9-BBN	Ph		46 : 54	c
c-C_6H_{11}	9-BBN	Ph		96 : 4	e
t-Bu	9-BBN	Ph		< 3 : 97	e
Et	c-C_6H_{11}	Ph		21 : 79	c
i-Pr	c-C_6H_{11}	Ph		< 3 : 97	e
c-C_6H_{11}	c-C_6H_{11}	Ph		< 1 : 99	e
t-Bu	c-C_6H_{11}	Ph		< 3 : 97	e
Et	2-BCOB	Ph		3 : 97	f
i-Pr	2-BCOB	Ph		< 3 : 97	f
c-C_6H_{11}	2-BCOB	Ph		< 3 : 97	f
t-Bu	2-BCOB	Ph		< 3 : 97	f
Et	n-C_4H_9	Ph	96 : 4	95 : 5	g
n-C_5H_{11}	n-C_4H_9	Ph	95 : 5	94 : 6	g
n-C_9H_{19}	n-C_4H_9	Ph	91 : 9	91 : 9	g
$PhCH_2$	n-C_4H_9	Ph	95 : 5	95 : 5	g
3-C_5H_{11}	n-C_4H_9	Ph	99 : 1	> 99 : 1	g
c-C_6H_{11}	n-C_4H_9	Ph	98 : 2	> 99 : 1	g
c-C_6H_{11}	n-C_4H_9	$PhCH_2CH_2$	98 : 2	> 98 : 2	g

a. From a more complete compilation by C. H. Heathcock, in *Asymmetric Synthesis*, Vol. 3, J. D. Morrison, ed., Academic Press, New York, 1984, Chapter 3.
b. 9-BBN = 9-borabicyclo[3.3.1]nonane; 2-BCOB = *bis*-(bicyclo[2.2.2]octyl)borane.
c. D. A. Evans, J. V. Nelson, E. Vogel, and T. R. Taber, *J. Am. Chem. Soc.* **103**:3099 (1981).
d. D. E. Van Horn and S. Masumune, *Tetrahedron Lett.* **1979**:2229.
e. Using dialkylboron chloride and (i-Pr)$_2$ NEt; H. C. Brown, R. K. Dhar, R. K. Bakshi, P. K. Pandjarajan, and P. Singaram, *J. Am. Chem. Soc.* **111**:3441 (1989); H. C. Brown, K. Ganesan, and R. K. Dhar, *J. Org. Chem.* **58**:147 (1993).
f. Using bis(bicyclo[2.2.2]octyl)boron chloride and Et_3N; H. G. Brown, K. Ganesan, and R. K. Dhar, *J. Org. Chem.* **57**:3767 (1992).
g. I. Kuwajima, M. Kato, and A. Mori, *Tetrahedron Lett.* **21**:4291 (1980).

cases, esters give *Z*-enolates which lead to *syn* adducts, but there are exceptions.

$CH_3CH_2CO_2C_2H_5$ —1) $Bu_2BO_3SCF_3$, *i*-Pr_2NEt; 2) $(CH_3)_2CHCH{=}O$→ $(CH_3)_2CH$–CH(OH)–CH(CH_3)–$CO_2C_2H_5$ 81% yield, 95:5 *syn:anti*

Ref. 23

$RCH_2CO_2C_2H_5$ —1) $(C_6H_{11})_2BI$, Et_3N; 2) PhCH=O→ Ph–CH(OH)–CH(R)–$CO_2C_2H_5$ or Ph–CH(OH)–CH(R)–$CO_2C_2H_5$

anti favored for R = *i*-Pr, *t*-Bu, Ph *syn* favored for R = CH_3, CH_2CH_3

2.1.3.3. Titanium, Tin, and Zirconium Enolates. Metals such as Ti, Sn, and Zr give enolates which are intermediate in structural character between the largely ionic Li^+ enolates and covalent boron enolates. The Ti, Sn, or Zr enolates provide oxygen–metal bonds that are largely covalent in character but can also accommodate additional ligands at the metal. Depending on the degree of substitution, both cyclic and acyclic transition states can be involved.

Titanium enolates can be prepared from lithium enolates by reaction with trialkoxytitanium(IV) chlorides, such as (isopropoxy)titanium chloride.[24] Titanium enolates can also be prepared directly from ketones by reaction with $TiCl_4$ and a tertiary amine.[25]

O —1) $TiCl_4$; 2) $(i\text{-}Pr)_2NEt$→ $OTiCl_3$ —$(CH_3)_2CHCH{=}O$→ O OH

Under these conditions, the *Z*-enolate is formed and the aldol adducts have *syn* stereochemistry. The addition can proceed through a cyclic transition state assembled around titanium.

Ti O R H R′ R^E O R^Z → Ti O R H R′ R^E O R^Z R^E = H *syn* R^Z = H *anti*

Titanium enolates can also be prepared from *N*-acyloxazolidinones. These enolates

23. A. Abiko, J.-F. Liu, and S. Masamune, *J. Org. Chem.* **61:**2590 (1996).
24. C. Siegel and E. Thornton, *J. Am. Chem. Soc.* **111:**5722 (1989).
25. D. A. Evans, D. L. Rieger, M. T. Bilodeau, and F. Urpi, *J. Am. Chem. Soc.* **113:**1047 (1991).

are considered to be chelated with the oxazolidinone carbonyl oxygen.[26]

$TiCl_4$, $(i\text{-}Pr)_2NEt$; $(CH_3)_2CHCH{=}O$

87% yield, 94:6 *syn:anti*

Trialkoxytitanium chlorides, which are somewhat less reactive, can also be used. Reactions of these enolates with aldehydes give mainly *syn* products, with the absolute stereochemistry being determined by the configuration of the oxazolidinone.[27]

1) LDA
2) $(i\text{-}PrO)_3TiCl$
3) $PhCH{=}O$

These results are explained on the basis of a transition state which is hexacoordinate at titanium. The oxazolidinone substituent dictates the approach of the aldehyde.

Procedures which are catalytic in titanium have been developed.[28] These reactions appear to exhibit the same stereoselectivity trends as other titanium-mediated additions.

\+ $PhCH{=}O$ — 5% TiF_4, CH_3CH_2CN → 86%

26. D. A. Evans, F. Urpi, T. C. Somers, J. S. Clark, and M. T. Bilodeau, *J. Am. Chem. Soc.* **112:**8215 (1990).
27. M. Nerz-Stormes and E. R. Thornton, *J. Org. Chem.* **56:**2489 (1991).
28. R. Mahrwald, *Chem.Ber.* **128:**919 (1995).

Tin enolates can be generated from ketones and $Sn(O_3SCF_3)_2$ in the presence of tertiary amines.[29,30] The subsequent aldol addition is *syn*-selective.[31]

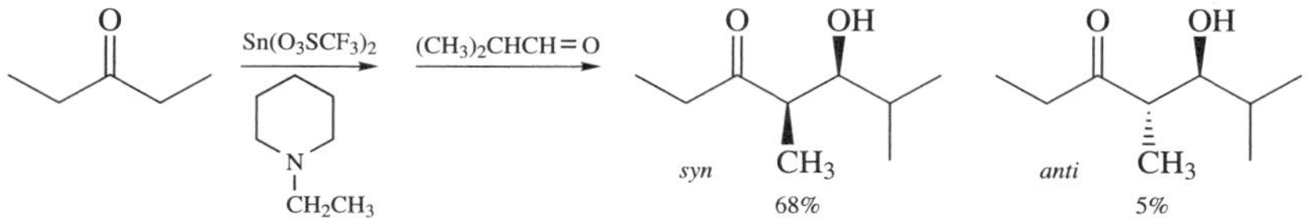

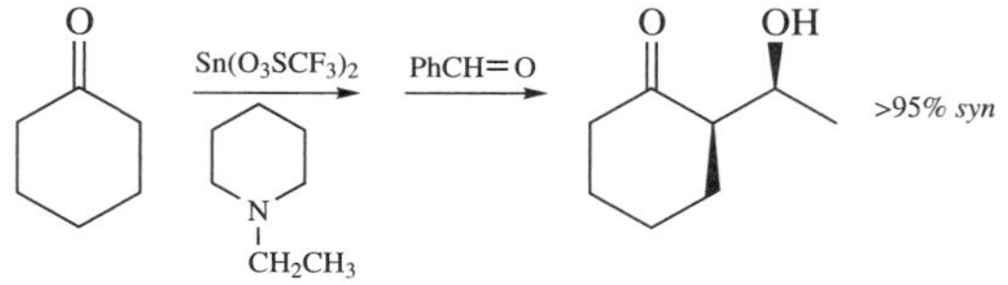

$Sn(O_3SCF_3)_2$ / N-ethylpiperidine; $(CH_3)_2CHCH{=}O$ → *syn* 68% + *anti* 5%

$Sn(O_3SCF_3)_2$ / N-ethylpiperidine; $PhCH{=}O$ → >95% *syn*

Tin(II) enolates prepared in this way also show good reactivity toward ketones as the carbonyl component.

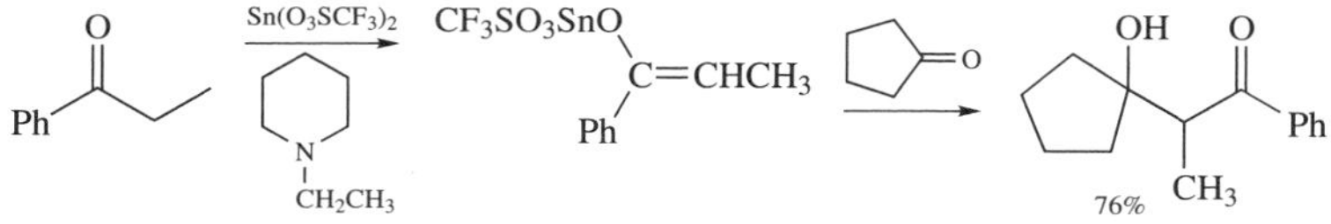

$Sn(O_3SCF_3)_2$ / N-ethylpiperidine → $CF_3SO_3SnO(Ph)C{=}CHCH_3$; cyclopentanone → 76% — Ref. 32

N-Acylthiazolinethiones are also useful enolate precursors under these conditions.

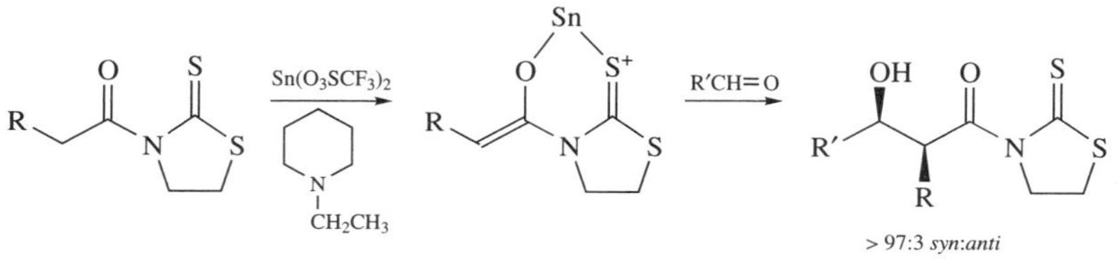

$Sn(O_3SCF_3)_2$ / N-ethylpiperidine; $R'CH{=}O$ → > 97:3 *syn:anti* — Ref. 33

Uncatalyzed additions of trialkylstannyl enolates to benzaldehyde show *anti* stereoselectivity, suggesting a cyclic transition state.[34]

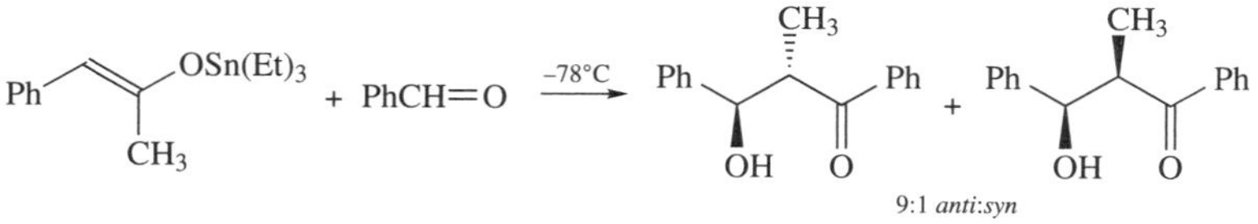

$PhCH{=}C(CH_3)OSn(Et)_3$ + $PhCH{=}O$ $\xrightarrow{-78°C}$ 9:1 *anti:syn*

Isolated tributylstannyl enolates react with benzaldehyde under the influence of metal salts

29. T. Mukaiyama and S. Kobayashi, *Org. React.* **46**:1 (1994).
30. T. Mukaiyama, N. Iwasawa, R. W. Stevens, and T. Haga, *Tetrahedron* **40**:1381 (1984); I. Shibata and A. Babu, *Org. Prep. Proc. Int.* **26**:85 (1994).
31. T. Mukaiyama, R. W. Stevens, and N. Iwasawa, *Chem. Lett.* **1982**:353.
32. R. W. Stevens, N. Iwasawa, and T. Mukaiyama, *Chem. Lett.* **1982**:1459.
33. T. Mukaiyama and N. Iwasawa, *Chem. Lett.* **1982**:1903; N. Iwasawa, H. Huang, and T. Mukaiyama, *Chem. Lett.* **1985**:1045.
34. S. S. Labadie and J. K. Stille, *Tetrahedron* **40**:2329 (1984).

including $Pd(O_3SCF_3)_2$, $Zn(O_3SCF_3)_2$, and $Cu(O_3SCF_3)_2$.[35] The *anti*:*syn* ratio depends on the catalyst.

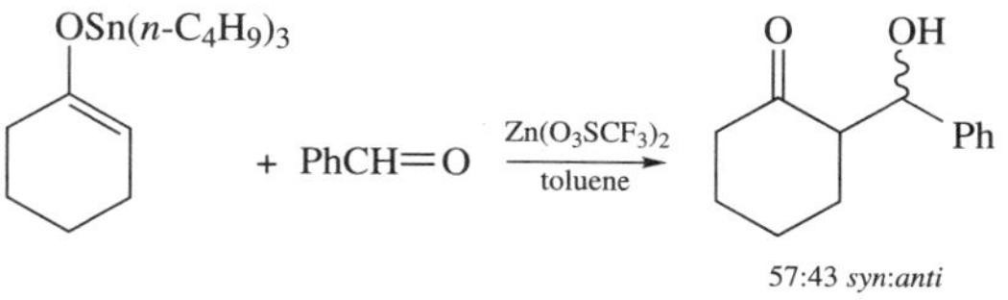

57:43 *syn*:*anti*

Zirconium enolates are prepared by reaction of lithium enolates with $(Cp)_2ZrCl_2$ ($C_p = \eta^5\text{-}C_5H_5$).[36] They act as nucleophiles in aldol addition reactions.

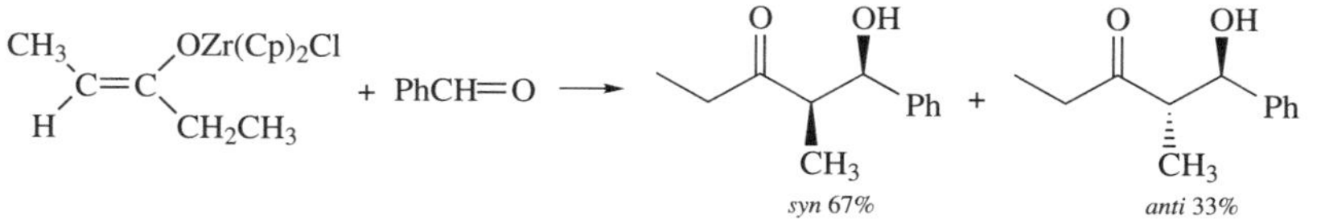

syn 67% *anti* 33% Ref. 37

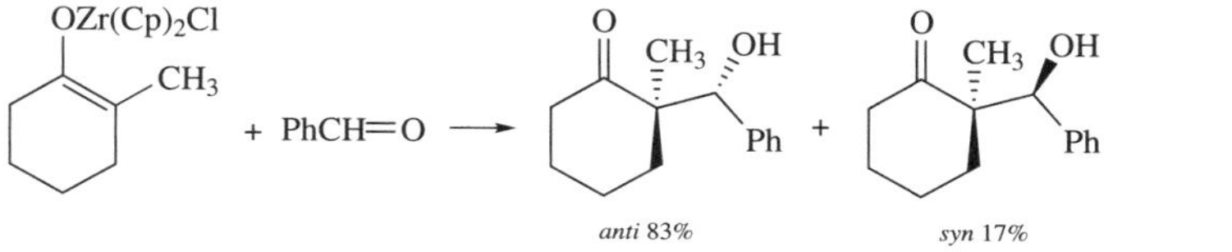

anti 83% *syn* 17% Ref. 38

Aldol additions of silyl enol ethers and ketene silyl acetals can be catalyzed by $(Cp)_2Zr^{2+}$ species, including $[(Cp)_2ZrO\text{-}t\text{-}Bu]^+$ and $(Cp)_2Zr(O_3SCF_3)_2$.[39]

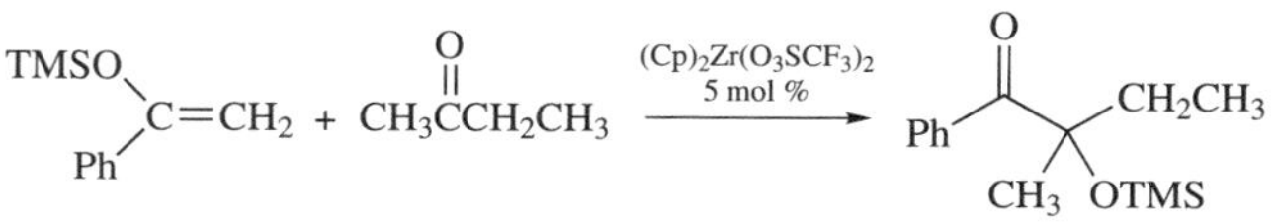

A comprehensive comparison of the *anti*:*syn* diastereoselectivity of the lithium, dibutylboron, and $(Cp)_2Zr$ enolates of 3-methyl-2-hexanone with benzaldehyde has been reported.[38] The order of stereoselectivity is $Bu_2B > (Cp)_2Zr > Li$. These results are consistent with reactions proceeding through a cyclic transition state.

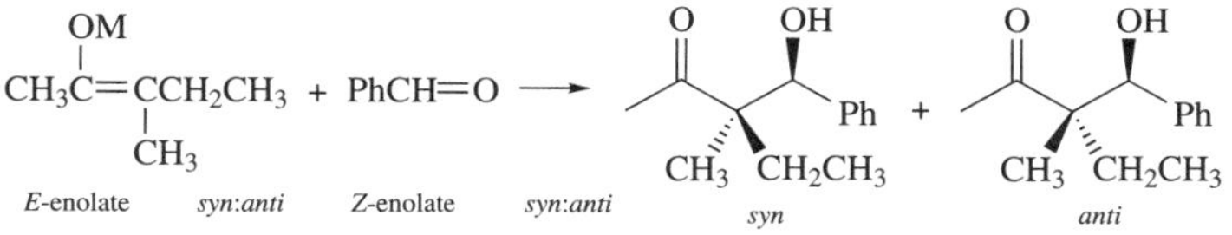

E-enolate	*syn*:*anti*	*Z*-enolate	*syn*:*anti*
Li	17:83	Li	45:55
Bu_2B	3:97	Bu_2B	94:6
$(Cp)_2ZrCl$	9:91	$(Cp)_2ZrCl$	86:14

35. A. Yanagisawa, K. Kimura, Y. Nakatsuka, and M. Yamamoto, *Synlett* **1998:**958.
36. (a) D. A. Evans and L. R. McGee, *Tetrahedron Lett.* **21:**3975 (1980); (b) M. Braun and H. Sacha, *Angew. Chem. Int. Ed. Engl.* **30:**1318 (1991); (c) S. Yamago, D. Machii, and E. Nakamura, *J. Org. Chem.* **56:**2098 (1991).
37. Y. Yamamoto and K. Maruyama, *Tetrahedron Lett.* **21:**4607 (1980).
38. (a) T. K. Hollis, N. P. Robinson, and B. Bosnich, *Tetrahedron Lett.* **33:**6423 (1992); (b) Y. Hong, D. J. Norris, and S. Collins, *J. Org. Chem.* **58:**3591 (1993).
39. S. Yamago, D. Machii, and E. Nakamura, *J. Org. Chem.* **56:**2098 (1991).

2.1.3.4. The Mukaiyama Reaction. The Mukaiyama reaction refers to Lewis acid-catalyzed aldol addition reactions of enol derivatives. The initial examples involved silyl enol ethers.[40] Silyl enol ethers do not react with aldehydes because the silyl enol ether is not a strong enough nucleophile. However, Lewis acids do cause reaction to occur by activating the ketone. The simplest mechanistic formulation of the Lewis acid catalysis is that complexation occurs at the carbonyl oxygen, activating the carbonyl group to nucleophilic attack.

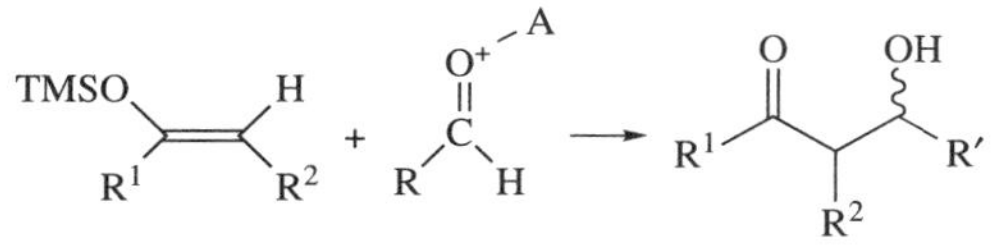

If there is no other interaction, such a reaction should proceed through an acyclic transition state, and steric factors should determine the amount of *syn* versus *anti* addition.[41] This seems to be the case with BF_3, where stereoselectivity increases with the steric bulk of the silyl enol ether substituent R^1.[42]

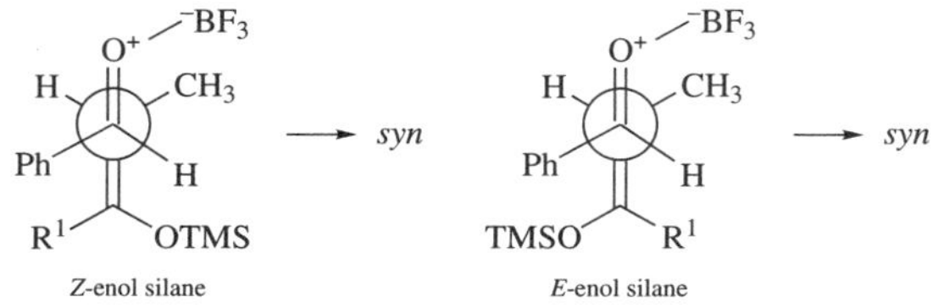

R^1	*Z*-enol silane *syn:anti*	*E*-enol silane *syn:anti*
Et	60:40	57:43
i-Pr	56:44	35:65
t-Bu	<5:95	– –
Ph	47:53	30:70

α-Substituted aldehydes show a preference for a *syn* relationship between the α-substituent and hydroxy group. This is consistent with a Felkin–Ahn transition state (see Section 3.10 to review the effect of α-substituents on carbonyl addition reactions.).[43]

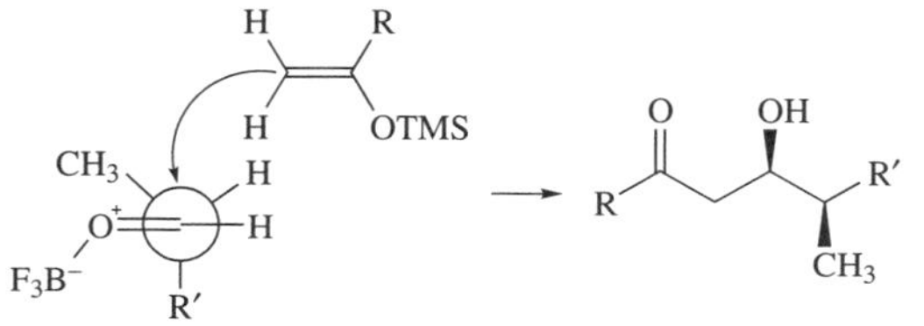

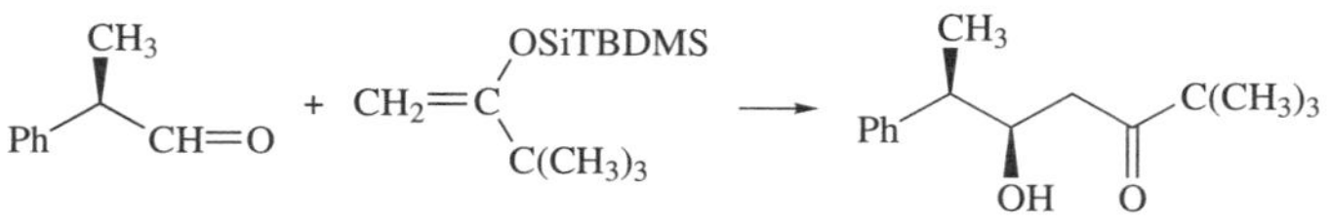

24:1 *syn:anti*

The results suggest that competition between *antiperiplanar* and *synclinal* transitions

40. T. Mukaiyama, K. Banno, and K. Narasaka, *J. Am. Chem. Soc.* **96:**7503 (1974).
41. S. Murata, M. Suzuki, and R. Noyori, *J. Am. Chem. Soc.* **102:**3248 (1980); Y. Yamamoto, H. Yatagai, Y. Naruta, and K. Maruyama, *J. Am. Chem. Soc.* **102:**7107 (1980).
42. C. H. Heathcock, K. T. Hug, and L. A. Flippin, *Tetrahedron Lett.* **25:**5973 (1984).
43. C. H. Heathcock and L. A. Flippin, *J. Am. Chem. Soc.* **105:**1667 (1983).

states is controlled by both steric and electrostatic effects.

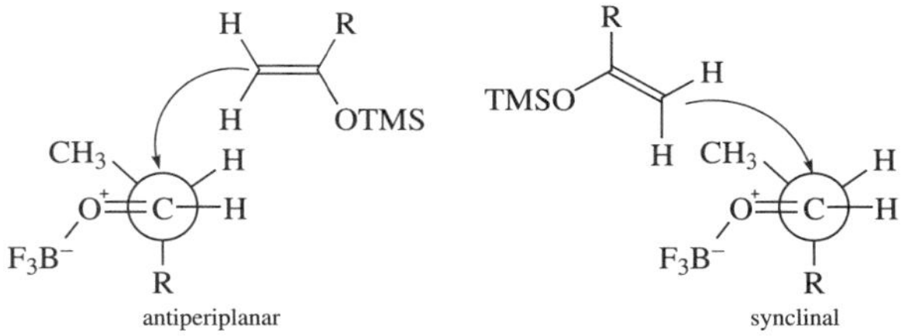

antiperiplanar

synclinal

The analysis of the transition-state effect on stereoselectivity has been extended to incorporate α,β-disubstituted systems.[44]

Quite a number of Lewis acids besides $TiCl_4$ and BF_3 can catalyze the Mukaiyama reaction, including $Bu_2Sn(O_3SCF_3)_2$,[45] Bu_3SnClO_4,[46] $Sn(O_3SCF_3)_2$,[47] $Zn(O_3SCF_3)_2$[48] and $LiClO_4$.[49] Triaryl perchlorate salts are also very active catalysts.[50] Examples of these reactions are included in Scheme 2.5.

Trialkylsilyl cations may play a key role in Lewis acid-catalyzed reactions. Trimethylsilyl triflate itself is not a good catalyst, but in combination with other Lewis acids it generates excellent catalytic activity.

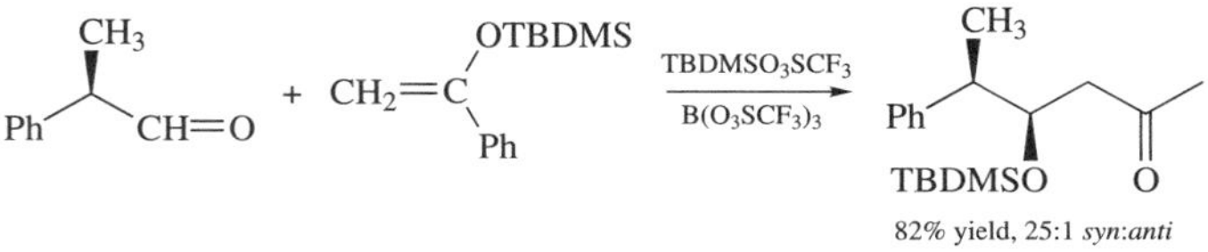

Ref. 51

82% yield, 25:1 *syn:anti*

Hindered bis-(phenoxy)aluminum derivatives are also powerful co-catalysts (see entry 15, Scheme 2.5). They are believed to act by sequestering the triflate anion.[52]

Silyl enol ethers react with formaldehyde and benzaldehyde in water–THF mixtures with the use of lanthanide triflates such as $Yb(O_3SCF_3)_3$ as catalysts. The catalysis reflects the strong affinity of lanthanides for carbonyl oxygen, even in aqueous solution.

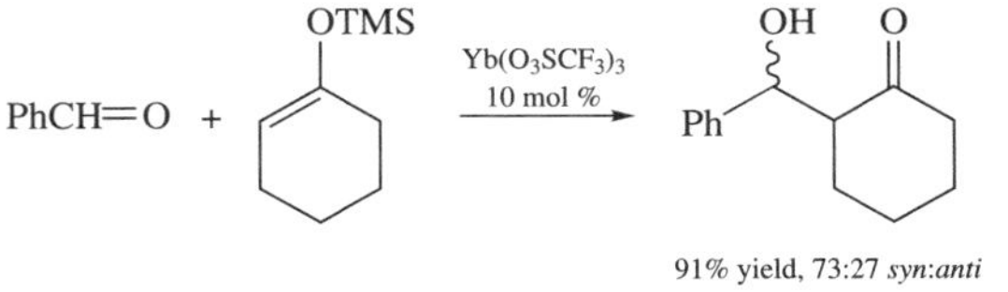

Ref. 53

91% yield, 73:27 *syn:anti*

Cerium, samarium, and other lanthanide halides promote addition of ketene silyl enol ethers to aldehydes.[54] Imines react with ketene silyl acetals in the presence of $Yb(O_3SCF_3)_3$. Preferential addition to the imine occurs even in the presence of aldehyde

44. D. A. Evans, M. J. Dart, J. L. Duffy, and M. G. Yang, *J. Am. Chem. Soc.* **118:**4322 (1996).
45. T. Sato, J. Otera, and H. Nozaki, *J. Am. Chem. Soc.* **112:**901 (1990).
46. J. Otera and J. Chen, *Synlett* **1996:**321.
47. T. Oriyama, K. Iwanami, Y. Miyauchi, and G. Koga, *Bull. Chem. Soc. Jpn.* **63:**3716 (1990).
48. M. Chini, P. Crotti, C. Gardelli, F. Minutolo, and M. Pineschi, *Gazz. Chim. Ital.* **123:**673 (1993).
49. M. T. Reetz and D. N. A. Fox, *Tetrahedron Lett.* **34:**1119 (1993).
50. T. Mukaiyama, S. Kobayashi, and M. Murakami, *Chem. Lett.* **1985:**447; T. Mukaiyama, S. Kobayashi, and M. Murakami, *Chem. Lett.* **1984:**1759; S. E. Denmark and C.-T. Chen, *Tetrahedron Lett.* **35:**4327 (1994).
51. A. P. Davis and S. J. Plunkett, *J. Chem. Soc., Chem. Commun.* **1995:**2173; A. P. Davis, J. E. Muir, and S. J. Plunkett, *Tetrahedron Lett.* **37:**9401 (1996).
52. M. Oishi, S. Aratake, and H. Yamamoto, *J. Am. Chem. Soc.* **120:**8271 (1998).
53. S. Kobayashi and I. Hachiya, *J. Org. Chem.* **59:**3590 (1994).
54. P. Van de Weghe and J. Collin, *Tetrahedron Lett.* **34:**3881 (1993); A. E. Vougioukas and H. B. Kagan, *Tetrahedron Lett.* **28:**5513 (1987).

Scheme 2.5. Mukaiyama Reactions

A. Reactions of silyl end ethers with aldehydes and ketones

1[a] (2-methyl-6-(trimethylsilyloxy)cyclohexene, OTMS, CH_3) + $PhCH{=}O$ $\xrightarrow[\text{2) } H_2O]{\text{1) } R_4N^+\ {}^-F}$ 2-methyl-6-(hydroxy(phenyl)methyl)cyclohexanone 68%

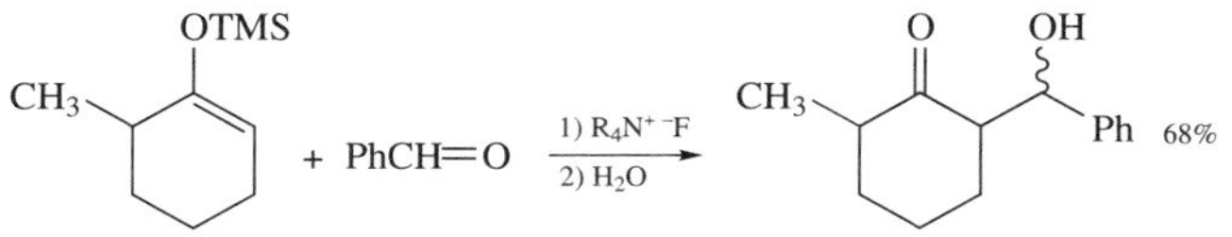

2[b] $PhCH(CH_3)CH{=}O$ + $CH_2{=}C(OTBDMS)CH_3$ $\xrightarrow{BF_3}$ Ph–CH(CH_3)–CH(OH)–CH_2–C(=O)–CH_3 75%

3[c] TMSO-pinene enol ether + $CH_3CH{=}O$ $\xrightarrow{TiCl_4}$ aldol product (CH_3CH(OH)– substituted pinanone, H) 96%

4[d] $PhC(OTMS){=}CH_2$ + $(CH_3)_2C{=}O$ $\xrightarrow{TiCl_4}$ $PhCOCH_2C(OH)(CH_3)_2$ 70–74%

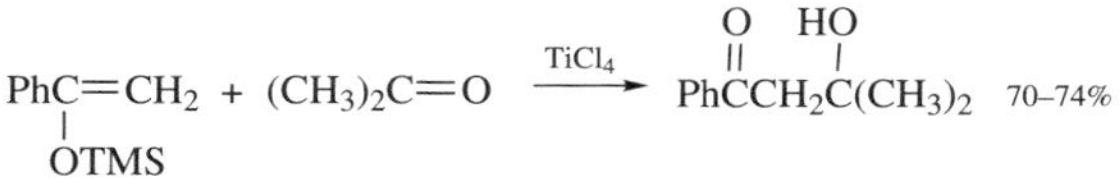

B. Reactions with acetals

5[e] $PhCH(CH_3)CH(OCH_3)_2$ + $CH_2{=}C(OTMS)C(CH_3)_3$ $\xrightarrow[-78^\circ C]{TiCl_4}$ Ph–CH(CH_3)–CH(OCH_3)–CH_2–C(=O)–$C(CH_3)_3$ (syn) + Ph–CH(CH_3)–CH(OCH_3)–CH_2–C(=O)–$C(CH_3)_3$ (anti)

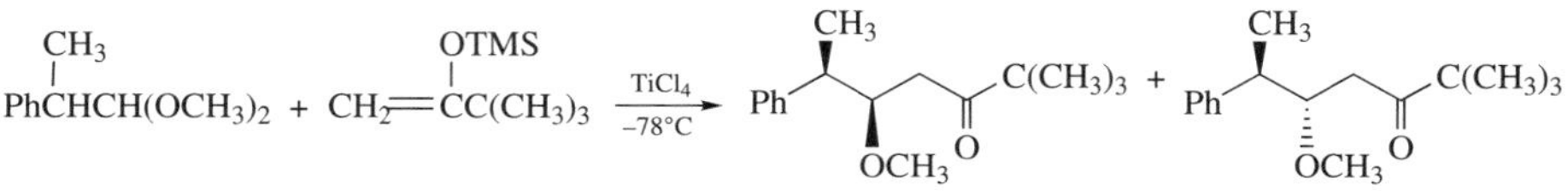

84% yield, 2.5:1 *syn*:*anti*

6[f] 1-(trimethylsilyloxy)cyclopentene (OTMS) + $(CH_3)_2C(OCH_3)_2$ $\xrightarrow[\text{5 mol \%}]{(CH_3)_3SiO_3SCF_3}$ 2-($C(CH_3)_2OCH_3$)cyclopentanone 87%

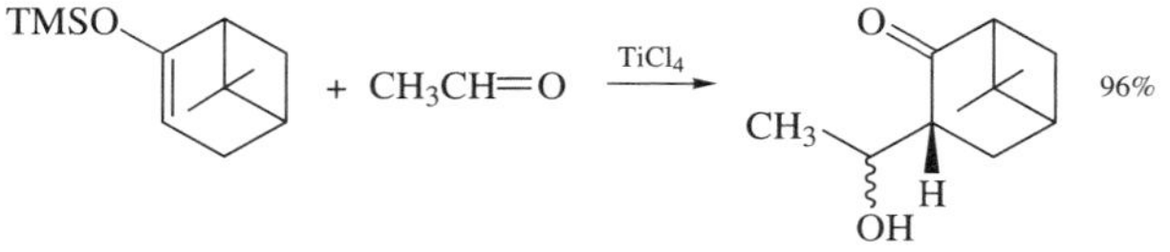

7[g] TMSO-cyclohexene (CH_3, CH_3, $C(CH_3)_2$) + 2-methoxytetrahydrofuran (OCH_3) $\xrightarrow{Ph_3C^+\ {}^-ClO_4}$ 2-(tetrahydrofuran-2-yl)-2-methylcyclohexanone (CH_3, $C(CH_3)_2$) 80%

8[h] $CH_3(CH_2)_3C(OCH_3)_2CH_3$ + $CH_2{=}C(OTMS)C(CH_3)_3$ $\xrightarrow[-78^\circ C]{Bu_2Sn(O_3SCF_3)_2\text{, 5 mol \%}}$ $CH_3(CH_2)_3C(CH_3)(OCH_3)CH_2COC(CH_3)_3$ 100%

9[i] cyclohexene (CH_3, CO_2CH_3, CH_3, CH_3) with 1,3-dioxolan-2-yl substituent + 3-methyl-3-vinyl-1-(OTMS)cyclohexene (CH_3) $\xrightarrow{TiCl_4}$ condensation product (CH_3, CO_2CH_3, CH_3, CH_3, H, CH_3, O) 96%

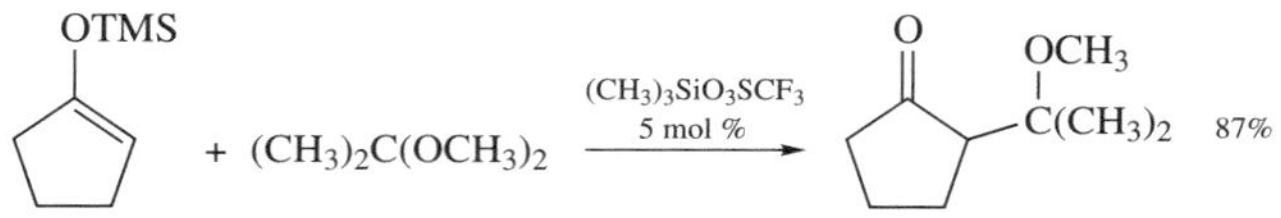

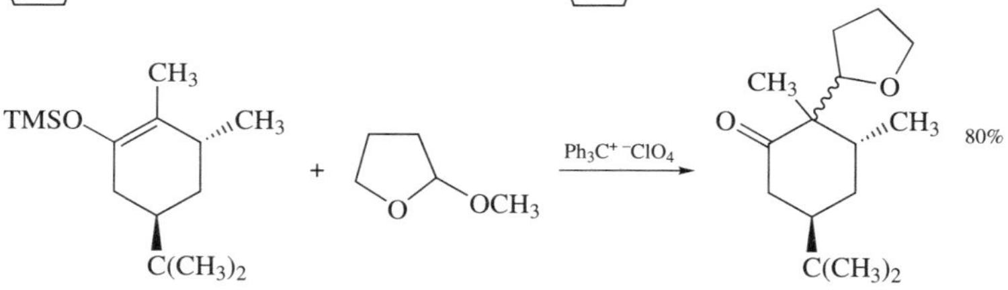

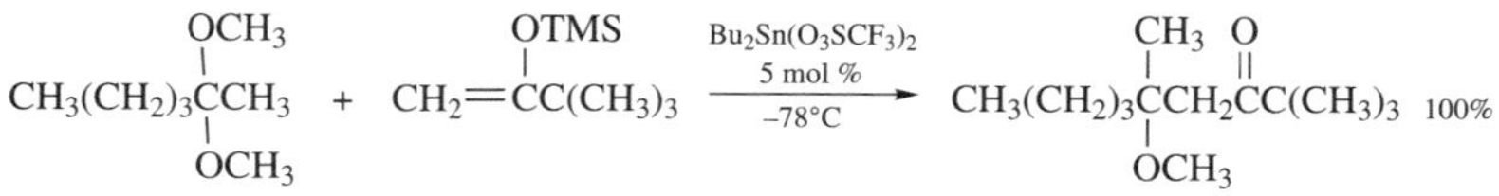

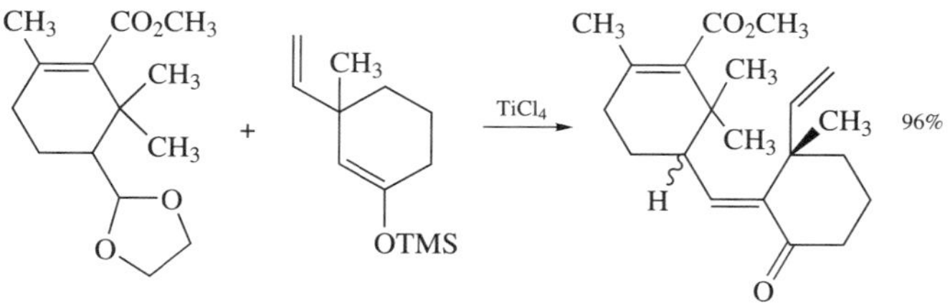

C. Catalytic Mukaiyama reactions

10[j] (acetonide aldehyde) + $(CH_3)_2C{=}C(OTMS)(OCH_3)$ $\xrightarrow{LiClO_4}$ TMSO adduct, CO_2CH_3 — >98% *syn*

11[k] $(CH_3)_2CHCH{=}O$ + $H_2C{=}C(OTMS)Ph$ $\xrightarrow{(Cp)_2Ti(O_3SCF_3)_2}$ $(CH_3)_2CHCH(OTMS)CH_2C(O)Ph$

12[l] $CH_3CH_2CH{=}O$ + $CH_3CH{=}C(OTMS)(OCH_3)$ $\xrightarrow{(Cp)_2Ti(O_3SCF_3)_2}$ OH, CH_3, CO_2CH_3 product

91% yield, 1:1.4 *syn*:*anti*

13[m] $CH_3CH(OCH_2Ph)CH{=}O$ + $H_2C{=}C(OTBDMS)(OCH_3)$ $\xrightarrow[-30°C]{LiClO_4\ 3\ mol\ \%}$ OCH_2Ph, OH, CO_2CH_3 product

92:8 *syn*:*anti*

14[n] $CH_3CH(OCH_2Ph)CH{=}O$ + $H_2C{=}C(OTMS)C(CH_3)_3$ $\xrightarrow{TiCl_4}$ $PhCH_2O$, OH, $C(O)C(CH_3)_3$ product >97% *syn*

15[o] cyclohexanone + $H_2C{=}C(OTMS)Ph$ $\xrightarrow[MABR\ 5\%]{TMSOTf}$ 1-hydroxycyclohexyl-$CH_2C(O)Ph$ 90%

MABR = *bis*(4-bromo-2,6-di-*tert* butylphenoxy) methyl aluminum

a. R. Noyori, K. Yokoyama, J. Sakata, I. Kuwajima, E. Nakamura, and M. Shimizu, *J. Am. Chem. Soc.* **99:**1265 (1977).
b. C. H. Heathcock and L. A. Flippin, *J. Am. Chem. Soc.* **105:**1667 (1983).
c. T. Yanami, M. Miyashita, and A. Yoshikoshi, *J. Org. Chem.* **45:**607 (1980).
d. T. Mukaiyama and K. Narasaka, *Org. Synth.* **65:**6 (1987).
e. I. Mori, K. Ishihara, L. A. Flippin, K. Nozaki, H. Yamamoto, P. A. Bartlett, and C. H. Heathcock,*J. Org. Chem.* **55:**6107 (1990).
f. S. Murata, M. Suzuki, and R. Noyori, *Tetrahedron* **44:**4259 (1988).
g. T. M. Meulmans, G. A. Stork, B. J. M. Jansen, and A. de Groot, *Tetrahedron Lett.***39:**6565 (1998).
h. T. Satay, J. Otera, and H. N. Zaki, *J. Am. Chem. Soc.* **112:**901 (1990).
i. A. S. Kende, S. Johnson, P. Sanfilippo, J. C. Hodges, and L. N. Jungheim, *J. Am. Chem. Soc.* **108:**3513 (1986).
j. J. Ipaktschi and A. Heydari, *Chem. Ber.* **126:**1905 (1993).
k. T. K. Hollis, N. Robinson, and B. Bosnich, *Tetrahedron Lett.* **33:**6423 (1992).
l. Y. Hong, D. J. Norris, and S. Collins, *J. Org. Chem.* **58:**3591 (1993).
m. M. T. Reetz and D. N. A. Fox, *Tetrahedron Lett.* **34:**1119 (1993).
n. M. T. Reetz, B. Raguse, C. F. Marth, H. M. Hugel, T. Bach, and D. N. A. Fox, *Tetrahedron* **48:**5731 (1992).
o. M. Oishi, S. Aratake, and H. Yamamoto, *J. Am. Chem. Soc.* **120:**8271 (1998).

and is attributed to coordination of the lanthanide at the imine nitrogen.[55]

$$CH_3(CH_2)_6CH{=}NCH_2Ph + TMSOC(OCH_3){=}C(CH_3)_2 \xrightarrow[20\ mol\ \%]{Yb(O_3SCF_3)_3} CH_3(CH_2)_6CH(NHCH_2Ph)C(CH_3)_2CO_2CH_3 \quad 86\%$$

In addition to aldehydes, acetals and ketals can serve as electrophiles in Mukaiyama reactions.[56]

$$RCH(OR')_2 + MX_n \longrightarrow RCH{=}O^+R' + [R'OMX_n]^-$$

$$RCH{=}O^+R' + R^2CH{=}C(OTMS)R^3 \longrightarrow RCH(OR')CH(R^2)C(=O)R^3$$

Effective catalysts include $TiCl_4$,[57] $SnCl_4$,[58] $(CH_3)_3SiO_3SCF_3$,[59] and $Bu_2SnO_3SCF_3$.[60] Indium trichloride catalyzes Mukaiyama additions in aqueous solution. The reaction is best conducted by preforming the aldehyde–$InCl_3$ complex and then adding the silyl enol ether and water.

$$ArCH{=}O + \text{1-(trimethylsilyloxy)cyclohexene (OTMS)} \xrightarrow{InCl_3,\ H_2O} \text{2-[ArCH(OH)]cyclohexanone} \quad 60\text{–}80\% \qquad \text{Ref. 61}$$

It has been proposed that there may be a single-electron-transfer mechanism for the Mukaiyama reaction.[62] For example, photolysis of benzaldehyde dimethylacetal and 1-trimethylsilyloxycyclohexene in the presence of a typical photoelectron acceptor, triphenylpyrylium cation, gives an excellent yield of the addition product.

$$PhCH(OMe)_2 + \text{1-(trimethylsilyloxy)cyclohexene (OTMS)} \xrightarrow[\text{2,4,6-triphenylpyrylium}]{h\nu} \text{2-[PhCH(OH)]cyclohexanone} \qquad \text{Ref. 63}$$

94% yield, 66:34 *syn:anti*

These reactions may operate by providing a source of trimethylsilyl cations, which act as the active catalyst.

55. S. Kobayashi and S. Nagayama, *J. Am. Chem. Soc.* **119:**10049 (1997); S. Kobayashi and S. Nagayama, *J. Org. Chem.* **62:**232 (1997).
56. T. Mukaiyama and M. Murakami, *Synthesis* **1987:**1043.
57. T. Mukaiyama and M. Hayashi, *Chem. Lett.* **1974:**15.
58. R. C. Cambie, D. S. Larsen, C. E. F. Rickard, P. S. Rutledge, and P. D. Woodgate, *Austr. J. Chem.* **39:**487 (1986).
59. S. Murata, M. Suzuki, and R. Noyori, *Tetrahedron* **44:**4259 (1988).
60. T. Sato, J. Otera, and H. Nozaki, *J. Am. Chem. Soc.* **112:**901 (1990).
61. T.-P. Loh, J. Pei, K. S.-V. Koh, G.-Q. Cao, and X.-R. Li, *Tetrahedron Lett.* **38:**3465, 3993 (1997).
62. T. Miura and Y. Masaki, *J. Chem. Soc., Perkin Trans. 1* **1994:**1659; T. Miura and Y.Masaki, *J. Chem. Soc., Perkin Trans. 1* **1995:**2155; J. Otera, Y. Fujita, N. Sakuta, M.Fujita, and S. Fukuzumi, *J. Org. Chem.* **61:**2951 (1996).
63. M. Kamata, S. Nagai, M. Kato, and E. Hasegawa, *Tetrahedron Lett.* **37:**7779 (1996).

2.1.3.5. Control of Enantioselectivity. In the previous sections, the most important factors in determining the *syn* or *anti* stereoselectivity of aldol and Mukaiyana reactions were identified as the nature of the transition state (cyclic versus acyclic) and the configuration (*E* or *Z*) of the enolate. Additional factors affect the enantioselectivity of aldol additions and related reactions. Nearby chiral centers in either the carbonyl compound or the enolate can impose facial selectivity. Chiral auxiliaries can achieve the same effect. Finally, use of chiral Lewis acids as catalysts can also achieve enantioselectivity. Although the general principles of control of the stereochemistry of aldol addition reactions have been developed for simple molecules, the application of the principles to more complex molecules and the selection of the optimum enolate system requires analysis of the individual cases.[64] Not infrequently, one of the enolate systems proves to be superior,[65] or a remote structural feature strongly influences the stereoselectivity.[66] The issues that need to be addressed in specific cases include the structure of the enolate, including its stereochemistry and potential sites for chelation, the organization of the transition state (cyclic versus acyclic), and the factors effecting the facial selectivity.

Up to this point, we have considered primarily the effect of enolate geometry on the stereochemistry of the aldol condensation and have considered achiral or racemic aldehydes and enolates. If the aldehyde is chiral, particularly when the chiral center is adjacent to the carbonyl group, the selection between the two diastereotopic faces of the carbonyl group will influence the stereochemical outcome of the reaction. Similarly, there will be a degree of selectivity between the two faces of the enolate when the enolate contains a chiral center. If both the aldehyde and enolate are chiral, mutual combinations of stereoselectivity will come into play. One combination should provide complementary, reinforcing stereoselection, whereas the alternative combination would result in opposing preferences and lead to diminished overall stereoselectivity. The combined interactions of chiral centers in both the aldehyde and the enolate determine the stereoselectivity. The result is called *double stereodifferentiation*.[67]

R-enolate ←favored→ *R*-aldehyde   or   *R*-enolate ←favored→ *S*-aldehyde

S-enolate ←favored→ *S*-aldehyde   *S*-enolate ←favored→ *R*-aldehyde

The analysis and prediction of the direction of preferred reaction depend on the same principles as for simple diastereoselectivity and are done by analysis of the attractive and repulsive interactions in the presumed transition state.

Analysis of results for α-substituted aldehydes with *E*- and *Z*-enolates indicates that the cyclic transition states shown below are favored with lithium and boron enolates.[64a]

64. (a) W. R. Roush, *J. Org. Chem.* **56:**4151 (1991); (b) C. Gennari, S. Vieth, A. Comotti, A. Vulpetti, J. M. Goodman, and I. Paterson, *Tetrahedron* **48:**4439 (1992); (c) D. A. Evans, M. J. Dart, J. L. Duffy, and M. G. Yang, *J. Am. Chem. Soc.* **118:**4322 (1996); (d) A. S. Franklin and I. Paterson, *Contemp. Org. Synth.* **1:**317 (1994).
65. E. J. Corey, G. A. Reichard, and R. Kania, *Tetrahedron Lett.* **34:**6977 (1993).
66. A. Balog, C. Harris, K. Savin, X.-G. Zhang, T. C. Chou, and S. J. Danishefsky, *Angew. Chem. Int. Ed. Engl.* **37:**2675 (1998).
67. S. Masamune, W. Choy, J. S. Petersen, and L. R. Sita, *Angew. Chem. Int. Ed. Engl.* **24:**1 (1985).

The larger α-substituent is aligned *anti* to the approaching enolate.

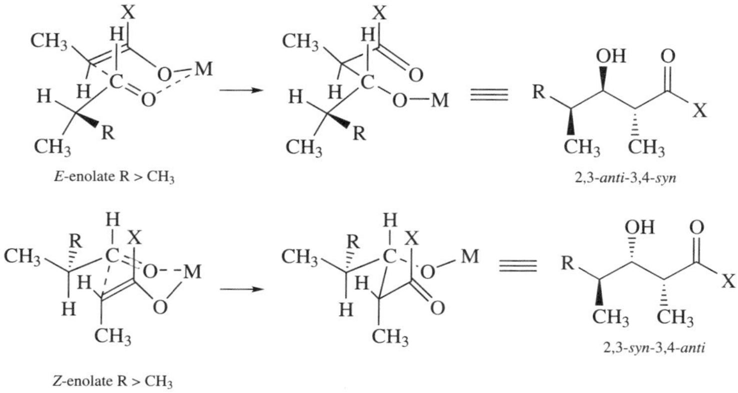

E-enolate R > CH_3 2,3-*anti*-3,4-*syn*

Z-enolate R > CH_3 2,3-*syn*-3,4-*anti*

The stereoselectivity resulting from interactions of chiral aldehydes and enolates has been useful in the construction of systems with several contiguous chiral centers.

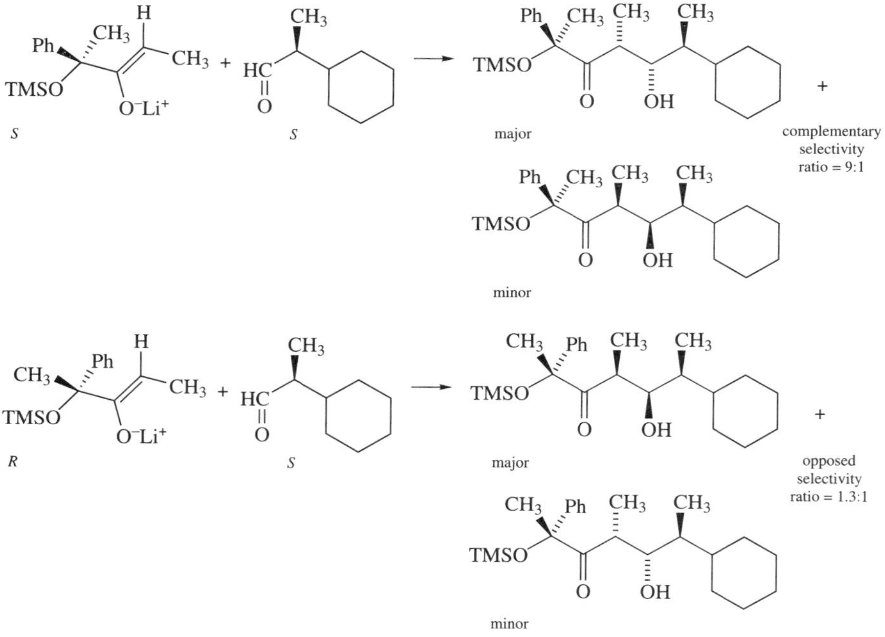

S *S* major minor complementary selectivity ratio = 9:1

Ref. 68

R *S* major minor opposed selectivity ratio = 1.3:1

Other structural features may influence the stereoselectivity of aldol condensations. One such factor is chelation by a donor substituent.[69] Several β-alkoxyaldehydes show a preference for *syn*-aldol products on reaction with *Z*-enolates. A chelated transition state can account for the observed stereochemistry.[70] The chelated aldehyde is most easily

68. S. Masamune, S. A. Ali, D. L. Snitman, and D. S. Garvey, *Angew. Chem. Int. Ed. Engl.* **19:**557 (1980).
69. M. T. Reetz, *Angew. Chem. Int. Ed. Engl.* **23:**556 (1984).
70. S. Masamune, J. W. Ellingboe, and W. Choy, *J. Am. Chem. Soc.* **104:**5526 (1982).

approached from the face opposite the methyl and R′ substituents.

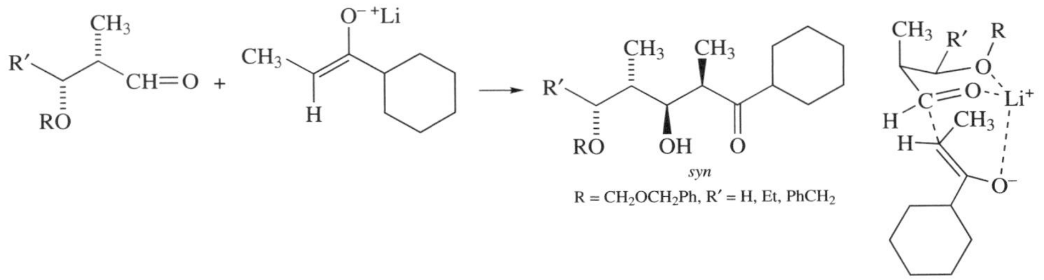

syn

R = CH_2OCH_2Ph, R′ = H, Et, $PhCH_2$

A similar stereoselectivity has been noted for the $TiCl_4$-mediated condensation of β-alkoxyaldehydes with silyl enol ethers.

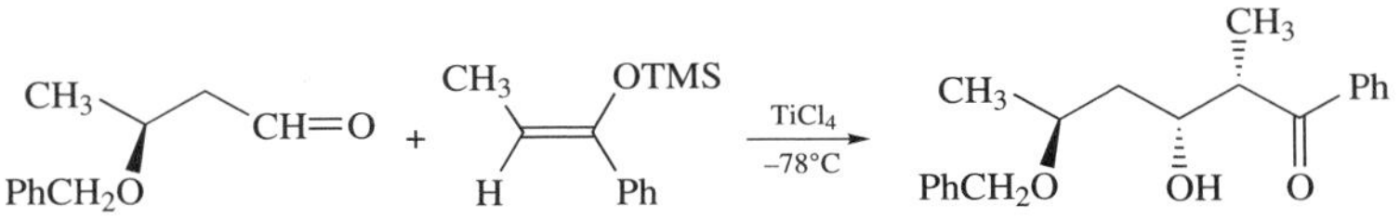

Ref. 71

The preceding reactions illustrate control of stereochemistry by aldehyde substituents. Substantial effort has also been devoted to use of chiral auxiliaries and chiral catalysts to effect enantioselective aldol reactions.[72] A very useful approach for enantioselective aldol condensations has been based on the oxazolidinones **1–3**, which are readily available in enantiomerically pure form.

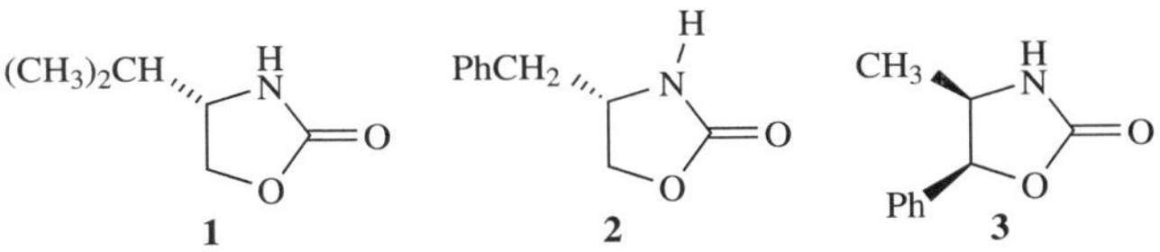

1 **2** **3**

These compounds can be acylated and converted to the lithium or boron enolates by the same methods applicable to ketones. The enolates are the *Z*-stereoisomers.[73]

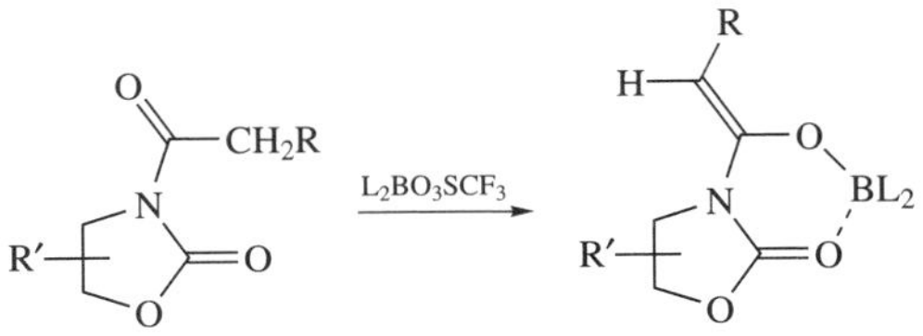

The oxazolinone substituents R′ then direct the approach of the aldehyde. Because of the differing steric encumbrance provided by **1** and **3**, the products have the opposite configuration at the new stereogenic sites. The acyl oxazolidinones are easily solvolyzed

71. M. T. Reetz and A. Jung, *J. Am. Chem. Soc.* **105:**4833 (1983).
72. M. Braun and H. Sacha, *J. Prakt. Chem.***335:**653 (1993).
73. D. A. Evans, J. Bartoli, and T. L. Shih, *J. Am. Chem. Soc.* **103:**2127 (1981).

in water or alcohols to give the enantiomeric β-hydroxy acid or ester.

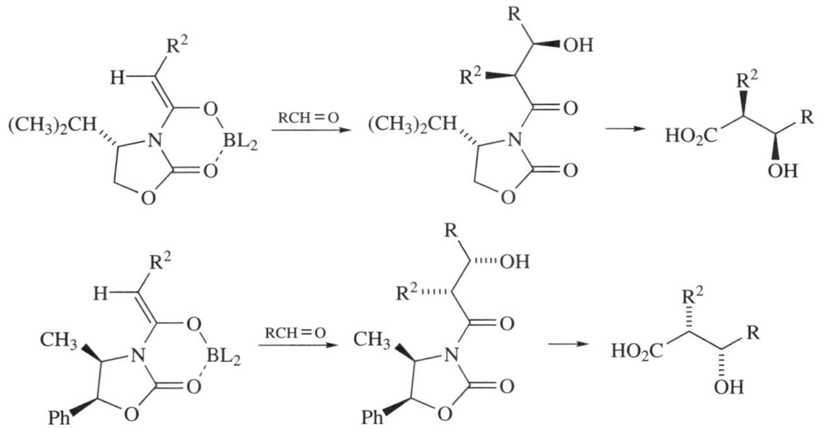

R^2 H $(CH_3)_2CH$ N O BL_2 O O RCH=O R OH R^2 O N O O HO_2C R^2 R OH

R^2 H CH_3 N O BL_2 O Ph O RCH=O R OH R^2 O CH_3 N O Ph O HO_2C R^2 R OH

1,3-Thiazoline-2-thiones are another useful type of chiral auxiliary. These can be used in conjunction with $Sn(O_3SCF_3)_2$,[74] $Bu_2BO_3SCF_3$,[75] or $TiCl_4$[76] for generation of enolates.

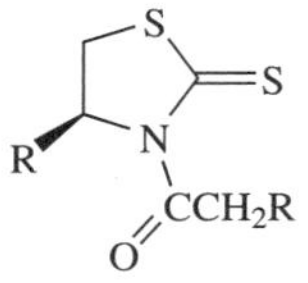

S S R N CCH_2R O

$R = C_3H_5, (CH_3)_3C, CO_2CH_3$

The chiral auxiliaries can be used under conditions where either cyclic or noncyclic transition states are involved. This frequently allows control of the *syn* or *anti* stereoselectivity. Scheme 2.6 gives some examples where good stereoselectivity has been achieved. The selectivity is believe to be determined by the cyclic or acyclic nature of the transition state.[77]

Enantioselectivity can also be induced by use of chiral boronates in the preparation of boron enolates. Both the (+) and (−) enantiomers of diisopinocamphylboron triflate have been used to generate *syn* adducts through a cyclic transition state.[78] The enantioselectivity was greater than 80% for most cases that were examined.

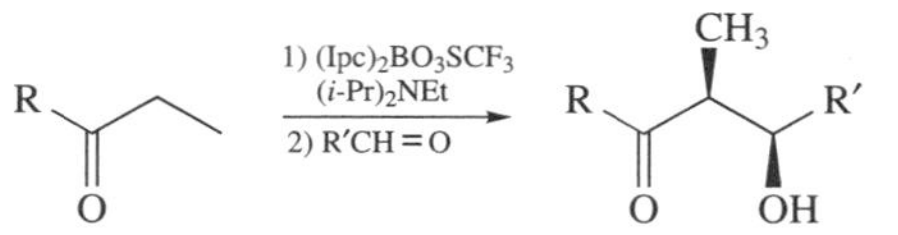

R O 1) $(Ipc)_2BO_3SCF_3$ $(i\text{-}Pr)_2NEt$ 2) $R'CH=O$ CH_3 R R′ O OH

R = Et, Ph, *i*-Pr   R′ = Me, *n*-Pr, *i*-Pr

74. Y. Nagao, Y. Hagiwara, T. Kumagai, M. Ochiai, T. Inoue, K. Hashimoto, and E. Fujita, *J. Org. Chem.* **51**:2391 (1986).
75. C.-N. Hsiao, L. Liu, and M. J. Miller, *J. Org. Chem.* **52**:2201 (1987).
76. D. A. Evans, S. J. Miller, M. D. Ennis, and P. L. Ornstein, *J. Org. Chem.* **57**:2067 (1992).
77. T. H. Yan, C. W. Tan, H.-C. Lee, H-C. Lo, and T. Y. Huang, *J. Am. Chem. Soc.* **115**:1613 (1993).
78. I. Paterson, J. M. Goodman, M. A. Lister, R. C. Schumann, C. K. McClure, and R. D. Norcross, *Tetrahedron* **46**:4663 (1990).

Scheme 2.6. Control of *syn* : *anti* Selectivity by Use of Alternate Reaction Conditions with Chiral Auxiliaries

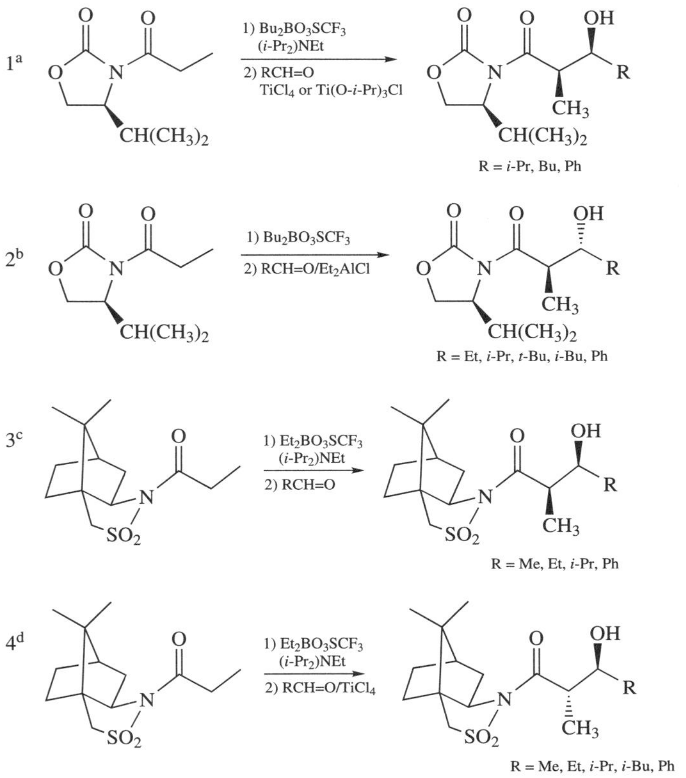

1[a] 1) $Bu_2BO_3SCF_3$, $(i\text{-}Pr_2)NEt$; 2) RCH=O, $TiCl_4$ or $Ti(O\text{-}i\text{-}Pr)_3Cl$

$CH(CH_3)_2$ … OH, R, CH_3, $CH(CH_3)_2$

R = *i*-Pr, Bu, Ph

2[b] 1) $Bu_2BO_3SCF_3$; 2) $RCH{=}O/Et_2AlCl$

$CH(CH_3)_2$ … OH, R, CH_3, $CH(CH_3)_2$

R = Et, *i*-Pr, *t*-Bu, *i*-Bu, Ph

3[c] 1) $Et_2BO_3SCF_3$, $(i\text{-}Pr_2)NEt$; 2) RCH=O

N, SO_2 … OH, R, CH_3

R = Me, Et, *i*-Pr, Ph

4[d] 1) $Et_2BO_3SCF_3$, $(i\text{-}Pr_2)NEt$; 2) $RCH{=}O/TiCl_4$

N, SO_2 … OH, R, CH_3

R = Me, Et, *i*-Pr, *i*-Bu, Ph

a. M. Nerz-Stormes and E. R. Thornton, *J. Org. Chem.* **56**:2489 (1991); D. A. Evans, F. Urpi, T. C. Somers, J. S. Clark, and M. T. Bilondeau, *J. Am. Chem. Soc.* **112**:8215 (1990).
b. M. A. Walker and C. H. Heathcock, *J. Org. Chem.* **56**:5747 (1991).
c. W. Oppolzer, J. Blagg, I. Rodriquez, and E. Walther, *J. Am. Chem. Soc.* **112**:2767 (1990).
d. W. Oppolzer and P. Lienhard, *Tetrahedron Lett.* **34**:4321 (1993).

Another promising boron enolate is derived from (−)-menthone. It gives *E*-boron enolates that give good enantioselectivity and result in formation of *anti* products.[79]

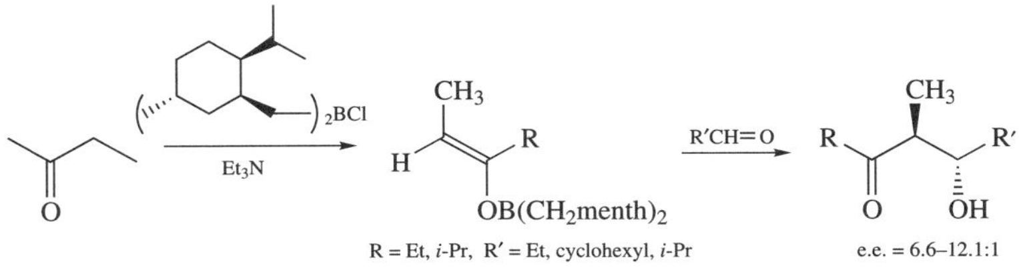

$)_2BCl$, Et_3N → CH_3, H, R, $OB(CH_2menth)_2$ → R′CH=O → CH_3, R, R′, O, OH

R = Et, *i*-Pr, R′ = Et, cyclohexyl, *i*-Pr

e.e. = 6.6–12.1:1

79. G. Gennari, C. T. Hewkin, F. Molinari, A. Bernardi, A. Comotti, J. M. Goodman, and I. Paterson, *J. Org. Chem.* **57**:5173 (1992).

The stereoselectivity in these cases has its origin in steric effects of the boron substituents.

Several hetereocyclic boron enolates with chirality installed at boron have been found to be useful for enantioselective additions. The diazaboridine below is an example.[80]

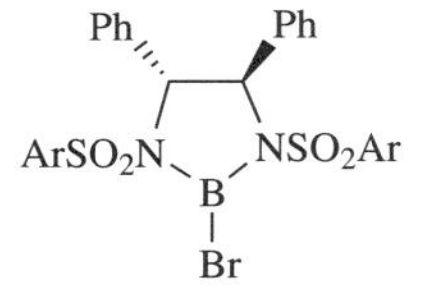

Ar = 3,5-di(trifluoromethyl)phenyl

Derivatives with various substituted sulfonamides have also been developed and used to form enolates from esters and thioesters.[81] An additional feature of these chiral auxiliaries is the ability to select for *syn* or *anti* products, depending upon choice of reagents and reaction conditions. The diastereoselectivity is determined by whether the *E*- or *Z*-enolate is formed.[82]

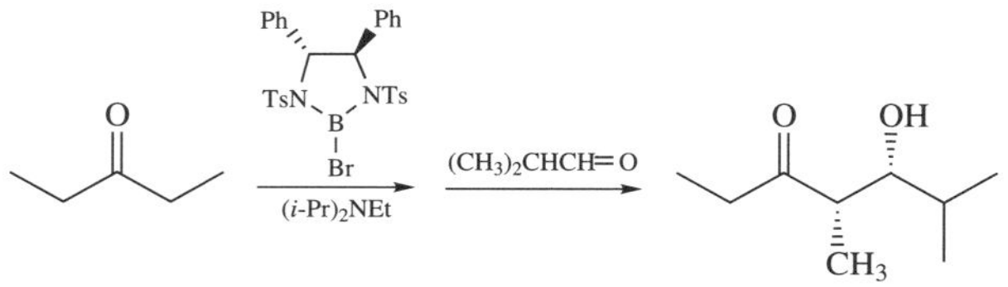

$(i\text{-Pr})_2NEt$; $(CH_3)_2CHCH{=}O$

85% yield, 98:2 *syn*:*anti*, 95% e.e.

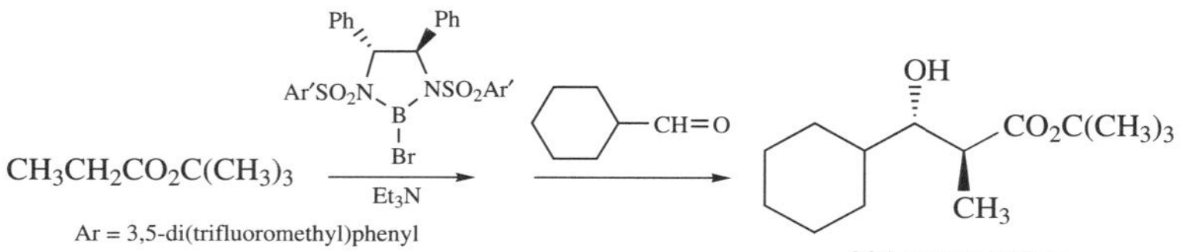

$CH_3CH_2CO_2C(CH_3)_3$; Et_3N

Ar = 3,5-di(trifluoromethyl)phenyl

96:4 *syn*:*anti*, 75% e.e.

Considerable effort has been devoted to finding Lewis acid or other catalysts that could induce high enantioselectivity in the Mukaiyama reaction. As with aldol addition reactions involving enolates, high diastereoselectivity and enantioselectivity requires involvement of a transition state with substantial facial selectivity with respect to the electrophilic reactant and a preferred orientation of the nucleophile. Scheme 2.4 shows some examples of enantioselective catalysts.

One example involves the addition of stannyl enol ethers to benzaldehyde in the presence of silver triflate and the chiral 2,2′-bis(diphenylphosphino)-1,1′-binaphthyl

80. E. J. Corey, R. Imwinkelried, S. Pikul, and Y. B. Xiang, *J. Am. Chem. Soc.* **111**:5493 (1989).
81. E. J. Corey and S. S. Kim, *J. Am. Chem. Soc.* **112**:4976 (1990).
82. E. J. Corey and D. H. Lee, *Tetrahedron Lett.* **34**:1737 (1993).

(BINAP) ligand. The observed enantioselectivity can be accounted for by a cyclic transition state.

PhCH=O + [1-(tributylstannyloxy)cyclohexene] $\xrightarrow{R\text{-(BINAP)-AgO}_3\text{SCF}_3}$ [product] 93% *anti* 94% e.e. Ref. 83

Scheme 2.7 gives some examples of chiral Lewis acids that have been used to catalyze aldol and Mukaiyama reactions. Scheme 2.8 shows some enantioselective aldol additions effected with these reagents.

2.1.4. Intramolecular Aldol Reactions and the Robinson Annulation

The aldol reaction can be applied to dicarbonyl compounds in which the two carbonyl groups are favorably disposed for intramolecular reaction. For formation of five- and six-membered rings, the use of a catalytic amount of a base is frequently satisfactory. With more complex structures, the special techniques required for directed aldol condensations are used. Scheme 2.9 illustrates intramolecular aldol condensations.

A particularly important example is the Robinson annulation, a procedure which constructs a new six-membered ring from a ketone.[84] The reaction sequence starts with conjugate addition of the enolate to methyl vinyl ketone or a similar enone. This is followed by cyclization involving an intramolecular aldol addition. Dehydration frequently occurs to give a cyclohexenone derivative. Scheme 2.10 shows some examples of Robinson annulation reactions.

$CH_3CCH{=}CH_2$ (with C=O) + cyclohexanone enolate $\xrightarrow{\text{conjugate addition}}$ 2-(3-oxobutyl)cyclohexanone $\xrightarrow{\text{aldol addition and dehydration}}$ octalone

A precursor of methyl vinyl ketone, 4-(trimethylamino)-2-butanone, was used as the reagent in the early examples of the reaction. This compound generates methyl vinyl ketone *in situ*, by β elimination. Other α,β-unsaturated enones can be used, but the reaction

83. A. Yanagisawa, Y. Matsumoto, H. Nakashima, K. Asakawa, and H. Yamamoto, *J. Am. Chem. Soc.* **119:**9319 (1997).
84. E. D. Bergmann, D. Ginsburg, and R. Pappo, *Org. React.* **10:**179 (1950); J. W. Cornforth and R. Robinson, *J. Chem. Soc.* **1949:**1855; R. Gawley, *Synthesis* **1976:**777; M. E. Jung, *Tetrahedron* **32:**3 (1976); B. P. Mundy, *J. Chem. Educ.* **50:**110 (1973).

Scheme 2.7. Chiral Catalysts for the Mukaiyama Reaction

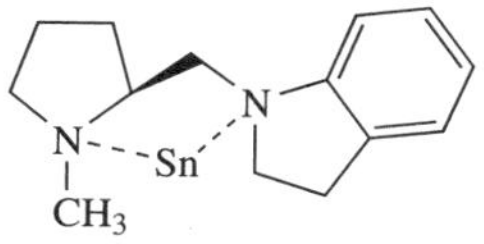

A[a] 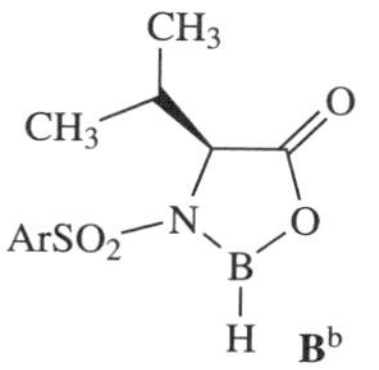 B[b] 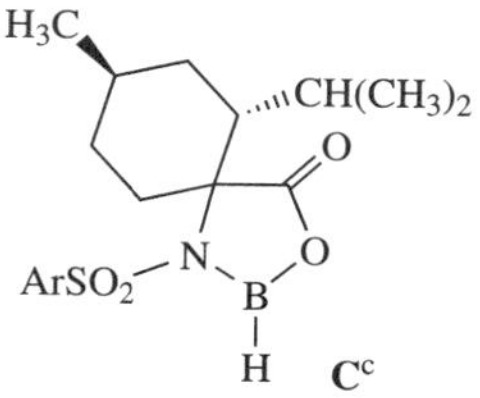 C[c] 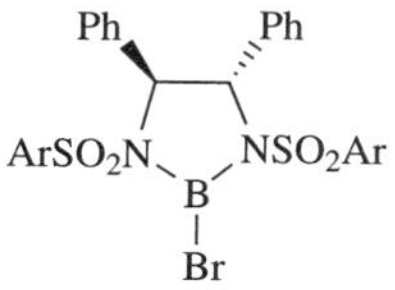 D[d] 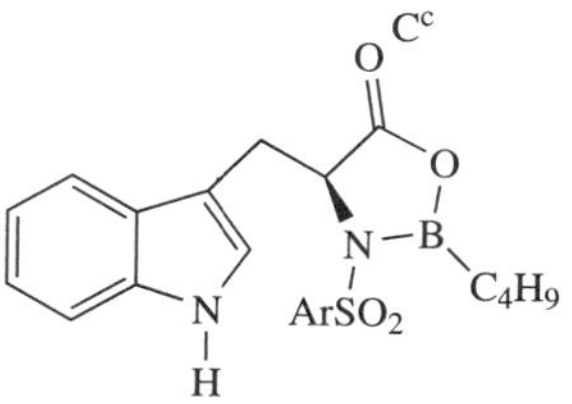 E[e] 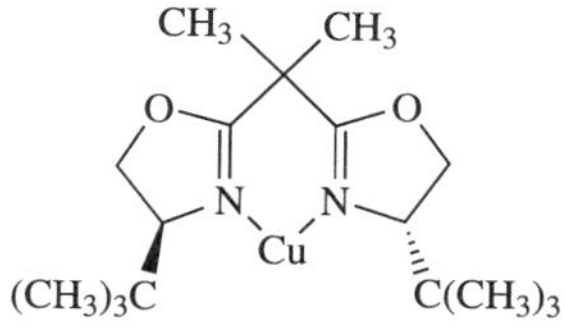 F[f] G[g] H[h]

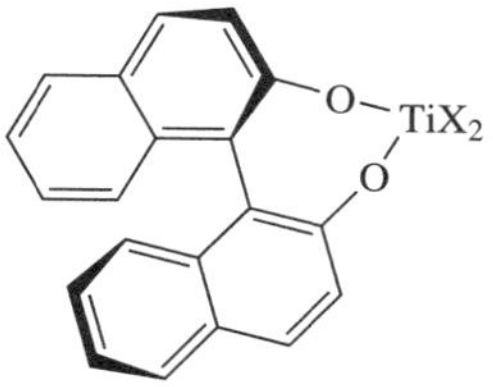

X = Cl or $OCH(CH_3)_2$

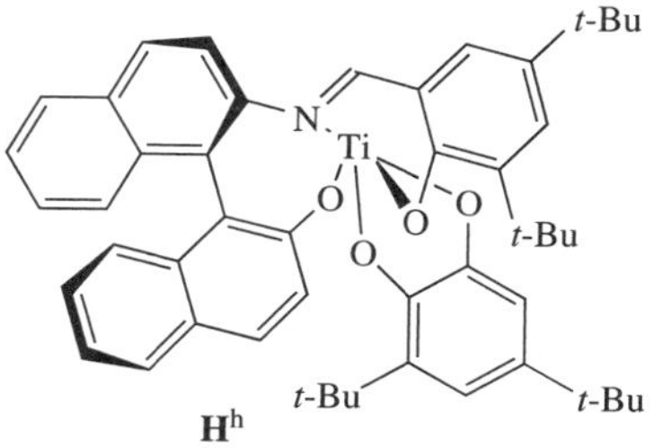

a. S. Kobayashi and M. Horibe, *Chem. Eur. J.* **3:**1472 (1997).
b. S. Kiyooka, Y. Kaneko, M. Komura, H. Matsuo, and M. Nakano, *J. Org. Chem.* **56:**2276 (1991).
c. E. R. Parmee, O. Tempkin, S. Masamune, and A. Akibo, *J. Am. Chem. Soc.* **113:**9365 (1991).
d. E. J. Corey, R. Imwinkelried, S. Pakul, and Y. B. Xiang, *J. Am. Chem. Soc.* **111:**5493 (1989).
e. E. J. Corey, C. L. Cywin, and T. D. Roper, *Tetrahedron Lett.* **33:**6907 (1992); E. J. Corey, D. Barnes-Seeman, and T. W. Lee, *Tetrahedron Lett.* **38:**1699 (1997).
f. D. A. Evans, J. A. Murry, and M. C. Kozlowski, *J. Am. Chem. Soc.* **118:**5814 (1996); D. A. Evans, M. C. Kozlowski, C. S. Burgey, and D. W. C. MacMillan, *J. Am. Chem. Soc.* **119:**7893 (1997); D. A. Evans, D. W. C. MacMillan, and K. R. Campos, *J. Am. Chem. Soc.* **119:**10859 (1997).
g. K. Mitami and S. Matsukawa, *J. Am. Chem. Soc.* **115:**7039 (1993); K. Mitami and S. Matsukawa, *J. Am. Chem. Soc.* **116:**4077 (1994); G. E. Keck and D. Krishnamurthy, *J. Am. Chem. Soc.* **117:**2363 (1995); G. E. Keck, D. Krishnamurthy, and M. C. Grier, *J. Org. Chem.* **58:**6543 (1993); G. E. Keck, X.-Y. Li, and D. Krishnamurthy, *J. Org. Chem.* **60:**5998 (1995).
h. E. M. Carreira, R. A. Singer, and W. Lee, *J. Am. Chem. Soc.* **116:**8837 (1994).

is somewhat sensitive to substitution at the β carbon, and adjustment of the reaction conditions is necessary.[85]

The original conditions developed for the Robinson annulation reaction are such that the ketone enolate composition is under thermodynamic control. This usually results in the formation of the more substituted enolate and gives a product with a substituent at a ring juncture when monosubstituted cyclohexanones are used as reactants. The alternative regiochemistry can be achieved by using an enamine of the ketone. As discussed in Section 1.9, the less substituted enamine is favored, so addition occurs at the less substituted position. Entry 4 of Scheme 2.10 illustrates this variation of the reaction.

85. C. J. V. Scanio and R. M. Starrett, *J. Am. Chem. Soc.* **93:**1539 (1971).

Scheme 2.8. Enantioselective Aldol and Mukaiyama Additions

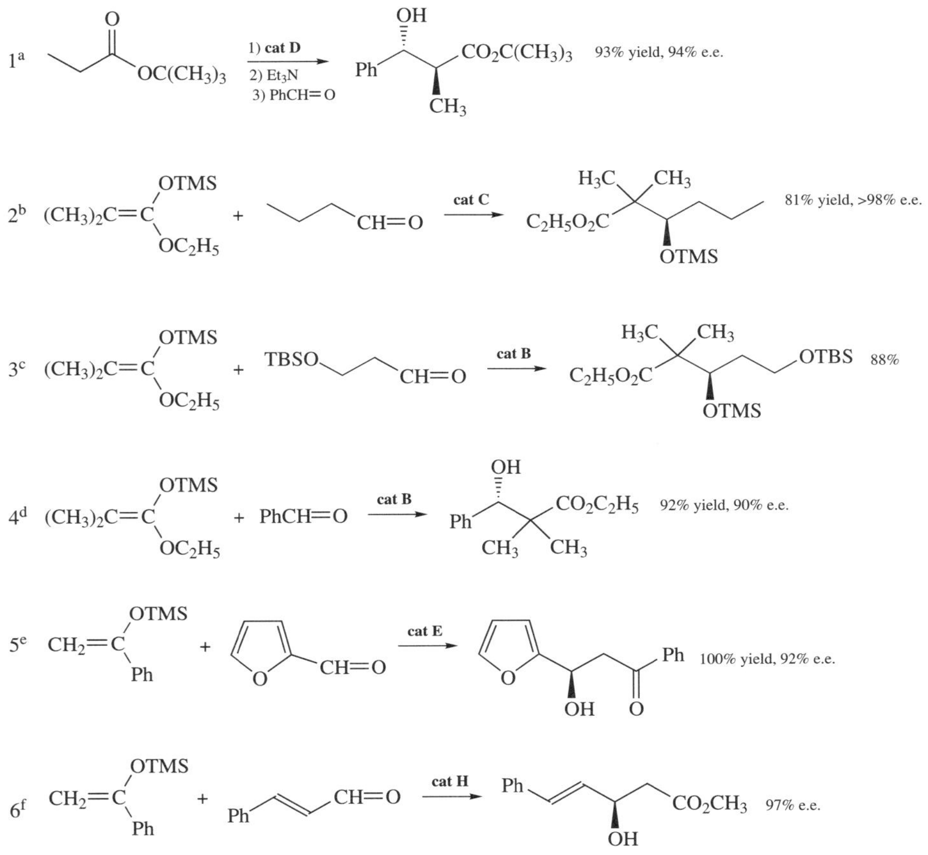
1[a]
1) cat D
2) Et_3N
3) PhCH=O
93% yield, 94% e.e.
2[b]
cat C
81% yield, >98% e.e.
3[c]
cat B
88%
4[d]
cat B
92% yield, 90% e.e.
5[e]
cat E
100% yield, 92% e.e.
6[f]
cat H
97% e.e.

a. E. J. Corey and S. S. Kim, *J. Am. Chem. Soc.* **112:**4976 (1990).
b. E. R. Parmee, O. Tempkin, S. Masamune, and A. Akibo, *J. Am. Chem. Soc.* **113:**9365 (1991).
c. J. Mulzer, A. J. Mantoulidis, and E. Ohler, *Tetrahedron Lett.* **39:**8633 (1998).
d. S. Kiyooka, Y. Kaneko, and K. Kume, *Tetrahedron Lett.* **33:**4927 (1992).
e. E. J. Corey, C. L. Cywin, and T. D. Roper, *Tetrahedron Lett.* **33:**6907 (1992).
f. E. M. Carreira, R. A. Singer, and W. Lee, *J. Am. Chem. Soc.* **116:**8837 (1994).

Robinson annulation can also be carried out using aluminum tris(2,6-diphenylphenoxide) to effect the conjugate addition and cyclization.

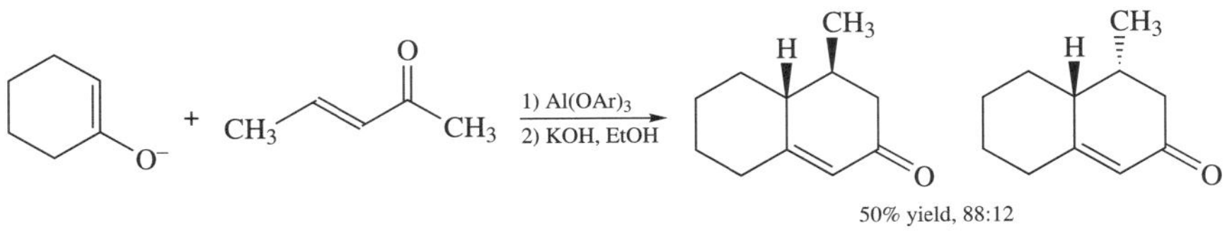
1) $Al(OAr)_3$
2) KOH, EtOH
50% yield, 88:12

Ref. 86

86. S. Saito, I. Shimada, Y. Takamori, M. Tanaka, K. Maruoka, and H. Yamamoto, *Bull. Chem. Soc. Jpn.* **70:**1671 (1997).

Scheme 2.9. Intramolecular Aldol and Mukaiyama Additions

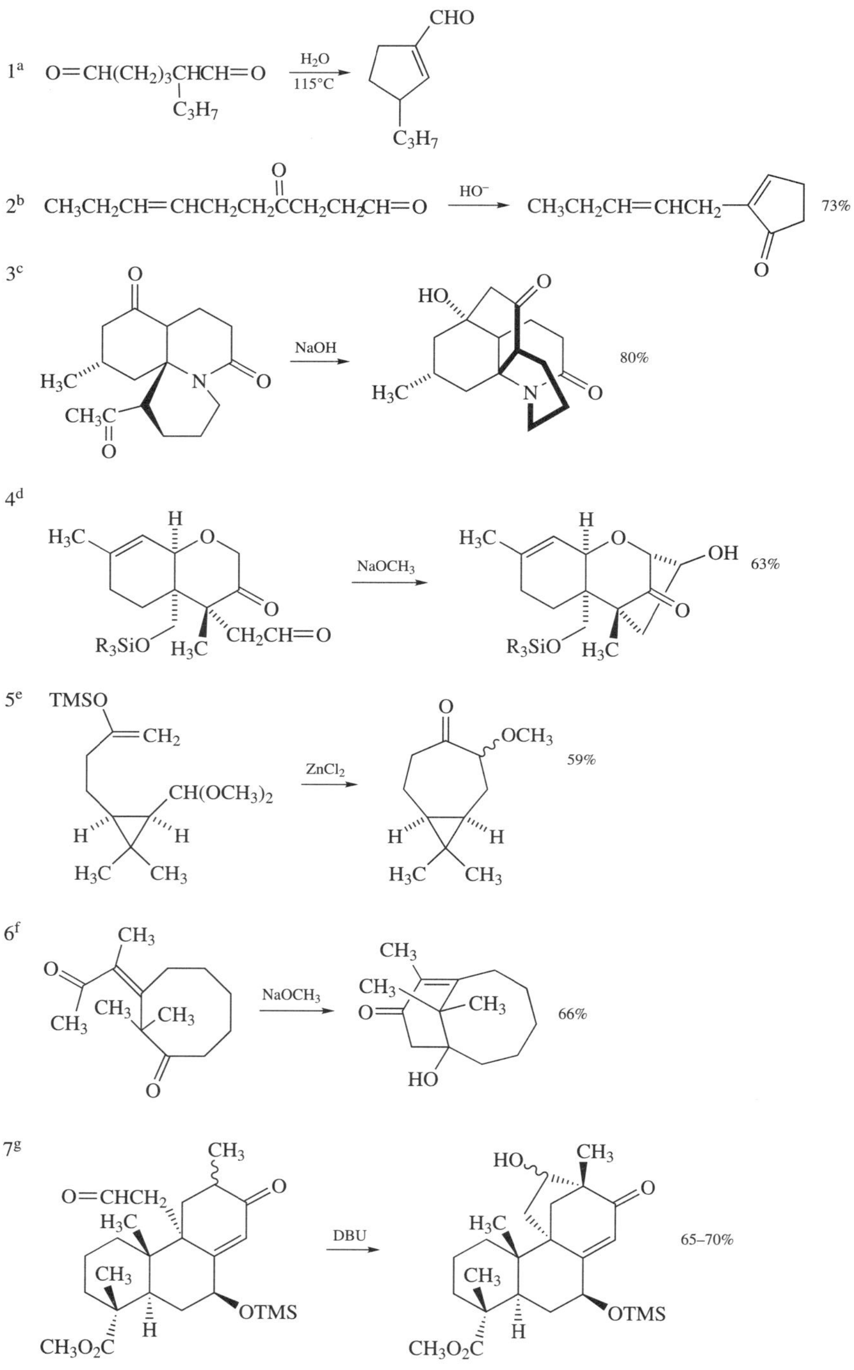

1[a] $O{=}CH(CH_2)_3CH(C_3H_7)CH{=}O$ —(H_2O, 115°C)→ 3-propylcyclopent-1-ene-1-carbaldehyde (CHO, C_3H_7)

2[b] $CH_3CH_2CH{=}CHCH_2CH_2C(O)CH_2CH_2CH{=}O$ —(HO^-)→ $CH_3CH_2CH{=}CHCH_2$–cyclopentenone 73%

3[c] —(NaOH)→ 80%

4[d] —($NaOCH_3$)→ 63%

5[e] —($ZnCl_2$)→ 59%

6[f] —($NaOCH_3$)→ 66%

7[g] —(DBU)→ 65–70%

a. J. English and G. W. Barber, *J. Am. Chem. Soc.* **71**:3310 (1949).
b. A. I. Meyers and N. Nazarenko, *J. Org. Chem.* **38**:175 (1973).
c. K. Wiesner, V. Musil, and K. J. Wiesner, *Tetrahedron Lett.* **1968**:5643.
d. G. A. Kraus, B. Roth, K. Frazier, and M. Shimagaki, *J. Am. Chem. Soc.* **104**:1114 (1982).
e. M. D. Taylor, G. Minaskanian, K. N. Winzenberg, P. Santone, and A. B. Smith III, *J. Org. Chem.* **47**:3960 (1982).
f. K. Yamada, H. Iwadare, and T. Mukaiyama, *Chem. Pharm. Bull.* **45**:1898 (1997).
g. J. R. Tagat, M. S. Puar, and S. W. McCombie, *Tetraheron Lett.* **37**:8463 (1996).

1[a] O, CH_3, O + $CH_2{=}CHCOCH_3$ $\xrightarrow{\text{KOH}}$ O, CH_3, $CH_2CH_2COCH_3$, O $\xrightarrow{\text{pyrrolidine}}$ O, CH_3, O 63–65%

2[b] $CO_2CH_2CH_3$, O + $CH_2{=}CHCOCH_2CH_3$ $\xrightarrow[\text{EtOH}]{\text{NaOEt}}$ $CO_2CH_2CH_3$, O, CH_3 59%

3[c] CH_3, O, CH_3O + $CH_3COCH_2CH_2\overset{+}{N}(CH_3)_3$ $\xrightarrow{^{-}\text{OEt}}$ O, H_3C, CH_3O 71%

4[d] CH_3, O → CH_3, N → 1) $CH_2{=}CHCOCH_3$ benzene, reflux 2) HOAc, NaOAc, H_2O, reflux → CH_3, O 45%

5[e] O, CH_3, O + $CH_2{=}CH\overset{O}{\overset{\|}{C}}CH_2CH_3$ → 1) DABCO 2) Et_3N, $PhCO_2H$ 140°C 24 h → O, CH_3, O, CH_3 75%

6[f] O, O, OCH_3, OCH_3, O → 1) LDA 2) $CH_3CH{=}\overset{O}{\overset{\|}{C}}CCH_3$ ($Si(CH_3)_3$) 3) MeO^- → O, O, OCH_3, OCH_3, O 62%

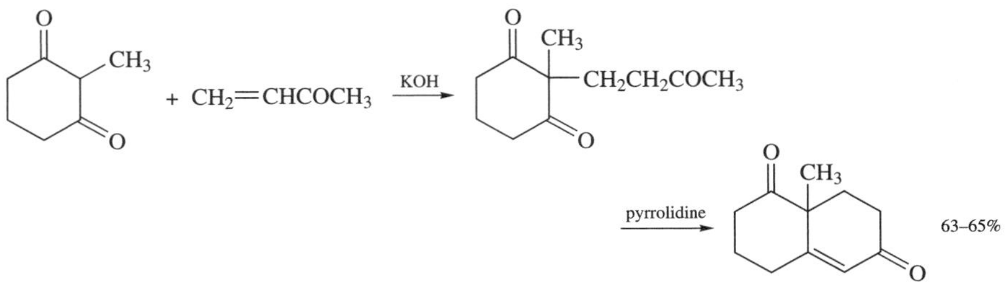

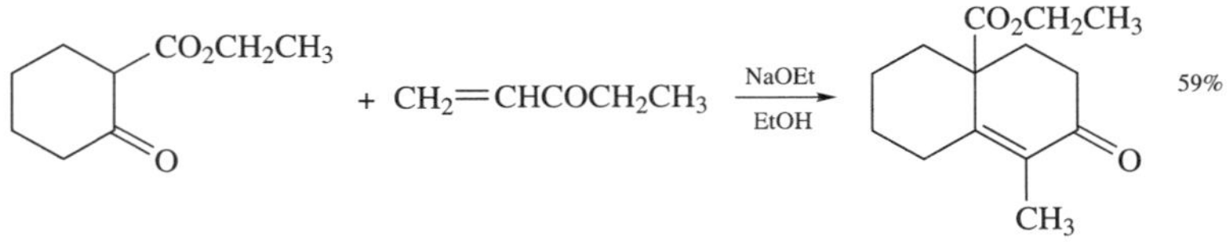

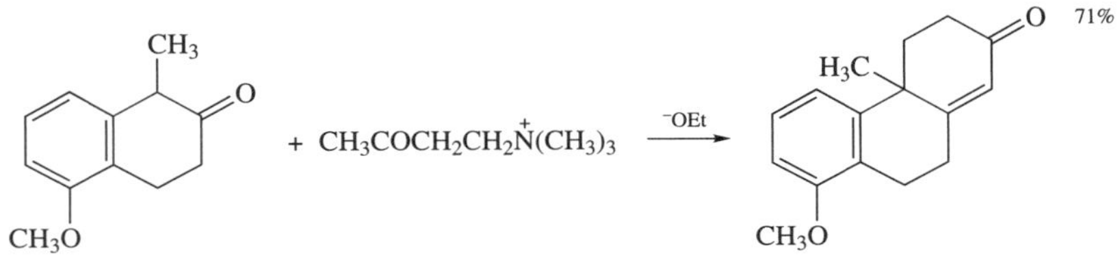

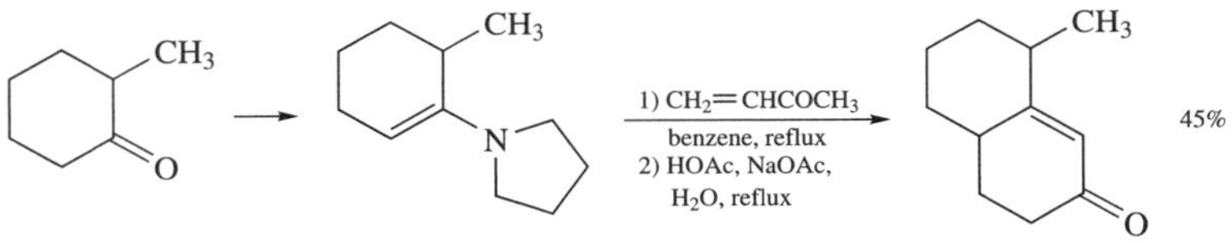

7[g] $O^{-}\,{}^{+}Li$ + $CH_2{=}\underset{SPh}{C}\overset{O}{\overset{\|}{C}}CH_3$ $\xrightarrow{-70°\text{C}}$ SPh, O, OH 80%

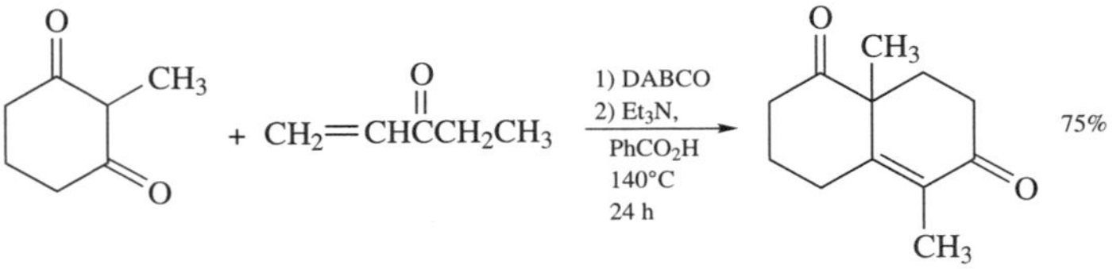

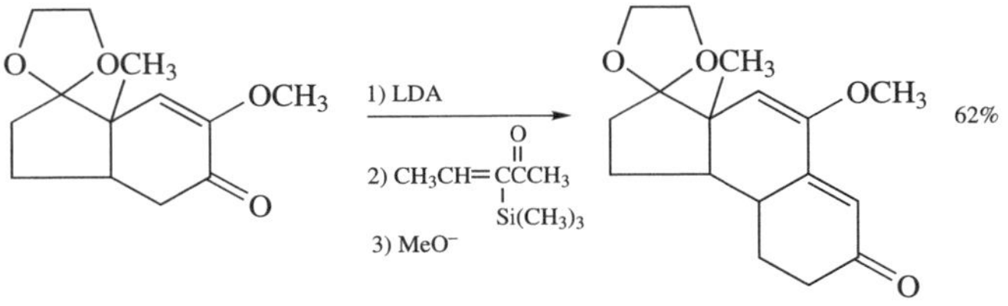

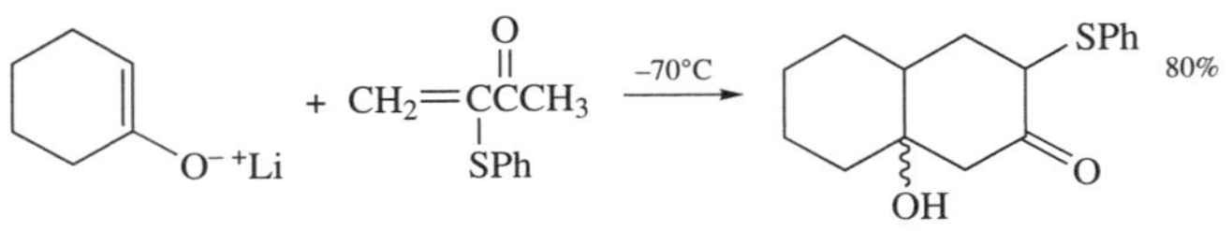

Scheme 2.10. (*continued*)

8[h] (cyclohexenyl OTMS) + $CH_2CH{=}CHCCH_3$ (C=O) $\xrightarrow{TiCl_4}$ 2-(1-methyl-3-oxobutyl)cyclohexanone 72% $\xrightarrow{KOH}$ octalone (CH_3)

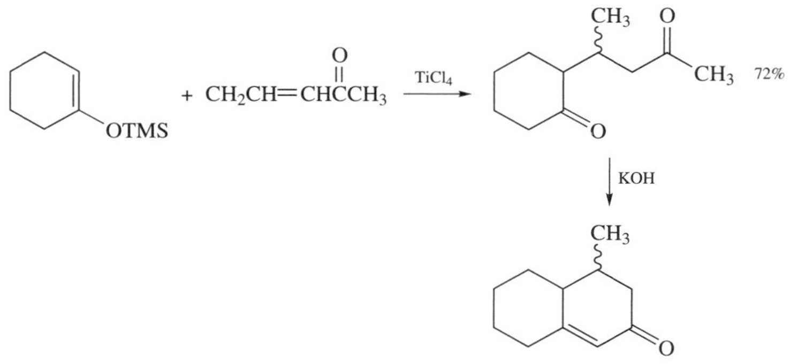

a. S. Ramachandran and M. S. Newman, *Org. Synth.* **41:**38 (1961).
b. D. L. Snitman, R. J. Himmelsbach, and D. S. Watt, *J. Org. Chem.* **43:**4578 (1978).
c. J. W. Cornforth and R. Robinson, *J. Chem. Soc.* **1949:**1855.
d. G. Stork, A. Brizzolara, H. Landesman, J. Szmuszkovicz, and R. Terrell, *J. Am. Chem. Soc.* **85:**207 (1963).
e. F. E. Ziegler, K.-J. Hwang, J. F. Kadow, S. I. Klein, U. K. Pati, and T.-F. Wang, *J. Org. Chem.* **51:**4573 (1986).
f. G. Stork, J. D. Winkler, and C. S. Shiner, *J. Am. Chem. Soc.* **104:**3767 (1982).
g. K. Takaki, M. Okada, M. Yamada, and K. Negoro, *J. Org. Chem.* **47:**1200 (1982).
h. J. W. Huffman, S. M. Potnis, and A. V. Satish, *J. Org. Chem.* **50:**4266 (1985).

Another version of the Robinson annulation procedure involves the use of methyl 1-trimethylsilylvinyl ketone. The reaction follows the normal sequence of conjugate addition, aldol cyclization, and dehydration.

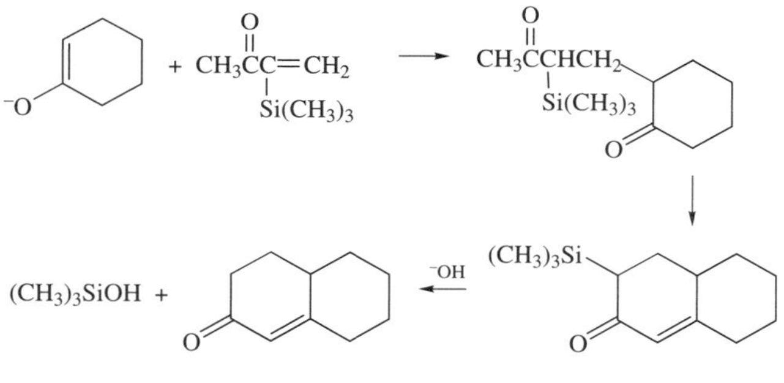

(cyclohexenolate, ^-O) + $CH_3CC{=}CH_2$ (C=O, $Si(CH_3)_3$) $\longrightarrow$ CH_3CCHCH_2–cyclohexanone (C=O, $Si(CH_3)_3$) $\longrightarrow$ $(CH_3)_3Si$–octalone $\xrightarrow{^-OH}$ $(CH_3)_3SiOH$ + octalone

Ref. 87

The role of the trimethylsilyl group is to stabilize the enolate formed in the conjugate addition. The silyl group is then removed during the dehydration step. The advantage of methyl 1-trimethylsilylvinyl ketone is that it can be used under aprotic conditions which are compatible with regiospecific methods for enolate generation. The direction of annulation of unsymmetrical ketones can therefore be controlled by the method of

87. G. Stork and B. Ganem, *J. Am. Chem. Soc.* **95:**6152 (1973); G. Stork and J. Singh, *J. Am. Chem. Soc.* **96:**6181 (1974).

enolate formation.

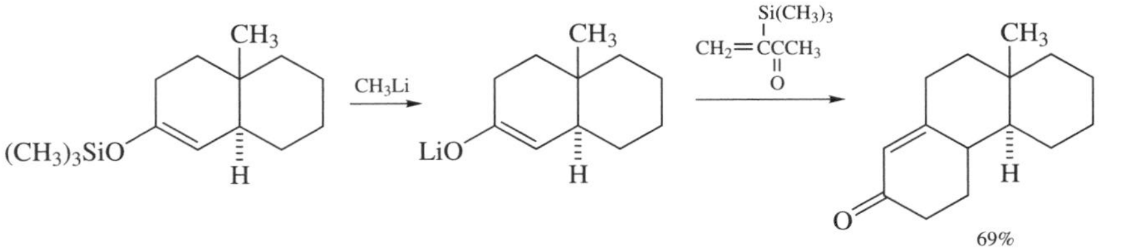

Ref. 88

Methyl 1-phenylthiovinyl ketones can also be used as enones in kinetically controlled Robinson annulation reactions, as illustrated by entry 7 in Scheme 2.10.

The product in entry 1 of Scheme 2.10 is commonly known as the Wieland–Miescher ketone and is a useful starting material for the preparation of steroids and terpenes. The Robinson annulation to prepare this ketone can be carried out enantioselectively by using the amino acid L-proline to form an enamine intermediate. The *S*-enantiomer of the product is obtained in high enantiomeric excess.[89] This compound and the corresponding product obtained from cyclopentane-1,3-dione[90] are key intermediates in the enantioselective synthesis of steroids.[91]

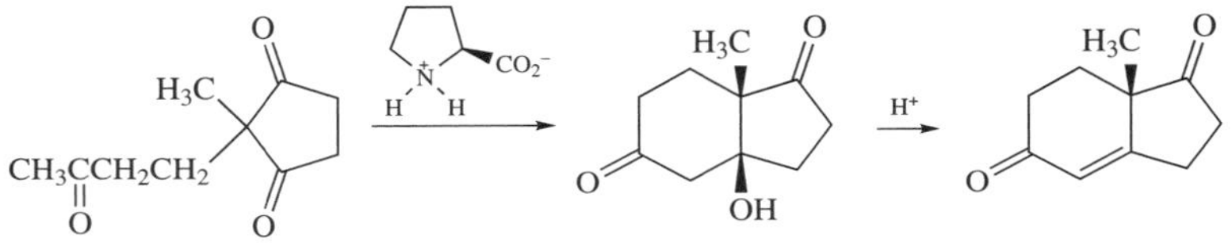

The detailed mechanism of this enantioselective transformation remains under investigation.[92] It is known that the acidic carboxylic group is crucial. The cyclization is believed to occur via the enamine derived from the catalyst and the exocyclic ketone. There is evidence that a second molecule of the catalyst is involved, and it has been suggested that this molecule participates in the proton-transfer step which completes the cyclization reaction.[93]

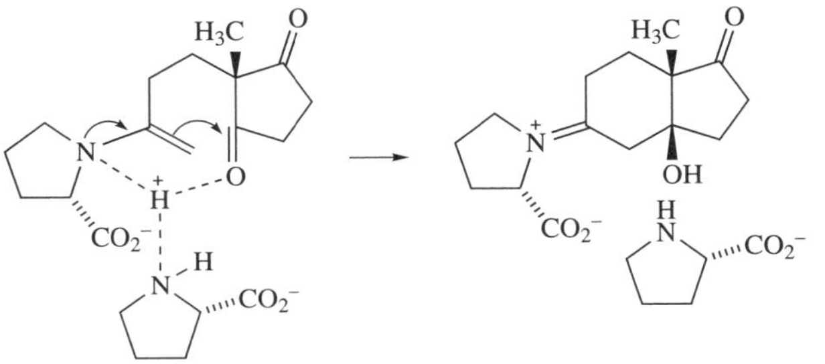

88. R. K. Boeckman, Jr., *J. Am. Chem. Soc.* **96:**6179 (1974).
89. J. Gutzwiller, P. Buchshacher, and A. Fürst, *Synthesis* **1977:**167; P. Buchshacher and A. Fürst, *Org. Synth.* **63:**37 (1984).
90. Z. G. Hajos and D. R. Parrish, *J. Org. Chem.* **39:**1615 (1974); U. Eder, G. Sauer, and R. Wiechert, *Angew. Chem. Int. Ed. Engl.* **10:**496 (1971). Z. G. Hajos and D. R. Parrish, *Org. Synth.* **63:**26 (1985).
91. N. Cohen, *Acc. Chem. Res.* **9:**412 (1976).
92. P. Buchschacher, J.-M. Cassal, A. Fürst, and W. Meier, *Helv. Chim. Acta* **60:**2747 (1977); K. L. Brown, L. Damm, J. D. Dunitz, A. Eschenmoser, R. Hobi, and C. Kratky, *Helv. Chim. Acta* **61:**3108 (1978); C. Agami, F. Meynier, C. Puchot, J. Guilhem, and C. Pascard, *Tetrahedron* **40:**1031 (1984).
93. C. Agami, J. Levisalles, and C. Puchot, *J. Chem. Soc., Chem. Commun.* **1985:**441; C. Agami, *Bull. Soc. Chim. Fr.* **1988:**499.

2.2. Addition Reactions of Imines and Iminium Ions

Imines and iminium ions are nitrogen analogs of carbonyl compounds, and they undergo nucleophilic additions like those involved in aldol condensations. The reactivity order is $C{=}NR < C{=}O < [C{=}NR_2]^+ < [C{=}OH]^+$. Because iminium ions are more reactive than imines, condensations involving imines are frequently run under acidic conditions where the imine is protonated.

2.2.1. The Mannich Reaction

The *Mannich reaction* is the condensation of an enolizable carbonyl compound with an iminium ion.[94] The reaction effects α-alkylation and introduces a dialkylaminomethyl substituent.

$$RCH_2\overset{O}{\overset{\|}{C}}R' + CH_2{=}O + HN(CH_3)_2 \rightarrow (CH_3)_2NCH_2\underset{R}{\underset{|}{C}}H\overset{O}{\overset{\|}{C}}R'$$

The electrophilic species is often generated *in situ* from the amine and formaldehyde.

$$CH_2{=}O + HN(CH_3)_2 \rightleftharpoons HOCH_2N(CH_3)_2 \overset{H^+}{\rightleftharpoons} H_2O + CH_2{=}\overset{+}{N}(CH_3)_2$$

The reaction is usually limited to secondary amines, because dialkylation can occur with primary amines. The dialkylation reaction can be used advantageously in ring closures.

$$CH_3O_2C\underset{}{\overset{CH_3CH_2}{\overset{|}{C}}}H-\overset{O}{\overset{\|}{C}}-\overset{CH_2CH_3}{\overset{|}{C}}HCO_2CH_3 + CH_2O + CH_3NH_2 \longrightarrow \text{3,5-diethyl-3,5-bis(methoxycarbonyl)-1-methyl-4-piperidone}$$

Ref. 95

Entries 1 and 2 in Scheme 2.11 show the preparation of "Mannich bases" from a ketone, formaldehyde, and a dialkylamine following the classical procedure. Alternatively, formaldehyde equivalents may be used, such as bis(dimethylamino)methane in entry 3. On treatment with trifluoroacetic acid, this aminal generates the iminium trifluoroacetate as a reactive electrophile.

N,N-Dimethylmethyleneammonium iodide is commercially available and is known as "Eschenmoser's salt".[96] This compound is sufficiently electrophilic to react directly with silyl enol ethers in neutral solution.[97] The reagent can be added to a solution of an enolate

94. F. F. Blicke, *Org. React.* **1:**303 (1942); J. H. Brewster and E. L. Eliel, *Org. React.* **7:**99 (1953); M. Tramontini and L. Angiolini, *Tetrahedron* **46:**1791 (1990); M. Tramontini and L. Angiolini, *Mannich Bases—Chemistry and Uses*, CRC Press, Boca Raton, Florida, 1994; M. Ahrend, B. Westerman, and N. Risch, *Angew. Chem. Int. Ed. Engl.* **37:**1045 (1998).
95. C. Mannich and P. Schumann, *Berichte* **69:** 2299 (1936).
96. J. Schreiber, H. Maag, N. Hashimoto, and A. Eschenmoser, *Angew. Chem. Int. Ed. Engl.* **10:**330 (1971).
97. S. Danishefsky, T. Kitahara, R. McKee, and P. F. Schuda, *J. Am. Chem. Soc.* **98:**6715 (1976).

Scheme 2.11. Synthesis and Utilization of Mannich Bases

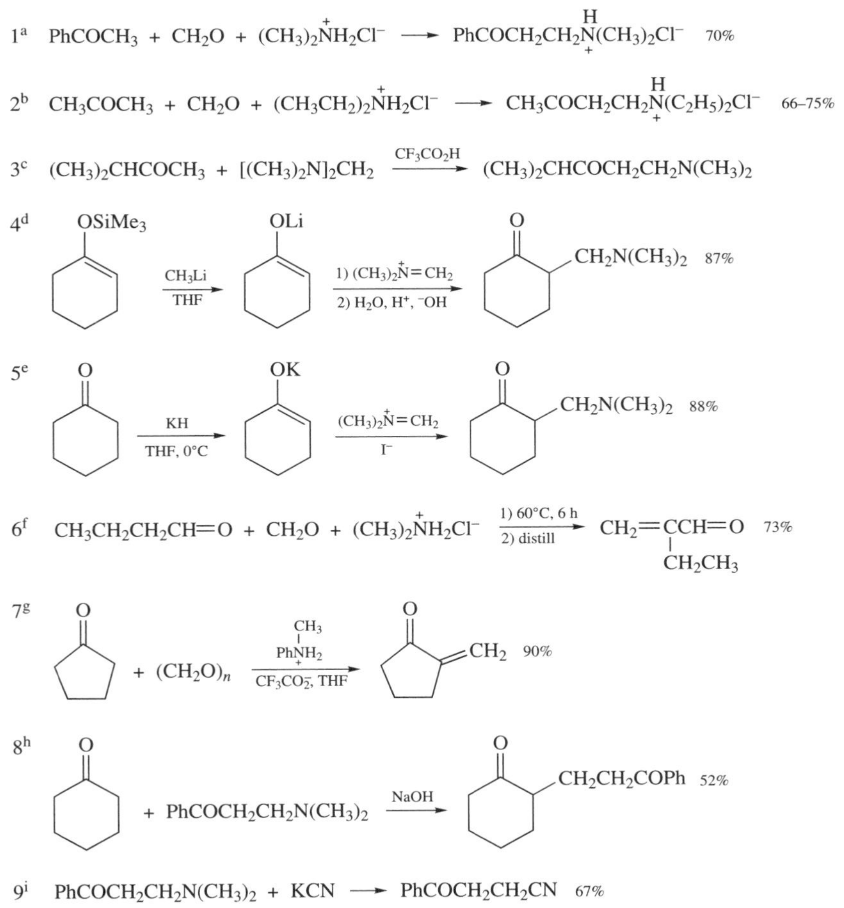

1[a] $PhCOCH_3 + CH_2O + (CH_3)_2\overset{+}{N}H_2Cl^- \longrightarrow PhCOCH_2CH_2\overset{+}{N}H(CH_3)_2Cl^-$ 70%

2[b] $CH_3COCH_3 + CH_2O + (CH_3CH_2)_2\overset{+}{N}H_2Cl^- \longrightarrow CH_3COCH_2CH_2\overset{+}{N}H(C_2H_5)_2Cl^-$ 66–75%

3[c] $(CH_3)_2CHCOCH_3 + [(CH_3)_2N]_2CH_2 \xrightarrow{CF_3CO_2H} (CH_3)_2CHCOCH_2CH_2N(CH_3)_2$

4[d] 1-(trimethylsilyloxy)cyclohexene $\xrightarrow[THF]{CH_3Li}$ lithium cyclohexenolate $\xrightarrow[2)\ H_2O,\ H^+,\ ^-OH]{1)\ (CH_3)_2\overset{+}{N}=CH_2}$ 2-($CH_2N(CH_3)_2$)cyclohexanone 87%

5[e] cyclohexanone $\xrightarrow[THF,\ 0°C]{KH}$ potassium cyclohexenolate $\xrightarrow[I^-]{(CH_3)_2\overset{+}{N}=CH_2}$ 2-($CH_2N(CH_3)_2$)cyclohexanone 88%

6[f] $CH_3CH_2CH_2CH=O + CH_2O + (CH_3)_2\overset{+}{N}H_2Cl^- \xrightarrow[2)\ distill]{1)\ 60°C,\ 6\ h} CH_2=C(CH_2CH_3)CH=O$ 73%

7[g] cyclopentanone $+ (CH_2O)_n \xrightarrow[CF_3CO_2^-,\ THF]{PhN(CH_3)H_2^+}$ 2-methylenecyclopentanone ($=CH_2$) 90%

8[h] cyclohexanone $+ PhCOCH_2CH_2N(CH_3)_2 \xrightarrow{NaOH}$ 2-(CH_2CH_2COPh)cyclohexanone 52%

9[i] $PhCOCH_2CH_2N(CH_3)_2 + KCN \longrightarrow PhCOCH_2CH_2CN$ 67%

a. C. E. Maxwell, *Org. Synth.* **III:**305 (1955).
b. A. L. Wilds, R. M. Nowak, and K. E. McCaleb, *Org. Synth.* **IV:**281 (1963).
c. M. Gaudry, Y. Jasor, and T. B. Khac, *Org. Synth.* **59:**153 (1979).
d. S. Danishefsky, T. Kitahara, R. McKee, and P. F. Schuda, *J. Am. Chem. Soc.* **98:**6715 (1976).
e. J. L. Roberts, P. S. Borromeo, and C. D. Poulter, *Tetrahedron Lett.* **1977:**1621.
f. C. S. Marvel, R. L. Myers, and J. H. Saunders, *J. Am. Chem. Soc.* **70:**1694 (1948).
g. J. L. Gras, *Tetrahedron Lett.* **1978:**2111, 2955.
h. A. C. Cope and E. C. Hermann, *J. Am. Chem. Soc.* **72:**3405 (1950).
i. E. B. Knott, *J. Chem. Soc.* **1947:**1190.

or enolate precursor, which permits the reaction to be carried out under nonacidic conditions. Entries 4 and 5 of Scheme 2.11 illustrate the preparation of Mannich bases with Eschenmoser's salt.

The dialkylaminomethyl ketones formed in the Mannich reaction are useful synthetic intermediates.[98] Thermal elimination of the amines or the derived quaternary salts provides

98. G. A. Gevorgyan, A. G. Agababyan, and O. L. Mndzhoyan, *Russ. Chem. Rev.* (Engl. Transl.) **54:**495 (1985).

α-methylene carbonyl compounds.

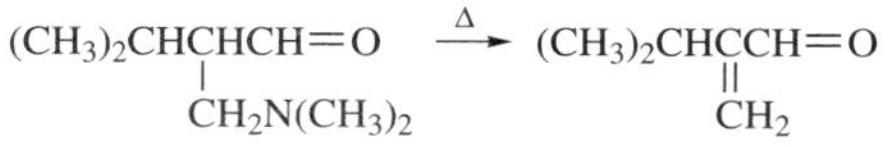

$$(CH_3)_2CHCH(CH_2N(CH_3)_2)CH{=}O \xrightarrow{\Delta} (CH_3)_2CHC(=CH_2)CH{=}O$$ Ref. 99

These α,β-unsaturated ketones and aldehydes are used as reactants in Michael additions (Section 1.10) and Robinson annulations (Section 2.1.4), as well as in a number of other reactions that we will encounter later. Entries 8 and 9 in Scheme 2.11 illustrate Michael reactions carried out by *in situ* generation of α,β-unsaturated carbonyl compounds from Mannich bases.

α-Methylene lactones are present in a number of natural products.[100] The reaction of ester enolates with *N,N*-dimethylmethyleneammonium trifluoroacetate,[101] or Eschenmoser's salt,[102] has been used for introduction of the α-methylene group in the synthesis of vernolepin, a compound with antileukemic activity.[103,104]

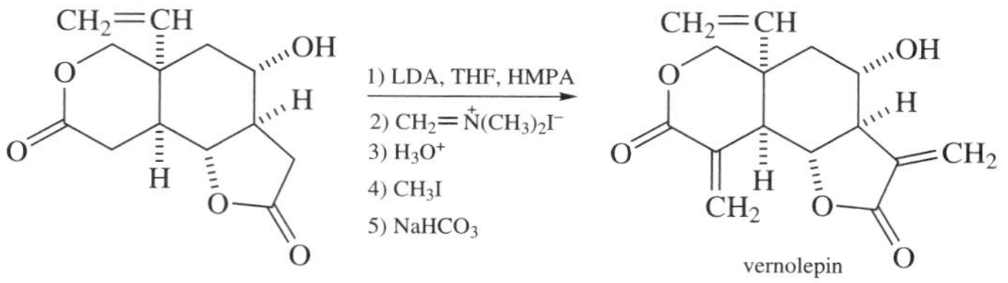

1) LDA, THF, HMPA
2) $CH_2{=}\overset{+}{N}(CH_3)_2I^-$
3) H_3O^+
4) CH_3I
5) $NaHCO_3$

vernolepin

Mannich reactions, or a close mechanistic analog, are important in the biosynthesis of many nitrogen-containing natural products. As a result, the Mannich reaction has played an important role in the synthesis of such compounds, especially in syntheses patterned after the mode of biosynthesis, i.e., *biogenetic-type synthesis*. The earliest example of the use of the Mannich reaction in this way was the successful synthesis of tropinone, a derivative of the alkaloid tropine, by Sir Robert Robinson in 1917.

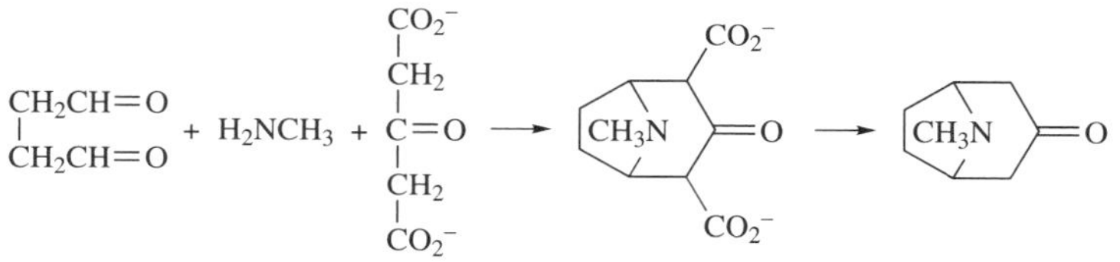

$$\begin{matrix} CH_2CH{=}O \\ | \\ CH_2CH{=}O \end{matrix} + H_2NCH_3 + {}^-O_2CCH_2C(=O)CH_2CO_2^- \longrightarrow \longrightarrow$$ Ref. 105

99. C. S. Marvel, R. L. Myers, and J. H. Saunders, *J. Am. Chem. Soc.* **70:**1694 (1948).
100. S. M. Kupchan, M. A. Eakin, and A. M. Thomas, *J. Med. Chem.* **14:**1147 (1971).
101. N. L. Holy and Y. F. Wang, *J. Am. Chem. Soc.* **99:**499 (1977).
102. J. L. Roberts, P. S. Borromes, and C. D. Poulter, *Tetrahedron Lett.* **1977:**1621.
103. S. Danishefsky, P. F. Schuda, T. Kitahara, and S. J. Etheredge, *J. Am. Chem. Soc.* **99:**6066 (1977).
104. For reviews of methods for the synthesis of α-methylene lactones, see R. B. Gammill, C. A. Wilson, and T. A. Bryson, *Synth. Commun.* **5:**245 (1975); J. C. Sarma and R. P. Sharma, *Heterocycles* **24:**441 (1986); N. Petragnani, H. M. C. Ferraz, and G. V. J. Silva, *Synthesis* **1986:**157.
105. R. Robinson, *J. Chem. Soc.* **1917:**762.

Even more reactive C=N bonds are present in *N-acyliminium ions.*[106]

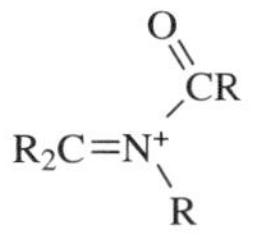

O
\\
CR
/
$R_2C{=}N^+$
\
R

These compounds are sufficiently electrophilic that they are usually prepared *in situ* in the presence of a potential nucleophile. There are several ways of generating acyliminium ions. Cyclic examples can be generated by partial reduction of imides.[107]

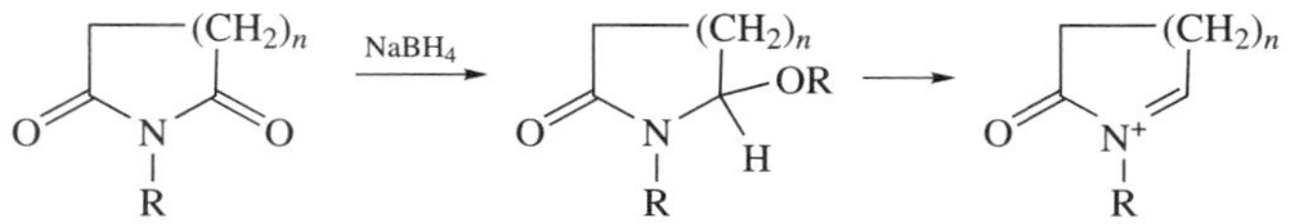

$(CH_2)_n$ $NaBH_4$ $(CH_2)_n$ OR $(CH_2)_n$
O N O O N H O N^+
R R R

Various oxidations of amines can also generate acyliminium ions. The methods most used in synthetic procedures involve electrochemical oxidation to form α-alkoxy amides and lactams, which then generate acyliminium ions.[108] Acyliminium ions are sufficiently electrophilic to react with enolate equivalents such as silyl enol ethers[109] and enol esters.[110]

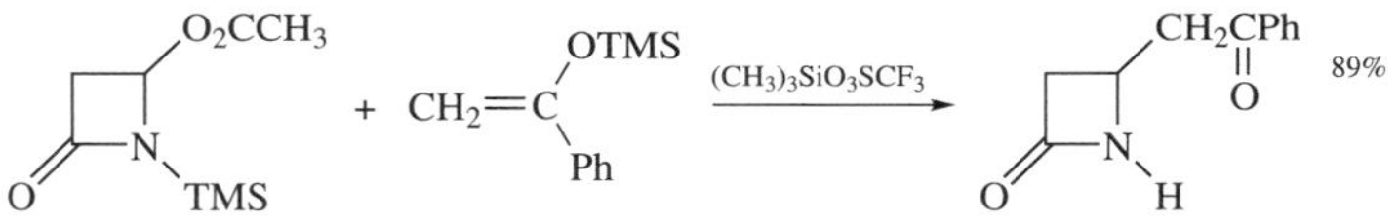

O_2CCH_3 OTMS CH_2CPh
\+ $CH_2{=}C$ $(CH_3)_3SiO_3SCF_3$ O 89%
N Ph N
O TMS O H

Acyliminium ions can be used in enantioselective additions with enolates having chiral auxiliaries, such as boron enolates or *N*-acylthiazolidinethiones.

106. H. Hiemstra and W. N. Speckamp, in *Comprehensive Organic Synthesis*, Vol. 2, B. Trost and I.Fleming, eds., 1991, pp. 1047–1082.
107. J. C. Hubert, J. B. P. A. Wijnberg, and W. Speckamp, *Tetrahedron* **31:**1437 (1975); H. Hiemstar, W. J. Klaver, and W. N. Speckamp, *J. Org. Chem.* **49:**1149 (1984); R. A. Pilli, L. C. Dias, and A. O. Maldaner, *J. Org. Chem.* **60:**717 (1995).
108. T. Shono, H. Hamaguchi, and Y. Matsumura, *J. Am. Chem. Soc.* **97:**4264 (1975); T. Shono, Y. Matsamura, K. Tsubata, Y. Sugihara, S Yamane, T. Kanazawa, and T. Aoki, *J. Am. Chem. Soc.* **104:**6697 (1982); T. Shono, *Tetrahedron* **40:**811 (1984).
109. R. P. Attrill, A. G. M. Barrett, P. Quayle, J. van der Westhuizen, and M. J. Betts, *J. Org. Chem.* **49:**1679 (1984); K. T. Wanner, A. Kartner, and E. Wadenstorfer, *Heterocycles* **27:**2549 (1988); M. A. Ciufolini, C. W. Hermann, K. H. Whitmire, and N. E. Byrne, *J. Am. Chem. Soc.* **111:**3473 (1989).
110. T. Shono, Y. Matsumura, and K. Tsubata, *J. Am. Chem. Soc.* **103:**1172 (1981).

CH_3CO_2 … O_2CCH_3, N–CH_2Ph + $PhCH_2OCH_2CH_2CH{=}C$ (O–B) → CH_3CO_2, H, N, CH_2Ph, OCH_2Ph Ref. 111

O_2CCH_3, N, TMS + CH_3, Sn, O, S, N, S → O, N, H, CH_3, O, S, N, S Ref. 112

2.2.2. Amine-Catalyzed Condensation Reactions

Iminium ions are intermediates in a group of reactions which form α,β-unsaturated compounds having structures corresponding to those formed by mixed aldol addition followed by dehydration. These reactions are catalyzed by amines or buffer systems containing an amine and an acid and are referred to as *Knoevenagel condensations*.[113] The general mechanism is believed to involve iminium ions as the active electrophiles, rather than the amine simply acting as a base for the aldol condensation. Knoevenagel condensation conditions frequently involve both an amine and a weak acid. The reactive electrophile is probably the protonated form of the imine, because this is a more reactive electrophile than the corresponding carbonyl compound.[114]

$$ArCH{=}O + C_4H_9NH_2 \rightleftharpoons ArCH{=}NC_4H_9$$

$$ArCH{=}NC_4H_9 \xrightarrow{H^+,\ {}^-CH_2NO_2} ArCH(CH_2NO_2)NHC_4H_9 \longrightarrow ArCH(-CHNO_2-H)-NHC_4H_9 \xrightarrow{H^+} ArCH{=}CHNO_2$$

The carbon nucleophiles in amine-catalyzed reaction conditions are usually rather acidic compounds containing two electron-attracting substituents. Malonic esters, cyanoacetic esters, and cyanoacetamide are examples of compounds which undergo condensation reactions under Knoevenagel conditions.[115] Nitroalkanes are also effective nucleophilic reactants. The single nitro group sufficiently activates the α hydrogens to permit deprotonation under the weakly basic conditions. Usually, the product that is isolated is

111. R. A. Pilli and D. Russowsky, *J. Org. Chem.* **61:**3187 (1996).
112. Y. Nagao, T. Kumagai, S. Tamai, T. Abe, Y. Kuramoto, T. Taga, S. Aoyagi, Y. Nagase, M. Ochiai, Y. Inoue, and E. Fujita, *J. Am. Chem. Soc.* **108:**4673 (1986).
113. G. Jones, *Org. React.* **15:**204 (1967); R. L. Reeves, in *The Chemistry of the Carbonyl Group*, S. Patai, ed., Interscience, New York, 1966, pp. 593–599.
114. T. I. Crowell and D. W. Peck, *J. Am. Chem. Soc.* **75:**1075 (1953).
115. A. C. Cope, C. M. Hofmann, C. Wyckoff, and E. Hardenbergh, *J. Am. Chem. Soc.* **63:**3452 (1941).

the "dehydrated," i.e., α,β-unsaturated, derivative of the original adduct.

$$R_2C(X)-CH(CO_2R)(CN) \xrightarrow{B} R_2C=C(CO_2R)(CN)$$

X = OH or NR$_2$

A relatively acidic proton in the nucleophile is important for two reasons. First, it permits weak bases, such as amines, to provide a sufficient concentration of the enolate for reaction. A highly acidic proton also facilitates the elimination step which drives the reaction to completion.

Malonic acid and cyanoacetic acid can also be used as the potential nucleophiles. The mechanism of the addition step is likely to involve iminium ions when secondary amines are used as catalysts. With malonic acid or cyanoacetic acid as reactant, the products usually undergo decarboxylation. This may occur as a concerted decomposition of the adduct.[116]

$$RC(=O)R + CH_2(CO_2H)_2 \longrightarrow R_2C(X)-CH(CO_2H)-C(=O)O^- \longrightarrow R_2C=CHCO_2H$$

X = OH or NR$_2$

Decarboxylative condensations of this type are sometimes carried out in pyridine. Pyridine can not form an imine intermediate, but it has been shown to catalyze the decarboxylation of arylidene malonic acids.[117] The decarboxylation occurs by concerted decomposition of the adduct of pyridine to the α,β-unsaturated diacid.

$$ArCH=C(CO_2H)_2 + C_5H_5NH^+ \longrightarrow ArCH(N^+C_5H_5)-CH(CO_2H)-C(=O)O-H \longrightarrow ArCH=CHCO_2H$$

Scheme 2.12 gives some examples of Knoevenagel condensation reactions.

2.3. Acylation of Carbanions

The reactions to be discussed in this section involve carbanion addition to carbonyl centers with a potential leaving group. The tetrahedral intermediate formed in the addition step then reacts by expulsion of the leaving group. The overall transformation results in the

116. E. J. Corey, *J. Am. Chem. Soc.* **74:**5897 (1952).
117. E. J. Corey and G. Fraenkel, *J. Am. Chem. Soc.* **75:**1168 (1953).

Scheme 2.12. Amine-Catalyzed Condensations of the Knoevenagel Type

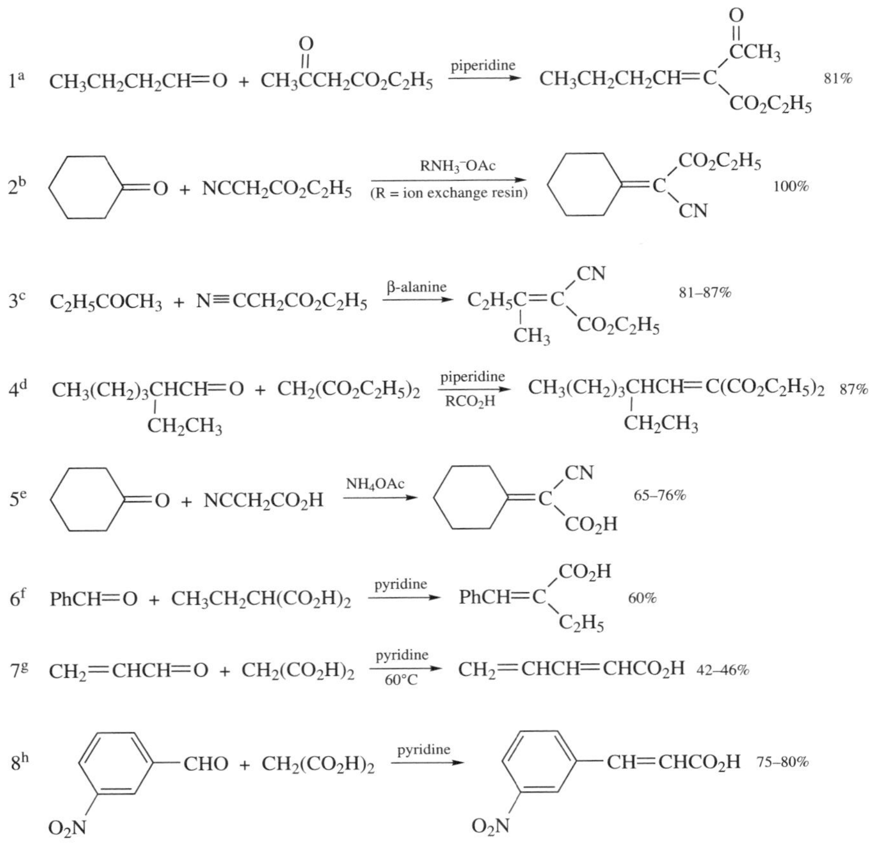

1[a] $CH_3CH_2CH_2CH{=}O + CH_3\overset{O}{\overset{\|}{C}}CH_2CO_2C_2H_5 \xrightarrow{\text{piperidine}} CH_3CH_2CH_2CH{=}C(\overset{O}{\overset{\|}{C}}CH_3)(CO_2C_2H_5)$ 81%

2[b] cyclohexanone ($C_6H_{10}{=}O$) $+ NCCH_2CO_2C_2H_5 \xrightarrow[\text{(R = ion exchange resin)}]{RNH_3^-OAc}$ $C_6H_{10}{=}C(CO_2C_2H_5)(CN)$ 100%

3[c] $C_2H_5COCH_3 + N{\equiv}CCH_2CO_2C_2H_5 \xrightarrow{\beta\text{-alanine}} C_2H_5C(CH_3){=}C(CN)(CO_2C_2H_5)$ 81–87%

4[d] $CH_3(CH_2)_3CH(CH_2CH_3)CH{=}O + CH_2(CO_2C_2H_5)_2 \xrightarrow[RCO_2H]{\text{piperidine}} CH_3(CH_2)_3CH(CH_2CH_3)CH{=}C(CO_2C_2H_5)_2$ 87%

5[e] cyclohexanone ($C_6H_{10}{=}O$) $+ NCCH_2CO_2H \xrightarrow{NH_4OAc}$ $C_6H_{10}{=}C(CN)(CO_2H)$ 65–76%

6[f] $PhCH{=}O + CH_3CH_2CH(CO_2H)_2 \xrightarrow{\text{pyridine}} PhCH{=}C(CO_2H)(C_2H_5)$ 60%

7[g] $CH_2{=}CHCH{=}O + CH_2(CO_2H)_2 \xrightarrow[60°C]{\text{pyridine}} CH_2{=}CHCH{=}CHCO_2H$ 42–46%

8[h] $m\text{-}O_2NC_6H_4CHO + CH_2(CO_2H)_2 \xrightarrow{\text{pyridine}} m\text{-}O_2NC_6H_4CH{=}CHCO_2H$ 75–80%

a. A. C. Cope and C. M. Hofmann, *J. Am. Chem. Soc.* **63**:3456 (1941).
b. R. W. Hein, M. J. Astle, and J. R. Shelton, *J. Org. Chem.* **26**:4874 (1961).
c. F. S. Prout, R. J. Hartman, E. P.-Y. Huang, C. J. Korpics, and G. R. Tichelaar, *Org. Synth.* **IV**:93 (1963).
d. E. F. Pratt and E. Werbie, *J. Am. Chem. Soc.* **72**:4638 (1950).
e. A. C. Cope, A. A. D'Addieco, D. E. Whyte, and S. A. Glickman, *Org. Synth.* **IV**:234 (1963).
f. W. J. Gensler and E. Berman, *J. Am. Chem. Soc.* **80**:4949 (1958).
g. P. J. Jessup, C. B. Petty, J. Roos, and L. E. Overman, *Org. Synth.* **59**:1 (1979).
h. R. H. Wiley and N. R. Smith, *Org. Synth.* **IV**:731 (1963).

acylation of the carbon nucleophile.

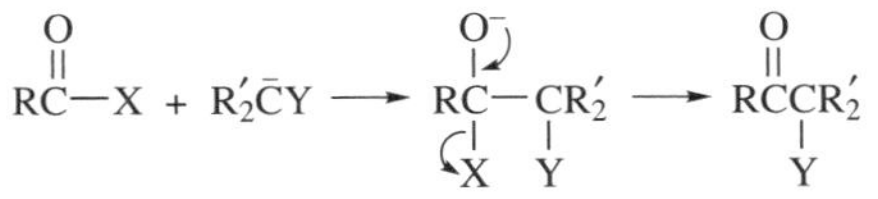

$$R\overset{O}{\overset{\|}{C}}{-}X + R'_2\bar{C}Y \longrightarrow R\overset{O^-}{\overset{|}{C}}(X){-}CR'_2(Y) \longrightarrow R\overset{O}{\overset{\|}{C}}CR'_2(Y)$$

An important group of these reactions involves esters, in which case the leaving group is alkoxy or aryloxy. The self-condensation of esters is known as the *Claisen condensation*.[118] Ethyl acetoacetate, for example, is prepared by Claisen condensation of ethyl

118. C. R. Hauser and B. E. Hudson, Jr., *Org. React.* **1**:266 (1942).

acetate. All of the steps in the mechanism are reversible.

$$CH_3CO_2CH_2CH_3 + CH_3CH_2O^- \rightleftharpoons {}^-CH_2CO_2CH_2CH_3 + CH_3CH_2OH$$

$$CH_3\overset{O}{\overset{\|}{C}}OCH_2CH_3 + {}^-CH_2CO_2CH_2CH_3 \rightleftharpoons CH_3\underset{CH_2CO_2CH_2CH_3}{\overset{O^-}{C}}OCH_2CH_3$$

$$CH_3\underset{CH_2CO_2CH_2CH_3}{\overset{O^-}{C}}-OCH_2CH_3 \rightleftharpoons CH_3\overset{O}{\overset{\|}{C}}CH_2CO_2CH_2CH_3 + CH_3CH_2O^-$$

$$CH_3\overset{O}{\overset{\|}{C}}CH_2CO_2CH_2CH_3 + CH_3CH_2O^- \longrightarrow CH_3\overset{O}{\overset{\|}{C}}\underset{-}{C}HCO_2CH_2CH_3 + CH_3CH_2OH$$

The final step drives the reaction to completion. Ethyl acetoacetate is more acidic than any of the other species present, and it is converted to its conjugate base in the final step. A full equivalent of base is needed to bring the reaction to completion. The β-ketoester product is obtained after neutralization and workup. As a practical matter, the alkoxide used as the base must be the same as the alcohol portion of the ester to prevent product mixtures resulting from ester interchange. Because the final proton transfer cannot occur when α-substituted esters are used, such compounds do not condense under the normal reaction conditions. This limitation can be overcome by use of a very strong base that converts the reactant ester completely to its enolate. Entry 2 of Scheme 2.13 illustrates the use of triphenylmethylsodium for this purpose.

Sodium hydride with a small amount of alcohol is frequently used as the base for ester condensation. It is likely that the reactive base is the sodium alkoxide formed by reaction of sodium hydride with the alcohol released in the condensation.

$$R'OH + NaH \longrightarrow R'ONa + H_2$$

The sodium alkoxide is also no doubt the active catalyst in procedures in which sodium metal is used, such as in entry 3 of Scheme 2.13. The alkoxide is formed by reaction of the alcohol that is formed as the reaction proceeds with sodium.

The intramolecular version of ester condensation is called the *Dieckmann condensation.*[119] It is an important method for the formation of five- and six-membered rings and has occassionally been used for formation of larger rings. Entries 3–6 in Scheme 2.13 are illustrative.

Because ester condensation is reversible, product structure is governed by thermodynamic control, and in situations in which more than one enolate may be formed, the product is derived from the most stable enolate. An example of this effect is the cyclization of the diester **4**.[120] Only **6** is formed, because **5** cannot be converted to a stable enolate. If **5**, synthesized by another method, is subjected to the conditions of the cyclization, it is isomerized to **6** by the reversible condensation mechanism:

$$\mathbf{5}\ (\text{2-}CO_2C_2H_5\text{-2-}CH_3\text{-cyclohexanone}) \;\not\longleftarrow\; C_2H_5O_2CCH_2(CH_2)_3\underset{}{\overset{CH_3}{C}}HCO_2C_2H_5\ (\mathbf{4}) \xrightarrow[\text{xylene}]{\text{NaOEt}} \mathbf{6}\ (O^-,\ CH_3,\ CO_2C_2H_5)$$

$$\mathbf{5} \xrightarrow[\text{xylene}]{\text{NaOEt}} \mathbf{6}$$

119. J. P. Schaefer and J. J. Bloomfield, *Org. React.* **15**:1 (1967).
120. N. S. Vul'fson and V. I. Zaretskii, *J. Gen. Chem. USSR* **29**:2704 (1959).

Scheme 2.13. Acylation of Nucleophilic Carbon by Esters

A. Intermolecular ester condensations

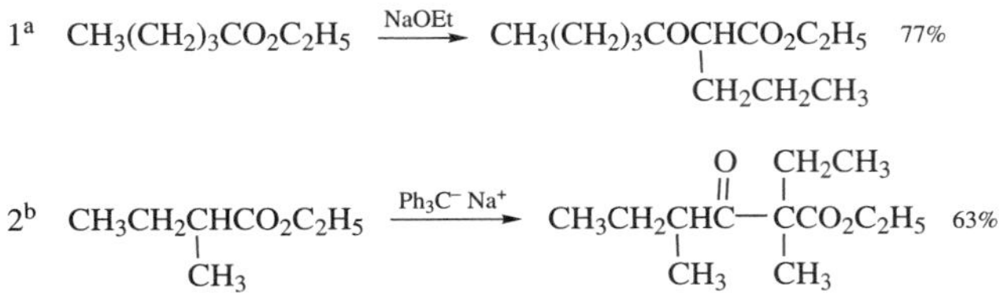
1a
NaOEt
77%
2b
Ph3C⁻ Na⁺
63%

B. Cyclization of diesters

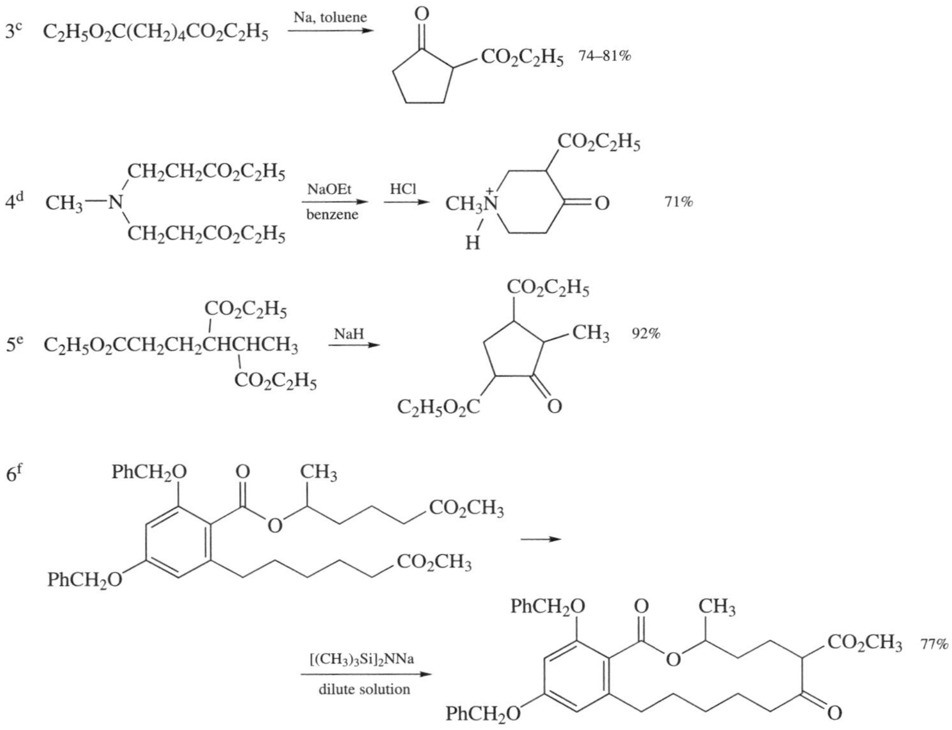
3c
Na, toluene
74–81%
4d
NaOEt
benzene
HCl
71%
5e
NaH
92%
6f
[(CH3)3Si]2NNa
dilute solution
77%

C. Mixed ester condensations

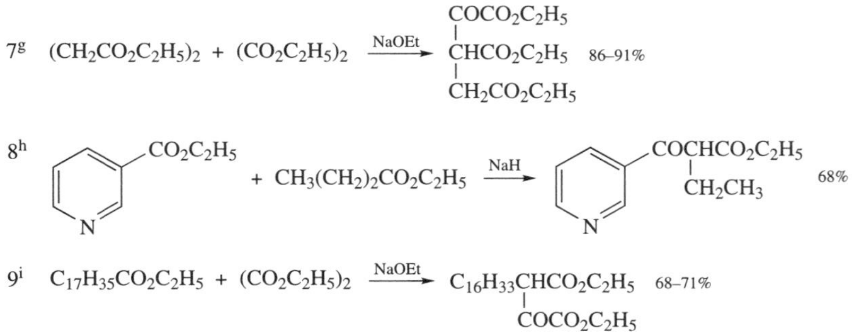
7g
NaOEt
86–91%
8h
NaH
68%
9i
NaOEt
68–71%

10[j] $C_6H_5CO_2C_2H_5 + CH_3CH_2CO_2C_2H_5 \xrightarrow{(i\text{-Pr})_2NMgBr} C_6H_5COCH(CH_3)CO_2C_2H_5$ 51%

a. R. R. Briese and S. M. McElvain, *J. Am. Chem. Soc.* **55**:1697 (1933).
b. B. E. Hudson, Jr., and C. R. Hauser, *J. Am. Chem. Soc.* **63**:3156 (1941).
c. P. S. Pinkney, *Org. Synth.* **II**:116 (1943).
d. E. A. Prill and S. M. McElvain, *J. Am. Chem. Soc.* **55**:1233 (1933).
e. M. S. Newman and J. L. McPherson, *J. Org. Chem.* **19**:1717 (1954).
f. R. N. Hurd and D. H. Shah, *J. Org. Chem.* **38**:390 (1973).
g. E. M. Bottorff and L. L. Moore, *Org. Synth.* **44**:67 (1964).
h. F. W. Swamer and C. R. Hauser, *J. Am. Chem. Soc.* **72**:1352 (1950).
i. D. E. Floyd and S. E. Miller, *Org. Synth.* **IV**:141 (1963).
j. E. E. Royals and D. G. Turpin, *J. Am. Chem. Soc.* **76**:5452 (1954).

Mixed condensations of esters are subject to the same general restrictions as outlined for mixed aldol condensations (Section 2.1.2) One reactant must act preferentially as the acceptor and another as the nucleophile for good yields to be obtained. Combinations which work most effectively involve one ester that cannot form an enolate but that is relatively reactive as an electrophile. Esters of aromatic acids, formic acid, and oxalic acid are especially useful. Some examples are shown in Section C of Scheme 2.13.

Acylation of ester enolates can also be carried out with more reactive acylating agents such as acid anhydrides and acyl chlorides. These reactions must be done in inert solvents to avoid solvolysis of the acylating agent. The preparation of diethyl benzoylmalonate (entry 1 in Scheme 2.14) is an example employing an acid anhydride. Entries 2–5 illustrate the use of acyl chlorides. Acylations with these more reactive compounds can be complicated by competing O-acylation. *N*-Methoxy-*N*-methylamides are also useful for acylation of ester enolates.

$$CH_3(CH_2)_4C(=O)N(OCH_3)CH_3 + CH_2{=}C(O^-Li^+)OC_2H_5 \xrightarrow[3)\ HCl]{1)\ -78°C\ \ 2)\ 25°C} CH_3(CH_2)_4C(=O)CH_2CO_2C_2H_5 \quad 82\%$$

Ref. 121

Magnesium enolates play an important role in C-acylation reactions. The magnesium enolate of diethyl malonate, for example, can be prepared by reaction with magnesium metal in ethanol. It is soluble in ether and undergoes C-acylation by acid anhydrides and acyl chlorides (entries 1 and 3 in Scheme 2.14). Monoalkyl esters of malonic acid react with Grignard reagents to give a chelated enolate of the malonate monoanion.

$$R'O_2CCH_2CO_2H + 2\ RMgX \longrightarrow R'OC(O^-)=CHC(=O)O^- \cdot Mg^{2+}$$

121. J. A. Turner and W. S. Jacks, *J. Org. Chem.* **54**:4229 (1989).

Scheme 2.14. Acylation of Ester Enolates with Acyl Halides, Anhydrides, and Imidazolides

A. Acylation with acyl halides and mixed anhydrides

1[a] $PhCOCOC_2H_5$ (O O above carbonyls) + $C_2H_5OMgCH(CO_2C_2H_5)_2$ → $PhCOCH(CO_2C_2H_5)_2$ 68–75%

2[b] $CH_3C(O^-){=}CHCO_2C_2H_5$ + PhCOCl → $[PhC(=O){-}\bar{C}(CCH_3{=}O)CO_2C_2H_5]$ → $PhC(=O)CH_2CO_2C_2H_5$ 68–71%

3[c] (2-nitrophenyl)COCl + $C_2H_5OMgCH(CO_2C_2H_5)_2$ → (2-nitrophenyl)$COCH(CO_2C_2H_5)_2$ 82–88%

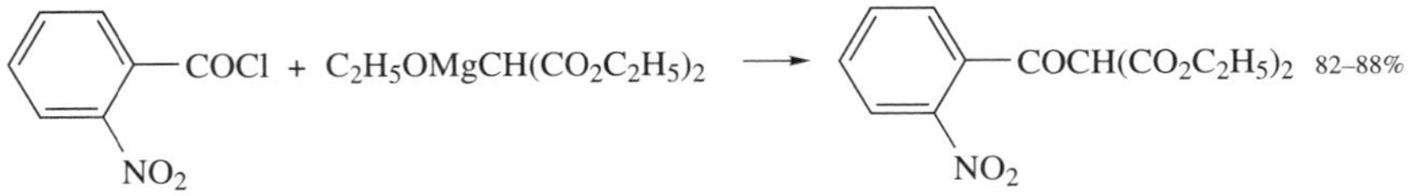

4[d] $CH_3C(O^-){=}CHCO_2C_2H_5$ + $ClC(=O)(CH_2)_3CO_2C_2H_5$ → $C_2H_5O_2C(CH_2)_3C(=O){-}CH(CCH_3{=}O)CO_2C_2H_5$ 61–66%

5[e] $CH_3CO_2C_2H_5$ $\xrightarrow{R_2NLi}$ $LiCH_2CO_2C_2H_5$ $\xrightarrow[-78°C]{(CH_3)_3CCCl (=O)}$ $(CH_3)_3CC(=O)CH_2CO_2C_2H_5$ 70%

6[f] $CH_3CO_2CH_3$ → (1) LDA; 2) $ClCO(CH_2)_{12}CH_3$; 3) H^+) → $CH_3O_2CCH_2C(=O)(CH_2)_{12}CH_3$ 83%

B. Acylation with imidazolides

7[g] (4-oxocyclopent-2-enyl)$CH_2C(=O)N$(imidazolyl) + $Mg(O_2CCH_2CO_2C_2H_5)_2$ → (1) 25°C; 2) H^+) → (4-oxocyclopent-2-enyl)$CH_2C(=O)CH_2CO_2C_2H_5$ 66%

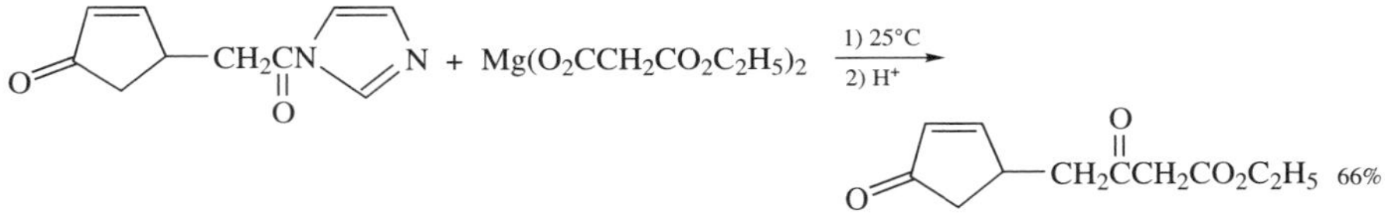

8[h] (imidazolide, CH_3) + $LiCH_2CO_2C(CH_3)_3$ $\xrightarrow[1\ h]{-78°C}$ $\xrightarrow{H^+}$ (product, CH_3, $C(=O)CH_2CO_2C(CH_3)_3$) 83%

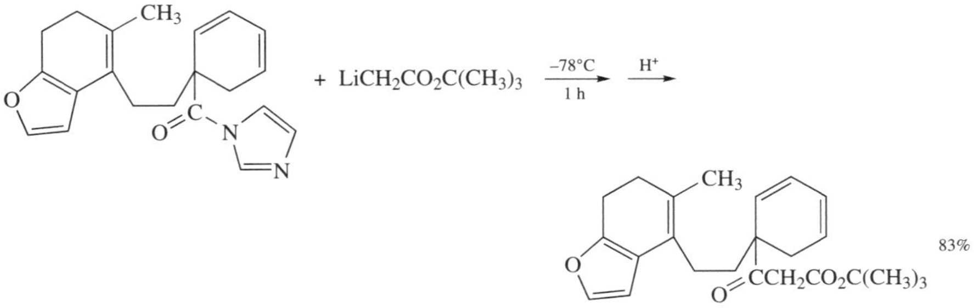

9[i] $O_2NCH_2^- + CH_3CN$-imidazole (C=O) $\xrightarrow[16\,h]{65°C}$ $\xrightarrow{H^+}$ $CH_3CCH_2NO_2$ (C=O) 80%

10[j] t-$BuO_2CNCHCO_2H$ (N–H, $CH(CH_3)_2$) $\xrightarrow[2)\ Mg\ salt\ of\ ethyl\ malonate\ (O_2CCH{=}C(OEt)O)]{1)\ imidazole\text{-}NCN\text{-}imidazole\ (C=O)}$ t-$BuO_2CNCHCCH_2CO_2C_2H_5$ (N–H, C=O, $CH(CH_3)_2$) 83%

a. J. A. Price and D. S. Tarbell, *Org. Synth.* **IV:**285 (1963).
b. J. M. Straley and A. C. Adams, *Org. Synth.* **IV:**415 (1963).
c. G. A. Reynolds and C. R. Hauser, *Org. Synth.* **IV:**708 (1963).
d. M. Guha and D. Nasipuri, *Org. Synth.* **V:**384 (1973).
e. M. W. Rathke and J. Deitch, *Tetrahedron Lett.* **1971:**2953.
f. D. F. Taber, P. B. Deker, H. M. Fales, T. H. Jones, and H. A. Lloyd, *J. Org. Chem.* **53:**2968 (1988).
g. A. Barco, S. Bennetti, G. P. Pollini, P. G. Baraldi, and C. Gandolfi, *J. Org. Chem.* **45:**4776 (1980).
h. E. J. Corey, G. Wess, Y. B. Xiang, and A. K. Singh, *J. Am. Chem. Soc.* **109:**4717 (1987).
i. M. E. Jung, D. D. Grove, and S. I. Khan, *J. Org. Chem.* **52:**4570 (1987).
j. J. Maibaum and D. H. Rich, *J. Org. Chem.* **53:**869 (1988).

These carbon nucleophiles react with acyl chlorides[122] or acyl imidazolides.[123] The initial products decarboxylate readily so the isolated products are β-ketoesters.

Mg^{2+} chelate of $R'O_2C\text{-}C(CH_3){=}C(O^-)\text{-}O^-$ + RCOCl or RCOIm $\longrightarrow$ $R'O_2CCHCR$ (C=O, CH_3)

Acyl imidazolides are more reactive than esters but not as reactive as acyl halides. β-Keto esters are formed by reaction of magnesium salts of monoalkyl esters of malonic acid with imidazolides.

$RC{-}N$-imidazole (C=O) + $Mg(O_2CCH_2CO_2R')_2$ $\longrightarrow$ $\xrightarrow[-CO_2]{H^+}$ $RCCH_2CO_2R'$ (C=O)

Acyl imidazolides have also been used for acylation of ester enolates and nitromethane anion, as illustrated by entries 9 and 10 in Scheme 2.14.

122. R. E. Ireland and J. A. Marshall, *J. Am. Chem. Soc.* **81:**2907 (1959).
123. J. Maibaum and D. H. Rich, *J. Org. Chem.* **53:**869 (1988); W. H. Moos, R. D. Gless, and H. Rapoport, *J. Org. Chem.* **46:**5064 (1981); D. W. Brooks, L. D.-L. Lu, and S. Masamune, *Angew. Chem. Int. Ed. Engl.* **18:**72 (1979).

Both diethyl malonate and ethyl acetoacetate can be acylated by acyl chlorides using magnesium chloride and triethylamine or pyridine.[124]

$$(C_2H_5O_2C)_2CH_2 \xrightarrow[Et_3N]{MgCl_2} \xrightarrow{RCOCl} (C_2H_5O_2C)_2CHCOR$$

$$C_2H_5O_2CCH_2COCH_3 \xrightarrow[pyridine]{MgCl_2} \xrightarrow{RCOCl} C_2H_5O_2CCH(COR)COCH_3$$

Rather similar conditions can be used to convert ketones to β-ketoacids by carboxylation.[125]

$$CH_3CH_2COCH_2CH_3 \xrightarrow[CH_3CN,\ CO_2,\ Et_3N]{MgCl_2,\ NaI} \xrightarrow{H^+} CH_3CH_2COCH(CO_2H)CH_3$$

Such reactions presumably involve formation of a magnesium chelate of the ketoacid. The β-keto acid is liberated when the reaction mixture is acidified during workup.

Carboxylation of ketones and esters can be achieved by using the magnesium salt of monomethyl carbonate:

$$C_6H_5COCH_3 + Mg(O_2COCH_3)_2 \xrightarrow[110°C]{DMF} \xrightarrow{H^+} C_6H_5COCH_2CO_2H \quad \text{Ref. 126}$$

$$\xrightarrow[2)\ H^+]{1)\ Mg(O_2COMe)_2} \quad 75\% \quad \text{Ref. 127}$$

The enolates of ketones can be acylated by esters and other acylating agents. The products of these reactions are all β-dicarbonyl compounds. They are all rather acidic and can be alkylated by the procedures described in Section 1.4. Reaction of ketone enolates

124. M. W. Rathke and P. J. Cowan, *J. Org. Chem.* **50:**2622 (1985).
125. R. E. Tirpak, R. S. Olsen, and M. W. Rathke, *J. Org. Chem.* **50:**4877 (1985).
126. M. Stiles, *J. Am. Chem. Soc.* **81:**2598 (1959).
127. W. L. Parker and F. Johnson, *J. Org. Chem.* **38:**2489 (1973).

with formate esters gives a β-keto aldehydes. Because these compounds exist in the enol form, they are referred to as *hydroxymethylene derivatives*. Product formation is under thermodynamic control so the structure of the product can be predicted on the basis of the stability of the various possible product anions.

$$RCH_2\overset{O}{\overset{\|}{C}}R' + HCO_2C_2H_5 \xrightarrow{NaOEt} R\overset{O}{\overset{\|}{C}}\underset{\underset{H-C-ONa}{\|}}{C}R' \xrightarrow{H^+} RC(=CHOH)-CR'=O$$

Ketones are converted to β-ketoesters by acylation with diethyl carbonate or diethyl oxalate, as illustrated by entries 5 and 6 in Scheme 2.15. Alkyl cyanoformate can be used as the acylating reagent under conditions where a ketone enolate has been formed under kinetic control.[128]

$$\text{2-methylcyclohexanone} \xrightarrow{LDA} \xrightarrow[\text{TMF, HMPA}]{EtO_2CCN} \xrightarrow{H_2O} \text{ethyl 3-methyl-2-oxocyclohexanecarboxylate } (CO_2C_2H_5)\quad 86\%$$

When this type of reaction is quenched with trimethylsilyl chloride, rather than by neutralization, a trimethylsilyl ether of the adduct is isolated. This result shows that the tetrahedral adduct is stable until the reaction mixture is hydrolyzed.

$$\text{cyclohexanone} \xrightarrow[2)\ EtO_2CCN]{1)\ LDA} \xrightarrow{(Me)_3SiCl} \text{2-[C(OSi(CH}_3)_3)(OC_2H_5)(CN)]\text{cyclohexanone} \qquad \text{Ref. 129}$$

β-Keto sulfoxides can be prepared by acylation of dimethyl sulfoxide ion with esters.[130]

$$R\overset{O}{\overset{\|}{C}}OR' + {}^{-}CH_2\overset{O}{\overset{\|}{S}}CH_3 \rightleftharpoons R\overset{O}{\overset{\|}{C}}\overset{-}{C}H\overset{O}{\overset{\|}{S}}CH_3 + R'OH$$

Mechanistically, this reaction is similar to ketone acylation. The β-keto sulfoxides have several synthetic applications. The sulfoxide substituent can be removed reductively, leading to methyl ketones:

$$CH_3O-C_6H_4-\overset{O}{\overset{\|}{C}}CH_2SOCH_3 \xrightarrow{Zn\ Hg} CH_3O-C_6H_4-\overset{O}{\overset{\|}{C}}CH_3 \qquad \text{Ref. 131}$$

128. L. N. Mander and S. P. Sethi, *Tetrahedron Lett.* **24:**5425 (1983).
129. F. E. Ziegler and T.-F. Wang, *Tetrahedron Lett.* **26:**2291 (1985).
130. E. J. Corey and M. Chaykovsky, *J. Am. Chem. Soc.* **87:**1345 (1965); H. D. Becker, G. J. Mikol, and G. A. Russell, *J. Am. Chem. Soc.* **85:**3410 (1963).
131. G. A. Russell and G. J. Mikol, *J. Am. Chem. Soc.* **88:**5498 (1966).

Scheme 2.15. Acylation of Ketones with Esters

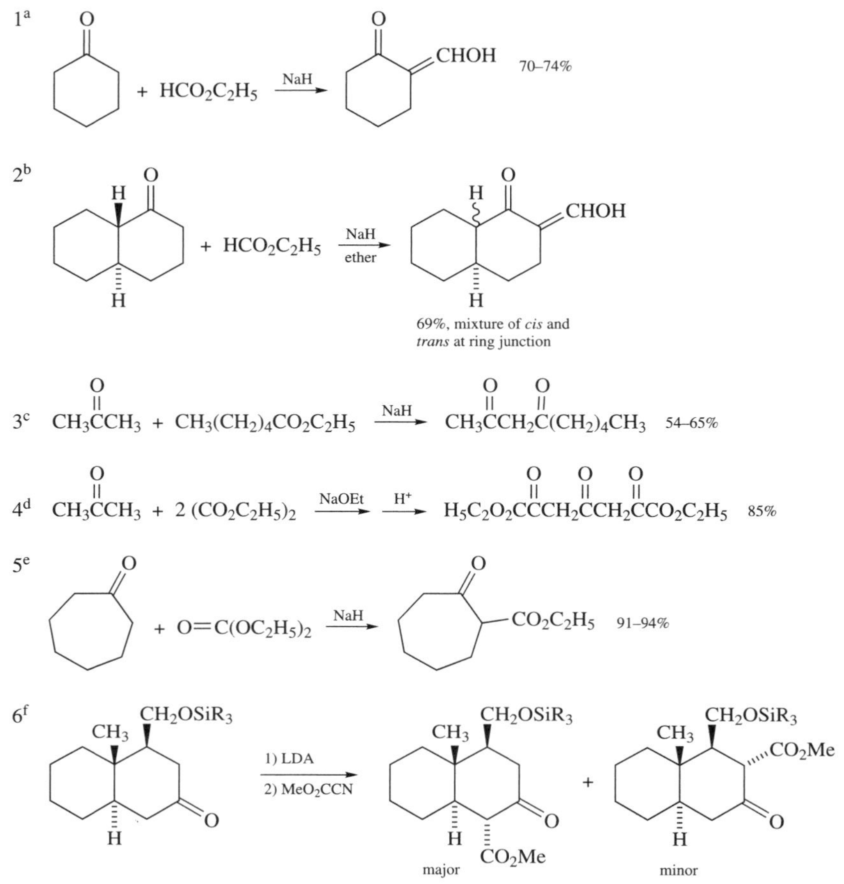

1[a] cyclohexanone + $HCO_2C_2H_5$ $\xrightarrow{\text{NaH}}$ 2-(hydroxymethylene)cyclohexanone (=CHOH) 70–74%

2[b] trans-1-decalone + $HCO_2C_2H_5$ $\xrightarrow[\text{ether}]{\text{NaH}}$ 2-(hydroxymethylene)-1-decalone (=CHOH)

69%, mixture of *cis* and *trans* at ring junction

3[c] $CH_3\overset{O}{\overset{\|}{C}}CH_3$ + $CH_3(CH_2)_4CO_2C_2H_5$ $\xrightarrow{\text{NaH}}$ $CH_3\overset{O}{\overset{\|}{C}}CH_2\overset{O}{\overset{\|}{C}}(CH_2)_4CH_3$ 54–65%

4[d] $CH_3\overset{O}{\overset{\|}{C}}CH_3$ + 2 $(CO_2C_2H_5)_2$ $\xrightarrow{\text{NaOEt}}$ $\xrightarrow{H^+}$ $H_5C_2O_2C\overset{O}{\overset{\|}{C}}CH_2\overset{O}{\overset{\|}{C}}CH_2\overset{O}{\overset{\|}{C}}CO_2C_2H_5$ 85%

5[e] cycloheptanone + $O{=}C(OC_2H_5)_2$ $\xrightarrow{\text{NaH}}$ 2-($CO_2C_2H_5$)cycloheptanone 91–94%

6[f] decalone bearing CH_3, CH_2OSiR_3, H $\xrightarrow[\text{2) } MeO_2CCN]{\text{1) LDA}}$ CO_2Me product (major) + CO_2Me product (minor)

a. C. Ainsworth, *Org. Synth.* **IV:**536 (1963).
b. P. H. Lewis, S. Middleton, M. J. Rosser, and L. E. Stock, *Aust. J. Chem.* **32:**1123 (1979).
c. N. Green and F. B. La Forge, *J. Am. Chem. Soc.* **70:**2287 (1948); F. W. Swamer and C. R. Hauser, *J. Am. Chem. Soc.* **72:**1352 (1950).
d. E. R. Riegel and F. Zwilgmeyer, *Org. Synth.* **II:**126 (1943).
e. A. P. Krapcho, J. Diamanti, C. Cayen, and R. Bingham, *Org. Synth.* **47:**20 (1967).
f. F. E. Ziegler, S. I. Klein, U. K. Pati, and T.-F. Wang, *J. Am. Chem. Soc.***107:**2730 (1985).

The β-keto sulfoxides can be alkylated via their anions. Inclusion of an alkylation step prior to the reduction provides a route to ketones with longer chains.

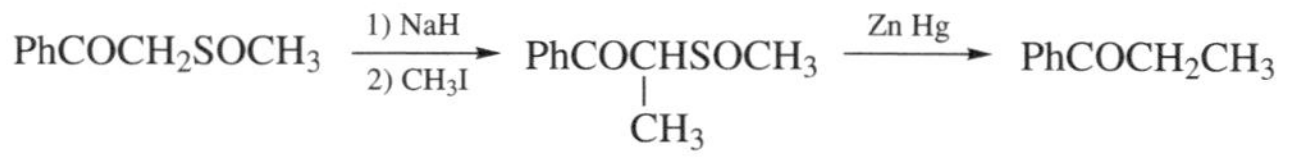

$$PhCOCH_2SOCH_3 \xrightarrow[\text{2) } CH_3I]{\text{1) NaH}} PhCO\underset{CH_3}{\underset{|}{C}}HSOCH_3 \xrightarrow{\text{Zn Hg}} PhCOCH_2CH_3$$ Ref. 132

Dimethyl sulfone can be subjected to similar reaction sequences.[133]

132. P. G. Gassman and G. D. Richmond, *J. Org. Chem.* **31:**2355 (1966).
133. H. O. House and J. K. Larson, *J. Org. Chem.* **33:**61 (1968).

2.4. The Wittig and Related Reactions of Phosphorus-Stabilized Carbon Nucleophiles

The *Wittig reaction* involves phosphorus ylides as the nucleophilic carbon species.[134] An *ylide* is a molecule that has a contributing Lewis structure with opposite charges on adjacent atoms, each of which has an octet of electrons. Although this definition includes other classes of compounds, the discussion here will be limited to ylides with the negative charge on carbon. Phosphorus ylides are stable, but usually quite reactive, compounds. They can be represented by two limiting resonance structures, which are sometimes referred to as the ylide and ylene forms. The ylene form is pentavalent at phosphorus and implies involvement of phosphorus 3*d* orbitals. Using $(CH_3)_3PCH_2$ (trimethylphosphonium methylide) as an example, the two forms are

$$\underset{\text{ylide}}{(CH_3)_3\overset{+}{P}-CH_2^-} \longleftrightarrow \underset{\text{ylene}}{(CH_3)_3P{=}CH_2}$$

Nuclear magnetic resonance (NMR) spectroscopic studies (^1H, ^{13}C, and ^{31}P), are consistent with the dipolar ylide structure and suggest only a minor contribution from the ylene structure.[135] Theoretical calculations support this view, also.[136]

The synthetic potential of phosphorus ylides was initially developed by G. Wittig and his associates at the University of Heidelberg. The reaction of a phosphorus ylide with an aldehyde or ketone introduces a carbon–carbon double bond in place of the carbonyl bond:

$$R_3\overset{+}{P}-\overset{-}{C}R'_2 + R''_2C{=}O \longrightarrow R''_2C{=}CR'_2 + R_3P{=}O$$

The mechanism proposed is an addition of the nucleophilic ylide carbon to the carbonyl group to yield a dipolar intermediate (a *betaine*), followed by elimination of a phosphine oxide. The elimination is presumed to occur after formation of a four-membered oxaphosphetane intermediate. An alternative mechanism might involve direct formation of the oxaphosphetane.[137] There have been several theoretical studies of these intermediates.[138] Oxaphosphetane intermediates have been observed by NMR studies at low temperature.[139] Betaine intermediates have been observed only under special conditions that retard the

134. For general reviews of the Wittig reaction, see A. Maercker, *Org. React.* **14:**270 (1965); I. Gosney and A. G. Rowley, in *Organophosphorus Reagents in Organic Synthesis*, J. I. G. Cadogan, ed., Academic Press, London, 1979, pp. 17–153; B. A. Maryanoff and A. B. Reitz, *Chem. Rev.* **89:**863 (1989); A. W. Johnson, *Ylides and Imines of Phosphorus*, John Wiley & Sons, New York, 1993; K. C. Nicolaou, M. W. Harter, J. L. Gunzer, and A. Nadin, *Liebigs Ann. Chem.* **1997:**1283.
135. H. Schmidbaur, W. Bucher, and D. Schentzow, *Chem. Ber.* **106:**1251 (1973).
136. A. Streitwieser, Jr., A. Rajca, R. S. McDowell, and R. Glaser, *J. Am. Chem. Soc.* **109:**4184 (1987); S. M. Bachrach, *J. Org. Chem.* **57:**4367 (1992).
137. E. Vedejs and K. A. J. Snoble, *J. Am. Chem. Soc.* **95:**5778 (1973); E. Vedejs and C. F. Marth, *J. Am. Chem. Soc.* **112:**3905 (1990).
138. R. Holler and H. Lischka, *J. Am. Chem. Soc.* **102:**4632 (1980); F. Volatron and O. Eisenstein, *J. Am. Chem. Soc.* **106:**6117 (1984); F. Mari, P. M. Lahti, and W. E. McEwen, *J. Am. Chem. Soc.* **114:**813 (1992); A. A. Restrepocossio, C. A. Gonzalez, and F. Mari, *J. Phys. Chem.* **102:**6993 (1998); H. Yamataka and S. Nagase, *J. Am. Chem. Soc.* **120:**7530 (1998).
139. E. Vedejs, G. P. Meier, and K. A. J. Snoble, *J. Am. Chem. Soc.* **103:**2823 (1981); B. E. Maryanoff, A. B. Reitz, M. S. Mutter, R. R. Inners, H. R. Almond, Jr., R. R. Whittle, and R. A. Olofson, *J. Am. Chem. Soc.* **108:**7684 (1986).

cyclization and elimination steps.[140]

$$R_3\overset{+}{P}-\overset{-}{C}R'_2 + R''_2C{=}O \longrightarrow \left[\begin{array}{c} R_3\overset{+}{P}-CR'_2 \\ | \\ \overset{-}{O}-CR''_2 \\ \text{(betaine intermediate)} \\ \downarrow \\ R_3P-CR'_2 \\ |\quad\ | \\ O-CR''_2 \\ \text{(oxaphosphetane intermediate)} \end{array} \right] \longrightarrow R_3P{=}O + R''_2C{=}CR'_2$$

Phosphorus ylides are usually prepared by deprotonation of phosphonium salts. The phosphonium salts most often used are alkyltriphenylphosphonium halides, which can be prepared by the reaction of triphenylphosphine and an alkyl halide:

$$Ph_3P + RCH_2X \longrightarrow Ph_3\overset{+}{P}-CH_2RX^-$$
$$X = I, Br, \text{ or } Cl$$
$$Ph_3\overset{+}{P}\overset{-}{C}H_2R \xrightarrow{\text{base}} Ph_3P{=}CHR$$

The alkyl halide must be one that is reactive toward S_N2 displacement. Alkyltriphenylphosphonium halides are only weakly acidic, and strong bases must be used for deprotonation. These include organolithium reagents, the sodium salt of dimethyl sulfoxide, amide ion, or substituted amide anions such as hexamethyldisilylamide (HMDS). The ylides are not normally isolated so the reaction is carried out either with the carbonyl compound present or it may be added immediately after ylide formation. Ylides with nonpolar substituents, for example, H, alkyl, or aryl, are quite reactive toward both ketones and aldehydes. Scheme 2.16 gives some examples of Wittig reactions.

When a hindered ketone is to be converted to a methylene derivative, the best results have been obtained when a potassium *t*-alkoxide is used as a base in a hydrocarbon solvent. Under these conditions, the reaction can be carried out at elevated temperature.[141] Entries 10 and 11 in Scheme 2.16 illustrate this procedure.

β-Ketophosphonium salts are condsiderably more acidic than alkylphosphonium salts and can be converted to ylides by relatively weak bases. The resulting ylides, which are stabilized by the carbonyl group, are substantially less reactive than unfunctionalized ylides. More vigorous conditions may be required to bring about reactions with ketones. Entries 6 and 7 in Scheme 2.16 involve stabilized ylides.

The stereoselectivity of the Wittig reaction depends strongly on both the structure of the ylide and the reaction conditions. The broadest generalization is that unstabilized ylides give predominantly the *Z*-alkene whereas stabilized ylides give mainly the *E*-alkene.[142] Use of sodium amide or sodium hexamethyldisilylamide as bases gives higher selectivity for *Z*-alkenes than is obtained when ylides are prepared with alkyllithium reagents as base (see entries 3 and 5 of Scheme 2.16). The dependence of the stereoselectivity on the nature of the base is attributed to complexes involving the lithium halide salt which is present when alkyllithium reagents are used as bases. Stabilized ylides

140. R. A. Neumann and S. Berger, *Eur. J. Org. Chem.* **1998:**1085.
141. J. M. Conia and J. C. Limasset, *Bull. Soc. Chim. Fr.* **1967:**1936; J. Provin, F. Leyendecker, and J. M. Conia, *Tetrahedron Lett.* **1975:**4053; S. R. Schow and T. C. Morris, *J. Org. Chem.* **44:**3760 (1979).
142. M. Schlosser, *Top. Stereochem.* **5:**1 (1970).

such as (carboethoxymethylidene)triphenylphosphorane (entries 6 and 7) react with aldehydes to give exclusively *trans* double bonds. Benzylidenetriphenylphosphorane (entry 8) gives a mixture of both *cis*- and *trans*-stilbene on reaction with benzaldehyde.

The stereoselectivity of the Wittig reaction is believed to be the result of steric effects which develop as the ylide and carbonyl compound approach one another. The three phenyl substituents on phosphorus impose large steric demands which govern the formation of the diastereomeric adducts.[143] Reactions of unstabilized phosphoranes are believed to proceed through an early transition state, and steric factors usually make such transition states selective for the *Z*-alkene.[144] The empirical generalization concerning the preference for *Z*-alkenes from unstabilized ylides under salt-free conditions and *E*-alkenes from stabilized ylides serves as a guide to predicting stereoselectivity.

The reaction of unstabilized ylides with aldehydes can be induced to yield *E*-alkenes with high stereoselectivity by a procedure known as the *Schlosser modification* of the Wittig reaction.[145] In this procedure, the ylide is generated as a lithium halide complex and allowed to react with an aldehyde at low temperature, presumably forming a mixture of diastereomeric betaine–lithium halide complexes. At the temperature at which the addition is carried out, fragmentation to an alkene and triphenylphosphine oxide does not occur. This complex is then treated with an equivalent of strong base such as phenyllithium to form a β-oxido ylide. Addition of *t*-butyl alcohol protonates the β-oxido ylide stereoselectively to give the more stable *syn*-betaine as a lithium halide complex. Warming the solution causes the *syn*-betaine–lithium halide complex to give the *E*-alkene by a *syn* elimination.

RCH—CHR′ (Li^+O^-, P^+Ph_3) —PhLi→ RCH—CR′(Li) (Li^+O^-, P^+Ph_3) —*t*-BuOH→ R, H, O^-Li^+, R′, H, P^+Ph_3 → H, R′, R, H

A useful extension of this method is one in which the β-oxido ylide intermediate, instead of being protonated, is allowed to react with formaldehyde. The β-oxido ylide and formaldehyde react to give, on warming, an allylic alcohol. Entry 12 in Scheme 2.16, is an example of this reaction. The reaction is valuable for the stereoselective synthesis of *Z*-allylic alcohols from aldehydes.[146]

$RCH(O^-)CH(R')—\overset{+}{P}Ph_3$ (betaine) —RLi, −25°C→ $RCH(O^-)C(R')=PPh_3$ (β-oxido ylide) —1) $CH_2=O$ 2) 25°C→ R, CH_2OH, H, R′

143. M. Schlosser and B. Schaub, *J. Am. Chem. Soc.* **104:**5821 (1982); H. J. Bestmann and O. Vostrowsky, *Top. Curr. Chem.* **109:**85 (1983); E. Vedejs, T. Fleck, and S. Hara, *J. Org. Chem.* **52:**4637 (1987).
144. E. Vedejs and C. F. Marth, *J. Am. Chem. Soc.* **110:**3948 (1988).
145. M. Schlosser and K.-F. Christmann, *Justus Liebigs Ann. Chem.* **708:**1 (1967); M. Schlosser, K.-F. Christmann, and A. Piskala, *Chem. Ber.* **103:**2814 (1970).
146. E. J. Corey and H. Yamamoto, *J. Am. Chem. Soc.* **92:**226 (1970); E. J. Corey, H. Yamamoto, D. K. Herron, and K. Achiwa, *J. Am. Chem. Soc.* **92:**6635 (1970); E. J. Corey and H. Yamamoto, *J. Am. Chem. Soc.* **92:**6636 (1970); E. J. Corey and H. Yamamoto, *J. Am. Chem. Soc.* **92:**6637 (1970); E. J. Corey, J. I. Shulman, and H. Yamamoto, *Tetrahedron Lett.* **1970:**447.

Scheme 2.16. The Wittig Reaction

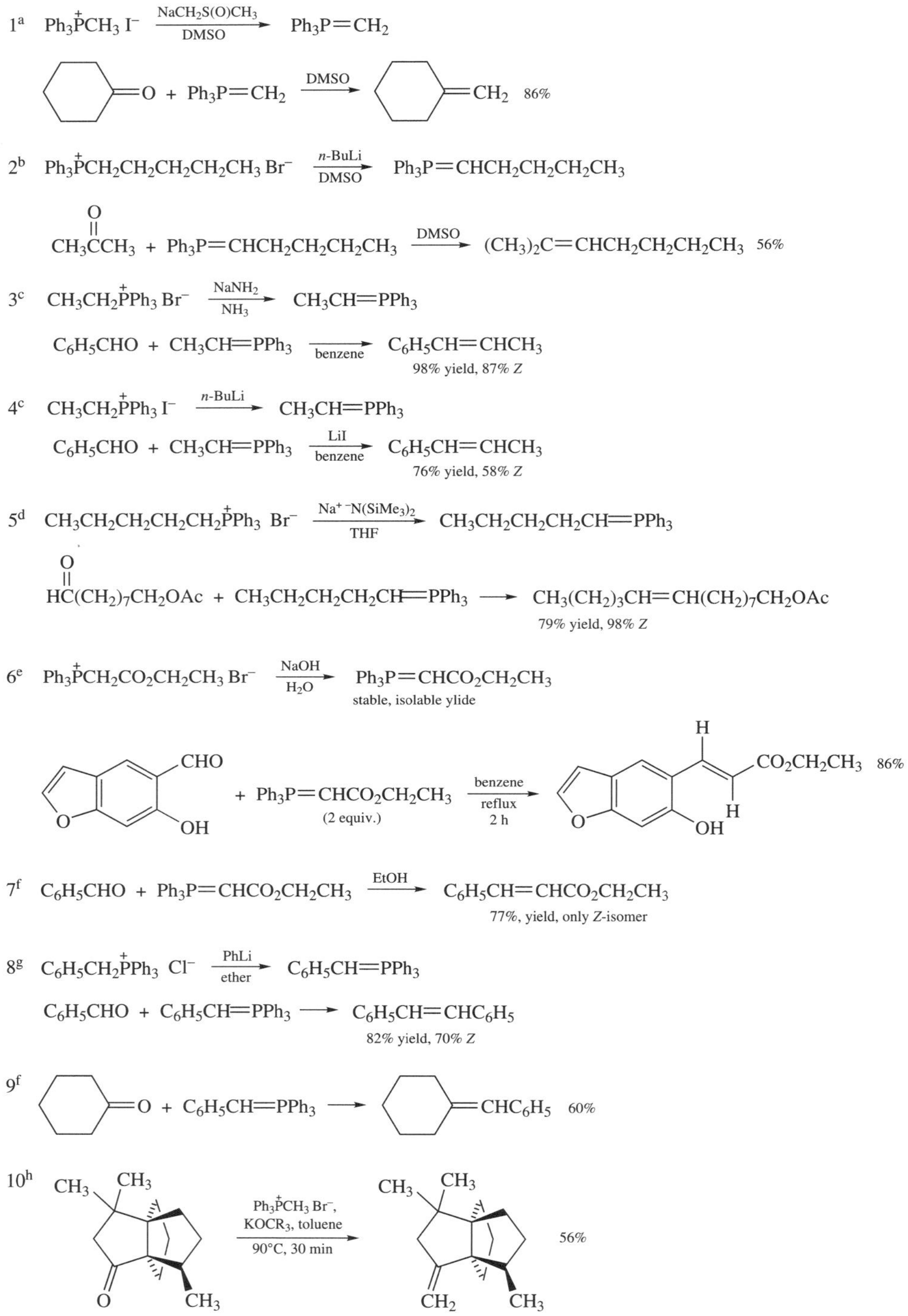

1[a] $Ph_3\overset{+}{P}CH_3\ I^- \xrightarrow[DMSO]{NaCH_2S(O)CH_3} Ph_3P{=}CH_2$

cyclohexanone $=O$ + $Ph_3P{=}CH_2 \xrightarrow{DMSO}$ cyclohexylidene $=CH_2$ 86%

2[b] $Ph_3\overset{+}{P}CH_2CH_2CH_2CH_2CH_3\ Br^- \xrightarrow[DMSO]{n\text{-BuLi}} Ph_3P{=}CHCH_2CH_2CH_2CH_3$

$CH_3\overset{O}{\overset{\|}{C}}CH_3 + Ph_3P{=}CHCH_2CH_2CH_2CH_3 \xrightarrow{DMSO} (CH_3)_2C{=}CHCH_2CH_2CH_2CH_3$ 56%

3[c] $CH_3CH_2\overset{+}{P}Ph_3\ Br^- \xrightarrow[NH_3]{NaNH_2} CH_3CH{=}PPh_3$

$C_6H_5CHO + CH_3CH{=}PPh_3 \xrightarrow[benzene]{} C_6H_5CH{=}CHCH_3$
98% yield, 87% *Z*

4[c] $CH_3CH_2\overset{+}{P}Ph_3\ I^- \xrightarrow{n\text{-BuLi}} CH_3CH{=}PPh_3$

$C_6H_5CHO + CH_3CH{=}PPh_3 \xrightarrow[benzene]{LiI} C_6H_5CH{=}CHCH_3$
76% yield, 58% *Z*

5[d] $CH_3CH_2CH_2CH_2CH_2\overset{+}{P}Ph_3\ Br^- \xrightarrow[THF]{Na^{+\ -}N(SiMe_3)_2} CH_3CH_2CH_2CH_2CH{=}PPh_3$

$H\overset{O}{\overset{\|}{C}}(CH_2)_7CH_2OAc + CH_3CH_2CH_2CH_2CH{=}PPh_3 \longrightarrow CH_3(CH_2)_3CH{=}CH(CH_2)_7CH_2OAc$
79% yield, 98% *Z*

6[e] $Ph_3\overset{+}{P}CH_2CO_2CH_2CH_3\ Br^- \xrightarrow[H_2O]{NaOH} Ph_3P{=}CHCO_2CH_2CH_3$
stable, isolable ylide

(benzofuran with CHO and OH) + $Ph_3P{=}CHCO_2CH_2CH_3$ (2 equiv.) $\xrightarrow[\text{reflux, 2 h}]{benzene}$ (benzofuran with OH and H/H trans $CH{=}CHCO_2CH_2CH_3$) 86%

7[f] $C_6H_5CHO + Ph_3P{=}CHCO_2CH_2CH_3 \xrightarrow{EtOH} C_6H_5CH{=}CHCO_2CH_2CH_3$
77%, yield, only *Z*-isomer

8[g] $C_6H_5CH_2\overset{+}{P}Ph_3\ Cl^- \xrightarrow[ether]{PhLi} C_6H_5CH{=}PPh_3$

$C_6H_5CHO + C_6H_5CH{=}PPh_3 \longrightarrow C_6H_5CH{=}CHC_6H_5$
82% yield, 70% *Z*

9[f] cyclohexanone $=O$ + $C_6H_5CH{=}PPh_3 \longrightarrow$ cyclohexylidene $=CHC_6H_5$ 60%

10[h] (tricyclic ketone: CH_3, CH_3, O, CH_3) $\xrightarrow[90°C,\ 30\ min]{Ph_3\overset{+}{P}CH_3\ Br^-,\ KOCR_3,\ toluene}$ (tricyclic alkene: CH_3, CH_3, CH_2, CH_3) 56%

11[i] Camphor $\xrightarrow[\text{100°C, 2 h}]{Ph_3\overset{+}{P}CH_3\ Br^-,\ KOCR_3}$ 2-methylene analog (91%)

12[b] $CH_3CH_2CH_2CH_2CHO + CH_3CH{=}PPh_3 \xrightarrow[\text{2) BuLi; 3) } CH_2O\text{, 25°C}]{\text{1) LiBr, THF, –78°C}}$ $CH_3CH_2CH_2CH_2(H)C{=}C(CH_2OH)CH_3$

13[j] OCH$_3$-substituted dienyl-$P^+Ph_3\ I^-$ + Boc-protected oxazolidine aldehyde ($CH{=}O$; CH_3, CH_3) $\xrightarrow{\text{LiHMDS, THF/HMPA}}$ Z-alkene product (H; O; N; Boc; CH_3; H_3C; OCH$_3$) 69%

14[k] NC–C$_6$H$_4$–CHO + $Ph_3P^+CH_2$–(1,3-dioxolan-2-yl) $\xrightarrow[\text{phase transfer}]{\text{satd. } K_2CO_3\text{, } CH_2Cl_2}$ NC–C$_6$H$_4$–CH=CH–(1,3-dioxolan-2-yl)

100% yield, 72:28 *Z*:*E*

15[l] ArO-substituted tetrahydrofuranol (OH) + $Ph_3P^+(CH_2)_4CO_2H \xrightarrow[\text{toluene, –78°C}]{\text{NaHMDS}}$ HO–CH$_2$–CH(OAr)–CH$_2$–CH=CH–(CH$_2$)$_3$–CO$_2$H 60%

Ar = 4-methoxybenzyl

a. R. Greenwald, M. Chaykovsky, and E. J. Corey, *J. Org. Chem.* **28:**1128 (1963).
b. U. T. Bhalerao and H. Rapoport, *J. Am. Chem. Soc.* **93:**4835 (1971).
c. M. Schlosser and K. F. Christmann, *Justus Liebigs Ann. Chem.* **708**:1 (1967).
d. H. J. Bestmann, K. H. Koschatzky, and O. Vostrowsky, *Chem. Ber.* **112:**1923 (1979).
e. Y. Y. Liu, E. Thom, and A. A. Liebman, *J. Heterocycl. Chem.* **16:**799 (1979).
f. G. Wittig and W. Haag, *Chem. Ber.* **88:**1654 (1955).
g. G. Wittig and U. Schöllkopf, *Chem. Ber.* **87:**1318 (1954).
h. A. B. Smith III and P. J. Jerris, *J. Org. Chem.* **47:**1845 (1982).
i. L. Fitjer and U. Quabeck, *Synth. Commun.* **15:**855 (1985).
j. J. D. White, T. S. Kim, and M. Nambu, *J. Am. Chem. Soc.* **119:**103 (1997).
k. N. Daubresse, C. Francesch, and G. Rolando, *Tetrahedron* **54:**10761 (1998).
l. D. J. Critcher, S. Connoll, and M. Wills, *J. Org. Chem.* **62:**6638 (1997).

The Wittig reaction can be extended to functionalized ylides.[147] Methoxymethylene and phenoxymethylene ylides lead to vinyl ethers, which can be hydrolyzed to aldehydes.[148]

$$\text{(bicyclic ketone, }OCH_2OCH_2CH_2OCH_3\text{)}=O \xrightarrow{Ph_3P=CHOMe} \text{(bicyclic, }OCH_2OCH_2CH_2OCH_3\text{)}=CHOCH_3$$

Ref. 149

2-(1,3-Dioxolanyl)methyl ylides can be used for the introduction of α,β-unsaturated aldehydes (see entry 14 in Scheme 2.16). Methyl ketones have been prepared by an analogous reaction.

$$CH_3(CH_2)_5CHO + CH_3OC(CH_3)=PPh_3 \xrightarrow[-40°C]{DME} CH_3(CH_2)_5CH=C(CH_3)OCH_3 \xrightarrow[CH_3OH,\ \Delta]{H_2O,\ HCl} CH_3(CH_2)_5CH_2C(=O)CH_3 \quad 57\%$$

Ref. 150

An important complement to the Wittig reaction is the reaction of phosphonate carbanions with carbonyl compounds.[151] The alkylphosphonate esters are made by the reaction of an alkyl halide, preferably primary, with a phosphite ester. Phosphonate carbanions are more nucleophilic than an analogous ylide, and even when R is a carbanion-stabilizing substituent, they react readily with aldehydes and ketones to give alkenes. Phosphonate carbanions are generated by treating alkylphosphonate esters with bases such as sodium hydride, *n*-butyllithium, or sodium ethoxide. Alumina coated with KF or KOH has also found use as the base.[152]

Reactions with phosphonoacetate esters are used frequently to prepare α,β-unsaturated esters. This is known as the *Wadsworth–Emmons* reaction. These reactions usually lead to the *E*-isomer. Scheme 2.17 gives a number of examples.

$$RCH_2X + P(OC_2H_5)_3 \longrightarrow RCH_2P(=O)(OC_2H_5)_2 + C_2H_5X$$

$$RCH_2P(=O)(OC_2H_5)_2 \xrightarrow{base} R\bar{C}HP(=O)(OC_2H_5)_2$$

$$R\bar{C}HP(=O)(OC_2H_5)_2 + R'_2C=O \longrightarrow R'_2C(O^-)-CHR-P(=O)(OC_2H_5)_2 \longrightarrow R'_2C=CHR + (C_2H_5O)_2P(=O)-O^-$$

147. S. Warren, *Chem. Ind. (London)* **1980:**824.
148. S. G. Levine, *J. Am. Chem. Soc.* **80:**6150 (1958); G. Wittig, W. Boll, and K. H. Kruck, *Chem. Ber.* **95:**2514 (1962).
149. M. Yamazaki, M. Shibasaki, and S. Ikegami, *J. Org. Chem.* **48:**4402 (1983).
150. D. R. Coulsen, *Tetrahedron Lett.* **1964:**3323.
151. For reviews of reactions of phosphonate carbanions with carbonyl compounds, see: J. Boutagy and R. Thomas, *Chem. Rev.* **74:**87 (1974); W. S. Wadsworth, Jr., *Org. React.* **25:**73 (1977); H. Gross and I. Keitels, *Z. Chem.* **22:**117 (1982).
152. F. Texier-Boullet, D. Villemin, M. Ricard, H. Moison, and A. Foucaud, *Tetrahedron* **41:**1259 (1985); M. Mikolajczyk and R. Zurawinski, *J. Org. Chem.* **63:**8894 (1998).

Three modified phosphonoacetate esters have been found to show selectivity for the Z-enoate product. Trifluoroethyl,[153] phenyl,[154] and 2,6-difluorophenyl[155] esters give good Z-stereoselectivity.

$$RCH{=}O + CH_3O_2CCH_2\overset{O}{\overset{\|}{P}}(OR')_2 \longrightarrow \underset{R}{\overset{H}{}}C{=}C\underset{CO_2CH_3}{\overset{H}{}}$$

$R' = CH_2CF_3$, phenyl, 2,6-difluorophenyl

An alternative procedure for effecting the condensation of phophonates is to carry out the reaction in the presence of lithium chloride and an amine such as N,N-diisopropyl-N-ethylamine or diazabicycloundecene (DBU). The lithium chelate of the substituted phosphonate is sufficiently acidic to be deprotonated by the amine.[156]

$$(R'O)_2P(=O\cdots Li^+)CH_2C(=O\cdots)OR \xrightarrow{R_3N} (R'O)_2P(=O\cdots Li^+)CH{=}C(O^-)OR \xrightarrow{R''CH=O} R''CH{=}CHCO_2R$$

Entries 10 and 11 of Scheme 2.17 also illustrate this procedure.

Intramolecular reactions have been used to prepare cycloalkenes.[157]

$$CH_3\underset{O}{\underset{\|}{C}}(CH_2)_3\underset{O}{\underset{\|}{C}}CH_2\underset{O}{\underset{\|}{P}}(OC_2H_5)_2 \xrightarrow{NaH} \text{3-methylcyclohex-2-enone}$$ Ref. 158

Intramolecular condensation of phosphonate carbanions with carbonyl groups carried out under conditions of high dilution has been utilized in macrocycle synthesis (entries 8 and 9 in Scheme 2.17)

Carbanions derived from phosphine oxides also add to carbonyl compounds. The adducts are stable but undergo elimination to form alkenes on heating with a base such as sodium hydride. This reaction is known as the *Horner–Wittig* reaction.[159]

$$Ph_2\overset{O}{\overset{\|}{P}}CH_2R \xrightarrow{RLi} Ph_2\overset{O}{\overset{\|}{P}}\underset{Li}{\underset{|}{C}}HR \xrightarrow{R'CH=O} Ph_2\overset{O}{\overset{\|}{P}}\underset{R}{\underset{|}{C}}H\underset{H}{\underset{|}{\overset{O^-}{\overset{|}{C}}}}R' \longrightarrow RCH{=}CHR'$$

The unique feature of the Horner–Wittig reaction is that the addition intermediate can be isolated and purified. This provides a means for control of the stereochemistry of the reaction. It is possible to separate the two diastereomeric adducts in order to prepare the pure alkenes. The elimination process is *syn* so that the stereochemistry of the alkene depends on the stereochemistry of the adduct. Usually, the *anti* adduct is the major product, so it is the Z-alkene which is favored. The *syn* adduct is most easily obtained by reduction of β-keto phosphine oxides.[160]

153. W. C. Still and C. Gennari, *Tetrahedron Lett.* **24:**4405 (1983).
154. K. Ando, *Tetrahedron Lett.* **36:**4105 (1995); K. Ando, *J. Org. Chem.* **63:**8411 (1998).
155. K. Kokin, J. Motoyoshiya, S. Hayashi, and H. Aoyama, *Synth. Commun.* **27:**2387 (1997).
156. M. A. Blanchette, W. Choy, J. T. Davis, A. P. Essenfeld, S. Masamune, W. R. Roush, and T. Sakai, *Tetrahedron Lett.* **25:**2183 (1984).
157. K. B. Becker, *Tetrahedron* **36:**1717 (1980).
158. P. A. Grieco and C. S. Pogonowski, *Synthesis* **1973:**425.
159. For a review, see J. Clayden and S. Warren, *Angew. Chem. Int. Ed. Engl.* **35:**241 (1996).
160. A. D. Buss and S. Warren, *J. Chem. Soc., Perkin Trans. 1* **1985:**2307.

Scheme 2.17. Carbonyl Olefination Using Phosphonate Carbanions

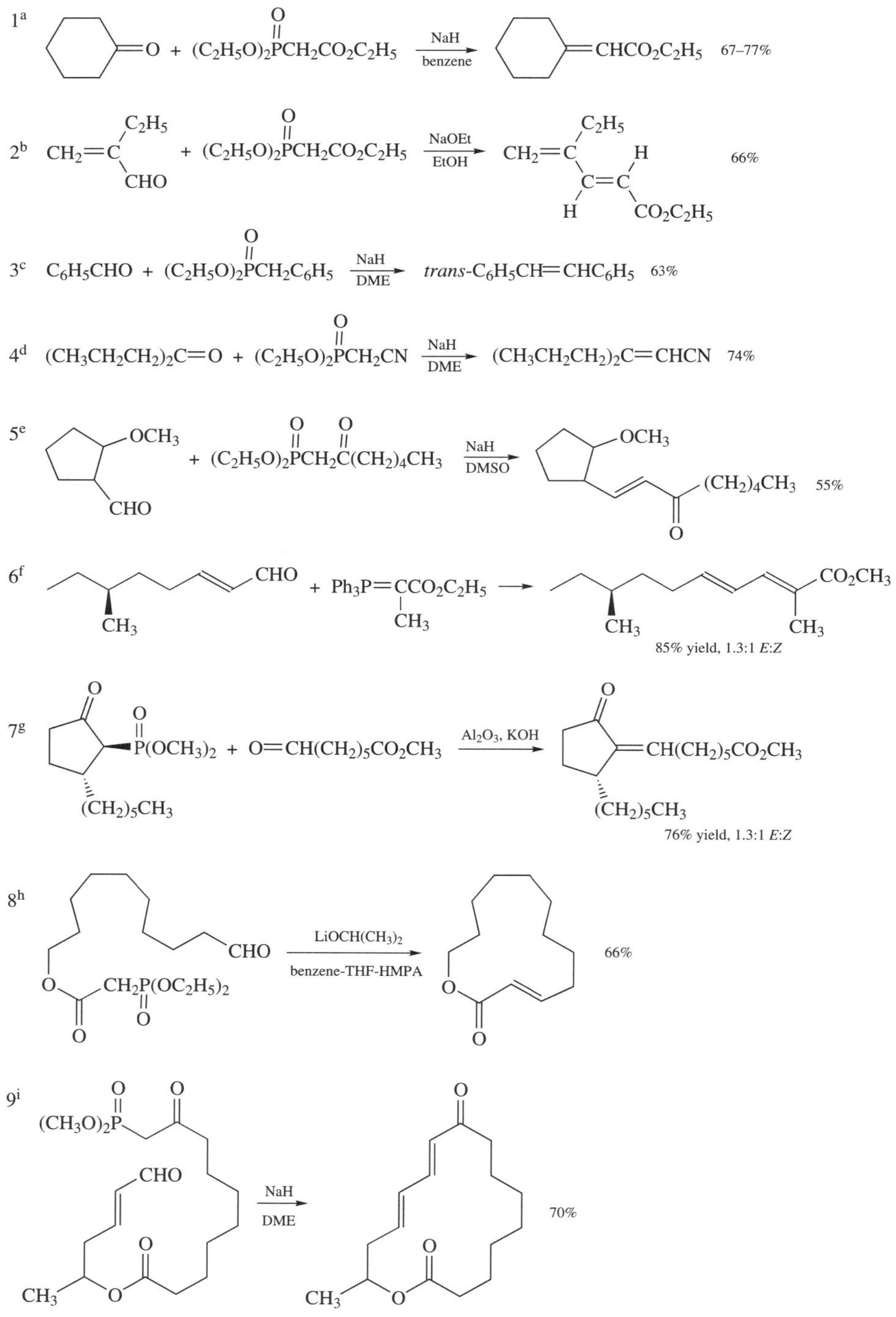

1[a] cyclohexanone + $(C_2H_5O)_2P(O)CH_2CO_2C_2H_5$ —NaH, benzene→ cyclohexylidene$=CHCO_2C_2H_5$ 67–77%

2[b] $CH_2=C(C_2H_5)CHO$ + $(C_2H_5O)_2P(O)CH_2CO_2C_2H_5$ —NaOEt, EtOH→ $CH_2=C(C_2H_5)CH=CHCO_2C_2H_5$ 66%

3[c] C_6H_5CHO + $(C_2H_5O)_2P(O)CH_2C_6H_5$ —NaH, DME→ *trans*-$C_6H_5CH=CHC_6H_5$ 63%

4[d] $(CH_3CH_2CH_2)_2C=O$ + $(C_2H_5O)_2P(O)CH_2CN$ —NaH, DME→ $(CH_3CH_2CH_2)_2C=CHCN$ 74%

5[e] 2-methoxycyclopentanecarbaldehyde (OCH_3, CHO) + $(C_2H_5O)_2P(O)CH_2C(O)(CH_2)_4CH_3$ —NaH, DMSO→ (OCH_3) enone $(CH_2)_4CH_3$ 55%

6[f] CHO (CH_3) + $Ph_3P=C(CH_3)CO_2C_2H_5$ → CO_2CH_3 (CH_3, CH_3) 85% yield, 1.3:1 *E*:*Z*

7[g] 2-oxocyclopentyl $P(O)(OCH_3)_2$, $(CH_2)_5CH_3$ + $O=CH(CH_2)_5CO_2CH_3$ —Al_2O_3, KOH→ $=CH(CH_2)_5CO_2CH_3$, $(CH_2)_5CH_3$ 76% yield, 1.3:1 *E*:*Z*

8[h] CHO, $OC(O)CH_2P(O)(OC_2H_5)_2$ —$LiOCH(CH_3)_2$, benzene-THF-HMPA→ macrolactone 66%

9[i] $(CH_3O)_2P(O)CH_2C(O)$…, CHO, CH_3, O —NaH, DME→ macrolide (CH_3) 70%

10[j]

PhCH$_2$O R$_3$SiO P(OCH$_3$)$_2$ O O + O=CH CO$_2$CH$_3$

LiCl, DBU CH$_3$CN 25°C, 2h

PhCH$_2$O R$_3$SiO O CO$_2$CH$_3$ 70%

11[k]

O CSC$_2$H$_5$ TBSO H$_3$C O O P(OC$_2$H$_5$)$_2$ O

OH CH$_3$CHCH$_2$CH=O LiCl, (IPr)$_2$) NEt

O CSC$_2$H$_5$ OH TBSO H$_3$C O O 72%

a. W. S. Wadsworth, Jr. and W. D. Emmons, *Org. Synth.* **45:**44 (1965).
b. R. J. Sundberg, P. A. Buckowick, and F. O. Holcombe, *J. Org. Chem.* **32:**2938 (1967).
c. W. S. Wadsworth, Jr. and W. D. Emmons, *J. Am. Chem. Soc.* **83:**1733 (1961).
d. J. A. Marshall, C. P. Hagan, and G. A. Flynn, *J. Org. Chem.* **40:**1162 (1975).
e. N. Finch, J. J. Fitt, and I. H. S. Hsu, *J. Org. Chem.* **40:**206 (1975).
f. A. G. M. Barrett, M. Pena, and J. A. Willardsen, *J. Org. Chem.* **61:**1082 (1996).
g. M. Mikolajczyk and R. Zurawinski, *J. Org. Chem.* **63:**8894 (1998).
h. G. Stork and E. Nakamura, *J. Org. Chem.* **44:**4010 (1979).
i. K. C. Nicolaou, S. P. Seitz, M. R. Pavia, and N. A. Petasis, *J. Org. Chem.* **44:**4010 (1979).
j. M. A. Blanchette, W. Choy, J. T. Davis, A. P. Essenfeld, S. Masamune, W. R. Roush, and T. Sakai, *Tetrahedron Lett.* **25:**2183 (1984).
k. G. E. Keck and J. A. Murry, *J. Org. Chem.* **56:**6606 (1991).

O Ph$_2$PCHCH$_2$CH$_2$Ph CH$_3$

1) BuLi 2) CH$_3$CH=O

PhCH$_2$CH$_2$ O CH$_3$ Ph$_2$P O CH$_3$

NaBH$_4$

H OH PhCH$_2$CH$_2$ CH$_3$ Ph$_2$P O CH$_3$ + HO H PhCH$_2$CH$_2$ CH$_3$ Ph$_2$P O CH$_3$

separate

NaH NaH

CH$_3$ CH$_2$CH$_2$Ph H CH$_3$ CH$_3$ CH$_2$CH$_2$Ph CH$_3$ H

Ref. 161

161. A. D. Buss and S. Warren, *Tetrahedron Lett.* **24:**111, 3931 (1983); A. D. Buss and S. Warren, *J. Chem. Soc., Perkin Trans. 1* **1985:**2307.

2.5. Reactions of Carbonyl Compounds with α-Trimethylsilylcarbanions

β-Hydroxyalkyltrimethylsilanes are converted to alkenes in either acidic or basic solution.[162] These eliminations provide a synthesis of alkenes that begins with the nucleophilic addition of an α-trimethylsilyl-substituted carbanion to an aldehyde or ketone. The reaction is sometimes called the Peterson reaction.[163] For example, the organometallic reagents derived from chloromethyltrimethylsilane adds to an aldehyde or ketone, and the intermediate can be converted to a teminal alkene by base.[164]

$$(CH_3)_3SiCH_2X \xrightarrow{n\text{-BuLi}} (CH_3)_3SiCH(Li)X \xrightarrow{R_2C=O} R_2C{=}CHX$$

Similarly, organolithium reagents of the type $(CH_3)_3SiCH(Li)X$, where X is a carbanion-stabilizing substituent, can be prepared by deprotonation of $(CH_3)_3SiCH_2X$ with *n*-butyllithium. These reagents usually react with aldehydes and ketones to give substituted alkenes directly. No separate elimination step is necessary because fragmentation of the intermediate occurs spontaneously under the reaction conditions.

In general, the elimination reactions are *anti* under acidic conditions and *syn* under basic conditions. This stereoselectivity is the result of a cyclic elimination mechanism under basic conditions, whereas under acidic conditions an acyclic β-elimination occurs.

The *anti* elimination can also be achieved by converting the β-silyl alcohols to trifluoroacetate esters.[165] Because the overall stereoselectivity of the Peterson olefination depends on the generation of pure *syn* or *anti* β-silyl alcohols, several strategies have been developed for their stereoselective preparation.[166] Several examples of synthesis of substituted alkenes in this way are given in Scheme 2.18.

162. P. F. Hudrlik and D. Peterson, *J. Am. Chem. Soc.* **97:**1464 (1975).
163. For reviews, see D. J. Ager, *Org. React.* **38:**1 (1990); D. J. Ager, *Synthesis* **1984:**384; A. G. M. Barrett, J. M. Hill, E. M. Wallace, and J. A. Flygare, *Synlett* **1991:**764.
164. D. J. Peterson, *J. Org. Chem.* **33:**780 (1968).
165. M. F. Connil, B. Jousseaume, N. Noiret, and A. Saux, *J. Org. Chem.* **59:**1925 (1994).
166. A. G. M. Barrett and J. A. Flygare, *J. Org. Chem.* **56:**638 (1991); L. Duhamel, J. Gralak, and A. Bouyanzer, *J. Chem. Soc., Chem. Commun.* **1993:**1763.

1[a] $Me_3SiCH(Li)CO_2C_2H_5$ + cyclododecanone → cyclododecylidene=$CHCO_2C_2H_5$ 94%

2[b] $Me_3SiCH(Li)CO_2Li$ + cyclopentanone → cyclopentylidene=$CHCO_2H$ 84%

3[c] $Me_3SiCH(Li)CN$ + $C_6H_5CH{=}CHCHO$ → $C_6H_5CH{=}CHCH{=}CHCN$ 95%

4[d] $Me_3SiCH(Li)SC_6H_5$ + $(CH_3)_3CC(O)CH_3$ → $C_6H_5SCH{=}C(CH_3)C(CH_3)_3$ 55%

5[e] $Me_3SiCH(Li)S(O)C_6H_5$ + $C_6H_5CH{=}CHCHO$ → $C_6H_5CH{=}CHCH{=}CHS(O)C_6H_5$ 70%

6[f] 2-Lithio-2-trimethylsilyl-1,3-dithiane + CH_3CH_2CHO → $CH_3CH_2CH{=}$(1,3-dithian-2-ylidene) 75%

7[d] $Me_3SiCH(Li)P(O)(OC_2H_5)_2$ + $(CH_3)_2CHCHO$ → $(CH_3)_2CHCH{=}CHP(O)(OC_2H_5)_2$ 92%

8[g] $Me_3SiC(Li)(SeC_6H_5)_2$ + C_6H_5CHO → $C_6H_5CH{=}C(SeC_6H_5)_2$ 75%

9[h] cycloheptanone + $Me_3SiCH(Li)OCH_3$ → $\xrightarrow{KH}$ cycloheptylidene=$CHOCH_3$ 51%

10[i] (CH_3O, H_3C, OTBDMS-substituted cyclohexyl)$CH(CH_3)CHO$ + $(C_2H_5)_3SiC(Li)(CH_3)CH{=}N$–cyclohexyl $\xrightarrow[\text{2) } CF_3CO_2H,\ 0°C]{\text{1) } -30°C}$ (CH_3O, H_3C, OTBDMS-substituted cyclohexyl)$CH(CH_3)CH{=}C(CH_3)CHO$ 91%

a. K. Shimoji, H. Taguchi, H. Yamamoto, K. Oshima and H. Hozaki, *J. Am. Chem. Soc.* **96**:1620 (1974).
b. P. A. Grieco, C. L. J. Wang, and S. D. Burke, *J. Chem. Soc., Chem. Commun.* **1975**:537.
c. I. Matsuda, S. Murata, and Y. Ishii, *J. Chem. Soc., Perkin Trans. 1* **1979**:26.
d. F. A. Carey and A. S. Court, *J. Org. Chem.* **37**:939 (1972).
e. F. A. Carey and O. Hernandez, *J. Org. Chem.* **38**:2670 (1973).
f. D. Seebach, M. Kolb, and B.-T. Grobel, *Chem. Ber.* **106**:2277 (1973).
g. B. T. Grobel and D. Seebach, *Chem. Ber.* **110**:852 (1977).
h. P. Magnus and G. Roy, *Organometallics* **1**:553 (1982).
i. S. F. Martin, J. A. Dodge, L. E. Burgess, and M. Hartmann, *J. Org. Chem.* **57**:1070 (1992).

2.6. Sulfur Ylides and Related Nucleophiles

Sulfur ylides are next to phosphorus ylides in importance as synthetic reagents.[167] Dimethylsulfonium methylide and dimethylsulfoxonium methylide are especially useful.[168] These sulfur ylides are prepared by deprotonation of the corresponding sulfonium salts, both of which are commercially available.

$$(CH_3)_2\overset{+}{S}CH_3\ I^- \xrightarrow[\text{DMSO}]{NaCH_2\overset{O}{\overset{\|}{S}}CH_3} (CH_3)_2\overset{+}{S}-CH_2^-$$

dimethylsulfonium methylide

$$(CH_3)_2\underset{+}{\overset{O}{\overset{\|}{S}}}CH_3\ I^- \xrightarrow[\text{DMSO}]{\text{NaH}} (CH_3)_2\underset{+}{\overset{O}{\overset{\|}{S}}}-CH_2^-$$

dimethylsulfoxonium methylide

There is an important difference between the reactions of these sulfur ylides and those of phosphorus ylides. Whereas phosphorus ylides normally react with carbonyl compounds to give alkenes, dimethylsulfonium methylide and dimethylsulfoxonium methylide yield epoxides. Instead of a four-center elimination, the adducts formed from the sulfur ylides undergo intramolecular displacement of the sulfur substituent by oxygen.

$$R_2C{=}O + (CH_3)_2\overset{+}{S}-\overset{-}{C}H_2 \longrightarrow R_2\overset{O^-}{\overset{|}{C}}-CH_2-\overset{+}{S}(CH_3)_2 \longrightarrow R_2C\overset{O}{\frown}CH_2 + (CH_3)_2S$$

$$R_2C{=}O + (CH_3)_2\underset{+}{\overset{O}{\overset{\|}{S}}}-\overset{-}{C}H_2 \longrightarrow R_2\overset{O^-}{\overset{|}{C}}-CH_2-\underset{+}{\overset{O}{\overset{\|}{S}}}(CH_3)_2 \longrightarrow R_2C\overset{O}{\frown}CH_2 + (CH_3)_2S{=}O$$

Examples of the use of dimethylsulfonium methylide and dimethylsulfoxonium methylide in the preparation of epoxides are listed in Scheme 2.19. Entries 1–4 illustrate epoxide formation with simple aldehydes and ketones.

Dimethylsulfonium methylide is both more reactive and less stable than dimethylsulfoxonium methylide, so it is generated and used at a lower temperature. A sharp distinction between the two ylides emerges in their reactions with α,β-unsaturated carbonyl compounds. Dimethylsulfonium methylide yields epoxides, whereas dimethylsulfoxonium methylide reacts by conjugate addition to give cyclopropanes (entries 5 and 6 in Scheme 2.19). It appears that the reason for the difference in their behavior lies in the relative rates of the two reactions available to the betaine intermediate: (a) reversal to starting materials or (b) intramolecular nucleophilic displacement.[169] Presumably, both reagents react most rapidly at the carbonyl group. In the case of dimethylsulfonium methylide, the intramolecular displacement step is faster than the reverse of the addition, and epoxide formation

167. B. M. Trost and L. S. Melvin, Jr., *Sulfur Ylides*, Academic Press, New York, 1975; E. Block, *Reactions of Organosulfur Compounds*, Academic Press, New York, 1978.
168. E. J. Corey and M. Chaykovsky, *J. Am. Chem. Soc.* **87:**1353 (1965).
169. C. R. Johnson, C. W. Schroeck, and J. R. Shanklin, *J. Am. Chem. Soc.* **95:**7424 (1973).

Scheme 2.19. Reactions of Sulfur Ylides

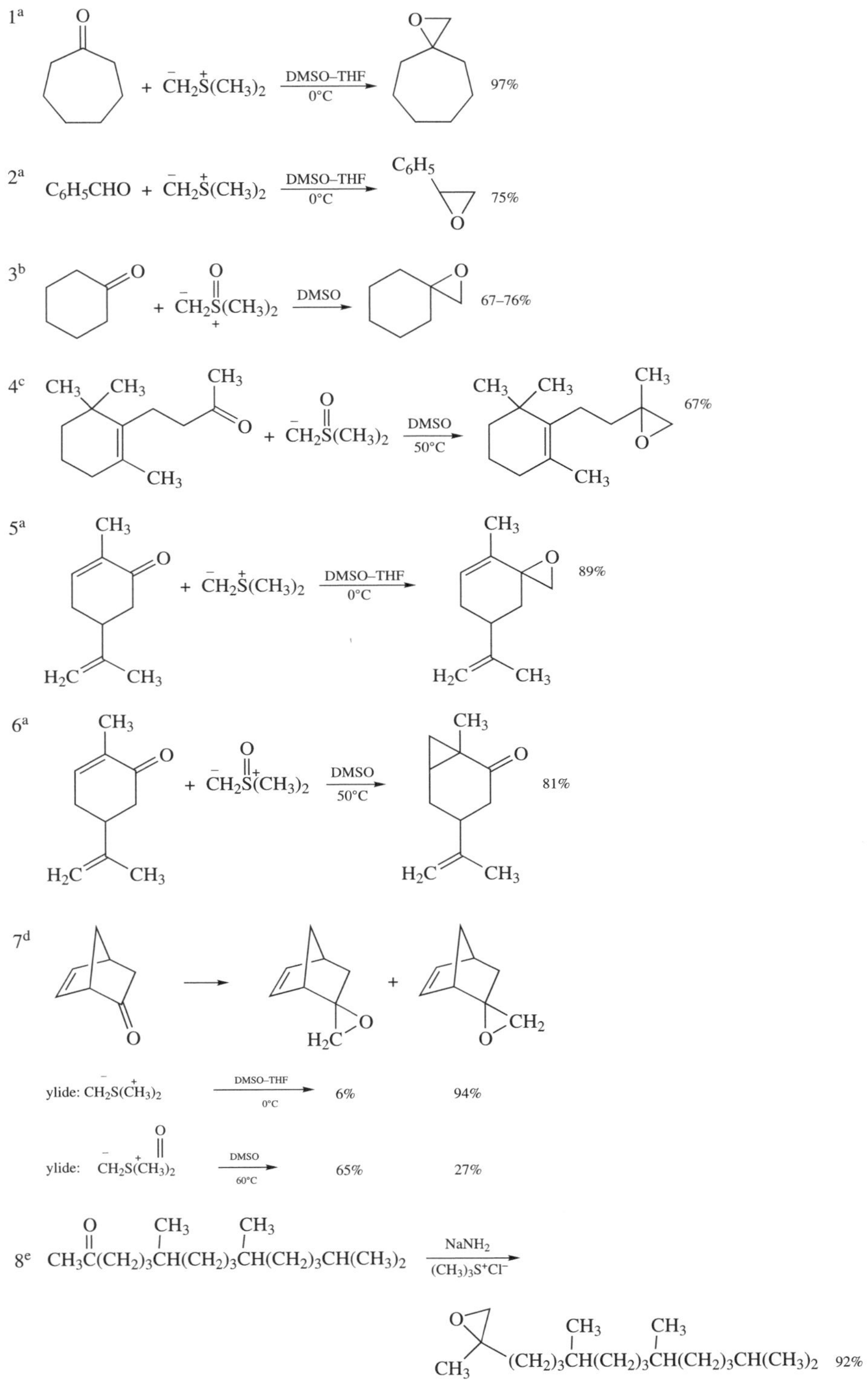

1[a] Cycloheptanone + $\bar{C}H_2\overset{+}{S}(CH_3)_2$ $\xrightarrow[0°C]{DMSO–THF}$ spiro-epoxide 97%

2[a] $C_6H_5CHO + \bar{C}H_2\overset{+}{S}(CH_3)_2$ $\xrightarrow[0°C]{DMSO–THF}$ C_6H_5-oxirane 75%

3[b] Cyclohexanone + $\bar{C}H_2\overset{+}{S}(O)(CH_3)_2$ $\xrightarrow{DMSO}$ spiro-epoxide 67–76%

4[c] CH_3, CH_3, CH_3 (ketone), CH_3 + $\bar{C}H_2\overset{+}{S}(O)(CH_3)_2$ $\xrightarrow[50°C]{DMSO}$ epoxide (CH_3, CH_3, CH_3, O, CH_3) 67%

5[a] CH_3, O, H_2C, CH_3 + $\bar{C}H_2\overset{+}{S}(CH_3)_2$ $\xrightarrow[0°C]{DMSO–THF}$ CH_3, O, H_2C, CH_3 89%

6[a] CH_3, O, H_2C, CH_3 + $\bar{C}H_2\overset{+}{S}(O)(CH_3)_2$ $\xrightarrow[50°C]{DMSO}$ CH_3, O, H_2C, CH_3 81%

7[d] O ⟶ (H_2C, O) + (O, CH_2)

ylide: $\bar{C}H_2\overset{+}{S}(CH_3)_2$ $\xrightarrow[0°C]{DMSO–THF}$ 6% 94%

ylide: $\bar{C}H_2\overset{+}{S}(O)(CH_3)_2$ $\xrightarrow[60°C]{DMSO}$ 65% 27%

8[e] $CH_3C(O)(CH_2)_3CH(CH_3)(CH_2)_3CH(CH_3)(CH_2)_3CH(CH_3)_2$ $\xrightarrow[(CH_3)_3S^+Cl^-]{NaNH_2}$

O, CH_3 (epoxide) $(CH_2)_3CH(CH_3)(CH_2)_3CH(CH_3)(CH_2)_3CH(CH_3)_2$ 92%

9[f] $(CH_3)_3S^+Cl^-$, NaOH → 87%

10[g] + $(CH_3)_2\bar{C}\overset{+}{S}Ph_2$ DME, 50°C → 82%

11[h] CH_3 ... CO_2CH_3 + $(CH_3)_2\bar{C}\overset{+}{S}Ph_2$ DME, −20°C → 72%

12[i] + cyclopropyl$^-$ $\overset{+}{S}Ph_2$ DMSO, 25°C → 75%

13[j] $CH_3C(O)(CH_2)_5CH_3$ + cyclopropyl$^-$ $\overset{+}{S}Ph_2$ DMSO, 25°C → 92%

14[j] Ph–CH=CH–C(O)N(OCH_3)CH_3 + $[(CH_3)_2CH]_2\overset{O}{S}{=}NSO_2Ar$ n-BuLi → 97%

a. E. J. Corey and M. Chaykovsky, *J. Am. Chem. Soc.* **87:**1353 (1965).
b. E. J. Corey and M. Chaykovsky, *Org. Synth.* **49:**78 (1969).
c. M. G. Fracheboud, O. Shimomura, R. K. Hill, and F. H. Johnson, *Tetrahedron Lett.* **1969:**3951.
d. R. S. Bly, C. M. DuBose, Jr., and G. B. Konizer, *J. Org. Chem.* **33:**2188 (1968).
e. G. L. Olson, H.-C. Cheung, K. Morgan, and G. Saucy, *J. Org. Chem.* **45:**803 (1980).
f. M. Rosenberger, W. Jackson, and G. Saucy, *Helv. Chim. Acta* **63:**1665 (1980).
g. E. J. Corey, M. Jautelat, and W. Oppolzer, *Tetrahedron Lett.* **1967:**2325.
h. E. J. Corey and M. Jautelat, *J. Am. Chem.Soc.* **89:**3112 (1967).
i. B. M. Trost and M. J. Bogdanowicz, *J. Am. Chem. Soc.* **95:**5307 (1973).
j. B. M. Trost and M. J. Bogdanowicz, *J. Am. Chem. Soc.* **95:**5311 (1973).
k. K. E. Rodrigues, *Tetrahedron Lett.* **32:**1275 (1991).

takes place.

With the more stable dimethylsulfoxonium methylide, the reversal is relatively more rapid, and product formation takes place only after conjugate addition.

Another difference between dimethylsulfonium methylide and dimethylsulfoxonium methylide concerns the stereoselectivity in formation of epoxides from cyclohexanones. Dimethylsulfonium methylide usually adds from the axial direction whereas dimethylsulfoxonium methylide favors the equatorial direction. This result may also be due to reversibility of addition in the case of the sulfoxonium methylide.[169] The product from the sulfonium ylide would be the result of the kinetic preference for axial addition by small nucleophiles (see Part A, Section 3.10). In the case of reversible addition of the sulfoxonium ylide, product structure would be determined by the rate of displacement, and this may be faster for the more stable epoxide.

ylide	conditions	axial-CH₂ epoxide	equatorial-CH₂ epoxide
$\bar{C}H_2\overset{+}{S}(CH_3)_2$	THF, 0°C	83%	17%
$\bar{C}H_2\overset{+}{S}(O)(CH_3)_2$	THF, 65°C	not formed	only product

Dimethylsulfonium methylide reacts with reactive alkylating reagents such as allylic and benzylic bromides to give terminal alkenes. A similar reaction occurs with primary alkyl bromides in the presence of LiI. The reaction probably involves alkylation of the ylide, followed by elimination.[170]

$$RCH_2{-}X + CH_2{=}S^+(CH_3)_2 \longrightarrow RCH_2CH_2S^+(CH_3)_2 \longrightarrow RCH{=}CH_2$$

Sulfur ylides are also available which allow transfer of substituted methylene units, such as isopropylidene (entries 10 and 11 in Scheme 2.19) or cyclopropylidene (entries 12 and 13). The oxaspiropentanes formed by reaction of aldehydes and ketones with diphenylsulfonium cyclopropylide are useful intermediates in a number of transformations such as acid-catalyzed rearrangement to cyclobutanones.[171]

CH3 (CH2)5CH3 C O —H+→ CH3 (CH2)5CH3 O= 92%

Aside from the methylide and cyclopropylide reagents, the sulfonium ylides are not very stable. A related group of reagents derived from sulfoximines offer greater versatility in alkylidene transfer reactions.[172] The preparation and use of this class of ylides is illustrated by the following sequence:

$$ArS(O)CH_2CH_3 \xrightarrow[H_2SO_4,\ CHCl_3]{NaN_3} ArS(O)(=NH)CH_2CH_3 \xrightarrow{(CH_3)_3O^+BF_4^-} ArS^+(O)(N(CH_3)_2)CH_2CH_3\ \bar{B}F_4$$

$$\xrightarrow{NaH,\ DMF} ArS^+(O)(N(CH_3)_2){-}\bar{C}HCH_3 \xrightarrow{C_6H_5CHO} C_6H_5CH{-}CHCH_3\ (\text{epoxide, O})\quad 67\%$$

Ar = p-$CH_3C_6H_4$–

A similar pattern of reactivity has been demonstrated for the anions formed by deprotonation of *S*-alkyl-*N*-*p*-toluenesulfoximines[173] (see entry 14 in Scheme 2.19).

CH_3–C_6H_4–$S^+(O)(NMe_2)$–$\bar{C}XY$
dimethylaminooxosulfonium ylide

CH_3–C_6H_4–$S(O)(NTs)$–$\bar{C}XY$
N-tosylsulfoximine anion

The sulfur atom in both these types of reagents is chiral. They have been utilized in the preparation of enantiomerically enriched epoxides and cyclopropanes.[174]

170. L. Alcaraz, J. J. Harnett, C. Mioskowski, J. P. Martel, T. LeGall, D.-S. Shin, and J. R. Falck, *Tetrahedron Lett.* **35:**5453 (1994).
171. B. M. Trost and M. H. Bogdanowicz, *J. Am. Chem. Soc.* **95:**5321 (1973).
172. C. R. Johnson, *Acc. Chem. Res.* **6:**341 (1973).
173. C. R. Johnson, R. A. Kirchoff, R. J. Reischer, and G. F. Katekar, *J. Am. Chem. Soc.* **95:**4287 (1973).
174. C. R. Johnson and E. R. Janiga, *J. Am. Chem. Soc.* **95:**7673 (1973).

The pattern of nucleophilic addition at a carbonyl group followed by intramolecular nucleophilic displacement of a leaving group present in the nucleophile can also be recognized in a much older synthetic technique, the *Darzens reaction*.[175] The first step in the reaction is addition of the enolate of the α-halo ester to the carbonyl compound. The alkoxide oxygen formed in the addition then effects nucleophilic attack, displacing the halide and forming an α,β-epoxy ester (also called a glycidic ester).

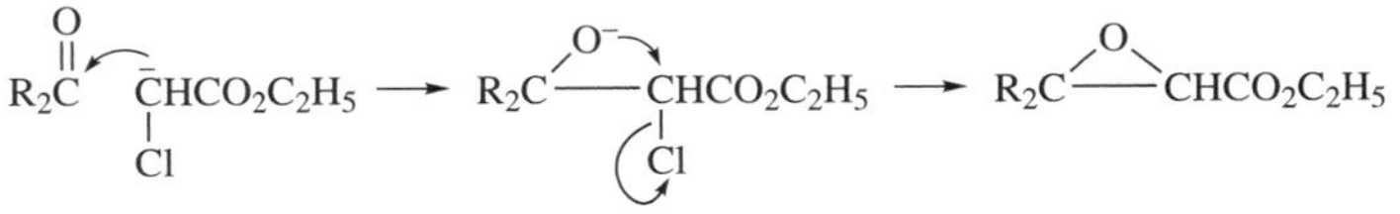

$$R_2C{=}O + {}^-CH(Cl)CO_2C_2H_5 \longrightarrow R_2C(O^-)\text{—}CH(Cl)CO_2C_2H_5 \longrightarrow R_2C\overset{O}{\text{—}}CHCO_2C_2H_5$$

Scheme 2.20 gives some examples of the Darzens reaction.

Trimethylsilylepoxides can be prepared by an addition–cyclization process. Reaction of chloromethyltrimethylsilane with *sec*-butyllithium at very low temperature gives an α-chloro lithium reagent which gives an epoxide on reaction with an aldehyde or ketone.[176]

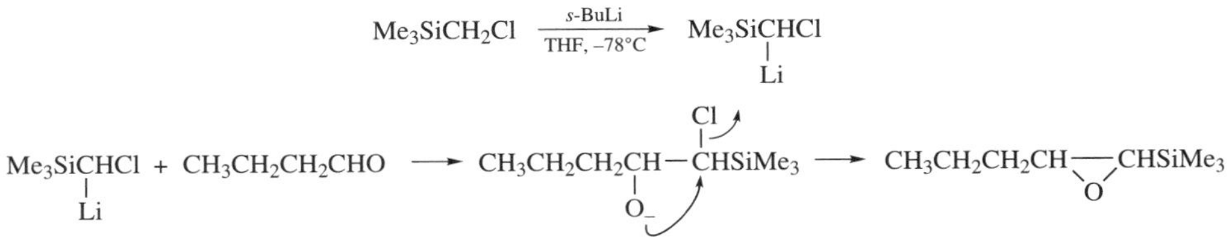

$$Me_3SiCH_2Cl \xrightarrow[\text{THF, }-78^\circ C]{s\text{-BuLi}} Me_3SiCH(Li)Cl$$

$$Me_3SiCH(Li)Cl + CH_3CH_2CH_2CHO \longrightarrow CH_3CH_2CH_2CH(O^-)\text{—}CH(Cl)SiMe_3 \longrightarrow CH_3CH_2CH_2CH\overset{O}{\text{—}}CHSiMe_3$$

Scheme 2.20. Darzens Condensation Reactions

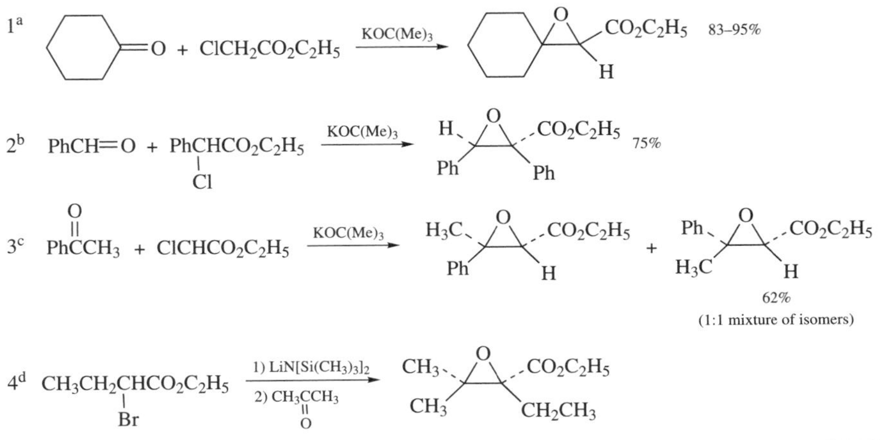

1[a] Cyclohexanone + $ClCH_2CO_2C_2H_5$ $\xrightarrow{KOC(Me)_3}$ spiro epoxide ($CO_2C_2H_5$, H) 83–95%

2[b] $PhCH{=}O + PhCH(Cl)CO_2C_2H_5$ $\xrightarrow{KOC(Me)_3}$ epoxide (H, Ph / $CO_2C_2H_5$, Ph) 75%

3[c] $PhC(O)CH_3 + ClCH_2CO_2C_2H_5$ $\xrightarrow{KOC(Me)_3}$ epoxide (H_3C, Ph / $CO_2C_2H_5$, H) + epoxide (Ph, H_3C / $CO_2C_2H_5$, H) 62% (1:1 mixture of isomers)

4[d] $CH_3CH_2CH(Br)CO_2C_2H_5$ $\xrightarrow[\text{2) } CH_3C(O)CH_3]{\text{1) } LiN[Si(CH_3)_3]_2}$ epoxide (CH_3, CH_3 / $CO_2C_2H_5$, CH_2CH_3)

a. R. H. Hunt, L. J. Chinn, and W. S. Johnson, *Org. Synth.* **IV:**459 (1963).
b. H. E. Zimmerman and L. Ahramjian, *J. Am. Chem. Soc.* **82:**5459 (1960).
c. F. W. Bachelor and R. K. Bansal, *J. Org. Chem.* **34:**3600 (1969).
d. R. F. Borch, *Tetrahedron Lett.* **1972:**3761.

175. M. S. Newman and B. J. Magerlein, *Org. React.* **5:**413 (1951).
176. C. Burford, F. Cooke, E. Ehlinger, and P. D. Magnus, *J. Am. Chem. Soc.* **99:**4536 (1977).

General References

Aldol Additions and Condensations

M. Braun in *Advances in Carbanion Chemistry*, Vol. 1. V. Snieckus, ed., JAI Press, Greenwich, Connecticut, 1992.
D. A. Evans, J. V. Nelson, and T. R. Taber, *Top. Stereochem.* **13:**1 (1982).
A. S. Franklin and I. Paterson, *Contemp. Org. Synth.* **1:**317 (1994).
C. H. Heathcock, in *Comprehensive Carbanion Chemistry*, E. Buncel and T. Durst, eds., Elsevier, Amsterdam, 1984.
C. H. Heathcock, in *Asymmetric Synthesis*, Vol 3, J. D. Morrison, ed., Academic Press, New York, 1984.
S. Masamune, W. Choy, J. S. Petersen, and L. R. Sita, *Angew. Chem. Int. Ed. Engl.* **24:**1 (1985).
T. Mukaiyama, *Org. React.* **28:**203 (1982).
A. T. Nielsen and W. T. Houlihan, *Org. React.* **16:**1 (1968).

Annulation Reactions

R. E. Gawley, *Synthesis* **1976:**777.
M. E. Jung, *Tetrahedron* **32:**3 (1976).

Mannich Reactions

F. F. Blicke, *Org. React.* **1:**303 (1942).
H. Bohme and M. Heake, in *Iminium Salts in Organic Chemistry*, H. Bohmne and H. G. Viehe, eds., Wiley-Interscience, New York, 1976, pp. 107–223.
M. Tramontini and L. Angiolini, *Mannich Bases—Chemistry and Uses*, CRC Press, Boca Raton, Florida, 1994.

Phosphorus-Stabilized Ylides and Carbanions

J. Boutagy and R. Thomas, *Chem. Rev.* **74:**87 (1974).
I. Gosney and A. G. Rowley in *Organophosphorus Reagents in Organic Synthesis*, J. I. G. Cadogan, ed., Academic Press, London, 1979, pp. 17–153.
A. W. Johnson, *Ylides and Imines of Phosphorus*, John Wiley & Sons, New York, 1993.
A. Maercker, *Org.React.* **14:**270 (1965).
W. S. Wadsworth, Jr., *Org. React.* **25:**73 (1977).

Problems

(*References for these problems will be found on page 924.*)

1. Predict the product formed in each of the following reactions:

(a) γ-butyrolactone + ethyl oxalate $\xrightarrow[\text{2) } H^+]{\text{1) } NaOCH_2CH_3}$

(b) 4-bromobenzaldehyde + ethyl cyanoacetate $\xrightarrow[\text{piperidine}]{\text{ethanol}}$

(c) $CH_3CH_2CH_2\overset{\overset{O}{\|}}{C}CH_3$ $\xrightarrow[\text{2) } CH_3CH_2CHO\text{, 15 min; 3) } H_2O]{\text{1) } LiN(i\text{-}Pr)_2\text{, }-78°C}$

(d) 2-Furaldehyde (C_4H_3O–CHO) + $PhCH_2\overset{O}{\overset{\|}{C}}CH_3$ $\xrightarrow{NaOH,\ H_2O}$

(e) $C_6H_5CH{=}C(OAc)CH_3$ $\xrightarrow[2)\ ZnCl_2\ \ 3)\ n\text{-}C_3H_7CHO]{1)\ CH_3Li,\ 2\ equiv.}$

(f) 2-Oxocyclohexyl–$CH_2\overset{+}{N}(CH_2CH_3)_2(CH_3)\ I^-$ + $CH_3\overset{O}{\overset{\|}{C}}CH_2CO_2CH_2CH_3$ $\xrightarrow[ethanol,\ \Delta]{NaOCH_2CH_3}$ $C_{10}H_{14}O$

(g) 2-Methylcyclopentanone + $HCO_2CH_2CH_3$ $\xrightarrow[ether]{Na}$

(h) Cyclohexyl–$\overset{O}{\overset{\|}{C}}CH_3$ + $(CH_3CH_2O)_2C{=}O$ $\xrightarrow[toluene]{NaNH_2}$

(i) $C_6H_5\overset{O}{\overset{\|}{C}}CH_3$ + $(CH_3CH_2O)_2\overset{O}{\overset{\|}{P}}CH_2CN$ $\xrightarrow[THF]{NaH}$

(j) $CH_3CH_2\overset{O}{\overset{\|}{C}}CH_2CH_2CO_2CH_2CH_3$ $\xrightarrow[xylene]{NaOCH_3}$

(k) 1-Cyclopentenyl–$\overset{O}{\overset{\|}{C}}{-}CH_3$ + $(CH_3)_2\overset{O}{\overset{\|}{S}}{=}CH_2$ $\longrightarrow$

(l) 3-Pyridyl–$CO_2C_2H_5$ + $CH_2{=}\overset{O^-}{\overset{|}{C}}CH{=}\overset{O^-}{\overset{|}{C}}OC_2H_5$ $\longrightarrow$ $\xrightarrow{H^+}$ $C_{10}H_7NO_3$

(m) Bicyclic enone (CH_3 at ring fusion, $C{=}O$, $CH(OMe)_2$, CH_3, $=CH_2$) $\xrightarrow[2)\ KH]{1)\ (CH_3)_3Si\overset{Li}{\overset{|}{C}}HOCH_3}$

2. Indicate reaction conditions or a series of reactions that could effect each of the following synthetic conversions.

(a) $CH_3CO_2C(CH_3)_3$ $\longrightarrow$ $(CH_3)_2\overset{OH}{\overset{|}{C}}CH_2CO_2C(CH_3)_3$

(b) 2-Tetrahydropyranyl–$O(CH_2)_3CH{=}O$ $\longrightarrow$ 2-Tetrahydropyranyl–$O(CH_2)_3CH{=}C(CH_3)CH_2OH$ (H and CH_3 cis)

(c) [cyclohexanone] → [2-(hydroxymethylene)cyclohexanone, =CHOH]

(d) $Ph_2C{=}O$ → [1-cyano-2,2-diphenylcyclopropane-1-carboxylate: Ph, Ph, CN, $CO_2C_2H_5$]

(e) [cyclododecanone] → [2-($CO_2C_2H_5$)cyclododecanone]

(f) [hexahydrobenzofuranone lactone] → [α-(CH_2OH) lactone]

(g) [γ-butyrolactone] → $HO(CH_2)_3\overset{O}{\overset{\|}{C}}CH_2\overset{O}{\overset{\|}{S}}CH_3$

(h) $H_2C{=}CHCH_2CH_2\overset{O}{\overset{\|}{C}}CH_3$ → $H_2C{=}CHCH_2CH_2\overset{O}{\overset{\|}{C}}CH_2CO_2C_2H_5$ → $H_2C{=}CHCH_2CH_2\overset{O}{\overset{\|}{C}}CH(CO_2C_2H_5)CH_2CH_2\overset{O}{\overset{\|}{C}}CH_3$ → [2-($H_2C{=}CHCH_2$)-3-CH_3-cyclohex-2-enone]

(i) $H_5C_2O_2CCH_2CH_2CO_2C_2H_5$ → [cyclohexane-1,4-dione]

(j) $C_6H_5\overset{O}{\overset{\|}{C}}CH_2\overset{O}{\overset{\|}{C}}CH_3$ → $C_6H_5\overset{O}{\overset{\|}{C}}CH_2\overset{O}{\overset{\|}{C}}CH_2\overset{O}{\overset{\|}{C}}-C_6H_4-OCH_3$

(k) $CH_3O-C_6H_4-\overset{O}{\overset{\|}{C}}CH_3$ → $CH_3O-C_6H_4-\overset{O}{\overset{\|}{C}}CH{=}CH_2$

(l) 3,4-$(CH_3O)_2C_6H_3-\overset{O}{\overset{\|}{C}}H$ → 3,4-$(CH_3O)_2C_6H_3-CH{=}CHCH{=}CH_2$

(m) [cis-1,5-dimethylbicyclo[3.3.0]octan-3-one, CH_3, CH_3] → [2-($=CHSCH_2CH_2CH_2CH_3$) derivative, H_3C, CH_3]

(n) O=HC (pyrrole, N–H) $CSCH_2CH_3$ (C=O) → CH_3CH=CH (pyrrole, N–H) $CSCH_2CH_3$ (C=O)

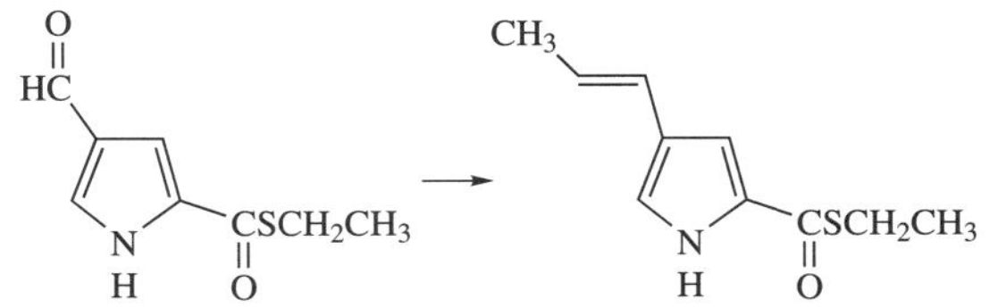

(o) CH_3, CH_3, O → CH_3, CH_3, O

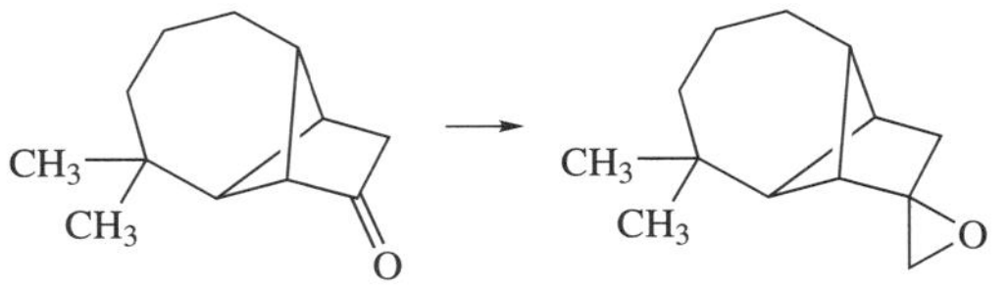

(p) 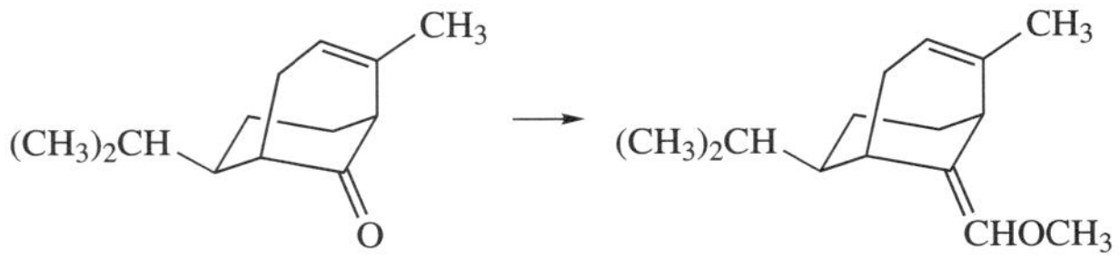 $(CH_3)_2CH$, CH_3, O → $(CH_3)_2CH$, CH_3, $CHOCH_3$

(q) 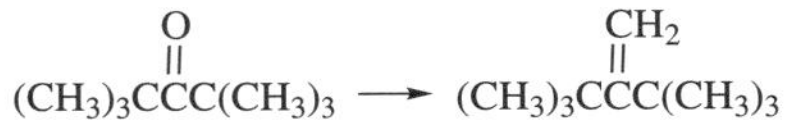 $(CH_3)_3CC(=O)C(CH_3)_3 \longrightarrow (CH_3)_3CC(=CH_2)C(CH_3)_3$

(r) 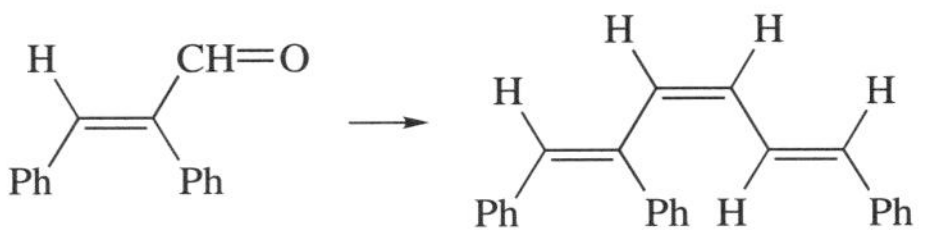 H, CH=O, Ph, Ph → H, H, H, H, Ph, Ph H, Ph

(s) 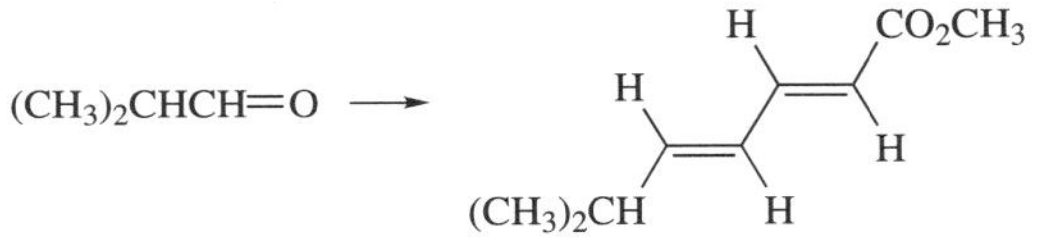 $(CH_3)_2CHCH{=}O \longrightarrow$ $(CH_3)_2CH$, H, H, H, CO_2CH_3, H

(t) 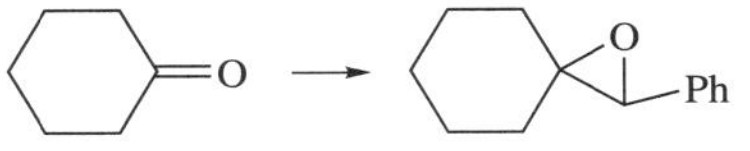 cyclohexanone (=O) → O, Ph (epoxide)

(u) 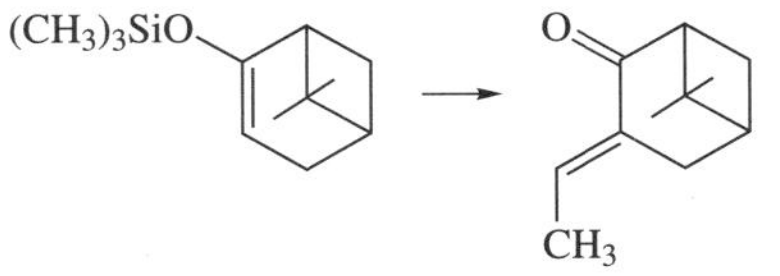 $(CH_3)_3SiO$ → O, CH_3

3. Step-by-step retrosynthetic analysis of each of the target molecules reveals that they can be efficiently prepared in a few steps from the starting material shown on the right. Show a retrosynthetic analysis and suggest reagents and conditions for carrying out the desired synthesis.

(a) 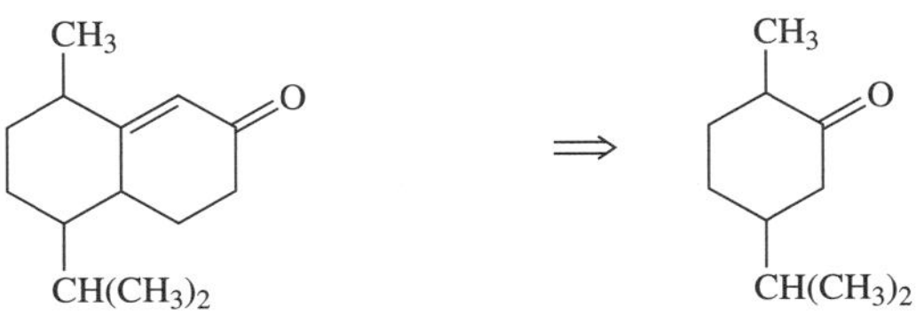 CH_3, O, $CH(CH_3)_2$ $\Rightarrow$ CH_3, O, $CH(CH_3)_2$

(b) O $CH(CH_3)_2$ O $\Longrightarrow$ OH $\overset{\text{O}}{\|}$ CCH_3

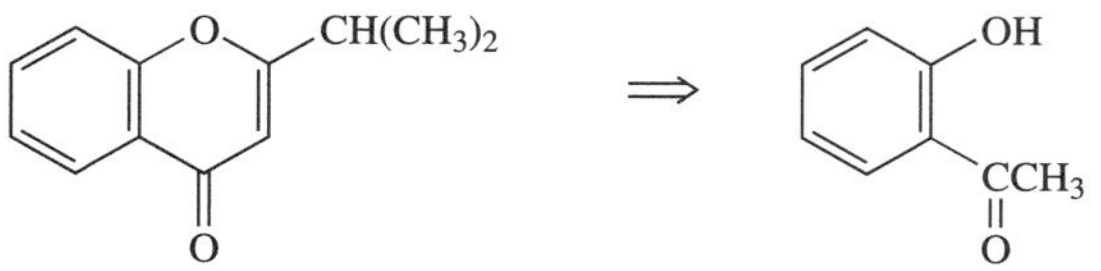

(c) $CH_2{=}\underset{CH_3}{C}CH{=}CH\underset{CH(CH_3)_2}{C}HCH_2CH_2CO_2CH_2CH_3 \Longrightarrow (CH_3)_2CHCH_2\overset{O}{\overset{\|}{C}}H$

(d) C_6H_5 O $\Longrightarrow$ $C_6H_5CH{=}CH\overset{O}{\overset{\|}{C}}H$

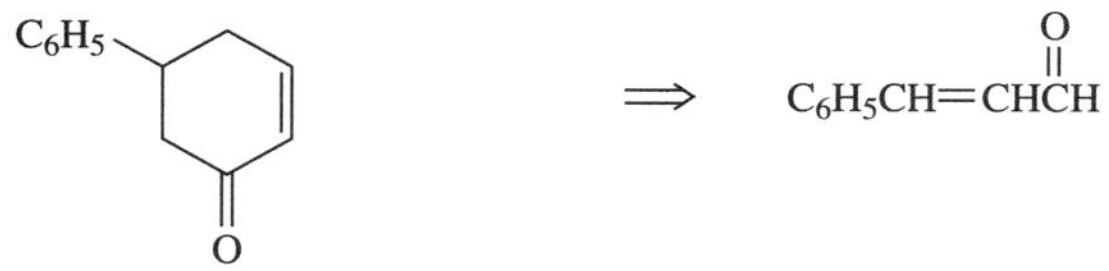

(e) CH_3 $\underset{O}{\underset{\|}{C}}CH_2CH_2\underset{CH_3}{C}{=}CH_2$ $\Longrightarrow$ CH_3 $\underset{O}{\underset{\|}{C}}CH_3$

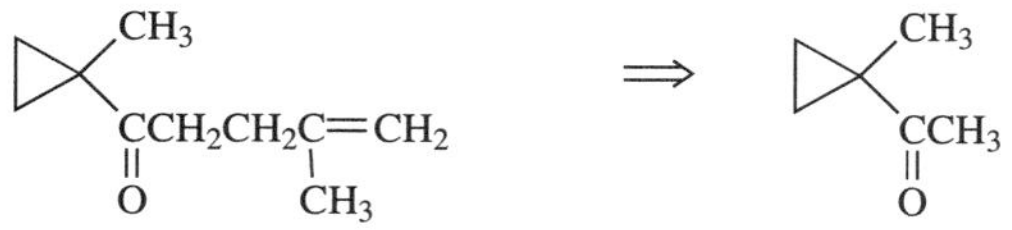

(f) O CH_3 CH_3 O $\Longrightarrow$ O O

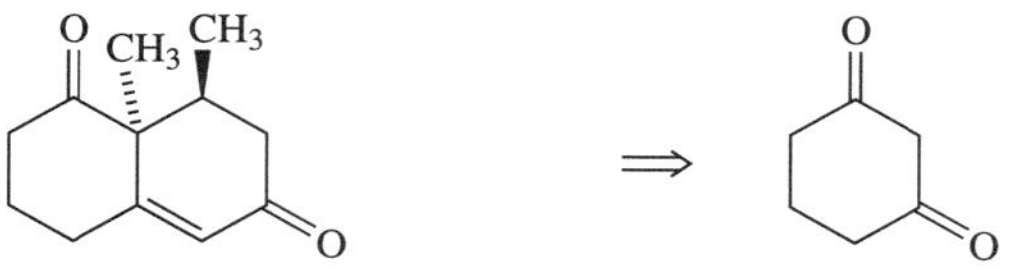

(g) O N $\Longrightarrow$ O

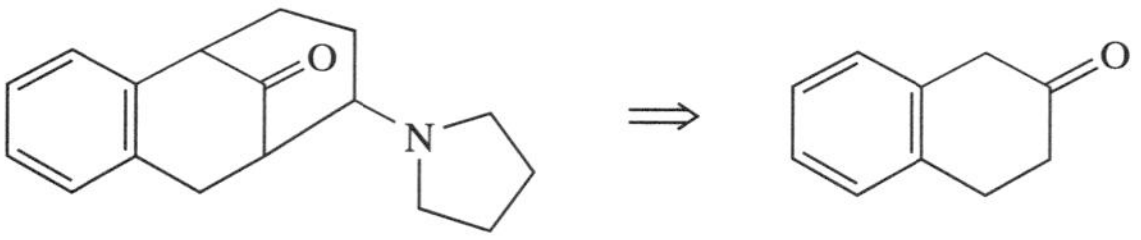

(h) $Ph_2C{=}CHCH{=}O \Longrightarrow Ph_2C{=}O$

(i) $CH_3CH_2\underset{CH_2}{\underset{\|}{C}}CH{=}CHCO_2C_2H_5 \Longrightarrow CH_3CH_2CH_2CH{=}O,\ ClCH_2CO_2C_2H_5$

(j) O N CH_3 $\Longrightarrow CH_3NH_2,\ CH_2{=}CHCO_2C_2H_5$

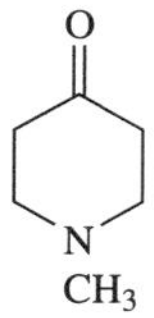

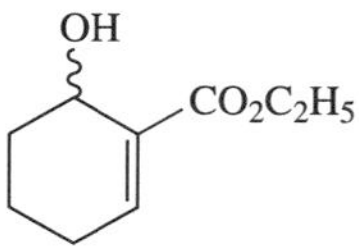

(k) OH $CO_2C_2H_5$ $\Longrightarrow O{=}CH(CH_2)_3CH{=}O$

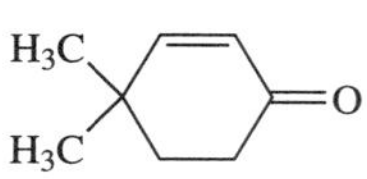

(l) H_3C H_3C O $\Longrightarrow (CH_3)_2CHCH{=}O,\ CH_2{=}CH\overset{O}{\overset{\|}{C}}CH_3$

(m) [structure: 2-bromophenyl–C(CH3)(OH)–CH2NH2] ⟹ [structure: 2-bromophenyl–CCH3 (C=O)]

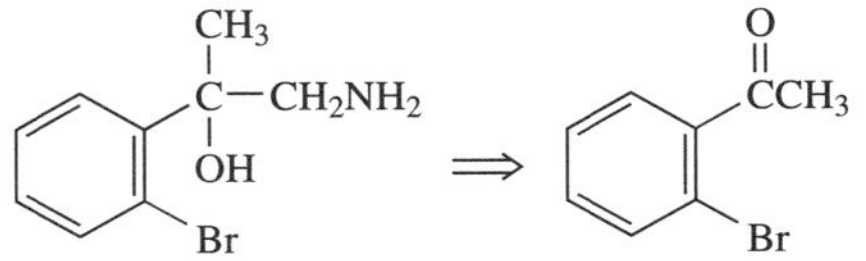

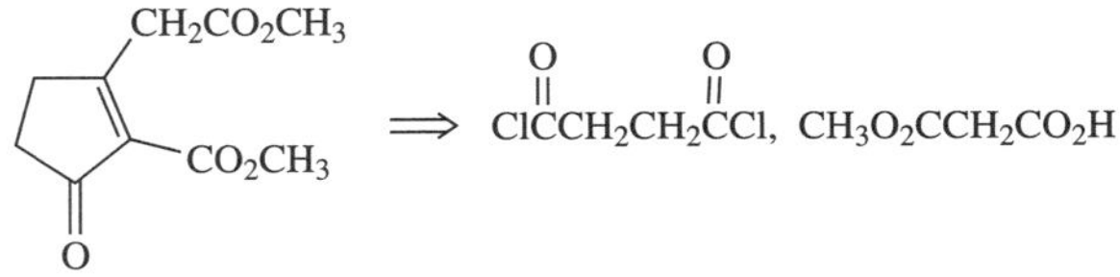

(n) [structure: cyclopentenone bearing $CH_2CO_2CH_3$ and CO_2CH_3] ⟹ $ClCCH_2CH_2CCl$ (two C=O), $CH_3O_2CCH_2CO_2H$

(o) [structure: bicyclic with CH_3O_2C, CH_2, $PhCH_2N$, CH_3O_2C] ⟹ [structure: 2,6-bis(CO_2CH_3)cyclohexanone]

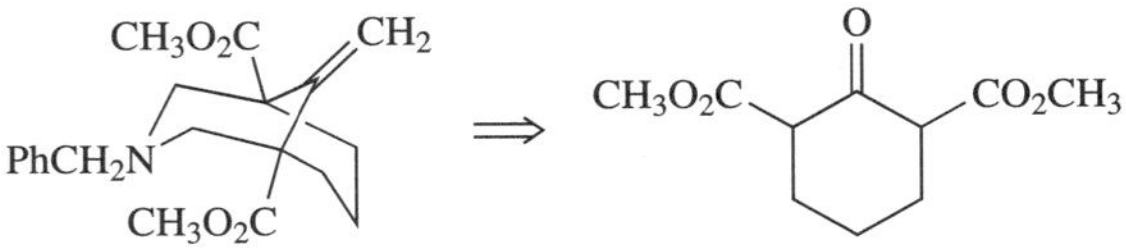

(p) [structure: HO_2C, CH_3O–C6H4–CH_2O, CO_2H] ⟹ [structure: CH_3O–C6H4–CH_2O–tetrahydrofuranol, OH]

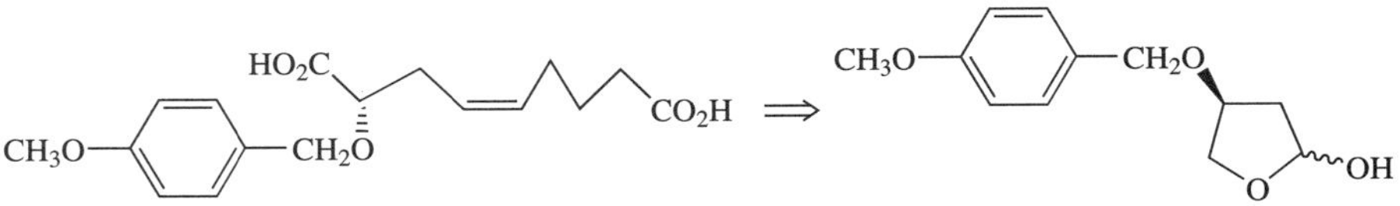

(q) [structure: pyrrolidinone with $CO_2C_2H_5$, $CH_2{=}CHCH_2$, N–CO_2CH_2Ph] ⟹ [structure: alkene chain with $CO_2C_2H_5$, $NHCO_2CH_2Ph$]

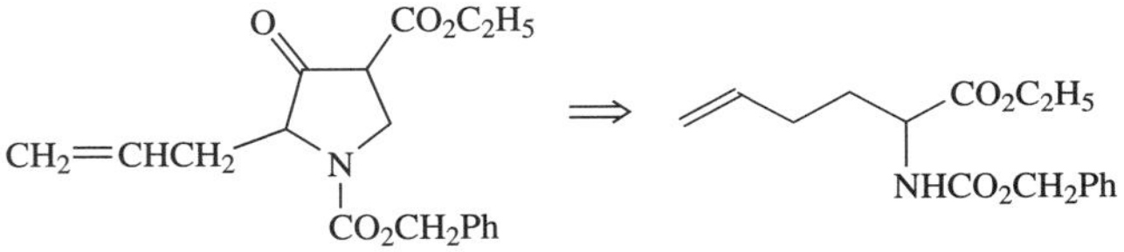

4. Offer a mechanism for each of the following transformations.

(a) [structure: dimethyl phthalate (CO_2CH_3, CO_2CH_3)] $\xrightarrow[\text{NaH, benzene}]{C_2H_5CC_2H_5 \text{ (C=O)}}$ [structure: 2-methylindane-1,3-dione, CH_3]

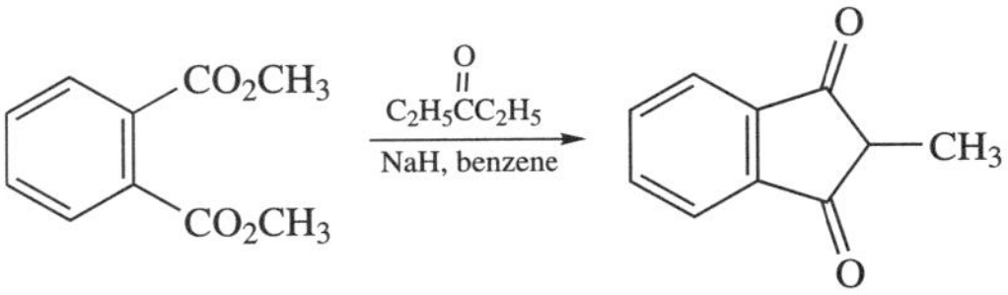

(b) [structure: O=CH, OH, $CH_2P(OCH_3)_2$ (P=O), $(CH_3)_3CO$, CH_3] $\xrightarrow[t\text{-BuOH}]{\text{KO-}t\text{-Bu}}$ [structure: H, =O, $(CH_3)_3CO$, CH_3]

(c) [structure: CH_3CH_2C(=O)–cyclopropane–CH_3, $OCCH_3$ (C=O)] $\xrightarrow{\text{NaOH, MeOH}}$ [structure: 2,3-dimethylcyclopent-2-enone, CH_3, CH_3]

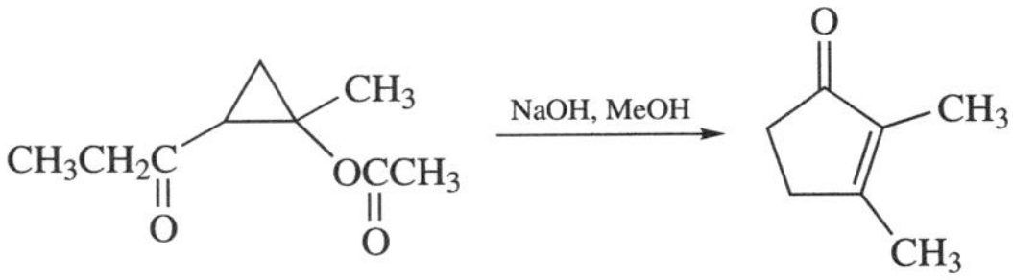

(d) $CH_3CH_2C(CH_3)(Ph)$—$C(OH)(Ph)CH_3$ $\xrightarrow[\text{dioxane, 150°C}]{\text{KOH, } H_2O}$ $CH_3CH_2CH(Ph)CH_3$ + $PhCCH_3$ (C=O)

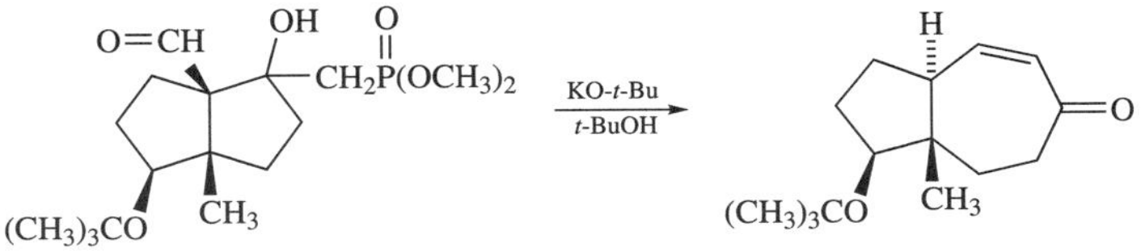

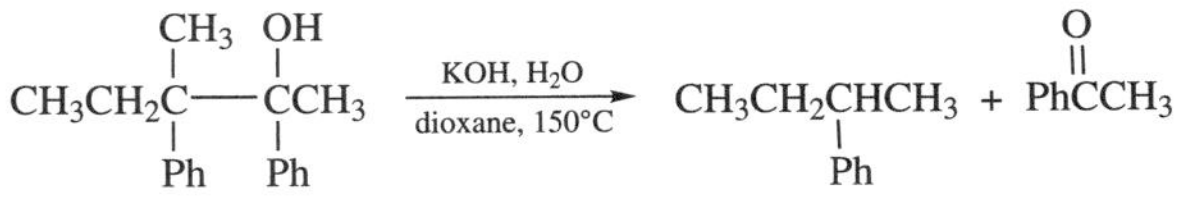

(e) + $CH_3CH{=}CHCH{=}PPh_3$ →

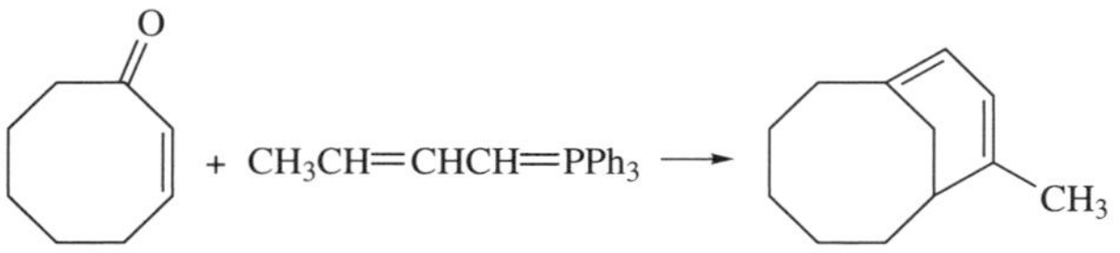

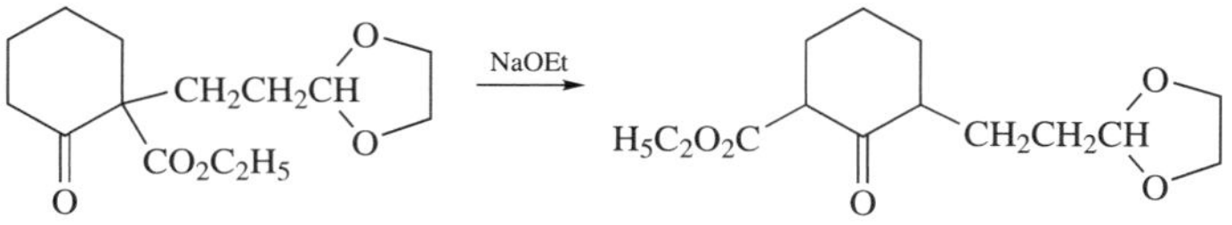

(f) + $H_2C{=}CHCCH_3$ (C=O) →

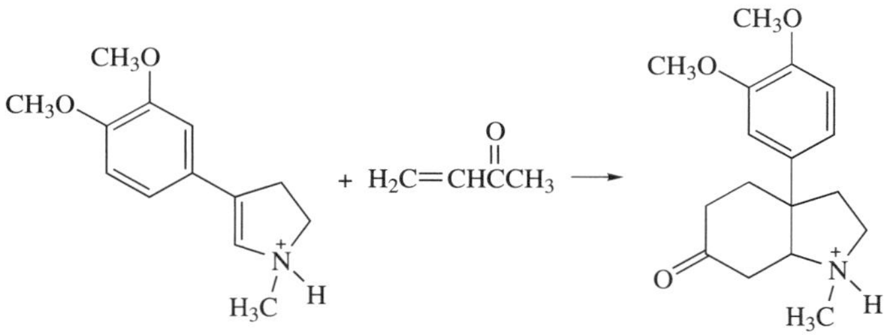

(g) NaOEt

(h) 1) LDA, 0°C 2) H^+

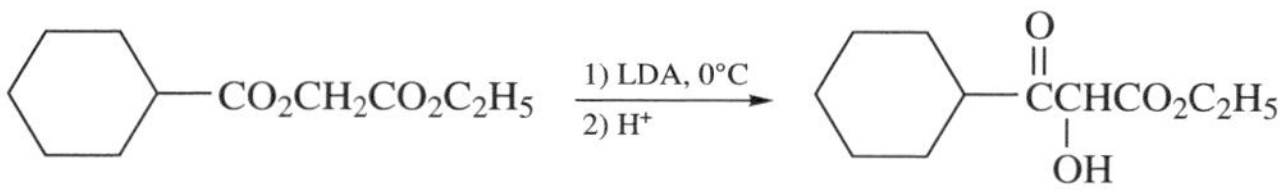

(i) +

1) $CsCO_3$ 2) H^+, 80°C

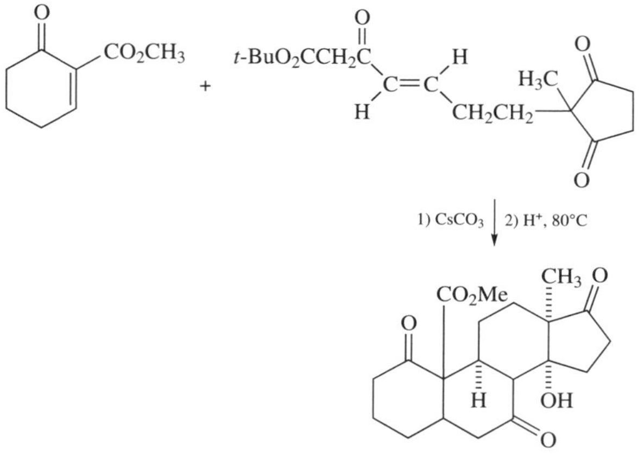

5. Tetraacetic acid (or a biological equivalent) has been suggested as an intermediate in the biosynthesis of phenolic natural products. Its synthesis has been described, as has its ready conversion to orsellinic acid. Suggest a mechanism for formation of orsellinic acid under the conditions specified.

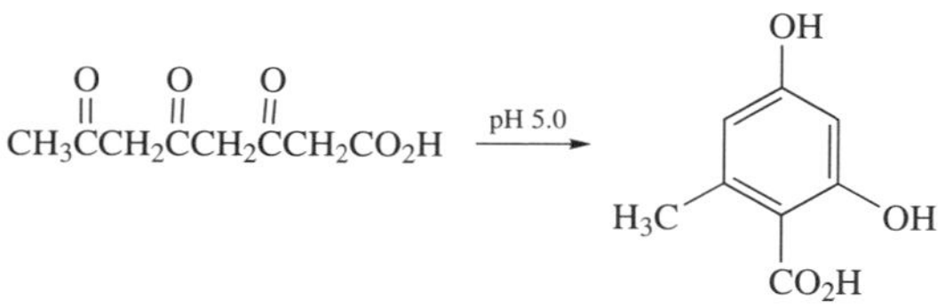

$CH_3CCH_2CCH_2CCH_2CO_2H$ (three C=O) $\xrightarrow{\text{pH 5.0}}$

orsellinic acid

6. (a) A stereospecific method for deoxygenating epoxides to alkenes involves reaction of the epoxide with the diphenylphosphide ion, followed by methyl iodide. The method results in overall inversion of the alkene stereochemistry. Thus, *cis*-cyclooctene epoxide gives *trans*-cyclooctene. Propose a mechanism for this process and discuss the relationship of the reaction to the Wittig reaction.

   (b) Reaction of the epoxide of *E*-4-octene (*trans*-2,3-di-*n*-propyloxirane) with trimethylsilylpotassium affords *Z*-4-octene as the only alkene in 93% yield. Suggest a reasonable mechanism for this reaction.

7. (a) A fairly general method for ring closure that involves vinyltriphenylphosphonium halides has been developed. Two examples are shown. Comment on the mechanism of the reaction and suggest two additional types of rings that could be synthesized using vinyltriphenylphosphonium salts.

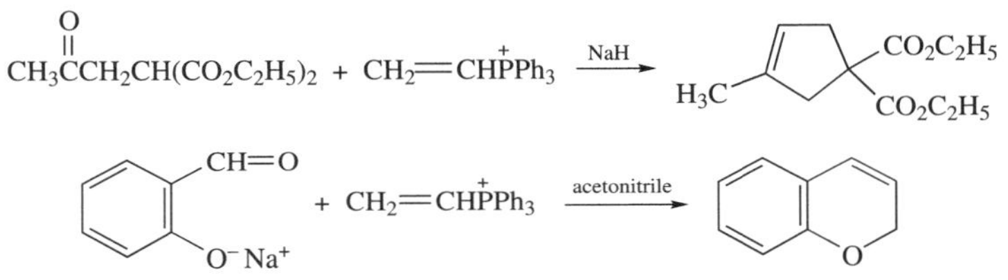

$$CH_3\overset{O}{\overset{\|}{C}}CH_2CH(CO_2C_2H_5)_2 + CH_2{=}CH\overset{+}{P}Ph_3 \xrightarrow{NaH}$$ (cyclopentene product bearing H_3C, $CO_2C_2H_5$, $CO_2C_2H_5$)

(salicylaldehyde sodium salt: CH=O, O^-Na^+) $+ CH_2{=}CH\overset{+}{P}Ph_3 \xrightarrow{acetonitrile}$ (2*H*-chromene, O)

   (b) The two phosphonium salts shown have both been used in syntheses of cyclohexadienes. Suggest appropriate co-reactants and catalysts that would be expected to lead to cyclohexadienes.

$$H_2C{=}CHCH_2\overset{+}{P}Ph_3 \qquad H_2C{=}CHCH{=}CH\overset{+}{P}Ph_3$$

   (c) The product shown below is formed by the reaction of vinyltriphenylphosphonium bromide, the lithium enolate of cyclohexanone, and 1,3-diphenyl-2-propen-1-one. Formulate a mechanism.

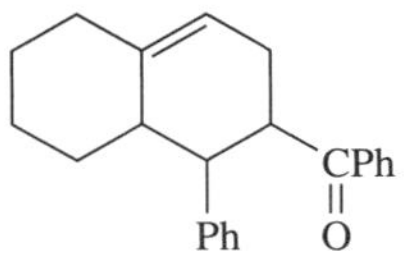

(bicyclic product bearing Ph and $\overset{O}{\overset{\|}{C}}Ph$)

8. Compounds **A** and **B** are key intermediates in one total synthesis of cholesterol. Rationalize their formation by the routes shown.

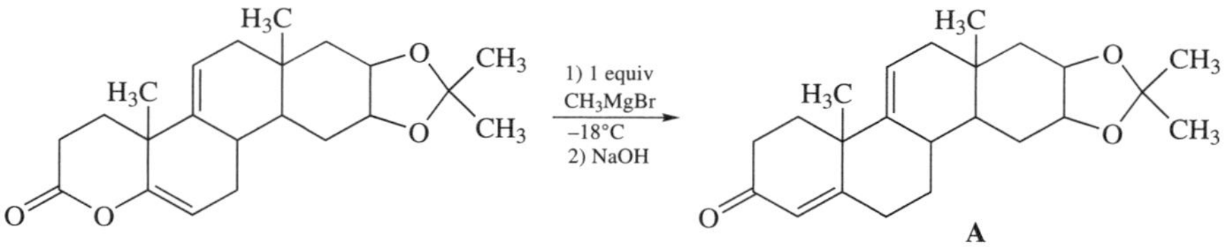

1) 1 equiv CH_3MgBr −18°C
2) NaOH

A

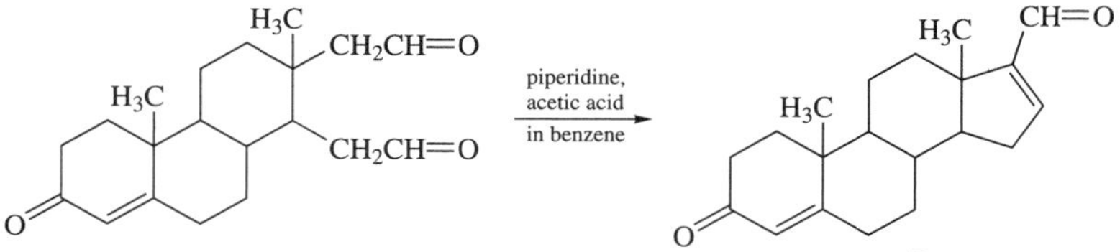

piperidine, acetic acid in benzene

B

9. The first few steps of a synthesis of a alkaloid conessine produce **D** from **C**. Suggest a sequence of reactions for effecting this conversion.

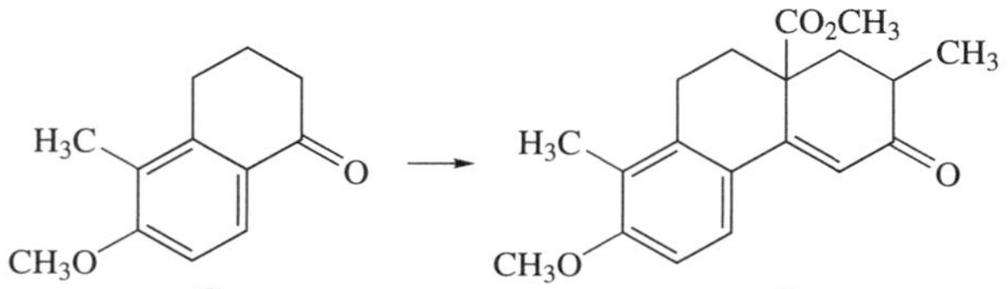

C

D

10. A substance known as elastase is involved in arthritis, various inflammations, pulmonary emphysema, and pancreatitis. Elastase activity can be inhibited by a compound known as elasnin, obtained from the culture broth of a particular microorganism. The structure of elasnin is shown. A synthesis of elasnin has been reported which utilized compound **E** as a key intermediate. Suggest a synthesis of compound **E** from methyl hexanoate and hexanal.

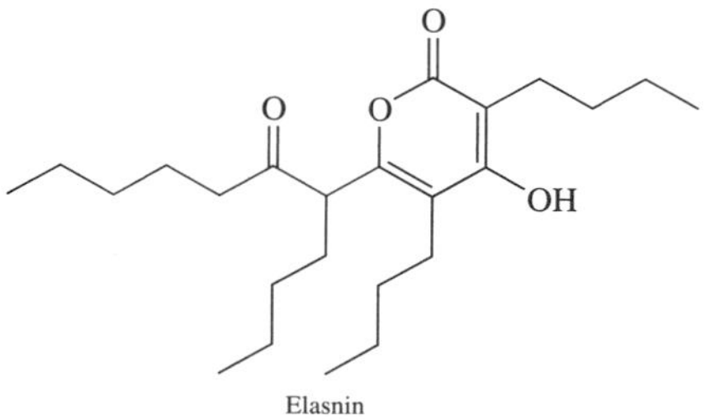

Elasnin

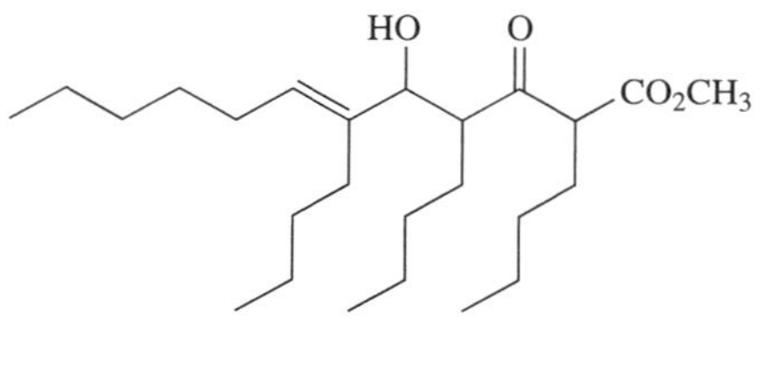

E

11. Treatment of compound **F** with lithium diisopropylamide followed by cyclohexanone gives either **G** or **H**. **G** is formed if the aldehyde is added at −78°C whereas **H** is formed if the aldehyde is added at 0°C. Furthermore, treatment of **G** with lithium diisopropylamide at 0°C gives **H**. Explain these results.

$CH_2{=}CHCH(CN)OCH_2CH_2OC_2H_5$

F

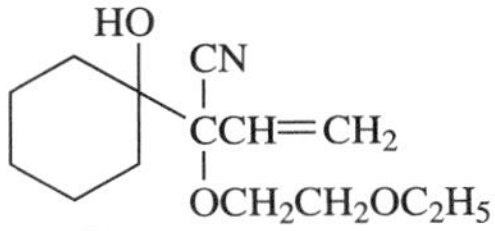

HO-cyclohexyl–$C(CN)(CH{=}CH_2)OCH_2CH_2OC_2H_5$

G

OH-cyclohexyl–$CH_2CH{=}C(CN)OCH_2CH_2OC_2H_5$

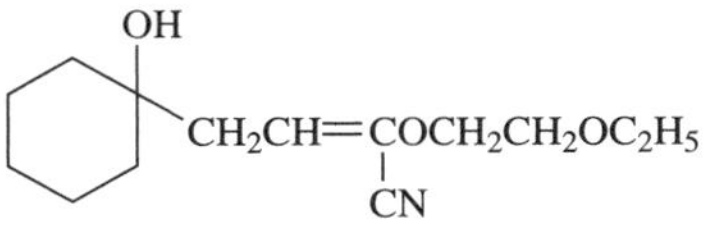

H

12. Dissect the following molecules into potential precursors by locating all bond connections which could be made by aldol-type reactions. Suggest the structure for potential precursors and conditions for performing the desired condensation.

(a) (b) CH_3 O CH_3 CH_3

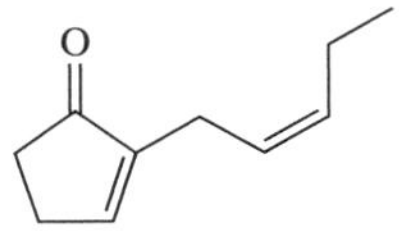

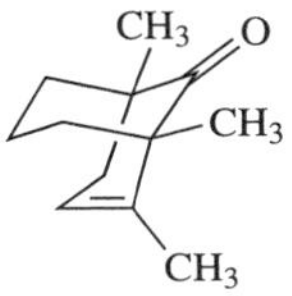

13. Mannich condensations permit one-step reactions to form the following substances from substantially less complex starting materials. By retrosynthetic analysis, identify a potential starting material which could give rise to the product shown in a single step under Mannich reaction conditions.

(a) N CO_2CH_3 O CO_2CH_3 (b) N $PhCH_2OCH_2CH_2CH_2$ CH_3 O

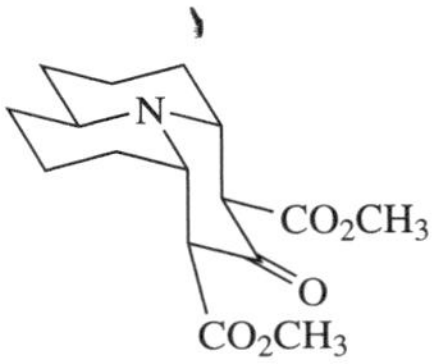

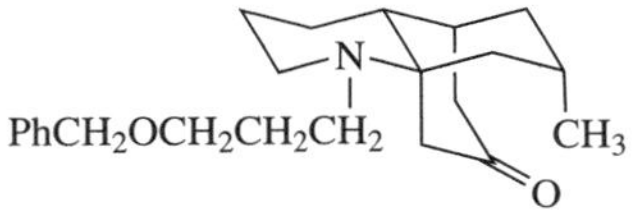

14. (a) The reagent **I** has found use in constructing rather complex molecules from simple precursors; for example, the enolate of 3-pentanone, treated first with **I**, then with benzaldehyde, gives **J** as a 2 : 1 mixture of stereoisomers. Explain the mechanism by which this synthesis occurs.

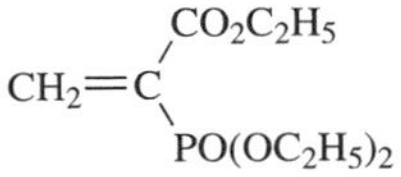

$$CH_2{=}C(CO_2C_2H_5)PO(OC_2H_5)_2$$

I

$$CH_3CH_2\overset{O}{\overset{\|}{C}}CH_2CH_3 \xrightarrow[2)\ \mathbf{I}]{1)\ LDA,\ -78°C} \xrightarrow[68°C,\ 45\ min]{PhCH{=}O} CH_3CH_2\overset{O}{\overset{\|}{C}}CH(CH_3)CH_2C(CO_2C_2H_5){=}CHPh \quad 74\%$$

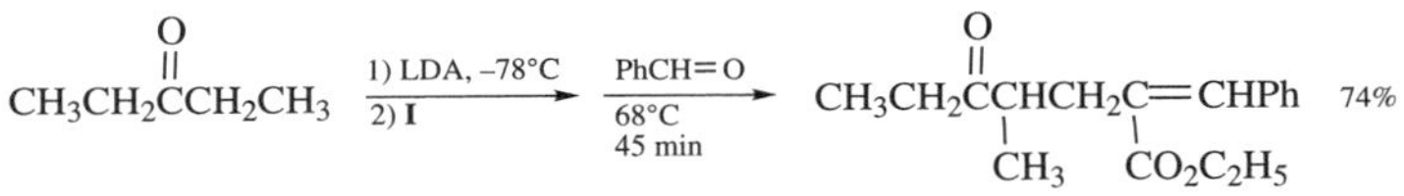

J

(b) The reagent **K** converts enolates of aldehydes into the cyclohexadienyl phosphonates **L**. What is the mechanism of this reaction? What alternative product might have been expected?

$$\underset{\mathbf{K}}{CH_2{=}CHCH{=}CHP(O)(OC_2H_5)_2} + R_2C{=}CH{-}O^- \longrightarrow \underset{\mathbf{L}}{\text{4,4-}R_2\text{-cyclohexa-1,5-dienyl}{-}P(O)(OC_2H_5)_2}$$

15. Indicate whether the aldol reactions shown below would be expected to exhibit high stereoselectivity. If high stereoselectivity is to be expected, show the relative configuration which is expected for the predominant product.

(a) $Ph_3CC(O)CH_2CH_3 \xrightarrow[\text{2) PhCH=O}]{\text{1) BuLi, }-50^\circ\text{C (enolate formation)}}$

(b) $CH_3CH_2C(O)CH(C_6H_{11})OSi(CH_3)_2C(CH_3)_3 \xrightarrow[\text{2) }Bu_2BOSO_2CF_3,\ -78^\circ\text{C}]{\text{1) }(i\text{-Pr})_2NC_2H_5} \xrightarrow{CH_3CH_2CH=O}$

(c) $CH_3CH_2C(O)CH_2CH_3 \xrightarrow[\text{2) }C_6H_5CH=O]{\text{1) LDA, THF, }-70^\circ\text{C}}$

(d) 1-$OSi(CH_3)_3$-2,6-dimethylcyclohexene $\xrightarrow[C_6H_5CH=O]{KF}$

(e) $PhC(O)CH_2CH_3 \xrightarrow[(i\text{-Pr})_2NC_2H_5,\ -78^\circ\text{C}]{\text{1) }Bu_2BOSO_2CF_3} \xrightarrow{\text{2) PhCH=O}}$

(f) $CH_3CH_2C(O){-}C(CH_3)_2OSi(CH_3)_3 \xrightarrow[\text{2) }(CH_3)_2CHCH=O]{\text{1) LDA, }-70^\circ\text{C}}$

(g) $CH_3CH_2CH{=}O \xrightarrow[\text{2) PhCH=O, }-78^\circ\text{C}]{\text{1) }Et_3N\text{ (2 equiv), }TiCl_4\text{ (2 equiv)}}$

(h) N-propionyl-4-methyl-5-phenyl-oxazolidin-2-one (Ph, CH_3 cis) $\xrightarrow[\text{2) }(C_2H_5)_3N]{\text{1) }n\text{-}Bu_2BO_3SCF_3} \xrightarrow{\text{furan-2-}CH{=}O}$

16. The stereoselectivity of several α-oxy derivatives of ketone **M** are given below. Suggest a transition state which accounts for the observed stereoselectivity.

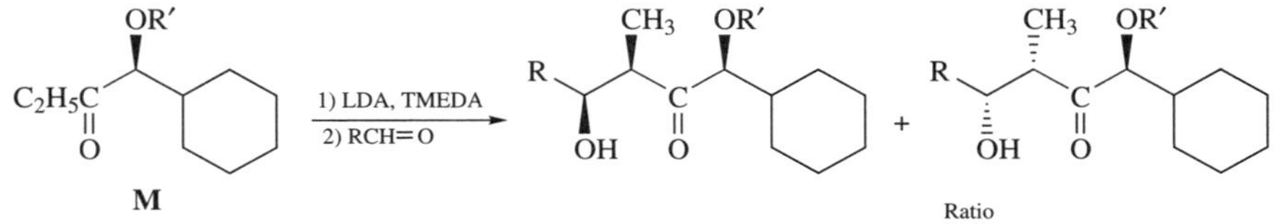

R′	Ratio
CH_2OCH_3	8:92
$Si(CH_3)_3$	9:91
$Si(CH_3)_2C(CH_3)_3$	83:17

17. Suggest a transition state which would account for the observed stereoselectivity of the following reaction sequence.

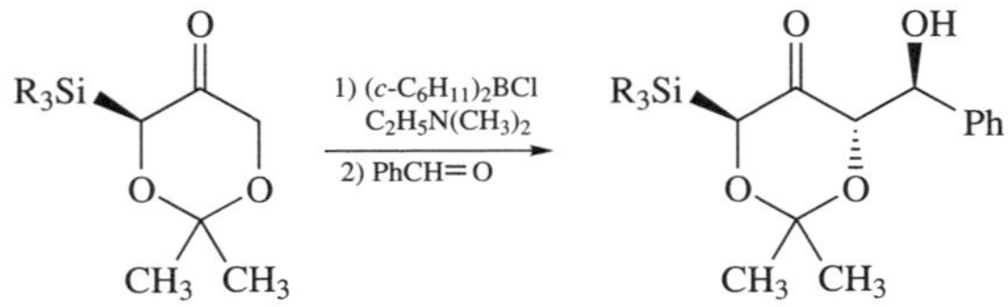

1) $(c\text{-}C_6H_{11})_2BCl$, $C_2H_5N(CH_3)_2$; 2) PhCH=O

$R_3Si = (C_2H_5)_2C(CH_3)Si(CH_3)_2$

18. Provide a mechanistic explanation for the influence of the Lewis acid in determining the stereoselectivity of addition of the silyl enol ether to the aldehyde

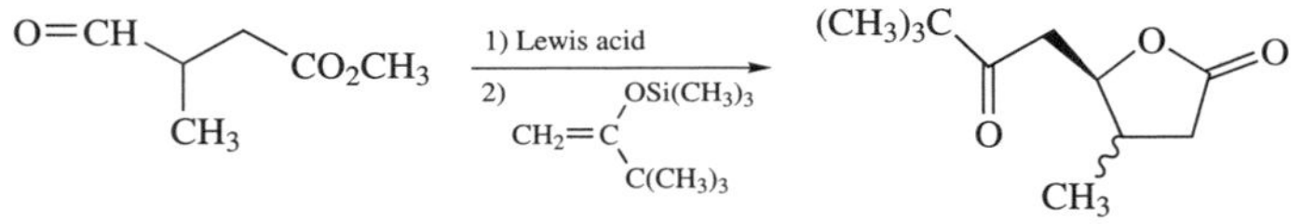

1) Lewis acid; 2) $CH_2{=}C(OSi(CH_3)_3)C(CH_3)_3$

	cis:trans
$TiCl_4$	29:71
BF_3	57:43
$SnCl_4$	66:34

19. The reaction of 3-benzyloxybutanal with the trimethylsilyl enol ether of acetophenone is stereoselective for the *anti* diasteromer.

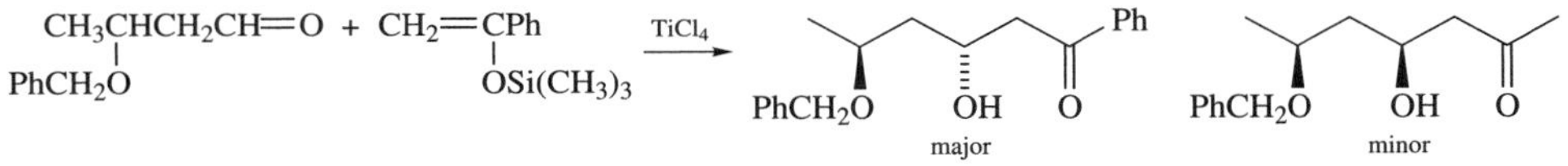

$CH_3CH(OCH_2Ph)CH_2CH{=}O + CH_2{=}C(OSi(CH_3)_3)Ph \xrightarrow{TiCl_4}$ major + minor

Propose a transition state which would account for the observed stereoselectivity.

3

Functional Group Interconversion by Nucleophilic Substitution

Introduction

The first two chapters dealt with formation of new carbon–carbon bonds by processes in which one carbon acts as the nucleophile and the other as the electrophile. In this chapter, we turn our attention to noncarbon nucleophiles. Nucleophilic substitution at both sp^3 and sp^2 centers is used in a variety of synthetic operations, particularly in the inverconversion of functional groups. The mechanistic aspects of nucleophilic substitutions were considered in Part A, Chapters 5 and 8.

3.1. Conversion of Alcohols to Alkylating Agents

3.1.1. Sulfonate Esters

Alcohols are a very important class of compounds for synthesis. However, because hydroxide is a very poor leaving group, they are not reactive as alkylating agents. The preparation of sulfonate esters from alcohols is an effective way of installing a reactive leaving group on an alkyl chain. The reaction is very general, and complications arise only if the resulting sulfonate ester is sufficiently reactive to require special precautions. *p*-Toluenesulfonate (tosylate) and methanesulfonate (mesylate) esters are the most frequently used groups for preparative work, but the very reactive trifluoromethanesulfonates (triflates) are useful when an especially good leaving group is required. The usual method for introducing tosyl or mesyl groups is to allow the alcohol to react with the sulfonyl chloride in pyridine at 0–25°C.[1] An alternative for preparing mesylates and tosylates is to convert the alcohol to a lithium salt, which is then allowed to react with the sulfonyl chloride.[2]

$$ROLi + ClSO_2\text{—}C_6H_4\text{—}CH_3 \longrightarrow ROSO_2\text{—}C_6H_4\text{—}CH_3$$

Trifluoromethanesulfonates of alkyl and allylic alcohols can be prepared by reaction with trifluoromethanesulfonic anhydride in halogenated solvents in the presence of pyridine.[3] Because the preparation of sulfonate esters does not disturb the C–O bond, problems of rearrangement or racemization do not arise in the ester formation step. However, sensitive sulfonate esters, such as allylic systems, may be subject to reversible ionization reactions, so that appropriate precautions must be taken to ensure structural and stereochemical integrity. Tertiary alkyl tosylates are not as easily prepared nor as stable as those from primary and secondary alcohols. Under the standard conditions, tertiary alcohols are likely to be converted to the corresponding alkene.

3.1.2. Halides

The prominent role of alkyl halides in formation of carbon–carbon bonds by nucleophilic substitution was evident in Chapter 1. The most common precursors for alkyl halides are the corresponding alcohols, and a variety of procedures have been developed for this transformation. The choice of an appropriate reagent is usually dictated by the sensitivity of the alcohol and any other functional groups present in the molecule. Unsubstituted primary alcohols can be converted to bromides with hot concentrated hydrobromic acid.[4] Alkyl chlorides can be prepared by reaction of primary alcohols with hydrochloric acid–zinc chloride.[5] These reactions proceed by an S_N2 mechanism, and elimination and rearrangements are not a problem for primary alcohols. Reactions with tertiary alcohols proceed by an S_N1 mechanism so these reactions are preparatively useful only when the carbocation intermediate is unlikely to give rise to rearranged product.[6] Because of the harsh conditions, these procedures are only applicable to very acid-stable molecules.

Another general method for converting alcohols to halides involves reactions with halides of certain non-metallic elements. Thionyl chloride, phosphorus trichloride, and phosphorus tribromide are the most common examples of this group of reagents. These reagents are suitable for alcohols that are neither acid-sensitive nor prone to structural rearrangements. The reaction of alcohols with thionyl chloride initially results in the formation of a chlorosulfite ester. There are two mechanisms by which the chlorosulfite

1. R. S. Tipson, *J. Org. Chem.* **9**:235 (1944); G. W. Kabalka, M. Varma, R. S. Varma, P. C. Srivastava and F. F. Knapp, Jr. *J. Org. Chem.* **51**:2386 (1986).
2. H. C. Brown, R. Bernheimer, C. J. Kim, and S. E. Scheppele, *J. Am. Chem. Soc.*, **89**:370 (1967).
3. C. D. Beard, K. Baum, and V. Grakauskas, *J. Org. Chem.* **38**:3673 (1973).
4. E. E. Reid, J. R. Ruhoff, and R. E. Burnett, *Org. Synth.* **II**:246 (1943).
5. J. E. Copenhaver and A. M. Wharley, *Org. Synth.* **I**:142 (1941).
6. J. F. Norris and A. W. Olmsted, *Org. Synth.* **I**:144 (1941); H. C. Brown and M. H. Rei, *J. Org. Chem.* **31**:1090 (1966).

can be converted to a chloride. In nucleophilic solvents, such as dioxane, the solvent participates and can lead to overall retention of configuration.[7]

$$ROH + SOCl_2 \longrightarrow ROS(=O)Cl + HCl$$

$$\text{dioxane} + ROS(=O)Cl \longrightarrow \text{dioxane-}\overset{+}{O}{-}R + SO_2 + Cl^- \longrightarrow R{-}Cl + \text{dioxane}$$

In the absence of solvent participation, chloride attack on the chlorosulfite ester leads to product with inversion of configuration.

$$ROH + SOCl_2 \longrightarrow ROS(=O)Cl + HCl$$

$$Cl^- \; R{-}OS(=O){-}Cl \longrightarrow R{-}Cl + SO_2 + Cl^-$$

Another method that provides chlorides from alcohols with retention of configuration involves conversion to a xanthate ester, followed by reaction with sulfuryl chloride. This method is thought to involve collapse of a chlorinated adduct of the xanthate ester. The reaction is useful for secondary alcohols, including sterically hindered structures.[8]

$$RCH(OH)R' \xrightarrow[2)\ SO_2Cl_2]{1)\ NaH,\ CS_2,\ CH_3I} RCH(Cl)R' \quad \text{retention}$$

$$RCH(R'){-}O{-}C(=S){-}SCH_3 \longrightarrow RCH(R'){-}O{-}C(SCl)(Cl){-}SCH_3 \longrightarrow RCH(R'){-}O{-}C(SSCH_3)(Cl){-}Cl \longrightarrow RCH(R'){-}Cl$$

The mechanism for the reactions with phosphorus halides can be illustrated using phosphorus tribromide. Initial reaction between the alcohol and phosphorus tribromide leads to a trialkyl phosphite ester by successive displacements of bromide. The reaction stops at this stage if it is run in the presence of an amine which neutralizes the hydrogen bromide that is formed.[9] If the hydrogen bromide is not neutralized, the phosphite ester is protonated, and each alkyl group is successively converted to the halide by nucleophilic substitution by bromide ion. The driving force for cleavage of the C—O bond is the

7. E. S. Lewis and C. E. Boozer, *J. Am. Chem. Soc.* **74:**308 (1952).
8. A. P. Kozikowski and J. Lee, *Tetrahedron Lett.* **29:**3053 (1988).
9. A. H. Ford-Moore and B. J. Perry, *Org. Synth.* **IV:**955 (1963).

formation of a strong phosphoryl double bond.

$$ROH + PBr_3 \longrightarrow (RO)_3P + 3\,HBr$$

$$(RO)_3P + HBr \longrightarrow RBr + O{=}P(H)(OR)_2$$

$$O{=}P(H)(OR)_2 + HBr \longrightarrow R{-}Br + O{=}P(H)(OH)OR$$

$$O{=}P(H)(OH){-}OR + HBr \longrightarrow RBr + O{=}P(H)(OH)_2$$

Because C–Br bond formation occurs by back-side attack, inversion of configuration at carbon is anticipated. However, both racemization and rearrangement can be observed as competing processes.[10] For example, conversion of enantiomerically pure 2-butanol to 2-butyl bromide with PBr_3 is accompanied by 10–13% racemization, and a small amount of *t*-butyl bromide is also formed.[11] The extent of rearrangement increases with increasing chain length and branching.

$$CH_3CH_2CH(OH)CH_2CH_3 \xrightarrow[\text{ether}]{PBr_3} \underset{85\text{–}90\%}{CH_3CH_2CH(Br)CH_2CH_3} + \underset{10\text{–}15\%}{CH_3CH_2CH_2CH(Br)CH_3} \qquad \text{Ref. 12}$$

$$(CH_3)_3CCH_2OH \xrightarrow[\text{quinoline}]{PBr_3} \underset{63\%}{(CH_3)_3CCH_2Br} + \underset{26\%}{(CH_3)_2C(Br)CH_2CH_3} + \underset{11\%}{CH_3CH(Br)CH(CH_3)_2} \qquad \text{Ref. 13}$$

Because of the very acidic solutions involved, these methods are limited to acid-stable molecules. Milder reagents are necessary for most functionally substituted alcohols. A very general and important method for activating alcohols toward nucleophilic substitution is by converting them to alkoxyphosphonium ions.[14] The alkoxyphosphonium ions are very reactive toward nucleophilic attack, with the driving force for substitution being formation of the strong phosphoryl bond.

$$R'_3P + \mathbf{E}{-}Y \longrightarrow R'_3P(\mathbf{E})(Y) \rightleftharpoons R'_3\overset{+}{P}{-}\mathbf{E} + Y^-$$

$$R'_3\overset{+}{P}{-}\mathbf{E} + ROH \longrightarrow R'_3\overset{+}{P}{-}OR + H\mathbf{E} \qquad R'_3\overset{+}{P}{-}OR + \mathbf{Nu}^- \longrightarrow R'_3P{=}O + R{-}\mathbf{Nu}$$

10. H. R. Hudson, *Synthesis* **1969:**112.
11. D. G. Goodwin and H. R. Hudson, *J. Chem. Soc.* B, **1968**:1333; E. J. Coulson, W. Gerrard, and H. R. Hudson, *J. Chem. Soc.* **1965**:2364.
12. J. Cason and J. S. Correia, *J. Org. Chem.* **26**:3645 (1961).
13. H. R. Hudson, *J. Chem. Soc.* **1968**:664.
14. B. P. Castro, *Org. React.* **29:**1 (1983).

A wide variety of species can function as the electrophile $\mathbf{E}^+$ in the general mechanism. The most useful synthetic procedures for preparation of halides are based on the halogens, positive halogen sources, and diethyl azodicarboxylate.

A 1 : 1 adduct formed by triphenylphosphine and bromine converts alcohols to bromides.[15] The alcohol displaces bromide ion from the pentavalent adduct, giving an alkoxyphosphonium intermediate. The phosphonium ion intermediate then undergoes nucleophilic attack by bromide ion, displacing triphenylphosphine oxide.

$$PPh_3 + Br_2 \longrightarrow Br_2PPh_3$$

$$Br_2PPh_3 + ROH \longrightarrow RO\overset{+}{P}Ph_3\ Br^- + HBr$$

$$Br^- + RO\overset{+}{P}Ph_3 \longrightarrow RBr + Ph_3P{=}O$$

Because the alkoxyphosphonium intermediate is formed by a reaction that does not break the C−O bond and the second step proceeds by back-side displacement on carbon, the stereochemistry of the overall process is inversion.

H3C C8H17 H3C HO Br2, Ph3P H3C C8H17 H3C Br Ref. 16

2,4,4,6-Tetrabromocyclohexa-2,5-dienone has been found to be a useful bromine source.

O O C12H25 OH Ph3P Br Br Br O Br O O C12H25 Br Ref. 17

Triphenylphosphine dichloride exhibits similar reactivity and has been used to prepare chlorides.[18] The most convenient methods for converting alcohols to chlorides are based on *in situ* generation of chlorophosphonium ions[19] by reaction of triphenylphosphine with various chlorine compounds such as carbon tetrachloride[20] and hexachloroacetone.[21]

$$Ph_3P + CCl_4 \longrightarrow Ph_3\overset{+}{P}{-}Cl + {}^-CCl_3$$

$$Ph_3P + Cl_3C\overset{O}{\overset{\|}{C}}CCl_3 \longrightarrow Ph_3\overset{+}{P}{-}Cl + {}^-CCl_2\overset{O}{\overset{\|}{C}}CCl_3$$

15. G. A. Wiley, R. L. Hershkowitz, B. M. Rein, and B. C. Chung, *J. Am. Chem. Soc.* **86:**964 (1964).
16. D. Levy and R. Stevenson, *J. Org. Chem.* **30:**2635 (1965).
17. A. Tanaka and T. Oritani, *Tetrahedron Lett.* **38:**1955 (1997).
18. L. Horner, H. Oediger, and H. Hoffmann, *Justus Liebigs Ann. Chem.* **626:**26 (1959).
19. R. Appel, *Angew. Chem. Int. Ed. Engl.* **14:**801 (1975).
20. J. B. Lee and T. J. Nolan, *Can. J. Chem.* **44:**1331 (1966).
21. R. M. Magid, O. S. Fruchey, W. L. Johnson, and T. G. Allen, *J. Org. Chem.* **44:**359 (1979).

The chlorophosphonium ion then reacts with the alcohol to give an alkoxyphosphonium ion, which is converted to the chloride:

$$Ph_3\overset{+}{P}-Cl + ROH \longrightarrow Ph_3\overset{+}{P}-OR + HCl$$

$$Ph_3\overset{+}{P}-OR + Cl^- \longrightarrow Ph_3P{=}O + R-Cl$$

Various modifications of halophosphonium ion-based procedures have been developed. The use of triphenylphosphine and imidazole in combination with iodine or bromine gives good conversion of alcohols to iodides or bromides.[22] An even more reactive system consists of chlorodiphenylphosphine, imidazole, and the halogen.[23] The latter system has the further advantage that the resulting phosphorus by-product, diphenylphosphinic acid, can be extracted with base during product workup.

$$Ph_2PCl + \text{H–N(imidazole)} + I_2 + ROH \longrightarrow RI + Ph_2\overset{O}{\overset{\|}{P}}H$$

A very mild procedure for converting alcohols to iodides uses triphenylphosphine, diethyl azodicarboxylate (DEAD) and methyl iodide.[24] This reaction occurs with clean inversion of stereochemistry.[25] The key intermediate is again an alkoxyphosphonium ion.

$$Ph_3P + ROH + C_2H_5O_2CN{=}NCO_2C_2H_5 \longrightarrow Ph_3\overset{+}{P}OR + C_2H_5O_2C\bar{N}NHCO_2C_2H_5$$

$$C_2H_5O_2C\bar{N}NHCO_2C_2H_5 + CH_3I \longrightarrow C_2H_5O_2CN(CH_3)NHCO_2C_2H_5 + I^-$$

$$Ph_3\overset{+}{P}OR + I^- \longrightarrow RI + Ph_3P{=}O$$

The role of the diethyl azodicarboxylate is to activate the triphenylphosphine toward nucleophilic attack by the alcohol. In the course of the reaction, the N=N double bond is reduced. As will be discussed subsequently, this method is applicable for activation of alcohols to attack by other nucleophiles in addition to halide ions. The activation of alcohols to nucleophilic attack by the triphenylphosphine–diethyl azodicarboxylate combination is called the *Mitsunobu* reaction.

There are a number of other useful methods for converting alcohols to halides. A very mild method which is useful for compounds that are prone to allylic rearrangement involves prior conversion of the alcohol to the mesylate, followed by nucleophilic displacement with halide ion:

$$(CH_3CH_2CH_2)_2C{=}CHCH_2OH \xrightarrow[\text{2) LiCl, DMF}]{\text{1) } CH_3SO_2Cl} (CH_3CH_2CH_2)_2C{=}CHCH_2Cl \quad 83\%$$ Ref. 26

22. P. J. Garegg, R. Johansson, C. Ortega, and B. Samuelsson, *J. Chem. Soc., Perkin Trans. 1* **1982:**681.
23. B. Classon, Z. Liu, and B. Samuelsson, *J. Org. Chem.* **53**: 6126 (1988).
24. O. Mitsunobu, *Synthesis* **1981:**1.
25. H. Loibner and E. Zbiral, *Helv. Chim. Acta* **59:**2100 (1976).
26. E. W. Collington and A. I. Meyers, *J. Org. Chem.* **36:**3044 (1971).

Another very mild procedure involves reaction of the alcohol with the heterocyclic 2-chloro-3-ethylbenzoxazolium cation.[27] The alcohol adds to the electrophilic heterocyclic ring, displacing chloride. The alkoxy group is thereby activated toward nucleophilic substitution, which forms a stable product, 3-ethylbenzoxazolinone.

C_2H_5 C_2H_5 C_2H_5

(benzoxazolium)–Cl + ROH ⟶ (benzoxazolium)–OR + Cl^- ⟶ (benzoxazolinone)=O + RCl

The reaction can be used for making either chlorides or bromides by using the appropriate tetraalkylammonium salt as a halide source.

Scheme 3.1 gives some examples of the various alcohol-to-halide conversions that have been discussed.

3.2. Introduction of Functional Groups by Nucleophilic Substitution at Saturated Carbon

The mechanistic aspects of nucleophilic substitution reactions were treated in detail in Chapter 5 of Part A. That mechanistic understanding has contributed to the development of nucleophilic substitution reactions as importantl synthetic processes. The S_N2 mechanism, because of its predictable stereochemistry and avoidance of carbocation intermediates, is the most desirable substitution process from a synthetic point of view. This section will discuss the role of S_N2 reactions in the preparation of several classes of compounds. First, however, the important role that solvent plays in S_N2 reactions will be reviewed. The knowledgeable manipulation of solvent and related medium effects has led to significant improvement of many synthetic procedures that proceed by the S_N2 mechanism.

3.2.1. General Solvent Effects

The objective in selecting the reaction conditions for a preparative nucleophilic substitution is to enhance the mutual reactivity of the leaving group and nucleophile so that the desired substitution occurs at a convenient rate and with minimal competition from other possible reactions. The generalized order of leaving-group reactivity $RSO_3^- > I^- > Br^- > Cl^-$ pertains for most S_N2 processes. (See Part A, Section 5.6, for more complete data). Mesylates, tosylates, iodides, and bromides are all widely used in synthesis. Chlorides usually react rather slowly, except in especially reactive systems, such as allylic and benzylic compounds. The overall synthetic objective normally governs the choice of the nucleophile. Optimization of reactivity, therefore, must be achieved by choice of the reaction conditions, particularly the solvent. Several generalizations about solvents can be made. Hydrocarbons, halogenated hydrocarbons, and ethers are usually unsuitable solvents for reactions involving metal-ion salts. Acetone and acetonitrile are somewhat more polar, but the solubility of most ionic compounds in these solvents is low. Solubility can be considerably improved by use of salts of cations having substantial nonpolar

27. T. Mukaiyama, S. Shoda, and Y. Watanabe, *Chem. Lett.* **1977**:383; T. Mukaiyama, *Angew. Chem. Int. Ed. Engl.* **18**:707 (1979).

Scheme 3.1. Preparation of Alkyl Halides

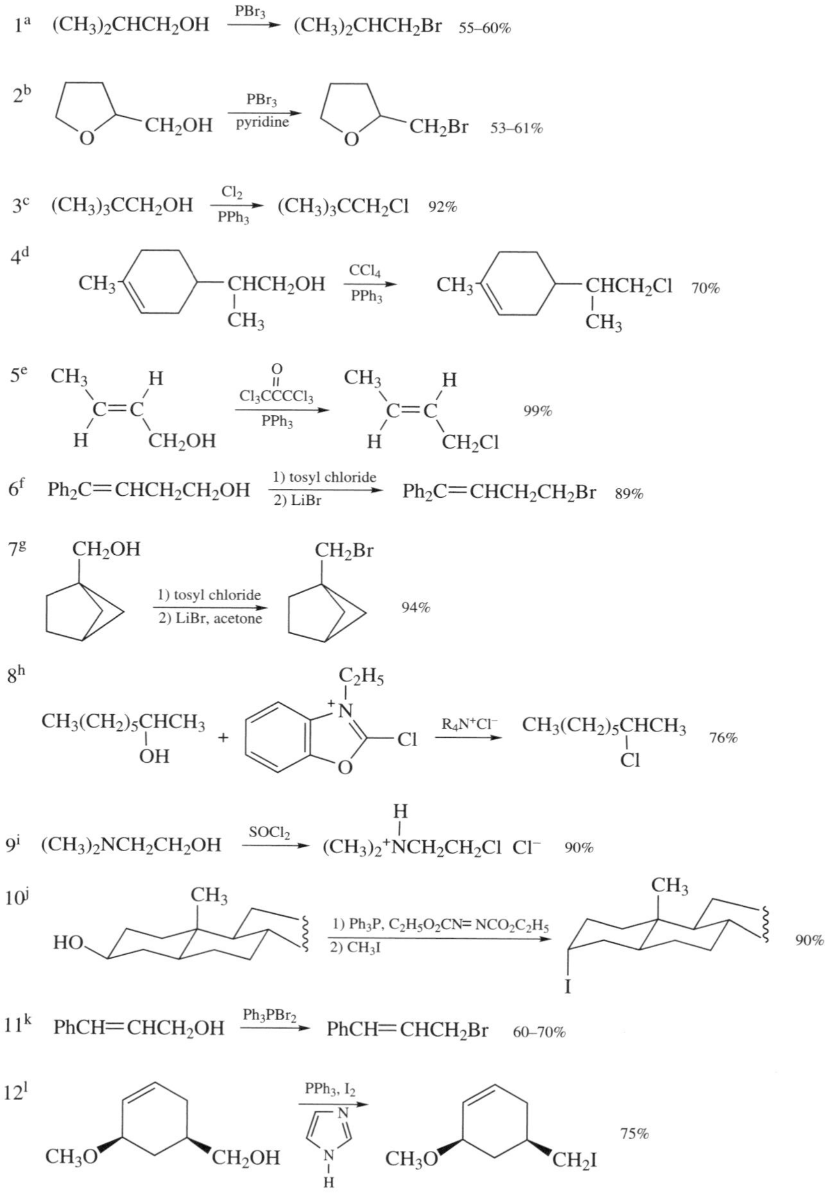

1[a] $(CH_3)_2CHCH_2OH \xrightarrow{PBr_3} (CH_3)_2CHCH_2Br$ 55–60%

2[b] (tetrahydrofuran-2-yl)$CH_2OH \xrightarrow[\text{pyridine}]{PBr_3}$ (tetrahydrofuran-2-yl)CH_2Br 53–61%

3[c] $(CH_3)_3CCH_2OH \xrightarrow[PPh_3]{Cl_2} (CH_3)_3CCH_2Cl$ 92%

4[d] 4-methylcyclohex-3-enyl–$CH(CH_3)CH_2OH \xrightarrow[PPh_3]{CCl_4}$ 4-methylcyclohex-3-enyl–$CH(CH_3)CH_2Cl$ 70%

5[e] $CH_3CH{=}CHCH_2OH$ (E) $\xrightarrow[PPh_3]{Cl_3CC(O)CCl_3}$ $CH_3CH{=}CHCH_2Cl$ (E) 99%

6[f] $Ph_2C{=}CHCH_2CH_2OH \xrightarrow[\text{2) LiBr}]{\text{1) tosyl chloride}} Ph_2C{=}CHCH_2CH_2Br$ 89%

7[g] bicyclic $CH_2OH \xrightarrow[\text{2) LiBr, acetone}]{\text{1) tosyl chloride}}$ bicyclic CH_2Br 94%

8[h] $CH_3(CH_2)_5CH(OH)CH_3$ + 3-ethyl-2-chlorobenzoxazolium (C_2H_5, N^+, Cl) $\xrightarrow{R_4N^+Cl^-} CH_3(CH_2)_5CH(Cl)CH_3$ 76%

9[i] $(CH_3)_2NCH_2CH_2OH \xrightarrow{SOCl_2} (CH_3)_2{}^+NHCH_2CH_2Cl\ Cl^-$ 90%

10[j] HO–steroid (CH_3) $\xrightarrow[\text{2) } CH_3I]{\text{1) } Ph_3P,\ C_2H_5O_2CN{=}NCO_2C_2H_5}$ I–steroid (CH_3) 90%

11[k] $PhCH{=}CHCH_2OH \xrightarrow{Ph_3PBr_2} PhCH{=}CHCH_2Br$ 60–70%

12[l] CH_3O–cyclohexene–$CH_2OH \xrightarrow{PPh_3,\ I_2,\ \text{imidazole}}$ CH_3O–cyclohexene–CH_2I 75%

a. C. R. Noller and R. Dinsmore, *Org. Synth.* **II:**358 (1943).
b. L. H. Smith, *Org. Synth.* **III:**793 (1955).
c. G. A. Wiley, R. L. Hershkowitz, B. M. Rein, and B. C. Chung, *J. Am. Chem. Soc.* **86:**964 (1964).
d. B. D. MacKenzie, M. M. Angelo, and J. Wolinsky, *J. Org. Chem.* **44:**4042 (1979).
e. R. M. Magid, O. S. Fruchy, W. L. Johnson, and T. G. Allen, *J. Org. Chem.* **44:**359 (1979).
f. M. E. H. Howden, A. Maereker, J. Burdon, and J. D. Roberts, *J. Am. Chem. Soc.* **88:**1732 (1966).
g. K. B. Wiberg and B. R. Lowry, *J. Am. Chem. Soc.* **85:**3188 (1963).
h. T. Mukaiyama, S. Shoda, and Y. Watanabe, *Chem. Lett.* **1977:**383.
i. L. A. R. Hall, V. C. Stephens, and J. H. Burckhalter, *Org. Synth.* **IV:**333 (1963).
j. H. Loibner and E. Zbiral, *Helv. Chim. Acta* **59:**2100 (1976).
k. J. P. Schaefer, J. G. Higgins, and P. K. Shenoy, *Org. Synth.* **V:**249 (1973).
l. R. G. Linde II, M. Egbertson, R. S. Coleman, A. B. Jones, and S. J. Danishefsky, *J. Org. Chem.* **55:**2771 (1990).

character, such as those containing tetraalkylammonium ions. Alcohols are reasonably good solvents for salts, but the nucleophilicity of hard anions is relatively low in alcohols because of extensive solvation. The polar aprotic solvents, particularly DMF and DMSO, are good solvents for salts, and, by virtue of selective cation solvation, anions usually show enhanced nucleophilicity in these solvents. The miscibility with water of these solvents and their high boiling points can sometimes cause problems in product separation and purification. HMPA, *N,N*-diethylacetamide, and *N*-methylpyrrolidinone are other examples of useful polar aprotic solvents.[28] In addition to enhancing reactivity, polar aprotic solvents also affect the order of reactivity of nucleophilic anions. In DMF the halides are all of comparable nucleophilicity,[29] whereas in hydroxylic solvents the order is $I^- > Br^- > Cl^-$ and the differences in reactivity are much greater.[30]

In addition to exploiting solvent effects on reactivity, there are two other valuable approaches to enhancing reactivity in nucleophilic substitutions. These are use of *crown ethers* as catalysts and the use of *phase-transfer conditions*. The crown ethers are a family of cyclic polyethers, three examples of which are shown below:

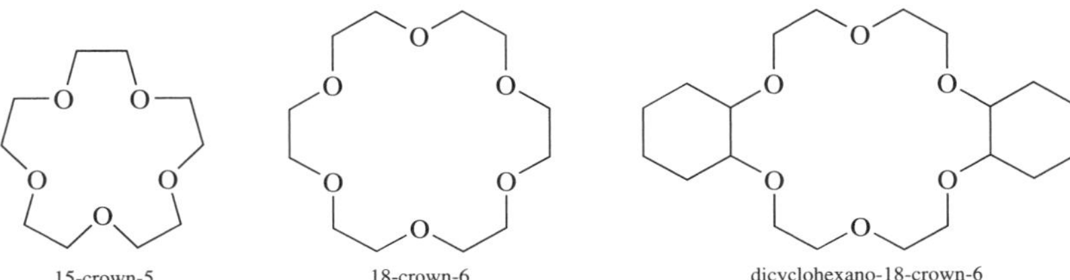

The first number designates the ring size, and the second number, the number of oxygen atoms in the ring. These materials have cation-complexing properties and catalyze nucleophilic substitution under many conditions. By complexing the cation in the cavity of the crown ether, these compounds solubilize many salts in nonpolar solvents. Once in solution, the anions are highly reactive as nucleophiles because they are weakly solvated. Tight ion-pairing is also precluded by the complexation of the cation by the nonpolar crown ether. As a result, nucleophilicity approaches or exceeds that observed in aprotic polar solvents.[31]

The second method for enhancing nucleophilic substitution processes is to use phase-transfer catalysts.[32] The phase-transfer catalysts are ionic substances, usually quaternary ammonium or phosphonium salts, in which the size of the hydrocarbon groups in the cation is large enough to convey good solubility of the salt in organic solvents. In other words, the cation must be highly lipophilic. Phase-transfer catalysis usually is done in a two-phase system. The organic reactant is dissolved in a water-immiscible solvent such as a hydrocarbon or halogenated hydrocarbon. The salt containing the nucleophile is dissolved in water. Even with vigorous mixing, such systems show little tendency to react, because the nucleophile and reactant remain separated in the water and organic

28. A. F. Sowinski and G. M. Whitesides, *J. Org. Chem.* **44:**2369 (1979).
29. W. M. Weaver and J. D. Hutchinson, *J. Am. Chem. Soc.* **86:**261 (1964).
30. R. G. Pearson and J. Songstad, *J. Org. Chem.* **32:**2899 (1967).
31. M. Hiraoka, *Crown Compounds. Their Characteristics and Application*, Elsevier, Amsterdam, 1982.
32. E. V. Dehmlow and S. S. Dehmlow, *Phase Transfer Catalysis*, 3rd ed., Verlag Chemie, Weinheim 1992; W. P. Weber and G. W. Gokel, *Phase Transfer Catalysis in Organic Synthesis*, Springer Verlag, New York, 1977; C. M. Stark, C. Liotta, and M. Halpern, *Phase Transfer Catalysis: Fundamentals, Applications and Industrial Perspective*, Chapman and Hall, New York, 1994.

phases, respectively. When a phase-transfer catalyst is added, the lipophilic cations are transferred to the nonpolar phase and, to maintain electrical neutrality in this phase, anions are transferred from the water to the organic phase. The anions are only weakly solvated in the organic phase and therefore exhibit enhanced nucleophilicity. As a result, the substitution reactions proceed under relatively mild conditions. The salts of the nucleophile are often used in high concentration in the aqueous solution, and in some procedures the solid salt is used.

3.2.2. Nitriles

The replacement of a halide or tosylate ion, extending the carbon chain by one atom and providing an entry to carboxylic acid derivatives, has been a reaction of synthetic importance since the early days of organic chemistry. The classical conditions for preparing nitriles involves heating a halide with a cyanide salt in aqueous alcohol solution:

$$C_6H_5CH_2Cl + NaCN \xrightarrow[\text{reflux 4 h}]{H_2O,\ C_2H_5OH} C_6H_5CH_2CN \quad \text{80–90\%} \qquad \text{Ref. 33}$$

$$ClCH_2CH_2CH_2Br + KCN \xrightarrow[\text{reflux 1.5 h}]{H_2O,\ C_2H_5OH} ClCH_2CH_2CH_2CN \quad \text{40–50\%} \qquad \text{Ref. 34}$$

These reactions proceed more rapidly in aprotic polar solvents. In DMSO, for example, primary alkyl chlorides are converted to nitriles in one hour or less at temperatures of 120–140°C.[35] Phase-transfer catalysis by hexadecyltributylphosphonium bromide permits conversion of 1-chlorooctane to octyl cyanide in 95% yield in 2 h at 105°C.[36]

$$CH_3CH_2CH_2CH_2Cl \xrightarrow[\text{DMSO, 90–160°C}]{NaCN} CH_3CH_2CH_2CH_2CN \quad 93\%$$

$$CH_3(CH_2)_6CH_2Cl \xrightarrow[H_2O,\ \text{decane},\ CH_3(CH_2)_{15}P^+(CH_2CH_2CH_2CH_3)_3,\ 105°C,\ 2\ h]{NaCN} CH_3(CH_2)_6CH_2CN \quad 95\%$$

Catalysis by 18-crown-6 of the reaction of solid potassium cyanide with a variety of chlorides and bromides has been demonstrated.[37] With primary bromides, yields are high and reaction times are 15–30 h at reflux in acetonitrile (83°C). Interestingly, the chlorides are more reactive and require reaction times of only 2 h. Secondary halides react more slowly, and yields drop because of competing elimination. Tertiary halides do not react satisfactorily because elimination processes dominate.

3.2.3. Azides

Azides are useful intermediates for synthesis of various nitrogen-containing compounds. They undergo cycloaddition reactions, as will be discussed in Section 6.2,

33. R. Adams and A. F. Thal, *Org. Synth.* **1**:101 (1932).
34. C. F. H. Allen, *Org. Synth.* **I**:150 (1932).
35. L. Friedman and H. Schecter, *J. Org. Chem.* **25**:877 (1960); R. A. Smiley and C. Arnold, *J. Org. Chem.* **25**:257 (1960).
36. C. M. Starks, *J. Am. Chem. Soc.* **93**:195 (1971); C. M. Starks and R. M. Owens, *J. Am. Chem. Soc.* **95**:3613 (1973).
37. F. L. Cook, C. W. Bowers, and C. L. Liotta, *J. Org. Chem.* **39**:3416 (1974).

and can also be easily reduced to primary amines. Azido groups are usually introduced into aliphatic compounds by nucleophilic substitution.[38] The most reliable procedures involve heating the appropriate halide with sodium azide in DMSO[39] or DMF.[40] Alkyl azides can also be prepared by reaction in high-boiling alcohols[41]:

$$CH_3(CH_2)_3CH_2I + NaN_3 \xrightarrow[H_2O]{CH_3CH_2OCH_2CH_2OCH_2CH_2OH} CH_3(CH_2)_3CH_2N_3 \quad 84\%$$

Phase-transfer conditions have also been used for the preparation of azides[42]:

$$CH_2{=}CH\text{–}CH(CH_3)\text{–}CH_2\text{–}CH(Br)\text{–}CO_2CH_3 \xrightarrow[R_4P^{+}\,^{-}Br,\ 4\ h,\ 25°C]{NaN_3} CH_2{=}CH\text{–}CH(CH_3)\text{–}CH_2\text{–}CH(N_3)\text{–}CO_2CH_3$$

Tetramethylguanidinium azide, an azide salt which is readily soluble in halogenated solvents, is a useful source of azide ions in the preparation of azides from reactive halides such as α-haloketones, α-haloamides, and glycosyl halides.[43]

There are also useful procedures for preparation of azides directly from alcohols. Reaction of alcohols with 2-fluoro-1-methylpyridinium iodide followed by reaction with lithium azide gives good yields of alkyl azides[44]:

$$ROH + \text{2-fluoro-1-methylpyridinium} \longrightarrow \text{2-OR-1-methylpyridinium} \xrightarrow{N_3^-} \text{1-methyl-2-pyridone} + RN_3$$

Diphenylphosphoryl azide reacts with alcohols in the presence of triphenylphosphine and diethyl azodicarboxylate.[45] Hydrazoic acid, HN_3, can also serve as the azide ion source under these conditions.[46] These reactions are examples of the Mitsunobu reaction discussed earlier.

$$ROH + Ph_3P + C_2H_5O_2CN{=}NCO_2C_2H_5 \longrightarrow RO\overset{+}{P}Ph_3 + C_2H_5O_2C\bar{N}NHCO_2C_2H_5$$

$$RO\overset{+}{P}Ph_3 + N_3^- \longrightarrow RN_3 + Ph_3P{=}O$$

38. M. E. C. Biffin, J. Miller and D. B. Paul, in *The Chemistry of the Azido Group*, S. Patai, ed., John Wiley & Sons, New York, 1971, Chapter 2.
39. R. Goutarel, A. Cave, L. Tan, and M. Leboeuf, *Bull. Soc. Chim. Fr.* **1962:**646.
40. E. J. Reist, R. R. Spencer, B. R. Baker, and L. Goodman, *Chem. Ind. (London)* **1962**:1794.
41. E. Lieber, T. S. Chao, and C. N. R. Rao, *J. Org. Chem.* **22:**238 (1957); H. Lehmkuhl, F. Rabet, and K. Hauschild, *Synthesis* **1977:**184.
42. W. P. Reeves and M. L. Bahr, *Synthesis* **1976:**823; B. B. Snider and J. V. Duncia, *J. Org. Chem.* **46:**3223 (1981).
43. Y. Pan, R. L. Merriman, L. R. Tanzer, and P. L. Fuchs, *Biomed. Chem. Lett.* **2:**967 (1992); C. Li, A. Arasappan, and P. L. Fuchs, *Tetrahedron Lett.* **34:**3535 (1993); D. A. Evans, T. C. Britton, J. A. Ellman, and R. L. Dorow, *J. Am. Chem. Soc.* **112:**4011 (1990).
44. K. Hojo, S. Kobayashi, K. Soai, S. Ikeda, and T. Mukaiyama, *Chem. Lett.* **1977:**635.
45. B. Lal, B. N. Pramanik, M. S. Manhas, and A. K. Bose, *Tetrahedron Lett.* **1977:**1977.
46. J. Schweng and E. Zbiral, *Justus Liebigs Ann. Chem.* **1978:**1089; M. S. Hadley, F. D. King, B. McRitchie, D. H. Turner, and E. A. Watts, *J. Med. Chem.* **28:**1843 (1985).

Diphenylphosphoryl azide also gives good conversion of primary alkyl and secondary benzylic alcohols to azides in the presence of the strong organic base diazabicycloundecene (DBU). These reactions proceed by O-phosphorylation followed by S_N2 displacement.[47]

$$\text{ArCH(OH)CH}_2\text{CH}_3 \xrightarrow[\text{DBU}]{(\text{PhO})_2\text{P(O)N}_3} \text{ArCH(N}_3)\text{CH}_2\text{CH}_3$$

This reaction can be extended to secondary alcohols with the more reactive bis(4-nitrophenyl)phosphorazidate.[48]

3.2.4. Oxygen Nucleophiles

The oxygen nucleophiles that are of primary interest in synthesis are the hydroxide ion (or water), alkoxide ions, and carboxylate anions, which lead, respectively, to alcohols, ethers, and esters. Because each of these nucleophiles can also act as a base, reaction conditions must be selected to favor substitution over elimination.

Usually, a given alcohol is more easily obtained than the corresponding halide so the halide-to-alcohol transformations is not extensively used for synthesis. The hydrolysis of benzyl halides to the corresponding alcohols proceeds in good yield. This can be a useful synthetic transformation, because benzyl halides are available either by side-chain halogenation or by the chloromethylation reaction (Section 11.1.3).

$$\text{NC-C}_6\text{H}_4\text{-CH}_2\text{Cl} \xrightarrow[\text{H}_2\text{O, 100°C, 2.5 h}]{\text{K}_2\text{CO}_3} \text{NC-C}_6\text{H}_4\text{-CH}_2\text{OH} \quad 85\%$$ Ref. 49

Ether formation for alkoxides and alkylating reagents is a reaction of wide synthetic importance. The conversion of phenols to methoxyaromatics, for example, is a very common reaction. Methyl iodide, methyl tosylate, or dimethyl sulfate can be used as the alkylating agent. The reaction proceeds in the presence of a weak base, such as Na_2CO_3 or K_2CO_3, which deprotonates the phenol. The conjugate bases of alcohols are considerably more basic than phenoxides, and therefore β elimination can become a problem. Phase-transfer conditions can be used in troublesome cases.[50] Fortunately, the most useful and commonly encountered ethers are methyl and benzyl ethers, where elimination is not a problem and the corresponding halides are especially reactive. Entries 13–16 in Scheme 3.2 provide some typical examples of ether preparations.

Two methods for converting carboxylic acids to esters fall into the mechanistic group under discussion. One of these methods is the reaction of carboxylic acids with diazo compounds, especially diazomethane. The second is alkylation of carboxylate anions by halides or sulfonates. The esterification of carboxylic acids with diazomethane is a very quick and clean reaction.[51] The alkylating agent is the extremely reactive methyldiazonium

47. A. S. Thompson, G. R. Humphrey, A. M. DeMarco, D. J. Mathre, and E. J. J. Grabowski, *J. Org. Chem.* **58**: 5886 (1993).
48. M. Mizuno and T. Shiori, *J. Chem. Soc., Chem. Commun.* **1997**:2165.
49. J. N. Ashley, H. J. Barber, A. J. Ewins, G. Newbery, and A. D. Self, *J. Chem. Soc.* **1942**:103.
50. F. Lopez-Calahorra, B. Ballart, F. Hombrados, and J. Marti, *Synth. Commun.* **28**:795 (1998).
51. T. H. Black, *Aldrichimica* **16**:3 (1983).

ion, which is generated by proton transfer from the carboxylic acid to diazomethane. The collapse of the resulting ion pair with loss of nitrogen is extremely rapid:

$$RCO_2H + CH_2N_2 \longrightarrow [RCO_2^- + CH_3\overset{+}{N}_2] \longrightarrow RCO_2CH_3 + N_2$$

The main drawback to this reaction is the toxicity of diazomethane and some of its precursors. One possible alternative is the use of alkyltriazenes as reactive alkylating agents.[52] Alkyltriazenes are readily prepared from primary amines and aryldiazonium salts.[53] The triazenes, on being protonated by the carboxylic acid, generate a reactive alkylating agent that is equivalent, if not identical, to the alkyldiazonium ions generated from diazoalkanes.

$$RCO_2H + Ar{-}N{=}NNHR' \rightleftharpoons Ar{-}\overset{+}{N}H_2{-}N{=}N{-}R' \;\; {}^-O_2CR \longrightarrow RCO_2R' + ArNH_2 + N_2$$

Especially for large-scale work, esters, may be more safely and efficiently prepared by reaction of carboxylate salts with alkyl halides or tosylates. Carboxylate anions are not very reactive nucleophiles so the best results are obtained in polar aprotic solvents[54] or with crown ether catalysts.[55] The reactivity for the salts is $Na^+ < K^+ < Rb^+ < Cs^+$. Cesium carboxylates are especially useful in polar aprotic solvents. The enhanced reactivity of the cesium salts is due both to high solubility and to the absence of ion pairing with the anion.[56] Acetone has been found to be a good solvent for reaction of carboxylate anions with alkyl iodides.[57] Cesium fluoride in DMF is another useful combination.[58] Carboxylate alkylation procedures have been particularly advantageous for preparation of hindered esters that can be relatively difficult to prepare by the acid-catalyzed esterification method (Fischer esterification) which will be discussed in Section 3.4.2. Sections F and G of Scheme 3.2 give some specific examples of ester synthesis by the reaction of carboxylic acids with diazomethane and by carboxylate alkylation.

In the course of synthesis, it is sometimes necessary to invert the configuration at an oxygen-substituted center. One of the best ways of doing this is to activate a hydroxyl group to substitution by a carboxylate anion. The activation is frequently done using the Mitsunobu reagents.[59] Hydrolysis of the resulting ester gives the alcohol of inverted

52. E. H. White, H. Maskill, D. J. Woodcock, and M. A. Schroeder, *Tetrahedron Lett.* **1969:**1713.
53. E. H. White and H. Scherrer, *Tetrahedron Lett.* **1961:**758.
54. P. E. Pfeffer, T. A. Foglia, P. A. Barr, I. Schmeltz, and L. S. Silbert, *Tetrahedron Lett.* **1972:**4063; J. E. Shaw, D. C. Kunerth, and J. J. Sherry, *Tetrahedron Lett.* **1973:**689; J. Grundy, B. G. James, and G. Pattenden, *Tetrahedron Lett.* **1972:**757.
55. C. L. Liotta, H. P. Harris, M. McDermott, T. Gonzalez, and K. Smith, *Tetrahedron Lett.* **1974:**2417.
56. G. Dijkstra, W. H. Kruizinga, and R. M. Kellog, *J. Org. Chem.* **52:**4230 (1987).
57. G. G. Moore, T. A. Foglia, and T. J. McGahan, *J. Org. Chem.* **44:**2425 (1979).
58. T. Sato, J. Otera, and H. Nozaki, *J. Org. Chem.* **57:**2166 (1992).
59. D. L. Hughes, *Org. React.* **42:**335 (1992); D. L. Hughes, *Org. Prep. Proced. Int.* **28:**127 (1996).

configuration:

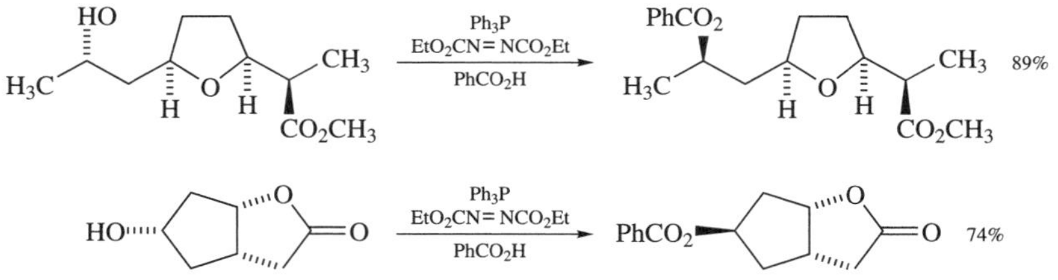

Ph_3P, $EtO_2CN{=}NCO_2Et$, $PhCO_2H$ — 89% Ref. 60

Ph_3P, $EtO_2CN{=}NCO_2Et$, $PhCO_2H$ — 74% Ref. 61

Carboxylate anions derived from somewhat stronger acids, such as *p*-nitrobenzoic acid and chloroacetic acid, seem to be particularly useful in this Mitsunobu inversion reaction.[62]

Sulfonate esters can also be prepared under Mitsunobu conditions. Use of zinc tosylate in place of the carboxylic acid gives a tosylate of inverted configuration:

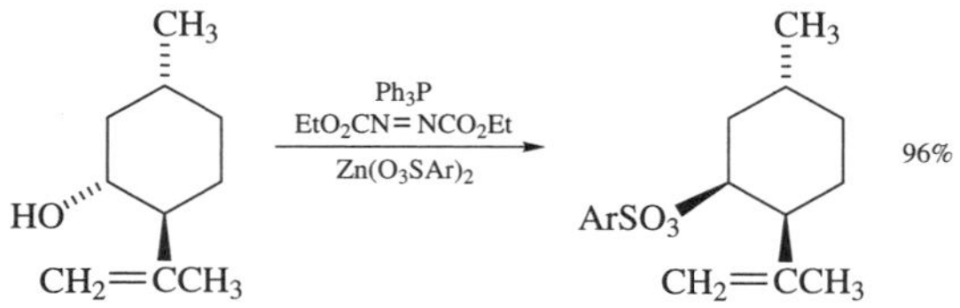

Ph_3P, $EtO_2CN{=}NCO_2Et$, $Zn(O_3SAr)_2$ — 96% Ref. 63

Entry 21 in Scheme 3.2 provides another example.

The Mitsunobu conditions can also be used to effect a variety of other important and useful nucleophilic substitution reactions, such as conversions of alcohols to mixed phosphite esters.[64] The active phosphitylating agent is believed to be a mixed phosphoramidite.

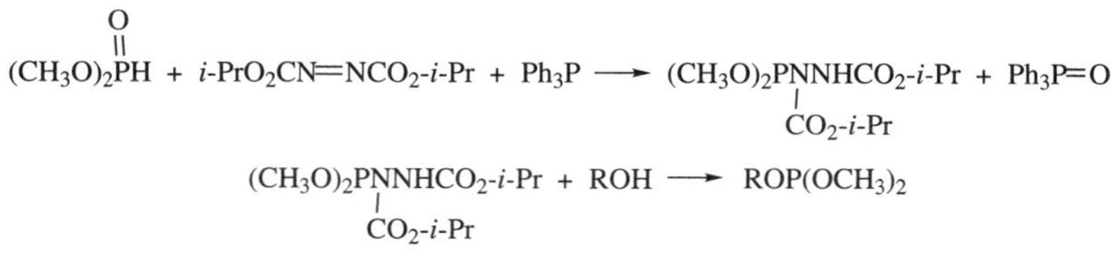

$$(CH_3O)_2P(=O)H + i\text{-}PrO_2CN{=}NCO_2\text{-}i\text{-}Pr + Ph_3P \longrightarrow (CH_3O)_2PN(CO_2\text{-}i\text{-}Pr)NHCO_2\text{-}i\text{-}Pr + Ph_3P{=}O$$

$$(CH_3O)_2PN(CO_2\text{-}i\text{-}Pr)NHCO_2\text{-}i\text{-}Pr + ROH \longrightarrow ROP(OCH_3)_2$$

60. M. J. Arco, M. H. Trammel, and J. D. White, *J. Org. Chem.* **41**:2075 (1976).
61. C.-T. Hsu, N.-Y. Wang, L. H. Latimer, and C. J. Sih, *J. Am. Chem. Soc.* **105**:593 (1983).
62. J. A. Dodge, J. I. Tujillo, and M. Presnell, *J. Org. Chem.* **59**:234 (1994); M. Saiah, M. Bessodes, and K. Antonakis, *Tetrahedron Lett.* **33**:4317 (1992); S. F. Martin and J. A. Dodge, *Tetrahedron Lett.* **32**:3017 (1991).
63. I. Galynker and W. C. Still, *Tetrahedron Lett.* **1982**:4461.
64. I. D. Grice, P. J. Harvey, I. D. Jenkins, M. J. Gallagher, and M. G. Ranasinghe, *Tetrahedron Lett.*, **37**:1087 (1996).

Mixed phosphonate esters can be prepared from alkylphosphonate monoesters, although here the activation is believed to occur at the alcohol.[65]

$$ROH + i\text{-}PrO_2CN{=}NCO_2\text{-}i\text{-}Pr + Ph_3P \longrightarrow ROP^+(Ph)_3$$

$$ROP^+(Ph)_3 + R'P(OCH_3)O_2^- \longrightarrow R'P(=O)(OCH_3)OR + Ph_3P{=}O$$

3.2.5. Nitrogen Nucleophiles

The alkylation of neutral amines by halides is complicated from a synthetic point of view because of the possibility of multiple alkylation which can proceed to the quaternary ammonium salt in the presence of excess alkyl halide:

$$RNH_2 + R'{-}X \longrightarrow R\overset{+}{N}H_2R' + X^-$$

$$R\overset{+}{N}H_2R' + RNH_2 \rightleftharpoons RNHR' + R\overset{+}{N}H_3$$

$$RNHR' + R'{-}X \longrightarrow R\overset{+}{N}HR'_2 + X^-$$

$$R\overset{+}{N}HR'_2 + RNH_2 \rightleftharpoons RNR'_2 + R\overset{+}{N}H_3$$

$$RNR'_2 + R'{-}X \longrightarrow R\overset{+}{N}R'_3 + X^-$$

Even with a limited amount of the alkylating agent, the equilibria between protonated product and the neutral starting amine are sufficiently fast that a mixture of products may be obtained. For this reason, when monoalkylation of amine is desired, the reaction is usually best carried out by *reductive amination*, a reaction which will be discussed in Chapter 5. If complete alkylation to the quaternary salt is desired, use of excess alkylating agent and a base to neutralize the liberated acid normally results in complete reaction.

Amides are only weakly nucleophilic and react very slowly with alkyl halides. The anions of amides are substantially more reactive. The classical Gabriel procedure for synthesis of amines from phthalimide is illustrative.[66]

$$\text{(phthalimide)}N^-K^+ + BrCH_2CH_2Br \longrightarrow \text{(phthalimide)}NCH_2CH_2Br \quad 70\text{–}80\%$$ Ref. 67

The enhanced acidity of the NH group in phthalimide permits formation of an anion which is readily aklylated by alkyl halides or tosylates. The amine can then be liberated by

65. D. A. Campbell, *J. Org. Chem.* **57:**6331 (1992); D. A. Campbell and J. C. Bermak, *J. Org. Chem.* **59:**658 (1994).
66. M. S. Gibson and R. W. Bradshaw, *Angew. Chem. Int. Ed. Engl.* **7:**919 (1968).
67. P. L. Salzberg and J. V. Supniewski, *Org. Synth.* **I:**119 (1932).

reaction of the substituted phthalimide with hydrazine:

$$CH_3O_2CCH(Br)CH_2CH(Br)CO_2CH_3 \longrightarrow CH_3O_2CCH(\text{phthal})CH_2CH(\text{phthal})CO_2CH_3 \xrightarrow[CH_3OH]{NH_2NH_2} \xrightarrow{HCl} HO_2CCH(NH_2)CH_2CH(NH_2)CO_2H$$

phthal = phthalimido

Ref. 68

Secondary amides can be alkylated on nitrogen by using sodium hydride for proton abstraction, followed by reaction with an alkyl halide[69]:

NH NaH, benzene / CH_3I NCH_3

Neutral tertiary and secondary amides react with very reactive alkylating agents, such as triethyloxonium tetrafluoroborate, to give O-alkylation.[70] The same reaction occurs, but more slowly, with tosylates and dimethyl sulfate. Neutralization of the resulting salt provides iminoethers:

$$RC(=O)NHR' \xrightarrow[2)\ ^-OH]{1)\ (CH_3O)_2SO_2} RC(OCH_3)=NR'$$

Sulfonamides are relatively acidic, and their anions can serve as nucleophiles.[71] Sulfonamido groups can be introduced at benzylic positions with a high level of inversion under Mitsunobu conditions.[72]

OH, OCH_2Ph, CH_3, OCH_2Ph, OCH_3 $\xrightarrow[TsNHCH_2CH(OCH_3)_2]{Ph_3P,\ C_2H_5O_2CN{=}NCO_2C_2H_5}$ $TsNHCH_2CH(OCH_3)_2$, OCH_2Ph, CH_3, OCH_2Ph, OCH_3

68. J. C. Sheehan and W. A. Bolhofer, *J. Am. Chem. Soc.* **72**:2786 (1950).
69. W. S. Fones, *J. Org. Chem.* **14**:1099 (1949); R. M. Moriarty, *J. Org. Chem.* **29**:2748 (1964).
70. L. Weintraub, S. R. Oles, and N. Kalish, *J. Org. Chem.* **33**:1679 (1968); H. Meerwein, E. Battenberg, H. Gold, E. Pfeil, and G. Willfang, *J. Prakt. Chem.* **154**:83 (1939).
71. T. Doornbos and J. Strating, *Org. Prep. Proced.* **2**:101 (1970).
72. T. S. Kaufman, *Tetrahedron Lett.* **37**:5329 (1996); D. Papaioannou, C. Athanassopoulos, V. Magafa, N. Karamanos, G. Stavropoulos, A. Napoli, G. Sindona, D. W. Aksnes, and G. W. Francis, *Acta Chem. Scand.* **48**:324 (1994).

The Mitsunobu conditions can be used for alkylation of 2-pyridones, as in the course of synthesis of analogs of the antitumor agent camptothecin.

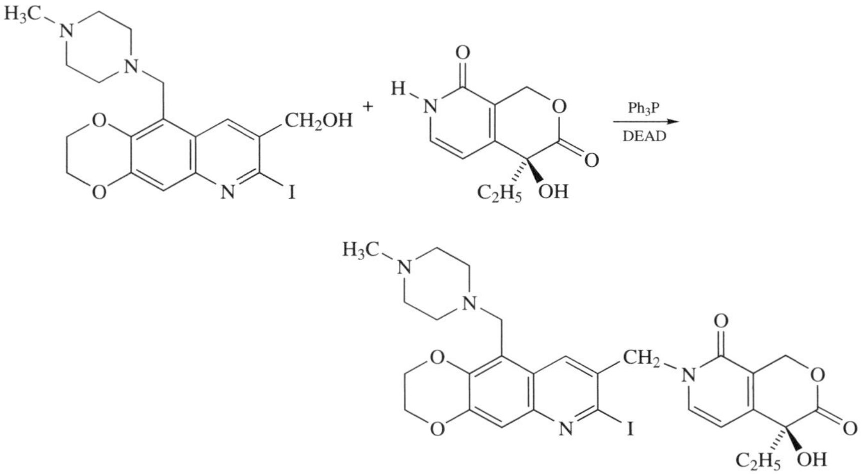

Ref. 73

Proline analogs can be obtained by cylization of δ-hydroxyalkylamino acid carbamates.

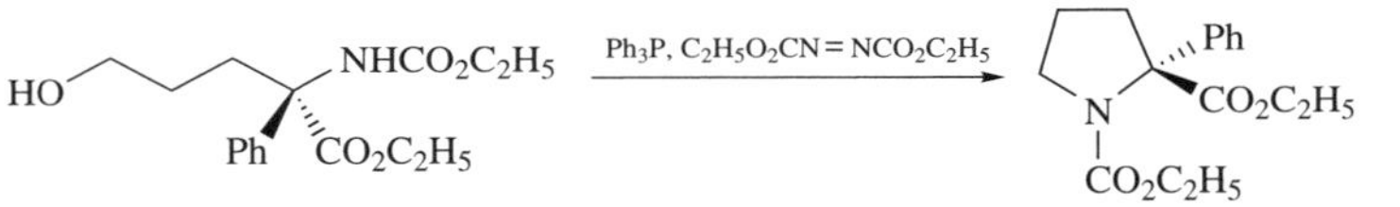

Ref. 74

Mitsunobu conditions are found effective for glycosylation of weak nitrogen nucleophiles, such as indoles.

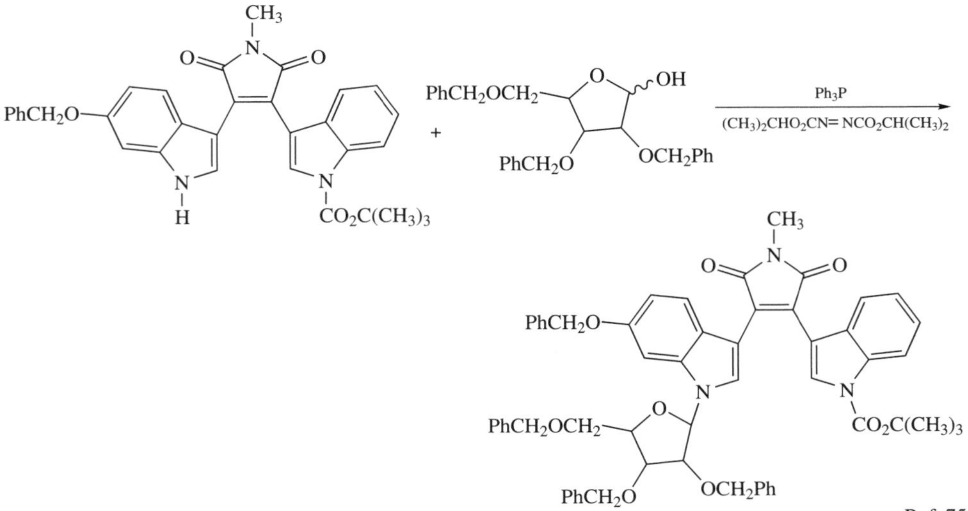

Ref. 75

73. F. G. Fang, D. D. Bankston, E. M. Huie, M. R. Johnson, M.-C. Kang, C. S. LeHoullier, G. C. Lewis, T. C. Lovelace, M. W. Lowery, D. L. McDougald, C. A. Meerholz, J. J. Partridge, M. J. Sharp, and S. Xie, *Tetrahedron* **53:**10953 (1997).
74. J. van Betsbrugge, D. Tourwe, B. Kaptein, H. Kierkels, and R. Broxterman, *Tetrahedron* **53:**9233 (1997).
75. M. Ohkubo, T. Nishimura, H. Jona, T. Honma, S. Ito, and H. Morishima, *Tetrahedron* **53:**5937 (1997).

3.2.6. Sulfur Nucleophiles

Anions derived from thiols are very nucleophilic and can easily be alkylated by halides.

$$CH_3S^-Na^+ + ClCH_2CH_2OH \xrightarrow{C_2H_5OH} CH_3SCH_2CH_2OH \quad 75\text{–}80\%$$ Ref. 76

Neutral sulfur compounds are also good nucleophiles. Sulfides and thioamides readily form salts with methyl iodide, for example:

$$(CH_3)_2S + CH_3I \xrightarrow[12\text{–}16\text{ h}]{25°C} (CH_3)_3\overset{+}{S}\ I^-$$ Ref. 77

$$\text{1-methylpyrrolidine-2-thione} + CH_3I \xrightarrow[12\text{ h}]{25°C} \text{1-methyl-2-(methylthio)-3,4-dihydro-2H-pyrrolium } (\overset{+}{N}\text{–}CH_3,\ SCH_3)$$ Ref. 78

Even sulfoxides, where nucleophilicity is decreased by the additional oxygen, can be alkylated by methyl iodide. These sulfoxinum salts have useful synthetic applications, as discussed in Section 2.6.

$$(CH_3)_2S{=}O + CH_3I \xrightarrow[72\text{ h}]{25°C} (CH_3)_3\overset{+}{S}{=}O\ I^-$$ Ref. 79

3.2.7 Phosphorus Nucleophiles

Both neutral and anionic phosphorus compounds are good nucleophiles toward alkyl halides. Examples of these reactions were already encountered in Chapter 2 in connection with the preparation of the valuable phosphorane and phosphonate intermediates used for Wittig reactions:

$$Ph_3P + CH_3Br \xrightarrow[\text{2 days}]{\text{room temp}} Ph_3\overset{+}{P}CH_3\ Br^-$$ Ref. 80

$$[(CH_3)_2CHO]_3P + CH_3I \longrightarrow [(CH_3)_2CHO]_2\overset{O}{\overset{\|}{P}}CH_3 + (CH_3)_2CHI$$ Ref. 81

The reaction with phosphite esters is known as the *Michaelis–Arbuzov reaction* and proceeds through an unstable trialkoxyphosphonium intermediate. The second stage in the reaction is another example of the great tendency of alkoxyphosphonium ions to react with nucleophiles to break the O–C bond, resulting in formation of phosphoryl P=O bond.

76. W. Windus and P. R. Shildneck, *Org. Synth.* **II:**345 (1943).
77. E. J. Corey and M. Chaykovsky, *J. Am. Chem. Soc.* **87:**1353 (1965).
78. R. Gompper and W. Elser, *Org. Synth.* **V:**780 (1973).
79. R. Kuhn and H. Trischmann, *Justus Liebigs Ann. Chem.* **611:**117 (1958).
80. G. Wittig and U. Schoellkopf, *Org. Synth.* **V:**75‘1 (1973).
81. A. H. Ford-Moore and B. J. Perry, *Org. Synth.* **IV:**325 (1963).

3.2.8. Summary of Nucleophilic Substitution at Saturated Carbon

In the preceding sections, some of the nucleophilic substitution reactions at sp^3 carbon which are most valuable for synthesis have been outlined. These reactions all fit into the general mechanistic patterns that were discussed in Chapter 5 of Part A. The order of reactivity of alkylating groups is benzyl ~ allyl > methyl > primary > secondary. Tertiary halides and sulfonates are generally not satisfactory because of the preference for ionization processes over S_N2 substitution. Because of their high reactivity toward nucleophilic substitution, α-haloesters, α-haloketones, and α-halonitriles are usually favorable reactants for substitution reactions. The reactivity of leaving groups is sulfonate > iodide > bromide > chloride. Steric hindrance greatly decreases the rate of nucleophilic substitution. Thus, projected synthetic steps involving nucleophilic substitution must be evaluated for potential steric problems. Scheme 3.2 gives some representative examples of nucleophlic substitution process drawn from *Organic Synthesis* and from recent synthetic efforts.

3.3. Nucleophilic Cleavage of Carbon–Oxygen Bonds in Ethers and Esters

The cleavage of carbon–oxygen bonds in ethers or esters by nucleophilic substitution is frequently a useful synthetic transformation

$$R{-}O{-}CH_3 + \mathbf{Nu^-} \longrightarrow RO^- + CH_3{-}\mathbf{Nu}$$

$$RC(=O){-}O{-}CH_3 + \mathbf{Nu^-} \longrightarrow RCO_2^- + CH_3{-}\mathbf{Nu}$$

The classical ether cleavage conditions involving concentrated hydrogen halides are much too strenuous for most polyfunctional molecules, so several milder reagents have been developed.[82] These reagents include boron tribromide,[83] dimethylboron bromide,[84] trimethyl iodide,[85] and boron trifluoride in the presence of thiols.[86] The mechanism for ether cleavage with boron tribromide involves attack of bromide ion on an adduct formed from the ether and the electrophilic boron reagent. The cleavage step can occur by either an S_N2 or S_N1 process, depending on the nature of the alkyl group.

$$R{-}O{-}R + BBr_3 \longrightarrow R{-}\overset{+}{O}(^-BBr_3){-}R \rightleftharpoons R{-}\overset{+}{O}(BBr_2){-}R + Br^-$$

$$R{-}\overset{+}{O}(BBr_2){-}R \;\; ^-Br \longrightarrow R{-}O{-}BBr_2 + RBr$$

$$R{-}O{-}BBr_2 + 3\,H_2O \longrightarrow ROH + B(OH)_3 + 2\,HBr$$

82. M. V. Bhatt and S. U. Kulkarni, *Synthesis* **1983:**249.
83. J. F. W. McOmie, M. L. Watts, and D. E. West, *Tetrahedron* **24:**2289 (1968).
84. Y. Guindon, M. Therien, Y, Girard, and C. Yoakim, *J. Org. Chem.* **52:**1680 (1987).
85. M. E. Jung and M. A. Lyster, *J. Org. Chem.* **42:**3761 (1977).
86. M. Node, H. Hori, and E. Fujita, *J. Chem. Soc., Perkin Trans.* 12237 (1976); K. Fuji, K. Ichikawa, M. Node, and E. Fujita, *J. Org. Chem.* **44:**1661 (1979).

Scheme 3.2. Transformation of Functional Groups by Nucleophilic Substitution

A. Nitriles

1[a] CH_3CHCH_2OH (on hexahydroanthracene) → 1) CH_3SO_2Cl, pyridine; 2) NaCN, DMF, 40–60°C, 3 h → CH_3CHCH_2CN 85%

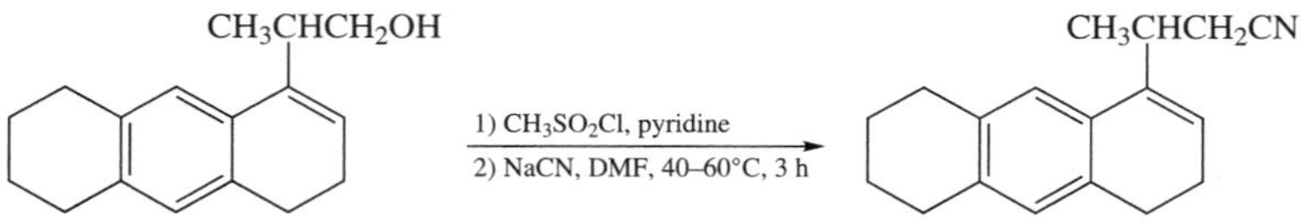

2[b] CH_3 / $CHCH_2OH$, CH_3 → 1) $ArSO_2Cl$; 2) NaCN, DMSO, 90°C, 5 h → CH_3 / $CHCH_2CN$, CH_3 80%

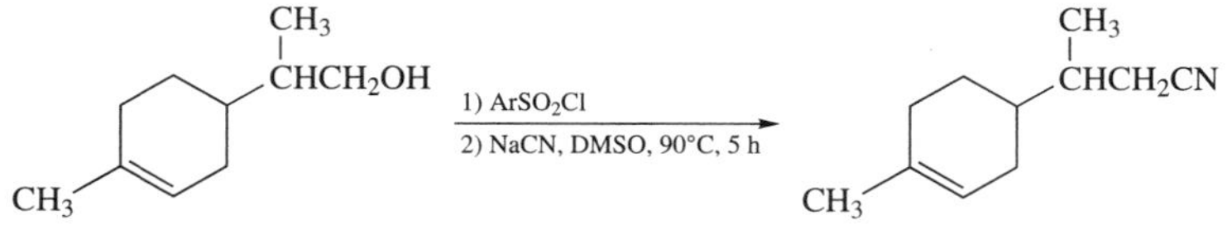

3[c] CH_2OH, CH_2OH → 1) $ArSO_2Cl$; 2) NaCN, DMSO → CH_2CN, CH_2CN

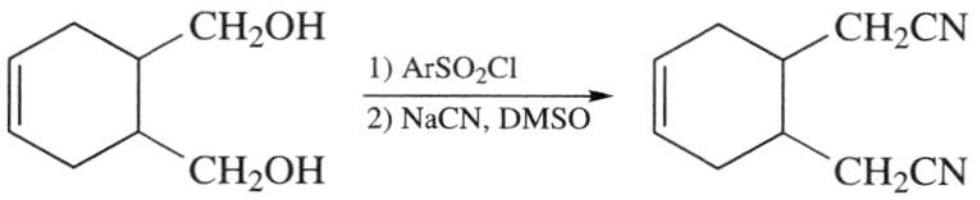

B. Azides

4[d] $CH_3CH_2CH_2CH_2Br + NaN_3 \xrightarrow[H_2O,\ 100°C,\ 6\ h]{R_4N^+Cl^-} CH_3CH_2CH_2CH_2N_3$ 97%

5[e] OH, CH_3 → 1) CH_3SO_2Cl, $(C_2H_5)_3N$; 2) NaN_3, HMPA → N_3, CH_3 57%

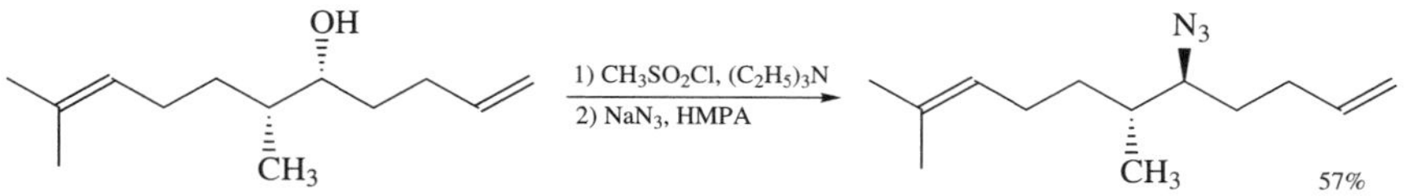

6[f] CH_3, N, OH, H → Ph_3P, $C_2H_5O_2CN{=}NCO_2C_2H_5$, $(PhO)_2P(O)N_3$ → CH_3, N, N_3, H 60%

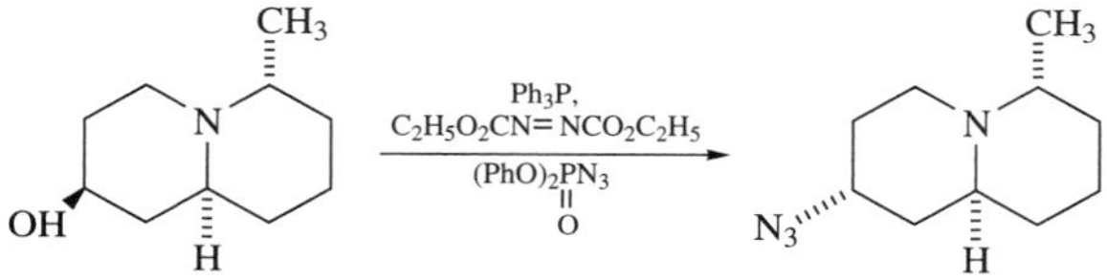

7[g] OH, O → $(PhO)_2P(O)N_3$, DBU → N_3, O 90%

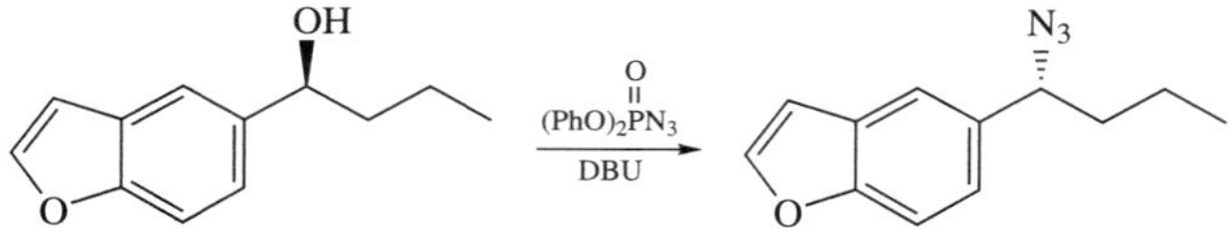

C. Amines and amides

8[h] pyrrolidine NH + $CH_3CHCO_2C_2H_5$ (Br) → $NCHCO_2C_2H_5$ (CH_3) 80–90%

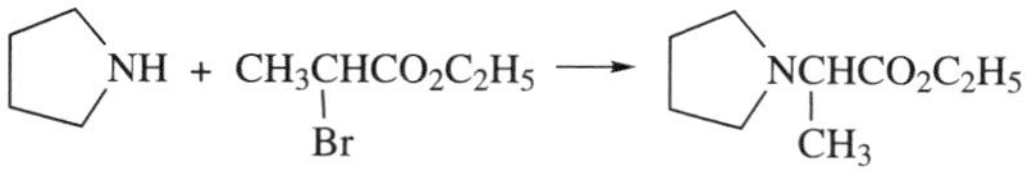

9[i] HN $\overset{+}{N}H_2$ (piperazinium) → $PhCH_2Cl$ → ^-OH → $PhCH_2N$ NH 65–75%

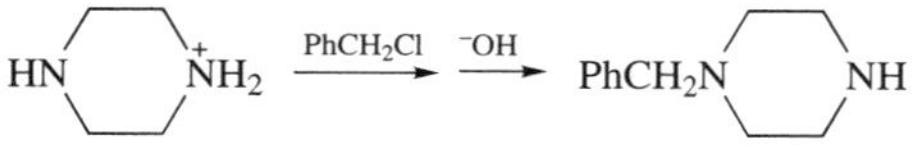

10[j] NH, =O (caprolactam) → $(CH_3O)_2SO_2$, benzene → K_2CO_3 → N, OCH_3 60–70%

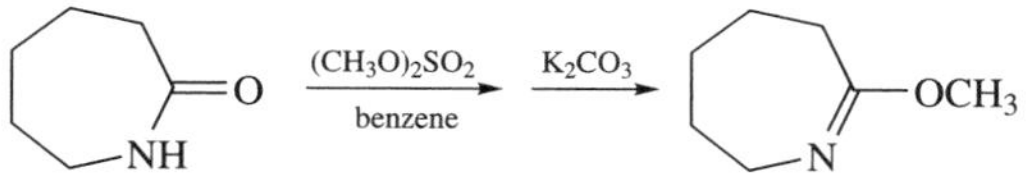

D. Hydrolysis by alkyl halides

11[k] $CH_3C_6H_4C(=O)CH(Cl)C_6H_5 \xrightarrow[4\ h,\ 25°C]{NaOH,\ H_2O} CH_3C_6H_4C(=O)CH(OH)C_6H_5$ 92%

12[l] $(CH_3O)_2C_6H_3CH(Br)CH(Br)CO_2CH_3 \xrightarrow[10\ min]{H_2O,\ 100°C} (CH_3O)_2C_6H_3CH(OH)CH(Br)CO_2CH_3$ 92%

E. Ethers by base-catalyzed alkylation

13[m] HO-substituted bis(isopropylidene) furanose $\xrightarrow{NaH,\ TMF,\ DMSO,\ PhCH_2Cl,\ heat,\ 3\ h}$ $PhCH_2O$-substituted product 95%

14[n] o-nitrophenol (OH, NO_2) + $CH_3CH_2CH_2CH_2Br \xrightarrow{K_2CO_3}$ o-$NO_2C_6H_4OCH_2CH_2CH_2CH_3$ 75–80%

15[o] $CH_3OC_6H_4CH_2Cl + HOCH_2CH_2C_6H_4NO_2 \xrightarrow[CH_2Cl_2]{Bu_4N^+HSO_4^-,\ 50\%\ aq.\ NaOH}$ $CH_3OC_6H_4CH_2OCH_2CH_2C_6H_4NO_2$ 88%

16[p] 2,4-dihydroxyacetophenone (COCH$_3$, OH, HO) $\xrightarrow[K_2CO_3]{CH_3I}$ 2-hydroxy-4-methoxyacetophenone (COCH$_3$, OH, CH_3O) 55–65%

F. Esterification by diazoalkanes

17[q] $c\text{-}C_3H_5CH_2CO_2H + CH_2N_2 \longrightarrow c\text{-}C_3H_5CH_2CO_2CH_3$ 79%

G. Esterification by nucleophilic substitution with carboxylate salts

18[r] $(CH_3)_3CCO_2^- + BrCH_2C(=O)C_6H_4Br \xrightarrow{18\text{-}crown\text{-}6} (CH_3)_3CCO_2CH_2C(=O)C_6H_4Br$ 95%

19[s] $2,4,6\text{-}(CH_3)_3C_6H_2CO_2^- + CH_3CH(I)(CH_2)_5CH_3 \xrightarrow[56°C]{acetone} 2,4,6\text{-}(CH_3)_3C_6H_2CO_2CH(CH_3)(CH_2)_5CH_3$ 100%

Scheme 3.2. *(continued)*

20[t] (acetonide-protected furanose carboxylic acid, CO_2H) $\xrightarrow[\text{18 h}]{CH_3I,\ KF,\ DMF,\ 25^\circ C}$ (corresponding methyl ester, CO_2CH_3) 84%

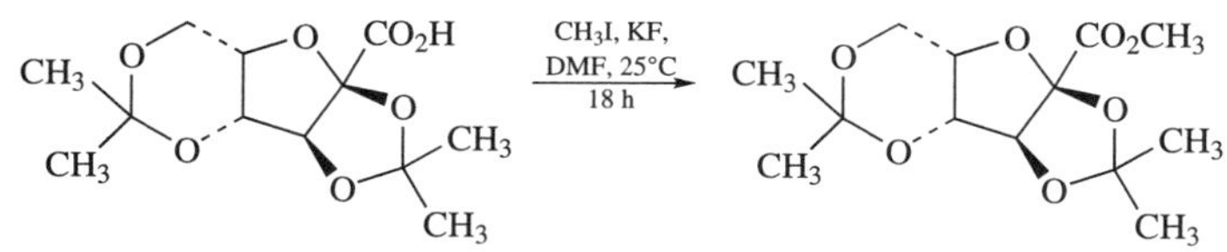

H. Sulfonate esters

21[u] N-benzoyl-4-hydroxyproline methyl ester (HO, CO_2CH_3, N–$C(=O)Ph$) $\xrightarrow[p\text{-toluenesulfonic acid, } (C_2H_5)_3N]{PPh_3,\ i\text{-}Pr\text{-}O_2CN{=}NCO_2\text{-}i\text{-}Pr}$ ($ArSO_3$, CO_2CH_3, N–$C(=O)Ph$) $Ar = p\text{-}CH_3C_6H_5$

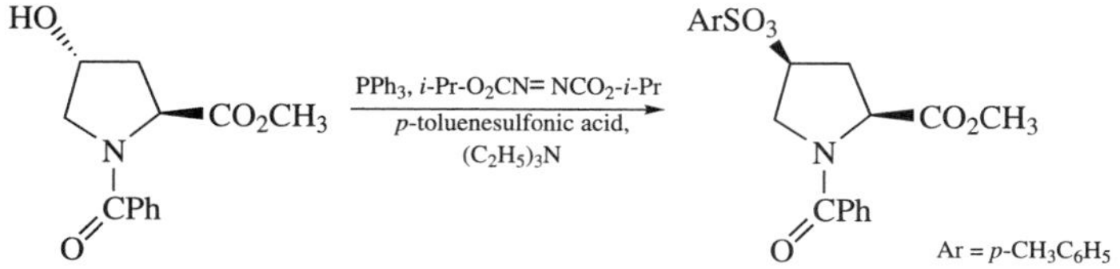

I. Phosphorus nucleophiles

22[v] $Ph_3P + BrCH_2CH_2OPh \longrightarrow Ph_3\overset{+}{P}CH_2CH_2OPh\ Br^-$

23[w] $[(CH_3)_2CHO]_3P + CH_3I \longrightarrow [(CH_3)_2CHO]_2\overset{O}{\overset{\|}{P}}CH_3 + (CH_3)_2CHI$ 85–90%

J. Sulfur nucleophiles

24[x] $CH_3(CH_2)_{10}CH_2Br + S{=}C(NH_2)_2 \longrightarrow \xrightarrow[H_2O]{NaOH} CH_3(CH_2)_{10}CH_2SH$ 80%

25[y] $Na^+\ ^-SCH_2CH_2S^-\ Na^+ + BrCH_2CH_2Br \longrightarrow$ 1,4-dithiane (S, S) 55–60%

26[z] 1-methylpyrrolidine-2-thione (N–CH_3, $=S$) $\xrightarrow[2)\ (CH_3)_3CO^-K^+]{1)\ CH_3I}$ 1-methyl-2-(methylthio)pyrroline (N–CH_3, SCH_3) 62%

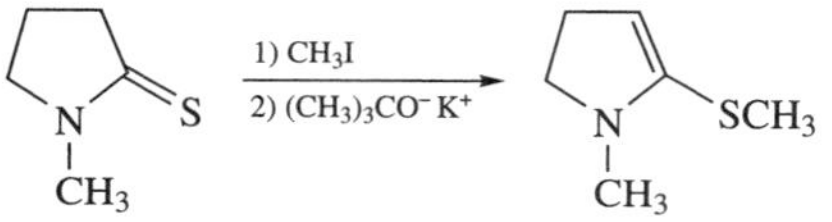

a. M. S. Newman and S. Otsuka, *J. Org. Chem.* **23**:797 (1958).
b. B. A. Pawson, H.-C. Cheung, S. Gurbaxani, and G. Saucy, *J. Am. Chem. Soc.* **92**:336 (1970).
c. J. J. Bloomfield and P. V. Fennessey, *Tetrahedron Lett.* **1964**:2273.
d. W. P. Reeves and M. L. Bahr, *Synthesis* **1976**:823.
e. D. F. Taber, M. Rahimizadeh, and K. K. You, *J. Org. Chem.* **60**:529 (1995).
f. M. S. Hadley, F. D. King, B. McRitchie, D. H. Turner, and E. A. Watts, *J. Med. Chem.* **28**:1843 (1985).
g. A. S. Thompson, G. G. Humphrey, A. M. De Marco, D. J. Mathre, and E. J. J. Grabowski, *J. Org. Chem.* **58**:5886 (1993).
h. R. B. Moffett, *Org. Synth.* **IV**:466 (1963).
i. J. C. Craig and R. J. Young, *Org. Synth.* **V**:88 (1973).
j. R. E. Benson and T. L. Cairns, *Org. Synth.* **IV**:588 (1963).
k. R. N. McDonald and P. A. Schwab, *J. Am. Chem. Soc.* **85**:4004 (1963).
l. E. Adler and K. J. Bjorkquist, *Acta Chem. Scand.* **5**:241 (1951).
m. C. H. Heathcock, C. T. White, J. J. Morrison, and D. VanDerveer, *J. Org. Chem.* **46**:1296 (1981).
n. E. S. West and R. F. Holden, *Org. Synth.* **III**:800 (1955).
o. F. Lopez-Calahorra, B. Ballart, F. Hombrados, and J. Marti, *Synth. Commun.* **28**:795 (1998).
p. G. N. Vyas and M. N. Shah, *Org. Synth.* **IV**:836 (1963).
q. L. I. Smity and S. McKenzie, Jr., *Org. Chem.* **15**:74 (1950); A. I. Vogel, *Practical Organic Chemistry*, third edition, Wiley (1956), p. 973.
r. H. D. Durst, *Tetrahedron Lett.*, 2421 (1974).
s. G. G. Moore, T. A. Foglia, and T. J. McGahan, *J. Org. Chem.* **44**:2425 (1979).
t. C. H. Heathcock, C. T. White, J. Morrison, and D. VanDerveer, *J. Org. Chem.* **46**:1296 (1981).
u. N. G. Anderson, D. A. Lust, K. A. Colapret, J. H. Simpson, M. F. Malley, and J. Z. Glougoutas, *J. Org. Chem.* **61**:7955 (1996).
v. E. E. Schweizer and R. D. Bach, *Org. Synth.* **V**:1145 (1973).
w. A. H. Ford-Moore and B. J. Perry, *Org. Synth.* **IV**:325 (1963).
x. G. G. Urquhart, J. W. Gates, Jr., and R. Conor, *Org. Synth.* **III**:363 (1965).
y. R. G. Gillis and A. B. Lacey, *Org. Synth.* **IV**:396 (1963).
z. R. Gompper and W. Elser, *Org. Synth.* **V**:780 (1973).

Good yields are generally observed, especially for methyl ethers. The combination of boron tribromide with dimethyl sulfide has been found to be particularly effective for cleaving aryl methyl ethers.[87] Trimethylsilyl iodide cleaves methyl ethers in a period of a few hours at room temperature.[85] Benzyl and *t*-butyl systems are cleaved very rapidly, whereas secondary systems require longer times. The reaction presumably proceeds via an initially formed silyl oxonium ion:

$$\mathrm{R{-}O{-}R} + \mathrm{(CH_3)_3SiI} \longrightarrow \mathrm{R{-}\overset{+}{O}(Si(CH_3)_3){-}R} + \mathrm{I^-} \longrightarrow \mathrm{R{-}O{-}Si(CH_3)_3} + \mathrm{RI}$$

The direction of cleavage in unsymmetrical ethers is determined by the relative ease of O–R bond breaking by either S_N2 (methyl, benzyl) or S_N1 (*t*-butyl) processes. Because trimethylsilyl iodide is rather expensive, alternative procedures that generate the reagent *in situ* have been devised:

$$\mathrm{(CH_3)_3SiCl} + \mathrm{NaI} \xrightarrow{\mathrm{CH_3CN}} \mathrm{(CH_3)_3SiI} + \mathrm{NaCl} \qquad \text{Ref. 88}$$

$$\mathrm{PhSi(CH_3)_3} + \mathrm{I_2} \longrightarrow \mathrm{(CH_3)_3SiI} + \mathrm{PhI} \qquad \text{Ref. 89}$$

Diiodosilane, SiH_2I_2, is an especially effective reagent for cleaving secondary alkyl ethers.[90]

Trimethylsilyl iodide also effects rapid cleavage of esters. The first products formed are trimethylsilyl esters, but these are hydrolyzed rapidly on exposure to water.[91]

$$\mathrm{R\overset{O}{\overset{\|}{C}}O{-}R'} + \mathrm{(CH_3)_3SiI} \longrightarrow \mathrm{R\overset{^+OSi(CH_3)_3}{\overset{\|}{C}}O{-}R'} + \mathrm{I^-} \longrightarrow \mathrm{R\overset{O}{\overset{\|}{C}}OSi(CH_3)_3} + \mathrm{R'I}$$

$$\mathrm{R\overset{O}{\overset{\|}{C}}OSi(CH_3)_3} + \mathrm{H_2O} \longrightarrow \mathrm{RCO_2H} + \mathrm{(CH_3)_3SiOH}$$

Benzyl, methyl, and *t*-butyl esters are rapidly cleaved, but secondary esters react more slowly. In the case of *t*-butyl esters, the initial silylation is followed by a rapid ionization to the *t*-butyl cation.

The boron trifluoride–alkylthiol combination (Entries 6–8, Scheme 3.3) also operates on the basis of nucleophilic attack on an oxonium ion generated by reaction of the ether

87. P. G. Williard and C. R. Fryhle, *Tetrahedron Lett.* **21:**3731 (1980).
88. T. Morita, Y. Okamoto, and H. Sakurai, *J. Chem. Soc., Chem. Commun.* **1978:**874; G. A. Olah, S. C. Narang, B. G. B. Gupta, and R. Malhotra, *Synthesis* **1979:**61.
89. T. L. Ho and G. A. Olah, *Synthesis* **1977:**417; A. Benkeser, E. C. Mozdzen, and C. L. Muth, *J. Org. Chem.* **44:**2185 (1979).
90. E. Keinan and D. Perez, *J. Org. Chem.* **52:**4846 (1987).
91. T. L. Ho and G. A. Olah, *Angew. Chem. Int. Ed. Engl.* **15:**774 (1976); M. E. Jung and M. A. Lyster, *J. Am. Chem. Soc.* **99:**968 (1977).

with boron trifluoride[92]:

$$\mathrm{R{-}O{-}R + BF_3 \longrightarrow R{-}\overset{+}{O}(^{-}BF_3){-}R}$$

$$\mathrm{R{-}\overset{+}{O}(^{-}BF_3){-}R + R'SH \longrightarrow RO\bar{B}F_3 + RSR' + H^+}$$

Ether cleavage can also be effected by reaction with acetic anhydride and Lewis acids such as BF_3, $FeCl_3$, and $MgBr_2$.[93] Mechanistic investigations have pointed to acylium ions generated from the anhydride and Lewis acid as the reactive electrophile:

$$\mathrm{(RCO)_2O + MX_n \longrightarrow RC{\equiv}\overset{+}{O} + [MX_nO_2CR]^-}$$

$$\mathrm{RC{\equiv}\overset{+}{O} + R'{-}O{-}R' \longrightarrow R'{-}\overset{+}{O}(R{-}C{=}O){-}R'}$$

$$\mathrm{R'{-}\overset{+}{O}(R{-}C{=}O){-}R' + X^- \longrightarrow R'{-}X + RCO_2R'}$$

Scheme 3.3 gives some specific examples of ether and ester cleavage reactions.

3.4. Interconversion of Carboxylic Acid Derivatives

The classes of compounds which are conveniently considered together as derivatives of carboxylic acids include the carboxylic acid anhydrides, acyl chlorides, esters, and amides. In the case of simple aliphatic and aromatic acids, synthetic transformations among these derivatives are usually a straightforward matter involving such fundamental reactions as ester saponification, formation of acyl chlorides, and the reactions of amines with acid anhydrides or acyl chlorides:

$$\mathrm{RCO_2CH_3 \xrightarrow[H_2O]{^-OH} RCO_2^- + CH_3OH}$$

$$\mathrm{RCO_2H + SOCl_2 \longrightarrow RCOCl + HCl + SO_2}$$

$$\mathrm{RCOCl + R'_2NH \longrightarrow RCONR'_2 + HCl}$$

When a multistep synthesis is being undertaken with other sensitive functional groups present in the molecule, milder reagents and reaction conditions may be necessary. As a result, many alternative methods for effecting intereconversion of the carboxylic acid derivatives have been developed, and some of the most useful reactions will be considered in the succeeding sections.

92. K. Fuji, K. Ichikawa, M. Node, and E. Fujita, *J. Org. Chem.* **44:**1661 (1979).
93. C. R. Narayanan and K. N. Iyer, *J. Org. Chem.* **30:**1734 (1965); B. Ganem and V. R. Small, Jr., *J. Org. Chem.* **39:**3728 (1974); D. J. Goldsmith, E. Kennedy, and R. G. Campbell, *J. Org. Chem.* **40:**3571 (1975).

1[a] CH_3O OCH_3 → BBr_3 → H_2O → HO OH 75–85%

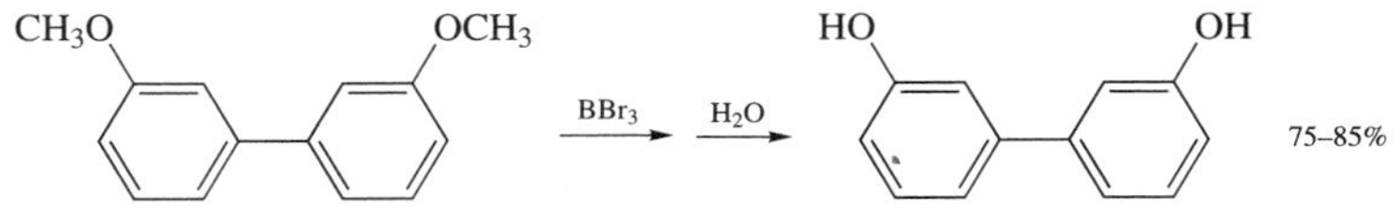

2[b] $CH_2{=}CH$ CH_2OCH_3 $CH_3O_2CCH_2$ H → BBr_3, CH_2Cl_2, −78°C → $CH{=}CH_2$ O O H 88%

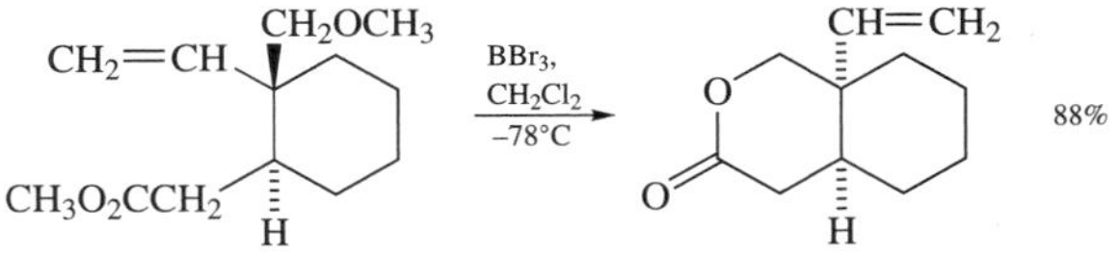

3[c] OCH_3 → $(CH_3)_3SiI$ → OH 83–89%

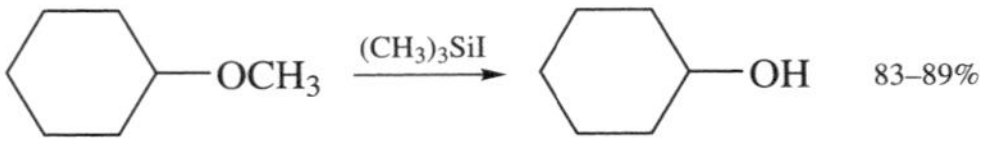

4[d] CO_2CH_3 → $(CH_3)_3SiCl$, NaI, CH_3CN → $CO_2Si(CH_3)_3$ + CH_3I 86%

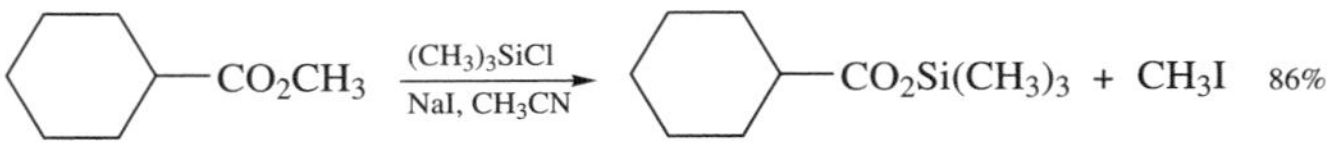

5[e] OCH_3 OCH_3 CH_3 → $(CH_3)_3SiI$ → H_2O → OCH_3 OH CH_3

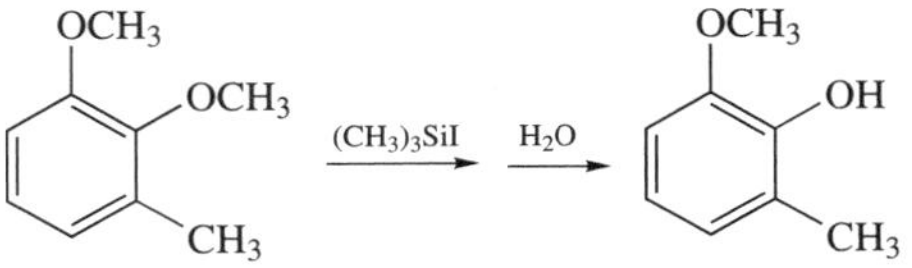

6[f] CH_3 O−CH_2 Br → BF_3, C_2H_5SH → CH_3 OH Br 90%

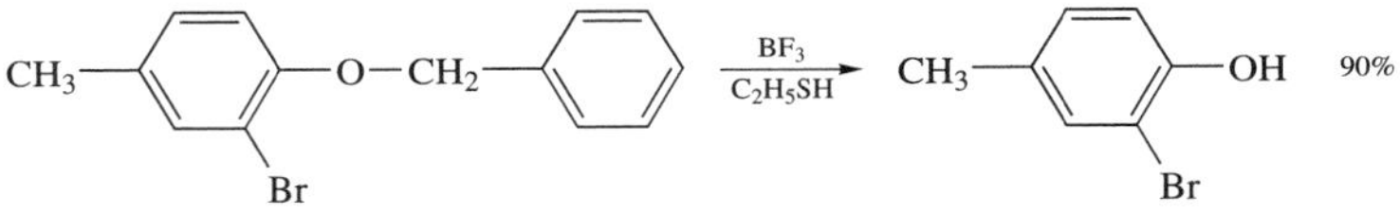

7[g] OCH_3 H_3C H H_3C H CH_3 → BF_3, C_2H_5SH → OH H_3C H H_3C H CH_3 75%

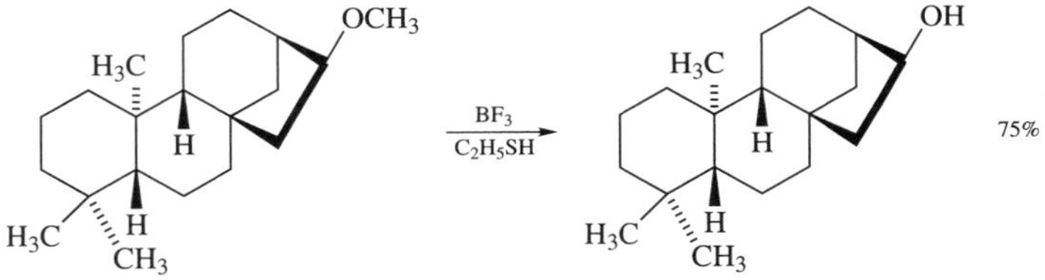

8[h] OH CH_3 CH_3 CH_3 $PhCH_2O$ H_2C → $BF_3{\cdot}OEt_2$, EtSH, NaOAc → OH CH_3 CH_3 CH_3 HO H_2C 61%

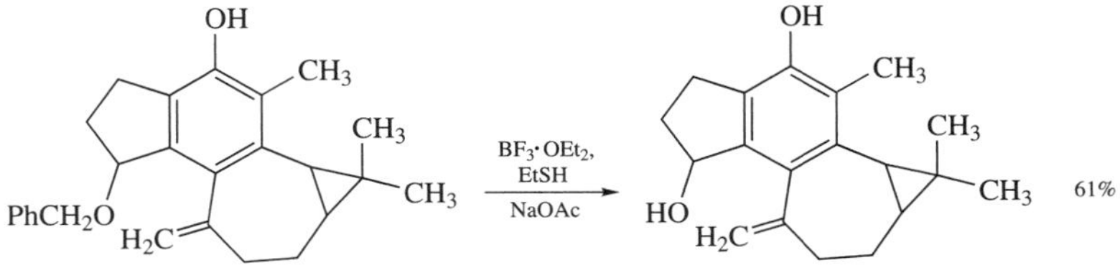

9[i] O → $(CH_3)_2BBr$ → Br OH 85%

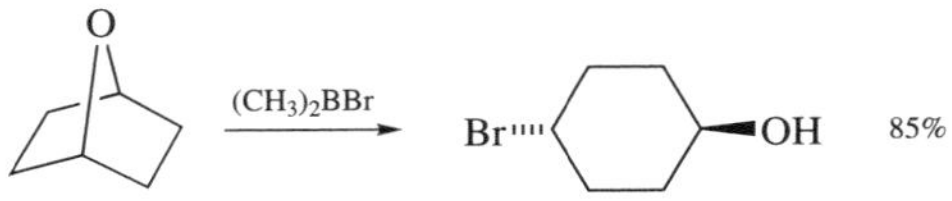

Scheme 3.3. *(continued)*

10[j] $(CH_3)_2CHOCH(CH_3)_2 \xrightarrow[(CH_3CO)_2O]{FeCl_3} (CH_3)_2CHCO_2CH_3$ 83%

a. J. F. W. McOmie and D. E. West, *Org. Synth.* **V**:412 (1973).
b. P. A. Grieco, K. Hiroi, J. J. Reap, and J. A. Noguez, *J. Org. Chem.* **40**:1450 (1975).
c. M. E. Jung and M. A. Lyster, *Org. Synth.* **59**:35 (1980).
d. T. Morita, Y. Okamoto, and H. Sakurai, *J. Chem. Soc., Chem. Commun.* **1978**:874.
e. E. H. Vickery, L. F. Pahler, and E. J. Eisenbraun, *J. Org. Chem.* **44**:4444 (1979).
f. K. Fuji, K. Ichikawa, M. Node, and E. Fujita, *J. Org. Chem.* **44**:1661 (1979).
g. M. Nobe, H. Hori, and E. Fujita, *J. Chem. Soc., Perkin Trans 1* **1976**:2237.
h. A. B. Smith III, N. J. Liverton, N. J. Hrib, H. Sivaramakrishnan, and K. Winzenberg, *J. Am. Chem. Soc.* **108**:3040 (1986).
i. Y. Guindon, M. Therien, Y. Girard, and C. Yoakim, *J. Org. Chem.* **52**:1680 (1987).
j. B. Ganem and V. R. Small, Jr., *J. Org. Chem.* **39**:3728 (1974).

3.4.1. Preparation of Reactive Reagents for Acylation

The traditional method for transforming carboxylic acids into reactive acylating agents capable of converting alcohols to esters or amines to amides is by formation of the acyl chloride. Molecules devoid of acid-sensitive functional groups can be converted to acyl chlorides with thionyl chloride or phosphorus pentachloride. When milder conditions are necessary, the reaction of the acid or its sodium salt with oxalyl chloride provides the acyl chloride. When a salt is used, the reaction solution remains essentially neutral.

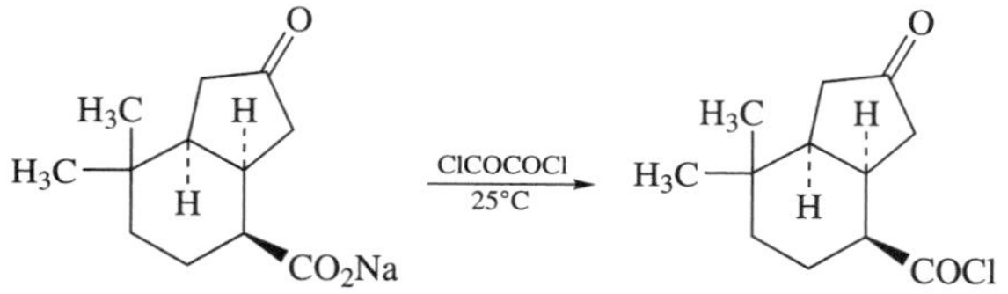

H3C, H3C, H, H, O, CO2Na — ClCOCOCl, 25°C → H3C, H3C, H, H, O, COCl   Ref. 94

Acyl chlorides are highly reactive acylating agents and react very rapidly with amines. For alcohols, preparative procedures often call for use of pyridine as a catalyst. Pyridine catalysis involves initial formation of an acylpyridinium ion, which then reacts with the alcohol. Pyridine is a better nucleophile than the neutral alcohol, but the acylpyridinium ion reacts more rapidly with the alcohol than the acyl chloride.[95]

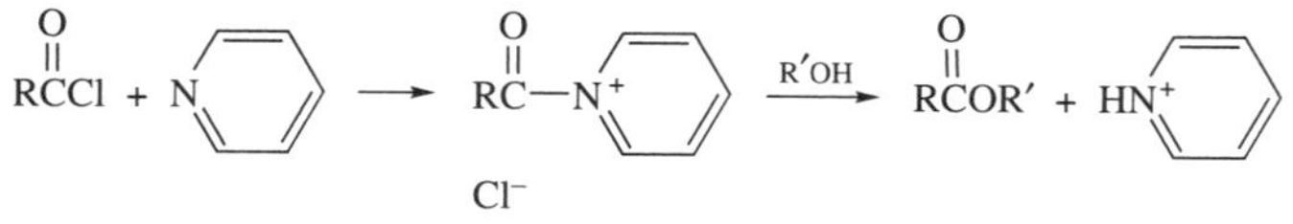

$RCCl(=O) + C_5H_5N \longrightarrow RC(=O)\text{—}N^+C_5H_5\ Cl^- \xrightarrow{R'OH} RCOR'(=O) + HN^+C_5H_5$

An even stronger catalytic effect is obtained when 4-dimethylaminopyridine (DMAP) is used as a nucleophilic catalyst.[96] The dimethylamino group acts as an electron-donor

94. M. Miyano and C. R. Dorn, *J. Org. Chem.* **37**:268 (1972).
95. A. R. Fersht and W. P. Jencks, *J. Am. Chem. Soc.* **92**:5432, 5442 (1970).
96. G. Hofle, W. Steglich, and H. Vorbruggen, *Angew. Chem. Int. Ed. Engl.* **17**:569 (1978); E. F. V. Scriven, *Chem. Soc. Rev.* **12**:129 (1983).

substituent, increasing both the nucleophilicity and the basicity of the pyridine nitrogen.

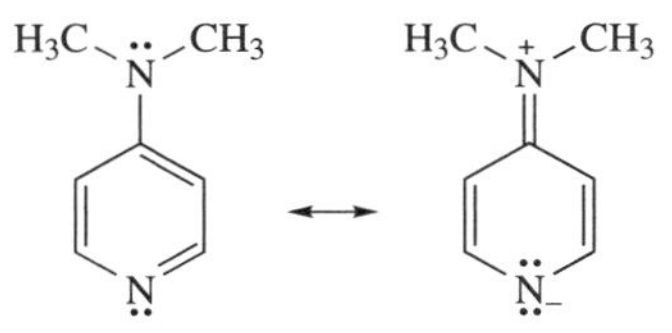

The inclusion of DMAP to the extent of 5–20 mol% in acylations by acid anhydrides and acyl chlorides increases acylation rates by up to four orders of magnitude and permits successful acylation of tertiary and other hindered alcohols.

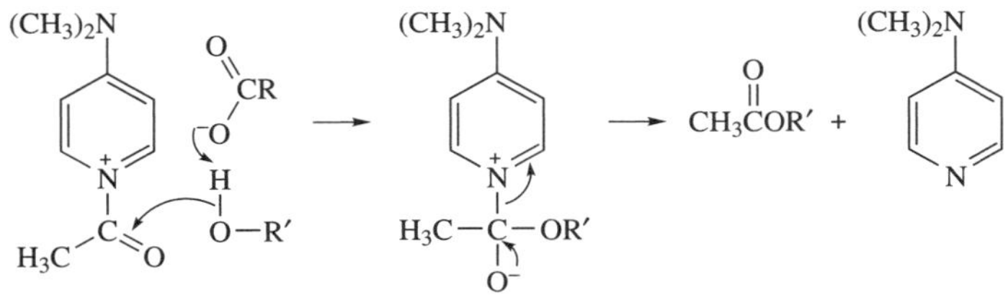

The reagent combination of an acid anhydride with $MgBr_2$ and a hindered tertiary amine, for example, $(i\text{-Pr})_2N(C_2H_5)$ or 1,2,2,6,6,-pentamethylpiperidine, gives an even more reactive acylation system which is useful for hindered and sensitive alcohols.[97] Another efficient catalyst for acylation is $Sc(O_3SCF_3)_3$. It can be used in combination with anhydrides[98] and other reactive acylating agents[99] and is a mild reagent for acylation of tertiary alcohols. Other lanthanide triflates have similar catalytic effects. $Yb(O_3SCF_3)_3$ and $Lu(O_3SCF_3)_3$, for example, were used in selective acylation of 10-deacetylbaccatin III, an important intermediate for preparation of the antitumor agent paclitaxel (taxol).[100]

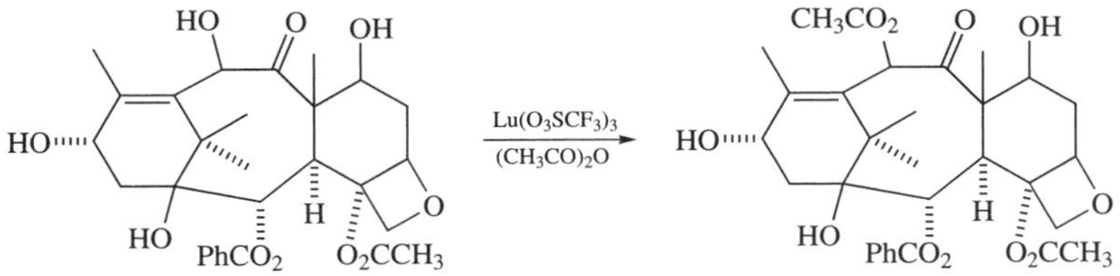

Trimethylsilyl triflate is also a powerful catalyst for acylations by anhydrides. Reactions of alcohols with a modest excess (1.5 equiv) of anhydride proceed in inert solvents at 0°C. Even tertiary alcohols react rapidly.[101] The active acylation reagent is

97. E. Vedejs and O. Daugulis, *J. Org. Chem.* **61**:5702 (1996).
98. K. Ishihara, M, Kubota, H., Kurihara, and H. Yamamoto, *J. Org. Chem.* **61**:4560 (1996); A. G. M. Barrett and D. C. Braddock, *J. Chem. Soc., Chem. Commun.* **1997**:351.
99. H. Zhao, A. Pendri, and R. B. Greenwald, *J. Org. Chem.* **63**:7559 (1998).
100. E. W. P. Damen, L. Braamer, and H. W. Scheeren, *Tetrahedron Lett.* **39**:6081 (1998).
101. P. A. Procopiou, S. P. D. Baugh, S. S. Flack, and G. G. A. Inglis, *J. Org. Chem.* **63**:2342 (1998).

presumably generated by O-silylation of the anhydride.

$$\text{1-methylcyclohexanol} \xrightarrow[\substack{(CH_3)_3SiO_3SCF_3 \\ 5\ \text{mol}\ \%}]{\substack{(CH_3CO)_2O \\ 5\ \text{equiv}}} \text{1-methylcyclohexyl acetate } (O_2CCH_3)$$

Tri-*n*-butylphosphine is also an effective catalyst for acylations by anhydrides. It is thought to act as a nucleophilic catalyst by generating an acylphosphonium ion.[102]

$$Bu_3P + RC(=O)OC(=O)R \longrightarrow Bu_3P^+{-}C(=O)R + RCO_2^-$$

There are other activation procedures which generate acyl halides *in situ* in the presence of the nucleophile. Refluxing a carboxylic acid, triphenylphosphine, bromotrichloromethane, and an amine gives rise to the corresponding amide[103].

$$RCO_2H + R'NH_2 \xrightarrow{PPh_3,\ CBrCl_3} RC(=O)NHR'$$

This reaction presumably proceeds via the acyl chloride, because it is known that triphenylphosphine and carbon tetrachloride convert acids to the corresponding acyl chloride.[104] Similarly, carboxylic acids react with the triphenylphosphine–bromine adduct to give acyl bromides.[105] Triphenylphosphine/*N*-bromosuccinimide also generates acyl bromides *in situ*.[106] Alcohols can be esterified by heating in excess ethyl formate or ethyl acetate and triphenylphosphine in carbon tetrabromide.[107] All these reactions are mechanistically analogous to the alcohol-to-halide conversions that were discussed in Section 3.1.2.

$$RCO_2H + Ph_3\overset{+}{P}Br \longrightarrow RC(=O){-}O{-}\overset{+}{P}Ph_3 + HBr$$

$$Br^- + RC(=O){-}O{-}\overset{+}{P}Ph_3 \longrightarrow RC(=O)Br + Ph_3P{=}O$$

In addition to acyl chlorides and acyl bromides, there are a number of milder and more selective acylating agents which can readily prepared from carboxylic acids. Imidazolides, the *N*-acyl derivatives of imidazole, are examples.[108] Imidazolides are isolable substances and can be prepared directly from the carboxylic acid by reaction with

102. E. Vedejs, N. S. Bennett, L. M. Conn, S. T. Diver, M. Gingras, S. Lin, P. A. Oliver, and M. J. Peterson, *J. Org. Chem.* **58**:7286 (1993); E Vedejs and S. T. Diver, *J. Am. Chem. Soc.* **115**:3358 (1993).
103. L. E. Barstow and V. J. Hruby, *J. Org. Chem.* **36**:1305 (1971).
104. J. B. Lee, *J. Am. Chem. Soc.* **88**:3440 (1966).
105. H. J. Bestmann and L. Mott, *Justus Liebigs Ann. Chem.* **693**:132 (1966).
106. K. Sucheta, G. S. R. Reddy, D. Ravi, and N. Rama Rao, *Tetrahedron Lett.* **35**:4415 (1994).
107. H. Hagiwara, K. Morohashi, H. Sakai, T. Suzuki, and M. Ando, *Tetrahedron* **54**:5845 (1998).
108. H. A. Staab and W. Rohr, *Newer Methods Prep. Org. Chem.* **5**:61 (1968).

$$RCO_2H + \text{N}\!\!\underset{}{\overset{}{\text{(imidazolyl)}}}\text{N–C(=O)–N(imidazolyl)N} \longrightarrow RC(=O)\text{–N(imidazolyl)N} + \text{HN(imidazolyl)N} + CO_2$$

Two factors are responsible for the high reactivity of the imidazolides as acylating reagents. One is the relative weakness of the "amide" bond. Because of the aromatic character of imidazole, there is little of the N $\rightarrow$ C=O delocalization that stabilizes normal amides. The reactivity of the imidazolides is also enhanced by protonation of the other imidazole nitrogen, which makes the imidazole ring a better leaving group.

$$\mathbf{Nu} + RC(=O)\text{–N(imidazolyl)N} \xrightarrow{H^+} \mathbf{Nu}\text{–}C(=O)R + \text{N(imidazolyl)NH}$$

Imidazolides can also be activated by N-alkylation with methyl triflate.[109] Imidazolides react with alcohols on heating to give esters and react at room temperature with amines to give amides. Imidazolides are particularly appropriate for acylation of acid-sensitive materials.

Dicyclohexylcarbodiimide (DCC) is another example of a reagent which converts carboxylic acids to reactive acylating agents. This compound has been particularly widely applied in the acylation step in the synthesis of polypeptides from amino acids.[110] (See also Section 13.6). The reactive species is an *O*-acyl isourea. The acyl group is highly reactive in this environment because the cleavage of the acyl–oxygen bond converts the carbon–nitrogen double bond of the isourea to a more stable carbonyl group.[111]

$$RCO_2H + RN{=}C{=}NR \longrightarrow RC(=O)\text{–O–}C(=NR)NHR$$

$$RC(=O)\text{–O–}C(=NR)NHR \;(\mathbf{Nu}^-\text{ attack}) \xrightarrow{H^+} RC(=O)\mathbf{Nu} + RNHC(=O)NHR$$

The combination of carboxyl activation by DCC and catalysis by DMAP provides a useful method for *in situ* activation of carboxylic acids for reaction with alcohols. The reaction proceeds at room temperature[112]:

$$Ph_2CHCO_2H + C_2H_5OH \xrightarrow[\text{DMAP}]{\text{DCC}} Ph_2CHCO_2C_2H_5$$

2-Chloropyridinium[113] and 3-chloroisoxazolium[114] cations also activate carbonxyl groups toward nucleophilic attack. In each instance, the halide is displaced from the

109. G. Ulibarri, N. Choret, and D. C. H. Bigg, *Synthesis* **1996:**1286.
110. F. Kurzer and K. Douraghi-Zadeh, *Chem. Rev.* **67:**107 (1967).
111. D. F. DeTar and R. Silverstein, *J. Am. Chem. Soc.* **88:**1013, 1020 (1966); D. F. DeTar, R. Silverstein, and F. F. Rogers, Jr., *J. Am. Chem. Soc.* **88:**1024 (1966).
112. A. Hassner and V. Alexanian, *Tetrahedron Lett.* **1978:**4475; B. Neises and W. Steglich, *Angew. Chem. Int. Ed. Engl.* **17:**522 (1978).
113. T. Mukaiyama, M. Usui, E. Shimada, and K. Saigo, *Chem. Lett.* **1975:**1045.
114. K. Tomita, S. Sugai, T. Kobayashi, and T. Murakami, *Chem. Pharm. Bull.* **27:**2398 (1979).

heterocycle by the carboxylate via an addition–elimination mechanism. Nucleophilic attack on the activated carbonyl group results in elimination of the heterocyclic ring with the departing oxygen being converted to an amide-like structure. The positive charge on the heterocylic ring accelerates both the initial addition step and subsequent elimination of the heterocycle.

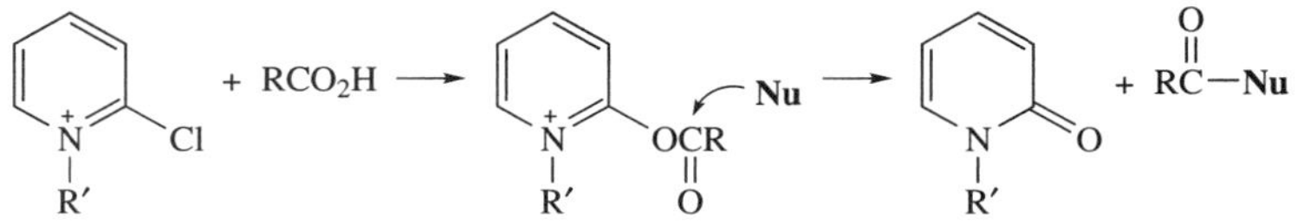

Carboxylic acid esters of thiols are considerably more reactive as acylating reagents than are the esters of alcohols. Particularly reactive are esters of pyridine-2-thiol because there is an additional driving force—the formation of the more stable pyridine-2-thione tuatomer:

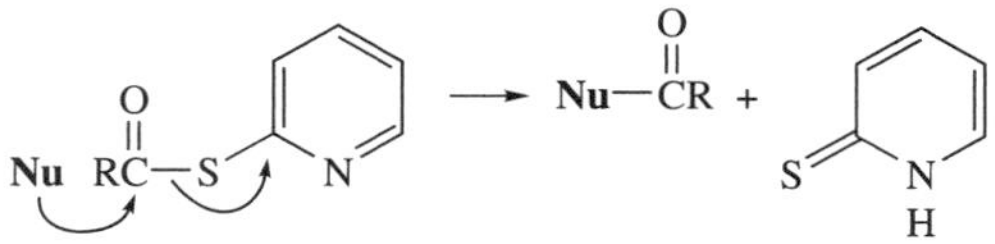

Additional acceleration of the rate of acylation can be obtained by inclusion of cupric salts that coordinate at the pyridine nitrogen. This modification is especially useful for preparation of highly hindered esters.[115] Pyridine-2-thiol esters can be prepared by reaction of the carboxylic acid with 2,2′-dipyridyl disulfide and triphenylphosphine[116] or directly from the acid and 2-pyridyl thiochloroformate.[117]

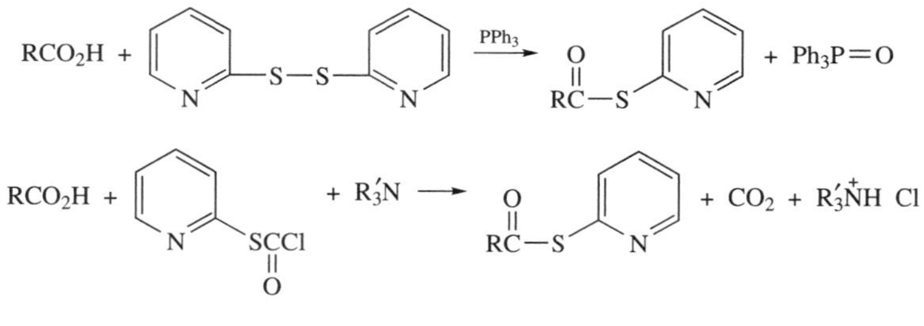

The 2-pyridyl and related 2-imidazolyl disulfides have found special use in the closure of large lactone rings.[118] This type of structure is encountered in a number of antibiotics which, because of the presence of numerous other sensitive functional groups, require mild conditions for cyclization. It has been suggested that the pyridyl and imidazoyl thioesters function by a mechanism in which the heterocyclic nitrogen acts as

115. S. Kim and J. I. Lee, *J. Org. Chem.* **49:**1712 (1984).
116. T. Mukaiyama, R. Matsueda, and M. Suzuki, *Tetrahedron Lett.* **1970:**1901.
117. E. J. Corey and D. A. Clark, *Tetrahedron Lett.* **1979:**2875.
118. E. J. Corey and K. C. Nicolaou, *J. Am. Chem. Soc.*, **96:**5614 (1974); K. C. Nicolaou, *Tetrahedron* **33:**683 (1977).

a base, deprotonating the alcohol group:

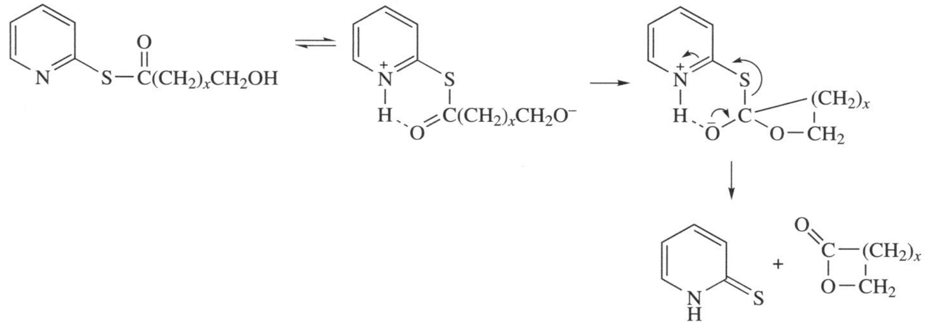

This provides a cyclic transition state in which hydrogen bonding can enhance the reactivity of the carbonyl group.[119] Excellent yields of large-ring lactones are achieved by this method.

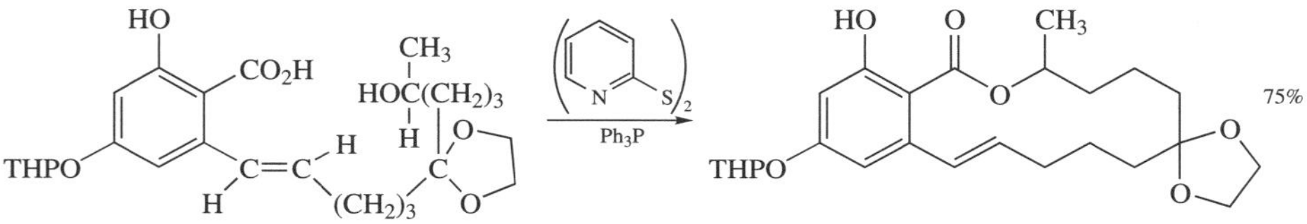

Ref. 118

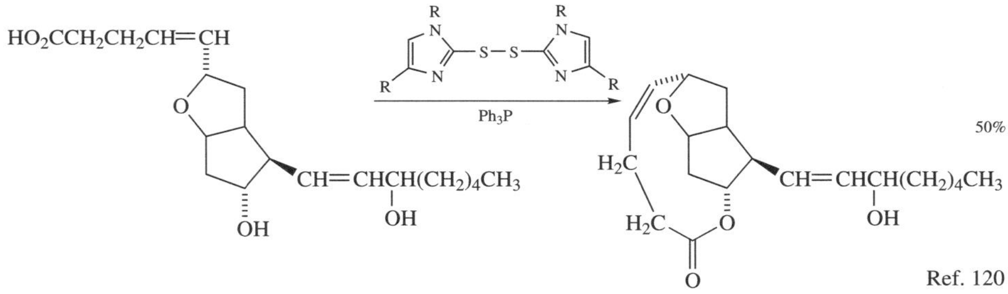

Ref. 120

Intramolecular lactonization can also be carried out with DCC and DMAP. As with other macrolactonizations, the reactions must be carried out in rather dilute solution to promote the intramolecular transformation in competition with intermolecular reaction, which leads to dimers or higher oligomers. A study with 15-hydroxypentadecanoic acid has demonstrated that a proton source is beneficial under these conditions and found the hydrochloride of DMAP to be convenient.[121]

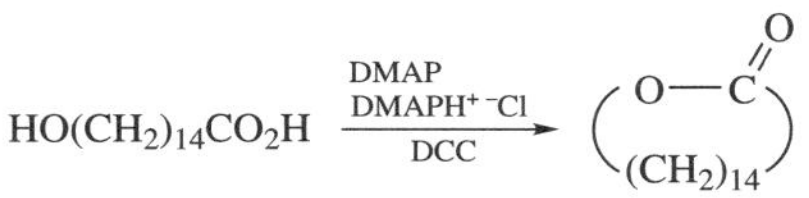

$$HO(CH_2)_{14}CO_2H \xrightarrow[\text{DCC}]{\text{DMAP, DMAPH}^{+}\ ^{-}\text{Cl}} \text{lactone } (CH_2)_{14}$$

119. E. J. Corey, K. C. Nicolaou, and L. S. Melvin, Jr., *J. Am. Chem. Soc.* **97:**654 (1975); E. J. Corey, D. J. Brunelle, and P. J. Stork, *Tetrahedron Lett.* **1976:**3405.
120. E. J. Corey, H. L. Pearce, I. Szekely, and M. Ishiguro, *Tetrahedron Lett.* **1978:**1023.
121. E. P. Boden and G. E. Keck, *J. Org. Chem.* **50:**2394 (1985).

Scheme 3.4 gives some typical examples of preparation and use of active acylating agents from carboxylic acids.

3.4.2. Preparation of Esters

As mentioned in the preceding section, one of the most general methods of synthesis of esters is by reaction of alcohols with an acyl chloride or other activated carboxylic acid derivative. Section 3.2.4 included a discussion of two other important methods, namely, reactions with diazoalkanes and reactions of carboxylate salts with alkyl halides or sulfonate esters. There remains to be mentioned the acid-catalyzed reaction of carboxylic acids with alcohols, which is frequently referred to as Fischer esterification:

$$RCO_2H + R'OH \xrightarrow{H^+} RCO_2R' + H_2O$$

This is an equilibrium process, and there are two techniques which are used to drive the reaction to completion. One is to use a large excess of the alcohol. This is feasible for simple and relatively inexpensive alcohols. The second method is to drive the reaction forward by irreversible removal of water. Azeotropic distillation is one method for doing this. Entries 1–4 in Scheme 3.5 are examples of acid-catalyzed esterifications. Entry 5 is the preparation of a diester starting with an anhydride. This is a closely related reaction in which the initial opening of the anhydride ring is followed by an acid-catalyzed esterification.

3.4.3. Preparation of Amides

By far the most common method for preparation of amides is the reaction of ammonia or a primary or secondary amine with one of the reactive reagents described in Section 3.4.1. When acyl halides are used, some provision for neutralizing the hydrogen halide is necessary, because it will otherwise react with the reagent amine to form the corresponding salt. Acid anhydrides give rapid acylation of most amines and are convenient if available. The Schotten–Bauman conditions, which involve shaking an amine with excess anhydride or acyl chloride and an alkaline aqueous solution, provide a very satisfactory method for preparation of simple amides.

$$\text{(piperidine) NH} + PhC(=O)Cl \xrightarrow{NaOH} \text{(piperidine) N}-C(=O)Ph \quad 90\%$$ Ref. 122

A great deal of work has been done on the *in situ* activation of carboxylic acids toward nucleophilic substitution by amines. This type of reaction forms the backbone of the methods for synthesis of peptides and proteins. (See also Section 13.6). DCC is very widely used for coupling carboxylic acids and amines to give amides. Because amines are better nucleophiles than alcohols, the leaving group in the acylation reagent need not be as reactive as is necessary for alcohols. The *p*-nitrophenyl[123] and 2,4,5-trichlorophenyl[124]

122. C. S. Marvel and W. A. Lazier, *Org. Synth.* **I**:99 (1941).
123. M. Bodanszky and V. DuVigneaud, *J. Am. Chem. Soc.* **81**:5688 (1959).
124. J. Pless and R. A. Boissonnas, *Helv. Chim. Acta* **46**:1609 (1963).

1[a] (Bicyclic acid with CH_3, CH_3, CH_3CO_2, CH_3 substituents) CO_2H $\xrightarrow[25°C]{ClCOCOCl}$ $COCl$

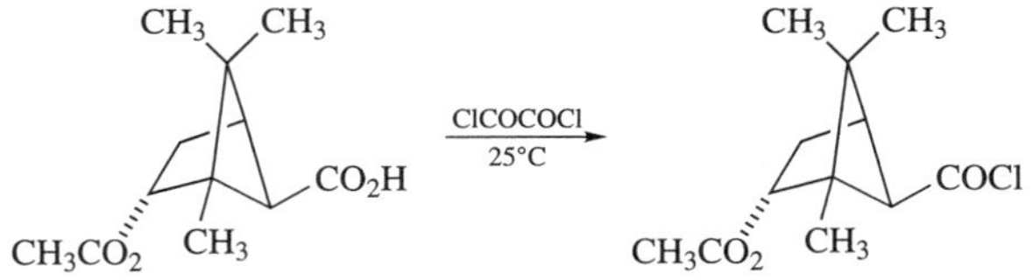

2[b] $(CH_3)_2C{=}CHCH_2CH_2C(CH_3){=}CHCH_2CH_2CH_2CO_2^-\ Na^+ \xrightarrow{ClCOCOCl} (CH_3)_2C{=}CHCH_2CH_2C(CH_3){=}CHCH_2CH_2CH_2COCl$

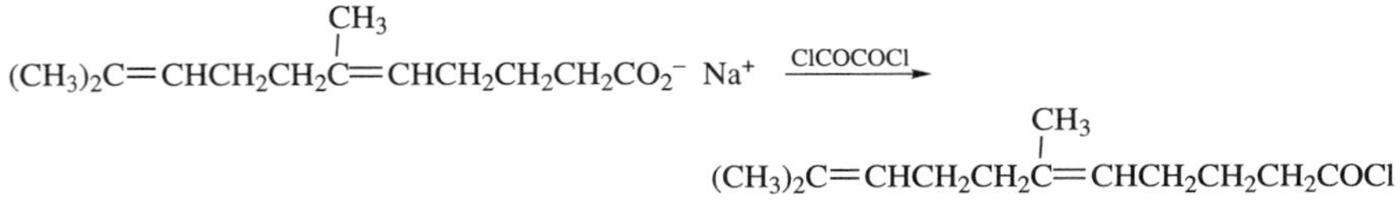

3[c] $PhCO_2H$ —(carbonyldiimidazole)→ —($CH_2{=}CHCH(OH)CH_2{-}N$ aziridine)→ $PhCO_2CH(CH{=}CH_2)CH_2{-}N$ (aziridine) 60%

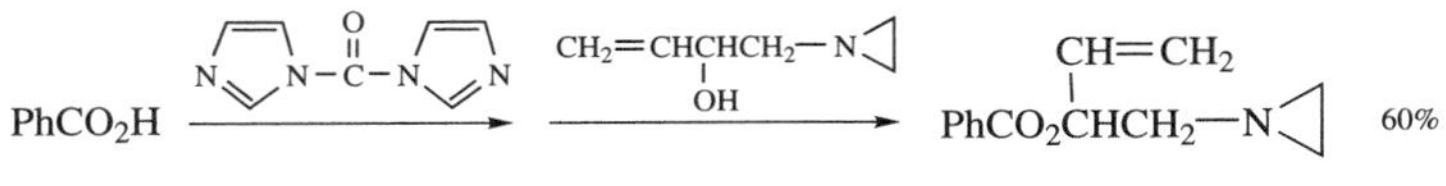

4[d] $C_6H_5CO_2H$ + HO–C_6H_4–NO_2 $\xrightarrow{DCC}$ $C_6H_5C(O){-}O{-}C_6H_4{-}NO_2$

5[e] $PhCH_2CO_2H$ —(2-chloro-1-methylpyridinium, CH_3, Cl)→ —($PhCH(CH_3)OH$)→ $PhCH_2C(O)OCH(CH_3)Ph$ 88%

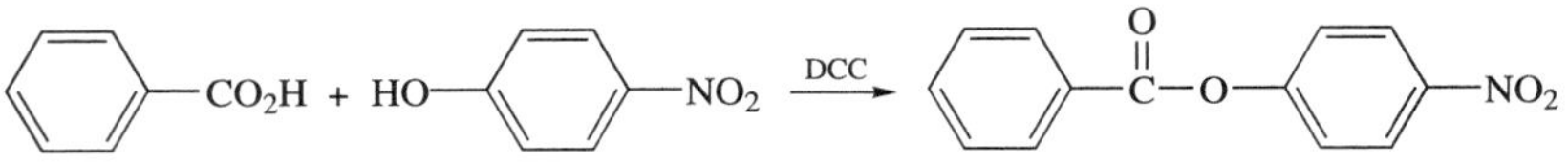

6[f] $CH_3CH{=}CHCH{=}CHCO_2H$ —(2-pyridyl-S-C(O)Cl)→ $CH_3CH{=}CHCH{=}CHC(O)S$–(2-pyridyl)

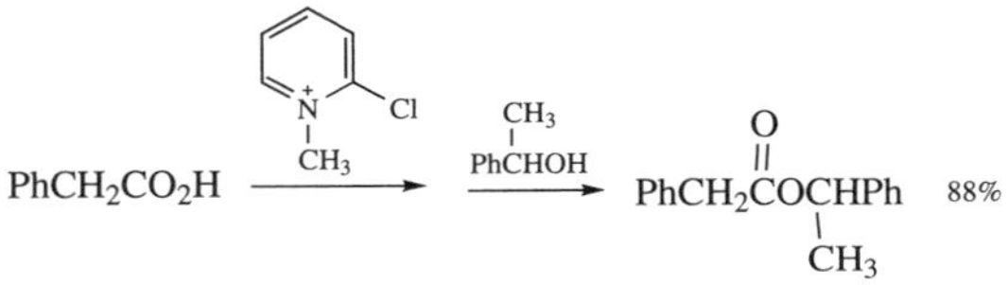

7[g] (HCO_2CH_2, $HCO_2(CH_2)_3CH(CH_3)$, OH, CH_2 lactone) $\xrightarrow[DMAP]{(CH_2{=}C(CH_3)CO)_2O}$ (HCO_2CH_2, $HCO_2(CH_2)_3CH(CH_3)$, $OC(O)C(CH_3){=}CH_2$, CH_2 lactone)

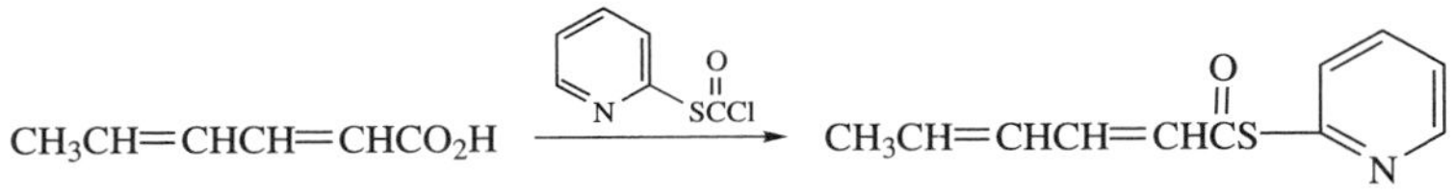

8[h] (R_3SiO, OCH_3, O, H, H, OH, H_3C, H) $\xrightarrow[DCC,\ DMAP]{CH_3CH(CO_2H)CH_2CH_3}$ (R_3SiO, OCH_3, O, H, H, $O_2CCH(CH_3)CH_2CH_3$, H_3C, H) 97%

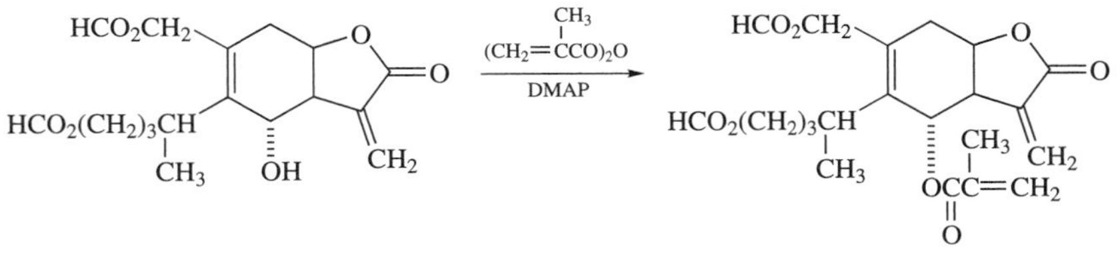

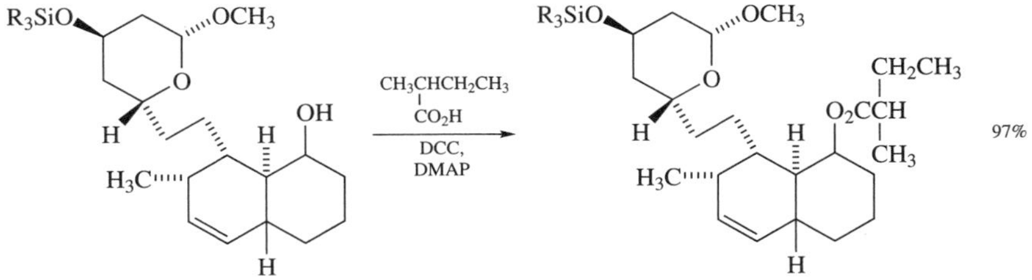

Scheme 3.4. *(continued)*

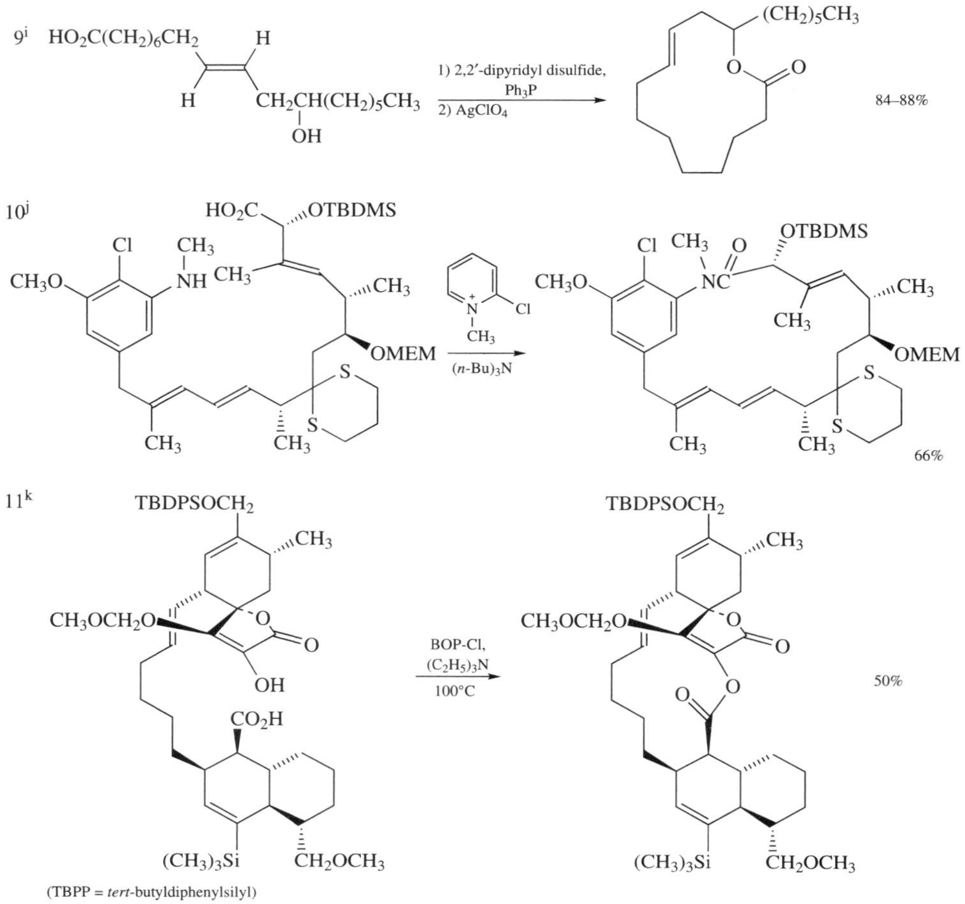

9[i] $HO_2C(CH_2)_6CH_2CH{=}CHCH_2CH(OH)(CH_2)_5CH_3$ → 1) 2,2′-dipyridyl disulfide, Ph_3P 2) $AgClO_4$ → 84–88%

10[j] → (2-chloro-1-methylpyridinium), $(n\text{-Bu})_3N$ → 66%

11[k] → BOP-Cl, $(C_2H_5)_3N$, 100°C → 50%

(TBPP = *tert*-butyldiphenylsilyl)

a. J. Meinwald, J. C. Shelton, G. L. Buchanan, and A. Courtain, *J. Org. Chem.* **33:**99 (1968).
b. U. T. Bhalerao, J. J. Plattner, and H. Rapaport, *J. Am. Chem. Soc.* **92:**3429 (1970).
c. H. A. Staab and Rohr, *Chem. Ber.* **95:**1298 (1962).
d. S. Neelakantan, R. Padmasani, and T. R. Seshadri, *Tetrahedron* **21:**3531 (1965).
e. T. Mukaiyama, M. Usui, E. Shimada, and K. Saigo, *Chem. Lett.* **1975:**1045.
f. E. J. Corey and D. A. Clark, *Tetrahedra Lett.* **1979:**2875.
g. P. A. Grieco, T. Oguri, S. Gilman, and G. DeTitta, *J. Am. Chem. Soc.* **100:**1616 (1978).
h. Y.-L. Yang, S. Manna, and J. R. Falck, *J. Am. Chem. Soc.* **106:**3811 (1984).
i. A. Thalman, K. Oertle, and H. Gerlach, *Org. Synth.* **63:**192 (1984).
j. M. Benechie and F. Khuong-Huu, *J. Org. Chem.* **61:**7133 (1996).
k. W. R. Rousch and R. J. Sciotti, *J. Am. Chem. Soc.* **120:**7411 (1998).

esters of amino acids are sufficiently reactive toward amines to be useful in peptide synthesis. Acyl derivatives of *N*-hydroxysuccinimide are also useful for synthesis of peptides and other types of amides.[125,126] Like the *p*-nitrophenyl esters, the acylated

125. G. W. Anderson, J. E. Zimmerman, and F. M. Callahan, *J. Am. Chem. Soc.* **86:**1839 (1964).
126. E. Wunsch and F. Drees, *Chem. Ber.* **99:**110 (1966); E. Wunsch, A. Zwick, and G. Wendlberger, *Chem. Ber.* **100:**173 (1967).

Scheme 3.5. Acid-Catalyzed Esterification

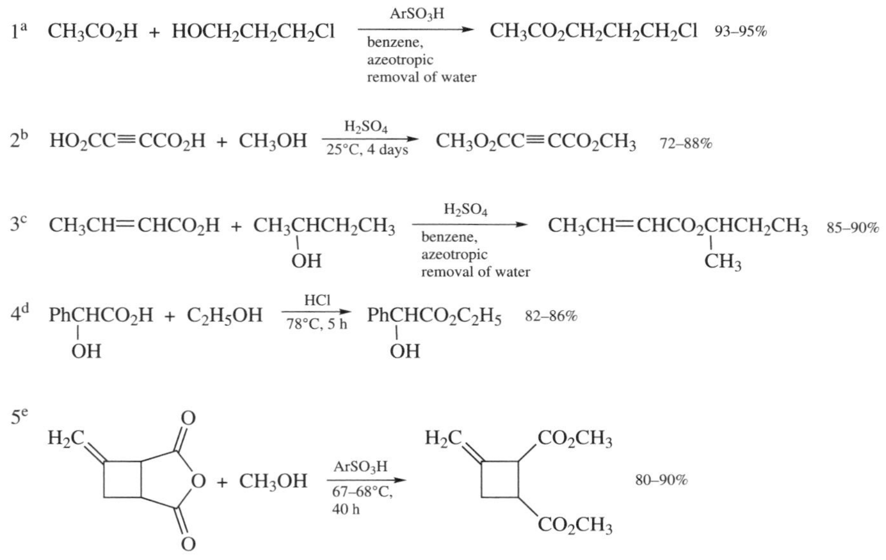

1[a] $CH_3CO_2H + HOCH_2CH_2CH_2CH_2Cl \xrightarrow[\text{benzene, azeotropic removal of water}]{ArSO_3H} CH_3CO_2CH_2CH_2CH_2CH_2Cl$ 93–95%

2[b] $HO_2CC{\equiv}CCO_2H + CH_3OH \xrightarrow[\text{25°C, 4 days}]{H_2SO_4} CH_3O_2CC{\equiv}CCO_2CH_3$ 72–88%

3[c] $CH_3CH{=}CHCO_2H + CH_3CH(OH)CH_2CH_3 \xrightarrow[\text{benzene, azeotropic removal of water}]{H_2SO_4} CH_3CH{=}CHCO_2CH(CH_3)CH_2CH_3$ 85–90%

4[d] $PhCH(OH)CO_2H + C_2H_5OH \xrightarrow[\text{78°C, 5 h}]{HCl} PhCH(OH)CO_2C_2H_5$ 82–86%

5[e] 3-methylenecyclobutane-1,2-dicarboxylic anhydride $+ CH_3OH \xrightarrow[\text{67–68°C, 40 h}]{ArSO_3H}$ dimethyl 3-methylenecyclobutane-1,2-dicarboxylate ($H_2C{=}$, CO_2CH_3, CO_2CH_3) 80–90%

a. C. F. H. Allen and F. W. Spangler, *Org. Synth.* **III**:203 (1955).
b. E. H. Huntress, T. E. Lesslie, and J. Bornstein, *Org. Synth.* **IV**:329 (1963).
c. J. Munch-Petersen, *Org. Synth.* **V**:762 (1973).
d. E. L. Eliel, M. T. Fisk, and T. Prosser, *Org. Synth.* **IV**:169 (1963).
e. H. B. Stevenson, H. N. Cripps, and J. K. Williams, *Org. Synth.* **V**:459 (1973).

N-hydroxysuccinimides can be isolated and purified, but they react rapidly with free amino groups.

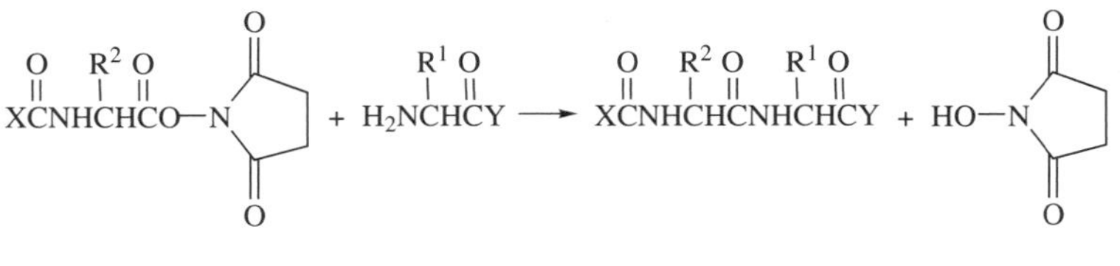

$$XC(=O)NHCH(R^2)C(=O)\text{–}N(\text{succinimide}) + H_2NCH(R^1)C(=O)Y \longrightarrow XC(=O)NHCH(R^2)C(=O)NHCH(R^1)C(=O)Y + HO\text{–}N(\text{succinimide})$$

The *N*-hydroxysuccinimide that is liberated is easily removed because of its solubility in dilute base. The relative stability of the anion of *N*-hydroxysuccinimide is also responsible for the acyl derivative being reactive toward nucleophilic attack by an amino group.

Esters of *N*-hydroxysuccinimide are also used to carry out chemical modification of peptides, proteins, and other biological molecules by acylation of nucleophilic groups in these molecules. For example, detection of estradiol antibodies can be accomplished using an estradiol analog to which a fluorescent label has been attached.

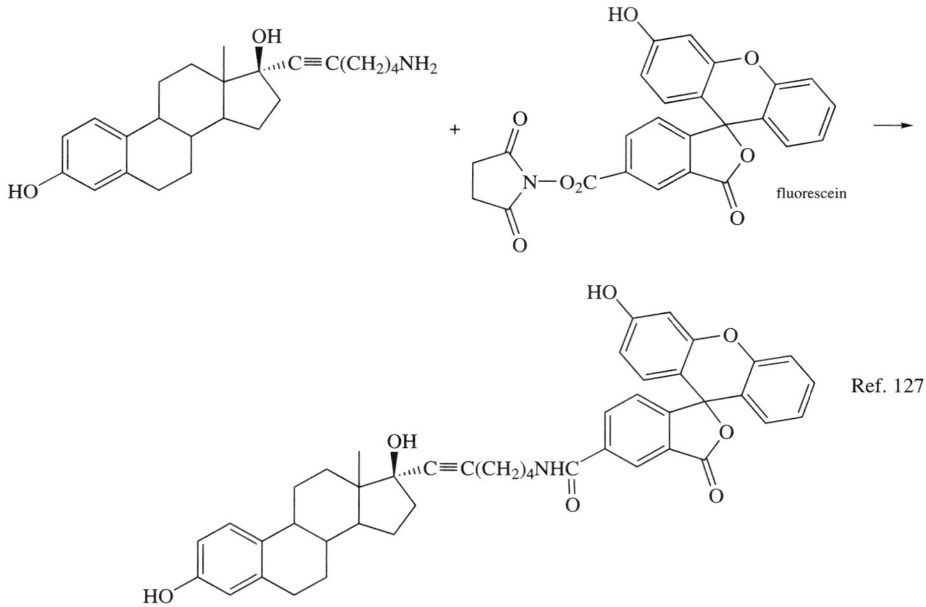

Ref. 127

Similarly, photolabels such as 4-azidobenzoylglycine can be attached to peptides and used to detect the peptide binding sites in proteins.[128]

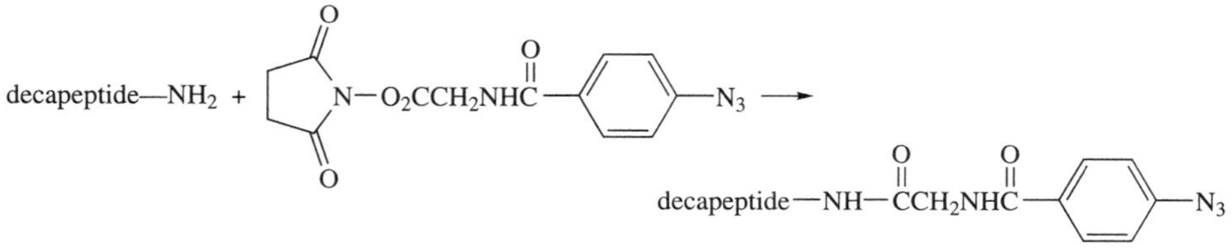

1-Hydroxybenzotriazole is also useful in conjunction with DCC.[129] For example, *t*-butoxycarbonyl (Boc)-protected leucine and the methyl ester of phenylalanine can be coupled in 88% yield with these reagents.

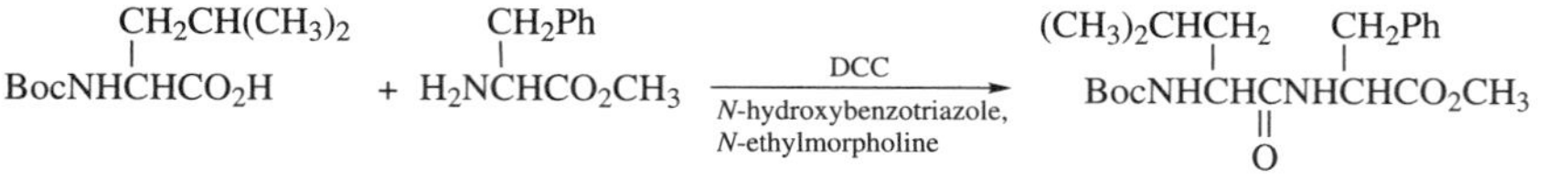

Ref. 130

Carboxylic acids can also be activated by formation of mixed anhydrides with various phosphoric acid derivatives. Diphenylphosphoryl azide, for example, is an effective

127. M. Adamczyk, Y.-Y. Chen, J. A. Moore, and P. G. Mattingly, *Biorg. Med. Chem. Lett.* **8:**1281 (1998); M. Adamczyk, J. R. Fishpaugh, and K. J. Heuser, *Bioconjug. Chem.* **8:**253 (1997).
128. G. C. Kundu, I. Ji, D. J. McCormick, and T. H. Ji, *J. Biol. Chem.* **271:**11063 (1996).
129. W. Konig and R. Geiger, *Chem. Ber.* **103:**788 (1970).
130. M. Bodanszky and A. Bodanszky, *The Prentice of Peptide Synthesis*, 2nd ed., Springer-Verlag, Berlin, 1994, pp. 119–120.

reagent for conversion of amines to amides.[131] The postulated mechanism involves formation of the acyl azide as a reactive intermediate:

$$RCO_2^- + (PhO)_2P(=O)N_3 \longrightarrow RC(=O)-O-P(=O)(OPh)_2 + N_3^-$$

$$RC(=O)-O-P(=O)(OPh)_2 + N_3^- \longrightarrow RC(=O)N_3 + {}^-O_2P(OPh)_2$$

$$RC(=O)N_3 + R'NH_2 \longrightarrow RC(=O)NHR' + HN_3$$

Another useful reagent for amide formation is compound **1**, known as BOP-Cl.[132] This reaction also proceeds via a mixed carboxylic phosphoric anhydride.

$$RCO_2^- + \mathbf{1} \longrightarrow RC(=O)-O-P(=O)(N\text{-oxazolidinonyl})_2$$

The preparation of amides directly from alkyl esters is also feasible but is usually too slow for preparative convenience. Entries 4 and 5 in Scheme 3.6 are successful examples. The reactivity of ethyl cyanoacetate (entry 4) is higher than that of unsubstituted aliphatic esters because of the inductive effect of the cyano group.

Another method for converting esters to amides involves aluminum amides, which can be prepared from trimethylaluminum and the amine. These reagents convert esters directly to amides at room temperature.[133]

$$C_6H_{11}CO_2CH_3 \xrightarrow{(CH_3)_2AlN(H)CH_2Ph} C_6H_{11}C(=O)NHCH_2Ph \quad 78\%$$

The driving force for this reaction is the strength of the aluminum–oxygen bond relative to the aluminum–nitrogen bond. This reaction provides a good way of making synthetically useful amides of *N*-methoxy-*N*-methylamine.[134] Trialkylamidotin and bis(hexamethyldi-

131. T. Shiori and S. Yamada, *Chem. Pharm. Bull.* **22:**849 (1974); T. Shioiri and S. Yamada, *Chem. Pharm. Bull.* **22:**855 (1974); T. Shioiri and S. Yamada, *Chem. Pharm. Bull.* **22:**859 (1974).
132. J. Diago-Mesequer, A. L. Palomo-Coll, J. R. Fernandez-Lizarbe, and A. Zugaza-Bilbao, *Synthesis* **1980:**547; R. D. Tung, M. K. Dhaon, and D. H. Rich, *J. Org. Chem.* **51:**3350 (1986); W. J. Colucci, R. D. Tung, J. A. Petri, and D. H. Rich, *J. Org. Chem.* **55:**2895 (1990); J. Jiang, W. R. Li, R. M. Przeslawski, and M. M. Joullie, *Tetrahedron Lett.* **34:**6705 (1993).
133. A. Basha, M. Lipton, and S. M. Weinreb, *Tetrahedron Lett.* **1977:**4171; A. Solladie-Cavallo and M. Benchegroun, *J. Org. Chem.* **57:**5831 (1992).
134. J. I. Levin, E. Turos, and S. M. Weinreb, *Synth. Commun.* **12:**989 (1982); T. Shimizu, K. Osako, and T. Nakata, *Tetrahedron Lett.* **38:**2685 (1997).

Scheme 3.6. Synthesis of Amides

A. From acyl chlorides and anhydrides

1[a] $(CH_3)_2CHCO_2H \xrightarrow[2)\ NH_3]{1)\ SOCl_2} (CH_3)_2CHC(=O)NH_2$ 70%

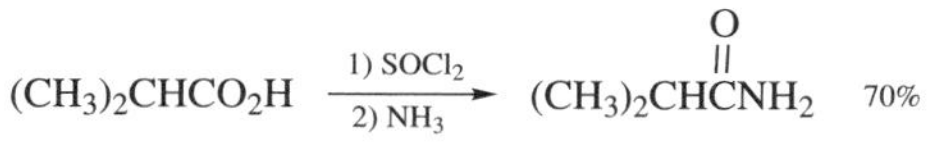

2[b] $C_6H_{11}CO_2H \xrightarrow[2)\ (CH_3)_2NH]{1)\ SOCl_2} C_6H_{11}C(=O)N(CH_3)_2$ 85–90%

3[c] $(CH_3CO)_2O + H_2NCH_2CO_2H \longrightarrow CH_3C(=O)NHCH_2CO_2H$ 90%

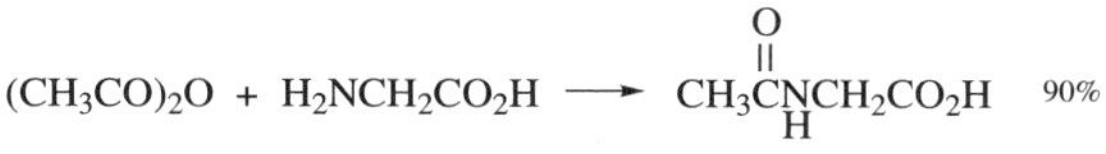

B. From esters

4[d] $NCCH_2CO_2C_2H_5 \xrightarrow{NH_4OH} NCCH_2C(=O)NH_2$

5[e] phenyl salicylate (CO_2Ph, OH) + o-toluidine (CH_3, H_2N) $\xrightarrow[185\text{–}200°C]{\text{trichlorobenzene}}$ N-(2-methylphenyl)salicylamide (OH, C(=O)–NH, CH_3) 75%

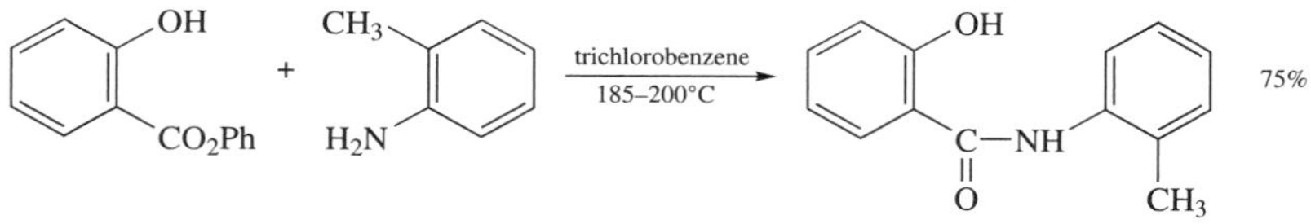

C. From carboxylic acids

6[f] methyl 4-azidopyrrolidine-2-carboxylate (N_3, CO_2CH_3, N–H) $\xrightarrow[EtN]{PhCO_2H,\ DCC}$ N-benzoyl product (N_3, CO_2CH_3, N–C(=O)Ph) 63%

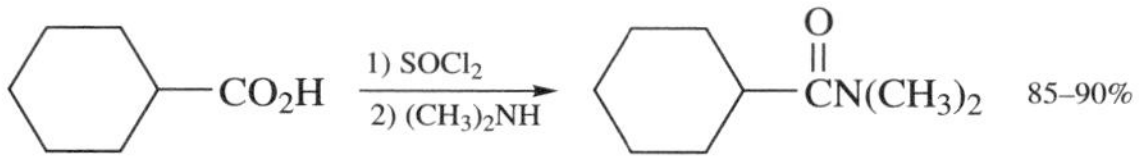

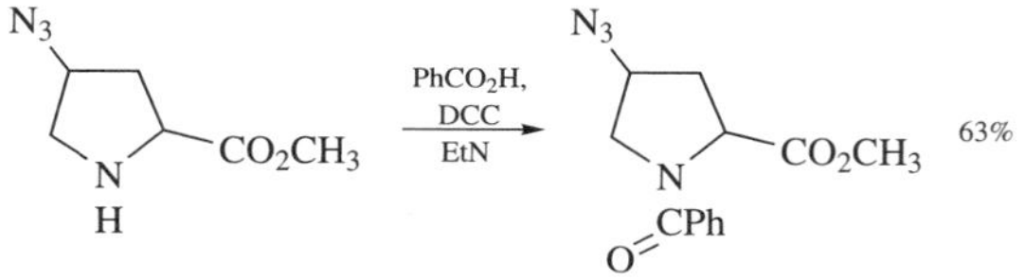

7[g] 2-(CH_2CN)benzoic acid (CO_2H) + aniline (NH_2) $\xrightarrow[\text{bis(2-oxo-3-oxazolidinyl)phosphinic chloride}]{Et_3N}$ 2-(CH_2CN)-C(=O)NH–Ph

8[h] 2,3-dimethoxybenzoic acid (OCH_3, OCH_3, CO_2H) + $H_2N(CH_2)_2CO_2H$ $\xrightarrow[\text{N-hydroxysuccinimide}]{DCC}$ 2,3-dimethoxy-$CONH(CH_2)_2CO_2H$ 82%

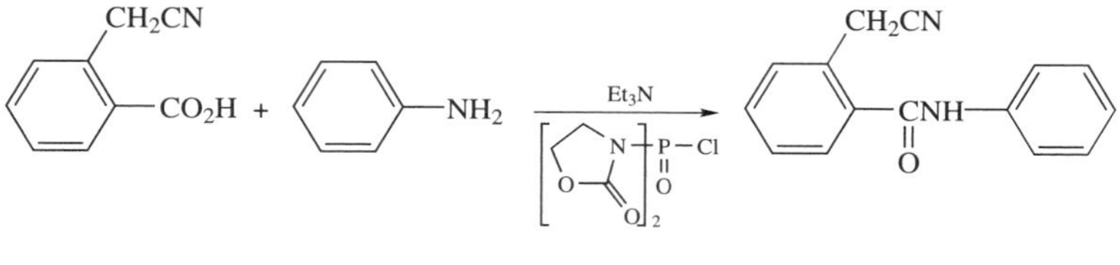

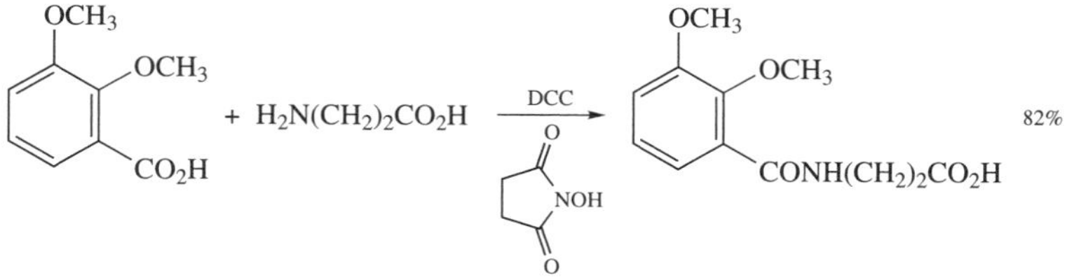

D. From nitriles

9[i] $PhCH_2CN \xrightarrow[40\text{–}50°C,\ 1\ h]{HCl,\ H_2O} PhCH_2C(=O)NH_2$ 80%

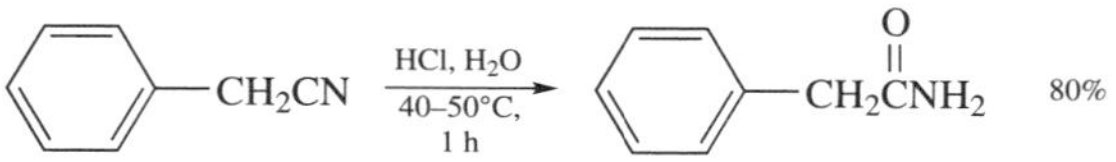

10[j] o-methylbenzonitrile $\xrightarrow[40\text{–}50°C,\ 4\ h]{30\%\ H_2O_2,\ NaOH}$ o-methylbenzamide (2-$CH_3C_6H_4CNH_2$, C=O) 90%

a. R. E. Kent and S. M. McElvain, *Org. Synth.* **III**:490 (1955).
b. A. C. Cope and E. Ciganek, *Org. Synth.* **IV**:339 (1963).
c. R. M. Herbst and D. Shemin, *Org. Synth.* **II**:11 (1943).
d. B. B. Corson, R. W. Scott, and C. E. Vose, *Org. Synth.* **I**:179 (1941).
e. C. F. H. Allen and J. Van Allan, *Org. Synth.* **III**:765 (1955).
f. D. J. Abraham, M. Mokotoff, L. Sheh, and J. E. Simmons, *J. Med. Chem.* **26**:549 (1983).
g. J. Diago-Mesenguer, A. L. Palamo-Coil, J. R. Fernandez-Lizarbe, and A. Zugaza-Bilbao, *Synthesis* **1980**:547.
h. R. J. Bergeron, S. J. Kline, N. J. Stolowich, K. A. McGovern, and P. S. Burton, *J. Org. Chem.* **46**:4524 (1981).
i. W. Wenner, *Org. Synth.* **IV**:760 (1963).
j. C. R. Noller, *Org. Synth.* **II**:586 (1943).

silylamido)tin amides as well as tetrakis(dimethylamino)titanium show similar reactivity.[135]

The cyano group is at the carboxylic acid oxidation level so nitriles are potential precursors of primary amides. Partial hydrolysis is sometimes possible.[136]

$$PhCH_2C{\equiv}N \xrightarrow[40\text{–}50°C,\ 1\ h]{HCl,\ H_2O} PhCH_2\overset{O}{\overset{\|}{C}}NH_2$$

A milder procedure involves the reaction of a nitrile with an alkaline solution of hydrogen peroxide.[137] The strongly nucleophilic hydrogen peroxide adds to the nitrile, and the resulting adduct gives the amide. There are several possible mechanisms for the subsequent decomposition of the peroxycarboximidic adduct.[138]

$$RC{\equiv}N + {}^{-}O_2H \longrightarrow R\overset{NH}{\overset{\|}{C}}OO^- \overset{H_2O}{\rightleftharpoons} R\overset{NH}{\overset{\|}{C}}OOH + H_2O_2 \longrightarrow R\overset{O}{\overset{\|}{C}}NH_2 + O_2 + H_2O$$

In all the mechanisms, the hydrogen peroxide is converted to oxygen and water, leaving the organic substrate hydrolyzed, but at the same oxidation level. Scheme 3.6 (Entries 9 and 10) includes two specific examples of conversion of nitriles to amides.

135. G. Chandra, T. A. George, and M. F. Lappert, *J. Chem. Soc., C* **1969**:2565; W.-B. Wang and E. J. Roskamp, *J. Org. Chem.* **57**:6101 (1992); W.-B. Wang, J. A. Restituyo, and E. J. Roskamp, *Tetrahedron Lett.* **34**:7217 (1993).
136. W. Wenner, *Org. Synth.* **IV**:760 (1963).
137. C. R. Noller, *Org. Synth.* **II**:586 (1943); J. S. Buck and W. S. Ide, *Org. Synth.* **II**:44 (1943).
138. K. B. Wiberg, *J. Am. Chem. Soc.* **75**:3961 (1953); *J. Am. Chem. Soc.* **77**:2519 (1955); J. E. McIsaac, Jr., R. E. Ball, and E. J. Behrman, *J. Org. Chem.* **36**:3048 (1971).

(References for these problems will be found on page 926.)

1. Give the products which would be expected to be formed under the specified reaction conditions. Be sure to specify all aspects of stereochemistry.

(a) CH_3CH_2–(γ-lactone, =O) $\xrightarrow[CH_3OH]{HCl}$

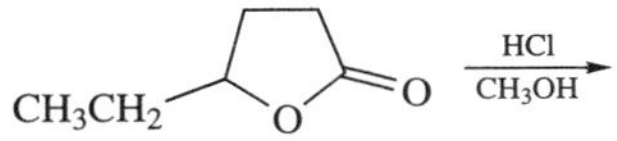

(b) $CH_3(CH_2)_4CH_2OH$ + $ClCH_2OCH_3$ $\xrightarrow[CH_2Cl_2]{ELH_2N(i\text{-}Pr)_2}$

(c) (*S*)-$CH_3(CH_2)_3CHCH_3$ (OH) + 2-chloro-3-ethylbenzoxazolium (C_2H_5, $\overset{+}{N}$, O, Cl) $\xrightarrow[Et_4N^+Cl^-]{Et_3N}$

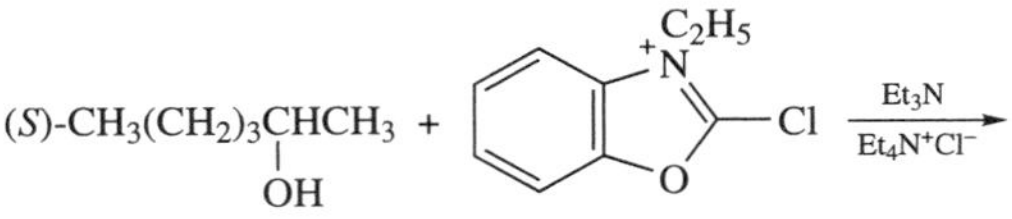

(d) $C_2H_5O_2CCH_2CHCO_2C_2H_5$ (OH) $\xrightarrow[2)\ C_2H_5O_2CN{=}NCO_2C_2H_5]{1)\ Ph_3P,\ HN_3}$

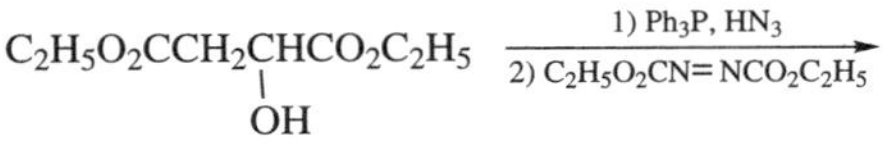

(e) N, N, N, N, $N(\overset{O}{\overset{\|}{C}}R)_2$, $HOCH_2$, O, RCO_2, O_2CR $\xrightarrow[DMF,\ 20°C,\ 10\ min]{(PhO)_3\overset{+}{P}CH_3\ I^-}$

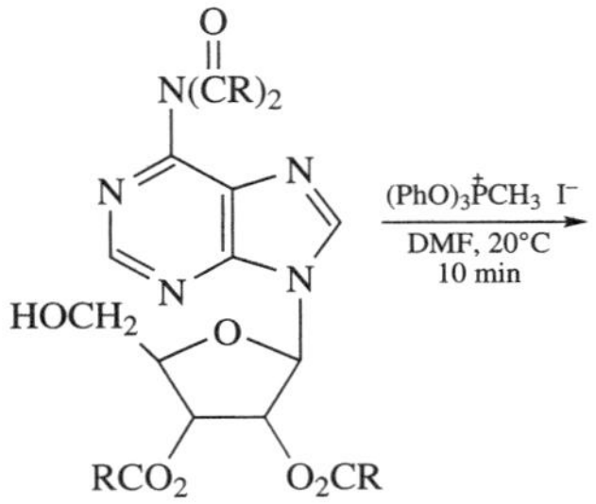

(f) CH_3, CH_3CHCH_2OH $\xrightarrow[CCl_4]{PPh_3}$

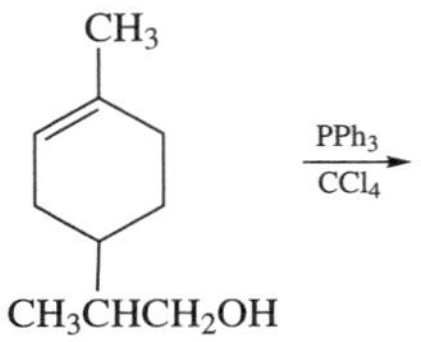

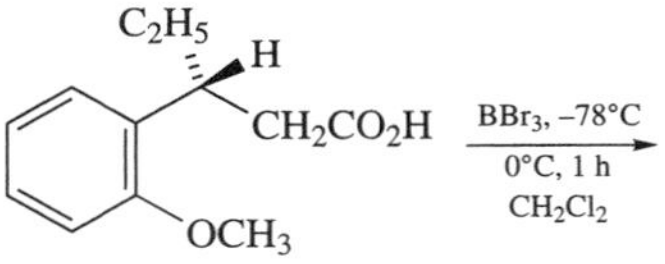

(g) C_2H_5, H, CH_2CO_2H, OCH_3 $\xrightarrow[0°C,\ 1\ h,\ CH_2Cl_2]{BBr_3,\ -78°C}$

(h) CO_2CH_3, Ph, H, OH $\xrightarrow[2)\ PhS^-\ Na^+]{1)\ p\text{-toluenesulfonyl chloride}}$

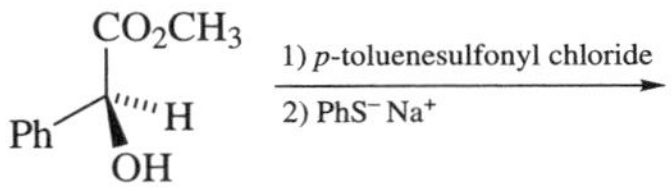

(i) $(C_6H_5)_2CHBr$ + $P(OCH_3)_3$ ⟶

(j) $CH_3SO_2OCH_2$, $CH_3SO_2OCH_2$ $\xrightarrow[HMPA]{Na_2S}$

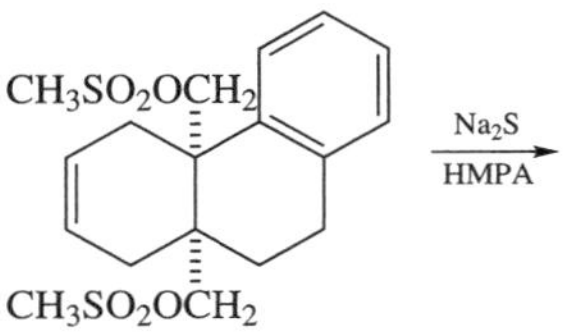

(k) CH_3O, CH_3O, O, $NCH_2C_6H_5$ $\xrightarrow[heat]{48\%\ HBr}$

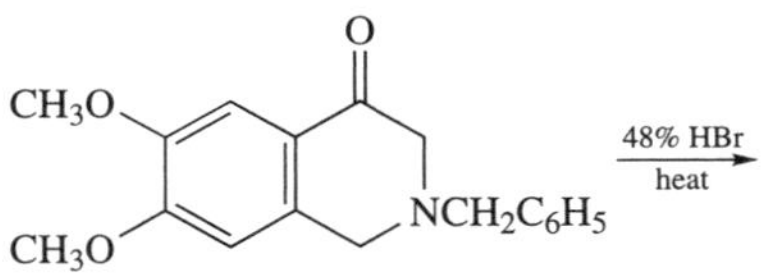

(l) H, CO_2H, C=C, $C_2H_5O_2C$, H $\xrightarrow[DCC,\ DMAP]{t\text{-BuOH}}$

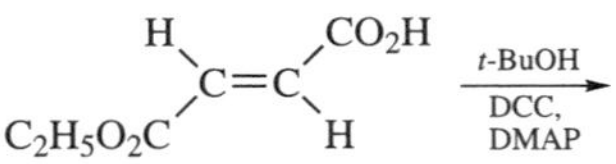

2. When (*R*)-(−)-5-hexen-2-ol was treated with triphenylphosphine in refluxing carbon tetrachloride, (+)-5-chloro-1-hexene was obtained. Conversion of (*R*)-(−)-5-hexen-2-ol to its *p*-bromobenzenesulfonate ester and subsequent reaction with lithium chloride

gave (+)-5-chloro-1-hexene. Reaction of (*S*)-(+)-5-hexen-2-ol with phosphorus pentachloride in ether gave (−)-5-chloro-1-hexene.

(a) Write chemical equations for each of the reactions described above and specify whether each one proceeds with net retention or inversion of configuration.
(b) What is the sign of rotation of (*R*)-5-chloro-1-hexene?

3. A careful investigation of the extent of isomeric products formed by reaction of several alcohols with thionyl chloride has been reported. The product compositions for several of the alcohols are given below. Show how each of the rearranged products arises and discuss the structural features which promote isomerization.

$$ROH \xrightarrow[100°C]{SOCl_2} RCl$$

R	Percent unrearranged RCl	Structure and amount of rearranged RCl
$CH_3CH_2CH_2CH_2$—	100	
$(CH_3)_2CHCH_2$—	99.7	$(CH_3)_2\underset{\underset{Cl}{\mid}}{C}HCH_3$ 0.3%
$(CH_3)_2CHCH_2CH_2$—	100	
$CH_3CH_2\underset{\underset{CH_3}{\mid}}{C}HCH_2$—	78	$CH_3\underset{\underset{Cl}{\mid}}{C}HCH_2CH_2CH_3$ 1%, $CH_3CH_2\underset{\underset{Cl}{\mid}}{C}HCH_2CH_3$ 11%, $CH_3CH_2\underset{\underset{Cl}{\mid}}{C}(CH_3)_2$ 10%
$(CH_3)_3CCH_2$—	2	$CH_3CH_2\underset{\underset{Cl}{\mid}}{C}(CH_3)_2$ 98%
$CH_3CH_2CH_2\underset{\mid}{C}HCH_3$	98	$CH_3CH_2\underset{\underset{Cl}{\mid}}{C}HCH_2CH_3$ 2%
$CH_3CH_2\underset{\mid}{C}HCH_2CH_3$	90	$CH_3CH_2CH_2\underset{\underset{Cl}{\mid}}{C}HCH_3$ 10%
$(CH_3)_2CH\underset{\mid}{C}HCH_3$	5	$CH_3CH_2\underset{\underset{Cl}{\mid}}{C}(CH_3)_2$ 95%

4. Give a reaction mechanism which will explain the following observations and transformations.

(a) Kinetic measurements reveal that solvolytic displacement is about 5×10^5 faster for **B** than for **A**.

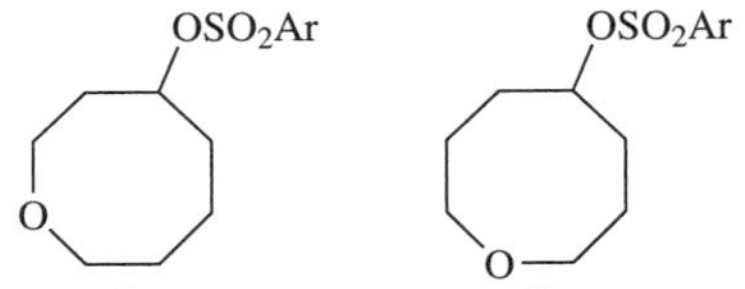

OSO_2Ar OSO_2Ar

O O

A **B**

(b) 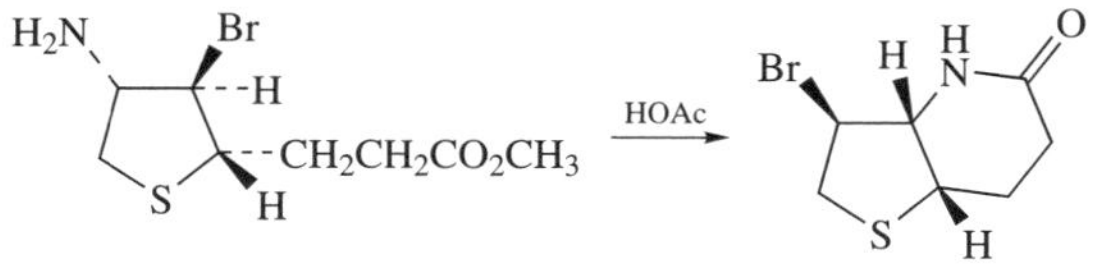 [scheme: bromo amino thiolane ester → HOAc → bicyclic lactam]

(c) 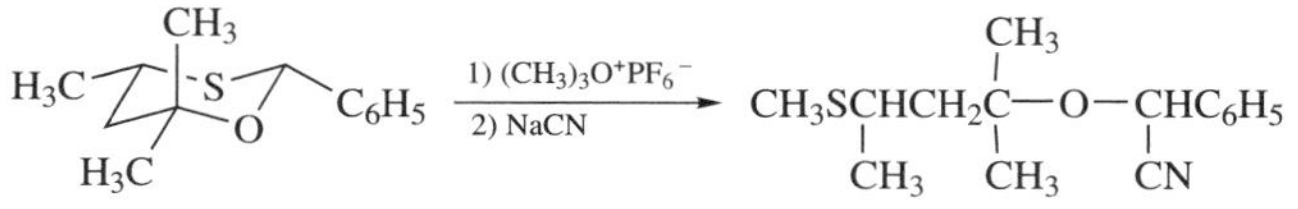 [scheme: oxathiane with C_6H_5 → 1) $(CH_3)_3O^+PF_6^-$ 2) NaCN → $CH_3SCH(CH_3)CH_2C(CH_3)_2{-}O{-}CH(CN)C_6H_5$]

(d) 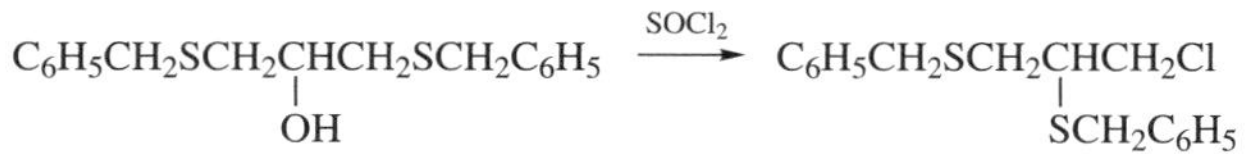 $C_6H_5CH_2SCH_2CH(OH)CH_2SCH_2C_6H_5 \xrightarrow{SOCl_2} C_6H_5CH_2SCH_2CH(SCH_2C_6H_5)CH_2Cl$

(e) 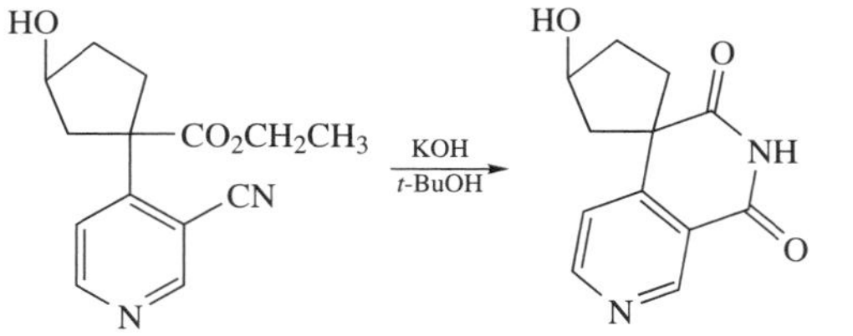 [scheme: hydroxycyclopentyl pyridine ester nitrile → KOH, t-BuOH → spiro imide]

(f) 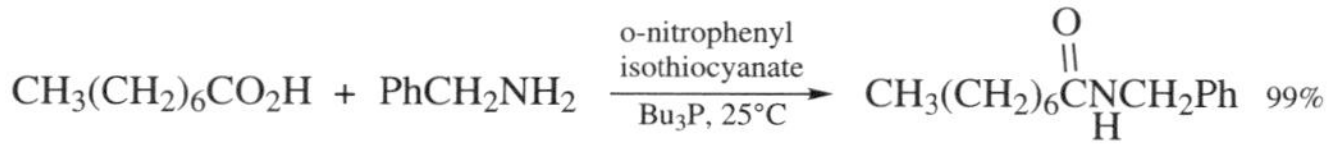 $CH_3(CH_2)_6CO_2H + PhCH_2NH_2 \xrightarrow[\text{Bu}_3\text{P, 25°C}]{\text{o-nitrophenyl isothiocyanate}} CH_3(CH_2)_6C(O)NHCH_2Ph$ 99%

(g) 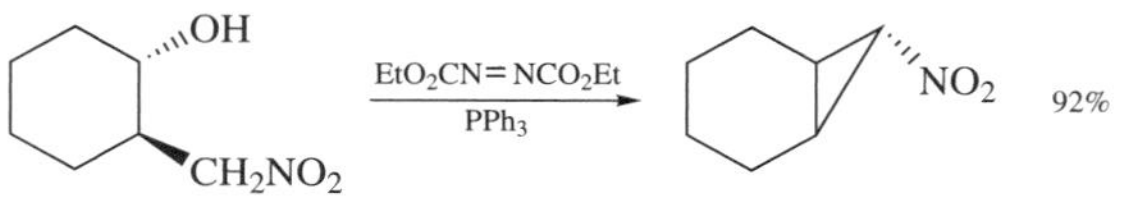 [scheme: 2-(nitromethyl)cyclohexanol → $EtO_2CN{=}NCO_2Et$, PPh_3 → nitro bicyclo[4.1.0]heptane] 92%

(h) Both **C** and **D** gave the same product when subjected to Mitsunobu conditions with phenol as the nucleophile.

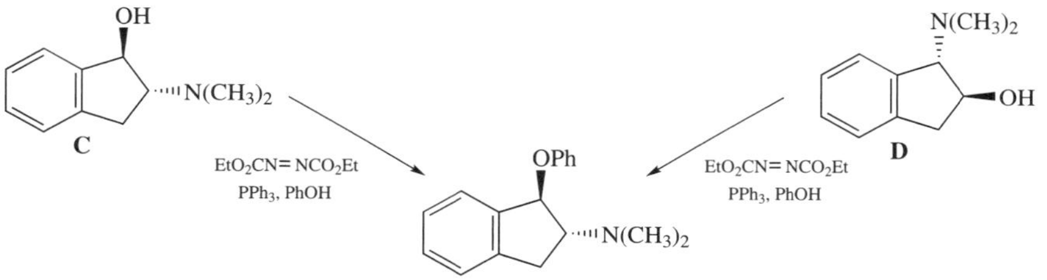

[scheme: **C** and **D** (dimethylamino indanols) → $EtO_2CN{=}NCO_2Et$, PPh_3, PhOH → same phenyl ether]

5. Substances such as carbohydrates, amino acids, and other small molecules available from natural sources are valuable starting materials for the synthesis of stereochemically defined substances. Suggest a sequence of reactions which could effect the following transformations, taking particular care to ensure that the product would be obtained stereochemically pure.

(a) 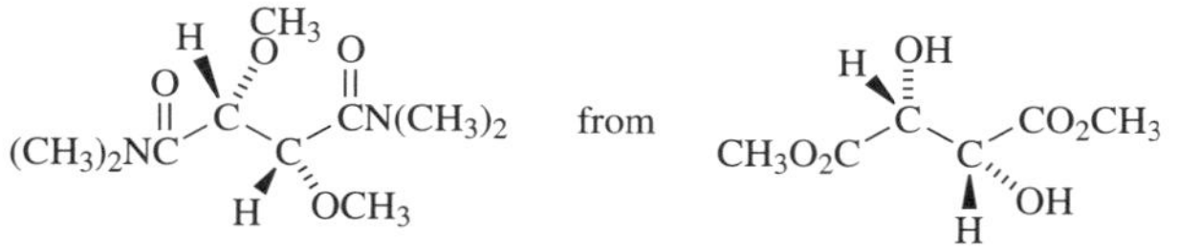 $(CH_3)_2NC(O)CH(OCH_3)CH(OCH_3)C(O)N(CH_3)_2$ from $CH_3O_2CCH(OH)CH(OH)CO_2CH_3$

(b) 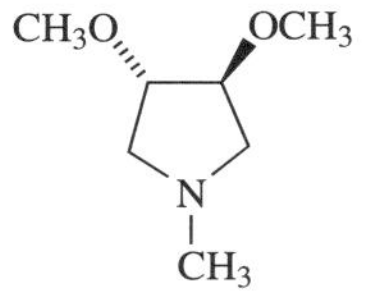 from 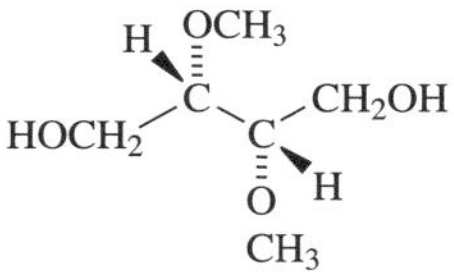

(c) 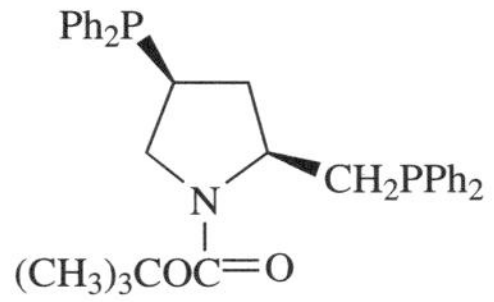 from 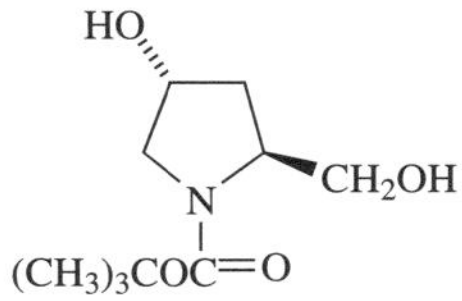

(d) 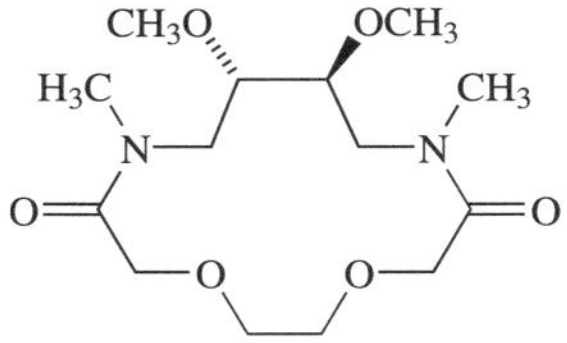 from 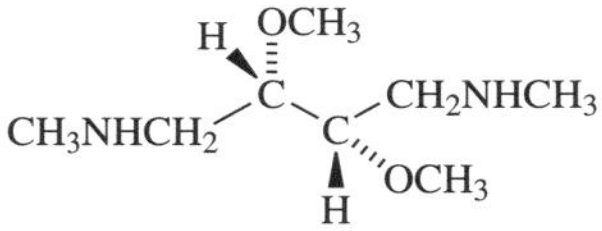

(e) 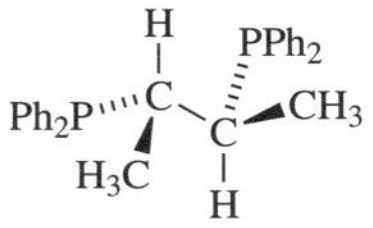 from 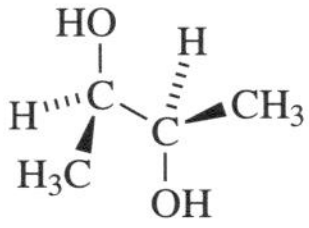

(f) 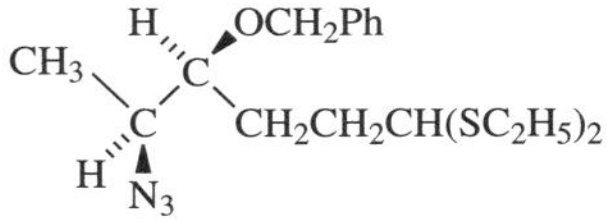 from 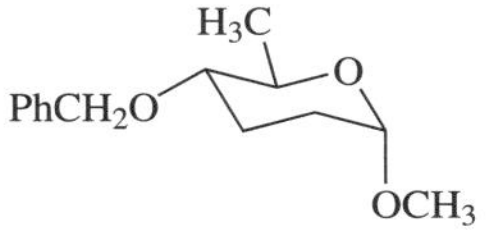

(g) 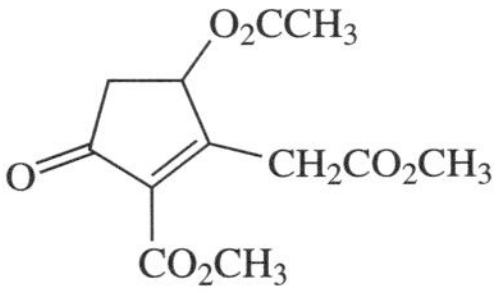 from 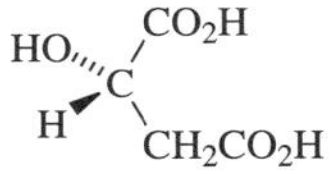

(h) 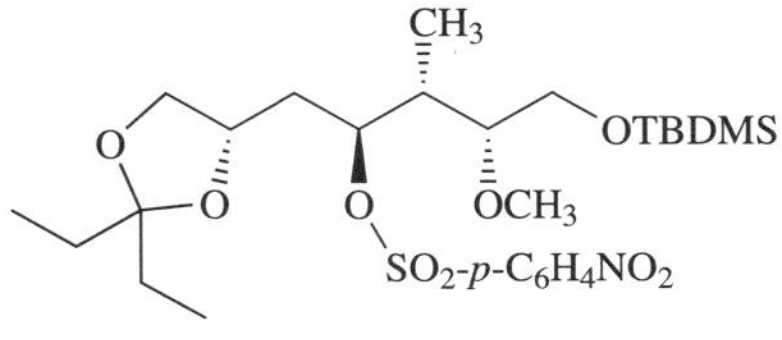 from 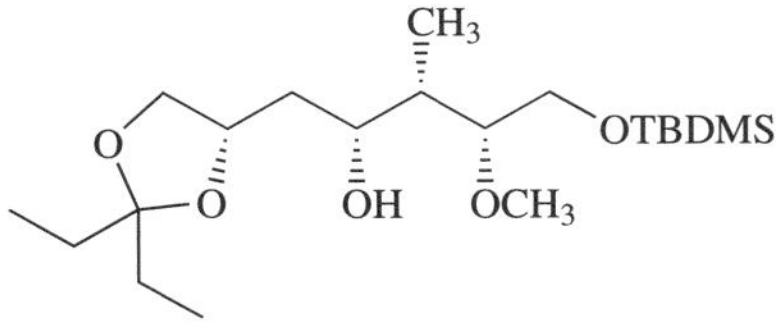

(i) 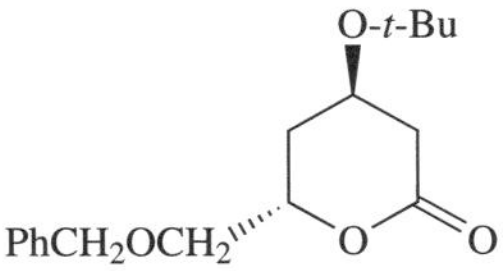 from 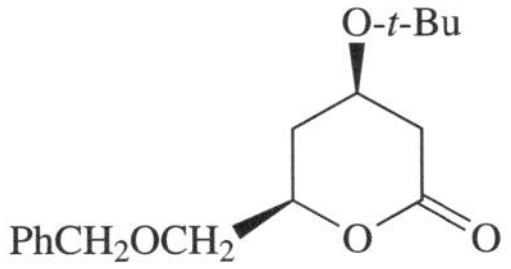

6. Suggest reagents and reaction conditions which could be expected to effect the following conversions.

(a) 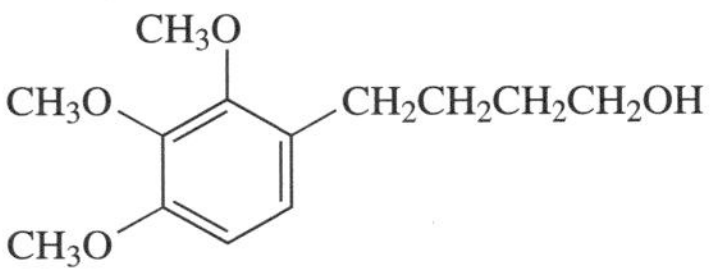 → 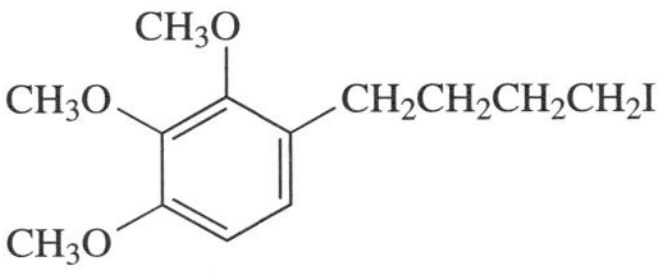

(b) 2,4,6-$(CH(CH_3)_2)_3C_6H_2CO_2H$ → 2,4,6-$(CH(CH_3)_2)_3C_6H_2CO_2CH_2CH{=}CH_2$

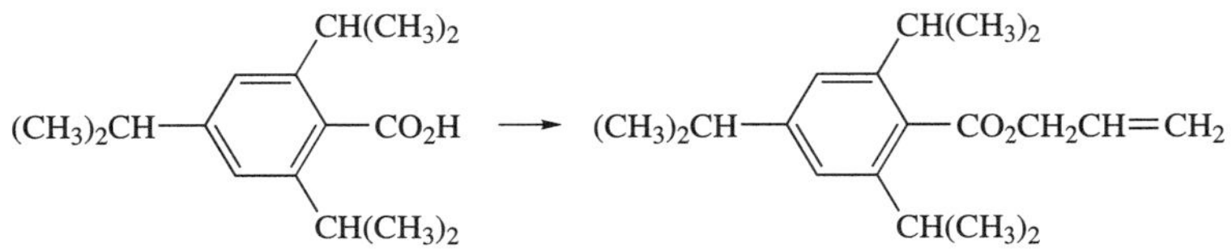

(c) CH_2OH (2,2-dimethyl-1,3-dioxolane) → CH_2CN

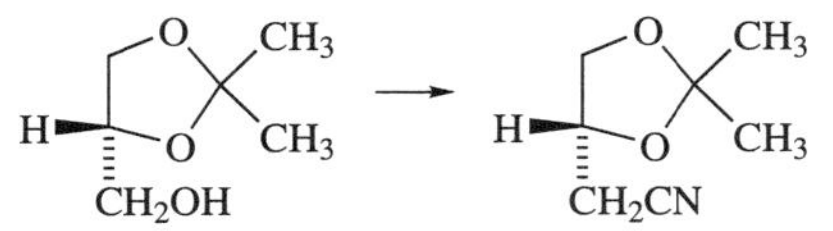

(d) 1,4-bis(CH_2OH)cyclohexane → 1,4-bis(CH_2SH)cyclohexane (more than one step is required)

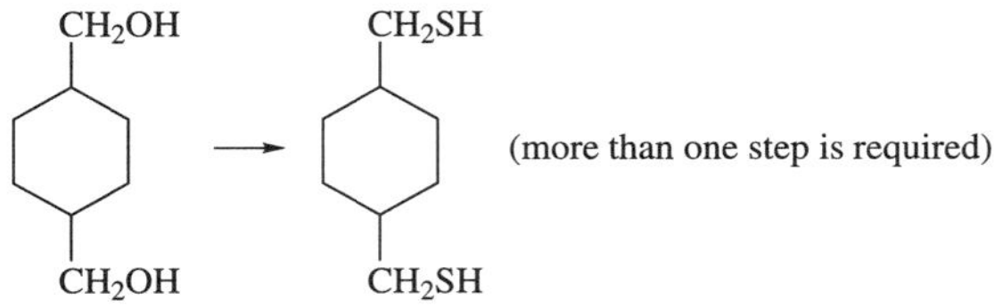

(e) $(CH_3)_3COCONHCH(CH_2CH_2CH_2CH_2OH)CO_2H$ → $(CH_3)_3COCONHCH(CH_2CH_2CH_2CH_2OH)CONHOCH_2C_6H_5$

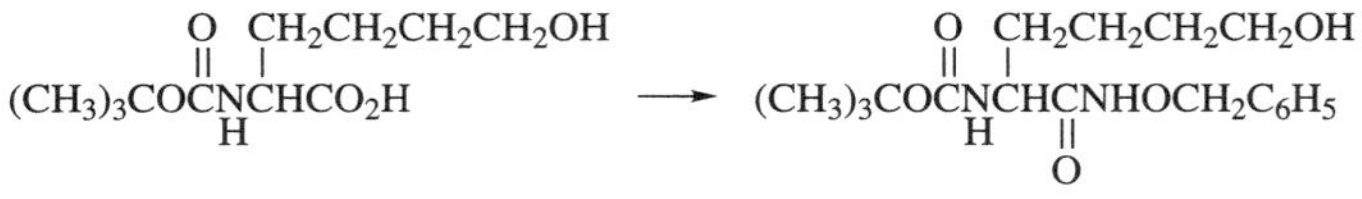

(f) HO, HO cyclopentane with $CH_2CH{=}CH(CH_2)_3CO_2CH_3$ and $CH{=}CHCH(OSiR_3)(CH_2)_4CH_3$ → Br, Br analog with $CH_2CH{=}CH(CH_2)_3CO_2CH_3$ and $CH{=}CHCH(OSiR_3)(CH_2)_4CH_3$

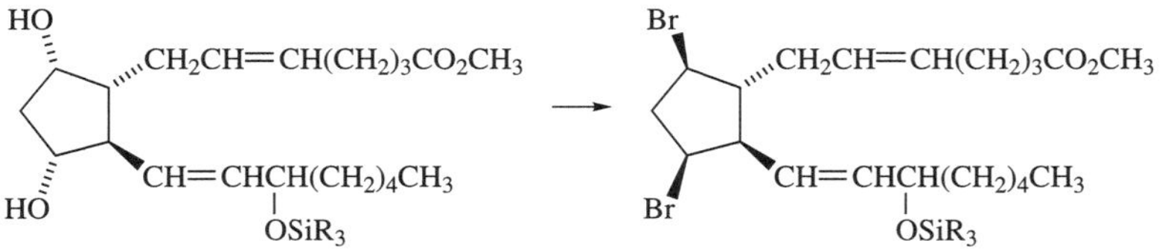

(g) $(CH_3)_2C(OH)CH_2CH(OH)CH_3$ → $(CH_3)_2C(Br)CH_2CH(OH)CH_3$

7. Provide a mechanistic interpretation for each of the following observations.

(a) A procedure for inverting the configuration of alcohols has been developed and demonstrated using cholesterol as a substrate:

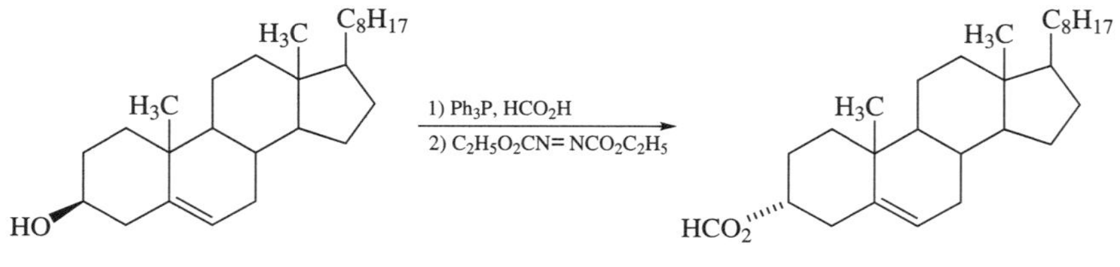

Cholesterol (HO, H_3C, H_3C, C_8H_{17}) —1) Ph_3P, HCO_2H; 2) $C_2H_5O_2CN{=}NCO_2C_2H_5$→ inverted formate ($HCO_2$, H_3C, H_3C, C_8H_{17})

Show the details of the mechanism of the key step which converts cholesterol to the inverted formate ester.

(b) It has been found that triphenylphosphine oxide reacts with trifluoromethylsulfonic anhydride to give an ionic substance with the composition of a simple 1 : 1 adduct. When this substance is added to a solution containing a carboxylic acid, followed by addition of an amine, amides are formed in good yield. Similarly, esters are formed by treating carboxylic acids first with the reagent and then with an alcohol. What is the likely structure for this ionic substance and how can it effect the activation of the carboxylic acids?

(c) Sulfonate esters having quarternary nitrogen substituents, such as **A** and **B**, show exceptionally high reactivity toward nucleophilic displacement reactions. Discuss factors which might contribute to the reactivity of these substances.

$ROSO_2CH_2CH_2\overset{+}{N}(CH_3)_3$ **A**

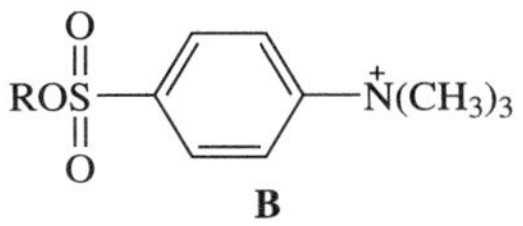

$RO\overset{O}{\underset{O}{S}}-C_6H_4-\overset{+}{N}(CH_3)_3$ **B**

(d) Alcohols react with hexachloroacetone in the presence of dimethylformamide to give alkyl trichloroacetates in high yield. Primary alcohols react fastest. Tertiary alcohols to not react. Suggest a reasonable mechanism for this reaction.

(e) The hydroxy amino acids serine and threonine can be converted to their respective bis(*O*-*t*-butyl) derivatives by reaction with isobutylene and sulfuric acid. Subsequent treatment with 1 equiv of trimethylsilyl triflate and then water cleaves the ester group but not the ether group. What is the basis for the selectivity?

$(CH_3)_3COCH(R)CH(NHCO_2CH_2Ph)CO_2C(CH_3)_3 \xrightarrow[2)\ H_2O]{1)\ (CH_3)_3SiO_3SCF_3,\ (C_2H_5)_3N} (CH_3)_3COCH(R)CH(NHCO_2CH_2Ph)CO_2H$

R = H or CH_3

8. Short synthetic sequences have been used to accomplish synthesis of the material at the left from that on the right. Suggest appropriate methods. No more than three separate steps should be required.

(a) $PhCH(OCH_3)C(=O)NHCH(CH_3)CH_2C_6H_5 \Longrightarrow PhCH(OCH_3)CO_2H$ (with retention of configuration)

(b) $(CH_3)_2CHCH_2CH{=}CHCH(CH_3)CH_2CO_2C_2H_5 \Longrightarrow$ 4-methyl-γ-butyrolactone (H_3C-substituted five-membered lactone)

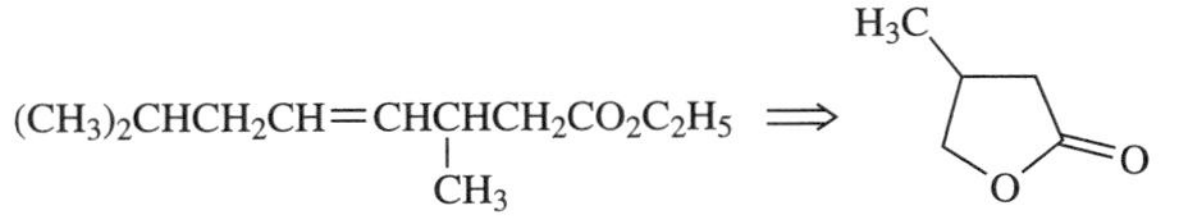

(c) TsO-substituted N-acetyl ($CH_3C{=}O$) pyrrolidine with CO_2CH_3 $\Longrightarrow$ HO-substituted pyrrolidine (N–H) with CO_2H

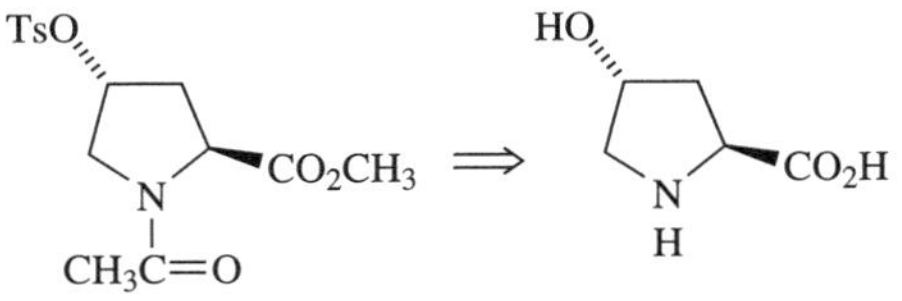

(d) $(CH_3)_2CH(H)C{=}C(H)\text{–}(H)C{=}C(CH_3)CH_2CN \Longrightarrow (CH_3)_2CH(H)C{=}C(H)\text{–}(H)C{=}C(CH_3)CH_2OH$

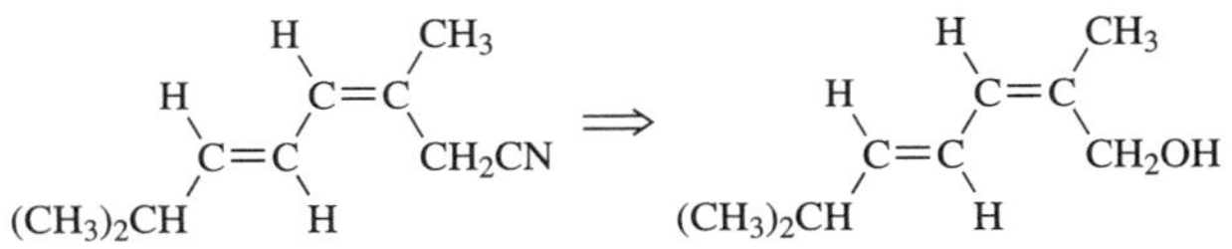

(e) 3,5-$(CH_3O)_2C_6H_3$–$CH(CH_3)CH(CO_2H)(CH_2)_4CH_3$ ⟹ 3,5-$(CH_3O)_2C_6H_3$–$CH(OH)CH_3$

9. Amino acids can be converted to epoxides in high enantiomeric purity by the following reaction sequence. Analyze the stereochemistry at each step of the reaction.

$$H_2N-C(CO_2H)(R)-H \xrightarrow[HCl]{NaNO_2} RCH(Cl)CO_2H \xrightarrow{LiAlH_4} RCH(Cl)CH_2OH \xrightarrow{KOH} \text{epoxide (H, R)}$$

10. A reagent which has been found to be useful for introduction of the benzyloxycarbonyl group onto amino groups of nucleosides is prepared by allowing benzyl chloroformate to react first with imidazole and then with trimethyloxonium tetrafluoroborate. What is the structure of the resulting reagent (a salt), and why is it an especially reactive acylating reagent?

11. (a) Write the equilibrium expression for phase transfer involving a tetraalkylammonium salt, $R_4N^+X^-$, NaOH, a water phase, and a nonaqueous phase.
(b) The concentration of ^-OH in the nonaqueous phase under phase-transfer conditions is a function of the anion X^-. What structural characteristics of X^- would be expected to influence the position of the equilibrium?
(c) It has been noted in a comparison of 15% aqueous NaOH versus 50% NaOH that the extent of transfer of ^-OH to the nonaqueous phase is less for 50% NaOH than for lower concentrations. What could be the cause of this?

12. The scope of the reaction of triphenylphosphine/hexachloroacetone with allylic alcohols has been studied. Primary and some secondary alcohols such as **1** and **2** give good yields of unrearranged halides. Certain other alcohols, such as **3** and **4**, give more complex mixtures. Discuss structural features which are probably important in determining how cleanly a given alcohol is converted to halide

1: $CH_3(H)C{=}C(H)CH_2OH$

2: $H_2C{=}C(H)CH(OH)CH_3$

3: $H_2C{=}C(H)C(CH_3)_2OH \xrightarrow[Cl_3CCOCCl_3]{Ph_3P} CH_2{=}CHC(Cl)(CH_3)_2$ (21%) + $ClCH_2CH{=}C(CH_3)_2$ (43%) + $CH_2{=}CHC(CH_3){=}CH_2$ (18%)

4: $H_2C{=}C(H)CH(OH)CH(CH_3)_2 \xrightarrow[Cl_3CCOCCl_3]{Ph_3P} CH_2{=}CHCH(Cl)CH(CH_3)_2$ (27%) + $ClCH_2CH{=}CHCH(CH_3)_2$ (15%) + $CH_2{=}CHCH{=}C(CH_3)_2$ (58%)

13. Two heterocyclic ring systems which have found some use in the formation of amides under mild conditions are *N*-alkyl-5-arylisoxazolium salts (structure **A**) and **N**-acyloxy-2-alkoxydihydroquinolines (structure **B**).

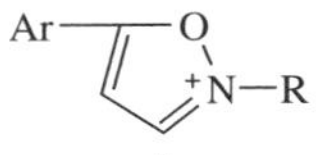
Ar
O
N—R

A

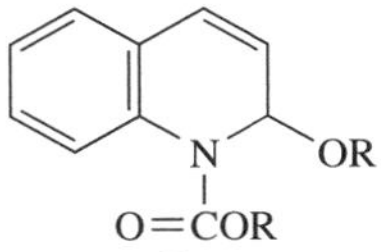
N
OR
O=COR

B

A typical set of reactions conditions is indicated below for each reagent. Consider mechanisms by which these heterocyclic molecules might function to activate the carboxylic acid group under these conditions, and outline the mechanisms you consider to be most likely.

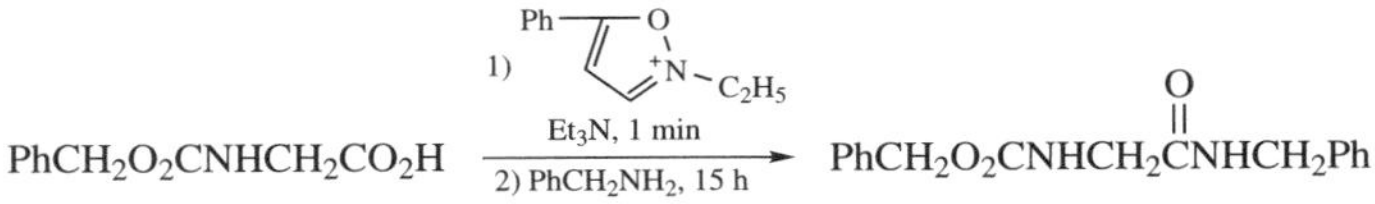
PhCH2O2CNHCH2CO2H
1)
Ph
O
N
C2H5
Et3N, 1 min
2) PhCH2NH2, 15 h
O
PhCH2O2CNHCH2CNHCH2Ph

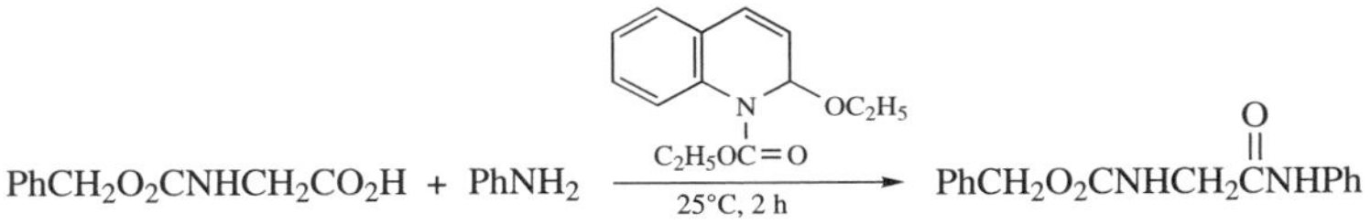
PhCH2O2CNHCH2CO2H + PhNH2
N
OC2H5
C2H5OC=O
25°C, 2 h
O
PhCH2O2CNHCH2CNHPh

14. Either because of potential interference with other functional groups present in the molecule or because of special structural features, the following reactions would require especially careful selection of reagents and reaction conditions. Identify the special requirements of each substrate and suggest appropriate conditions for effecting the desired transformation.

(a)

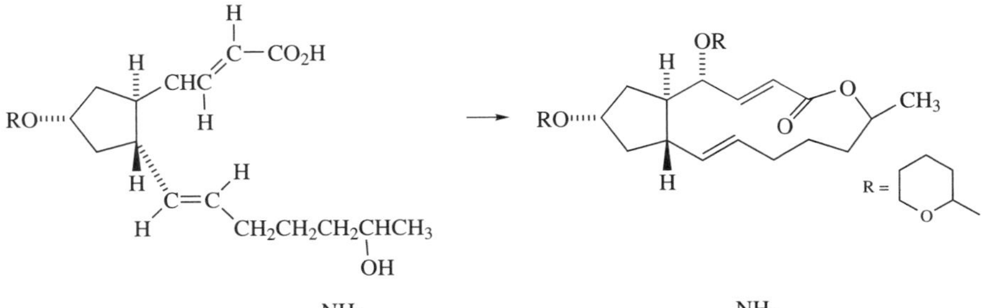
H
C—CO2H
H
CHC
H
RO
H
H
C=C
H
CH2CH2CH2CHCH3
OH
OR
H
O
CH3
RO
O
H
R =
O

(b)

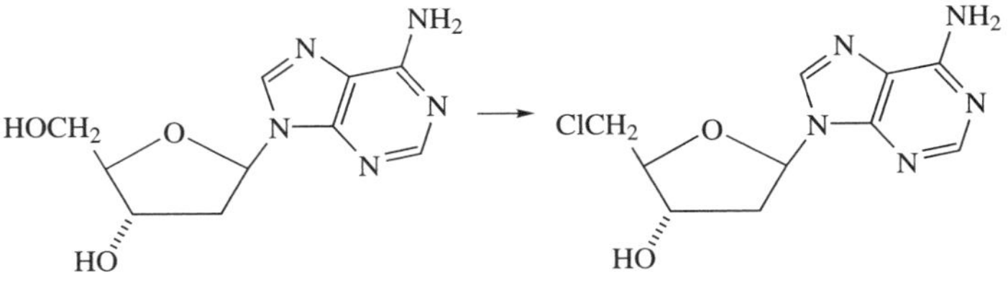
NH2
N
N
HOCH2
O
N
N
HO
NH2
N
N
ClCH2
O
N
N
HO

(c)

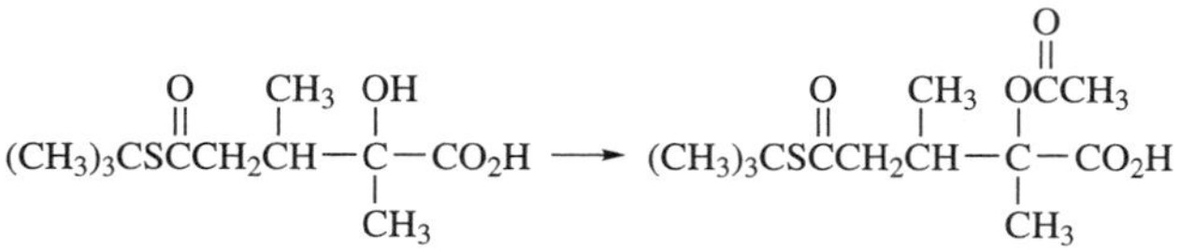

$$(CH_3)_3CSC(=O)CH_2CH(CH_3)-C(OH)(CH_3)-CO_2H \longrightarrow (CH_3)_3CSC(=O)CH_2CH(CH_3)-C(OC(=O)CH_3)(CH_3)-CO_2H$$

(d) 2,6-bis$((CH_3)_2CH)$-3,5-bis$(N(CH_3)_2)$-4-CO_2H benzene → 2,6-bis$((CH_3)_2CH)$-3,5-bis$(N(CH_3)_2)$-4-$CO_2C_2H_5$ benzene

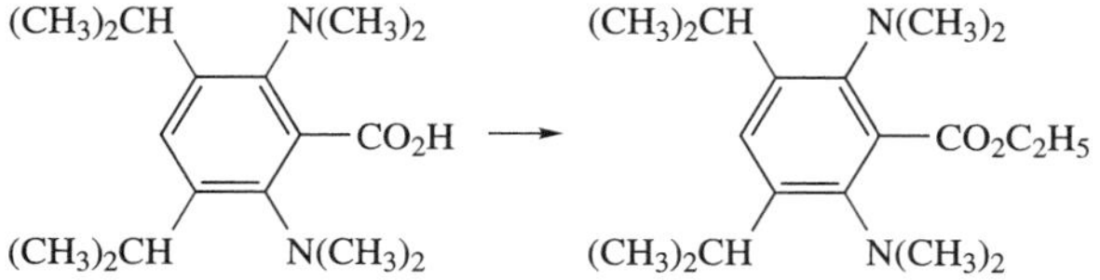

15. The preparation of nucleosides by reaction of carbohydrates and heterocyclic bases is fundamental to the study of the important biological activity of such substances. Several methods have been developed for accomplishing this reaction.

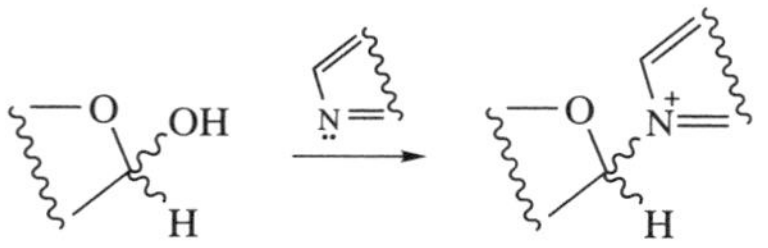

OH, H ⟶ (with N-heterocycle) N⁺, H

Application of 2-chloro-3-ethylbenzoxazolium chloride to this problem has been investigated using 2,3,4,6-tetra-*O*-acetyl-*β*-D-glucopyranose as the carbohydrate derivative. Good yields were observed, and, furthermore, the process was stereoselective, giving the *β*-nucleoside. Suggest a mechanism and explain the stereochemistry.

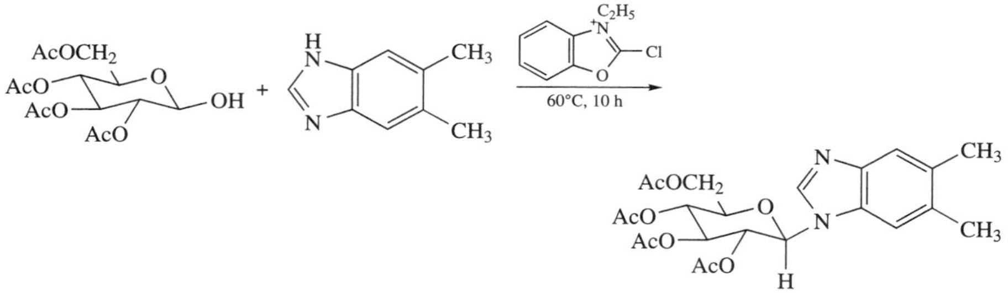

AcOCH₂, AcO, AcO, AcO, O, OH + 5,6-dimethylbenzimidazole (H, N, N, CH₃, CH₃) → (2-chloro-3-ethylbenzoxazolium, C_2H_5, $\overset{+}{N}$, O, Cl; 60°C, 10 h) → AcOCH₂, AcO, AcO, AcO, O, N, N, CH₃, CH₃, H

16. A route to α-glycosides has been described in which 2,3,4,6-tetra-*O*-benzyl-α-D-glucopyranosyl bromide is treated with an alcohol and tetraethylammonium bromide and diisopropylethylamiine in dichloromethane.

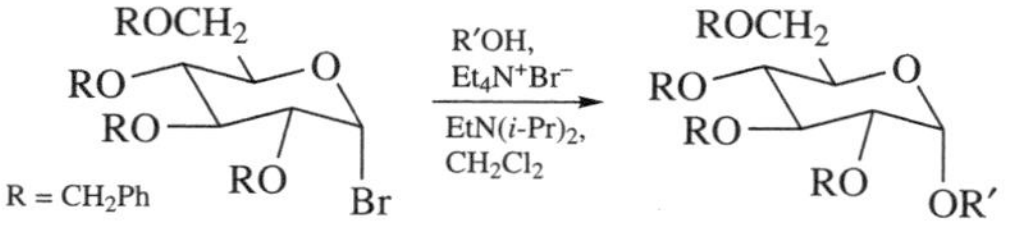

ROCH₂, RO, RO, RO, O, Br; R = CH₂Ph → ($R'OH$, $Et_4N^+Br^-$, $EtN(i\text{-}Pr)_2$, CH_2Cl_2) → ROCH₂, RO, RO, RO, O, OR′

Suggest an explanation for the stereochemical course of this reaction.

17. Write mechanisms for the formation of 2-pyridylthio esters by the following reactions.

(a) RCO_2H + (2-pyridyl)–S–S–(2-pyridyl) + PPh_3 ⟶ $RC(=O)$–S–(2-pyridyl) + $Ph_3P{=}O$

(b) RCO_2H + (2-pyridyl)–S–C(=O)Cl + R'_3N ⟶ $RC(=O)$–S–(2-pyridyl) + CO_2 + $R_3\overset{+}{N}H$ Cl^-

4

Electrophilic Additions to Carbon–Carbon Multiple Bonds

Introduction

One of the most general and useful reactions of alkenes and alkynes for synthetic purposes is the addition of electrophilic reagents. This chapter is restricted to reactions which proceed through polar intermediates or transition states. Several other classes of addition reactions are also of importance, and these are discussed elsewhere. Nucleophilic additions to electrophilic alkenes were covered in Chapter 1, and cycloadditions involving concerted mechanisms will be encountered in Chapter 6. Free-radical addition reactions are considered in Chapter 10.

4.1. Addition of Hydrogen Halides

Hydrogen chloride and hydrogen bromide react with alkenes to give addition products. In early work, it was observed that addition usually takes place to give the product in which the halogen atom is attached to the more substituted carbon of the double bond. This behavior was sufficiently general that the name *Markownikoff's rule* was given to the statement describing this mode of addition. A rudimentary picture of the reaction mechanism reveals the basis of Markownikoff's rule. The addition involves either protonation or a transition state involving a partial transfer of a proton to the double bond. The relative stability of the two possible carbocations from an unsymmetrical alkenes favors formation of the more substituted cationic intermediate. Addition is completed when the carbocation reacts with a halide anion.

$$R_2C{=}CH_2 + HX \longrightarrow R_2\overset{+}{C}{-}CH_3 + X^- \longrightarrow R_2CXCH_3$$

A more complete discussion of the mechanism of ionic addition of hydrogen halides to alkenes is given in Chapter 6 of Part A. In particular, the question of whether or not discrete carbocations are always involved is considered there.

The term *regioselective* is used to describe addition reactions that proceed selectively in one direction with unsymmetrical alkenes.[1] Markownikoff's rule describes a specific case of regioselectivity that is based on the stabilizing effect that alkyl and aryl substituents have on carbocations.

$$^{+}CH_2CH_2R \quad CH_3\overset{+}{C}HR \quad CH_3\overset{+}{C}HAr \quad CH_3\overset{+}{C}R_2 \quad CH_3\overset{+}{C}(Ar)_2$$

—— increasing stability ——→

Terminal and disubstituted internal alkenes react very slowly with HCl. The rate is greatly accelerated in the presence of silica or alumina in noncoordinating solvents such as dichloromethane or chloroform. Preparatively convenient conditions have been developed in which HCl is generated *in situ* from $SOCl_2$ or ClCOCOCl.[2] These heterogeneous reaction systems give Markownikoff addition. The mechanism is thought to involve interaction of the silica or alumina surface with HCl.

Another convenient procedure for hydrochlorination involves adding trimethylsilyl chloride to a mixture of an alkene and water. Good yields of HCl addition products (Markownikoff orientation) are obtained.[3] These conditions presumably involve generation of HCl from the silyl chloride, but it is unclear if the silicon plays any further role in the reaction.

$$CH_3CH{=}C(CH_3)CH_2CH_3 \xrightarrow[H_2O]{(CH_3)_3SiCl} CH_3CH_2C(CH_3)(Cl)CH_2CH_3 \quad 98\%$$

In nucelophilic solvents, products that arise from reaction of the solvent with the cationic intermediate may be encountered. For example, reaction of cyclohexene with hydrogen bromide in acetic acid gives cyclohexyl acetate as well as cyclohexyl bromide.

1. A. Hassner, *J. Org. Chem.* **33:**2684 (1968).
2. P. J. Kropp, K. A. Daus, M. W. Tubergen, K. D. Kepler, V. P. Wilson, S. L. Craig, M. M. Baillargeon, and G. W. Breton, *J. Am. Chem. Soc.* **115:**3071 (1993).
3. P. Boudjouk, B.-K. Kim, and B.-H. Han, *Synth. Commun.* **26:**3479 (1996); P. Boudjouk, B.-K. Kim, and B.-H. Han, *J. Chem. Ed.* **74:**1223 (1997).

This occurs because acetic acid acts as a nucleophile in competition with the bromide ion.

cyclohexene + HBr $\xrightarrow[40°C]{AcOH}$ bromocyclohexane (85%) + cyclohexyl acetate (15%) Ref. 4

Since carbocations are involved as intermediates, carbon skeleton rearrangement can occur during electrophilic addition reactions. Reaction of *t*-butylethylene with hydrogen chloride in acetic acid gives both rearranged and unrearranged chloride.[5]

$$(CH_3)_3CCH{=}CH_2 \xrightarrow[HCl]{AcOH} (CH_3)_3CCHCH_3(Cl) + (CH_3)_2CCH(CH_3)_2(Cl) + (CH_3)_3CCHCH_3(OAc)$$

35–40% 40–50% 15–20%

The stereochemistry of addition of hydrogen halides to alkenes is dependent on the structure of the alkene and also on the reaction conditions. Addition of hydrogen bromide to cyclohexene and to *E*- and *Z*-2-butene is *anti*.[6] The addition of hydrogen chloride to 1-methylcyclopentene is entirely *anti* when carried out at 25°C in nitromethane.[7]

[Scheme: deuterium-labeled 1-methylcyclopentene → *anti* addition product with Me, Cl, H, D labels]

1,2-Dimethylcyclohexene is an example of an alkene for which the stereochemistry of hydrogen chloride addition is dependent on the solvent and temperature. At −78°C in dichloromethane, 88% of the product is the result of *syn* addition, whereas at 0°C in ether, 95% of the product results from *anti* addition.[8] *Syn* addition is particulary common with alkenes having an aryl substituent. Table 4.1 lists examples of several alkenes for which the stereochemistry of addition of hydrogen chloride or hydrogen bromide has been studied.

The stereochemistry of addition, depends on the details of the mechanism. The addition can proceed through an ion-pair intermediate formed by an initial protonation step.

$$RCH{=}CH_2 + HCl \longrightarrow R\overset{+}{C}HCH_3\ Cl^- \longrightarrow RCHCH_3(Cl)$$

Most alkenes, however, react via a transition state that involves the alkene, hydrogen halide, and a third species which delivers the nucleophile. This termolecular mechanism is generally pictured as a nucleophilic attack on the alkene–hydrogen halide complex. This

4. R. C. Fahey and R. A. Smith, *J. Am. Chem. Soc.* **86:**5035 (1964).
5. R. C. Fahey and C. A. MckPherson, *J. Am. Chem. Soc.* **91:**3865 (1969).
6. D. J. Pasto, G. R. Meyer, and S. Kang, *J. Am. Chem. Soc.* **91:**4205 (1969).
7. Y. Pocker and K. D. Stevens, *J. Am. Chem. Soc.* **91:**4205 (1969).
8. K. B. Becker and C. A. Grob, *Synthesis* **1973:**789.

Table 4.1. Stereochemistry of Addition of Hydrogen Halides to Alkenes

Alkene	Hydrogen halide	Stereochemistry	Reference
1,2-Dimethylcyclohexene	HBr	*anti*	a
1,2-Dimethylcyclohexene	HCl	solvent- and temperature-dependent	a
Cyclohexene	HBr	*anti*	b
Z-2-Butene	DBr	*anti*	c
E-2-Butene	DBr	*anti*	c
1,2-Dimethylcyclopentene	HBr	*anti*	d
1-Methylcyclopentene	HCl	*anti*	e
Norbornene	HBr	*syn* and rearrangement	f
Norbornene	HCl	*syn* and rearrangement	g
E-1-Phenylpropene	HBr	*syn* (9 : 1)	h
Z-1-Phenylpropene	HBr	*syn* (8 : 1)	h
Bicyclo[3.1.0]hex-2-ene	DCl	*syn*	i
1-Phenyl-4-*t*-butylcyclohexene	DCl	*syn*	j

a. G. S. Hammond and T. D. Nevitt, *J. Am. Chem. Soc.* **76:**4121 (1954); R. C. Fahey and C. A. McPherson, *J. Am. Chem. Soc.* **93:**2445 (1971); K. B. Becker and C. A. Grob, *Synthesis* **1973:**789.
b. R. C. Fahey and R. A. Smith, *J. Am. Chem. Soc.* **86:**5035 (1964).
c. D. J. Pasto, G. R. Meyer, and B. Lepeska, *J. Am. Chem. Soc.* **96:**1858 (1974).
d. G. S. Hammond and C. H. Collins, *J. Am. Chem. Soc.* **82:**4323 (1960).
e. Y. Pocker and K. D. Stevens, *J. Am. Chem. Soc.* **91:**4205 (1969).
f. H. Kwart and J. L. Nyce. *J. Am. Chem. Soc.* **86:**2601 (1964).
g. J. K. Stille, F. M. Sonnenberg, and T. H. Kinstle, *J. Am. Chem. Soc.* **88:**4922 (1966).
h. M. J. S. Dewar and R. C. Fahey, *J. Am. Chem. Soc.* **85:**3645 (1963).
i. P. K. Freeman, F. A. Raymond, and M. F. Grostic, *J. Org. Chem.* **32:**24 (1967).
j. K. D. Berlin, R. O. Lyerla, D. E. Gibbs, and J. P. Devlin, *J. Chem. Soc., Chem. Commun.* **1970:**1246.

mechanism bypasses a discrete carbocation.

$$\text{RCH{=}CHR} + \text{HX} \rightleftharpoons \underset{\text{H—X}}{\overset{\text{H—X}}{\text{RCH{=}CHR}}} \longrightarrow \text{RCH(X)—CH(H)R} + \text{HX}$$

The major factor in determining which mechanism is followed is the stability of the carbocation intermediate. Alkenes that can give rise to a particularly stable carbocation are likely to react via the ion-pair mechanism. The ion-pair mechanism would not be expected to be stereospecific, because the carbocation intermediate permits loss of stereochemistry relative to the reactant alkene. It might be expected that the ion-pair mechanism would lead to a preference for *syn* addition, since at the instant of formation of the ion pair, the halide is on the same side of the alkene as the proton being added. Rapid collapse of the ion-pair intermediate leads to *syn* addition. If the lifetime of the ion pair is longer and the ion pair dissociates, a mixture of *syn* and *anti* addition products is formed. The termolecular mechanism is expected to give *anti* addition. Attack by the nucleophile occurs at the opposite side of the double bond from proton addition.

H—Cl

C—C

Nu:

Section 6.1 of Part A gives further discussion of the structural features that affect the competition between the two possible mechanisms.

4.2. Hydration and Other Acid-Catalyzed Additions of Oxygen Nucleophiles

Other nucleophilic species can be added to double bonds under acidic conditions. A fundamental example is the hydration of alkenes in strongly acidic aqueous solution:

$$R_2C{=}CH_2 + H^+ \longrightarrow R_2\overset{+}{C}CH_3 \xrightarrow{H_2O} R_2C(\overset{+}{O}H_2)CH_3 \xrightarrow{-H^+} R_2C(OH)CH_3$$

Addition of a proton occurs to give the more substituted carbocation and addition is regioselective, in accord with Markownikoff's rule. A more detailed discussion of the reaction mechanism is given in Section 6.2 of Part A. The reaction is occasionally applied to the synthesis of tertiary alcohols:

$$(CH_3)_2C{=}CHCH_2CH_2\overset{O}{\overset{\|}{C}}CH_3 \xrightarrow[H_2O]{H_2SO_4} (CH_3)_2C(OH)CH_2CH_2CH_2\overset{O}{\overset{\|}{C}}CH_3 \qquad \text{Ref. 9}$$

Because of the strongly acidic and rather vigorous conditions required to effect hydration of most alkenes, these conditions are only applicable to molecules that have no acid-sensitive functional groups. Also, because of the involvement of cationic intermediates, rearrangements can occur in systems where a more stable cation would result by aryl, alkyl, or hydrogen migration. A much milder and more general procedure for alkene hydration is discussed in the next section.

Addition of nucleophilic solvents such as alcohols and carboxylic acids can be effected by use of strong acids as catalysts[10]:

$$(CH_3)_2C{=}CH_2 + CH_3OH \xrightarrow{HBF_4} (CH_3)_3COCH_3$$

$$CH_3CH{=}CH_2 + CH_3CO_2H \xrightarrow{HBF_4} (CH_3)_2CHO_2CCH_3$$

Trifluoroacetic acid is a sufficiently strong acid to react with alkenes under relatively mild conditions.[11] The addition is regioselective in the direction predicted by Markownikoff's rule.

$$ClCH_2CH_2CH_2CH{=}CH_2 \xrightarrow[\Delta]{CF_3CO_2H} ClCH_2CH_2CH_2CH(O_2CCF_3)CH_3$$

9. J. Meinwald, *J. Am. Chem. Soc.* **77:**1617 (1955).
10. R. D. Morin and A. E. Bearse, *Ind. Eng. Chem.* **43:**1596 (1951); D. T. Dalgleish, D. C. Nonhebel, and P. L. Pauson, *J. Chem. Soc., C* **1971:**1174.
11. P. E. Peterson, R. J. Bopp, D. M. Chevli, E. L. Curran, D. E. Dillard, and R. J. Kamat, *J. Am. Chem. Soc.* **89:**5902 (1967).

Ring strain enhances alkene reactivity. Norbornene, for example, undergoes rapid addition at 0°C.[12]

4.3. Oxymercuration

The addition reactions which were discussed in Sections 4.1 and 4.2 are initiated by interaction of a proton with the alkene, which causes nucleophilic attack on the double bond. The role of the initial electrophile can be played by metal cations as well. Mercuric ion is the reactive electrophile in several synthetically valuable procedures.[13] The most commonly used reagent is mercuric acetate, but the trifluoroacetate, trifluoromethanesulfonate, or nitrate salts are preferable in some applications. A general mechanism depicts a mercurinium ion as an intermediate.[14] Such species can be detected by physical measurements when alkenes react with mercuric ions in nonnucleophilic solvents.[15] Depending on the structure of the particular alkene, the mercurinium ion may be predominantly bridged or open. The addition is completed by attack of a nucleophile at the more substituted carbon:

$$RCH{=}CH_2 + Hg(II) \longrightarrow R\overset{Hg^{2+}}{CH{\cdots}}CH_2 \text{ or } RCH{-}\overset{\overset{+}{Hg}}{CH_2} \xrightarrow{Nu^-} R\underset{Nu}{C}HCH_2{-}Hg^+$$

The nucleophiles that are used for synthetic purposes include water, alcohols, carboxylate ions, hydroperoxides, amines, and nitriles. After the addition step is complete, the mercury is usually reductively removed by sodium borohydride. The net result is the addition of hydrogen and the nucleophile to the alkene. The regioselectivity is excellent and is in the same sense as is observed for proton-initiated additions.[16] Scheme 4.1 includes examples of these reactions. Electrophilic attack by mercuric ion can affect cyclization by intramolecular capture of a nucleophilic functional group, as illustrated by entries 9–11. Inclusion of triethylboron in the reduction has been found to improve yields (entry 9).[17]

The reductive replacement of mercury using sodium borohydride is a free-radical process.[18]

$$RHgX + NaBH_4 \longrightarrow RHgH$$
$$RHgH \longrightarrow R\cdot + Hg^{(I)}H$$
$$R\cdot + RHgH \longrightarrow RH + Hg^{(II)} + R\cdot$$

12. H. C. Brown, J. H. Kawakami, and K.-T. Liu, *J. Am. Chem. Soc.* **92:**5536 (1979).
13. R. C. Larock, *Angew. Chem. Int. Ed. Engl.* **17:**27 (1978); W. Kitching, *Organomet, Chem. Rev.* **3:**61 (1968).
14. S. J. Cristol, J. S. Perry, Jr., and R. S. Beckley, *J. Org. Chem.* **41:**1912; D. J. Pasto and J. A. Gontarz, *J. Am. Chem. Soc.* **93:**6902 (1971).
15. G. A. Olah and P. R. Clifford, *J. Am. Chem. Soc.* **95:**6067 (1973); G. A. Olah and S. H. Yu, *J. Org. Chem.* **40:**3638 (1975).
16. H. C. Brown and P. J. Geoghegan, Jr., *J. Org. Chem.* **35:**1844 (1970); H. C. Brown, J. T. Kurek, M.-H. Rei, and K. L. Thompson, *J. Org. Chem.* **49:**2551 (1984); H. C. Brown, J. T. Kurek, M.-H. Rei, and K. L. Thompson, *J. Org. Chem.* **50:**1171 (1985).
17. S. H. Kang, J. H. Lee, and S. B. Lee, *Tetrahedron Lett.* **39:**59 (1998).
18. C. L. Hill and G. M. Whitesides, *J. Am. Chem. Soc.* **96:**870 (1974).

A. Alcohols

1[a] $(CH_3)_3CCH{=}CH_2$ —1) $Hg(OAc)_2$; 2) $NaBH_4$→ $(CH_3)_3CCH(OH)CH_3$ (97%) + $(CH_3)_3CCH_2CH_2OH$ (3%)

2[b] 3-[$(CH_2)_8CH{=}CH_2$]phthalide —1) $Hg(OAc)_2$; 2) $NaBH_4$→ 3-[$(CH_2)_8CH(OH)CH_3$]phthalide 80%

3[c] 2-methylenenorbornane ($=CH_2$) —1) $Hg(OAc)_2$; 2) $NaBH_4$→ 2-methyl-2-norbornanol (OH, CH_3) 99.5%

B. Ethers

4[d] cyclohexene —1) $Hg(O_2CCF_3)_2$, $(CH_3)_2CHOH$; 2) $NaBH_4$→ cyclohexyl–$OCH(CH_3)_2$ 98%

5[e] $CH_3(CH_2)_3CH{=}CH_2$ —$Hg(O_2CCF_3)_2$, EtOH→ $CH_3(CH_2)_3CH(OC_2H_5)CH_3$ 97%

C. Amides

6[f] $CH_3(CH_2)_3CH{=}CH_2$ —1) $Hg(NO_3)_2$, CH_3CN; 2) $NaBH_4$, H_2O→ $CH_3CH_2CH_2CH_2CH(HNCOCH_3)CH_3$ 92%

D. Peroxides

7[g] $CH_3(CH_2)_4CH{=}CHCH_3$ —1) $Hg(OAc)_2$, *t*-BuOOH; 2) $NaBH_4$→ $CH_3(CH_2)_4CH(OOC(CH_3)_3)CH_2CH_3$ 40%

E. Amines

8[h] 3-CH_3O-C_6H_4-$CH_2CH{=}CH_2$ + $PhCH_2NH_2$ —1) $Hg(ClO_4)_2$; 2) $NaBH_4$→ 3-CH_3O-C_6H_4-$CH_2CH(HNCH_2Ph)CH_3$ 70%

F. Cyclizations

9[i] $CH_2{=}CHCH_2CH(Ph)CO_2H$ —1) $Hg(O_2CCF_3)_2$, K_2CO_3; 2) $(C_2H_5)_3B$; 3) $NaBH_4$→ 5-CH_3-3-Ph-dihydrofuran-2-one (γ-lactone, =O) 93%

10[j] $PhCH_2OCH_2CH{=}CH\!-\!CH_2\!-\!OCH_2N(H)CO_2CH_2Ph$ (i.e. $PhCH_2OCH_2CH{=}CHCH_2OCH_2NHCO_2CH_2Ph$) —1) $Hg(NO_3)_2$; 2) $NaBH_4$→ 4-($PhCH_2OCH_2CH_2$)-3-($PhCH_2O_2C$)-oxazolidine 81%

Scheme 4.1. *(continued)*

11[k] CbzNH–CH(CH$_3$)–CH$_2$CH$_2$CH=CH$_2$ $\xrightarrow{\text{1) Hg(OAc)}_2\text{; 2) NaBr, NaHCO}_3\text{; 3) O}_2\text{, NaBH}_4}$ H$_3$C-pyrrolidine(N-Cbz)-CH$_2$OH 67%

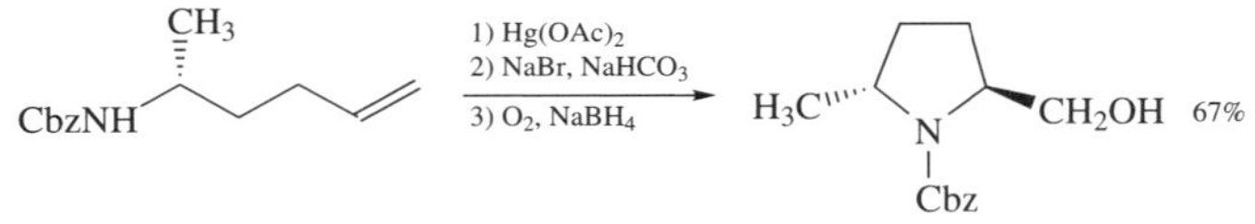

a. H. C. Brown and P. J. Geoghegan, Jr., *J. Org. Chem.* **35:**1844 (1970).
b. H. L. Wehrmeister and D. E. Roberston, *J. Org. Chem.* **33:**4173 (1968).
c. H. C. Brown and W. J. Hammar, *J. Am. Chem. Soc.* **89:**1524 (1967).
d. H. C. Brown and M.-H. Rei, *J. Am. Chem. Soc.* **91:**5646 (1969).
e. H. C. Brown, J. T. Kurek, M.-H. Rei, and K. L. Thompson, *J. Org. Chem.* **50:**1171 (1985).
f. H. C. Brown and J. T. Kurek, *J. Am. Chem. Soc.* **91:**5647 (1969).
g. D. H. Ballard and A. J. Bloodworth, *J. Chem. Soc. C.* **1971:**945.
h. R. C. Griffith, R. J. Gentile, T. A. Davidson, and F. L. Scott, *J. Org. Chem.* **44:**3580 (1979).
i. S. H. Kang, J. H. Lee, and S. B. Lee, *Tetrahedron Lett.* **39:**59 (1998).
j. K. E. Harding and D. R. Hollingsworth, *Tetrahedron Lett.* **29:**3789 (1988).

The evidence for this mechanism includes the fact that the course of the reaction can be diverted by oxygen, an efficient radical scavenger. In the presence of oxygen, the mercury is replaced by a hydroxy group. Also consistent with occurrence of a free-radical intermediate is the formation of cyclic products when hex-5-enylmercury compounds are reduced with sodium borohydride.[19] In the presence of oxygen, no cyclic product is formed, indicating that O_2 trapping of the radical is much faster than cyclization.

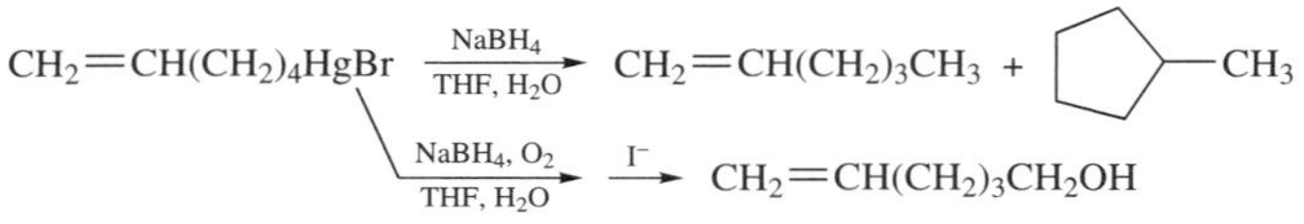

$$CH_2{=}CH(CH_2)_4HgBr \xrightarrow[\text{THF, H}_2\text{O}]{\text{NaBH}_4} CH_2{=}CH(CH_2)_3CH_3 + \text{(methylcyclopentane) } CH_3$$

$$CH_2{=}CH(CH_2)_4HgBr \xrightarrow[\text{THF, H}_2\text{O}]{\text{NaBH}_4\text{, O}_2} \xrightarrow{I^-} CH_2{=}CH(CH_2)_3CH_2OH$$

The trapping of the radical intermediate by oxygen has been exploited as a method for introduction of a hydroxyl substituent. The example below and entry 11 in Scheme 4.1 illustrate this reaction.

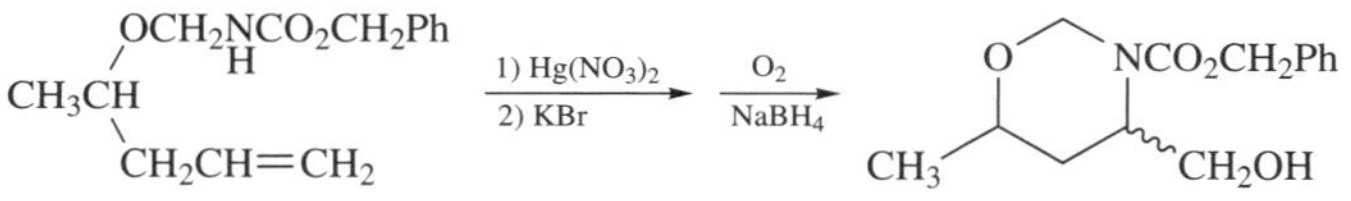

$$CH_3CH(OCH_2NHCO_2CH_2Ph)CH_2CH{=}CH_2 \xrightarrow{\text{1) Hg(NO}_3)_2\text{; 2) KBr}} \xrightarrow[\text{NaBH}_4]{\text{O}_2} \text{(CH}_3\text{, CH}_2\text{OH-substituted 1,3-oxazinane, NCO}_2\text{CH}_2\text{Ph)} \quad 80\%$$ Ref. 20

An alternative reagent for demercuration is sodium amalgam in a protic solvent. Here the evidence is that free radicals are not involved and the mercury is replaced with complete retention of configuration[21]:

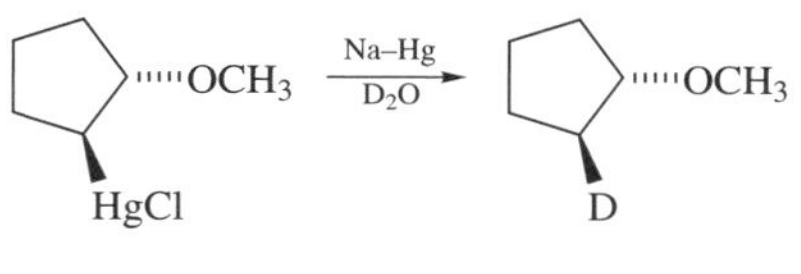

$$\text{(cyclopentane: OCH}_3\text{, HgCl)} \xrightarrow[\text{D}_2\text{O}]{\text{Na–Hg}} \text{(cyclopentane: OCH}_3\text{, D)}$$

19. R. P. Quirk and R. E. Lea, *J. Am. Chem. Soc.* **98:**5973 (1976).
20. K. E. Harding, T. H. Marman and D. Nam, *Tetrahedron Lett.* **29:**1627 (1988).
21. F. R. Jensen, J. J. Miller, S. J. Cristol, and R. S. Beckley, *J. Org. Chem.* **37:**434 (1972); R. P. Quirk, *J. Org. Chem.* **37:**3554 (1972); W. Kitching, A. R. Atkins, G. Wickham, and V. Alberts, *J. Org. Chem.* **46:**563 (1981).

The stereochemistry of oxymercuration has been examined in a number of systems. Conformationally biased cyclic alkenes such as 4-*t*-butylcyclohexene and 4-*t*-butyl-1-methylcyclohexene give exclusively the product of *anti* addition, which is consistent with a mercurinium ion intermediate.[16,22]

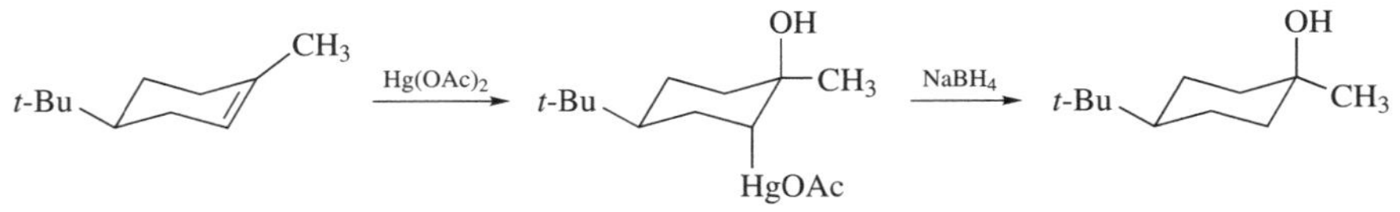

The reactivity of different alkenes toward mercuration spans a considerable range and is governed by a combination of steric and electronic factors.[23] Terminal double bonds are more reactive than internal ones. Disubstituted terminal alkenes, however, are more reactive than monosubstituted ones, as would be expected for electrophilic attack. The differences in relative reactivities are large enough that selectivity can be achieved in certain dienes:

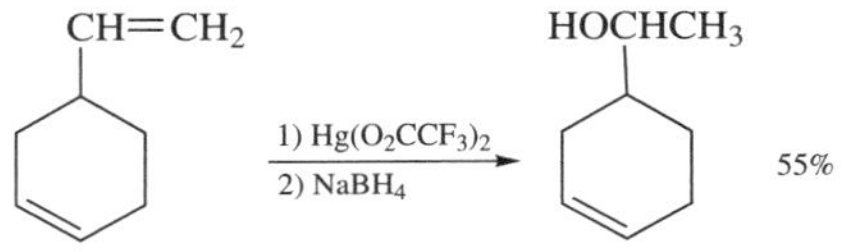

55% Ref. 23b

The relative reactivity data for some pentene derivatives are given in Table 4.2.

Diastereoselectivity has been observed in oxymercuration of alkenes with nearby oxygen substituents. Terminal allylic alcohols show a preference for formation of the *anti*

Table 4.2. Relative Reactivity of Some Alkenes in Oxymercuration

Alkene	Relative reactivity[a]
1-Pentene	6.6
2-Methyl-1-pentene	48
Z-2-Pentene	0.56
E-2-Pentene	0.17
2-Methyl-2-pentene	1.24

a. Relative to cyclohexene; data from H. C. Brown and P. J. Geoghegan, Jr., *J. Org. Chem.* **37:**1937 (1972).

22. H. C. Brown, G. J. Lynch, W. J. Hammar, and L. C. Liu, *J. Org. Chem.* **44:**1910 (1979).
23. H. C. Brown and J. P. Geoghegan, Jr., *J. Org. Chem.* **37:**1937 (1972); H. C. Brown, P. J. Geoghegan, Jr., G. J. Lynch, and J. T. Kurek, *J. Org. Chem.* **37:**1941 (1972); H. C. Brown, P. J. Geoghegan, Jr., and J. T. Kurek, *J. Org. Chem.* **46:**3810 (1981).

2,3-diols.

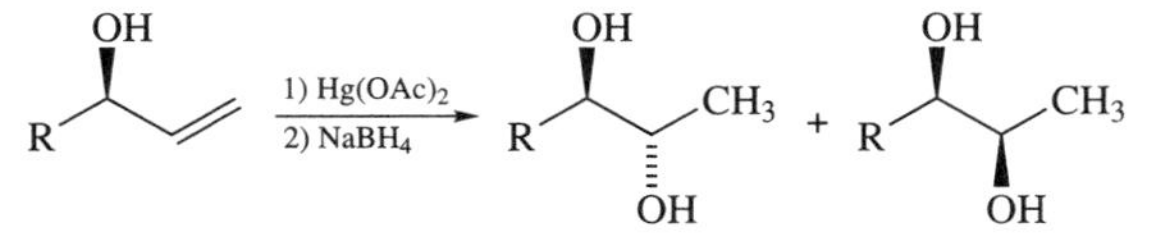

R	*anti*	*syn*
Et	76	24
i-Pr	80	20
t-Bu	98	2
Ph	88	12

This result can be explained in terms of a steric preference for transition state **A** over **B**. The approach of the mercuric ion is directed by the hydroxyl group. The selectivity increases with the size of the substituent R.[24]

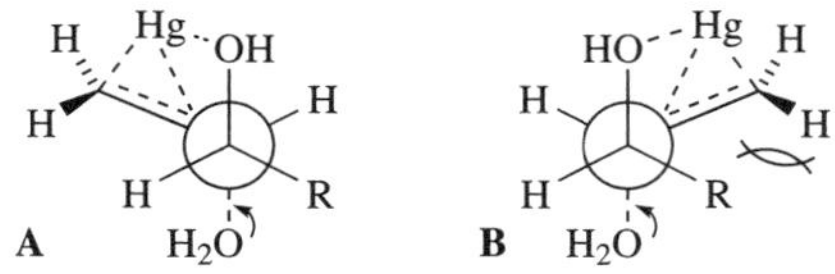

When the hydroxyl group is acetylated, the *syn* isomer is preferred. This result is attributed to direct nucleophilic participation by the carbonyl oxygen of the ester.

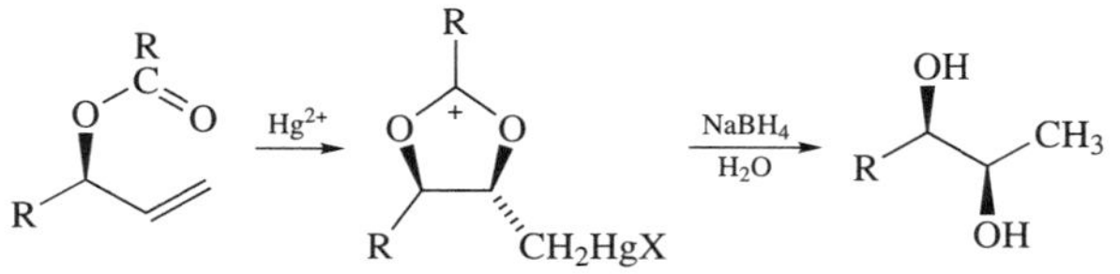

4.4. Addition of Halogens to Alkenes

The addition of chlorine or bromine to alkenes is a very general reaction. Considerable insight has been gained into the mechanism of halogen addition by studies on the stereochemistry of the reaction. Most types of alkenes are known to add bromine in a stereospecific manner, giving the product of *anti* addition. Among the alkenes that are known to give *anti* addition products are maleic and fumaric acid, *Z*-2-butene, *E*-2-butene, and a number of cycloalkenes.[25] Cyclic, positively charged bromonium ion intermediates provide an explanation for the observed stereospecificity.

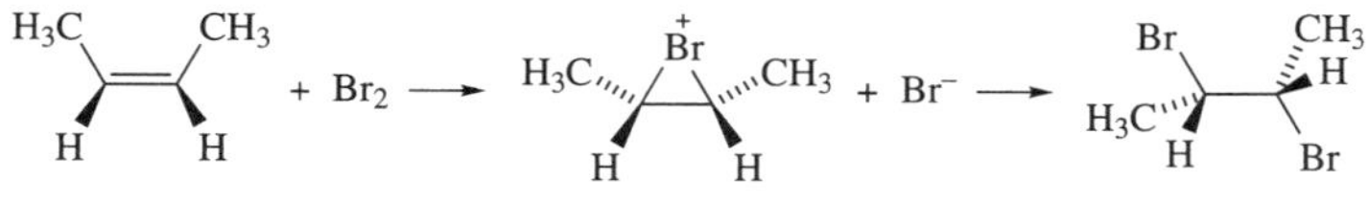

The bridging by bromine prevents rotation about the remaining bond and back-side nucleophilic opening of the bromonium ion by bromide ion leads to the observed *anti*

24. B. Giese and D. Bartmann, *Tetrahedron Lett.* **26:** 1197 (1985).
25. J. H. Rolston and K. Yates, *J. Am. Chem. Soc.* **91:**1469, 1477 (1969).

addition. Direct evidence for the existence of bromonium ions has been obtained from NMR measurements.[26] A bromonium ion salt (with Br_3^- as the counterion) has been isolated from the reaction of bromine with the very hindered alkene adamantylideneadamantane.[27] (See Part A, Section 6.3, for further mechanistic discussion.)

Substantial amounts of *syn* addition have been observed for *cis*-1-phenylpropene (27–80% *syn* addition), *trans*-1-phenylpropene (17–29% *syn* addition), and *cis*-stilbene (up to 90% *syn* addition in polar solvents).

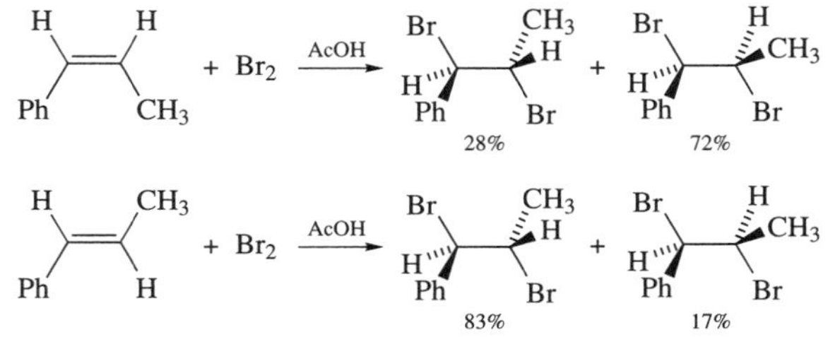

H H Ph CH3 + Br2 →(AcOH) Br CH3 H H Ph Br 28% + Br H CH3 H Ph Br 72%

H CH3 Ph H + Br2 →(AcOH) Br CH3 H H Ph Br 83% + Br H CH3 H Ph Br 17%

Ref. 28

A common feature of the compounds that give extensive *syn* addition is the presence of at least one phenyl substituent on the double bond. The presence of a phenyl substituent diminishes the strength of bromonium ion bridging by stabilizing the cationic center. A weakly bridged structure in equilibrium with an open benzylic cation can account for the loss in stereospecificity.

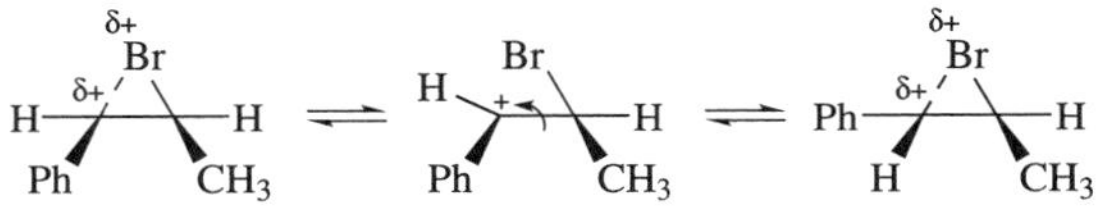

δ+ Br δ+ H H Ph CH3 ⇌ Br H + H Ph CH3 ⇌ δ+ Br δ+ Ph H H CH3

The diminished stereospecificity is similar to that noted for hydrogen halide addition to phenyl-substituted alkenes.

Although chlorination of aliphatic alkenes usually gives *anti* addition, *syn* addition is often dominant for phenyl-substituted alkenes[29]:

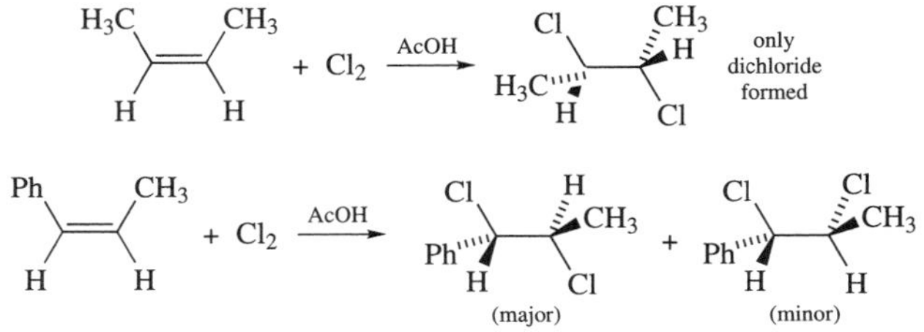

H3C CH3 H H + Cl2 →(AcOH) Cl CH3 H H3C H Cl only dichloride formed

Ph CH3 H H + Cl2 →(AcOH) Cl H CH3 Ph H Cl (major) + Cl Cl CH3 Ph H H (minor)

These results, too, reflect a difference in the extent of bridging in the intermediates. With

26. G. A. Olah, J. M. Bollinger, and J. Brinich, *J. Am. Chem. Soc.* **90**:2587 (1968); G. A. Olah, P. Schilling, P. W. Westerman, and H. C. Lin, *J. Am. Chem. Soc.* **96**:3581 (1974).
27. J. Strating, J. H. Wierenga, and H. Wynberg, *J. Chem. Soc., Chem. Commun.* **1969**:907.
28. J. H. Rolston and K. Yates, *J. Am. Chem. Soc.* **91**:1469, 1477 (1969).
29. M. L. Poutsma, *J. Am. Chem. Soc.* **87**:2161, 2172 (1965); R. C. Fahey, *J. Am. Chem. Soc.* **88**:4681 (1966); R. C. Fahey and C. Shubert, *J. Am. Chem. Soc.* **87**:5172 (1965).

unconjugated alkenes, there is strong bridging and high *anti* stereospecificity. Phenyl substitution leads to greater cationic character at the benzylic site, and there is more *syn* addition. Because of its smaller size and lesser polarizability, chlorine is not as effective as bromine in maintaining bridging for any particular alkene. Bromination therefore generally gives a higher degree of *anti* addition than chlorination, all other factors being the same.[30]

Chlorination can be accompanied by other reactions that are indicative of carbocation intermediates. Branched alkenes can give products that are the result of elimination of a proton from a cationic intermediate.

$$(CH_3)_2C{=}CH_2 \xrightarrow{Cl_2} (CH_3)_2\overset{+}{C}{-}CH_2Cl \longrightarrow H_2C{=}C(CH_3){-}CH_2Cl \quad 80\%$$

$$(CH_3)_2C{=}C(CH_3)_2 \xrightarrow{Cl_2} (CH_3)_2\overset{+}{C}{-}C(Cl)(CH_3)_2 \longrightarrow H_2C{=}C(CH_3)C(Cl)(CH_3)_2 \quad 99\%$$

Skeletal rearrangements have also been observed in systems that are prone toward migration.

$$(CH_3)_3CCH{=}CHC(CH_3)_3 \xrightarrow{Cl_2} CH_3C({=}CH_2)CH(CH_3)CH(Cl)C(CH_3)_3 \qquad \text{Ref. 31}$$

$$Ph_3CCH{=}CH_2 \xrightarrow{Br_2} Ph_3CCH(Br)CH_2Br + Ph_2C{=}C(Ph)CH_2Br \qquad \text{Ref. 32}$$

Because halogenation involves electrophilic attack, substituents on the double bond that increase electron density increase the rate of reaction, whereas electron-withdrawing substituents have the opposite effect. Bromination of simple alkenes is an extremely fast reaction. Some specific rate data are tabulated and discussed in Section 6.3 of Part A.

In nucleophilic solvents, the solvent can compete with halide ion for the cationic intermediate. For example, the bromination of styrene in acetic acid leads to substantial amounts of the acetoxybromo derivative.

$$PhCH{=}CH_2 + Br_2 \xrightarrow{AcOH} \underset{80\%}{PhCH(Br)CH_2Br} + \underset{20\%}{PhCH(OAc)CH_2Br} \qquad \text{Ref. 33}$$

30. R. J. Abraham and J. R. Monasterios, *J. Chem. Soc., Perkin Trans. 1* **1973:**1446.
31. M. L. Poutsma, *J. Am. Chem. Soc.* **87:**4285 (1965).
32. R. O. C. Norman and C. B. Thomas, *J. Chem. Soc. B* **1967:**598.
33. J. H. Rolston and K. Yates, *J. Am. Chem. Soc.* **91:**1469 (1969).

The acetoxy group is introduced exclusively at the benzylic carbon, in accord with the intermediate being a weakly bridged species or a benzylic cation.

$$\text{H}\,\overset{\delta+}{\ }\ \overset{\delta+}{\text{Br}}\text{-bridged Ph intermediate} + \text{AcO–H} \longrightarrow \text{PhCH(OAc)–CH}_2\text{Br}$$

The addition of bromide salts to the reaction mixture diminishes the amount of acetoxy compound formed by tipping the competition between acetic acid and bromide ion for the electrophile in favor of the bromide ion. Chlorination in nucleophilic solvents can also lead to solvent incorporation, as, for example, in the chlorination of phenylpropene in methanol:[34]

$$\text{PhCH{=}CHCH}_3 + \text{Cl}_2 \xrightarrow{\text{CH}_3\text{OH}} \underset{82\%}{\text{PhCH(OCH}_3\text{)CH(Cl)CH}_3} + \underset{18\%}{\text{PhCH(Cl)–CH(Cl)CH}_3}$$

From a synthetic point of view, the participation of water in brominations, leading to bromohydrins, is the most important example of nucleophilic participation by solvent. In the case of unsymmetrical alkenes, water reacts at the more substituted carbon, which is the carbon with the greatest cationic character. To favor introduction of water, it is necessary to keep the concentration of the bromide ion as low as possible. One method for accomplishing this is to use *N*-bromosuccinimide (NBS) as the brominating reagent.[35,36] High yields of bromohydrins are obtained by use of NBS in aqueous DMSO. The reaction is a stereospecific *anti* addition. As in bromination, a bromonium ion intermediate can explain the *anti* stereospecificity. It has been shown that the reactions in DMSO involve initial nucleophilic attack by the sulfoxide oxygen. The resulting intermediate reacts with water to give the bromohydrin.

$$\text{RCH{=}CH}_2 \xrightarrow{\text{Br}^+} \text{R–CH–CH}_2\ (\text{Br}^+\text{ bridged}) \xrightarrow{(\text{CH}_3)_2\text{S{=}O}} (\text{CH}_3)_2\overset{+}{\text{S}}\text{–O–CH(R)CH}_2\text{Br} \xrightarrow{\text{H}_2\text{O}} \text{HOCH(R)CH}_2\text{Br} + \text{H}^+$$

In accord with the Markownikoff rule, the hydroxyl group is introduced at the carbon best able to support positive charge:

$$(\text{CH}_3)_3\text{C(CH}_3\text{)C{=}CH}_2 \xrightarrow[\text{DMSO, H}_2\text{O}]{\text{NBS}} (\text{CH}_3)_3\text{CC(CH}_3\text{)(OH)–CH}_2\text{Br}\quad 60\% \qquad \text{Ref. 37}$$

$$\text{PhCH}_2\text{CH{=}CH}_2 \xrightarrow[\text{DMSO, H}_2\text{O}]{\text{NBS}} \text{PhCH}_2\text{CH(OH)CH}_2\text{Br}\quad 89\% \qquad \text{Ref. 38}$$

34. M. L. Poutsma and J. L. Kartch, *J. Am. Chem. Soc.* **89**:6595 (1967).
35. A. J. Sisti and M. Meyers, *J. Org. Chem.* **38**:4431 (1973).
36. C. O. Guss and R. Rosenthal, *J. Am. Chem. Soc.* **77**:2549 (1955).
37. D. R. Dalton, V. P. Dutta, and D. C. Jones, *J. Am. Chem. Soc.* **90**:5498 (1968).
38. A. W. Langman and D. R. Dalton, *Org. Synth.* **59**:16 (1979).

Another procedure which is useful for the preparation of both bromohydrins and iodohydrins involves *in situ* generation of the hypohalous acid from $NaBrO_3$ and $NaIO_4$.[39]

$NaBrO_3$, $NaHSO_3$, H_2O, CH_3CN → trans-2-bromocyclohexanol 75%

HIO_4, $NaHSO_3$, H_2O, CH_3CN → trans-2-iodocyclohexanol 80%

These reactions show the same regioselectivity and stereoselectivity as other reactions which proceed through halonium ion intermediates.

Because of its high reactivity, special precautions must be used in reactions of fluorine, and its use is somewhat specialized.[40] Nevertheless, there is some basis for comparison with the less reactive halogens. Addition of fluorine to *Z*- and *E*-1-propenylbenzene is not stereospecific, but *syn* addition is somewhat favored.[41] This result suggests formation of a cationic intermediate.

$$PhCH{=}CHCH_3 \xrightarrow{F_2} PhCHFCHFCH_3 \text{ (two diastereomers)}$$

In methanol, the solvent incorporation product is formed, as would be expected for a cationic intermediate.

$$PhCH{=}CHCH_3 \xrightarrow[MeOH]{F_2} PhCH(OMe)CH(F)CH_3$$

These results are consistent with the expectation that fluorine would not be an effective bridging atom. There are other reagents, such as CF_3OF and CH_3CO_2F, which appear to transfer an electrophile fluorine to double bonds and form an ion pair that collapses to an addition product.

$$PhCH{=}CHPh + CF_3OF \longrightarrow PhCH(OCF_3)CH(F)Ph$$ Ref. 42

$$CH_3(CH_2)_9CH{=}CH_2 + CH_3CO_2F \longrightarrow CH_3(CH_2)_9CH(O_2CCH_3)CH_2F \quad 30\%$$ Ref. 43

39. H. Masuda, K. Takase, M. Nishio, A. Hasegawa, Y. Nishiyama, and Y. Ishii, *J. Org. Chem.* **59:**5550 (1994).
40. H. Vypel, *Chimia* **39:**305 (1985).
41. R. F. Merritt, *J. Am. Chem. Soc.* **89:**609 (1967).
42. D. H. R. Barton, R. H. Hesse, G. P. Jackman, L. Ogunkoya, and M. M. Pechet, *J. Chem. Soc., Perkin Trans. 1* **1974:**739.
43. S. Rozen, O. Lerman, M. Kol, and D. Hebel, *J. Org. Chem.* **50:**4753 (1985).

The stability of hypofluorites is improved in derivatives having electron-withdrawing substituents, such as 2,2-dichloropropanoyl hypofluorite.[44]

Various other fluorinating agents have been developed and used. These include *N*-fluoropyridinium salts such as the triflate[45] and heptafluorodiborate.[46] The reactivity of these reagents can be "tuned" by variation of the pyridine ring substituents. In contrast to the hypofluorites, these reagents are storable.[47] In nucleophilic solvents such as acetic acid or alcohols, the reagents give addition products whereas in nonnucleophilic solvents, alkene substitution products resulting from a carbocation intermediate are formed.

$$PhC(CH_3)=CH_2 \xrightarrow[(CH_3)_2CHOH]{\text{2,6-dichloro-}N\text{-fluoropyridinium}} PhC(CH_3)(OCH(CH_3)_2)CH_2F \quad 70\%$$

$$PhC(CH_3)=CH_2 \xrightarrow[\text{2,6-dichloro-}N\text{-fluoropyridinium}]{CH_2Cl_2} PhC(=CH_2)CH_2F \quad 73\%$$

Addition of iodine to alkenes can be accomplished by a photochemically initiated reaction. Elimination of iodine is catalyzed by excess iodine radicals, but the diiodo compounds can be obtained if unreacted iodine is removed.[48]

$$RCH{=}CHR + I_2 \rightleftharpoons RCH(I){-}CH(I)R$$

The diiodo compounds are very sensitive to light and have not been used very often in synthesis.

Iodine is a very good electrophile for effecting intramolecular nucleophilic addition to alkenes, as exemplified by the iodolactonization reaction.[49] Reaction of iodine with carboxylic acids having carbon–carbon double bonds placed to permit intramolecular reaction results in formation of iodolactones.[50] The reaction shows a preference for formation of five-membered rings over six-membered ones[51] and is a strictly *anti* stereospecific addition when carried out under basic conditions.

$$\text{cyclopentene with } CH_2CO_2H \text{ and } CH{=}CH_2 \xrightarrow[NaHCO_3]{I_2,\ I^-} \text{iodolactone bearing } CH{=}CH_2$$ Ref. 52

44. S. Rozen and D. Hebel, *J. Org. Chem.* **55:**2621 (1990).
45. T. Umemoto, S. Fukami, G. Tomizawa, K. Harasawa, K. Kawada, and K. Tomita, *J. Am. Chem. Soc.* **112:**8563 (1990).
46. A. J. Poss, M. Van Der Puy, D. Nalewajek, G. A. Shia, W. J. Wagner, and R. L. Frenette, *J. Org. Chem.* **56:**5962 (1991).
47. T. Umemoto, K. Tomita, and K. Kawada, *Org. Synth.* **69:**129 (1990).
48. P. S. Skell and R. R. Pavlis, *J. Am. Chem. Soc.* **86:**2956 (1964); R. L. Ayres, C. J. Michejda, and E. P. Rack, *J. Am. Chem. Soc.* **93:**1389 (1971).
49. G. Cardillo and M. Orena, *Tetrahedron* **46:**3321 (1990).
50. M. D. Dowle and D. I. Davies, *Chem. Soc. Rev.* **8:**171 (1979).
51. S. Ranganathan, D. Ranganathan, and A. K. Mehrota, *Tetrahedron* **33:**807 (1977); C. V. Ramana, K. R. Reddy, and M. Nagarajan, *Ind. J. Chem. B* **35:**534 (1996).

The *anti* addition is kinetically controlled and results from irreversible back-side opening of an iodonium ion intermediate by the carboxylate nucleophile.

When iodolactonization is carried out under nonbasic conditions, the addition step becomes reversible and the product is then the thermodynamically favored one.[53] This usually results in the formation of the stereoisomeric lactone which has adjacent substituents *trans* with respect to one another.

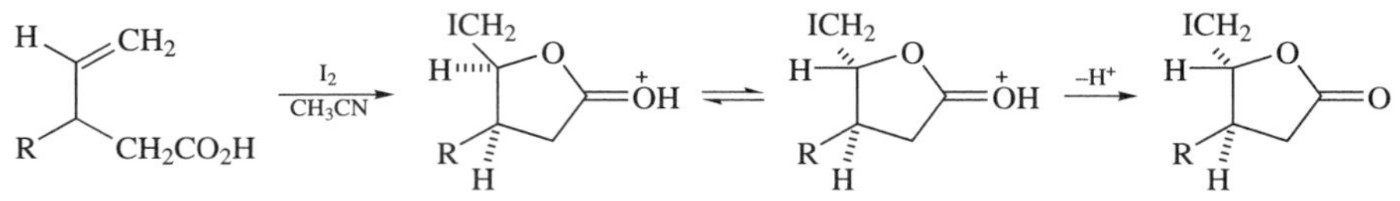

Several other nucleophilic functional groups can be induced to participate in iodocyclization reactions. *t*-Butyl carbonate esters cyclize to diol carbonates[54]:

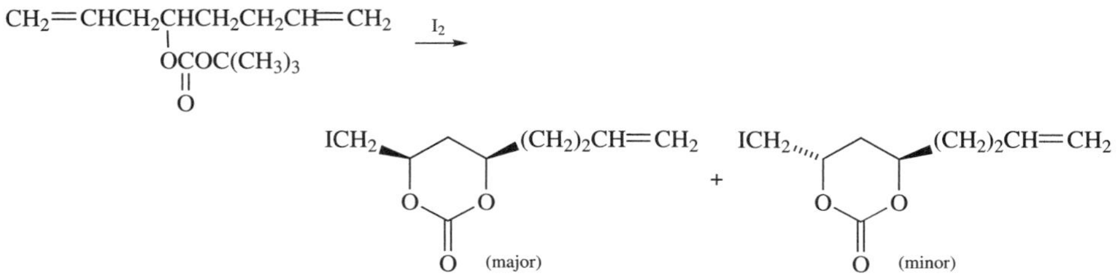

Enhanced stereoselectivity has been found using IBr, which reacts at a lower temperature.[55] Lithium salts of carbonate monoesters can also be prepared and cyclized.[56]

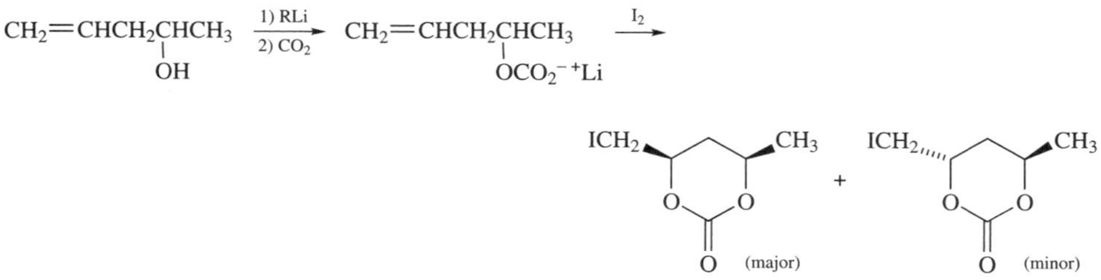

52. L. A. Paquette, G. D. Crouse, and A. K. Sharma, *J. Am. Chem. Soc.* **102:**3972 (1980).
53. P. A. Bartlett and J. Myerson, *J. Am. Chem. Soc.* **100:**3950 (1978).
54. P. A. Bartlett, J. D. Meadows, E. G. Brown, A. Morimoto, and K. K. Jernstedt, *J. Org. Chem.* **47:**4013 (1982).
55. J. J.-W. Duan and A. B. Smith III, *J. Org. Chem.* **58:**3703 (1993).
56. A. Bogini, G. Cardillo, M. Orena, G. Ponzi, and S. Sandri, *J. Org. Chem.* **47:**4626 (1982).

Because the iodocyclization products have a potentially nucleophilic oxygen substituent β to the iodide, they are useful in stereospecific synthesis of epoxides and diols:

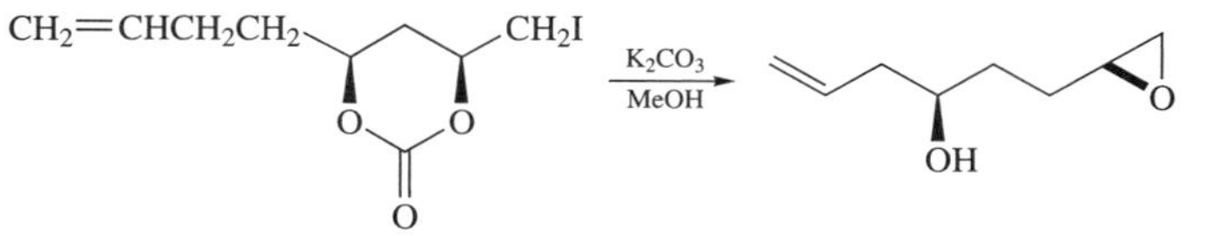

$CH_2{=}CHCH_2CH_2$ ... CH_2I $\xrightarrow[\text{MeOH}]{K_2CO_3}$ ... OH ... Ref. 54

HO_2C ... OH $\xrightarrow{I_2}$ ... $\xrightarrow[\text{MeOH}]{Na_2CO_3}$ MeO_2C ... OH ... Ref. 57

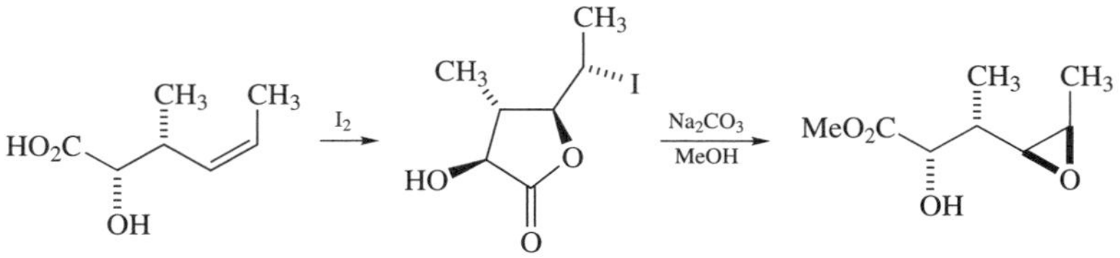

Iodolactones can also be obtained form *N*-pentenoyl amides. These reactions occur by O-alkylation, followed by hydrolysis of the iminoether intermediate.[58]

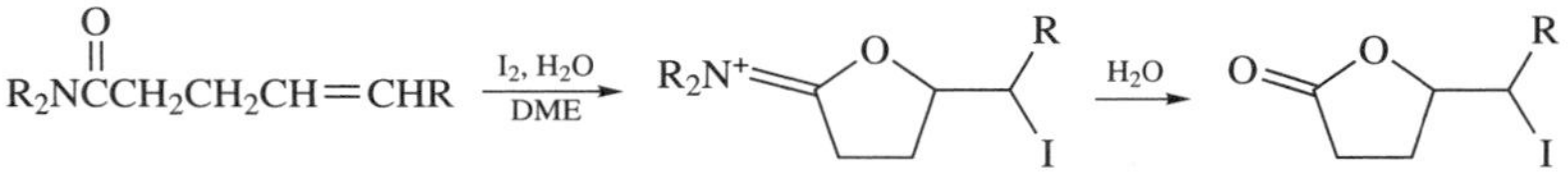

$R_2NCCH_2CH_2CH{=}CHR \xrightarrow[\text{DME}]{I_2,\ H_2O} R_2N^+{=}$ ... $\xrightarrow{H_2O}$ O= ...

Use of a chiral amide can promote enantioselective cyclization.[59]

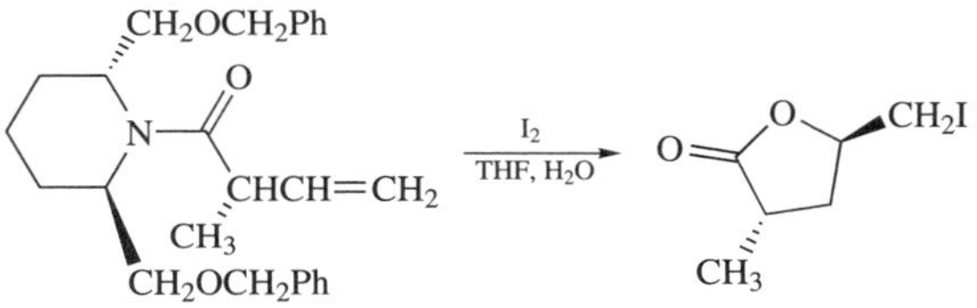

CH_2OCH_2Ph ... $CHCH{=}CH_2$, CH_3, CH_2OCH_2Ph $\xrightarrow[\text{THF, } H_2O]{I_2}$ ... CH_2I ... CH_3

Lactams can be obtained by iodolactonization of *O,N*-trimethylsilyl imidates[60]:

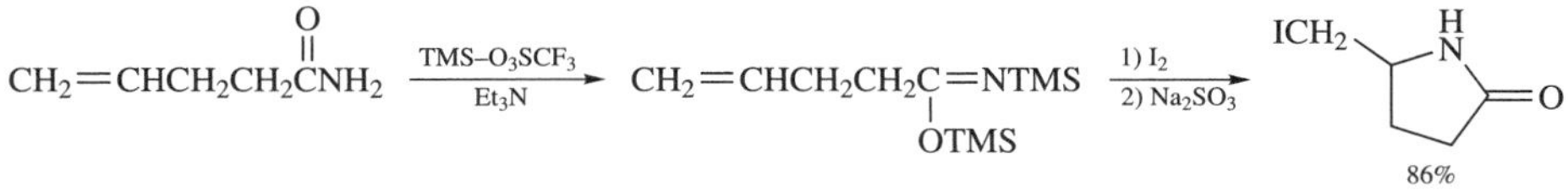

$CH_2{=}CHCH_2CH_2CNH_2 \xrightarrow[Et_3N]{TMS\text{–}O_3SCF_3} CH_2{=}CHCH_2CH_2C(OTMS){=}NTMS \xrightarrow[2)\ Na_2SO_3]{1)\ I_2}$ ICH_2 ... 86%

As compared with amides, where oxygen is the most nucleophilic atom, the silyl imidates are more nucleophilic at nitrogen.

Examples of iodolactonization and related iodocyclizations can be found in Scheme 4.2.

The elemental halogens are not the only source of electrophilic halogen atoms, and, for some synthetic purposes, other "positive halogen" compound may be preferable

57. C. Neukome, D. P. Richardson, J. H. Myerson, and P. A. Bartlett, *J. Am. Chem. Soc.* **108:**5559 (1986).
58. Y. Tamaru, M. Mizutani, Y. Furukawa, S. Kawamura, Z. Yoshida, K. Yanagi, and M. Minobe, *J. Am. Chem. Soc.* **106:**1079 (1984).
59. S. Najda, D. Reichlin, and M. J. Kurth, *J. Org. Chem.* **55:**6241 (1990).
60. S. Knapp, K. E. Rodriquez, A. T. Levorse, and R. M. Ornat, *Tetrahedron Lett.* **26:**1803 (1985).

Scheme 4.2. Iodolactonization and Other Cyclizations Induced by Iodine

1[a] CH_2CO_2H, $CH{=}CH_2$ — I_2, $NaHCO_3$ → O, O, I, $CH{=}CH_2$

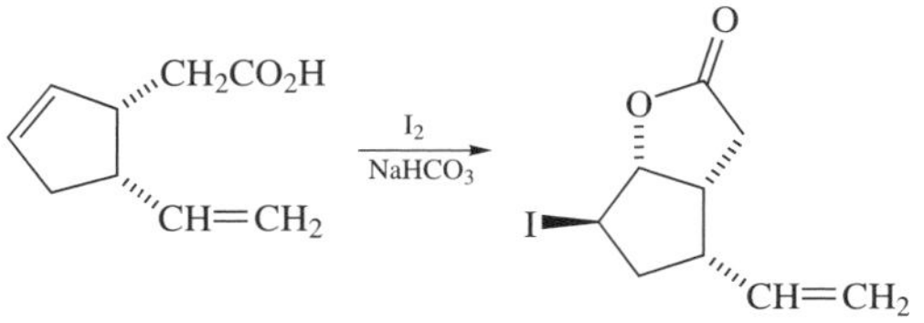

2[b] $OSiR_3$, CH_3, CH_2CO_2H — I_2, $NaHCO_3$ → CH_3, $OSiR_3$, O, O, I

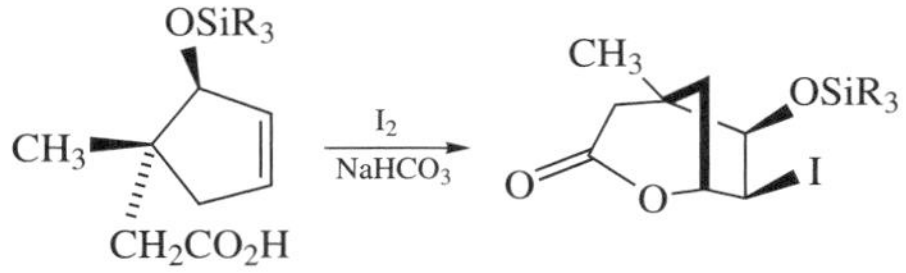

3[c] CH_3, HO_2C — 1) I_2, CH_3CN 2) $NaHCO_3$ → CH_3, O, O, CH_2I 85%

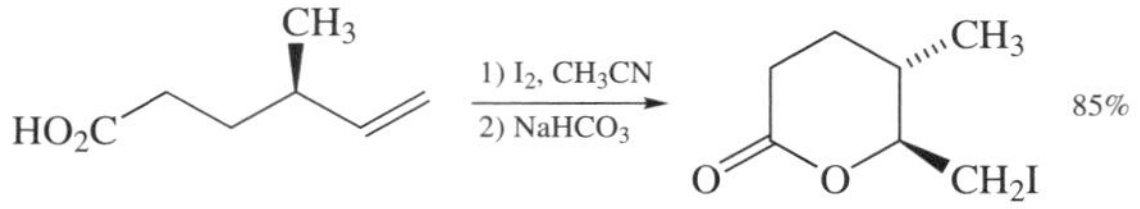

4[d] $OCO_2C(CH_3)_3$ — 1) I_2, CH_3CN 2) $NaHCO_3$ → ICH_2, O, O, O major (68%); ICH_2, O, O, O minor

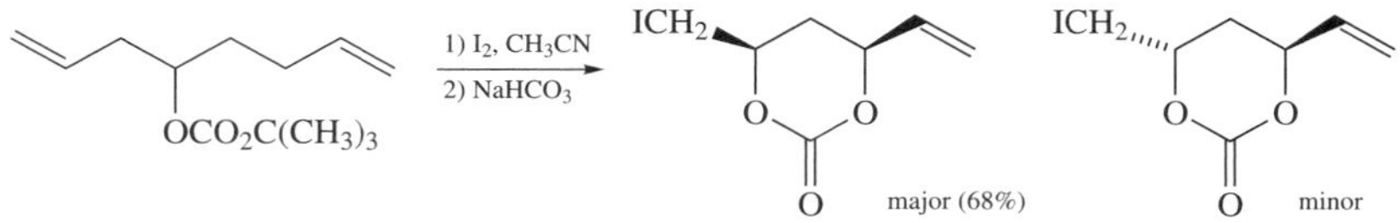

5[e] $O_2COC(CH_3)_3$ — IBr, –80°C → ICH_2, O, O, O major + ICH_2, O, O, O minor 95% (25.8:1)

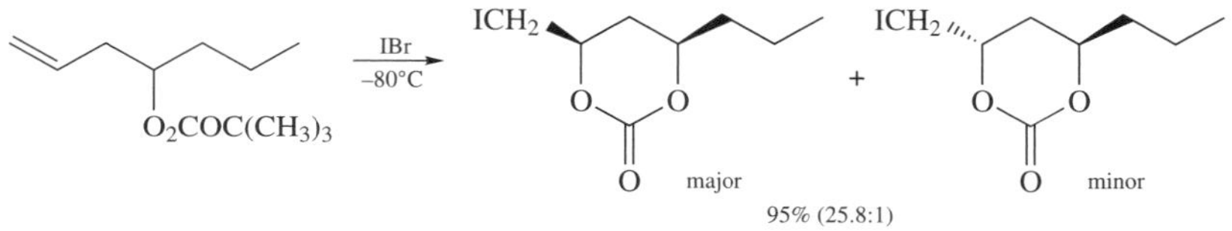

6[f] CH_2OH — I_2, $NaHCO_3$ → O, CH_2I 92%

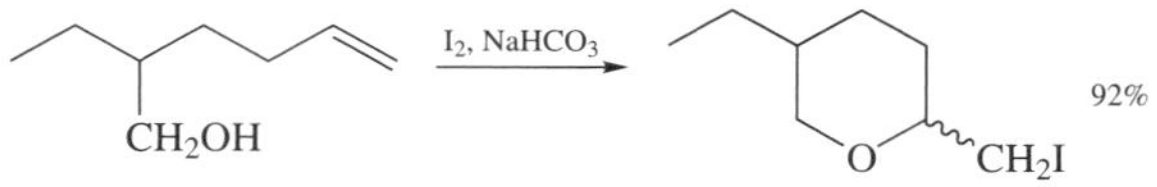

7[g] OH — 1) RLi, CO_2 2) I_2 → ICH_2, CH_3, O, O, O major (80%); ICH_2, CH_3, O, O, O minor

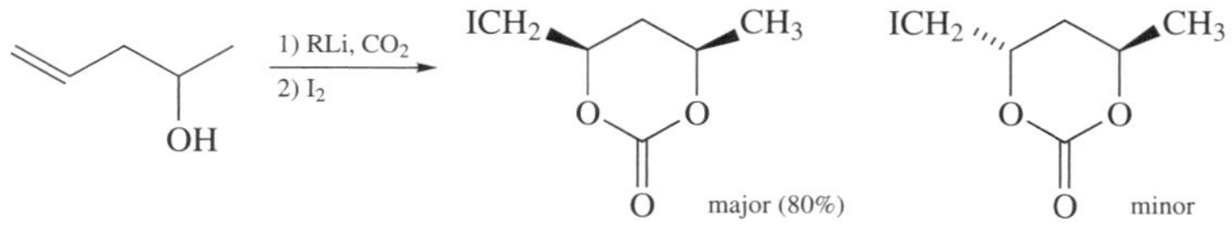

8[h] CH_2, HO, OH, H_3C, CH_3 — *N*-iodosuccinimide → O, CH_2I, OH, H_3C, CH_3

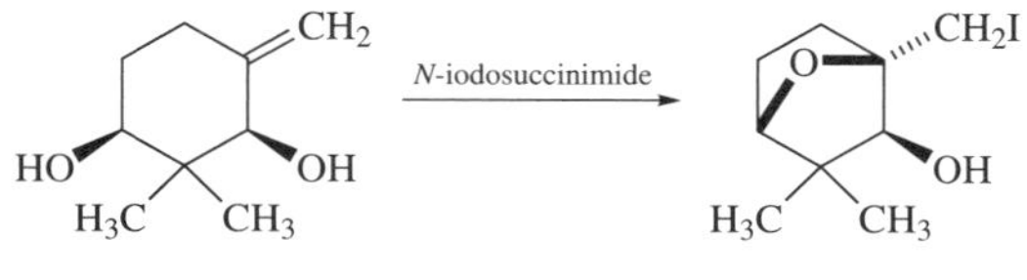

9[i]

CH_2CNH_2 (=O) $\xrightarrow[Et_3N]{Me_3SiO_3SCF_3}$ $CH_2C(NSiMe_3)(OSiMe_3)$ $\xrightarrow[2)\ Na_2SO_3]{1)\ I_2,\ THF}$ 88%

a. L. A. Paquette, G. D. Crouse, and A. K. Sharma, *J. Am. Chem. Soc.* **102**:3972 (1980).
b. A. J. Pearson and S.-Y. Hsu, *J. Org. Chem.* **51**:2505 (1986).
c. A. G. M. Barrett, R. A. E. Carr, S. V. Attwood, G. Richardson, and N. D. A. Walshe, *J. Org. Chem.* **51**:4840 (1986).
d. P. A. Bartlett, J. D. Meadows, E. G. Brown, A. Morimoto, and K. K. Jernstedt, *J. Org. Chem.* **47**:4013 (1982).
e. J. J.-W. Duan and A. B. Smith, III, *J. Org. Chem.* **58**:3703 (1993).
f. L. F. Tietze and C. Schneider, *J. Org. Chem.* **56**:2476 (1991).
g. A. Bongini, G. Cardillo, M. Orena, G. Porzi, and S. Sandri, *J. Org. Chem.* **47**:4626 (1982).
h. A. Murai, N. Tanimoto, N. Sakamoto, and T. Masamune, *J. Am. Chem. Soc.* **110**:1985 (1988).
i. S. Knapp and A. T. Levorse, *J. Org. Chem.* **53**:4006 (1988).

sources of the desired electrophile. The utility of N-bromosuccinimide in formation of bromohydrins was mentioned earlier. Other compounds which are useful for specific purposes are indicated in Table 4.3. Pyridinium hydrotribromide (pyridinium hydrobromide perbromide), benzyltrimethyl ammonium tribromide, and dioxane–bromine complex of examples of complexes of bromine in which its reactivity is somewhat attenuated, resulting in increased selectivity. *N*-Chlorosuccinimide and *N*-bromosuccinimide transfer electrophile halogen, with the succinimide anion acting as the leaving group. This anion is subsequently protonated to give the weak nucleophile succinimide. These reagents therefore favor nucleophilic additions by solvent and cyclization reactions, because there is no competition from a nucleophilic anion. In tetrabromocyclohexadienone, the leaving group is 2,4,6-tribromophenoxide ion. This reagent is a very mild and selective source of electrophilic bromine.

Br, Br, Br, Br, =O → "Br^{+}" + Br—(Br, Br)—O^-

Electrophilic iodine reagents have also been employed in iodocyclization. Several salts of pyridine complexes with I^+ such as bis(pyridinium)iodonium tetrafluoroborate and bis(collidine)iodonium hexafluorophosphate have proven especially effective.[61] γ-Hydroxy- and δ-hydroxyalkenes can be cyclized to tetrahydrofuran and tetrahydropyran derivatives, respectively, by positive halogen reagents.[62] (see entries 6 and 8 in Scheme 4.2).

4.5. Electrophilic Sulfur and Selenium Reagents

Compounds in which sulfur and selenium atoms are bound to more electronegative elements react with alkenes to give addition products. The mechanism is similar to that in

61. Y. Brunel and G. Rousseau, *J. Org. Chem.* **61**:5793 (1996).
62. A. B. Reitz, S. O. Nortey, B. E. Maryanoff, D. Liotta, and R. Monahan III, *J. Org. Chem.* **52**:4191 (1987).

Table 4.3. Other Sources of Positive Halogen

Source	Synthetic applications[a]
A. Chlorinating agents	
Sodium hypochlorite solution	Formation of chlorohydrins from alkenes
N-Chlorosuccinimide	Chlorination with solvent participation and cyclization
Antimony pentachloride	Controlled chlorination of acetylenes
B. Brominating agents	
Pyridinium hydrotribromide (pyridinium hydrobromide perbromide)	Substitute for bromine when increased selectivity or mild reaction conditions are required
Dioxane–bromine complex	Same as for pyridinium hydrotribromide
N-Bromosuccinimide	Substitute for bromine when low Br^- concentration is required
2,4,4,6-Tetrabromocyclohexadienone	Selective bromination of polyolefins and cyclization induced by Br^+
Benzyltrimethylammonium tribromide[b]	Selective bromination of alkenes and carbonyl compounds
C. Iodinating Agents	
Bis(pyridine)iodonium tetrafluoroborate[c]	Selective iodination and iodocyclization

a. For specific examples, consult M. Fieser and L. F. Fieser, *Reagents for Organic Synthesis*, Vols 1–8, John Wiley & Sons, New York, 1979.
b. S. Kajgaeshi and T. Kakinami, *Ind. Chem. Libr.* **7**:29 (1995).
c. J. Barluenga, J. M. Gonzalez, M. A. Garcia-Martin, P. J. Campos, and G. Asensio, *J. Org. Chem.* **58**:2058 (1993).

halogenation with a bridged cationic intermediate being involved.

$$R'S{-}Cl + RCH{=}CHR \longrightarrow RCH\overset{\overset{R'}{|}}{\underset{S^+}{\wedge}}CHR \xrightarrow{Cl^-} RCH(SR')CH(Cl)R$$

$$R'Se{-}Cl + RCH{=}CHR \longrightarrow RCH\overset{\overset{R'}{|}}{\underset{^+Se}{\wedge}}CHR \xrightarrow{Cl^-} R{-}CH(SeR'){-}CH(Cl)R$$

In many synthetic applications, the sulfur or selenium substituent is subsequently removed by elimination, as will be discussed in Chapter 6. Arenesulfenyl halides, ArSCl, are the most commonly used of the sulfur reagents. A variety of electrophilic selenium reagents have been employed, and several examples are given in Scheme 4.3.

Mechanistic studies have been most thorough with the sulfenyl halides.[63] The reactions show moderate sensitivity to alkene structure, with electron-releasing groups on the alkene accelerating the reaction. The addition can occur in either the Markownikoff or anti-Markownikoff sense.[64] The variation in regioselectivity can be understood by

63. W. A. Smit, N. S. Zefirov, I. V. Bodrikov, and M. Z. Krimer, *Acc. Chem. Res.* **12**:282 (1979); G. H. Schmid and D. G. Garratt, *The Chemistry of Double-Bonded Functional Groups*, S. Patai, ed., John Wiley & Sons, New York, 1977, Chapter 9; G. A. Jones, C. J. M. Stirling, and N. G. Bromby, *J. Chem. Soc., Perkin Trans. 2* **1983**:385.
64. W. H. Mueller and P. E. Butler, *J. Am. Chem. Soc.* **90**:2075 (1968); G. H. Schmid and D. I. Macdonald, *Tetrahedron Lett.* **25**:157 (1984).

Scheme 4.3. Sulfur and Selenium Reagents for Electrophile Addition

	Reagent	Product
1[a,b]	CH_3SCl	CH_3S–RCHCHR–Cl
2[a]	PhSCl	PhS–RCHCHR–Cl
3[c]	PhSeCl	PhSe–RCHCHR–Cl
4[d]	$PhSeO_2CCF_3$	PhSe–RCHCHR–O_2CCF_3
5[e]	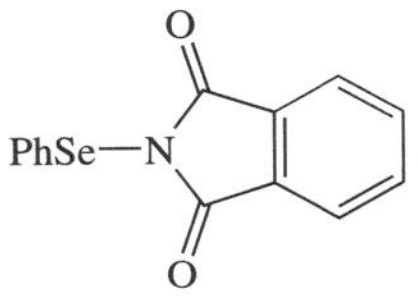 N-(phenylseleno)phthalimide (PhSe—N), H_2O	PhSe–RCHCHR–OH
6[f]	$PhSeO_2H$, H_3PO_2	PhSe–RCHCHR–OH
7[g]	PhSeCN, Cu(II), R′OH	PhSe–RCHCHR–OR′
8[h]	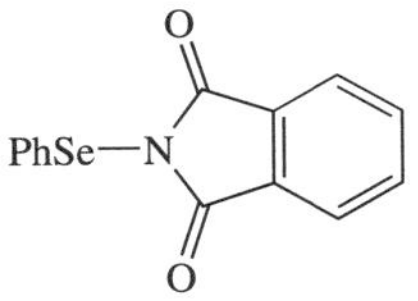 N-(phenylseleno)phthalimide (PhSe—N), $(CH_3)_3SiN_3$	PhSe–R_2CCHR–N_3
9[i]	PhSeCl, $AgBF_4$, $H_2NCO_2C_2H_5$	PhSe–RCHCHR–$HNCO_2C_2H_5$

a. W. M. Mueller and P. E. Butler, *J. Am. Chem. Soc.* **90:**2075 (1968).
b. W. A. Thaler, *J. Org. Chem.* **34:**871 (1969).
c. K. B. Sharpless and R. F. Lauer, *J. Org. Chem.* **39:**429 (1974); D. Liotta and G. Zima, *Tetrahedron Lett.* **1978:**4977.
d. H. J. Reich, *J. Org. Chem.* **39:**428 (1974); A. G. Kulateladze, J. L. Kice, T. G. Kutateladze, N. S. Zefirov, and N. V. Zyk, *Tetrahedron Lett.* **33:**1949 (1992).
e. K. C. Nicolaou, D. A. Claremon, W. E. Barnette, and S. P. Seitz, *J. Am. Chem. Soc.* **101:**3704 (1979).
f. D. Labar, A. Krief, and L. Hevesi, *Tetrahedron Lett.* **1978:**3967.
g. A. Toshimitsu, T. Aoai, S. Uemura, and M. Okano, *J. Org. Chem.* **45:**1953 (1980).
h. R. M. Giuliano and F. Duarte, *Synlett* **1992:**419.
i. C. G. Francisco, E. I. León, J. A. Salazar, and E. Suárez, *Tetrahedron Lett.* **27:**2513 (1986).

focusing attention on the sulfur-bridged intermediate, which may range from being a sulfonium ion to being a less electrophilic chlorosulfurane.

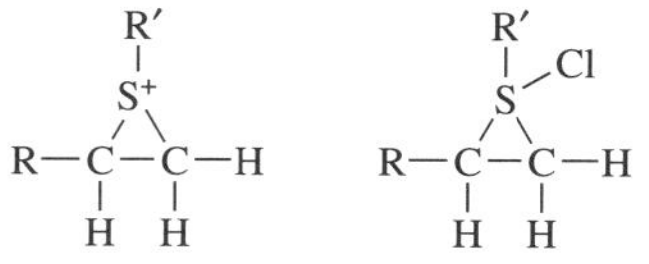

R′–S⁺ bridged (R–CH–CH–H) ; R′–S(Cl) bridged (R–CH–CH–H)

Compared to the C–Br bonds in a bromonium ion, the C–S bonds are stronger and the transition state for nucleophilic addition will be reached later. Steric interactions that dictate access by the nucleophile become a more important factor in determining the direction of addition. For reactions involving phenylsulfenyl chloride or methylsulfenyl chloride, the intermediate is a fairly stable species, and ease of approach by the nucleophile is the major factor in determining the direction of ring opening. In these cases, the product has the anti-Markownikoff orientation.[65]

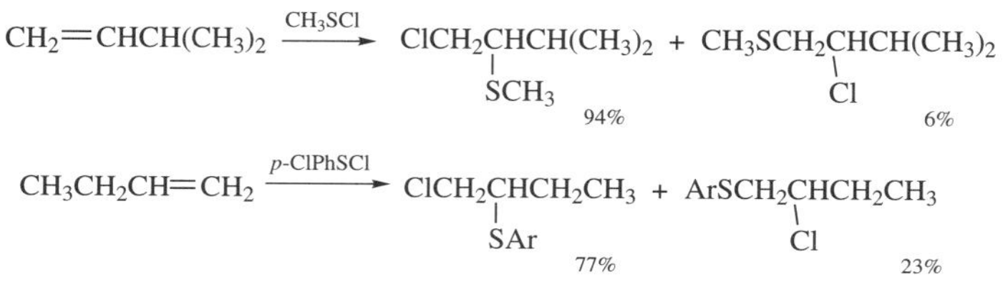

$$CH_2{=}CHCH(CH_3)_2 \xrightarrow{CH_3SCl} ClCH_2CH(SCH_3)CH(CH_3)_2\ (94\%) + CH_3SCH_2CH(Cl)CH(CH_3)_2\ (6\%)$$ Ref. 66

$$CH_3CH_2CH{=}CH_2 \xrightarrow{p\text{-}ClPhSCl} ClCH_2CH(SAr)CH_2CH_3\ (77\%) + ArSCH_2CH(Cl)CH_2CH_3\ (23\%)$$ Ref. 67

The stereospecific *anti* addition of phenylsulfenyl chloride to norbornene is a particularly interesting example of the stability of the intermediate. Neither rearrangement nor *syn* addition products, which are observed with many of the other electrophilic reagents, are formed.[63] This result indicates that the intermediate must be quite stable and reacts only by nucleophilic attack.[64]

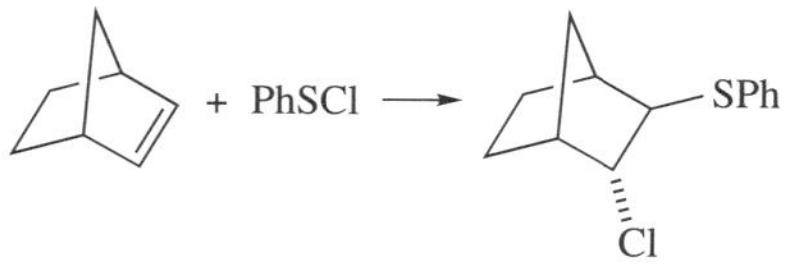

norbornene + PhSCl ⟶ (SPh, Cl anti-substituted norbornane)

When nonnucleophilic salts, for example $LiClO_4$, are included in the reaction medium, products indicative of a more reactive intermediate with carbocationic character are

65. G. H. Schmid, M. Strukelj, S. Dalipi, and M. D. Ryan, *J. Org. Chem.* **52**:2403 (1987).
66. W. H. Mueller and P. E. Butler, *J. Am. Chem. Soc.* **90**:2075 (1968).
67. G. H. Schmid, C. L. Dean, and D. G. Garratt, *Can. J. Chem.* **54**:1253 (1976).

observed.

ArSCl, $LiClO_4$, CH_3CO_2H → SAr + SAr, O_2CCH_3 + ArS, O_2CCH_3 Ref. 68

Ar = O_2N–(2-NO_2-phenyl)–

These contrasting results can be interpreted in terms of a relatively unreactive species, perhaps a chlorosulfurane, being the main intermediate in the absence of the salt. The presence of the lithium cation gives rise to a more reactive species such as the episulfonium ion, as the result of ion pairing with the chloride ion.

$$\text{(Ar)(Cl)S} + \text{Li}^+ \rightleftharpoons \overset{+}{\text{S}}\text{–Ar} + \text{Li}^{+\,-}\text{Cl}$$

Terminal alkenes react with selenenyl halides with anti-Markownikoff regioselectivity.[69] However, the β-selenenyl halide addition product can readily rearrange to isomeric products[70]:

$$R_2C{=}CH_2 + ArSeX \longrightarrow R_2\underset{X}{C}CH_2SeAr \rightleftharpoons R_2\overset{ArSe}{\underset{X}{C}}CH_2X$$

When reactions with phenylselenenyl chloride are carried out in aqueous acetonitrile solution β-hydroxyselenides are formed as a result of solvolysis of the chloride.[71]

Electrophilic selenium reagents are very effective in promoting cyclization of unsaturated molecules containing potentially nucleophilic substituents.[72] Unsaturated carboxylic acids, for example, give selenolactones, and this reaction has been termed *selenolactonization*[73]:

CH_2CO_2H —PhSeCl→ PhSe, O, O 93%

68. N. S. Zefirov, N. K. Sadovaja, A. M. Maggerramov, I. V. Bodrikov, and V. E. Karstashov, *Tetrahedron* **31:**2949 (1975); see also S. Dalipi and G. H. Schmid, *J. Org. Chem.* **47:**5027 (1982); N. S. Zefirov and I. V. Bodrikov, *J. Org. Chem. USSR Engl. Trans.* **1983:**1940.
69. D. Liotta and G. Zima, *Tetrahedron Lett.* **1978:**4977; P. T. Ho and R. J. Holt, *Can. J. Chem.* **60:**663 (1982).
70. S. Raucher, *J. Org. Chem.* **42:**2950 (1977).
71. A. Toshmitsu, T. Aoai, H. Owada, S. Uemura, and M. Okano, *Tetrahedron* **41:**5301 (1985).
72. K. Fujita, *Rev. Hetereoatom. Chem.* **16:**101 (1997).
73. K. C. Nicolaou, S. P. Seitz, W. J. Sipio, and J. F. Blount, *J. Am. Chem. Soc.* **101:**3884 (1979).

N-Phenylselenenophthalimide is an excellent reagent for this process and permits the formation of large-ring lactones.[74] The advantage of the reagent in this particular application is the low nucleophilicity of the phthalimide anion, which does not compete with the remote internal nucleophile.

The reaction of phenylselenenyl chloride of *N*-phenylselenenophthalimide with unsaturated alcohols leads to formation of β-phenylselenenyl ethers:

CH_2CH_2OH + PhSeN → PhSe Ref. 75

Another useful reagent for selenoncyclization is phenylselenenyl sulfate. This reagent is capable of cyclizing unsaturated acids[76] and alcohols.[77] This reagent can be prepared *in situ* by oxidation of diphenyl diselenide with ammonium peroxydisulfate.[78]

$CH_3CH(OH)CH_2CH{=}C(CH_3)_2 \xrightarrow[(NH_4^+)_2S_2O_8^{2-}]{(PhSe)_2}$ $C(CH_3)_2$(SePh) 90%

Chiral selenenylating reagents have been developed and shown to be capable of effecting enantiselective additions and cyclizations. For example the reagent show below (SeAr*) achieves $>90\%$ enantioselectivity in typical reactions.[79]

Ph, CH_3 → SeAr*, Ph, CH_3, OCH_3 95% d.e.

Ph, CH_2OH → Ar*Se, Ph, O 94% d.e.

CO_2H → SeAr*, O, =O 95% d.e.

N, Se^+, PF_6^-, O, Ph

Scheme 4.4 gives some examples of cyclizations induced by selenium electrophiles.

74. K. C. Nicolaou, D. A. Claremon, W. E. Barnette, and S. P. Seitz, *J. Am. Chem. Soc.* **101:**3704 (1979).
75. K. C. Nicolaou, R. L. Magolda, W. J. Sipio, W. E. Barnette, Z. Lysenko, and M. M. Joullie, *J. Am. Chem. Soc.* **102:**3784 (1980).
76. S. Murata and T. Suzuki, *Chem. Lett.* **1987:**849.
77. A. G. Kutateladze, J. L. Kice, T. G. Kutateladze, N. S. Zefirov, and N. V. Zyk, *Tetrahedron Lett.* **33:**1949 (1992).
78. M. Tiecco, L. Testaferri, M. Tingoli, D. Bartoli, and R. Balducci, *J. Org. Chem.* **55:**429 (1990).
79. K. Fujita, K. Murata, M. Iwaoka, and S. Tomoda, *Tetrahedron* **53:**2029 (1997); K. Fujita, *Rev. Heteroatom Chem.* **16:**101 (1997); T. Wirth, *Tetrahedron* **55:**1 (1999).

1[a] $CH_2{=}CH(CH_2)_4OH$ —PhSCl, $(i\text{-}Pr)_2NEt$→ PhSCH$_2$ (tetrahydropyran) 85%

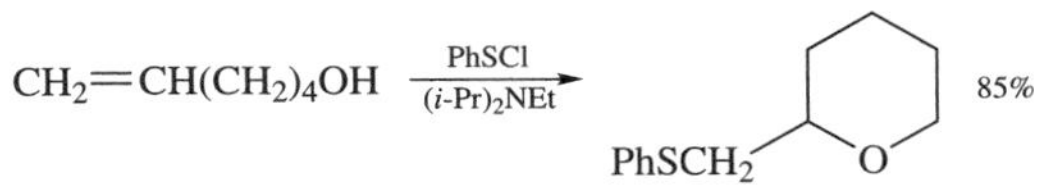

2[b] PhSCl → CH_2SPh 35%

3[c] $CH_2{=}C(CH_3)CH_2N(CH(CH_3)_2)CO_2C_2H_5$ —PhSCl→ H_3C, $PhSCH_2$, $CH(CH_3)_2$ 42%

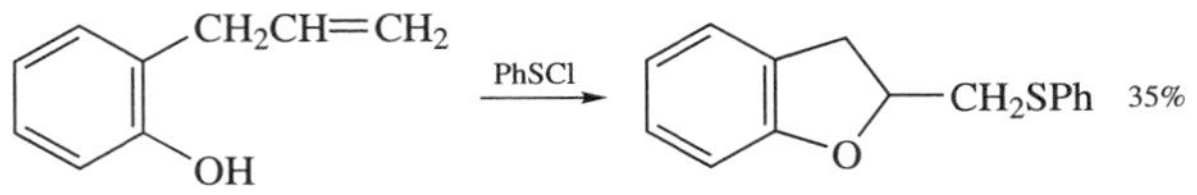

4[d] PhCH$_2$, HO, H, CH_2CO_2H —PhSeCl→ PhSe, HO, H, PhCH$_2$ 70–75%

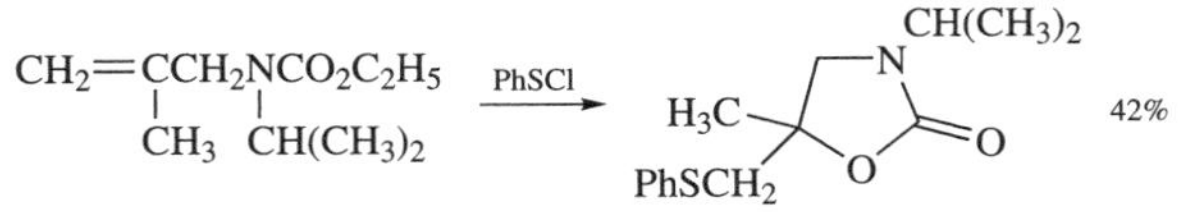

5[e] H, H_3C, OAc, H, CH_2OAc, $HOCH_2$ —PhSeCl→ SePh, OAc, H_3C, H, CH_2OAc 52%

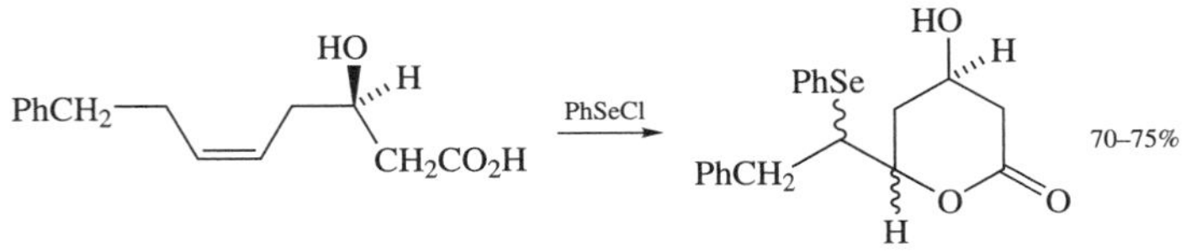

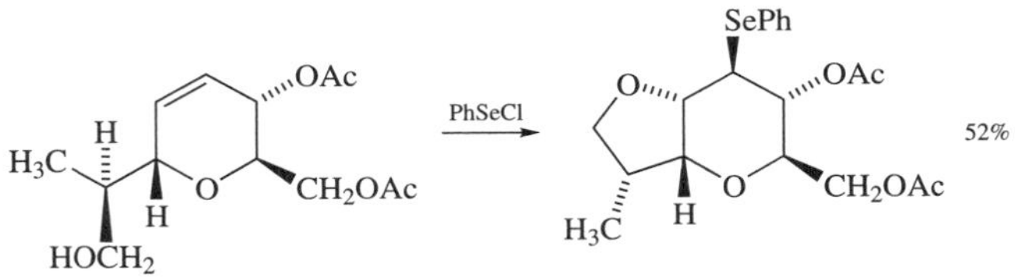

6[f] OH, CH_3 —PhSe–N (phthalimide)→ H_3C, SePh, H_3C 82%

7[g] CH_2OH, CH_3 —$PhSeO_2CCF_3$→ CH_3, SePh

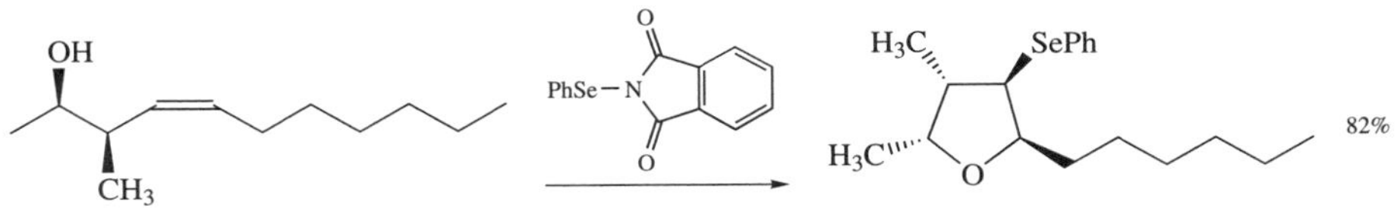

8[h] $CH_3O_2CCH_2C(=O)CH_2CH_2CH{=}CH_2$ —$(PhSe)_2$, $(NH_4^+)_2S_2O_5^{2-}$→ CH_2SePh, CH_3O_2C 58%

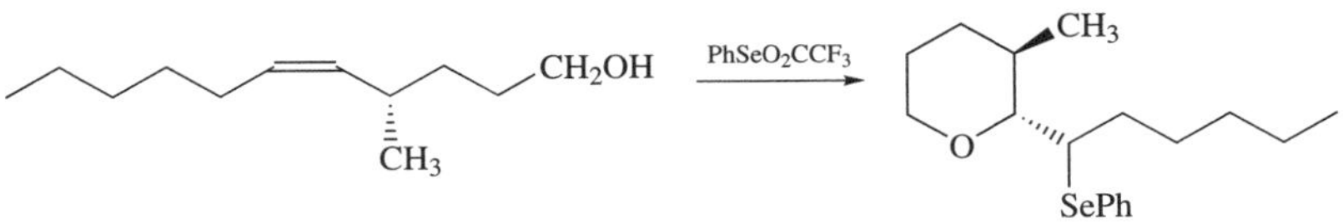

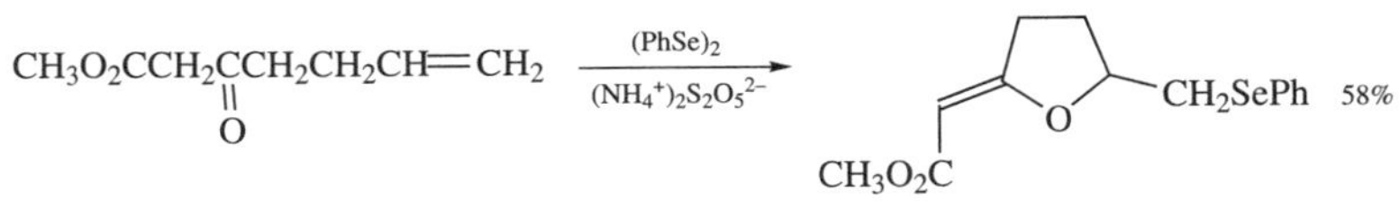

9[i] $HOCH_2$, CH_3 —PhSeCN, $Cu(O_3SCF_3)_2$→ CH_3, CH_2SePh 95%

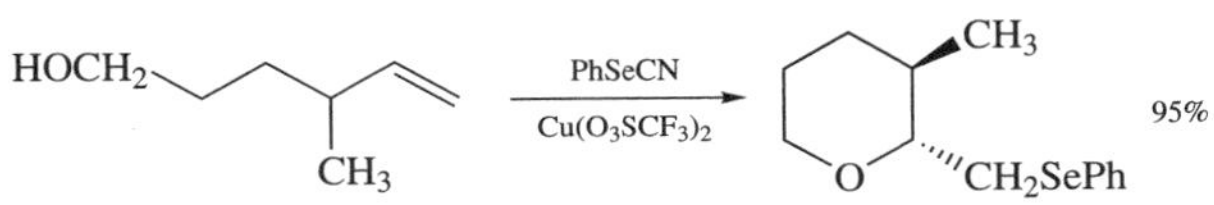

Scheme 4.4. (*continued*)

10[j] $CH_2{=}CHCH_2CH(CH_2CH_3)C(=O)NHPh$ $\xrightarrow{PhSeCl}$ $PhSeCH_2$–(2-(N-phenylimino)-3-ethyltetrahydrofuran-5-yl) 85%

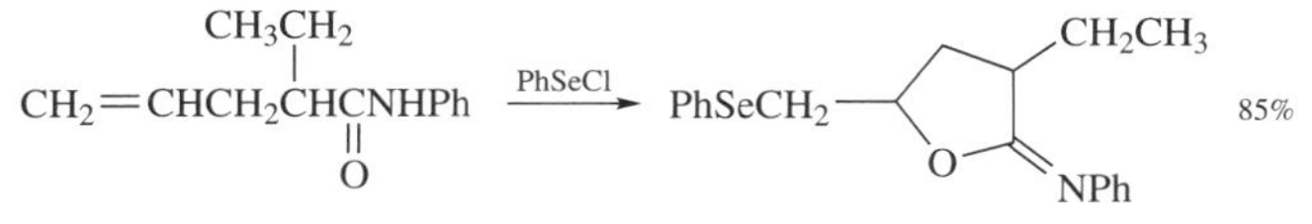

11[k] 5-($CH_2CH_2CH{=}CH_2$)-pyrrolidin-2-one $\xrightarrow{PhSeBr}$ 3-(CH_2SePh)-pyrrolizidin-5-one 68%

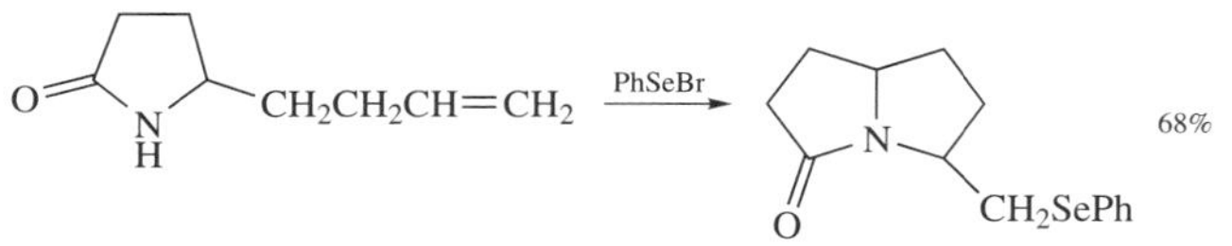

a. S. M. Tuladhar and A. G. Fallis, *Tetrahedron Lett.* **28:**523 (1987).
b. M. Muehlstaedt, C. Schubert, and E. Kleinpeter, *J. Prakt, Chem.* **327:**270 (1985).
c. M. Muehlstaedt, R. Widera, and B. Olk, *J. Prakt. Chem.* **324:**362 (1982).
d. F. Bennett and D. W. Knight, *Tetrahedron Lett.* **29:**4625 (1988).
e. S. J. Danishefsky, S. DeNinno, and P. Lartey, *J. Am. Chem. Soc.* **109:**2082 (1987).
f. E. D. Mihelich and G. A. Hite, *J. Am. Chem. Soc.* **114:**7318 (1992).
g. G. Li and W. C. Still, *J. Org. Chem.* **56:**6964 (1991).
h. M. Ticocco, L. Testaferri, M. Tingoli, D. Bartoli, and R. Balducci, *J. Org. Chem.* **55:**429 (1990).
i. H. Inoue and S. Murata, *Heterocycles* **45:**847 (1997).
j. A. Toshimitsu, K. Terao, and S. Uemura, *J. Org. Chem.* **52:**2018 (1987).
k. A. Toshimitsu, K. Terao, and S. Uemura, *J. Org. Chem.* **51:**1724 (1986).

4.6. Addition of Other Electrophilic Reagents

Many other halogen-containing compounds react with alkenes to give addition products by mechanisms similar to halogenation. A complex is generated, and the halogen is transferred to the alkene to generate a cationic intermediate. This may be a symmetrically bridged ion or an unsymmetrically bridged species, depending on the ability of the reacting carbon atoms of the alkene to accommodate positive charge. The direction of opening of the bridged intermediate is usually governed by electronic factors. That is, the addition is completed by attack of the nucleophile at the more positive carbon atom of the bridged intermediate. The orientation of addition therefore follows Markownikoff's rule. The stereochemistry of addition is usually *anti*, because of the involvement of a bridged halonium intermediate.[80] Several reagents of this type are listed in Scheme 4.5.

In the case of thiocyanogen chloride and thiocyanogen, the formal electrophile is $[NCS]^+$. The presumed intermediate is a cyanosulfonium ion. The thiocyanate anion is an ambident nucleophile, and both carbon–sulfur and carbon–nitrogen bond formation can be observed, depending upon the reaction conditions (see entry 9 in Scheme 4.5).

4.7. Electrophilic Substitution Alpha to Carbonyl Groups

Although the reaction of ketones and other carbonyl compounds with electrophiles such as bromine leads to substitution rather than addition, the mechanism of the reaction is closely related to that of electrophilic additions to alkenes. An enol or enolate derived from the carbonyl compound is the reactive species, and the electrophilic attack by the halogen is analogous to the attack on alkenes. The reaction is completed by deprotonation and restoration of the carbonyl bond, rather than by addition of a nucleophile. The acid- and

80. A. Hassner and C. Heathcock, *J. Org. Chem.* **30:**1748 (1965).

Scheme 4.5. Addition Reactions of Other Electrophilic Reagents

	Reagent	Preparation	Product
1[a]	I—N=C=O	AgCNO, I_2	RCH(I)—CHR(NCO)
2[b]	$Br—N=\overset{+}{N}=N^-$	HN_3, Br_2	RCH(Br)—CHR(N_3)
3[c]	$I—N=\overset{+}{N}=N^-$	NaN_3, ICl	RCH(I)—CHR(N_3)
4[d]	I—S=C≡N	$(NCS)_2$, I_2	RCH(I)—CHR(S—C≡N)
5[e]	I—ONO_2	$AgNO_3$, ICl	RCH(I)—CHR(ONO_2)
6[f]	O=N—Cl		RC(=NOH)—CHR(Cl)
7[g]	O=N—CO_2H	$(CH_3)_2CHCH_2CH_2ON=O$, HCO_2H	RC(=NOH)—CHR(O_2CH)
8[h]	Cl—SCN	$Pb(SCN)_2$, Cl_2	RCH(Cl)—CHR(SCN)
9[i]	N≡CS—SC≡N	$Pb(SCN)_2$, Br_2	RCH(SC≡N)—CHR(SC≡N) and RCH(SC≡N)—CHR(N=C=S)

a. A. Hassner, R. P. Hoblitt, C. Heathcock, J. E. Kropp, and M. Lorber, *J. Am. Chem. Soc.* **92:**1326 (1970); A. Hassner, M. E. Lorber, and C. Heathcock, *J. Org. Chem.* **32:**540 (1967).
b. A. Hassner, F. P. Boerwinkle, and A. B. Levy, *J. Am. Chem. Soc.* **92:**4879 (1970).
c. F. W. Fowler, A. Hassner, and L. A. Levy, *J. Am. Chem. Soc.* **89:** 2077 (1967).
d. R. J. Maxwell and L. S. Silbert, *Tetrahedron Lett.* **1978:**4991.
e. J. W. Lown and A. V. Joshua, *J. Chem. Soc., Perkin Trans. 1* **1973:**2680.
f. J. Meinwald, Y. C. Meinwald, and T. N. Baker III, *J. Am. Chem. Soc.* **86:**4074 (1964).
g. H. C. Hamann and D. Swern, *J. Am. Chem. Soc.* **90:**6481 (1968).
h. R. G. Guy and I. Pearson, *J. Chem. Soc., Perkin Trans. 1* **1973:**281; *J. Chem. Soc., Perkin Trans. 2* **1973:**1359.
i. R. J. Maxwell, L. S. Silbert, and J. R. Russell, *J. Org. Chem.* **42:**1510 (1977).

base-catalyzed halogenation of ketones, which were discussed briefly in Part A, Chapter 7, are the most studied examples of the reaction.

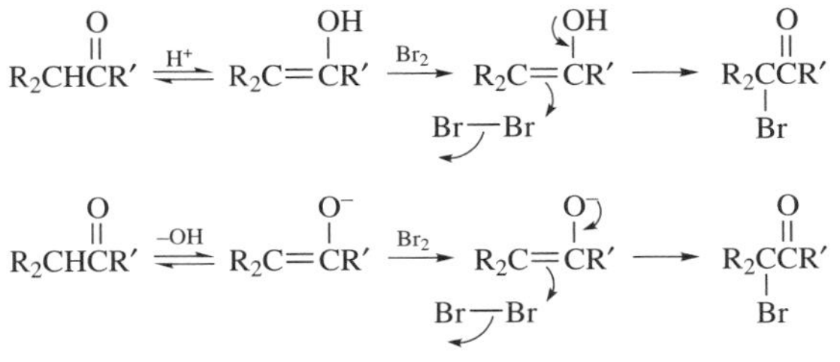

$$R_2CHC(=O)R' \underset{}{\overset{H^+}{\rightleftharpoons}} R_2C=C(OH)R' \xrightarrow{Br_2} R_2C=C(OH)R' \;(Br—Br) \longrightarrow R_2C(Br)C(=O)R'$$

$$R_2CHC(=O)R' \underset{}{\overset{^-OH}{\rightleftharpoons}} R_2C=C(O^-)R' \xrightarrow{Br_2} R_2C=C(O^-)R' \;(Br—Br) \longrightarrow R_2C(Br)C(=O)R'$$

The most common preparative procedures involve use of the halogen, usually bromine, in acetic acid. Other suitable halogenating agents include *N*-bromosuccinimide, sulfuryl chloride and tetrabromocyclohexadienone.

Br–C₆H₄–C(=O)CH₃ $\xrightarrow[CH_3CO_2H]{Br_2}$ Br–C₆H₄–C(=O)CH₂Br 69–72% Ref. 81

Cycloheptanone $\xrightarrow[CCl_4]{N\text{-bromosuccinimide}}$ 2-bromocycloheptanone Ref. 82

2-Methylcyclohexanone $\xrightarrow{SO_2Cl_2}$ 2-chloro-2-methylcyclohexanone 83–85% Ref. 83

C₆H₅–CH=CHC(=O)CH₃ $\xrightarrow{\text{tetrabromocyclohexadienone}}$ C₆H₅–CH=CHC(=O)CH₂Br 91% Ref. 84

The reactions involving bromine or chlorine generate hydrogen halide and are autocatalytic. Reactions with *N*-bromosuccinimide or tetrabromocyclohexadienone form no hydrogen bromide, and these reagents may therefore be preferable in the case of acid-sensitive compounds.

As was pointed out in Part A, Section 7.3, under many conditions halogenation is faster than enolization. When this is true, the position of substitution in unsymmetrical ketones is governed by the relative rates of formation of the isomeric enols. In general, mixtures are formed with unsymmetrical ketones. The presence of a halogen substituent decreases the rate of acid-catalyzed enolization and therefore retards the introduction of a second halogen at the same site. Monohalogenation can therefore usually be carried out satisfactorily. A preparatively useful procedure for monohalogenation of ketones involves reaction with cupric chloride or cupric bromide.[85]

Cyclodecanone $\xrightarrow[HCCl_3,\ CH_3CO_2C_2H_5]{CuBr_2}$ 2-bromocyclodecanone Ref. 86

In contrast, in basic solution halogenation tends to proceed to polyhalogenated products. This is because the inductive effect of a halogen accelerates base-catalyzed

81. W. D. Langley, *Org. Synth.* **1:**122 (1932).
82. E. J. Corey, *J. Am. Chem. Soc.* **75:**2301 (1953).
83. E. W. Warnhoff, D. G. Martin, and W. S. Johnson, *Org. Synth.* **IV:**162 (1963).
84. V. Calo, L. Lopez, G. Pesce, and P. E. Todesco, *Tetrahedron* **29:**1625 (1973).
85. E. M. Kosower, W. J. Cole, G-S. Wu, D. E. Cardy, and G. Meisters, *J. Org. Chem.* **28:**630 (1963); E. M. Kosower and G.-S. Wu, *J. Org. Chem.* **28:**633 (1963).
86. D. P. Bauer and R. S. Macomber, *J. Org. Chem.* **40:**1990 (1975).

enolization. With methyl ketones, base-catalyzed reaction with iodine or bromine leads eventually to cleavage to a carboxylic acid.[87] The reaction can also be effected with hypochlorite ion.

$$(CH_3)_2C{=}CHC(=O)CH_3 + {}^-OCl \longrightarrow (CH_3)_2C{=}CHCO_2H \quad 49\text{–}53\%$$ Ref. 88

Instead of direct halogenation of ketones, reactions with more reactive ketone derivatives such as silyl enol ethers and enamines have advantages in certain cases.

1-(trimethylsilyloxy)cyclohexene —[1) I_2, AgOAc; $R_4N^+F^-$]→ 2-iodocyclohexanone 84% Ref. 89

1-(1-pyrrolidinyl)cyclohexene —[Cl_2, −78°C]→ 2-chlorocyclohexylidene pyrrolidinium —[H_2O]→ 2-chlorocyclohexanone 65% Ref. 90

There are also procedures in which the enolate is generated and allowed to react with a halogenating agent. Among the sources of halogen that have been used under these conditions are bromine,[91] *N*-chlorosuccinimide,[92] trifluoromethanesulfonyl chloride,[93] and hexachloroethane.[94]

2-methyl-5-(1-methoxy-1-methylethyl)cyclohex-2-enone —[1) LDA; 2) CF_3SO_2Cl]→ 6-chloro-2-methyl-5-(1-methoxy-1-methylethyl)cyclohex-2-enone

α-Fluoroketones have been made primarily by reactions of enol acetates or silyl enol ethers with fluorinating agents such as CF_3OF,[95] XeF_2,[96] and dilute F_2.[97] Other fluorinating reagents which can be used include *N*-fluoropyridinium salts,[98] 1-fluoro-4-

87. S. J. Chakabartty, in *Oxidations in Organic Chemistry*, Part C, W. Trahanovsky, ed., Academic Press, New York, 1978, Chapter V.
88. L. J. Smith, W. W. Prichard, and L. J. Spillane, *Org. Synth.* **III:**302 (1955).
89. G. M. Rubottom and R. C. Mott, *J. Org. Chem.* **44:**1731 (1979); G. A. Olah, L. Ohannesian, M. Arvanaghi, and G. K. S. Prakash, *J. Org. Chem.* **49:**2032 (1984).
90. W. Seufert and F. Effenberger, *Chem. Ber.* **112:**1670 (1979).
91. T. Woolf, A. Trevor, T. Baille, and N. Castagnoli, *Jr.,* J. Org. Chem. **49:**3305 (1984).
92. A. D. N. Vaz and G. Schoellmann, *J. Org. Chem.* **49:**1286 (1984).
93. P. A. Wender and D. A. Holt, *J. Am. Chem. Soc.* **107:**7771 (1985).
94. M. B. Glinski, J. C. Freed, and T. Durst, *J. Org. Chem.* **52:**2749 (1987).
95. W. J. Middleton and E. M. Bingham, *J. Am. Chem. Soc.* **102:**4845 (1980).
96. B. Zajac and M. Zupan, *J. Chem. Soc., Chem. Commun.* **1980:**759.
97. S. Rozen and Y. Menahem, *Tetrahedron Lett.* **1979:**725.
98. T. Umemoto, M. Nagayoshi, K. Adachi, and G. Tomizawa, *J. Org. Chem.* **63:**3379 (1998).

hydroxy-1,4-diazabicyclo[2.2.2]octane,[99] and 1,4-difluoro-1,4-diazabicyclooctane.[100] These reagents fluorinate readily enolizable carbonyl compounds and silyl enol ethers.

$$PhC(=O)CH_2CH_3 + F{-}N^{+}(CH_2CH_2)_3N^{+}OH \longrightarrow PhC(=O)CH(F)CH_3 \quad 88\%$$ Ref. 101

Another example of α-halogenation which has synthetic utility is the α-halogenation of acyl chlorides. The mechanism is presumed to be similar to that of ketone halogenation and to proceed through an enol. The reaction can be effected in thionyl chloride as solvent to give α-chloro, α-bromo, or α-iodo acyl chlorides using, respectively, *N*-chlorosuccinimide, *N*-bromosuccinimide, or molecular iodine as the halogenating agent.[102] Because thionyl chloride rapidly converts carboxylic acids to acyl chlorides, the acid can be used as the starting material.

$$CH_3(CH_2)_3CH_2CO_2H \xrightarrow[SOCl_2]{N\text{-chlorosuccinimide}} CH_3(CH_2)_3CH(Cl)COCl \quad 87\%$$

$$PhCH_2CH_2CO_2H \xrightarrow[SOCl_2]{I_2} PhCH_2CH(I)COCl \quad 95\%$$

The α-sulfenylation[103] and α-selenation[104] of carbonyl compounds have become very important reactions, because these derivatives can subsequently be oxidized to sulfoxides and selenoxides. The sulfoxides and selenoxides readily undergo elimination (see Section 6.8.3), generating the corresponding α,β-unsaturated carbonyl compound. Sulfenylations and selenations are usually carried out under conditions in which the enolate of the carbonyl compound is the reactive species. Scheme 4.6 gives some specific examples of these types of reactions. The most general procedure involves generating the enolate by deprotonation, or one of the alternative methods, followed by reaction with the sulfenylation or selenation reagent, Disulfides are the most common sulfenylation reagents, whereas diselenides or selenenyl halides are used for selenation. As entries 7 and 8 in Scheme 4.6 indicate, the selenation of ketones can also be effected by reactions of enol acetates or silyl enol ethers. If a specific enolate is generated by one of the methods described in Chapter 1, the position of sulfenylation of selenation can be controlled.[105]

99. S. Stavber, M. Zupan, A. J. Poss, and G. A. Shia, *Tetrahedron Lett.* **36:**6769 (1995).
100. T. Umemoto and M. Nagayoshi, *Bull. Chem. Soc. Jpn.* **69:**2287 (1996).
101. S. Stavber and M. Zupan, *Tetrahedron Lett.* **37:**3591 (1996).
102. D. N. Harpp, L. Q. Bao, C. J. Black, J. G. Gleason, and R. A. Smith, *J. Orgn. Chem.* **40:**3420 (1975); Y. Ogata, K. Adachi, and F.-C. Chen, *J. Org. Chem.* **48:**4147 (1983).
103. B. M. Trost, *Chem. Rev.* **78:**363 (1978).
104. H. J. Reich, *Acc. Chem. Res.* **12:**22 (1979); H. J. Reich, J. M. Renga, and I. L. Reich, *J. Am. Chem. Soc.* **97:**5434 (1975).
105. P. G. Gassman, D. P. Gilbert, and S. M. Cole, *J. Org. Chem.* **42:**3233 (1977).

Scheme 4.6. α-Sulfenylation and α-Selenenylation of Carbonyl Compounds

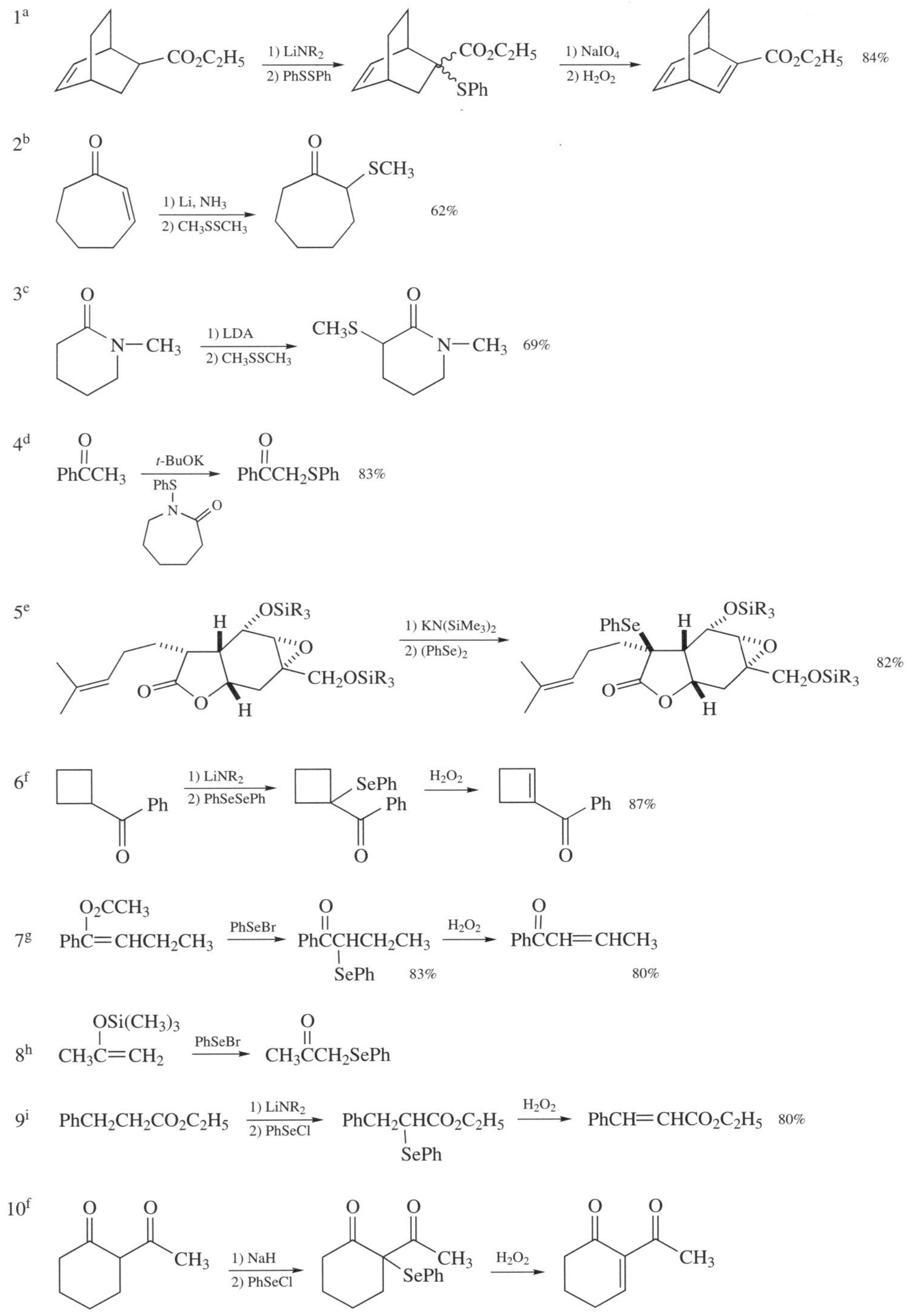

1[a] $CO_2C_2H_5$ —1) $LiNR_2$ 2) PhSSPh→ $CO_2C_2H_5$, SPh —1) $NaIO_4$ 2) H_2O_2→ $CO_2C_2H_5$ 84%

2[b] O —1) Li, NH_3 2) CH_3SSCH_3→ O, SCH_3 62%

3[c] O, N—CH_3 —1) LDA 2) CH_3SSCH_3→ CH_3S, O, N—CH_3 69%

4[d] $PhCCH_3$ (O) —t-BuOK, PhS–N(caprolactam)→ $PhCCH_2SPh$ (O) 83%

5[e] H, $OSiR_3$, O, CH_2OSiR_3, O, O, H —1) $KN(SiMe_3)_2$ 2) $(PhSe)_2$→ PhSe, H, $OSiR_3$, O, CH_2OSiR_3, O, O, H 82%

6[f] Ph, O —1) $LiNR_2$ 2) PhSeSePh→ SePh, Ph, O —H_2O_2→ Ph, O 87%

7[g] $PhC(O_2CCH_3){=}CHCH_2CH_3$ —PhSeBr→ $PhCCH(SePh)CH_2CH_3$ (O) 83% —H_2O_2→ $PhCCH{=}CHCH_3$ (O) 80%

8[h] $CH_3C(OSi(CH_3)_3){=}CH_2$ —PhSeBr→ CH_3CCH_2SePh (O)

9[i] $PhCH_2CH_2CO_2C_2H_5$ —1) $LiNR_2$ 2) PhSeCl→ $PhCH_2CH(SePh)CO_2C_2H_5$ —H_2O_2→ $PhCH{=}CHCO_2C_2H_5$ 80%

10[f] O, O, CH_3 —1) NaH 2) PhSeCl→ O, O, CH_3, SePh —H_2O_2→ O, O, CH_3

Scheme 4.6. (*continued*)

11[j] 1) $LiNR_2$ 2) PhSeBr 3) H_2O_2 82%

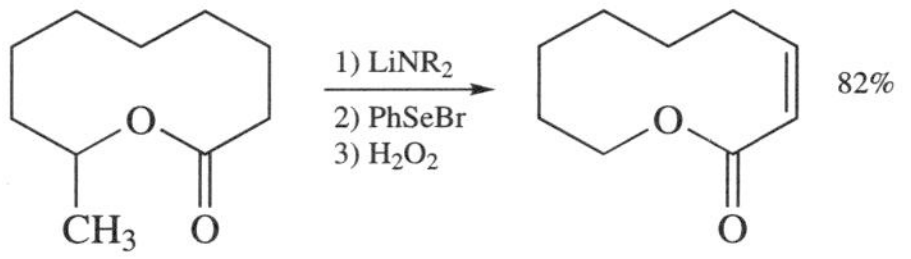

a. B. M. Trost, T. N. Salzmann, and K. Hiroi, *J. Am. Chem. Soc.* **98:**4887 (1976).
b. P. G. Gassman, D. P. Gilbert, and S. M. Cole, *J. Org. Chem.* **42:**3233 (1977).
c. P. G. Gassman and R. J. Balchunis, *J. Org. Chem.* **42:**3236 (1977).
d. G. Foray, A. Penenory, and A. Rossi, *Tetrahedron Lett.* **38:**2035 (1997).
e. A. B. Smith III and R. E. Richmond, *J. Am. Chem. Soc.* **105:**575 (1983).
f. H. J. Reich, J. M. Renga, and I. L. Reich, *J. Am. Chem. Soc.* **97:**5434 (1975).
g. H. J. Reich, I. L. Reich, and J. M. Renga, *J. Am. Chem. Soc.* **95:**5813 (1973).
h. I. Ryu, S. Murai, I. Niwa, and N. Sonoda, *Synthesis* **1977:**874.
i. J. M. Renga and H. J. Reich, *Org. Synth.* **59:**58 (1979).
j. T. Wakamatsu, K. Akasaka, and Y. Ban, *J. Org. Chem.* **44:**2008 (1979).

4.8. Additions to Allenes and Alkynes

Both allenes[106] and alkynes[107] require special consideration with regard to mechanisms of electrophilic addition. The attack by a proton on allene can conceivably lead to the allyl cation or the 2-propenyl cation:

$$^{+}CH_2{-}CH{=}CH_2 \xleftarrow{H^+} CH_2{=}C{=}CH_2 \xrightarrow{H^+} CH_3{-}\overset{+}{C}{=}CH_2$$

An immediate presumption that the more stable allyl ion will be formed overlooks the stereoelectronic aspects of the reaction. Protonation at the center carbon without rotation of one of the terminal methylene groups leads to a primary carbocation which is not stabilized by resonance, because the the adjacent π bond is orthogonal to the empty p orbital.

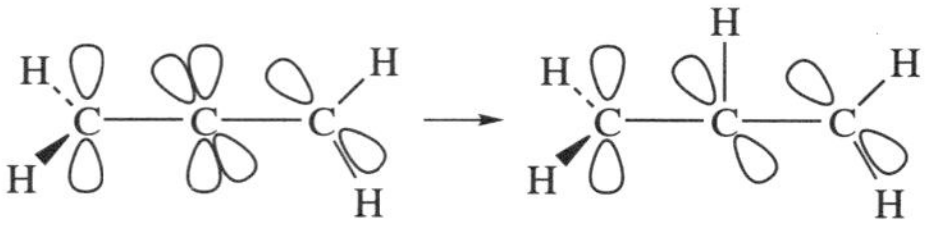

The addition of HCl, HBr, and HI to allene has been studied in some detail.[108] In each case, a 2-halopropene is formed, corresponding to protonatin at a terminal carbon. The initial product can undergo a second addition, giving rise to 2,2-dihalopropanes. Dimers are also formed, but we will not consider them.

$$CH_2{=}C{=}CH_2 + HX \longrightarrow CH_3\overset{X}{\overset{|}{C}}{=}CH_2 + H_3C\overset{X}{\underset{X}{C}}CH_3$$

106. H. F. Schuster and G. M. Coppola, *Allenes in Organic Synthesis*, John Wiley & Sons, New York.
107. W. Drenth, in: *The Chemistry of Triple Bonded Functional Groups*, Supplement C2, Vol. 2, S. Patai, ed., John Wiley & Sons, New York, 1994, pp. 873–915.
108. K. Griesbaum, W. Naegele, and G. G. Wanless, *J. Am. Chem. Soc.*, **87:**3151 (1965).

The presence of a phenyl group results in the formation of products from protonation at the *sp* carbon[109]:

$$PhCH{=}C{=}CH_2 \xrightarrow[HOAc]{HCl} PhCH{=}CHCH_2Cl$$

Two alkyl substituents, as in 1,1-dimethylallene, also lead to protonation at the *sp* carbon[110]:

$$(CH_3)_2C{=}C{=}CH_2 \longrightarrow (CH_3)_2C{=}CHCH_2Cl$$

These substituent effects are due to the stabilization of the carbocation resulting from protonation at the center carbon. Even if allylic conjugation is not available in the transition state, the aryl and alkyl substituents make the terminal carbocation more stable than the alternative, a secondary vinyl cation.

Alkynes, although not as prevalent as alkenes, have a number of important uses in synthesis. In general, alkynes are somewhat less reactive than alkenes toward many electrophiles. A major reason for this difference in reactivity is the substantially higher energy of the vinyl cation intermediate that is formed by an electrophilic attack on an alkyne. It is estimated that vinyl cations are about 10 kcal/mol less stable than an alkyl cation with similar substitution. The observed differences in rate of addition in direct comparisons between alkenes and alkynes depend upon the specific electrophile and the reaction conditions.[111] Table 4.4 summarizes some illustrative rate comparisons. A more complete discussion of the mechanistic aspects of addition to alkynes can be found in Section 6.5 of Part A.

Acid-catalyzed additions to alkynes follow the Markownikoff rule.

$$CH_3(CH_2)_6C{\equiv}CH \xrightarrow{Et_4N^+HBr_2^-} CH_3(CH_2)_6C(Br){=}CH_2 \quad 77\%$$ Ref. 112

The rate and selectivity of the reactions can be considerably enhanced by using an added quaternary bromide salt in 1 : 1 trifluoroacetic acid (TFA) : CH_2Cl_2. Clean formation of the *anti* addition product occurs under these conditions.[113]

$$CH_3CH_2CH_2C{\equiv}CCH_2CH_2CH_3 \xrightarrow[1:4\ TFA:CH_2Cl_2,\ 144\ h]{1.0\ M\ Bu_4N^+Br^-} CH_3CH_2CH_2CH{=}C(Br)CH_2CH_2CH_3 \quad 100\%$$

$$HC{\equiv}C(CH_2)_5CH_3 \xrightarrow[1:4\ TFA:CH_2Cl_2,\ 336\ h]{1.0\ M\ Bu_4N^+Br^-} CH_2{=}C(Br)(CH_2)_5CH_3 \quad 98\%$$

109. T. Okuyama, K. Izawa, and T. Fueno, *J. Am. Chem. Soc.* **95:**6749 (1973).
110. T. L. Jacobs and R. N. Johnson, *J. Am. Chem. Soc.* **82:**6397 (1960).
111. K. Yates, G. H. Schmid, T. W. Regulski, D. G. Garratt, H. W. Leung, and R. McDonald, *J. Am. Chem. Soc.* **95:**160 (1973).
112. J. Cousseau, *Synthesis* **1980:**805.
113. H. W. Weiss and K. M. Touchette, *J. Chem. Soc., Perkin Trans. 2* **1998:**1523.

Table 4.4. Relative Reactivity of Alkenes and Alkynes[a]

	Ratio of second-order rate constants (alkene/alkyne)		
Alkene and alkyne	Bromination, acetic acid	Chlorination, acetic acid	Acid-catalyzed hydration, water
$CH_3CH_2CH_2CH_2CH{=}CH_2$ $CH_3CH_2CH_2CH_2C{\equiv}CH$	1.8×10^5	5.3×10^5	3.6
trans-$CH_3CH_2CH{=}CHCH_2CH_3$ $CH_3CH_2C{\equiv}CCH_2CH_3$	3.4×10^5	$\sim 1 \times 10^5$	16.6
$PhCH{=}CH_2$ $PhC{\equiv}CH$	2.6×10^3	7.2×10^2	0.65

a. From data tabulated in Ref. 111.

Surface-mediated addition of HCl or HBr can be carried out in the presence of silica or alumina.[114] The hydrogen halides can be generated from thionyl chloride, oxalyl chloride, oxalyl bromide, phosphorus tribromide, or acetyl bromide.

$$PhC{\equiv}CCH_3 \xrightarrow[SiO_2]{SOCl_2} \underset{Ph}{\overset{Cl}{}}C{=}C\underset{CH_3}{\overset{H}{}} \longrightarrow \underset{Ph}{\overset{Cl}{}}C{=}C\underset{H}{\overset{CH_3}{}}$$

The kinetic products from HCl results from *syn* addition, but isomerization to the more stable *Z*-isomer occurs on continued exposure to the acid halide.

The initial products of addition to alkynes are not always stable. Addition of acetic acid, for example, results in the formation of enol acetates, which are easily converted to the corresponding ketone under the reaction conditions[115]:

$$H_5C_2C{\equiv}CC_2H_5 \xrightarrow[CH_3CO_2H]{H^+} H_5C_2\underset{O_2CCH_3}{\underset{|}{C}}{=}CHCH_2CH_3 \longrightarrow H_5C_2\underset{O}{\underset{\|}{C}}CH_2CH_2CH_3$$

The most synthetically valuable method for converting alkynes to ketones is by mercuric ion-catalyzed hydration. Terminal alkynes give methyl ketones, in accordance with the Markownikoff rule. Internal alkynes will give mixtures of ketones unless some structural feature promotes regioselectivity. Reactions with $Hg(OAc)_2$ in other nucleophilic solvents such as acetic acid or methanol proceed to β-acetoxy- or β-methoxyalkenylmercury intermediates.[116] These intermediates can be reduced to alkenyl acetates or solvolyzed to ketones. The regiochemistry is indicative of a mercurinium ion intermediate which is opened by nucleophilic attack at the more positive carbon; that is, the additions follow the Markownikoff rule. Scheme 4.7 gives some examples of alkyne addition reactions.

114. P. J. Kropp and S. D. Crawford, *J. Org. Chem.* **59:**3102 (1994).
115. R. C. Fahey and D.-J. Lee, *J. Am. Chem. Soc.* **90:**2124 (1968).
116. S. Uemura, H. Miyoshi, and M. Okano, *J. Chem. Soc., Perkin Trans. 1* **1980:**1098; R. D. Bach, R. A. Woodard, T. J. Anderson, and M. D. Glick, *J. Org. Chem.* **47:**3707 (1982); M. Bassetti, B. Floris, and G. Spadafora, *J. Org. Chem.* **54:**5934 (1989).

Scheme 4.7. Ketones by Hydration of Alkynes

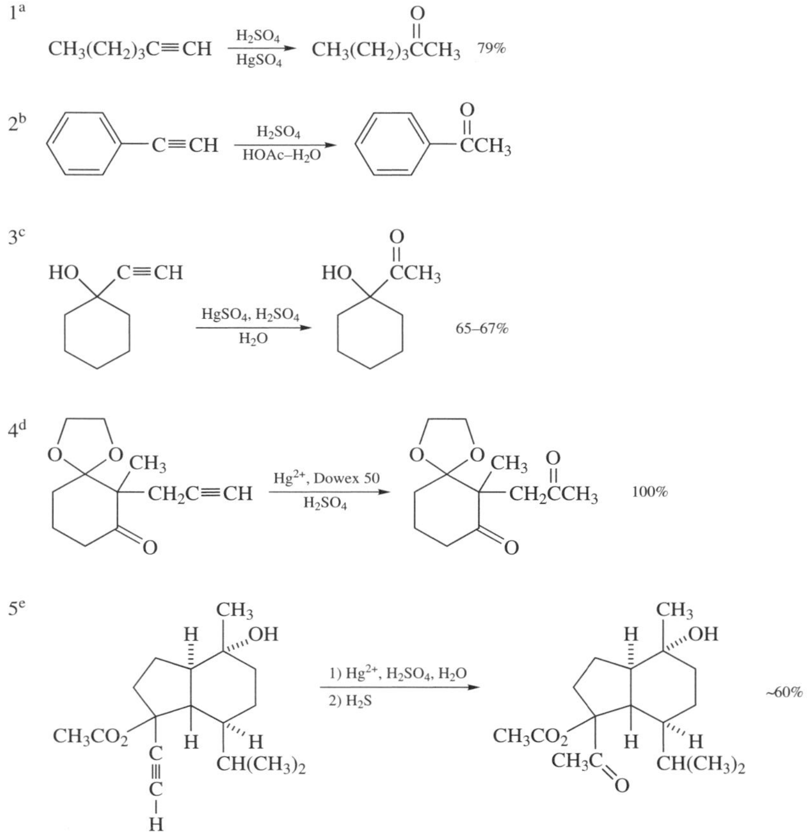

1[a] $CH_3(CH_2)_3C{\equiv}CH \xrightarrow[HgSO_4]{H_2SO_4} CH_3(CH_2)_3\overset{O}{\overset{\|}{C}}CH_3$ 79%

2[b] $C_6H_5C{\equiv}CH \xrightarrow[HOAc\text{–}H_2O]{H_2SO_4} C_6H_5\overset{O}{\overset{\|}{C}}CH_3$

3[c] 1-ethynylcyclohexanol (HO, C≡CH) $\xrightarrow[H_2O]{HgSO_4,\ H_2SO_4}$ 1-acetylcyclohexanol (HO, $\overset{O}{\overset{\|}{C}}CH_3$) 65–67%

4[d] (O, O ketal; CH_3; $-CH_2C{\equiv}CH$; =O) $\xrightarrow[H_2SO_4]{Hg^{2+},\ Dowex\ 50}$ (O, O ketal; CH_3; $-CH_2\overset{O}{\overset{\|}{C}}CH_3$; =O) 100%

5[e] (CH_3, OH, H, CH_3CO_2, C≡C–H, H, H, $CH(CH_3)_2$) $\xrightarrow[2)\ H_2S]{1)\ Hg^{2+},\ H_2SO_4,\ H_2O}$ (CH_3, OH, H, CH_3CO_2, $CH_3C{=}O$, H, H, $CH(CH_3)_2$) ~60%

a. R. J. Thomas, K. N. Campbell, and G. F. Hennion, *J. Am. Chem. Soc.* **60:**718 (1938).
b. R. W. Bott, C. Eaborn, and D. R. M. Walton, *J. Chem. Soc.* **1965:**384.
c. G. N. Stacy and R. A. Mikulec, *Org. Synth.* **IV:**13 (1963).
d. W. G. Dauben and D. J. Hart, *J. Org. Chem.* **42:**3787 (1977).
e. D. Caine and F. N. Tuller, *J. Org. Chem.* **38:**3663 (1973).

Addition of chlorine to 1-butyne is slow in the absence of light. When addition is initiated by light, the major product is *E*-1,2-dichlorobutene when butyne is in large excess[117]:

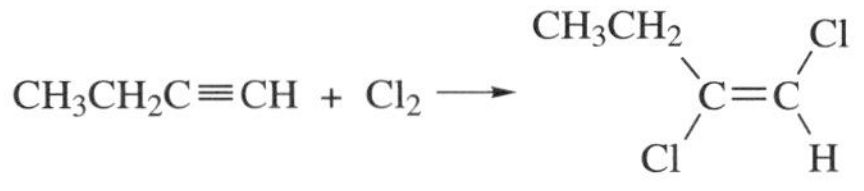

$$CH_3CH_2C{\equiv}CH + Cl_2 \longrightarrow \underset{Cl}{\overset{CH_3CH_2}{}}C{=}C\underset{H}{\overset{Cl}{}}$$

In acetic acid, both 1-pentyne and 1-hexyne give the *syn* addition product. With 2-butyne and 3-butyne, the major products are β-chlorovinyl acetates of *E*-configuration.[118] Some of the dichloro compounds are also formed, with more of the *E*- than the *Z*-isomer being

117. M. L. Poutsma and J. L. Kartch, *Tetrahedron* **22:**2167 (1966).
118. K. Yates and T. A. Go, *J. Org. Chem.* **45:**2385 (1980).

observed.

$$RC{\equiv}CR \xrightarrow[CH_3CO_2H]{Cl_2} (R)(Cl)C{=}C(O_2CCH_3)(R) + (R)(Cl)C{=}C(R)(Cl) + (R)(Cl)C{=}C(Cl)(R)$$

The reactions of the internal alkynes are considered to involve a cyclic halonium ion intermediate, whereas the terminal alkynes seem to react by a rapid collapse of a vinyl cation.

Alkynes react with bromine via an electrophilic addition mechanism. A bridged bromonium ion intermediate has been postulated for alkyl-substituted acetylenes, while vinyl cations are suggested for aryl-substituted examples.[119] 1-Phenylpropyne gives mainly the *anti* addition product in acetic acid, but some of the *syn* isomer is formed.[120] The proportion of dibromide formed and stereoselectivity are enhanced when lithium bromide is added to the reaction mixture.

$$PhC{\equiv}CCH_3 \xrightarrow[CH_3CO_2H]{Br_2} (Br)(Ph)C{=}C(CH_3)(Br) + (Br)(Ph)C{=}C(Br)(CH_3) + PhC(AcO){=}C(Br)CH_3$$

(both isomers)

	anti dibromide	syn dibromide	AcO/Br adduct
no LiBr	59%	14%	21%
LiBr added	98%	0.2%	1.5%

Some of the most useful reactions of alkynes are with organometallic reagents. These reactions, which can lead to carbon–carbon bond formation, will be discussed in Chapter 8.

4.9. Addition at Double Bonds via Organoborane Intermediates

4.9.1. Hydroboration

Borane, BH_3, is an avid electron-pair acceptor, having only six valence electrons on boron. Pure borane exists as a dimer in which two hydrogens bridge the borons. In aprotic solvents that can act as electron donors such as ethers, tertiary amines, and sulfides, borane forms Lewis acid–base adducts.

$$R_2\overset{+}{O}{-}\overset{-}{B}H_3 \qquad R_3\overset{+}{N}{-}\overset{-}{B}H_3 \qquad R_2\overset{+}{S}{-}\overset{-}{B}H_3$$

Borane dissolved in THF or dimethyl sulfide undergoes addition reactions rapidly with most alkenes. This reaction, which is known as hydroboration, has been extensively studied, and a variety of useful synthetic processes have been developed, largely through the work of H. C. Brown and his associates.

Hydroboration is highly regioselective and is stereospecific. The boron becomes bonded primarily to the *less substituted* carbon atom of the alkene. A combination of steric and electronic effects work together to favor this orientation. Borane is an electrophilic reagent. The reaction with substituted styrenes exhibits a weakly negative ρ value

119. G. H. Schmid, A. Modro, and K. Yates, *J. Org. Chem.* **45**:665 (1980).
120. J. A. Pinock and K. Yates, *J. Am. Chem. Soc.* **90**:5643 (1968).

(−0.5).[121] Compared with bromination ($\rho^+ = -4.3$),[122] this is a small substituent effect, but it does favor addition of the electrophilic boron at the less substituted end of the double bond. In contrast to the case of addition of protic acids to alkenes, it is the boron atom, not hydrogen, which is the more electrophilic atom. This electronic effect is reinforced by steric factors. Hydroboration is usually done under conditions in which the borane eventually reacts with three alkene molecules to give a trialkylborane. The second and third alkyl groups would result in severe steric repulsion if the boron were added at the internal carbon.

$[(CH_3)_2CH(CH_3)C]_3B$ (each boron-bound carbon carrying two CH_3 groups; B bonded to three $C(CH_3)_2$ carbons)

severe nonbonded repulsions

$[(CH_3)_2CHCH_2]_3B$ ($H_3C{-}CH(CH_3){-}CH_2{-}B{-}CH_2{-}CH(CH_3){-}CH_3$, with third $CH_2CH(CH_3)_2$ group on B)

nonbonded repulsions reduced

Table 4.5 provides some data on the regioselectivity of addition of diborane and several of its derivatives to representative alkenes. The table includes data for some mono- and dialkylboranes which show even higher regioselectivity than diborane itself. These derivatives have been widely used in synthesis and are frequently referred to by the shortened names shown with the structures.

$[(CH_3)_2CHCH(CH_3){-}]_2BH$

disiamylborane
bis(3-methyl-2-butyl)borane

$(CH_3)_2CHC(CH_3)_2{-}BH_2$

thexylborane
1,1,2-trimethylpropylborane

BH (bicyclic)

9-BBN
9-borabicyclo[3.3.1]nonane

These reagents are prepared by hydroboration of the appropriate alkene, using control of stoichiometry to terminate the hydroboration at the desired degree of alkylation:

$$2\,(CH_3)_2C{=}CHCH_3 + BH_3 \longrightarrow [(CH_3)_2CHCH(CH_3){-}]_2BH$$

$$(CH_3)_2C{=}C(CH_3)_2 + BH_3 \longrightarrow (CH_3)_2CHC(CH_3)_2{-}BH_2$$

Hydroboration is a sterospecific *syn* addition. The addition occurs through a four-center transition state with essentially simultaneous bonding to boron and hydrogen. Both the new C−B and C−H bonds are, therefore, formed from the same side of the double bond. In molecular orbital terms, the addition is viewed as taking place by interaction of the filled alkene π orbital with the empty p orbital on boron, accompanied by concerted

121. L. C. Vishwakarma and A. Fry, *J. Org. Chem.* **45**:5306 (1980).
122. J. A. Pincock and K. Yates, *Can. J. Chem.* **48**:2944 (1970).

Table 4.5. Regioselectivity of Diborane and Alkylboranes toward Representative Alkenes

	Percent of boron added at less substituted carbon			
		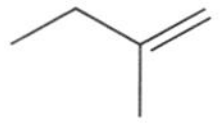	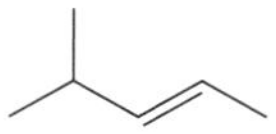	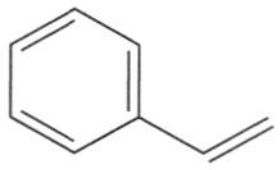
Hydroborating reagent	1-Hexene	2-Methyl-1-butene	4-Methyl-2-pentene	Styrene
Diborane[a]	94	99	57	80
Chloroborane–dimethyl sulfide[b]	99	99.5	—	98
Disiamylborane[a]	99	—	97	98
Thexylborane[c]	94	—	66	95
Thexylchloroborane–dimethyl sulfide[d]	99	99	97	99
9-BBN[e]	99.9	99.8[f]	99.8	98.5

a. G. Zweifel and H. C. Brown, *Org. React.* **13**:1 (1963).
b. H. C. Brown, N. Ravindran, and S. U. Kulkari, *J. Org. Chem.* **44**:2417 (1969); H. C. Brown and U. S. Racherla, *J. Org. Chem.* **51**:895 (1986).
c. H. C. Brown and G. Zweifel, *J. Am. Chem. Soc.* **82**:4708 (1960).
d. H. C. Brown, J. A. Sikorski, S. U. Kulkarni, and H. D. Lee, *J. Org. Chem.* **45**:4540 (1980).
e. H. C. Brown, E. F. Knights, and C. G. Scouten, *J. Am. Chem. Soc.* **96**:7765 (1974).
f. Data for 2-methyl-1-pentene.

C–H bond formation.[123]

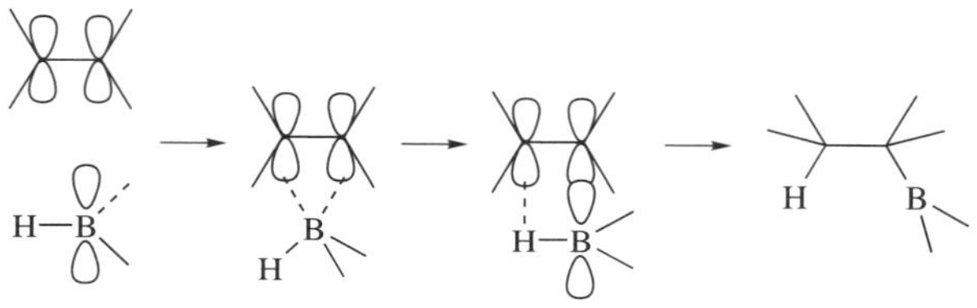

As is true for most reagents, there is a preference for approach of the borane from the less hindered side of the molecule. Because diborane itself is a relatively small molecule, the stereoselectivity is not high for unhindered molecules. Table 4.6 gives some data comparing the direction of approach for three cyclic alkenes. The products in all cases result from *syn* addition, but the mixtures result both from the low regioselectivity and from addition to both faces of the double bond. Even the quite hindered 7,7-dimethylnorbornene shows only modest preference for *endo* addition with diborane. The selectivity is enhanced with the bulkier reagent 9-BBN.

The haloboranes BH_2Cl, BH_2Br, $BHCl_2$, and $BHBr_2$ are also useful hydroborating reagents.[124] These compounds are somewhat more regioselective than borane itself but otherwise show similar reactivity. The most useful aspects of the chemistry of the haloboranes is their application in sequential introduction of substituents at boron. The

123. D. J. Pasto, B. Lepeska, and T.-C. Cheng, *J. Am. Chem. Soc.* **94**:6083 (1972); P. R. Jones, *J. Org. Chem.* **37**:1886 (1972); S. Nagase, K. N. Ray, and K. Morokuma, *J. Am. Chem. Soc.* **102**:4536 (1980); X. Wang, Y. Li, Y.-D. Wu, M. N. Paddon-Row, N. G. Rondan, and K. N. Houk, *J. Org. Chem.* **55**:2601 (1990); N. J. R. van Eikema Hommes and P. v. R. Schleyer, *J. Org. Chem.* **56**:4074 (1991).
124. H. C. Brown and S. U. Kulkarni, *J. Organomet. Chem.* **239**:23 (1982).

Table 4.6. Stereoselectivity of Hydroboration of Cyclic Alkenes[a]

Hydroborating reagent	Product composition[b]								
	3-Methylcyclopentene			3-Methylcyclohexene				7,7-Dimethylnorbornene	
	trans-2	*cis*-3	*trans*-3	*cis*-2	*trans*-2	*cis*-3	*trans*-3	*exo*	*endo*
Borane	45	55 (*cis*-3 + *trans*-3)		16	34	18	32	22	78[c]
Disiamylborane	40	60 (*cis*-3 + *trans*-3)		18	30	27	25	—	—
9-BBN	25	50	25	0	20	40	40	3	97

a. Data from H. C. Brown, R. Liotta, and L. Brener, *J. Am. Chem. Soc.* **99:**3427 (1977), except where noted otherwise.
b. Product composition refers to methylcycloalkanol formed by subsequent oxidation.
c. H. C. Brown, J. H. Kawakami, and K.-T. Liu, *J. Am. Chem. Soc.* **95:**2209 (1973).

halogens can be replaced by alkoxide or by hydride. When halogen is replaced by hydride, a second hydroboration step can be carried out.

$$R_2BX + NaOR' \longrightarrow R_2BOR'$$
$$R_2BX + LiAlH_4 \longrightarrow R_2BH$$
$$RBX_2 + LiAlH_4 \longrightarrow RBH_2$$
$$X = Cl, Br$$

Application of these transformations will be discussed in Chapter 9, where carbon–carbon bond-forming reactions of organoboranes are covered.

Catecholborane and pinacoloborane, in which the boron has two oxygen substituents, are much less reactive hydroborating reagents than alkyl- or haloboranes. Nevertheless, they are useful reagents for certain applications. The reactivity of catecholborane has been found to be substantially enhanced by addition of 10–20% of *N,N*-dimethylacetamide to CH_2Cl_2.[125] Hydroboration by catecholborane and pinacolborane is also catalyzed by transition metals.[126]

catecholborane　　pinacolborane

One frequently used catalyst is Wilkinson's catalyst $Rh(PPh_3)_3Cl$.[127] The general mechanism for catalysis is believed to involve addition of the borane to the metal by oxidative addition[128] (see Section 8.2.3.3). Catalyzed hydroboration has proven to be valuable in

125. C. E. Garrett and G. C. Fu, *J. Org. Chem.* **61:**3224 (1996).
126. I. Beletskaya and A. Pelter, *Tetrahedron* **53:**4957 (1997); H. Wadepohl, *Angew. Chem. Int. Ed. Engl.* **36:**2441 (1997); K. Burgess and M. J. Ohlmeyer, *Chem. Rev.* **91:**1179 (1991).
127. D. A. Evans, G. C. Fu, and A. H. Hoveyda, *J. Am. Chem. Soc.* **110:**6917 (1988); D. Männig and H. Nöth, *Angew. Chem. Int. Ed. Engl.* **24:**878 (1985).
128. D. A. Evans, G. C. Fu, and B. A. Anderson, *J. Am. Chem. Soc.* **114:**6679 (1992).

controlling the stereoselectivity of hydroboration of functionalized alkenes.[129]

Several other catalysts have been described, including, for example, dimethyltitanocene.[130]

$$RCH{=}CH_2 \xrightarrow[(Cp)_2Ti(CH_3)_2,\ (Cp = \eta^5\text{-}C_5H_5)]{\text{catecholborane}} RCH_2CH_2{-}B(\text{catechol}) \xrightarrow[H_2O_2]{NaOH} RCH_2CH_2OH$$

The use of chiral ligands in catalysis can lead to enantioselective hydroboration. Rh-BINAP[131] and the related structure **D**[132] have shown good enantioselectivity in the hydroboration of styrene and related compounds.

C **D**

	styrene	indene	
C	96% e.e.	13% e.e.	(Ref. 131)
D	67% e.e.	84% e.e.	(Ref. 132)

Hydroboration is thermally reversible. At 160°C and above, B–H moieties are eliminated from alkylboranes, but the equilibrium is still in favor of the addition products. This provides a mechanism for migration of the boron group along the carbon chain by a

129. D. A. Evans, G. C. Fu, and A. H. Hoveyda, *J. Am. Chem. Soc.* **114:**6671 (1992).
130. X. He and J. F. Hartwig, *J. Am. Chem. Soc.* **118:**1696 (1996).
131. T. Hayashi, Y. Matsumoto, and Y. Ito *Tetrahedron Asymmetry* **2:**601 (1991).
132. J. M. Valk, G. A. Whitlock, T. P. Layzell, and J. M. Brown, *Tetrahedron Asymmetry* **6:**2593 (1995).

series of eliminations and additions.

Migration cannot occur past a quaternary carbon, however, since the required elimination is blocked. At equilibrium, the major trialkylborane is the least substituted terminal isomer that is accessible, because this is the isomer which minimizes unfavorable steric interactions.

Ref. 133

$CH_3(CH_2)_{13}CH{=}CH(CH_2)_{13}CH_3 \xrightarrow[2)\ 80°C,\ 14\ h]{1)\ B_2H_6} [CH_3(CH_2)_{29}]_3B$ Ref. 134

More bulky substituents on boron faciliate the migration. Bis-Bicyclo[2.2.2]octanylborane, in which there are no complications from migrations in the bicylic substituent, have been found to be particularly useful.

Ref. 135

There is also evidence that boron migration can occur intramolecularly.[136] A transition state that could describe this process has been located computationally.[137] It involves an electron-deficient π-complex about 20–25 kcal above the trialkylborane.

133. G. Zweifel and H. C. Brown, *J. Am. Chem. Soc.* **86:**393 (1964).
134. K. Maruyama, K. Terada, and Y. Yamamoto, *J. Org. Chem.* **45:**737 (1980).
135. H. C. Brown and U. S. Racherla, *J. Am. Chem. Soc.* **105:**6506 (1983).
136. S. E. Wood and B. Rickborn, *J. Org. Chem.* **48:**555 (1983).
137. N. J. R. van Eikema Hommes and P. v. R. Schleyer, *J. Org. Chem.* **56:**4074 (1991).

Migration of boron to terminal positions is observed under much milder conditions in the presence of transition metal catalysts. For example, catalytic hydroboration of 2-methyl-3-hexene by pinacolborane leads to the terminal boronic ester.

$$(CH_3)_2CHCH{=}CHCH_2CH_3 \xrightarrow[Rh(PPh_3)_3Cl]{\text{H–B(pinacolato)}} (CH_3)_2CH(CH_2)_4{-}B(\text{pinacolato})$$ Ref. 138

4.9.2. Reactions of Organoboranes

The organoboranes have proven to be very useful intermediates in organic synthesis. In this section, we will discuss methods by which the boron atom can efficiently be replaced by hydroxyl, halogen, or amino groups. There are also important processes which use alkyboranes in the formation of new carbon–carbon bonds. These reactions will be discussed in Section 9.1.

The most widely used reaction of organoboranes is the oxidation to alcohols. Alkaline hydrogen peroxide is the reagent usually employed to effect the oxidation. The mechanism is outlined below.

$$R_3B + HOO^- \longrightarrow R_2\overset{-}{B}(R){-}O{-}OH \longrightarrow R_2B{-}OR + {}^-OH$$

$$R_2BOR + HOO^- \longrightarrow (R{-}O)R\overset{-}{B}(R){-}O{-}O{-}H \longrightarrow R{-}B(RO)_2 + {}^-OH$$

$$(RO)_2BR + HOO^- \longrightarrow (RO)_2\overset{-}{B}(R){-}O{-}O{-}H \longrightarrow (RO)_3B + {}^-OH$$

$$(RO)_3B + 3\,H_2O \longrightarrow 3\,ROH + B(OH)_3$$

The R–O–B bonds are hydrolysed in the alkaline aqueous solution, generating the alcohol. The oxidation mechanism involves a series of B-to-O migrations of the alkyl groups. The stereochemical outcome is replacement of the C–B bond by a C–O bond with *retention of configuration*. In combination with the stereospecific *syn* hydroboration, this allows the structure and stereochemistry of the alcohols to be predicted with confidence. The preference for hydroboration at the least substituted carbon of a double bond results in the alcohol being formed with regiochemistry which is complementary to that observed in the case of direct hydration or oxymercuration, that is, anti-Markownikoff.

138. S. Pereira and M. Srebnik, *J. Am. Chem. Soc.* **118:**909 (1996); S. Pereira and M. Srebnik, *Tetrahedron Lett.* **37:**3283 (1996).

Conditions that permit oxidation of organoboranes to alcohols using molecular oxygen,[139] sodium peroxycarbonate,[140] or amine oxides[141] as oxidants have also been developed. The reaction with molecular oxygen is particularly effective in perfluoroalkane solvents.[142]

1) $HB(C_2H_5)_2$
2) O_2, $Br(CF_2)_7CF_3$
OH 82%

The oxidation by amine oxides provides a basis for selection among non-equivalent groups on boron. In acyclic organoboranes, the order of reaction is *tertiary > secondary > primary.* In cyclic boranes, stereoelectronic factors dominate. With 9-BBN derivatives, for example, preferential migration of a C−B bond which is part of the bicylic ring structure occurs.

$+ (CH_3)_3\bar{N}{-}O^-$ → $^+N(CH_3)_3$

This is attributed to the unfavourable steric interactions which arise in the transition state that is required for antiperiplanar migration of the exocyclic substituent.[143] Some examples of synthesis of alcohols by hydroboration–oxidation are included in Scheme 4.8.

More vigorous oxidizing agents such as Cr(VI) reagents effect replacement of boron and oxidation to the carbonyl level.[144]

1) B_2H_6
2) $K_2Cr_2O_7$

An alternative procedure for oxidation to ketones involves treatment of the alkylborane with a quaternary ammonium perruthenate salt and an amine oxide.[145] (see entry 6, in Scheme 4.8). Use of the dibromoborane–dimethyl sulfide complex for hydroboration of terminal alkenes, followed by hydrolysis and Cr(VI) oxidation, gives carboxylic acids.[146]

$$RCH{=}CH_2 \xrightarrow[2)\ H_2O]{1)\ BHBr_2S(CH_3)_2} RCH_2CH_2B(OH)_2 \xrightarrow[HOAc,\ H_2O]{Cr(VI)} RCH_2CO_2H$$

139. H. C. Brown, and M. M. Midland, and G. W. Kalbalka, *J. Am. Chem. Soc.* **93:**1024 (1971).
140. G. W. Kabalka, P. P. Wadgaonkar, and T. M. Shoup, *Tetrahedron Lett.* **30:**5103 (1989).
141. G. W. Kabalka and H. C. Hedgecock, Jr., *J. Org. Chem.* **40:**1776 (1975); R. Koster and Y. Monta, *Justus Liebigs Ann. Chem.* **704:**70 (1967).
142. I. Klement and P. Knochel, *Synlett* **1996:**1004.
143. J. A. Soderquist and M. R. Najafi, *J. Org. Chem.* **51:**1330 (1986).
144. H. C. Brown and C. P. Garg, *J. Am. Chem. Soc.* **83:**2951 (1961); H. C. Brown, C. Rao and S. Kulkarni, *J. Organomet. Chem.* **172:**C20 (1979).
145. M. H. Yates, *Tetrahedron Lett.* **38:**2813 (1997).
146. H. C. Brown, S. V. Kulkarni, V. V. Khanna, V. D. Patil, and U. S. Racherla, *J. Org. Chem.* **57:**6173 (1992).

Scheme 4.8. Alcohols, Ketones, Aldehydes, and Amines from Organoboranes

A. Alcohols

1[a] CH_3 — 1) B_2H_6 2) H_2O_2, ^-OH → H_3C, H, OH 85%

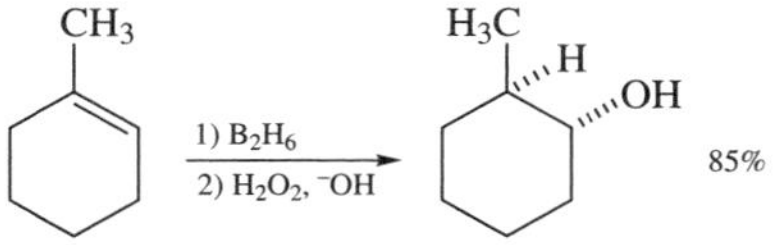

2[b] CH_2, CH_3, CH_3 — 1) B_2H_6 2) H_2O_2, ^-OH → H, CH_2OH, CH_3, CH_3 76%

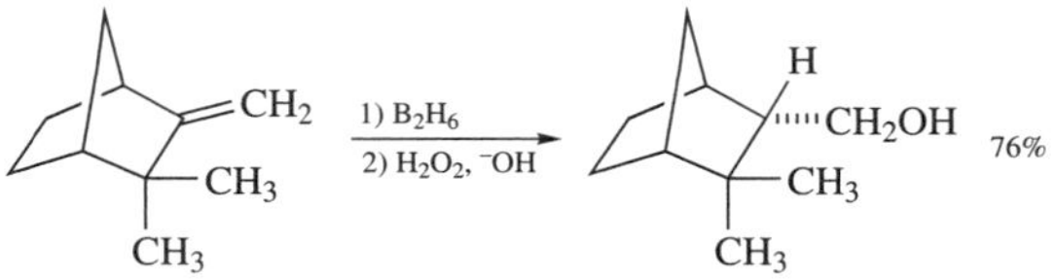

3[c] CH_3 — 1) B_2H_6 2) H_2O_2, ^-OH → CH_3, H, OH 85%

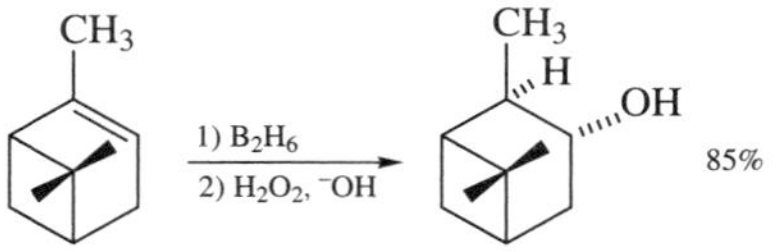

4[d] H, CH_2OCH_2Ph, C=C, CH_3, O, C, H, CH_3 — 1) B_2H_6, THF 2) H_2O_2, ^-OH → OH, H, C, CH_2OCH_2Ph, O, C, C, H, H, CH_3, CH_3 85%

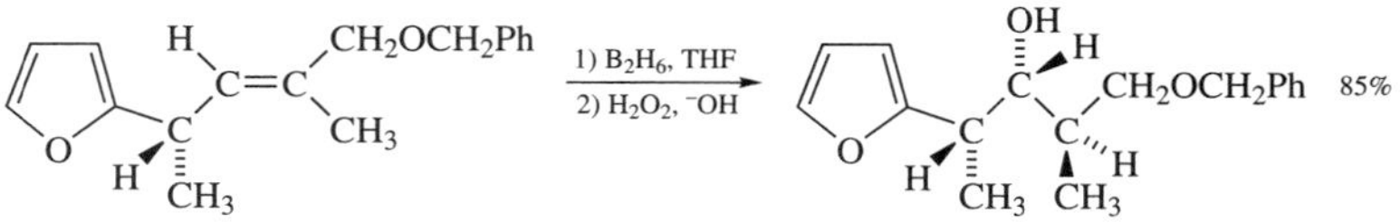

B. Ketones and aldehydes

5[e] H_3C, H_3C, Ph, C, H — 1) B_2H_6 2) CrO_3 → H_3C, H_3C, Ph, O 50%

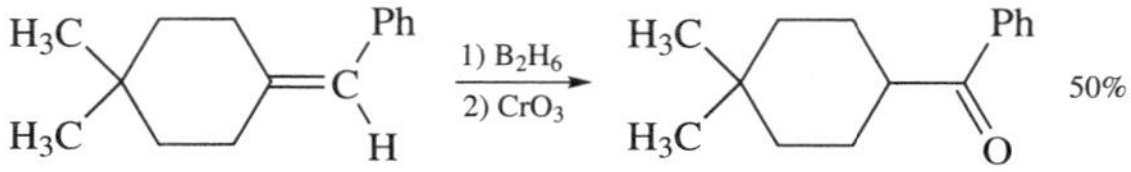

6[f] 1) $BH_3/S(CH_3)_2$ 2) *N*-methylmorpholine-*N*-oxide, $R_4N^+RuO_4^-$ → O

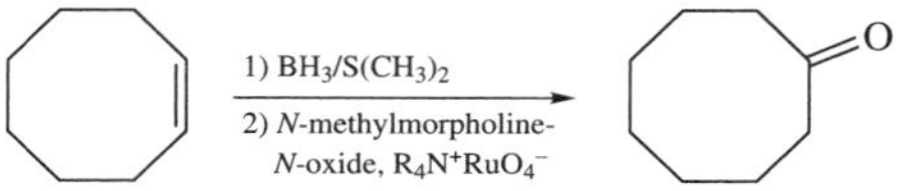

7[g] H_3C, H, CH_2OAc, $CH{=}CH_2$, H — disiamylborane, pyridinium chlorochromate → H_3C, H, CH_2OAc, $CH_2CH{=}O$, H 80%

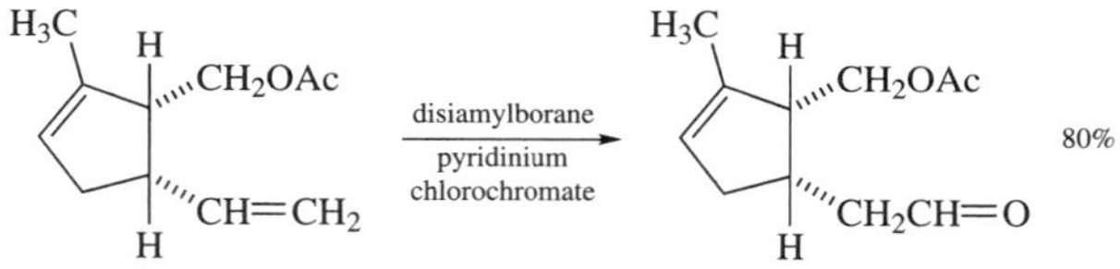

C. Amines

8[h] CH_3 — 1) B_2H_6 2) H_2NOSO_3H → CH_3, NH_2 42%

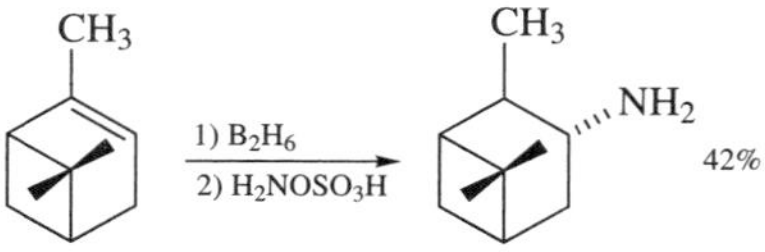

9[i] cyclopentene $\xrightarrow[\text{2) } PhN_3, H_2O]{\text{1) } BHCl_2}$ cyclopentyl–NHPh 84%

a. H. C. Brown and G. Zweifel, *J. Am. Chem. Soc.* **83:**2544 (1961).
b. R. Dulou, Y. Chretien-Bessiere, *Bull. Soc. Chim. France*, 1362 (1959).
c. G. Zweifel and H. C. Brown, *Org. Synth.* **52:**59 (1972).
d. G. Schmid, T. Fukuyama, K. Akasaka, and Y. Kishi, *J. Am. Chem. Soc.* **101:**259 (1979).
e. W. B. Farnham, *J. Am. Chem. Soc.* **94:**6857 (1972).
f. M. H. Yates, *Tetrahedron Lett.* **38:**2813 (1997).
g. H. C. Brown, S. U. Kulkarni, and C. G. Rao, *Synthesis*, 151 (1980); T. H. Jones and M. S. Blum, *Tetrahedron Lett.* **22:**4373 (1981).
h. M. W. Rathke and A. A. Millard, *Org. Synth.* **58:**32 (1978).
i. H. C. Brown, M. M. Midland, and A. B. Levy, *J. Am. Chem. Soc.* **95:**2394 (1973).

The boron atoms can also be replaced by an amino group.[147] The reagents that effect this conversion are chloramine or hydroxylamine-*O*-sulfonic acid. The mechanism of these reactions is very similar to that or the hydrogen peroxide oxidation of organoboranes. The nitrogen-containing reagent initially reacts as a nucleophile by adding at boron, and then rearrangement with expulsion of chloride or sulfate ion follows. As in the oxidation, the migration step occurs with retention of configuration. The amine is freed by hydrolysis.

$$R_3B + NH_2X \longrightarrow R_2\bar{B}(R){-}NH{-}X \longrightarrow R_2B{-}NH(R) \xrightarrow{H_2O} RNH_2$$

$X = Cl$ or OSO_3

Secondary amines are formed by reaction of trisubstituted boranes with alkyl or aryl azides. The most efficient borane intermediates to use are monoalkyldichloroboranes, which are generated by reaction of an alkene with $BHCl_2 \cdot Et_2O$.[148] The entire sequence of steps and the mechanism of the final stages are summarized by the equations below.

$$BHCl_2 \cdot Et_2O + RCH{=}CH_2 \longrightarrow RCH_2CH_2BCl_2$$

$$RCH_2CH_2BCl_2 + R'{-}N_3 \longrightarrow Cl_2B^{-}(CH_2CH_2R){-}N(R'){-}\overset{+}{N}{\equiv}N \longrightarrow Cl_2BN(R')CH_2CH_2R$$

$$Cl_2BN(R')CH_2CH_2R \xrightarrow{H_2O} R'NHCH_2CH_2R$$

Secondary amines can also be made using the *N*-chloro derivatives of primary amines[149]:

$$(CH_3CH_2)_3B + ClN H(CH_2)_7CH_3 \longrightarrow CH_3CH_2NH(CH_2)_7CH_3 \quad 90\%$$

147. M. W. Rathke, N. Inoue, K. R. Varma, and H. C. Brown, *J. Am. Chem. Soc.* **88:**2870 (1966); G. W. Kabalka, K. A. R. Sastry, G. W. McCollum, and H. Yoshioka, *J. Org. Chem.* **46:**4296 (1981).
148. H. C. Brown, M. M. Midland, and A. B. Levy, *J. Am. Chem. Soc.* **95:**2394 (1973).
149. G. W. Kabalka, G. W. McCollum, and S. A. Kunda, *J. Org. Chem.* **49:**1656 (1984).

Organoborane intermediates can also be used to synthesize alkyl halides. Replacement of boron by iodine is rapid in the presence of base.[150] The best yields are obtained with sodium methoxide in methanol.[151] If less basic conditions are desirable, the use of iodine monochloride and sodium acetate gives good yields.[152] As is the case in hydroboration–oxidation, the regioselectivity of hydroboration–halogenation is opposite to that observed for direct ionic addition of hydrogen halides to alkenes. Terminal alkenes give primary halides.

$$RCH{=}CH_2 \xrightarrow[\text{2) } Br_2\text{, NaOH}]{\text{1) } B_2H_6} RCH_2CH_2Br$$

4.9.3. Enantioselective Hydroboration

Several alkylboranes are available in enantiomerically enriched or enantiomerically pure form, and they can be used to prepare enantiomerically enriched alcohols and other compounds available via organoborane intermediates.[153] One route to enantiopure boranes is by hydroboration of readily available terpenes that occur naturally in enantiomerically enriched or enantiomerically pure form. The most thoroughly investigated of these is bis(isopinocampheyl)borane [$(Ipc)_2BH$], which can be prepared in 100% enantiomeric purity from the readily available terpene α-pinene.[154] Both enantiomers are available.

$$+ BH_3 \longrightarrow [\;\;]_2BH$$

Other examples of chiral organoboranes derived from terpenes are **E**, **F**, and **G**, which are derived from longifolene,[155] 2-carene,[156] and limonene,[157] respectively.

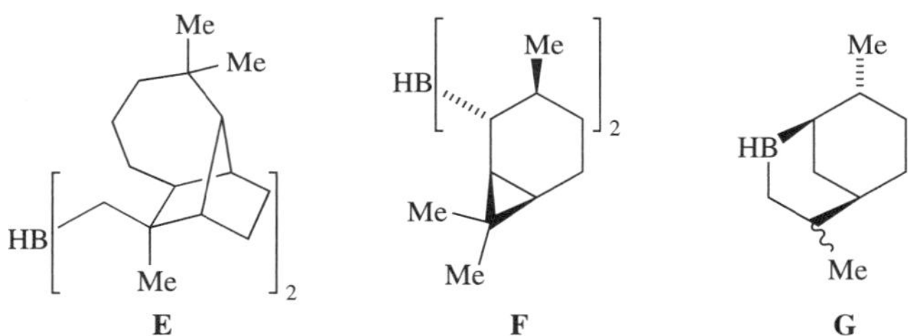

150. H. C. Brown, M. W. Rathke, and M. M. Rogic, *J. Am. Chem. Soc.* **90:**5038 (1968).
151. N. R. De Lue and H. C. Brown, *Synthesis* **1976:**114.
152. G. W. Kabalka and E. E. Gooch III, *J. Org. Chem.* **45:**3578 (1980).
153. H. C. Brown and B. Singaram, *Acc. Chem. Res.* **21:**287 (1988); D. S. Matteson, *Acc. Chem. Res.* **21:**294 (1988).
154. H. C. Brown, P. K. Jadhav, and A. K. Mandal, *Tetrahedron* **37:**3547 (1981); H. C. Brown and P. K. Jadhav, in *Asymmetric Synthesis*, Vol. 2, J. D. Morrison, ed., Academic Press, New York, 1983, Chapter 1.
155. P. K. Jadhav and H. C. Brown, *J. Org. Chem.* **46:**2988 (1981).
156. H. C. Brown, J. V. N. Vara Prasad, and M. Zaidlewics, *J. Org. Chem.* **53:**2911 (1988).
157. P. K. Jadhav and S. U. Kulkarni, *Heterocycles* **18:**169 (1982).

$(Ipc)_2BH$ adopts a conformation which minimizes steric interactions. This conformation results in transition states **H** and **I**, where the S, M, and L substituents are, respectively, the 3-H, 4-CH_2, and 2-$CHCH_3$ groups of the carbocyclic structure. The steric environment at boron in this conformation is such that *Z*-alkenes encounter less steric encumbrance in transition state **I** than in **H**.

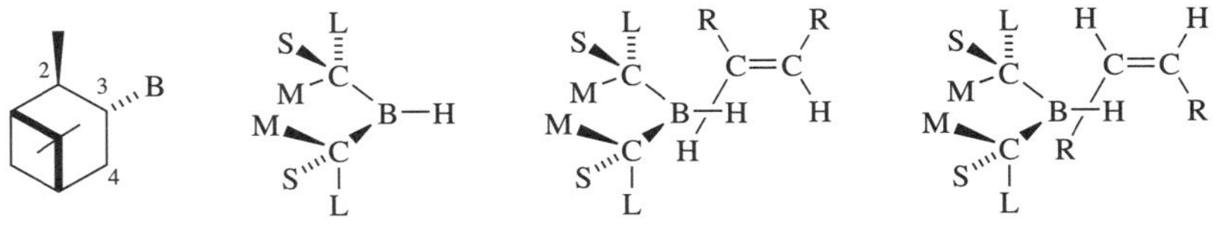

The degree of enantionselectivity of $(Ipc)_2BH$ is not high for all simple alkenes. *Z*-Disubstituted alkenes give good enantioselectivity (75–90%), but *E*-alkenes and simple cycloalkenes give low enantioselectivity (5–30%).

Monoisopinocampheylborane ($IpcBH_2$) can be prepared in enantiomerically pure form by purification of a TMEDA adduct.[158] When this monoalkylborane reacts with a prochiral alkene, one of the diastereomeric products is normally formed in excess and can be obtained in high enatiomeric purity by an appropriate separation.[159] Oxidation of the borane then provides the corresponding alcohol in the same enantiomeric purity achieved for the borane.

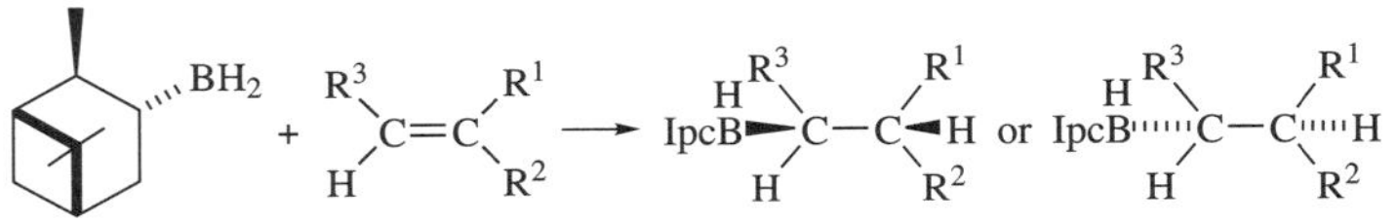

Because oxidation also converts the original chiral terpene-derived group to an alcohol, it is not directly reusable as a chiral auxillary. Although this is not a problem with inexpensive materials, the overall efficiency of generation of enantiomerically pure product is improved by procedures that can regenerate the original terpene. This can be done by heating the dialkylborane intermediate with acetaldehyde. The α-pinene is released and a diethoxyborane is produced.[160]

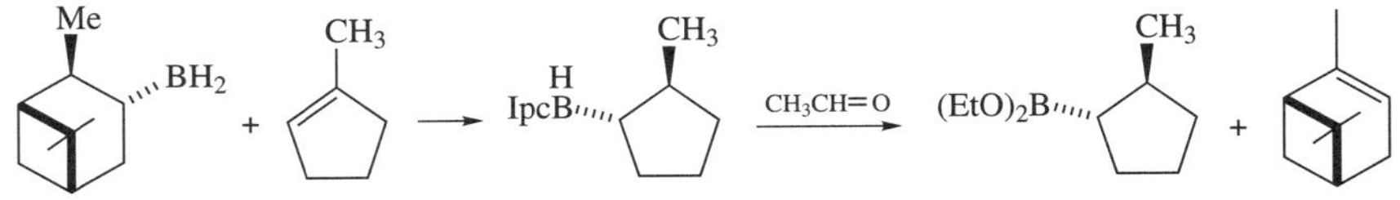

The usual oxidation conditions then convert this boronate ester to an alcohol.[161]

The corresponding haloboranes are also useful for enantioselective hydroboration. Isopinocampheylchloroborane can achieve 45–80% e.e. with representative alkenes.[162] The corresponding dibromoborane achieves 65–85% enantioselectivity with simple

158. H. C. Brown, J. R. Schwier, and B. Singaram, *J. Org. Chem.* **43:**4395 (1978); H. C. Brown, A. K. Mandal, N. M. Yoon, B. Singaram, J. R. Schwier, and P. K. Jadhav, *J. Org. Chem.* **47:**5069 (1982).
159. H. C. Brown and B. Singaram, *J. Am. Chem. Soc.* **106:**1797 (1984); H. C. Brown, P. K. Jadhav, and A. K. Mandal, *J. Org. Chem.* **47:**5074 (1982).
160. H. C. Brown, B. Singaram, and T. E. Cole, *J. Am. Chem. Soc.* **107:**460 (1985); H. C. Brown, T. Imai, M. C. Desai, and B. Singaram, *J. Am. Chem. Soc.* **107:**4980 (1985).
161. D. S. Matteson and K. M. Sadhu, *J. Am. Chem. Soc.* **105:**2077 (1983).
162. U. P. Dhokte, S. V. Kulkarni, and H. C. Brown, *J. Org. Chem.* **61:**5140 (1996).

alkenes when used at −78°C.[163]

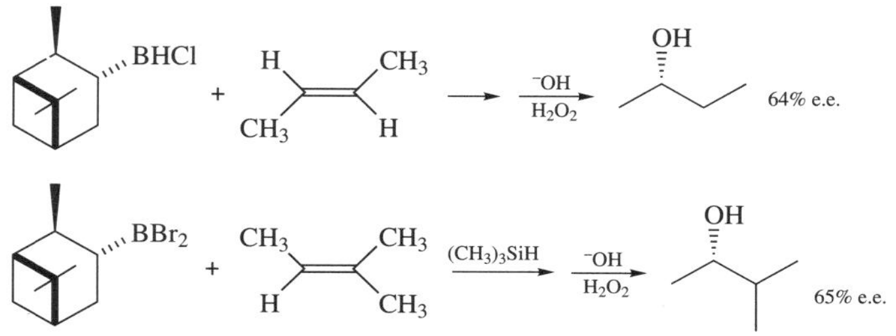

Procedures for synthesis of chiral amines[164] and halides[165] based on chiral alkylboranes have been developed by applying the methods discussed earlier to the homochiral organoborane intermediates. For example, enantiomerically pure terpenes can be converted to trialkylboranes and then aminated with hydroxylaminesulfonic acid.

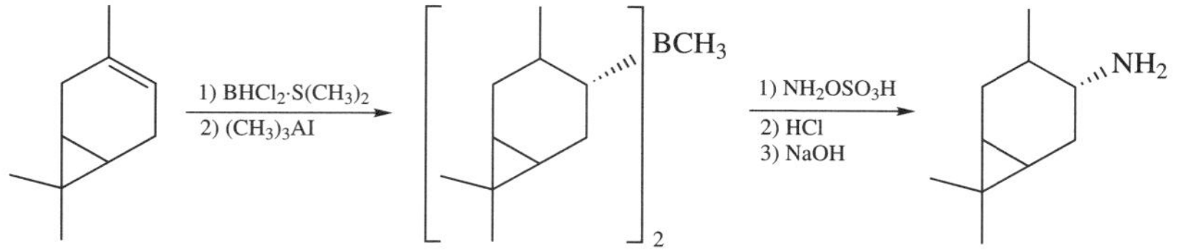

Ref. 166

Combining catalytic enantioselective hydroboration (see p. 230) with amination has provided certain amines with good enantioselectivity.

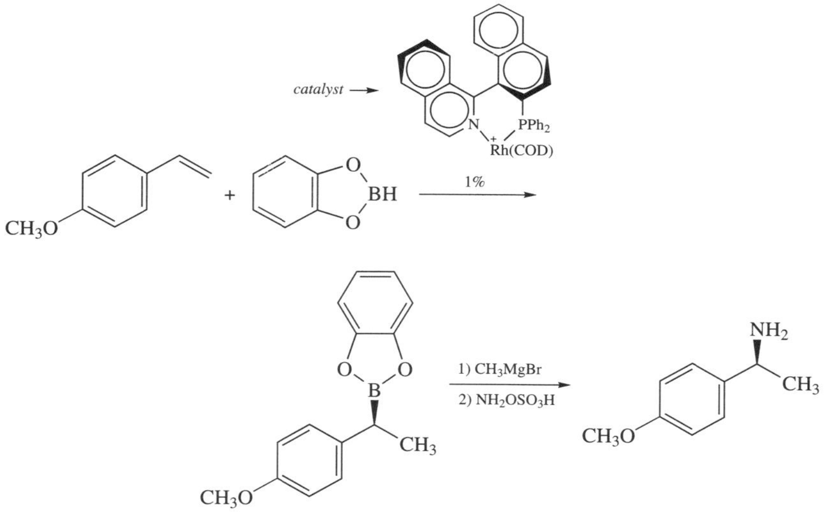

Ref. 167

163. U. P. Dhokte and H. C. Brown, *Tetrahedron Lett.* **37:**9021 (1996).

164. L. Verbit and P. J. Heffron, *J. Org. Chem.* **32:**3199 (1967); H. C. Brown, K.-W. Kim, T. E. Cole, and B. Singaram, *J. Am. Chem. Soc.* **108:**6761 (1986); H. C. Brown, A. M. Sahinke, and B. Singaram, *J. Org. Chem.* **56:**1170 (1991).

165. H. C. Brown, N. R. De Lue, G. W. Kabalka, and H. C. Hedgecock, Jr., *J. Am. Chem. Soc.* **98:**1290 (1976).

166. H. C. Brown, S. V. Malhotra, and P. V. Ramachandran, *Tetrahedron Asymmetry* **7:**3527 (1996).

167. E. Fernandez, M. W. Hooper, F. I. Knight, and J. M. Brown, *J. Chem. Soc., Chem. Commun.* **1997:**173.

Alkynes are reactive toward hydroboration reagents. The most useful procedures involve addition of a disubstituted borane to the alkyne. This avoids the complications which occur with borane that lead to polymeric structures. Catecholborane is a particularly useful reagent for hydroboration of alkynes.[168] Protonolysis of the adduct with acetic acid results in reduction of the alkyne to the corresponding *Z*-alkene. Oxidative workup with hydrogen peroxide gives ketones via enol intermediates.

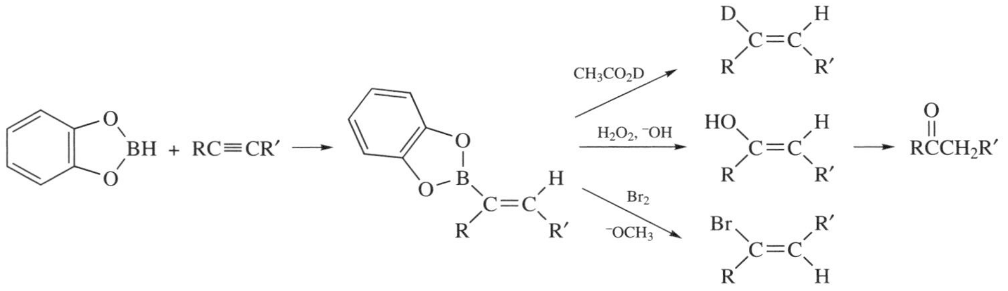

Treatment of the vinylborane with bromine and base leads to vinyl bromides. The reaction occurs with net *anti* addition. The stereoselectivity is explained on the basis of *anti* addition of bromine followed by a second *anti* elimination of bromide and boron:

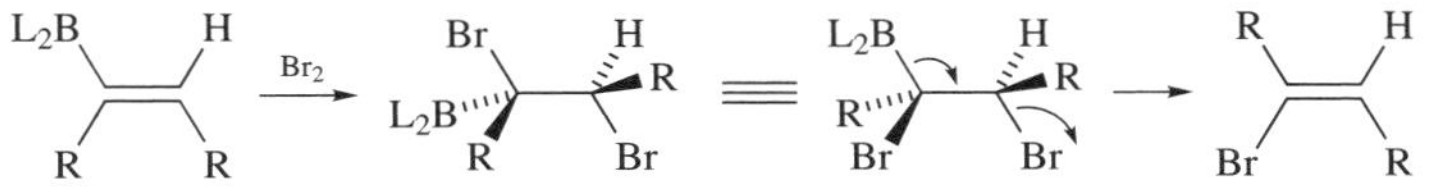

Exceptions to this stereoselectivity have been noted.[169]

The adducts derived from catecholborane are hydrolysed to vinylboronic acids. These materials are useful intermediates for preparation of terminal vinyl iodides. Because the hydroboration is a *syn* addition and the iodinolysis occurs with retention of the alkene geometry, the iodides have the *E*-configuration.[170]

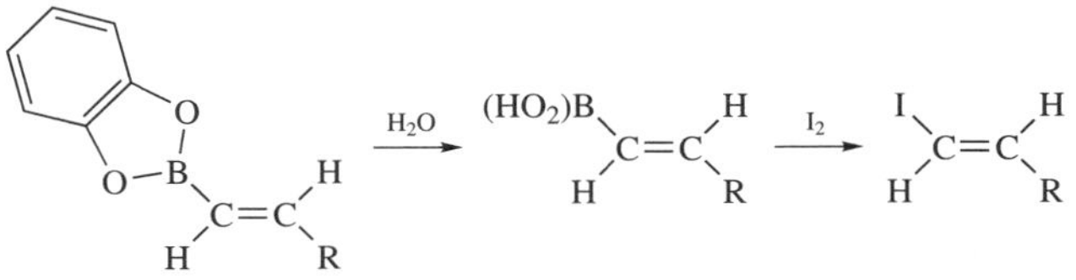

168. H. C. Brown, T. Hamaoka, and N. Ravindran, *J. Am. Chem. Soc.* **95:**6456 (1973); C. F. Lane and G. W. Kabalka, *Tetrahedron* **32:**981 (1976).
169. J. R. Wiersig, N. Waespe-Sarcevic, and C. Djerassi, *J. Org. Chem.* **44:**3374 (1979).
170. H. C. Brown, T. Hamaoka, and N. Ravindran, *J. Am. Chem. Soc.* **95:**5786 (1973).

The dimethyl sulfide complex of dibromoborane[171] and pinacolborane[172] are also useful for synthesis of E-vinyl iodides from terminal alkynes.

$$Br_2\bar{B}H-\overset{+}{S}(CH_3)_2 + HC\equiv CR \longrightarrow (Br_2B)(H)C=C(H)(R) \xrightarrow[2)\ ^-OH,\ I_2]{1)\ OH,\ H_2O} (I)(H)C=C(H)(R)$$

Other disubstituted boranes have also been used for selective hydroboration of alkynes. 9-BBN can be used to hydroborate internal alkynes. Protonlysis can be carried out with methanol, and this provides a convenient method for formation of a disubstituted Z-alkene[173].

$$R-C\equiv C-R + \text{9-BBN} \longrightarrow (R)(H)C=C(R)(B) \xrightarrow{MeOH} (R)(H)C=C(R)(H)$$

A large number of procedures which involve carbon–carbon bond formation have developed around organoboranes. These reactions are considered in Chapter 9.

General References

Addition of Hydrogen Halide, Halogens, and Related Electrophiles

P. B. De la Mare and R. Bolton, *Electrophilic Addition to Unsaturated Systems*, Elsevier, Amsterdam, 1982.
R. C. Fahey, *Top Stereochem.* **2:**237 (1968).

Solvomercuration

W. Kitching, *Organomet. Chem. Rev.* **3:**61 (1968).
R. C. Larock, *Angew. Chem. Int. Ed. Engl.* **12:**27 (1978).

Addition of Sulfur and Selenium Reagents

D. J. Clive, *Tetrahedron* **34:**1049 (1978).
D. Liotta, ed., *Organoselenium Chemistry*, John Wiley & Sons, New York, 1987.
S. Patai and Z. Rappoport, ed., *The Chemistry of Organic Selenium and Tellurium Compounds*, Johnn Wiley & Sons, New York, 1986.
C. Paulmier, *Selenium Reagents and Intermediates in Organic Synthesis*, Pergamon, Oxford, 1986.

Additions to Acetylenes and Allenes

T. F. Rutledge, *Acetylenes and Allenes*, Reinhold, New York, 1969.
G. H. Schmid, in *The Chemistry of the Carbon–Carbon Triple Bond*, S. Patai, ed., John Wiley & Sons, New York, 1978, Chapter 8.
L. Brandsma, *Preparative Acetylenic Chemistry*, 2nd ed. Elsevier, Amsterdam, 1988.

171. H. C. Brown and J. B. Campbell, Jr., *J. Org. Chem.* **45:**389 (1980); H. C. Brown, T. Hamaoka, N. Ravindran, C. Subrahmanyam, V. Somayaji, and N. G. Bhat, *J. Org. Chem.* **54:**6075 (1989).
172. C. E. Tucker, J. Davidson, and P. Knochel, *J. Org. Chem.* **57:**3482 (1992).

H. C. Brown, *Organic Synthese via Boranes*, John Wiley & Sons, New York, 1975.
G. Cragg, *Organoboranes in Organic Synthesis*, Marcel Dekker, New York, 1973.
A. Pelter, K. Smith, and H. C. Brown, *Borane Reagents*, Academic Press, New York, 1988.

Problems

(*References for these problems will be found on page 927.*)

1. Predict the direction of addition and structure of the product for each of the following reactions.

(a) $CH_3CH{=}CH_2$ + O_2N–$C_6H_3(NO_2)$–SCl $\longrightarrow$ (2,4-dinitrobenzenesulfenyl chloride; NO_2 ortho to SCl)

(b) $(CH_3)_2C(OH)C{\equiv}COCH_2CH_3 \xrightarrow[\text{2) KOH}]{\text{1) H}^+\text{, H}_2\text{O}}$

(c) cyclohexane$=CH_2 \xrightarrow{\text{HOBr}}$

(d) $(CH_3)_2C{=}CHCH_3 \xrightarrow[\text{2) H}_2\text{O}_2\text{, HO}^-]{\text{1) disiamylborane}}$

(e) $CH_3CH_2CH_2CH_2CH{=}CH_2 \xrightarrow{\text{IN}_3}$

(f) $(CH_3)_3CCH{=}CHCH_3 \xrightarrow{\text{IN}_3}$

(g) $C_6H_5CH{=}CHCH(OCH_3)_2 \xrightarrow{\text{IN}_3}$

(h) cyclohexenyl–$OSi(CH_3)_3 \xrightarrow[\text{ether, }-78°\text{C}]{\text{PhSeBr}} \xrightarrow[\text{H}_2\text{O}]{\text{NaHCO}_3}$

(i) $HC{\equiv}CCH_2CH_2CO_2H \xrightarrow[\text{CH}_2\text{Cl}_2]{\text{Hg(OAc)}_2} \xrightarrow[\text{10 min}]{\text{H}_2\text{O, NaHCO}_3}$

(j) $H_2C{=}CHCH_2CH_2CH_2CH_2OH \xrightarrow[\text{2) NaBH}_4]{\text{1) Hg(OAc)}_2}$

(k) cyclohexene $\xrightarrow[\text{CHCl}_3\text{, crown ether}]{\text{I}_2\text{, NaN}_3}$

(l) $\Delta^{9,10}$-octalin $\xrightarrow{\text{NOCl}}$

(m) cyclohexene $\xrightarrow[\text{LiClO}_4\text{, CH}_3\text{CN}]{\text{PhSCl, Hg(OAc)}_2}$

(n) $PhCH(CH{=}CH_2)CH_2CO_2H \xrightarrow[\text{CH}_3\text{CN}]{\text{I}_2}$

2. Bromination of 4-*t*-butylcyclohexene in methanol gives a 45:55 mixture of two compounds, each of compositions $C_{11}H_{21}BrO$. Predict the structure and stereochemistry of these two products. How would you confirm your prediction?

3. Hydroboration–oxidation of $PhCH{=}CHOC_2H_5$ gives **A** as the major product if the hydroboration step is of short duration (7 s), but **B** is the major product if the hydroboration is allowed to proceed for a longer time (2 h). Explain.

$PhCH(OH)CH_2OC_2H_5$ (**A**) $\qquad$ $PhCH_2CH_2OH$ (**B**)

4. Oxymercuration of 4-*t*-butylcyclohexene, followed by $NaBH_4$ reduction, gives *cis*-4-*t*-butylcyclohexanol and *trans*-3-*t*-butylcyclohexanol in approximately equal amounts. 1-Methyl-4-*t*-butylcyclohexene under similar conditions gives only *cis*-4-*t*-butyl-1-methylcyclohexanol. Formulate a mechanism for the oxymercuration–reduction process that is consistent with this stereochemical result.

5. Treatment of compound C with *N*-bromosuccinimide in acetic acid containing sodium acetate gives a product $C_{13}H_{19}BrO_3$. Propose the structure, including stereochemistry, of the product and explain the basis for your proposal.

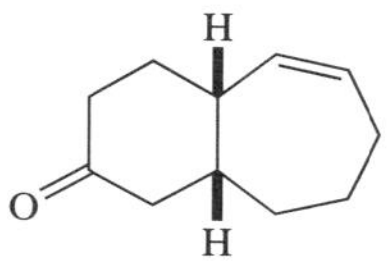

H H O

C

6. The hydration of 5-undecyn-2-one with mercuric sulfate and sulfuric acid in methanol is regioselective, giving 2,5-undecadione in 85% yield. Suggest an explanation for the high selectivity.

7. A procedure for the preparation of allylic alcohols has been devised in which the elements of phenylselenenic acid are added to an alkene, and then the reaction mixture is treated with *t*-butyl hydroperoxide. Suggest a mechanistic rationale for this process.

$$CH_3CH_2CH_2CH{=}CHCH_2CH_2CH_3 \xrightarrow[\text{2) } t\text{-BuOOH}]{\text{1) “}C_6H_5SeOH\text{”}} CH_3CH_2CH_2CH(OH)CH{=}CHCH_2CH_3 \quad 88\%$$

8. Suggest synthetic sequences that could accomplish each of the following transformations.

(a) O, $CO_2C_2H_5$ → O

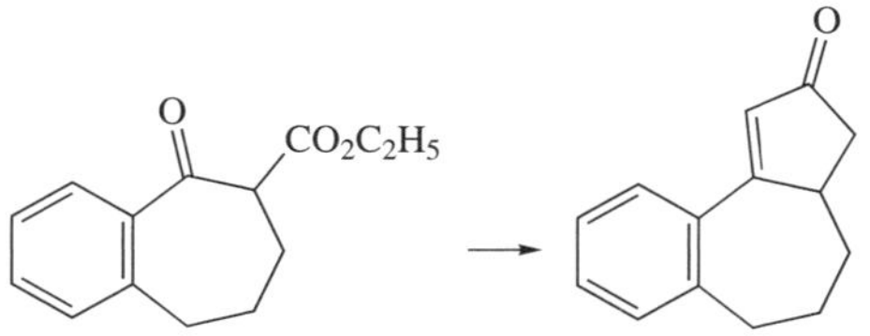

(b) CH_3, OTHP, O, CH_3 → CH_3, OTHP, O, CH_3C, CH_3

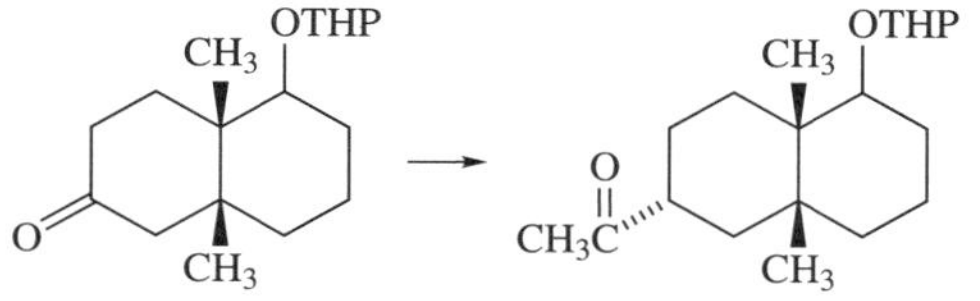

(c) CH_3, $C(CH_3)_2$ → CH_3, CH_3, COH, H, CH_3

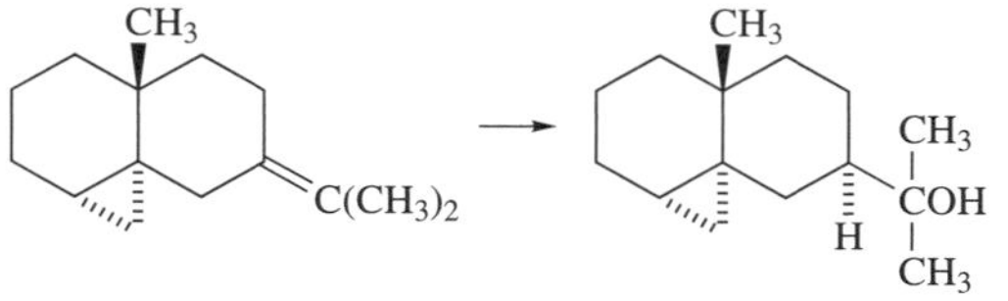

(d) CH_3 → OH, CH_3

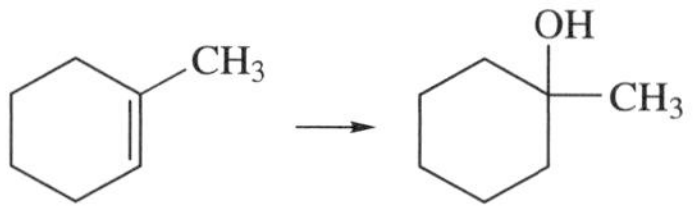

(e) $CH_3CH_2CH_2CH_2C{\equiv}CH$ → $CH_3CH_2CH_2CH_2$, H, C=C, H, I

(f) CH_3, O, H_3C, C, CH_3 → CH_3, O, $HC(CH_3)_2$

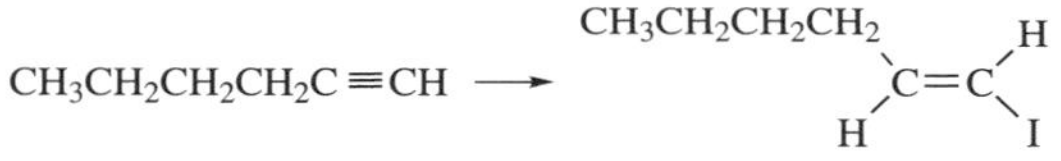

(g) $CH_3\overset{O}{\overset{\|}{C}}CH_2CH_2CH{=}C(CH_3)_2$ → O, $(CH_3)_2CH$

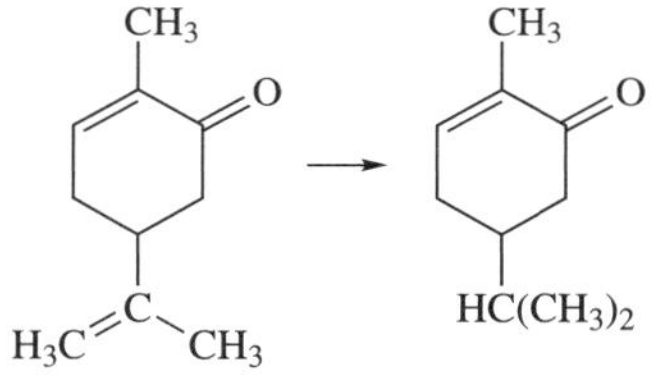

(h) H_3C, $C{=}CH_2$, $CH{=}CH_2$, CH_3 → H_3C, $C{=}CH_2$, CH_2CH_2OH, CH_3

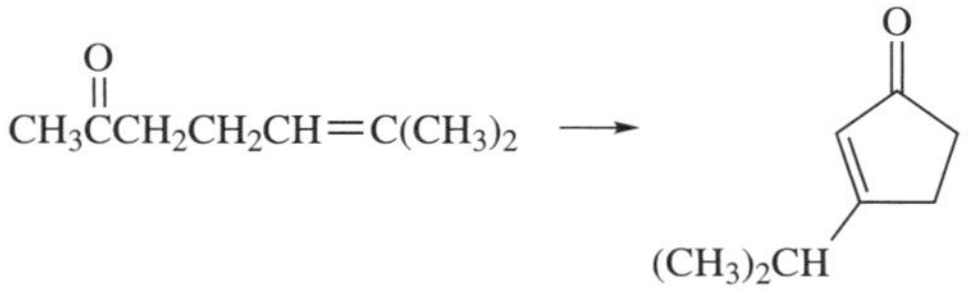

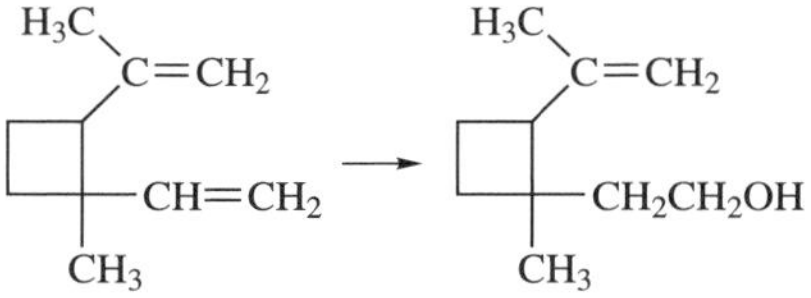

(i) O, O, $CH_2CH{=}CH_2$ → O, O, $(CH_2)_3I$

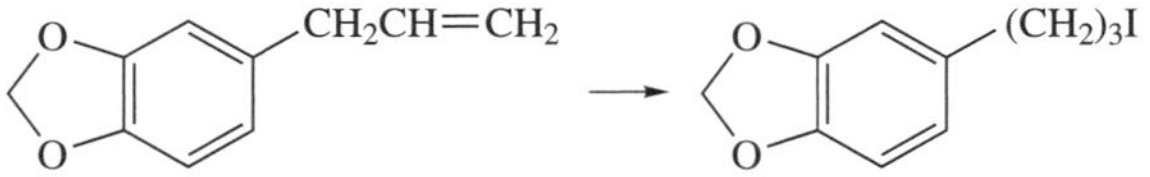

(j) CH_3, O, $C{\equiv}CH$ → CH_3, O, $CH_2CH{=}O$

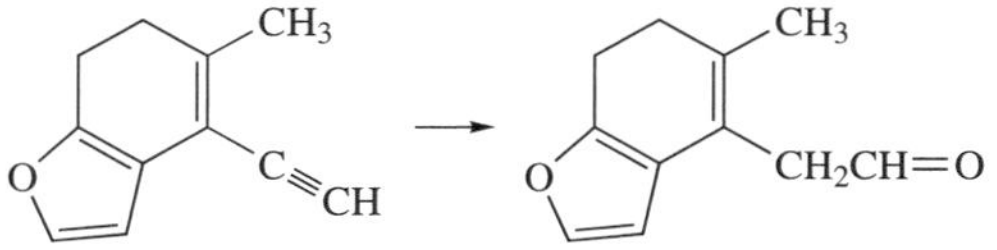

(k) [structure: MeO-steroid with OH and C≡CH] → [structure: MeO-steroid with OH and CH=CHI]

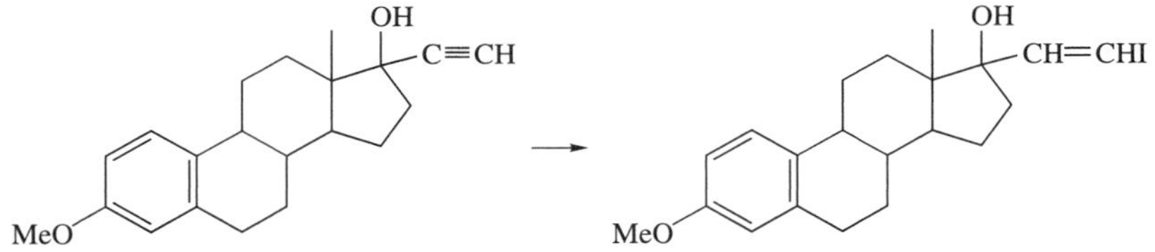

(l) $CH_3CH_2CH(HC(OCH_3)_2)CH_2CH{=}CH_2 \longrightarrow CH_3CH_2CH(HC(OCH_3)_2)CH_2CH_2CH_2Br$

(m) $CH_3(CH_2)_5C{\equiv}CH \longrightarrow CH_3(CH_2)_5CH{=}CHBr$ (H, H; Br)

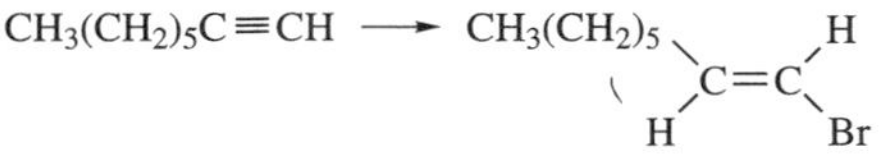

(n) [structure: NHCO₂C(CH₃)₃] → [structure: HOCH₂, N–CO₂C(CH₃)₃]

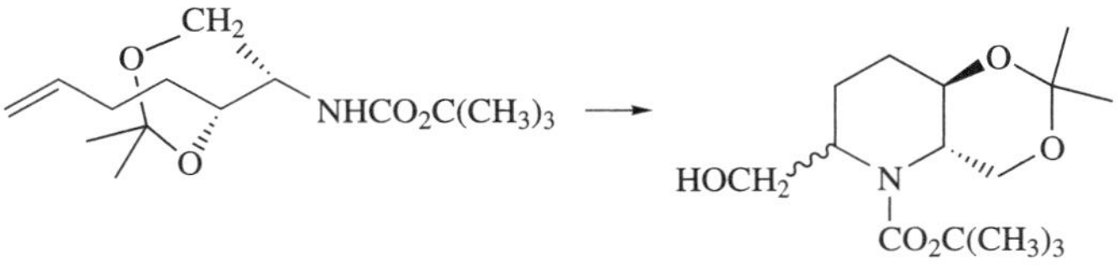

(o) [structure: CH₂OCH₃, H₃C–N, HOCH₂] → [structure: CH₂OCH₃, H₃C–N]

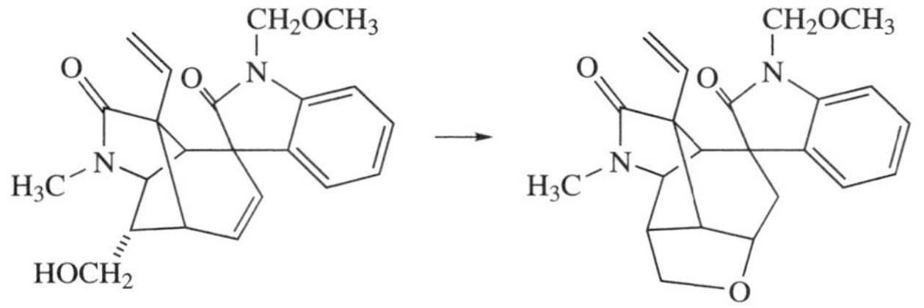

9. Three methods for the preparation of nitroalkenes are outlined as shown. Describe in mechanistic terms how each of these transformations might occur.

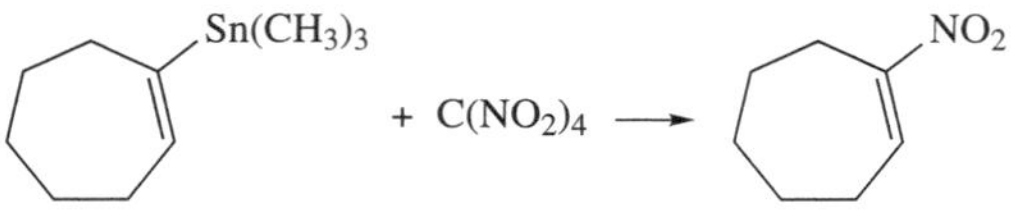

(a) cyclohexene $\xrightarrow[\text{2) NaOH}]{\text{1) HgCl}_2\text{, NaNO}_2}$ 1-nitrocyclohexene (NO_2)

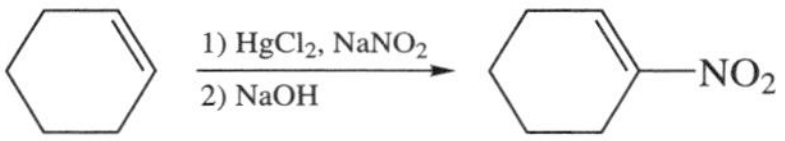

(b) 1-(trimethylstannyl)cycloheptene ($Sn(CH_3)_3$) + $C(NO_2)_4$ $\longrightarrow$ 1-nitrocycloheptene (NO_2)

(c) cyclohexene $\xrightarrow{NO_2^+BF_4^-}$ 1-nitrocyclohexene (NO_2)

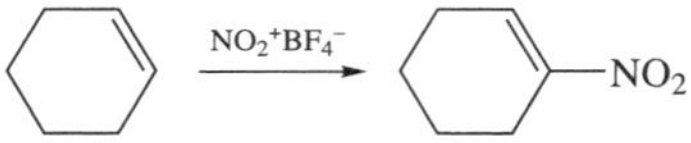

10. Hydroboration–oxidation of 1,4-di-*t*-butylcyclohexene gave three alcohols: C (77%), D (20%), and E (3%). Oxidation of C gave ketone F, which was readily converted in either acid or base to an isomeric ketone G. Ketone G was the only oxidation product of alcohols D and E. What are the structures of compounds C–G?

11. Show how, using enolate chemistry and organoselenium reagents, you could convert 2-phenylcyclohexanone regiospecifically to either 2-phenyl-2-cyclohexen-1-one or 6-phenyl-2-cyclohexen-1-one.

12. On the basis of the mechanistic picture of oxymercuration involving a mercurinium ion, predict the structure and stereochemistry of the major alcohols to be expected by application of the oxymercuration–demercuration sequence to each of the following substituted cyclohexenes.

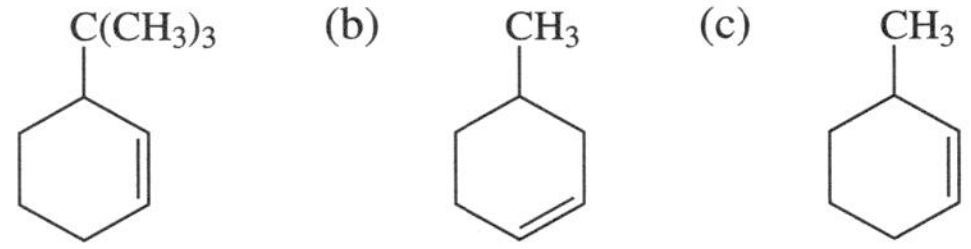

(a) $C(CH_3)_3$ (b) CH_3 (c) CH_3

13. Reaction of the unsaturated acid **A** and I_2 in acetonitrile (no base) gives rise in 89% yield to a 20 : 1 mixture of two stereoisomeric iodolactones. Formulate the complete stereochemistry of both the major and the minor product to be expected under these conditions.

A $H_3C{=}CH$, CH_2, CO_2H, H_3C, H H, CH_3

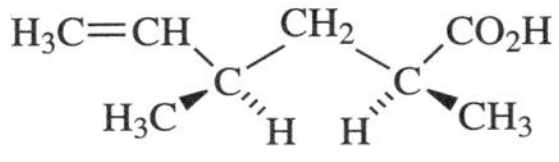

14. Give the structure, including stereochemistry, of the expected product.

(a) 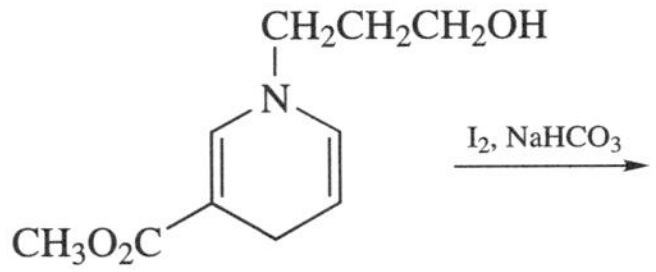 $CH_2CH_2CH_2OH$, N, CH_3O_2C $\xrightarrow{I_2,\ NaHCO_3}$

(b) 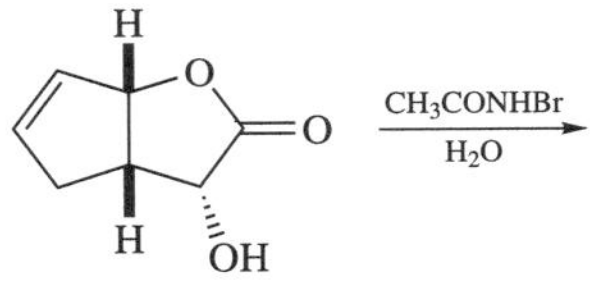 H, O, =O, H, OH $\xrightarrow[H_2O]{CH_3CONHBr}$

(c) 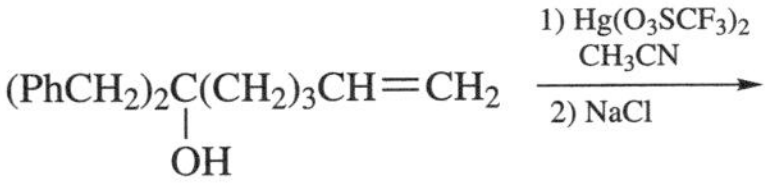 $(PhCH_2)_2C(OH)(CH_2)_3CH{=}CH_2$ $\xrightarrow[2)\ NaCl]{1)\ Hg(O_3SCF_3)_2,\ CH_3CN}$

(d) 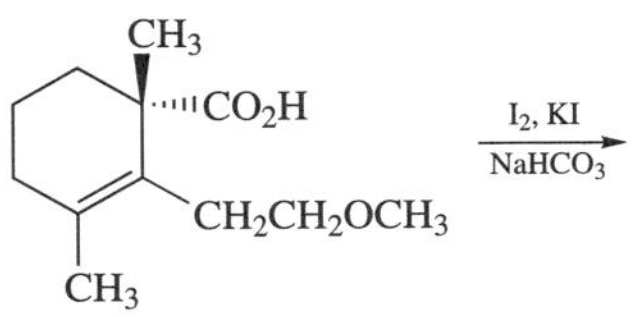 CH_3, CO_2H, $CH_2CH_2OCH_3$, CH_3 $\xrightarrow[NaHCO_3]{I_2,\ KI}$

(e) 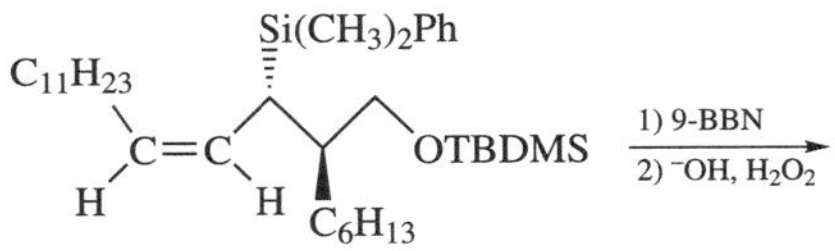 $Si(CH_3)_2Ph$, $C_{11}H_{23}$, C=C, H, H, OTBDMS, C_6H_{13} $\xrightarrow[2)\ ^-OH,\ H_2O_2]{1)\ 9\text{-BBN}}$

(f) $H_2C{=}C(CH_2OC_{16}H_{33})(CH_2OCH_3)$ $\xrightarrow[\text{2) }^{-}\text{OH, }H_2O_2]{\text{1) (+)-(Ipc)}_2\text{BH}}$

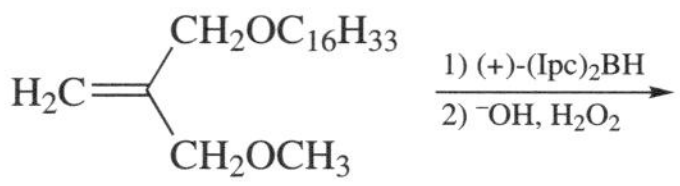

15. Some synthetic transformations are shown in the retrosynthetic format. Propose a short series of reactions (no more than three steps should be necessary) which could effect the synthetic conversion.

(a) [structure: oxazolidinone, N-acyl C(NH₂)CH(CH₃)₂, CH₂Ph] ⟹ [structure: oxazolidinone, N-acyl CH₂CH(CH₃)₂, CH₂Ph] (enantioselective)

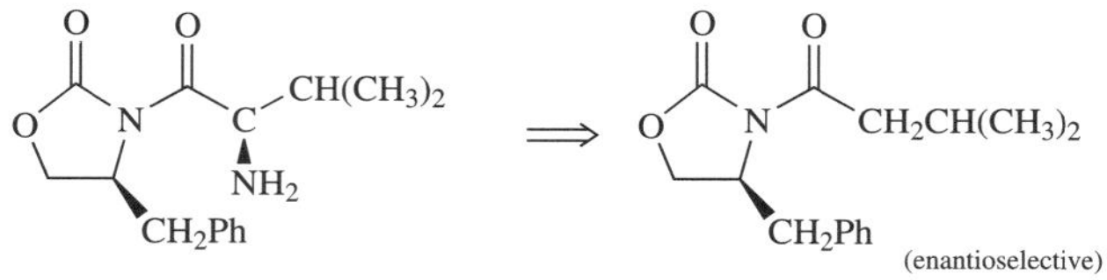

(b) [structure: bicyclic ether] ⟹ [structure: cyclohexene with CH₂OH]

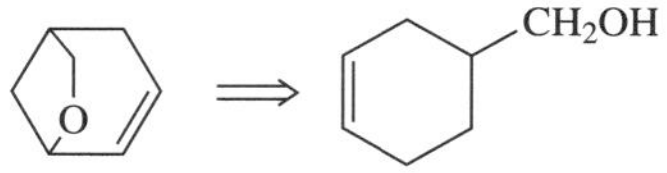

(c) I—CHCH₂CH₂CH₂CO₂CH₃ [structure with S ring, RO, H, C=C, H, CH(CH₂)₄CH₃, OR] ⟹ HO [structure with CH₂, C=C, H, CH₂CH₂CH₂CO₂CH₃, H, H, C=C, CH(CH₂)₄CH₃, RO, H, OR]

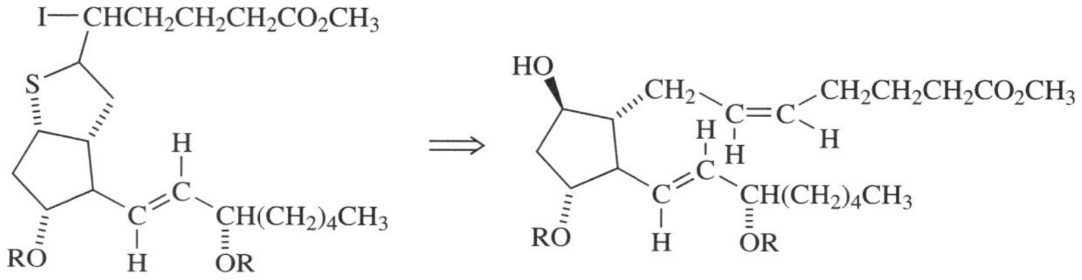

(d) [structure: CH₃, PhCHN, O, bicyclic lactone, Br] ⟹ [structure: dihydropyran with C(=O)NHCH(CH₃)Ph]

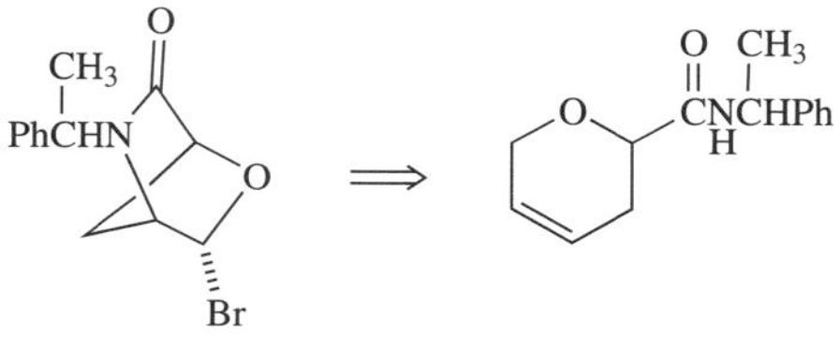

(e) [structure: CH₃-cyclopentene, CH₂O₂CCH₃, C(=CH₂)CH₃] ⟹ [structure: CH₃-cyclopentene, CH₂O₂CCH₃, CH₃CHCH=O]

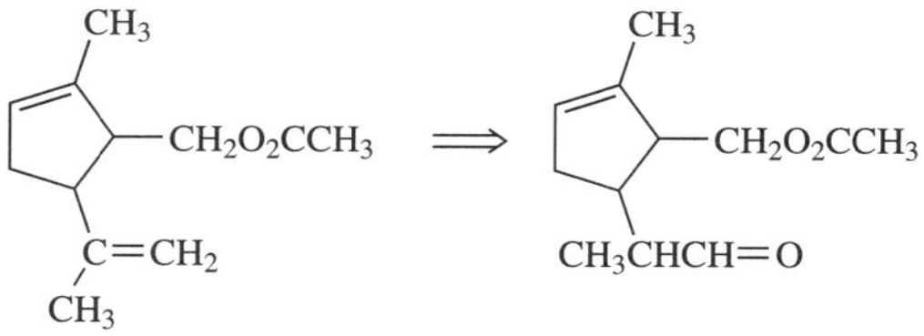

16. Write detailed mechanisms for the following reactions.

(a) [structure: 5,5-dimethylcyclohexane-1,3-dione, H₃C, H₃C, O, O] $\xrightarrow{\text{NaOCl}}$ $HO_2CCH_2C(CH_3)_2CH_2CO_2H$

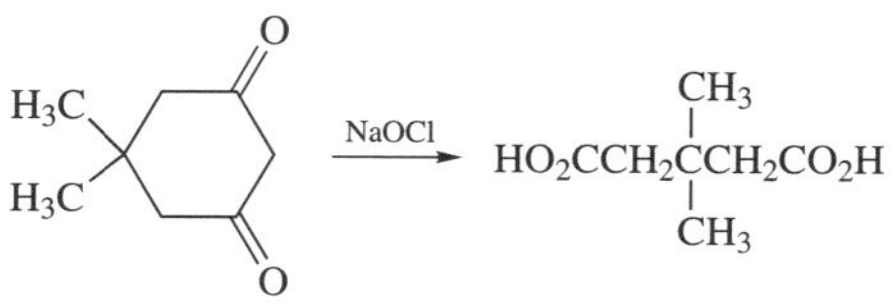

(b) $CH_2{=}CHCH_2CH(CH_3)OCH_2NHCO_2CH_2Ph$ $\xrightarrow[2)\ KBr]{1)\ Hg(NO_3)_2}$ $\xrightarrow[O_2]{NaBH_4}$ (product: NCO_2CH_2Ph, H_3C, CH_2OH) 3:1 *cis:trans*

(c) (CO_2H, N, Ph, CH_3) $\xrightarrow[2)\ NaOCH_3]{1)\ NBS}$ (O, N, CO_2CH_3, Ph, CH_3) >90% enantiomerically pure

17. It has been observed that 4-pentenyl amides such as **1** cyclize to lactams **2** on reaction with phenylselenenyl bromide. The 3-butenyl compound **3**, on the other hand, cyclizes to an imino ether, **4**. What is the basis for the different course of these reactions?

$CH_2{=}CHCH_2CH(R)CH_2NHC(O)CH_3$ (**1**) $\xrightarrow{PhSeBr}$ **2** ($PhSeCH_2$, R, N, $O{=}CCH_3$)

$CH_2{=}CHCH_2CH_2NHC(O)CH_3$ (**3**) $\xrightarrow{PhSeBr}$ **4** ($PhSeCH_2$, O, N, CH_3)

18. Procedures for enantioselective synthesis of derivatives of α-bromoacids based on reaction of compounds **A** and **B** with *N*-bromosuccinimide have been developed. Predict the absolute configuration at the halogenated carbon in each product. Explain the basis of your prediction.

A: $(C_6H_{13})_2NSO_2CH_2$, O, C, R, H, OTMS $\xrightarrow{NBS}$

B: Bu, Bu, B, O, O, O, N, R, H, CH_2Ph $\xrightarrow{NBS}$

19. The stereochemical outcome of the hydroboration–oxidation of 1,1′-bicyclohexenyl depends on the amount of diborane used in the hydroboration. When 1.1 equiv is used, the product is a 3 : 1 mixture of **C** and **D**. When 2.1 equiv is used, **C** is formed nearly exclusively. Offer an explanation of these results.

$\xrightarrow[2)\ ^{-}OH,\ H_2O_2]{1)\ B_2H_6}$ **C** (H H, OH HO) + **D** (H H, OH HO)

20. Predict the absolute configuration of the product obtained from the following reactions based on enantioselective hydroboration.

(a) 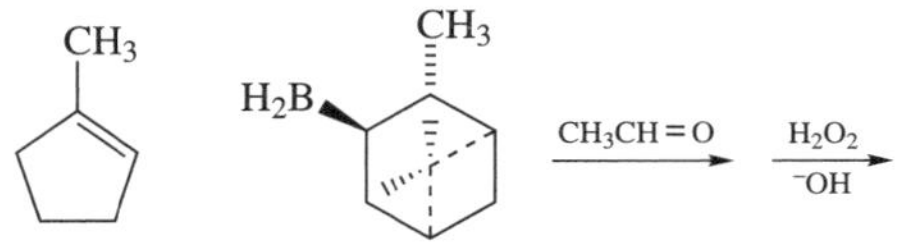 CH_3, H_2B, CH_3, $CH_3CH{=}O$, H_2O_2, ^-OH

(b) 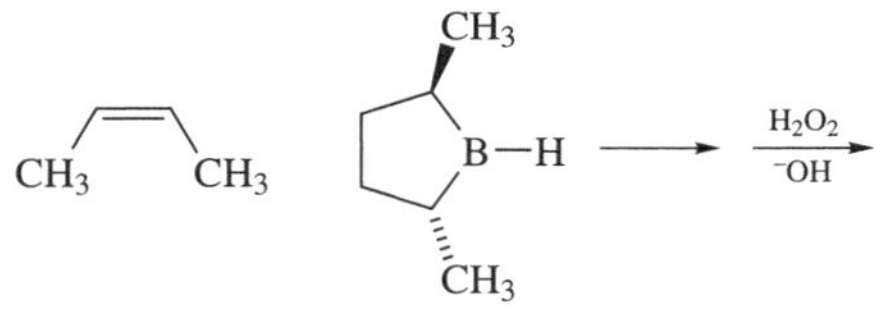 CH_3, CH_3, CH_3, B—H, CH_3, H_2O_2, ^-OH

21. The regioselectivity and stereoselectivity of electrophilic additions to 2-benzyl-2-azabicyclo[2.2.1]hept-5-en-3-one are quite dependent on the specific electrophile. Discuss the factors which could influence the differing selectivity patterns and compare this system to norbornene.

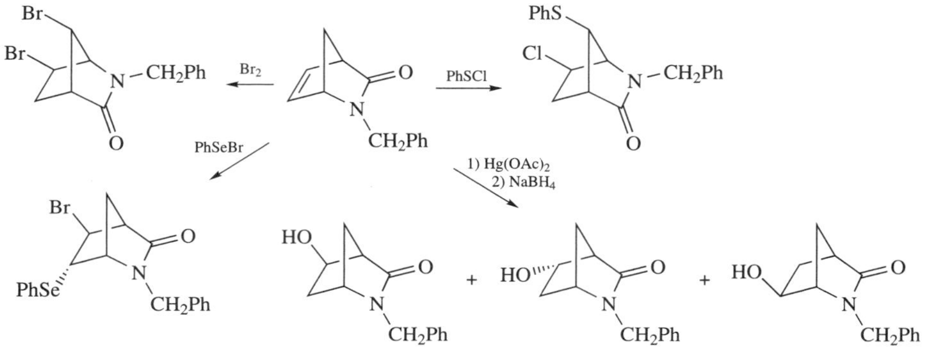

Br, Br, N–CH_2Ph, O, Br_2, O, N, CH_2Ph, PhSCl, PhS, Cl, N–CH_2Ph, O, PhSeBr, 1) $Hg(OAc)_2$ 2) $NaBH_4$, Br, O, N, PhSe, CH_2Ph, HO, O, N, CH_2Ph, +, HO, O, N, CH_2Ph, +, HO, O, N, CH_2Ph

22. Offer a mechanistic explanation of the following observations.

(a) In the cyclization shown, **A** is the preferred product for R=H, but *endo* cyclization to **B** is preferred for R=phenyl or methyl.

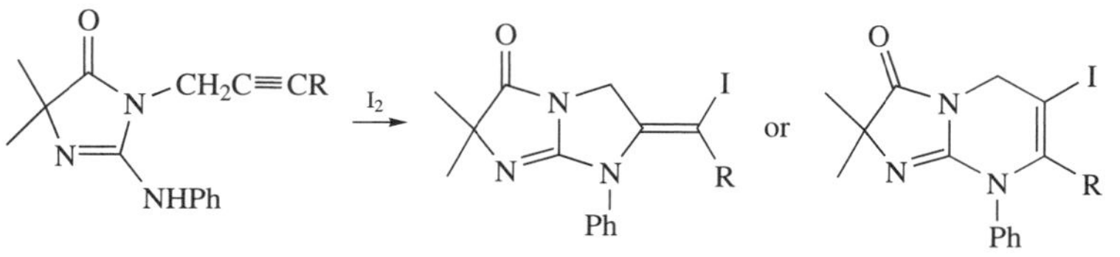

O, N–$CH_2C{\equiv}CR$, N, NHPh, I_2, O, N, N, N, Ph, I, R, **A**, or, O, N, N, N, Ph, I, R, **B**

(b) The pent-4-enoyl group has been developed as a protecting group for amines. The conditions for cleavage involve treatment with iodine and an mixed aqueous solution with THF or acetonitrile. Give a mechanism which accounts for the mild deprotection under these conditions.

5

Reduction of Carbonyl and Other Functional Groups

Introduction

The topic of this chapter is reduction reactions that are especially important in synthesis. Reduction can be accomplished by several broad methods, including addition of hydrogen and/or electrons to a molecule or removal of oxygen or other electronegative substituents. The most important reducing agents from a synthetic point of view are molecular hydrogen and hydride derivatives of boron and aluminum. Other important procedures use metals such as lithium, sodium, or zinc as electron donors. Certain reductions that proceed via a free-radical mechanism involve hydrogen atom donors such as the trialkyl tin hydrides. Reductive removal of oxygen from functional groups such as alcohols, benzylic carbonyls, α-oxycarbonyls, and diols is also important in synthesis, since these reactions provide important methods for interconversion of functional groups. There are also reductive procedures which involve formation of carbon–carbon bonds. Most of these begin with an electron transfer that generates a radical intermediate which then undergoes a coupling or addition reaction.

5.1. Addition of Hydrogen

5.1.1. Catalytic Hydrogenation

The most widely used method for adding the elements of hydrogen to carbon–carbon double bonds is catalytic hydrogenation. Except for very sterically hindered alkenes, this reaction usually proceeds rapidly and cleanly. The most common catalysts are various forms of transition metals, particularly platinum, palladium, rhodium, ruthenium, and nickel. Both the metals, as finely dispersed solids or adsorbed on inert supports such as carbon or alumina, and certain soluble complexes of these metals exhibit catalytic activity. Depending upon conditions and catalyst, other functional groups are also subject to catalytic hydrogenation.

$$RCH{=}CHR + H_2 \xrightarrow{\text{catalyst}} RCH_2CH_2R$$

The mechanistic description of alkene hydrogenation is somewhat vague, partly because the reactive sites on the metal surface are not as easily described as small-molecule reagents in solution. As understanding of the chemistry of soluble hydrogenation catalysts has developed, it has become possible to extrapolate some mechanistic concepts to heterogeneous catalysts. It is known that hydrogen is adsorbed onto the metal surface, presumably forming metal–hydrogen bonds similar to those in transition-metal hydride complexes. Alkenes are also adsorbed on the catalyst surface, and at least three types of intermediates have been implicated in the process of hydrogenation. The initially formed intermediate is pictured as attached at both carbon atoms of the double bond by π-type bonding, as shown in **A**. The bonding is regarded as an interaction between the alkene π and π^* orbitals and acceptor and donor orbitals of the metal. A hydrogen can be added to the adsorbed group, leading to **B**, which involves a σ-type carbon–metal bond. This species can react with another hydrogen to give the alkane, which is desorbed from the surface. A third intermediate species, shown as **C**, accounts for double-bond isomerization and the exchange of hydrogen which sometimes accompanies hydrogenation. This intermediate is equivalent to an allyl group bound to the metal surface by π bonds. It can be formed from adsorbed alkene by abstraction of an allylic hydrogen atom by the metal. In Chapter 8, the reactions of transition metals with organic compounds will be discussed. There are well-characterized examples of structures corresponding to each of the intermediates **A**, **B**, and **C** that are involved in hydrogenation.

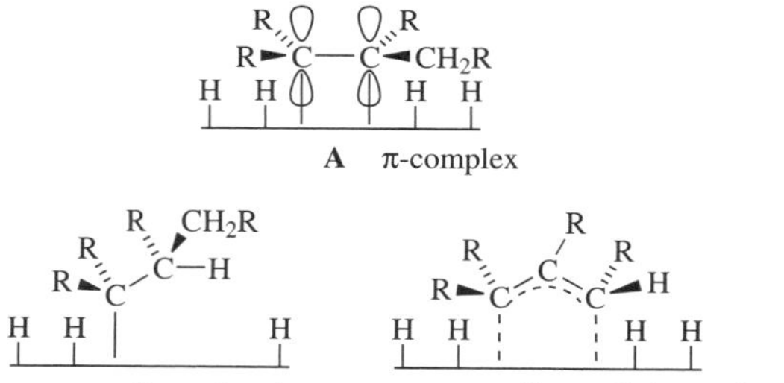

A π-complex

B σ-bond

C π-allyl complex

In most cases, both hydrogen atoms are added to the same side of the reactant (*syn* addition). If hydrogenation occurs by addition of hydrogen in two steps, as implied by the mechanism above, the intermediate must remain bonded to the metal surface in such a way that the stereochemical relationship is maintained. Adsorption to the catalyst surface normally involves the less sterically congested side of the double bond, and, as a result, hydrogen is added from the less hindered face of the double bond. Scheme 5.1 illustrates some hydrogenations in which the *syn* addition from the less hindered side is observed. Some exceptions are also included. There are many hydrogenations in which hydrogen addition is not entirely *syn*, and independent corroboration of the stereochemistry is normally necessary.

Scheme 5.1. Stereochemistry of Hydrogenation of Some Alkenes

A. Examples of preferential *syn* addition from less hindered side

1[a] Pt, H_2: 70%, 30%

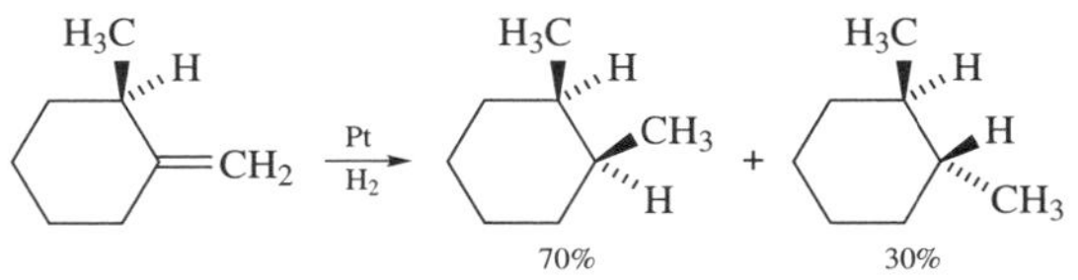

2[b] Pt, H_2: 70–85%, 15–30%

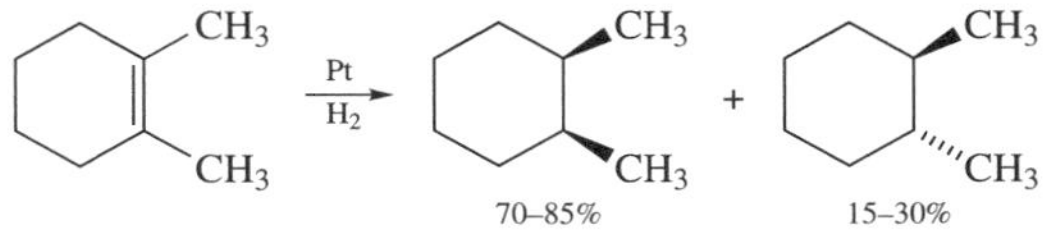

3[c]

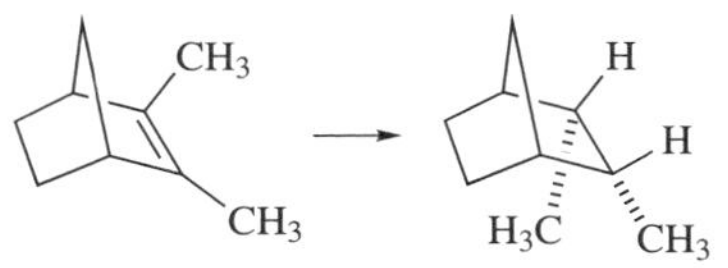

4[b] $Pt(BH_4^-)$—C, H_2

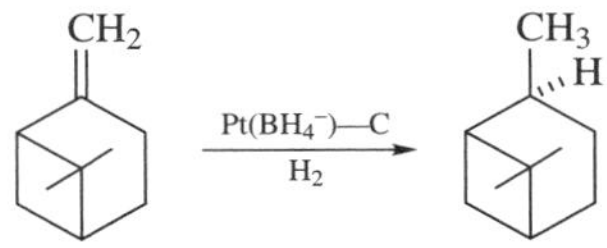

B. Exceptions

5[a] Pd, H_2: 75%, 25%

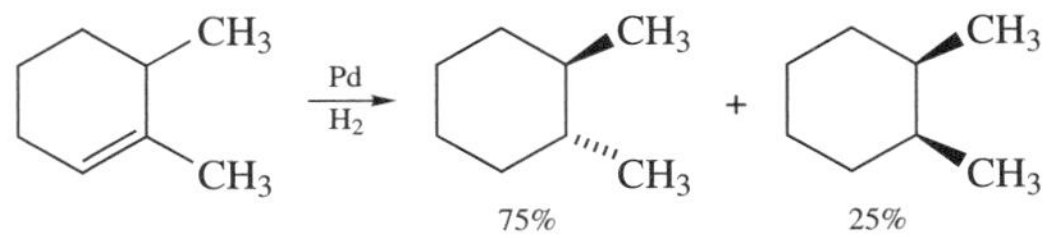

6[d] Pt, H_2

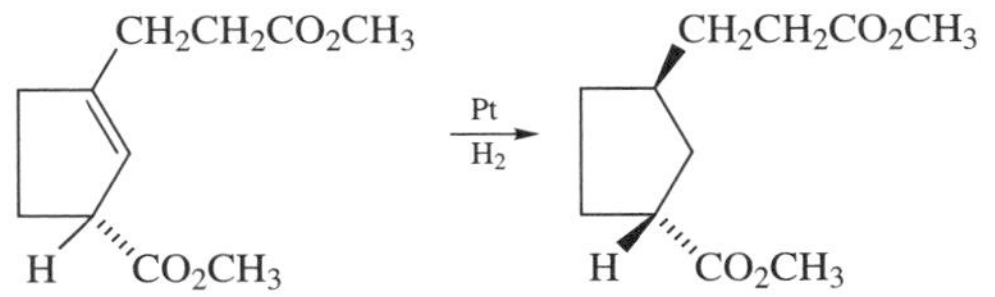

7[e] Ni, H_2: 95%, 5%

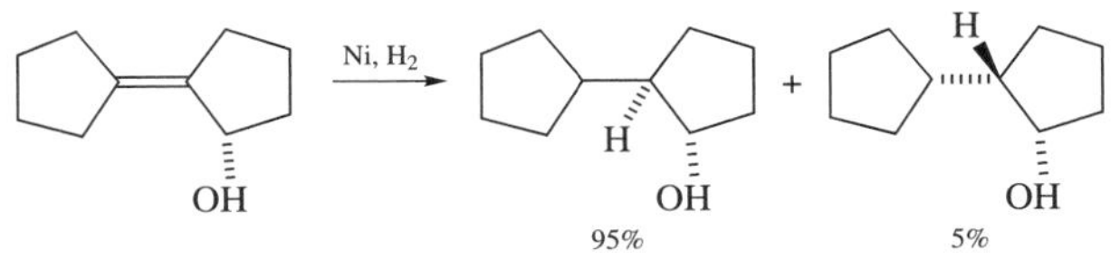

8[f] Pt, H_2, acetic acid: 80%, 20%

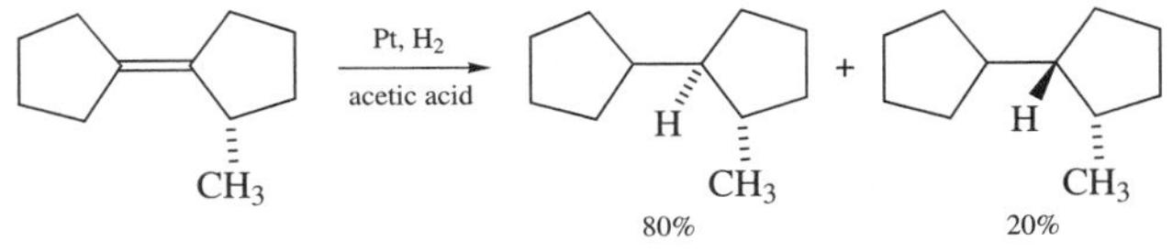

a. S. Siegel and G. V. Smith, *J. Am. Chem. Soc.* **82:**6082, 6087 (1960).
b. C. A. Brown, *J. Am. Chem. Soc.* **91:**5901 (1969).
c. K. Alder and W. Roth, *Chem. Ber.* **87:**161 (1954).
d. J. P. Ferris and N. C. Miller, *J. Am. Chem. Soc.* **88:**3522 (1966).
e. S. Mitsui, Y. Senda, and H. Saito, *Bull. Chem. Soc. Jpn.* **39:**694 (1966).
f. S. Siegel and J. R. Cozort, *J. Org. Chem.* **40:**3594 (1975).

The facial stereoselectivity of hydrogenation is affected by the presence of polar functional groups that can influence the mode of adsorption to the catalyst surface. For instance, there are many examples where the presence of a hydroxyl group results in the hydrogen being introduced from the side of the molecule occupied by the hydroxyl group. This implies that the hydroxyl group is involved in the interaction with the catalyst surface. This behavior can be illustrated with the alcohol **1a** and the ester **1b**.[1] Although the overall shapes of the two molecules are similar, the alcohol gives mainly the product with a *cis* ring juncture (**2a**), whereas the ester gives a product with *trans* stereochemistry (**3b**). The stereoselectivity of hydroxyl-directed hydrogenation is a function of solvent and catalyst. The *cis* isomer is the main product in hexane. This suggests that the hydroxyl group directs the molecule to the catalyst surface. In ethanol, the competing interaction of the solvent molecules evidently swamps out the effect of the hydroxymethyl group in **4**.

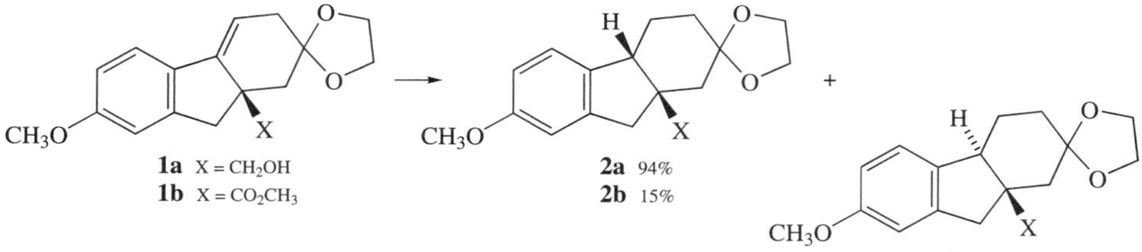

CH_3O **1a** X = CH_2OH **1b** X = CO_2CH_3 → **2a** 94% **2b** 15% + **3a** 6% **3b** 85%

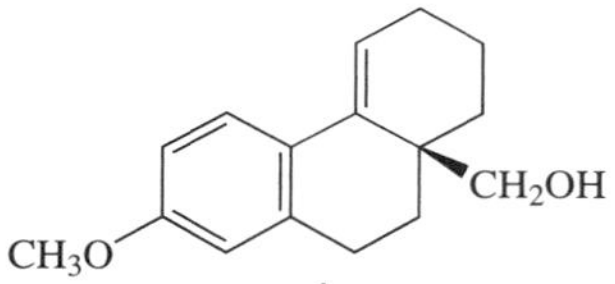

CH_3O **4** CH_2OH

Solvent	% *cis*	% *trans*
Hexane	61	39
DME	20	80
EtOH	6	94

Catalytic hydrogenations are usually extremely clean reactions with little by-product formation, unless reduction of other groups is competitive. Careful study, however, sometimes reveals that double-bond migration can take place in competition with reduction. For example, hydrogenation of 1-pentene over Raney nickel is accompanied by some isomerization to both *E*- and *Z*-2-pentene.[2] The isomerized products are converted to pentane, but at a slower rate than 1-pentene. Exchange of hydrogen atoms between the reactant and adsorbed hydrogen can be detected by exchange of deuterium for hydrogen. Allylic positions undergo such exchange particularly rapidly.[3] Both the isomerization and allylic hydrogen exchange can be explained by the intervention of the π-allyl intermediate **C** in the general mechanism for hydrogenation. If this intermediate adds a hydrogen at the alternative end of the allyl system, an isomeric alkene is formed. Hydrogen exchange occurs if a hydrogen from the metal surface, rather than the original hydrogen, is transferred prior to desorption.

1. H. W. Thompson, *J. Org. Chem.* **36**:2577 (1971); H. W. Thompson, E. McPherson, and B. L. Lences, *J. Org. Chem.* **41**:2903 (1976).
2. H. C. Brown and C. A. Brown, *J. Am. Chem. Soc.* **85**:1005 (1963).
3. G. V. Smith and J. R. Swoap, *J. Org. Chem.* **31**:3904 (1966).

Besides solid transition metals, certain soluble transition-metal complexes are active hydrogenation catalysts.[4] The most commonly used example is tris(triphenylphosphine)-chlororhodium, which is known as *Wilkinson's catalyst*.[5] This and related homogeneous catalysts usually minimize exchange and isomerization processes. Hydrogenation by homogeneous catalysts is believed to take place by initial formation of a π-complex, followed by transfer of hydrogen from rhodium to carbon.

$$\mathrm{Rh{-}H} + \mathrm{RCH{=}CHR} \longrightarrow \underset{\mathrm{RCH{=}CHR}}{\mathrm{Rh{-}H}} \longrightarrow \underset{\mathrm{RCH_2CHR}}{\mathrm{Rh}} \xrightarrow{H_2} \underset{\mathrm{RCH_2CHR}}{\mathrm{Rh{-}H}} \longrightarrow \mathrm{RCH_2CHR}$$

The phosphine ligands serve both to provide a stable soluble, complex and to adjust the reactivity of the metal center. Scheme 5.2 gives some examples of hydrogenations carried out with homogeneous catalysts. One potential advantage of homogeneous catalysts is the ability to achieve a high degree of selectivity among different functional groups. Entries 3 and 5 in Scheme 5.2 are examples of such selectivity.

The stereochemistry of reduction by homogeneous catalysts is often controlled by functional groups in the reactant. Homogeneous iridium catalysts have been found to be influenced not only by hydroxyl groups, but also by amide, ester, and ether substituents.[6]

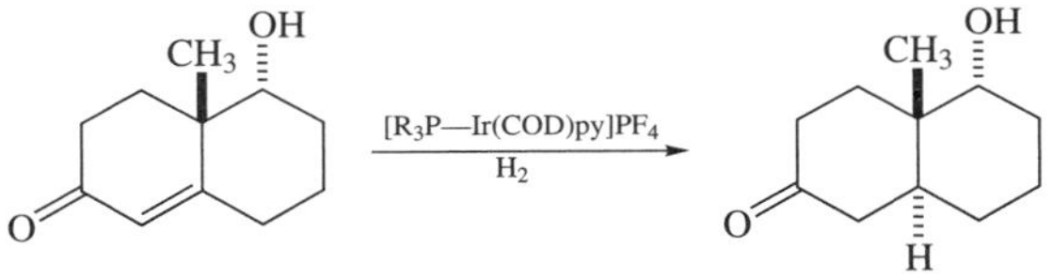

$\xrightarrow[H_2]{[R_3P{-}Ir(COD)py]PF_4}$ Ref. 7

$\xrightarrow[H_2]{[R_3P{-}Ir(COD)py]PF_4}$ Ref. 8

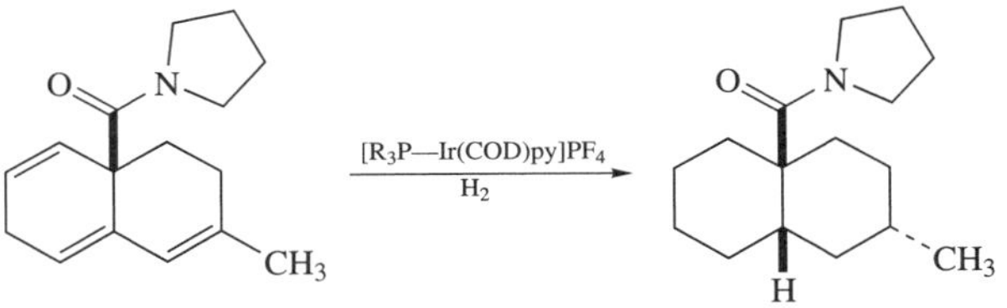

Delivery of hydrogen occurs *syn* to the polar functional group. Presumably, the stereoselectivity is the result of coordination of iridium by the functional group. The crucial property required for a catalyst to be stereodirective is that it be able to coordinate with both the directive group and the double bond and still accommodate the metal hydride bond necessary for hydrogenation. In the iridium catalyst illustrated above, the cyclooctadiene (COD) ligand in the catalyst is released upon coordination of the reactant.

A number of chiral ligands, especially phosphines, have been explored in order to develop enantioselective hydrogenation catalysts.[9] Some of the most successful catalysts

4. A. J. Birch and D. H. Williamson, *Org. React.* **24:**1 (1976); B. R. Jones, *Homogeneous Hydrogenation*, John Wildey & Sons, New York, 1973.
5. J. A. Osborn, F. H. Jardine, J. F. Young, and G. Wilkinson, *J. Chem. Soc. A.* **1966:**1711.
6. R. H. Crabreee and M. W. Davis, *J. Org. Chem.* **51:**2655 (1986); P. J. McCloskey and A. G. Schultz, *J. Org. Chem.* **53:**1380 (1988).
7. G. Stork and D. E. Kahne, *J. Am. Chem. Soc.* **105:**1072 (1983).
8. A. G. Schultz and P. J. McCloskey, *J. Org. Chem.* **50:**5905 (1985).
9. B. Bosnich and M. D. Fryzuk, *Top. Stereochem.* **12:**119 (1981); W. S. Knowles, W. S. Chrisopfel, K. E. Koenig, and C. F. Hobbs, *Adv. Chem. Ser.* **196:**325 (1982); W. S. Knowles, *Acc. Chem. Res.* **16:**106 (1983).

Scheme 5.2. Homogeneous Catalytic Hydrogenation

1[a] $(Ph_3P)_3RhBr$, D_2; H^+, H_2O — 56%

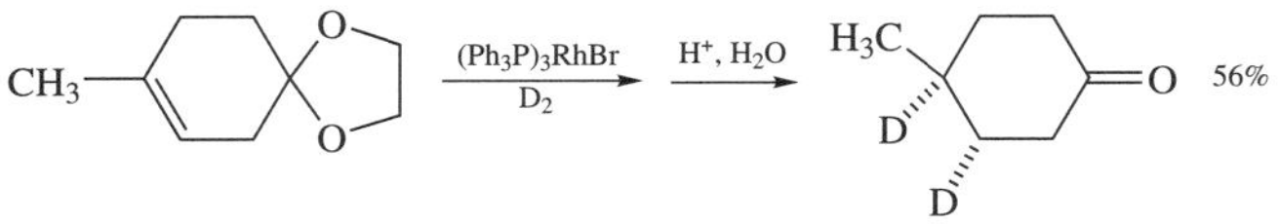

2[b] $(Ph_3P)_3RhCl$, H_2 — 90%

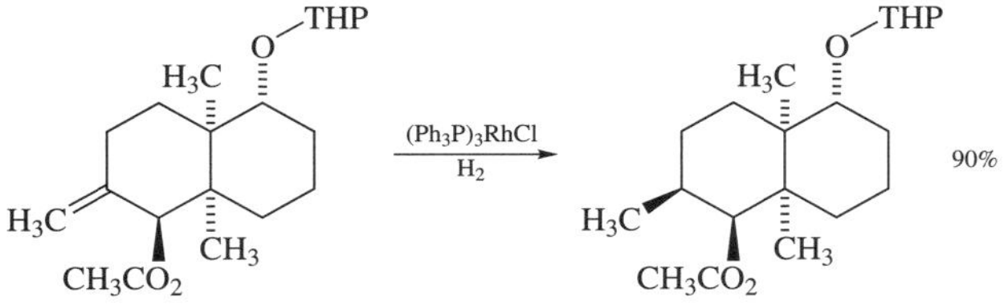

3[c] $(Ph_3P)_3RhCl$, H_2 — 94%

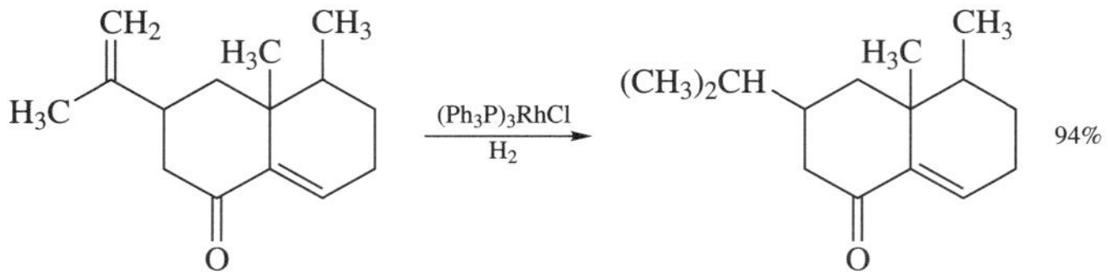

4[d] $CH_3O-C_6H_4-CH{=}CHNO_2 \xrightarrow[H_2]{(Ph_3P)_3RhCl} CH_3O-C_6H_4-CH_2CH_2NO_2$ 90%

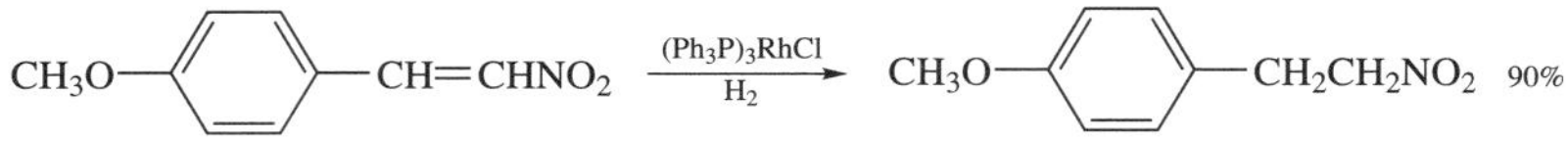

5[e] $(Ph_3P)_3RhCl$, H_2 — 90–94%

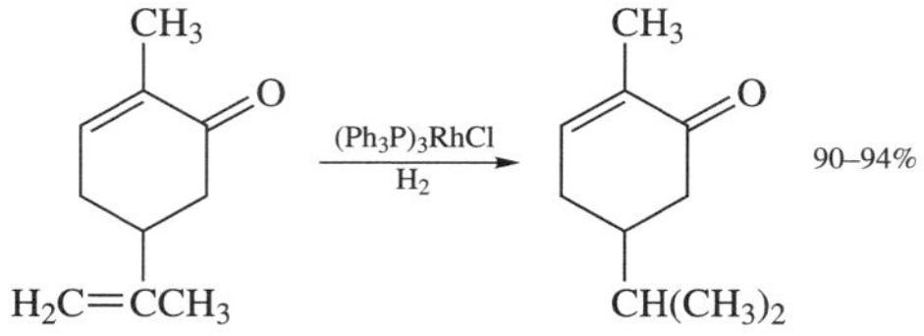

6[f] $[R_3P{-}Ir(COD)py]BF_4$ — 67%

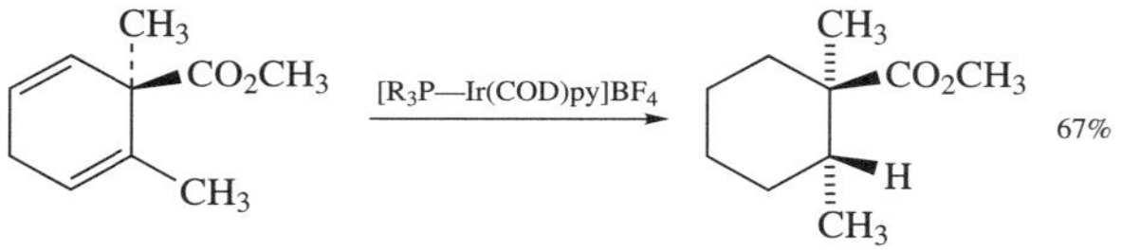

7[g] $[Rh(NBD)(diphosp)]BF_4$ — 95%

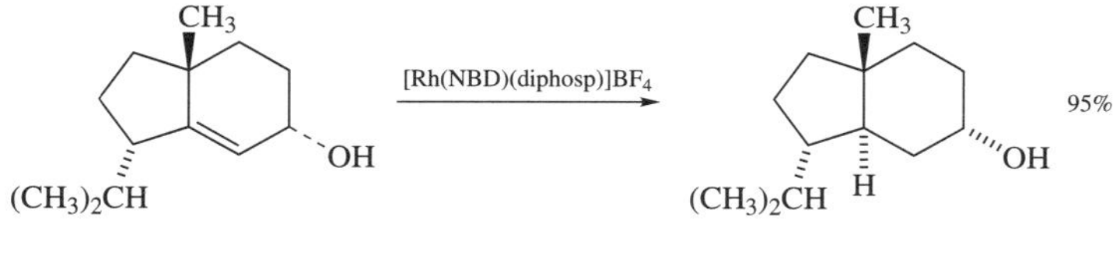

8[h] $[R_3P{-}Ir(COD)py]PF_6$ — 100%

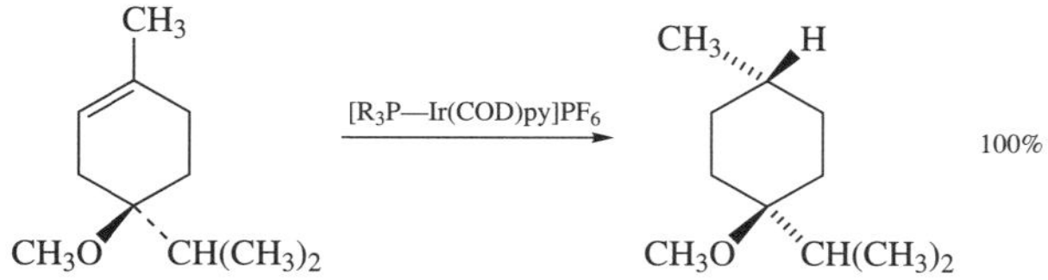

a. W. C. Agosta and W. L. Schreiber, *J. Am. Chem. Soc.* **93:**3947 (1971).
b. E. Piers, W. de Waal, and R. W. Britton, *J. Am. Chem. Soc.* **93:**5113 (1971).
c. M. Brown and L. W. Piszkiewicz, *J. Org. Chem.* **32:**2013 (1967).
d. R. E. Harmon, J. L. Parsons, D. W. Cooke, S. K. Gupta, and J. Schoolenberg, *J. Org. Chem.* **34:**3684 (1969).
e. R. E. Ireland and P. Bey, *Org. Synth.* **53:**63 (1973).
f. A. G. Schulz and P. J. McCloskey, *J. Org. Chem.* **50:**5905 (1985).
g. D. A. Evans and M. M. Morrissey, *J. Am. Chem. Soc.* **106:**3866 (1984).
h. R. H. Crabtree and M. W. Davies, *J. Org. Chem.* **51:**2655 (1986).

are derived from chiral 1,1′-binaphthyldiphosphines (BINAP).[10] These ligands are chiral by virtue of the sterically restricted rotation of the two naphthyl rings.

PPh_2
PPh_2

Ruthenium complexes containing this phosphine ligand are able to reduce a variety of double bonds with enantiomeric excesses above 95%. In order to achieve high enantioselectivity, the compound to be reduced must show a strong preference for a specific orientation when complexed with the catalyst. This ordinarily requires the presence of a functional group that can coordinate with the metal. The ruthenium binaphthyldiphosphine catalyst has been used successfully with unsaturated amides,[11] allylic and homoallylic alcohols,[12] and unsaturated carboxylic acids.[13]

OH $\xrightarrow{\text{Ru}(S\text{-BINAP})(\text{OAc})_2}$ OH 99% e.e. Ref. 12

An especially important case is the enantioselective hydrogenation of α-amidoacrylic acids, which leads to α-amino acids.[14] A particularly detailed study has been carried out on the mechanism of reduction of methyl *Z*-α-acetamidocinnamate by a rhodium catalyst with a chiral disphosphine ligand.[15] It has been concluded that the reactant can bind reversibly to the catalysts to give either of two complexes. Addition of hydrogen at rhodium then leads to a reactive rhodium hybride and eventually to product. Interestingly, the addition of hydrogen occurs most rapidly in the minor isomeric complex, and the enantioselectivity is

10. R. Noyori and H. Takaya, *Acc. Chem. Res.* **23:**345 (1990).
11. R. Noyori, M. Ohta, Y. Hsiao, M. Kitamura, T. Ohta, and H. Takaya, *J. Am. Chem. Soc.* **108:**7117 (1986).
12. H. Takaya, T. Ohta, N. Sayo, H. Kumobayashi, S. Akutagawa, S. Inoue, I. Kasahara, and R. Noyori, *J. Am. Chem. Soc.* **109:**1596 (1987).
13. T. Ohta, H. Takaya, M. Kitamura, K. Nagai, and R. Noyori, *J. Org. Chem.* **52:**3176 (1987).
14. A. Pfaltz and J. M. Brown, in *Stereoselective Synthesis*, G. Helmchen, R. W. Hoffmann, J. Mulzer, and E. Schauman, eds., Thieme, New York, 1996, Part D, Sect. 2.5.1.2; U. Nagel and J. Albrecht, *Top. Catal.* **5:**3 (1998).
15. C. R. Landis and J. Halpern, *J. Am. Chem. Soc.* **109:**1746 (1987).

due to this kinetic preference.

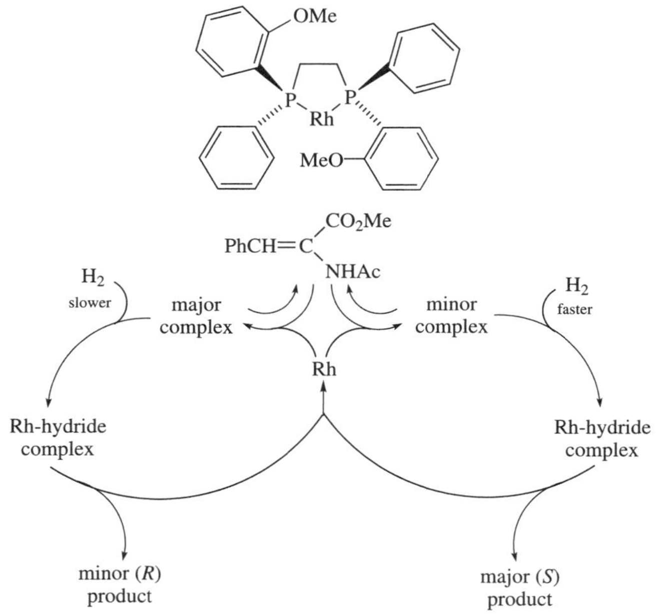

α,β-Unsaturated acids can be reduced enantioselectively with ruthenium and rhodium catalysts having chiral phosphine ligands. The mechanism of such reactions using $Ru(BINAP)(O_2CCH_3)_2$ has been studied and is consistent with the idea that coordination of the carboxy group establishes the geometry at the metal ion.[16]

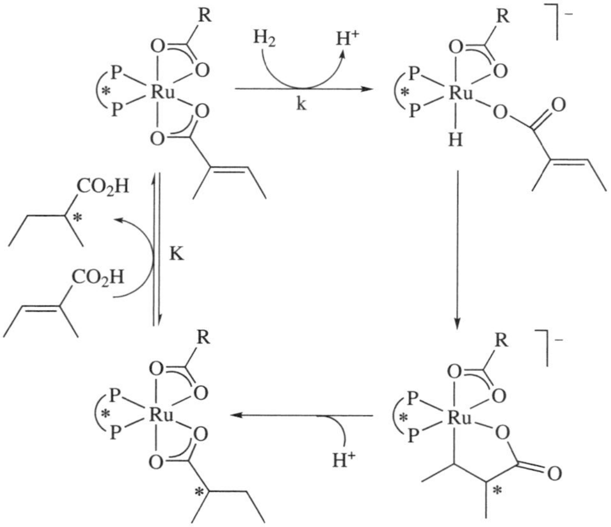

Table 5.1 gives the enantioselectivity of some hydrogenations of substituted acrylic acids.

16. M. T. Ashby and J. T. Halpern, *J. Am. Chem. Soc.* **113:**589 (1991).

Table 5.1. Enantiomeric Excess (e.e.) for Asymmetric Catalytic Hydrogenation of Substituted Acrylic Acids

Substrate	Catalyst	Product	Configuration	% e.e.	Reference
$CH_2{=}C(CO_2H)NHC(=O)CH_3$		$CH_3CH(NHC(=O)CH_3)CO_2H$	*R*	90	a
$(H)(Ph)C{=}C(CO_2H)NHC(=O)CH_3$	Same as above	$PhCH_2CH(NHC(=O)CH_3)CO_2H$	*R*	95	a
$(H)(Ph)C{=}C(CO_2H)NHC(=O)CH_3$		$PhCH_2CH(NHC(=O)CH_3)CO_2H$	*R*	94	b
$(H)(Ph)C{=}C(CO_2H)NHC(=O)Ph$		$PhCH_2CH(NHC(=O)Ph)CO_2H$	*S*	100	c

(*continued*)

Table 5.1. *(continued)*

Substrate	Catalyst	Product	Configuration	% e.e.	Reference
H(Ph)C=C($CO_2C_2H_5$)(O_2CCH_3)	CH_3O … PH, Rh, PH … OCH_3	$PhCH_2CH(O_2CCH_3)CO_2C_2H_5$	*S*	90	b
$CH_2{=}C(CO_2CH_3)CH_2CO_2CH_3$	Same as above	$CH_3CH(CH_2CO_2CH_3)CO_2CH_3$	*R*	88	d
$CH_3CH{=}CHCH{=}C(CO_2CH_3)NHCCH_3$ (C=O)	H_5C_2, C_2H_5, P, Rh, P, H_5C_2, C_2H_5	$CH_3CH{=}CHCH_2CH(CO_2CH_3)NHCCH_3$ (C=O)	*R*	99.2	e
$(CH_3)_2CHCH{=}C(CO_2CH_3)CH_2CO_2H$	Same as above	$(CH_3)_2CHCH_2CH(CO_2CH_3)CH_2CO_2H$	*R*	99	f

Substrate	Catalyst	Product	Config.	% e.e.	Ref.
CH_3O_2C, HC, Ph, CO_2H	Rh, PPh₂, Ph_2PCH_2, N, CNHPh, O	$PhCH_2$, HO_2C, CO_2H	*S*	>95	g
CH_2, CH_3O_2C, CO_2H	$(C_6H_{11})_2$, P, Rh, P, $(C_6H_{11})_2$, CH_3, CH_3	CH_3, HO_2C, CO_2H	*S*	96	h

a. M. D. Fryzuk and B. Bosnich, *J. Am. Chem. Soc.* **99:**6262 (1977).
b. B. D. Vineyard, W. S. Knowles, M. J. Sabacky, G. L. Bachman, and D. J. Weinkauff, *J. Am. Chem. Soc.* **99:**5946 (1977).
c. A. Miyashita, H. Takaya, T. Souchi, and R. Noyori, *Tetrahedron* **40:**1245 (1984).
d. W. C. Christopfel and B. D. Vineyard, *J. Am. Chem. Soc.* **101:**4406 (1979).
e. M. J. Burk, J. G. Allen, and W. F. Kiesman, *J. Am. Chem. Soc.* **120:**657 (1998).
f. M. J. Burk, F. Bienewald, M. Harris, and A. Zanotti-Gerosa, *Angew. Chem. Int. Ed. Engl.* **37:**1931 (1998).
g. H. Jendralla, *Tetrahedron Lett.* **32:**3671 (1991).
h. T. Chiba, A. Miyashita, H. Nohira, and H. Takaya, *Tetrahedron Lett.* **32:**4745 (1991).

Partial reduction of alkynes to *Z*-alkenes is another important application of selective hydrogenation catalysts. The transformation can be carried out under heterogeneous or homogeneous conditions. Among heterogeneous catalysts, the one which is most successful is *Lindlar's catalyst*, which is a lead-modified palladium–$CaCO_3$ catalyst.[17] A nickel–boride catalyst prepared by reduction of nickel salts with $NaBH_4$ is also useful.[18] Rhodium catalysts have also been reported to show good selectivity.[19]

Many other functional groups are also reactive under conditions of catalytic hydrogenation. The reduction of nitro compounds to amines, for example, usually proceeds very rapidly. Ketones, aldehydes, and esters can all be reduced to alcohols, but in most cases these reactions are slower than alkene reductions. For most synthetic applications, the hydride-transfer reagents to be discussed in Section 5.2 are used for reduction of carbonyl groups. Amides and nitriles can be reduced to amines. Hydrogenation of amides requires extreme conditions and is seldom used in synthesis, but reduction of nitriles is quite useful. Table 5.2 gives a summary of the approximate conditions for catalytic hydrogenation of some common functional groups.

Certain functional groups can be entirely removed and replaced by hydrogen. This is called *hydrogenolysis*. For example, aromatic halogen substituents are frequently removed by hydrogenation over transition-metal catalysts. Aliphatic halogens are somewhat less reactive, but hydrogenolysis is promoted by base.[20] The most useful type of hydrogeolysis reactions involves removal of functional groups at benzylic and allylic positions.[21]

$$C_6H_5CH_2OR \xrightarrow{H_2,\ Pd} C_6H_5CH_3 + HOR$$

Hydrogenolysis of halides and benzylic groups presumably involves intermediates formed by *oxidative addition* to the active metal catalysts to generate intermediates similar to those involved in hydrogenation.

$$C_6H_5CH_2X + Pd^0 \longrightarrow C_6H_5CH_2PdX \xrightarrow{H_2} C_6H_5CH_2PdH_2X \longrightarrow C_6H_5CH_3 + Pd^0$$

Many other examples of this pattern of reactivity will be discussed in Chapter 8.

The facile cleavage of the benzyl–oxygen bond has made the benzyl group a useful "protecting group" in multistep synthesis. A particularly important example is the use of the carbobenzyloxy group in peptide synthesis. The protecting group is removed by hydrogenolysis. The substituted carbamic acid generated by the hydrogenolysis decarboxylates spontaneously to provide the amine.

$$PhCH_2OC(=O)NHR \longrightarrow PhCH_3 + HOC(=O)NHR \longrightarrow CO_2 + H_2NR$$

17. H. Lindlar and R. Dubuis, *Org. Synth.* **V**:880 (1973).
18. H. C. Brown and C. A. Brown, *J. Am. Chem. Soc.* **85**:1005 (1963); E. J. Corey, K. Achiwa, and J. A. Katzenellenbogen, *J. Am. Chem. Soc.* **91**:4318 (1969).
19. R. R. Schrock and J. A. Osborn, *J. Am. Chem. Soc.* **98**:2143 (1976); J. M. Tour, S. L. Pendalwar, C. M. Kafka, and J. P. Cooper, *J. Org. Chem.* **57**:4786 (1992).
20. A. R. Pinder, *Synthesis* **1980**:425.
21. W. H. Hartung and R. Simonoff, *Org. React.* **7**:263 (1953); P. N. Rylander, *Catalytic Hydrogenation over Platinum Metals*, Academic Press, New York, 1967, Chapter 25; P. N. Rylander, *Catalytic Hydrogenation in Organic Synthesis*, Academic Press, New York, 1979, Chapter 15; P. N. Rylander, *Hydrogenation Methods*, Academic Press, Orlando, Florida, 1985, Chapter 13.

Table 5.2. Conditions for Catalytic Reduction of Various Functional Groups[a]

Functional group	Reduction product	Common catalysts	Typical reaction conditions
$>C=C<$	$-\overset{\mid}{\underset{H}{C}}-\overset{\mid}{\underset{H}{C}}-$	Pd, Pt, Ni, Ru, Rh	Rapid at room temperature (R.T.) and 1 atm except for highly substituted or hindered cases
$-C\equiv C-$	$>C=C<$ (cis, H H)	Lindlar	R. T. and low pressure, quinoline or lead added to deactivate catalyst
C_6H_5-	$C_6H_{11}-$	Rh, Pt	Moderate pressure (5–10 atm), 50–100°C
C_6H_5-	$C_6H_{11}-$	Ni, Pd	High pressure (100–200 atm), 100–200°C
RCR (C=O)	$RCHR$ (OH)	Pt, Ru	Moderate rate at R. T. and 1–4 atm. acid-catalyzed
RCR (C=O)	$RCHR$ (OH)	Cu–Cr, Ni	High pressure, 50–100°C
C_6H_5CR (C=O) or C_6H_5CHR (OR)	$C_6H_5CH_2R$	Pd	R. T., 1–4 atm. acid-catalyzed
C_6H_5CHR (NR_2)	$C_6H_5CH_2R$	Pd, Ni	50–100°C, 1–4 atm
$RCCl$ (C=O)	RCH (C=O)	Pd	R. T., 1 atm. quinoline or other catalyst moderator used
$RCOH$ (C=O)	RCH_2OH	Pd, Ni, Ru	Very strenuous conditions required
$RCOR$ (C=O)	RCH_2OH	Cu–Cr, Ni	200°C, high pressure
$RC\equiv N$	RCH_2NH_2	Ni, Rh	50–100°C, usually high pressure, NH_3 added to increase yield of primary amine
$RCNH_2$ (C=O)	RCH_2NH_2	Cu–Cr	Very strenuous conditions required
RNO_2	RNH_2	Pd, Ni, Pt	R. T., 1–4 atm
RCR (C=NR)	R_2CHNHR	Pd, Pt	R. T., 4–100 atm
R—Cl, R—Br, R—I	R—H	Pd	Order of reactivity: 1 > Br > Cl > F, bases promote reactions for R = alkyl
$>C\overset{O}{—}C<$ (epoxide)	$-\overset{H}{C}-\overset{OH}{C}-$	Pt, Pd	Proceeds slowly at R. T., 1–4 atm, acid-catalyzed

a. General references: M. Freifelder, *Catalytic Hydrogenation in Organic Synthesis: Procedures and Commentary*, John Wiley & Sons, New York, 1978; P. N. Rylander, *Hydrogenation Methods*, Academic Press, Orlando, Florida, 1985.

5.1.2. Other Hydrogen-Transfer Reagents

Catalytic hydrogenation transfers the elements of molecular hydrogen through a series of complexes and intermediates. Diimide, HN=NH, an unstable hydrogen donor that can only be generated *in situ*, finds some specialized application in the reduction of carbon–carbon double bonds. Simple alkenes are reduced efficiently by diimide, but other easily reduced functional groups, such as nitro and cyano, are unaffected. The mechanism of the reaction is pictured as a transfer of hydrogen via a nonpolar cyclic transition state.

HN=NH + >C=C< → [>C–C< / H H / N=N] → >C–C< / H H + N≡N

In agreement with this mechanism is the fact that the stereochemistry of addition is *syn*.[22] The rate of reaction with diimide is influenced by torsional and angle strain in the alkene. More strained double bonds react more rapidly.[23] For example, the more strained *trans* double bond is selectively reduced in *Z*,*E*-1,5-cyclodecadiene.

$\xrightarrow[Cu^{2+},\ O_2]{NH_2NH_2}$ Ref. 24

Diimide selectively reduces terminal over internal double bonds in polyunsaturated systems.[25] There are several methods for generation of diimide and they are illustrated in Scheme 5.3.

5.2. Group III Hydride-Donor Reagents

5.2.1. Reduction of Carbonyl Compounds

Most reductions of carbonyl compounds are done with reagents that transfer a hydride from boron or aluminum. The numerous reagents of this type that are available provide a considerable degree of chemoselectivity and stereochemical control. Sodium borohydride and lithium aluminum hydride are the most widely used of these reagents. Sodium borohydride is a mild reducing agent that reacts rapidly with aldehydes and ketones but quite slowly with esters. Lithium aluminum hydride is a much more powerful hydride-donor reagent. It will rapidly reduce esters, acids, nitriles, and amides, as well as aldehydes and ketones. Neither sodium borohydride nor lithium aluminum hydride reacts with isolated carbon–carbon double bonds. The reactivity of these reagents and some related reducing reagents is summarized in Table 5.3.

22. E. J. Corey, D. J. Pasto, and W. L. Mock, *J. Am. Chem. Soc.* **83:**2957 (1961).
23. E. W. Garbisch, Jr., S. M. Schildcrout, D. B. Patterson, and C. M. Sprecher, *J. Am. Chem. Soc.* **87:**2932 (1965).
24. J. G. Traynham, G. R. Franzen, G. A. Kresel, and D. J. Northington, Jr., *J. Org. Chem.* **32:**3285 (1967).
25. E. J. Corey, H. Yamamoto, D. K. Herron, and K. Achiwa, *J. Am. Chem. Soc.* **92:**6635 (1970); E. J. Corey and H. Yamamoto, *J. Am. Chem. Soc.* **92:**6636, 6637 (1970).

Scheme 5.3. Reductions with Diimide

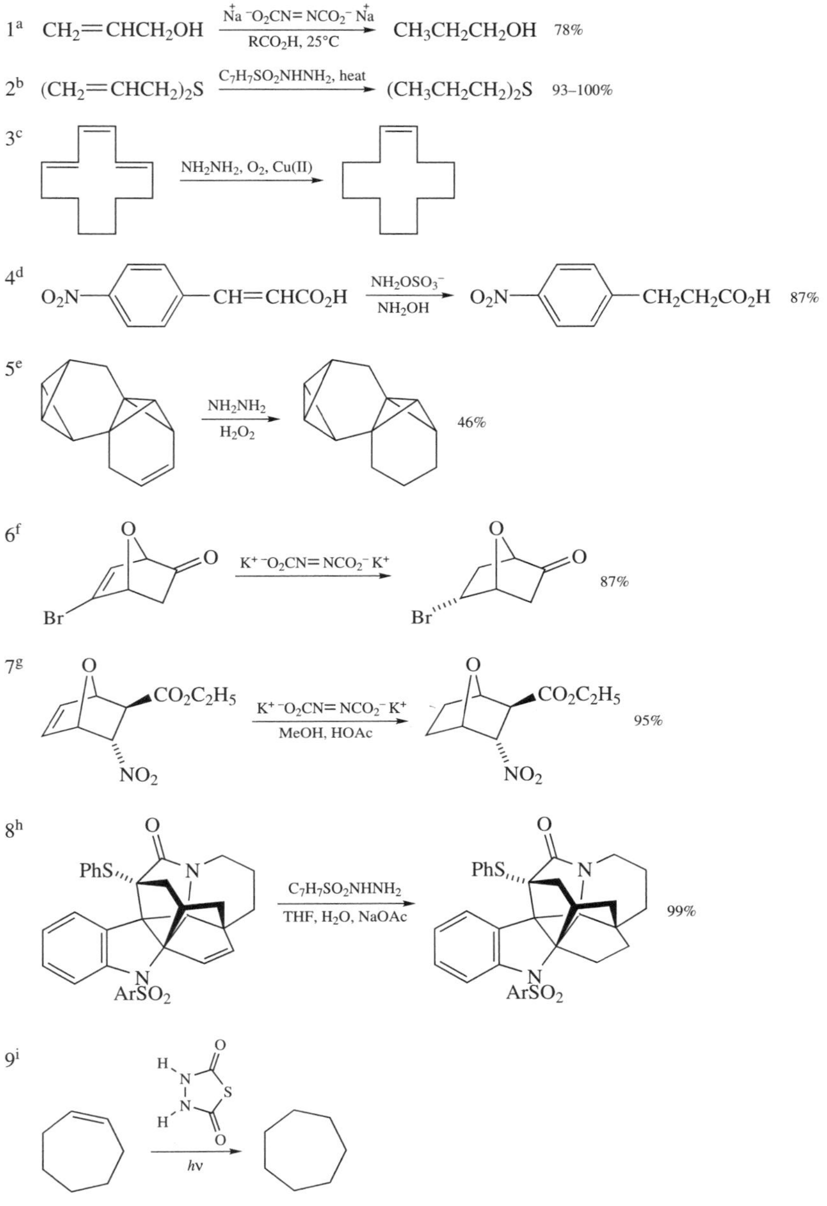

1[a] $CH_2{=}CHCH_2OH \xrightarrow[RCO_2H,\ 25°C]{\overset{+}{Na}\ ^-O_2CN{=}NCO_2^-\ \overset{+}{Na}} CH_3CH_2CH_2OH$ 78%

2[b] $(CH_2{=}CHCH_2)_2S \xrightarrow{C_7H_7SO_2NHNH_2,\ heat} (CH_3CH_2CH_2)_2S$ 93–100%

3[c] NH_2NH_2, O_2, Cu(II)

4[d] $O_2N-C_6H_4-CH{=}CHCO_2H \xrightarrow[NH_2OH]{NH_2OSO_3^-} O_2N-C_6H_4-CH_2CH_2CO_2H$ 87%

5[e] $\xrightarrow[H_2O_2]{NH_2NH_2}$ 46%

6[f] Br; $K^+\ ^-O_2CN{=}NCO_2^-\ K^+$; 87%

7[g] $CO_2C_2H_5$, NO_2; $\xrightarrow[MeOH,\ HOAc]{K^+\ ^-O_2CN{=}NCO_2^-\ K^+}$ 95%

8[h] PhS, N, ArSO₂; $\xrightarrow[THF,\ H_2O,\ NaOAc]{C_7H_7SO_2NHNH_2}$ 99%

9[i] $\xrightarrow{h\nu}$

a. E. E. van Tamelen, R. S. Dewey, and R. J. Timmons, *J. Am. Chem. Soc.* **83:**3725 (1961).
b. E. E. van Tamelen, R. S. Dewey, M. F. Lease, and W. H. Pirkle, *J. Am. Chem. Soc.* **83:**4302 (1961).
c. M. Ohno and M. Okamoto, *Org. Synth.* **49:**30 (1969).
d. W. Durckheimer, *Justus Liebigs Ann. Chem.* **721:**240 (1969).
e. L. A. Paquette, A. R. Browne, E. Chamot, and J. F. Blount, *J. Am. Chem. Soc.* **102:**643 (1980).
f. J.-M. Durgnat and P. Vogel, *Helv. Chim. Acta* **76:**222 (1993).
g. P. A. Grieco, R. Lis, R. E. Zelle, and J. Finn, *J. Am. Chem. Soc.* **108:**5908 (1986).
h. P. Magnus, T. Gallagher, P. Brown, and J. C. Huffman, *J. Am. Chem. Soc.* **106:**2105 (1984).
i. M. Squillacote, J. DeFelipppis, and Y. L. Lai, *Tetrahedron Lett.* **34:**4137 (1993).

Table 5.3. Relative Reactivity of Hydride-Donor Reducing Agents

	Reduction products[a]						
Hydride donor	Iminium ion	Acyl halide	Aldehyde	Ketone	Ester	Amide	Carboxylate salt
$LiAlH_4$[b]	Amine	Alcohol	Alcohol	Alcohol	Alcohol	Amine	Alcohol
$LiAlH_2(OCH_2CH_2OCH_3)_2$[c]		Alcohol	Alcohol	Alcohol	Alcohol	Amine	Alcohol
$LiAlH[(OC(CH_3)]_3$[d]		Aldehyde[e]	Alcohol	Alcohol	Alcohol[f]	Aldehyde[f]	
$NaBH_4$[b]	Amine		Alcohol	Alcohol	Alcohol[f]		
$NaBH_3CN$[g]	Amine		Alcohol				
B_2H_6[h]			Alcohol	Alcohol		Amine	Alcohol[f]
AlH_3[j]		Alcohol	Alcohol	Alcohol	Alcohol	Amine	Alcohol
$[(CH_3)_2CHCH(CH_3)]_2BH$[k]			Alcohol	Alcohol		Aldehyde[e]	
$[(CH_3)_2CHCH_2]_2AlH$[l]			Alcohol	Alcohol	Aldehyde[e]	Aldehyde[e]	Alcohol

a. Products shown are the usual products of synthetic operations. Where no entry is given, the combination has not been studied or is not of major synthetic utility.
b. See the general references at the end of the chapter.
c. J. Malék, *Org. React.* **34:**1 (1985); **36:**249 (1989).
d. H. C. Brown and R. F. McFarlin, *J. Am. Chem. Soc.* **78:**752 (1956); **80:**5372 (1958); H. C. Brown and B. C. Subba Rao, *J. Am. Chem. Soc.* **80:**5377 (1958); H. C. Brown and A. Tsukamoto, *J. Am. Chem. Soc.* **86:**1089 (1964).
e. Reaction must be controlled by use of a stoichiometric amount of reagent and low temperature.
f. Reaction occurs slowly.
g. C. F. Lane, *Synthesis* **1975:**135.
h. H. C. Brown, P. Heim, and N. M. Yoon, *J. Am. Chem. Soc.* **92:**1637 (1970); N. M. Yoon, C. S. Park, H. C. Brown, S. Krishnamurthy, and T. P. Stocky, *J. Org. Chem.* **38:**2786 (1973); H. C. Brown and P. Heim, *J. Org. Chem.* **38:**912 (1973).
i. Reaction occurs via the triacyl borate.
j. H. C. Brown and N. M. Yoon, *J. Am. Chem. Soc.* **88:**1464 (1966).
k. H. C. Brown, D. B. Bigley, S. K. Arora, and N. M. Yoon, *J. Am. Chem. Soc.* **92:**7161 (1970); H. C. Brown and V. Varma, *J. Org. Chem.* **39:**1631 (1974).
l. E. Winterfeldt, *Synthesis* **1975:**617; H. Reinheckel, K. Haage, and D. Jahnke, *Organomet. Chem. Res.* **4:**47 (1969); N. M. Yoon and Y. S. Gyoung, *J. Org. Chem.* **50:**2443 (1985).

The mechanism by which the group III hydrides effect reduction involves nucleophilic transfer of hydride to the carbonyl group. Activation of the carbonyl group by coordination with a metal cation is probably involved under most conditions. As reduction proceeds and hydride is transferred, the Lewis acid character of boron and aluminum can also be involved.

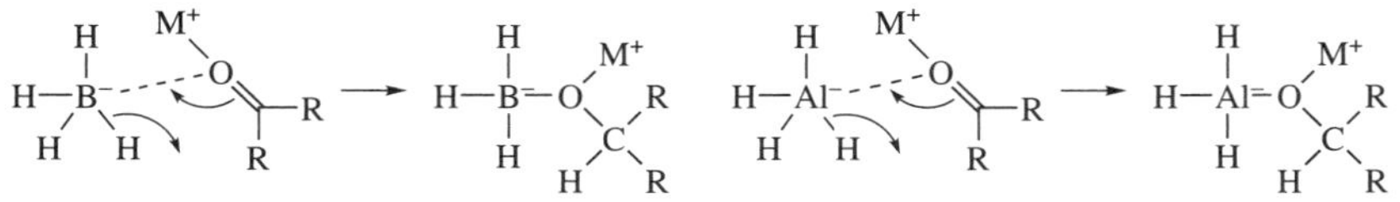

Because all four of the hydrides can eventually be transferred, there are actually several distinct reducing agents functioning during the course of the reaction.[26] Although this somewhat complicates interpretation of rates and stereoselectivity, it does not detract from the synthetic utility of these reagents. Reduction with $NaBH_4$ is usually done in aqueous or

26. B. Rickborn and M. T. Wuesthoff, *J. Am. Chem. Soc.* **92:**6894 (1970).

alcoholic solution, and the alkoxyboranes formed as intermediates are rapidly solvolyzed.

$$BH_4^- + R_2CO \longrightarrow R_2CHO\bar{B}H_3$$
$$R_2CHO\bar{B}H_3 + R_2CO \longrightarrow [R_2CHO]_2\bar{B}H_2$$
$$[R_2CHO]_2\bar{B}H_2 + R_2CO \longrightarrow [R_2CHO]_3\bar{B}H$$
$$[R_2CHO]_3\bar{B}H + R_2CO \longrightarrow [R_2CHO]_4\bar{B}$$
$$[R_2CHO]_4B^- + 4\ SOH \longrightarrow 4\ R_2CHOH + B(OS)_4^-$$

The mechanism for reduction by $LiAlH_4$ is very similar. However, because $LiAlH_4$ reacts very rapidly with protic solvents to form molecular hydrogen, reductions with this reagent must be carried out in aprotic solvents, usually ether or THF. The products are liberated by hydrolysis of the aluminum alkoxide at the end of the reaction.

Hydride reduction of esters to alcohols involves elimination steps, in addition to hydride transfer.

$$RC(=O)OR\cdots\bar{A}lH_3 \longrightarrow RCH(OR)O-\bar{A}lH_3 \longrightarrow RCH=O + RO\bar{A}lH_3$$

$$RCH=O\cdots\bar{A}lH_2OR \longrightarrow RCH_2O-\bar{A}lH_2OR \xrightarrow{H_2O} RCH_2OH$$

Amides are reduced to amines because the nitrogen is a poorer leaving group than oxygen at the intermediate stage of the reduction. Primary and secondary amides are rapidly deprotonated by the strongly basic $LiAlH_4$, so the addition step involves the conjugate base.

$$RC(=O)\bar{N}H\cdots\bar{A}lH_3 \longrightarrow RCH(O-\bar{A}lH_3)\bar{N}H \longrightarrow RHC=NH + H-AlH_2O^- \longrightarrow RCH_2N(H)\bar{A}lH_2O^- \xrightarrow{H_2O} RCH_2NH_2$$

Reduction of amides by $LiAlH_4$ is an important method for synthesis of amines:

$$C_6H_{11}-CON(CH_3)_2 \xrightarrow[\text{ether, }35^\circ C,\ 15\ h]{LiAlH_4} C_6H_{11}-CH_2N(CH_3)_2 \quad 88\%$$ Ref. 27

$$\text{5,5-dimethyl-2-pyrrolidinone} \xrightarrow[\text{THF, }65^\circ C,\ 8\ h]{LiAlH_4} \text{2,2-dimethylpyrrolidine} \quad 67\text{–}79\%$$ Ref. 28

Several factors affect the reactivity of the boron and aluminum hydrides. These include the metal cation present and the ligands, in addition to hydride, in the metallo

27. A. C. Cope and E. Ciganek, *Org. Synth.* **IV:**339 (1963).
28. R. B. Moffett, *Org. Synth.* **IV:**354 (1963).

hydride. Some of these effects can be illustrated by considering the reactivity of ketones and aldehydes toward various hydride-transfer reagents. Comparison of $LiAlH_4$ and $NaAlH_4$ has shown the former to be more reactive.[29] This can be attributed to the greater Lewis acid strength and hardness of the lithium cation. Both $LiBH_4$ and $Ca(BH_4)_2$ are more reactive than sodium borohydride. This enhanced reactivity is due to the greater Lewis acid strength of Li^+ and Ca^{2+}, compared with Na^+. Both of these reagents can reduce esters and lactones efficiently.

$CO_2C_2H_5$ / CN (benzene ring) $\xrightarrow{Ca(BH_4)_2}$ CH_2OH / CN (benzene ring) 70% Ref. 30

C_7H_{15} (γ-lactone) $\xrightarrow{LiBH_4}$ $C_7H_{15}CH(OH)CH_2CH_2CH_2OH$ 45% Ref. 31

Zinc borohydride is also a useful reagent.[32] It is prepared by reaction of $ZnCl_2$ with $NaBH_4$ in THF. Because of the stronger Lewis acid character of Zn^{2+}, $Zn(BH_4)_2$ is more reactive than $NaBH_4$ toward esters and amides and reduces them to alcohols and amines, respectively.[33] The reagent also smoothly reduces α-amino acids to β-amino alcohols.[34]

$$PhCH(NH_2)CO_2H + Zn(BH_4)_2 \longrightarrow PhCH(NH_2)CH_2OH \quad 87\%$$

An extensive series of aluminum hydrides in which one or more of the hydrides is replaced by an alkoxide ion can be prepared by addition of the correct amount of the appropriate alcohol.

$$LiAlH_4 + 2\,ROH \longrightarrow LiAlH_2(OR)_2 + 2\,H_2$$
$$LiAlH_4 + 3\,ROH \longrightarrow LiAlH(OR)_3 + 3\,H_2$$

These reagents generally show increased solubility, particularly at low temperatures, in organic solvents and are useful in certain selective reductions.[35] Lithium tri-*t*-butoxyaluminum hydride and lithium or sodium bis(2-methoxyethoxy)aluminum hydride (Red-Al)[36] are examples of these types of reagents which have wide synthetic use. Their reactivity toward typical functional groups is included in Table 5.3. Sodium cyanoborohydride[37] is a useful derivative of sodium borohydride. The electron-attracting cyano substituent reduces reactivity, and only iminum groups are rapidly reduced by this reagent.

29. E. C. Ashby and J. R. Boone, *J. Am. Chem. Soc.* **98:**5524 (1976); J. S. Cha and H. C. Brown, *J. Org. Chem.* **58:**4727 (1993).
30. H. C. Brown, S. Narasimhan, and Y. M. Choi, *J. Org. Chem.***47:**4702 (1982).
31. K. Soai and S. Ookawa, *J. Org. Chem.* **51:**4000 (1986).
32. S. Narasimhan and R. Balakumar, *Aldrichimica Acta* **31:**19 (1998).
33. S. Narasimhan, S. Madhavan, R. Balakumar, and S. Swarnalakshmi, *Synth. Commun.* **27:**391 (1997).
34. S. Narasimhan, S. Madhavan, and K. G. Prasad, *Synth. Commun.* **26:**703 (1996).
35. J. Malek and M. Cerny, *Synthesis* **1972:**217; J. Malek, *Org. React.* **34:**1 (1985).
36. Red-Al is a trademark of Aldrich Chemical Company.
37. C. F. Lane, *Synthesis* **1975:**135.

Alkylborohydrides are also used as reducing agents. These compounds have greater steric demands than the borohydride ion and therefore are more stereoselective in situations in which steric factors are controlling.[38] They are prepared by reaction of trialkylboranes with lithium, sodium, or potassium hydride.[39] Several of the compounds are available commercially under the trade name Selectrides.[40]

$Li^+H\bar{B}(-CH(CH_3)CH_2CH_3)_3$ — L-selectride

$Li^+H\bar{B}[-CH(CH_3)CH(CH_3)_2]_3$ — LS-selectride

$Na^+H\bar{B}(-CH(CH_3)CH_2CH_3)_3$ — N-selectride

$K^+H\bar{B}(-CH(CH_3)CH_2CH_3)_3$ — K-selectride

Closely related to, but distinct from, the anionic boron and aluminum hydrides are the neutral boron (borane, BH_3) and aluminum (alane, AlH_3) hydrides. These molecules also contain hydrogen that can be transferred as hydride. Borane and alane differ from the anionic hydrides in being electrophilic species by virtue of a vacant *p* orbital at the metal. Reduction by these molecules occurs by an intramolecular hydride transfer in a Lewis acid–base complex of the reactant and reductant.

$$R_2MH + R_2C{=}O \longrightarrow R_2C{=}\overset{+}{O}{-}\overset{-}{M}HR_2 \longrightarrow R_2CH{-}O{-}MR_2$$

Alkyl derivatives of borane and alane can function as reducing agents in a similar fashion. Two reagents of this group, disiamylborane and diisobutylaluminum hydride (DIBAlH), are included in Table 5.3. The latter is an especially useful reagent.

In synthesis, the principal factors affecting the choice of a reducing agent are selectivity among functional groups (chemoselectivity) and stereoselectivity. Chemoselectivity can involve two issues. It may be desired to effect a *partial reduction* of a particular functional group, or it may be necessary to *reduce one group in preference to another*. The reagents in Table 5.3 are arranged in approximate order of decreasing reactivity as hydride donors.[41] The relative ordering of reducing agents with respect to particular functional groups can permit selection of the appropriate reagent.

One of the more difficult partial reductions to accomplish is the conversion of a carboxylic acid derivative to an aldehyde without over-reduction to the alcohol. Aldehydes are inherently more reactive than acids or esters so the challenge is to stop the reduction at the aldehyde stage. Several approaches have been used to achieve this objective. One is to replace some of the hydrogens in a group III hydride with more bulky groups, thus modifying reactivity by steric factors. Lithium tri-*t*-butoxyaluminum hydride is an example of this approach.[42] Sodium tri-*t*-butoxyaluminum hydride can also be used to reduce acyl chlorides to aldehydes without over-reduction to the alcohol.[43] The excellent solubility of sodium bis(2-methoxyethoxy)aluminum hydride makes it a useful reagent for selective

38. H. C. Brown and S. Krishnamurthy, *J. Am. Chem. Soc.* **94:**7159 (1972); S. Krishnamurthy and H. C. Brown, *J. Am. Chem. Soc.* **98:**3383 (1976).
39. H. C. Brown, S. Krishnamurthy, and J. L. Hubbard, *J. Am. Chem. Soc.* **100:**3343 (1978).
40. Selectride is a trade name of the Aldrich Chemical Company.
41. For more complete discussion of functional group selectivity of hydride reducing agents, see E. R. H. Walter, *Chem. Soc. Rev.* **5:**23 (1976).
42. H. C. Brown and B. C. SubbaRao, *J. Am. Chem. Soc.* **80:**5377 (1958).
43. J. S. Cha and H. C. Brown, *J. Org. Chem.* **58:**4732 (1993).

reductions. The reagent is soluble in toluene even at −70°C. Selectivity is enhanced by the low temperature. It is possible to reduce esters to aldehydes and lactones to lactols with this reagent.

CH_3O–C_6H_4–$CH_2CH_2CO_2CH_3$ → ($NaAlH_2(OCH_2CH_2OCH_3)_2$, N-methylpiperazine) → CH_3O–C_6H_4–$CH_2CH_2CH{=}O$ Ref. 44

Lactone (THPO, $(CH_2)_4CO_2C(CH_3)_3$, OTHP) → ($NaAlH_2(OCH_2CH_2OCH_3)_2$) → lactol (OH, THPO, $(CH_2)_4CO_2C(CH_3)_3$, OTHP) Ref. 45

Probably the most widely used reagent for partial reduction of esters and lactones at the present time is diisobutylaluminum hydride.[46] By use of a controlled amount of the reagent at low temperature, partial reduction can be reliably achieved. The selectivity results from the relative stability of the hemiacetal intermediate that is formed. The aldehyde is not liberated until the hydrolytic workup and is therefore not subject to overreduction. At higher temperatures, at which the intermediate undergoes elimination, diisobutylaluminum hydride reduces esters to primary alcohols.

Benzoquinolizidine (CH_3O, CH_3O, C_2H_5, $CH_2CO_2C_2H_5$) → ($(i\text{-Bu})_2AlH$, toluene, −60°C) → (CH_3O, CH_3O, C_2H_5, $CH_2CH{=}O$) 83% Ref. 47

(CH_3SCH_2O, $CO_2C_2H_5$) → ($(i\text{-Bu})_2AlH$, −90°C) → (CH_3SCH_2O, $CH{=}O$) Ref. 48

Selective reduction to aldehydes can also be achieved using *N*-methoxy-*N*-methylamides.[49] Lithium aluminum hydride and diisobutylaluminum hydride have both been used as the hydride donor. The partial reduction is believed to be the result of the stability of the initial reduction product. The *N*-methoxy substituent permits a chelated structure which is

44. R. Kanazawa and T. Tokoroyama, *Synthesis* **1976:**526.
45. H. Disselnkötter, F. Liob, H. Oedinger, and D. Wendisch, *Liebigs Ann. Chem.* **1982:**150.
46. F. Winterfeldt, *Synthesis* **1975:**617; N. M. Yoon and Y. G. Gyoung, *J. Org. Chem.* **50:**2443 (1985).
47. C. Szantay, L. Toke, and P. Kolonits, *J. Org. Chem.* **31:**1447 (1966).
48. G. E. Keck, E. P. Boden, and M. R. Wiley, *J. Org. Chem.* **54:**896 (1989).
49. S. Nahm and S. M. Weinreb, *Tetrahedron Lett.* **22:**3815 (1981).

stable until acid hydrolysis occurs during workup.

$$RC(=O)N(CH_3)OCH_3 + M-H \longrightarrow \text{[R(H)C(O-M)(OCH}_3\text{)N(CH}_3\text{)]} \xrightarrow[H_2O]{H^+} RCH=O$$

Another useful approach to aldehydes is by partial reduction of nitriles to imines. The imines are then hydrolyzed to the aldehyde. Diisobutylaluminum hydride seems to be the best reagent for this purpose.[50,51] The reduction stops at the imine stage because of the low electrophilicity of the deprotonated imine intermediate.

$$CH_3CH=CHCH_2CH_2CH_2C\equiv N \xrightarrow[2)\ H^+,\ H_2O]{1)\ (i\text{-}Bu)_2AlH} CH_3CH=CHCH_2CH_2CH_2CH=O \quad 64\%$$

A second type of chemoselectivity arises in the context of the need to reduce one functional group in the presence of another. If the group to be reduced is more reactive than the one to be left unchanged, it is simply a matter of choosing a reducing reagent with the appropriate reactivity. Sodium borohydride, for example, is very useful in this respect because it reduces ketones and aldehydes much more rapidly than esters. Sodium cyanoborohydride is used to reduce imines to amines. This reagent is only reactive toward protonated imines. At pH 6–7, $NaBH_3CN$ is essentially unreactive toward carbonyl groups. When an amine and a ketone are mixed together, equilibrium is established with the imine. At mildly acidic pH, $NaBH_3CN$ is reactive only toward the protonated imine.[52]

$$R_2C=O + R'NH_2 + H^+ \rightleftharpoons R_2C=\overset{+}{N}HR'$$

$$R_2C=\overset{+}{N}HR' + BH_3CN^- \longrightarrow R_2CHNHR'$$

Reductive animation by $NaBH_3CN$ can also be carried out in the presence of $Ti(O\text{-}i\text{-}Pr)_4$. These conditions are especially useful for situations in which it is not practical to use the amine in excess (as is typically the case under acid-catalyzed conditions) or for acid-sensitive compounds. The $Ti(O\text{-}i\text{-}Pr)_4$ may act as a Lewis acid in generation of a tetrahedral adduct, which then may be reduced directly or via a transient iminium intermediate.[53]

$$R_2C=O + HNR'_2 \xrightarrow{Ti(O\text{-}i\text{-}Pr)_4} R_2C(OTi(O\text{-}i\text{-}Pr)_3)-NR'_2 \rightleftharpoons R_2C=N^+R'_2 \xrightarrow{NaBH_3CN} R_2CHNR'_2$$

Sodium triacetoxyborohhydride is an alternative to $NaBH_3CN$ for reductive amination. This reagent can be used with a wide variety of aldehydes and ketones mixed with primary and secondary amines, including aniline derivatives.[54] This reagent has been used

50. N. A. LeBel, M. E. Post, and J. J. Wang, *J. Am. Chem. Soc.* **86**:3759 (1964).
51. R. V. Stevens and J. T. Lai, *J. Org. Chem.* **37**:2138 (1972); S. Trofimenko, *J. Org. Chem.* **29**:3046 (1964).
52. R. F. Borch, M. D. Bernstein, and H. D. Durst, *J. Am. Chem. Soc.* **93**:2897 (1971).
53. R. J. Mattson, K. M. Pham, D. J. Leuck, and K. A. Cowen, *J. Org. Chem.* **55**:2552 (1990).
54. A. F. Abdel-Magid, K. G. Carson, B. D. Harris, C. A. Maryanoff, and R. D. Shah, *J. Org. Chem.* **61**:3849 (1996).

successfully to alkylate amino acid esters.[55]

$$R_2C{=}O + HNR'_2 \xrightarrow{NaBH(OAc)_3} R_2CHNR'_2$$

$$PhCH_2CH(NH_3^+Cl^-)CO_2CH_3 + CH_3(CH_2)_4CH{=}O \xrightarrow{NaBH(OAc)_3} PhCH_2CH(NH(CH_2)_5CH_3)CO_2CH_3 \quad 79\%$$

Zinc borohydride has been found to effect very efficient reductive amination in the presence of silica. The amine and the carbonyl compound are mixed with silica, and the powder is then treated with a solution of $Zn(BH_4)_2$. Excellent yields are reported and the procedure works well for unsaturated aldehydes and ketones.[56]

$$\text{(trimethylcyclohexenone)}{=}O + H_2N\text{–cyclohexyl} \xrightarrow[2)\ Zn(BH_4)_2]{1)\ SiO_2} \text{–NH–cyclohexyl} \quad 80\%$$

Aromatic aldehydes can be reductively aminated with the combination $Zn(BH_4)_2$–$ZnCl_2$.[57]

$$F\text{–}C_6H_4\text{–}CH{=}O + H\text{–}N\text{(pyrrolidine)} \xrightarrow[2)\ Zn(BH_4)_2]{1)\ ZnCl_2} F\text{–}C_6H_4\text{–}CH_2\text{–}N\text{(pyrrolidine)} \quad 77\%$$

The $ZnCl_2$ assists in imine formation in this procedure.

Diborane also has a useful pattern of selectivity. It reduces carboxylic acids to primary alcohols under mild conditions which leave esters unchanged.[58] Nitro and cyano groups are also relatively unreactive toward diborane. The rapid reaction between carboxylic acids and diborane is the result of formation of triacyloxyborane intermediate by protonolysis of the B–H bonds. This compound is essentially a mixed anhydride of the carboxylic acid and boric acid in which the carbonyl groups have enhance reactivity

$$3\ RCO_2H + BH_3 \longrightarrow (RCO_2)_3B + 3\ H_2$$

$$RC({=}O){-}O{-}B(O_2CR)_2 \longleftrightarrow RC({=}O){-}\overset{+}{O}{=}\overset{-}{B}(O_2CR)_2$$

Diborane is also a useful reagent for reducing amides. Tertiary and secondary amides are easily reduced, but primary amides react only slowly.[59] The electrophilicity of diborane is involved in the reduction of amides. The boron coordinates at the carbonyl oxygen,

55. J. M. Ramanjulu and M. M. Joullie, *Synth. Commun.* **26:**1379 (1996).
56. B. C. Ranu, A. Majee, and A. Sarkar, *J. Org. Chem.* **63:**370 (1998).
57. S. Bhattacharyya, A. Chatterjee, and J. S. Williamson, *Synth. Commun.* **27:**4265 (1997).
58. M. N. Yoon, C. S. Pak, H. C. Brown, S. Krishnamurthy, and T. P. Stocky, *J. Org. Chem.* **38:**2786 (1973).
59. H. C. Brown and P. Heim, *J. Org. Chem.* **38:**912 (1973).

enhancing the reactivity of the carbonyl center.

Amides require vigorous reaction conditions for reduction by $LiAlH_4$ so that little selectivity can be achieved with this reagent. Diborane, however, permits the reduction of amides in the presence of ester and nitro groups.

Alane is also a useful group for reducing amides, and it too can be used to reduce amides to amines in the presence of ester groups.

Ref. 60

Again, the electrophilicity of alane is the basis for the selective reaction with the amide group. Alane is also useful for reducing azetidinones to azetidines. Most nucleophilic hydride reducing agents lead to ring-opened products. DiBAlH, AlH_2Cl, and $AlHCl_2$ can also reduce azetidinones to azetidines.[61]

Ref. 62

Another approach to reduction of an amide group in the presence of more easily reduced groups is to convert the amide to a more reactive species. One such method is conversion of the amide to an O-alkyl imidate with a positive charge on nitrogen.[63] This method has proven successful for tertiary and secondary, but not primary, amides. Other compounds which can be readily derived from amides and that are more reactive than amides toward hydride reducing agents are α-alkylthioimmonium ions[64] and α-chloroimmonium ions.[65]

$$RC(=O)NR_2 + Et_3O^+ \longrightarrow RC(OEt)=\overset{+}{N}R_2$$

$$RC(OEt)=\overset{+}{N}R_2 + NaBH_4 \longrightarrow RCH_2NR_2$$

60. S. F. Martin, H. Rüeger, S. A. Williamson, and S. Grejszczak, *J. Am. Chem. Soc.* **109:**6124 (1987).
61. I. Ojima, M. Zhao, T. Yamato, K. Nakahashi, M. Yamashita, and R. Abe, *J. Org. Chem.* **56:**5263 (1991).
62. M. B. Jackson, L. N. Mander, and T. M. Spotswood, *Aust. J. Chem.* **36:**779 (1983).
63. R. F. Borch, *Tetrahedron Lett.* **1968:**61.
64. S. Raucher and P. Klein, *Tetrahedron Lett.* **1980:**4061; R. J. Sundberg, C. P. Walters, and J. D. Bloom, *J. Org. Chem.* **46:**3730 (1981).
65. M. E. Kuehne and P. J. Shannon, *J. Org. Chem.* **42:**2082 (1977).

An important case of chemoselectivity arises in the reduction of α,β-unsaturated carbonyl compounds. Reduction can occur at the carbonyl group, giving an allylic alcohol, or at the double bond, giving a saturated ketone. If a hydride is added at the β position, the initial product is an enolate. In protic solvents, this leads to the ketone, which can be reduced to the saturated alcohol. If hydride is added at the carbonyl group, the allylic alcohol is usually not susceptible to further reduction. These alternative reaction modes are called 1,2- and 1,4-reduction, respectively. Both $NaBH_4$ and $LiAlH_4$ have been observed to give both types of product, although the extent of reduction to saturated alcohol is usually greater with $NaBH_4$.[66]

1,2-reduction

$$R_2C{=}CHC(=O)R' + [H^-] \longrightarrow R_2C{=}CHC(O^-)(H)R' \xrightarrow{H^+} R_2C{=}CHCH(OH)R'$$

1,4-reduction leading to saturated alcohol

$$R_2C{=}CHC(=O)R' + [H^-] \longrightarrow R_2CH{-}CH{=}C(O^-)R' \xrightarrow{H^+} R_2CHCH_2C(=O)R'$$

$$R_2CHCH_2C(=O)R' + [H^-] \longrightarrow R_2CHCH_2CH(O^-)R' \xrightarrow{H^+} R_2CHCH_2CH(OH)R'$$

Several reagents have been developed which lead to exclusive 1,2- or 1,4-reduction. Use of $NaBH_4$ in combination with cerium chloride results in clean 1,2-reduction.[67] Diisobutylaluminum hydride[68] and the dialkylborane 9-BBN[69] also give exclusive carbonyl reduction. In each case, the reactivity of the carbonyl group is enhanced by a Lewis acid complexation at oxygen. Selective reduction of the carbon–carbon double bond can usually be achieved by catalytic hydrogenation. A series of reagents prepared from a hydride reducing agent and copper salts also give primarily the saturated ketone.[70] Similar reagents have been shown to reduce α,β-unsaturated esters[71] and nitriles[72] to the corresponding saturated compounds. The mechanistic details are not known with certainty, but it is likely that "copper hydrides" are the active reducing agents and that they form an organocopper intermediate by conjugate addition.

$$\text{"H}{-}\overset{|}{\underset{|}{Cu}}{-}\text{H"} + RCH{=}CHC(=O)R \longrightarrow {-}\overset{|}{\underset{|}{Cu}}{-}CH(R){-}CH_2C(=O)R \longrightarrow RCH_2CH_2C(=O)R$$

Combined use of cobalt(II) acetylacetonate $[Co(acac)_2]$ and DiBAlH also gives selective 1,4-reduction for α,β-unsaturated ketones, esters, and amides.[73]

66. M. R. Johnson and B. Richborn, *J. Org. Chem.* **35:**1041 (1970); W. R. Jackson and A. Zurqiyah, *J. Chem. Soc.*, 5280 (1965).
67. J.-L. Luche, *J. Am. Chem. Soc.*, **100:**2226 (1978); J.-L. Luche, L. Rodriquez-Hahn, and P. Crabbe, *J. Chem. Soc. Chem. Commun.* **1978:**601.
68. K. E. Wilson, R. T. Seidner, and S. Masamune, *J. Chem. Soc., Chem. Commun.* **1970:**213.
69. K. Krishnamurthy and H. C. Brown, *J. Org. Chem.* **42:**1197 (1977).
70. S. Masamune, G. S. Bates, and P. E. Georghiou, *J. Am. Chem. Soc.* **96:**3686 (1974); E. C. Ashby, J.-J. Lin, and R. Kovar, *J. Org. Chem.* **41:**1939 (1976); E. C. Ashby, J.-J. Lin, and A. B. Goel, *J. Org. Chem.* **43:**183 (1978); W. S. Mahoney, D. M. Brestensky, and J. M. Stryker, *J. Am. Chem. Soc.* **110:**291 (1988); D. S. Brestensky, D. E. Huseland, C. McGettigan, and J. M. Stryker, *Tetrahedron Lett.* **29:**3749 (1988); T. M. Koenig, J. F. Daeuble, D. M. Brestensky, and J. M. Stryker, *Tetrahedron Lett.* **31:**3237 (1990).
71. M. F. Semmelhack, R. D. Stauffer, and A. Yamashita, *J. Org. Chem.* **42:**3180 (1977).
72. M. E. Osborn, J. F. Pegues, and L. A. Paquette, *J. Org. Chem.* **45:**167 (1980).
73. T. Ikeno, T. Kimura, Y. Ohtsuka, and T. Yamada, *Synlett* **1999:**96.

Another reagent combination that selectively reduces the carbon–carbon double bond is Wilkinson's catalyst and triethylsilane. The initial product is the silyl enol ether.[74]

$$(CH_3)_2C{=}CH(CH_2)_2\overset{CH_3}{\overset{|}{C}}{=}CHCH{=}O \xrightarrow[(Ph_3P)_3RhCl]{Et_3SiH} (CH_3)_2C{=}CH(CH_2)_2\overset{CH_3}{\overset{|}{C}}HCH{=}CHOSiEt_3$$

$$\xrightarrow{H_2O} (CH_3)_2C{=}CH(CH_2)_2\overset{CH_3}{\overset{|}{C}}HCH_2CH{=}O$$

Unconjugated double bonds are unaffected by this reducing system.[75]

The enol ethers of β-dicarbonyl compounds are reduced to α,β-unsaturated ketones by $LiAlH_4$, followed by hydrolysis.[76] Reduction stops at the allylic alcohol, but subsequent acid hydrolysis of the enol ether and dehydration lead to the isolated product. This reaction is a useful method for synthesis of substituted cyclohexenones.

$$\text{(OC}_2\text{H}_5\text{, Ph, Ph, O)} \xrightarrow{LiAlH_4} \text{(OC}_2\text{H}_5\text{, Ph, Ph, }^-\text{O, H)} \xrightarrow{H^+} \text{(O, Ph, Ph)}$$

5.2.2. Stereoselectivity of Hydride Reduction

A very important aspect of reductions by hydride-transfer reagents is their stereoselectivity. The stereochemistry of hydride reduction has been studied most thoroughly with conformationally biased cyclohexanone derivatives. Some reagents give predominantly axial cyclohexanols whereas others give the equatorial isomer. Axial alcohols are likely to be formed when the reducing agent is a sterically hindered hydride donor. This is because the equatorial direction of approach is more open and is preferred by bulky reagents. This is called *steric approach control*.[77]

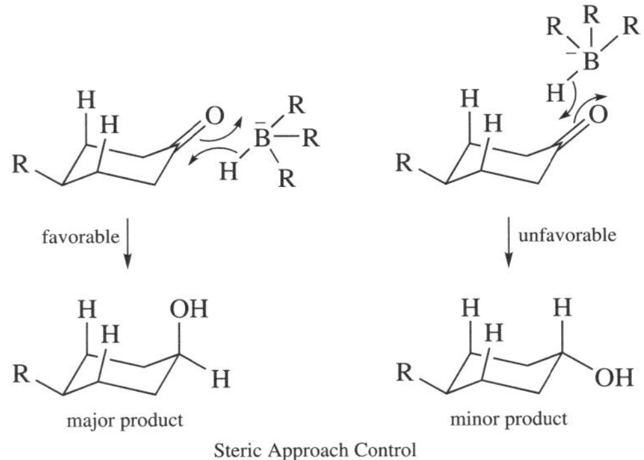

Steric Approach Control

74. I. Ojima, T. Kogure, and Y. Nagai, *Tetrahedron Lett.* **1972:**5035; I. Ojima, M. Nihonyanagi, T. Kogure, M. Kumagai, S. Horiuchi, K. Nakatsugawa, and Y. Nogai, *J. Organomet. Chem.* **94:**449 (1973).
75. H.-J. Liu and E. N. C. Browne, *Can. J. Chem.* **59:**601 (1981); T. Rosen and C. H. Heathcock, *J. Am. Chem. Soc.* **107:**3731 (1985).
76. H. E. Zimmerman and D. I. Schuster, *J. Am. Chem. Soc.* **84:**4527 (1962); W. F. Gannon and H. O. House, *Org. Synth.* **40:**14 (1960).
77. W. G. Dauben, G. J. Fonken, and D. S. Noyce, *J. Am. Chem. Soc.* **78:**2579 (1956).

With less hindered hydride donors, particularly $NaBH_4$ and $LiAlH_4$, cyclohexanones give predominantly the equatorial alcohol. The equatorial alcohol is normally the more stable of the two isomers. However, hydride reductions are exothermic reactions with low activation energies. The transition state should resemble starting ketone, so product stability should not control the stereoselectivity. One explanation of the preference for formation of the equatorial isomer involves the torsional strain that develops in formation of the axial alcohol.[78]

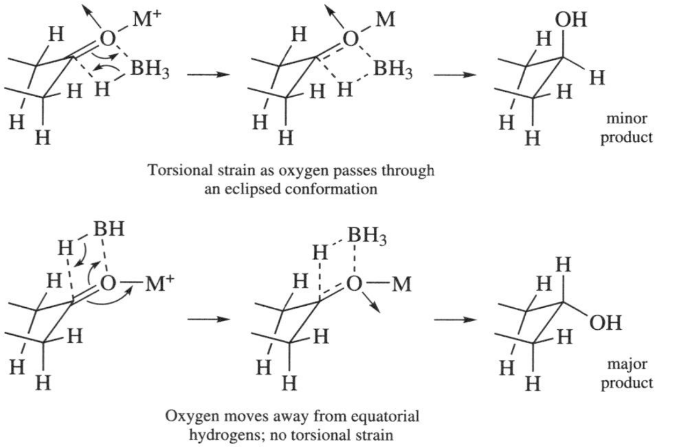

minor product

Torsional strain as oxygen passes through an eclipsed conformation

major product

Oxygen moves away from equatorial hydrogens; no torsional strain

An alternative suggestion is that the carbonyl group π-antibonding orbital which acts as the lowest unoccupied molecular orbital (LUMO) in the reaction has a greater density on the axial face.[79] It is not entirely clear at the present time how important such orbital effects are. Most of the stereoselectivities which have been reported can be reconciled with torsional and steric effects being dominant.[80] See Section 3.10 of Part A for further discussion of this issue.

When a ketone is relatively hindered, as for example in the bicyclo[2.2.1]heptan-2-one system, steric factors govern stereoselectivity even for small hydride donors.

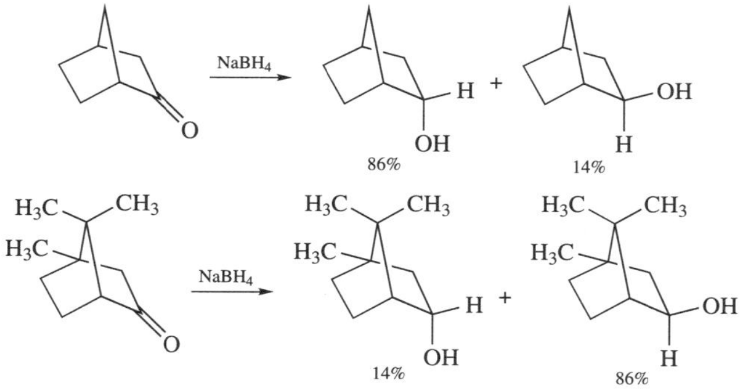

$NaBH_4$: 86% + 14%

$NaBH_4$: 14% + 86%

78. M. Cherest, H. Felkin, and N. Prudent, *Tetrahedron Lett.* **1968:**2205; M. Cherest and H. Felkin, *Tetrahedron Lett.* **1971:**383.
79. J. Klein, *Tetrahedron Lett.* **1973:**4307; N. T. Ahn, O. Eisenstein, J.-M. Lefour, and M. E. Tran Huu Dau, *J. Am. Chem. Soc.* **95:**6146 (1973).
80. W. T. Wipke and P. Gund, *J. Am. Chem. Soc.* **98:**8107 (1976); J.-C. Perlburger and P. Müller, *J. Am. Chem. Soc.* **99:**6316 (1977); D. Mukherjee, Y.-D. Wu, F. R. Fornczek, and K. N. Houk, *J. Am. Chem. Soc.* **110:**3328 (1988).

Table 5.4. Stereoselectivity of Hydride Reducing Agents[a]

	Percentage of the alcohol favored by steric approach control				
Reducing agent	% axial	% axial	% axial	% *endo*	% *exo*
$NaBH_4$	20[b]	25[c]	58[c]	86[d]	86[d]
$LiAlH_4$	8	24	83	89	92
$LiAl(OMe)_3H$	9	69		98	99
$LiAl(t\text{-}BuO)_3H$	9[e]	36[f]	95	94[f]	94[f]
$[CH_3CH_2CH(CH_3)]_3\bar{B}HLi^+$	93[g]	98[g]	99.8[g]	99.6[g]	99.6[g]
$[(CH_3)_2CHCH(CH_3)]_3\bar{B}HLi^+$	>99[h]	>99[h]		>99[h]	NR[h]

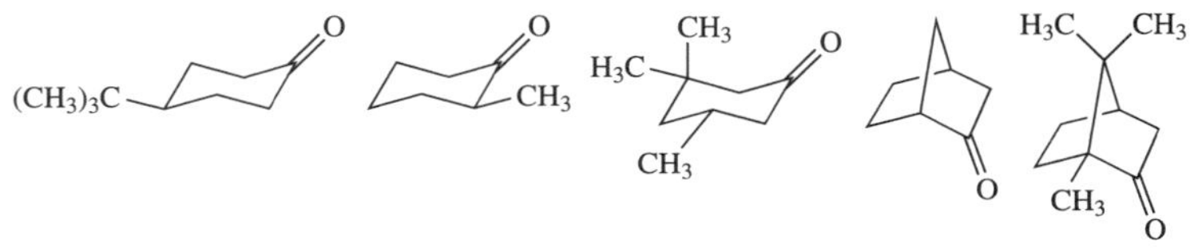

a. Except where otherwise noted, data are those given by H. C. Brown and W. D. Dickason, *J. Am. Chem. Soc.* **92:**709 (1970). Data for many other cyclic ketones and reducing agents are given by A. V. Kamernitzky and A. A. Akhrem, *Tetrahedron* **18:**705 (1962) and W. T. Wipke and P. Gund, *J. Am. Chem. Soc.* **98:**8107 (1976).
b. P. T. Lansbury and R. E. MacLeay, *J. Org. Chem.* **28:**1940 (1963).
c. B. Rickborn and W. T. Wuesthoff, *J. Am. Chem. Soc.* **92:**6894 (1970).
d. H. C. Brown and J. Muzzio, *J. Am. Chem. Soc.* **88:**2811 (1966).
e. J. Klein, E. Dunkelblum, E. L. Eliel, and Y. Senda, *Tetrahedron Lett.* **1968:**6127.
f. E. C. Ashby, J. P. Sevenair, and F. R. Dobbs, *J. Org. Chem.* **36:**197 (1971).
g. H. C. Brown and S. Krishnamurthy, *J. Am. Chem. Soc.* **94:**7159 (1972).
h. S. Krishnamurthy and H. C. Brown, *J. Am. Chem. Soc.* **98:**3383 (1976).

A large amount of data has been accumulated on the stereoselectivity of reduction of cyclic ketones.[81] Table 5.4 compares the stereochemistry of reduction of several ketones by hydride donors of increasing seric bulk. The trends in the table illustrate the increasing importance of steric approach control as both the hydride reagent and the ketone become more highly substituted. The alkyl-substituted borohydrides have especially high selectivity for the least hindered direction of approach.

The stereochemistry of reduction of acylic aldehydes and ketones is a function of the substitution on the adjacent carbon atom and can be predicted on the basis of a conformational model of the transition state.[78]

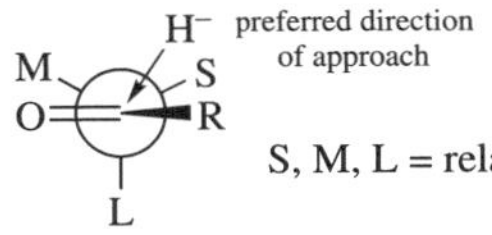

H⁻ preferred direction of approach

S, M, L = relative size of substituents

This model is rationalized by a combination of steric and stereoelectronic effects. From a purely steric standpoint, an approach from the direction of the smallest substituent, involving minimal steric interaction with the groups L and M, is favorable. The stereoelectronic effect involves the interaction between the approaching hydride ion and the LUMO of the carbonyl group. This orbital, which accepts the electrons of the incoming

81. D. C. Wigfield, *Tetrahedron* **35:**449 (1979); D. C. Wigfield and D. J. Phelps, *J. Org. Chem.* **41:**2396 (1976).

nucleophile, is stabilized when the group L is perpendicular to the plane of the carbonyl group.[82] This conformation permits a favourable interaction between the LUMO and the antiboding σ^* orbital associated with the C–L bond.

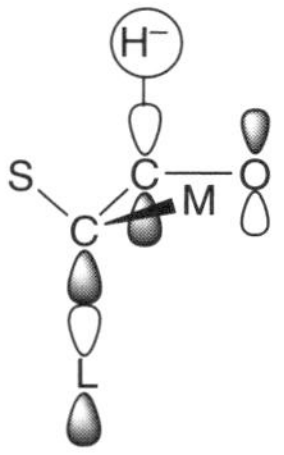

Steric factors arising from groups which are more remote from the center undergoing reduction can also influence the stereochemical course of reduction. Such steric factors are magnified with the use of bulky reducing agents. For example, a 4.5 : 1 preference for stereoisomer **E** over **F** is achieved by using the trialkylborohydride **D** as the reducing agent in the reduction of a prostaglandin intermediate.[83]

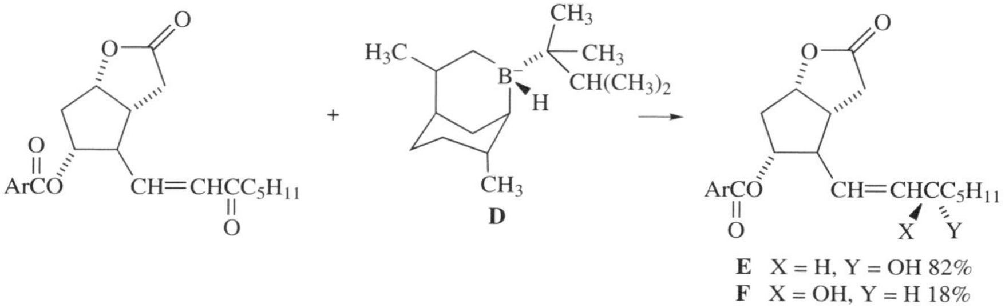

E X = H, Y = OH 82%
F X = OH, Y = H 18%

The stereoselectivity of reduction of carbonyl groups is effected by the same combination of steric and stereoelectronic factors which control the addition of other nucleophiles, such as enolates and organometallic reagents to carbonyl groups. A general discussion of these factors on addition of hydride is given in Section 3.10 of Part A.

The stereoselectivity of reduction of carbonyl groups can also be controlled by chelation effects when there is a nearby donor substituent. In the presence of such a group, specific complexation between the substituent, the carbonyl oxygen, and the Lewis acid can establish a preferred conformation for the reactant which then controls reduction. Usually, hydride is then delivered from the less sterically hindered face of the chelate.

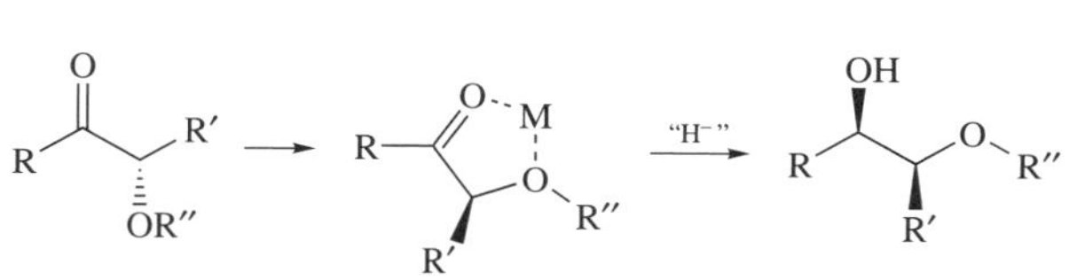

α-Hydroxy ketones[84] and α-alkoxy ketones[85] are reduced to *anti* 1,2-diols by $Zn(BH_4)_2$,

82. N. T. Ahn, *Top. Curr. Chem.* **88:**145 (1980).
83. E. J. Corey, S. M. Albonico, U. Koelliker, T. K. Shaaf, and R. K. Varma, *J. Am. Chem. Soc.* **93:**1491 (1971).
84. T. Nakata, T. Tanaka, and T. Oishi, *Tetrahedron Lett.* **24:**2653 (1983).
85. G. J. McGarvey and M. Kimura, *J. Org. Chem.* **47:**5420 (1982).

which reacts through a chelated transition state. This stereoselectivity is consistent with the preference for transition state **G** over **H**. The stereoselectivity increases with the bulk of substituent R^2.

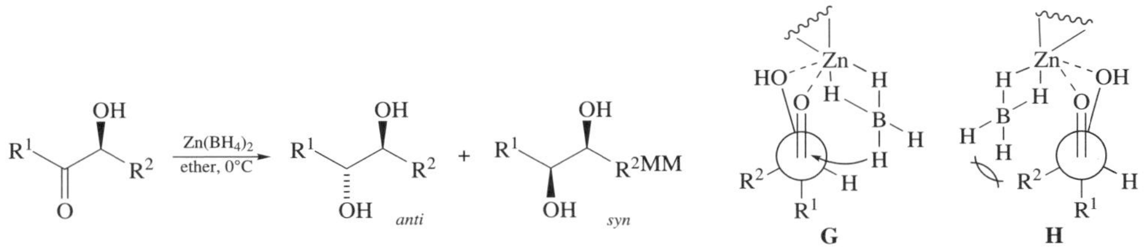

R^1	R^2	$Zn(BH_4)_2$ *anti* : *syn*	$LiAlH_4$ *anti* : *syn*
n-C_5H_{11}	CH_3	77 : 23	64 : 36
CH_3	n-C_5H_{11}	85 : 15	70 : 30
i-C_3H_7	CH_3	85 : 15	58 : 42
CH_3	i-C_3H_7	96 : 4	73 : 27
Ph	CH_3	98 : 2	87 : 13
CH_3	Ph	90 : 10	80 : 20

Reduction of β-hydroxyketones through chelated transitions states fovors *syn*-1,3-diols. Boron chelates have been exploited to achieve this stereoselectivity.[86] One procedure involves *in situ* generation of diethylmethoxyboron, which then forms a chelate with the β-hydroxy ketone. Reduction with $NaBH_4$ leads to the *syn* diol.[87]

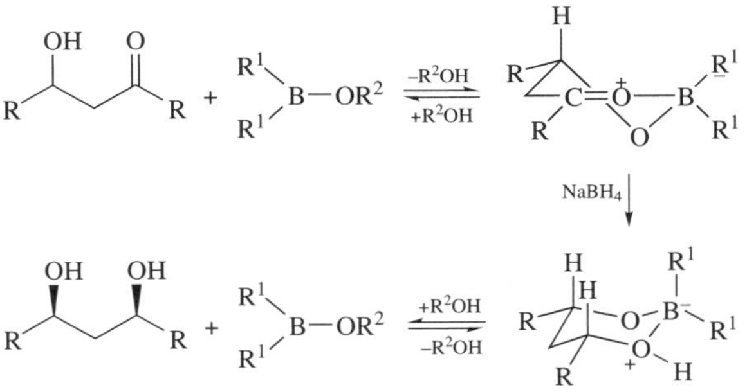

β-Hydroxy ketones also give primarily *syn* 1,3-diols when chelates prepared with BCl_3 are reduced with quaternary ammonium salts of $BH_4{}^-$ or BH_3CN^-.[88]

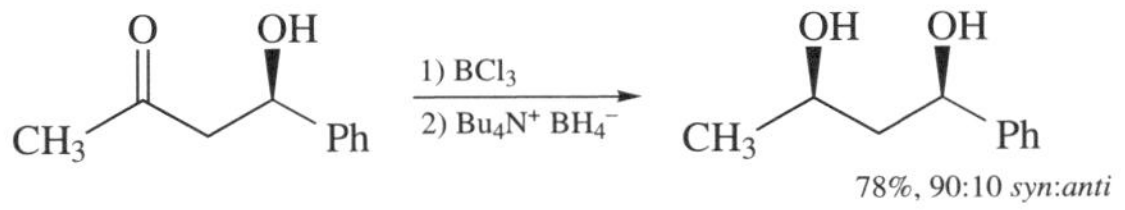

78%, 90:10 *syn*:*anti*

86. K. Narasaka and F.-C. Pai, *Tetrahedron* **40:**2233 (1984); K.-M. Chen, G. E. Hardtmann, K. Prasad, O. Repic, and M. J. Shapiro, *Tetrahedron Lett.* **28:**155 (1987).
87. K.-M. Chen, K. G. Gunderson, G. E. Hardtmann, K. Prasad, O. Repic, and M. J. Shapiro, *Chem. Lett.* **1987:**1923.
88. C. R. Sarko, S. E. Collibee, A. L. Knorr, and M. DiMare, *J. Org. Chem.* **61:**868 (1996).

Similar results are obtained with β-methoxyketones using $TiCl_4$ as the chelating reagent.[89]

A survey of several alkylborohydrides found that $LiBu_3BH$ in ether–pentane gave the best ratio of chelation-controlled reduction products from α- and β-alkoxyketones.[90] In this case, the Li^+ cation must act as the Lewis acid. The alkylborohydride provides an added increment of steric discrimination.

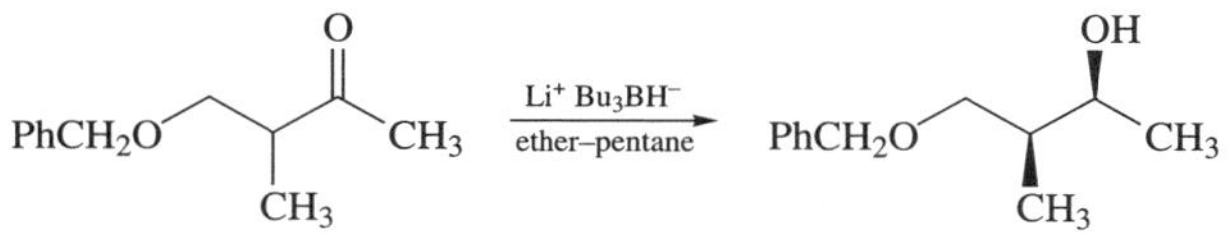

$$PhCH_2OCH_2CH(CH_3)COCH_3 \xrightarrow[\text{ether–pentane}]{Li^+\ Bu_3BH^-} PhCH_2OCH_2CH(CH_3)CH(OH)CH_3$$

Syn 1,3-diols also can be obtained from β-hydroxyketones using $LiI{-}LiAlH_4$ at low temperatures.[91]

The reduction of an unsymmetrical ketone creates a new stereo center. Because of the importance of hydroxy groups both in synthesis and in relation to the properties of molecules, including biological activity, there has been a great deal of effort directed toward enantioselective reduction of ketones. One approach is to use chiral borohydride reagents.[92] Boranes derived from chiral alkenes can be converted to borohydrides, and there has been much study of the enantioselectivity of these reagents. Several of the reagents are commercially available.

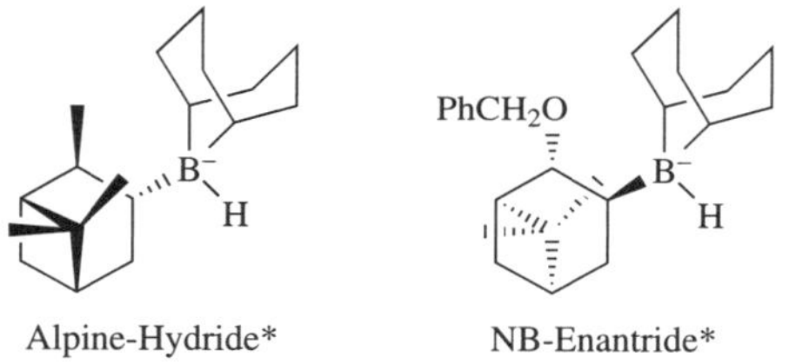

Alpine-Hydride*

$PhCH_2O$ … B^- H

NB-Enantride*

Chloroboranes have also been found to be useful for enantioselective reduction. Diisopinocampheylchloroborane,[93] $(Ipc)_2BCl$, and *t*-butylisopinocampheylchloroborane[94] achieve high enantioselectivity for aryl and hindered dialkyl ketones. Diiso-2-ethylapopinocampheylchloroborane,[95] $(Eap)_2BCl$, shows good enantioselectivity with a wider range

* The names are trademarks of Aldrich Chemical Company.

89. C. R. Sarko, I. C. Guch, and M. DiMare, *J. Org. Chem.* **59:**705 (1994); G. Bartoli, M. C. Bellucci, M. Bosco, R. Dalpozzo, E. Marcantoni, and L. Sambri, *Tetrahedron Lett.* **40:**2845 (1999).
90. A.-M. Faucher, C. Brochu, S. R. Landry, I. Duchesne, S. Hantos, A. Roy, A. Myles, and C. Legault, *Tetrahedron Lett.* **39:**8425 (1998).
91. Y. Mori, A. Takeuchi, H. Kageyama, and M. Suzuki, *Tetrahedron Lett.* **29:**5423 (1988).
92. M. M. Midland, *Chem. Rev.* **89:**1553 (1989).
93. H. C. Brown, J. Chandrasekharan, and P. V. Ramachandran *J. Am. Chem. Soc.* **110:**1539 (1988); M. Zhao, A. O. King, R. D. Larsen, T. R. Verhoeven, and P. J. Reider, *Tetrahedron Lett.* **38:**2641 (1997).
94. H. C. Brown, M. Srebnik, and P. V. Ramachandran, *J. Org. Chem.* **54:**1577 (1989).
95. H. C. Brown, P. V. Ramachandran, A. V. Teodorovic, and S. Swaminathan, *Tetrahedron Lett.* **32:**6691 (1991).

Table 5.5. Enantioselective Reduction of Ketones

Reagent	Ketone	% e.e.	Config.	Reference
Alpine-Borane[a]	3-Methyl-2-butanone	62	*S*	b
NB-Enantride[a]	2-Octanone	79	*S*	c
$(Ipc)_2BCl$	2-Acetylnaphthalene	94	*S*	d
(IpcB(*t*-Bu)Cl	Acetophenone	96	*R*	e
$(Ipc)_2BCl$	2,2-Dimethycyclohexanone	91	*S*	f
$(Eap)_2BCl$	3-Methyl-2-butanone	95	*R*	g

a. Trademark of Aldrich Chemical Company.
b. H. C. Brown and G. G Pai, *J. Org. Chem.* **50**:1384 (1985).
c. M. M. Midland and A. Kozubski, *J. Org. Chem.* **47**:2495 (1982).
d. M. Zhao, A. O. King, R. D. Larsen, T. R. Verhoeven, and A. J. Reider, *Tetrahedron Lett.* **38**:2641 (1997).
e. H. C. Brown, M. Srebnik, and P. V. Ramachandran, *J. Org. Chem.* **54**:1577 (1989).
f. H. C. Brown, J. Chandrasekharan, and P. V. Ramachandran, *J. Am. Chem. Soc.* **110**:1539 (1988).
g. H. C. Brown, P. V. Ramachandran, A. V. Teodorovic, and S. Swaminathan, *Tetrahedron Lett.* **32**:6691 (1991).

of alcohols.

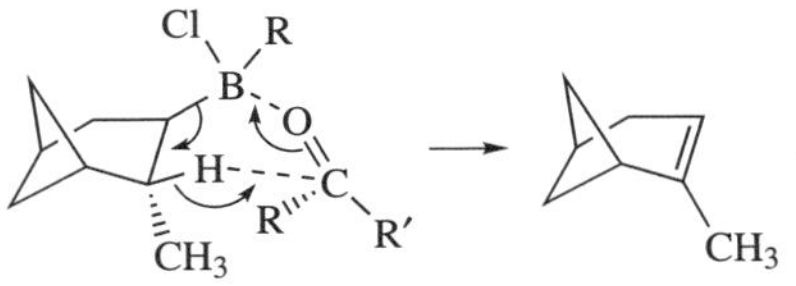

Table 5.5 give some typical results for enantioselective reduction of ketones.

An even more efficient approach to enantioselective reduction is to use a chiral catalyst. One of the most promising is the oxazaborolidine **I**, which is ultimately derived from the amino acid proline.[96] The enantiomer is also available. A catalytic amount (5–20 mol %) of this reagent along with BH_3 as the reductant can reduce ketones such as acetophone and pinacolone in >95% e.e. An adduct of borane and **I** is the active reductant.

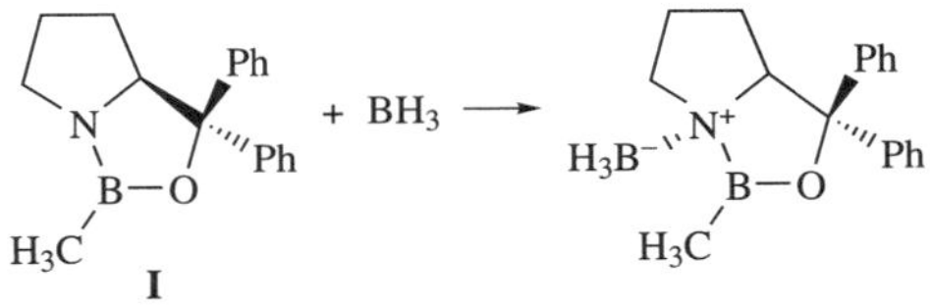

This adduct can be prepared, stored, and used as a stoichiometric reagent if so desired.[97] Catecholborane can also be used as the reductant.[98]

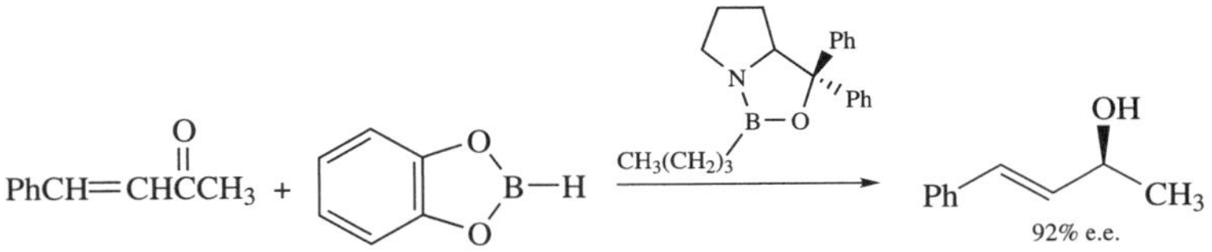

96. E. J. Corey, R. K. Bakshi, S. Shibata, C. P. Chen, and V. K. Singh, *J. Am. Chem. Soc.* **109**:7925 (1987); E. J. Corey and C. J. Helal, *Angew, Chem. Int. Ed. Engl.* **37**:1987 (1998).
97. D. J. Mathre, A. S. Thompson, A. W. Douglas, K. Hoogsteen, J. D. Carroll, E. G. Corley, and E.-J. J. Grabowski, *J. Org. Chem.* **58**:2880 (1993).
98. E. J. Corey and R. K. Bakshi, *Tetrahedron Lett.* **31**:611 (1990).

The enantioselectivity and reactivity of these catalysts can be modified by changes in substituent groups to optimize selectivity toward a particular ketone.[99]

The enantioselectivity in these reductions is proposed to arise from a chairlike transition state in which the governing steric interaction is with the alkyl substituent on boron.[100] There are data indicating that the steric demand of this substituent influences enantioselectivity.[101]

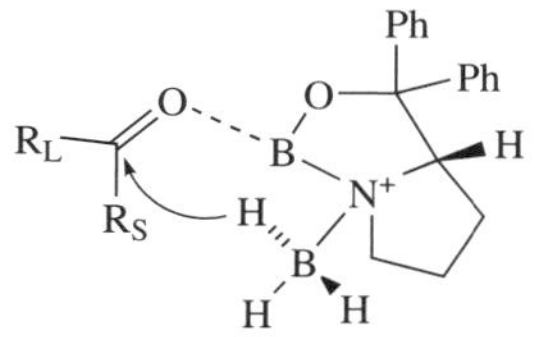

Scheme 5.4 shows some examples of enantioselective reduction of ketones using **I**. Adducts of borane with several other chiral β-aminoalcohols are being explored as chiral catalyst for reduction of ketones.[102] Table 5.6 shows the enantioselectivity of several of these catalysts toward acetophenone.

5.2.3. Reduction of Other Functional Groups by Hydride Donors

Although reductions of the common carbonyl and carboxylic acid derivatives are the most prevalent uses of hydride donors, these reagents can reduce a number of other groups in ways that are of synthetic utility. Scheme 5.5 illustrates some of these other applications of the hydride donors. Halogen and sulfonate leaving groups can undergo replacement by hydride. Both aluminum and boron hydrides exhibit this reactivity. Lithium trialkylborohydrides are especially reactive.[103] The reduction is particularly rapid and efficient in polar aprotic solvents such as DMSO, DMF, and HMPA. Table 5.7 gives some indication of the reaction conditions. The normal factors in susceptibility to nucleophilic attack govern reactivity, with the order of reactivity being I > Br > Cl in terms of the leaving group and benzyl ~ allyl > primary > secondary > tertiary in terms of the substitution site.[104] For alkyl groups, it is likely that the reaction proceeds by an S_N2 mechanism. However, the range of halides that can be reduced includes aryl halides and bridgehead halides, which cannot react by the S_N2 mechanism.[105] There is loss of stereochemical integrity in the reduction of vinyl halides, suggesting the involvement of radical intermediates.[106] Formation and subsequent dissociation of a radical anion by one-electron transfer is a likely mechanism for reductive dehalogenation of compounds that cannot react

99. A. W. Douglas, D. M. Tschaen, R. A. Reamer, and Y.-J. Shi, *Tetrahedron Asymmetry* **7:**1303 (1996).
100. D. K. Jones, D. C. Liotta, I. Shinkai, and D. J. Mathre, *J. Org. Chem.* **58:**799 (1993).
101. E. J. Corey and R. K. Bakshi, *Tetrahedron Lett.* **31:**611 (1990); T. K. Jones, J. J. Mohan, L. C. Xavier, T. J. Blacklock, D. J. Mathre, P. Sohar, E. T. T. Jones, R. A. Beamer, F. E. Roberts, and E. J. J. Grabowski, *J. Org. Chem.* **56:**763 (1991).
102. G. J. Qaullich and T. M. Woodall, *Tetrahedron Lett.* **34:**4145 (1993); J. Martens, C. Dauelsberg, W. Behnen, and S. Wallbaum, *Tetrahedron Asymmetry* **3:**347 (1992); Z. Shen, W. Huang, J. W. Feng, and Y. W. Zhang, *Tetrahedron Asymmetry* **9:**1091 (1998); N. Hashimot, T. Ishizuko, and T. Kunieda, *Heterocycles* **46:**189 (1997).
103. S. Krishnamurthy and H. C. Brown, *J. Org. Chem.* **45:**849 (1980).
104. S. Krishnamurthy and H. C. Brown, *J. Org. Chem.* **47:**276 (1982).
105. C. W. Jefford, D. Kirkpatrick, and F. Delay, *J. Am. Chem. Soc.* **94:**8905 (1972).
106. S.-K. Chung, *J. Org. Chem.* **45:**3513 (1980).

1[a]

CH_3O_2C ... $\overset{O}{\|}CCH_2Br$ → (BH_3) CH_3O_2C ... $\overset{OH}{CHCH_2Br}$ 80%

2[b]

Ar = 4-biphenyl ... C_5H_{11} → (BH_3) ... 90% d.e.

3[c]

... + (H_3B ... CH_3) → ... 98.8% e.e.

4[d]

... → (BH_3, $S(CH_3)_2$) ... 98% e.e.

5[e]

$CH_3O_2C(CH_2)_3\overset{O}{\|}C$–CH=CH–$Sn(C_4H_9)_3$ → ($CH_2Si(CH_3)_3$) $CH_3O_2C(CH_2)_3\overset{OH}{CH}$–CH=CH–$Sn(C_4H_9)_3$ 90%

6[f]

$Ph\overset{O}{\|}CCH_2OSi[CH(CH_3)_2]_3$ → (BH_3) $Ph\overset{OH}{CH}CH_2OSi[CH(CH_3)_2]_3$ 99% e.e.

a. K. G. Hull, M. Visnick, W. Tautz, and A. Sheffron, *Tetrahedron* **53:**12405 (1997).
b. E. J. Corey, R. K. Bakshi, S. Shibata, C.-P. Chen, and V. K. Singh, *J. Am. Chem. Soc.*, **109:**7925 (1987).
c. D. J. Mathre, A. S. Thompson, A. W. Douglas, K. Hoogsteen, J. D. Carroll, E. G. Corley, and E. J. J. Grabowski, *J. Org. Chem.* **58:**2880 (1993).
d. T. K. Jones, J. J. Mohan, L. C. Xavier, T. J. Blacklock, D. J. Mathre, P. Sohar, E. T. T. Jones, R. A. Reamer, F. E. Roberts, and E. J. J. Grabowski, *J. Org. Chem.* **56:**763 (1991).
e. E. J. Corey, A Guzman-Perez, and S. E. Lazerwith, *J. Am. Chem. Soc.* **119:**11769 (1997).
f. B. T. Cho and Y. S. Chun, *J. Org. Chem.* **63:**5280 (1998).

Table 5.6. Catalysts for Enantioselective Reduction of Acetophenone

Catalyst	Reductant	% e.e.	Config.	Reference
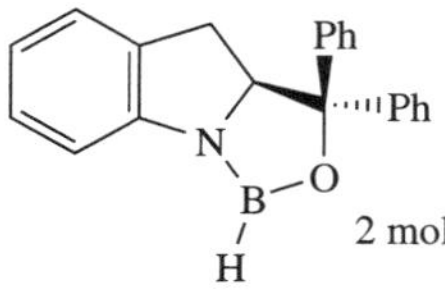 Ph, Ph, N, B, O, H, 2 mol%	BH_3	93%	*R*	a
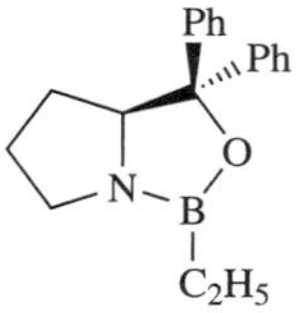 Ph, Ph, O, N, B, C_2H_5	BH_3	96%	*R*	b
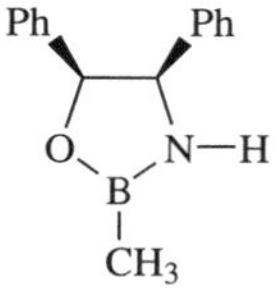 Ph, Ph, O, N—H, B, CH_3	BH_3	92%	*R*	c
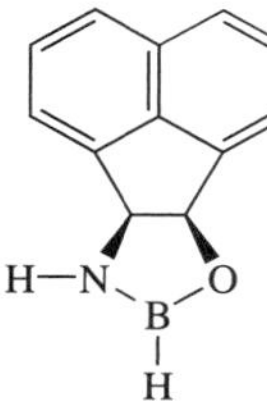 H—N, O, B, H, 10 mol%	BH_3	95%	*R*	d
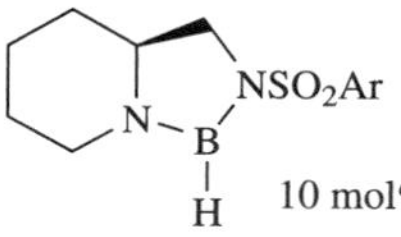 NSO_2Ar, N, B, H, 10 mol%	BH_3	72%	*R*	e
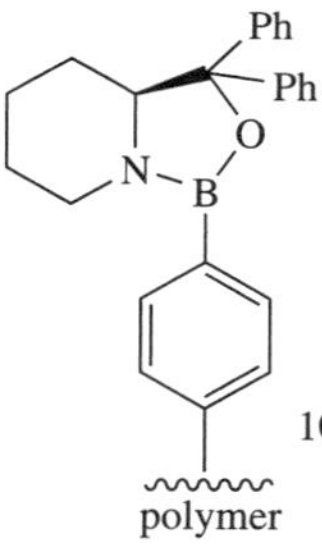 Ph, Ph, O, N, B, 10 mol%, polymer	BH_3	93–98	–	f

a. J. Martens, C. Dauelsburg, W. Behnen, and S. Wallbaum, *Tetrahedron Asymmetry* **3:**347 (1992).
b. E. J. Corey and J. O. Link, *Tetrahedron Lett.* **33:**4141 (1992).
c. G. J. Quallich and T. M. Woodall, *Tetrahedron Lett.* **34:**4145 (1993).
d. A. Sudo, M. Matsumoto, Y. Hashimoto, and K. Saigo, *Tetrahedron Asymmetry* **6:**1853 (1995).
e. O. Froelich, M. Bonin, J.-C. Quirion, and H.-P. Husson, *Tetrahedron Asymmetry* **4:**2335 (1993).
f. C. Franot, G. B. Stone, P. Engeli, C. Spondlin, and E. Waldvogel, *Tetrahedron Asymmetry* **6:**2755 (1995).

Table 6.7. Reaction Conditions for Reductive Replacement of Halogen and Tosylate by Hydride Donors

	Approximate conditions for complete reduction	
Hydride donor	Halides	Tosylates
$NaBH_3CN$[a]	1-Iodododecane, HMPA, 25°C, 4 h	1-Dodecyl tosylate, HMPA, 70°C, 8 h
$NaBH_4$[b]	1-Bromododecane, DMSO, 85°C, 1.5 h	1-Dodecyl tosylate, DMSO, 85°C, 2 h
$LiAlH_4$[c,d]	1-Bromooctane, THF, 25°C, 1 h	1-Octyl tosylate, DME, 25°C, 6 h
$LiB(C_2H_5)_3H$[c]	1-Bromooctane, THF, 25°C, 3 h	

a. R. O. Hutchins, D. Kandasamy, C. A. Maryanoff, D. Masilamani, and B. E. Maryanoff, *J. Org. Chem.* **42:**82 (1977).
b. R. O. Hutchins, D. Kandasamy, F. Dux III, C. A. Maryanoff, D. Rotstein, B. Goldsmith, W. Burgoyne, F. Cistone, J. Dalessandro, and J. Puglis, *J. Org. Chem.* **43:**2259 (1978).
c. S. Krishnamurthy and H. C. Brown, *J. Org. Chem.* **45:**849 (1980).
d. S. Krishnamurthy, *J. Org. Chem.* **45:**25250 (1980).

by an S_N2 mechanism.

$$R{-}X + e^- \longrightarrow R{-}X^{\dot{-}}$$
$$R{-}X^{\dot{-}} \longrightarrow R\cdot + X^-$$
$$R\cdot + X^- \longrightarrow R{-}H + e^-$$

One experimental test for the involvement of radical intermediates is to study 5-hexenyl systems and look for the characteristic cyclization to cyclopentane derivatives (see Section 12.2 of Part A). When 5-hexenyl bromide or iodide reacts with $LiAlH_4$, no cyclizataion products are observed. However, the more hindered 2,2-dimethyl-5-hexenyl iodide gives mainly cyclic product.[107]

$$CH_2{=}CH(CH_2)_3CH_2I + LiAlH_4 \xrightarrow[1\ h]{24°C} CH_2{=}CH(CH_2)_3CH_3 \quad 94\%$$

$$CH_2{=}CH(CH_2)_2C(CH_3)_2CH_2I + LiAlH_4 \xrightarrow[1\ h]{24°C} CH_2{=}CH(CH_2)_2C(CH_3)_2CH_3 \quad 3\% \;+\; \text{1,1,3-trimethylcyclopentane} \quad 81\%$$

Some cyclization also occurs with the bromide but not with the chloride or the tosylate. The secondary iodide 6-iodo-1-heptene gives a mixure of cyclic and acyclic product in THF.[108]

$$CH_2{=}CH(CH_2)_3CH(I)CH_3 \xrightarrow[THF]{LiAlH_4} CH_2{=}CH(CH_2)_3CH_2CH_3 \;(21\%) + \text{1,2-dimethylcyclopentane} \;(72\%,\ 3.7{:}1\ \textit{cis:trans})$$

107. E. C. Ashby, R. N. DePriest, A. B. Goel, B. Wenderoth, and T. N. Pham, *J. Org. Chem.* **49:**3545 (1984).
108. E. C. Ashby, T. N. Pham, and A. Amrollah-Madjadabadi, *J. Org. Chem.* **56:**1596 (1991).

The occurrence of a radical intermediate is also indicated in the reduction of 2-octyl iodide by $LiAlD_4$ since, in contrast to the other halides, extensive racemization accompanies reduction.

The presence of transition-metal ions has a catalytic effect on reduction of halides and tosylates by $LiAlH_4$.[109] Various "copper hydride" reducing agents are effective for removal of halide and tosylate groups.[110] The primary synthetic value of these reductions is for the removal of a hydroxyl function after conversion to a halide or tosylate. Entry 6 in Scheme 5.5 is an example of the use of the reaction in synthesis.

Epoxides are converted to alcohols by $LiAlH_4$. The reaction occurs by nucleophilic attack, and hydride addition at the least hindered carbon of the epoxide is usually observed.

$$PhCH(-O-)CH_2 + LiAlH_4 \longrightarrow PhCH(OH)CH_3$$

Cyclohexene epoxides are preferentially reduced by an axial approach of the nucleophile.[111]

$(CH_3)_3C$–cyclohexene epoxide (H, H, O) $\xrightarrow{LiAlH_4}$ $(CH_3)_3C$–cyclohexanol (H, OH)

$(CH_3)_3C$–cyclohexene epoxide (O, H) $\xrightarrow{LiAlH_4}$ $(CH_3)_3C$–cyclohexane (H, H)

Lithium triethylborohydride is a superior reagent for reduction of epoxides that are relatively unreactive or prone to rearrangement.[112]

Alkynes are reduced to *E*-alkenes by $LiAlH_4$.[113] This stereochemistry is complementary to that of partial hydrogenation, which gives *Z*-isomers. Alkyne reduction by $LiAlH_4$ is greatly accelerated by a nearby hydroxyl group. Typically, propargylic alcohols react in ether or THF over a period of several hours,[114] whereas forcing conditions are required for isolated triple bonds.[115] (Compare entries 8 and 9 in Scheme 5.5.) This is presumably the result of coordination of the hydroxyl group at aluminum and formation of cyclic intermediate. The involvement of intramolecular Al–H addition has been demonstrated by use of $LiAlD_4$ as the reductant. When reduction by $LiAlD_4$ is followed by quenching with normal water, propargylic alcohol gives 3-^{2}H-prop-2-enol. Quenching

109. E. C. Ashby and J. J. Lin, *J. Org. Chem.* **43:**1263 (1978).
110. S. Masamune, G. S. Bates, and P. E. Georghiou, *J. Am. Chem. Soc.* **96:**3686 (1974); E. C. Ashby, J. J. Lin, and A. B. Goel, *J. Org. Chem.* **43:**183 (1978).
111. B. Rickborn and J. Quartucci, *J. Org. Chem.* **29:**3185 (1964); B. Rickborn and W. E. Lamke II, *J. Org. Chem.* **32:**537 (1967); D. K. Murphy, R. L. Alumbaugh, and B. Rickborn, *J. Am. Chem. Soc.* **91:**2649 (1969).
112. H. C. Brown, S. C. Kim, and S. Krishnamurthy, *J. Org. Chem.* **45:**1 (1980); H. C. Brown, S. Narasimhan, and V. Somayaji, *J. Org. Chem.* **48:**3091 (1983).
113. E. F. Magoon and L. H. Slaugh, *Tetrahedron* **23:**4509 (1967).
114. N. A. Porter, C. B. Ziegler, Jr., F. F. Khouri, and D. H. Roberts, *J. Org. Chem.* **50:**2252 (1985).
115. H. C. Huang, J. K. Rehmann, and G. R. Gray, *J. Org. Chem.* **47:**4018 (1982).

Scheme 5.5. Reduction of Other Functional Groups by Hydride Donors

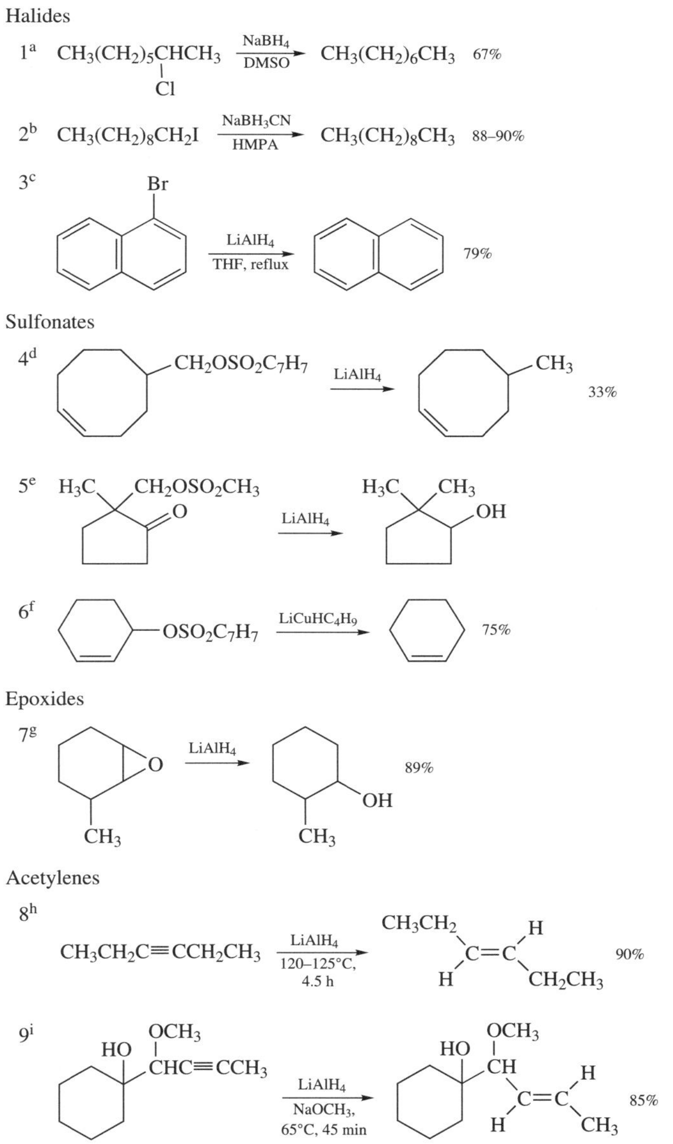

Halides

1[a] $CH_3(CH_2)_5CH(Cl)CH_3 \xrightarrow[\text{DMSO}]{NaBH_4} CH_3(CH_2)_6CH_3$ 67%

2[b] $CH_3(CH_2)_8CH_2I \xrightarrow[\text{HMPA}]{NaBH_3CN} CH_3(CH_2)_8CH_3$ 88–90%

3[c] 1-Bromonaphthalene $\xrightarrow[\text{THF, reflux}]{LiAlH_4}$ naphthalene 79%

Sulfonates

4[d] Cyclooctenyl-$CH_2OSO_2C_7H_7$ $\xrightarrow{LiAlH_4}$ cyclooctenyl-CH_3 33%

5[e] 2-($CH_2OSO_2CH_3$)-2-(H_3C)cyclopentanone $\xrightarrow{LiAlH_4}$ 2-(CH_3)-2-(H_3C)cyclopentanol (OH)

6[f] Cyclohexenyl-$OSO_2C_7H_7$ $\xrightarrow{LiCuHC_4H_9}$ cyclohexene 75%

Epoxides

7[g] Methylcyclohexene oxide (CH_3, O) $\xrightarrow{LiAlH_4}$ 2-methylcyclohexanol (CH_3, OH) 89%

Acetylenes

8[h] $CH_3CH_2C{\equiv}CCH_2CH_3 \xrightarrow[\text{120–125°C, 4.5 h}]{LiAlH_4}$ $(CH_3CH_2)(H)C{=}C(H)(CH_2CH_3)$ 90%

9[i] 1-[$CH(OCH_3)C{\equiv}CCH_3$]cyclohexanol (HO) $\xrightarrow[\text{NaOCH}_3\text{, 65°C, 45 min}]{LiAlH_4}$ 1-[$CH(OCH_3)CH{=}CHCH_3$]cyclohexanol (HO, H, H, CH_3) 85%

a. R. O. Hutchins, D. Hoke, J. Keogh, and D. Koharski, *Tetrahedron Lett.* **1969:**3495; H. M. Bell, C. W. Vanderslice, and A. Spehar, *J. Org. Chem.* **34:**3923 (1969).
b. R. O. Hutchins, C. A. Milewski, and B. E. Maryanoff, *Org. Synth.* **53:**107 (1973).
c. H. C. Brown and S. Krishnamurthy, *J. Org. Chem.* **34:**3918 (1969).
d. A. C. Cope and G. L. Woo, *J. Am. Chem. Soc.* **85:**3601 (1963).
e. A. Eschenmoser and A. Frey, *Helv. Chim. Acta* **35:**1660 (1952).
f. S. Masamune, G. S. Bates, and P. E. Geoghiou, *J. Am. Chem. Soc.* **96:**3686 (1974).
g. B. Rickborn and W. E. Lamke II, *J. Org. Chem.* **32:**537 (1967).
h. E. F. Magoon and L. H. Slaugh, *Tetrahedron* **23:**4509 (1967).
i. D. A. Evans and J. V. Nelson, *J. Am. Chem. Soc.* **102:**774 (1980).

with D_2O gives 2-^{2}H-3-^{2}H-prop-2-enol indicating overall *anti* addition.[116]

$$D_3\bar{A}lOCH_2C{\equiv}CH \longrightarrow \text{(cyclic D}_2\text{Al}^- \text{ vinyl intermediate)} \xrightarrow{D_2O} HOCH_2(D)C{=}C(D)H$$
$$\xrightarrow{H_2O} HOCH_2(D)C{=}CH_2$$

The efficiency and stereospecificity of reduction are improved by using a 1 : 2 $LiAlH_4$–$NaOCH_3$ mixture as the reducing agent.[117] The mechanistic basis of this effect has not been explored in detail.

5.3. Group IV Hydride Donors

Both Si–H and C–H compounds can function as hydride donors under certain circumstances. The silicon–hydrogen bond is capable of transferring a hydride to carbocations. Alcohols that can be ionized in trifluoroacetic acid are reduced to hydrocarbons in the presence of a silane.

$$\xrightarrow[CF_3CO_2H]{Ph_3SiH} \quad 92\%$$

Ref. 118

Aromatic aldehydes and ketones are reduced to alkylaromatics.[119]

$$ArC(=O)R + H^+ \rightleftharpoons Ar\overset{+}{C}R(OH) \xrightarrow{R_3SiH} ArCHR(OH) \rightleftharpoons Ar\overset{+}{C}HR + H_2O$$

$$Ar\overset{+}{C}HR + R_3SiH \longrightarrow ArCH_2R$$

Aliphatic ketones can be reduced to hydrocarbons by triethylsilane and gaseous BF_3.[120] The BF_3 is a sufficiently strong Lewis acid to promote formation of a carbocation from the

116. J. E. Baldwin and K. A. Black, *J. Org. Chem.* **48:**2778 (1983).
117. E. J. Corey, J. A. Katzenellenbogen, and G. H. Posner, *J. Am. Chem. Soc.* **89:**4245 (1967); B. B. Molloy and K. L. Hauser, *J. Chem. Soc., Chem. Commun.* **1968:**1017.
118. F. A. Carey and H. S. Tremper, *J. Org. Chem.* **36:**758 (1971).
119. C. T. West, S. J. Donnelly, D. A. Kooistra, and M. P. Doyle, *J. Org. Chem.* **38:**2675 (1973); M. P. Doyle, D. J. DeBruyn, and D. A. Kooistra, *J. Am. Chem. Soc.* **94:**3659 (1972); M. P. Doyle and C. T. West, *J. Org. Chem.* **40:**3821 (1975).
120. J. L. Frey, M. Orfanopoulos, M. G. Adlington, W. R. Dittman, Jr., and S. B. Silverman, *J. Org. Chem.* **43:**374 (1978).

intermediate alcohol.

$$R\overset{\overset{+}{O}-\bar{B}F_3}{\overset{\|}{C}}R \xrightarrow{Et_3SiH} R-\underset{H}{\overset{O\bar{B}F_3}{C}}-R \longrightarrow \underset{H}{\overset{R\quad R}{\overset{+}{C}}} \xrightarrow{Et_3SiH} RCH_2R$$

Aryl ketones have also been reduced with triethylsilane and $TiCl_4$. This method was used to prepare γ-aryl amino acids.[121]

$$ArCCH_2\underset{CO_2H}{C}HNHCO_2CH_3 \;(\text{C=O}) \xrightarrow[2)\,(C_2H_5)_3SiH,\ TiCl_4]{1)\,TMSCl,\ Et_3N} ArCH_2CH_2\underset{CO_2H}{C}HNHCO_2CH_3$$

All of these reactions involve formation of oxonium and carbocation intermediates that can abstract hydride from the silane donor.

There is also a group of reactions in which hydride is transferred from carbon. The carbon–hydrogen bond has little intrinsic tendency to act as a hydride donor so especially favorable circumstances are required to observe this reactivity. Frequently, these reactions proceed through a cyclic transition state in which a new C–H bond is formed simultaneously with the C–H cleavage. Hydride transfer is facilitated by high electron density at the carbon atom. Aluminum alkoxides catalyze transfer of hydride from an alcohol to a ketone. This is generally an equlibrium process, and the reaction can be driven to completion if the ketone is removed from the system by distillation, for example. This process is called the Meerwein–Pondorff–Verley reduction.[122]

$$3\,R_2C{=}O + Al[OCH(CH_3)_2]_3 \longrightarrow [R_2CHO]_3Al + 3\,CH_3\underset{O}{\underset{\|}{C}}CH_3$$

The reaction proceeds via a cyclic transition state involving coordination of both the alcohol and ketone oxygens to the aluminum. Hydride donation usually takes place from the less hindered face of the carbonyl group.[123]

$$\text{(cyclic transition state: Al coordinated to O of } (H_3C)_2C(H){-}O \text{ and O}{=}CR_2\text{; H transferred from C to C)}$$

Certain lanthanide alkoxides, such as t-BuOSmI_2, have also been found to catalyze hydride exchange between alcohols and ketones.[124] Isopropanol can serve as the reducing agent for aldehydes and ketones that are thermodynamically better hydride acceptors than

121. M. Yato, K. Homma, and A. Ishida, *Heterocycles* **49**:233 (1998).
122. A. L. Wilds, *Org. React.* **2**:178 (1944); C. F. de Graauw, J. A. Peters, H. van Bekkum and J. Huskens, *Synthesis*, 1007 (1994).
123. F. Nerdel, D. Frank, and G. Barth, *Chem. Ber.* **102**:395 (1969).
124. J. L. Namy, J. Souppe, J. Collins, and H. B. Kagan, *J. Org. Chem.* **49**:2045 (1984).

acetone.

$$O_2N\text{-}C_6H_4\text{-}CH{=}O \xrightarrow[t\text{-BuOSmI}_2]{CH_3CH(OH)CH_3} O_2N\text{-}C_6H_4\text{-}CH_2OH \quad 94\%$$

Like the Meerwein–Pondorff–Verley reduction, these reactions are believed to proceed under thermodynamic control, and the more stable stereoisomer is the main product.[125]

Another reduction process, catalysed by iridium chloride and characterized by very high axial : equatorial product ratios for cyclohexanones, apparently involves hydride transfer from isopropanol.[126]

$$(CH_3)_3C\text{-cyclohexanone} \xrightarrow[(CH_3)_2CHOH]{IrCl_4,\ HCl,\ (CH_3O)_3P,\ H_2O} (CH_3)_3C\text{-cyclohexanol (axial OH)}$$

Formic acid can also act as a donor of hydrogen. The driving force in this case is the formation of carbon dioxide.

$$RNH_2 + CH_2{=}O + HCO_2H \longrightarrow RN(CH_3)_2 + CO_2$$

A useful application is the Clark–Eschweiler reductive alkylation of amines. Heating a primary or secondary amine with formaldehyde and formic acid results in complete methylation to the tertiary amine.[127] The hydride acceptor is the iminium ion resulting from condensation of the amine with formaldehyde.

$$R_2\overset{+}{N}{=}CH_2 \quad H\text{-}O\text{-}C(H){=}O$$

5.4. Hydrogen-Atom Donors

Reduction by hydrogen-atom donors involves free-radical intermediates. Tri-*n*-butyltin hydride is the most prominent example of this type of reducing agent. It is able to reductively replace halogen by hydrogen in organic compounds. Mechanistic studies have indicated a free-radical chain mechanism.[128] The order of reactivity for the haldes is RI > RBr > RCl > RF, which reflects the relative ease of the halogen-atom abstraction.[129]

$$In\cdot + Bu_3SnH \longrightarrow In{-}H + Bu_3Sn\cdot \quad (In\cdot = \text{initiator})$$
$$Bu_3Sn\cdot + R{-}X \longrightarrow R\cdot + Bu_3SnX$$
$$R\cdot + Bu_3SnH \longrightarrow RH + Bu_3Sn\cdot$$

Tri-*n*-butyltin hydride shows substantial selectivity toward polyhalogenated compounds, permitting partial dehalogenation. The reason for the greater reactivity of

125. D. A. Evans, S. W. Kaldor, T. K. Jones, J. Clardy, and T. J. Stout, *J. Am. Chem. Soc.* **112:**7001 (1990).
126. E. L. Eliel, T. W. Doyle, R. O. Hutchins, and E. C. Gilbert, *Org. Synth.* **50:**13 (1970).
127. M. L. Moore, *Org. React.* **5:**301 (1949); S. H. Pine and B. L. Sanchez, *J. Org. Chem.* **36:**829 (1971).
128. L. W. Menapace and H. G. Kuivila, *J. Am. Chem. Soc.* **86:**3047 (1964).
129. H. G. Kuivila and L. W. Menapace, *J. Org. Chem.* **28:**2165 (1963).

Scheme 5.6. Dehalogenations with Stannanes

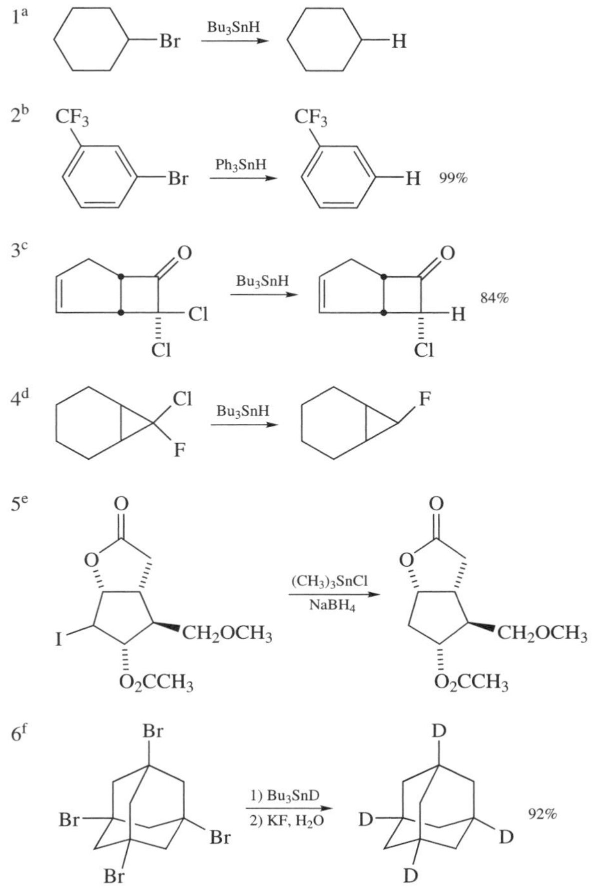

1[a] Bromocyclohexane $\xrightarrow{Bu_3SnH}$ cyclohexane (H)

2[b] 3-Bromobenzotrifluoride (CF_3, Br) $\xrightarrow{Ph_3SnH}$ benzotrifluoride (H) 99%

3[c] Dichlorobicyclic ketone (Cl, Cl, O) $\xrightarrow{Bu_3SnH}$ monochloro ketone (H, Cl, O) 84%

4[d] Chlorofluorobicycloheptane (Cl, F) $\xrightarrow{Bu_3SnH}$ fluorobicycloheptane (F)

5[e] Iodolactone (I, CH_2OCH_3, O_2CCH_3) $\xrightarrow[NaBH_4]{(CH_3)_3SnCl}$ lactone (CH_2OCH_3, O_2CCH_3)

6[f] Tetrabromoadamantane (Br, Br, Br, Br) $\xrightarrow[2)\ KF,\ H_2O]{1)\ Bu_3SnD}$ tetradeuterioadamantane (D, D, D, D) 92%

a. H. G. Kuivila, L. W. Menapace, and C. R. Warner, *J. Am. Chem. Soc.* **84:**3584 (1962).
b. D. H. Lorenz, P. Shapiro, A. Stern, and E. I. Becker, *J. Org. Chem.* **28:**2332 (1963).
c. W. T. Brady and E. F. Hoff, Jr. *J. Org. Chem.* **35:**3733 (1970).
d. T. Ando, F. Namigata, H. Yamanaka, and W. Funasaka, *J. Am. Chem. Soc.* **89:**5719 (1967).
e. E. J. Corey and J. W. Suggs, *J. Org. Chem.* **40:**2554 (1975).
f. J. E. Leibner and J. Jacobson, *J. Org. Chem.* **44:**449 (1979).

more highly halogented carbons toward reduction lies in the stabilizing effect that the remaining halogen has on the radical intermediate. This selectivity has been used, for example, to reduce dihalocyclopropanes to monohalocyclopropanes as in entry 4 of Scheme 5.6. A procedure which is catalytic in Bu_3SnH and uses $NaBH_4$ as the stoichiometric reagent has been developed.[130] This procedure has advantages in the isolation and purification of product. Entry 5 in Scheme 5.6 is an example of this procedure.

130. E. J. Corey and J. W. Suggs, *J. Org. Chem.* **40:**2554 (1975).

Tri-*n*-butyltin hydride also serves as a hydrogen-atom donor in radical-mediated methods for reductive deoxygenation of alcohols.[131] The alcohol is converted to a thiocarbonyl derivative. These thioesters undergo a radical reaction with tri-*n*-butyltin hydride.

$$R{-}O\overset{S}{\overset{\|}{C}}X + Bu_3Sn\cdot \longrightarrow RO\overset{S-SnBu_3}{\overset{|}{\dot{C}}}X \longrightarrow R\times + X\overset{O}{\overset{\|}{C}}S{-}SnBu_3$$

$$R\times + Bu_3SnH \longrightarrow R{-}H + Bu_3Sn\cdot$$

This procedure gives good yields from secondary alcohols and, by appropriate adjustment of conditions, can also be adapted to primary alcohols.[132] Scheme 5.7 illustrates some of the conditions which have been developed for the reductive deoxygenation of alcohols.

Because of the expense, toxicity, and purification problems associated with use of stoichiometric amounts of tin hydrides, there has been interest in finding other hydrogen-atom donors. The trialkylboron–oxygen system for radical initiation has been used with tris(trimethylsily)silane or diphenylsilane as a hydrogen-donor system.[133]

$$c\text{-}C_{12}H_{23}O\overset{S}{\overset{\|}{C}}O{-}C_6H_4{-}F \xrightarrow[(Ph)_2SiH_2]{Et_3B,\ O_2} c\text{-}C_{12}H_{24} \quad 96\%$$

$$(C_2H_5)_3B + O_2 \longrightarrow C_2H_5\cdot$$

$$C_2H_5\cdot + R_3SiH \longrightarrow C_2H_6 + R_3Si\cdot$$

$$R_3Si\cdot + R'O\underset{S}{\underset{\|}{C}}OR' \longrightarrow R'O\underset{SSiR_3}{\underset{|}{\dot{C}}}OR'$$

$$R'O\underset{SSiR_3}{\underset{|}{\dot{C}}}OR' \longrightarrow R'\cdot + RO_2CSSiR_3$$

$$R'\cdot + R_3SiH \longrightarrow R_3Si\cdot$$

The alcohol derivatives that have been successfully deoxygenated include thiocarbonates and xanthates.[134] Peroxides can also be used as initiators.[135] Dialkyl phosphites can also be used as hydrogen donors.[136] (see Entry 4, Scheme 5.7)

5.5. Dissolving-Metal Reductions

Another group of synthetically useful reductions employs a metal as the reducing agent. The organic substrate under these conditions accepts one or more electrons from the metal. The subsequent course of the reaction depends on the structure of the reactant and reaction conditions. Three broad classes of reactions can be recognized, and these will be discussed separately. These include reactions in which the overall change involves (a) net

131. D. H. R. Barton and S. W. McCombie, *J. Chem. Soc., Perkin I Trans, 1* **1975:**1574; for reviews of this method, see W. Hartwig, *Tetrahedron* **39:**2609 (1983); D. Crich and L. Quintero, *Chem. Rev.* **89:**1413 (1989).
132. D. H. R. Barton, W. B. Motherwell, and A. Stange, *Synthesis* **1981:**743.
133. D. H. R. Barton, D. O. Jang, and J. C. Jaszberenyi, *Tetrahedron Lett.* **31:**4681 (1990).
134. J. N. Kirwan, B. P. Roberts, and C. R. Willis, *Tetrahedron Lett.* **31:**5093 (1990).
135. D. H. Barton, D. O. Jang, and J. C. Jaszberenyi, *Tetrahedron Lett.* **32:**7187 (1991).
136. D. H. R. Barton, D. O. Jang, and J. C. Jaszberenyi, *Tetrahedron Lett.* **33:**2311 (1992).

1[a]

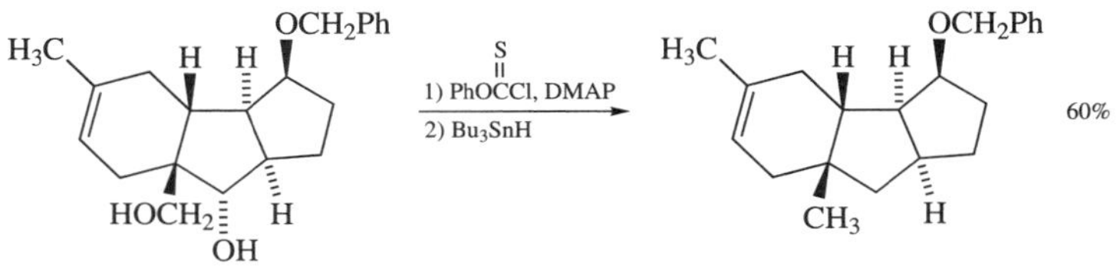

2[b]

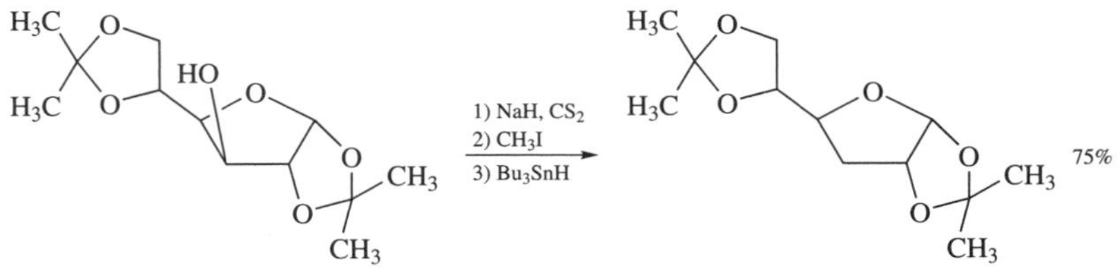

3[c]

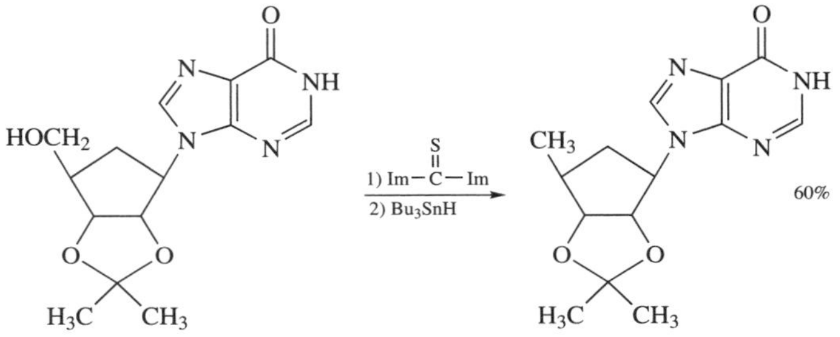

4[d]

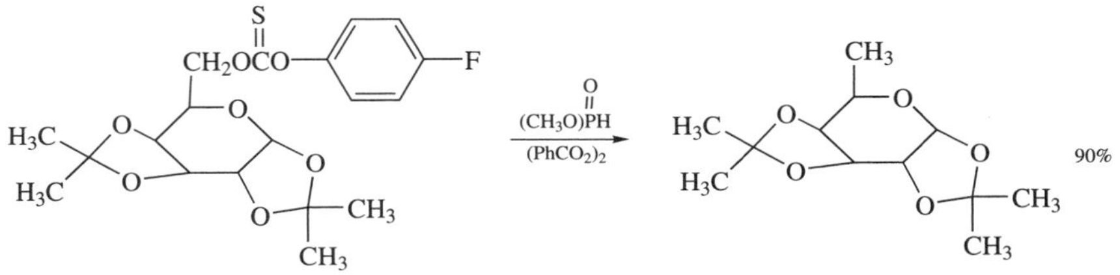

5[e]

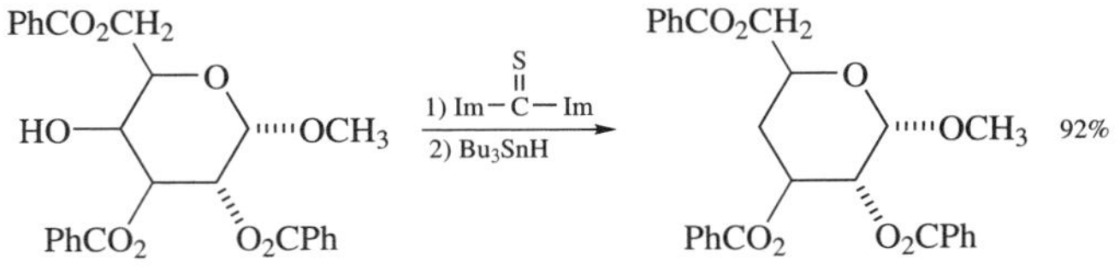

6[f]

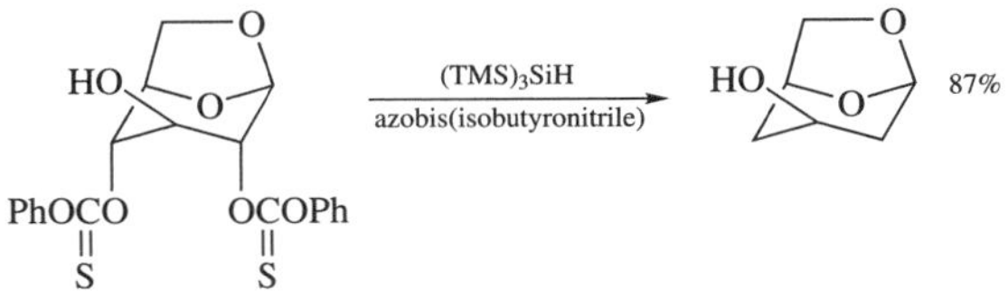

a. H. J. Liu and M. G. Kulkarni, *Tetrahedron Lett.* **26:**4847 (1985).
b. S. Iacono and J. R. Rasmussen, *Org. Synth.* **64:**57 (1985).
c. O. Miyashita, F. Kasahara, T. Kusaka, and R. Marumoto, *J. Antibiot.* **38:**981 (1985).
d. D. H. R. Barton, D. O. Jang, and J. C. Jaszberenyi, *Tetrahedron Lett.* **33:**2311 (1992).
e. J. R. Rasmussen, C. J. Slinger, R. J. Kordish, and D. D. Newman-Evans, *J. Org. Chem.* **46:**4843 (1981).
f. D. H. R. Barton, D. O. Jang, and J. C. Jaszberenyi, *Tetrahedron Lett.* **33:**6629 (1992).

addition of hydrogen, (b) reductive removal of a functional group, and (c) formation of carbon–carbon bonds.

5.5.1. Addition of Hydrogen

Although the method has been supplanted for synthetic purposes by the use of hydride donors, the reduction of ketones to alcohols by alkali metals in ammonia or alcohols provides some mechanistic insight into dissolving-metal reductions. The outcome of the reaction of ketones with metal reductants is determined by the fate of the initial ketyl intermediate formed by a single-electron transfer. The intermediate, depending on its structure and the reaction medium, may be protonated, disproportionate, or dimerize.[137] In hydroxylic solvents such as liquid ammonia or in the presence of an alcohol, the protonation process dominates over dimerization. As will be discussed in Section 5.5.3, dimerization may become the dominant process under other conditions.

$$RCH_2-C(=O)-R' \xrightarrow{e^-} RCH_2-\dot{C}(O^-)-R' \text{ (ketyl)}$$

$$\xrightarrow[S-H]{\text{protonation}} RCH_2-\dot{C}(OH)-R' \xrightarrow[S-H]{e^-} RCH_2C(O^-)(H)-R'$$

$$\xrightarrow{\text{dimerization}} RCH_2-C(O^-)(R')-C(O^-)(R')-CH_2R$$

$$\xrightarrow{\text{disproportionation}} RCH_2-CH(O^-)R' + RCH{=}C(O^-)R' \;\xrightarrow[SH]{\text{protonation}}\; RCH_2-CH(OH)R'$$

α,β-Unsaturated carbonyl compounds are cleanly reduced to the enolate of the corresponding saturated ketone on reduction with lithium in ammonia.[138] Usually, an alcohol is added to the reduction solution to serve as the proton source.

$$R_2C{=}CH-C(=O)-R \xrightarrow{e^-} R_2\dot{C}-CH{=}C(O^-)-R \xrightarrow[S-H]{e^-} R_2CH-CH{=}C(O^-)-R$$

As mentioned in Chapter 1, this is one of the best methods for generating a specific enolate of a ketone. The enolate generated by conjugate reduction can undergo the characteristic alkylation and addition reactions which were discussed in Chapters 1 and 2. When this is the objective of the reduction, it is important to use only one equivalent of the proton donor. Ammonia, being a weaker acid than an aliphatic ketone, does not protonate the enolate, and it remains available for reaction. If the saturated ketone is the desired product, the enolate is protonated either by use of excess proton donor during the reduction or on

137. V. Rautenstrauch and M. Geoffroy, *J. Am. Chem. Soc.* **99:**6280 (1977); J. W. Huffman and W. W. McWhorter, *J. Org. Chem.* **44:**594 (1979); J. W. Huffman, P. C. Desai, and J. E. LaPrade, *J. Org. Chem.* **48:**1474 (1983).
138. D. Cain, *Org. React.* **23:**1 (1976).

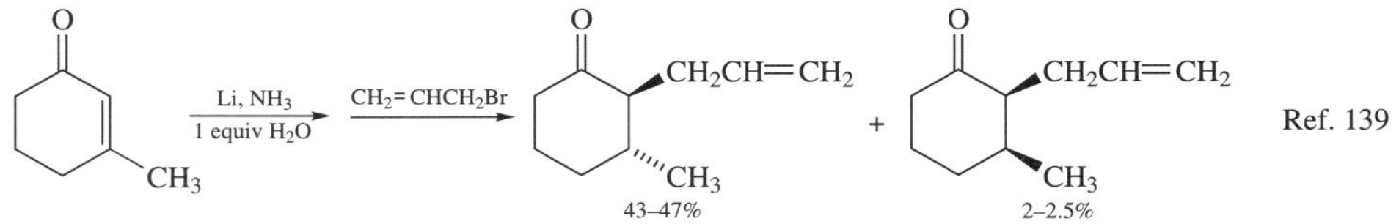

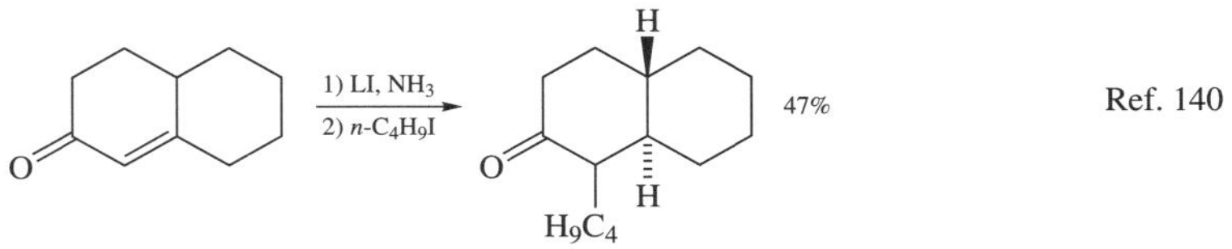

Ref. 139

Ref. 140

The stereochemistry of conjugate reduction is established by the proton transfer to the β carbon. In the well-studied case of $\Delta^{1,9}$-2-octalones, the ring junction is usually *trans*.[141]

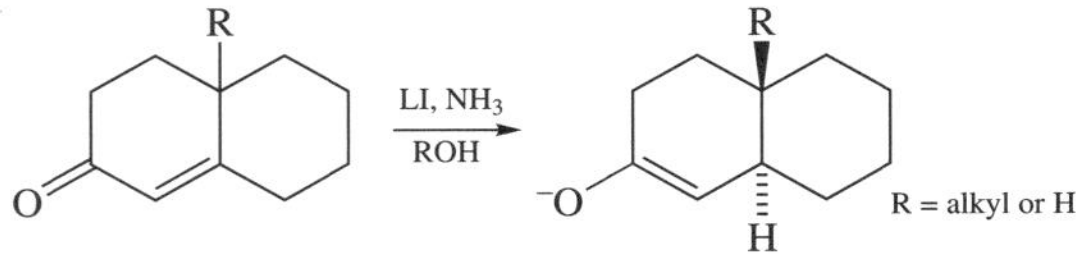

The stereochemistry is controlled by a stereoelectronic preference for protonation perpendicular to the enolate system, and, given that this requirement is met, the stereochemistry will normally correspond to protonation of the most stable conformation of the dianion intermediate from its least hindered side.

Dissolving-metal systems constitute the most general method for partial reduction of aromatic rings. The reaction is called the *Birch reduction*.[142] The usual reducing medium is lithium or sodium in liquid ammonia. The reaction occurs by two successive electron-transfer/protonation steps.

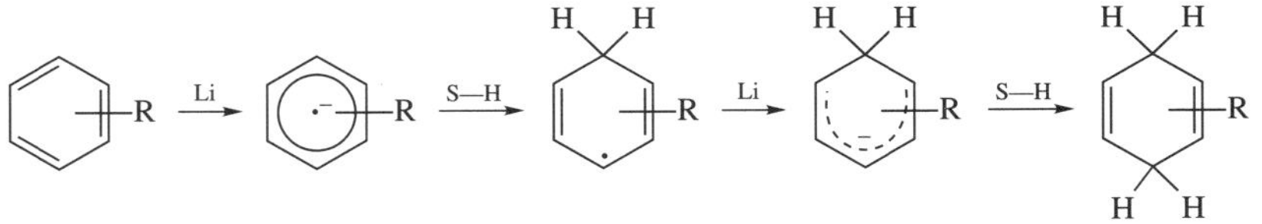

The isolated double bonds in the dihydro product are much less easily reduced than the conjugated ring, so the reduction stops at the dihydro stage. Alkyl and alkoxy aromatics, phenols, and benzoate anions are the most useful reactants for Birch reduction. In aromatic ketones and nitro compounds, the substituents are reduced in preference to the aromatic ring. Substituents also govern the position of protonation, Alkyl and alkoxy aromatics

139. D. Caine, S. T. Chao, and H. A. Smith, *Org. Synth.* **56:**52 (1977).
140. G. Stork, P. Rosen, and N. L. Goldman, *J. Am. Chem. Soc.* **83:**2965 (1961).
141. G. Stork, P. Rosen, N. Goldman, R. V. Coombs, and J. Tsuji, *J. Am. Chem. Soc.* **87:**275 (1965); M. J. T. Robinson, *Tetrahedron* **21:**2475 (1965).
142. A. J. Birch and G. Subba Rao, *Adv. Org. Chem.* **8:**1 (1972); R. G. Harvey, *Synthesis* **1980:**161; J. M. Hook and L. N. Mander, *Nat. Prod. Rep.* **3:**35 (1986); P. W. Rabideau, *Tetrahedron* **45:**1599 (1989); A. J. Birch, *Pure Appl. Chem.* **68:**553 (1996).

normally give the 2,5-dihydro derivative. Benzoate anions give 1,4-dihydro derivatives.

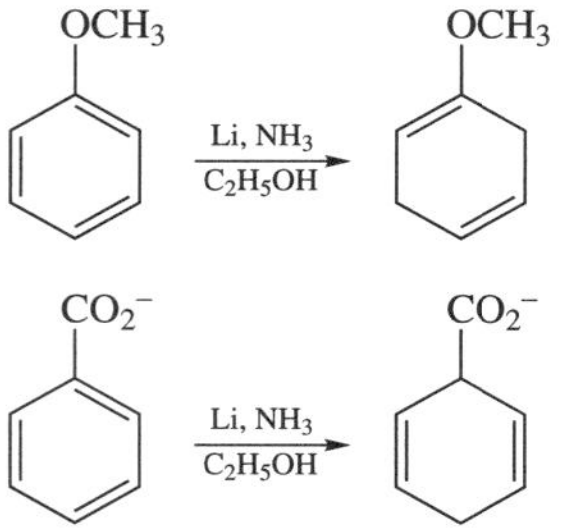

The structure of the products is determined by the site of protonation of the radical-anion intermediate formed after the first electron-transfer step. In general, electron-releasing substituents favor protonation at the *ortho* position, whereas electron-attracting groups favor protonation at the *para* position.[143] Addition of a second electron gives a pentadienyl anion, which is protonated at the center carbon. As a result, 2,5-dihydro products are formed with alkyl or alkoxy substituents, and 1,4-products are formed from aromatics with electron-attracting substituents. The preference for protonation of the central carbon of the pentadienyl anion is believed to be the result of the greater 1,2 and 4,5 bond order and a higher concentration of negative charge at the 3-carbon.[144] The reduction of methoxybenzenes is of importance in the synthesis of cyclohexenones via hydrolysis of the intermediate enol ethers:

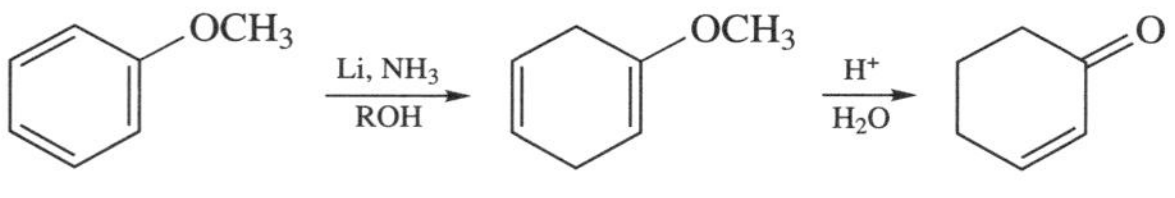

The anionic intermediates formed in Birch reductions can be used in tandem reactions.

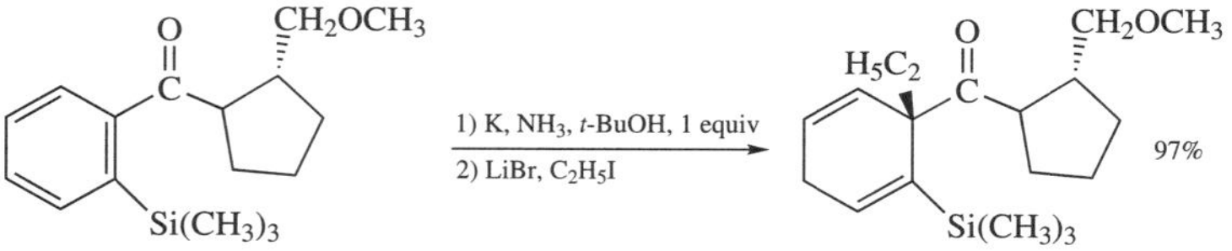

Ref. 145

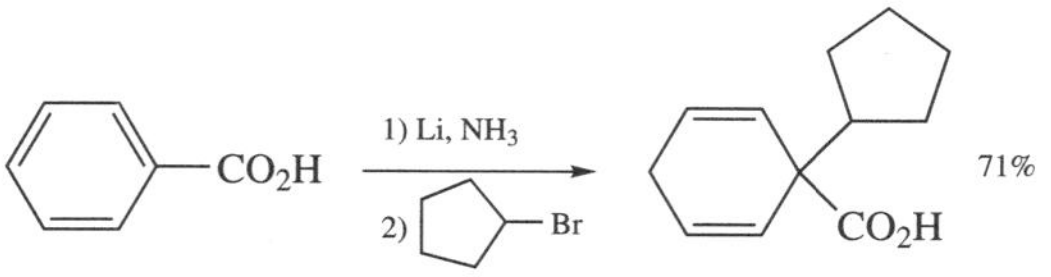

Ref. 146

Scheme 5.8 lists some examples of the use of the Birch reduction.

143. A. J. Birch, A. L. Hinde, and L. Radom, *J. Am. Chem. Soc.* **102:**2370 (1980); H. E. Zimmerman and P. A. Wang, *J. Am. Chem. Soc.* **112:**1280 (1990).
144. P. W. Rabideau and D. L. Huser, *J. Org. Chem.* **48:**4266 (1983); H. E. Zimmerman and P. A. Wang, *J. Am. Chem. Soc.* **115:**2205 (1993).
145. P. A. Baguley and J. C. Walton, *J. Chem. Soc., Perkin Trans. 1* **1998:**2073.
146. A. G. Schultz and L. Pettus, *J. Org. Chem.* **62:**6855 (1997).

Scheme 5.8. Birch Reduction of Aromatic Rings

1[a] 4-tert-butylanisole $\xrightarrow{\text{Li, NH}_3}$ 1-methoxy-4-tert-butyl-1,4-cyclohexadiene 63%

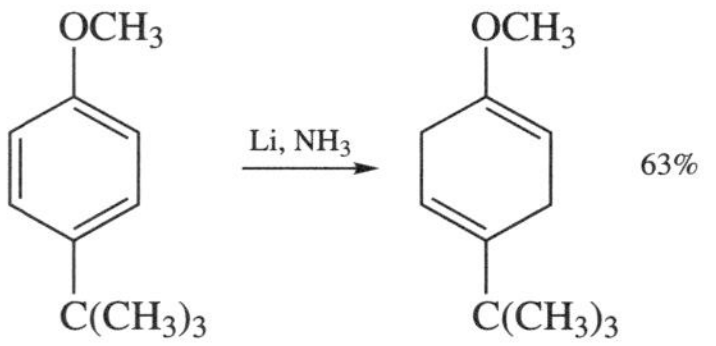

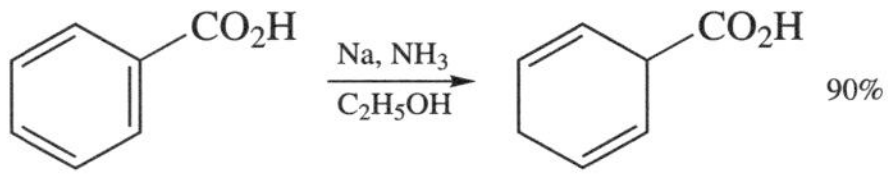

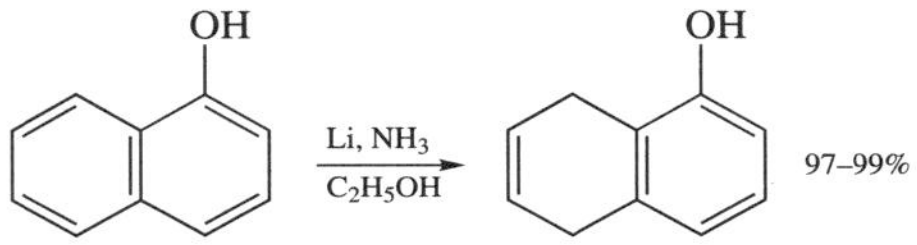

2[b] 1,4-di-tert-butylbenzene $\xrightarrow[\text{C}_2\text{H}_5\text{NH}_2]{\text{Li}}$ 1,4-di-tert-butyl-1,4-cyclohexadiene 56%

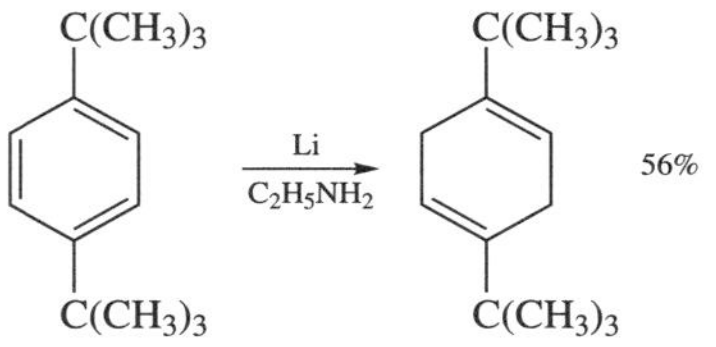

3[c] 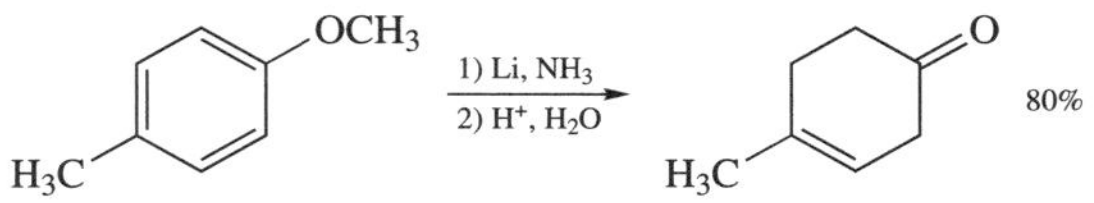 4-methylanisole $\xrightarrow[\text{2) H}^+\text{, H}_2\text{O}]{\text{1) Li, NH}_3}$ 4-methylcyclohex-3-enone 80%

4[d] $C_6H_5CO_2H$ $\xrightarrow[\text{C}_2\text{H}_5\text{OH}]{\text{Na, NH}_3}$ 2,5-cyclohexadiene-1-carboxylic acid 90%

5[e] 1-naphthol $\xrightarrow[\text{C}_2\text{H}_5\text{OH}]{\text{Li, NH}_3}$ 5,6,7,8-tetrahydro-1-naphthol 97–99%

6[f] 2-ethoxynaphthalene $\xrightarrow[\text{C}_2\text{H}_5\text{OH}]{\text{Na}}$ 2-ethoxy-3,4-dihydronaphthalene

a. D. A. Bolton, *J. Org. Chem.* **35:**715 (1970).
b. H. Kwart and R. A. Conley, *J. Org. Chem.* **38:**2011 (1973).
c. E. A. Braude, A. A. Webb, and M. U. S. Sultanbawa, *J. Chem. Soc.* **1958:**3328;
W. C. Agosta and W. L. Schreiber, *J. Am. Chem. Soc.* **93:**3947 (1971).
d. M. E. Kuehne and B. F. Lambert, *Org. Synth.* **V:**400 (1973).
e. C. D. Gutsche and H. H. Peter, *Org. Synth.* **IV:**887 (1963).
f. M. D. Soffer, M. P. Bellis, H. E. Gellerson, and R. A. Stewart, *Org. Synth.* **IV:**903 (1963).

Reduction of alkynes with sodium in ammonia,[147] lithium in low-molecular-weight amines,[148] or sodium in hexamethylphosphoric triamide containing *t*-butanol as a proton source[149] leads to the corresponding *E*-alkene. The reaction is assumed to involve successive electron-transfer and proton-transfer steps.

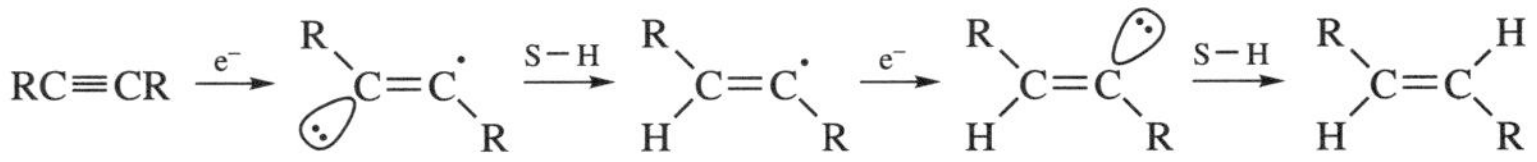

$$RC{\equiv}CR \xrightarrow{e^-} R(\ddot{\ })C{=}\dot{C}R \xrightarrow{S-H} R(H)C{=}\dot{C}R \xrightarrow{e^-} R(H)C{=}C(\ddot{\ })R \xrightarrow{S-H} R(H)C{=}C(H)R$$

147. K. N. Campbell and T. L. Eby, *J. Am. Chem. Soc.* **63:**216, 2683 (1941); A. L. Henne and K. W. Greenlee, *J. Am. Chem. Soc.* **65:**2020 (1943).
148. R. A. Benkeser, G. Schroll, and D. M. Sauve, *J. Am. Chem. Soc.* **77:**3378 (1955).
149. H. O. House and E. F. Kinloch, *J. Org. Chem.* **39:**747 (1974).

5.5.2. Reductive Removal of Functional Groups

The reductive removal of halogen can be accomplished with lithium or sodium. Tetrahydrofuran containing *t*-butanol is a useful reaction medium. Good results have also been achieved with polyhalogenated compounds by using sodium in ethanol.

Cl Cl Cl Cl Cl Cl O_2CCH_3 $\xrightarrow{Na,\ C_2H_5OH}$ OH 70% Ref. 150

An important synthetic application of this reaction is in dehalogenation of dichloro- and dibromocyclopropanes. The dihalocyclopropanes are accessible via carbene addition reactions (see Section 10.2.3). Reductive dehalogenation can also be used to introduce deuterium at a specific site. Some examples of these types of reactions are given in Scheme 5.9. The mechanism of the reaction presumably involves electron transfer to form a radical anion, which then fragments with loss of a halide ion. The resulting radical is reduced to a carbanion by a second electron transfer and subsequently protonated.

$$\text{R—X} \xrightarrow{e^-} \text{R—}\dot{\text{X}}^- \xrightarrow{-X^-} \text{R}\cdot \xrightarrow{e^-} \bar{\text{R}}: \xrightarrow{\text{S—H}} \text{R—H}$$

Phosphate groups can also be removed by dissolving-metal reduction. Reductive removal of vinyl phosphate groups is one of the better methods for conversion of a carbonyl compound to an alkene.[151] The required vinyl phosphate esters are obtained by phosphorylation of the enolate with diethyl phosphorochloridate or *N*,*N*,*N*′,*N*′-tetramethyldiamidophosphorochloridate.[152]

$$\text{RCH}_2\overset{\text{O}}{\overset{\|}{\text{C}}}\text{R}' \xrightarrow[\text{(X)}_2\text{POCl}]{\text{LiNR}_2} \text{RCH}{=}\overset{\text{OPO(X)}_2}{\overset{|}{\text{C}}}\text{R}' \xrightarrow[t\text{-BuOH}]{\text{Li, RNH}_2} \text{RCH}{=}\text{CHR}'$$

X = OEt or NMe_2

Reductive removal of oxygen from aromatic rings can also be achieved by reductive cleavage of aryl diethyl phosphate esters.

CH_3–C₆H₃(OCH₃)–$OP(O)(OC_2H_5)_2$ $\xrightarrow{K,\ NH_3}$ CH_3–C₆H₄–OCH_3 77% Ref. 153

There are also examples where phosphate esters of saturated alcohols are reductively deoxygenated.[154] Mechanistic studies of the cleavage of aryl dialkyl phosphates have

150. B. V. Lap and M. N. Paddon-Row, *J. Org. Chem.* **44:**4979 (1979).
151. R. E. Ireland and G. Pfister, *Tetrahedron Lett.* **1969:**2145.
152. R. E. Ireland, D. C. Muchmore, and U. Hengartner, *J. Am. Chem. Soc.* **94:**5098 (1972).
153. R. A. Rossi and J. F. Bunnett, *J. Org. Chem.* **38:**2314 (1973).
154. R. R. Muccino and C. Djerassi, *J. Am. Chem. Soc.* **96:**556 (1974).

A. Dehalogenation

1[a] PhCl → (Mg, i-PrOH) → PhH

2[b] CH3O OCH3 / Cl Cl Cl Cl Cl → (Na, t-BuOH, THF) → CH3O OCH3 40%

3[c] Cl Cl Cl Cl Cl Cl → (Na, t-BuOH, THF) → 69%

4[d] Ph, Cl, Cl → (C_2H_5MgBr, Ti(O-i-Pr)$_4$) → Ph, Cl 10%

B. Deoxygenation

5[e] $(CH_3)_2C=$ / O / OP(OC_2H_5)$_2$ / CH_3 / CH_3 / $(CH_3O)_2CH$ → (Li, $C_2H_5NH_2$) → $(CH_3)_2C=$ / H / CH_3 / CH_3 / $(CH_3O)_2CH$

6[f] $(CH_3)_2CH$—C6H4—OH → (ClP(O)(OC_2H_5)$_2$) → (Ti(0)) → $(CH_3)_2CH$—C6H5 92%

7[g] O / OP(OC_2H_5)$_2$ → (Li, NH_3) → 85%

a. D. Bryce-Smith and B. J. Wakefield, *Org. Synth.* **47:**103 (1967).
b. P. G. Gassman and J. L. Marshall, *Org. Synth.* **48:**68 (1968).
c. B. V. Lap and M. N. Paddon-Row, *J. Org. Chem.* **44**:4979 (1979).
d. J. R. Al Dulayymi, M. S. Baird, I. G. Bolesov, V. Tversovsky, and M. Rubin, *Tetrahedron Lett.* **37:**8933 (1996).
e. S. C. Welch and T. A. Valdes, *J. Org. Chem.* **42;**2108 (1977).
f. S. S. Welch and M. E. Walter, *J. Org. Chem.* **43:**4797 (1978).
g. M. R. Detty and L. A. Paquette, *J. Am. Chem. Soc.* **99:**821 (1977).

indicated that the crucial C–O cleavage occurs after transfer of two electrons.[155]

$$\mathrm{ArOP(=O)(OC_2H_5)_2} \xrightarrow{2e^-} [\mathrm{ArOPO(OEt)_2}]^{2-} \longrightarrow \mathrm{Ar^-} + \mathrm{(EtO)_2PO_2^-}$$

For preparative purposes, titanium metal can be used in place of sodium or lithium in liquid ammonia for both the vinyl phosphate[156] and aryl phosphate[157] cleavages. The titanium metal is generated *in situ* from $TiCl_3$ by reduction with potassium metal in THF.

Both metallic zinc and aluminum amalgam are milder reducing agents than the alkali metals. These reductants selectively remove oxygen and sulfur functional groups α to carbonyl groups. The mechanistic picture which seems most generally applicable is a net two-electron reduction with expulsion of the oxygen or sulfur substituent as an anion. The reaction seems to be a concerted process because the isolated functional groups are not reduced under these conditions.

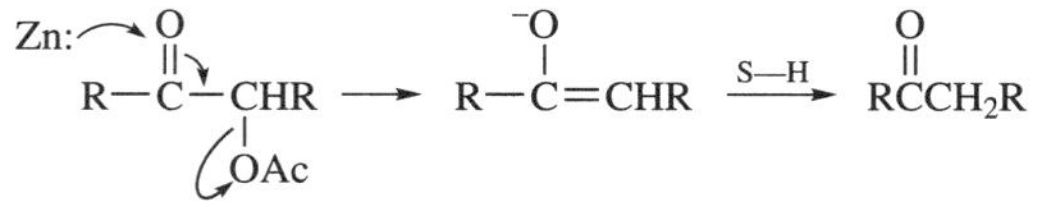

Another useful reagent for reduction of α-acetoxyketones and similar compounds is samarium diiodide.[158] SmI_2 is a strong one-electron reducing agent, and it is believed that the reductive elimination occurs after a net two-electron reduction of the carbonyl group.

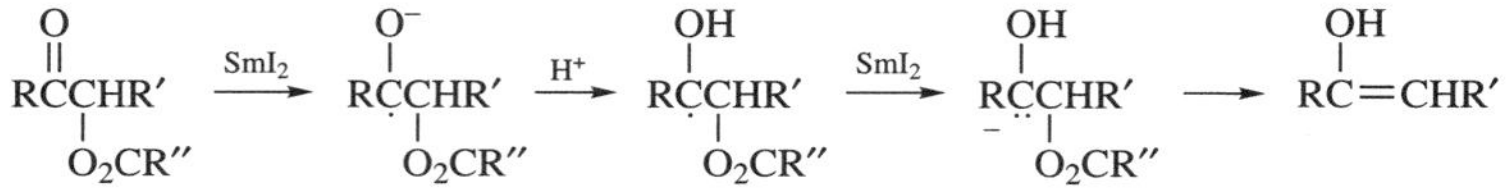

These conditions were used, for example, in the preparation of the anticancer compound 10-deacetoxytaxol.

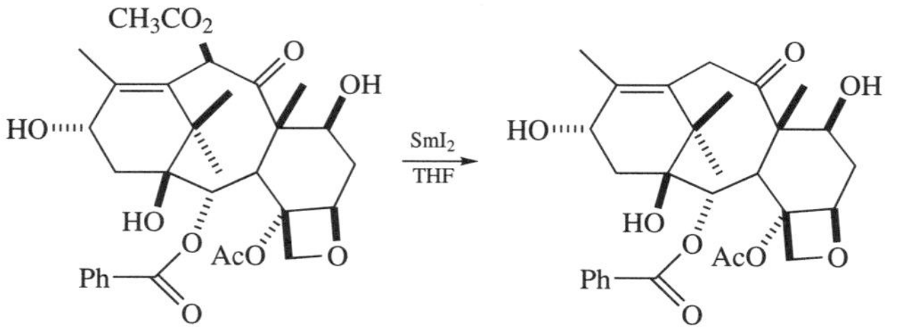

Ref. 159

The reaction is also useful for deacetoxylaton or dehydroxylation of α-oxygenated lactones derived from carbohydrates.[160] (See entires 9 and 10 in Scheme 5.10.) Some other examples of this type of reaction are given in Scheme 5.10. Vinylogous oxygen

155. S. J. Shafer, W. D. Closson, J. M. F. vanDijk, O. Piepers, and H. M. Buck, *J. Am. Chem. Soc.* **99:**5118 (1977).
156. S. C. Welch and M. E. Walters, *J. Org. Chem.* **43:**2715 (1978).
157. S. C. Welch and M. E. Walters, *J. Org. Chem.* **43:**4797 (1978).
158. G. A. Molander and G. Hahn, *J. Org. Chem.* **51:**1135 (1986).
159. R. A. Holton, C. Somoza, and K.-B. Chai, *Tetrahedron Lett.* **35:**1665 (1994).
160. S. Hanessian, C. Girard, and J. L. Chiara, *Tetrahedron Lett.* **33:**573 (1992).

substituents are also subject to reductive elimination by zinc or aluminum amalgam (see entry 8 in Scheme 5.10).

5.5.3. Reductive Carbon–Carbon Bond Formation

Because reductions by metals often occur as one-electron processes, radicals are involved as intermediates. When the reaction conditions are adjusted so that coupling competes favorably with other processes, the formation of a carbon–carbon bond can occur. The reductive coupling of acetone to form 2,3-dimethyl-2,3-butanediol (pinacol) is an example of such a process.

$$(CH_3)_2CO \xrightarrow{\text{Mg–Hg}} (CH_3)_2C(OH)-C(OH)(CH_3)_2 \qquad \text{Ref. 161}$$

Reduced forms of titanium are currently the most versatile and dependable reagents for reductive coupling of carbonyl compounds. Depending on the reagent used, either diols or alkenes can be formed.[162] One reagent for effecting diol formation is a combination of $TiCl_4$ and magnesium amalgam.[163] The active reductant is presumably titanium metal formed by reduction of $TiCl_4$.

Mg–Hg, $TiCl_4$ 95%

$$CH_3COCH_2CH_2COCH_3 \xrightarrow[\text{TiCl}_4]{\text{Mg–Hg}} \text{1,2-dimethylcyclobutane-1,2-diol} \quad 81\%$$

Good yields of pinacols from aromatic aldehydes and ketones are obtained by adding catechol to the $TiCl_3$–Mg reagent prior to the coupling.[164]

$$PhCOCH_3 \xrightarrow[\text{THF, catechol}]{\text{TiCl}_3\text{, Mg}} PhC(OH)(CH_3)-C(OH)(CH_3)Ph \quad 95\%$$

Pinacols are also obtained using $TiCl_3$ in conjunction with Zn–Cu as the reductant.[165] This reagent is capable of forming normal, medium, and large rings with comparable efficiency. The macrocyclization has proven useful in the formation of a number of natural products.[166] (See entry 3 in Scheme 5.11.)

161. R. Adams and E. W. Adams, *Org. Synth.* **I:**448 (1932).
162. J. E. McMurry, *Chem. Rev.* **89:**1513 (1989).
163. E. J. Corey, R. L. Danheiser, and S. Chandrashekaran, *J. Org. Chem.* **41:**260 (1976).
164. N. Balu, S. K. Nayak, and A. Banerji, *J. Am. Chem. Soc.* **118:**5932 (1996).
165. J. E. McMurry and J. G. Rico, *Tetrahedron Lett.* **30:**1169 (1989).
166. J. E. McMurry, J. G. Rico, and Y. Shih, *Tetrahedron Lett.* **30:**1173 (1989); J. E. McMurry and R. G. Dushin, *J. Am. Chem. Soc.* **112:**6942 (1990).

Scheme 5.10. Reductive Removal of Functional Groups from α-Substituted Carbonyl Compounds

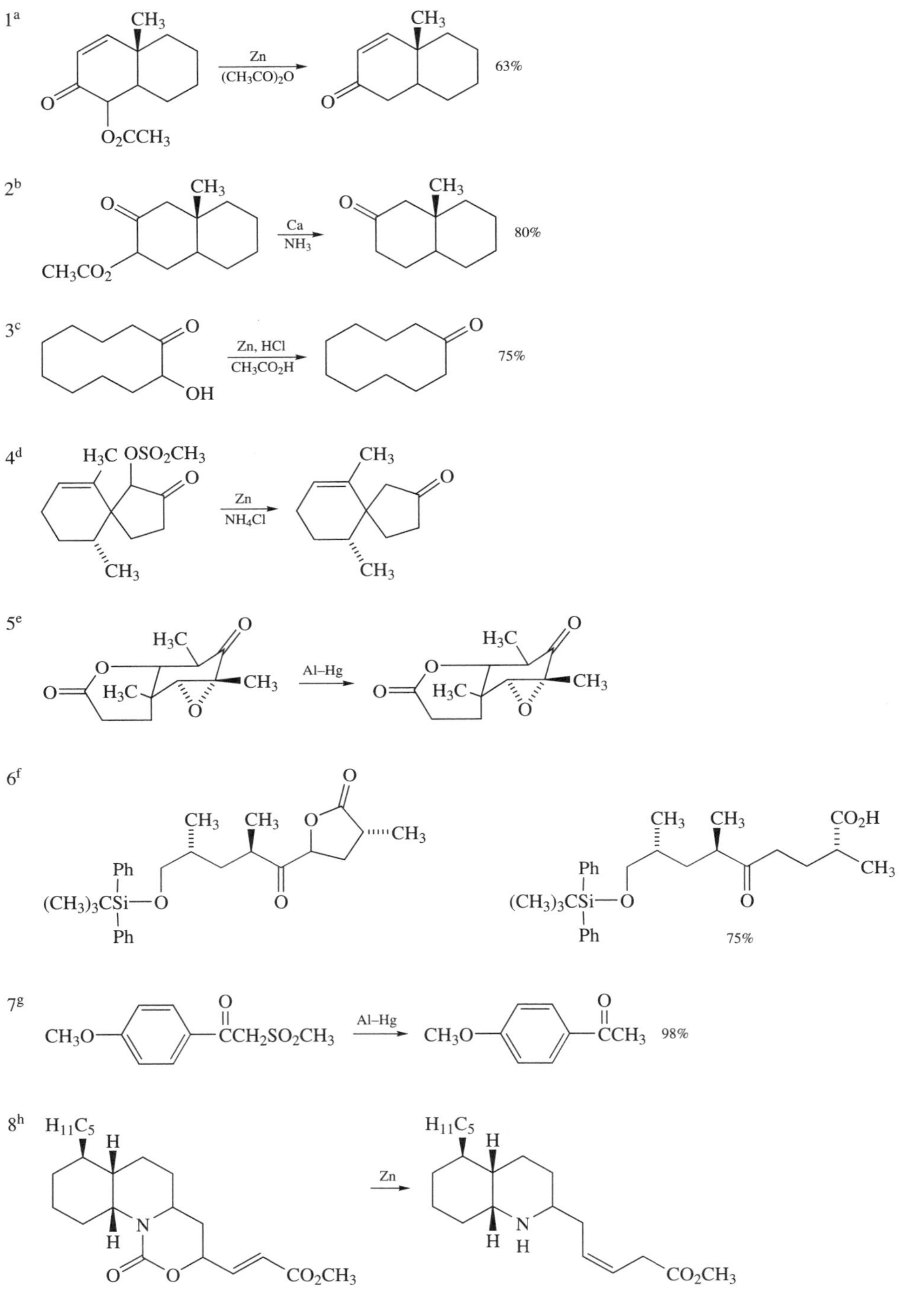

1[a] CH_3 O_2CCH_3 Zn, $(CH_3CO)_2O$ → CH_3 63%

2[b] CH_3 CH_3CO_2 Ca, NH_3 → CH_3 80%

3[c] O OH Zn, HCl, CH_3CO_2H → O 75%

4[d] H_3C OSO_2CH_3 O CH_3 Zn, NH_4Cl → CH_3 O CH_3

5[e] H_3C O O H_3C CH_3 O Al–Hg → H_3C O O H_3C CH_3 O

6[f] CH_3 CH_3 O O CH_3 Ph $(CH_3)_3CSi$—O Ph O → CH_3 CH_3 CO_2H CH_3 Ph $(CH_3)_3CSi$—O Ph O 75%

7[g] CH_3O—C₆H₄—$\overset{O}{\parallel}CCH_2SO_2CH_3$ Al–Hg → CH_3O—C₆H₄—$\overset{O}{\parallel}CCH_3$ 98%

8[h] $H_{11}C_5$ H H N O O CO_2CH_3 Zn → $H_{11}C_5$ H H N H CO_2CH_3

9[i] $\xrightarrow{SmI_2}$ 90%

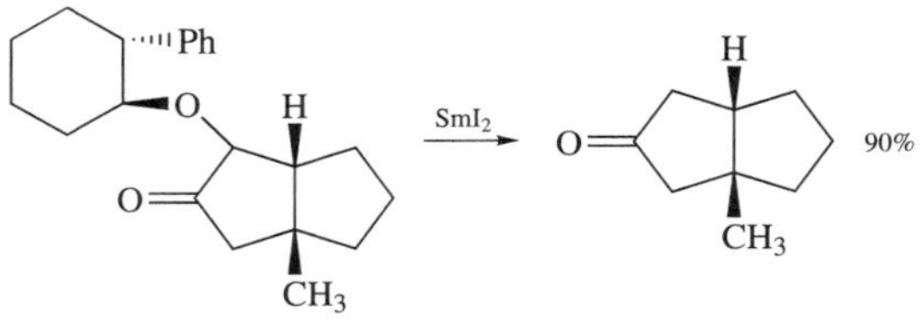

10[j] $\xrightarrow{SmI_2}$

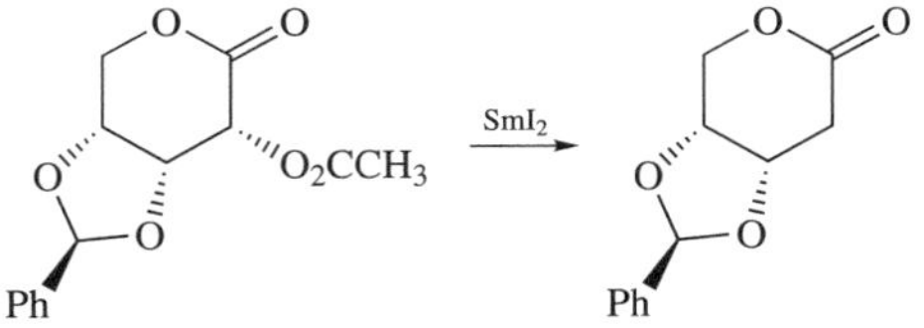

a. R. B. Woodward, F. Sondheimer, D. Taub, K. Heusler, and W. M. McLamore, *J. Am. Chem. Soc.* **74:**4223 (1952).
b. J. A. Marshall and H. Roebke, *J. Org. Chem.* **34:**4188 (1969).
c. A. C. Cope, J. W. Barthel, and R. D. Smith, *Org. Synth.* **IV:**218 (1963).
d. T. Ibuka, K. Hayashi, H. Minakata, and Y. Inubushi, *Tetrahedron Lett.* **1979:**159.
e. E. J. Corey, E. J. Trybulski, L. S. Melvin, Jr., K. C. Nicolaou, J. A. Secrist, R. Lett, P. W. Sheldrake, J. R. Falck, D. J. Brunelle, M. F. Haslanger, S. Kim, and S. Yoo, *J. Am. Chem. Soc.* **100:**4618 (1978).
f. P. A. Grieco, E. Williams, H. Tanaka, and S. Gilman, *J. Org. Chem.* **45:**3537 (1980).
g. E. J. Corey and M. Chaykovsky, *J. Am. Chem. Soc.* **86:**1639 (1964).
h. L. E. Overman and C. Fukaya, *J. Am. Chem. Soc.* **102:**1454 (1980).
i. J. Castro, H. Sorensen, A. Riera, C. Morin, C. Morin, A. Moyano, M. A. Percias, and A. E. Green, *J. Am. Chem. Soc.* **112:**9338 (1990).
j. S. Hanessian, C. Girard, and J. L. Chiara, *Tetrahedron Lett.* **33:**573 (1992).

Titanium metal generated by stronger reducing agents, such as $LiAlH_4$, or lithium or potassium metal, results in complete removal of oxygen with formation of an alkene.[167] A particularly active form of Ti is obtained by reducing $TiCl_3$ with lithium metal and then treating the reagent with 25 mol% of I_2.[168] This reagent is especially reliable when prepared form $TiCl_3$ purified as a DME complex.[170] A version of titanium-mediated reductive coupling in which $TiCl_3$–Zn–Cu serves as the reductant is efficient in closing large rings.

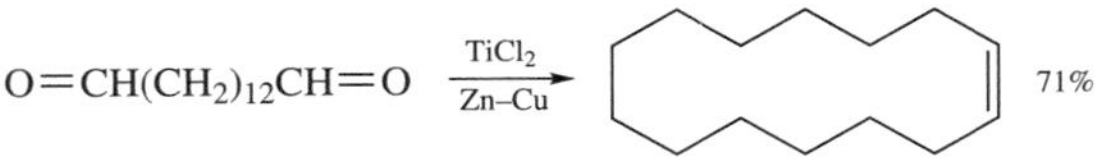

$$O{=}CH(CH_2)_{12}CH{=}O \xrightarrow[Zn\text{–}Cu]{TiCl_2}$$ 71% Ref. 171

Alkenes as large as 36-membered macrocycles have been prepared using the $TiCl_3$–Zn–Cu combination.[169]

Alkene formation can also be achieved using potassium/graphite (C_8K) or sodium naphthalenide for reduction.[172] The reductant prepared in this way is more efficient at

167. J. E. McMurry and M. P. Fleming; *J. Org. Chem.* **41:**896 (1976); J. E. McMurry and L. R. Krepski, *J. Org. Chem.* **41:**3929 (1976); J. E. McMurry, M. P. Fleming, K. L. Kees, and L. R. Krepski, *J. Org. Chem.* **43:**3255 (1978); J. E. McMurry, *Acc. Chem. Res.* **16:**405 (1983).
168. S. Talukdar, S. K. Nayak, and A. Banerji, *J. Org. Chem.* **63:**4925 (1998).
169. T. Eguchi, T. Terachi, and K. Kakinuma, *Tetrahedron Lett.* **34:**2175 (1993).
170. J. E. McMurry, T. Lectka, and J. G. Rico, *J. Org. Chem.* **54:**3748 (1989).
171. J. E. McMurry, J. R. Matz, K. L. Kees, and P. A. Bock, *Tetrahedron Lett.* **23:**1777 (1982).
172. D. L. J. Clive, C. Zhang, K. S. K. Murthy, W. D. Hayward, and S. Daigneault, *J. Org. Chem.* **56:**6447 (1991).

Scheme 5.11. Reductive Carbon–Carbon Bond Formation

A. Pinacol formation

1[a] H; O; $CH_2CH{=}O$ —— 1) Mg–Al/$(CH_3)_2SiCl_2$, 2) ^-OH ——> H; HO; HO 75%

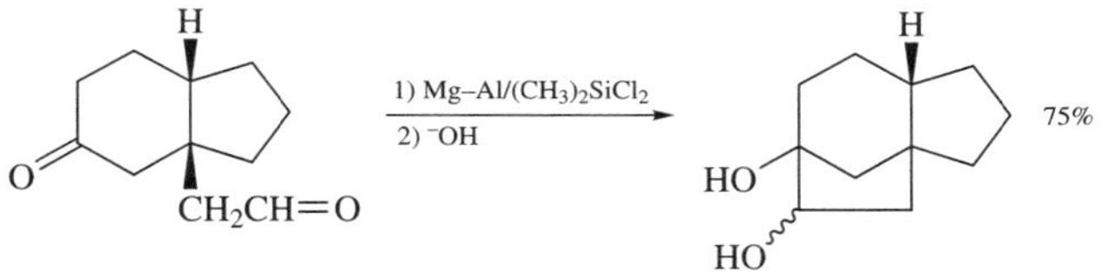

2[b] O —— Mg–Hg, $TiCl_4$ ——> HO; OH 93%

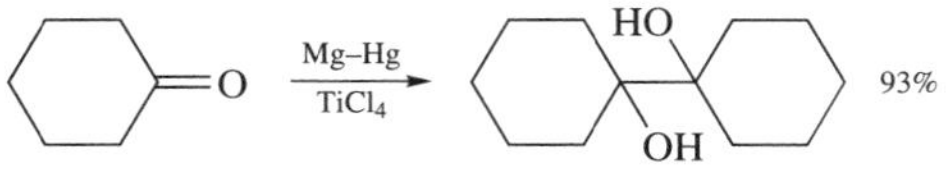

3[c] O; O=CHCH$_2$; O; O —— $TiCl_3$, Zn–Cu ——> OH; OH; O; O 58%

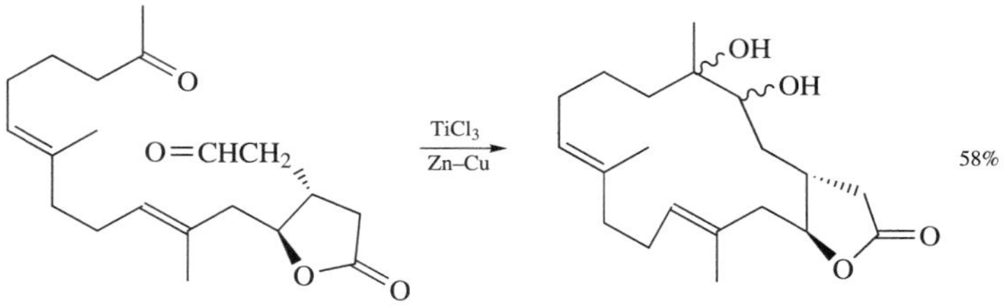

B. Alkene formation

4[d] O —— $TiCl_3$, K ——> 86%

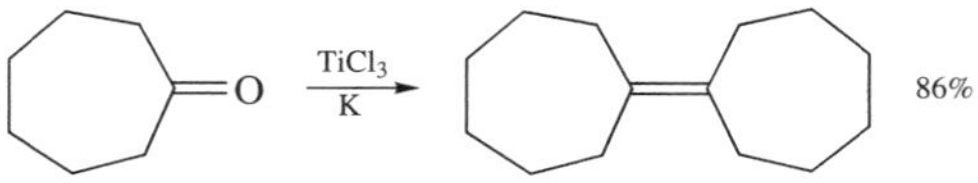

5[d] Ph; O; $(CH_2)_3CH{=}O$ —— $TiCl_3$, Zn–Cu ——> Ph 80%

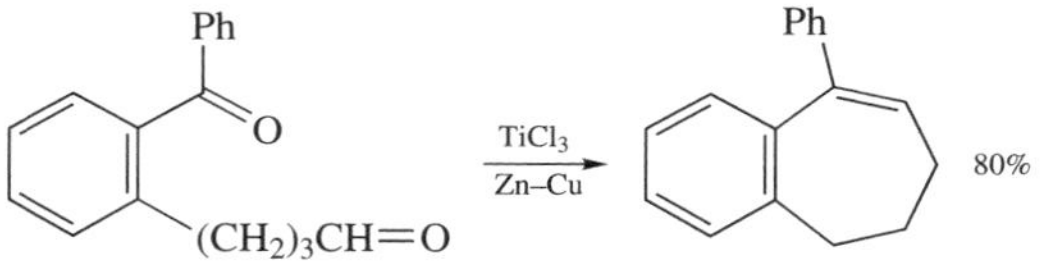

C. Acyloin formation

6[f] $CH_3O_2C(CH_2)_8CO_2CH_3$ —— 1) Na, xylene, 2) CH_3CO_2H ——> O; OH 70%

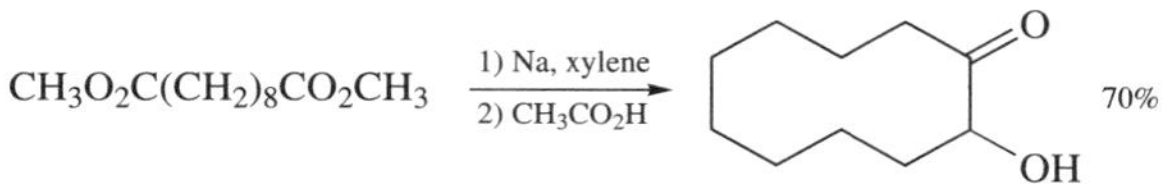

7[g] $C_2H_5O_2CH_2CH_2CO_2C_2H_5$ —— 1) Na, $(CH_3)_3SiCl$, 2) CH_3OH ——> O; OH 85%

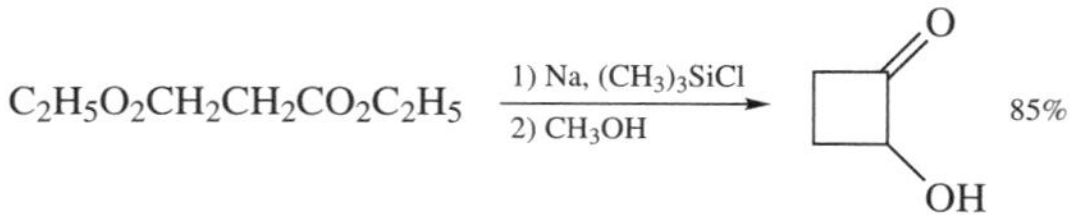

8[h] $CH_3(CH_2)_6CO_2C_2H_5$ —— Na/NaCl, benzene ——> $CH_3(CH_2)_6CH(OH)C(=O)(CH_2)_6CH_3$ 78%

a. E. J. Corey and R. L. Carney, *J. Am. Chem. Soc.* **93:**7318 (1971).
b. E. J. Corey, R. L. Danheiser, and S. Chandrasekaran, *J. Org. Chem.* **41:**260 (1976).
c. J. E. McMurry and R. G. Dushin, *J. Am. Chem. Soc.* **112:**6942 (1990).
d. J. E. McMurry, M. P. Fleming, K. L. Kees, and L. R. Krepski, *J. Org. Chem.* **43:**3255 (1978).
e. C. B. Jackson and G. Pattenden, *Tetrahedron Lett.* **26:**3393 (1985).
f. N. L. Allinger, *Org. Synth.* **IV:**840 (1963).
g. J. J. Bloomfield and J. M. Nelke, *Org. Synth.* **57:**1 (1977).
h. M. Makosza and K. Grela, *Synlett* **1997:**267.

coupling reactants with several oxygen substituents.

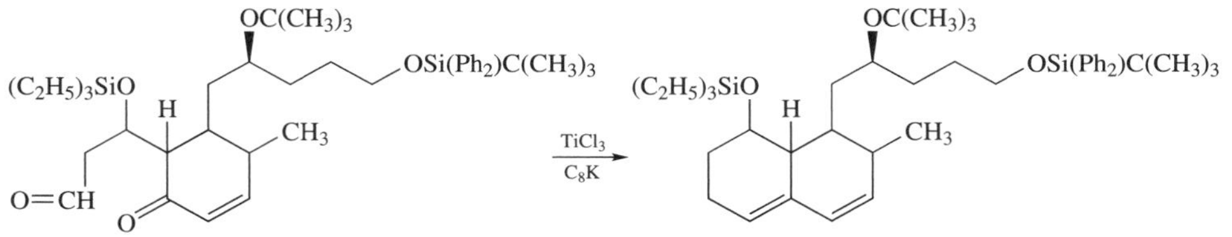

Both unsymmetrical diols and alkenes can be prepared by applying these methods to mixtures of two different carbonyl compounds. An excess of one component can be used to achieve a high conversion of the more valuable reactant. A mixed reductive deoxygenation using $TiCl_4/Zn$ has been used to prepare 4-hydroxytamoxifen, the active anti-estrogenic metabolite of tamoxifen.

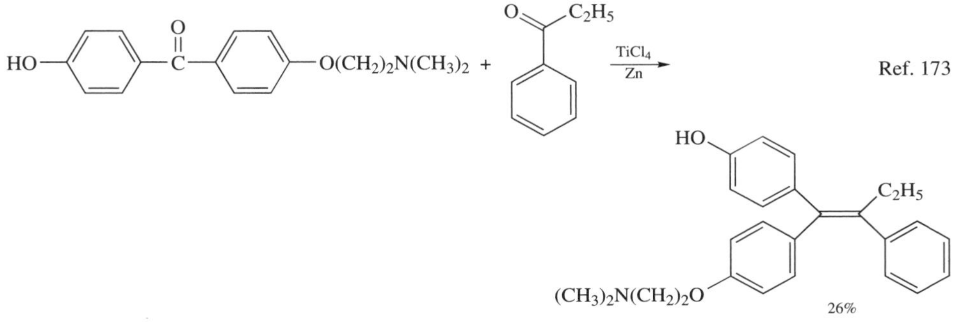

Ref. 173

The mechanism of the titanium-mediated reductive couplings is presumably similar to that of reduction by other metals, but titanium is uniquely effective in reductive coupling of carbonyl compounds. The strength of Ti–O bonds is probably the basis for this efficiency. Titanium-mediated reductive couplings are normally heterogeneous, and it is likely that the reaction takes place at the metal surface.[174] The partially reduced intermediates are probably bound to the metal surface, and this may account for the effectiveness of the reaction in forming medium and large rings.

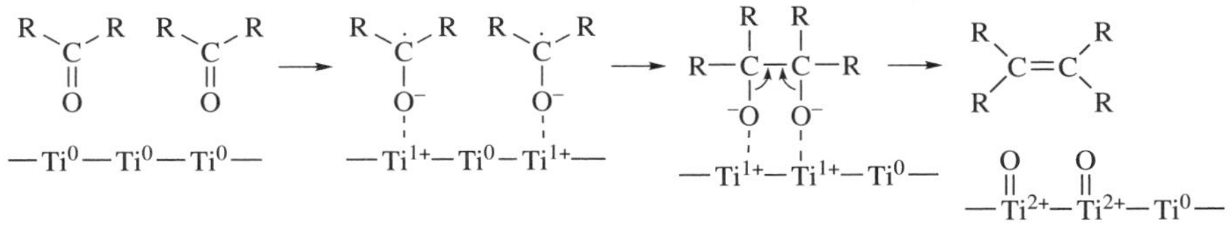

Samarium diiodide is another powerful one-electron reducing agent that can effect carbon–carbon bond formation under appropriate conditions.[175] Aromatic aldehydes and

173. S. Gauthier, J. Mailhot, and F. Labrie, *J. Org. Chem.* **61**:3890 (1996).
174. R. Dams, M. Malinowski, I. Westdrop, and H. Y. Geise, *J. Org. Chem.* **47**:248 (1982).
175. G. A. Molander, *Org. React.* **46**:211 (1994); J. L. Namy, J. Souppe, and H. B. Kagan, *Tetrahedron Lett.* **24**:765 (1983); A. Lebrun, J.-L. Namy, and H. B. Kagan, *Tetrahedron Lett.* **34**:2311 (1993); H. Akane, T. Hatano, H. Kusui, Y. Nishiyama, and Y. Ishii, *J. Org. Chem.* **59**:7902 (1994).

aliphatic aldehydes and ketones undergo pinacol-type coupling, with SmI_2 or $SmBr_2$.

$$RCR'(=O) \xrightarrow{SmI_2} RC(R')(OH)-C(R')(OH)R$$

$$ArCH{=}O \xrightarrow{SmI_2} ArCH(OH)-CH(OH)Ar$$

δ-Ketoaldehydes and δ-diketones are reduced to *cis*-cyclopentanediols.[176] ε-Diketo compounds can be cyclized to cyclohexanediols, again with a preference for *cis*-diols.[177] These reactions are believed to occur through successive one-electron transfer, radical coupling, and a second electron transfer with Sm^{2+} serving as a template and Lewis acid.

Many of the compounds used have additional functional groups, including ester, amide, ether, and acetal. These groups may be involved in coordination to samarium and thereby influence the stereoselectivity of the reaction.

The ketyl intermediates in SmI_2 reductions can also be trapped by carbon–carbon double bonds, leading to cyclization of δ,ε-enones to cyclopentanols.

$CH_2{=}CH(CH_2)_3C(R)(COCH_3)CO_2C_2H_5 \xrightarrow{SmI_2}$ (HO, CH_3, H_3C, $CO_2C_2H_5$, R substituted cyclopentanol) Ref. 178

(2-oxocyclopentyl)$(CH_2)_2CH{=}CHCO_2CH_3 \xrightarrow{SmI_2}$ (OH, $CH_2CO_2CH_3$, H bicyclic product) 87% Ref. 179

SmI_2 has also been used to form cyclooctanols by cyclization of 7,8-enones.[180] These alkene addition reactions all presumably proceed by addition of the ketyl radical to the

176. G. A. Molander and C. Kemp, *J. Am. Chem. Soc.* **111:**8236 (1989); J. Uenishi, S. Masuda, and S. Wakabashi, *Tetrahedron Lett.* **32:**5097 (1991).
177. J. L. Chiara, W. Cabri, and S. Hanessian, *Tetrahedron Lett.* **32:**1125 (1991); J. P. Guidok, T. Le Gall, and C. Mioskowski, *Tetrahedron Lett.* **35:**6671 (1994).
178. G. Molander and C. Kenny, *J. Am. Chem. Soc.* **111:**8236 (1989).
179. E. J. Enholm and A. Trivellas, *Tetrahedron Lett.* **30:**1063 (1989).
180. G. A. Molander and J. A. McKie, *J. Org. Chem.* **59:**3186 (1994).

double bond, followed by a second electron transfer.

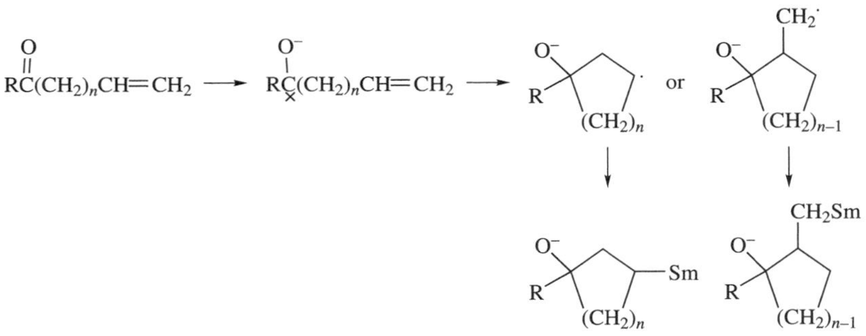

The initial products of such additions under aprotic conditions are organosamarium reagents, and further (tandem) transformations are possible, including addition to ketones, anhydrides, and carbon dioxide.

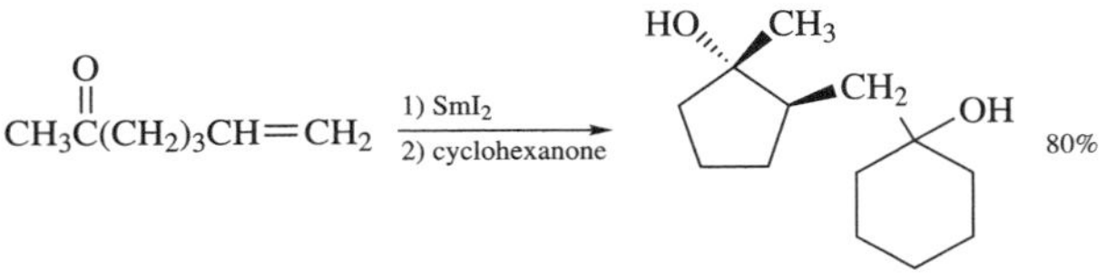

Ref. 181

Another reagent which has found use in pinacolic coupling is prepared from VCl_3 and zinc dust.[182] This reagent is selective for aldehydes that can form chelated intermediates, such as β-formyl amides, α-amido aldehydes, α-phosphinoyl aldehydes,[183] and γ-keto aldehydes.[184] It can be used for both homodimerization and heterodimerization. In the latter case, the more reactive aldehyde is added to an excess of the second aldehyde. Under these conditions, the ketal formed from the chelated aldehyde reacts with the second aldehyde.

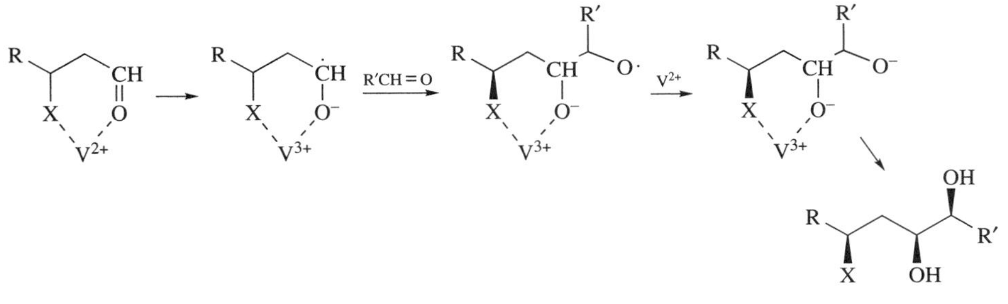

Another important reductive coupling is the conversion of esters to α-hydroxyketones (acyloins).[185] This reaction is usually carried out with sodium metal in an inert solvent.

181. G. A. Molander and J. A. McKie, *J. Org. Chem.* **57**:3132 (1992).
182. J. H. Freudenberg, A. W. Konradi, and S. F. Pedersen, *J. Am. Chem. Soc.* **111**:8014 (1989).
183. J. Park and S. F. Pedersen, *J. Org. Chem.* **55**:5924 (1990).
184. A. S. Raw and S. F. Pederson, *J. Org. Chem.* **56**:830 (1991).
185. J. J. Bloomfield, D. C. Owsley, and J. M. Nelke, *Org. React.* **23**:259 (1976).

Good results have also been reported for sodium metal dispersed on solid supports.[186] Diesters undergo intramolecular reactions, and this is also an important method for preparation of medium and large carbocyclic rings.

$$CH_3O_2C(CH_2)_8CO_2CH_3 \xrightarrow[\text{2) } CH_3CO_2H]{\text{1) Na}}$$ Ref. 187

There has been considerable discussion of the mechanism of the acyloin condensation. A simple formulation of the mechanism envisages coupling of radicals generated by one-electron transfer.

An alternative mechanism bypasses the postulated α-diketone intermediate since its involvement is doubtful.[188]

Regardless of the details of the mechanism, the product prior to neutralization is the dianion of the final α-hydroxy ketone, namely, an enediolate. It has been found that the overall yields are greatly improved if trimethylsilyl chloride is present during the reduction to trap these dianions as trimethylsilyl ethers.[189] These derivatives are much more stable under the reaction conditions than the enediolates. Hydrolysis during workup gives the acyloin product. This modified version of the reaction has been applied to cyclizations leading to small, medium, and large rings, as well as to intermolecular couplings.

A few examples of acyloin formation from esters are given in Scheme 5.11.

186. M. Makosza and K. Grela, *Synlett.* **1997:**267; M. Makosza, P. Nieczypor, and K. Grela, *Tetrahedron* **54:**10827 (1998).
187. N. Allinger, *Org. Synth.* **IV:**840 (1963).
188. J. J. Bloomfield, D. C. Owsley, C. Ainsworth, and R. E. Robertson, *J. Org. Chem.* **40:**393 (1975).
189. K. Ruhlmann, *Synthesis* **1971:**236.

Several methods are available for reductive removal of carbonyl groups form organic molecules. Complete reduction to methylene groups or conversion to alkenes can be achieved. Some examples of both types of reactions are given in Scheme 5.12.

Zinc and hydrochloric acid is a classical reagent combination for conversion of carbonyl groups to methylene groups. The reaction is known as the *Clemmensen reduction*.[190] The corresponding alcohols are not reduced under the conditions of the reaction, so they are evidently not intermediates. The Clemmensen reaction works best for aryl ketones and is less reliable with unconjugated ketones. The mechanism is not known in detail, but it most likely involves formation of carbon–zinc bonds at the metal surface.[191] The reaction is commonly carried out in hot concentrated hydrochloric acid with ethanol as a co-solvent. These conditions preclude the presence of acid-sensitive or hydrolyzable functional groups. A modification in which the reaction is run in ether saturated with dry hydrogen chloride gave good results in the reduction of steroidal ketones.[192]

The *Wolf–Kishner reaction*[193] is the reduction of carbonyl groups to methylene groups by base-catalyzed decomposition of the hydrazone of the carbonyl compound. Alkyldiimides are believed to be formed and then collapse with loss of nitrogen.

$$R_2C{=}N{-}NH_2 + \bar{O}H \rightleftharpoons R_2C{=}N{-}\bar{N}H \longrightarrow R_2\underset{H}{C}{-}N{=}N{-}H \xrightarrow{-N_2} R_2CH_2$$

The reduction of tosylhydrazones by $LiAlH_4$ or $NaBH_4$ also converts carbonyl groups to methylene groups.[194] It is believed that a diimide is involved, as in the Wolff–Kishner reaction.

$$R_2C{=}NNHSO_2Ar \xrightarrow{NaBH_4} R_2CHN\overset{H}{-}\overset{H}{N}{-}SO_2Ar \longrightarrow R_2CHN{=}NH \longrightarrow R_2CH_2$$

Excellent yields can also be obtained by using $NaBH_3CN$ as the reducing agent.[195] The $NaBH_3CN$ can be added to a mixture of the carbonyl cmpound and *p*-toluensulfonylhydrazide. Hydrazone formation is faster than reduction of the carbonyl group by $NaBH_3CN$, and the tosylhydrazone is reduced as it is formed. Another reagent which can reduce tosylhydrazones to give methylene groups is $CuBH_4(PPh_3)_2$.[196]

Reduction of tosylhydrazones of α,β-unsaturated ketones by $NaBH_3CN$ gives alkenes with double bond located between the former carbonyl carbon and the α carbon.[197] This reaction is believed to proceed by an initial conjugate reduction, followed by decomposi-

190. E. Vedejs, *Org. React.* **22:**401 (1975).
191. M. L. Di Vona and V. Rosnati, *J. Org. Chem.* **56:**4269 (1991).
192. M. Toda, M. Hayashi, Y. Hirata, and S. Yamamura, *Bull. Chem. Soc. Jpn.* **45:**264 (1972).
193. D. Todd, *Org. React.* **4:**378 (1948); Huang-Minlon, *J. Am. Chem. Soc.* **68:**2487 (1946).
194. L. Caglioti, *Tetrahedron* **22:**487 (1966).
195. R. O. Hutchins, C. A. Milewski, and B. E. Maryanoff, *J. Am. Chem. Soc.* **95:**3662 (1973).
196. B. Milenkov and M. Hesse, *Helv. Chim. Acta* **69:**1323 (1986).
197. R. O. Hutchins, M. Kacher, and L. Rua, *J. Org. Chem.* **40:**923 (1975).

Scheme 5.12. Carbonyl-to=Methylene Reductions

A. Clemmensen

1[a] OH, OCH3, CH=O → Zn (Hg), HCl → OH, OCH3, CH_3 60–67%

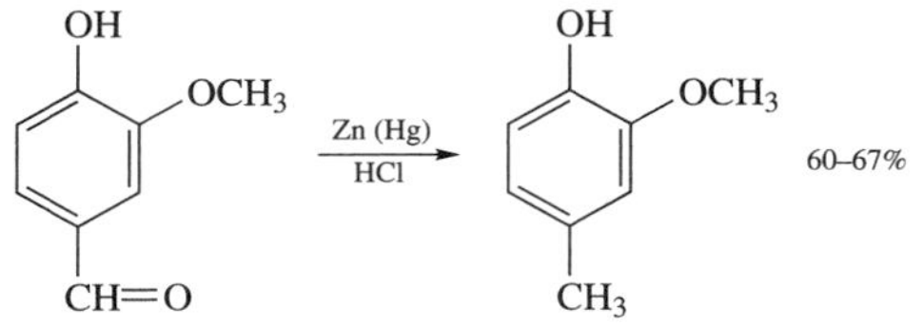

2[b] OH, $CO(CH_2)_5CH_3$ → Zn (Hg), HCl → OH, $CH_2(CH_2)_5CH_3$ 81–86%

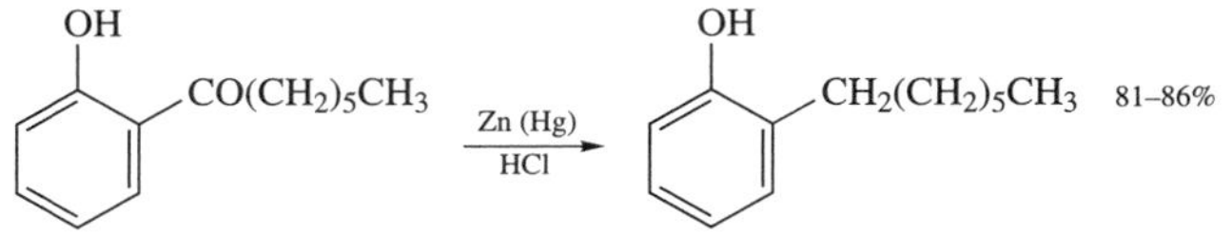

B. Wolff–Kishner

3[c] $HO_2C(CH_2)_4CO(CH_2)_4CO_2H$ $\xrightarrow[KOH]{NH_2NH_2}$ $HO_2C(CH_2)_9CO_2H$ 87–93%

4[d] Ph—C(=NNH_2)—Ph $\xrightarrow[DMSO]{KOC(CH_3)_3}$ $PhCH_2Ph$ 90%

C. Tosylhydrazone reduction

5[e] $CH{=}NNHSO_2C_7H_7$ → $LiAlH_4$ → CH_3 70%

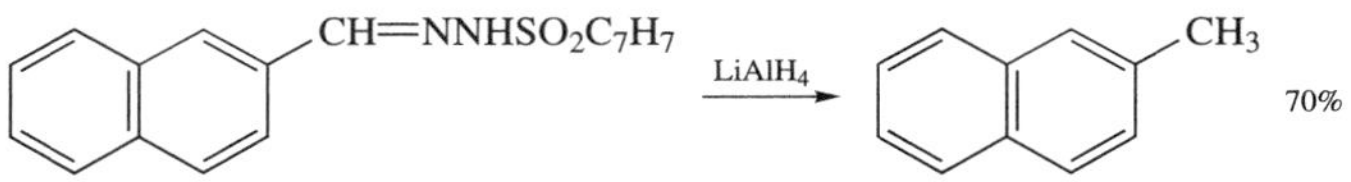

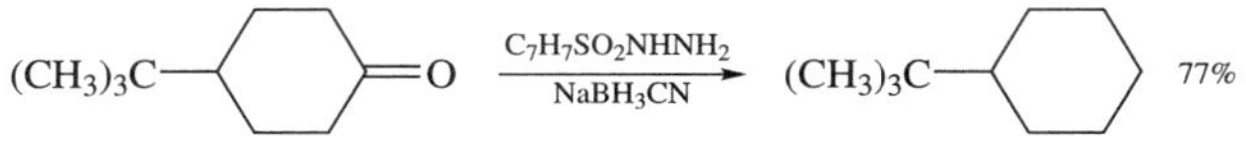

6[f] $(CH_3)_3C$—⟨cyclohexanone⟩=O $\xrightarrow[NaBH_3CN]{C_7H_7SO_2NHNH_2}$ $(CH_3)_3C$—⟨cyclohexane⟩ 77%

7[g] $CO_2C_2H_5$, TsNHN → (catecholborane, O, BH, O) → $CO_2C_2H_5$ 67%

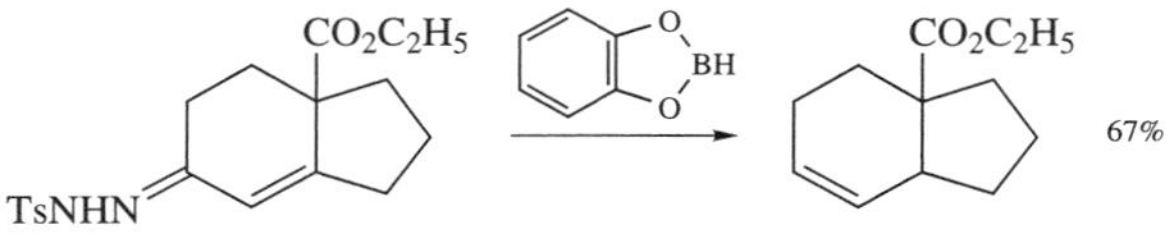

D. Thioketal desulfurization

8[h] $H_5C_2O_2C$—(S, S, S, S)—$CO_2C_2H_5$ → Raney Ni → $H_5C_2O_2C$—$CO_2C_2H_5$

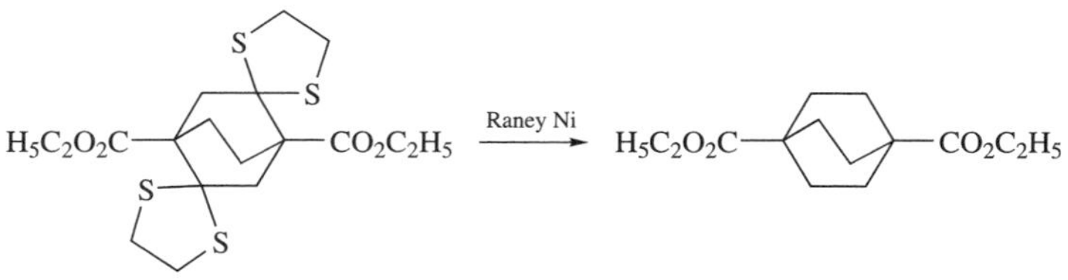

9[i] H_3C, O, $(CH_3)_2CH$, CH_3 → 1) $HSCH_2CH_2SH$, BF_3 2) Raney Ni → CH_3, $(CH_3)_2CH$, CH_3 58%

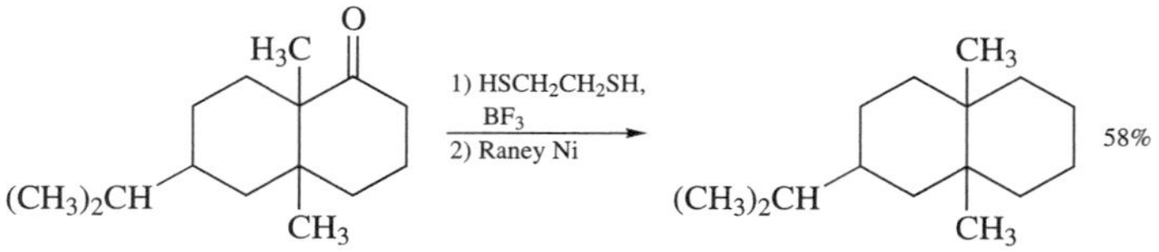

a. R. Schwarz and H. Hering, *Org. Synth.* **IV**:203 (1963).
b. R. R. Read and J. Wood Jr., *Org. Synth.* **III**:444 (1955).
c. L. J. Durham, D. J. McLeod, and J. Cason, *Org. Synth.* **IV**:510 (1963).
d. D. J. Cram, M. R. V. Sahyun, and G. R. Knox, *J. Am. Chem. Soc.* **84**:1734 (1962).
e. L. Caglioti and M. Magi, *Tetrahedron* **19**:1127 (1963).
f. R. O. Hutchins, B. E. Maryanoff, and C. A. Milewski, *J. Am. Chem. Soc.* **93**:1793 (1971).
g. M. N. Greco and B. E. Maryanoff, *Tetrahedron Lett.* **33**:5009 (1992).
h. J. D. Roberts and W. T. Moreland Jr., *J. Am. Chem. Soc.* **75**:2167 (1953).
i. P. N. Rao, *J. Org. Chem.* **36**:2426 (1971).

tion of the resulting vinylhydrazine to a vinyldiimide.

$$RCH{=}CH\overset{\overset{NNHSO_2Ar}{\parallel}}{C}R' \xrightarrow{NaBH_3CN} RCH_2CH{=}\overset{\overset{NHNHSO_2Ar}{|}}{C}R' \longrightarrow RCH_2CH{=}\overset{\overset{N=NH}{|}}{C}R' \xrightarrow{-N_2} RCH_2CH{=}CHR'$$

Catecholborane or sodium borohydride in acetic acid can also be used as reducing reagents in this reaction.[198]

Ketones can also be reduced to alkenes via enol triflates. The use of $Pd(OAc)_2$, triphenylphosphine as the catalyst, and tertiary amines as the hydrogen donors is effective.[199]

H_3C–N, CO_2CH_3, O_3SCF_3 $\xrightarrow[(C_2H_5)_3N,\ HCO_2H]{Pd(O_2CCH_3)_2,\ PPh_3}$ H_3C–N, CO_2CH_3 Ref. 200

Carbonyl groups can be converted to methylene groups by desulfurization of thioketals. The cyclic thioketal from ethanedithiol is commonly used. Reaction with excess Raney nickel causes hydrogenolysis of both C–S bonds.

H_3C, H_3C, H_3C, O $\xrightarrow[HSCH_2CH_2SH]{BF_3}$ H_3C, H_3C, H_3C, S, S $\xrightarrow{Ni}$ H_3C, H_3C, H_3C 81% Ref. 201

Other reactive forms of nickel including nickel boride[202] and nickel alkoxide complexes[203] can also be used for desulfurization. Tri-*n*-butyltin hydride is an alternative reagent for desulfurization.[204]

The conversion of ketone *p*-toluenesulfonylhydrazones to alkenes takes place on treatment with strong bases such as an alkyllithium or lithium dialkylamide.[205] This is known as the *Shapiro reaction*.[206] The reaction proceeds through the anion of a vinyldiimide, which decomposes to a vinyllithium reagent. Contact of this intermediate

198. G. W. Kabalka, D. T. C. Yang, and J. D. Baker, Jr., *J. Org. Chem.* **41:**574 (1976); R. O. Hutchins and N. R. Natale, *J. Org. Chem.* **43:**2299 (1978).
199. W. J. Scott and J. K. Stille, *J. Am. Chem. Soc.* **108:**3033 (1986); L. A. Paquette, P. G. Meiser, D. Friedrich, and D. R. Sauer, *J. Am. Chem. Soc.* **115:**49 (1993).
200. K. I. Keverline, P. Abraham, A. H. Lewin, and F. I. Carroll, *Tetrahedron Lett.* **36:**3099 (1995).
201. F. Sondheimer and S. Wolfe, *Can. J. Chem.* **37:**1870 (1959).
202. W. E. Truce and F. M. Perry, *J. Org. Chem.* **30:**1316 (1965).
203. S. Becker, Y. Fort, and P. Caubere, *J. Org. Chem.* **55:**6194 (1990).
204. C. G. Guiterrez, R. A. Stringham, T. Nitasaka, and K. G. Glasscock, *J. Org. Chem.* **45:**3393 (1980).
205. R. H. Shapiro and M. J. Health, *J. Am. Chem. Soc.* **89:**5734 (1967).
206. R. H. Shapiro, *Org. React.* **23:**405 (1976); R. M. Adington and A. G. M. Barrett, *Acc. Chem. Res.* **16:**53 (1983); A. R. Chamberlin and S. H. Bloom, *Org. React.* **39:**1 (1990).

with a proton source gives the alkene.

$$\underset{\text{RCCH}_2\text{R}'}{\overset{\text{NNHSO}_2\text{Ar}}{\|}} \xrightarrow{2\text{ RLi}} \text{Li}^+\,\underset{\underset{\text{Li}}{|}}{\underset{\text{RCCHR}'}{\overset{\text{NNSO}_2\text{Ar}}{\|}}} \xrightarrow{-\text{LiSO}_2\text{Ar}} \underset{\text{RC}=\text{CHR}'}{\overset{\text{N}=\text{N}^-\text{Li}^+}{|}} \xrightarrow{-\text{N}_2} \underset{\text{RC}=\text{CHR}'}{\overset{\text{Li}}{|}}$$

The Shapiro reaction has been particularly useful for cyclic ketones, but the scope of the reaction also includes acyclic systems. In the case of unsymmetrical acyclic ketones, questions of both regiochemistry and stereochemistry arise. 1-Octene is the exclusive product from 2-octanone.[207]

$$\underset{\text{CH}_3\text{C(CH}_2)_5\text{CH}_3}{\overset{\text{C}_7\text{H}_7\text{SO}_2\overset{\text{H}}{\text{N}}\text{N}}{\|}} \xrightarrow{2\text{ LiNR}_2} \text{CH}_2{=}\text{CH(CH}_2)_5\text{CH}_3$$

This regiospecificity has been shown to depend on the stereochemistry of the C=N bond in the starting hydrazone. There is evidently a strong preference for abstracting the proton *syn* to the arenesulfonyl group, probably because this permits chelation with the lithium ion.

$$\underset{\text{CH}_3\text{CCH}_2\text{R}}{\overset{\text{ArSO}_2\text{N}^-\!\diagdown\text{N}}{\|}} \longrightarrow \text{ArSO}_2\text{N}^-\text{(Li)}\;\underset{\text{CH}_2\text{CCH}_2\text{R}}{\overset{\text{N}}{\|}} \longrightarrow \xrightarrow{\text{H}^+} \text{CH}_2{=}\text{CHCH}_2\text{R}$$

The Shapiro reaction converts the *p*-toluenesulfonylhydrazones of α,β-unsaturated ketones to dienes (see entries 3–5 in Scheme 5.13).[208]

5.7. Reductive Elimination and Fragmentation

The placement of a potential leaving group β to the site of carbanionic character usually leads to β elimination.

$$2e^- \; \text{X}\frown\frown\text{Y} \longrightarrow \text{X}^- + {=} + \text{Y}^-$$

Similarly, carbanionic character δ to a leaving group can lead to β,γ-fragmentation.

$$2e^- \; \text{X}\frown\frown\frown\frown\text{Y} \longrightarrow \text{X}^- + {=} + {=} + \text{Y}^-$$

In some useful synthetic procedures, the carbanionic character results from a reductive process. A classical example of the β-elimination reaction is the reductive debromination of vicinal dibromides. Zinc metal is the traditional reducing agent.[209] A multitude of other

207. K. J. Kolonko and R. H. Shapiro, *J. Org. Chem.* **43:**1404 (1978).
208. W. G. Dauben, G. T. Rivers, and W. T. Zimmerman, *J. Am. Chem. Soc.* **99:**3414 (1977).
209. J. C. Sauer, *Org. Synth.* **IV**:268 (1965).

Scheme 5.13. Conversion of Ketones to Alkenes via Sulfonylhydrazones

1[a] H_3C CH_3 CH_3 $NNHSO_2C_7H_7$ —CH_3Li→ H_3C CH_3 CH_3 98–99%

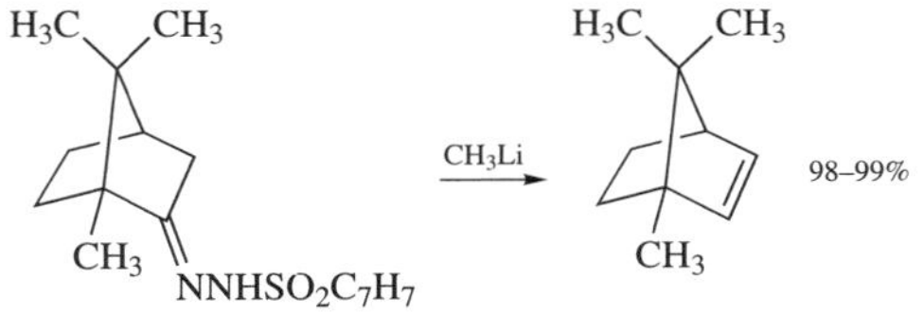

2[b] H $NNHSO_2C_7H_7$ O O H CH_3 —CH_3Li→ H O O H CH_3

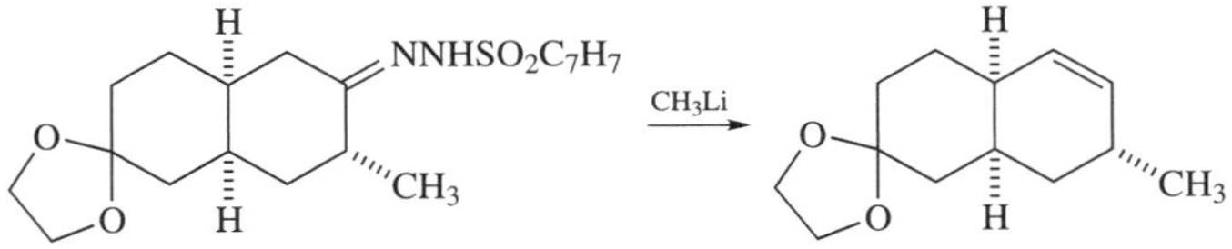

3[c] O H_3C CH_3 CH_3 —1) $C_7H_7SO_2NHNH_2$ 2) CH_3Li→ H_3C CH_3 CH_3 100%

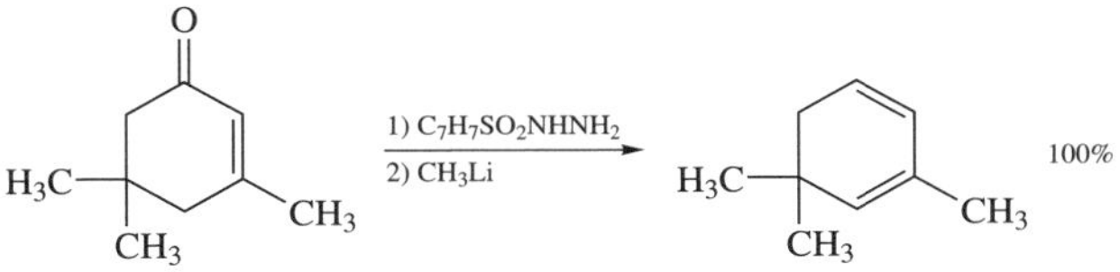

4[d] $NNHSO_2C_7H_7$ H_3C CH_3 —CH_3Li→ H_3C CH_3 80% + H_3C CH_2 9%

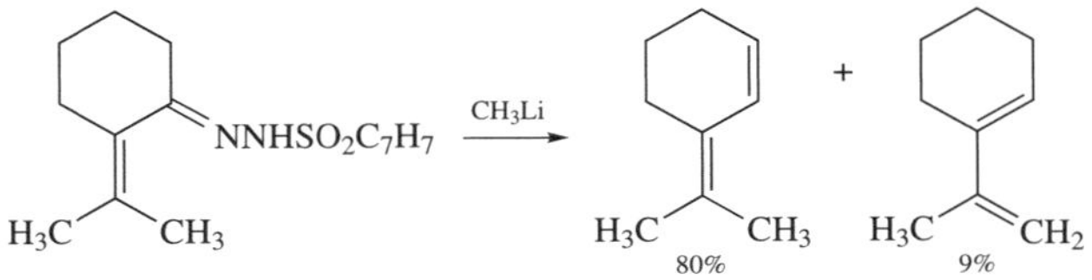

5[e] $PhCH_2O$ CH_3 O H CH_3 —1) $C_7H_7SO_2NHNH_2$ 2) $LiN(i\text{-}Pr)_2$→ $PhCH_2O$ CH_3 H CH_3 98%

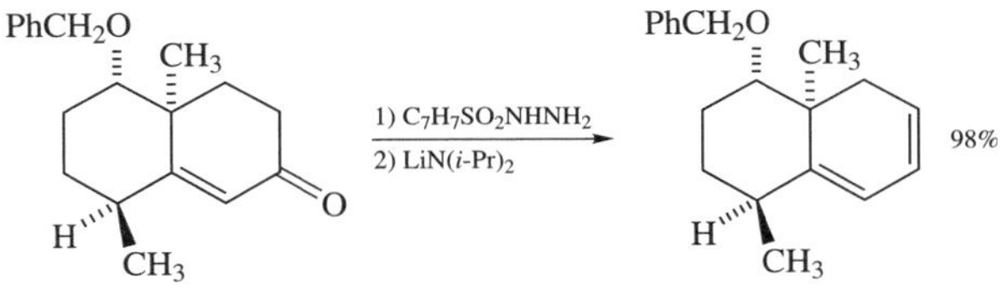

6[f] H $NNSO_2C_7H_7$ —$Li^+\bar{N}$→ 35–55%

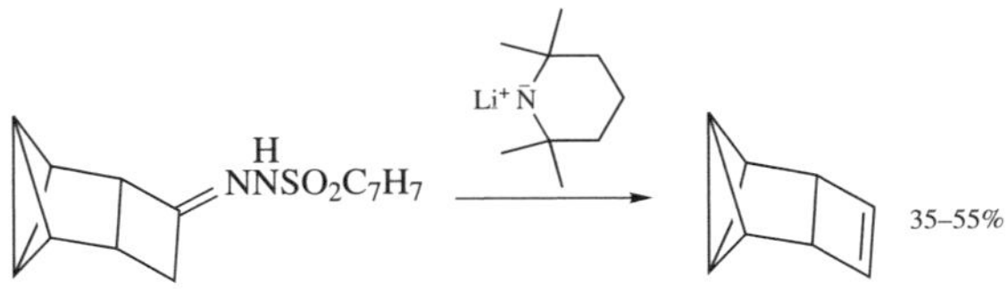

a. R. H. Shapiro and J. H. Duncan, *Org. Synth.* **51**:66 (1971).
b. W. L. Scott and D. A. Evans, *J. Am. Chem. Soc.* **94**:4779 (1972).
c. W. G. Dauben, M. E. Lorber, N. D. Vietmeyer, R. H. Shapiro, J. H. Duncan, and K. Tomer, *J. Am. Chem. Soc.* **90**:4762 (1968).
d. W. G. Dauben, G. T. Rivers and W. T. Zimmerman, *J. Am. Chem. Soc.* **99**:3414 (1977).
e. P. A. Grieco, T. Oguri, C.-L. J. Wang, and E. Williams, *J. Org. Chem.* **42**:4113 (1977).
f. L. R. Smith, G. R. Gream, and J. Meinwald, *J. Org. Chem.* **42**:927 (1977).

Table 5.8. Reagents for Reductive Dehalogenation

Reagent	*Anti* stereoselectivity	Reference
Zn, cat. $TiCl_4$	Yes	a
Zn, H_2NCSNH_2	?	b
$SnCl_2$, DIBAlH	?	c
Sm, CH_3OH	No	d
Fe, graphite	Yes	e
C_2H_5MgBr, cat. Ni(dppe)Cl_2	No	f

a. F. Sato, T. Akiyama, K. Iida, and M. Sato, *Synthesis* **1982:**1025.
b. R. N. Majumdar and H. J. Harwood, *Synth. Commun.* **11:**901 (1981).
c. T. Oriyama and T. Mukaiyama, *Chem. Lett.* **1984:**2069.
d. R. Yanada, N. Negoro, K. Yanada, and T. Fujita, *Tetrahedron Lett.* **37:**9313 (1996).
e. D. Savoia, E. Tagliavini, C. Trombini, and A. Umani-Ronchi, *J. Org. Chem.* **47:**876 (1982).
f. C. Malanga, L. A. Aronica, and L. Lardicci, *Tetrahedron Lett.* **36:**9189 (1995).

reducing agents have been found to give this and similar reductive eliminations. Some examples are given in Table 5.8. Some of the reagents exhibit *anti* stereospecificity while others do not. A *stringent* test for *anti* stereoselectivity is the extent of *Z*-alkene formation from a *syn* precursor.

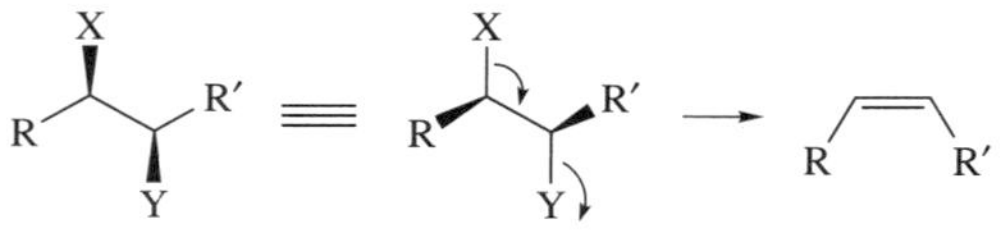

Anti stereospecificity is associated with a concerted reductive elimination, whereas single-electron transfer–fragmentation leads to loss of stereospecificity.

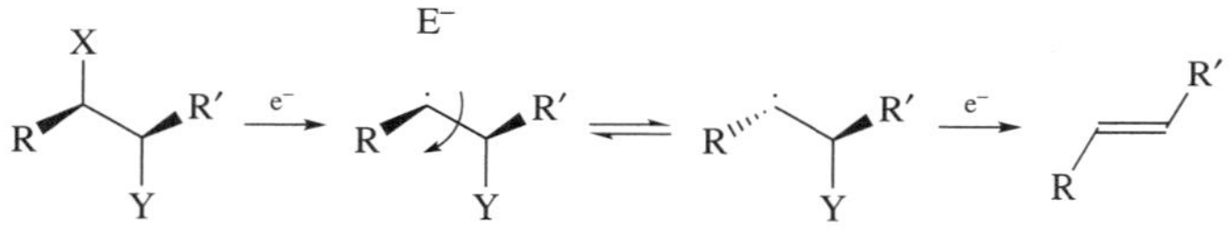

Because vicinal dibromides are usually made by bromination of alkenes, their utility for synthesis is limited, except for temporary masking of a double bond. Much more frequently, it is desirable to convert a diol to an alkene. Several useful procedures have been developed. The reductive deoxygenation of diols via thiocarbonates was developed by Corey and co-workers.[210] Triethyl phosphite is useful for many cases, but the more reactive reductant 1,3-dimethyl-2-phenyl-1,3,2-diazaphospholidine can be used when milder conditions are required.[211] The reaction presumably occurs by initial P–S bonding

210. E. J. Corey and R. A. E. Winter, *J. Am. Chem. Soc.* **85:**2677 (1963); E. J. Cory, F. A. Cary, and R. A. E. Winter, *J. Am. Chem. Soc.* **87:**934 (1965).
211. E. J. Corey and P. B. Hopkins, *Tetrahedral Lett.* **23:**1979 (1982).

followed by a concerted elimination of carbon dioxide and the thiophosphoryl compound.

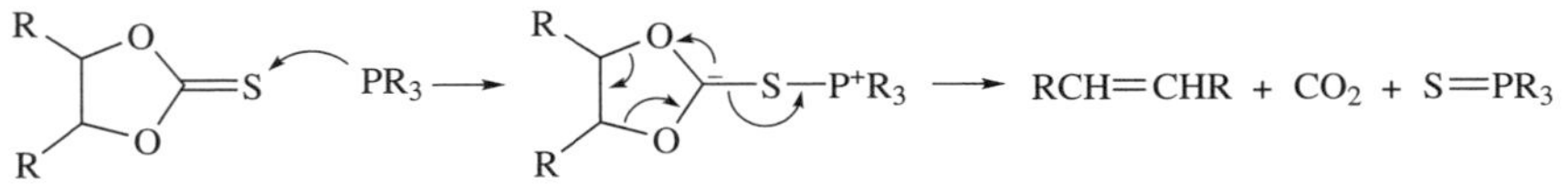

$$RCH{=}CHR + CO_2 + S{=}PR_3$$

Diols can also be deoxygenated via bis-sulfonate esters using sodium naphthalenide.[212] Cyclic sulfate esters are also cleanly reduced by lithium naphthalenide.[213]

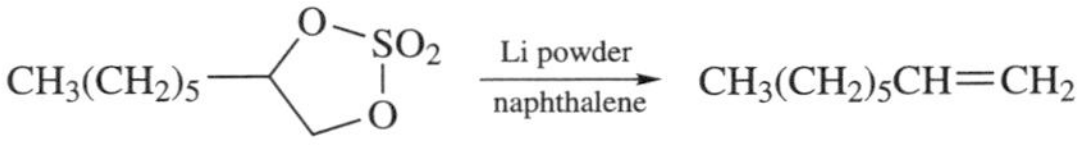

$$\xrightarrow[\text{naphthalene}]{\text{Li powder}} CH_3(CH_2)_5CH{=}CH_2$$

This reaction, using sodium naphthalenide, has been used to prepare unsaturated nucleosides.

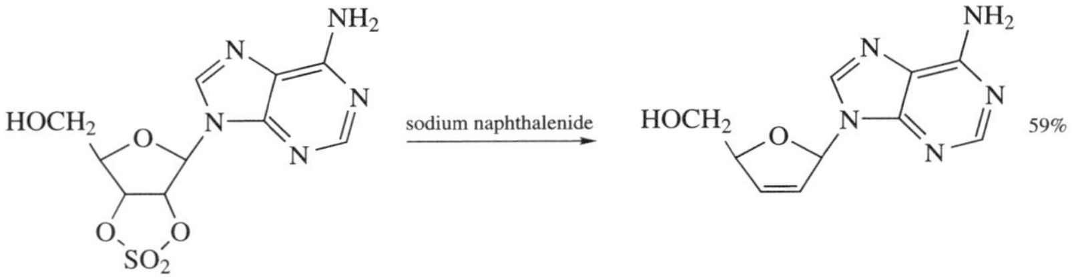

sodium naphthalenide 59% Ref. 214

It is not entirely clear whether these reactions involve a redox reaction at sulfur or proceed via organometallic intermediates.

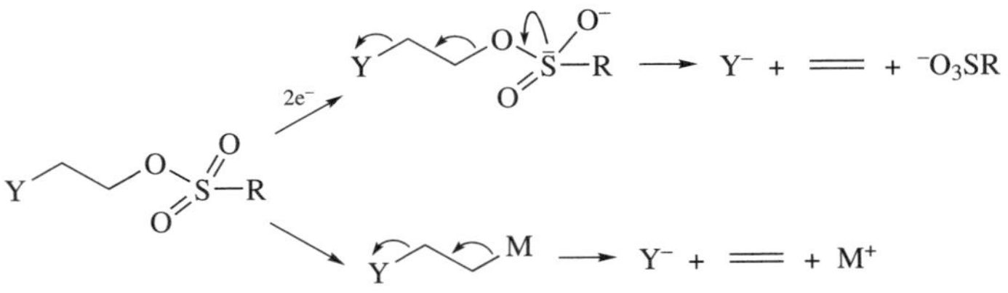

Iodination reagents combined with aryl phosphines and imidazole can also effect reductive conversion of diols to alkenes. One such combination is 2,4,5-triiodoimidazole, imidazole, and triphenylphosphine.[215] These reagent combinations are believed to give oxyphosphonium intermediates which then serve as leaving groups, forming triphenylphosphine oxide as in the Mitsunobu reaction (see Section 3.2.4). The iodide serves as both a

212. J. C. Carnahan, Jr., and W. D. Closson, *Tetrahedron Lett.* **1972:**3447; R. J. Sundberg and R. J. Cherney, *J. Org. Chem.* **55:**6028 (1990).
213. D. Guijarro, B. Mancheno, and M. Yus, *Tetrahedron Lett.* **33:**5597 (1992).
214. M. J. Robbins, E. Lewandowska, and S. F. Wnuk, *J. Org. Chem.* **63:**7375 (1998).
215. P. J. Garegg and B. Samuelsson, *Synthesis* **1979:**813; Y. Watanabe, M. Mitani, and S. Ozaki, *Chem. Lett.* **1987:**123.

nucleophile and a reductant.

$$\mathrm{RCH(OH)-CHR(OH)} \longrightarrow \mathrm{RCH(OP^+Ph_3)-CHR(OP^+Ph_3)} \text{ or } \mathrm{RCH(OP^+Ph_3)-CHR(I)} \xrightarrow{I^-} \mathrm{RCH{=}CHR}$$

In a related procedure, chlorodiphenylphosphine, imidazole, iodine, and zinc cause reductive elimination of diols.[216] β-Iodophosphinate esters can be shown to be intermediates in some cases.

HO, HO, OCH_2Ph, O, O, O $\xrightarrow[\text{imidazole}]{Ph_2PCl,\ I_2}$ I, Ph_2PO_2, OCH_2Ph, O, O, O $\xrightarrow{Zn}$ $CH_2{=}CH$, OCH_2Ph, O, O, O

Another alternative to conversion of diols to alkenes is the use of the Barton radical fragmentation conditions (see Section 5.4) with a silane hydrogen-atom donor.[217]

$$\mathrm{CH_3SC(=S)O-CHR-CHR-OC(=S)SCH_3} \xrightarrow[(PhCO_2)_2]{Et_3SiH} \mathrm{RCH{=}CHR}$$

The reductive elimination of β-hydroxysulfones is the final step in the Julia–Lythgoe olefin synthesis.[218] The β-hydroxysulfones are normally obtained by an aldol addition.

$$\mathrm{RCH{=}O} + \mathrm{R'CH_2SO_2R''} \xrightarrow{\text{base}} \mathrm{RCH(HO)-CHR'(SO_2R'')} \xrightarrow{2e^-} \mathrm{RCH{=}CHR''}$$

Several reducing agents have been used for the elimation, including sodium amalgam[219] and samarium diiodide.[220] The elimination can also be done by converting the hydroxy group to a xanthate or thiocarbonate and using radical fragmentation.[221]

Reductive elimination from 2-ene-1,4-diol derivatives has been used to generate 1,3-dienes. Low-valent titanium generated from $TiCl_3/LiAlH_4$ can be used directly with the

216. Z. Liu, B. Classon, and B. Samuelsson, *J. Org. Chem.* **55:**4273 (1990).
217. D. H. R. Barton, D. O. Jang, and J. C. Jaszberenyi, *Tetrahedron Lett.* **32:**2569 (1991); D. H. R. Barton, D. O. Jang and J. C. Jaszberenyi, *Tetrahedron Lett.* **32:**7187 (1991).
218. P. Kocienski, *Phosphorus Sulfur* **24:**97 (1985).
219. P. J. Kocienski, B. Lythgoe, and I. Waterhouse, *J. Chem. Soc., Perkin Trans. 1* **1980**:1045; A. Armstrong, S. V. Ley, A. Madin, and S. Mukherjee, *Synlett* **1990:**328; M. Kageyama, T. Tamura, M. H. Nantz, J. C. Roberts, P. Somfai, D. C. Whritenour, and S. Masamune, *J. Am. Chem. Soc.* **112:**7407 (1990).
220. A. S. Kende and J. S. Mendoza, *Tetrahedron Lett.* **31:**7105 (1990); I. E. Marko, F. Murphy, and S. Dolan, *Tetrahedron Lett.* **37:**2089 (1996); G. E. Keck, K. A. Savin, and M. A. Weglarz, *J. Org. Chem.* **60:**3194 (1995).
221. D. H. R. Barton, J. C. Jaszberenyi, and C. Tachdjian, *Tetrahedron Lett.* **32:**2703 (1991).

diols. This reaction has been used successfully to create extended polyene conjugation.[222]

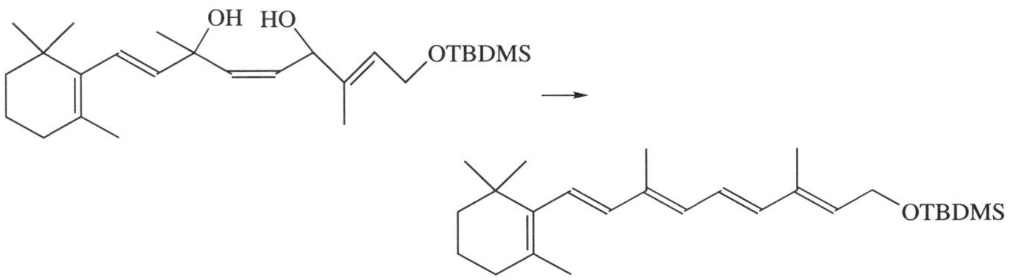

Benzoate esters of 2-ene-1,4-diols undergo reductive elimination with sodium amalgam.[223]

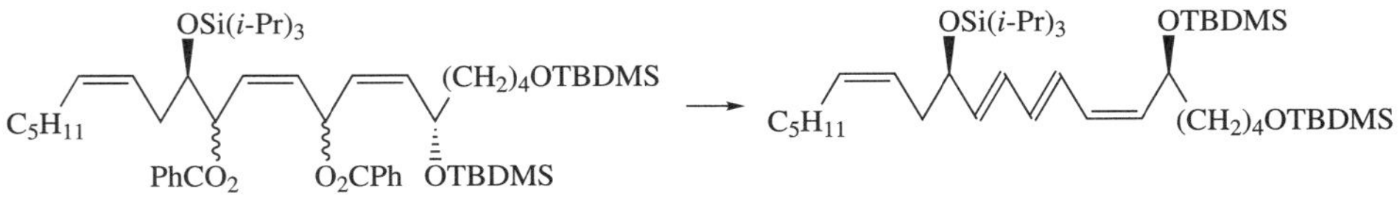

The β,γ fragmentation is known as *Grob fragmentation*. Its synthetic application is usually in the construction of medium-sized rings by fragmentation of fused ring systems.

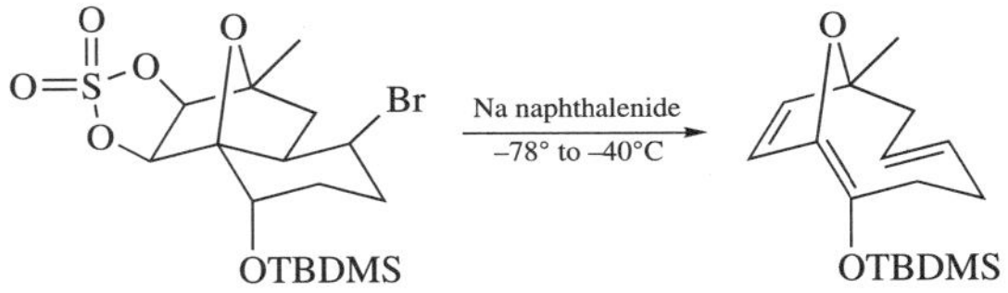

Na naphthalenide, −78° to −40°C

Ref. 224

General References

R. L. Augustine, ed., *Reduction Techniques and Applications in Organic Synthesis*, Marcel Dekker, New York, 1968.

M. Hudlicky, *Reductions in Organic Chemistry*, Halstead Press, New York, 1984.

Catalytic Reduction

M. Freifelder, *Catalytic Hydrogenation in Organic Synthesis, Procedures and Commentary*, John Wiley & Sons, New York, 1978.

B. R. James, *Homogeneous Hydrogenation*, John Wiley & Sons, New York, 1973.

P. N. Rylander, *Hydrogenation Methods*, Academic Press, Orlando, Florida, 1985.

P. N. Rylander, *Hydrogenation in Organic Synthesis*, Academic Press, New York, 1979.

222. G. Solladie, A. Givardin, and G. Lang, *J. Org. Chem.* **54:**2620 (1989); G. Solladie and V. Berl, *Tetrahedron Lett.* **33:**3477 (1992).
223. G. Solladie, A. Urbana, and G. B. Stone, *Tetrahedron Lett.* **34:**6489 (1993).
224. W. B. Wang and E. J. Roskamp, *Tetrahedral Lett.* **33:**7631 (1992).

Metal Hydrides

A. Hajos, *Complex Hydrides and Related Reducing Agents in Organic Synthesis*, Elsevier, New York, 1979.

J. Malek, *Org. React.* **34**:1 (1985); **36**:249 (1988).

J. Sayden-Penne, *Reductions by the Alumino- and Borohydrides in Organic Synthesis*, VCH Publishers, New York, 1991.

Dissolving-Metal Reductions

A. A. Akhrem, I. G. Rshetova, and Y. A. Titov, *Birch Reduction of Aromatic Compounds*, IFGI/Plenum, New York, 1972.

Problems

(*References for these problems will be found on page 929.*)

1. Give the product(s) to be expected from the following reactions. Be sure to specify all facets of stereochemistry.

(a) $(CH_3)_2CHCH{=}CHCH{=}CHCO_2CH_3 \xrightarrow{(i\text{-Bu})_2AlH}$

(b)

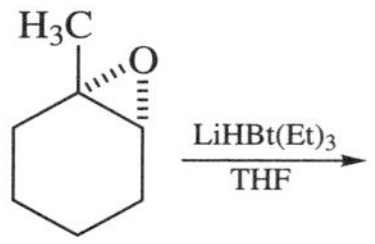

(c)

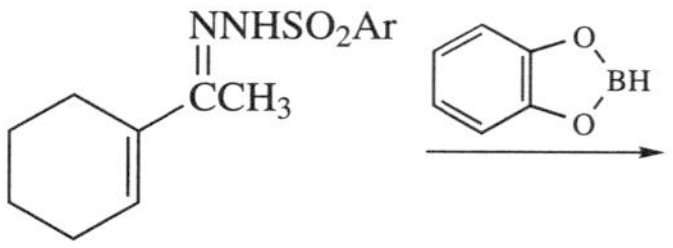

(d)

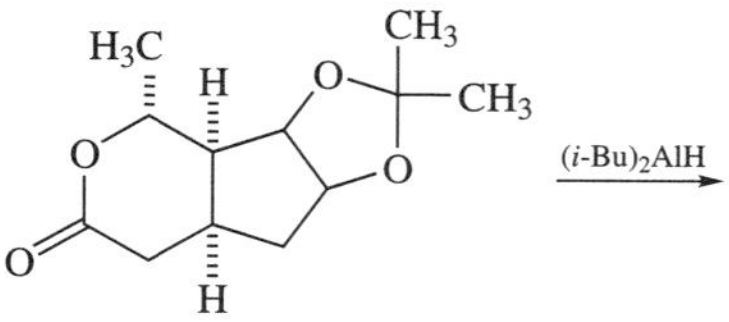

(e)

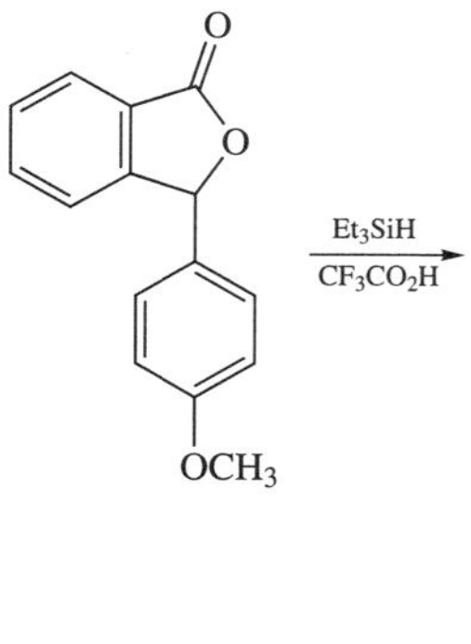

(f) 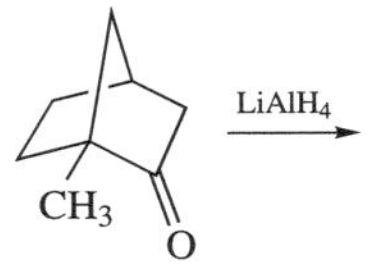 CH3, O — $\xrightarrow{LiAlH_4}$

(g) 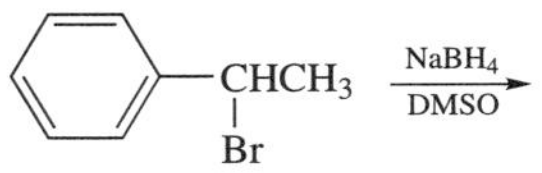 C_6H_5–CHCH3, Br — $\xrightarrow[DMSO]{NaBH_4}$

(h) 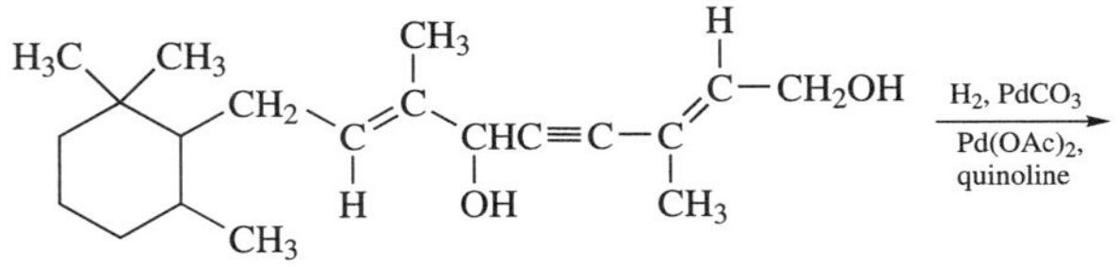 H3C, CH3, CH2, CH3, C, H, CH3, CHC≡C–C, OH, H, C–CH2OH, CH3 — $\xrightarrow[\text{Pd(OAc)}_2,\ \text{quinoline}]{H_2,\ PdCO_3}$

(i) 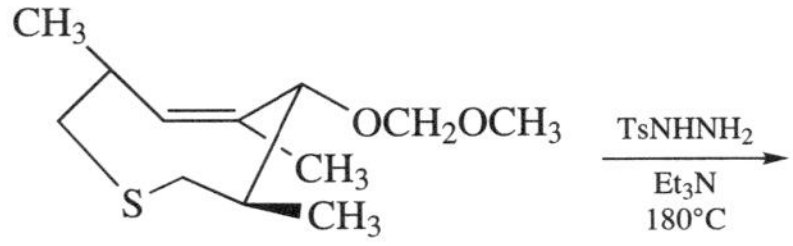 CH3, S, OCH2OCH3, CH3, CH3 — $\xrightarrow[\text{Et}_3\text{N},\ 180°\text{C}]{TsNHNH_2}$

(j) 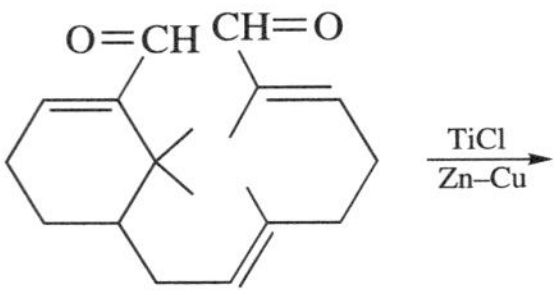 O=CH CH=O — $\xrightarrow[Zn–Cu]{TiCl}$

2. Indicate the stereochemistry of the major alcohol that would be formed by sodium borohydride reduction of each of the cyclohexanone derivatives shown:

(a) 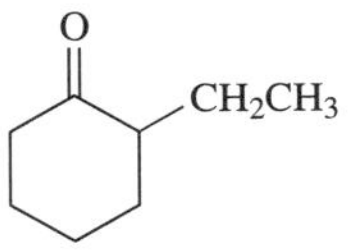 O, CH2CH3

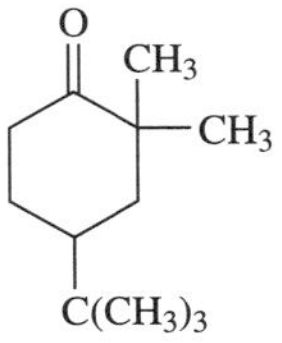

(b) 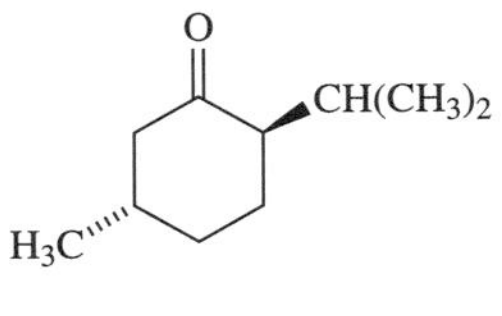 O, CH(CH3)2, H3C

(c) O, CH3, CH3, C(CH3)3

(d) 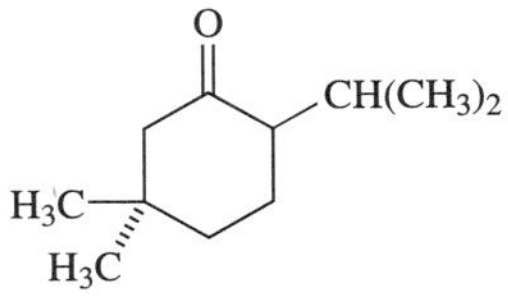 O, CH(CH3)2, H3C, H3C

3. Indicate reaction conditions that would accomplish each of the following transformations in one step.

(a) 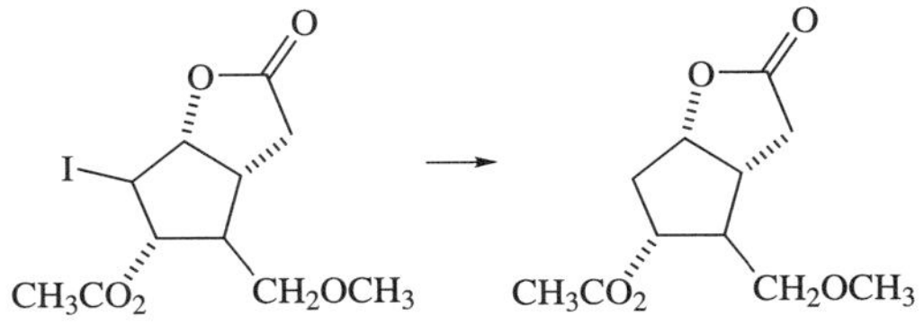 I, O, O, CH3CO2, CH2OCH3 → O, O, CH3CO2, CH2OCH3

(b)

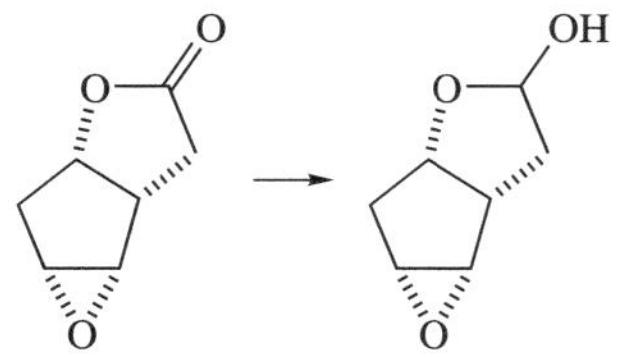

(c)

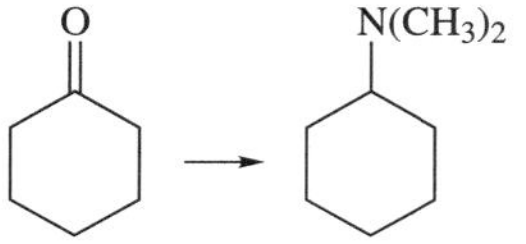

(d)

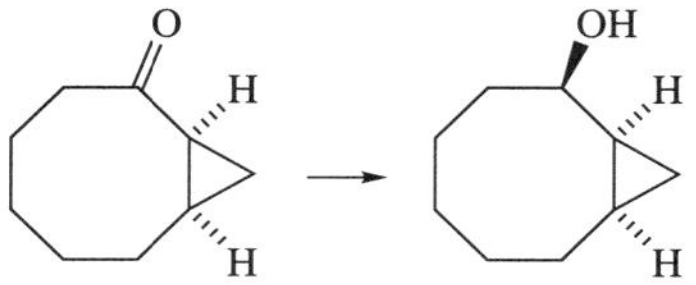

(e)

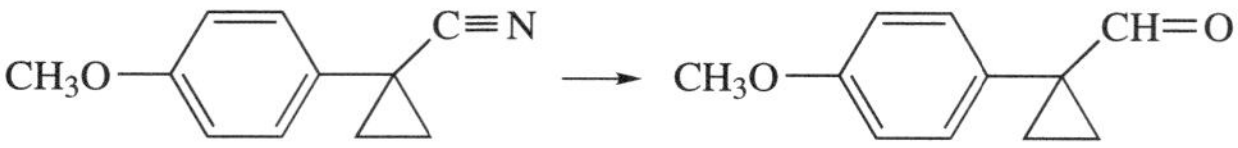

(f)

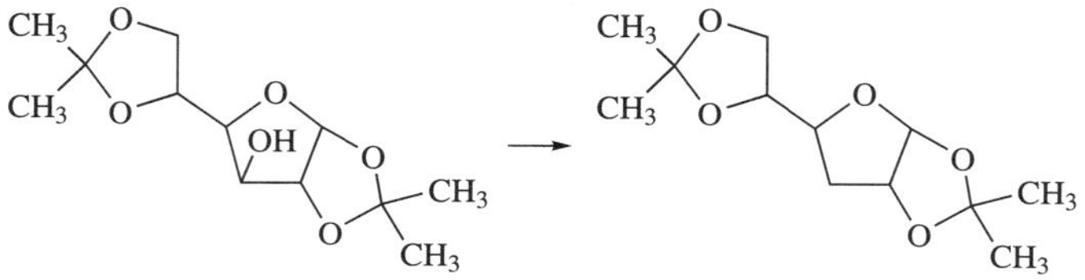

(g)

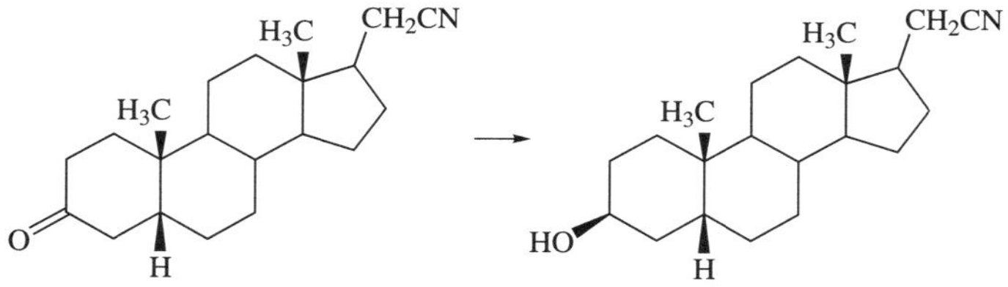

(h)

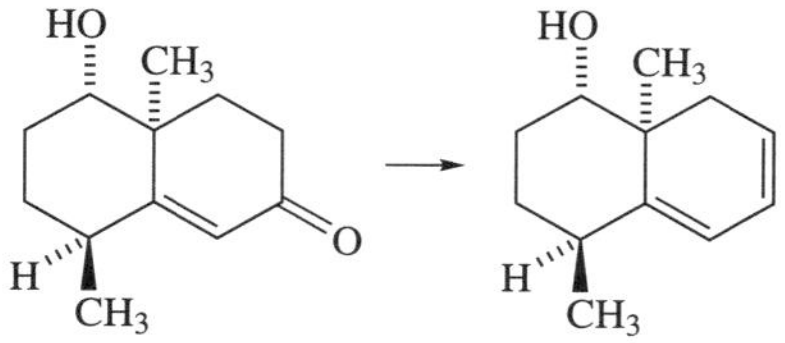

(i)

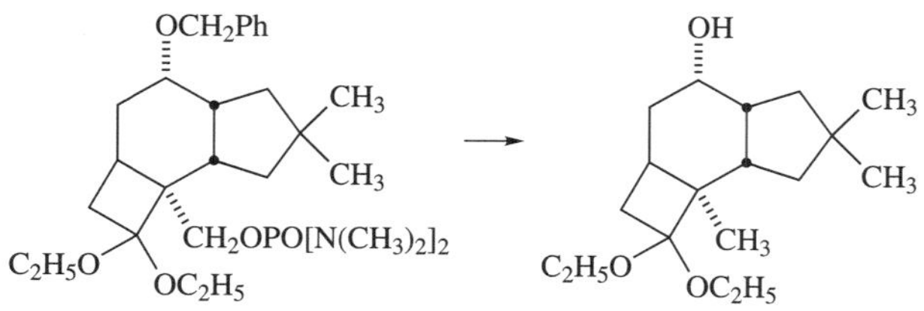

(j)

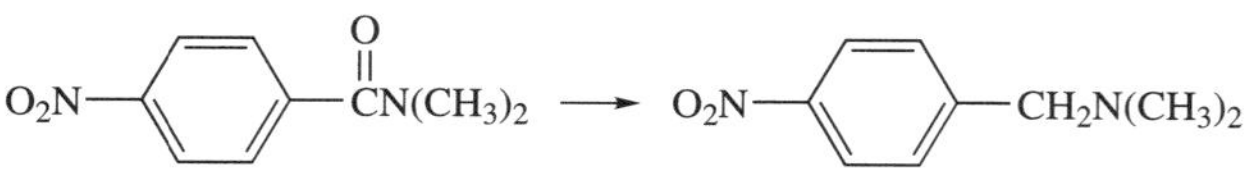

(k) H3C, CH3, CH3, O → H3C, CH3, CH3

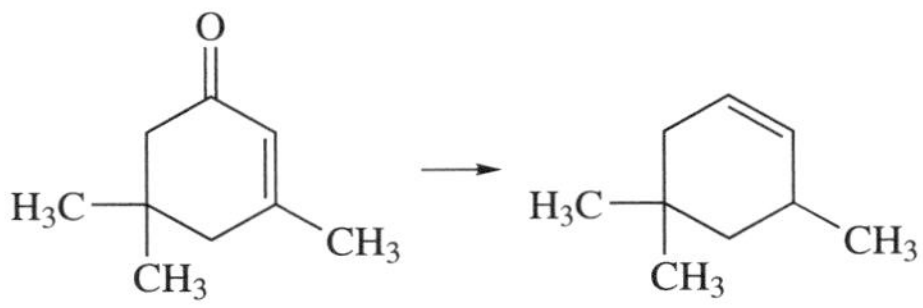

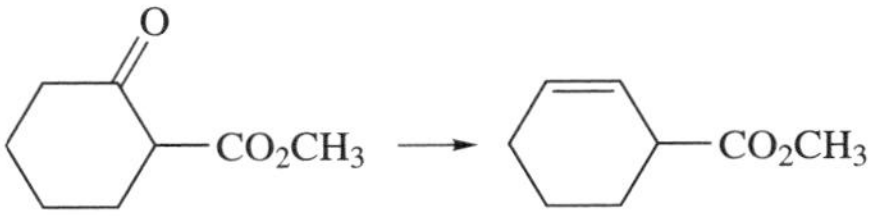

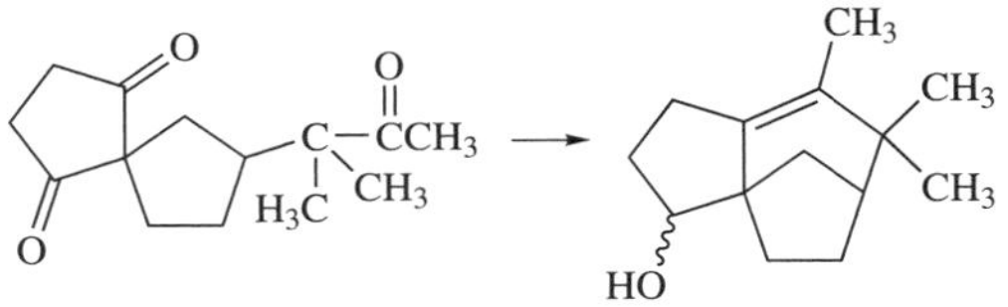

(l) →

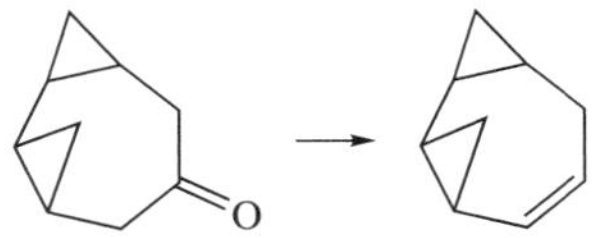

(m) CO_2CH_3 → CO_2CH_3

(n) O, O, O, C—CCH3, H3C CH3 → CH3, CH3, CH3, HO

(o) O O, H3C (CH2)3C≡C(CH2)3OTHP → O O, H3C (CH2)3, H, C=C, (CH2)3OTHP, H

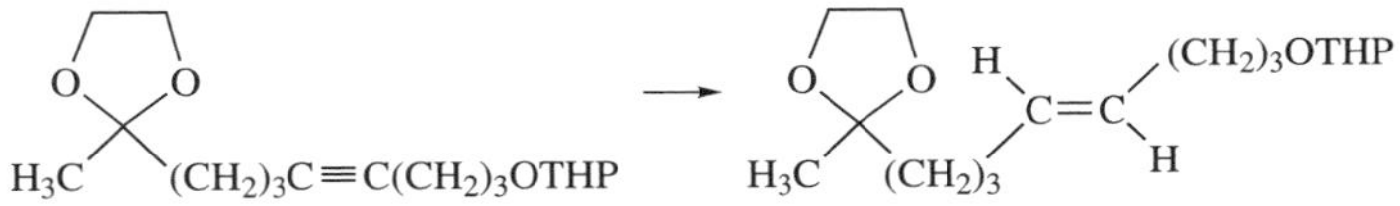

4. Predict the stereochemistry of the products from the following reactions and justify your prediction.

(a) O, O, O, O, CH3, CH3 $\xrightarrow[H_2O]{KBH_4}$

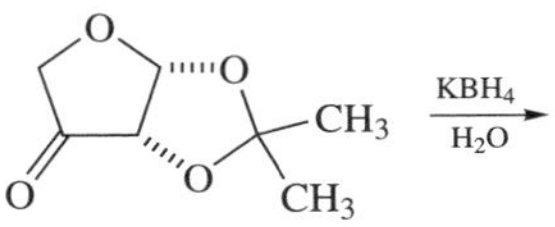

(b) H, CH3, C—C, O, Ph, Ph $\xrightarrow[Et_2O]{LiAlH_4}$

(c) H, O, H $\xrightarrow{LiBH(Et)_3}$

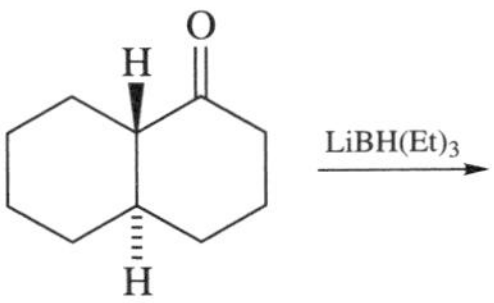

(d) $(CH_3)_2CH$, HO, CH_3 $\xrightarrow[ethanol]{Pt, H_2}$

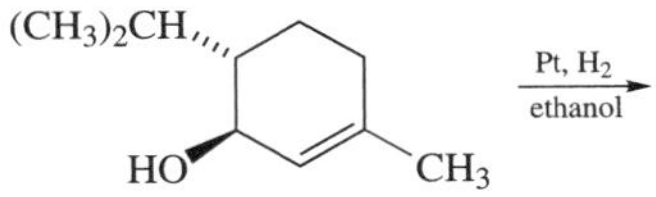

(e) O, O, H3C, H3C, CH3, O_2CCH_3 $\xrightarrow[(PPh_3)_3RhCl]{H_2}$

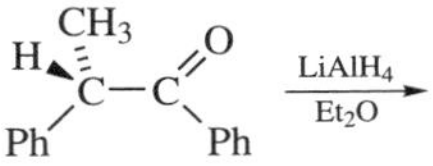

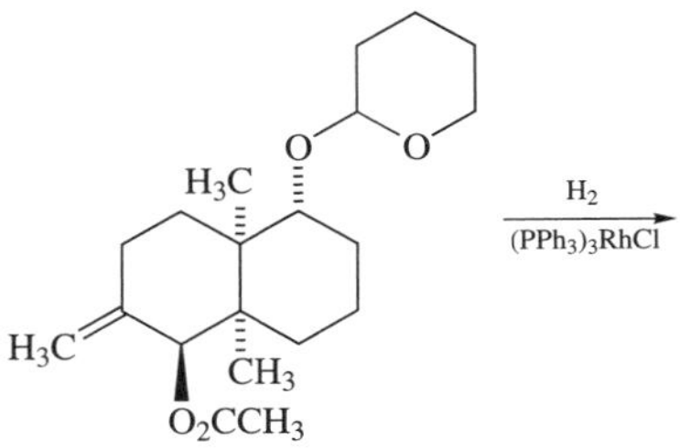

(f) $C(CH_3)_3$, H_3C, CH_3, OH $\xrightarrow[Rh/Al_2O_3]{H_2}$

(g) H_2C, O, HO, OCH_3, $HNCCH_3$, O $\xrightarrow{H_2, Pd}$

(h) CH_3, O, O, O, CH_3 $\xrightarrow{H_2/Pd\text{–}C}$

(i) OCH_2OCH_3, $PhCH_2O$, CH_3OCH_2O, O, OMe $\xrightarrow{Zn(BH_4)_2}$

(j) CH_3, OH $\xrightarrow{[NBD-Rh(P\;P)]^{1+}}$

(k) $CH_3(CH_2)_4C{\equiv}CCH_2OH$ $\xrightarrow[ether]{LiAlH_4}$

(l) CH_3, OH, CH_3 $\xrightarrow{[R_3P-Ir(COD)py]^{1+}}$

(m) $PhCHCCH_2CH_3$ (O above C; CH_3 below CH) $\xrightarrow{L\text{-}Selectride}$

(n) H, N, N–H, O, CH_3CO_2 $\xrightarrow[H_2, CH_2Cl_2]{[(C_6H_{11})_3P-Ir(COD)py]PF_6}$

(o) O_2N, OCH_3, OCH_3, CH_3O, TBDMSO, O, $NOCH_3$, CH_3, O, CH_3, CH_3, CH_3, CH_3, OCH_3 $\xrightarrow{ZnBH_4}$

(p) $\xrightarrow[H_2,\ CH_2Cl_2]{[(C_6H_{11})_3P-Ir(COD)py]PF_6}$

CH₃ O ... CH₂Ph CH₂OCH₃

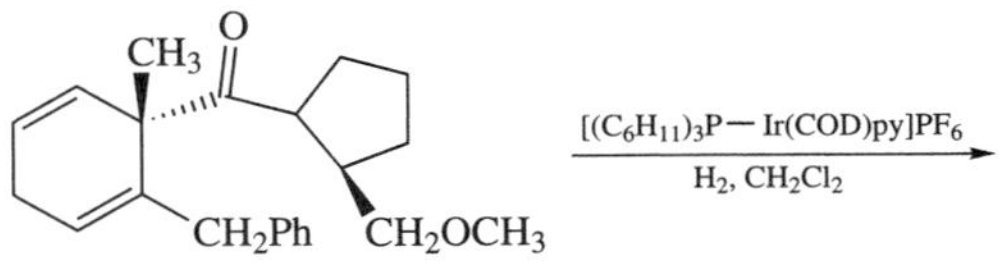

(q) $PhCH(CH_3)C(=O)CH_2CH_3 \xrightarrow{\text{L-Selectride}}$

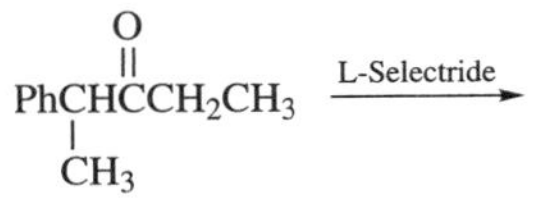

5. Suggest a convenient method for carrying out the following syntheses. The compound on the left is to be synthesized from the one on the right (retrosynthetic notation). No more than three steps should be necessary.

(a) $\Rightarrow$ C=O

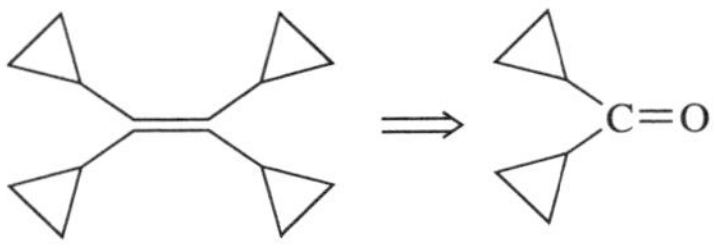

(b) CO_2CH_3 $\Rightarrow$ CO_2CH_3, O

(c) H_3C, O, O, H_3C, H, CH_2OH $\Rightarrow$ H_3C, O, O, CH_3

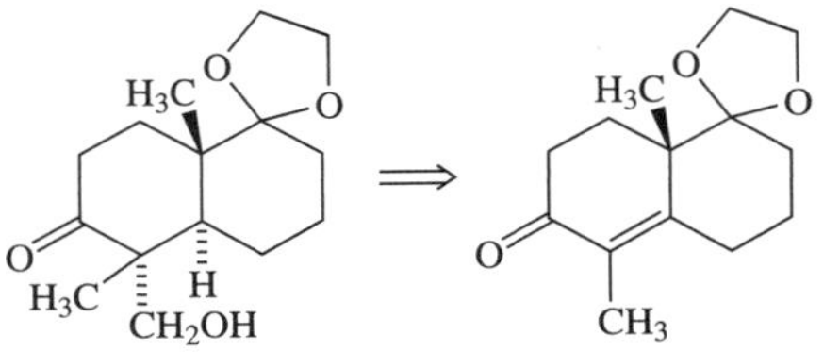

(d) HO H OH H, H_3C, C, C, C, C, CH_2OH, HO H H OH $\Rightarrow$ $HOCH_2$, O, OH, HO, OH, HO

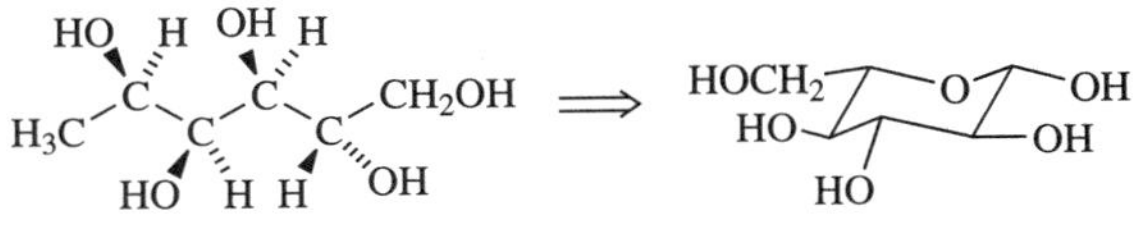

(e) CH_3 $\Rightarrow$ O, CH_3

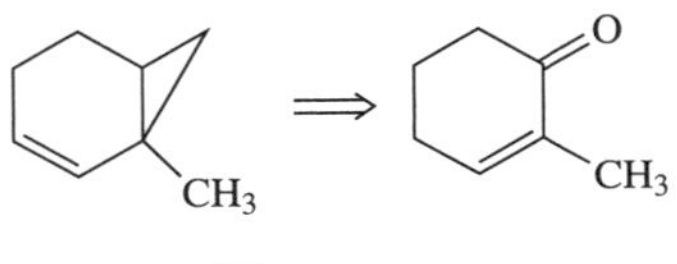

(f) CH_3, $HOCH_2$, OCH_3, OCH_3 $\Rightarrow$ HO_2C, OCH_3, OCH_3

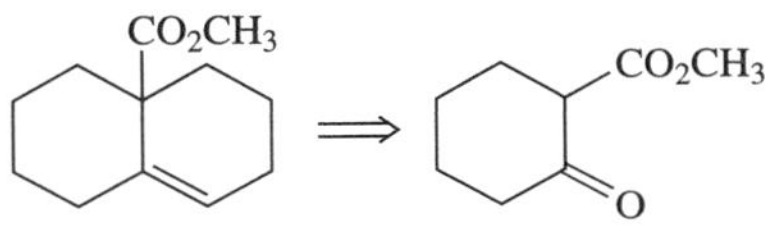

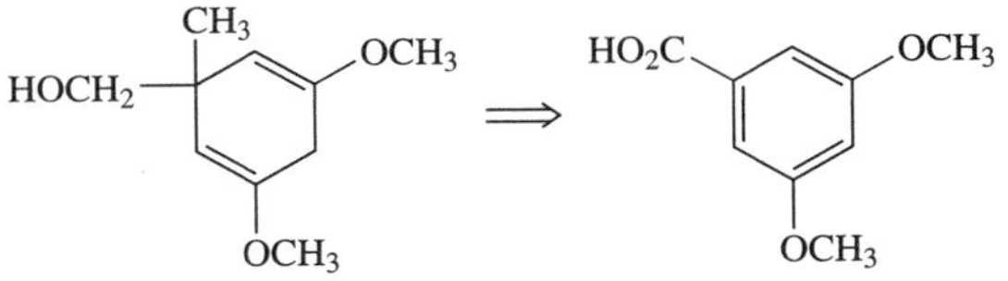

(g) CH_3, =O, CH_3 $\Rightarrow$ O=, CH_3, =O, CH_3

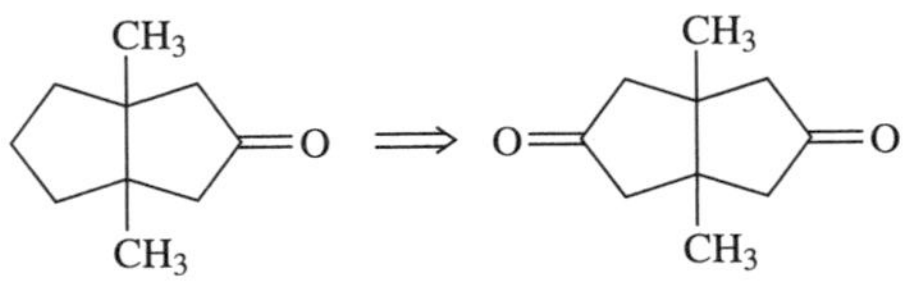

(h) *meso*-$(CH_3)_2CHCH(OH)CH(OH)CH(CH_3)_2 \Rightarrow (CH_3)_2CHCO_2CH_3$

(i) 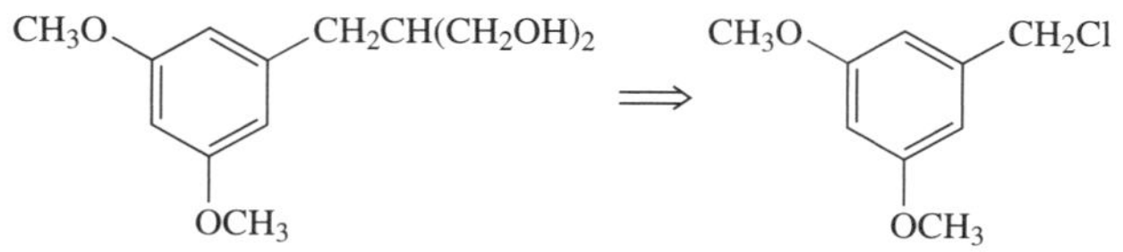 CH_3O–C_6H_3(–OCH_3)–$CH_2CH(CH_2OH)_2$ $\Longrightarrow$ CH_3O–C_6H_3(–OCH_3)–CH_2Cl

(j) 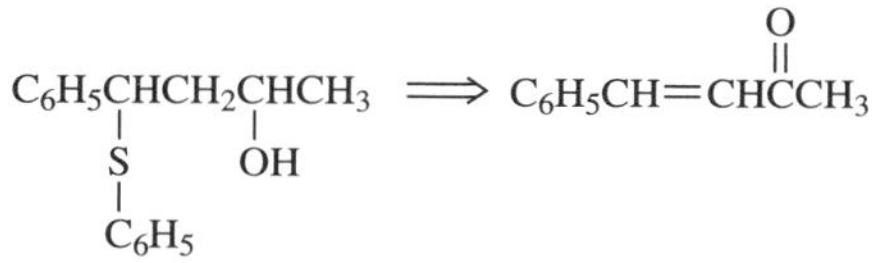 $C_6H_5CH(SC_6H_5)CH_2CH(OH)CH_3 \Longrightarrow C_6H_5CH{=}CHC(O)CH_3$

6. Offer an explanation to account for the observed differences in rate which are described.

   (a) $LiAlH_4$ reduces the ketone camphor about 30 times faster than does $NaAlH_4$.
   (b) The rate of reduction of camphor by $LiAlH_4$ is decreased by a factor of about 4 when a crown ether is added to the reaction mixture.
   (c) For reduction of cyclohexanones by lithium tri-*t*-butyoxyaluminum hydride, the addition of one methyl group at C-3 has little effect on the rate, but a second group has a large effect. The addition of a third methyl group at C-5 has no effect. The effect of a fourth group is also rather small.

	Rate
Cyclohexanone	439
3-Methylcyclohexanone	280
3,3-Dimethylcyclohexanone	17.5
3,3,5-Trimethylcyclohexanone	17.4
3,3,5,5-Tetramethylcyclohexanone	8.9

7. Suggest reaction conditions appropriate for stereoselectively converting the octalone shown to each of the diastereomeric decalones.

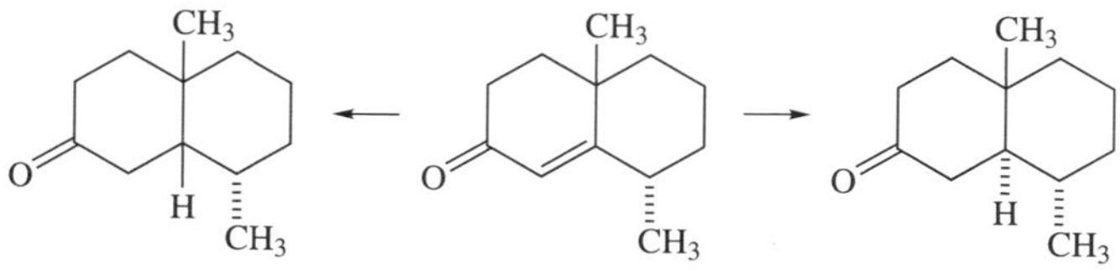

8. The fruit of a shrub which grows in Sierra Leone is very toxic and has been used as a rat poison. The toxic principle has been identified as *Z*-18-fluoro-9-octadecenoic acid. Suggest a synthesis for this material from 8-fluorooctanol, 1-chloro-7-iodo-heptane, acetylene, and any other necessary organic or inorganic reagents.

9. Each of the following molecules contains more than one potentially reducible group. Indicate a reducing agent which would be suitable for effecting the desired selective

reduction. Explain the basis for the expected selectivity.

(a)

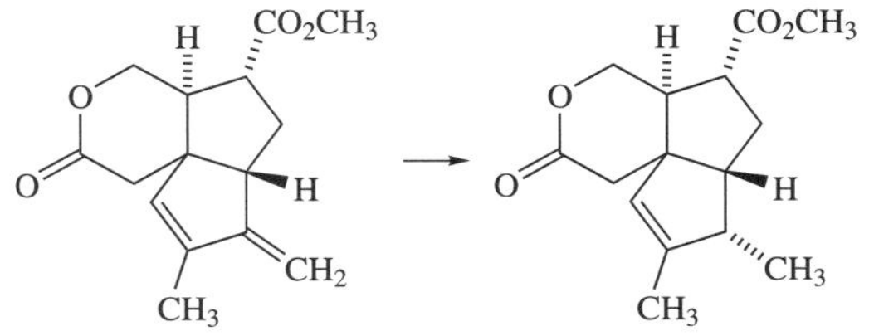

(b)

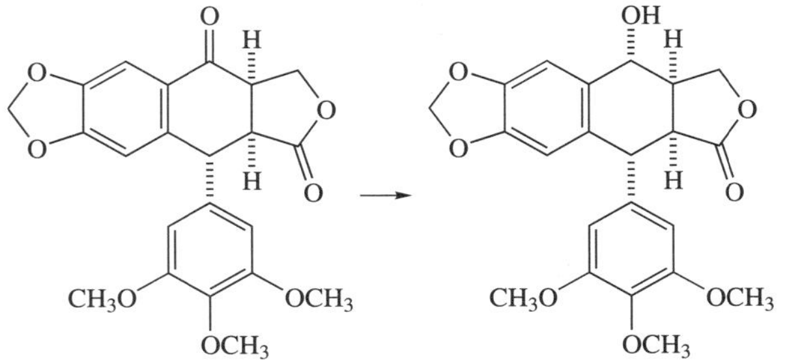

(c)

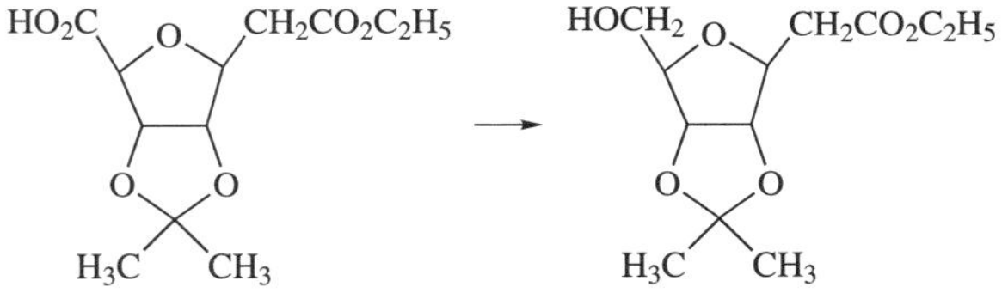

(d) $CH_3CH_2C{\equiv}CCH_2C{\equiv}CCH_2OH \longrightarrow CH_3CH_2C{\equiv}CCH_2CH{=}CHCH_2OH$

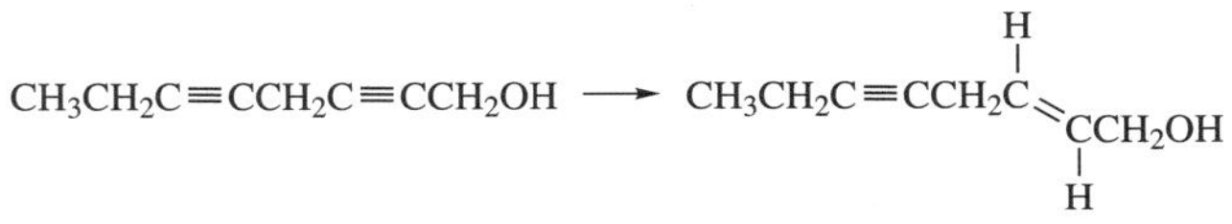

(e)

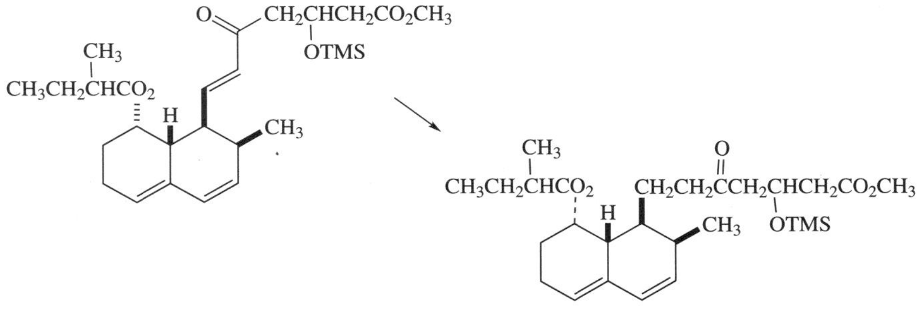

(f) $CH_3(CH_2)_3C(=O)(CH_2)_4C(=O)Cl \longrightarrow CH_3(CH_2)_3C(=O)(CH_2)_4CH{=}O$

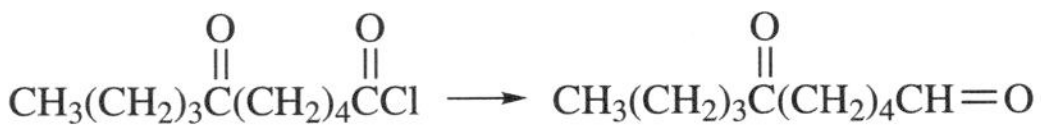

(g)

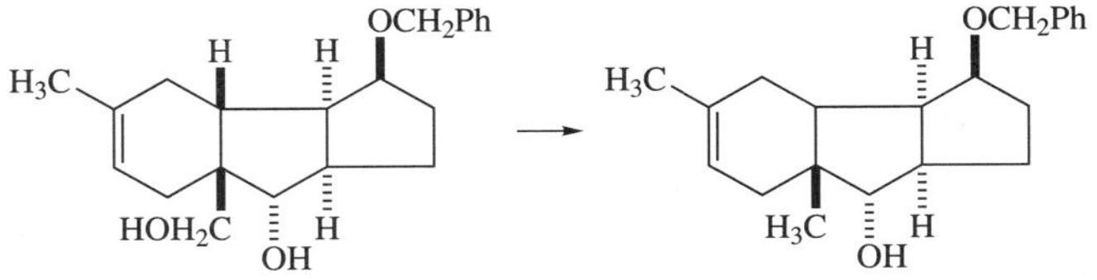

(h) $CH_3(CH_2)_2C(=O)(CH_2)_4C(=O)Cl \longrightarrow CH_3(CH_2)_2C(=O)(CH_2)_4C(=O)H$

(i) Ph, CH_3O, O, $NOCH_3$, CH_3, CH_3

(j) CH_3, CH_3, O, O, H, CH_3, $OSiR_3$ → O, HC, H, CH_3, HO, $OSiR_3$

10. Explain the basis of the observed stereoselectivity for the following reductions.

(a) O, CH_3 —(H–B⁻)→ OH, CH_3

(b) H, Br, Br, H —(Bu_3SnH)→ H, Br, H, H

(c) H_3C, OCH_3, O —(Li, NH_3, EtOH)→ H_3C, OCH_3, O, H

11. A valuable application of sodium cyanoborohydride is in the synthesis of amines by reductive amination. What combination of carbonyl and amine components would you choose to prepare the following amines by this route? Explain your choices.

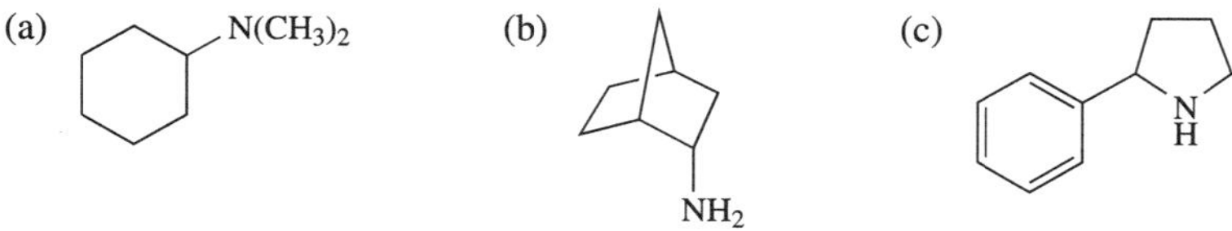

12. The reduction of *o*-bromophenyl allyl ether by $LiAlH_4$ has been studied in several solvents. In ether, two products are formed. The ratio **A** : **B** increases with increasing $LiAlH_4$ concentration. When $LiAlD_4$ is used as the reductant about half of the product **B** is a monodeuterated derivative. Provide a mechanistic rationale for these results.

What is the most likely location of the deuterium atom in the deuterated product? Why is the product not completely deuterated.

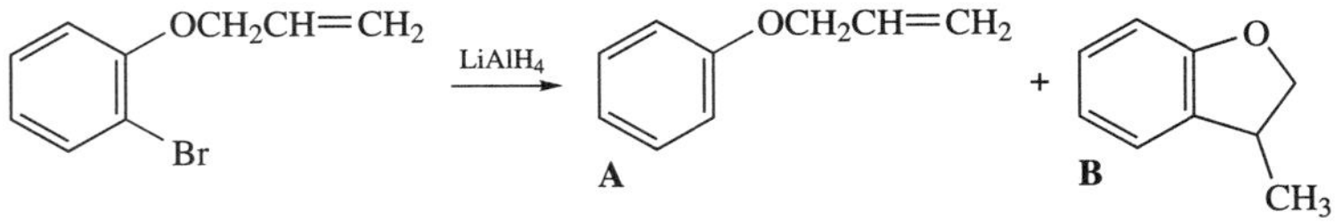

$$\text{o-BrC}_6\text{H}_4\text{OCH}_2\text{CH}{=}\text{CH}_2 \xrightarrow{\text{LiAlH}_4} \underset{\mathbf{A}}{\text{C}_6\text{H}_5\text{OCH}_2\text{CH}{=}\text{CH}_2} + \underset{\mathbf{B}}{\text{3-methyl-2,3-dihydrobenzofuran}}$$

13. A simple synthesis of 2-substituted cyclohexenones has been developed. Although the yields are only 25–30%, it is carried out as a "one-pot" process using the sequence of reactions shown below. Explain the mechanistic basis of this synthesis and identify the intermediate present after each stage of the reaction.

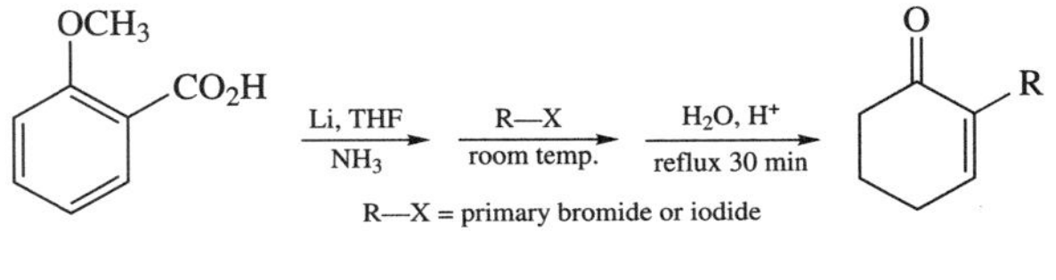

$$\text{2-methoxybenzoic acid (OCH}_3\text{, CO}_2\text{H)} \xrightarrow[\text{NH}_3]{\text{Li, THF}} \xrightarrow[\text{room temp.}]{\text{R—X}} \xrightarrow[\text{reflux 30 min}]{\text{H}_2\text{O, H}^+} \text{2-R-cyclohex-2-enone}$$

R—X = primary bromide or iodide

14. Birch reduction of 3,4,5-trimethoxybenzoic acid gives in 94% yield a dihydrobenzoic acid which bears only *two* methoxy substituents. Suggest a plausible structure for this product based on the mechanism of the Birch reduction.

15. The cyclohexenone **C** has been prepared in a one-pot process beginning with 4-methylpent-3-en-2-one. The reagents which are added in succession are 4-methoxyphenyllithium, Li, and NH_3, followed by acidic workup. Show the intermediate steps that are involved in this process.

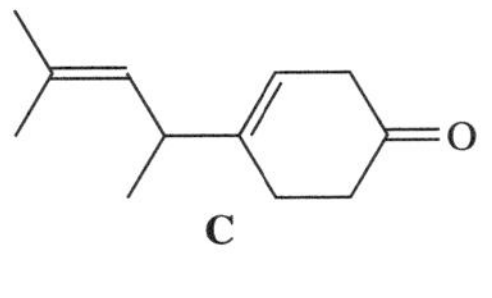

C

16. Ketones can be converted to nitriles by the following sequence of reagents. Indicate the intermediate stages of the reaction.

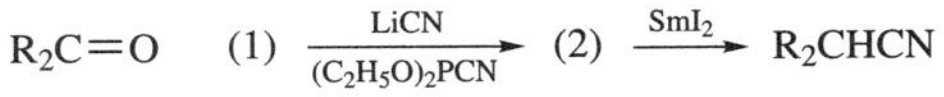

$$R_2C{=}O \quad (1) \xrightarrow[(C_2H_5O)_2PCN]{LiCN} \quad (2) \xrightarrow{SmI_2} R_2CHCN$$

17. Provide a mechanistic rationale for the outcome, including stereoselectivity, of the following reactions.

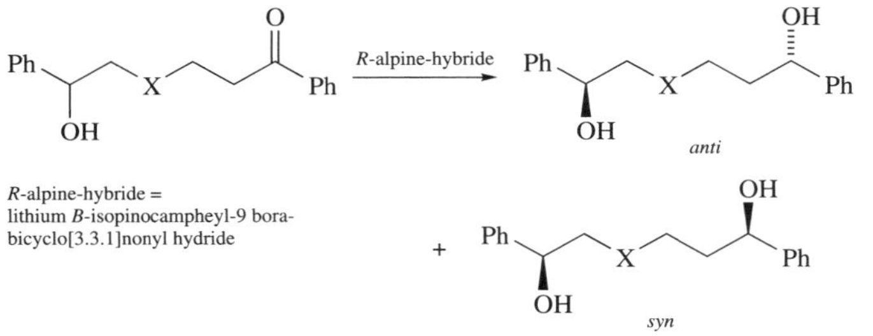
Ph
X
O
Ph
OH
R-alpine-hybride
OH
anti
syn
+
R-alpine-hybride =
lithium B-isopinocampheyl-9 bora-
bicyclo[3.3.1]nonyl hydride

X	*anti*:*syn*
CH_2	1.2:1
S	1.3:1
NCH_2Ph	12:1

18. In a multistep synthetic sequence, it was necessary to remove selectively one of two secondary hydroxyl groups.

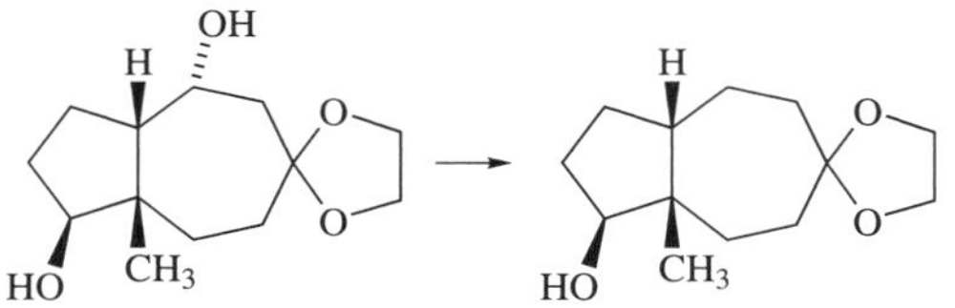
OH
H
O
O
HO
CH3

Consider several (at least three) methods by which this transformation might be accomplished. Discuss the relative merits of the various possibilities and recommend one as the most likely to succeed or be most convenient. Explain your choice.

19. In the synthesis of fluorinated analogs of an acetylcholinesterase inhibitor, huperzine A, it was necessary to accomplish reductive elimination of the diol **D** to **E**. Of the methods for diol reduction, which seem most compatible with the other functional groups in the molecule?

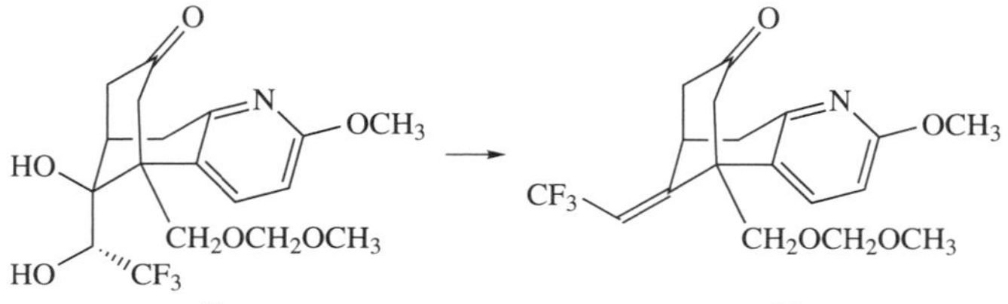
O
N
OCH3
HO
HO
CF3
CH2OCH2OCH3

D **E**

20. Wolff–Kishner reduction of ketones that bear other functional groups sometimes give products other than the corresponding methylene compound. Some examples are

given. Indicate a mechanism for each of the reactions.

(a) $(CH_3)_3CCCH_2OPh$ (with C=O) → $(CH_3)_3CCH{=}CH_2$

(b) [structure] → [structure]

H₃C, CH₃, CH₃, O → H₃C, CH₃, CH₃, OH

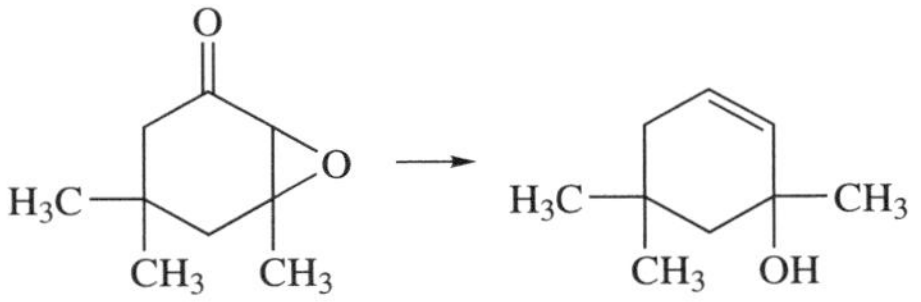

(c) [structure] → [structure]

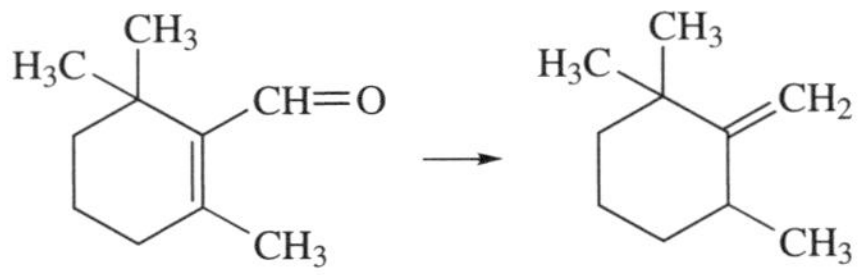

(d) $PhCH{=}CHCH{=}O$ → Ph–[pyrazoline] H H

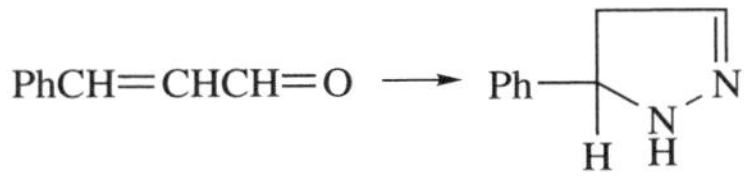

21. Suggest reagents and reaction conditions that would be suitable for each of the following selective or partial reductions.

(a) $HO_2C(CH_2)_4CO_2C_2H_5$ → $HOCH_2(CH_2)_4CO_2C_2H_5$

(b) [decalin with CH_3, $CH(CH_3)_2$, $CH_2CN(CH_3)_2$ (C=O)] → [decalin with CH_3, $CH(CH_3)_2$, $CH_2CH{=}O$]

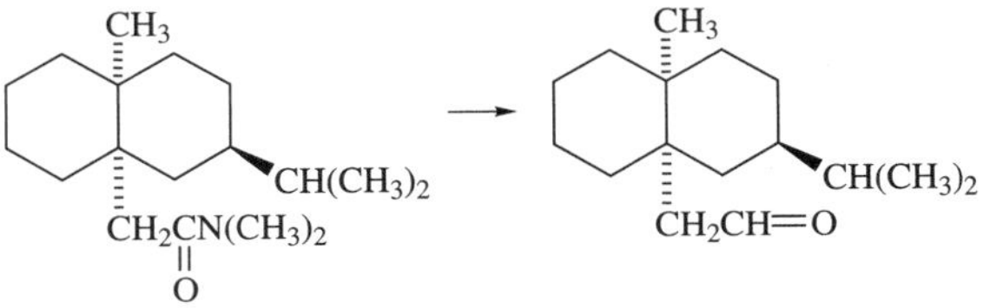

(c) $CH_3C(CH_2)_2CO_2C_8H_{17}$ (C=O) → $CH_3(CH_2)_3CO_2C_8H_{17}$

(d) CH_3CNH–C_6H_4–CO_2CH_3 (C=O) → CH_3CH_2NH–C_6H_4–CO_2CH_3

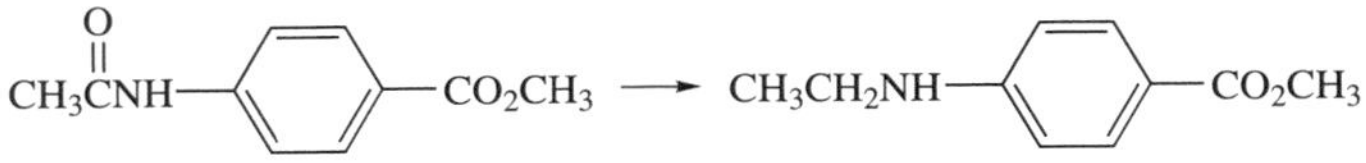

(e) O_2N–C_6H_4–C(=O)–C_6H_5 → O_2N–C_6H_4–CH_2–C_6H_5

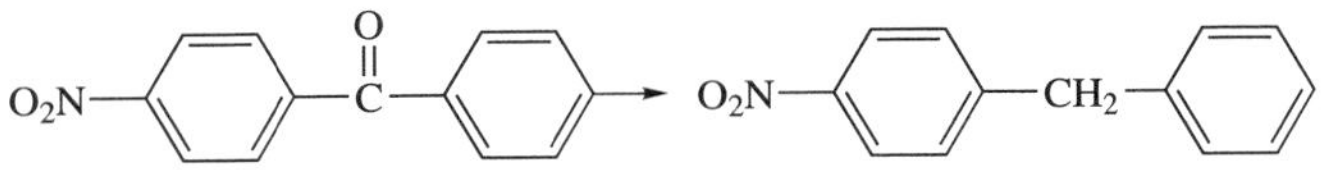

(f) 3-methylcyclohex-2-enone (CH_3, =O) → 3-methylcyclohex-2-enol (CH_3, OH)

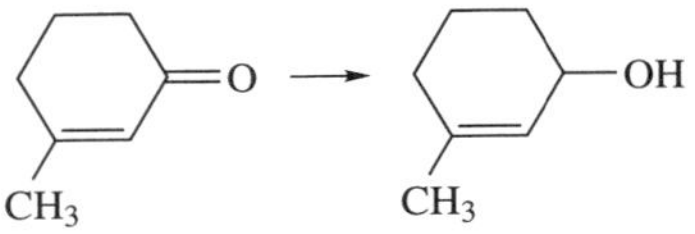

(g) [structure with O–O and O] → [structure with O–O and O]

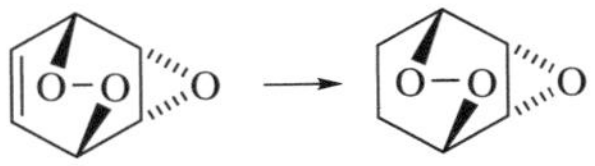

22. In the reduction of the ketone **F**, product **G** is favored with *increasing* stereoselectivity in the order $NaBH_4$ < $LiAlH_2(OCH_2CH_2OCH_3)_2$ < $Zn(BH_4)_2$. With L-Selectride, stereoisomer **H** is favored. Account for the dependence of the stereoselectivity on the various reducing agents.

F → **G** + **H**

Ar = 4-methoxyphenyl
R = benzyl
MOM = methoxymethyl

23. The following reducing agents effect enantioselective reduction of ketones. Propose a mechanism and transition-state structure which would be in accord with the observed enantioselectivity.

(a) $PhCOCO_2CH_3$ + (pinanyl borane) → *R*-α-hydroxyester in 90% e.e.

(b) $PhCOCH_2CH_3$ + BH_3 (0.6 equiv) + (oxazaborolidine, Ph, Ph, N–B–Me) (0.1 equiv) → *R*-alcohol in 97% e.e.

(c) 4-Br-$C_6H_4COCH_3$ + (pinanyl$)_2$BCl → *S*-alcohol in 97% e.e.

24. Devise a sequence of reactions which would accomplish the following synthesis:

(target, MeO) from (MeO ketone)

25. A group of topologically unique molecules known as "betweenanenes" have been synthesized. Successful synthesis of such molecules depends on effective means of closing large rings. Suggest an overall strategic approach (details are not required) to synthesize such molecules. Suggest reaction types which might be considered for formation of the large rings.

26. Give the products expected from the following reactions of Sm(II) reagents.

(a) PhCH$_2$O OCH$_2$Ph O=CH CH=O PhCH$_2$O OCH$_2$Ph $\xrightarrow{SmI_2}$

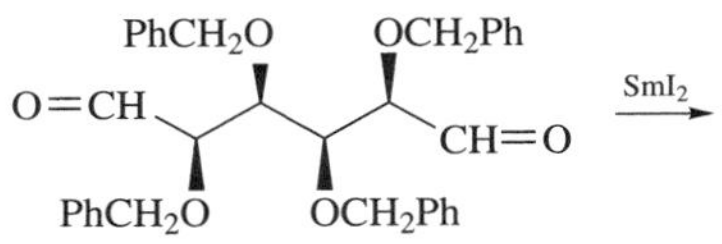

(b) (CH$_2$)$_3$CH=CHCO$_2$CH$_3$ CH$_3$ O $\xrightarrow{SmI_2}$

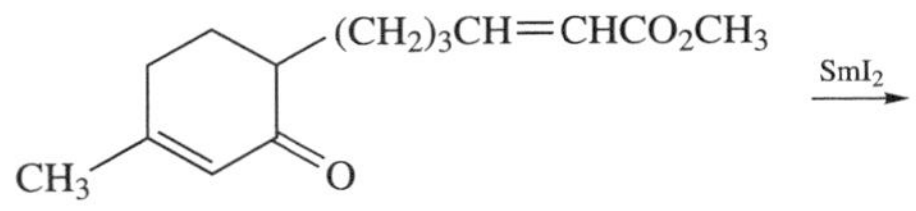

(c) CH$_3$ H O OH O CH$_2$CCH$_3$ O $\xrightarrow[HMPA]{SmI_2}$

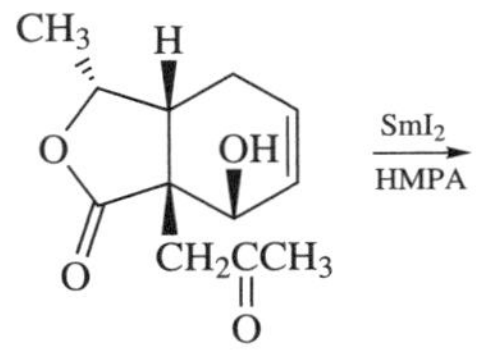

(d) TBDMSO CO$_2$CH$_3$ CH$_3$ O CO$_2$CH$_3$ CH$_3$ O CH$_3$ $\xrightarrow{SmI_2}$

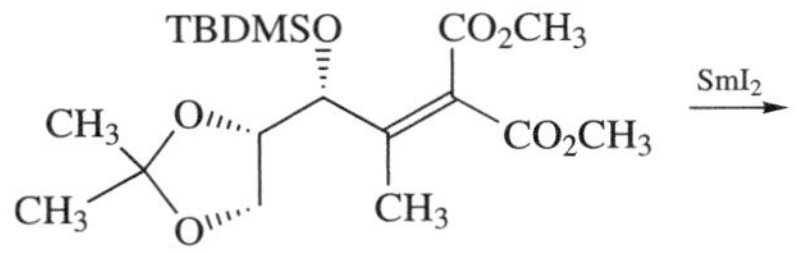

(e) OTBDMS CH=O CO$_2$CH$_3$ O O CH$_3$ CH$_3$ $\xrightarrow{\text{1) SmI}_2\text{ 2) cyclohexyl-CH=O}}$

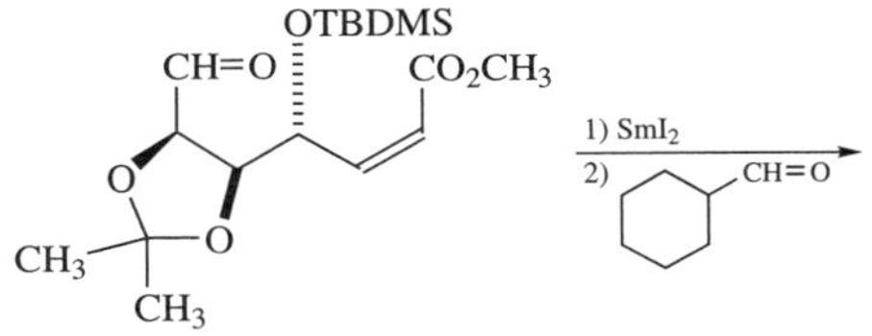

(f) CO$_2$Ph (CH$_3$)$_3$SiC≡CCH$_2$CH$_2$N CH CH$_2$OTBDMS $\xrightarrow{SmI_2}$

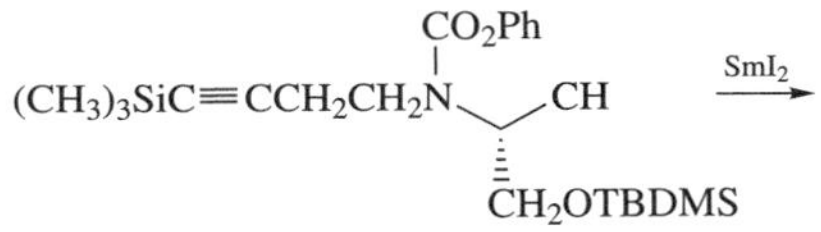

6

Cycloadditions, Unimolecular Rearrangements, and Thermal Eliminations

Introduction

Most of the reactions described in the preceding chapters involve polar or polarizable reactants and proceed through polar intermediates or transition states. One reactant can be identified as nucleophilic, and the other as electrophilic. Carbanion alkylations, nucleophilic additions to carbonyl groups, and electrophilic additions to alkenes are examples of such reactions. The reactions to be examined in this chapter, on the other hand, occur by a reorganization of valence electrons through activated complexes that are not much more polar than the reactants. These reactions usually proceed through cyclic transition states, and little separation of charge occurs during these processes. The energy necessary to attain the transition state is usually provided by thermal or photochemical excitation of the reactant(s), and frequently no other reagents are involved. Many of the transformations fall into the category of *concerted pericyclic reactions*, and the transition states are stabilized by favorable orbital interactions, as discussed in Chapter 11 of Part A. We will also discuss some reactions which effect closely related transformations but which, on mechanistic scrutiny, are found to proceed through discrete intermediates.

6.1. Cycloaddition Reactions

Cycloaddition reactions result in the formation of a new ring from two reacting molecules. A concerted mechanism requires that a single transition state, and therefore no intermediate, lie on the reaction path between reactants and adduct. Two important examples of cycloadditions that usually occur by concerted mechanisms are the *Diels–Alder reaction*,

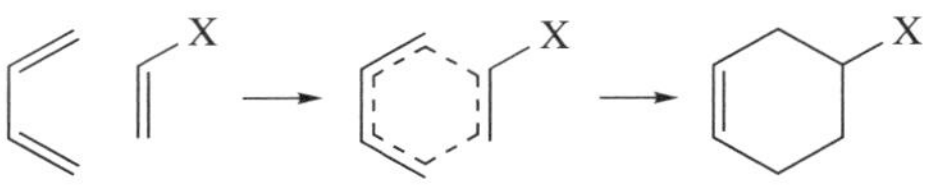

and *1,3-dipolar cycloaddition*:

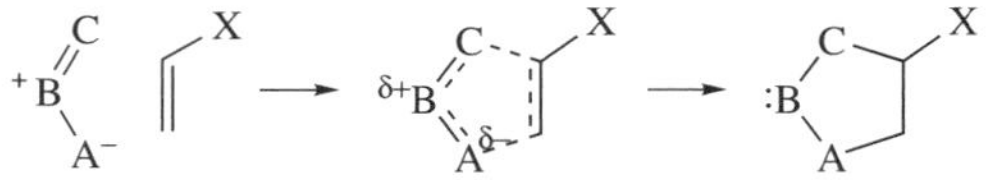

A firm understanding of concerted cycloaddition reactions developed as a result of the formulation of the mechanisms within the framework of molecular orbital (MO) theory. Consideration of the molecular orbitals of reactants and products revealed that, in many cases, a smooth transformation of the orbitals of the reactants to those of products is possible. In other cases, reactions that might appear feasible if no consideration is given to the symmetry and spatial orientation of the orbitals are found to require high-energy transition states when the orbitals are considered in detail. (Review Section 11.3 of Part A for a discussion of the orbital symmetry analysis of cycloaddition reactions.) The relationships between reactant and transition-state orbitals permit description of potential cycloaddition reactions as "allowed" or "forbidden" and permit conclusions as to whether specific reactions are likely to be energetically feasible. In this chapter, the synthetic applications of cycloaddition reactions will be emphasized. The same orbital symmetry relationships that are informative as to the feasibility of a reaction are often predictive of the regiochemistry and stereochemistry. This predictability is an important feature for synthetic purposes. Another attractive feature of cycloaddition reactions is the fact that *two* new bonds are formed in a single reaction. This can enhance the efficiency of a synthetic process.

6.1.1. The Diels–Alder Reaction: General Features

The cycloaddition of alkenes and dienes is a very useful method for forming substituted cyclohexenes. This reaction is known as the *Diels–Alder reaction*.[1] The concerted nature of the mechanism was generally accepted and the stereospecificity of the reaction was firmly established before the importance of orbtial symmetry was recognized. In the terminology of orbital symmetry classification, the Diels–Alder reaction is a $[4\pi_s + 2\pi_s]$ cycloaddition, an allowed process. The transition state for a concerted reaction requires that the diene adopt the *s-cis* conformation. The diene and substituted alkene (which is called the *dienophile*) approach each other in approximately parallel planes. The symmetry properties of the π orbitals permit stabilizing interations between C-1 and C-4 of the diene and the dienophile. Usually, the strongest interaction is between the highest occupied molecular orbital (HOMO) of the diene and the lowest unoccupied molecular orbital (LUMO) of the dienophile. The interaction between the frontier orbitals is depicted in Fig. 6.1.

1. L. W. Butz and A. W. Rytina, *Org. React.* **5**:136 (1949); M. C. Kloetzel, *Org. React.* **4**:1 (1948); A. Wasserman, *Diels–Alder Reactions*, Elsevier, New York, 1965; R. Huisgen, R. Grashey, and J. Sauer, in *Chemistry of Alkenes*, S. Patai, ed., John Wiley & Sons, New York, 1964, pp. 878–928.

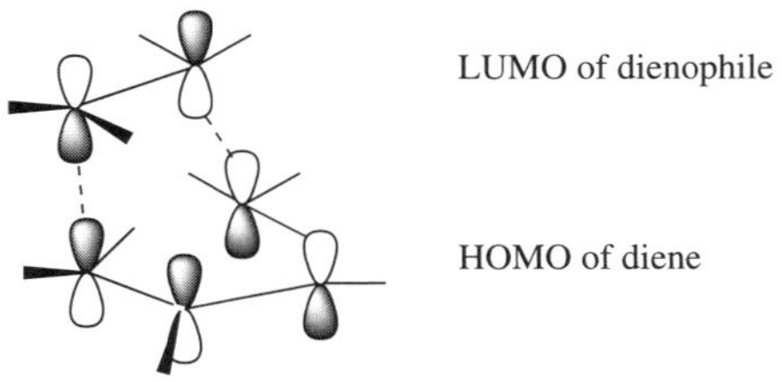

LUMO of dienophile

HOMO of diene

Fig. 6.1. Cycloaddition of an alkene and a diene, showing interaction of LUMO of alkene with HOMO of diene.

There is a strong electronic substituent effect on the Diels–Alder addition. The alkenes that are most reactive toward simple dienes are those with electron-attracting groups. Thus, among the most reactive dienophiles are quinones, maleic anhydride, and nitroalkenes. α,β-Unsaturated aldehydes, esters, ketones, and nitriles are also effective dienophiles. It is significant that if an electron-poor diene is utilized, the preference is reversed and electron-rich alkenes, such as vinyl ethers, are the best dienophiles. Such reactions are called *inverse electron demand* Diels–Alder reactions. These relationships are readily understood in terms of frontier orbital theory. Electron-rich dienes have high-energy HOMOs and interact strongly with the LUMOs of electron-poor dienophiles. When the substituent pattern is reversed and the diene is electron-poor, the strongest interaction is between the dienophile HOMO and the diene LUMO.

A question of regioselectivity arises when both the diene and the alkene are unsymmetrically substituted. Generally, there is a preference for the "*ortho*" and "*para*" orientations, respectively, as in the examples shown.[2]

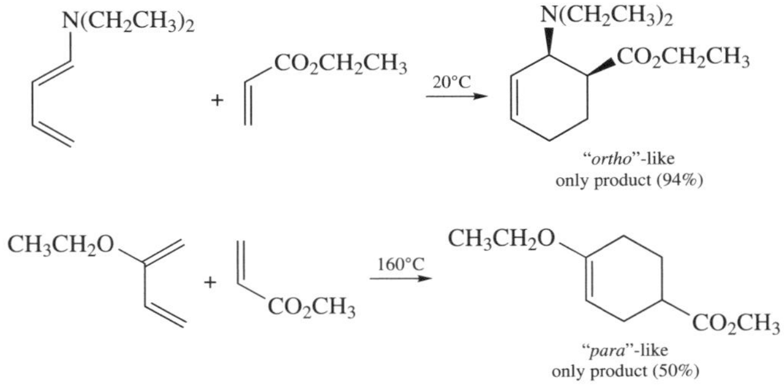

$N(CH_2CH_3)_2$ + $CO_2CH_2CH_3$ $\xrightarrow{20°C}$ $N(CH_2CH_3)_2$ $CO_2CH_2CH_3$

"*ortho*"-like
only product (94%)

CH_3CH_2O + CO_2CH_3 $\xrightarrow{160°C}$ CH_3CH_2O CO_2CH_3

"*para*"-like
only product (50%)

This preference can also be understood in terms of frontier orbital theory.[3] When the dienophile bears an electron-withdrawing substituent and the diene an electron-releasing one, the strongest interaction is between the HOMO of the diene and the LUMO of the dienophile. The reactants are oriented so that the carbons having the highest coefficients of the two frontier orbitals begin the bonding process. This is illustrated in Fig. 6.2 and leads to the observed regiochemical preference.

2. J. Sauer, *Angew. Chem. Int. Ed. Engl.* **6**:16 (1967).
3. K. N. Houk, *Acc. Chem. Res.* **8**:361 (1975); I. Fleming, *Frontier Orbitals and Organic Chemical Reactions*, Wiley-Interscience, New York, 1976; O. Eisenstein, J. M. LeFour, N. T. Anh, and R. F. Hudson, *Tetrahedron* **33**:523 (1977).

(a) Coefficient of C-2 is higher than coefficient of C-1 in LUMO of dienophile bearing an electron-withdrawing substituent.

EWG is a π acceptor such as $-C(O)R$, $-NO_2$, $-CN$

(b) Coefficient of C-4 is higher than coefficient of C-1 in HOMO of diene bearing an electron-releasing substituent at C-1.

ERG is π donor such as $-OR$, $-SR$, $-OSiMe_3$

(c) Coefficient of C-1 is higher than coefficient of C-4 in HOMO of diene bearing an electron-releasing substituent at C-2.

(d) Regioselectivity of Diels–Alder addition corresponds to that given by matching carbon atoms having the largest coefficients in the frontier orbitals.

"*ortho*"-like orientation:

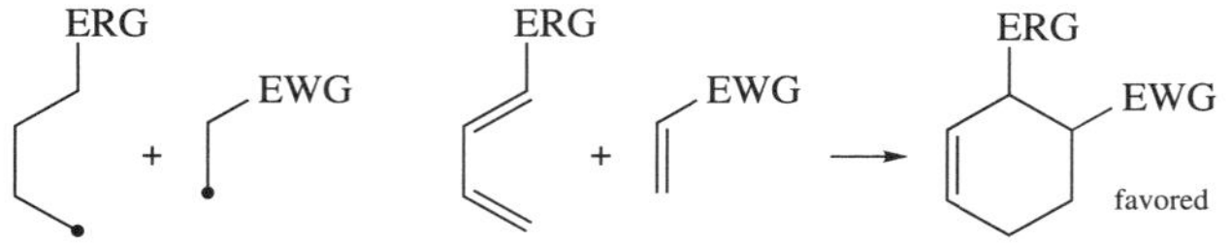

"*para*"-like orientation:

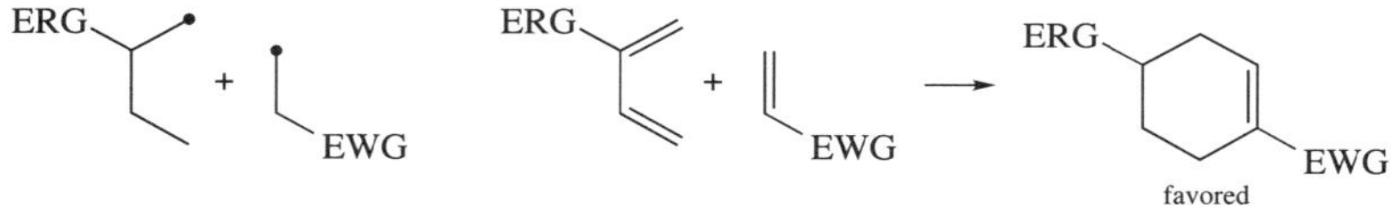

Fig. 6.2. HOMO–LUMO interactions rationalize regioselectivity of Diels–Alder cycloaddition reactions.

For an unsymmetrical dienophile, there are two possible stereochemical orientations with respect to the diene. The two possible orientations are called *endo* and *exo*, as illustrated in Fig. 6.3. In the *endo* transition state, the reference substituent on the dienophile is oriented toward the π orbitals of the diene. In the *exo* transition state, the substituent is oriented away from the π system. For many substituted butadiene derivatives, the two transition states lead to two different stereoisomeric products. The *endo* mode of addition is usually preferred when an electron-attracting substituent such as a carbonyl group is present on the dienophile. The empirical statement which describes this preference is called the *Alder rule*. Frequently, a mixture of both stereoisomers is formed, and sometimes the *exo* product predominates, but the Alder rule is a useful initial guide to prediction of the stereochemistry of a Diels–Alder reaction. The *endo* product is often the more sterically congested. The preference for the *endo* transition state

(a)

(b)

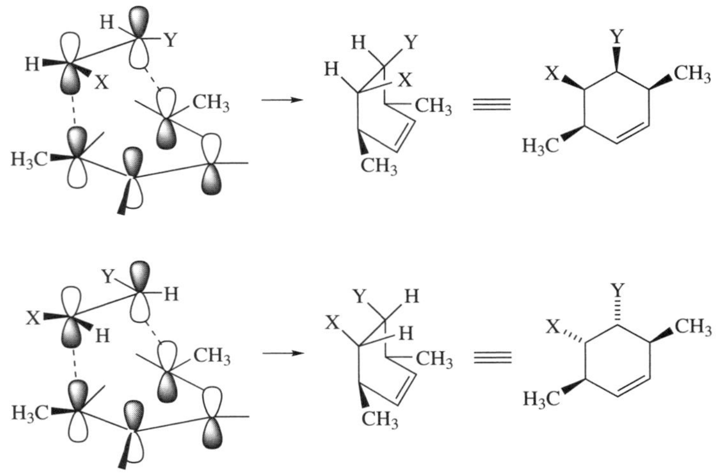

Fig. 6.3. *Endo* (a) and *exo* (b) addition in a Diels–Alder reaction.

is the result of interaction between the dienophile substituent and the π electrons of the diene. Dipolar attractions and van der Waals attractions may also be involved.[4]

Diels–Alder cycloadditions are sensitive to steric effects of two major types. Bulky substituents on the dienophile or on the termini of the diene can hinder approach of the two components to each other and decrease the rate of reaction. This effect can be seen in the relative reactivity of 1-substituted butadienes toward maleic anhydride.[5]

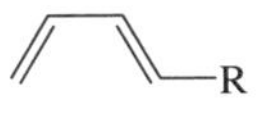

R	k_{rel} (25°C)
$-H$	1
$-CH_3$	4.2
$-C(CH_3)_3$	< 0.05

Substitution of hydrogen by a methyl group results in a slight rate *increase*, as a result of the electron-releasing effect of the methyl group. A *t*-butyl substituent produces a large rate *decrease*, because the steric effect is dominant.

The other type of steric effect has to do with interactions between diene substituents. Adoption of the *s-cis* conformation of the diene in the transition state brings the *cis*-oriented 1- and 4-substituents on the diene close together. *trans*-1,3-Pentadiene is 10^3 times more reactive than 4-methyl-1,3-pentadiene toward the very reactive dienophile tetracyanoethylene. This is because of the unfavorable interaction between the additional methyl substituent and the C-1 hydrogen in the *s-cis* conformation.[6]

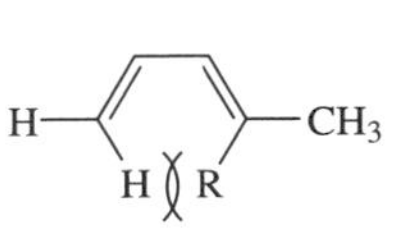

R	k_{rel}
$-H$	1
$-CH_3$	10^{-3}

4. Y. Kobuke, T. Sugimoto, J. Furukawa, and T. Fueno, *J. Am. Chem. Soc.* **94:**3633 (1972); K. L. Williamson and Y.-F. L. Hsu, *J. Am. Chem. Soc.* **92:**7385 (1970).
5. D. Craig, J. J. Shipman, and R. B. Fowler, *J. Am. Chem. Soc.* **83:**2885 (1961).
6. C. A. Stewart, Jr., *J. Org. Chem.* **28:**3320 (1963).

Relatively small substituents at C-2 and C-3 of the diene exert little steric influence on the rate of Diels–Alder addition. 2,3-Dimethylbutadiene reacts with maleic anhydride about 10 times faster than butadiene, and this is because of the electronic effect of the methyl groups. 2-*t*-Butyl-1,3-butadiene is 27 times more reactive than butadiene. This is because the *t*-butyl substituent favors the *s-cis* conformation, because of the steric repulsions in the *s-trans* conformation.

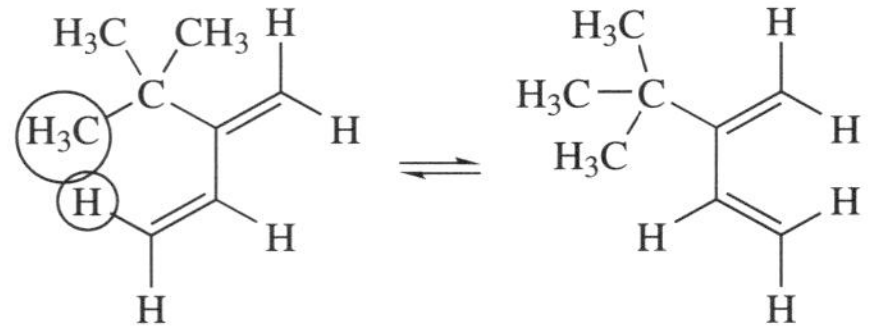

The presence of a *t*-butyl substituent on *both* C-2 and C-3, however, prevents attainment of the *s-cis* conformation, and Diels–Alder reactions of 2,3-di(*t*-butyl)-1,3-butadiene have not been observed.[7]

Lewis acids such as zinc chloride, boron trifluoride, aluminum chloride, and diethylaluminum chloride catalyze Diels–Alder reactions.[8] The catalytic effect is the result of coordination of the Lewis acid with the dienophile. The complexed dienophile is more electrophilic and more reactive toward electron-rich dienes. The mechanism of the cycloaddition is still believed to be concerted, and high stereoselectivity is observed.[9] Lewis acid catalysts also usually increase the regioselectivity of the reaction.

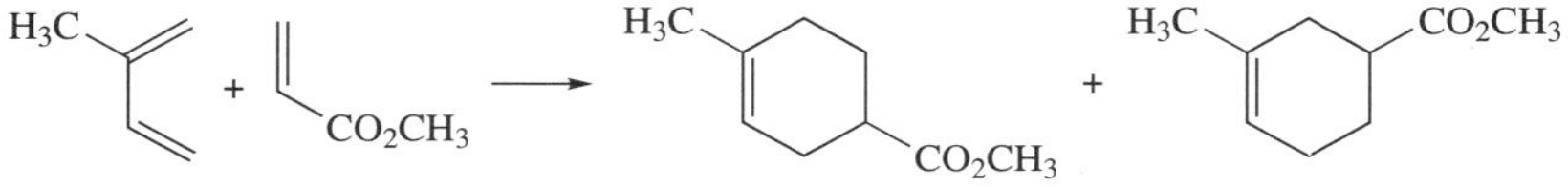

Ref. 10

	"*para*"-like	"*meta*"-like
	Product ratio	
Uncatalyzed reaction: 120°C, 6 h	70%	30%
Aluminum chloride catalyzed: 20°C, 3 h	95%	5%

The stereoselectivity of any particular reaction depends on the details of the structure of the transition state. The structures of several enone–Lewis acid complexes have been determined by X-ray crystallography.[11] The site of complexation is the carbonyl oxygen, which maintains a trigonal geometry, but with somewhat expanded angles (130–140°). The Lewis acid is normally *anti* to the larger carbonyl substituent. Boron trifluoride

7. H. J. Backer, *Rec. Trav. Chim. Pays-Bas* **58:**643 (1939).
8. P. Yates and P. Eaton, *J. Am. Chem. Soc.* **82:**4436 (1960); T. Inukai and M. Kasai, *J. Org. Chem.* **30:**3567 (1965); T. Inukai and T. Kojima, *J. Org. Chem.* **32:**869, 872 (1967); F. Fringuelli, F. Pizzo, A. Taticchi, and E. Wenkert, *J. Org. Chem.* **48:**2802 (1983); F. K. Brown, K. N. Houk, D. J. Burnell, and Z. Valenta, *J. Org. Chem.* **52:**3050 (1987).
9. K. N. Houk, *J. Am. Chem. Soc.* **95:**4094 (1973).
10. T. Inukai and T. Kojima, *J. Org. Chem.* **31:**1121 (1966).
11. S. Shambayati, W. E. Crowe, and S. L. Schreiber, *Angew. Chem. Int. Ed. Engl.* **29:**256 (1990).

complexes are tetrahedral, but Sn(IV) and Ti(IV) complexes can be trigonal bipyramidal or octahedral. The structure of the 2-methylpropenal–BF_3 complex is illustrative.[12]

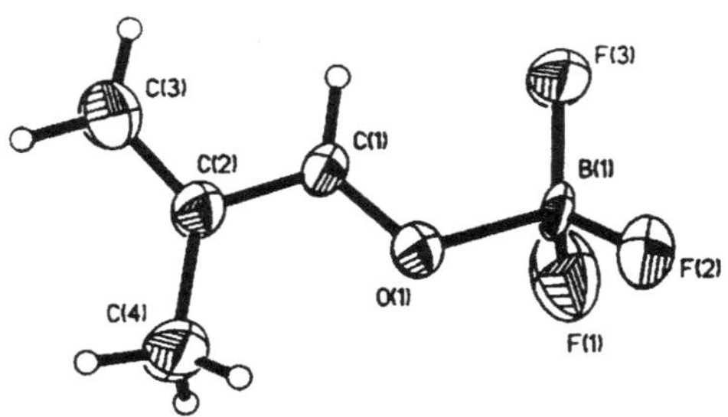

Chelation can favor a particular structure. For example, *O*-acryloyl lactates adopt a chelated structure with $TiCl_4$.[13]

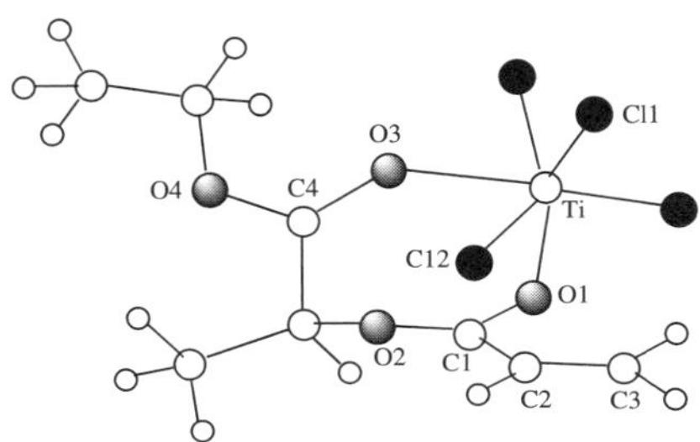

Theoretical calculations (6-31G*) have been used to compare the energies of four possible transition states for Diels–Alder reaction of the BF_3 complex of methyl acrylate with 1,3-butadiene. The results are summarized in Fig. 6.4. The *endo* transition state with the *s-trans* conformation of the dienophile is preferred to the others by about 2 kcal/mol.[14]

Some Diels–Alder reactions are also catalyzed by high concentrations of $LiClO_4$ in ether.[15] This catalysis may be a reflection of Lewis acid complexation of Li^+ with the dienophile.[16] Other cations can catalyze Diels–Alder reactions of certain dienophiles. For

12. Structure reprinted from E. J. Corey, T.-P. Loh, S. Sarshar, and M. Azimioara, *Tetrahedron Lett.* **33:**6945, Copyright 1992, with permission from Elsevier Science.
13. Structure reprinted from T. Poll, J. O. Metter, and G. Helmchen, *Angew. Chem. Int. Ed. Engl.* **24:**112 (1985), with permission.
14. (a) J. I. Garcia, J. A. Mayoral, and L. Salvatella, *J. Am. Chem. Soc.* **118:**11680 (1996); (b) J. I. Garcia, J. A. Mayoral, and L. Salvatella, *Tetrahedron* **53:**6057 (1997).
15. P. A. Grieco, J. J. Nunes, and M. D. Gaul, *J. Am. Chem. Soc.* **112:**4595 (1990).
16. M. A. Forman and W. P. Dailey, *J. Am. Chem. Soc.* **113:**2761 (1991).

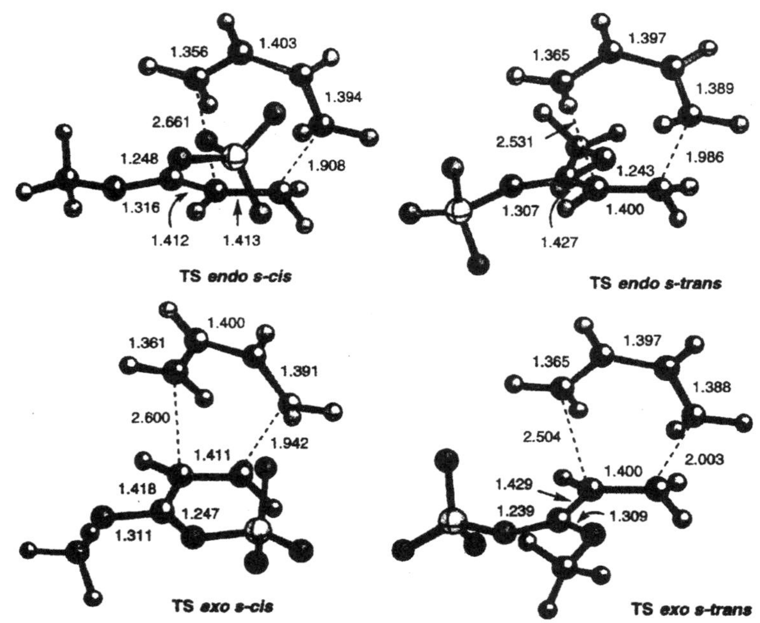

Structure	RHF/6-31G*// RHF/6-31G*		RHF/6-311++G**// RHF/6-31G*		B3LYP/6-311+G(2d,p)// RHF/6-31G*	
TS *endo s-cis*–BF_3	−782.749754	2.84	−782.959815	2.51	−787.274886	2.23
TS *endo s-trans*–BF_3	−782.754290	0.00	−782.963810	0.00	−787.278442	0.00
TS *exo s-cis*–BF_3	−782.751645	1.66	−782.960503	2.07	−787.277135	0.82
TS *exo s-trans*–BF_3	−782.751519	1.74	−782.961505	1.45	−787.277118	0.83

Fig. 6.4. Transition structures of the reaction between 1,3-butadiene and methyl acrylate, calculated at the *ab initio* RHF/6-31G* level. Total energies are in hartrees and relative energies in kcal/mol. (Reprinted from Ref. 14b, Copyright 1997, with permission from Elsevier Science.)

example, Cu^{2+} strongly catalyzes addition reactions of 2-pyridyl styryl ketones, presumably through a chelate with the carbonyl oxygen and pyridine nitrogen.[17]

Rate (M^{-1} s^{-1})	Relative rate	Solvent
1.3×10^{-5}	1	Acetonitrile
3.8×10^{-5}	2.9	Ethanol
4.0×10^{-3}	310	Water
3.25	250,000	Water + 0.01 *M* $Cu(NO_3)_2$

17. S. Otto and J. B. F. N. Engberts, *Tetrahedron Lett.* **36**:2645 (1995).

Lanthanide salts have also been found to catalyze Diels–Alder reactions. For example, with 10 mol % $Sc(O_3SCF_3)_3$ added, isoprene and methyl vinyl ketone react to give the expected adduct in 91% yield after 13 h at 0°C.[18]

$$CH_3\overset{O}{\overset{\|}{C}}CH{=}CH_2 + CH_2{=}\overset{CH_3}{\overset{|}{C}}CH{=}CH_2 \xrightarrow{10\ \text{mol}\ \%\ Sc(O_3SCF_3)_3}$$

Among the unique features of $Sc(O_3SCF_3)_3$ is its ability to function as a catalyst in hydroxylic solvents. Other dienophiles, including *N*-acryloyloxazolinones (see page 349), also are subject to catalysis by $Sc(O_3SCF_3)_3$.

The solvent also has an important effect on the rate of Diels–Alder reactions. The traditional solvents have been nonpolar organic solvents such as aromatic hydrocarbons. However, water and other highly polar solvents, such as ethylene glycol and formamide, have been found to accelerate a number of Diels–Alder reactions.[19] The accelerating effect of water is attributed to "enforced hydrophobic interactions." That is, the strong hydrogen-bonding network in water tends to exclude nonpolar solutes and force them together, resulting in higher effective concentrations and also relative stabilization of the developing transition state.[20] More specific hydrogen bonding with the transition state also contributes to the rate acceleration.[21]

6.1.2. The Diels–Alder Reaction: Dienophiles

Examples of some compounds which exhibit a high level of reactivity as dienophiles are collected in Table 6.1. Scheme 6.1 presents some typical Diels–Alder reactions. Each of the reactive dienophiles has at least one strongly electron-attracting substituent on the double or triple carbon–carbon bond. Ethylene, acetylene, and their alkyl derivatives are poor dienophiles and react only under extreme conditions.

Diels–Alder reactions have long played an important role in synthetic organic chemistry. The reaction of a substituted benzoquinone and 1,3-butadiene, for example, was the first step in one of the early syntheses of steroids. The angular methyl group is introduced from the quinone, and the other functional groups were used for further structural elaboration.

benzene, 100°C — 86% — Ref. 22

18. S. Kobayashi, I. Hachiya, M. Araki, and H. Ishitami, *Tetrahedron Lett.* **34:**3755 (1993); S. Kobayahsi, *Eur. J. Org. Chem.* **1999:**15.
19. D. Rideout and R. Breslow, *J. Am. Chem. Soc.* **102:**7816 (1980); R. Breslow and T. Guo, *J. Am. Chem. Soc.* **110:**5613 (1988); T. Dunams, W. Hoekstra, M. Pentaleri, and D. Liotta, *Tetrahedron Lett.* **29:**3745 (1988).
20. R. Breslow and C. J. Rizzo, *J. Am. Chem. Soc.* **113:**4340 (1991).
21. W. Blokzijl, M. J. Blandamer, and J. B. F. N. Engberts, *J. Am. Chem. Soc.* **113:**4241 (1991); W. Blokzijl and J. B. F. N. Engberts, *J. Am. Chem. Soc.* **114:**5440 (1992); S. Otto, W. Blokzijl, and J. B. F. N. Engberts, *J. Org. Chem.* **59:**5372 (1994); A. Meijer, S. Otto, and J. B. F. N. Engberts, *J. Org. Chem.* **63:**8989 (1998).
22. R. B. Woodward, F. Sondheimer, D. Taub, K. Heusler, and W. M. McLamore, *J. Am. Chem. Soc.* **74:**4223 (1952).

Table 6.1. Representative Dienophiles

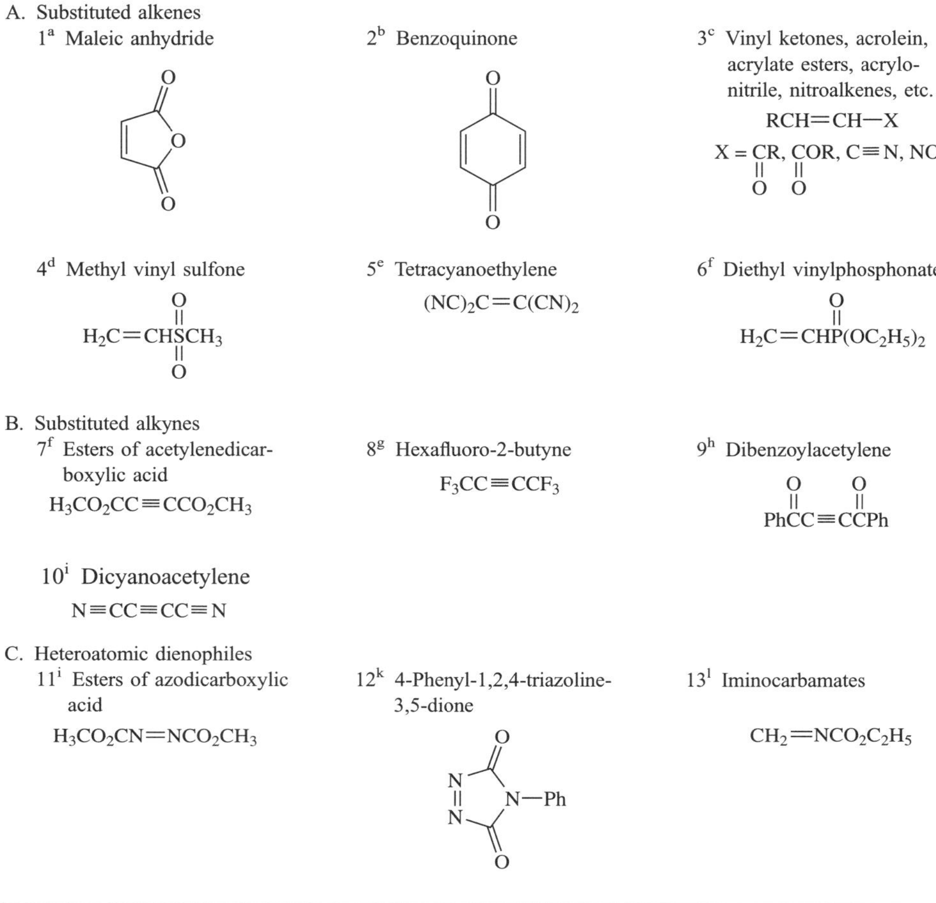

A. Substituted alkenes

1[a] Maleic anhydride	2[b] Benzoquinone	3[c] Vinyl ketones, acrolein, acrylate esters, acrylonitrile, nitroalkenes, etc. RCH=CH—X; X = CR(=O), COR(=O), C≡N, NO_2
4[d] Methyl vinyl sulfone $H_2C{=}CHS(O_2)CH_3$	5[e] Tetracyanoethylene $(NC)_2C{=}C(CN)_2$	6[f] Diethyl vinylphosphonate $H_2C{=}CHP(O)(OC_2H_5)_2$

B. Substituted alkynes

7[f] Esters of acetylenedicarboxylic acid $H_3CO_2CC{\equiv}CCO_2CH_3$	8[g] Hexafluoro-2-butyne $F_3CC{\equiv}CCF_3$	9[h] Dibenzoylacetylene PhC(O)C≡CC(O)Ph
10[i] Dicyanoacetylene N≡CC≡CC≡N		

C. Heteroatomic dienophiles

11[i] Esters of azodicarboxylic acid $H_3CO_2CN{=}NCO_2CH_3$	12[k] 4-Phenyl-1,2,4-triazoline-3,5-dione	13[l] Iminocarbamates $CH_2{=}NCO_2C_2H_5$

a. M. C. Kloetzel, *Org. React.* **4:**1 (1948).
b. L. W. Butz and A. W. Rytina, *Org. React.* **5:**136 (1949).
c. H. L. Holmes, *Org. React.* **4:**60 (1948).
d. J. C. Philips and M. Oku, *J. Org. Chem.* **37:**4479 (1972).
e. W. J. Middleton, R. E. Heckert, E. L. Little, and C. G. Krespan, *J. Am. Chem. Soc.* **80:**2783 (1958); E. Ciganek, W. J. Linn, and O. W. Webster, *The Chemistry of the Cyano Group*, Z. Rappoport, ed., John Wiley & Sons, New York, 1970, pp. 423–638.
f. W. M. Daniewski and C. E. Griffin, *J. Org. Chem.* **31:**3236 (1966).
g. R. E. Putnam, R. J. Harder, and J. E. Castle, *J. Am. Chem. Soc.* **83:**391 (1961); C. G. Krespan, B. C. McKusick, and T. L. Cairns, *J. Am. Chem. Soc.* **83:**3428 (1961).
h. J. D. White, M. E. Mann, H. D. Kirshenbaum, and A. Mitra, *J. Org. Chem.* **36:**1048 (1971).
i. C. D. Weis, *J. Org. Chem.* **28:**74 (1963).
j. B. T. Gillis and P. E. Beck, *J. Org. Chem.* **28:**3177 (1963).
k. B. T. Gillis and J. D. Hagarty, *J. Org. Chem.* **32:**330 (1967).
l. M. P. Cava, C. K. Wilkins, Jr., D. R. Dalton, and K. Bessho, *J. Org. Chem.* **30:**3772 (1965); G. Krow, R. Rodebaugh, R. Carmosin, W. Figures, H. Pannella, G. De Vicaris, and M. Grippi, *J. Am. Chem. Soc.* **95:**5273 (1973).

The synthetic utility of the Diels–Alder reaction can be significantly expanded by the use of dienophiles that contain *masked functionality* and are the *synthetic equivalents* of unreactive or inaccessible species (see Section 13.2 for a more complete discussion of the concept of synthetic equivalents). For example, α-chloroacrylonitrile shows satisfactory reactivity as a dienophile. The α-chloronitrile functionality in the adduct can be hydrolyzed to a carbonyl group. Thus, α-chloroacrylonitrile can function as the equivalent of ketene,

Scheme 6.1. Diels–Alder Reactions of Some Representative Dienophiles

1[a] Maleic anhydride

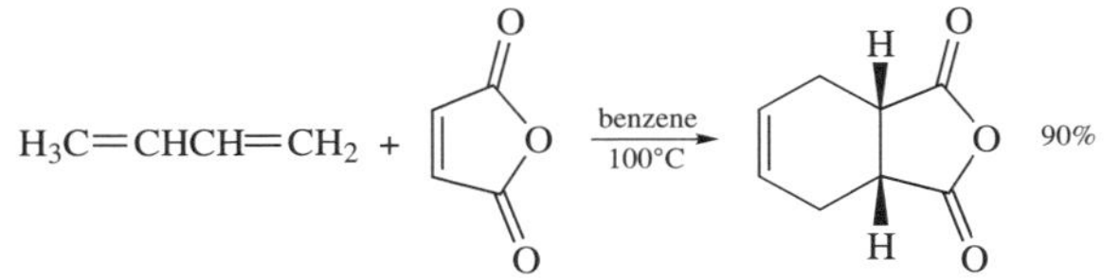

$H_3C{=}CHCH{=}CH_2$ + (maleic anhydride) —benzene, 100°C→ (product) 90%

2[b] Benzoquinone

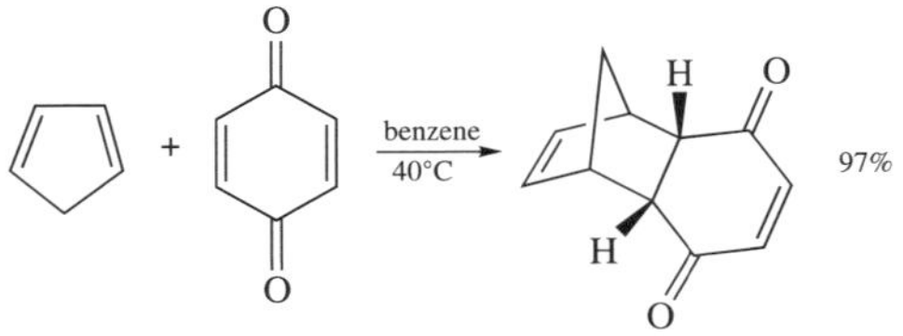

(cyclopentadiene) + (benzoquinone) —benzene, 40°C→ (product) 97%

3[c] Methyl vinyl ketone

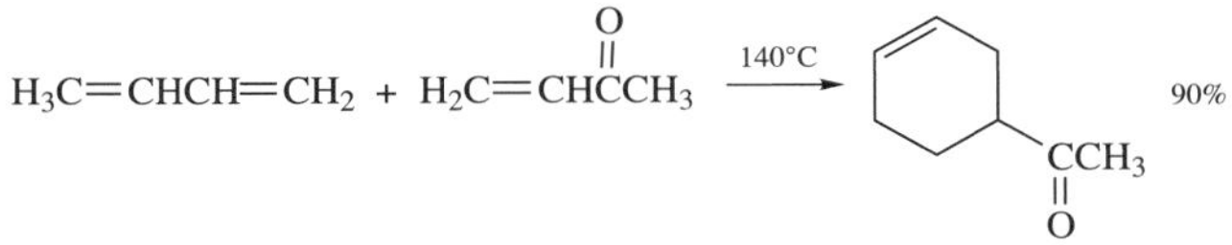

$H_3C{=}CHCH{=}CH_2 + H_2C{=}CHCCH_3$ (C=O) —140°C→ (cyclohexenyl CCH_3, C=O) 90%

4[d] Methyl acrylate

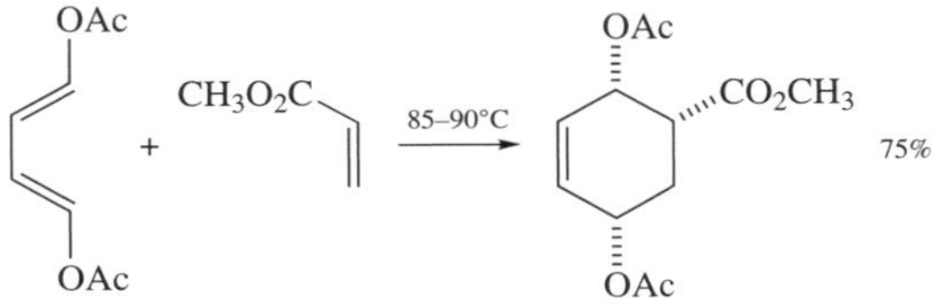

(1,4-diacetoxybutadiene, OAc, OAc) + CH_3O_2C (vinyl) —85–90°C→ (OAc, CO_2CH_3, OAc product) 75%

5[e] Acrolein

$H_3C{=}C(OCH_3)CH{=}CH_2 + H_2C{=}CHCH{=}O$ → (CH_3O-cyclohexene-$CH{=}O$) 75%

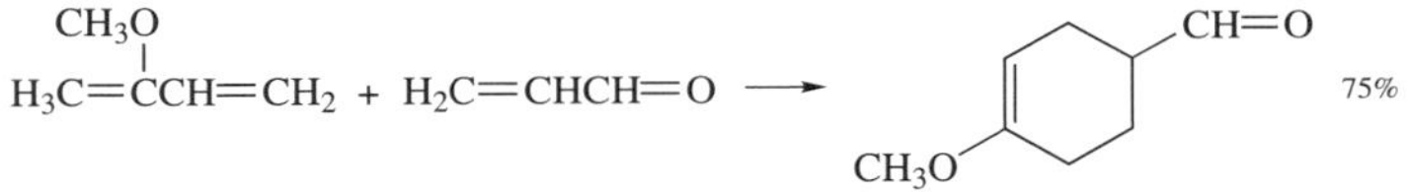

6[f] Tetracyanoethylene

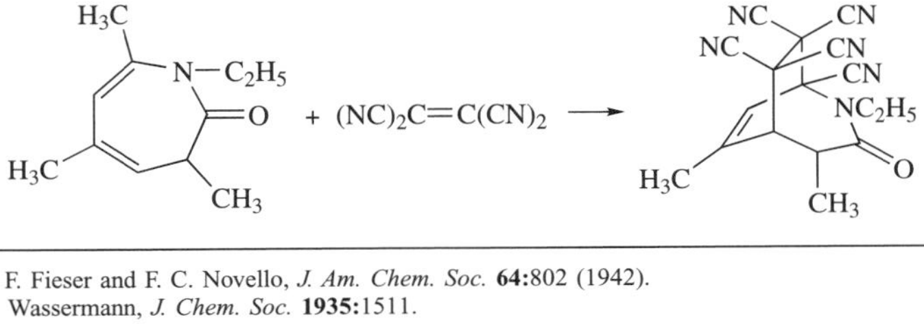

(H_3C, $N{-}C_2H_5$, =O, H_3C, CH_3 azepinone) + $(NC)_2C{=}C(CN)_2$ → (NC, NC, CN, CN, NC_2H_5, O, H_3C, CH_3 adduct)

a. L. F. Fieser and F. C. Novello, *J. Am. Chem. Soc.* **64:**802 (1942).
b. A. Wassermann, *J. Chem. Soc.* **1935:**1511.
c. W. K. Johnson, *J. Org. Chem.* **24:**864 (1959).
d. R. McCrindle, K. H. Overton, and R. A. Raphael, *J. Chem. Soc.* **1960:**1560; R. K. Hill and G. R. Newkome, *Tetrahedron Lett.* **1968:**1851.
e. J. I. DeGraw, L. Goodman, and B. R. Baker, *J. Org. Chem.* **26:**1156 (1961).
f. L. A. Paquette, *J. Org. Chem.* **29:**3447 (1964).

$CH_2{=}C{=}O$.[23] Ketene is not a suitable dienophile because it has a tendency to react with dienes by [2 + 2] cycloaddition, rather than in the desired [4 + 2] fashion.

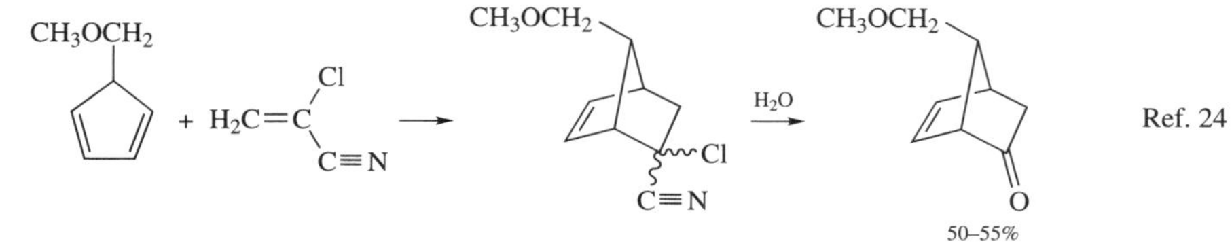

Ref. 24

Nitroalkenes are good dienophiles, and the variety of transformations that are available for nitro groups make them versatile intermediates.[25] Nitro groups can be converted to carbonyl groups by reductive hydrolysis, so nitroethylene can be used as a ketene equivalent.[26]

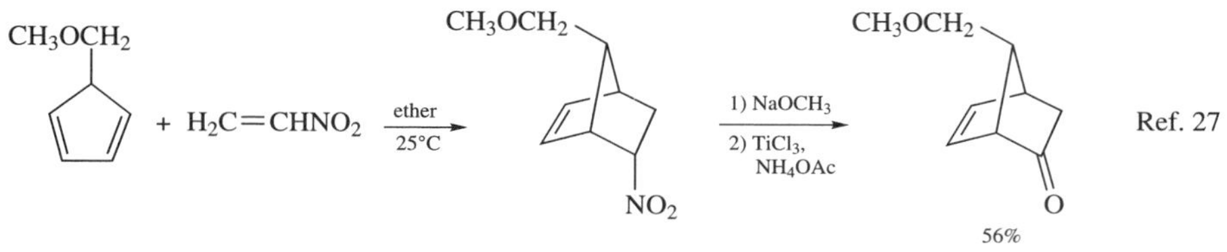

Ref. 27

Vinyl sulfones are reactive as dienophiles. The sulfonyl group can be removed reductively with sodium amalgam (see Section 5.5.2). In this two-step reaction sequence, the vinyl sulfone functions as an ethylene equivalent. The sulfonyl group also permits alkylation of the Diels–Alder adduct, via the carbanion. This three-step sequence allows the vinyl sulfone to serve as the synthetic equivalent of a terminal alkene.[28]

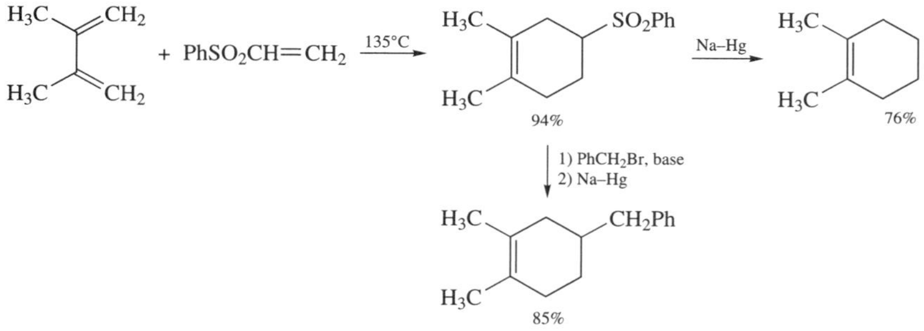

Phenyl vinyl sulfoxide is a useful acetylene equivalent. Its Diels–Alder adducts can

23. V. K. Aggarwal, A. Ali, and M. P. Coogan, *Tetrahedron* **55**:293 (1999).
24. E. J. Corey, N. M. Weinshenker, T. K. Schaaf, and W. Huber, *J. Am. Chem. Soc.* **91**:5675 (1969).
25. D. Ranganathan, C. B. Rao, S. Ranganathan, A. K. Mehrotra, and R. Iyengar, *J. Org. Chem.* **45**:1185 (1980).
26. For a review of ketene equivalents, see S. Ranganathan, D. Ranganathan, and A. K. Mehrotra, *Synthesis* **1977**:289.
27. S. Ranganathan, D. Ranganathan, and A. K. Mehrotra, *J. Am. Chem. Soc.* **96**:5261 (1974).
28. R. V. C. Carr and L. A. Paquette, *J. Am. Chem. Soc.* **102**:853 (1980); R. V. C. Carr, R. V. Williams, and L. A. Paquette, *J. Org. Chem.* **48**:4976 (1983); W. A. Kinney, G. O. Crouse, and L. A. Paquette, *J. Org. Chem.* **48**:4986 (1983).

undergo elimination of benzenesulfenic acid.

$$\text{C}_5\text{Cl}_6 + \text{PhS(O)CH{=}CH}_2 \xrightarrow{100^\circ\text{C}} \text{adduct (S(=O)Ph)} \xrightarrow{100^\circ\text{C}} \text{hexachloronorbornadiene (83\%)}$$

Ref. 29

Cis- and *trans*(bisbenzenesulfonyl)ethene are also acetylene equivalents. The two sulfonyl groups undergo reductive elimination on reaction with sodium amalgam.

$$\text{cyclohexadiene} + \text{PhSO}_2\text{CH{=}CHSO}_2\text{Ph} \longrightarrow \text{bicyclo[2.2.2]octene-(SO}_2\text{Ph)}_2 \xrightarrow[\text{MeOH}]{\text{Na–Hg}} \text{bicyclo[2.2.2]octadiene (69\%)}$$

Ref. 30

Vinylphosphonium salts are reactive as dienophiles as a result of the electron-withdrawing capacity of the phosphonium substituent. The Diels–Alder adducts can be deprotonated to give ylides which undergo the Wittig reaction to introduce an exocyclic double bond. This sequence of reactions corresponds to a Diels–Alder reaction employing allene as the dienophile.[31]

$$\text{cyclohexadiene} + \text{H}_2\text{C{=}CH}\overset{+}{\text{P}}\text{Ph}_3 \longrightarrow \text{adduct-}\overset{+}{\text{P}}\text{Ph}_3\ (96\%) \xrightarrow{1)\ \text{LiNR}_2\quad 2)\ \text{CH}_2\text{=O}} \text{adduct=CH}_2\ (50\%)$$

The use of 2-vinyldioxolane, the ethylene glycol acetal of acrolein, as a dienophile illustrates application of the masked functionality concept in a different way. The acetal itself would not be expected to be a reactive dienophile, but in the presence of a catalytic amount of acid, the acetal is in equilibrium with the highly reactive oxonium ion.

$$\text{CH}_2\text{=CH–(2-dioxolanyl)} + \text{H}^+ \rightleftharpoons \text{CH}_2\text{=CH–CH=}\overset{+}{\text{O}}\text{–CH}_2\text{CH}_2\text{OH}$$

Diels–Alder addition occurs through this cationic intermediate at room temperature.[32]

29. L. A. Paquette, R. E. Moerck, B. Harirchian, and P. D. Magnus, *J. Am. Chem. Soc.* **100:**1597 (1978).
30. O. DeLucchi, V. Lucchini, L. Pasquato, and G. Modena, *J. Org. Chem.* **49:**596 (1984).
31. R. Bonjouklian and R. A. Ruden, *J. Org. Chem.* **42:**4095 (1977).
32. P. G. Gassman, D. A. Singleton, J. J. Wilwerding, and S. P. Chavan, *J. Am. Chem. Soc.* **109:**2182 (1987).

Similar reactions occur with substituted alkenyldioxolanes.

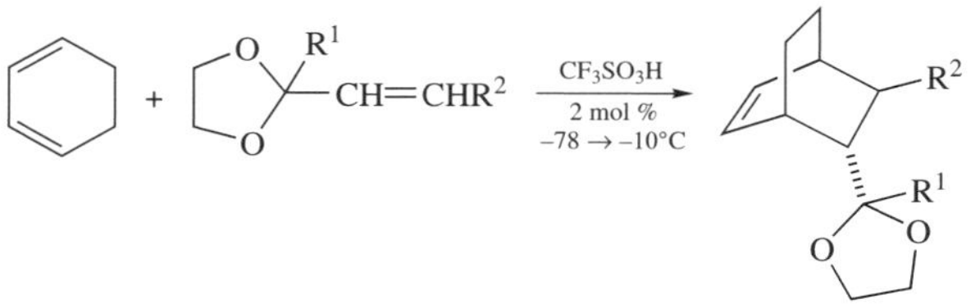

CF_3SO_3H, 2 mol %, $-78 \rightarrow -10°C$

Alkenyl- and alkynylboranes also function as dienophiles. The electron-deficient boron is responsible for the electronic effect.

$$CH_2{=}CH{-}BR_2 \longleftrightarrow \overset{+}{C}H_2{-}CH{=}\overset{-}{B}R_2$$

Alkenylboranes are less sensitive to substituents on the diene than are carbonyl-activated dienophiles.[33] Relatively hindered dialkylboranes, such as *B*-vinyl-9-BBN, show steric effects which lead to a preference for the "*meta*" regioisomer and reduced *endo* : *exo* ratios.[34]

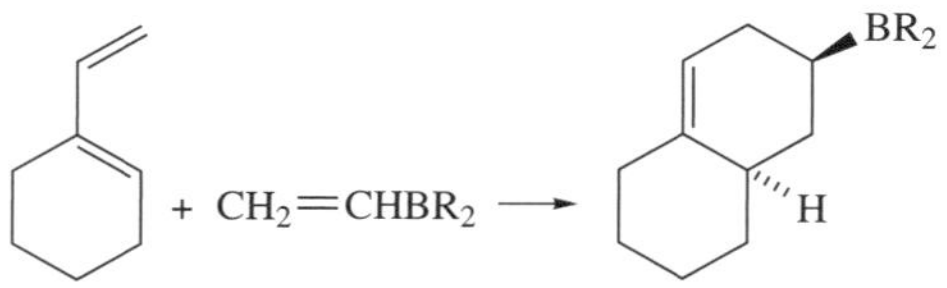

$+ \; CH_2{=}CHBR_2 \longrightarrow$

β-Trimethylsilylvinyl-9-BBN shows a preference for the "*meta*" adduct with both isoprene and 2-(*t*-butyldimethylsilyloxy)butadiene.[35]

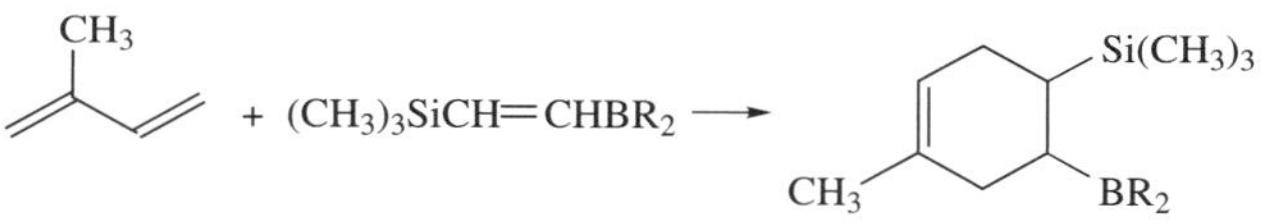

$+ \; (CH_3)_3SiCH{=}CHBR_2 \longrightarrow$

The characterisitics of the vinylboranes as dienophiles can be rationalized in terms of a strong interaction of the diene with the empty π orbital at boron. Molecular orbital calculations show a strong interaction between B and C-1 in the transition state, and the transition state shows little charge separation, accounting for the relative insensitivity to substituent effects. As for regiochemistry, the "*para*-like" selectivity would also be expected to be reduced because the LUMO of the dienophile is nearly equally distributed between B and C-2.[36]

33. Y. Singleton, J. P. Martinez, and J. V. Watson, *Tetrahedron Lett.* **33:**1017 (1992).
34. D. A. Singleton and J. P. Martinez, *J. Am. Chem. Soc.* **112:**7423 (1990).
35. D. A. Singleton and S.-W. Leung, *J. Org. Chem.* **57:**4796 (1992).
36. D. A. Singleton, *J. Am. Chem. Soc.* **114:**6563 (1992).

Simple dienes react readily with good dienophiles in Diels–Alder reactions. Functionalized dienes are also important in organic synthesis. One example which illustrates the versatility of such reagents is 1-methoxy-3-trimethylsilyloxy-1,3-butadiene (*Danishefsky's diene*).[37] Its Diels–Alder adducts are trimethylsilyl enol ethers which can be readily hydrolyzed to ketones. The β-methoxy group is often eliminated during hydrolysis.

OCH$_3$ Me$_3$SiO + H$_2$C=CCHO (CH$_3$) —benzene, Δ→ OCH$_3$ CHO CH$_3$ Me$_3$SiO —H$_3$O$^+$→ O CHO CH$_3$ 72%

Related transformations of the adduct with dimethyl acetylenedicarboxylate lead to dimethyl 4-hydroxyphthalate.

OCH$_3$ Me$_3$SiO + CH$_3$OC(O)C≡CC(O)OCH$_3$ —benzene, Δ→ OCH$_3$ CO$_2$CH$_3$ Me$_3$SiO CO$_2$CH$_3$ —H$_3$O$^+$→ HO CO$_2$CH$_3$ CO$_2$CH$_3$

The corresponding enamine shows a similar reactivity pattern.[38]

N(CH$_3$)$_2$ TBDMSO + CH$_3$ CH=O → (CH$_3$)$_2$N CH$_3$ CH=O TBDMSO → O CH$_3$ CH=O

Unstable dienes can also be generated *in situ* in the presence of a dienophile. Among the most useful examples of this type of diene are the quinodimethanes. These compounds are exceedingly reactive as dienes because the cycloaddition reestablishes a benzenoid ring and results in aromatic stabilization.[39]

CH$_2$ CH$_2$ + X → X

quinodimethane

37. S. Danishefsky and T. Kitahara, *J. Am. Chem. Soc.* **96:**7807 (1974).
38. S. A. Kozmin and V. H. Rawal, *J. Org. Chem.* **62:**5252 (1997).
39. W. Oppolzer, *Angew. Chem. Int. Ed. Engl.* **16:**10 (1977); T. Kametani and K. Fukumoto, *Heterocycles* **3:**29 (1975); J. J. McCullogh, *Acc. Chem. Res.* **13:**270 (1980); W. Oppolzer, *Synthesis* **1978:**73; J. L. Charlton and M. M. Alauddin, *Tetrahedron* **43:**2873 (1987); H. N. C. Wong, K.-L. Lau and K. F. Tam, *Top. Curr. Chem.* **133:**85 (1986); P. Y. Michellys, H. Pellissier, and M. Santelli, *Org. Prep. Proced. Int.* **28:**545 (1996).

There are several general routes to quinodimethanes. One is pyrolysis of benzocyclobutenes.[40]

heat

CH_2

CH_2

Eliminations from α,α'-*ortho*-disubstituted benzenes with various potential leaving groups can be carried out. Benzylic silyl substituents can serve as the carbanion precursors.

CH_3 $CHSi(CH_3)_3$ $CH\overset{+}{N}(CH_3)_3$ CH_3

F^-, 50°C

CH_3O_2C, H, C=C, H, CO_2CH_3

CH_3 CO_2CH_3 CO_2CH_3 CH_3

100%

Ref. 41

Several procedures have been developed for obtaining quinodimethane intermediates from *ortho*-substituted benzylstannanes. The reactions occur by generating an electrophilic center at the adjacent benzylic position, which triggers a 1,4-elimination.

R E^+ X SnR_3 → R + XE SnR_3 → R XE

$=\!\!-$X = CHOH, =O, =CH_2

Specific examples include treatment of *o*-stannyl benzyl alcohols with TFA,[42] reactions of ketones and aldehydes with Lewis acids,[43] and selenation of styrenes.[44]

CH=O $CH_2Sn(C_4H_9)_3$ + CH_3O_2C H CO_2CH_3 H $\xrightarrow{MgBr_2}$ OH CO_2CH_3 CO_2CH_3 69%

CH=CH_2 $CH_2Sn(C_4H_9)_3$ + CH_2=$CHCO_2CH_3$ (O, N−SePh, O) → CH_2SePh CO_2CH_3 59%

40. M. P. Cava and M. J. Mitchell, *Cyclobutadiene and Related Compounds*, Academic Press, New York, 1967, Chapter 6; I. L. Klundt, *Chem. Rev.* **70**:471 (1970); R. P. Thummel, *Acc. Chem. Res.* **13**:70 (1980).
41. Y. Ito, M. Nakatsuka, and T. Saegusa, *J. Am. Chem. Soc.* **104**:7609 (1982).
42. H. Sano, H. Ohtsuka, and T. Migita, *J. Am. Chem. Soc.* **110**:2014 (1988).
43. S. H. Woo, *Tetrahedron Lett.* **35**:3975 (1994).
44. S. H. Woo, *Tetrahedron Lett.* **34**:7587 (1993).

o-(Dibromomethyl)benzenes can be converted to quinodimethanes with reductants such as zinc, nickel, chromous ion, and tri-n-butylstannide.[45]

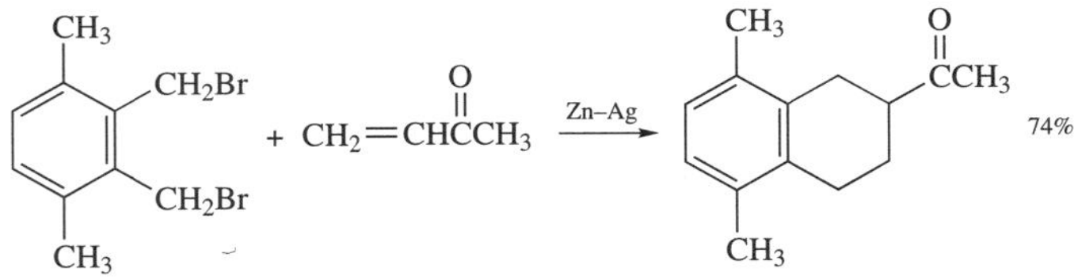

Quinodimethanes have been especially useful in intramolecular Diels–Alder reactions, as will be illustrated in Section 6.1.5.

Another group of dienes with extraordinarily high reactivity are derivatives of benzo[c]furan (isobenzofuran).[46]

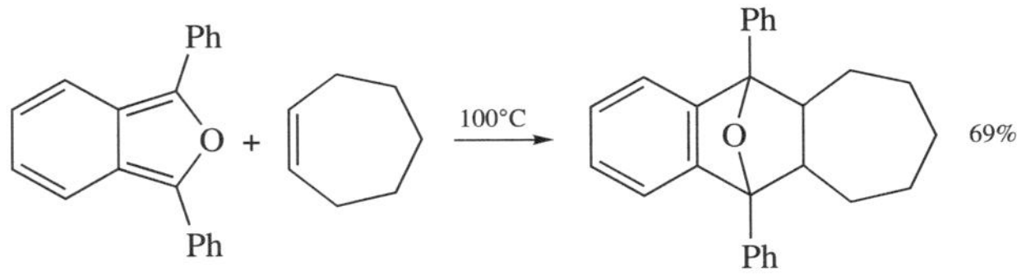

Ref. 47

Here again, the high reactivity can be traced to the gain in aromatic stabilization of the adduct.

Polycyclic aromatic hydrocarbons are moderately reactive as the diene component of Diels–Alder reactions. Anthracene forms adducts with a number of reactive dienophiles. The addition occurs at the center ring. There is no net loss of resonance stabilization, because the anthracene ring (resonance energy = 1.60 eV) is replaced by two benzenoid rings (total resonance energy = $2 \times 0.87 = 1.74$ eV).[48]

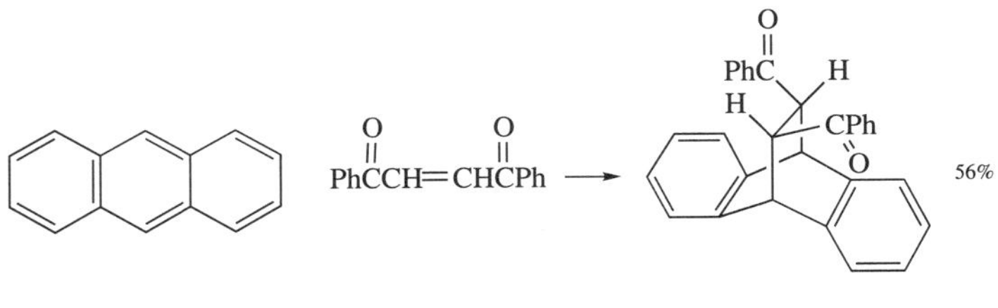

Ref. 49

The naphthalene ring is much less reactive. Polymethylnaphthalenes are more reactive than the parent molecule, and 1,2,3,4-tetramethylnaphthalene gives an adduct with maleic anhydride in 82% yield. Reaction occurs exclusively in the substituted ring.[50] This is because the steric repulsions between the methyl groups, which are relieved in the nonplanar adduct, exert an accelerating effect.

45. G. M. Rubottom and J. E. Wey, *Synth. Commun.* **14:**507 (1984); S. Inaba, R. M. Wehmeyer, M. W. Forkner, and R. D. Rieke, *J. Org. Chem.* **53:**339 (1988); D. Stephan, A. Gorgues, and A. LeCoq, *Tetrahedron Lett.* **25:**5649 (1984); H. Sato, N. Isono, K. Okamura, T. Date, and M. Mori, *Tetrahedron Lett.* **35:**2035 (1994).
46. M. J. Haddadin, *Heterocycles* **9:**865 (1978); W. Friedrichsen, *Adv. Heterocycl. Chem.* **26:**135 (1980).
47. G. Wittig and T. F. Burger, *Justus Liebigs Ann. Chem.* **632:**85 (1960).
48. M. J. S. Dewar and D. de Llano, *J. Am. Chem. Soc.* **91:**789 (1969).
49. D. M. McKinnon and J. Y. Wong, *Can. J. Chem.* **49:**3178 (1971).
50. A. Oku, Y. Ohnishi, and F. Mashio, *J. Org. Chem.* **37:**4264 (1972).

Diels–Alder addition of simple benzene derivatives is difficult and occurs only with very reactive dienophiles. Formation of an adduct between benzene and dicyanoacetylene in the presence of $AlCl_3$ has been reported, for example.[51]

+ N≡CC≡CC≡N $\xrightarrow{AlCl_3}$ (adduct) C≡N, C≡N

Pyrones are a useful type of diene. Although they are not particularly reactive dienes, the adducts have the potential for elimination of carbon dioxide, resulting in the formation of an aromatic ring.

CO_2Et, CH_3, CH_3, O, O + $(MeO)_2C{=}CH_2$ ⟶ EtO_2C, CH_3, CH_3, O, O=C, OMe, OMe $\xrightarrow[-MeOH]{-CO_2}$ EtO_2C, CH_3, CH_3, OMe Ref. 52

O, O + $CH_2{=}C(OCH_3)_2$ $\xrightarrow{110°C}$ OCH_3 84% Ref. 53

Vinyl ethers are frequently used as dienophiles with pyrones. These reactions can be catalyzed by Lewis acids such as bis(alkoxy)titanium dichlorides[54] and lanthanide salts.[55]

CO_2CH_3, O, O + $CH_2{=}CHOC_4H_9$ $\xrightarrow{Yb(hfc)_3}$ O, O, OC_4H_9 94%

CO_2CH_3, O, O + $CH_2{=}CHO{-}c{-}C_6H_{11}$ $\xrightarrow[R\text{-BINOL},\ i\text{-}C_3H_7N(C_2H_5)_2]{Yb(O_3SCF_3)_3}$ O, O, O-c-C_6H_{11} >95%

Another use of special dienes, the polyaza benzene heterocyclics, such as triazines and tetrazines, will be discussed in Section 6.8.2.

51. E. Ciganek, *Tetrahedron Lett.* **1967:**3321.
52. M. E. Jung and J. A. Hagenah, *J. Org. Chem.* **52:**1889 (1987).
53. D. L. Boger and M. D. Mullican, *Org. Synth.* **65:**98 (1987).
54. G. H. Posner, J.-C. Carry, J. K. Lee, D. S. Bull, and H. Dai, *Tetrahedron Lett.* **35:**1321 (1994); G. H. Posner, H. Dai, D. S. Bull, J.-K. Lee, F. Eydoux, Y. Ishihara, W. Welsh, N. Pryor, and S. Petr, Jr., *J. Org. Chem.* **61:**671 (1996).
55. G. H. Posner, J.-C. Carry, T. E. N. Anjeh, and A. N. French, *J. Org. Chem.* **57:**7012 (1992).

The highly ordered cyclic transition state of the Diels–Alder reaction permits design of reaction parameters which lead to a preference between the transition states leading to diastereomeric or enantiomeric adducts. (See Part A, Section 2.3, to review the principles of diastereoselectivity and enantioselectivity.) One way to achieve this is to install a chiral auxiliary.[56] The cycloaddition proceeds to give two diastereomeric products which can be separated and purified. Because of the lower temperature required and the greater stereoselectivity observed in Lewis acid-catalyzed reactions, the best enantioselectivity is often observed in catalyzed reactions. Chiral esters and amides of acrylic acid are particularly useful because the chiral auxiliary can be easily recovered upon hydrolysis of the adduct to give the enantiomerically pure carboxylic acid.

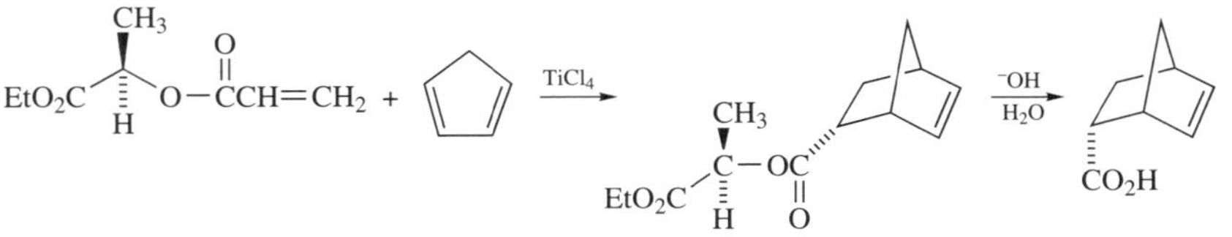

Ref. 57

Prediction and analysis of diastereoselectivity is based on steric, stereoelectronic, and complexing interactions in the transition state.[58]

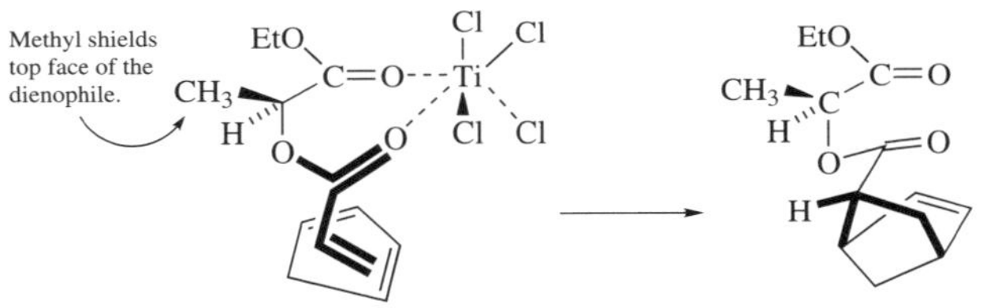

Methyl shields top face of the dienophile.

α,β-Unsaturated derivatives of chiral oxazolinones have proven to be especially useful for enantioselective Diels–Alder additions. Reaction occurs at low temperatures in the presence of such Lewis acids as $SnCl_4$, $TiCl_4$ and $(C_2H_5)_2AlCl$.[59]

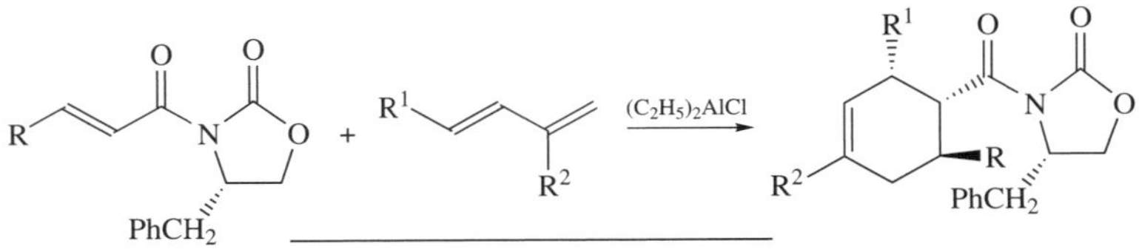

R	R^1	R^2	Yield	dr
H	H	CH_3	85%	95:5
H	CH_3	H	84%	>100:1
CH_3	H	CH_3	83%	94:6
CH_3	CH_3	H	77%	95:5

56. W. Oppolzer, *Angew. Chem. Int. Ed. Engl.* **23:**876 (1984); M. J. Tascher, in *Organic Synthesis, Theory and Applications*, Vol. 1, T. Hudlicky, ed., JAI Press, Greenwich, Connecticut, 1989, pp. 1–101; H. B. Kagan and O. Riant, *Chem. Rev.* **92:**1007 (1992); K. Narasaka, *Synthesis* **1991:**1.
57. T. Poll, G. Helmchen, and B. Bauer, *Tetrahedron Lett.* **25:**2191 (1984).
58. For example, see T. Poll, A. Sobczak, H. Hartmann, and G. Helmchen, *Tetrahedron Lett.* **26:**3095 (1985).
59. D. A. Evans, K. T. Chapman, and J. Bisaha, *J. Am. Chem. Soc.* **110:**1238 (1988).

Scheme 6.2 gives some other examples of use of chiral auxiliaries in Diels–Alder reactions.[60]

The alkenyl oxonium ion dienophiles generated from dioxolanes have been made enantioselective by use of chiral diols. For example, dioxolanes derived from *syn*-1,2-diphenylethane-1,2-diol react with dienes such as cyclopentadiene and isoprene, but the stereoselectivity is very modest in most cases.

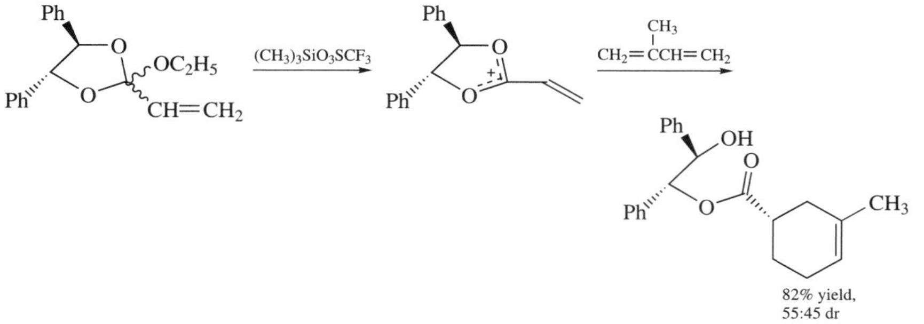
Ph
OC2H5
CH=CH2
(CH3)3SiO3SCF3
CH3
CH2=CCH=CH2
OH
82% yield,
55:45 dr

Ref. 61

Acetals derived from *anti*-pentane-2,4-diol react with dienes under the influence of $TiCl_4/Ti(i\text{-}OPr)_4$ to give adducts with stereoselectivity ranging from 3 : 1 to 15 : 1.

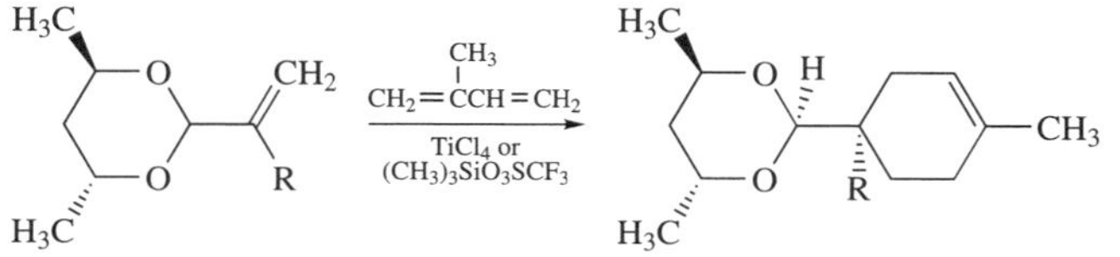
H3C
CH2
R
CH3
CH2=CCH=CH2
TiCl4 or
(CH3)3SiO3SCF3

Ref. 62

Enantioselectivity can also be achieved with chiral catalysts. For example, additions of *N*-acryloyloxazolinones can be made enantioselective using $Sc(O_3SCF_3)_3$ in the presence of a BINOL ligand.[63] Optimized conditions involved use of 5–20 mol % of the catalyst along with a hindered amine such as *cis*-1,2,6-trimethylpiperidine. A hexacoordinate transition state in which the amine is hydrogen-bonded to the BINOL has been proposed.

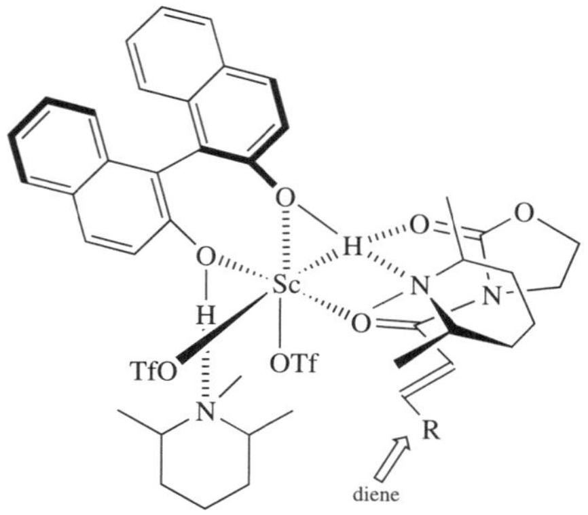
Sc
TfO
OTf
R
diene

60. For additional examples see W. Oppolzer, *Tetrahedron* **43:**1969, 4057 (1987).
61. A. Haudrechy, W. Picoul, and Y. Langlois, *Tetrahedron Asymmetry* **8:**139 (1997).
62. T. Sammakia and M. A. Berliner, *J. Org. Chem.* **59:**6890 (1994).
63. S. Kobayashi, M. Araki, and I. Hachiya, *J. Org. Chem.* **59:**3758 (1994).

Scheme 6.2. Diels–Alder Reactions with Chiral Auxiliaries

Entry	Dienophile	Diene	Catalyst	Yield (%)	dr
1[a]	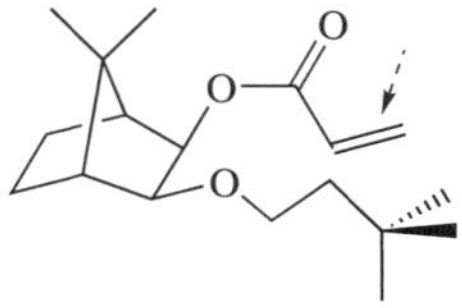 O, O, O		$TiCl_2(i\text{-}OPr)_2$, –20°C	90	>99:1
2[b]	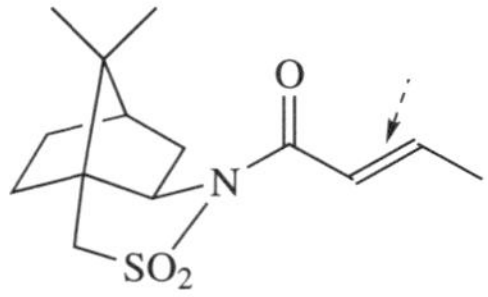 O, N, SO_2		$(C_2H_5)_2AlCl$, –78°C	88	99:1
3[c]	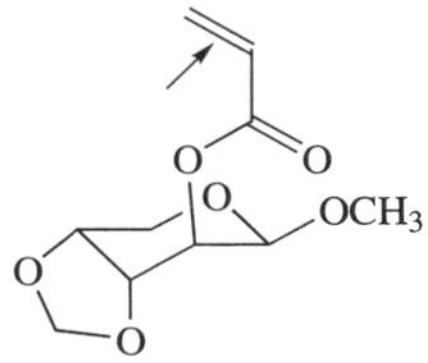 O, O, O, OCH_3, O, O		$SnCl_4$, –78°C	93	96:4
4[d]	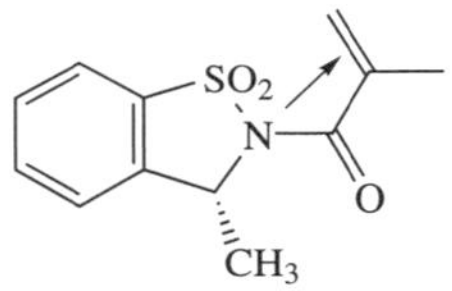 SO_2, N, O, CH_3	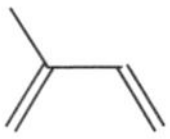	$(C_2H_5)_2AlCl$, –40°C	94	98:2
5[e]	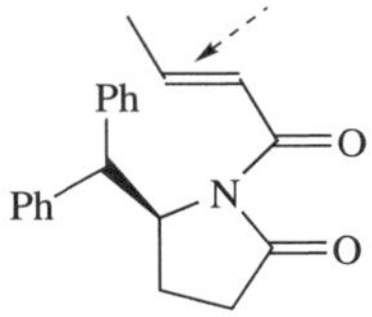 Ph, Ph, N, O, O	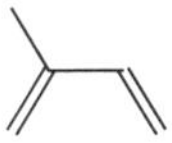	$ZnCl_4$, –78°C	86	>99:1
6[f]	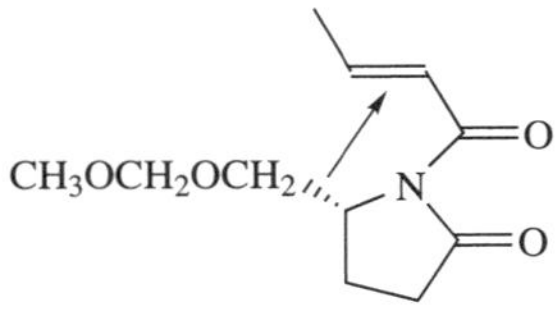 $CH_3OCH_2OCH_2$, N, O, O		$(C_2H_5)_2AlCl$, –40°C	62	97:3
7[g]	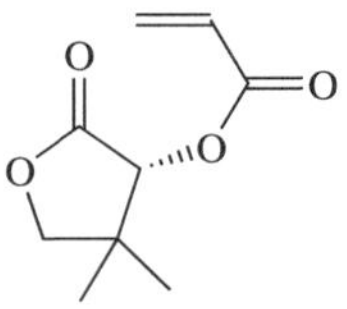 O, O, O, O	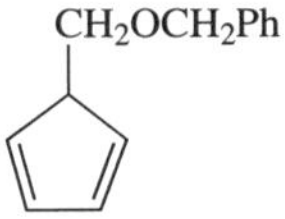 CH_2OCH_2Ph	$TiCl_4$, –55 to –20°C	79	96:2

a. W. Oppolzer, C. Chapuis, D. Dupuis, and M. Guo, *Helv. Chim. Acta* **68:**2100 (1985).
b. W. Oppolzer, C. Chapuis, and G. Bernardinelli, *Helv. Chim. Acta* **67:**1397 (1984); M. Vanderwalle, J. Van der Eycken, W. Oppolzer, and C. Vullioud, *Tetrahedron* **42:**4035 (1986).
c. R. Nougier, J.-L. Gras, B. Giraud, and A. Virgili, *Tetrahedron Lett.* **32:**5529 (1991).
d. W. Oppolzer, B. M. Seletsky, and G. Bernardinelli, *Tetrahedron Lett.* **35:**3509 (1994).
e. M. P. Sibi, P. K. Deshpande, and J. Ji, *Tetrahedron Lett.* **36:**8965 (1995).
f. N. Ikota, *Chem. Pharm. Bull.* **37:**2219 (1989).
g. K. Miyaji, Y. Ohara, Y. Takahashi, T. Tsuruda, and K. Arai, *Tetrahedron Lett.* **32:**4557 (1991).

The chiral oxazaborolidines introduced in Section 2.1.3.5 as enantioselective aldol addition catalysts have also been found to be useful in Diels–Alder reactions. The tryptophan-derived catalyst **A**, for example, can achieve 99% enantioselectivity in the Diels–Alder reaction between 5-benzyloxymethyl-1,3-cyclopentadiene and 2-bromopropenal. The adduct is an important intermediate in the synthesis of prostaglandins.[64]

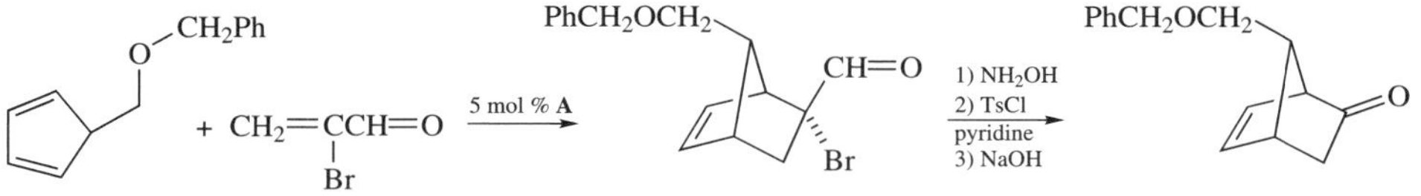

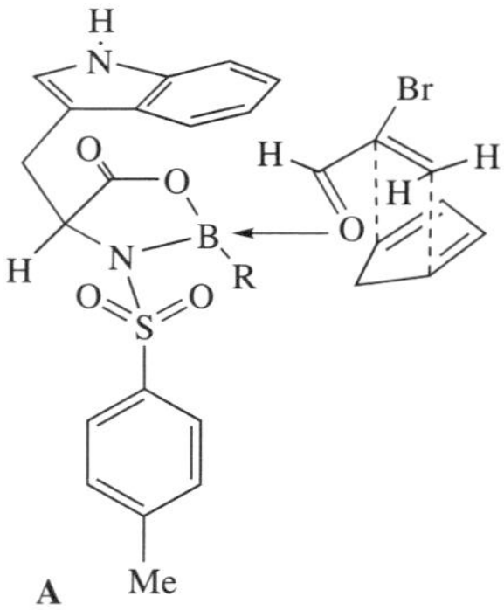

Enantioselective Diels–Alder reactions of acrolein are also catalyzed by 3-(2-hydroxy-3-phenyl) derivatives of BINOL in the presence of an aromatic boronic acid. The optimum boronic acid is 3,5-di(trifluoromethyl)benzeneboronic acid, with which >95% e.e. can be achieved. The transition state is believed to involve Lewis acid complexation of the boronic acid at the carbonyl oxygen and hydrogen bonding with the hydroxyl substituent. In this transition state, π,π-interactions between the dienophile and the hydroxybiphenyl substitutent can also help to align the dienophile.[65]

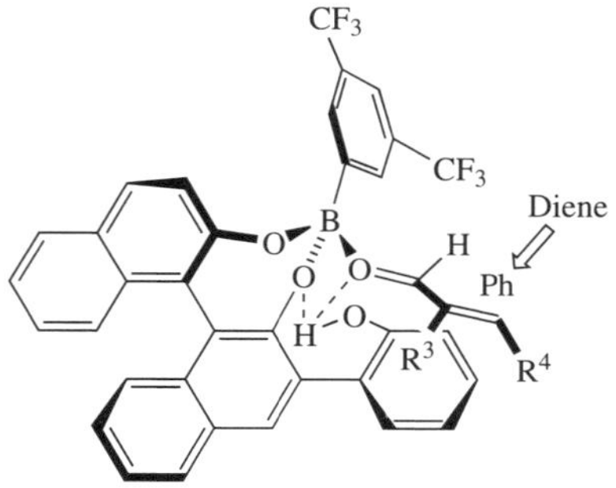

Dienophile	Yield (%)	*exo*:*endo*	e.e. (%)
$CH_2{=}CHCH{=}O$	84	3:97	95
$CH_2{=}C(Br)CH{=}O$	99	90:10	>99
E-$CH_3CH{=}CHCH{=}O$	94	10:90	95
E-$PhCH{=}CHCH{=}O$	94	26:74	80

64. E. J. Corey and T. P. Loh, *J. Am. Chem. Soc.* **113:**8966 (1991).
65. K. Ishihara, H. Kurihara, M. Matsumoto, and H. Yamamoto, *J. Am. Chem. Soc.* **120:**6920 (1998).

Another useful group of catalysts are Cu^{2+} chelates of bis-oxazolines.

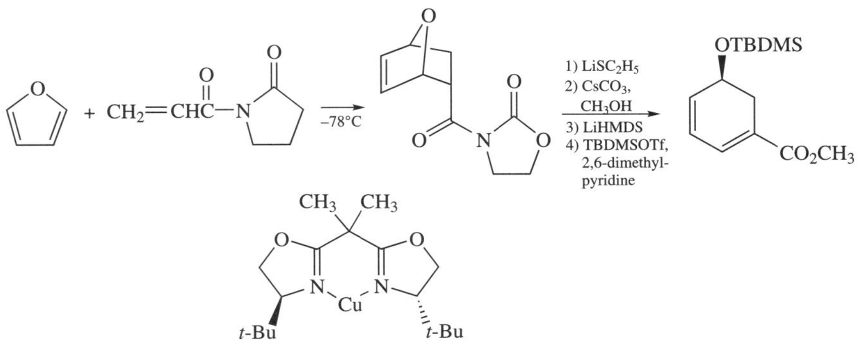

Ref. 66

Several other examples of catalytic enantioselective Diels–Alder reactions are given in Scheme 6.3.

6.1.5. Intramolecular Diels–Alder Reactions

Intramolecular Diels–Alder reactions have proven very useful in the synthesis of polycylic compounds.[67] Some examples are given in Scheme 6.4.

In entry 1 of Scheme 6.4, the dienophilic portion bears a carbonyl substituent, and cycloaddition occurs easily. Two stereoisomeric products are formed in a 90:10 ratio, but both have the *cis* ring fusion. This is the stereochemistry expected for an *endo* transition state.

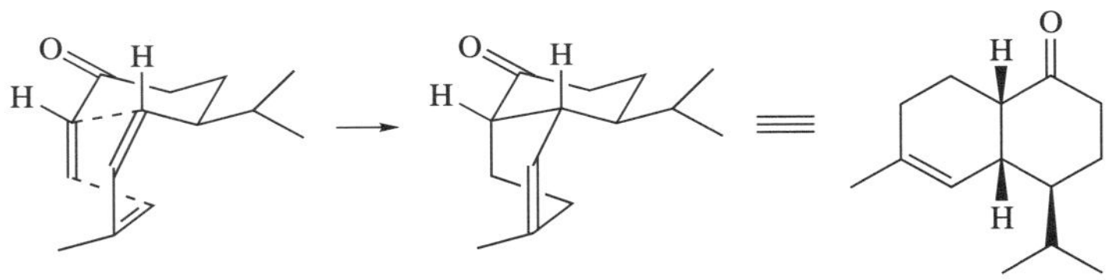

In entry 2, a similar triene that lacks the activating carbonyl group undergoes reaction, but a much higher temperature is required. In this case, the ring junction is *trans*. This corresponds to an *exo* transition state and presumably reflects the absence of an important secondary orbital interaction between the diene and dienophile.

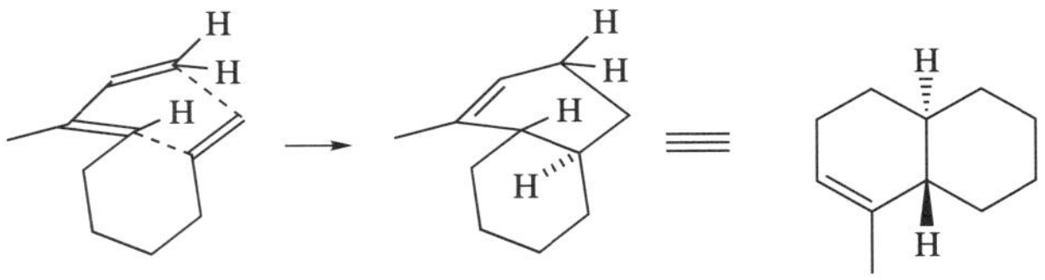

In entry 3, the dienophilic double bond bears an electron-withdrawing group, but a higher temperature than for entry 1 is required because the connecting chain contains one

66. D. A. Evans and D. M. Barnes, *Tetrahedron Lett.* **38:**57 (1997).
67. W. Oppolzer, *Angew. Chem. Int. Ed. Engl.* **16:**10 (1977); G. Brieger and J. N. Bennett, *Chem. Rev.* **80:**63 (1980); E. Ciganek, *Org. React.* **32:**1 (1984); D. F. Taber, *Intramolecular Diels–Alder and Alder Ene Reactions*, Springer-Verlag, Berlin, 1984.

Scheme 6.3. Catalytic Enantioselective Diels–Alder Reactions

Entry	Dienophile	Diene	Catalyst	Yield (%)	e.e.
1[a]	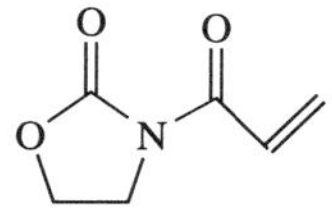		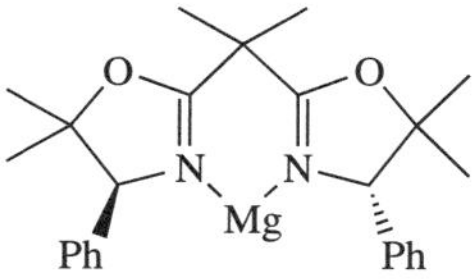	82	95
2[b]	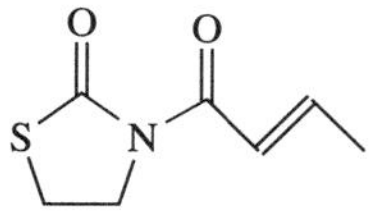		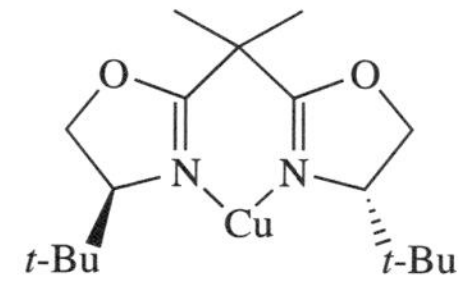	79	94
3[c]	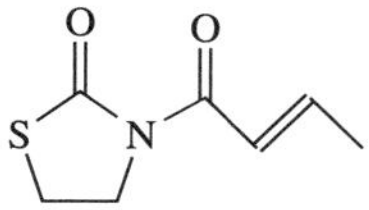		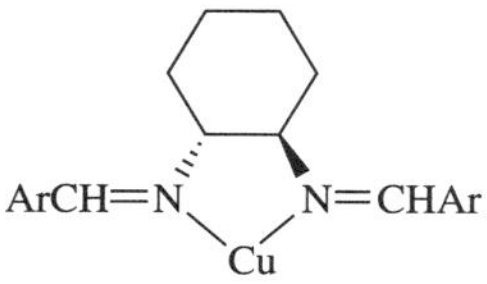 Ar = 2,6-dichlorophenyl	79	91
4[d]	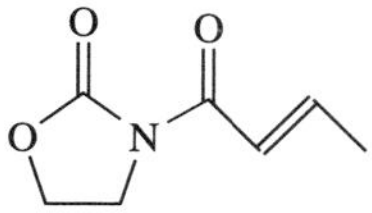		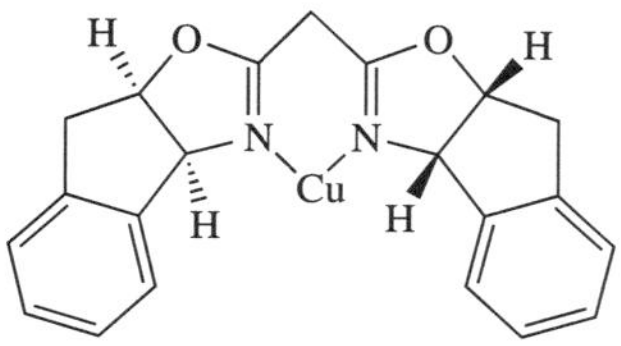	88	84
5[e]	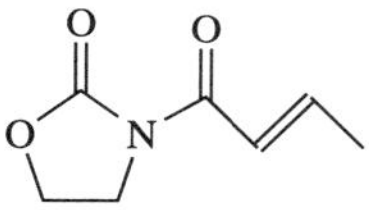		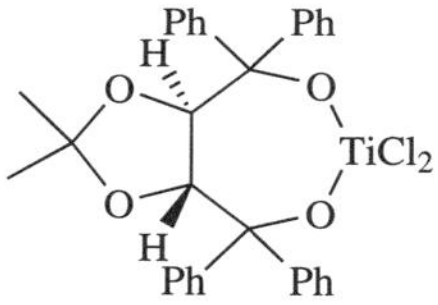	79	91
6[f]	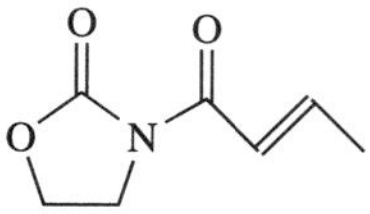		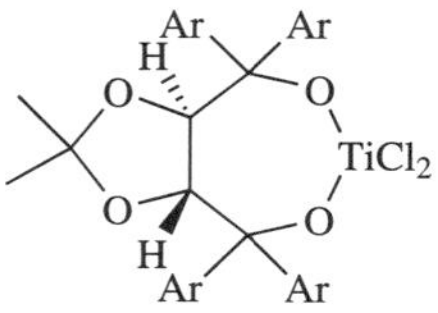 Ar = 2,6-dimethylphenyl	92	93
7[g]	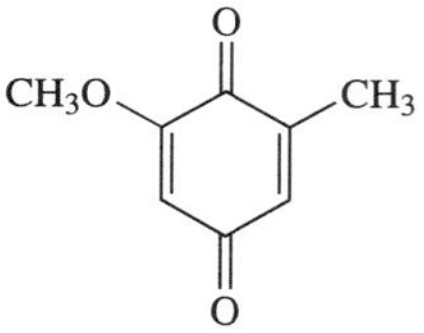	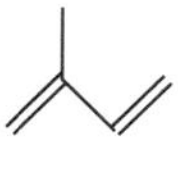	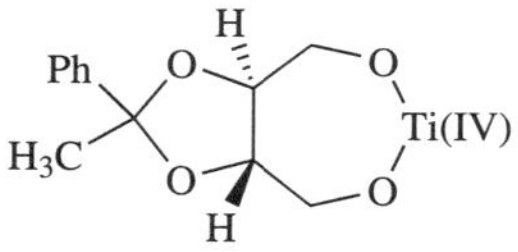	94	80

8[h]		OCH_3	Ar = 3,5-dimethylphenyl	98	93
9[i]	$CH_2{=}C(Br)CH{=}O$		Ar = 3-indolyl		>99.5

a. E. J. Corey and K. Ishihara, *Tetrahedron Lett.* **33:**6807 (1992).
b. D. A. Evans, S. J. Miller, and T. Lectka, *J. Am. Chem. Soc.* **115:**6460 (1993).
c. D. A. Evans, T. Lectka, and S. J. Miller, *Tetrahedron Lett.* **34:**7027 (1993).
d. A. K. Ghosh, H. Cho, and J. Cappiello, *Tetrahedron Asymmetry* **9:**3687 (1998).
e. K. Narasaka, N. Iwasawa, M. Inoue, T. Yamada, M. Nakashima, and J. Sugimori, *J. Am. Chem. Soc.* **111:**5340 (1989).
f. E. J. Corey and Y. Matsumura, *Tetrahedron Lett.* **32:**6289 (1991).
g. T. A. Engler, M. A. Letavic, K. O. Lynch, Jr., and F. Takusagawa, *J. Org. Chem.* **59:**1179 (1994).
h. E. J. Corey, S. Sarshar, and D.-H. Lee, *J. Am. Chem. Soc.* **116:**12089 (1994).
i. E. J. Corey, T.-P. Loh, T. D. Roper, M. D. Azimioara, and M. C. Noe, *J. Am. Chem. Soc.* **114:**8290 (1992).

less methylene group and this leads to a more strained transition state. A mixture of stereoisomers is formed, reflecting a conflict between the Alder rule, which favors *endo* addition, and conformational factors that favor the *exo* transition state. The stereoselectivity of a number of intramolecular Diels–Alder reactions has been analyzed, and conformational factors in the transition state seem to play the dominant role in determining product structure.[68]

Lewis acid catalysis usually substantially improves the stereoselectivity of intramolecular Diels–Alder reactions, just as it does in intermolecular cases. For example, the thermal cyclization of **1** at 160°C gives a 50 : 50 mixture of two stereoisomers, but the use of Et_2AlCl as a catalyst permits the reaction to proceed at room temperature, and *endo* addition is favored by 8 : 1.[69]

CO_2CH_3 **1** → MeO_2C H H MeO_2C H H

	endo T.S.	*exo* T.S.
thermal (160°C)	50%	50%
Et_2AlCl (23°C)	88%	12%

68. W. R. Roush, A. I. Ko, and H. R. Gillis, *J. Org. Chem.* **45:**4264 (1980); R. K. Boeckman, Jr. and S. K. Ko, *J. Am. Chem. Soc.* **102:**7146 (1980); W. R. Roush and S. E. Hall, *J. Am. Chem. Soc.* **103:**5200 (1981); K. A. Parker and T. Iqbal, *J. Org. Chem.* **52:**4369 (1987).
69. W. R. Roush and H. R. Gillis, *J. Org. Chem.* **47:**4825 (1982).

Scheme 6.4. Intramolecular Diels–Alder Reactions

1[a] H_3C, O, $CH(CH_3)_2$ —0°C→ H, O, H_3C, H, $CH(CH_3)_2$ 87%

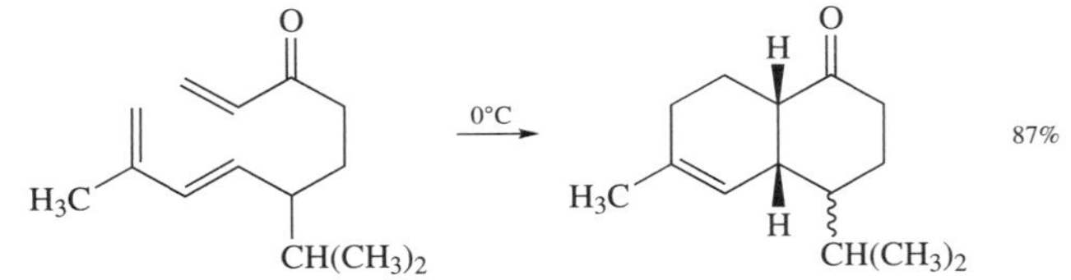

2[b] CH_3 —160°C→ H, H_3C, H 95%

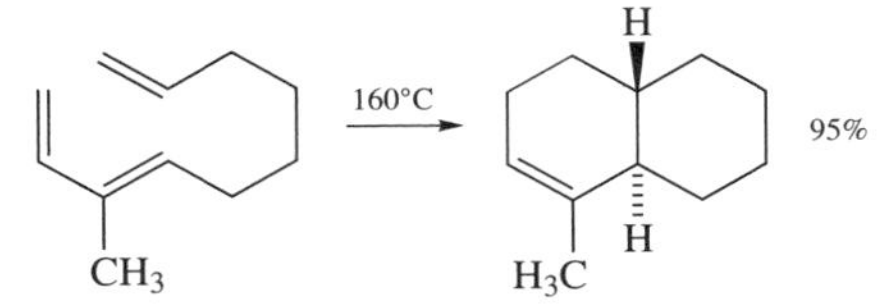

3[c] CH_3O_2C, $(CH_3)_2CH$, OH —150°C→ CH_3O_2C, H, $(CH_3)_2CH$, H, OH 60%

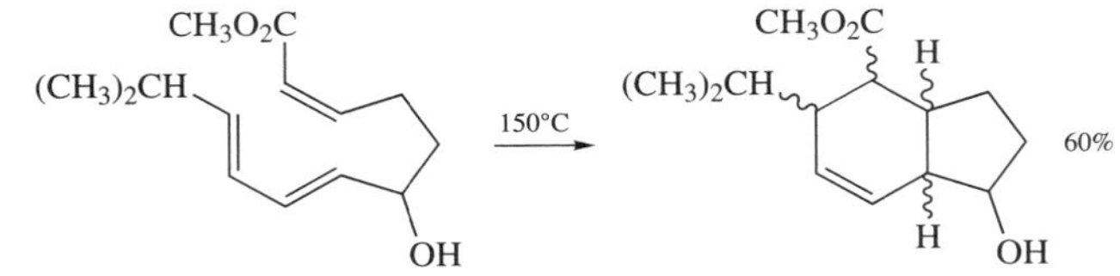

mixture of stereoisomers

4[d] H, H_3C, N, $CH_2CH_2CH_3$, O, $CH(CH_3)_2$ —230°C, 16 h→ H_3C, H, H, N, $CH_2CH_2CH_3$, O, $CH(CH_3)_2$ 60%

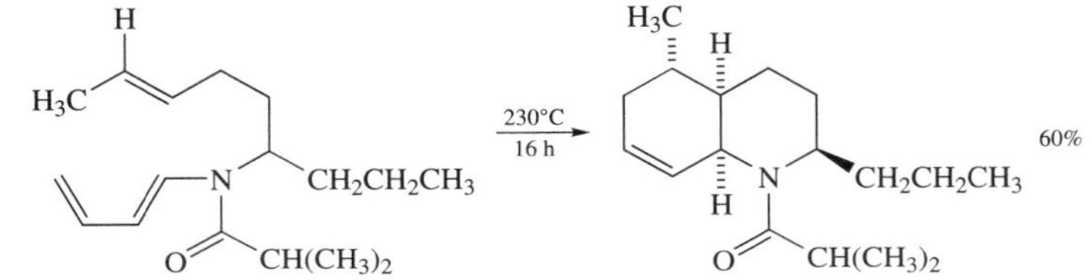

5[e] O=CH, $OSiR_3$ —Et_2AlCl→ H, H, O=CH, $OSiR_3$ 62%

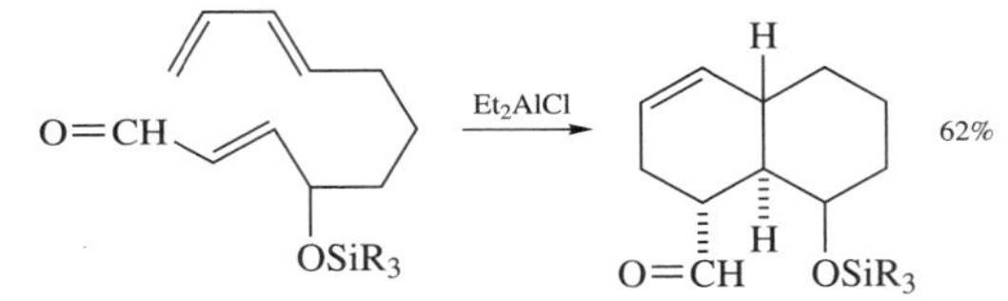

6[f] H_3C, OCH_3, HO, H —195°C, 4.5 h→ OCH_3, H_3C, H, H, HO, H

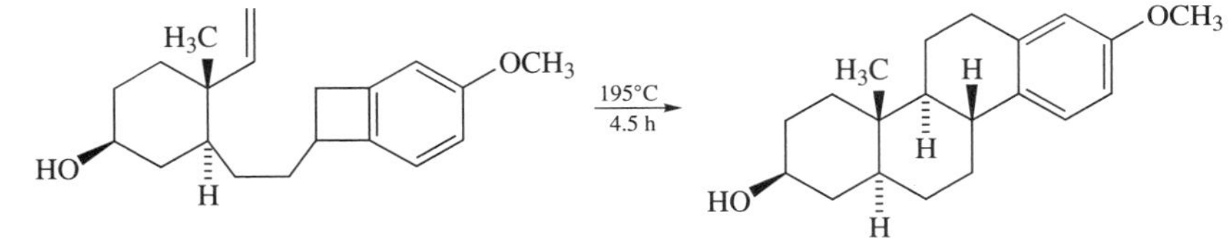

7[g] H_3C, OH, CH_3O, H —200°C, 7.5 h→ H_3C, OH, H, H, H, CH_3O 91%

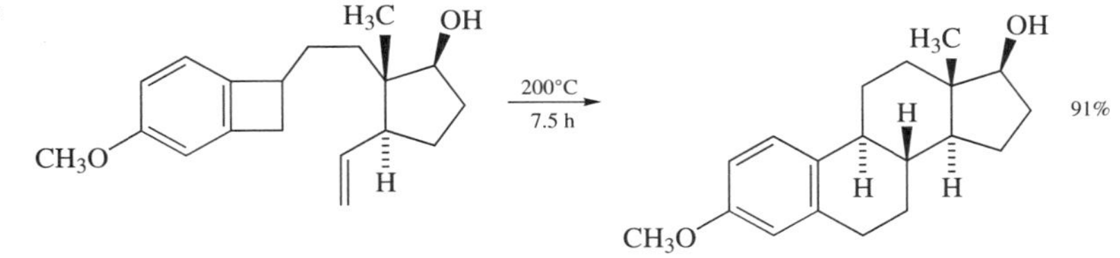

a. D. F. Taber and B. P. Gunn, *J. Am. Chem. Soc.* **101:**3992 (1979).
b. S. R. Wilson and D. T. Mao, *J. Am. Chem. Soc.* **100:**6289 (1978).
c. W. R. Roush, *J. Am. Chem. Soc.* **102:**1390 (1980).
d. W. Oppolzer and E. Flaskamp, *Helv. Chim. Acta* **60:**204 (1977).
e. J. A. Marshall, J. E. Audia, and J. Grote, *J. Org. Chem.* **49:**5277 (1984).
f. T. Kametani, K. Suzuki, and H. Nemoto, *J. Org. Chem.* **45:**2204 (1980); *J. Am. Chem. Soc.* **103:**2890 (1981).
g. P. A. Grieco, T. Takigawa, and W. J. Schillinger, *J. Org. Chem.* **45:**2247 (1980).

It has also been noted in certain systems that the stereoselectivity is a function of the activating substituent on the double bond, both for thermal and Lewis acid-catalyzed reactions.[70] The general trend in these systems is consistent with frontier orbital interactions and conformational effects being the main factors in determining stereoselectivity. Because the conformational interactions depend on the substituent pattern in each specific case, no general rules regarding stereoselectivity can be put forward. Molecular modeling can frequently identify the controlling structural features.[71]

As in intermolecular reactions, enantioselectivity can be enforced in intramolecular Diels–Alder additions by use of chiral structures. For example, the dioxolane rings in **2** and **3** result in transition-state structures that lead to enantioselective reactions.[72]

2

3 OCH_2OCH_3 CH_3OCH_2O CH_3OCH_2O

Chiral catalysts (see Section 6.1.4) can also achieve enantioselectivity in intramolecular Diels–Alder reactions.

TBDMSO t-Bu Cu i-Bu TBDMSO Ref. 73

96% e.e.

70. J. A. Marshall, J. E. Audia, and J. Grote, *J. Org. Chem.* **49:**5277 (1984); W. R. Roush, A. P. Essenfeld, and J. S. Warmus, *Tetrahedron Lett.* **28:**2447 (1987); T.-C. Wu and K. N. Houk, *Tetrahedron Lett.* **26:**2293 (1985).
71. K. J. Shea, L. D. Burke, and W. P. England, *J. Am. Chem. Soc.* **110:**860 (1988); L. Raimondi, F. K. Brown, J. Gonzalez, and K. N. Houk, *J. Am. Chem. Soc.* **114:**4796 (1992); D. P. Dolata and L. M. Harwood, *J. Am. Chem. Soc.* **114:**10738 (1992); F. K. Brown, U. C. Singh, P. A. Kollman, L. Raimondi, K. N. Houk, and C. W. Bock, *J. Org. Chem.* **57:**4862 (1992); J. D. Winkler, H. S. Kim, S. Kim, K. Ando, and K. N. Houk, *J. Org. Chem.* **62:**2957 (1997).
72. T. Wong, P. D. Wilson, S. Woo, and A. G. Fallis, *Tetrahedron Lett.* **38:**7045 (1997).
73. D. A. Evans and J. S. Johnson, *J. Org. Chem.* **62:**786 (1997).

The favorable kinetics of intramolecular Diels–Alder additions can be exploited by temporary links (tethers) between the diene and dienophile components.[74] After the addition reaction, the tether can be broken. Siloxy derivatives have been used in this way, because silicon–oxygen bonds can be broken by solvolysis or by fluoride ion.[75]

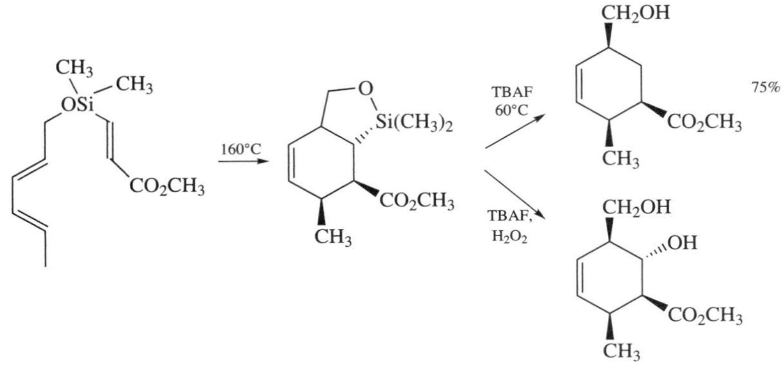

Ref. 75a

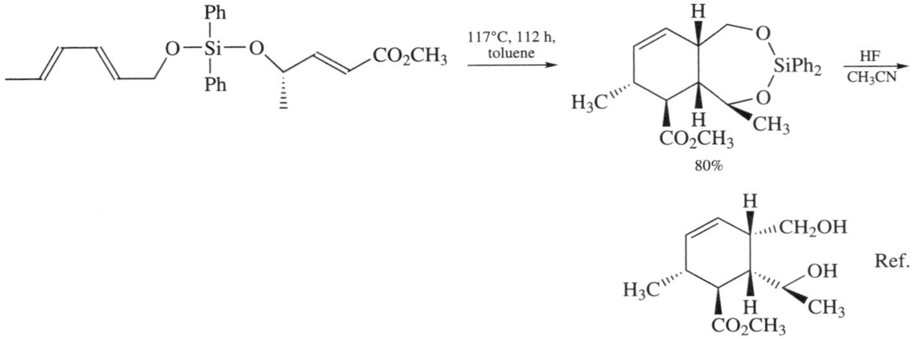

Ref. 75d

Acetals have also been used as removable tethers.

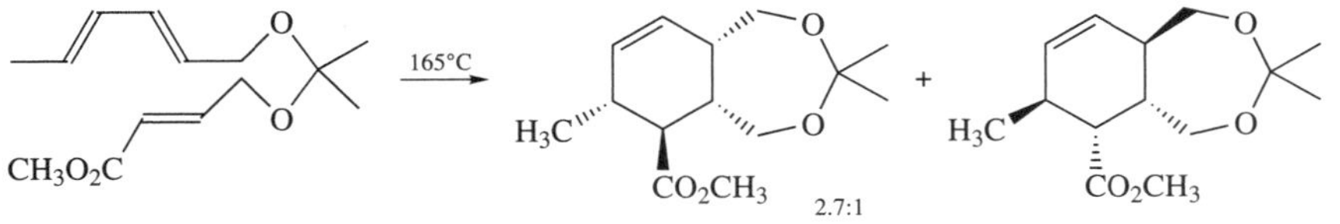

Ref. 76

The activating capacity of boronate groups can be combined with the ability for facile transesterification at boron in such a way as to permit intramolecular reactions between

74. L. Fensterbank, M. Malacria, and S. McN. Sieburth, *Synthesis* **1997:**813; M. Bols and T. Skrydstrup, *Chem. Rev.* **95:**1253 (1995).
75. (a) G. Stork, T. Y. Chan, and G. A. Breault, *J. Am. Chem. Soc.* **114:**7578 (1992); (b) S. McN. Sieburth, and L. Fensterbank, *J. Org. Chem.* **57:**5279 (1992); (c) J. W. Gillard, R. Fortin, E. L. Grimm, M. Maillard, M. Tjepkema, M. A. Bernstein, and R. Glaser, *Tetrahedron Lett.* **32:**1145 (1991); (d) D. Craig and J. C. Reader, *Tetrahedron Lett.* **33:**4073 (1992).
76. P. J. Ainsworth, D. Craig, A. J. P. White, and D. J. Williams, *Tetrahedron* **52:**8937 (1996).

vinylboronates and 2,4-dienols.

Ref. 77

6.2. Dipolar Cycloaddition Reactions

In Chapter 11 of Part A, the mechanistic classification of 1,3-dipolar cycloadditions as a type of concerted cycloadditions was developed. Dipolar cycloaddition reactions are useful both for the synthesis of heterocyclic compounds and for carbon–carbon bond formation. Table 6.2 lists some of the types of molecules that are capable of dipolar cycloaddition. These molecules, which are called *1,3-dipoles*, have π-electron systems that are isoelectronic with allyl anion, consisting of two filled and one empty orbital. Each molecule has at least one charge-separated resonance structure with opposite charges in a 1,3-relationship. It is this structural feature that leads to the name *1,3-dipolar cycloadditions* for this class of reactions.[78] The other reactant in a dipolar cycloaddition, usually an alkene or alkyne, is referred to as the *dipolarophile*. Other multiply bonded functional groups such as imine, azo, and nitroso groups can also act as dipolarophiles. The transition states for 1,3-dipolar cycloadditions involve four π electrons from the 1,3-dipole and two from the dipolarophile. As in the Diels–Alder reaction, the reactants approach one another in parallel planes.

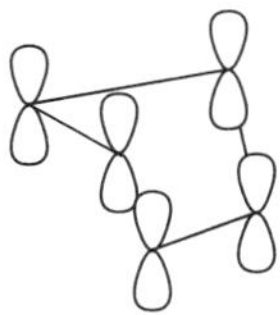

Mechanistic studies have shown that the transition state for 1,3-dipolar cycloaddition is not very polar. The rate of reaction is not strongly sensitive to solvent polarity. In most

77. R. A. Batey, A. N. Thadani, and A. J. Lough, *J. Am. Chem. Soc.* **121:**450 (1999).
78. For comprehensive reviews of 1,3-dipolar cycloaddition reactions, see G. Bianchi, C. DeMicheli, and R. Gandolfi, in *The Chemistry of Double Bonded Functional Groups*, *Part I*, *Supplement A*, S. Patai, ed., John Wiley & Sons, New York, 1977, pp. 369–532; A. Padwa, ed., *1,3-Dipolar Cycloaddition Chemistry*, John Wiley & Sons, New York, 1984. For a review of intramolecular 1,3-dipolar cycloaddition reactions, see A. Padwa, *Angew. Chem. Int. Ed. Engl.* **15:**123 (1976).

Table 6.2. 1,3-Dipolar Compounds

$:\overset{+}{N}{=}\ddot{N}{-}\overset{-}{\ddot{C}}R_2 \longleftrightarrow :N{\equiv}\overset{+}{N}{-}\overset{-}{\ddot{C}}R_2$	Diazoalkane
$:\overset{+}{N}{=}\ddot{N}{-}\overset{-}{\ddot{N}}R \longleftrightarrow :N{\equiv}\overset{+}{N}{-}\overset{-}{\ddot{N}}R$	Azide
$R\overset{+}{C}{=}\ddot{N}{-}\overset{-}{\ddot{C}}R_2 \longleftrightarrow RC{\equiv}\overset{+}{N}{-}\overset{-}{\ddot{C}}R_2$	Nitrile ylide
$R\overset{+}{C}{=}\ddot{N}{-}\overset{-}{\ddot{N}}R \longleftrightarrow RC{\equiv}\overset{+}{N}{-}\overset{-}{\ddot{N}}R$	Nitrile imine
$R\overset{+}{C}{=}\ddot{N}{-}\overset{-}{\ddot{O}}: \longleftrightarrow RC{\equiv}\overset{+}{N}{-}\overset{-}{\ddot{O}}:$	Nitrile oxide
$R_2\overset{+}{C}{-}\ddot{N}(R){-}\overset{-}{\ddot{C}}R_2 \longleftrightarrow R_2C{=}\overset{+}{N}(R){-}\overset{-}{\ddot{C}}R_2$	Azomethine ylide
$R_2\overset{+}{C}{-}\ddot{N}(R){-}\overset{-}{\ddot{O}}: \longleftrightarrow R_2C{=}\overset{+}{N}(R){-}\overset{-}{\ddot{O}}:$	Nitrone
$R_2\overset{+}{C}{-}\ddot{O}{-}\ddot{O}:^{-} \longleftrightarrow R_2C{=}\overset{+}{O}{-}\ddot{O}:^{-}$	Carbonyl oxide

cases, the reaction is a concerted $[2\pi_s + 4\pi_s]$ cycloaddition.[79] The destruction of charge separation that is implied is more apparent than real, because most 1,3-dipolar compounds are not highly polar. The polarity implied by any single structure is balanced by other contributing structures.

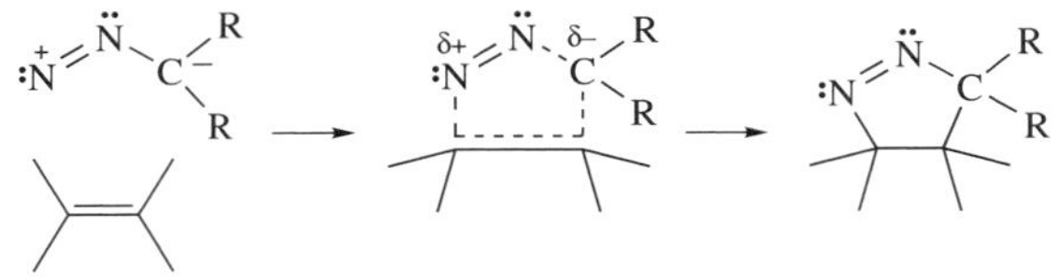

Two questions are of immediate interest for predicting the structure of 1,3-dipolar cycloaddition products: (1) What is the regioselectivity? and (2) what is the stereoselectivity? Many specific examples demonstrate that 1,3-dipolar cycloaddition is a stereospecific *syn* addition with respect to the dipolarophile. This is what would be expected for a concerted process.

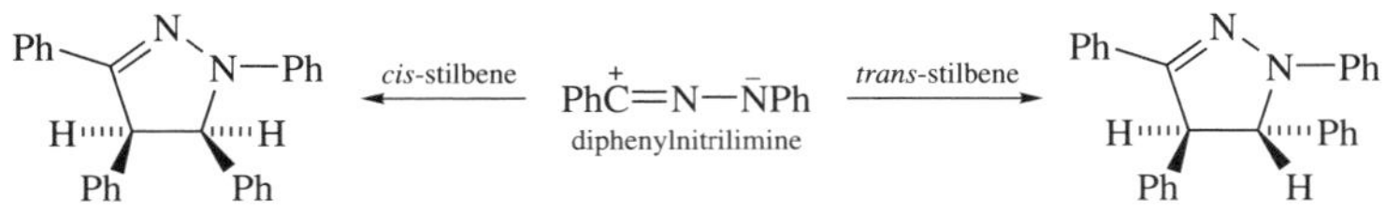

cis-stilbene ← $Ph\overset{+}{C}{=}N{-}\overset{-}{N}Ph$ diphenylnitrilimine → *trans*-stilbene — Ref. 80

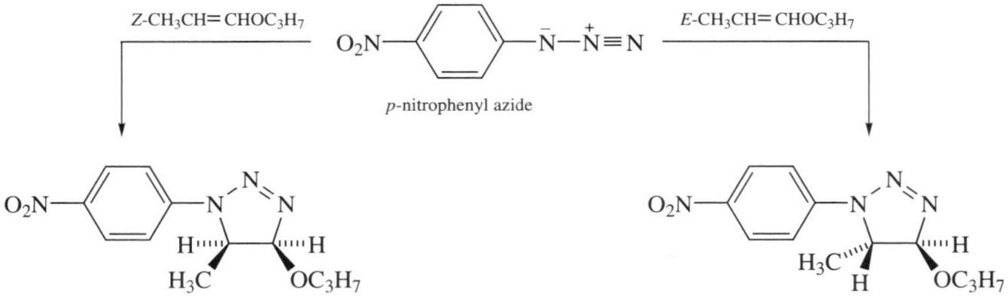

Z-$CH_3CH{=}CHOC_3H_7$ ← $O_2N{-}C_6H_4{-}\overset{-}{N}{-}\overset{+}{N}{\equiv}N$ *p*-nitrophenyl azide → *E*-$CH_3CH{=}CHOC_3H_7$ — Ref. 81

79. P. K. Kadaba, *Tetrahedron* **25**:3053 (1969); R. Huisgen, G. Szeimes, and L. Mobius, *Chem. Ber.* **100**:2494 (1967); P. Scheiner, J. H. Schomaker, S. Deming, W. J. Libbey, and G. P. Nowack, *J. Am. Chem. Soc.* **87**:306 (1965).
80. R. Huisgen, M. Seidel, G. Wallibillich, and H. Knupfer, *Tetrahedron* **17**:3 (1962).
81. R. Huisgen and G. Szeimies, *Chem. Ber.* **98**:1153 (1965).

With some 1,3-dipoles, two possible stereoisomers can be formed by *syn* addition. These result from two differing orientations of the reacting molecules, which are analogous to the *endo* and *exo* transition states in Diels–Alder reactions. Diazoalkanes, for example, can add to unsymmetrical dipolarophiles to give two diastereomers.

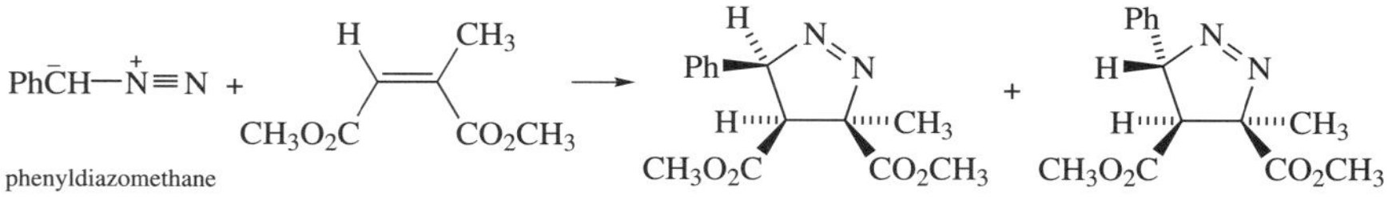

Ref. 82

Each 1,3-dipole exhibits a characteristic regioselectivity toward different types of dipolarophiles. The dipolarophiles can be grouped, as were dienophiles, depending upon whether they have electron-donating or electron-withdrawing substituents. The regioselectivity can be interpreted in terms of frontier orbital interactions. Depending on the relative orbital energies in the 1,3-dipole and dipolarophile, the strongest interaction may be between the HOMO of the dipole and the LUMO of the dipolarophile or vice versa. Usually, for dipolarophiles with electron-withdrawing groups, the dipole-HOMO/dipolarophile-LUMO interaction is dominant. The reverse is true for dipolarophiles with donor substituents. In some circumstances, the magnitudes of the two interactions may be comparable.[83]

The prediction of regiochemistry requires estimation or calculation of the energies of the orbitals that are involved, which permits identification of the frontier orbitals. The energies and orbital coefficients for the most common dipoles and dipolarophiles have been summarized.[83] Figure 11.14 in Part A gives the orbital coefficients of some representative 1,3-dipoles. Regioselectivity is determined by the preference for the orientation that results in bond formation between the atoms having the largest coefficients in the two frontier orbitals. This analysis is illustrated in Fig. 6.5.

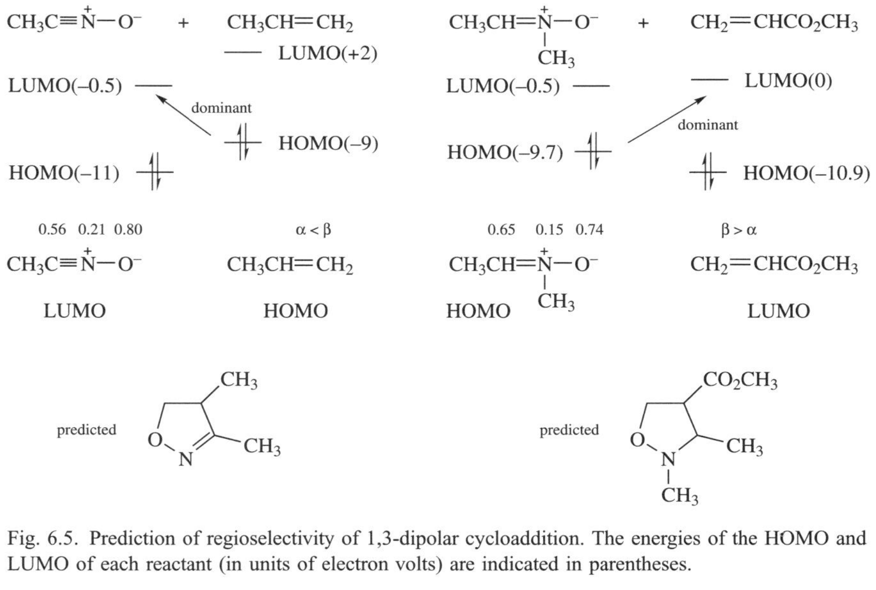

Fig. 6.5. Prediction of regioselectivity of 1,3-dipolar cycloaddition. The energies of the HOMO and LUMO of each reactant (in units of electron volts) are indicated in parentheses.

82. R. Huisgen and P. Eberhard, *Tetrahedron Lett.* **1971**:4343.
83. K. N. Houk, J. Sims, B. E. Duke, Jr., R. W. Strozier, and J. K. George, *J. Am. Chem. Soc.* **95**:7287 (1973); I. Fleming, *Frontier Orbitals and Organic Chemical Reactions*, John Wiley & Sons, New York, 1977; K. N. Houk, in *Pericyclic Reactions*, Vol. II, A. P. Marchand and R. E. Lehr, eds., Academic Press, New York, 1977, pp. 181–271.

Table 6.3. Relative Reactivity of Substituted Alkenes toward Some 1,3-Dipoles[a,b]

Substituted alkene	Ph_2CN_2	PhN_3	$Ph\overset{+}{N}{=}N{-}\overset{-}{N}Ph$	$PhC{\equiv}\overset{+}{N}{-}\overset{-}{O}$	$PhCH{=}\overset{+}{N}(O^-){-}CH_3$
Dimethyl fumarate	100	31	283	94	18.3
Dimethyl maleate	27.8	1.25	7.9	1.61	6.25
Norbornene	1.15	700	3.1	97	0.13
Ethyl acrylate	28.8	36.5	48	66	11.1[c]
Butyl vinyl ether	—	1.5	—	15	—
Styrene	0.57	1.5	1.6	9.3	0.32
Ethyl crotonate	1.0	1.0	1.0	1.0	1.0
Cyclopentene	—	6.9	0.13	1.04	0.022
Terminal alkene	—	0.8[d]	0.15[d]	2.6[e]	0.072[d]
Cyclohexene	—	—	0.011	0.055	—

a. Data are selected from those compiled by R. Huisgen, R. Grashey, and J. Sauer, in *Chemistry of Alkenes*, S. Patai, ed., John Wiley & Sons, New York, 1964, pp. 806–977.
b. Conditions such as solvent and temperature vary for each 1,3-dipole, so comparison from dipole to dipole is not possible. Following Huisgen, Grashey, and Sauer,[a] ethyl crotonate is assigned reactivity = 1.0 for each 1,3-dipole.
c. Methyl ester
d. Heptene
d. Hexene

In addition to the role of substituents in determining regioselectivity, several other structural features affect the reactivity of dipolarophiles. Strain increases reactivity. Norbornene, for example, is consistently more reactive than cyclohexene in 1,3-dipolar cycloadditions. Conjugated functional groups also usually increase reactivity. This increased reactivity has most often been demonstrated with electron-attracting substituents, but for some 1,3-dipoles, enamines, enol ethers, and other alkenes with donor substituents are also quite reactive. Some reactivity data for a series of alkenes with a few 1,3-dipoles are given in Table 6.3. Scheme 6.5 gives some examples of 1,3-dipolar cycloaddition reactions.

Dipolar cycloadditions are an important means of synthesis of a wide variety of heterocyclic molecules, some of which are useful intermediates in multistage synthesis. Pyrazolines, which are formed from alkenes and diazo compounds, for example, can be pyrolyzed or photolyzed to give cyclopropanes.

$PhCH{=}CH_2$ + $N_2CHCH(OMe)_2$ → [pyrazoline: Ph, $CH(OMe)_2$, N=N] $\xrightarrow{h\nu}$ [cyclopropane: Ph, $CH(OMe)_2$] Ref. 84

TBDPSO– [butenolide] $\xrightarrow{CH_2N_2}$ TBDPSO– [pyrazoline-fused lactone, N=N] $\xrightarrow{h\nu}$ TBDPSO– [cyclopropane-fused lactone] Ref. 85

84. P. Carrie, *Heterocycles* **14:**1529 (1980).
85. M. Martin-Vila, N. Hanafi, J. M. Jiminez, A. Alvarez-Larena, J. F. Piniella, V. Branchadell, A. Oliva, and R. M. Ortuno, *J. Org. Chem.* **63:**3581 (1998).

Scheme 6.5. Typical 1,3-Dipolar Cycloaddition Reactions

A. Intermolecular cycloaddition

1[a] $O_2N-C_6H_4-N=\overset{+}{N}=\overset{-}{N}$ + norbornene → 92%

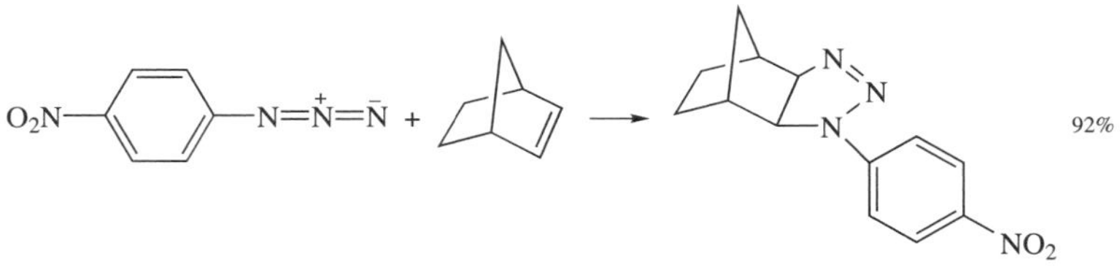

2[b] $C_6H_5-N=\overset{+}{N}=\overset{-}{N}$ + $H_3CO_2CC{\equiv}CCO_2CH_3$ → 87%

3[c] CH_2N_2 + $H_2C{=}CH$–(1,3-dioxolan-2-yl) → 80%

4[d] $PhCH{=}\overset{+}{N}(O^-)CH_3$ + $H_2C{=}CHC{\equiv}N$ → 91%

5[e] $\xrightarrow[Et_3N]{PhNCO}$ $C{\equiv}\overset{+}{N}-O^-$ $\xrightarrow{C_5H_{11}C{\equiv}CH}$ 60%

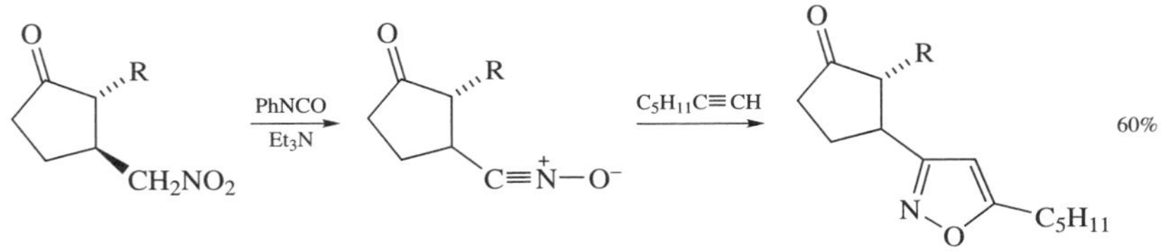

$R = -(CH_2)_6CO_2(CH_2)_3CH_3$

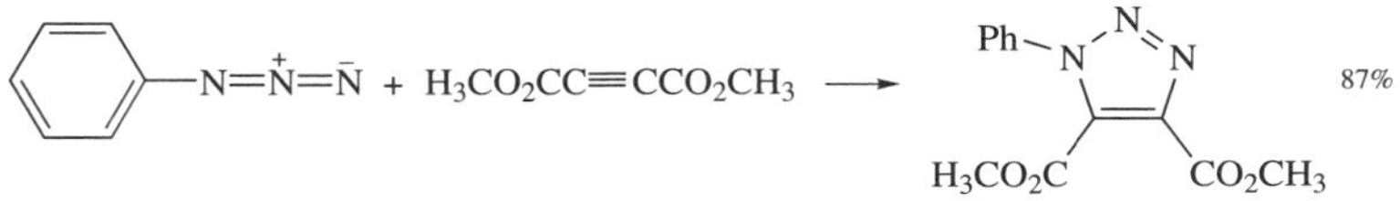

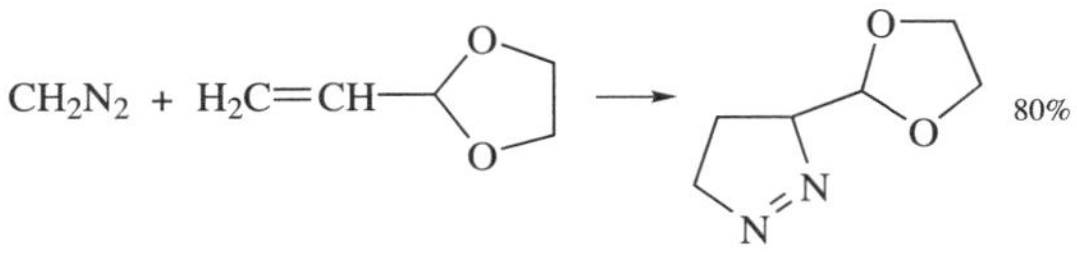

B. Intramolecular cycloaddition

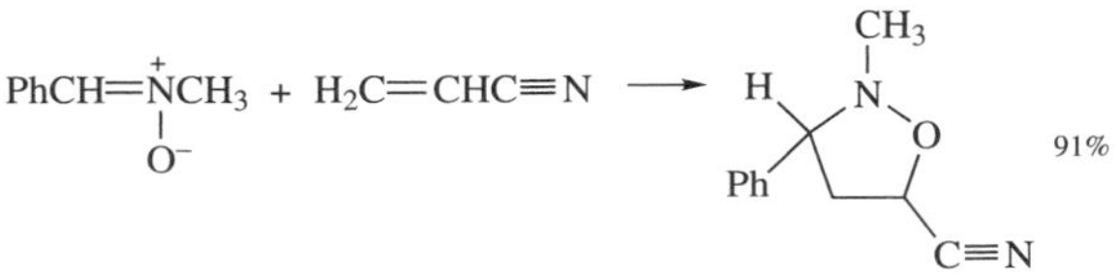

6[f] $CH_3CH_2CH_2CH(NHOH)CH_2CH_2CH{=}CHCH_3$ + $PhSO_2(CH_2)_3CH{=}O$ → 74%

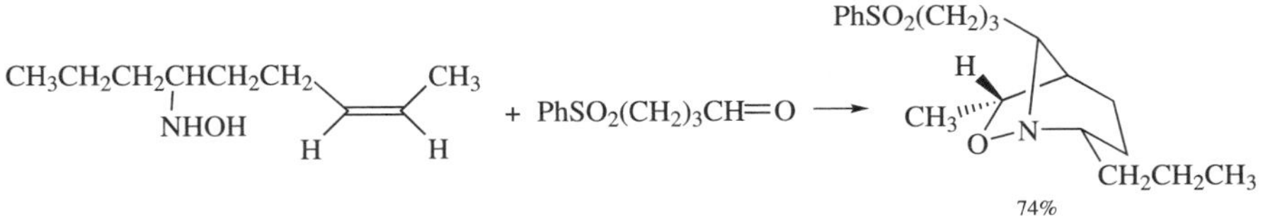

7[g] $(CH_3)_2C{=}CHCH_2CH_2CH(CH_3)CH_2CH{=}O$ $\xrightarrow[NaOCH_3\ toluene,\ \Delta]{CH_3NHOH{\cdot}HCl}$ 64–67%

8[h] $\xrightarrow[\Delta]{toluene}$ $\xrightarrow[2)\ CH_2O,\ HCO_2H]{1)\ H_2,\ Pd/C}$

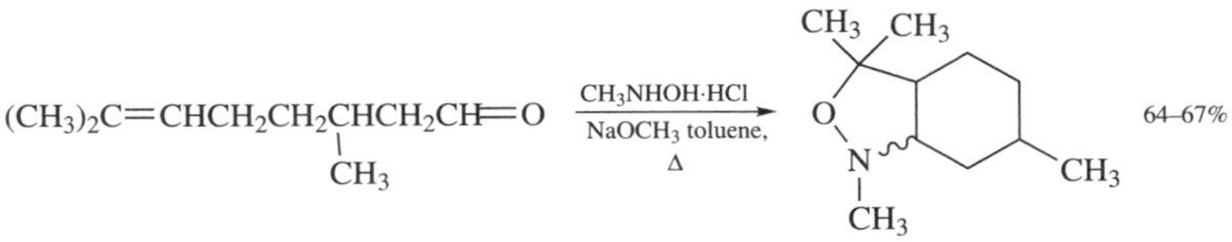

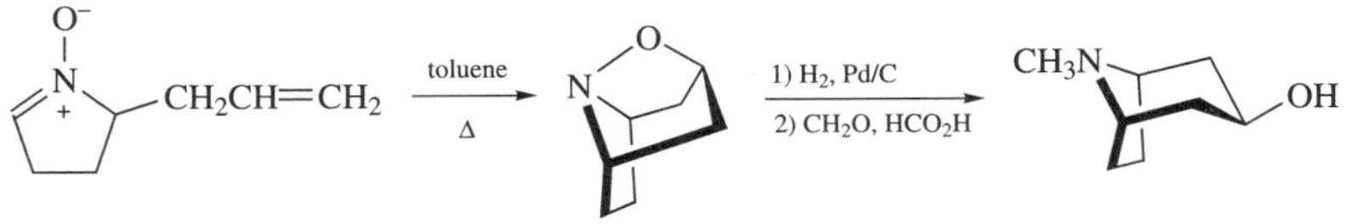

(*continued*)

Scheme 6.5. Typical 1,3-Dipolar Cycloaddition Reactions

9[i] O=$CHCH_2S$– (cycloheptenyl) $\xrightarrow{PhCH_2NHOH, \text{ acetonitrile}}$ $PhCH_2N$–O (66%) $\xrightarrow{Zn, HOAc, H_2O}$ $PhCH_2NH$ HO (96%)

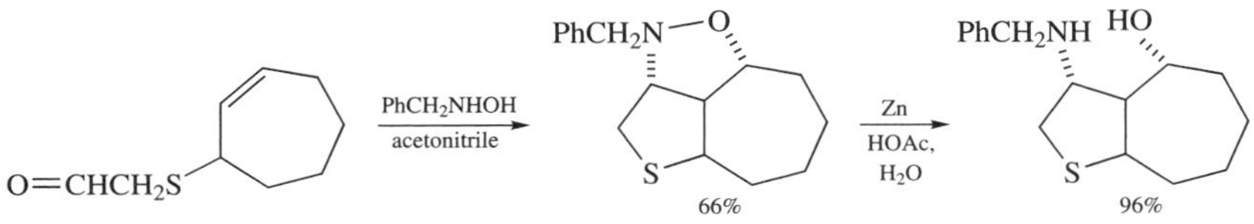

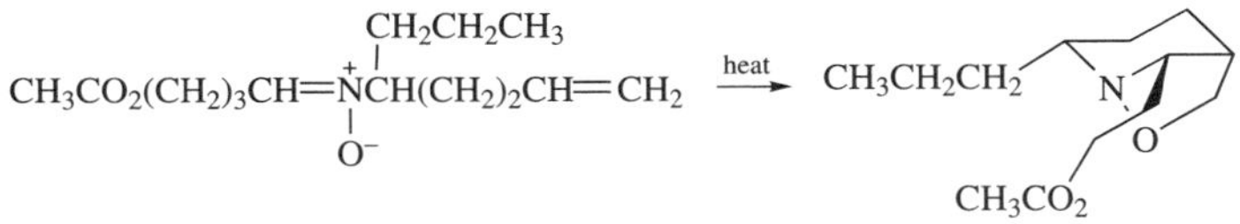

10[j] $CH_3CO_2(CH_2)_3CH{=}\overset{+}{N}(O^-)CH(CH_2CH_2CH_3)(CH_2)_2CH{=}CH_2$ $\xrightarrow{heat}$ $CH_3CH_2CH_2$, N, O, CH_3CO_2

11[k] CH_3O_2C, CH_3, $OC(CH_3)_3$, CH_3, $(CH_2)_2CH{=}NOH$ $\xrightarrow{NaOCl}$ CH_3O_2C, CH_3, $OC(CH_3)_3$, CH_3, N–O 96%

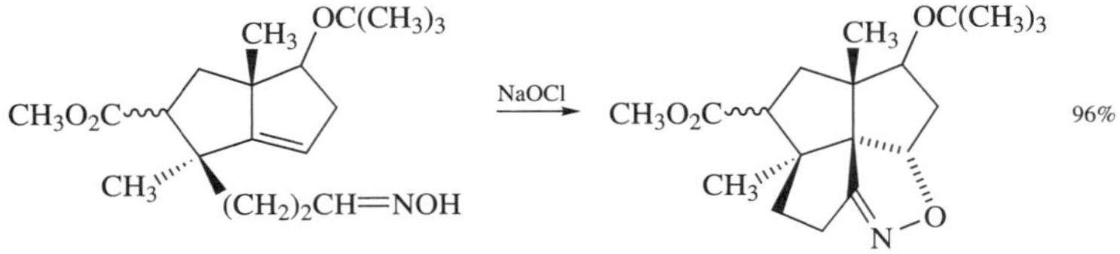

a. P. Scheiner, J. H. Schomaker, S. Deming, W. J. Libbey, and G. P. Nowack, *J. Am. Chem. Soc.* **87:**306 (1965).
b. R. Huisgen, R. Knorr, L. Mobius, and G. Szeimies, *Chem. Ber.* **98:**4014 (1965).
c. J. M. Stewart, C. Carlisle, K. Kem, and G. Lee, *J. Org. Chem.* **35:**2040 (1970).
d. R. Huisgen, H. Hauck, R. Grashey, and H. Seidl, *Chem. Ber.* **101:**2568 (1968).
e. A. Barco, S. Benetti, G. P. Pollini, P. G. Baraldi, M. Guarneri, D. Simoni, and C. Gandolfi, *J. Org. Chem.* **46:**4518 (1981).
f. N. A. LeBel and N. Balasubramanian, *J. Am. Chem. Soc.* **111:**3363 (1989).
g. N. A. LeBel and D. Hwang, *Org. Synth.* **58:**106 (1978).
h. J. J. Tufariello, G. B. Mullen, J. J. Tegeler, E. J. Trybulski, S. C. Wong, and S. A. Ali, *J. Am. Chem. Soc.* **101:**2435 (1979).
i. P. N. Confalone, G. Pizzolato, D. I. Confalone, and M. R. Uskokovic, *J. Am. Chem. Soc.* **102:**1954 (1980).
j. A. L. Smith, S. F. Williams, A. B. Holmes, L. R. Hughes, Z. Lidert, and C. Swithenbank, *J. Am. Chem. Soc.* **110:**8696 (1988).
k. M. Ihara, Y. Tokunaga, N. Taniguchi, K. Fukumoto, and C. Kabuto, *J. Org. Chem.* **56:**5281 (1991).

Intramolecular 1,3-dipolar cycloadditions have proven to be especially useful in synthesis. The addition of nitrones to alkenes serves both to form a carbon–carbon bond and to introduce oxygen and nitrogen functionality.[86] Entry 7 in Scheme 6.5 is an example. The nitrone **B** is generated by condensation of the aldehyde group with *N*-methylhydroxylamine and then goes on to product by intramolecular cycloaddition.

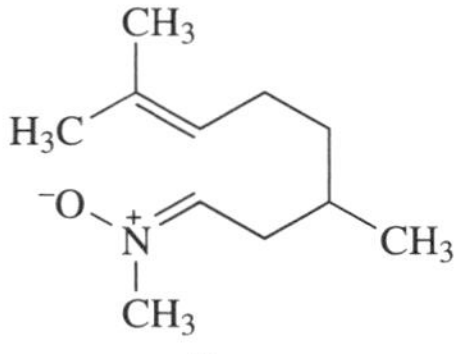

CH_3, H_3C, ^-O, $\overset{+}{N}$, CH_3, CH_3

B

The products of nitrone–alkene cycloadditions are isoxazolines, and the oxygen–nitrogen bond can be cleaved by reduction, leaving both an amino and a hydroxy function in place.

86. For reviews of nitrone cycloadditions, see D. St. C. Black, R. F. Crozier, and V. C. Davis, *Synthesis* **1975:**205; J. J. Tufariello, *Acc. Chem. Res.* **12:**396 (1979); P. N. Confalone and E. M. Huie, *Org. React.* **36:**1 (1988).

A number of clever syntheses have employed this strategy. Entry 8 in Scheme 6.5 shows the final steps in the synthesis of the alkaloid pseudotropine. The proper stereochemical orientation of the hydroxyl group is ensured by the structure of the isoxazoline from which it is formed by reduction. Entry 9 portrays the early stages of a synthesis of the biologically important molecule biotin.

Nitrile oxides, which are formed by dehydration of nitroalkanes or by oxidation of oximes with hypochlorite,[87] are also useful 1,3-dipoles. They are highly reactive and must be generated *in situ*.[88] They react with both alkenes and alkynes. Entry 5 in Scheme 6.5 is an example in which the cycloaddition product (an isoxazole) was eventually converted to a prostaglandin derivative.

As with the Diels–Alder reaction, it is possible to achieve enantioselective cycloaddition in the presence of chiral catalysts.[89] The Ti(IV) catalyst **C** with chiral diol ligands leads to moderate to high enantioselectivity in nitrone–alkene cycloadditions.[90]

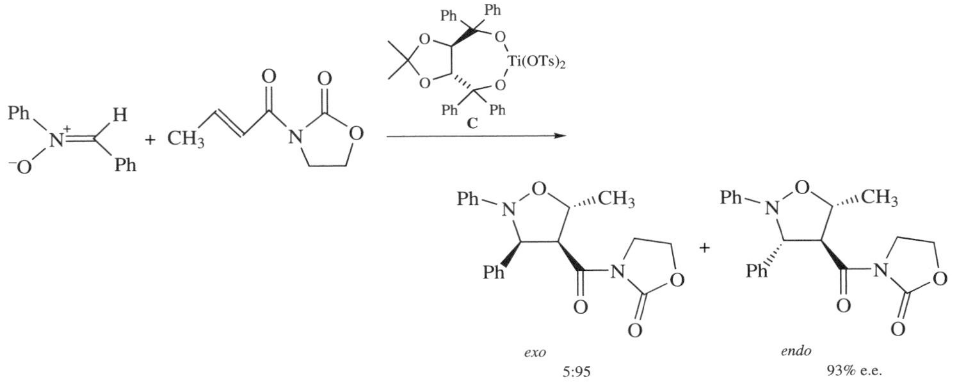

exo 5:95 endo 93% e.e.

Other effective catalysts include $Yb(O_3SCF_3)_3$[91] with BINOL, Mg^{2+}-bis-oxazolines,[92] and oxaborolidines.[93] Intramolecular nitrone cycloadditions can be facilitated by Lewis acids such as $ZnCl_2$.[94] The catalysis can be understood as resulting from a lowering of the LUMO energy of the 1,3-dipole, reasoning which is analogous to that employed to account for the Lewis acid catalysis of Diels–Alder reactions. The more organized transistion state, incorporating the metal ion and associated ligands, then enforces a

87. G. A. Lee, *Synthesis* **1982:**508.
88. K. Torssell, *Nitrile Oxides, Nitrones and Nitronates in Organic Synthesis*, VCH Publishers, New York, 1988.
89. K. V. Gothelf and K. A. Jorgensen, *Chem. Rev.* **98:**863 (1998); M. Frederickson, *Tetrahedron* **53:**403 (1997).
90. K. V. Gothelf and K. A. Jorgensen, *Acta Chem. Scand.* **50:**652 (1996); K. B. Jensen, K. V. Gothelf, R. G. Hazell, and K. A. Jorgensen, *J. Org. Chem.* **62:**2471 (1997); K. B. Jensen, K. V. Gothelf, and K. A. Jorgensen, *Helv. Chim. Acta* **80:**2039 (1997).
91. M. Kawamura and S. Kobayashi, *Tetrahedron Lett.* **40:**3213 (1999).
92. G. Desimoni, G. Faita, A. Mortoni, and P. Righetti, *Tetrahedron Lett.* **40:**2001 (1999); K. V. Gothelf, R. G. Hazell, and K. A. Jorgensen, *J. Org. Chem.* **63:**5483 (1998).
93. J. P. G. Seerden, M. M. M.Boeren, and H. W. Scheeren, *Tetrahedron* **53:**11843 (1997).
94. J. Marcus, J. Brussee, and A. van der Gen, *Eur. J. Org. Chem.* **1998:**2513.

preferred orientation of the reagents.

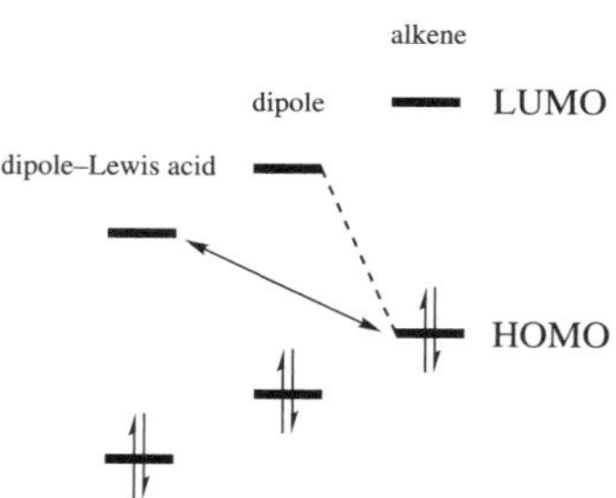

The change in frontier orbitals by coordination of a Lewis acid to the dipole

An interesting variation of the 1,3-dipolar cycloaddition involves generation of 1,3-dipoles from three-membered rings. As an example, aziridines **4** and **6** give adducts derived from apparent formation of 1,3-dipoles **5** and **7**, respectively.[95]

The evidence for the involvement of 1,3-dipoles as discrete intermediates includes the observation that the reaction rates are independent of dipolarophile concentration. This fact indicates that the ring opening is the rate-determining step in the reaction. Ring opening is most facile for aziridines that have an electron-attracting substituent to stabilize the carbanion center in the dipole.

Cyclopropanones are also reactive toward certain types of cycloadditions. Theoretical modeling indicates that a dipolar species resulting from reversible cleavage of the cyclopropanone ring is the reactive species.[96] *cis*-Disubstituted cyclopropanes with bulky substituents exhibit NMR features that indicate a barrier of 10–13 kcal/mol for

95. R. Huisgen and H. Mader, *J. Am. Chem. Soc.* **93:**1777 (1971).
96. D. Lim, D. A. Hrovat, W. T. Borden, and W. L. Jorgenson, *J. Am. Chem. Soc.* **116:**3494 (1994); B. A. Hess, Jr., U. Eckart, and J. Fabian, *J. Am. Chem. Soc.* **120:**12310 (1998).

the ring-opening process.[97]

$\Delta G^{\ddagger}$ = 10–13 kcal/mol for R = $(CH_3)_2CCH_2CH_3$, $(CH_3)_2CCH(CH_3)_2$, $(CH_3)_2CC(CH_3)_3$

The ring-opened intermediates, which are known as oxyallyl cations, can also be generated by a number of other reaction processes.[98]

$(C_2H_5)_3N$, H_2O — 30:70, 76% total yield — Ref. 99

$OSi(CH_3)_3$, H_2C, $CH(OCH_3)_2$ — $(CH_3)_3SiO_3SCF_3$ 5 mol % — OCH_3 67% — Ref. 100

6.3. [2 + 2] Cycloadditions and Other Reactions Leading to Cyclobutanes

[2 + 2] Cycloadditions of ketenes and alkenes have been shown to have synthetic utility for the preparation of cyclobutanones.[101] The stereoselectivity of ketene–alkene cycloaddition can be analyzed in terms of the Woodward–Hoffmann rules.[102] To be an allowed process, the $[2\pi + 2\pi]$ cycloaddition must be suprafacial in one component and antarafacial in the other. An alternative description of the transition state is a $[2\pi_s + (2\pi_s + 2\pi_s)]$ addition.[103] Figure 6.6 illustrates these transition states. The ketene, utilizing its low-lying LUMO, is the antarafacial component and interacts with the HOMO of the alkene. The stereoselectivity of ketene cycloadditions can be rationalized in terms of steric effects in this transition state. Minimization of interaction between the substituents R and R′ leads to a cyclobutanone in which these substituents are *cis*. This is the

97. T. S. Sorensen and F. Sun, *J. Chem. Soc., Perkin Trans. 2* **1998:**1053.
98. N. J. Turro, S. S. Edelson, J. R. Williams, T. R. Darling, and W. B. Hammond, *J. Am. Chem. Soc.* **91:**2283 (1969); S. S. Edelson and N. J. Turro, *J. Am. Chem. Soc.* **92:**2770 (1970); N. J. Turro, *Acc. Chem. Res.* **2:**25 (1969); J. Mann, *Tetrahedron* **42:**4611 (1986).
99. A. Lubineau and G. Bouchain, *Tetrahedron Lett.* **38:**8031 (1997).
100. D. H. Murray and K. F. Albizati, *Tetrahedron Lett.* **31:**4109 (1990).
101. For reviews, see W. T. Brady in *The Chemistry of Ketenes, Allenes, and Related Compounds*, S. Patai, ed., John Wiley & Sons, New York, 1980, Chapter 8; W. T. Brady, *Tetrahedron* **37:**2949 (1981).
102. R. B. Woodward and R. Hoffmann, *Angew. Chem. Int. Ed. Engl.* **8:**781 (1969).
103. E. Valenti, M. A. Pericas, and A. Moyano, *J. Org. Chem.* **55:**3582 (1990).

stereochemistry usually observed in these reactions.

H(C₂H₅)C=C=O + cyclopentadiene → bicyclic cyclobutanone (H, H, H, C₂H₅, O) Ref. 104

Ketenes are especially reactive in [2 + 2] cycloadditions, and an important reason is that they offer a low degree of steric interactions in the transition state. Another reason is the electrophilic character of the ketene LUMO. The best yields are obtained in reactions in which the ketene has an electronegative substituent, such as halogen. Simple ketenes are not very stable and usually must be generated *in situ*. The most common method for generating ketenes for synthesis is by dehydrohalogenation of acyl chorides. This is usually done with an amine such as triethylamine.[105] Other activated carboxylic acid derivatives, such as acyloxypyridinium ions, have also been used as ketene precursors[106]

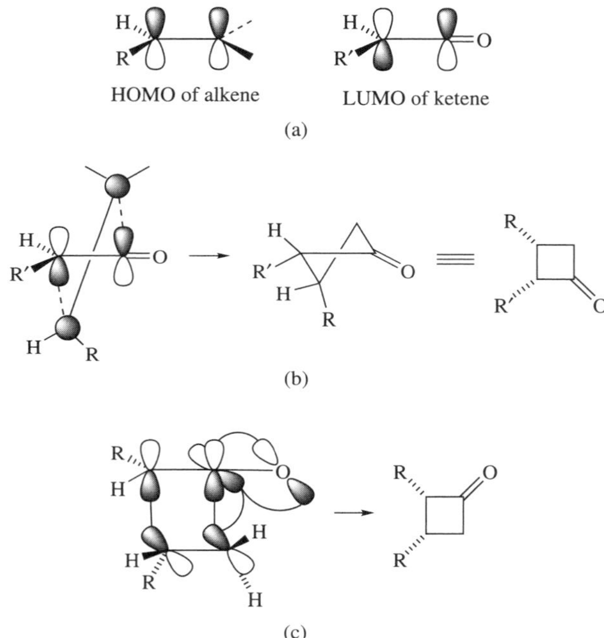

Fig. 6.6. HOMO–LUMO interactions in the [2 + 2] cycloaddition of an alkene and a ketene. (a) Frontier orbitals of alkene and ketene. (b) $[2\pi_s + 2\pi_a]$ Transition state required for suprafacial addition to alkene and antarafacial addition to ketene, leading to R and R′ in *cis* orientation in cyclobutanone products. (c) $[2\pi_s + (2\pi_s + 2\pi_s)]$ alternative transition state.

104. M. Rey, S. M. Roberts, A. S. Dreiding, A. Roussel, H. Vanlierde, S. Toppert, and L. Ghosez, *Helv. Chim. Acta* **65:**703 (1982).
105. K. Shishido, T. Azuma, and M. Shibuya, *Tetrahedron Lett.* **31:**219 (1990).
106. R. L. Funk, P. M. Novak, and M. M. Abelman, *Tetrahedron Lett.* **29:**1493 (1988).

Scheme 6.6. [2 + 2] Cycloadditions of Ketenes

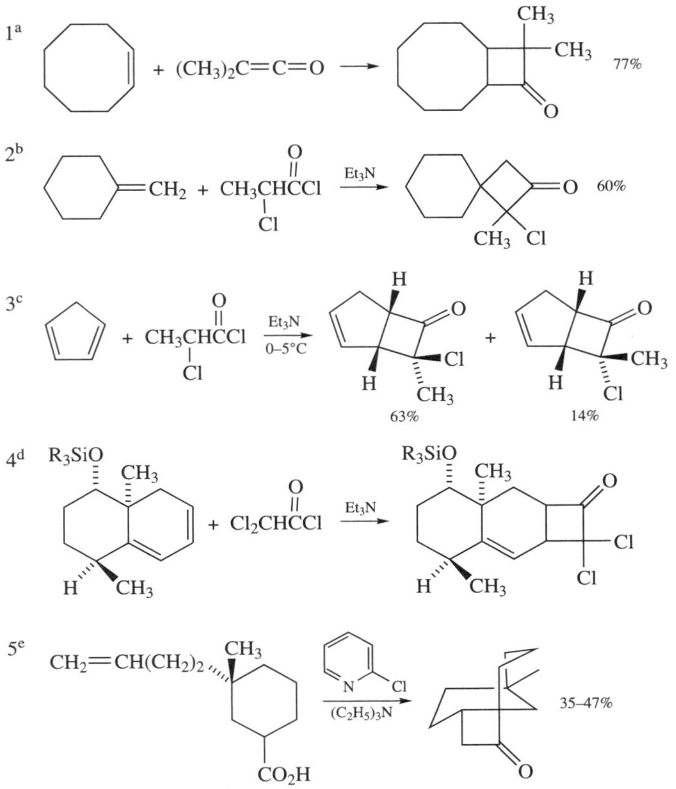

1[a] + $(CH_3)_2C{=}C{=}O$ → 77%

2[b] $=CH_2$ + $CH_3CHClCOCl$ $\xrightarrow{Et_3N}$ 60%

3[c] + $CH_3CHClCOCl$ $\xrightarrow[0-5°C]{Et_3N}$ 63% + 14%

4[d] + $Cl_2CHCOCl$ $\xrightarrow{Et_3N}$

5[e] $CH_2{=}CH(CH_2)_2$ ... CO_2H $\xrightarrow{\text{2-chloropyridine}, (C_2H_5)_3N}$ 35–47%

a. A. P. Krapcho and J. H. Lesser, *J. Org. Chem.* **31:**2030 (1966).
b. W. T. Brady and A. D. Patel, *J. Org. Chem.* **38:**4106 (1973).
c. W. T. Brady and R. Roe, *J. Am. Chem. Soc.* **93:**1662 (1971).
d. P. A. Grieco, T. Oguri, and S. Gilman, *J. Am. Chem. Soc.* **102:**5886 (1980).
e. R. L. Funk, P. M. Novak, and M. M. Abraham, *Tetrahedron Lett.* **29:**1493 (1988).

(see entry 5 in Scheme 6.6). Ketene itself and certain alkyl derivatives can be generated by pyrolysis of carboxylic anhydrides.[107] Scheme 6.6 gives some specific examples of ketene–alkene cycloadditions.

Intramolecular ketene cycloadditions are possible if the ketene and alkene functionalities can achieve an appropriate orientation.[108]

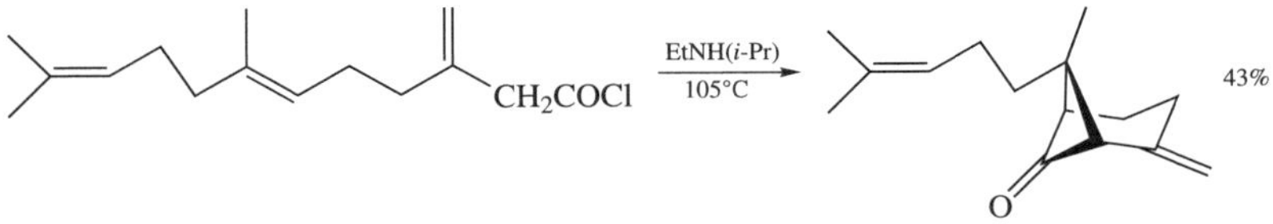

CH_2COCl $\xrightarrow[105°C]{EtNH(i\text{-}Pr)}$ 43% Ref. 109

107. G. J. Fisher, A. F. MacLean, and A. W. Schnizer, *J. Org. Chem.* **18:**1055 (1953).
108. B. B. Snider, R. A. H. F. Hui, and Y. S. Kulkarni, *J. Am. Chem. Soc.* **107:**2194 (1985); B. B. Snider and R. A. H. F. Hui, *J. Org. Chem.* **50:**5167 (1985); W. T. Brady and Y. F. Giang, *J. Org. Chem.* **50:**5177 (1985).
109. E. J. Corey and M. C. Desai, *Tetrahedron Lett.* **26:**3535 (1985).

Cyclobutanes can also be formed by nonconcerted processes involving zwitterionic intermediates. The combination of an electron-rich alkene (enamine, enol ether) and a very electrophilic one (nitro- or polycyanoalkene) is required for such processes.

ERG = electron releasing group (–OR, –NR_2)
EWG = electron withdrawing group (–NO_2, –C≡N)

Two examples of this reaction type are:

$CH_3CH_2CH{=}CHN$(pyrrolidine) + $PhCH{=}CHNO_2$ → 100% Ref. 110

$H_3C{=}CHOCH_3$ + $(NC)_2C{=}C(CN)_2$ → 90% Ref. 111

The stereochemistry of these reactions depends on the lifetime of the dipolar intermediate, which, in turn, is influenced by the polarity of the solvent. In the reactions of enol ethers with tetracyanoethylene, the stereochemistry of the enol ether portion is retained in nonpolar solvents. In polar solvents, cycloaddition is nonstereospecific, as a result of a longer lifetime for the zwitterionic intermediate.[112]

6.4. Photochemical Cycloaddition Reactions

Photochemical cycloadditions provide a method that is often complementary to thermal cycloadditions with regard to the types of compounds that can be prepared. The theoretical basis for this complementary relationship between thermal and photochemical modes of reaction lies in orbital symmetry relationships, as discussed in Chapter 13 of Part A. The reaction types permitted by photochemical excitation that are particularly useful for synthesis are [2 + 2] additions between two carbon–carbon double bonds and [2 + 2] additions of alkenes and carbonyl groups to form oxetanes. Photochemical cycloadditions are often not concerted processes because in many cases the reactive excited state is a triplet. The initial adduct is a triplet 1,4-diradical, which must undergo spin inversion before product formation is complete. Stereospecificity is lost if the intermediate 1,4-

110. M. E. Kuehne and L. Foley, *J. Org. Chem.* **30:**4280 (1965).
111. J. K. Williams, D. W. Wiley, and B. C. McKusick, *J. Am. Chem. Soc.* **84:**2210 (1962).
112. R. Huisgen, *Acc. Chem. Res.* **10:**117, 199 (1977).

diradical undergoes bond rotation faster than ring closure.

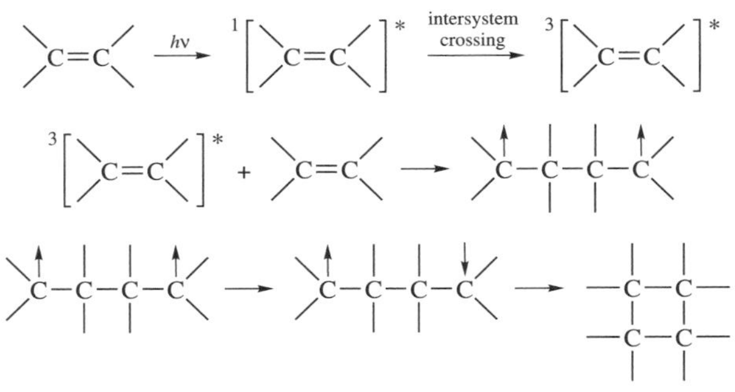

Intermolecular photocycloadditions of alkenes can be carried out by photosensitization with mercury or directly with short-wavelength light.[113] Relatively little preparative use has been made of this reaction for simple alkenes. Dienes can be photosensitized using benzophenone, butane-2,3-dione, or acetophenone.[114] The photodimerization of derivatives of cinnamic acid was among the earliest photochemical reactions to be studied.[115] Good yields of dimers are obtained when irradiation is carried out in the crystalline state. In solution, *cis–trans* isomerization is the dominant reaction.

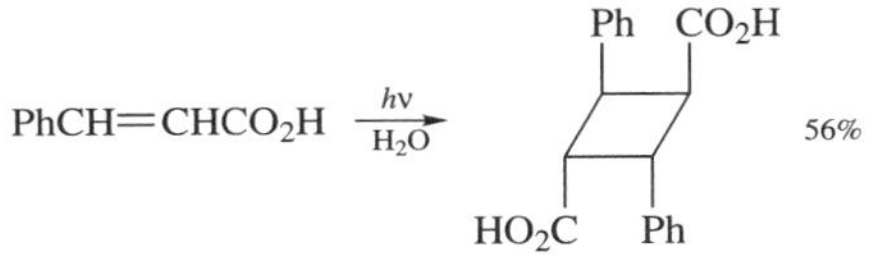

$PhCH{=}CHCO_2H \xrightarrow[H_2O]{h\nu}$ 56%

The presence of Cu(I) salts promotes intermolecular photocycloaddition of simple alkenes. Copper(I) triflate is especially effective.[116] It is believed that the photoreactive species is a 2 : 1 alkene : Cu(I) complex in which the two alkene molecules are brought together prior to photoexcitation.[117]

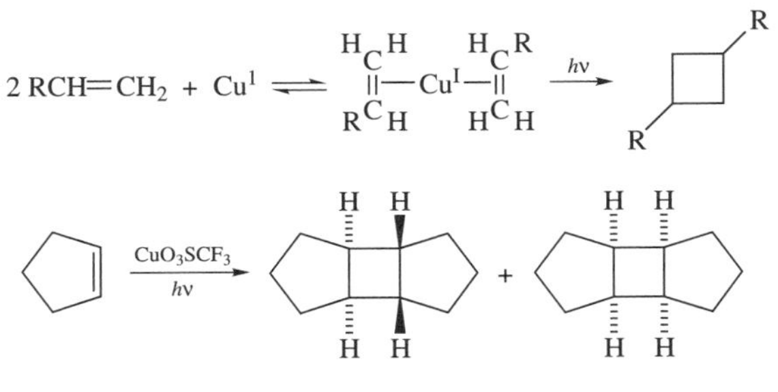

$2\,RCH{=}CH_2 + Cu^I \rightleftharpoons$ complex $\xrightarrow{h\nu}$

$\xrightarrow[h\nu]{CuO_3SCF_3}$

113. H. Yamazaki and R. J. Cvetanovic, *J. Am. Chem. Soc.* **91:**520 (1969).
114. G. S. Hammond, N. J. Turro, and R. S. H. Liu, *J. Org. Chem.* **28:**3297 (1963).
115. A. Mustafa, *Chem. Rev.* **51:**1 (1962).
116. R. G. Salomon, *Tetrahedron* **39:**485 (1983); R. G. Salomon and S. Ghosh, *Org. Synth.* **62:**125 (1984).
117. R. G. Salomon, K. Folking, W. E. Streib, and J. K. Kochi, *J. Am. Chem. Soc.* **96:**1145 (1974).

Intramolecular [2 + 2] photocycloaddition of dienes is an important method of formation of bicyclic compounds containing four-membered rings.[118] Direct irradiation of simple nonconjugated dienes leads to cyclobutanes.[119] Strain makes the reaction unfavorable for 1,4-dienes, but when the alkene units are separated by at least two carbon atoms, cycloaddition becomes possible.

Ref. 120

The most widely exploited photochemical cycloadditions involve irradiation of dienes in which the two double bonds are fairly close and result in formation of polycyclic cage compounds. Some examples are given in Scheme 6.7. Copper(I) triflate facilitates these intramolecular additions, as was the case for intermolecular reactions.

OH, $CH{=}CH_2$, $CH_2CH{=}CH_2$, H — CuO_3SCF_3, $h\nu$ → OH, H 51%

Ref. 121

Another class of molecules that undergo photochemical cycloadditions is α,β-unsaturated ketones.[122] The reactive excited state is either an n–π^* or a π–π^* triplet. The reaction is most successful with cyclopentenones and cyclohexenones. The excited states of acyclic enones and larger ring compounds are rapidly deactivated by *cis–trans* isomerization and do not readily add to alkenes. Photoexcited enones can also add to alkynes.[123] Unsymmetrical alkenes can undergo two regioisomeric modes of addition. It is generally observed that alkenes with donor groups are oriented such that the substituted carbon becomes bound to the β carbon, whereas with acceptor substituents the other orientation is preferred.[124] Selectivity is low for alkenes without strong donor or acceptor substituents.[125]

O + X $h\nu$ → O, X or O, X

favored for X = electron donor

favored for X = electron acceptor

118. P. deMayo, *Acc. Chem. Res.* **4:**41 (1971).
119. R. Srinivasan, *J. Am. Chem. Soc.* **84:**4141 (1962); R. Srinivasan, *J. Am. Chem. Soc.* **90:**4498 (1968).
120. J. Meinwald and G. W. Smith, *J. Am. Chem. Soc.* **89:**4923 (1967); R. Srinivasan and K. H. Carlough, *J. Am. Chem. Soc.* **89:**4932 (1967).
121. K. Avasthi and R. G. Salomon, *J. Org. Chem.* **51:**2556 (1986).
122. A. C. Weedon, in *Synthetic Organic Photochemistry*, W. M. Horspool, ed., Plenum, New York, 1984, Chapter 2; D. I. Schuster, G. Lem, and N. A. Kaprinidis, *Chem. Rev.* **93:**3 (1993); M. T. Crimmins and T. L. Reinhold, *Org. React.* **44:**297 (1993).
123. R. L. Cargill, T. Y. King, A. B. Sears, and M. R. Willcott, *J. Org. Chem.* **36:**1423 (1971); W. C. Agosta and W. W. Lowrance, *J. Org. Chem.* **35:**3851 (1970).
124. E. J. Corey, J. D. Bass, R. Le Mahieu, and R. B. Mitra, *J. Am. Chem. Soc.* **86:**5570 (1984).
125. J. D. White and D. N. Gupta, *J. Am. Chem. Soc.* **88:**5364 (1966); P. E. Eaton, *Acc. Chem. Res.* **1:**50 (1968).

Scheme 6.7. Intramolecular [2 + 2] Photochemical Cycloaddition Reactions of Dienes

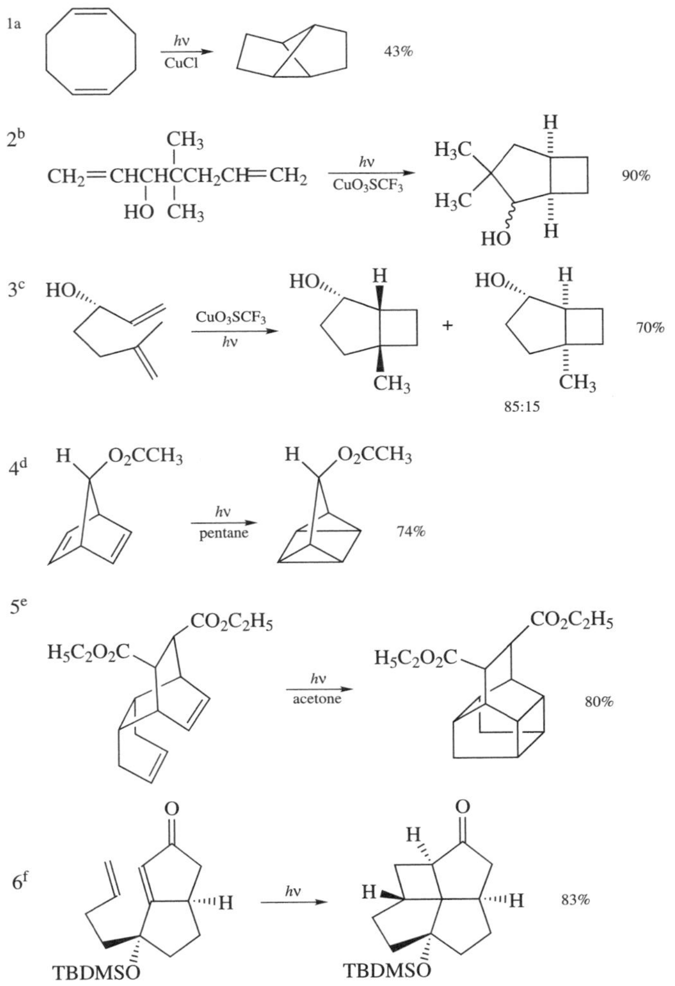

1a — $h\nu$, CuCl — 43%

2[b] $CH_2{=}CHCH(OH)C(CH_3)_2CH_2CH{=}CH_2$ — $h\nu$, CuO_3SCF_3 — 90%

3[c] — CuO_3SCF_3, $h\nu$ — 85:15 — 70%

4[d] — $h\nu$, pentane — 74%

5[e] — $h\nu$, acetone — 80%

6[f] TBDMSO — $h\nu$ — 83%

a. P. Srinivasan, *Org. Photochem. Synth.* **1:**101 (1971); *J. Am. Chem. Soc.* **86:**3318 (1964).
b. R. G. Salomon and S. Ghosh, *Org. Synth.* **62:**125 (1984).
c. K. Langer and J. Mattay, *J. Org. Chem.* **60:**7256 (1995).
d. P. G. Gassman and D. S. Patton, *J. Am. Chem. Soc.* **90:**7276 (1968).
e. B. M. Jacobson, *J. Am. Chem. Soc.* **95:**2579 (1973).
f. M. Thommen and R. Keese, *Synlett.* **1997:**231.

The cycloadditions are believed to proceed through 1,4-diradical intermediates. Trapping experiments with hydrogen-atom donors indicated that the initial bond formation can take place at either the α or β carbon of the enone. The final product ratio reflects both the rate of formation of the diradical and the efficiency of ring closure.[126]

126. D. I. Schuster, G. E. Heibel, P. B. Brown, N. J. Turro, and C. V. Kumar, *J. Am. Chem. Soc.* **110:**8261 (1988); D. Andrew, D. J. Hastings, and A. C. Weedon, *J. Am. Chem. Soc.* **116:**10870 (1994).

Intramolecular enone–alkene cycloadditions are also possible.

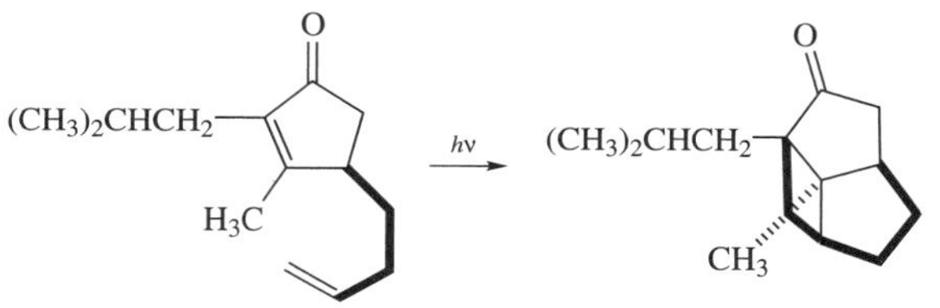

Ref. 127

In the case of β-(5-pentenyl) substituents, there is a general preference for *exo*-type cyclization to form a five-membered ring.[127,128] This is consistent with the general pattern for radical cyclizations and implies initial bonding at the β carbon of the enone.

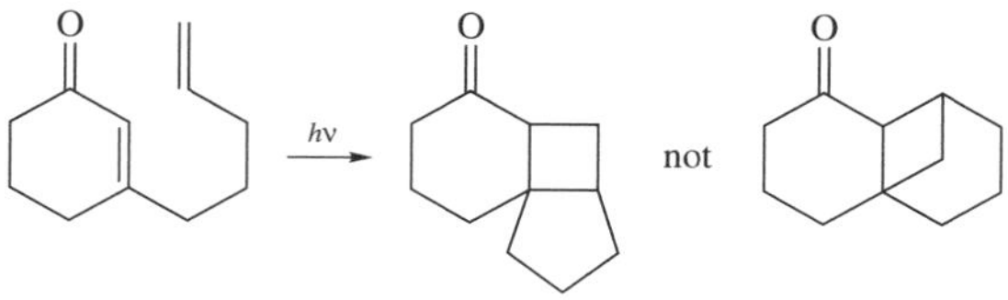

Some examples of photochemical enone–alkene cycloadditions are given in Scheme 6.8.

With many other ketones and aldehydes, reaction between the photoexcited carbonyl chromophore and alkene can result in formation of four-membered cyclic ethers (oxetanes). This reaction is often referred to as the *Paterno–Büchi reaction*.[129]

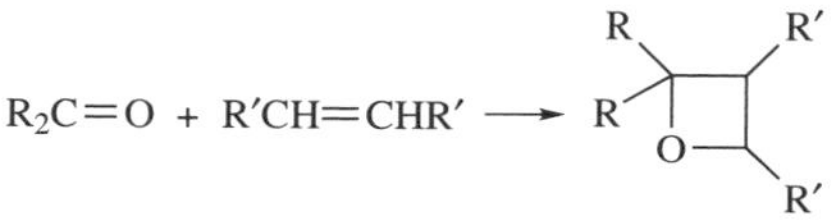

The reaction is stereospecific for at least some aliphatic ketones but not for aromatic carbonyl compounds.[130] This result suggests that the reactive excited state is a singlet for aliphatics and a triplet for aromatics. With aromatic aldehydes and ketones, the regioselecitivity of addition can usually be predicted on the basis of formation of the more stable of the two possible diradical intermediates by bond formation between oxygen and the alkene.

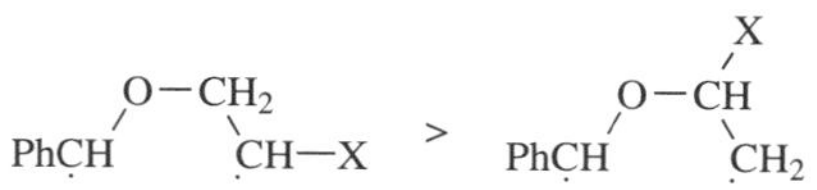

127. P. J. Connolly and C. H. Heathcock, *J. Org. Chem.* **50:**4135 (1985).
128. W. C. Agosta and S. Wolff, *J. Org. Chem.* **45:**3139 (1980); M. C. Pirrung, *J. Am. Chem. Soc.* **103:**82 (1981).
129. D. R. Arnold, *Adv. Photochem.* **6:**301 (1968); H. A. J. Carless, in *Synthetic Organic Photochemistry*, W. M. Horspool, ed., Plenum, New York, 1984, Chapter 8.
130. N. C. Yang and W. Eisenhardt, *J. Am. Chem. Soc.* **93:**1277 (1971); D. R. Arnold, R. L. Hinman, and A. H. Glick, *Tetrahedron Lett.* **1964:**1425; N. J. Turro and P. A. Wriede, *J. Am. Chem. Soc.* **90:**6863 (1968); J. A. Barltrop and H. A. J. Carless, *J. Am. Chem. Soc.* **94:**8761 (1972).

Scheme 6.8. Photochemical Cycloaddition Reactions of Enones with Alkenes and Alkynes

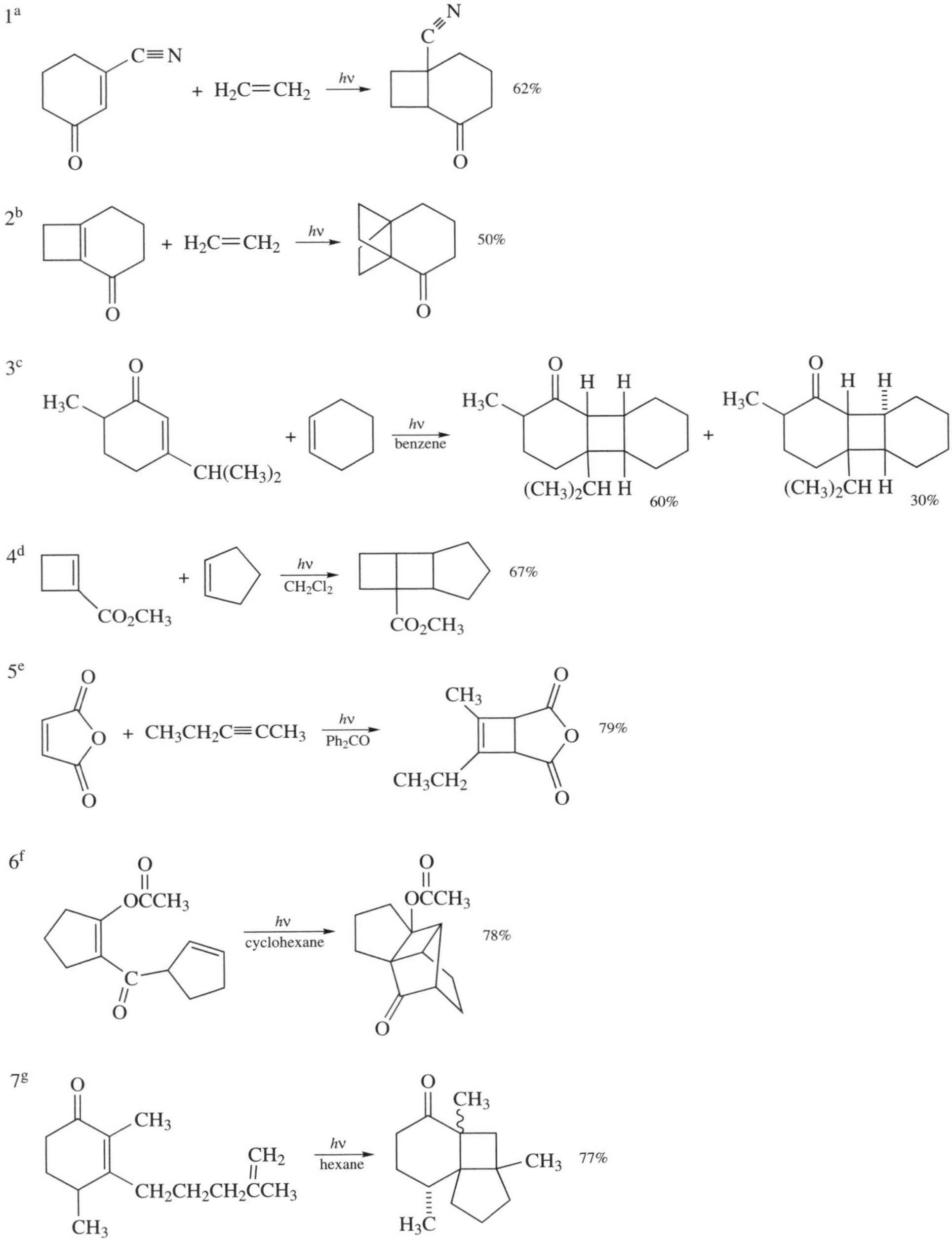

1[a] C≡N; + $H_2C{=}CH_2$ —hν→ C≡N; 62%

2[b] + $H_2C{=}CH_2$ —hν→ 50%

3[c] H_3C; O; $CH(CH_3)_2$ + —hν, benzene→ O; H H; H_3C; $(CH_3)_2CH$ H; 60% + O; H H; H_3C; $(CH_3)_2CH$ H; 30%

4[d] CO_2CH_3 + —hν, CH_2Cl_2→ CO_2CH_3; 67%

5[e] O + $CH_3CH_2C{\equiv}CCH_3$ —hν, Ph_2CO→ CH_3; CH_3CH_2; O; 79%

6[f] $OCCH_3$ (O); C=O —hν, cyclohexane→ $OCCH_3$ (O); O; 78%

7[g] O; CH_3; CH_2; $CH_2CH_2CH_2CCH_3$; CH_3 —hν, hexane→ O; CH_3; CH_3; H_3C; 77%

a. W. C. Agosta and W. W. Lowrance, Jr., *J. Org. Chem.* **35**:3851 (1970).
b. P. E. Eaton and K. Nyi, *J. Am. Chem. Soc.* **93**:2786 (1971).
c. P. Singh, *J. Org. Chem.* **36**:3334 (1971).
d. P. A. Wender and J. C. Lechleiter, *J. Am. Chem. Soc.* **99**:267 (1977).
e. R. M. Scarborough, Jr., B. H. Toder, and A. B. Smith III, *J. Am. Chem. Soc.* **102**:3904 (1980).
f. W. Oppolzer and T. Godel, *J. Am. Chem. Soc.* **100**:2583 (1978).
g. M. C. Pirrung, *J. Am. Chem. Soc.* **101**:7130 (1979).

Stereochemistry can also be interpereted in terms of conformational effects in the 1,4-diradical intermediates.[131] Vinyl enol ethers and enamides add to benzaldehyde to give 3-substituted oxetanes, usually with the *cis* isomer preferred.[132–135]

$PhCH{=}O + CH_2{=}C(OTMS)C(CH_3)_3 \xrightarrow{h\nu}$ 2-Ph-3-OTMS-3-$C(CH_3)_3$-oxetane 59% Ref. 132

$PhCH{=}O +$ 5-methyl-2,3-dihydrofuran $\xrightarrow{h\nu}$ Ph-substituted bicyclic oxetane (H_3C) 32% Ref. 133

$PhCH{=}O +$ N-acetyl-2,3-dihydropyrrole ($N{-}COCH_3$) $\xrightarrow{h\nu}$ Ph-substituted bicyclic oxetane ($N{-}COCH_3$) Ref. 134

Some other examples of Paterno–Büchi reactions are given in Scheme 6.9.

6.5. [3,3] Sigmatropic Rearrangements

The mechanistic basis of sigmatropic rearrangements was introduced in Chapter 11 of Part A. The sigmatropic process that is most widely applied in synthesis is the [3,3] sigmatropic rearrangement. The principles of orbital symmetry establish that concerted [3,3] sigmatropic rearrangements are allowed processes. Stereochemical predictions and analyses are based on the cyclic transition state implied by a concerted reaction mechanism. Some of the various [3,3] sigmatropic rearrangements that are used in synthesis are presented in outline form in Scheme 6.10.[136]

6.5.1. Cope Rearrangements

The Cope rearrangement is the conversion of a 1,5-hexadiene derivative to an isomeric 1,5-hexadiene by the [3,3] sigmatropic mechanism. The reaction is both stereospecific and stereoselective. It is stereospecific in that a Z or E configurational relationship at either double bond is maintained in the transition state and governs the stereochemical relationship at the newly formed single bond in the product.[137] However, the relationship depends upon the conformation of the transition state. When a chair transition state is favored, the E,E- and Z,Z-dienes lead to *anti*-3,4-diastereomers whereas the E,Z and Z,E-isomers give the 3,4-*syn* product. Transition-state conformation also

131. A. G. Griesbeck and S. Stadtmüller, *J. Am. Chem. Soc.* **113:**6923 (1991).
132. T. Bach, *Tetrahedron Lett.* **32:**7037 (1991).
133. A. G. Griesbeck and S. Stadtmüller, *J. Am. Chem. Soc.* **113:**6923 (1991).
134. T. Bach, *Liebigs Ann. Chem.* **1997:**1627.
135. T. Bach, *Synthesis* **1998:**683.
136. For reviews of synthetic application of [3,3] sigmatropic rearrangements, see G. B. Bennett, *Synthesis* **1977:**58; F. E. Ziegler, *Acc. Chem. Res.* **10:**227 (1977).
137. W. v. E. Doering and W. R. Roth, *Tetrahedron* **18:**67 (1962).

Scheme 6.9. Photochemical Cycloaddition Reactions of Carbonyl Compounds with Alkenes

1[a] PhCH=O + (cyclohexene) $\xrightarrow{h\nu}$ (oxetane: O, Ph, H) 38%

2[b] (norbornene) + $Ph_2CH=O$ $\xrightarrow[\text{benzene}]{h\nu}$ (oxetane: O, Ph, Ph) 81%

3[c] (norbornenyl methyl ketone: H, CCH_3, O) $\xrightarrow[\text{benzene}]{h\nu}$ (O, CH_3) 83%

4[d] PhCH=O + $PhCH=CH_2$ $\xrightarrow{h\nu}$ (oxetane: O, Ph, Ph) 31%

a. J. S. Bradshaw, *J. Org. Chem.* **31**:237 (1966).
b. D. R. Arnold, A. H. Glick, and V. Y. Abraitys, *Org. Photochem. Synth.* **1**:51 (1971).
c. R. R. Sauers, W. Schinski, and B. Sickles, *Org. Photochem. Synth.* **1**:76 (1971).
d. H. A. J. Carless, A. K. Maitra, and H. S. Trivedi, *J. Chem. Soc., Chem. Commun.* **1979**:984.

determines the stereochemistry of the new double bond. If both *E*- and *Z*-stereoisomers are possible for the product, the product ratio will normally reflect product (and transition-state) stability. Thus, an *E* arrangement is normally favored for the newly formed double bonds. The stereochemical aspects of the Cope rearrangements for relatively simple reactants are consistent with a chairlike transition state in which the larger substituent at C-3 (or C-4) adopts an equatorial-like conformation.

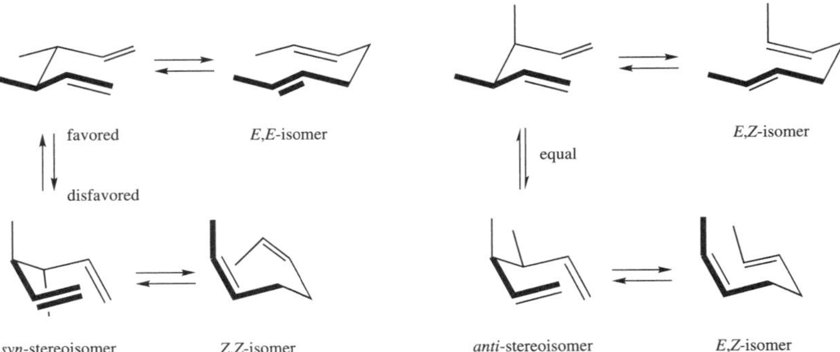

Scheme 6.10. [3,3] Sigmatropic Rearrangements

1[a] Cope rearrangement

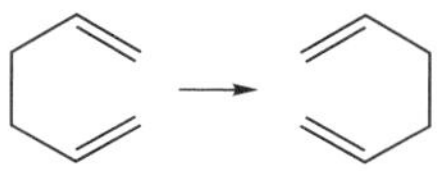

2[b] Oxy-Cope rearrangement

HO HO O

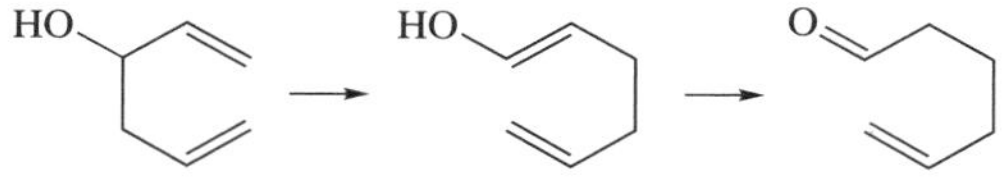

3[c] Anionic oxy-Cope rearrangement

$\bar{O}$ $\bar{O}$ H^+ O

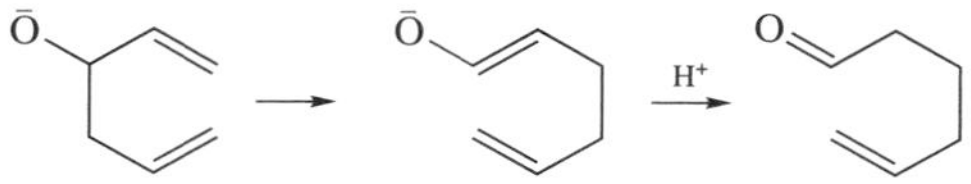

4[d] Claisen rearrangement of allyl vinyl ethers

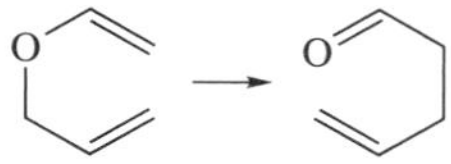

O O

5[d] Claisen rearrangement of allyl phenyl ethers

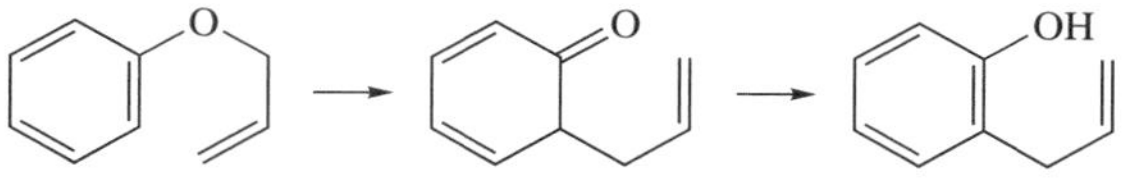

O O OH

6[e] Ortho ester Claisen rearrangement

RO OR O –ROH OR O OR O

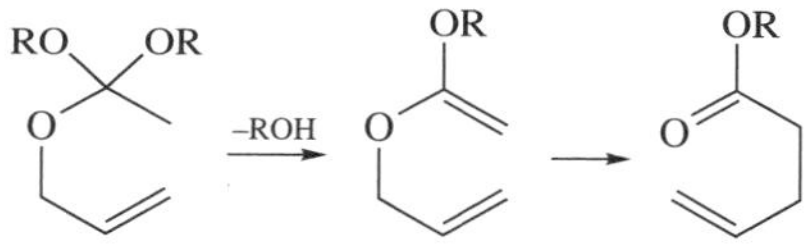

7[f] Claisen rearrangement of *O*-allyl-*O*′-trimethylsilyl ketene acetals

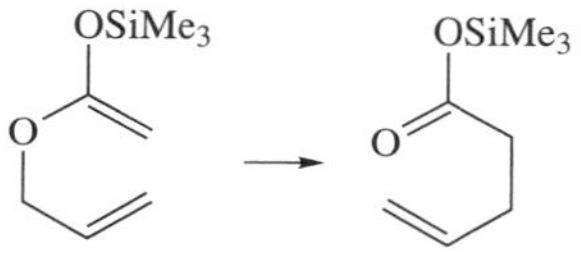

$OSiMe_3$ O $OSiMe_3$ O

8[g] Ester enolate Claisen rearrangement

O^- O O^- O

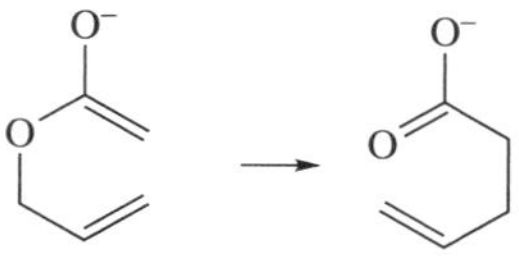

9[h] Claisen rearrangement of *O*-allyl-*N*,*N*-dialkyl ketene aminals

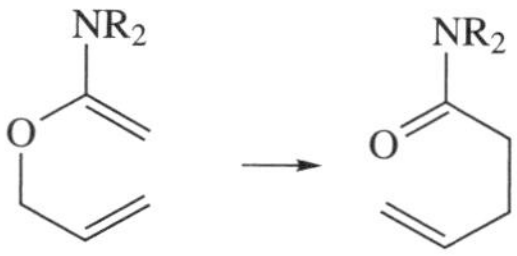

NR_2 O NR_2 O

Scheme 6.10. *(continued)*

10[i] Aza-Claisen rearrangement of *O*-allyl imidates

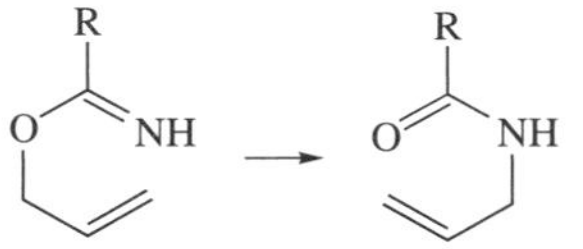

a. S. J. Rhoads and N. R. Raulins, *Org. React.* **22:**1 (1975).
b. J. A. Berson and M. Jones, Jr., *J. Am. Chem. Soc.* **86:**5019 (1964).
c. D. A. Evans and A. M. Golob, *J. Am. Chem. Soc.* **97:**4765 (1975).
d. D. S. Tarbell, *Org. React.* **2:**1 (1944).
e. W. S. Johnson, L. Werthemann, W. R. Bartlett, T. J. Brocksom, T. Li, D. J. Faulkner, and M. R. Petersen, *J. Am. Chem. Soc.* **92:**741 (1970).
f. R. E. Ireland and R. H. Mueller, *J. Am. Chem. Soc.* **94:**5898 (1972).
g. R. E. Ireland, R. H. Mueller, and A. K. Willard, *J. Am. Chem. Soc.* **98:**2868 (1976).
h. D. Felix, K. Gschwend-Steen, A. E. Wick, and A. Eschenmoser, *Helv. Chim. Acta* **52:**1030 (1969).
i. L. E. Overman, *Acc. Chem. Res.* **13:**218 (1980).

Because of the concerted mechanism, chirality at C-3 (or C-4) leads to enantiospecific formation of new chiral centers at C-1 (or C-6).[138] These relationships are illustrated in the example below. Both the configuration of the new chiral center and that of the new double bond are those expected on the basis of a chairlike transition state. Because there are two stereogenic centers, the double bond and the asymmetric carbon, there are four possible stereoisomers of the product. Only two are formed. The *E*-double-bond isomer has the *S*-configuration at C-4 whereas the *Z*-isomer has the *R*-configuration. These are the products expected for a chair transition state. The stereochemistry of the new double bond is determined by the relative stability of the two chair transition states. Transition state **B** is less favorable than **A** because of the axial placement of the larger phenyl substituent.

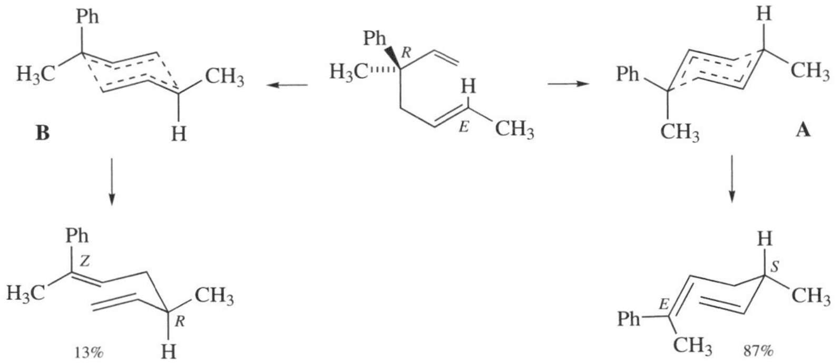

B 13% A 87%

The products corresponding to boatlike transition states are usually not observed for acyclic dienes. However, the boatlike transition state is allowed, and if steric factors make a

138. R. K. Hill and N. W. Gilman, *J. Chem. Soc., Chem. Commun.* **1967:**619; R. K. Hill, in *Asymmetric Synthesis*, Vol. 4, J. D. Morrison, ed., Academic Press, New York, 1984, pp. 503–572.

boat transition state preferable to a chair, reaction will proceed through a boat.

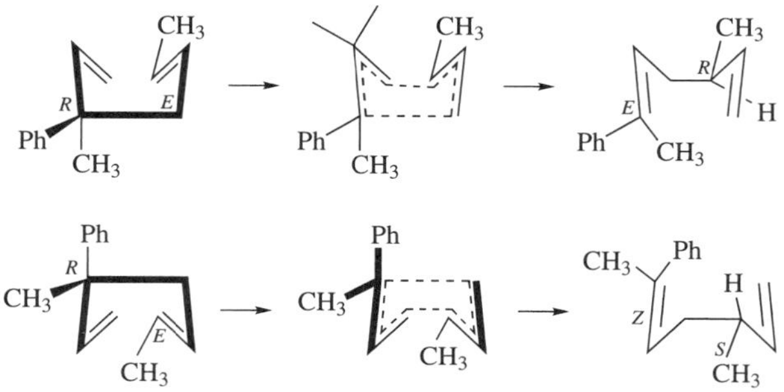

Cope rearrangements are reversible reactions, and, because there are no changes in the number or types of bonds as a result of the reaction, to a first approximation the total bond energy is unchanged. The position of the final equilibrium is governed by the relative stability of the starting material and the product. In the example just cited, the equilibrium is favorable for product formation because the product is stabilized by conjugation of the alkene with the phenyl ring. Some other examples of Cope rearrangements are given in Scheme 6.11. In entry 1, the equilibrium is biased toward product by the fact that the double bonds in the product are more highly substituted, and therefore more stable, than those in the reactant. In entry 2, a gain in conjugation is offset by the formation of a less highly substituted double bond, and the equilibrium mixture contains both dienes.

When ring strain is relieved, Cope rearrangements can occur at much lower temperatures and with complete conversion to ring-opened products. A striking example of such a process is the conversion of *cis*-divinylcyclopropane to 1,4-cycloheptadiene, a reaction which occurs readily at temperatures below −40°C.[139]

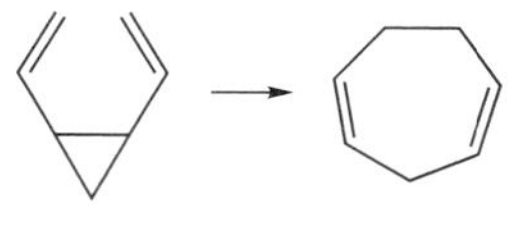

Entry 3 in Scheme 6.11 illustrates the application of a *cis*-divinylcyclopropane rearrangement in the prepartion of an intermediate for the synthesis of pseudoguaiane-type natural products.

Several transition-metal species, especially Pd(II) salts, have been found to catalyze Cope rearrangements.[140] The catalyst that has been adopted for synthetic purposes is $PdCl_2(CH_3CN)_2$. With this catalyst, the rearrangement of **8** to **9** and **10** occurs at room temperature, as contrasted to 240°C in its absence.[141] The catalyzed reaction shows

139. W. v. E. Doering and W. R. Roth, *Tetrahedron* **19:**715 (1963).
140. R. P. Lutz, *Chem. Rev.* **84:**205 (1984).
141. L. E. Overman and F. M. Knoll, *J. Am. Chem. Soc.* **102:**865 (1980).

Scheme 6.11. Cope Rearrangements of 1,5-Dienes

A. Thermal

1[a] 350°C, 1 h; CH_2 CH_2 → CH_2-CH_2 100%

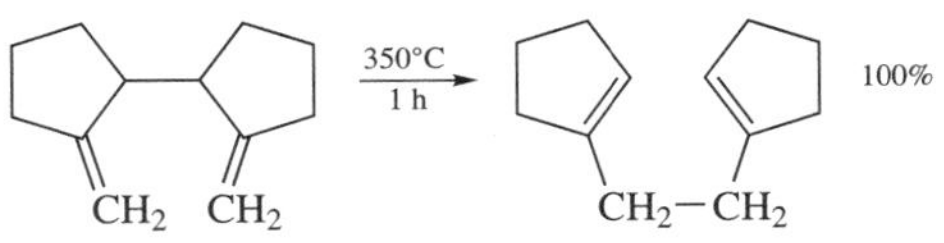

2[b] H_3C, H_3C, $CO_2C_2H_5$ ⇌ ($K = 0.25$, 275°C) CH_3, CH_3, $CO_2C_2H_5$

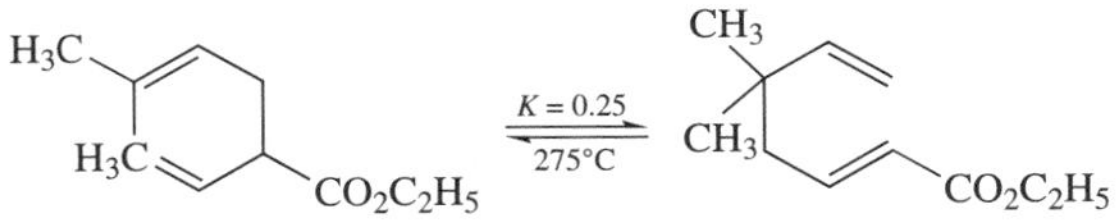

3[c] H, H, CH_3, O → (98°C) CH_3, CH_3, O 80–90%

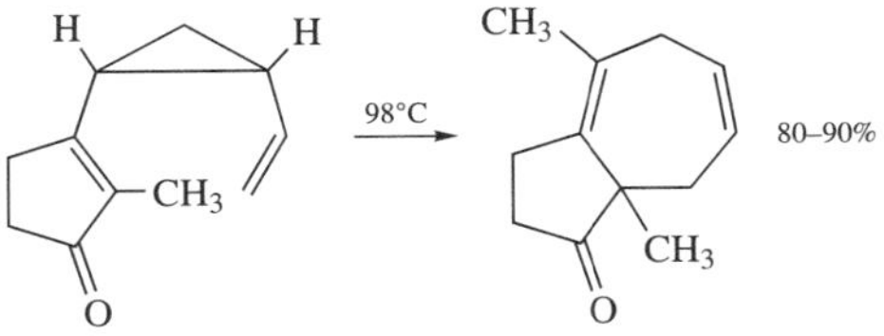

4[d] OH, $CH=CH_2$, $CH=CH_2$ → (320°C) O 90%

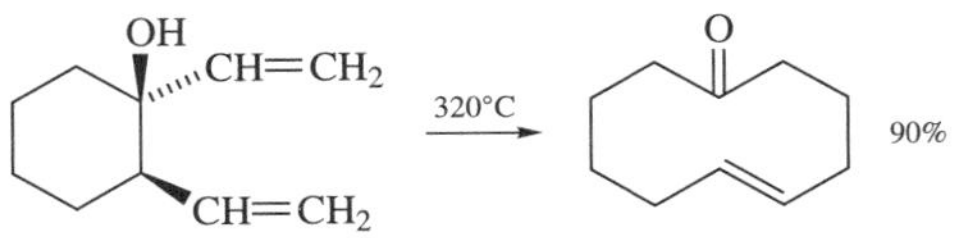

B. Anionic oxy-Cope

5[e] OH → (KH, THF, reflux, 18 h) H, H, O 98%

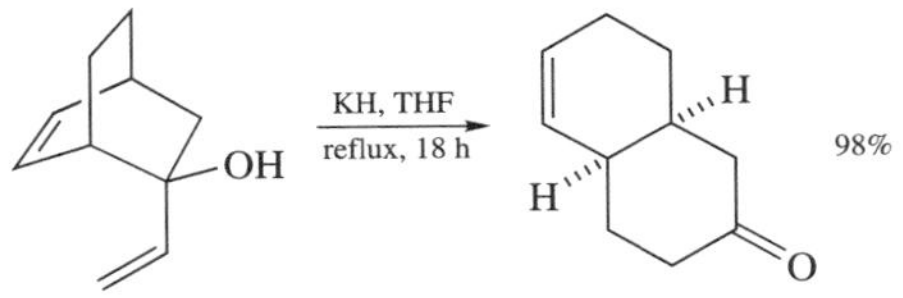

6[f] CH_3, $C=CH_2$, H_3C, $CH=CH_2$, OH → (KH, 18-crown-6, 25°C, 18 h) CH_3, H_3C, O 75%

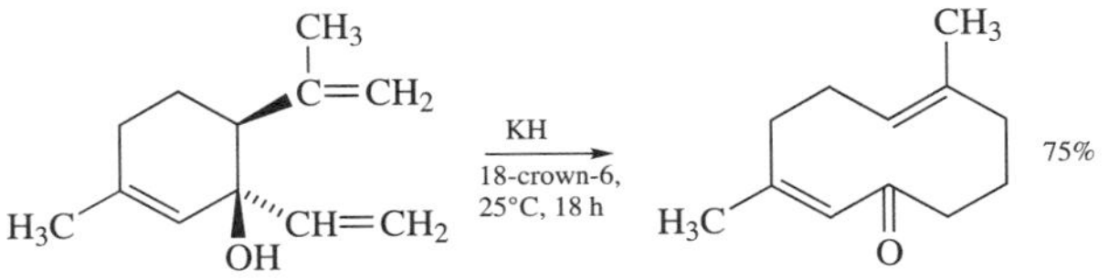

7[g] CH_3O, OCH_3, OH, C_2H_5, CH_3 → (KH, THF, 25°C) H, H, CH_3, C_2H_5, H, CH_3O, OCH_3, O

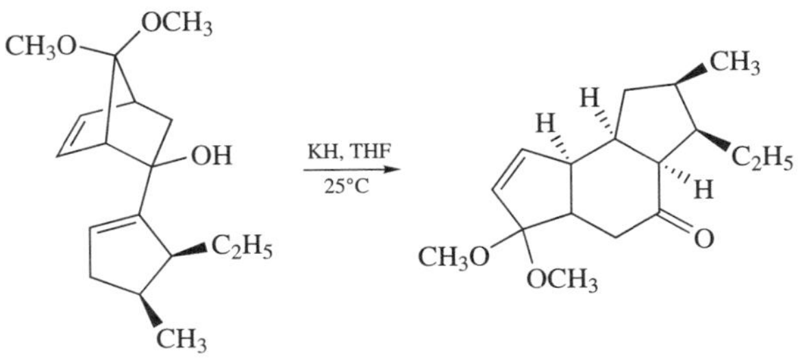

a. K. J. Shea and R. B. Phillips, *J. Am. Chem. Soc.* **102:**3156 (1980).
b. F. E. Zeigler and J. J. Piwinski, *J. Am. Chem. Soc.* **101:**1612 (1979).
c. P. A. Wender, M. A. Eissenstat, and M. P. Filosa, *J. Am. Chem. Soc.* **101:**2196 (1979).
d. E. N. Marvell and W. Whalley, *Tetrahedron Lett.* **1970:**509.
e. D. A. Evans, A. M. Golob, N. S. Mandel, and G. S. Mandel, *J. Am. Chem. Soc.* **100:**8170 (1978).
f. W. C. Still, *J. Am. Chem. Soc.* **99:**4186 (1977).
g. L. A. Paquette, K. S. Learn, J. L. Romine, and H.-S. Lin, *J. Am. Chem. Soc.* **110:**879 (1988); L. A. Paquette, J. L. Romine, H.-S. Lin, and J. Wright, *J. Am. Chem. Soc.* **112:**9284 (1990).

enhanced stereoselectivity and is consistent with a chairlike transition-state structure.

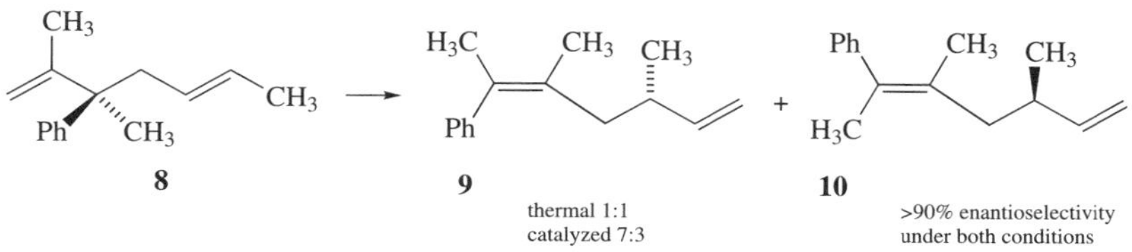

8 → 9 + 10

thermal 1:1
catalyzed 7:3

>90% enantioselectivity under both conditions

The mechanism for catalysis is formulated as a stepwise process in which the electrophilic character of Pd(II) facilitates the reaction.[142]

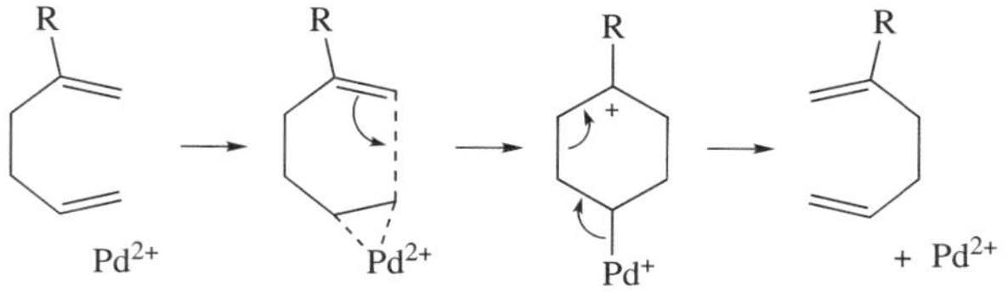

Pd^{2+} → Pd^{2+} → Pd^{+} → + Pd^{2+}

When there is a hydroxyl substituent at C-3 of the diene system, the Cope rearrangement product is an enol, which is subsequently converted to the corresponding carbonyl compound. This is called the *oxy-Cope rearrangement*.[143] The formation of the carbonyl compound provides a net driving force for the reaction.[144]

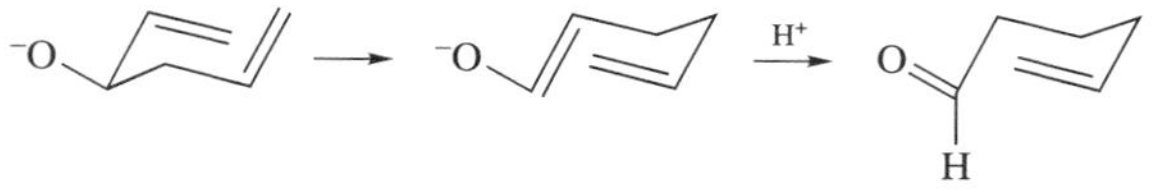

Entry 4 in Scheme 6.11 illustrates the use of the oxy-Cope rearrangement in formation of a medium-sized ring.

An important improvement in the oxy-Cope reaction was made when it was found that the reactions are markedly catalyzed by base.[145] When the C-3 hydroxyl group is converted to its alkoxide, the reaction is accelerated by factors of 10^{10}–10^{17}. These base-catalyzed reactions are called *anionic oxy-Cope rearrangements*. The rates of anionic oxy-Cope rearrangements depend on the degree of cation coordination at the oxy anion. The reactivity trend is $K^+ > Na^+ > Li^+$. Catalytic amounts of tetra-*n*-butylammonium salts lead to accelerated rates in some cases. This presumably results from the dissociation of less reactive ion-pair species promoted by the tetra-*n*-butylammonium ion.[146] Entries 5, 6, and 7 in Scheme 6.11 illustrate the mild conditions under which rearrangement occurs.

Silyl ethers of vinyl allyl alcohols can also be used in oxy-Cope rearrangements. This methodology has been used in connection with *syn*-selective aldol additions in stereo-

142. L. E. Overman and A. F. Renaldo, *J. Am. Chem. Soc.* **112:**3945 (1990).
143. S. R. Wilson, *Org. React.* **43:**93 (1993); L. A. Paquette, *Angew. Chem. Int. Ed. Engl.* **29:**609 (1990); L. A. Paquette, *Tetrahedron* **53:**13971 (1997).
144. A. Viola, E. J. Iorio, K. K. Chen, G. M. Glover, U. Nayak, and P. J. Kocienski, *J. Am. Chem. Soc.* **89:**3462 (1967).
145. D. A. Evans and A. M. Golob, *J. Am. Chem. Soc.* **97:**4765 (1975); D. A. Evans, D. J. Balillargeon, and J. V. Nelson, *J. Am. Chem. Soc.* **100:**2242 (1978).
146. M. George, T.-F. Tam, and B. Fraser-Reid, *J. Org. Chem.* **50:**5747 (1985).

selective synthesis.[147] The use of the silyloxy group prevents reversal of the aldol addition, which would otherwise occur under anionic conditions. The reactions proceed at convenient rates at 140–180°C.

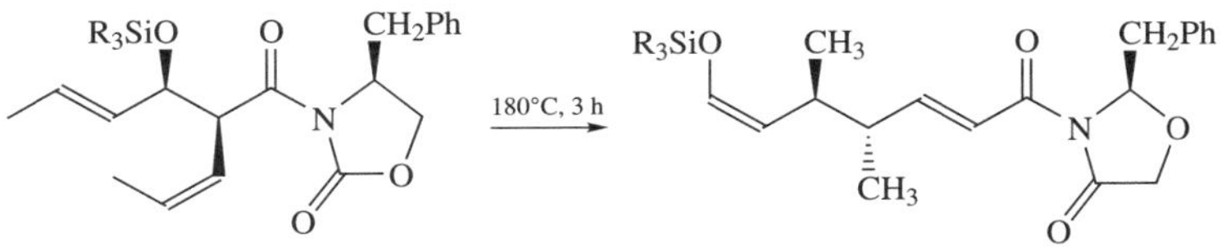

R_3SiO O CH_2Ph → (180°C, 3 h) R_3SiO CH_3 O CH_2Ph CH_3 Ref. 148

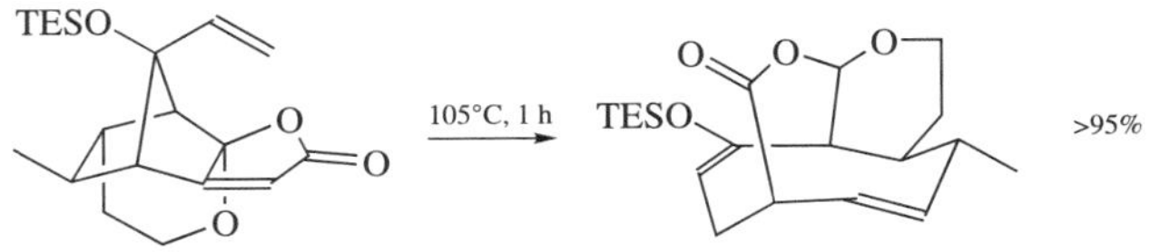

TESO → (105°C, 1 h) TESO >95% Ref. 149

(TES = triethylsilyl)

6.5.2. Claisen Rearrangements

The [3,3] sigmatropic rearrangement of allyl vinyl ethers leads to γ,δ-enones and is known as the *Claisen rearrangement*.[150] The reaction is mechanistically analogous to the Cope rearrangement. Because the product is a carbonyl compound, the equilbrium is usually favorable for product formation. The reactants can be made from allylic alcohols by mercuric ion-catalyzed exchange with ethyl vinyl ether.[151] The allyl vinyl ether need not be isolated but is usually prepared under conditions which lead to its rearrangement. The simplest of all Claisen rearrangements, the conversion of allyl vinyl ether to 4-pentenal, typifies this process.

$$CH_2{=}CHCH_2OH + CH_2{=}CHOCH_2CH_3 \xrightarrow[\Delta]{Hg(OAc)_2} [CH_2{=}CHCH_2OCH{=}CH_2] \longrightarrow \underset{96\%}{CH_2{=}CHCH_2CH_2CH{=}O}$$ Ref. 152

Acid-catalyzed exchange can also be used to prepare the vinyl ethers.

$$RCH{=}CHCH_2OH + CH_3CH_2OCH{=}CH_2 \xrightarrow{H^+} RCH{=}CHCH_2OCH{=}CH_2$$ Ref. 153

Allyl vinyl ethers can also be generated by thermal elimination reactions. For example, base-catalyzed conjugate addition of allyl alcohols to phenyl vinyl sulfone generates 2-

147. C. Schneider and M. Rehfeuter, *Synlett* **1996:**212; C. Schneider and M. Rehfeuter, *Tetrahedron* **53:**133 (1997); W. C. Black, A. Giroux, and G. Greidanus, *Tetrahedron Lett.* **37:**4471 (1996).
148. C. Schneider, *Eur. J. Org. Chem.* **1998:**1661.
149. M. M. Bio and J. L. Leighton, *J. Am. Chem. Soc.* **121:**890 (1999).
150. F. E. Ziegler, *Chem. Rev.* **88:**1423 (1988).
151. W. H. Watanabe and L. E. Conlon, *J. Am. Chem. Soc.* **79:**2828 (1957); D. B. Tulshian, R. Tsang, and B. Fraser-Reid, *J. Org. Chem.* **49:**2347 (1984).
152. S. E. Wilson, *Tetrahedron Lett.* **1975:**4651.
153. G. Saucy and R. Marbet, *Helv. Chim. Acta* **50:**2091 (1967); R. Marbet and G. Saucy, *Helv. Chim. Acta* **50:**2095 (1967).

(phenylsulfinyl)ethyl ethers, which can undergo elimination at 200°C.[154] The sigmatropic rearrangement proceeds under these conditions. Allyl vinyl ethers can also be prepared by Wittig reactions using ylides generated from allyloxymethylphosphonium salts.[155]

$$RCH{=}CHCH_2OH + CH_2{=}CH\overset{O}{\overset{\|}{S}}Ph \xrightarrow{NaH} RCH{=}CHCH_2OCH_2CH_2SPh \xrightarrow{200°C} RCH{=}CHCH_2OCH{=}CH_2$$

$$R_2C{=}O + Ph_3P^+CH_2OCH_2CH{=}CH_2 \xrightarrow{K^+\ ^-O\text{-}t\text{-}Bu} R_2C{=}CHOCH_2CH{=}CH_2$$

Catalysis of Claisen rearrangements has been achieved using highly hindered bis(phenoxy)methylaluminum as a Lewis acid.[156] Reagents of this type also have the ability to control the $E{:}Z$ ratio of the products. Very bulky catalysts tend to favor the Z-isomer by forcing the α substituent of the allyl group into an axial conformation.

R R R R O $O^+{-}{}^-AlR_3$ O $O^+{-}{}^-AlR_3$ O R

Z-isomer *E*-isomer

Some representative Claisen rearrangements are shown in Scheme 6.12. Entry 1 illustrates the application of the Claisen rearrangement in introduction of a substituent at the junction of two six-membered rings. Introduction of a substituent at this type of position is frequently necessary in the synthesis of steroids and terpenes. In entry 2, rearrangement of a 2-propenyl ether leads to formation of a methyl ketone. Entry 3 illustrates the use of 3-methoxyisoprene to form the allylic ether. The rearrangement of this type of ether leads to introduction of isoprene structural units into the reaction product.

There are several variations of the Claisen rearrangement that make it a powerful tool for the synthesis of γ,δ-unsaturated carboxylic acids. The ortho ester modification of the Claisen rearrangement allows carboalkoxymethyl groups to be introduced at the γ-position of allylic alcohols.[157] A mixed ortho ester is formed as an intermediate and undergoes sequential elimination and sigmatropic rearrangement.

$$RCH{=}CHCH_2OH + CH_3C(OCH_3)_3 \rightleftharpoons RCH{=}CHCH_2O\underset{OCH_3}{\overset{OCH_3}{C}}CH_3 \rightleftharpoons RCH{=}CHCH_2O\underset{OCH_3}{C}{=}CH_2 \longrightarrow R\underset{}{\overset{CH_2CO_2CH_3}{C}}HCH{=}CH_2$$

154. T. Mandai, S. Matsumoto, M. Kohama, M. Kawada, J. Tsuji, S. Saito, and T. Moriwake, *J. Org. Chem.* **55:**5671 (1990); T. Mandai, M. Ueda, S. Hagesawa, M. Kawada, J. Tsuji, and S. Saito, *Tetrahedron Lett.* **31:**4041 (1990).
155. M. G. Kulkarni, D. S. Pendharkar, and R. M. Rasne, *Tetrahedron Lett.* **38:**1459 (1997).
156. K. Nonoshita, H. Banno, K. Maruoka, and H. Yamamoto, *J. Am. Chem. Soc.* **112:**316 (1990).
157. W. S. Johnson, L. Werthemann, W. R. Bartlett, T. J. Brocksom, T. Li, D. J. Faulkner, and M. R. Petersen, *J. Am. Chem. Soc.* **92:**741 (1970).

A. Rearrangements of allyl vinyl ethers

1[a] (bicyclic allyl vinyl ether with $OCH{=}CH_2$) $\xrightarrow{195°C}$ (bicyclic product with angular $CH_2CH{=}O$) 87%

2[b] $(CH_3)_2C(OH)CH{=}CH_2 + H_2C{=}C(CH_3)OCH_3 \xrightarrow[125°C]{H^+} (CH_3)_2C{=}CHCH_2CH_2CCH_3$ (C=O) 94%

3[c] $CH_2{=}C(CH_3)CH(OH)CH_2CH_3 + CH_2{=}C(CH_3)C(OCH_3){=}CH_2 \xrightarrow[H^+]{110°C} CH_2{=}C(CH_3)C(O)CH_2CH_2C(CH_3){=}CHCH_2CH_3$ ~70%

4[d] $CH_3CH{=}CHCH_2OC(CH_3){=}C(H)CO_2CH_2CH{=}CHCH_3 \xrightarrow{140–145°C} CH_3C(O)CH(CH(CH_3)CH{=}CH_2)CO_2CH_2CH{=}CHCH_3$ 61%

5[e] (cyclohexene with H_3C, H_3C, $CH_2CH_2CH_2CN$, CH_2OH) $\xrightarrow[Hg(O_2CF_3)_2]{CH_2{=}CHOCH_2CH{=}CH_2}$ (cyclohexene with H_3C, H_3C, $CH_2CH_2CH_2CN$, $CH_2OCH{=}CH_2$) 73% $\xrightarrow{200°C}$ (cyclohexane with H_3C, H_3C, $CH_2CH_2CH_2CN$, $=CH_2$, $CH_2CH{=}O$) 95%

6[f] (geraniol, OH) $\xrightarrow[POCl_3]{CH_2{=}C(OCH_3)CH_3} \xrightarrow[0°C]{(i\text{-Bu})_3Al}$ (product, OH) 89%

B. Rearrangements via ortho esters

7[g] $H_2C{=}CHCH_2CH_2CH(OH)C(CH_3){=}CH_2 \xrightarrow[H^+, 140°C]{CH_3C(OC_2H_5)_3} H_2C{=}CHCH_2CH_2C(H){=}C(CH_3)CH_2CH_2CO_2C_2H_5$ 83–88%

8[h] $CH_2{=}C(C_2H_5)CH(OH)CH_2CH_2C(CH_3){=}C(H)CO_2CH_3 \xrightarrow[110°C]{CH_3C(OCH_3)_3} CH_3O_2CCH_2CH_2C(C_2H_5){=}C(H)CH_2CH_2C(CH_3){=}C(H)CO_2CH_3$ 85%

(continued)

Scheme 6.12. (*continued*)

9[i] HO, CH_2Ph, N, CH_2CH_3 → $CH_3C(OC_2H_5)_3$, 140°C → CH_2Ph, N, CH_2CH_3, $CH_2CO_2C_2H_5$ 74%

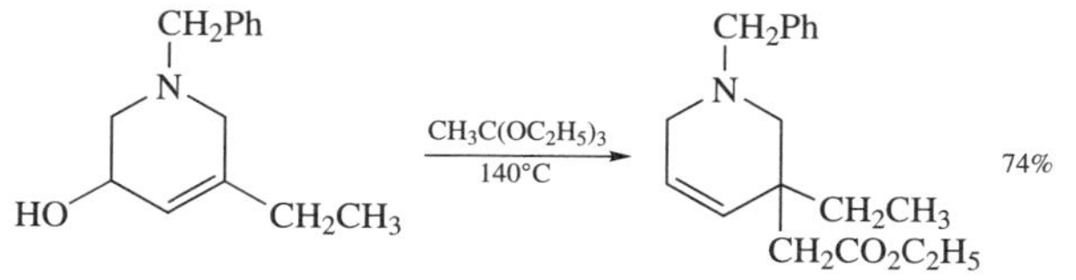

10[j] CH_3, N, O, CH_2OH → $CH_3CH_2CH_2C(OCH_3)_3$, 145°C → CH_3, N, O, CH_2, $CH_3CH_2CHCO_2CH_3$ 96%

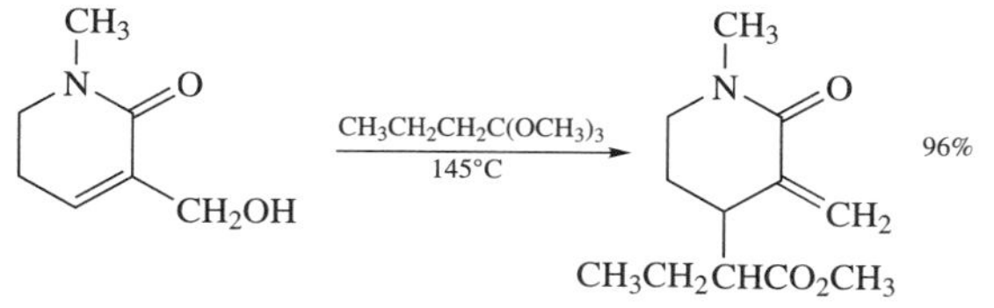

11[k] $PhthNCH_2$, C=C, CH_3, H, $CHCH_3$, OH (Phth = phthaloyl) → $CH_3C(OC_2H_5)_3$, $CH_3CH_2CO_2H$, heat → $PhthNCH_2CHCH_2CO_2C_2H_5$, H, C=C, CH_3, CH_3 68%

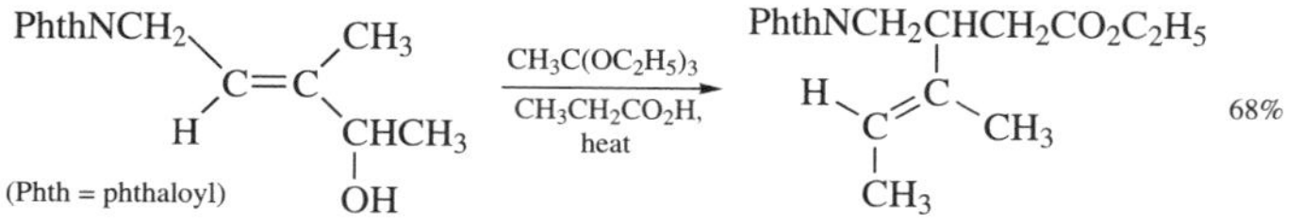

12[l] H, N, H, C, CH_2OH, Ph, C, CH_2, CH_3 → $CH_3C(OC_2H_5)_3$, $CH_3CH_2CO_2H$, 155°C → H, N, H, C, $CH_2CH_2CO_2CH_3$, Ph, C, CH_2, CH_3 93%

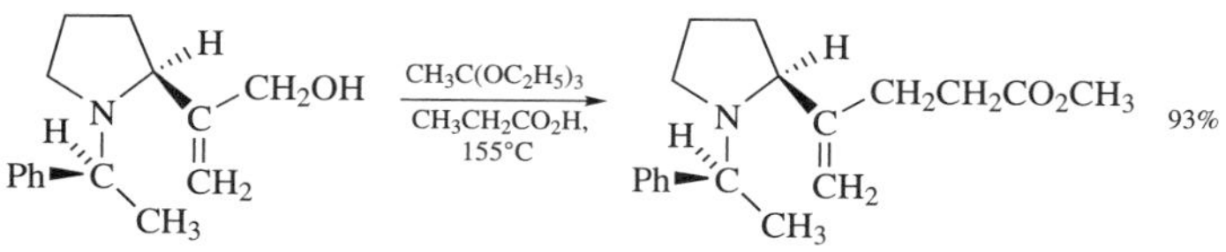

13[m] OH, S, Cl → $CH_3C(OC_2H_5)_3$, $CH_3CH_2CO_2H$, 120–130°C → $CH_2CO_2C_2H_5$, S, Cl

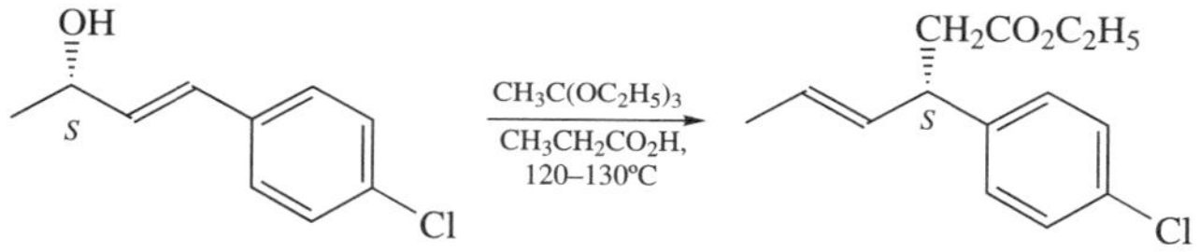

C. Rearrangements of ester enolates and silyl enol ethers

14[n] CH_3, H, C=C, H, $CH_2OC{=}CH_2$, $OSiMe_3$ → 1) 67°C, 2) CH_3OH, 3) HO^- → $CH_2{=}CHCHCH_2CO_2H$, CH_3 70%

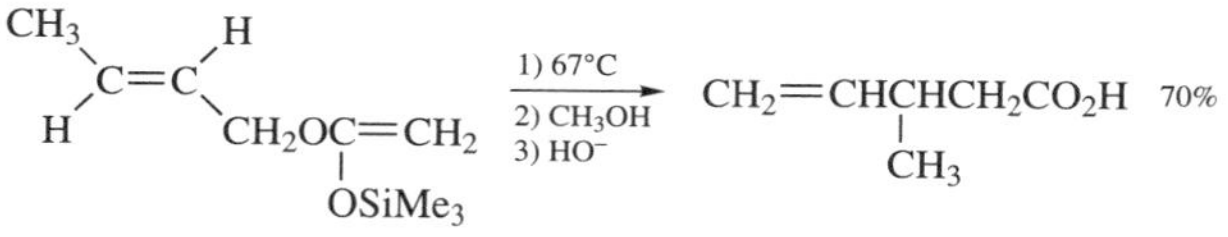

15[o] CH_3, $CH_3(CH_2)_5CHC{=}CH_2$, O, $C{=}CH_2$, $OSiMe_3$ → 1) 70°C, 2) H_3O^+ → $CH_3(CH_2)_5$, C=C, CH_3, H, $CH_2CH_2CO_2H$ 53%

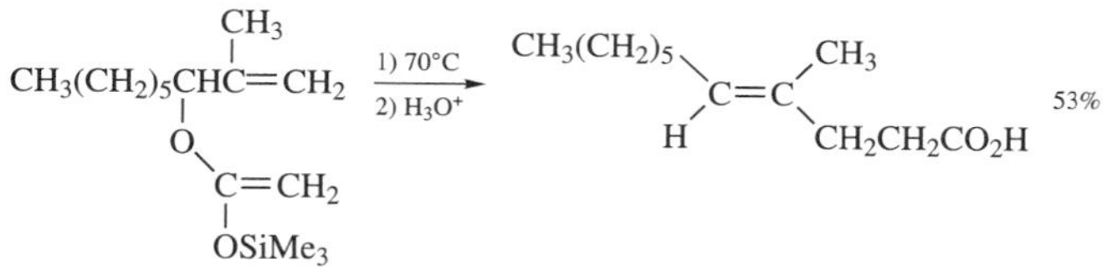

16[n] CH_3, C, $(CH_2)_5CH_3$, H_2C, C, H, O, C, CH_2CH_3, O → 1) $Li^+[(CH_3)_2CH\bar{N}C_6H_{11}]$, 2) 25°C, 3 h → CH_3, $CH_2{=}CCH(CH_2)_5CH_3$, CH_3CHCO_2H 71%

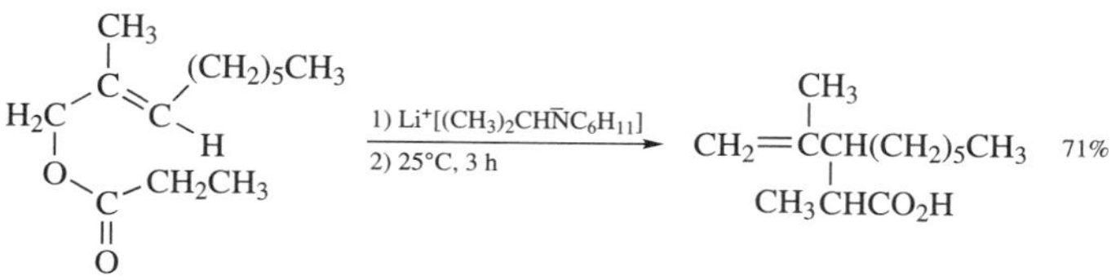

17[p] H_3C CH_3 O O PhCH$_2$O H O OCH$_2$ H CH_3 C=C H — 1) LDA, TMS–Cl 2) CH_2N_2 → H_3C CH_3 O O CH_3 PhCH$_2$O O CH=CH$_2$ CO_2CH_3 80%

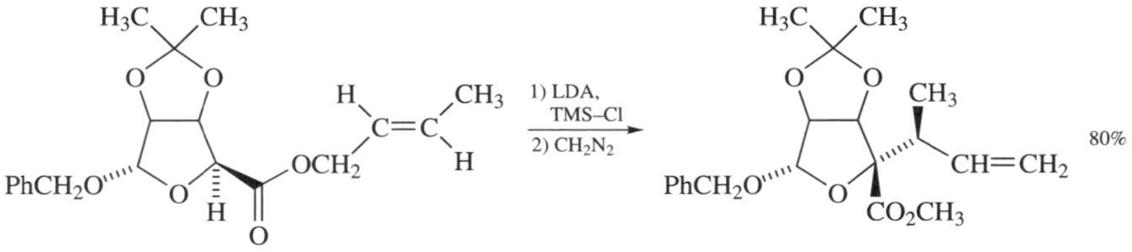

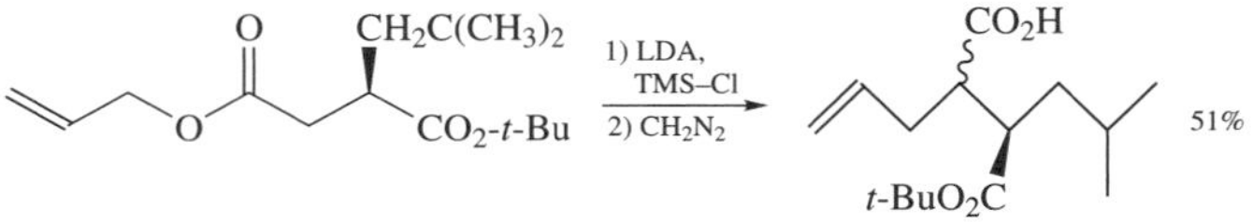

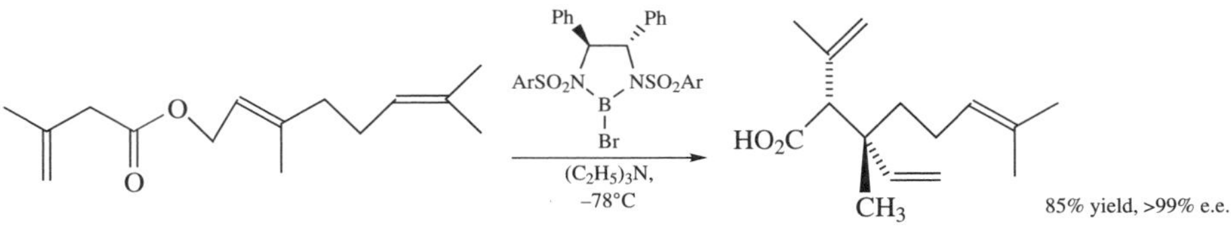

18[q] O $CH_2C(CH_3)_2$ O CO_2-*t*-Bu — 1) LDA, TMS–Cl 2) CH_2N_2 → CO_2H *t*-BuO$_2$C 51%

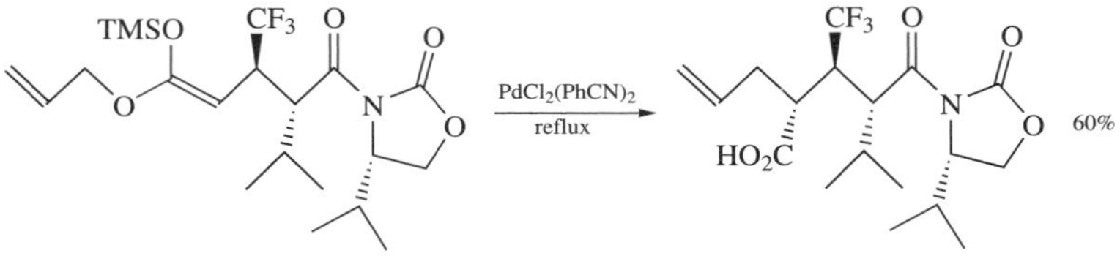

19[r] O O — Ph Ph ArSO$_2$N B NSO$_2$Ar Br ($C_2H_5)_3N$, –78°C → HO$_2$C CH_3 85% yield, >99% e.e.

20[s] TMSO CF_3 O O O N O — PdCl$_2$(PhCN)$_2$ reflux → CF_3 O O HO$_2$C N O 60%

21[t] O O NHCOCF$_3$ — 1) 4.5 equiv LHDMS, 2 equiv quinine 1.2 equiv Mg(OC$_2$H$_5$)$_2$ –78° to 0°C → CH_3 CO_2H NHCOCF$_3$

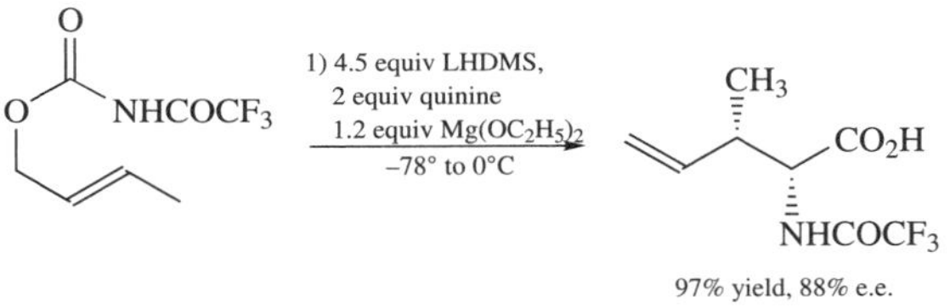

97% yield, 88% e.e.

22[u] OCH$_2$OCH$_3$ F$_2$C O$_2$CCH$_2$NHCO$_2$C(CH$_3$)$_3$ Ph — 1) 3 LDA –78°C 2) ZnCl$_2$ → CH$_3$OCH$_2$O NHCO$_2$C(CH$_3$)$_3$ Ph CO$_2$H F F 92%

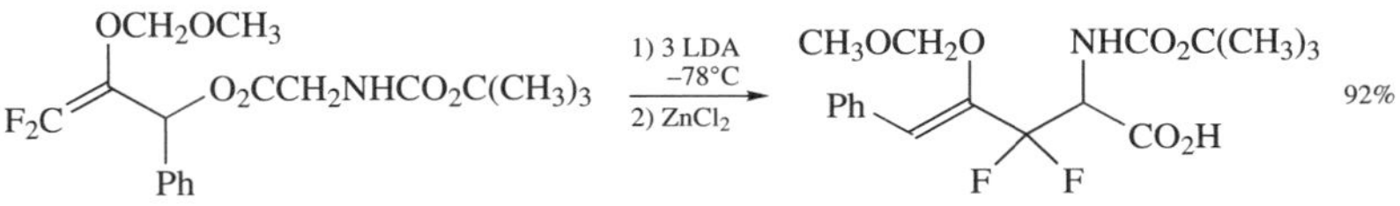

D. Rearrangement of ortho amides

23[i] CH$_2$Ph N HO CH$_2$CH$_3$ — (CH$_3$)$_2$NC(OCH$_3$)$_2$ CH$_3$ diglyme, 160°C → CH$_2$Ph N CH$_2$CH$_3$ CH$_2$CN(CH$_3$)$_2$ O 45%

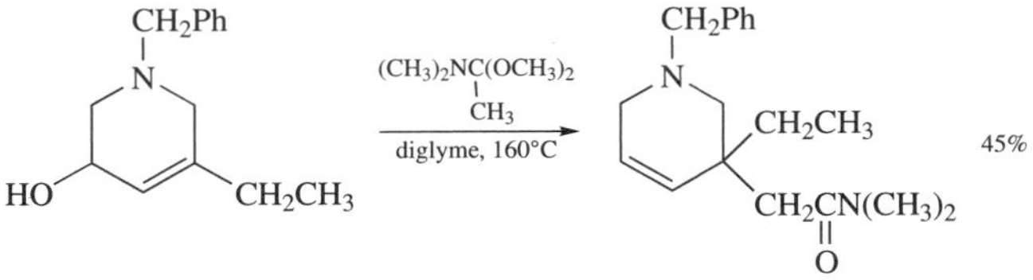

(*continued*)

Scheme 6.12. *(continued)*

24[v] PhCH$_2$O ... CH$_3$, OH — $(CH_3O)_2CHN(CH_3)_2$, 160°C, 48 h → PhCH$_2$O ... $CH_2CON(CH_3)_2$, CH$_3$ 92%

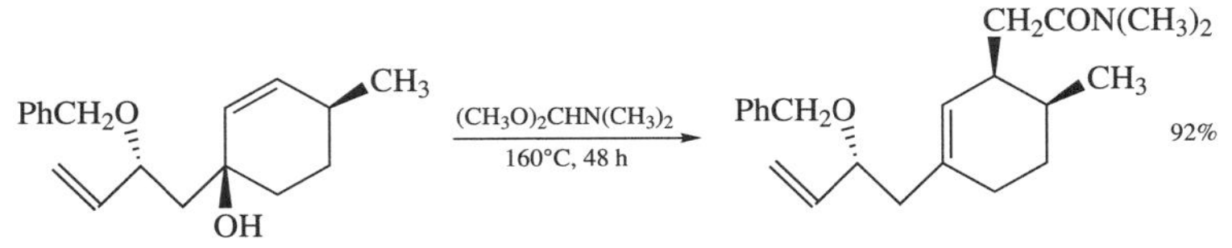

a. A. W. Burgstahler and I. C. Nordin, *J. Am. Chem. Soc.* **83:**198 (1961).
b. G. Saucy and R. Marbet, *Helv. Chim. Acta* **50:**2091 (1967).
c. D. J. Faulkner and M. R. Petersen, *J. Am. Chem. Soc.* **95:**553 (1973).
d. J. W. Ralls, R. E. Lundin, and G. F. Bailey, *J. Org. Chem.* **28:**3521 (1963).
e. L. A. Paquette, T.-Z. Wang, S. Wang, and C. M. G. Philippo, *Tetrahedron Lett.* **34:**3523 (1993).
f. S. D. Rychnovsky and J. L. Lee, *J. Org. Chem.* **60:**4318 (1995).
g. R. I. Trust and R. E. Ireland, *Org. Synth.* **53:**116 (1973).
h. C. A. Hendrick, R. Schaub, and J. B. Siddall, *J. Am. Chem. Soc.* **94:**5374 (1972).
i. F. E. Ziegler and G. B. Bennett, *J. Am. Chem. Soc.* **95:**7458 (1973).
j. J. J. Plattner, R. D. Glass, and H. Rapoport, *J. Am. Chem. Soc.* **94:**8614 (1972).
k. L. Serfass and P. J. Casara, *Biorg. Med. Chem. Lett.* **8:**2599 (1998).
l. D. N. A. Fox, D. Lathbury, M. F. Mahon, K. C. Molloy, and T. Gallagher, *J. Am. Chem. Soc.* **113:**2652 (1991).
m. E. Brenna, N. Caraccia, C. Fuganti, and P. Grasselli, *Tetrahedron Asymmetry* **8:**3801 (1997).
n. R. E. Ireland, R. H. Mueller, and A. K. Willard, *J. Am. Chem. Soc.* **98:**2868 (1976).
o. J. A. Katzenellenbogen and K. J. Christy, *J. Org. Chem.* **39:**3315 (1974).
p. R. E. Ireland and D. W. Norbeck, *J. Am. Chem. Soc.* **107:**3279 (1985).
q. L. M. Pratt, S. A. Bowler, S. F. Courney, C. Hidden, C. N. Lewis, F. M. Martin, and R. S. Todd, *Synlett* **1998:**531.
r. E. J. Corey, B. E. Roberts, and B. R. Dixon, *J. Am. Chem. Soc.* **117:**193 (1995).
s. T. Yamazaki, N. Shinohara, T. Ktazume, and S. Sato, *J. Org. Chem.* **60::**8140 (1995).
t. A. Kazmaier and A. Krebs, *Tetrahedron Lett.* **40:**479 (1999).
u. J. M. Percy, M. E. Prime, and M. J. Broadhurst, *J. Org. Chem.* **63:**8049 (1998).
v. A. R. Daniewski, P. M. Waskulich, and M. R. Uskokovic, *J. Org. Chem.* **57:**7133 (1992).

Both the exchange and elimination are catalyzed by addition of a small amount of a weak acid, such as propionic acid. Entries 7–13 in Scheme 6.12 are representative examples.

The mechanism and stereochemistry of the ortho ester Claisen rearrangement are analogous to those of the Cope rearrangement. The reaction is stereospecific with respect to the double bond present in the initial allylic alcohol. In acyclic molecules, the stereochemistry of the product can usually be predicted on the basis of a chairlike transition state.[158] When steric effects or ring geometry preclude a chairlike structure, the reaction can proceed through a boatlike transition state.[159]

High levels of enantiospecificity have been observed in the rearrangement of chiral reactants. This method can be used to establish the configuration of the newly formed carbon–carbon bond on the basis of the chirality of the C–O bond in the starting allylic alcohol. Treatment of (2*R*,3*E*)-3-penten-2-ol with ethyl orthoacetate gives the ethyl ester of (3*R*,4*E*)-3-methyl-4-hexenoic acid in 90% enantiomeric purity.[160] The configuration of the new chiral center is that predicted by a chairlike transition state with the methyl group occupying a pseudoequatorial position.

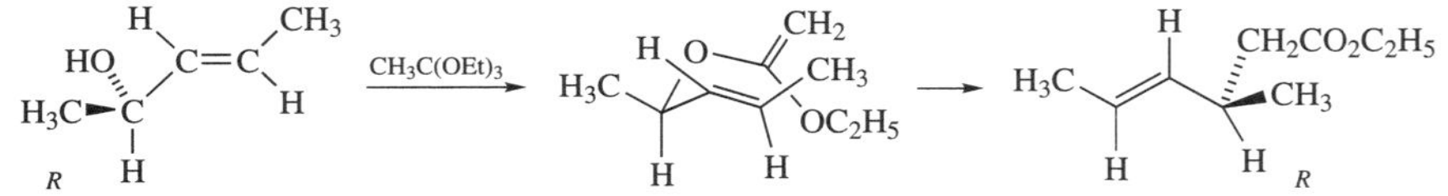

HO, H_3C, C, H, *R*, C=C, H, CH_3, H — $CH_3C(OEt)_3$ → H_3C, H, O, CH_2, CH_3, OC_2H_5, H, H → H_3C, H, H, $CH_2CO_2C_2H_5$, CH_3, H, *R*

158. G. W. Daub, J. P. Edwards, C. R. Okada, J. W. Allen, C. T. Maxey, M. S. Wells, A. S. Goldstien, M. J. Dibley, C. J. Wang, D. P. Ostercamp, S. Chung, P. S. Cunningham, and M. A. Berliner, *J. Org. Chem.* **62:**1976 (1997).
159. R. J. Cave, B. Lythgoe, D. A. Metcalf, and I. Waterhouse, *J. Chem. Soc., Perkin Trans. 1* **1997:**1218; G. Büchi and J. E. Powell, Jr., *J. Am. Chem. Soc.* **92:**3126 (1970); J. J. Gajewski and J. L. Jiminez, *J. Am. Chem. Soc.* **108:**468 (1986).
160. R. K. Hill, R. Soman, and S. Sawada, *J. Org. Chem.* **37:**3737 (1972); **38:**4218 (1973).

Esters of allylic alcohols can be rearranged to γ,δ-unsaturated carboxylic acids via the *O*-trimethylsilyl ether of the ester enolate.[161] This rearrangement takes place under much milder conditions than the ortho ester method. The reaction occurs at or slightly above room temperature. Entries 14 and 15 of Scheme 6.12 are examples. The example in entry 16 is a rearrangement of the enolate without intervention of the silyl enol ether.

The stereochemistry of the silyl enol ether Claisen rearrangement is controlled not only by the stereochemistry of the double bond in the allylic alcohol but also by the stereochemistry of the silyl enol ether. For the chair transition state, the configuration at the newly formed C–C bond is predicted to be determined by the *E*- or *Z*-configuration of the silyl enol ether.

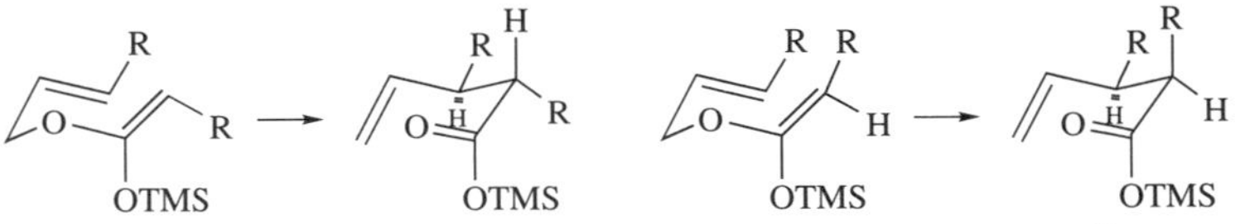

Z-silyl ether → *syn* isomer; E-silyl ether → *anti* isomer

The stereochemistry of the silyl enol ether can be controlled by the conditions of preparation. The base that is usually used for enolate formation is LDA. If the enolate is prepared in pure THF, the *E*-enolate is generated, and this stereochemistry is maintained in the silylated derivative. The preferential formation of the *E*-enolate can be explained in terms of a cyclic transition state in which the proton is abstracted from the stereoelectronically preferred orientation.

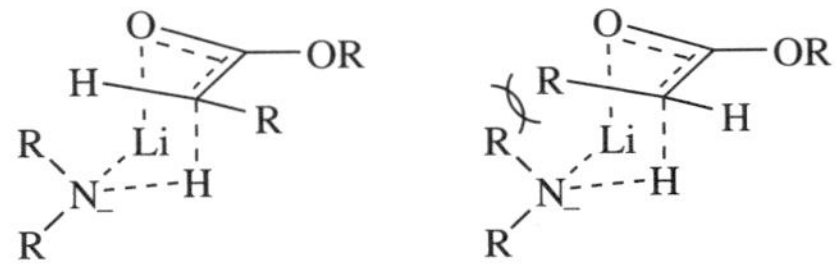

transition state for *E*-enolate

transition state for *Z*-enolate

If HMPA is included in the solvent, the *Z*-enolate predominates.[162] DMPU also favors the *Z*-enolate. The switch to the *Z*-enolate with HMPA or DMPU can be attributed to a loose, perhaps acyclic, transition state being favored as the result of strong solvation of the lithium ion by HMPA or DMPU. The steric factors favoring the *E* transition state are therefore diminished.[163] These general principles of solvent control of enolate stereochemistry are applicable to other systems.[164]

A number of steric effects on the rate of rearrangement have been observed and can be accommodated by the chairlike transition-state model.[165] The *E*-silyl enol ethers

161. R. E. Ireland, R. H. Mueller, and A. K. Willard, *J. Am. Chem. Soc.* **98:** 2868 (1976); S. Pereira and M. Srebnik, *Aldrichimica Acta* **26:**17 (1993).
162. R. E. Ireland and A. K. Willard, *Tetrahedron Lett.* **1975:**3975; R. E. Ireland, P. Wipf, and J. D. Armstrong III, *J. Org. Chem.* **56:**650 (1991).
163. C. H. Heathcock, C. T. Buse, W. A. Kleschick, M. C. Pirrung, J. E. Sohn, and J. Lamp, *J. Org. Chem.* **45:**1066 (1980).
164. J. Corset, F. Froment, M.-F. Lutie, N. Ratovelomanana, J. Seyden-Penne, T. Strzalko, and M. C. Roux-Schmitt, *J. Am. Chem. Soc.* **115:**1684 (1993).
165. C. S. Wilcox and R. E. Babston, *J. Am. Chem. Soc.* **108:**6636 (1986).

rearrange somewhat more slowly than the corresponding *Z*-isomers. This is interpreted as resulting from the pseudoaxial placement of the methyl group in the *E* transition state.

Z-isomer *E*-isomer

The size of the substituent R also influences the rate, with the rate increasing somewhat for both isomers as R becomes larger. It is believed that steric interactions with R are relieved as the C–O bond stretches. The rate acceleration would reflect the higher ground-state energy resulting from these steric interactions.

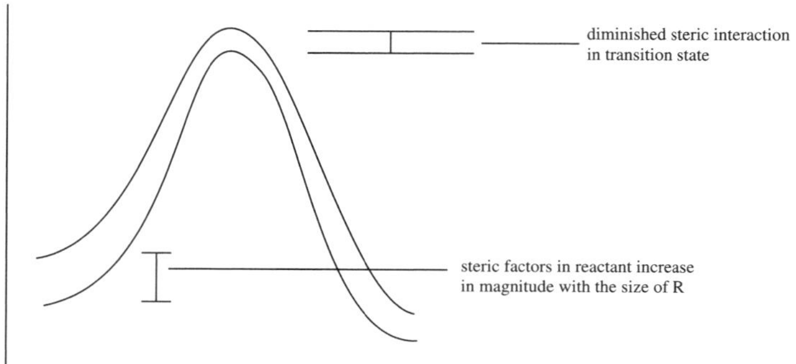

The silyl ketene acetal rearrangement can also be carried out by reaction of the ester with a silyl triflate and tertiary amine, without formation of the ester enolate. Optimum results have been obtained with bulky silyl triflates and amines, for example, *t*-butyldimethylsilyl triflate and *N*-methyl-*N*,*N*-dicyclohexylamine. Under these conditions, the reaction is stereoselective for the *Z*-silyl ketene acetal, and the stereochemistry of the allylic double bond determines the *syn* or *anti* configuration.

Ref. 166

The possibility of using chiral auxiliaries or chiral catalysts to achieve enantioselective Claisen rearrangements has been explored.[167] One approach is to use boron enolates with chirality installed at the boron atom. For example, enolates prepared with L^*_2BBr led

166. M. Kobayashi, K. Matsumoto, E. Nakai, and T. Nakai, *Tetrahedron Lett.* **37**:3005 (1996).
167. D. Enders, M. Knopp, and R. Schiffers, *Tetrahedron Asymmetry* **7**:1847 (1996).

to rearranged products of >95% enantiomeric excess.[168]

L*$_2$BBr, $(C_2H_5)_3N$ → OBL*$_2$ → CO_2H 65% yield, 96% e.e.

$(i\text{-Pr})_2NC_2H_5$, L*$_2$BBr → OBL*$_2$ → CO_2H 75% yield, >97% e.e.

L*$_2$BBr = (Ph, Ph, $ArSO_2N$, NSO_2Ar, B, Br) Ar = 3,5-bis(trifluoromethyl)phenyl

As with other ester enolate rearrangements, the presence of chiral ligands can render the reaction enantioselective. Use of quinine or quinidine with the chelating metal leads to enantioselectivity (see entry 21 in Scheme 6.12).

The stereoselectivity of ester enolate Claisen rearrangements can also be controlled by specific intramolecular interactions.[169] The enolates of α-alkoxy esters give the *Z*-silyl derivatives because of chelation by the alkoxy substituent.

$ROCH_2COCH_2CH{=}CHR'$ —LDA→ (Li^+ chelate: RO, O^-, C=C, H, $OCH_2CH{=}CHR'$) —$ClSiR_3$→ (RO, $OSiR_3$, C=C, H, $OCH_2CH{=}CHR'$) *Z*-isomer

The configuration at the newly formed C—C bond is then controlled by the stereochemistry of the double bond in the allylic alcohol. The *E*-isomer gives a *syn* orientation whereas the *Z*-isomer gives rise to *anti* stereochemistry.[170]

(*E*: O, R′, OR, $OSiR_3$ → O, H, OR, R′, $OSiR_3$ ≡ OR, CO_2SiR_3, R′)

(*Z*: O, OR, $OSiR_3$, R′ → O, H, OR, R_3SiO, R′ ≡ OR, CO_2SiR_3, R′)

168. E. J. Corey and D.-H. Lee, *J. Am. Chem. Soc.* **113**:4026 (1991); E. J. Corey, B. E. Roberts, and B. R. Dixon, *J. Am. Chem. Soc.* **117**:193 (1995).
169. H. Frauenrath, in *Stereoselective Synthesis*, G. Helmchen, R. W. Hoffmann, J. Mulzer, and E. Schaumann, eds., Georg Thieme Verlag, Stuttgart, 1996.
170. T. J. Gould, M. Balestra, M. D. Wittman, J. A. Gary, L. T. Rossano, and J. Kallmerten, *J. Org. Chem.* **52**:3889 (1987); S. D. Burke, W. F. Fobare, and G. J. Pacofsky, *J. Org. Chem.* **48**:5221 (1983); P. A. Bartlett, D. J. Tanzella, and J. F. Barstow, *J. Org. Chem.* **47**:3941 (1982).

Similar chelation effects appear to be present in α-alkoxymethyl derivatives. Magnesium enolates give predominantly the Z-enolate as a result of this chelation. The corresponding trimethylsilyl enol ethers give E/Z mixtures because of a relatively weak steric differentiation between the ethyl and alkoxymethyl substituents.[171]

$R = CH_3$ or CH_2OCH_3 — C_2H_5NMgBr — 85% yield, >95% Z

Enolates of allyl esters of α-amino acids are also subject to chelation-controlled Claisen rearrangement.[172]

2.5 equiv LDA, 1.1 equiv $ZnCl_2$

Various salts can promote chelation, but $ZnCl_2$ and $MgCl_2$ are suitable for most cases. The rearrangement is a useful reaction for preparing amino acid analogs and has also been applied to modified dipeptides.[173]

1) 4 equiv LDA 2) $MnCl_2$ 3) CH_2N_2 — 90% yield, 62:38 mixture

A reaction which is related to the ortho ester Claisen rearrangement utilizes an amide acetal, such as dimethylacetamide dimethyl acetal, rather than an ortho ester in the exchange reaction with allylic alcohols.[174] The stereochemistry of the reaction is analogous to that of the other variants of the Claisen rearrangement.[175]

$$RCH{=}CHCH_2OH + (CH_3)_2NC(OCH_3)_2CH_3 \rightleftharpoons (CH_3)_2NC(OCH_3)(CH_3)OCH_2CH{=}CHR \rightleftharpoons (CH_3)_2NC(=CH_2)OCH_2CH{=}CHR \rightarrow (CH_3)_2NC(=O)CH_2CH(R)CH{=}CH_2$$

171. M. E. Krafft, S. Jarrett, and O. A. Dasse, *Tetrahedron Lett.* **34:**8209 (1993).
172. U. Kazmaier, *Liebigs Ann. Chem.* **1997:**285; U. Kazmaier, *J. Org. Chem.* **61:**3694 (1996); U. Kazmaier and S. Maier, *Tetrahedron* **52:**941 (1996).
173. U. Kazmaier and S. Maier, *J.. Chem. Soc., Chem. Commun.* **1998:**2535.
174. A. E. Wick, D. Felix, K. Steen, and A. Eschenmoser, *Helv. Chim. Acta* **47:**2425 (1964); D. Felix, K. Gschwend-Steen, A. E. Wick, and A. Eschenmoser, *Helv. Chim. Acta* **52:**1030 (1969).
175. W. Sucrow, M. Slopianka, and P. P. Calderia, *Chem. Ber.* **108:**1101 (1975).

O-Allyl imidate esters undergo [3,3] sigmatropic rearrangements to *N*-allyl amides. Trichloroacetimidates can be easily made from allylic alcohols by reaction with trichloroacetonitrile. The rearrangement then provides trichloroacetamides of *N*-allylamines.[176]

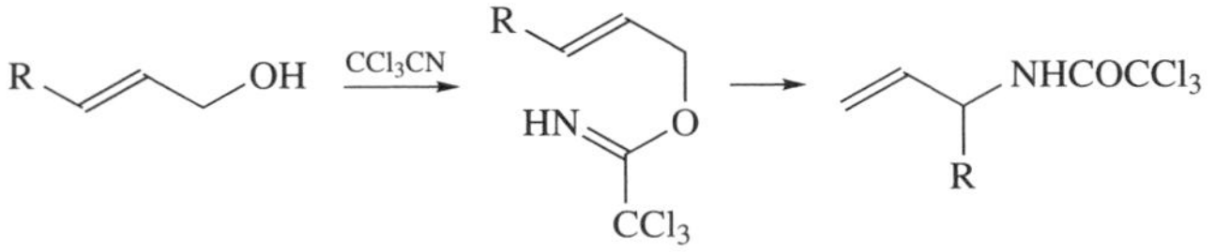

Yields in the reaction are sometimes improved by inclusion of K_2CO_3 in the reaction mixture.[177]

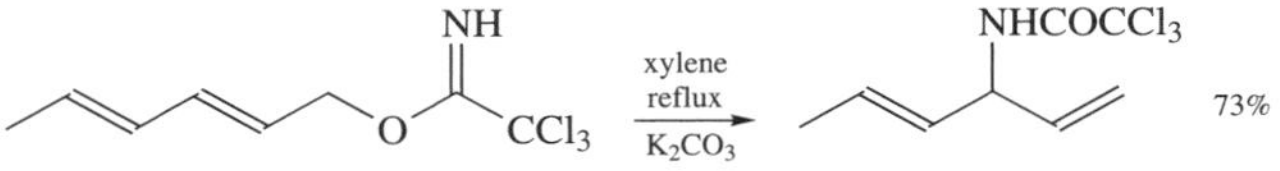

Trifluoroacetimidates show similar reactivity.[178] Imidate rearrangements are catalyzed by palladium salts.[179] The mechanism is presumably similar to that for the Cope rearrangement (see p. 382).

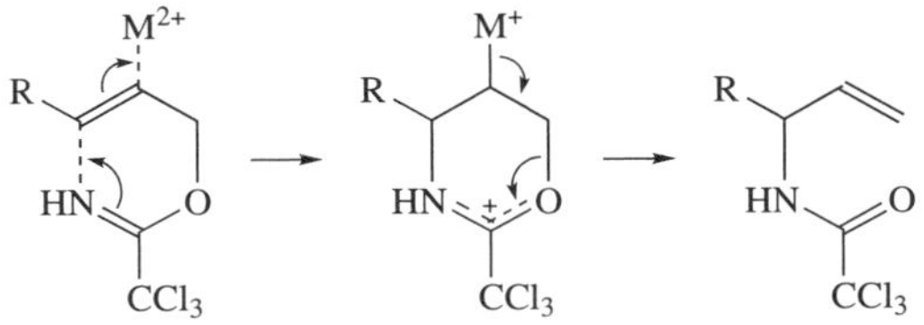

Imidate esters can also be generated by reaction of imidoyl chlorides and allylic alcohols. The anions of these imidates, prepared using lithium diethylamide, rearrange at around 0°C. When a chiral amine is used, this reaction can give rise to enantioselective formation of γ,δ-unsaturated amides. Good results were obtained with a chiral binaphthylamine.[180] The methoxy substitutent is believed to play a role as a Li^+ ligand in the reactive enolate.

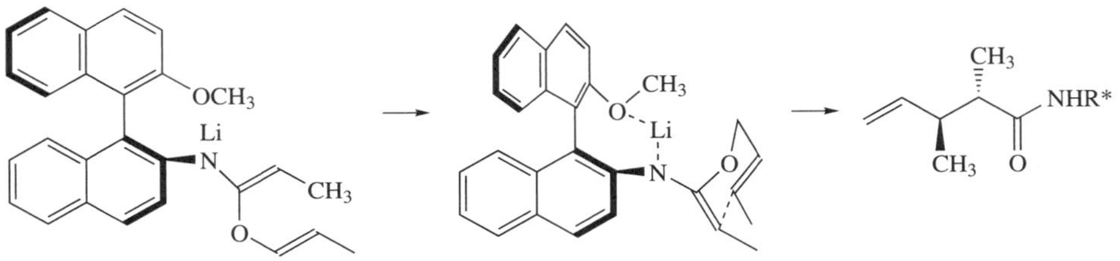

176. L. E. Overman, *J. Am. Chem. Soc.* **98:**2901 (1976); L. E. Overman, *Acc. Chem. Res.* **13:**218 (1980).
177. T. Nishikawa, M. Asai, N. Ohyabu, and M. Isobe, *J. Org. Chem.* **63:**188 (1998).
178. A. Chen, I. Savage, E. J. Thomas, and P. D. Wilson, *Tetrahedron Lett.* **34:**6769 (1993).
179. L. E. Overman, *Angew. Chem. Int. Ed. Engl.* **23:**579 (1984); T. G. Schenck and B. Bosnich, *J. Am. Chem. Soc.* **107:**2058 (1985).
180. P. Metz and B. Hungerhoff, *J. Org. Chem.* **62:**4442 (1997).

Aryl allyl ethers can also undergo [3,3] sigmatropic rearrangement. Claisen rearrangements of allyl phenyl ethers to *ortho*-allyl phenols were the first [3,3] sigmatropic rearrangements to be thoroughly studied.[181] The reaction proceeds through a cyclohexadienone that enolizes to the stable phenol.

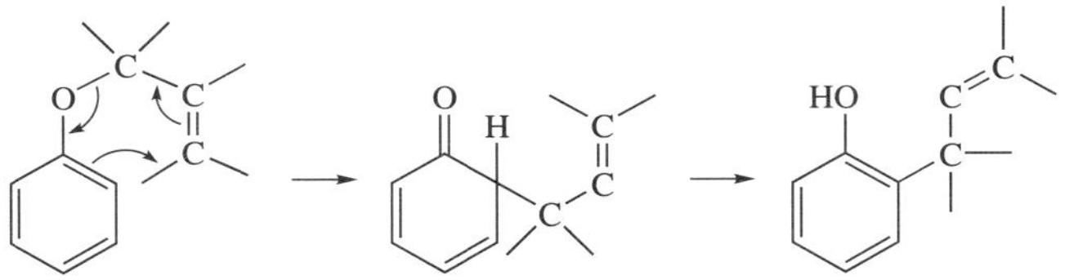

If both *ortho* positions are substituted, the allyl group undergoes a second sigmatropic migration, giving the *para*-substituted phenol:

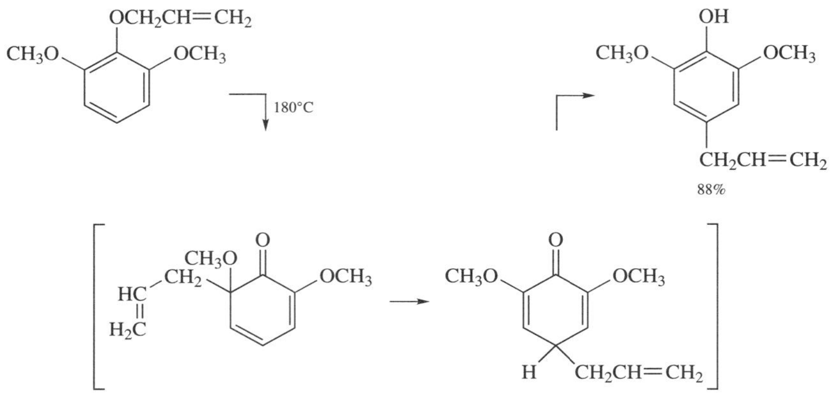

180°C

88%

Ref. 182

6.6. [2,3] Sigmatropic Rearrangements

The [2,3] sigmatropic class of rearrangements is represented by two generic charge types:

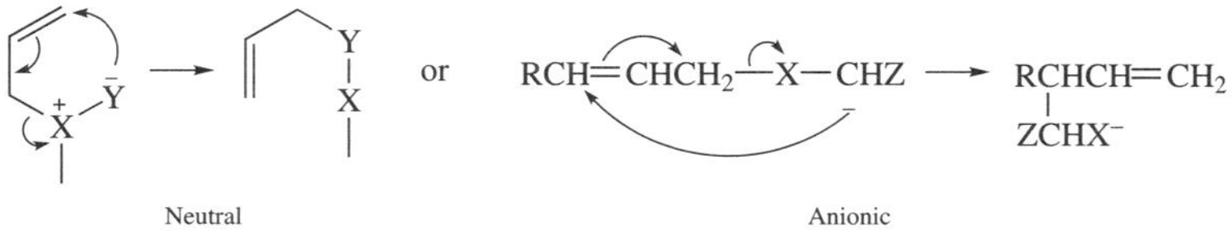

Neutral

Anionic

The rearrangements of allylic sulfoxides, selenoxides, and nitrones are the most useful examples of the first type whereas rearrangements of carbanions of allyl ethers are the major examples of the anionic type.

181. S. J. Rhoads, in *Molecular Rearrangements*, Vol. 1, P. de Mayo, ed., Interscience, New York, 1963, pp. 655–684.

The sigmatropic rearrangement of allylic sulfoxides to allylic sulfenates first received study in connection with the mechanism of racemization of allyl aryl sulfoxides.[183] Although the allyl sulfoxide structure is strongly favored at equilibrium, rearrangement through the achiral allyl sulfenate provides a low-energy pathway for racemization.

$$R-\overset{+}{S}(O^-)-CH_2-CH=CH_2 \rightleftharpoons RS-O-CH_2-CH=CH_2$$

The synthetic utility of the allyl sulfoxide–allyl sulfenate rearrangement is as a method of preparation of allylic alcohols.[184] The reaction is carried out in the presence of a reagent, such as phenylthiolate or trimethyl phosphite, which reacts with the sulfenate to cleave the S–O bond:

$$(CH_3)_3C\text{-}C_6H_9{=}CHCH_2\overset{+}{S}(O^-)Ph \longrightarrow (CH_3)_3C\text{-}C_6H_{10}(OSPh)(CH{=}CH_2) \xrightarrow{PhS^-} (CH_3)_3C\text{-}C_6H_{10}(OH)(CH{=}CH_2)\ \ 95\%$$

Ref. 185

An analogous transposition occurs with allylic selenoxides when they are generated *in situ* by oxidation of allylic seleno ethers.[186]

$$PhCH_2CH_2CH(SePh)CH{=}CHCH_3 \xrightarrow{H_2O_2} PhCH_2CH_2CH{=}CHCH(OH)CH_3$$

Allylic sulfonium ylides readily undergo [2,3] sigmatropic rearrangement.[187]

$$(CH_3)_2C{=}CHCH_2\overset{+}{S}(CH_3)-\overset{-}{C}HCH{=}C(CH_3)_2 \longrightarrow (CH_3)_2C{=}CHCH(SCH_3)C(CH_3)_2CH{=}CH_2\ \ 95\%$$

This reaction results in carbon–carbon bond formation. It has found synthetic application in ring-expansion sequences for generation of medium-sized rings. The reaction proceeds best when the ylide has a carbanion-stabilizing substituent. Part A of Scheme 6.13 shows some examples of the reaction.

The corresponding nitrogen ylides can also be generated when one of the nitrogen substituents has an anion-stabilizing group on the α carbon. For example, quaternary salts

182. I. A. Pearl, *J. Am. Chem. Soc.* **70:**1746 (1948).
183. R. Tang and K. Mislow, *J. Am. Chem. Soc.* **92:**2100 (1970).
184. D. A. Evans and G. C. Andrews, *Acc. Chem. Res.* **7:**147 (1974).
185. D. A. Evans, G. C. Andrews, and C. L. Sims, *J. Am. Chem. Soc.* **93:**4956 (1971).
186. H. J. Reich, *J. Org. Chem.* **40:**2570 (1975); D. L. J. Clive, G. Chittatu, N. J. Curtis, and S. M. Menchen, *J. Chem. Soc., Chem. Commun.* **1978:**770.
187. J. E. Baldwin, R. E. Hackler, and D. P. Kelly, *J. Chem. Soc., Chem. Commun.* **1968:**537.

Scheme 6.13. Carbon–Carbon Bond Formation via [2,3] Sigmatropic Rearrangements of Sulfur and Nitrogen Ylides

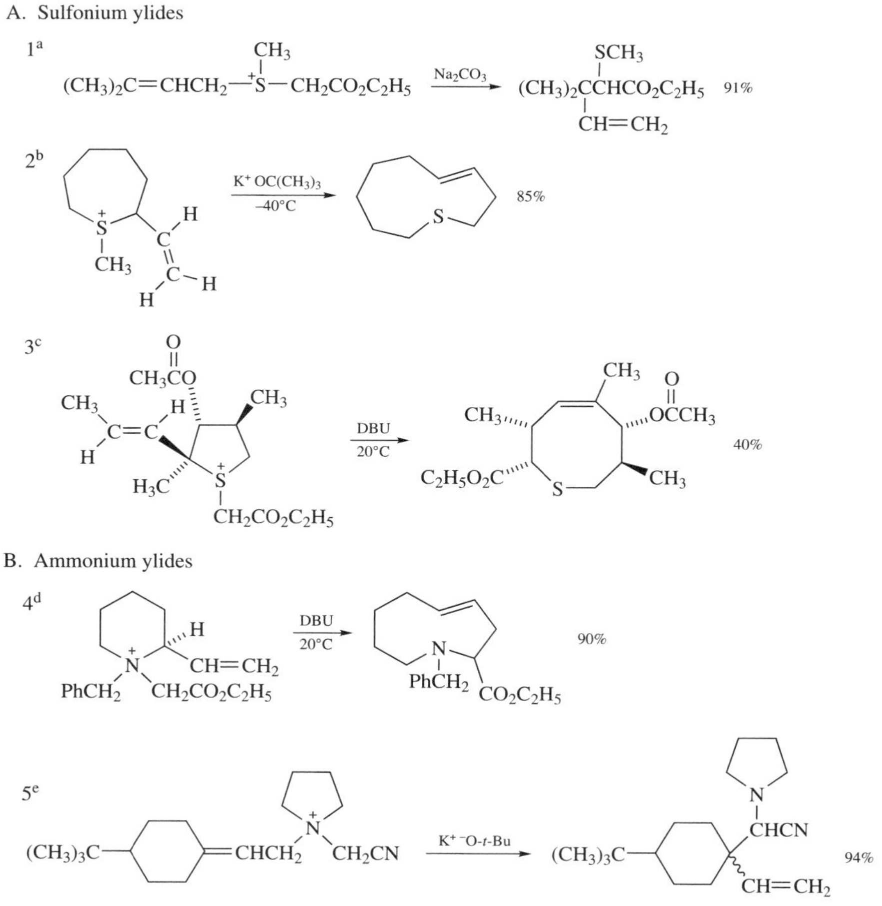

A. Sulfonium ylides

1[a] $(CH_3)_2C{=}CHCH_2{-}\overset{+}{S}(CH_3){-}CH_2CO_2C_2H_5 \xrightarrow{Na_2CO_3} (CH_3)_2C(SCH_3)CHCO_2C_2H_5$ with $CH{=}CH_2$ 91%

2[b] $\xrightarrow[-40°C]{K^+\,OC(CH_3)_3}$ 85%

3[c] $\xrightarrow[20°C]{DBU}$ 40%

B. Ammonium ylides

4[d] $\xrightarrow[20°C]{DBU}$ 90%

5[e] $(CH_3)_3C$–cyclohexylidene$=CHCH_2\overset{+}{N}$(pyrrolidinium)$CH_2CN \xrightarrow{K^+\,{}^-O\text{-}t\text{-}Bu} (CH_3)_3C$–cyclohexyl–$CH(CN)N$(pyrrolidine), $CH{=}CH_2$ 94%

a. K. Ogura, S. Furukawa, and G. Tsuchihashi, *J. Am. Chem. Soc.* **102:**2125 (1980).
b. V. Cere, C. Paolucci, S. Pollicino, E. Sandri, and A. Fava, *J. Org. Chem.* **43:**4826 (1978).
c. E. Vedejs and M. J. Mullins, *J. Org. Chem.* **44:**2947 (1979).
d. E. Vedejs, M. J. Arco, D. W. Powell, J. M. Renga, and S. P. Singer, *J. Org. Chem.* **43:**4831 (1978).
e. L. N. Mander and J. V. Turner, *Aust. J. Chem.* **33:**1559 (1980).

of N-allyl α-aminoesters readily rearrange to α-allyl products.[188]

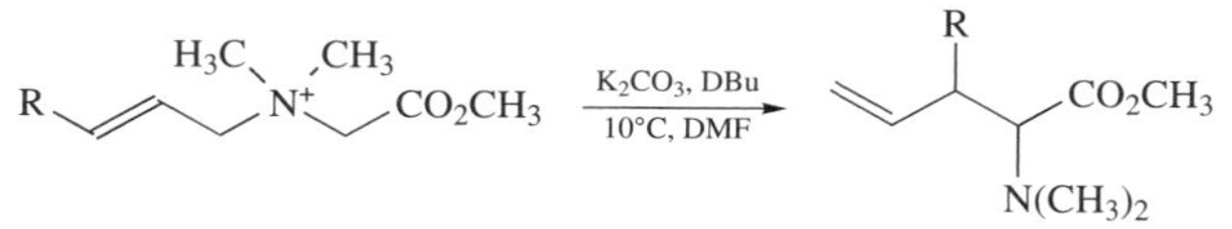

$$R\text{-}CH{=}CH\text{-}CH_2\overset{+}{N}(CH_3)_2CH_2CO_2CH_3 \xrightarrow[10°C,\ DMF]{K_2CO_3,\ DBu} CH_2{=}CH\text{-}CH(R)\text{-}CH(N(CH_3)_2)CO_2CH_3$$

Entries 4 and 5 in Scheme 6.13 are other examples. Entry 4 illustrates the use of the reaction for ring expansion.

188. I. Coldham, M. L. Middleton, and P. L. Taylor, *J. Chem. Soc., Perkin Trans. 1* **1997:**2951; I. Coldham, M. L. Middleton and P. L. Taylor, *J. Chem. Soc., Perkin Trans. 1* **1998:**2817.

N-Allylamine oxides possess the general structure pattern for [2,3] sigmatropic rearrangement where X = N and Y = O^-. The rearrangement proceeds readily to provide *O*-allyl hydroxylamine derivatives.

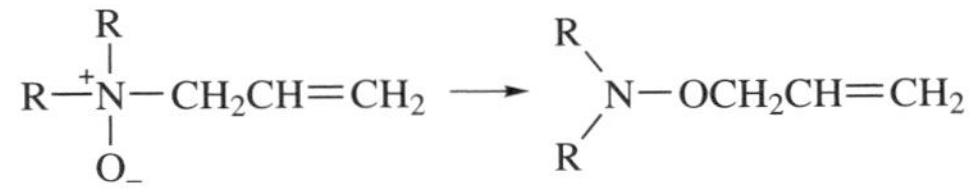

$$R_2\overset{+}{N}(O^-)-CH_2CH{=}CH_2 \longrightarrow R_2N-OCH_2CH{=}CH_2$$

A useful method for *ortho*-alkylation of aromatic amines is based on [2,3] sigmatropic rearrangement of *S*-anilinosulfonium ylides. These ylides are generated from anilinosulfonium ions, which can be prepared from *N*-chloroanilines and sulfides.[189]

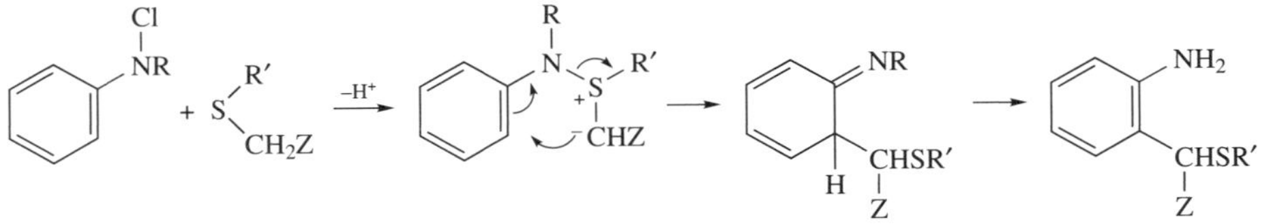

This method is the basis for synthesis of nitrogen-containing heterocyclic compounds when Z is a carbonyl-containing group.[190]

The [2,3] sigmatropic rearrangement pattern is also observed with anionic species. The most important case for synthetic purposes is the *Wittig rearrangement*, in which a strong base converts allylic ethers to α-allyl alkoxides.[191]

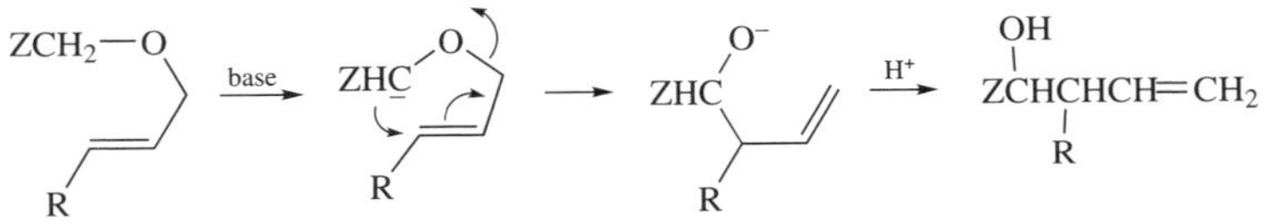

Because the deprotonation at the α′ carbon must compete with deprotonation of the α carbon in the allyl group, most examples involve a conjugated or electron-withdrawing substituent Z.[192]

The stereochemistry of the Wittig rearrangement can be predicted in terms of a cyclic five-membered transition state in which the α substituent prefers an equatorial orientation.[193]

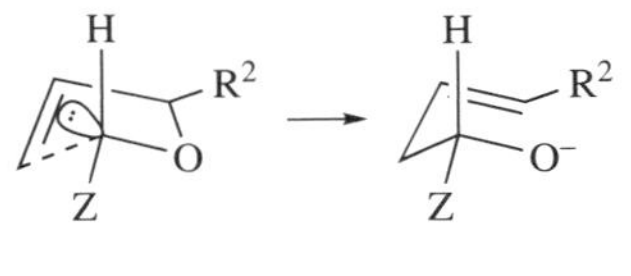

189. P. G. Gassman and G. D. Gruetzmacher, *J. Am. Chem. Soc.* **96:**5487 (1974); P. G. Gassman and H. R. Drewes, *J. Am. Chem. Soc.* **100:**7600 (1978).
190. P. G. Gassman, T. J. van Bergen, D. P. Gilbert, and B. W. Cue, Jr., *J. Am. Chem. Soc.* **96:**5495 (1974); P. G. Gassman and T. J. van Bergern, *J. Am. Chem. Soc.* **96:**5508 (1974); P. G. Gassman, G. Gruetzmacher, and T. J. van Bergen, *J. Am. Chem. Soc.* **96:**5512 (1974).
191. J. Kallmarten, in *Stereoselective Synthesis*, Houben Weyl Methods in Organic Chemistry Vol. E21d, R. W. Hoffmann, J. Mulzer, and E. Schaumann, eds., G. Thieme Verlag, Stuttgart, 1995, pp. 3810.
192. For reviews of [2,3] sigmatropic rearrangement of allyl ethers, see T. Nakai and K. Mikami, *Chem. Rev.* **86:**885 (1986).
193. R. W. Hoffmann, *Angew. Chem. Int. Ed. Engl.* **18:**563 (1979). K. Mikami, Y. Kimura, N. Kishi, and T. Nakai, *J. Org. Chem.* **48:**279 (1983); K. Mikami, K. Azuma, and T. Nakai, *Tetrahedron* **40:**2303 (1984); Y.-D. Wu, K. N. Houk, and J. A. Marshall, *J. Org. Chem.* **55:**1421 (1990).

A consistent feature of the observed stereochemistry is a preference for *E*-stereochemistry at the newly formed double bond. The reaction can also show stereoselectivity at the newly formed single bond. This stereoselectivity has been carefully studied for the case in which the substituent Z is an acetylenic group.

E-isomer → *anti*-isomer

Z-isomer → *syn*-isomer

The preferred stereochemistry arises from the transition state that minimizes interaction between the ethynyl and isopropyl substituents. This stereoselectivity is revealed in the rearrangement of **11** to **12**.

11 → **12** Ref. 194

There are other means of generating the anions of allyl ethers. For synthetic purposes, one of the most important involves lithium–tin exchange on stannylmethyl ethers.[195]

$$R\text{CH=CHCH}_2\text{OCH}_2\text{SnR}_3 \xrightarrow{\text{RLi}} R\text{CH=CHCH}_2\text{OCH}_2\text{Li} \longrightarrow R\text{CH(CH}_2\text{OLi)CH=CH}_2$$

Another method involves reduction of allylic acetals of aromatic aldehydes by SmI_2.[196]

$$\text{ArCH(OCH}_2\text{CH=CHR)}_2 \xrightarrow{3\text{ equiv SmI}_2} \text{Ar}\underline{\text{C}}\text{HOCH}_2\text{CH=CHR} \longrightarrow \text{ArCH(OH)CH(R)CH=CH}_2$$

194. M. M. Midland and J. Gabriel, *J. Org. Chem.* **50:**1143 (1985).
195. W. C. Still and A. Mitra, *J. Am. Chem. Soc.* **100:**1927 (1978).
196. H. Hioki, K. Kono, S. Tani, and M. Kunishima, *Tetrahedron Lett.* **39:**5229 (1998).

[2,3] Sigmatropic rearrangements of anions of *N*-allylamines have also been observed and are known as aza-Wittig rearrangements.[197] The reaction requires anion-stabilizing substituents and is favored by *N*-benzyl and by silyl or sulfenyl substituents on the allyl group.[198] The trimethylsilyl substituents also can influence the stereoselectivity of the reaction. The steric interactions between the benzyl group and allyl substituent govern the stereoselectivity, and which is markedly higher in the trimethylsilyl derivatives.[199]

R	X	*anti*:*syn*
CH_3	H	3:2
C_2H_5	H	1:1
$(CH_3)_2CH$	H	4:3
CH_3	$Si(CH_3)_3$	<1:20
C_2H_5	$Si(CH_3)_3$	1:18
$(CH_3)_2CH$	$Si(CH_3)_3$	1:11

The [2,3] Wittig rearrangement has proven useful for ring contraction in the synthesis of a number of medium-ring unsaturated structures, as illustrated by entry 3 in Scheme 6.14.

6.7. Ene Reactions

Certain electrophilic carbon–carbon and carbon–oxygen double bonds can undergo an addition reaction with alkenes in which an allylic hydrogen is transferred to the electrophile. This process is called the *ene reaction*, and the electrophile is called an *enophile*.[200]

ene + enophile
(EWG = electron-withdrawing group)

197. C. Vogel, *Synthesis* **1997**:497.
198. J. C. Anderson, S. C. Smith, and M. E. Swarbrick, *J. Chem. Soc., Perkin Trans 1* **1997**:1517.
199. J. C. Anderson, D. C. Siddons, S. C. Smith, and M. E. Swarbrick, *J. Org. Chem.* **61**:4820 (1996).
200. For review of the ene reaction, see H. M. R. Hoffmann, *Angew. Chem. Int. Ed. Engl.* **8**:856 (1969); W. Oppolzer, *Pure Appl. Chem.* **53**:1181 (1981).

Scheme 6.14. [2,3] Wittig Rearrangements

1[a] $(CH_3)_2CH$, H, C=C, H, CH_3, $OCH_2C{=}CH_2$, CH_3 → (n-BuLi, $K^{+}\,{}^{-}O$-t-Bu) → $(CH_3)_2CH$–C(H)=C(H)–C(OH)–C(CH_3)–C(=CH_2)CH_3 95%

2[b] OCH_2SiMe_3, $O(CH_2)_2OTMS$ → (1) BuLi, −78 → 0°C; 2) NH_4Cl) → CH_2OH, $O(CH_2)_2OTMS$ 45%

3[c] O → (n-BuLi, HMPA) → OH 60%

4[d] H_3C, $OCH_2C{\equiv}CH$, OMe → (2 n-BuLi, −78°C) → H_3C, OH, $C{\equiv}CH$, OMe 75%

5[e] O, N, O → (n-BuLi, −78°C) → OH, O, N

a. D. J.-S. Tsai and M. M. Midland, *J. Am. Chem. Soc.* **107:**3915 (1985).
b. T. Sugimura and L. A. Paquette, *J. Am. Chem. Soc.* **109:**3017 (1987).
c. J. A. Marshall, T. M. Jenson, and B. S. De Hoff, *J. Org. Chem.* **51:**4316 (1986).
d. K. Mikami, K. Kawamoto, and T. Nakai, *Tetrahedron Lett.* **26:**5799 (1985).
e. M. H. Kress, B. F. Kaller, and Y. Kishi, *Tetrahedron Lett.* **34:**8047 (1993).

The concerted mechanism is allowed by the Woodward–Hoffmann rules. The transition state involves the π electrons of the alkene and enophile and the σ electrons of the C–H bond.

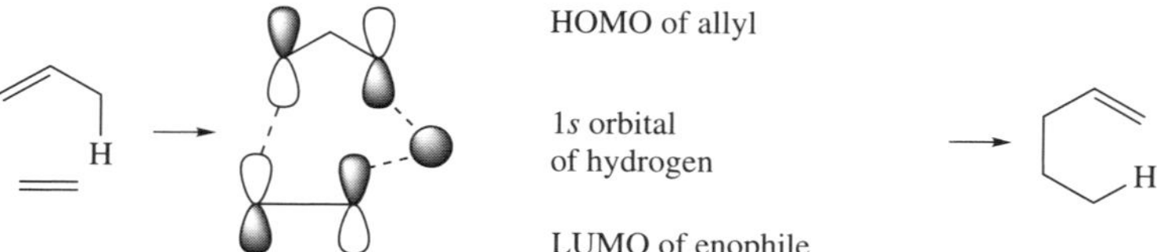

A concerted ene reaction corresponds to the interaction of a hydrogen atom with the HOMO of an allyl radical and the LUMO of the enophile and is allowed.

Ene reactions have relatively high activation energies and intermolecular reaction is observed only for strongly electrophilic enophiles. Some examples are given in Scheme 6.15.

The thermal ene reaction of carbonyl compounds generally requires electron-attracting substituents. Glyoxalate and oxomalonate esters are useful reagents for the ene reaction.[201,202] Mechanistic studies have been designed to determine if the concerted cyclic transition state is a good representation of the mechanism. The reaction is only moderately sensitive to electronic effects. The ρ value for reaction of diethyl oxomalonate with a series of 1-arylcyclopentenes is -1.2, which would indicate there is little charge development in the transition state. The reaction shows a primary kinetic isotope effect indicative of C−H bond-breaking in the rate-determining step.[203] These observations are consistent with a concerted process.

The ene reaction is strongly catalyzed by Lewis acids such as aluminum chloride and diethylaluminum chloride.[204] Coordination by the aluminum at the carbonyl group increases the electrophilicity of the conjugated system and allows reaction to occur below room temperature, as illustrated in Entry 6. Intramolecular ene reactions can be carried out under either thermal (Entry 3) or catalyzed (Entry 7) conditions.[205] Formaldehyde in acidic solution can form allylic alcohols, as in entry 1. Other carbonyl ene reactions are carried out with Lewis acid catalysts. Aromatic aldehydes and acrolein undergo the ene reaction with activated alkenes such as enol ethers in the presence of $Yb(fod)_3$.[206] $Sc(O_3SCF_3)_3$ has also been used to catalyze ene reactions.[207]

$=CH_2$ + $ArCH{=}O$ $\xrightarrow{Sc(O_3SCF_3)_3}$ $-CH_2CHAr$ / O_2CCF_3

With chiral catalysts, the reaction becomes enantioselective. Among the successful catalysts are diisopropoxyTi(IV) BINOL and copper bis-oxazoline complexes.

$(CH_3)_2C{=}CH_2 + O{=}CHCO_2CH_3 \xrightarrow{(i\text{-PrO})_2Ti/BINOL}$ OH, CO_2CH_3 Ref. 208

72% yield, 95% e.e.

$=CH_2 + O{=}CHCO_2C_2H_5$ → (O, N, Cu, N, O, t-Bu, t-Bu) $CO_2C_2H_5$, OH Ref. 209

95% yield, 96% e.e.

201. K. Mikami and M. Shimizu, *Chem. Rev.* **92:**1020 (1992).
202. M. F. Salomon, S. N. Pardo, and R. G. Salomon, *J. Org. Chem.* **49:**2446 (1984); M. F. Salomon, S. N. Pardo, and R. G. Salomon, *J. Am. Chem. Soc.* **106:**3797 (1984).
203. O. Achmatowicz and J. Szymoniak, *J. Org. Chem.* **45:**4774 (1980); H. Kwart and M. Brechbiel, *J. Org. Chem.* **47:**3353 (1982).
204. B. B. Snider, *Acc. Chem. Res.* **13:**426 (1980).
205. W. Oppolzer and V. Snieckus, *Angew. Chem. Int. Ed. Engl.* **17:**476 (1978).
206. M. A. Ciufolini, M. V. Deaton, S. Zhu, and M. Chen, *Tetrahedron* **53:**16299 (1997); M. A. Ciufolini and S. Zhu, *J. Org. Chem.* **63:**1668 (1998).
207. V. K. Aggarawal, G. P. Vennall, P. N. Davey, and C. Newman, *Tetrahedron Lett.* **39:**1997 (1998).
208. K. Mikami, M. Terada, and T. Nakai, *J. Am. Chem. Soc.* **112:**3949 (1990).
209. D. A. Evans, C. S. Burgey, N. A. Paras, T. Vojkovsky, and S. W. Tegay, *J. Am. Chem. Soc.* **120:**5824 (1998).

Scheme 6.15. Ene Reactions

1[a] CH_3; $CH_3C{=}CH_2$ + CH_2O $\xrightarrow[Ac_2O,\ CH_2Cl_2]{BF_3}$ CH_3; $H_2C{=}C(CH_2CH_2OCCH_3)$ (O) 84%

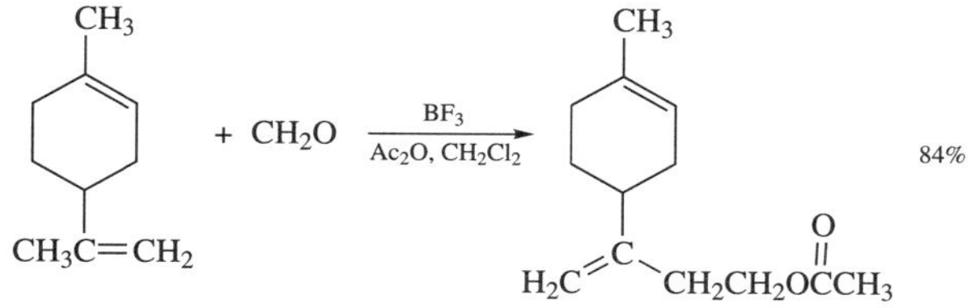

2[b] $(CH_3)_2C{=}CH_2$ + $HC{\equiv}CCO_2CH_3$ $\xrightarrow[25°C]{AlCl_3}$ $CH_2{=}C(CH_3)CH_2CH{=}CHCO_2CH_3$ (H_3C) 61%

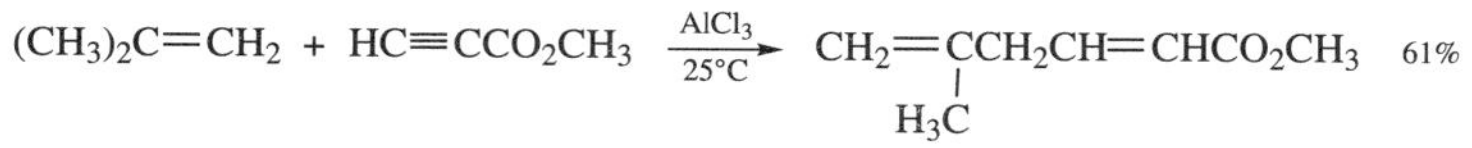

3[c] $CO_2C_2H_5$; $CHCH_2CH_2CH{=}CH_2$ $\xrightarrow{280°C}$ $CO_2C_2H_5$; CH_3 68%

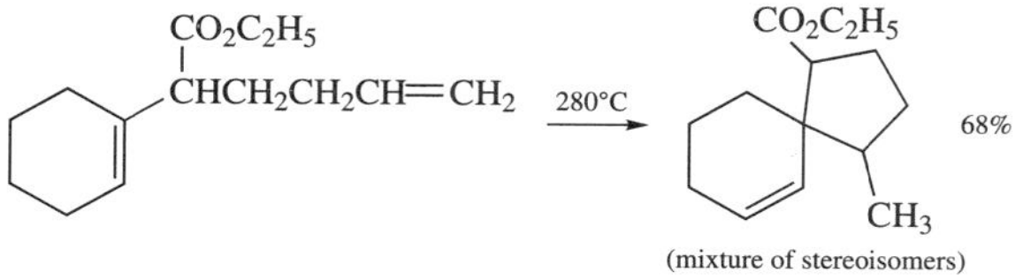

(mixture of stereoisomers)

4[d] $(CH_3)_2C{=}CHCH_2$; O; $NCCF_3$; EtO_2C; $C(CO_2Et)_2$; C=C; H; H $\xrightarrow[-78°C]{Et_2AlCl}$ $O{=}CCF_3$; N; CO_2Et; CO_2Et; CH_3; C; CH_2; CH_2CO_2Et 90%

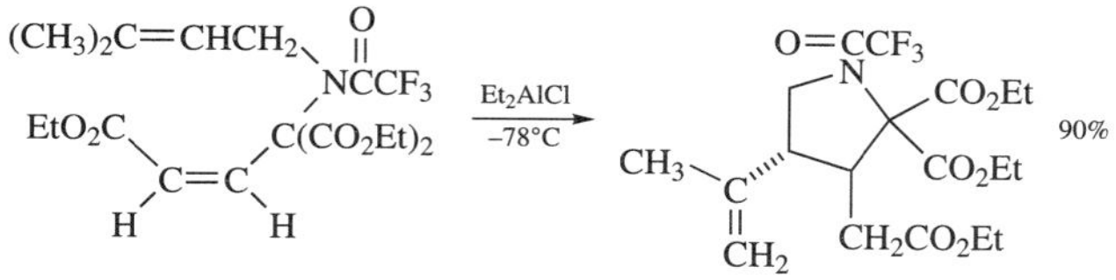

5[e] $(C_2H_5)_2C{=}CH_2$ + $O{=}CHCO_2C(CH_3)_3$ $\xrightarrow{(i\text{-PrO})_2TiCl_2}$ CH_3; C_2H_5; OH; $CO_2C(CH_3)_3$

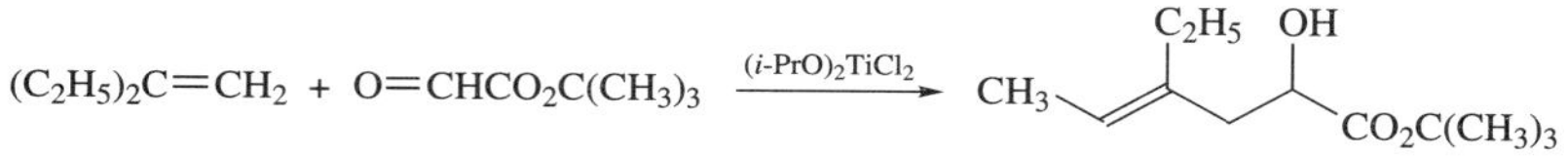

6[f] CH_2; OTBDMS + $O{=}CHC{\equiv}CCO_2CH_3$ $\xrightarrow[0°C]{20\ mol\ \%(i\text{-PrO})_2TiCl_2,\ BINOL}$ OH; CO_2CH_3; OTBDMS

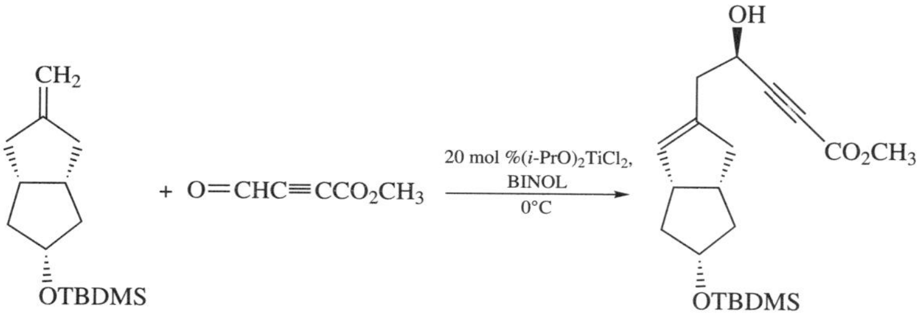

7[g] TBDPSO; $CO_2C_2H_5$; $CO_2C_2H_5$ $\xrightarrow{ZnBr_2}$ TBDPSO; $CH(CO_2C_2H_5)_2$ 90%

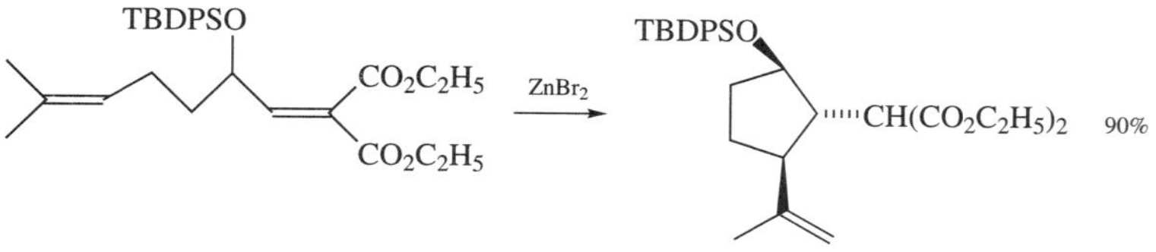

a. A. T. Blomquist and R. J. Himics, *J. Org. Chem.* **33**:1156 (1968).
b. B. B. Snider, D. J. Rodini, R. S. E. Conn, and S. Sealfon, *J. Am. Chem. Soc.* **101**:5283 (1979).
c. W. Oppolzer, K. K. Mahalanabis, and K. Battig, *Helv. Chim. Acta* **60**:2388 (1977).
d. W. Oppolzer and C. Robbiani, *Helv. Chim. Acta* **63**:2010 (1980).
e. M. A. Brimble and M. K. Edmonds, *Synth. Commun.* **26**:243 (1996).
f. K. Mikami, A. Yoshida, and Y. Matsumoto, *Tetrahedron Lett.* **37**:8515 (1996).
g. T. K. Sarkar, B. K. Ghorai, S. K. Nandy, B. Mukherjee, and A. Banerji, *J. Org. Chem.* **62**:6006 (1997).

Intramolecular ene reactions can also be carried out with Lewis acid catalysts. Several examples are included in Scheme 6.15. Mechanistic analysis of Lewis acid-catalyzed reactions indicates they may more closely resemble an electrophilic substitution process related to reactions which will be discussed in Section 10.1.1.[210]

6.8. Unimolecular Thermal Elimination Reactions

This section will describe reactions in which elimination to form a double bond or a new ring occurs as a result of thermal activation. There are several such thermal elimination reactions which find use in synthesis. Some of these are concerted processes. The transition-state energy requirements and stereochemistry of concerted elimination processes can be analyzed in terms of orbital symmetry considerations. We will also consider an important group of unimolecular β-elimination reactions in Section 6.8.3.

6.8.1. Cheletropic Elimination

Cheletropic processes are defined as reactions in which two bonds are broken at a single atom. Concerted cheletropic reactions are subject to orbital symmetry restrictions in the same way that cycloadditions and sigmatropic processes are.

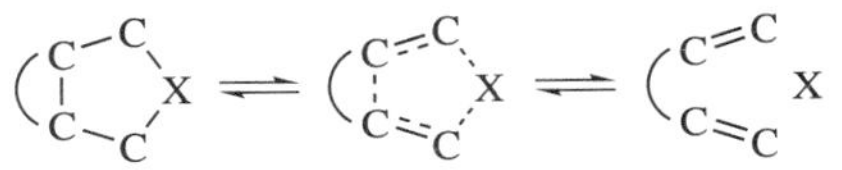

In the elimination processes of interest here, the atom X is normally bound to other atoms in such a way that elimination will give rise to a stable molecule. The most common examples involve five-membered rings.

A good example of a concerted cheletropic elimination is the reaction of 3-pyrroline with N-nitrohydroxylamine, which gives rise to a diazene, that then undergoes elimination of nitrogen.

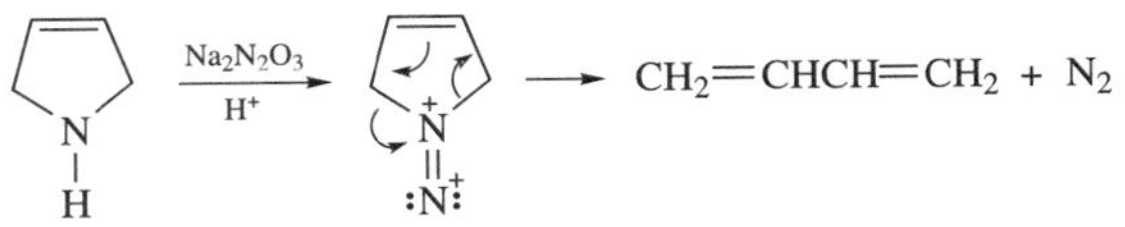

$$\xrightarrow[H^+]{Na_2N_2O_3} \longrightarrow CH_2{=}CHCH{=}CH_2 + N_2$$

Use of substituted systems has shown that the reaction is completely stereospecific.[211] The groups on C-2 and C-5 of the pyrroline ring rotate in the disrotatory mode on going to

210. B. B. Snider, D. M. Roush, D. J. Rodini, D. M. Gonzalez, and D. Spindell, *J. Org. Chem.* **45:**2773 (1980); J. V. Duncia, P. T. Lansbury, Jr., T. Miller, and B. B. Snider, *J. Org. Chem.* **47:**4538 (1982); B. B. Snider and G. B. Phillips, *J. Org. Chem.* **48:**464 (1983); B. B. Snider and E. Ron, *J. Am. Chem. Soc.* **107:**8160 (1985); O. Achmatowicz and E. Bialecka-Florjanczyk, *Tetrahedron* **52:**8827 (1996).
211. D. M. Lemal and S. D. McGregor, *J. Am. Chem. Soc.* **88:**1335 (1966).

product. This stereochemistry is consistent with conservation of orbital symmetry.

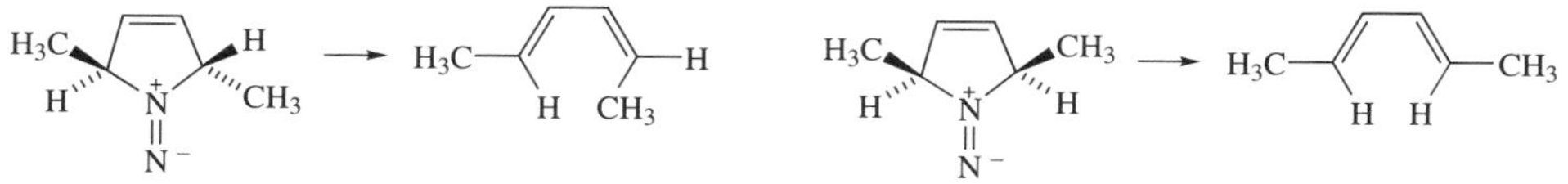

The most synthetically useful cheletropic elimination involves 2,5-dihydrothiophene-1,1-dioxides (sulfolene dioxides). At elevated temperatures, they fragment to give dienes and sulfur dioxide.[212] The reaction is stereospecific. For example, the dimethyl derivatives **13** and **14** give the *E,E*- and *Z,E*-isomers of 2,4-hexadiene, respectively, at temperatures of 100–150°C.[213] This sterespecificity corresponds to disrotatory elimination.

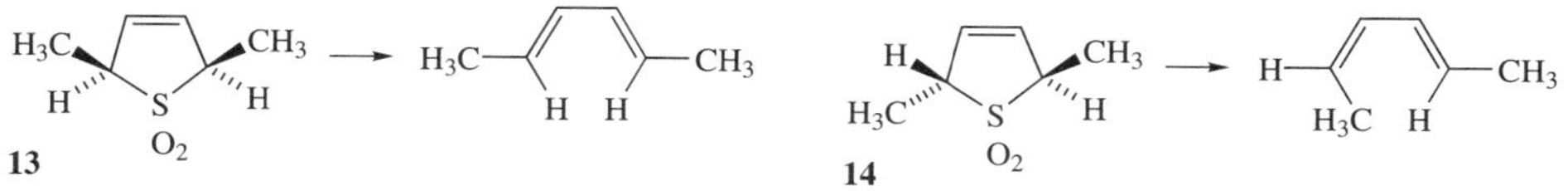

13 **14**

Elimination of sulfur dioxide has proven to be a useful method for generating dienes which can undergo subsequent Diels–Alder addition.

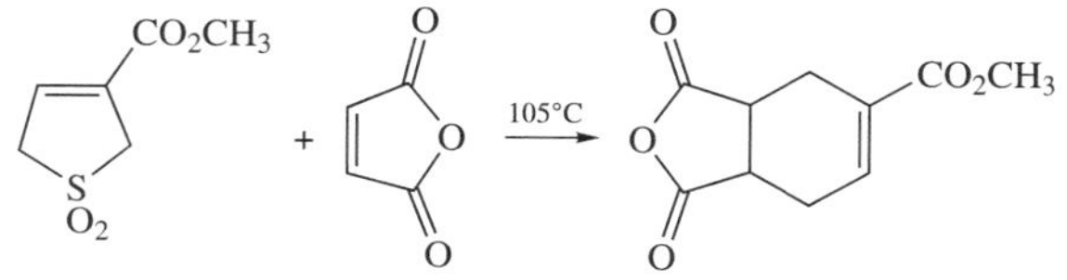

Ref. 214

The method is particularly useful in formation of *o*-quinodimethanes.

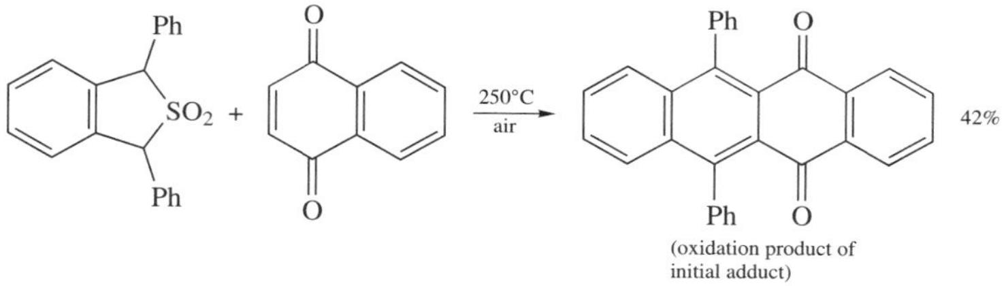

(oxidation product of initial adduct)

42% Ref. 215

85% Ref. 216

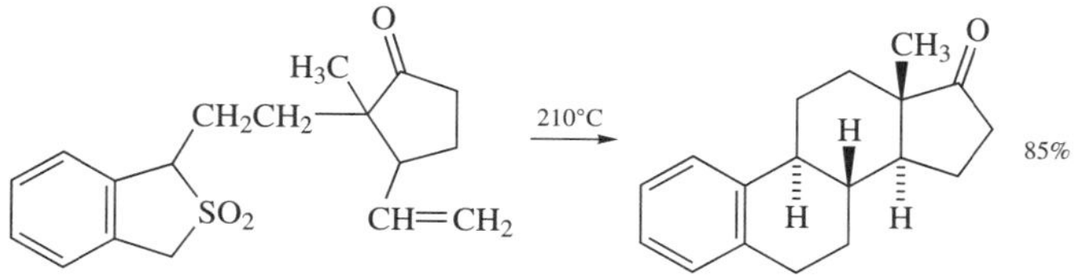

The elimination of carbon monoxide can occur by a concerted process in some cyclic ketones. The elimination of carbon monoxide from bicyclo[2.2.1]heptadien-7-ones is very

212. W. L. Mock, in *Pericyclic Reactions*, Vol. II, A. P. Marchand and R. E. Lehr, eds., Academic Press, New York, 1977, Chapter 3.
213. W. L. Mock, *J. Am. Chem. Soc.* **88:**2857 (1966); S. D. McGregor and D. M. Lemal, *J. Am. Chem. Soc.* **88:**2858 (1966).
214. J. M. McIntosh and R. A. Sieler, *J. Org. Chem.* **43:**4431 (1978).
215. M. P. Cava, M. J. Mitchell, and A. A. Deana, *J. Org. Chem.* **25:**1481 (1960).
216. K. C. Nicolaou, W. E. Barnette, and P. Ma, *J. Org. Chem.* **45:**1463 (1980).

facile. In fact, generation of bicyclo[2.2.1]heptadien-7-ones is usually accompanied by spontaneous elimination.

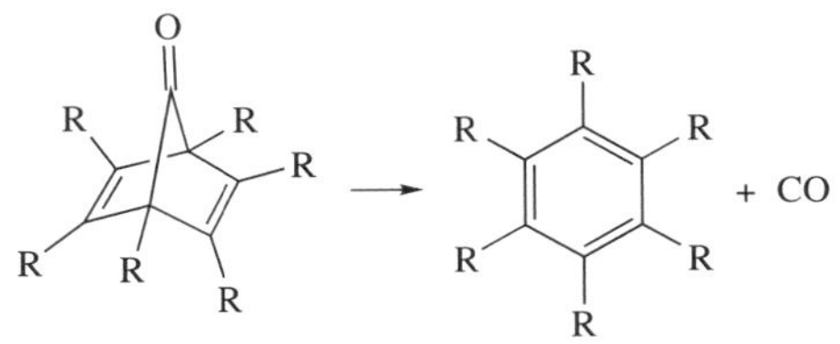

The ring system can be generated by Diels–Alder addition of a substituted cyclopentadienone and an alkyne. A reaction sequence involving addition followed by CO elimination can be used for the synthesis of highly substituted benzene rings.[217]

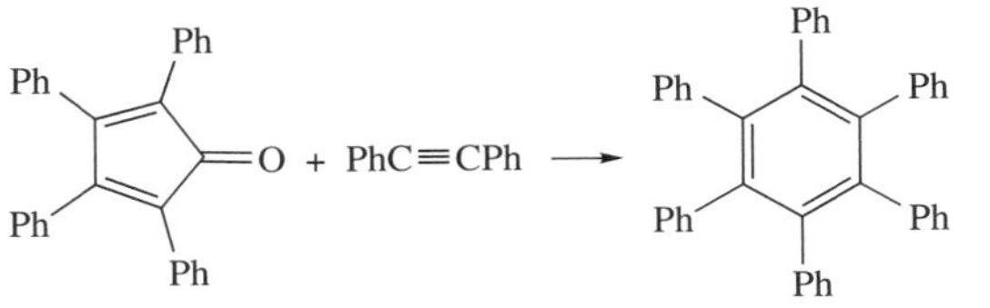

Ref. 218

Exceptionally facile elimination of CO also takes place from **15**, in which homoaromaticity can stabilize the transition state:

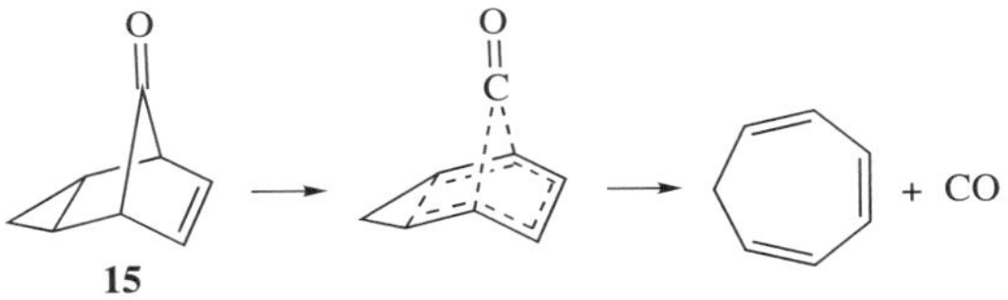

Ref. 219

6.8.2. Decomposition of Cyclic Azo Compounds

Another significant group of elimination reactions involves processes in which a small molecule is eliminated from a ring system and the two reactive sites that remain react to re-form a ring.

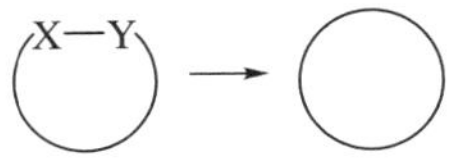

The most widely studied example is decomposition of azo compounds, where –X–Y– is –N=N–.[220] The elimination of nitrogen from cyclic azo compounds can be carried out either photochemically or thermally. Although the reaction generally does not proceed by a concerted mechanism, there are some special cases in which concerted elimination is possible. We will consider some of these cases first and then consider the more general case.

217. M. A. Ogliaruso, M. G. Romanelli, and E. I. Becker, *Chem. Rev.* **65**:261 (1965).
218. L. F. Fieser, *Org. Synth.* **V**:604 (1973).
219. B. A. Halton, M. A. Battiste, R. Rehberg, C. L. Deyrup, and M. E. Brennan, *J. Am. Chem. Soc.* **89**:5964 (1967).
220. P. S. Engel, *Chem. Rev.* **80**:99 (1980).

An interesting illustration of the importance of orbital symmetry effects is the contrasting stability of azo compounds **16** and **17**. Compound **16** decomposes to norbornene and nitrogen only above 100°C. In contrast **17** eliminates nitrogen immediately on preparation, even at −78°C.[221]

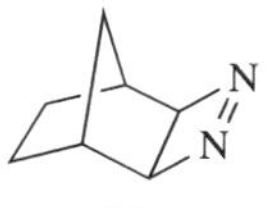

16 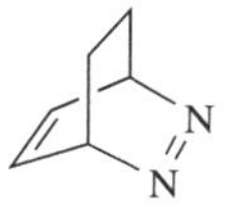 **17**

The reason for this difference is that if **16** were to undergo a concerted elimination, it would have to follow the forbidden (high-energy) $[2\pi_s + 2\pi_s]$ pathway. For **17**, the elimination can take place by the allowed $[2\pi_s + 4\pi_s]$ pathway. Thus, these reactions are the reverse of, respectively, the [2 + 2] and [4 + 2] cycloadditions, and only the latter is an allowed concerted process. The temperature at which **16** decomposes is fairly typical for strained azo compounds, and the decomposition presumably proceeds by a nonconcerted diradical mechanism. Because a C−N bond must be broken without concomitant compensation by carbon–carbon bond formation, the activation energy is much higher than for a concerted process.

Although the concerted mechanism is available only to those azo compounds with appropriate orbital arrangements, the nonconcerted mechanism occurs at low enough temperatures to be synthetically useful. The elimination can also be carried out photochemically. These reactions presumably occur by stepwise elimination of nitrogen.

$$\text{R—N=N—R}' \xrightarrow{\text{slow}} \text{R—N=N}\cdot + \cdot\text{R}' \xrightarrow{\text{fast}} \text{R}\cdot\ \text{N}\equiv\text{N}\ \cdot\text{R}' \longrightarrow \text{R—R}'$$

The stereochemistry of the nonconcerted reaction has been a topic of considerable study and discussion. Frequently, there is only partial randomization, indicating a short-lived diradical intermediate. The details vary from case to case, and both preferential inversion and retention of relative stereochemistry have been observed.

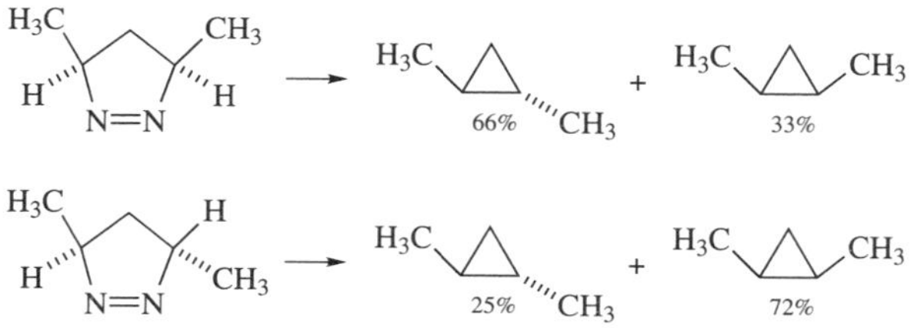

66% 33%

Ref. 222

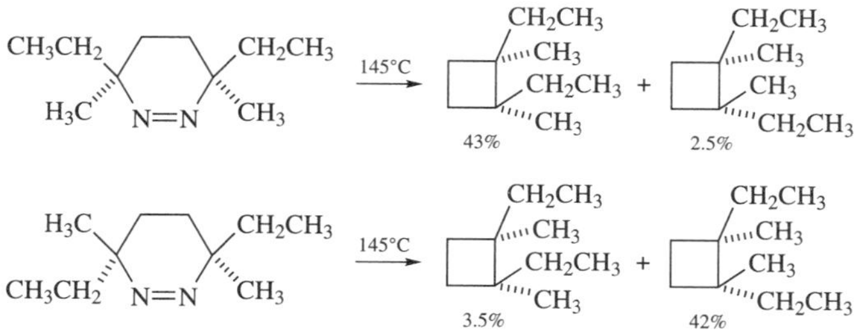

25% 72%

145°C

43% 2.5%

Ref. 223

145°C

3.5% 42%

These results can be interpreted in terms of competition between recombination of the

221. N. Rieber, J. Alberts, J. A. Lipsky, and D. M. Lemal, *J. Am. Chem. Soc.* **91**:5668 (1969).
222. R. J. Crawford and A. Mishra, *J. Am. Chem. Soc.* **88**:3963 (1966).
223. P. D. Bartlett and N. A. Porter, *J. Am. Chem. Soc.* **90**:5317 (1968).

diradical intermediate and conformational equilibration which would destroy the stereochemical relationships present in the azo compound. The main synthetic application of azo compound decomposition is in the synthesis of cyclopropanes and other strained ring systems. Some of the required azo compounds can be made by dipolar cycloadditions of diazo compounds.

Elimination of nitrogen from Diels–Alder adducts of certain heteroaromatic rings has been useful in the synthesis of substituted aromatic compounds.[224] Pyridazines, triazines, and tetrazines react with electron-rich dienophiles in inverse-electron-demand cycloadditions. The adducts then rearomatize with loss of nitrogen and the dienophile substituent.[225]

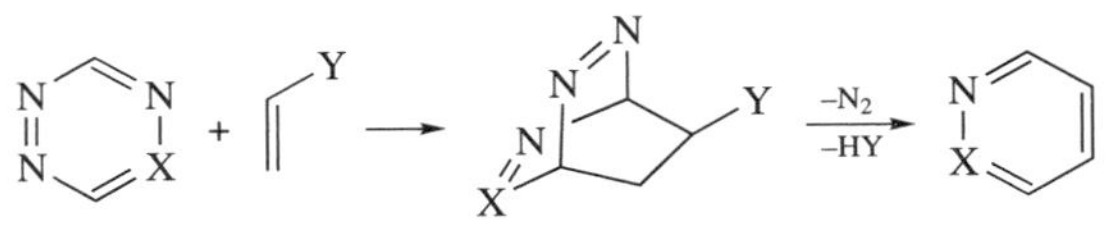

Pyridazine-3,6-dicarboxylate esters react with electron-rich alkenes to give adducts that undergo subsequent elimination to give benzene derivatives.[226]

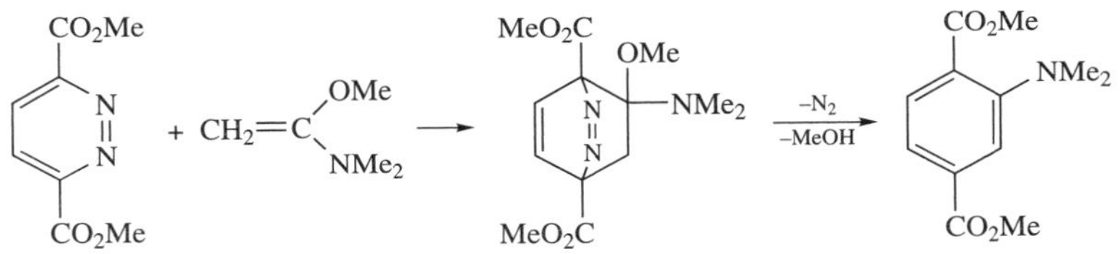

Similar reactions have been developed for 1,2,4-triazines and 1,2,4,5-tetrazines.

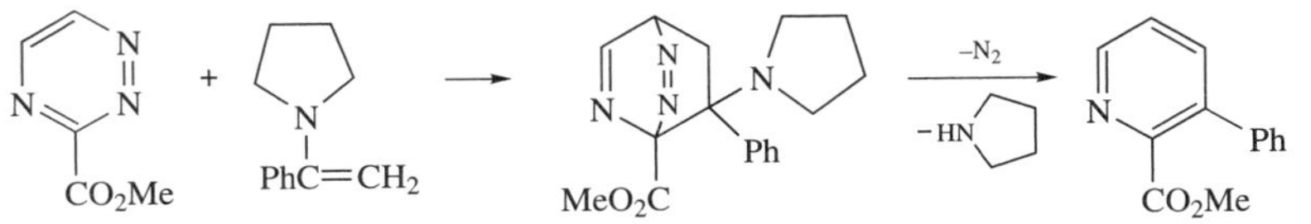

Ref. 227

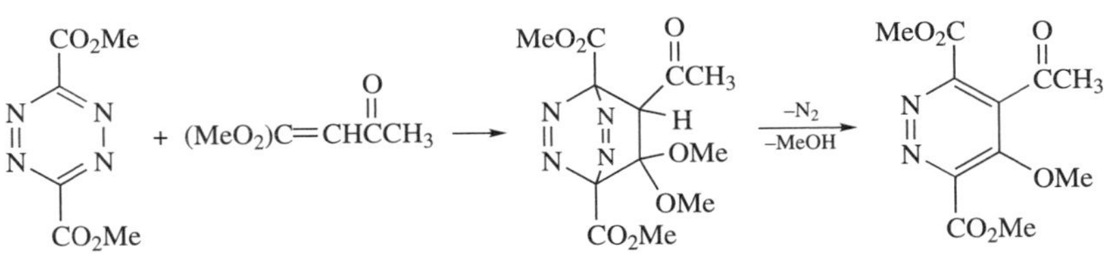

Ref. 228

224. D. L. Boger, *Chem. Rev.* **86:**781 (1986).
225. D. L. Boger, *J. Heterocycl Chem.* **33:**1519 (1996).
226. H. Neunhoeffer and G. Werner, *Justus Liebigs Ann. Chem.* **1973:** 1955.
227. D. L. Boger and J. S. Panek, *J. Am. Chem. Soc.* **107:**5745 (1985).
228. D. L. Boger and R. S. Coleman, *J. Am. Chem. Soc.* **109:**2717 (1987).

The heterocycles frequently carry substituents such as chloro, methylthio, or alkoxycarbonyl.

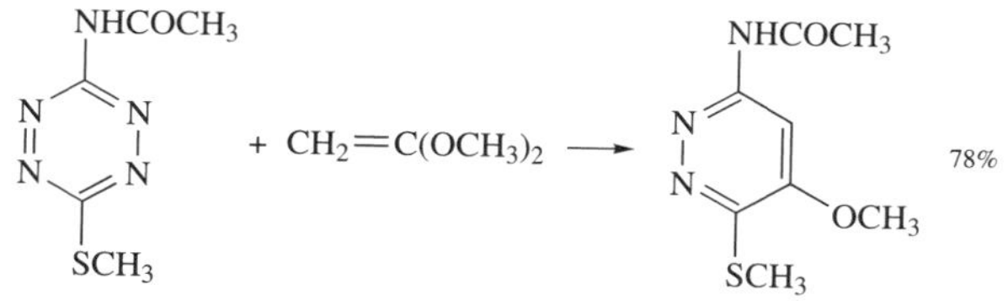

Ref. 229

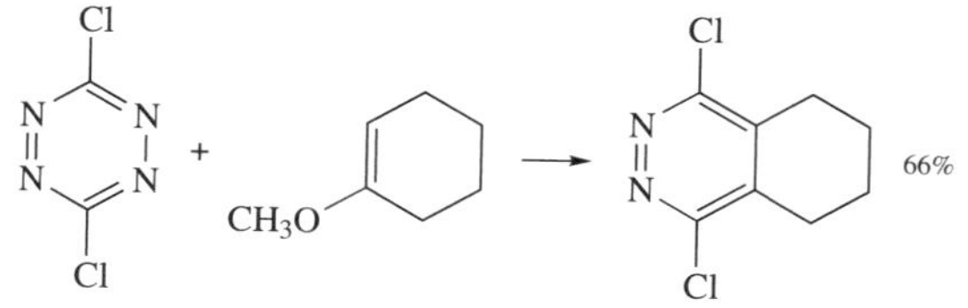

Ref. 230

Acetylenic dienophiles lead directly to aromatic adducts on loss of nitrogen.

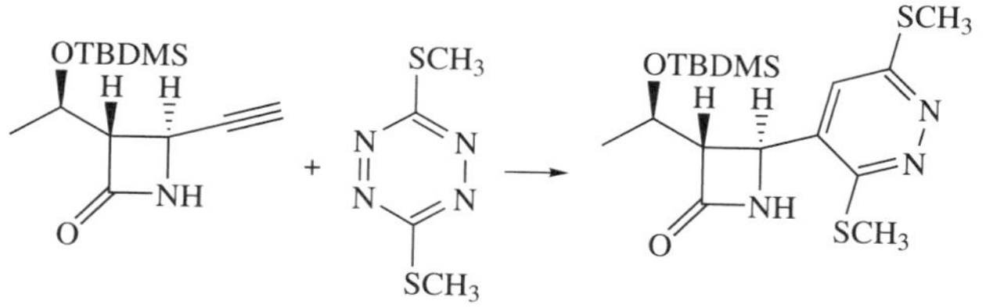

Ref. 231

6.8.3. β Eliminations Involving Cyclic Transition States

Another important family of elimination reactions has as the common mechanistic feature cyclic transition states in which an intramolecular proton transfer accompanies elimination to form a new carbon–carbon double bond. Scheme 6.16 depicts examples of the most important of these reaction types. These reactions are thermally activated unimolecular reactions that normally do not involve acidic or basic catalysts. There is, however, a wide variation in the temperature at which elimination proceeds at a convenient rate. The cyclic transition states dictate that elimination occurs with *syn* stereochemistry. At least in a formal sense, all the reactions can proceed by a concerted mechanism. The reactions, as a group, are referred to as *thermal syn eliminations*.

Amine oxide pyrolysis occurs at temperatures of 100–150°C. The reaction can proceed at room temperature in DMSO.[232] If more than one type of β hydrogen can attain the eclipsed conformation of the cyclic transition state, a mixture of alkenes will be formed. The product ratio parallels the relative stability of the competing transition states. Usually, more of the E-alkene is formed because of the additional eclipsed interactions present in the transition state leading to the Z-alkene. The selectivity is usually not high,

229. D. L. Boger, R. P. Schaum, and R. M. Garbaccio, *J. Org. Chem.* **63:**6329 (1998).
230. T. J. Sparey and T. Harrison, *Tetrahedron Lett.* **39:**5873 (1998).
231. S. M. Sakya, T. W. Strohmeyer, S. A. Lang, and Y.-I. Lin, *Tetrahedron Lett.* **38:**5913 (1997).
232. D. J. Cram, M. R. V. Sahyun, and G. R. Knox, *J. Am. Chem. Soc.* **84:**1734 (1962).

Scheme 6.16. Elimination via Cyclic Transition States

Reactant	Transition state	Product	Temperature range
1[a] R—CH(H)—CHR(O⁻—N⁺(CH₃)₂)	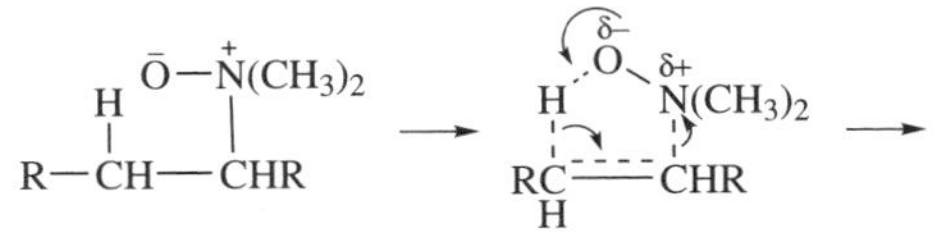 cyclic (δ−O···δ+N(CH₃)₂)	RCH=CHR + $HON(CH_3)_2$	100–150°C
2[b] R—CH(H)—CHR(O⁻—Se⁺R′)	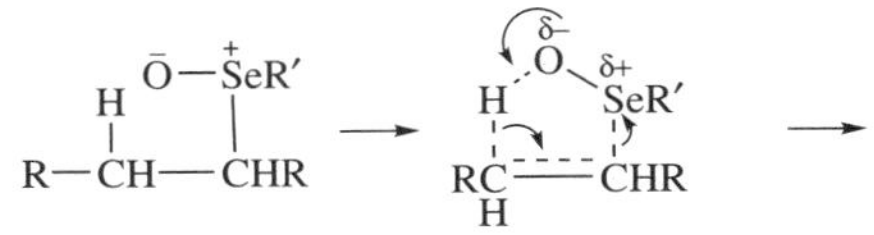 cyclic (δ−O···δ+SeR′)	RCH=CHR + HOSeR′	0–100°C
3[c] R—CH(H)—CHR—O—C(=O)CH_3	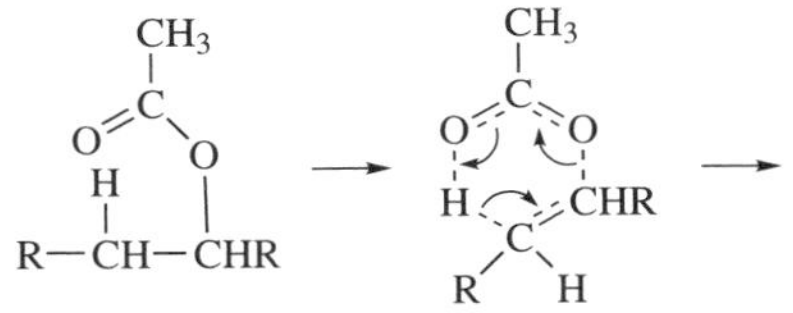 cyclic six-membered	RCH=CHR + CH_3CO_2H	400–600°C
4[d] R—CH(H)—CHR—O—C(=S)SCH_3	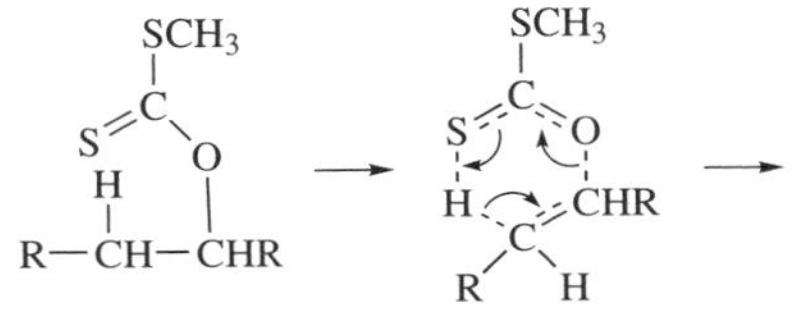 cyclic six-membered	RCH=CHR + CH_3SH + SCO	150–250°C

a. A. C. Cope and E. R. Trumbull, *Org. React.* **11**:317 (1960).
b. D. L. J. Clive, *Tetrahedron* **34**:1049 (1978).
c. C. H. De Puy and R. W. King, *Chem. Rev.* **60**:431 (1960).
d. H. R. Nace, *Org. React.* **12**:57 (1962).

however.

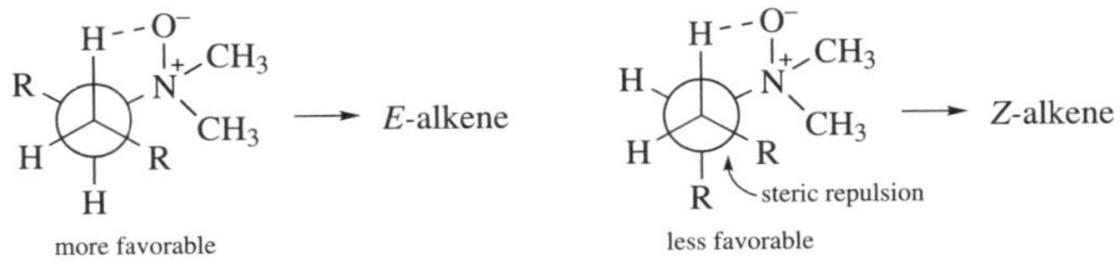

H- -O⁻, N⁺ CH_3, CH_3, R, H, R, H → *E*-alkene (more favorable)

H- -O⁻, N⁺ CH_3, CH_3, H, H, R, R → *Z*-alkene (less favorable; steric repulsion)

In cyclic systems, conformational effects and the requirement for a cyclic transition state determine the product composition. This effect can be seen in the product ratios from pyrolysis of *N,N*-dimethyl-2-phenylcyclohexylamine *N*-oxide.

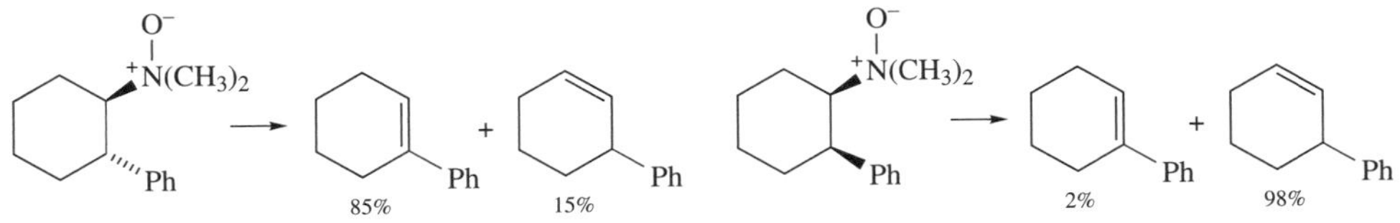

O⁻ ⁺N$(CH_3)_2$, Ph → 1-phenylcyclohexene (Ph) 85% + 3-phenylcyclohexene (Ph) 15%

O⁻ ⁺N$(CH_3)_2$, Ph → 1-phenylcyclohexene (Ph) 2% + 3-phenylcyclohexene (Ph) 98%

Elimination to give a double bond conjugated with an aromatic ring is especially favorable. This presumably reflects both the increased acidity of the proton α to the phenyl ring and the stabilizing effect of the developing conjugation at the transition state. Amine oxides can be readily prepared from amines by oxidation with hydrogen peroxide or a peroxycarboxylic acid. Some typical examples are given in section A of Scheme 6.17.

Selenoxides are even more reactive than amine oxides toward β elimination. In fact, many selenoxides react spontaneously when generated at room temperature. Synthetic procedures based on selenoxide eliminations usually involve synthesis of the corresponding selenide followed by oxidation and *in situ* elimination. We have already discussed examples of these procedures in Section 4.7, where the conversion of ketones and esters to their α,β-unsaturated derivatives was considered. Selenides can also be prepared by electrophilic addition of selenenyl halides and related compounds to alkenes (see Section 4.5). Selenide anions are powerful nucleophiles that can displace halides or tosylates and open epoxides.[233] Selenide substituents stabilize an adjacent carbanion so that α-selenenyl carbanions can be prepared. One versatile procedure involves conversion of a ketone to a bis-selenoketal which can then be cleaved by n-butyllithium.[234] The carbanions in turn add to ketones to give β-hydroxyselenides.[235] Elimination gives an allylic alcohol.

$$RCH_2C(R')=O + 2\,PhSeH \longrightarrow RCH_2C(R')(SePh)_2 \xrightarrow{BuLi} RCH_2C(Li)(R')SePh \xrightarrow{R''CH=O} RCH_2C(R')(SePh)-CH(OH)R'' \xrightarrow{[O]} RCH=C(R')-CH(OH)R''$$

Alcohols can be converted to o-nitrophenylselenides by reaction with o-nitrophenyl selenocyanate and tri-n-butylphosphine.[236]

$$RCH_2OH + o\text{-}O_2NC_6H_4SeCN \xrightarrow{Bu_3P} RCH_2Se\text{-}C_6H_4\text{-}o\text{-}NO_2$$

The selenides prepared by any of these methods can be converted to selenoxides by such oxidants as hydrogen peroxide, sodium metaperiodate, peroxycarboxylic acids, t-butyl hydroperoxide, or ozone.

Like amine oxide eliminations, selenoxide eliminations normally favor formation of the E-isomer in acyclic structures. In cyclic systems, the stereochemical requirements of the cyclic transition state govern the product structure. Section B of Scheme 6.17 gives some examples of selenoxide eliminations.

A third category of *syn* eliminations involves pyrolytic decomposition of esters with elimination of a carboxylic acid. The pyrolysis of acetate esters normally requires temperatures above 400°C. The pyrolysis is usually a vapor-phase reaction. In the

233. D. L. J. Clive, *Tetrahedron* **34:**1049 (1978).
234. W. Dumont, P. Bayet, and A. Krief, *Angew. Chem. Int. Ed. Engl.* **13:**804 (1974).
235. D. Van Ende, W. Dumont, and A. Krief, *Angew. Chem. Int. Ed. Engl.* **14:**700 (1975); W. Dumont, and A. Krief, *Angew. Chem. Int. Ed. Engl.* **14:**350 (1975).
236. P. A. Grieco, S. Gilman, and M. Nishizawa, *J. Org. Chem.* **41:**1485 (1976); A. Krief and A.-M. Laval, *Bull. Soc. Chim. Fr.* **134:**869 (1997).

Scheme 6.17. Thermal Eliminations via Cyclic Transition States

A. Amine oxide pyrolyses

1[a] $PhCH(CH_3)CH(CH_3)\overset{+}{N}(CH_3)_2(O^-)$ → $PhC(CH_3){=}CHCH_3$ 92% + $PhCH(CH_3)CH{=}CH_2$ 8%

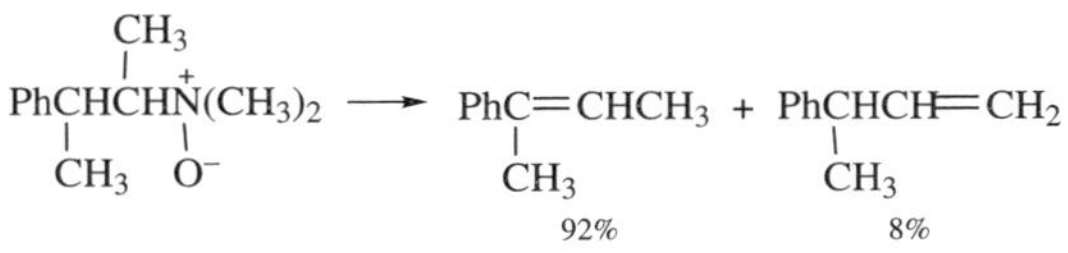

2[b] $C_6H_{11}CH_2\overset{+}{N}(CH_3)_2(O^-)$ → methylenecyclohexane ($=CH_2$) 85%

3[c] $\overset{+}{N}(CH_3)_2$, O^- → 67%

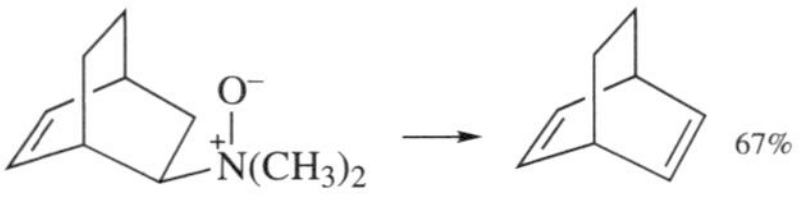

B. Selenoxide elimination

4[d] $CH_2CH_2OSO_2Ar$, O, O, H —1) $PhSe^-$, 2) O_3→ $CH_2CH_2\overset{O}{\overset{\|}{Se}}Ph$, O, O, H —77°C, CCl_4, 10 min→ $CH{=}CH_2$, O, O, H 60%

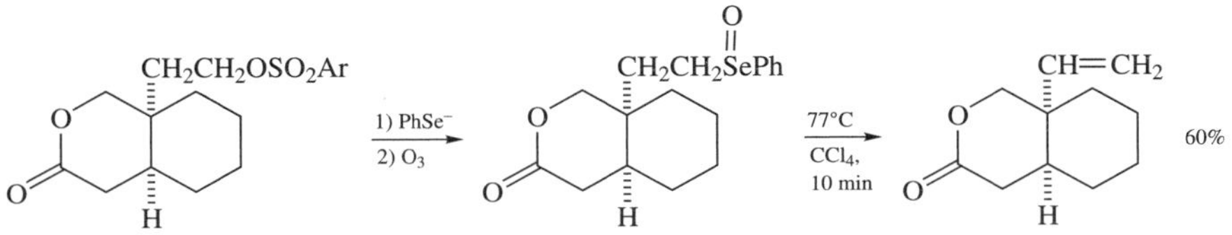

5[e] O, O —1) $PhSe^-$, 2) O_3→ CO_2H, $CH_2\overset{}{Se}Ph$ (=O) —3) pyridine, 4) H^+→ CO_2H, CH_2

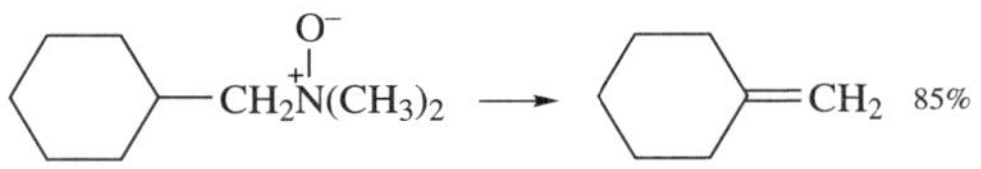

6[f] CH_2CO_2H —$PhSeCl$, Et_3N, CH_2Cl_2→ O, O, PhSe 93% —H_2O_2, THF→ O, O 92%

7[g] O, OH, H —1) NO_2, SeCN, Ph_3P; 2) O_3→ O, Se, O, O_2N —DBU→ O 5%

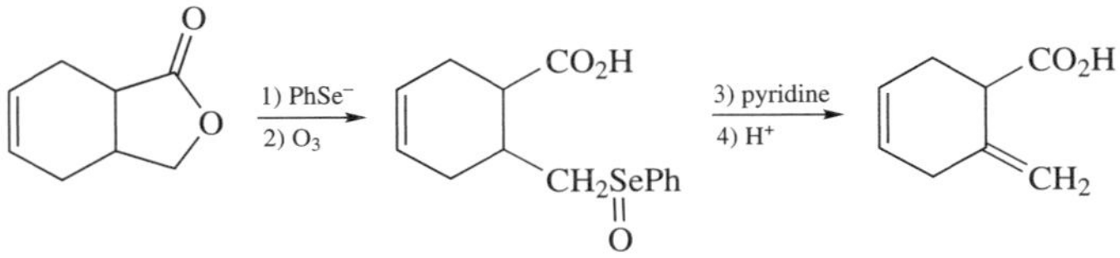

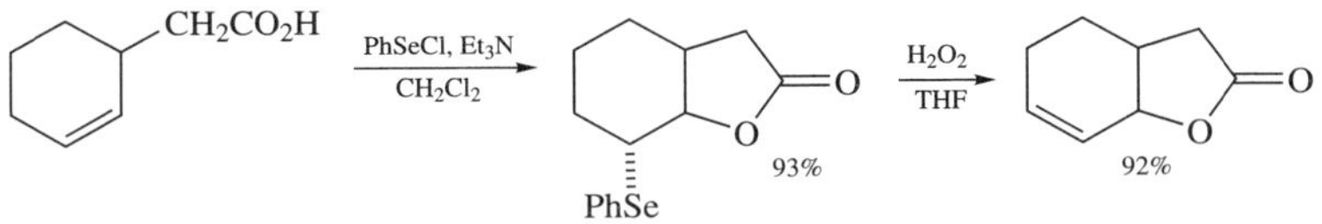

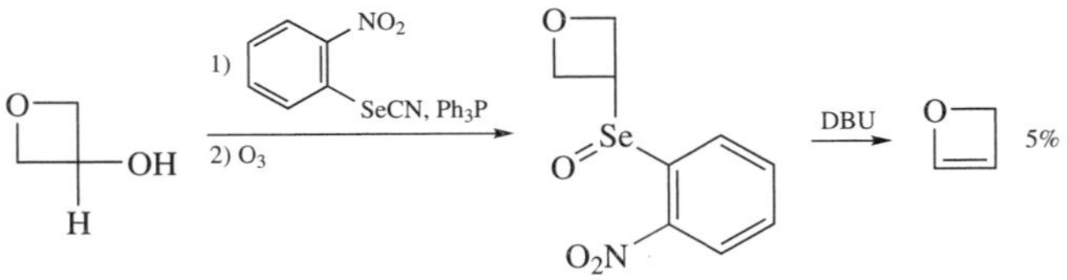

C. Acetate pyrolyses

8[h] $N{\equiv}C{-}C_6H_4{-}CH(O_2CCH_3)CH_3$ —575–600°C→ $N{\equiv}C{-}C_6H_4{-}CH{=}CH_2$ 76%

9[i] $CH_2O_2CCH_3$, $CH_2O_2CCH_3$ —555°C→ CH_2, CH_2 61% + CH_2, $CH_2O_2CCH_3$ 24%

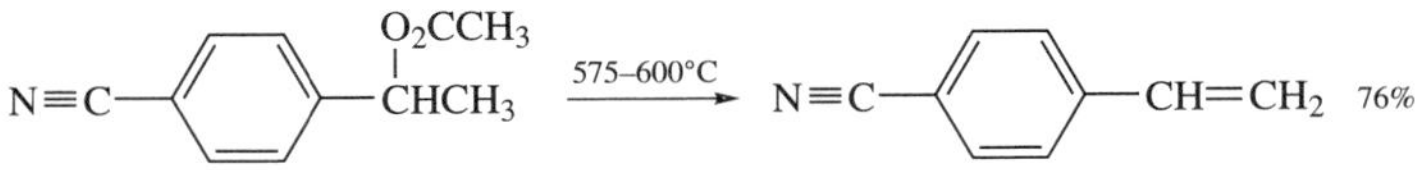

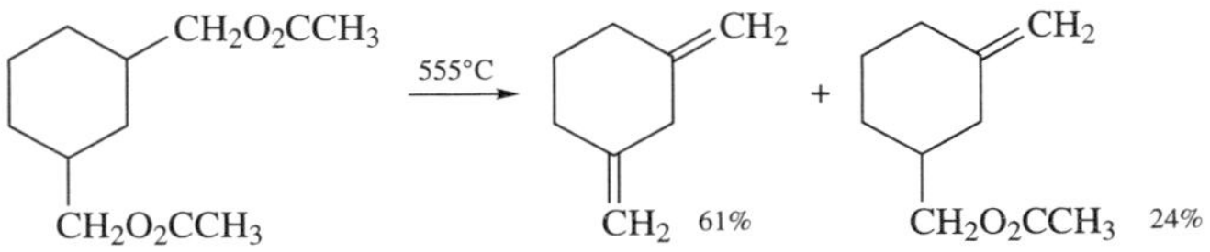

(continued)

Scheme 6.17. *(continued)*

10[j] (keto-ester) $\xrightarrow{400°C}$ (alkene); CH_3, CH_3, CH_3, $CH_2O_2CCH_3$, O → CH_3, CH_3, CH_3, CH_2, O

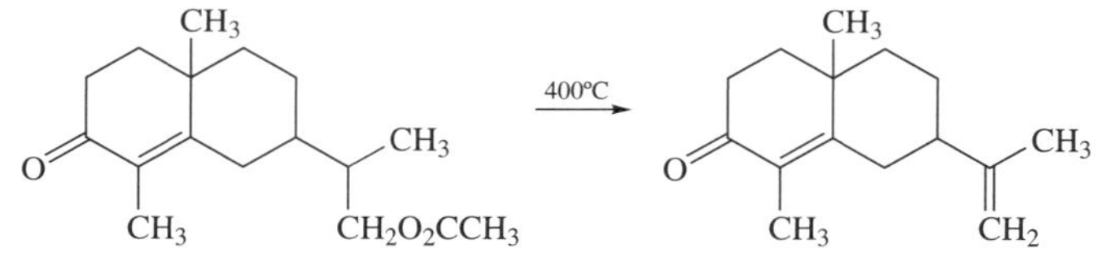

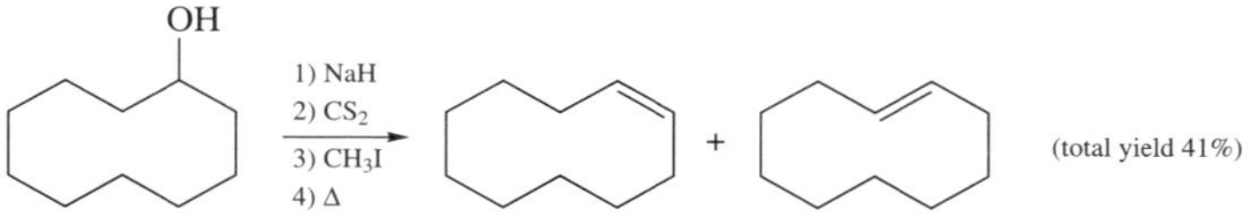

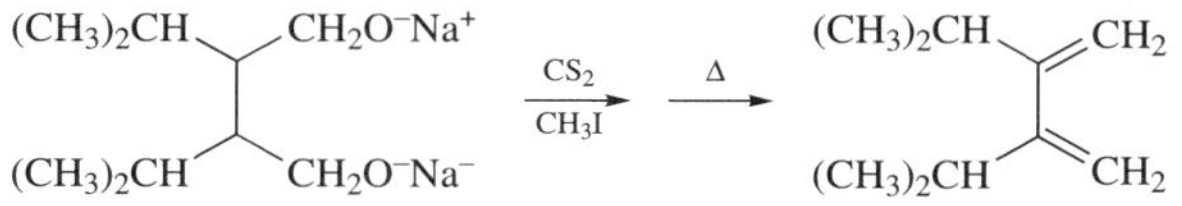

D. Xanthate ester pyrolyses

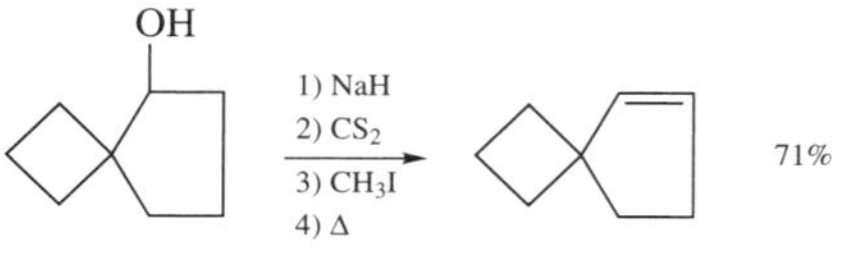

11[k] $PhCH(CH_3)CH(OH)CH_3$ $\xrightarrow{\text{1) K; 2) } CS_2\text{; 3) } CH_3I\text{; 4) }\Delta}$ $PhC(CH_3){=}CHCH_3$ 91%

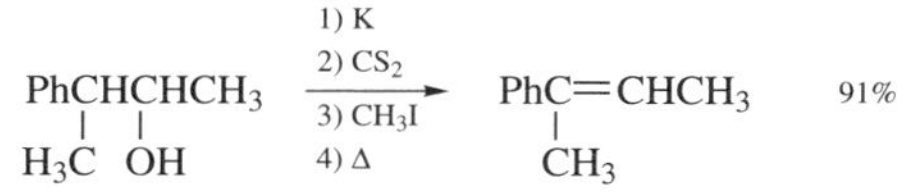

12[l] cyclodecanol (OH) $\xrightarrow{\text{1) NaH; 2) } CS_2\text{; 3) } CH_3I\text{; 4) }\Delta}$ cyclodecene + cyclodecene (total yield 41%)

13[m] $(CH_3)_2CH$, $CH_2O^-Na^+$, $(CH_3)_2CH$, $CH_2O^-Na^-$ $\xrightarrow{CS_2,\ CH_3I}$ $\xrightarrow{\Delta}$ $(CH_3)_2CH$, CH_2, $(CH_3)_2CH$, CH_2

14[n] OH (spiro[3.4]octanol) $\xrightarrow{\text{1) NaH; 2) } CS_2\text{; 3) } CH_3I\text{; 4) }\Delta}$ (spiro alkene) 71%

15[o] (bicyclic ketone, OH) $\xrightarrow{\text{1) NaH, } CS_2\text{, 2) } CH_3I\text{; 3) }\Delta}$ (bicyclic ketone alkene, O) 60%

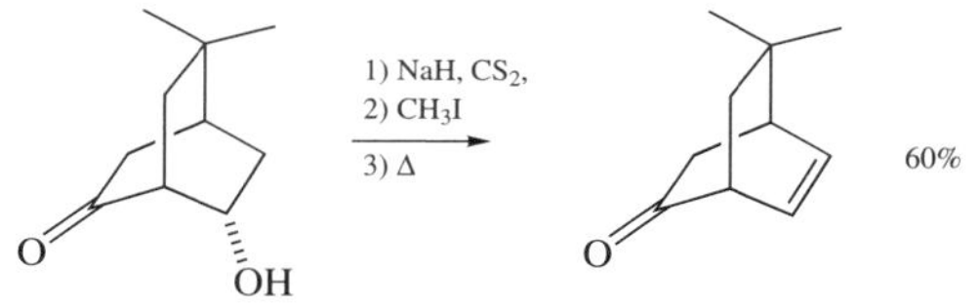

a. D. J. Cram and J. E. McCarty, *J. Am. Chem. Soc.* **76**:5740 (1954).
b. A. C. Cope, E. Ciganek, and N. A. LeBel, *J. Am. Chem. Soc.* **81**:2799 (1959).
c. A. C. Cope and C. L. Bumgardner, *J. Am. Chem. Soc.* **78**:2812 (1956).
d. R. D. Clark and C. H. Heathcock, *J. Org. Chem.* **41**:1396 (1976).
e. D. Liotta and H. Santiesteban, *Tetrahedron Lett.* **1977**:4369; R. M. Scarborough, Jr. and A. B. Smith III, *Tetrahedron Lett.* **1977**:4361.
f. K. C. Nicolaou and Z. Lysenko, *J. Am. Chem. Soc.* **99**:3185 (1977).
g. L. E. Friedrich and P. Y. S. Lam, *J. Org. Chem.* **46**:306 (1981).
h. C. G. Overberger and R. E. Allen, *J. Am. Chem. Soc.* **68**:722 (1946).
i. W. J. Bailey and J. Economy, *J. Org. Chem.* **23**:1002 (1958).
j. E. Piers and K. F. Cheng, *Can. J. Chem.* **46**:377 (1968).
k. D. J. Cram, *J. Am. Chem. Soc.* **71**:3883 (1949).
l. A. T. Blomquist and A. Golstein, *J. Am. Chem. Soc.* **77**:1001 (1955).
m. A. de Groot, B. Evenhuis, and H. Wynberg, *J. Org. Chem.* **33**:2214 (1968).
n. C. F. Wilcox, Jr. and C. G. Whitney, *J. Org. Chem.* **32**:2933 (1967).
o. L. A. Paquette and H.-C. Tsui, *J. Org. Chem.* **61**:142 (1996).

laboratory, this can be carried out by using a glass tube in the heating zone of a small furnace. The vapors of the reactant are swept through the hot chamber by an inert gas and into a cold trap. Similar reactions occur with esters derived from long-chain acids. If the boiling point of the ester is above the decomposition temperature, the reaction can be carried out in the liquid phase.

Ester pyrolysis has been shown by use of deuterium labels to be a *syn* elimination in the case of formation of stilbene.[237]

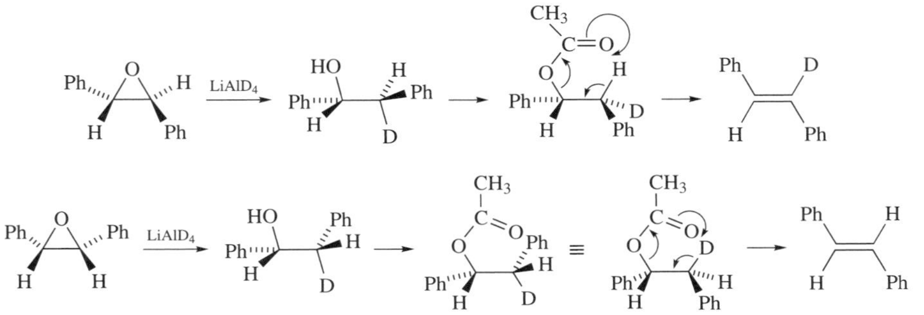

Although the existence of the concerted cyclic mechanism is recognized, it has been proposed that most preparative pyrolyses proceed as surface-catalyzed reactions.[238]

Mixtures of alkenes are formed when more than one type of β hydrogen is present. In acyclic compounds, the product composition often approaches that expected on a statistical basis from the number of each type of hydrogen. The *E*-alkene usually predominates over the *Z*-alkene for a given isomeric pair. In cyclic structures, elimination is in the direction in which the cyclic mechanism can operate most favorably.

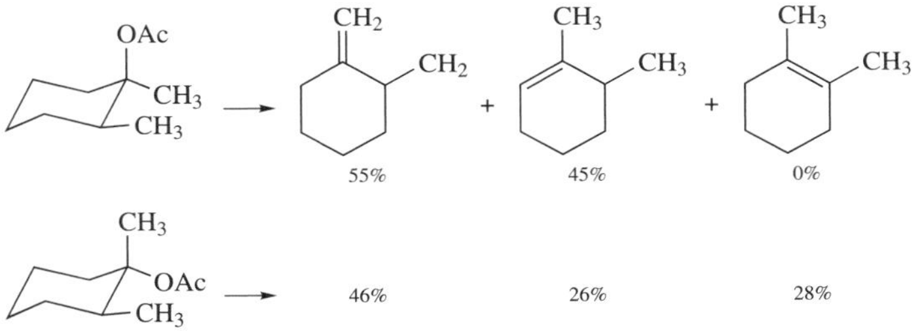

55% 45% 0% Ref. 239

46% 26% 28%

Alcohols can be dehydrated via xanthate esters at temperatures that are much lower than those required for acetate pyrolysis. The preparation of xanthate esters involves reaction of the alkoxide with carbon disulfide. The resulting salt is alkylated with methyl iodide.

$$RO^- Na^+ + CS_2 \longrightarrow ROC(=S)S^- Na^+ \xrightarrow{CH_3I} ROC(=S)SCH_3$$

The elimination is often effected simply by distillation.

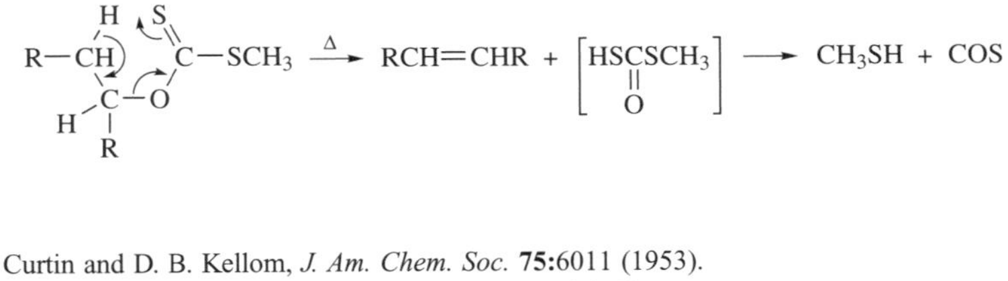

$$RCH_2CH(R)OC(=S)SCH_3 \xrightarrow{\Delta} RCH{=}CHR + [HSC(=O)SCH_3] \longrightarrow CH_3SH + COS$$

237. D. Y. Curtin and D. B. Kellom, *J. Am. Chem. Soc.* **75**:6011 (1953).
238. D. H. Wertz and N. L. Allinger, *J. Org. Chem.* **42**:698 (1977).
239. D. H. Froemsdorf, C. H. Collins, G. S. Hammond, and C. H. DePuy, *J. Am. Chem. Soc.* **81**:643 (1959).

Product mixtures are observed when more than one type of β hydrogen can participate in the reaction. As with the other *syn* thermal eliminations, there are no intermediates that are prone to skeletal rearrangement.

General References

W. Carruthers, *Cycloaddition Reactions in Organic Synthesis*, Pergamon Press, Oxford, 1990.
A. P. Marchand and R. E. Lehr (eds.), *Pericyclic Reactions*, Vols. I and II, Academic Press, New York, 1977.
A. Williams, *Concerted Organic and Bro-Organic Mechanisms*. CRC Press, Boca Raton, FL, 2000.
R. B. Woodward and R. Hoffmann, *The Conservation of Orbital Symmetry*, Academic Press, New York, 1970.

Diels–Alder Reactions

G. Brieger and J. N. Bennett, *Chem. Rev.* **80:**63 (1980).
E. Ciganek, *Org. React.* **32:**1 (1984).
W. Oppolzer, *Angew Chem. Int. Ed. Engl.* **23:**876 (1984).
W. Oppolzer, *Synthesis* **1978:**793.

Cycloaddition Reactions

A. Padwa (ed.), *1,3-Dipolar Cycloaddition Chemistry*, Wiley, New York, 1984.

Sigmatropic Rearrangements

E. Block, *Reactions of Organosulfur Compounds*. Academic Press, New York, 1978, Chapter 7.
H.-J. Hansen, in *Mechanisms of Molecular Migrations*, Vol. 3, B. S. Thyagarajan (ed.), Wiley-Interscience, New York, 172, pp. 177–236.
R. K. Hill, in *Asymmetric Synthesis*, Vol. 3, J. D. Morrison (ed.), Academic Press, New York, 1984, Chapter 8.
S. J. Rhoads and N. R. Raulins, *Org. React.* **22:**1 (1975).
T. S. Stevens and W. E. Watts, *Selected Molecular Rearrangements*, Van Nostrand Reinhold, London, 1973, Chapter 8.
B. M. Trost and L. S. Melvin, Jr., *Sulfur Ylides*, Academic Press, New York, 1975, Chapter 7.

Elimination Reactions

W. H. Saunders, Jr. and A. F. Cockerill, *Mechanisms of Elimination Reactions*, Wiley, New York, 1973, Chapter VIII.

Problems

(*References for these problems will be found on page 931.*)

1. Predict the product of each of the following reactions, clearly showing stereochemistry where relevant.

(a) OAc

CH_2CH_3

+ $CH_2{=}CHCHO$ $\xrightarrow[\text{toluene, }-10^\circ C]{BF_3,\ Et_2O}$

(b) $OSiMe_3$ CH_3 CH_3

+ $CH_2{=}CHCHO$ $\longrightarrow$

(c) $NHCO_2C_2H_5$

+ (*E*)-$CH_3CH{=}CHCHO$ $\xrightarrow{110^\circ C}$

(d) $H_2C{=}CH$ H H $\xrightarrow{60^\circ C}$

(e) CH_3 CH_3 CH_3 H O O $\xrightarrow{200^\circ C}$

(f) H_3C O $C(CH_3){=}C(H)CO_2CH_3$ $\xrightarrow{230^\circ C}$

(g) $CH_3\overset{O}{\overset{\|}{C}}CH_2CH_2CH_2CH{=}CH_2$ + $CH_3NHOH{\cdot}HCl$ $\longrightarrow$

(h) OH H_3C CH_3 $\xrightarrow[\text{2) }210^\circ C]{\text{1) }C_2H_5OCH{=}CH_2,\ Hg^{2+}}$

(i) OH, $CCH_2CH{=}CH_2$, CH_3 $\xrightarrow[\text{dimethoxyethane, }80^\circ C]{KH}$

(j) O CH_3 $CH_2CH_2CH_2C(CH_3){=}CH_2$ CH_3 $\xrightarrow{h\nu}$

(k) $C_6H_5CH(SeCH_3)_2$ → 1) *n*-BuLi; 2) 1,2-epoxybutane; 3) H_2O_2

(l) [cyclonona-2,4,7-trien-1-ol structure: H, OH] → KH, THF, 25°C

(m) [structure: $CH_3CH{=}CH{-}C(H)(HO){-}CH_2CH(CH_3)_2$] → $(CH_3)_2N\overset{CH_3}{C}(OCH_3)_2$, Δ

(n) [cyclic allyl ether structure with two CH_3 groups, O] → 100°C

(o) [same cyclic ether structure, CH_3, CH_3, O] → Li⁺ ⁻N(2,2,6,6-tetramethylpiperidide)

(p) [macrocyclic lactone structure: CH_3OCH_2O, H_3C, H, O, O] → 1) LDA, HMPA; 2) *t*-$BuMe_2SiCl$; 3) 105°C

(q) $CH_2{=}CH(CH_2)_3$–[dioxinone fused cyclopentene structure: H_3C, CH_3, O, O, O] → $h\nu$

(r) [anthraquinone structure: O, OMe, CH_2Br, CH_2Br, OMe, O] + $CH_2{=}CH\overset{O}{\overset{\|}{C}}CH_3$ → Zn

(s) $CH_2{=}\underset{CH_3}{C}CH_2CH_2CH_2OCH_2CO_2H$ → 1) ClCOCOCl; 2) Et_3N

(t) [structure: vinyl–CH(SPh)–C(CH_3)_2–CH_2CN] → 1) MCPBA; 2) $(C_2H_5)_2NH$

(u) [structure: Ph–CH_2–CH(N(CH_2Ph)_2)–CH=CH–CO_2C_2H_5] → MCPBA

(v) 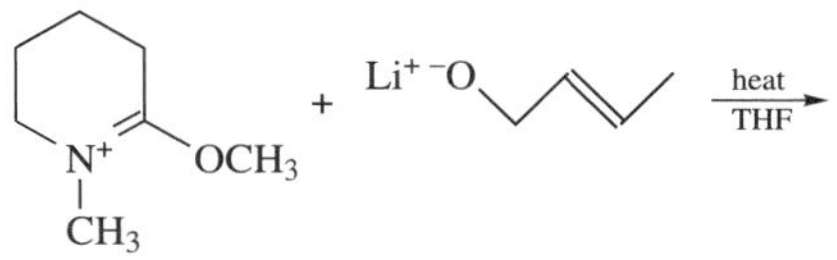 $Li^+ \ ^-O$ heat THF $\longrightarrow$

N^+ OCH_3 CH_3

(w) 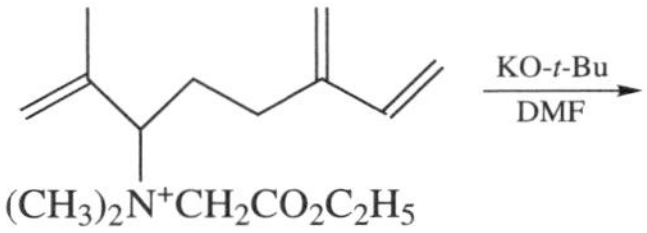 KO-*t*-Bu DMF $\longrightarrow$

$(CH_3)_2N^+CH_2CO_2C_2H_5$

(x) 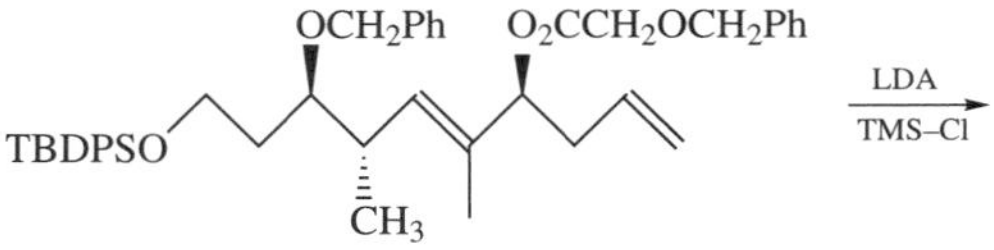 OCH_2Ph $O_2CCH_2OCH_2Ph$ LDA TMS–Cl $\longrightarrow$

TBDPSO CH_3

(y) 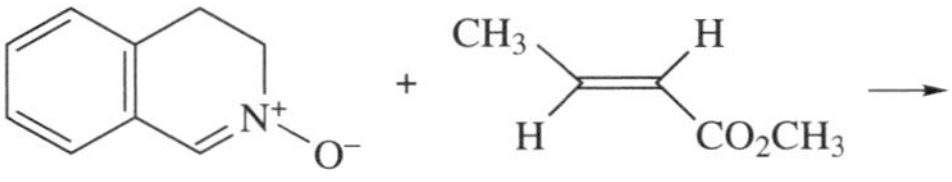 CH_3 H + H CO_2CH_3 $\longrightarrow$

N^+ O^-

2. Intramolecular cycloaddition reactions occur under the reaction conditions specified for each of the following reactants. Show the structure of the product, including all aspects of its stereochemistry, and indicate the structures of any intermediates which are involved in the reactions.

(a) 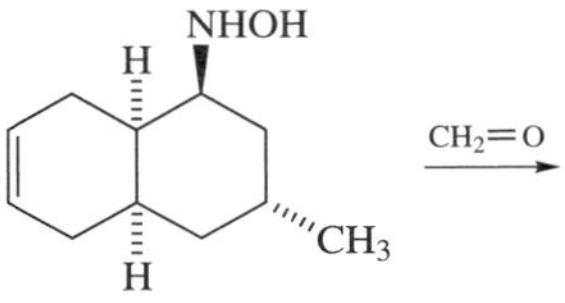 NHOH H H CH_3 $CH_2{=}O$ $\longrightarrow$

(b) C_6H_5 N CH_3 $CH_2CH_2CH_2CH{=}CH_2$ $h\nu$ $\longrightarrow$

(c) O= H N N H S S $CH_2\overset{O}{\overset{\|}{C}}NHNHCH_3$ $C_6H_5CH{=}O$ 80°C, 1 h $\longrightarrow$

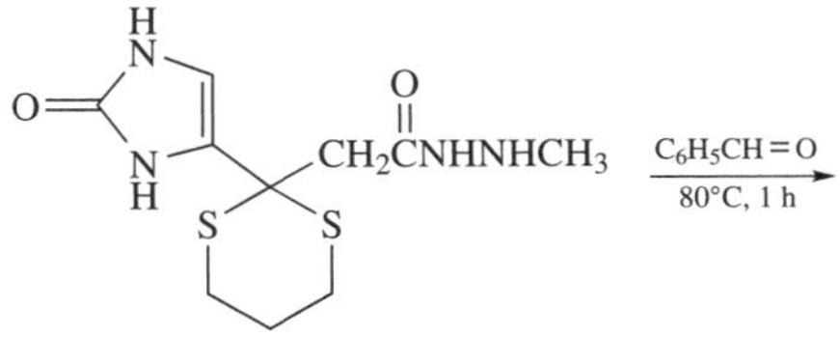

(d) 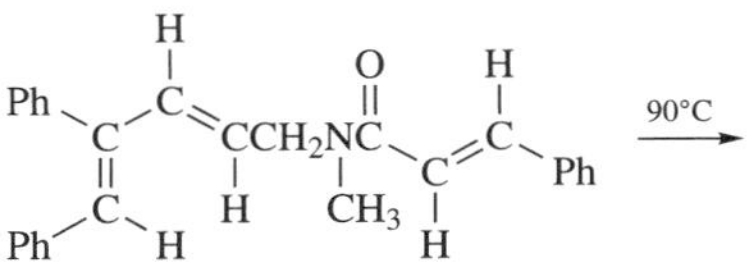 Ph Ph H H O H H CCH_2NC H CH_3 Ph 90°C $\longrightarrow$

(e) 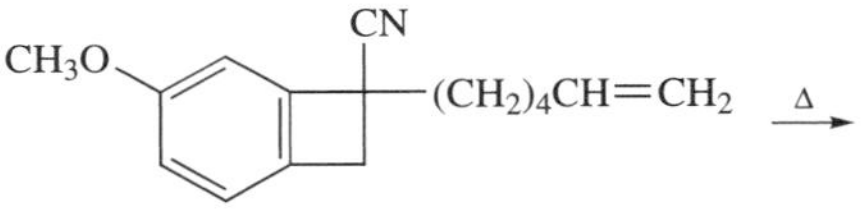 CH_3O CN $-(CH_2)_4CH{=}CH_2$ Δ $\longrightarrow$

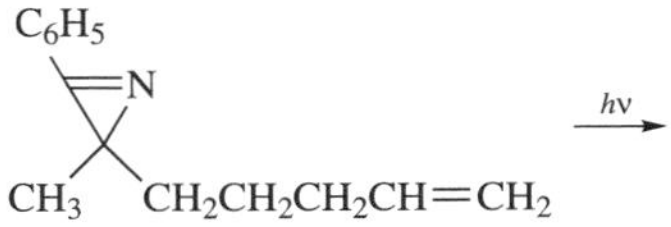

(f) 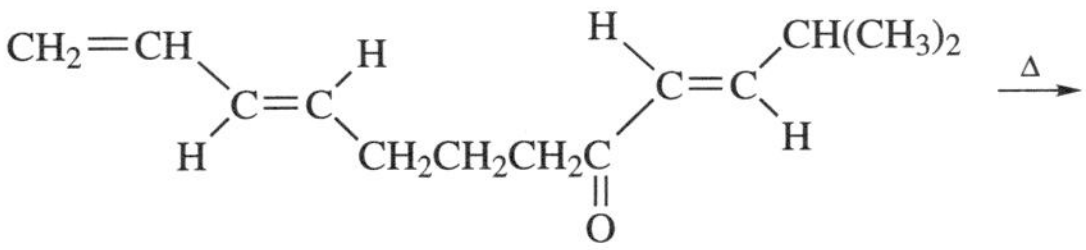 $CH_2{=}CH(H)C{=}C(H)CH_2CH_2CH_2C(=O)(H)C{=}C(H)CH(CH_3)_2 \xrightarrow{\Delta}$

(g) 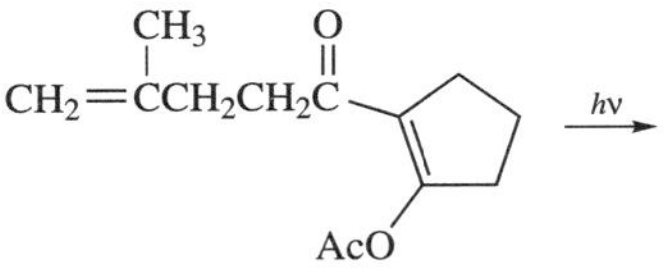 $CH_2{=}C(CH_3)CH_2CH_2C(=O)$–(2-acetoxycyclopent-1-enyl) $\xrightarrow{h\nu}$

3. Indicate the mechanistic type to which each of the following reactions belongs.

(a) 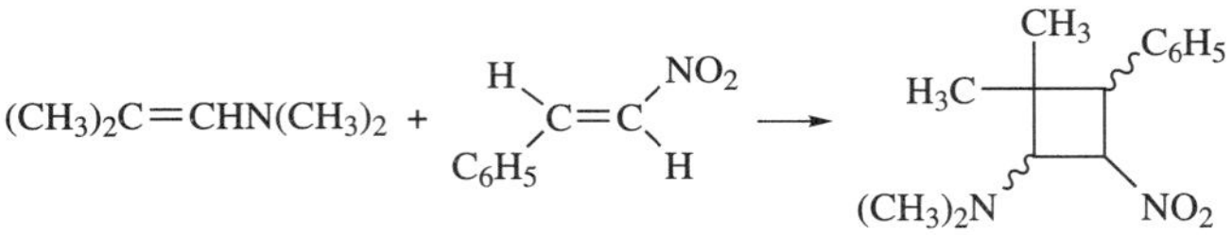 $(CH_3)_2C{=}CHN(CH_3)_2$ + $H(C_6H_5)C{=}C(H)NO_2$ ⟶ cyclobutane bearing CH_3, H_3C, C_6H_5, $(CH_3)_2N$, NO_2

(b) $CH_3CH{=}CHCH_2Br$ + $\xrightarrow[\text{2) } K_2CO_3]{\text{1) } CH_3SCH_2CPh(=O)}$ $CH_3CHCH{=}CH_2$ / $CH_3SCHCPh(=O)$

(c) N-benzyl-N-($CH_2CO_2C_2H_5$)-2-vinylpiperidinium (PhCH$_2$, $CH_2CO_2C_2H_5$, H) $\xrightarrow[20°C]{\text{DBU}}$ N-PhCH$_2$ azacyclodecene bearing $CO_2C_2H_5$

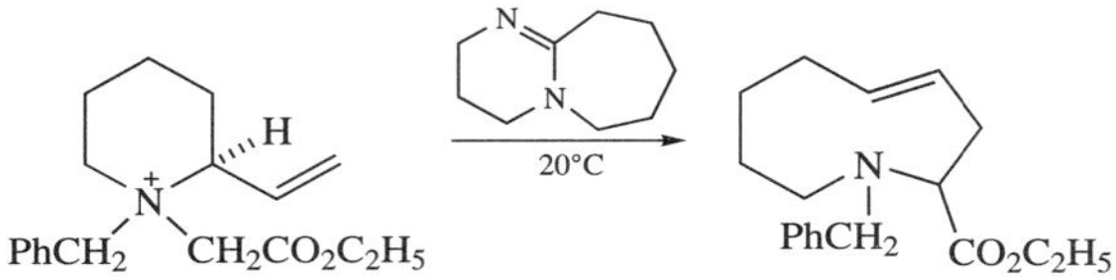

(d) 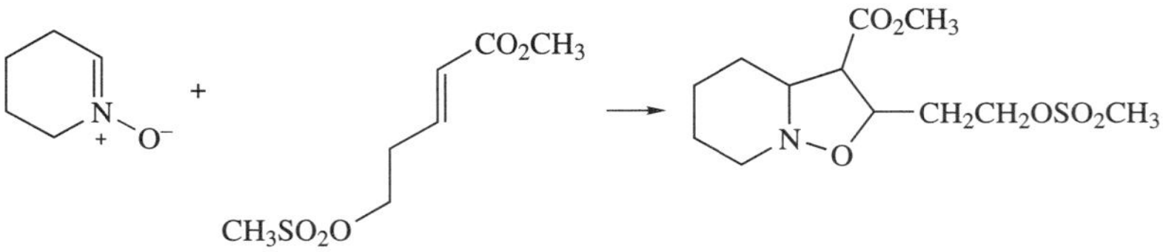 tetrahydropyridine N-oxide (N^+–O^-) + $CH_3SO_2O(CH_2)_3CH{=}CHCO_2CH_3$ ⟶ bicyclic isoxazolidine (N–O) bearing CO_2CH_3 and $CH_2CH_2OSO_2CH_3$

(e) 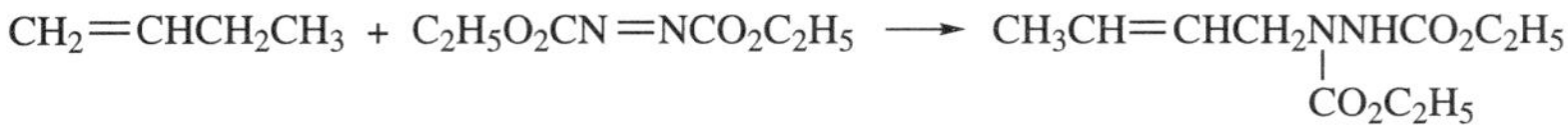 $CH_2{=}CHCH_2CH_3$ + $C_2H_5O_2CN{=}NCO_2C_2H_5$ ⟶ $CH_3CH{=}CHCH_2N(CO_2C_2H_5)NHCO_2C_2H_5$

(f) 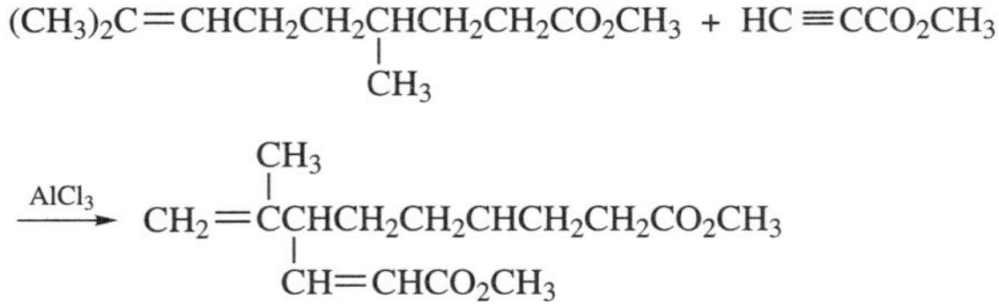 $(CH_3)_2C{=}CHCH_2CH_2CH(CH_3)CH_2CH_2CO_2CH_3$ + $HC{\equiv}CCO_2CH_3$

$\xrightarrow{AlCl_3}$ $CH_2{=}C(CH_3)CH(CH{=}CHCO_2CH_3)CH_2CH_2CHCH_2CH_2CO_2CH_3$ (with CH_3 branch)

(g) 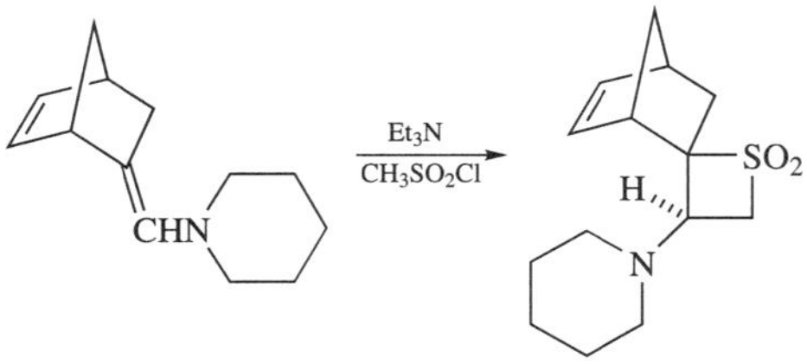 norbornenylidene-CHN(piperidine) $\xrightarrow[CH_3SO_2Cl]{Et_3N}$ spiro thietane dioxide (SO_2) bearing H and piperidin-1-yl (N)

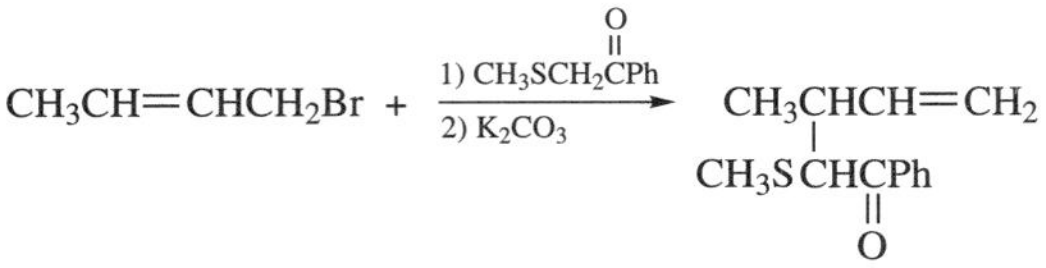

(h) PhN_3 + (piperidinyl)NCH=CHPh ⟶ (triazoline product: Ph–N–N=N ring with piperidinyl N and Ph)

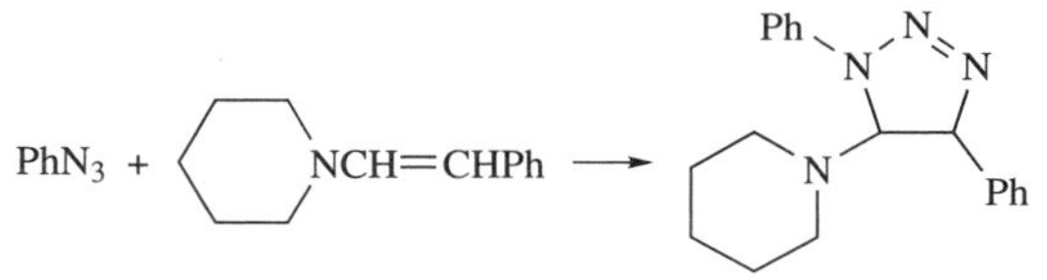

(i) $(CH_3)_2CHCH(NC)OCH_2CH{=}C(CH_3)_2 \xrightarrow[\text{2) } H_3O^+]{\text{1) } LiNR_2} (CH_3)_2CHC(=O){-}C(CH_3)_2CH{=}CH_2$

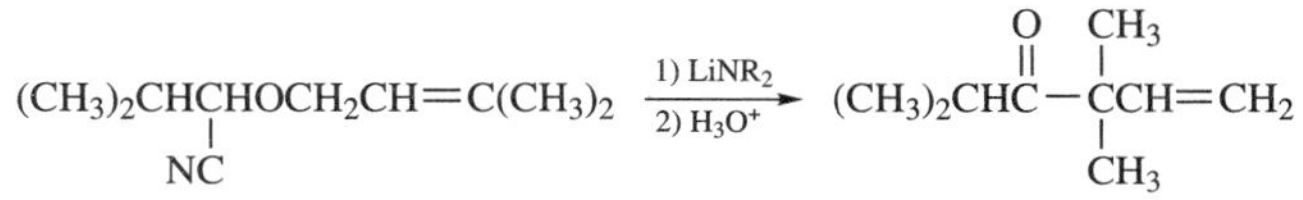

(j) (bicyclic pyranone with CO_2CH_3) + $CH_2{=}C(OCH_3)_2 \xrightarrow{110°C}$ (tetrahydronaphthalene with OCH_3 and CO_2CH_3)

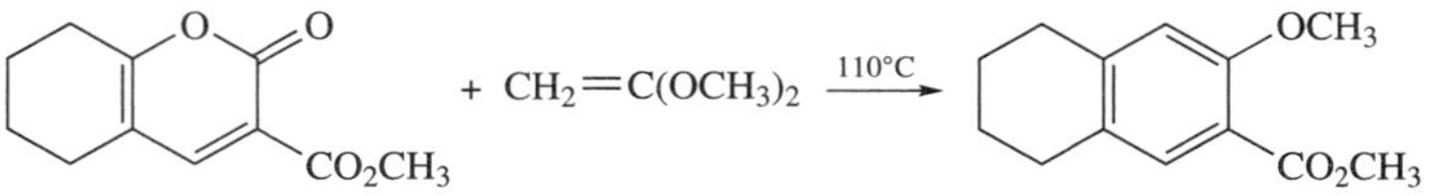

4. By applying the principles of retrosynthetic analysis, show how each of the indicated target molecules could be prepared from the starting material(s) given. No more than three separate transformations are necessary in any of the syntheses.

(a) (polycyclic target with CH_3O, OCH_3, Cl, Cl, Cl, Cl) ⟹ (tetrachloro-5,5-dimethoxycyclopentadiene: Cl, Cl, Cl, Cl, CH_3O, OCH_3) + (cyclooctadiene)

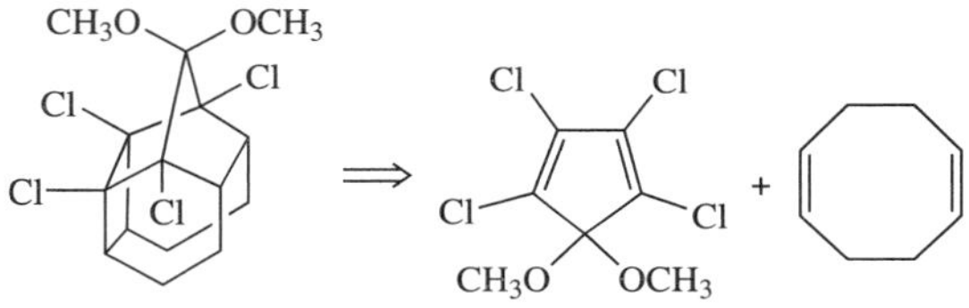

(b) (oxa-bridged target with O, CH_3, CO_2CH_3, CH_3, CO_2CH_3) ⟹ H_3C–(furan)–CH_3 + dimethyl acetylenedicarboxylate

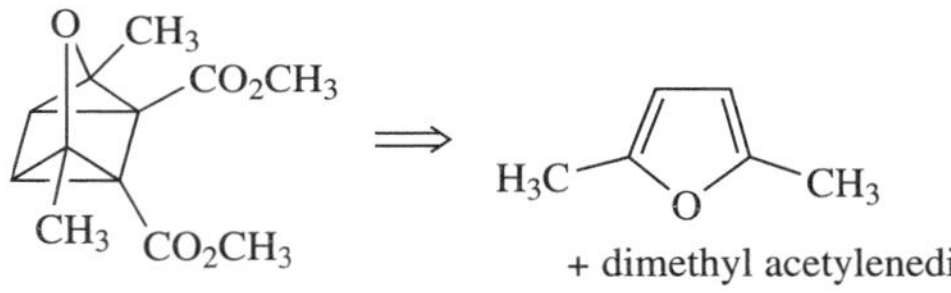

(c) (cyclohexene with $N(C_2H_5)_2$, C(O)Ph, C(O)Ph) ⟹ crotonaldehyde, diethylamine, and trans-1,2-dibenzoylethylene

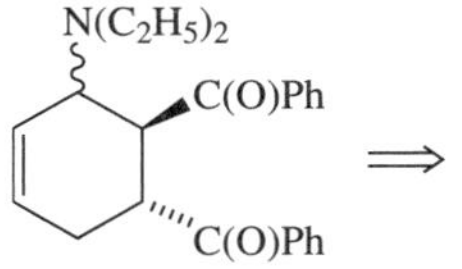

(d) $CH_3CH_2C{\equiv}CCH{=}CHCH_2CH_2CO_2CH_3 \Longrightarrow$
$CH_3CH_2C{\equiv}CH + CH_2{=}CHCHO + CH_3C(OCH_3)_3$

(e) (benzene ring with OH, HO, HO, $CH_2CH{=}CH_2$) ⟹ (benzodioxolone O= with OH) + $H_2C{=}CHCH_2Br$

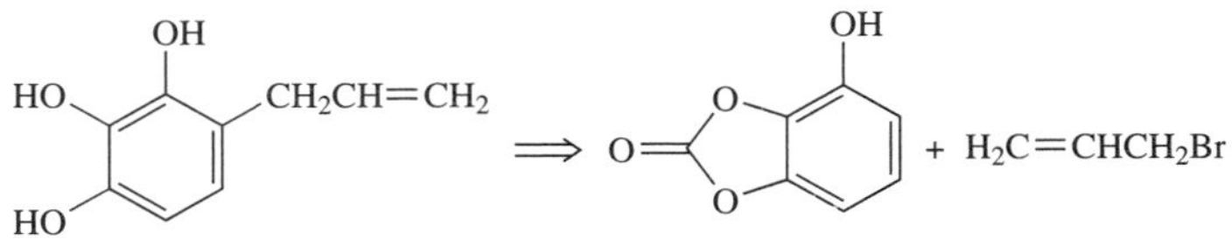

(f) (cyclobutane with H_3C, H_3C, H, Ph, $CO_2C_2H_5$, $CO_2C_2H_5$, Ph, H) ⟹ *trans*-stilbene, diethyl malonate, and acetone

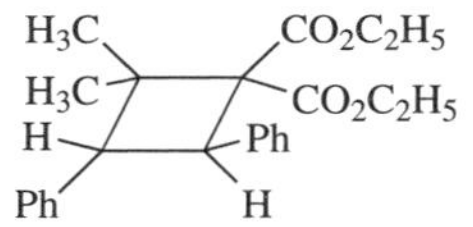

(g) $\Longrightarrow$ and any other necessary reagents

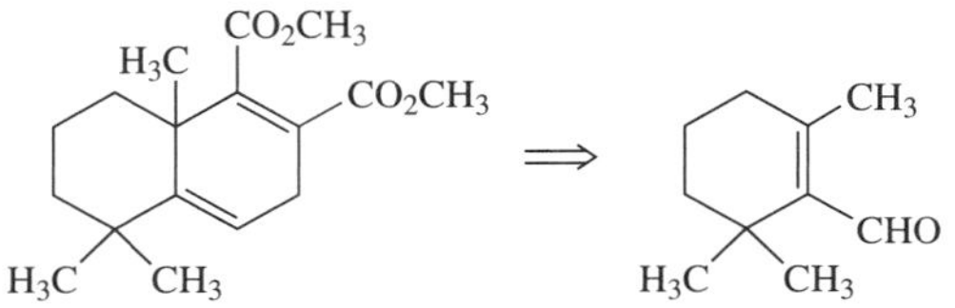

(h) $\Longrightarrow$ (*E*)-$O_2NCH{=}CHCO_2CH_3$ and any other necessary reagents

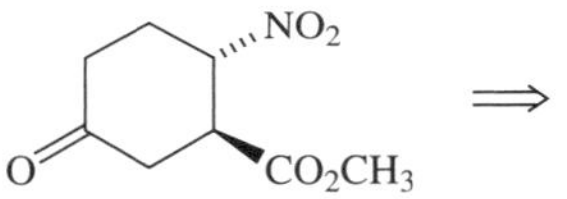

(i) $\Longrightarrow$ and any other necessary reagents

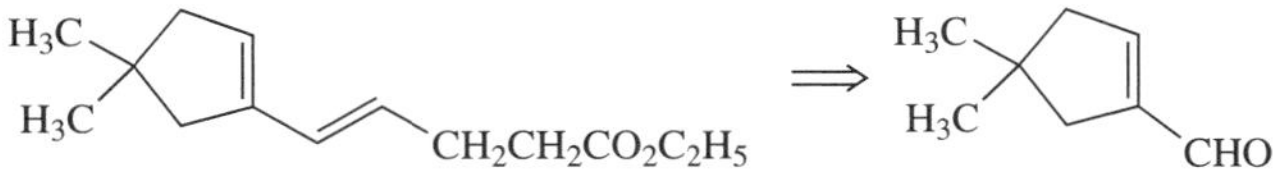

(j) $\Longrightarrow$

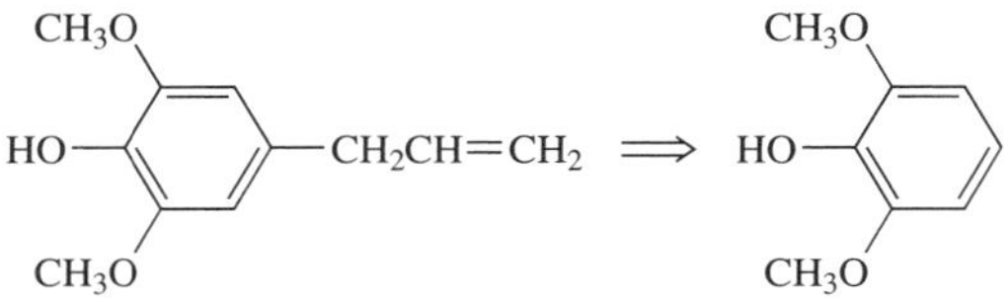

(k) $\Longrightarrow$

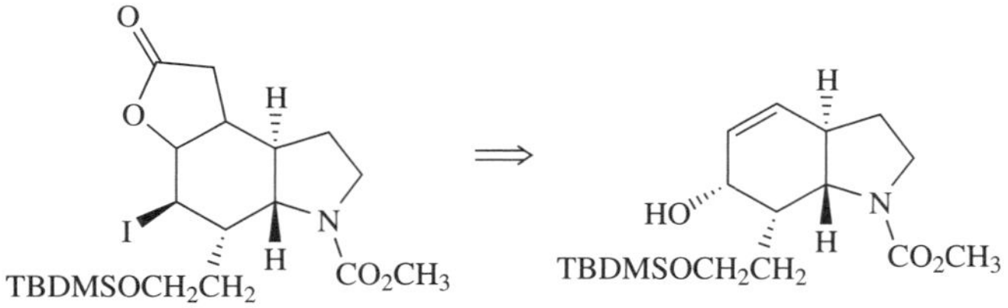

(l) $\Longrightarrow$

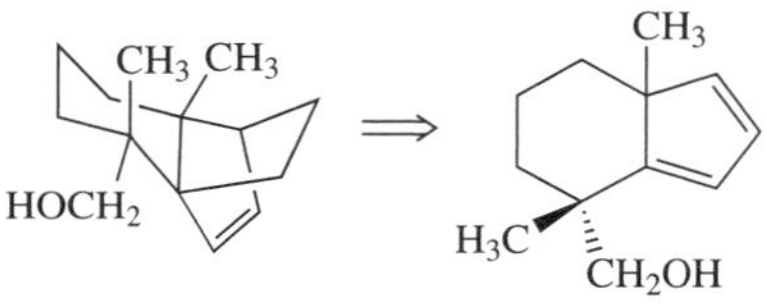

(m) $\Longrightarrow$

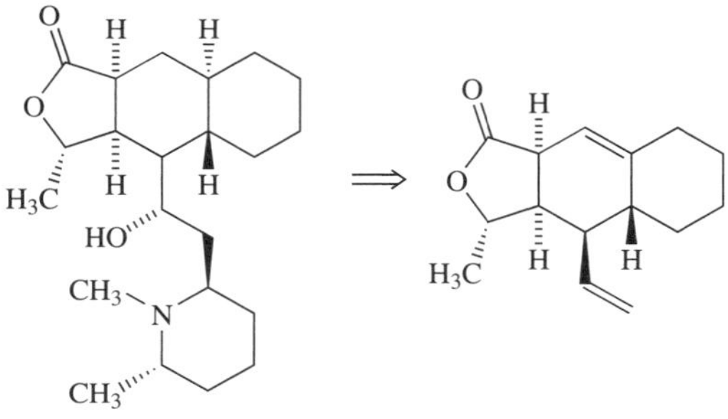

5. Reaction of α-pyrone (**A**) with methyl acrylate at reflux for extended periods gives a mixture of stereoisomers of **B**. Account for the formation of this product.

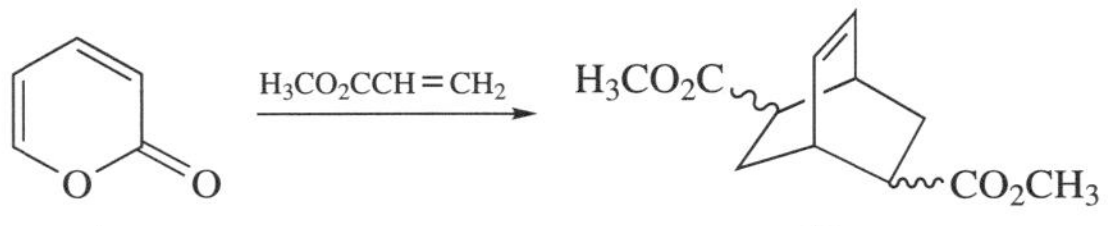

A **B**

6. When 2-methylpropene and acrolein are heated at 300°C under pressure, 3-methylenecyclohexanol and 6,6-dimethyldihydropyran are formed. Explain the formation of these products.

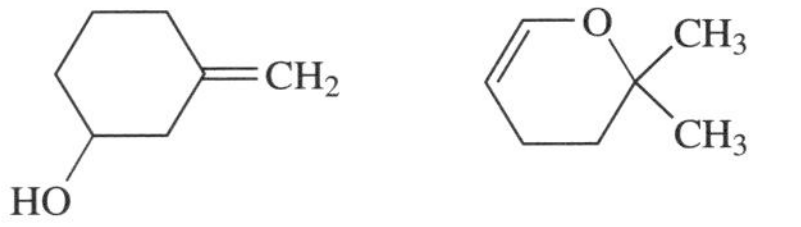

3-methylenecyclohexanol 6,6-dimethyldihydropyran

7. Vinylcyclopropane, when irradiated with benzophenone or benzaldehyde, gives a mixture of two types of products. Suggest the mechanism by which product of type **C** is formed.

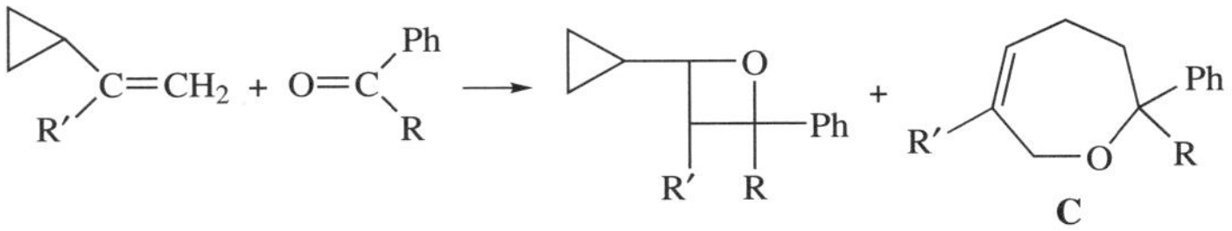

C

8. The addition reaction of tetracyanoethylene and ethyl vinyl ether in acetone gives 94% of the 2 + 2 adduct and 6% of an adduct having the composition tetracyanoethylene + ethyl vinyl ether + acetone. If the 2 + 2 adduct is kept in contact with acetone for several days, it is completely converted to the minor product. Suggest a structure for this product, and indicate its mode of formation (a) in the initial reaction and (b) on standing in acetone.

9. A convenient preparation of 2-allylcyclohexanone involves simply heating the diallylketal of cyclohexanone in toluene containing a trace of *p*-toluenesulfonic acid and collecting a distillate consisting of toluene and allyl alcohol. Distillation of the residue gives a 90% yield of 2-allylcyclohexanone. Outline the mechanism of this reaction.

10. The preparation of a key intermediate in an imaginative synthesis of prephenic acid is depicted below. Write a series of equations showing the important steps and intermediates in this process. Indicate the reagents required to bring about the desired

transformation where other than thermal reactions are involved.

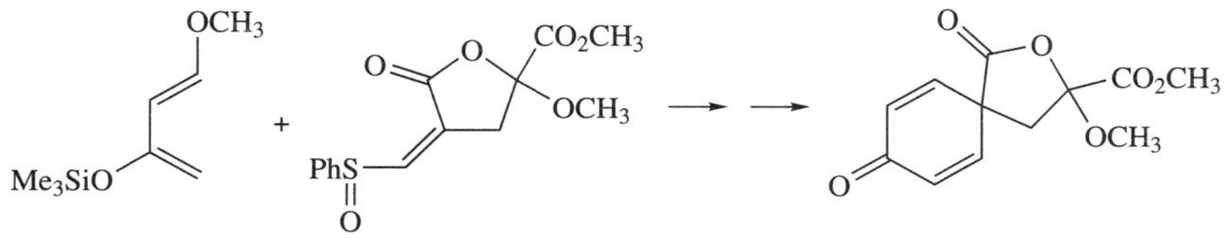

11. A route to hasubanan alkaloids has been described involved reaction of 1-butadienyl phenyl sulfoxide with the tetrahydrobenzindole **D**. Treatment of the resulting adduct with sodium sulfide in refluxing methanol gave **F**. Suggest a structure for **E**, and rationalize the formation of **F** from **E**.

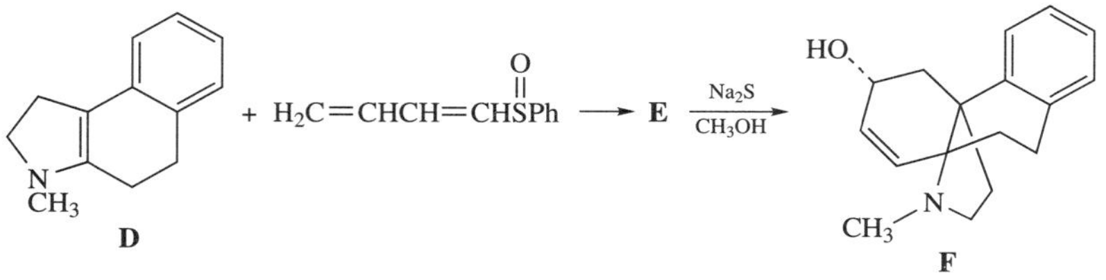

$$\mathbf{D} + H_2C{=}CHCH{=}CH\overset{O}{\overset{\|}{S}}Ph \longrightarrow \mathbf{E} \xrightarrow[CH_3OH]{Na_2S} \mathbf{F}$$

12. A solution of 2-butenal, 2-acetoxypropene, and dimethyl acetylenedicarboxylate refluxed in the presence of a small amount of an acidic catalyst gives an 80% yield of dimethyl phthalate. Explain the course of this reaction.

13. Irradiation of the dienone shown generates three isomeric saturated ketones, all of which contain cyclobutane rings. Postulate reasonable structures.

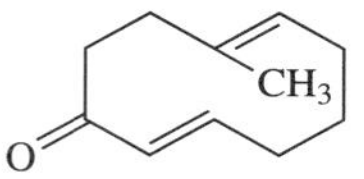

14. Irradiation of *o*-methylbenzaldehyde in the presence of maleic anhydride give **G**. The same compound is obtained when **H** is heated with maleic anhydride. Both reactions give only the stereoisomer shown. Formulate a mechanism.

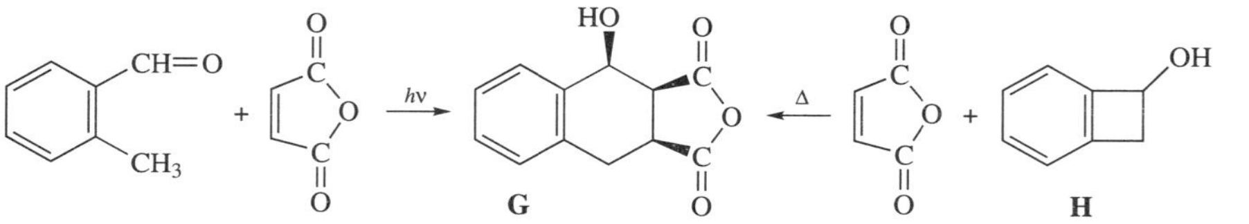

15. The ester **I** gives alternative stereoisomers when subjected to Claisen rearrangement as the lithium enolate or as the trimethylsilyl enol ether. Analyze the respective transition states and develop a rationale for this observation.

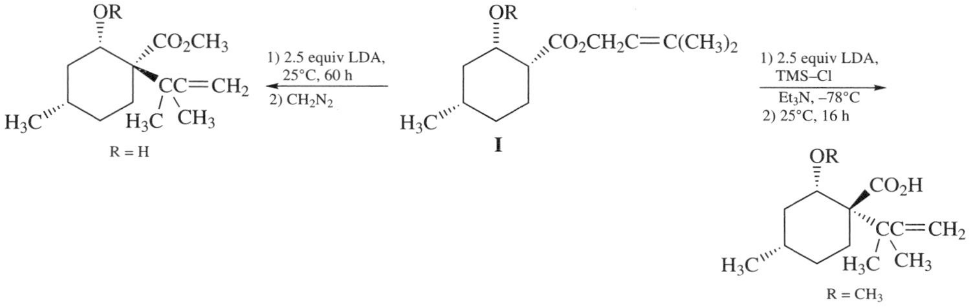

16. Photolysis of **J** gives an isomeric compound **K** in 83% yield. Alkaline hydrolysis of **K** affords a hydroxy carboxylic acid **L**, $C_{25}H_{32}O_4$. Treatment of **K** with silica gel in hexane yields **M**, $C_{24}H_{28}O_2$. **M** is converted by sodium periodate–potassium permanganate to a mixture of **N** and **O**. What are the structures of **K**, **L**, and **M**?

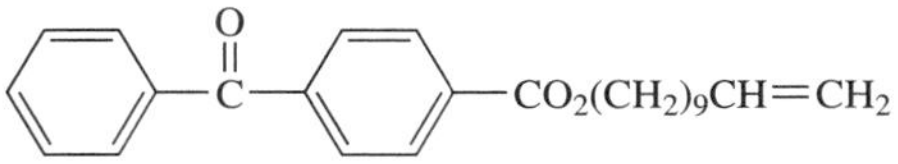

J

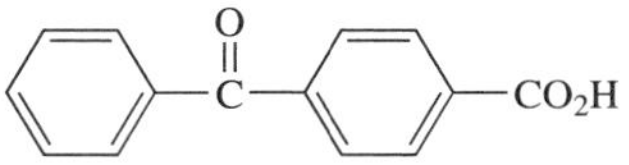

N

$HO(CH_2)_9CO_2H$

O

17. (a) 1,2,4,5-Tetrazines react with alkenes to give dihydropyridazines, as illustrated in the equation below. Suggest a mechanism.

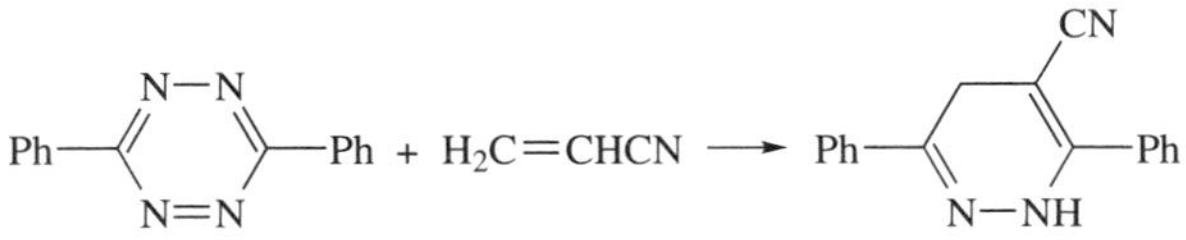

(b) Compounds **P** and **Q** are both unstable toward loss of nitrogen at room temperature. Both compounds give **R** as the product of decomposition. Account for the formation of **R**.

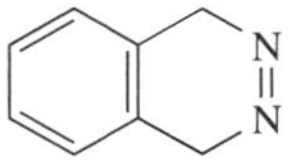

P

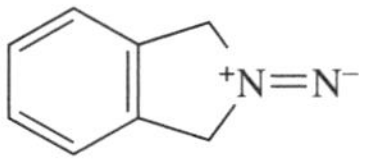

Q

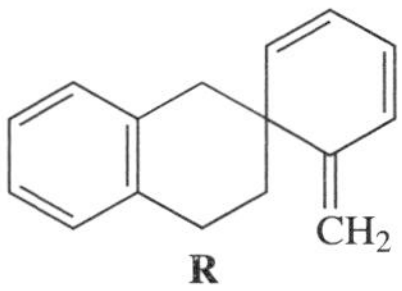

R

18. For each of the following reactions, develop a transition-state structure using molecular models which would account for the observed stereoselectivity. Identify important conformational and/or steric features of the proposed transition state.

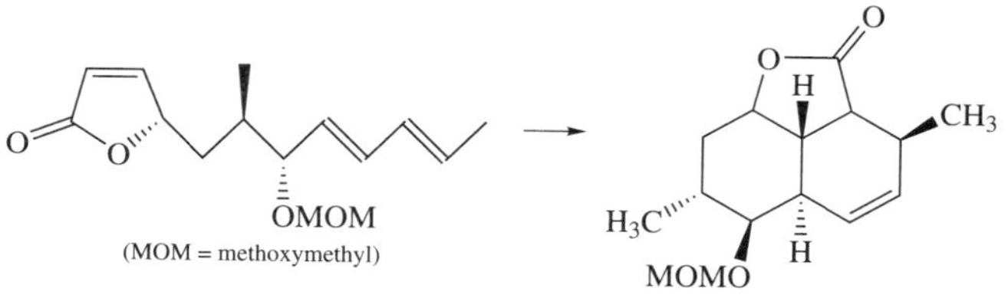

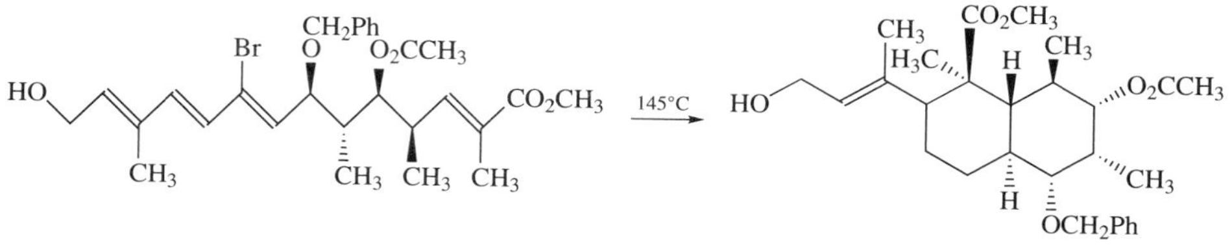

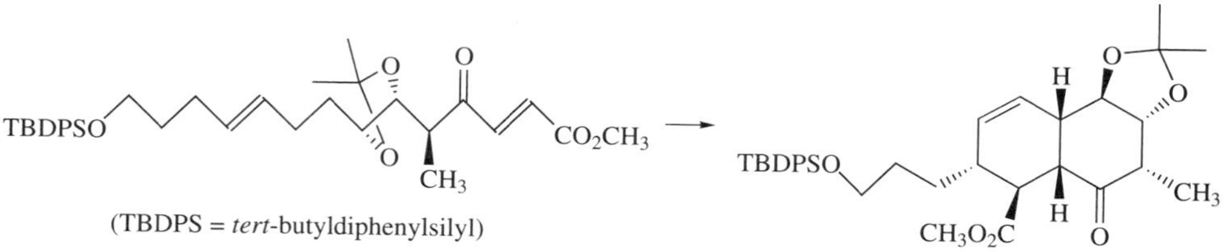

OMOM

(MOM = methoxymethyl)

H_3C MOMO H H CH_3

HO Br CH_2Ph O O_2CCH_3 CO_2CH_3 CH_3 CH_3 CH_3 CH_3 145°C

HO CH_3 H_3C CO_2CH_3 H CH_3 O_2CCH_3 CH_3 H OCH_2Ph

TBDPSO O O O CO_2CH_3 CH_3

(TBDPS = *tert*-butyldiphenylsilyl)

TBDPSO H O O H CH_3 CH_3O_2C O

19. Provide a detailed mechanistic explanation for each of the following synthetically useful transformations.

(a) CH_2OH — 1) p-$O_2NC_6H_4SeCN$ 2) Bu_3P 3) H_2O_2 → CH_2, OH

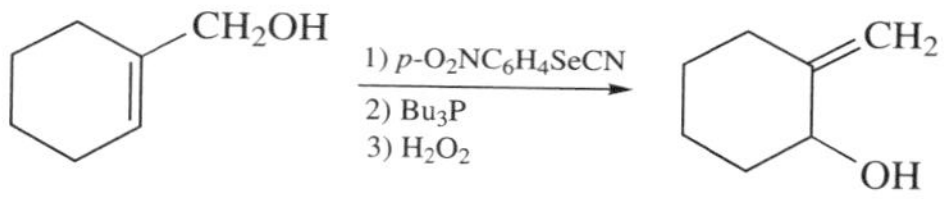

(b) $HC{\equiv}CCH(NHCCH_3)CH(OH)CH{=}CH_2$ (O) — 1) Cl_3CCN 2) Δ → $HC{\equiv}CCH(NHCCH_3)CH{=}CHCH_2NHCCCl_3$ (O)

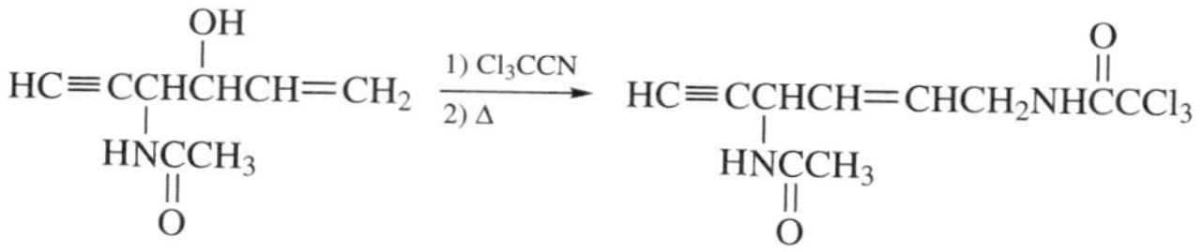

(c) O, CH_2SPh — $Me_3SiCH_2O_3SCF_3$, CsF, PhCH=O → O, CH_2, CH_2SPh + PhCH—CH_2 (O)

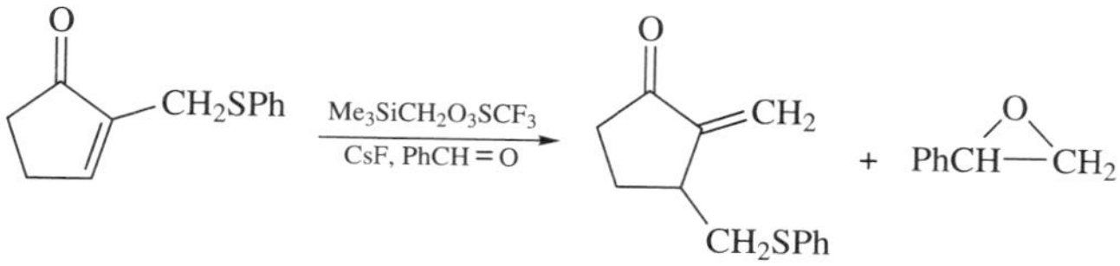

(d) OH — KH, 25°C, 16 h → OH

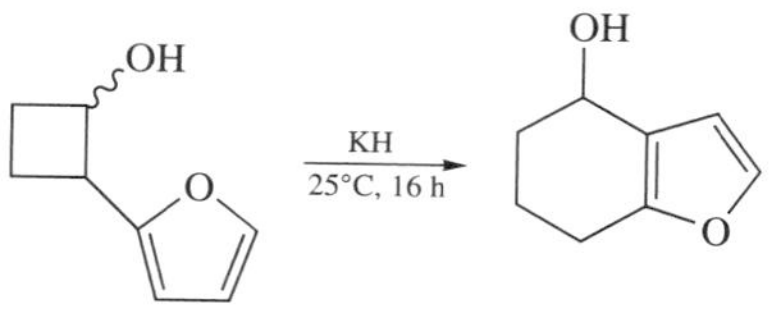

(e) $CH_2{=}C(CH_3)CH_2SePh$ + $H_2NCH_2CO_2C_2H_5$ — N-Cl (N-chlorosuccinimide) → $CH_2{=}C(CH_3)CH_2NHCH_2CO_2C_2H_5$

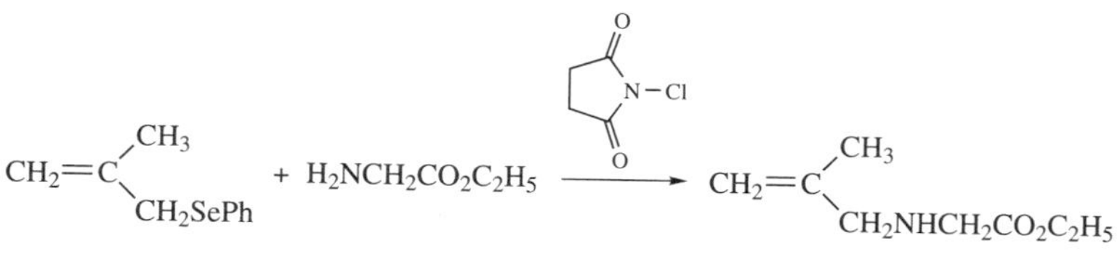

(f) $CH_2{=}CHCH_2CH_2$ H$_3$C CO_2H 1) N-methyl-2-fluoropyridinium, Me, F 2) Et_3N

(g) H O^- $CH{=}\overset{+}{N}C(CH_3)_3$ 1) CH_3O_2CCl 2) Et_3N H_2O OCO_2CH_3 $CH{=}O$

(h) CH_3 H C=C H CH_2Br + $CH_3SCH_2\overset{O}{\overset{\|}{C}}Ph$ K_2CO_3 $Ph\overset{O}{\overset{\|}{C}}CH\overset{CH_3}{\overset{|}{C}H}CH{=}CH_2$ SCH_3

(i) OTMS $CH_2CH_2\overset{O}{\overset{\|}{C}}NCH_2O_2CCH_3$ H $CH_3CH_2CH_2$ SO_2 380°C OTMS $CH_3CH_2CH_2$ N O

(j) OH $C{\equiv}CCHC(CH_3)_2CH_2CH{=}CH_2$ CH_3 PhSCl Et_3N, 25°C, 38 h O=SPh CH_3 CH_3 CH_3

(k) $CH_2Si(CH_3)_3$ CH_3 H H CH_2Cl C=C H $PhCH_2$ OAc N O CH_3C O S H O NaI, K_2CO_3 CH_3CN $CH_2Si(CH_3)_3$ CH_3 H H $PhCH_2$ OAc N O CH_3C O S H O

(l) $CH_2CO_2C(CH_3)_3$ N $C(CH_3)_3$ LDA $(CH_3)_3CO_2C$ N H $C(CH_3)_3$

(m) $CH{=}CH_2$ N CH_3 + $CH_3O_2CC{\equiv}CCO_2CH_3$ TsOH, 3 mol %, $CDCl_3$, 41 h N CH_3 CO_2CH_3 CO_2CH_3

(n) OH $CH_2CCO_2CH_3$ CCO_2CH_3 O 2 equiv $SnCl_4$ $EtN(i\text{-}Pr)_2$ OH CO_2CH_3 H OH CO_2CH_3 65% + OH CO_2CH_3 OH CO_2CH_3 67%

(o) [structure: $CH_2=C(Si(CH_3)_3)CH_2CH_2CH_2CH_2OCH(CH_3)OC_2H_5$] $\xrightarrow{SnCl_2}$ [eight-membered cyclic ether bearing $Si(CH_3)_3$ on the ring double bond and CH_3 adjacent to O]

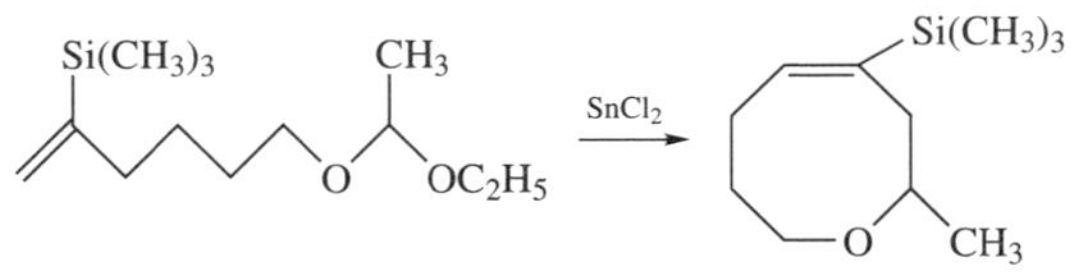

20. In each part below, the molecule shown has been employed as a synthetic equivalent in cycloaddition reactions. Show the sequence of reactions by which the adduct could be converted to the adduct that cannot be obtained directly.

(a) $RC(NO_2)=CHSO_2Ph$ as an acetylene equivalent

(b) $PhSO_2CH=CHSiMe_3$ as an acetylene equivalent

(c) $CH_2=C(O_2CCH_3)CN$ as a ketene equivalent

(d) $CH_2=CHNO_2$ as a ketene equivalent

21. Suggest sequences of reactions for accomplishing each of the following synthetic transformations.

(a) Squalene from succinaldehyde, isopropyl bromide, and 3-methoxy-2-methyl-1,3-butadiene.

(b) [bicyclic structure bearing $(CH_3)_2N$ and OH] from [cyclohex-3-enecarbaldehyde, $CH=O$]

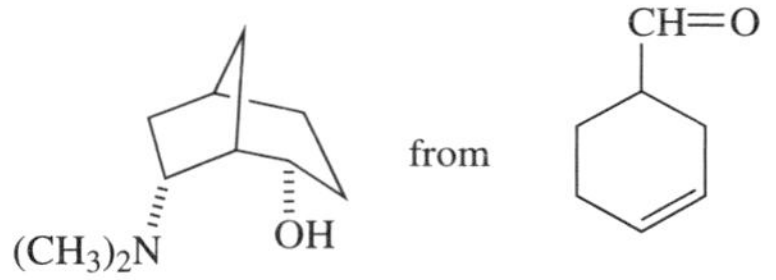

(c) $H_2C=CH(H_3C)C=C(H)CH_2CH=C(CH_3)CH_2CH_2CH=C(CH_3)CH=O$ (with H on the trisubstituted alkene carbons)

from $H_2C=CH(H_3C)C=C(H)CH_2CH(C(CH_3)=CH_2)C(CH_3)(CH=CH_2)CH=O$

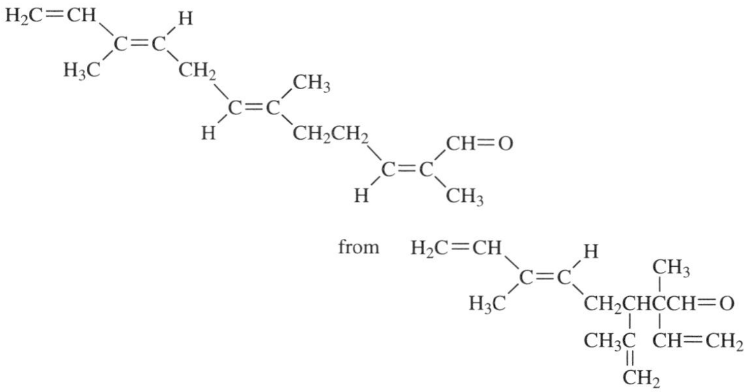

(d) $H_5C_2O_2CCH_2CH_2C(C_2H_5)=CHCH_2CH_2Cl$ from $H_2C=C(C_2H_5)CH(OH)CH_2CH_2Cl$

(e) [cyclopentenone with H, C_5H_{11}, CH_3 and H substituents] from $H_5C_2CO_2CH_2C(OC_2H_5)=CHC_5H_{11}$

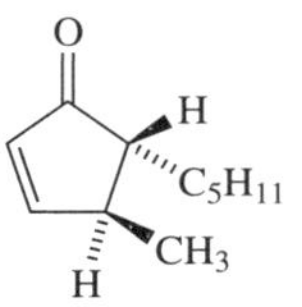

(f) 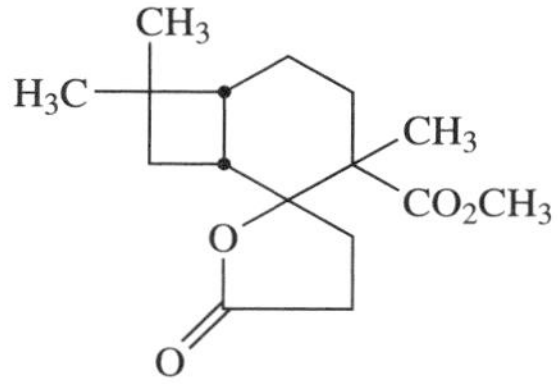 [spiro lactone structure with CH_3, H_3C, CH_3, CO_2CH_3] from [cyclohexenone]

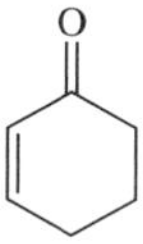

(g) 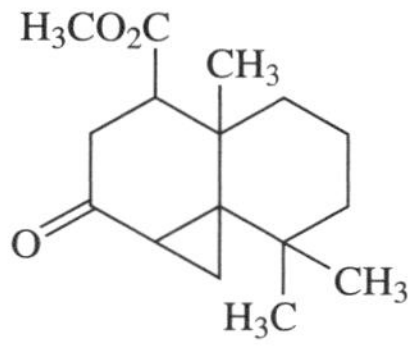 [structure with H_3CO_2C, CH_3, CH_3, H_3C] from $HO_2CCH_2CH(CO_2CH_3)$-[cyclohexane with CH_3, $H_2C=$, H_3C, CH_3]

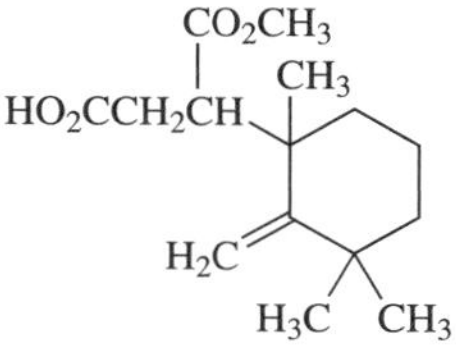

(h) $(CH_3)_2C{=}CHCH(CH{=}CH_2)C(CH_3){=}CH_2$ from $(CH_3)_2C{=}CHCH{=}CHCH_2OH$

(i) $(CH_3)_2CHCH_2C(=O)CH_2C(CH_3){=}CH_2$ from $H_2C{=}CHOCH_2CH{=}CHCH_3$ and $(CH_3)_2CHCH_2Br$

(j) 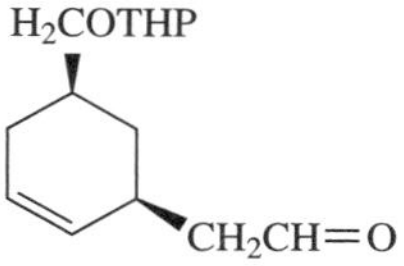 [cyclohexene with H_2COTHP and $CH_2CH{=}O$] from [cyclohexene with H_2COTHP and HO]

(k) $(CH_3)_3C$-[cyclohexane with CH_2CH_3 and $CH_2CO_2C_2H_5$] from $(CH_3)_3C$-[cyclohexanone]

(l) [bicyclic lactone with $CH{=}CH_2$] from [cyclooctatrienol, OH]

(m) [tricyclic ketone with H_3C, Me_2N, CH_3, CH_3, H] from [3,3-dimethylcyclopentanone with CH_3, CH_3, O, CH_2-(2-methylcyclopentenyl), H]

(n) [decalone with AcO, H, CH_2OSiR_3, CH_3, HO, H, O] from [cyclohexene with AcO, H, $CH(CH_2OSiR_3)C(CH_3)$-$(CH_2)_4NO_2$]

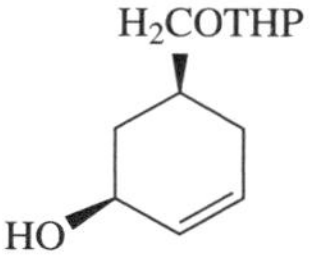

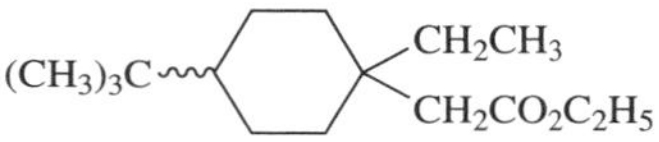

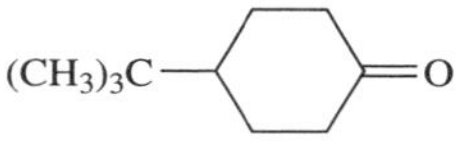

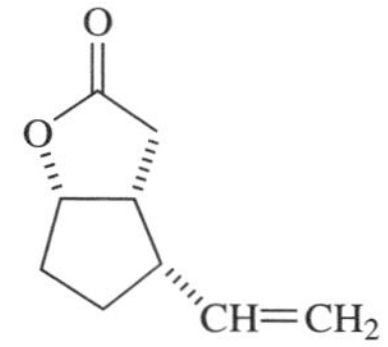

(o) [cyclopentanone with $CO_2C_2H_5$ and $CH_2{-}CH{=}C(CH_3)CH_2OH$] from [cyclopentanone with $CO_2C_2H_5$] and $BrCH_2C(S(O)Ph){=}C(CH_3)_2$

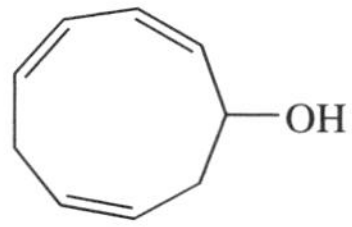

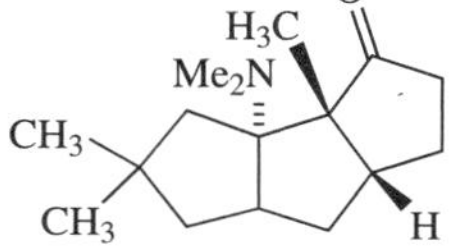

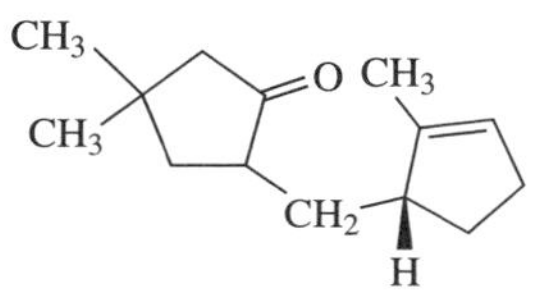

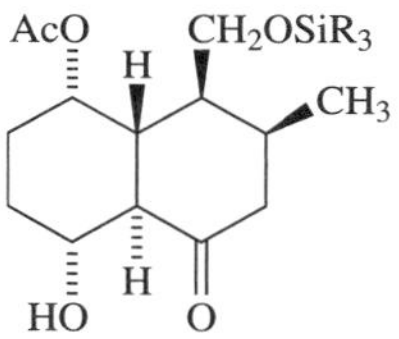

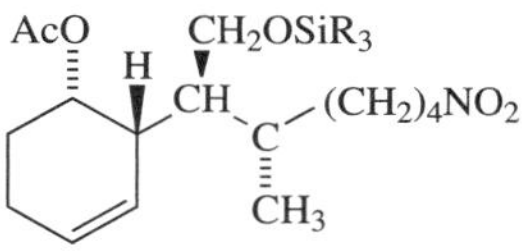

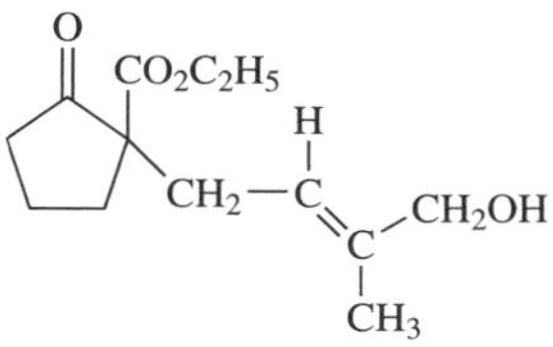

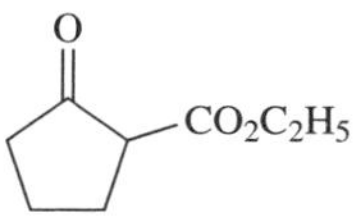

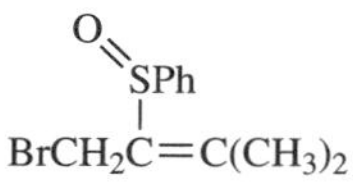

(p) from and

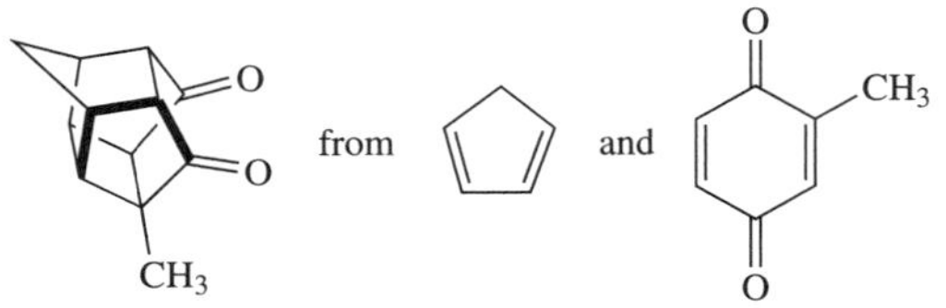

(q) $PhCH_2O$ $O{=}CH$ H_2C from $PhCH_2O$ $O{=}CH$ $(CH_3)_2C{=}CHCH_2CH_2Br$, and $MeO_2CCH_2PO(OEt)_2$

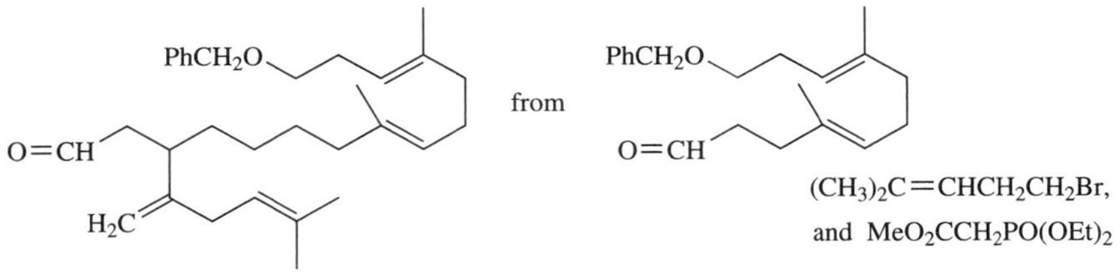

(r) C_6H_{13} NH OH $PhOCH_2$ from PhO NHOH C_6H_{13}

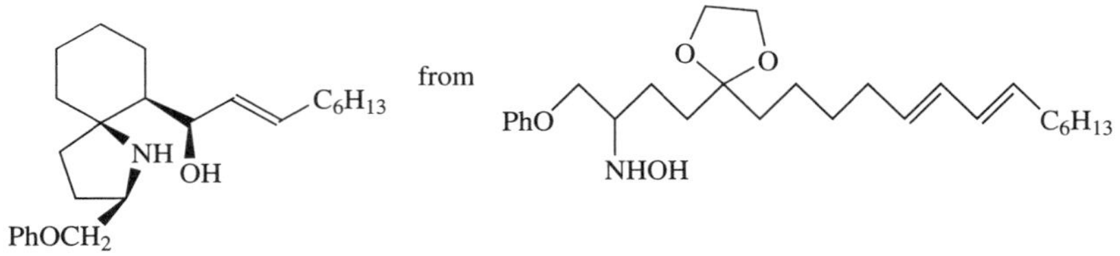

(s) $CO_2C_2H_5$ from $CO_2C_2H_5$

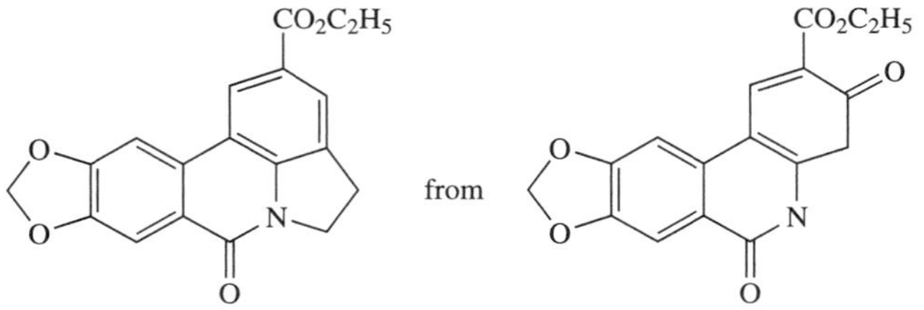

22. Identify a precursor which could provide the desired product by a single pericyclic reaction. Indicate approximate reaction conditions.

(a) H H OCH_2Ph CH_3O

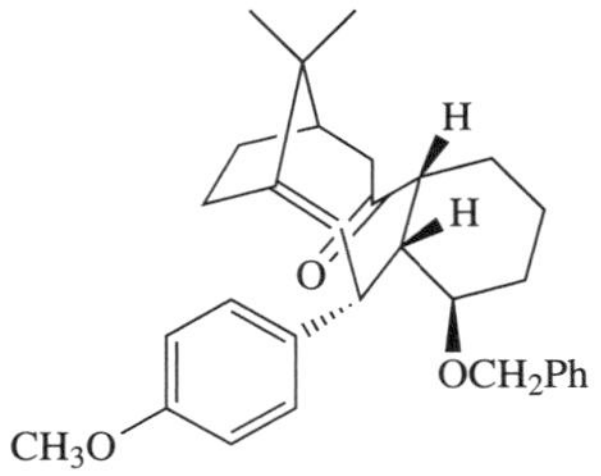

(b) H CH_3 H CH_3O_2C HO_2C O CH_2

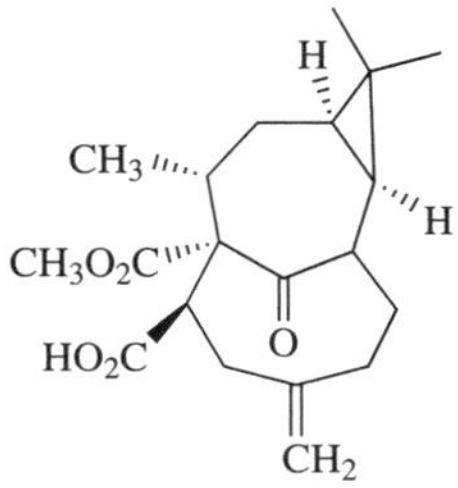

23. Predict the major product and its stereochemistry for each of the following reactions. Provide a structure for the transition state, and indicate the features that control the stereochemistry.

(a) [structure: 2,5-bis($PhCH_2CH_2$)-3-azabicyclo[3.2.0]heptane, N–H] $\xrightarrow{\text{1) } ONOSO_3H \text{; 2) } LiAlH_4 \text{; 3) } HgO}$

(b) $4{-}CH_3C_6H_4S$, CH_2OAc, OC_2H_5, 2,4,6-triMe—$PhCO_2$ $\xrightarrow{\text{MCPBA}}$

(c) $OSiR_3$, $O{-}CH_2$, NC, CH_3 $\xrightarrow{165°C}$

(d) $CH_3CH_2CO_2$, CH_3, O, CH_2OTMS $\xrightarrow{\text{1) LDA; 2) TMS–Cl; 3) 50°C; 4) } CH_2N_2}$

(e) $CH_2O_2CCH_2CH_3$, $C(CH_3)_3$ $\xrightarrow{\text{1) LDA, –78°C, THF; 2) } t\text{-}BuMe_2SiCl\text{, HMPA; 3) 50°C}}$

(f) $(CH_3)_2CH$, H, H, C=C, CH_3, CH, CH_3, $CH_2{=}CCH_2O$ $\xrightarrow[KOC(CH_3)_3]{n\text{-BuLi}}$

(g) $CH_2{=}CCH_3$, S, CH_3, OCH_2Ph, CH_3 $\xrightarrow{\text{1) } EtO_2C\overset{CH_3}{C}HO_3SCF_3\text{; 2) DBU}}$

(h) [structure] versus [structure] in intramolecular Diels–Alder reaction

(i) CH_3O_2C, O, O, OTBDMS $\xrightarrow[\text{Diels–Alder}]{\text{intramolecular}}$

(j) TBDMSO, O $\xrightarrow[-20°C, 45 min]{n\text{-BuLi}}$

24. Intramolecular Diels–Alder reactions occur when **Ta-c** are treated with $LiClO_4$ and TFA in ether. Indicate the mechanism of the reaction and analyze the stereoselectivity.

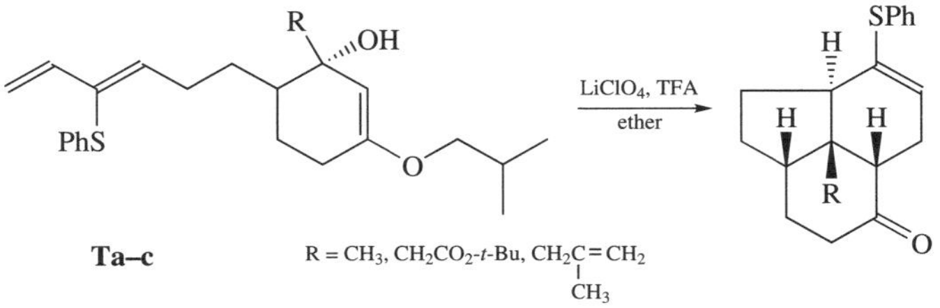
R
OH
PhS
O
LiClO4, TFA
ether
SPh
H
H
H
R
O
Ta–c
R = CH3, CH2CO2-t-Bu, CH2C=CH2
CH3

25. Oxepin is in equilibrium with benzene oxide by a [3,3] sigmatropic shift. Suggest a way to take advantage of this equilibrium to develop a short synthesis of barrelene.

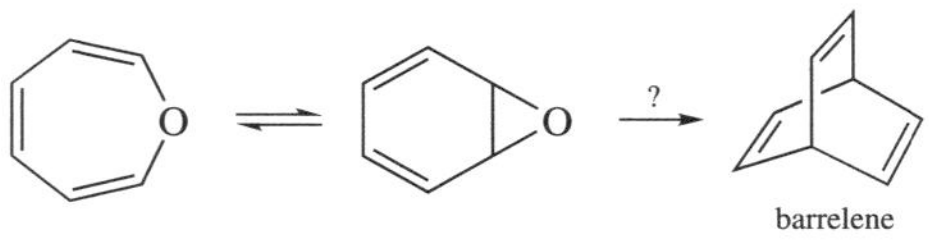
O
O
?
barrelene

26. Provide a detailed mechanistic description for each of the following transformations.

(a)

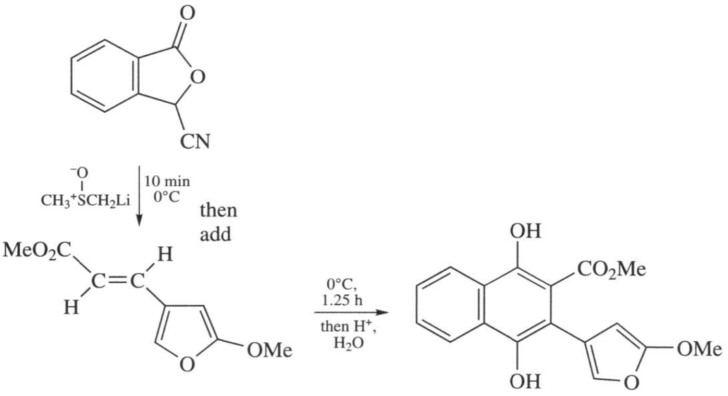
O
O
CN
-O
CH3+SCH2Li
10 min
0°C
then
add
MeO2C
H
C=C
H
O
OMe
0°C,
1.25 h
then H+,
H2O
OH
CO2Me
OH
O
OMe

(b)

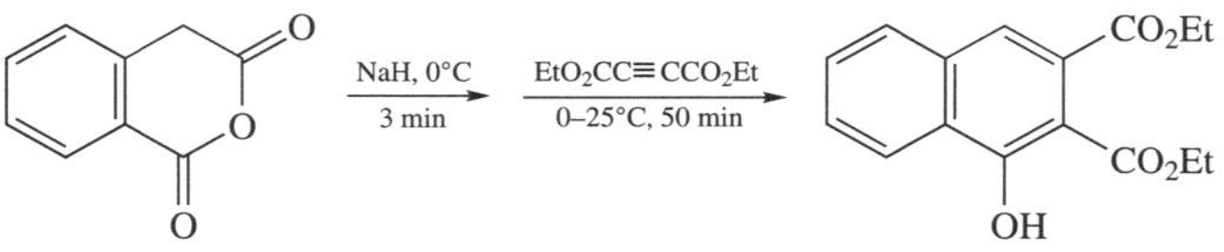
O
O
O
NaH, 0°C
3 min
EtO2CC≡CCO2Et
0–25°C, 50 min
CO2Et
CO2Et
OH

27. When the lactone silyl enol ether **U** is heated to 135°C, a mixture of four stereoisomers is obtained. Although the major one is that expected for a [3,3] sigmatropic rearrangement, lesser amounts of the other possible C-4a and C-5 epimers are also formed. When the reactant is heated to 100°C, partial conversion to the same mixture of stereoisomers is observed, but most of the product at this temperature is an acyclic triene ester.

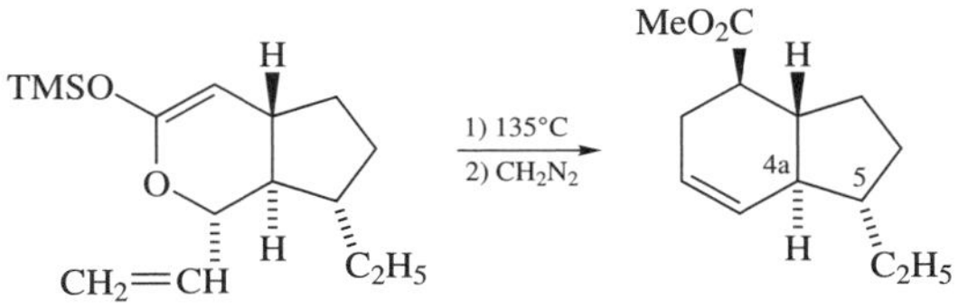

TMSO, O, H, H, $CH_2{=}CH$, C_2H_5, **U** —1) 135°C, 2) CH_2N_2→ MeO_2C, H, 4a, 5, H, C_2H_5

Suggest a structure for the triene ester, and show how it could be formed. Discuss the significance of the observation of the triene ester for the incomplete stereospecificity in the rearrangement.

28. Each of the following cycloaddition reactions exhibits a good degree of diastereoselectivity. Provide a rationalization of the observed diastereoselectivity in terms of a preferred transition-state structure.

(a)

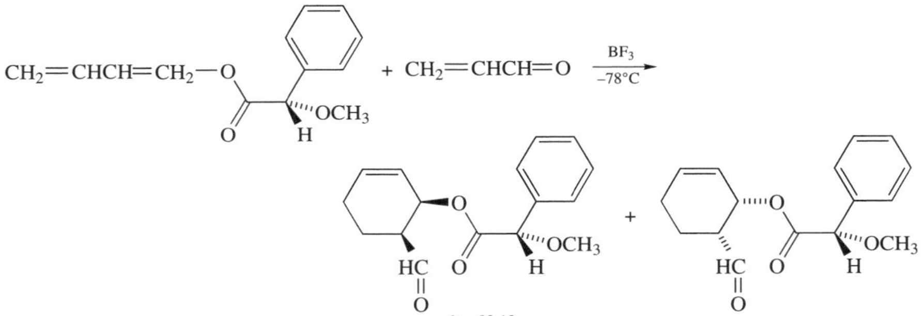

$CH_2{=}CHCH{=}CH_2{-}O$ ... O, OCH_3, H + $CH_2{=}CHCH{=}O$ —BF_3, −78°C→

dr = 82:18

(b)

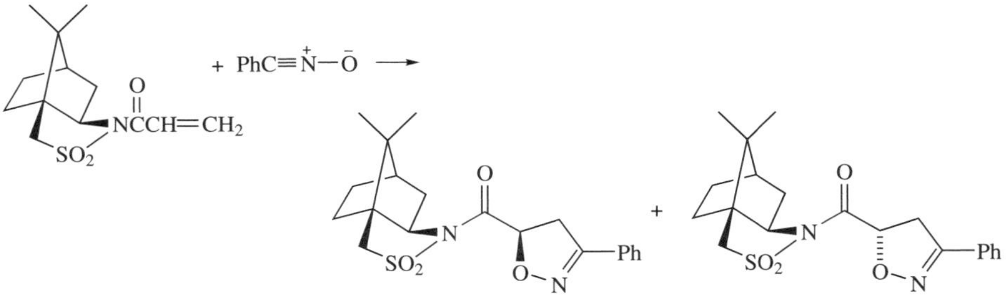

+ $Ph\overset{+}{C}{\equiv}N{-}\bar{O}$ ⟶ $NCCH{=}CH_2$, SO_2

dr = 95:5

(c)

O, $CH{=}CH_2$, N, $PhCH_2O_2C$ + cyclopentadiene —$TiCl_4$, 10°C→

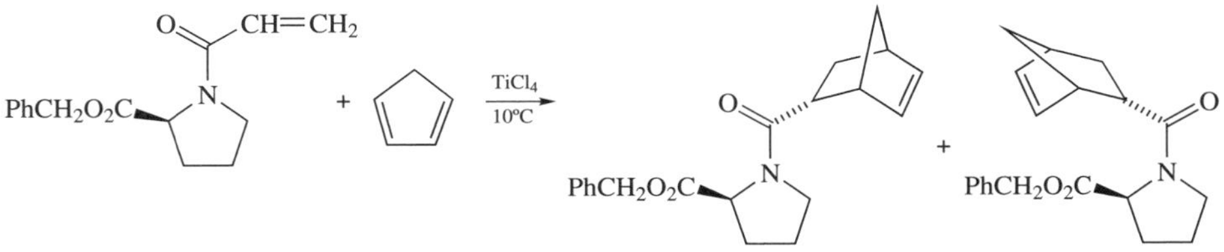

dr = 94.6

7

Organometallic Compounds of the Group I, II, and III Metals

Introduction

The use of organometallic reagents in organic synthesis had its beginning around 1900 when Victor Grignard discovered that alkyl and aryl halides react with magnesium metal to give homogeneous solutions containing organomagnesium compounds. The "Grignard reagents" proved to be highly reactive carbon nucleophiles and have remained very useful synthetic reagents since that time. Organolithium reagents came into synthetic use somewhat later. This chapter will focus primarily on Grignard reagents and organolithium compounds. These reagents are highly reactive carbon nucleophiles with fundamental applications in organic synthesis. We will also consider zinc, cadmium, mercury, indium, and lanthanide organometallics. Each of these classes of compounds has more specialized places in synthetic methodology. Certain of the transition metals, such as copper, palladium, and nickel, are important in synthetic methodology and will be discussed in Chapter 8.

7.1. Preparation and Properties

The compounds of lithium and magnesium are the most important of the group IA and IIA organometallics from a synthetic perspective. The metals in these two groups are the most electropositive of the elements. The polarity of the metal–carbon bond is such as to place high electron density on carbon. This electronic distribution is responsible for the strong nucleophilicity and basicity of these compounds.

The reaction of magnesium metal with an alkyl or aryl halide in diethyl ether is the classical method for synthesis of Grignard reagents.

$$RX + Mg \longrightarrow RMgX$$

The order of reactivity of the halides is RI > RBr > RCl. Solutions of Grignard reagents such as methylmagnesium bromide, ethylmagnesium bromide, and phenylmagnesium bromide are available commercially. Some Grignard reagents are formed in tetrahydrofuran more rapidly than in ether. This is true of vinylmagnesium bromide, for example.[1] The solubility of Grignard reagents in ethers is the result of strong Lewis acid–base complex formation between the ether molecules and the magnesium ion. A number of Grignard reagents have been subjected to X-ray structure determination.[2] Ethylmagnesium bromide has been observed in both monomeric and dimeric forms in crystal structures.[3] Figure 7.1a shows the crystal structure of the monomer with two diethyl ether molecules coordinated to magnesium. Figure 7.1b shows a dimeric structure with one diisopropyl ether molecule per magnesium.

Organic halides that are unreactive toward magnesium shavings can often be induced to react by using an extremely reactive form of magnesium that is obtained by reducing magnesium salts with sodium or potassium metal.[4] Even alkyl fluorides, which are normally unreactive, form Grignard reagents under these conditions. Sonication or mechanical pretreatment can also be used to activate magnesium.[5]

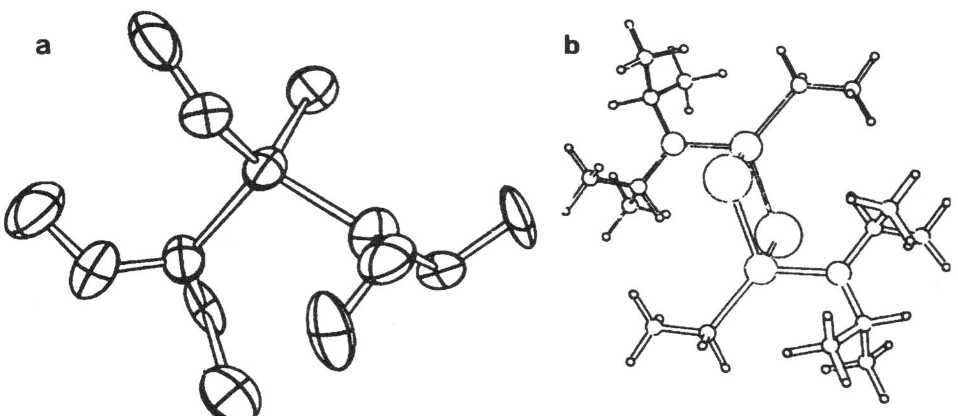

Fig. 7.1. Crystal structures of ethylmagnesium bromide. (a) Monomeric $C_2H_5MgBr[O(C_2H_5)_2]_2$. Reproduced with permission from Ref. 3a. Copyright 1968 American Chemical Society, (b) Dimeric $C_2H_5MgBr\ [O(i\text{-}C_3H_7)_2]$. Reproduced from Ref. 3b, Copyright 1974, with permission from Elsevier Science.)

1. D. Seyferth and F. G. A. Stone, *J. Am. Chem. Soc.* **79:**515 (1957); H. Normant, *Adv. Org. Chem.* **2:**1 (1960).
2. C, E. Holloway and M. Melinik, *Coord. Chem. Rev.* **135:**287 (1994); H. L. Uhm, in *Handbook of Grignard Reagents*, G. S. Silverman and P. E. Rakita, eds., Marcel Dekker, New York, 1996, pp. 117–144.
3. (a) L. J. Guggenberger and R. E. Rundle, *J. Am. Chem. Soc.* **90:**5375 (1968); (b) A. L. Spek, P. Voorbergen, G. Schat, C. Blomberg, and F. Bickelhaupt, *J. Organomet. Chem.*, **77:**147 (1974).
4. R. D. Rieke and S. E. Bales, *J. Am. Chem. Soc.* **96:**1775 (1974); R. D. Rieke, *Acc. Chem. Res.* **10:**301 (1977).
5. K. V. Baker, J. M. Brown, N. Hughes, A. J. Skarnulis, and A. Sexton, *J. Org. Chem.* **56:**698 (1991); J.-L. Luche and J.-C. Damiano, *J. Am. Chem. Soc.* **102:**7926 (1980).

The formation of Grignard reagents takes place at the metal surface. The reaction appears to begin at discrete sites.[6] Reaction commences with an electron transfer and decomposition of the radical ion, followed by rapid combination of the organic group with a magnesium ion.[7]

$$\mathrm{R{-}Br + Mg \longrightarrow R{-}Br^{\cdot -} + Mg(I)}$$
$$\mathrm{R{-}Br^{\cdot -} \longrightarrow R\cdot + Br^-}$$
$$\mathrm{R\cdot + Mg(I) + Br^- \longrightarrow R{-}Mg{-}Br}$$

One test for the involvement of radical intermediates is to determine if cyclization occurs in the 6-hexenyl system, in which radical cyclization is rapid (see Section 12.2.2 in Part A). Small amounts of cyclized products are formed upon preparation of the Grignard reagent from 5-hexenyl bromide.[8] This indicates that cyclization of the intermediate radical competes to a small extent with combination of the radical with the metal. A point of considerable discussion is whether the radicals generated are "free" or associated with the metal surface.[9]

The preparation of Grignard reagents from alkyl halides normally occurs with stereochemical randomization at the site of the reaction. Stereoisomeric halides give rise to organomagnesium compounds of identical composition.[10] The main exceptions to this generalization are cyclopropyl and alkenyl systems, which can be prepared with partial retention of configuration.[11] Once formed, secondary alkylmagnesium compounds undergo stereochemical inversion only slowly. *Endo*- and *exo*-norbornylmagnesium bromide, for example, require one day at room temperature to reach equilibrium.[12] NMR studies have demonstrated that inversion of configuration is quite slow, on the NMR time scale, even up to 170°C.[13] In contrast, the inversion of configuration of primary alkylmagnesium halides is very fast.[14] This difference between the primary and secondary systems may be the result of a mechanism for inversion that involves exchange of alkyl

6. C. E. Teerlinck and W. J. Bowyer, *J. Org. Chem.* **61:**1059 (1996).
7. H. R. Rogers, C. L. Hill, Y. Fujuwara, R. J. Rogers, H. L. Mitchell, and G. M. Whitesides, *J. Am. Chem. Soc.* **102:**217 (1980); J. F. Garst, J. E. Deutch, and G. M. Whitesides, *J. Am. Chem. Soc.* **108:**2490 (1986); E. C. Ashby and J. Oswald, *J. Org. Chem.* **53:**6068 (1988); H. M. Walborsky, *Acc. Chem. Res.* **23:**286 (1990); H. M. Walborsky and C. Zimmermann, *J. Am. Chem. Soc.* **114:**4996 (1992); C. Hamdouchi, M. Topolski, V. Goedken, and H. M. Walborsky, *J. Org. Chem.* **58:**3148 (1993); C. Hamdouchi and H. M. Walborsky, in *Handbook of Grignard Reagents*, G. S. Silverman and P. E. Rakita, eds., Marcel Dekker, New York, 1996, pp. 145-218.
8. R. C. Lamb, P. W. Ayers, and M. K. Toney, *J. Am. Chem. Soc.* **85:**3483 (1963); R. C. Lamb and P. W. Ayers, *J. Org. Chem.* **27:**1441 (1962); C. Walling and A. Cioffari, *J. Am. Chem. Soc.* **92:**6609 (1970); H. W. H. J. Bodewitz, C. Blomberg, and F. Bickelhaupt, *Tetrahedron* **31:**1053 (1975); J. F. Garst and B. L. Swift, *J. Am. Chem. Soc.* **111:**241 (1989).
9. C. Walling, *Acc. Chem. Res.* **24:**255 (1991); J. F. Garst, F. Ungvary, R. Batlaw, and K. E. Lawrence, *J. Am. Chem. Soc.* **113:**5392 (1991).
10. N. G. Krieghoff and D. O. Cowan, *J. Am. Chem. Soc.* **88:**1322 (1966).
11. T. Yoshino and Y. Manabe, *J. Am. Chem. Soc.* **85:**2860 (1963); H. M. Walborsky and A. E. Young, *J. Am. Chem. Soc.* **86:**3288 (1964); H. M. Walborsky and B. R. Banks, *Bull. Soc. Chim. Belg.* **89:**849 (1980).
12. F. R. Jensen and K. L. Nakamaye, *J. Am. Chem. Soc.* **88:**3437 (1966); N. G. Krieghoff and D. O. Cowan, *J. Am. Chem. Soc.* **88:**1322 (1966).
13. E. Pechold, D. G. Adams, and G. Fraenkel, *J. Org. Chem.* **36:**1368 (1971).
14. G. M. Whitesides, M. Witanowski, and J. D. Roberts, *J. Am. Chem. Soc.* **87:**2854 (1965); G. M. Whitesides and J. D. Roberts, *J. Am. Chem. Soc.* **87:**4878 (1965); G. Fraenkel and D. T. Dix, *J. Am. Chem. Soc.* **88:**979 (1966).

groups between magnesium atoms:

$$\begin{array}{c} \text{X}-\text{Mg}-\text{CR}_2\text{H} \\ \text{HR}_2\text{C}-\text{Mg}-\text{X} \end{array} \rightleftharpoons \text{R}_2\text{HC}-\text{Mg}(\mu\text{-X})_2\text{Mg}-\text{CHR}_2$$

If such bridged intermediates are involved, the larger steric bulk of secondary systems would retard the reaction. Steric restrictions may be further enhanced by the fact that organomagnesium reagents are often present as clusters (see below).

The usual designation of Grignard reagents as RMgX is a useful but incomplete representation of the composition of the compounds in ether solution. An equilbrium exists with magnesium bromide and the dialkylmagnesium.

$$2\ \text{RMgX} \rightleftharpoons \text{R}_2\text{Mg} + \text{MgX}_2$$

The position of the equilibrium depends upon the solvent and the identity of the specific organic group but lies far to the left in ether for simple aryl-, alkyl-, and alkenylmagnesium halides.[15] Solutions of organomagnesium compounds in diethyl ether contain aggregated species.[16] Dimers predominate in ether solutions of alkylmagnesium chlorides.

$$2\ \text{RMgCl} \rightleftharpoons \text{R}-\text{Mg}(\mu\text{-Cl})_2\text{Mg}-\text{R}$$

The corresponding bromides and iodides show concentration-dependent behavior, and in very dilute solutions they exist as monomers. In tetrahydrofuran, there is less tendency to aggregate, and several alkyl and aryl Grignard reagents have been found to be monomeric in this solvent.

Most simple organolithium reagents can be prepared by reaction of an appropriate halide with lithium metal.

$$\text{R}-\text{X} + 2\ \text{Li} \longrightarrow \text{RLi} + \text{LiX}$$

As with organomagnesium reagents, there is usually loss of stereochemical integrity at the site of reaction during the preparation of alkyllithium compounds.[17] Alkenyllithium reagents can usually be prepared with retention of configuration of the double bond.[18]

For some halides, it is advantageous to use finely powdered lithium and a catalytic amount of an aromatic hydrocarbon, ususally naphthalene or 4,4′-di-*t*-butylbiphenyl (DTBB).[19] These reactions may involve anions generated by reduction of the aromatic ring (see Section 5.5.1) which then convert the halide to a radical anion. Several useful

15. G. E. Parris and E. C. Ashby, *J. Am. Chem. Soc.* **93:**1206 (1971); P. E. M. Allen, S. Hagias, S. F. Lincoln, C. Mair, and E. H. Williams, *Ber. Bunsenges. Phys. Chem.* **86:**515 (1982).
16. E. C. Ashby and M. B. Smith, *J. Am. Chem. Soc.* **86:**4363 (1964); F. W. Walker and E. C. Ashby, *J. Am. Chem. Soc.* **91:**3845 (1969).
17. W. H. Glaze and C. M. Selman, *J. Org. Chem.* **33:**1987 (1968).
18. J. Millon, R. Lorne, and G. Linstrumelle, *Synthesis* **1975:**434.
19. M. Yus, *Chem. Soc. Rev.* **25:**155 (1996) D. J. Ramon and M. Yus, *Tetrahedron* **52:**13739 (1996).

functionalized lithium reagents have been prepared by this method.

$$\mathrm{ClCH{=}C(OC_2H_5)_2 \xrightarrow[\text{DTBB}]{\text{Li, 5 equiv}} LiCH{=}C(OC_2H_5)_2}$$ Ref. 20

Ref. 21

$$\mathrm{[(CH_3)_2CH]_2NC(O)Cl + PhCH{=}O \xrightarrow[\text{naphthalene}]{\text{Li}} [(CH_3)_2CH]_2NC(O)CH(OH)Ph}$$ 79% Ref. 22

Alkyllithium reagents can also be generated by reduction of sulfides.[23] This technique is especially useful for the preparation of α-lithio ethers, sulfides, and silanes.[24] The lithium radical anion of naphthalene, DTBB, or dimethylaminonaphthalene (LDMAN) is used as the reducing agent.

$$\mathrm{PhSC(CH_3)_2Si(CH_3)_3 \xrightarrow{\text{LDMAN}} LiC(CH_3)_2Si(CH_3)_3}$$

Alkenyllithium and substituted alkyllithium reagents can be prepared from sulfides.[25] The method can also be applied to unsubstituted alkyllithium reagents, although in this case there is no special advantage over the conventional procedure.

$$\mathrm{PhCH_2CH_2SPh \xrightarrow[\text{Li^+Naph^-}]{\text{Li or}} PhCH_2CH_2Li}$$ Ref. 26

Sulfides can also be converted to lithium reagents by the catalytic electron-transfer process described earlier for halides.[27]

20. M. Si-Fodil, H. Ferrreira, J. Gralak, and L. Duhamel, *Tetrahedron Lett.* **39:**8975 (1998).
21. A. Bachki, F. Foubelo, and M. Yus, *Tetrahedron.* **53:**4921 (1997).
22. A. Guijarro, B. Mandeno, J. Ortiz, and M. Yus, *Tetrahedron* **52:**1643 (1993).
23. T. Cohen and M. Bhupathy, *Acc. Chem. Res.* **22:**152 (1989).
24. T. Cohen and J. R. Matz, *J. Am. Chem. Soc.* **102:**6900 (1980); T. Cohen, J. P. Sherbine, J. R. Matz, R. R. Hutchins, B. M. McHenry, and P. R. Wiley, *J. Am. Chem. Soc.* **106:**3245 (1984); S. D. Rychnovsky, K. Plzak, and D. Pickering, *Tetrahedron Lett.* **35:**6799 (1994); S. D. Rychnovsky and D. J. Skalitzky, *J. Org. Chem.* **57:**4336 (1992).
25. T. Cohen and M. D. Doubleday, *J. Org. Chem.* **55:**4784 (1990); D. J. Rawson and A. I Meyers, *Tetrahedron Lett.* **32:**2095 (1991); H. Liu and T. Cohen, *J. Org. Chem.* **60:**2022 (1995).
26. C. G. Screttas and M. Micha-Screttas, *J. Org. Chem.* **43:**1064 (1978); C. G. Screttas and M. Micha-Screttas, *J. Org. Chem.* **44:**713 (1979).
27. F. Foubelo, A. Gutierrez, and M. Yus, *Synthesis* **1999:**503.

The simple alkyllithium reagents exist mainly as hexamers in hydrocarbon solvents.[28] In ethers, tetrameric structures are usually dominant.[29] The tetramers, in turn, are solvated by ether molecules.[30] Phenyllithium is tetrameric in cyclohexane and a mixture of monomer and dimer in tetrahydrofuran.[31] Chelating ligands such as tetramethylethylenediamine (TMEDA) reduce the degree of aggregation.[32] Strong donor molecules such as hexamethylphosphorictriamide (HMPA) and *N,N*-dimethylpropyleneurea (DMPU) also lead to more dissociated and more reactive organolithium reagents.[33] NMR studies on phenyllithium show that TMEDA, other polyamine ligands, HMPA, and DMPU favor monomeric solvated species.[34]

$O{=}P[N(CH_3)_2]_3$

HMPA

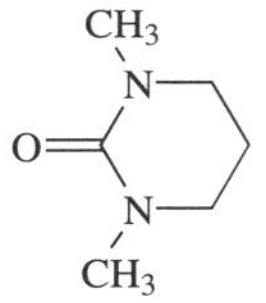

DMPU

The crystal structures of many organolithium compounds have been determined.[35] Phenyllithium has been crystallized as an ether solvate. The structure is tetrametic, with lithium and carbon atoms at alternating corners of a strongly distorted cube. Each carbon is 2.33 Å from the three neighboring lithium atoms. An ether molecule is coordinated to each lithium atom. Figure 7.2a shows the Li–C cluster, and Fig. 7.2b shows the complete array of atoms, except for hydrogen.[36] Section 7.1 of Part A provides additional information on the structure of organolithium compounds.

There are two other general methods that are very useful for preparing organolithium reagents. The first of these is hydrogen–metal exchange or metalation. This reaction is the usual method for preparing alkynylmagnesium and alkynyllithium reagents. The reaction proceeds readily because of the relative acidity of the hydrogen bound to *sp* carbon.

$$H{-}C{\equiv}C{-}R + R'MgBr \longrightarrow BrMgC{\equiv}C{-}R + R'{-}H$$

$$H{-}C{\equiv}C{-}R + R'Li \longrightarrow LiC{\equiv}C{-}R + R'{-}H$$

Although of limited utility for other types of Grignard reagents, metalation is an important means of preparing a variety of organolithium compounds. The position of lithiation is determined by the relative acidity of the available hydrogens and the directing effect of

28. G. Fraenkel, W. E. Beckenbaugh, and P. P. Yang, *J. Am. Chem. Soc.* **98:**6878 (1976); G. Fraenkel, M. Henrichs, J. M. Hewitt, B. M. Su, and M. J. Geckle, *J. Am. Chem. Soc.* **102:**3345 (1980).
29. H. L. Lewis and T. L. Brown, *J. Am. Chem. Soc.* **92:**4664 (1970); P. West and R. Waack, *J. Am. Chem. Soc.* **89:**4395 (1967); J. F. McGarrity and C. A. Ogle, *J. Am. Chem. Soc.* **107:**1085 (1985); D. Seebach, R. Hässig, and J. Gabriel, *Helv. Chim. Acta* **66:**308 (1983); T. L. Brown, *Adv. Organomet. Chem.* **3:**365 (1965); W. N. Setzer and P. v. R. Schleyer, *Adv. Organomet. Chem.* **24:**354 (1985); W. Bauer, T. Clark, and P. v. R. Schleyer, *J. Am. Chem. Soc.* **109:**970 (1987).
30. P. D. Bartlett, C. V. Goebel, and W. P. Weber, *J. Am. Chem. Soc.* **91:**7425 (1969).
31. L. M. Jackman and L. M. Scarmoutzos, *J. Am. Chem. Soc.* **106:**4627 (1984); O. Eppers and H. Gunther, *Helv. Chim. Acta* **75:**2553 (1992).
32. W. Bauer and C. Griesinger, *J. Am. Chem. Soc.* **115:**10871 (1993); D. Hofffmann and D. B. Collum, *J. Am. Chem. Soc.* **120:**5810 (1998).
33. H. J. Reich and D. P. Green, *J. Am. Chem. Soc.* **111:**8729 (1989).
34. H. J. Reich, D. P. Green, M. A. Medina, W. S. Goldenberg, B. O. Gudmundsson, R. R. Dykstra, and N. H. Phillips, *J. Am. Chem. Soc.* **120:**7201 (1998).
35. E. Weiss, *Angew. Chem. Int. Ed. Engl.* **32:**1501 (1993).
36. H. Hope and P. P. Power, *J. Am. Chem. Soc.* **105:**5320 (1983).

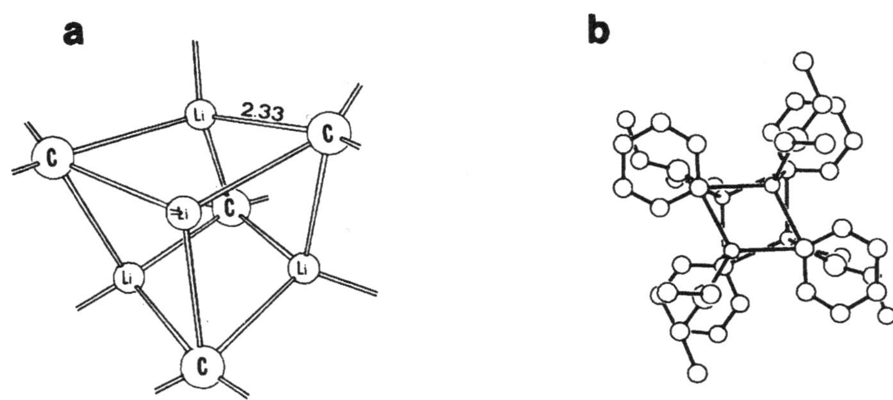

Fig. 7.2. Crystal structure of tetrameric phenyllithium etherate. (a) Tetrameric cluster. (b) Complete structure except for hydrogens. (Reproduced with permission from Ref. 36. Copyright 1983 American Chemical Society.)

substituent groups. Benzylic and allylic hydrogens are relatively reactive toward lithiation because of the resonance stabilization of the resulting anions.[37] Substituents which can coordinate to the lithium atom, such as alkoxy, amido, sulfoxide, and sulfonyl, have a powerful influence on the position and rate of lithiation of aromatic compounds.[38] Some substituents, such as *t*-butoxycarbonylamido and carboxy, undergo deprotonation during the lithiation process.[39] The methoxymethoxy substituent is particularly useful among the alkoxy directing groups. It can provide selective lithiation and, being an acetal, is readily removed by hydrolysis.[40] In heteroaromatic compounds, the preferred site for lithiation is usually adjacent to the heteroatom. The features which characterize the activating groups include a donor pair that can coordinate lithium and an electron-withdrawing functional group that can stabilize anionic character.[41] If competing nucleophilic attack is a possibility, as in tertiary amides, steric bulk is also an important factor. Scheme 7.1 gives some examples of the preparation of organolithium compounds by lithiation.

Reaction conditions can be modified to accelerate the rate of lithiation when necessary. Addition of tertiary amines, especially TMEDA, accelerates lithiation[42] by

37. R. D. Clark and A. Jahangir, *Org. React.* **47:**1 (1995).
38. D. W. Slocum and C. A. Jennings, *J. Org. Chem.* **41:**3653 (1976); J. M. Mallan and R. C. Rebb, *Chem. Rev.* **69:**693 (1969); H. W. Gschwend and H. R. Rodriguez, *Org. React.* **26:**1 (1979); V. Snieckus, *Chem. Rev.* **90:**879 (1990); C. Quesnelle, T. Iihama, T. Aubert, H. Perrier, and V. Snieckus, *Tetrahedron Lett.* **33:**2625 (1992); M. Iwao, T. Iihama, K. K. Mahalandabis, H. Perrier, and V. Snieckus, *J. Org. Chem.* **54:**24 (1989); L. A. Spangler, *Tetrahedron Lett.* **37:**3639 (1996).
39. J. M. Muchowski and M. C. Venuti, *J. Org. Chem.* **45:**4798 (1980); P. Stanetty, H. Koller, and M. Mihovilovic, *J. Org. Chem.* **57:**6833 (1992); J. Mortier, J. Moyroud, B. Bennetau, and P. A. Cain, *J. Org. Chem.* **59:**4042 (1994).
40. C. A. Townsend and L. M. Bloom, *Tetrahedron Lett.* **22:**3923 (1981); R. C. Ronald and M. R. Winkle, *Tetrahedron* **39:**2031 (1983); M. R. Winkle and R. C. Ronald, *J. Org. Chem.* **47:**2101 (1982).
41. N. J. R. van Eikema Hommes and P. v. R. Schleyer, *Angew. Chem. Int. Ed. Engl.* **31:**755 (1992); N. J. R. van Eikema Hommes and P. v. R. Schleyer, *Tetrahedron* **50:**5903 (1994).
42. G. G. Eberhardt and W. A. Butte, *J. Org. Chem.* **29:**2928 (1964); R. West and P. C. Jones, *J. Am. Chem. Soc.* **90:**2656 (1968); S. Akiyama and J. Hooz, *Tetrahedron Lett.* **1973:**4115; D. W. Slocum, R. Moon, J. Thompson, D. S. Coffey, J. D. Li, M. G. Slocum, A. Siegel, and R. Gayton-Garcia, *Tetrahedron Lett.* **35:**385 (1994); M. Khaldi, F. Chretien, and Y. Chapleur, *Tetrahedron Lett.* **35:**401 (1994); D. B. Collum, *Acc. Chem. Res.* **25:**448 (1992).

Scheme 7.1. Organolithium Compounds by Metalation

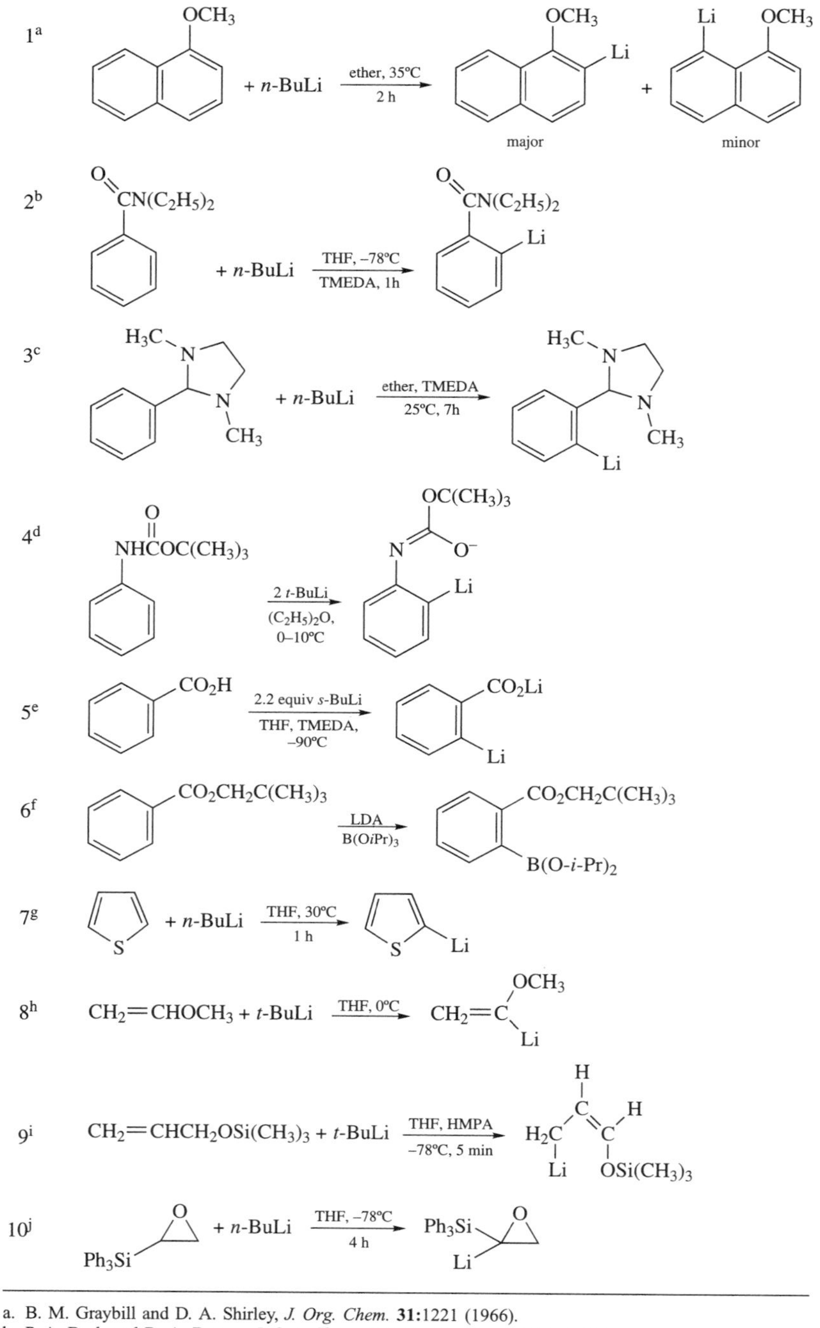

1[a] 1-methoxynaphthalene + n-BuLi → (ether, 35°C, 2 h) 2-lithio-1-methoxynaphthalene (major) + 8-lithio-1-methoxynaphthalene (minor)

2[b] $C_6H_5CON(C_2H_5)_2$ + n-BuLi → (THF, –78°C, TMEDA, 1h) o-Li-$C_6H_4CON(C_2H_5)_2$

3[c] 1,3-dimethyl-2-phenylimidazolidine + n-BuLi → (ether, TMEDA, 25°C, 7h) 2-(2-lithiophenyl)-1,3-dimethylimidazolidine

4[d] $C_6H_5NHCOC(CH_3)_3$ (O over C) → (2 t-BuLi, $(C_2H_5)_2O$, 0–10°C) o-Li-$C_6H_4N{=}C(O^-)OC(CH_3)_3$

5[e] $C_6H_5CO_2H$ → (2.2 equiv s-BuLi, THF, TMEDA, –90°C) o-Li-$C_6H_4CO_2Li$

6[f] $C_6H_5CO_2CH_2C(CH_3)_3$ → (LDA, $B(OiPr)_3$) o-$(B(O\text{-}i\text{-}Pr)_2)C_6H_4CO_2CH_2C(CH_3)_3$

7[g] thiophene + n-BuLi → (THF, 30°C, 1 h) 2-lithiothiophene

8[h] $CH_2{=}CHOCH_3$ + t-BuLi → (THF, 0°C) $CH_2{=}C(OCH_3)Li$

9[i] $CH_2{=}CHCH_2OSi(CH_3)_3$ + t-BuLi → (THF, HMPA, –78°C, 5 min) $LiH_2C\text{–}CH{=}CH\text{–}OSi(CH_3)_3$

10[j] Ph_3Si-oxirane + n-BuLi → (THF, –78°C, 4 h) 2-lithio-2-(Ph_3Si)oxirane

a. B. M. Graybill and D. A. Shirley, *J. Org. Chem.* **31:**1221 (1966).
b. P. A. Beak and R. A. Brown, *J. Org. Chem.* **42:**1823 (1977); **44:**4463 (1979).
c. T. D. Harris and G. P. Roth, *J. Org. Chem.* **44:**2004 (1979).
d. P. Stanetty, H. Koller, and M. Mihovilovic, *J. Org. Chem.* **57:**6833 (1992).
e. B. Bennetau, J. Mortier, J. Moyroud, and J.-L. Guesnet, *J. Chem. Soc., Perkin Trans. 1* **1995:**1265.
f. S. Caron and J. M. Hawkins, *J. Org. Chem.* **63:**2054 (1998).
g. E. Jones and I. M. Moodie, *Org. Synth.* **50:**104 (1970).
h. J. E. Baldwin, G. A. Höfle, and O. W. Lever, Jr., *J. Am. Chem. Soc.* **96:**7125 (1974).
i. W. C. Still and T. L. Macdonald, *J. Org. Chem.* **41:**3620 (1976).
j. J. J. Eisch and J. E. Galle, *J. Am. Chem. Soc.* **98:**4646 (1976).

chelation at the lithium, which promotes dissociation of aggregated structures. Kinetic and spectroscopic evidence indicates that, in the presence of TMEDA, lithiation of anisole involves the solvated dimeric species $(BuLi)_2(TMEDA)_2$.[43] The reaction shows an isotope effect for the *ortho*-hydrogen, establishing that proton abstraction is rate-determining.[44] It is likely that there is a precomplexation between the anisole and organometallic dimer.

Lithiation of alkyl groups is also possible, and again a combination of donor chelation and dipolar stabilization of anionic character are the requirements. Amides and carbamates can be lithiated α to the nitrogen.

$CH_3CH_2N(CH_3)CO_2C(CH_3)_3$ —1) *s*-BuLi, TMEDA; 2) $(CH_3)_3SiCl$→ $CH_3CH_2N(CH_2Si(CH_3)_3)CO_2C(CH_3)_3$ Ref. 45

CH_3 (N-$CO_2C(CH_3)_3$ piperidine) —1) *s*-BuLi, TMEDA; 2) E^+→ CH_3, E (N-$CO_2C(CH_3)_3$ piperidine); intermediate: $(CH_3)_3CO$, N, O--Li, CH_3 Ref. 46

Formamidines can also be lithiated.[47]

H, H, N, H, $NC(CH_3)_3$ —*t*-BuLi→ H, H, N, Li, H, $NC(CH_3)_3$

Tertiary amides with carbanion stabilization at the β carbon give β-lithiation.[48]

$CH_3CH(CH_2R)CON[(CH(CH_3)_2]_2$ —*s*-BuLi, TMEDA→ $CH_3CH(LiCHR)CON[(CH(CH_3)_2]_2$ —E^+→ $CH_3CH(ECHR)CON[(CH(CH_3)_2]_2$

R = Ph, PhS, $CH_2 = CH$

$E^+ = RCH_2I$, $RCH = O$

β-Lithiation has also been observed for deprotonated secondary amides of 3-phenylpropanoic acid.

$PhCH_2CH_2CONHR$ —2 *s*-BuLi, –78°C→ Li-----O^-, Ph, NR Ref. 49

R = CH_3, $CH(CH_3)_2$

43. R. A. Reynolds, A. J. Maliakal, and D. B. Collum, *J. Am. Chem. Soc.* **120:**421 (1998).
44. M. Stratakis, *J. Org. Chem.* **62:**3024 (1997).
45. V. Snieckus, M. Rogers-Evans, P. Beak, W. K. Lee, E. K. Yum, and J. Freskos, *Tetrahedron Lett.* **35:**4067 (1994).
46. P. Beak and W. K. Lee, *J. Org. Chem.* **58:**1109 (1993).
47. A. I. Meyers and G. Milot, *J. Am. Chem. Soc.* **115:**6652 (1993).
48. P. Beak, J. E. Hunter, Y. M. Jun, and A. P. Wallin, *J. Am. Chem. Soc.* **109:**5403 (1987); G. P. Lutz, A. P. Wallin, S. T. Kerrick, and P. Beak, *J. Org. Chem.* **56:**4938 (1991).
49. G. P. Lutz, H. Du, D. J. Gallagher and P. Beak, *J. Org. Chem.* **61:**4542 (1996).

The mechanism of directed lithiation appears to involve an association between the amide substituent and the lithiating agent.[50]

Hydrocarbons lacking directing substituents are not very reactive toward metalation, but it has been found that a mixture of *n*-butyllithium and potassium *t*-butoxide[51] is sufficiently reactive to give allyl anions from alkenes such as isobutene.[52]

$$CH_2{=}C(CH_3)_2 \xrightarrow[KOC(CH_3)_3]{n\text{-BuLi}} CH_2{=}C(CH_3)CH_2Li$$

Metal–halogen exchange is also an important method for preparation of organolithium reagents. This reaction proceeds in the direction of forming the more stable organolithium reagent, that is, the one derived from the more acidic compound. Thus, by use of the very basic alkyllithium compounds, such as *n*-butyl- or *t*-butyllithium, halogen substituents at carbons where the anion is stabilized are readily exchanged to give the corresponding lithium compound. Halogen–metal exchange is particularly useful for converting aryl and alkenyl halides to the corresponding lithium compounds. The driving force of the reaction is the greater stability of sp^2 carbanions in comparison with sp^3 carbanions. Scheme 7.2 gives some examples of these reactions.

$$\text{(H)(Ph)C{=}C(H)(Br)} \xrightarrow[\text{pentane-THF-Et}_2\text{O}]{t\text{-BuLi, }-120^\circ\text{C}} \text{(H)(Ph)C{=}C(H)(Li)}$$ Ref. 53

$$m\text{-MeOC}_6\text{H}_4\text{Br} \xrightarrow[-78^\circ\text{C}]{n\text{-BuLi}} m\text{-MeOC}_6\text{H}_4\text{Li}$$ Ref. 54

Metal–halogen exchange is a very fast reaction and is usually carried out at −60 to −120°C. This makes it possible to prepare aryllithium compounds containing functional groups, such as cyano and nitro, which would react under the conditions required for preparation from lithium metal. Entries 6 and 7 in Scheme 7.2 are examples. For alkyl halides, halogen–metal exchange is restricted by competing reactions, but primary alkyllithium reagents can be prepared from iodides under carefully controlled conditions.[55]

Retention of configuration is often observed when organolithium compounds are prepared by metal–halogen exchange. The degree of retention is low for exchange of most

50. W. Bauer and P. v. R. Schleyer, *J. Am. Chem. Soc.* **111:**7191 (1989); P. Beak, S. T. Kerrick, and D. J. Gallagher, *J. Am. Chem. Soc.* **115:**10628 (1993).
51. L. Lochmann, J. Pospisil, and D. Lim, *Tetrahedron Lett.* **1966:**257.
52. M. Schlosser and J. Hartmann, *Angew. Chem. Int. Ed. Engl.* **12:**508 (1973); J. J. Bahl, R. B. Bates, and B. Gordon, III, *J. Org. Chem.* **44:**2290 (1979); M. Schlosser and G. Rauchshwalbe, *J. Am. Chem. Soc.* **100:**3258 (1978).
53. N. Neumann and D. Seebach, *Tetrahedron Lett.* **1976:**4839.
54. T. R. Hoye, S. J. Martin, and D. R. Peck, *J. Org. Chem.* **47:**331 (1982).
55. W. F. Bailey and E. R. Punzalan, *J. Org. Chem.* **55:**5404 (1990); E. Negishi, D. R. Swanson, and C. J. Rousset, *J. Org. Chem.* **55:**5406 (1990).

1[a] $H_3C(H)C{=}C(H)Br$ + t-BuLi —(−120°C)→ $H_3C(H)C{=}C(H)Li$

2[b] 1-bromocyclopentene + n-BuLi —(−70°C)→ 1-lithiocyclopentene

3[c] $CH_3(CH_2)_3CH{=}CHI$ —(2 equiv t-BuLi, hexane, 25°C)→ $CH_3(CH_2)_3CH{=}CHLi$

4[d] 1-bromo-2-methoxycyclopropane (CH_3O, Br) + t-BuLi ⟶ (CH_3O, Li) cyclopropane

5[e] $C_4H_9(H)C{=}C(Si(CH_3)_3)Br$ + s-BuLi —(−70°C)→ $C_4H_9(H)C{=}C(Si(CH_3)_3)Li$

6[f] 3-bromobenzonitrile ($N{\equiv}C$, Br) + n-BuLi —(−100°C)→ ($N{\equiv}C$, Li)

7[g] 2,5-dibromonitrobenzene (Br, NO_2, Br) + n-BuLi —(−100°C)→ (Br, NO_2, Li)

8[h] 2-chloropyridine (N, Cl) —(Li naphthalenide, −78°C)→ 2-lithiopyridine (N, Li)

9[i] $CH_3(CH_2)_5$ / SPh / O / O / $CH(CH_3)_2$ —(Li naphthalenide, −78°C)→ $CH_3(CH_2)_5$ / Li / O / O / $CH(CH_3)_2$ —(−20°C)→ $CH_3(CH_2)_5$ / Li / O / O / $CH(CH_3)_2$

a. H. Neuman and D. Seebach, *Tetrahedron Lett.* **1976:**4839.
b. J. Millon, R. Lorne, and G. Linstrumelle, *Synthesis* **1975:**434.
c. M. A. Peterson and R. Polt, *Synth. Commun.* **22:**477 (1992).
d. E. J. Corey and P. Ulrich, *Tetrahedron Lett.* **1975:**3685.
e. R. B. Miller and G. McGarvey, *J. Org. Chem.* **44:**4623 (1979).
f. W. E. Parham and L. D. Jones, *J. Org. Chem.* **41:**1187 (1976).
g. W. E. Parham and R. M. Piccirilli, *J. Org. Chem.* **42:**257 (1977).
h. Y. Kondo, N. Murata, and T. Sakamoto, *Heterocycles* **37:**1467 (1994).
i. S. D. Rychnovsky and D. J. Skalitsky, *J. Org. Chem.* **57:**4336 (1992).

alkyl systems,[56] but it is normally high for cyclopropyl and vinyl halides.[57] Once formed, both cyclopropyl- and vinyllithium reagents retain their configuration at room temperature.

Another useful method of preparing organolithium reagents involves metal–metal exchange. The reaction between two organometallic compounds proceeds in the direction

56. R. L. Letsinger, *J. Am. Chem. Soc.* **72:**4842 (1950); D. Y. Curtin and W. J. Koehl, Jr., *J. Am. Chem. Soc.* **84:**1967 (1962).
57. H. M. Walborsky, F. J. Impastato, and A. E. Young, *J. Am. Chem. Soc.* **86:**3283 (1964); D. Seyferth and L. G. Vaughan, *J. Am. Chem. Soc.* **86:**883 (1964); M. J. S. Dewar and J. M. Harris, *J. Am. Chem. Soc.* **91:**3652 (1969); E. J. Corey and P. Ulrich, *Tetrahedron Lett.* **1975:**3685; N. Neumann and D. Seebach, *Tetrahedron Lett.* **1976:**4839; R. B. Miller and G. McGarvey, *J. Org. Chem.* **44:**4623 (1979).

of placing the more electropositive metal at the more stable carbanion position. Exchanges between vinyltin reagents and alkyllithium reagents are particularly significant from a synthetic point of view.

$$HC\equiv CCH_2OTHP \xrightarrow{Bu_3SnH} (H)(Bu_3Sn)C{=}C(CH_2OTHP)(H) \xrightarrow{n\text{-}BuLi} (H)(Li)C{=}C(CH_2OTHP)(H)$$ Ref. 58

$$RCH(OR')SnBu_3 + n\text{-}BuLi \xrightarrow{-78^\circ C} RCH(OR')Li$$ Ref. 59

$$R_2NCH_2SnBu_3 + n\text{-}BuLi \xrightarrow{0^\circ C} R_2NCH_2Li$$ Ref. 60

The α-tri-*n*-butylstannyl derivatives needed for the latter two examples are readily available.

$$RCH{=}O + Bu_3SnLi \longrightarrow RCH(SnBu_3){-}O^- \xrightarrow{R'X} RCH(OR')SnBu_3$$

$$R_2NCH_2SPh + Bu_3SnLi \longrightarrow R_2NCH_2SnBu_3$$

The exchange reactions of α-alkoxystannanes occur with retention of configuration at the carbon–metal bond.[61]

$$RCH_2C(H)(OCH_2OR')SnBu_3 \xrightarrow{RLi} RCH_2C(H)(OCH_2OR')Li$$

Alkenyllithium compounds are intermediates in the Shapiro reaction, which was discussed in Section 5.6. The reaction can be run in such a way that the organolithium compound is generated in high yield and subsequently allowed to react with a variety of electrophiles.[62] This method provides a route to vinyllithium compounds starting with a ketone.

2-ethyl-3-methylcyclopentanone $\xrightarrow{TsNHNH_2}$ tosylhydrazone (NNHTs) $\xrightarrow[TMEDA]{2\ n\text{-}BuLi}$ 1-lithio-5-ethyl-4-methylcyclopentene (Li, C_2H_5, H_3C) Ref. 63

58. E. J. Corey and R. H. Wollenberg, *J. Org. Chem.* **40:**2265 (1975).
59. W. C. Still, *J. Am. Chem. Soc.* **100:**1481 (1978).
60. D. J. Peterson, *J. Am. Chem. Soc.* **93:**4027 (1971).
61. W. C. Still and C. Sreekumar, *J. Am. Chem. Soc.* **102:**1201 (1980); J. S. Sawyer, A. Kucerovy, T. L. Macdonald, and G. J. McGarvey, *J. Am. Chem. Soc.* **110:**842 (1988).
62. F. T. Bond and R. A. DiPietro, *J. Org. Chem.* **46:**1315 (1981); T. H. Chan, A. Baldassarre, and D. Massuda, *Synthesis* **1976:**801. B. M. Trost and T. N. Nanninga, *J. Am. Chem. Soc.* **107:**1293 (1985).
63. W. Barth and L. A. Paquette, *J. Org. Chem.* **50:**2438 (1985).

7.2.1. Reactions with Alkylating Agents

The organometallic compounds of the group I and II metals are strongly basic and nucleophilic. The main limitation on alkylation reactions is complications from electron-transfer processes which can lead to radical reactions. Methyl and other primary iodides usually give good results in alkylation reactions. HMPA can accelerate the reaction and improve yields when electron transfer is a complication.[64]

$$\text{2-lithio-N-}(CH{=}NC(CH_3)_3)\text{azepane} \xrightarrow[\text{HMPA}]{n\text{-}C_7H_{15}I} \text{2-}(CH_2)_6CH_3\text{-N-}(CH{=}NC(CH_3)_3)\text{azepane}$$

Organolithium reagents in which the carbanion is delocalized are less subject to competing electron-transfer processes. Allyllithium and benzyllithium reagents can be alkylated by secondary alkyl bromides, and a high degree of inversion of configuration is observed.[65]

$$PhCH_2Li + CH_3CH_2C(H)(CH_3)Br \longrightarrow PhCH_2{-}C(H)(CH_2CH_3)CH_3 \quad \begin{array}{l}\text{58\% yield}\\ \text{100\% inversion}\end{array}$$

Alkenyllithium reagents can be alkylated in good yields by alkyl iodides and bromides.[66]

$$(CH_3)(Br)C{=}C(CH_3)(H) \xrightarrow[\text{2) } CH_3(CH_2)_3I]{\text{1) Li}} (CH_3)(CH_3(CH_2)_3)C{=}C(CH_3)(H)$$

Alkylation by allylic halides is usually a satisfactory reaction. The reaction in this case may proceed through a cyclic mechanism.[67] For example, when [1-^{14}C]-allyl chloride reacts with phenyllithium, about three-fourths of the product has the labeled carbon at the terminal methylene group.

$$H_2C{=}CH{-}\overset{*}{C}H_2Cl + PhLi \longrightarrow PhCH_2CH{=}\overset{*}{C}H_2$$

Intramolecular reactions have been useful for forming small rings. The reaction of 1,3-, 1,4-, and 1,5-diiodides with *t*-butyllithium is an effective means of ring closure,

64. A. I. Meyers, P. D. Edwards, W. F. Rieker, and T. R. Bailey, *J. Am. Chem. Soc.* **106**:3270 (1984); A. I. Meyers and G. Milot, *J. Am. Chem. Soc.* **115**:6652 (1993).
65. L. H. Sommer and W. D. Korte, *J. Org. Chem.* **35**:22 (1970).
66. J. Millon, R. Lorne, and G. Linstrumelle, *Synthesis* **1975**:434.
67. R. M. Magid and J. G. Welch, *J. Am. Chem. Soc.* **90**:5211 (1968); R. M. Magid, E. C. Nieh, and R. D. Gandour, *J. Org. Chem.* **36**:2099 (1971); R. M. Magid, and E. C. Nieh, *J. Org. Chem.* **36**:2105 (1971).

but 1,6-diiodides give very little cyclization.[68]

CH_2I CH_2I —t-BuLi→ 97%

Both trialkylsilyl and trialkylstannyl halides usually give high yields of substitution products with organolithium reagents, and this is an important route to silanes and stannanes.

Grignard reagents are somewhat less reactive toward alkylation but can be of synthetic value, especially when methyl, allyl, or benzyl halides are involved.

H_3C CH_3 Br H_3C CH_3 —1) Mg 2) $CH_2{=}CHCH_2Br$→ H_3C CH_3 $CH_2CH{=}CH_2$ H_3C CH_3 79% Ref. 69

Synthetically useful alkylation of Grignard reagents can also be carried out with alkyl sulfonates and sulfates.

$PhCH_2MgCl + CH_3CH_2CH_2CH_2OSO_2C_7H_7 \longrightarrow PhCH_2CH_2CH_2CH_2CH_3$ 50–59% Ref. 70

CH_3 H_3C—MgBr + $(CH_3O)_2SO_2$ ⟶ H_3C—CH_3 CH_3 CH_3 CH_3 52–60% Ref. 71

7.2.2. Reactions with Carbonyl Compounds

The most important type of reactions of Grignard reagents for synthesis involves addition to carbonyl groups. The transition state for addition of Grignard reagents is often represented as a cyclic array containing the carbonyl group and two molecules of the Grignard reagent. There is considerable evidence favoring this mechanism involving a termolecular complex.[72]

$R'_2C{=}O\cdots Mg(R)X \longrightarrow$ [cyclic transition state: R′, R′, O, Mg—R, X, R, Mg, X] $\longrightarrow R'_2C(R){-}O{-}Mg{-}R + MgX_2$

When the carbonyl carbon is substituted with a leaving group, the tetrahedral adduct can break down to regenerate a C=O bond, and a second addition step can occur. Esters,

68. W. F. Bailey, R. P. Gagnier, and J. J. Patricia, *J. Org. Chem.* **49**:2098 (1984).
69. J. Eustache, J.-M. Bernardon, and B. Shroot, *Tetrahedron Lett.* **28**:4681 (1987).
70. H. Gilman and J. Robinson, *Org. Synth.* **II**:47 (1943).
71. L. I. Smith, *Org. Synth.* **II**:360 (1943).
72. E. C. Ashby, R. B. Duke, and H. M. Neuman, *J. Am. Chem. Soc.* **89**:1964 (1967); E. C. Ashby, *Pure Appl. Chem.* **52**:545 (1980).

for example, usually are converted to tertiary alcohols, rather than ketones, in reactions with Grignard reagents.

$$RMgX + R'COR'' \longrightarrow R{-}C(OMgX)(R'){-}OR'' \longrightarrow RCR' (=O) + R''OMgX$$

$$RCR'(=O) + RMgX \xrightarrow{\text{fast}} R_2CR'(OMgX)$$

The addition of Grignard reagents to aldehydes, ketones, and esters is the basis for synthesis of a wide variety of alcohols. A number of examples are given in Scheme 7.3.

Grignard additions are sensitive to steric effects, and with hindered ketones a competing process involving reduction of the carbonyl group is observed. A cyclic transition state is involved.

The extent of this reaction increases with the steric bulk of the ketone and Grignard reagent. For example, no addition occurs between diisopropyl ketone and isopropylmagnesium bromide, and the reduction product diisopropylcarbinol is formed in 70% yield.[73] Competing reduction can be minimized in troublesome cases by using benzene or toluene as the solvent.[74] Alkyllithium compounds are much less prone to reduction and are preferred for the synthesis of highly substituted alcohols. This is illustrated by the comparison of the reaction of ethyllithium and ethylmagnesium bromide with adamantanone. A 97% yield of the tertiary alcohol is obtained with ethyllithium, whereas the Grignard reagent gives mainly the reduction product.[75]

C_2H_5MgBr / C_2H_5Li 97%

Enolization of the ketone is also sometimes a competing reaction. Because the enolate is unreactive toward nucleophilic addition, the ketone is recovered unchanged after hydrolysis. Enolization has been shown to be especially important when a considerable portion of the Grignard reagent is present as an alkoxide.[76] Alkoxides are formed as the

73. D. O. Cowan and H. S. Mosher, *J. Org. Chem.* **27**:1 (1962).
74. P. Canonne, G. B. Foscolos, and G. Lemauy, *Tetrahedron Lett.* **1979**:4383.
75. S. Landa, J. Vais, and J. Burkhard, *Coll. Czech. Chem. Commun.* **32**:570 (1967).
76. H. O. House and D. D. Traficante, *J. Org. Chem.* **28**:355 (1963).

Scheme 7.3. Synthetic Procedures Involving Grignard Reagents

A. Primary alcohols from formaldehyde

1[a] $C_6H_{11}MgCl + CH_2O \xrightarrow[H^+]{H_2O} C_6H_{11}CH_2OH$ 64–69%

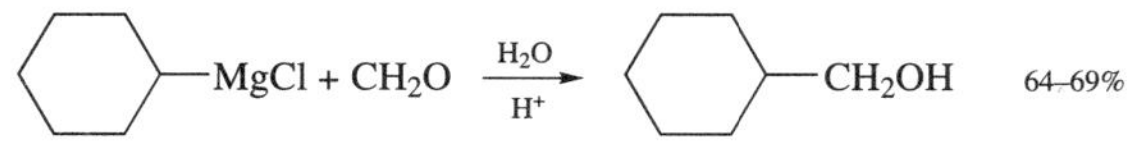

B. Primary alcohols from ethylene oxide

2[b] $CH_3(CH_2)_3MgBr + H_2C\text{—}CH_2$ (epoxide, O) $\longrightarrow \xrightarrow[H^+]{H_2O} CH_3(CH_2)_5OH$ 60–62%

C. Secondary alcohols from aldehydes

3[c] $PhCH{=}CHCH{=}O + HC{\equiv}CMgBr \longrightarrow \xrightarrow[H^+]{H_2O} HC{\equiv}CCH(OH)CH{=}CHPh$ 58–69%

4[d] 3-chlorophenyl-MgBr $+ CH_3CH{=}O \longrightarrow \xrightarrow{H_2O}$ 3-chlorophenyl-CHOHCH$_3$ 82–85%

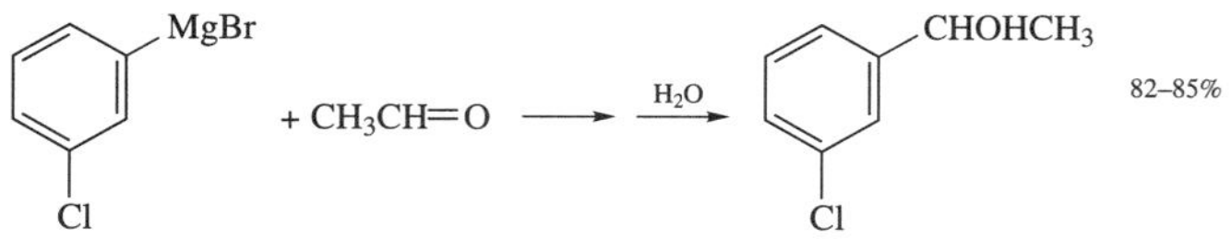

5[e] $CH_3CH{=}CHCH{=}O + CH_3MgCl \longrightarrow \xrightarrow{H_2O} CH_3CH{=}CHCH(OH)CH_3$ 81–86%

6[f] $(CH_3)_2CHMgBr + CH_3CH{=}O \longrightarrow (CH_3)_2CHCH(OH)CH_3$ 53–54%

D. Secondary alcohols from formate esters

7[g] $2\ CH_3(CH_2)_3MgBr + HCO_2C_2H_5 \longrightarrow \xrightarrow[H^+]{H_2O} (CH_3CH_2CH_2CH_2)_2CHOH$ 83–85%

E. Tertiary alcohols from ketones, esters, and lactones

8[h] 1,3-dioxolan-2-yl-CH_2MgBr + (isopropylidene-protected keto sugar: CH$_3$, CH$_3$, O, O, C=O, CH$_3$, CH$_3$) $\xrightarrow{LiBr}$ tertiary alcohol (dioxolanyl-CH_2, OH) 89%

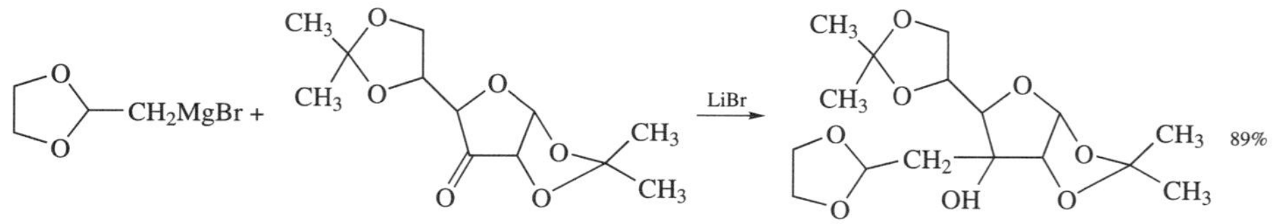

9[i] $3\ C_2H_5MgBr + (C_2H_5O)_2CO \longrightarrow \xrightarrow[NH_4Cl]{H_2O} (C_2H_5)_3COH$ 82–88%

10[j] $2\ PhMgBr + PhCO_2C_2H_5 \longrightarrow \xrightarrow{H_2O} Ph_3COH$ 89–93%

11[k] $CH_3(CH_2)_4$-(γ-lactone, O, C=O) $+ 2\ CH_3MgBr \longrightarrow \xrightarrow[H^+]{H_2O} CH_3(CH_2)_4CH(OH)(CH_2)_2C(OH)(CH_3)_2$ 57%

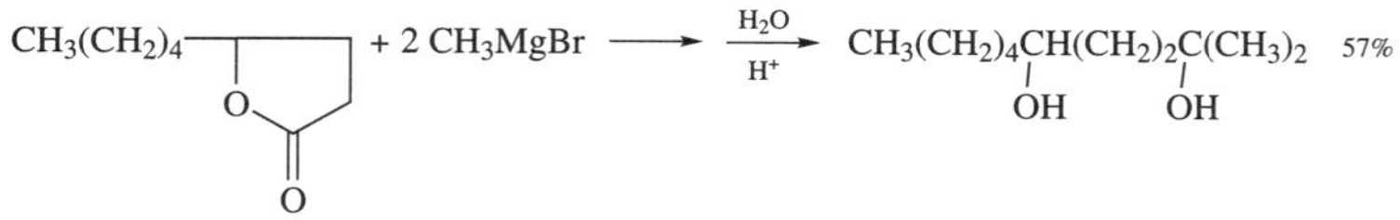

F. Aldehydes from triethyl orthoformate

12[l] 9-phenanthryl-MgBr $+ HC(OC_2H_5)_3 \longrightarrow \xrightarrow[H^+]{H_2O}$ 9-phenanthryl-CH=O 40–42%

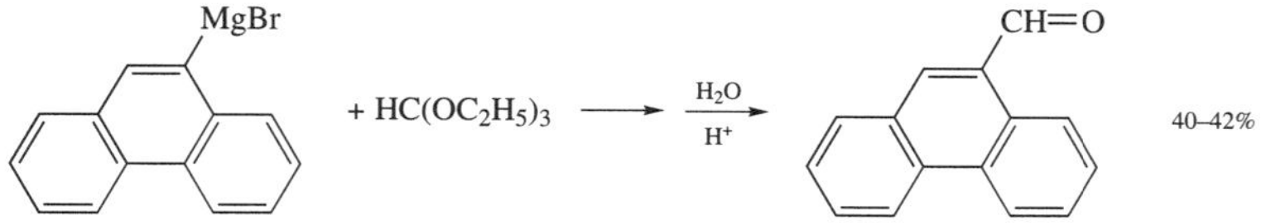

13[m] $CH_3(CH_2)_4MgBr + HC(OC_2H_5)_3 \longrightarrow \xrightarrow[H^+]{H_2O} CH_3(CH_2)_4CH{=}O$ 45–50%

G. Ketones from nitriles, thioesters, amides, and anhydrides

14[n] 9-cyanophenanthrene + $CH_3MgI \longrightarrow \xrightarrow[HCl]{H_2O}$ 9-acetylphenanthrene ($\overset{O}{\overset{\|}{C}}CH_3$) 52–59%

15[o] $CH_3OCH_2C{\equiv}N + PhMgBr \longrightarrow \xrightarrow[HCl]{H_2O} Ph\overset{O}{\overset{\|}{C}}CH_2OCH_3$ 71–78%

16[p] 2-methyl-1,3-dithiolane-2-$CH_2CH_2\overset{O}{\overset{\|}{C}}S$-(2-pyridyl) + (Z)-$BrMgCH_2CH{=}CHCH_2CH_3$ $\longrightarrow$ 2-methyl-1,3-dithiolane-2-$CH_2CH_2\overset{O}{\overset{\|}{C}}CH_2CH{=}CHCH_2CH_3$ (Z) 93%

17[q] $PhMgBr + ClCH_2\overset{O}{\overset{\|}{C}}N(OCH_3)CH_3 \longrightarrow Ph\overset{O}{\overset{\|}{C}}CH_2Cl$ 92%

18[r] $HC{\equiv}CCH_2CH_2\overset{O}{\overset{\|}{C}}N(OCH_3)CH_3 + CH_2{=}CHMgBr \longrightarrow HC{\equiv}CCH_2CH_2\overset{O}{\overset{\|}{C}}CH{=}CH_2$

19[s] $PhC{\equiv}CMgBr + (CH_3CO)_2O \longrightarrow PhC{\equiv}C\overset{O}{\overset{\|}{C}}CH_3$ 80%

H. Carboxylic acids by carbonation

20[t] 2,4,6-trimethylphenyl-MgBr (H_3C, CH_3, CH_3) $+ CO_2 \longrightarrow \xrightarrow[H^+]{H_2O}$ 2,4,6-trimethylphenyl-CO_2H 86–87%

21[u] $CH_3CH_2CH(MgBr)CH_3 + CO_2 \longrightarrow \xrightarrow[H^+]{H_2O} CH_3CH_2CH(CO_2H)CH_3$ 76–86%

22[v] 1-chloronorbornane $\xrightarrow{\text{1) active Mg; 2) } CO_2\text{; 3) } H^+, H_2O}$ norbornane-1-CO_2H 60–70%

I. Amines from imines

23[w] $PhCH{=}NCH_3 + PhCH_2MgCl \longrightarrow \xrightarrow{H_2O} PhCH(NHCH_3)CH_2Ph$ 96%

Scheme 7.3. *(continued)*

J. Alkenes after dehydration of intermediate alcohols

24[x] $PhCH{=}CHCH{=}O + CH_3MgBr \longrightarrow \xrightarrow{H_2SO_4} PhCH{=}CHCH{=}CH_2$ 75%

25[y] $2\ PhMgBr + CH_3CO_2C_2H_5 \longrightarrow \xrightarrow[H_2O]{H^+} Ph_2C{=}CH_2$ 67–70%

a. H. Gilman and W. E. Catlin, *Org. Synth.* **I:**182 (1932).
b. E. E. Dreger, *Orth. Synth.* **I:**299 (1932).
c. L. Skattebøl, E. R. H. Jones, and M. C. Whiting, *Org. Synth.* **IV:**792 (1963).
d. C. G. Overberger, J. H. Saunders, R. E. Allen, and R. Gander, *Org. Synth.* **III:**200 (1955).
e. E. R. Coburn, *Org. Synth.* **III:**696 (1955).
f. N. L. Drake and G. B. Cooke, *Org. Synth.* **II:**406 (1943).
g. G. H. Coleman and D. Craig, *Org. Synth.* **II:**179 (1943).
h. M. Schmeichel and H. Redlich, *Synthesis* **1996:**1002.
i. W. W. Moyer and C. S. Marvel, *Org. Synth.* **II:**602 (1943).
j. W. E. Bachman and H. P. Hetzner, *Org. Synth.* **III:**839 (1955).
k. J. Colonge and R. Marey, *Org. Synth.* **IV:**601 (1963).
l. C. A. Dornfeld and G. H. Coleman, *Org. Synth.* **III:**701 (1955).
m. G. B. Bachman, *Org. Synth.* **II:**323 (1943).
n. J. E. Callen, C. A. Dornfeld, and G. H. Coleman, *Org. Synth.* **III:**26 (1955).
o. R. B. Moffett and R. L. Shriner, *Org. Synth.* **III:**562 (1955).
p. T. Mukaiyama, M. Araki, and H. Takei, *J. Am. Chem. Soc.* **95:**4763 (1973); M. Araki, S. Sakata, H. Takei, and T. Mukaiyama, *Bull. Chem. Soc. Jpn.* **47:**1777 (1974).
q. R. Tillyer, L. F. Frey, D. M. Tschaen, and U.-H. Dolling, *Synlett* **1996:**225.
r. B. M. Trost and Y. Shi, *J. Am. Chem. Soc.* **115:**942 (1993).
s. A. Zanka, *Org. Process Res. Dev.* **2:**60 (1998).
t. D. M. Bowen, *Org. Synth.* **III:**553 (1955).
u. H. Gilman and R. H. Kirby, *Org. Synth.* **I:**353 (1932).
v. R. D. Rieke, S. E. Bales, P. M. Hudnall, and G. S. Poindexter, *Org. Synth.* **59:**85 (1977).
w. R. B. Moffett, *Org. Synth.* **IV:**605 (1963).
x. O. Grummitt and E. I. Beckeer, *Org. Synth.* **IV:**771 (1963).
y. C. F. H. Allen and S. Converse, *Org. Synth.* **I:**221 (1932).

addition reaction proceeds. They also can be present as the result of oxidation of some of the Grignard reagent by oxygen during preparation or storage. As with reduction, enolization is most seriously competitive in cases in which addition is retarded by steric factors.

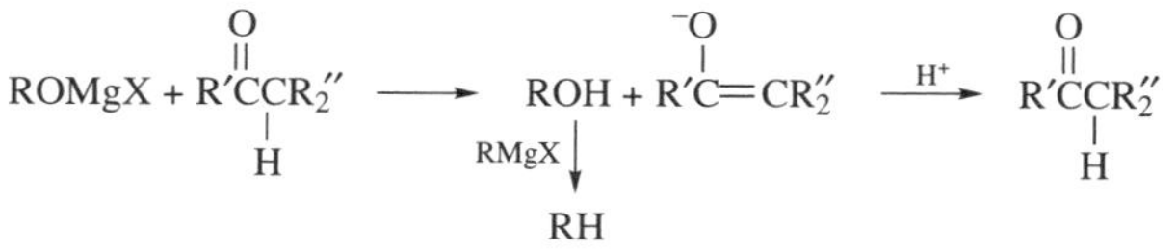

$$ROMgX + R'\overset{O}{\overset{\|}{C}}\underset{H}{\underset{|}{C}}R''_2 \longrightarrow ROH + R'\overset{^-O}{\overset{|}{C}}{=}CR''_2 \xrightarrow{H^+} R'\overset{O}{\overset{\|}{C}}\underset{H}{\underset{|}{C}}R''_2$$

$$ROMgX \xrightarrow{RMgX} RH$$

Grignard reagents are quite restricted in the types of functional groups that can be present either in the organometallic or in the carbonyl compound. Alkene, ether, and acetal functionality usually causes no difficulty, but unprotected OH, NH, SH, or carbonyl groups cannot be present, and CN and NO_2 groups cause problems in many cases.

Grignard reagents add to nitriles, and, after hydrolysis of the reaction mixture, a ketone is obtained. Hydrocarbons are the preferred solvent for this reaction.[77]

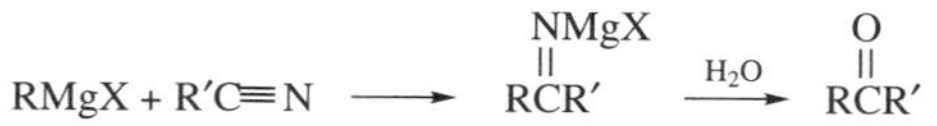

$$RMgX + R'C{\equiv}N \longrightarrow R\overset{NMgX}{\overset{\|}{C}}R' \xrightarrow{H_2O} R\overset{O}{\overset{\|}{C}}R'$$

77. P. Canonne, G. B. Foscolos, and G. Lemay, *Tetrahedron Lett.* **1980:**155.

Ketones can also be prepared from acyl chlorides by reaction at low temperature using an excess of the acyl chloride. Tetrahydrofuran is the preferred solvent.[78] The reaction conditions must be controlled to prevent formation of tertiary alcohol by addition of Grignard reagent to the ketone as it is formed.

$$CH_3(CH_2)_5MgBr + CH_3CH_2CH_2\overset{O}{\overset{\|}{C}}Cl \xrightarrow{-30^\circ C} CH_3(CH_2)_5\overset{O}{\overset{\|}{C}}(CH_2)_2CH_3 \quad 92\%$$

2-Pyridinethiolate esters, which are easily prepared from acyl chlorides, also react with Grignard reagents to give ketones (see entry 16 in Scheme 7.3).[79] *N*-Methoxy-*N*-methylamides are also converted to ketones by Grignard reagents (see entries 17 and 18).

Aldehydes can be obtained by reaction of Grignard reagents with triethyl orthoformate. The addition step is preceded by elimination of one of the alkoxy groups to generate an electrophilic carbon. The elimination is promoted by the magnesium ion acting as a Lewis acid.[80] The acetals formed by the addition are stable under the reaction conditions but are hydrolyzed to aldehydes on contact with aqueous acid (see entries 12 and 13).

$$(C_2H_5O)_2CH{-}OC_2H_5 \cdots RMgX \longrightarrow HC^{+}(OC_2H_5)_2 \xrightarrow{RMgX} RCH(OC_2H_5)_2 + C_2H_5OMgR + X^-$$

Aldehydes can also be obtained from Grignard reagents by reaction with formamides, such as *N*,*N*-dimethylformamide, *N*-methylformanilide and *N*-formylpiperidine.

$$PhCH_2CH_2MgCl + H\overset{O}{\overset{\|}{C}}{-}N(CH_2)_5 \longrightarrow PhCH_2CH_2CH{=}O \quad 66\text{–}76\% \qquad \text{Ref. 81}$$

Carboxylic acids are obtained from Grignard reagents by reaction with carbon dioxide. Scheme 7.3 includes some specific examples of procedures described in *Organic Syntheses*.

Structural rearrangements are not encountered with saturated Grignard reagents. Allylic and homoallylic systems can give products resulting from isomerization. NMR studies indicate that allylmagnesium bromide exists as a σ-bonded structure in which there is rapid equilibration of the two terminal carbons.[82] 2-Butenylmagnesium bromide and 1-methylpropenylmagnesium bromide are in equilibrium in solution. Addition products are derived from the latter compound, although it is the minor component at

78. F. Sato, M. Inoue, K. Oguro, and M. Sato, *Tetrahedron Lett.* **1979:**4303.
79. T. Mukaiyama, M. Araki, and H. Takei, *J. Am. Chem. Soc.* **95:**4763 (1973); M. Araki, S. Sakata, H. Takei, and T. Mukaiyama, *Bull. Chem. Soc. Japan* **47:**1777 (1974).
80. E. L. Eliel and F. W. Nader, *J. Am. Chem. Soc.* **92:**584 (1970).
81. G. A. Olah and M. Arvanaghi, *Org. Synth.* **64:**114 (1985).
82. M. Schlosser and N. Stähle, *Angew. Chem. Int. Ed. Engl.* **19:**487 (1980); M. Stähle and M. Schlosser, *J. Organomet. Chem.* **220:**277 (1981).

equilibrium.[83] Addition is believed to occur through a cyclic process that leads to an allylic shift.

$$R_2C{=}O \cdots MgBr\text{—}CH_2CH{=}CHCH_3 \longrightarrow R_2C(OMgBr)CH(CH_3)CH{=}CH_2$$

3-Butenylmagnesum bromide is in equilibrium with a small amount of cyclopropylmethylmagnesium bromide. The existence of the mobile equilibrium has been established by deuterium-labeling techniques.[84] Cyclopropylmethylmagnesium bromide[85] (and cyclopropylmethyllithium[86]) can be prepared by working at low temperature. At room temperature, the ring-opened 3-butenyl reagents are formed.

$$CH_2{=}CHCH_2CD_2MgBr \rightleftharpoons \underset{CD_2}{\overset{CH_2}{|}}\!\!>CHCH_2MgBr \rightleftharpoons BrMgCH_2CD_2CH{=}CH_2$$

When the double bond is further removed, as in 5-hexenylmagnesium bromide, there is no evidence of a similar equilibrium.[87]

$$CH_2{=}CHCH_2CH_2CH_2CH_2MgBr \not\rightleftharpoons BrMgCH_2\text{—}(\text{cyclopentyl})$$

The corresponding lithium reagent remains uncyclized at −78°C but cyclizes on warming.[88] In the case of γ-, δ-, and ε-alkynyllithium reagents, *exo*-cyclization to α-cycloalkylidene isomers occurs.[89] Anion-stabilizing substituents are required for the three- and four-membered rings, but not for the *exo*-5 cyclization.

$$Li(CH_2)_nC{\equiv}C\text{—}X \longrightarrow (CH_2)_n\,C{=}C(Li)X$$

X = Ph, TMS; n = 2,3

$$Li(CH_2)_4C{\equiv}C(CH_2)_3CH_3 \longrightarrow (\text{cyclopentylidene}){=}C(Li)(CH_2)_3CH_3$$

83. R. A. Benkeser, W. G. Young, W. E. Broxterman, D. A. Jones, Jr., and S. J. Piaseczynski, *J. Am. Chem. Soc.* **91:**132 (1969).
84. M. E. H. Howden, A. Maercker, J. Burdon, and J. D. Roberts, *J. Am. Chem. Soc.* **88:**1732 (1966).
85. D. J. Patel, C. L. Hamilton, and J. D. Roberts, *J. Am. Chem. Soc.* **87:**5144 (1965).
86. P. T. Lansbury, V. A. Pattison, W. A. Clement, and J. D. Sidler, *J. Am. Chem. Soc.* **86:**2247 (1964).
87. R. C. Lamb, P. W. Ayers, M. K. Toney, and J. F. Garst, *J. Am. Chem. Soc.* **88:**4261 (1966).
88. W. F. Bailey, J. J. Patricia, V. C. Del Gobbo, R. M. Jarrett, and P. J. Okarma, *J. Org. Chem.* **50:**1999 (1985); W. F. Bailey, T. T. Nurmi, J. L. Patricia, and W. Wang, *J. Am. Chem. Soc.* **109:**2442 (1987); W. F. Bailey, A. D. Khanolkar, K. Gavaskar, T. V. Oroska, K. Rossi, Y. Thiel, and K. B. Wiberg, *J. Am. Chem. Soc.* **113:**5720 (1991).
89. W. F. Bailey and T. V. Ovaska, *J. Am. Chem. Soc.* **115:**3080 (1993).

The reactivity of organolithium reagents toward carbonyl compounds is generally similar to that of Grignard reagents. The lithium reagents are less likely to undergo the competing reduction reaction with ketones, however. Organolithium compounds can add to α,β-unsaturated ketones by either 1,2- or 1,4-addition. The most synthetically important version of the 1,4-addition involves organocopper intermediates and will be discussed in Chapter 8. However, 1,4-addition is observed under some conditions even in the absence of copper catalysts. Highly reactive organolithium reagents usually react by 1,2-addition. The addition of small amounts of HMPA has been found to favor 1,4-addition. This is attributed to solvation of the lithium ion, which attenuates its Lewis acid character toward the carbonyl oxygen.[90]

O- -Li   R   O Li^+HMPA R^-

One reaction that is quite efficient for lithium reagents but poor for Grignard reagents is the synthesis of ketones from carboxylic acids.[91] The success of the reaction depends upon the stability of the dilithio adduct that is formed. This intermediate does not break down until hydrolysis, at which point the ketone is liberated. Some examples of this reaction are shown in Section C of Scheme 7.4.

$$\mathrm{RLi} + \mathrm{R'C(=O)O^-Li^+} \longrightarrow \mathrm{R'C(R)(O^-Li^+)O^-Li^+} \xrightarrow[\mathrm{H_2O}]{\mathrm{H^+}} \mathrm{R'C(R)(OH)OH} \longrightarrow \mathrm{RC(=O)R'}$$

A study aimed at optimizing yields in this reaction found that carbinol formation was a major competing process if the reaction was not carried out in such a way that all of the lithium compound had been consumed prior to hydrolysis.[92] Any excess lithium reagent that is present reacts extremely rapidly with the ketone as it is formed by hydrolysis. Another way to avoid the problem of carbinol formation is to quench the reaction mixture with trimethylsilyl chloride.[93] This procedure generates the disilyl acetal, which is stable until hydrolysis.

$$\text{cyclohexyl-}\mathrm{CO_2H} \xrightarrow[\text{3) } \mathrm{H_2O, H^+}]{\text{1) 4 equiv MeLi; 2) TMS-Cl}} \text{cyclohexyl-}\mathrm{C(=O)CH_3} \quad 92\%$$

90. H. J. Reich and W. H. Sikorski, *J. Org. Chem.* **64**:14 (1999).
91. M. J. Jorgenson, *Org. React.* **18**:1 (1971).
92. R. Levine, M. J. Karten, and W. M. Kadunce, *J. Org. Chem.* **40**:1770 (1975).
93. G. M. Rubottom and C. Kim, *J. Org. Chem.* **48**:1550 (1983).

Scheme 7.4. Synthetic Procedures Involving Organolithium Reagents

A. Alkylation

1[a] 3,5-dimethoxyphenyl-$OSiR_3$ $\xrightarrow[\text{2) } BrCH_2CH{=}C(CH_3)_2]{\text{1) } n\text{-BuLi}}$ $(CH_3)_2C{=}CHCH_2$-(2,6-dimethoxy-4-$OSiR_3$-phenyl) 60%

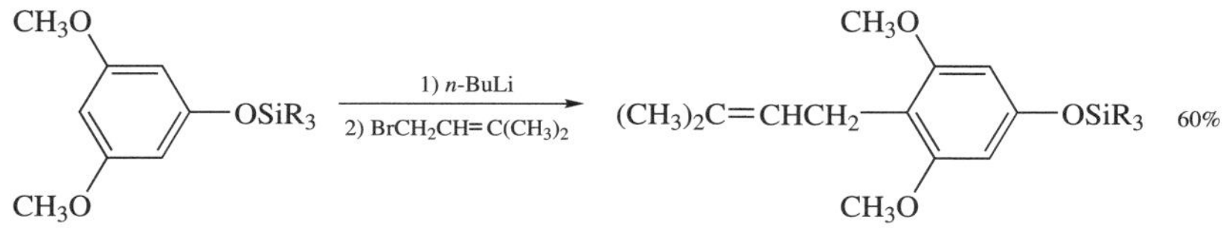

2[b] $(CH_3)_3COCH(Li)CH{=}CH_2 + CH_3(CH_2)_5I \longrightarrow (CH_3)_3CO(H)C{=}C(H)(CH_2)_6CH_3$ 83%

3[c] (bicyclic CH_2Br, H_3C, CH_3, H_3C, CH_3) + Li-aryl (CH_3O, CH_2OSiR_3, THPO, CH_2OSiR_3) $\longrightarrow$ coupled product (CH_3O, CH_2OSiR_3, CH_2OSiR_3, CH_2, OTHP, H_3C, CH_3, H_3C, CH_3) 65%

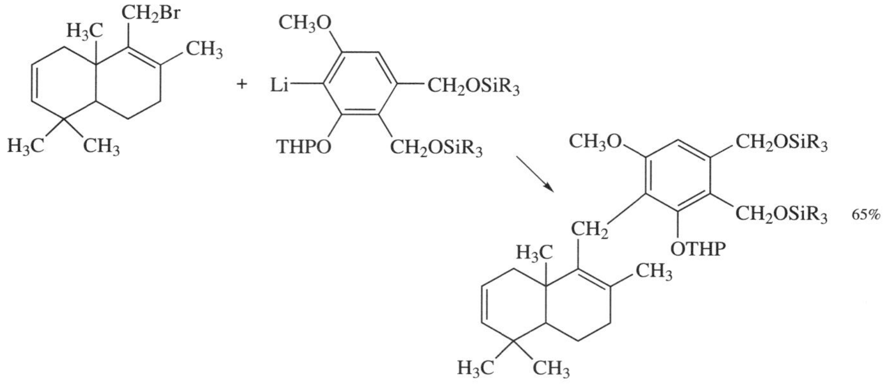

4[d] cyclohexene oxide (O) + C_4H_9Li/BF_3 (3 equiv) $\longrightarrow$ trans-2-$(CH_2)_3CH_3$-cyclohexanol (OH) 97%

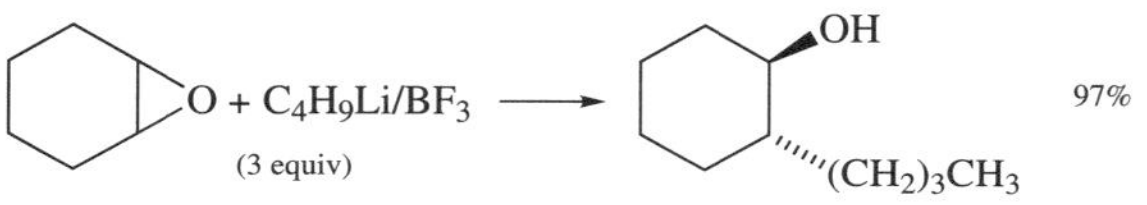

B. Reactions with aldehydes and ketones to give alcohols

5[e] $CH_2{=}CHCH_2Li + CH_3C(O)CH_2CH(CH_3)_2 \longrightarrow CH_2{=}CHCH_2C(OH)(CH_3)CH_2CH(CH_3)_2$ 70–72%

6[f] C_4H_9Li + cyclohexanone (O) $\longrightarrow$ 1-$CH_2CH_2CH_2CH_3$-cyclohexanol (OH) 89%

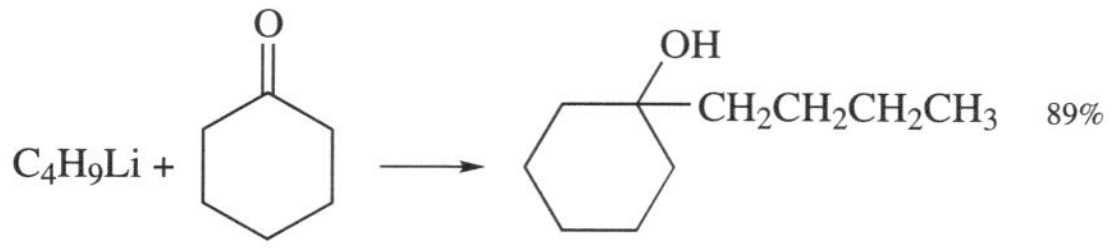

7[g] 2-CH_3-pyridine (N) $\xrightarrow{PhLi}$ 2-CH_2Li-pyridine (N) $\xrightarrow{CH_3CH{=}O}$ 2-$CH_2CH(OH)CH_3$-pyridine (N) 44–50%

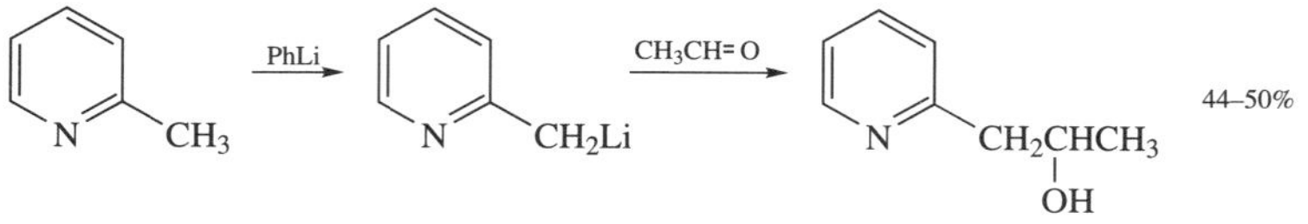

8[h] $CH_3CH{=}CHBr \xrightarrow[-120°C]{2\ t\text{-BuLi}} CH_3CH{=}CHLi \xrightarrow{PhCH{=}O} CH_3CH{=}CHCH(OH)Ph$ 72%

9[i] 3,5-(CH_3O)$_2C_6H_3C(=O)N(C_2H_5)_2$ $\xrightarrow{s\text{-BuLi}}$ 2-Li-3,5-(CH_3O)$_2C_6H_2C(=O)N(C_2H_5)_2$ $\xrightarrow{\text{2,5-(CH}_3\text{O)}_2\text{-4-CH}_3\text{C}_6\text{H}_2\text{CH=O}}$ product (OH, $(C_2H_5)_2NC=O$, OH_3C, CH_3O, OCH_3, CH_3, OCH_3) 63%

10[j] CH_3OCH_2Cl + cyclohexanone $\xrightarrow[\text{5 mol\% DTBB}]{\text{Li}}$ 1-(methoxymethyl)cyclohexanol (HO, CH_3OCH_2) 90%

C. Reactions with carboxylic acids, acyl chlorides, acid anhydrides, and *N*-methoxyamides to give ketones

11[k] cyclohexyl-CO_2Li + CH_3Li $\longrightarrow$ $\xrightarrow{H_2O}$ cyclohexyl-$\overset{O}{\overset{\|}{C}}CH_3$ 91%

12[l] $(CH_3)_3CCO_2H$ + 2 PhLi $\longrightarrow$ $\xrightarrow{H_2O}$ $C_6H_5\overset{O}{\overset{\|}{C}}C(CH_3)_3$ 65%

13[m] (H, Ph, CO_2H, CH_3 cyclopropane) $\xrightarrow{2\ CH_3Li}$ (H, Ph, $\overset{O}{\overset{\|}{C}}CH_3$, CH_3 cyclopropane) 90%

14[n] CH_3O–C_6H_3(Br)(CO_2H) $\xrightarrow[-100°C]{2\ n\text{-BuLi}}$ CH_3O–C_6H_3(Li)(CO_2Li) $\xrightarrow{3\text{-}CH_3OC_6H_4COCl}$ CH_3O–C_6H_3(CO_2H)–$\overset{O}{\overset{\|}{C}}$–$C_6H_4$–$OCH_3$ 78%

15[o] NC–C_6H_4–Br $\xrightarrow[\text{2) phthalic anhydride}]{\text{1) n-BuLi, −100°C}}$ NC–C_6H_4–C(=O)–C_6H_4–CO_2H 71%

(*continued*)

Scheme 7.4. *(continued)*

16[p] PhCH₂CH(OCH₃)C(=O)N(CH₃)OCH₃ + $CH_3C{\equiv}CLi$ → PhCH₂CH(OCH₃)C(=O)C≡CCH₃ 89%

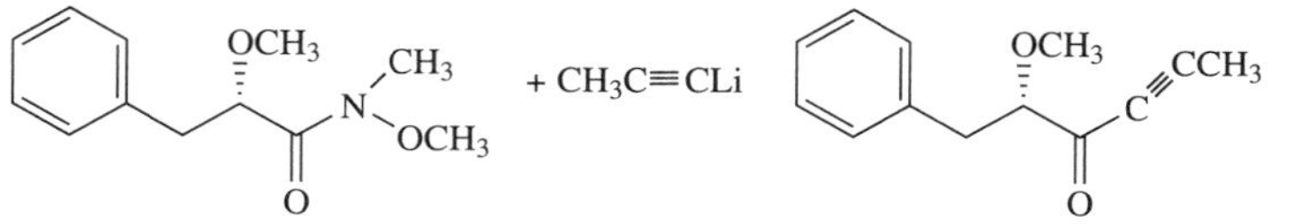

D. Reactions with carbon dioxide to give carboxylic acids

17[q] 3-(OCH_2OCH_3)-1-H_3C-benzene $\xrightarrow[\text{2) } CO_2]{\text{1) } t\text{-BuLi, }-100^\circ C}$ H_3C–C₆H₃(OCH₂OCH₃)–CO_2H 90%

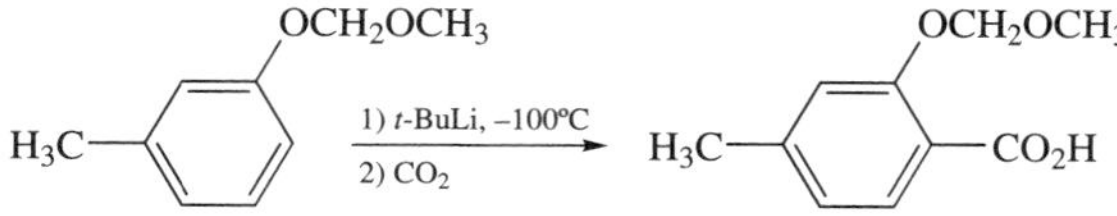

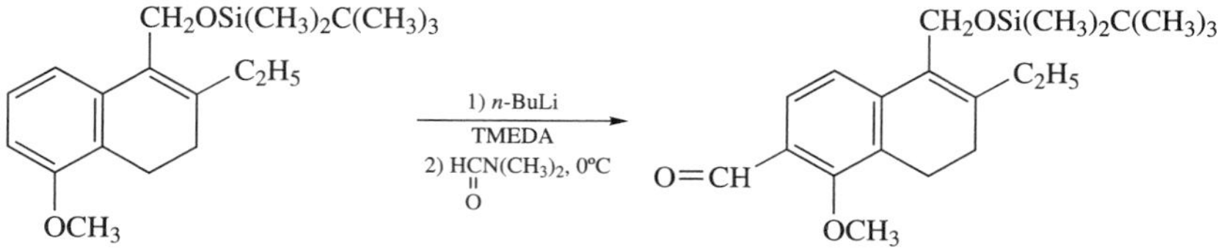

18[r] 2-methylpyridine $\xrightarrow[\text{2) } CO_2]{\text{1) PhLi}}$ 2-pyridyl–CH_2CO_2Li

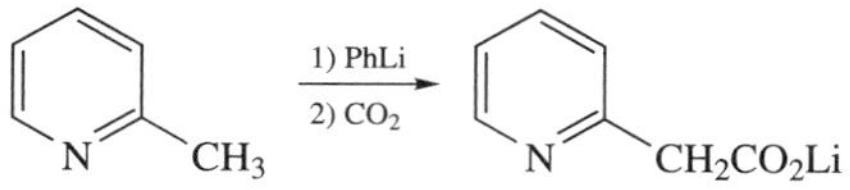

E. Other reactions

19[s] 5-methoxy-1-($CH_2OSi(CH_3)_2C(CH_3)_3$)-2-(C_2H_5)-3,4-dihydronaphthalene $\xrightarrow[\text{2) } HCN(CH_3)_2\text{ (C=O), } 0^\circ C]{\text{1) } n\text{-BuLi, TMEDA}}$ O=CH–(6-formyl derivative, OCH_3, $CH_2OSi(CH_3)_2C(CH_3)_3$, C_2H_5) 80%

20[t] $(CH_3)_2C{=}CHC(O)N(i\text{-}Pr)_2$ $\xrightarrow[\text{2) } CH_3I]{\text{1) } LiN(i\text{-}Pr)_2}$ $H_2C{=}C(CH_3)CH(CH_3)C(O)N(i\text{-}Pr)_2$ 98%

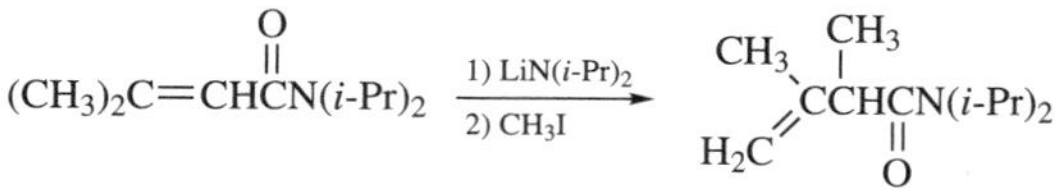

a. T. L. Shih, M. J. Wyvratt, and H. Mrozik, *J. Org. Chem.* **52**:2029 (1987).
b. D. A. Evans, G. C. Andrews, and B. Buckwalter, *J. Am. Chem. Soc.* **96**:5560 (1974).
c. J. E. McMurry and M. D. Erion, *J. Am. Chem. Soc.* **107**:2712 (1985).
d. M. J. Eis, J. E. Wrobel, and B. Ganem, *J. Am. Chem. Soc.* **106**:3693 (1984).
e. D. Seyferth and M. A. Weiner, *Org. Synth.* **V**:452 (1973).
f. J. D. Buhler, *J. Org. Chem.* **38**:904 (1973).
g. L. A. Walker, *Org. Synth.* **III**:757 (1955).
h. H. Neumann and D. Seebach, *Tetrahedron Lett.* **1976**:4839.
i. S. O. diSilva, M. Watanabe, and V. Snieckus, *J. Org. Chem.* **44**:4802 (1979).
j. A. Guijarro, B. Mandeno, J. Ortiz, and M. Yus, *Tetrahedron* **52**:1643 (1993).
k. T. M. Bare and H. O. House, *Org. Synth.* **49**:81 (1969).
l. R. Levine and M. J. Karten, *J. Org. Chem.* **41**:1176 (1976).
m. C. H. DePuy, F. W. Breitbeil, and K. R. DeBruin, *J. Am. Chem. Soc.* **88**:3347 (1966).
n. W. E. Parham, C. K. Bradsher, and K. J. Edgar, *J. Org. Chem.* **46**:1057 (1981).
o. W. E. Parham and R. M. Piccirilli, *J. Org. Chem.* **41**:1268 (1976).
p. F. D'Aniello, A. Mann, and M. Taddei, *J. Org. Chem.* **61**:4870 (1996).
q. R. C. Ronald, *Tetrahedron Lett.* **1975**:3973.
r. R. B. Woodward and E. C. Kornfeld, *Org. Synth.* **III**:413 (1955).
s. A. S. Kende and J. R. Rizzi, *J. Am. Chem. Soc.* **103**:4247 (1981).
t. M. Majewski, G. B. Mpango, M. T. Thomas, A. Wu, and V. Snieckus, *J. Org. Chem.* **46**:2029 (1981).

The synthesis of unsymmetrical ketones can be carried out in a tandem one-pot process by successive addition of two different alkyllithium reagents.[94]

$$RLi + CO_2 \longrightarrow RCO_2^-Li^+ \xrightarrow{R'Li} \xrightarrow[H^+]{H_2O} RC(=O)R'$$

N-Methyl-*N*-methoxyamides are also useful starting materials for preparation of

94. G. Zadel and E. Breitmaier, *Angew. Chem. Int. Ed. Engl.* **31**:1035 (1992).

ketones (see entry 16 in Scheme 7.4). Again, the reaction depends upon stabilizing the tetrahedral intermediate against elimination and a second addition step. In this case, chelation with the *N*-methoxy substituent is responsible for the stability of the intermediate.

$$RC(=O)N(OCH_3)CH_3 + R'Li \longrightarrow RR'C(O^-\cdots Li^+)N(CH_3)OCH_3 \xrightarrow{H^+, H_2O} RC(=O)R'$$

Scheme 7.4 illustrates some of the important synthetic reactions in which organolithium reagents act as nucleophiles. In addition to this type of reactivity, the lithium reagents have enormous importance in synthesis as bases and as lithiating reagents. The commerically available methyl, *n*-butyl, *s*-butyl, and *t*-butyl reagents are used most frequently in this context.

The stereochemistry of the addition of organomagnesium and organolithium compounds to cyclohexanones is similar.[95] With unhindered ketones, the stereoselectivity is not high, but there is generally a preference for attack from the equatorial direction to give the axial alcohol. This preference for the equatorial approach increases with the size of the alkyl group. With alkyllithium reagents, added salts improve the stereoselectivity. For example, one equivalent of $LiClO_4$, enhances the proportion of the axial alcohol in the addition of methyllithium to 4-*t*-butylcyclohexanone.[96]

4-*t*-Bu-cyclohexanone $\xrightarrow{CH_3Li}$ axial OH (equatorial CH_3) + equatorial OH (axial CH_3)

	axial OH	equatorial OH
no $LiClO_4$	65%	35%
1 equiv $LiClO_4$	92%	8%

Bicyclic ketones react with organometallic reagents to give the products of addition from the less hindered face of the carbonyl group.

The stereochemistry of addition of organometallic reagents to acyclic carbonyl compounds parallels the behavior of the hydride reducing agents, as discussed in Section 5.2.1. Organometallic compounds were included in the early studies that established the preference for addition according to Cram's rule.[97]

$$(L)(M)(S)C\text{–}C(=O)R \xrightarrow{R'MgX} (L)(M)(S)C\text{–}C(R')(OMgX)R$$

S, M, L = relative size of substituents

The interpretation of the basis for this stereoselectivity can be made in terms of the steric, torsional, and stereoelectronic effects discussed in connection with reduction by hydrides. It has been found that crown ethers enhance stereoselectivity in the reactions of both Grignard reagents and alkyllithium compounds.[98]

95. E. C. Ashby and J. T. Laemmle, *Chem. Rev.* **75:**521 (1975).
96. E. C. Ashby and S. A. Noding, *J. Org. Chem.* **44:**4371 (1979).
97. D. J. Cram and F. A. A. Elhafez, *J. Am. Chem. Soc.* **74:**5828 (1952).
98. Y. Yamamoto and K. Maruyama, *J. Am. Chem. Soc.* **107:**6411 (1985).

For ketones and aldehydes in which adjacent substituents permit chelation with the metal ion in the transition state, the stereochemistry can often be interpreted in terms of the steric requirements of the chelated transition state. In the case of α-alkoxyketones, for example, an assumption that both the alkoxy and carbonyl oxygens will be coordinated with the metal ion and that addition will occur from the less hindered side of this structure correctly predicts the stereochemistry of addition. The predicted product dominates by as much as 100:1 for several Grignard reagents.[99] Further supporting the importance of chelation is the correlation between rate and stereoselectivity. Groups which facilitate chelation cause an increase in both rate and stereoselectivity.[100]

$$\text{R(H)(R'O)C–C(=O)CH}_3 + \text{C}_4\text{H}_9\text{MgBr} \xrightarrow[-78^\circ\text{C}]{\text{THF}} \text{R(H)(R'O)C–C(OH)(CH}_3)\text{C}_4\text{H}_9$$

$R = C_7H_{15}$ $R' = CH_2OCH_3$, $CH_2OCH_2CH_2OCH_3$, CH_2Ph, CH_2OCH_2Ph

An alternative to preparation of organometallic reagents and then carrying out reaction with a carbonyl compound is to generate the organometallic intermediate *in situ* in the presence of the carbonyl compound. The organometallic compound then reacts immediately with the carbonyl compound. This procedure is referred to as the *Barbier reaction*.[101] This technique has no advantage over the conventional one for most cases. However, when the organometallic reagent is very unstable, it can be a useful method. Allylic halides, which are difficult to convert to Grignard reagents in good yield, frequently give excellent results in the Barbier procedure. Because solid metals are used, one of the factors affecting the rate of the reaction is the physical state of the metal. Ultrasonic irradiation has been found to have a favorable effect on the Barbier reaction, presumably by accelerating the generation of reactive sites on the metal surface.[102]

$$(CH_3)_2CHCH_2CH{=}O + CH_2{=}C(CH_3)CH_2Cl \xrightarrow[\text{ether}]{\text{Mg}} (CH_3)_2CHCH_2CH(OH)CH_2C(CH_3){=}CH_2 \quad 92\%$$

7.3. Organic Derivatives of Group IIB and Group IIIB Metals

In this section, we will discuss organometallic derivatives of zinc, cadmium, mercury, and indium. The group IIB and IIIB metals have the d^{10} electronic configuration in the 2+ and 3+ oxidation states, respectively. Because of the filled d level, the 2+ or 3+ oxidation states are quite stable, and reactions of the organometallics usually do not involve changes in oxidation level. This property makes the reactivity patterns of these organometallics more similar to those of derivatives of the group IA and IIA metals than to those of derivatives of transition metals with vacancies in the d levels. The IIB metals, however, are

99. W. C. Still and J. H. McDonald III, *Tetrahedron Lett.* **1980:**1031.
100. X. Chen, E. R. Hortelano, E. L. Eliel, and S. V. Frye, *J. Am. Chem. Soc.* **112:**6130 (1990).
101. C. Blomberg and F. A. Hartog, *Synthesis* **1977:**18.
102. J.-L. Luche and J.-C. Damiano, *J. Am. Chem. Soc.* **102:**7926 (1980).

much less electropositive than the IA and IIA metals, and the nucleophilicity of the organometallics is less than for organolithium or organomagnesium compounds. Many of the synthetic applications of these organometallics are based on this attenuated reactivity.

7.3.1. Organozinc Compounds

Organozinc compounds can be prepared by reaction of Grignard or organolithium reagents with zinc salts. A one-pot process in which the organic halide, magnesium metal, and zinc chloride are sonicated has proven to be a convenient method for the preparation.[103] Organozinc compounds can also be prepared from organic halides by reaction with highly reactive zinc metal.[104] Simple alkylzinc compounds, which are distillable liquids, can also be prepared from alkyl halides and Zn–Cu couple.[105]

When prepared *in situ* from $ZnCl_2$ and Grignard reagents, organozinc reagents add to carbonyl compounds to give carbinols.[106] This must reflect activation of the carbonyl group through Lewis acid catalysis by magnesium ion, because ketones are much less reactive toward pure dialkylzinc reagents and tend to react by reduction rather than addition.[107] The addition of alkylzinc reagents is also promoted by trimethylsilyl chloride, and this leads to isolation of silyl ethers of the alcohol products.[108]

$$(C_2H_5)_2Zn + PhC(=O)CH_3 \xrightarrow{(CH_3)_3SiCl} PhC(OSi(CH_3)_3)(CH_3)CH_2CH_3 \quad 93\%$$

One attractive feature of organozinc reagents is that many functional groups which would interfere with organomagnesium or organolithium reagents can be present in organozinc reagents.[109,110] Functionalized reagents can be prepared by halogen–metal exchange reactions with diethylzinc.[111] The reaction equilibrium is driven to completion by use of excess diethylzinc and removal of the ethyl halide by distillation. The pure organozinc reagent can be obtained by removal of the excess diethylzinc under vacuum.

$$2\,X(CH_2)_nI + (C_2H_5)_2Zn \longrightarrow X(CH_2)_nZnC_2H_5 \rightleftharpoons [X(CH_2)_n]_2Zn + 2\,C_2H_5I$$

$$n = 2, 3, 4, 5;\; X = CH_3CO_2, (CH_3)_3COCO_2, N{\equiv}C, Cl$$

These exchange reactions are subject to catalysis by certain transition-metal ions, and, with small amounts of $MnBr_2$ or CuCl, the reaction proceeds satisfactorily with alkyl

103. J. Boersma, *Comprehensive Organometallic Chemistry*, G. Wilkinson, ed., Vol. 2, Pergamon Press, Oxford, 1982, Chapter 16; G. E. Coates and K. Wade, *Organometallic Compounds*, Vol. 1, 3rd ed., Methuen, London, 1967, pp. 121–128.
104. R. D. Rieke, P. T.-J. Li, T. P. Burns, and S. T. Uhm, *J. Org. Chem.* **46:**4323 (1981).
105. C. R. Noller, *Org. Synth.* **II:**184 (1943).
106. P. R. Jones, W. J. Kauffman, and E. J. Goller, *J. Org. Chem.* **36:**186 (1971); P. R. Jones, E. J. Goller, and W. J. Kaufmann, *J. Org. Chem.* **36:**3311 (1971).
107. G. Giacomelli, L. Lardicci, and R. Santi, *J. Org. Chem.* **39:**2736 (1974).
108. C. Alvisi, S. Casolari, A. L. Costa, M. Ritiani, and E. Tagliavini, *J. Org. Chem.* **63:**1330 (1998).
109. P. Knochel, J. J. A. Perea, and P. Jones, *Tetrahedron* **54:**8275 (1998).
110. P. Knochel and R. D. Singer, *Chem. Rev.* **93:**2117 (1993).
111. M. J. Rozema, A. R. Sidduri, and P. Knochel, *J. Org. Chem.* **57:**1956 (1992); A. Boudier, L. O. Bromm, M. Lotz, and P. Knochel, *Angew. Chem. Int. Ed.* **39**: 4415 (2000).

bromides.[112]

$$X(CH_2)_nBr + (C_2H_5)_2Zn \xrightarrow[3\%\ CuCl]{5\%\ MnBr_2} X(CH_2)_nZnBr$$

$$n = 3, 4;\ X = C_2H_5O_2C,\ N{\equiv}C,\ Cl$$

Another effective catalyst is $Ni(acac)_2$.[113]

Organozinc reagents can also be prepared from trialkylboranes by exchange with dimethylzinc.[114]

CH_2 1) $HB(C_2H_5)_2$ 2) $(CH_3)_2Zn$ ($CH_2)_2Zn$

This route can be used to prepare enantiomerically enriched organozinc reagents using enantioselective hydroboration (see Section 4.9.3), followed by exchange with diisopropylzinc. Trisubstituted cycloalkenes such as 1-methyl- or 1-phenylcyclohexene give enantiomeric purity exceeding 95%. The exchange reaction takes place with retention of configuration.[115]

CH_3 $IpcBH_2$ CH_3 BHIpc 1) $(C_2H_5)_2BH$ 2) $(i\text{-}Pr)_2Zn$ CH_3 $ZnCH(CH_3)_2$ 94% e.e.

Exchange with boranes can also be used to prepare alkenylzinc reagents.[116]

$$[CH_3(CH_2)_3CH{=}C((CH_2)_2CH_3){-}]_3B + (C_2H_5)_2Zn \longrightarrow [CH_3(CH_2)_3CH{=}C((CH_2)_2CH_3){-}]_2Zn$$

Alkenylzinc reagents can also be made from alkynes.[117]

$$CH_3(CH_2)_3C{\equiv}C(CH_2)_3CH_3 \xrightarrow[(Cp)_2TiCl_2\ 10\ mol\ \%]{ZnI_2,\ LiH} \begin{matrix} CH_3(CH_2)_3 & & (CH_2)_3CH_3 \\ & C{=}C & \\ H & & ZnI \end{matrix}$$

112. I. Klement, P. Knochel, K. Chau, and G. Cahiez, *Tetrahedron Lett.* **35:**1177 (1994).
113. S. Vettel, A. Vaupel, and P. Knochel, *J. Org. Chem.* **61:**7473 (1996).
114. F. Langer, J. Waas, and P. Knochel, *Tetrahedron Lett.* **34:**5261 (1993); L. Schwink and P. Knochel, *Tetrahedron Lett.* **35:**9007 (1994); F. Langer, A. Devasagayraj, P.-Y. Chavant, and P. Knochel, *Synlett* **1994:**410; F. Langer, L. Schwink, A. Devasagayraj, P.-Y. Chavant, and P. Knochel, *J. Org. Chem.* **61:**8229 (1996).
115. A. Boudier, F. Flachsmann, and P. Knochel, *Synlett* **1998:**1438.
116. M.Srebnik, *Tetrahedron Lett.* **32:**2449 (1991); K.A. Agrios and M. Srebnik, *J. Org. Chem.* **59:**5468 (1994).
117. Y. Gao, K. Harada, T. Hata, H. Urabe, and F. Sato, *J. Org. Chem.* **60:**290 (1995).

Arylzinc reagents can be made from aryl halides with activated zinc[118] or from Grignard reagents by metal–metal exchange with zinc salts.[119]

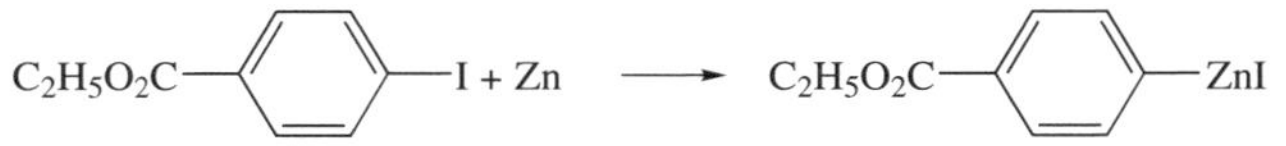
C2H5O2C
I + Zn
C2H5O2C
ZnI

$$2\ PhMgBr + ZnCl_2 \longrightarrow Ph_2Zn + 2\ MgBrCl$$

High degrees of enantioselectivity have been observed when alkylzinc reagents react with aldehydes in the presence of chiral ligands.[120] Among several compounds that have been used as ligands are *exo*-(dimethylamino)norborneol (**A**)[121] and diphenyl(1-methylpyrrolin-2-yl)methanol (**B**)[122] as well as ephedrine derivatives **C**[123] and **D**.[124]

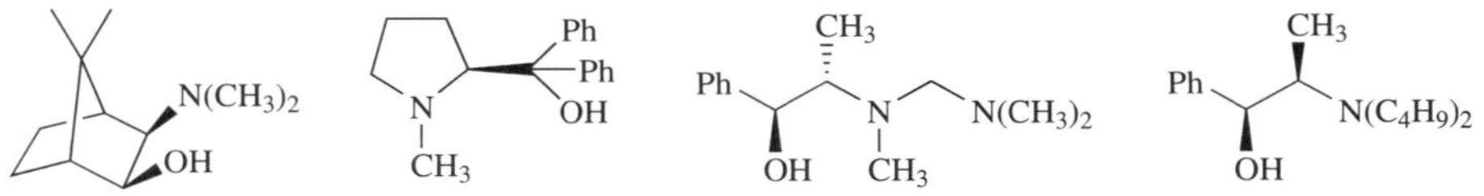
N(CH3)2
OH
Ph
Ph
OH
N
CH3
CH3
Ph
N
N(CH3)2
OH
CH3
CH3
Ph
N(C4H9)2
OH

A **B** **C** **D**

The bis-trifluoromethanesulfonamide of *trans*-cyclohexane-1,2-diamine also leads to enantioselective additions in 80% or greater enantiomeric excess.[125]

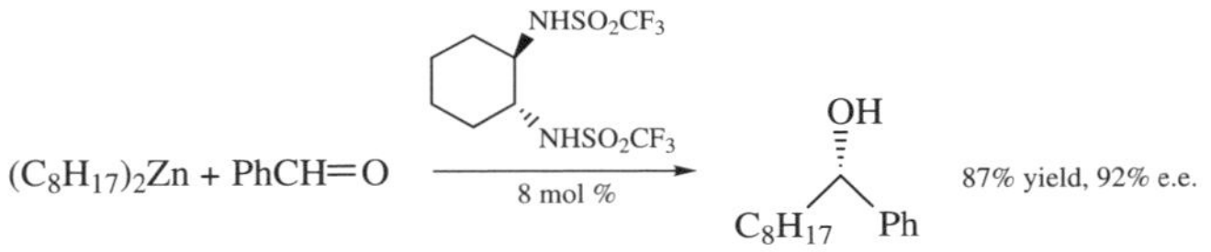
NHSO2CF3
NHSO2CF3
(C8H17)2Zn + PhCH=O
8 mol %
OH
C8H17
Ph
87% yield, 92% e.e.

The enantioselectivity is the result of chelation of the zinc by the chiral ligand. The transition states of the additions are generally believed to involve two zinc atoms. The proposed transition state for ligand **A**, for example, is:[126]

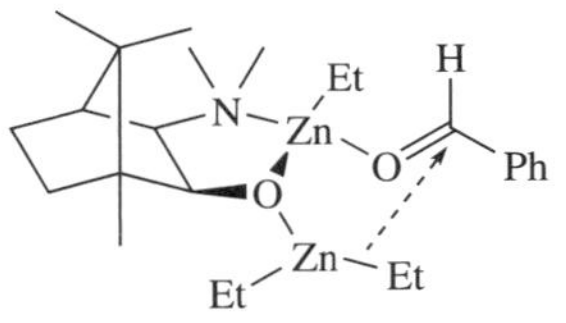
N
Zn
Et
H
O
Ph
O
Zn
Et
Et

118. L. Zhu, R. M. Wehmeyer, and R. D. Rieke, *J. Org. Chem.* **56**:1445 (1991); T. Sakamoto, Y. Kondo, N. Murata, and H. Yamanaka, *Tetrahedron Lett.* **33**:5373 (1992).
119. K. Park, K. Yuan, and W. J. Scott, *J. Org. Chem.* **58**:4866 (1993).
120. K. Soai, A. Ookawa, T. Kaba, and K. Ogawa, *J. Am. Chem. Soc.* **109**:7111 (1987); M. Kitamura, S. Suga, K. Kawai, and R. Noyori, *J. Am. Chem. Soc.* **108**:6071 (1986); W. Oppolzer and R. N. Rodinov, *Tetrahedron Lett.* **29**:5645 (1988).
121. M. Kitamura, S. Suga, K. Kawai, and R. Noyori, *J. Am. Chem. Soc.* **108**:6071 (1986); M. Kitamura, H.Oka, and R. Noyori, *Tetrahedron* **55**:3605 (1999).
122. K. Soai, A. Ookawa, T. Kaba, and E. Ogawa, *J. Am. Chem. Soc.* **109**:7111 (1987).
123. E. J. Corey and F. J. Hannon, *Tetrahedron Lett.* **28**:5233 (1987).
124. K. Soai, S. Yokoyama, and T. Hayasaka, *J. Org. Chem.* **56**:4264 (1991).
125. F. Langer, L. Schwink, A. Devasagayaraj, P.-Y. Chavant, and P. Knochel, *J. Org. Chem.* **61**:8229 (1996); C. Lutz and P. Knochel, *J. Org. Chem.* **62**:7895 (1997).
126. D. A. Evans, *Science*, **240**:420 (1988); E. J. Corey, P.-W. Yuen, F. J. Hannon, and D. A. Wierda, *J. Org. Chem.* **55**:784 (1990); B. Goldfuss and K. N. Houk, *J. Org. Chem.* **63**:8998 (1998).

Additions to aldehydes are also catalyzed by Lewis acids, especially $Ti(i\text{-}OPr)_4$ and trimethylsilyl chloride. These additions can be carried out in the presence of other chiral ligands that induce enantioselectivity.[127]

A frequently used reaction involving zinc is the *Reformatsky reaction*, in which zinc, an α-haloester, and a carbonyl compound react to give a β-hydroxyester.[128] The zinc and α-haloester react to form an organozinc reagent. Because the ester group can stabilize the carbanionic character, the product is essentially the zinc enolate of the dehalogenated ester.[129] The enolate can then carry out a nucleophilic attack on the carbonyl group.

$$C_2H_5O_2CCH_2Br + Zn \longrightarrow C_2H_5O\overset{O^-Zn^{2+}}{\overset{|}{C}}{=}CH_2 + Br^-$$

$$+ \text{ cyclohexanone} \longrightarrow \text{1-hydroxy-1-}(CH_2CO_2C_2H_5)\text{cyclohexane}$$

Several techniques have been used to "activate" the zinc metal and improve yields. For example, pretreatment of zinc dust with a solution of copper acetate gives a more reactive zinc–copper couple.[130] Exposure to trimethylsilyl chloride also activates the zinc.[131] Scheme 7.5 gives some examples of the Reformatsky reaction.

Zinc enolates prepared from α-haloketones can be used as nucleophiles in mixed aldol condensations (see Section 2.1.3). Entry 7 in Scheme 7.5 is an example. This reaction can be conducted in the presence of the Lewis acid diethylaluminum chloride, in which case addition occurs at $-20°C$.[132]

The reagent combination $Zn/CH_2Br_2/TiCl_4$ gives rise to an organometallic reagent called *Lombardo's reagent*. It converts ketones to methylene groups.[133] The active reagent is presumed to be a dimetalated species which adds to the ketone under the influence of the Lewis acidity of titanium. β-Elimination then generates the methylene group.

$$R{-}\overset{O\cdots Ti}{\overset{\|}{C}}{-}R \;+\; {-}Zn{-}CH_2{-}Zn{-} \longrightarrow R{-}\overset{O-Ti}{\overset{|}{C}}{-}R \text{ } (\underset{CH_2-Zn}{|}) \longrightarrow \overset{R\quad R}{C}{=}CH_2$$

Use of esters and 1,1-dibromoalkanes as reactants gives enol ethers:[134]

$$C_4H_9CO_2CH_3 + (CH_3)_2CHCHBr_2 \xrightarrow[\text{TiCl}_4,\ \text{TMEDA}]{\text{Zn}} \overset{C_4H_9}{\underset{CH_3O}{}}C{=}C\overset{H}{\underset{CH(CH_3)_2}{}} \quad 95\%$$

127. D. Seebach, D. A. Plattner, A. K. Beck, Y. M. Wang, D. Hunziker, and W. Petter, *Helv. Chim. Acta* **75:**2171 (1992).
128. R. L. Shriner, *Org. React.*, **1:**1 (1942); M. W. Rathke, *Org. React.* **22:**423 (1975).
129. W. R. Vaughan and H. P. Knoess, *J. Org. Chem.* **35:**2394 (1970).
130. E. Le Goff, *J. Org. Chem.* **29:**2048 (1964); L. R. Krepski, L. E. Lynch, S. M. Heilmann, and J. K. Rasmussen, *Tetrahedron Lett.* **26:**981 (1985).
131. G. Picotin and P. Miginiac, *J. Org. Chem.* **52:**4796 (1987).
132. K. Maruoka, S. Hashimoto, Y. Kitagawa, H. Yamamoto, and H. Nozaki, *J. Am. Chem. Soc.* **99:**7705 (1977).
133. K. Oshima, K. Takai, Y. Hotta, and H. Nozaki, *Tetrahedron Lett.* **1978:**2417; L. Lombardo, *Tetrahedron Lett.* **23:**4293 (1982); L. Lombardo, *Org. Synth.* **65:**81 (1987).
134. T. Okazoe, K. Takai, K. Oshima, and K. Utimoto, *J. Org. Chem.* **52:**4410 (1987).

Scheme 7.5. Condensation of α-Halocarbonyl Compounds Using Zinc—The Reformatsky Reaction

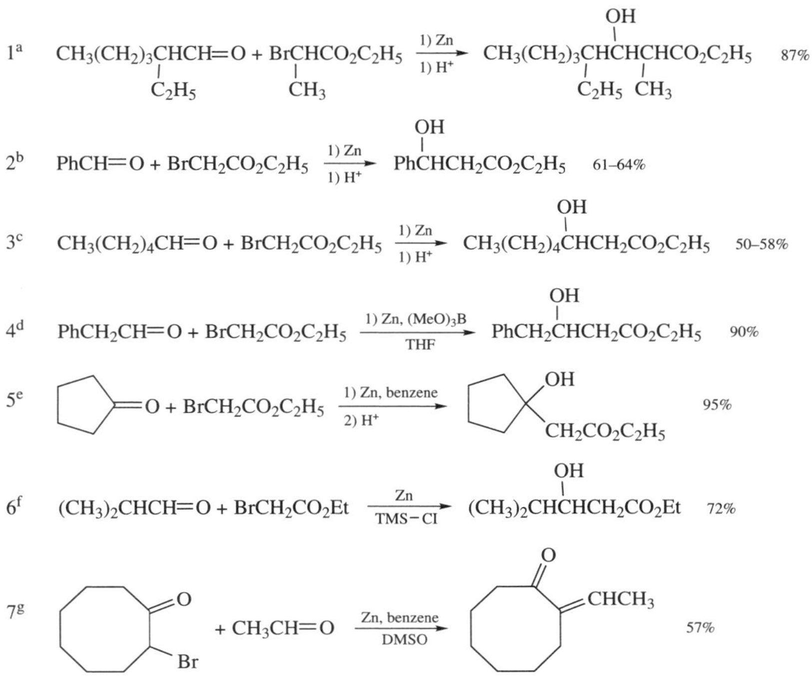

1[a] $CH_3(CH_2)_3CH(C_2H_5)CH{=}O + BrCH(CH_3)CO_2C_2H_5 \xrightarrow[1)\ H^+]{1)\ Zn} CH_3(CH_2)_3CH(C_2H_5)CH(OH)CH(CH_3)CO_2C_2H_5$ 87%

2[b] $PhCH{=}O + BrCH_2CO_2C_2H_5 \xrightarrow[1)\ H^+]{1)\ Zn} PhCH(OH)CH_2CO_2C_2H_5$ 61–64%

3[c] $CH_3(CH_2)_4CH{=}O + BrCH_2CO_2C_2H_5 \xrightarrow[1)\ H^+]{1)\ Zn} CH_3(CH_2)_4CH(OH)CH_2CO_2C_2H_5$ 50–58%

4[d] $PhCH_2CH{=}O + BrCH_2CO_2C_2H_5 \xrightarrow[THF]{1)\ Zn,\ (MeO)_3B} PhCH_2CH(OH)CH_2CO_2C_2H_5$ 90%

5[e] cyclopentanone $+ BrCH_2CO_2C_2H_5 \xrightarrow[2)\ H^+]{1)\ Zn,\ benzene}$ 1-hydroxy-1-($CH_2CO_2C_2H_5$)cyclopentane 95%

6[f] $(CH_3)_2CHCH{=}O + BrCH_2CO_2Et \xrightarrow[TMS-Cl]{Zn} (CH_3)_2CHCH(OH)CH_2CO_2Et$ 72%

7[g] 2-bromocyclooctanone $+ CH_3CH{=}O \xrightarrow[DMSO]{Zn,\ benzene}$ 2-ethylidenecyclooctanone ($=CHCH_3$) 57%

a. K. L. Rinehart, Jr., and E. G. Perkins, *Org. Synth.* **IV:**444 (1963).
b. C. R. Hauser and D. S. Breslow, *Org. Synth.* **III:**408 (1955).
c. J. W. Frankenfield and J. J. Werner, *J. Org. Chem.* **34:**3689 (1969).
d. M. W. Rathke and A. Lindert, *J. Org. Chem.* **35:**3966 (1970).
e. J. F. Ruppert and J. D. White, *J. Org. Chem.* **39:**269 (1974).
f. G. Picotin and P. Miginiac, *J. Org. Chem.* **52:**4796 (1987).
g. T. A. Spencer, R. W. Britton, and D. S. Watt, *J. Am. Chem. Soc.* **89:**5727 (1967).

A similar procedure starting with trimethylsilyl esters generates trimethylsilyl enol ethers.[135]

$$PhCO_2Si(CH_3)_3 + CH_3CHBr_2 \xrightarrow[TMEDA]{Zn,\ TiCl_4} PhC(OSi(CH_3)_3){=}CHCH_3$$

Organozinc reagents are also used in conjunction with palladium in a number of carbon–carbon bond-forming processes which will be discussed in Section 8.2.

7.3.2. Organocadmium Compounds

Organocadmium compounds can be prepared from Grignard reagents or organolithium compounds by reaction with Cd(II) salts.[136] Organocadmium compounds can also

135. K. Takai, Y. Kataoka, T. Okazoe, and K. Utimoto, *Tetrahedron Lett.* **29:**1065 (1988).
136. P. R. Jones and P. J. Desio, *Chem. Rev.* **78:**491 (1978).

be prepared directly from alkyl, benzyl, and aryl halides by reaction with highly reactive cadmium metal generated by reduction of Cd(II) salts.[137]

$$NC-C_6H_4-CH_2Br \xrightarrow{Cd} NC-C_6H_4-CH_2CdBr$$

The reactivity of these reagents is similar to that of the corresponding organozinc compounds.

The most common application of organocadmium compounds has been in the preparation of ketones by reaction with acyl chlorides. A major disadvantage of the use of organocadmium reagents is the toxicity and environmental problems associated with use of cadmium.

$$[(CH_3)_2CHCH_2CH_2]_2Cd + ClCCH_2CH_2CO_2CH_3 \longrightarrow (CH_3)_2CHCH_2CH_2COCH_2CH_2CO_2CH_3 \quad 73\text{–}75\%$$

Ref. 138

(bicyclic acid chloride with two H_3C groups and COCl) + $(CH_3)_2Cd$ ⟶ (bicyclic ketone with two H_3C groups and $\overset{O}{\parallel}CCH_3$) 60% Ref. 139

7.3.3. Organomercury Compounds

There are several useful means for preparation of organomercury compounds. The general metal–metal exchange reaction between mercury(II) salts and organolithium or magnesium compounds is applicable. The oxymercuration reaction discussed in Section 4.3 provides a means of acquiring certain functionalized organomercury reagents. Organomercury compounds can also be obtained by reaction of mercuric salts with trialkylboranes, although only primary alkyl groups react readily.[140] Other organoboron compounds, such as boronic acids and boronate esters, also react with mercuric salts.

$$R_3B + 3\,Hg(O_2CCH_3)_2 \longrightarrow 3\,RHgO_2CCH_3$$

$$RB(OH)_2 + Hg(O_2CCH_3)_2 \longrightarrow RHgO_2CCH_3$$

$$RB(OR')_2 + Hg(O_2CCH_3)_2 \longrightarrow RHgO_2CCH_3$$

137. E. R. Burkhardt and R. D. Rieke, *J. Org. Chem.* **50:**416 (1985).
138. J. Cason and F. S. Prout, *Org. Synth.* **III:**601 (1955).
139. M. Miyano and B. R. Dorn, *J. Org. Chem.* **37:**268 (1972).
140. R. C. Larock and H. C. Brown, *J. Am. Chem. Soc.* **92:**2467 (1970); J. J. Tufariello and M. M. Hovey, *J. Am. Chem. Soc.* **92:**3221 (1970).

Alkenylmercury compounds, for example, can be prepared by hydroboration of an alkyne with catecholborane, followed by reaction with mercuric acetate.[141]

$$RC{\equiv}CR + HB(O_2C_6H_4) \longrightarrow RCH{=}C(R)B(O_2C_6H_4) \xrightarrow{Hg(O_2CCH_3)_2} RCH{=}C(R)HgO_2CCH_3$$

The organomercury compounds can be used *in situ*, or they can be isolated as organomercuric halides.

Organomecury compounds are very weakly nucleophilic and react only with very reactive electrophiles. They readily undergo electrophilic substitution by halogens.

$$CH_3(CH_2)_6CH{=}CH_2 \xrightarrow[\text{2) } Hg(O_2CCH_3)_2 \text{ 3) } Br_2]{\text{1) } B_2H_6} CH_3(CH_2)_8Br \quad 69\%$$ Ref. 140

$$\text{2-(HgCl)C}_6\text{H}_4\text{OH} + I_2 \longrightarrow \text{2-IC}_6\text{H}_4\text{OH}$$ Ref. 142

Organomercury reagents do not react with ketones or aldehydes, but Lewis acids cause reaction with acyl chlorides.[143] With alkenylmercury compounds, the reaction probably proceeds by electrophilic attack on the double bond, with the regiochemistry being directed by the stabilization of the β carbocation by the mercury (see Section 6.10, Part A).[144]

$$RCH{=}CH{-}HgCl + R'CCl(=O) \xrightarrow{AlCl_3} R\overset{+}{C}H{-}CH(CR'{=}O){-}HgCl \longrightarrow RCH{=}CHCR'(=O)$$

The majority of the synthetic applications of organomercury compounds are in transition-metal-catalyzed processes in which the organic substituent is transferred from mercury to the transition metal in the course of the reaction. Examples of this type of reaction will be considered in Chapter 8.

7.3.4. Organoindium Reagents

Indium is a group IIIA metal and is a congener of aluminum. Considerable interest has developed recently in the synthetic application of organoindium reagents.[145] One of the properties that makes them useful is that the first oxidation potential is less than that of

141. R. C. Larock, S. K. Gupta, and H. C. Brown, *J. Am. Chem. Soc.* **94:**4371 (1972).
142. F. C. Whitmore and E. R. Hanson, *Org. Synth.* **I:**326 (1941).
143. A. L. Kurts, I. P. Beletskaya, I. A. Savchenko, and O. A. Reutov, *J. Organomet. Chem.* **17:**P21 (1969).
144. R. C. Larock and J. C. Bernhardt, *J. Org. Chem.* **43:**710 (1978).
145. P. Cintas, *Synlett* **1995:**1087.

zinc and even less than that of magnesium, making indium quite reactive as an electron donor to halides. Indium metal reacts with allylic halides in the presence of aldehydes to give the corresponding carbinols.

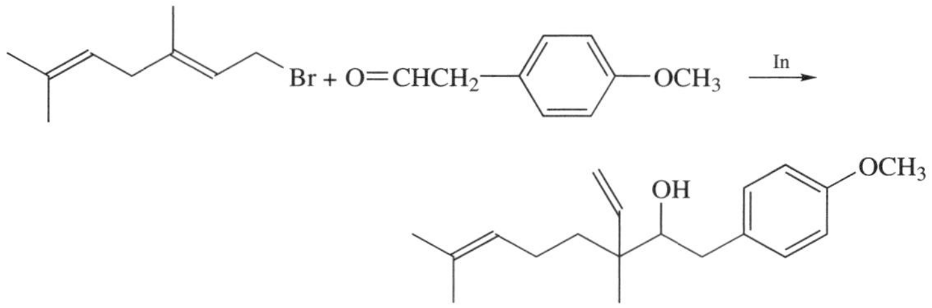

Br + O=CHCH$_2$–C$_6$H$_4$–OCH$_3$ $\xrightarrow{\text{In}}$

OH, OCH$_3$ 85% Ref. 146

The reaction is believed to proceed through a cyclic transition state, and the nucleophile is believed to be an In(I) species.[147]

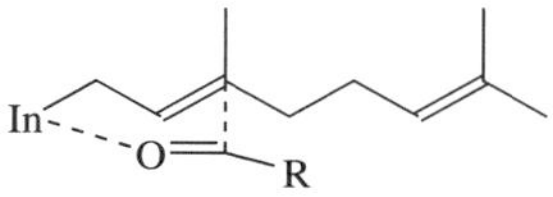

In, O, R

A striking feature of the reactions of indium and allylic halides is that they can be carried out in aqueous solution.[148] The aldehyde traps the organometallic intermediate as it is formed.

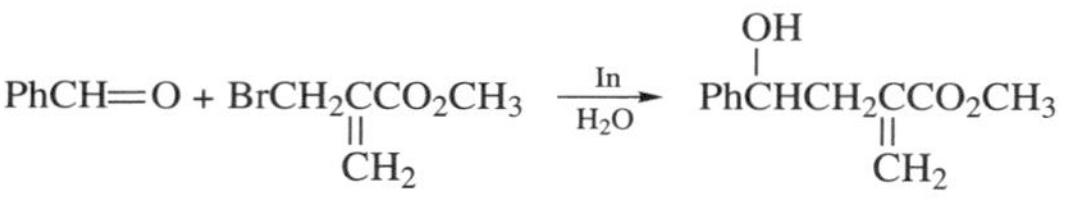

$$\mathrm{PhCH{=}O + BrCH_2C(=CH_2)CO_2CH_3 \xrightarrow[H_2O]{In} PhCH(OH)CH_2C(=CH_2)CO_2CH_3}\quad 96\%$$

The reaction has been found to be applicable to functionalized allylic halides and aldehydes.

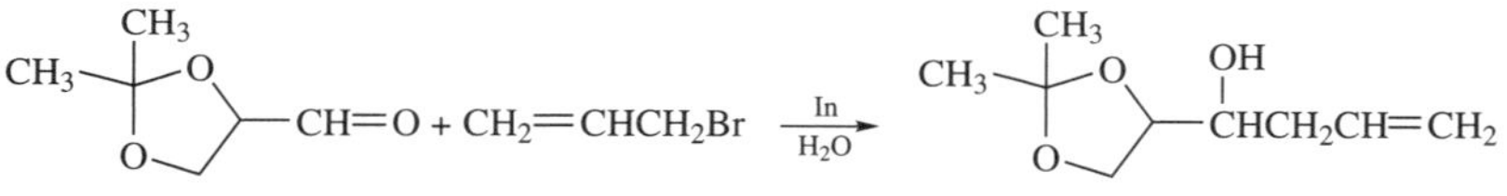

CH_3, CH_3, O, O –CH=O + CH_2=CHCH$_2$Br $\xrightarrow[H_2O]{In}$ CH_3, CH_3, O, O –CH(OH)CH$_2$CH=CH$_2$ 83%

Ref. 149

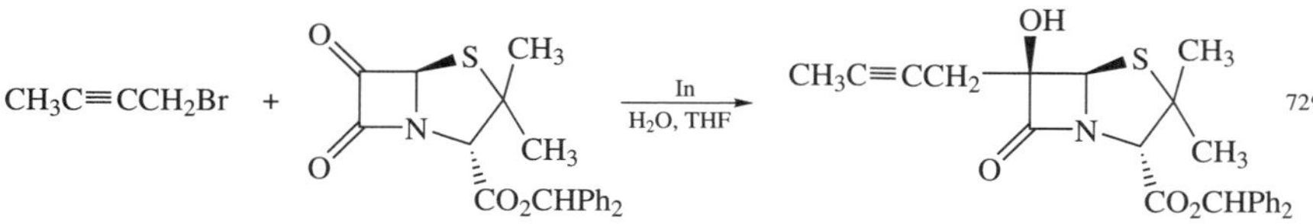

$CH_3C{\equiv}CCH_2Br$ + O, O, N, S, CH_3, CH_3, CO_2CHPh_2 $\xrightarrow[H_2O,\ THF]{In}$ $CH_3C{\equiv}CCH_2$–, OH, O, N, S, CH_3, CH_3, CO_2CHPh_2 72%

Ref. 150

146. S. Araki and Y. Bustugan, *J. Chem. Soc., Perkin Trans. 1* **1991:**2395.
147. T. H. Chan and Y. Yang, *J. Am. Chem. Soc.* **121:**3228 (1999).
148. C.-J. Li and T. H. Chan, *Tetrahedron Lett.* **32:**7017 (1991); C.-J. Li, *Tetrahedron* **52:**5643 (1996).
149. L. A. Paquette and T. M. Mitzel, *J. Am. Chem. Soc.* **118:**1931 (1996); L. A. Paquette and R. R. Rothhaar, *J. Org. Chem.* **64:**217 (1999).
150. Y. S. Cho, J. E. Lee, A. N. Pae, K. I. Choi, and H. Y. Koh, *Tetrahedron Lett.* **40:**1725 (1999).

Aldehydes with α-hydroxy and similar donor substituents react through chelated transition states, which convey good stereoselectivity to the addition.

7.4. Organolanthanide Reagents

The lanthanides are congeners of the group IIIB metals, with the 3+ oxidation state usually being the most stable. Recent years have seen the development of synthetic procedures involving lanthanide metals such as cerium.[151] In the synthetic context, one of the most important applications is the conversion of organolithium compounds to organocerium derivatives by $CeCl_3$.[152] The precise details of preparation of the $CeCl_3$ and its reaction with the organolithium compound can be important to the success of individual reactions.[153] The organocerium compounds are useful for addition to carbonyl compounds that are prone to enolization or highly sterically hindered.[154] The organocerium reagents retain strong nucleophilicity but show a much reduced tendency to effect deprotonation. For example, in the addition of TMS-methyllithium to relatively acidic ketones such as 2-indanone, the yield was substantially increased by use of the organocerium intermediate[155]:

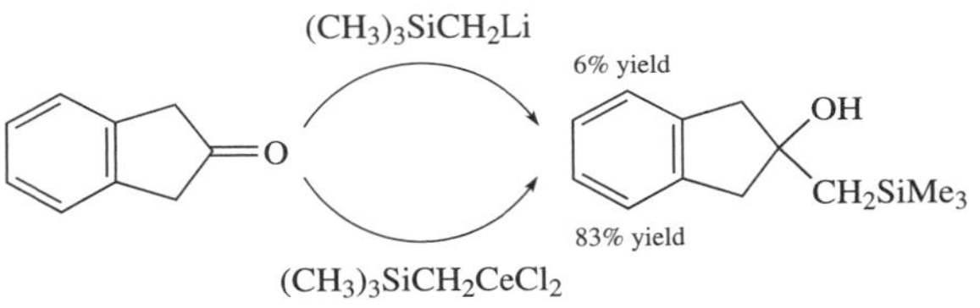

An organocerium reagent gave better yields than either the lithium or Grignard reagents in addition to carbonyl at the 17-position on steroids.[156]

Additions of both Grignard and organolithium reagents can be catalyzed by 5–10 mol% of $CeCl_3$.

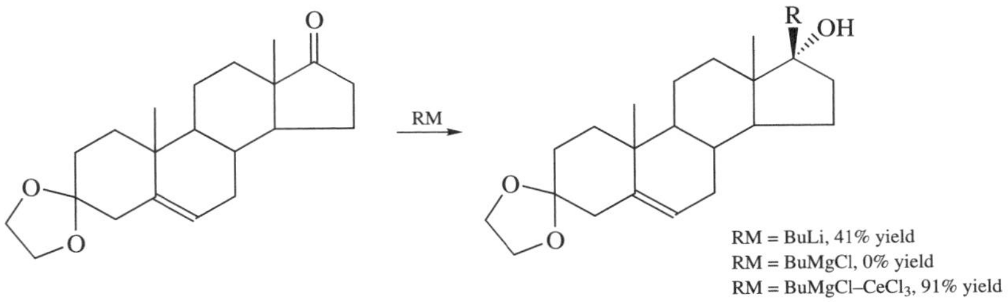

151. H.-J. Liu, K.-S. Shia, X. Shange, and B.-Y. Zhu, *Tetrahedron* **55:**3803 (1999).
152. T. Imamoto, T. Kusumoto, Y. Tawarayama, Y. Sugiura, T. Mita, Y. Hatanaka, and M. Yokoyama, *J. Org. Chem.* **49:**3904 (1984).
153. D. L. J. Clive, Y. Bo, Y. Tao, S. Daigneault, Y.-J. Wu, and G. Meignan, *J. Am. Chem. Soc.* **120:**10332 (1998); W. J. Evans, J. D. Feldman, and T. W. Ziller, *J. Am. Chem. Soc.* **118:**4581 (1996); V. Dimitrov, K. Kostova, and M. Genov, *Tetrahedron Lett.* **37:**6787 (1996).
154. T. Imamoto, N. Takiyama, K. Nakumura, T. Hatajima, and Y. Kamiya, *J. Am. Chem. Soc.* **111:**4392 (1989).
155. C. R. Johnson and B. D. Tait, *J. Org. Chem.* **52:**281 (1987).
156. V. Dimitrov, S. Bratovanov, S. Simova, and K. Kostova, *Tetrahedron Lett.* **35:**6713 (1994); X. Li, S. M. Singh, and F. Labrie, *Tetrahedron Lett.* **35:**1157 (1994).

Cerium reagents have also been found to give improved yields in the reaction of organolithium reagents with carboxylate salts to give ketones.

$$CH_3(CH_2)_2Li + CH_3(CH_2)_4CO_2Li \xrightarrow{\text{2 equiv } CeCl_3} CH_3(CH_2)_2C(=O)(CH_2)_4CH_3 \quad 83\%$$ Ref. 157

Organocerium reagents also show excellent reactivity toward nitriles and imines.[158] Organocerium compounds were also found to be the preferred organometallic reagent for addition to hydrazones in an enantioselective synthesis of amines[159]:

$$RLi \xrightarrow{CeCl_3} RCeCl_2 \xrightarrow{R'CH_2CH{=}NN(\text{2-}CH_2OCH_3\text{-pyrrolidine})} \xrightarrow{ClCO_2CH_3} R'CH_2CH(R){-}N(CO_2CH_3){-}N(\text{2-}CH_2OCH_3\text{-pyrrolidine}) \xrightarrow{H_2,\ \text{Raney Ni}} R'CH_2CH(R)NH_2$$

General References

E. Erdik, *Organozinc Reagents in Organic Synthesis*, CRC Press, Boca Raton, Florida, 1996.
P. Knochel and P. Jones, editors, *Organic Zinc Reagents. A Practical Approach*, Oxford University Press, Oxford, UK, 1999.
P. R. Jenkins, *Organometallic Reagents in Synthesis*, Oxford University Press, Oxford, UK, 1992.
R. C. Larock, *Organomercury Compounds in Organic Synthesis*, Springer Verlag, Berlin, 1985.
M. Schlosser, editor, *Organometallic in Synthesis; A Manual*, Wiley, New York, 1994.
G. S. Silverman and P. E. Rakita, editors, *Handbook of Grignard Reagents*, Marcel Dekker, New York, 1996.
B. J. Wakefield, *The Chemistry of Organolithium Compounds*, Pergamon, Oxford, 1974.
B. J. Wakefield, *Organolithium Methods*, Academic Press, Orlando, Florida, 1988.
B. J. Wakefield, *Organomagnesium Methods in Organic Synthesis*, Academic Press, London, 1995.

Problems

(*References for these problems will be found on page 934.*)

1. Predict the product of each of the following reactions. Be sure to specify all elements of stereochemistry.

157. Y. Ahn and T. Cohen, *Tetrahedron Lett.* **35:**203 (1994).
158. E. Ciganek, *J. Org. Chem.* **57:**4521 (1992).
159. S. E. Denmark, T. Weber, and D. W. Piotrowski, *J. Am. Chem. Soc.* **109:**2224 (1987).

(a) $H_3C(H)C=C(H)Br$ (trans) $\xrightarrow[\text{THF/ether/pentane, } -120^\circ C]{2\ t\text{-BuLi}}$ $\xrightarrow{PhCH=O}$

(b) Cyclohexyl–MgBr + $(CH_3)_2CHCN$ $\xrightarrow[25^\circ C]{\text{benzene}}$

(c) 3,5-$(CH_3O)_2C_6H_3$–$OSi(CH_3)_2C(CH_3)_3$ $\xrightarrow[2)\ BrCH_2C=C(CH_3)_2]{1)\ n\text{-BuLi}}$

(d) $THPO(CH_2)_3C\equiv CH$ $\xrightarrow[2)\ CH_3C\equiv CCH_2Br]{1)\ EtMgBr}$

(e) CH_3O–C_6H_4–CO_2H $\xrightarrow[2)\ 20\text{ equiv TMS}-Cl]{1)\ 8\text{ equiv MeLi, } 0^\circ C}$ $\xrightarrow{H^+, H_2O}$

(f) 1-I-2-$(ICH_2CH_2)C_6H_4$ $\xrightarrow{n\text{-BuLi}}$

(g) $PhCO_2CH_3 + CH_3CHBr_2$ $\xrightarrow[\text{TMEDA, } 25^\circ C]{Zn,\ TiCl_4}$

(h) $CH_3(CH_2)_4CH=O + BrCH_2CO_2Et$ $\xrightarrow[\text{benzene}]{\text{Zn dust}}$

(i) C_6H_5–CH_2Br $\xrightarrow{\text{active Cd}}$ $\xrightarrow{PhCOCl}$

(j) $PhCH_2C(H)(NHCO_2C(CH_3)_3)CH=O$ + $CH_2=CHCH_2MgBr$ $\longrightarrow$

2. Reaction of the epoxide of 1-butene with methyllithium gives 3-pentanol in 90% yield. In contrast, methylmagnesium bromide under similar conditions gives the array of products shown below. Explain the difference in the reactivity of the two organometallic compounds toward this epoxide.

$$CH_3CH_2CH\overset{O}{-}CH_2 \xrightarrow{CH_3MgBr} \underset{5\%}{(CH_3CH_2)_2CHOH} + \underset{15\%}{CH_3CH_2CH_2CH(OH)CH_3} + \underset{7\%}{CH_3CH_2C(OH)(CH_3)_2} + \underset{63\%}{CH_3CH_2CH(OH)CH_2Br}$$

3. Devise an efficient synthesis for the following organometallic compounds from the starting material specified.

(a) 1-Li-1-(OCH_2OCH_3)cyclohexane from cyclohexanone

(b) $(CH_3)_2C(OCH_3)Li$ from $(CH_3)_2C(OCH_3)_2$

(c) $CH_3OCH_2OCH_2Li$ from Bu_3SnCH_2OH

(d) [bicyclic structure: $H_2C=$, CH_3, Li] from [bicyclic ketone: $H_2C=$, CH_3, O]

(e) $LiCH_2C(OSi(CH_3)_3)=NSi(CH_3)_3$ from $CH_3C(=O)NH_2$

(f) [structure: CH_3, Li, $PhCH_2OCH_2OCH_2$, OCH_2OCH_3] from $PhCH_2OCH_2OCH_2CH(CH_3)CH=O$

(g) $(CH_3)_3Si(H)C=C(H)Li$ from $(CH_3)_3SiC\equiv CH$

4. Each of the following compounds gives products in which one or more lithium atoms have been introduced under the conditions specified. Predict the structure of the lithiated product on the basis of structural features known to promote lithiation and/or stabilization of lithiated species. The number of lithium atoms introduced is equal to the number of moles of lithium reagent used in each case.

(a) $H_2C=C(CH_3)C(=O)NHC(CH_3)_3$ $\xrightarrow[\text{TMEDA, THF, }-20^\circ C]{2\ n\text{-BuLi}}$

(b) $(CH_3)_2C=CH_2$ $\xrightarrow[\text{TMEDA, hexane}]{n\text{-BuLi}}$

(c) $C_6H_5CH_2N(CH_3)_2$ $\xrightarrow[\text{ether, }25^\circ C\text{, 24h}]{n\text{-BuLi}}$

(d) 1-methoxynaphthalene (OCH_3) $\xrightarrow[\text{ether, }38^\circ C\text{, 20 h}]{n\text{-BuLi}}$

(e) $HC\equiv CCO_2CH_3$ $\xrightarrow[\text{THF/pentane/ether}]{n\text{-BuLi, }-120^\circ C}$

(f) $C_6H_5NHC(=O)C(CH_3)_3$ $\xrightarrow[\text{THF, }0^\circ C\text{, 2 h}]{2\ n\text{-BuLi}}$

(g) $(CH_3)_2CH-C_6H_4-OCH_3$ $\xrightarrow[\text{TMEDA, ether}]{n\text{-BuLi}}$

(h) $Ph(H)C=C(H)CN$ $\xrightarrow[-113^\circ C]{\text{LDA}}$

(i) $CH_2=C(CH_3)CH_2OH$ $\xrightarrow[2\ n\text{-BuLi, }0^\circ C]{2\ K^+\ ^-O\text{-}t\text{-Bu}}$

(j) N-($PhSO_2$)indole $\xrightarrow[-5^\circ C]{2\ t\text{-BuLi}}$

(k) $Ph_2NCH_2CH=CH_2$ $\xrightarrow{n\text{-BuLi}}$

5. Each of the following compounds can be prepared from reactions of organometallic reagents and readily available starting materials. Identify the appropriate organome-

tallic reagent in each case, and show how it would be prepared. Show how the desired product would be made.

(a) $H_2C{=}CHCH_2CH_2CH_2OH$

(b) $H_2C{=}C(CH_3)C(OH)(CH_2CH_2CH_2CH_3)_2$

(c) $PhC(OH)(CH_2OCH_3)_2$

(d) 2-$N(CH_3)_2$-5-CH_3-C_6H_3-CPh_2OH

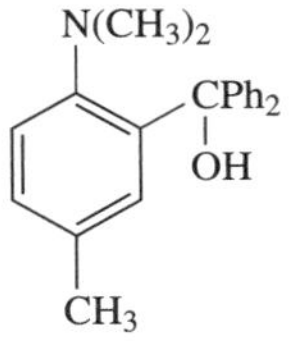

(e) $(CH_3)_3CCH(CO_2C_2H_5)_2$

(f) $H_2C{=}CHCH{=}CHCH{=}CH_2$

6. Identify an organometallic reagent system which will permit formation of the product on the left of each equation from the specified starting material in a "one-pot" process.

(a) [structure: CH_2, H, O, O, H, CH_3, OTBDMS] $\Longrightarrow$ [structure: O, H, O, O, H, CH_3, OTBDMS]

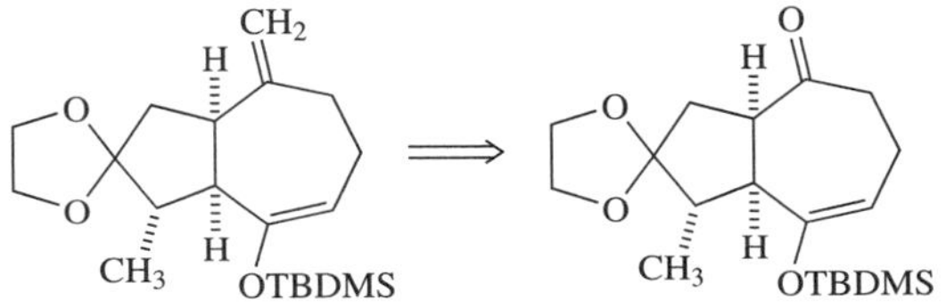

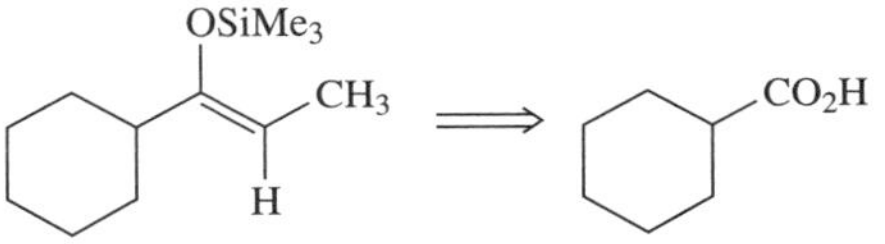

(b) [structure: $OSiMe_3$, CH_3, H] $\Longrightarrow$ [structure: CO_2H]

(c) $PhCCH_2CH_2CO_2C_2H_5$ (C=O) $\Longrightarrow$ $PhCOCl$

(d) $(CH_3)_2CH(CH_2)_2C(CH_2)_6CO_2C_2H_5$ (C=O) $\Longrightarrow$ $ClC(CH_2)_6CO_2C_2H_5$ (C=O)

7. The solvomercuration reaction (Section 4.3) provides a convenient source of such organomercury compounds as **A** and **B**. How could these be converted to functionalised lithium reagents such as **C** and **D**?

A	**B**	**C**	**D**
$HOCH(R)CH_2HgBr$	$PhN(H)CH_2CH_2HgBr$	$LiOCH(R)CH_2Li$	$PhN(Li)CH_2CH_2Li$

Would the procedure you have suggested also work for the following transformation? Explain your reasoning.

$$CH_3OCH(R)CH_2HgBr \longrightarrow CH_3OCH(R)CH_2Li$$

8. Predict the stereochemical outcome of the following reactions and indicate the basis of your predictions.

(a) 2-(OCH_2OCH_2Ph)cyclohexanone $\xrightarrow{CH_3MgCl}$

(b) $CH_3(CH_2)_6C(=O)C(H)(OCH_2OCH_2CH_2OCH_3)CH_3$ $\xrightarrow[\text{THF}]{n\text{-BuMgBr}}$

(c) trans-1-decalone (ring-fusion H's trans) $\xrightarrow{CH_3MgI}$

9. Tertiary amides **E**, **F**, and **G** are lithiated at the β carbon, rather than the α carbon, when treated with s-BuLi/TMEDA. It is estimated, however, that the intrinsic acidity of the α position exceeds that of the β position by ~ 9 pK units. What would cause the β deprotonation to be kinetically preferred?

$$CH_3CH(CH_2R)C(=O)N(i\text{-Pr})_2 \longrightarrow CH_3CH(CHRLi)C(=O)N(i\text{-Pr})_2$$

E R = Ph
F R = CH=CH$_2$
G R = SPh

10. The following reaction sequence converts esters to bromomethyl ketones. Show the intermediates that are involved in each of the steps in the sequence.

$$CH_2Br_2 \xrightarrow[-90^\circ C]{\text{LDA}} \xrightarrow[-90^\circ C]{RCO_2Et} \xrightarrow[-90^\circ C]{n\text{-BuLi}} \xrightarrow[-78^\circ C]{H^+} RC(=O)CH_2Br$$

11. Normally, the reaction of an ester with one equivalent of a Grignard reagent leads to formation of a mixture of tertiary alcohol, ketone, and unreacted ester. However, if one equivalent of LDA is present along with the Grignard reagent, good yields of ketones are obtained. What is the role of LDA in this process?

12. Several examples of intramolecular additions to carbonyl groups by organolithium reagents generated by halogen-metal exchange have been reported, such as the examples shown below. What relative reactivity relationships must hold in order for

such procedures to succeed?

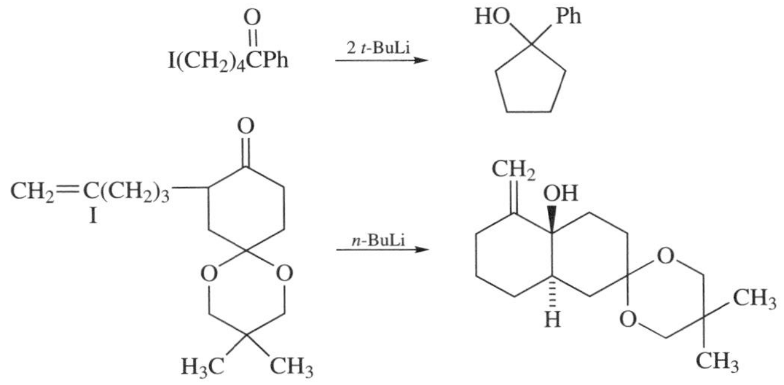

$I(CH_2)_4C(=O)Ph \xrightarrow{2\ t\text{-BuLi}}$ 1-phenylcyclopentanol (HO, Ph)

$CH_2{=}C(I)(CH_2)_3$–(spiroketal cyclohexanone, H_3C, CH_3) $\xrightarrow{n\text{-BuLi}}$ (CH_2, OH, H, O, O, CH_3, CH_3)

13. Short synthetic sequences (three steps or less) involving functionally substituted organometallic compounds as key reagents can effect the following synthetic transformations. Suggest reaction sequences which would be effective for each transformation. Indicate how the required organometallic reagent could be prepared.

(a) $CH_3(CH_2)_4$–(γ-lactone, O, =O) ⟶ $CH_3(CH_2)_4$–(spiroketal, O, O, CH_3, CH_3)

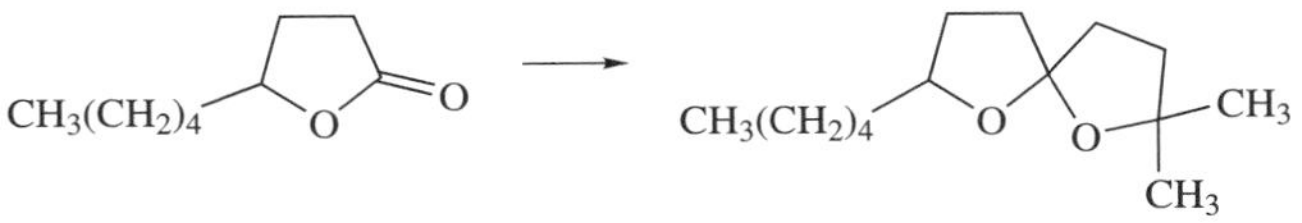

(b) $CH_3CH{=}O$ ⟶ (bicyclic ketal: CH_3, O, O, CH_3)

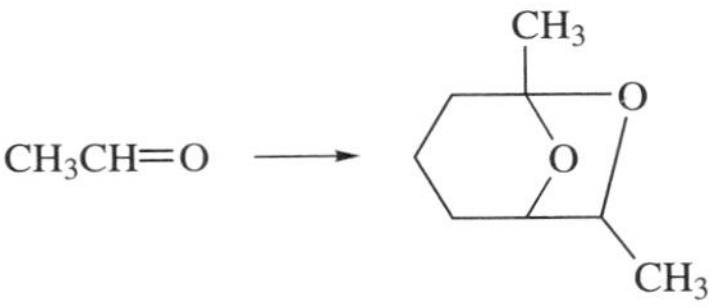

(c) MeO–C_6H_4–CH=O ⟶ (3-aryl isochromanone: O, =O, OMe)

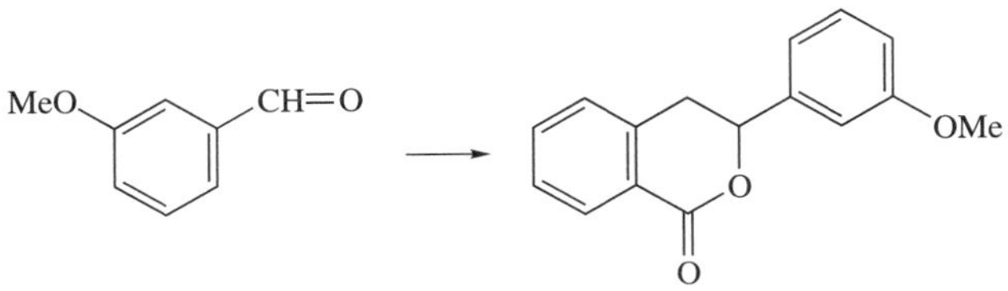

(d) 3-furyl–CH=O ⟶ 3-furyl–$CH(CH_3)OCH(CH_3)OCH_2CH_3$

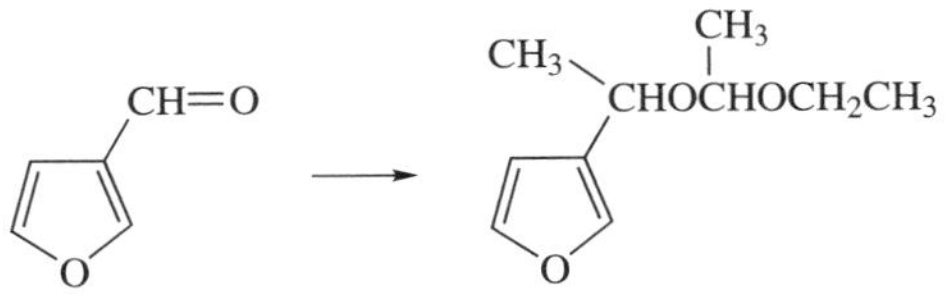

(e) $THPOCH_2CH_2C{\equiv}CH$ ⟶ $(H)(THPOCH_2CH_2)C{=}C(CH_2CH{=}CH_2)(H)$

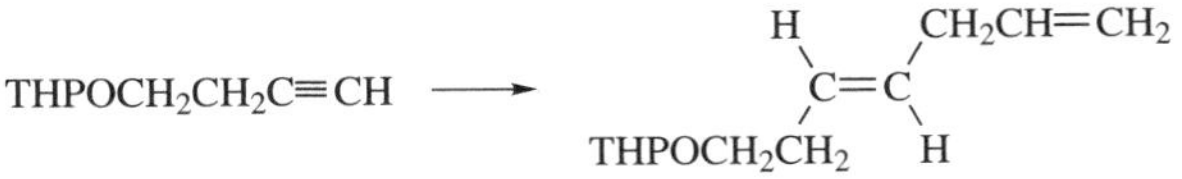

(f) $C_4H_9C(=O)OCH_3$ ⟶ $(C_4H_9)(CH_3O)C{=}C(H)(C_5H_{11})$

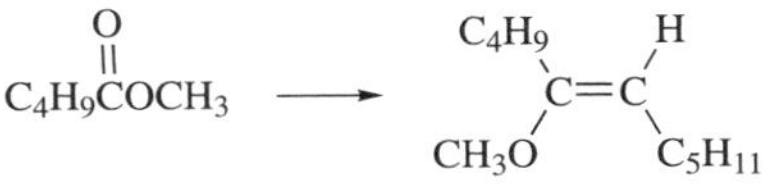

(g) H_3C, CH_3 (epoxide) → H_3C, CH_3, O, O (lactone)

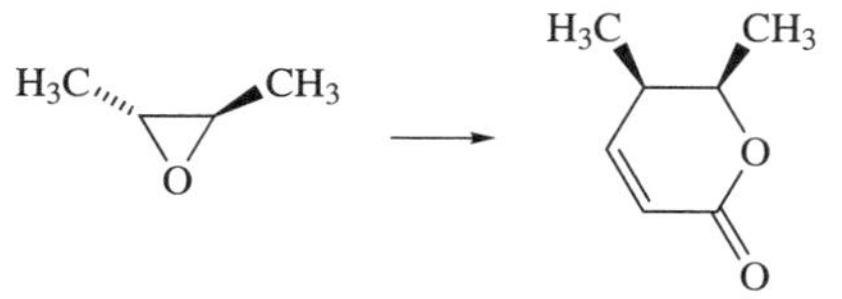

(h) CH=O, OCH_3, CH_3O, CH_3 → CH_3O, O, O, H, CH_3O, OCH_3, CH_3O, CH_3

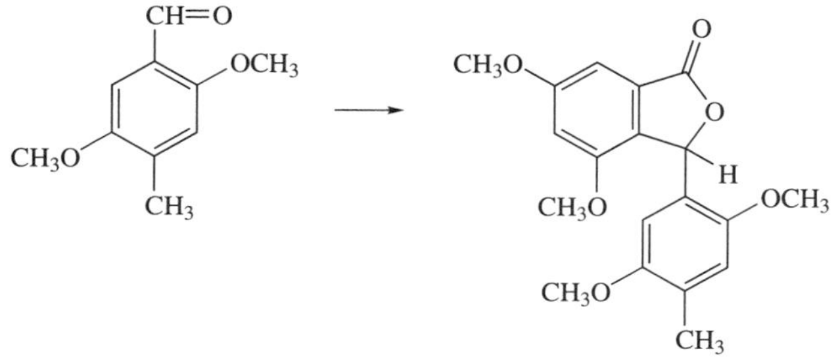

14. Chiral aminoalcohols both catalyze reactions of simple dialkylzinc reagents with aldehydes and also induce a high degree of enantioselectivity, even when used in only catalytic amounts. Two examples are given below. Indicate how the aminoalcohols can have a catalytic effect. Suggest transition states for the examples show which would be in accord with the observed enantioselectivity.

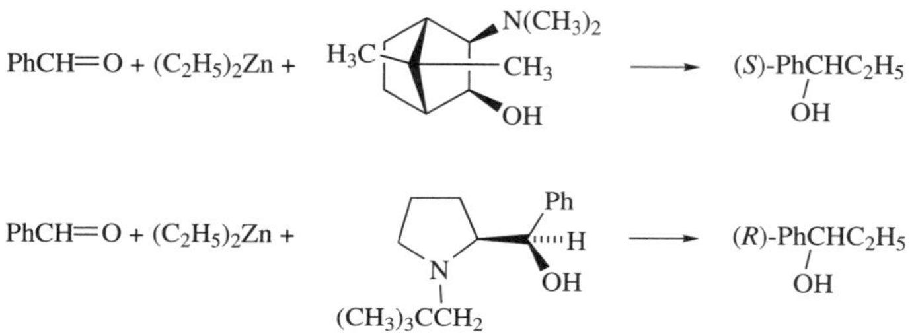

$PhCH{=}O + (C_2H_5)_2Zn +$ [H_3C, $N(CH_3)_2$, CH_3, OH] → (*S*)-$PhCH(OH)C_2H_5$

$PhCH{=}O + (C_2H_5)_2Zn +$ [Ph, H, OH, N, $(CH_3)_3CCH_2$] → (*R*)-$PhCH(OH)C_2H_5$

15. Both of the steps outlined in the retrosynthesis below can be achieved by use of organometallic reagents. Devise a sequence of reactions which would achieve the desired synthesis.

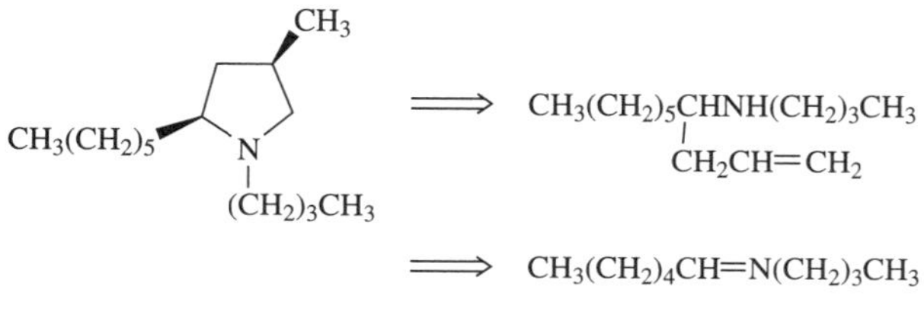

[CH_3, $CH_3(CH_2)_5$, N, $(CH_2)_3CH_3$] ⟹ $CH_3(CH_2)_5CH(CH_2CH{=}CH_2)NH(CH_2)_3CH_3$

⟹ $CH_3(CH_2)_4CH{=}N(CH_2)_3CH_3$

16. When simple 4-substituted 2,2-dimethyl-1,3-dioxolanes react with Grignard reagents, the bond which is broken is the one at the less substituted oxygen.

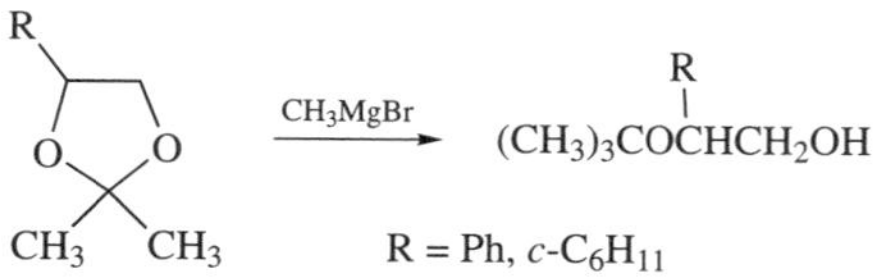

[R, O, O, CH_3, CH_3] $\xrightarrow{CH_3MgBr}$ $(CH_3)_3COCH(R)CH_2OH$

R = Ph, *c*-C_6H_{11}

However, with **H** and **I** the regioselectivity is reversed.

H CH_3 CH_3 O O O O CH_3 CH_3 $\xrightarrow{CH_3MgBr}$ $(CH_3)_3COCH_2CH(OH)$— O O CH_3 CH_3

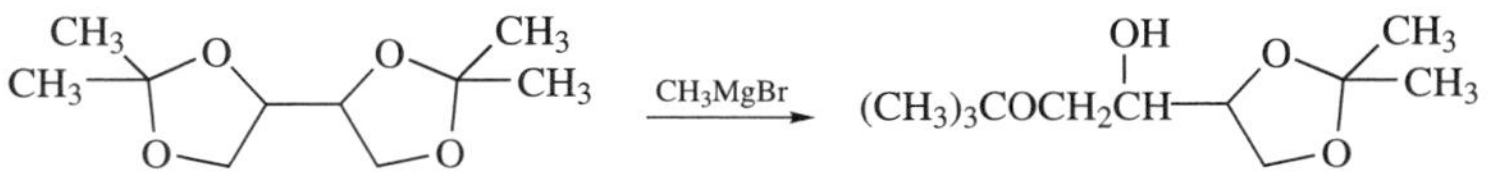

I $(CH_3)_3COCH_2CH(NH_2)$— O O CH_3 CH_3 $\xrightarrow{CH_3MgBr}$ $(CH_3)_3COCH_2CH(NH_2)CH(OH)CH_2OC(CH_3)_3$

What factors might lead to the reversal of regioselectivity?

8

Reactions Involving the Transition Metals

Introduction

While the group I and II elements magnesium and lithium were the first metals to have a prominent role in organic synthesis, several of the transition metals are also very important. In this chapter, we will discuss reactions which are important in synthetic organic chemistry that involve transition-metal compounds and intermediates. In contrast to reactions involving lithium and magnesium, in which the organometallic reagents are used in stoichiometric quantity, many of the transition-metal reactions are catalytic processes. Another distinguishing feature of transition-metal reactions is that they frequently involve oxidation state changes at the metal atom.

8.1. Organocopper Intermediates

8.1.1. Preparation and Structure of Organocopper Reagents

The synthetic application of organocopper compounds received a major impetus from the study of the catalytic effect of copper salts on reactions of Grignard reagents with α,β-unsaturated ketones.[1] Whereas Grignard reagents normally add to conjugated enones to give the 1,2-addition product, the presence of catalytic amounts of Cu(I) results in

1. H. O. House, W. L. Respess, and G. M. Whitesides, *J. Org. Chem.*, **31**:3128 (1966).

conjugate addition. Mechanistic study of this effect pointed to a very fast reaction by an organocopper intermediate.

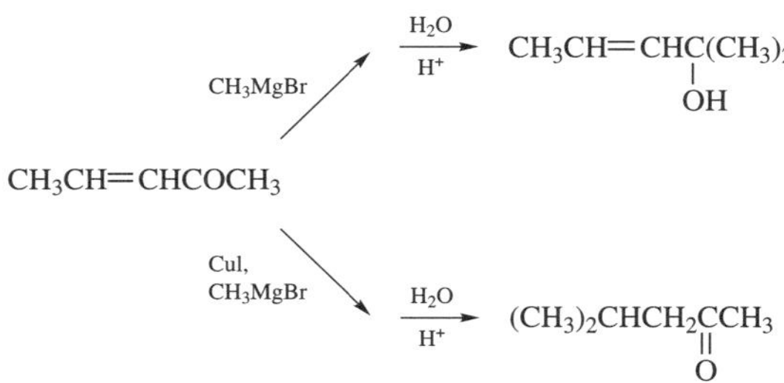

Many subsequent studies have led to the characterization of several types of organocopper compounds that result from reaction of organolithium reagents with copper salts.[2]

$$RLi + Cu(I) \longrightarrow [RCu] + Li^+$$
$$2\,RLi + Cu(I) \longrightarrow [R_2CuLi] + Li^+$$
$$3\,RLi + Cu(I) \longrightarrow [R_3CuLi_2] + Li^+$$

The 2 : 1 species are known as cuprates and are the most important as synthetic reagents. In solution, lithium dimethylcuprate exists as a dimer, $[LiCu(CH_3)_2]_2$.[3] The compound is often represented as four methyl groups attached to a tetrahedral cluster of lithium and copper atoms. However, in the presence of LiI, the compound seems to be a monomer of composition $(CH_3)_2CuLi$.[4]

Discrete diarylcuprate anions have been observed in crystals in which the lithium cation is complexed by crown ethers.[5] Both tetrahedral Ph_4Cu_4 and $[Ph_2Cu]^-$ units have been observed in complex cuprates containing $(CH_3)_2S$ as a ligand. $[Ph_3Cu]^{2-}$ units have also been observed as parts of larger aggregates.[6]

Cuprates with two different copper substituents have been developed. These compounds can have important advantages in cases where one of the substituents is derived from a valuable synthetic intermediate. Table 8.1 presents some of these mixed cuprate reagents.

2. E. C. Ashby and J. J. Lin, *J. Org. Chem.* **42:**2805 (1977); E. C. Ashby and J. J. Watkins, *J. Am. Chem. Soc.*, **99:**5312 (1977).
3. R. G. Pearson and C. D. Gregory, *J. Am. Chem. Soc.* **98:**4098 (1976); B. H. Lipshutz, J. A. Kozlowski and C. M. Breneman, *J. Am. Chem. Soc.* **107:**3197 (1985).
4. A. Gerold, J. T. B. H. Jastrezebski, C. M. P. Kronenburg, N. Krause, and G. Van Koten, *Angew. Chem. Int. Ed. Engl.* **36:**755 (1997).
5. H. Hope, M. M. Olmstead, P. P. Power, J. Sandell, and X. Xu, *J. Am. Chem. Soc.* **107:**4337 (1985).
6. M. M. Olmstead and P. P. Power, *J. Am. Chem. Soc.* **112:**8008 (1990).

Table 8.1. Mixed Culprate Reagents

Mixed cuprate	Reactivity and properties	Reference(s)
$[RC{\equiv}C{-}Cu{-}R]Li$	Conjugate addition to α,β-unsaturated ketones and certain esters	a
$[ArS{-}Cu{-}R]Li$	Nucleophilic substitution and conjugate addition to unsaturated ketones; ketones from acyl chlorides	b, c
$[(CH_3)_3CO{-}Cu{-}R]Li$	Nucleophilic substitution and conjugate addition to α,β-unsaturated ketones	b
$[(c\text{-}C_6H_{11})_2N{-}Cu{-}R]Li$	Normal range of nucleophilic reactivity, improved thermal stability	d
$[Ph_2P{-}Cu{-}R]Li$	Normal range of nucleophilic reactivity, improved thermal stability	d
$[MeS(=O)CH_2{-}Cu{-}R]Li$	Normal range of nucleophilic reactivity, ease of preparation, thermal stability	e
$[N{\equiv}C{-}Cu{-}R]Li$	Efficient opening of epoxides	f
$[R_2CuCN]^{2-}$	Nucleophilic substitution and conjugate addition	g
R—Cu⁻—(2-thienyl)	Nucleophilic substitution, conjugate addition, and epoxide ring opening	h
$R{-}Cu^{-}{-}CH_2C(CH_3)_3$	Conjugate addition	i
$[RCuCH_2Si(CH_3)_3]^{-}$	High reactivity, thermal stability	j
$\{RCuN[Si(CH_3)_3]_2\}^{-}$	High reactivity, thermal stability	j
$[R{-}Cu{-}BF_3]$	Conjugate addition including addition to acrylate esters and acrylonitrile; S_N2' displacement of allylic halides	k

a. H. O. House and M. J. Umen, *J. Org. Chem.* **38**:3893 (1973); E. J. Corey, D. Floyd, and B. H. Lipshutz, *J. Org. Chem.* **43**:3418 (1978).
b. G. H. Posner, C. E. Whitten, and J. J. Sterling, *J. Am. Chem. Soc.* **95**:7788 (1973).
c. G. H. Posner and C. E. Whitten, *Org. Synth.* **55**:122 (1975).
d. S. H. Bertz, G. Dabbagh, and G. M. Villacorta, *J. Am. Chem. Soc.* **104**:5824 (1982).
e. C. R. Johnson and D. S. Dhanoa, *J. Org. Chem.* **52**:1885 (1987).
f. R. D. Acker, *Tetrahedron Lett.* **1977**:3407; J. P. Marino and N. Hatanaka, *J. Org. Chem.* **44**:4467 (1979).
g. B. H. Lipshutz and S. Sengupta, *Org. React.* **41**:135 (1992).
h. H. Malmberg, M. Nilsson, and C. Ullenius, *Tetrahedron Lett.* **23**:3823 (1982); B. H. Lipshutz, M. Koernen, and D. A. Parker, *Tetrahedron Lett.* **28**:945 (1987).
i. C. Lutz, P. Jones, and P. Knochel, *Synthesis* **1999**:312.
j. S. H. Bertz, M. Eriksson, G. Miao, and J. P. Snyder, *J. Am. Chem. Soc.* **118**:10906 (1996).
k. Maruyama and Y. Yamamoto, *J. Am. Chem. Soc.* **99**:8068 (1977); Y. Yamamoto and K. Maruyama, *J. Am. Chem. Soc.* **100**:3240 (1978).

An important type of mixed cuprates is prepared from a 2:1 ratio of an alkyllithium and CuCN.[7] These compounds are called *higher-order cyanocuprates*. The composition is $R_2CuCNLi_2$ in THF solution, but it is thought that most of the molecules probably are present as dimers. The cyanide does not seem to be bound directly to the copper, but instead to the lithium cation.[8]

$$2\,RLi + CuCN \longrightarrow [R_2Cu]^- + [Li_2CN]^+ \rightleftharpoons [R_2Cu]_2^-[Li_2CN]_2^+$$

7. B. H. Lipshutz, R. S. Wilhelm, and J. Kozlowski, *Tetrahedron* **40**:5005 (1984); B. H. Lipshutz, *Synthesis* **1987**:325.
8. T. M. Barnhart, H. Huang, and J. E. Penner-Hahn, *J. Org. Chem.* **60**:4310 (1995); J. P. Snyder and S. H. Bertz, *J. Org. Chem.* **60**:4312 (1995); T. L. Semmler, T. M. Barnhart, J. E. Penner-Hahn, C. E. Tucker, P. Knochel, M. Böhme, and G. Frenking, *J. Am. Chem. Soc.* **117**:12489 (1995); S. H. Bertz, G. Miao, and M. Eriksson, *J. Chem. Soc., Chem. Commun.* **1996**:815.

These reagents are qualitatively similar to other cuprates in reactivity, but they are more stable than the dialkylcuprates. Because cyanocuprate reagents usually transfer only one of the two organic groups, it is useful to incorporate a group which normally does not transfer. The 2-thienyl group has been used for this purpose.[9] In a mixed alkyl–thienyl cyanocuprate, only the alkyl substituent is normally transferred as a nucleophile.

$$[\text{2-thienyl}-Cu(CN)(CH_2)_3CH_3]^{2-} + \text{cyclohexenone} \longrightarrow \text{3-}(CH_2)_3CH_3\text{-cyclohexanone}$$

Another type of mixed cyanocuprate has both methyl and vinyl groups attached to copper. Interestingly, these reagents selectively transfer the alkenyl group in conjugate addition reactions.[10] These reagents can be prepared from alkynes via hydrozirconation, followed by metal–metal exchange.[11]

$$CH_3(CH_2)_7C{\equiv}CH \xrightarrow[\text{3) } (CH_3)_2Cu(CN)Li_2,\ -78^\circ C]{\text{1) } (Cp)_2ZrHCl \quad \text{2) } CH_3Li,\ -78^\circ C} CH_3(CH_2)_7CH{=}CH{-}Cu(CN)Li_2(CH_3)$$

Alkenylcyanocuprates can also be made by metal–metal exchange from alkenylstannanes.[12]

$$\text{1-cyclohexenyl}{-}SnBu_3 + (CH_3)_2Cu(CN)Li_2 \longrightarrow \text{1-cyclohexenyl}{-}Cu(CN)Li_2(CH_3)$$

The 1:1 organocopper reagents can be prepared directly from the halide and highly reactive copper metal prepared by reducing Cu(I) salts with lithium naphthalenide.[13] This method of preparation is advantageous for organocuprates containing substituents that are incompatible with organolithium compounds. For example, nitrophenyl and cyanophenyl copper reagents can be prepared in this way. Alkylcopper reagents having ester and cyano substituents have also been prepared.[14] Allylic chlorides and acetates can also be converted

9. B. H. Lipshutz, J. A. Kozlowski, D. A. Parker, S. L. Nguyen, and K. E. McCarthy, *J. Organomet. Chem.* **285**:437 (1985); B. H. Lipshutz, M. Koerner, and D. A. Parker, *Tetrahedron Lett.* **28**:945 (1987).
10. B. H. Lipshutz, R. S. Wilhelm, and J. A. Kozlowski, *J. Org. Chem.* **49**:3938 (1984).
11. B. H. Lipshutz and E. L. Ellsworth, *J. Am. Chem. Soc.* **112**:7440 (1990).
12. J. R. Behling, K. A. Babiak, J. S. Ng, A. L. Campbell, R. Moretti, M. Koerner, and B. H. Lipshutz, *J. Am. Chem. Soc.* **110**:2641 (1988).
13. G. W. Ebert and R. D. Rieke, *J. Org. Chem.* **49**:5280 (1984); G. W. Ebert and R. D. Rieke, *J. Org. Chem.* **53**:4482 (1988); G. W. Ebert, J. W. Cheasty, S. S. Tehrani, and E. Aouad, *Organometallics* **11**:1560 (1992); G. W. Ebert, D. R. Pfennig, S. D. Suchan, and T. J. Donovan, Jr., *Tetrahedron Lett.* **34**:2279 (1993).
14. R. M. Wehmeyer and R. D. Rieke, *J. Org. Chem.* **52**:5056 (1987); T.-C. Wu, R. M. Wehmeyer, and R. D. Rieke, *J. Org. Chem.* **52**:5059 (1987); R. M. Wehmeyer and R. D. Rieke, *Tetrahedron Lett.* **29**:4513 (1988).

to cyanocuprates by reaction with lithium naphthalenide in the presence of CuCN and LiCl.[15]

$$(CH_3)_2C{=}CHCH_2Cl \xrightarrow[\text{CuCN, LiCl}]{\text{Li naphthalenide}} [((CH_3)_2C{=}CHCH_2)_2CuCN]Li_2$$

There has been much study on the effect of solvents and other reaction conditions on the stability and reactivity of organocuprate species.[16] These studies have found, for example, that $(CH_3)_2S{-}CuBr$, a readily prepared and purified complex of CuBr, is an especially reliable source of Cu(I) for cuprate preparation.[17] Copper(I) cyanide and iodide are also generally effective and are preferable in some cases.[18]

8.1.2. Reactions Involving Organocopper Reagents and Intermediates

The most important reactions of organocuprate reagents are nucleophilic displacements on halides and sulfonates, epoxide ring opening, conjugate additions to α,β-unsaturated carbonyl compounds, and additions to alkynes.[19] Scheme 8.1 gives some examples of each of these reaction types, and they are discussed in more detail in the following paragraphs.

Corey and Posner discovered that lithium dimethylcuprate could replace iodine or bromine by methyl in a wide variety of compounds, including aryl, alkenyl, and alkyl derivatives. This halogen displacement reaction is more general and gives higher yields than displacements with Grignard or lithium reagents.[20]

$$C_6H_5I + (CH_3)_2CuLi \longrightarrow C_6H_5CH_3 \quad 90\%$$

$$PhCH{=}CHBr + (CH_3)_2CuLi \longrightarrow PhCH{=}CHCH_3 \quad 81\%$$

Secondary bromides and tosylates react with inversion of stereochemistry, as in the classical S_N2 substitution reaction.[21] Alkyl iodides, however, lead to racemized product. Aryl and alkenyl halides are reactive, even though the direct displacement mechanism is not feasible. With there halides the mechanism probably consists of two steps. The addition of halides to transition-metal species with low oxidation states is a common reaction in transition-metal chemistry and is called *oxidative addition*. An oxidative addition to the copper occurs in the first step of the mechanism, and the formal oxidation

15. D. E. Stack, B. T. Dawson, and R. D. Rieke, *J. Am. Chem. Soc.* **114:**5110 (1992).
16. R. H. Schwartz and J. San Filippo, Jr., *J. Org. Chem.* **44:**2705 (1979).
17. H. O. House, C.-Y. Chu, J. M. Wilkins, and M. J. Umen, *J. Org. Chem.* **40:**1460 (1975).
18. B. H. Lipshutz, R. S. Wilhelm, and D. M. Floyd, *J. Am. Chem. Soc.* **103:**7672 (1981); S. H. Bertz, C. P. Gibson, and G. Dabbagh, *Tetrahedron Lett.* **28:**4251 (1987); B. H. Lipshutz, S. Whitney, J. A. Kozlowski, and C. M. Breneman, *Tetrahedron Lett.* **27:**4273 (1986).
19. For reviews of the reactions of organocopper reagents see G. H. Posner, *Org. React.* **19:**1 (1972); G. H. Posner, *Org. React.* **22:**253 (1975); G. H. Posner, *An Introduction to Synthesis Using Organocopper Reagents*, John Wiley & Sons, New York, 1980; N. Krause and A. Gerold, *Angew. Chem. Int. Ed. Engl.* **36:** 187 (1997).
20. E. J. Corey and G. H. Posner, *J. Am. Chem. Soc.* **89:**3911 (1967).
21. C. R. Johnson and G. A. Dutra, *J. Am. Chem. Soc.* **95:**7783 (1973); B. H. Lipshutz and R. S. Wilhelm, *J. Am. Chem. Soc.* **104:**4696 (1982); E. Hebert, *Tetrahedron Lett.* **23:**415 (1982).

Scheme 8.1. Reactions of Organocopper Intermediates

A. Conjugate addition reactions

1[a] =O + Me_2CuLi → =O 98%
CH_3 H_3C CH_3

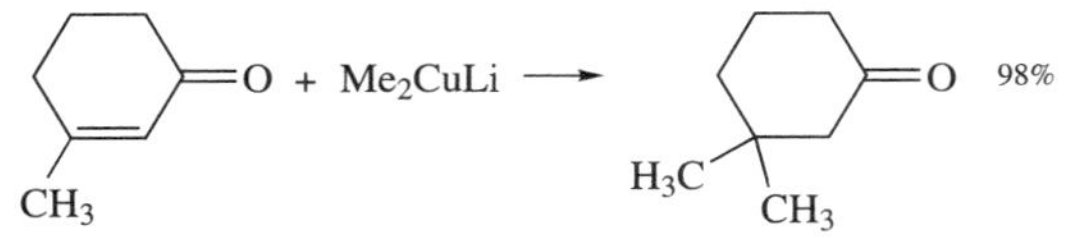

2[b] CH_3 $\overset{O}{\parallel}$ $OCCH_3$ + Me_2CuLi → H_3C CH_3 $\overset{O}{\parallel}$ $OCCH_3$ 55%
O H O H

3[c] O $(CH_2)_6CO_2CH_3$ + $LiCu(CH{=}CH_2)_2$ → O $(CH_2)_6CO_2CH_3$ 66%
$CH{=}CH_2$

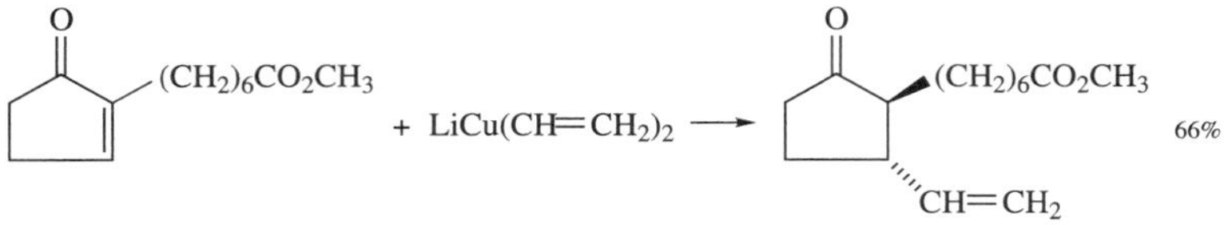

4[d] $CH_3(CH_2)_3Li$ $\xrightarrow[PPh_3]{CuI}$ $CH_3(CH_2)_3Cu$ + O → O 82%
$CH_2CH_2CH_2CH_3$

5[e] CH_3—N O $\xrightarrow{Ph_2CuLi}$ Ph CH_3—N O H 75%

6[f] O + $[(CH_3)_2C—]_2Cu(CN)Li_2$ CH_3OCH_2O → 1) TMS–Cl 2) Et_3N 3) $TiCl_4$ → O O CH_3 CH_3 73%

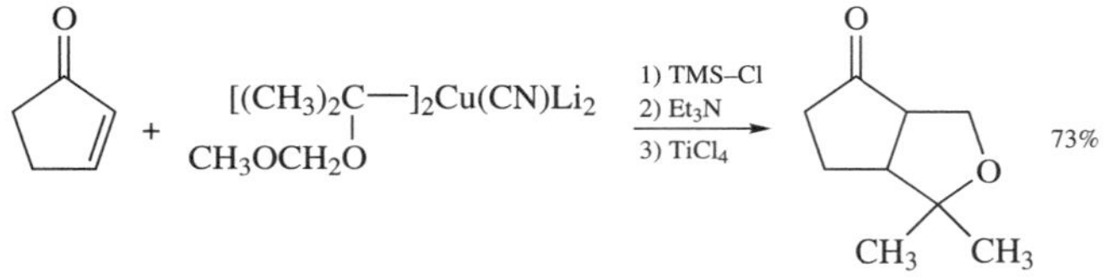

7[g] O CH_3 CH_3 CH_3 + $Ph_2Cu(CN)Li_2$ + BF_3 → O CH_3 CH_3 CH_3 Ph 95%

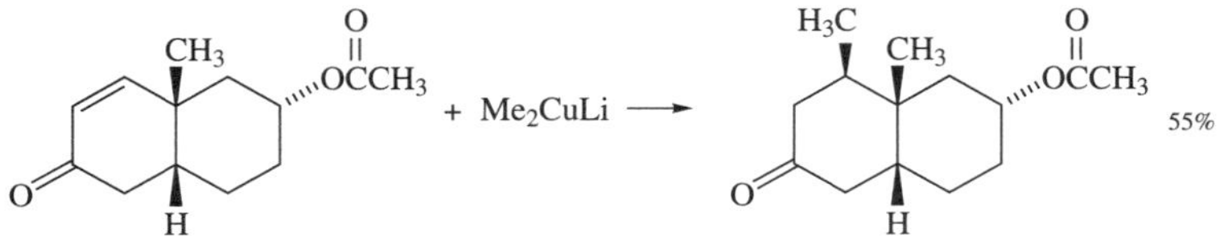

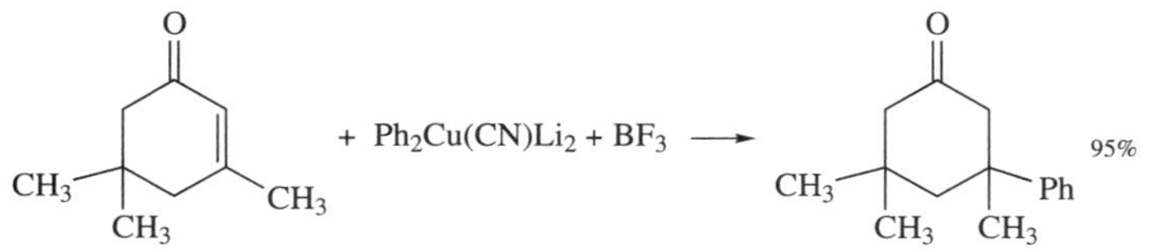

8[h] O + $CH_3(CH_2)_7CH{=}CHCu(CN)Li_2$ → O 86%
CH_3 $CH{=}CH(CH_2)_7CH_3$

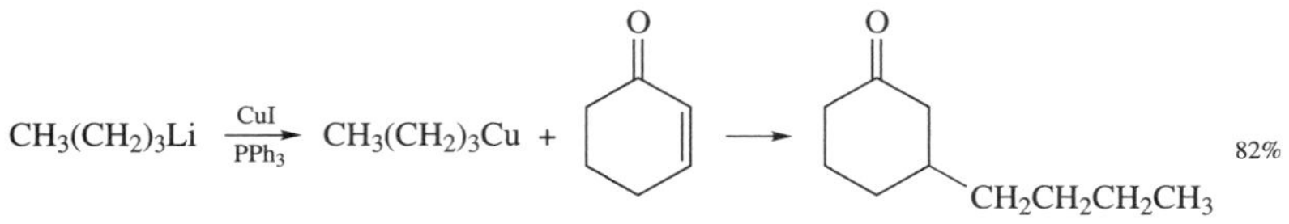

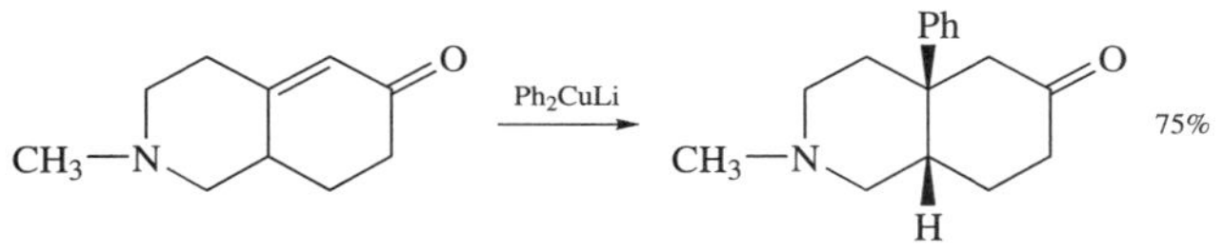

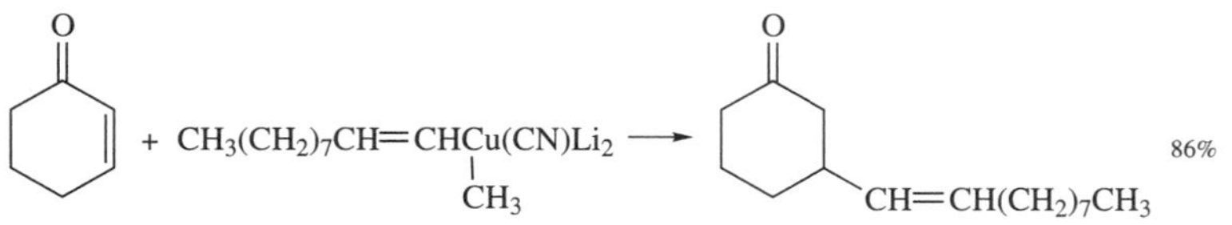

9[i] + $CH_2{=}CHCH_2Cu{\cdot}LiCl$ $\xrightarrow{(CH_3)_3SiCl}$ 87%

O, $CH(CH_3)_2$; O, $CH_2CH{=}CH_2$, $CH(CH_3)_2$

10[j] CH_3O, H, O, O, O $\xrightarrow[\text{2) } NH_4Cl,\ H_2O]{\text{1) 9 equiv } t\text{-BuCu(CN)Li, 18 equiv TMS–Cl}}$ CH_3O, $(CH_3)_3C$, H, O, O, O 87%

11[k] CH_3, CH_3, O, O, O, CH_3, CH_3, C_2H_5, CH_3, H, CH_3, O, O, CH_3, CH_3 $\xrightarrow[BF_3,\ -78°C]{Me_2CuLi}$ CH_3, CH_3, O, O, O, CH_3, CH_3, C_2H_5, CH_3, CH_3, H, CH_3, O, O, CH_3, CH_3 80%

12[l] O, CH_3, H, CH_2OCH_3, O, O $\xrightarrow[-78°C]{CH_2{=}CHCu,\ LiI,\ Bu_3P}$ O, CH_3, $CH{=}CH_2$, H, CH_2OCH_3, O, O 95%

B. Halide substitution

13[m] Br, Br + Me_2CuLi ⟶ CH_3, CH_3 65%

14[n] Br, Br + Me_2CuLi ⟶ H_3C, CH_3 95%

15[o] C_2H_5, I, $CH_3C{=}C(H)CH_2CH_2C{=}C(H)CH_2OH$ $\xrightarrow{Et_2CuLi}$ C_2H_5, C_2H_5, $CH_3C{=}C(H)CH_2CH_2C{=}C(H)CH_2OH$ 65%

Scheme 8.1. *(continued)*

16[p] (Cl-substituted diene, CH_2OCPh_3) $\xrightarrow{Me_2CuLi}$ (triene with C_2H_5, C_2H_5, CH_3 substituents, CH_2OCPh_3)

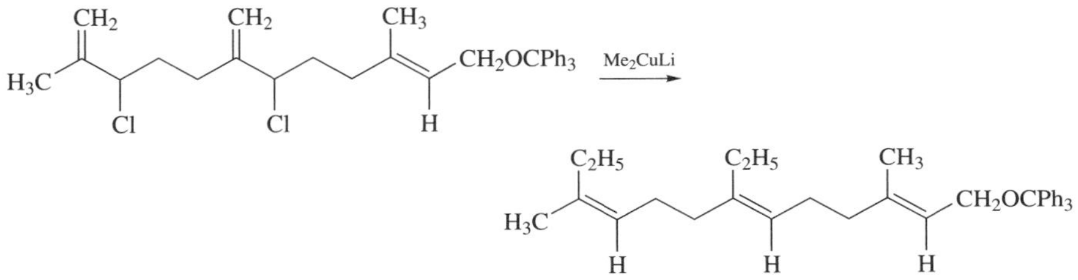

17[q] $CH_3CH{=}CHCl \xrightarrow{Li\ (1\%\ Na)} CH_3CH{=}CHLi \xrightarrow[CH_3(CH_2)_6CH_2I]{CuI,\ -78^\circ C} CH_3CH{=}CHCH_2(CH_2)_6CH_3$ 90–93%

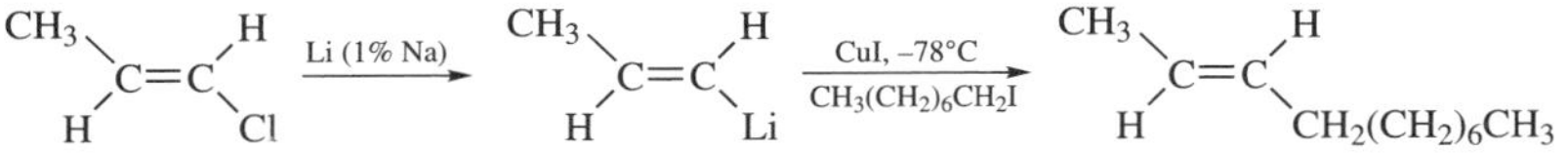

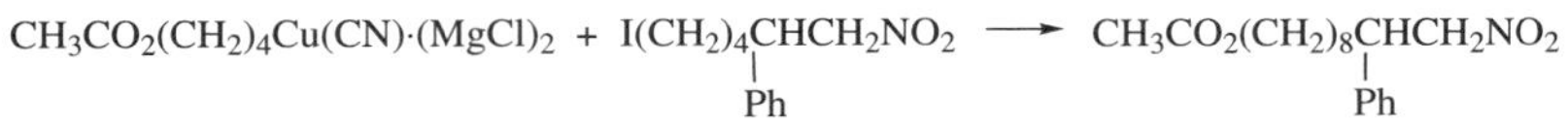

18[r] $CH_3CO_2(CH_2)_4Cu(CN){\cdot}(MgCl)_2 + I(CH_2)_4CH(Ph)CH_2NO_2 \longrightarrow CH_3CO_2(CH_2)_8CH(Ph)CH_2NO_2$ 83%

19[s] (OTBDMS, OTs, CO_2CH_3 substrate) + $Me_2Cu(CN)Li_2{\cdot}BF_3 \longrightarrow$ (OTBDMS, CH_3, CO_2CH_3 product) 96%

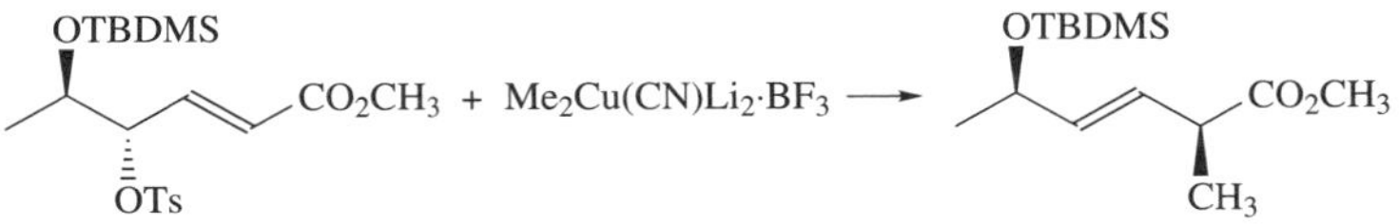

C. Displacement of allylic acetates

20[t] $CH_3C(=CH_2)CH(O_2CCH_3)CH_2CH_2C(CH_3){=}CHCH_2CH_2C(CH_3){=}CHCO_2CH_3 \xrightarrow[ether,\ -10^\circ C]{Me_2CuLi}$

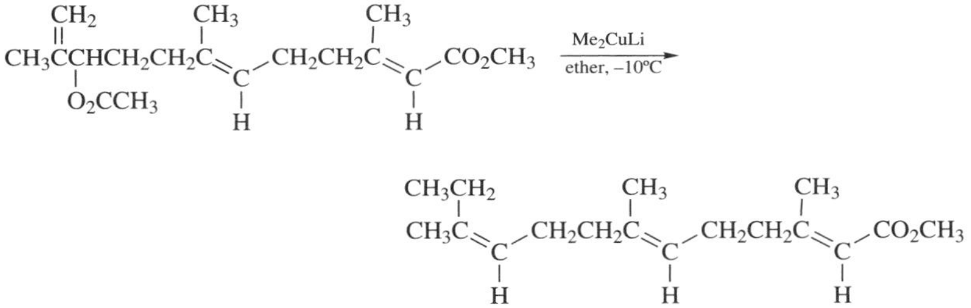

$CH_3CH_2(CH_3)C{=}CHCH_2CH_2C(CH_3){=}CHCH_2CH_2C(CH_3){=}CHCO_2CH_3$ 87%

21[u] (CH_3, O_2CCH_3 cyclohexene) $\xrightarrow{Me_2CuLi}$ (CH_3, CH_3 cyclohexene) 90–95%

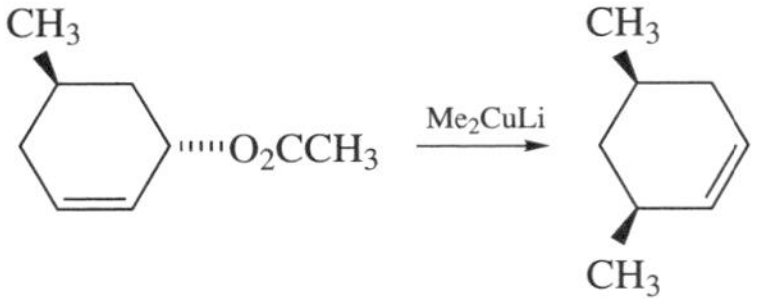

D. Epoxide ring-opening reactions

22[v] (cyclohexene oxide, O) + $(n\text{-}C_4H_9)_2Cu(CN)Li \longrightarrow$ (OH, $(CH_2)_3CH_3$ cyclohexane) 88%

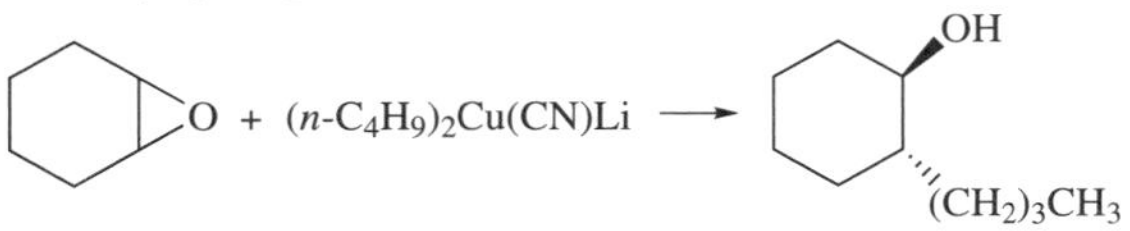

23[w] ($HOCH_2$, CH_3, O, OCH_2Ph) $\xrightarrow{Et_2CuLi}$ ($HOCH_2$, C_2H_5, CH_3, OH, OCH_2Ph)

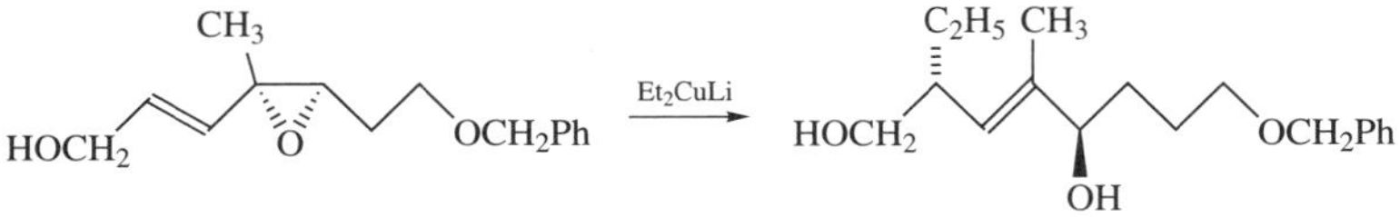

24[x] $PhCH_2O$–(epoxy alcohol, CH_3, CH_2OH) $\xrightarrow{Me_2Cu(CN)Li_2}$ $PhCH_2O$–(CH$_3$, CH$_3$, OH, CH_2OH) 80%

E. Ketones from acyl chlorides

25[y] $PhCCl$ (O) + $[(CH_3)_3CCuSPh]Li$ ⟶ $PhCC(CH_3)_3$ (O) 84–87%

26[z] $ClCCH_2CH_2$(O)–(steroid: H_3C, H_3C, $OC(CH_3)_3$, O) $\xrightarrow{Et_2CuLi}$ $C_2H_5CCH_2CH_2$(O)–(steroid: H_3C, H_3C, $OC(CH_3)_3$, O) 65%

a. H. O. House, W. L. Respess, and G. M. Whitesides, *J. Org. Chem.* **31**:3128 (1966).
b. J. A. Marshall and G. M. Cohen, *J. Org. Chem.* **36**:877 (1971).
c. F. S. Alvarez, D. Wren, and A. Prince, *J. Am. Chem. Soc.* **94**:7823 (1972).
d. M. Suzuki, T. Suzuki, T. Kawagishi, and R. Noyori, *Tetrahedron Lett.* **21**:1247 (1980).
e. N. Finch, L. Blanchard, R. T. Puckett, and L. H. Werner, *J. Org. Chem.* **39**:1118 (1974).
f. R. J. Linderman and A. Godfrey, *J. Am. Chem. Soc.* **110**:6249 (1988).
g. B. H. Lipshutz, D. A. Parker, J. A. Kozlowski, and S. L. Nguyen, *Tetrahedron Lett.* **25**:5959 (1984).
h. B. H. Lipshutz and E. L. Ellsworth, *J. Am. Chem. Soc.* **112**:7440 (1990).
i. B. H. Lipshutz, E. L. Ellsworth, S. H. Dimock, and R. A. J. Smith, *J. Am. Chem. Soc.* **112**:4404 (1990).
j. E. J. Corey and K. Kamiyama, *Tetrahedron Lett.* **31**:3995 (1990).
k. B. Delpech and R. Lett, *Tetrahedron Lett.* **28**:4061 (1987).
l. T. Kawabata, P. Grieco, H. L. Sham, H. Kim, J. Y. Jaw, and S. Tu, *J. Org. Chem.* **52**:3346 (1987).
m. E. J. Corey and G. H. Posner, *J. Am. Chem. Soc.* **89**:3911 (1967).
n. W. E. Konz, W. Hechtl, and R. Huisgen, *J. Am. Chem. Soc.* **92**:4104 (1970).
o. E. J. Corey, J. A. Katzenellenbogen, N. W. Gilman, S. A. Roman, and B. W. Erickson, *J. Am. Chem. Soc.* **90**:5618 (1968).
p. E. E. van Tabelen and J. P. McCormick, *J. Am. Chem. Soc.* **92**:737 (1970).
q. G. Linstrumelle, J. K. Krieger, and G. M. Whitesides, *Org. Synth.* **55**:103 (1976).
r. C. E. Tucker and P. Knochel, *J. Org. Chem.* **58**:4781 (1993).
s. T. Ibuka, T. Nakao, S. Nishii, and Y. Yamamoto, *J. Am. Chem. Soc.* **108**:7420 (1986).
t. R. J. Anderson, C. A. Henrick, J. B. Siddall, and R. Zurfluh, *J. Am. Chem. Soc.* **94**:5379 (1972).
u. H. L. Goering and V. D. Singleton, Jr., *J. Am. Chem. Soc.* **98**:7854 (1976).
v. B. H. Lipshutz, J. Kozlowski, and R. S. Wilhelm, *J. Am. Chem. Soc.* **104**:2305 (1982).
w. J. A. Marshall, T. D. Crute III, and J. D. Hsi, *J. Org. Chem.* **57**:115 (1992).
x. A. B. Smith III, B. A. Salvatore, K. G. Hull, and J. J.-W. Duan, *Tetrahedron Lett.* **32**:4859 (1991).
y. G. Posner and C. E. Whitten, *Org. Synth.* **55**:122 (1976).
z. W. G. Dauben, G. Ahlgren, T. J. Leitereg, W. C. Schwarzel, and M. Yoshioko, *J. Am. Chem. Soc.* **94**:8593 (1972).

state of copper after this addition step is 3+. This step is followed by combination of two of the alkyl groups from copper. This process, which is also very common for transition-metal intermediates, is called *reductive elimination*.

$$R{-}X + R'_2Cu \longrightarrow R{-}\overset{R'}{\underset{R'}{Cu}}X \longrightarrow R{-}R' + R'CuX$$

Allylic halides usually give both S_N2 products and products of substitution with an allylic shift (S_N2' products) although the mixed organocopper reagent $RCu{-}BF_3$ is reported to give mainly the S_N2' product.[22] Allylic acetates undergo displacement with an allylic shift (S_N2' mechanism).[23] The allylic substitution process may involve initial

22. K. Maruyama and Y. Yamamoto, *J. Am. Chem. Soc.* **99**:8068 (1977).
23. R. J. Anderson, C. A. Henrick, and J. B. Siddall, *J. Am. Chem. Soc.* **92**:735 (1970); E. E. van Tamelen and J. P. McCormick, *J. Am. Chem. Soc.* **92**:737 (1970).

coordination with the double bond.[24]

$$[R_2Cu]^- + CH_2{=}CHCH_2X \rightleftharpoons \text{R–Cu}^-\text{–R}\cdots CH_2{=}CH{-}CH_2{-}X \longrightarrow \text{R–Cu–R (CH}_2CH{=}CH_2) \longrightarrow RCH_2CH{=}CH_2 + RCu$$

The reaction shows a preference for *anti* stereochemistry in cyclic systems.[25]

$[Me_2Cu]Li$

It has been suggested that the preference for the *anti* stereochemistry is the result of simultaneous overlap of a *d* orbital on copper with both the π^* and σ^* orbitals of the allyl system.[26]

Propargylic acetates, halides, and sulfonates also react with a double-bond shift to give allenes.[27] Some direct substitution product can be formed, as well. A high ratio of allenic product is usually found with $CH_3Cu{-}LiBr{-}MgBrI$, which is prepared by addition of methylmagnesium bromide to a 1 : 1 LiBr–CuI mixture.[28]

$$C_6H_{11}{-}C{\equiv}CCH(O_2CCH_3)C_5H_{11} + CH_3Cu\text{-}LiBr\text{-}MgBrI \longrightarrow (C_6H_{11})(H_3C)C{=}C{=}C(H)C_5H_{11} \quad 100\%$$

Coupling of Grignard reagents with primary halides and tosylates can by catalyzed by Li_2CuCl_4.[29] This method, for example, was used to synthesize the long-chain carboxylic acid **1** in >90% yield.

$$CH_2{=}CH(CH_2)_9MgCl + Br(CH_2)_{11}CO_2MgBr \xrightarrow[2)\ H^+]{1)\ Li_2CuCl_4} \underset{\mathbf{1}}{CH_2{=}CH(CH_2)_{20}CO_2H} \quad \text{Ref. 30}$$

24. H. L. Goering and S. S. Kantner, *J. Org. Chem.* **49:**422 (1984).
25. H. L. Goering and V. D. Singleton, Jr., *J. Am. Chem. Soc.* **98:**7854 (1976); H. L. Goering and C. C. Tseng, *J. Org. Chem.* **48:**3986 (1983).
26. E. J. Corey and N. W. Boaz, *Tetrahedron Lett.* **25:**3063 (1984).
27. P. Rona and P. Crabbé, *J. Am. Chem. Soc.* **90:**4733 (1968); R. A. Amos and J. A. Katzenellenbogen, *J. Org. Chem.* **43:**555 (1978); D. J. Pasto, S.-K. Chou, E. Fritzen, R. H. Shults, A. Waterhouse, and G. F. Hennion, *J. Org. Chem.* **43:**1389 (1978).
28. T. L. Macdonald, D. R. Reagan, and R. S. Brinkmeyer, *J. Org. Chem.* **45:**4740 (1980).
29. M. Tamura and J. Kochi, *Synthesis*, **1971:**303; T. A. Baer and R. L. Carney, *Tetrahedron Lett.* **1976:**4697.
30. S. B. Mirviss, *J. Org. Chem.* **54:**1948 (1989).

Another excellent catalyst for coupling is a mixture of $CuBr-S(CH_3)_2$, LiBr, and LiSPh. This catalyst can effect coupling of a wide variety of Grignard reagents with tosylates and mesylates and is superior to Li_2CuCl_4 in coupling with secondary sulfonates.[31]

$$CH_3(CH_2)_9MgBr + CH_3\underset{\displaystyle O_3SCH_3}{\underset{|}{C}}HCH_2CH_3 \xrightarrow{\text{catalyst}} CH_3(CH_2)_9\underset{\displaystyle CH_3}{\underset{|}{C}}HCH_2CH_3$$

Catalyst	Yield
Li_2CuCl_4	17%
CuBr/HMPA	30%
$CuBr–S(CH_3)_2$, LiBr, LiSPH	62%

Halogens α to carbonyl groups can be successfully coupled with organocopper reagents. For example, 3,9-dibromocamphor can be selectively arylated α to the carbonyl.

$BrCH_2$ (3,9-dibromocamphor, Br, O) + (OCH_3, OCH_3 aryl $)_2CuLi$ ⟶ $BrCH_2$ (camphor, O) aryl (OCH_3, CH_3O) 79% Ref. 32

Saturated epoxides are opened in good yield by lithium dimethylcuprate.[33] The methyl group is introduced at the less hindered carbon of the epoxide ring.

$$CH_3CH_2\text{-epoxide (O)} + (CH_3)_2CuLi \longrightarrow CH_3CH_2\underset{\displaystyle OH}{\underset{|}{C}}HCH_2CH_3 \quad 88\%$$

Epoxides with alkenyl substituents undergo alkylation at the double bond with a double-bond shift accompanying ring opening[34].

$$(CH_3)_2CuLi + H_2C{=}\overset{\displaystyle CH_3}{\overset{|}{C}}\text{-epoxide(O)-}CH_3 \longrightarrow CH_3CH_2\overset{\displaystyle CH_3}{\overset{|}{C}}{=}CH\underset{\displaystyle OH}{\underset{|}{C}}HCH_3$$

All of the types of mixed cuprate reagents described in Table 8.1 react with conjugated enones. A comparison of various Cu(I) salts suggests that $CuBr-S(CH_3)_2$ and CuCN are the best.[35] A number of improvements in methodology for carrying out the

31. D. H. Burns, J. D. Miller, H.-K. Chan, and M. O. Delaney, *J. Am. Chem. Soc.* **119:**2125 (1997).
32. V. Vaillancourt and K. F. Albizati, *J. Org. Chem.* **57:**3627 (1992).
33. C. R. Johnson, R. W. Herr, and D. M. Wieland, *J. Org. Chem.* **38:**4263 (1973).
34. R. J. Anderson, *J. Am. Chem. Soc.* **92:**4978 (1970); R. W. Herr and C. R. Johnson, *J. Am. Chem. Soc.* **92:**4979 (1970).
35. S. H. Bertz, C. P. Gibson, and G. Dabbagh, *Tetrahedron Lett.* **28:**4251 (1987).

conjugate addition reactions have been introduced. The addition is accelerated by trimethylsilyl chloride or a combination of trimethylsilyl chloride and HMPA.[36] Under these conditions, the initial product is a silyl enol ether. The rate enhancement is attributed to trapping of a reversibly formed complex between the enone and cuprate.[37]

$$R_2CuLi + R'CH{=}CHCR'' (C{=}O) \rightleftharpoons [R_2Cu^-]\cdots\text{complex} \xrightarrow[\text{fast}]{TMS-Cl} \text{silyl enol ether (OTMS)}; \quad \xrightarrow{\text{slow}} \text{enolate } (^-O)$$

This technique also greatly improves yields of conjugate addition of cuprates to α,β-unsaturated esters and amides.[38] Trimethylsilyl cyanide also accelerates conjugate addition.[39]

Another useful reagent is prepared from a 1 : 1 : 1 ratio of organolithium reagent, CuCN, and $BF_3 \cdot O(C_2H_5)_2$.[40] The BF_3 appears to interact with the cyanocuprate reagent, giving a more reactive species.[41] The efficiency of the reaction is improved by the addition of trialkylphosphines to the reaction mixture.[42] Even reagents prepared from a 1 : 1 ratio of organocopper and organolithium compounds are reactive in the presence of phosphines.[43]

$$(CH_3)_2CHCH{=}CHCCH_3 (C{=}O) + PhCu{\cdot}LiI \xrightarrow{(n\text{-}C_4H_9)_3P} (CH_3)_2CHCH(Ph)CH_2CCH_3 (C{=}O) \quad 84\%$$

The conjugate addition reactions probably occur by a mechanism similar to that for substitution on allylic halides. There is probably an initial complex between the cuprate and enone.[44] The key intermediate for formation of the new carbon–carbon bond is an adduct formed between the enone and the organocopper reagent. The adduct is formulated as a Cu(III) species which then undergoes reductive elimination. The lithium ion also plays a key role, presumably by Lewis acid coordination at the carbonyl oxygen.[45] Solvent

36. S. H. Bertz and G. Dabbagh, *Tetrahedron* **45:**425 (1989); S. H. Bertz and R. A. J. Smith, *Tetrahedron* **46:**4091 (1990); K. Yamanaka, H. Ogura, J. Jaukuta, H. Inoue, K. Hamada, Y. Sugiyama, and S. Yamada, *J. Org. Chem.* **63:**4449 (1998); M. Kanai, Y. Nakagawa, and K. Tomioka, *Tetrahedron* **55:**3831 (1999).
37. E. J. Corey and N. W. Boaz, *Tetrahedron Lett.* **26:**6019 (1985); E. Nakamura, S. Matsuzawa, Y. Horiguchi, and I. Kuwajima, *Tetrahedron Lett.* **27:**4029 (1986); C. R. Johnson and T. J. Marren, *Tetrahedron Lett.* **28:**27 (1987).
38. A. Alexakis, J. Berlan, and Y. Besace, *Tetrahedron Lett.* **27:**1047 (1986).
39. B. H. Lipshutz and B. James, *Tetrahedron Lett.* **34:**6689 (1993).
40. T. Ibuka, N. Akimoto, M. Tanaka, S. Nishii, and Y. Yamamoto, *J. Org. Chem.* **54:**4055 (1989).
41. B. H. Lipshutz, E. L. Ellsworth, and T. J. Siahaan, *J. Am. Chem. Soc.* **111:**1351 (1989); B. H. Lipshutz, E. L. Ellsworth, and S. H. Dimock, *J. Am. Chem. Soc.* **112:**5869 (1990).
42. M. Suzuki, T. Suzuki, T. Kawagishi, and R. Noyori, *Tetrahedron Lett.* **1980:**1247.
43. T. Kawabata, P. A. Grieco, H.-L. Sham, H. Kim, J. Y. Jaw, and S. Tu, *J. Org. Chem.* **52:**3346 (1987).
44. S. R. Krauss and S. G. Smith, *J. Am. Chem. Soc.* **103:**141 (1981); E. J. Corey and N. W. Boaz, *Tetrahedron Lett.* **26:**6015 (1985); E. J. Corey and F. J. Hannon, *Tetrahedron Lett.* **31:**1393 (1990).
45. H. O. House, *Acc. Chem. Res.* **9:**59 (1976); H. O. House and P. D. Weeks, *J. Am. Chem. Soc.* **97:**2770, 2778 (1975); H. O. House and K. A. J. Snoble, *J. Org. Chem.* **41:**3076 (1976); S. H. Bertz, G. Dabbagh, J. M. Cook, and V. Honkan, *J. Org. Chem.* **49:**1739 (1984).

molecules also affect the reactivity of the complex.[46] Thus, the mechanism can be outlined as occurring in three steps.

$$R_2Cu^- + R'CH{=}CH\overset{\overset{O}{\|}}{C}Z \longrightarrow \underset{R'CH{=}CH\overset{\|}{C}Z}{R_2Cu^- \;\; O^{\cdot\cdot}Li^+} \longrightarrow R_2Cu^{III}\underset{\underset{R'}{|}}{C}H{-}CH{=}\overset{\overset{O^-Li^+}{|}}{C}Z$$

$$\longrightarrow R\underset{\underset{R'}{|}}{C}H{-}CH{=}\overset{\overset{O^-Li^+}{|}}{C}Z + RCu^I$$

Isotope effects indicate that the collapse of the adduct by reductive elimination is the rate-determining step.[47] Theoretical treatements of the mechanism suggest similar intermediates.[48]

There is a correlation between the reduction potential of the carbonyl compounds and the ease of reaction with cuprate reagents.[49] The more easily reduced, the more reactive is the compound toward cuprate reagents. Compounds such as α,β-unsaturated esters and nitriles, which are not as easily reduced as the corresponding ketones, do not react as readily with dialkylcuprates, even though they are good Michael acceptors in classical Michael reactions with carbanions. α,β-Unsaturated esters are borderline in terms of reactivity toward standard dialkylcuprate reagents. β-Substitution retards reactivity. The $RCu{-}BF_3$ reagent combination is more reactive toward conjugated esters and nitriles.[50] Additions to hindered α,β-unsaturated ketones are also accelerated by BF_3.[51]

Prior to protonolysis, the products of conjugate addition to unsaturated carbonyl compounds are enolates and, therefore, potential nucleophiles. A useful extension of the conjugate addition method is to combine it with an alkylation step that adds a substituent at the α position.[52] Several examples of this tandem conjugate addition/alkylation procedure are given in Scheme 8.2.

The preparation of organozinc reagents was discussed in Section 7.3.1. Many of these reagents can be converted to mixed copper–zinc organometallics that have useful synthetic applications.[53] A virtue of these reagents is that they can contain a number of functional groups that are not compatible with the organolithium route to cuprate reagents. The mixed copper–zinc reagents are not very basic and can be prepared and allowed to react in the presence of weakly acidic functional groups that would be deprotonated by more basic organometallic reagents. For example, reagents containing secondary amide or indole groups can be prepared.[54] The mixed copper–zinc reagents are mild nucleophiles and are especially useful in conjugate addition. Mixed copper–zinc reagents can be prepared by

46. C. J. Kingsbury and R. A. J. Smith, *J. Org. Chem.* **62:**4629, 7637 (1997).
47. D. E. Frantz, D. A. Singleton, and J. P. Snyder, *J. Am. Chem. Soc.* **119:**3383 (1997).
48. E. Nakamura, S. Mori, and K. Morokuma, *J. Am. Chem. Soc.* **119:**4900 (1997); S. Mori and E. Nakamura, *Chem. Eur. J.* **5:**1534 (1999).
49. H. O. House and M. J. Umen, *J. Org. Chem.* **38:**3893 (1973); B. H. Lipshutz, R. S. Wilhelm, S. T. Nugent, R. D. Little, and M. M. Baizer, *J. Org. Chem.* **48:**3306 (1983).
50. Y. Yamamoto and K. Maruyama, *J. Am. Chem. Soc.* **100:**3240 (1978); Y. Yamamoto, *Angew. Chem. Int. Ed. Engl.* **25:**947 (1986).
51. A. B. Smith, III and P. J. Jerris, *J. Am. Chem. Soc.* **103:**194 (1981).
52. For a review of such reactions, see R. J. K. Taylor, *Synthesis* **1985:**364.
53. P. Knochel and R. D. Singer, *Chem. Rev.* **93:**2117 (1993); P. Knochel, *Synlett* **1995:**393.
54. H. P. Knoess, M. T. Furlong, M. J. Rozema, and P. Knochel, *J. Org. Chem.* **56:**5974 (1991).

Scheme 8.2. Tandem Reactions Involving Alkylation of Enolates Generated by Conjugate Addition of Organocopper Reagents

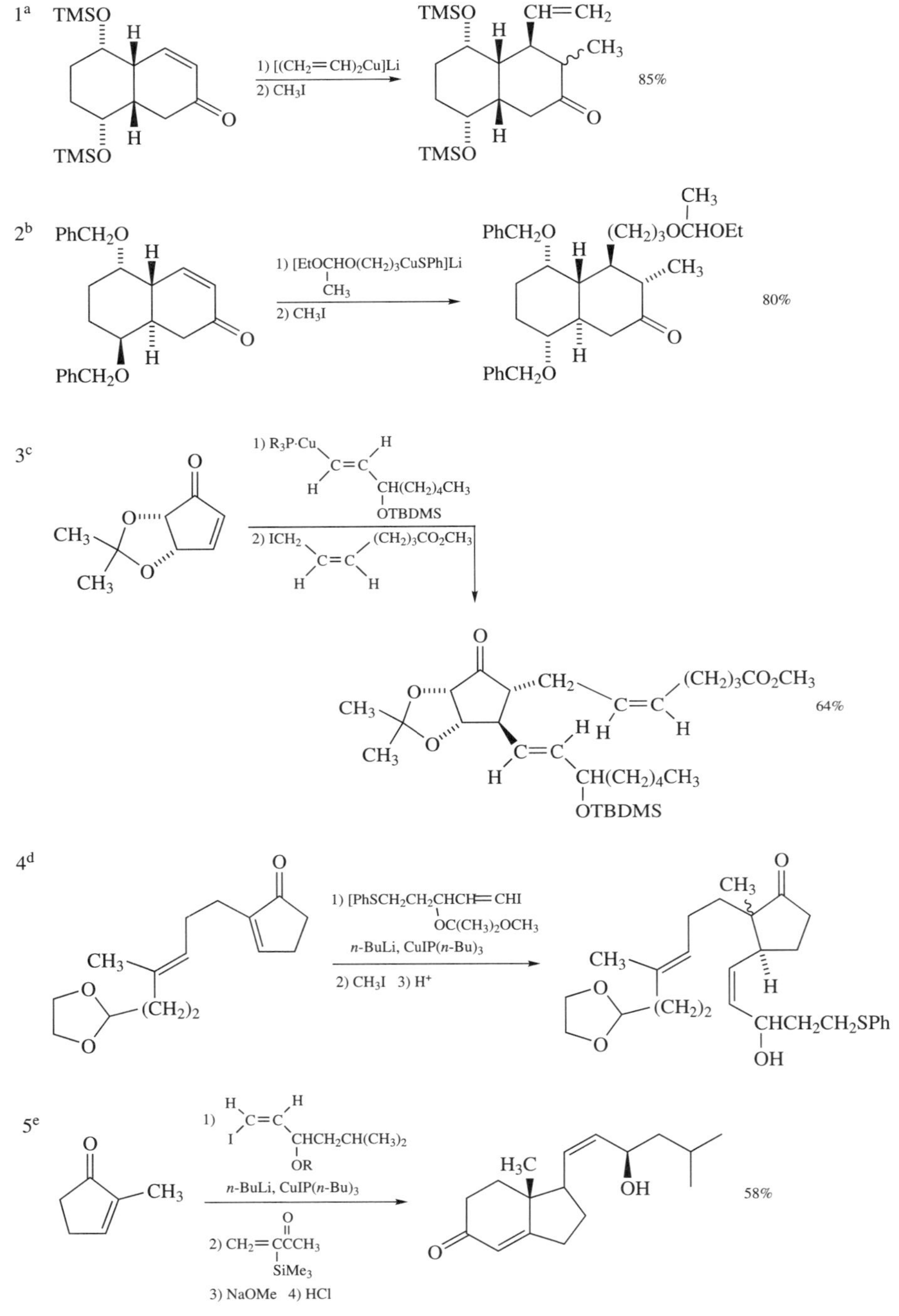

1[a] TMSO, H, H, TMSO, O → 1) $[(CH_2{=}CH)_2Cu]Li$ 2) CH_3I → TMSO, H, $CH{=}CH_2$, CH_3, O, H, TMSO 85%

2[b] $PhCH_2O$, H, O, H, $PhCH_2O$ → 1) $[EtOCHO(CH_2)_3CuSPh]Li$ (CH_3) 2) CH_3I → $PhCH_2O$, H, $(CH_2)_3OCHOEt$ (CH_3), CH_3, O, H, $PhCH_2O$ 80%

3[c] O, O, CH_3, CH_3, O → 1) $R_3P{\cdot}Cu$, H, C=C, H, $CH(CH_2)_4CH_3$, OTBDMS 2) ICH_2, C=C, $(CH_2)_3CO_2CH_3$, H, H → O, O, CH_2, C=C, $(CH_2)_3CO_2CH_3$, H, H, CH_3, CH_3, O, H, C=C, H, $CH(CH_2)_4CH_3$, OTBDMS 64%

4[d] O, CH_3, O, $(CH_2)_2$, O → 1) $[PhSCH_2CH_2CHCH{=}CHI$ ($OC(CH_3)_2OCH_3$) n-BuLi, $CuIP(n\text{-}Bu)_3$ 2) CH_3I 3) H^+ → O, CH_3, CH_3, O, $(CH_2)_2$, O, H, $CHCH_2CH_2SPh$, OH

5[e] O, CH_3 → 1) H, H, C=C, I, $CHCH_2CH(CH_3)_2$, OR n-BuLi, $CuIP(n\text{-}Bu)_3$ 2) $CH_2{=}CCCH_3$ (O, $SiMe_3$) 3) NaOMe 4) HCl → H_3C, OH, O 58%

R = $C(CH_3)_2OCH_3$

a. N. N. Girotra, R. A. Reamer, and N. L. Wendler, *Tetrahedron Lett.* **25:**5371 (1984).
b. N.-Y. Wang, C.-T. Hsu, and C. J. Sih, *J. Am. Chem. Soc.* **103:**6538 (1981).
c. C. R. Johnson and T. D. Penning, *J. Am. Chem. Soc.* **110:**4726 (1988).
d. T. Takahashi, K. Shimizu, T. Doi, and J. Tsuji, *J. Am. Chem. Soc.* **110:**2674 (1988).
e. T. Takahashi, H. Okumoto, J. Tsuji, and N. Harada, *J. Org. Chem.* **49:**948 (1984).

addition of CuCN to organozinc iodides.[55] These reagents are analogous to the cyanocuprates prepared from alkyllithium and CuCN, but with Zn^{2+} in place of Li^+. These reagents react with enones, nitroalkenes, and allylic halides.[56] In the presence of BF_3, they add to aldehydes.[57] Several specific examples are given in Scheme 8.3. Note, in particular, the regiospecific S_N2' reaction with allylic halides.

In addition to use of stoichiometric amounts of cuprate or cyanocuprate reagents for conjugate addition, there are also procedures which require only a catalytic amount of copper and use another organometallic reagent as the stoichiometric reagent. Some of the most useful examples involve organozinc reagents.[58] As discussed in Chapter 7, organozinc reagents which incorporate common functional groups can be prepared. In the presence of LiI, TMS–Cl, and a catalytic amount of $(CH_3)_2Cu(CN)Li_2$, conjugate addition of organozinc reagents occurs in good yield.

2-Cyclohexenone + $CH_3Zn(CH_2)_4CPh$ (LiI, C=O) $\xrightarrow[(CH_3)_2Cu(CN)Li_2\ 5\ mol\ \%,\ -78°C]{(CH_3)_3SiCl}$ 3-$(CH_2)_4CPh$(=O)-cyclohexanone 85% Ref. 59

Simple organozinc reagents, such as diethylzinc, undergo conjugate addition with 0.5 mol % CuO_3SCF_3 as catalyst in the presence of phoshines or phosphites.

2-Cyclohexenone + $(C_2H_5)_2Zn$ $\xrightarrow[P(OC_2H_5)_3,\ 1.0\ mol\ \%]{CuO_3SCF_3,\ 0.5\ mol\ \%}$ 3-C_2H_5-cyclohexanone 100% Ref. 60

CuI or CuCN (10 mol %) in conjunction with BF_3 and TMS–Cl catalyzes addition of alkylzinc bromides to enones.

$(CH_3)_3CBr$ $\xrightarrow{Zn^0}$ $(CH_3)_3CZnBr$ $\xrightarrow[\text{CuI, 0.1 mol \%, 1.5 equiv } BF_3\text{, 2.0 equiv TMS-Cl}]{\text{2-cyclohexenone}}$ 3-$C(CH_3)_3$-cyclohexanone 96% Ref. 61

Conjugate addition reactions involving organocopper intermediates can be made enantioselective by using chiral ligands.[62] Several mixed cuprate reagents containing chiral ligands have been explored to determine the degree of enantioselectivity that can be

55. P. Knochel, J. J Almena Perea, and P. Jones, *Tetrahedron* **54:**8275 (1998).
56. P. Knochel, M. C. P. Yeh, S. C. Berk, and J. Talbert, *J. Org. Chem.* **53:**2390 (1988); M. C. P. Yeh and P. Knochel, *Tetrahedron Lett.* **29:**2395 (1988); S. C. Berk, P. Knochel, and M. C. P. Yeh, *J. Org. Chem.* **53:**5789 (1988); T.-S. Chen and P. Knochel, *J. Org. Chem.* **55:**4791 (1990).
57. M. C. P. Yeh, P. Knochel, and L. E. Santa, *Tetrahedron Lett.* **29:**3887 (1988).
58. B. H. Lipshutz, *Acc. Chem. Res.* **30:**277 (1997).
59. B. H. Lipshutz, M. R. Wood, and R. J. Tirado, *J. Am. Chem. Soc.* **117:**6126 (1995).
60. A. Alexakis, J. Vastra, and P. Mageney, *Tetrahedron Lett.* **38:**7745 (1997).
61. R. D. Rieke, M. V. Hanson, J. D. Brown, and Q. J. Niu, *J. Org. Chem.* **61:**2726 (1996).
62. N. Krause and A. Gerold, *Angew. Chem. Int. Ed. Engl.* **36:**186 (1997); N. Krause, *Angew. Chem. Int. Ed. Engl.* **37:**283 (1998).

Scheme 8.3. Conjugate Addition and Alkylation Reactions of Mixed Copper–Zinc Reagents

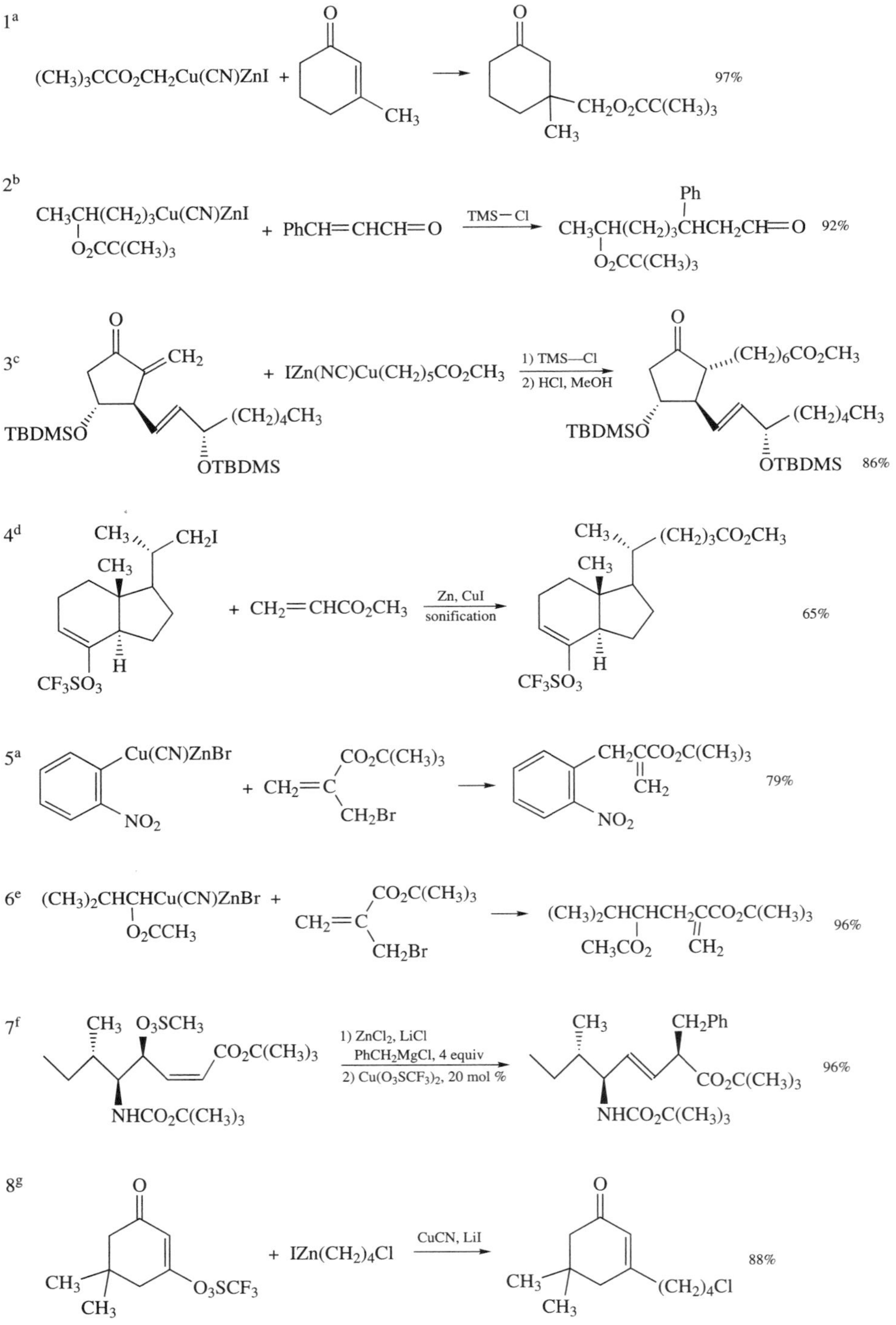

1[a] $(CH_3)_3CCO_2CH_2Cu(CN)ZnI$ + 3-methylcyclohex-2-enone → 3-methyl-3-($CH_2O_2CC(CH_3)_3$)cyclohexanone 97%

2[b] $CH_3CH(O_2CC(CH_3)_3)(CH_2)_3Cu(CN)ZnI$ + $PhCH{=}CHCH{=}O$ $\xrightarrow{\text{TMS-Cl}}$ $CH_3CH(O_2CC(CH_3)_3)(CH_2)_3CH(Ph)CH_2CH{=}O$ 92%

3[c] (TBDMSO, CH_2, $(CH_2)_4CH_3$, OTBDMS) + $IZn(NC)Cu(CH_2)_5CO_2CH_3$ $\xrightarrow{\text{1) TMS—Cl; 2) HCl, MeOH}}$ ($(CH_2)_6CO_2CH_3$, $(CH_2)_4CH_3$, TBDMSO, OTBDMS) 86%

4[d] (CH_3, CH_2I, CH_3, H, CF_3SO_3) + $CH_2{=}CHCO_2CH_3$ $\xrightarrow{\text{Zn, CuI; sonification}}$ (CH_3, $(CH_2)_3CO_2CH_3$, CH_3, H, CF_3SO_3) 65%

5[a] 2-nitrophenyl-$Cu(CN)ZnBr$ + $CH_2{=}C(CO_2C(CH_3)_3)CH_2Br$ → 2-nitrophenyl-$CH_2C({=}CH_2)CO_2C(CH_3)_3$ 79%

6[e] $(CH_3)_2CHCH(O_2CCH_3)Cu(CN)ZnBr$ + $CH_2{=}C(CO_2C(CH_3)_3)CH_2Br$ → $(CH_3)_2CHCH(CH_3CO_2)CH_2C({=}CH_2)CO_2C(CH_3)_3$ 96%

7[f] (CH_3, O_3SCH_3, $CO_2C(CH_3)_3$, $NHCO_2C(CH_3)_3$) $\xrightarrow{\text{1) ZnCl}_2\text{, LiCl, PhCH}_2\text{MgCl, 4 equiv; 2) Cu(O}_3\text{SCF}_3)_2\text{, 20 mol \%}}$ (CH_3, CH_2Ph, $CO_2C(CH_3)_3$, $NHCO_2C(CH_3)_3$) 96%

8[g] 5,5-dimethyl-3-(O_3SCF_3)cyclohex-2-enone + $IZn(CH_2)_4Cl$ $\xrightarrow{\text{CuCN, LiI}}$ 5,5-dimethyl-3-($(CH_2)_4Cl$)cyclohex-2-enone 88%

9[h] $CH_3(CH_2)_2C(CH_3)_2Br$ $\xrightarrow[\text{2) CuCN, 10 mol \%, LiBr}]{\text{1) Zn}}$ $\xrightarrow{\text{PhCOCl}}$ $CH_3(CH_2)_2C(CH_3)_2$—C(=O)Ph 86%

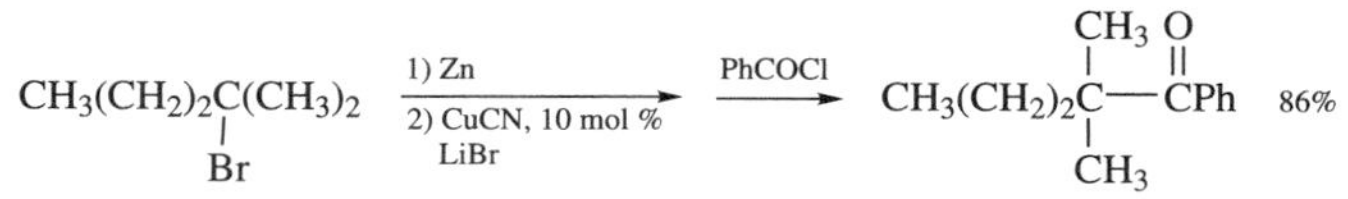

a. P. Knochel, T. S. Chou, C. Jubert, and D. Rajagopal, *J. Org. Chem.* **58:**588 (1993).
b. M. C. P. Yeh, P. Knochel, and L. E. Santa, *Tetrahedron Lett.* **29:**3887 (1988).
c. H. Tsujiyama, N. Ono, T. Yoshino, S. Okamoto, and F. Sato, *Tetrahedron Lett.* **31:**4481 (1990).
d. J. P. Sestalo, J. L. Mascarenas, L. Castedo, and A. Mourina, *J. Org. Chem.* **58:**118 (1993).
e. C. Tucker, T. N. Majid, and P. Knochel, *J. Am. Chem. Soc.* **114:**3983 (1992).
f. N. Fujii, K. Nakai, H. Habashita, H. Yoshizawa, T. Ibuka, F. Garrido, A. Mann, Y. Chounan, and Y. Yamamot, *Tetrahedron Lett.* **34:**4227 (1993).
g. B. H. Lipshutz and R. W. Vivian, *Tetrahedron Lett.* **40:**2871 (1999).
h. R. D. Rieke, M. V. Hanson, J. C. Brown, and Q. J. Niu, *J. Org. Chem.* **61:**2726 (1966).

achieved in conjugate addition. Several amide and phosphine ligands have been explored.

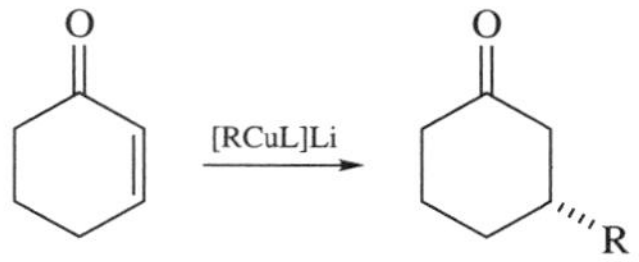

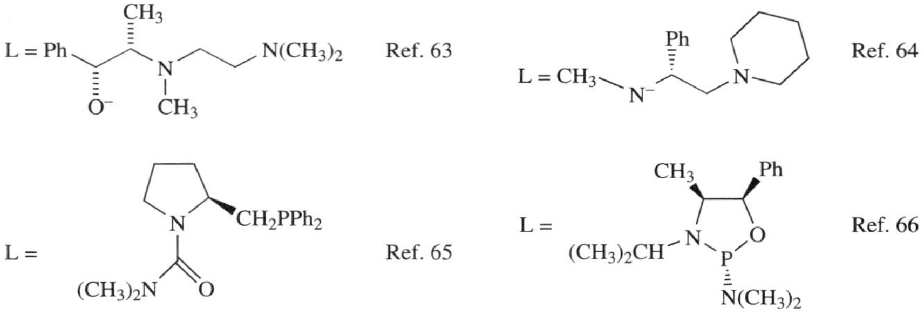

[RCuL]Li

L = Ph, O⁻, CH₃, N, CH₃, N(CH₃)₂ Ref. 63

L = CH₃, N⁻, Ph, N Ref. 64

L = N, CH₂PPh₂, (CH₃)₂N, O Ref. 65

L = CH₃, Ph, (CH₃)₂CH, N, O, P, N(CH₃)₂ Ref. 66

Enantioselectivity can also be observed under catalytic conditions. Scheme 8.4 shows some examples.

Conjugated acetylenic esters react readily with cuprate reagents, with *syn* addition being kinetically preferred.[67]

$$(C_4H_9)_2CuLi + CH_3C{\equiv}CCO_2CH_3 \longrightarrow \xrightarrow{H^+} (CH_3)(C_4H_9)C{=}C(CO_2CH_3)(H) \quad 86\%$$

The intermediate adduct can be substituted at the α position by a variety of electrophiles, including acyl chlorides, epoxides, aldehydes, and ketones.[68]

63. E. J. Corey, R. Naef, and F. J. Hannon, *J. Am. Chem. Soc.* **108:**7114 (1986).
64. N. M. Swingle, K. V. Reddy, and B. E. Rossiter, *Tetrahedron* **50:**4455 (1994); G. Miao and B. E. Rossiter, *J. Org. Chem.* **60:**8424 (1995).
65. M. Kanai and K. Tomioka, *Tetrahedron Lett.* **35:**895 (1994); **36:**4273, 4275 (1995).
66. A. Alexakis, J. Frutos, and P. Mangeney, *Tetrahedron Asymmetry* **4:**2427 (1993).
67. R. J. Anderson, V. L. Corbin, G. Cotterrell, G. R. Cox, C. A. Henrick, F. Schaub, and J. B. Siddall, *J. Am. Chem. Soc.* **97:**1197 (1975).
68. J. P. Marino and R. G. Lindeman, *J. Org. Chem.* **48:**4621 (1983).

Scheme 8.4. Enantioselective Conjugate Addition to Cyclohexenone

	Reagent	Catalyst	Chiral ligand	Yield (%)	e.e.
1[a]	n-C_4H_9MgCl	CuI	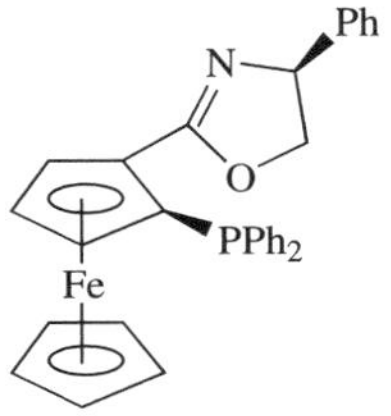	97	83
2[b]	$(C_2H_5)_2Zn$	$Cu(O_3SCF_3)_2$	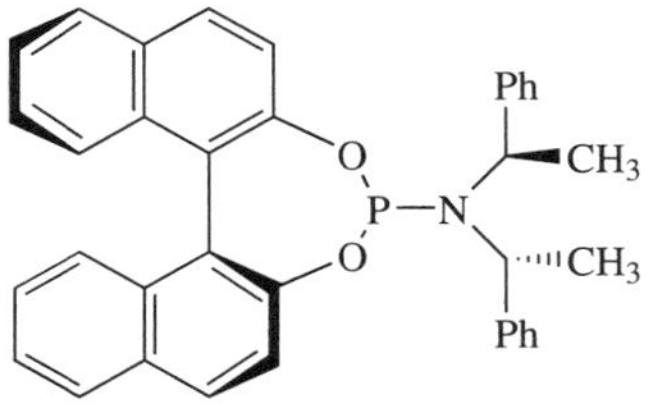	94	>98
3[c]	$(C_2H_5)_2Zn$	$Cu(O_3SCF_3)_2$	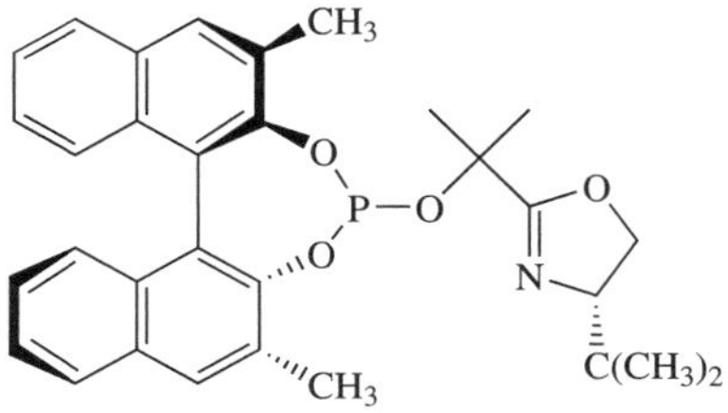	96	90
4[d]	$(C_2H_5)_2Zn$	$Cu(O_3SCF_3)_2$	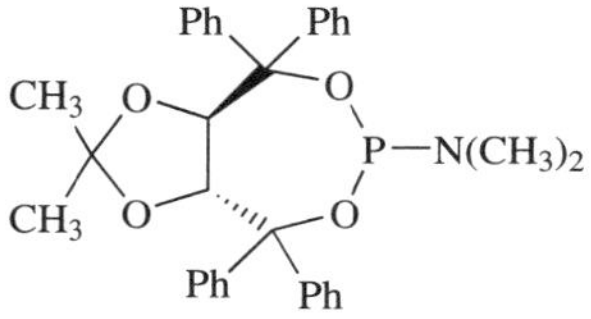	90	71
5[e]	n-C_4H_9MgCl	CuI	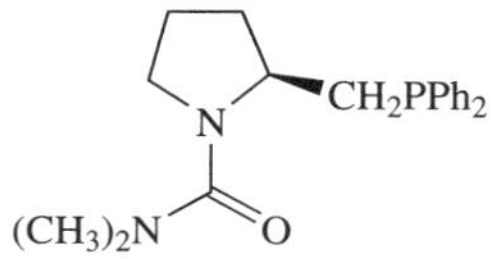	92	90

a. E. L. Stangeland and T. Sammakia, *Tetrahedron* **53:**16503 (1997).
b. B. L. Feringa, M. Pineschi, L. A. Arnold, R. Imbos, and A. H. M. de Vries, *Angew. Chem. Int. Ed. Engl.* **36:**2620 (1997).
c. A. K. H. Knöbel, I. H. Escher, and A. Pflatz, *Synlett* **1997:**1429.
d. E. Keller, J. Maurer, R. Naasz, T. Schader, A. Meetsma, and B. L. Feringa, *Tetrahedron Asymmetry* **9:**2409 (1998).
e. M. Kanai, Y. Nakagawa, and K. Tomioka, *Tetrahedron* **55:**3843 (1999).

The cuprate reagents that have been discussed in the preceding paragraphs are normally prepared by reaction of an organolithium reagent with a copper(I) salt, using a 2:1 ratio of lithium reagent to Cu(I). There are also valuable synthetic procedures which involve organocopper intermediates that are generated in the reaction system by use of only a catalytic amount of a copper(I) species.[69] Conjugate addition to α,β-unsaturated

69. For a review, see E. Erdiky, *Tetrahedron* **40:**641 (1984).

esters can often be effected by copper-catalyzed reaction with a Grignard reagent. Other reactions, such as epoxide ring opening, can also be carried out under catalytic conditions. Some examples of catalyzed additions and alkylations are given in Scheme 8.5.

Mixed copper–magnesium reagents analogous to the lithium cuprates can be prepared.[70] These compounds are often called *Normant reagents*. The precise structural nature of these compounds has not been determined. Individual species with differing Mg:Cu ratios may be in equilibrium.[71] These reagents undergo addition to terminal alkynes to generate alkenylcopper reagents. The addition is stereospecifically *syn*.

$$C_2H_5MgBr + CuBr \longrightarrow C_2H_5CuMgBr_2$$

$$C_2H_9CuMgBr_2 + CH_3C{\equiv}CH \longrightarrow (C_2H_5)(CH_3)C{=}C(CuMgBr_2)(H) \xrightarrow{H_2O} (C_2H_5)(CH_3)C{=}C(H)(H)$$

The alkenylcopper adducts can be worked up by protonolysis, or they can be subjected to further elaboration by alkylation or nucleophilic addition. Some examples are given in Scheme 8.6.

The mixed copper–zinc reagents also react with alkynes to give alkenylcopper species which can undergo subsequent electrophilic substitution.

$$Z(CH_2)_nCu(CN)ZnI \xrightarrow{(CH_3)_2Cu(CN)Li} Z(CH_2)_nCu(CN)Li{\cdot}Zn(CH_3)_2$$

$$\xrightarrow{PhC{\equiv}CH} (Z(CH_2)_n)(Ph)C{=}C(Cu(CN){\cdot}Zn(CH_3)_2)(H) \xrightarrow{CH_2{=}CHCH_2Br} (Z(CH_2)_n)(Ph)C{=}C(CH_2CH{=}CH_2)(H)$$ Ref. 72

Organocopper intermediates are also involved in several procedures for coupling of two organic reactants to form a new carbon–carbon bond. A classical example of this type of reaction is the *Ullman coupling* of aryl halides, which is done by heating an aryl halide with a copper–bronze alloy.[73] Good yields by this method are limited to halides with electron-attracting substituents.[74] Mechanistic studies have established the involvement of arylcopper intermediates. Soluble Cu(I) salts, particularly the triflate, effect coupling of aryl halides at much lower temperatures and under homogeneous conditions.[75]

$$\text{2-}O_2NC_6H_4Br \xrightarrow[\text{24 h, 25°C}]{CuO_3SCF_3,\ NH_3} \text{2,2'-dinitrobiphenyl}$$

70. J. F. Normant and M. Bourgain, *Tetrahedron Lett.* **1971:**2583; J. F. Normant, G. Cahiez, M. Bourgain, C. Chuit, and J. Villieras, *Bull. Soc. Chim. Fr.* **1974:**1656; H. Westmijze, J. Meier, H. J. T. Bos, and P. Vermeer, *Rec. Trav. Chim. Pays-Bas* **95:**299, 304 (1976).
71. E. C. Ashby, R. S. Smith, and A. B. Goel, *J. Org. Chem.* **46:**5133 (1981); E. C. Ashby and A. B. Goel, *J. Org. Chem.* **48:**2125 (1983).
72. S. A. Rao and P. Knochel, *J. Am. Chem. Soc.* **113:**5735 (1991).
73. P. E. Fanta, *Chem.Rev.* **64:**613 (1964); P. E. Fanta, *Synthesis* **1974:**9.
74. R. C. Fuson and E. A. Cleveland, *Org. Synth.* **III:**339 (1955).
75. T. Cohen and I. Cristea, *J. Am. Chem. Soc.* **98:**748 (1976).

Scheme 8.5. Copper-Catalyzed Reactions of Grignard Reagents

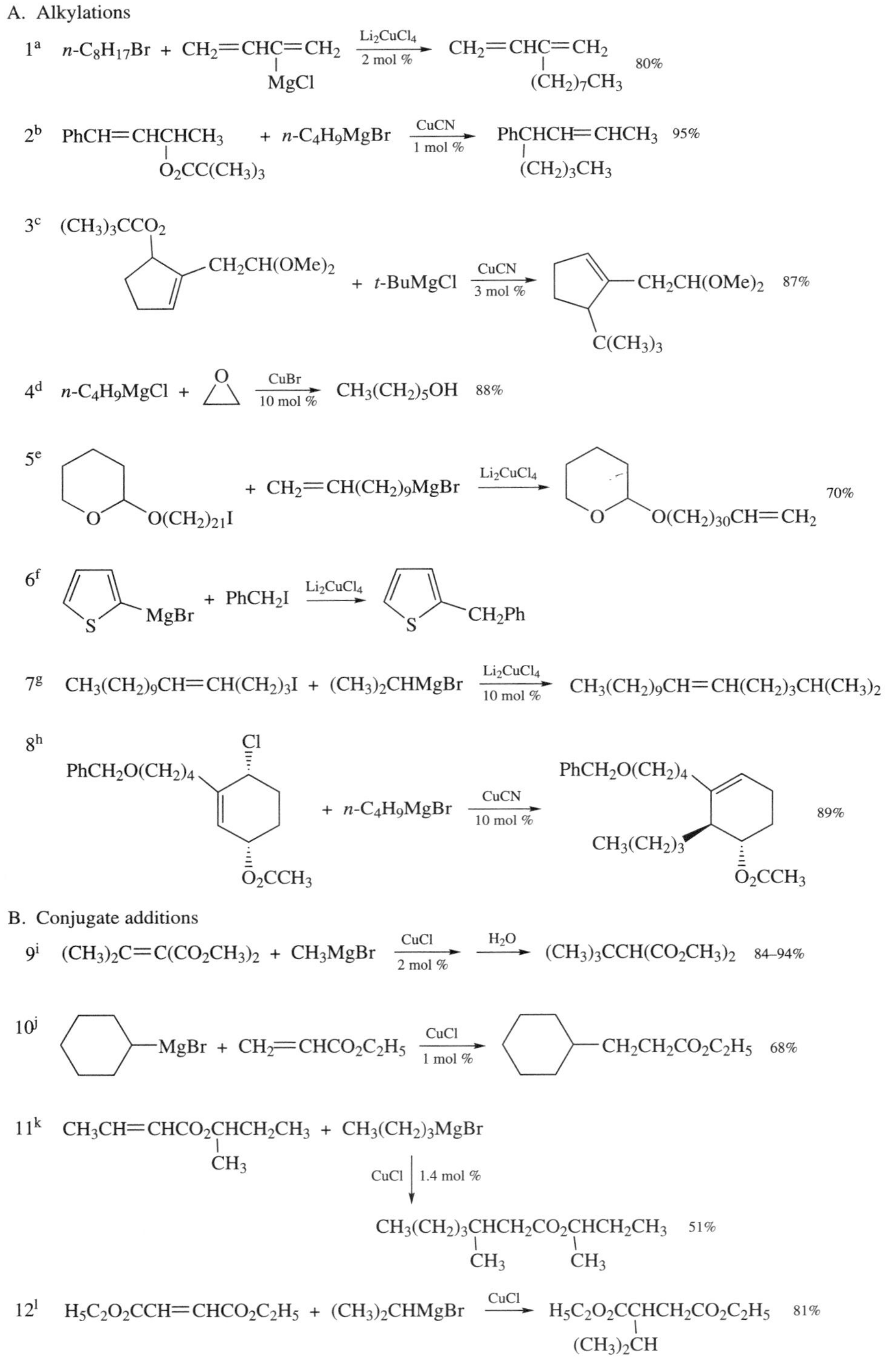

A. Alkylations

1[a] $n\text{-}C_8H_{17}Br + CH_2{=}CHC(MgCl){=}CH_2 \xrightarrow[\text{2 mol \%}]{Li_2CuCl_4} CH_2{=}CHC((CH_2)_7CH_3){=}CH_2$ 80%

2[b] $PhCH{=}CHCH(O_2CC(CH_3)_3)CH_3 + n\text{-}C_4H_9MgBr \xrightarrow[\text{1 mol \%}]{CuCN} PhCH((CH_2)_3CH_3)CH{=}CHCH_3$ 95%

3[c] 2-(2,2-dimethoxyethyl)-2-cyclopentenyl pivalate [$(CH_3)_3CCO_2$; $CH_2CH(OMe)_2$] + $t\text{-}BuMgCl \xrightarrow[\text{3 mol \%}]{CuCN}$ 1-[$CH_2CH(OMe)_2$]-5-[$C(CH_3)_3$]cyclopentene 87%

4[d] $n\text{-}C_4H_9MgCl$ + ethylene oxide $\xrightarrow[\text{10 mol \%}]{CuBr} CH_3(CH_2)_5OH$ 88%

5[e] 2-[$O(CH_2)_{21}I$]tetrahydropyran + $CH_2{=}CH(CH_2)_9MgBr \xrightarrow{Li_2CuCl_4}$ 2-[$O(CH_2)_{30}CH{=}CH_2$]tetrahydropyran 70%

6[f] 2-thienyl-MgBr + $PhCH_2I \xrightarrow{Li_2CuCl_4}$ 2-thienyl-CH_2Ph

7[g] $CH_3(CH_2)_9CH{=}CH(CH_2)_3I + (CH_3)_2CHMgBr \xrightarrow[\text{10 mol \%}]{Li_2CuCl_4} CH_3(CH_2)_9CH{=}CH(CH_2)_3CH(CH_3)_2$

8[h] $PhCH_2O(CH_2)_4$-substituted cyclohexene (Cl, O_2CCH_3) + $n\text{-}C_4H_9MgBr \xrightarrow[\text{10 mol \%}]{CuCN}$ $PhCH_2O(CH_2)_4$-substituted cyclohexene ($CH_3(CH_2)_3$, O_2CCH_3) 89%

B. Conjugate additions

9[i] $(CH_3)_2C{=}C(CO_2CH_3)_2 + CH_3MgBr \xrightarrow[\text{2 mol \%}]{CuCl} \xrightarrow{H_2O} (CH_3)_3CCH(CO_2CH_3)_2$ 84–94%

10[j] cyclohexyl-MgBr + $CH_2{=}CHCO_2C_2H_5 \xrightarrow[\text{1 mol \%}]{CuCl}$ cyclohexyl-$CH_2CH_2CO_2C_2H_5$ 68%

11[k] $CH_3CH{=}CHCO_2CH(CH_3)CH_2CH_3 + CH_3(CH_2)_3MgBr \xrightarrow{CuCl\ 1.4\ \text{mol \%}} CH_3(CH_2)_3CH(CH_3)CH_2CO_2CH(CH_3)CH_2CH_3$ 51%

12[l] $H_5C_2O_2CCH{=}CHCO_2C_2H_5 + (CH_3)_2CHMgBr \xrightarrow{CuCl} H_5C_2O_2CCH(CH(CH_3)_2)CH_2CO_2C_2H_5$ 81%

13[m] 3-methyl-4-($CH_2CO_2C(CH_3)_3$)cyclohex-2-enone + $(CH_3)_2C{=}CHMgBr$ $\xrightarrow[CuBr \cdot S(CH_3)_2]{(CH_3)_3SiCl}$ 3-methyl-3-($CH{=}C(CH_3)_2$)-4-($CH_2CO_2C(CH_3)_3$)cyclohexanone 78%

14[n] $CH_3O_2C(CH_2)_4COCl + CH_3(CH_2)_3MgBr \xrightarrow{CuI,\ 5\ mol\ \%} CH_3O_2C(CH_2)_4C(=O)(CH_2)_3CH_3$ 85%

15[o] $C_6H_5{-}C{\equiv}N + (CH_3)_3CMgCl \xrightarrow[2\ mol\ \%]{CuBr} C_6H_5{-}C(=NH)C(CH_3)_3$ 95%

a. S. Nunomoto, Y. Kawakami, and Y. Yamashita, *J. Org. Chem.* **48:**1912 (1983).
b. C. C. Tseng, S. D. Paisley, and H. L. Goering, *J. Org. Chem.* **51:**2884 (1986).
c. E. J. Corey and A. V. Gavai, *Tetrahedron Lett.* **29:**3201 (1988).
d. G. Huynh, F. Derguini-Boumechal, and G. Linstrumelle, *Tetrahedron Lett.* **1979:**1503.
e. U. F. Heiser and B. Dobner, *J. Chem. Soc., Perkin Trans. 1* **1997:**809.
f. Y.-T. Ku, R. R. Patel, and D. P. Sawick, *Tetrahedron Lett.* **37:**1949 (1996).
g. E. Keinan, S. C. Sinha, A. Sinha-Bagchi, Z.-M. Wang, X.-L. Zhang, and K. B. Sharpless, *Tetrahedron Lett.* **33:**6411 (1992).
h. D. Tanner, M. Sellen, and J. Bäckvall, *J. Org. Chem.* **54:**3374 (1989).
i. E. L. Eliel, R. O. Hutchins, and M. Knoeber, *Org. Synth.* **50:**38 (1970).
j. S.-H. Liu, *J. Org. Chem.* **42:**3209 (1977).
k. T. Kindt-Larsen, V. Bitsch, I. G. K. Andersen, A. Jart, and J. Munch-Petersen, *Acta Chem. Scand.* **17:**1426 (1963).
l. V. K. Andersen and J. Munch-Petersen, *Acta Chem. Scand.* **16:**947 (1962).
m. Y. Horiguchi, E. Nakamura, and I. Kuwajima, *J. Am. Chem. Soc.* **111:**6257 (1989).
n. T. Fujisawa and T. Sato, *Org. Synth.* **66:**116 (1988).
o. F. J. Weiberth and S. S. Hall, *J. Org. Chem.* **52:**3901 (1987).

Arylcopper intermediates can be generated from organolithium compounds as in the preparation of cuprates.[76] These compounds react with a second aryl halide to provide unsymmetrical biaryls. This reaction is essentially a variant of the cuprate alkylation process discussed earlier in the chapter (p. 481). An alternative procedure involves generation of a mixed diarylcyanocuprate by sequential addition of two different aryllithium reagents to CuCN. The second addition must be carried out at very low temperature to prevent equilibration with the symmetrical diarylcyanocuprates. These unsymmetrical diarylcyanocuprates then undergo decomposition to biaryls on exposure to oxygen.[77]

$$Ar'Li + CuCN \longrightarrow Ar'Cu(CN)Li$$

$$Ar''Li + Ar'Cu(CN)Li \longrightarrow Ar'Cu(CN)Li(Ar'')$$

$$Ar'Cu(CN)Li(Ar'') \xrightarrow{O_2} Ar'{-}Ar''$$

76. F. E. Ziegler, I. Chliwner, K. W. Fowler, S. J. Kanfer, S. J. Kuo, and N. D. Sinha, *J. Am. Chem. Soc.* **102:**790 (1980).
77. B. H. Lipshutz, K. Siegmann, and E. Garcia, *Tetrahedron* **48:**2579 (1992); B. H. Lipshutz, K. Siegmann, E. Garcia, and F. Kayser, *J. Am. Chem. Soc.* **115:**9276 (1993).

Scheme 8.6. Generation and Reactions of Alkenylcopper Reagents by Additions to Alkynes

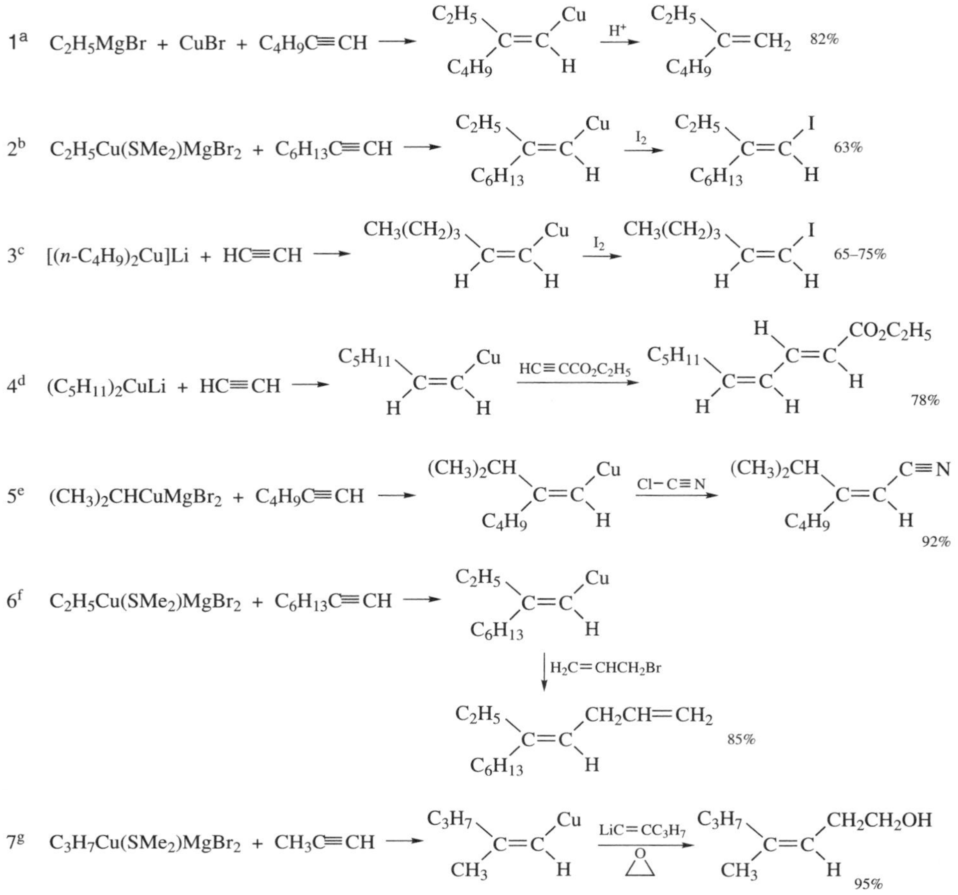
1a C2H5MgBr + CuBr + C4H9C≡CH
C2H5 Cu C=C C4H9 H
H+
C2H5 C=CH2 C4H9
82%
2b C2H5Cu(SMe2)MgBr2 + C6H13C≡CH
I2
63%
3c [(n-C4H9)2Cu]Li + HC≡CH
CH3(CH2)3
65–75%
4d (C5H11)2CuLi + HC≡CH
HC≡CCO2C2H5
CO2C2H5
78%
5e (CH3)2CHCuMgBr2 + C4H9C≡CH
(CH3)2CH
Cl−C≡N
C≡N
92%
6f C2H5Cu(SMe2)MgBr2 + C6H13C≡CH
H2C=CHCH2Br
CH2CH=CH2
85%
7g C3H7Cu(SMe2)MgBr2 + CH3C≡CH
LiC≡CC3H7
CH2CH2OH
95%

a. J. F. Normant, G. Cahiez, M. Bourgain, C. Chuit, and J. Villieras, *Bull. Chim. Soc. Fr.* **1974:**1656.
b. N. J. LaLima, Jr., and A. B. Levy, *J. Org. Chem.* **43:**1279 (1978).
c. A. Alexakis, G. Cahiez, and J. F. Normant, *Org. Synth.* **62:**1 (1984).
d. A. Alexakis, J. Normant, and J. Villieras, *Tetrahedron Lett.* **1976:**3461.
e. H. Westmijze and P. Vermeer, *Synthesis* **1977:**784.
f. R. S. Iyer and P. Helquist, *Org. Synth.* **64:**1 (1985).
g. P. R. McGuirk, A. Marfat, and P. Helquist, *Tetrahedron Lett.* **1978:**2465.

Intramolecular variations of this reaction have been achieved.

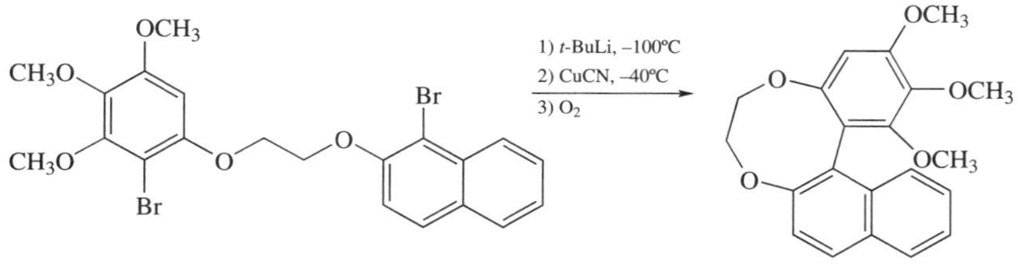
OCH3
CH3O
CH3O
Br
O
O
Br
1) t-BuLi, –100°C
2) CuCN, –40°C
3) O2
OCH3
OCH3
OCH3

Ref. 78

78. B. H. Lipshutz, F. Kayser, and N. Maullin, *Tetrahedron Lett.* **35:**815 (1994).

Organopalladium intermediates have become very important in synthetic organic chemistry. Usually, organic reactions involving palladium do not involve the preparation of stoichiometric organopalladium reagents. Instead, organopalladium species are generated *in situ* in the course of the reaction. Indeed, in the most useful processes only a *catalytic amount* of palladium is used. Catalytic processes have both economic and environmental advantages. Because, in principle, the catalyst is not consumed, it can be used to make product without generating by-products. Some processes use *solid-phase catalysts*, which further improves the economic and environmental advantages of catalyst recovery. Furthermore, processes that involve use of chiral catalysts can generate enantiomerically enriched or pure materials from achiral starting materials. In this section, we will focus on carbon–carbon bond formation, but in Chapter 11, we will see that palladium can also catalyze aromatic substitution reactions.

Three types of organopalladium intermediates are of primary importance in the reactions that have found synthetic application. Alkenes react with Pd(II) to give π-complexes which are subject to nucleophilic attack. These reactions are closely related to the solvomercuration reactions discussed in Section 4.3. The products that are derived from the resulting intermediates depend upon specific reaction conditions. The palladium can be replaced by hydrogen under reductive conditions (path a). In the absence of a reducing agent, an elimination of Pd(0) and a proton occurs, leading to net substitution of a vinyl hydrogen by the nucleophile (path b). We will return to specific examples of these reactions shortly.

$$RCH{=}CH_2 + Pd(II) \rightleftharpoons RCH{=}CH_2 \cdot\cdot\cdot Pd^{2+}$$

$$Nu + RCH{=}CH_2(\cdot\cdot\cdot Pd^{2+}) \longrightarrow Nu{-}CH(R)CH_2Pd^{2+}$$

$$Nu{-}CH(R)CH_2Pd^{2+} \xrightarrow{[H]} Nu{-}CH(R)CH_3 \quad \text{(path a)}$$

$$Nu{-}CH(R)CH_2Pd^{2+} \xrightarrow[-H^+]{-Pd(0)} Nu{-}C(R){=}CH_2 \quad \text{(path b)}$$

A second type of organopalladium intermediates are π-allyl complexes. These complexes can be obtained from Pd(II) salts and allylic acetates and other compounds with potential leaving groups in an allylic position.[79] The same type of π-allyl complexes can be prepared from alkenes by reaction with $PdCl_2$ or $Pd(O_2CCF_3)_2$.[80] The reaction occurs by electrophilic attack on the π electrons followed by loss of a proton. The proton loss probably proceeds via an unstable species in which the hydrogen is bound to

79. R. Huttel, *Synthesis* **1970:**225; B. M. Trost, *Tetrahedron* **33:**2615 (1977).
80. B. M. Trost and P. J. Metzner, *J. Am. Chem. Soc.* **102:**3572 (1980); B. M. Trost, P. E. Strege, L. Weber, T. J. Fullerton, and T. J. Dietsche, *J. Am. Chem. Soc.* **100:**3407 (1978).

palladium.[81]

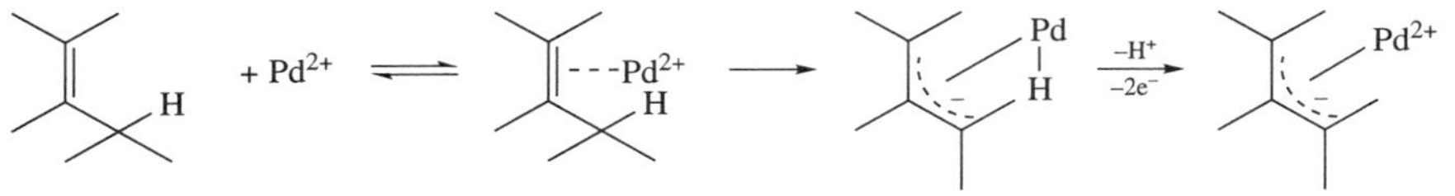

The π-allyl complexes can be isolated as halide-bridged dimers.

These π-allyl complexes are electrophilic in character and undergo reaction with a variety of nucleophiles. After nucleophilic addition occurs, the resulting organopalladium intermediate usually breaks down by elimination of Pd(0) and H^+.

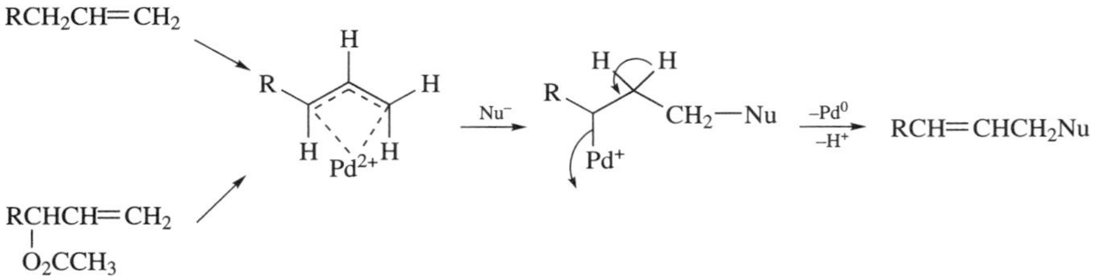

The third general process involves the reaction of Pd(0) species with halides or sulfonates by oxidative addition, generating reactive intermediates having the organic group attached to Pd(II) by a σ bond. The oxidative addition reaction is very useful for aryl and alkenyl halides, but the products from saturated alkyl halides usually decompose by elimination. The σ-bonded species formed by oxidative addition can react with alkenes and other unsaturated compounds to form new carbon–carbon bonds. The σ-bound species also react with a variety of organometallic reagents to give coupling products.

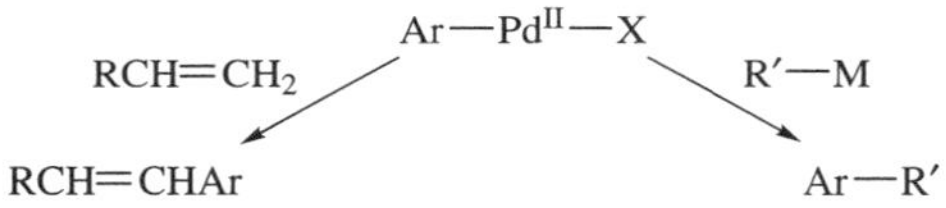

Specific examples of these reactions will be discussed below.

In considering the mechanisms involved in organopalladium chemistry, several general points should be kept in mind. Frequently, reactions involving organopalladium intermediates are done in the presence of phosphine ligands. These ligands coordinate at palladium and play a key role in the reaction by influencing the reactivity. Another general point concerns the relative weakness of the C—Pd bond and, especially, the instability of alkylpalladium species in which there is a β hydrogen. The final stage in many palladium-mediated reactions is the elimination of Pd(0) and H^+ to generate a carbon–carbon double bond. This tendency toward elimination distinguishes organopalladium species from most of the organometallic species that we have discussed to this point. Finally, organopalladium(II) species with two organic substituents show the same tendency to decompose with recombination of the organic groups by reductive elimination that is exhibited by copper(III) intermediates.

81. D. R. Chrisope, P. Beak, and W. H. Saunders, Jr., *J. Am. Chem. Soc.* **110:**230 (1988).

An important industrial process based on Pd–alkene complexes is the *Wacker reaction*, a catalytic method for conversion of ethylene to acetaldehyde. The first step is addition of water to the Pd-activated alkene. The addition intermediate undergoes the characteristic elimination of Pd(0) and H^+ to generate the enol of acetaldehyde.

$$CH_2{=}CH_2 + Pd(II) \longrightarrow \underset{\vdots\,Pd^{2+}}{CH_2{=}CH_2} \xrightarrow{H_2O} HO{-}CH_2CH_2{-}Pd^+ \longrightarrow (HO)(H)C{=}CH_2 + Pd^0 + H^+$$

$$(HO)(H)C{=}CH_2 \longrightarrow CH_3CH{=}O$$

The reaction is run with only a catalytic amount of Pd. The co-reagents $CuCl_2$ and O_2 serve to reoxidize the Pd(0) to Pd(II). The net reaction consumes only alkene and oxygen.

When the Wacker condition are applied to terminal alkenes, methyl ketones are formed.[82]

$$CH_2{=}CHCH_2C(CH_3)_2CH{=}O \xrightarrow[H_2O,\ DMF,\ O_2]{CuCl_2,\ PdCl_2} CH_3C(=O)CH_2C(CH_3)_2CH{=}O \quad 78\%$$

π-Allyl palladium species are involved in a number of useful reactions which result in allylation of nucleophiles.[83] This reaction can be applied to carbon–carbon bond formation with relatively stable carbanions, such as those derived from malonate esters and β-sulfonyl esters.[84] The π-allyl complexes are usually generated *in situ* by reaction of an allylic acetate with a catalytic amount of tetrakis(triphenylphosphine)palladium.[85] The

82. (a) J. Tsuji, I. Shimizu, and K. Yamamoto, *Tetrahedron Lett.* **1976:**2975; (b) J. Tsuji, H. Nagashima, and H. Nemoto, *Org. Synth.* **62:**9 (1984); (c) D. Pauley, F. Anderson, and T. Hudlicky, *Org. Synth.* **67:**121 (1988); (d) K. Januszkiewicz and H. Alper, *Tetrahedron Lett.* **24:**5159 (1983); (e) K. Januszkiewicz and D. J. H. Smith, *Tetrahedron Lett.* **26:**2263 (1985).
83. G. Consiglio and R. M. Waymouth, *Chem. Rev.* **89:**257 (1989).
84. B. M. Trost, W. P. Conway, P. E. Strege, and T. J. Dietsche, *J. Am. Chem. Soc.* **96:**7165 (1974); B. M. Trost, L. Weber, P. E. Strege, T. J. Fullerton, and T. J. Dietsche, *J. Am. Chem. Soc.* **100:**3416 (1978); B. M. Trost, *Acc. Chem. Res.* **13:**385 (1980).
85. B. M. Trost and T. R. Verhoeven, *J. Am. Chem. Soc.* **102:**4730 (1980).

Scheme 8.7. Enantioselective Alkylation of Diethyl Malonate

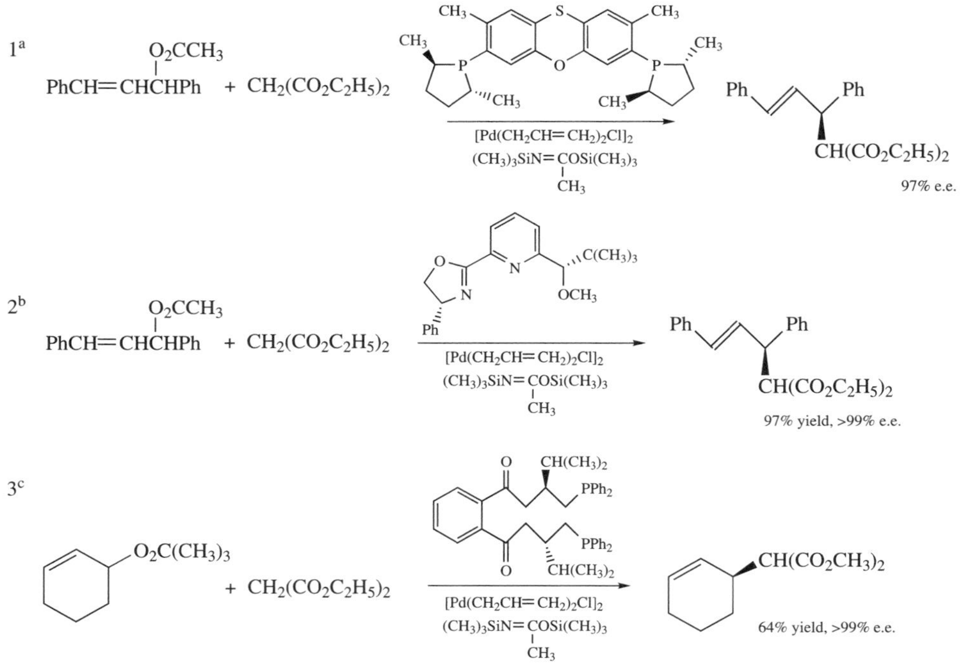

1[a] PhCH=CHCH(O_2CCH_3)Ph + $CH_2(CO_2C_2H_5)_2$ → ($[Pd(CH_2CH=CH_2)_2Cl]_2$, $(CH_3)_3SiN=C(CH_3)OSi(CH_3)_3$) → PhCH=CHCH(Ph)$CH(CO_2C_2H_5)_2$ 97% e.e.

2[b] PhCH=CHCH(O_2CCH_3)Ph + $CH_2(CO_2C_2H_5)_2$ → ($[Pd(CH_2CH=CH_2)_2Cl]_2$, $(CH_3)_3SiN=C(CH_3)OSi(CH_3)_3$) → PhCH=CHCH(Ph)$CH(CO_2C_2H_5)_2$ 97% yield, >99% e.e.

3[c] cyclohexenyl-$O_2C(CH_3)_3$ + $CH_2(CO_2C_2H_5)_2$ → ($[Pd(CH_2CH=CH_2)_2Cl]_2$, $(CH_3)_3SiN=C(CH_3)OSi(CH_3)_3$) → cyclohexenyl-$CH(CO_2CH_3)_2$ 64% yield, >99% e.e.

a. P. Dierkes, S. Ramdeehul, L. Barley, A. DeCian, J. Fischer, P. C. J. Kramer, P. W. N. M. van Leeuwen, and J. A. Osborne, *Angew. Chem. Int. Ed. Engl.* **37:**3116 (1998).
b. K. Nordstrom, E. Macedo, and C. Moberg, *J. Org. Chem.* **62:**1604 (1997); U. Bremberg, F. Rahm, and C. Moberg, *Tetrahedron Asymmetry* **9:**3437 (1998).
c. A. Saitoh, M. Misawa, and T. Morimoto, *Tetrahedron Asymmetry* **10:**1025 (1999).

reactive Pd(0) species is regenerated in an elimination step.

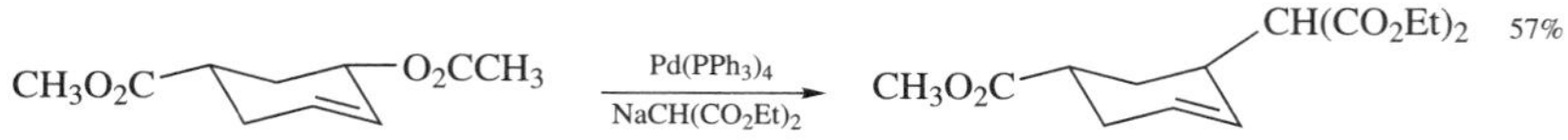

CH_3O_2C-cyclohexenyl-O_2CCH_3 → ($Pd(PPh_3)_4$, $NaCH(CO_2Et)_2$) → CH_3O_2C-cyclohexenyl-$CH(CO_2Et)_2$ 57% Ref. 86

Allylation reactions can be made highly enantioselective by use of various chiral ligands.[87] Examples are included in Scheme 8.7.

The allylation reaction has also been used to form rings. β-Sulfonyl esters have proven particularly useful in this application for formation of both medium and large rings.[88] In some cases, medium-sized rings are formed in preference to six- and seven-

86. B. M. Trost and P. E. Strege, *J. Am. Chem. Soc.* **99:**1649 (1977).
87. S. J. Sesay and J. M. J. Williams, in *Advances in Asymmetric Synthesis* Vol. 3, A. Hassner, ed., JAI Press, Stamford, Connecticut, 1998, pp. 235–271; G. Helmchen, *J. Organmet. Chem.* **576:**203 (1999).
88. B. M. Trost, *Angew. Chem. Int. Ed. Engl.* **28:**1173 (1989).

membered rings.[89]

$$PhSO_2CH(CO_2CH_3)CH_2C(=O)OCH_2CH_2CH{=}CHCH_2O_2CCH_3 \xrightarrow[\text{NaH}]{Pd(PPh_3)_4}$$ 54% (+5% of *E*- isomer)

$\xrightarrow{Pd(PPh_3)_4}$ 60%

The sulfonyl substituent can be removed by reduction after the ring closure (see Section 5.5.2).

8.2.2. The Heck Reaction

A third important type of reactivity of palladium, namely, oxidative addition to Pd(0), is the foundation for several methods of forming carbon–carbon bonds. Aryl[90] and alkenyl[91] halides react with alkenes in the presence of catalytic amounts of palladium to give net substitution of the halide by the alkenyl group. The reaction is called the *Heck reaction*. The reaction is quite general and has been observed for simple alkenes, aryl-substituted alkenes, and electrophilic alkenes such as acrylate esters and *N*-vinylamides.[92] Many procedures use $Pd(OAc)_2$ or other Pd(II) salts as catalysts, with the catalytically active Pd(0) being generated *in situ*. The reactions are usually carried out in the presence of a phosphine ligand, with tri-*o*-tolylphosphine being preferred in many cases. Several chelating diphosphines, shown below with their common abbreviations, are also effective.

89. B. M. Trost and T. R. Verhoeven, *J. Am. Chem. Soc.* **102:**4743 (1980); B. M. Trost and S. J. Brickner, *J. Am. Chem. Soc.* **105:**568 (1983); B. M. Trost, B. A. Vos, C. M. Brzezowski, and D. P. Martina, *Tetrahedron Lett.* **33:**717 (1992).
90. H. A. Dieck and R. F. Heck, *J. Am. Chem. Soc.* **96:**1133 (1974); R. F. Heck, *Acc. Chem. Res.* **12:**146 (1979); R. F. Heck, *Org. React.* **27:**345 (1982).
91. B. A. Patel and R. F. Heck, *J. Org. Chem.* **43:**3898 (1978); B. A. Patel, J. I. Kim, D. D. Bender, L. C. Kao, and R. F. Heck, *J. Org. Chem.* **46:**1061 (1981); J. I. Kim, B. A. Patel, and R. F. Heck, *J. Org. Chem.* **46:**1067 (1981).
92. C. B. Ziegler, Jr., and R. F. Heck, *J. Org. Chem.* **43:**2941 (1978); W. C. Frank, Y. C. Kim, and R. F. Heck, *J. Org. Chem.* **43:**2947 (1978); C. B. Ziegler, Jr., and R. F. Heck, *J. Org. Chem.* **43:**2949 (1978); H. A. Dieck and R. F. Heck, *J. Am. Chem. Soc.* **96:**1133 (1974); C. A. Busacca, R. E. Johnson, and J. Swestock, *J. Org. Chem.* **58:**3299 (1993).

Tri-2-furylphosphine is also used frequently. Phosphites are also good ligands.[93]

$Ph_2PCH_2CH_2PPh_2$
dppe

$Ph_2P(CH_2)_3PPh_2$
dppp

$Ph_2P(CH_2)_4PPh_2$
dppb

dppf

chiraphos

DINAP

The reaction is initiated by oxidative addition of the halide to a palladium(0) species generated *in situ* from the Pd(II) catalyst. The arylpalladium(II) intermediate then forms a complex with the alkene, which rearranges to a σ complex with carbon–carbon bond formation. The σ-complex decomposes with regeneration of Pd(0) by β-elimination.

$PhBr + Pd(PPh_3)_2 \longrightarrow PhPd(PPh_3)_2Br \xrightarrow{CH_2=CH-Y}$ Ph–Pd(PPh_3)$_2$Br···CH_2=CHY

$\longrightarrow PhCH_2-CHY-Pd(PPh_3)_2Br \longrightarrow PhCH=CHY + Pd(PPh_3)_2 + HBr$

At least two different Pd(0) species can be involved in both the oxidative addition and π-coordination steps, depending on the anions and ligands present. High halide concentration promotes formation of the anionic species $[PdL_2X]^-$ by addition of a halide ligand. Use of trifluoromethanesulfonate anions promotes dissociation of the anion from the Pd(II) adduct and accelerates complexation with electron-rich alkenes.

$R-Pd^{II}L(CH_2CH_2R')$ → (L; R–CH=CHR′) $H-PdL_2$ → (B; BH^+) $[Pd^0L_2]$ ⇌ (X^-) $[PdL_2X]^-$ → (RX; X^-) $[R-Pd^{II}L_2X]$ ⇌ (-X^-) $[R-Pd^{II}L_2]^+$ → (R′CH=CH_2) $R-Pd^{II}LX$···R′CH=CH_2 → $R-Pd^{II}L(CH_2CH_2R')$

93. M. Beller and A. Zapf, *Synlett* **1998:**792.

A number of modified reaction conditions have been developed. One involves addition of silver salts, which activate the halide toward displacement.[94] Use of sodium bicarbonate or sodium carbonate in the presence of a phase-transfer catalyst permits especially mild conditions to be used for many systems.[95] Tetraalkylammonium salts often accelerate reaction.[96] Solid-phase catalysts in which the palladium is complexed by polymer-bound phosphine groups have also been developed.[97] Aryl chlorides are not very reactive under normal Heck reaction conditions, but reaction can be achieved by inclusion of triphenylphosphonium salts with $Pd(OAc)_2$ or $PdCl_2$ as the catalyst.[98]

$$C_6H_5{-}Cl + C_6H_5{-}CH{=}CH_2 \xrightarrow[\text{NMP}]{Pd(CH_3CN)_2Cl_2,\ 2\ \text{mol}\ \%,\ NaO_2CCH_3,\ Ph_4P^+Cl^-} C_6H_5CH{=}CHC_6H_5 \quad 79\%$$

Pretreatment with nickel bromide also causes normally unreactive aryl chlorides to undergo Pd-catalyzed substitution.[99] Aryl and vinyl triflates have also been found to be excellent substrates for Pd-catalyzed vinylations.[100] Scheme 8.8 illustrates some of these reaction conditions.

Heck reactions can result in regioisomers depending on whether migration occurs to the α or β carbon of the alkene. Alkenes having electron-withdrawing substituents normally result in β arylation. However, alkenes with donor substituents give a mixture of α and β regioisomers. The regiochemistry can be controlled to some extent by specific reaction conditions. With vinyl ethers and *N*-vinylamides, it is possible to promote α arylation by use of bidentate phosphine ligands such as dppe and dppp, using aryl triflates as reactants. These reactions are believed to occur through a more electrophilic form of Pd(II) generated by dissociation of the triflate anion.[101] Electronic factors favor migration of the aryl group to the α carbon.

$$Ar{-}Pd^{II}(P\frown P)\cdots H_2C{=}CHX \longrightarrow (P\frown P)Pd^{II}{-}CH_2{-}CH(X)Ar \longrightarrow CH_2{=}C(Ar)X$$

Favored for X = OR, O_2CCH_3, CH_2OH, N-(2-oxopyrrolidinyl)

Allylic silanes show a pronounced tendency to react at the α carbon.[102] This regiochemistry is attributed to the stabilization of cationic character at the β carbon by the silyl

94. M. M. Abelman, T. Oh, and L. E. Overman, *J. Org. Chem.* **52**:4130 (1987); M. M. Abelman and L. E. Overman, *J. Am. Chem. Soc.* **110**:2328 (1988).
95. T. Jeffery, *J. Chem. Soc., Chem. Commun.*, **1984**:1287; T. Jeffery, *Tetrahedron Lett.* **26**:2667 (1985); T. Jeffery, *Synthesis* **1987**:70; R. C. Larock and S. Babu, *Tetrahedron Lett.* **28**:5291 (1987).
96. A. de Meijere and F. E. Meyer, *Angew. Chem. Int. Ed. Engl.* **33**:2379 (1994); R. Grigg, *J. Heterocycl. Chem.* **31**:631 (1994); T. Jeffery, *Tetrahedron* **52**:10113 (1996).
97. C.-M. Andersson, K. Karabelas, A. Hallberg, and C. Andersson, *J. Org. Chem.* **50**:3891 (1985).
98. M. T. Reetz, G. Lehmer, and R. Schwickard, *Angew. Chem. Int. Ed. Engl.* **37**:481 (1998).
99. J. J. Bozell and C. E. Vogt, *J. Am. Chem. Soc.* **110**:2655 (1988).
100. A. M. Echavarren and J. K. Stille, *J. Am. Chem. Soc.* **109**:5478 (1987); K. Karabelas and A. Hallberg, *J. Org. Chem.* **53**:4909 (1988).
101. W. Cabri, I. Candiani, A. Bedeschi, and R. Santi, *J. Org. Chem.* **55**:3654 (1990); W. Cabri, I. Candiani, A. Bedeschi, and R. Santi, *J. Org. Chem.* **57**:3558 (1992); W. Cabri, I. Candiani, A. Bedeschi, A. Penco, and R. Santi, *J. Org. Chem.* **57**:1481 (1992).
102. K. Olofsson, M. Larhed, and A. Hallberg, *J. Org. Chem.* **63**:5076 (1998).

Scheme 8.8. Palladium-Catalyzed Vinylation of Aryl and Alkenyl Halides

1[a] 2-Bromoiodobenzene + $CH_2{=}CHCO_2H$ $\xrightarrow[Et_3N]{Pd(OAc)_2}$ 2-$BrC_6H_4CH{=}CHCO_2H$ 82%

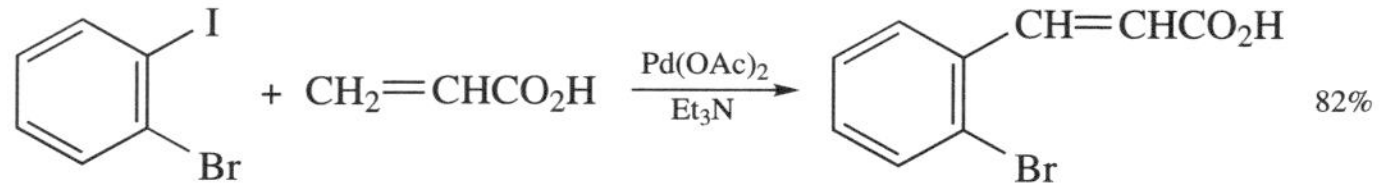

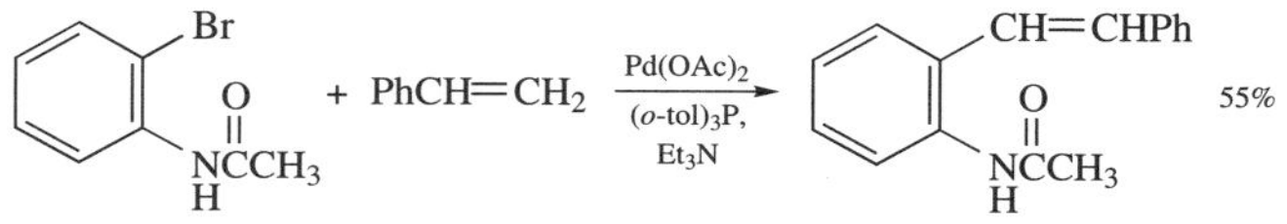

2[b] 2-Bromoacetanilide ($NHCOCH_3$) + $PhCH{=}CH_2$ $\xrightarrow[(o\text{-tol})_3P,\ Et_3N]{Pd(OAc)_2}$ 2-($CH{=}CHPh$)acetanilide 55%

3[c] $(H_3C)(CH_3O_2C)C{=}C(Br)H$ + cyclohexene $\xrightarrow[(o\text{-tol})_3P,\ Et_3N]{Pd(OAc)_2}$ cyclohexenyl–$CH{=}C(CH_3)CO_2CH_3$ 57%

4[d] PhI + cyclohexene $\xrightarrow[R_4N^+Cl^-,\ DMF]{Pd(OAc)_2,\ KOAc}$ 3-phenylcyclohexene 70%

5[e] cyclohexenyl–O_3SCF_3 + $CH_2{=}CHCH(NHCO_2CH_2Ph)CH_2O_2CCH_3$ $\xrightarrow[n\text{-}Bu_4N^+\ {}^-O_3SCF_3,\ K_2CO_3]{Pd(OAc)_2,\ 10\ mol\ \%}$ cyclohexenyl–$CH{=}CHCH(NHCO_2CH_3Ph)CH_2O_2CCH_3$ 80%

6[f] OTBDMS, CH_3, O, OCH_3, H, CH_3, I + $CH_2{=}CHCO_2CH_3$ $\xrightarrow[(C_2H_5)_3N,\ DMF]{Pd(PPh_3)_4,\ 10\ mol\ \%}$ OTBDMS, CH_3, O, OCH_3, H, CH_3, CO_2CH_3 69%

7[g] HO, $CH_2OTBDMS$, OH, CH_3, CH_3, I + $CH_3CH{=}CHCO_2C(CH_3)_3$ $\xrightarrow[(C_2H_5)_3N,\ AgCO_3]{Pd(OAc)_2,\ 10\ mol\ \%}$ HO, $CH_2OTBDMS$, OH, CH_3, CH_3, CH_3, $CO_2C(CH_3)_3$ 84%

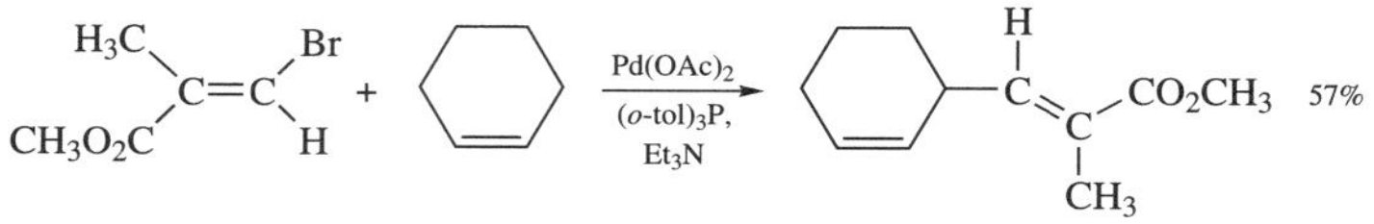

8[h] 2-Br-C_6H_4–$N(CH_3)$–$C(=O)$–cyclopentenyl $\xrightarrow[Et_3N]{Pd(OAc)_2,\ PPh_3}$ spiro(cyclopentene)-N-CH_3-oxindole (N, CH_3, O) 85%

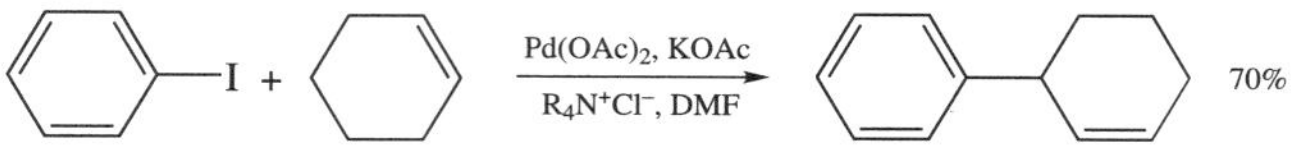

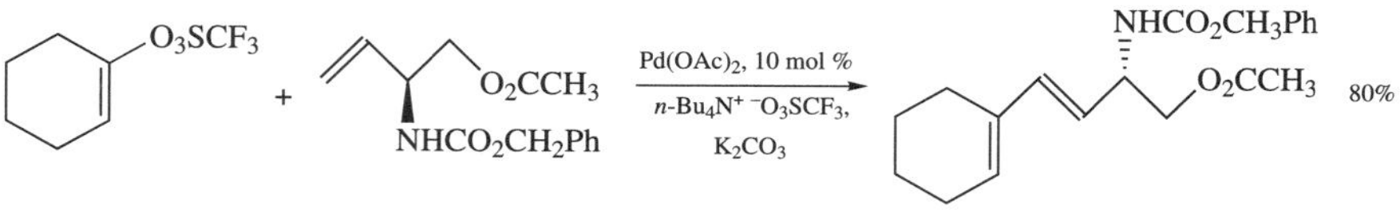

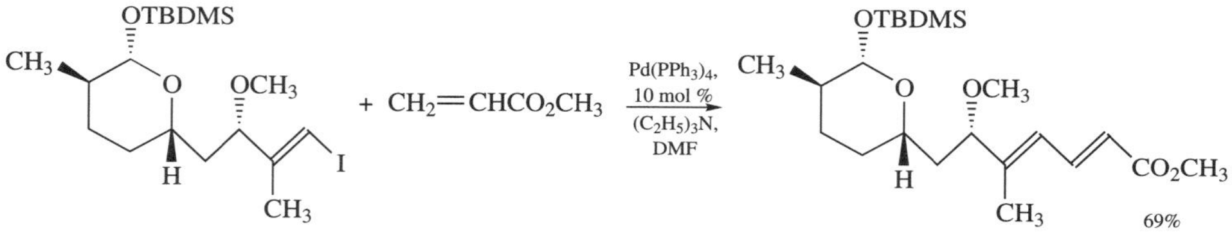

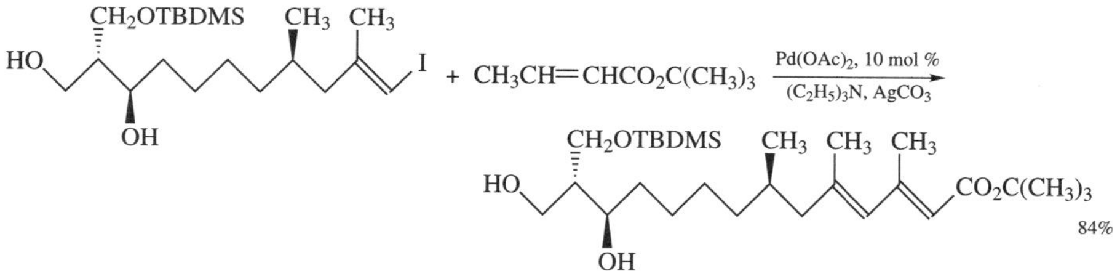

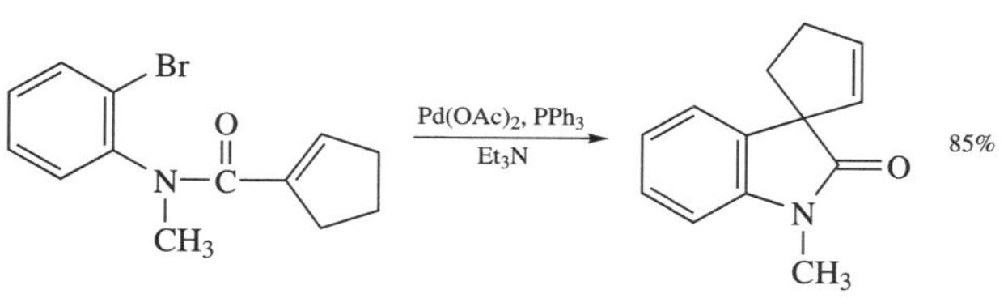

9[i] 3-iodo-2-methoxy-4-[$CH_2OCH_2CH{=}CHCH_3$]pyridine $\xrightarrow[n\text{-}Bu_4N^+Cl^-,\ DMF]{Pd(OAc)_2,\ K_2CO_3}$ product with OCH_3 and CH_2CH_3 substituents, 79%

a. J. E. Plevyak, J. E. Dickerson, and R. F. Heck, *J. Org. Chem.* **44:**4078 (1979).
b. P. de Mayo, L. K. Sydnes, and G. Wenska, *J. Org. Chem.* **45:**1549 (1980).
c. J.-I. I. Kim, B. A. Patel, and R. F. Heck, *J. Org. Chem.* **46:**1067 (1981).
d. R. C. Larock and B. E. Baker, *Tetrahedron Lett.* **29:**905 (1988).
e. G. T. Crisp and M. G. Gebauer, *Tetrahedron* **52:**12465 (1996).
f. L. Harris, K. Jarowicki, P. Kocienski, and R. Bell, *Synlett* **1996:**903.
g. P. M. Wovkulich, K. Shankaran, J. Kiegel, and M. R. Uskokovic, *J. Org. Chem.* **58:**832 (1993); T. Jeffery and J.-C. Galland, *Tetrahedron Lett.* **35:**4103 (1994).
h. M. M. Abelman, T. Oh, and L. E. Overman, *J. Org. Chem.* **52:**4130 (1987).
i. F. G. Fang, S. Xie, and M. W. Lowery, *J. Org. Chem.* **59:**6142 (1994).

substituent.

$$PhO_3SCF_3 + CH_2{=}CHCH_2Si(CH_3)_3 \xrightarrow[dppf]{Pd(OAc)_2} CH_2{=}C(Ph)CH_2Si(CH_3)_3 \quad 67\%$$

95:5 β:γ

8.2.3. Palladium-Catalyzed Cross Coupling

8.2.3.1. Coupling with Organometallic Reagents. Palladium can catalyze carbon–carbon bond formation between aryl and vinyl halides and sulfonates and a wide range of organometallic reagents. These are called *cross-coupling reactions.*[103] The organometallic reagents can be organomagnesium and organozinc, mixed cuprate, stannane, or organoboron compounds. The reaction is quite general for formation of sp^2-sp^2 and sp^2-sp bonds in biaryls, dienes and polyenes, and enynes. There are also some conditions which can couple alkyl organometallic reagents, but these reactions are less general because of the tendency of alkylpalladium intermediates to decompose by β elimination.

The basic steps in the cross-coupling reaction include oxidative addition of the aryl or vinyl halide (or sulfonate) to Pd(0), followed by transfer of an organic ligand from the organometallic to the resulting Pd(II) intermediate. The disubstituted Pd(II) intermediate then undergoes reductive elimination, which gives the product by carbon bond formation and regenerates the catalytically active Pd(0) oxidation level. Ligands and anions play a crucial role in determining the rates and equilibria of the various steps by controlling the detailed coordination environment at palladium.[104]

103. F. Diederich and P. J. Stang, *Metal-Catalyzed Cross-Coupling Reactions*, Wiley-VCH, New York, 1998; S. P. Stanforth, *Tetrahedron* **54:**263 (1998).
104. P. J. Stang, M. H. Kowalski, M. D. Schiavelli, and D. Longford, *J. Am. Chem. Soc.* **111:**3347 (1989); P. J. Stang and M. H. Kowalski, *J. Am. Chem. Soc.* **111:**3356 (1989); M.Portnoy and D. Milstein, *Organometallics* **12:**1665 (1993).

Tetrakis(triphenylphosphine)palladium catalyzes coupling of alkenyl halides with Grignard reagents and organolithium reagents.

$$\underset{C_6H_{13}}{H}\!\!\begin{array}{c}\\C{=}C\\\end{array}\!\!\underset{I}{H} + \underset{H}{BrMg}\!\!\begin{array}{c}\\C{=}C\\\end{array}\!\!\underset{H}{H} \xrightarrow{Pd(PPh_3)_4} \underset{C_6H_{13}}{H}\!\!\begin{array}{c}\\C{=}C\\\end{array}\!\!\underset{CH=CH_2}{H} \quad 75\%$$

Ref. 105

$$\underset{C_4H_9}{H}\!\!\begin{array}{c}\\C{=}C\\\end{array}\!\!\underset{Br}{H} + C_4H_9Li \xrightarrow{Pd(PPh_3)_4} \underset{C_4H_9}{H}\!\!\begin{array}{c}\\C{=}C\\\end{array}\!\!\underset{C_4H_9}{H} \quad 63\%$$

Ref. 106

Organozinc compounds are also useful in palladium-catalyzed coupling with aryl and alkenyl halides. Procedures for arylzinc,[107] alkenylzinc,[108] and alkylzinc[109] reagents have been developed. Ferrocenyldiphosphine (dppf) has been found to be an especially good Pd ligand for these reactions.[110]

$$2\text{-}CH_3C_6H_4ZnCl + Br\text{-}C_6H_4\text{-}NO_2 \xrightarrow{Pd(PPh_3)_4} 2\text{-}CH_3C_6H_4\text{-}C_6H_4\text{-}NO_2 \quad 78\%$$

Ref. 107

$$CH_2{=}CH(CH_2)_2ZnCl + \underset{H}{I}\!\!\begin{array}{c}\\C{=}C\\\end{array}\!\!\underset{(CH_2)_3CH_3}{CH_3} \xrightarrow{Pd(PPh_3)_4} \underset{H}{CH_2{=}CHCH_2CH_2}\!\!\begin{array}{c}\\C{=}C\\\end{array}\!\!\underset{(CH_2)_3CH_3}{CH_3} \quad 81\%$$

Ref. 109

$$\underset{H_5C_2}{Ph}\!\!\begin{array}{c}\\C{=}C\\\end{array}\!\!\underset{Ph}{Br} + MeO\text{-}C_6H_4\text{-}ZnCl \xrightarrow{Pd(PPh_3)_4} \underset{H_5C_2}{Ph}\!\!\begin{array}{c}\\C{=}C\\\end{array}\!\!\underset{Ph}{C_6H_4\text{-}OMe}$$

Ref. 111

Other examples of Pd-catalyzed cross-coupling of organometallic reagents are given in Scheme 8.9.

A promising recent development is the extension of Pd-catalyzed cross coupling to simple enolates and enolate equivalents. This provides an important way of arylating enolates, which is normally a difficult transformation to accomplish. Use of tri-*t*-butylphosphine with a catalytic amount of $Pd(OAc)_2$ results in phenylation of the enolates of

105. M. P. Dang and G. Linstrumelle, *Tetrahedron Lett.* **1978:**191.
106. M. Yamamura, I. Moritani, and S. Murahashi, *J. Organomet.Chem.* **91:**C39 (1975).
107. E. Negishi, A. O. King, and N. Okukado, *J. Org. Chem.* **42:**1821 (1977); E. Negishi, T. Takahashi, and A. O. King, *Org. Synth.* **66:**67 (1987).
108. U. H. Lauk, P. Skrabal, and H. Zollinger, *Helv. Chim. Acta* **68:**1406 (1985); E. Negishi, T. Takahashi, S. Baba, D. E. Van Horn, and N. Okukado, *J. Am. Chem. Soc.* **109:**2393 (1987); J.-M. Duffault, J. Einhorn, and A. Alexakis, *Tetrahedron Lett.* **32:**3701 (1991).
109. E. Negishi, L. F. Valente, and M. Kobayashi, *J. Am. Chem. Soc.* **102:**3298 (1980).
110. T. Hayashi, M. Konishi, Y. Kobori, M. Kumada, T. Higuchi, and K. Hirotsu, *J. Am. Chem. Soc.* **106:**158 (1984).
111. R. B. Miller and M. I. Al-Hassan, *J. Org. Chem.* **50:**2121 (1985).

Scheme 8.9. Palladium-Catalyzed Cross Coupling of Organomagnesium and Organozinc Reagents

1[a] 2-methylphenyl-ZnCl + Br–C$_6$H$_4$–NO$_2$ $\xrightarrow{Pd(PPh_3)_4,\ 1\ mol\ \%}$ 2-CH$_3$-C$_6$H$_4$–C$_6$H$_4$–NO$_2$ 78%

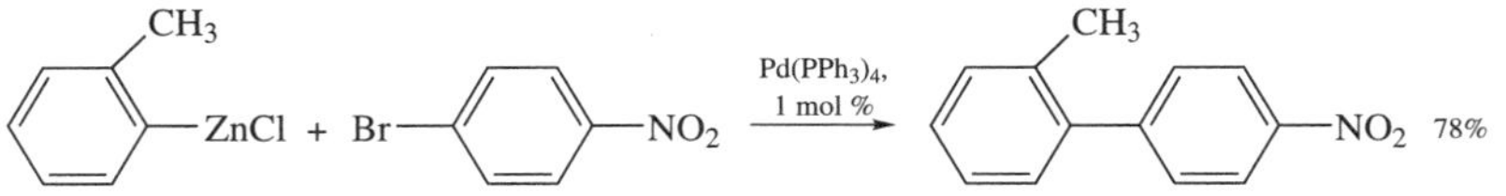

2[b] 2-biphenylyl-O$_3$SCF$_3$ + PhMgBr $\xrightarrow{PdCl_2}$ (ligand: CH$_3$CH(N(CH$_3$)$_2$)CH$_2$PPh$_2$) o-terphenyl 95%

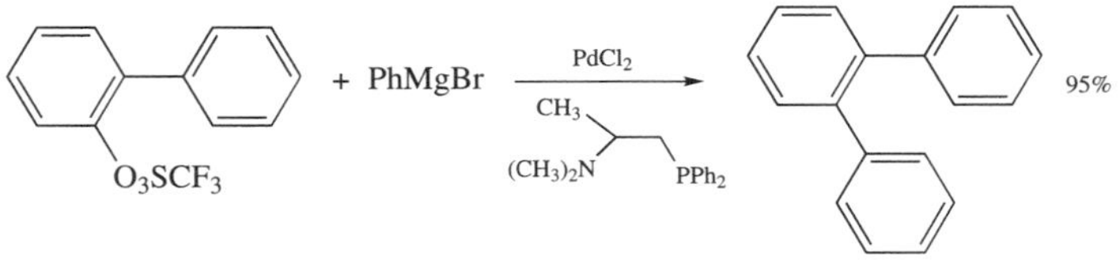

3[c] CH$_3$ / O$_3$SC$_4$H$_9$ biphenyl + BrZn–C$_6$H$_4$–Cl $\xrightarrow[dppf,\ 2\ mol\ \%]{Pd(dba)_2,\ 2\ mol\ \%}$ CH$_3$ … Cl terphenyl 96%

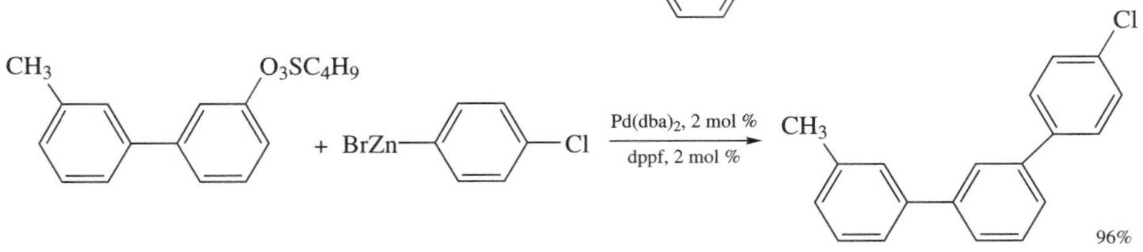

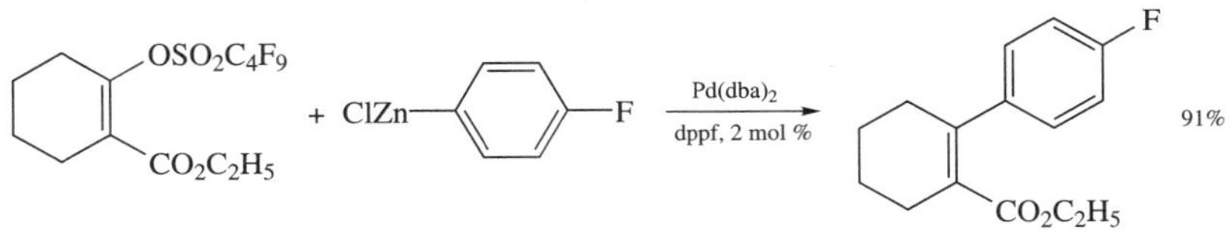

(dba = dibenzylideneacetone)

4[d] 3-methylindene (CH$_3$) $\xrightarrow{1)\ IpcBH_2\ 2)\ (C_2H_5)_2BH\ 3)\ (i\text{-}Pr)_2Zn}$ CH$_3$, H, ZnCH(CH$_3$)$_2$ $\xrightarrow[Pd(dba)_2,\ 2\ mol\ \%;\ (o\text{-}Tol)_3P,\ 4\ mol\ \%]{ICH=CH(CH_2)_3CH_3}$ CH$_3$, H, CH=CH(CH$_2$)$_3$CH$_3$ 35%

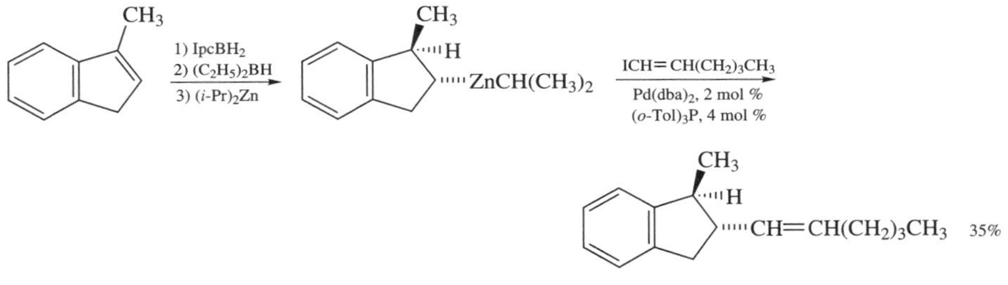

5[e] OSO$_2$C$_4$F$_9$ / CO$_2$C$_2$H$_5$ cyclohexene + ClZn–C$_6$H$_4$–F $\xrightarrow[dppf,\ 2\ mol\ \%]{Pd(dba)_2}$ F-C$_6$H$_4$ / CO$_2$C$_2$H$_5$ cyclohexene 91%

6[f] bornenyl-O$_3$SCF$_3$ + PhZnCl $\xrightarrow{Pd(PPh_3)_4\ 2\ mol\ \%}$ bornenyl-Ph 55%

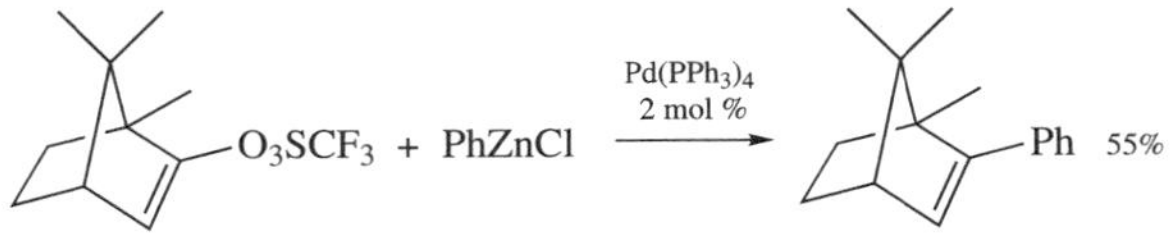

7[g] PhZnCl + Cl, Cl vinyl, CH$_3$ CH$_3$, CO$_2$C$_2$H$_5$ cyclopropane $\xrightarrow{Pd(dppb)Cl_2,\ 14\ mol\ \%}$ Ph, Cl vinyl, CH$_3$ CH$_3$, CO$_2$C$_2$H$_5$ 86%

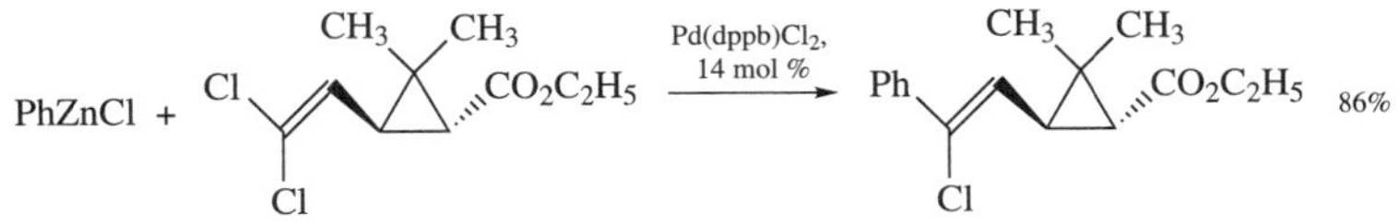

a. E. Negishi, T. Takahashi, and A. O. King, *Org. Synth.* **66:**67 (1987).
b. T. Kamikawa and T. Hayashi, *Synlett* **1997:**163.
c. M. Rottländer and P. Knochel, *J. Org. Chem.* **63:**203 (1998).
d. A. Boudier and P. Knochel, *Tetrahedron Lett.* **40:**687 (1999).
e. F. Bellina, D. Ciucci, R. Rossi, and P. Vergamini, *Tetrahedron* **55:**2103 (1999).
f. G. Stork and R. C. A. Isaacs, *J. Am. Chem. Soc.* **112:**7399 (1990).
g. A. Minato, *J. Org. Chem.* **56:**4052 (1991).

aromatic ketones and diethyl malonate.[112]

$$CH_3CH{=}C(O^-)Ph + PhBr \xrightarrow[25^\circ C,\ 2\ h]{Pd(O_2CCH_3)_2,\ 1\ mol\ \%;\ P(t\text{-}Bu)_3,\ 1\ mol\ \%} CH_3CH(Ph)C(=O)Ph \quad 96\%$$

$$^-CH(CO_2C_2H_5)_2 \xrightarrow[20^\circ C]{Pd(O_2CCH_3)_2,\ 2\ mol\ \%;\ P(t\text{-}Bu)_3,\ 2\ mol\ \%} PhCH(CO_2C_2H_5)_2 \quad 86\%$$

Arylation has also been observed with the diphosphine ligand BINAP.

$$CH_3CH_2CPh(=O) + Br\text{-}C_6H_4\text{-}OCH_3 \xrightarrow[NaOtBu]{Pd_2(dba)_3,\ 1.5\ mol\ \%;\ BINAP,\ 3\ mol\ \%} CH_3O\text{-}C_6H_4\text{-}CH(CH_3)C(=O)Ph \quad 91\%$$

Ref. 113

Arylacetate esters have been generated by coupling aryl bromides with enolates generated from O-silyl ketene acetals in the presence of tributylstannyl fluoride.

$$ArBr + CH_2{=}C(OTBDMS)(OC(CH_3)_3) \xrightarrow[2\ Bu_3SnF]{Pd(o\text{-}tol_3P)_2Cl_2\ (cat)} ArCH_2CO_2C(CH_3)_3$$

Ref. 114

A combination of $Pd(PPh_3)_4$ and Cu(I) effects coupling of terminal alkynes with vinyl or aryl halides.[115] The alkyne is presumably converted to the copper acetylide. The halide reacts with Pd(0) by oxidative addition. Transfer of the acetylide group to Pd results in reductive elimination and formation of the observed product.

$$HC{\equiv}CR \xrightarrow[R_3N]{Cu(I)} CuC{\equiv}CR$$

$$R'X + Pd^0 \longrightarrow R'Pd^{II}X$$

$$CuC{\equiv}CR + R'Pd^{II}X \longrightarrow R'Pd^{II}(C{\equiv}CR) \longrightarrow R'C{\equiv}CR + Pd^0$$

Use of alkenyl halides in this reaction has proven to be an effective method for the synthesis of enynes.[116] The reaction can be carried out directly with the alkyne, using amines for deprotonation.

$$CH_3(CH_2)_4CH{=}CHI + HC{\equiv}C(CH_2)_2OH \xrightarrow[pyrrolidine]{Pd(PPh_3)_4,\ 5\ mol\ \%;\ CuI,\ 10\ mol\ \%} CH_3(CH_2)_4CH{=}CHC{\equiv}C(CH_2)_2OH \quad 90\%$$

Ref. 117

112. M. Kawatsura and J. F. Hartwig, *J. Am. Chem. Soc.* **121:**1473 (1999).
113. M. Palucki and S. L. Buchwald, *J. Am. Chem. Soc.* **119:**11108 (1997).
114. F. Agnelli and G. A. Sulikowski, *Tetrahedron Lett.* **39:**8807 (1998).
115. K. Sonogashira, Y. Tohda, and N. Hagihara, *Tetrahedron Lett.* **1975:**4467.
116. V. Ratovelomana and G. Linstrumelle, *Synth. Commun.* **11:**917 (1981); L. Crombie and M. A. Horsham, *Tetrahedron Lett.* **28:**4879 (1987); G. Just and B. O'Connor, *Tetrahedron Lett.* **29:**753 (1988); D. Guillerm and G. Linstrumelle, *Tetrahedron Lett.* **27:**5857 (1986).
117. M. Alami, F. Ferri, and G. Linstrumelle, *Tetrahedron Lett.* **34:**6403 (1993).

8.2.3.2. Coupling with Stannanes. Another important group of cross-coupling reactions uses aryl and alkenyl stannanes as the organometallic component. These are called *Stille reactions.*[118] The reaction has proven to be very general with respect to both the halides and the types of stannnanes that can be used. Benzylic, aryl, alkenyl, and allylic halides all can be used.[119] The groups that can be transferred from tin include alkyl, alkenyl, aryl, and alkynyl. The approximate order of effectiveness of transfer of groups from tin is alkynyl > alkenyl > aryl > methyl > alkyl, so unsaturated groups are normally transferred selectively.[120] Subsequent studies have found improved ligands, including tri-2-furylphosphine[121] and triphenylarsine.[122] Aryl–aryl coupling rates are increased by the presence of Cu(I) co-catalyst.[123] These improvements have led to a simplified protocol in which Pd/C catalyst, along with CuI and Ph_3As, gives excellent yields of biaryls.

+ $(n\text{-}C_4H_9)_3Sn$—

Pd/C, 0.5 mol % Pd
CuI, 10 mol %
Ph_3As, 20 mol %
NMP, 80°C, 16 h

77%

Ref. 124

The basic mechanism of the Stille reaction involves transmetallation, either directly or via an organocopper intermediate, with a Pd(II) intermediate generated by oxidative addition from the aryl halide or triflate.

$$\mathrm{ArPd(L)_nX + Ar'Cu} \; (\rightleftharpoons \mathrm{Ar'SnR_3}) \longrightarrow [\mathrm{ArPd(Ar')(L)_yX}]^- \longrightarrow \mathrm{Ar{-}Ar' + Pd^0(L)_y + X^-}$$

In addition to aryl–aryl coupling, the Stille reaction can be used with alkenylstannanes and alkenyl halides and triflates.[125] The reactions occur with retention of configuration at both the halide and the stannane. These reactions have become very useful in stereospecific construction of dienes and polyenes, as illustrated by some of the examples in Scheme 8.10.

The coupling reaction is very general with respect to the functionality which can be carried both in the halide and in the tin reagent. Groups such as ester, nitrile, nitro, cyano, and formyl can be present. This permits applications involving “masked functionality.” For example, when the coupling reaction is applied to 1-alkoxy-2-butenylstannanes, the

118. J. K. Stille, *Angew. Chem. Int. Ed. Engl.* **25:**508 (1986); T. N. Mitchell, *Synthesis* **1992:**803.
119. F. K. Sheffy, J. P. Godschalx, and J. K. Stille, *J. Am. Chem. Soc.* **106:**4833 (1984); I. P. Beltskaya, *J. Organomet. Chem.* **250:**551 (1983); J. K. Stille and B. L. Grot, *J. Am. Chem. Soc.* **109:**813 (1987).
120. J. W. Labadie and J. K. Stille, *J. Am. Chem. Soc.* **105:**6129 (1983).
121. V. Farina and B. Krishnan, *J. Am. Chem. Soc.* **113:**9585 (1991).
122. V. Farina, B. Krishnan, D. R. Marshall, and G. P. Roth, *J. Org. Chem.* **58:**9434 (1993).
123. V. Farina, S. Kapadia, B. Krishnan, C. Wang, and L. S. Liebeskind, *J. Org. Chem.* **59:**5905 (1994).
124. G. P. Roth, V. Farina, L. S. Liebeskind, and E. Pena-Cabrera, *Tetrahedron Lett.* **36:**2191 (1995).
125. W. J. Scott and J. K. Stille, *J. Am. Chem. Soc.* **108:**3033 (1986).

Scheme 8.10. Palladium-Catalyzed Coupling of Stannanes with Halides and Sulfonates

A. Aryl halides

1[a] $CH_3O-C_6H_4-Br + CH_2=CHCH_2Sn(n\text{-}Bu)_3 \xrightarrow[120°C,\ 20\ h]{Pd(PPh_3)_4} CH_3O-C_6H_4-CH_2CH=CH_2$ 96%

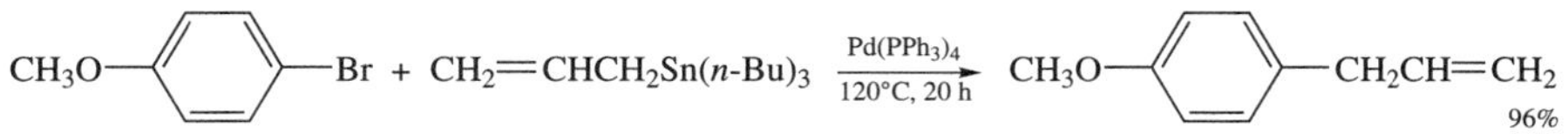

2[b] $O_2N-C_6H_4-Br + CH_2=CHSn(n\text{-}Bu)_3 \xrightarrow[105°C,\ 4\ h]{Pd(PPh_3)_4} O_2N-C_6H_4-CH=CH_2$ 80%

3[c] $CH_3O_2C-C_6H_4-I$ + 2-pyridyl-$Sn(CH_3)_3$ $\xrightarrow{Pd(PPh_3)_4Cl_2}$ 2-pyridyl-$C_6H_4-CO_2CH_3$ 95%

4[d] 2-pyridyl-$Sn(n\text{-}Bu)_3$ + 4-bromopyridine (Br) $\xrightarrow[AgO,\ 1\ equiv,\ DMF,\ 100°C]{Pd(PPh_3)_4,\ 10\ mol\ \%}$ 2,4'-bipyridine 70%

5[e] 3-$OCH(CH_3)_2$-4-$Sn(n\text{-}Bu)_3$-cyclobutenedione + $I-C_6H_4-OCH_3$ $\xrightarrow[DMF]{PhCH_2Pd(PPh_3)_2Cl,\ 5\ mol\ \%;\ CuI,\ 10\ mol\ \%}$ 3-$OCH(CH_3)_2$-4-(C_6H_4-OCH_3)-cyclobutenedione 80%

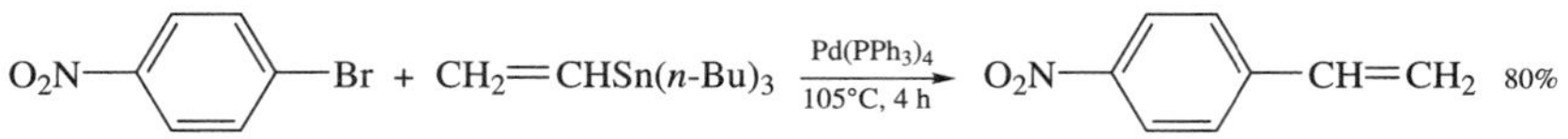

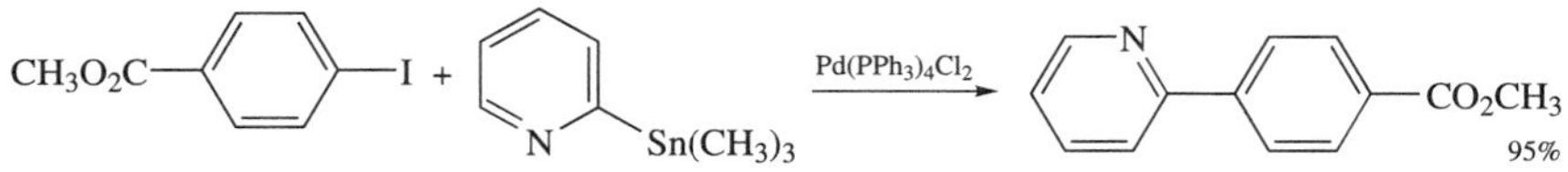

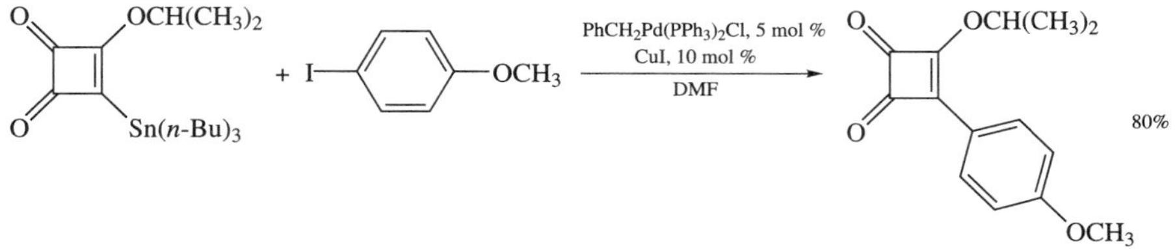

B. Alkenyl halides and sulfonates

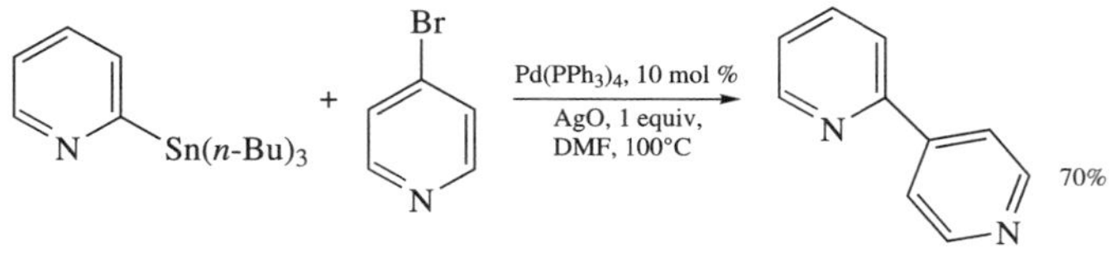

6[f] $PhCH=CHI + CH_2=CHSn(n\text{-}Bu)_3 \xrightarrow[25°C,\ 0.1\ h]{PdCl_2(CH_3CN)_2} PhCH=CHCH=CH_2$ 85%

7[g] 2,5,5-trimethylcyclopentenyl-OSO_2CF_3 (CH_3, CH_3, CH_3) + $(CH_3)_3Sn(H)C=C(H)Si(CH_3)_3$ $\xrightarrow{Pd(PPh_3)_4}$ 2,5,5-trimethylcyclopentenyl-$C(H)=C(H)Si(CH_3)_3$ 100%

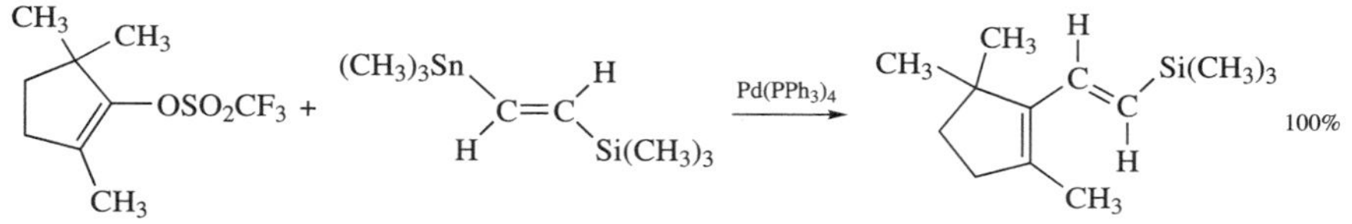

8[h] 2-iodo-3-methylcyclopent-2-enone (O, I, CH_3) + $(n\text{-}Bu)_3Sn-C_6H_4-CH_3$ $\xrightarrow[Ph_3As,\ 10\ mol\ \%;\ NMP,\ 100°C]{Pd(CH_3CN)_2Cl_2,\ 5\ mol\ \%}$ 2-(C_6H_4-CH_3)-3-methylcyclopent-2-enone

9[i] tetrahydropyranyl-CH(OTBDMS)-CH=CH-$Sn(n\text{-}Bu)_3$ + I-CH=C(CH_3)-CH_2CH_2-CH(CH_3)-CH_2OH $\xrightarrow[THF,\ DMF]{Pd(PPh_3)_4}$ tetrahydropyranyl-CH(OTBDMS)-CH=CH-CH=C(CH_3)-CH_2CH_2-CH(CH_3)-CH_2OH 42%

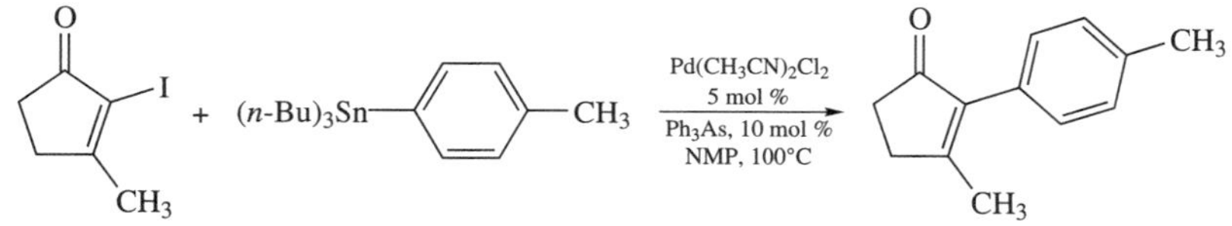

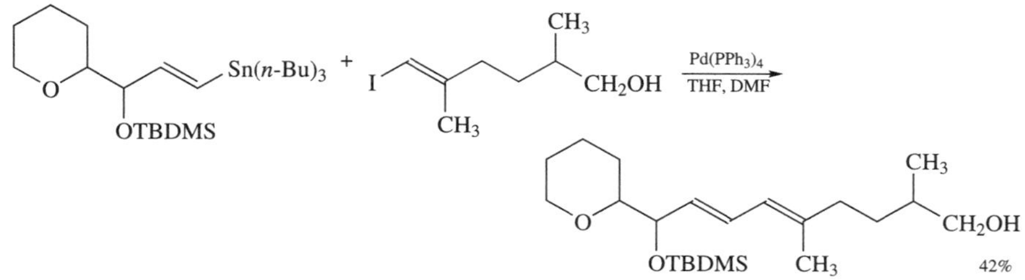

10[j] $Pd(CH_3CN)_2Cl_2$, 2.5 mol %, DMF — 82%

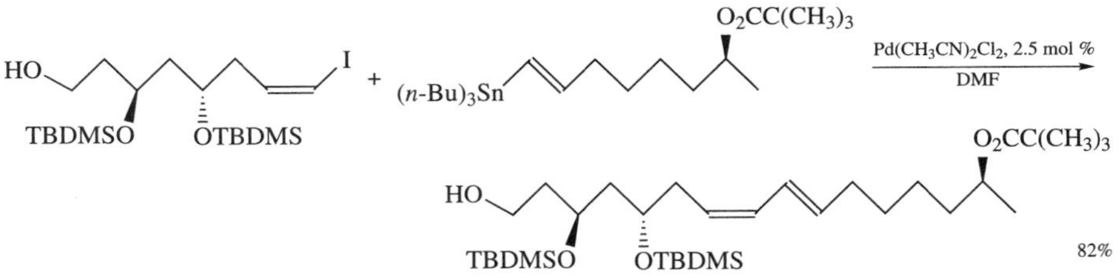

HO, TBDMSO, OTBDMS, I + $(n\text{-}Bu)_3Sn$, $O_2CC(CH_3)_3$

11[k] $Pd_2(dba)_3$, 5 mol %, Ph_3As, 10 mol %, DMF — 85%

Ph, O_3SCF_3 + $(n\text{-}Bu)_3Sn$, CO_2CH_3, $NHCO_2C(CH_3)_3$

12[l] $Pd(CH_3CN)_2Cl_2$ — 94%

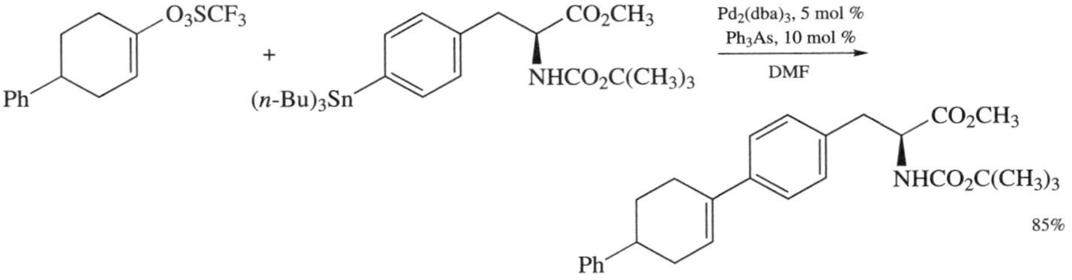

CH_3, N, O, $Sn(n\text{-}Bu)_3$ + I, CH_3, O_3SCH_3, OTBDMS

13[m] $Pd(dba)_2$, trifurylphosphine — 53%

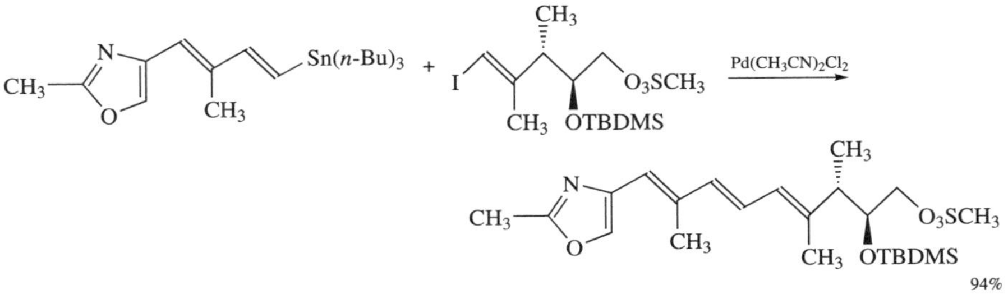

$C_2H_5O_2C$, S, N, Br, O_2CCH_3 + $(n\text{-}Bu)_3Sn$, CH_3, $CH_2N(CH_3)_2$

14[n] $Pd(PPh_3)_4$, LiCl, Cl, N, NH_2

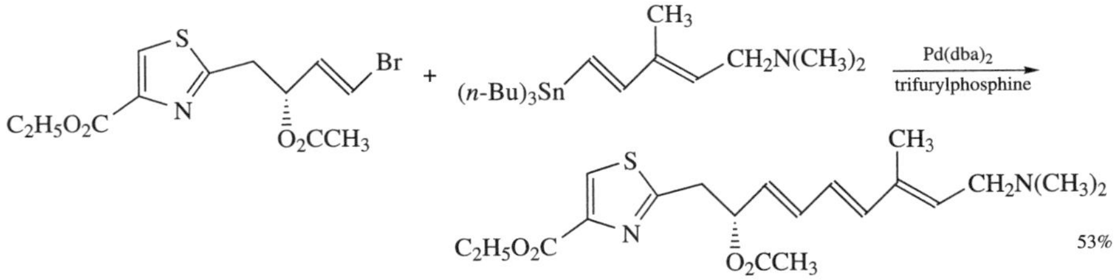

CH_3, OTBDMS, O, $CH(CH_3)_2$, O_3SCF_3 + $(n\text{-}Bu)_3SnCH_2O$, O, O_2CCH_3

CH_2O

(continued)

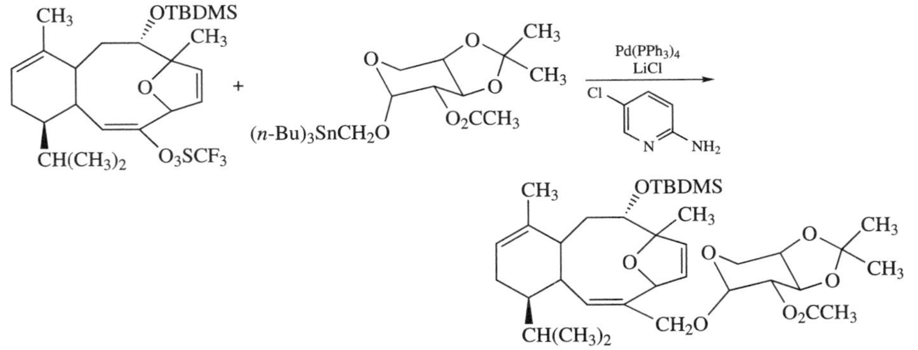

Scheme 8.10. *(continued)*

15[o] $(n\text{-Bu})_3Sn$ ... Br, HO, CH_3, CH_3, OTMS $\xrightarrow{Pd(PPh_3)_4,\ 5\ mol\ \%}$ HO, CH_3, CH_3, OTMS

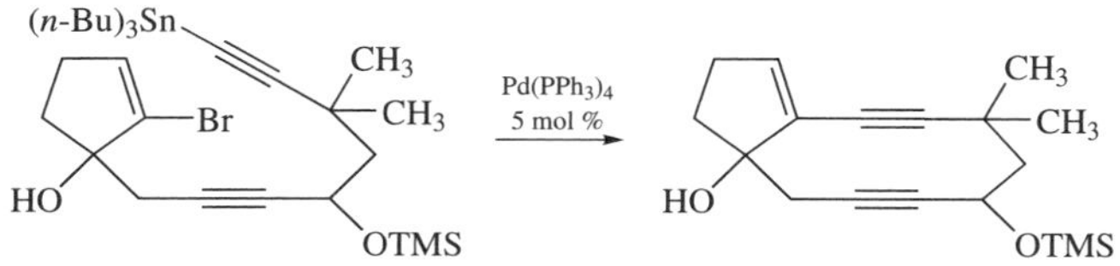

C. Allylic and benzylic halides

16[p] $CH_3CH{=}CHSn(n\text{-Bu})_3$ + $BrCH_2C(OCH_3){=}CHCO_2CH_3$ $\xrightarrow{Pd(dba)_2,\ PPh_3}$ $CH_3CH{=}CHCH_2C(OCH_3){=}CHCO_2CH_3$ 86%

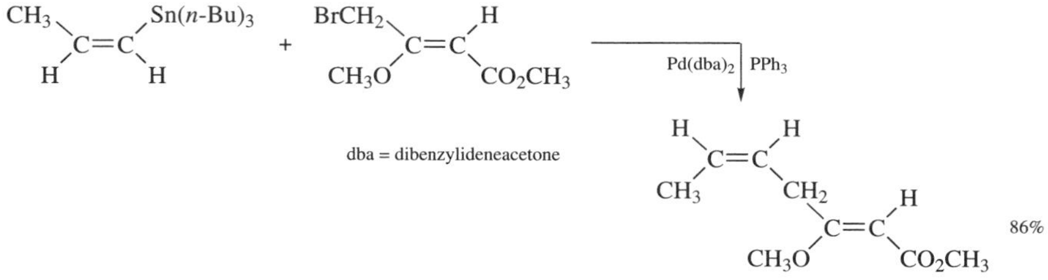

dba = dibenzylideneacetone

17[q] $PhCH_2Br$ + $(n\text{-Bu})_3SnC(H){=}C(CH_3)(CH_2)_2OCH_2Ph$ $\xrightarrow{Pd(PPh_3)_4}$ $PhCH_2CH{=}C(CH_3)(CH_2)_2OCH_2Ph$ 81%

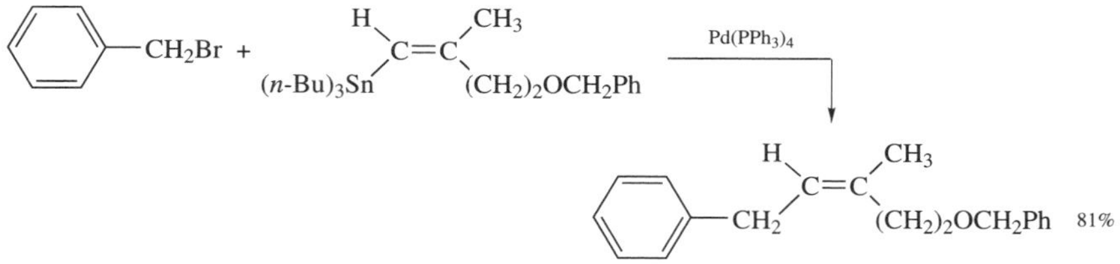

a. M. Kosugi, K. Sasazawa, Y. Shimizu, and T. Migata, *Chem. Lett.* **1977**:301.
b. D. R. McKean, G. Parrinello, A. F. Renaldo, and J. K. Stille, *J. Org. Chem.* **52**:422 (1987).
c. T. R. Bailey, *Tetrahedron Lett.* **27**:4407 (1986).
d. J. Malm, P. Bjork, S. Gronowitz, and A.-B. Hörnfeldt, *Tetrahedron Lett.* **33**:2199 (1992).
e. L. S. Liebeskind and R. W. Fengl, *J. Org. Chem.* **55**:5359 (1990).
f. J. K. Stille and B. L. Groh, *J. Am. Chem. Soc.* **109**:813 (1987).
g. W. J. Scott, G. T. Crisp, and J. K. Stille, *J. Am. Chem. Soc.* **106**:4630 (1984).
h. C. R. Johnson, J. P. Adams, M. P. Braun, and C. B. W. Senanayake, *Tetrahedron Lett.* **33**:919 (1992).
i. E. Claus and M. Kalesse, *Tetrahedron Lett.* **40**:4157 (1999).
j. A. B. Smith III and G. R. Ott, *J. Am. Chem. Soc.* **120**:3935 (1998).
k. E. Morera and G. Ortar, *Synlett* **1997**:1403.
l. J. D. White, M. A. Holoboski, and N. J. Green, *Tetrahedron Lett.* **38**:7333 (1997).
m. D. Romo, R. M. Rzasa, H. E. Shea, K. Park, J. M. Langenhan, L. Sun, A. Akhiezer, and J. O. Liu, *J. Am. Chem. Soc.* **120**:12237 (1998).
n. X.-T. Chen, B. Zhou, S. K. Bhattaharya, C. E. Gutteridge, T. R. R. Pettus, and S. Danishefsky, *Angew. Chem. Int. Ed. Engl.* **37**:789 (1999).
o. M. Hirama, K. Fujiwara, K. Shigematu, and Y. Fukazawa, *J. Am. Chem. Soc.* **111**:4120 (1989).
p. F. K. Sheffy, J. P. Godschalx, and J. K. Stille, *J. Am. Chem. Soc.* **106**:4833 (1984).
q. J. Hibino, S. Matsubara, Y. Morizawa, K. Oshima, and H. Nozaki, *Tetrahedron Lett.* **25**:2151 (1984).

double-bond shift leads to a vinyl ether, which can be hydrolyzed to an aldehyde.

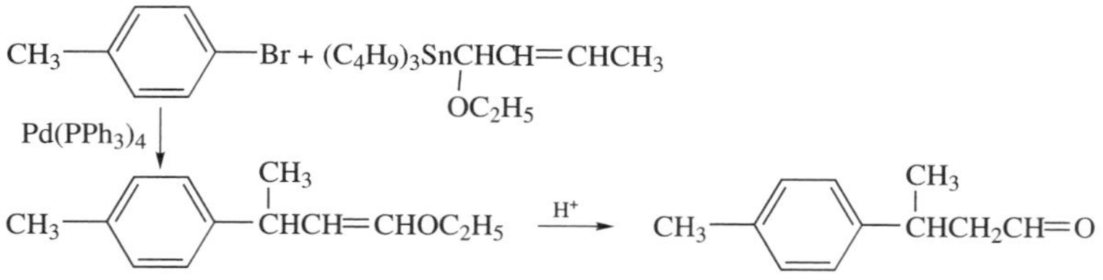

$$CH_3C_6H_4Br + (C_4H_9)_3SnCH(OC_2H_5)CH{=}CHCH_3 \xrightarrow{Pd(PPh_3)_4} CH_3C_6H_4CH(CH_3)CH{=}CHOC_2H_5 \xrightarrow{H^+} CH_3C_6H_4CH(CH_3)CH_2CH{=}O$$

Ref. 126

126. A. Duchene and J.-P. Quintard, *Synth. Commun.* **15**:873 (1987).

The versatility of Pd-catalyzed coupling of stannanes has been extended by the demonstration that alkenyl triflates are also reactive.[127]

$$\text{2-CH}_3\text{-cyclohexenyl-OSO}_2\text{CF}_3 + (CH_3)_3Sn\text{CH=CHSi}(CH_3)_3 \xrightarrow{Pd(PPh_3)_4} \text{2-CH}_3\text{-cyclohexenyl-CH=CH-Si}(CH_3)_3$$

The alkenyl triflates can be prepared from ketones.[128] Methods for regioselective preparation of alkenyl triflates from unsymmetrical ketones are available.[129]

$$\text{2-methylcyclohexanone} \xrightarrow[2)\ (CF_3SO_2)_2NPh]{1)\ LDA} \text{6-methylcyclohex-1-enyl-OSO}_2CF_3$$

Some examples of Pd-catalyzed coupling of organostannanes with halides and triflates are included in Scheme 8.10

8.2.3.3. Coupling with Organoboranes. The *Suzuki reaction* is a cross-coupling reaction in which the organometallic component is an aryl or vinyl boron compound.[130] The organoboron compounds that undergo coupling include boronic acids,[131] boronate esters,[132] and boranes.[133] Scheme 8.11 illustrates some of the successful reaction conditions, which include the absence of phosphine ligands (entry 7) and use of solid-phase reactions (entry 10).

The overall mechanism is closely related to that of the other cross-coupling methods. The aryl halide or triflate reacts with the Pd(0) catalyst by oxidative addition. The organoboron compound serves as the source of the second organic group by transmetalation. The disubstituted Pd(II) intermediate then undergoes reductive elimination. It appears that either the oxidative addition or the transmetalation can be rate-determining, depending on reaction conditions.[134] With boronic acids as reactants, base catalysis is normally required and is believed to involve the formation of the more reactive boronate anion.[135]

$$ArX + Pd^0 \longrightarrow Ar\text{-}Pd^{II}\text{-}X$$

$$Ar'B(OH)_2 + {}^-OH \rightleftharpoons [Ar'B(OH)_3]^-$$

$$[Ar'B(OH)_3]^- + Ar\text{-}Pd^{II}\text{-}X \longrightarrow Ar\text{-}Pd^{II}\text{-}Ar' + B(OH)_3 + X^-$$

$$Ar\text{-}Pd^{II}\text{-}Ar' \longrightarrow Ar{-}Ar' + Pd^0$$

127. W. J. Scott, G. T. Crisp, and J. K. Stille, *J. Am. Chem. Soc.* **106:**4630 (1984); W. J. Scott and J. E. McMurry, *Acc. Chem. Res.* **21:**47 (1988).
128. P. J. Stang, M. Hanack, and L. R.Subramanian, *Synthesis* **1982:**85.
129. J. E. McMurry and W. J. Scott, *Tetrahedron Lett.* **24:**979 (1983).
130. N. Miyaura, T. Yanagi, and A. Suzuki, *Synth. Commun.* **11:**513 (1981); A. Miyaura and A. Suzuki, *Chem. Rev.* **95:**2457 (1995). A. Suzuki, *J. Organomet. Chem.* **576:**147 (1999).
131. W. R. Roush, K. J. Moriarty, and B. B. Brown, *Tetrahedron Lett.* **31:**6509 (1990); W. R. Roush, J. S.Warmus, and A. B. Works, *Tetrahedron Lett.* **34:**4427 (1993); A. R. de Lera, A. Torrado, B. Iglesias, and S. Lopez, *Tetrahedron Lett.* **33:**6205 (1992).
132. T. Oh-e, N. Miyaura, and A. Suzuki, *Synlett*, **1990**:221; J. Fu, B. Zhao, M. J. Sharp, and V. Sniekus, *J. Org. Chem.* **56:**1683 (1991).
133. T. Oh-e, N. Miyaura, and A. Suzuki, *J. Org. Chem.* **58:**2201 (1993); Y. Kobayashi, T. Shimazaki, H. Taguchi, and F. Sato, *J. Org. Chem.* **55:**5324 (1990).
134. G. B. Smith, G. C. Dezeny, D. L. Hughes, A. O. King, and T. R. Verhoeven, *J. Org. Chem.* **59:**8151 (1994).
135. K. Matos and J. A. Soderquist, *J. Org. Chem.* **63:**461 (1998).

Scheme 8.11. Palladium-Catalyzed Cross Couplings of Organoboron Reagents

A. Biaryl formation

1[a] CH_3–C6H4–Cl + $(HO)_2B$–Ph → ($Pd_2(dba)_3$, 1.5 mol %; $P(t\text{-}Bu)_3$, 3.6 mol %; 1.2 equiv Cs_2CO_3; dioxane, 80°C) → CH_3–C6H4–Ph 87%

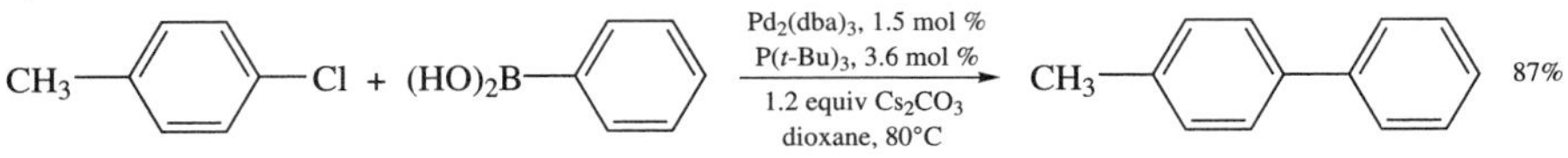

2[b] O_2N–C6H4–I + $(HO)_2B$–Ph → ($Pd(OAc)_2$, 0.2 mol %; K_2CO_3; acetone, water) → O_2N–C6H4–Ph 97%

3[c] CH_3O–C6H4–Br + $(HO)_2B$–C6H4–CF_3 → ($Pd(OAc)_2$, 2 mol %; $Bu_4N^+ Br^-$; 2.5 equiv K_2CO_3) → CH_3O–C6H4–C6H4–CF_3 95%

4[d] 2-CH_3–C6H4–N_2^+ + $(HO)_2B$–Ph → ($Pd_2(OAc)_2$; CH_3OH) → 2-CH_3–C6H4–Ph 90%

5[e] CH_3O–C6H4–N_2^+ + $(HO)_2B$–Ph → ($Pd(OAc)_2$, 5 mol %) → CH_3O–C6H4–Ph 79%

6[f] 2,3,4-$(CH_3O)_3$C6H2–$B(OH)_2$ + Br–C6H2(CONHC2H5)(OCH_3)2 → ($Pd(PPh_3)_4$; K_2CO_3, DME) → biaryl with $C_2H_5NHC(=O)$, CH_3O, CH_3O, OCH_3, OCH_3, OCH_3 substituents 77%

7[b] CF_3–C6H4–B(OCH2CH2O) + Br–C6H4–OCH_3 → ($Pd(OAc)_2$, 2 mol %; K_2CO_3) → CF_3–C6H4–C6H4–OCH_3 97%

8[g] C6H5–$CONHC_2H_5$ → (1) s-BuLi; 2) $B(OMe)_3$; 3) H^+) → 2-$B(OH_2)$–C6H4–$CONHC_2H_5$ + 2-bromothiophene (Br, S) → ($Pd(PPh_3)_4$) → 2-(2-thienyl)–C6H4–$CONHC_2H_5$ 92%

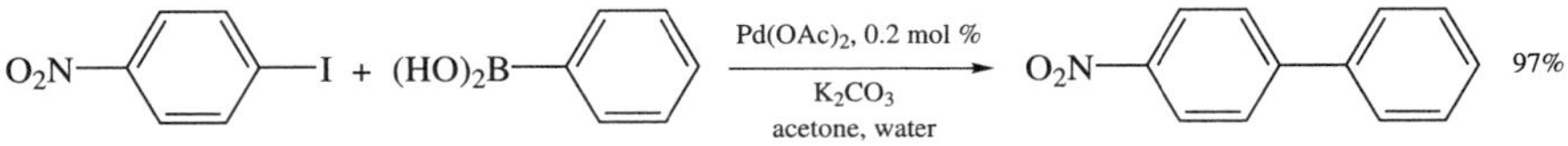

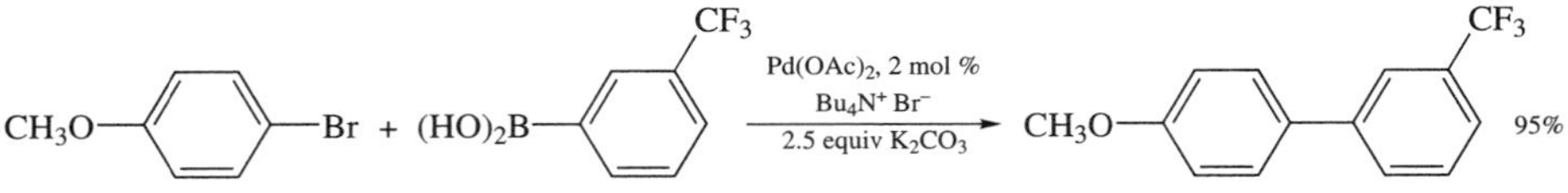

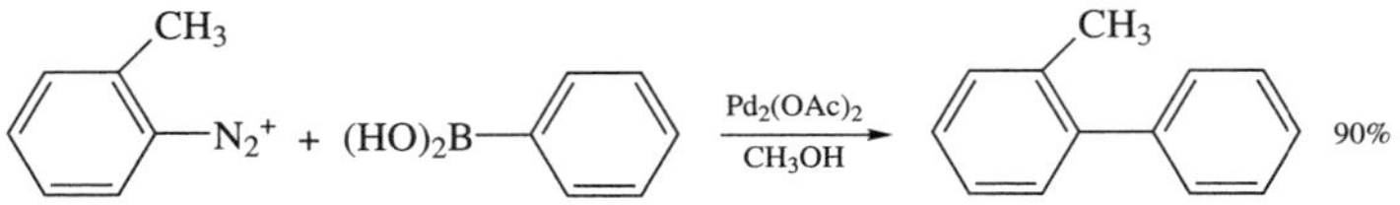

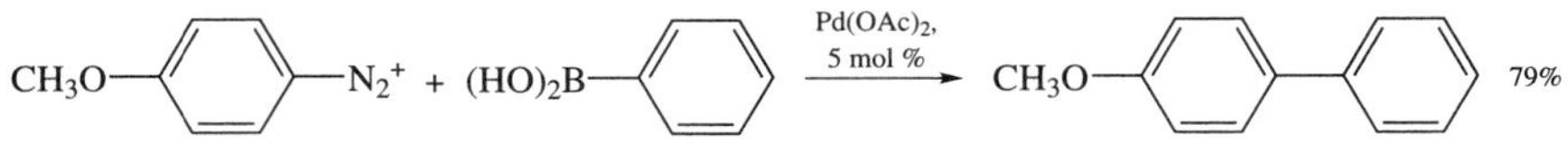

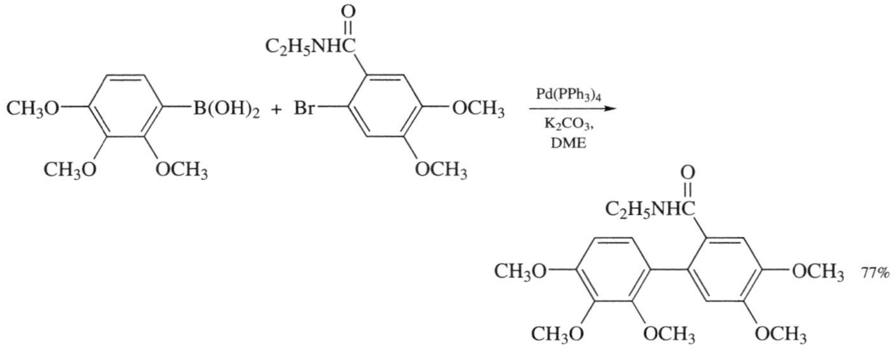

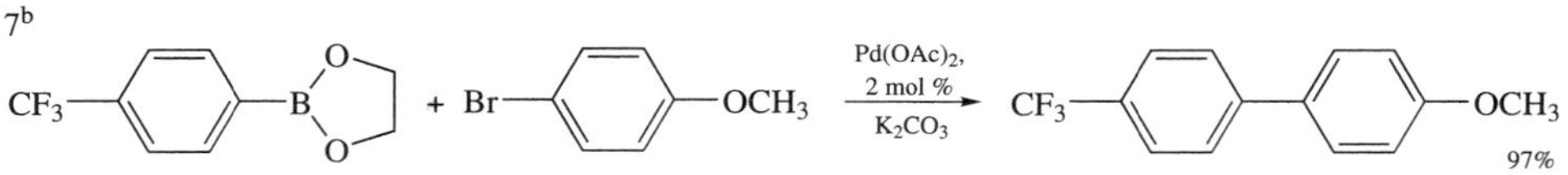

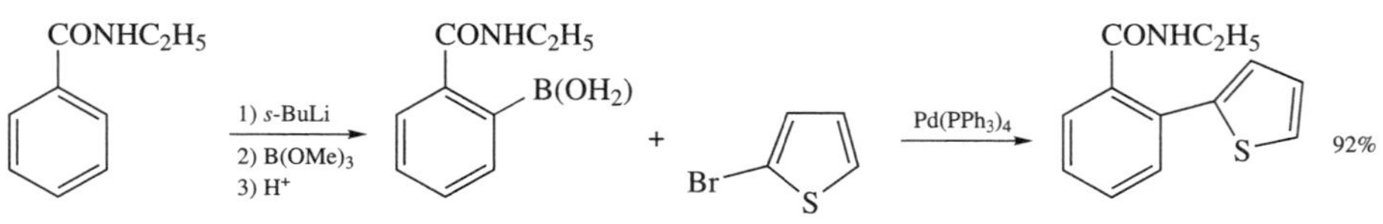

9[h] 4-bromopyridine $\xrightarrow[\text{2) } Et_2BOMe]{\text{1) } n\text{-BuLi}}$ 4-$B(C_2H_5)_2$-pyridine + 4-bromoanisole (OCH_3) $\xrightarrow[\text{KOH, } Bu_4NBr]{Pd(PPh_3)_4\text{, 5 mol \%}}$ 4-(4-OCH_3-phenyl)pyridine 75%

10[i] polystyrene—O_2C—C_6H_4—I + $(HO)_2B$—(2-thienyl) $\xrightarrow[\text{2) TFA/}CH_2Cl_2]{\text{1) } Pd(dba)_3\text{, } K_2CO_3}$ HO_2C—C_6H_4—(2-thienyl) 91%

B. Alkenylboranes and alkylboronic acids

11[j] $CH_3(CH_2)_3CH{=}CH$–B(catechol) + $BrCH{=}CHPh$ $\xrightarrow[NaOC_2H_5]{Pd(PPh_3)_4}$ $CH_3(CH_2)_3CH{=}CH{-}CH{=}CHPh$ 86%

12[k] $CH_3(CH_2)_5CH{=}CHB(O\text{-}i\text{-}Pr)_2$ + I—C_6H_5 $\xrightarrow[NaOEt]{Pd(PPh_3)_4}$ $CH_3(CH_2)_5CH{=}CHPh$ 98%

13[l] $CH_3CH_2CH{=}CHBR_2$ + $BrCH{=}CH(CH_2)_8OTHP$ $\xrightarrow[NaOC_2H_5]{Pd(PPh_3)_4}$ $CH_3CH_2CH{=}CH{-}CH{=}CH(CH_2)_8OTHP$ 73%

14[m] CH_3OCH_2, R_3SiO, R_3SiO, R_3SiO, O, OH, I + $(HO)_2B$, $OSiR_3$, R_3SiO, $OSiR_3$, $OSiR_3$ $\xrightarrow[TlOH]{Pd(PPh_3)_4}$

CH_3OCH_2, R_3SiO, O, R_3SiO, R_3SiO, OH, R_3SiO, $OSiR_3$, $OSiR_3$, $OSiR_3$

15[n] CH_3, CH_3, CH_3, $B(OH)_2$ + I, CH_3, $CO_2C_2H_5$ $\xrightarrow[TlOH]{Pd(PPh_3)_4\text{, 7 mol \%}}$

CH_3, CH_3, CH_3, CH_3, $CO_2C_2H_5$ 67%

Scheme 8.11. *(continued)*

16[o] $\xrightarrow[\text{NaCO}_3,\ \text{LiCl}]{\text{Pd(PPh}_3)_4}$ 80%

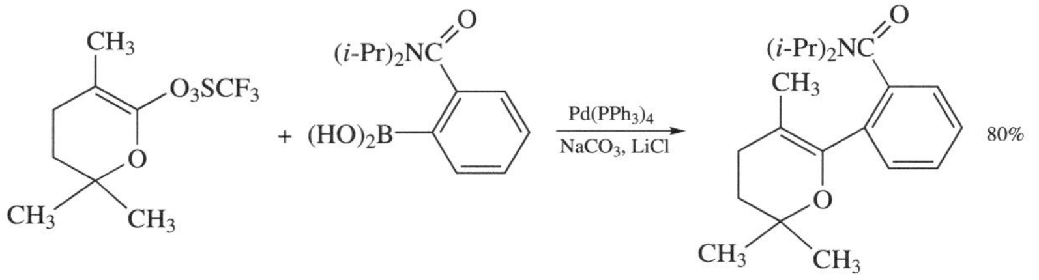

17[p] $\xrightarrow[\text{NaOH}]{\text{Pd(PPh}_3)_4,\ 4\ \text{mol}\ \%}$ 62%

18[q] R = 3-methyl-2-butyl $\xrightarrow{\text{Pd(PPh}_3)_4,\ 7\ \text{mol}\ \%}$ 76%

19[r] $\xrightarrow[\text{Cs}_2\text{CO}_3]{\text{Pd(dppf)}_2,\ \text{Ph}_3\text{As}}$ 72%

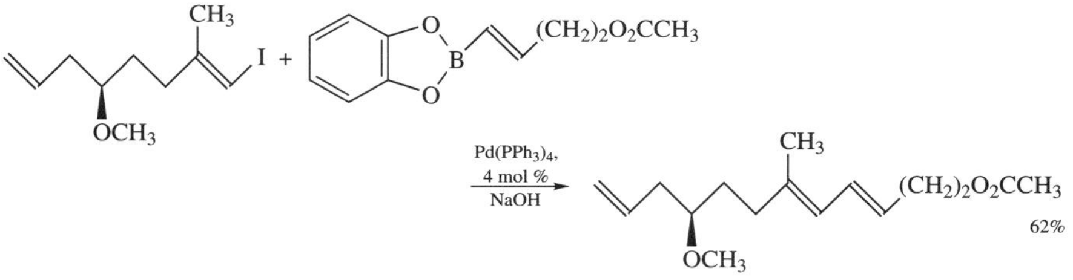

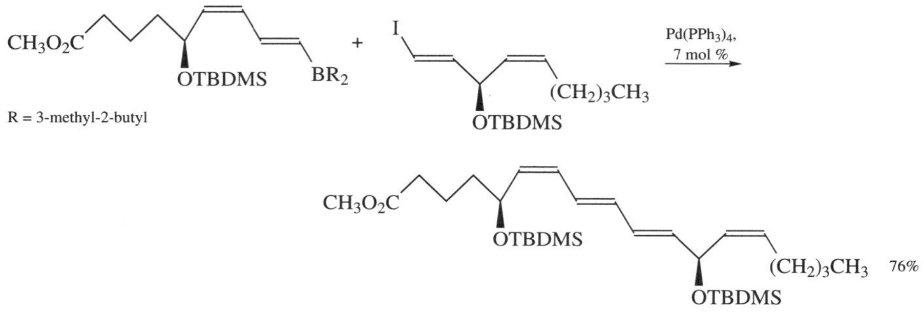

20[s] $\xrightarrow[\text{TlOH}]{\text{Pd(PPh}_3)}$

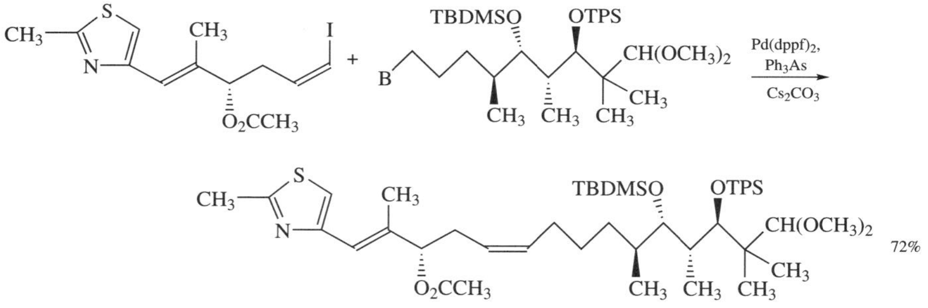

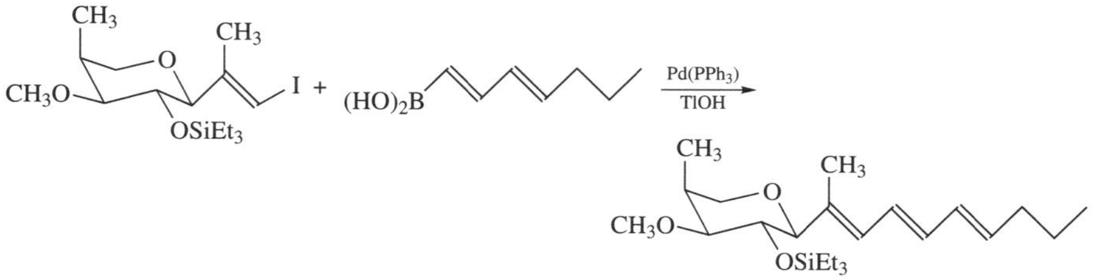

C. Alkyl–aryl coupling

21[t] $PhO(CH_2)_3B$ + CF_3SO_3–C_6H_4–OCH_3 → ($Pd(PPh_3)_4$, 2 mol %, K_3PO_4) $PhO(CH_2)_3$–C_6H_4–OCH_3 92%

22[u] $CH_3(CH_2)_7B$ + I–(benzodioxole) → ($Pd(dppf)Cl_2$, 3 mol %, $NaOCH_3$) $CH_3(CH_2)_7$–(benzodioxole) 78%

23[v] $CH_3(CH_2)_7B$ + OCH_3/I-benzene → ($Pd(dppf)Cl_2$, NaOH) OCH_3/$(CH_2)_7CH_3$-benzene 90%

24[w] C_6H_5–$B(OH)_2$ + $BrCH_2CH{=}CH$–C_6H_5 → ($Pd_2(dba)_3$, K_2CO_3) C_6H_5–$CH_2CH{=}CH$–C_6H_5 73%

a. F. Little and G. C. Fu, *Angew. Chem. Int. Ed. Engl.* **37:**3387 (1998).
b. T. L. Wallow and B. M. Novak, *J. Org. Chem.* **59:**5034 (1994).
c. D. Badone, M. B. R. Cardamone, A. Ielimini, and U. Guzzi, *J. Org. Chem.* **62:**7170 (1997).
d. S. Darses, T. Jeffery, J.-L. Brayer, J.-P. Demoute, and J.-P. Genet, *Bull. Soc. Chim. Fr.* **133:**1095 (1996); S. Sengupta and S. Bhattacharyya, *J. Org. Chem.* **62:**3405 (1997).
e. S. Darses, T. Jeffery, T.-P. Genet, J.-L. Brayer, and J.-P. Demoute, *Tetrahedron Lett.* **37:**3857 (1996).
f. B. I. Alo, A. Kandil, P. A. Patil, M. J. Sharp, M. A. Siddiqui, and V. Snieckus, *J. Org. Chem.* **56:**3763 (1991).
g. J. Sharp and V. Snieckus, *Tetrahedron Lett.* **26:**5997 (1985).
h. M. Ishikura, T. Ohta, and M. Terashima, *Chem. Pharm. Bull.* **33:**4755 (1985).
i. J. W. Guiles, S. G. Johnson, and W. V. Murray, *J. Org. Chem.* **61:**5169 (1996).
j. N. Miyaura, K. Yamada, H. Suginome, and A. Suzuki, *J. Am. Chem. Soc.* **107:**972 (1985).
k. N. Miyaura, M. Satoh, and A. Suzuki, *Tetrahedron Lett.* **27:**3745 (1986).
l. F. Björkling, T. Norin, C. R. Unelius, and R. B. Miller, *J. Org. Chem.* **52:**292 (1987).
m. J. Uenishi, J.-M. Beau, R. W. Armstrong, and Y. Kishi, *J. Am. Chem. Soc.* **109:**4756 (1987).
n. A. R. de Lera, A. Torrado, B. Iglesias, and S. Lopez, *Tetrahedron Lett.* **33:**6205 (1992).
o. M. A. F. Brabdao, A. B. de Oliveira, and V. Snieckus, *Tetrahedron Lett.* **34:**2437 (1993).
p. J. D. White, T. S. Kim, and M. Nambu, *J. Am. Chem. Soc.* **119:**103 (1997).
q. Y. Kobayashi, T. Shimazaki, H. Taguchi, and F. Sato, *J. Org. Chem.* **55:**5324 (1990).
r. D. Meng, P. Bertinato, A. Balog, D.-S. Su, T. Kamenecka, E. J. Sorensen, and S. J. Danishefsky, *J. Am. Chem. Soc.* **119:**103 (1997).
s. A. G. M. Barrett, A. J. Bennett, S. Merzer, M. L. Smith, A. J. P. White, and P. J. Williams, *J. Org. Chem.* **64:**162 (1999).
t. T. Oh-e, N. Miyaura, and A. Suzuki, *J. Org. Chem.* **58:**2201 (1993).
u. N. Miyaura, T. Ishiyama, H. Sasaki, M. Ishikawa, M. Satoh, and A. Suzuki, *J. Am. Chem. Soc.* **111:**314 (1989).
v. N. Miyaura, T. Ishiyama, M. Ishikawa, and A. Suzuki, *Tetrahedron Lett.* **27:**6369 (1986).
w. M. Moreno-Manas, F. Pajuelo, and R. Pleixarts, *J. Org. Chem.* **60:**2396 (1995).

In some synthetic applications, specific bases such as Cs_2CO_3[136] or TlOH[137] have been found preferable to NaOH. Conditions for effecting Suzuki coupling in the absence of phosphine ligands have been developed.[138] One of the potential advantages of the Suzuki reaction, especially when boronic acids are used, is that the by-product boric acid is more innocuous than the tin by-products generated in Stille-type couplings.

136. A. F. Littke and G. C. Fu, *Angew. Chem. Int. Ed. Engl.* **37:**3387 (1998).
137. J. Uenishi, J.-M. Beau, R. W. Armstrong, and Y. Kishi, *J. Am. Chem. Soc.* **109:**4756 (1987); J. C. Anderson, H. Namli, and C. A. Roberts, *Tetrahedron* **53:**15123 (1997).
138. T. L. Wallow and B. M. Novak, *J. Org. Chem.* **59:**5034 (1994); D. Badone, M. Baroni, R. Cardamore, A. Ielmini, and U. Guzzi, *J. Org. Chem.* **62:**7170 (1997).

In addition to aryl halides and triflates, aryldiazonium ions can be the source of the electrophilic component in coupling with arylboronic acids[139] (entries 4 and 5 in Scheme 8.11).

Alkenylboronic acids, alkenyl boronate esters, and alkenylboranes can be coupled with alkenyl halides by palladium catalysts to give dienes.[140]

$$RCH{=}CHBX_2 + R'CH{=}CHY \xrightarrow{Pd(PPh_3)_4} RCH{=}CH{-}CH{=}CHR'$$

X = OH, OR, R
Y = Br, I

These reactions proceed with retention of double-bond configuration in both the boron derivative and the alkenyl halide. The basic steps involve oxidative addition by the alkenyl halide, transfer of an alkenyl group from boron to palladium, and reductive elimination.

$$R'CH{=}CHY \xrightarrow{Pd^0} R'CH{=}CHPd^{II} + RCH{=}CHBX_2 \longrightarrow R'CH{=}CH{-}Pd^{II}{-}CH{=}CHR \longrightarrow R'CH{=}CH{-}CH{=}CHR$$

Alkyl substituents on boron in 9-BBN derivatives can be coupled with both vinyl and aryl halides through the use of Pd catalysts[141] (Entries 21–23, Scheme 8.11). This is an especially interesting reaction because of its ability to effect coupling of saturated alkyl groups. Palladium-catalyzed couplings of alkyl groups by most other methods fail because of the tendency for β elimination.

$$\begin{matrix} Ar{-}X \\ \text{or} \\ R'CH{=}CHX \end{matrix} + RBL_2 \xrightarrow[NaOMe]{Pd} \begin{matrix} Ar{-}R \\ \text{or} \\ R'CH{=}CHR \end{matrix}$$

Both β-alkenylcatecholboranes and alkenyl disiamylboranes couple stereospecifically with alkenyl bromides.[142]

$$THPO(CH_2)_8CH{=}CHBr + (siam)_2BCH{=}CHC_2H_5 \xrightarrow{Pd(PPh_3)_4} THPO(CH_2)_8CH{=}CH{-}CH{=}CHC_2H_5$$

$$THPO(CH_2)_8CH{=}CHBr + \text{(catecholboryl)}CH{=}CHC_2H_5 \xrightarrow{Pd(PPh_3)_4} THPO(CH_2)_8CH{=}CH{-}CH{=}CHC_2H_5$$

139. S. Darses, T. Jeffery, J.-P. Genet, J.-L. Brayer, and J.-P. Demoute, *Tetrahedron Lett.* **37**:3857 (1996); S. Darses, T. Jeffery, J.-L. Brayer, J.-P. Demoute, and J.-P. Genet, *Bull. Soc. Chim. Fr.* **133**:1095 (1996); S. Sengupta and S. Bhatacharyya, *J. Org. Chem.* **62**:3405 (1997).
140. N. Miyaura, K. Yamada, H. Suginome, and A. Suzuki, *J. Am. Chem. Soc.* **107**:972 (1985); N. Miyaura, M. Satoh, and A. Suzuki, *Tetrahedron Lett.* **27**:3745 (1986); F. Bjorkling, T. Norin, C. R. Unelius, and R. B. Miller, *J. Org. Chem.* **52**:292 (1987).
141. N. Miyaura, T. Ishiyama, M. Ishikawa, and A. Suzuki, *Tetrahedron Lett.* **27**:6369 (1986).
142. N. Miyaura, K. Yamada, H. Suginome, and A. Suzuki, *J. Am. Chem. Soc.* **107**:972 (1985); F. Björkling, T. Norin, C. R. Unelius, and R. B. Miller, *J. Org. Chem.* **52**:292 (1987); Y. Satoh, H. Serizawa, N. Miyaura, S. Hara, and A. Suzuki, *Tetrahedron Lett.* **29**:1811 (1988).

When vinylboronic acids are used as reactants, bases, especially $Tl(OH)_3$, can accelerate the reaction.[143] Scheme 8.11 gives a number of examples of coupling using organoborane reagents.

8.2.4. Carbonylation Reactions

Carbonylation reactions have been observed using both Pd(II)–alkene complexes and σ-bonded Pd(II) species. A catalytic process that includes copper(II) results in concomitant addition of nucleophilic solvent. The copper(II) reoxidizes Pd(0) to the Pd(II) state.[144]

2 Cu^{1+} 2 Cu^{2+} Pd^{2+} $CH_2{=}CHR$ Pd^0 $MeOCCH_2CHR(=O)(OMe)$ Pd^{2+} $CH_2{=}CHR$ $Pd^+{-}C(=O)CH_2CHR(OMe)$ MeOH MeOH $Pd^+{-}CH_2CHR(OMe)$ CO

Organopalladium(II) intermediates generated from halides or triflates by oxidative addition react with carbon monoxide in the presence of alcohols to give carboxylic acids[145] or esters.[146]

$$(C_2H_5)(H)C{=}C(C_2H_5)(I) + CO \xrightarrow[n\text{-BuOH}]{Pd(PPh_3)_2I_2} (C_2H_5)(H)C{=}C(C_2H_5)(CO_2C_4H_9) \quad 74\%$$

The carbonyl insertion step takes place by migration of the organic group from the metal to the coordinated carbon monoxide.

$$R{-}Pd{-}C{\equiv}O^+ \longrightarrow Pd^+{-}\overset{O}{\overset{\|}{C}}{-}R \xrightarrow{R'OH} Pd^0 + R'O{-}\overset{O}{\overset{\|}{C}}{-}R + H^+$$

The detailed mechanisms of such reactions have been shown to involve addition and elimination of phosphine ligands. The efficiency of individual reactions can often be improved by careful study of the effect of added ligands.

143. J. Uenishi, J.-M. Beau, R. W. Armstrong, and Y. Kishi, *J. Am. Chem. Soc.* **109:**4756 (1987).
144. D. E. James and J. K. Stille, *J. Am. Chem. Soc.* **98:**1810 (1976).
145. S. Cacchi and A. Lupi, *Tetrahedron Lett.* **33:**3939 (1992).
146. A. Schoenberg, I. Bartoletti, and R. F. Heck, *J. Org. Chem.* **39:**3318 (1974); S. Cacchi, E. Morera, and G. Ortar, *Tetrahedron Lett.* **26:**1109 (1985).

Application of the carbonylation reaction to halides with appropriately placed hydroxyl groups leads to lactone formation. In this case, the acylpalladium intermediate is trapped intramolecularly.

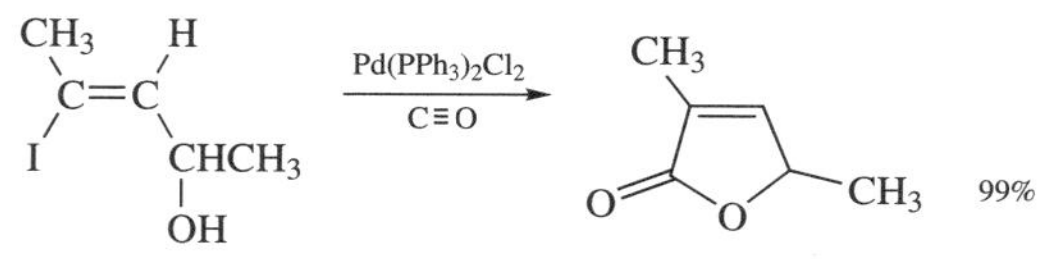
Pd(PPh3)2Cl2
C≡O
99%

Ref. 147

Coupling of organometallic reagents with halides in a carbon monoxide atmosphere leads to ketones by incorporation of a carbonylation step.[148] These reactions involved a migration of one of the organic subsituents to the carbonyl carbon, followed by reductive elimination. These reactions can be carried out with stannanes[149] or boronic acids[150] as the nucleophilic component.

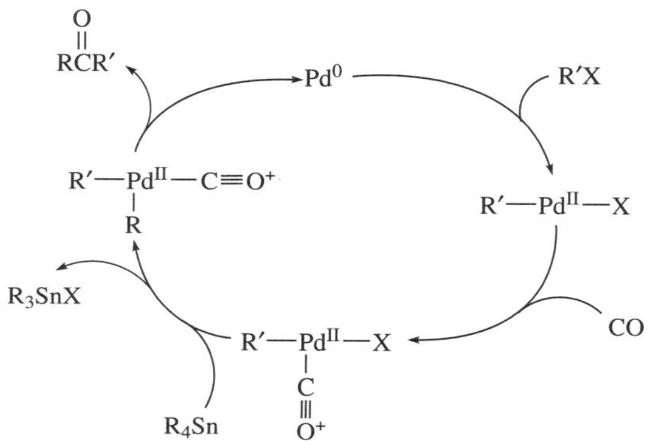
RCR′
Pd0
R′X
R′—PdII—C≡O+
R
R′—PdII—X
R3SnX
CO
R′—PdII—X
R4Sn

This method can also be applied to alkenyl triflates.

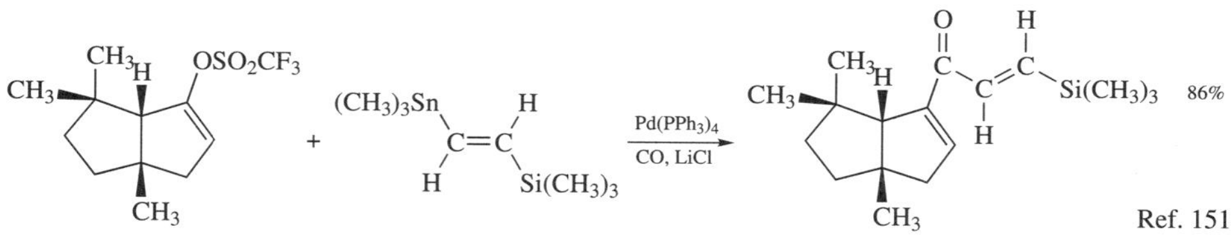
Pd(PPh3)4
CO, LiCl
86%

Ref. 151

Carbonylation can also be carried out as a tandem reaction in intramolecular Heck

147. A. Cowell and J. K. Stille, *J. Am. Chem. Soc.* **102:**4193 (1980).
148. M. Tanaka, *Tetrahedron Lett.* **1979:**2601.
149. A. M. Echavarren and J. K. Stille, *J. Am. Chem. Soc.* **110:**1557 (1988).
150. T. Ishiyama, H. Kizaki, N. Miyaura, and A. Suzuki, *Tetrahedron Lett.* **34:**7595 (1993); T. Ishiyama, H. Kizaki, T. Hayashi, A. Suzuki, and N. Miyaura, *J. Org. Chem.* **63:**4726 (1998).
151. G. T. Crisp, W. J. Scott, and J. K. Stille, *J. Am. Chem. Soc.* **106:**7500 (1984).

Scheme 8.12. Synthesis of Ketones, Esters, Acids, and Amides by Palladium-Catalyzed Acylation and Carbonylation

A. Ketones from acyl halides

1[a] $O_2N-C_6H_4-COCl + (CH_3)_3Sn-C_6H_5 \xrightarrow[18\ h]{PhCH_2PdCl/PPh_3} O_2N-C_6H_4-C(=O)-C_6H_5$ 97%

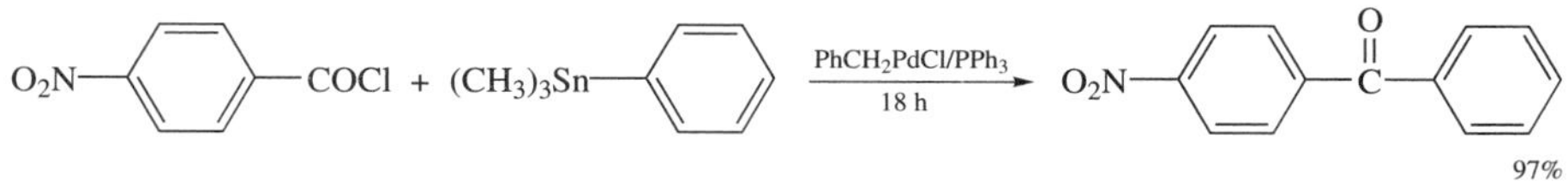

2[a] $C_6H_5-COCl + (CH_3)_3SnC\equiv CC_3H_7 \xrightarrow[23\ h]{PhCH_2PdCl/PPh_3} C_6H_5-C(=O)C\equiv CC_3H_7$ 70%

3[b] $(CH_3)_2C{=}CHCOCl + (n\text{-Bu})_3Sn\text{-cyclopentenyl} \xrightarrow{PdCl_2,\ PPh_3} (CH_3)_2C{=}CHC(=O)\text{-cyclopentenyl}$ 85%

4[c] $CH_3C(=O)NHCH(CO_2C_2H_5)(CH_2)_5COCl + (CH_2{=}CH)_4Sn \xrightarrow{PhCH_2PdCl/PPh_3} CH_3C(=O)NHCH(CO_2C_2H_5)(CH_2)_5C(=O)CH{=}CH_2$ 70%

5[d] $O_2N-C_6H_4-C(=O)Cl + (n\text{-Bu})_3SnCH{=}CHCO_2C_2H_5 \xrightarrow[CO]{PhCH_2Pd(PPh_3)_2Cl,\ 0.7\ mol\ \%} O_2N-C_6H_4-C(=O)CH{=}CHCO_2C_2H_5$ 80%

6[e] $PhCH(O_2CCH_3)Sn(n\text{-Bu})_3 + PhCOCl \xrightarrow[76°C]{Pd(PPh_3)_2Cl_2,\ CuCN} PhCH(O_2CCH_3)C(=O)Ph$ 78%

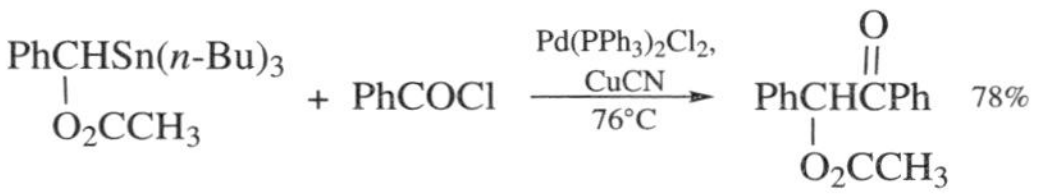

B. Ketones by carbonylation

7[f] $Br-C_6H_4-I + (HO)_2B-C_6H_5 \xrightarrow[CO,\ K_2CO_3]{Pd(PPh_3)_2Cl_2,\ 3\ mol\ \%} Br-C_6H_4-C(=O)-C_6H_5$ 86%

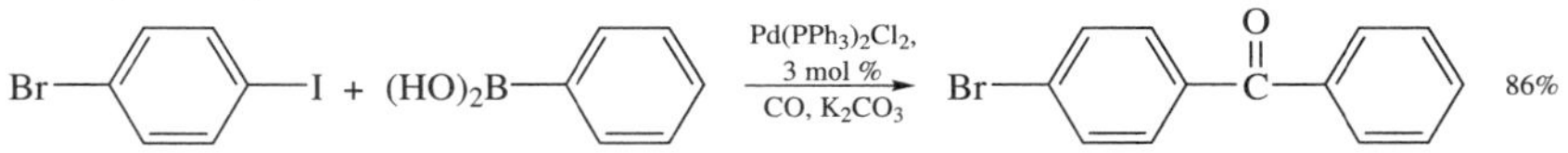

8[b] $\text{cyclohexenyl}-I + (n\text{-Bu})_3SnCH{=}CH_2 \xrightarrow[CO,\ 50°C]{PhCH_2PdCl/PPh_3} \text{cyclohexenyl}-C(=O)CH{=}CH_2$ 93%

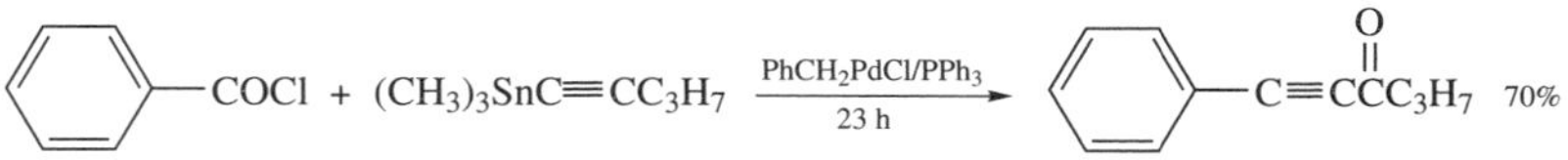

9[g] $(CH_3)_2C{=}CHCH_2Cl + \text{3-(Sn(CH}_3)_3\text{)furan} \xrightarrow[CO]{PhCH_2PdCl/PPh_3} \text{3-furyl}-C(=O)CH_2CH{=}C(CH_3)_2$ 75%

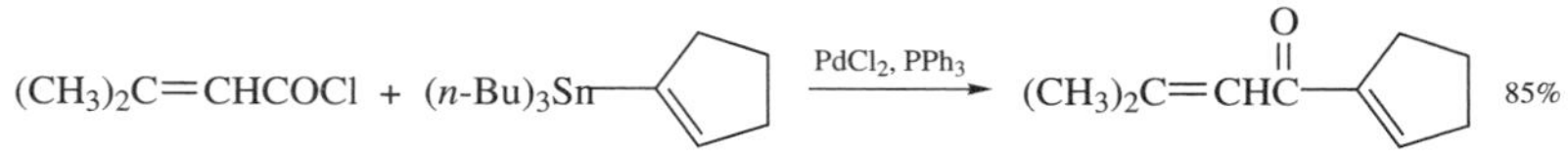

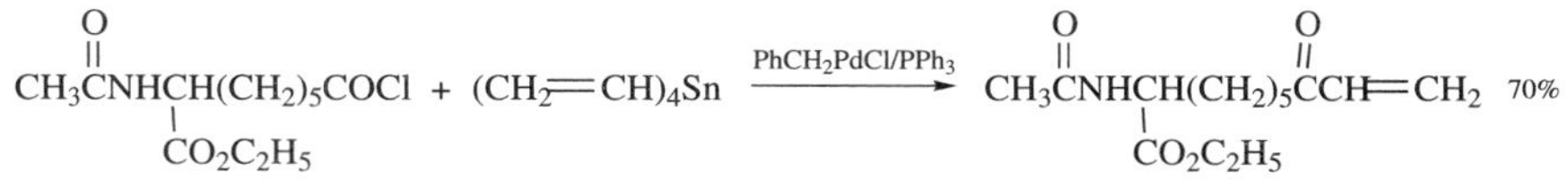

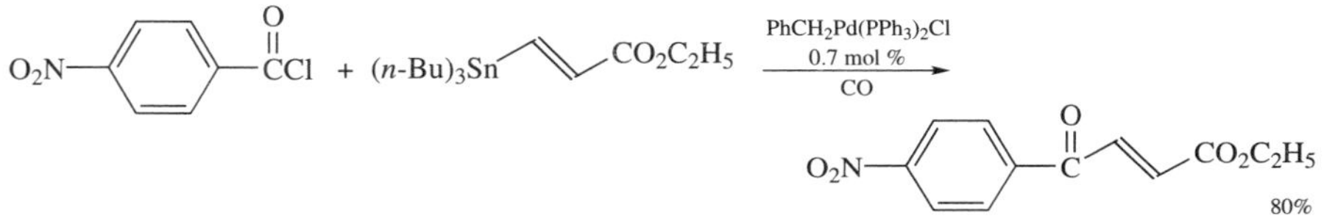

10[h] 2-iodophenyl-substituted triazinone (N–CH₃, N–CH₃, C=O) + (CH₃)₃Sn-cyclopentenyl bearing C(CH₂OTIPS)=CHCH₃ $\xrightarrow[CO,\ LiCl,\ THF]{Pd_2(dba)_3,\ 2.5\ mol\ \%;\ Ph_3As,\ 2.2\ mol\ \%}$ coupled aryl-cyclopentenyl product (CH₂OTIPS, CH₃; triazinone with CH₃, CH₃, O) 85%

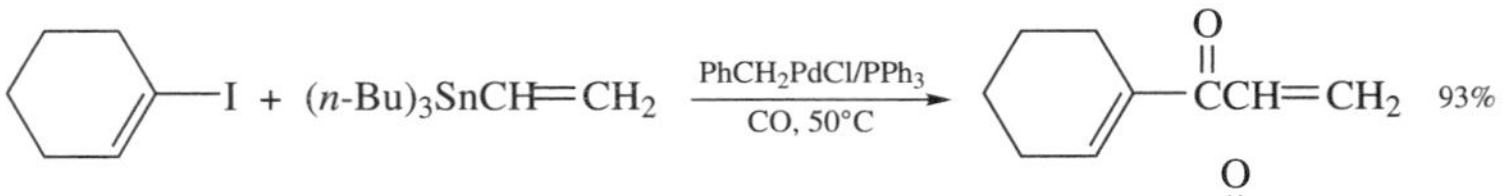

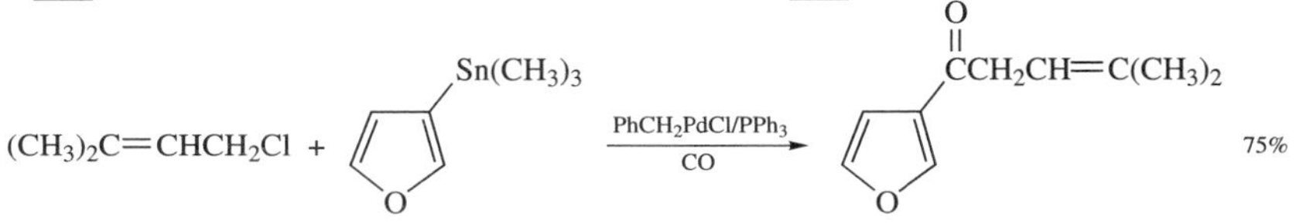

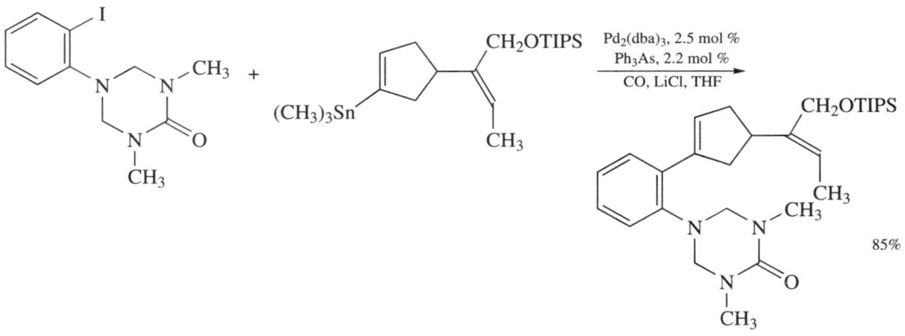

(*continued*)

Scheme 8.12. *(continued)*

C. Esters, acids and amides

11[i] Ph–(cyclohexenyl)–O_3SCF_3 $\xrightarrow{Pd(PPh_3)_2(OAc)_2,\ CO,\ NaOAc}$ Ph–(cyclohexenyl)–CO_2H 82%

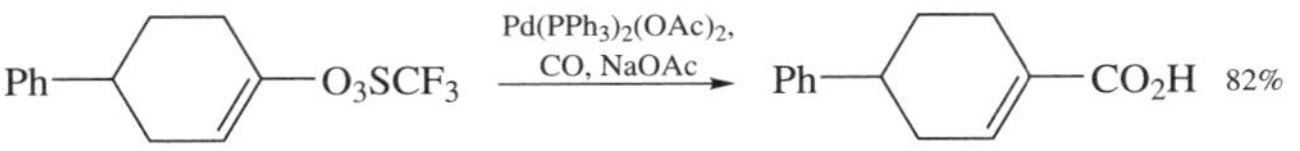

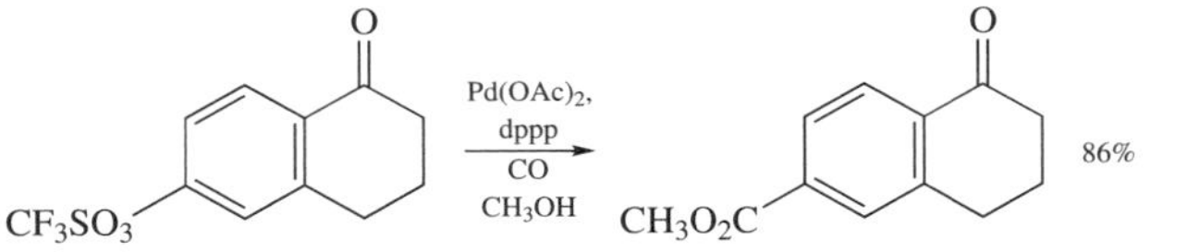

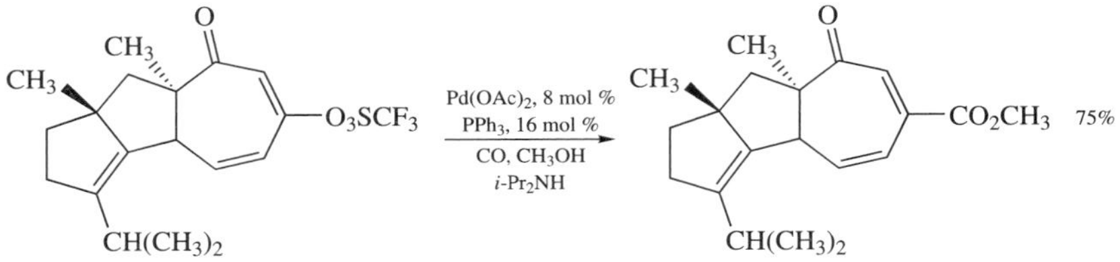

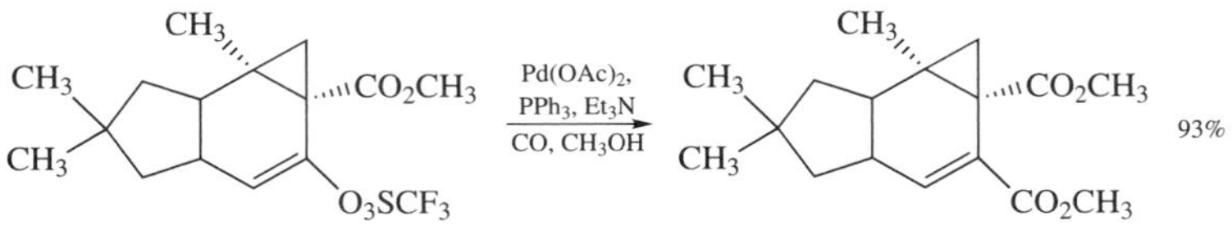

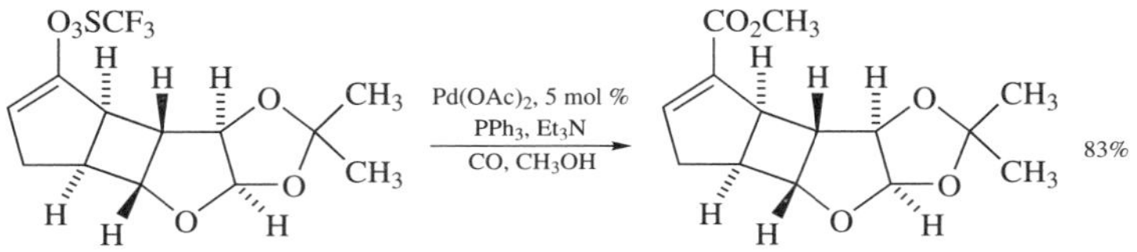

12[j] CF_3SO_3–(tetralone) $\xrightarrow{Pd(OAc)_2,\ dppp,\ CO,\ CH_3OH}$ CH_3O_2C–(tetralone) 86%

13[k] $\xrightarrow{Pd(OAc)_2,\ 8\ mol\ \%,\ PPh_3,\ 16\ mol\ \%,\ CO,\ CH_3OH,\ i\text{-}Pr_2NH}$ (O_3SCF_3 → CO_2CH_3; CH_3, CH_3, $CH(CH_3)_2$) 75%

14[l] $\xrightarrow{Pd(OAc)_2,\ PPh_3,\ Et_3N,\ CO,\ CH_3OH}$ (O_3SCF_3 → CO_2CH_3; CH_3, CH_3, CH_3, CO_2CH_3) 93%

15[m] $\xrightarrow{Pd(OAc)_2,\ 5\ mol\ \%,\ PPh_3,\ Et_3N,\ CO,\ CH_3OH}$ (O_3SCF_3 → CO_2CH_3; H, H, H, H, H, H, O, O, O, CH_3, CH_3) 83%

16[n] CH_3O–C_6H_4–I $\xrightarrow{Pd(PPh_3)_2Cl_2,\ [(CH_3)_3Si]_2NH,\ CO,\ DMF,\ 80°C}$ CH_3O–C_6H_4–$CONH_2$

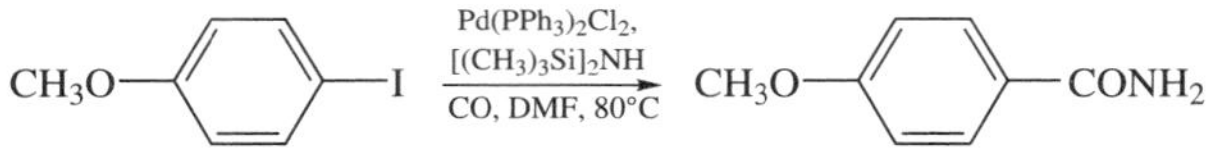

a. J. W. Labadie, D. Tueting, and J. K. Stille, *J. Org. Chem.* **48:**4634 (1983).
b. W. F. Goure, M. E. Wright, P. D. Davis, S. S. Labadie, and J. K. Stille, *J. Am. Chem. Soc.* **106:**6417 (1984).
c. D. H. Rich, J. Singh, and J. H. Gardner, *J. Org. Chem.* **48:**432 (1983).
d. A. F. Renaldo, J. W. Labadie, and J. K. Stille, *Org. Synth.* **67:**86 (1988).
e. J. Ye, R. K. Bhatt, and J. R. Falck, *J. Am. Chem. Soc.* **116:**1 (1994).
f. T. Ishiyama, H. Kizaki, T. Hayashi, A. Suzuki, and N. Miyaura, *J. Org. Chem.* **63:**4726 (1998).
g. F. K. Sheffy, J. P. Godschalx, and J. K. Stille, *J. Am. Chem. Soc.* **106:**4833 (1984).
h. S. R. Angle, J. M. Fevig, S. D. Knight, R. W. Marquis, Jr., and L. E. Overman, *J. Am. Chem. Soc.* **115:**3966 (1993).
i. S. Cacchi and A. Lupi. *Tetrahedron Lett.* **33:**3939 (1992).
j. U. Gerlach and T. Wollmann, *Tetrahedron Lett.* **33:**5499 (1992).
k. B. B. Snider, N. H. Vo, and S. V. O'Neil, *J. Org. Chem.* **63:**4732 (1998).
l. S. K. Thompson and C. H. Heathcock, *J. Org. Chem.* **55:**3004 (1990).
m. A. B. Smith III, G. A. Sulikowski, M. M. Sulikowski, and K. Fujimoto, *J. Am. Chem. Soc.* **114:**2567 (1992).
n. E. Morera and G. Ortar, *Tetrahedron Lett.* **39:**2835 (1998).

$$\xrightarrow[\text{CO, } CH_3OH]{Pd(PPh_3)_2Cl_2\text{, 5 mol \%, 3 equiv TlOAc}}$$ 86% Ref. 152

Procedures for synthesis of ketones based on coupling of organostannanes with acyl chlorides have also been developed.[153] The catalytic cycle is similar to that involved in the coupling with alkyl or aryl halides. The scope of compounds to which the procedure can be applied is wide and includes successful results with tetra-*n*-butylstannane. This example implies that the reductive elimination step in the mechanism can compete successfully with β-elimination.

RCR′ (=O) ← RC(=O)—Pd^{II}—R′ → Pd^0; Pd^0 + RCCl (=O) → RC(=O)—Pd^{II}—Cl; RC(=O)—Pd^{II}—Cl + R'_4Sn → RC(=O)—Pd^{II}—R′ + R'_3SnCl

Scheme 8.12 gives some examples of these palladium-based ketone syntheses.

Carbonylation can also be carried out via *in situ* generation of other types of electrophiles. For example, good yields of *N*-acyl α-amino acids are obtained in a process in which an amide and aldehyde combine to generate a carbinolamide and, presumably, an acyliminium ion. The organopalladium intermediate is then carbonylated.[154]

$$RCH{=}O + CH_3CONH_2 \xrightarrow[\text{NMP}]{Pd(PPh_3)_2Br_2,\ H^+,\ LiBr,\ CO} RCH(NHCOCH_3)CO_2H$$

8.3. Reactions Involving Organonickel Compounds

The original synthetic processes using organonickel compounds involved the coupling of halides. Allylic halides react with nickel carbonyl, $Ni(CO)_4$, to give π-allyl

152. R. Grigg, P. Kennewell, and A. J. Teasdale, *Tetrahedron Lett.* **33:**7789 (1992).
153. D. Milstein and J. K. Stille, *J. Org. Chem.* **44:**1613 (1979); J. W. Labadie and J. K. Stille, *J. Am. Chem. Soc.* **105:**6129 (1983).
154. M. Beller, M. Eckert, F. M. Vollmüller, S. Bogdanovic, and H. Geissler, *Angew. Chem. Int. Ed. Engl.* **36:**1494 (1997); M. Beller, W. A. Maradi, M. Eckert, and H. Neumann, *Tetrahedron Lett.* **40:**4523 (1999).

complexes. These complexes react with a variety of halides to give coupling products.[155]

$$2\ CH_2{=}CHCH_2Br + 2\ Ni(CO)_4 \longrightarrow [(\eta^3\text{-allyl})Ni(\mu\text{-Br})]_2$$

$CH_2{=}CHBr + [(CH_2{=}CH{=}CH_2)NiBr]_2 \longrightarrow CH_2{=}CHCH_2CH{=}CH_2$ 70% Ref. 156

$C_6H_{11}I + [(CH_2{=}CH{=}CH_2)NiBr]_2 \longrightarrow C_6H_{11}CH_2CH{=}CH_2$ 91%

These coupling reactions are believed to involve Ni(I) and Ni(III) intermediates in a chain process which is initiated by formation of a small amount of a Ni(I) species.[157]

Ni^{II} e$^-$ —Ni^I— Ni^I RCH$_2$CH=CH$_2$ R Ni^{III} R—X

Nickel carbonyl effects coupling of allylic halides when the reaction is carried out in very polar solvents such as DMF or DMSO. This coupling reaction has been used intramolecularly to bring about cyclization of bis-allylic halides and was found useful in the preparation of large rings.

$$BrCH_2CH{=}CH(CH_2)_{12}CH{=}CHCH_2Br \xrightarrow{Ni(CO)_4} \text{cycloalkadiene}\quad 76\text{–}84\%$$

Ref. 158

$BrCH_2CH{=}CHCH_2CH_2C(=O)OCH_2CH_2CH_2CH{=}CHCH_2Br \xrightarrow{Ni(CO)_4}$ macrocyclic lactone 70–75% Ref. 159

Nickel carbonyl is an extremely toxic compound, and a number of other nickel reagents with generally similar reactivity can be used in its place. The Ni(0) complex of 1,5-

155. M. F. Semmelhack, *Org. React.* **19:**115 (1972).
156. E. J. Corey and M. F. Semmelhack, *J. Am. Chem. Soc.* **89:**2755 (1967).
157. L. S. Hegedus and D. H. P. Thompson, *J. Am. Chem. Soc.* **107:**5663 (1985).
158. E. J. Corey and E. K. W. Wat, *J. Am. Chem. Soc.* **89:**2757 (1967).
159. E. J. Corey and H. A. Kirst, *J. Am. Chem. Soc.* **94:**667 (1972).

cyclooctadiene, $Ni(COD)_2$, has been found to bring about coupling of allylic, alkenyl, and aryl halides.

$$PhCH{=}CHBr \xrightarrow{Ni(COD)_2} PhCH{=}CH{-}CH{=}CHPh \quad 46\% \qquad \text{Ref. 160}$$

$$N{\equiv}C{-}C_6H_4{-}Br \xrightarrow{Ni(COD)_2} N{\equiv}C{-}C_6H_4{-}C_6H_4{-}C{\equiv}N \quad 81\% \qquad \text{Ref. 161}$$

Tetrakis(triphenylphoshphine)nickel(0) is also an effective reagent for coupling aryl halides.[162] Medium-sized rings can be formed in intramolecular reactions.

$$\text{Ref. 163} \quad Ni(PPh_3)_4$$

The coupling of aryl halides and triflates can be made catalytic in nickel by using zinc as a reductant for *in situ* regeneration of the active Ni(0) species.

$$O{=}CH{-}C_6H_4{-}Cl \xrightarrow[\text{NiCl}_2\text{ (5 mol \%)},\ \text{PPh}_3\text{ (5 mol \%)}]{\text{Zn, NaBr}} O{=}CH{-}C_6H_4{-}C_6H_4{-}CH{=}O \quad 62\% \qquad \text{Ref. 164}$$

$$CH_3O{-}C_6H_4{-}O_3SCF_3 \xrightarrow[\text{Zn, KI}]{\text{Ni(dppe)Cl}_2\text{, 10 mol \%}} CH_3O{-}C_6H_4{-}C_6H_4{-}OCH_3 \qquad \text{Ref. 165}$$

Mechanistic study of the aryl couplings has revealed the importance of the changes in redox state which are involved in the reaction.[166] Ni(I), Ni(II), and Ni(III) states are believed to be involved. Changes in the degree of coordination by phosphine ligands are also believed to be involved, but these have been omitted in the mechanism shown here. The detailed kinetics of the reaction are inconsistent with a mechanism involving only

160. M. F. Semmelhack, P. M. Helquist, and J. D. Gorzynski, *J. Am. Chem. Soc.* **94:**9234 (1972).
161. M. F. Semmelhack, P. M. Helquist, and L. D. Jones, *J. Am. Chem. Soc.* **93:**5908 (1971).
162. A. S. Kende, L. S. Liebseskind, and D. M. Braitsch, *Tetrahedron Lett.* **1975:**3375.
163. S. Brandt, A. Marfat, and P. Helquist, *Tetrahedron Lett.* **1979:**2193.
164. M. Zembayashi, K. Tamao, J. Yoshida, and M. Kumada, *Tetrahedron Lett.* **1977:**4089; I. Colon and D. R. Kelly, *J. Org. Chem.* **51:**2627 (1986).
165. A. Jutand and A. Mosleh, *J. Org. Chem.* **62:**261 (1997).
166. T. T. Tsou and J. K. Kochi, *J. Am. Chem. Soc.* **101:**7547 (1979); C. Amatore and A. Jutland, *Organometallics* **7:**2203 (1988).

formation and decomposition of a biarylnickel(II) intermediate.

initiation by electron transfer $ArNi(II)X + ArX \longrightarrow ArNi(III)X^+ + Ar\cdot + X^-$

propagation $ArNi(III)X^+ + ArNi(II)X \longrightarrow Ar_2Ni(III)X + Ni^{2+} + X^+$

$Ar_2Ni(III)X \longrightarrow Ar\text{-}Ar + Ni(I)X$

$Ni(I)X + ArX \longrightarrow ArNi(III)X^+ + X^-$

The key aspects of the mechanism are (1) the reductive elimination which occurs via a diaryl Ni(III) intermediate and (2) the oxidative addition which involves a Ni(I) species.

Nickel(II) salts are able to catalyze the coupling of Grignard reagents with alkenyl and aryl halides. A soluble bis-phosphine comples, $Ni(dppe)_2Cl_2$, is a particularly effective catalyst.[167] The main distinction between this reaction and Pd-catalyzed cross coupling is that the nickel reaction can be more readily applied to saturated alkyl groups because of a reduced tendency for β-elimination.

m-Dichlorobenzene (Cl, Cl) + $CH_3CH_2CH_2CH_2MgBr$ $\xrightarrow{Ni(dppe)_2Cl_2}$ m-di($CH_2CH_2CH_2CH_3$)benzene 94%

The reaction has been applied to the synthesis of cyclophane-type structures by use of dihaloarenes and Grignard reagents from α,ω-dihalides.

m-Dichlorobenzene (Cl, Cl) + $BrMg(CH_2)_{12}MgBr$ $\xrightarrow{Ni(dppe)_2Cl_2}$ cyclophane (CH_2–$(CH_2)_{10}$–CH_2) 18%

When secondary Grignard reagents are used, the coupling product sometimes is derived from the corresponding primary alkyl group.[169] This transformation can occur by reversible formation of a nickel–alkene complex from the σ-bonded alkyl group. Re-formation of the σ-bonded structure will be preferred at the less hindered primary position.

$(R_3P)_2Ni(Ph)Cl + (CH_3)_2CHMgX \longrightarrow (R_3P)_2Ni(Ph)\text{—}CH(CH_3)_2 \rightleftharpoons (R_3P)_2Ni(Ph)(H)(CH_2{=}CH\text{—}CH_3)$

$\rightleftharpoons R_3P\text{—}Ni(PR_3)(Ph)\text{—}CH_2CH_2CH_3 \longrightarrow PhCH_2CH_2CH_3$

167. K. Tamao, K. Sumitani, and M. Kumada, *J. Am. Chem. Soc.* **94**:4374 (1972).
168. K. Tamao, S. Kodama, T. Nakatsuka, Y. Kiso, and A. Kumada, *J. Am. Chem. Soc.* **97**:4405 (1975).
169. K. Tamao, Y. Kiso, K. Sumitani, and M. Kumada, *J. Am. Chem. Soc.* **94**:9268 (1972).

Nickel acetylacetonate, $Ni(acac)_2$, in the presence of a styrene derivative promotes coupling of primary alkyl iodides with organozinc reagents. The added styrene serves to stabilize the active catalytic species, and among the styrene derivatives examined, *m*-trifluoromethylstyrene was the best.[170]

$$\text{(piperidin-1-yl)}-\overset{O}{\overset{\|}{C}}CH_2CH_2I + (n\text{-}C_5H_{11})_2Zn \xrightarrow[m\text{-}CF_3C_6H_4CH{=}CH_2]{Ni(acac)_2} \text{(piperidin-1-yl)}-\overset{O}{\overset{\|}{C}}(CH_2)_6CH_3 \quad 70\%$$

This method can extend Ni-catalyzed cross coupling to functionalized organometallic reagents.

Nickel can also be used in place of Pd in Suzuki-type couplings of boronic acids. The main advantage of nickel in this application is that it reacts more readily with aryl chlorides[171] and methanesulfonates[172] than does the Pd system. These reactants may be more economical than iodides or triflates in large-scale syntheses.

$$CH_3-C_6H_4-B(OH)_2 + CH_3SO_3-C_6H_4-CN \xrightarrow[3\text{ equiv }K_2CO_3]{Ni(dppf)_2Cl_2\ 4\text{ mol }\%} CH_3-C_6H_4-C_6H_4-CN \quad 97\%$$

Similarly, nickel catalysis permits the extension of cross coupling to vinyl phosphates, which are in some cases more readily obtained and handled than vinyl triflates.[173]

$$\text{1-}OPO(OPh)_2\text{-cycloheptadiene} + PhMgBr \xrightarrow[1\text{ mol }\%]{Ni(dppe)_2Cl_2} \text{1-Ph-cycloheptadiene} \quad 92\%$$

8.4. Reactions Involving Rhodium and Cobalt

Rhodium and cobalt participate in several reactions which are of value in organic synthesis. Rhodium and cobalt are active catalysts for the reaction of alkenes with hydrogen and carbon monoxide to give aldehydes. This reaction is called *hydro-*

170. R. Giovannini, T. Studemann, G. Dussin, and P. Knochel, *Angew. Chem. Int. Ed. Engl.* **37:**2387 (1998); R. Giovannini, T. Studemann, A. Devasagayaraj, G. Dussin, and P. Knochel, *J. Org. Chem.* **64:**3544 (1999).
171. S. Saito, M. Sakai, and N. Miyaura, *Tetrahedron Lett.* **37:**2993 (1996); S. Saito, S. Oh-tani, and N. Miyaura, *J. Org. Chem.* **62:**8024 (1997).
172. V. Percec, J.-Y. Bae, and D. H. Hill, *J. Org. Chem.* **60:**1060 (1995); M. Ueda, A. Saitoh, S. Oh-tani, and N. Miyaura, *Tetrahedron* **54:**13079 (1998).
173. A. Sofia, E. Karlström, K. Itami, and J.-E. Bäckvall, *J. Org. Chem.* **64:**1745 (1999).

formylation.[174]

cyclohexene + $CO + H_2$ $\xrightarrow[\text{100°C, 50–150 atm}]{Rh_2O_3}$ cyclohexyl–$CH{=}O$ 82–84% Ref. 175

cyclohexyl–$CH{=}CH_2$ $\xrightarrow[\text{2,6-dimethyl-}B\text{-cyclodextrin}]{[Rh(acac)(CO)_2],\ P(C_6H_5SO_3Na)_3}$ cyclohexyl–$CH_2CH_2CH{=}O$ + cyclohexyl–$CH(CH_3)CH{=}O$ 100% yield, 3.2:1 ratio Ref. 176

The key steps in the reaction are addition of hydridorhodium to the double bond of the alkene and migration of the alkyl group to the complexed carbon monoxide.

cyclohexene + HRh(CO) → cyclohexyl–$Rh{-}C{\equiv}O^+$ ⟶ $Rh{-}C({=}O){-}$cyclohexyl $\xrightarrow{H_2}$ $Rh{-}H + HC({=}O){-}$cyclohexyl

Carbonylation can also be carried out under conditions in which the acylrhodium intermediate is trapped by internal nucleophiles.

$CH_3CH(NH_2)CH_2CH{=}CH_2$ $\xrightarrow[\text{CO, }H_2,\ C_2H_5OH]{Rh(OAc)_2,\ PPh_3}$ 6-methyl-2-piperidone (CH_3, N–H, =O) + 3-methyl-2-pyrrolidone (CH_3, N–H, =O) 80% yield, 70:30 ratio Ref. 177

The steps in the hydroformylation reaction are closely related to those that occur in the *Fischer-Tropsch process*. The Fischer–Tropsch process is the reductive conversion of carbon monoxide to alkanes. It occurs by a repetitive series of carbonylation, migration, and reduction steps which can build up a hydrocarbon chain.

$$M + CO \longrightarrow M{-}CO \xrightarrow{+H_2} M{-}CH_3 \xrightarrow{+CO} OC{-}M{-}CH_3$$

$$OC{-}M{-}CH_3 \longrightarrow M{-}\overset{O}{\overset{\|}{C}}{-}CH_3 \xrightarrow{+H_2} M{-}CH_2CH_3 \quad \text{etc.}$$

The Fischer–Tropsch process is of great economic interest because it is the basis of conversion of carbon monoxide to synthetic hydrocarbon fuels, and extensive work has been done on optimization of catalyst systems.

174. R. L. Pruett, *Adv. Organometal. Chem.* **17:**1 (1979); H. Siegel and W. Himmele, *Angew. Chem. Int. Ed. Engl.* **19:**178 (1980); J. Falbe, *New Syntheses with Carbon Monoxide*, Springer-Verlag, Berlin, 1980.
175. P. Pino and C. Botteghi, *Org. Synth.* **57:**11 (1977).
176. E. Monflier, S. Tilloy, G. Fremey, Y. Castanet, and A. Mortreaux, *Tetrahedron Lett.* **36:**9481 (1995); E. Monflier, G. Fremy, Y. Castanet, and A. Mortreaux, *Angew. Chem. Int. Ed. Engl.* **34:**2269 (1995).
177. D. Anastasiou and W. R. Jackson, *Tetrahedron Lett.* **31:**4795 (1990).

The carbonylation step which is involved in both hydroformylation and the Fischer–Tropsch reaction can be reversible. Under appropriate conditions, rhodium catalyst can be used for the decarbonylation of aldehydes[178] and acyl chlorides.[179]

$$RC(=O)H + Rh(PPh_3)_3Cl \longrightarrow RH \qquad RC(=O)Cl + Rh(PPh_3)_3Cl \longrightarrow RCl$$

An acylrhodium intermediated is involved in both cases. The elimination of the hydrocarbon or halide occurs by reductive elimination.[180]

$$RC(=O)X + Rh(PPh_3)_3Cl \longrightarrow RC(=O){-}Rh(Cl)(X)(PPh_3)_2 \longrightarrow R{-}Rh(Cl)(X)(PPh_3)_2 + CO$$

$$R{-}Rh(Cl)(X)(PPh_3)_2 \longrightarrow R{-}X + Rh(PPh_3)_2Cl \qquad X = H, Cl$$

Although the very early studies of transition-metal-catalyzed coupling of organometallic reagents included Co salts, the use of cobalt for synthetic purposes is quite limited. Vinyl bromides and iodides couple with Grignard reagents in good yield, but a good donor solvent such as NMP or DMPU is required as a co-catalyst.

$$PhCH{=}CHBr + C_6H_{11}MgCl \xrightarrow[\text{THF, 4 equiv NMP}]{Co(acac)_2\text{, 3 mol \%}} PhCH{=}CH{-}C_6H_{11} \quad 87\% \qquad \text{Ref. 181}$$

$Co(acac)_2$ also catalyzes cross coupling of organozinc reagents under these conditions.[182]

$$CH_3(CH_2)_5CH{=}CHI + CH_3(CH_2)_3ZnI \xrightarrow[\text{THF, NMP}]{Co(acac)_2\text{, 20 mol \%}} CH_3(CH_2)_5CH{=}CH(CH_2)_3CH_3 \quad 80\%$$

8.5. Organometallic Compounds with π Bonding

The organometallic intermediates discussed in the previous sections have in most cases involved carbon–metal σ bonds, although examples of π bonding with alkenes and allyl groups were also encountered. The compounds which are emphasized in this section involve organic groups that are bound to the metal through delocalized π systems. Among the classes of organic compounds that serve as π ligands are alkenes, allyl groups, dienes,

178. J. A. Kampmeier, S. H. Harris, and D. K. Wedgaertner, *J. Org. Chem.* **45**:315 (1980); J. M. O'Connor and J. Ma, *J. Org. Chem.* **57**:5074 (1992).
179. J. K. Stille and M. T. Regan, *J. Am. Chem. Soc.* **96**:1508 (1974); J. K. Stille and R. W. Fries, *J. Am. Chem. Soc.* **96**:1514 (1974).
180. J. E. Baldwin, T. C. Barden, R. L. Pugh, and W. C. Widdison, *J. Org. Chem.* **52**:3303 (1987).
181. G. Cahiez and H. Avedissian, *Tetrahedron Lett.* **39**:6159 (1998).
182. H. Avedissian, L. Berillon, G. Cahiez, and P. Knochel, *Tetrahedron Lett.* **39**:6163 (1998).

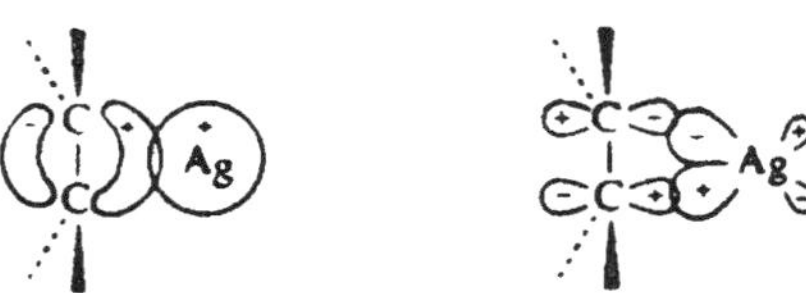

Fig. 8.1. Representation of π bonding in alkene–transition-metal complexes.

the cyclopentadienide anion, and aromatic compounds. There are many such organometallic compounds, and we can illustrate only a few examples.

The bonding in π complexes of alkenes is the result of two major contributions. The filled π orbital acts as an electron donor to empty d orbitals of the metal ion. There is also a contribution to bonding, called "back-bonding," from a filled metal orbital interacting with the alkene π^* orbital. These two types of bonding are represented in Fig. 8.1.

These same general bonding concepts apply to all the other π organometallics. The details of structure and reactivity of the individual compound depend on such factors as (a) the number of electrons that can be accommodated by the metal orbitals, (b) the oxidation level of the metal, and (c) the electronic character of other ligands on the metal.

Alkene–metal complexes are usually prepared by a process in which some other ligand is dissociated from the metal. Both thermal and photochemical reactions are used.

$$(C_6H_5CN)_2PdCl_2 + 2\ RCH{=}CH_2 \longrightarrow [(RCH{=}CH_2)PdCl(\mu\text{-}Cl)]_2$$ Ref. 183

$$[(OC)_2Rh(\mu\text{-}Cl)]_2 + 2\ \text{cyclooctadiene} \longrightarrow [(\text{COD})Rh(\mu\text{-}Cl)]_2$$ Ref. 184

π-Allyl complexes of nickel can be prepared either by oxidative addition on Ni(0) or by transmetalation of a Ni(II) salt.

$$2\ CH_2{=}CHCH_2Br + 2\ Ni(CO)_4 \longrightarrow [(\pi\text{-allyl})Ni(\mu\text{-}Br)]_2 + 8\ CO$$ Ref. 185

$$2\ CH_2{=}CHCH_2MgBr + NiBr_2 \longrightarrow (\pi\text{-allyl})_2Ni + 2\ MgBr_2$$ Ref. 186

Organic ligands with a cyclic array of four carbon atoms have been of particular interest in connection with the chemistry of cyclobutadiene. Organometallic compounds containing cyclobutadiene as a ligand were first prepared in 1965.[187] The carbocyclic ring

183. M. S. Kharasch, R. C. Seyler, and F. R. Mayo, *J. Am. Chem. Soc.* **60:**882 (1938).
184. J. Chatt and L. M. Venanzi, *J. Chem. Soc.* **1957:**4735.
185. E. J. Corey and M. F. Semmelhack, *J. Am. Chem. Soc.* **89:**2755 (1967).
186. D. Walter and G. Wilke, *Angew. Chem. Int. Ed. Engl.* **5:**151 (1966).
187. G. F. Emerson, L. Watts, and R. Pettit, *J. Am. Chem. Soc.* **87:**131 (1965); R. Pettit and J. Henery, *Org. Synth.* **50:**21 (1970).

Scheme 8.13. Reactions of Cyclobutadiene

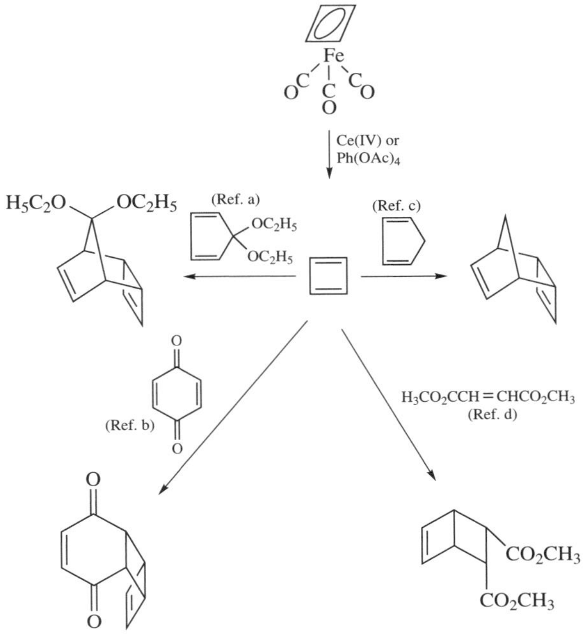

Ce(IV) or $Ph(OAc)_4$

(Ref. a) (Ref. b) (Ref. c) (Ref. d)

$H_3CO_2CCH{=}CHCO_2CH_3$

a. J. C. Barborak and R. Pettit, *J. Am. Chem. Soc.* **89:**3080 (1967).
b. J. C. Barborak, L. Watts, and R. Pettit, *J. Am. Chem. Soc.* **88:**1328 (1966).
c. L. Watts, J. D. Fitzpatrick, and R. Pettit. *J. Am. Chem. Soc.* **88:**623 (1966).
d. P. Reeves, J. Henery, and R. Pettit, *J. Am. Chem. Soc.* **91:**5889 (1969).

in the cyclobutadiene–iron tricarbonyl complex reacts as an aromatic ring and can undergo electrophilic substitutions.[188] Subsequent studies provided evidence that oxidative decomposition of the complex could liberate cyclobutadiene and that it could be trapped by appropriate reactants.[189] Some examples of these reactions are given in Scheme 8.13.

One of the best known of the π-organometallic compounds is ferrocene. It is a neutral compound that can be readily prepared from cyclopentadienide anion and iron(II).[190]

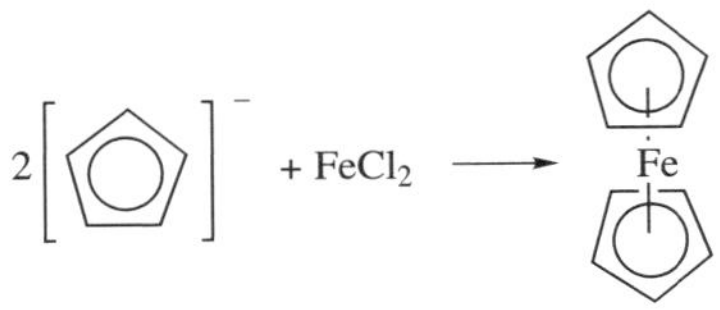

$2[C_5H_5]^- + FeCl_2 \longrightarrow Fe(C_5H_5)_2$

Numerous chemical reactions have been carried out on ferrocene and its derivatives. The molecule behaves as an electron-rich aromatic system, and electrophilic substitution reactions occur readily. Reagents that are relatively strong oxidizing agents, such as the halogens, effect oxidation at iron and destroy the compound.

Many other π-organometallic compounds have been prepared. In the most stable compounds, the total number of electrons contributed by the ligands (six for each

188. J. D. Fitzpatrick, L. Watts, G. F. Emerson, and R. Pettit, *J. Am. Chem. Soc.* **87:**3254 (1965).
189. R. H. Grubbs and R. A. Grey, *J. Am. Chem. Soc.* **95:**5765 (1973).
190. G. Wilkinson, *Org. Synth.* **IV:**473, 476 (1963).

cyclopentadiene ion) plus the valence shell electrons on the metal atom or ion usually totals 18, to satisfy the effective atomic number rule.[191]

	Mn	Ni	Ti
Metal	6	9	2
Ligands	12	9	16
Total	18	18	18

One of the most useful types of π-complexes of aromatic compounds from the synthetic point of view are chromium complexes obtained by heating benzene or other aromatics with $Cr(CO)_6$.

$$C_6H_6 + Cr(CO)_6 \longrightarrow C_6H_6Cr(CO)_3$$ Ref. 192

$$C_6H_5Cl + Cr(CO)_6 \longrightarrow (C_6H_5Cl)Cr(CO)_3$$ Ref. 193

The $Cr(CO)_3$ unit in these compounds is strongly electron-withdrawing and activates the ring to nucleophilic attack. Reactions with certain carbanions results in arylation.[194]

$$(OC)_3Cr(C_6H_5Cl) + (CH_3)_2\bar{C}CN \longrightarrow (OC)_3Cr(C_6H_5C(CH_3)_2C{\equiv}N) \longrightarrow C_6H_5C(CH_3)_2C{\equiv}N \quad 78\%$$

In compounds in which the aromatic ring does not have a leaving group addition occurs, and the intermediate can by oxidized by I_2.

$$(OC)_3Cr(C_6H_6) + LiC(CH_3)_2CO_2C(CH_3)_3 \longrightarrow [H,\ C(CH_3)_2CO_2C(CH_3)_3\text{-cyclohexadienyl}]Cr(CO)_3 \xrightarrow{I_2} C_6H_5C(CH_3)_2CO_2C(CH_3)_3 \quad 91\%$$ Ref. 195

191. M. Tsutsui, M. N. Levy, A. Nakamura, M. Ichikawa, and K. Mori, *Introduction to Metal π-Complex Chemistry*, Plenum Press, New York, 1970, pp. 44–45; J. P. Collman, L. S. Hegedus, J. R. Norton, and R. G. Finke, *Principles and Applications of Organotransition Metal Chemistry*, University Science Books, Mill Valley, California, 1987, pp. 166–173.
192. W. Strohmeier, *Chem. Ber.* **94:**2490 (1961).
193. J. F. Bunnett and H. Hermann, *J. Org. Chem.* **36:**4081 (1971).
194. M. F. Semmelhack and H. T. Hall, *J. Am. Chem. Soc.* **96:**7091 (1974).
195. M. F. Semmelhack, H. T. Hall, M. Yoshifuji, and G. Clark, *J. Am. Chem. Soc.* **97:**1247 (1975); M. F. Semmelhack, H. T. Hall, Jr., R. Farina, M. Yoshifuji, G. Clark, T. Bargar, K. Hirotsu, and J. Clardy, *J. Am. Chem. Soc.* **101:**3535 (1979).

Existing substituent groups such as CH_3, OCH_3, and $^+N(CH_3)_3$ exert a directive effect, often resulting in a major amount of the *meta* substitution product.[196] The intermediate adducts can be converted to cyclohexadiene derivatives if the adduct is protonolyzed.[197]

$$CH_3O\text{-}C_6H_5\text{-}Cr(CO)_3 + LiC(CH_3)_2C{\equiv}N \longrightarrow [\text{adduct: } C(CH_3)_2C{\equiv}N, H;\ Cr(CO)_3] \xrightarrow{CF_3CO_2H} CH_3O\text{-cyclohexadienyl-}C(CH_3)_2C{\equiv}N$$

Not all carbon nucleophiles will add to arene chromiumtricarbonyl complexes. For example, alkyllithium reagents and simple ketone enolates do not give adducts.[198]

Organometallic chemistry is a very large active field of research, and new types of compounds, new reactions, and catalysts are being discovered at a rapid rate. These developments have had a major impact on organic synthesis, and developments can be expected to continue.

General References

J. P. Collman, L. S. Hegedus, J. R. Norton, and R. G. Finke, *Principles and Applications of Organotransition Metal Chemistry*, University Science Books, Mill Valley, California, 1987.
H. M. Colquhoun, J. Holton, D. J. Thomson, and M. V. Twigg, *New Pathways for Organic Synthesis*, Plenum, New York, 1984.
S. G. Davies, *Organo-transition Metal Chemistry: Applications in Organic Synthesis*, Pergamon, Oxford, 1982.
F. Diederich and P. J. Stang, *Metal-Catalyzed Cross-Coupling Reactions*, Wiley-VCH, New York, 1998.
J. K. Kochi, *Organometallic Mechanisms and Catalysis*, Academic Press, New York, 1979.
E. Negishi, *Organometallics in Organic Synthesis*, John Wiley & Sons, New York, 1980.
M. Schlosser, ed., *Organometallics in Synthesis: A Manual*, John Wiley & Sons, Chichester, U.K., 1994.

Organocopper Reactions

G. Posner, *Org. React.* **19**:1 (1972).
G. Posner, *Org. React.* **22**:253 (1975).
G. Posner, *An Introduction to Synthesis Using Organocopper Reagents*, Wiley-Interscience, New York, 1975.
R. J. K. Taylor, ed., *Organnocopper Reagents*, Oxford University Press, Oxford, 1995.

Organopalladium Reactions

R. F. Heck, *Palladium Reagents in Organic Synthesis*, Academic Press, Orlando, Florida, 1985.
R. F. Heck, *Org. React.* **27**:345 (1982).
J. Tsuji, *Palladium Reagents and Catalysts: Innovations in Organic Synthesis*, John Wiley & Sons, New York, 1996.

196. M. F. Semmelhack, G. R. Clark, R. Farina, and M. Saeman, *J. Am. Chem. Soc.* **101**:217 (1979).
197. M. F. Semmelhack, J. J. Harrison, and Y. Thebtaranonth, *J. Org. Chem.* **44**:3275 (1979).
198. R. J. Card and W. S. Trahanovsky, *J. Org. Chem.* **45**:2555, 2560 (1980).

Problems

(References for these problems will be found on page 935.)

1. Predict the product of the following reactions. Be sure to specify all elements of stereochemistry.

(a) $H_2C{=}C(CH_3)MgBr$ + $CH_3{-}CH{-}CH_2$ (epoxide, O) $\xrightarrow{Cu(I)}$

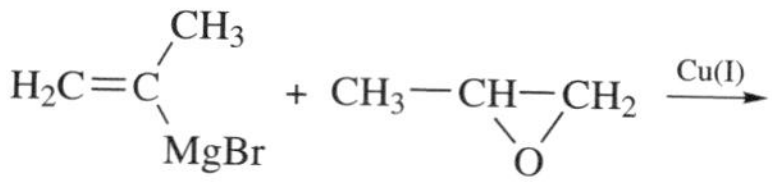

(b) C_2H_5MgBr $\xrightarrow[\text{2) } C_6H_{13}C\equiv CH \text{ 3) } I_2]{\text{1) } CuBr{-}S(CH_3)_2,\ -45°C}$

(c) (2-(hydroxymethyl)benzyl bromide: CH_2Br, CH_2OH) $\xrightarrow[CO]{Pd(PPh_3)_2Cl_2}$

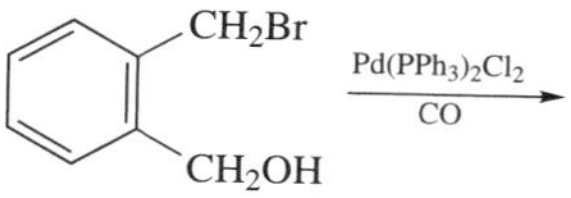

(d) $H_2C{=}CHCH_2MgBr$ + (diketene: $HC{=}C$, O, O) $\xrightarrow{Cu(I)}$

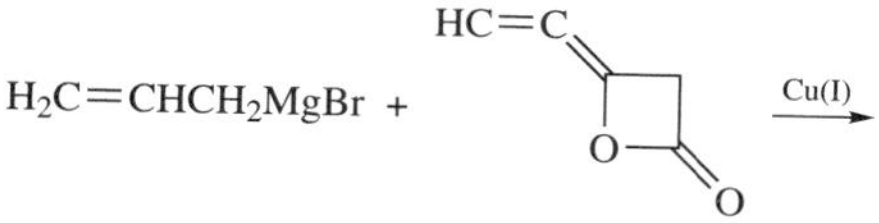

(e) CH_3CH_2MgBr + (H)(C$_6$H$_{13}$)C=C(H)(I) $\xrightarrow{Pd(PPh_3)_4}$

(f) (2-methylcyclopentane-1,3-dione: O, O, CH_3) + $H_2C{=}CHCH_2O_2CCH_3$ $\xrightarrow[80°C,\ DBU]{Pd(PPh_3)_4}$

(g) (1-chloro-1-ethynylcyclohexane: Cl, $C\equiv CH$) + $[CH_3(CH_2)_3]_2CuLi$ $\longrightarrow$

(h) (CH_3, O epoxide; $C(CH_3)_3$) + $[(C_2H_5)_2CuCN]Li_2$ $\longrightarrow$

(i) ($PhCH_2OCH_2$, CH_3, =O, O lactone) + $(CH_3)_2CuLi$ $\longrightarrow$

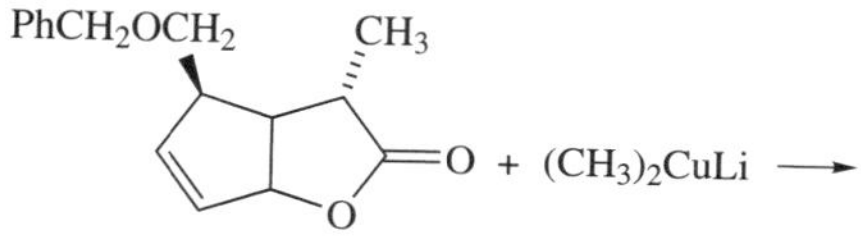

(j) (1,3-dioxolane, O, O; CH_2CH_2; $CH_2CH{=}CH_2$; =O; CH_3, CH_3) $\xrightarrow[O_2,\ DMF,\ H_2O]{PdCl_2,\ CuCl_2}$

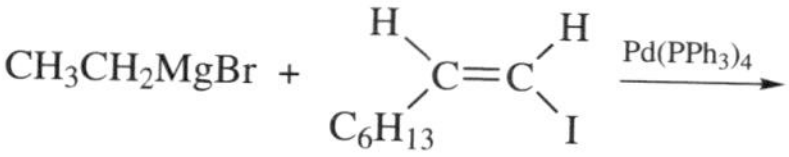

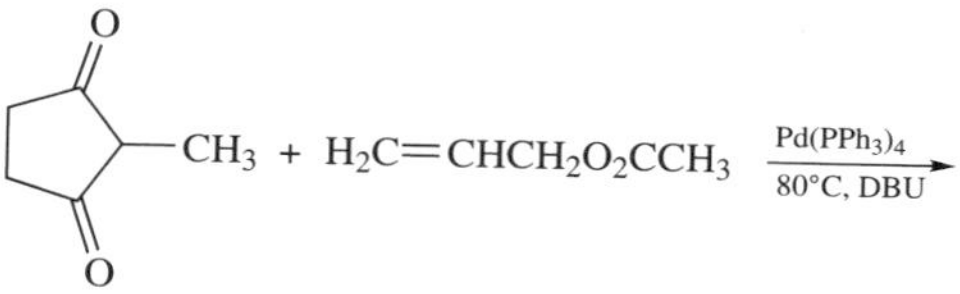

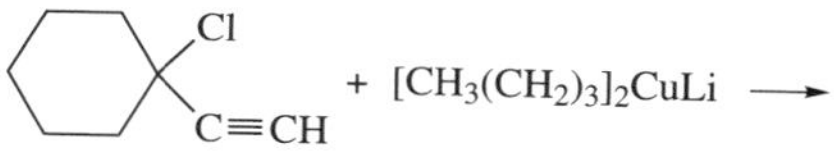

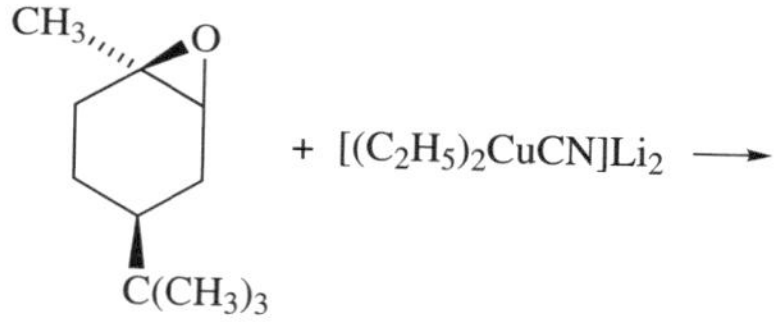

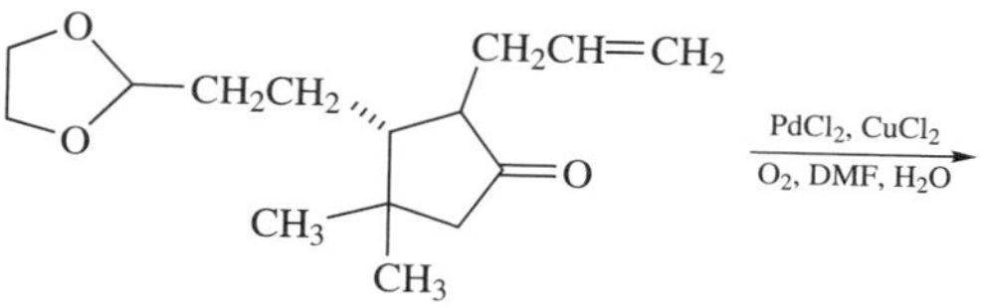

(k) C_4H_9Li $\xrightarrow[\text{2) } HC\equiv CH \text{ 3) } I_2]{\text{1) } CuBr{\cdot}SMe}$

(l) $PhOCH_2CH{=}CHCH_2CH_2\overset{O}{\overset{\|}{C}}CH_2CO_2CH_3 \xrightarrow[PPh_3]{Pd(OAc)_2}$

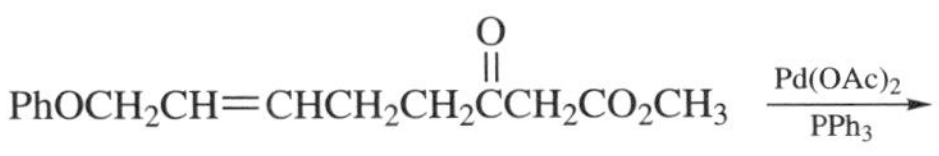

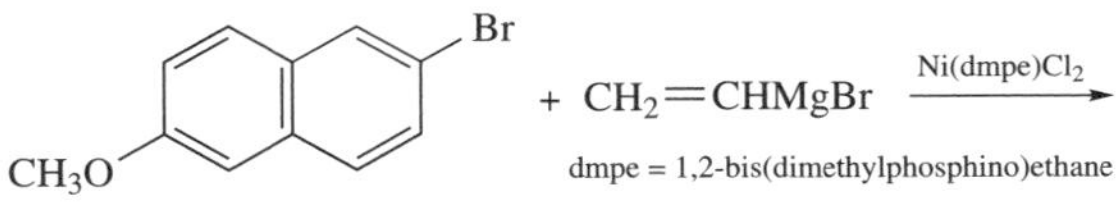

(m) 2-bromo-6-methoxynaphthalene (CH_3O, Br) + $CH_2{=}CHMgBr \xrightarrow{Ni(dmpe)Cl_2}$

dmpe = 1,2-bis(dimethylphosphino)ethane

(n) PhN (bicyclic triazolinedione adduct, O, N–N, 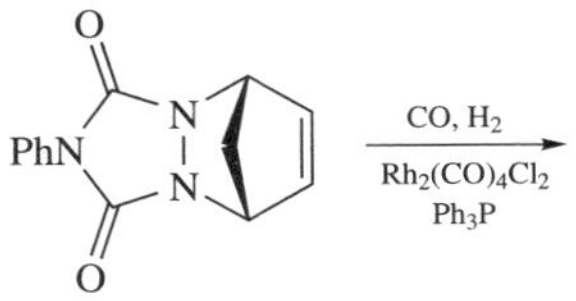 O) $\xrightarrow[Rh_2(CO)_4Cl_2,\ Ph_3P]{CO,\ H_2}$

(o) 2-(3-bromophenyl)oxazoline (N, O, Br) + $CH_3CH_2CH_2MgBr \xrightarrow{NiCl_2(dppe)}$

dppe = 1,2-bis(diphenylphosphino)ethane

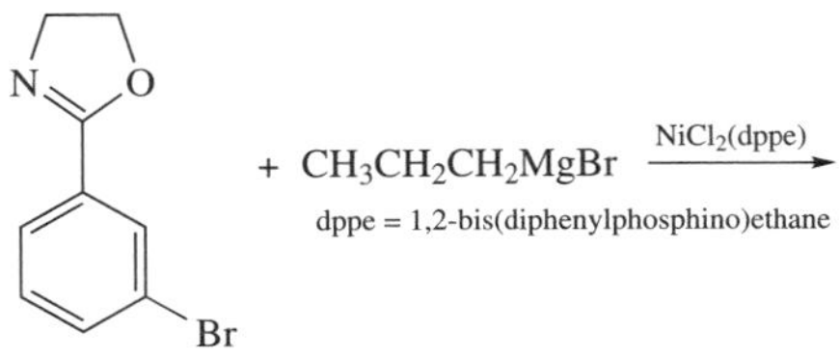

(p) CH_3O, $CON(C_2H_5)_2$ (benzene ring) $\xrightarrow[3)\ CH_2{=}CHCH_2Br]{1)\ s\text{-BuLi, TMEDA}\quad 2)\ CuI{\cdot}S(CH_3)_2}$

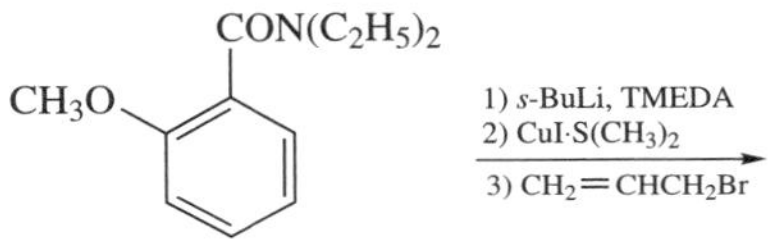

(q) $CH_3CH_2CH_2CH_2(H)C{=}C(H)$–B(catechol, O, O) + Br(H)C=C(H)Ph $\xrightarrow[NaOC_2H_5]{Pd(PPh_3)_4}$

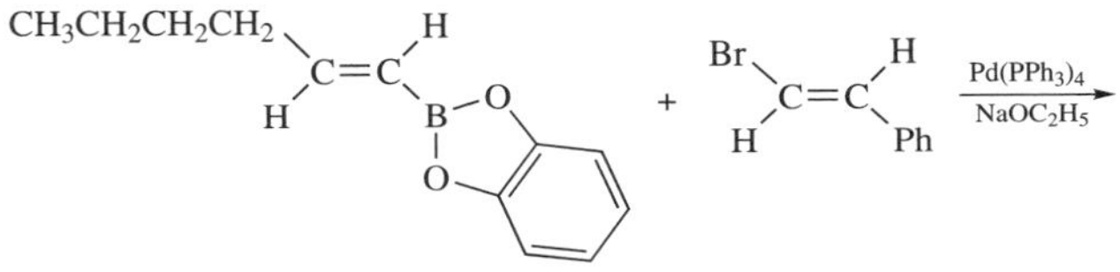

(r) $(C_4H_9)_3Sn(H)C{=}C(H)CH_2OTHP$ + $CH_3O(BrCH_2)C{=}C(H)CO_2C_2H_5 \xrightarrow[bis(dba)Pd]{CO,\ 55\ psi}$

dba = dibenzylideneacetone

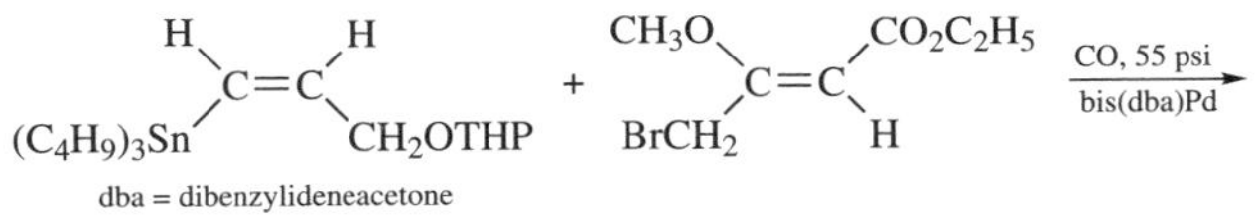

2. Give the products to be expected from each of the following reactions involving mixed or higher-order cuprate reagents.

(a) cyclohexenone (O) + [2-thienyl(S)–Cu$(CH_2)_3CH_3$ (CN)] Li_2

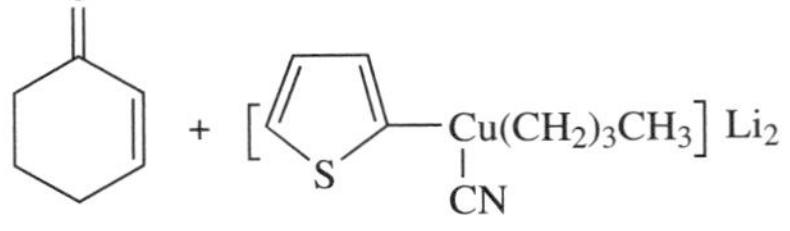

(b) cyclohexyl–I + $[(CH_3CH_2CH_2CH_2)_2CuCN]Li_2$

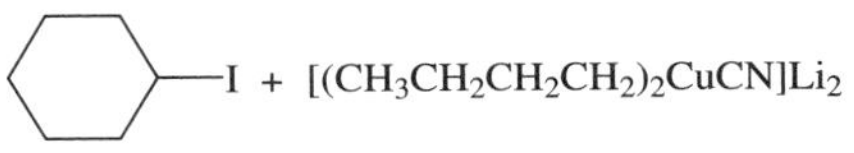

(c) cyclopentadiene monoepoxide (O) + $[(CH_3)_3CCuCN]Li \longrightarrow$

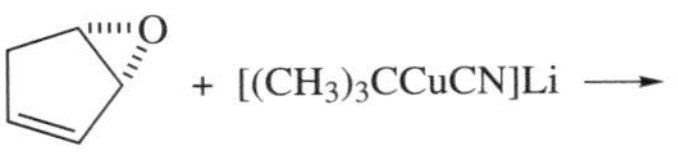

(d) [epoxycyclohexane bearing $C{\equiv}CCH_3$] + $[(CH_3)_3CCuCH_2\overset{O}{\overset{\|}{S}}CH_3]Li$ $\longrightarrow$

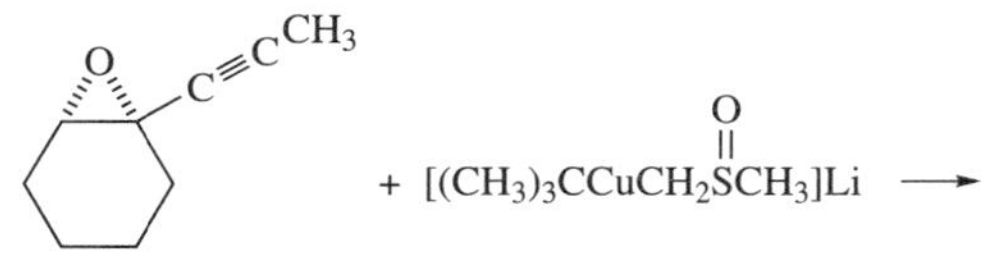

3. Write a mechanism for each of the following transformations which accounts for the observed product and is in accord with other information which is available concerning the reaction.

(a) $CH_3(CH_2)_5CH{=}CH_2 + CO + (CH_3CO)_2O \xrightarrow[CuCl_2]{PdCl_2,\ O_2} CH_3(CH_2)_5\underset{CH_3CO_2}{CH}CH_2\overset{O}{\overset{\|}{C}}O\overset{O}{\overset{\|}{C}}CH_3$

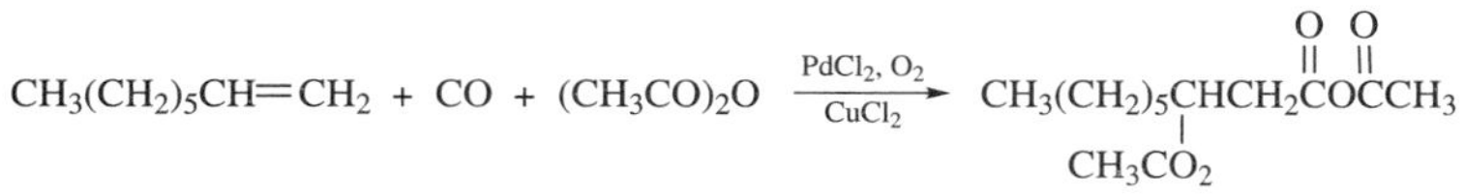

(b) $CH_2{=}CHCH_2CH_2\overset{OSi(CH_3)_3}{C}{=}CH_2 \xrightarrow[10\ h,\ 25°C]{Pd(OAc)_2}$ 3-methylcyclopent-2-enone (H_3C, O) $+ Pd^0$

(c) $PhCH_2O$–(CH$_3$, CH$_3$, OH substituted alkene) $\xrightarrow[Rh_2(OAc)_2,\ PPh_3,\ 100°C]{CO,\ H_2}$ $PhCH_2O$–(CH$_3$, H, CH$_3$ substituted δ-lactone, O, O)

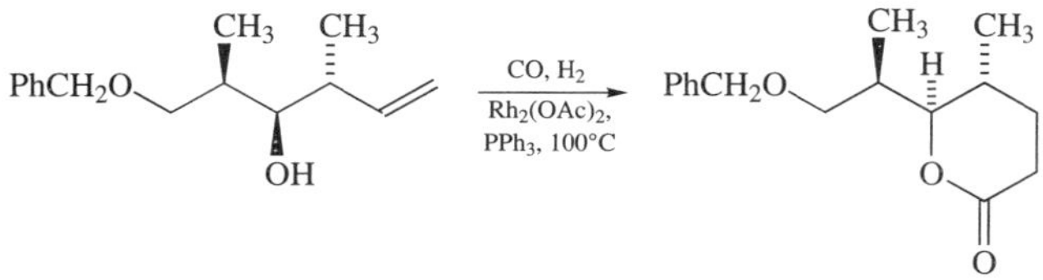

(d) $PhCOCl + CH_3(CH_2)_5C{\equiv}CH \xrightarrow[PPh_3]{[RhCl(COD)]_2}$ $CH_3(CH_2)_5(Cl)C{=}CH(Ph)$

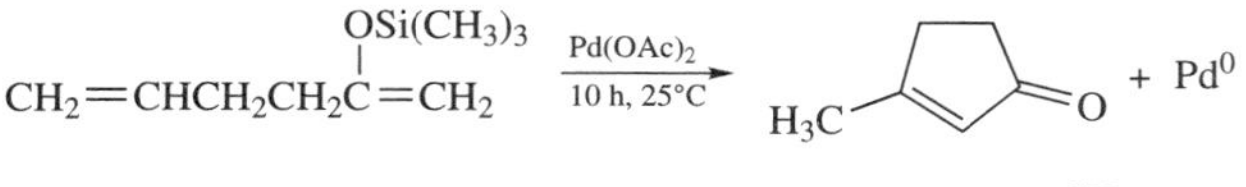

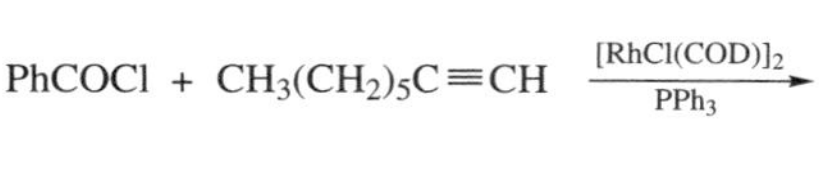

4. Indicate appropriate conditions and reagents for effecting the following transformations. "One-pot" processes are possible in all cases.

(a) $(CH_3CH_2)_2C{=}CHCH_2CH_2Br \longrightarrow (CH_3CH_2)_2C{=}CHCH_2CH_2\overset{CH_3}{C}{=}\underset{H}{C}CO_2CH_3$

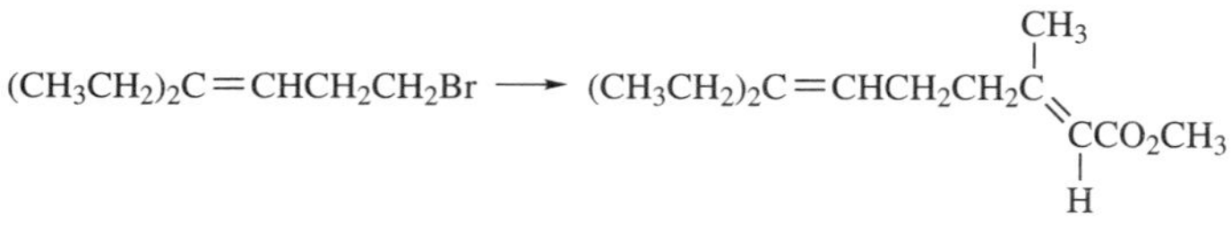

(b) o-bromoacetanilide (Br, $NH\overset{O}{\overset{\|}{C}}CH_3$) $\longrightarrow$ ortho-substituted benzene ($CH{=}CHCN$, $NH\overset{O}{\overset{\|}{C}}CH_3$)

(c) $CH_3CH_2CH_2CH_2Br + CH_3O_2CC{\equiv}CCO_2CH_3 \longrightarrow$ $(CH_3O_2C)(CH_3(CH_2)_3)C{=}C(CO_2CH_3)(H)$

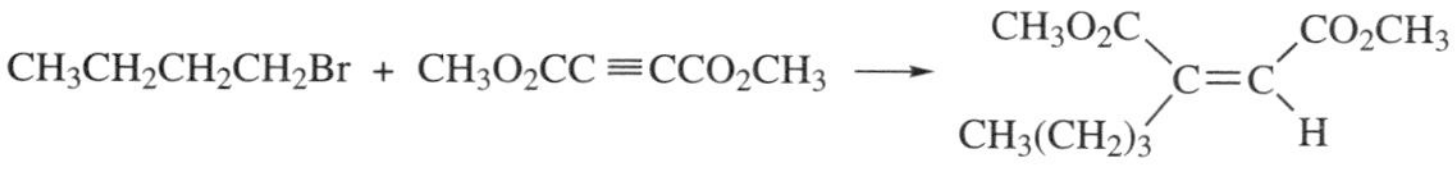

(d) bicyclic enone (O, H_3C, CH_3) $\longrightarrow$ bicyclic ketone (O, H_3C, CH_3, $CH_2CH_2CH{=}CH_2$)

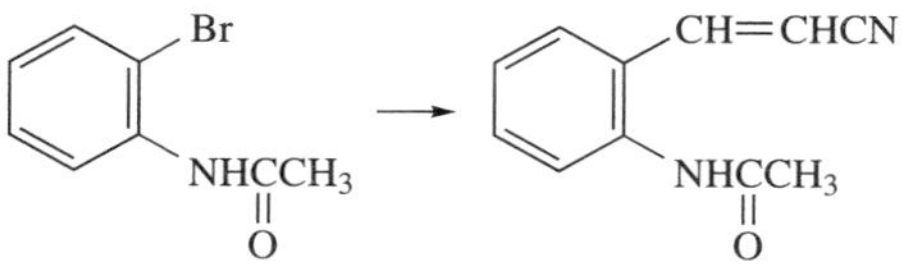

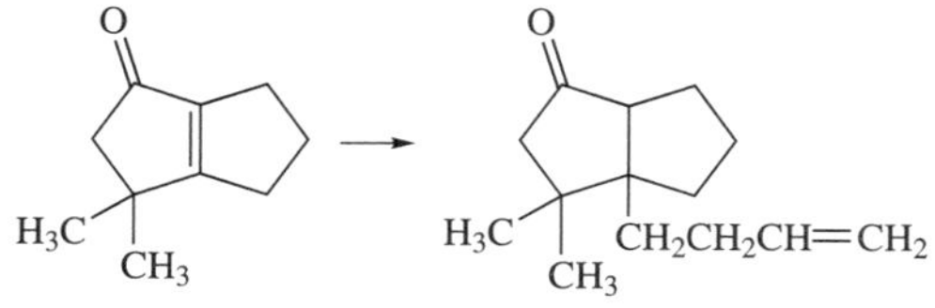

(e) $(CH_3)_2CHCH(H)C=C(H)CH(O_2CCH_3)CH_2CH_2C(H)=C(CH_3)CH_2Br$ with $CH_2CH_2C(CH_3)=C(H)CH_2Br$ → macrocycle bearing O_2CCH_3, $(CH_3)_2CH$, CH_3, H_3C

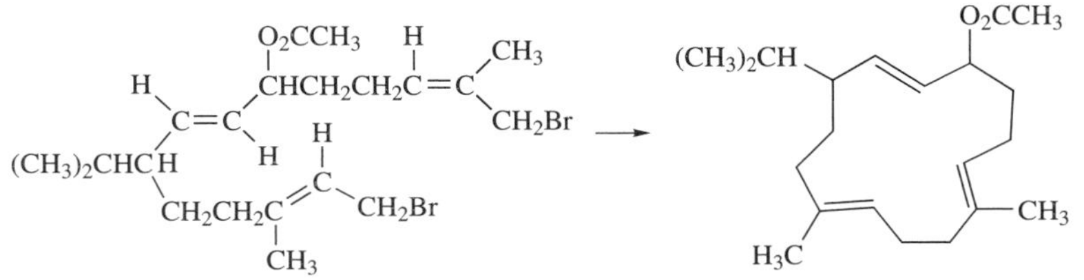

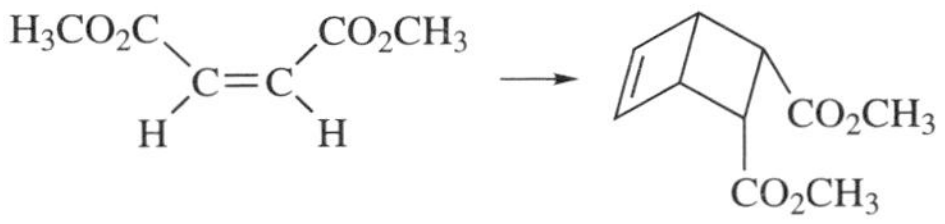

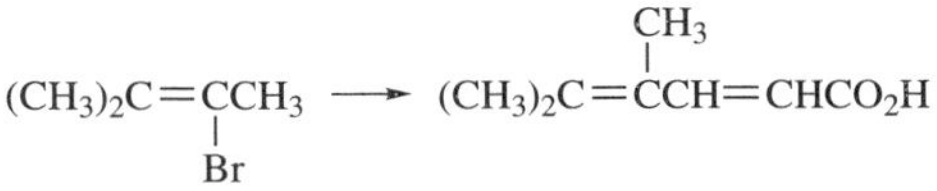

(f) $H_3CO_2C(H)C=C(H)CO_2CH_3$ → bicyclic diester (CO_2CH_3, CO_2CH_3)

(g) $(CH_3)_2C=C(Br)CH_3$ → $(CH_3)_2C=C(CH_3)CH=CHCO_2H$

(h) bicyclic enone (O) → O, H, CH_3

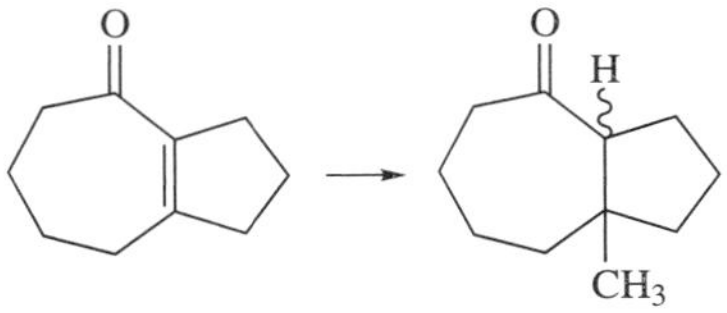

(i) 2-bromo-3-($OCH_2OCH_2CH_2Si(CH_3)_3$)pyridine (N, Br) → bipyridine with $OCH_2OCH_2CH_2Si(CH_3)_3$ and $(CH_3)_3SiCH_2CH_2OCH_2O$ (N, N)

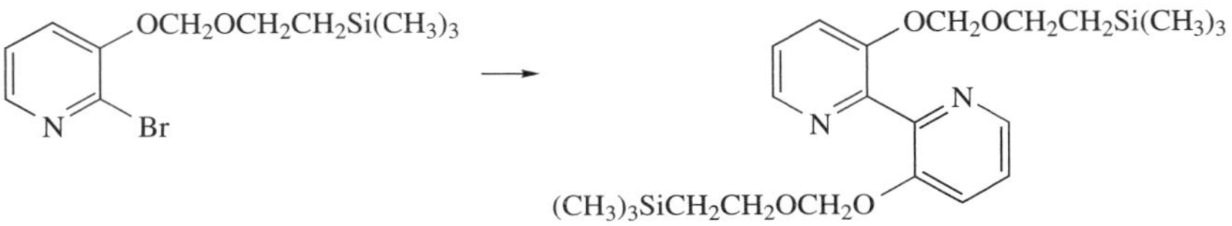

(j) 3-pyridyl diene iodide (N, I) → 3-pyridyl tetraene CH_2OH (N)

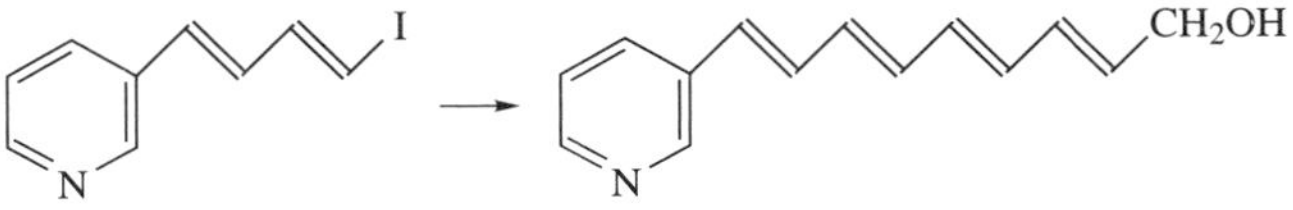

(k) H_2C, $SnBu_3$, CH_3, O, O, CH_3, CH_3 → $HOCH_2$, CH_3, CH_3, CH_2, CH_3, O, O, CH_3, CH_3

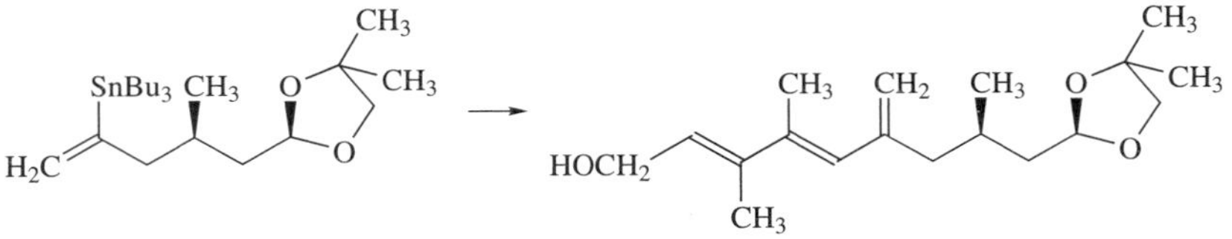

(l) Br, OCH_2Ph, CH_3O, O, O → furyl, O, OCH_2Ph, CH_3O, O, O

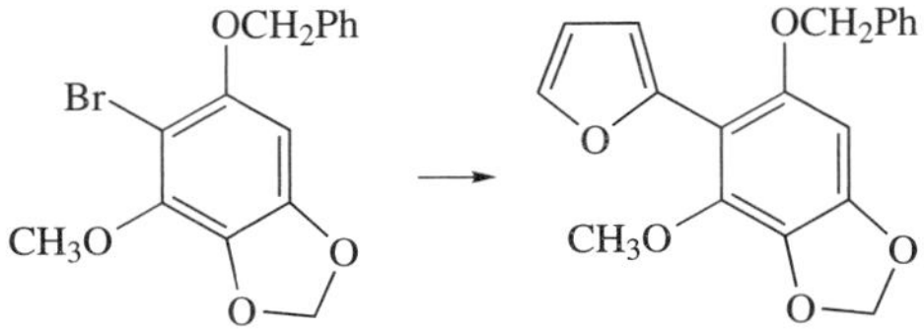

(m) CH_3, CH_3, CH_3, $SnBu_3$, CH_3 → CH_3, CH_3, CH_3, CH_3, CO_2H, CH_3

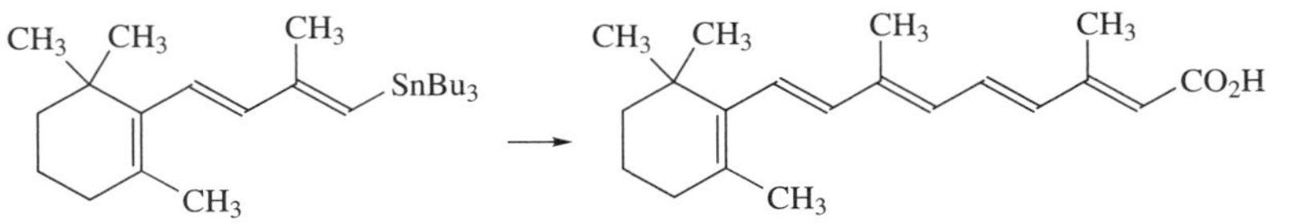

5. Vinyl triphenylphosphonium ion has been found to react with cuprate reagents by nucleophilic addition, generating an ylide structure. This intermediate can then be

treated with an aldehyde to give an alkene by the Wittig reaction. Show how organocuprate intermediates could be used in conjunction with vinyl triphenylphosphonium ion to generate the following products from the specified starting material.

(a) $CH_3(CH_2)_3CH{=}CHCH_2CH{=}CHPh$ (cis: H, H on same side) from $CH_3(CH_2)_3CH{=}CHI$ (cis: H, H on same side)

(b) $PhCH_2CH{=}CH(CH_2)_3CH_3$ from C_6H_5I (iodobenzene)

6. It has been observed that the reaction of $[(C_2H_5)_2Cu]Li$ or $[(C_2H_5)_2CuCN]Li_2$ with 2-iodooctane proceeds with racemization in both cases. On the other hand, the corresponding bromide reacts with nearly complete inversion of configuration with both reagents. When 6-halo-2-heptenes are used in similar reactions with dimethylcuprate, the iodide gives mainly the cyclic product 1-ethyl-2-methylcyclopentane whereas the bromide gives mainly 6-methyl-1-heptene. Provide a mechanism which accounts for the different behavior of the iodides as compared with the bromides.

7. Short synthetic sequences involving no more than three steps can be used to prepare the compound shown on the left from the potential starting material on the right. Suggest an appropriate series of reactions for each transformation.

(a) 3-acetylcyclohexanone (ring C=O, with $\underset{\|}{C}CH_3$ over O substituent) $\Longrightarrow$ 2-cyclohexenone and $H_2C{=}CHOCH_3$

(b) $C_6H_5CH_2O$–cyclohexane bearing $CH{=}CH_2$ and $CH_2\underset{\|}{C}CH_3$ (=O) on the same carbon $\Longrightarrow$ $C_6H_5CH_2O$–cyclohexanone ($=O$)

(c) O= cyclopentenone fused to cyclooctene $\Longrightarrow$ 2-($CH_2{=}CHCH_2$)cyclooctenone (O)

(d) 2-(CO_2CH_3)-3-($CH{=}CH_2$)cyclopentanone (O) $\Longrightarrow$ CH_3O_2C–$CH_2CH{=}CHCH_2$–CO_2CH_3

(e) OTBDMS, THPO cyclopentane bearing allene chain $=C=$ $(CH_2)_2CO_2CH_3$ and side chain with OTHP, OPh $\Longrightarrow$ OTBDMS, THPO cyclopentane bearing $CH_2CH(Br)C{\equiv}CH$ chain and side chain with OTHP, OPh

(f) 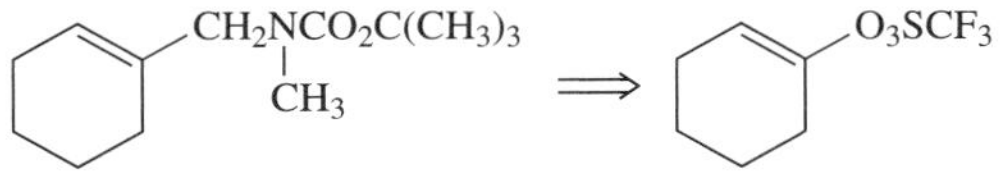 [structure: $CH_2NCO_2C(CH_3)_3$ / CH_3 substituted cyclohexene] $\Longrightarrow$ [structure: cyclohexenyl O_3SCF_3]

(g) 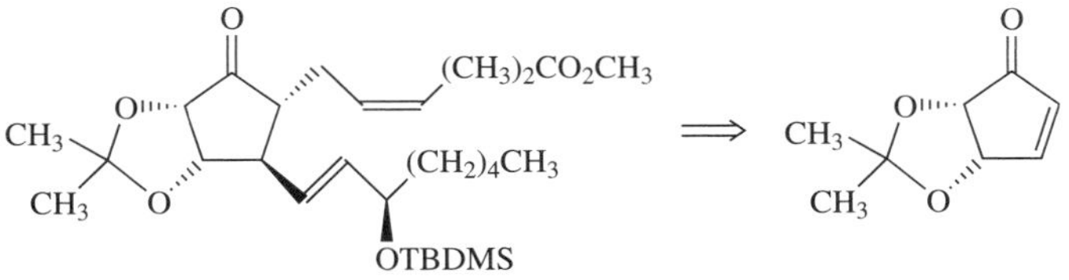 [structure: O; $(CH_3)_2CO_2CH_3$; CH_3; CH_3; O; O; $(CH_2)_4CH_3$; OTBDMS] $\Longrightarrow$ [structure: O; CH_3; CH_3; O; O]

(h) 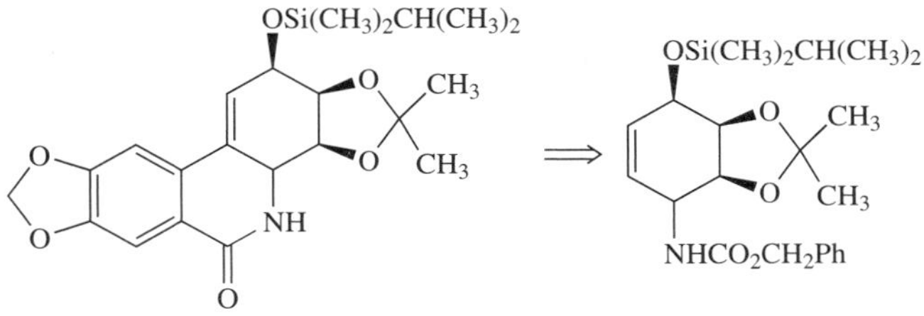 [structure: $OSi(CH_3)_2CH(CH_3)_2$; O; CH_3; O; CH_3; O; O; NH; O] $\Longrightarrow$ [structure: $OSi(CH_3)_2CH(CH_3)_2$; O; CH_3; O; CH_3; $NHCO_2CH_2Ph$]

8. The conversions shown were carried out in multistep, but "one-pot," synthetic processes in which none of the intermediates needs to be isolated. Show how you could perform the transformation by suggesting a sequence of organic and inorganic reagents to be employed and the approximate reaction conditions.

(a) 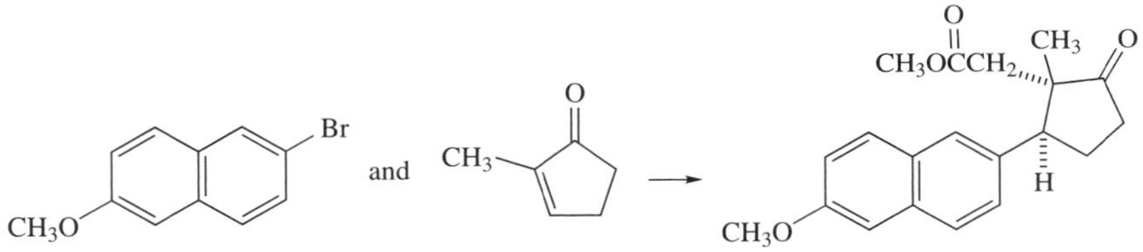 [structure: CH_3O-naphthalene-Br] and [structure: CH_3-cyclopentenone] $\longrightarrow$ [structure: $CH_3O\overset{O}{\overset{\|}{C}}CH_2$, CH_3, O, H, CH_3O-naphthyl cyclopentanone]

(b) 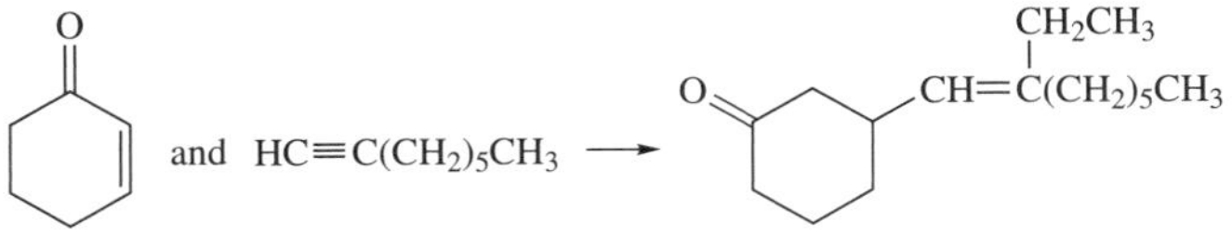 [structure: cyclohexenone, O] and $HC{\equiv}C(CH_2)_5CH_3$ $\longrightarrow$ [structure: O=cyclohexyl-$CH{=}\overset{CH_2CH_3}{C}(CH_2)_5CH_3$]

(c) [structure: CH_3, CH_2, =O, CH_3 bicyclic ketone] and $BrCH_2C{\equiv}CSi(CH_3)_3$ $\longrightarrow$ [structure: CH_3, CH_3, $CH_2CH_2C{\equiv}CSi(CH_3)_3$, =O, CH_3]

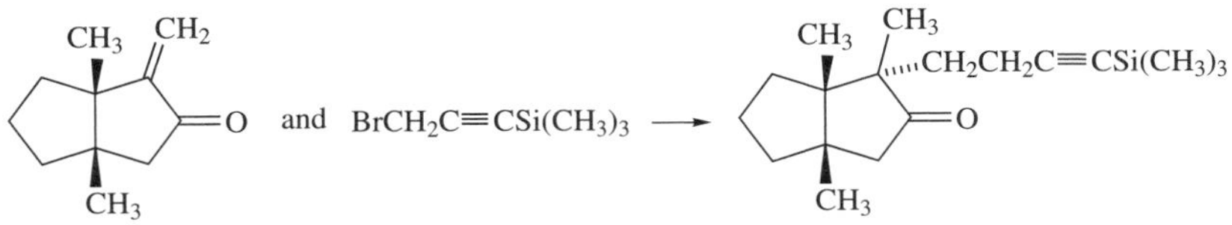

(d) 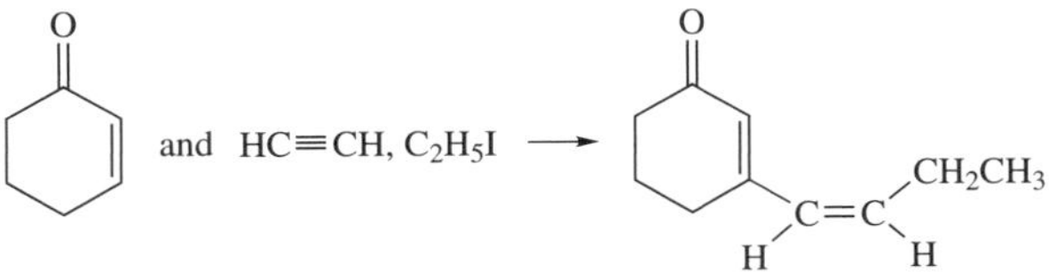 [structure: cyclohexenone, O] and $HC{\equiv}CH$, C_2H_5I $\longrightarrow$ [structure: O, cyclohexenone with $C{=}C$, H, H, CH_2CH_3]

(e) [Structures: 4-bromo-1-fluoro-2-methylbenzene and 2-bromo-6-(CH=CHCN)-toluene derivative (Br, CH=CHCN, CH_3, CH_3) → biaryl with F, H_3C, NCCH=CH, CH_3, CH_3]

(f) $PhCH{=}CHCH_2OC_2H_5 \longrightarrow PhCH{=}CHCH_2CH(CO_2C_2H_5)C(=O)CH_3$

9. A number of syntheses of medium- and large-ring compounds which involve transition-metal reagents have been described. Suggest an organometallic reagent or metal complex which could bring about each of the following conversions:

(a) $BrCH_2CH{=}C(CH_3)CH_2CH{=}CHC(CH_3)_2CH_2CH{=}C(CH_3)CH_2Br \longrightarrow$ [cyclic triene with H_3C, CH_3, CH_3, CH_3]

(b) $CH_3CO_2CH_2CH{=}C(CH_3)CH_2CH_2CH{=}C(CH_3)CH_2CH_2C(=O)CH(CH_3)CO_2CH_3 \longrightarrow$ [cyclic dienone with CH_3, CH_3, CH_3, =O, CO_2CH_3]

(c) $(PhSO_2)_2CH(CH_2)_{10}CO_2(CH_2)_{10}$–(epoxide)–$CH{=}CH_2 \longrightarrow$ [macrocyclic lactone: CO, $(CH_2)_{10}$, $(CH_2)_{10}$, $(PhSO_2)_2C$–CH=CH–CH(OH)]

(d) [bis(2-iodo-5-methoxyphenyl) compound: MeO, OMe, $CH_2CH_2N(CH_3)CH_2CH_2$, I, I] $\longrightarrow$ [bridged biaryl: CH_3–N, MeO, OMe]

(e) $TBDMSOCH_2C(=CH_2)$–(epoxide)–$(CH_2)_8CH(SO_2Ph)CO_2CH_3 \longrightarrow$ [macrocycle with PhO_2S, CH_3O_2C, OTBDMS, OH]

10. The cyclobutadiene complex **1** can be prepared in enantiomerically pure form. When the complex is reacted with an oxidizing agent and a compound capable of trapping cyclobutadienes, the products are racemic. When the reaction is carried only to partial completion, the recovered complex remains enantiomerically pure. Discuss the relevance of these results to the following questions: "In oxidative decomposition of cyclobutadiene complexes, is the cyclobutadiene liberated from the complex before or after it has reacted with the trapping reagent?"

CH_3 CH_2OCH_3 $Fe(CO)_3$ **1** —Ce(IV), $(NC)_2C{=}C(CN)_2$→ NC H NC NC NC H CH_3 CH_2OCH_3

11. When the isomeric acetates **A** and **B** react with dialkylcuprates, both give a give a very similar product mixture containing mainly **C** with small amounts of **D**. Only trace amounts of the corresponding *Z*-isomers are found. Suggest a mechanism to account for the formation of essentially the same product mixture from both reactants.

Ph H C=C H CH CH_3 | O_2CCH_3 **A** or PhCH | CH_3CO_2 C=C H H CH_3 **B** —R_2CuLi→ Ph H C=C H $CHCH_3$ | R **C** + PhCH | R C=C H H CH_3 **D**

12. The compound shown below is a constitutent of the pheromone of the codling moth. It has been synthesized using *n*-propyl bromide, propyne, 1-pentyne, ethylene oxide, and CO_2 as the source of the carbon atoms. Devise a route for such a synthesis. Hint: Extensive use of the chemistry of organocopper reagents is the basis for the existing synthesis.

$CH_3CH_2CH_2$ CH_3 C=C CH_2CH_2 H —C($CH_2CH_2CH_3$)=C—H | CH_2OH

13. (*S*)-3-Hydroxy-2-methylpropanoic acid, **E**, can be obtained enantiomerically pure from isobutyric acid by a microbiological oxidation. The aldehyde **F** is available from a natural product, pulegone, also in enantiomerically pure form.

CH_3 H $HOCH_2$—C CO_2H **E**

CH_3 H O=$CHCH_2$—C $CH_2CH_2CH_2CH(CH_3)_2$ **F**

Devise a synthesis of enantiomerically pure **G**, a compound of interest as a starting material for the synthesis of α-tocopherol (vitamin E).

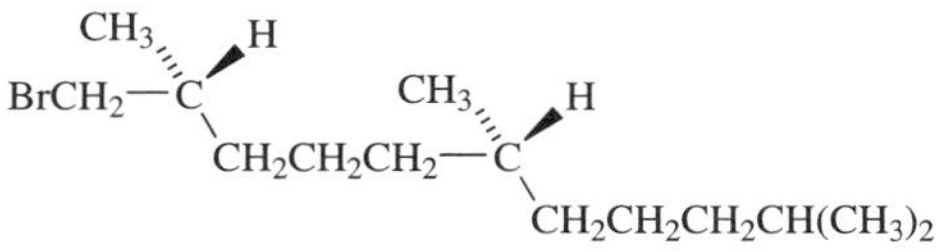

CH_3 H
$BrCH_2$—C
CH_3 H
$CH_2CH_2CH_2$—C
$CH_2CH_2CH_2CH(CH_3)_2$
G

14. Each of the following transformations can be carried out in good yield under optimized conditions. Consider the special factors in each case, and discuss the most appropriate reagent and reaction conditions to obtain good yields.

(a)

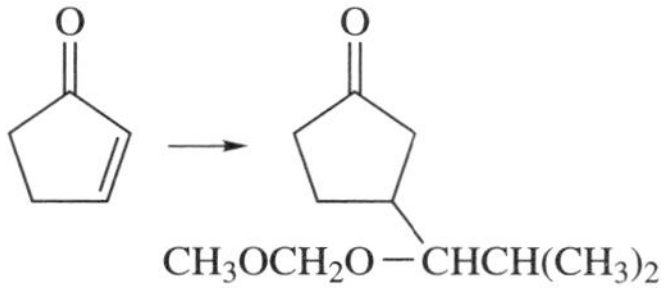

O → O
$CH_3OCH_2O-CHCH(CH_3)_2$

(b)

O → O
H_3C CH_3 H_3C CH_3 $CH_2CH{=}CH_2$

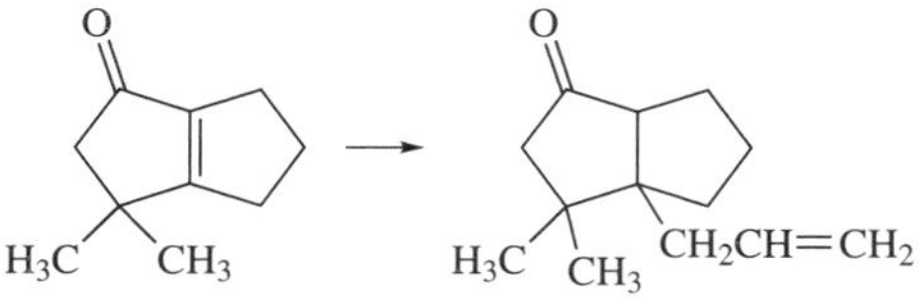

(c) $(CH_3)_2C{=}CHCO_2C_2H_5$ → $CH_3(CH_2)_3C(CH_3)_2CH_2CO_2C_2H_5$

(d)

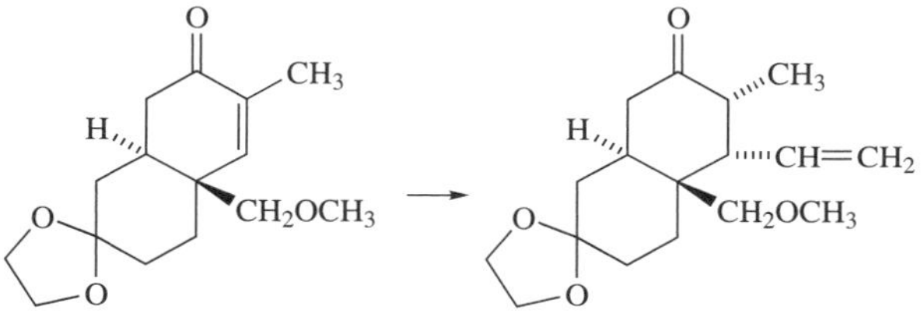

O CH_3 H O O CH_2OCH_3 → O CH_3 H $CH{=}CH_2$ O O CH_2OCH_3

15. Each of the following synthetic transformations can be accomplished by use of organometallic reagents and/or catalysts. Indicate a sequence of reactions which would permit each of the syntheses to be completed.

(a)

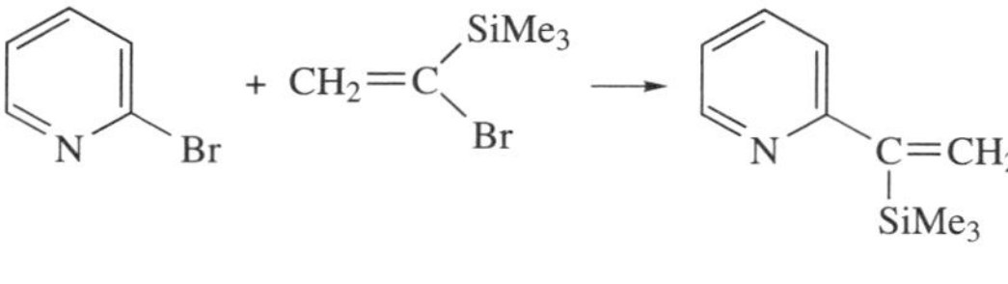

N Br + $CH_2{=}C$ $SiMe_3$ Br → N $C{=}CH_2$ $SiMe_3$

(b)

O O Br + MeO I MeO MeO $CH{=}N$— → O O HC N MeO OMe OMe

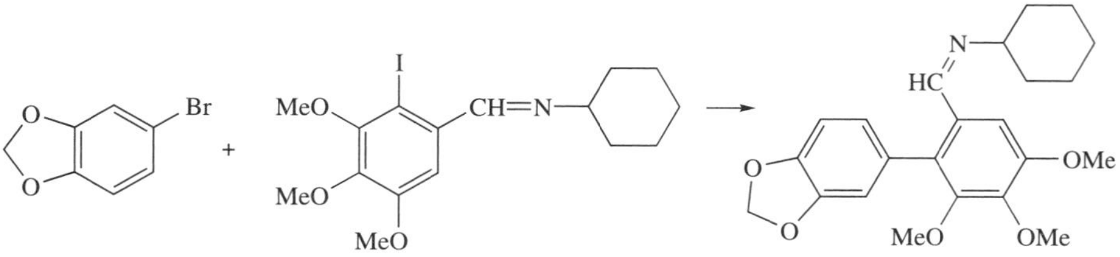

(c)

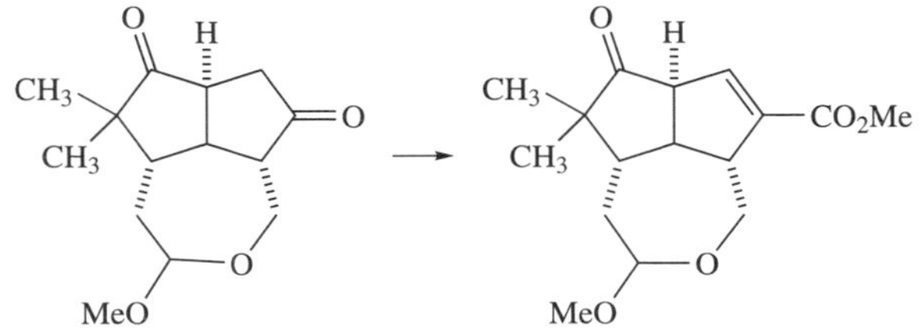
O
H
CH3
CH3
O
MeO
O
H
CH3
CH3
CO2Me
O
MeO

(d)

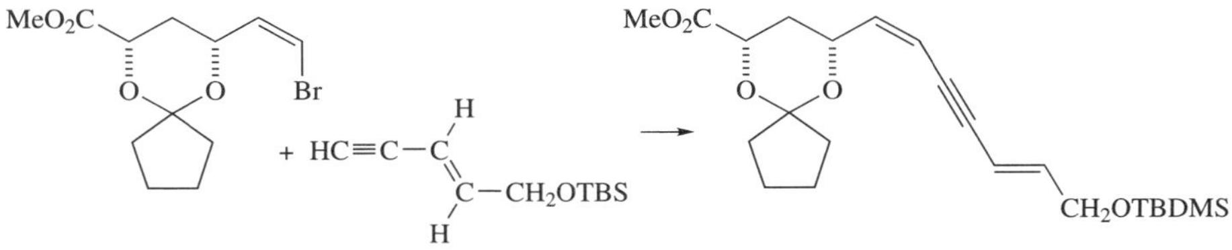
MeO2C
O
O
Br
+ HC≡C−C
H
C−CH2OTBS
H
MeO2C
O
O
CH2OTBDMS

(e)

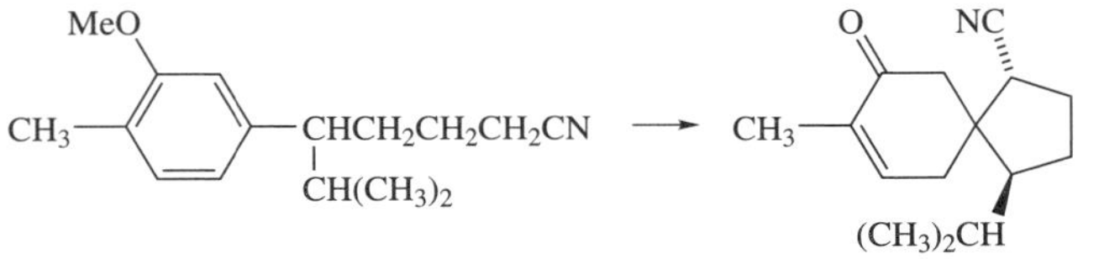
MeO
CH3
CHCH2CH2CH2CN
CH(CH3)2
O
NC
CH3
(CH3)2CH

(f)

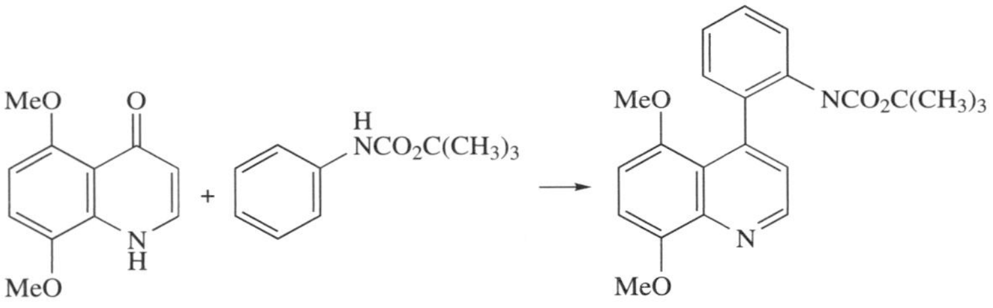
MeO
O
N
H
MeO
+
H
NCO2C(CH3)3
MeO
NCO2C(CH3)3
N
MeO

16. Each of the following reactions is accomplished with a palladium reagent or catalyst. Write a detailed mechanism for each reaction. The number of equivalents of each reagent which is used is given in parentheses. Be sure your mechanism accounts for the regeneration of a catalytically active species in those reactions which are catalytic palladium.

(a)

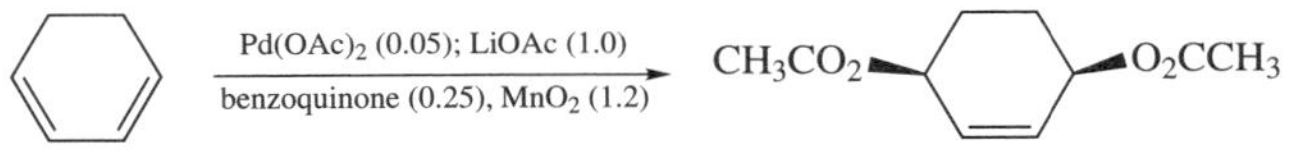
Pd(OAc)2 (0.05); LiOAc (1.0)
benzoquinone (0.25), MnO2 (1.2)
CH3CO2
O2CCH3

(b)

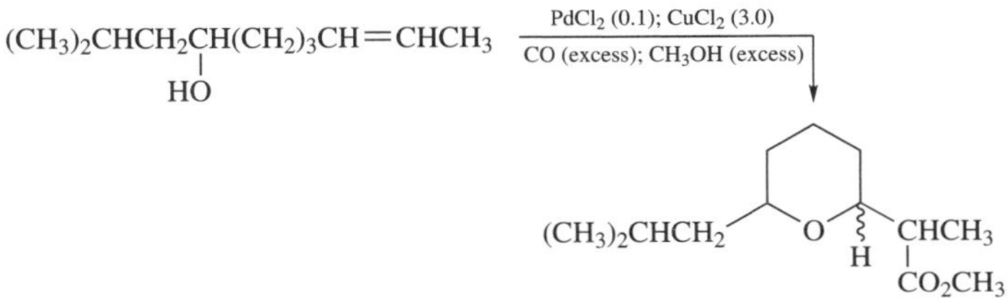
(CH3)2CHCH2CH(CH2)3CH=CHCH3
HO
PdCl2 (0.1); CuCl2 (3.0)
CO (excess); CH3OH (excess)
(CH3)2CHCH2
O
H
CHCH3
CO2CH3

(c)

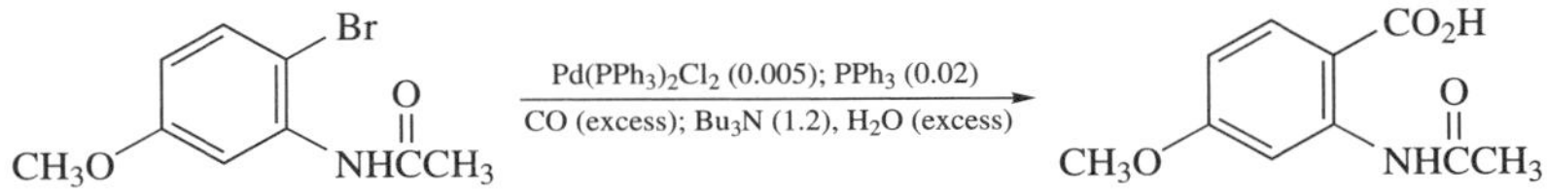
Br
CH3O
O
NHCCH3
Pd(PPh3)2Cl2 (0.005); PPh3 (0.02)
CO (excess); Bu3N (1.2), H2O (excess)
CO2H
CH3O
O
NHCCH3

(d) $CH_2{=}CHCH_2CH_2$, CH_3, CH_3, $CH_2CO_2C_2H_5$, $OSi(CH_3)_3$ $\xrightarrow{Pd(OAc)_2\ (1.0)}$ CH_3, CH_3, O, CH_2, CO_2CH_3, CH_2 + CH_3, CH_3, O, CH_2, CO_2CH_3, CH_3

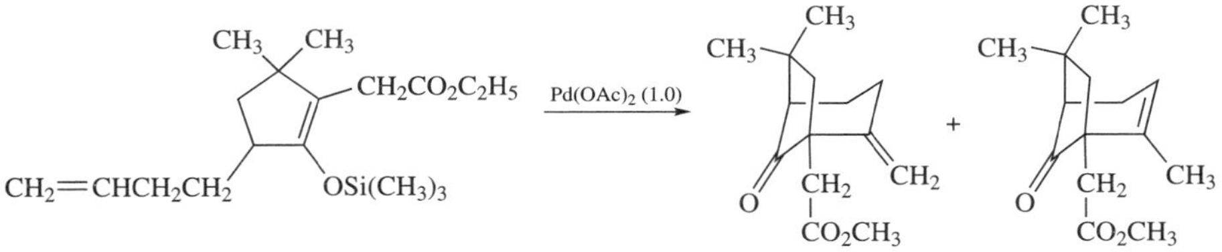

17. The reaction of lithium dimethylcuprate with **H** shows considerable 1,4-diastereoselectivity. Offer an explanation in the form of a transition-state model.

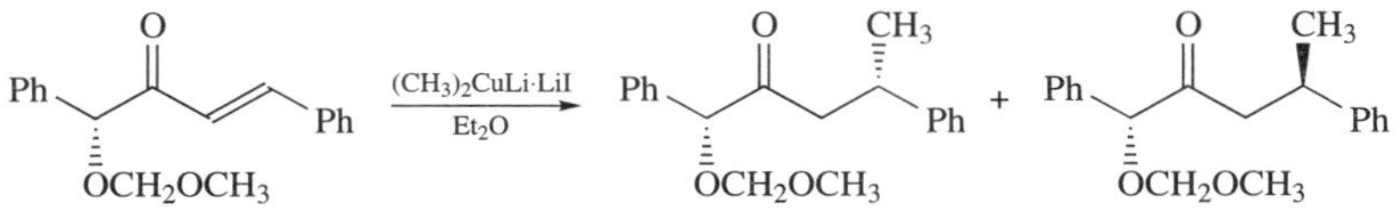

Ph, O, OCH_2OCH_3, Ph **H** $\xrightarrow[Et_2O]{(CH_3)_2CuLi{\cdot}LiI}$ Ph, O, CH_3, Ph, OCH_2OCH_3 + Ph, O, CH_3, Ph, OCH_2OCH_3

13:1 ratio

18. The following transformations have been carried out *enantiospecifically* by synthetic sequences involving organometallic reagents. Devise a strategy by which each desired material could be prepared in high enantiomeric purity from the specified starting material.

(a) $CH_2{=}CHCH_2$, $CH_2CH_2CH{=}CH_2$, C, H, OH $\Longrightarrow$ CH_3O_2C, $CH_2CO_2CH_3$, C, H, OH

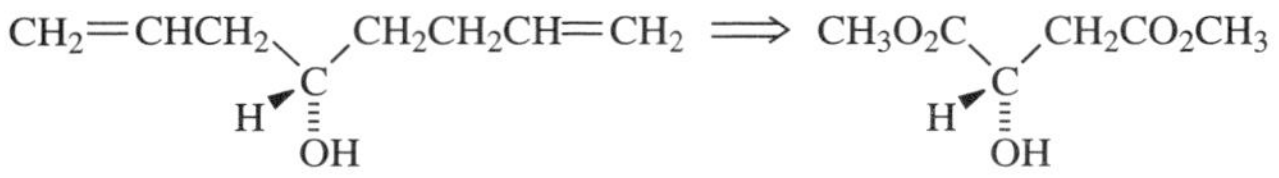

(b) $(CH_3)_2C{=}CHCH_2$, $CO_2C_2H_5$, C, H, OH $\Longrightarrow$ $HOCH_2$, CO_2H, C, H, NH_2

(c) N, Ts, CH_3O_2C $\Longrightarrow$ TsNH, CO_2CH_3, O_2CCH_3

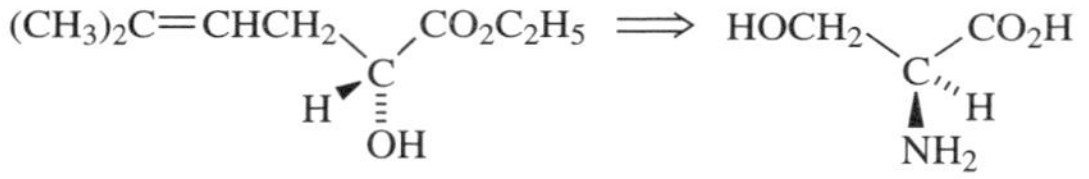

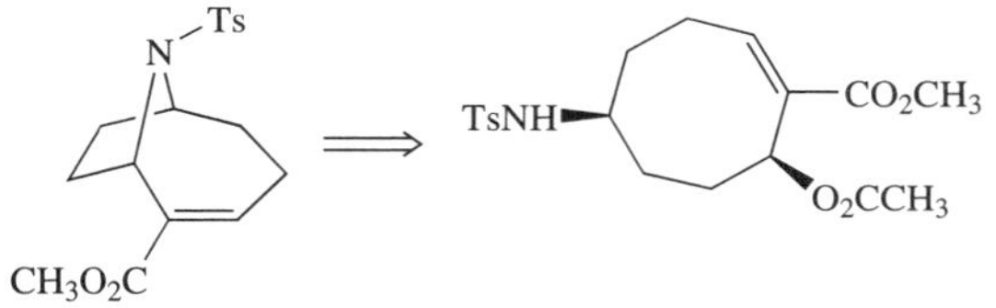

9

Carbon–Carbon Bond-Forming Reactions of Compounds of Boron, Silicon, and Tin

Introduction

In this chapter, we will discuss the use of boron, silicon, and tin compounds to form carbon–carbon bonds. These elements are at the boundary of the metals and nonmetals, with boron being the most and tin the least electronegative of the three. The neutral alkyl derivatives of boron have the formula R_3B, whereas for silicon and tin they are R_4Si and R_4Sn, respectively. These compounds are relatively volatile, nonpolar substances that exist as discrete molecules and in which the carbon–metal bonds are largely covalent. The boranes are Lewis acids, whereas the silanes and stannanes are not, unless substituted by a leaving group. The synthetically important reactions of these compounds involve transfer of a carbon substituent with one (radical equivalent) or two (carbanion equivalent) electrons to a reactive carbon center. This chapter will emphasize the nonradical reactions. In contrast to the transition metals, which often undergo a change in oxidation level at the metal during the reaction, there is usually no oxidation level change for boron, silicon, and tin compounds. We have already discussed one important aspect of boron and tin chemistry in the transmetalation reactions involved in Pd-catalyzed cross-coupling reactions discussed in Section 8.2.3.

9.1. Organoboron Compounds

9.1.1. Synthesis of Organoboranes

The most widely used route to organoboranes is hydroboration, which was discussed in Section 4.9.1. Hydroboration provides access to both alkyl- and alkenylboranes. Aryl-, methyl- and benzylboranes cannot be prepared by hydroboration. One route to methyl and aryl derivatives is by reaction of a dialkylborane, such as 9-BBN, with a cuprate reagent.[1]

$$\text{BH} + R_2CuLi \longrightarrow \text{B—R} + [RCuH]^-Li^+$$

These reactions occur by oxidative addition at copper, followed by decomposition of the Cu(III) intermediate.

$$R'_2B\text{—}H + {}^-Cu^IR_2 \longrightarrow R'_2B\text{—}Cu^{III}(H)(R)R \longrightarrow R'_2B\text{—}R + [RCu^IH]^-$$

Two successive reactions with different organocuprates can convert thexylborane to an unsymmetrical trialkylborane.[2]

$$\text{ThexylBH}_2 \xrightarrow{R^1_2CuLi} \xrightarrow{R^2_2CuLi} \text{ThexylB}(R^1)(R^2)$$

Alkyl, aryl, and allyl derivatives of boron can be prepared directly from the corresponding halides, BF_3, and magnesium metal. This process presumably involves *in situ* generation of a Grignard reagent, which then displaces fluoride from boron.[3]

$$3\ R\text{—}X + BF_3 + 3\ Mg \longrightarrow R_3B + 3\ MgXF$$

Organometallic displacement reactions on haloboranes provide another route to boranes that cannot be obtained by hydroboration.[4]

$$[\ \ ]_2BCl \xrightarrow{CH_2{=}CHCHMgBr} [\ \ ]_2BCH_2CH{=}CH_2$$

Alkoxy groups can be displaced from boron by both alkyl- and aryllithium reagents. The reaction of diisopropoxyboranes with an organolithium reagent, for example, provides good yields of unsymmetrically disubstituted isopropoxyboranes.[5]

$$RB(O\text{—}i\text{-}Pr)_2 + R'Li \longrightarrow RR'B\text{—}O\text{—}i\text{-}Pr$$

Alkoxyboron compounds are usually named as esters. Compounds with one alkoxy group are esters of borinic acids and are called borinates. Compounds with two alkoxy groups are

1. C. G. Whiteley, *J. Chem. Soc., Chem. Commun.* **1981**:5.
2. C. G. Whiteley, *Tetrahedron Lett.* **25**:5563 (1984).
3. H. C. Brown and U. S. Racherla, *J. Org. Chem.* **51**:427 (1986).
4. H. C. Brown and P. K. Jadhar, *J. Am. Chem. Soc.* **105**:2092 (1983).
5. H. C. Brown, T. E. Cole, and M. Srebnik, *Organometallics* **4**:1788 (1985).

called boronate esters.

R_2BOH	R_2BOR'	$RB(OH)_2$	$RB(OR')_2$
borinic acid	borinate ester	boronic acid	boronate ester

9.1.2. Carbon–Carbon Bond-Forming Reactions of Organoboranes

9.1.2.1. Carbonylation and Other One-Carbon Homologation Reactions. The reactions of organoboranes that were discussed in Chapter 4 are valuable methods for introducing functional groups into alkenes. In this section, we will discuss carbon–carbon-bond forming reactions of organoboranes.[6] Trivalent organoboranes are not very nucleophilic, but they are moderately reactive Lewis acids. Most reactions in which carbon–carbon bonds are formed involve a tetracoordinate intermediate with a negative charge on boron. Adduct formation weakens the boron–carbon bonds and permits a transfer of a carbon substituent with its electrons. The general mechanistic pattern is:

$$R_3B + :\mathbf{Nu}^- \longrightarrow [R_3B{-}\mathbf{Nu}]^-$$

$$[R_3B{-}\mathbf{Nu}]^- + \mathbf{E}^+ \longrightarrow R_2B{-}\mathbf{Nu} + R{-}\mathbf{E}$$

The electrophilic center is sometimes generated from the Lewis base by formation of the adduct.

$$R_3B + :\mathbf{Nu}{=}X \longrightarrow R_3\overset{-}{B}{-}\overset{+}{\mathbf{Nu}}{=}X \qquad R_2B{-}\overset{+}{\mathbf{Nu}}(R){-}X^-$$

An important group of reactions of this type are the reactions of organoboranes with carbon monoxide. Carbon monoxide forms Lewis acid–base complexes with organoboranes. In these adducts, the boron bears a formal negative charge and carbon is electrophilic because of the triple bond to oxygen bearing a formal positive charge. The adducts undergo boron-to-carbon migration of the boron substituents. The reaction can be controlled so that it results in the migration of one, two, or all three of the boron substituents.[7]

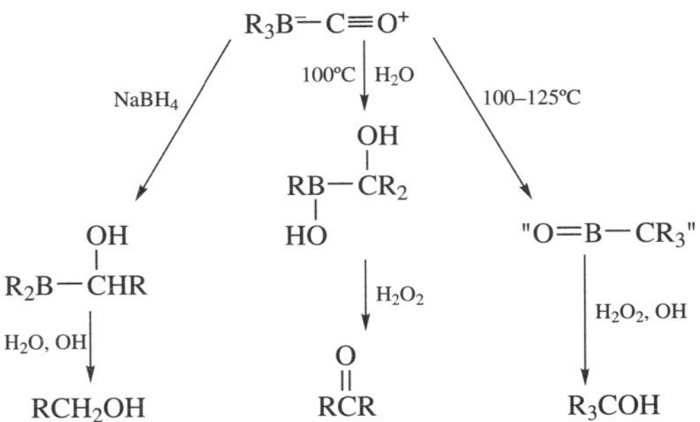

If the organoborane is heated with carbon monoxide to 100–125°C, all of the groups migrate and a tertiary alcohol is obtained after workup by oxidation. The presence of water

6. For a review of this topic see E. Negishi and M. Idacavage, *Org. React.* **33**:1 (1985).
7. H. C. Brown and M. W. Rathke, *J. Am. Chem. Soc.* **89**:2737 (1967).

in the reaction mixture causes the reaction to cease after migration of two groups from boron to carbon. Oxidation of the reaction mixture at this stage gives a ketone.[8] Primary alcohols are obtained when the carbonylation is carried out in the presence of sodium borohydride or lithium borohydride.[9] The product of the first migration step is reduced, and subsequent hydrolysis gives a primary alcohol.

In this synthesis of primary alcohols, only one of the three groups in the organoborane is converted to product. This disadvantage can be overcome by using a dialkylborane, particularly 9-BBN, in the initial hydroboration. After carbonylation and B $\rightarrow$ C migration, the reaction mixture can be processed to give an aldehyde, an alcohol, or the homologated 9-alkyl-BBN.[10] The utility of 9-BBN in these procedures is the result of the minimal tendency of the bicyclic ring to undergo migration.

H
B

RCH=CH$_2$

B—CH$_2$CH$_2$R

CO | KBH(O-*i*-Pr)$_3$

$^-$OH

HOCH$_2$CH$_2$CH$_2$R

OH

B—CHCH$_2$CH$_2$R

LiAlH
–20°C

B—CH$_2$CH$_2$CH$_2$R

H$_2$O$_2$

O=CHCH$_2$CH$_2$R

Several alternative procedures have been developed in which other reagents replace carbon monoxide as the migration terminus. Perhaps the most generally applicable of these methods involves the use of cyanide ion and trifluoroacetic anhydride as illustrated by entries 9 and 10 in Scheme 9.1. In this reaction, the borane initially forms an adduct with cyanide ion. The migration is induced by N-acylation of the cyano group by trifluoroacetic anhydride.

$$R_3B + {}^-CN \longrightarrow R_3\bar{B}{-}C{\equiv}N \xrightarrow{(CF_3CO)_2O} R_3\bar{B}{-}C{\equiv}\overset{+}{N}{-}\overset{O}{\overset{\|}{C}}CF_3$$

$$R_3\bar{B}{-}C{\equiv}\overset{+}{N}{-}\overset{O}{\overset{\|}{C}}CF_3 \longrightarrow R_2B{-}\underset{R}{\underset{|}{C}}{=}N{-}\overset{O}{\overset{\|}{C}}CF_3 \longrightarrow R{-}B\text{(O—CCF}_3\text{, N, C(R)R ring)} \xrightarrow{H_2O_2} R\overset{O}{\overset{\|}{C}}R$$

Another useful reagent for introduction of the carbonyl carbon is dichloromethyl methyl ether. In the presence of a hindered alkoxide base, it is deprotonated and acts as a

8. H. C. Brown and M. W. Rathke, *J. Am. Chem. Soc.* **89:**2738 (1967).
9. M. W. Rathke and H. C. Brown, *J. Am. Chem. Soc.* **89:**2740 (1967).
10. H. C. Brown, E. F. Knights, and R. A. Coleman, *J. Am. Chem. Soc.* **91:**2144 (1969); H. C. Brown, T. M. Ford, and J. L. Hubbard, *J. Org. Chem.* **45:**4067 (1980).

nucleophile toward boron. Rearrangement then ensues.

$$R_3B + :\bar{C}Cl_2OCH_3 \longrightarrow R_3B^{-}\text{—}CCl_2OCH_3$$

$$R_3B^{-}\text{—}CCl_2OCH_3 \longrightarrow R_2B\text{—}\underset{R}{CClOCH_3} \longrightarrow R\underset{Cl}{\overset{Cl}{B}}\text{—}\underset{R}{\overset{R}{C}}OCH_3$$

$$R\text{—}\underset{Cl}{\overset{Cl}{B}}\text{—}\underset{R}{\overset{R}{C}}OCH_3 \xrightarrow{H_2O_2} R_2C{=}O$$

Entries 11 and 12 in Scheme 9.1 illustrate other methods which proceed by a generally similar mechanism involving adduct formation and a B → C rearrangement. Problem 9.3 deals with the mechanisms of these reactions.

Unsymmetrical ketones can be made by using either thexylborane or thexylchloroborane.[11] Thexylborane works well when one of the desired carbonyl substituents is derived from a moderately hindered alkene. Under these circumstances, a clean monoalkylation of thexylborane can be accomplished. This is followed by reaction with a second alkene and carbonylation.

$$(CH_3)_2CH\underset{CH_3}{\overset{CH_3}{C}}\text{—}BH_2 \xrightarrow{RCH{=}CHR} \xrightarrow{R'CH{=}CH_2} \xrightarrow[\text{2) } H_2O,\ 100^\circ C \quad \text{3) } H_2O_2]{\text{1) CO}} RCH_2\underset{R}{CH}\overset{O}{\overset{\|}{C}}CH_2CH_2R'$$

Thexylchloroborane can be alkylated and then converted to a dialkylborane by a reducing agent such as $KBH[OCH(CH_3)_2]_3$. This approach is preferred for terminal alkenes.

$$(CH_3)_2CH\underset{CH_3}{\overset{CH_3}{C}}\text{—}BHCl \xrightarrow[\text{2) } KBH[OCH(CH_3)_2]_3]{\text{1) } CH_3CH_2CH{=}CH_2} (CH_3)_2CH\underset{CH_3}{\overset{CH_3}{C}}\text{—}BH\text{—}CH_2CH_2CH_2CH_3$$

$$\downarrow \text{ (cyclohexenyl)}CH{=}CH_2$$

$$(CH_3)_2CH\underset{CH_3}{\overset{CH_3}{C}}H\text{—}B(CH_2CH_2CH_2CH_3)(CH_2CH_2\text{—cyclohexenyl}) \xrightarrow[\text{2) } (CF_3CO)_2O,\ -78^\circ C \quad \text{3) NaOH, } H_2O_2]{\text{1) NaCN}} \text{cyclohexenyl—}CH_2CH_2\overset{O}{\overset{\|}{C}}CH_2CH_2CH_2CH_3 \quad 67\%$$

The success of both of these approaches depends upon the thexyl group being noncompetitive with the other groups in the migration steps.

The formation of unsymmetrical ketones can also be done starting with $IpcBCl_2$. Sequential reduction and hydroboration is carried out with two different alkenes. The first

11. H. C. Brown and E. Negishi, *J. Am. Chem. Soc.* **89:**5285 (1967); S. U. Kulkarni, H. D. Lee, and H. C. Brown, *J. Org. Chem.* **45:**4542 (1980).
12. H. C. Brown, S. V. Kulkarni, U. S. Racherla, and U. P. Dhokte, *J. Org. Chem.* **63:**7030 (1998).

Scheme 9.1. Homologation of Organoboranes by Carbon Monoxide and Other One-Carbon Donors

A. Formation of alcohols

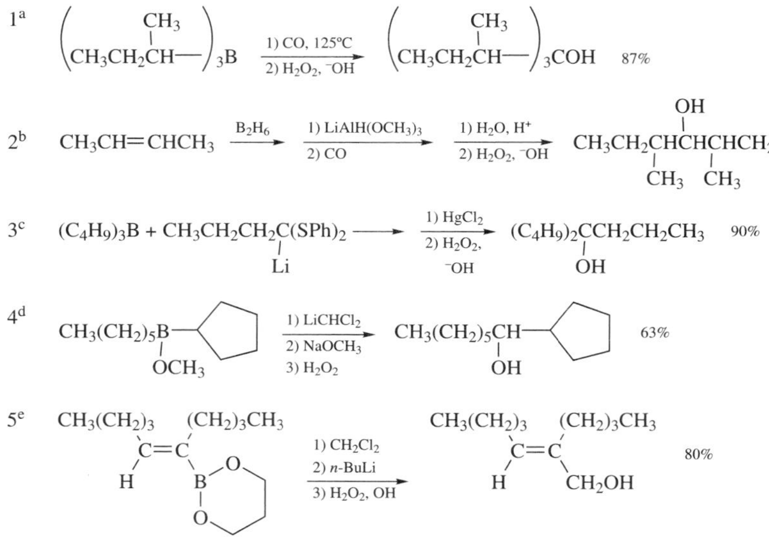

1[a] $(CH_3CH_2CH(CH_3)-)_3B$ —1) CO, 125°C; 2) H_2O_2, ^-OH→ $(CH_3CH_2CH(CH_3)-)_3COH$ 87%

2[b] $CH_3CH{=}CHCH_3$ —B_2H_6→ —1) $LiAlH(OCH_3)_3$; 2) CO→ —1) H_2O, H^+; 2) H_2O_2, ^-OH→ $CH_3CH_2CH(CH_3)CH(OH)CH(CH_3)CH_2CH_3$ 82%

3[c] $(C_4H_9)_3B + CH_3CH_2CH_2C(Li)(SPh)_2$ ⟶ —1) $HgCl_2$; 2) H_2O_2, ^-OH→ $(C_4H_9)_2C(OH)CH_2CH_2CH_3$ 90%

4[d] $CH_3(CH_2)_5B(OCH_3)$–cyclopentyl —1) $LiCHCl_2$; 2) $NaOCH_3$; 3) H_2O_2→ $CH_3(CH_2)_5CH(OH)$–cyclopentyl 63%

5[e] $CH_3(CH_2)_3(H)C{=}C((CH_2)_3CH_3)$–B(1,3,2-dioxaborinane) —1) CH_2Cl_2; 2) n-BuLi; 3) H_2O_2, OH→ $CH_3(CH_2)_3(H)C{=}C((CH_2)_3CH_3)CH_2OH$ 80%

B. Formation of ketones

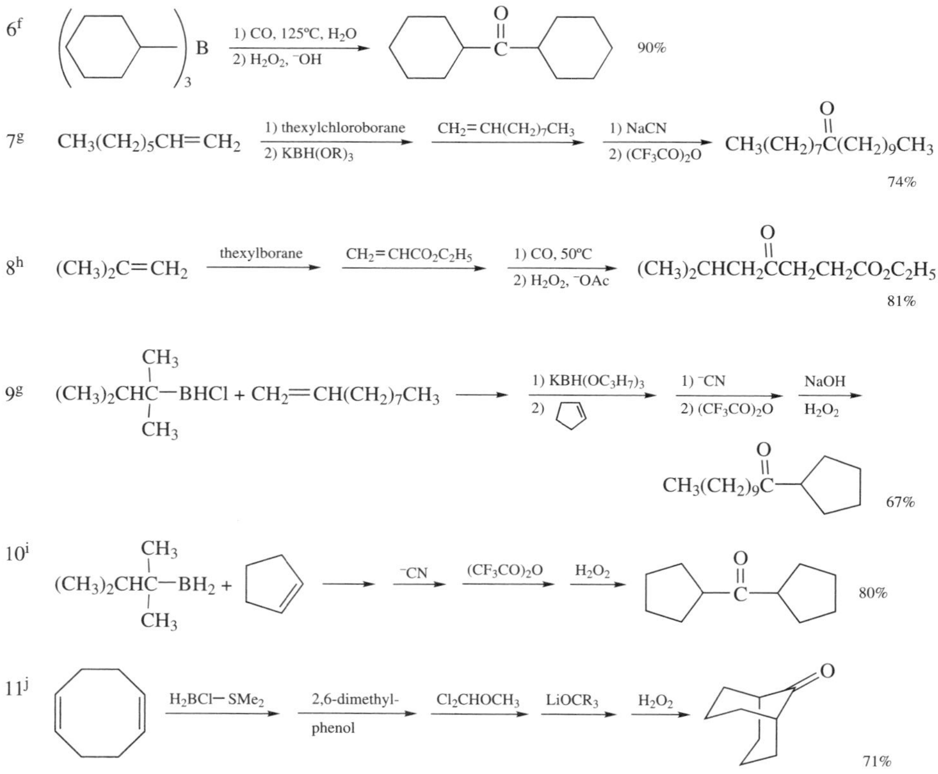

6[f] (cyclohexyl)$_3$B —1) CO, 125°C, H_2O; 2) H_2O_2, ^-OH→ cyclohexyl–C(=O)–cyclohexyl 90%

7[g] $CH_3(CH_2)_5CH{=}CH_2$ —1) thexylchloroborane; 2) $KBH(OR)_3$→ —$CH_2{=}CH(CH_2)_7CH_3$→ —1) NaCN; 2) $(CF_3CO)_2O$→ $CH_3(CH_2)_7C(=O)(CH_2)_9CH_3$ 74%

8[h] $(CH_3)_2C{=}CH_2$ —thexylborane→ —$CH_2{=}CHCO_2C_2H_5$→ —1) CO, 50°C; 2) H_2O_2, ^-OAc→ $(CH_3)_2CHCH_2C(=O)CH_2CH_2CO_2C_2H_5$ 81%

9[g] $(CH_3)_2CHC(CH_3)_2$–BHCl + $CH_2{=}CH(CH_2)_7CH_3$ ⟶ —1) $KBH(OC_3H_7)_3$; 2) cyclopentene→ —1) ^-CN; 2) $(CF_3CO)_2O$→ —NaOH, H_2O_2→ $CH_3(CH_2)_9C(=O)$–cyclopentyl 67%

10[i] $(CH_3)_2CHC(CH_3)_2$–BH_2 + cyclopentene ⟶ —^-CN→ —$(CF_3CO)_2O$→ —H_2O_2→ cyclopentyl–C(=O)–cyclopentyl 80%

11[j] 1,5-cyclooctadiene —$H_2BCl-SMe_2$→ —2,6-dimethylphenol→ —Cl_2CHOCH_3→ —$LiOCR_3$→ —H_2O_2→ bicyclo[3.3.1]nonan-9-one 71%

12[k] 1) CH_3Li; 2) CH_3OH, HCl; 1) Cl_2CHOCH_3, $(CH_3)_3CO^-K^+$; 2) H_2O_2

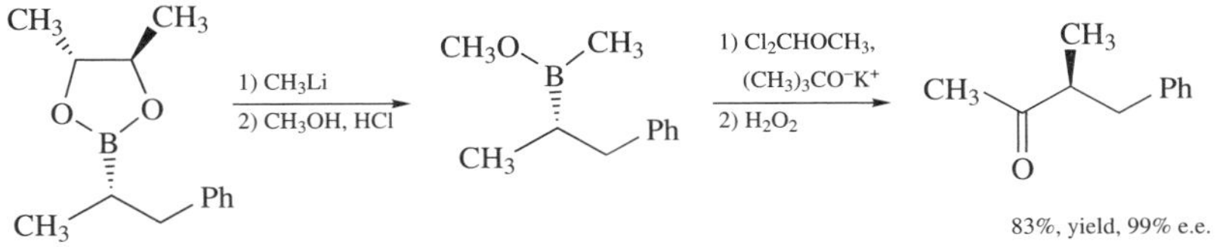

83%, yield, 99% e.e.

13[l] 1) siamylborane; 2) CO, 50°C; 3) H_2O_2

53%

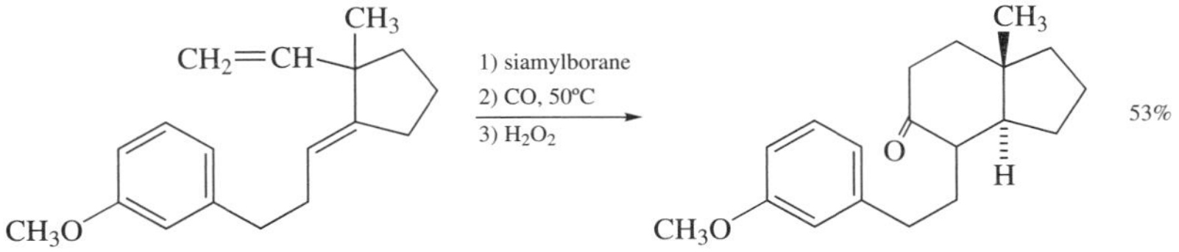

C. Formation of aldehydes

14[m] 9-BBN; 1) $KBH(O\text{-}i\text{-}Pr)_3$; 2) CO; 3) H_2O_2, ^-OH

96%

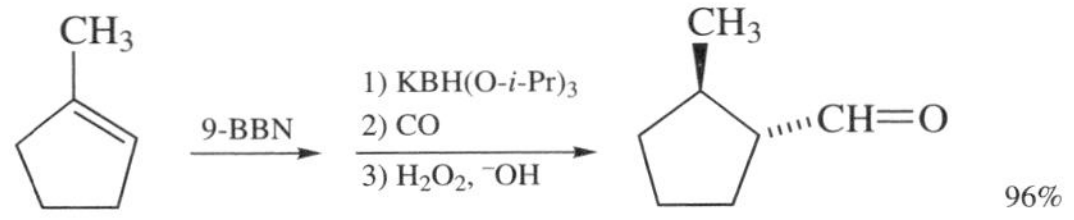

a. H. C. Brown and M. W. Rathke, *J. Am. Chem. Soc.* **89:**2737 (1967).
b. J. L. Hubbard and H. C. Brown, *Synthesis* **1978:**676.
c. R. J. Hughes, S. Ncube, A. Pelter, K. Smith, E. Negishi, and T. Yoshida, *J. Chem. Soc., Perkin Trans. 1* **1977:**1172; S. Ncube, A. Pelter, and K. Smith, *Tetrahedron Lett.* **1979:**1895.
d. H. C. Brown, T. Imai, P. T. Perumal, and B. Singaram, *J. Org. Chem.* **50:**4032 (1985).
e. H. C. Brown, A. S. Phadke, and N. G. Bhat, *Tetrahedron Lett.* **34:**7845 (1993).
f. H. C. Brown and M. W. Rathke, *J. Am. Chem. Soc.* **89:**2738 (1967).
g. S. U. Kulkarni, H. D. Lee, and H. C. Brown, *J. Org. Chem.* **45:**4542 (1980).
h. H. C. Brown and E. Negishi, *J. Am. Chem. Soc.* **89:**5285 (1967).
i. A. Pelter, K. Smith, M. G. Hutchings, and K. Rowe, *J. Chem. Soc., Perkin Trans. 1* **1975:**129.
j. H. C. Brown and S. U. Kulkarni, *J. Org. Chem.* **44:**2422 (1979).
k. M. V. Rangaishenvi, B. Singaran, and H. C. Brown, *J. Org. Chem.* **56:**3286 (1991).
l. T. A. Bryson and W. E. Pye, *J. Org. Chem.* **42:**3214 (1977).
m. H. C. Brown, J. L. Hubbard, and K. Smith, *Synthesis* **1979:**701.

reduction can be done with $(CH_3)_3SiH$, but the second stage requires $LiAlH_4$. In this procedure, dichloromethyl methyl ether is used as the source of the carbonyl carbon.[12]

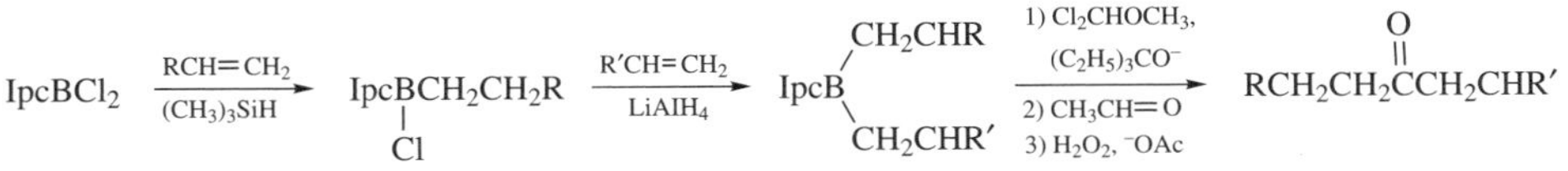

$$\mathrm{IpcBCl_2} \xrightarrow[(CH_3)_3SiH]{RCH=CH_2} \mathrm{IpcB(Cl)CH_2CH_2R} \xrightarrow[LiAlH_4]{R'CH=CH_2} \mathrm{IpcB(CH_2CHR)(CH_2CHR')} \xrightarrow[\text{2) } CH_3CH=O \text{ 3) } H_2O_2,\ ^-OAc]{\text{1) } Cl_2CHOCH_3,\ (C_2H_5)_3CO^-} \mathrm{RCH_2CH_2C(=O)CH_2CHR'}$$

As can be judged from the preceding discussion, organoboranes are versatile intermediates for formation of carbon–carbon bonds. An important aspect of all of these synthetic procedures involving boron-to-carbon migration is that they involve *retention of the configuration* of the migrating group. Because effective procedures for enantioselective hydroboration have been developed (see Section 4.9.3), these reactions offer the opportunity for enantioselective synthesis. A sequence for enantioselective formation of ketones starts with hydroboration of monoisopinocampheylborane ($IpcBH_2$), which can be obtained in high enantiomeric purity.[13] The hydroboration of a prochiral alkene establishes

13. H. C. Brown, P. K. Jadhav, and A. K. Mandal, *J. Org. Chem.* **47:**5074 (1982).

a new stereocenter. A third alkyl group can be introduced by a second hydroboration step.

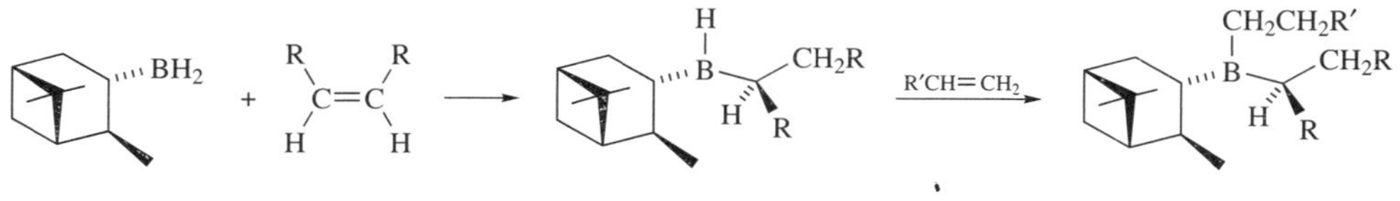

The trialkylborane can be transformed to a dialkyl(ethoxy)borane by heating with acetaldehyde, which releases the original chiral α-pinene. Finally, application of one of the carbonylation procedures outlined in Scheme 9.1 gives a chiral ketone.[14] The enantiomeric excess observed for ketones prepared in this way ranges from 60 to 90%.

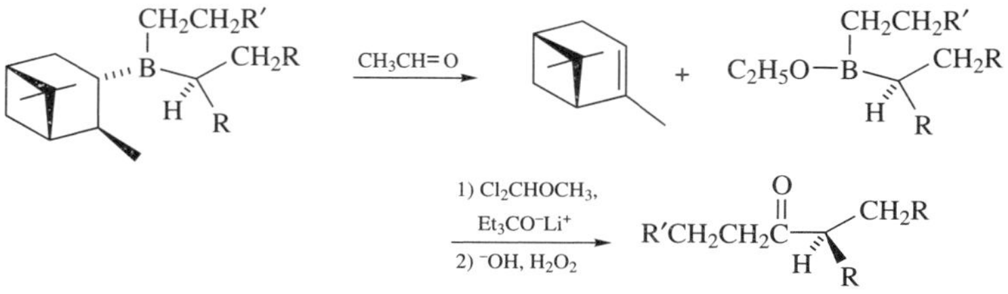

Higher enantiomeric purity can be obtained by a modified procedure in which the monoalkylborane intermediate is prepared.[15]

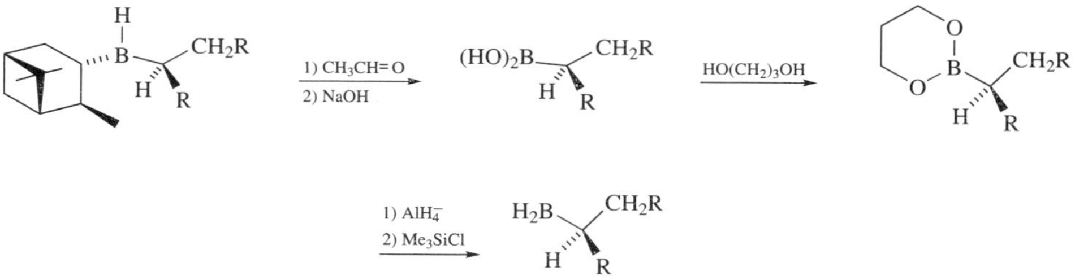

Subsequent steps involve introduction of a thexyl group and then the second ketone substituent. Finally, the ketone is formed by the cyanide–trifluoroacetic anhydride method.

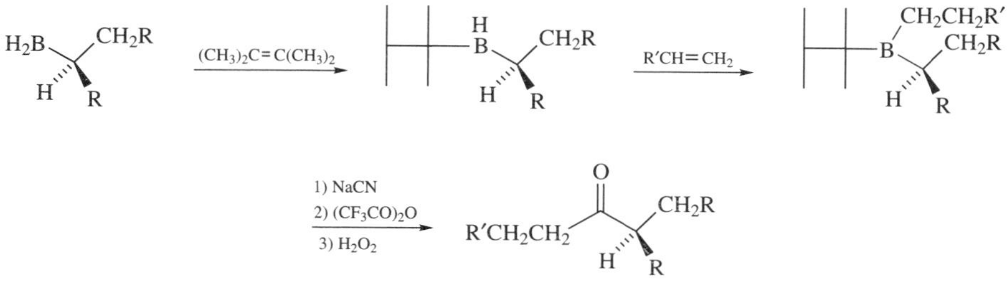

14. H. C. Brown, R. K. Jadhav, and M. C. Desai, *Tetrahedron* **40:**1325 (1984).
15. H. C. Brown, R. K. Bakshi, and B. Singaram, *J. Am. Chem. Soc.* **110:**1529 (1988); H. C. Brown, M. Srebnik, R. K. Bakshi, and T. E. Cole, *J. Am. Chem. Soc.* **109:**5420 (1987).

By starting with enantiomerically enriched IpcBHCl, it is possible to construct chiral cyclic ketones. For example, stepwise hydroboration of 1-allylcyclohexene and ring construction provides *trans*-1-decalone in >99% e.e.[16]

IpcBHCl + [1-allylcyclohexene] → (0.25 equiv $LiAlH_4$) [cyclohexenyl-$CH_2CH_2CH_2$-BHIpc]

1) $CH_3CH{=}O$ 2) Cl_2CHOCH_3 3) $(CH_3)_3CO^-K^+$ 4) H_2O_2 → *trans*-1-decalone, >99% e.e.

An efficient process for one-carbon homologation to aldehydes is based on cyclic boronate esters.[17] These can be prepared by hydroboration of an alkene with dibromoborane, followed by conversion of the dibromoborane to the cyclic ester. The homologation step is carried out by addition of methoxy(phenylthio)methyllithium to the boronate ester. The migration step is induced by mercuric ion. Use of enantioenriched boranes and boronates leads to products containing the groups of retained configuration.[18]

$$RCH{=}CH_2 + HBBr_2 \longrightarrow RCH_2CH_2BBr_2 \xrightarrow{Me_3SiO(CH_2)_3OSiMe_3} RCH_2CH_2B(OCH_2CH_2CH_2O)$$

$$RCH_2CH_2B(OCH_2CH_2CH_2O) + LiCH(SPh)OCH_3 \longrightarrow RCH_2CH_2B^-(OCH_2CH_2CH_2O)CH(SPh)OCH_3 \xrightarrow{Hg^{2+}} RCH_2CH_2CH(OCH_3)B(OCH_2CH_2CH_2O) \xrightarrow{H_2O_2,\ pH\ 8} RCH_2CH_2CH{=}O$$

9.1.2.2. Homologation via α-Haloenolates. Organoboranes can also be used to construct carbon–carbon bonds by several other types of reactions that involve migration of a boron substituent to carbon. One such reaction involves α-halocarbonyl compounds.[19] For example, ethyl bromoacetate reacts with trialkylboranes in the presence of base to give alkylated acetic acid derivatives in excellent yield. The reaction is most efficiently carried out with a 9-BBN derivative. These reactions can also be effected with B-alkenyl derivatives of 9-BBN to give *β,γ*-unsaturated esters.[20]

$$\text{9-BBN–R} + BrCH_2CO_2R' \xrightarrow{^-OC(CH_3)_3} RCH_2CO_2R'$$

16. H. C. Brown, V. K. Mahindroo, and U. P. Dhokte, *J. Org. Chem.* **61:**1906 (1996); U. P. Dhokte, P. M. Pathare, V. K. Mahindroo, and H. C. Brown, *J. Org. Chem.* **63:**8276 (1998).
17. H. C. Brown and T. Imai, *J. Am. Chem. Soc.* **105:**6285 (1983).
18. M. V. Rangashenvi, B. Singaram, and H. C. Brown, *J. Org. Chem.* **56:**3286 (1991).
19. H. C. Brown, M. M. Rogić, M. W. Rathke, and G. W. Kabalka, *J. Am. Chem. Soc.* **90:**818 (1968); H. C. Brown and M. M. Rogić, *J. Am. Chem. Soc.* **91:**2146 (1969).
20. H. C. Brown, N. G. Bhat, and J. B. Cambell, Jr., *J. Org. Chem.* **51:**3398 (1986).

The mechanism of these alkylations involves a tetracoordinate boron intermediate formed by addition of the enolate of the α-bromoester to the organoborane. The migration then occurs with displacement of bromide ion. In agreement with this mechanism, retention of configuration of the migrating group is observed.[21]

$$R_3B + \bar{C}H(Br)CO_2C_2H_5 \longrightarrow R_2\bar{B}(R)-CH(Br)CO_2C_2H_5 \longrightarrow R_2B-CH(R)CO_2C_2H_5 \xrightarrow{RO^-} RCH_2CO_2C_2H_5$$

α-Haloketones and α-halonitriles undergo similar reactions.[22]

A closely related reaction employs α-diazoesters or α-diazoketones.[23] With these compounds, molecular nitrogen acts as the leaving group in the migration step. The best results are achieved using dialkylchloroboranes or monoalkyldichloroboranes.

$$RBCl_2 + N_2CHCO_2CH_3 \longrightarrow RCH_2CO_2CH_3$$

A number of these alkylation reactions are illustrated in Scheme 9.2.

9.1.2.3. Stereoselective Alkene Synthesis. Terminal alkynes can also be alkylated by organoboranes. Adducts are formed between a lithium acetylide and a trialkylborane. Reaction with iodine induces migration and results in the formation of the alkylated alkyne.[24]

$$(\text{cyclopentyl})_3\bar{B}-C\equiv C(CH_2)_3CH_3\ Li^+ \xrightarrow[-78^\circ C]{I_2} \text{cyclopentyl}-C\equiv C(CH_2)_3CH_3 \quad 100\%$$

The mechanism involves electrophilic attack by iodine at the triple bond, which induces migration of an alkyl group from boron. This is followed by elimination of dialkyliodoborane.

$$Li^+\ R_3B^--C\equiv C-R' \xrightarrow{I_2} (R_2B)(R)C=C(I)(R') \longrightarrow R-C\equiv C-R' + R_2BI$$

Related procedures have been developed for the synthesis of both *Z*- and *E*-alkenes. Treatment of alkenyldialkylboranes with iodine results in the formation of the *Z*-alkene.[25]

$$(R_2B)(H)C=C(H)(R) \xrightarrow{I_2} \text{iodonium ion} \longrightarrow I^-\ \ RB(I)-CH(R)-CH(R)(I) \equiv \ldots \longrightarrow (H)(R)C=C(H)(R)$$

21. H. C. Brown, M. M. Rogić, M. W. Rathke, and G. W. Kabalka, *J. Am. Chem. Soc.* **91:**2151 (1969).
22. H. C. Brown, M. M. Rogić, H. Nambu, and M. W. Rathke, *J. Am. Chem. Soc.* **91:**2147 (1969); H. C. Brown, H. Nambu, and M. M. Rogić, *J. Am. Chem. Soc.* **91:**6853, 6855 (1969).
23. H. C. Brown, M. M. Midland, and A. B. Levy, *J. Am. Chem. Soc.* **94:**3662 (1972); J. Hooz, J. N. Bridson, J. G. Calzada, H. C. Brown, M. M. Midland, and A. B. Levy, *J. Org. Chem.* **38:**2574 (1973).
24. A. Suzuki, N. Miyaura, S. Abiko, H. C. Brown, J. A. Sinclair, and M. M. Midland, *J. Am. Chem. Soc.* **95:**3080 (1973); A. Suzuki, N. Miyaura, S. Abiko, M. Itoh, M. M. Midland, J. A. Sinclair, and H. C. Brown, *J. Org. Chem.* **51:**4507 (1986).
25. G. Zweifel, H. Arzoumanian, and C. C. Whitney, *J. Am. Chem. Soc.* **89:**3652 (1967); G. Zweifel, R. P. Fisher, J. T. Snow, and C. C. Whitney, *J. Am. Chem. Soc.* **93:**6309 (1971).

Scheme 9.2. Alkylation of Trialkylboranes by α-Halocarbonyl and Related Compounds

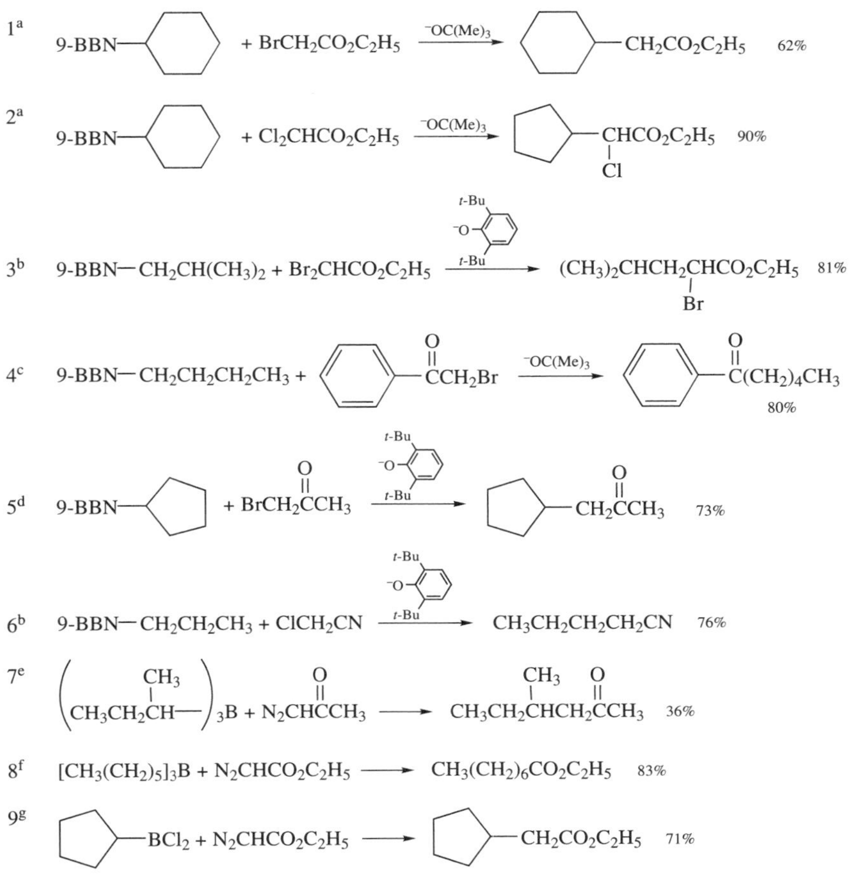

1[a] 9-BBN–cyclohexyl + $BrCH_2CO_2C_2H_5$ $\xrightarrow{^-OC(Me)_3}$ cyclohexyl–$CH_2CO_2C_2H_5$ 62%

2[a] 9-BBN–cyclohexyl + $Cl_2CHCO_2C_2H_5$ $\xrightarrow{^-OC(Me)_3}$ cyclopentyl–$CH(Cl)CO_2C_2H_5$ 90%

3[b] 9-BBN–$CH_2CH(CH_3)_2$ + $Br_2CHCO_2C_2H_5$ $\xrightarrow{\text{2,6-di-}t\text{-Bu-phenoxide}}$ $(CH_3)_2CHCH_2CH(Br)CO_2C_2H_5$ 81%

4[c] 9-BBN–$CH_2CH_2CH_2CH_3$ + $C_6H_5COCH_2Br$ $\xrightarrow{^-OC(Me)_3}$ $C_6H_5CO(CH_2)_4CH_3$ 80%

5[d] 9-BBN–cyclopentyl + $BrCH_2COCH_3$ $\xrightarrow{\text{2,6-di-}t\text{-Bu-phenoxide}}$ cyclopentyl–CH_2COCH_3 73%

6[b] 9-BBN–$CH_2CH_2CH_3$ + $ClCH_2CN$ $\xrightarrow{\text{2,6-di-}t\text{-Bu-phenoxide}}$ $CH_3CH_2CH_2CH_2CN$ 76%

7[e] $(CH_3CH_2CH(CH_3)-)_3B$ + $N_2CHCOCH_3$ $\longrightarrow$ $CH_3CH_2CH(CH_3)CH_2COCH_3$ 36%

8[f] $[CH_3(CH_2)_5]_3B$ + $N_2CHCO_2C_2H_5$ $\longrightarrow$ $CH_3(CH_2)_6CO_2C_2H_5$ 83%

9[g] cyclopentyl–BCl_2 + $N_2CHCO_2C_2H_5$ $\longrightarrow$ cyclopentyl–$CH_2CO_2C_2H_5$ 71%

a. H. C. Brown and M. M. Rogić *J. Am. Chem. Soc.* **91:**2146 (1969).
b. H. C. Brown, H. Nambu, and M. M. Rogić *J. Am. Chem. Soc.* **91:**6855 (1969).
c. H. C. Brown, M. M. Rogić, H. Nambu, and M. W. Rathke, *J. Am. Chem. Soc.* **91:**147 (1969).
d. H. C. Brown, H. Nambu, and M. M. Rogić, *J. Am. Chem. Soc.* **91:**6853 (1969).
e. J. Hooz and S. Linke, *J. Am. Chem. Soc.* **90:**5936 (1968).
f. J. Hooz and S. Linke, *J. Am. Chem. Soc.* **90:**6891 (1968).
g. J. Hooz, J. N. Bridson, J. G. Caldaza, H. C. Brown, M. M. Midland, and A. B. Levy, *J. Org. Chem.* **38:**2574 (1973).

Similarly, alkenyllithium reagents add to dimethyl boronates to give adducts which decompose to *Z*-alkenes on treatment with iodine.[26]

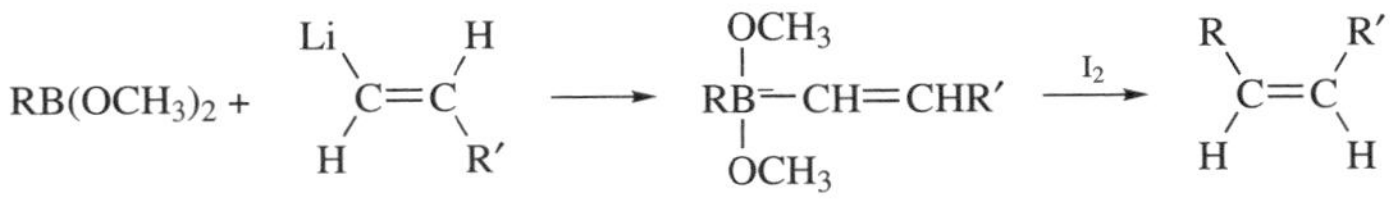

$$RB(OCH_3)_2 + \underset{H}{\overset{Li}{}}C{=}C\underset{R'}{\overset{H}{}} \longrightarrow R\overset{OCH_3}{\underset{OCH_3}{B^-}}{-}CH{=}CHR' \xrightarrow{I_2} \underset{H}{\overset{R}{}}C{=}C\underset{H}{\overset{R'}{}}$$

26. D. A. Evans, T. C. Crawford, R. C. Thomas, and J. A. Walker, *J. Org. Chem.* **41:**3947 (1976).

The synthesis of *Z*-alkenes can also be carried out by starting with an alkylbromoborane, in which case migration presumably follows replacement of the bromide by methoxide.[27]

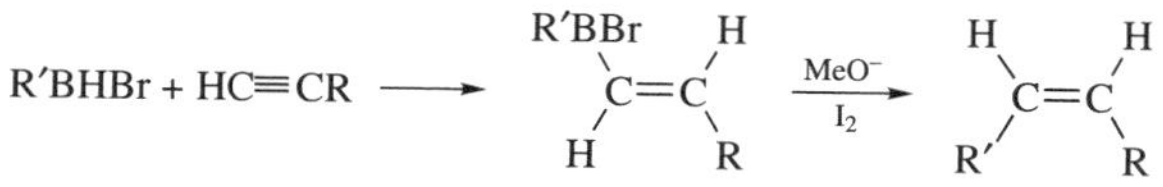

The stereoselectivity of these reactions arises from a base-induced *anti* elimination after the migration. The elimination is induced by addition of methoxide to the boron, generating an anionic center.

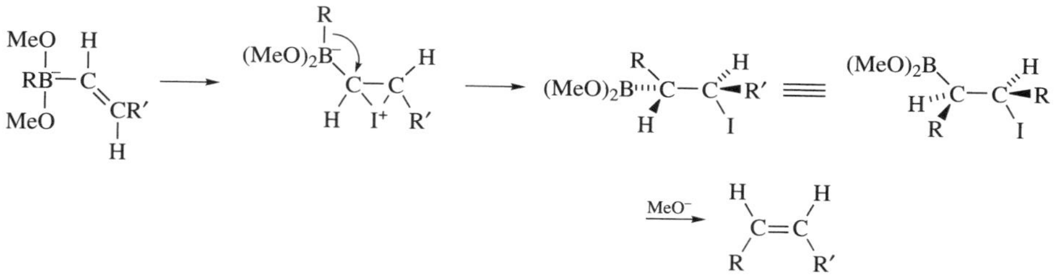

E-Alkenes can be prepared by several related reactions.[28] Hydroboration of a bromoalkyne generates an α-bromoalkenylborane. On treatment with methoxide ion, these intermediates undergo B → C migration to give an alkyl alkenylborinate. Protonolysis generates an *E*-alkene.

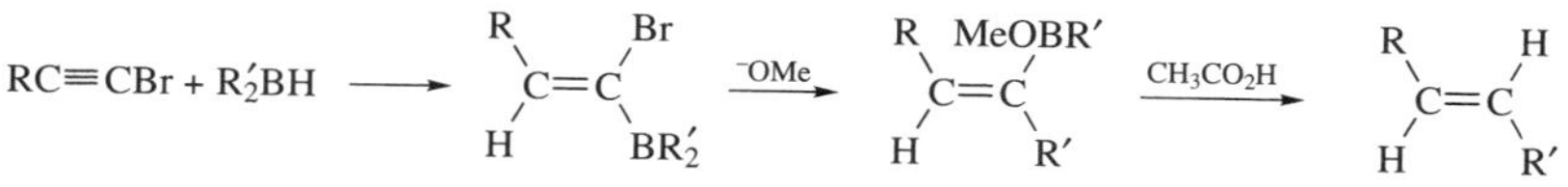

The dialkylboranes can be prepared from thexylchloroborane. The thexyl group does not normally migrate.

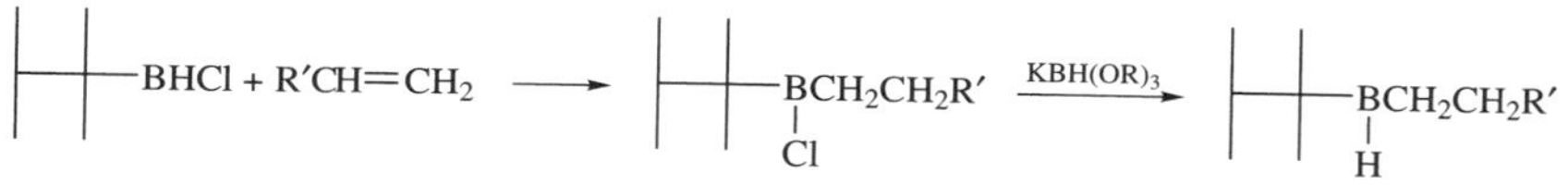

A similar strategy involves initial hydroboration by $BrBH_2$.[29]

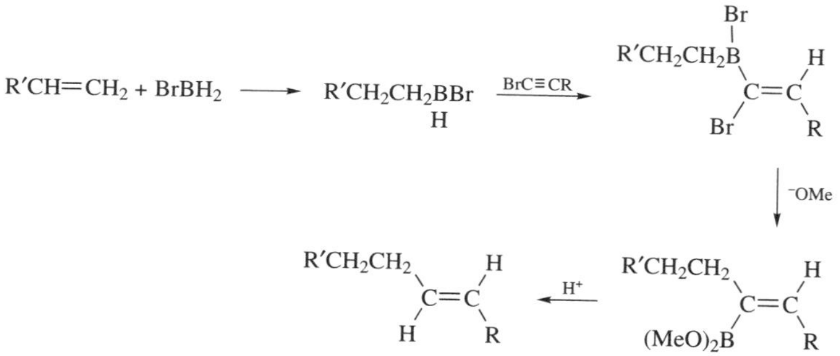

27. H. C. Brown, D. Basavaiah, S. U. Kulkarni, N. G. Bhat, and J. V. N. Vara Prasad, *J. Org. Chem.* **53:**239 (1988).
28. H. C. Brown, D. Basavaiah, S. U. Kulkarni, H. P. Lee, E. Negishi, and J.-J. Katz, *J. Org. Chem.* **51:**5270 (1986).
29. H. C. Brown, T. Imai, and N. G. Bhat, *J. Org. Chem.* **51:**5277 (1986); H. C. Brown, D. Basavaiah, and S. U. Kulkarni, *J. Org. Chem.* **47:**3808 (1982).

Stereoselective syntheses of trisubstituted alkenes are based on *E*- and *Z*-alkenyldioxaborinanes. Reaction with an alkyllithium reagent forms an "ate" adduct which rearranges on treatment with iodine in methanol.[30]

$$\text{(R)(H)C=C(R')(B-dioxaborinane)} \xrightarrow[\text{3) NaOH}]{\text{1) R''Li, 2) I}_2\text{, CH}_3\text{OH}} \text{(R)(H)C=C(R')(R'')} \quad \text{or} \quad \text{(R)(H)C=C(R')(B-dioxaborinane)} \xrightarrow[\text{3) NaOH}]{\text{1) R''Li, 2) I}_2\text{, CH}_3\text{OH}} \text{(R)(H)C=C(R'')(R')}$$

The B → C migration can also be induced by other types of electrophiles. Trimethylsilyl chloride or trimethylsilyl triflate induces a stereospecific migration to form β-trimethylsilyl alkenylboranes with the silicon and boron substituents *cis*.[31] It has been suggested that this stereospecificity arises from a silicon-bridged intermediate.

$$R_3\bar{B}-C{\equiv}C-R' + (CH_3)_3Si-X \longrightarrow R_2\bar{B}(R)C{=}C(R')\cdots Si(CH_3)_3\cdots X \longrightarrow (R)(R_2B)C{=}C(R')(Si(CH_3)_3)$$

Tributyltin chloride also induces migration and also gives the product in which the C–Sn bond is *cis* to the C–B bond. Protonolysis of both the C–Sn and C–B bonds by acetic acid gives the *Z*-alkene.[32]

$$R_3\bar{B}-C{\equiv}C-R' + ClSnR''_3 \longrightarrow (R)(R_2B)C{=}C(R')(SnR''_3) \xrightarrow{CH_3CO_2H} (R)(H)C{=}C(R')(H)$$

9.1.2.4. Nucleophilic Addition by Allylboron Derivatives. Allylboranes such as 9-allyl-9-BBN react with aldehydes and ketones to give allylic carbinols. Bond formation takes place at the γ carbon of the allyl group, and the double bond shifts.[33]

$$R_2C{=}O + CH_2{=}CHCH_2B< \longrightarrow [R_2C{=}O\cdots B(CH_2)\ H_2C{=}C(H)] \longrightarrow R_2C(CH_2CH{=}CH_2)-O-B< \xrightarrow{HO(CH_2)_2NH_2} R_2C(OH)CH_2CH{=}CH_2$$

This reaction begins by Lewis acid–base coordination at the carbonyl oxygen, which both increases the electrophilicity of the carbonyl group and weakens the C–B bond of the allyl

30. H. C. Brown and N. G. Bhat, *J. Org. Chem.* **53:**6009 (1988).
31. P. Binger and R. Köster, *Synthesis* **1973:**309; E. J. Corey and W. L. Seibel, *Tetrahedron Lett.* **27:**905 (1986).
32. K. K. Wang and K.-H. Chu, *J. Org. Chem.* **49:**5175 (1984).
33. G. W. Kramer and H. C. Brown, *J. Org. Chem.* **42:**2292 (1977).

borane. The dipolar adduct then reacts through a cyclic transition state. After the reaction is complete, the carbinol product is liberated from the borinate ester by displacement with ethanolamine. Yields for a series of aldehydes and ketones were usually above 90% with 9-allyl-9-BBN.

The cyclic mechanism would predict that the addition reaction would be stereospecific with respect to the geometry of the double bond in the allylic group. This has been demonstrated to be the case. The *E*- and *Z*-2-butenyl cyclic boronate esters **1** and **2** were synthesized and allowed to react with aldehydes. The *E*-boronate gave the carbinol having *anti* stereochemistry whereas the *Z*-boronate gave the *syn* product.[34]

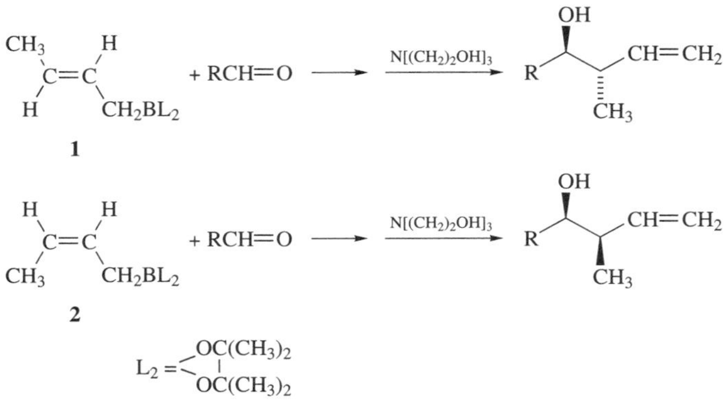

This stereochemistry is that predicted on the basis of a cyclic transition state in which the aldehyde substituent occupies an equatorial postion.

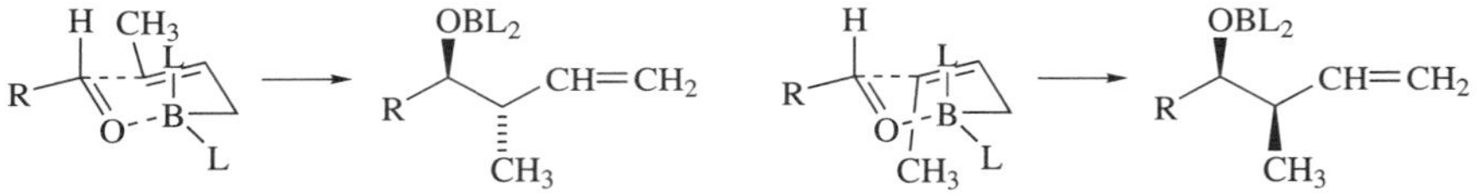

The diastereoselectivity observed in simple systems led to investigation of enantiomerically pure aldehydes. It was found that the *E*- and *Z*-2-butenylboronates both exhibit high *syn–anti* diastereoselectivity with chiral α-substituted aldehydes. However, only the *Z*-isomer also exhibited high selectivity toward the diastereotopic faces of the aldehyde.[35]

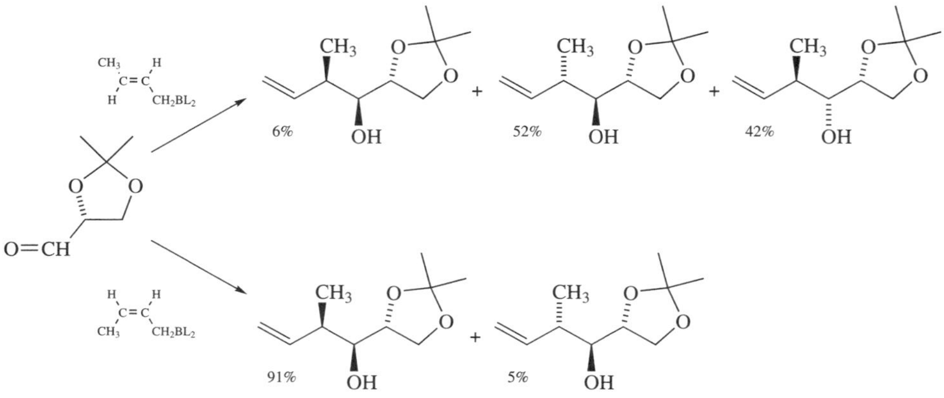

34. R. W. Hoffmann and H.-J. Zeiss, *J. Org. Chem.* **46:**1309 (1981); K. Fujita and M. Schlosser, *Helv. Chim. Acta* **65:**1258 (1982).
35. W. R. Roush, M. A. Adam, A. E. Walts, and D. J. Harris, *J. Am. Chem. Soc.* **108:**3422 (1986).

Addition reactions of allylic boron compounds have proven to be quite general and useful. Several methods for synthesis of allylic boranes and boronate esters have been developed.[36] The reaction has found some application in the stereoselective synthesis of complex structures.

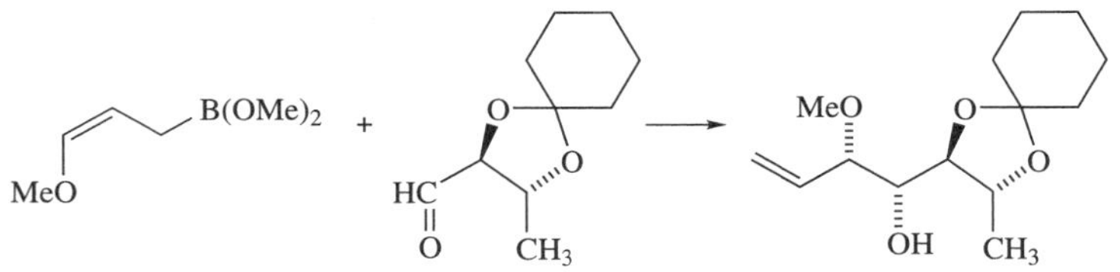

Ref. 37

The allylation reaction has also been extended to enantiomerically pure allylic boranes and borinates. For example, the 3-methyl-2-butenyl derivative of $(Ipc)_2BH$ reacts with aldehydes to give carbinols of >90% enantiomeric excess in most cases.[38]

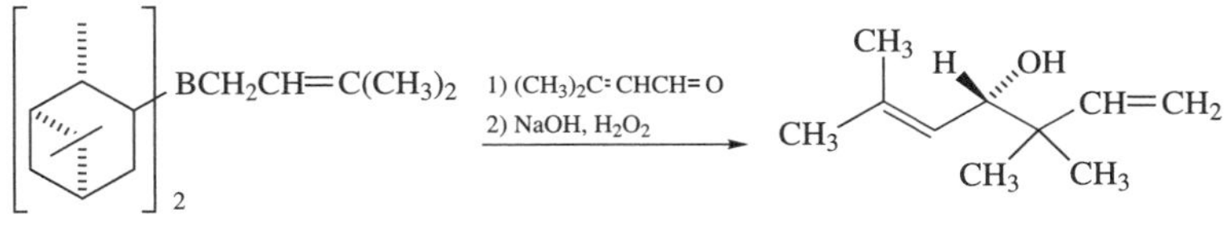

$BCH_2CH{=}C(CH_3)_2$ 1) $(CH_3)_2C{=}CHCH{=}O$ 2) NaOH, H_2O_2

85% yield
96% e.e.

B-Allyl-bis(isopinocampheyl)borane exhibits high stereoselectivity in reactions with chiral α-substituted aldehydes.[39]

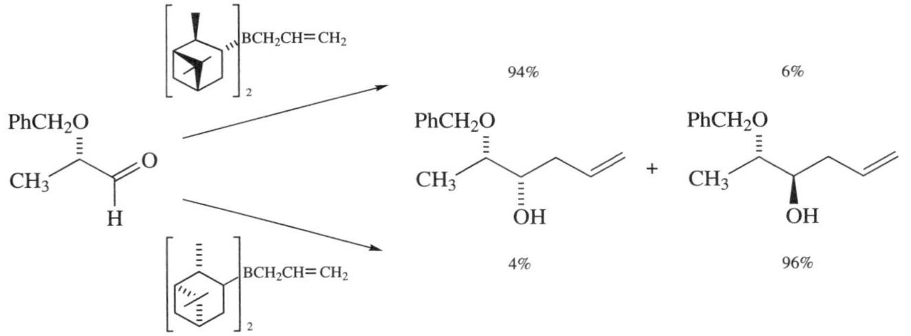

94% 6%

4% 96%

The most extensively developed allylboron reagents for enantioselective synthesis are derived from tartrate esters.[40]

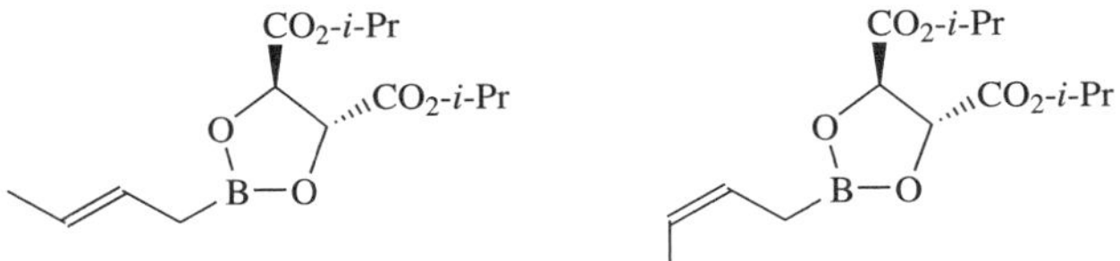

E-boronate *Z*-boronate

36. P. G. M. Wuts, P. A. Thompson, and G. R. Callen, *J. Org. Chem.* **48:**5398 (1983); E. Moret and M. Schlosser, *Tetrahedron Lett.* **25:**4491 (1984).
37. W. R. Roush, M. R. Michaelides, D. F. Tai, and W. K. M. Chong, *J. Am. Chem. Soc.* **109:**7575 (1987).
38. H. C. Brown and P. K. Jadhav, *Tetrahedron Lett.* **25:**1215 (1984); H. C. Brown, P. K. Jadhav, and K. S. Bhat, *J. Am. Chem. Soc.* **110:**1535 (1988).
39. H. C. Brown, K. S. Bhat, and R. S. Randad, *J. Org. Chem.* **52:**319 (1987); H. C. Brown, K. S. Bhat, and R. S. Randad, *J. Org. Chem.* **54:**1570 (1989).
40. W. R. Roush, K. Ando, D. B. Powers, R. L. Halterman, and A. Palkowitz, *Tetrahedron Lett.* **29:**5579 (1988); W. R. Roush, L. Banfi, J. C. Park, and L. K. Hoong, *Tetrahedron Lett.* **30:**6457 (1989).

With unhindered aldehydes such as cyclohexanecarboxaldehyde, the stereoselectivity is >95%, with the *E*-boronate giving the *anti* adduct and the *Z*-boronate giving the *syn* adduct. Enantioselectivity is about 90% for the *E*-boronate and 80% for the *Z*-boronate. With more hindered aldehydes, such as pivaldehyde, the diastereoselectivity is maintained but the enantioselectivity drops somewhat. These reagents also give excellent double stereodifferentiation when used with chiral aldehydes. For example, the aldehydes **3** and **4** give at least 90% enantioselection with both the *E*- and *Z*-boronates.[41]

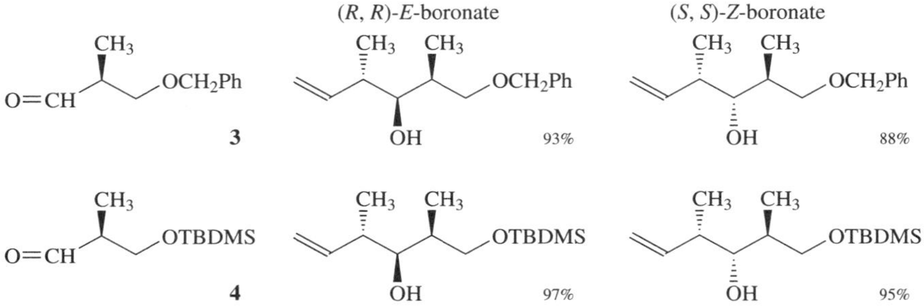

(*R*, *R*)-*E*-boronate (*S*, *S*)-*Z*-boronate

CH_3 OCH_2Ph $O{=}CH$ **3** — CH_3 CH_3 OCH_2Ph OH 93% — CH_3 CH_3 OCH_2Ph OH 88%

CH_3 OTBDMS $O{=}CH$ **4** — CH_3 CH_3 OTBDMS OH 97% — CH_3 CH_3 OTBDMS OH 95%

These reagents have proven to be very useful in stereoselective synthesis of polypectide natural products which frequently contain arrays of alternating methyl and oxygen substituents.[42]

The enantioselectivity is consistent with cyclic transition states. The key element determining the orientation of the aldehyde within the transition state is the interaction of the aldehyde group with the tartrate ester substituents.

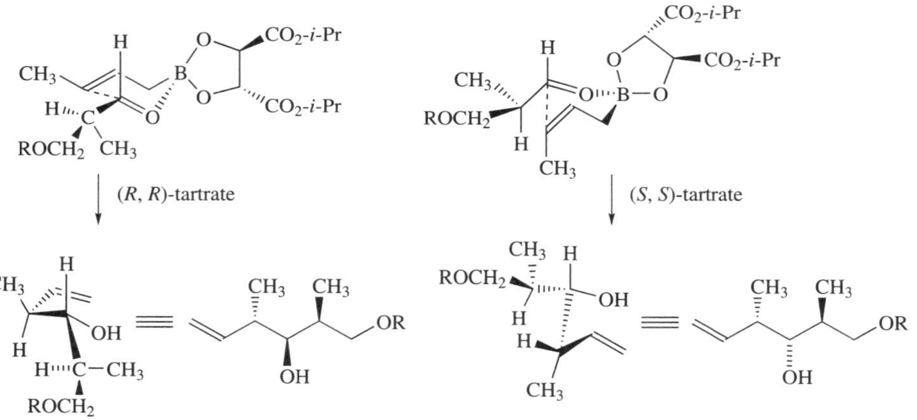

H CO_2-*i*-Pr CH_3 B O O CO_2-*i*-Pr H C O $ROCH_2$ CH_3 — H CO_2-*i*-Pr O CO_2-*i*-Pr CH_3 O B O $ROCH_2$ H CH_3

(*R*, *R*)-tartrate (*S*, *S*)-tartrate

H CH_3 OH H H C CH_3 $ROCH_2$ ≡ CH_3 CH_3 OR OH — CH_3 H $ROCH_2$ OH H H CH_3 ≡ CH_3 CH_3 OR OH

The preferred orientation results from the greater repulsive interaction between the carbonyl groups of the aldehyde and ester in the disfavored orientation.[43] This orientation

41. W. R. Roush, A. D. Palkowitz, and M. A. J. Palmer, *J. Org. Chem.* **52**:316 (1987); W. R. Roush, K. Ando, D. B. Powers, A. D. Palkowitz, and R. L. Halterman, *J. Am. Chem. Soc.* **112**:6339 (1990); W. R. Roush, A. D. Palkowitz, and K. Ando, *J. Am. Chem. Soc.* **112**:6348 (1990).
42. W. R. Roush and A. D. Palkowitz, *J. Am. Chem. Soc.* **109**:953 (1987).
43. W. R. Roush, A. E. Walts, and L. K. Hoong, *J. Am. Chem. Soc.* **107**:8186 (1985); W. R. Roush, L. K. Hoong, M. A. J. Palmer, and J. C. Park, *J. Org. Chem.* **55**:4109 (1990); W. R. Roush, C. K. Hoong, M. A. J. Palmer, J. A. Straub, and A. D. Palkowitz, *J. Org. Chem.* **55**:4117 (1990).

and the *E*- or *Z*-configuration of the allylic group as part of a chair cyclic transition state determine the stereochemistry of the product.

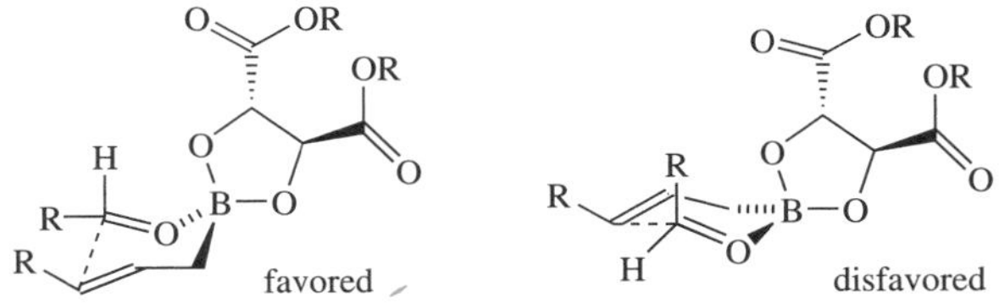

favored disfavored

Another useful chiral allylboron reagent is derived from *N*,*N*-bis(*p*-toluenesulfonyl)-1,2-diphenyl-1,2-ethanediamine. This reagent gives homoallylic alcohols with >90% e.e. with typical aldehydes.[44]

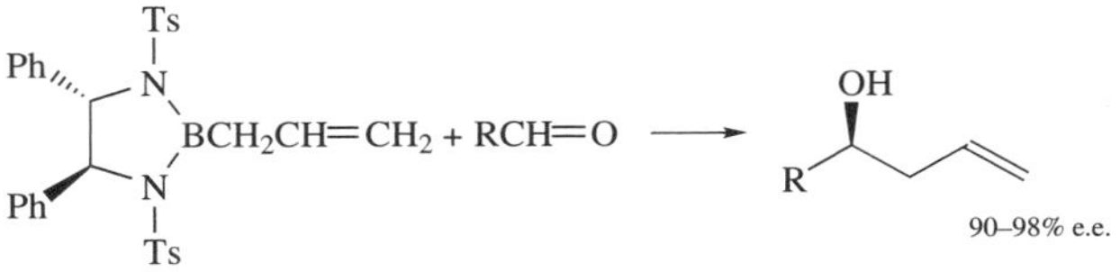

$$\text{(Ts-N, Ph)}_2\text{-BCH}_2\text{CH}{=}\text{CH}_2 + \text{RCH}{=}\text{O} \longrightarrow \text{RCH(OH)CH}_2\text{CH}{=}\text{CH}_2 \quad \text{90–98\% e.e.}$$

$R = n\text{-}C_5H_{11}$, Ph, cyclohexyl, CH=CHPh

Scheme 9.3 illustrates some examples of synthesis of allylic carbinols via allylic boranes and boronate esters.

B-Alkynyl derivatives of 9-BBN act as mild sources of nucleophilic acetylenic groups. Reaction occurs with both aldehydes and ketones, but the rate is at least 100 times faster for aldehydes.[45]

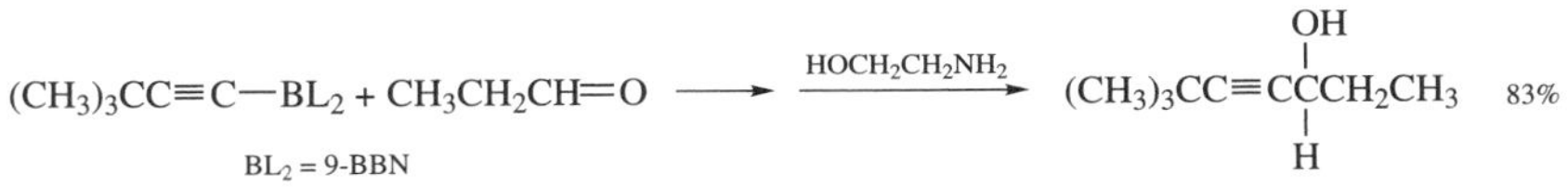

$$(CH_3)_3CC{\equiv}C{-}BL_2 + CH_3CH_2CH{=}O \longrightarrow \xrightarrow{HOCH_2CH_2NH_2} (CH_3)_3CC{\equiv}CCH(OH)CH_2CH_3 \quad 83\%$$

BL_2 = 9-BBN

Ethanolamine is used to displace the adduct from the borinate ester. The facility with which the transfer of acetylenic groups occurs is associated with the relative stability of the *sp*-hybridized carbon. This reaction is an alternative to the more common addition of magnesium or lithium salts of acetylides to aldehydes.

9.2. Organosilicon Compounds

9.2.1. Synthesis of Organosilanes

The two most general means of synthesis of organosilanes are nucleophilic displacement of halogen from a halosilane by an organometallic reagent and addition of silanes at double or triple bonds (*hydrosilation*). Organomagnesium and organolithium compounds

44. E. J. Corey, C.-M. Yu, and S. S. Kim, *J. Am. Chem. Soc.* **111**:5495 (1989).
45. H. C. Brown, G. A. Molander, S. M. Singh, and U. S. Racherla, *J. Org. Chem.* **50**:1577 (1985).

Scheme 9.3. Addition Reactions of Allylic Boranes with Carbonyl Compounds

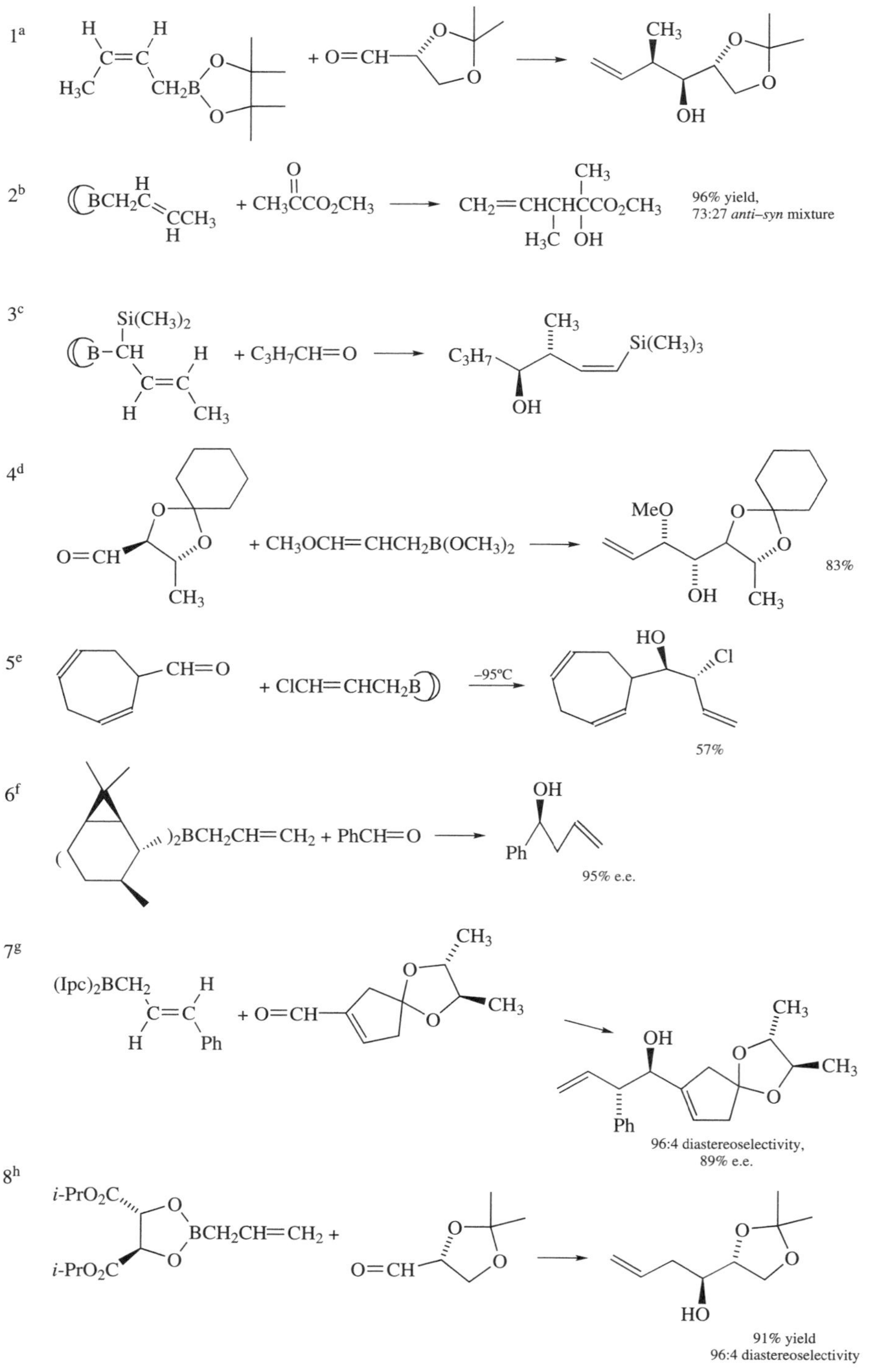

1[a] + O=CH— →

2[b] + $CH_3CCO_2CH_3$ → $CH_2=CHCHCCO_2CH_3$ (with CH_3, H_3C, OH) 96% yield, 73:27 *anti–syn* mixture

3[c] + $C_3H_7CH=O$ → C_3H_7 … $Si(CH_3)_3$ (with CH_3, OH)

4[d] O=CH … CH_3 + $CH_3OCH=CHCH_2B(OCH_3)_2$ → MeO … OH … CH_3 83%

5[e] CH=O + $ClCH=CHCH_2B$ $\xrightarrow{-95°C}$ HO … Cl 57%

6[f] (… $)_2BCH_2CH=CH_2$ + PhCH=O → OH, Ph 95% e.e.

7[g] $(Ipc)_2BCH_2$ … H, H, Ph + O=CH— … CH_3, CH_3 → OH, Ph, CH_3, CH_3 96:4 diastereoselectivity, 89% e.e.

8[h] i-PrO_2C … i-PrO_2C … $BCH_2CH=CH_2$ + O=CH— → HO 91% yield 96:4 diastereoselectivity

9[i]

CO_2-*i*-Pr

CO_2iPr

B—O

+ O=CH$(CH_2)_4CH_3$

OH

$(CH_2)_4CH_3$ 73% e.e.

CH_3 C_2H_5

10[j] TBDPSO CH=O +

$CO_2C_2H_5$

$CO_2C_2H_5$

OH

TBDPSO

95% yield, >96% e.e.

11[k]

OTBDMS

$PhCH_2O$ CH=O + TBDMSO

OCH_3

TBDMSO OH

$PhCH_2O$ OTBDMS

OCH_3 CH=CH_2

85% total yield, major isomer

12[k]

$(Ipc)_2B$ OCH_3 + O=$CHCH_2CH_2OCH_2Ph$ ⟶

OH

OCH_2Ph 67%

OCH_3

13[l]

$(Ipc)_2B$ OCH_3 + O=$CHCH(CH_3)_2$ —ethanolamine→

OH

CH_3O

57% yield, 100% *anti*, 88% e.e.

14[m] H_3C CH_2—B

C=C

H H

+ PhCH=O —−78°C→

OH

Ph

CH_3

diastereosolectivity > 95%

15[m]

H CH_2—B

C=C

H_3C H

+ PhCH=O —−78°C→

OH

Ph

CH_3

diastereosolectivity > 95%

(*continued*)

Scheme 9.3. *(continued)*

16[n] $(Ipc)_2BCH_2CH{=}CH_2$ + O=CH–(CH₂)₂C(CH₃)=CHCH₂SO₂Ph —(−100°C)→ CH₂=CHCH₂CH(OH)(CH₂)₂C(CH₃)=CHCH₂SO₂Ph

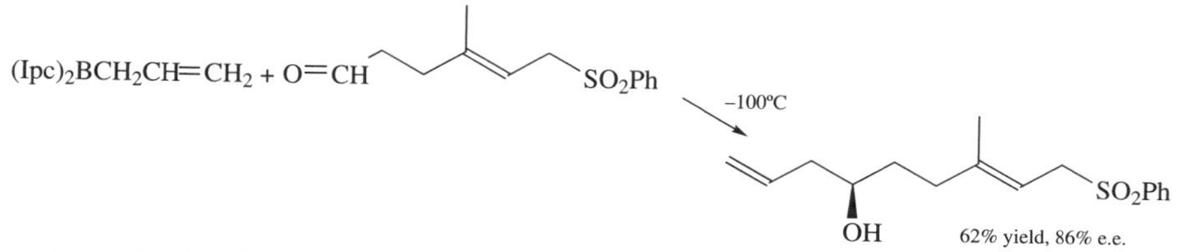

62% yield, 86% e.e.

a. W. R. Roush and A. E. Walts, *Tetrahedron Lett.* **26:**3427 (1985); W. R. Rousch, M. A. Adam, and D. J. Harris, *J. Org. Chem.* **50:**2000 (1985).
b. Y. Yamamoto, K. Maruyama, T. Komatsu, and W. Ito, *J. Org. Chem.* **51:**886 (1986).
c. Y. Yamamoto, H. Yatagai, and K. Maruyama, *J. Am. Chem. Soc.* **103:**3229 (1981).
d. W. R. Roush, M. R. Michaelides, D. F. Tai, and W. K. M. Chong, *J. Am. Chem. Soc.* **109:**7575 (1987).
e. C. Hertweck and W. Boland, *Tetrahedron* **53:**14651 (1997).
f. H. C. Brown, R. S. Randad, K. S. Bhat, M. Zaidlewicz, and U. S. Racherla, *J. Am. Chem. Soc.* **112:**2389 (1990).
g. L. K. Truesdale, D. Swanson, and R. C. Sun, *Tetrahedron Lett.* **26:**5009 (1985).
h. W. R. Roush, A. E. Walts, and L. K. Hoong, *J. Am. Chem. Soc.* **107:**8186 (1985).
i. Y. Yamamoto, S. Hara, and A. Suzuki, *Synlett* **1996:**883.
j. W. R. Roush, J. A. Straub, and M. S. VanNieuwenhze, *J. Org. Chem.* **56:**1636 (1991).
k. P. G. M. Wuts and S. S. Bigelow, *J. Org. Chem.* **53:**5023 (1988).
l. H. C. Brown, P. K. Jadhav, and K. S. Bhat, *J. Am. Chem. Soc.* **110:**1535 (1988).
m. R. W. Hoffmann and H.-J. Zeiss, *J. Org. Chem.* **46:**1309 (1981).
n. M. Z. Hoemann, K. A. Agrios, and J. Aube, *Tetrahedron* **53:**11087 (1997).

react with trimethylsilyl chloride to give the corresponding tetrasubstituted silanes.

$$CH_2{=}CHMgBr + (CH_3)_3SiCl \longrightarrow CH_2{=}CHSi(CH_3)_3 \qquad \text{Ref. 46}$$

$$CH_2{=}C(Li)OC_2H_5 + (CH_3)_3SiCl \longrightarrow CH_2{=}C(Si(CH_3)_3)OC_2H_5 \qquad \text{Ref. 47}$$

The carbon–silicon bond is quite strong (≈75 kcal/mol), and trimethylsilyl groups are stable under many of the reaction conditions which are typically used in organic synthesis. Thus, much of the repertoire of synthetic organic chemistry can be used for elaboration of organosilanes.[48]

Silicon substituents can also be introduced into alkenes and alkynes by hydrosilation.[49] This reaction, in contrast to hydroboration, does not occur spontaneously, but it can be carried out in the presence of catalysts, the most common of which is hexachloroplatinic acid, H_2PtCl_6. Other catalysts are also available.[50] Silanes that have one or

46. R. K. Boeckman, Jr., D. M. Blum, B. Ganem, and N. Halvey, *Org. Synth.* **58:**152 (1978).
47. R. F. Cunico and C.-P. Kuan, *J. Org. Chem.* **50:**5410 (1985).
48. L. Birkofer and O. Stuhl in *The Chemistry of Organic Silicon Compounds*, S. Patai and Z. Rappoport, eds., Wiley Interscience, 1989, New York, Chapter 10.
49. J. L. Speier, *Adv. Organomet. Chem.*, **17:**407 (1979); E. Lukenvics, *Russ. Chem. Rev., Engl. Transl.*, **46:**264 (1977).
50. A. Onopchenko and E. T. Sabourin, *J. Org. Chem.* **52:**4118 (1987); H. M. Dickens, R. N. Hazeldine, A. P. Mather, and R. V. Parish, *J. Organomet. Chem.* **161:**9 (1978); A. J. Cornish and M. F. Lappert, *J. Organomet. Chem.* **271:**153 (1984).

more halogen substituents are more reactive than trialkylsilanes.[51]

$$\text{methylenecyclohexane}\ (=CH_2) \xrightarrow[H_2PtCl_6]{CH_3SiCl_2H} \text{cyclohexyl}-CH_2SiCl_2CH_3$$

With more substituted alkenes, reaction under these conditions is often accompanied by double-bond migrations which eventually lead to the formation of an alkyltrichlorosilane with a primary alkyl group.[52]

Alkenylsilanes can be made by Lewis acid-catalyzed hydrosilation of alkynes.[53] The reaction proceeds by net *anti* addition, giving the *Z*-silane.

$$PhCH_2C{\equiv}CH + (C_2H_5)_3SiH \xrightarrow{AlCl_3} (PhCH_2)HC{=}CH(Si(C_2H_5)_3)$$

Catalysis of hydrosilation by rhodium gives *E*-alkenylsilanes from 1-alkynes.[54]

$$RC{\equiv}CH + (C_2H_5)_3SiH \xrightarrow[Ph_3P]{Rh(COD)_2BF_4} (R)(H)C{=}C(H)(Si(C_2H_5)_3)$$

Syn addition of alkyl and trimethylsilyl groups can be accomplished with dialkylzinc and trimethylsilyl iodide in the presence of a Pd^0 catalyst.[55]

$$RC{\equiv}CH + R'_2Zn + (CH_3)_3SiI \xrightarrow{Pd(PPh_3)_4} (R)(R')C{=}C(H)(Si(CH_3)_3)$$

9.2.2. Carbon–Carbon Bond-Forming Reactions

The carbon–silicon bond to saturated alkyl groups is not very reactive because there are no high-energy electrons in the sp^3-sp^3 bonds. Most of the valuable synthetic procedures based on organosilanes involve either alkenyl or allylic silicon substituents. The dominant reactivity pattern involves attack by an electrophilic carbon intermediate at the double bond.

$$\geq C^{\delta+} \cdots (R_3Si)(H)C{=}CHR \longrightarrow H{-}C(SiR_3)(C{\leq}){-}\overset{+}{C}HR \longrightarrow (\geq C)(H)C{=}CHR$$

$$\geq C^{\delta+} \cdots CH_2{=}CHCH_2SiR_3 \longrightarrow \geq C{-}CH_2\overset{+}{C}HCH_2{-}SiR_3 \longrightarrow \geq C{-}CH_2CH{=}CH_2$$

51. T. G. Selin and R. West, *J. Am. Chem. Soc.* **84:**1863 (1962).
52. R. A. Benkeser, S. Dunny, G. S. Li, P. G. Nerlekar, and S. D. Work, *J. Am. Chem. Soc.* **90:**1871 (1968).
53. N. Asao, T. Sudo, and Y. Yamamoto, *J. Org. Chem.* **61:**7654 (1996).
54. R. Takeuchi, S. Nitta, and D. Watanabe, *J. Org. Chem.* **60:**3045 (1995).
55. N. Chatani, N. Amishiro, T. Morii, T. Yamashita, and S. Murai, *J. Org. Chem.* **60:**1834 (1995).

Attack on alkenylsilanes takes place at the α carbon and results in replacement of the silicon substituent by the electrophile. Attack on allylic groups is at the γ carbon and results in replacement of the silicon substituent and an allylic shift of the double bond. The crucial influence on the reactivity pattern in both cases is the *very high stabilization which silicon provides for carbocationic character at the β-carbon atom*. This stabilization is attributed primarily to a hyperconjugation with the C–Si bond.[56]

or

The most useful electrophiles from a synthetic standpoint are carbonyl compounds, iminium ions, and electrophilic alkenes.

Most reactions of alkenylsilanes require strong carbon electrophiles, and Lewis acid catalysts are often involved. Reaction with acyl chlorides is catalyzed by aluminum chloride or stannic chloride.[57]

$$RCH{=}CHSi(CH_3)_3 + RCOCl \xrightarrow[\text{or } SnCl_4]{AlCl_3} RCH{=}CHC(=O)R$$

Titanium tetrachloride induces reaction with dichloromethyl methyl ether.[58]

$$RCH{=}CHSi(CH_3)_3 + Cl_2CHOCH_3 \xrightarrow{TiCl_4} RCH{=}CHCH{=}O$$

Similar conditions are used to effect reactions of allylsilanes with acyl halides.[59]

$$PhC(=O)Cl + CH_2{=}CHCH_2Si(CH_3)_3 \xrightarrow{AlCl_3} PhC(=O)CH_2CH{=}CH_2$$

Several examples of the reactions of alkenyl and allylic silanes are given in Scheme 9.4.

A variety of electrophilic catalysts promote the addition of allylic silanes to carbonyl compounds.[60] The original catalysts included typical Lewis acids such as $TiCl_4$ and BF_3.[61]

$$CH_2{=}CHCH_2SiR_3 + R_2C{=}O \xrightarrow[\text{or } BF_3]{TiCl_4} R_2C(OH)CH_2CH{=}CH_2$$

56. S. G. Wierschke, J. Chandrasekhar, and W. L. Jorgensen, *J. Am. Chem. Soc.* **107:**1496 (1985); J. B. Lambert, G. Wang, R. B. Finzel, and D. H. Teramura, *J. Am. Chem. Soc.* **109:**7838 (1987).
57. I. Fleming and A. Pearce, *J. Chem. Soc., Chem. Commun.* **1975:**633; W. E. Fristad, D. S. Dime, T. R. Bailey, and L. A. Paquette, *Tetrahedron Lett.* **1979:**1999.
58. K. Yamamoto, O. Nunokawa, and J. Tsuji, *Synthesis* **1977:**721.
59. J.-P. Pillot, G. Déléris, J. Dunoguès, and R. Calas, *J. Org. Chem.* **44:**3397 (1979); R. Calas, J. Dunogues, J.-P. Pillot, C. Biran, F. Pisciotti, and B. Arreguy, *J. Organomet. Chem.* **85:**149 (1975).
60. A. Hosomi, *Acc. Chem. Res.*, **21:**200 (1988); I. Fleming, J. Dunoquès, and R. Smithers, *Org. React.* **37:**57 (1989).
61. A. Hosomi and H.Sakurai, *Tetrahedron Lett.* **1976:**1295.

A. Reactions with carbonyl compounds

1[a] Reagents: CH_3CCl (C=O), $TiCl_4$; products: $CH{=}CH_2$, CCH_3 (C=O); starting material $CH_2Si(CH_3)_3$. 82%

2[b] $CH_2Si(CH_3)_3$ → CH_3CCl (C=O), $AlCl_3$, 90°C → CH_2, CCH_3 (C=O). 50%

3[c] $Si(CH_3)_3$ → $CH_3CH_2CH_2CH{=}O$, $TiCl_4$ → OH, $CHCH_2CH_2CH_3$ 78%

4[d] $CH_2{=}CHCH_2Si(CH_3)_3$ + $O{=}CH$ ... OCH_3 → BF_3 → $CH_2{=}CHCH_2$, OH, OCH_3 80%

B. Reactions with acetals and related compounds

5[e] $CH_2{=}CHCH_2Si(CH_3)_3$ + CH_3O, H_3C, CH_3 → $ZnBr_2$ → $CH_2{=}CHCH_2$, H_3C, CH_3 99%

6[f] $CH_2{=}CHCH_2Si(CH_3)_3$ + $AcOCH_2$, AcO, AcO, AcO, OAc → BF_3, 4°C, CH_3CN → $AcOCH_2$, AcO, AcO, AcO, $CH_2CH{=}CH_2$ 81%

7[g] $CH_2{=}CHCH_2Si(CH_3)_3$ + $ROCH_2$, OR, RO, RO, OAc, OCH_3; R = CH_2Ph → $Me_3SiO_3SCF_3$ → $ROCH_2$, OR, RO, RO, OAc, $CH_2CH{=}CH_2$ 87%

8[h] OCH_3, $CH(OCH_3)_2$, OCH_3 + OCH_3, CO_2CH_3, $Si(CH_3)_2Ph$ → $(CH_3)_3SiO_3SCF_3$ → OCH_3, OCH_3, OCH_3, CO_2CH_3, CH_3, OCH_3

92% yields, 95% e.e.

(*continued*)

Scheme 9.4. *(continued)*

C. Iminium ions

9[i] O=C(pyrrolidinone) N–$CH_2CH_2CH{=}CHSi(CH_3)_3$, OH $\xrightarrow{CF_3CO_2H}$ bicyclic product 91%

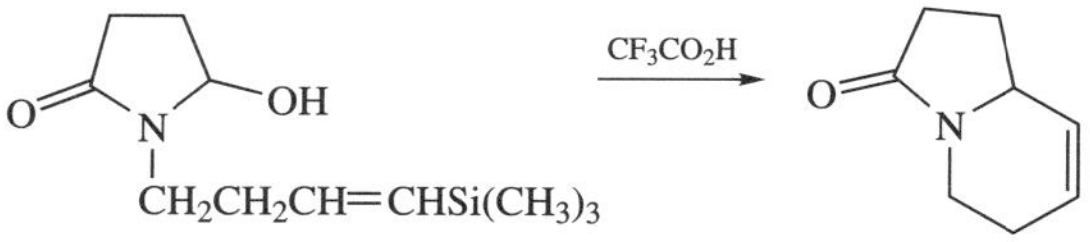

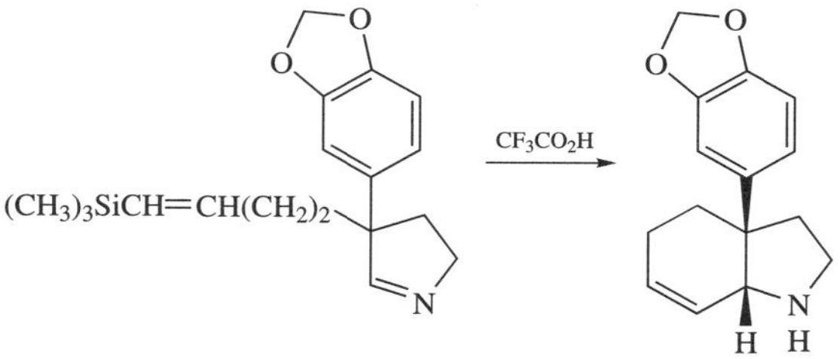

10[j] $(CH_3)_3SiCH{=}CH(CH_2)_2$– (benzodioxolyl)pyrroline $\xrightarrow{CF_3CO_2H}$ H H, N 73%

11[k] $C_9H_{19}CH(Si(CH_3)_3)CH{=}CH_2$ + C_2H_5O–piperidine(OTIPS, OTIPS), N–$CO_2C(CH_3)_3$ $\xrightarrow{BF_3}$ $C_9H_{19}CH{=}CHCH_2$–piperidine(OTIPS, OTIPS), N–$CO_2C(CH_3)_3$

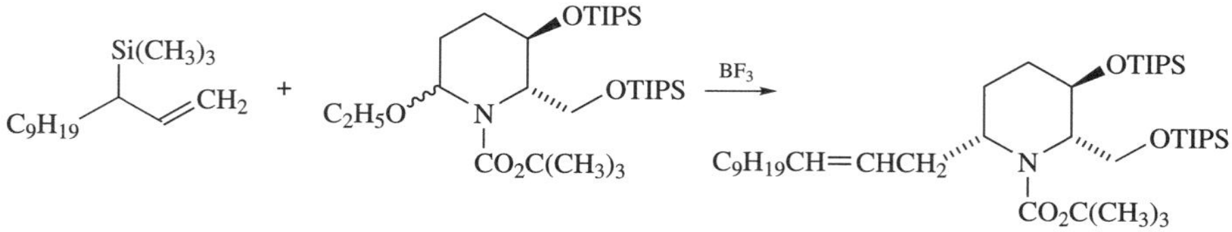

a. I. Fleming and I. Paterson, *Synthesis* **1979:**446.
b. J. P. Pillot, G. Déléris, J. Dunoguès, and R. Calas, *J. Org. Chem.* **44:**3397 (1979).
c. I. Ojima, J. Kumagai, and Y. Miyazawa, *Tetrahedron Lett.* **1977:**1385.
d. S. Danishefsky and M. De Ninno, *Tetrahedron Lett.* **26:**823 (1985).
e. H. Suh and C. S. Wilcox, *J. Am. Chem. Soc.* **110:**470 (1988).
f. A. Giannis and K. Sanshoff, *Tetrahedron Lett.* **26:**1479 (1985).
g. A. Hosomi, Y. Sakata, and H. Sakurai, *Tetrahedron Lett.* **25:**2383 (1984).
h. J. S. Panek and M. Yang, *J. Am. Chem. Soc.* **113:**6594 (1991).
i. C. Flann, T. C. Malone, and L. E. Overman, *J. Am. Chem. Soc.* **109:**6097 (1987).
j. L. E. Overman and R. M. Burk, *Tetrahedron Lett.* **25:**5739 (1984).
k. I. Ojima and E. S. Vidal, *J. Org. Chem.* **63:**7999 (1998).

These reactions involve activation of the carbonyl group by the Lewis acid. A nucleophile, either a ligand from the Lewis acid or the solvent, assists in the desilylation step.

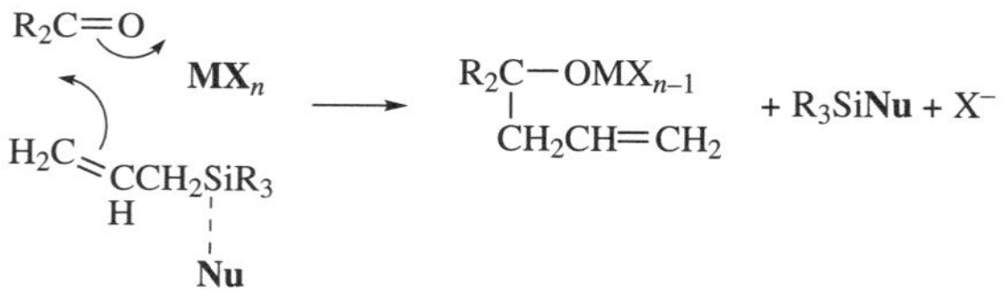

$$R_2C{=}O \;\; \mathbf{MX}_n \;\; H_2C{=}C(H)CH_2SiR_3 \cdots \mathbf{Nu} \longrightarrow R_2C(-OMX_{n-1})CH_2CH{=}CH_2 + R_3Si\mathbf{Nu} + X^-$$

Lanthanide salts, such as $Sc(O_3SCF_3)_3$ are also effective catalysts.[62] Conventional silylating reagents such as TMS–I and TMS triflate have only a modest catalytic effect, but a still more powerful silylating reagent, $(CH_3)_3SiB(O_3SCF_3)_2$, does induce addition to aldehydes.[63]

$$RCH{=}O + CH_2{=}CHCH_2Si(CH_3)_3 \xrightarrow{(CH_3)_3SiB(O_3SCF_3)_2} RCH(OSi(CH_3)_3)CH_2CH{=}CH_2$$

62. V. K. Aggarwal and G. P. Vennall, *Tetrahedron Lett.* **37:**3745 (1996).
63. A. P. Davis and M. Jaspars, *Angew. Chem. Int. Ed. Engl.* **31:**470 (1992).

Although the allylation reaction is formally analogous to the addition of allylboranes to carbonyl derivatives, it does not appear to occur through a cyclic transition state. This is because, in contrast to the boron in allyl boranes, the silicon in allylic silanes has no Lewis acid character and would not be expected to coordinate at the carbonyl oxygen. The stereochemistry of addition of allylic silanes to carbonyl compounds is consistent with an acyclic transition state. Both the *E*- and *Z*-stereoisomers of 2-butenyl(trimethyl)silane give the product in which the newly formed hydroxyl group is *syn* to the methyl substituent.[64] The preferred orientation of approach by the silane minimizes interaction between the aldehyde substituent R and the methyl group.

LA = Lewis acid

preferred T.S. for *E*-allylsilane

syn product

destabilized T.S. for *E*-allylsilane

preferred T.S. for *Z*-allylsilane

syn product

destabilized T.S. for *Z*-allylsilane

When chiral aldehydes such as **5** are used, there is a modest degree of diastereoselectivity in the direction predicted by Cram's rule.

5 + $CH_2{=}CHCH_2Si(CH_3)_3$ $\xrightarrow{TiCl_4}$

major

minor

86% yield, ratio = 1.6:1

Aldehydes with donor substituents that can form a chelate with the Lewis acid catalyst react by approach from the less hindered side of the chelate structure. The best electrophile

64. T. Hayashi, K. Kabeta, I. Hamachi, and M. Kumada, *Tetrahedron Lett.* **24**:2865 (1983).

in this case is $SnCl_4$.[65]

$$\text{PhCH}_2\text{O(Sn chelate)CH}_2\text{CH(CH}_3\text{)CHO} + CH_2{=}CHCH_2Si(CH_3)_3 \longrightarrow \text{PhCH}_2\text{OCH}_2\text{CH(CH}_3\text{)CH(OH)CH}_2\text{CH}{=}\text{CH}_2 \text{ (major)} + \text{(minor)}$$

preferred approach

Both ketals[66] and enol ethers[67] can be used in place of aldehydes with the selection of appropriate catalysts. Trimethylsilyl iodide causes addition to occur.[68] The trimethylsilyl iodide can be used in catalytic quantity because it is regenerated by recombination of iodide ion with silicon in the desilation step.

$$R_2C(OMe)_2 + CH_2{=}CHCH_2Si(CH_3)_3 \xrightarrow{\text{TMS-I}} R_2C{=}\overset{+}{O}Me \;\; CH_2{=}CH{-}CH_2{-}Si(CH_3)_3 \; I^-$$

$$\longrightarrow R_2C(OMe)CH_2CH{=}CH_2$$

This reaction has been used for the extension of the carbon chain of protected carbohydrate acetals.[69]

$$\text{ROCH}_2\text{-furanose(RO, OR)-O}_2\text{CCH}_3 \xrightarrow{CH_2{=}CHCH_2SiMe_3,\ BF_3\cdot OEt_2} \text{ROCH}_2\text{-furanose(RO, OR)-CH}_2\text{CH}{=}\text{CH}_2 \text{ (major)} + \text{(minor)}$$

Reaction of allylic silanes with enantiomerically pure 1,3-dioxanes has been found to proceed with high enantioselectivity.[70] The enantioselectivity is dependent on several reaction variables, including the Lewis acid and the solvent. The observed stereoselectivity appears to reflect differences in the precise structure of the electrophilic species that is generated. Mild Lewis acids tend to react with inversion of configuration at the reaction site, whereas very strong Lewis acids cause loss of enantioselectivity. These trends, and related effects of solvent and other experimental variables, determine the nature of the electrophile. With mild Lewis acids, a tight ion pair favors inversion, whereas stronger

65. C. H. Heathcock, S. Kiyooka, and T. Blumenkopf, *J. Org. Chem.* **49:**4214 (1984).
66. T. K. Hollis, N. P. Robinson, J. Whelan, and B. Bosnich, *Tetrahedron Lett.* **34:**4309 (1993).
67. T. Yokozawa, K. Furuhashi, and H. Natsume, *Tetrahedron Lett.* **36:**5243 (1995).
68. H. Sakurai, K. Sasaki, and A. Hosmoni, *Tetrahedron Lett.* **22:**745 (1981).
69. A. P. Kozikowski, K. Sorgi, B. C. Wang, and Z. Xu, *Tetrahedron Lett.* **24:**1563 (1983).
70. P. A. Bartlett, W. S. Johnson, and J. D. Elliott, *J. Am. Chem. Soc.* **105:**2088 (1983).

Lewis acids cause complete dissociation to an acyclic species. These two species represent extremes of behavior, and intermediate levels of enantioselectivity are also observed.[71]

inversion of configuration in tight ion-pair intermediate

loss of enantioselectivity in dissociated acyclic species

The homoallylic alcohol can be liberated by oxidation followed by base-catalyzed β-elimination. The alcohols obtained in this way are formed in $70 \pm 5\%$ e.e. A similar reaction occurs with alkynylsilanes to give propargylic alcohols in 70–90% e.e.[72]

$$+ R'C{\equiv}CSi(CH_3)_3 \xrightarrow{TiCl_4} \xrightarrow[2)\ ^-OH]{1)\ oxid} RCH(OH)C{\equiv}CR'$$

Section B of Scheme 9.4 gives some additional examples of Lewis acid-mediated reactions of allylic silanes with aldehydes and acetals.

Reaction of allylic silanes with aldehydes and ketones can also be induced by fluoride ion, which is usually supplied by the THF-soluble salt tetrabutylammonium fluoride (TBAF). Fluoride adds at silicon to form a hypervalent anion with much enhanced nucleophilicity.[73] An alternative reagent to TBAF is tetrabutylammonium triphenyldifluorosilicate.[74]

$$CH_2{=}CHCH_2SiR_3 + F^- \longrightarrow CH_2{=}CHCH_2\bar{Si}{-}F$$

Unsymmetrical allylic anions generated in this way react with ketones at their less substituted terminus.

$$CH_2{=}CHCH_2Si(CH_3)_3 + F^- \longrightarrow CH_2{=}CHCH_2\bar{Si}F(CH_3)_3 \xrightarrow{RCH=O} RCH(OH)CH_2CH{=}CH_2$$

$$(CH_3)_2C{=}CHCH_2Si(CH_3) + Ph_2C{=}O \xrightarrow{TBAF} \xrightarrow{H_2O} (CH_3)_2C{=}CHCH_2CPh_2OH \quad 87\%$$

71. S. E. Denmark and N. G. Almstead, *J. Am. Chem. Soc.* **113:**8089 (1991).
72. W. S. Johnson, R. Elliott, and J. D. Elliott, *J. Am. Chem. Soc.* **105:**2904 (1983).
73. A. Hosomi, A. Shirahata, and H. Sakurai, *Tetrahedron Lett.* **1978:**3043; G. G. Furin, O. A. Vyazankina, B. A. Gostevsky, and N. S. Vyazankin, *Tetrahedron* **44:**2675 (1988).
74. A. S. Pilcher and P. De Shong, *J. Org. Chem.* **61:**6901 (1996).

An allylic silane of this type serves as a reagent for introduction of isoprenoid structures.[75]

$$(CH_3)_2C{=}CHCH{=}O + (CH_3)_3SiCH_2\underset{\underset{CH_2}{\|}}{C}CH{=}CH_2 \xrightarrow{R_4N^+F^-} (CH_3)_2C{=}CH\underset{\underset{OH}{|}}{C}HCH_2\underset{\underset{CH_2}{\|}}{C}CH{=}CH_2 \quad 70\%$$

Fluoride-induced desilation has also been used to effect ring closures.[76]

$$\text{H, O, } CH_2{=}C(CH_2Si(CH_3)_3)CH_2\text{-}, SO_2Ph, \text{H} \xrightarrow{R_4N^+F^-} \text{HO, H, } {=}CH_2, SO_2Ph, \text{H} \quad 94\%$$

Iminium compounds are reactive electrophiles toward both alkenyl and allylic silanes. Useful techniques for closing nitrogen-containing rings are based on *in situ* generation of iminium ions from amines and formaldehyde.[77]

$$PhCH_2\underset{H}{N}CH_2CH_2\overset{\overset{CH_2}{\|}}{C}CH_2Si(CH_3)_3 \xrightarrow[H_2C=O]{CF_3CO_2H} PhCH_2\underset{\underset{CH_2}{\|}}{\overset{+}{N}}CH_2CH_2\underset{\underset{CH_2}{\|}}{C}CH_2Si(CH_3)_3 \longrightarrow PhCH_2N\text{(piperidine)}{=}CH_2 \quad 73\%$$

When primary amines are employed, the initially formed 3-butenylamine undergoes a further reaction forming 4-piperidinols.[78]

$$PhCH_2\overset{+}{N}H_3 + CH_2{=}CHCH_2Si(CH_3)_3 + CH_2{=}O \longrightarrow PhCH_2N\text{(piperidine)}\text{—}OH$$

Reactions of this type can also be observed with 4-(trimethylsilyl)-3-alkenylamines.[79]

$$R'\underset{H}{N}CH_2CH_2CH{=}\underset{\underset{Si(CH_3)_3}{|}}{C}R^3 \xrightarrow[H^+]{CH_2=O} \text{(tetrahydropyridine: } R^3, N, R')$$

Mechanistic investigation in this case has shown that there is an equilibration between an alkenyl silane and an allylic silane by a rapid [3,3] sigmatropic process. The cyclization occurs through the more reactive allylic silane.

$$R'\underset{H}{N}(CH_2)_2CH{=}\underset{\underset{Si(CH_3)_3}{|}}{C}R_3 + CH_2{=}O \longrightarrow R'\overset{+}{N}{=}CH_2\text{, } C\text{—}R^3\text{, } (CH_3)_3Si \rightleftharpoons R'\overset{+}{N}{=}CH_2\text{, } (CH_3)_3Si\text{, } R^3 \longrightarrow R'N\text{(tetrahydropyridine)}R^3$$

75. A. Hosomi, Y. Araki, and H. Sakurai, *J. Org. Chem.* **48:**3122 (1983).
76. B. M. Trost and J. E. Vincent, *J. Am. Chem. Soc.* **102:**5680 (1980); B. M. Trost and D. P. Curran, *J. Am. Chem. Soc.* **103:**7380 (1981).
77. P. A. Grieco and W. F. Fobare, *Tetrahedron Lett.* **27:**5067 (1986).
78. S. D. Larsen, P. A. Grieco, and W. F. Fobare, *J. Am. Chem. Soc.* **108:**3512 (1986).
79. C. Flann, T. C. Malone, and L. E. Overman, *J. Am. Chem. Soc.* **109:**6097 (1987).

N-Acyliminium ions are even more reactive toward alkenyl and allylic silanes. *N*-Acyliminium ions are usually obtained from imides by partial reduction. The partially reduced *N*-acylcarbinolamines can then generate acyliminium ions. Intramolecular examples of such reactions have been observed.

$CH_2(CH_2)_2CH{=}CHCH_2Si(CH_3)_3$ $\xrightarrow{CF_3CO_2H}$ CH=CH$_2$ Ref. 80

$(CH_3)_3Si$ $\xrightarrow[2)\ CF_3CO_2H]{1)\ NaBH_4}$ Ref. 81

Section C of Scheme 9.4 give some other examples of cyclization involving iminium ions as electrophiles in reactions with unsaturated silanes.

Allylic silanes act as nucleophilic species toward α,β-unsaturated ketones in the presence of Lewis acids such as $TiCl_4$.[82]

$(CH_3)_3SiCH_2CH{=}CH_2$ + $\xrightarrow[-78^\circ C]{TiCl_4}$ $CH_2{=}CHCH_2$ 85%

The stereochemistry of this reaction in cyclic systems is in accord with expectations for stereoelectronic control. The allylic group approaches from a trajectory that is appropriate for interaction with the LUMO of the conjugated system.[83]

=O- -$TiCl_4$ R H

The stereoselectivity then depends on the conformation of the enone and the location of substituents which establish a steric bias for one of the two potential directions of approach. In the ketone **6**, the preferred approach is from the β face, because this permits a chair conformation to be maintained as the reaction proceeds.[84]

$(CH_3)_3SiCH{=}CH_2$ O- -$TiCl_4$ $(CH_2)_3CH_3$ H_3C **6** ⟶ $CH_2{=}CHCH_2$ O $(CH_2)_3CH_3$ H_3C

80. H. Hiemstra, M. H. A. M. Sno, R. J. Vijn, and W. N. Speckamp, *J. Org. Chem.* **50:**4014 (1985).
81. G. Kim, M. Y. Chu-Moyer, S. J. Danishefsky, and G. K. Schulte, *J. Am. Chem. Soc.* **115:**30 (1993).
82. A. Hosomi and H. Sakurai, *J. Am. Chem. Soc.* **99:**1673 (1977).
83. T. A. Blumenkopf and C. H. Heathcock, *J. Am. Chem. Soc.* **105:**2354 (1983).
84. W. R. Roush and A. E. Walts, *J. Am. Chem. Soc.* **106:**721 (1984).

Intramolecular conjugate addition of allylic silanes can also be used to construct new rings, as illustrated by entry 6 in Scheme 9.5.

Conjugate addition can also be carried out by fluoride-mediated desilylation. A variety of α,β-unsaturated esters and amides have been found to undergo this reaction.[85]

$$PhCH{=}CHCO_2C_2H_5 + CH_2{=}CHCH_2Si(CH_3)_3 \xrightarrow{F^-} PhCH(CH_2CH{=}CH_2)CH_2CO_2C_2H_5$$

With unsaturated aldehydes, 1,2-addition occurs, and with ketones both the 1,2- and 1,4-products are formed.

$$PhCH{=}CHC({=}O)CH_3 + CH_2{=}CHCH_2Si(CH_3)_3 \xrightarrow[HMPA]{TBAF} \underset{25\%}{PhCH(CH_2CH{=}CH_2)CH_2C({=}O)CH_3} + \underset{50\%}{PhCH{=}CHC(OH)(CH_2CH{=}CH_2)CH_3}$$

9.3. Organotin Compounds

9.3.1. Synthesis of Organostannanes

The readily available organotin compounds include trisubstituted tin hydrides (stannanes) and chlorides. Trialkylstannanes can be added to carbon–carbon double and triple bonds. The reaction is normally carried out by a radical-chain process.[86] Addition is facilitated by the presence of radical-stabilizing substituents.

$$(C_2H_5)_3SnH + CH_2{=}CHCN \longrightarrow (C_2H_5)_3SnCH_2CH_2CN \qquad \text{Ref. 87}$$

$$(C_4H_9)_3SnH + CH_2{=}C(CO_2CH_3)Ph \longrightarrow (C_4H_9)_3SnCH_2CH(Ph)CO_2CH_3 \qquad \text{Ref. 88}$$

With terminal alkynes, the stannyl group is added at the unsubstituted carbon, and the Z-stereoisomer is initially formed but is isomerized to the E-isomer.[89]

$$HC{\equiv}CCH_2OTHP \xrightarrow[AIBN]{(C_4H_9)SnH} \underset{(Z)}{(C_4H_9)_3Sn(H)C{=}C(H)CH_2OTHP} \longrightarrow \underset{(E)}{(C_4H_9)_3Sn(H)C{=}C(H)CH_2OTHP}$$

The reaction with internal alkynes leads to a mixture of regioisomers and stereoisomers.[90] Lewis acid-catalyzed hydrostannylation has also been observed using $ZrCl_4$. With terminal

85. G. Majetich, A. Casares, D. Chapman, and M. Behnke, *J. Org. Chem.* **51**:1745 (1986).
86. H. G. Kuivila, *Adv. Organomet. Chem.* **1**:47 (1964).
87. A. J. Leusinsk and J. G. Noltes, *Tetrahedron Lett.* **1966**:335.
88. I. Fleming and C. J. Urch, *Tetrahedron Lett.* **24**:4591 (1983).
89. E. J. Corey and R. H. Wollenberg, *J. Org. Chem.* **40**:2265 (1975).
90. H. E. Ensley, R. R. Buescher, and K. Lee, *J. Org. Chem.* **47**:404 (1982).

1[a] $PhCH{=}CHC(O)CH_3 + (CH_3)_3SiCH_2CH{=}CH_2 \xrightarrow{TiCl_4} PhCH(CH_2CH{=}CH_2)CH_2C(O)CH_3$ 80%

2[b] $(CH_3)_2C{=}CHC(O)CH_3 + CH_2{=}CHCH_2Si(CH_3)_3 \xrightarrow{TiCl_4} CH_2{=}CHCH_2C(CH_3)_2CH_2C(O)CH_3$ 87%

3[c] 1-acetylcyclopentene + $CH_2{=}CHCH_2Si(CH_3)_3 \xrightarrow{TiCl_4}$ 1-acetyl-2-($CH_2CH{=}CH_2$)cyclopentane

4[d] (bicyclic enone with $=C(H)CH_3$) + $CH_2{=}CHCH_2Si(CH_3)_3 \xrightarrow{TiCl_4}$ $CH_2{=}CHCH_2CH(CH_3)$-(bicyclic ketone) 89%

5[e] $PhC(CH_3){=}CHCO_2C_2H_5 + CH_2{=}CHCH_2Si(CH_3)_3 \xrightarrow{F^-} PhC(CH_3)(CH_2CH{=}CH_2)CH_2CO_2C_2H_5$ 47%

6[f] 3-methyl-4-[$(CH_2)_2CH{=}CHCH_2Si(CH_3)_3$]cyclohex-2-enone $\xrightarrow[0°C]{EtAlCl_2}$ bicyclic ketone with CH_3 and $CH{=}CH_2$ substituents

90% yield, 2:1 mixture of stereoisomers

a. H. Sakurai, A. Hosomi, and J. Hayashi, *Org. Synth.* **62**:86 (1984).
b. D. H. Hua, *J. Am. Chem. Soc.* **108**:3835 (1986).
c. H. O. House, P. C. Gaa, and D. VanDerveer, *J. Org. Chem.* **48**:1661 (1983).
d. T. Yanami, M. Miyashita, and A. Yoshikoshi, *J. Org. Chem.* **45**:607 (1980).
e. G. Majetich, A. Casares, D. Chapman, and M. Behnke, *J. Org. Chem.* **51**:1745 (1986).
f. D. Schinzer, S. Sólymom, and M. Becker, *Tetrahedron Lett.* **26**:1831 (1985).

alkynes, the *Z*-alkenylstannane is formed.[91]

$$RC{\equiv}CH + (n\text{-}C_4H_9)_3SnH \xrightarrow{ZrCl_4} (Z)\text{-}RCH{=}CHSn(n\text{-}C_4H_9)_3$$

Palladium-catalyzed procedures have also been developed for addition of stannanes to alkynes.[92]

$$PhC{\equiv}CCH_3 + (C_4H_9)SnH \xrightarrow{PdCl_2\text{-}PPh_3} (C_4H_9)_3Sn(Ph)C{=}C(CH_3)H$$

91. N. Asao, J.-X. Liu, T. Sudoh, and Y. Yamamoto, *J. Org. Chem.* **61**:4568 (1996).
92. H. X. Zhang, F. Guibe and G. Balavoine, *Tetrahedron Lett.* **29**:619 (1988); M. Bénéchie, T. Skrydstrup, and F. Khuong-Huu, *Tetrahedron Lett.* **32**:7535 (1991).

Hydrostannylation of terminal alkynes can also be achieved by reaction with stannylcyanocuprates.

$$HOCH_2CH_2C{\equiv}CH + (CH_3)_3Sn\underset{\displaystyle CH_3}{\underset{|}{C}}u(CN)Li_2 \longrightarrow \underset{H}{\overset{HOCH_2CH_2}{}}C{=}C\underset{Sn(CH_3)_3}{\overset{H}{}} \qquad \text{Ref. 93}$$

$$(C_2H_5O)_2CHC{\equiv}CH + (n\text{-}C_4H_9)_3Sn\underset{\displaystyle C_4H_9}{\underset{|}{C}}u(CN)Li_2 \longrightarrow \underset{H}{\overset{(C_2H_5O)_2CH}{}}C{=}C\underset{Sn(n\text{-}C_4H_9)_3}{\overset{H}{}} \qquad \text{Ref. 94}$$

These reactions proceed via a *syn* addition followed by protonolysis.

$$RC{\equiv}CH + R'_3Sn\underset{\displaystyle R'}{\underset{|}{C}}u(CN)Li_2 \longrightarrow \underset{-Cu-}{\overset{R}{}}C{=}C\underset{SnR'_3}{\overset{H}{}} \xrightarrow{SOH} \underset{H}{\overset{R}{}}C{=}C\underset{SnR'_3}{\overset{H}{}}$$

Terminal alkenylstannanes can be prepared after homologation of aldehydes via the Corey–Fuchs reaction.[95]

$$RCH{=}O + CBr_4 \xrightarrow[Zn]{P(Ph)_3} RCH{=}CBr_2 \xrightarrow[2)\ H_2O]{1)\ n\text{-BuLi}} RC{\equiv}CH \xrightarrow[AIBN]{(n\text{-}C_4H_9)_3SnH} RCH{=}CHSn(n\text{-}C_4H_9)_3$$

Another sequence involves a dibromomethyl(trialkyl)stannane as an intermediate which undergoes addition to the aldehyde, followed by reductive elimination.[96]

$$RCH{=}O + R'_3SnCHBr_2 \xrightarrow[LiI]{CrCl_2} RCH{=}CHSnR'_3$$

Deprotonated trialkylstannanes are potent nucleophiles. Addition to carbonyl groups or iminium intermediates provides routes to α-alkoxy- and α-aminoalkylstannanes.

$$RCH{=}O + (C_4H_9)_3SnLi \longrightarrow R\overset{O^-}{\overset{|}{C}}HSn(C_4H_9)_3 \xrightarrow{R'X} R\overset{OR'}{\overset{|}{C}}HSn(C_4H_9)_3 \qquad \text{Ref. 97}$$

$$R_2NCH_2SPh + (C_4H_9)_3SnLi \longrightarrow R_2NCH_2Sn(C_4H_9)_3 \qquad \text{Ref. 98}$$

93. I. Beaudet, J.-L. Parrain, and J.-P. Quintard, *Tetrahedron Lett.* **32:**6333 (1991).
94. A. C. Oehlschlager, M. W. Hutzinger, R. Aksela, S. Sharma, and S. M. Singh, *Tetrahedron Lett.* **31:**165 (1990).
95. E. J. Corey and P. L. Fuchs, *Tetrahedron Lett.* **1972**: 3769.
96. M.D. Cliff and S. G. Pyne, *Tetrahedron Lett.* **36:**763 (1995); D. M. Hodgson, *Tetrahedron Lett.* **33:**5603 (1992).
97. W. C. Still, *J. Am. Chem. Soc.* **100:**1481 (1978).
98. D. J. Peterson, *J. Am. Chem. Soc.* **93:**4027 (1971).

α-Silyoxystannanes can be prepared directly from aldehydes and tri-n-butyl(trimethylsilyl)stannane.[99]

$$\mathrm{RCH{=}O} + (n\text{-}\mathrm{C_4H_9})_3\mathrm{SnSi(CH_3)_3} \xrightarrow{\mathrm{R'_4N^+CN^-}} \mathrm{RCH(OSi(CH_3)_3)Sn}(n\text{-}\mathrm{C_4H_9})_3$$

Addition of tri-n-butylstannyllithium to aldehydes, followed by iodination and dehydrohalogenation, gives primarily E-alkenylstannanes.[100]

$$\mathrm{RCH_2CH{=}O} + (n\text{-}\mathrm{C_4H_9})_3\mathrm{SnLi} \xrightarrow{\mathrm{Ph_3P.I_2}} \mathrm{RCH_2CH(I)Sn}(n\text{-}\mathrm{C_4H_9})_3 \xrightarrow{\mathrm{DBU}} (E)\text{-}\mathrm{RCH{=}CHSn}(n\text{-}\mathrm{C_4H_9})_3$$

Another major route for synthesis of stannanes is reaction of an organometallic reagent with a trisubstituted halostannane. This is the normal route for preparation of arylstannanes.

$$\mathrm{CH_3O{-}C_6H_4{-}MgBr} + \mathrm{BrSn(CH_3)_3} \longrightarrow \mathrm{CH_3O{-}C_6H_4{-}Sn(CH_3)_3} \quad \text{Ref. 101}$$

$$\mathrm{(H)(Ph)C{=}C(Li)(OCH_3)} + \mathrm{(CH_3)_3SnCl} \longrightarrow \mathrm{(H)(Ph)C{=}C(Sn(CH_3)_3)(OCH_3)} \quad \text{Ref. 102}$$

9.3.2. Carbon–Carbon Bond-Forming Reactions

As with the silanes, some of the most useful synthetic procedures involve electrophilic attack on alkenyl and allylic stannanes. The stannanes are considerably more reactive than the corresponding silanes because there is more anionic character on carbon in the C–Sn bond and it is a weaker bond.[103] There are also useful synthetic procedures in which organotin compounds act as carbanion donors in palladium-catalyzed reactions, as discussed in Section 8.2.3 Organotin compounds are also very important in free-radical reactions, which will be discussed in Chapter 10.

Tetrasubstituted organotin compounds are not sufficiently reactive to add directly to aldehydes and ketones, although reactions with aldehydes do occur with heating.

$$\mathrm{Cl{-}C_6H_4{-}CH{=}O} + \mathrm{CH_2{=}CHCH_2Sn(C_2H_5)_3} \xrightarrow[4\,\mathrm{h}]{100^\circ\mathrm{C}} \mathrm{Cl{-}C_6H_4{-}CH(OSn(C_2H_5)_3)CH_2CH{=}CH_2} \quad 90\% \quad \text{Ref. 104}$$

99. R. K. Bhatt, J. Ye, and J. R. Falck, *Tetrahedron Lett.* **35:**4081 (1994).
100. J. M. Chong and S. B. Park, *J. Org. Chem.* **58:**523 (1993).
101. C. Eaborn, A. R. Thompson, and D. R. M. Walton, *J. Chem. Soc. C* **1967:**1364; C. Eaborn, H. L. Hornfeld, and D. R. M. Walton, *J. Chem. Soc. B* **1967:**1036.
102. J. A. Soderquist and G. J.-H. Hsu, *Organometallics* **1:**830 (1982).
103. J. Burfeindt, M. Patz, M. Müller, and H. Mayr, *J. Am. Chem. Soc.* **120:**3629 (1998).
104. K. König and W. P. Neumann, *Tetrahedron Lett.* **1967:**495.

Use of Lewis acid catalysts allows allylic stannanes to react under mild conditions. As was the case with allylic silanes, a double-bond shift occurs in conjunction with destannylation.[105]

$$\mathrm{PhCH{=}O} + \underset{\mathrm{H}\qquad\mathrm{H}}{\overset{\mathrm{CH_3}\qquad\mathrm{CH_2Sn(C_4H_9)_3}}{\mathrm{C{=}C}}} \xrightarrow{\mathrm{BF_3}} \mathrm{Ph}\underset{\mathrm{OH}}{\mathrm{CH}}\overset{\mathrm{CH_3}}{\mathrm{CH}}\mathrm{CH{=}CH_2} \quad 92\%$$

Use of di-*n*-butylstannyl dichloride along with an acyl or silyl halide leads to addition of allylstannanes to the aldehydes.[106] Reaction is also promoted by butylstannyl trichloride.[107] Both $SnCl_4$ and $SnCl_2$ also catalyze this kind of addition. Reactions of tetraallylstannane with aldehydes catalyzed by $SnCl_4$ also appear to involve a halostannane intermediate. It can be demonstrated by NMR that there is a rapid redistribution of the allyl group.[108]

$$\mathrm{RCH{=}O} + \mathrm{CH_2{=}CHCH_2Sn}(n\text{-}\mathrm{C_4H_9})_3 \xrightarrow[\mathrm{RCOCl\ or\ (CH_3)_3SiCl}]{(n\text{-}\mathrm{C_4H_9})_2\mathrm{SnCl_2}} \mathrm{R}\overset{\mathrm{OX}}{\mathrm{C}}\mathrm{HCH_2CH{=}CH_2}$$

$$\mathrm{X = RCO\ or\ (CH_3)_3Si}$$

Various allylhalostannanes can transfer allyl groups to carbonyl compounds. In this case, the reagent acts both as a Lewis acid and as the source of the nucleophilic allyl group. Reactions with halostannanes are believed to proceed through cyclic transition states.

$$\mathrm{PhCH_2CH_2}\overset{\mathrm{O}}{\overset{\|}{\mathrm{C}}}\mathrm{CH_3} + \mathrm{(CH_2{=}CHCH_2)_2SnBr_2} \longrightarrow \mathrm{PhCH_2CH_2}\underset{\mathrm{CH_3}}{\overset{\mathrm{OH}}{\mathrm{C}}}\mathrm{CH_2CH{=}CH_2} \qquad \text{Ref. 109}$$

The halostannanes can also be generated *in situ* by reactions of allylic halides with tin metal or with stannous halides.

$$\mathrm{PhCH{=}O} + \mathrm{CH_2{=}CHCH_2I} \xrightarrow{\mathrm{Sn}} \xrightarrow{\mathrm{H_2O}} \mathrm{Ph}\overset{\mathrm{OH}}{\mathrm{C}}\mathrm{HCH_2CH{=}CH_2} \qquad \text{Ref. 109}$$

$$\mathrm{PhCH{=}CHCH{=}O} + \mathrm{CH_2{=}CHCH_2I} \xrightarrow{\mathrm{SnF_2}} \mathrm{PhCH{=}CH}\overset{\mathrm{OH}}{\mathrm{C}}\mathrm{HCH_2CH{=}CH_2} \qquad \text{Ref. 110}$$

105. H. Yatagai, Y. Yamamoto, and K. Maruyama, *J. Am. Chem. Soc.* **102:**4548 (1980); Y. Yamamoto, *Acc. Chem. Res.* **20:**243 (1987); Y. Yamamoto and N. Asao, *Chem. Rev.* **93:**2207 (1993).
106. J. K. Whitesell and R. Apodaca, *Tetrahedron Lett.* **37:**3955 (1996); M. Yasuda, Y. Sugawa, A. Yamomoto, I. Shibata, and A. Baba, *Tetrahedron Lett.* **37:**5951 (1996).
107. H. Miyake and F. Yamamura, *Chem. Lett.* **1992:**1369.
108. S. E. Denmark, T. Wilson, and T. M. Wilson, *J. Am. Chem. Soc.* **110:**984 (1988); G. E. Keck, M. B. Andrus, and S. Castellino, *J. Am. Chem. Soc.* **111:**8136 (1989).
109. T. Mukaiyama and T. Harada, *Chem. Lett.* **1981:**1527.
110. T. Mukaiyama, T. Harada, and S. Shoda, *Chem. Lett.* **1980:**1507.

The allylation reaction can be adapted to the synthesis of terminal dienes by using 1-bromo-3-iodopropene and stannous chloride.

$$\mathrm{PhCH{=}O + ICH_2CH{=}CHBr \xrightarrow{SnCl_2} PhCH(OH)CH(Br)CH{=}CH_2 \longrightarrow PhCH{=}CHCH{=}CH_2}$$ Ref. 111

The elimination step is a reductive elimination of the type discussed in Section 6.10 of Part A. Excess stannous chloride acts as the reducing agent.

The stereoselectivity of addition to aldehydes and ketones has been of considerable interest.[112] With benzaldehyde, the addition of 2-butenylstannanes catalyzed by BF_3 gives the *syn* isomer, irrespective of the stereochemistry of the butenyl group.[113]

H_3C, H / C=C / H, $CH_2Sn(C_4H_9)_3$ $\xrightarrow{PhCH=O}$ Ph–CH(OH)–CH(CH_3)–CH=CH_2 $\xleftarrow{PhCH=O}$ H, H / C=C / H_3C, $CH_2Sn(C_4H_9)_3$

The stereoselectivity is higher for the *E*-stannane.[114] This stereochemistry is the same as that observed for allylic silanes and can be interpreted in terms of an acyclic transition state. (See page 571). Either an *anti* or *gauche* conformation can lead to the preferred *syn* product. An electronic π interaction between the stannane HOMO and the carbonyl LUMO is thought to favor the gauche conformation.[115]

CH_2SnBu_3, H, R, H_3C, H, O^+, F_3B^- — *anti*

CH_3, H, R, H, O^+, CH_2SnBu_3, F_3B^- — *gauche*

When $TiCl_4$ is used as the catalyst, the stereoselectivity depends on the order of addition of the reagents. When *E*-2-butenylstannane is added to a $TiCl_4$–aldehyde mixture, *syn* stereoselectivity is observed. When the aldehyde is added to a premixed solution of the

111. J. Auge, *Tetrahedron Lett.* **26:**753 (1985).
112. Y. Yamomoto, *Acc. Chem. Res.* **20:**243 (1987).
113. Y. Yamamoto, H. Yatagi, H. Ishihara, and K. Maruyama, *Tetrahedron* **40:**2239 (1984).
114. G. E. Keck, K. A. Savin, E. N. K. Cressman, and D. E. Abbott, *J. Org. Chem.* **59:**7889 (1994).
115. S. E. Denmark, E. J. Weber, T. Wilson, and T. M. Willson, *Tetrahedron* **45:**1053 (1989).

2-butenylstannane and $TiCl_4$, the *anti* isomer predominates.[116]

$TiCl_4$ + cyclohexyl–CH=O → (add $CH_3CH{=}CHCH_2SnBu_3$) → product (OH, CH_3)

$TiCl_4 + CH_3CH{=}CHCH_2SnBu_3$ → (add cyclohexyl–CH=O) → product (OH, CH_3)

The formation of the *anti* stereoisomer is attributed to involvement of a butenyltitanium intermediate formed by rapid exchange with the butenylstannane. This intermediate then reacts through a cyclic transition state.

$CH_3CH{=}CHCH_2SnBu_3 + TiCl_4 \rightleftharpoons CH_3CH{=}CHCH_2TiCl_3$

When an aldehyde subject to "chelation control" is used, the *syn* stereoisomer dominates with $MgBr_2$ as the Lewis acid.[117]

Allylic stannanes with γ-oxygen substituents have been used to build up polyoxygenated carbon chains. For example, **7** reacts with the stannane **8** to give a high preference for the stereoisomer in which the two oxygen substituents are *anti*. This stereoselectivity is

116. G. E. Keck, D. E. Abbott, E. P. Boden, and E. J. Enholm, *Tetrahedron Lett.* **25**:3927 (1984).
117. G. E. Keck and E. P. Boden, *Tetrahedron Lett.* **25**:265 (1984); G. E. Keck, D. E. Abbott, and M. R. Wiley, *Tetrahedron Lett.* **28**:139 (1987); G. E. Keck, K. A. Savin, E. N. K. Cressman, and D. E. Abbott, *J. Org. Chem.* **59**:7889 (1994).

consistent with chelation control.

PhCH$_2$O O H CH$_3$ **7** + H H C=C TBDMSO CH$_2$SnBu$_3$ **8** ⟶ PhCH$_2$O OH CH$_3$ OTBDMS

Mg PhCH$_2$O O H CH$_3$

preferred attack from side away from the methyl group

Allylstannane additions to aldehydes can be made enantioselective by use of chiral catalysts. A catalyst prepared from the chiral binaphthols *R*- and *S*-BINOL and Ti(O-*i*-*Pr*)$_4$ achieved 80–95% enantioselectivity.[118]

$$\mathrm{PhCH{=}O + CH_2{=}CHCH_2SnR_3 \xrightarrow[\text{Ti(O}i\text{-Pr)}_4]{R\text{-BINOL}} PhCH(OH)CH_2CH{=}CH_2} \quad \text{87–96\% e.e.}$$

Lewis acid-mediated ionization of acetals also generates electrophilic carbon intermediates that react readily with allylic stannanes.[119] Dithioacetals are activated by the sulfonium salt $[(CH_3)_2SSCH_3]^+BF_4^-$.[120]

$$\mathrm{PhCH_2CH_2CH(OCH_3)_2 + CH_2{=}CHCH_2Sn(CH_3)_3 \xrightarrow{(R_2AlO)_2SO_2} PhCH_2CH_2CH(OCH_3)CH_2CH{=}CH_2}$$

$$\mathrm{PhCH_2CH(OCH_3)_2 + Sn(CH_2CH{=}CH_2)_4 \xrightarrow[CH_3OH]{CF_3CO_2H,\ SiO_2} PhCH_2CH(OH)CH_2CH{=}CH_2}$$

$$\mathrm{CH_3(CH_2)_4C(CH_3)(SCH_3)_2 + CH_2{=}CHCH_2Sn(C_4H_9)_3 \xrightarrow{[(CH_3)_2SSCH_3]^+BF_4^-} CH_3(CH_2)_4C(SCH_3)(CH_3)CH_2CH{=}CH_2}$$

Scheme 9.6 gives some other examples of Lewis acid-catalyzed reactions of allylic stannanes with carbonyl compounds.

118. G. E. Keck, K. H. Tarbet, and L. S. Geraci, *J. Am. Chem. Soc.* **115:**8467 (1993); A. L. Costa, M. G. Piazza, E. Tagliavini, C. Trombini, and A. Umani-Ronchi, *J. Am. Chem. Soc.* **115:**7001 (1993); G. E. Keck and L.S. Geraci, *Tetrhahedron Lett.* **34:**7827 (1993). G. E. Keck, D. Krishnamurthy, and M. C. Grier, *J. Org. Chem.* **58:**6543 (1993).
119. A. Hosomi, H. Iguchi, M. Endo, and H. Sakurai, *Chem. Lett.* **1979:**977.
120. B. M. Trost and T. Sato, *J. Am. Chem. Soc.* **107:**719 (1985).

Scheme 9.6. Reactions of Allylic Stannanes with Carbonyl Compounds

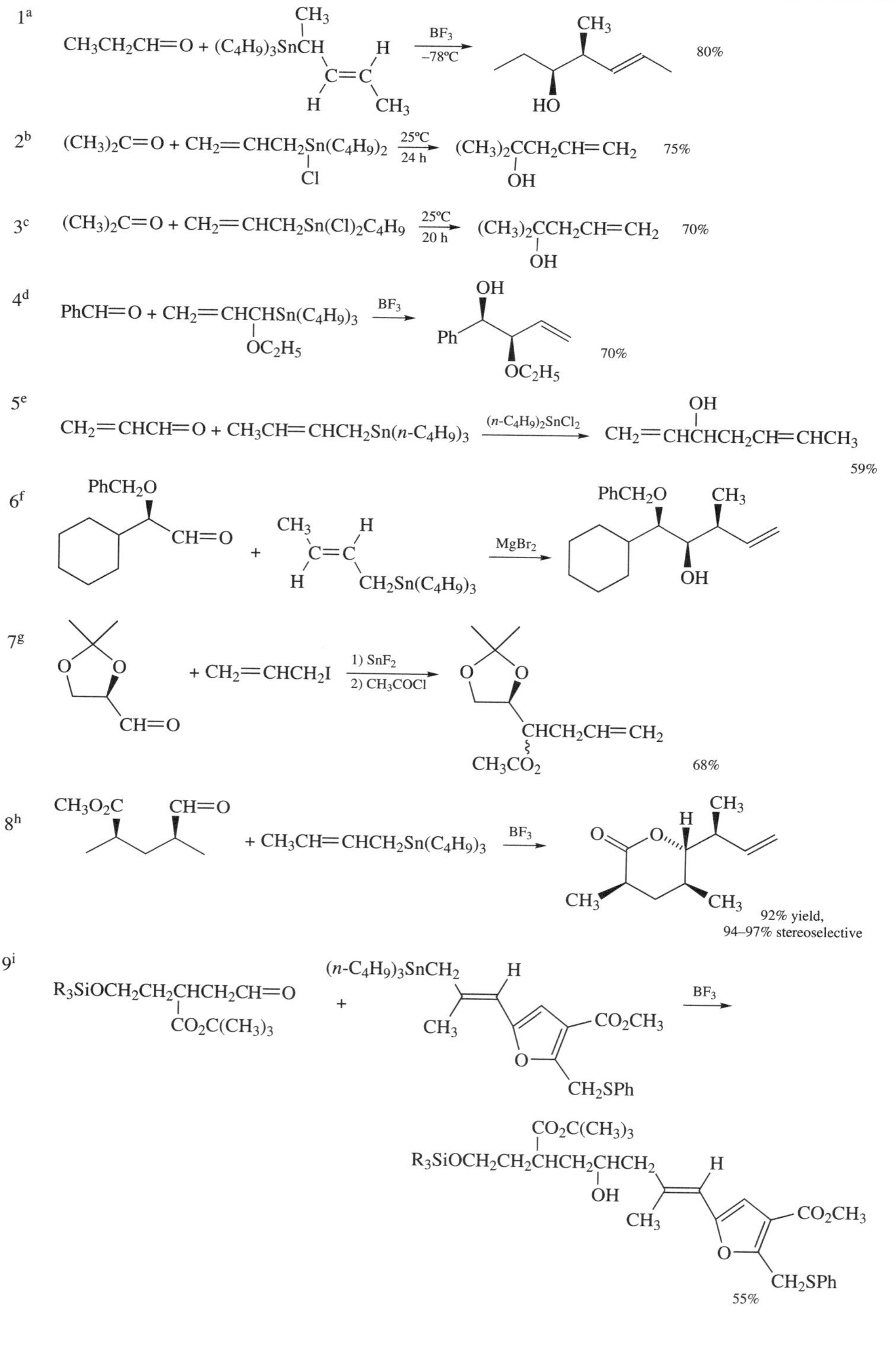

1[a] $CH_3CH_2CH{=}O + (C_4H_9)_3SnCH(CH_3)CH{=}CHCH_3$ (E) $\xrightarrow[-78°C]{BF_3}$ 4-methyl-5-hepten-3-ol (HO, CH_3) 80%

2[b] $(CH_3)_2C{=}O + CH_2{=}CHCH_2Sn(Cl)(C_4H_9)_2 \xrightarrow[24\ h]{25°C} (CH_3)_2C(OH)CH_2CH{=}CH_2$ 75%

3[c] $(CH_3)_2C{=}O + CH_2{=}CHCH_2Sn(Cl)_2C_4H_9 \xrightarrow[20\ h]{25°C} (CH_3)_2C(OH)CH_2CH{=}CH_2$ 70%

4[d] $PhCH{=}O + CH_2{=}CHCH(OC_2H_5)Sn(C_4H_9)_3 \xrightarrow{BF_3}$ PhCH(OH)CH(OC_2H_5)CH=CH_2 70%

5[e] $CH_2{=}CHCH{=}O + CH_3CH{=}CHCH_2Sn(n\text{-}C_4H_9)_3 \xrightarrow{(n\text{-}C_4H_9)_2SnCl_2} CH_2{=}CHCH(OH)CH_2CH{=}CHCH_3$ 59%

6[f] cyclohexyl–CH(OCH_2Ph)CH=O + (Z)-$CH_3CH{=}CHCH_2Sn(C_4H_9)_3$ $\xrightarrow{MgBr_2}$ cyclohexyl–CH(OCH_2Ph)CH(OH)CH(CH_3)CH=CH_2

7[g] 2,2-dimethyl-1,3-dioxolane-4-CH=O + $CH_2{=}CHCH_2I$ $\xrightarrow{1)\ SnF_2;\ 2)\ CH_3COCl}$ 2,2-dimethyl-1,3-dioxolan-4-yl–CH(CH_3CO_2)$CH_2CH{=}CH_2$ 68%

8[h] CH_3O_2C–CH(CH_3)CH_2CH(CH_3)–CH=O + $CH_3CH{=}CHCH_2Sn(C_4H_9)_3 \xrightarrow{BF_3}$ lactone (O, H, CH_3, CH_3, CH_3) 92% yield, 94–97% stereoselective

9[i] $R_3SiOCH_2CH_2CH(CO_2C(CH_3)_3)CH_2CH{=}O$ + $(n\text{-}C_4H_9)_3SnCH_2C(CH_3){=}CH$–furan($CO_2CH_3$, CH_2SPh) $\xrightarrow{BF_3}$ $R_3SiOCH_2CH_2CH(CO_2C(CH_3)_3)CH_2CH(OH)CH_2C(CH_3){=}CH$–furan($CO_2CH_3$, CH_2SPh) 55%

Scheme 9.6. (*continued*)

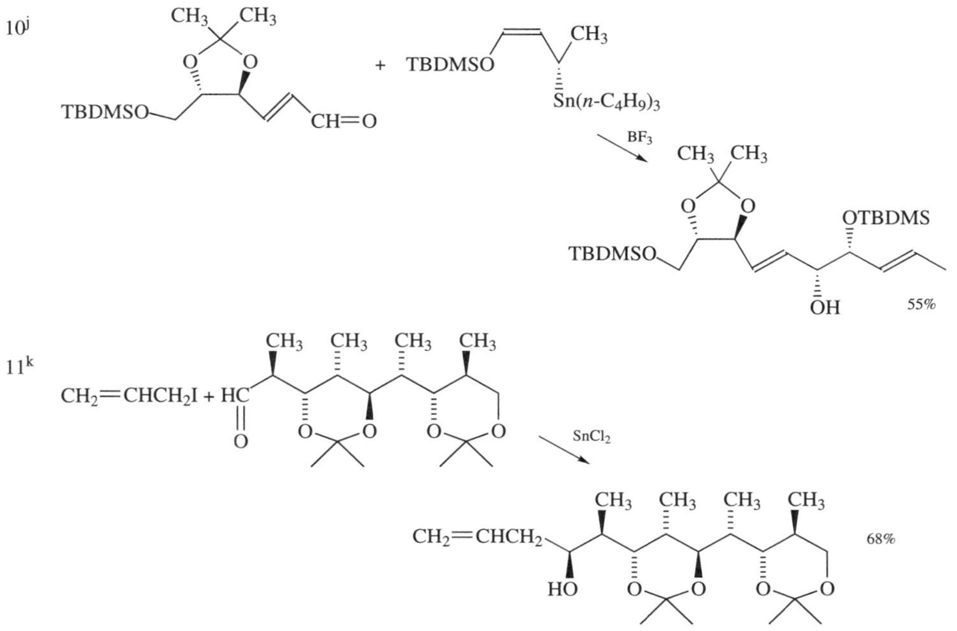

10[j]

CH_3 CH_3 O O TBDMSO CH=O + TBDMSO CH_3 $Sn(n\text{-}C_4H_9)_3$ $\xrightarrow{BF_3}$ CH_3 CH_3 O O TBDMSO OTBDMS OH 55%

11[k]

$CH_2{=}CHCH_2I$ + HC(=O) CH_3 CH_3 CH_3 CH_3 O O O O $\xrightarrow{SnCl_2}$ $CH_2{=}CHCH_2$ CH_3 CH_3 CH_3 CH_3 HO O O O O 68%

a. M. Koreeda and Y. Tanaka, *Chem. Lett.* 1297 (1982).
b. V. Peruzzo and G. Tagliavini, *J. Organomet. Chem.* **162:**37 (1978).
c. A. Gambaro, V. Peruzzo, G. Plazzogna, and G. Tagliavini, *J. Organomet. Chem.* **197:**45 (1980).
d. D.-P. Quintard, B. Elissondo, and M. Pereyre, *J. Org. Chem.* **48:** 1559 (1983).
e. L. A. Paquette and G. D. Maynard, *J. Am. Chem. Soc.* **114:**5018 (1992).
f. G. E. Keck and E. P. Boden, *Tetrahedron Lett.* **25:**1879 (1984).
g. T. Harada and T. Mukaiyama, *Chem. Lett.*, 1109 (1981).
h. K. Maruyama, Y. Ishihara, and Y. Yamamoto, *Tetrahedron Lett.* **22:**4235 (1981).
i. L. A. Paquette and P. C. Astles, *J. Org. Chem.* **58:**165 (1993).
j. J. A. Marshall, S. Beaudoin, and K. Lewinski, *J. Org. Chem.* **58:**5876 (1993).
k. H. Nagaoka and Y. Kishi, *Tetrahedron* **37:**3873 (1981).

General References

Organoborane Compounds

H. C. Brown, *Organic Synthesis via Boranes*, John Wiley & Sons, New York, 1975.
E. Negishi and M. Idacavage, *Org. React.* **33:**1 (1985).
A. Pelter, K. Smith, and H. C. Brown, *Borane Reagents*, Academic Press, New York, 1988.
A. Pelter, in *Rearrangements in Ground and Excited States*, Vol. 2, P. de Mayo, ed., Academic Press, New York, 1980, Chapter 8.
B. M. Trost, ed., *Stereodirected Synthesis with Organoboranes*, Springer, Berlin, 1995.

Organosilicon Compounds

T. H. Chan and I. Fleming, *Synthesis* **1979:**761.
E. W. Colvin, *Silicon Reagents in Organic Synthesis*, Academic Press, London, 1988.
I. Fleming, J. Dunogues, and R. Smithers, *Org. React.* **37:**57 (1989).
W. Weber, *Silicon Reagents for Organic Synthesis*, Springer, Berlin, 1983.

Organotin Compounds

A. G. Davies, *Organotin Chemistry*, VCH, Weinheim, 1997.

S. Patai, ed., *The Chemistry of Organic Germanium, Tin and Lead Compounds*, Wiley-Interscience, New York, 1995.

M. Pereyre, J.-P. Quintard, and A. Rahm, *Tin in Organic Synthesis*, Butterworths, London, 1983.

Problems

(*References for these problems will be found on page 936.*)

1. Give the expected product(s) for the following reactions.

(a) $(CH_3)_3CC{\equiv}CCH_3 \xrightarrow[2)\ H_2C{=}CHC(O)CH_3,\ 65°C]{1)\ 9\text{-BBN},\ 0°C,\ 16\ h} \xrightarrow[H_2O_2]{NaOH}$

(b) $PhCH{=}CHCOCl + (CH_3)_4Sn \xrightarrow[HMPA]{PhCH_2PdCl/Ph_3P}$

(c) $[CH_3CO_2(CH_2)_5]_3B + LiC{\equiv}C(CH_2)_3CH_3 \longrightarrow \xrightarrow{I_2}$

(d) PhCH=O + $(CH_3)(H)C{=}C(H)(CH_2Sn(C_4H_9)_3)$ $\xrightarrow{BF_3}$

(e) $CH_3O{-}C_6H_4{-}CH{=}CHCH_2Si(CH_3)_3$ + $CH_3CH_2CH(OCH_3){-}C_6H_4{-}OCH_3$ $\xrightarrow{TiCl_4}$

(f) $(CH_3O_2C)_2C(CH_3)CH(Si(CH_3)_3)CH{=}CHSi(CH_3)_3 + PhCH{=}O \xrightarrow{F^-}$

(g) 2-methylcyclopentyl (CH₃ wedge, B hashed) 1,3,2-dioxaborinane $\xrightarrow[2)\ HgCl_2;\ 3)\ H_2O_2,\ pH\ 8]{1)\ LiCH(SPh)OCH_3}$

(h) 9-BBN–CH=CH(CH₂)₂CH₃ (trans; B–C(H)=C(H)C(CH₂)₂CH₃) + $ClCH_2CN$ $\xrightarrow{\text{potassium 2,6-di-}t\text{-butylphenoxide}}$

2. Starting with an alkene $RCH{=}CH_2$, indicate how an organoborane intermediate could be used for each of the following synthetic transformations:

(a) $RCH{=}CH_2 \longrightarrow RCH_2CH_2CH_2C(O){-}C_6H_5$

(b) $RCH{=}CH_2 \longrightarrow RCH_2CH_2CH{=}O$

(c) $RCH{=}CH_2 \longrightarrow RCH_2CH_2\overset{CH_3}{\overset{|}{C}}HCH_2\overset{O}{\overset{\|}{C}}CH_3$

(d) $RCH{=}CH_2 \longrightarrow$ (RCH$_2$CH$_2$)(H)C=C(CH$_3$)(H)

(e) $RCH{=}CH_2 \longrightarrow RCH_2CH_2\overset{O}{\overset{\|}{C}}CH_2CH_2R$

(f) $RCH{=}CH_2 \longrightarrow RCH_2CH_2CH_2CO_2C_2H_5$

3. In Scheme 9.1 there are described reactions of organoboranes with cyanide ion, lithiodichloromethane, and dichloromethyl methyl ether. Compare the structures of these reagents and the final reaction products from each of these reagents. Develop a general mechanistic outline with encompasses these reactions, and discuss the structural features which these reagents have in common with one another and with carbon monoxide.

4. Give appropriate reagents, other organic reactants, and approximate reaction conditions for effecting the following syntheses in a "one-pot" process.

(a) (CH$_3$)(H)C=C(H)((CH$_2$)$_5$O$_2$CCH$_3$) from $IC{\equiv}CCH_2CH_3$ and $CH_2{=}CH(CH_2)_3O_2CCH_3$

(b) $CH_3(CH_2)_3C{\equiv}C(CH_2)_7CH_3$ from $CH_2{=}CH(CH_2)_5CH_3$ and $HC{\equiv}C(CH_2)_3CH_3$

(c) $CH_3(CH_2)_{11}\overset{O}{\overset{\|}{C}}$–cyclopentyl from $CH_2{=}CH(CH_2)_9CH_3$ and cyclopentene

(d) Bicyclic ketone (H, CH$_3$, TBSO, CH$_3$, =O) from cyclopentane derivative (H, C(CH$_3$)=CH$_2$, CH=CH$_2$, TBSO, CH$_3$)

5. Give a mechanism for each of the following reactions.

(a) $C_6H_5\text{–}Br + CH_2{=}CHSn(C_4H_9)_3 \xrightarrow{Pd(PPh_3)_4} C_6H_5\text{–}CH{=}CH_2$

(b) $CH_3(CH_2)_6\overset{OSi(CH_3)_3}{\overset{|}{C}}{=}CH_2 + PhBr + Bu_3SnF \xrightarrow[\text{3 mol \%}]{PdCl_2[P(tol)_3]} CH_3(CH_2)_6\overset{O}{\overset{\|}{C}}CH_2Ph$

(c) $(\text{cyclopentyl})_3\bar{B}\text{–}C{\equiv}C(CH_2)_5CH_3 \xrightarrow[\text{2) } H_2O_2,\ ^{-}OAc]{\text{1) } ICH_2CN} \text{cyclopentyl–}\overset{O}{\overset{\|}{C}}\text{–}\underset{CH_2CN}{\underset{|}{C}}H(CH_2)_5CH_3$

(d) $(CH_3)_3\bar{B}C{\equiv}C(CH_2)_3CH_3 \xrightarrow[\text{ether/THF}]{2.5\text{ equiv }CH_3SO_3H} HO{-}C(CH_3)_2(CH_2)_4CH_3$

(e) $PhCH_2NCH_2CH_2C{\equiv}CSi(CH_3)_3 \xrightarrow[NaN_3]{CH_2{=}O}$ PhCH₂N (tetrahydropyridine with Si(CH₃)₃ and N₃)

(f) B–Cl (9-BBN-Cl) → 1) 2,6-dimethylphenol 2) Cl_2CHOCH_3 3) $(C_2H_5)_3CO^-Li^+$ 4) H_2O_2, ^-OH → bicyclic ketone

6. Offer a detailed mechanistic explanation for the following observations.

(a) When the *E*- and *Z*-isomers of the 2-butenyl boradioxolane **A** are caused to react with the aldehyde **B**, the *Z*-isomer gives the *syn* product **C** with >90% stereoselectivity. The *E*-isomer, however, gives a nearly 1 : 1 mixture of **D** and **E**.

$CH_3CH{=}CHCH_2B$ (dioxaborolane) **A** + O=CH (dioxolane) **B**

Z-isomer → **C**: $CH_2{=}CH$, CH_3, OH

E-isomer → **D**: $CH_2{=}CH$, CH_3, OH + **E**: $CH_2{=}CH$, CH_3, OH

(b) The reaction of a variety of $\Delta^{2,3}$-pyranyl acetates with allysilanes under the influence of Lewis acid catalysts gives 2-allyl-$\Delta^{3,4}$-pyrans. The stereochemistry of the methyl group is dependent on whether the *E*- or *Z*-allylsilane is used. Predict which stereoisomer is formed in each case, and explain the mechanistic basis of your prediction.

(pyran with O_2CCH_3, O_2CCH_3, $CH_2O_2CCH_3$) + $CH_3CH{=}CHCH_2Si(CH_3)_3$ ⟶ $CH_2{=}CHCH(H_3C)$–(pyran with H, O_2CCH_3, $CH_2O_2CCH_3$)

(c) When trialkylboranes react with α-diazoketones or α-diazoesters in D_2O, the resulting products are monodeuterated. Formulate the reaction mechanism in sufficient detail to account for this fact.

$$R_3B + N_2CHCR' \xrightarrow{D_2O} RCHDCR'$$

(with C=O on the carbonyl carbons: $N_2CHC(=O)R'$ → $RCH(D)C(=O)R'$)

(d) It is observed that the stereoselectivity of condensation of aminosilane **F** with aldehydes depends on the steric bulk of the amino substituent. Offer an explanation for this observation in terms of the transition state for the addition reaction.

$$\mathbf{F}\ (Si(CH_3)_3,\ CH_2CH_2NHR) + PhCH_2CH{=}O \xrightarrow[CH_3OH]{ZnI_2,\ 5\ mol\%} \text{product } (R\text{–}N,\ H,\ CH_2Ph)$$

R	Yield (%)	*trans:cis* ratio
CH_3[a]	68	20:80
$PhCH_2$	88	58:42
Ph_2CH	73	>99:1
Dibenzocycloheptyl	67	>99:1

[a] $Ph(CH_3)_2Si$ instead of $(CH_3)_3Si$.

7. A number of procedures for stereoselective syntheses of alkenes involving alkenylboranes have been developed. For each of the procedures given below, give the structures of the intermediates and describe the mechanism in sufficient detail to account for the observed stereoselectivity.

(a) $R^1C{\equiv}CBr \xrightarrow[2)\ HO(CH_2)_3OH]{1)\ BHBr_2\cdot SMe_2} A \xrightarrow{R^2Li} B \xrightarrow[2)\ I_2,\ MeOH;\ 3)\ NaOH]{1)\ R^3Li} (R^1)(H)C{=}C(R^2)(R^3)$

(b) $R^1C{\equiv}CBr \xrightarrow[2)\ HO(CH_2)_3OH]{1)\ BHBr_2\cdot SMe_2} A \xrightarrow{R^2Li} B \xrightarrow[2)\ NaOMe,\ MeOH]{1)\ Br_2} C \xrightarrow[2)\ (i\text{-}PrO)_3B;\ 3)\ HCl]{1)\ s\text{-}BuLi} D$

$D \xrightarrow[HO(CH_2)_3OH]{H_2O} E \xrightarrow[2)\ I_2,\ MeOH;\ 3)\ NaOH]{1)\ R^3Li} (R^1)(H)C{=}C(R^3)(R^2)$

(c) $R^2_2BCl \xrightarrow[R^1C{\equiv}CH]{LiAlH_4} A \xrightarrow[I_2]{NaOCH_3} (R^1)(H)C{=}C(R^2)(H)$

(d) $R^1C{\equiv}CH \xrightarrow[2)\ HO(CH_2)_2OH]{1)\ BHBr_2\cdot SMe_2} A \xrightarrow[2)\ I_2,\ MeOH;\ 3)\ NaOH]{1)\ R^2Li} (R^1)(H)C{=}C(R^2)(H)$

8. Suggest reagents which could be effective for the following cyclization reactions.

(a) [structure: acyclic alkynyl aldehyde (–C≡C–CH=O) bearing OCH_2OCH_3 and $SnBu_3$ on an allylic carbon] → [structure: macrocyclic alkyne with OH and $CH=CHOCH_3$ substituents]

(b) [structure: lactone with O–CH_2–C(=CH_2)–CH_2SnBu_3 side chain and $C(SCH_3)_2$ ethyl-substituted chain; labels $SnBu_3$, CH_3S, SCH_3] → [structure: macrocycle with H_2C= exocyclic methylene, C_2H_5 and SCH_3 substituents]

9. Show how the following silanes and stannanes could be synthesized from the suggested starting material.

(a) 1-methyl-2-(trimethylsilyl)cyclopentene [H_3C, $Si(CH_3)_3$] from 2-methylcyclopentanone [H_3C, O]

(b) $(C_2H_5)((CH_3)_3Sn)C{=}C(H)CO_2C_2H_5$ from $CH_3CH_2C{\equiv}CCO_2C_2H_5$

(c) $Bu_3SnCH{=}CHSnBu_3$ from $HC{\equiv}CH$, Bu_3SnCl, and Bu_3SnH

(d) 2-(trimethylstannyl)pyridine [N, $Sn(CH_3)_3$] from 2-chloropyridine [N, Cl]

(e) $RC({=}CH_2)CH_2Si(CH_3)_3$ from RCOCl or RCO_2R'

(f) $CH_2{=}C(CH_2Si(CH_3)_3)CH{=}CH_2$ from $CH_2{=}C(Cl)CH{=}CH_2$

10. Each of the cyclic amines shown below has been synthesized by reaction of an amino-substituted allylic silane with formaldehyde in the presence of trifluoroacetic acid. Identify the appropriate precursor of each amine and suggest a method for its synthesis.

(a) N-benzyl-3-vinylpiperidine [CH_2Ph, N, $CH{=}CH_2$]

(b) N-benzyl-3-vinylpyrrolidine [CH_2Ph, N, $CH{=}CH_2$]

(c) N-benzyl-4-methylenepiperidine [CH_2Ph, N, CH_2]

11. Both the *E*- and *Z*-stereoisomers of the terpene γ-bisabolene can be isolated from natural sources. Recently, stereoselective syntheses of these compounds were developed which rely heavily on borane intermediates.

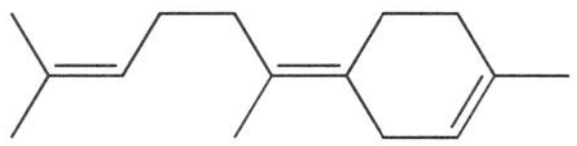

E-γ-bisabolene 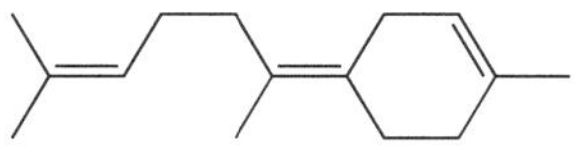 *Z*-γ-bisabolene

The syntheses of the *E*-isomer, proceeds from **G** through **H** as a key intermediate. Show how **H** could be obtained from **G** and converted to *E*-γ-bisabolene.

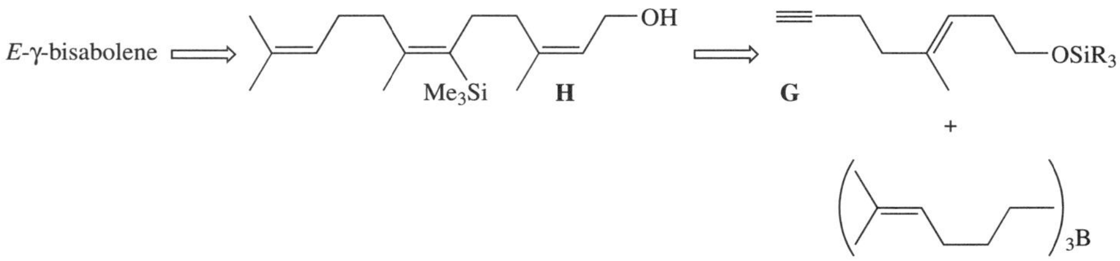

E-γ-bisabolene ⟹ OH Me_3Si **H** ⟹ $OSiR_3$ **G** + $_3B$

The synthesis of the *Z*-isomer employs the bromide **I** and the borane **J** as starting materials. Devise a method for synthesis of the *Z*-isomer from **I** and **J**.

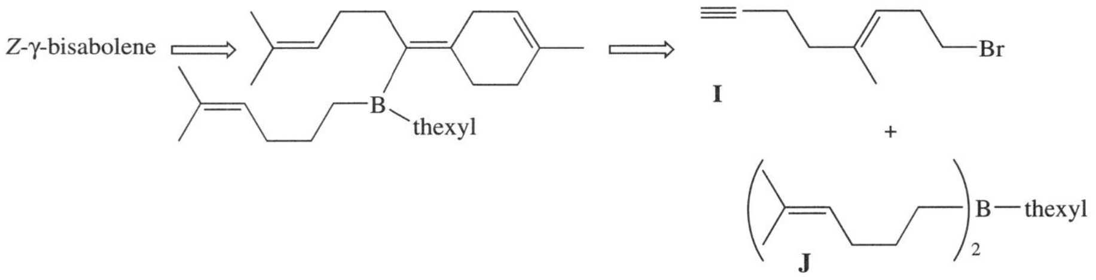

Z-γ-bisabolene ⟹ B thexyl ⟹ Br **I** + B—thexyl 2 **J**

The crucial element of these syntheses is control of the stereochemistry of the exocyclic double bond on the basis of the stereochemistry of migration of a substituent from boron to carbon. Discuss the requirements for a stereoselective synthesis, and suggest how these requirements might be met.

12. Devise a sequence of reactions which would provide the desired compound from the suggested starting material(s).

(a) OH CHOMOM from

$HC{\equiv}CCH_2CH_2C(H){=}C(CH_3)CH_2CH_2C(H){=}C(CH_3)CH_2CH_2C(H){=}C(H)CH{=}O$

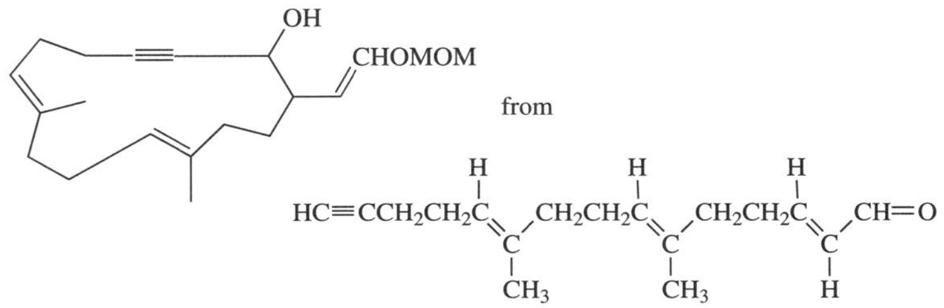

(b) from $CH_3(CH_2)_3CH{=}O$ and

(c) from and

(d) from $(C_4H_9)_3SnCH_2C(=CH_2)CH_2OH$ and $CH_3CH_2C(SCH_3)_2(CH_2)_9CO_2H$

(e) from

(f) from

(g) from , $BrCH_2C(CH_3){=}CHCO_2CH_3$ and $(C_4H_9)_3SnCH{=}CHSn(C_4H_9)_3$

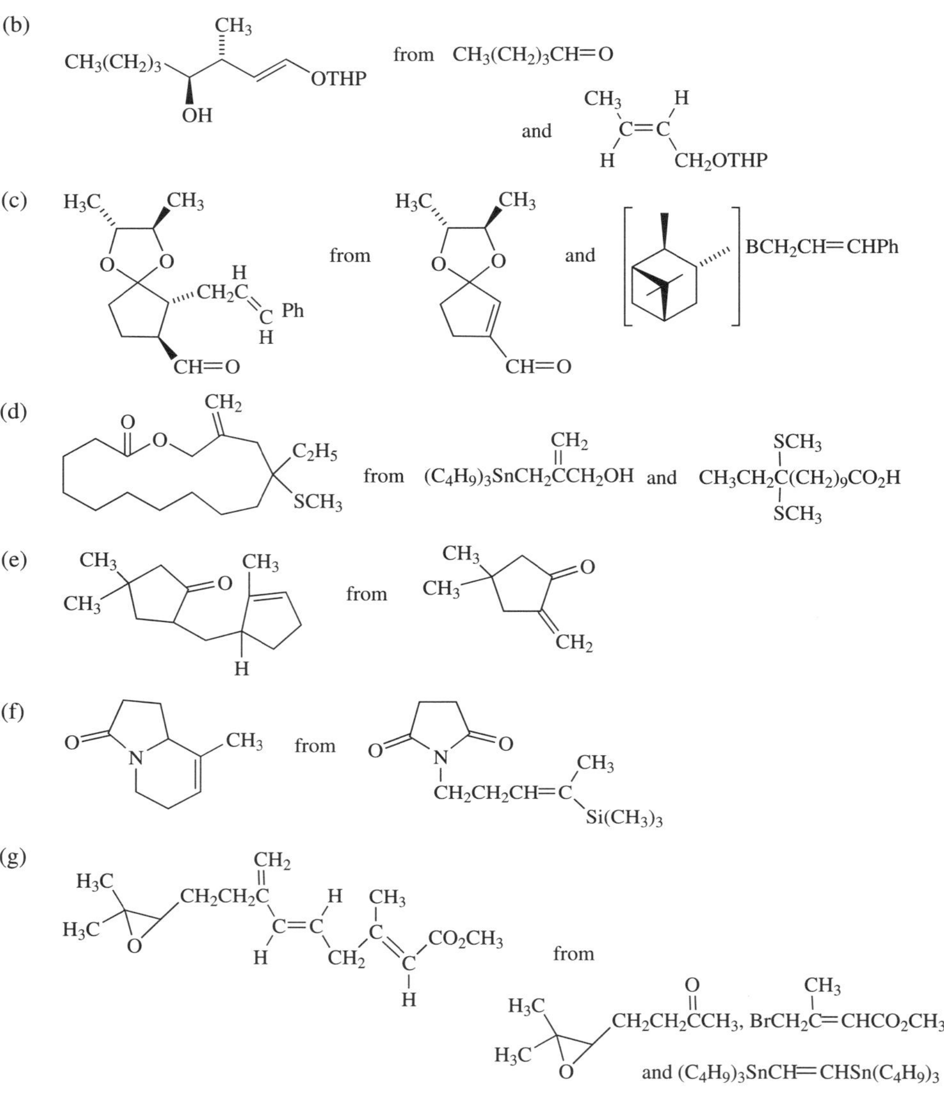

13. Show how the following compounds can be prepared in high enantiomeric purity using enantiopure boranes as reagents

(a)

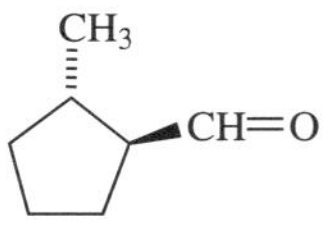

(b)

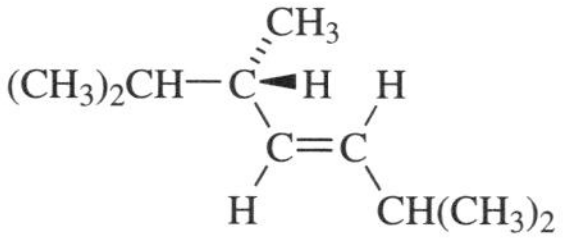

(c)

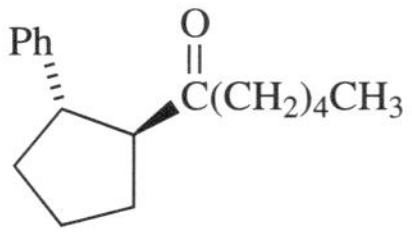

(d)

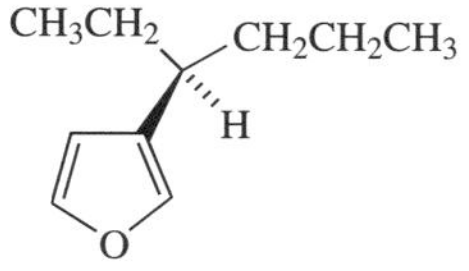

14. Show how organoborane intermediates could be used to synthesize the gypsy moth pheromone E,Z-$CH_3CO_2(CH_2)_4CH{=}CH(CH_2)_2CH{=}CH(CH_2)_2CH_3$ from hept-6-ynyl acetate, allyl bromide, and 1-hexyne.

15. Predict the major stereoisomer which will be formed in the following reactions

(a) TBDPSO–(epoxide)–CH=O + $CH_2{=}CHCH_2{-}B$(dioxaborolane with two $CO_2C_2H_5$ groups) →

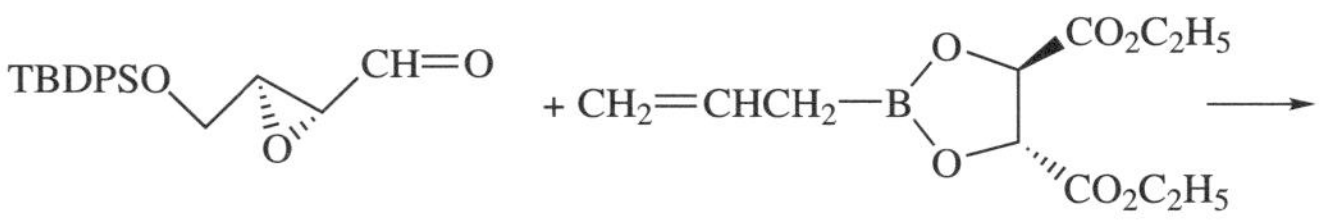

(b) $CH_2{=}CHCH{=}O + CH_3CH{=}CHCH_2Sn(n\text{-}C_4H_9)_3 \xrightarrow{(n\text{-}C_4H_9)_2SnCl_2}$

(c) $PhCH{=}O$ + (C_2H_5, H)C=C($CH_2Sn(n\text{-}C_4H_9)_3$, $C(CH_3)_3$) $\xrightarrow{BF_3}$

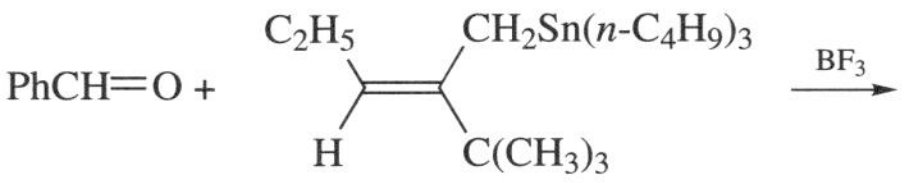

(d) $SiCH_3)_2Ph$ (cyclohexenyl) CH_2CH_2N–(dibenzosuberyl) + ($CH_2CH{=}O$, I, OCH_2Ph, OCH_3 substituted benzene) $\xrightarrow[C_2H_5OH]{ZnI_2}$

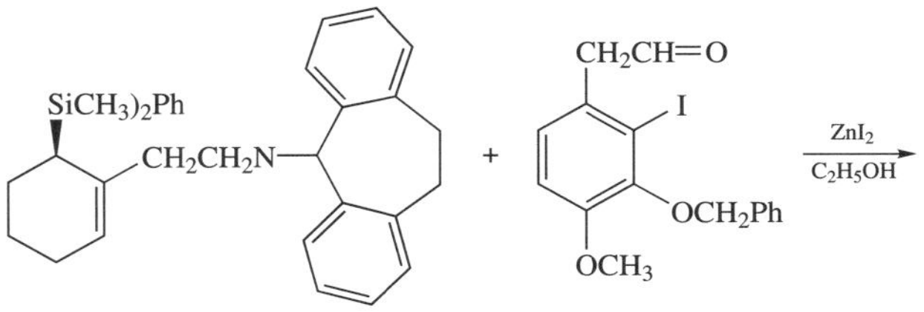

(e) (benzene-1,2-dicarbaldehyde: CH=O, CH=O) + 2.5 equiv (pinanyl$)_2BCH_2CH{=}CH_2$

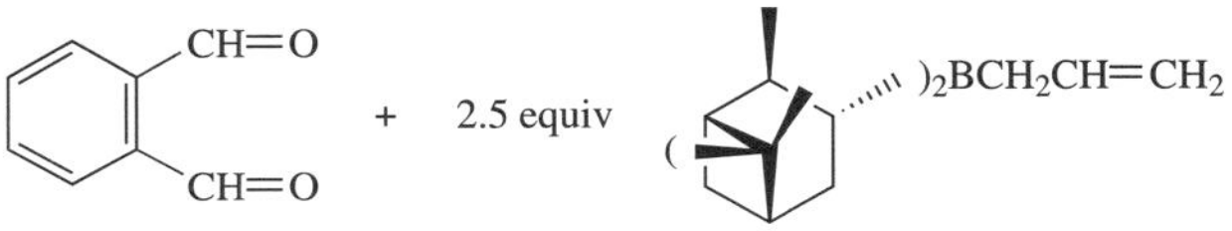

(f) $PhCH_2O$ (tetrahydropyran) $OCH_2CH{=}O$ + $CH_2{=}CHCH_2SnBu_3 \xrightarrow{\text{4 equiv, } MgBr_2}$

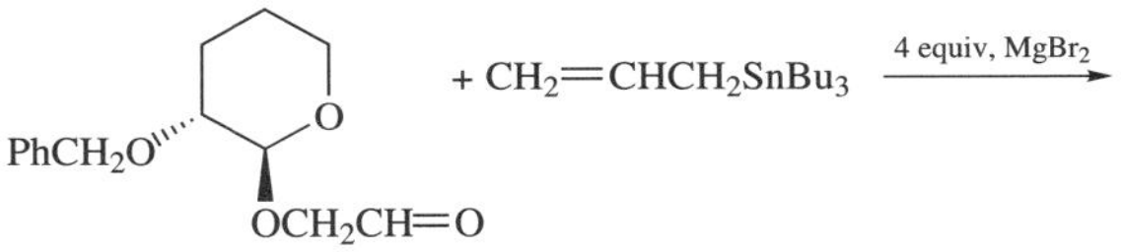

(g) $PhCH_2O_2C$–(chain with CH_3CO_2, CH_3, OCH_2Ph, CH_3)–CH=O + $CH_2{=}CHCH_2Si(CH_3)_3 \xrightarrow[\text{2) TIPSOTf}]{\text{1) } TiCl_4}$

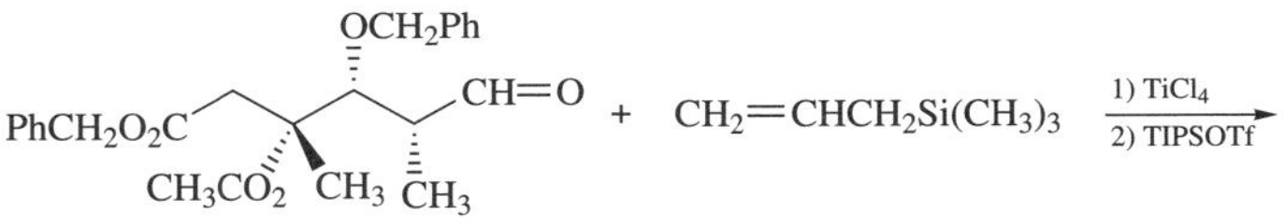

10

Reactions Involving Carbocations, Carbenes, and Radicals as Reactive Intermediates

Introduction

Trivalent carbocations, carbanions, and radicals are the most fundamental classes of reactive intermediates. Discussion of carbanion intermediates began in Chapter 1 and has continued in several other chapters. The focus in this chapter will be on *electron-deficient reactive intermediates*. Carbocations are the most fundamental example, but carbenes and nitrenes also play a significant role in synthetic reactions. Each of these intermediates has a carbon or nitrogen atom with *six* valence electrons, and they are therefore electron-deficient and electrophilic in character. Because of their electron deficiency, carbocations and carbenes have the potential for skeletal rearrangements. We will also discuss the use of carbon radicals to form carbon–carbon bonds. Radicals, too, are electron-deficient but react through homolytic bond-breaking and bond-forming reactions. A common feature of all of these intermediates is that they are of high energy, relative to structures with filled bonding orbitals. Their lifetimes are usually very short. Reaction conditions designed to lead to synthetically useful outcomes must take this high reactivity into account.

10.1. Reactions Involving Carbocation Intermediates

In this section, we will emphasize carbocation reactions that modify the carbon skeleton, including carbon–carbon bond formation, rearrangements, and fragmentations.

The fundamental structural and reactivity characteristics of carbocations, especially toward nucleophilic substitution, were introduced in Chapter 5 of Part A.

10.1.1. Carbon–Carbon Bond Formation Involving Carbocations

The formation of carbon–carbon bonds by electrophilic attack on the π system is an important reaction in aromatic chemistry, with both Friedel–Crafts alkylation and acylation following this pattern (see Chapter 11). There also are valuable synthetic procedures in which carbon–carbon bond formation results from electrophilic attack by a carbocation on an alkene. The reaction of a carbocation with an alkene to form a new carbon–carbon bond is both kinetically accessible and thermodynamically favorable because a π bond is replaced by a stronger σ bond.

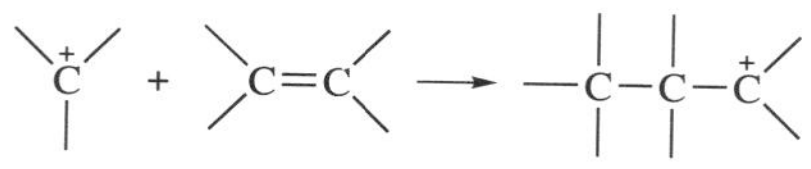

There are, however, important problems that must be overcome in the application of this reaction to synthesis. The product is a new carbocation which can react further. Repetitive addition to alkene molecules leads to polymerization. Indeed, this is the mechanism of acid-catalyzed polymerization of alkenes. There is also the possibility of rearrangement. A key requirement for adapting the reaction of carbocations with alkenes to the synthesis of small molecules is control of the reactivity of the newly formed carbocation intermediate. Synthetically valuable carbocation–alkene reactions require a suitable termination step. We have already encountered one successful strategy in the reaction of alkenyl and allylic silanes and stannanes with electrophilic carbon (see Chapter 8). In those reactions, the silyl or stannyl substituent is eliminated and a stable alkene is formed.

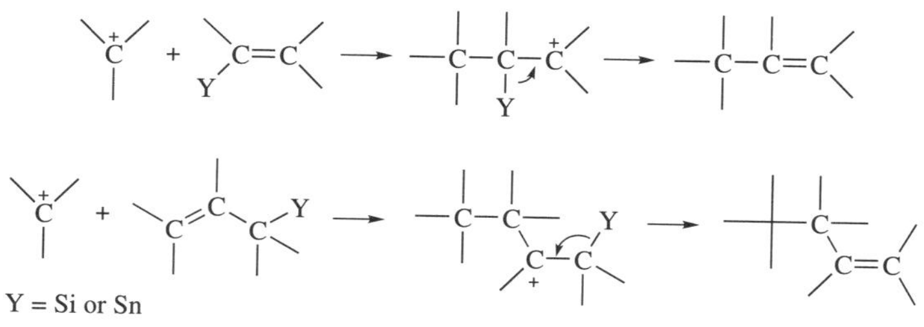

Y = Si or Sn

The increased reactivity of the silyl- and stannyl-substituted alkenes enhances the synthetic utility of carbocation–alkene reactions.

Silyl enol ethers offer both enhanced reactivity and an effective termination step. Electrophilic attack is followed by desilylation to give an α-substituted carbonyl compound. The carbocations can be generated from tertiary chlorides and a Lewis acid, such as $TiCl_4$. This reaction provides a method for introducing tertiary alkyl groups α to a carbonyl, a transformation which cannot be achieved by base-catalyzed alkylation because of the strong tendency for tertiary halides to undergo elimination.

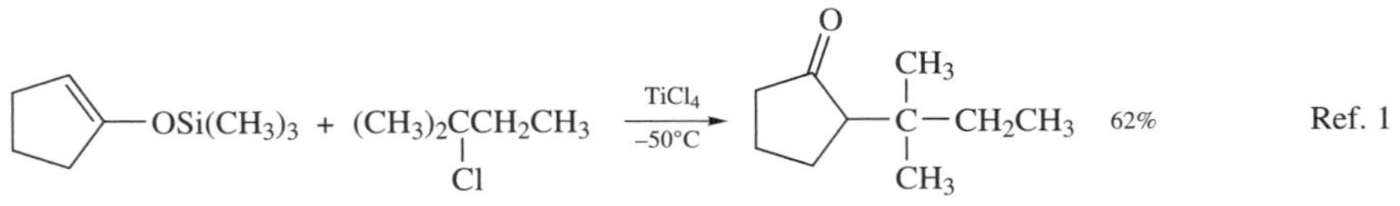

Ref. 1

Secondary benzylic bromides, allylic bromides, and α-chloro ethers can undergo analogous reactions with the use of $ZnBr_2$ as the catalyst.[2] Primary iodides react with silyl enol ethers in the presence of AgO_2CCF_3.[3]

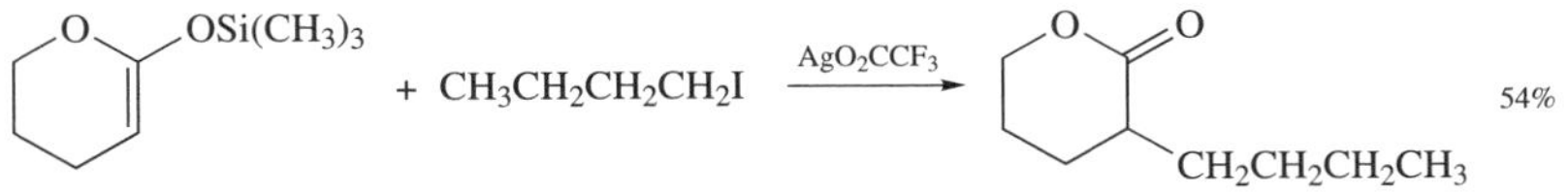

Alkylations by allylic cations have been observed with the use of $LiClO_4$ to promote ionization.[4]

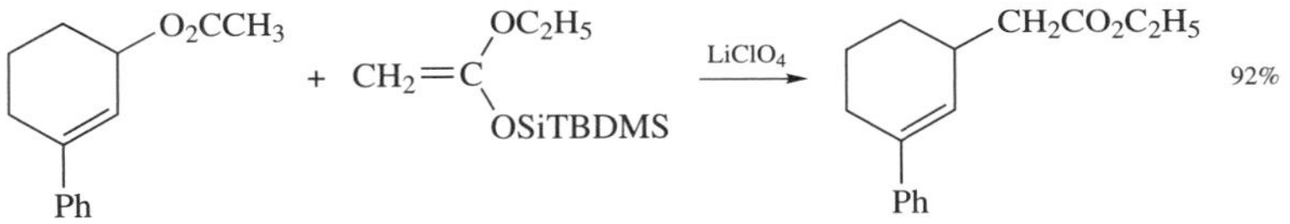

Alkenes react with acyl halides or acid anhydrides in the presence of a Lewis acid catalyst. The reaction works better with cyclic alkenes than with acyclic ones.

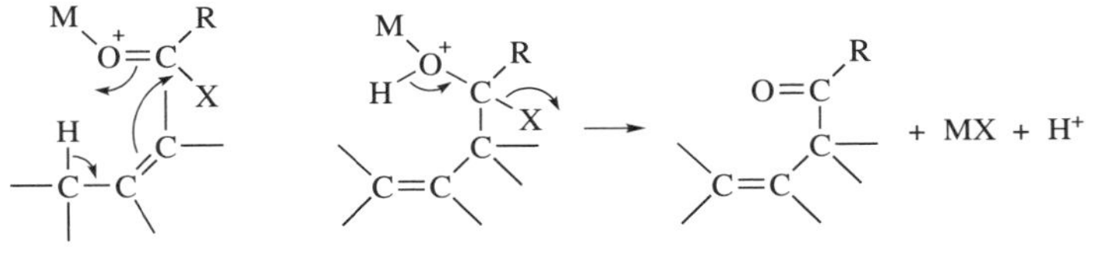

A mechanistically significant feature of this reaction is the kinetic preference for formation of β,γ-unsaturated ketones. It has been suggested that this regiochemistry results from an intramolecular deprotonation, as shown in the mechanism above.[5] A related reaction occurs between alkenes and acylium ions, as in the reaction between 2-methylpropene and

1. M. T. Reetz, I. Chatziiosifidis, U. Löwe, and W. F. Maier, *Tetrahedron Lett.* **1979:**1427; M. T. Reetz, I. Chatziiosifidis, F. Hübner, and H. Heimbach, *Org. Synth.* **62:**95 (1984).
2. I. Paterson, *Tetrahedron Lett.* **1979:**1519.
3. C. W. Jefford, A. W. Sledeski, P. Lelandais, and J. Bolukouvalas, *Tetrahedron Lett.* **33:**1855 (1992).
4. W. H. Pearson and J. M. Schkeryantz, *J. Org. Chem.* **57:**2986 (1992).
5. P. Beak and K. R. Berger, *J. Am. Chem. Soc.* **102:**3848 (1980).

the acetylium ion.[6] The reaction leads regiospecifically to β,γ-enones. A concerted "ene reaction" mechanism has been suggested (see Section 6.7).

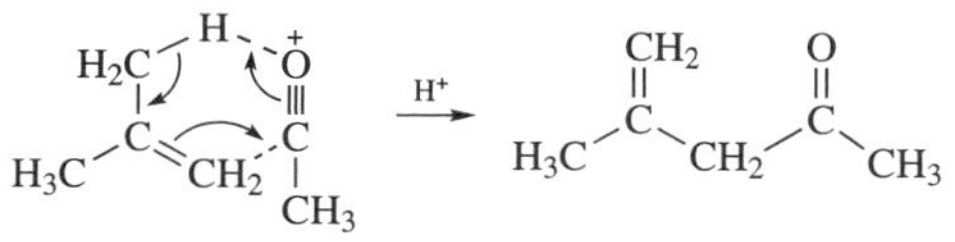

A variety of other reaction conditions have been examined for acylation of alkenes by acyl chlorides. With the use of Lewis acid catalysts, reaction typically occurs to give both enones and β-halo ketones.[7] The latter reaction has been most synthetically useful in intramolecular cyclizations. The following reactions are illustrative.

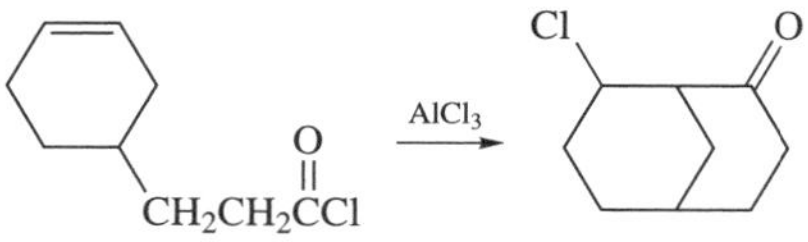

Ref. 8

Ref. 9

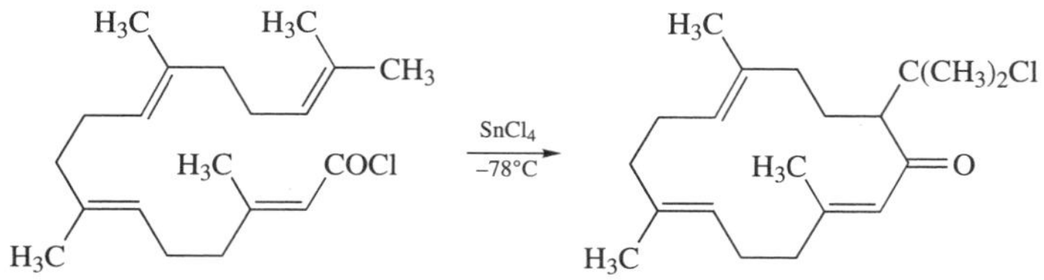

Lewis acid-catalyzed cyclization of unsaturated aldehydes is also an effective reaction. Stannic chloride is the usual reagent for this cyclization.

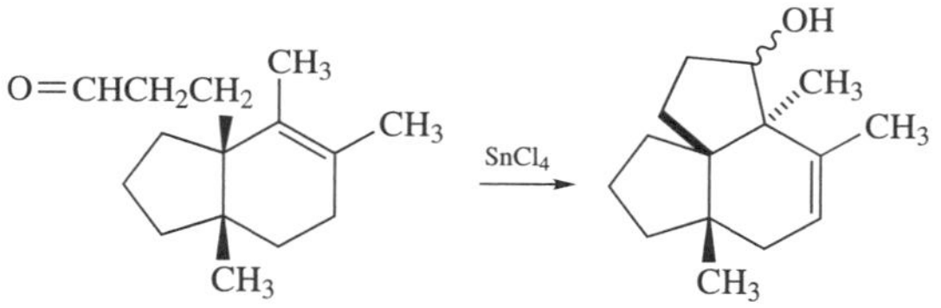

Ref. 10

In those cases in which the molecular geometry makes it possible, the proton-transfer step is intramolecular.[11]

Perhaps most useful from a synthetic point of view are reactions of polyenes having two or more double bonds positioned in such a way that successive bond-forming steps can occur. This process, called *polyene cyclization*, has proven to be an effective way of making polycyclic compounds containing six- and, in some cases, five-membered rings. The reaction proceeds through an electrophilic attack and requires that the double bonds

6. H. M. R. Hoffman and T. Tsushima, *J. Am. Chem. Soc.* **99**:6008 (1977).
7. For example, T. S. Cantrell, J. M. Harless, and B. L. Strasser, *J. Org. Chem.* **36**:1191 (1971); L. Rand and R. J. Dolinski, *J. Org. Chem.* **31**:3063 (1966).
8. E. N. Marvell, R. S. Knutson, T. McEwen, D. Sturmer, W. Federici, and K. Salisbury, *J. Org. Chem.* **35**:391 (1970).
9. T. Kato, M. Suzuki, T. Kobayashi, and B. P. Moore, *J. Org. Chem.* **45**:1126 (1980).
10. L. A. Paquette and Y.-K. Han, *J. Am. Chem. Soc.* **103**:1835 (1981).
11. N. H. Andersen and D. W. Ladner, *Synth. Commun.* **1978**:449.

which participate in the cyclization be properly positioned. For example, compound **1** is converted quantitatively to **2** on treatment with formic acid. The reaction is initiated by protonation and ionization of the allylic alcohol. It is terminated by nucleophilic capture of the secondary carbocation.

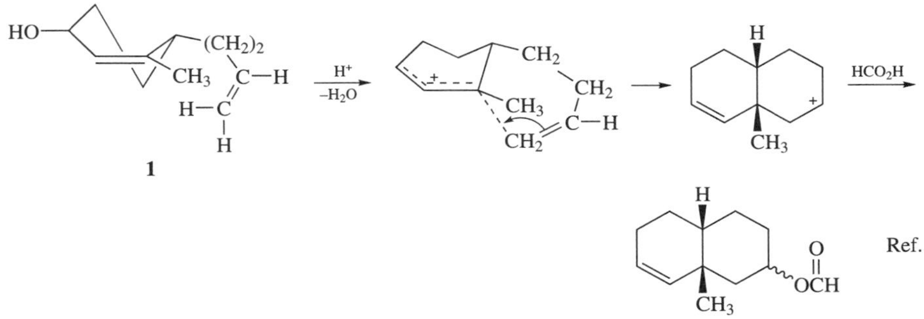

Ref. 12

More extended polyenes can cyclize to tricylic systems:

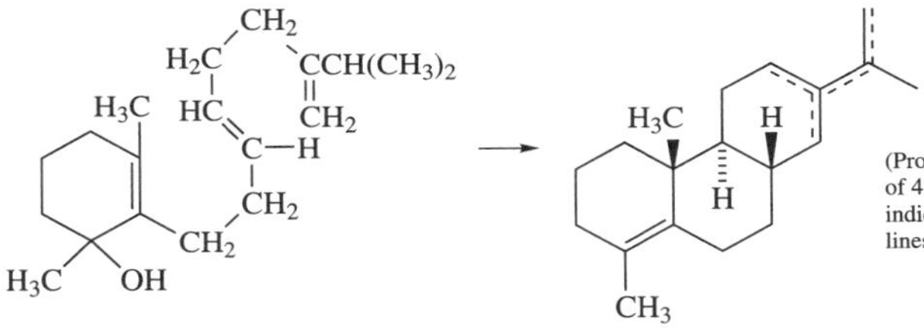

(Product is a mixture of 4 diene isomers indicated by dotted lines)

Ref. 13

These cyclizations are usually highly stereoselective, with the stereochemical outcome being predictable on the basis of reactant conformation.[14] The stereochemistry of cyclization products in the decalin family can be predicted by assuming that cyclizations will occur through conformations which resemble chair cyclohexane rings. The stereochemistry at ring junctures is that expected for *anti* attack at the participating double bonds:

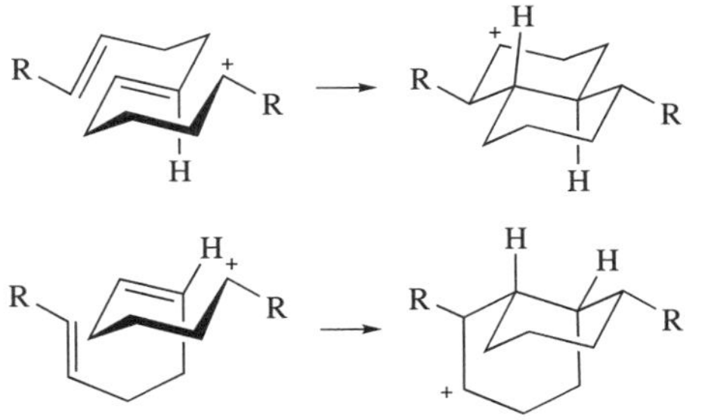

12. W. S. Johnson, P. J. Neustaedter, and K. K. Schmiegel, *J. Am. Chem. Soc.* **87:**5148 (1965).
13. W. J. Johnson, N. P. Jensen, J. Hooz, and E. J. Leopold, *J. Am. Chem. Soc.* **90:**5872 (1968).
14. W. S. Johnson, *Acc. Chem. Res.* **1:**1 (1968); P. A. Bartlett, in *Asymmetric Synthesis*, Vol. 3, J. D. Morrison, eds Academic Press, New York, 1984, Chapter 5.

To be of maximum synthetic value, the generation of the cationic site that initiates cyclization must involve mild reaction conditions. Formic acid and stannic chloride have proved to be effective reagents for cyclization of polyunsaturated allylic alcholos. Acetals generate α-alkoxy carbocations in acidic solution and can also be used to initiate the cyclization of polyenes[15]:

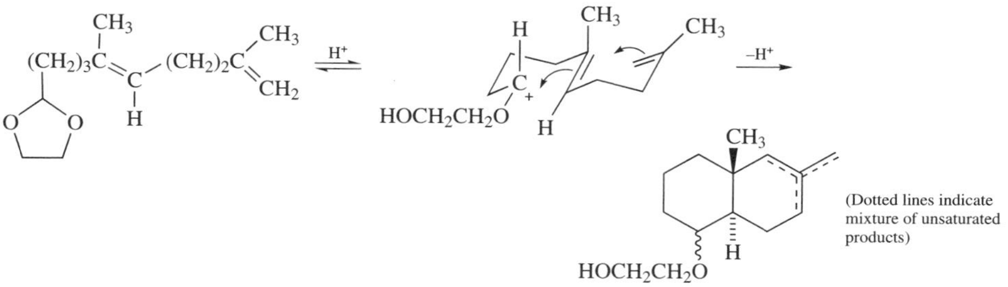

(Dotted lines indicate mixture of unsaturated products)

Another significant method for generating the electrophilic site is acid-catalyzed epoxide ring opening.[16] Lewis acids such as BF_3, $SnCl_4$, CH_3AlCl_2, or $TiCl_3$(O-*i*-Pr) can also be used,[17] as indicated by entries 4–6 in Scheme 10.1.

Mercuric ion has been found to be capable of inducing cyclization of polyenes.

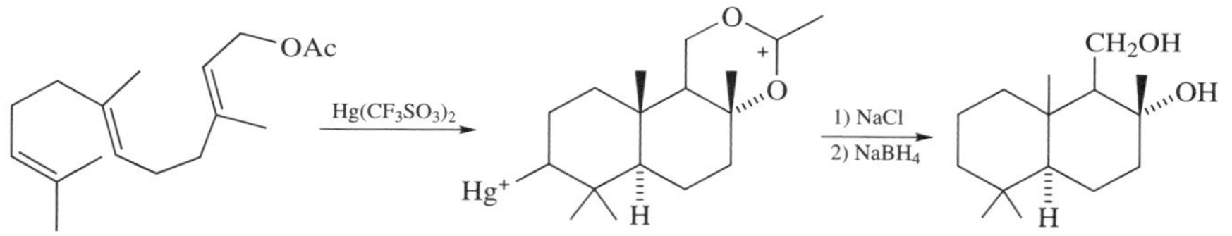

Ref. 18

The particular example shown also has a special mechanism for stabilization of the cyclized carbocation. The adjacent acetoxy group is captured to form a stabilized dioxanylium cation. After reductive demercuration (see Section 4.3) and hydrolysis, a diol is isolated.

Because the immediate product of the polyene cyclization is a carbocation, the reaction often yields a mixture of closely related compounds resulting from the competing modes of reaction of the carbocation. The products result from capture of the carbocation by solvent or other nucleophile or by deprotonation to form an alkene. Polyene cyclization can be carried out on reactants which have special structural features that facilitate transformation of the carbocation to a stable product. Allylic silanes, for example, are

15. A van der Gen, K. Wiedhaup, J. J. Swoboda, H. C. Dunathan, and W. S. Johnson, *J. Am. Chem. Soc.* **95:**2656 (1973).
16. E. E. van Tamelen and R. G. Nadeau, *J. Am. Chem. Soc.* **89:**176 (1967).
17. E. J. Corey and M. Sodeoka, *Tetrahedron Lett* **32**:7005 (1991); P. V. Fish, A. R. Sudhakar, and W. S. Johnson, *Tetrahedron Lett.* **34:**7849 (1993).
18. M. Nishizawa, H. Takenaka, and Y. Hayashi, *J. Org. Chem.* **51:**806 (1986); E. J. Corey, J. G. Reid, A. G. Myers, and R. W. Hahl, *J. Am. Chem. Soc.* **109:**918 (1987).

stabilized by desilylation.[19]

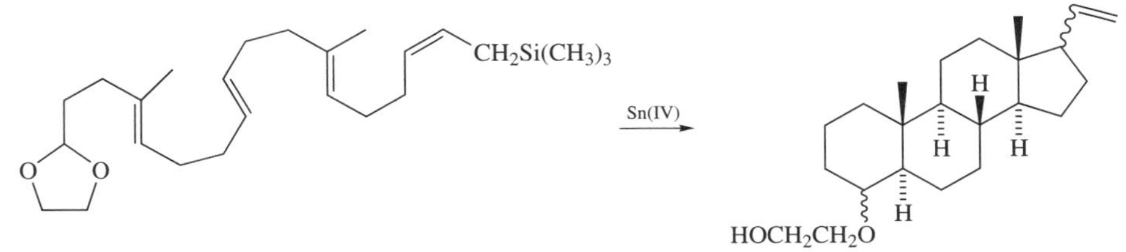

With terminating alkynyl groups, vinyl cations are formed. Capture of water leads to formation of a ketone.[20]

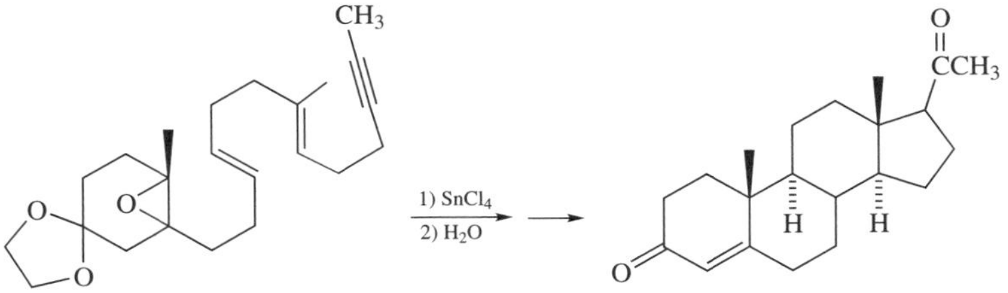

Polyene cyclizations have been of substantial value in the synthesis of polycyclic natural products of the terpene type. These syntheses resemble the processes by which terpenoid and steroidal compounds are assembled in nature. The most dramatic example of biological synthesis of a polycyclic skeleton from a polyene intermediate is the conversion of squalene oxide to the steroid lanosterol. In the biological reaction, the enzyme presumably functions not only to induce the cationic cyclization but also to bind the substrate in a conformation corresponding to the stereochemistry of the polycyclic product.[21]

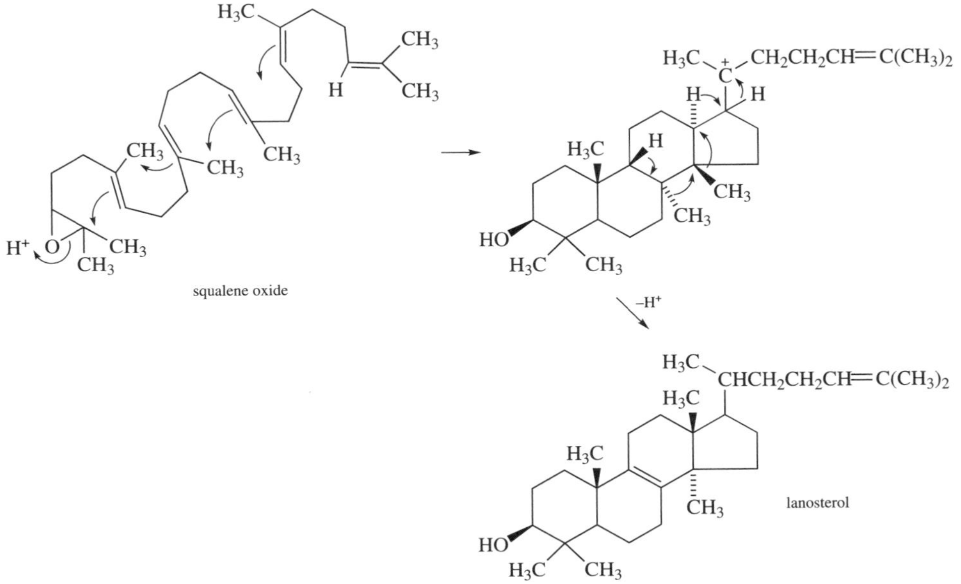

19. W. S. Johnson, Y.-Q Chen, and M. S. Kellogg, *J. Am. Chem. Soc.* **105:**6653 (1983).
20. E. E. van Tamelen and J. R. Hwu, *J. Am. Chem. Soc.* **105:**2490 (1983).
21. D. Cane, *Chem. Rev.* **90:**1089 (1990); I. Abe, M. Rohmer, and G. D. Prestwich, *Chem. Rev.*, **93:**2189 (1993).

Scheme 10.1 gives some representative laboratory syntheses involving polyene cyclization.

10.1.2. Rearrangement of Carbocations

Carbocations can be stabilized by the migration of hydrogen, alkyl, or aryl groups, and, occasionally, functional groups. A mechanistic discussion of these reactions is given in Section 5.11 of Part A. Reactions involving carbocations can be complicated by competing rearrangement pathways. Rearrangements can be highly selective and, therefore, reliable synthetic reactions when the structural situation favors a particular pathway. One example is the reaction of carbocations having a hydroxyl group on an adjacent carbon, which can lead to the formation of a carbonyl group.

$$\mathrm{R{-}C(OH)(R){-}\overset{+}{C}R_2 \longrightarrow RC({=}O)CR_3}$$

A reaction that follows this pattern is the acid-catalyzed conversion of diols to ketones, which is known as the *pinacol rearrangement*.[22] The classic example of this reaction is the conversion of 2,3-dimethylbutane-2,3-diol (pinacol) to methyl *t*-butyl ketone (pinacolone)[23]:

$$\mathrm{(CH_3)_2C(OH){-}C(OH)(CH_3)_2 \xrightarrow{H^+} CH_3C({=}O)C(CH_3)_3} \quad 67\text{–}72\%$$

The acid-catalyzed mechanism involves carbocation formation and substituent migration assisted by the hydroxyl group:

$$\mathrm{R_2C(OH){-}CR_2(OH) \xrightarrow{H^+} [RC(\overset{\delta+}{O}H){\cdots}R^{\delta+}{\cdots}CR_2] \longrightarrow RC({=}\overset{+}{O}H){-}CR_3 \longrightarrow RC({=}O)CR_3 + H^+}$$

Under acidic conditions, the more easily ionized C—O bond generates the carbocation, and migration of one of the groups from the adjacent carbon ensues. Both stereochemistry and "migratory aptitude" can be factors in determining the extent of migration of the different groups. Vinyl groups, for example, have a relatively high tendency toward migration.[24] This tendency is further enhanced by donor substituents, and selective migration of

22. C. J. Collins, *Q. Rev.* **14**:357 (1960).
23. G. A. Hill and E. W. Flosdorf, *Org. Synth.* **I**:451 (1932).
24. K. Nakamura and Y. Osamura, *J. Am. Chem. Soc.* **115**:9112 (1993).

Scheme 10.1. Polyene Cyclizations

1[a] HCO_2H >50%

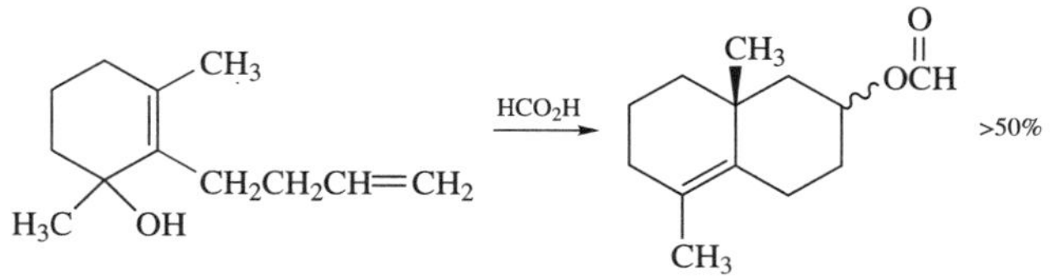

2[b] 1) CF_3CO_2H 2) $LiAlH_4$ 52%

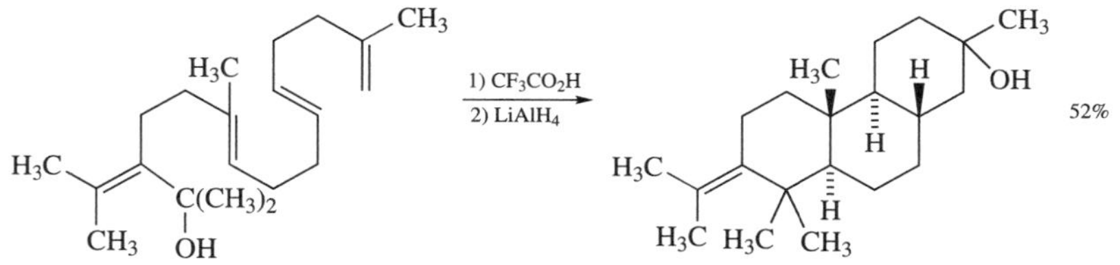

3[c] CF_3CO_2H, ethylene carbonate, HCF_2CH_3, –25°C 65%

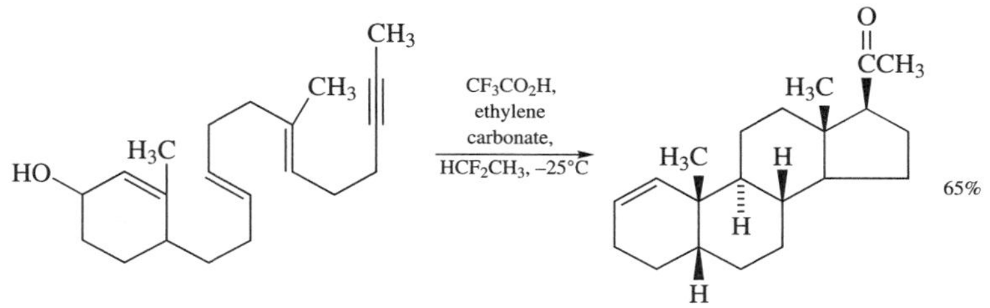

4[d] $SnCl_4$ / CH_3NO_2

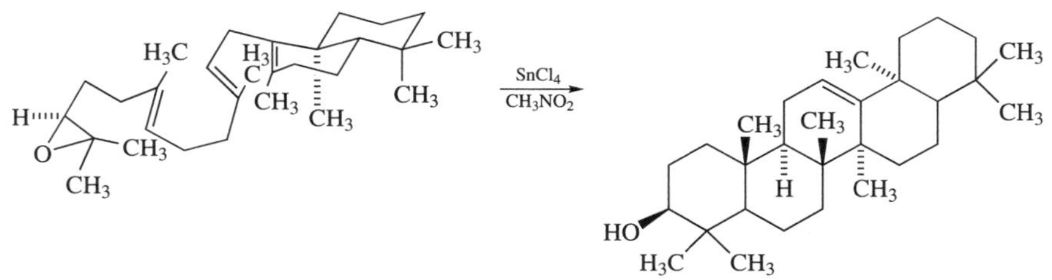

5[e] $BF_3 \cdot OEt_2$ / Et_3N, –78°C 25–35%

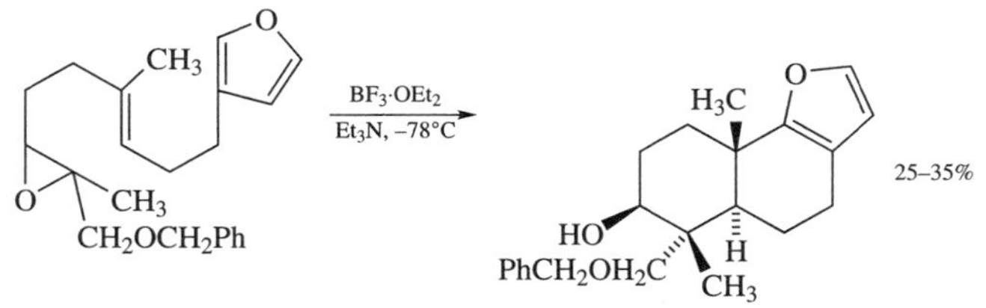

6[f] CH_3AlCl_2 / –94°C 84%

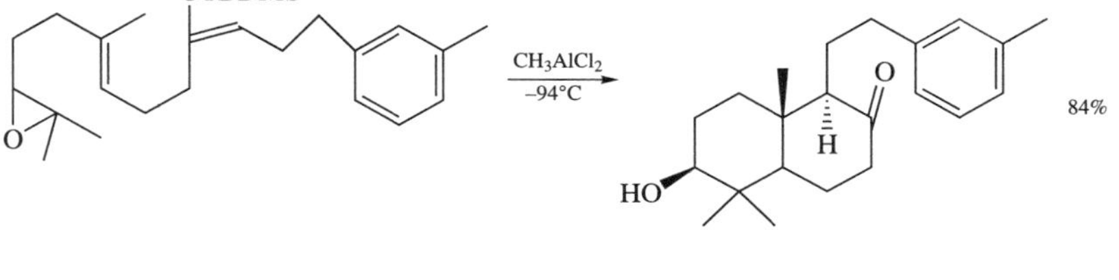

7[g]

$\xrightarrow[\text{CH}_2\text{Cl}_2,\ -70^\circ\text{C}]{\text{CF}_3\text{CO}_2\text{H}}$

65–70%

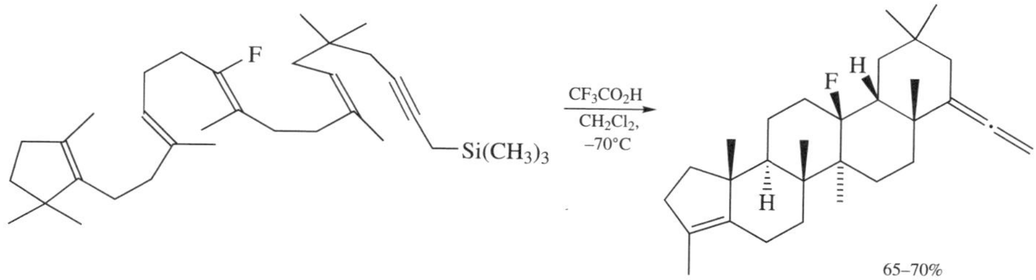

a. J. A. Marshall, N. Cohen, and A. R. Hochstetler, *J. Am. Chem. Soc.* **88:**3408 (1966).
b. W. S. Johnson and T. K. Schaaf, *J. Chem. Soc., Chem. Commun.* **1969:**611.
c. B. E. McCarry, R. L. Markezich, and W. S. Johnson, *J. Am. Chem. Soc.* **95:**4416 (1973).
d. E. E. van Tamelen, R. A. Holton, R. E. Hopla, and W. E. Konz, *J. Am. Chem. Soc.* **94:**8228 (1972).
e. S. P. Tanis, Y.-H. Chuang, and D. B. Head, *J. Org. Chem.* **53:**4929 (1988).
f. E. J. Corey, G. Luo, and L. S. Lin, *Angew. Chem. Int. Ed. Engl.* **37:**1126 (1998).
g. W. S. Johnson, M. S. Plummer, S. P. Reddy, and W. R. Bartlett, *J. Am. Chem. Soc.* **115:**515 (1993).

trimethylsilylvinyl groups has been exploited in pinacol rearrangements.[25]

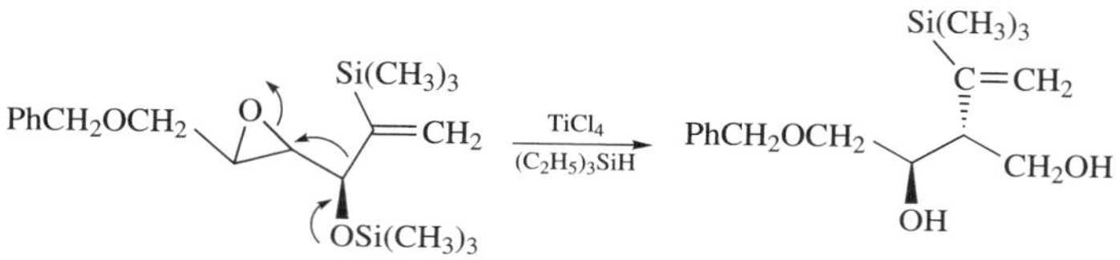

$\xrightarrow[(\text{C}_2\text{H}_5)_3\text{SiH}]{\text{TiCl}_4}$

Another method for carrying out the same net rearrangement involves synthesis of a glycol monosulfonate ester. These compounds rearrange under the influence of base.

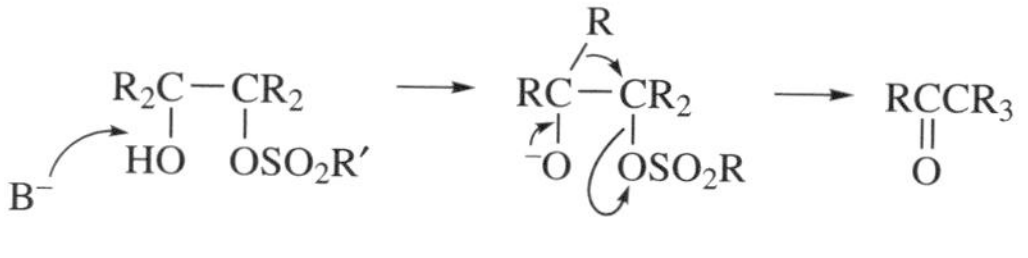

$$\text{R}_2\text{C(OH)}-\text{CR}_2(\text{OSO}_2\text{R}') \xrightarrow{\text{B}^-} \text{RC}(\text{R})(\text{O}^-)-\text{CR}_2(\text{OSO}_2\text{R}) \longrightarrow \text{RC(=O)CR}_3$$

Rearrangements of monosulfonates permit greater control over the course of the rearrangement, because ionization can take place only at the sulfonylated alcohol. These reactions have been of value in the synthesis of ring systems, especially terpenes, as illustrated by entries 3 and 4 in Scheme 10.2. Entry 7 in Scheme 10.2 illustrates the use of a Lewis acid, $(C_2H_5)_2AlCl$, to promote rearrangements. Under these conditions, the reaction was shown to proceed with *inversion* of configuration at the migration terminus, as would be implied

25. K. Suzuki, T. Ohkuma, and G. Tsuchihashi, *Tetrahedron Lett.* **26:**861 (1985); K. Suzuki, M. Shimazaki, and G. Tsuchihashi, *Tetrahedron Lett.* **27:**6233 (1986); M. Shimazaki, M. Morimoto, and K. Suzuki, *Tetrahedron Lett.* **31:**3335 (1990).

A. Pinacol-type rearrangements

1[a] OH, COH, Ph, Ph; H_2SO_4; O, Ph, Ph 99%

2[b] O_2CH, OH; H_2SO_4; O 69–81%

3[c] HO, OSO_2Ar, H, CH_3, H_3C, CH_3; $K^+ {}^-OC(CH_3)_3$; O, H, H, CH_3, H_3C, CH_3 85%

4[d] CH_3SO_2O, OH, CH_3, CH_3, H_3CO_2C; pyridine, Et_3N; H, O, CH_3, CH_3, H_3CO_2C 91%

5[e] $CF_3SO_3CH_2$, CH_3, CH_3, CH_3, O, O; H_2SO_4, CF_3CH_2OH, H_2O, 80°C; HO, CH_3, CH_3, CH_3, O, O 100%

6[f] OH, $O_2C_6H_4NO_2$, OH; $HC(OCH_3)_3$, $SnCl_4$, 20 mol %; O, $O_2C_6H_4NO_2$ 97%

7[g] H, OH, OH; 1) CH_3SO_2Cl 2) $(C_2H_5)_2AlCl$; O 37%

Scheme 10.2. *(continued)*

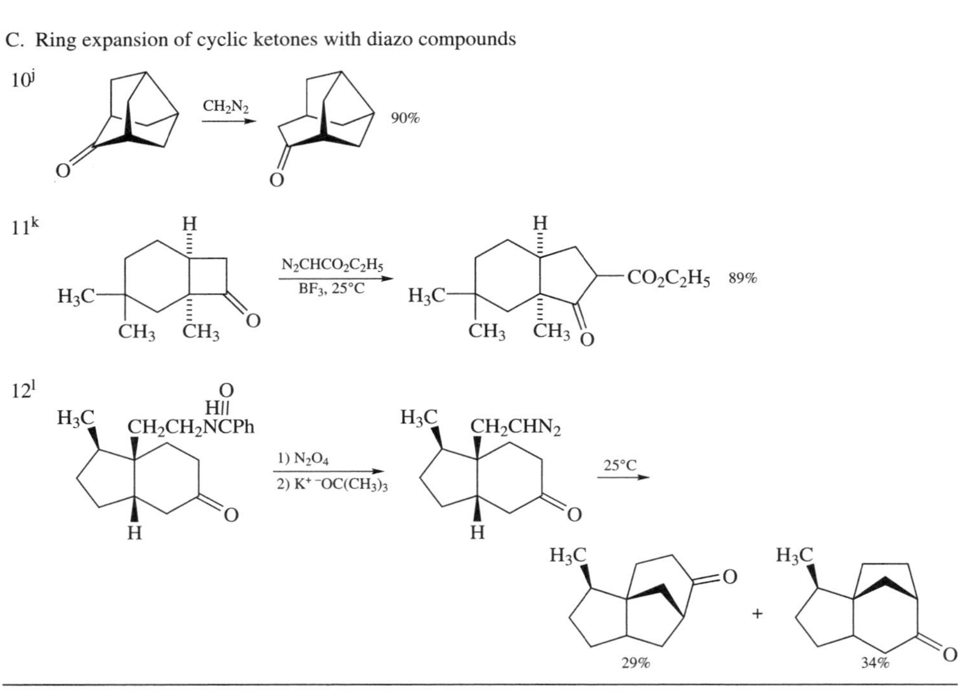

a. H. E. Zaugg, M. Freifelder, and B. W. Horrom, *J. Org. Chem.* **15:**1191 (1950).
b. J. E. Horan and R. W. Schiessler, *Org. Synth.* **41:**53 (1961).
c. G. Büchi, W. Hofheinz, and J. V. Paukstelis, *J. Am. Chem. Soc.* **88:**4113 (1966).
d. D. F. MacSweeney and R. Ramage, *Tetrahedron* **27:**1481 (1971).
e. P. Magnus, C. Diorazio, T. J. Donohoe, M. Giles, P. Pye, J. Tarrant, and S. Thom, *Tetrahedron* **52:**14147 (1996).
f. Y. Kita, Y. Yoshida, S. Mihara, D.-F. Fang, K. Higuchi, A. Furukawa, and H. Fujioka, *Tetrahedron Lett.* **38:**8315 (1997).
g. J. H. Rigby and K. R. Fales, *Tetrahedron Lett.* **39:**1525 (1998).
h. R. B. Woodward, J. Gosteli, I. Ernest, R. J. Friary, G. Nestler, H. Raman, R. Sitrin, C. Suter, and J. K. Whitesell, *J. Am. Chem. Soc.* **95:**6853 (1973).
i. E. G. Breitholle and A. G. Fallis, *J. Org. Chem.* **43:**1964 (1978).
j. Z. Majerski, S. Djigas, and V. Vinkovic, *J. Org. Chem.* **44:**4064 (1979).
k. H. J. Liu and T. Ogino, *Tetrahedron Lett.* **1973:**4937.
l. P. R. Vettel and R. M. Coates, *J. Org. Chem.* **45:**5430 (1980).

by a concerted mechanism.[26]

In cyclic systems which enforce structural rigidity or conformational bias, the course of the rearrangement is controlled by stereoelectronic factors. The carbon substituent that is *anti* to the leaving group is the one which undergoes migration. In cyclic systems such as **3**, for example, selective migration of the ring fusion bond occurs because of this stereoelectronic effect.

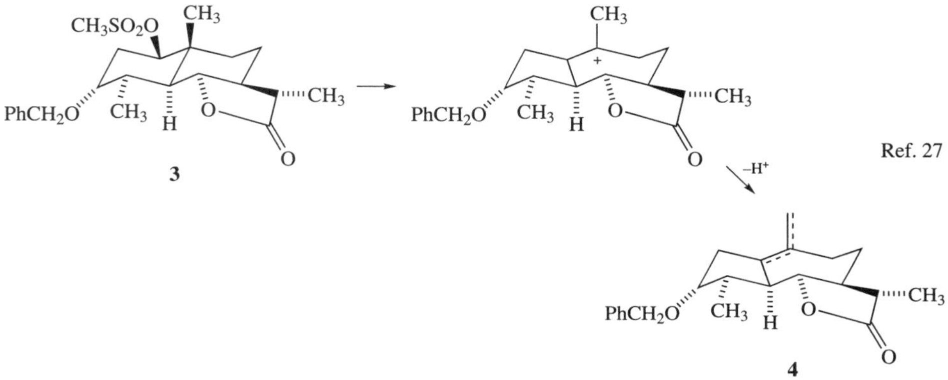

Ref. 27

Similarly, **5** gives **6** by antiperiplanar migration:

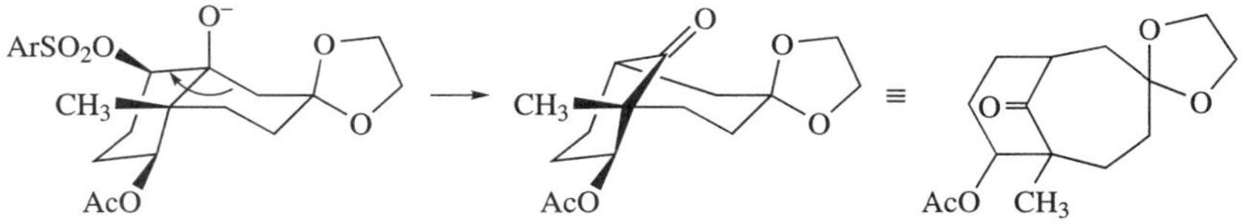

Ref. 28

There is kinetic evidence that the migration step in these base-catalyzed rearrangements is concerted with ionization. Thus, in cyclopentane derivatives, the rate of reaction depends on the nature of the *trans* substituent R, which implies that the migration is part of the rate-determining step.[29]

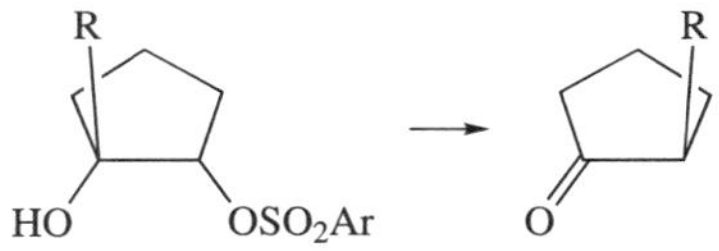

rate: R = H > Ph > alkyl

26. G. Tsuchihashi, K. Tomooka, and K. Suzuki, *Tetrahedron Lett.* **25:**4253 (1984).
27. M. Ando, A. Akahane, H. Yamaoka, and K. Takase, *J. Org. Chem.* **47:**3909 (1982).
28. C. H. Heathcock, E. G. Del Mar, and S. L. Graham, *J. Am. Chem. Soc.* **104:**1907 (1982).
29. E. Wistuba and C. Rüchardt, *Tetrahedron Lett.* **22:**4069 (1981).

Aminomethyl carbinols yield ketones when treated with nitrous acid. This reaction has been used to form ring-expanded cyclic ketones, a procedure which is called the *Tiffeneau–Demjanov reaction*.[30]

$$R_2C(OH)CH_2NH_2 \xrightarrow{HONO} R_2C(OH)CH_2\overset{+}{N}\equiv N \longrightarrow RC(OH)(R)-\overset{+}{C}H_2 \longrightarrow RC(=O)CH_2R$$

Ref. 31

1-(Aminomethyl)cycloheptanol (HO, CH_2NH_2) $\xrightarrow{HONO}$ cyclooctanone 61%

The diazotization reaction generates the same type of β-hydroxy carbocation that is involved in the pinacol rearrangement. (See Section 5.6 in Part A for a discussion of the formation of carbocations from diazo compounds.)

The reaction of ketones with diazoalkanes sometimes leads to a ring-expanded ketone in synthetically useful yields.[32] The reaction occurs by addition of the diazoalkane, followed by elimination of nitrogen and migration:

$$\text{cyclic ketone }(CH_2)_x + CH_2N_2 \longrightarrow {}^{-}O,\ CH_2\overset{+}{N}\equiv N\ \text{adduct }(CH_2)_x \longrightarrow \text{cyclic ketone }(CH_2)_{x+1}$$

The rearrangement proceeds via essentially the same intermediate that is involved in the Tiffeneau–Demjanov reaction. Because the product is also a ketone, subsequent addition of diazomethane can lead to higher homologs. The best yields are obtained when the starting ketone is substantially more reactive than the product. For this reason, strained ketones work especially well. Higher diazoalkanes can also be used in place of diazomethane. The reaction is found to be acclerated by alcoholic solvents. This effect probably involves the hydroxyl group being hydrogen-bonded to the carbonyl oxygen and serving as a proton donor in the addition step.[33]

$$R-C(=O\cdots H-O-R)(R) + :CH_2N_2 \longrightarrow R-C(OH)(R)-CH_2\overset{+}{N}\equiv N$$

30. P. A. S. Smith and D. R. Baer, *Org. React.* **11:**157 (1960).
31. F. F. Blicke, J. Azuara, N. J. Dorrenbos, and E. B. Hotelling, *J. Am. Chem. Soc.* **75:**5418 (1953).
32. C. D. Gutsche, *Org. React.* **8:**364 (1954).
33. J. N. Bradley, G. W. Cowell, and A. Ledwith, *J. Chem. Soc.* **1964:**4334.

Trimethylaluminum also promotes ring expansion by diazoalkanes.[34]

$$\text{cyclohexanone} + CH_3(CH_2)_4CHN_2 \xrightarrow{(CH_3)_3Al} \text{2-}(CH_2)_4CH_3\text{-cycloheptanone} \quad 88\%$$

Ketones react with esters of diazoacetic acid in the presence of Lewis acids such as BF_3 or $SbCl_5$.[35]

$$\text{cyclohexanone} + N_2CHCO_2C_2H_5 \xrightarrow{SbCl_5} \text{2-}CO_2C_2H_5\text{-cycloheptanone}$$

Triethyloxonium tetrafluoroborate also effects ring expansion of cyclic ketones by ethyl diazoacetate.[36]

$$\text{cyclooctanone} + N_2CHCO_2C_2H_5 \xrightarrow{(C_2H_5)_3O^+BF_4^-} \text{2-}CO_2C_2H_5\text{-cyclononanone}$$

These reactions involve addition of the diazoester to an adduct of the carbonyl compound and the Lewis acid. Elimination of nitrogen then triggers migration. Sections B and C of Scheme 10.2 give some additional examples of pinacol rearrangements involving diazo and diazonium intermediates.

10.1.3. Related Rearrangements

α-Halo ketones when treated with base undergo a skeletal change that is similar to the pinacol rearrangement. The most commonly used bases are alkoxide ions, which lead to esters as the reaction products:

$$RCH_2\overset{O}{\overset{\|}{C}}\underset{X}{\underset{|}{C}}HR' \xrightarrow{CH_3O^-} CH_3O\overset{O}{\overset{\|}{C}}\underset{CH_2R}{\underset{|}{C}}HR' + X^-$$

34. K. Maruoka, A. B. Concepcion, and H. Yamamoto, *J. Org. Chem.* **59:**4725 (1994).
35. H. J. Liu and T. Ogino, *Tetrahedron Lett.* **1973:**4937; W. T. Tai and E. W. Warnhoff, *Can. J. Chem.* **42:**1333 (1964); W. L. Mock and M. E. Hartman, *J. Org. Chem* **42:**459 (1977); V. Dave and E. W. Warnhoff, *J. Org. Chem.* **48:**2590 (1983).
36. L. J. MacPherson, E. K. Bayburt, M. P. Capparelli, R. S. Bohacek, F. H. Clarke, R. D. Ghai, Y. Sakane, C. J. Berry, J. V. Peppard, and A. J. Trapani, *J. Med. Chem.* **36:**3821 (1993).

This reaction is known as the *Favorskii rearrangement*.[37] If the ketone is cyclic, a ring contraction occurs.

2-chlorocyclohexanone $\xrightarrow{Na^+\ ^-OMe}$ cyclopentyl-CO_2Me Ref. 38

There is considerable evidence that the rearrangement involves cyclopropanones and/or the 1,3-dipolar isomers of cyclopropanone as reaction intermediates.[39]

$$RCH_2C(=O)CHR'X \underset{}{\overset{^-OR''}{\rightleftharpoons}} R\overset{-}{C}HC(=O)CHR'X \longrightarrow R\overset{-}{C}HC(=O)\overset{+}{C}HR' \longleftrightarrow R\overset{+}{C}HC(=O)\overset{-}{C}HR'$$

$$\longrightarrow \text{cyclopropanone (RHC—CHR', C=O)} \xrightarrow{^-OR''} \text{RHC—CHR' ring with } C(O^-)(OR'') \longrightarrow RCH(CO_2R'')CH_2R' + RCH_2CH(CO_2R'')R'$$

There is also a related mechanism that can operate in the absence of an acidic α hydrogen, which is called the "semibenzilic" rearrangement.

$$RC(=O)CHR'X \xrightarrow{R''O^-} RC(O^-)(OR'')CHR'X \longrightarrow R''O{-}C(=O){-}CHR'R$$

The net structural change is the same for both mechanisms. The energy requirements of the cyclopropanone and semibenzilic mechanisms may be fairly closely balanced.[40] Cases of operation of the semibenzilic mechanism have been reported even for compounds with a hydrogen available for enolization.[41] Among the evidence that the cyclopropanone mechanism usually operates is the demonstration that a symmetrical intermediate is involved. The isomeric chloroketones **7** and **8**, for example, lead to the same ester.

$$\underset{\mathbf{7}}{PhCH(Cl)C(=O)CH_3} \xrightarrow{CH_3O^-} \underset{\mathbf{9}}{PhCH_2CH_2CO_2CH_3} \xleftarrow{CH_3O^-} \underset{\mathbf{8}}{PhCH_2C(=O)CH_2Cl}$$ Ref. 39

37. A. S. Kende, *Org. React.* **22:**261 (1960); A. A. Akhrem, T. K. Ustynyuk, and Y. A. Titov, *Russ. Chem. Rev.* (English transl.) **39:**732 (1970).
38. D. W. Goheen and W. R. Vaughan, *Org. Synth.* **IV:**594 (1963).
39. F. G. Bordwell, T. G. Scamehorn and W. R. Springer, *J. Am. Chem. Soc.* **91:**2087 (1969); F. G. Bordwell and J. G. Strong, *J. Org. Chem.* **38:**579 (1973).
40. V. Moliner, R. Castillo, V. S. Safont, M. Oliva, S. Bohn, I. Tunon, and J. Andres, *J. Am. Chem. Soc.* **119:**1941 (1997).
41. E. W. Warnhoff, C. M. Wong, and W. T. Tai, *J. Am. Chem. Soc.* **90:**514 (1968).

The occurrence of a symmetrical intermediate has also been demonstrated by ^{14}C labeling in the case of α-chlorocyclohexanone.[42]

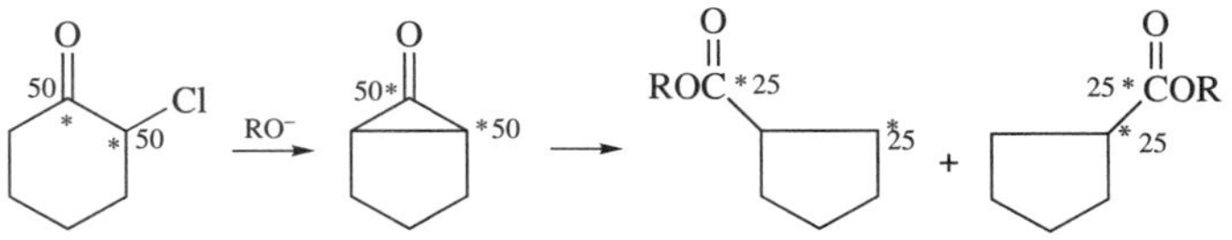

$* = ^{14}C$ label

Numbers refer to percentage of label at each carbon.

Because of the cyclopropanone mechanism, the structure of the ester product cannot be predicted directly from the structure of the reacting halo ketone. Instead, the identity of the product is governed by the direction of ring opening of the cyclopropanone intermediate. The dominant mode of ring opening would be expected to be the one which forms the more stable of the two possible ester homoenolates. For this reason, a phenyl substituent favors breaking the bond to the substituted carbon, but an alkyl group directs the cleavage to the less substituted carbon.[43] That both **7** and **8** above give the same ester, **9**, is illustrative of the directing effect that the phenyl group can have on the ring-opening step.

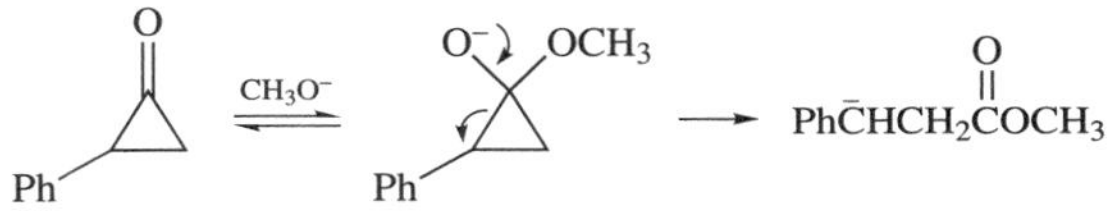

The Favorskii reaction has been used to effect ring contraction in the synthesis of strained ring compounds. Entry 4 in Scheme 10.3 illustrates this application of the reaction. With α,α'-dihalo ketones, the rearrangement is accompanied by dehydrohalogenation to yield an α,β-unsaturated ester, as illustrated by entry 3 in Scheme 10.3.

α-Halo sulfones undergo a related rearrangment known as the *Ramberg–Bäcklund reaction*.[44] The carbanion formed by deprotonation gives an unstable thiirane dioxide which decomposes with elimination of sulfur dioxide.

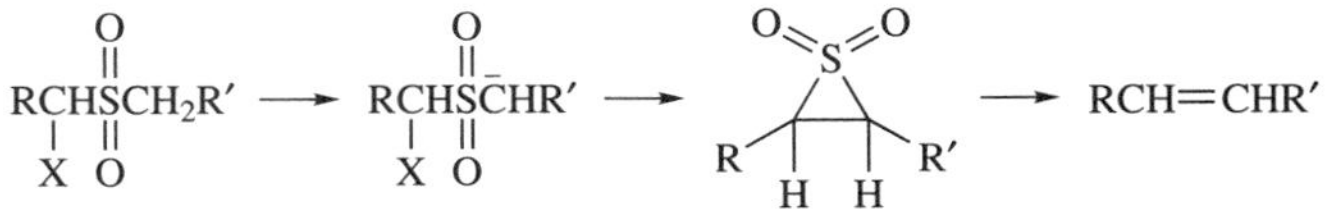

The reaction is useful for the synthesis of certain types of strained alkenes.

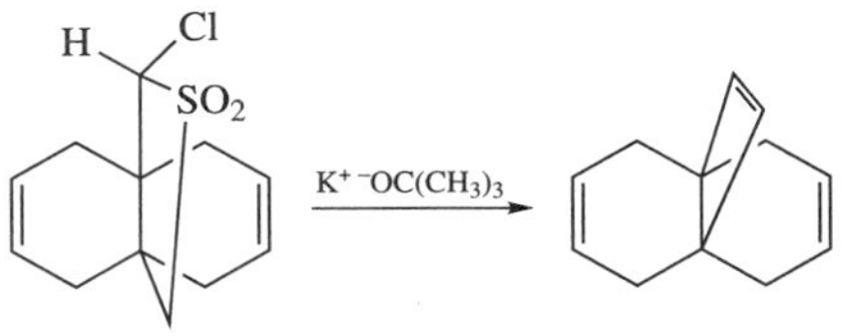

Ref. 45

42. R. B. Loftfield, *J. Am. Chem. Soc.* **73**:4707 (1951).
43. C. Rappe, L. Knutsson, N. J. Turro, and R. B. Gagosian, *J. Am. Chem. Soc.* **92**:2032 (1970).
44. L. A. Paquette, *Acc. Chem. Res.* **1**:209 (1968); L. A. Paquette, in *Mechanisms of Molecular Migrations*, Vol. 1, B. S. Thyagarajan, ed., Wiley-Interscience, New York, 1968, Chapter 3; L. A. Paquette, *Org. React.* **25**:1 (1977); R. J. K. Taylor, *Chem. Commun.* **1999**:217.
45. L. A. Paquette, J. C. Philips, and R. E. Wingard, Jr., *J. Am. Chem. Soc.* **93**:4516 (1971).

Scheme 10.3. Base-Catalyzed Rearrangements of α-Halo-ketones

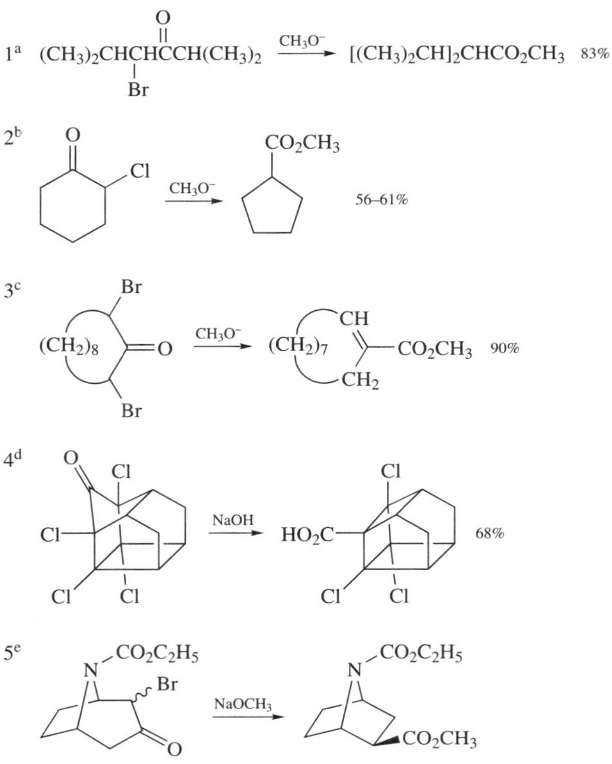

1[a] $(CH_3)_2CHCH(Br)C(=O)CH(CH_3)_2 \xrightarrow{CH_3O^-} [(CH_3)_2CH]_2CHCO_2CH_3$ 83%

2[b] 2-chlorocyclohexanone $\xrightarrow{CH_3O^-}$ cyclopentane-CO_2CH_3 56–61%

3[c] 2,10-dibromocyclodecanone [$(CH_2)_8$, Br, Br, =O] $\xrightarrow{CH_3O^-}$ $(CH_2)_7$ ring with CH=C(CO_2CH_3)–CH_2 90%

4[d] polychlorinated cage ketone (O, Cl, Cl, Cl, Cl) $\xrightarrow{NaOH}$ HO_2C– cage (Cl, Cl, Cl) 68%

5[e] N-$CO_2C_2H_5$ bicyclic bromo ketone (Br, O) $\xrightarrow{NaOCH_3}$ N-$CO_2C_2H_5$ bicycle with CO_2CH_3

a. S. Sarel and M. S. Newman, *J. Am. Chem. Soc.* **78**:416 (1956).
b. D. W. Goheen and W. R. Vaughan, *Org. Synth.* **IV**:594 (1963).
c. E. W. Garbisch, Jr., and J. Wohllebe, *J. Org. Chem.* **33**:2157 (1968).
d. R. J. Stedman, L. S. Miller, L. D. Davis, and J. R. E. Hoover, *J. Org. Chem.* **35**:4169 (1970).
e. D. Bai, R. Xu, G. Chu, and X. Zhu, *J. Org. Chem.* **61**:4600 (1996).

10.1.4. Fragmentation Reactions

The classification "fragmentation" applies to reactions in which a carbon–carbon bond is broken. A structural feature that permits fragmentation to occur readily is the presence of a carbon β to a developing electron deficiency that can accommodate carbocationic character. This type of reaction occurs particulary readily when the γ atom is a heteroatom, such as nitrogen or oxygen, having an unshared electron pair that can stabilize the new cationic center.[46] This is called the *Grob fragmentation*.

$$\underset{\gamma}{Y}-\underset{\beta}{C}-\underset{\alpha}{C}-A-X \longrightarrow \overset{+}{Y}=C + C=A + X^-$$

46. C. A. Grob, *Angew. Chem. Int. Ed. Engl.* **8**:535 (1969).

The fragmentation can be concerted or stepwise. The concerted mechanism is restricted to molecular geometry that is appropriate for continuous overlap of the participating orbitals. An example is the solvolysis of 4-chloropiperidine, which is more rapid than the solvolysis of chlorocyclohexane and occurs by fragmentation of the C(2)–C(3) bond[47]:

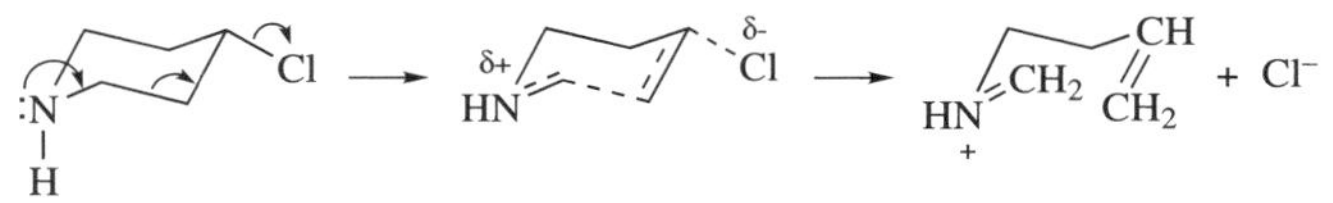

Diols or hydroxy ethers in which the two oxygen substituents are in a 1,3-relationship are particularly useful substrates for fragmentation. If a diol or hydroxy ether is converted to a monotosylate, the remaining hydroxyl group can serve to promote fragmentation.

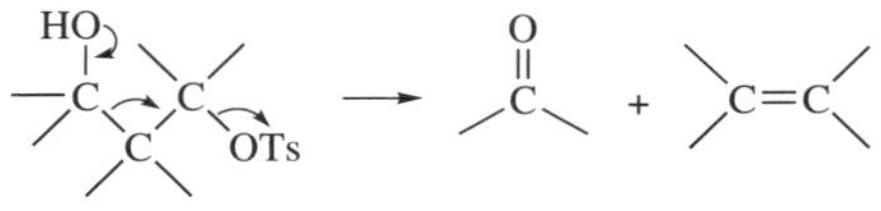

A similar reaction pattern can be seen in a fragmentation used to construct the ring structure found in the taxane group of diterpenes.

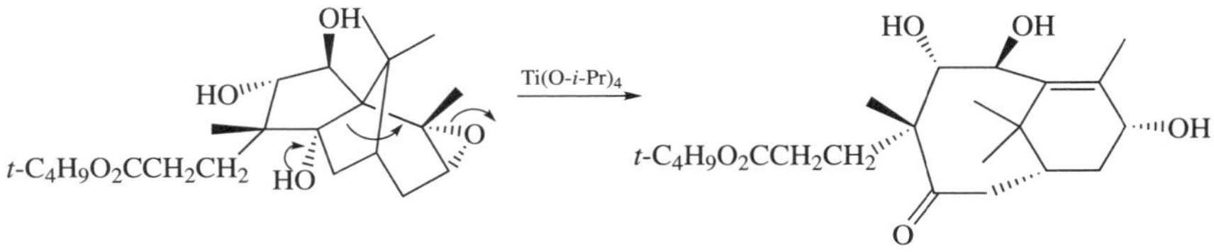

Ref. 48

Similarly, a carbonyl group at the fifth carbon from a leaving group, reacting as the enolate, promotes fragmentation with formation of an enone.[49]

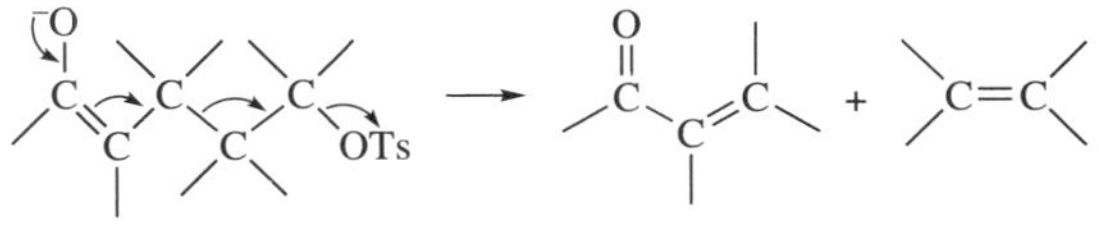

Organoboranes have been shown to undergo fragmentation if a good leaving group is present on the δ carbon.[50] The electron donor is the tetrahedral species formed by addition

47. R. D'Arcy, C. A. Grob, T. Kaffenberger, and V. Krasnobajew, *Helv. Chim. Acta* **49:**185 (1966).
48. R. A. Holton, R. R. Juo, H. B. Kim, A. D. Williams, S. Harusawa, R. E. Lowenthal, and S. Yogai, *J. Am. Chem. Soc.* **110:**6558 (1988).
49. J. M. Brown, T. M. Cresp, and L. N. Mander, *J. Org. Chem.* **42:**3984 (1977); D. A. Clark and P. J. Fuchs, *J. Am. Chem. Soc.* **101:**3567 (1979).
50. J. A. Marshall, *Synthesis* **1971:**229; J. A. Marshall and G. L. Bundy, *J. Chem. Soc., Chem. Commun.*, **1967:**854; P. S. Wharton, C. E. Sundin, D. W. Johnson, and H. C. Kluender, *J. Org. Chem.*, **37:**34 (1972).

of hydroxide ion at boron.

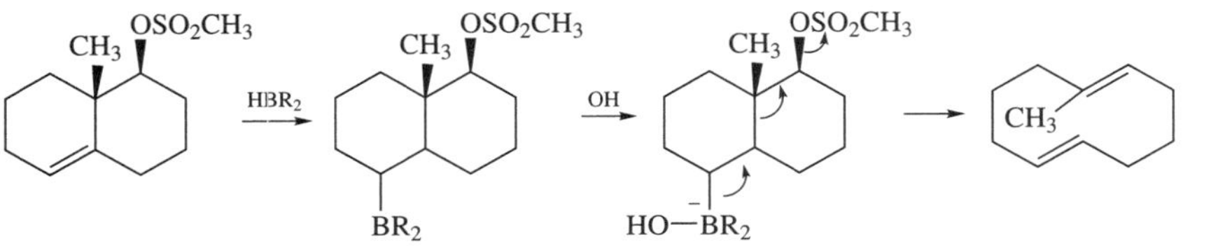

Ref. 51

The usual synthetic objective of a fragmentation reaction is the construction of a medium-sized ring from a fused ring system. Furthermore, because the fragmentation reactions of the type being discussed are usually concerted stereoselective processes, the stereochemistry is predictable. In 3-hydroxy tosylates, the fragmentation is most favorable for a geometry in which the carbon–carbon bond being broken is in an antiperiplanar relationship to the leaving group.[52] Other stereochemical relationships in the molecule are retained during the concerted fragmentation. In the case below, for example, the newly formed double bond has the E-configuration.

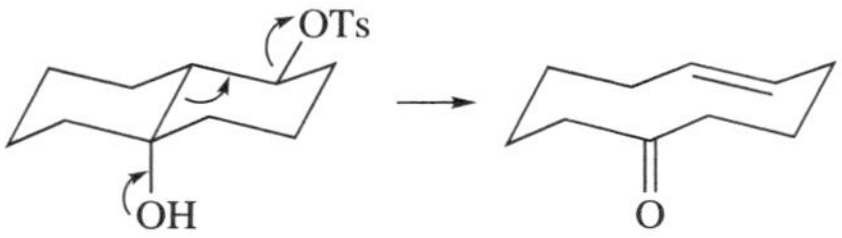

Scheme 10.4 provides some additional examples of fragmentation reaction that have been employed in a synthetic context.

10.2. Reactions Involving Carbenes and Nitrenes

Carbenes are neutral divalent derivatives of carbon. Carbenes can be included with carbanions, carbocations, and carbon-centered radicals as among the fundamental intermediates in the reactions of carbon compounds. Depending on whether the nonbonding electrons are of the same of opposite spin, they can be triplet or singlet species.

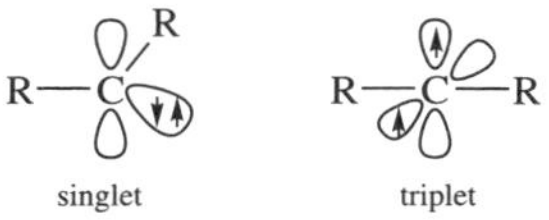

singlet triplet

As would be expected from their electron-deficient nature, carbenes are highly reactive.

Carbenes can be generated by α-elimination reactions.

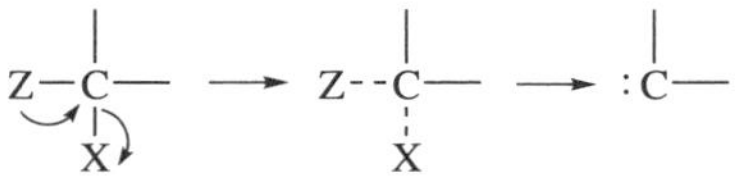

51. J. A. Marshall and G. L. Bundy, *J. Am. Chem. Soc.* **88**:4291 (1966).
52. P. S. Wharton and G. A. Hiegel, *J. Org. Chem.* **30**:3254 (1965); C. H. Heathcock and R. A. Badger, *J. Org. Chem.* **37**:234 (1972).

A. Heteroatom-promoted fragmentation

1[a] CH_3SO_2O, O^-, CH_3 → $K^+ {}^-OC(CH_3)_3$ → $H_2C{=}CH$, O, CH_3 64%

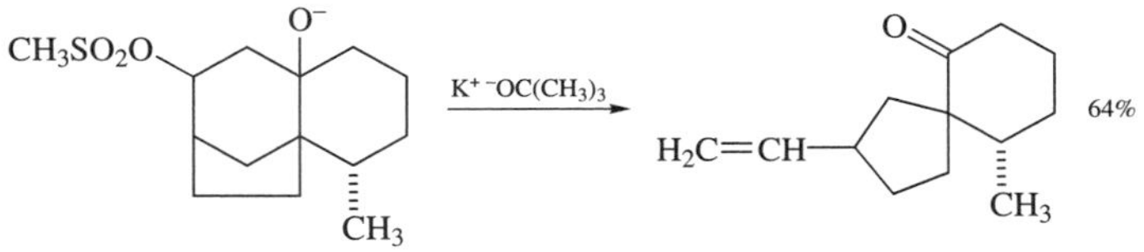

2[b] H_3C, OSO_2CH_3, OH → $LiAlH_4$ → H_3C, OH 71%

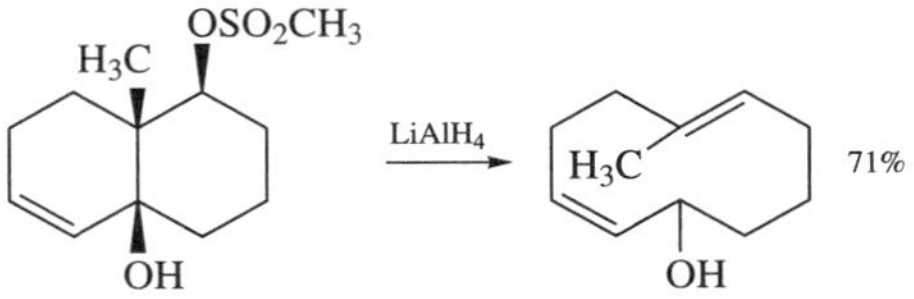

3[c] OCH_3, CCH_3, O → H^+, CH_3CO_2H, 25°C, 0.5 h → O, $CH_2CH_2CCH_3$, O 70%

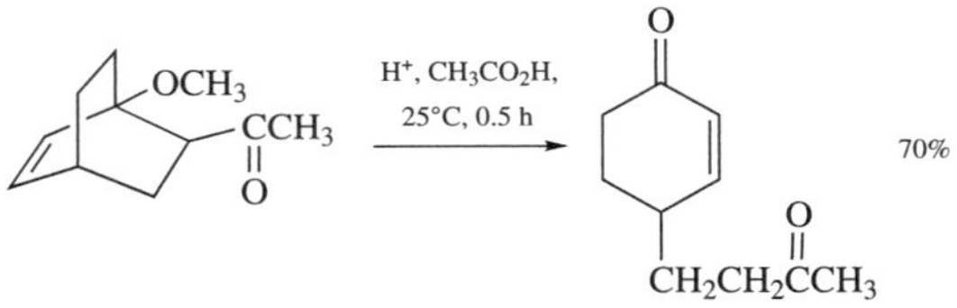

4[d] H, OSO_2Ar, N, H, CH_2Ph → solvolysis in the presence of $NaBH_4$ → N, CH_2Ph 44–58%

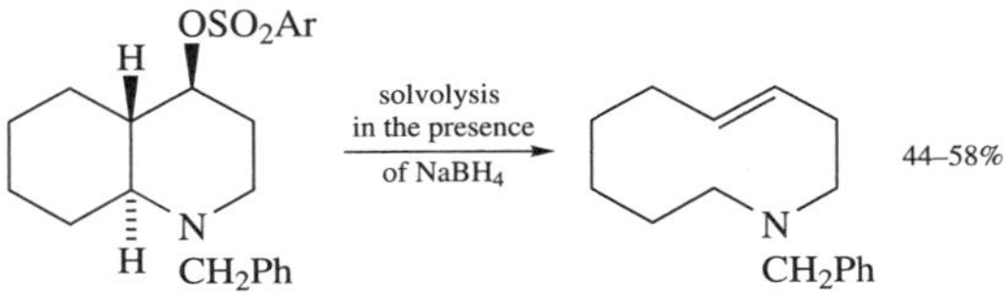

5[e] CH_3SO_3, $CH_2OCH_2OCH_3$, $OSi(CH_3)_3$ → 1) n-$Bu_4N^+F^-$ 2) NaH, 15-crown-5 → $CH_2OCH_2OCH_3$, O ≡ CH_2, $CH_2OCH_2OCH_3$, O, CH_3

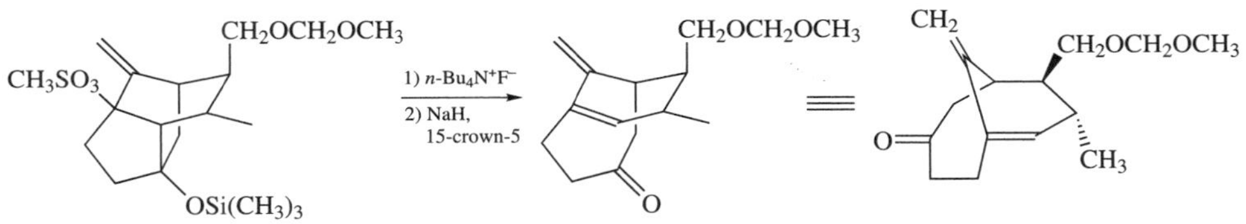

6[f] OH, CH_3, O, O, CH_3, CH_3, H, OCH_2Ph, O_3SCH_3 → $K^+ {}^-OC(CH_3)_3$ → $CH{=}O$, CH_3, O, O, CH_3, CH_3, H, OCH_2Ph, CH_2 81%

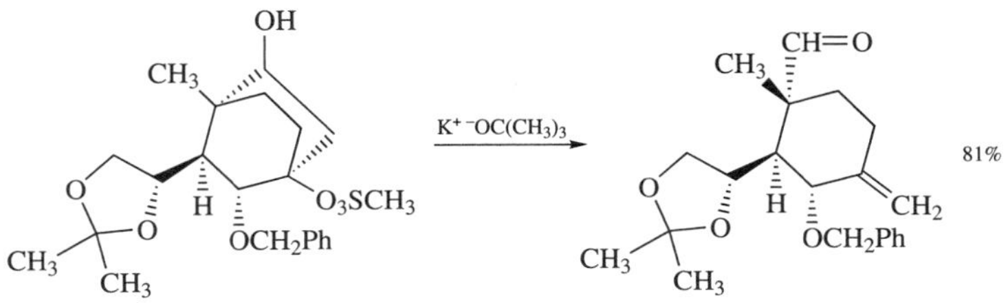

B. Boronate fragmentation

7[g] OSO_2CH_3, CH_3, H_3C, CH_3 → 1) B_2H_6 2) H_2O_2, ^-OH → H_3C, CH_3, CH_3 70%

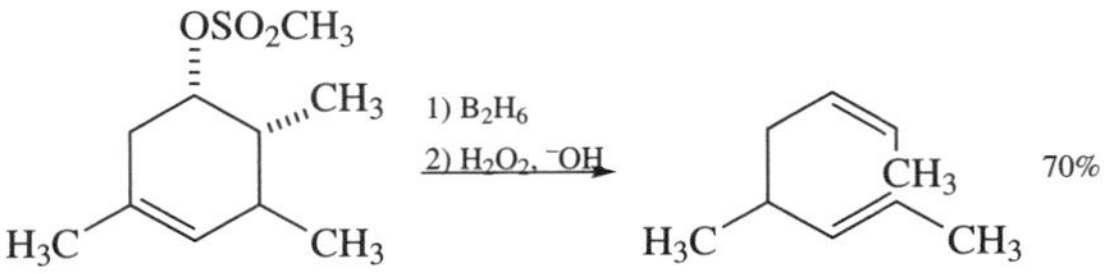

Scheme 10.4. (*continued*)

C. δ-Tosyloxy fragmentation

8[h] (bicyclic ketone with CH_3, CH_3 and $OSO_2C_6H_5$) → 1) $LiNR_2$, 2) R_2AlH, 25°C, 2 h → (OH, CH_3, CH_3, H) 93%

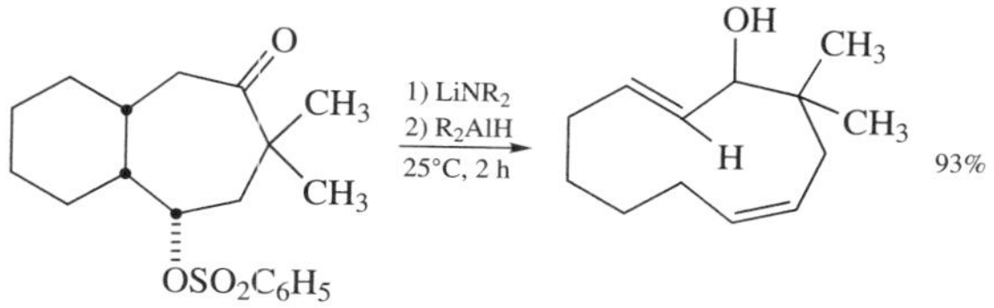

a. J. A. Marshall and S. F. Brady, *J. Org. Chem.* **35:**4068 (1970).
b. J. A. Marshall, W. F. Huffman, and J. A. Ruth, *J. Am. Chem. Soc.* **94:**4691 (1972).
c. A. J. Birch and J. S. Hill, *J. Chem. Soc. C* **1966:**419.
d. J. A. Marshall and J. H. Babler, *J. Org. Chem.* **34:**4186 (1969).
e. T. Yoshimitsu, M. Yanagiya, and H. Nagaoka, *Tetrahedron Lett.* **40:**5215 (1999).
f. Y. Hirai, T. Suga, and H. Nagaoka, *Tetrahedron Lett.* **38:**4997 (1997).
g. J. A. Marshall and J. H. Babler, *Tetrahedron Lett.* **1970:**3861.
h. D. A. Clark and P. L. Fuchs, *J. Am. Chem. Soc.* **101:**3567 (1979).

Under some circumstances, the question arises as to whether the carbene has a finite lifetime, and in some cases a completely free carbene structure is never attained. When a reaction involves a species that reacts as expected for a carbene, but must still be at least partially bound to other atoms, the term *carbenoid* is applied. Some reactions that proceed by carbene-like processes involve transition-metal ions. In many of these reactions, the divalent carbene is bound to the transition metal. Some compounds of this type are stable whereas others exist only as transient intermediates.

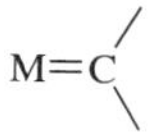

M=C

metal-bound carbene

Carbenes and carbenoids can add to double bonds to form cyclopropanes or insert into C−H bonds. These reactions have *very* low activation energies when the intermediate is a "free" carbene. Intermolecular insertion reactions are inherently nonselective. The course of intramolecular reactions is frequently controlled by the proximity of the reacting groups.[53]

Nitrenes are neutral monovalent nitrogen analogs of carbenes. The term *nitrenoid* is applied to nitrene-like intermediates that transfer a monosubstituted nitrogen fragment.

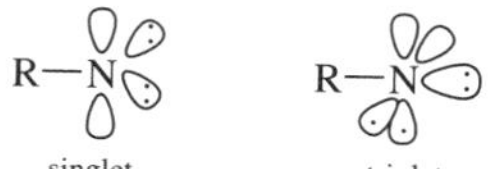

R—N　　R—N

singlet nitrene　　triplet nitrene

We will also consider a number of rearrangement reactions that probably do not involve carbene or nitrene intermediates but give overall transformations that correspond to those characteristic of a carbene or nitrene.

53. S. D. Burke and P. A. Grieco, *Org. React.* **26:**361 (1979).

10.2.1. Structure and Reactivity of Carbenes

Depending upon the mode of generation, a carbene can be formed in either the singlet or the triplet state, no matter which is lower in energy. The two electronic configurations have different geometry and reactivity. A rough picture of the bonding in the singlet assumes sp^2 hybridization at carbon, with the two unshared electrons in an sp^2 orbital. The p orbital is unoccupied. The R–C–R angle would be expected to be contracted slightly from the normal 120° because of the electronic repulsions between the unshared electron pair and the electrons in the two bonding σ orbitals. The bonds in the corresponding triplet carbene structure are formed from sp orbitals, with the unpaired electrons being in two orthogonal p orbitals. A linear structure would be predicted for this bonding arrangement.

Both theoretical and experimental studies have provided more detailed information about carbene structure. MO calculations lead to the prediction of H–C–H angles for methylene of ~135° for the triplet and ~105° for the singlet. The triplet is calculated to be about 8 kcal/mol lower in energy than the singlet.[54] Experimental determinations of the geometry of CH_2 tend to confirm the theoretical results. The H–C–H angle of the triplet state, as determined from the electron paramagnetic resonance (EPR) spectrum, is 125–140°. The H–C–H angle of the singlet state is found to be 102° by electronic spectroscopy. The available evidence is consistent with the triplet being the ground-state species.

Substituents perturb the relative energies of the singlet and triplet state. In general, alkyl groups resemble hydrogen as a substituent, and dialkylcarbenes have triplet ground-states. Substituents that act as electron-pair donors stabilize the singlet state more than the triplet state by delocalization of an electron pair into the empty p orbital.[55,56]

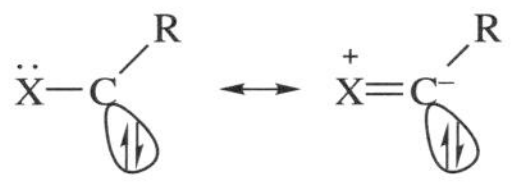

X = F, Cl, OR, NR_2

The presence of more complex substituent groups complicates the description of carbene structure. Furthermore, because carbenes are high-energy species, structural entities that would be unrealistic for more stable species must be considered. As an example, one set of MO calculations[57] arrives at structure **A** as a better description of carbomethoxycarbene than the conventional structure **B**.

A **B**

From the point of view of both synthetic and mechanistic interest, much attention has

54. J. F. Harrison, *Acc. Chem. Res.* **7**:378 (1974); P. Saxe, H. F. Shaefer, and N. C. Hardy, *J. Phys. Chem.* **85**:745 (1981); C. C. Hayden, M. Neumark, K. Shobatake, R. K. Sparks, and Y. T. Lee, *J. Chem. Phys.* **76**:3607 (1982); R. K. Lengel and R. N. Zare, *J. Am. Chem. Soc.* **100**:739 (1978); C. W. Bauschlicher, Jr., and I. Shavitt, *J. Am. Chem. Soc.* **100**:739 (1978); A. R. W. M. Kellar, P. R. Bunker, T. J. Sears, K. M. Evenson, R. Saykally and S. R. Langhoff, *J. Chem. Phys.* **79**:5251 (1983).
55. N. C. Baird and K. F. Taylor, *J. Am. Chem. Soc.* **100**:1333 (1978).
56. J. F. Harrison, R. C. Liedtke, and J. F. Liebman, *J. Am. Chem. Soc.* **101**:7162 (1979).
57. R. Noyori and M. Yamanaka, *Tetrahedron Lett.* **1980**:2851.

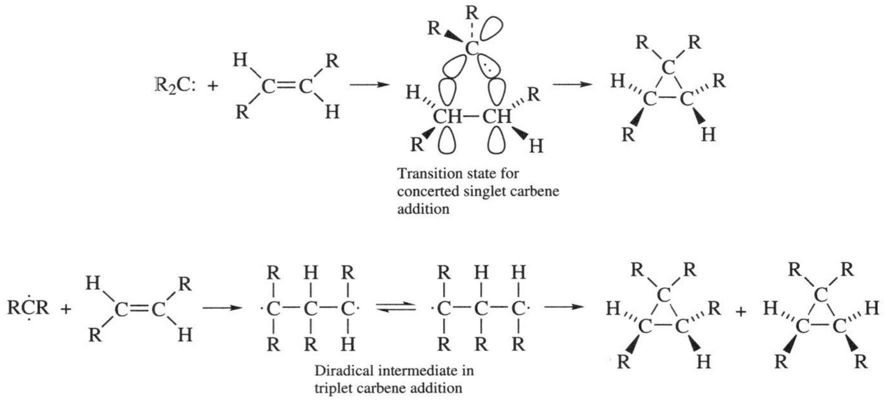

Transition state for concerted singlet carbene addition

Diradical intermediate in triplet carbene addition

Fig. 10.1. Mechanisms for addition of singlet and triplet carbenes to alkenes.

been focused on the addition reaction between carbenes and alkenes to give cyclopropanes. Characterization of the reactivity of substituted carbenes in addition reactions has emphasized stereochemistry and selectivity. The reactivity of singlet and triplet states is expected to be different. The triplet state is a diradical and should exhibit a selectivity similar to that of free radicals and other species with unpaired electrons. The singlet state, with its unfilled *p* orbital, should be electrophilic and exhibit reactivity similar to that of other electrophiles. Also, a triplet addition process must go through an intermediate that has two unpaired electrons of the same spin. In contrast, a singlet carbene can go to a cyclopropane in a single concerted step[58] (see Fig. 10.1). As a result, it was predicted[59] that additions of singlet carbenes would be stereospecific whereas those of triplet carbenes would not be. This expectation has been confirmed, and the stereoselectivity of addition reactions with alkenes has come to be used as a test for the involvement of the singlet versus the triplet carbene in specific reactions.[60]

The radical versus electrophilic character of triplet and singlet carbenes also shows up in relative reactivity patterns shown in Table 10.1. The relative reactivity of singlet dibromocarbene toward alkenes is more similar to that of electrophiles (bromination, epoxidation) than to that of radicals ($\cdot CCl_3$). Carbene reactivity is strongly affected by substituents.[61] Various singlet carbenes have been characterized as nucleophilic, ambiphilic, and electrophilic as shown in Table 10.2. This classification is based on relative reactivity toward a series of different alkenes containing both nucleophilic alkenes, such as tetramethylethylene, and electrophilic ones, such as acrylonitrile. The principal structural feature that determines the reactivity of the carbene is the ability of the substituents to act as electron donors. For example, dimethoxycarbene is devoid of electrophilicity toward

58. A. E. Keating, S. R. Merrigan, D. A. Singleton, and K. N. Houk, *J. Am. Chem. Soc.* **121**:3933 (1999).
59. P. S. Skell and A. Y. Garner, *J. Am. Chem. Soc.* **78**:5430 (1956).
60. R. C. Woodworth and P. S. Skell, *J. Am. Chem. Soc.* **81**:3383 (1959); P. S. Skell, *Tetrahedron* **41**:1427 (1985).
61. A comprehensive review of this topic is given by R. A. Moss, in *Carbenes*, M. Jones, Jr., and R. A. Moss, eds., John Wiley & Sons, New York, 1973, pp. 153–304; more recent work is reviewed in the series *Reactive Intermediates*, R. A. Moss and M. Jones, Jr., eds., John Wiley & Sons, New York; R. A. Moss, *Acc. Chem. Res.* **22**:15 (1989).

Table 10.1. Relative Rates of Addition to Alkenes[a]

Alkene	$\cdot CCl_3$	$:CBr_2$	Br_2	Epoxidation
2-Methylpropene	1.00	1.00	1.00	1.00
Styrene	>19	0.4	0.6	0.1
2-Methyl-2-butene	0.17	3.2	1.9	13.5

a. P. S. Skell and A. Y. Garner, *J. Am. Chem. Soc.* **78:**5430 (1956).

Table 10.2. Classification of Carbenes on the Basis of Reactivity toward Alkenes[a]

Nucleophilic	Ambiphilic	Electrophilic
$CH_3O\underset{\cdot\cdot}{C}OCH_3$	$CH_3O\underset{\cdot\cdot}{C}Cl$	$Cl\underset{\cdot\cdot}{C}Cl$
$CH_3O\underset{\cdot\cdot}{C}N(CH_3)_2$	$CH_3O\underset{\cdot\cdot}{C}F$	$Ph\underset{\cdot\cdot}{C}Cl$
		$CH_3\underset{\cdot\cdot}{C}Cl$
		$Br\underset{\cdot\cdot}{C}CO_2C_2H_5$

a. R. A. Moss and R. C. Munjal, *Tetrahedron Lett.* **1979:**4721; R. A. Moss, *Acc. Chem. Res.* **13:**58 (1980); R. A. Moss, *Acc. Chem. Res.* **22:**15 (1989).

alkenes[62] because of electron donation by the methoxy groups.

$$CH_3-\ddot{\underset{\cdot\cdot}{O}}-\underset{\cdot\cdot}{C}-\ddot{\underset{\cdot\cdot}{O}}-CH_3 \longleftrightarrow CH_3-\overset{+}{\underset{\cdot\cdot}{O}}=\overset{-}{\underset{\cdot\cdot}{C}}-\ddot{\underset{\cdot\cdot}{O}}-CH_3 \longleftrightarrow CH_3-\ddot{\underset{\cdot\cdot}{O}}-\overset{-}{\underset{\cdot\cdot}{C}}=\overset{+}{\underset{\cdot\cdot}{O}}-CH_3$$

π-Delocalization involving divalent carbon in conjugated cyclic systems has been studied in the interesting species cyclopropenylidene (**C**)[63] and cycloheptatrienylidene (**D**).[64] In these molecules, the empty p orbital on the carbene carbon can be part of the aromatic π system and delocalized over the entire ring. Currently available data indicate that the ground-state structure for both **C** and **D** is a singlet, but for **D**, the most advanced theoretical calculations indicate that the most stable singlet structure has an electronic configuration in which only one of the nonbonding electrons is in the π orbital.[65]

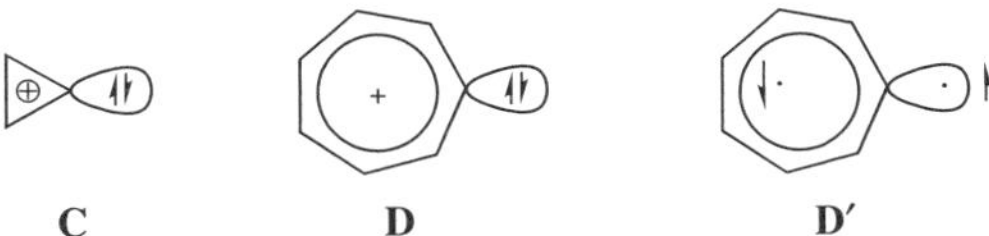

C **D** **D′**

62. D. M. Lemal, E. P. Gosselink, and S. D. McGregor, *J. Am. Chem. Soc.* **88:**582 (1966).
63. H. P. Reisenauer, G. Maier, A. Reimann, and R. W. Hoffmann, *Angew. Chem. Int. Ed. Engl.* **23:**641 (1984); T. J. Lee, A. Bunge, and H. F. Schaefer III, *J. Am. Chem. Soc.* **107:**137 (1985); J. M. Bofill, J. Farras, S. Olivella, A. Sole, and J. Vilarrasa, *J. Am. Chem. Soc.* **110:**1694 (1988).
64. R. J. McMahon and O. L. Chapman, *J. Am. Chem. Soc.* **108:**1713 (1986); M. Kusaz, H. Lüerssen, and C. Wentrup *Angew. Chem. Int. Ed. Engl.* **25:**480 (1986); C. L. Janssen and H. F. Schaefer III, *J. Am. Chem. Soc.* **109:**5030 (1987); M. W. Wong and C. Wentrup, *J. Org. Chem.* **61:**7022 (1996).
65. S. Matzinger, T. Bally, E. V. Patterson, and R. J. McMahon, *J. Am. Chem. Soc.* **118:**1535 (1996); P. R. Schreiner, W. L. Karney, P. v. R. Schleyer, W. T. Borden, T. P. Hamilton, and H. F. Schaefer III, *J. Org. Chem.* **61:**7030 (1996).

Table 10.3. General Methods for Generation of Carbenes

Precursor	Conditions	Products	Reference
Diazoalkanes $R_2C{=}\overset{+}{N}{=}N^-$	Photolysis, thermolysis, or metal-ion catalysis	$R_2C{:} + N_2$	a
Salts of sulfonylhydrazones $R_2C{=}N{-}\bar{N}SO_2Ar$	Photolysis or thermolysis; diazoalkanes are intermediates	$R_2C{:} + N_2 + ArSO_2^-$	b
Diazirines (R₂C<N=N ring)	Photolysis	$R_2C{:} + N_2$	c
Halides $R_2CH{-}X$	Strong base or organometallic compounds	$R_2C{:} + BH + X^-$	d
α-Halomercury compounds $R_2C(X)HgR'$	Thermolysis	$R_2C{:} + R'HgX$	e

a. W. J. Baron, M. R. DeCamp, M. E. Hendrick, M. Jones, Jr., R. H. Levin, and M. B. Sohn, in *Carbenes*, M. Jones, Jr., and R. A. Moss, eds, John Wiley & Sons, New York, 1973, pp. 1–151.
b. W. R. Bamford and T. S. Stevens, *J. Chem. Soc.* **1952**:4735.
c. H. M. Frey, *Adv. Photochem.* **4**:225 (1966); R. A. G. Smith and J. R. Knowles, *J. Chem. Soc., Perkin Trans. 2* **1975**:686.
d. W. Kirmse, *Carbene Chemistry*, Academic Press, New York, 1971, pp. 96–109, 129–149.
e. D. Seyferth, *Acc. Chem. Res.* **5**:65 (1972).

10.2.2. Generation of Carbenes

There are several ways of generating carbene intermediates. Some of the most general routes are summarized in Table 10.3 and will be discussed in the succeeding paragraphs.

Decomposition of diazo compounds to carbenes is a quite general reaction. Examples include the decomposition of diazomethane and other diazoalkanes, diazoalkenes, and diazo compounds with aryl and acyl substituents. The main restrictions on this method are limitations on the synthesis of the diazo compounds and their modest stability. The lower diazoalkenes are toxic and potentially explosive. They are usually prepared immediately before use. The most general synthetic route involves base-catalyzed decomposition of N-nitroso derivatives of amides or sulfonamides. These reactions are illustrated by several methods used for the preparation of diazomethane.

$$CH_3N(N{=}O){-}C({=}NH)NHNO_2 \xrightarrow{KOH} CH_2N_2$$ Ref. 66

$$CH_3N(N{=}O){-}C({=}O)NH_2 \xrightarrow[RO^-]{HO^-} CH_2N_2$$ Ref. 67

$$CH_3N(N{=}O){-}C({=}O){-}C_6H_4{-}C({=}O){-}N(N{=}O)CH_3 \xrightarrow{NaOH} CH_2N_2$$ Ref. 68

$$CH_3N(N{=}O){-}SO_2Ph \xrightarrow{KOH} CH_2N_2$$ Ref. 69

66. M. Neeman and W. S. Johnson, *Org. Synth.* **V**:245 (1973).
67. F. Arndt, *Org. Synth.* **II**:165 (1943).
68. Th. J. de Boer and H. J. Backer, *Org. Synth.* **IV**:250 (1963).
69. J. A. Moore and D. E. Reed, *Org. Synth.* **V**:351 (1973).

The details of the base-catalyzed decompositions vary somewhat, but the mechanisms involve two essential steps.[70] The initial reactants undergo a base-catalyzed elimination to form an alkyl diazoate. This is followed by a deprotonation of the α carbon and elimination of the oxygen:

$$\mathrm{RCH_2N(N{=}O)C({=}X)Z} \xrightarrow{^-\mathrm{OH}} \mathrm{RCH_2N(N{=}O)\!-\!C(X^-)(OH)\!-\!Z} \longrightarrow \mathrm{RCH(H)\!-\!N{=}N\!-\!O\!-\!H}\ (\mathrm{HO^-}) \xrightarrow{-\mathrm{H_2O}} \mathrm{RCH{=}\overset{+}{N}{=}\ddot{N}^-}$$

Diazo compounds can also be obtained by oxidation of the corresponding hydrazone. This route is employed most frequently when one of the substituents is an aromatic ring:

$$\mathrm{Ph_2C{=}NNH_2} \xrightarrow{\mathrm{HgO}} \mathrm{Ph_2C{=}\overset{+}{N}{=}\ddot{N}^-} \qquad \text{Ref. 71}$$

The higher diazoalkanes can be made by $Pb(O_2CCH_3)_4$ oxidation of hydrazones.[72]

α-Diazoketones are especially useful in synthesis.[73] There are several methods of preparation. Reaction of diazomethane with an acyl chloride results in formation of a diazomethyl ketone:

$$\mathrm{RC({=}O)Cl} + \mathrm{H_2C{=}\overset{+}{N}{=}N^-} \longrightarrow \mathrm{RC({=}O)CH{=}\overset{+}{N}{=}N^-}$$

The HCl generated in this reaction destroys one equivalent of diazomethane. This can be avoided by including a base, such as triethylamine, to neutralize the acid.[74] Cyclic α-diazoketones, which are not available from acyl chlorides, can be prepared by reaction of an enolate equivalent with a sulfonyl azide. This reaction is called *diazo transfer*.[75] Various arenesulfonyl azides[76] and methanesulfonyl azide[77] are used most frequently. Several types of compounds can act as the carbon nucleophile. These include the anion of the hydroxymethylene derivative of the ketone[78] or the dialkylaminomethylene derivative of

70. W. M. Jones, D. L. Muck, and T. K. Tandy, Jr., *J. Am. Chem. Soc.* **88:**3798 (1966); R. A. Moss, *J. Org. Chem.* **31:**1082 (1966); D. E. Applequist and D. E. McGreer, *J. Am. Chem. Soc.* **82:**1965 (1960); S. M. Hecht and J. W. Kozarich, *J. Org. Chem.* **38:**1821 (1973); E. H. White, J. T. DePinto, A. J. Polito, I. Bauer, and D. F. Roswell, *J. Am. Chem. Soc.* **110:**3708 (1988).
71. L. I. Smith and K. L. Howard, *Org. Synth.*, **III:**351 (1955).
72. T. L. Holton and H. Shechter, *J. Org. Chem.* **60:**4725 (1995).
73. T. Ye and M. A. McKervey, *Chem. Rev.* **94:**1091 (1994).
74. M. S. Newman and P. Beall III, *J. Am. Chem. Soc.* **71:**1506 (1949); M. Berebom and W. S. Fones, *J. Am. Chem. Soc.* **71:**1629 (1949); L. T. Scott and M. A. Minton, *J. Org. Chem.* **42:**3757 (1977).
75. F. W. Bollinger and L. D. Tuma, *Synlett* **1996:**407.
76. J. B. Hendrickson and W. A. Wolf, *J. Org. Chem.* **33:**3610 (1968); J. S. Baum, D. A. Shook, H. M. L. Davies, and H. D. Smith, *Synth Commun.* **17:**1709 (1987); L. Lombardo and L. N. Mander, *Synthesis* **1980:**368.
77. D. F. Taber, R. E. Ruckle, and M. J. Hennessy, *J. Org. Chem.* **51:**4077 (1986); R. L. Danheiser, D. S. Casebier, and F. Firooznia, *J. Org. Chem.* **60:**8341 (1995).
78. M. Regitz and G. Heck, *Chem. Ber.* **97:**1482 (1964); M. Regitz, *Angew. Chem. Int. Ed. Engl.* **6:**733 (1967).

the ketone.[79]

$$\mathrm{RC(=O)C(R')=CHY} + \mathrm{ArSO_2N_3} \longrightarrow \mathrm{RC(=O)C(R')=\overset{+}{N}=N^-}$$

$Y = O^-$ or NR_2

α-Trifluoroacetyl derivatives of ketones are also good substrates for diazo transfer.[80]

1) LiHMDS, $CF_3CO_2CH_2CF_3$
2) $CH_3SO_2N_3$, $(C_2H_5)_3N$

α-Diazoketones can also be made by first converting the ketone to an α-oximino derivative by nitrosation and then allowing the oximino ketone to react with chloramine.[81]

1) RONO, $KOC(CH_3)_3$
2) NH_3, CaOCl

70% Ref. 82

The driving force for decomposition of diazo compounds to carbenes is the formation of the very stable nitrogen molecule. Activation energies for decomposition of diazoalkanes in the gas phase are in the neighborhood of 30 kcal/mol. The requisite energy can also be supplied by photochemical excitation. It is often possible to control the photochemical process to give predominantly singlet or triplet carbene. Direct photolysis leads to the singlet intermediate when the dissociation of the excited diazoalkene is faster than intersystem crossing to the triplet state. The triplet carbene is the principal intermediate in photosensitized decomposition of diazoalkanes. (See Chapter 13, Part A, to review photosensitization.)

Reaction of diazo compounds with a variety of transition metal compounds leads to evolution of nitrogen and formation of products of the same general type as those formed by thermal and photochemical decomposition of diazoalkanes. These transition metal-catalyzed reactions in general appear to involve carbenoid intermediates in which the carbene becomes bound to the metal.[83] The metals which have been used most frequently in synthesis are copper and rhodium.

The second method listed in Table 10.3, thermal or photochemical decomposition of salts of arenesulfonylhydrazones, is actually a variation of the diazoalkane method, since diazo compounds are intermediates. The conditions of the decomposition are usually such

79. M. Rosenberger, P. Yates, J. B. Hendrickson, and W. Wolf, *Tetrahedron Lett.* **1964:**2285; K. B. Wiberg, B. L. Furtek, and L. K. Olli, *J. Am. Chem. Soc.* **101:**7675 (1979).
80. R. L. Danheiser, R. F. Miller, R. G. Brisbois, and S. Z. Park, *J. Org. Chem.* **55:**1959 (1990).
81. T. N. Wheeler and J. Meinwald, *Org. Synth.* **52:**53 (1972).
82. T. Sasaki, S. Eguchi, and Y. Hirako, *J. Org. Chem.* **42:**2981 (1977).
83. W. R. Moser, *J. Am. Chem. Soc.* **91:**1135, 1141 (1969); M. P. Doyle, *Chem. Rev.* **86:**919 (1986); M. Brookhart, *Chem. Rev.* **87:**411 (1987).

that the diazo compound reacts immediately on formation.[84] The nature of the solvent plays an important role in the outcome of sulfonylhydrazone decompositions. In protic solvents, the diazoalkane can be diverted to a carbocation by protonation.[85] Aprotic solvents favor decomposition via the carbene pathway.

$$RC(=O)R + NH_2NHSO_2Ar \longrightarrow R_2C{=}N{-}NHSO_2Ar \xrightarrow{\text{base}} R_2C{=}N{-}\bar{N}SO_2Ar$$

$$R_2C{=}N{-}\bar{N}SO_2Ar \xrightarrow[\text{or } \Delta]{h\nu} R_2C{=}\overset{+}{N}{=}N^-$$

$$R_2C{=}\overset{+}{N}{=}N^- \xrightarrow{\text{XOH}} R_2CH{-}\overset{+}{N}{\equiv}N \longrightarrow R_2\overset{+}{C}H + N_2$$

The diazirine precursors of carbenes (entry 3 in Table 10.3) are cyclic isomers of diazo compounds. The strain of the small ring and the potential for formation of nitrogen make them highly reactive on photoexcitation. They are, in general, somewhat less easily available than diazo compounds or arenesulfonylhydrazones. However, there are several useful synthetic routes.[86]

Ref. 87

R = TMS

1) NH_3, NH_2OSO_3H; 2) I_2

Ref. 88

$$CH_3O_2C{-}C_6H_4{-}C(=NH)NH_2 \xrightarrow{\text{NaOCl, LiCl}} CH_3O_2C{-}C_6H_4{-}\text{(3-chlorodiazirin-3-yl)}$$

The α elimination of hydrogen halide induced by strong base (entry 4, in Table 10.3) is restricted to reactants that do not have β hydrogens, because dehydrohalogenation by β elimination dominates when it can occur. The classic example of this method is the generation of dichlorocarbene by base-catalyzed decomposition of chloroform.[89]

$$HCCl_3 + {}^-OR \rightleftharpoons :\bar{C}Cl_3 \longrightarrow :CCl_2 + Cl^-$$

Both phase-transfer and crown ether catalysis have been used to promote α-elimination reactions of chloroform and other haloalkanes.[90] The carbene is trapped by alkenes to form halocyclopropanes.

84. G. M. Kaufman, J. A. Smith, G. G. Vander Stouw, and H. Schechter, *J. Am. Chem. Soc.* **87**:935 (1965).
85. J. H. Bayless, L. Friedman, F. B. Cook, and H. Schechter, *J. Am. Chem. Soc.* **90**:531 (1968).
86. For reviews of synthesis of diazirines, see E. Schmitz, *Dreiringe mit Zwei Heteroatomen*, Springer-Verlag, Berlin, 1967, pp. 114–121; E. Schmitz, *Adv. Heterocycl. Chem.* **24**:63 (1979); H. W. Heine, in *Chemistry of Heterocyclic Compounds*, Vol. 42, Pt. 2, A. Hassner, ed. Wiley-Interscience, New York, 1983, pp. 547–628.
87. G. Kurz, J. Lehmann, and R. Thieme, *Carbohyd. Res.* **136**:125 (1983).
88. D. F. Johnson and R. K. Brown, *Photochem. Photobiol.* **43**:601 (1986).
89. J. Hine, *J. Am. Chem. Soc.* **72**:2438 (1950); J. Hine and A. M. Dowell, Jr., *J. Am. Chem. Soc.* **76**:2688 (1954).
90. W. P. Weber and G. W. Gokel, *Phase Transfer Catalysis in Organic Synthesis*, Springer-Verlag, New York, 1977, Chapters 2–4.

$$Ph_2C{=}CH_2 + CHCl_3 \xrightarrow[\text{50\% NaOH}]{PhCH_2\overset{+}{N}(C_2H_5)_3} \text{1,1-dichloro-2,2-diphenylcyclopropane}$$ Ref. 91

Dichlorocarbene can also be generated by sonication of a solution of chloroform with powdered KOH.[92]

α-Elimination also occurs in the reaction of dichloromethane and benzyl chlorides with alkyllithium reagents. The carbanion stabilization provided by the chloro and phenyl groups make the lithiation feasible

$$H_2CCl_2 + RLi \longrightarrow RH + LiCHCl_2 \longrightarrow :CHCl + LiCl$$ Ref. 93

$$ArCH_2X + RLi \longrightarrow RH + Ar\overset{Li}{C}HX \longrightarrow Ar\ddot{C}H + LiX$$ Ref. 94

The reactive intermediates under some conditions may be the carbenoid α-haloalkyllithium compounds or carbene-lithium halide complexes.[95] In the case of the trichloromethyllithium → dichlorocarbene conversion, the equilibrium lies heavily to the side of trichloromethyllithium at $-100°$.[96] The addition reaction with alkenes seems to involve dichlorocarbene, however, since the pattern of reactivity towards different alkenes is identical to that observed for the free carbene in the gas phase.[97]

Hindered lithium dialkylamides can generate aryl-substituted carbenes from benzyl halides.[98] Reaction of α,α-dichlorotoluene or α,α-dibromotoluene with potassium *t*-butoxide in the presence of 18-crown-6 generates the corresponding α-halophenylcarbene.[99] The relative reactivity data for carbenes generated under these latter conditions suggest that they are "free." The potassium cation would be expected to be strongly solvated by the crown ether and it is evidently not involved in the carbene-generating step.

A method that provides an alternative route to dichlorocarbene is the decarboxylation of trichloroacetic acid.[100] The decarboxylation generates the trichloromethyl anion which decomposes to the carbene. Treatment of alkyl trichloroacetates with an alkoxide salt also generates dichlorocarbene.

91. E. V. Dehmlow and J. Schönefeld, *Liebigs Ann. Chem.* **744:**42 (1971).
92. S. L. Regen and A. Singh, *J. Org. Chem.* **47:**1587 (1982).
93. G. Köbrich, H. Trapp, K. Flory, and W. Drischel, *Chem. Ber.* **99:**689 (1966); G. Köbrich and H. R. Merkle, *Chem. Ber.* **99:**1782 (1966).
94. G. L. Gloss and L. E. Closs, *J. Am. Chem. Soc.* **82:**5723 (1960).
95. G. Köbrich, *Angew. Chem. Int. Ed. Engl.* **6:**41 (1967).
96. W. T. Miller, Jr., and D. M. Whalen, *J. Am. Chem. Soc.* **86:**2089 (1964); D. F. Hoeg, D. I. Lusk, and A. L. Crumbliss, *J. Am. Chem. Soc.* **87:**4147 (1965).
97. P. S. Skell and M. S. Cholod, *J. Am. Chem. Soc.* **91:**6035, 7131 (1969); P. S. Skell and M. S. Cholod, *J. Am. Chem. Soc.* **92**:3522 (1970).
98. R. A. Olofson and C. M. Dougherty, *J. Am. Chem. Soc.* **95:**581 (1973).
99. R. A. Moss and F. G. Pilkiewicz, *J. Am. Chem. Soc.* **96:**5632 (1974).
100. W. E. Parham and E. E. Schweizer, *Org. React.* **13:**55 (1963).

$$^{-}O-\overset{O}{\overset{\|}{C}}-CCl_3 \xrightarrow{-CO_2} :\bar{C}Cl_3 \longleftarrow Cl_3C-\overset{O^-}{\underset{OR'}{\overset{|}{\underset{|}{C}}}}OR \longleftarrow Cl_3\overset{O}{\overset{\|}{C}}COR + ^{-}OR'$$

$$:\bar{C}Cl_3 \longrightarrow :CCl_2 + Cl^-$$

The applicability of these methods is restricted to polyhalogenated compounds, because the inductive effect of the halogen atoms is necessary for facilitating formation of the carbanion.

The α-elimination mechanism is also the basis of the use of organomercury compounds for carbene generation (entry 5, in Table 10.3). The carbon–mercury bond is much more covalent than the C–Li bond, however, so the mercury reagents are generally stable at room temperature and can be isolated. They then decompose to the carbene on heating.[101] Addition reactions occur in the presence of alkenes. The decomposition rate is not greatly influenced by the alkene, which implies that a free carbene is generated from the organomercury precursor in the rate-determining step.[102]

$$PhHg-\overset{Cl}{\underset{Cl}{\overset{|}{\underset{|}{C}}}}-Br \longrightarrow :CCl_2 + PhHgBr$$

A variety of organomercury compounds that can serve as precursors of substituted carbenes have been synthesized. For example, carbenes with carbomethoxy or trifluoromethyl substituents can be generated in this way.[103]

$$PhHg-\overset{Cl}{\underset{CF_3}{\overset{|}{\underset{|}{C}}}}-Br \longrightarrow Cl\ddot{C}CF_3$$

$$PhHgCCl_2CO_2CH_3 \longrightarrow Cl\ddot{C}CO_2CH_3$$

The addition reactions of alkenes and phenylmercuric bromide typically occur at about 80°C. Phenylmercuric iodides are somewhat more reactive and may be advantageous in reactions with relatively unstable alkenes.[104]

10.2.3. Addition Reactions

Addition reactions with alkenes to form cyclopropanes are the best studied reactions of carbene intermediates, both from the point of view of understanding carbene mechanisms and for synthetic applications. A concerted mechanism is possible for singlet carbenes. As a result, the stereochemistry present in the alkene is retained in the cyclopropane. With triplet carbenes, an intermediate diradical is involved. Closure to cyclopropane requires spin inversion. The rate of spin inversion is slow relative to that of

101. D. Seyferth, J. M. Burlitch, R. J. Minasz, J. Y.-P. Mui, H. D. Simmons, Jr., A. J. H. Treiber, and S. R. Dowd, *J. Am. Chem. Soc.* **87:**4259 (1965).
102. D. Seyferth, J. Y.-P. Mui, and J. M. Burlitch, *J. Am. Chem. Soc.* **89:**4953 (1967).
103. D. Seyferth, D. C. Mueller, and R. L. Lambert, Jr., *J. Am. Chem. Soc.* **91:**1562 (1969).
104. D. Seyferth and C. K. Haas, *J. Org. Chem.* **40:**1620 (1975).

rotation about single bonds, so that mixtures of the two possible stereoisomers are obtained from either alkene stereoisomer.

singlet mechanism

triplet mechanism

Reactions involving free carbenes are very exothermic because two new σ bonds are formed and only the alkene π bond is broken. The reactions are very fast, and, in fact, theoretical treatment of the addition of singlet methylene to ethylene suggests that there is no activation barrier.[105] The addition of carbenes to alkenes is an important method for the synthesis of many types of cyclopropanes, and several of the methods for carbene generation listed in Table 10.3 have been adapted for use in synthesis.

A very effective means for conversion of alkenes to cyclopropanes by transfer of a CH_2 unit involves reaction with methylene iodide and zinc–copper couple, commonly referred to as the *Simmons–Smith reagent.*[106] The active species is iodomethylzinc iodide.[107] The transfer of methylene occurs stereospecifically. Free $:CH_2$ is not an intermediate. Entries 1–3 in Scheme 10.5 are typical examples.

A modified version of the Simmons–Smith reaction uses dibromomethane and *in situ* generation of the Cu–Zn couple.[108] Sonication is used in this procedure to promote reaction at the metal surface.

CH_2Br_2, Zn–Cu, sonication — 50% — Ref. 109

Another useful reagent combination involves diethylzinc and diiodomethane or chloroiodomethane.

$(C_2H_5)_2Zn$, $ClCH_2I$ — Ref. 110

105. B. Zurawski and W. Kutzelnigg, *J. Am. Chem. Soc.* **100:**2654 (1978).
106. H. E. Simmons and R. D. Smith, *J. Am. Chem. Soc.* **80:**5323 (1958); H. E. Simmons and R. D. Smith, **81**: 4256 (1959); H. E. Simmons, T. L. Cairns, S. A. Vladuchick, and C. M. Hoiness, *Org. Chem.* **20:**1 (1973).
107. A. B. Charette and J.-F. Marcoux, *J. Am. Chem. Soc.* **118:**4539 (1996).
108. E. C. Friedrich, J. M. Demek, and R. Y. Pong, *J. Org. Chem.* **50:**4640 (1985).
109. S. Sawada and Y. Inouye, *Bull. Chem. Soc. Jpn.* **42:**2669 (1969); N. Kawabata, T. Nakagawa, T. Nakao, and S. Yamashita, *J. Org. Chem.* **42:**3031 (1977); J. Furukawa, N. Kawabata, and J. Nishimura, *Tetrahedron* **24:**53 (1968).
110. S. Miyano and H. Hashimoto, *Bull. Chem. Soc. Jpn.* **46:**892 (1973); S. E. Denmark and J. P. Edwards, *J. Org. Chem.* **56:**6974 (1991).

A. Cyclopropanes by methylene transfer

1[a] + CH_2I_2 $\xrightarrow[CuCl]{Zn\ dust}$ 92%

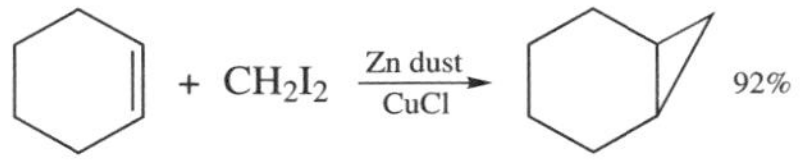

2[b] OH H + CH_2I_2 $\xrightarrow{Cu-Zn}$ H OH H H 66%

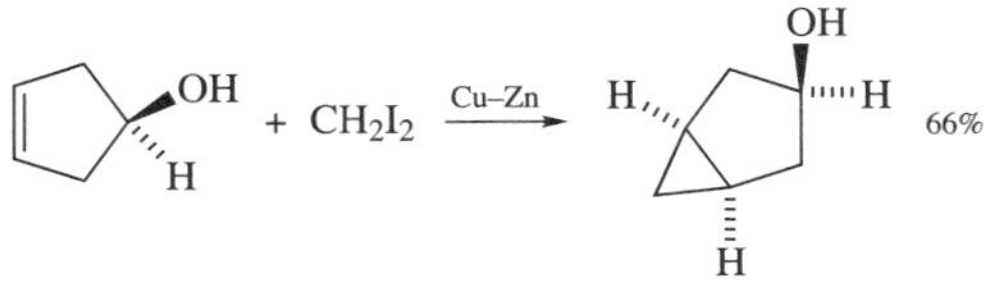

3[c] HO CH_3 O O $\xrightarrow[Cu-Zn]{CH_2I_2}$ HO CH_3 O O 76%

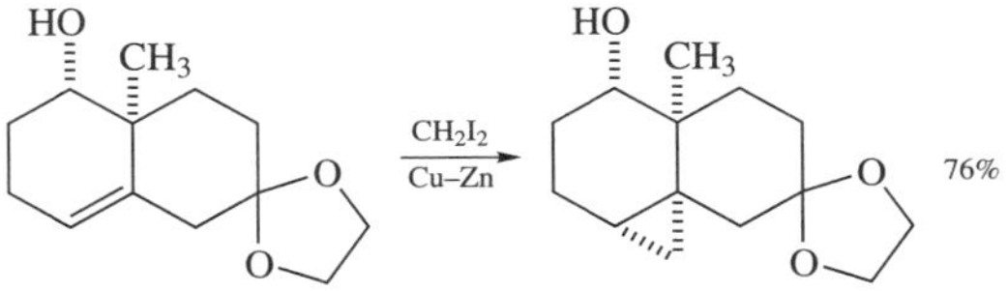

4[d] OH CH_3O CH_3 $\xrightarrow[CH_2I_2]{(C_2H_5)_2Zn}$ OH CH_3O CH_3 99%

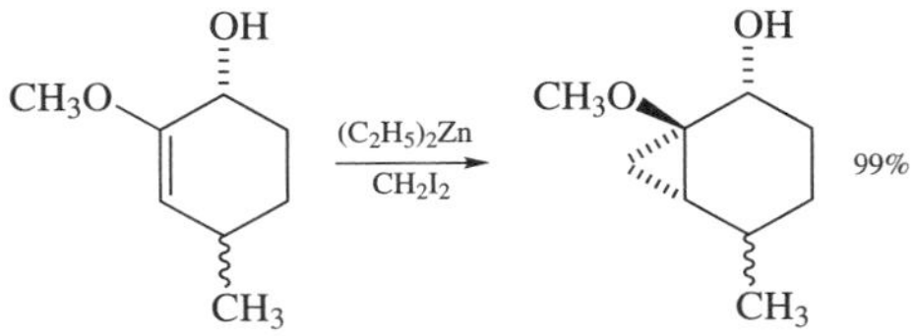

B. Catalytic cyclopropanation by diazo products and metal salts

5[e] + $N_2CHCO_2C_2H_5$ $\xrightarrow{CuCN}$ $CO_2C_2H_5$ 58%

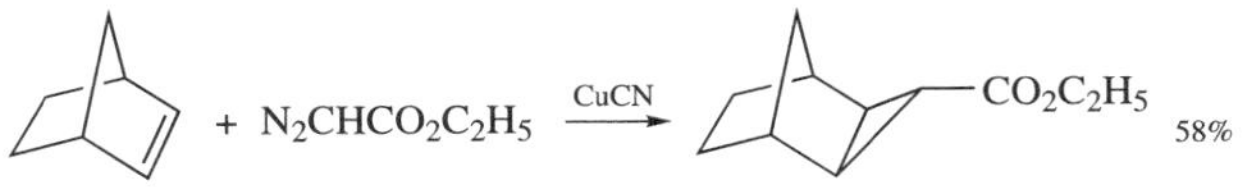

6[f] H_3C H C=C H CH_3 + $N_2CHCO_2C_2H_5$ $\xrightarrow{CuO_3SCF_3}$ $CO_2C_2H_5$ CH_3 H_3C 51%

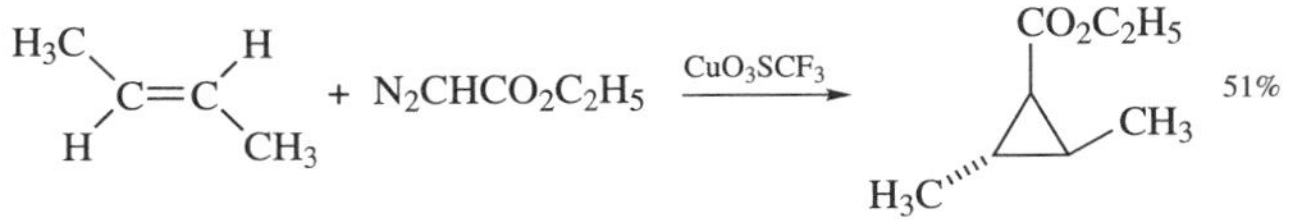

7[g] $H_2C{=}CHO_2CCH_3$ + $N_2CHCO_2C_2H_5$ $\xrightarrow{Rh(O_2CCH_3)_4}$ $CO_2C_2H_5$ O_2CCH_3 77%

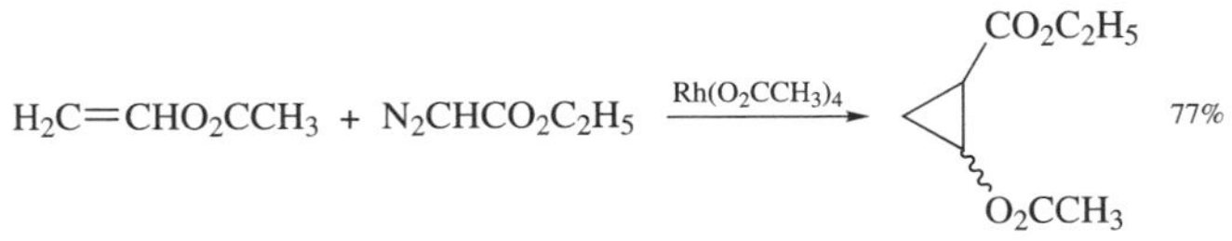

8[h] $(CH_3)_3C$ $=CH_2$ + $N_2CH\overset{O}{\overset{\|}{C}}CO_2C_2H_5$ $\xrightarrow{Cu(acac)_2}$ $(CH_3)_3C$ $\overset{O}{\overset{\|}{C}}CO_2C_2H_5$ 45%

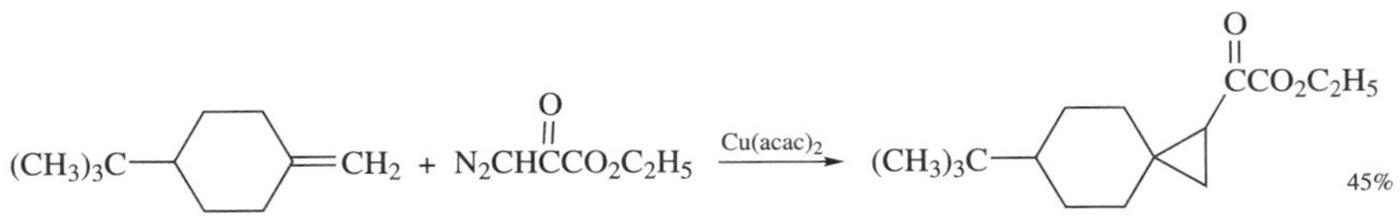

9[i] $PhCH{=}CH_2$ + $\xrightarrow{\textbf{cat 3}}$ Ph CO_2-t-C_4H_9 + Ph CO_2-t-C_4H_9

61% yield, 96% e.e. 14% yield, 93% e.e.

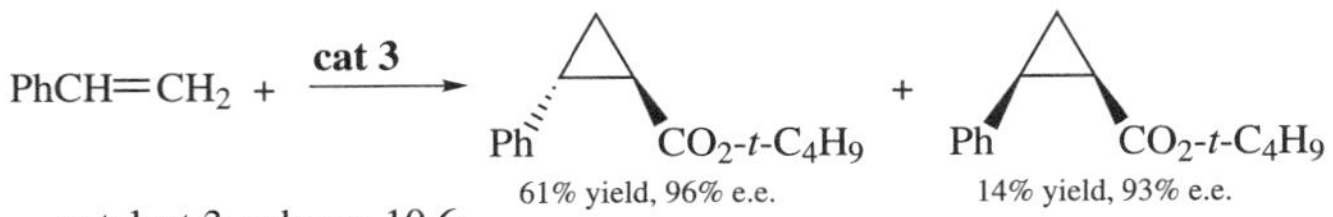

catalyst 3, scheme 10.6

10[j] CH_2, H, O, CH_3 + CH_2N_2 $\xrightarrow{Pd(O_2CCH_3)_2}$ H, O, CH_3 96%

C. Cyclopropane formation using haloalkylmercurials

11[k] + $PhHgC(Br)_2CO_2CH_3$ ⟶ CO_2CH_3, Br + Br, CO_2CH_3

50% total yield

12[l] $(CH_3)_2C{=}C(CH_3)_2$ + $PhHgC(CF_3)_2Br$ ⟶ Cl, CF_3, H_3C, CH_3, H_3C, CH_3 58%

D. Reactions of carbenes generated by α elimination

13[m] CH_3 + $HCBr_3$ + $K^+\,^-OC(CH_3)_3$ ⟶ Br, Br, CH_3 79%

14[n] CH_3O–C6H4–$CHBr_2$ + H_3C, CH_3, C=C, H, H $\xrightarrow{n\text{-BuLi}}$ H_3C, H_3C, OCH_3 55%

15[o] $(CH_3)_2C{=}CHCH_3$ + $CFBr_3$ $\xrightarrow{n\text{-BuLi}}$ H_3C, Br, F, H_3C, CH_3 55%

16[p] + $CHCl_3$ $\xrightarrow[\text{50\% NaOH, benzene}]{PhCH_2\overset{+}{N}(C_2H_5)_3Cl^-}$ Cl, Cl

E. Intramolecular addition reactions

17[q] $AcOCH_2$, CH_2OAc, $O{=}CCHN_2$ $\xrightarrow{Rh_2(OAc)_4}$ CH_2OAc, $AcOCH_2$, O 37%

18[r] N_2CH ... CO_2CH_3 $\xrightarrow{h\nu}$ CO_2CH_3 50%

19[s] CH_3O_2C CHN_2 $\xrightarrow{\text{cat 5}}$ CH_3 CO_2CH_3

catalyst 5, scheme 10.6

77% yield, 90% e.e.

a. R. J. Rawson and I. T. Harrison, *J. Org. Chem.* **35**:2057 (1970).
b. S. Winstein and J. Sonnenberg, *J. Am. Chem. Soc.* **83**:3235 (1961).
c. P. A. Grieco, T. Oguir, C.-L. J. Wang, and E. Williams, *J. Org. Chem.* **42**:4113 (1977).
d. R. C. Gadwood, R. M. Lett, and J. E. Wissinger, *J. Am. Chem. Soc.* **108**:6343 (1986).
e. R. R. Sauers and P. E. Sonnett, *Tetrahedron* **20**:1029 (1964).
f. R. G. Salomon and J. K. Kochi, *J. Am. Chem. Soc.* **95**:3300 (1973).
g. A. J. Anciaux, A. J. Hubert, A. F. Noels, N. Petiniot, and P. Teyssie, *J. Org. Chem.* **45**:695 (1980).
h. M. E. Alonso, P. Jano, and M. I. Hernandez, *J. Org. Chem.* **45**:5299 (1980).
i. D. A. Evans, K. A. Woerpel, M. M. Hinman, and M. M. Faul, *J. Am. Chem. Soc.* **113**:726 (1991).
j. L. Strekowski, M. Visnick, and M. A. Battiste, *J. Org. Chem.* **51**:4836 (1986).
k. D. Seyferth, D. C. Mueller, and R. L. Lambert, Jr., *J. Am. Chem. Soc.* **91**:1562 (1969).
l. D. Seyferth and D. C. Mueller, *J. Am. Chem. Soc.* **93**:3714 (1971).
m. L. A. Paquette, S. E. Wilson, R. P. Henzel, and G. R. Allen, Jr., *J. Am. Chem. Soc.* **94**:7761 (1972).
n. G. L. Closs and R. A. Moss, *J. Am. Chem. Soc.* **86**:4042 (1964).
o. D. J. Burton and J. L. Hahnfeld, *J. Org. Chem.* **42**:828 (1977).
p. T. T. Sasaki, K. Kanematsu, and N. Okamura, *J. Org. Chem.* **40**:3322 (1975).
q. P. Dowd, P. Garner, R. Schappert, H. Irngartinger, and A. Goldman, *J. Org. Chem.* **47**:4240 (1982).
r. B. M. Trost, R. M. Cory, P. H. Scudder, and H. B. Neubold, *J. Am. Chem. Soc.* **95**:7813 (1973).
s. T. G. Grant, M. C. Noe, and E. J. Corey, *Tetrahedron Lett.* **36**:8745 (1995).

Several modifications of the Simmons–Smith procedure have been developed in which an electrophile or Lewis acid is included. Inclusion of acetyl chloride accelerates the reaction and permits the use of dibromomethane.[111] Titanium tetrachloride has a similar effect on the reactions of unfunctionalized alkenes.[112] Reactivity can be induced by inclusion of a small amount of trimethylsilyl chloride.[113] The Simmons–Smith reaction has also been found to be sensitive to the purity of the zinc used. Electrolytically prepared zinc is much more reactive than zinc prepared by metallurgic smelting. This difference has been traced to small amounts of lead in the latter material.

In molecules with hydroxyl groups, the CH_2 unit is selectivity introduced on the side of the double bond *syn* to the hydroxyl group. This indicates that the reagent is complexed to the hydroxyl group and that the complexation directs the addition. Entries 2, 3 and 4 in Scheme 10.5 illustrates the stereodirective effect of the hydroxyl group.

The directive effect of allylic hydroxyl groups can be used in conjunction with chiral catalysts to achieve enantioselective cyclopropanation. The chiral ligand used is a boronate

111. E. C. Friedrich and E. J. Lewis, *J. Org. Chem.* **55**:2491 (1990).
112. E. C. Friedrich, S. E. Lunetta, and E. J. Lewis, *J. Org. Chem.* **54**:2385 (1989).
113. K. Takai, T. Kakiuchi, and K. Utimoto, *J. Org. Chem.* **59**:2671 (1994).

ester derived from the *N,N,N′,N′*-tetramethyl amide of tartaric acid.[114]

$(CH_3)_2NCO$ $CON(CH_3)_2$ O B O n-C_4H_9

Ph–CH=CH–CH$_2$OH → [$Zn(CH_2I)_2$·DME, CH_2Cl_2] → Ph (cyclopropyl) OH 93% e.e.

These conditions have been used to make natural products containing several successive cyclopropane rings.[115]

U-106305

The transition-metal-catalyzed decomposition of diazo compounds is a very useful reaction for formation of substituted cyclopropanes. The reaction has been carried out with several copper salts.[116] Both Cu(I) and Cu(II) triflate are useful.[117] Several Cu(II)salen complexes, such as the *N*-*t*-butyl derivative $Cu(TBS)_2$, have become popular catalysts.[118]

Ph ... O_2CHN_2 → [$Cu(TBS)_2$] → Ph ... H H O O $(CH_3)_3C$ N Cu O$^-$)$_2$ $Cu(TBS)_2$ Ref. 119

Catalysts of the copper–imine class are enantioselective when chiral imines are used. Some of the structures are shown in Scheme 10.6.

A wide variety of other transition-metal complexes are also useful, including rhodium,[120] palladium,[121] and molybdenum[122] compounds. The catalytic cycle can be

114. A. B. Charette and H. Juteau, *J. Am. Chem. Soc.* **116:**2651 (1994); A. B. Charette, S. Prescott, and C. Brochu, *J. Org. Chem.* **60:**1081 (1995).
115. A. B. Charette and H. Lebel, *J. Am. Chem. Soc.* **118:**10327 (1996).
116. W. von E. Doering and W. R. Roth, *Tetrahedron* **19:**715 (1963); J. P. Chesick, *J. Am. Chem. Soc.* **84:**3250 (1962); H. Nozaki, H. Takaya, S. Moriuti, and R. Noyori, *Tetrahedron* **24:**3655 (1968); R. G. Salomon and J. K. Kochi, *J. Am. Chem. Soc.* **95:**3300 (1973); M. E. Alonso, P. Jano, and M. I. Hernandez, *J. Org. Chem.* **45:**5299 (1980); T. Hudlicky, F. J. Koszyk, T. M. Kutchan, and J. P. Sheth, *J. Org. Chem.* **45:**5020 (1980); M. P. Doyle and M. L. Truell, *J. Org. Chem.* **49:**1196 (1984); E. Y. Chen, *J. Org. Chem.* **49:**3245 (1984).
117. R. T. Lewis and W. B. Motherwell, *Tetrahedron Lett.* **29:**5033 (1988).
118. E. J. Corey and A. G. Myers, *Tetrahedron Lett.* **25:**3559 (1984); J. D. Winkler and E. Gretler, *Tetrahedron Lett.* **32:**5733 (1991).
119. S. F. Martin, R. E. Austin, and C. J. Oalmann, *Tetrahedron Lett.* **31:**4731 (1990).
120. S. Bien and Y. Segal, *J. Org. Chem.* **42:**1685 (1977); A. J. Anciaux, A. J. Hubert, A. F. Noels, N. Petiniot, and P. Teyssie, *J. Org. Chem.* **45:**695 (1980); M. P. Doyle, W. H. Tamblyn, and V. Baghari, *J. Org. Chem.* **46:**5094 (1981); D. F. Taber and R. E. Ruckle, Jr., *J. Am. Chem. Soc.* **108:**7686 (1986).
121. R. Paulissen, A. J. Hubert, and P. Teyssie, *Tetrahedron Lett.* **1972:**1465; U. Mende, B. Radüchel, W. Skuballa, and H. Vorbrüggen, *Tetrahedron Lett.* **1975:**629; M. Suda, *Synthesis* **1981**:714; M. P. Doyle, L. C. Wang, and K.-L. Loh, *Tetrahedron Lett.* **25:**4087 (1984); L. Strekowski, M. Visnick, and M. A. Battiste, *J. Org. Chem.* **51:**4836 (1986).
122. M. P. Doyle and J. G. Davidson, *J. Org. Chem.* **45:**1538 (1980); M. P. Doyle, R. L. Dorow, W. E. Buhro, J. H. Tamblyn, and M. L. Trudell, *Organometallics* **3:**44 (1984).

Scheme 10.6. Chiral Copper Catalysts

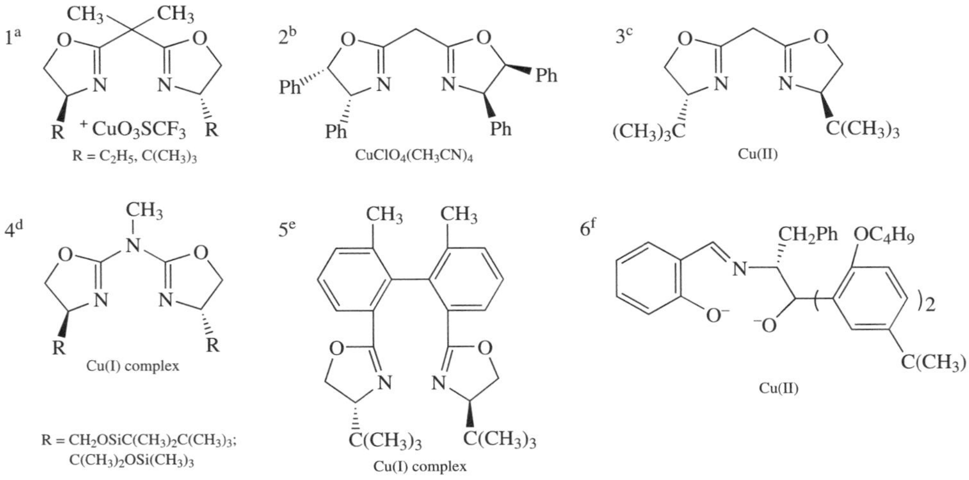

1[a] $^{+}CuO_3SCF_3$; R = C_2H_5, $C(CH_3)_3$

2[b] $CuClO_4(CH_3CN)_4$

3[c] Cu(II)

4[d] Cu(I) complex; R = $CH_2OSiC(CH_3)_2C(CH_3)_3$; $C(CH_3)_2OSi(CH_3)_3$

5[e] Cu(I) complex

6[f] Cu(II)

a. D. A. Evans, K. A. Woerpel, M. M. Hinman, and M. M. Faul *J. Am. Chem. Soc.* **113:**726 (1991); D. A. Evans, K. A. Woerpel, and M. I. Scott, *Angew. Chem. Int. Ed. Engl.* **31:**430 (1992).
b. R. E. Lowenthal and S. Masamune, *Tetrahedron Lett.* **32:**7373 (1991).
c. R. E. Lowenthal, A. Abiko, and S. Masamune, *Tetrahedron Lett.* **31:**6005 (1990).
d. A. Pfaltz, *Acc. Chem. Res.* **26:**339 (1993).
e. T. G. Tant, M. C. Noe, and E. J. Corey, *Tetrahedron Lett.* **36:**8745 (1995).
f. T. Aratani, Y. Yoneyoshi, and T. Nagase, *Tetrahedron Lett.* **1982:**685.

generally represented as below[123]:

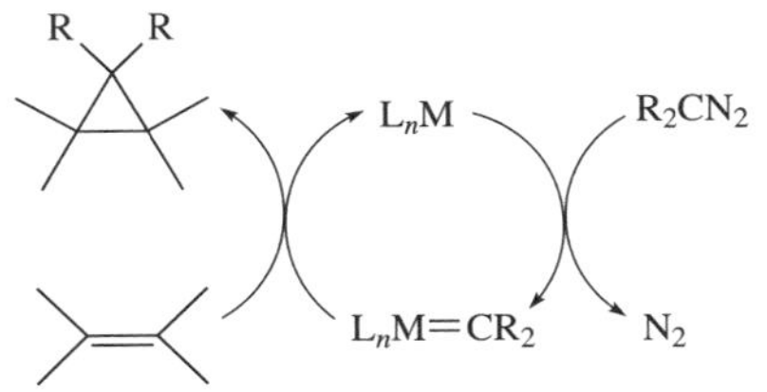

$$L_nM + R_2CN_2 \longrightarrow L_nM{=}CR_2 + N_2; \quad L_nM{=}CR_2 + \text{alkene} \longrightarrow \text{cyclopropane} + L_nM$$

The metal–carbene complexes are electrophilic in character. They can, in fact, be represented as metal-stabilized carbocations.

$$\ddot{M} + R_2\bar{C}{-}\overset{+}{N}{\equiv}N \xrightarrow{-N_2} \ddot{M}{-}C^{+}R_2 \longleftrightarrow M{=}CR_2$$

In most transition-metal-catalyzed reactions, one of the carbene substituents is a carbonyl group, which further enhances the electrophilicity of the intermediate. There are two general mechansism that can be considered for cyclopropane formation. One involves formation of a four-membered ring intermediate that incorporates the metal. The alternative represents an electrophilic attack giving a polar species which undergoes 1,3-bond

123. M. P. Doyle, *Chem. Rev.* **86**:919 (1986).

formation.

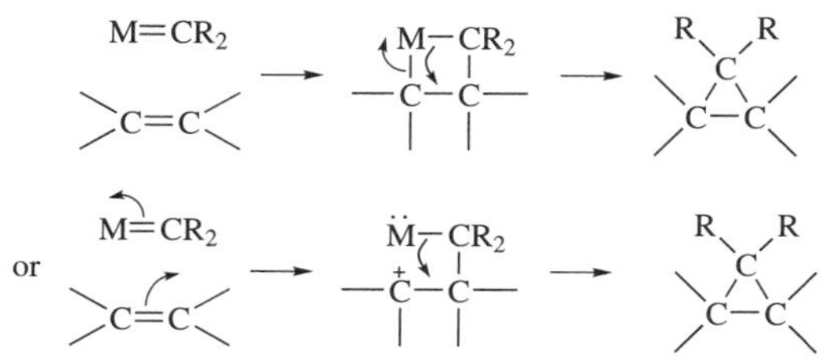

Because the additions are normally stereospecific with respect to the alkene, if an open-chain intermediate is involved, it must collapse to product at a rate more rapid than that of single-bond rotations, which would destroy the stereoselectivity. Entries 5–10 in Scheme 10.5 are examples of transition-metal-catalyzed carbene addition reactions.

In recent years, much attention has been focused on rhodium-mediated carbenoid reactions. The goal has been to understand how the rhodium ligands control reactivity and selectivity, especially in cases in which both addition and insertion reactions are possible. These catalysts contain Rh–Rh bonds but function by mechanisms similar to other transition-metal catalysts.

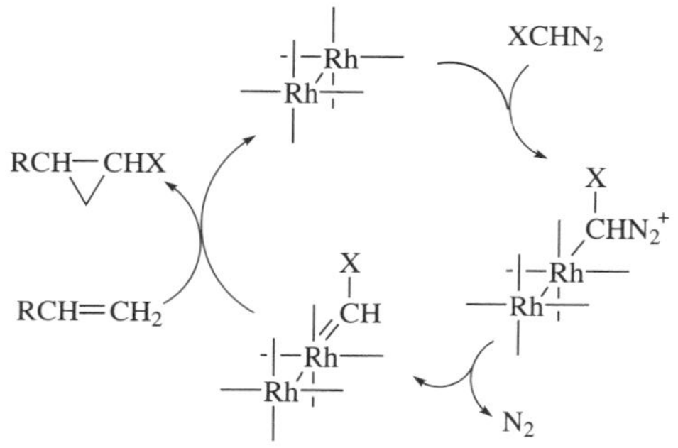

The original catalyst used was $Rh_2(O_2CCH_3)_4$, but other carboxylates such as nonafluorobutanoate and amide anions also have good catalytic activity.[124] The ligands adjust the electrophilicity of the catalyst, with the nonafluorobutanoate being more electrophilic and the amide ligands less electrophilic than the acetate. For example, $Rh_2(O_2C_4F_9)_4$ was found to favor aromatic substitution over cyclopropanation, whereas Rh_2(caprolactamate)$_4$

124. M. P. Doyle, V. Bagheri, T. J. Wandless, N. K. Harn, D. B. Brinker, C. T. Eagle, and K.-L. Loh, *J. Am. Chem. Soc.* **112**:1906 (1990).

was selective for cyclopropanation.[125]

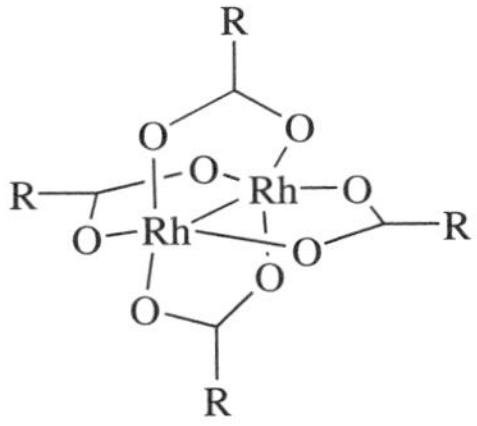

R = CH_3, $(CF_2)_3CF_3$

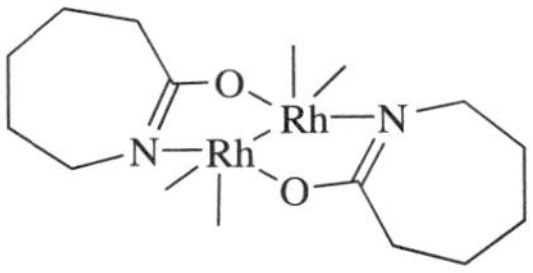

Rh_2(caprolactamate)$_4$ (two ligands not shown)

In competition between tertiary alkyl insertion versus cyclopropanation, the order favoring cyclopropanation is also Rh_2(caprolactamate)$_4$ > $Rh_2(O_2CCH_3)_4$ > $Rh_2(O_2CC_4F_9)_4$.

Various chiral amide ligands have been found to lead to enantioselective reactions.[126] For example, the lactamate of pyroglutamic acid gives enantioselective cyclopropanation reactions.

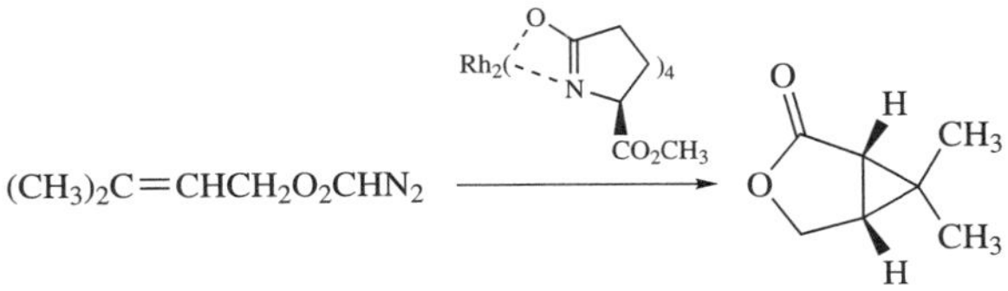

$(CH_3)_2C{=}CHCH_2O_2CHN_2$ → (catalyst Rh_2(lactamate with CO_2CH_3)$_4$) 82% yield, 92% e.e.

Various substituted analogs, some of which give improved results, have been described.

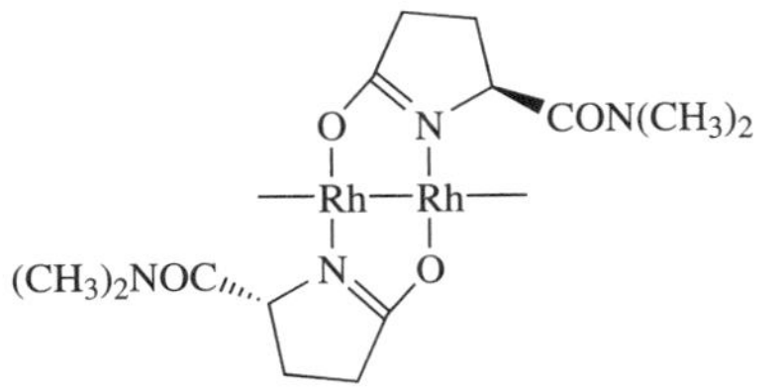

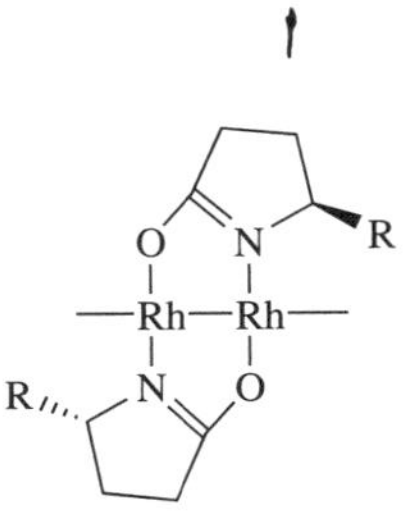

R = CH_3, Ph, CH_2Ph

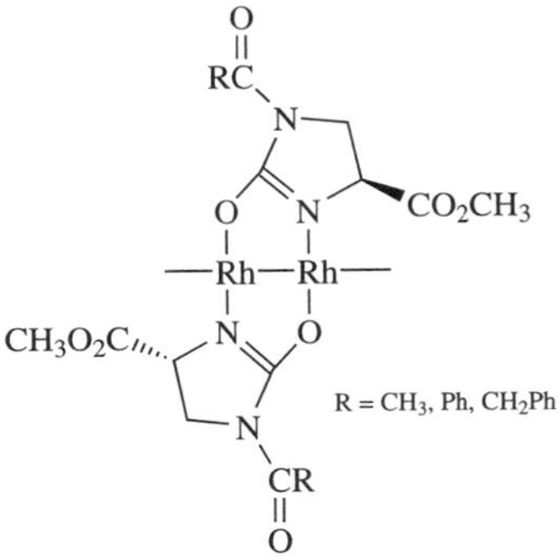

R = CH_3, Ph, CH_2Ph

Haloalkylmercury compounds are also useful in synthesis. The addition reactions are usually carried out by heating the organomercury compound with the alkene. Two typical examples are given in section C of Scheme 10.5.

125. A. Padwa, D. J. Austin, A. T. Price, M. A. Semones, M. P. Doyle, M. N. Protopova, W. R. Winchester, and A. Tran, *J. Am. Chem. Soc.* **115:**8669 (1993).

126. M. P. Doyle, R. E. Austin, A. S. Bailey, M. P. Dwyer, A. B. Dyatkin, A. V. Kalinin, M. M. Y. Kwan, S. Liras, C. J. Oalmann, R. J. Pieters, M. N. Protopopova, C. E. Raab, G. H. P. Roos, Q. L. Zhou, and S. F. Martin, *J. Am. Chem. Soc.* **117:**5763 (1995); M. P. Doyle, A. B. Dyatkin, M. N. Protopopova, C. I. Yang, G. S. Miertschin, W. R. Winchester, S. H. Simonsen, V. Lynch, and R. Ghosh, *Rec. Trav. Chim. Pays-Bas* **114:**163 (1995); M. P. Doyle, *Pure Appl. Chem.* **70:**1123 (1998); M. P. Doyle and M. N. Protopopova, *Tetrahedron* **54:**7919 (1998); M. P. Doyle and D. C. Forbes, *Chem. Rev.* **98:**911 (1998).

The addition of dichlorocarbene, generated from chloroform, to alkenes is a useful synthesis of dichlorocyclopropanes. The procedures based on lithiated halogen compounds have been less generally used in synthesis. Section D of Scheme 10.5 gives a few examples of addition reactions of carbenes generated by α elimination.

Intramolecular carbene addition reactions have a special importance in the synthesis of strained ring compounds. Because of the high reactivity of carbene or carbenoid species, the formation of highly strained bonds is possible. The strategy for synthesis is to construct a potential carbene precursor, such as diazo compounds or di- or trihalo compounds, which can undergo intramolecular addition to the desired structure. Section E of Scheme 10.5 gives some representative examples.

The high reactivity of carbenes is also essential to the addition reactions that occur with aromatic compounds.[127] The resulting adducts are in thermal equilibrium with the corresponding cycloheptatrienes. The position of the equilibrium depends on the nature of the substituent (see Section 11.1 of Part A).

+ CH_2N_2 $\xrightarrow{h\nu}$ Ref. 128

+ $(N{\equiv}C)_2CN_2$ $\xrightarrow{80°C}$ C≡N C≡N Ref. 129

+ $H_5C_2O_2CHN_2$ $\xrightarrow{\text{heat}}$ H $CO_2C_2H_5$ Ref. 130

10.2.4. Insertion Reactions

Insertion reactions are processes in which a reactive intermediate, in this case a carbene, interposes itself into an existing bond. In terms of synthesis, this usually involves C–H bonds. Many singlet carbenes are sufficiently reactive that this insertion can occur as a one-step process.

$$CH_3{-}CH_2{-}CH_3 + :CH_2 \longrightarrow CH_3{-}\underset{\displaystyle CH_3}{\underset{|}{CH}}{-}CH_3$$

127. E. Ciganek, *J. Am. Chem. Soc.* **93:**2207 (1971).
128. G. A. Russell and D. G. Hendry, *J. Org. Chem.* **28:**1933 (1963).
129. E. Ciganek, *J. Am. Chem. Soc.* **89:**1454 (1967).
130. J. E. Baldwin and R. A. Smith, *J. Am. Chem. Soc.* **89:**1886 (1967).

The same products can be formed by a two-step hydrogen abstraction and recombination involving a triplet carbene.

$$CH_3{-}CH_2{-}CH_3 + \cdot\dot{C}H_2 \longrightarrow CH_3{-}\underset{\cdot}{C}H{-}CH_3 + CH_3\cdot \longrightarrow CH_3{-}\underset{\substack{|\\ CH_3}}{CH}{-}CH_3$$

It is sometimes difficult to distinguish clearly between these mechanisms, but determination of reaction stereochemistry provides one approach. The one-step insertion must occur with complete *retention* of configuration. The results for the two-step process will depend on the rate of recombination in competition with stereorandomization of the radical intermediate.

Because of the very high reactivity of the intermediates that are involved, intermolecular carbene insertion reactions are not very selective. The distribution of products from the photolysis of diazomethane in heptane, for example, is almost exactly that which would be expected on a statistical basis.[131]

$$CH_3CH_2CH_2CH_2CH_2CH_2CH_3 \xrightarrow[h\nu]{CH_2N_2} \underset{38\%}{CH_3(CH_2)_6CH_3} + \underset{25\%}{CH_3\underset{\substack{|\\ CH_3}}{CH}(CH_2)_4CH_3}$$

$$+ \underset{24\%}{CH_3CH_2\underset{\substack{|\\ CH_3}}{CH}(CH_2)_3CH_3} + \underset{13\%}{(CH_3CH_2CH_2)_2CHCH_3}$$

There is some increase in selectivity with functionally substituted carbenes, but the selectivity is still not high enough to prevent formation of mixtures. Phenylchlorocarbene gives a relative reactivity ratio of 2.1 : 1 : 0.09 in insertion reactions with isopropylbenzene, ethylbenzene, and toluene.[132] For cycloalkanes, tertiary positions are about 15 times more reactive than secondary positions toward phenylchlorocarbene.[133] Carbethoxycarbene inserts at tertiary C–H bonds about three times as fast as at primary C–H bonds in simple alkanes.[134] Because of low selectivity, intermolecular insertion reactions are seldom useful in synthesis. Intramolecular insertion reaction are of considerably more use. Intramolecular insertion reactions usually occur at the C–H bond that is closest to the carbene, and good yields can frequently be obtained. Intramolecular insertion reactions can provide routes to highly strained structures that would be difficult to obtain in other ways.

Rhodium carboxylates have been found to be effective catalysts for intramolecular C–H insertion reactions of α-diazo ketones and esters.[135] In flexible systems, five-membered rings are formed in preference to six-membered ones. Insertion into a methine hydrogen is preferred to insertion into a methylene hydrogen. Intramolecular insertion can be competitive with intramolecular addition. Product preferences can to some extent be

131. D. B. Richardson, M. C. Simmons, and I. Dvoretzky, *J. Am. Chem. Soc.* **83:**1934 (1961).
132. M. P. Doyle, J. Taunton, S.-M. Oon, M. T. H. Liu, N. Soundararajan, M. S. Platz, and J. E. Jackson, *Tetrahedron Lett.* **29:**5863 (1988).
133. R. M. Moss and S. Yan, *Tetrahedron Lett.* **39:**9381 (1998).
134. W. v. E. Doering and L. H. Knox, *J. Am. Chem. Soc.* **83:**1989 (1961).
135. D. F. Taber and E. H. Petty, *J. Org. Chem.* **47:**4808 (1982); D. F. Taber and R. E. Ruckle, Jr., *J. Am. Chem. Soc.* **108:**7686 (1986).

controlled by the specific rhodium catalyst that is used.[136]

$Rh_2(X^-)_4$	Yield (%)	Ratio
$Rh_2(O_2CCH_3)_4$	99	67:33
$Rh_2(O_2CC_4F_9)_4$	95	0:100
Rh_2(caprolactamate)$_4$	72	100:0

The insertion reaction can be used to form lactones from α-diazo-β-ketoesters.

$CH_3COC(N_2)CO_2(CH_2)_7CH_3$ $\xrightarrow[80°C, benzene]{Rh_2(O_2CCH_3)_4}$ lactone, 92%

When the structure provides more than one kind of hydrogen for insertion, the catalyst can influence selectivity. For example, whereas $Rh_2(acam)_4$ (acam = acetamido) gives exclusively insertion at a tertiary position, $Rh_2(O_2CC_4F_9)_4$ leads to almost a statistical mixture.[137]

$CH_3COC(N_2)CO_2C[CH(CH_3)_2]_2CH_3$ →

$Rh_2(X^-)_4$	Ratio
$Rh_2(O_2CCH_3)_4$	90:10
$Rh_2(O_2CC_4F_9)_4$	39:61
$Rh_2(NHCOCH_3)_4$	>99:1

Stereoselectivity is also influenced by the catalysts. For example, **10** can lead to either *cis* or *trans* products. Whereas $Rh_2(O_2CCH_3)_4$ is unselective, the lactamate catalyst **E** is

136. A. Padwa and D. J. Austin, *Angew. Chem. Int. Ed. Engl.* **33:**1797 (1994).
137. M. P. Doyle, L. J. Westrum, W. N. E. Wolthuis, M. M. See, W. P. Boone, V. Bagheri, and M. M. Pearson, *J. Am. Chem. Soc.* **115:**958 (1993).

selective for the *cis* isomer and also gives excellent enantioselectivity in the major product.[138]

Certain sterically hindered rhodium catalysts also lead to improved selectivity. For example, rhodium triphenylacetate improves the selectivity for **11** over **12** from 5 : 1 to 99 : 1.[139]

Scheme 10.7 gives some additional examples of intramolecular insertion reactions.

10.2.5. Generation and Reactions of Ylides by Carbenoid Decomposition

Compounds in which a carbonyl or other nucleophilic functional group is close to a carbenoid carbon can react to give intermediates by intramolecular bonding.[140] One example is the formation of carbonyl ylides which go on to react by 1,3-dipolar addition.

138. M. P. Doyle, A. B. Dyatkin, G. H. P. Roos, F. Canas, D. A. Pierson, and A. van Basten, *J. Am. Chem. Soc.* **116:**4507 (1994).
139. S. Hashimoto, N. Watanabe, and S. Ikegami, *J. Chem. Soc., Chem. Commun.* **1992:**1508; S. Hashimoto, N. Watanabe, and S. Ikegami, *Tetrahedron Lett.* **33:**2709 (1992).
140. A. Padwa and S. F. Hornbuckle, *Chem. Rev.* **91:**263 (1991).

Scheme 10.7. Intramolecular Carbene-Insertion Reactions

1[a] H_3C, CH_3, CH_3, =$NNHSO_2Ar$ —MeO^-, diglyme→ H_3C, CH_3, H_3C 97%

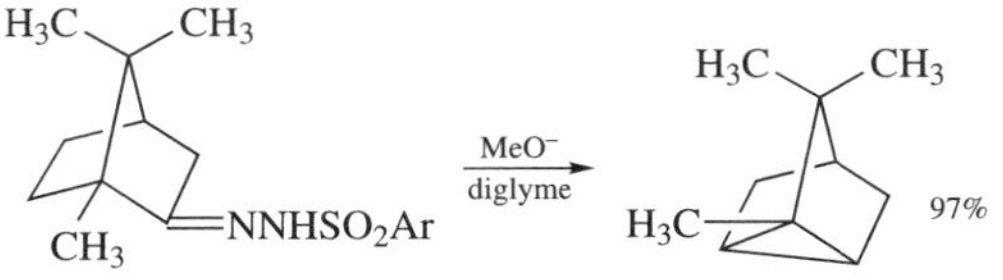

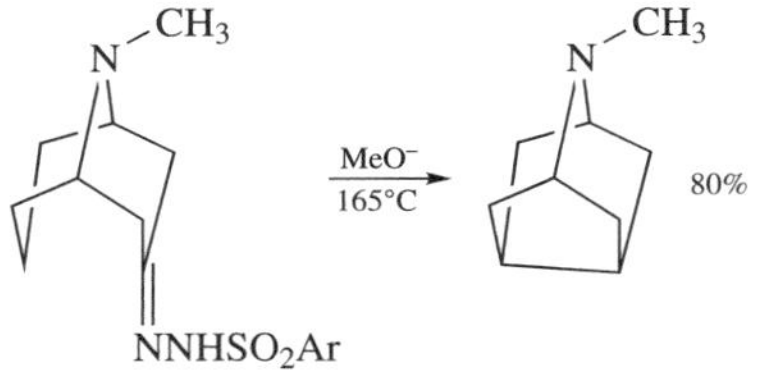

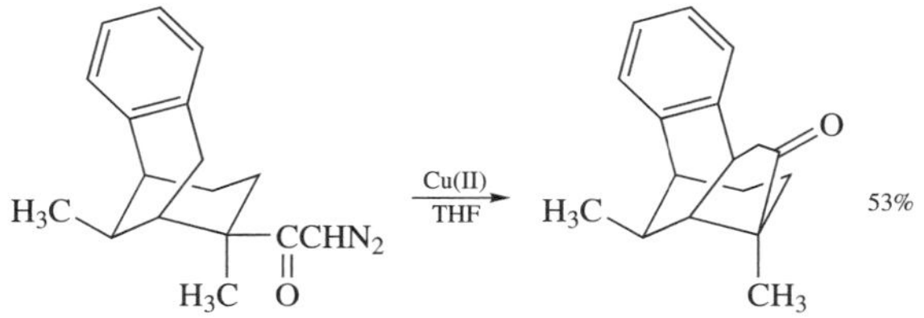

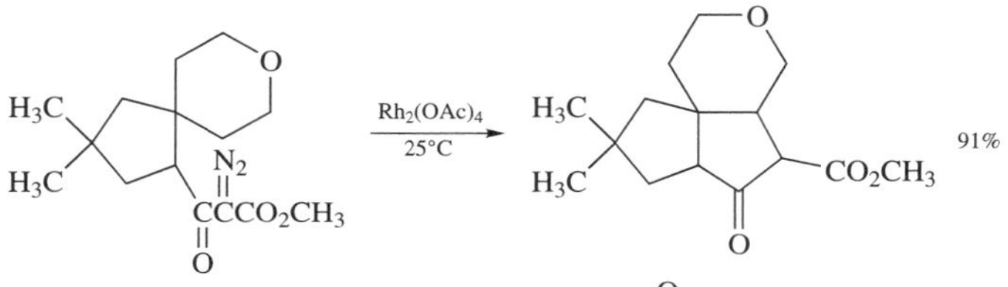

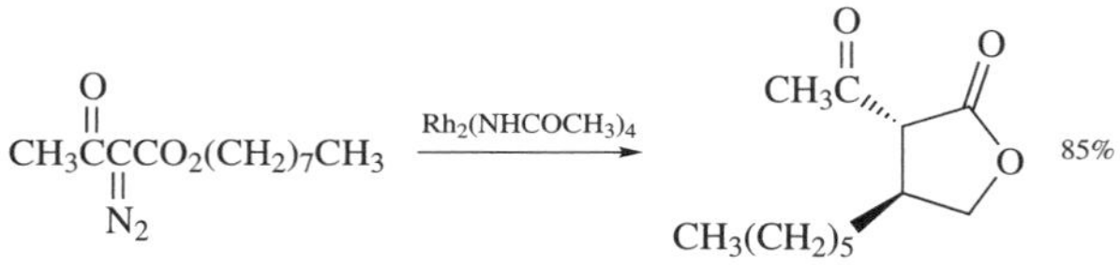

2[b] N–CH_3, $NNHSO_2Ar$ —MeO^-, 165°C→ N–CH_3 80%

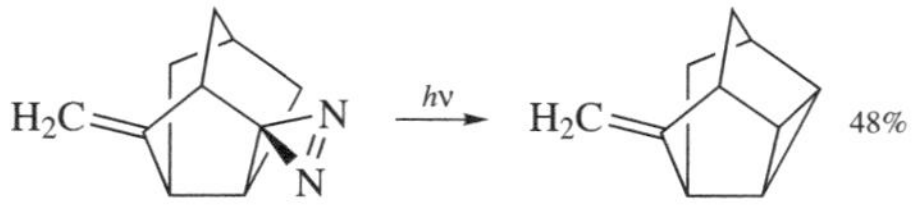

3[c] H_3C, $CCHN_2$ (C=O), H_3C —Cu(II), THF→ H_3C, O, CH_3 53%

4[d] H_3C, H_3C, O, N_2, $CCCO_2CH_3$ (C=O) —$Rh_2(OAc)_4$, 25°C→ H_3C, H_3C, O, CO_2CH_3, O 91%

5[e] $CH_3CCCO_2(CH_2)_7CH_3$ (O, N_2) —$Rh_2(NHCOCH_3)_4$→ CH_3C(=O), O, O, $CH_3(CH_2)_5$ 85%

6[f] H_2C=, N=N —$h\nu$→ H_2C= 48%

7[g] Br, Br, CH_3, H_3C —CH_3Li→ H_3C, H_3C 27%

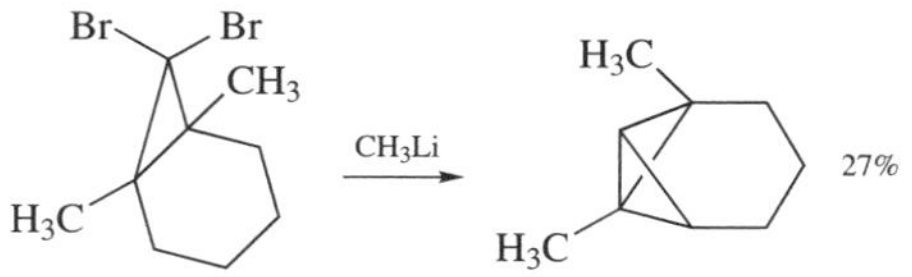

a. R. H. Shapiro, J. H. Duncan, and J. C. Clopton, *J. Am. Chem. Soc.* **89**:1442 (1967).
b. T. Sasaki, S. Eguchi, and T. Kiriyama, *J. Am. Chem. Soc.* **91**:212 (1969).
c. U. R. Ghatak and S. Chakrabarty, *J. Am. Chem. Soc.* **94**:4756 (1972).
d. D. F. Taber and J. L. Schuchardt, *J. Am. Chem. Soc.* **107**:5289 (1985).
e. M. P. Doyle, V. Bagheri, M. M. Pearson, and J. D. Edwards, *Tetrahedron Lett.* **30**:7001 (1989).
f. Z. Majerski, Z. Hamersak, and R. Sarac-Arneri, *J. Org. Chem.* **53**:5053 (1988).
g. L. A. Paquette, S. E. Wilson, R. P. Henzel, and G. R. Allen, Jr., *J. Am. Chem. Soc.* **94**:7761 (1972).

$Rh_2(O_2CCH_3)_4$ Ref. 141

$Rh_2(O_2CCH_3)_4$, $CH_3O_2CC{\equiv}CCO_2CH_3$ Ref. 140

$CH_2{=}CH(CH_2)_3C(O)(CH_2)_2C(O)CHN_2 \xrightarrow{Rh_2(O_2CCH_3)_4}$ Ref. 142

Allylic ethers and acetals can react with carbenoid reagents to generate oxonium ylides which undergo 2,3-sigmatropic shifts.[143]

$PhCH{=}CHCH_2OCH_3 + N_2CHC(O)Ph \xrightarrow{Rh_2(O_2CCH_3)_4} PhCH{=}CHCH_2O^+(CH_3)\,{}^-CHC(O)Ph \longrightarrow PhCH(CH{=}CH_2)CH(OCH_3)C(O)Ph$

10.2.6. Rearrangement Reactions

The most common rearrangement reaction of alkyl carbenes is the shift of hydrogen, generating an alkene. This mode of stabilization predominates to the exclusion of most intermolecular reactions of aliphatic carbenes and often competes with intramolecular insertion reactions. For example, the carbene generated by decomposition of the tosylhydrazone of 2-methylcyclohexanone gives mainly 1- and 3-methylcyclohexene rather than the intramolecular insertion product:

2-methylcyclohexanone $NNHSO_2Ar$ $\xrightarrow[180^\circ C]{NaOCH_3}$ 1-methylcyclohexene (38%) + 3-methylcyclohexene (16%) + bicyclo[4.1.0]heptane (trace) Ref. 144

141. A. Padwa, S. P. Carter, H. Nimmesgern, and P. D. Stull, *J. Am. Chem. Soc.* **110**:2894 (1988).
142. A. Padwa, S. F. Hornbuckle, G. E. Fryxell, and P. D. Stull, *J. Org. Chem.* **54**:819 (1989).
143. M. P. Doyle, V. Bagheri, and N. K. Harn, *Tetrahedron Lett.* **29**:5119 (1988).
144. J. W. Wilt and W. J. Wagner, *J. Org. Chem.* **29**:2788 (1964).

Carbenes can also be stabilized by migration of alkyl or aryl groups. 2-Methyl-2-phenyl-1-diazopropane provides a case in which both phenyl and methyl migration, as well as intramolecular insertion, are observed.

$$PhC(CH_3)_2CHN_2 \xrightarrow{60°C} (CH_3)_2C{=}CHPh \; (50\%) + PhC(CH_3){=}CHCH_3 \; (9\%) + Ph{-}C(CH_3)(CH_2CH_2) \; (41\%)$$

Ref. 145

Bicyclo[3.2.2]non-1-ene, a strained bridgehead alkene, is generated by rearrangement when bicyclo[2.2.2]octyldiazomethane is photolyzed.[146]

N_2CH $\xrightarrow{h\nu}$ $:CH$ $\longrightarrow$

Carbene centers adjacent to double bonds (vinyl carbenes) usually cyclize to cyclopropenes.[147]

$$CH_3CH_2(CH_3)C{=}C(CH_3)CH{=}NNHSO_2Ar \longrightarrow$$ H, CH_2CH_3, CH_3, CH_3

Ref. 148

Cyclopropylidenes undergo ring opening to give allenes. Reactions that would be expected to generate a cyclopropylidene therefore lead to allene, often in preparatively useful yields.

Ph, Ph, $N{=}O$, $NCNH_2$, O $\xrightarrow{LiOC_2H_5}$ [Ph, Ph] $\longrightarrow$ $Ph(H)C{=}C{=}C(H)Ph$ 79%

Ref. 149

H_3C, Cl, Cl, $CH_3CH_2CH_2$ $\xrightarrow{BuLi}$ $CH_3CH{=}C{=}CHCH_2CH_2CH_3$

Ref. 150

145. H. Philip and J. Keating, *Tetrahedron Lett.* **1961**:523.
146. M. S. Gudipati, J. G. Radziszewski, P. Kaszynski, and J. Michl, *J. Org. Chem.* **58**:3668 (1993).
147. G. L. Closs, L. E. Closs, and W. A Böll, *J. Am. Chem. Soc.* **85**:3796 (1963).
148. E. J. York, W. Dittmar, J. R. Stevenson, and R. G. Bergman, *J. Am. Chem. Soc.* **95**:5680 (1973).
149. W. M. Jones, J. W. Wilson, Jr., and F. B. Tutwiler, *J. Am. Chem. Soc.* **85**:3309 (1963).
150. W. R. Moore and H. R. Ward, *J. Org. Chem.* **25**:2073 (1960).

10.2.7. Related Reactions

There are several transformations that are conceptually related to carbene reactions but do not involve carbene, or even carbenoid, intermediates. Usually, these are reactions in which the generation of a carbene is circumvented by a concerted rearrangement process. An important example of this type of reaction is the thermal and photochemical reactions of α-diazoketones. When α-diazoketones are decomposed thermally or photochemically, they usually rearrange to ketenes. This reaction is known as the *Wolff rearrangement*.

$$\mathrm{R{-}C({=}O){-}\overset{-}{C}H{-}\overset{+}{N}{\equiv}N \longrightarrow O{=}C{=}CHR} \quad \text{concerted mechanism}$$

$$\mathrm{R{-}C({=}O){-}CH{=}\overset{+}{N}{=}\ddot{N}^{-} \longrightarrow R{-}C({=}O){-}\ddot{C}H \longrightarrow O{=}C{=}CHR} \quad \text{carbene mechanism}$$

$$\mathrm{R{-}C({=}O){-}\ddot{C}H \rightleftharpoons} \text{ oxirene}$$

If this reaction proceeds in a concerted fashion, a carbene intermediate is avoided. Mechanistic studies have been aimed at determining if migration is concerted with loss of nitrogen. The conclusion that has emerged is that a carbene is generated in photochemical reactions but that the reaction can be concerted under thermal conditions. A related issue is whether the carbene, when it is involved, is in equilibrium with a ring-closed isomer, an oxirene.[151] This aspect of the reaction has been probed by isotopic labeling. If a symmetrical oxirene is formed, the label should be distributed to both the carbonyl carbon and the α carbon. A concerted reaction or a carbene intermediate that did not equilibrate with the oxirene should have label only in the carbonyl carbon. The extent to which the oxirene is formed depends on the structure of the diazo compound. For diazoacetaldehyde, photolysis leads to only 8% migration of label, which would correspond to formation of 16% of the product through the oxirene.[152]

$$\mathrm{H{-}\overset{*}{C}({=}O){-}CHN_2 \longrightarrow H{-}\overset{*}{C}({=}O){-}\ddot{C}H \rightleftharpoons H\overset{*}{C}{=}CH\ (oxirene) \rightleftharpoons H\ddot{\overset{*}{C}}{-}C({=}O){-}H}$$

$$\mathrm{CH_2{=}\overset{*}{C}{=}O \xrightarrow{ROH} CH_3CO_2R \xleftarrow{ROH} {}^{*}CH_2{=}C{=}O}$$

8% 92%
distribution of label

151. M. Torres, E. M. Lown, H. E. Gunning, and O. P. Strausz, *Pure Appl. Chem.* **52:**1623 (1980); E. G. Lewars, *Chem. Rev.* **83:**519 (1983); A. P. Scott, R. H. Nobes, H. F. Schaeffer III, and L. Radom, *J. Am. Chem. Soc.* **116:**10159 (1994).
152. K.-P. Zeller, *Tetrahedron Lett.* **1977:**707.

The diphenyl analog shows about 20–30% rearrangement.[153] α-Diazocyclohexanone gives no evidence of an oxirene intermediate, since all the label remains at the carbonyl carbon.[154]

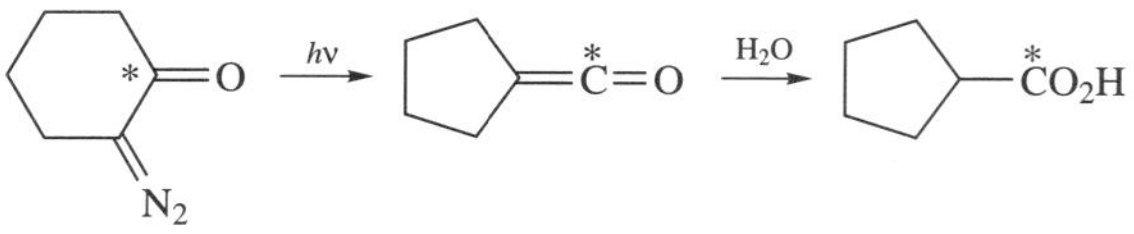

The main synthetic application of the Wolff rearrangement is for the one-carbon homologation of carboxylic acids.[155] In this procedure, a diazomethyl ketone is synthesized from an acyl chloride. The rearrangement is then carried out in a nucleophilic solvent which traps the ketene to form a carboxylic acid (in water) or an ester (in alcohols). Silver oxide is often used as a catalyst, because it seems to promote the rearrangement over carbene formation.[156]

The photolysis of cyclic α-diazoketones results in ring contraction to a ketene, which is usually isolated as the corresponding ester.

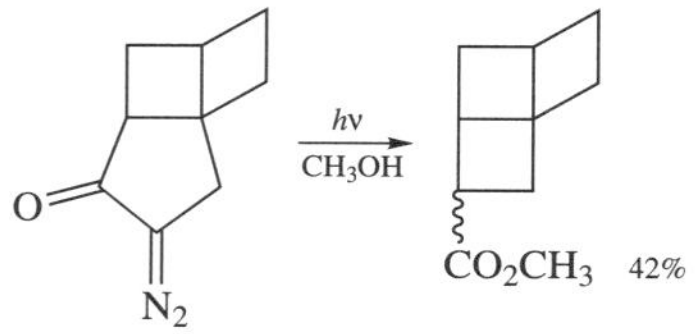

Ref. 157

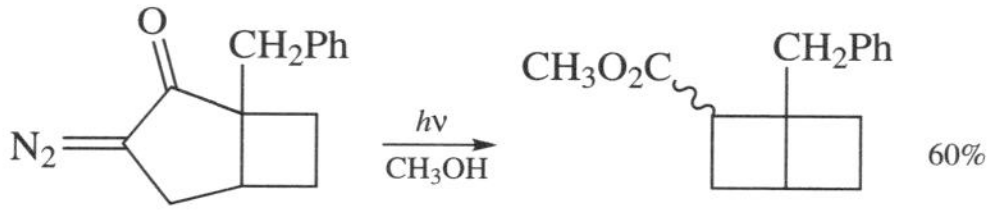

Ref. 158

Scheme 10.8 gives some other examples of Wolff rearrangement reactions.

10.2.8 Nitrenes and Related Intermediates

The nitrogen analogs of carbenes are called nitrenes. As with carbenes, both singlet and triplet electronic states are possible. The triplet state is usually the ground state for simple structures, but either species can be involved in reactions. The most common

153. K.-P. Zeller, H. Meier, H. Kolshorn, and E. Müller, *Chem. Ber.* **105:**1875 (1972).
154. U. Timm, K.-P. Zeller, and H. Meier, *Tetrahedron* **33:**453 (1977).
155. W. E. Bachmann and W. S. Stuve, *Org. React.* **1:**38 (1942); L. L. Rodina and I. K. Korobitsyna, *Russ. Chem. Rev.* (Engl. Transl.) **36:**260 (1967); W. Ando, in *Chemistry of Diazonium and Diazo Groups*, S. Patai, ed., John Wiley & Sons, New York, 1978, pp. 458–475; H. Meier and K.-P. Zeller, *Angew. Chem. Int. Ed. Engl.* **14:**32 (1975).
156. T. Hudlicky and J. P. Sheth, *Tetrahedron Lett.* **1979:**2667.
157. K. B. Wiberg, L. K. Olli, N. Golembeski, and R. D. Adams, *J. Am. Chem. Soc.* **102:**7467 (1980).
158. K. B. Wiberg, B. L. Furtek, and L. K. Olli, *J. Am. Chem. Soc.* **101:**7675 (1979).

Scheme 10.8. Wolff Rearrrangement of α-Diazoketones

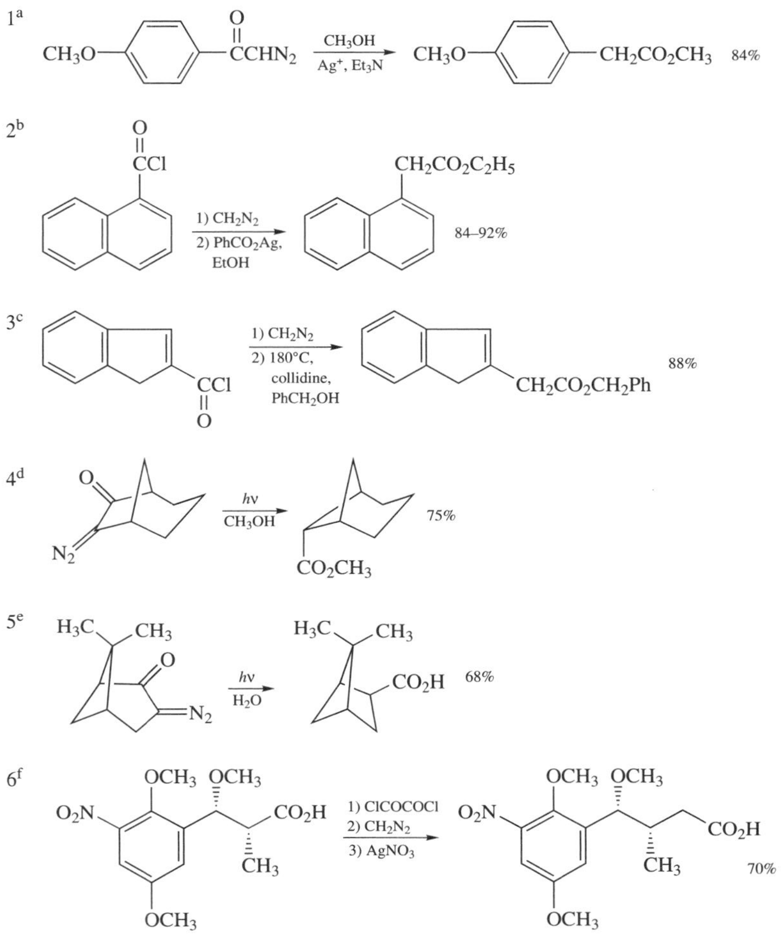

1[a] $CH_3O-C_6H_4-C(=O)CHN_2$ $\xrightarrow[\text{Ag}^+,\ Et_3N]{CH_3OH}$ $CH_3O-C_6H_4-CH_2CO_2CH_3$ 84%

2[b] 1-naphthyl-CCl(=O) $\xrightarrow[\text{2) PhCO}_2\text{Ag, EtOH}]{\text{1) CH}_2\text{N}_2}$ 1-naphthyl-$CH_2CO_2C_2H_5$ 84–92%

3[c] inden-3-yl-CCl(=O) $\xrightarrow[\text{2) 180°C, collidine, PhCH}_2\text{OH}]{\text{1) CH}_2\text{N}_2}$ inden-3-yl-$CH_2CO_2CH_2Ph$ 88%

4[d] $\xrightarrow[CH_3OH]{h\nu}$ CO_2CH_3 75%

5[e] $\xrightarrow[H_2O]{h\nu}$ CO_2H 68%

6[f] $\xrightarrow[\text{3) AgNO}_3]{\text{1) ClCOCOCl, 2) CH}_2\text{N}_2}$ 70%

a. M. S. Newman and P. F. Beal III, *J. Am. Chem. Soc.* **72**:5163 (1950).
b. V. Lee and M. S. Newman, *Org. Synth.* **50**:77 (1970).
c. E. D. Bergmann and E. Hoffmann, *J. Org. Chem.* **26**:3555 (1961).
d. K. B. Wilberg and B. A. Hess, Jr., *J. Org. Chem.* **31**:2250 (1966).
e. J. Meinwald and P. G. Gassman, *J. Am. Chem. Soc.* **82**:2857 (1960).
f. D. A. Evans, S. J. Miller, M. D. Ennis, and P. L. Ornstein *J. Org. Chem.* **57**:1067 (1992); D. A. Evans, S. J. Miller, and M. D. Ennis, *J. Org. Chem.* **58**:471 (1993).

method for generating nitrene intermediates is by thermolysis or photolysis of azides.[159] This method is analogous to formation of carbenes from diazo compounds.

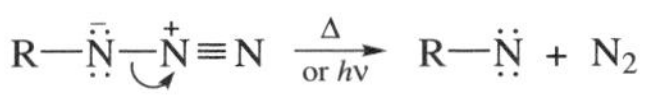

$$R-\ddot{N}^{-}-\overset{+}{N}\equiv N \xrightarrow[\text{or } h\nu]{\Delta} R-\ddot{N} + N_2$$

159. E. F. V. Scriven, ed. *Azides and Nitrenes; Reactivity and Utility*, Academic Press, Orlando, Florida, 1984.

The types of azides that have been used for generation of nitrenes include alkyl,[160] aryl,[161] acyl,[162] and sulfonyl[163] derivatives.

The characteristic reactions of an alkyl nitrene is migration of one of the substituents to nitrogen, giving an imine:

$$R_3C-\ddot{N}^{-}-\overset{+}{N}\equiv N \xrightarrow[\text{or } h\nu]{\Delta} R_2C=N-R$$

R = H or alkyl

Intramolecular insertion and addition reactions are almost unknown for alkyl nitrenes. In fact, it is not clear that the nitrenes are formed as discrete species. The migration may be concerted with elimination, as in the case of the thermal Wolff rearrangement.[164]

Aryl nitrenes also generally rearrange rather than undergo addition or insertion reactions.[165]

Nu = HNR_2, etc.

A few intramolecular insertion reactions, especially in aromatic systems, proceed in high yield.[166]

The nitrenes that most consistently give addition and insertion reactions are carbo-alkoxynitrenes generated from alkyl azidoformates.

$$RO-\overset{O}{\overset{\|}{C}}-N_3 \xrightarrow[\text{or } h\nu]{\Delta} RO-\overset{O}{\overset{\|}{C}}-N:$$

160. F. D. Lewis, and W. H. Saunders, Jr., in *Nitrenes*, W. Lwowski, ed., Interscience, New York, 1970, pp. 47–98; E. P. Kyba, in *Azides and Nitrenes*; *Reactivity and Utility*, E. F. V. Scriven, ed., Academic Press, Orlando, Florida, 1984, pp. 2–34.
161. P. A. Smith, in *Nitrenes*, W. Lwowski, ed., Interscience, New York, 1970, pp. 99–162; P. A. S. Smith, in *Azides and Nitrenes*: *Reactivity and Utility*, E. F. V. Scriven, editor, Academic Press, Orlando, Florida, 1984, pp. 95–204.
162. W. Lwowski, in *Nitrenes*, W. Lwowski, ed., Interscience, New York, 1970, pp. 185–224; W. Lwowski, in *Azides and Nitrenes*: *Reactivity and Utility*, E. F. V. Scriven, ed., Academic Press, Orlando, Florida, 1984, pp. 205–246.
163. D. S. Breslow, in *Nitrenes*, W. Lwowski, ed., Interscience, New York, 1970, pp. 245–303; R. A. Abramovitch and R. G. Sutherland, *Fortschr. Chem. Forsch.* **16:**1 (1970).
164. R. M. Moriarty and R. C. Reardon, *Tetrahedron* **26:**1379 (1970); R. A. Abramovitch and E. P. Kyba, *J. Am. Chem. Soc.* **93:**1537 (1971); R. M. Moriarty and P. Serridge, *J. Am. Chem. Soc.* **93:**1534 (1971).
165. O. L. Chapman and J.-P. LeRoux, *J. Am. Chem. Soc.* **100:**282 (1978); O. L. Chapman, R. S. Sheridan, and J.-P. LeRoux, *Rec. Trav. Chim. Pays-Bas* **98:**334 (1979); R. J. Sundberg, S. R. Suter, and M. Brenner, *J. Am. Chem. Soc.* **94:**573 (1972).
166. P. A. S. Smith and B. B. Brown, *J. Am. Chem. Soc.* **73:**2435, 2438 (1951); J. S. Swenton, T. J. Ikeler, and B. H. Williams, *J. Am. Chem. Soc.* **92:**3103 (1970).

These intermediates undergo addition reactions with alkenes and aromatic compounds and insertion reactions with saturated hydrocarbons.[167]

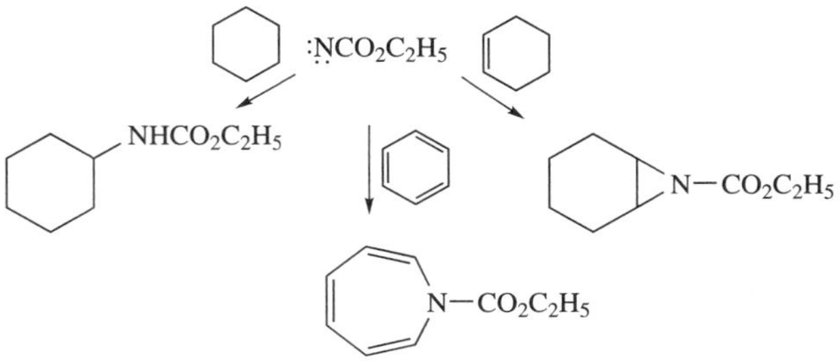

Carboalkoxynitrenes are somewhat more selective than the corresponding carbenes, showing selectivities of roughly 1 : 10 : 40 for the primary, secondary, and tertiary positions in 2-methylbutane in insertion reactions.

Sulfonylnitrenes are formed by thermal decomposition of sulfonyl azides. Insertion reactions occur with saturated hydrocarbons.[168] With aromatic rings, the main products are formally insertion products, but they are believed to be formed through addition intermediates.

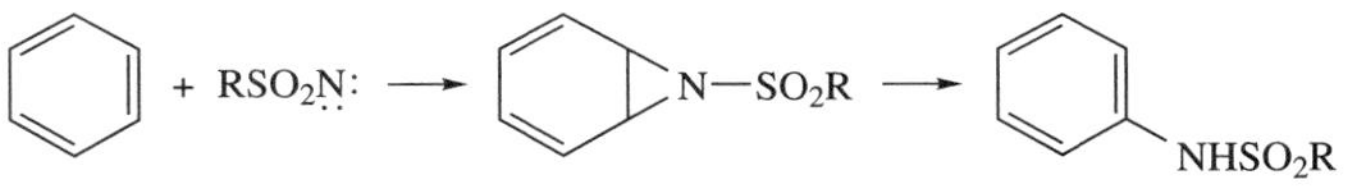

Ref. 169

Aziridination of alkenes can be carried out using *N-p*-toluenesulfonyliminophenyliodinane and copper triflate or other copper salts.[170] These reactions are mechanistically analogous to metal-catalyzed cyclopropanation. Rhodium acetate also acts as a catalyst.[171] Other arenesulfonyliminoiodinanes can be used,[172] as can chloramines T[173] and bromamine T.[174] The range of substituted alkenes which react includes acrylate esters.[175]

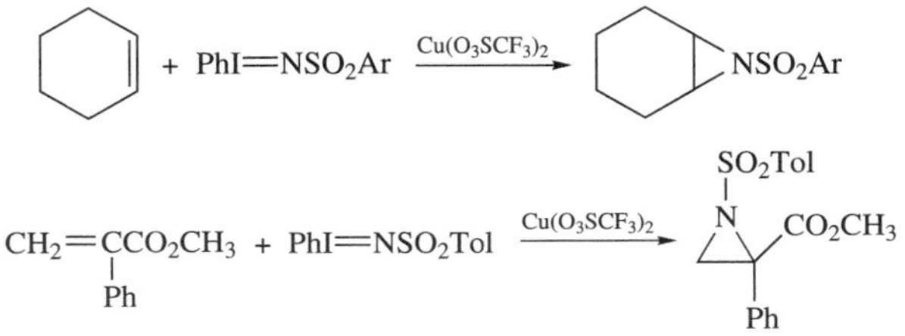

167. W. Lwowski, *Angew. Chem. Int. Ed. Engl.* **6:**897 (1967).
168. D. S. Breslow, M. F. Sloan, N. R. Newburg, and W. B. Renfrow, *J. Am. Chem. Soc.* **91:**2273 (1969).
169. R. A. Abramovitch, G. N. Knaus, and V. Uma, *J. Org. Chem.* **39:**1101 (1974).
170. D. A. Evans, M. M. Faul, and M. T. Bilodeau, *J. Am. Chem. Soc.* **116**:2742 (1994).
171. P. Müller, C. Baud, and Y. Jacquier, *Tetrahedron* **52:**1543 (1996).
172. M. J. Södergren, D. A. Alonso, and P. G. Andersson, *Tetrahedron Asymmetry* **8:**3563 (1991); M. J. Södergren, D. A. Alonso, A. V. Bedekar, and P. G. Andersson, *Tetrahedron Lett.* **38:**6897 (1997).
173. D. P. Albone, P. S. Aujla, P. C. Taylor, S. Challenger, and A. M. Derrick, *J. Org. Chem.* **63:**9569 (1998).
174. R. Vyas, B. M. Chanda, and A. V. Bedekar, *Tetrahedron Lett.* **39:**4715 (1998).
175. P. Dauban and R. H. Dodd, *Tetrahedron Lett.* **39:**5739 (1998).

10.2.9. Rearrrangements to Electron-Deficient Nitrogen

In contrast to the somewhat limited synthetic utility of nitrenes, there is an important group of reactions in which migration occurs to electron-deficient nitrogen. One of the most useful of these reactions is the *Curtius rearrangement*.[176] This reaction has the same relationship to acylnitrene intermediates that the Wolff rearrangement does to acylcarbenes. The initial product is an isocyanate, which can be isolated or trapped by a nucleophilic solvent.

$$\mathrm{RC(=O)N^-{-}N^+{\equiv}N} \longrightarrow [\mathrm{R\cdots C(=O){=}N{:}\cdots N{\equiv}N}] \longrightarrow \mathrm{O{=}C{=}N{-}R} \begin{cases} \xrightarrow{H_2O} [\mathrm{R{-}NH{-}C(=O){-}OH}] \longrightarrow \mathrm{RNH_2} + \mathrm{CO_2} \\ \xrightarrow{R'OH} \mathrm{R{-}NH{-}C(=O){-}OR'} \end{cases}$$

This reaction is considered to be a concerted process in which migration accompanies loss of nitrogen.[177] The migrating group retains its stereochemical configuration. The temperature required for reaction is in the vicinity of 100°C.

The acyl azide intermediates are prepared either by reaction of sodium azide with a reactive acylating agent or by diazotization of an acyl hydrazide. An especially convenient version of the former process is to treat the carboxylic acid with ethyl chloroformate to form a mixed anhydride, which then reacts with azide ion.[178]

$$\mathrm{RCO_2H} \xrightarrow{\mathrm{ClC(=O)OEt}} \mathrm{RC(=O)OC(=O)OEt} \xrightarrow{N_3^-} \mathrm{RC(=O)N_3}$$

$$\mathrm{RC(=O)NHNH_2} \xrightarrow[H^+]{\mathrm{NaNO_2}} \mathrm{RC(=O)N_3}$$

The reaction can also be carried out on the acid using diphenyl phosphoryl azide.[179]

$$\mathrm{RCO_2H} + \mathrm{(PhO)_2P(=O)N_3} \longrightarrow \mathrm{RC(=O)N_3} \xrightarrow{R'OH} \mathrm{RNHC(=O)OR'}$$

Some examples of the Curtius reaction are given in Scheme 10.9.

Another reaction that can be used for conversion of carboxylic acids to the corresponding amines with loss of carbon dioxide is the *Hofmann rearrangement*. The reagent is hypobromite ion, which reacts to form an *N*-bromoamide intermediate. Like the

176. P. A. S. Smith, *Org. React.* **3:**337 (1946).
177. S. Linke, G. T. Tisue, and W. Lwowski, *J. Am. Chem. Soc.* **89:**6308 (1967).
178. J. Weinstock, *J. Org. Chem.* **26:**3511 (1961).
179. D. Kim and S. M. Weinreb, *J. Org. Chem.* **43:**125 (1978).

A. Curtius rearrangement reactions

1[a] $CH_3(CH_2)_{10}CCl$ (C=O) → 1) NaN_3; 2) benzene, 70°C → $CH_3(CH_2)_{10}N{=}C{=}O$

2[b] $H_5C_2O_2C(CH_2)_4CO_2C_2H_5$ → 1) N_2H_4; 2) HNO_2; 3) Δ; 4) H^+, H_2O → $Cl^-\ H_3\overset{+}{N}(CH_2)_4\overset{+}{N}H_3\ Cl^-$

3[c] 1-phenylcyclopentanecarboxylic acid → 1) EtOCCl (C=O); 2) NaN_3; 3) heat; 4) H^+, H_2O → 1-phenylcyclopentylammonium chloride ($\overset{+}{N}H_3$ Cl^-) 76–81%

4[d] $(H_3C)(CH_3CH_2)(Ph)C{-}CO_2H$ → 1) $SOCl_2$, pyridine; 2) NaN_3, xylene → $(H_3C)(CH_3CH_2)(Ph)C{-}NH_2$ 66%

5[e] pyridine with CH_3O_2C, CH_3, Ph, CO_2H, CH_3O_2C substituents → 1) $(PhO)_2PN_3$ (P=O), 80°C; 2) MeOH → pyridine with CH_3O_2C, CH_3, Ph, $NHCO_2CH_3$, CH_3O_2C substituents 100%

B. Schmidt reactions

6[f] $PhCH_2CO_2H$ → NaN_3, polyphosphoric acid → $PhCH_2NH_2$

7[g] 3,5-di-tert-butylbenzoic acid (CO_2H; $(CH_3)_3C$, $C(CH_3)_3$) → H_2SO_4, NaN_3 → 3,5-di-tert-butylanilinium ($^+NH_3$) 93%

8[h] adamantanone → NaN_3, CF_3CO_2H → lactam (NH, C=O) 59%

C. Beckmann rearrangement reactions

9[i] $PhC(=NOH)C(CH_3)_3$ → HCl, CH_3CO_2H → $(CH_3)_3CC(=O)NH{-}Ph$ 94%

10[j] decalin with $H_3C{-}C{=}NOH$ at ring junction, H → p-toluenesulfonyl chloride, pyridine → decalin with $NHC(=O)CH_3$ at ring junction, H 92%

Scheme 10.9. *(continued)*

11[k] H_3C ... NOH —polyphosphoric acid→ H_3C ... NH 92%

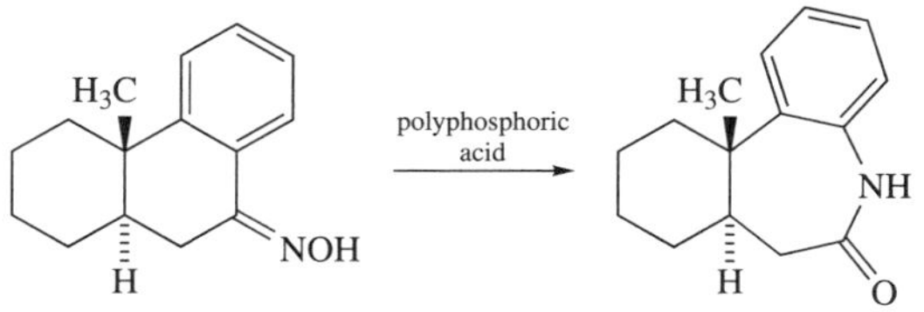

12[l] O =NOH —p-toluenesulfonyl chloride, pyridine→ O, O, NH 91%

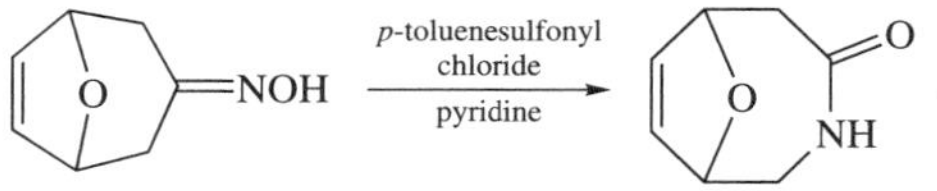

a. C. F. H. Allen and A. Bell, *Org. Synth.* **III**:846 (1955).
b. P. A. S. Smith, *Org. Synth.* **IV**:819 (1963).
c. C. Kaiser and J. Weinstock, *Org. Synth.* **51**:48 (1971).
d. D. J. Cram and J. S. Bradshaw, *J. Am. Chem. Soc.* **85**:1108 (1963).
e. D. Kim and S. M. Weinreb, *J. Org. Chem.* **43**:125 (1978).
f. R. M. Palmere and R. T. Conley, *J. Org. Chem.* **35**:2703 (1970).
g. J. W. Elder and R. P. Mariella, *Can. J. Chem.* **41**:1653 (1963).
h. T. Sasaki, S. Eguchi, and T. Toru, *J. Org. Chem.* **35**:4109 (1970).
i. R. F. Brown, N. M. van Gulick, and G. H. Schmid, *J. Am. Chem. Soc.* **77**:1094 (1955).
j. R. K. Hill and O. T. Chortyk, *J. Am. Chem. Soc.* **84**:1064 (1962).
k. R. A. Barnes and M. T. Beachem, *J. Am. Chem. Soc.* **77**:5388 (1955).
l. S. R. Wilson, R. A. Sawicki, and J. C. Huffman, *J. Org. Chem.* **46**:3887 (1981).

Curtius reaction, the rearrangement is believed to be a concerted process.

$$RC(=O)NH_2 + {}^-OBr \longrightarrow RC(=O)NHBr + {}^-OH \rightleftharpoons RC(=O)\bar{N}Br + H_2O$$

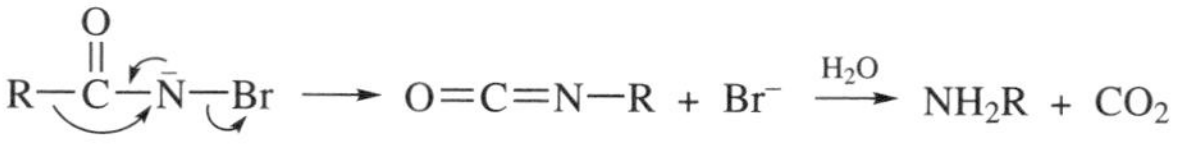

$$R-C(=O)-\bar{N}-Br \longrightarrow O{=}C{=}N-R + Br^- \xrightarrow{H_2O} NH_2R + CO_2$$

The reaction has been useful in the conversion of aromatic carboxylic acids to aromatic amines.

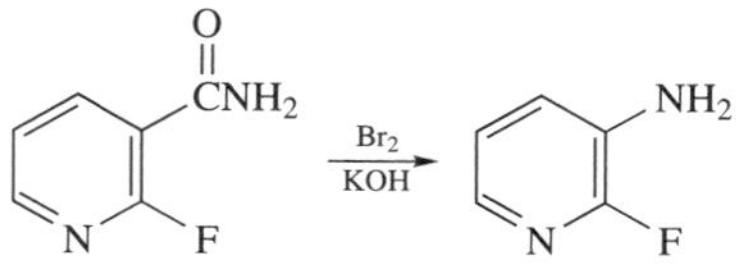

2-fluoro-3-pyridinecarboxamide (CNH_2 with C=O, N, F) $\xrightarrow[KOH]{Br_2}$ 3-amino-2-fluoropyridine (NH_2, N, F) Ref. 180

Use of *N*-bromosuccinimide in methanol in the presence of sodium methoxide or DBU as a base traps the isocyanate intermediate as a carbamate.[181]

$$RC(=O)NH_2 \xrightarrow[CH_3OH,\ NaOCH_3]{NBS} RNHCO_2CH_3$$

180. G. C. Finger, L. D. Starr, A. Roe, and W. J. Link, *J. Org. Chem.* **27**:3965 (1962).
181. X. Huang and J. W. Keillor, *Tetrahedron Lett.* **38**:313 (1997); X. Huang, M. Seid, and J. W. Keillor, *J. Org. Chem.* **62**:7495 (1997).

Direct oxidation of amides also can lead to Hofmann-type rearrangement with formation of amines or carbamates. One reagent that is used is $Pb(O_2CCH_3)_4$.

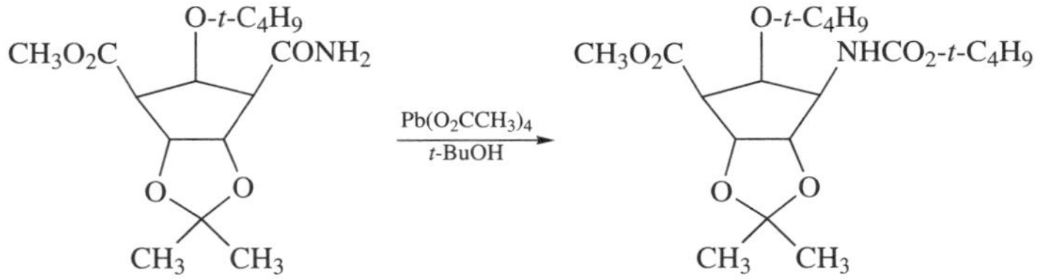

Ref. 182

Ref. 183

Phenyliodonium diacetate,[184] phenyliodonium trifluoroacetate,[185] and iodosobenzene diacetate[186] are also useful oxidants for amides.

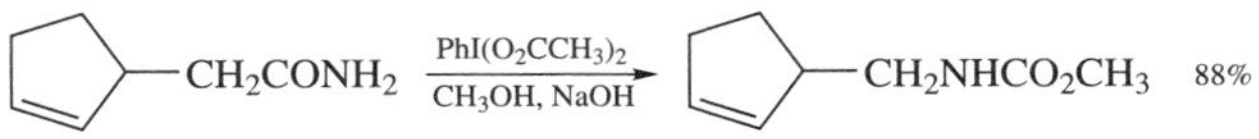

Carboxylic acids and esters can also be converted to amines with loss of the carbonyl group by reaction with hydrazoic acid, HN_3. This is known as the *Schmidt reaction*.[187] The mechanism is related to that of the Curtius reaction. An azido intermediate is generated by addition of hydrazoic acid to the carbonyl group. The migrating group retains its stereochemical configuration.

The reaction with hydrazoic acid converts ketones to amides.

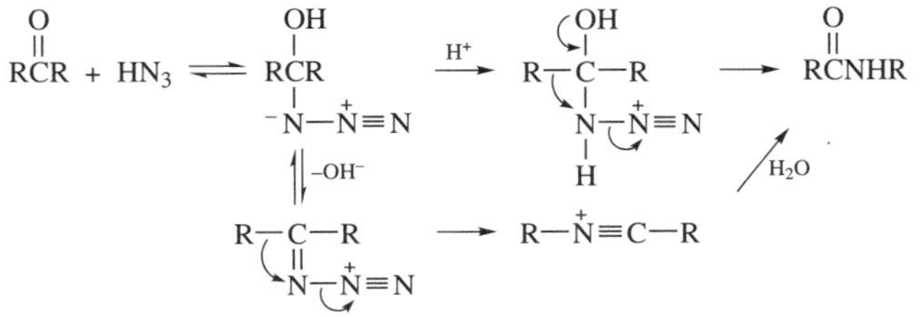

182. A. Ben Cheikh, L. E. Craine, S. G. Recher, and J. Zemlicka, *J. Org. Chem.* **53:**929 (1988).
183. R. W. Dugger, J. L. Ralbovsky, D. Bryant, J. Commander, S. S. Massett, N. S. Sage, and J. R. Selvidio, *Tetrahedron Lett.* **33:**6763 (1992).
184. R. M. Moriarty, C. J. Chany II, R. K. Vaid, O. Prakash, and S. M. Tuladhar, *J. Org. Chem.* **58:**2478 (1993).
185. G. M. Loudon, A. S. Radhakrishna, M. R. Almond, J. K. Blodgett, and R. H. Boutin, *J. Org. Chem.* **49:**4272 (1984).
186. L. Zhang, G. S. Kaufman, J. A. Pesti, and J. Yin, *J. Org. Chem.* **62:**6918 (1997).
187. H. Wolff, *Org. React.* **3:**307 (1946); P. A. S. Smith, in *Molecular Rearrangements*, P. de Mayo, ed., Vol. 1, Interscience, New York, 1963, pp. 507–522.

Unsymmetrical ketones can give mixtures of products because it is possible for either group to migrate.

$$\mathrm{RC(=O)R'} \xrightarrow{HN_3} \mathrm{RC(=O)NHR'} + \mathrm{RNHC(=O)R'}$$

Both inter- and intramolecular variants of the Schmidt reaction in which an alkyl azide effects overall insertion have been observed.

Cyclohexanone + PhN_3 $\xrightarrow{TiCl_4}$ N-phenyl caprolactam 80% Ref. 188

2-($(CH_2)_4N_3$)cyclohexanone $\xrightarrow{TiCl_4}$ bicyclic lactam 91% Ref. 189

These reactions are especially favorable for β- and γ-hydroxy azides which can proceed through a hemiketal intermediate.

Cyclopentanone + $HO(CH_2)_nN_3$ ⇌ HO, $O(CH_2)_nN_3$ hemiketal → $N-N_2^+$ → → lactam with N-$(CH_2)_nOH$

Ref. 190

Section B of Scheme 10.9 includes some examples of the Schmidt reaction.

Another important reaction involving migration to electron-deficient nitrogen is the *Beckmann rearrangement*, in which oximes are converted to amides:[191]

$$\mathrm{R{-}C({=}N{-}OH){-}R'} \longrightarrow \mathrm{R{-}NH{-}C({=}O){-}R'}$$

A variety of protic acids, Lewis acids, acid anhydrides, and acyl halides can cause the reaction to occur. The mechanism involves conversion of the oxime hydroxyl group to a leaving group. Ionization and migration then occur as a concerted process, with the group that is *anti* to the oxime leaving group migrating. The migration results in formation of a

188. J. Aube, G. L. Milligan and C. J. Mossman, *J. Org. Chem.* **57:**1635 (1992).
189. J. Aube and G. L. Milligan, *J. Am. Chem. Soc.* **113:**8965 (1991).
190. V. Gracias, K. E. Frank, G. L. Milligan, and J. Aube, *Tetrahedron* **53:**16241 (1997).
191. L. G. Donaruma and W. Z. Heldt, *Org. React.* **11:**1 (1960); P. A. S. Smith, *Open Chain Nitrogen Compounds*, Vol. II, W. A. Benjamin, New York, 1966, pp. 47–54; P. A. S. Smith, in *Molecular Rearrangements*, Vol. 1, P. de Mayo, (ed.), Interscience, New York, 1963, pp. 483–507; G. R. Krow, *Tetrahedron* **37:**1283 (1981); R. E. Gawley, *Org. React.* **35:**1 (1988).

nitrilium ion, which captures a nucleophile. Eventually, hydrolysis leads to the amide:

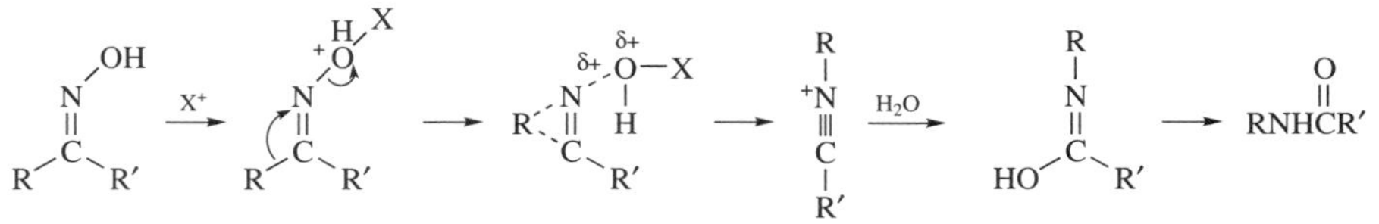

The migrating group retains its configuration. Some reaction conditions can lead to *syn–anti* isomerization occurring at a rate exceeding that of rearrangement. When this occurs, a mixture of products will be formed. The reagents that have been found least likely to cause competing isomerization are phosphorus pentachloride and *p*-toluenesulfonyl chloride.[192]

A fragmentation reaction occurs if one of the oxime substituents can give rise to a relatively stable carbocation. Fragmentation is very likely to occur if X is a nitrogen, oxygen, or sulfur atom.

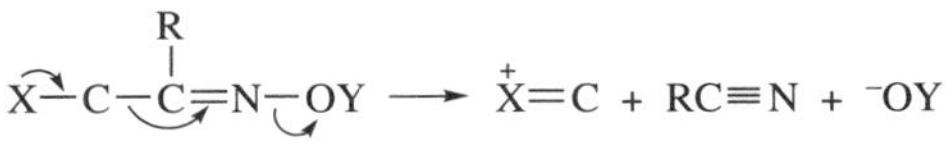

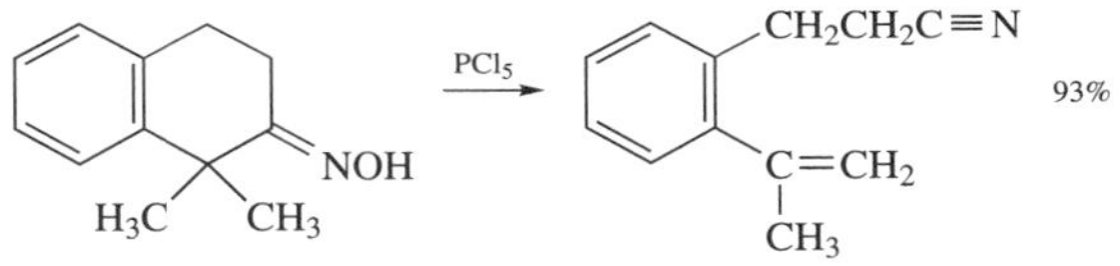

Ref. 193

Scheme 10.9 provides some examples of the Beckmann rearrangement.

10.3. Reactions Involving Free-Radical Intermediates

The fundamental mechanisms of free-radical reactions were considered in Chapter 12 of Part A. Several mechanistic issues are crucial in development of free-radical reactions for synthetic applications.[194] Successful free-radical reactions are usually chain processes. The lifetimes of the intermediate radicals are very short. To meet the synthetic requirements of high selectivity and efficiency, all steps in a desired process must be fast in comparison with competing reactions. Because of the requirement that all steps be quite fast, only steps that are exothermic or very slightly endothermic can participate in chain processes. Comparison of two sets of radical processes can illustrate this point. Let us compare addition of a radical to a carbon–carbon double bond with addition to a carbonyl group:

192. R. F. Brown, N. M. van Gulick, and G. H. Schmid, *J. Am. Chem. Soc.* **77**:1094 (1955); J. C. Craig and A. R. Naik, *J. Am. Chem. Soc.* **84**:3410 (1962).
193. R. T. Conley and R. J. Lange, *J. Org. Chem.* **28**:210 (1963).
194. C. Walling, *Tetrahedron* **41**:3887 (1985).

$$-\overset{|}{\underset{|}{C}}\cdot + {>}C{=}C{<} \longrightarrow -\overset{|}{\underset{|}{C}}-\overset{|}{\underset{|}{C}}-C\cdot{<}$$

$$\Delta H = (C{-}C) - (C^{\pi}{-}C^{\pi}) = -81 - (-64) = -17$$

$$-\overset{|}{\underset{|}{C}}\cdot + {>}C{=}O \longrightarrow -\overset{|}{\underset{|}{C}}-\overset{|}{\underset{|}{C}}-O\cdot$$

$$\Delta H = (C{-}C) - (C^{\pi}{-}O^{\pi}) = -81 - (-94) = +13$$

This comparison suggests that of these two similar reactions, only the former is likely to be a part of an efficient radical-chain reaction. Addition to a carbonyl group, in contrast, is endothermic. Radical additions to carbon–carbon double bonds can be further facilitated by radical-stabilizing groups. A similar comparison can be made for abstraction of hydrogen from carbon as opposed to oxygen:

$$-\overset{|}{\underset{|}{C}}\cdot + H-\overset{|}{\underset{|}{C}}- \longrightarrow -\overset{|}{\underset{|}{C}}-H + \cdot\overset{|}{\underset{|}{C}}-$$

$$\Delta H = 0$$

$$-\overset{|}{\underset{|}{C}}\cdot + H-O-\overset{|}{\underset{|}{C}}- \longrightarrow -\overset{|}{\underset{|}{C}}-H + \cdot O-\overset{|}{\underset{|}{C}}-$$

$$\Delta H = (C{-}H) - (O{-}H) = -98 - (-109) = +11$$

The reaction endothermicity establishes a *minimum* for the activation energy, and while abstraction of a hydrogen atom from carbon may be a feasible step in a chain process, abstraction of a hydrogen atom from a hydroxyl group is less favorable. Homolytic cleavage of an O–H bond is likely only if the resulting oxygen radical is highly stabilized in some way.

10.3.1. Sources of Radical Intermediates

A discussion of some of the radical sources used for mechanistic studies was given in Section 12.1.4 of Part A. Some of the reactions discussed there, particularly the use of azo compounds and peroxides as reaction initiators, are also important in synthetic chemistry.

One of the most useful sources of free radicals in preparative chemistry is the reaction of halides with stannyl radical

$$\text{initiation} \quad In\cdot + R'_3SnH \longrightarrow R'_3Sn\cdot + In{-}H$$

$$\text{propagation} \quad R{-}X + R'_3Sn\cdot \longrightarrow R\times + R'_3Sn{-}X$$

$$R\cdot + X{=}Y \longrightarrow R{-}X{-}Y\cdot$$

$$R{-}X{-}Y\cdot + R'_3Sn{-}H \longrightarrow R{-}X{-}Y{-}H + R'_3Sn\cdot$$

This generalized reaction sequence consumes the halide, the stannane, and the reactant X=Y and effects addition of the organic radical and a hydrogen atom to the X=Y bond. The order of reactivity of organic halides toward stannyl radicals is iodides > bromides > chlorides.

The esters of *N*-hydroxypyridine-2-thione are another versatile source of radicals.[195] The radical is formed by decarboxylation of an adduct formed by attack at sulfur by the

195. D. Crich, *Aldrichimica Acta* **20:**35 (1987); D. H. R. Barton, *Aldrichimica Acta* **23:**3 (1990).

chain-carrying radical.[196] The generalized chain sequence is as follows:

$$R\cdot + X{-}Y \longrightarrow R{-}Y + X\cdot$$

When X–Y is $R_3Sn{-}H$, the net reaction is decarboxylation and reduction of the resulting alkyl radical. When X–Y is $Cl_3C{-}Cl$, the final product is a chloride.[197] Use of $Cl_3C{-}Br$ gives the corresponding bromide.[198]

$$\xrightarrow{CCl_4} R{-}Cl + \quad + CO_2$$

The precise reaction conditions for optimal yields depend upon the specific reagents, and both thermal[199] and photochemical[200] conditions for obtaining good yields have been developed.

Selenenyl groups can be abstracted from acyl selenides to generate radicals on reaction with stannyl radicals.[201] Normally, some type of stabilization of the potential reaction site is necessary. Among the types of selenides that are generated by selenenyl abstraction are α-selenenyl cyanides[202] and α-selenenyl phosphates.[203]

Alkyl radicals can be generated from alkyl iodides in a chain process initiated by a trialkylborane and oxygen.[204] The alkyl radicals are generated by breakdown of a borane–oxygen adduct.

$$R_3B + O_2 \longrightarrow \cdot O{-}OBR_2 + R\cdot$$
$$R\cdot + O_2 \longrightarrow RO_2\cdot$$
$$RO_2\cdot + R_3B \longrightarrow RO_2BR_2 + R\cdot$$

196. D. H. R. Barton, D. Crich, and W. B. Motherwell, *Tetrahedron* **41:**3901 (1985); D. H. R. Barton, D. Crich, and G. Kretzschmar, *J. Chem. Soc., Trans. 1* **1986:**39 D. H. R. Barton, D. Bridson, I. Fernandez-Picot, and S. Z. Zard, *Tetrahedron* **43:**2733 (1987).
197. D. H. R. Barton, D. Crich, and W. B. Motherwell, *Tetrahedron Lett.* **24:**4979 (1983).
198. D. H. R. Barton, R. Lacher, and S. Z. Zard, *Tetrahedron Lett.* **26:**5939 (1985).
199. D. H. R. Barton, J. C. Jaszberenyi, and D. Tang, *Tetrahedron Lett.* **34:**3381 (1993).
200. J. Bouivin, E. Crepon, and S. Z. Zard, *Tetrahedron Lett.* **32:**199 (1991).
201. J. Pfenninger, C. Heuberger, and W. Graf, *Helv. Chim. Acta* **63:**2328 (1980); D. L. Boger and R. J. Mathvink, *J. Org. Chem.* **53:**3377 (1988); D. L. Boger and R. J. Mathvink, *J. Org. Chem.* **57:**1429 (1992).
202. D. L. J. Clive, T. L. B. Boivin, and A. G. Angoh, *J. Org. Chem.* **52:**4943 (1987).
203. P. Balczewski, W. M. Pietrzykowski, and M. Mikolajczyk, *Tetrahedron* **51:**7727 (1995).
204. H. C. Brown and M. M. Midland, *Angew. Chem. Int. Ed. Engl.* **11:**692 (1972); K. Nozaki, K. Oshima, and K. Utimoto, *Tetrahedron Lett.* **29:**1041 (1988).

For example, addition of alkyl radicals to alkynes can be accomplished under these conditions.

$$\text{C}_6\text{H}_{11}\text{–I} + \text{HC}\equiv\text{CSi(CH}_3)_3 \xrightarrow{(\text{C}_2\text{H}_5)_3\text{B}} \text{C}_6\text{H}_{11}\text{CH}=\text{C(I)Si(CH}_3)_3$$

Ref. 205

These reactions result in *iodine-atom transfer* and introduce a potential functional group into the product. This method of radical generation can also be used in conjunction with either tri-*n*-butylstannane or tris(trimethylsilyl)silane, in which case the reaction is terminated by hydrogen-atom transfer.

The reductive decomposition of alkylmercury compounds is also a useful source of radicals.[206] The organomercury compounds are available by oxymercuration (Section 4.3) or from an organometallic compound as a result of metal–metal exchange (Section 7.3.3). The mercuric hydride formed by reduction undergoes chain decomposition to generate alkyl radicals.

$$\text{RHgX} + \text{NaBH}_4 \longrightarrow \text{RHgH}$$
$$\text{RHgH} \longrightarrow \text{R}\cdot + \text{HgH}$$
$$\text{propagation} \quad \text{R}\cdot + \text{RHgH} \longrightarrow \text{R–H} + \text{RHg}$$
$$\text{RHg} \longrightarrow \text{R}\cdot + \text{Hg}^0$$

10.3.2. Introduction of Functionality by Radical Reactions

The introduction of halogen substituents by free-radical substitution was discussed in Section 12.3 of Part A. Halogenation is a fairly general method for functionalization, but the synthetic utility is dependent on the selectivity which can be achieved. Selective bromination of tertiary positions is usually possible. Halogenations at benzylic and allylic positions are also useful synthetic reactions. The high reactivity of free radicals toward molecular oxygen can also be exploited for the introduction of oxygen functionality. For example, when tri-*n*-butylstannane-mediated dehalogenation is done in aerated solution, the products are alcohols.

$$(\text{CH}_3)_2\text{CHCH(O}_2\text{CCH}_3)\text{CH}=\text{CHCH}_2\text{Br} \xrightarrow[\text{air, 0°C}]{n\text{-Bu}_3\text{SnH}} (\text{CH}_3)_2\text{CHCH(O}_2\text{CCH}_3)\text{CH}=\text{CHCH}_2\text{OH} \quad 71\%$$

Ref. 207

In this section, we will focus on intramolecular functionalization. Such reactions normally achieve selectivity on the basis of proximity of the reacting centers. In acylic molecules, intramolecular functionalization normally involves hydrogen-atom abstraction via a six-membered cyclic transition state. The net result is introduction of functionality at

205. Y. Ichinose, S. Matsunaga, K. Fugami, K. Oshima, and K. Utimoto, *Tetrahedron Lett.* **30**:3155 (1989).
206. G. A. Russell, *Acc. Chem. Res.* **22**:1 (1989).
207. E. Nakamura, T. Inubushi, S. Aoki, and D. Machii, *J. Am. Chem. Soc.* **113**:8980 (1991).

the δ atom in relation to the radical site:

$$\text{C-C-C-C(H)-C-C-X}\cdot \longrightarrow \text{C-C-C-}\dot{\text{C}}\text{-C-C-X-H} \xrightarrow{\text{Y-Z}} \text{C-C-C-C(Y)-C-C-X-H} + \text{Z}\cdot$$

One example of this type of reaction is the photolytically initiated decomposition of N-chloroamines in acidic solution, which is known as the *Hofmann–Löffler reaction*.[208] The initial products are δ-chloroamines, but these are usually converted to pyrrolidines by intramolecular nucleophilic substitution.

$$\text{RCH}_2\text{CH}_2\text{CH}_2\text{CH}_2\overset{+}{\text{N}}(\text{Cl})\text{HCH}_3 \xrightarrow{h\nu} \text{RCH}_2\text{CH}_2\text{CH}_2\text{CH}_2\overset{+}{\dot{\text{N}}}\text{HCH}_3 + \text{Cl}\cdot$$

$$\text{RCH}_2\text{CH}_2\text{CH}_2\text{CH}_2\overset{+}{\dot{\text{N}}}\text{HCH}_3 \longrightarrow \text{R}\dot{\text{C}}\text{HCH}_2\text{CH}_2\text{CH}_2\overset{+}{\text{N}}\text{H}_2\text{CH}_3$$

$$\text{R}\dot{\text{C}}\text{HCH}_2\text{CH}_2\text{CH}_2\overset{+}{\text{N}}\text{H}_2\text{CH}_3 + \text{RCH}_2\text{CH}_2\text{CH}_2\text{CH}_2\overset{+}{\text{N}}(\text{Cl})\text{HCH}_3 \longrightarrow$$

$$\text{RCH(Cl)CH}_2\text{CH}_2\text{CH}_2\overset{+}{\text{N}}\text{H}_2\text{CH}_3 + \text{RCH}_2\text{CH}_2\text{CH}_2\text{CH}_2\overset{+}{\dot{\text{N}}}\text{HCH}_3$$

$$\text{RCH(Cl)CH}_2\text{CH}_2\text{CH}_2\overset{+}{\text{N}}\text{H}_2\text{CH}_3 \xrightarrow{\text{NaOH}} \text{2-R-1-methylpyrrolidine (N-CH}_3\text{)}$$

A closely related procedure results in formation of γ-lactones. Amides are converted to N-iodoamides by reaction with iodine and t-butyl hypochlorite. Photolysis of the N-iodoamides gives lactones via iminolactone intermediates.[209]

$$\text{RCH}_2(\text{CH}_2)_2\overset{\text{O}}{\overset{\|}{\text{C}}}\text{NHI} \xrightarrow{h\nu} \text{RCH(I)(CH}_2)_2\overset{\text{O}}{\overset{\|}{\text{C}}}\text{NH}_2 \longrightarrow \text{5-R-2-iminotetrahydrofuran (=}\overset{+}{\text{N}}\text{H}_2\ \text{I}^-\text{)} \xrightarrow{\text{H}_2\text{O}} \text{5-R-}\gamma\text{-lactone (C=O)}$$

208. M. E. Wolff, *Chem. Rev.* **63**:55 (1963).
209. D. H. R. Barton, A. L. J. Beckwith, and A. Goosen, *J. Chem. Soc.* **1965**:181.

Steps similar to the Hofmann–Löffler reaction are also involved in cyclization of *N*-alkylmethanesulfonamides by oxidation with $Na_2S_2O_4$ in the presence of cupric ion.[210]

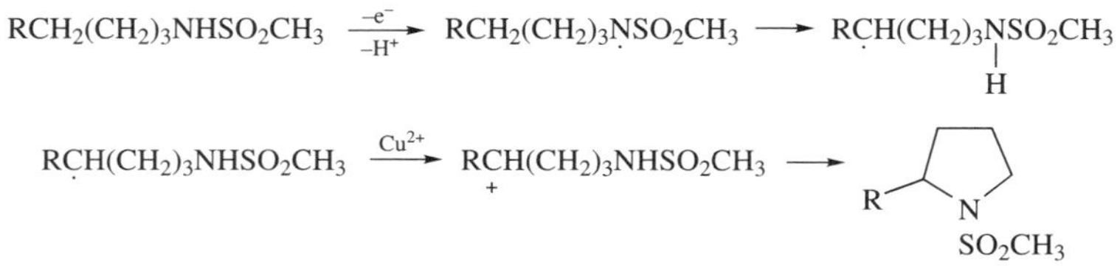

$$RCH_2(CH_2)_3NHSO_2CH_3 \xrightarrow[-H^+]{-e^-} RCH_2(CH_2)_3\dot{N}SO_2CH_3 \longrightarrow R\dot{C}H(CH_2)_3NSO_2CH_3 \text{ (N–H)}$$

$$R\dot{C}H(CH_2)_3NHSO_2CH_3 \xrightarrow{Cu^{2+}} R\overset{+}{C}H(CH_2)_3NHSO_2CH_3 \longrightarrow \text{2-R-1-}(SO_2CH_3)\text{pyrrolidine}$$

There are also useful intramolecular functionalization methods which involve hydrogen-atom abstraction by oxygen radicals. The conditions that were originally developed involved thermal or photochemical dissociation of alkoxy derivatives of Pb(IV) generated by exchange with $Pb(OAc)_4$.[211]

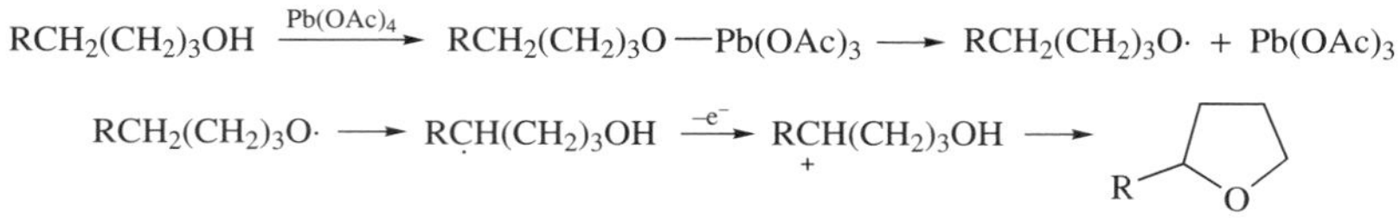

$$RCH_2(CH_2)_3OH \xrightarrow{Pb(OAc)_4} RCH_2(CH_2)_3O{-}Pb(OAc)_3 \longrightarrow RCH_2(CH_2)_3O\cdot + Pb(OAc)_3$$

$$RCH_2(CH_2)_3O\cdot \longrightarrow R\dot{C}H(CH_2)_3OH \xrightarrow{-e^-} R\overset{+}{C}H(CH_2)_3OH \longrightarrow \text{2-R-tetrahydrofuran}$$

The subsequent oxidation of the radical to a carbocation is effected by Pb(IV) or Pb(III). Current procedures include iodine and are believed to involve a hypoiodite intermediate.[212]

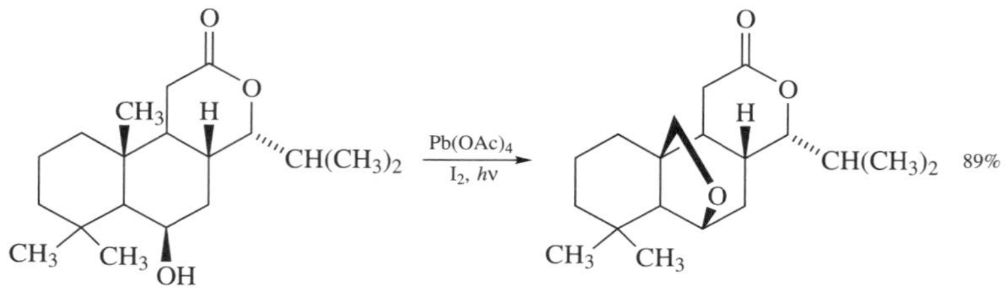

$\xrightarrow[I_2,\ h\nu]{Pb(OAc)_4}$ 89% Ref. 213

Alkoxy radicals are also the active hydrogen-abstracting species in a procedure which involves photolysis of nitrite esters. This reaction was originally developed as a method for functionalization of methyl groups in steroids.[214]

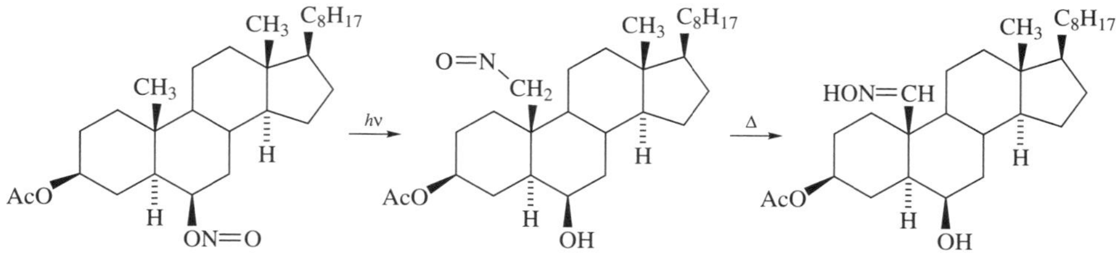

$\xrightarrow{h\nu}$ $\xrightarrow{\Delta}$

210. G. I. Nikishin, E. I. Troyansky, and M. Lazareva, *Tetrahedron Lett.* **26:**1877 (1985).
211. K. Heusler, *Tetrahedron Lett.* **1964:**3975.
212. K. Heusler, P. Wieland, and C. Meystre, *Org. Synth.* **V:**692 (1973); K. Heusler and J. Kalvoda, *Angew. Chem. Int. Ed. Engl.* **3:**525 (1964).
213. S. D. Burke, L. A. Silks III, and S. M. S. Strickland, *Tetrahedron Lett.* **29:**2761 (1988).
214. D. H. R. Barton, J. M. Beaton, L. E. Geller, and M. M. Pechet, *J. Am. Chem. Soc.* **83:**4076 (1961).

It has found other synthetic applications.

OH, $(CH_2)_3CH_3$ → 1) NOCl, 2) $h\nu$ → HO–N=, OH, $(CH_2)_3CH_3$ Ref. 215

10.3.3. Addition Reactions of Radicals to Substituted Alkenes

The most general method for formation of new carbon–carbon bonds via radical intermediates involves addition of the radical to an alkene. The addition reaction generates a new radical which can propagate a chain reaction. The preferred alkenes for trapping alkyl radicals are ethylene derivatives with electron-attracting groups, such as cyano, ester, or other carbonyl substituents.[216] There are two factors that make such compounds particularly useful: (1) alkyl radicals are relatively nucleophilic, and they react at enhanced rates with alkenes having electron-withdrawing substituents and (2) alkenes with such substituents exhibit a good degree of regioselectivity, resulting from a combination of steric and radical-stabilizing effects of the substituent. The "nucleophilic" versus "electrophilic" character of radicals can be understood in terms of the MO description of substituent effects on radicals. The three most important cases are outlined in Fig. 10.2.

Radicals for addition reactions can be generated by halogen-atom abstraction by stannyl radicals. The chain mechanism for alkylation of alkyl halides by reaction with a substituted alkene is outlined below. There are three reactions in the propagation cycle of this chain mechanism, shown in Fig. 10.3. The rates of each of these steps must exceed those of competing chain termination reactions in order for good yields to be obtained. The most important competitions are between the addition step k_1 and reaction of the intermediate R· with Bu_3SnH and between the H-abstraction step k_2 and addition to another molecule of the alkene. If the addition step k_1 is not fast enough, the radical R· will abstract H from the stannane, and the overall reaction will simply be dehalogenation. If

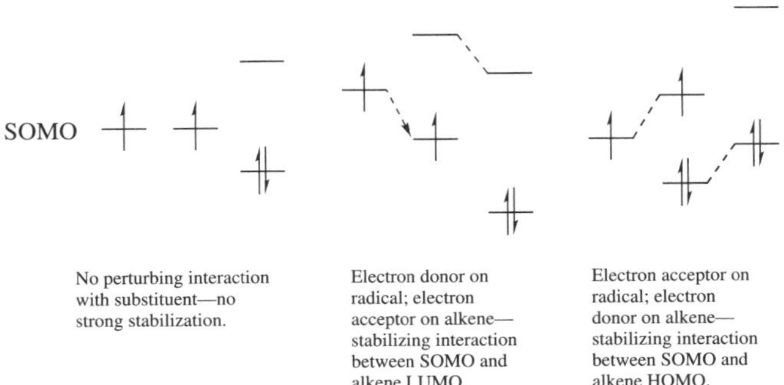

Fig. 10.2. Frontier orbital interpretation of radical substituent effects. SOMO, Singley occupied molecular orbitals.

215. E. J. Corey, J. F. Arnett, and G. N. Widiger, *J. Am. Chem. Soc.* **97**:430 (1975).
216. B. Giese, *Angew. Chem. Int. Ed. Engl.* **22**:753 (1983); B. Giese, *Angew. Chem. Int. Ed. Engl.* **24**:553 (1985).

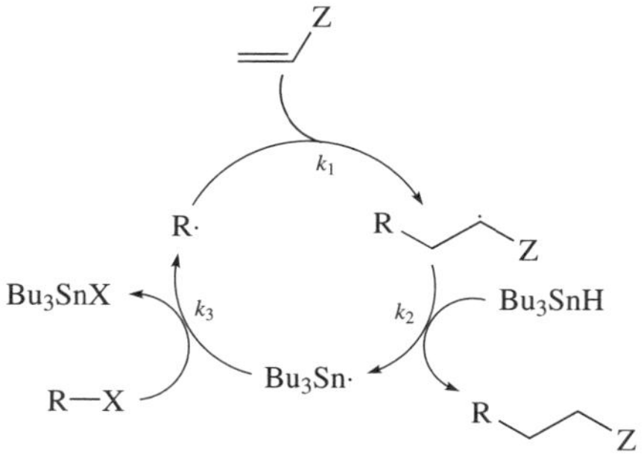

Fig. 10.3. Chain mechanism for radical addition reactions mediated by trialkylstannyl radicals.

step k_2 is not fast relative to a successive addition step, formation of oligomers containing several alkene units will occur. For good yields, R· must be more reactive toward the substituted alkene than is $RCH_2\dot{C}HZ$, and $RCH_2\dot{C}HZ$ must be more reactive toward Bu_3SnH than is R·. These requirements are met when Z is an electron-attracting group. Yields are also improved if the concentration of Bu_3SnH is kept low to minimize reductive dehalogenation. This can be done by adding the stannane slowly as the reaction proceeds. Another method is to use only a small amount of the trialkyltin hydride along with a reducing agent, such as $NaBH_4$ or $NaBH_3CN$, which can regenerate the reactive stannane.[217]

Radicals formed by fragmentation of xanthate and related thiono esters can also be trapped by reactive alkenes.[217] The mechanism of radical generation from thiono esters was discussed in connection with the Barton deoxygenation method in Section 5.4.

Although most radical reactions involving chain propagation by hydrogen-atom transfer have been done using trialkylstannanes, several silanes have been investigated as alternatives.[218] Tris(trimethylsilyl)silane reacts with alkyl radicals at a rate of about one-tenth of that at which tri-*n*-butylstannane reacts. The tris(trimethylsily)silyl radical is reactive toward iodides, sulfides, selenides, and thiono esters, permitting chain reaction. Thus, it is possible to substitute tris(trimethylsilyl)silane for tri-*n*-butylstannane in reactions such as dehalogenations, cyclizations, and radical additions.

$$CH_3(CH_2)_{15}I \xrightarrow[\text{AIBN}]{[(CH_3)_3Si]_3SiH} CH_3(CH_2)_{14}CH_3 \qquad \text{Ref. 219}$$

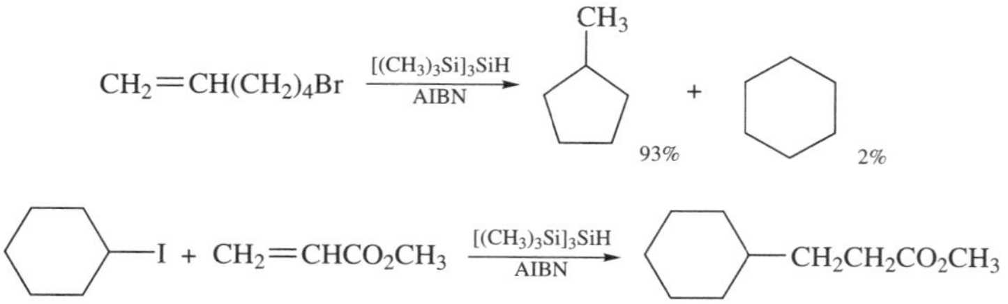

$$CH_2{=}CH(CH_2)_4Br \xrightarrow[\text{AIBN}]{[(CH_3)_3Si]_3SiH} \text{methylcyclopentane (93\%)} + \text{cyclohexane (2\%)} \qquad \text{Ref. 220}$$

$$\text{cyclohexyl–I} + CH_2{=}CHCO_2CH_3 \xrightarrow[\text{AIBN}]{[(CH_3)_3Si]_3SiH} \text{cyclohexyl–}CH_2CH_2CO_2CH_3 \qquad \text{Ref. 220}$$

A virtue of the silane donors is that they avoid the by-products of stannane reactions, which frequently cause purification problems.

217. B. Giese, J. A. Gonzalez-Gomez, and T. Witzel, *Angew. Chem. Int. Ed. Engl.* **23:**69 (1984).
218. C. Chatgilialoglu, *Acc. Chem. Res.* **25:**188 (1991).
219. C. Chatgilialoglu, A. Guerrini, and G. Seconi, *Synlett* **1990:**219.
220. B. Giese, B. Kopping, and C. Chatgilialoglu, *Tetrahedron Lett.* **30:**681 (1989).

Dialkyl phosphates and hypophosphorous acid (H_3PO_2) are also excellent hydrogen-atom donors. These compounds have weak H–P bonds which react with alkyl radicals.

$$\underset{\text{dialkyl phosphite}}{HP(=O)(OR)_2} \qquad \underset{\text{hypophosphorous acid}}{H_2P(=O)OH}$$

Reactions with thiono esters, iodides, bromides, and selenides proceed efficiently with dimethyl phosphite or with hypophosphorous acid in the presence of a tertiary amine and AIBN.[221]

$$R{-}X + HP(OCH_3)_2 \xrightarrow{\text{AIBN}} R{-}H$$

Organomercury compounds are also sources of alkyl radicals. Organomercurials can be prepared either by solvomercuration (Section 4.3) or from organometallic reagents (Section 7.3.3).

$$RCH{=}CH_2 + HgX_2 \xrightarrow{SH} RCH(S)CH_2HgX \qquad \text{(SH = solvent)}$$

$$RLi + HgX_2 \longrightarrow RHgX + LiX$$

Radicals are generated by reduction of the organomercurial with $NaBH_4$ or a similar reductant. These techniques have been applied to β-hydroxy,[222] β-alkoxy-[223] and β-amido-[224] alkylmercury derivatives.

Alkylmercury reagents can also be prepared from alkylboranes.

$$R_3B + 3\,Hg(OAc)_2 \longrightarrow 3\,RHgOAc \qquad \text{Ref. 225}$$

α-Acetoxyalkylmercury compounds can be prepared from hydrazones by reaction with mercuric oxide and mercuric acetate.

$$R_2C{=}O \xrightarrow{H_2NNH_2} R_2C{=}N{-}NH_2 \xrightarrow[HgO]{Hg(OAc)_2} R_2C(OAc){-}HgOAc \xrightarrow[CH_2=CHZ]{NaBH_4} R_2C(OAc)CH_2CH_2Z \qquad \text{Ref. 226}$$

There are also reactions in which electrophilic radicals react with relatively nucleophilic alkenes. These reactions are represented by a group of procedures in which a radical intermediate is formed by oxidation of the enol of a readily enolized compound. This reaction was initially developed for β-ketoacids.[227] The method has been extended to β-diketones, malonic acids, and cyanoacetic acid.[228]

$$HO_2CCH_2CN + Mn^{3+} \longrightarrow HO_2C\dot{C}HCN$$

$$HO_2C\dot{C}HCN + CH_2{=}CHR \longrightarrow HO_2CCH(CN)CH_2\dot{C}HR \xrightarrow{Mn^{3+}} HO_2CCH(CN)CH_2\overset{+}{C}HR \longrightarrow \text{(3-cyano-5-R-}\gamma\text{-butyrolactone)}$$

221. D. H. R. Barton, D. O. Jang, and J. C. Jaszberenyi, *J. Org. Chem.* **58:**6838 (1993).
222. A. P. Kozikowski, T. R. Nieduzak, and J. Scripko, *Organometallics* **1:**675 (1982).
223. B. Giese and K. Heuck, *Chem. Ber.* **112:**3759 (1979); B. Giese and U. Lüning, *Synthesis* **1982:**735.
224. A. P. Kozikowski and J. Scripko, *Tetrahedron Lett.* **24:**2051 (1983).
225. R. C. Larock and H. C. Brown, *J. Am. Chem. Soc.* **92:**2467 (1970).
226. B. Giese and U. Erfort, *Chem. Ber.* **116:**1240 (1983).
227. E. Heiba and R. M. Dessau, *J. Org. Chem.* **39:**3456 (1974).
228. E. J. Corey and M. C. Kang, *J. Am. Chem. Soc.* **106:**5384 (1984); E. J. Corey and A. W. Gross, *Tetrahedron Lett.* **26:**4291 (1985); W. E. Fristad and S. S. Hershberger, *J. Org. Chem.* **50:**1026 (1985).

The radicals formed by the addition step are rapidly oxidized to cations, which give rise to the final product by intramolecular capture of a nucleophilic carboxylate group.

Scheme 10.10 illustrates the addition reaction of radicals with alkenes using a variety of methods for radical generation.

Another class of compounds which undergo addition reactions with alkyl radicals are allylstannanes. The chain is propagated by elimination of the trialkylstannyl radical.[229]

$$R{-}X + Bu_3Sn\cdot \longrightarrow R\cdot + Bu_3SnX$$

$$R\times + CH_2{=}CHCH_2SnBu_3 \longrightarrow RCH_2\dot{C}HCH_2SnBu_3 \longrightarrow RCH_2CH{=}CH_2 + \cdot SnBu_3$$

$$\text{(cyclohexyl bromide)} \quad Bu_3SnCH_2CH{=}CH_2 \longrightarrow \text{(cyclohexyl)}{-}CH_2CH{=}CH_2$$

The radical source must have some functional group X that can be abstracted by trialkylstannyl radicals. In addition to halides, both thiono esters[230] and selenides[231] are reactive. Allyl tris(trimethylsilyl)silane can also react similarly.[232] Scheme 10.11 illustrates allylation by reaction of radical intermediates with allylstannanes.

10.3.4. Cyclization of Free-Radical Intermediates

Cyclization reactions of radical intermediates have become an important method for ring synthesis.[233] The key step in these procedures involves addition of a radical center to an unsaturated functional group. The radical formed by the cyclization must then give rise to a new radical that can propagate the chain. Many of these reactions involve halides as the source of the radical intermediate. The radicals are normally generated by halogen-atom abstraction using a trialkylstannane as the reagent and AIBN as the initiator. The cyclization step must be fast relative to hydrogen abstraction from the stannane. The chain is propagated when the cyclized radical abstracts hydrogen from the stannane.

$$In\cdot + Bu_3Sn{-}H \longrightarrow In{-}H + Bu_3Sn\cdot$$

$$Bu_3Sn\cdot + X{-}CH_2 \frown CH{=}CH_2 \longrightarrow Bu_3Sn{-}X + \cdot CH_2 \frown CH{=}CH_2 \longrightarrow CH_2{-}CH{-}CH_2\cdot$$

$$CH_2{-}CH{-}CH_2\cdot + Bu_3Sn{-}H \longrightarrow CH_2CH{-}CH_3 + Bu_3Sn\cdot$$

From a synthetic point of view, the regioselectivity and stereoselectivity of the cyclization are of paramount importance. As discussed in Section 12.2.2 of Part A, there is usually a preference for ring formation in the order $5 > 6 > 7$ because of stereoelectronic factors.[234] The other major influence on the direction of cyclization is the presence of substituents. Attack at a less hindered position is favored, both by steric effects and by the stabilizing effect that most substituents have on a radical center. For relatively rigid cyclic

229. G. E. Keck and J. B. Yates, *J. Am. Chem. Soc.* **104**:5829 (1982).
230. G. E. Keck, D. F. Kachensky, and E. J. Enholm, *J. Org. Chem.* **49**:1462 (1984).
231. R. R. Webb and S. Danishefsky, *Tetrahedron Lett.* **24**:1357 (1983); T. Toru, T. Okumura, and Y. Ueno, *J. Org. Chem.* **55**:1277 (1990).
232. C. Chatgilialoglu, C. Ferreri, M. Ballestri, and D. P. Curran, *Tetrahedron Lett.* **37**:6387 (1996).
233. D. P. Curran, *Synthesis* **1988**:417, 489; D. P. Curran, *Synlett* **1991**:63; C. P. Jasperse, D. P. Curran, and T. L. Fevig, *Chem. Rev.* **91**:1237 (1991).
234. A. L. J. Beckwith, C. J. Easton, T. Lawrence, and A. K. Serelis, *Aust. J. Chem.* **36**:545 (1983).

Scheme 10.10. Alkylation of Alkyl Radicals by Reaction with Alkenes

A. With radical generation using trisubstituted stannanes

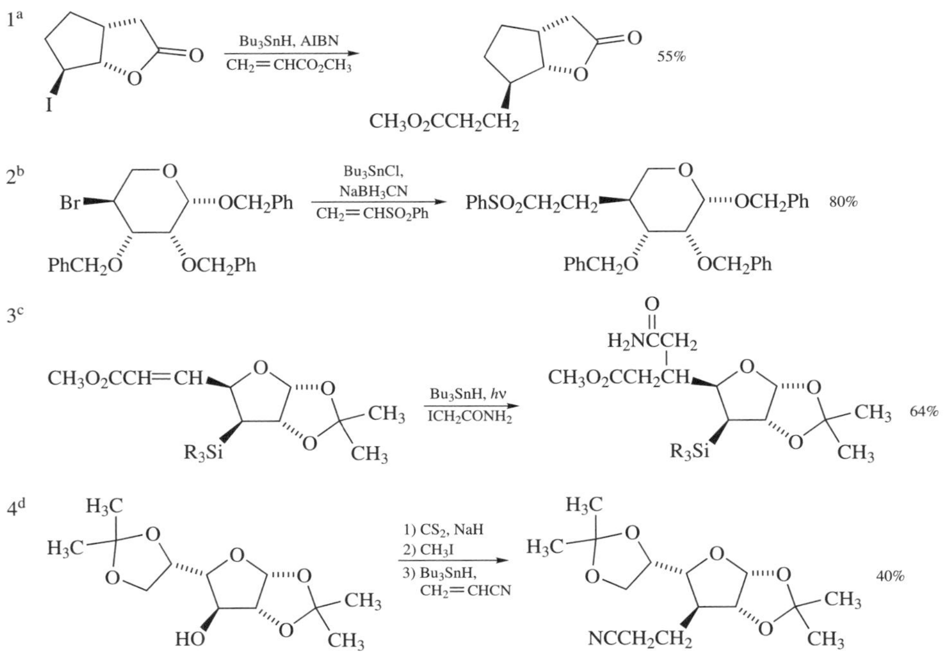

1[a] $\xrightarrow[CH_2=CHCO_2CH_3]{Bu_3SnH,\ AIBN}$ $CH_3O_2CCH_2CH_2$ 55%

2[b] Br, OCH_2Ph, $PhCH_2O$, OCH_2Ph $\xrightarrow[CH_2=CHSO_2Ph]{Bu_3SnCl,\ NaBH_3CN}$ $PhSO_2CH_2CH_2$, OCH_2Ph, $PhCH_2O$, OCH_2Ph 80%

3[c] $CH_3O_2CCH=CH$, R_3Si, CH_3, CH_3 $\xrightarrow[ICH_2CONH_2]{Bu_3SnH,\ h\nu}$ H_2NCCH_2 (C=O), $CH_3O_2CCH_2CH$, R_3Si, CH_3, CH_3 64%

4[d] H_3C, H_3C, HO, CH_3, CH_3 $\xrightarrow{1)\ CS_2,\ NaH;\ 2)\ CH_3I;\ 3)\ Bu_3SnH,\ CH_2=CHCN}$ H_3C, H_3C, $NCCH_2CH_2$, CH_3, CH_3 40%

B. Using other methods of radical generation

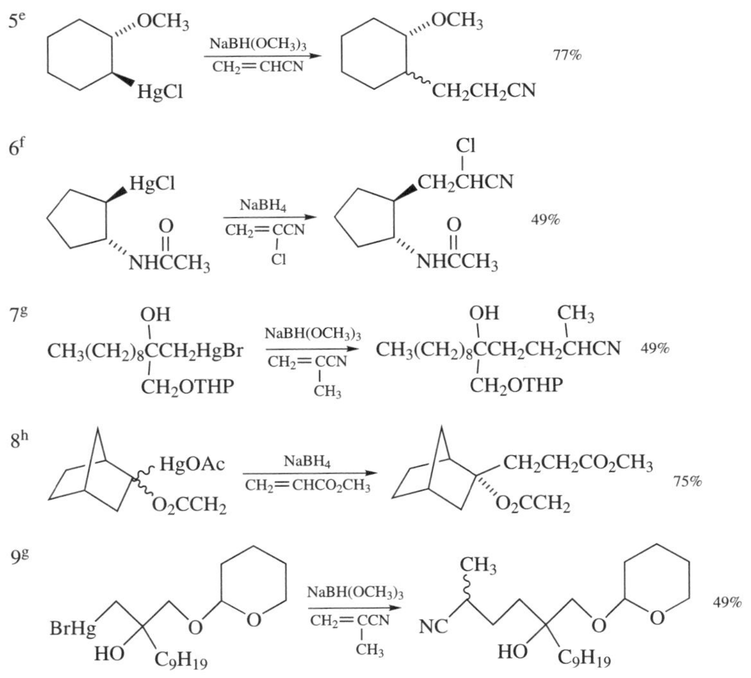

5[e] OCH_3, HgCl $\xrightarrow[CH_2=CHCN]{NaBH(OCH_3)_3}$ OCH_3, CH_2CH_2CN 77%

6[f] HgCl, $NHCCH_3$ (C=O) $\xrightarrow[CH_2=C(Cl)CN]{NaBH_4}$ CH_2CHCN (Cl), $NHCCH_3$ (C=O) 49%

7[g] $CH_3(CH_2)_8C(OH)(CH_2OTHP)CH_2HgBr$ $\xrightarrow[CH_2=C(CH_3)CN]{NaBH(OCH_3)_3}$ $CH_3(CH_2)_8C(OH)(CH_2OTHP)CH_2CH_2CH(CH_3)CN$ 49%

8[h] HgOAc, O_2CCH_2 $\xrightarrow[CH_2=CHCO_2CH_3]{NaBH_4}$ $CH_2CH_2CO_2CH_3$, O_2CCH_2 75%

9[g] BrHg, HO, C_9H_{19} $\xrightarrow[CH_2=C(CH_3)CN]{NaBH(OCH_3)_3}$ NC, CH_3, HO, C_9H_{19} 49%

Scheme 10.10. *(continued)*

10[i] N-(CO_2CH_2Ph)-2-($CH_2HgO_2CCH_3$)pyrrolidine $\xrightarrow[CH_2{=}CHCO_2CH_3]{NaBH(OCH_3)_3}$ N-(CO_2CH_2Ph)-2-($(CH_2)_3CO_2CH_3$)pyrrolidine 64%

11[j] $PhC(=O)SePh + CH_2{=}CHCO_2CH_3 \xrightarrow[AIBN]{Bu_3SnH,\ 1.3\ equiv} PhC(=O)CH_2CH_2CO_2CH_3$ 58%

a. S. D. Burke, W. B. Fobare, and D. M. Arminsteadt, *J. Org. Chem.* **47**:3348 (1982).
b. M. V. Rao and M. Nagarajan, *J. Org. Chem.* **53**:1432 (1988).
c. G. Sacripante, C. Tan, and G. Just, *Tetrahedron Lett.* **26**:5643 (1985).
d. B. Giese, J. A. Gonzalez-Gomez, and T. Witzel, *Angew. Chem. Int. Ed. Engl.* **23**:69 (1984).
e. B. Giese and K. Heuck, *Chem. Ber.* **112**:3759 (1979).
f. R. Henning and H. Urbach, *Tetrahedron Lett.* **24**:5343 (1983).
g. A. P. Kozikowski, T. R. Nieduzak, and J. Scripko, *Organometallics*, **1**:675 (1982).
h. B. Giese and U. Erfort, *Chem. Ber.* **116**:1240 (1983).
i. S. Danishefsky, E. Taniyama, and R. P. Webb, II, *Tetrahedron Lett* **24**:11 (1983).
j. D. L. Boger and R. J. Mathvink, *J. Org. Chem.* **57**:1429 (1992).

structures, proximity factors determined by the specific geometry of the ring system are a major factor. Theoretical analysis of radical addition indicates that the major interaction of the attacking radical is with the alkene LUMO.[235] The preferred direction of attack is not perpendicular to the π system but, instead, at an angle of about 110°.

Five-membered rings usually are fused onto the other rings in a *cis* manner in order to minimize strain. When cyclization is followed by hydrogen abstraction, the hydrogen atom is normally delivered from the less hindered side of the molecule. The example below illustrates these principles. The initial tetrahydrofuran ring closure gives the *cis*-fused ring. The subsequent hydrogen abstraction is from the less hindered axial direction.[236]

[Scheme: n-Bu_3SnH, AIBN; structures with H_3C, OCH_3, CH_2Br, CH_3, O, H, CH_2 labels]

235. A. L. J. Beckwith and C. H. Schiesser, *Tetrahedron* **41**:3925 (1985); D. C. Spellmeyer and K. N. Houk, *J. Org. Chem.* **52**:959 (1987).
236. M. J. Begley, H. Bhandal, J. H. Hutchinson, and G. Pattenden, *Tetrahedron Lett.* **28**:1317 (1987).

1[a] $CH_2=CHCH_2SnBu_3$, $h\nu$ — 80–93%

2[b] $CH_2=CHCH_2SnBu_3$, $h\nu$ — 82%

3[c] $CH_2=CHCH_2SnBu_3$, AIBN — 88%

4[d] $CH_2=CHCH_2SnBu_3$, AIBN — 73%

5[e] + C_2H_5I + $CH_2=CHCH_2SnBu_3$ $\xrightarrow[(C_2H_5)_3B,\ O_2]{MgBr_2}$

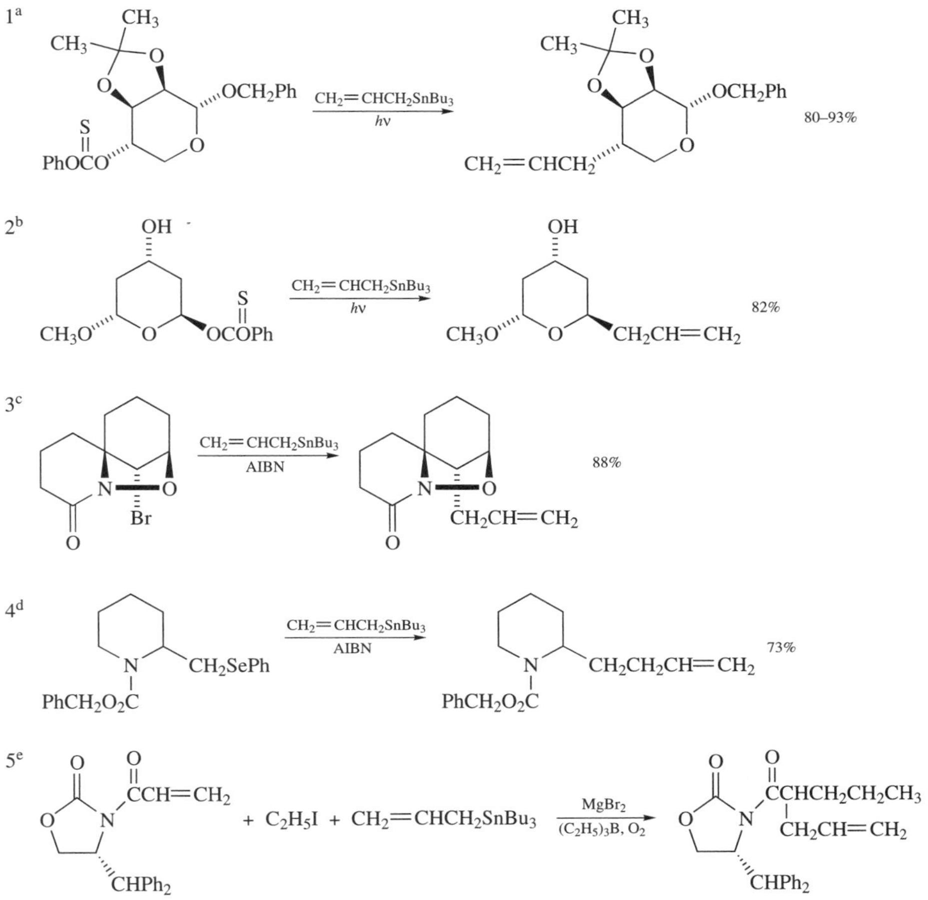

a. G. E. Keck, D. F. Kachensky, and E. J. Enholm, *J. Org. Chem.* **50**:4317 (1985).
b. G. E. Keck and D. F. Kachensky, *J. Org. Chem.* **51**:2487 (1986).
c. G. E. Keck and J. B. Yates, *J. Org. Chem.* **47**:3590 (1982).
d. R. R. Webb II and S. Danishefsky, *Tetrahedron Lett.* **24**:1357 (1983).
e. M. P. Sibi and J. Ji, *J. Org. Chem.* **61**:6090 (1996).

Reaction conditions have been developed in which the cyclized radical can react in some manner other than hydrogen-atom abstraction. One such reaction is abstraction of an iodine atom. The cyclization of 2-iodo-2-methyl-6-heptyne is a structurally simple example.

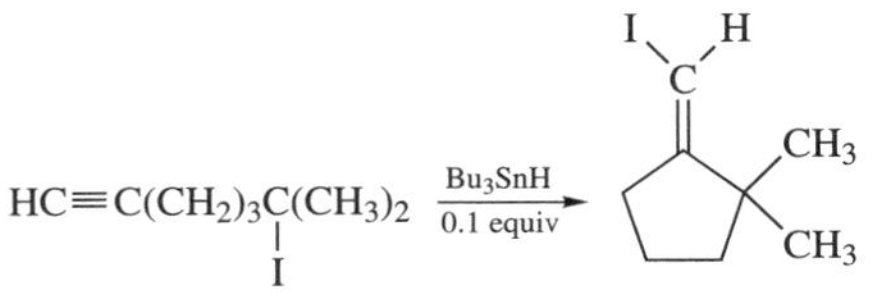

$$HC\equiv C(CH_2)_3C(CH_3)_2I \xrightarrow[0.1\ equiv]{Bu_3SnH}$$ Ref. 237

237. D. P. Curran, M.-H. Chen, and D. Kim, *J. Am. Chem. Soc.* **108**:2489 (1986); D. P. Curran, M.-H. Chen, and D. Kim, *J. Am. Chem. Soc.* **111**:6265 (1989).

In this reaction, the trialkylstannane serves to initiate the chain sequence, but it is present in low concentration to minimize the rate of H-atom abstraction from the stannane. Under these conditions, the chain is propagated by iodine-atom abstraction.

initiation $Bu_3Sn\cdot + I{-}C(CH_3)_2CH_2CH_2CH_2C{\equiv}CH \longrightarrow Bu_3SnI + (CH_3)_2\dot{C}CH_2CH_2CH_2C{\equiv}CH$

propagation $(CH_3)_2\dot{C}CH_2CH_2CH_2C{\equiv}CH \longrightarrow$ 2,2-dimethyl-1-(·CH=)cyclopentane (vinyl radical)

vinyl radical $+ I{-}C(CH_3)_2CH_2CH_2CH_2C{\equiv}CH \longrightarrow$ 2,2-dimethyl-1-(ICH=)cyclopentane $+ (CH_3)_2\dot{C}CH_2CH_2CH_2C{\equiv}CH$

The fact that the cyclization is directed to an acetylenic group and leads to formation of an alkenyl radical is significant. Formation of a saturated iodide would be expected to lead to a more complex product mixture because the cyclized product could undergo iodine abstraction and proceed to add to a second unsaturated center. Vinyl iodides are much less reactive, and the reaction product is stable to iodine-atom abstraction. Because of the potential for competition from reduction by the stannane, other reaction conditions have been developed to promote cyclization. Hexabutylditin is frequently used.[238]

cyclohex-2-enyl$-O_2CCH_2I \xrightarrow[80°C]{(n\text{-}Bu_3Sn)_2,\ 10\ \text{mol}\ \%}$ iodo-substituted hexahydrobenzofuran-2-one (H, H, I, =O)

83% yield, 6:1 *trans:cis*

An alternative system for initiating radical cyclization uses triethylborane and oxygen. Under these conditions, tris(trimethylsilyl)silane is an effective hydrogen atom donor.[239]

$CH_3O_2CC{\equiv}C(CH_2)_3CH(I)CH_3 \xrightarrow[[(CH_3)_3Si]H]{(C_2H_5)_3B,\ O_2}$ 2-methyl-1-($=CHCO_2CH_3$)cyclopentane 72%

238. D. P. Curran and J. Tamine, *J. Org. Chem.* **56:**2746 (1991).
239. T. B. Lowinger and L. Weiler, *J. Org. Chem.* **57:**6099 (1992); P. A. Evans and J. D. Roseman, *J. Org. Chem.* **61:**2252 (1996).

These cyclizations can also be carried out in the presence of ethyl iodide, in which case the chain is propagated by iodine-atom transfer.[240]

Intramolecular additions have also been accomplished with xanthate and thionocarbonates.

$CH_2{=}CH(CH_2)_2OCH(SC(=S)OC_2H_5)CO_2CH_3 \xrightarrow{(CH_3)_3COOC(CH_3)_3}$ cyclopentane with $CH_2SC(=S)OC_2H_5$ and CO_2CH_3 (67% yield, 56:44 *cis:trans*) + tetrahydropyran with $SC(=S)OC_2H_5$ and CO_2CH_3 (3%) Ref. 241

Cyclic thionocarbonate with $CH_2CH(OTBDMS)CH_2C{\equiv}CH \xrightarrow[AIBN]{Bu_3SnH}$ cyclopentane with OTBDMS, $HOCH_2$ and $=CH_2$ Ref. 242

Scheme 10.12 gives other examples of cyclization, including examples in which radicals are generated by sulfuryl and selenyl abstraction and by $Mn(O_2CCH_3)_3$ oxidation.

The use of vinyl radicals in cyclizations of this type is particularly promising. Addition of a vinyl radical to a double bond is usually thermodynamically favorable because a more stable alkyl radical results. The vinyl radical can be generated by dehalogenation of vinyl bromides or iodides. An early study provided examples of both five- and six-membered rings being formed.[243]

$(CH_3O_2C)_2C$ bearing $CH_2C(Br){=}CH_2$ and $CH_2C(R){=}CHR'$ $\xrightarrow{Bu_3SnH}$ **F** (methylenecyclopentane with CH_2R' and R) + **G** (methylenecyclohexane with R and R')

R	R′	Product ratio **F**:**G**
H	H	3:1
CH_3	H	**G** exclusively
H	CH_3	2:1

240. T. J. Woltering and H. M. R. Hoffmann, *Tetrahedron* **51:**7389 (1995).
241. J. H. Udding, J. P. M. Giesselink, H. Hiemstra, and W. N. Speckamp, *J. Org. Chem.* **59:**6671 (1994).
242. F. E. Ziegler, C. A. Metcalf III, and G. Schulte, *Tetrahedron Lett.* **33:**3117 (1992).
243. G. Stork and N. H. Baine, *J. Am. Chem. Soc.* **104:**2321 (1982).

Cyclizations of both alkyl and acyl radicals generated by selenide abstraction have also been observed.

HO, C≡CPh, CN, $CH_2CHSePh$ → (Bu_3SnH) → OH, H, CPh, CN 91% Ref. 244

O, CH_2CH_2CSePh → (Bu_3SnH) → H, H, O Ref. 245

Several functional groups in addition to carbon–carbon double and triple bonds can participate in radical cyclizations. Among these are oxime ethers, imines, and hydrazones. Cyclization at these functional groups leads to amino-substituted products.

Br, $CH_2CH_2CH{=}NOCH_2Ph$ → (Bu_3SnH, AIBN) → $PhCH_2ONH$ 72% Ref. 246

$BrCH_2CH_2O_2CCH{=}NNPh_2$ → (Bu_3SnH, AIBN) → $NHNPh_2$, O, O Ref. 247

$CH_2CH_2N{=}CHPh$, Br → (Bu_3SnH, AIBN) → NH, Ph 70% Ref. 248

244. D. L. J. Clive, T. L. B. Boivin, and A. G. Angoh, *J. Org. Chem.* **52:**4943 (1987).
245. D. L. Boger and R. J. Mathvink, *J. Org. Chem.* **53:**3377 (1988).
246. J. W. Grissom, D. Klingberg, S. Meyenburg, and B. L. Stallman, *J. Org. Chem.* **59:**7876 (1994).
247. D. L. J. Clive and J. Zhang, *Chem. Commun.* **1997:**549.
248. M. J. Tomaszewski and J. Warkentin, *Tetrahedron Lett.* **33:**2123 (1992).

A radical cyclization of this type was used to synthesize the 3-amino-4-hydroxyhexahydroazepine group found in the protein kinase C inhibitor balanol.

$O{=}CH(CH_2)_3N(CO_2\text{-}t\text{-}C_4H_9)CH_2CH{=}NOCH_2Ph \xrightarrow{Bu_3SnH}$ [azepine: OH, $NHOCH_2Ph$, $N\text{-}CO_2\text{-}t\text{-}C_4H_9$] 50% 1:2:6 *cis*:*trans* Ref. 249

Vinyl radicals generated by addition of trialkylstannyl radicals to terminal alkynes can also undergo cyclization with a nearby double bond.

$HC{\equiv}CCH_2C(CO_2CH_3)_2CH_2CH{=}C(CH_3)_2 \xrightarrow[AIBN]{Bu_3SnH}$ [cyclopentane: CH_3O_2C, CO_2CH_3, $=CHSnBu_3$, $CH(CH_3)_2$] 90% Ref. 250

Vinyl radicals generated by intramolecular addition can add to a nearby double bond in a tandem cyclization process.

[1-methylcyclohexene with $CH_2CH_2C{\equiv}CCH_2OCH(OCH_2CH_2Cl)CH_2Br$] $\xrightarrow[80°C]{1.1\text{ equiv } Bu_3SnH,\ AIBN}$ [bicyclic product: CH_3, H, tetrahydrofuranylidene with OCH_2CH_2Cl] 75% Ref. 251

Scheme 10.12 gives some additional examples of cyclization reactions involving radical intermediates.

Radicals formed by intramolecular addition can be further extended by trapping the intermediate cyclic radical with an electrophilic alkene.

[cyclohexenyl $OCH(OC_2H_5)CH_2I$] $\xrightarrow[CH_2{=}CHCN]{0.2\text{ equiv } Bu_3SnCl,\ NaBH_3CN,\ h\nu}$ [bicyclic product: OC_2H_5, CH_2CH_2CN] Ref. 252

As with carbocation-initiated polyene cyclizations, radical cyclizations can proceed through several successive steps if the steric and electronic properties of the reactant

249. H. Miyabe, M. Torieda, K. Inoue, K. Tajiri, T. Kiguchi, and T. Naito, *J. Org. Chem.* **63:**4397 (1998).
250. G. Stork and R. Mook, Jr., *J. Am. Chem. Soc.* **109:**2829 (1987).
251. G. Stork and R. Mook, Jr., *J. Am. Chem. Soc.* **105:**3720 (1983).
252. G. Stork and P. M. Sher, *J. Am. Chem. Soc.* **108:**303 (1986).

Scheme 10.12. Radical Cyclizations

A. Cyclizations of halides terminated by hydrogen-atom abstraction or halogen-atom transfer

1[a]

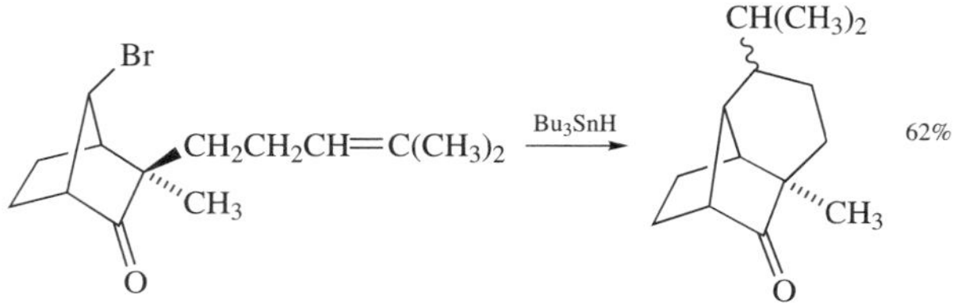

Br; $CH_2CH_2CH{=}C(CH_3)_2$; CH_3; O → (Bu_3SnH) → $CH(CH_3)_2$; CH_3; O — 62%

2[b]

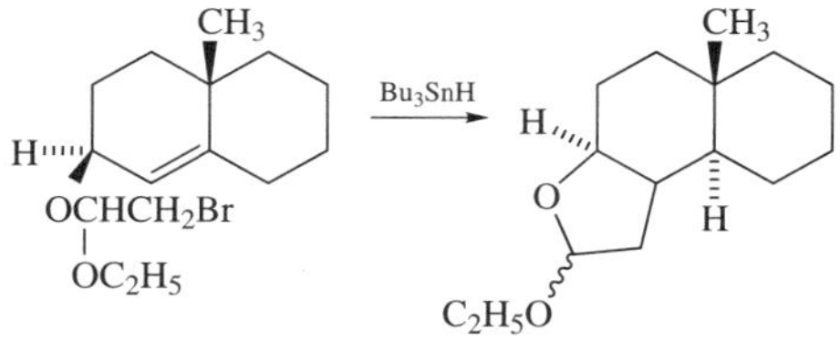

CH_3; H; $OCHCH_2Br$; OC_2H_5 → (Bu_3SnH) → CH_3; H; O; H; C_2H_5O

3[c]

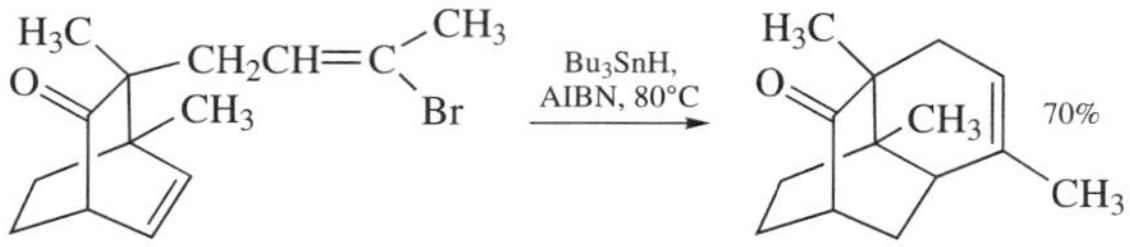

H_3C; O; $CH_2CH{=}C$; CH_3; Br; CH_3 → (Bu_3SnH, AIBN, 80°C) → H_3C; O; CH_3; CH_3 — 70%

4[d]

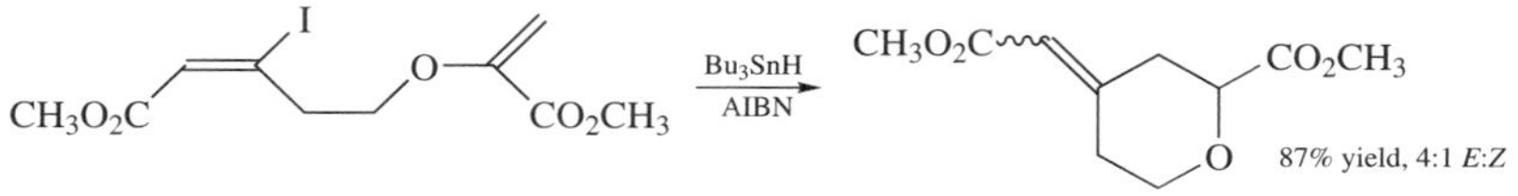

I; CH_3O_2C; O; CO_2CH_3 → (Bu_3SnH, AIBN) → CH_3O_2C; CO_2CH_3; O — 87% yield, 4:1 *E*:*Z*

5[e]

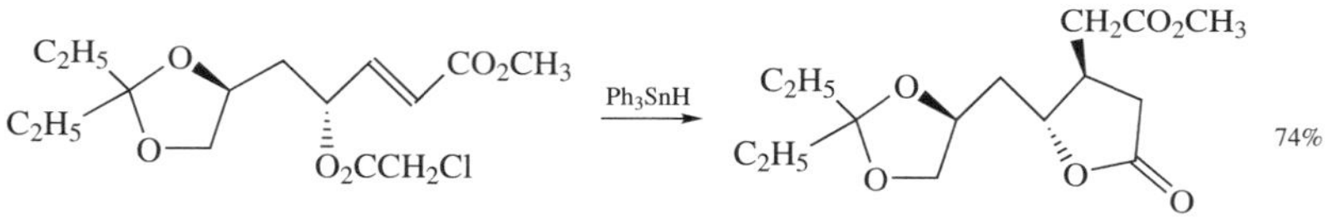

C_2H_5; C_2H_5; O; O; CO_2CH_3; O_2CCH_2Cl → (Ph_3SnH) → $CH_2CO_2CH_3$; C_2H_5; C_2H_5; O; O; O; O — 74%

6[f]

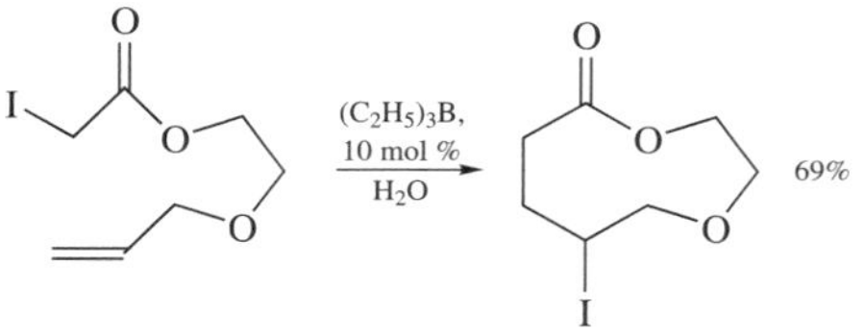

O; I; O; O → ($(C_2H_5)_3B$, 10 mol %, H_2O) → O; O; O; I — 69%

7[g]

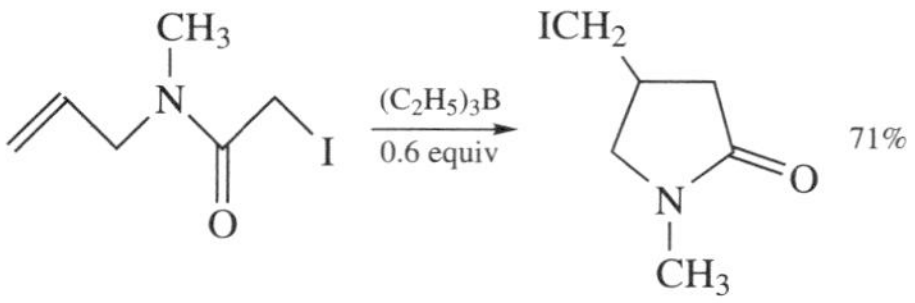

CH_3; N; I; O → ($(C_2H_5)_3B$ 0.6 equiv) → ICH_2; N; O; CH_3 — 71%

O; I; CH_3; O; $(CH_3)_2CH$; $CH_2CH_2C{\equiv}CSi(CH_3)_3$; OCH_3 → (Bu_3SnH, AIBN) → O; CH_3; $CH{\sim}Si(CH_3)_3$; O; $(CH_3)_2CH$; H; OCH_3 — 69% yield, 1:9 *E*:*Z*

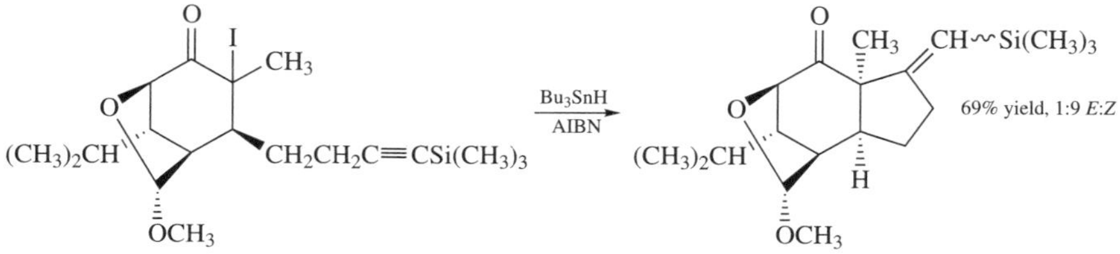

B. Cyclization of thioesters, sulfides, and selenides

9[i] Bu_3SnH

10[j] Bu_3SnH, AIBN — 88%

11[k] CH_2OCH_2SePh — Bu_3SnH, AIBN — 80%

12[l] Bu_3SnH, AIBN — 76%

13[m] Bu_3SnH, 1.1 equiv, AIBN — 68%

14[n] $(C_2H_5)_3B$, $[(CH_3)_3Si]_3SiH$ — 94% yield, 5.7:1 *cis*:*trans*

15[o] Bu_3SnH, 1.2 equiv, AIBN — 82%, 62:28 *trans*:cis

16[p] Bu_3SnH, AIBN — 73%

17[q] HO, C≡CPh, CH_2CHCN, SePh; Ph_3SnH, 15 equiv, AIBN; OH, CHPh, CN

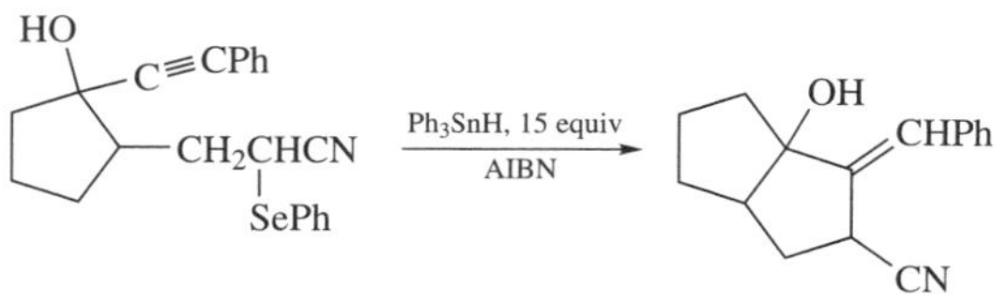

C. Oxidative cyclization with $Mn(O_2CCH_3)_3$

18[r] O, $CH_2CH{=}CH_2$; $Mn(O_2CCH_3)_3$, 2 equiv, $Cu(O_2CCH_3)_2$, 1 equiv, 80°C; O; 51%

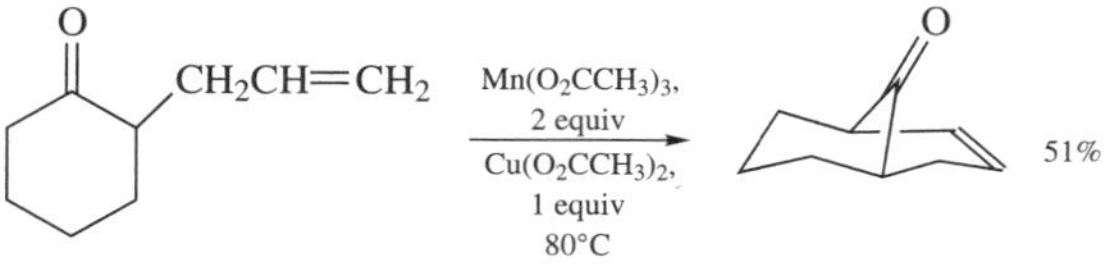

19[s] O, CH_3, $(CH_2)_2C{\equiv}CSi(CH_3)_3$; $Mn(O_2CCH_3)_3$, 15 equiv, 90°C; O, CH_3, $CHSi(CH_3)_3$; 58% yield, 1:1.4 *E*:*Z*

20[t] $CH_3CCH_2CH_2CCHCH_2CH{=}CH_2$ (with $=CH_2$, O, CO_2CH_3); $Mn(O_2CCH_3)_3$, $Cu(O_2CCH_3)_2$; O, CO_2CH_3, CH_3, CH_2; 86%

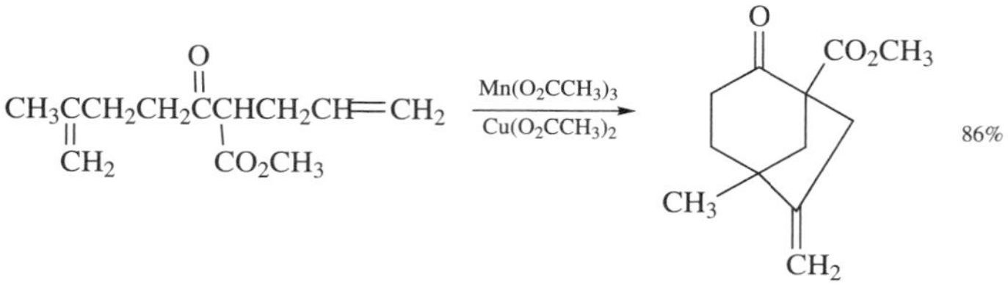

D. Additions to C=N bonds

21[u] $PhCH_2ON{=}CH$, HO, O, O, CH_3, CH_3; Ph_3SnH, $(C_2H_5)_3B$; $CHSnPh_3$, $PhCH_2ONH$, OH, O, O, CH_3, CH_3; 91% 1:14 *E*:*Z*

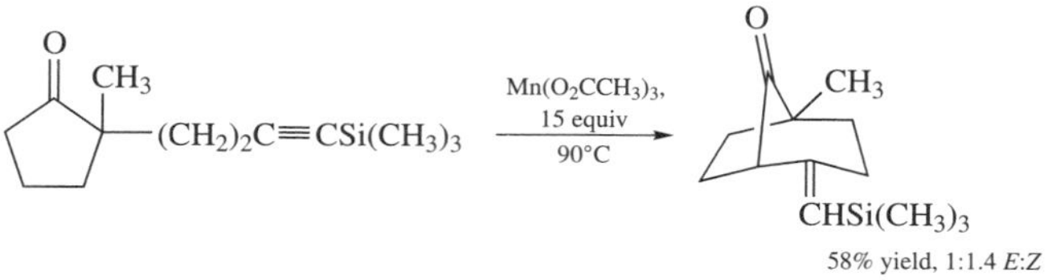

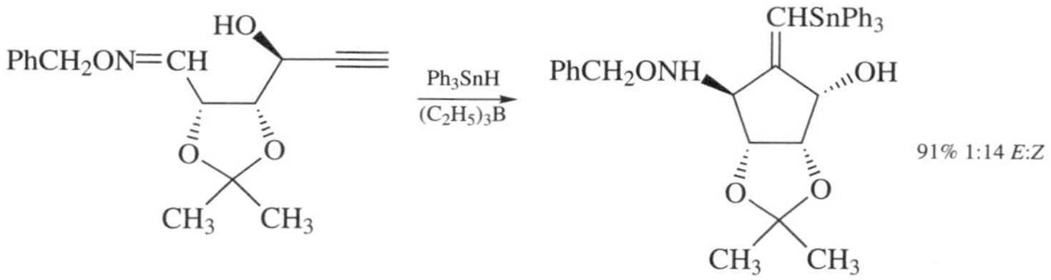

22[v] $PhSeCH_2$, O, $O_2CCH{=}NNPh_2$, SePh, OH; Ph_3SnH, AIBN; O, $NHNPh_2$, CH_3, O, OH, O; 79% 1:1.1 *trans*:*cis*

23[w] $O{=}CH(CH_2)_2NCH_2CH{=}NOCH_3$, CO_2CH_2Ph; Bu_3SnH, 2 equiv; OH, $NHOCH_3$, N, CO_2CH_2Ph; 62% yield, 1:1.3 *cis*:*trans*

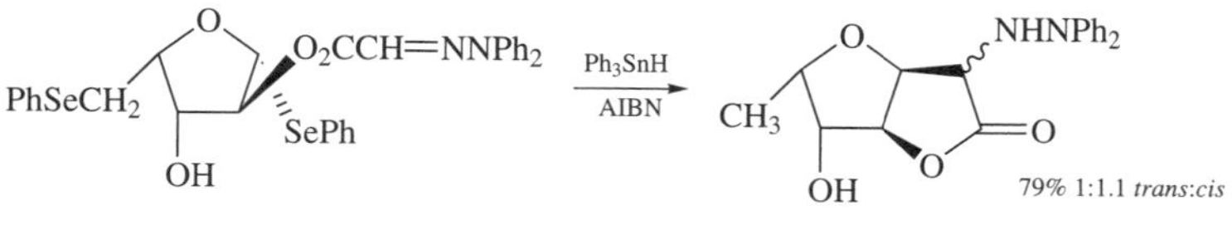

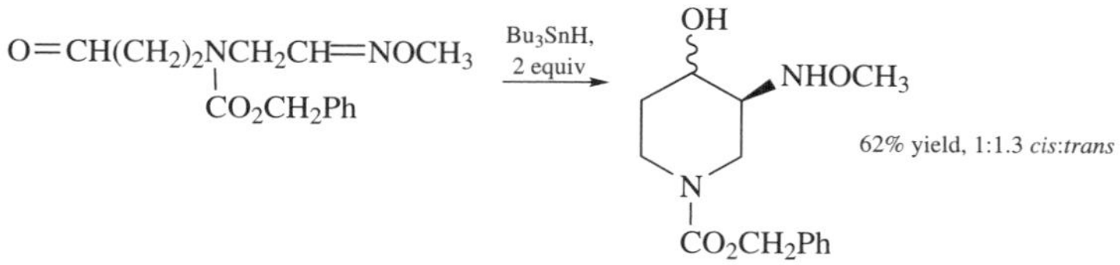

24[x] Bu_3SnH, AIBN 72%

E. Cyclization with tandem alkylation

25[y] Bu_3SnCl, $NaBH_4$, $CH_2{=}C(Si(CH_3)_3)C(O)(CH_2)_4CH_3$ 60%

26[z] Bu_3SnH 64%

27[aa] Bu_3SnH, AIBN 71%

28[bb] Ph_3SnH 61% + 26%

29[cc] Bu_3SnH 93%

30[dd]

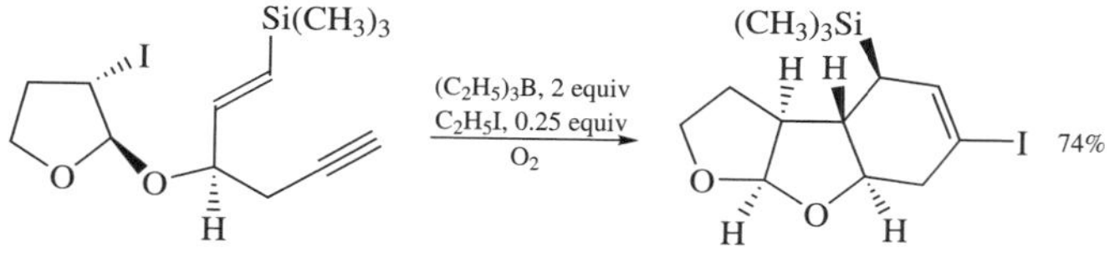

$Si(CH_3)_3$, I, O, H

$(C_2H_5)_3B$, 2 equiv
C_2H_5I, 0.25 equiv
O_2

$(CH_3)_3Si$, H H, O, H, O, H, —I 74%

31[ee]

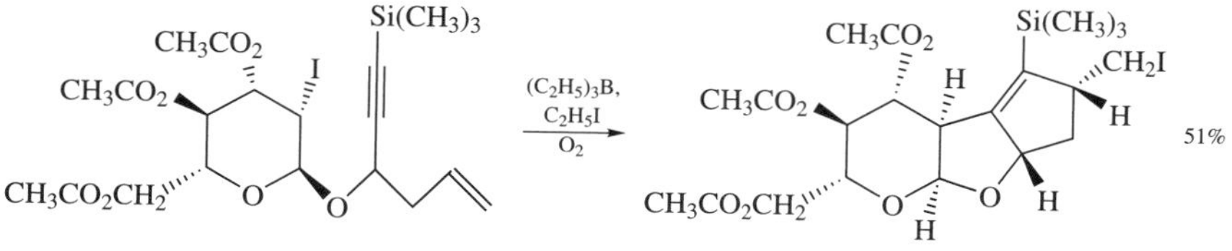

CH_3CO_2, CH_3CO_2, I, $Si(CH_3)_3$, $CH_3CO_2CH_2$, O, O

$(C_2H_5)_3B$,
C_2H_5I
O_2

CH_3CO_2, $Si(CH_3)_3$, CH_2I, CH_3CO_2, H, H, $CH_3CO_2CH_2$, O, H, O, H 51%

32[ff]

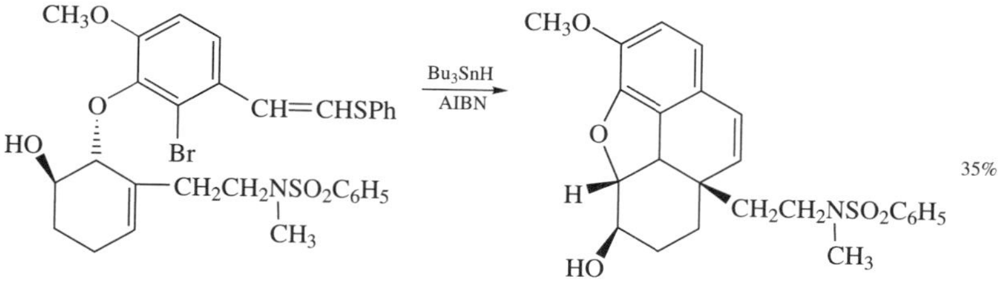

CH_3O, O, HO, Br, CH=CHSPh, $CH_2CH_2NSO_2C_6H_5$, CH_3

Bu_3SnH
AIBN

CH_3O, O, H, HO, $CH_2CH_2NSO_2C_6H_5$, CH_3 35%

33[gg]

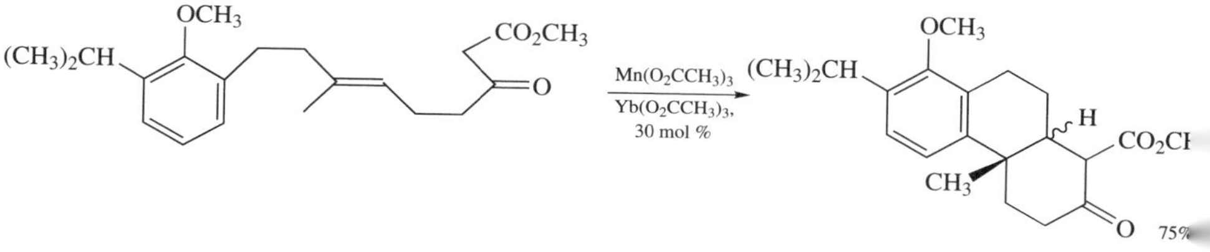

OCH_3, $(CH_3)_2CH$, CO_2CH_3, O

$Mn(O_2CCH_3)_3$
$Yb(O_2CCH_3)_3$,
30 mol %

OCH_3, $(CH_3)_2CH$, H, CO_2CH_3, CH_3, O 75%

34[hh]

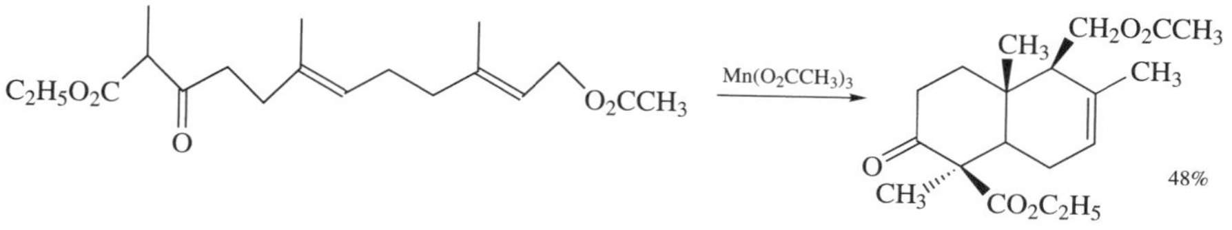

$C_2H_5O_2C$, O, O_2CCH_3

$Mn(O_2CCH_3)_3$

CH_3, $CH_2O_2CCH_3$, CH_3, O, CH_3, $CO_2C_2H_5$ 48%

35[ii]

Ph, O, O, O, CH_3, NC, OCH_3

1) Bu_3SnH
2) H_2O, H^+

Ph, O, O, O, CH_3, O, OCH_3, $SnBu_3$ 65%

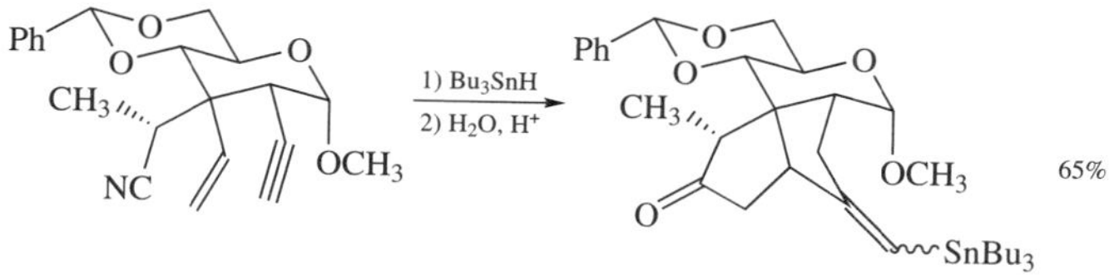

36[jj]

PhSH, $h\nu$ — 90%

a. P. Bakuzis, O. O. S. Campos, and M. L. F. Bakuzis, *J. Org. Chem.* **41**:3261 (1976).
b. G. Stork and M. Kahn, *J. Am. Chem. Soc.* **107**:500 (1985).
c. G. Stork and N. H. Baine, *Tetrahedron Lett.* **26**:5927 (1985).
d. R. J. Maguire, S. P. Munt, and E. J. Thomas, *J. Chem. Soc., Perkin Trans. 1* **1998**:2853.
e. S. Hanessian, R. DiFabio, J.-F. Marcoux, and M. Prud'homme, *J. Org. Chem.* **55**:3436 (1990).
f. H. Yorimitsu, T. Nakamura, H. Shinokubo, and K. Oshima, *J. Org. Chem.* **63**:8604 (1998).
g. M. Ikeda, H. Teranishi, K. Nozaki, and H. Ishibashi, *J. Chem. Soc., Perkin Trans. 1* **1998**:1691.
h. C.-K. Sha, R.-T. Chiu, C.-F. Yang, N.-T. Yao, W.-H. Tseng, F.-L. Liao, and S.-L. Wang, *J. Am. Chem. Soc.* **119**:4130 (1997).
i. T. V. Rajan Babu, *J. Org. Chem.* **53**:4522 (1988).
j. J.-K. Choi, D.-C. Ha, D. J. Hart, C.-S. Lee, S. Ramesh, and S. Wu, *J. Org. Chem.* **54**:279 (1989).
k. V. H. Rawal, S. P. Singh, C. Dufour, and C. Michoud, *J. Org. Chem.* **56**:5245 (1991).
l. D. L. J. Clive and V. S. C. Yeh, *Tetrahedron Lett.* **39**:4789 (1998).
m. S. Knapp and F. S. Gibson, *J. Org. Chem.* **57**:4802 (1992).
n. P. A. Evans and J. D. Roseman, *J. Org. Chem.* **61**:2252 (1996).
o. D. L. Boger and R. J. Mathvink, *J. Org. Chem.* **57**:1429 (1992).
p. A. K. Singh, R. K. Bakshi, and E. J. Corey, *J. Am. Chem. Soc.* **109**:6187 (1987).
q. D. L. J. Clive, T. L. B. Boivin, and A. G. Angoh, *J. Org. Chem.* **52**:4943 (1987).
r. B. McC. Cole, L. Han, and B. B. Snider, *J. Org. Chem.* **61**:7832 (1996).
s. S. V. O'Neil, C. A. Quickley, and B. B. Snider, *J. Org. Chem.* **62**:1970 (1997).
t. M. A. Dombroski, S. A. Kates, and B. B. Snider, *J. Am. Chem. Soc.* **112**:2759 (1990).
u. J. Marco-Contelles, C. Destabel, P. Gallego, J. L. Chiara, and M. Bernabe, *J. Org. Chem.* **61**:1354 (1996).
v. J. Zhang and D. L. J. Clive, *J. Org. Chem.* **64**:1754 (1999).
w. T. Naito, K. Nakagawa, T. Nakamura, A. Kasei, I. Ninomiya, and T. Kiguchi, *J. Org. Chem.* **64**:2003 (1999).
x. G. E. Keck, S. F. McHardy, and J. A. Murry, *J. Org. Chem.* **64**:4465 (1999).
y. G. Stork, P. M. Sher, and H.-L. Chen, *J. Am. Chem. Soc.* **108**:6384 (1986).
z. D. P. Curran and D. W. Rakewicz, *Tetrahedron* **41**:3943 (1985).
aa. S. Iwasa, M. Yamamoto, S. Kohmoto, and K. Yamada, *J. Org. Chem.* **56**:2849 (1991).
bb. S. R. Baker, A. F. Parsons, J.-F. Pons, and M. Wilson, *Tetrahedron Lett.* **39**:7197 (1998); S. R. Baker, K. I. Burton, A. F. Parsons, J.-F. Pons, and M. Wilson, *J. Chem. Soc., Perkin Trans. 1* **1999**:427.
cc. T. Takahashi, S. Tomida, Y. Sakamoto, and H. Yamada, *J. Org. Chem.* **62**:1912 (1997).
dd. M. Breithor, U. Herden, and H. M. R. Hoffmann, *Tetrahedron* **53**:8401 (1997).
ee. T. J. Woltering and H. M. R. Hoffmann, *Tetrahedron* **51**:7389 (1995).
ff. K. A. Parker and D. Fokas, *J. Am. Chem. Soc.* **114**:9688 (1992).
gg. D. Yang, X.-Y. Ye, S. Gu, and M. Xu, *J. Am. Chem. Soc.* **121**:5579 (1999).
hh. B. B. Snider, R. Mohan, and S. A. Kates, *Tetrahedron Lett.* **28**:841 (1987).
ii. H. Pak, I. I. Canalda, and B. Fraser-Reid, *J. Org. Chem.* **55**:3009 (1990).
jj. G. E. Keck, T. T. Wager, and J. F. Duarte Rodriquez, *J. Am. Chem. Soc.* **121**:5176 (1999).

provide potential reaction sites. These kinds of reactions are referred to as *tandem reactions*. Cyclization may be followed by a second intramolecular step or by an intermolecular addition or alkylation. Intermediate radicals can also be constructed so that hydrogen-atom transfer can occur as part of the overall process. For example, 2-halohexenes with radical-stabilizing substituents at C-6 can undergo cyclization after a hydrogen-atom transfer step.[253]

$Bu_3Sn\cdot$

E = CO_2CH_3; X,Y = TBDMSO, H; Ph, H; CO_2CH_3, H; CO_2CH_3, CO_2CH_3; 2-dioxolanyl

253. D. P. Curran, D. Kim, H. T. Liu, and W. Shen, *J. Am. Chem. Soc.* **110**:5900 (1988).

The success of such reactions depends on the intramolecular hydrogen transfer being faster than hydrogen-atom abstraction from the stannane reagent. In the example show, this is favored by the thermodynamic driving force of radical stabilization, by the intramolecular nature of the hydrogen transfer, and by the steric effects of the central quaternary carbon. This substitution pattern often favors intramolecular reactions as a result of conformational effects. Scheme 10.12 gives some other examples of tandem radical reactions.

10.3.5 Fragmentation and Rearrangement Reactions

Fragmentation is the reverse of radical addition. Fragmentation of radicals is often observed to be fast when the overall transformation is exothermic.

$$\cdot Y-\overset{|}{\underset{|}{C}}-\overset{|}{\underset{|}{C}}-X \longrightarrow Y{=}C\langle \;+\; \cdot C-X$$

Rearrangement of radicals frequently occurs by a series of addition–fragmentation steps. The fragmentation of alkoxyl radicals is especially common because the formation of a carbonyl bond makes such reactions exothermic. The following two reaction sequences are examples of radical rearrangements proceeding through addition–elimination.

$(CH_2)_n$ $CO_2C_2H_5$ $(CH_2)_4I$ $\xrightarrow{Bu_3Sn\cdot}$ $(CH_2)_n$ O· $CO_2C_2H_5$ $\longrightarrow$ $(CH_2)_n$ O $CO_2C_2H_5$ $\xrightarrow{Bu_3SnH}$ $(CH_2)_n$ O $CO_2C_2H_5$

Ref. 254

Br, $CH_2CH(CO_2CH_3)CCH_3$ (=O) $\xrightarrow{Bu_3Sn\cdot}$ $CH_2CH(CO_2CH_3)C(=O)CH_3$ $\longrightarrow$ O·, CH_3, CO_2CH_3 Ref. 255

$\downarrow$

$C(=O)CH_3$, $CH_2\dot{C}HCO_2CH_3$ $\xrightarrow{Bu_3SnH}$ $C(=O)CH_3$, $CH_2CH_2CO_2CH_3$

Both of these transformations feature addition of a carbon-centered radical to a carbonyl group, followed by fragmentation to a more stable radical. The rearranged radicals then abstract hydrogen from the co-reactant n-Bu_3SnH. The addition step must be fast relative to hydrogen abstraction, since if this were not the case, simple reductive dehalogenation would occur. The fragmentation step is usually irreversible. There are two reasons: (1) the reverse addition is endothermic; and (2) the stabilized radical substituted by electron-

254. P. Dowd and S.-C. Choi, *J. Am. Chem. Soc.* **109:**6548 (1987).
255. A. L. J. Beckwith, D. M. O'Shea, and S. W. Westwood, *J. Am. Chem. Soc.* **110:**2565 (1988).

withdrawing alkoxycarbonyl radicals is unreactive to addition to carbonyl bonds.

The two reactions above are examples of a more general reactivity pattern[255,256]:

The unsaturated group X=Y that is "transferred" by the rearrangement process can be C=C, C=O, C=N, or other groups that fulfill the following general criteria: (1) the addition step (a) must be fast relative to other potentially competing reactions: and (2) the group Z must stabilize the product radical so that the overall process is energetically favorable.

A direct comparison of the ease with which unsaturated groups migrate by cyclization–fragmentation has been made for the case of net 1,2-migration:

In this system, the overall driving force is the conversion of a primary radical to a tertiary one ($\Delta H \approx 5$ kcal/mol), and the activation barrier incorporates strain associated with formation of the three-membered ring. Rates and activation energies for several migrating groups have been determined.[257]

	X=Y				
	$HC{=}CH_2$	$(CH_3)_3C(C{=}O)$	phenyl	$-C{\equiv}CC(CH_3)_3$	$C{\equiv}N$
k_r (s^{-1})	10^7	1.7×10^5	7.6×10^3	93	0.9
E_a (kcal/mol)	5.7	7.8	11.8	12.8	16.4

Among the most useful radical fragmentation reactions from a synthetic point of view are decarboxylations and fragmentations of alkoxyl radicals. The use of *N*-hydroxythiopyridine esters for decarboxylation is quite widespread. Several procedures and reagents are available for preparation of the esters[258] and the reaction conditions are compatible with many functional groups.[259] *t*-Butyl mercaptan and thiophenol can serve as hydrogen-atom donors.

256. A. L. J. Beckwith, D. M. O'Shea, and S. W. Westwood, *J. Am. Chem. Soc.* **110**:2565 (1988); R. Tsang, J. K. Pickson, Jr., H. Pak, R. Walton, and B. Fraser-Reid, *J. Am. Chem. Soc.* **109**:3484 (1987).
257. D. A. Lindsay, J. Lusztyk, and K. U. Ingold, *J. Am. Chem. Soc.* **106**:7087 (1984).
258. F. J. Sardina, M. H. Howard, M. Morningstar, and H. Rapoport, *J. Org. Chem.* **55**:5025 (1990); D. Bai, R. Xu, G. Chu, and X. Zhu, *J. Org. Chem.* **61**:4600 (1996).
259. D. H. R. Barton, D. Crich, and W. B. M. Motherwell, *Tetrahedron* **41**:3901 (1985).

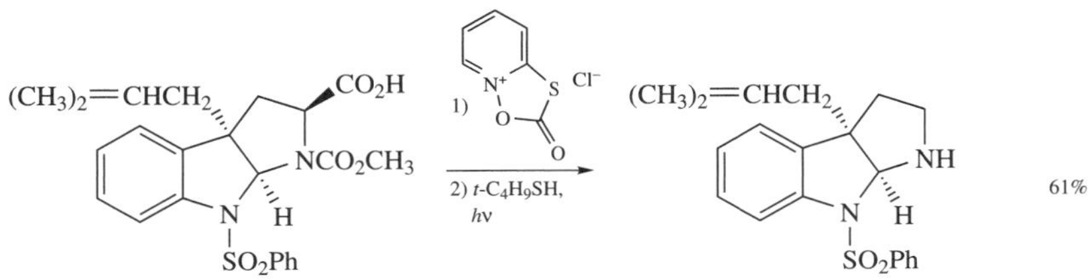

Ref. 260

Esters of N-hydroxyphthalimide can also be used for decarboxylation. Photolysis in the presence of an electron donor and a hydrogen-atom donor leads to decarboxylation.

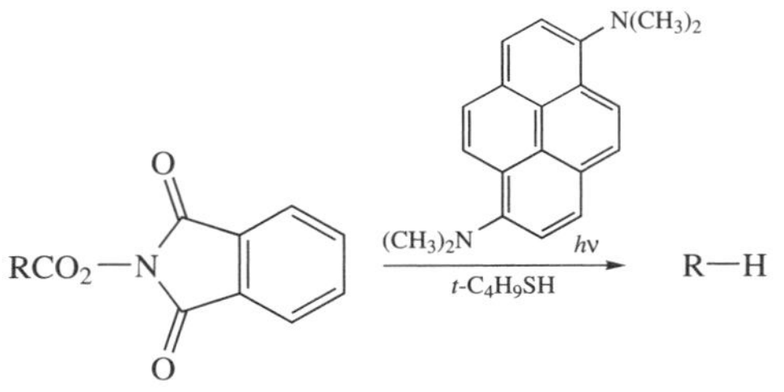

Ref. 261

Carboxyl radicals are formed from one-electron reduction of the phthalimide ring.

Fragmentation of cyclopropylcarbinyl radicals has been incorporated into several synthetic schemes.[262] For example, 2-dienyl-1,1-bis(methoxycarbonyl)cyclopropanes undergo ring expansion to cyclopentenes.

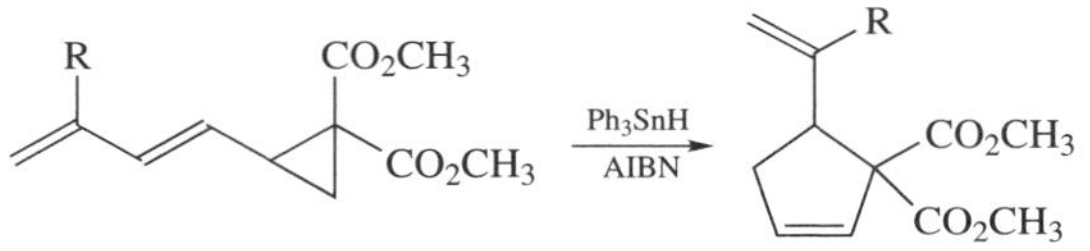

Ref. 263

These reactions presumably involve terminal addition of the chain-carrying radical, fragmentation, and recyclization.

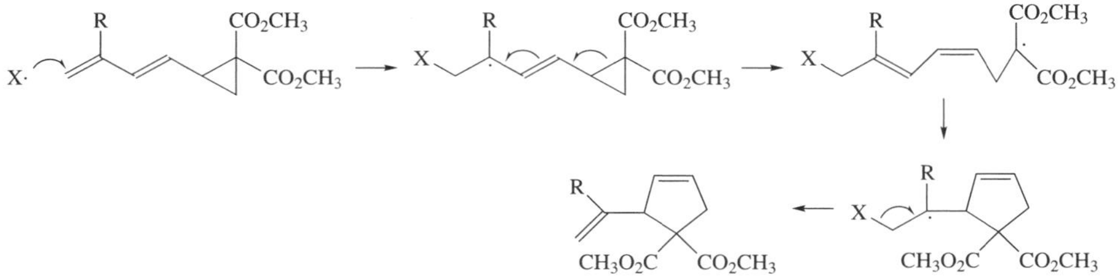

Other intramolecular cyclizations can follow generation and fragmentation of cyclopro-

260. M. Bruncko, D. Crich, and R. Samy, *J. Org. Chem.* **59:**5543 (1994).
261. K. Okada, K. Okamoto, and M. Oda, *J. Am. Chem. Soc.* **110:**8736 (1988).
262. P. Dowd and W. Zhang, *Chem. Rev.* **93:**2091 (1993).
263. K. Miura, K. Fugami, K. Oshima, and K. Utimoto, *Tetrahedron Lett.* **29:**1543 (1988).

pylcarbinyl radicals.

$\xrightarrow[\text{AIBN}]{\text{Bu}_3\text{SnH}}$ 81% Ref. 264

Cyclic α-halomethyl or α-phenylselenenylmethyl β-keto esters undergo one-carbon ring expansion via transient cyclopropylalkoxy radicals.[265]

X = Br, I, SePh; n = 1–3

Comparable cyclization–fragmentation sequences have been developed for acyclic and heterocyclic systems.

$CH_3COC(CH_3)(CO_2C_2H_5)CH_2Br \xrightarrow[\text{AIBN}]{\text{Bu}_3\text{SnH}} CH_3COCH_2CH(CH_3)CO_2C_2H_5$ 64% Ref. 266

$\xrightarrow[\text{AIBN}]{\text{Bu}_3\text{SnH}}$ 84% Ref. 267

264. R. A. Batey, J. D. Harling, and W. B. Motherwell, *Tetrahedron* **48:**8031 (1992).
265. P. Dowd and S.-C. Choi, *Tetrahedron* **45:**77 (1989); A. L. J. Beckwith, D. M. O'Shea, and S. W. Westwood, *J. Am. Chem. Soc.* **110:**2565 (1988), P. Dowd and S.-C. Choi, *Tetrahedron* **48:**4773 (1992).
266. P. Dowd and S.-C. Choi, *Tetrahedron* **45:**77 (1989).
267. Z. B. Zheng and P. Dowd, *Tetrahedron Lett.* **34:**7709 (1993); P. Dowd and S.C. Choi, *Tetrahedron* **47:**4847 (1991).

Fragmentation of alkoxy radicals finds use in construction of medium-size rings. One useful reagent is iodosobenzene diacetate and iodine.[268]

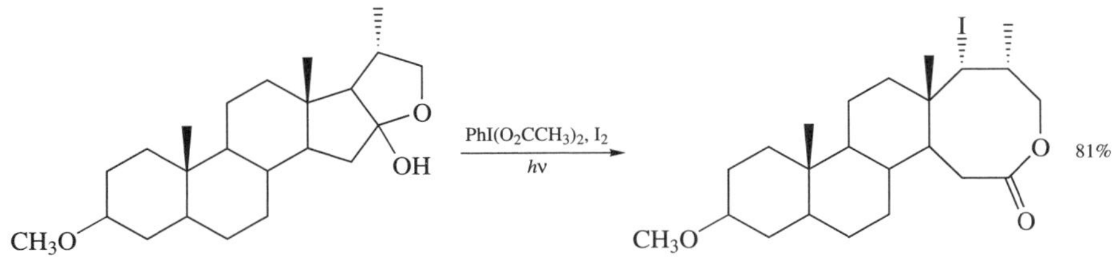

$PhI(O_2CCH_3)_2$, I_2, $h\nu$ — 81%

This reagent also can cleave the C(1)–C(2) bond in carbohydrates.

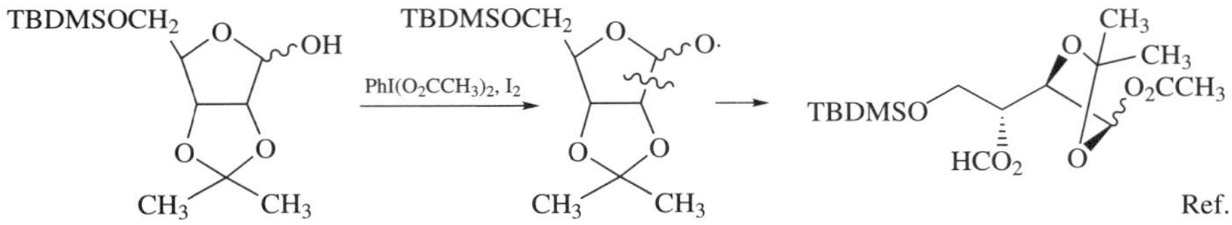

$PhI(O_2CCH_3)_2$, I_2

Ref. 269

When the 6-hydroxyl group is unprotected, it can capture the fragmented intermediate.[270]

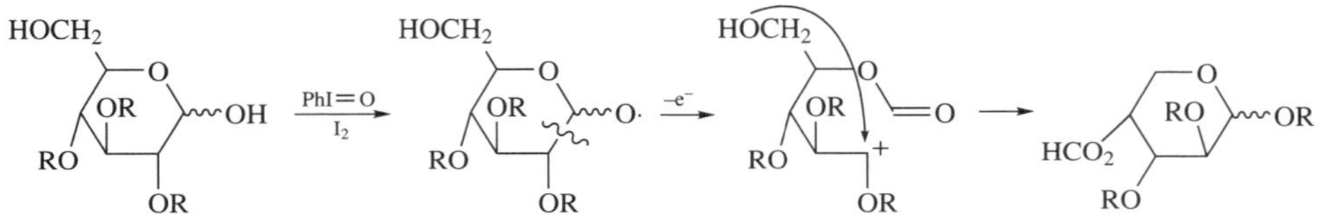

$PhI{=}O$, I_2; $-e^-$

Iodosobenzene diacetate and iodine convert pentanol to 2-methyltetrahydrofuran by a similar mechanism. The secondary radical is most likely captured by iodine or oxidized to the carbocation prior to cyclization.[271]

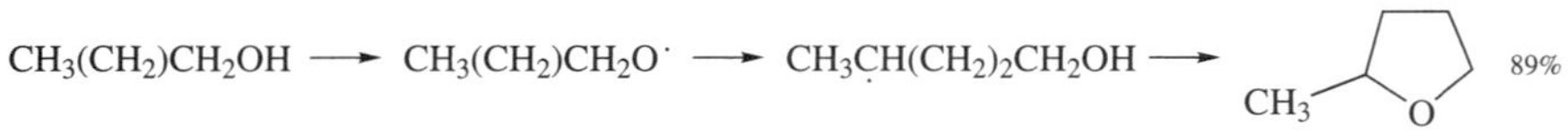

$CH_3(CH_2)CH_2OH \longrightarrow CH_3(CH_2)CH_2O^{\cdot} \longrightarrow CH_3\dot{C}H(CH_2)_2CH_2OH \longrightarrow$ 2-methyltetrahydrofuran 89%

Bicyclic lactols afford monocyclic iodolactones.

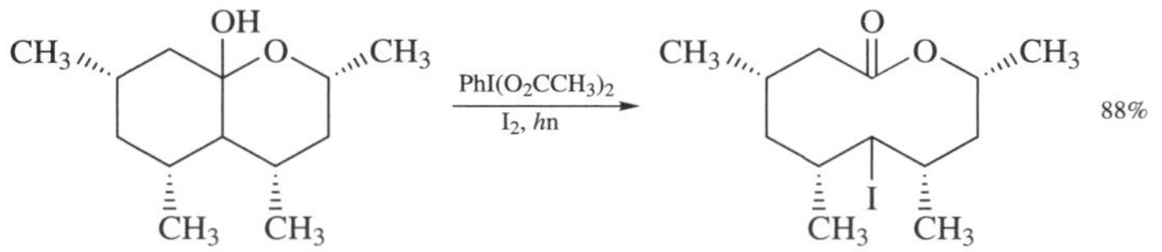

$PhI(O_2CCH_3)_2$, I_2, hn — 88%

Ref. 272

268. R. Freire, J. J. Marrero, M. S. Rodriquez, and E. Suarez, *Tetrahedron Lett.* **27:**383 (1986); M. T. Arencibia, R. Freire, A. Perales, M. S. Rodriguez, and E. Suarez, *J. Chem. Soc., Perkin Trans. 1* **1991:**3349.
269. P. de Armas, C. G. Francisco, and E. Suarez, *Angew. Chem. Intl. Ed. Engl.* **31:**772 (1992).
270. P. de Armas, C. G. Francisco, and E. Suarez, *J. Am. Chem. Soc.* **115:**8865 (1993).
271. J. L. Courtneidge, J. Lusztyk, and D. Page, *Tetrahedron Lett.* **35:**1003 (1994).

Similarly, bicyclic hemiacetals fragment to medium-size lactones.

$PhI(O_2CCH_3)_2$, I_2, $h\nu$ — Ref. 273

These reactions are believed to proceed through hypoiodite intermediates.

Alkoxy radical fragmentation is also involved in ring expansion of 3- and 4-haloalkyl cyclohexanones. The radical formed by halogen-atom abstraction adds to the carbonyl group, and fragmentation to the carboethoxy-stabilized radical then occurs.[274]

$(CH_2)_4I$, $CO_2C_2H_5$ $\xrightarrow{Bu_3SnH,\ AIBN}$ $CO_2C_2H_5$ ⟶ $CO_2C_2H_5$ 71% + $(CH_2)_3CH_3$, $CO_2C_2H_5$ 25%

The by-product results from competing reduction of the radical by hydrogen-atoms abstraction.

General References

S. P. McManus, ed., *Organic Reactive Intermediates*, Academic Press, New York, 1973.

Carbocation Cyclizations and Rearrangements

P. A. Bartlett, in *Asymmetric Synthesis*, Vol. 3, J. D. Morrison, ed., Academic Press, New York, 1984, Chapter 5.
W. S. Johnson, *Angew. Chem. Int. Ed. Engl.* **15:**9 (1976).
D. Redmore and C. D. Gutsche, *Adv. Alicyclic Chem.* **3:**1 (1971).
B. S. Thyagarajan, ed., *Mechanisms of Molecular Migrations*, Vols. 1–4, Wiley-Interscience, New York, 1968–1971.

Carbenes, Nitrenes, and Related Electron-Deficient Intermediates

G. L'Abbe, *Chem. Rev.* **69:**345 (1969).
R. A. Abramovitch and E. P. Kyba, in *The Chemistry of the Azido Group*, S. Patai, ed., John Wiley & Sons, New York, 1971, pp. 331–395.
D. Bethell, *Adv. Phys. Org. Chem.* **7:**153 (1968).
R. E. Gawley, *Org. React.* **35:**1 (1988).
M. Jones, Jr., and R. A. Moss, eds., Vols. I and II, John Wiley & Sons, New York, 1973, 1975.

272. M. Kaino, Y. Naruse, K. Ishihara, and H. Yamamoto, *J. Org. Chem.*, **55:**5814 (1990).
273. J. L. Courtneidge, J. Lusztyk, and D. Page, *Tetrahedron Lett.* **35:**1003 (1994).
274. P. Dowd and S.-C. Choi, *Tetrahedron* **45:**77 (1989); P. Dowd and S.-C. Choi, *J. Am. Chem. Soc.* **109:**6548 (1987).

W. Kirmse, *Carbene Chemistry*, Academic Press, New York, 1971.
W. Lwowski, ed., *Nitrenes*, Wiley-Interscience, New York, 1970.
E. F. V. Scriven, ed., *Azides and Nitrenes*, Academic Press, New York, 1984.

Free Radicals

A. L. J. Beckwith and K. U. Ingold, in *Rearrangements in Ground and Excited States*, Vol. 1, P. de Mayo, ed., Academic Press, 1980, Chapter 4.
B. Giese, *Radicals in Organic Synthesis: Formation of Carbon–Carbon Bonds*, Pergamon, Oxford, 1986.
B. Giese, *Angew. Chem. Int. Ed. Engl.* **24**:553 (1985).
W. Motherwell and D. Crich, *Free Radical Chain Reactions in Organic Synthesis*, Academic Press, London, 1992.
J. M. Tedder, *Angew. Chem. Int. Ed. Engl.* **21**:401 (1982).

Problems

(References for these problems will be found on page 938.)

1. Indicate the major product to be expected in the following reactions.

(a) cyclohexene + $CHCl_3$ $\xrightarrow[\text{NaOH, } H_2O]{PhCH_2\overset{+}{N}(C_2H_5)_3Cl}$

(b) adamantane + $CH_3OC(=O)N_3$ $\xrightarrow{\Delta}$

(c) $(CH_3)_2C{=}C(CH_3)_2$ + $CFCl_3$ $\xrightarrow[-120°C]{n\text{-BuLi}}$

(d) cyclooctene + $PhHgCF_3$ $\xrightarrow[12\ h]{80°C}$

(e) $(CH_3)_2C{=}C(CH_3)CH{=}NNHTs$ $\xrightarrow{NaOCH_3}$

(f) 1,3-cyclohexadiene + $N_2CHC(=O)C(=O)OC(CH_3)_3$ $\xrightarrow{Rh_2(OAc)_4}$

(g) $PhCH_2C(=O)CH(Br)CH_3 + CH_3CH_2O^- \longrightarrow$

(h) 2-furyl–C(=O)Cl $\xrightarrow{CH_2N_2}$

(i) $N{=}N$ (diazirine), H, $C(CH_3)_3$ $\xrightarrow[\text{nitrobenzene}]{\Delta}$

(j) H_5C_2, OH, $ArSO_2O$, H_5C_2, OH, CH_3 $\xrightarrow{NaH}$

(k) $(CH_3)_2C{=}CHCH_2CH_2(CH_3)C{=}CH CH_2CH_2 (CH_3)C{=}CH CH_2O_2CCH_3$ $\xrightarrow[\text{3) } NaBH_4]{\text{1) } Hg(O_3SCF_3)_2/PhN(CH_3)_2 \text{ 2) NaCl}}$

(l) $CH_3C(=O)O{-}N{=}$ (cyclohexylidene), C_5H_{11}, $Si(CH_3)_3$ $\xrightarrow{(CH_3)_3SiO_3SCF_3}$ $C_{11}H_{19}N$

(m) $ArSO_2O$, CH_3, OH $\xrightarrow{K^+ \ {}^-OC(CH_3)_3}$

(n) $(CH_3)_2CHCH_2CH_2C(=O)C(=N_2)CO_2CH_3$ $\xrightarrow{Rh_2(O_2CCH_3)_4}$

(o) CH_3, N, N, OH, CH_3 $\xrightarrow{PCl_5}$ $C_{14}H_{16}N_2$

(p) H, O, O_2CCH_3, O, CH_3, H, CO_2CH_3, CH_2OH $\xrightarrow[I_2,\ h\nu]{Pb(OAc)_4}$

(q) OH, CH_3O, O, $CH_2OC(=S)Ph$ $\xrightarrow[\text{AIBN}]{CH_2{=}CHCH_2SnBu_3,}$

(r) [structure] $\xrightarrow{h\nu}$

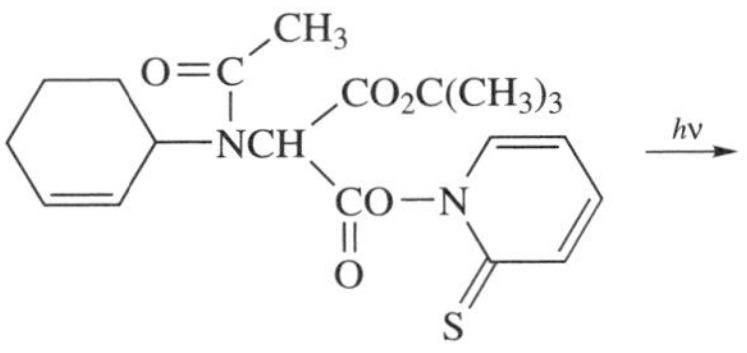

(s) [structure] $\xrightarrow{n\text{-}Bu_3SnH}$

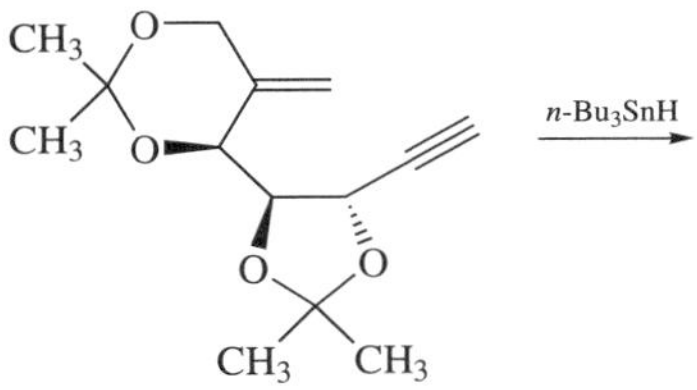

(t) [structure] $\xrightarrow{TiCl_4}$

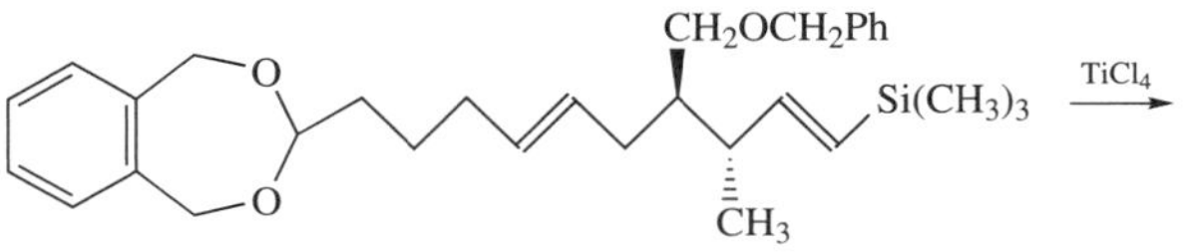

(u) [structure] $\xrightarrow{TiCl_4}$

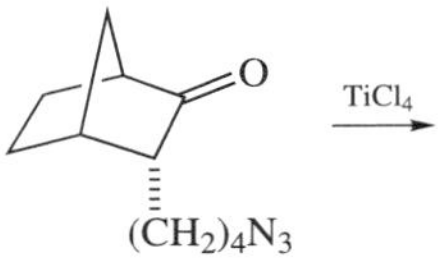

2. Indicate appropriate reagents and conditions or a short reaction sequence which could be expected to effect the following transformations.

(a) [structure] → [structure]

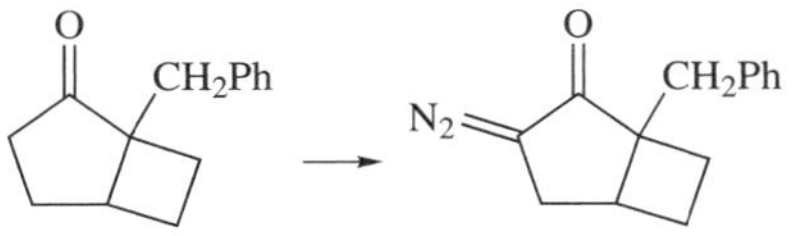

(b) [structure] → [structure]

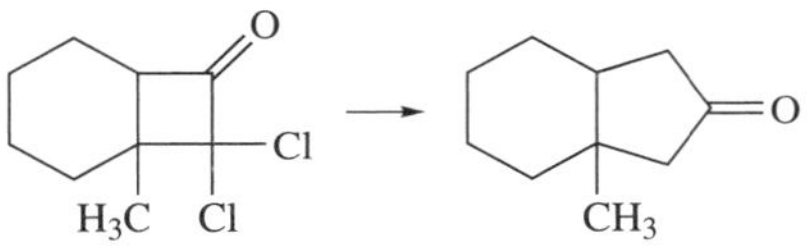

(c) [structure] → [structure]

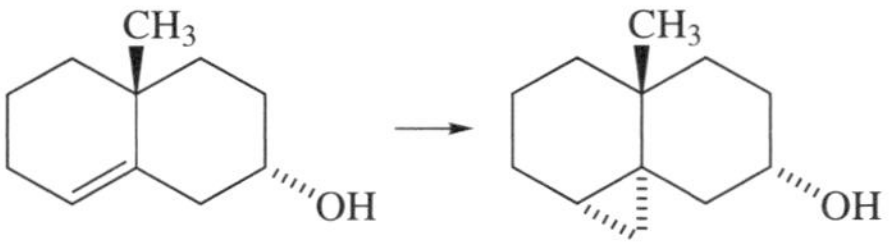

(d) [structure] → [structure]

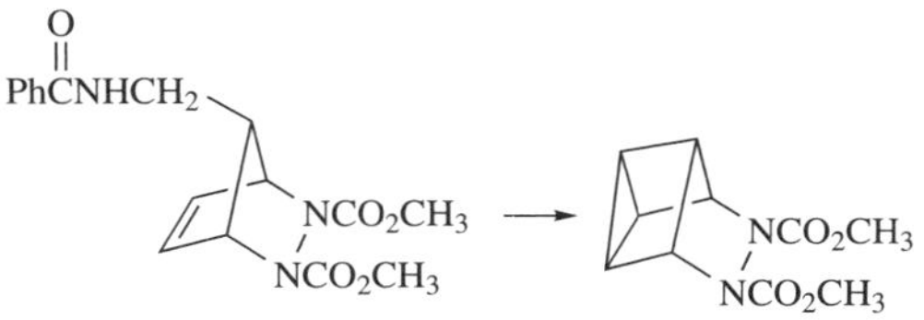

(e) $(H)(Ph)C{=}C(CO_2C_2H_5)_2$ → $(H)(Ph)C{=}C(NHCO_2CH_2Ph)(CO_2C_2H_5)$

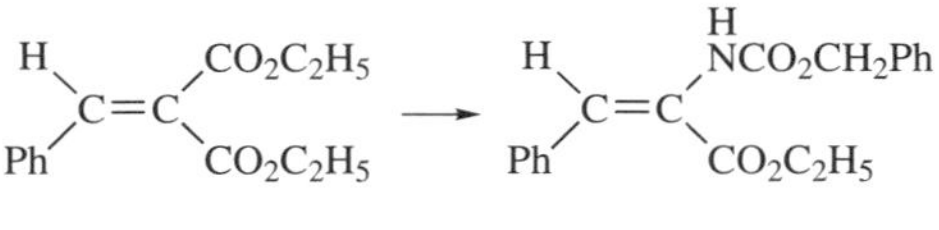

(f) [structures: 1-methyl-4-isopropylcyclohex-3-enecarboxylic acid → bicyclic ketone]

CH_3, CO_2H, $(CH_3)_2CH$ → CH_3, O, $(CH_3)_2C$

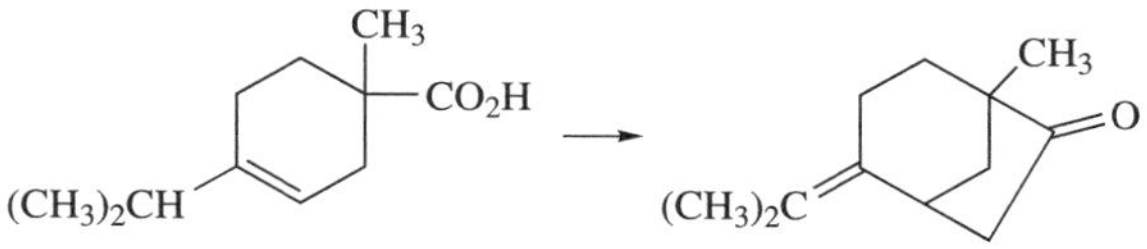

(g) $CH_2{=}CHCH{=}CHCO_2H$ → $CH_2{=}CHCH{=}CH\overset{H}{N}CO_2CH_2Ph$

(h) CH_3, O, H_3C → CH_3, H_3C

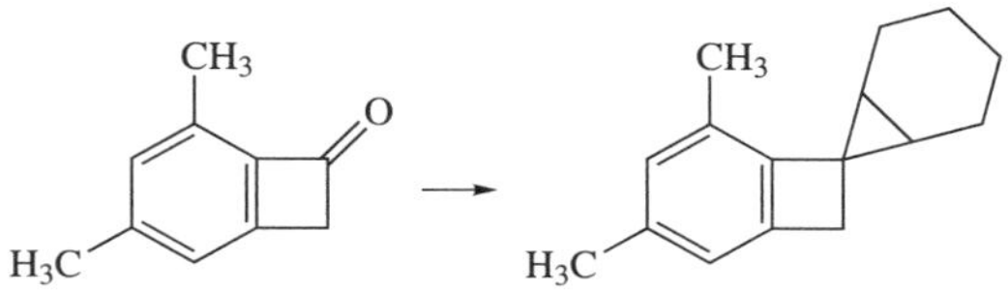

(i) CH_3, OCH_3 → $CO_2C_2H_5$, H_3C, OCH_3

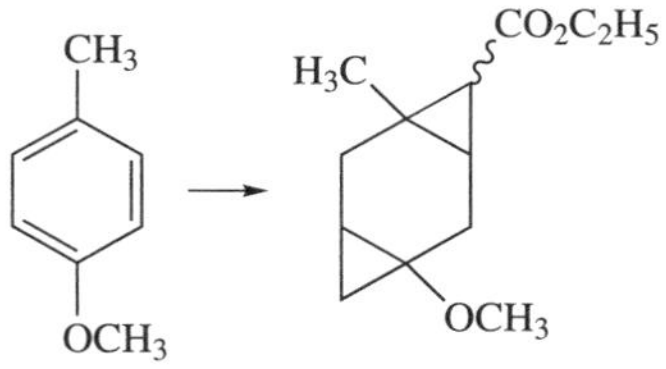

(j) H, OH, H, CH_2CH_2, C, C, H, CH_3O, $CHCH_2$, CH_3, CH_3 → CH_3, CH_3O, CH_3

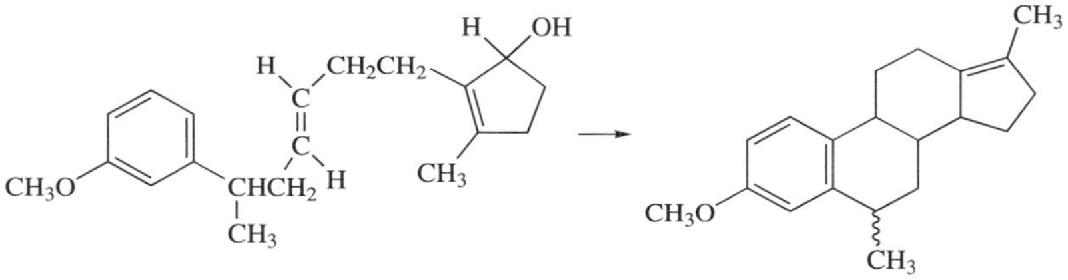

(k) OH, CH_3O, CH_3 → H, O, H, CH_3

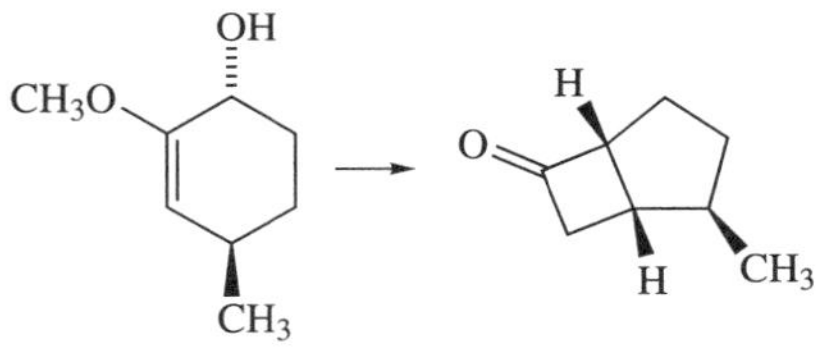

(l) O, CO_2CH_3, CH_3O_2C, $CH_2CH_2C{=}CH_2$, CH_3 → O, O, O, CH_3

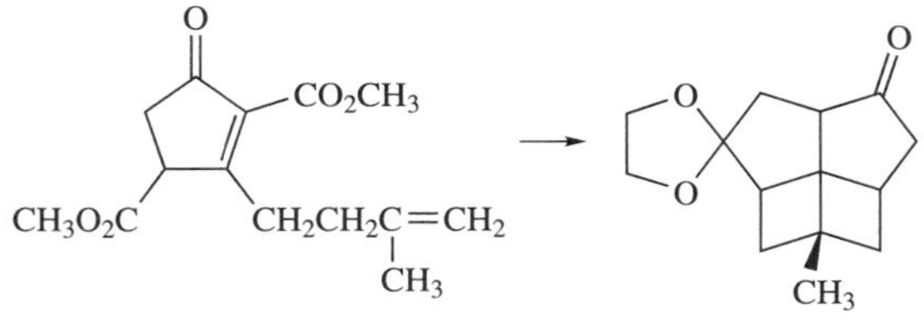

(m) $AcOCH_2$, O, AcO, I, AcO, OAc → $AcOCH_2$, O, O, NC_2H_5, AcO, O, AcO, OAc

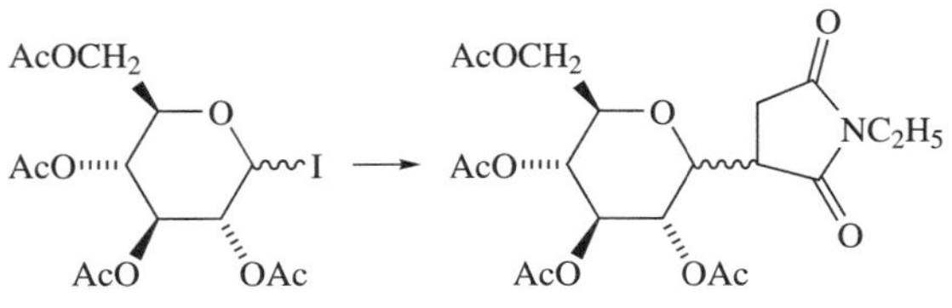

(n) $CH_3(CH_2)_7CO_2H$ → $CH_3(CH_2)_9CN$

(o) OC_2H_5, $OCHCH_2I$, R_3SiO → OC_2H_5, R_3SiO, CN

(p) $CH_2CCH_2CO_2H$ (O) → H, O

(q) CH_3, CH_3, O, O, OCH_2Ph, HO, O → CH_3, CH_3, O, O, OCH_2Ph, $CH_2{=}CHCH_2$, O

(r) $AcOCH_2$, O, AcO, OAc → $AcOCH_2$, O, O, AcO, AcO, CH_2

(s) HO, H_3C, H, $CH_2OTBDMS$ → H_3C, H, $CH_2OTBDMS$, O, O

(t) CH_3O, CH_3O, N, CH_2CH_2SePh, O → CH_3O, CH_3O, N, H, O

3. Each of the following carbenes has been predicted to have a singlet ground state, either as the result of qualitative structural considerations or on the basis of theoretical calculations. Indicate what structural feature in each case might lead to stabilization of the singlet state.

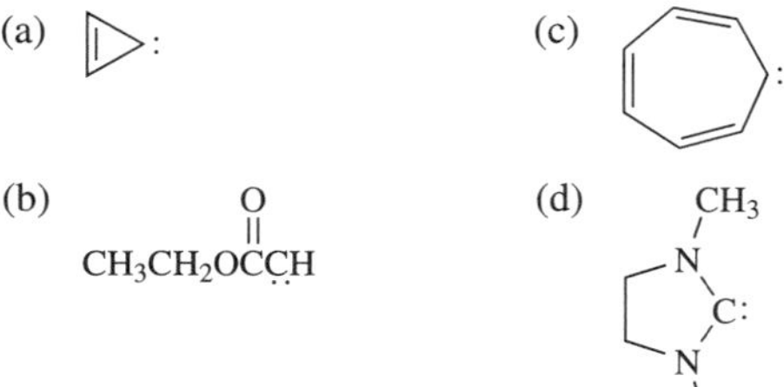

4. The hydroxyl group in *trans*-cycloocten-3-ol determines the stereochemistry of reaction of this compound with the Simmons–Smith reagent. By examining a model, predict the stereochemistry of the resulting product.

5. Discuss the significance of the relationships between substrate stereochemistry and product composition exhibited by the reactions shown below.

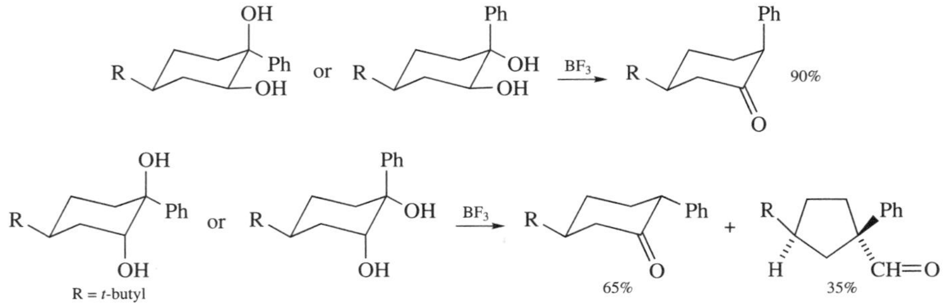

OH, Ph, OH or Ph, OH, OH $\xrightarrow{BF_3}$ Ph, O 90%

OH, Ph, OH or Ph, OH, OH $\xrightarrow{BF_3}$ Ph, O 65% + R, H, Ph, CH=O 35%

R = *t*-butyl

6. Suggest a mechanistic rationalization for the following reactions. Point out the structural features which contribute to the unusual or abnormal course of the reaction. What product might have been expected if the reaction followed a "normal" course?

(a) NOH, S $\xrightarrow{SOCl_2}$ C≡N, CH_2SCH_2Cl

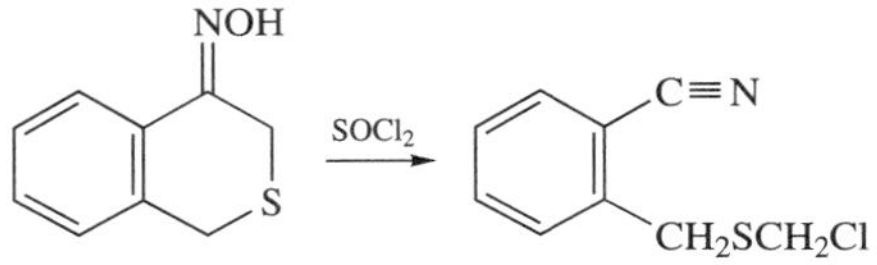

(b) H_3C, CH_3, C, $CCHN_2$, O $\xrightarrow[CH_3OH]{Cu(acac)_2}$ $C(CH_3)_2$, $CH_2CO_2CH_3$

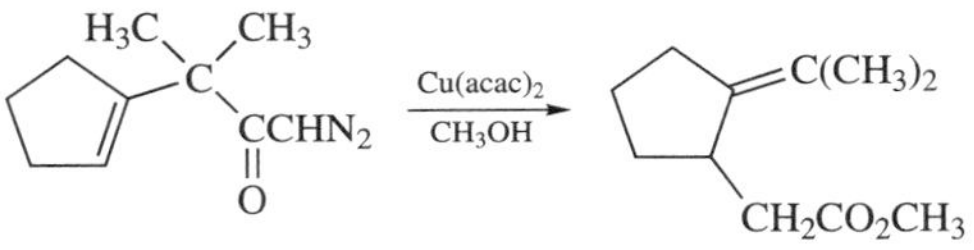

(c) O, $CCCO_2C_2H_5$, N_2, $OCH_2CH{=}CH_2$ $\xrightarrow{Rh_2(O_2CCH_3)_4}$ O, $CO_2C_2H_5$, O, $CH_2CH{=}CH_2$

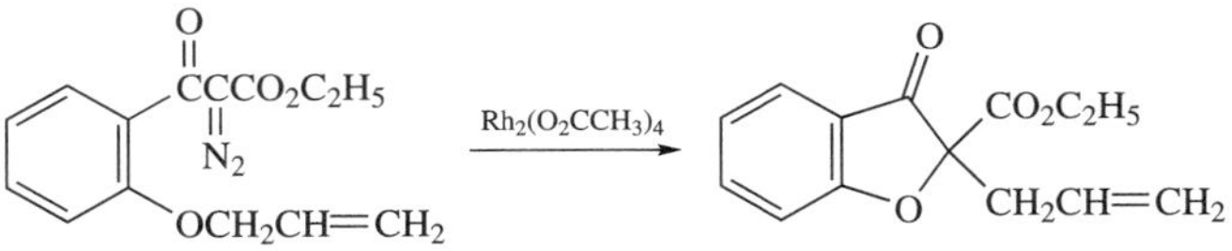

7. Give the structure of the expected Favorskii rearrangement product of compound **A**.

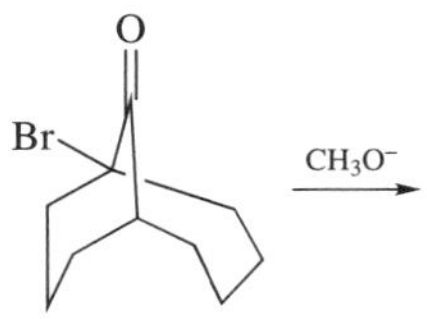

O, Br $\xrightarrow{CH_3O^-}$

A

Experimentally, it has been found that the above ketone can rearrange by either the cyclopropanone or the semibenzilic mechanism, depending on the reaction conditions. Devise two experiments that would permit you to determine which mechanism was operating under a given set of circumstances.

8. Predict the major product of the following reactions.

(a) [structure: bicyclic ketone bearing $CCHN_2$ (O=) and H] $\xrightarrow[\text{toluene, 105°C, 2 h}]{CuSO_4}$

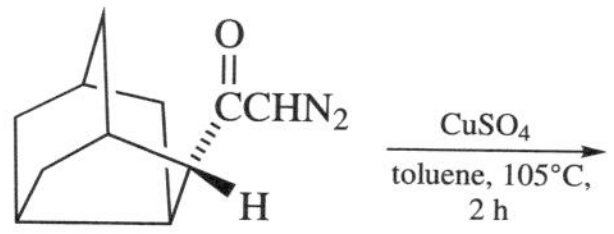

(b) [structure: polycyclic ketone bearing OSO_2Ar] $\xrightarrow[\text{2) } H^+]{\text{1) }^-OH\text{, 110°C}}$

(c) [structure: decalin bearing CH_3, H, H_3C, CH_3O, OTs, OH] $\xrightarrow{t\text{-AmO}^-}$

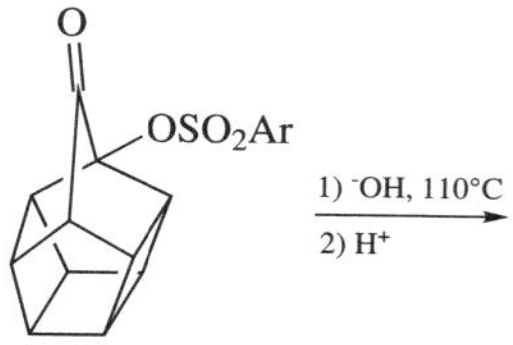

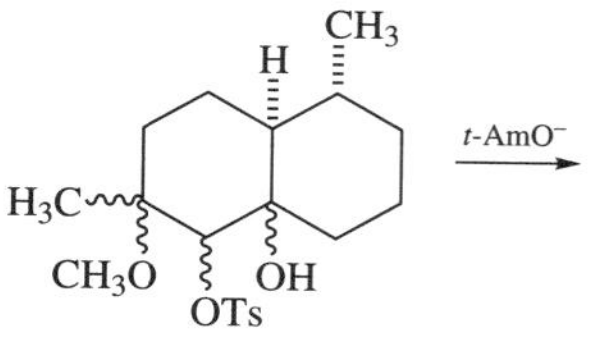

(d) [structure: H_3C, H_3C-cyclopentene with $CH{=}CHCH_2CH_2C(=O)C(=N_2)CH_3$ side chain] $\xrightarrow[\text{benzene, 80°C, 12 h}]{Cu(acac)_2}$

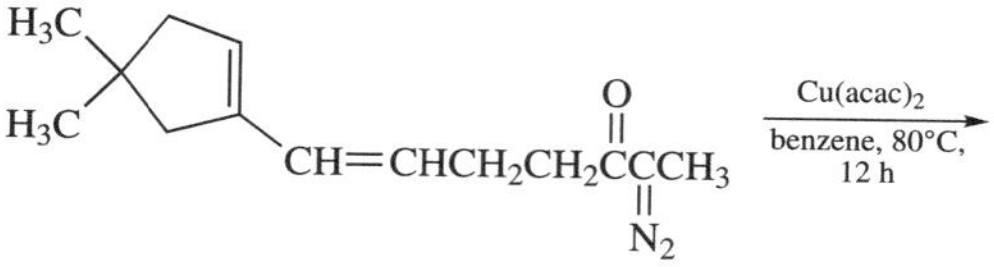

(e) [structure: $O{=}CHCH_2CH_2$-substituted bicyclic alkene bearing CH_3, CH_3, CH_3] $\xrightarrow[\text{benzene, 10°C,}]{SnCl_4}$

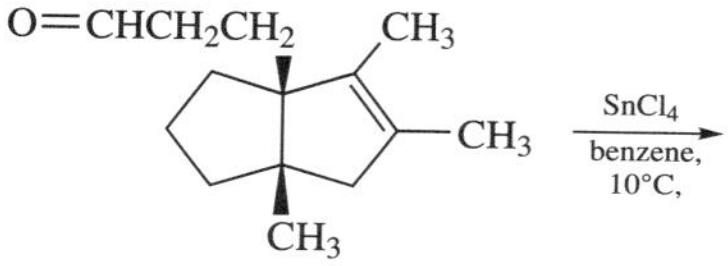

(f) [structure: bicyclic sulfone ($O{=}S{=}O$) bearing Cl] $\xrightarrow[H_2O\text{, 100°C,}]{KOH}$

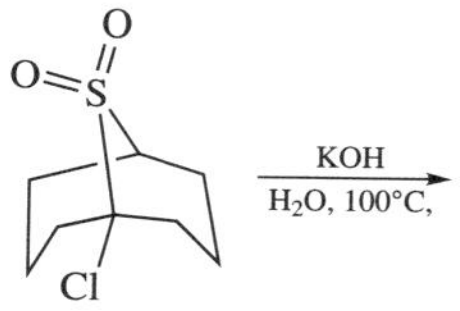

(g) [structure: polycyclic diol bearing CH_3, H, CH_3O, H, CH_3, OH, CH_3, CH_3, CH_3, CH_3, OH] $\xrightarrow[\text{2) } K^{+\,-}O\text{-}t\text{-}C_4H_9]{\text{1) } CH_3SO_2Cl\text{, } (C_2H_5)_3N}$

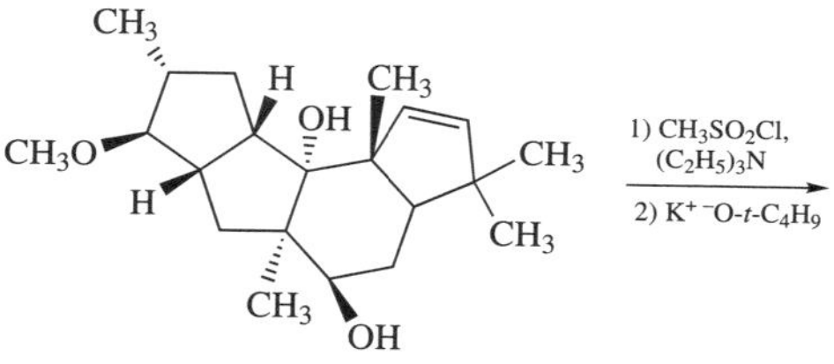

9. Short reaction series can effect formation of the desired material on the left from the starting material on the right. Devise an appropriate reaction sequence.

(a) [structure: bicyclo[4.1.0]heptane bearing $OSi(CH_3)_3$] $\Longrightarrow$ [structure: cyclohexanone]

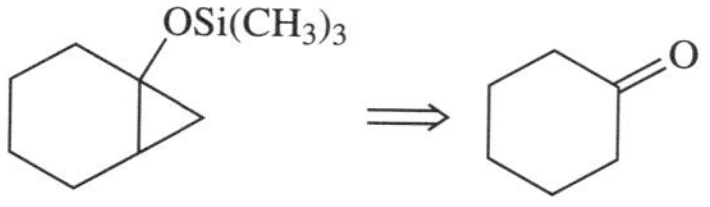

(b) OH ⟹ CH_2OH

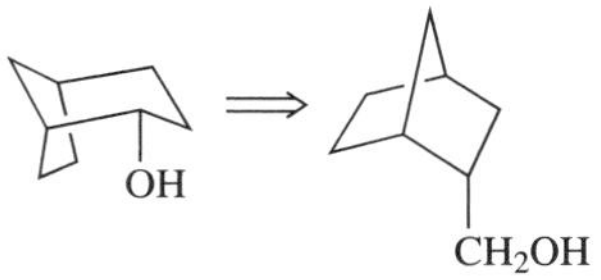

(c) $H_2C{=}CH$, H, OCH_3, O, $O{=}CHCH_2$, H, CO_2CH_3 ⟹ $HOCH_2$, H, OCH_3, CH_3CO_2, O, H, CO_2CH_3

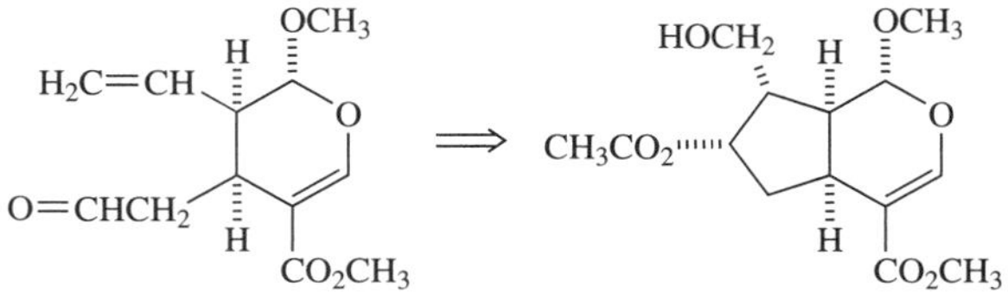

(d) H_3C, H_3C, CH_3, OH, CH_3 ⟹ H_3C, H_3C, CH_3

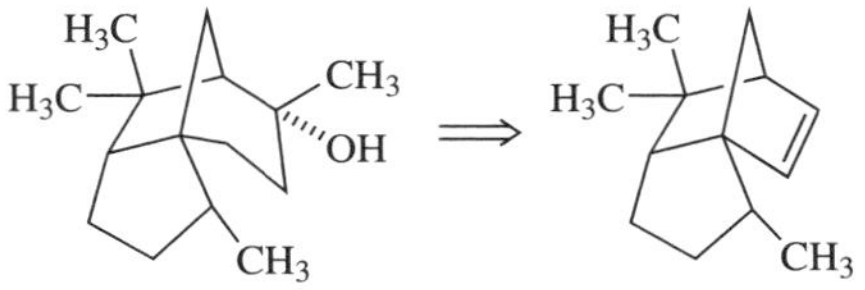

(e) CH_3, O, CO_2CH_3, CH_3 ⟹ CH_3, CH_2OH, CH_3

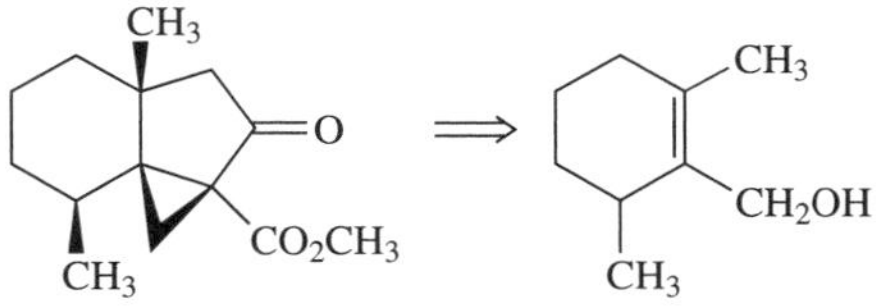

(f) H, H, H, O, H, H, H, O ⟹ H, H, O, H, H

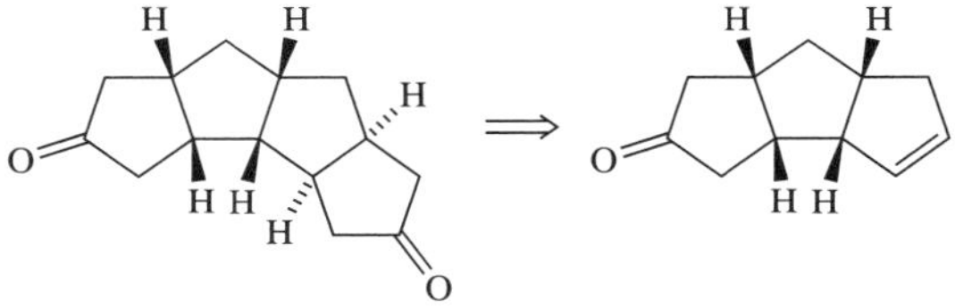

(g) CH_2OCH_3, O, H, OCH_2Ph, Ph, O ⟹ O, OH, O, H, Ph, O, OCH_2Ph

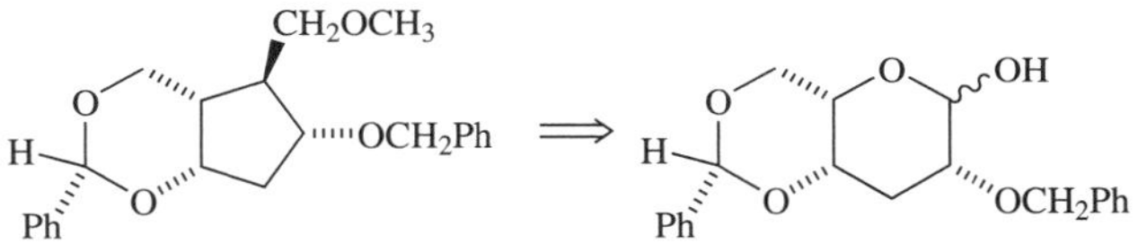

(h) ⟹ O, O, O

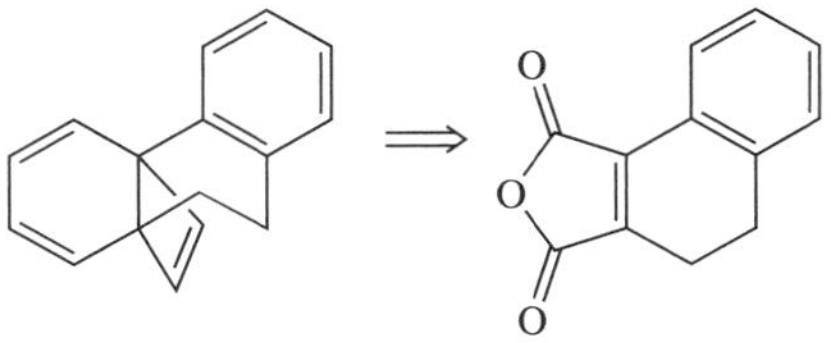

(i) HO, CH_2, H_3C, $CH_2CH_2OCH_2Ph$, CH_2OH, H, OH ⟹ H_2C, CH_3, C, $CH{=}O$, H_3C, $CHCH_2CH_2OCH_2Ph$, O, CH_3, H, O, CH_3

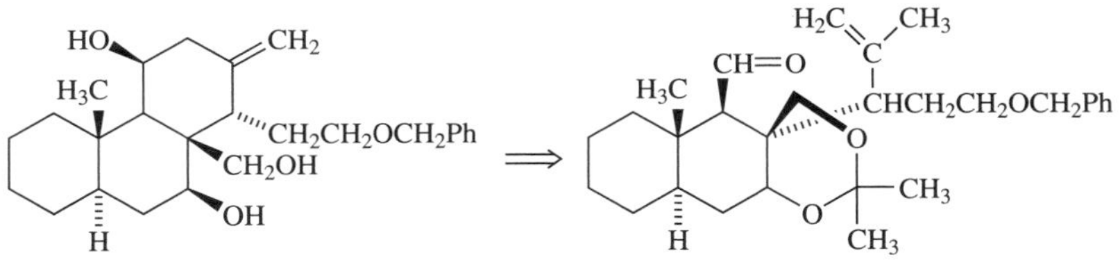

(j) O, H, C, C, CH_3, H ⟹ H, $CH_2CH_2CH_2CO_2H$, H, C=C, C=C, H, CH_3, H

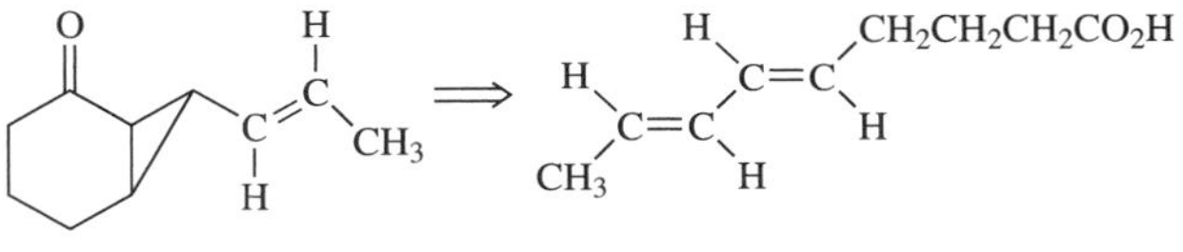

(k) [structure: ketal-protected tricyclic ketone with CH_2, H_3C, H labels] $\Longrightarrow$ [structure: spiro ketone with $CH_2Si(CH_3)_3$, H_3C, O, H labels]

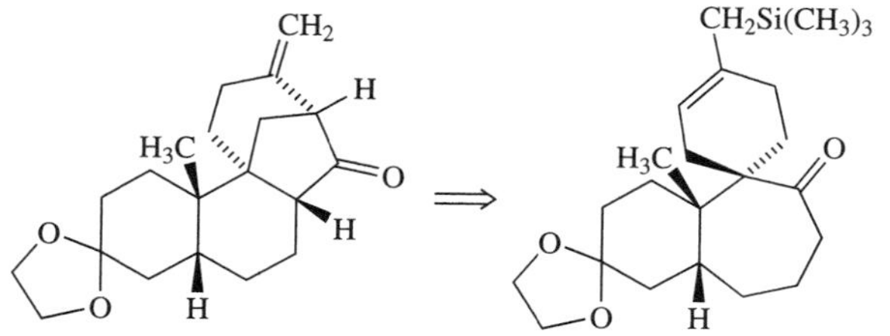

(l) [structure with CH_3, H, H, $CH_2CO_2C(CH_3)_3$, O labels] $\Longrightarrow$ [structure: 3-methylbenzoic acid, CH_3, CO_2H]

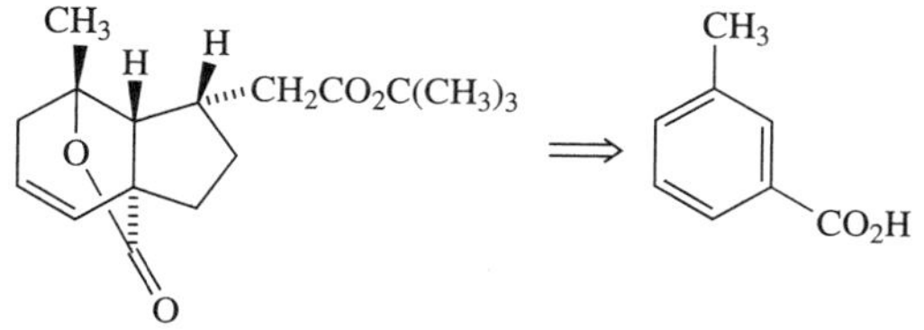

10. Formulate mechanisms for the following reactions.

(a) $(CH_3)_2C{=}CHCH_2CH_2$–[tricyclic ketone with CH_3, OSO_2CH_3] $\xrightarrow{NaNH_2}$ $(CH_3)_2C{=}CHCH_2CH_2$–[bicyclic with $\overset{O}{\overset{\|}{C}}NH_2$, CH_3]

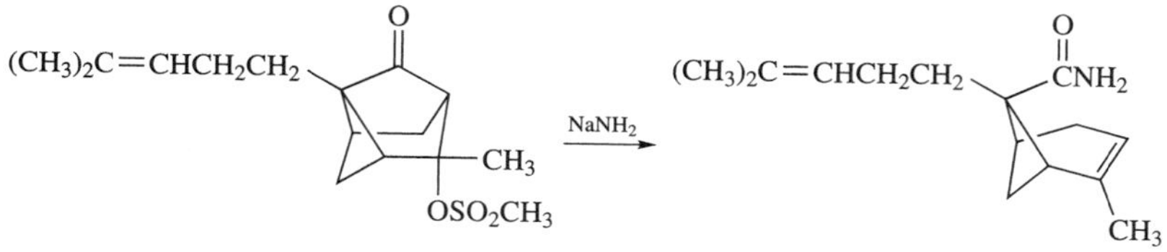

(b) [Cl-substituted tricyclic ketone] $\xrightarrow[2)\ H^+]{1)\ KOH}$ [tricyclic CO_2H]

(c) [Cl-substituted bicyclic ketone] $\xrightarrow[2)\ H^+]{1)\ KOH}$ $H_2C{=}CHCH_2\underset{\underset{CH_2}{\|}}{C}CO_2H$

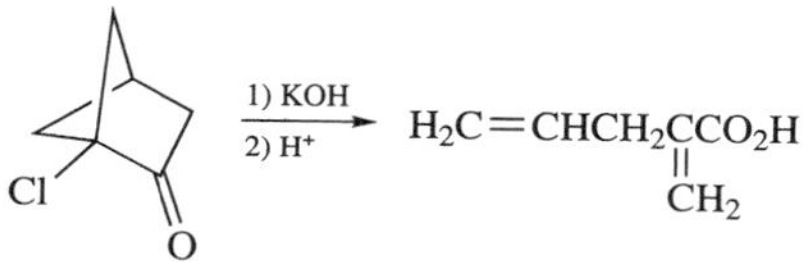

(d) $CH_2{=}CHCH{=}CH_2 + N_2CH\overset{O}{\overset{\|}{C}}CO_2C_2H_5 \xrightarrow{Rh_2(OAc)_4}$ [cyclopropane with $CH_2{=}CH$, $CO_2C_2H_5$, H] + [oxepine ring with $CO_2C_2H_5$]

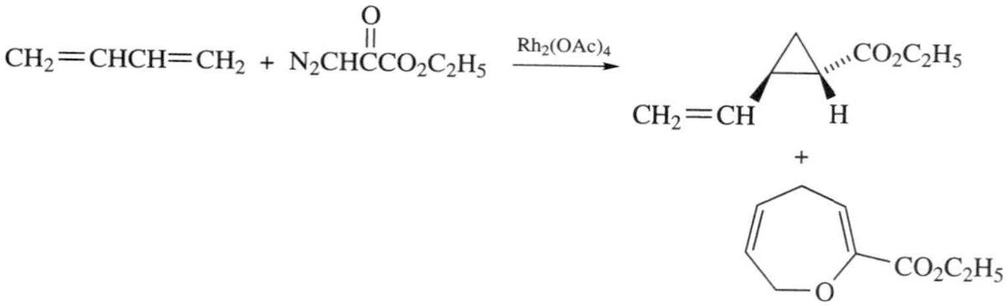

(e) [H, CH_3, C=C, H] $CH_2CH_2CH_2\overset{O}{\overset{\|}{C}}\underset{CO_2CH_3}{\underset{|}{C}}HCH_2CH{=}CH_2 \xrightarrow[Cu(OAc)_2]{Mn(OAc)_3}$ [bicyclic ketone with O, CO_2CH_3, $=CH_2$, H, H, CH_3]

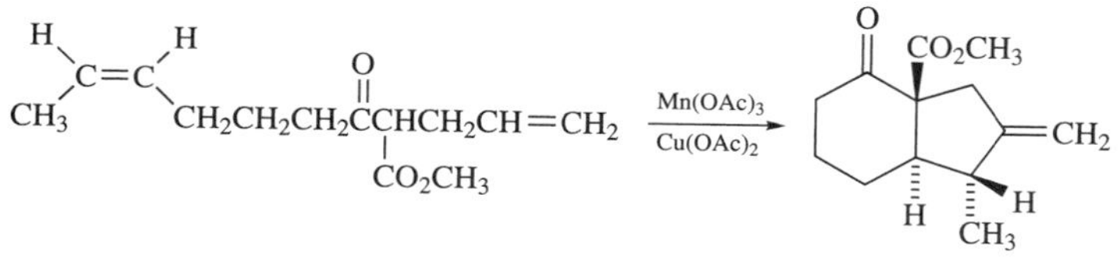

(f) $CH_3\underset{I}{\underset{|}{C}}HCO_2SnBu_3 + C_4H_9CH{=}CH_2 \xrightarrow{AIBN}$ [lactone with CH_3, O, O, C_4H_9]

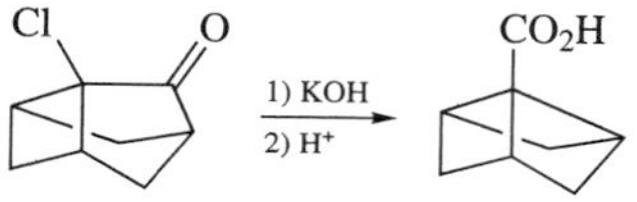

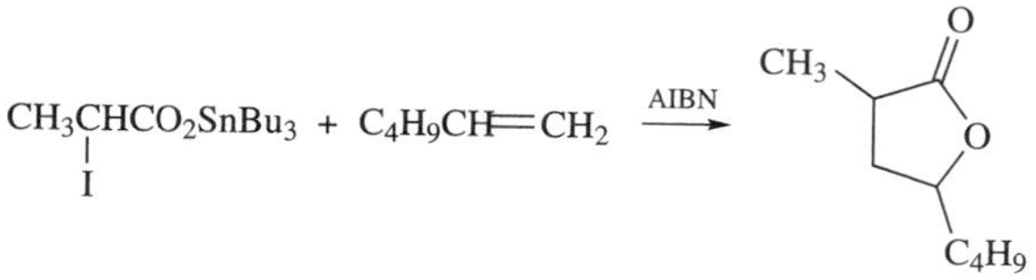

(g) Bu_3SnH, AIBN; $(CH_2)_3Ph$; CH_2; H; Ph; OH

(h) $(CH_3)_3Si$; Ph; $C{=}CH_2$; $(CH_2)_2CH(OCH_3)_2$; $(CH_3)_3SiO_3SCF_3$, 2,6-di-*t*-butylpyridine; O; Ph; OCH_3; H

90% yield, 2:1 mixture of stereoisomers

(i) Bu_3SnH, AIBN; OC(=S)–N (imidazole); CH_3; CH_3; $CH(CH_3)_2$; 53%; CH_3; CH_2CH_3; CH_3; CH_3; 14%

(j) $CH_2{=}CH$; Ph; Ph; + $CH_2{=}CHCO_2C(CH_3)_3$; cat PhSH; O; $CH_2{=}CH$; $CO_2C(CH_3)_3$; 51%

(k) Br; $CO_2C_2H_5$; Bu_3SnH, $(C_2H_5)_3B$; $CH_2CO_2C_2H_5$; O; H; 67%

(l) $PhCH_2$; N_2; N; $CO_2C_2H_5$; CH_3; O; O; $Rh_2(O_2CC_4F_9)_4$; O; O; $CO_2C_2H_5$; N; $PhCH_2$; CH_3

(m) CH_3CO_2; O; PhCONH; O; CH_3; OH; OTs; Ph; OTBDMS; HO; $PhCO_2$; O_2CCH_3; n-$Bu_4N^+F^-$, THF

CH_3CO_2; O; PhCONH; O; CH_3; $CH{=}O$; Ph; OTBDMS; HO; $PhCO_2$; H; OH; $CH_2O_2CCH_3$

(n) 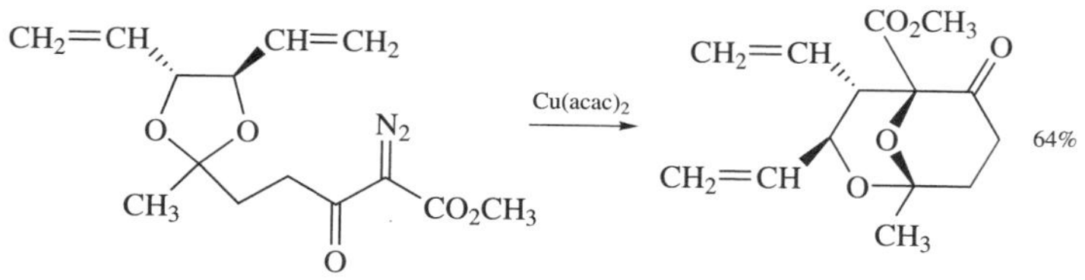 [Scheme: diazo keto ester with dioxolane bearing two $CH_2{=}CH$ groups and CH_3, N_2, CO_2CH_3 → $Cu(acac)_2$ → bicyclic ketone product with CO_2CH_3, two $CH_2{=}CH$ groups, CH_3; 64%]

11. A sequence of reactions for converting acyclic and cyclic ketones α,β-unsaturated ketones with an additional $=CCH_3$ unit has been developed.

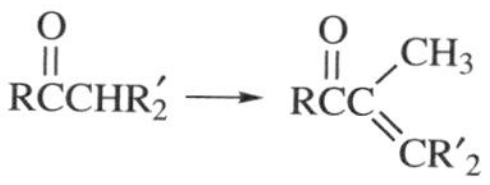

$$RCCHR'_2 \;(C{=}O) \longrightarrow RC(=O)C(CH_3){=}CR'_2$$

The method utilizes a carbenoid reagent, 1-lithio-1,1-dichloroethane, as a key reagent. The overall sequence involves three steps, one of them before and one of them after the carbenoid reaction. Atttempt, by analysis of the bond changes and with your knowledge of carbene chemistry, to devise such a reaction sequence.

12. The synthesis of globulol from the octalin derivative shown proceeds in four stages. These include, not necessarily in sequence, addition of a carbene, fragmentation, and an acid-catalyzed cyclization of a cyclodeca-2,7-dienol. The final stage of the process converts a dibromocyclopropane to a dimethylcyclopropane using dimethylcuprate. Working back from globulol, attempt to discover the appropriate sequence of reactions and suggest appropriate reagents for each step.

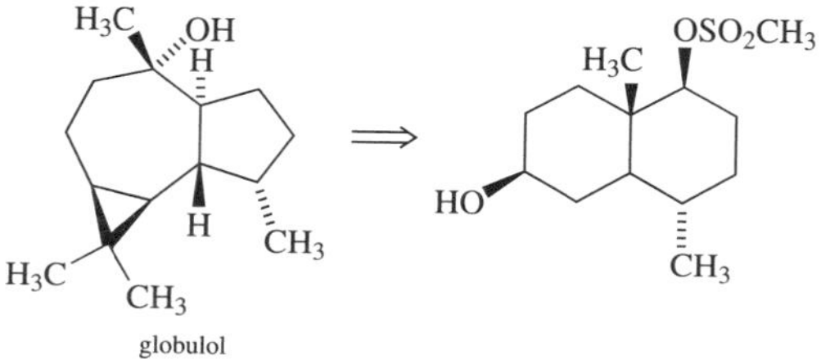

[Scheme: globulol (H_3C, OH, H, H, CH_3, H_3C, CH_3) ⟹ decalin derivative with OSO_2CH_3, H_3C, HO, CH_3]

globulol

13. Both the *E* and *Z* isomers of vinylsilane **B** have been subjected to polyene cyclization under the influence of $TiCl_4/Ti(O\text{-}i\text{-}Pr)_4$. While the *Z*-isomer gives an 85–90% yield, the *E*-isomer affords only a 30–40% yield. Offer an explanation.

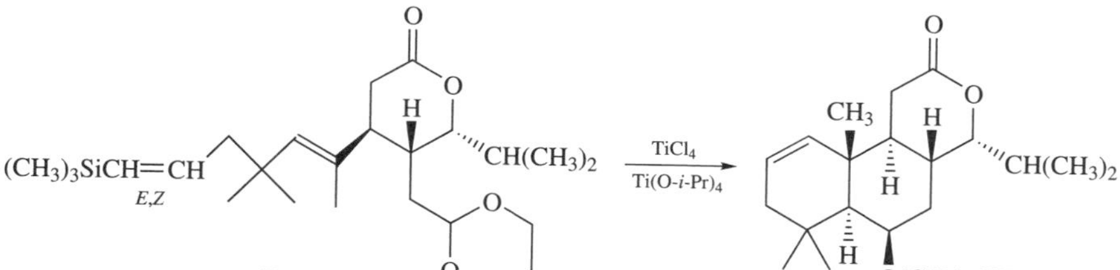

[Scheme: **B**, $(CH_3)_3SiCH{=}CH$ (*E,Z*) polyene lactone with $CH(CH_3)_2$ and dioxane acetal → $TiCl_4$, $Ti(O\text{-}i\text{-}Pr)_4$ → tricyclic lactone with CH_3, H, H, H, $CH(CH_3)_2$, $O(CH_2)_3OH$]

14. The three decahydroquinolines shown below each give a different product composition on solvolysis. One gives 9-methylamino-*trans*-5-nonenal, one gives 9-methylamino-*cis*-nonenal, and the third gives a mixture of the two quinoline derivatives **F** and **G**. Deduce which compound gives rise to which product. Explain your reasoning.

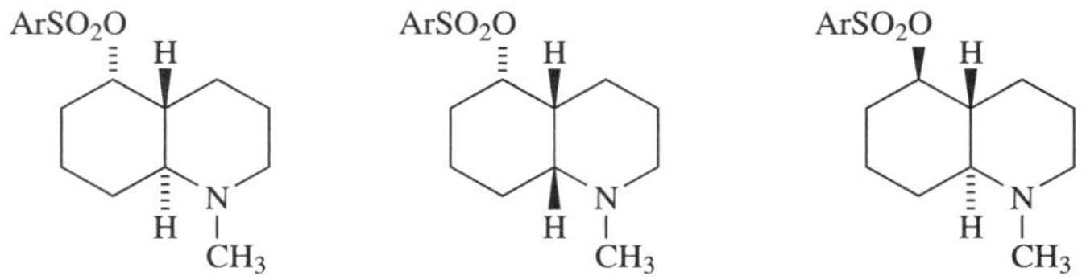
ArSO2O
H
N
H
CH3
ArSO2O
H
N
H
CH3
ArSO2O
H
N
H
CH3

C **D** **E**

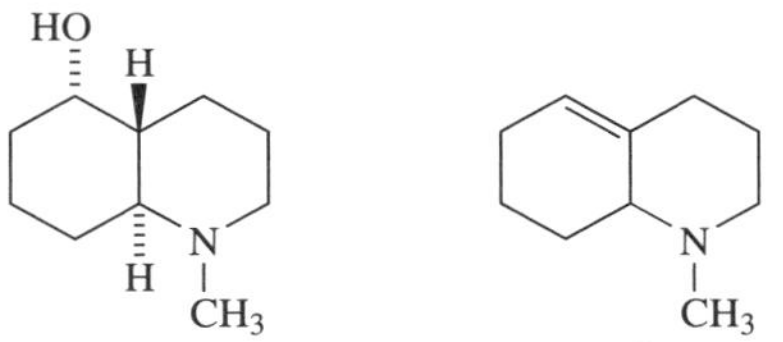
HO
H
N
H
CH3
N
CH3

F **G**

15. Normally, the dominant reaction between acyl diazo compounds and simple α,β-unsaturated carbonyl compounds is a cycloaddition.

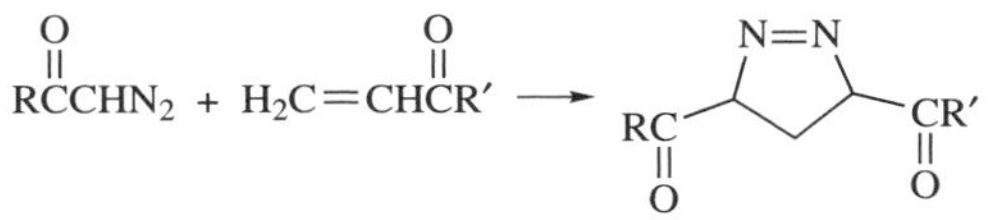
RCCHN2 + H2C=CHCR'
N=N
RC
CR'

If, however, the reaction is run in the presence of a Lewis acid, particularly antimony pentafluoride, the reaction takes a different course, giving a diacyl cyclopropane.

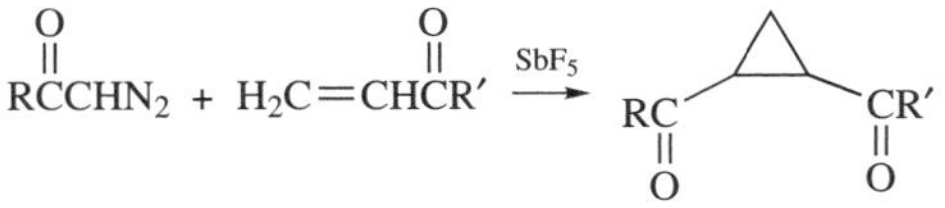
RCCHN2 + H2C=CHCR'
SbF5
RC
CR'

Formulate a mechanism to account for the altered course of the reaction in the presence of SbF_5.

16. Compound **H** on reaction with Bu_3SnH in the presence of AIBN gives **I** rather than **J**. How is **I** formed? Why is **J** not formed? What relationship do these results have to the rate data given on p. 675?

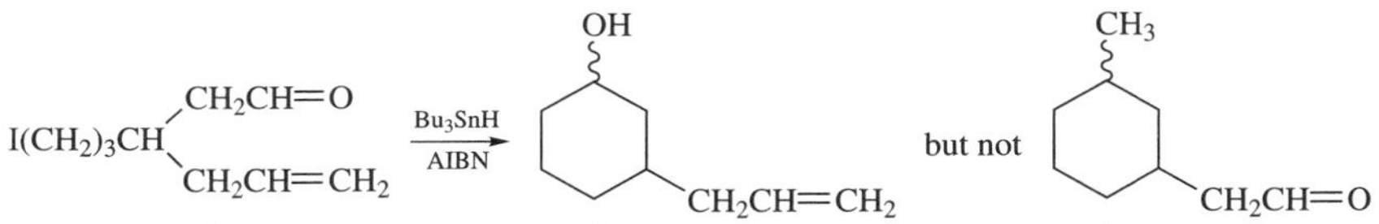
CH2CH=O
I(CH2)3CH
CH2CH=CH2
Bu3SnH
AIBN
OH
CH2CH=CH2
but not
CH3
CH2CH=O

H **I** **J**

11

Aromatic Substitution Reactions

Introduction

This chapter is concerned with reactions that introduce or replace substituent groups on aromatic rings. The most important group of reactions is electrophilic aromatic substitution. The mechanism of electrophile aromatic substitution has been studied in great detail, and much information is available about structure–reactivity relationships. There are also important reactions which occur by nucleophilic substitution, including reactions of diazonium ion intermediates and metal-catalyzed substitution. The mechanistic aspects of these reactions were discussed in Chapter 10 of Part A. In this chapter, the synthetic aspects of aromatic substitution will be emphasized.

11.1. Electrophilic Aromatic Substitution

11.1.1. Nitration

Nitration is the most important method for introduction of nitrogen functionality on aromatic rings. The nitro compounds can easily be reduced to the corresponding amino derivatives, which can provide access to diazonium ions. There are several reagent systems that are useful for nitration. A major factor in the choice of reagent is the reactivity of the ring to be nitrated. Concentrated nitric acid can effect nitration, but it is not as reactive as a mixture of nitric acid with sulfuric acid. The active nitrating species in both media is the nitronium ion, NO_2^+. The NO_2^+ ion is formed by protonation and dissociation of nitric acid. The concentration of NO_2^+ is higher in the more acidic sulfuric acid than in nitric acid.

$$HNO_3 + 2\,H^+ \rightleftharpoons H_3O^+ + NO_2^+$$

Nitration can also be carried out in organic solvents, with acetic acid and nitromethane being common examples. In these solvents, the formation of the NO_2^+ ion is often the rate-controlling step[1]:

$$2\,HNO_3 \rightleftharpoons H_2NO_3^+ + NO_3^-$$
$$H_2NO_3^+ \xrightarrow{\text{slow}} NO_2^+ + H_2O$$
$$ArH + NO_2^+ \xrightarrow{\text{fast}} ArNO_2 + H^+$$

Another useful medium for nitration is a solution prepared by dissolving nitric acid in acetic anhydride. This generates acetyl nitrate:

$$HNO_3 + (CH_3CO)_2O \rightleftharpoons CH_3\overset{O}{\overset{\|}{C}}ONO_2 + CH_3CO_2H$$

This reagent tends to give high *ortho* : *para* ratios for some nitrations.[2] Another convenient procedure involves reaction of the aromatic compound in chloroform or dichloromethane with a nitrate salt and trifluoroacetic anhydride.[3] Presumably, trifluoroacetyl nitrate is generated under these conditions.

$$NO_3^- + (CF_3CO)_2O \longrightarrow CF_3\overset{O}{\overset{\|}{C}}ONO_2 + CF_3CO_2^-$$

Nitration can be catalysed by lanthanide salts. For example, the nitration of benzene, toluene, and naphthalene by aqueous nitric acid proceeds in good yield in the presence of $Yb(O_3SCF_3)_3$.[4] The catalysis presumably results from an oxyphilic interaction of nitrate ion with the cation, which generates or transfers the NO_2^+ ion.[5] This catalytic procedure uses a stoichiometric amount of nitric acid and avoids the excess strong acid associated with conventional nitration conditions.

$$Ln^{3+}\text{---}O\text{–}N(O^-)(=O) \longrightarrow [O{=}N^+{=}O]$$

Salts containing the nitronium ion can be prepared, and they are reactive nitrating agents. The tetrafluoroborate salt has been used most frequently,[6] but the trifluoromethan-

1. E. D. Hughes, C. K. Ingold, and R. I. Reed, *J. Chem. Soc.* **1950:**2400; J. G. Hoggett, R. B. Moodie, and K. Schofield, *J. Chem. Soc., B* **1969:**1; K. Schofield, *Aromatic Nitration*, Cambridge University Press, Cambridge, 1980, Chapter 2.
2. A. K. Sparks, *J. Org. Chem.* **31:**2299 (1966).
3. J. V. Crivello, *J. Org. Chem.* **46:**3056 (1981).
4. F. J. Waller, A. G. M. Barrett, D. C. Braddock, and D. Ramprasad, *Chem. Commun.* **1997:**613.
5. F. J. Waller, A. G. M. Barrett, D. C. Braddock, R. M. McKinnell, and D. Ramprasad, *J. Chem. Soc., Perkin Trans. 1*, **1999:**867.
6. S. J. Kuhn and G. A. Olah, *J. Am. Chem. Soc.* **83:**4564 (1961); G. A. Olah and S. J. Kuhn, *J. Am. Chem. Soc.* **84:**3684 (1962); G. A. Olah, S. C. Narang, J. A. Olah, and K. Lammertsma, *Proc. Natl. Acad. Sci. U.S.A.* **79:**4487 (1982); C. L. Dwyer and C. W. Holzapfel, *Tetrahedron* **54:**7843 (1998).

sulfonate can also be prepared readily.[7] Nitrogen heterocycles like pyridine and quinoline form N-nitro salts on reaction with NO_2BF_4.[8] These N-nitro heterocycles can, in turn, act as nitrating reagents. This is called "transfer nitration." (See entry 9 in Scheme 11.1.)

Another nitration procedure uses ozone and nitrogen dioxide.[9] With aromatic hydrocarbons and activated derivatives, this nitration is believed to involve the radical cation of the aromatic reactant

$$NO_2 + O_3 \longrightarrow NO_3 + O_2$$

$$ArH + NO_3 \longrightarrow [ArH\cdot^+] + NO_3^-$$

$$[ArH]^{\cdot+} + NO_2 \longrightarrow [Ar\langle^{H}_{NO_2}]^+ \longrightarrow ArNO_2 + H^+$$

When this mechanism operates, position selectively is governed by the SOMO distribution of the radical cation. Compounds such as phenylacetate esters and phenylethyl ethers that have oxygen substituents that can serve as directing groups show high *ortho* : *para* ratios.

$$C_6H_5(CH_2)_2OCH_3 \xrightarrow{O_3,\ NO_2} O_2N\text{-}C_6H_4(CH_2)_2OCH_3 \quad o{:}m{:}p = 79{:}2{:}19$$

Nitration is a very general reaction, and satisfactory conditions can normally be developed for both activated and deactivated aromatic compounds. Because each successive nitro group reduces the reactivity of the ring, it is easy to control conditions to obtain a mononitration product. If polynitration is desired, more vigorous conditions are used. Scheme 11.1 gives some examples of nitration reactions.

11.1.2. Halogenation

The introduction of the halogens onto aromatic rings by electrophilic substitution is an important synthetic procedure. Chlorine and bromine are reactive toward aromatic hydrocarbons, but Lewis acid catalysts are normally needed to achieve desirable rates. Elemental fluorine reacts very exothermally, and very careful control of conditions is required. Iodine can effect substitution only on very reactive aromatics, but a number of more reactive iodination reagents have been developed.

Rate studies show that chlorination is subject to acid catalysis, although the kinetics are frequently complex.[10] The proton is believed to assist Cl–Cl bond breaking in a reactant–Cl_2 complex. Chlorination is much more rapid in polar than in nonpolar

7. C. L. Coon, W. G. Blucher, and M. E. Hill, *J. Org. Chem.* **38:**4243 (1973).
8. G. A. Olah, S. C. Narang, J. A. Olah, R. L. Pearson, and C. A. Cupas, *J. Am. Chem. Soc.* **102:**3507 (1980).
9. H. Suzuki and T. Mori, *J. Chem. Soc., Perkin Trans. 2* **1996:**677.
10. L. M. Stock and F. W. Baker, *J. Am. Chem. Soc.* **84:**1661 (1962); L. J. Andrews and R. M. Keefer, *J. Am. Chem. Soc.* **81:**1063 (1959); R. M. Keefer and L. J. Andrews, *J. Am. Chem. Soc.* **82:**4547 (1960); L. J. Andrews and R. M. Keefer, *J. Am. Chem. Soc.* **79:**5169 (1957).

Scheme 11.1. Aromatic Nitration

1[a] $C_6H_5CH_2CN$ $\xrightarrow[H_2SO_4]{HNO_3}$ $p\text{-}O_2NC_6H_4CH_2CN$ 50–54%

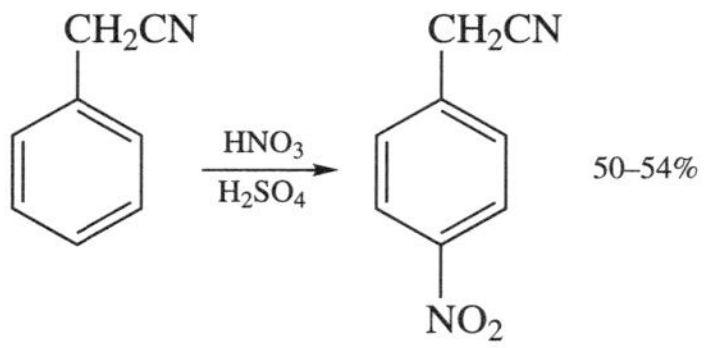

2[b] $C_6H_5N(CH_3)_2$ $\xrightarrow[HNO_3]{H_2SO_4}$ $m\text{-}O_2NC_6H_4N(CH_3)_2$ 56–63%

3[c] $C_6H_5CO_2H$ $\xrightarrow[HNO_3]{H_2SO_4}$ $3,5\text{-}(O_2N)_2C_6H_3CO_2H$ 54–58%

4[d] 3,4-$(CH_3O)_2C_6H_3CH{=}O$ $\xrightarrow{HNO_3}$ 4,5-$(CH_3O)_2$-2-$NO_2C_6H_2CH{=}O$ 73–79%

5[e] $C_6H_5CH{=}CHCH{=}O$ $\xrightarrow[Ac_2O]{HNO_3}$ $o\text{-}O_2NC_6H_4CH{=}CHCH{=}O$ 36–46%

6[f] $C_6H_5CO_2H$ $\xrightarrow[(CF_3CO)_2O]{NH_4^+NO_3^-}$ $m\text{-}O_2NC_6H_4CO_2H$ 94%

7[g] $p\text{-}HOC_6H_4CH_2CH(NHCO_2CH_3)CO_2CH_3$ $\xrightarrow[HCl]{La(NO_3)_3,\ NaNO_3}$ 4-HO-3-$O_2NC_6H_3CH_2CH(NHCO_2CH_3)CO_2CH_3$ 85%

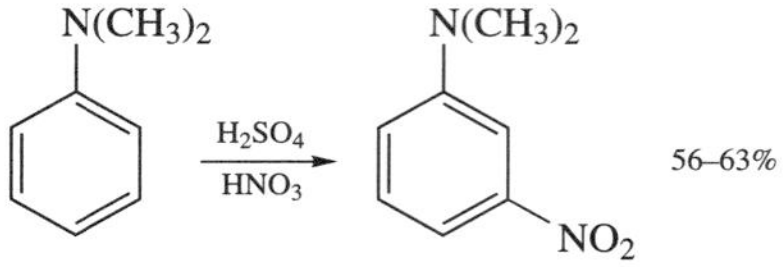

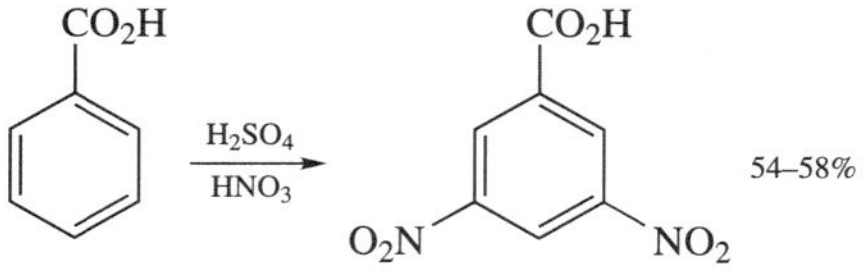

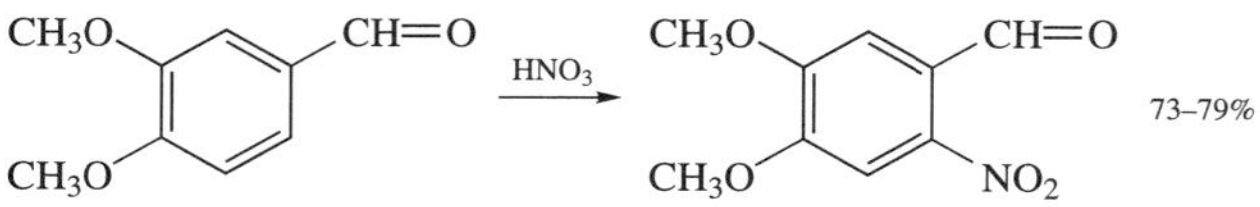

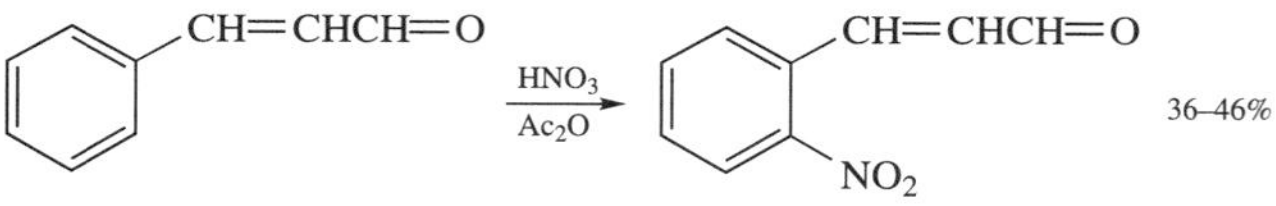

8[h] 1,2,4-$(CH_3O)_3C_6H_3$ $\xrightarrow[-50°C]{NO_2BF_4}$ 2,4,5-$(CH_3O)_3C_6H_2NO_2$ 81%

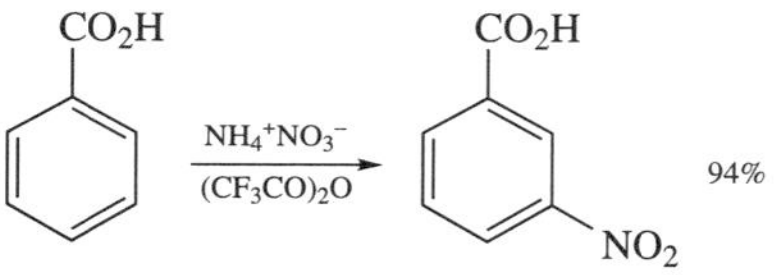

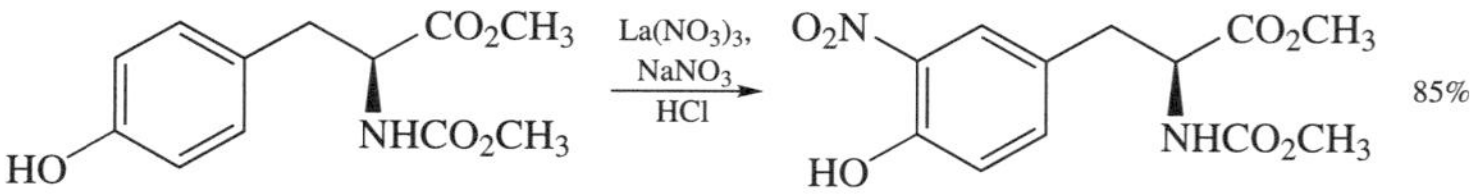

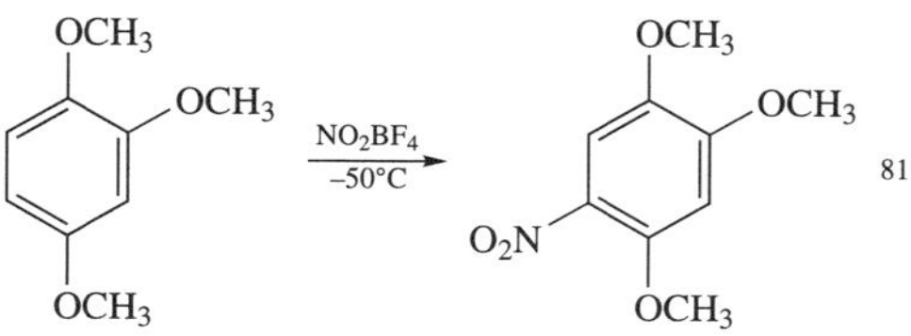

9[i] toluene + N-nitro-2-methylpyridinium → nitrotoluene ($o{:}m{:}p$ 100%, 64:3:33)

a. G. R. Robertson, *Org. Synth.* **I**:389 (1932).
b. H. M. Fitch, *Org. Synth.* **III**:658 (1955).
c. R. Q. Brewster, B. Williams, and R. Phillips, *Org. Synth.* **III**:337 (1955).
d. C. A. Fetscher, *Org. Synth.* **IV**:735 (1963).
e. R. E. Buckles and M. P. Bellis, *Org. Synth.* **IV**:722 (1963).
f. J. V. Crivello, *J. Org. Chem.* **46**:3056 (1981).
g. D. Ma and W. Tang, *Tetrahedron Lett.* **39**:7369 (1998).
h. C. L. Dwyer and C. W. Holzapel, *Tetrahedron* **54**:7843 (1998).
i. C. A. Cupas and R. L. Pearson, *J. Am. Chem. Soc.* **90**:4742 (1968).

solvents.[11] Bromination exhibits similar mechanistic features.

$$C_6H_6 + Cl{-}Cl + H{-}A \longrightarrow [C_6H_6Cl]^+ \longrightarrow \text{product} \quad + HCl + A^-$$

For preparative reactions, Lewis acid catalysts are used. Zinc chloride or ferric chloride can be used in chlorination, and metallic iron, which generates ferric bromide, is often used in bromination. The Lewis acid facilitates cleavage of the halogen–halogen bond.

$$MX_n + X_2 \rightleftharpoons X{-}\overset{\delta+}{X}\text{- -}\overset{\delta-}{M}X_n$$

$$C_6H_6 + X{-}\overset{\delta+}{X}\text{- -}\overset{\delta-}{M}X_n \longrightarrow [C_6H_6X]^+ \longrightarrow C_6H_5X + H^+$$

N-Bromosuccinimide (NBS) and *N*-chlorosuccinimide (NCS) are alternative halogenating agents. Activated aromatics, such as 1,2,4-trimethoxybenzene, are brominated by NBS at room temperature.[12] Both NCS and NBS can halogenate moderately active aromatics in nonpolar solvents with the use of $HClO_4$ as a catalyst.[13] Many other “positive halogen” compounds act as halogenating agents.

A wide variety of aromatic compounds can be brominated. Highly reactive ones, such as anilines and phenols, may undergo bromination at all activated positions. More selective reagents such as pyridinium bromide perbromide or tetraalkylammonium trihalides can be used in such cases.[14] Moderately reactive compounds such as anilides, haloaromatics, and

11. L. M. Stock and A. Himoe, *J. Am. Chem. Soc.* **83**:4605 (1961).
12. M. C. Carreno, J. L. Garcia Ruano, G. Sanz, M. A. Toledo, and A. Urbano, *J. Org. Chem.* **60**:5328 (1995).
13. Y. Goldberg and H. Alper, *J. Org. Chem.* **58**:3072 (1993).
14. W. P. Reeves and R. M. King II, *Synth. Commun.* **23**:855 (1993); J. Berthelot, C. Guette, P.-L. Desbene, and J.-J. Basselier, *Can. J. Chem.* **67**:2061 (1989); S. Kajigaeshi, T. Kakinami, T. Inoue, M. Kondo, H. Nakamura, M. Fujikawa, and T. Okamoto, *Bull. Chem. Soc. Jpn.* **61**:597 (1988); S. Kajigaeshi, T. Kakinami, H. Yamasaki, S. Fujisaki, and T. Okamoto, *Bull. Chem. Soc. Jpn.* **61**:2681 (1988); S. Gervat, E. Leonel, J.-Y. Barraud, and V. Ratovelomanana, *Tetrahedron Lett.* **34**:2115 (1993).

hydrocarbons can be readily brominated, and the usual directing effects control the regiochemistry. Use of Lewis acid catalysts permits bromination of rings with deactivating substituents, such as nitro and cyano.

Halogenations are strongly catalysed by mercuric acetate or trifluoroacetate. These conditions generate acyl hypohalites, which are the reactive halogenating agents. The trifluoroacetyl hypohalites are very reactive reagents. Even nitrobenzene, for example is readily brominated by trifluoroacetyl hypobromite.[15]

$$Hg(O_2CR)_2 + X_2 \rightleftharpoons HgX(O_2CR) + RCO_2X$$

A solution of bromine in CCl_4 containing sulfuric acid and mercuric oxide is also a reactive brominating agent[16] (see entry 9 in Scheme 11.2).

Fluorination can be carried out using fluorine diluted with an inert gas. However, great care is necessary to avoid uncontrolled reaction.[17] Several other reagents have been devised which are capable of aromatic fluorination.[18] Acetyl hypofluorite can be prepared *in situ* from fluorine and sodium acetate.[19] This reagent effects fluorination of activated aromatics. Although this procedure does not avoid the special precautions necessary for manipulation of elemental fluorine, it does provide a system with much greater selectivity. Acetyl hypofluorite shows a strong preference for *o*-fluorination of alkoxy- and acetamido-substituted rings. *N*-Fluoro-bis(trifluoromethansulfonyl)amine displays similar reactivity. It can fluorinate benzene and activated aromatics.[20]

$$CH_3O\text{–}C_6H_5 + (CF_3SO_2)_2NF \longrightarrow \underset{69\%}{o\text{-}CH_3O\text{–}C_6H_4\text{–}F} + \underset{24\%}{p\text{-}CH_3O\text{–}C_6H_4\text{–}F}$$

Iodinations can be carried out with mixtures of iodine and various oxidants such as periodic acid,[21] I_2O_5,[22] NO_2,[23] and $Ce(NH_3)_2(NO_3)_6$.[24] A mixture of cuprous iodide and a cupric salt can also effect iodination.[25]

$$p\text{-}CH_3\text{–}C_6H_4\text{–}CH_3 + CuI + CuCl_2 \longrightarrow 2\text{-iodo-}1,4\text{-}(CH_3)_2C_6H_3 \quad \sim 70\%$$

15. J. R. Barnett, L. J. Andrews, and R. M. Keefer, *J. Am. Chem. Soc.* **94:**6129 (1972).
16. S. A. Khan, M. A. Munawar, and M. Siddiq, *J. Org. Chem.* **53:**1799 (1988).
17. F. Cacace, P. Giacomello, and A. P. Wolf, *J. Am. Chem. Soc.* **102:**3511 (1980).
18. S. T. Purrington, B. S. Kagan, and T. B. Patrick, *Chem. Rev.* **86:**997 (1986).
19. O. Lerman, Y. Tor, and S. Rozen, *J. Org. Chem.* **46:**4629 (1981); O. Lerman, Y. Tor, D. Hebel, and S. Rozen, *J. Org. Chem.* **49:**806 (1984).
20. S. Singh, D. D. DesMarteau, S. S. Zuberi, M. Whitz, and H.-N. Huang, *J. Am. Chem. Soc.* **109:**7194 (1987).
21. H. Suzuki, *Org. Synth.* **VI:**700, (1988).
22. L. C. Brazdil and C. J. Cutler, *J. Org. Chem.* **61:**9621 (1996).
23. Y. Noda and M. Kashima, *Tetrahedron Lett.* **38:**6225 (1997).
24. T. Sugiyama, *Bull. Chem. Soc. Jpn.* **54:**2847 (1981).
25. W. C. Baird, Jr., and J. H. Surridge, *J. Org. Chem.* **35:**3436 (1970).

Iodination of moderately reactive aromatics can be effected by mixtures of iodine and silver or mercuric salts.[26] Hypoiodites are presumably the active iodinating species. Bis(pyridine)iodonium salts can iodinate benzene and activated derivatives in the presence of strong acids such as HBF_4 or CF_3SO_3H.[27] Scheme 11.2 gives some specific examples of aromatic halogenation reactions.

11.1.3. Friedel–Crafts Alkylations and Acylations

Friedel–Crafts reactions are among the most important methods for introducing carbon substituents on aromatic rings. The reactive electrophiles can be either discrete carbocations or acylium ions or polarized complexes that still contain the leaving group. Various combinations of reagents can be used to generate alkylating species. Alkylations usually involve alkyl halides and Lewis acids or reactions of alcohols or alkenes with strong acids.

$$R{-}X + AlCl_3 \rightleftharpoons R{-}\overset{+}{X}{-}\overset{-}{Al}Cl_3 \rightleftharpoons R^+ + X\overset{-}{Al}Cl_3$$

$$R{-}OH + H^+ \rightleftharpoons R{-}\overset{+}{O}H(H) \rightleftharpoons R^+ + H_2O$$

$$RCH{=}CH_2 + H^+ \rightleftharpoons R\overset{+}{C}HCH_3$$

Because of the involvement of carbocations, Friedel–Crafts alkylations are often accompanied by rearrangement of the alkylating group. For example, isopropyl groups are often introduced when *n*-propyl reactants are used.[28]

$$C_6H_6 + CH_3CH_2CH_2Cl \xrightarrow{AlCl_2} C_6H_5CH(CH_3)_2$$

Under a variety of reaction conditions, alkylation of benzene with either 2-chloro- or 3-chloropentane gives rise to a mixture of both 2-pentyl and 3-pentylbenzene.[29]

Rearrangement can also occur after the initial alkylation. The reaction of 2-chloro-2-methylbutane with benzene under Friedel–Crafts conditions is an example of this behavior.[30] With relatively mild Friedel–Crafts catalysis such as BF_3 or $FeCl_3$, the main

26. Y. Kobayashi, I. Kumadaki, and T. Yoshida, *J. Chem. Res. (Synopses)* **1977:**215; R. N. Hazeldine and A. G. Sharpe, *J. Chem. Soc.* **1952:**993; W. Minnis, *Org. Synth.* **II:**357 (1943); D. E. Janssen and C. V. Wilson, *Org. Synth.* **IV**: 547 (1963); W.-W. Sy and B. A. Lodge, *Tetrahedron Lett.* **30:**3769 (1989).
27. J. Barluenga, J. M. Gonzalez, M. A. Garcia-Martin, P. J. Campos, and G. Asensio, *J. Org. Chem.* **58:**2058 (1993).
28. S. H. Sharman, *J. Am. Chem. Soc.* **84:**2945 (1962).
29. R. M. Roberts, S. E. McGuire, and J. R. Baker, *J. Org. Chem.* **41:**659 (1976).
30. A. A. Khalaf and R. M. Roberts, *J. Org. Chem.* **35:**3717 (1970); R. M. Roberts and S. E. McGuire, *J. Org. Chem.* **35:**102 (1970).

Scheme 11.2. Aromatic Halogenation

A. Chlorination

1[a] F; Cl_2, $AlCl_3$ → F, Cl (25%) + F, Cl (73%) + F, Cl (2%)

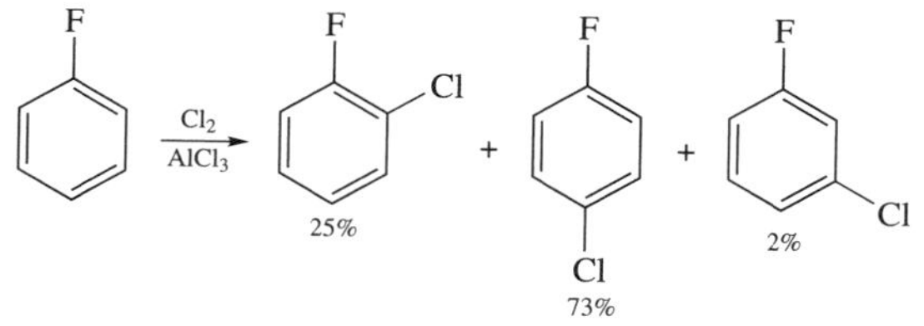

2[b] $\overset{O}{\|}$ CCl; $FeCl_3$, Cl_2 → $\overset{O}{\|}$ CCl, Cl

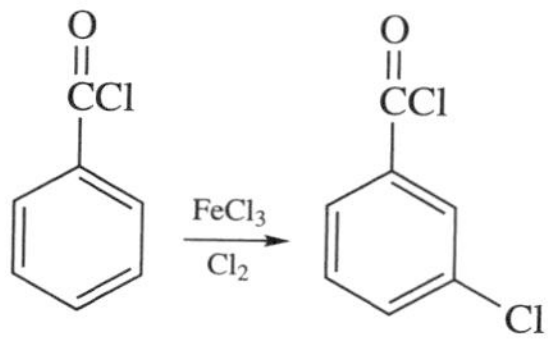

3[c] CH_3O, CH_3, CH_3; NCS, $HClO_4$, CCl_4 → CH_3O, CH_3, CH_3, Cl 94%

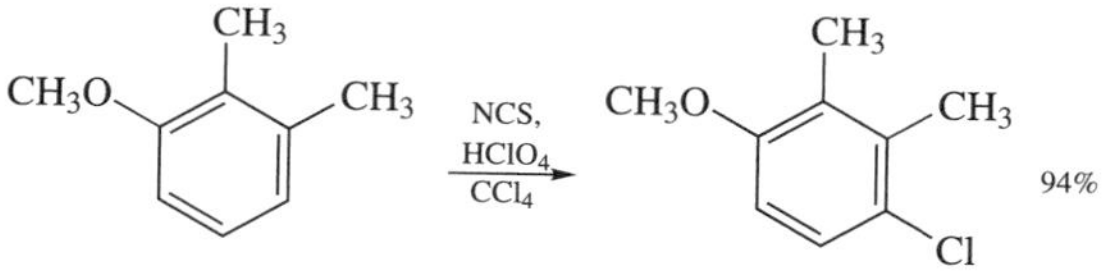

4[d] $NH\overset{O}{\|}CCH_3$; $(CH_3)_3COCl$ → $NH\overset{O}{\|}CCH_3$, Cl 92%

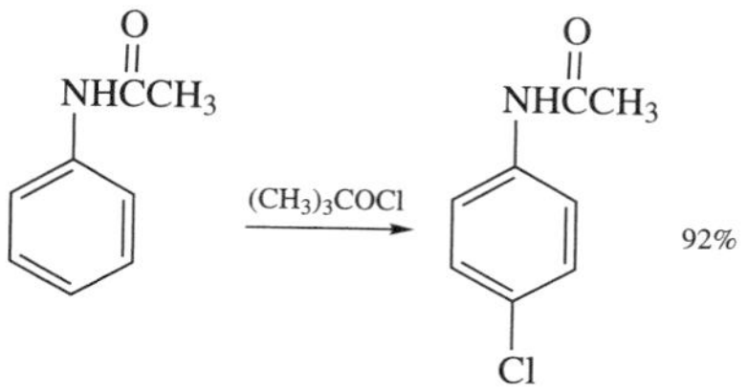

B. Bromination

5[e] NO_2, CH_3; Fe, Br_2 → NO_2, Br, CH_3 86–90%

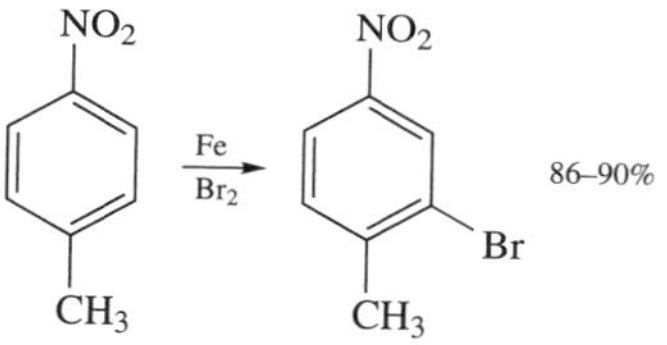

6[f] CO_2H, NH_2; HCl, Br_2 → CO_2H, Br, Br, NH_2, Br

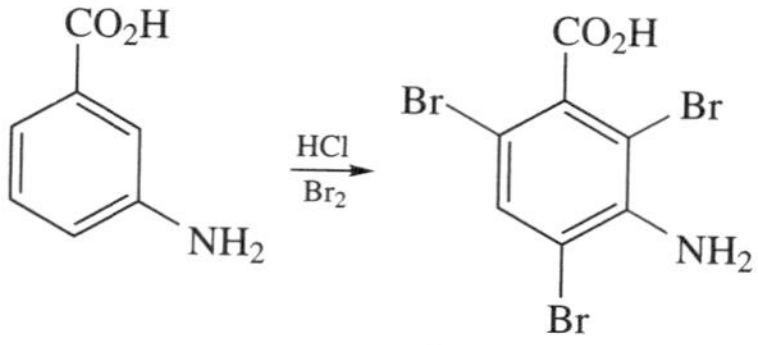

7[g] CH_3, NO_2, CO_2H; Br–N, N–Br, O, O, O, N–H (dibromoisocyanuric acid), H_2SO_4 → Br, CH_3, NO_2, CO_2H 98%

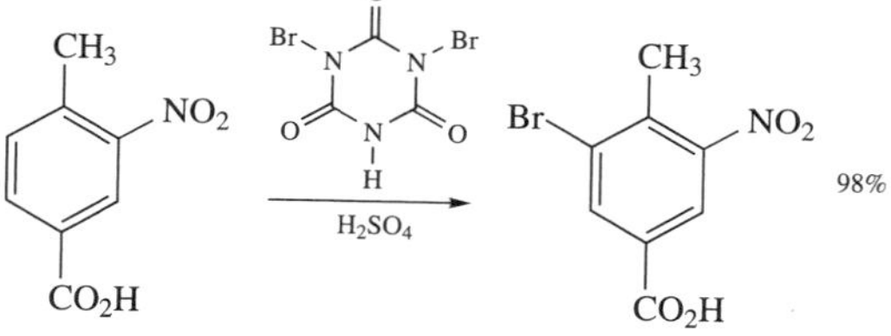

8[h] 1-methoxynaphthalene $\xrightarrow[CH_3CN]{NBS}$ 4-bromo-1-methoxynaphthalene 94%

9[i] $C_6H_5CO_2CH_3$ $\xrightarrow[H^+]{Br_2,\ HgO}$ 3-$BrC_6H_4CO_2CH_3$ 80%

10[j] $(CH_3)_2N$–C_6H_5 $\xrightarrow[CH_3OH]{Br_2,\ (C_2H_5)_4N^+Cl^-}$ $(CH_3)_2N$–C_6H_4–Br

C. Iodination

11[k] 2-$H_2NC_6H_4CO_2H$ $\xrightarrow{ICl}$ 5-iodo-2-aminobenzoic acid (CO_2H, NH_2, I) 76–84%

12[l] 3,4,5-$(CH_3O)_3C_6H_2CH_2OH$ $\xrightarrow[Hg(O_2CCH_3)_2]{I_2}$ 2-iodo-3,4,5-trimethoxybenzyl alcohol (CH_3O, CH_3O, OCH_3, CH_2OH, I) 76%

13[m] 1,2-$(OCH_3)_2C_6H_4$ $\xrightarrow[AgO_2CCF_3]{I_2}$ 4-iodo-1,2-dimethoxybenzene 85–91%

14[n] 1,2,4,5-tetramethylbenzene (H_3C, CH_3, H_3C, CH_3) $\xrightarrow[HIO_4]{I_2}$ 3-iodo-1,2,4,5-tetramethylbenzene 80–91%

15[o] 1,3-benzodioxole $\xrightarrow{I_2,\ NO_2}$ 5-iodo-1,3-benzodioxole 92%

16[p] $C_6H_5OCH_3$ $\xrightarrow[Ce(NH_3)_2(NO_3)_6]{n\text{-}Bu_4N^+I^-}$ 4-$IC_6H_4OCH_3$ 84%

Scheme 11.2. (*continued*)

17[q] Br–C6H5 → ($I^+(pyridine)_2$, CF_3CO_2H) Br–C6H4–I 90%

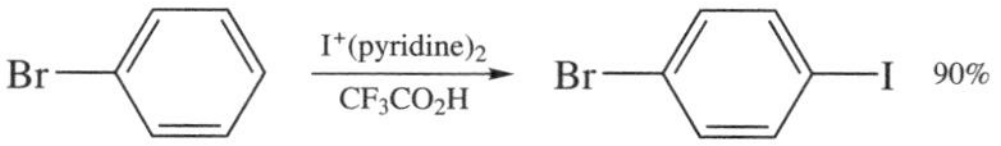

18[r] → (N-fluoropyridinium $CH_3SO_3^-$) 60%

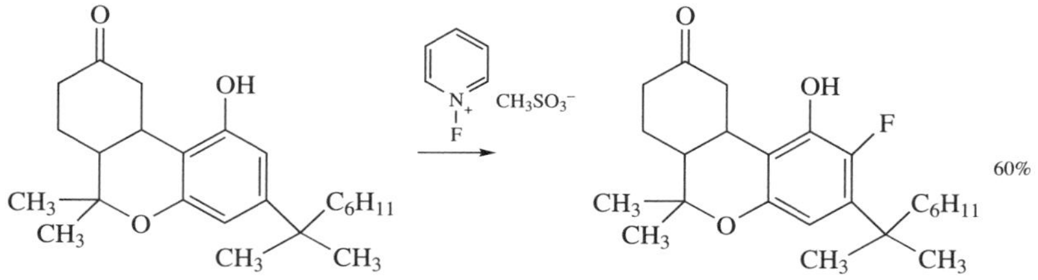

a. G. A. Olah, S. J. Kuhn, and B. A. Hardie, *J. Am. Chem. Soc.* **86:** 1055 (1964).
b. E. Hope and G. F. Riley, *J. Chem. Soc.* **121:**2510 (1922).
c. V. Goldberg and H. Alper, *J. Org. Chem.* **58:**3072 (1993).
d. I. Lengyel, V. Cesare, and R. Stephani, *Synth. Commun.* **28:**1891 (1998).
e. M. M. Robison and B. L. Robison, *Org. Synth.* **IV:**947 (1963).
f. W. A. Wisansky and S. Ansbacher, *Org. Synth.* **III:**138 (1955).
g. A. R. Leed, S. D. Boettger, and B. Ganem, *J. Org. Chem.* **45:**1098 (1980).
h. M. C. Carreno, J. L. Garcia Ruano, G. Sanz, M. A. Toledo, and A. Urbano, *J. Org. Chem.* **60:**5328 (1995).
i. S. A. Khan, M. A. Munawar, and M. Siddiq, *J. Org. Chem.* **53:**1799 (1988).
j. S. Gervat, E. Leonel, J.-Y. Barraud, and V. Ratovelomanana, *Tetrahedron Lett.* **34:**2115 (1993).
k. V. H. Wallingford and P. A. Krueger, *Org. Synth.* **II:**349 (1943).
l. F. E. Ziegler and J. A. Schwartz, *J. Org. Chem.* **43:**985 (1978).
m. D. E. Janssen and C. V. Wilson, *Org. Synth.* **IV:**547 (1963).
n. H. Suzuki, *Org. Synth.* **51:**94 (1971).
o. Y. Noda and M. Kashima, *Tetrahedron Lett.* **38:**6225 (1997).
p. T. Sugiyama, *Bull. Chem. Soc. Jpn.* **54:** 2847 (1981).
q. J. Barluenga, J. M. Gonzalez, M. A. Garcia-Martin, P. J. Campos, and G. Asensio, *J. Org. Chem.* **58:**2058 (1993).
r. M. A. Tius, J. K. Kawakami, W. A. G. Hill, and A. Makriyannis, *J. Chem. Soc., Chem. Commun.* **1996:**2085.

product is **A**. With $AlCl_3$, equilibrium of **A** and **B** occurs and the equilibrium favors **B**. The rearrangement is the result of product equilibration via reversibly formed carbocations.

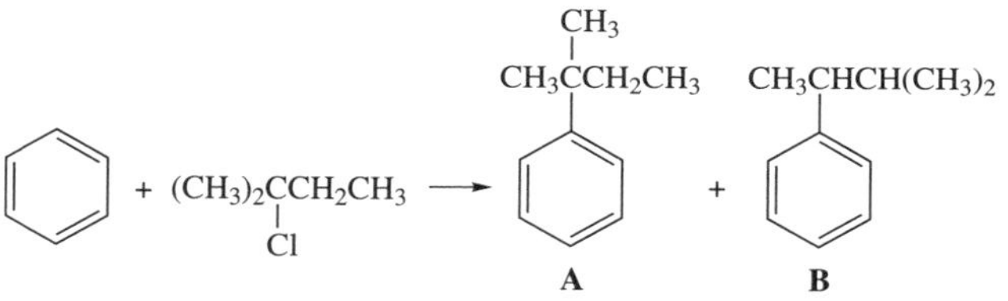

$$C_6H_6 + (CH_3)_2CClCH_2CH_3 \longrightarrow C_6H_5C(CH_3)_2CH_2CH_3\ (\mathbf{A}) + C_6H_5CH(CH_3)CH(CH_3)_2\ (\mathbf{B})$$

Alkyl groups can also migrate from one position to another on the ring.[31] Such migrations are also thermodynamically controlled and proceed in the direction of minimizing steric interactions between substituents.

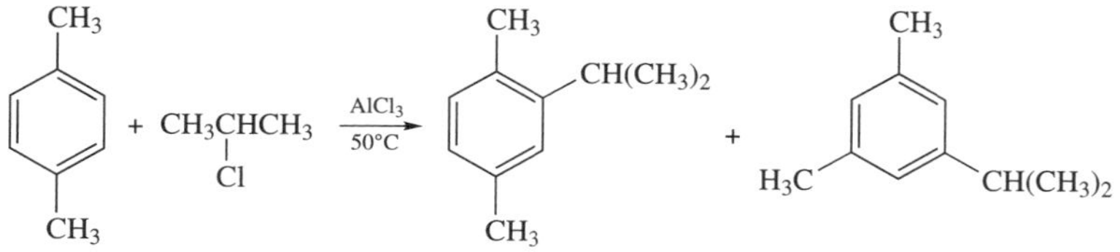

$$p\text{-}CH_3C_6H_4CH_3 + CH_3CHClCH_3 \xrightarrow[50°C]{AlCl_3} \text{2-isopropyl-1,4-dimethylbenzene} + \text{1-isopropyl-3,5-dimethylbenzene}$$

31. R. M. Roberts and D. Shiengthong, *J. Am. Chem. Soc.* **86:**2851 (1964).

Table 11.1. Relative Activity of Friedel–Crafts Catalysts

Very active	Moderately active	Mild
$AlCl_3$, $AlBr_3$, $GaCl_3$, $GaCl_2$, SbF_5, $MoCl_5$	$InCl_3$, $LnBr_3$, $SbCl_5$, $FeCl_3$, $AlCl_3-CH_3NO_2$, $SbF_5-CH_3NO_2$	BCl_3, $SnCl_4$, $TiCl_4$, $TiBr_4$, $FeCl_2$

a. G. A. Olah, S. Kobayashi, and M. Tashiro, *J. Am. Chem. Soc.* **94:**7448 (1972).

The relative reactivities of Friedel–Crafts catalysts have not been described in a quantitative way, but comparative studies using a series of benzyl halides have resulted in the qualitative groupings shown in Table 11.1. Proper choice of catalyst can minimize subsequent product equilibrations.

The Friedel–Crafts alkylation reaction usually does not proceed successfully with aromatic substrates having electron-attracting groups. Another limitation is that each alkyl group that is introduced increases the reactivity of the ring toward further substitution, so polyalkylation can be a problem. Polyalkylation can be minimized by using the aromatic reactant in excess.

As mentioned above, besides the alkyl halide–Lewis acid combination, two other sources of carbocations are often used in Friedel–Crafts reactions. Alcohols can serve as carbocation precursors in strong acids such as sulfuric or phosphoric acid. Alkylation can also be effected by alcohols in combination with BF_3 and $AlCl_3$.[32] Alkenes can also serve as alkylating agents when protic acids, especially H_2SO_4, H_3PO_4, or HF, or Lewis acids, such as BF_3 and $AlCl_3$, are used as catalysts.[33]

Stabilized carbocations can be generated from allylic and benzylic alcohols by reaction with $Sc(O_3SCF_3)_3$ and result in formation of alkylation products from benzene and activated derivatives.[34]

+ $CH_3CH_2CHCH{=}CH_2$ (OH) $\xrightarrow{Sc(O_3SCF_3)_3}$ $-CH_2CH{=}CHCH_2CH_3$

64% yield, 94:6 *E:Z*

This kind of reaction has been used to synthesize α-tocopherol.

HO, CH_3, CH_3, CH_3, OH + OH, CH_3, CH_3, H, 3 $\xrightarrow{Sc(O_3SCF_3)_3,\ 1\ mol\ \%}$ Ref. 35

HO, CH_3, CH_3, CH_3, O, CH_3, CH_3, H, 3 96%

32. A. Schriesheim, in *Friedel–Crafts and Related Reactions*, Vol. II, G. Olah, ed., Interscience, New York, 1964, Chapter XVIII.
33. S. H. Patinkin and B. S. Friedman, in *Friedel–Crafts and Related Reactions*, Vol. II, G. Olah, ed., Interscience, New York, 1964, Chapter XIV.
34. T. Tsuchimoto, K. Tobita, T. Hiyama, and S. Fukuzawa, *Synlett* **1996:**557; T. Tsuchimoto, K. Tobita, T. Hiyama, and S. Fukuzawa, *J. Org. Chem.* **62:**6997 (1997).
35. M. Matsui, N. Karibe, K. Hayashi, and H. Yamamoto, *Bull. Chem. Soc. Jpn.* **68:**3569 (1995).

Methanesulfonate esters of secondary alcohols also give Friedel–Crafts products in the presence of $Sc(O_3SCF_3)_3$.

benzene + cyclopentyl–O_3SCH_3 $\xrightarrow{Sc(O_3SCF_3)_3}$ cyclopentylbenzene 87% Ref. 36

Friedel–Crafts alkylation can occur intramolecularly to form a fused ring. It is somewhat easier to form six-membered rings than five-membered ones in such reactions. Thus, whereas 4-phenyl-1-butanol gives a 50% yield of cyclized product in phosphoric acid, 3-phenyl-1-propanol is mainly dehydrated to alkene.[37]

$PhCH_2(CH_2)_3CH_2OH$...

Ph–$(CH_2)_3CH_2OH$ $\xrightarrow{H_3PO_4}$ tetralin 50%

Ph–$(CH_2)_2CH_2OH$ $\xrightarrow{H_3PO_4}$ Ph–$CH{=}CHCH_3$ + Ph–$CH_2CH{=}CH_2$

If a potential carbocation intermediate can undergo a hydride or alkyl shift, this shift will occur in preference to closure of the five-membered ring.

Ph–$CH_2CH_2C(CH_3)(OH)CH(CH_3)_2$ $\xrightarrow{H_2SO_4}$ 1,1,2-trimethyltetralin (H_3C, CH_3, CH_3) 58% Ref. 38

This reflects a rather general tendency for the formation of six-membered rings in preference to five- and seven-membered rings in ring closure by intramolecular Friedel–Crafts reactions.[38,39] Intramolecular Friedel–Crafts reactions provide an important method for constructing polycyclic hydrocarbon frameworks. Entries 5–7 in Scheme 11.3 are examples of this type of reaction.

Friedel–Crafts acylation generally involves reaction of an acyl halide and Lewis acid such as $AlCl_3$, SbF_5, or BF_3. Bismuth(III) triflate is also a very active acylation catalyst.[40] Acid anhydrides can also be used in some cases. A combination of hafnium(IV) triflate

36. H. Kotsuki, T. Ohishi, and M. Inoue, *Synlett* **1998:**255; H. Kotsuki, T. Ohishi, M. Inoue, and T. Kojima, *Synthesis* **1999:**603.
37. A. A. Khalaf and R. M. Roberts, *J. Org. Chem.* **34:**3571 (1969).
38. A. A. Khalaf and R. M. Roberts, *J. Org. Chem.* **37:**4227 (1972).
39. R. J. Sundberg and J. P. Laurino, *J. Org. Chem.* **49:**249 (1984); S. R. Angle and M. S. Louie, *J. Org. Chem.* **56:**2853 (1991).
40. J.-R. Desmurs, M. Labrouillere, C. Le Roux, H. Gaspard, A. Laporterie, and J. Dubac, *Tetrahedron Lett.* **38:**8871 (1997); S. Repichet, and C. LeRoux, J. Dubac, and J.-R. Desmurs, *Eur. J. Org. Chem.* **1998:**2743.

A. Intermolecular reactions

1[a] $PhCH(Br)C(=O)CH_3$ + benzene $\xrightarrow{AlCl_3}$ $Ph_2CHC(=O)CH_3$ 53–57%

2[b] $CH_3C(CH_2Cl)=CH_2$ + benzene $\xrightarrow{H_2SO_4}$ $PhC(CH_3)_2CH_2Cl$ 70–73%

3[c] $PhCH(Br)CH(Br)CO_2H$ + benzene $\xrightarrow{AlCl_3}$ $Ph_2CHCH(Ph)CO_2H$ 66–78%

4[d] toluene + cyclohexene $\xrightarrow{p\text{-TsOH}}$ cyclohexyltoluene

98% yield, *o:m:p* = 29:18:53

B. Intramolecular Friedel–Crafts cyclizations

5[e] HO, CH_3, CH_3, H, O, CH_3O $\xrightarrow{\text{polyphosphoric acid}}$ H_3C, H, O, CH_3, CH_3O 87%

6[f] CH_3O–C_6H_4–$CH(OH)CH_2CH_2CH_2$–Ph $\xrightarrow[-78°C]{TiCl_4}$ CH_3O–C_6H_4–(tetrahydronaphthyl) 94%

7[g] $(CH_3O)_2C_6H_3(CH_2)_2N(CH_3)CH_2CH(OH)Ph$ $\xrightarrow[CH_2Cl_2]{HBF_4}$ CH_3O, CH_3O, Ph, N–CH_3

a. E. M. Shultz and S. Mickey, *Org. Synth.* **III:**343 (1955).
b. W. T. Smith, Jr., and J. T. Sellas, *Org. Synth.* **IV:**702 (1963).
c. C. P. Krimmol, L. E. Thielen, E. A. Brown, and W. J. Heidtke, *Org. Synth.* **IV:**960 (1963).
d. M. P. D. Mahindaratne and K. Wimalasena, *J. Org. Chem.* **63:**2858 (1998).
e. R. E. Ireland, S. W. Baldwin, and S. C. Welch, *J. Am. Chem. Soc.* **94:**2056 (1972).
f. S. R. Angle and M. S. Louie, *J. Org. Chem.* **56:**2853 (1991).
g. S. J. Coote, S. G. Davies, D. Middlemiss, and A. Naylor, *Tetrahedron Lett.* **30:**3581 (1989).

and $LiClO_4$ in nitromethane catalyzes acylation of moderately reactive aromatics.

$$\text{m-xylene} + (CH_3C(=O))_2O \xrightarrow[\text{LiClO}_4,\ CH_3NO_2]{\text{Hf(O}_3\text{SCF}_3)_4,\ 5\text{ mol \%}} \text{2,4-dimethylacetophenone}\quad 91\%$$ Ref. 41

Mixed anhydrides with trifluoroacetic acid are particularly reactive acylating agents.[42] For example, entry 5 in Scheme 11.4 shows the use of a mixed anhydride in the synthesis of the anticancer agent tamoxifen.

As in the alkylation reaction, the reactive intermediate can be a dissociated acylium ion or a complex of the acyl chloride and Lewis acid.[43] Recent mechanistic studies have indicated that with benzene and slightly deactivated derivatives, it is the protonated acylium ion that is the kinetically dominant electrophile.[44]

$$RC(=O)X + MX_n \longrightarrow R{-}C{\equiv}\overset{+}{O} + (MX_{n+1})^-$$

$$R{-}C{\equiv}\overset{+}{O} + H^+ \rightleftharpoons R\overset{+}{C}{=}\overset{+}{O}H$$

$$C_6H_6 + R\overset{+}{C}{=}\overset{+}{O}H \longrightarrow [C_6H_6^+(H)(C(=O)R)] + H^+ \longrightarrow C_6H_5{-}C(=O)R + H^+$$

or

$$RC(=O)X + MX_n \longrightarrow R{-}C(={}^{+}O{-}\overline{M}X_n){-}X \longrightarrow R{-}C{\equiv}O^+ + \overline{M}X_{n+1}$$

$$C_6H_6 + RC{\equiv}O^+ \longrightarrow [C_6H_6^+(H)(C(=O)R)] \longrightarrow C_6H_5{-}C(=O)R + H^+$$

Regioselectivity in Friedel–Crafts acylations can be quite sensitive to the reaction solvent and other procedural variables.[45] In general, *para* attack predominates for alkylbenzenes.[46] The percentage of *ortho* attack increases with the electrophilicity of the acylium ion, and as much as 50% *ortho* product is observed with the formylium and 2,4-dinitrobenzoylium ions.[47] Rearrangement of the acyl group is not a problem in

41. I. Hachiya, M. Moriwaki, and S. Kobayashi, *Tetrahedron Lett.* **36:**409 (1995); A. Kawada, S. Mitamura, and S. Kobayashi, *Chem. Commun.* **1996:**183; I. Hachiya, M. Moriwaki, and S. Kobayashi, *Bull. Chem. Soc. Jpn.* **68:**2053 (1995).
42. E. J. Bourne, M. Stacey, J. C. Tatlow, and J. M. Teddar, *J. Chem. Soc.* **1951:**719; C. Galli, *Synthesis* **1979:**303; B. C. Ranu, K. Ghosh, and U. Jana, *J. Org. Chem.* **61:**9546 (1996).
43. F. R. Jensen and G. Goldman, in *Friedel–Crafts and Related Reactions*, Vol. III, G. Olah, ed., Interscience, New York, 1964, Chapter XXXVI.
44. Y. Sato, M. Yato, T. Ohwada, S. Saito, and K. Shudo, *J. Am. Chem. Soc.* **117:**3037 (1995).
45. For example, see L. Friedman and R. J. Honour, *J. Am. Chem. Soc.* **91:**6344 (1969).
46. H. C. Brown, G. Marino, and L. M. Stock, *J. Am. Chem. Soc.* **81:**3310 (1959); H. C. Brown and G. Marino, *J. Am. Chem. Soc.* **81:**5611 (1959); G. A. Olah, M. E. Moffatt, S. J. Kuhn, and B. A. Hardie, *J. Am. Chem. Soc.* **86:**2198 (1964).
47. G. A. Olah and S. Kobayashi, *J. Am. Chem. Soc.* **93:**6964 (1971).

Friedel–Craft acylation. Neither is polyacylation, because the first acyl group serves to deactivate the ring to further attack.

Intramolecular acylations are very common. The normal procedure involving an acyl halide and a Lewis acid can be used. One useful alternative is to dissolve the carboxylic acid in polyphosphoric acid (PPA) and heat to effect cyclization. This procedure probably involves formation of a mixed phosphoric–carboxylic anhydride.[48]

$(CH_2)_3CO_2H$ PPA O

Cyclizations can also be carried out with an esterified oligomer of phosphoric acid called "polyphosphate ester." This material is soluble in chloroform.[49] (See entries 11 and 12 in Scheme 11.4.) Another reagent of this type is trimethylsilyl polyphosphate.[50] Neat methanesulfonic acid is also an effective reagent for intramolecular Friedel–Crafts acylation.[51] (See entry 13 in Scheme 11.4.)

A classical procedure for fusing a six-membered ring to an aromatic ring uses succinic anhydride or a derivative. An intermolecular acylation is followed by reduction and an intramolecular acylation. The reduction is necessary to provide a more reactive ring for the second acylation.

CH_3 CH_3 + H_3C O O O $\xrightarrow{AlCl_3}$ CH_3 O CH_3 CCH_2CHCO_2H CH_3

Pd, H_2 Ref. 52

CH_3 CH_3 O CH_3 $\xleftarrow{PPA}$ CH_3 CH_3 $(CH_2)_2CHCO_2H$ CH_3

Scheme 11.4 shows some other representative Friedel–Crafts acylation reactions.

48. W. E. Bachmann and W. J. Horton, *J. Am. Chem. Soc.* **69:**58 (1947).
49. Y. Kanaoka, O. Yonemitsu, K. Tanizawa, and Y. Ban, *Chem. Pharm. Bull.* **12:**773 (1964); T. Kametani, S. Takano, S. Hibino, and T. Terui, *J. Heterocycl. Chem.* **6:**49 (1969).
50. E. M. Berman and H. D. H. Showalter, *J. Org. Chem.* **54:**5642 (1989).
51. V. Premasagar, V. A. Palaniswamy, and E. J. Eisenbraun, *J. Org. Chem.* **46:**2974 (1981).
52. E. J. Eisenbraun, C. W. Hinman, J. M. Springer, J. W. Burnham, T. S. Chou, P. W. Flanagan, and M. C. Hamming, *J. Org. Chem.* **36:**2480 (1971).

Scheme 11.4. Friedel–Crafts Acylation Reactions

A. Intermolecular reactions

1[a] Bromobenzene + $(CH_3CO)_2O$ $\xrightarrow{AlCl_3}$ 4-Br-C_6H_4-$C(=O)CH_3$ ($O{=}CCH_3$) 69–79%

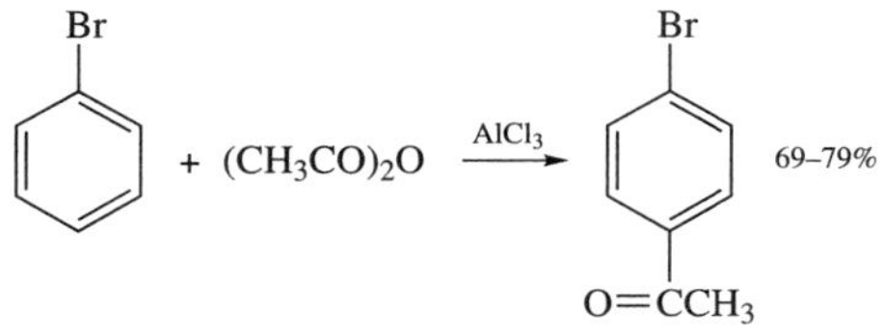

2[b] 4-$CH(CH_3)_2$-toluene + CH_3CCl (O) $\xrightarrow{AlCl_3}$ 2-acetyl product (CH_3, $\overset{O}{\|}CCH_3$, $CH(CH_3)_2$) 50–55%

3[c] Benzene + maleic anhydride $\xrightarrow{AlCl_3}$ $PhCCH{=}CHCO_2H$ (C=O) 80–85%

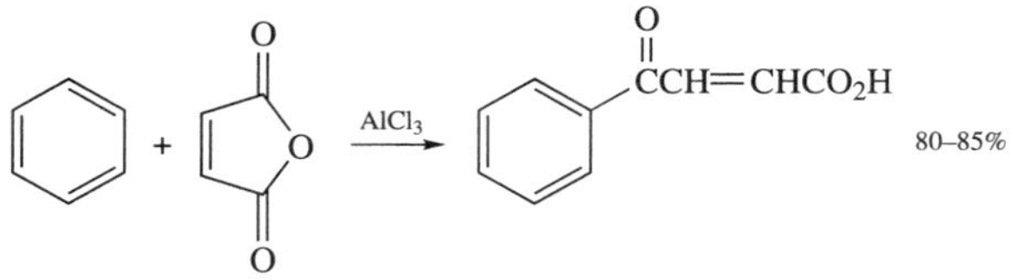

4[d] $NHCCH_3$ (O) benzene + $ClCCH_2Cl$ (O) $\xrightarrow{AlCl_3}$ 4-$NHCCH_3$-C_6H_4-$O{=}CCH_2Cl$ 80–85%

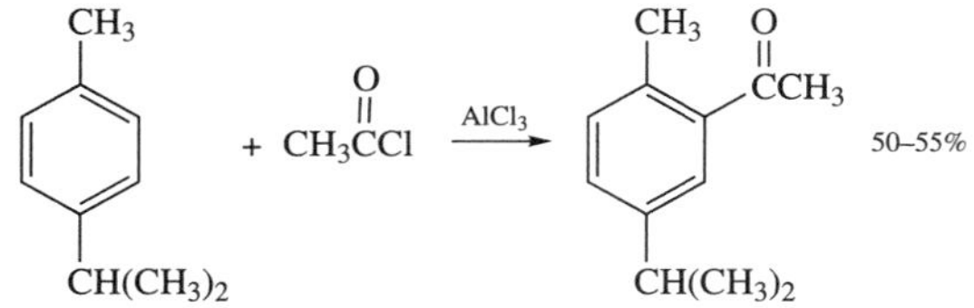

5[e] $OCH_2CH_2N(CH_3)_2$ benzene + $PhCHCH_2CH_3$ (CO_2H) $\xrightarrow[H_3PO_4]{(CF_3CO)_2O}$ $(CH_3)_2NCH_2CH_2O$-C_6H_4-C(=O)-CH(Ph)-C_2H_5 96%

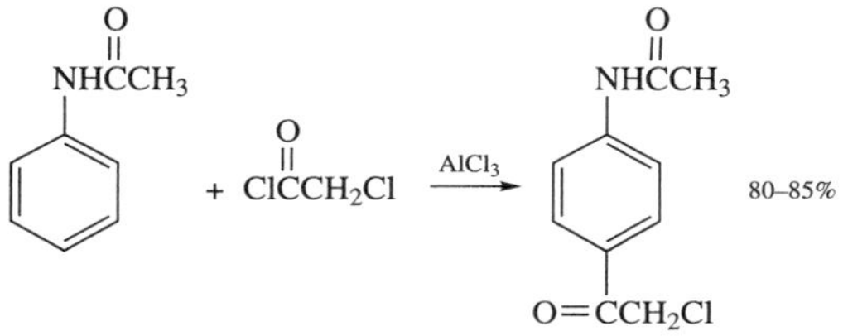

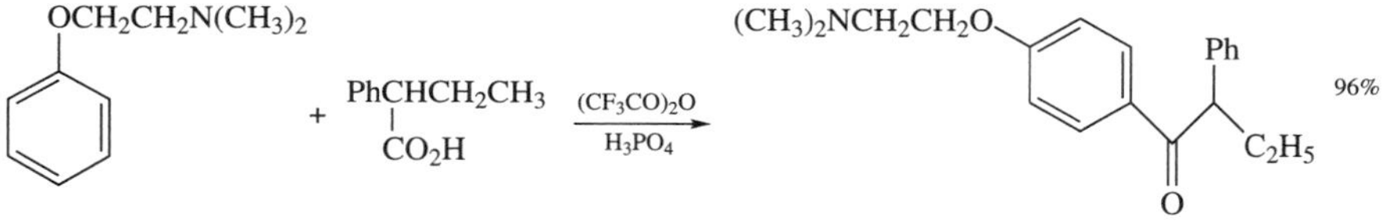

6[f] F-C_6H_5 + $PhCCl$ (O) $\xrightarrow{Bi(O_3SCF_3)_3,\ 10\ mol\ \%}$ F-C_6H_4-C(=O)Ph 86%

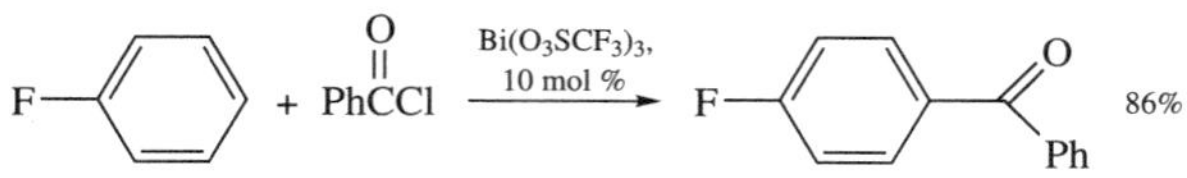

B. Intramolecular Friedel–Crafts acylations

7[g] $C_6H_5(CH_2)_3CO_2H$ $\xrightarrow{PPA}$ 1-tetralone (O) 75–86%

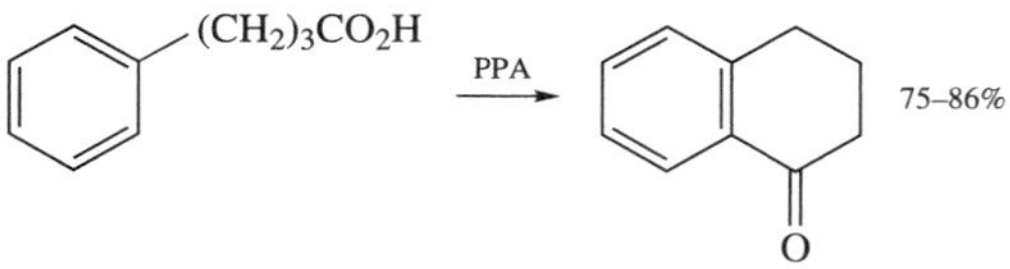

8[h] $(CH_2)_3CCl$ (O) $\xrightarrow{AlCl_3}$ 74–91%

9[i] + $\xrightarrow{AlCl_3}$ 91–96%

10[j] Cl, O; $CHCH_2CCl$; OCH_3 $\xrightarrow{\text{1) } AlCl_3 \text{ 2) Pyridine}}$ OCH_3 + CH_3O O
85% total yield

11[k] CH_3O, CH_3O, CH_3, $CH_2CHCH_2CO_2H$, $CH(CH_3)_2$ $\xrightarrow{\text{polyphosphate ester}, CH_2Cl_2}$ CH_3O, CH_3O, O, CH_3, $CH(CH_3)_2$

12[l] CH_3, $(CH_2)_3CO_2H$ $\xrightarrow{\text{polyphosphate ester}}$ CH_3, O 87%

13[m] $(CH_2)_3CO_2H$ $\xrightarrow{CH_3SO_3H, \ 90–95°C}$ O 95%

a. R. Adams and C. R. Noller, *Org. Synth.* **I:**109 (1941).
b. C. F. H. Allen, *Org. Synth.* **II:**3 (1943).
c. O. Grummitt, E. I. Becker, and C. Miesse, *Org. Synth.* **III:**109 (1955).
d. J. L. Leiserson and A. Weissberger, *Org. Synth.* **III:**183 (1955).
e. T. P. Smyth and B. W. Corby, *Org. Process Res. Dev.* **1:**264 (1977).
f. J. R. Desmurs, M. Labrouillere, C. Le Roux, H. Gaspard, A. Laporterie, and J. Dubac, *Tetrahedron Lett.* **38:**8871 (1997).
g. L. Arsenijevic, V. Arsenijevic, A. Horeau, and J. Jacques, *Org. Synth.* **53:**5 (1973).
h. E. L. Martin and L. F. Fieser, *Org. Synth.* **II:**569 (1943).
i. C. E. Olson and A. F. Bader, *Org. Synth.* **IV:**898 (1963).
j. M. B. Floyd and G. R. Allen, Jr., *J. Org. Chem.* **35:**2647 (1970).
k. M. C. Venuti, *J. Org. Chem.* **46:**3124 (1981).
l. G. Esteban, M. A. Lopez-Sanchez, E. Martinez, and J. Plumet, *Tetrahedron* **54:**197 (1998).
m. V. Premasagar, V. A. Palaniswamy, and E. J. Eisenbraun, *J. Org. Chem.* **46:**2974 (1981).

A special case of aromatic acylation is the *Fries rearrangement*, which is the conversion of an ester of a phenol to an *o*-acyl phenol by a Lewis acid.

OCH₃ / OCH₃ O₂CC₂H₅ —BF₃→ OCH₃ / OCH₃ OH C₂H₅ O 92% Ref. 53

O₂CC₂H₃ / CH₃ —ZrCl₄→ OH O CH₃ / CH₃ 95% Ref. 54

Lanthanide triflates are also good catalysts for Fries rearrangements.[55]

There are a number of other variations of the Friedel–Crafts reaction which are useful in synthesis. The introduction of chloromethyl substituents is brought about by reaction with formaldehyde in concentrated hydrochloric acid and a Lewis acid, especially zinc chloride.[56] The reaction proceeds with benzene and derivatives with electron-releasing groups. The reactive electrophile is probably the chloromethylium ion.

$$CH_2{=}O + HCl + H^+ \rightleftharpoons H_2\overset{+}{O}CH_2Cl \longrightarrow CH_2{=}Cl^+$$

$$C_6H_6 + CH_2{=}Cl^+ \longrightarrow C_6H_5CH_2Cl$$

Chloromethylation can also be carried out by using various chloromethyl ethers and $SnCl_4$.[57]

CH₃ / CH₃ —ClCH₂O(CH₂)₄OCH₂Cl, SnCl₄→ CH₃ CH₂Cl / CH₃

Carbon monoxide, hydrogen cyanide, and nitriles also react with aromatic compounds in the presence of strong acids or Lewis acid catalysts to introduce formyl or acyl substituents. The active electrophiles are believed to be *dications* resulting from diprotonation of CO, HCN, or the nitrile.[58] The general outlines of the mechanisms of

53. Y. Naruta, Y. Nishigaichi, and K. Maruyama, *J. Org. Chem.* **53:**1192 (1988).
54. D. C. Harrowven and R. F. Dainty, *Tetrahedron Lett.* **37:**7659 (1996).
55. S. Kobayashi, M. Moriwaki, and J. Hachiya, *Bull. Chem. Soc. Jpn.* **70:**267 (1997).
56. R. C. Fuson and C. H. McKeever, *Org. React.* **1:**63 (1942); G. A. Olah and S. H. Yu, *J. Am. Chem. Soc.* **97:**2293 (1975).
57. G. A. Olah, D. A. Beal, and J. A. Olah, *J. Org. Chem.* **41:**1627 (1976); G. A. Olah, D. A. Bell, S. H. Yu, and J. A. Olah, *Synthesis* **1974:**560.
58. M. Yato, T. Ohwada, and K. Shudo, *J. Am. Chem. Soc.* **113:**691 (1991); Y. Sato, M. Yato, T. Ohwada, S. Saito, and K. Shudo, *J. Am. Chem. Soc.* **117:**3037 (1995).

these reactions are given below:

a. Formylation with carbon monoxide:

$$\bar{C}{\equiv}\overset{+}{O} + H^+ \rightleftharpoons H{-}C{\equiv}\overset{+}{O} \overset{H^+}{\rightleftharpoons} H{-}\overset{+}{C}{=}\overset{+}{O}{-}H$$

$$ArH + H\overset{+}{C}{=}\overset{+}{O}{-}H \longrightarrow ArCH{=}O + 2\,H^+$$

b. Formylation with hydrogen cyanide:

$$H{-}C{\equiv}N + H^+ \rightleftharpoons H{-}C{\equiv}\overset{+}{N}{-}H \overset{H^+}{\rightleftharpoons} H\overset{+}{C}{=}\overset{+}{N}H_2$$

$$ArH + H\overset{+}{C}{=}\overset{+}{N}H_2 \longrightarrow Ar\underset{H}{\underset{|}{C}}{=}\overset{+}{N}H_2 \xrightarrow{H_2O} ArCH{=}O$$

c. Acylation with nitriles:

$$R{-}C{\equiv}N + H^+ \rightleftharpoons R{-}C{\equiv}\overset{+}{N}{-}H \overset{H^+}{\rightleftharpoons} R\overset{+}{C}{=}\overset{+}{N}H_2$$

$$ArH + R\overset{+}{C}{=}\overset{+}{N}H_2 \longrightarrow R{-}\underset{Ar}{\underset{|}{C}}{=}\overset{+}{N}H_2 \xrightarrow{H_2O} Ar\overset{O}{\overset{\|}{C}}R$$

Many specific examples of these reactions can be found in reviews in the *Organic Reactions* series.[59] Dichloromethyl ethers are also precursors of the formyl group via alkylation catalyzed by $SnCl_4$ or $TiCl_4$.[60] The dichloromethyl group is hydrolyzed to a formyl group.

$$Ar{-}H \xrightarrow[SnCl_4]{Cl_2CHOR} ArCHCl_2 \xrightarrow{H_2O} ArCH{=}O$$

Another useful method for introducing formyl and acyl groups is the *Vilsmeier–Haack reaction*.[61] An *N*,*N*-dialkylamide reacts with phosphorus oxychloride or oxalyl chloride[62] to give a chloroiminium ion, which is the reactive electrophile.

$$R\overset{O}{\overset{\|}{C}}N(CH_3)_2 + POCl_3 \longrightarrow R\overset{Cl}{\overset{|}{C}}{=}\overset{+}{N}(CH_3)_2$$

This species acts as an electrophile in the absence of any added Lewis acid, but only rings with electron-releasing substituents are reactive. Scheme 11.5 gives some examples of these acylation reactions.

11.1.4. Electrophilic Metalation

Aromatic compounds react with mercuric salts to give arylmercury compounds.[63] The reaction shows substituent effects that are characteristic of electrophilic aromatic

59. N. N. Crounse, *Org. React.* **5**:290 (1949); W. E. Truce, *Org. React.* **9**:37 (1957); P. E. Spoerri and A. S. DuBois, *Org. React.* **5**:387 (1949); see also G. A. Olah, L. Ohannesian, and M. Arvanaghi, *Chem. Rev.* **87**:671 (1987).
60. P. E. Sonnet, *J. Med. Chem.* **15**:97 (1972); C. H. Hassall and B. A. Morgan, *J. Chem. Soc., Perkin Trans. 1* **1973**:2853; R. Halterman and S.-T. Jan, *J. Org. Chem.* **56**:5253 (1991).
61. G. Martin and M. Martin, *Bull. Soc. Chim. Fr.* **1963**:1637; S. Seshadri, *J. Sci. Ind. Res.* **32**:128 (1973); C. Just, in *Iminium Salts in Organic Chemistry*, H. Böhme and H. G. Viehe, eds., Vol. 9 in *Advances in Organic Chemistry: Methods and Results*, Wiley-Interscience, New York, 1976, pp. 225–342.
62. J. N. Freskos, G. W. Morrow, and J. S. Swenton, *J. Org. Chem.* **50**:805 (1985).
63. W. Kitching, *Organomet. Chem. Rev.* **3**:35 (1968).

Scheme 11.5. Other Electrophilic Aromatic Substitutions Related to Friedel–Crafts Reactions

A. Chloromethylation

1[a]

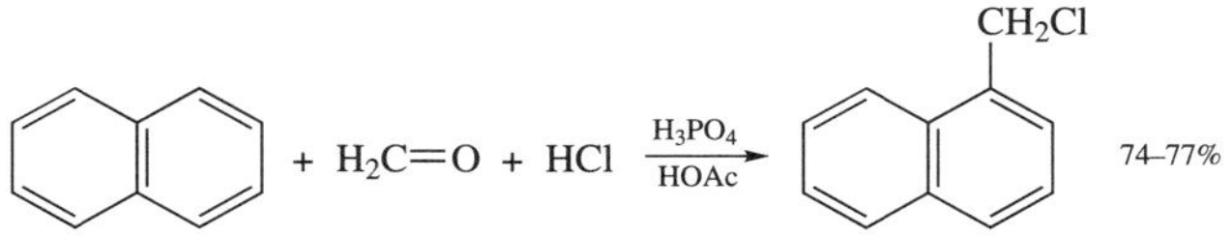

naphthalene + $H_2C{=}O$ + HCl $\xrightarrow[\text{HOAc}]{H_3PO_4}$ 1-(chloromethyl)naphthalene (CH_2Cl) 74–77%

B. Formylation

2[b]

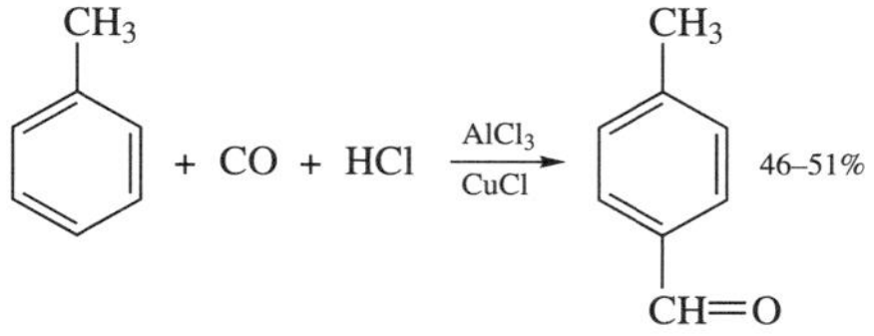

toluene (CH_3) + CO + HCl $\xrightarrow[\text{CuCl}]{AlCl_3}$ 4-methylbenzaldehyde (CH_3, $CH{=}O$) 46–51%

3[c]

methyl 3,5-dimethoxybenzoate (OCH_3, CH_3O, CO_2CH_3) $\xrightarrow[TiCl_4]{(ClCH_2)_2O}$ product with OCH_3, $CH{=}O$, CH_3O, CO_2CH_3 99%

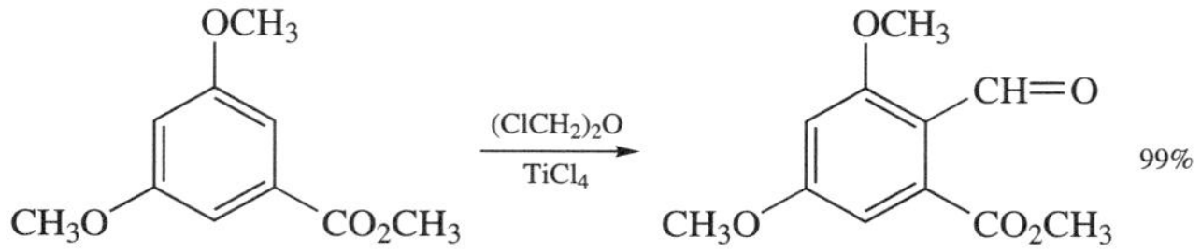

C. Acylation with cyanide and nitriles

4[d]

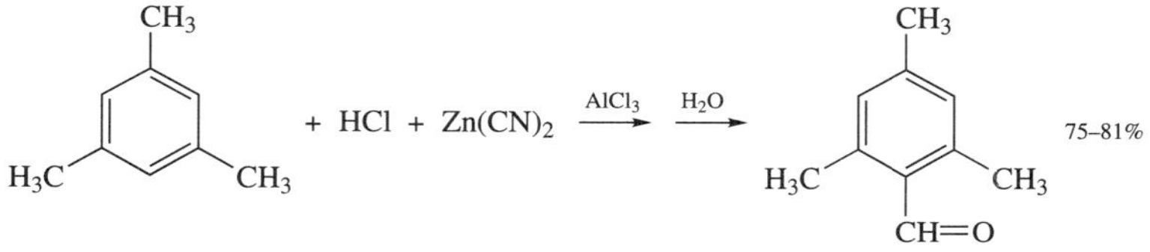

mesitylene (CH_3, H_3C, CH_3) + HCl + $Zn(CN)_2$ $\xrightarrow{AlCl_3}$ $\xrightarrow{H_2O}$ 2,4,6-trimethylbenzaldehyde (CH_3, H_3C, CH_3, $CH{=}O$) 75–81%

5[e]

phloroglucinol (OH, HO, OH) + CH_3CN $\xrightarrow[Zn(CN)_2]{HCl}$ $\xrightarrow{H_2O}$ 2,4,6-trihydroxyacetophenone (OH, $\overset{O}{\|}{C}CH_3$, HO, OH) 74–87%

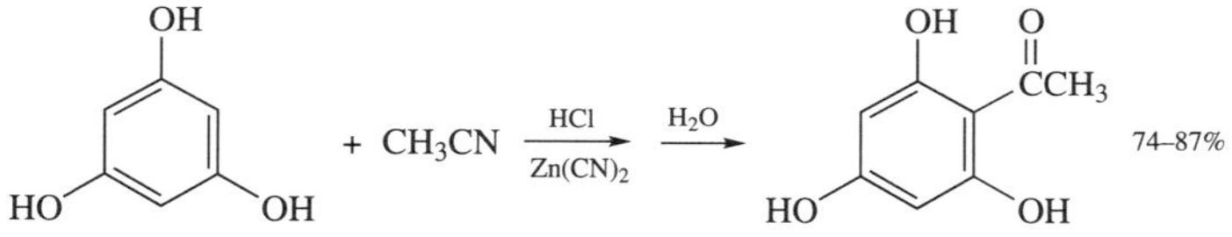

D. Vilsmeier–Haack acylation

6[f]

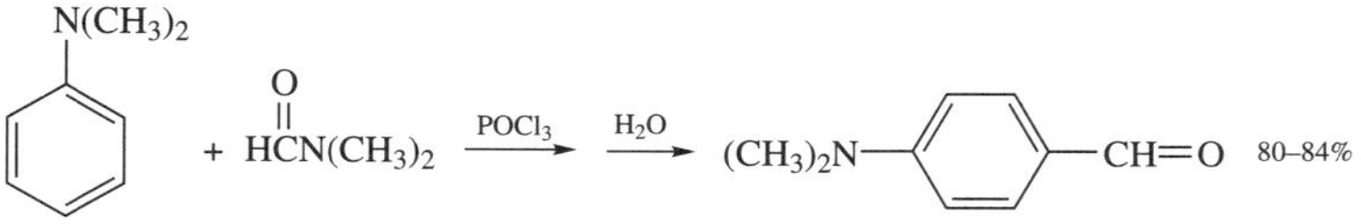

$N(CH_3)_2$-benzene + $H\overset{O}{\|}{C}N(CH_3)_2$ $\xrightarrow{POCl_3}$ $\xrightarrow{H_2O}$ $(CH_3)_2N-C_6H_4-CH{=}O$ 80–84%

7[g]

$N(CH_3)_2$-benzene + $Ph\overset{O}{\|}{C}NHPh$ $\xrightarrow{POCl_3}$ $\xrightarrow{H_2O}$ $(CH_3)_2N-C_6H_4-\overset{O}{\|}{C}-C_6H_5$ 72–77%

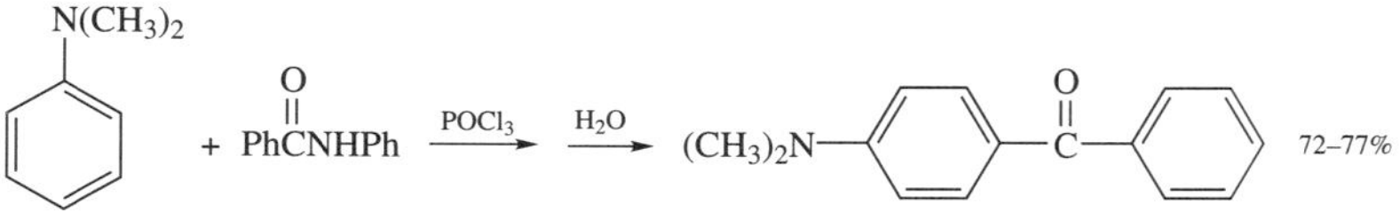

8[h] 2-ethoxynaphthalene (OC_2H_5) + HCNHPh (O=C) $\xrightarrow{POCl_3}$ $\xrightarrow{H_2O}$ 1-formyl-2-ethoxynaphthalene (CH=O, OC_2H_5) 74–84%

a. O. Grummitt and A. Buck, *Org. Synth.* **III:**195 (1955).
b. G. H. Coleman and D. Craig, *Org. Synth.* **II:**583 (1943).
c. C. H. Hassall and B. A. Morgan, *J. Chem. Soc., Perkin Trans. 1* **1973:**2853.
d. R. C. Fuson, E. C. Horning, S. P. Rowland, and M. L. Ward, *Org. Synth.* **III:**549 (1955).
e. K. C. Gulati, S. R. Seth, and K. Venkataraman, *Org. Syth.* **II:**522 (1943).
f. E. Campaigne and W. L. Archer, *Org. Synth.* **IV:**331 (1963).
g. C. D. Hurd and C. N. Webb, *Org. Synth.* **I:**217 (1941).
h. J. H. Wood and R. W. Bost, *Org. Synth.* **III:**98 (1955).

substitution.[64] Mercuration is one of the few electrophilic aromatic substitutions in which proton loss from the σ-complex is rate-determining. Mercuration of benzene shows an isotope effect $k_H/k_D = 6$,[65] which indicates that the σ-complex must be reversibly formed.

$$C_6H_6 + Hg^{2+} \rightleftharpoons [\sigma\text{-complex (H, Hg)}]^{+} \xrightarrow{slow} C_6H_5Hg^{+} + H^{+}$$

Mercuric acetate and mercuric trifluoroacetate are the usual reagents.[66] The synthetic utility of the mercuration reaction derives from subsequent transformations of the arylmercury compounds. As indicated in Section 7.3.3, arylmercury compounds are only weakly nucleophilic, but the carbon–mercury bond is reactive toward various electrophiles. The nitroso group can be introduced by reaction with nitrosyl chloride[67] or nitrosonium tetrafluoroborate[68] as the electrophile. Arylmercury compounds are also useful in certain palladium-catalyzed reactions, as discussed in Section 8.2.

Thallium(III), particularly as the trifluoroacetate salt, is also a reactive electrophilic metalating species, and a variety of synthetic schemes based on arylthallium intermediates have been devised.[69] Arylthallium compounds are converted to chlorides or bromides by reaction with the appropriate cupric halide.[70] Reaction with potassium iodide gives aryl iodides.[71] Fluorides are prepared by successive treatment with potassium fluoride and

64. H. C. Brown and C. W. McGary, Jr., *J. Am. Chem. Soc.* **77:**2300, 2310 (1955); A. J. Kresge and H. C. Brown, *J. Org. Chem.* **32:**756 (1967); G. A. Olah, I. Hashimoto, and H. C. Lin, *Proc. Natl. Acad. Sci. U.S.A.* **74:**4121 (1977).
65. C. Perrin and F. H. Westheimer, *J. Am. Chem. Soc.* **85:**2773 (1963); A. J. Kresge and J. F. Brennan, *J. Org. Chem.* **32:**752 (1967); C. W. Fung, M. Khorramdel-Vahad, R. J. Ranson, and R. M. G. Roberts, *J. Chem. Soc., Perkin Trans. 2* **1980:**267.
66. A. J. Kresge, M. Dubeck, and H. C. Brown, *J. Org. Chem.* **32:**745 (1967); H. C. Brown and R. A. Wirkkala, *J. Am. Chem. Soc.* **88:**1447, 1453, 1456 (1966).
67. L. I. Smith and F. L. Taylor, *J. Am. Chem. Soc.* **57:**2460 (1935); S. Terabe, S. Kuruma, and R. Konaka, *J. Chem. Soc., Perkin Trans. 2* **1973:**1252.
68. L. M. Stock and T. L. Wright, *J. Org. Chem.* **44:**3467 (1979).
69. E. C. Taylor and A. McKillop, *Acc. Chem. Res.* **3:**338 (1970).
70. S. Uemura, Y. Ikeda, and K. Ichikawa, *Tetrahedron* **28:**5499 (1972).
71. A. McKillop, J. D. Hunt, M. J. Zelesko, J. S. Fowler, E. C. Taylor, G. McGillivray, and F. Kienzle, *J. Am. Chem. Soc.* **93:**4841 (1971); M. L. dos Santos, G. C. de Magalhaes, and R. Braz Filho, *J. Organomet. Chem.* **526:**15 (1996).

boron trifluoride.[72] Procedures for converting arylthallium compounds to nitriles and phenols have also been described.[73]

The thallium intermediates can be useful in directing substitution to specific positions when the site of thallation can be controlled in an advantageous way. The two principal means of control are chelation and the ability to effect thermal equilibration of arylthallium intermediates. Oxygen-containing groups normally direct thallation to the *ortho* position by a chelation effect. The thermodynamically favored position is normally the *meta* position, and heating the thallium derivatives of alkybenzenes gives a predominance of the *meta* isomer.[74] Both mercury and thallium compounds are very toxic, so great care is needed in their manipulation.

11.2. Nucleophilic Aromatic Substitution

Many synthetically important substitutions of aromatic compounds are effected by nucleophilic reagents. There are several general mechanisms for substitution by nucleophiles. Unlike nucleophilic substitution at saturated carbon, aromatic nucleophilic substitution does not occur by a single-step mechanism. The broad mechanistic classes that can be recognized include addition–elimination, elimination–addition, metal-catalyzed, and radical or electron-transfer processes. (See Sections 10.5, 10.6, 12.8, and 12.9, Part A to review these mechanisms.)

11.2.1. Aryl Diazonium Ions as Synthetic Intermediates

The most broadly useful intermediates for nucleophilic aromatic substitution are the aryl diazonium salts. Aryl diazonium ions are usually prepared by reaction of an aniline with nitrous acid, which is generated *in situ* from a nitrite salt.[75] Unlike aliphatic diazonium ions, which decompose very rapidly to molecular nitrogen and a carbocation (see Section 10.1), aryl diazonium are stable enough to exist in solution at room temperature and below. They can also be isolated as salts with nonnucleophilic anions, such as tetrafluoroborate or trifluoroacetate.[76] The steps in forming a dizonium ion are addition of the nitrosonium ion, ^+NO, to the amino group, followed by elimination of water.

$$ArNH_2 + HONO \xrightarrow{H^+} ArN(H){-}N{=}O + H_2O$$

$$ArN(H){-}N{=}O \longrightarrow ArN{=}N{-}OH \xrightarrow{H^+} Ar\overset{+}{N}{\equiv}N + H_2O$$

72. E. C. Taylor, E. C. Bigham, and D. K. Johnson, *J. Org. Chem.* **42**:362 (1977).
73. S. Uemura, Y. Ikeda, and K. Ichikawa, *Tetrahedron* **28**:3025 (1972); E. C. Taylor, H. W. Altland, R. H. Danforth, G. McGillivray, and A. McKillop, *J. Am. Chem. Soc.* **92**:3520 (1970).
74. A. McKillop, J. D. Hunt, M. J. Zelesko, J. S. Fowler, E. C. Taylor, G. McGillivray, and F. Kienzle, *J. Am. Chem. Soc.* **93**:4841 (1971).
75. H. Zollinger, *Azo and Diazo Chemistry*, Interscience, New York, 1961; S. Patai, ed. *The Chemistry of Diazonium and Diazo Groups*, John Wiley & Sons, New York, 1978, Chapters 8, 11, and 14; H. Saunders and R. L. M. Allen, *Aromatic Diazo Compounds*, 3rd ed., Edward Arnold, London, 1985.
76. C. Colas and M. Goeldner, *Eur. J. Org. Chem.* **1999**:1357.

In alkaline solution, diazonium ions are converted to diazoate anions, which are in equilibrium with diazoxides.[77]

$$\mathrm{Ar\overset{+}{N}{\equiv}N + 2\,^{-}OH \longrightarrow ArN{=}N{-}O^{-} + H_2O}$$

$$\mathrm{ArN{=}N{-}O^{-} + Ar\overset{+}{N}{\equiv}N \rightleftharpoons ArN{=}N{-}O{-}N{=}NAr}$$

In addition to the classical techniques for diazotization in aqueous solution, diazonium ions can be generated in organic solvents by reaction with alkyl nitrites.

$$\mathrm{RO{-}N{=}O + ArNH_2 \longrightarrow ArN(H){-}N{=}O + ROH}$$

$$\mathrm{ArN(H){-}N{=}O \rightleftharpoons ArN{=}N{-}OH \xrightarrow{H^+} Ar\overset{+}{N}{\equiv}N + H_2O}$$

Diazonium ions form stable adducts with certain nucleophiles such as secondary amines and sulfides.[78] These compounds can be used as *in situ* precursors of diazonium ion intermediates.

$$\mathrm{Ar\overset{+}{N}{\equiv}N + HNR_2 \longrightarrow ArN{=}NNR_2}$$

$$\mathrm{Ar\overset{+}{N}{\equiv}N + {}^{-}SR \longrightarrow ArN{=}NSR}$$

The great usefulness of aryl diazonium ions as synthetic intermediates results from the excellence of N_2 as a leaving group. There are at least three general mechanisms by which substitution can occur. One involves unimolecular thermal decomposition of the diazonium ion, followed by capture of the resulting aryl cation by a nucleophile. The phenyl cation is very unstable (see Part A, Section 5.4) and therefore highly unselective.[79] Either the solvent or an anion can act as the nucleophile.

$$\mathrm{C_6H_5{-}\overset{+}{N}{\equiv}N \longrightarrow C_6H_5^{+} + N_2}$$

$$\mathrm{C_6H_5^{+} + X^{-} \longrightarrow C_6H_5{-}X}$$

Another possible mechanism for substitution is adduct formation followed by collapse of the adduct with loss of nitrogen.

$$\mathrm{C_6H_5{-}\overset{+}{N}{\equiv}N + X^{-} \rightleftharpoons C_6H_5{-}N{=}N{-}X \longrightarrow C_6H_5{-}X + N_2}$$

77. E. S. Lewis and M. P. Hanson, *J. Am. Chem. Soc.* **89:**6268 (1967).
78. M. L. Gross, D. H. Blank, and W. M. Welch, *J. Org. Chem.* **58:**2104 (1993); S. A. Haroutounian, J. P. DiZio, and J. A. Katzenellenbogen, *J. Org. Chem.* **56:**4993 (1991).
79. C. G. Swain, J. E. Sheats, and K. G. Harbison, *J. Am. Chem. Soc.* **97:**783 (1975).

The third mechanism involves redox processes.[80] This mechanism is particularly likely to operate in reactions in which copper salts are used as catalysts.[81]

$$Ar\overset{+}{N}{\equiv}N + [Cu(I)X_2]^- \longrightarrow Ar{-}Cu(III)X_2 + N_2$$
$$Ar{-}Cu(III)X_2 \longrightarrow ArX + Cu(I)X$$

Examples of the three mechanistic types are, respectively: (a) hydrolysis of diazonium salts to phenols;[82] (b) reaction with azide ion to form aryl azides;[83] and (c) reaction with cuprous halides to form aryl chlorides or bromides.[84] In the paragraphs which follow, these and other synthetically useful reactions of diazonium intermediates are considered. The reactions are organized on the basis of the group which is introduced, rather than on the mechanism involved. It will be seen that the reactions that are discussed fall into one of the three general mechanistic types.

Replacement of a nitro or amino group by hydrogen is sometimes required as a sequel to a synthetic operation in which the substituent has been used to control the regioselectively of a prior transformation. The best reagents for reductive dediazonation are hypophosphorous acid, H_3PO_2,[85] and $NaBH_4$.[86] The reduction by H_3PO_2 is substantially improved by catalysis by cuprous oxide.[87] The reduction by H_3PO_2 proceeds by one-electron reduction followed by loss of nitrogen and formation of the phenyl radical.[88]

initiation $$Ar\overset{+}{N}{\equiv}N + e^- \longrightarrow Ar\cdot + N_2$$

propagation $$Ar\cdot + H_3PO_2 \longrightarrow Ar{-}H + [H_2PO_2\cdot]$$
$$Ar\overset{+}{N}{\equiv}N + [H_2PO_2\cdot] \longrightarrow Ar\cdot + N_2 + [H_2PO_2^+]$$
$$[H_2PO_2^+] + H_2O \longrightarrow H_3PO_3 + H^+$$

An alternative method for reductive dediazonation involves *in situ* diazotization by an alkyl nitrite in dimethylformamide.[89] This reduction is a chain reaction in which the solvent acts as a hydrogen-atom donor.

initiation $$Ar\overset{+}{N}{\equiv}N + e^- \longrightarrow Ar\cdot + N_2$$

propagation $$Ar\cdot + HC({=}O)N(CH_3)_2 \longrightarrow Ar{-}H + \cdot C({=}O)N(CH_3)_2$$
$$Ar\overset{+}{N}{\equiv}N + \cdot C({=}O)N(CH_3)_2 \longrightarrow Ar\cdot + N_2 + C{\equiv}O + CH_3\overset{+}{N}H{=}CH_2$$

80. C. Galli, *Chem. Rev.* **88:**765 (1988).
81. T. Cohen, R. J. Lewarchi, and J. Z. Tarino, *J. Am. Chem. Soc.* **97:**783 (1975).
82. E. S. Lewis, L. D. Hartung, and B. M. McKay, *J. Am. Chem. Soc.* **91:**419 (1969).
83. C. D. Ritchie and D. J. Wright, *J. Am. Chem. Soc.*, **93:**2429 (1971); C. D. Ritchie and P. O. I. Virtanen, *J. Am. Chem. Soc.* **94:**4966 (1972).
84. J. K. Kochi, *J. Am. Chem. Soc.* **79:**2942 (1957); S. C. Dickerman, K. Weiss, and A. K. Ingberman, *J. Am. Chem. Soc.* **80:**1904 (1958).
85. N. Kornblum, *Org. React.* **2:**262 (1944).
86. J. B. Hendrickson, *J. Am. Chem. Soc.* **83:**1251 (1961).
87. S. Korzeniowski, L. Blum, and G. W. Gokel, *J. Org. Chem.* **42:**1469 (1977).
88. N. Kornblum, G. D. Cooper, and J. E. Taylor, *J. Am. Chem. Soc.* **72:**3013 (1950).
89. M. P. Doyle, J. F. Dellaria, Jr., B. Siegfried, and S. W. Bishop, *J. Org. Chem.* **42:**3494 (1977); J. H. Markgraf, R. Chang, J. R. Cort, J. L. Durant, Jr., M. Finkelstein, A. W. Gross, M. H. Lavyne, W. M. Moore, R. C. Petersen, and S. D. Ross, *Tetrahedron* **53:**10009 (1997).

This reaction can also be catalysed by $FeSO_4$.[90] (See entry 5 in Scheme 11.6.)

Aryl diazonium ions can be converted to phenols by heating in water. Under these conditions, formation of a phenyl cation probably occurs.

$$Ar\overset{+}{N}{\equiv}N \longrightarrow Ar^+ + N_2 \xrightarrow{H_2O} ArOH + H^+$$

By-products from capture of nucleophilic anions may be observed.[79] Phenols can be formed under milder conditions by an alternative redox mechanism.[91] The reaction is initiated by cuprous oxide, which effects reduction and decomposition to an aryl radical. The reaction is run in the presence of Cu(II) salts. The radical is captured by Cu(II) and oxidized to the phenol. This procedure is very rapid and gives good yields of phenols over a range of structural types.

$$Ar\overset{+}{N}{\equiv}N + Cu(I) \longrightarrow Ar\cdot + N_2 + Cu(II)$$

$$Ar\cdot + Cu(II) \longrightarrow [Ar{-}Cu]^{2+} \xrightarrow{H_2O} ArOH + Cu(I) + H^+$$

Replacement of diazonium groups by halide is a valuable alternative to direct halogenation for preparation of aryl halides. Aryl bromides and chlorides are usually prepared by reaction of aryl diazonium salts with the appropriate Cu(I) salt, a process which is known as the *Sandmeyer reaction*. Under the classic conditions, the diazonium salt is added to a hot acidic solution of the cuprous halide.[92] The Sandmeyer reaction is formulated as proceeding by an oxidative addition reaction of the diazonium ion with Cu(I) and halide transfer from the Cu(III) intermediate.

$$Ar\overset{+}{N}{\equiv}N + [CuX_2]^- \longrightarrow Ar{-}CuX_2 + N_2$$

$$Ar{-}CuX_2 \longrightarrow ArX + CuX$$

It is also possible to covert anilines to aryl halides by generating the diazonium ion *in situ*. Reaction of anilines with alkyl nitrites and Cu(II) halides in acetonitrile gives good yields of aryl chlorides and bromides.[93] Examples of these reactions are given in section C of Scheme 11.6.

Diazonium salts can also be converted to halides by processes involving aryl free radicals. In basic solutions, aryl diazonium ions are converted to radicals via diazoxides.[94]

$$2\,Ar\overset{+}{N}{\equiv}N + 2\,^-OH \longrightarrow ArN{=}N{-}O{-}N{=}NAr + H_2O$$

$$ArN{=}N{-}O{-}N{=}NAr \longrightarrow ArN{=}N{-}O\cdot + Ar\cdot + N_2$$

The reaction can be carried out efficiently using aryl diazonium tetrafluoroborates with crown ethers, polyethers, or phase-transfer catalysts.[95] In solvents that can act as halogen-

90. F. W. Wassmundt and W. F. Kiesman, *J. Org. Chem.* **60:**1713 (1995).
91. T. Cohen, A. G. Dietz, Jr., and J. R. Miser, *J. Org. Chem.* **42:**2053 (1977).
92. W. A. Cowdrey and D. S. Davies, *Q. Rev. Chem. Soc.* **6:**358 (1952); H. H. Hodgson, *Chem. Rev.* **40:**251 (1947).
93. M. P. Doyle, B. Siegfried, and J. F. Dellaria, Jr., *J. Org. Chem.* **42:**2426 (1977).
94. C. Rüchardt and B. Freudenberg, *Tetrahedron Lett.* **1964:**3623; C. Rüchardt and E. Merz, *Tetrahedron Lett.* **1964:**2431.
95. S. H. Korzeniowski and G. W. Gokel, *Tetrahedron Lett.* **1977:**1637.

Scheme 11.6. Aromatic Substitution via Diazonium Ions

A. Replacement by hydrogen

1[a] 2,4,6-Tribromoaniline —1) HONO; 2) C_2H_5OH→ 1,3,5-tribromobenzene 74–77%

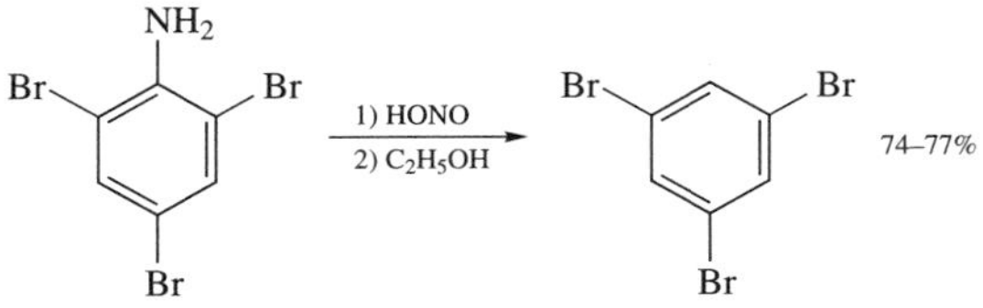

2[b] H_2N–(3-methylphenyl)–(3-methylphenyl)–NH_2 (H_3C, CH_3) —1) HONO; 2) H_3PO_2→ 3,3'-dimethylbiphenyl (H_3C, CH_3) 76–82%

3[c] 2,4-Dichlorobenzenediazonium $\overset{+}{N}_2BF_4^-$ —H_3PO_2, Cu_2O→ 1,3-dichlorobenzene (Cl, Cl) 97%

4[d] 4,5-Dichloro-2-nitroaniline (Cl, Cl, NH_2, NO_2) —$(CH_3)_3CONO$, DMF→ 1,2-dichloro-4-nitrobenzene (Cl, Cl, NO_2) 68%

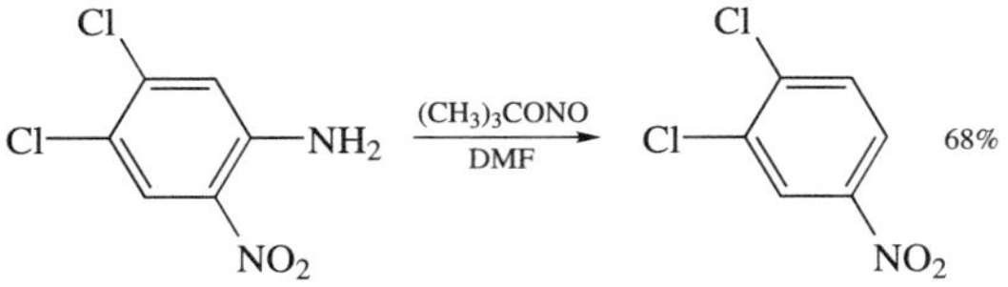

5[e] 2,4,6-Trimethylaniline (CH_3, CH_3, CH_3, NH_2) —1) $NaNO_2$, CH_3CO_2H; 2) $FeSO_4$, DMF→ 1,3,5-trimethylbenzene (CH_3, CH_3, CH_3) 76%

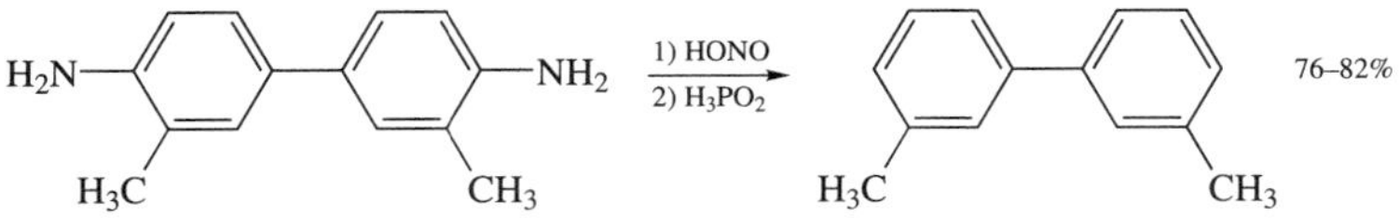

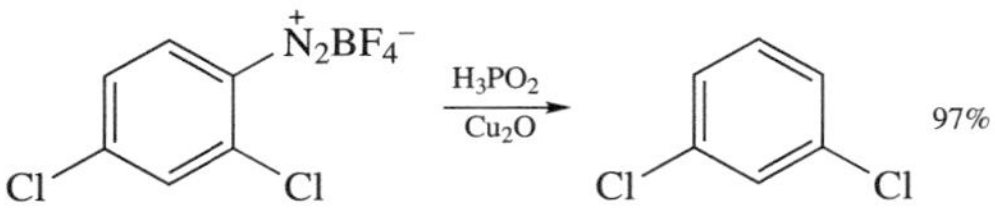

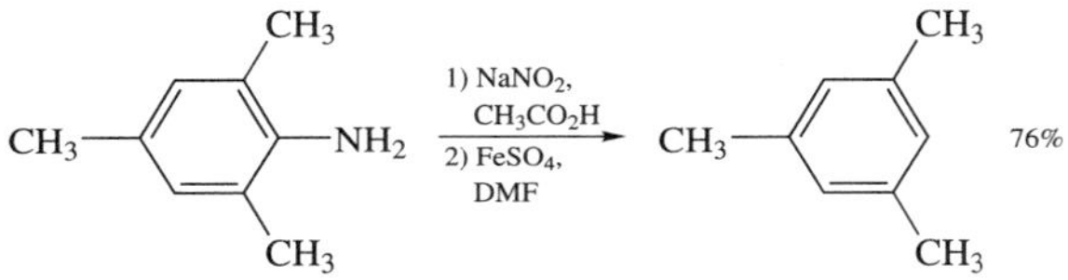

6[f] OH, CO_2H, $NHCO_2C(CH_3)_3$, NH_2 —1) C_2H_5ONO; 2) $NaBH_4$; 3) HCl→ OH, CO_2H, NH_3^+ 72%

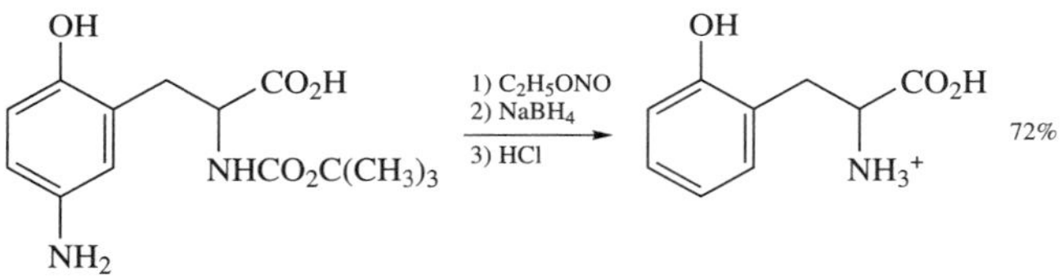

A. Replacement by hydroxyl

7[g] CH_3, Br, NH_2 —1) HONO; 2) H_2O, Δ→ CH_3, Br, OH 80–92%

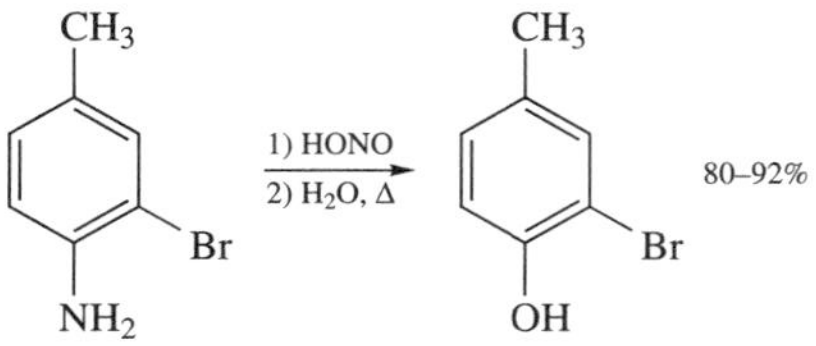

8[h] CH_3, NH_2, NO_2 —1) HONO; 2) $Cu(NO_3)_2$, CuO→ CH_3, OH, NO_2 95%

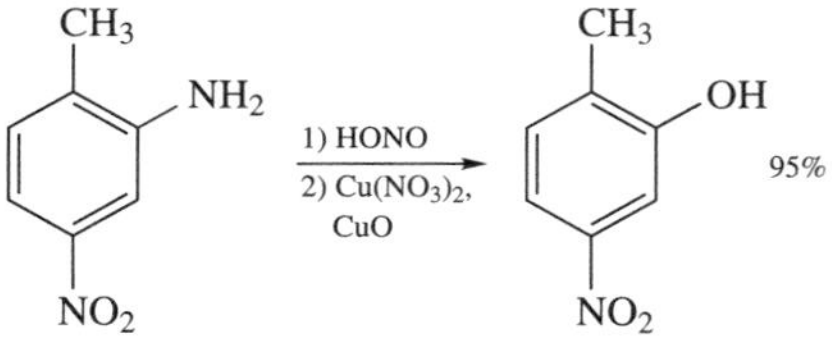

C. Replacement by halogen

9[i] 3-aminobenzaldehyde (CH=O, NH_2) $\xrightarrow{\text{1) HONO; 2) } Cu_2Cl_2}$ 3-chlorobenzaldehyde (CH=O, Cl) 75–79%

10[j] 2,6-dinitroaniline (O_2N, NH_2, NO_2) $\xrightarrow{\text{1) HONO; 2) } Cu_2Cl_2}$ 2-chloro-1,3-dinitrobenzene (O_2N, Cl, NO_2) 71–74%

11[k] $H_3C-C_6H_4-NH_2$ $\xrightarrow[CuBr_2]{RONO}$ $H_3C-C_6H_4-Br$ 76%

12[l] 3-Cl-$C_6H_4-\overset{+}{N}_2BF_4^-$ $\xrightarrow[BrCCl_3]{\text{NaOAc, 18-crown-6}}$ 3-Cl-C_6H_4-Br 88%

13[m] 2-bromoaniline (Br, NH_2) $\xrightarrow{\text{1) HONO; 2) KI}}$ 1-bromo-2-iodobenzene (Br, I) 72–83%

14[n] 2-bromoaniline (Br, NH_2) $\xrightarrow{\text{1) HONO; 2) } HPF_6\text{; 3) } \Delta}$ 1-bromo-2-fluorobenzene (Br, F) 73–75%

15[o] $H_2N-C_6H_4-C_6H_4-NH_2$ $\xrightarrow{\text{1) HONO; 2) } HBF_4\text{; 3) } \Delta}$ $F-C_6H_4-C_6H_4-F$ 54–56%

(*continued on next page*)

atom donors, the radicals react to give aryl halides.

$$Ar\cdot + S-X \longrightarrow ArX + S\cdot$$

Diiodomethane is used for iodides.[96] Bromotrichloromethane gives aryl bromides, and methyl iodide gives iodides.[97] The diazonium ions can also be generated by *in situ* methods. Under these conditions, bromoform and bromotrichloromethane have been used as bromine donors, and carbon tetrachloride is the best chlorine donor.[98] This method was

96. W. B. Smith and O. C. Ho, *J. Org. Chem.* **55:**2543 (1990).
97. S. H. Korzeniowski and G. W. Gokel, *Tetrahedron Lett.* **1977:**3519; R. A. Bartsch and I. W. Wang, *Tetrahedron Lett.* **1979:**2503.
98. J. I. G. Cadogan, D. A. Roy, and D. M. Smith, *J. Chem. Soc., C* **1966:**1249.

Scheme 11.6. (*continued*)

16[p]

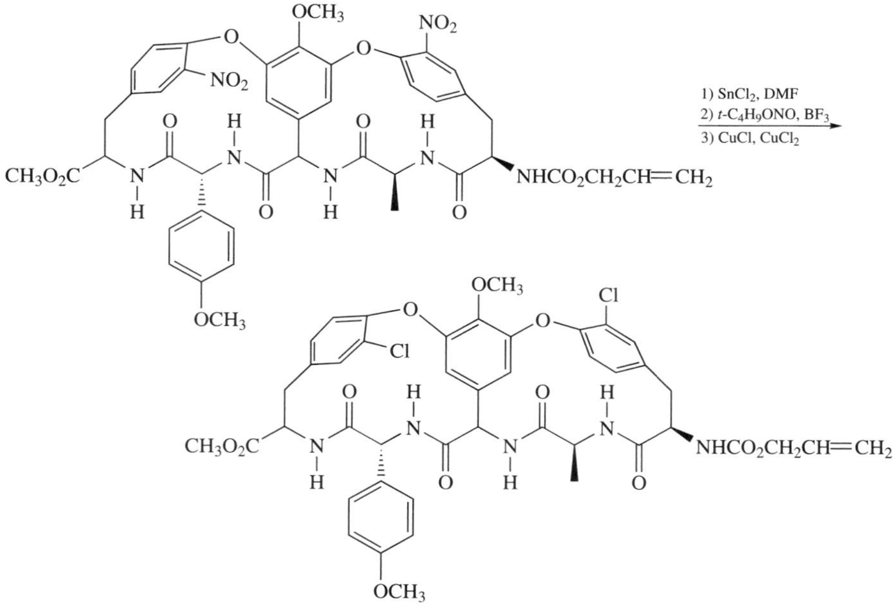

OCH3 NO2 O O NO2 O H O H CH3O2C N N N N NHCO2CH2CH=CH2 H O H O OCH3

1) $SnCl_2$, DMF
2) t-C_4H_9ONO, BF_3
3) CuCl, $CuCl_2$

OCH3 Cl O O Cl O H O H CH3O2C N N N N NHCO2CH2CH=CH2 H O H O OCH3

D. Replacement by other anions

17[q] NH_2, CH_3 → 1) HONO 2) CuCN → C≡N, CH_3 64–70%

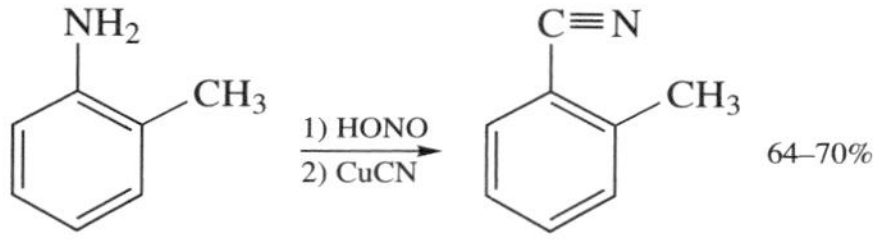

18[r] NH_2 → 1) HONO 2) NaN_3 → N_3 88%

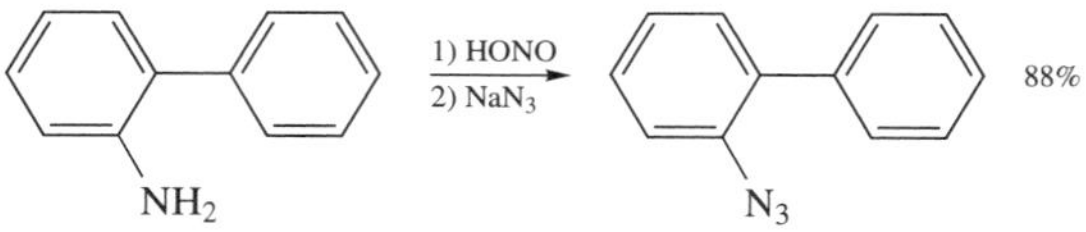

a. G. H. Coleman and W. F. Talbot, *Org. Synth.* **II:**592 (1943).
b. N. Kornblum, *Org. Synth.* **III:**295 (1955).
c. S. H. Korzeniowski, L. Blum, and G. W. Gokel, *J. Org. Chem.* **42:**1469 (1977).
d. M. P. Doyle, J. F. Dellaria, Jr., B. Siegfried, and S. W. Bishop, *J. Org. Chem.* **42:**3494 (1977).
e. F. W. Wassmundt and W. F. Kiesman, *J. Org. Chem.* **60:**1713 (1995).
f. C. Dugave, *J. Org. Chem.* **60:**601 (1995).
g. H. E. Ungnade and E. F. Orwoll, *Org. Synth.* **III:**130 (1955).
h. T. Cohen, A. G. Dietz, Jr., and J. R. Miser, *J. Org. Chem.* **42:**2053 (1977).
i. J. S. Buck and W. S. Ide, *Org. Synth.* **II:**130 (1943).
j. F. D. Gunstone and S. H. Tucker, *Org. Synth.* **IV:**160 (1963).
k. M. P. Doyle, B. Siegfried, and J. F. Dellaria, Jr., *J. Org. Chem.* **42:**2426 (1977).
l. S. H. Korzeniowski and G. W. Gokel, *Tetrahedron Lett.* **1977:**3519.
m. H. Heaney and I. T. Millar, *Org. Synth.* **40:**105 (1960).
n. K. G. Rutherford and W. Redmond, *Org. Synth.* **43:**12 (1963).
o. G. Schiemann and W. Winkelmuller, *Org. Synth.* **II:**188 (1943).
p. C. Vergne, M. Bois-Choussy, and J. Zhu, *Synlett* **1998:**1159.
q. H. T. Clarke and R. R. Read, *Org. Synth.* **I:**514 (1941).
r. P. A. S. Smith and B. B. Brown, *J. Am. Chem. Soc.* **73:**2438 (1951).

used successfully for a challenging chlorodeamination in the vancomycin system (entry 16 in Scheme 11.6).

Fluorine substituents can also be introduced via diazonium ions. One procedure is to isolate aryl diazonium tetrafluoroborates. These decompose thermally to give aryl fluorides.[99] This reaction probably involves formation of an aryl cation which abstracts fluoride ion from the tetrafluoroborate anion.[100]

$$\mathrm{Ar\overset{+}{N}{\equiv}N + BF_4^- \longrightarrow ArF + N_2 + BF_3}$$

Hexafluorophosphate salts behave similarly.[101] The diazonium tetrafluoroborates can be prepared either by precipitation from an aqueous solution by fluoroboric acid[102] or by anhydrous diazotization in ether, THF, or acetonitrile using *t*-butyl nitrite and boron trifluoride.[103] Somewhat milder conditions can be achieved by reaction of aryldiazo sulfide adducts with pyridine–HF in the presence of AgF or $AgNO_3$.

$$n\text{-}C_4H_9\text{-}C_6H_4\text{-}N{=}NSPh \xrightarrow[\text{AgNO}_3,\ 90^\circ\text{C}]{\text{pyridine–HF}} n\text{-}C_4H_9\text{-}C_6H_4\text{-}F \quad 39\%$$

Ref. 104

Aryl diazonium ions are converted to iodides in high yield by reaction with iodide salts. This reaction is initiated by reduction of the diazonium ion by iodide. The aryl radical then abstracts iodine from either I_2 or I_3^-. A chain mechanism then proceeds which consumes I^- and ArN_2^+.[105] Evidence for the involvement of radicals includes the isolation of cyclized products from *o*-allyl derivatives.

$$\mathrm{Ar\overset{+}{N}{\equiv}N + I^- \longrightarrow Ar\cdot + N_2 + I\cdot \qquad 2\,I\cdot \longrightarrow I_2}$$

$$\mathrm{Ar\cdot + I_3^- \longrightarrow ArI + I_2^{\dot{-}}}$$

$$\mathrm{Ar\overset{+}{N}{\equiv}N + I_2^{\dot{-}} \longrightarrow Ar\cdot + N_2 + I_2 \qquad I_2 + I^- \longrightarrow I_3^-}$$

Cyano and azido groups are also readily introduced via diazonium intermediates. The former process involves a copper-catalyzed reaction analogous to the Sandmeyer reaction. Reaction of diazonium salts with azide ion gives adducts which smoothly decompose to nitrogen and the aryl azide.[83]

$$\mathrm{Ar\overset{+}{N}{\equiv}N + {}^-N{=}\overset{+}{N}{=}N^- \longrightarrow ArN{=}N{-}N{=}\overset{+}{N}{=}N^- \longrightarrow ArN{=}\overset{+}{N}{=}N^- + N_2}$$

Aryl thiolates react with aryl diazonium ions to give diaryl sulfides. This reaction is believed to be a radical-chain process, similar to the mechanism for reaction of diazonium

99. A. Roe, *Org. React.* **5**:193 (1949).
100. C. G. Swain and R. J. Rogers, *J. Am. Chem. Soc.* **97**:799 (1975).
101. M. S. Newman and R. H. B. Galt, *J. Org. Chem.* **25**:214 (1960).
102. E. B. Starkey, *Org. Synth.* **II**:225 (1943); G. Schiemann and W. Winkelmuller, *Org. Synth.* **II**:299 (1943).
103. M. P. Doyle and W. J. Bryker, *J. Org. Chem.* **44**:1572 (1979).
104. S. A. Haroutounian, J. P. Dizio, and J. A. Katzenellenbogen, *J. Org. Chem.* **56**:4993 (1991).
105. P. R. Singh and R. Kumar, *Aust. J. Chem.* **25**:2133 (1972); A. Abeywickrema and A. L. J. Beckwith, *J. Org. Chem.* **52**:2568 (1987).

ions with iodide ion.[106]

$$\text{initiation} \quad Ar\overset{+}{N}{\equiv}N + PhS^- \longrightarrow ArN{=}NSPh$$
$$ArN{=}NSPh \longrightarrow Ar\cdot + N_2 + PhS\cdot$$
$$\text{propagation} \quad Ar\cdot + PhS^- \longrightarrow Ar\dot{\bar{S}}Ph$$
$$Ar\dot{\bar{S}}Ph + Ar\overset{+}{N}{\equiv}N \longrightarrow ArSPh + Ar\cdot + N_2$$

Scheme 11.6 gives some examples of the various substitution reactions of aryl diazonium ions.

Aryl diazonium ions can also be used to form certain types of carbon–carbon bonds. The copper-catalyzed reaction of diazonium ions with conjugated alkenes results in arylation of the alkene. This is known as the *Meerwein arylation reaction*.[107] The reaction sequence is initiated by reduction of the diazonium ion by Cu(I). The aryl radical adds to the alkene to give a new β-aryl radical. The final step is an oxidation/ligand transfer which takes place in the copper coordination sphere. An alternative course is oxidation/deprotonation, which gives a styrene derivative.

$$PhN_2^+ + Cu(I) \longrightarrow Ph\cdot + N_2 + Cu(II) \xrightarrow{H_2C=CHZ} PhCH_2{-}\dot{C}HZ$$

$$PhCH_2\dot{C}HZ + CuCl_2 \longrightarrow PhCH_2CH(Cl)Z + CuCl$$

The reaction gives better yields with dienes, styrenes, or alkenes substituted with electron-withdrawing groups than with simple alkenes. These groups increase the rate of capture of the aryl radical. The standard conditions for the Meerwein arylation employ aqueous solutions of diazonium ions prepared in the usual way. Conditions for *in situ* diazotization by *t*-butyl nitrite in the presence of $CuCl_2$ and acrylonitrile or styrene are also effective.[108]

Reduction of aryl diazonium ions by Ti(III) in the presence of α,β-unsaturated ketones and aldehydes leads to β arylation and formation of the saturated ketone or aldehyde. The early steps in this reaction parallel the copper-catalyzed reaction. However, rather than being oxidized, the radical formed by the addition step is reduced by Ti(III).[109] Scheme 11.7 illustrates some typical examples of arylation of alkenes by diazonium ions.

$$Ar\overset{+}{N}{\equiv}N + Ti(III) \longrightarrow Ar\cdot + N_2 + Ti(IV)$$

$$Ar\cdot + RCH{=}CH\overset{O}{\overset{\|}{C}}R \longrightarrow Ar\overset{R}{\overset{|}{C}}H\dot{C}H\overset{O}{\overset{\|}{C}}R \xrightarrow[H^+]{Ti(III)} Ar\overset{R}{\overset{|}{C}}HCH_2\overset{O}{\overset{\|}{C}}R$$

11.2.2. Substitution by the Addition–Elimination Mechanism

The addition of a nucleophile to an aromatic ring, followed by elimination of a substituent, results in nucleophilic substitution. The major energetic requirement for this

106. A. N. Abeywickrema and A. L. J. Beckwith, *J. Am. Chem. Soc.* **108:**8227 (1986).
107. C. S. Rondestvedt, Jr., *Org. React.* **11:**189 (1960); C. S. Rondesvedt, Jr., *Org. React.* **24:**225 (1976); A. V. Dombrovskii, *Russ. Chem. Rev.*, (Engl. Transl.) **53:**943 (1984).
108. M. P. Doyle, B. Siegfried, R. C. Elliot, and J. F. Dellaria, Jr., *J. Org. Chem.* **42:**2431 (1977).
109. A. Citterio and E. Vismara, *Synthesis* **1980:**191; A. Citterio, A. Cominelli, and F. Bonavoglia, *Synthesis* **1986:**308.

1[a] $O_2N-C_6H_4-N_2^+Cl^- + H_2C{=}CHCH{=}CH_2 \rightarrow O_2N-C_6H_4-CH_2CH{=}CHCH_2Cl$

2[b] $Cl-C_6H_4-N_2^+$ + N-isopropylmaleimide ($NCH(CH_3)_2$) $\xrightarrow[\text{pH 3}]{CuCl_2}$ 3-(4-chlorophenyl)-N-isopropylmaleimide ($NCH(CH_3)_2$) 51%

3[c] $O_2N-C_6H_4-N_2^+ + H_2C{=}CHCN \xrightarrow{CuCl_2} O_2N-C_6H_4-CH_2CH(Cl)CN$ 48%

4[d] $m\text{-}F-C_6H_4-NH_2 + CH_2{=}CHCO_2CH_3 \xrightarrow[\text{2) CuCl}]{\text{1) NaNO}_2\text{, HCl}} m\text{-}F-C_6H_4-CH{=}CHCO_2CH_3$ 93%

5[e] $Cl-C_6H_4-NH_2 + H_2C{=}CHCN \xrightarrow[CuCl_2]{t\text{-BuONO}} Cl-C_6H_4-CH_2CH(Cl)CN$ 71%

6[f] $Cl-C_6H_4-N_2^+ + CH_3CH{=}CHC(O)CH_3 \xrightarrow{Ti^{3+}} Cl-C_6H_4-CH(CH_3)CH_2C(O)CH_3$ 65–75%

a. G. A. Ropp and E. C. Coyner, *Org. Synth.* **IV**:727 (1963).
b. C. S. Rondestvedt, Jr., and O. Vogl, *J. Am. Chem. Soc.* **77**:2313 (1955).
c. C. F. Koelsch, *J. Am. Chem. Soc.* **65**:57 (1943).
d. G. Theodoridis and P. Malamus, *J. Heterocycl. Chem.* **28**:849 (1991).
e. M. P. Doyle, B. Siegfried, R. C. Elliott, and J. F. Dellaria, Ur., *J. Org. Chem.* **42**:2431 (1977).
f. A. Citterio and E. Vismara, *Synthesis*, **1980**:291; A. Citterio, *Org. Synth.* **62**:67 (1984).

mechanism is formation of the addition intermediate. The addition step is greatly facilitated by strongly electron-attracting substituents, so that nitroaromatics are the best substrates for nucleophilic aromatic substitution. Other electron-attracting groups such as cyano, acetyl, and trifluoromethyl also enhance reactivity.

$$O_2N-C_6H_4-X + Y^- \rightarrow (^-O)_2N{=}C_6H_4(X)(Y) \rightarrow O_2N-C_6H_4-Y + X^-$$

Nucleophilic substitution occurs when there is a potential leaving group present at the carbon at which addition occurs. Although halides are the most common leaving groups, alkoxy, cyano, nitro, and sulfonyl groups can also be displaced. The leaving-group ability does not necessarily parallel that found for nucleophilic substitution at saturated carbon. As a particularly striking example, fluoride is often a better leaving group than the other halogens in nucleophilic aromatic substitution. The relative reactivity of the *p*-halonitrobenzenes toward sodium methoxide at 50°C is F(312) $\gg$ Cl(1) $>$ Br(0.74) $>$ I(0.36).[110]

A principal reason for the order I > Br > Cl > F in S_N2 reactions is the carbon–halogen bond strength, which increases from I to F. The carbon–halogen bond strength is not so important a factor in nucleophilic aromatic substitution because bond breaking is not ordinarily part of the rate-determining step. Furthermore, the highly electronegative fluorine favors the addition step more than the other halogens.

There are not many successful examples of arylation of carbanions by nucleophilic aromatic substitution. A major limitation is the fact that aromatic nitro compounds often react with carbanions by electron-transfer processes.[111] However, such substitution can be carried out under the conditions of the $S_{RN}1$ reaction (see Section 11.4).

2-Halopyridines and other π-deficient nitrogen heterocycles are excellent reactants for nucleophilic aromatic substitution.[112] Substitution reactions also occur readily for other heterocyclic systems, such as 2-haloquinolines and 1-haloisoquinolines, in which a potential leaving group is adjacent to a pyridine-type nitrogen. 4-Halopyridines and related heterocyclic compounds can also undergo substitution by nucleophilic addition–elimination but are somewhat less reactive.

N Cl + $NaOC_2H_5$ → N OC_2H_5 Ref. 113

Scheme 11.8 gives some examples of nucleophilic aromatic substitution reactions.

11.2.3. Substitution by the Elimination–Addition Mechanism

The elimination–addition mechanism involves a highly unstable intermediate, which is called *dehydrobenzene* or *benzyne*.[114] (See Part A, Section 10.6, for a discussion of the structure of benzyne.)

X H + base → → ($^-$Nu, H^+) H **Nu**

A unique feature of this mechanism is that the entering nucleophile does not necessarily become bound to the carbon to which the leaving group was attached:

Y X H → Y → ($^-$Nu, H^+) Y **Nu** H + Y H **Nu**

110. G. P. Briner, J. Mille, M. Liveris, and P. G. Lutz, *J. Chem. Soc.* **1954:**1265.
111. R. D. Guthrie, in *Comprehensive Carbanion Chemistry, Part A*, E. Buncel and T. Durnst, eds., Elsevier, Amsterdam, 1980, Chapter 5.
112. H. E. Mertel, in *Heterocyclic Compounds*, Vol. 14, Part 2, E. Klingsberg, ed., Interscience, New York, 1961; M. M. Boudakian, in *Heterocyclic Compounds*, Vol. 14, Part 2, Supplement, R. A. Abramovitch, ed., Wiley-Interscience, New York, 1974, Chapter 6; B. C. Uff, in *Comprehensive Heterocyclic Chemistry*, Vol. 2A, A. J. Boulton and A. McKillop, eds., Pergamon Press, Oxford, U.K., 1984, Chapter 2.06.
113. N. Al-Awadi, J. Ballam, R. R. Hemblade, and R. Taylor, *J. Chem. Soc., Perkin Trans. 2* **1982:**1175.
114. R. W. Hoffmann, *Dehydrobenzene and Cycloalkynes*, Academic Press, New York, 1967.

1[a] 2-chloronitrobenzene + piperidine $\xrightarrow[16\ h]{120°C}$ 1-(2-nitrophenyl)piperidine 94%

2[b] 4-fluoronitrobenzene + 2-methylpiperidine $\xrightarrow[5\ days]{100°C}$ O_2N–C_6H_4–N(2-methylpiperidine) 85%

3[c] F–C_6H_4–CCH_3(=O) + $(CH_3)_2NH$ $\xrightarrow[5\ h]{80°C}$ $(CH_3)_2N$–C_6H_4–CCH_3(=O) 96%

4[d] 3,5-dinitrobenzonitrile + CH_3O^- $\xrightarrow{DMSO}$ 3-methoxy-5-nitrobenzonitrile

5[e] O_2N–C_6H_4–Cl + C_6H_5–OH $\xrightarrow[Cu,\ 150°C]{KOH}$ O_2N–C_6H_4–O–C_6H_5 80–82%

6[f] 1-(pyrrolidin-1-yl)cyclohexene + Cl–$C_6H_3(NO_2)$–NO_2 (2,4-dinitrochlorobenzene) $\xrightarrow[25°C]{(C_2H_5)_3N}$ 2-(2,4-dinitrophenyl)cyclohexanone 92%

7[g] $N{\equiv}C\underline{C}HCO_2C_2H_5$ + Cl–$C_6H_3(NO_2)$–NO_2 (2,4-dinitrochlorobenzene) $\xrightarrow[18\ h]{25°C}$ $C_2H_5O_2CCH(C{\equiv}N)$–$C_6H_3(NO_2)$–NO_2 80%

a. S. D. Ross and M. Finkelstein, *J. Am. Chem. Soc.* **85:**2603 (1963).
b. F. Pietra and F. Del Cima, *J. Org. Chem.* **33:**1411 (1968).
c. H. Bader, A. R. Hansen, and F. J. McCarty, *J. Org. Chem.* **31:**2319 (1966).
d. E. J. Fendler, J. H. Fendler, N. I. Arthur, and C. E. Griffin, *J. Org. Chem.* **37:**812 (1972).
e. R. O. Brewster and T. Groening, *Org. Synth.* **II:**445 (1943).
f. M. E. Kuehne, *J. Am. Chem. Soc.* **84:**837 (1962).
g. H. R. Snyder, E. P. Merica, C. G. Force, and E. G. White, *J. Am. Chem. Soc.* **80:**4622 (1958).

The elimination–addition mechanism is facilitated by electronic effects that favor removal of a hydrogen from the ring as a proton. Relative reactivity also depends on the halide. The order Br > I > Cl ≫ F has been established in the reaction of aryl halides with KNH_2 in liquid ammonia.[115] This order has been interpreted as representing a balance of two effects. The polar order favoring proton removal would be F > Cl > Br > I, but this is largely overwhelmed by the ease of bond breaking, which is in the reverse order, I > Br > Cl > F. Under these conditions, carbon–halogen bond breaking must be part of the rate-determining step. With organolithium reagents in aprotic solvents, the order of reactivity is F > Cl > Br > I, which indicates that the acidity of the ring hydrogen is the dominant factor governing reactivity.[116]

Addition of nucleophiles such as ammonia or alcohols, or their conjugate bases, to benzynes takes place very rapidly. The addition is believed to involve capture of the nucleophile by benzyne, followed by protonation to give the substitution product.[117] Electronegative groups tend to favor addition of the nucleophile at the more distant end of the "triple bond," because this permits maximum stabilization of the developing negative charge. Selectivity is usually not high, however, and formation of both possible products from monosubstituted benzynes is common.[118]

EWG + **Nu**:⁻ → EWG, **Nu**

There are several methods for generation of benzyne in addition to base-catalyzed elimination of hydrogen halide from a halobenzene, and some of these are more generally applicable for preparative work. Benzyne can also be generated from *o*-dihaloaromatics. Reaction with lithium amalgam or magnesium results in the formation of transient organometallic compounds that decompose with elimination of lithium halide. *o*-Fluorobromobenzene is the usual starting material in this procedure.[119]

F, Br —Li–Hg→ [F, Li] →

Probably the most useful method is diazotization of *o*-aminobenzoic acids.[120] Loss of nitrogen and carbon dioxide follows diazotization and generates benzyne. This method permits generation of benzyne in the presence of a number of molecules with which it can

115. F. W. Bergstrom, R. E. Wright, C. Chandler, and W. A. Gilkey, *J. Org. Chem.* **1:**170 (1936).
116. R. Huisgen and J. Sauer, *Angew. Chem.* **72:**91 (1960).
117. J. F. Bunnett, D. A. R. Happer, M. Patsch, C. Pyun, and H. Takayama, *J. Am. Chem. Soc.* **88:**5250 (1966); J. F. Bunnett and J. K. Kim, *J. Am. Chem. Soc.* **95:**2254 (1973).
118. E. R. Biehl, E. Nieh, and K. C. Hsu, *J. Org. Chem.* **34:**3595 (1969).
119. G. Wittig and L. Pohmer, *Chem. Ber.* **89:**1334 (1956); G. Wittig, *Org. Synth.* **IV:**964 (1963).
120. M. Stiles, R. G. Miller, and U. Burckhardt, *J. Am. Chem. Soc.* **85:**1792 (1963); L. Friedman and F. M. Longullo, *J. Org. Chem.* **34:**3089 (1969).

react.

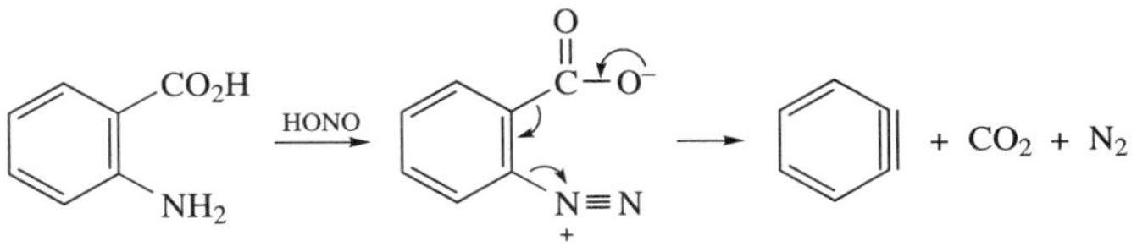

Oxidation of 1-aminobenzotriazole also serves as a source of benzyne under mild conditions. An oxidized intermediate decomposes with loss of two molecules of nitrogen.[121]

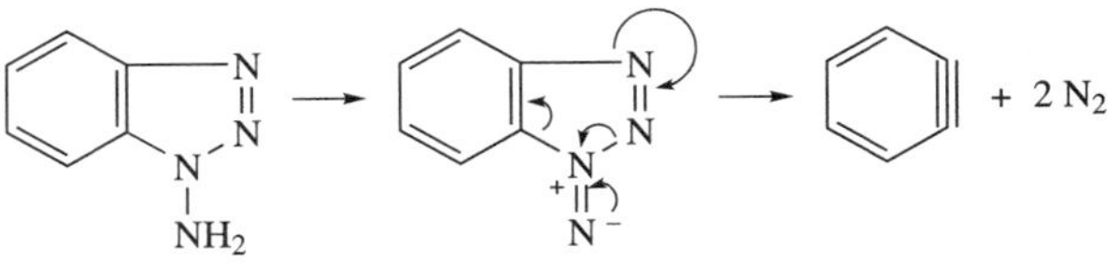

Another heterocyclic molecule that can serve as a benzyne precursor is benzothiadiazole-1,1-dioxide, which decomposes with elimination of nitrogen and sulfur dioxide.[122]

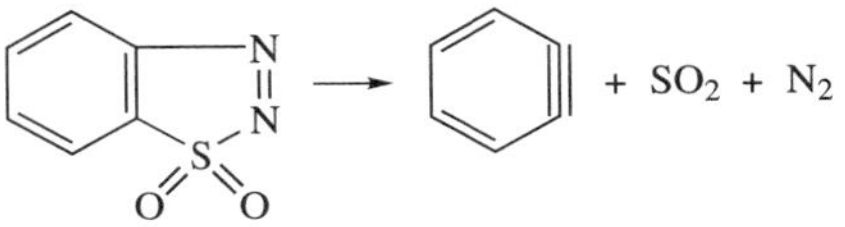

When benzyne is generated in the absence of another reactive molecule, it dimerizes to biphenylene.[123] In the presence of dienes, benzyne is a very reactive dienophile, and $[4+2]$ cycloaddition products are formed.

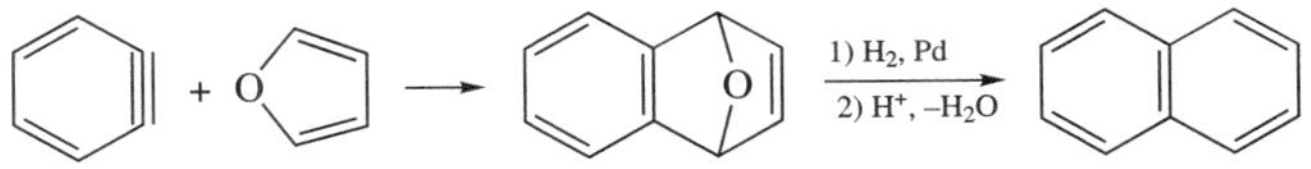

Ref. 124

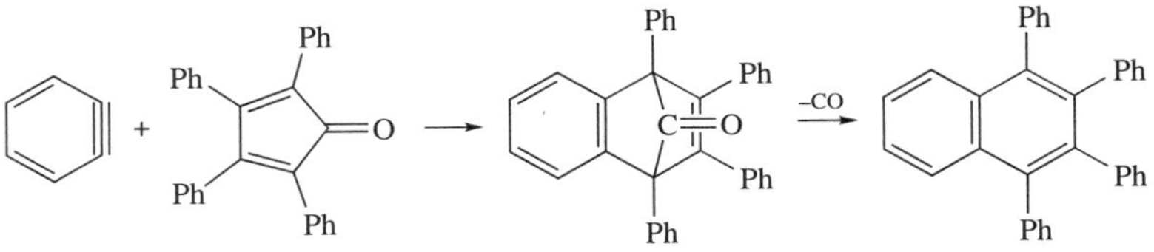

Ref. 125

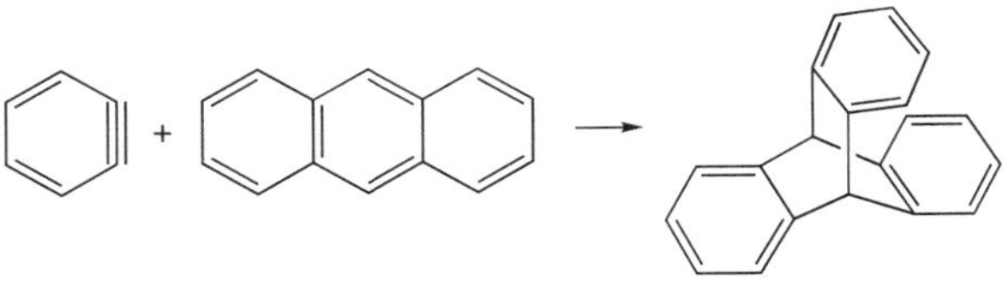

Ref. 126

121. C. D. Campbell and C. W. Rees, *J. Chem. Soc. C* **1969:**742, 752; S. E. Whitney and B. Rickborn, *J. Org. Chem.* **53:**5595 (1988); D. Ok and H. Hart, *J. Org. Chem.* **52:**3835 (1987).
122. G. Wittig and R. W. Hoffmann, *Org. Synth.* **47:**4 (1967); G. Wittig and R. W. Hoffmann, *Chem. Ber.* **95:**2718, 2729 (1962).
123. F. M. Logullo, A. H. Seitz, and L. Friedman, *Org. Synth.* **V:**54 (1973).
124. G. Wittig and L. Pohmer, *Angew. Chem.* **67:**348 (1955).
125. L. F. Fieser and M. J. Haddadin, *Org. Synth.* **V:**1037 (1973).
126. L. Friedman and F. M. Logullo, *J. Org. Chem.* **34:**3089 (1969).

Benzyne gives both [2 + 2] cycloaddition and ene reaction products with simple alkenes.[127]

major   minor

Scheme 11.9 illustrates some of the types of compounds that can be prepared via benzyne intermediates.

11.2.4. Transition-Metal-Catalyzed Substitution Reactions

Nucleophilic substitution of aromatic halides lacking activating substituents is generally difficult. It has been known for a long time that the nucleophilic substitution of aromatic halides is often catalysed by the presence of copper salts.[128] Synthetic procedures based on this observation are used to prepare aryl nitriles by reaction of aryl bromides with Cu(I)CN. The reaction is usually carried out at elevated temperature in DMF or a similar solvent.

CH_3 Br + CuCN $\xrightarrow[\Delta]{DMF}$ CH_3 CN 93% Ref. 129

Br + CuCN $\xrightarrow[200°C]{N\text{-methylpyrrolidone}}$ CN 95% Ref. 130

A general mechanistic description of the copper-promoted nucleophilic substitution pictures an oxidative addition of the aryl halide at Cu(I) followed by collapse of the arylcopper intermediate with a ligand transfer (reductive elimination).[131]

$$\text{Ar—X} + \text{Cu(I)Z} \longrightarrow \text{Ar—}\underset{\text{X}}{\underset{|}{\text{Cu(III)}}}\text{—Z} \longrightarrow \text{Ar—Z} + \text{CuX}$$

X = halide
Z = nucleophile

Many other kinds of nucleophiles can be arylated by copper-catalyzed substitution.[132] Among the reactive nucleophiles are carboxylate ions,[133] alkoxide ions,[134] amines,[135]

127. P. Crews and J. Beard, *J. Org. Chem.* **38:**522 (1973).
128. J. Lindley, *Tetrahedron* **40:**1433 (1984).
129. L. Friedman and H. Shechter, *J. Org. Chem.* **26:**2522 (1961).
130. M. S. Newman and H. Bode, *J. Org. Chem.* **26:**2525 (1961).
131. T. Cohen, J. Wood, and A. G. Dietz, *Tetrahedron Lett.* **1974:**3555.
132. For a review of this reaction, see Ref. 128.
133. T. Cohen and A. H. Lewin, *J. Am. Chem. Soc.* **88:**4521 (1966).
134. R. G. R. Bacon and S. C. Rennison, *J. Chem. Soc., C* **1969:**312.
135. A. J. Paine, *J. Am. Chem. Soc.* **109:**1496 (1987).

Scheme 11.9. Some Syntheses via Benzyne Intermediates

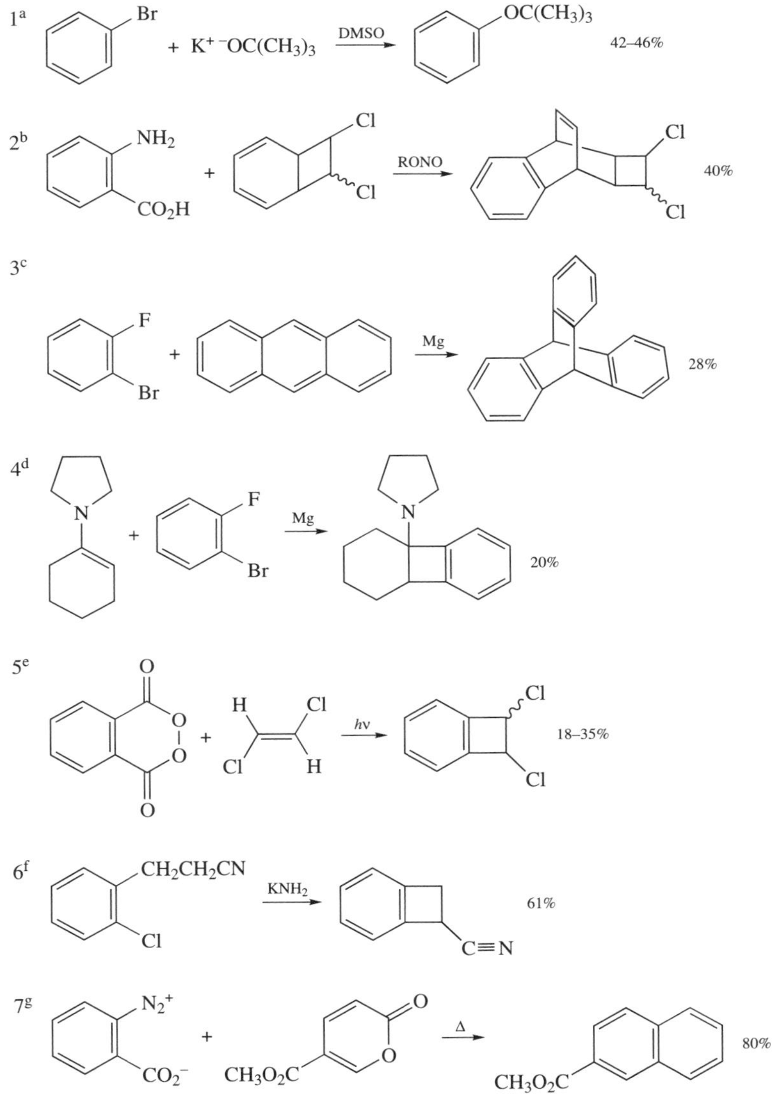

1[a] Br + $K^{+}\,{}^{-}OC(CH_3)_3$ —DMSO→ $OC(CH_3)_3$ 42–46%

2[b] NH_2, CO_2H + Cl, Cl —RONO→ Cl, Cl 40%

3[c] F, Br + —Mg→ 28%

4[d] N + F, Br —Mg→ N 20%

5[e] O, O, O, O + H, Cl, Cl, H —$h\nu$→ Cl, Cl 18–35%

6[f] CH_2CH_2CN, Cl —KNH_2→ $C{\equiv}N$ 61%

7[g] N_2^{+}, CO_2^{-} + CH_3O_2C, O, O —Δ→ CH_3O_2C 80%

a. M. R. Sahyun and D. J. Cram, *Org. Synth.* **45:**89 (1965).
b. L. A. Paquette, M. J. Kukla, and J. C. Stowell, *J. Am. Chem. Soc.* **94:**4920 (1972).
c. G. Wittig, *Org. Synth.* **IV:**964 (1963).
d. M. E. Kuehne, *J. Am. Chem. Soc.* **84:**837 (1962).
e. M. Jones, Jr., and M. R. DeCamp, *J. Org. Chem.* **36:**1536 (1971).
f. J. F. Bunnett and J. A. Skorcz, *J. Org. Chem.* **27:**3836 (1962).
g. S. Escudero, D. Perez, E. Guitian, and L. Castedo, *Tetrahedron Lett.* **38:**5375 (1997).

phthalimide anions,[136] thiolate anions,[137] and acetylides.[138] In some of these reactions, there is a competitive reduction of the aryl halide to the dehalogenated arene, which is attributed to protonolysis of the arylcopper intermediate. Traditionally, most of these reactions have been carried out at high temperature under heterogeneous conditions using copper powder or copper bronze as the catalyst. The general mechanism would suggest that these catalysts act as sources of Cu(I) ions. Homogeneous reactions have been carried out using soluble Cu(I) salts, particularly Cu(I)O_3SCF_3.[139] The range and effectiveness of coupling aryl halides and phenolates to give diaryl ethers has been further improved by use of Cs_2CO_3.[140]

CuO_3SCF_3, 2.5 mol %; Cs_2CO_3 — 79%

It has been found that palladium–phosphine combinations are even more effective catalysts for these nucleophilic substitution reactions. For example, conversion of aryl iodides to nitriles can be done under much milder conditions.

$Pd(PPh_3)_4$, $(C_2H_5)_3N$, $(CH_3)_3SiCN$ — 89% Ref. 141

A great deal of effort has been devoted to finding effective catalysts for substitution by oxygen and nitrogen nucleophiles.[142] These studies have led to optimization of the catalysis with ligands such as triarylphosphines,[143] bis-phosphines such as BINAP,[144] dppf,[145] and phosphines with additional chelating substituents.[146] Among the most effective catalysts are highly hindered trialkylphosphines such as tri-*t*-butyl- and tricyclohexylphosphine.[147] In addition to bromides and iodides, the reaction has been successfully

136. R. G. R. Bacon and A. Karim, *J. Chem. Soc., Perkin Trans. 1* **1973:**272.
137. H. Suzuki, H. Abe, and A. Osuka, *Chem. Lett.* **1980:**1303; R. G. R. Bacon and H. A. O. Hill, *J. Chem. Soc.* **1964:**1108.
138. C. E. Castro, R. Havlin, V. K. Honwad, A. Malte, and S. Moje, *J. Am. Chem. Soc.* **91:**6464 (1969).
139. T. Cohen and J. G. Tirpak, *Tetrahedron Lett.* **1975:**143.
140. J.-F. Marcoux, S. Doye, and S. L. Buchwald, *J. Am. Chem. Soc.* **119:**10539 (1997).
141. N. Chatani and T. Hanafusa, *J. Org. Chem.* **51:**4714 (1986).
142. A. S. Guram, R. A. Rennels and S. L. Buchwald, *Angew. Chem. Int. Ed. Engl.* **34:**1348 (1995); J. F. Hartwig, *Synlett* **1997:**329; J. F. Hartwig, *Angew. Chem. Int. Ed. Engl.* **37:**2047 (1998); J. P. Wolfe, S. Wagaw, J.-F. Marcoux, and S. L. Buchwald, *Acc. Chem. Res.* **31:**805 (1998); J. F. Hartwig, *Acc. Chem. Res.* **31:**852 (1998); B. H. Yang and S. L. Buchwald, *J. Organomet. Chem.* **576:**125 (1999).
143. J. P. Wolfe and S. L. Buchwald, *J. Org. Chem.* **61:**1133 (1996); J. Louie and J. F. Hartwig, *Tetrahedron Lett.* **36:**3609 (1995).
144. J. P. Wolfe, S. Wagaw, and S. L. Buchwald, *J. Am. Chem. Soc.* **118:**7215 (1996).
145. M. S. Driver and J. F. Hartwig, *J. Am. Chem. Soc.* **118:**7217 (1996).
146. D. W. Old, J. P. Wolfe, and S. L. Buchwald, *J. Am. Chem. Soc.* **120:**9722 (1998); B. C. Hamann and J. F. Hartwig, *J. Am. Chem. Soc.* **120:**7369 (1998); S. Vyskocil, M. Smrcina, and P. Kocovsky, *Tetrahedron Lett.* **39:**9289 (1998).
147. M. Nishiyama, T. Yamamoto, and Y. Koie, *Tetrahedron Lett.* **39:**617 (1998); N. P. Reddy and M. Tanaka, *Tetrahedron Lett.* **38:**4807 (1997).

extended to chlorides[148] and triflates.[149] These reaction conditions now permit substitution on both electron-poor and electron-rich aryl systems by a variety of nitrogen nucleophiles, including alkyl and aryl amines and heterocycles. These reactions proceed via a catalytic cycle involving Pd(0) and Pd(II) intermediates.

Ar—N(R′)CH$_2$R
$L_nPd(0)$
Ar—X
L_nPd^{II}(N(R′)CH$_2$R)(Ar)
L_nPd^{II}(X)(Ar)
HN(R′)CH$_2$R

Similar conditions have been used for substitution by alkoxide and phenoxide nucleophiles. Some examples are given in Scheme 11.10.

11.3. Aromatic Radical Substitution Reactions

Aromatic rings are moderately reactive toward addition of free radicals (see Part A, Section 12.2), and certain synthetically useful substitution reactions involve free-radical substitution. One example is the synthesis of biaryls.[150]

X–C$_6$H$_4$· + C$_6$H$_6$ ⟶ X–C$_6$H$_4$–C$_6$H$_6$· (H) ⟶ X–C$_6$H$_4$–C$_6$H$_5$

There are some inherent limits to the usefulness of such reactions. Radical substitutions are only moderately sensitive to substituent directing effects, so that substituted reactants usually give a mixture of products. This means that the practical utility is limited to symmetrical reactants, such as benzene, where the position of attack is immaterial. The best sources of aryl radicals for the reaction are aryl diazonium ions and *N*-nitrosoacetanilides. In the presence of base, diazonium ions form diazoxides, which decompose to aryl radicals.[151]

$$\mathrm{Ar\overset{+}{N}{\equiv}N + 2\ {}^{-}OH \longrightarrow ArN{=}N{-}O{-}N{=}NAr + H_2O}$$

$$\mathrm{ArN{=}N{-}O{-}N{=}NAr \longrightarrow Ar\cdot + N_2 + \cdot O{-}N{=}NAr}$$

148. X. Bei, A. S. Guram, H. W. Turner, and W. H. Weinberg, *Tetrahedron Lett.* **40:**1237 (1999).
149. J. P. Wolfe and S. L. Buchwald, *J. Org. Chem.* **62:**1264 (1997); J. Louie, M. S. Driver, B. C. Hamann, and J. F. Hartwig, *J. Org. Chem.* **62:**1268 (1997).
150. W. E. Bachmann and R. A. Hoffman, *Org. React.* **2:**224 (1944); D. H. Hey, *Adv. Free Radical Chem.* **2:**47 (1966).
151. C. Rüchardt and B. Freudenberg, *Tetrahedron Lett.* **1964:**3623; C. Rüchardt and E. Merz, *Tetrahedron Lett.* **1964:**2431; C. Galli, *Chem. Rev.* **88:**765 (1988).

Scheme 11.10. Copper- and Palladium-Catalyzed Aromatic Substitution

A. Copper-catalyzed substitution

1[a] CH_3O–C_6H_4–I + imidazole (HN, N) —(CuO_3SCF_3, 5 mol %, dba, phenanthroline / $CsCO_3$)→ CH_3O–C_6H_4–N-imidazole (N) 96%

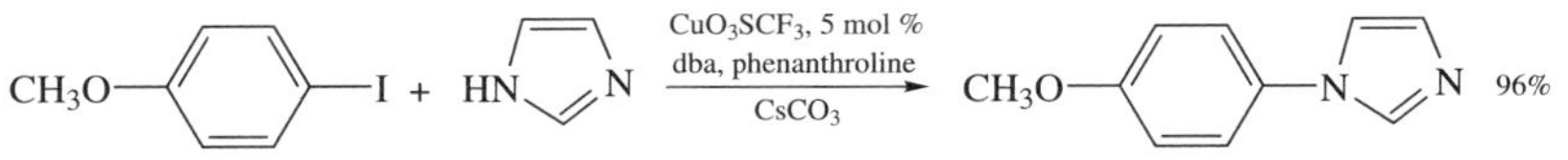

2[b] 4-(3-cyclopentyloxy-4-methoxyphenyl)pyrrolidin-2-one (CH_3O, O, O, NH) + I–C_6H_3(3,5-OCH_3)$_2$ —(Cu, K_2CO_3 / DMF)→ 1-(3,5-dimethoxyphenyl)-4-(3-cyclopentyloxy-4-methoxyphenyl)pyrrolidin-2-one (CH_3O, O, O, N, OCH_3, OCH_3) 61%

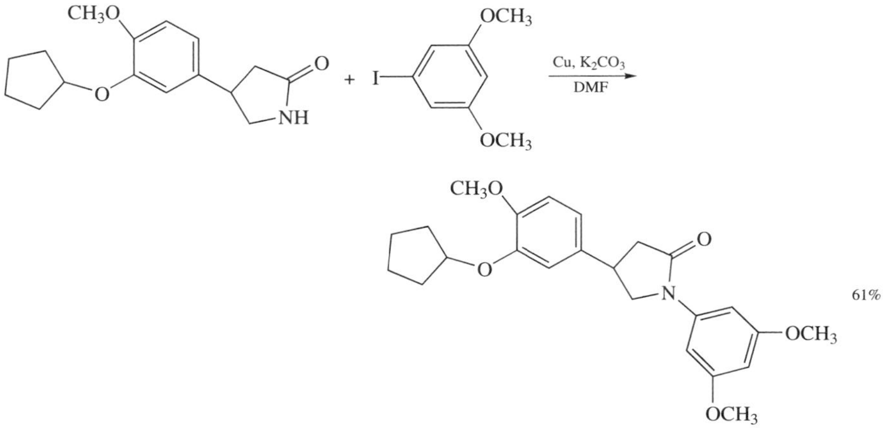

B. Palladium-catalyzed substitution

3[c] C_6H_5–Cl + HN(piperazine)NCH_3 —($Pd[P(C_6H_{11})_3]_2Cl_2$, 1 mol % / NaO-*t*-Bu, 120°C)→ C_6H_5–N(piperazine)NCH_3 88%

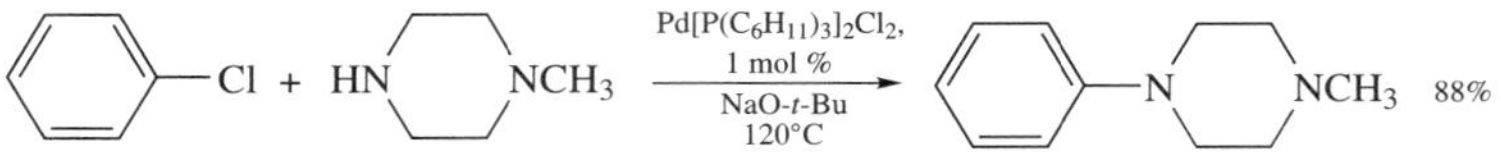

4[d] C_6H_5–Br + PhNH–C_6H_4–CH_3 —($Pd(O_2CCH_3)_2$, $P(t\text{-}C_4H_9)_3$ / NaO-*t*-Bu)→ Ph_2N–C_6H_4–CH_3

5[e] CH_3–C_6H_4–I + H_2N–C_6H_4–Cl —($Pd(dppf)Cl_2$, 5 mol % / 100°C)→ CH_3–C_6H_4–NH–C_6H_4–Cl 84%

6[f] $PhCH_2O$–$C_6H_2Cl_2$–Br + HN(piperazine)NH —($Pd(dba)_2$, 2 mol %, BINAP / NaO-*t*-C_4H_9)→ $PhCH_2O$–$C_6H_2Cl_2$–N(piperazine)NH 94%

7[g] 2-CH_3-C_6H_4–Br + $(H_2NCH_2CH_2)_2NH$ —($Pd(dba)_2$, 1.5 mol % / NaO-*t*-C_4H_9)→ (2-CH_3-C_6H_4–$NHCH_2CH_2)_2NH$ 95%

8[h] piperidin-2-one (N–H, O) + BrPh —($Pd(O_2CCH_3)_2$, 5 mol %, dppf / NaO-*t*-C_4H_9)→ 1-phenylpiperidin-2-one (N–Ph, O) 95%

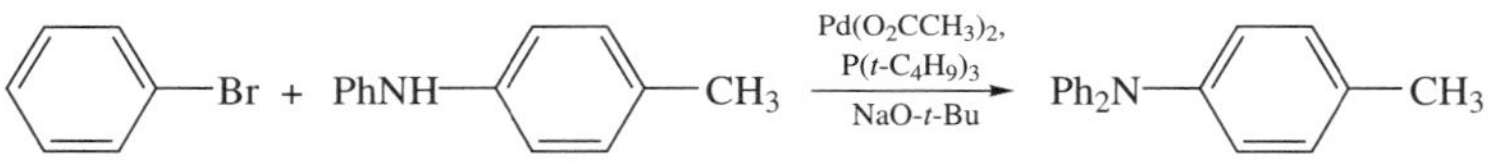

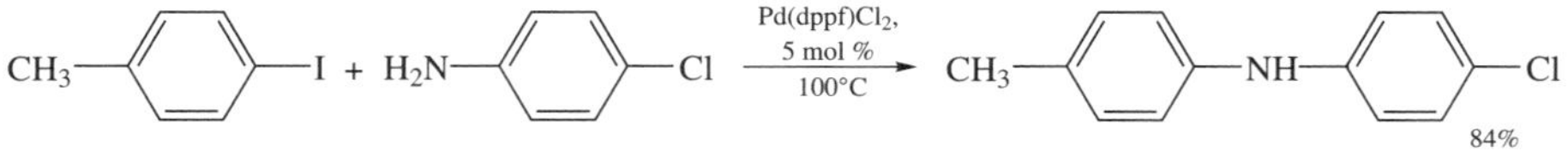

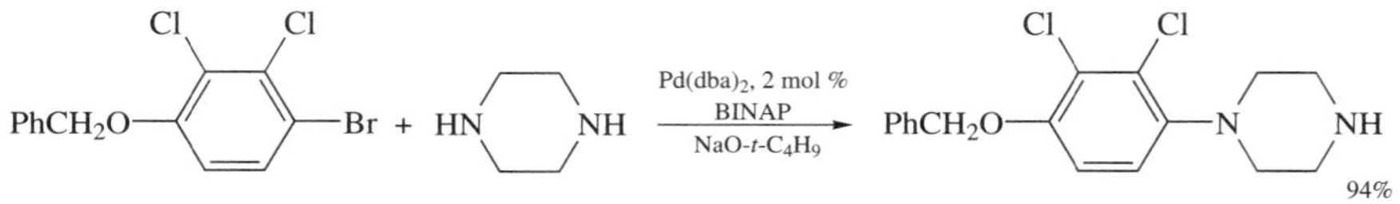

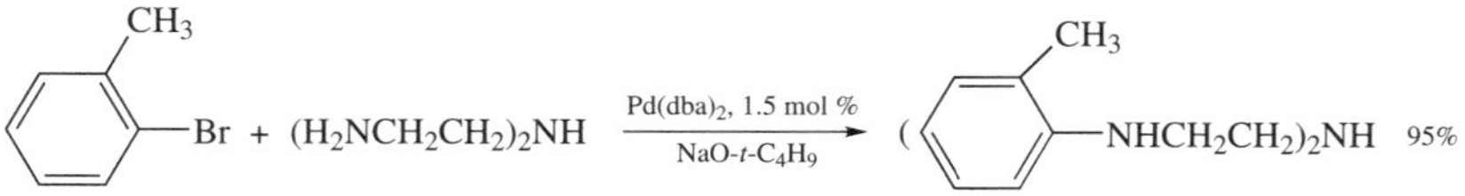

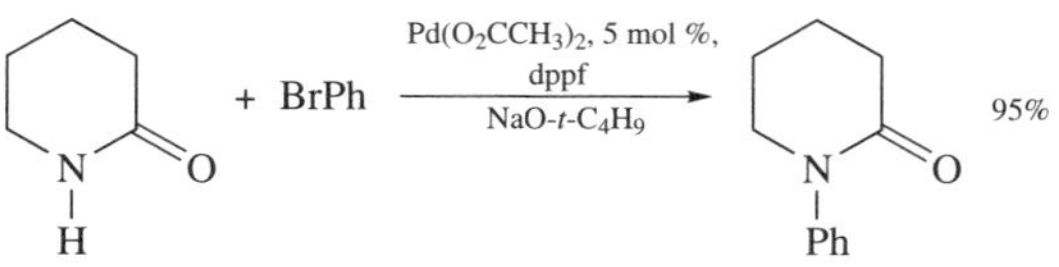

9[i] $CH_3O-C_6H_4-O_3SCF_3 + H_2NPh \xrightarrow[\text{dppf}]{\text{Pd(dba)}_2\text{, 1.5 mol \%,}} CH_3O-C_6H_4-NHPh$ 92%

10[j] $CH_3O-C_6H_4-O_3SCF_3 + CH_3NHPh \xrightarrow[\text{BINAP, CsCO}_3]{\text{Pd(O}_2\text{CCH}_3)_2\text{, 3 mol \%}} CH_3O-C_6H_4-N(CH_3)Ph$ 88%

11[k] $NC-C_6H_4-Br + NaO\text{-}t\text{-}C_4H_9 \xrightarrow[\text{120°C}]{\text{Pd(O}_2\text{CCH}_3)_2\text{, dppf}} NC-C_6H_4-OC(CH_3)_3$

12[l] $2\text{-}CH_3C_6H_4-Br + {}^-O-C_6H_4-OCH_3 \xrightarrow{\text{Pd(dba)}_2\text{, di-}t\text{-Budppf}} 2\text{-}CH_3C_6H_4-O-C_6H_4-OCH_3$ 85%

13[m] $CH_3O_2C-C_6H_4-Br + HOPh \xrightarrow[\text{K}_3\text{PO}_4]{\text{Pd(O}_2\text{CCH}_3)_2\text{, 2 mol \%, biPhP(}t\text{-Bu)}_2\text{, 3 mol \%}} CH_3O_2C-C_6H_4-OPh$ 89%

a. A. Kiyomori, J.-F. Marcoux, and S. L. Buchwald, *Tetrahedron Lett.* **40:**2657 (1999).
b. E. Aebischer, E. Bacher, F. W. J. Demnitz, T. H. Keller, M. Kurzmeyer, M. L. Ortiz, E. Pombo-Villar, and H.-P. Weber, *Heterocycles* **48:**2225 (1998).
c. N. P. Reddy and M. Tanaka, *Tetrahedron Lett.* **38:**4807 (1997).
d. T. Yamamoto, M. Nishiyama, and Y. Koie, *Tetrahedron Lett.* **39:**2367 (1998).
e. M. S. Driver and J. F. Hartwig, *J. Am. Chem. Soc.* **118:**7217 (1996).
f. S. Morita, K. Kitano, J. Matsubara, T. Ohtani, Y. Kawano, K. Otsubo, and M. Uchida, *Tetrahedron* **54:**4811 (1998).
g. Y. Hong, C. H. Senanayake, T. Xiang, C. P. Vandenbossche, G. J. Tanoury, R. P. Bakale, and S. A. Wald, *Tetrahedron Lett.* **39:**3121 (1998).
h. W. C. Shakespeare, *Tetrahedron Lett.* **40:**2035 (1999).
i. J. Louie, M. S. Driver, B. C. Hamann, and J. F. Hartwig, *J. Org. Chem.* **62:**1268 (1997).
j. J. Ahman and S. L. Buchwald, *Tetrahedron Lett.* **38:**6363 (1997).
k. G. Mann and J. F. Hartwig, *J. Org. Chem.* **62:**5413 (1997).
l. G. Mann, C. Incarvito, A. L. Rheingold, and J. F. Hartwig, *J. Am. Chem. Soc.* **121:**3224 (1999).
m. A. Aranyos, D. W. Old, A. Kiyomori, J. P. Wolfe, J. P. Sadighi, and S. L. Buchwald, *J. Am. Chem. Soc.* **121:**4369 (1999).

In the classical procedure, base is added to a two-phase mixture of the aqueous diazonium salt and an excess of the aromatic that is to be substituted. Improved yields have been obtained by using polyethers or phase-transfer catalysts with solid aryl diazonium tetrafluoroborate salts in an excess of the aromatic reactant.[152] Another source of aryl radicals is *N*-nitrosoacetanilides, which rearrange to diazonium acetates and give rise to aryl radicals via diazooxides.[153]

$$ArN(N{=}O)C(=O)CH_3 \longrightarrow ArN{=}N{-}OC(=O)CH_3$$

$$2\ ArN{=}N{-}OC(=O)CH_3 \longrightarrow ArN{=}N{-}O{-}N{=}NAr + (CH_3CO)_2O$$

A procedure for arylation involving *in situ* diazotization has also been developed.[154] Scheme 11.11 gives some representative preparative methods.

152. J. R. Beadle, S. H. Korzeniowski, D. E. Rosenberg, G. J. Garcia-Slanga, and G. W. Gokel, *J. Org. Chem.* **49:**1594 (1984).
153. J. I. G. Cadogan, *Acc. Chem. Res.* **4:**186 (1971); J. I. G. Cadogan, *Adv. Free Radical Chem.* **6:**185 (1980).
154. J. I. G. Cadogan, *J. Chem. Soc.*, 4257 (1962).

Scheme 11.11. Biaryls by Radical Substitution

1[a] Br–C$_6$H$_4$–N_2^+ + C$_6$H$_6$ $\xrightarrow{NaOH}$ Br–C$_6$H$_4$–C$_6$H$_5$ 35%

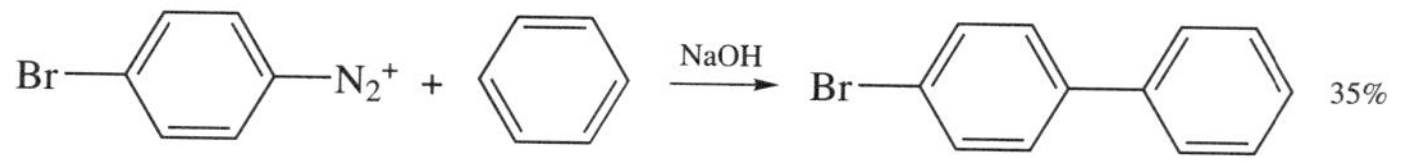

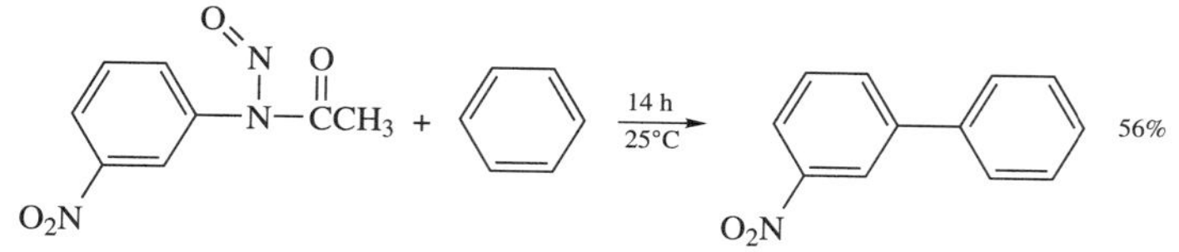

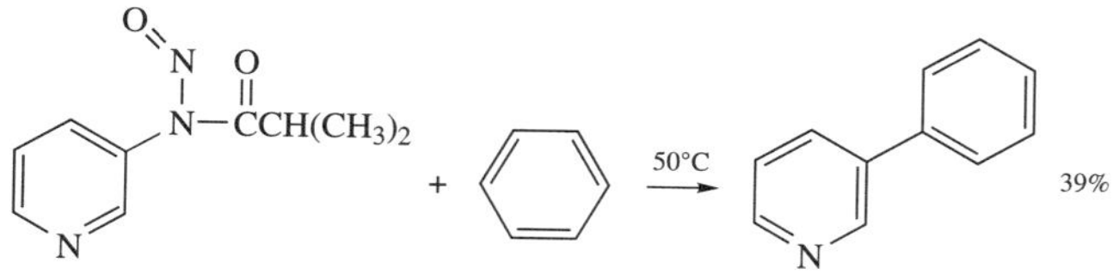

2[b] CH_3O–C$_6$H$_4$–N_2^{+} $^{-}BF_4$ + C$_6$H$_6$ $\xrightarrow[KO_2CCH_3]{18\text{-crown-6}}$ CH_3O–C$_6$H$_4$–C$_6$H$_5$ 80%

3[c] 3-O_2N–C$_6$H$_4$–N(N=O)–C(O)CH_3 + C$_6$H$_6$ $\xrightarrow[25°C]{14\ h}$ 3-O_2N–C$_6$H$_4$–C$_6$H$_5$ 56%

4[d] 3-pyridyl–N(N=O)–C(O)CH$(CH_3)_2$ + C$_6$H$_6$ $\xrightarrow{50°C}$ 3-phenylpyridine 39%

5[e] Cl–C$_6$H$_4$–NH_2 + C$_6$H$_6$ $\xrightarrow{C_5H_{11}ONO}$ Cl–C$_6$H$_4$–C$_6$H$_5$ 45%

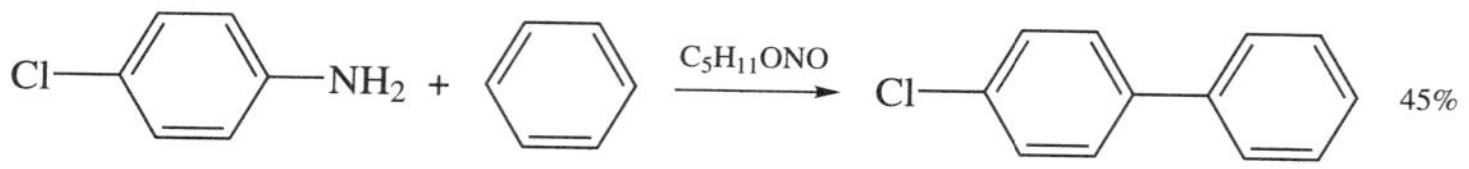

a. M. Gomberg and W. E. Bachmann, *Org. Synth.* **I:**113 (1941).
b. S. H. Korzeniowski, L. Blum, and G. W. Gokel, *Tetrahedron Lett.* **1977:**1871; J. R. Beadle, S. H. Korzeniowski, D. E. Rosenberg, B. J. Garcia-Slanga, and G. W. Gokel, *J. Org. Chem.* **49:**1594 (1984).
c. W. E. Bachmann and R. A. Hoffman, *Org. React.* **2:**249 (1944).
d. H. Rapoport, M. Lock, and G. J. Kelly, *J. Am. Chem. Soc.* **74:**6293 (1952).
e. J. I. G. Cadogan, *J. Chem. Soc.* **1962:**4257.

11.4. Substitution by the $S_{RN}1$ Mechanism

The mechanistic aspects of the $S_{RN}1$ reaction were discussed in Part A, Section 12.9. The distinctive feature of the $S_{RN}1$ mechanism is an electron transfer between the nucleophile and the aryl halide.[155] The overall reaction is normally a chain process.

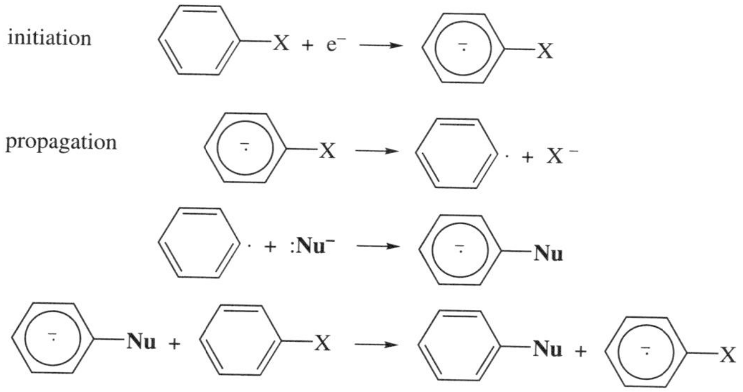

initiation: Ar–X + e$^-$ ⟶ [Ar–X]$^{\cdot -}$

propagation: [Ar–X]$^{\cdot -}$ ⟶ Ar$\cdot$ + X$^-$

Ar$\cdot$ + :Nu$^-$ ⟶ [Ar–**Nu**]$^{\cdot -}$

[Ar–**Nu**]$^{\cdot -}$ + Ar–X ⟶ Ar–**Nu** + [Ar–X]$^{\cdot -}$

155. J. F. Bunnett, *Acc. Chem. Res.* **11:**413 (1978); R. A. Rossi and R. H. de Rossi, *Aromatic Substitution by the $S_{RN}1$ Mechanism*, ACS Monograph Series, No. 178, American Chemical Society, Washington, D.C., 1983.

1[a] Bromobenzene + $H_2C{=}C(O^-)CH_3$ $\xrightarrow[hv]{NH_3}$ $C_6H_5CH_2C(=O)CH_3$ 86%

2[b] Bromobenzene + $H_2C{=}C(O^-)CH(CH_3)_2$ $\xrightarrow[hv]{NH_3}$ $C_6H_5CH_2C(=O)CH(CH_3)_2$ 79%

3[c] 2,4,6-Trimethyl-1-bromobenzene (H_3C, CH_3, CH_3, Br) + $H_2C{=}C(O^-)CH{=}C(O^-)CH_3$ $\xrightarrow[hv]{NH_3}$ 2,4,6-trimethylphenyl–$CH_2C(=O)CH_2C(=O)CH_3$

4[d] CH_3O–C_6H_4–I + $^-OP(OC_2H_5)_2$ $\xrightarrow[hv]{NH_3}$ CH_3O–C_6H_4–$P(=O)(OC_2H_5)_2$ 65%

5[e] 2-Chloropyridine (N, Cl) + $H_2C{=}C(O^-)CH_3$ $\xrightarrow[hv]{NH_3}$ 2-pyridyl–$CH_2C(=O)CH_3$ 84%

a. R. A. Rossi and J. F. Bunnett, *J. Org. Chem.* **38:**1407 (1973).
b. M. F. Semmelhack and T. Bargar, *J. Am. Chem. Soc.* **102:**7765 (1980).
c. J. F. Bunnett and J. E. Sundberg, *J. Org. Chem.* **41:**1702 (1976).
d. J. F. Bunnett and X. Creary, *J. Org. Chem.* **39:**3612 (1974).
e. A. P. Komin and J. F. Wolfe, *J. Org. Chem.* **42:**2481 (1977).

The potential advantage of the $S_{RN}1$ mechanism is that it is not particularly sensitive to the nature of other aromatic ring substituents, although electron-attracting substituents favor the nucleophilic addition step. For example, chloropyridines and chloroquinolines are also excellent reactants.[156] A variety of nucleophiles undergo the reaction, although not always in high yield. The nucleophiles that have been found to participate in $S_{RN}1$ substitution include ketone enolates,[157] 2,4-pentanedione dianion,[158] amide enolates,[159] pentadienyl and indenyl carbanions,[160] phenolates,[161] diethyl phosphite anion,[162] phosphides,[163] and thiolates.[164] The reactions are frequently initiated by light, which accelerates the initiation

156. J. V. Hay, T. Hudlicky, and J. F. Wolfe, *J. Am. Chem. Soc.* **97:**374 (1975); J. V. Hay and J. F. Wolfe, *J. Am. Chem. Soc.* **97:**3702 (1975); A. P. Komin and J. F. Wolfe, *J. Org. Chem.* **42:**2481 (1977); R. Beugelmans, M. Bois-Choussy, and B. Boudet, *Tetrahedron* **24:**4153 (1983).
157. M. F. Semmelhack and T. Bargar, *J. Am. Chem. Soc.* **102:**7765 (1980).
158. J. F. Bunnett and J. E. Sundberg, *J. Org. Chem.* **41:**1702 (1976).
159. R. A. Rossi and R. A. Alonso, *J. Org. Chem.* **45:**1239 (1980).
160. R. A. Rossi and J. F. Bunnett, *J. Org. Chem.* **38:**3020 (1973).
161. A. B. Pierini, M. T. Baumgartner, and R. A. Rossi, *Tetrahedron Lett.* **29:**3429 (1988).
162. J. F. Bunnett and X. Creary, *J. Org. Chem.* **39:**3612 (1974); A. Boumekouez, E. About-Jaudet, N. Collignon, and P. Savignac, *J. Organomet. Chem.* **440:**297 (1992).
163. E. Austin, R. A. Alonso, and R. A. Rossi, *J. Org. Chem.* **56:**4486 (1991).
164. J. F. Bunnett and X. Creary, *J. Org. Chem.* **39:**3173, 3611 (1974); J. F. Bunnett and X. Creary, *J. Org. Chem.* **40:**3740 (1975).

step. As for other radical-chain processes, the reaction is sensitive to substances that can intercept the propagation intermediates. Scheme 11.12 provides some examples of the preparative use of the $S_{RN}1$ reaction.

General References

Electrophilic Aromatic Substitution

J. G. Hoggett, R. B. Moodie, J. R. Penton, and K. S. Schofield, *Nitration and Aromatic Reactivity*, Cambridge University Press, Cambridge, 1971.

R. O. C. Norman and R. Taylor, *Electrophilic Substitution in Benzenoid Compounds*, Elsevier, Amsterdam, 1965.

G. A. Olah, *Friedel–Crafts Chemistry*, Wiley-Interscience, New York, 1973.

G. A. Olah, ed., *Friedel–Crafts and Related Reactions*, Vols. I–IV, Wiley-Interscience, New York, 1962–1964.

R. M. Roberts and A. A. Khalaf, *Friedel–Crafts Alkylation Chemistry*, Marcel Dekker, New York, 1984.

K. Schofield, *Aromatic Nitration*, Cambridge University Press, Cambridge, 1980.

L. M. Stock, *Atomatic Substitution Reactions*, Prentice-Hall, Englewood Cliffs, New Jersey, 1968.

R. Taylor, *Electrophilic Aromatic Substitution*, John Wiley & Sons, New York, 1990.

Nucleophilic Aromatic Substitution

R. W. Hoffmann, *Dehydrobenzene and Cycloalkynes*, Academic Press, New York, 1967.

J. Miller, *Aromatic Nucleophilic Substitution*, Elsevier, Amsterdam, 1968.

S. Patai, ed. *The Chemistry of Diazonium and Diazo Groups*, John Wiley & Sons, New York, 1978.

K. H. Saunders and R. L. M. Allen, *Aromatic Diazo Compounds*, Edward Arnold, London, 1985.

H. Zollinger, *Azo and Diazo Chemistry*, Interscience, New York, 1961.

Problems

(References for these problems will be found on page 939.)

1. Give reaction conditions that would accomplish each of the following transformations. Multistep schemes are not necessary. Be sure to choose conditions that would afford the desired isomer as the principal product.

(a) $H_3C-C_6H_4-Br \longrightarrow H_3C-C_6H_4-C{\equiv}N$

(b) $C_6H_5CO_2CH_3 \longrightarrow$ 2-$IC_6H_4CO_2CH_3$ (ortho-iodo)

(c) $(CH_3)_3C-C_6H_4-C(CH_3)_3 \longrightarrow (CH_3)_3C-C_6H_4-\overset{O}{\overset{\|}{C}}CH_3$

(d) CH_3O, CH_3O, CH_3O ... $CH_2CHCO_2C_2H_5$, $C(=O)CO_2C_2H_5$ → CH_3O, CH_3O, CH_3O ... $CO_2C_2H_5$, $CO_2C_2H_5$

(e) $CH(CH_3)_2$ → $CH(CH_3)_2$, I

(f) NH_2, NO_2 → $HC{=}CHCH{=}CH_2$, NO_2

(g) O_2CCH_3, O, O_2CCH_3 → O_2CCH_3, O, $CH_3C(=O)$, O_2CCH_3

(h) CH_3O, CH_3O, NH_2 → CH_3O, CH_3O, F

(i) $CH_3OC(=O)$ ... NH_2 → $CH_3OC(=O)$ ... OCH_3

2. Suggest a short series of reactions which could be expected to transform the material on the right into the desired product shown on the left.

(a) CH_3, CH_3, F ⟹ CH_3, CH_3

(b) H, H ⟹

(c) Cl, H_3C, $CCH_2CH_2CO_2H$ (C=O) ⟹ H_3C

(d) [2,4-dinitrophenyl phenyl ether (OC_6H_5, O_2N, NO_2) ⟹ chlorobenzene (Cl)]

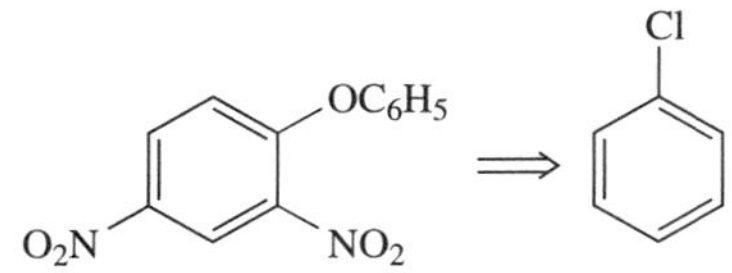

(e) [1,1-dichlorobenzocyclobutene (Cl, Cl) ⟹ anthranilic acid (CO_2H, NH_2)]

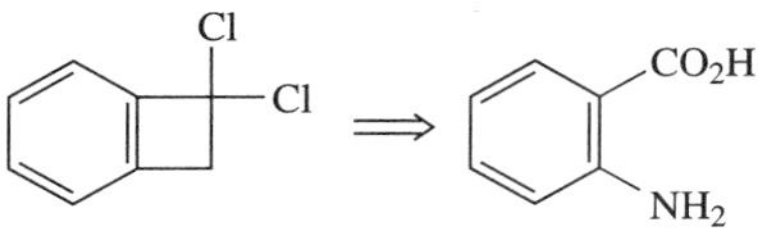

3. Write mechanisms that would account for the following reactions.

(a) [4-bromoanisole (OCH_3, Br) $\xrightarrow[Ac_2O]{HNO_3}$ 4-bromo-2-nitroanisole (OCH_3, NO_2, Br) + 4-nitroanisole (OCH_3, NO_2)]

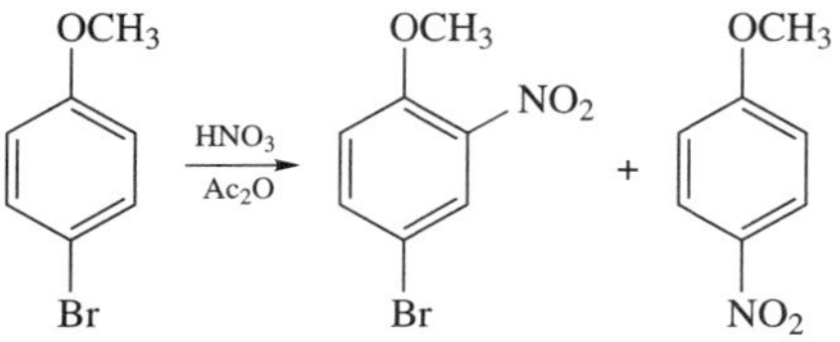

(b) [4-bromoanisole (OCH_3, Br) + Na + $CH_3\overset{O}{\overset{\|}{C}}CH_2^-$ $\xrightarrow{NH_3(l)}$ 4-methoxyphenylacetone (OCH_3, $CH_2\underset{O}{\underset{\|}{C}}CH_3$)]

(c) [benzenediazonium-2-carboxylate (CO_2^-, N_2^+) + H_3C–CH=CH–CH=CH–CH_3 (H, H, H, H) ⟶ cis-1,4-dimethyl-1,4-dihydronaphthalene (H, CH_3, H, CH_3) ~74% + H_2C=C(H)–CH(Ph)–C(H)=C(H)–CH_3 ~6%]

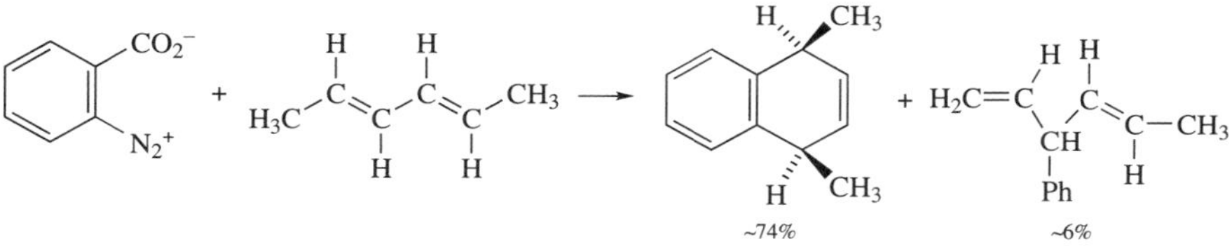

(d) [1,5-dimethoxy-4-naphthyl propanoate (OCH_3, CH_3O, $O_2CC_2H_5$) $\xrightarrow{BF_3}$ (OCH_3, CH_3O, OH, $\underset{O}{\underset{\|}{C}}C_2H_5$)]

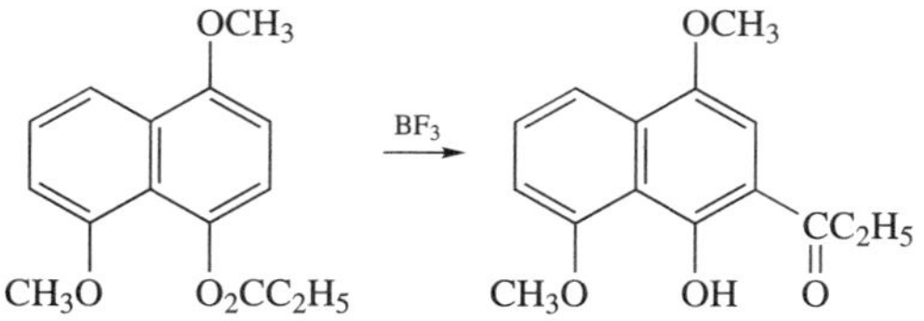

(e) [3-cyanophthalide (CN, O, O) $\xrightarrow[-78°C]{LDA}$ $\xrightarrow[-40 \rightarrow 25°C]{\text{2-bromo-1,4-dimethoxybenzene } (CH_3O, Br, CH_3O)}$ 1,4-dimethoxyanthraquinone (O, OCH_3, O, OCH_3)]

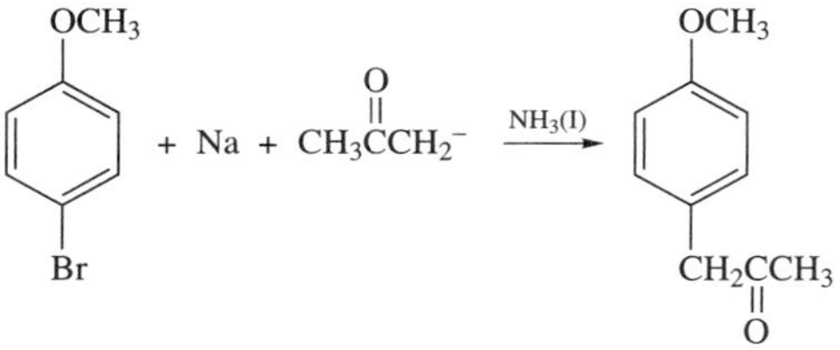

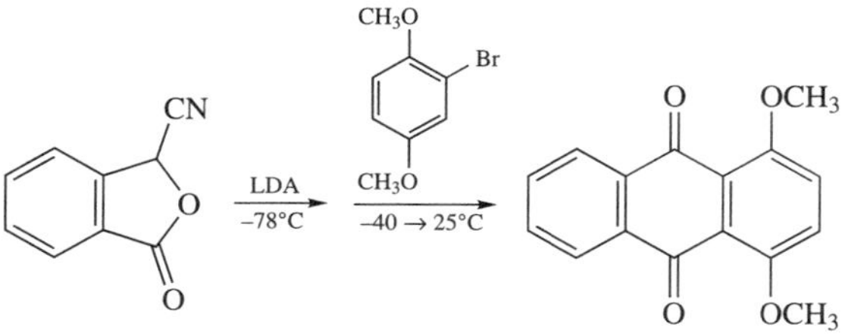

(f)

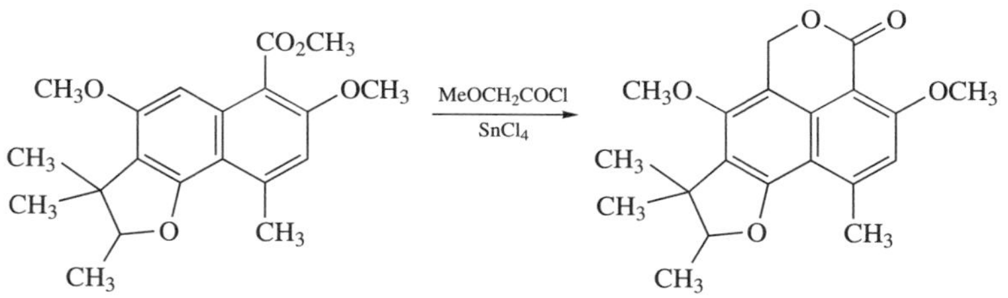

$\xrightarrow[SnCl_4]{MeOCH_2COCl}$

4. Predict the product(s) of the following reactions. If more than one product is expected, indicate which will be major and which will be minor.

(a) $\xrightarrow[DMF,\ 65°C]{(CH_3)_3CONO}$

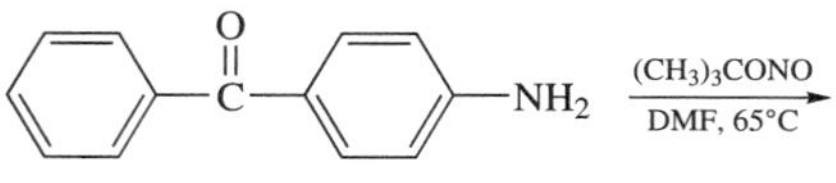

(b) 1) H_2SO_4, $NaNO_2$, 0°C; 2) $Cu(NO_3)_2$, CuO, H_2O

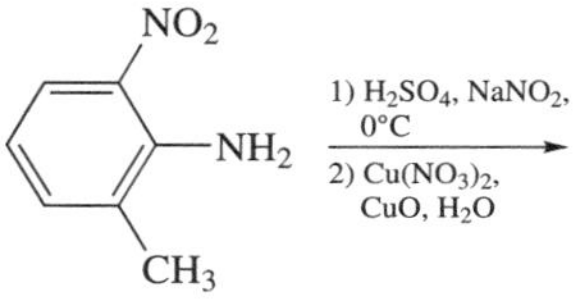

(c) Cl–C_6H_4–NH_2 + PhCH=CH_2 $\xrightarrow[CuCl]{(CH_3)_3CONO}$

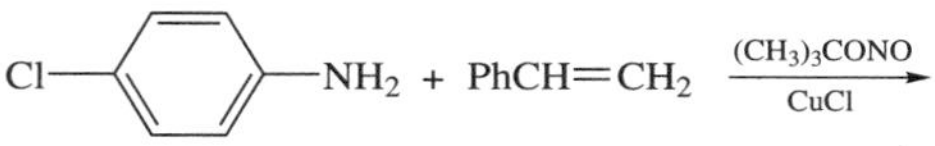

(d) $\xrightarrow[HgSO_4]{H_2SO_4,\ H_2O}$

(e) $\xrightarrow[0°C]{HNO_3,\ AcOH}$

(f) $\xrightarrow[CH_3CN]{(CH_3)_3CONO,\ CuCl}$

(g) + $HSCH_2CO_2CH_3$ $\xrightarrow[HMPA]{LiOH}$

(h) CH_3CCl (O), $AlCl_3$, $CHCl_3$

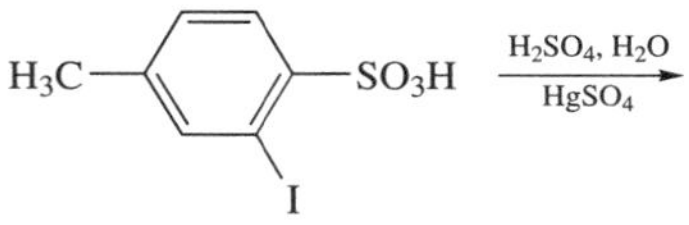

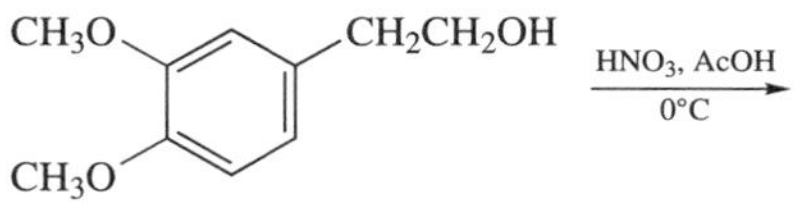

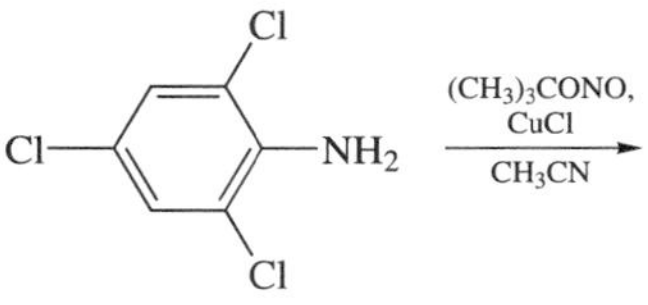

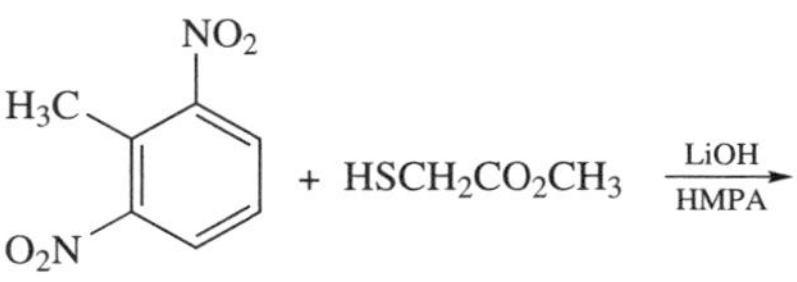

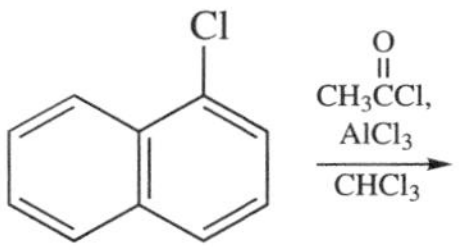

(i) $CH_3\overset{O}{\overset{\|}{C}}CH_2$– (on benz[a]anthracene) $\xrightarrow{CH_3SO_3H}$

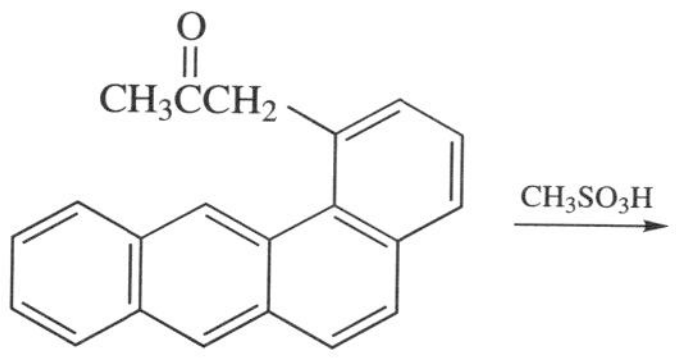

(j) $(CH_3)_3Si$–C_6H_4–$CH_2\underset{}{C}HCO_2CH_3$ (with $NH\overset{O}{\overset{\|}{C}}CH_3$) $\xrightarrow{I_2,\ AgBF_4}$

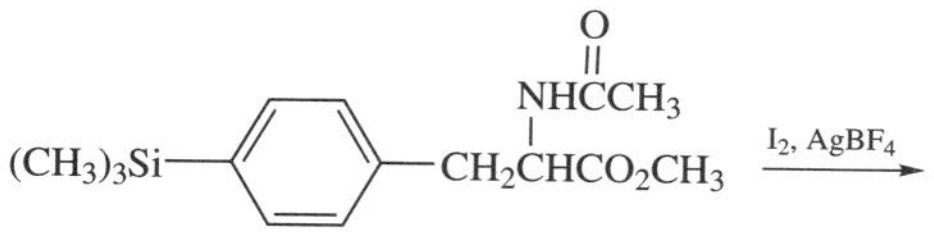

(k) Br–C_6H_5 $\xrightarrow[H^+]{Br_2,\ HgO}$

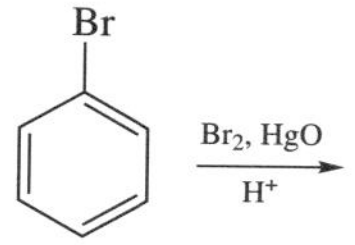

(l) C_6H_5–CN + F–C_6H_4–N_2^+ $\xrightarrow[KOAc]{18\text{-crown-6}}$

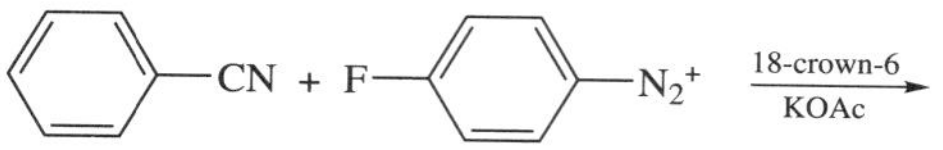

5. Suggest efficient syntheses of *o*-, *m*-, and *p*-fluoropropiophenone from benzene and any other necessary organic or inorganic reagents.

6. Treatment of compound **A** in dibromomethane with one equivalent of aluminum bromide yielded **B** as the only product in 78% yield. When three equivalents of aluminum bromide were used, however, compounds **C** and **D** were obtained in a combined yield of 97%. Suggest an explanation for these observations.

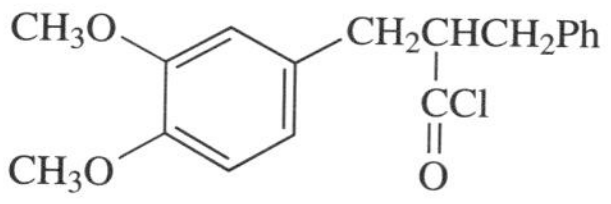

A: CH_3O, CH_3O, CH_2CHCH_2Ph, CCl, O

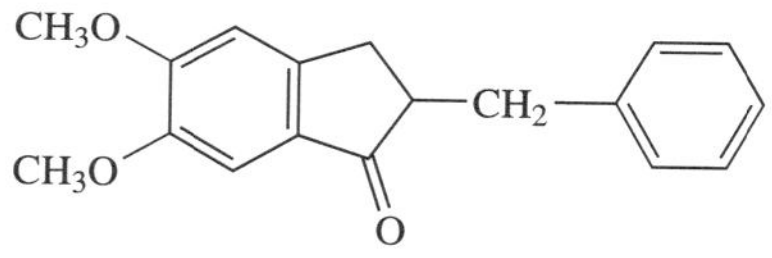

B: CH_3O, CH_3O, CH_2, O

RO, R′O, CH_2, O

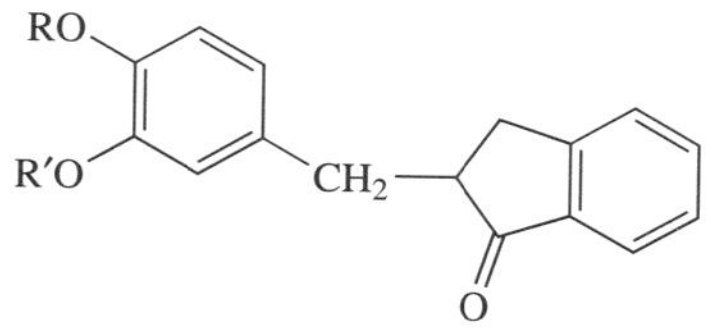

C $R = CH_3, R' = H$
D $R = H, R' = CH_3$

7. Some data for the alkylation of naphthalene by isopropyl bromide under various conditions are given.

Reaction medium A: $AlCl_3-CS_2$
Reaction medium B: $AlCl_3-CH_3NO_2$

Reaction time (min)	$\alpha:\beta$ Ratio A	$\alpha:\beta$ Ratio B
5	4:96	83:17
15	2.5:97.5	74:26
45	2:98	70:30

What factors are responsible for the difference in the product ratio for the two reaction media, and why might the ratio change with reaction time?

8. Addition of a solution of bromine and potassium bromide to a solution of the carboxylate salt **E** results in the precipitation of a neutral compund having the formula $C_{11}H_{13}BrO_3$. Various spectroscopic data show that the compound is non-aromatic. Suggest a structure and discuss the significance of the formation of this product.

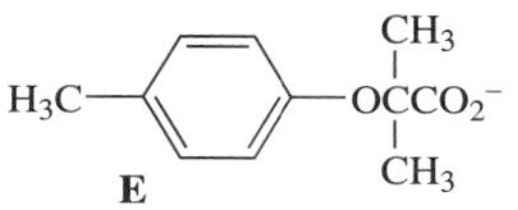

$H_3C-C_6H_4-OC(CH_3)_2CO_2^-$

E

9. Benzaldehyde, benzyl methyl ether, benzoic acid, methyl benzoate, and phenylacetic acid all undergo thallation initially in the *ortho* position. Explain this observation.

10. Reaction of 3,5,5-trimethyl-2-cyclohexenone with $NaNH_2$ (3 equiv) in THF generates its enolate. When bromobenzene is then added and this solution stirred for 4 h, the product **F** is isolated in 30% yield. Formulate a mechanism for this transformation

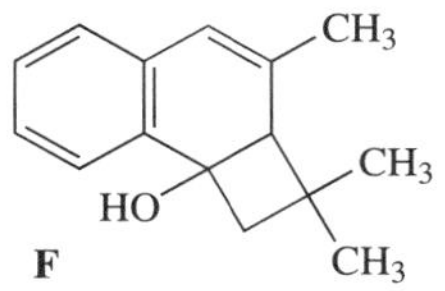

CH3 CH3 HO CH3

F

11. When phenylacetonitrile is converted to its anion in the presence of an excess of LDA and then allowed to react with a brominated aromatic ether such as 2-bromo-4-methylmethoxybenzene, the product is the result of both cyanation and benzylation.

Propose a mechanism for this reaction.

2-bromo-4-methylanisole (OCH_3, Br, CH_3) $\xrightarrow{PhCH_2CN \text{ and } >3 \text{ equiv LDA}}$ product with OCH_3, CN, CH_2Ph, CH_3

12. Suggest a reaction sequence that would permit synthesis of the following aromatic compounds from the starting material indicated on the right.

(a) 1,5-diamino-2,4-dinitrobenzene (H_2N, NH_2, O_2N, NO_2) $\Longrightarrow$ 1,3-dichlorobenzene (Cl, Cl)

(b) 2,6-dinitroaniline (NH_2, O_2N, NO_2) $\Longrightarrow$ chlorobenzene (Cl)

(c) 3,4-dichlorobiphenyl (Cl, Cl) $\Longrightarrow$ 3,4-dichlorobenzoic acid (Cl, Cl, CO_2H)

(d) $CH_3O-C_6H_4-CH_2C{\equiv}N \Longrightarrow CH_3O-C_6H_5$

(e) 1-bromo-2-fluorobenzene (Br, F) $\Longrightarrow$ 2-bromoaniline (Br, NH_2)

(f) 2,4,6-tribromobenzoic acid (CO_2H, Br, Br, Br) $\Longrightarrow$ benzoic acid (CO_2H)

(g) 1,2-bis(butylthio)benzene ($S(CH_2)_3CH_3$, $S(CH_2)_3CH_3$) $\Longrightarrow$ 1,2-dibromobenzene (Br, Br)

(h) $CH_3C(=O)-C_6H_4-CH_2CH(Br)CO_2H \Longrightarrow CH_3C(=O)-C_6H_4-NH_2$

13. Aromatic substitution reactions are key steps in multistep synthetic sequences that effect the following transformations. Suggest reaction sequences that might accomplish the desired syntheses.

(a) H_3C O H_3C H CH_3 CH_3O from H_3C O O , CH_3O O $(CH_2)_3\overset{\|}{C}CH{=}CH_2$

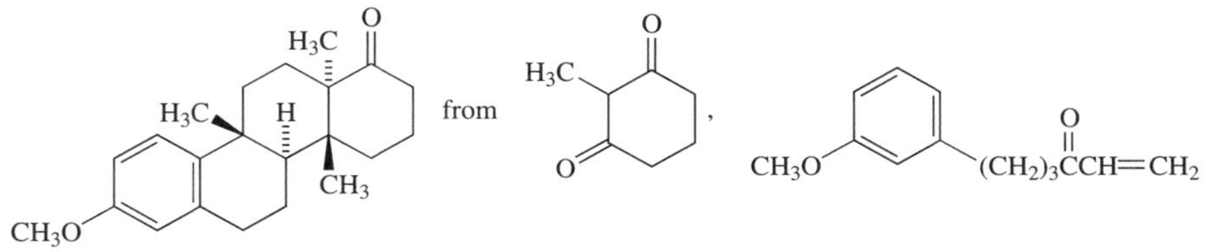

(b) Cl $CH_2CH_2CO_2CH_3$ CH_3O O from CO_2CH_3 CH_3O $CH_2{=}CHCO_2CH_3$, $H_3CO_2CCH_2CH_2CO_2CH_3$

(c) CH_3O CN CH_3O O from OCH_3 $CH{=}O$ OCH_3 $CO_2C_2H_5$

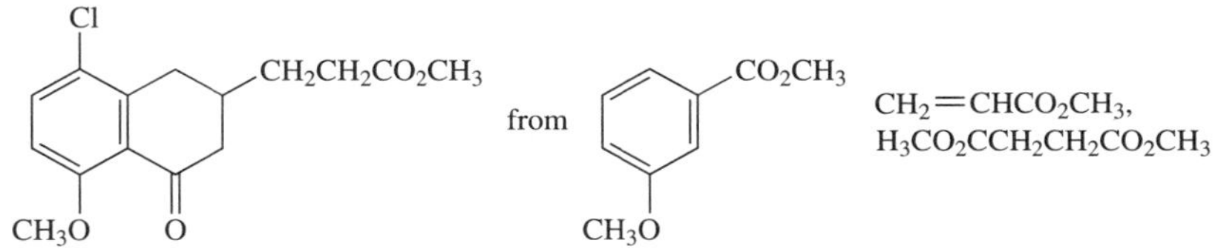

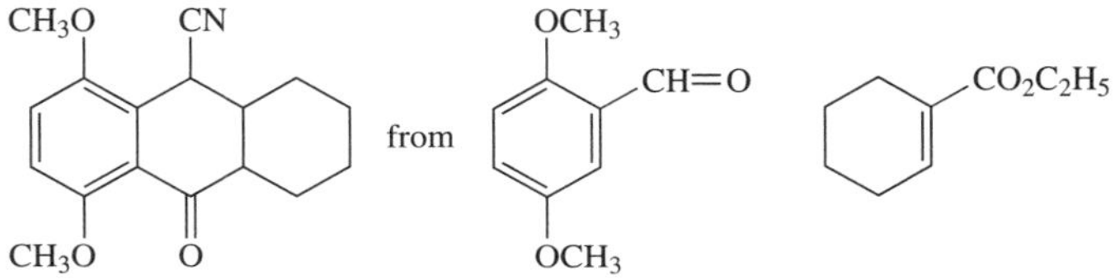

(d) CH_3O H N O CH_3 CH_3O N O H C=C H CH_3 from CO_2CH_3 CH_3O OCH_3 CH_3 H C=C CH_3 H $H_5C_2O_2C$ N H

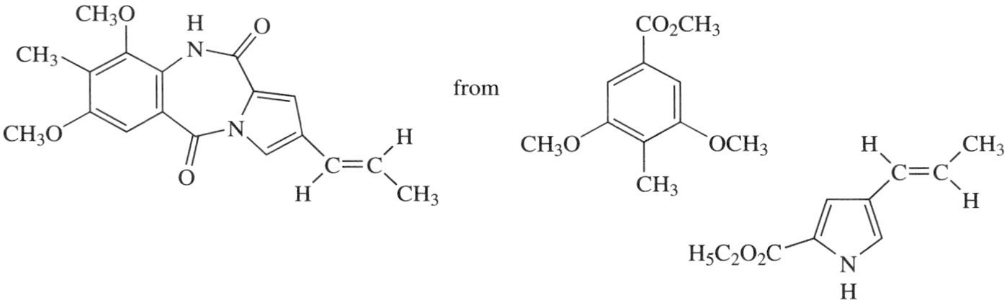

(e) O $\overset{\|}{C}CH_3$ CH_3 CH_3 from

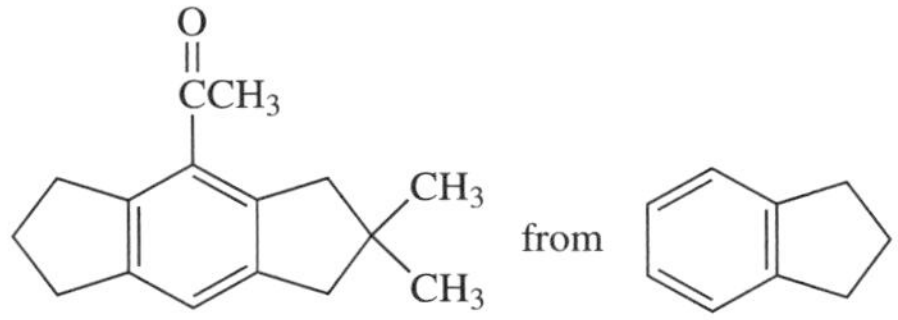

(f) CH_3O CH_3O $CO_2C_2H_5$ CH_3 from CH_3O OCH_3 $CH{=}O$

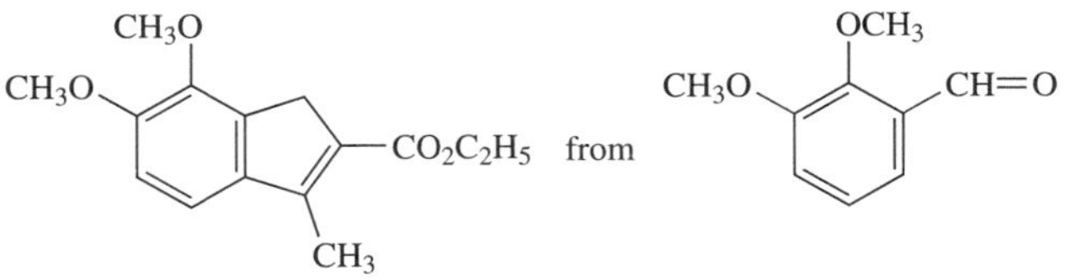

14. In the cyclization of **G** by an intramolecular Friedel–Crafts acylation, it is observed that the product formed depends on the amount of Lewis acid catalyst used in the

reaction. Offer a mechanistic explanation of this effect

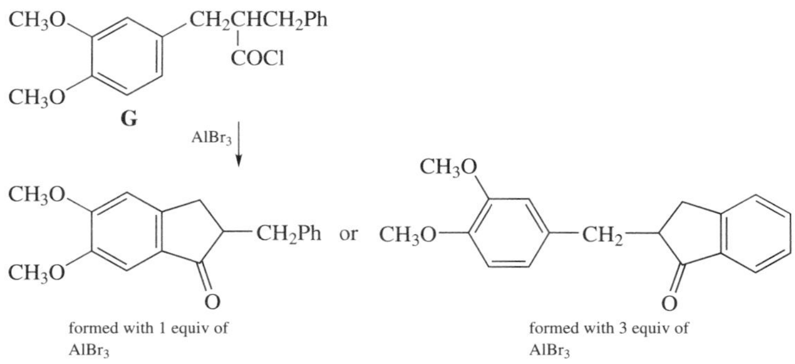

CH_3O CH_2CHCH_2Ph COCl
CH_3O
G
$AlBr_3$
CH_3O CH_3O
CH_2Ph or CH_3O CH_2
CH_3O
O O
formed with 1 equiv of $AlBr_3$
formed with 3 equiv of $AlBr_3$

15. The use of aryltrimethylsilanes as intermediates has been found to be a useful complement to direct thallation in the preparation of arylthallium intermediates. The advantages are (a) position specificity, because the thallium always replaces the silane substituent, and (b) improved ability to effect thallation of certain deactivated rings, such as those having trifluoromethyl substituents. What role does the silyl substituent play in these reactions?

16. The Pschorr reaction is a method of synthesis of phenanthrenes from diazotized *cis*-2-aminostilbene derivatives. A traditional procedure involves heating with a copper catalyst. Improved yields are frequently observed, however, if the diazonium salt is allowed to react with iodide ion. What might be the mechanism of the iodide-catalyzed reaction?

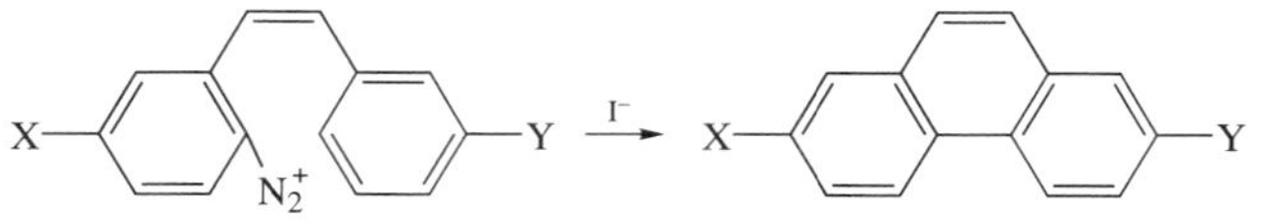

X N_2^+ Y $\xrightarrow{I^-}$ X Y

17. When compound **H** is dissolved in FSO_3H at $-78°C$, NMR spectroscopy shows that a carbocation is formed. If the solution is then allowed to warm to $-10°C$, a different ion forms. The first ion gives compound **J** when quenched with base, while the second gives **K**. What are the structures of the two carbocations, and why do they give different products on quenching?

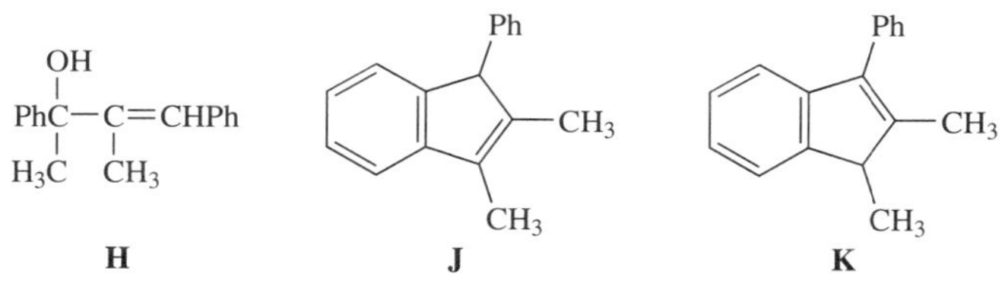

OH
PhC—C=CHPh
H_3C CH_3
H

Ph CH_3 CH_3
J

Ph CH_3 CH_3
K

18. Various phenols can be selectively hydroxymethylated at the *ortho* position by heating with paraformaldehyde and phenylboronic acid.

CH_3 OH
L
$(CH_2O)_n$, $PhB(OH)_2$, CH_3CO_2H → L → H_2O_2, Δ
CH_3 OH CH_2OH

An intermediate **L**, having the formula $C_{14}H_{13}O_2B$ for the case above, can be isolated after the first step. Postulate a structure for the intermediate, and comment on its role in the reaction.

19. The electrophilic cyclization of **M** and **N** gives two stereoisomers, but with the unsubstituted starting material **O**, only a single stereoisomer is formed. Explain the origin of the two stereoisomers and the absence of the isomer in the case of **O**.

M X = I
N X = Br
O X = H

Tf_2O, 2,6-lutidene, or 2,6-di-*t*-butylpyridine, CH_2Cl_2, RT, 1–3 h

X = I (42%)
X = Br (50%)

\+

X = I (21%)
X = Br (16%)

X = H (73%)

20. Entry 5 of Scheme 11.4 is a step in the synthesis of the anticancer drug tamoxifen. Explain why the 2-phenylbutanoyl group is introduced in preference to a trifluoroacetyl group.

12

Oxidations

Introduction

This chapter is concerned with reactions which transform a functional group to a more highly oxidized derivative. There are a very large number of oxidation methods, and the reactions have been chosen for discussion on the basis of their utility in organic synthesis. As the reactions are considered, it will become evident that the material in this chapter spans a wider variety of mechanistic patterns than that in most of the earlier chapters. Because of this range in mechanisms, the chapter has been organized by the functional group transformation that is accomplished. This organization facilitates comparison of the methods available for effecting a given synthetic transformation. The oxidants are grouped into three classes: transition-metal derivatives; oxygen, ozone and peroxides; and other reagents.

12.1. Oxidation of Alcohols to Aldehydes, Ketones, or Carboxylic Acids

12.1.1. Transition-Metal Oxidants

The most widely employed transition-metal oxidants are based on Cr(VI). The specific reagents are generally prepared from chromic trioxide (CrO_3) or a dichromate ($[Cr_2O_7]^{2-}$) salt. The form of Cr(VI) in aqueous solution depends upon concentration and pH. In dilute solution, the monomeric acid chromate ion is present; as concentration increases, the dichromate ion dominates. The extent of protonation of these ions depends on the pH.

$$2\,HO{-}Cr(=O)_2{-}O^- \rightleftharpoons {}^-O{-}Cr(=O)_2{-}O{-}Cr(=O)_2{-}O^- + H_2O$$

In acetic acid, Cr(VI) is present as mixed anhydrides of acetic acid and chromic acid.[1]

$$2CH_3CO_2H + CrO_3 \rightleftharpoons CH_3CO_2Cr(=O)_2{-}OH \rightleftharpoons CH_3CO_2Cr(=O)_2O_2CCH_3 + H_2O$$

In pyridine, an adduct involving Cr–N bonding is formed.

$$C_5H_5N + CrO_3 \longrightarrow C_5H_5N^+{-}Cr(=O)_2{-}O^-$$

The oxidation state of Cr in each of these species is VI, and they are powerful oxidants. The precise reactivity depends on the solvent and the chromium ligands, so substantial selectivity can be achieved by choice of the particular reagent and conditions.

The conversion of an alcohol to the corresponding ketone or aldehyde is often effected with CrO_3-based oxidants. The general mechanism of alcohol oxidation is outlined below:

$$R_2CHOH + HCrO_4^- + H^+ \longrightarrow R_2CHOCrO_3H + H_2O$$

$$R_2C(H){-}O{-}CrO_3H \xrightarrow{\text{slow}} R_2C{=}O + HCrO_3^- + H^+$$

An important piece of evidence pertinent to identification of the rate-determining step is the fact that a large isotope effect is observed when the α-H is replaced by deuterium.[2] The Cr(IV) that is produced in the initial step is not stable, and this species is capable of a further one-electron oxidation step. It is believed that Cr(IV) is reducd to Cr(II), which is then oxidized by Cr(VI), generating Cr(V). This mechanism accounts for the overall stoichiometry of the reaction.[3]

$$\begin{aligned}
R_2CHOH + Cr(VI) &\longrightarrow R_2C{=}O + Cr(IV) + 2\,H^+\\
R_2CHOH + Cr(IV) &\longrightarrow R_2C{=}O + Cr(II) + 2\,H^+\\
Cr(II) + Cr(VI) &\longrightarrow Cr(III) + Cr(V)\\
R_2CHOH + Cr(V) &\longrightarrow R_2C{=}O + Cr(III) + 2\,H^+\\
\hline
3\,R_2CHOH + 2\,Cr(VI) &\longrightarrow 3\,R_2C{=}O + 2\,Cr(III) + 6\,H^+
\end{aligned}$$

A variety of experimental conditions have been used for oxidations of alcohols by Cr(VI) on a laboratory scale. For simple unfunctionalized alcohols, oxidation can be done by addition of an acidic aqueous solution containing chromic acid (known as *Jones' reagent*) to an acetone solution of the alcohol. Oxidation normally occurs rapidly, and overoxidation is minimal. In acetone solution, the reduced chromium salts precipitate, and the reaction solution can be decanted. Entries 2, 3 and 4 in Scheme 12.1 are examples of this technique.

The chromium trioxide–pyridine complex is useful in situations in which other

1. K. B. Wiberg, *Oxidation in Organic Chemistry, Part A*, Academic Press, New York, 1965, pp. 69–72.
2. F. H. Westheimer and N. Nicolaides, *J. Am. Chem. Soc.* **71**:25 (1949).
3. S. L. Scott, A. Bakac, and J. H. Espenson, *J. Am. Chem. Soc.* **114**:4205 (1992).

Scheme 12.1. Oxidations with Cr(VI)

A. Chromic acid solutions

1[a] $CH_3CH_2CH_2OH \xrightarrow[H_2O]{H_2CrO_4} CH_3CH_2CH{=}O$ 45–49%

2[b] cyclooctanol $\xrightarrow[\text{acetone}]{H_2CrO_4}$ cyclooctanone 92–96%

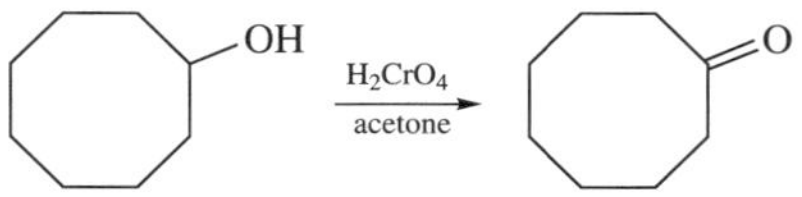

3[c] $\xrightarrow[\text{acetone}]{H_2CrO_4}$ 84%

4[d] $\xrightarrow[\text{acetone}]{H_2CrO_4}$ 79–88%

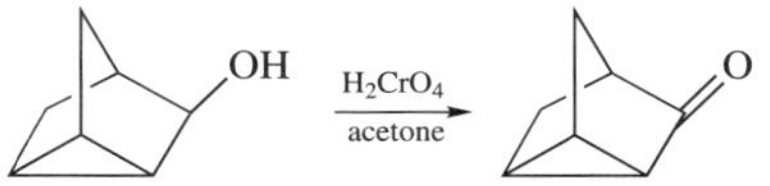

B. Chromium trioxide–pyridine

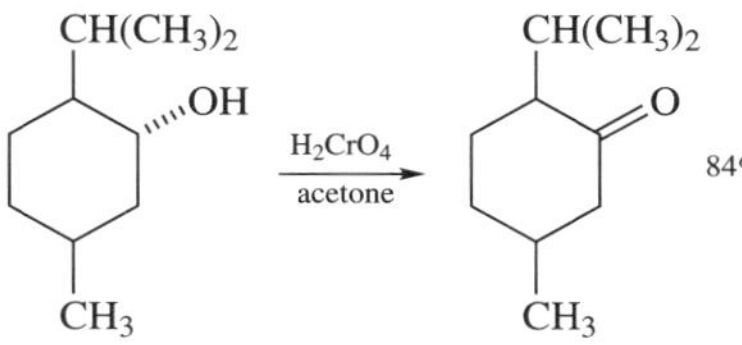

5[e] $CH_3(CH_2)_5CH_2OH \xrightarrow[CH_2Cl_2]{CrO_3\text{–pyridine}} CH_3(CH_2)_5CH{=}O$ 70–84%

6[f] $CH_3CH_2CH(CH_3)(CH_2)_4CH_2OH \xrightarrow[CH_2Cl_2]{CrO_3\text{–pyridine}} CH_3CH_2CH(CH_3)(CH_2)_4CH{=}O$ 69%

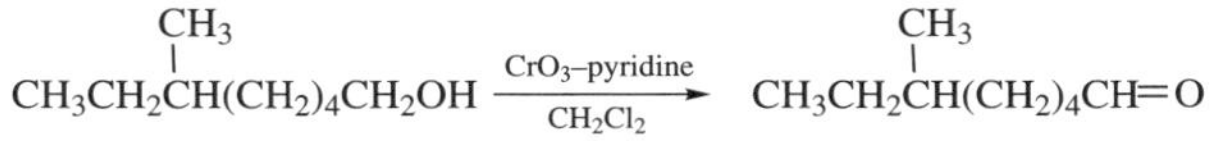

7[g] $\xrightarrow[CH_2Cl_2]{CrO_3\text{–pyridine}}$ 95%

8[h] $\xrightarrow[CH_2Cl_2]{CrO_3\text{–pyridine}}$

9[i] $\xrightarrow[CH_3CO_2H]{CrO_3\text{–pyridine}}$ 96%

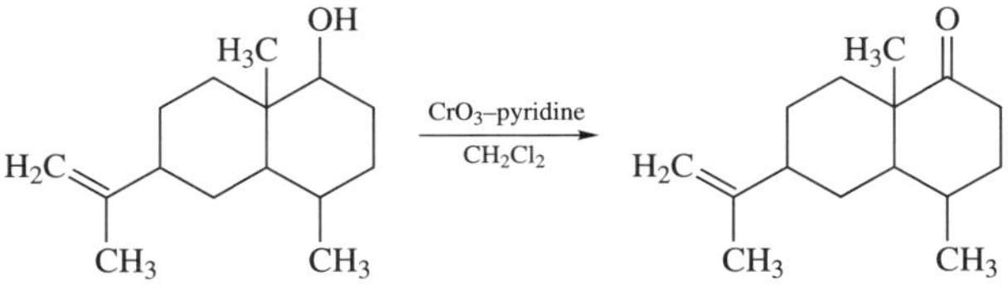

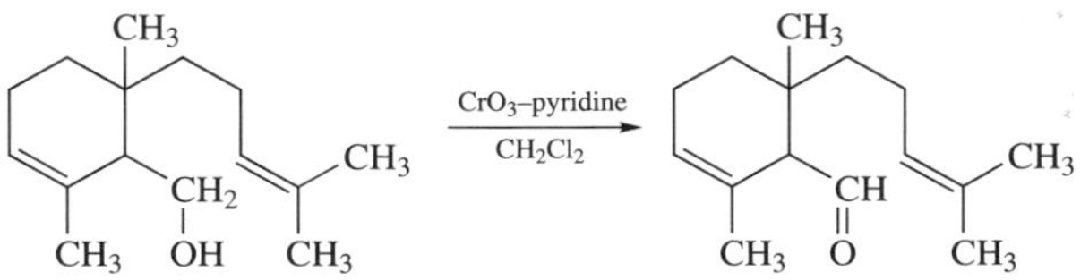

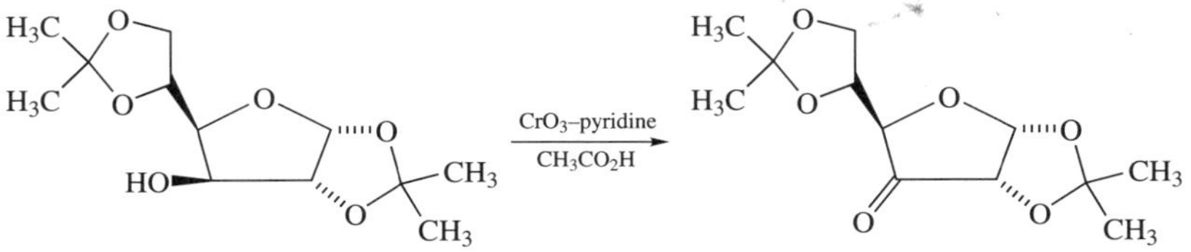

Scheme 12.1. (*continued*)

C. Pyridinium chlorochromate

10[j] $(CH_3)_2C{=}CHCH_2CH_2CH(CH_3)CH_2CH_2OH \xrightarrow{PCC} (CH_3)_2C{=}CHCH_2CH_2CH(CH_3)CH_2CH{=}O$ 82%

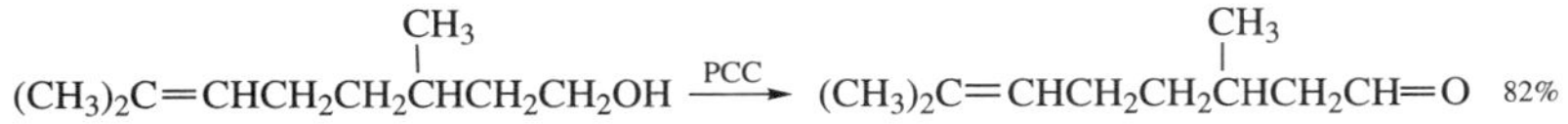

11[k] $HOCH_2CH_2C(CH_3)_2CH_2CH{=}CHCO_2CH_3 \xrightarrow{PCC} O{=}CHCH_2C(CH_3)_2CH_2CH{=}CHCO_2CH_3$ 83%

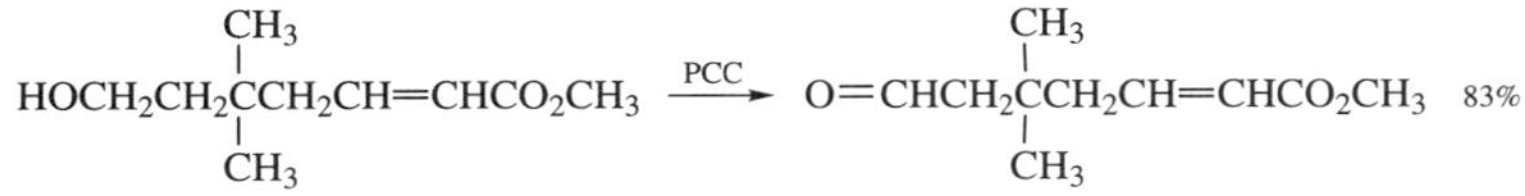

a. C. D. Hurd and R. N. Meinert, *Org. Synth.* **11:**541 (1943).
b. E. J. Eisenbraun, *Org. Synth.* **IV:**310 (1973).
c. H. C. Brown, C. P. Garg, and K.-T. Liu, *J. Org. Chem.* **36:**387 (1971).
d. J. Meinwald, J. Crandall, and W. E. Hymans, *Org. Synth.* **45:**77 (1965).
e. J. C. Collins and W. W. Hess, *Org. Synth.* **52:**5 (1972).
f. J. I. DeGraw and J. O. Rodin, *J. Org. Chem.* **36:**2902 (1971).
g. R. Ratcliffe and R. Rodehorst, *J. Org. Chem.* **35:**4000 (1970).
h. M. A. Schwartz, J. D. Crowell, and J. H. Musser, *J. Am. Chem. Soc.* **94:**4361 (1972).
i. C. Czernecki, C. Georgoulis, C. L. Stevens, and K. Vijayakumaran, *Tetrahedron Lett.* **26:**1699 (1985).
j. E. J. Corey and J. W. Suggs, *Tetrahedron Lett.* **1975:**2647.
k. R. D. Little and G. W. Muller, *J. Am. Chem. Soc.* **103:**2744 (1981).

functional groups might be susceptible to oxidation or when the molecule is sensitive to strong acid.[4] A procedure for utilizing the CrO_3–pyridine complex, which was originated by Collins,[5] has been quite widely adopted. The CrO_3–pyridine complex is isolated and dissolved in dichloromethane. With an excess of the reagent, oxidation of simple alcohols is complete in a few minutes, giving the aldehyde or ketone in good yield. A procedure that avoids isolation of the complex can further simplify the experimental operations.[6] Chromium trioxide is added to pyridine in dichloromethane. Subsequent addition of the alcohol to this solution results in oxidation in high yield. Other modifications for use of the CrO_3–pyridine complex have been developed.[7] Entries 5–9 in Scheme 12.1 demonstrate the excellent results that have been reported using the CrO_3–pyridine complex in dichloromethane.

Another very useful Cr(VI) reagent is pyridinium chlorochromate (PCC), which is prepared by dissolving CrO_3 in hydrochloric acid and adding pyridine to obtain a solid reagent having the composition $CrO_3ClpyrH^+$.[8] This reagent can be used in close to the stoichiometric ratio. Entries 10 and 11 in Scheme 12.1 are examples of the use of this reagent. Reaction of pyridine with CrO_3 in a small amount of water gives pyridinium dichromate (PDC), which is also a useful oxidant.[9] As a solution in DMF or a suspension in dichloromethane, this reagent oxidizes secondary alcohols to ketones. Allylic primary alcohols give the corresponding aldehydes. Depending upon the conditions, saturated

4. G. I. Poos, G. E. Arth, R. E. Beyler, and L. H. Sarett, *J. Am. Chem. Soc.* **75:**422 (1953); W. S. Johnson, W. A. Vredenburgh, and J. E. Pike, *J. Am. Chem. Soc.* **82:**3409 (1960); W. S. Allen, S. Bernstein, and R. Little, *J. Am. Chem. Soc.* **76:**6116 (1954).
5. J. C. Collins, W. W. Hess, and F. J. Frank, *Tetrahedron Lett.* **1968:**3363.
6. R. Ratcliffe and R. Rodehorst, *J. Org. Chem.* **35:**4000 (1970).
7. J. Herscovici, M.-J. Egron, and K. Antonakis, *J. Chem. Soc., Perkin Trans. 1* **1982:**1967; E. J. Corey and G. Schmidt, *Tetrahedron Lett.* **1979:**399; S. Czernecki, C. Georgoulis, C. L. Stevens, and K. Vijayakumaran, *Tetrahedron Lett.* **26:**1699 (1985).
8. E. J. Corey and J. W. Suggs, *Tetrahedron Lett.* **1975:**2647; G. Piancatelli, A. Scettri, and M. D'Auria, *Synthesis* **1982:**245.
9. E. J. Corey and G. Schmidt, *Tetrahedron Lett.* **1979:**399.

primary alcohols give either an aldehyde or the corresponding carboxylic acid.

$$CH_3(CH_2)_8CH_2OH \xrightarrow[\text{DMF, 25°C}]{\text{PDC}} CH_3(CH_2)_8CH{=}O \quad 98\%$$

Although Cr(VI) oxidants are very versatile and efficient, they have one drawback which becomes especially serious in larger-scale work. That is the toxicity and environmental hazards associated with chromium compounds. One possible alternative oxidant that has recently been investigated is an Fe(VI) species, potassium ferrate (K_2FeO_4) in a form supported on montmorillonite clay.[10] This reagent gives clean, high-yielding oxidation of benzylic and allylic alcohols, but saturated alcohols are less reactive.

$$PhCH_2OH \xrightarrow[\text{K10 montmorillonite clay}]{K_2FeO_4} PhCH{=}O$$

Potassium permanganate ($KMnO_4$) is another powerful transition-metal oxidant. Potassium permanganate has found relatively little application in the oxidation of alcohols to ketones and aldehydes. The reagent is less selective than Cr(VI), and overoxidation is a problem. On the other hand, manganese(IV) dioxide is quite useful.[11] This reagent is selective for allylic and benzylic hydroxyl groups. Manganese dioxide is prepared by reaction of $MnSO_4$ with $KMnO_4$ and sodium hydroxide. The precise reactivity of MnO_2 depends on its mode of preparation and the extent of drying.[12] Scheme 12.2 illustrates various types of alcohols that are most susceptible to MnO_2 oxidation.

A catalytic system that extends the reactivity of MnO_2 to saturated secondary alcohols has been developed.[13] This consists of a Ru(II) salt, $RuCl_2(p\text{-cymene})_2$, and 2,6-di-*t*-butylbenzoquinone.

$$\text{cyclohexanol} + MnO_2 \xrightarrow[\text{2,6-di-}t\text{-butylbenzoquinone, 20 mol \%}]{MnO_2,\ RuCl_2(p\text{-cymene})_2,\ 1\text{ mol \%}} \text{cyclohexanone}$$

The Ru and benzoquinone are believed to function as intermediary hydride-transfer agents.

R_2CHOH → $R_2C{=}O$; Ru ⇄ $Ru(H)_2$; 2,6-di-*t*-Bu-hydroquinone (OH, OH) ⇄ 2,6-di-*t*-Bu-benzoquinone (O, O) ; Mn(IV) → Mn(II)

10. L. Delaude and P. Laszlo, *J. Org. Chem.* **61:**6360 (1996).
11. D. G. Lee, in *Oxidation*, Vol. 1, R. L. Augustine, ed., Marcel Dekker, New York, 1969, pp. 66–70; A. J. Fatiadi, *Synthesis* **1976:**65, 133.
12. J. Attenburrow, A. F. B. Cameron, J. H. Chapman, R. M. Evans, A. B. A. Jansen, and T. Walker, *J. Chem. Soc.* **1952:**1094; I. M. Goldman, *J. Org. Chem.* **34:**1979 (1969).
13. U. Karlsson, G.-Z. Wang, and J.-E. Bäckvall, *J. Org. Chem.* **59:**1196 (1994).

Scheme 12.2. Oxidations of Alcohols with Manganese Dioxide

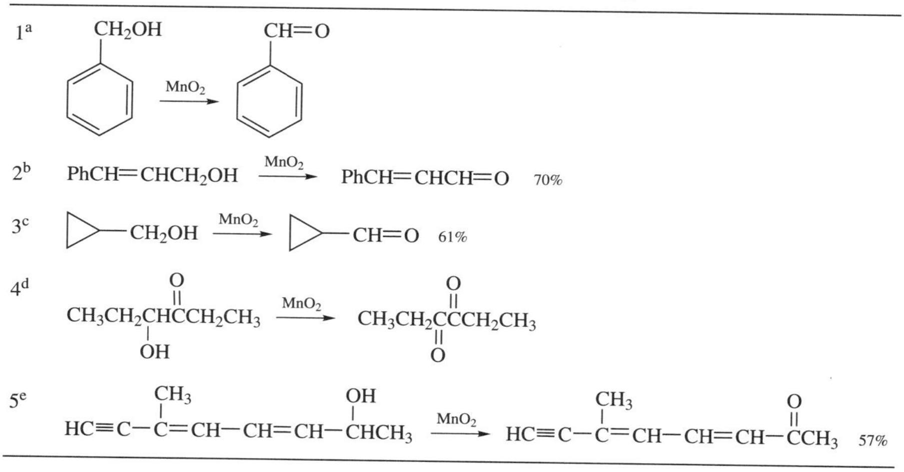

1[a] $C_6H_5CH_2OH \xrightarrow{MnO_2} C_6H_5CH{=}O$

2[b] $PhCH{=}CHCH_2OH \xrightarrow{MnO_2} PhCH{=}CHCH{=}O$ 70%

3[c] cyclopropyl–$CH_2OH \xrightarrow{MnO_2}$ cyclopropyl–$CH{=}O$ 61%

4[d] $CH_3CH_2CH(OH)C(=O)CH_2CH_3 \xrightarrow{MnO_2} CH_3CH_2C(=O)C(=O)CH_2CH_3$

5[e] $HC{\equiv}C{-}C(CH_3){=}CH{-}CH{=}CH{-}CH(OH)CH_3 \xrightarrow{MnO_2} HC{\equiv}C{-}C(CH_3){=}CH{-}CH{=}CH{-}C(=O)CH_3$ 57%

a. E. F. Pratt and J. F. Van De Castle, *J. Org. Chem.* **26:**2973 (1961).
b. I. M. Goldman, *J. Org. Chem.* **34:**1979 (1969).
c. L. Crombie and J. Crossley, *J. Chem. Soc.* **1963:**4983.
d. E. P. Papadopoulos, A. Jarrar, and C. H. Issidorides, *J. Org. Chem.* **31:**615 (1966).
e. J. Attenburrow, A. F. B. Cameron, J. H. Chapman, R. M. Evans, B. A. Hems, A. B. A. Jansen, and T. Walker, *J. Chem. Soc.* **1952:**1094.

Another reagent that finds application in oxidations of alcohols to ketones is ruthenium tetroxide. For example, the oxidation of **1** to **2** was successfully achieved with this reagent after a number of other methods failed.

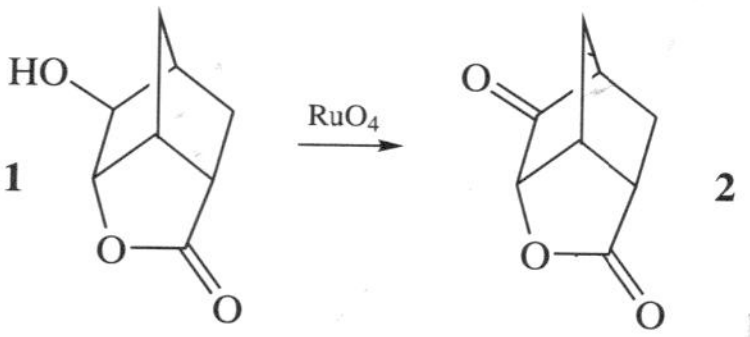

1 $\xrightarrow{RuO_4}$ **2** Ref. 14

This compound is a potent oxidant, however, and it readily attacks carbon–carbon double bonds.[15] Procedures for *in situ* generation of RuO_4 from RuO_2 using periodate or hypochlorite as oxidants are available.[16]

12.1.2. Other Oxidants

A very useful group of procedures for oxidation of alcohols to ketones have been developed which involve DMSO and any one of several electrophilic reagents, such as dicyclohexylcarbodiimide, acetic anhydride, trifluoroacetic anhydride, oxalyl chloride, or

14. R. M. Moriarty, H. Gopal, and T. Adams, *Tetrahedron Lett.* **1970:**4003.
15. J. L. Courtney and K. F. Swansborough, *Rev. Pure Appl. Chem.* **22:**47 (1972); D. G. Lee and M. van den Engh, in *Oxidation, Part B*, W. S. Trahanovsky, ed., Academic Press, New York, 1973, Chapter IV.
16. P. E. Morris, Jr. and D. E. Kiely, *J. Org. Chem.* **52:**1149 (1987).

sulfur trioxide.[17] The initial procedure involved DMSO and dicyclohexylcarbodiimide.[18] The main utility of the DMSO methods is for oxidation of molecules that are sensitive to the transition-metal oxidants. The mechanism of the oxidation involves formation of intermediate **A** by nucleophilic attack by DMSO on the carbodiimide, followed by reaction of the intermediate with the alcohol.[19]

$$RN{=}C{=}NR + {}^-O{-}\overset{+}{S}(CH_3)_2 \xrightarrow{H^-} \underset{\mathbf{A}}{RNH{-}C({=}NR){-}O{-}\overset{+}{S}(CH_3)_2} \xrightarrow{R_2CHOH} \underset{\mathbf{B}}{R_2CH{-}O{-}S(CH_3)(CH_2{-}H)\cdots O{-}C(NHR){=}NR} \longrightarrow \underset{\mathbf{C}}{R_2C(H){-}O{-}\overset{+}{S}(CH_3){-}\overset{-}{C}H_2}$$

$$\mathbf{C} \longrightarrow R_2C{=}O + (CH_3)_2S + RNHCNHR \ (\text{C}{=}\text{O})$$

The activation of DMSO toward the addition step can be accomplished by other electrophiles. All of these reagents are believed to form a sulfoxonium species by electrophilic attack at the sulfoxide oxygen. The addition of the alcohol and the departure of the sulfoxide oxygen as part of a leaving group generate an intermediate comparable to **C** in the above mechanism.

$$(CH_3)_2\overset{+}{S}{-}O^- + X^+ \longrightarrow (CH_3)_2\overset{+}{S}{-}O{-}X$$

$$R_2CHOH + (CH_3)_2\overset{+}{S}{-}O{-}X \longrightarrow R_2CHO{-}S(CH_3)_2{-}O{-}X \longrightarrow R_2CHO{-}\overset{+}{S}(CH_3)_2 + {}^-OX$$

$$R_2CHO{-}\overset{+}{S}(CH_3)_2 \longrightarrow R_2C{=}O + (CH_3)_2S + H^+$$

Preparatively useful procedures based on acetic anhydride,[20] trifluoroacetic anhydride,[21] and oxalyl chloride[22] have been developed. The latter method, known as *Swern oxidation*, is currently the most popular and is frequently preferred to Cr(VI) oxidation. Scheme 12.3 gives some representative examples of these methods. Entry 4 is an example of the use of a water-soluble carbodiimide as the activating reagent. The modified carbodiimide facilitates product purification by providing for easy removal of the urea by-product.

Oxidation of alcohols under extremely mild conditions can be achieved by a procedure that is mechanistically related to the DMSO methods. Dimethyl sulfide is converted to a chlorosulfonium ion by reaction with *N*-chlorosuccinimide. This sulfonium ion reacts with alcohols, generating the same kind of alkoxysulfonium ion that is involved

17. A. J. Mancuso and D. Swern, *Synthesis* **1981:**165; T. T. Tidwell, *Synthesis* **1990:**857.
18. K. E. Pfitzner and J. G. Moffatt, *J. Am. Chem. Soc.* **87:**5661, 5670 (1965).
19. J. G. Moffatt, *J. Org. Chem.* **36:**1909 (1971).
20. J. D. Albright and L. Goldman, *J. Am. Chem. Soc.* **89:**2416 (1967).
21. J. Yoshimura, K. Sato, and H. Hashimoto, *Chem. Lett.* **1977:**1327; K. Omura, A. K. Sharma, and D. Swern, *J. Org. Chem.* **41:**957 (1976); S. L. Huang, K. Omura, and D. Swern, *J. Org. Chem.* **41:**3329 (1976).
22. A. J. Mancuso, S.-L. Huang, and D. Swern, *J. Org. Chem.* **43:**2480 (1978).

Scheme 12.3. Oxidation of Alcohols Using Dimethyl Sulfoxide

1[a] DMSO, dicyclohexylcarbodimide — CH_2OH → $CH{=}O$ 84%

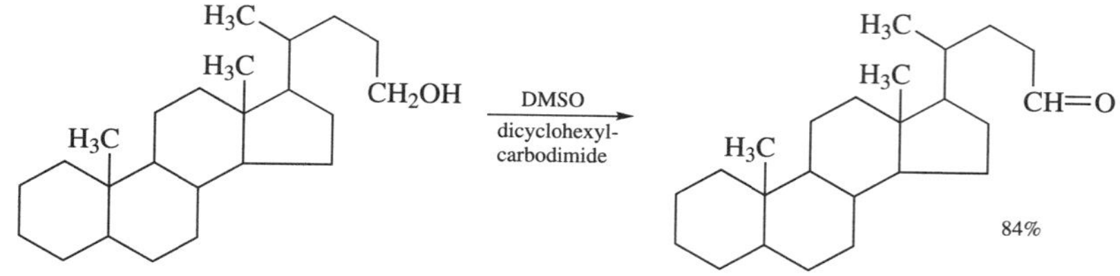

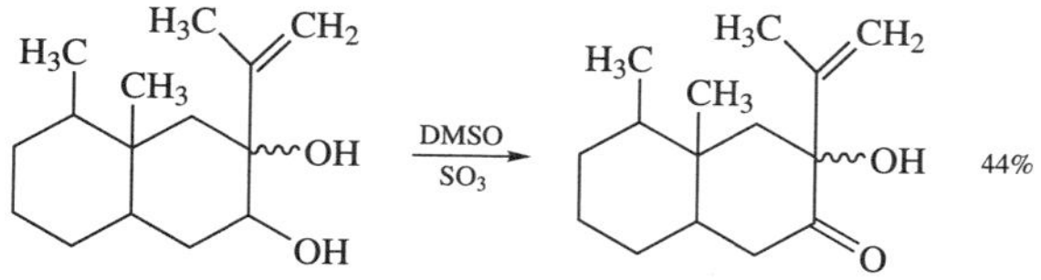

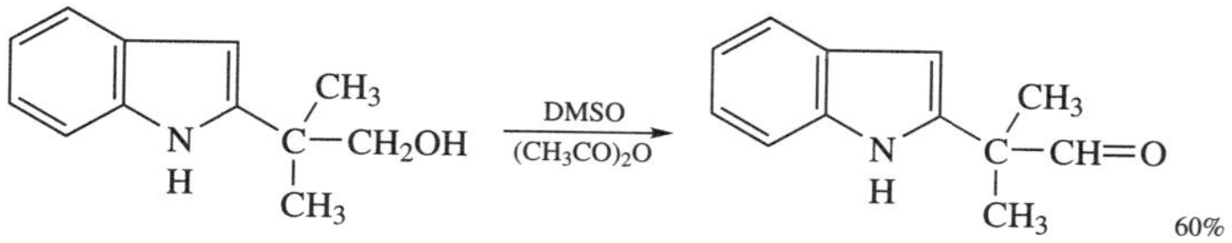

2[b] DMSO, SO_3 — 44%

3[c] $C(CH_3)_2CH_2OH$ → DMSO, $(CH_3CO)_2O$ → $C(CH_3)_2CH{=}O$ 60%

4[d] $PhSCH_2ON{=}$ ... CH_2OH → DMSO, $N{=}C{=}NCH_2CH_2N^+$ (CH_3) → $CH{=}O$ 98%

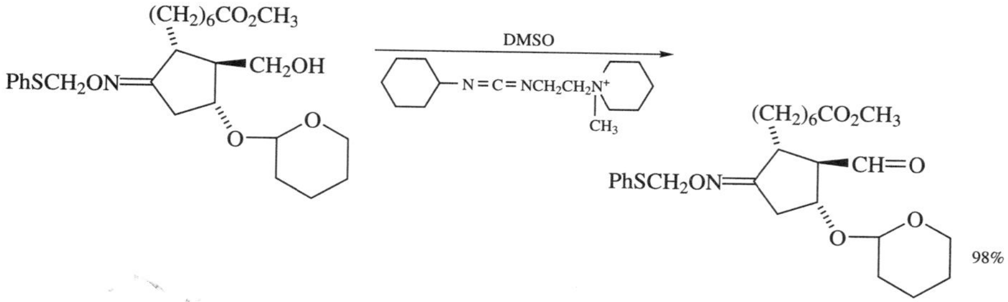

5[e] $(CH_3)_2CHCH{=}CHCH{=}CHCH_2OH$ → DMSO, ClCOCOCl → $(CH_3)_2CHCH{=}CHCH{=}CHCH{=}O$ 93%

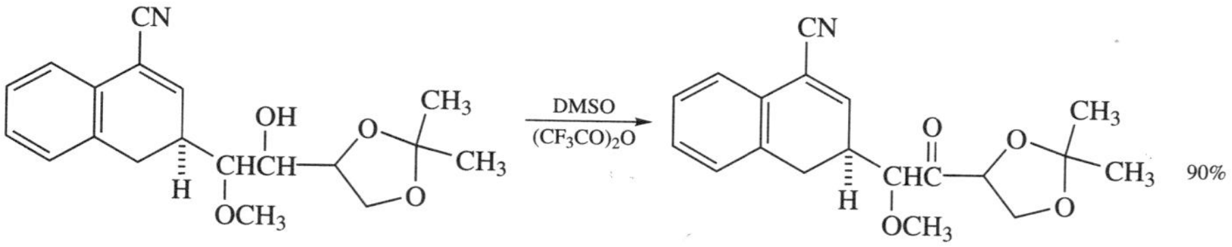

6[f] $CO_2C(CH_3)_3$ ... CH_2OH → DMSO, ClCOCOCl, $(i\text{-}C_3H_7)_2NC_2H_5$ → $CH{=}O$ 99%

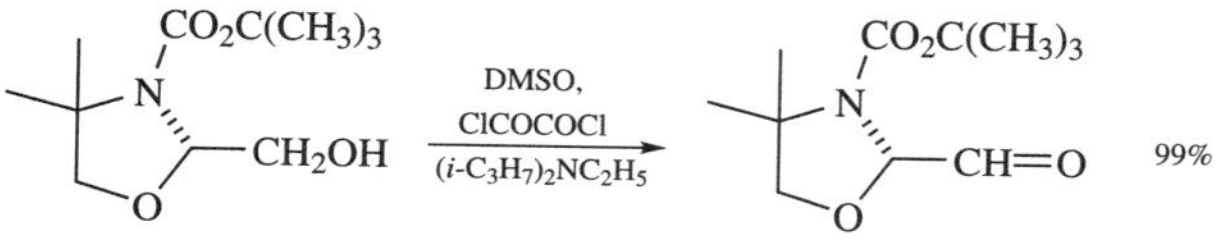

7[g] CN ... OH, $\overset{|}{C}HCH$, OCH_3 → DMSO, $(CF_3CO)_2O$ → $\overset{O}{\overset{\|}{CHC}}$ 90%

8[h] $CH_3CH{=}CHCH_2CH_2CH(OH)CH_2CO_2CH_3$ → DMSO, P_2O_5, Et_3N → $CH_3CH{=}CHCH_2CH_2C(O)CH_2CO_2CH_3$ 85%

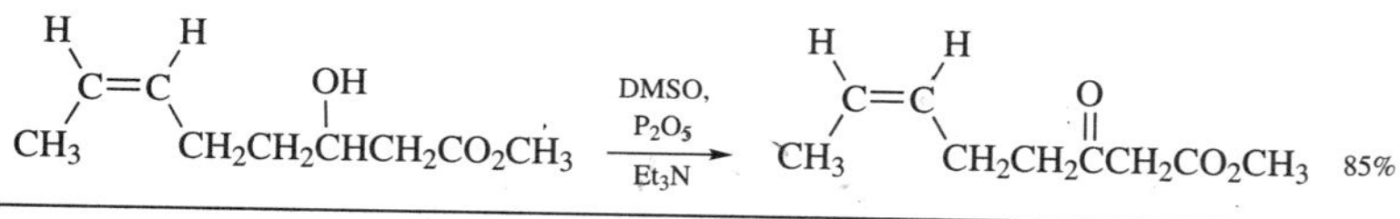

a. J. G. Moffat, *Org. Synth.* **47**:25 (1967).
b. J. A. Marshall and G. M. Cohen, *J. Org. Chem.* **36**:877 (1971).
c. E. Houghton and J. E. Saxton, *J. Chem. Soc. C* **1969**:595.
d. N. Finch, L. D. Vecchia, J. J. Fitt, R. Stephani, and I. Vlattas, *J. Org. Chem.* **38**:4412 (1973).
e. W. R. Roush, *J. Am. Chem. Soc.* **102**:1390 (1980).
f. A. Dondoni and D. Perrone, *Synthesis* **1997**:527.
g. R. W. Franck and T. V. John, *J. Org. Chem.* **45**:1170 (1980).
h. D. F. Taber, J. C. Amedio, Jr., and K.-Y. Jung, *J. Org. Chem.* **52**:5621 (1987).

in the DMSO procedures. In the presence of a mild base, elimination of dimethyl sulfide completes the oxidation.[23]

$$\text{N-chlorosuccinimide} + (CH_3)_2S \longrightarrow (CH_3)_2\overset{+}{S}{-}Cl \xrightarrow{R_2CHOH} R_2C(H){-}O{-}\overset{+}{S}(CH_3)_2 \;(B{:}) \longrightarrow R_2C{=}O + (CH_3)_2S$$

Similarly, reaction of chlorine and DMSO at low temperature gives an adduct that reacts with alcohols to give the ketone and DMSO.[24]

$$(CH_3)_2S{=}O \xrightarrow{Cl_2} (CH_3)_2\overset{+}{S}(O){-}Cl \xrightarrow{R_2CHOH} (CH_3)_2\overset{+}{S}(O){-}O{-}CR_2(H) \xrightarrow{Et_3N} (CH_3)_2S{=}O + R_2C{=}O$$

Another reagent which has become important for laboratory synthesis, known as the *Dess–Martin reagent*,[25] is a hypervalent iodine(V) compound.[26] The reagent is used in inert solvents such as chloroform or acetonitrile and gives rapid oxidation of primary and secondary alcohols. The by-product, *o*-iodosobenzoic acid, can be extracted with base and recycled.

$$R_2CHOH + \text{(Dess–Martin periodinane, } I(O_2CCH_3)_3\text{)} \longrightarrow R_2C{=}O + o\text{-}(I{=}O)C_6H_4CO_2H$$

The mechanism of the Dess–Martin oxidation involves exchange of the alcohol for acetate, followed by proton removal.[27]

$$\text{ArI}(O_2CCH_3)_3 + R_2CHOH \longrightarrow \text{ArI}(O_2CCH_3)_2(O{-}CHR_2)\;({:}B) \longrightarrow \text{ArI}(O_2CCH_3)_2 + O{=}CR_2$$

Several examples of oxidations by the Dess–Martin reagent are shown in Scheme 12.4.

23. E. J. Corey and C. U. Kim, *J. Am. Chem. Soc.* **94:**7586 (1972).
24. E. J. Corey and C. U. Kim, *Tetrahedron Lett.* **1973:**919.
25. D. B. Dess and J. C. Martin, *J. Org. Chem.* **48:**4155 (1983); R. E. Ireland and L. Liu, *J. Org. Chem.* **58:**2899 (1993); S. D. Meyer and S. L. Schreiber, *J. Org. Chem.* **59:**7549 (1994).
26. T. Wirth and U. H. Hirt, *Synthesis* **1999:**1271.
27. S. De Munari, M. Frigerio, and M. Santagostino, *J. Org. Chem.* **61:**9272 (1996).

Scheme 12.4. Oxidations by Dess–Martin Reagent

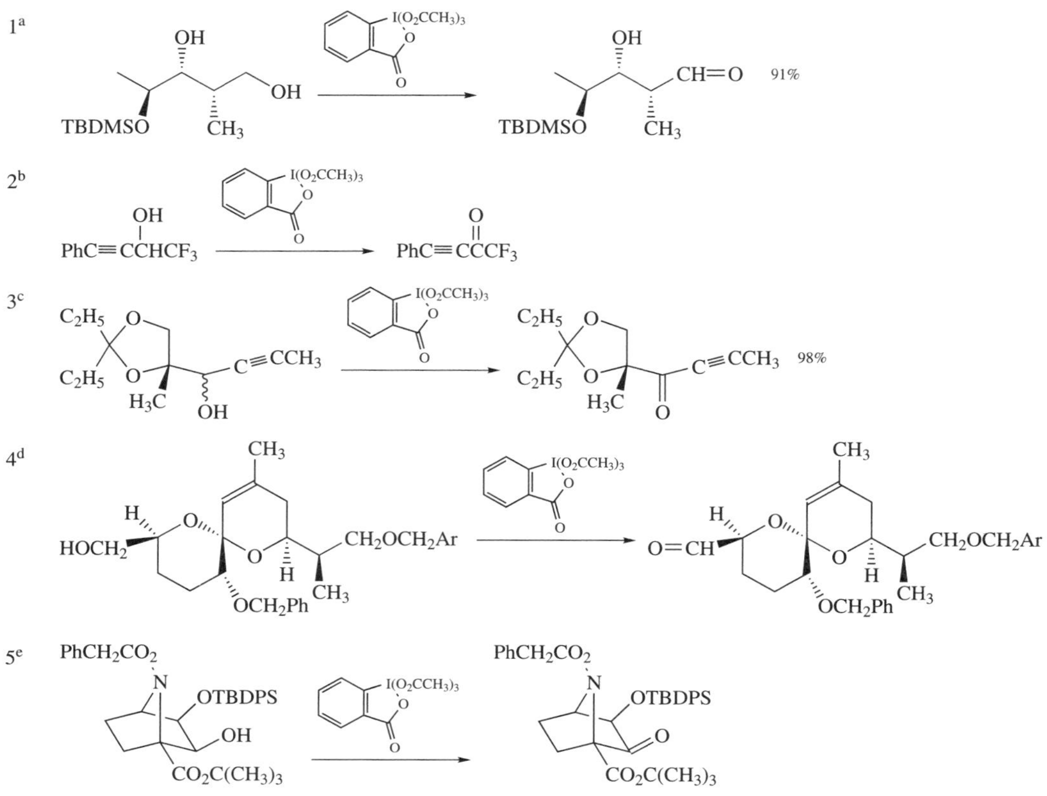

1[a] TBDMSO … OH … CH_3 … OH → TBDMSO … OH … CH=O, CH_3 91%

2[b] $PhC{\equiv}CCH(OH)CF_3$ → $PhC{\equiv}CC(=O)CF_3$

3[c] … $C{\equiv}CCH_3$ 98%

4[d] $HOCH_2$ … CH_2OCH_2Ar, OCH_2Ph, CH_3 → O=CH … CH_2OCH_2Ar

5[e] $PhCH_2CO_2$, N, OTBDPS, OH, $CO_2C(CH_3)_3$ → $PhCH_2CO_2$, N, OTBDPS, O, $CO_2C(CH_3)_3$

Reagent: $I(O_2CCH_3)_3$

a. P. R. Blakemore, P. J. Kocienski, A. Morley, and K. Muir, *J. Chem. Soc., Perkins Trans. 1* **1999:**955.
b. R. J. Linderman and D. M. Graves, *Tetrahedron Lett.* **28:**4259 (1987).
c. S. D. Burke, J. Hong, J. R. Lennox, and A. P. Mongin, *J. Org. Chem.* **63:**6952 (1998).
d. S. F. Sabes, R. A. Urbanek, and C. J. Forsyth, *J. Am. Chem. Soc.* **120:**2534 (1998).
e. B. P. Hart and H. Rapoport, *J. Org. Chem.* **64:**2050 (1999).

Another oxidation procedure uses an oxoammonium ion, usually derived from the stable nitroxide tetramethylpiperidine nitroxide (TEMPO) as the active reagent. It is regenerated in a catalytic cycle using hypochlorite ion[28] or NCS[29] as the stoichiometric oxidant. These reactions involve an intermediate adduct of the alcohol and the oxoammonium ion.

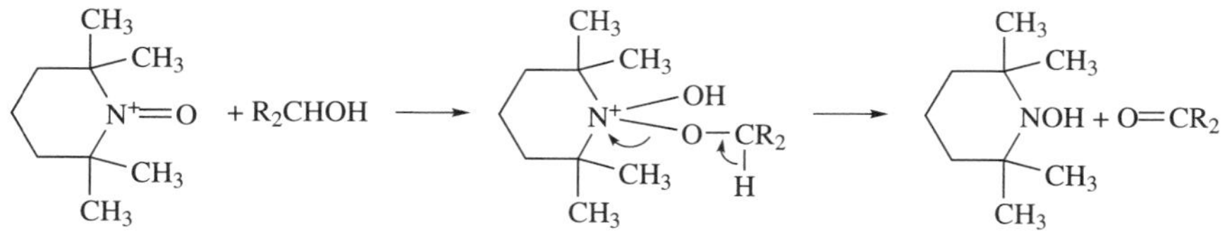

$N^+{=}O + R_2CHOH \longrightarrow N^+(OH)\text{–}O\text{–}CR_2H \longrightarrow NOH + O{=}CR_2$

28. R. Siedlecka, J. Skarzewski, and J. Mlochowski, *Tetrahedron Lett.* **31:**2177 (1990); T. Inokuchi, S. Matsumoto, T. Nishiyama, and S. Torii, *J. Org. Chem.* **55:**462 (1990); P. L. Anelli, S. Banfi, F. Montanari, and S. Quici, *J. Org. Chem.* **54:**2970 (1989); M. R. Leanna, T. J. Sowin, and H. E. Morton, *Tetrahedron Lett.* **33:**5029 (1992).
29. J. Einhorn, C. Einhorn, F. Ratajczak, and J.-L. Pierre, *J. Org. Chem.* **61:**7452 (1996).

Scheme 12.5. Oxidations with TEMPO

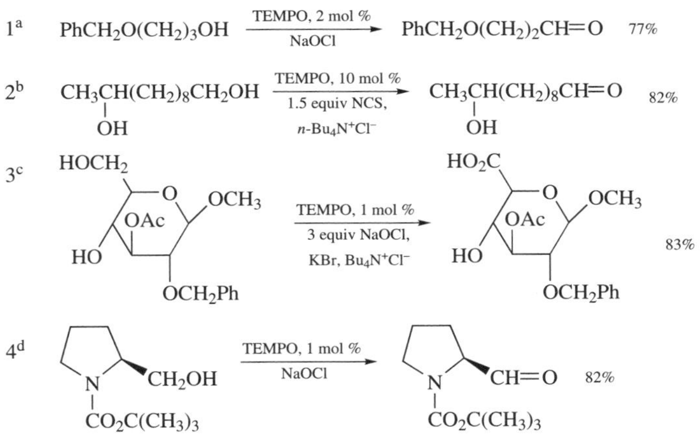

1[a] $PhCH_2O(CH_2)_3OH \xrightarrow[\text{NaOCl}]{\text{TEMPO, 2 mol \%}} PhCH_2O(CH_2)_2CH{=}O$ 77%

2[b] $CH_3CH(OH)(CH_2)_8CH_2OH \xrightarrow[\text{1.5 equiv NCS, } n\text{-}Bu_4N^+Cl^-]{\text{TEMPO, 10 mol \%}} CH_3CH(OH)(CH_2)_8CH{=}O$ 82%

3[c] (HOCH$_2$ sugar with OAc, OCH$_3$, HO, OCH$_2$Ph) $\xrightarrow[\text{3 equiv NaOCl, KBr, } Bu_4N^+Cl^-]{\text{TEMPO, 1 mol \%}}$ (HO$_2$C analogue) 83%

4[d] N-CO$_2$C(CH$_3$)$_3$-2-pyrrolidine-CH$_2$OH $\xrightarrow[\text{NaOCl}]{\text{TEMPO, 1 mol \%}}$ N-CO$_2$C(CH$_3$)$_3$-2-pyrrolidine-CH=O 82%

a. B. G. Szczepankiewicz and C. H. Heathcock, *Tetrahedron* **53:**8853 (1997).
b. J. Einhorn, C. Einhorn, F. Ratajczak, and J.-L. Pierre, *J. Org. Chem.* **61:**7452 (1996).
c. N. J. Davis and S. L. Flitsch, *Tetrahedron Lett.* **34:**1181 (1993).
d. M. R. Leanna, T. J. Sowin, and H. E. Morton, *Tetrahedron Lett.* **33:**5029 (1992).

One feature of this oxidation system is that it can selectively oxidize primary alcohols in preference to secondary alcohols. (See entry 2, Scheme 12.5) The reagent can also be used to oxidize primary alcohols to carboxylic acids by a subsequent oxidation with sodium chlorite.[30] Entry 3 in Scheme 12.5 shows the selective oxidation of a primary alcohol in a carbohydrate to a carboxylic acid without affecting the secondary alcohol group.

12.2. Addition of Oxygen at Carbon–Carbon Double Bonds

12.2.1. Transition-Metal Oxidants

Compounds of certain transition metals, in their higher oxidation states, particularly permanganate ion and osmium tetroxide, are effective reagents for addition of two oxygen atoms at a carbon–carbon double bond. Under carefully controlled reaction conditions, potassium permanganate can effect conversion of alkenes to glycols. This oxidant is, however, capable of further oxidizing the glycol with cleavage of the carbon–carbon bond. A cyclic manganese ester is an intermediate in these oxidations. Because of the cyclic nature of this intermediate, the glycols are formed by *syn* addition.

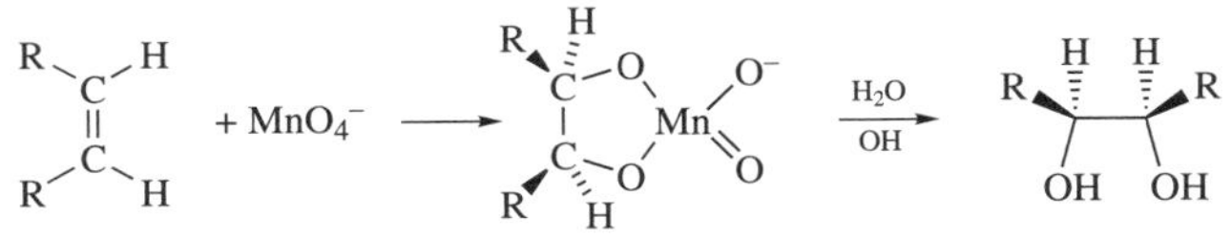

$RCH{=}CHR + MnO_4^- \longrightarrow$ (cyclic manganate ester) $\xrightarrow[\text{OH}]{H_2O}$ RCH(OH)–CH(OH)R

30. P. M. Wovkulich, K. Shankaran, J. Kiegiel, and M. R. Uskokovic, *J. Org. Chem.* **58:**832 (1993).

Ketols are also observed as products of permanganate oxidation of alkenes. The ketols are believed to be formed as a result of oxidation of the cyclic intermediate.[31]

$$\mathrm{R{-}CH{-}O{-}Mn(=O)(O^-){-}O{-}CH{-}R} \xrightarrow{-e^-} \mathrm{R{-}CH{-}O{-}Mn(=O){-}O{-}C(H){-}R} \longrightarrow \mathrm{R{-}CH(OH){-}C(=O){-}R} + MnO_2$$

Permanganate ion can be used to oxidize acetylenes to diones.

$$\mathrm{PhC{\equiv}CCH_2CH_2CH_3} \xrightarrow[R_4N^+,\ CH_2Cl_2]{KMnO_4} \underset{81\%}{\mathrm{PhC(=O){-}C(=O)CH_2CH_2CH_3}} \qquad \text{Ref. 32}$$

A mixture of $NaIO_4$ and RuO_2 in a heterogeneous solvent system is also effective.

$$\mathrm{PhC{\equiv}CCH_3} \xrightarrow[NaIO_4]{RuO_2} \underset{80\%}{\mathrm{PhC(=O)C(=O)CH_3}} \qquad \text{Ref. 33}$$

Osmium tetroxide is a highly selective oxidant which gives glycols by a stereospecific *syn* addition.[34] The reaction occurs through a cyclic osmate ester which is formed by a $[3 + 2]$ cycloaddition.[35]

$$\mathrm{RCH{=}CHR} + OsO_4 \longrightarrow \text{cyclic osmate ester} \longrightarrow \mathrm{RCH(OH)CH(OH)R}$$

The reagent is quite expensive, but this disadvantage can be minimized by procedures which use only a catalytic amount of osmium tetroxide. A very useful procedure involves an amine oxide such as morpholine-*N*-oxide as the stoichiometric oxidant.[36]

$$\mathrm{RCH{=}CHR} \xrightarrow{OsO_4} \text{osmate ester} \xrightarrow[H_2O]{R_3\overset{+}{N}{-}O^-} \mathrm{RCH(OH){-}CH(OH)R} + OsO_4 + R_3N$$

31. S. Wolfe, C. F. Ingold, and R. U. Lemieux, *J. Am. Chem. Soc.* **103:**938 (1981); D. G. Lee and T. Chen, *J. Am. Chem. Soc.* **111:**7534 (1989).
32. D. G. Lee and V. S. Chang, *J. Org. Chem.* **44:**2726 (1979).
33. R. Zibuck and D. Seebach, *Helv. Chim. Acta* **71:**237 (1988).
34. M. Schröder, *Chem. Rev.* **80:**187 (1980).
35. A. J. DelMonte, J. Haller, K. N. Houk, K. B. Sharpless, D. A. Singleton, T. Strassner, and A. A. Thomas, *J. Am. Chem. Soc.* **119:**9907 (1997); U. Pidun, C. Boehme, and G. Frenking, *Angew. Chem. Int. Ed. Engl.* **35:**2817 (1997).
36. V. Van Rheenen, R. C. Kelly, and D. Y. Cha, *Tetrahedron Lett.* **1976:**1973.

Scheme 12.6. *syn*-Dihydroxylation of Alkenes

A. Potassium permanganate

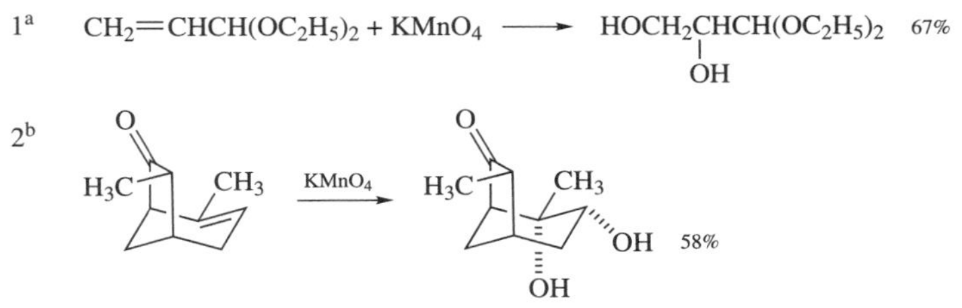
1[a]
CH2=CHCH(OC2H5)2 + KMnO4 → HOCH2CHCH(OC2H5)2 67%
OH
2[b]
O
H3C
CH3
KMnO4
O
H3C
CH3
OH 58%
OH

B. Osmium tetroxide

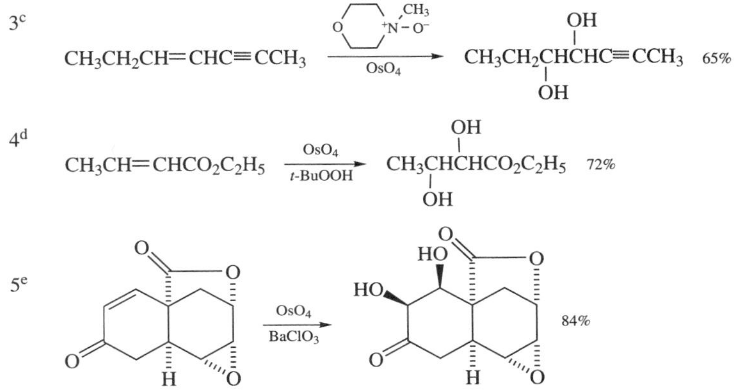
3[c]
CH3CH2CH=CHC≡CCH3
CH3
O
+N−O−
OsO4
OH
CH3CH2CHCHC≡CCH3 65%
OH
4[d]
CH3CH=CHCO2C2H5
OsO4
t-BuOOH
OH
CH3CHCHCO2C2H5 72%
OH
5[e]
O
O
O
H
O
OsO4
BaClO3
O
HO
O
HO
O
H
O
84%

a. E. J. Witzeman, W. L. Evans, H. Haas, and E. F. Schroeder, *Org. Synth.* **II**:307 (1943).
b. S. D. Larsen and S. A. Monti, *J. Am. Chem. Soc.* **99**:8015 (1977).
c. E. J. Corey, P. B. Hopkins, S. Kim, S. Yoo, K. P. Nambiar, and J. R. Falck, *J. Am. Chem. Soc.* **101**:7131 (1979).
d. K. Akashi, R. E. Palermo, and K. B. Sharpless, *J. Org. Chem.* **43**:2063 (1978).
e. S. Danishefsky, P. F. Schuda, T. Kitahara, and S. J. Etheredge, *J. Am. Chem. Soc.* **99**:6066 (1977).

t-Butyl hydroperoxide,[37] barium chlorate,[38] or potassium ferricyanide[39] can also be used as oxidants in catalytic procedures. Scheme 12.6 provides some examples of oxidations of alkenes to glycols by permanganate and by osmium tetroxide.

Osmium tetroxide hydroxylations can be highly enantioselective in the presence of chiral ligands. The most highly developed ligands are derived from the cinchona alkaloids dihydroquinine and dihydroquinidine.[40] The most effective ligands are dimeric derivatives

37. K. B. Sharpless and K. Akashi, *J. Am. Chem. Soc.* **98**:1986 (1976); K. Akashi, R. E. Palermo, and K. B. Sharpless, *J. Org. Chem.* **43**:2063 (1978).
38. L. Plaha, J. Weichert, J. Zvacek, S. Smolik, and B. Kakac, *Collect. Czech. Chem. Commun.* **25**:237 (1960); A. S. Kende, T. V. Bentley, R. A. Mader, and D. Ridge, *J. Am. Chem. Soc.* **96**:4332 (1974).
39. M. Minato, K. Yamamoto and J. Tsuji, *J. Org. Chem.* **55**:766 (1990); K. B. Sharpless, W. Amberg, Y. L. Bennani, G. A. Crispino, J. Hartung, K.-S. Jeong, H.-L. Kwong, K. Morikawa, Z.-M. Wang, D. Xu, and X.-L. Zhang, *J. Org. Chem.* **57**:2768 (1992); J. Eames, H. J. Mitchell, A. Nelson, P. O'Brien, S. Warren, and P. Wyatt, *Tetrahedron Lett.* **36**:1719 (1995).
40. H. C. Kolb, M. S. VanNieuwenhze, and K. B. Sharpless, *Chem. Rev.* **94**:2483 (1994).

of these alkaloids.[41] These ligands not only induce high enantioselectivity, but they also accelerate the reaction.[42] Optimization of the reaction conditions permits rapid and predictable dihydroxylation of many types of alkenes.[43] The premixed catalysts are available commercially and are referred to by the trade names AD-mix™. Several heterocyclic compounds including phthalazine (PHAL), pyrimidine (PYR), pyridazine (PYDZ) and diphenylpyrimidine (DPPYR) have been used in conjunction with the alkaloids.

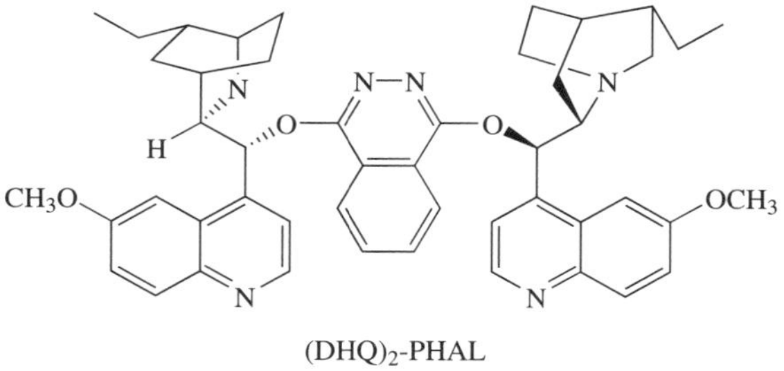

$(DHQ)_2$-PHAL

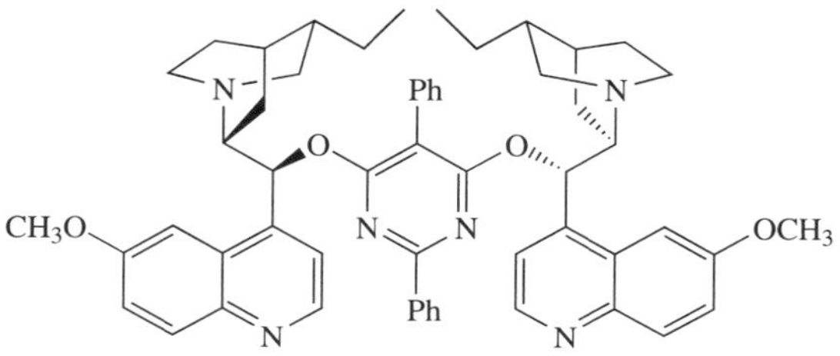

$(DHQD)_2$-DPPYR

Scheme 12.7 gives some examples of enantioselective hydroxylations using these reagents.

Various other chiral diamines have also been explored for use with OsO_4, and Scheme 12.8 illustrates some of them.

Other transition-metal oxidants can convert alkenes to epoxides. The most useful procedures involve *t*-butyl hydroperoxide as the stoichiometric oxidant in combination with vanadium, molybdenum, or titanium compounds. The most reliable substrates for oxidation are allylic alcohols. The hydroxyl group of the alcohol plays both an activating and a stereodirecting role in these reactions. *t*-Butyl hydroperoxide and a catalytic amount of $VO(acac)_2$ convert allylic alcohols to the corresponding epoxides in good yields.[44] The reaction proceeds through a complex in which the allylic alcohol is coordinated to

41. G. A. Crispino, K.-S. Jeong, H. C. Kolb, Z.-M. Wang, D. Xu, and K. B. Sharpless, *J. Org. Chem.* **58:**3785 (1993); G. A. Crispino, A. Makita, Z.-M. Wang, and K. B. Sharpless, *Tetrahedron Lett.* **35:**543 (1994); K. B. Sharpless, W. Amberg, Y. L. Bennani, G. A. Crispino, J. Hartung, K. S. Jeong, H.-L. Kwong, K. Morikawa, Z.-M. Wang, D. Xu, and X.-L. Zhang, *J. Org. Chem.* **57:**2768 (1992); W. Amberg, Y. L. Bennani, R. K. Chadha, G. A. Crispino, W. D. Davis, J. Hartung, K.-S. Jeong, Y. Ogino, T. Shibata, and K. B. Sharpless, *J. Org. Chem.* **58:**844 (1993); H. Becker, S. B. King, M. Taniguchi, K. P. M. Vanhessche, and K. B. Sharpless, *J. Org. Chem.* **60:**3940 (1995).
42. P. G. Andersson and K. B. Sharpless, *J. Am. Chem. Soc.* **115:**7047 (1993).
43. H.-L. Kwong, C. Sorato, Y. Ogino, H. Chen, and K. B. Sharpless, *Tetrahedron Lett.* **31:**2999 (1990); T. Göbel and K. B. Sharpless, *Angew. Chem. Int. Ed. Engl.* **32:**1329 (1993).
44. K. B. Sharpless and R. C. Michaelson, *J. Am. Chem. Soc.* **95:**6136 (1973).

1[a] $PhOCH_2CH{=}CH_2$ — $K_2OsO_2(OH)_2$, $(DHQD)_2$-PHAL, $K_3Fe(CN)_6$, K_2CO_3 → PhO–CH(OH)–CH_2OH 88% e.e.

2[b] 1-vinylcyclohexanol (HO, $CH{=}CH_2$) — $K_2OsO_2(OH)_2$, 1 mol %, $(DHQD)_2$-PYR, $K_3Fe(CN)_6$, K_2CO_3 → HO, OH, CH_2OH 88% yield, 90% e.e.

3[c] E-$C_2H_5O_2CCH_2CH_2CH{=}CH(CH_2)_{11}CH_3$ — $K_2OsO_2(OH)_2$, DHQ-PHAL, $K_3Fe(CN)_6$, K_2CO_3, $CH_3SO_2NH_2$ → lactone with OH, $(CH_2)_{11}CH_3$ 82% yield, 95% e.e.

4[d] stilbene — $K_2OsO_2(OH)_4$, 0.2 mol %, $(DHQD)_2$-PHAL, N-methylmorpholine-N-oxide → diol (OH, OH) 76% yield, 99% e.e.

5[e] geranyl acetate (O_2CCH_3) — K_2OsO_2, 0.01 mol %, $(DHQ)_2$-PYDZ, 1 mol %, $K_3Fe(CN)_6$, K_2CO_3 → OH, OH, O_2CCH_3 76% yield, >95% e.e.

6[f] Ph–CH=CH–CH$_2$–CO–N(CH_3)CH_2Ph — $K_2OsO_2(OH)_4$, 1 mol %, $(DHQD)_2$-PHAL, $CH_3SO_2NH_2$, $K_3Fe(CN)_6$ → Ph, OH, OH, O, CH_3, NCH_2Ph 97% yield, 98% e.e.

7[g] $(CH_3)_2CHCH_2$–C$_6$H$_4$–C(CH_3)$=CH_2$ — $K_2OsO_2(OH)_4$, 1 mol %, $(DHQ)_2$-PHAL, $K_3Fe(CN)_6$ → $(CH_3)_2CHCH_2$–C$_6$H$_4$–C(CH_3)(OH)CH_2OH 99%

8[h] E-$CH_3CH{=}CHCH_2CO_2CH_3$ — $K_2OsO_2(OH)_4$, 1 mol %, $(DHQ)_2$-PHAL, $K_3Fe(CN)_6$ → HO, CH_3 lactone (O, O) 48% yield, 80% e.e.

9[i] C_2H_5-furyl–CH=CH–CO_2CH_3 — $K_2OsO_2(OH)_4$, 1 mol %, $(DHQ)_2$-PHAL, $K_3Fe(CN)_6$ → C_2H_5, O, OH, OH, CO_2CH_3 93% yield, 97.5% e.e.

a. Z.-M. Wang, X.-L. Zhang, and K. B. Sharpless, *Tetrahedron Lett.* **34:**2267 (1993).
b. Z.-M. Wang and K. B. Sharpless, *Tetrahedron Lett.* **34:**8225 (1993).
c. Z.-M. Wang, X.-L. Zhang, K. B. Sharpless, S. C. Sinha, A. Sinha-Bagchi, and E. Keinan, *Tetrahedron Lett.* **33:**6407 (1992).
d. H. T. Chang and K. B. Sharpless, *J. Org. Chem.* **61:**6456 (1996).
e. E. J. Corey, M. C. Noe, and W.-C. Shieh, *Tetrahedron Lett.* **34:**5995 (1993).
f. Y. L. Bennani and K. B. Sharpless, *Tetrahedron Lett.* **34:**2079 (1993).
g. H. Ishibashi, M. Maeki, J. Yagi, M. Ohba, and T. Kanai, *Tetrahedron* **55:**6075 (1999).
h. T. Berkenbusch and R. Brückner, *Tetrahedron* **54:**11461 (1998).
i. T. Taniguchi, M. Takeuchi, and K. Ogasawara, *Tetrahedron Asymmetry* **9:**1451 (1998).

Scheme 12.8. Enantioselective Hydroxylation Using Other Chiral Amines

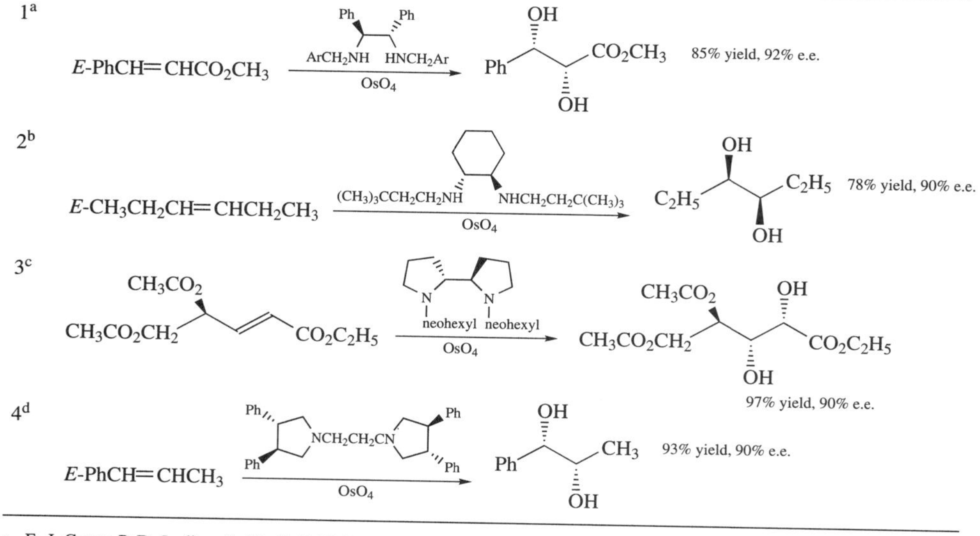
1[a]
E-PhCH=CHCO2CH3
Ph
Ph
ArCH2NH HNCH2Ar
OsO4
OH
Ph
CO2CH3
OH
85% yield, 92% e.e.
2[b]
E-CH3CH2CH=CHCH2CH3
(CH3)3CCH2CH2NH
NHCH2CH2C(CH3)3
OsO4
OH
C2H5
C2H5
OH
78% yield, 90% e.e.
3[c]
CH3CO2
CH3CO2CH2
CO2C2H5
N
N
neohexyl neohexyl
OsO4
CH3CO2
OH
CH3CO2CH2
CO2C2H5
OH
97% yield, 90% e.e.
4[d]
E-PhCH=CHCH3
Ph
Ph
NCH2CH2CN
Ph
Ph
OsO4
OH
Ph
CH3
OH
93% yield, 90% e.e.

a. E. J. Corey, P. D. Jardine, S. Virgil, P.-W. Yuen, and R. D. Connell, *J. Am. Chem. Soc.* **111**:9243 (1989).
b. S. Hanessian, P. Meffre, M. Girard, S. Beaudoin, J.-Y. Sanceau, and Y. Bennani, *J. Org. Chem.* **58**:1991 (1993).
c. T. Oishi, K. Iida, and M. Hirama, *Tetrahedron Lett.* **34**:3573 (1993).
d. K. Tomioka, M. Nakajima, and K. Koga, *Tetrahedron Lett.* **31**:1741 (1990).

vanadium by the hydroxyl group. In cyclic alcohols, this results in epoxidation *cis* to the hydroxyl group. In acyclic alcohols, the observed stereochemistry is consistent with a transition state in which the double bond is oriented at an angle of about 50° to the coordinated hydroxy group. This transition state leads to formation of an alcohol having *syn* stereochemistry. This stereochemistry is observed for both *cis* and *trans*-disubstituted allylic alcohols. If there is a substituent *cis* to the allylic carbon, the *syn-cis* isomer is the major product.[45]

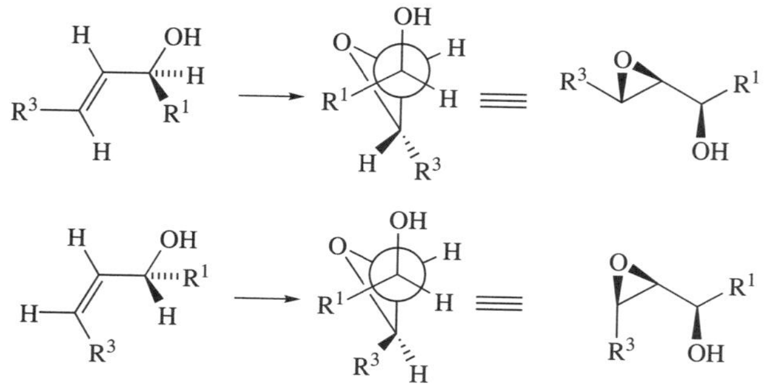
H
OH
H
R3
R1
H
OH
O
H
R1
H
H
R3
R3
O
R1
OH
H
OH
R1
H
H
R3
OH
O
H
R1
H
R3
H
O
R1
R3
OH

The epoxidation of allylic alcohols can also be effected by *t*-butyl hydroperoxide and titanium tetraisopropoxide. When enantiomerically pure tartrate esters are included in the system, the reaction is highly enantioselective. This reaction is called the *Sharpless*

45. E. D. Mihelich, *Tetrahedron Lett.* **1979**:4729; B. E. Rossiter, T. R. Verhoeven, and K. B. Sharpless, *Tetrahedron Lett.* **1979**:4733.

asymmetric epoxidation.[46] Either the (+) or (−) tartrate ester can be used so either enantiomer of the desired product can be obtained.

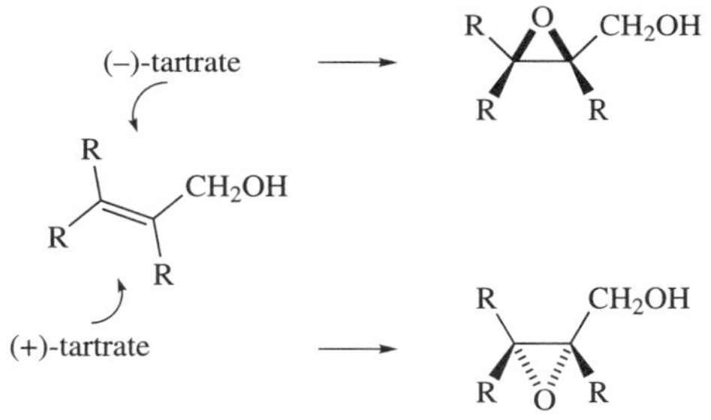

(−)-tartrate → R, O, CH$_2$OH, R, R

R, CH$_2$OH, R, R

(+)-tartrate → R, CH$_2$OH, R, O, R

The mechanism by which the enantionselective oxidation occurs is generally similar to that for the vanadium-catalyzed oxidations. The allylic alcohol serves to coordinate the reactant to titanium. The tartrate esters are also coordinated at titanium, which creates a chiral environment. The active catalyst is believed to be a dimeric species. The mechanism involves rapid exchange of the allylic alcohol and *t*-butyl hydroperoxide at the titanium atom.

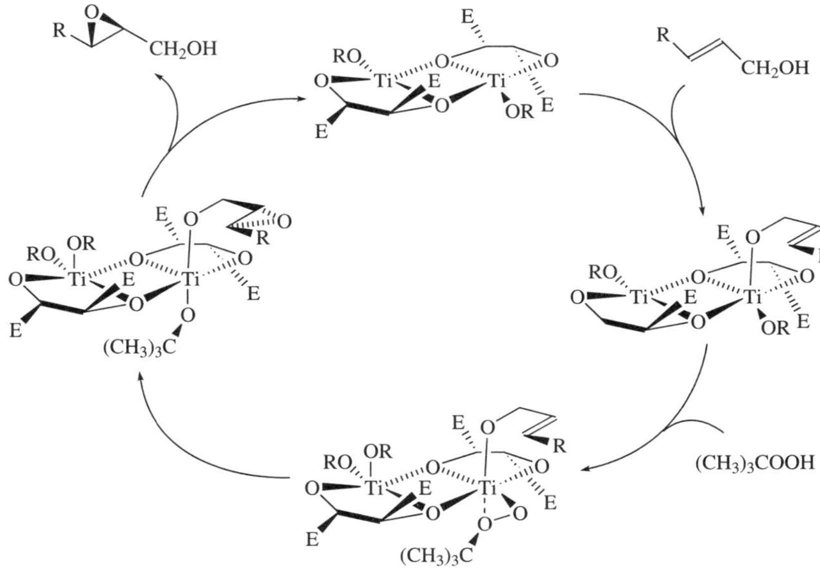

R, O, CH$_2$OH; RO, O, Ti, E, O, Ti, O, OR, E, E; R, CH$_2$OH; (CH$_3$)$_3$COOH; (CH$_3$)$_3$C

The orientation of the reactants is governed by the chirality of the tartrate ester. In the transition state, an oxygen atom from the peroxide is transferred to the double bond. The

46. For reviews, see A. Pfenninger, *Synthesis* **1986:**89; R. A. Johnson and K. B. Sharpless, in *Catalytic Asymmetric Synthesis*, I. Ojima, ed., VCH Publishers, New York, 1993, pp. 103–158.

enantioselectivity is consistent with a transition state such as that shown below.[47]

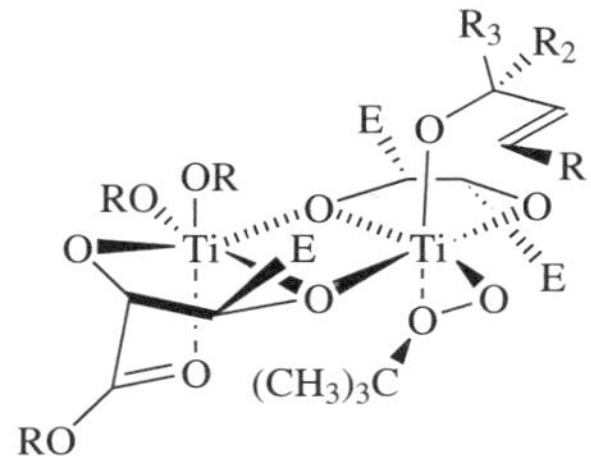

This method has proven to be an extremely useful means of synthesizing enantiomerically enriched compounds. Various improvements in the methods for carrying out the Sharpless oxidation have been developed.[48] The reaction can be done with catalytic amounts of titanium isopropoxide and the tartrate ester.[49] This procedure uses molecular sieves to sequester water, which has a deleterious effect on both the rate and enantioselectivity of the reaction. Scheme 12.9 gives some examples of enantioselective epoxidation of allylic alcohols.

Because of the importance of the allylic hydroxyl group in coordinating the reactant to the titanium, the structural relationship between the double bond and the hydroxyl group is crucial. Homoallylic alcohols can be oxidized, but the degree of enantioselectivity is reduced. Interestingly, the facial selectivity is reversed from that observed with allylic alcohols.[50] Compounds lacking a coordinating hydroxyl group are not reactive under these conditions.

Several catalysts that can effect enantioselective epoxidation of unfunctionalized alkenes have been developed, most notably manganese complexes of diimines such as **D** and **E** derived from salicylaldehyde and chiral diamines.[51]

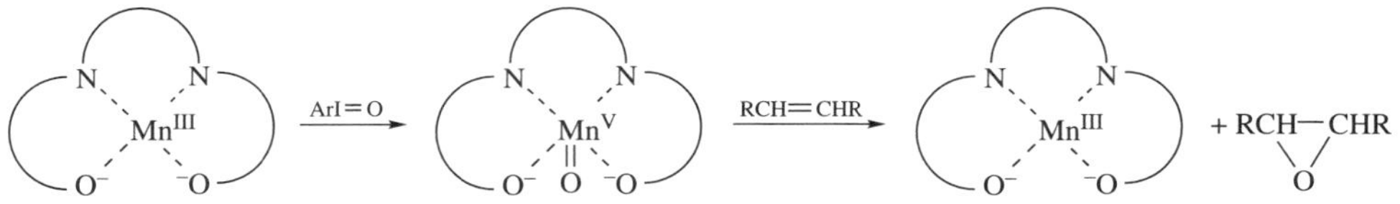

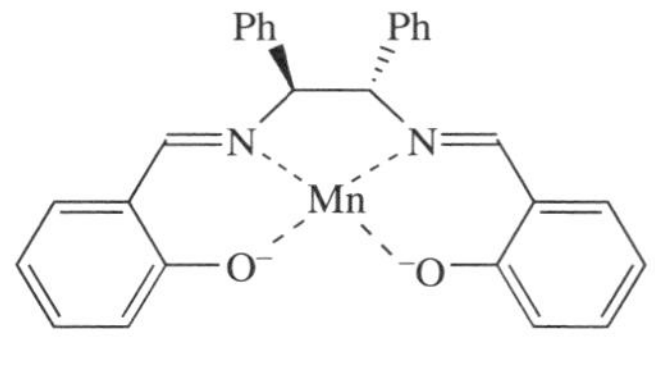

D

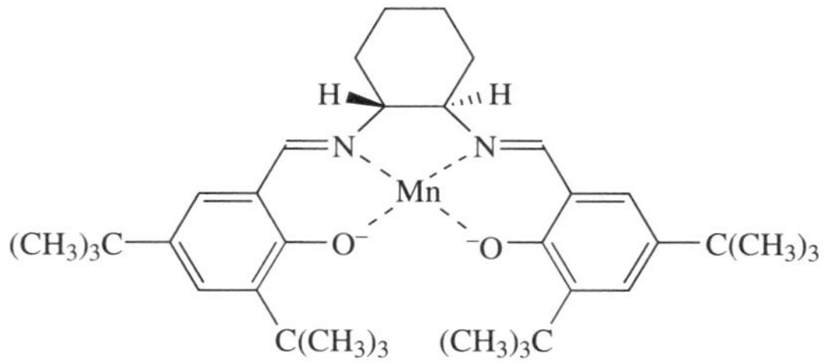

E

47. V. S. Martin, S. S Woodard, T. Katsuki, Y. Yamada, M. Ikeda, and K. B. Sharpless, *J. Am. Chem. Soc.* **103:**6237 (1981); K. B. Sharpless, S. S. Woodard, and M. G. Finn, *Pure Appl. Chem.* **55:**1823 (1983); M. G. Finn and K. B. Sharpless, in *Asymmetric Synthesis*, Vol. 5, J. D. Morrison, ed., Academic Press, New York, 1985, Chapter 8; M. G. Finn, and K. B. Sharpless, *J. Am. Chem. Soc.* **113:**113 (1991); B. H. McKee, T. H. Kalantar, and K. B. Sharpless, *J. Org. Chem.* **56:**6966 (1991); for an alternative description of the origin of enantioselectivity, see E. J. Corey, *J. Org. Chem.* **55:**1693 (1990).
48. J. G. Hill, B. E. Rossiter, and K. B. Sharpless, *J. Org. Chem.* **48:**3607 (1983); L. A. Reed III, S. Masamune, and K. B. Sharpless, *J. Am. Chem. Soc.* **104:**6468 (1982).
49. R. M. Hanson and K. B. Sharpless, *J. Org. Chem.* **51:**1922 (1986); Y. Gao, R. M. Hanson, J. M. Klunder, S. Y. Ko, H. Masamune, and K. B. Sharpless, *J. Am. Chem. Soc.* **109:**5765 (1987).
50. B. E. Rossiter and K. B. Sharpless, *J. Org. Chem.* **49:**3707 (1984).
51. W. Zhang, J. L. Loebach, S. R. Wilson, and E. N. Jacobsen, *J. Am. Chem. Soc.* **112:**2801 (1990); E. N. Jacobsen, W. Zhang, A. R. Muci, J. R. Ecker, and L. Deng, *J. Am. Chem. Soc.* **113:**7063 (1991).

1[a] (+)-diethyl tartrate, Ti(O-*i*-Pr)$_4$, *t*-BuOOH — 78% yield, 97% e.e.

2[b] (+)-diisopropyl tartrate, Ti(O-*i*-Pr)$_4$, *t*-BuOOH — 80% yield, 95% e.e.

3[c] (+)-diethyl tartrate, Ti(O-*i*-Pr)$_4$, *t*-BuOOH — 77% yield, 93% e.e.

4[d] (+)-diethyl tartrate, Ti(O-*i*-Pr)$_4$, *t*-BuOOH — 77% yield, 94 e.e.

5[e] (+)-diethyl tartrate (0.07 equiv), Ti(O-*i*-Pr)$_4$ (0.05 equiv), *t*-BuOOH — 95% yield, 91% e.e.

6[f] (−)-diethyl tartrate, Ti(O-*i*-Pr)$_4$, *t*-BuOOH — 77% yield

7[g] (−)-diisopropyl tartrate, Ti(O-*i*-Pr)$_4$, *t*-BuOOH — 85% yield; Ar = 4-methoxyphenyl

a. J. G. Hill and K. B. Sharpless, *Org. Synth.* **63:**66 (1985).
b. B. E. Rossiter, T. Katsuki, and K. B. Sharpless, *J. Am. Chem. Soc.* **103:**464 (1981).
c. Y. Gao, R. M. Hanson, J. M. Klunder, S. Y. Ko, H. Masamune, and K. B. Sharpless, *J. Am. Chem. Soc.* **109:**5765 (1987).
d. D. A. Evans, S. L. Bender, and J. Morris, *J. Am. Chem. Soc.* **110:**2506 (1988).
e. R. M. Hanson and K. B. Sharpless, *J. Org. Chem.* **51:**1922 (1986).
f. A. K. Ghosh and Y. Wang, *J. Org. Chem.* **64:**2789 (1999).
g. J. A. Marshall, Z.-H. Lu, and B. A. Johns, *J. Org. Chem.* **63:**817 (1998).

These catalysts are used in conjunction with a stoichiometric amount of an oxidant, and the active oxidant is believed to be an oxo Mn(V) species. The stoichiometric oxidants that have been used include NaOCl,[52] periodate,[53] and amine oxides.[54] Various other chiral salen-type ligands have also been explored.[55] These epoxidations are not always

52. W. Zhang and E. N. Jacobsen, *J. Org. Chem.* **56:**2296 (1991); B. D. Brandes and E. N. Jacobsen, *J. Org. Chem.* **59:**4378 (1994).
53. P. Pietikäinen, *Tetrahedron Lett.* **36:**319 (1995).
54. M. Palucki, P. J. Pospisil, W. Zhang, and E. N. Jacobsen, *J. Am. Chem. Soc.* **116:**9333 (1994).
55. N. Hosoya, R. Irie, and T. Katsuki, *Synlett* **1993:**261; S. Chang, R. M. Heid, and E. N. Jacobsen, *Tetrahedron Lett.* **35:**669 (1994).

stereospecific with respect to the alkene geometry. This is attributed to an electron-transfer mechanism which involves a radical intermediate.

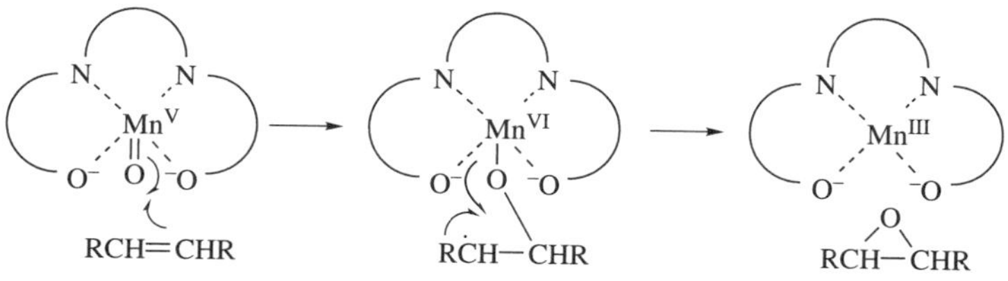
Mn V
RCH=CHR
Mn VI
RCH—CHR
Mn III
RCH—CHR

Scheme 12.10 gives some examples of these oxidations.

Scheme 12.10. Enantioselective Epoxidation with a Chiral Mn Catalyst[a]

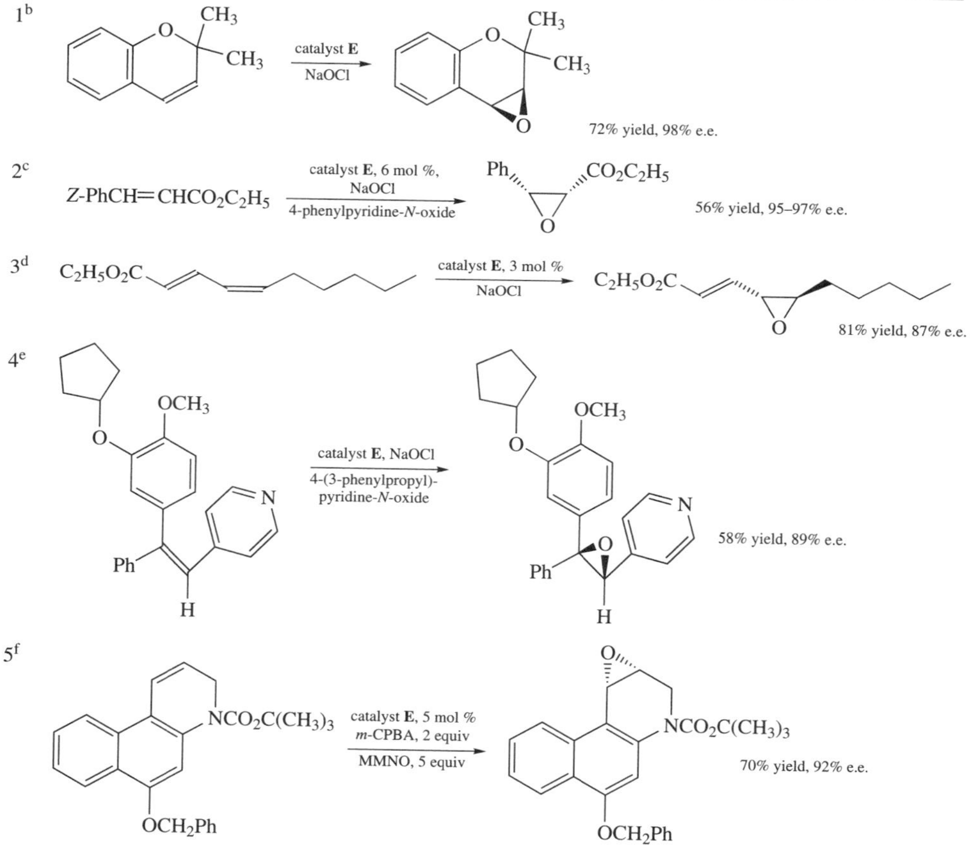
1b
catalyst E
NaOCl
72% yield, 98% e.e.
2c
Z-PhCH=CHCO2C2H5
catalyst E, 6 mol %,
NaOCl
4-phenylpyridine-N-oxide
56% yield, 95–97% e.e.
3d
catalyst E, 3 mol %
NaOCl
81% yield, 87% e.e.
4e
catalyst E, NaOCl
4-(3-phenylpropyl)-
pyridine-N-oxide
58% yield, 89% e.e.
5f
catalyst E, 5 mol %
m-CPBA, 2 equiv
MMNO, 5 equiv
70% yield, 92% e.e.

a. The structure of catalyst **E** is shown on p. 764.
b. E. N. Jacobsen, W. Zhang, A. R. Muci, J. R. Ecker, and L. Deng, *J. Am. Chem. Soc.* **113:**7063 (1991).
c. L. Deng and E. N. Jacobsen, *J. Org. Chem.* **57:**4320 (1992).
d. S. Chang, N. H. Lee, and E. N. Jacobsen, *J. Org. Chem.* **58:**6939 (1993).
e. J. E. Lynch, W.-B. Choi, H. R. O. Churchill, R. P. Volante, R. A. Reamer, and R. G. Ball, *J. Org. Chem.* **62:**9223 (1997).
f. D. L. Boger, J. A. McKie, and C. W. Boyce, *Synlett.* **1997:**515.

The most general reagents for conversion of simple alkenes to epoxides are peroxycarboxylic acids.[56] *m*-Chloroperoxybenzoic acid[57] (MCPBA) is a particularly convenient reagent, but it is not commercially available at the present time. The magnesium salt of monoperoxyphthalic acid has been recommended as a replacement.[58] Potassium hydrogen peroxysulfate, which is sold commercially as "oxone," is a convenient reagent for epoxidations that can be done in aqueous methanol.[59] Peroxyacetic acid, peroxybenzoic acid, and peroxytrifluoroacetic acid have also been used frequently for epoxidation. All of the peroxycarboxylic acids are potentially hazardous materials and require appropriate precautions.

It has been demonstrated that ionic intermediates are not involved in the epoxidation reaction. The reaction rate is not very sensitive to solvent polarity.[60] Stereospecific *syn* addition is consistently observed. The oxidation is therefore believed to be a concerted process. A representation of the transition state is shown below.

The rate of epoxidation of alkenes is increased by alkyl groups and other electron-donating substituents, and the reactivity of the peroxy acids is increased by electron-accepting substituents.[61] These structure–reactivity relationships demonstrate that the peroxyacid acts as an electrophile in the reaction. Very low reactivity is exhibited by double bonds that are conjugated with strongly electron-attracting substituents and very reactive peroxy acids, such as peroxytrifluoroacetic acid, are required for oxidation of such compounds.[62] Electron-poor alkenes can also be epoxidized by alkaline solutions of hydrogen peroxide or *t*-butyl hydroperoxide. A quite different mechanism, involving conjugate nucleophilic addition, operates in this case:[63]

$$\mathrm{RC(=O)CH{=}CHCH_3} + {}^{-}\mathrm{OOH} \longrightarrow \mathrm{RC(-O^-){=}C(H)-CHCH_3(O{-}OH)} \longrightarrow \text{epoxide} + \mathrm{OH^-}$$

The stereoselectivity of epoxidation with peroxycarboxylic acids has been well studied. Addition of oxygen occurs preferentially from the less hindered side of the

56. D. Swern, *Organic Peroxides*, Vol. II, Wiley-Interscience, New York, 1971, pp. 355–533; B. Plesnicar, in *Oxidation in Organic Chemistry, Part C*, W. Trahanovsky, ed., Academic Press, New York, 1978, pp. 211–253.
57. R. N. McDonald, R. N. Steppel, and J. E. Dorsey, *Org. Synth.* **50:**15 (1970).
58. P. Brougham, M. S. Cooper, D. A. Cummerson, H. Heaney, and N. Thompson, *Synthesis* **1987:**1015.
59. R. Bloch, J. Abecassis, and D. Hassan, *J. Org. Chem.* **50:**1544 (1985).
60. N. N. Schwartz and J. N. Blumbergs, *J. Org. Chem.* **29:**1976 (1964).
61. B. M. Lynch and K. H. Pausacker, *J. Chem. Soc.* **1955:**1525.
62. W. D. Emmons and A. S. Pagano, *J. Am. Chem. Soc.* **77:**89 (1955).
63. C. A. Bunton and G. J. Minkoff, *J. Chem. Soc.* **1949:**665.

molecule. Norbornene, for example, gives a 96 : 4 *exo* : *endo* ratio.[64] In molecules in which two potential modes of approach are not greatly different, a mixture of products is to be expected. For example, the unhindered exocyclic double bond in 4-*t*-butylmethylenecyclohexane gives both stereoisomeric products.[65]

m-chloroperoxybenzoic acid, CH_2Cl_2 → 69% + 31%

Hydroxyl groups exert a directive effect on epoxidation and favor approach from the side of the double bond closest to the hydroxyl group.[66] Hydrogen bonding between the hydroxyl group and the reagent evidently stabilizes this transition state.

peroxybenzoic acid →

This strong directing effect can exert stereochemical control even when opposed by steric effects. Several examples of epoxidation reactions are given in Scheme 12.11. Entries 4 and 5 illustrate the hydroxyl directing effect. Other substituents capable of hydrogen bonding, in particular amides, also can exert a *syn*-directing effect.[67]

A process that is effective for epoxidation and which avoids acidic conditions involves reaction of an alkene, a nitrile, and hydrogen peroxide.[68] The nitrile and hydrogen peroxide react, forming a peroxyimidic acid, which epoxidizes the alkene, presumably by a mechanism similar to that for peroxyacids. An important contribution to the reactivity of the peroxyimidic acids comes from the formation of the stable amide carbonyl group.

$$R'C{\equiv}N + H_2O_2 \longrightarrow R'C(=NH){-}O{-}OH$$

$$R'C(=NH){-}O{-}OH + R_2C{=}CR_2 \longrightarrow R'C(=O)NH_2 + \text{epoxide}$$

CH_3CN, H_2O_2, $KHCO_3$, CH_3OH → Ref. 69

64. H. Kwart and T. Takeshita, *J. Org. Chem.* **28:**670 (1963).
65. R. G. Carlson and N. S. Behn, *J. Org. Chem.* **32:**1363 (1967).
66. H. B. Henbest and R. A. L. Wilson, *J. Chem. Soc.* **1957:**1958.
67. F. Mohamadi and M. M. Spees, *Tetrahedron Lett.* **30:**1309 (1989); P. G. M. Wuts, A. R. Ritter, and L. E. Pruitt, *J. Org. Chem.* **57:**6696 (1992); A. Jenmalm, W. Bets, K. Luthman, I. Csöregh, and U. Hacksell, *J. Org. Chem.* **60:**1026 (1995); P. Kocovsky and I. Stary, *J. Org. Chem.* **55:**3236 (1990); A. Armstrong, P. A. Barsanti, P. A. Clarke, and A. Wood, *J. Chem. Soc., Perkin Trans 1* **1996:**1373.
68. G. B. Payne, *Tetrahedron* **18:**763 (1962); R. D. Bach and J. W. Knight, *Org. Synth.* **60:**63 (1981); L. A. Arias, S. Adkins, C. J. Nagel, and R. D. Bach, *J. Org. Chem.* **48:**888 (1983)
69. W. C. Frank, *Tetrahedron Asymmetry* **9:**3745 (1998).
70. R. D. Bach, M. W. Klein, R. A. Ryntz, and J. W. Holubka, *J. Org. Chem.* **44:**2569 (1979).

Scheme 12.11. Synthesis of Epoxides from Alkenes

A. Oxidation of alkenes with peroxy acids

1[a]

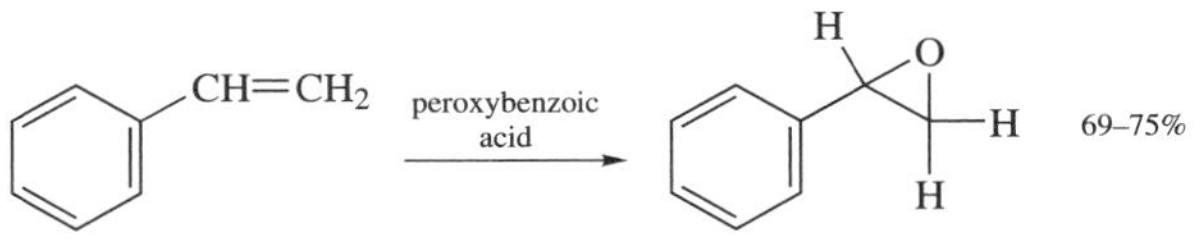
CH=CH2
peroxybenzoic acid
H
O
H
H
69–75%

2[b]

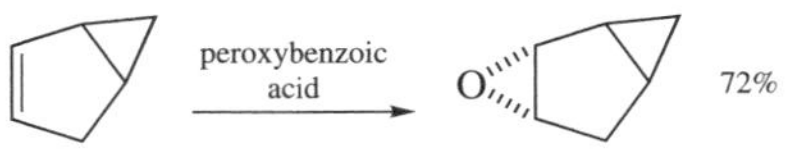
peroxybenzoic acid
O
72%

3[c]

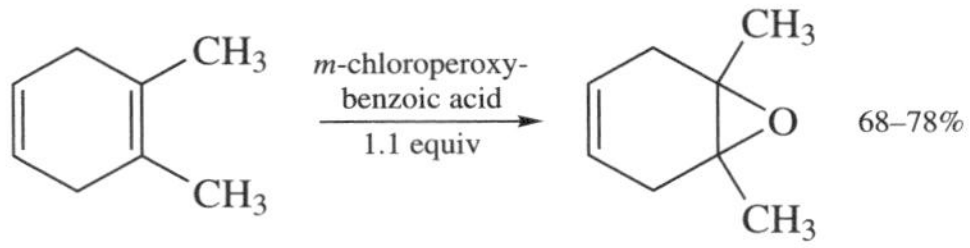
CH3
CH3
m-chloroperoxybenzoic acid
1.1 equiv
CH3
O
CH3
68–78%

4[d]

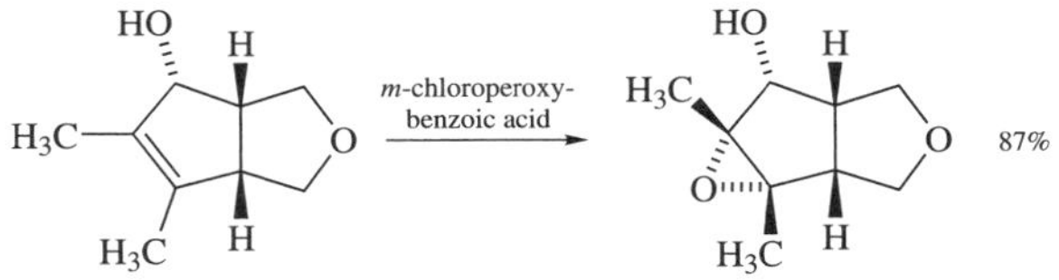
HO
H
H3C
O
H3C
H
m-chloroperoxybenzoic acid
HO
H
H3C
O
O
H3C
H
87%

5[e]

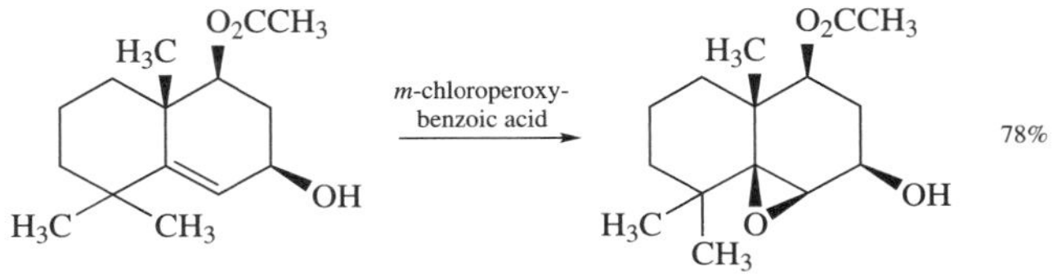
O2CCH3
H3C
OH
H3C
CH3
m-chloroperoxybenzoic acid
O2CCH3
H3C
OH
H3C
O
CH3
78%

6[f]

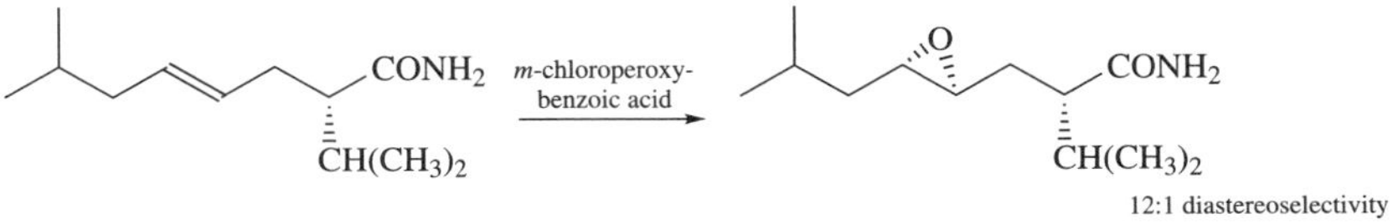
CONH2
CH(CH3)2
m-chloroperoxybenzoic acid
O
CONH2
CH(CH3)2
12:1 diastereoselectivity

B. Epoxidation of electrophilic alkenes

7[g]

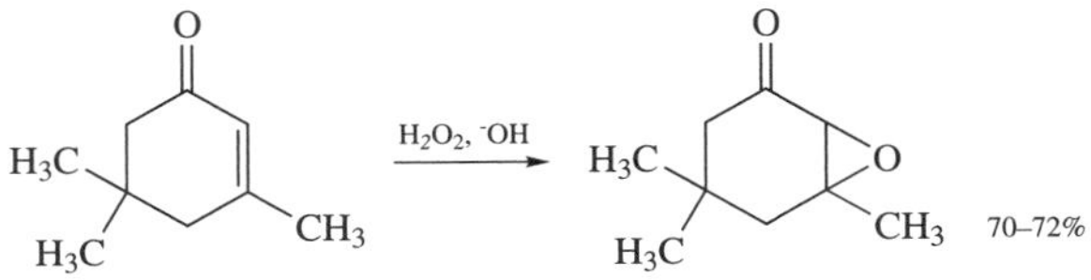
O
H3C
H3C
CH3
H2O2, -OH
O
H3C
H3C
O
CH3
70–72%

8[h]

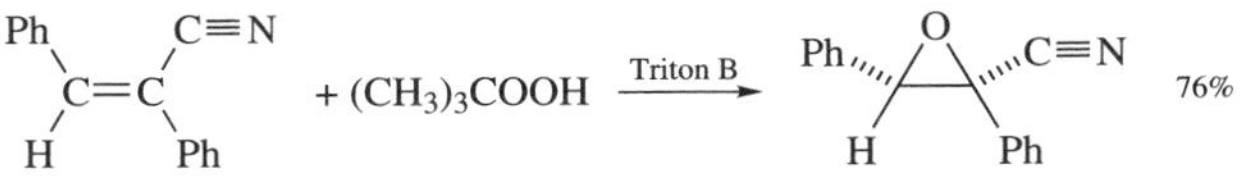
Ph
C≡N
C=C
H
Ph
+ (CH3)3COOH
Triton B
Ph
O
C≡N
H
Ph
76%

9[i]

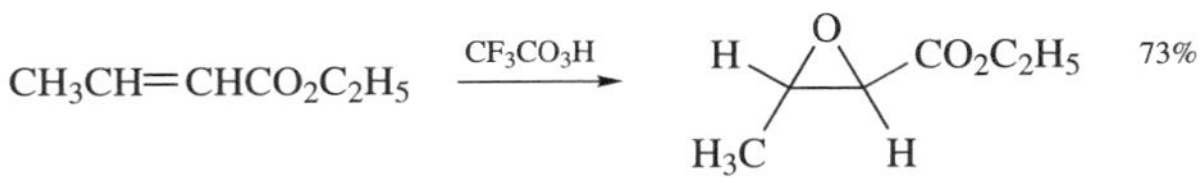
CH3CH=CHCO2C2H5
CF3CO3H
H
O
CO2C2H5
H3C
H
73%

Scheme 12.11. *(continued)*

10[j] cyclooctene $\xrightarrow[CH_3OH,\ KHCO_3]{H_2O_2,\ CH_3CN}$ cyclooctene oxide 60%

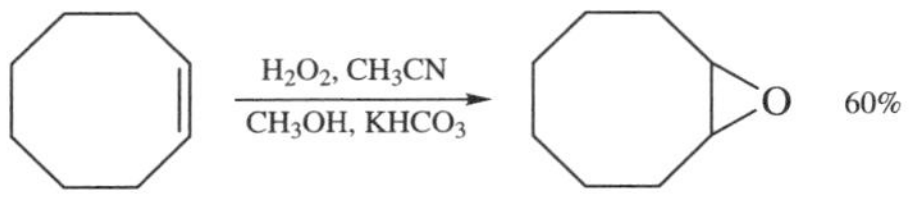

a. H. Hibbert and P. Burt, *Org. Synth.* **1:**481 (1932).
b. E. J. Corey and R. L. Dawson, *J. Am. Chem. Soc.* **85:**1782 (1963).
c. L. A. Paquette and J. H. Barrett, *Org. Synth.* **49:**62 (1969).
d. R. M. Scarborough, Jr., B. H. Toder, and A. B. Smith III, *J. Am. Chem. Soc.* **102:**3904 (1980).
e. M. Miyashita and A. Yoshikoshi, *J. Am. Chem. Soc.* **96:**1917 (1974).
f. P. G. M. Wuts, A. R. Ritter, and L. E. Pruitt, *J. Org. Chem.* **57:**6696 (1992).
g. R. L. Wasson and H. O. House, *Org. Synth.* **IV:**552 (1963).
h. G. B. Payne and P. H. Williams, *J. Org. Chem.* **26:**651 (1961).
i. W. D. Emmons and A. S. Pagano, *J. Am. Chem. Soc.* **77:**89 (1955).
j. R. D. Bach and J. W. Knight, *Org. Synth.* **60:**63 (1981).

A variety of other reagents have been examined with the goal of activating H_2O_2 to generate a good epoxidizing agent. In principle, any species which can convert one of the hydroxyl groups to a good leaving group will generate a reactive epoxidizing reagent:

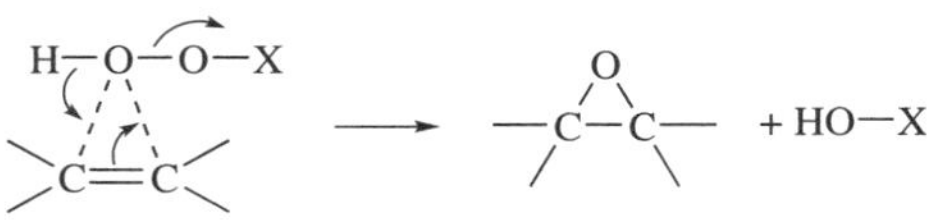

$$H{-}O{-}O{-}X + {>}C{=}C{<} \longrightarrow {-}C{-}C{-}\ (\text{epoxide}) + HO{-}X$$

In practice, promising results have been obtained for several systems. For example, fair-to-good yields of epoxides are obtained when a two-phase system containing an alkene and ethyl chloroformate is stirred with a buffered basic solution of hydrogen peroxide. The active oxidant is presumed to be *O*-ethyl peroxycarbonic acid.[70]

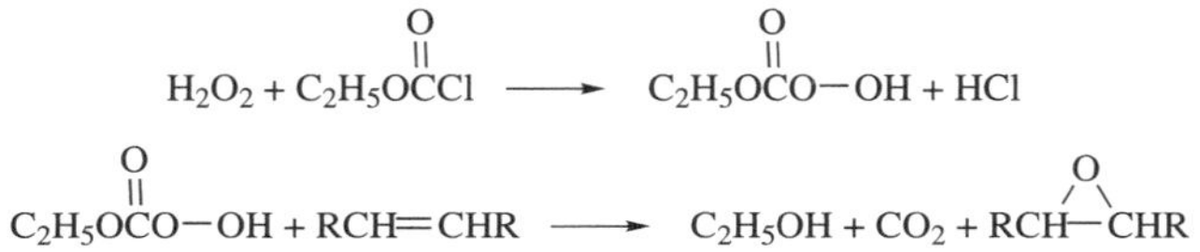

$$H_2O_2 + C_2H_5OCCl\ (C{=}O) \longrightarrow C_2H_5OCO{-}OH\ (C{=}O) + HCl$$

$$C_2H_5OCO{-}OH + RCH{=}CHR \longrightarrow C_2H_5OH + CO_2 + RCH{-}CHR\ (\text{epoxide})$$

Although none of these reagent combinations have been as generally useful as the peroxycarboxylic acids, they serve to illustrate that epoxidizing activity is not unique to the peroxy acids.

Another useful epoxidizing agent is dimethyldioxirane (DMDO).[71] This reagent is generated by *in situ* reaction of acetone and peroxymonosulfate in buffered aqueous solution. Distillation gives an $\sim 0.1\,M$ solution of DMDO in acetone.[72]

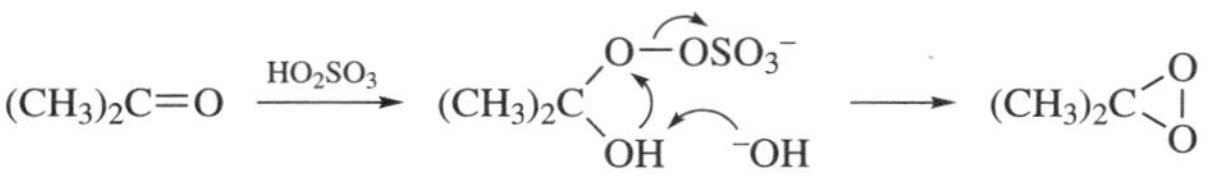

$$(CH_3)_2C{=}O \xrightarrow{HO_2SO_3} (CH_3)_2C(O{-}OSO_3^-)(OH)\ \ {}^-OH \longrightarrow (CH_3)_2C(O)_2$$

71. R. W. Murray, *Chem. Rev.* **89:**1187 (1989); W. Adam and L. P. Hadjiarapoglou, *Top. Curr. Chem.* **164:**45 (1993); W. Adam, A. K. Smerz, and C. G. Zhao, *J. Prakt. Chem., Chem. Zeit.* **339:**295 (1997).
72. R. W. Murray and R. Jeyaraman, *J. Org. Chem.* **50:**2847 (1985); W. Adam, J. Bialas, and L. Hadjiarapoglou, *Chem. Ber.* **124:**2377 (1991).

Higher concentrations of DMDO can be obtained by extraction of a 1 : 1 aqueous dilution of the distillate by CH_2Cl_2, $CHCl_3$, or CCl_4.[73] Another method involves *in situ* generation of DMDO under phase-transfer conditions.[74]

$$CH_3CH{=}CH(CH_2)_3OCH_2Ph \xrightarrow[\text{pH 7.8 buffer, } n\text{-}Bu_4N^+\ HSO_4^-]{CH_3C(=O)CH_3,\ KOSO_2OOH} CH_3\overset{O}{CH{-}CH}(CH_2)_3OCH_2Ph$$

The yields and rates of oxidation by DMDO under these *in situ* conditions depend on pH and other reaction conditions.[75] Various computational models of the transition state agree that the reaction occurs by a concerted mechanism.[76] Kinetics and isotope effects are consistent with this mechanism.[77]

Similarly to peroxycarboxylic acids, DMDO is subject to *cis* or *syn* stereoselectivity by hydroxy and other hydrogen-bonding functional groups.[78] For other substituents, both steric and dipolar factors seem to have an influence, and several complex reactants have shown good stereoselectivity, although the precise origins of the stereoselectivity are not always evident.[79]

Other ketones besides acetone can be used for *in situ* generation of dioxiranes by reaction with peroxysulfuric acid or another suitable peroxide. More electrophilic ketones give more reactive dioxiranes. 3-Methyl-3-trifluoromethyldioxirane is a more reactive analog of DMDO.[80] This reagent, which is generated *in situ* from 1,1,1-trifluoroacetone, is

73. M. Gilbert, M. Ferrer, F. Sanchez-Baeza, and A. Messequer, *Tetrahedron* **53:**8643 (1997).
74. S. E. Denmark, D. C. Forbes, D. S. Hays, J. S. DePue, and R. G. Wilde, *J. Org. Chem.* **60:**1391 (1995).
75. M. Frohn, Z.-X. Wang, and Y. Shi, *J. Org. Chem.* **63:**6425 (1998); A. O'Connell, T. Smyth, and B. K. Hodnett, *J. Chem. Technol. Biotechnol.* **72:**60 (1998).
76. R. D. Bach, M. N. Glukhovtsev, C. Gonzalez, M. Marquez, C. M. Estevez, A. G. Baboul, and H. B. Schlegel, *J. Phys. Chem.* **101:**6092 (1997); M. Freccero, R. Gandolfi, M. Sarzi-Amade, and A. Rastelli, *Tetrahedron* **54:**6123 (1998); J. Liu, K. N. Houk, A. Dinoi, C. Fusco, and R. Curci, *J. Org. Chem.* **63:**8565 (1998).
77. W. Adam, R. Paredes, A. K. Smerz, and L. A. Veloza, *Liebigs Ann. Chem.* **1997:**547; A. L. Baumstark, E. Michalenabaez, A. M. Navarro, and H. D. Banks, *Heterocycl. Commun.* **3:**393 (1997); Y. Angelis, X. Zhang, and M. Orfanopoulos, *Tetrahedron Lett.* **37:**5991 (1996).
78. R. W. Murray, M. Singh, B. L. Williams, and H. M. Moncrief, *J. Org. Chem.* **61:**1830 (1996); G. Asensio, C. Boix-Bernardini, C. Andreu, M. E. Gonzalez-Nunez, R. Mello, J. O. Edwards, and G. B. Carpenter, *J. Org. Chem.* **64:**4705 (1999).
79. R. C. Cambie, A. C. Grimsdale, P. S. Rutledge, M. F. Walker, and A. D. Woodgate, *Aust. Chem.* **44:**1553 (1991); P. Boricelli and P. Lupattelli, *J. Org. Chem.* **59:**4304 (1994); R. Curci, A. Detomaso, T. Prencipe, and G. B. Carpenter, *J. Am. Chem. Soc.* **116:**8112 (1994); T. C. Henninger, M. Sabat, and R. J. Sundberg, *Tetrahedron* **52:**14403 (1996).
80. R. Mello, M. Fiorentino, O. Sciacevolli, and R. Curci, *J. Org. Chem.* **53:**3890 (1988).

capable of oxidizing less reactive compounds such as methyl cinnamate.

$$\mathrm{PhCH{=}CHCO_2CH_3} \xrightarrow[\mathrm{CH_3CN,\ H_2O}]{\mathrm{CF_3C(O)CH_3,\ KOSO_2OOH}} \mathrm{PhCH{-}CHCO_2CH_3}\ \text{(epoxide)}\quad 97\%$$

Ref. 81

Hexafluoroacetone and hydrogen peroxide in buffered aqueous solution epoxidize alkenes and allylic alcohols.[82] *N,N*-Dialkylpiperidin-4-one salts are also good catalysts for epoxidation.[83] The quaternary nitrogen enhances the reactivity of the ketone toward nucleophilic addition and also makes the dioxirane intermediate more reactive.

$$\mathrm{PhCH{=}CHCH_2OH} \xrightarrow[\mathrm{KOSO_2OOH}]{\text{1-}C_{12}H_{25}\text{-1-}CH_3\text{-piperidin-4-one}^{+}} \mathrm{PhCH{-}CHCH_2OH}\ \text{(epoxide)}\quad 83\%$$

The use of chiral ketones can lead to enantioselective epoxidation.[84]

$$\mathrm{Ph{-}CH{=}CH{-}CH_2OH} \xrightarrow[\mathrm{KHSO_5,\ K_2CO_3}]{\text{chiral difluoro ketone (}CH_3,\ CH_3,\ F,\ F,\ =O\text{)}} \text{Ph-epoxide-}CH_2OH\quad 93\%\ \text{yield, } 89\%\ \text{e.e.}$$

Scheme 12.12 gives some example of epoxidations involving dioxirane reagents and intermediates.

12.2.3. Transformations of Epoxides

Epoxides are useful synthetic intermediates, and the conversion of an alkene to an epoxide is often part of a more extensive molecular transformation.[85] In many instances, advantage is taken of the high reactivity of the epoxide ring to introduce additional functionality. Because epoxide ring opening is usually stereospecific, such reactions can be used to establish stereochemical relationships between adjacent substituents. Such two- or three-step operations can accomplish specific oxidative transformations of an alkene that may not easily be accomplished in a single step. Scheme 12.13 provides a preview of the type of reactivity to be discussed.

Epoxidation may be preliminary to solvolytic or nucleophilic ring opening in synthetic sequences. In acidic aqueous solution, epoxides are opened to give diols by an

81. D. Yang, M.-K. Wong, and Y.-C. Yip, *J. Org. Chem.* **60:**3887 (1995).
82. R. P. Heggs and B. Ganem, *J. Am. Chem. Soc.* **101:**2484 (1979); A. J. Biloski, R. P. Hegge, and B Ganem, *Synthesis* **1980:**810; W. Adam, H.-G. Degen, and C. R. Saha-Möller, *J. Org. Chem.* **64:**1274 (1999).
83. S. E. Denmark, D. C. Forbes, D. S. Hays, J. S. DePue, and R. G. Wilde, *J. Org. Chem.* **60:**1391 (1995).
84. S. E. Denmark and Z. C. Wu, *Synlett* **1999:**847.
85. J. G. Smith, *Synthesis* **1984:**629.

Scheme 12.12. Epoxidation by Dioxiranes

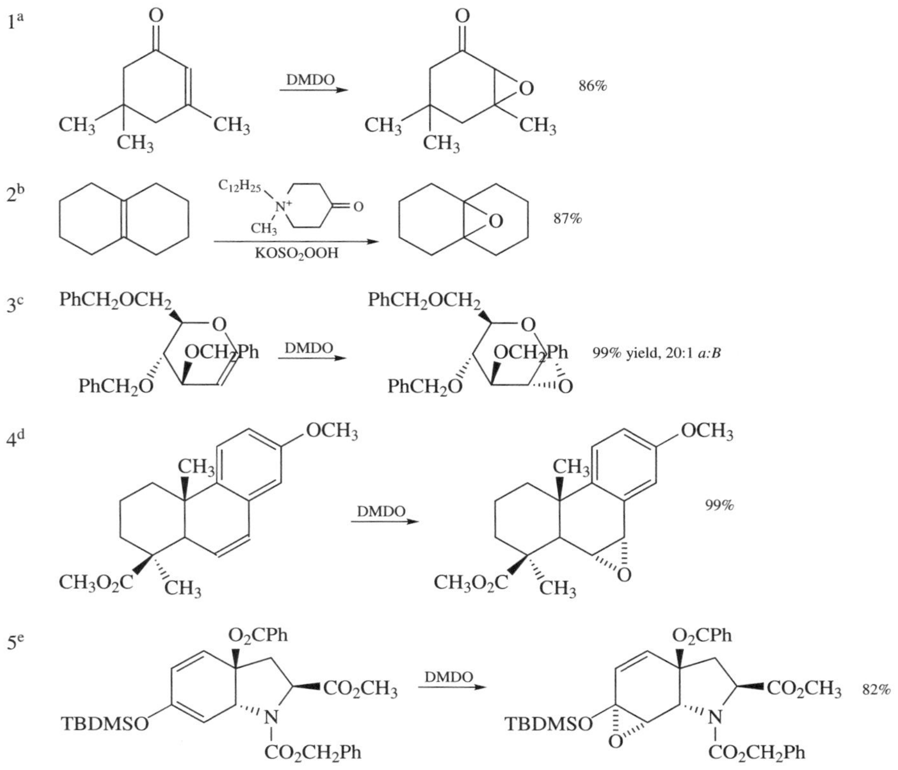

1[a] DMDO, 86%

2[b] $C_{12}H_{25}N^+(CH_3)$-piperidin-4-one, $KOSO_2OOH$, 87%

3[c] DMDO, 99% yield, 20:1 a:B

4[d] DMDO, 99%

5[e] DMDO, 82%

a. W. Adam, L. Hadjarapaglou, and B. Nestler, *Tetrahedron Lett.* **31:**331 (1990).
b. S. E. Denmark, D. C. Forbes, D. S. Hays, J. S. DePue, and R. G. Wilde, *J. Org. Chem.* **60:**1391 (1995).
c. R. L. Halcomb and S. J. Danishefsky, *J. Am. Chem. Soc.* **111:**6661 (1989).
d. R. C. Cambie, A. C. Grimsdale, P. S. Rutledge, M. F. Walker, and P. D. Woodgate, *Aust. J. Chem.* **44:**1553 (1991).
e. T. C. Henninger, M. Sabat, and R. J. Sundberg, *Tetrahedron* **52:**14403 (1996).

anti addition process. In cyclic systems, ring opening gives the diaxial diol.

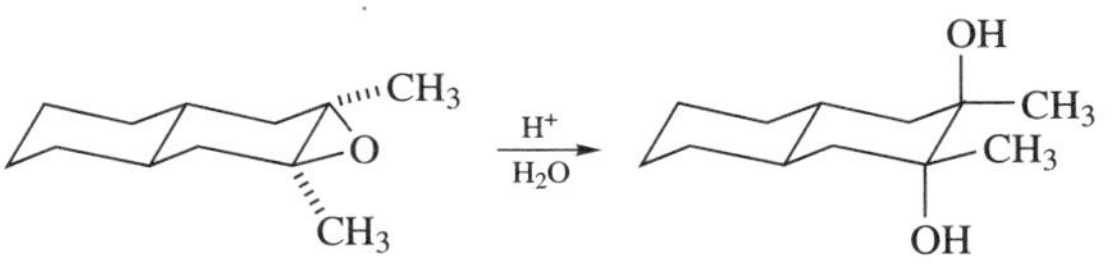

H^+, H_2O — Ref. 86

Base-catalyzed reactions, in which the nucleophile provides the driving force for ring opening, usually involve breaking the epoxide bond at the less substituted carbon, since this is the position most accessible to nucleophilic attack.[87] The situation in acid-catalyzed reactions is more complex. The bonding of a proton to the oxygen weakens the C–O

86. B. Rickborn and D. K. Murphy, *J. Org. Chem.* **34:**3209 (1969).
87. R. E. Parker and N. S. Isaacs, *Chem. Rev.* **59:**737 (1959).

Scheme 12.13. Multistep Synthetic Transformations via Epoxides

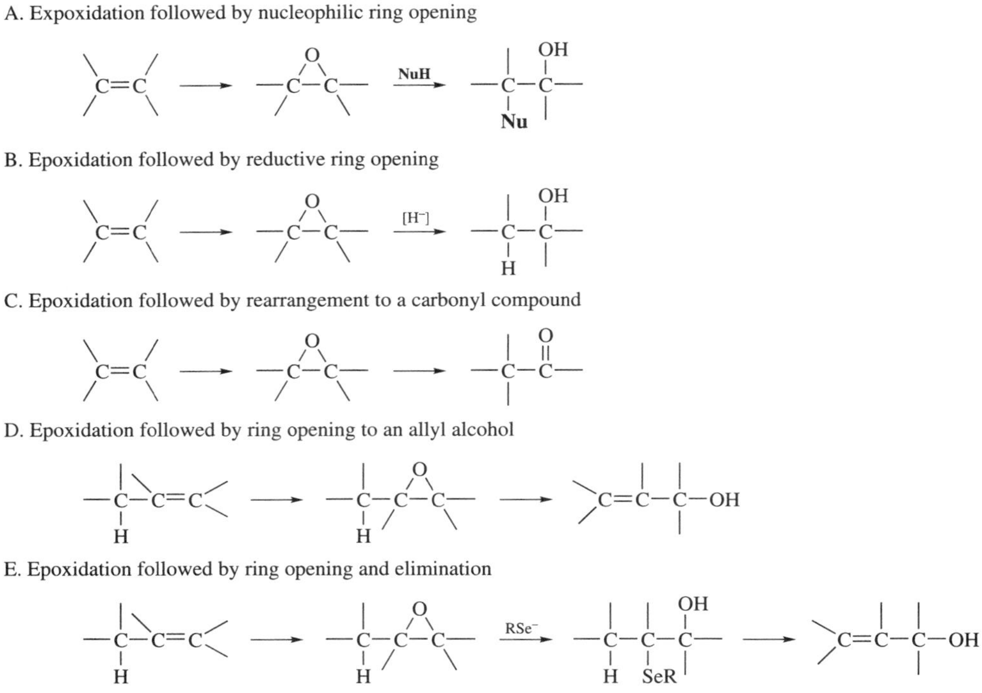
A. Expoxidation followed by nucleophilic ring opening
NuH
OH
Nu
B. Epoxidation followed by reductive ring opening
[H⁻]
C. Epoxidation followed by rearrangement to a carbonyl compound
D. Epoxidation followed by ring opening to an allyl alcohol
E. Epoxidation followed by ring opening and elimination
RSe⁻
SeR

bonds and facilitates rupture by weak nucleophiles. If the C–O bond is largely intact at the transition state, the nucleophile will become attached to the less substituted position for the same steric reasons that were cited for nucleophilic ring opening. If, on the other hand, C–O rupture is more complete when the transition state is reached, the opposite orientation is observed. This change in regiochemistry results from the ability of the more substituted carbon to stabilize the developing positive charge.

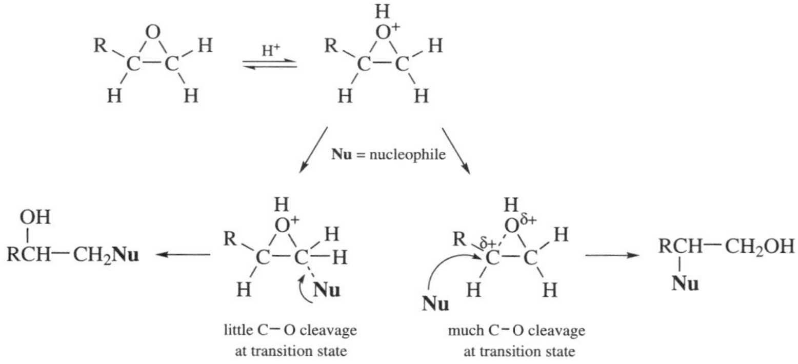
H⁺
Nu = nucleophile
RCH–CH₂Nu
RCH–CH₂OH
little C–O cleavage at transition state
much C–O cleavage at transition state

When simple aliphatic epoxides such as methyloxirane react with hydrogen halides, the dominant mode of reaction introduces halide at the less substituted primary carbon.[88]

$$H_3C\text{-oxirane} \xrightarrow[H_2O]{HBr} \underset{76\%}{CH_3CH(OH)CH_2Br} + \underset{24\%}{CH_3CH(Br)CH_2OH}$$

Substituents that further stabilize a carbocation intermediate lead to reversal of the mode of addition.[89] The case of styrene oxide hydrolysis has been carefully examined. Under acidic conditions, the bond breaking is exclusively at the benzylic position. Under basic conditions, ring opening occurs at both epoxide carbons.[90] Styrene oxide also undergoes highly regioselective ring opening in the presence of Lewis acids. For example, methanolysis is catalyzed by $SnCl_4$ and occurs with >95% attack at the benzyl carbon and with high inversion.[91] The stereospecificity indicates a concerted nucleophilic opening of the complexed epoxide.

$$\text{Ph-oxirane} \xrightarrow[CH_3OH]{SnCl_4} PhCH(OCH_3)CH_2OH$$

Synthetic procedures for epoxide ring opening can be based on nucleophilic or protic/Lewis acid-mediated electrophilic ring opening. Recently, a number of procedures which feature the oxyphilic Lewis acid character of metal ions, including lanthanides, have been developed. $LiClO_4$, LiO_3SCF_3, $Mg(ClO_4)_2$, $Zn(O_3SCF_3)_2$, and $Yb(O_3SCF_3)_3$ have all been shown to catalyze epoxide ring opening.[92] The cations catalyze *anti* addition of amines at the less substituted carbon, indicating a Lewis acid-assisted nucleophilic ring opening.

$$M^{n+} \cdots \text{R-oxirane} + :NHR'_2 \longrightarrow RCH(OH)CH_2NR'_2$$

Styrene oxide gives mixtures of products of C-2 and C-3 attack, as a result of competition between the activated benzylic site and the primary site.

$$\text{Ph-oxirane} + (C_2H_5)_2NH \xrightarrow{Yb(O_3SCF_3)_3} \underset{45\%}{PhCH(OH)CH_2N(C_2H_5)_2} + \underset{55\%}{PhCH(N(C_2H_5)_2)CH_2OH}$$

88. C. A. Stewart and C. A. VanderWerf, *J. Am. Chem. Soc.* **76:**1259 (1954).
89. S. Winstein and L. L. Ingraham, *J. Am. Chem. Soc.* **74:**1160 (1952).
90. R. Lin and D. L. Whalen, *J. Org. Chem.* **59:**1638 (1994); J. J. Blumenstein, V. C. Ukachukwu, R. S. Mohan, and D. Whalen, *J. Org. Chem.* **59:**924 (1994).
91. C. Moberg, L. Rakos, and L. Tottie, *Tetrahedron Lett.* **33:**2191 (1992).
92. M. Chini, P. Crotti, and F. Macchia, *Tetrahedron Lett.* **31:**4661 (1990); M. Chini, P. Crotti, L. Favero, F. Macchia, and M. Pineschi, *Tetrahedron Lett.* **35:**433 (1994); J. Auge and F. Leroy, *Tetrahedron Lett.* **37:**7715 (1996).

The same salts can be used to catalyze ring opening by other nucleophiles such as azide ion[93] and cyanide ion.[94] A number of successful reaction conditions have been developed for nucleophilic ring opening by cyanide.[95] Heating an epoxide with acetone cyanohydrin (which serves as the cyanide source) and triethylamine leads to ring opening.

$$CH_3(CH_2)_3\overset{O}{CH-CH_2} \xrightarrow[(C_2H_5)_3N]{(CH_3)_2C(CN)OH} CH_3(CH_2)_3CH(OH)CH_2CN \quad 74\% \qquad \text{Ref. 96}$$

Trimethylsilyl cyanide, in conjunction with KCN and a crown ether, also gives nucleophilic ring opening by cyanide.

$$CH_2{=}CH(CH_2)_2\overset{O}{CH-CH_2} \xrightarrow[\text{KCN, 18-crown-6}]{(CH_3)_3SiCN} CH_2{=}CH(CH_2)_2CH(OH)CH_2CN \quad 80\% \qquad \text{Ref. 97}$$

Diethylaluminum cyanide can also be used for preparation of β-hydroxynitriles.

$$\text{epoxide}-CH_2OSO_2Ar \xrightarrow{(C_2H_5)_2AlCN} NC-CH_2-CH(OH)-CH_2-OSO_2Ar \quad 96\% \qquad \text{Ref. 98}$$

Scheme 12.14 gives some examples of both acid-catalyzed and nucleophilic ring openings of epoxides.

Epoxides can also be reduced to saturated alcohols. Lithium aluminum hydride acts as a nucleophilic reducing agent, and the hydride is added at the less substituted carbon atom of the epoxide ring. Lithium triethylborohydride is more reactive than $LiAlH_4$ and is superior for epoxides that are resistant to reduction.[99] Reductions by dissolving metals, such as lithium in ethylenediamine,[100] also give good yields. Diisobutylaluminum hydride reduces epoxides. 1,2-Epoxyoctane gives 2-octanol in excellent yield, whereas styrene oxide gives a 1 : 6 mixture of the secondary and primary alcohols.[101]

$$R\overset{O}{CH-CH_2} \xrightarrow[\text{hexane}]{(i\text{-Bu})_2AlH} RCH(OH)CH_3 + RCH_2CH_2OH$$

$R = C_6H_{13}$ 100 : 0

$R = C_6H_5$ 14 : 86

93. M. Chini, P. Crotti, and F. Macchia, *Tetrahedron Lett.* **31:**5641 (1990); P. Van de Weghe and J. Collin, *Tetrahedron Lett.* **36:**1649 (1995).
94. M. Chini, P. Crotti, L. Favera, and F. Macchia, *Tetrahedron Lett.* **32:**4775 (1991).
95. R. A. Smiley and C. J. Arnold, *J. Org. Chem.* **25:**257 (1960); J. A. Ciaccio, C. Stanescu, and J. Bontemps, *Tetrahedron Lett.* **33:**1431 (1992).
96. D. Mitchell and T. M. Koenig, *Tetrahedron Lett.* **33:**3281 (1992).
97. M. B. Sassaman, G. K. Surya Prakash, and G. A. Olah, *J. Org. Chem.* **55:**2016 (1990).
98. J. M. Klunder, T. Onami, and K. B. Sharpless, *J. Org. Chem.* **54:**1295 (1989).
99. S. Krishnamurthy, R. M. Schubert, and H. C. Brown, *J. Am. Chem. Soc.* **95:**8486 (1973).
100. H. C. Brown, S. Ikegami, and J. H. Kawakami, *J. Org. Chem.* **35:**3243 (1970).
101. J. J. Eisch, Z.-R. Liu, and M. Singh, *J. Org. Chem.* **57:**1618 (1992).

Scheme 12.14. Nucleophilic and Solvolytic Ring Opening of Epoxides

A. Epoxidation with solvolysis of the intermediate epoxide

1[a] cyclohexene $\xrightarrow[HCO_2H]{H_2O_2}$ trans-1,2-cyclohexanediol 65–73%

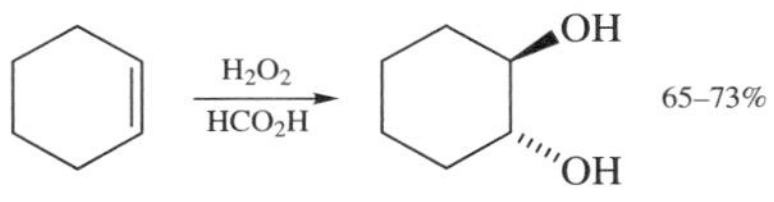

2[b] $\xrightarrow[2)\ NaOH]{1)\ HCO_2H,\ H_2O_2}$

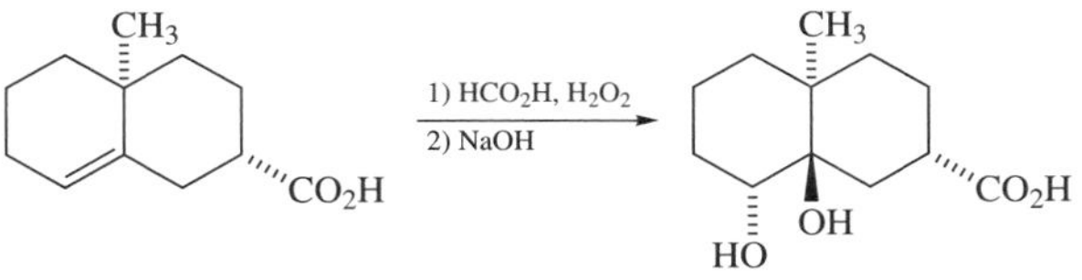

B. Acid-catalyzed solvolytic ring opening

3[c] $\xrightarrow[MeOH]{H_2SO_4}$ $(CH_3)_2C(OCH_3)-CH(OH)CH_3$ 76%

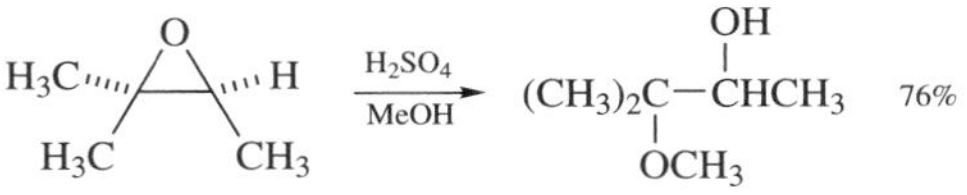

4[d] $\xrightarrow[benzene]{HCl}$ 93%

5[e] $\xrightarrow[H_2O]{HClO_4}$ 100%

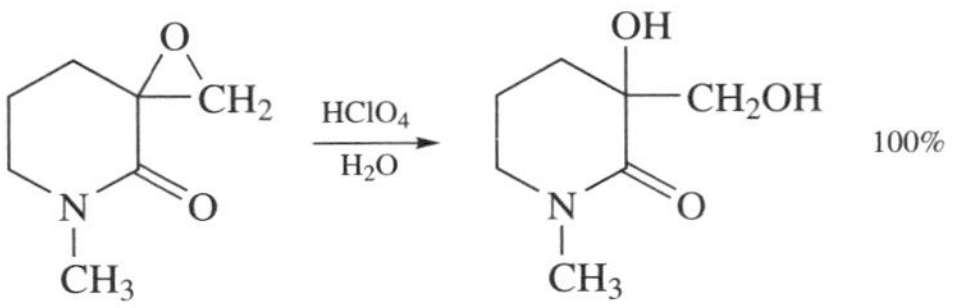

C. Nucleophilic ring-opening reactions

6[c] $+ CH_3O^- \longrightarrow (CH_3)_2C(OH)CH(OCH_3)CH_3$ 53%

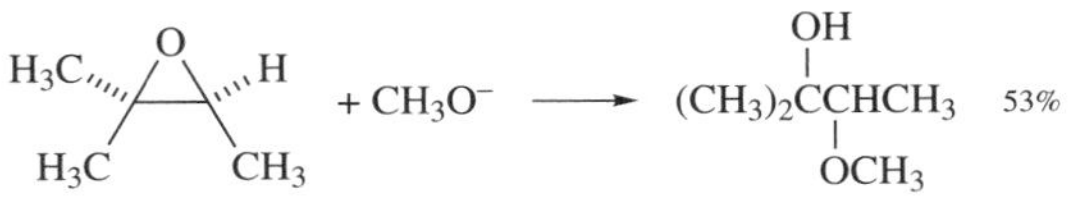

7[f] $+ HN$ (piperidine) $\longrightarrow (CH_3)_2C(OH)CH(N)CH_2CH_3$ 100%

8[g] CH_3CH_2–epoxide + HN(morpholine) $\xrightarrow[CH_3CN]{LiO_3SCF_3}$ $CH_3CH_2CH(OH)CH_2-N$(morpholine) 83%

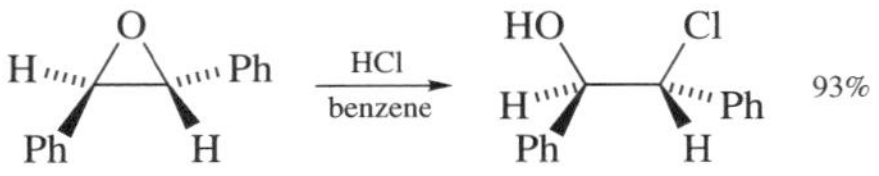

9[h] $PhOCH_2$–epoxide $+ NaN_3 \xrightarrow{Zn(O_3SCF_3)_2} PhOCH_2CH(OH)CH_2N_3$ 88%

10[i] epoxide–$CH_2N(C_2H_5)_2 + {}^-SH \longrightarrow HSCH_2CH(OH)CH_2N(C_2H_5)_2$ 63%

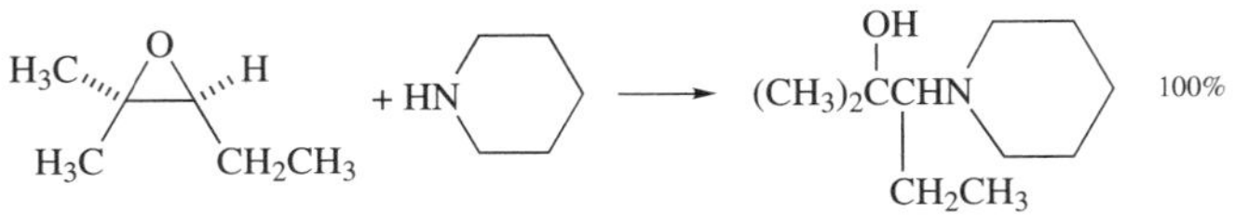

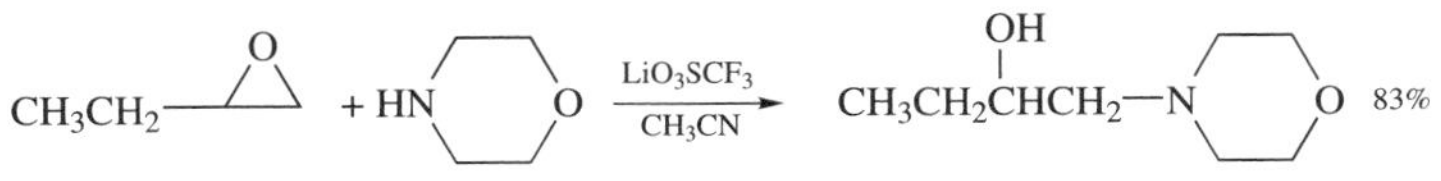

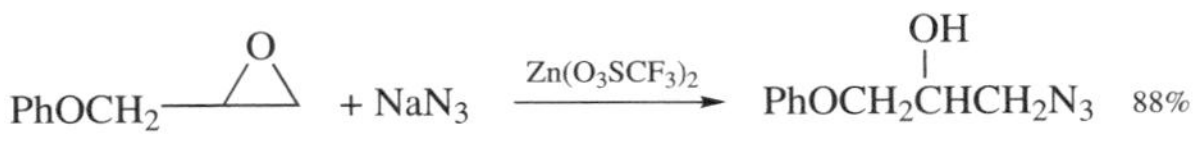

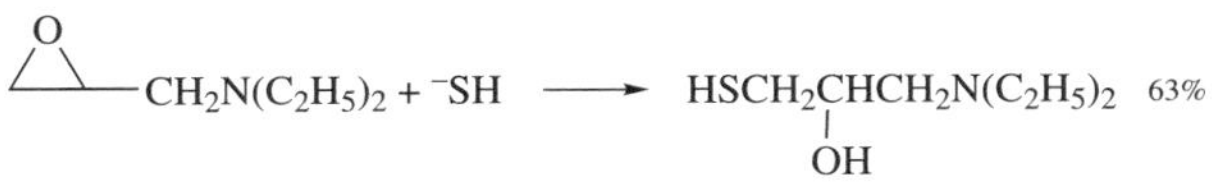

Scheme 12.14 (*continued*)

11[j]

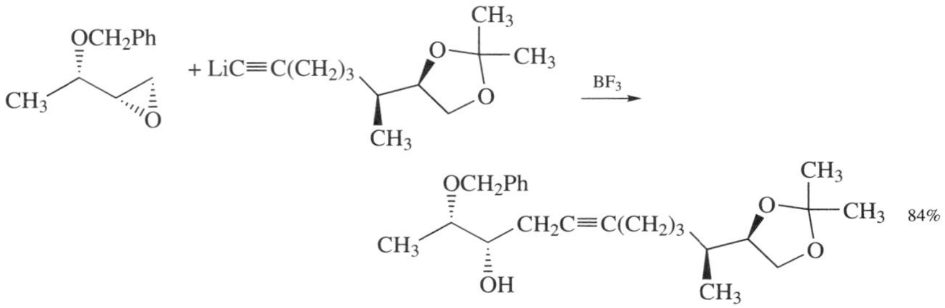

a. A. Roebuck and H. Adkins, *Org. Synth.* **III**:217 (1955).
b. T. R. Kelly, *J. Org. Chem.* **37**:3393 (1972).
c. S. Winstein and L. L. Ingrahm, *J. Am. Chem. Soc.* **74**:1160 (1952).
d. G. Berti, F. Bottari, P. L. Ferrarini, and B. Macchia, *J. Org. Chem.* **30**:4091 (1965).
e. M. L. Rueppel and H. Rapoport, *J. Am. Chem. Soc.* **94**:3877 (1972).
f. T. Colclough, J. I. Cunneen, and C. G. Moore, *Tetrahedron* **15**:187 (1961).
g. J. Auge and F. Leroy, *Tetrahedron Lett.* **37**:7715 (1996).
h. M. Chini, P. Crotti, and F. Macchia, *Tetrahedron Lett.* **31**:5641 (1990).
i. D. M. Burness and H. O. Bayer, *J. Org. Chem.* **28**:2283 (1963).
j. Z. Liu, C. Yu, R.-F. Wang, and G. Li, *Tetrahedron Lett.* **39**:5261 (1998).

Diborane in THF reduces epoxides, but the yields are low, and other products are formed by pathways that result from the electrophilic nature of diborane.[102] Better yields are obtained when BH_4^- is included in the reaction system, but the electrophilic nature of diborane is still evident because the dominant product results from addition of the hydride at the more substituted carbon:[103]

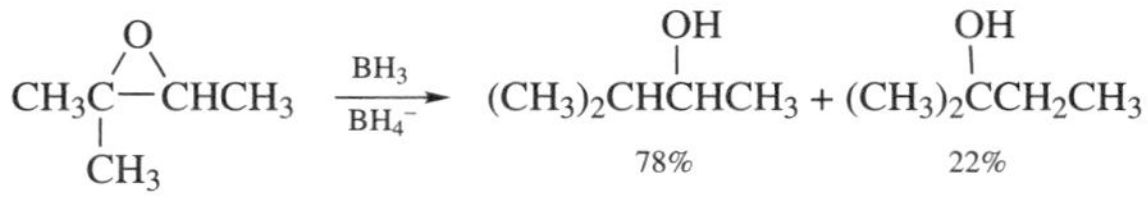

The overall transformation of alkenes to alcohols that is accomplished by epoxidation and reduction corresponds to alkene hydration. This reaction sequence is therefore an alternative to the methods discussed in Chapter 4 for converting alkenes to alcohols.

Epoxides can be isomerized to carbonyl compounds by Lewis acids.[104] Boron trifluoride is frequently used as the reagent. Carbocation intermediates appear to be involved, and the structure and stereochemistry of the product are determined by the factors which govern substituent migration in the carbocation. Clean, high-yield reactions can be expected only where structural or conformational factors promote a selective rearrangement.

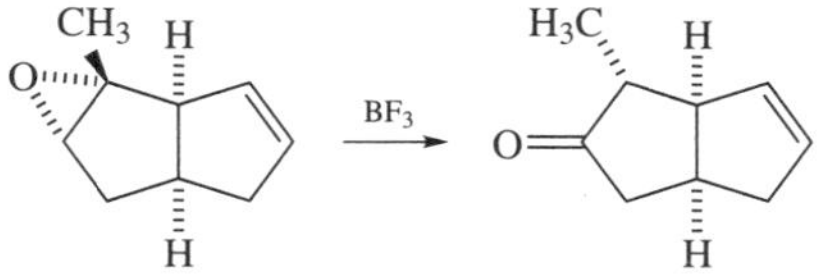

Ref. 105

102. D. J. Pasto, C. C. Cumbo, and J. Hickman, *J. Am. Chem. Soc.* **88**:2201 (1966).
103. H. C. Brown and N. M. Yoon, *J. Am. Chem. Soc.* **90**:2686 (1968).
104. J. N. Coxon, M. P. Hartshorn, and W. J. Rae, *Tetrahedron* **26**:1091 (1970).
105. J. K. Whitesell, R. S. Matthews, M. A. Minton, and A. M. Helbling, *J. Am. Chem. Soc.* **103**:3468 (1981).

Bulky diaryloxymethylaluminum reagents are also effective for this transformation.

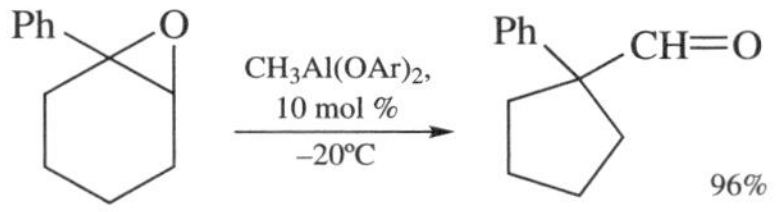

$CH_3Al(OAr)_2$, 10 mol %, –20°C; 96%; Ref. 106

Ar = 2,6-di-*t*-butyl-4-bromophenyl

Double bonds having oxygen and halogen substituents are susceptible to epoxidation, and the reactive epoxides that are generated serve as intermediates in some useful synthetic transformations. Vinyl chlorides furnish haloepoxides, which can rearrange to α- haloketones:

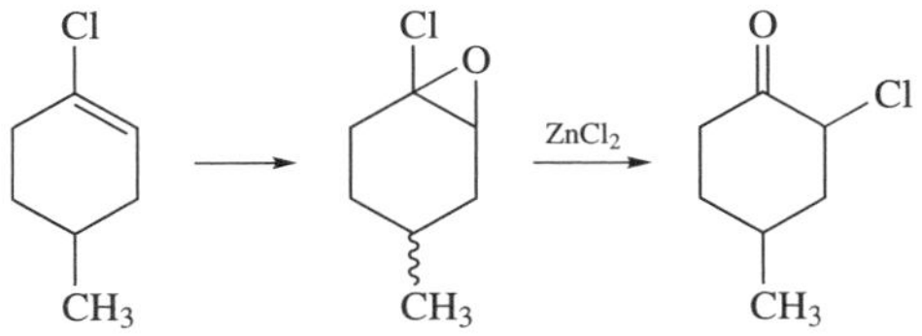

$ZnCl_2$; Ref. 107

Enol acetates form epoxides which can rearrange to α-acetoxy ketones:

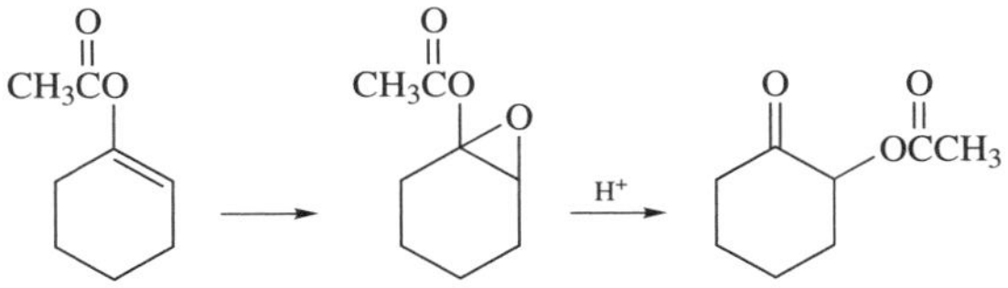

H^+; Ref. 108

The stereochemistry of the rearrangement of the acetoxy epoxides involves inversion at the carbon to which the acetoxy group migrates.[109] The reaction probably proceeds through a cyclic transition state:

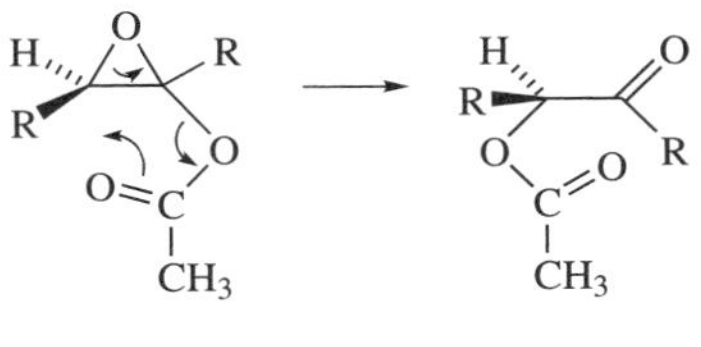

A more synthetically reliable version of this reaction involves epoxidation of trimethylsilyl enol ethers. Epoxidation of the silyl enol ethers, followed by aqueous

106. K. Maruoka, S. Nagahara, T. Ooi, and H. Yamamoto, *Tetrahedron Lett.* **30:**5607 (1989).
107. R. N. McDonald and T. E. Tabor, *J. Am. Chem. Soc.* **89:**6573 (1967).
108. K. L. Williamson, J. I. Coburn, and M. F. Herr, *J. Org. Chem.* **32:**3934 (1967).
109. K. L. Williamson and W. S. Johnson, *J. Org. Chem.* **26:**4563 (1961).

workup, gives α-hydroxyketones and α-hydroxyaldehydes.[110]

$$PhCH(CH_3)CH{=}O \longrightarrow Ph(CH_3)C{=}CH(OSi(CH_3)_3) \xrightarrow[2)\ H_2O,\ HCO_3^-]{1)\ RCO_3H} PhC(OH)(CH_3)CH{=}O \quad 85\%$$

$$CH_3C(O)C(CH_3)_3 \longrightarrow CH_2{=}C(OSi(CH_3)_3)C(CH_3)_3 \xrightarrow{RCO_3H} (CH_3)_3SiOCH_2C(O)C(CH_3)_3 \quad 73\%$$

The oxidation of silyl enol ethers with the osmium tetroxide–amine oxide combination also leads to α-hydroxyketones in generally good yields.[111]

Epoxides derived from vinylsilanes are converted under mildly acidic conditions into ketones or aldehydes.[112]

$$(CH_3)_3Si(H)C\overset{O}{-}CR_2 \xrightarrow{H^+,\ H_2O} R_2CHCH{=}O$$

The regioselective ring opening of the silyl epoxides is facilitated by the stabilizing effect that silicon has on a positive charge in the β position. This facile transformation permits vinylsilanes to serve as the equivalent of carbonyl groups in multistep synthesis.[113]

$$(CH_3)_3Si(R)C\overset{O^+H}{-}C(R)R \longrightarrow (CH_3)_3Si{-}C(R)(OH){-}\overset{+}{C}R_2 \longrightarrow RC(OH){=}CR_2 \longrightarrow RC(O)CHR_2$$

Base-catalyzed ring opening of epoxides constitutes a route to allylic alcohols.[114]

$$RCH_2CH\overset{O}{-}CH_2 \xrightarrow{B:^-} RCH{=}CHCH_2OH$$

Strongly basic reagents, such as lithium dialkylamides, are required to promote the reaction. The stereochemistry of the ring opening has been investigated by deuterium labeling. A proton *cis* to the epoxide ring is selectively removed.[115]

$$\text{4-}t\text{-butylcyclohexene oxide (D, H)} \xrightarrow{LiN(Et)_2} \text{HO-cyclohexene (H), } C(CH_3)_3$$

110. A. Hassner, R. H. Reuss, and H. W. Pinnick, *J. Org. Chem.* **40:**3427 (1975).
111. J. P. McCormick, W. Tomasik, and M. W. Johnson, *Tetrahedron Lett.* **1981:**607.
112. G. Stork and E. Colvin, *J. Am. Chem. Soc.* **93:**2080 (1971).
113. G. Stork and M. E. Jung, *J. Am. Chem. Soc.* **96:**3682 (1974).
114. J. K. Crandall and M. Apparu, *Org. React.* **29:**345 (1983).
115. R. P. Thummel and B. Rickborn, *J. Am. Chem. Soc.* **92:**2064 (1970).

A transition state represented by structure **F** can account for this stereochemistry. Such an arrangement would be favored by ion pairing that would bring the amide anion and lithium cation into close proximity. Simultaneous coordination of the lithium ion at the epoxide would result in a *syn* elimination.

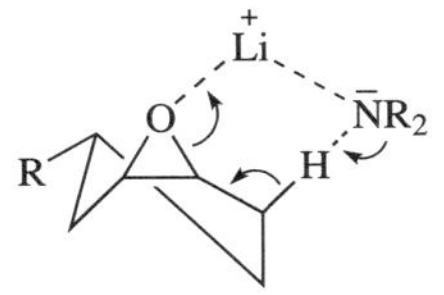

Among other reagents which effect epoxide ring opening are diethylaluminum 2,2,6,6-tetramethylpiperidide and bromomagnesium *N*-cyclohexyl-*N*-(2-propyl)amide.

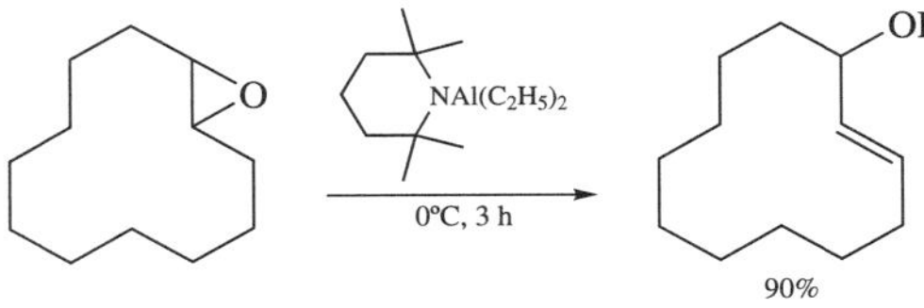

Ref. 116

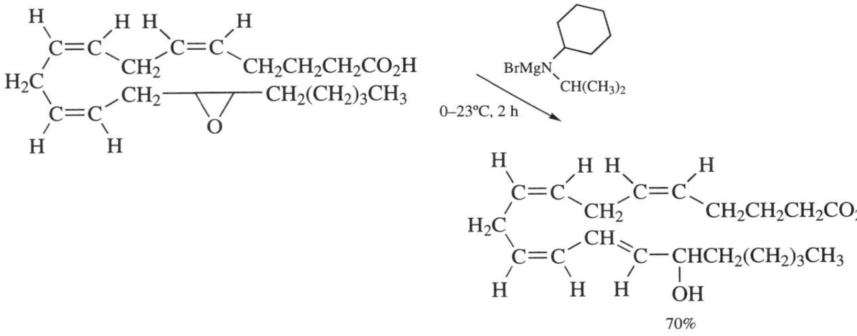

Ref. 117

These latter reagents are appropriate even for very sensitive molecules. Their efficacy is presumably due to the Lewis acid effect of the aluminum and magnesium ions. The hindered nature of the amide bases minimizes competition from nucleophilic ring opening.

Allylic alcohols can also be obtained from epoxides by ring opening with a selenide anion followed by elimination via the selenoxide (see Section 6.8.3 for discussion of selenoxide elimination). The elimination occurs regiospecifically away from the hydroxy group.[118]

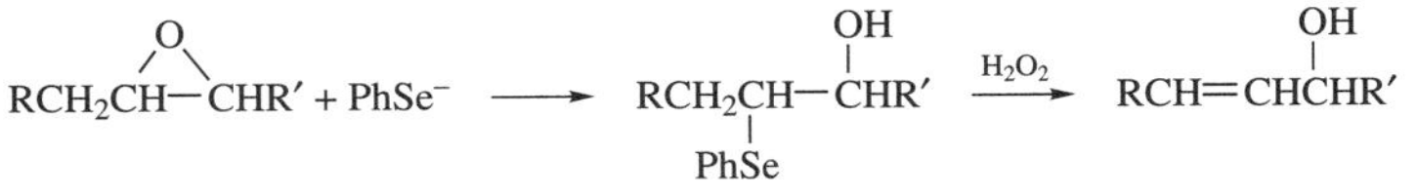

$$\mathrm{RCH_2CH\overset{O}{-}CHR'} + \mathrm{PhSe^-} \longrightarrow \mathrm{RCH_2\underset{PhSe}{CH}-\overset{OH}{CHR'}} \xrightarrow{H_2O_2} \mathrm{RCH{=}CH\overset{OH}{CHR'}}$$

116. A. Yasuda, S. Tanaka, K. Oshima, H. Yamamoto, and H. Nozaki, *J. Am. Chem. Soc.* **96:**6513 (1974).
117. E. J. Corey, A. Marfat, J. R. Falck, and J. O. Albright, *J. Am. Chem. Soc.* **102:**1433 (1980).
118. K. B. Sharpless and R. F. Lauer, *J. Am. Chem. Soc.* **95:**2697 (1973).

Epoxides can also be converted to allylic alcohols using electrophilic reagents. The treatment of epoxides with trisubstituted silyl iodides and an organic base such as DBN gives the silyl ether of the corresponding allylic alcohols.[119]

1) $(CH_3)_2SiC(CH_3)_3$ with I; DBN → $OSi(CH_3)_2C(CH_3)_3$ cycloheptenyl ether, 70–80%

Similar ring openings have been achieved using TMS triflate and 2,6-di-*t*-butylpyridine.[120]

Each of these procedures for epoxidation and ring opening is the equivalent of an allylic oxidation of a double bond with migration of the double bond:

$$R_2CHCH{=}CHR' \longrightarrow R_2C{=}CH{-}CH(OH)R'$$

In Section 12.2.4, alternative means of effecting this transformation will be described.

12.2.4. Reaction of Alkenes with Singlet Oxygen

Also among the oxidants that add oxygen at carbon–carbon double bonds is singlet oxygen.[121] For most alkenes, this reaction proceeds with the specific removal of an allylic hydrogen and shift of the double bond to provide an allylic hydroperoxide as the initial product.

O=O H → O—OH

Singlet oxygen is usually generated from oxygen by dye-sensitized photoexcitation. A number of alternative methods of generating singlet oxygen are summarized in Scheme 12.15.

Singlet oxygen decays to the ground-state triplet oxygen at a rate which is strongly dependent on the solvent.[122] Measured half-lives range from about 700 μs in carbon

119. M. R. Detty, *J. Org. Chem.* **45:**924 (1980); M. R. Detty and M. D. Seiler, *J. Org. Chem.* **46:**1283 (1981).
120. S. F. Martin and W. Li, *J. Org. Chem.* **56:**642 (1991).
121. H. H. Wasserman and R. W. Murray, eds., *Singlet Oxygen*, Academic Press, New York, 1979; A. A. Frimer, *Chem. Rev.* **79:**359 (1979); A. Frimer, ed., *Singlet Oxygen*, CRC Press, Boca Raton, Florida, 1985; C. S. Foote and E. L. Clennan, in *Active Oxygen in Chemistry*, C. S. Foote, J. S. Valentine, A. Greenberg, and J. F. Liebman, eds., Blackie Academic & Professional, London, 1995, pp. 105–140.
122. P. B. Merkel and D. R. Kearns, *J. Am. Chem. Soc.* **94:**1029, 7244 (1972); P. R. Ogilby and C. S. Foote, *J. Am. Chem. Soc.* **105:**3423 (1983); J. R. Hurst, J. D. McDonald, and G. B. Schuster, *J. Am. Chem. Soc.* **104:**2065 (1982).

Scheme 12.15. Generation of Singlet Oxygen

1[a] Photosensitizer + $h\nu$ ⟶ ^{1}[Photosensitizer]*

^{1}[Photosensitizer]* ⟶ ^{3}[Photosensitizer]*

^{3}[Photosensitizer]* + $^3O_2$ ⟶ $^1O_2$ + Photosensitizer

2[b] $H_2O_2 + {}^-OCl \longrightarrow {}^1O_2 + H_2O + Cl^-$

3[c] $(RO)_3P + O_3 \longrightarrow (RO)_3P(O_3) \longrightarrow (RO)_3P{=}O + {}^1O_2$

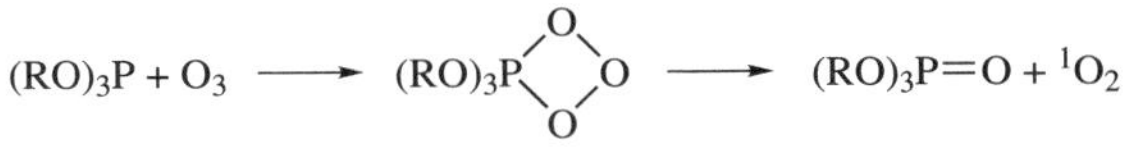

4[d] 9,10-diphenylanthracene endoperoxide ⟶ 9,10-diphenylanthracene + $^1O_2$

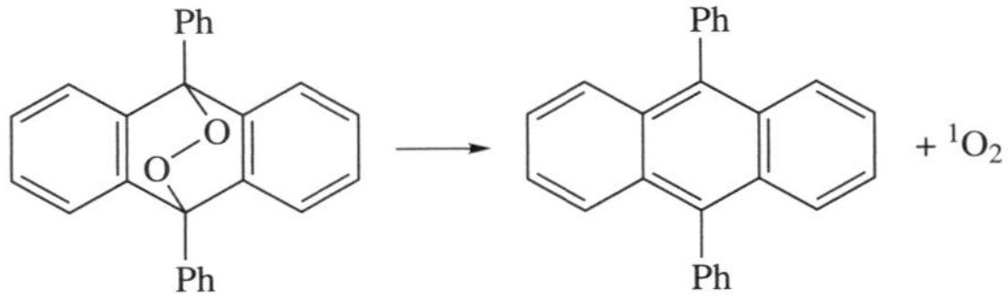

5[e] $(C_2H_5)_3SiH + O_3 \longrightarrow (C_2H_5)_3SiOOOH \longrightarrow (C_2H_5)_3SiOH + {}^1O_2$

a. C. S. Foote and S. Wexler, *J. Am. Chem. Soc.* **86:**3880 (1964).
b. C. S. Foote and S. Wexler, *J. Am. Chem. Soc.* **86:**3879 (1964).
c. R. W. Murray and M. L. Kaplan, *J. Am. Chem. Soc.* **90:**537 (1968).
d. H. H. Wasserman, J. R. Scheffler, and J. L. Cooper, *J. Am. Chem. Soc.* **94:**4991 (1972).
e. E. J. Corey, M. M. Mehotra, and A. U. Khan, *J. Am. Chem. Soc.* **108:**2472 (1986).

tetrachloride to 2 μs in water. The choice of solvent can, therefore, have a pronounced effect on the efficiency of oxidation; the longer the excited-state lifetime, the more likely it is that reaction with the alkene can occur.

The reactivity order of alkenes is that expected for attack by an electrophilic reagent. Reactivity increases with the number of alkyl substituents on the alkene.[123] Terminal alkenes are relatively inert. Steric effects govern the direction of approach of the oxygen, so the hydroperoxy group is usually introduced on the less hindered face of the double bond. The main mechanistic issue in singlet-oxygen oxidations is whether the reaction is a concerted process or involves an intermediate formulated as a "perepoxide." Most of the available evidence points to the perepoxide mechanism.[124]

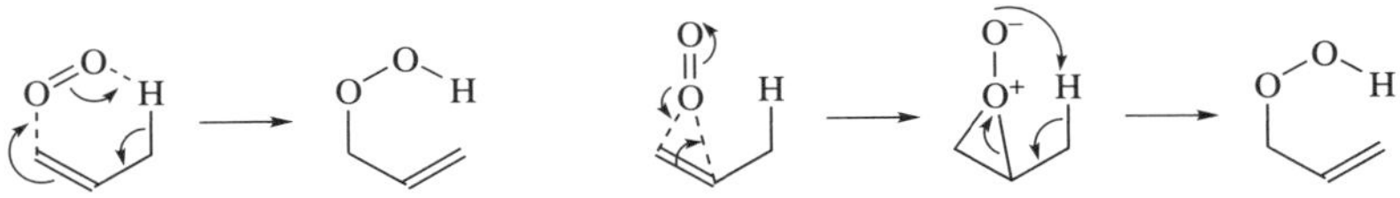

concerted mechanism

perepoxide-intermediate mechanism

Many alkenes present several different allylic hydrogens, and in this type of situation

123. K. R. Kopecky and H. J. Reich, *Can. J. Chem.* **43:**2265 (1965); C. S. Foote and R. W. Denny, *J. Am. Chem. Soc.* **93:**5162 (1971); A. Nickon and J. F. Bagli, *J. Am. Chem. Soc.* **83:**1498 (1961).
124. M. Orfanopoulos, I. Smonou, and C. S. Foote, *J. Am. Chem. Soc.* **112:**3607 (1990); M. Stratakis, M. Orfanopoulos, J. S. Chen, and C. S. Foote, *Tetrahedron Lett.* **37:**4105 (1996).

it is important to be able to predict the degree of selectivity. A useful generalization is that *there is a preference for removal of a hydrogen from the more congested side of the double bond.*[125]

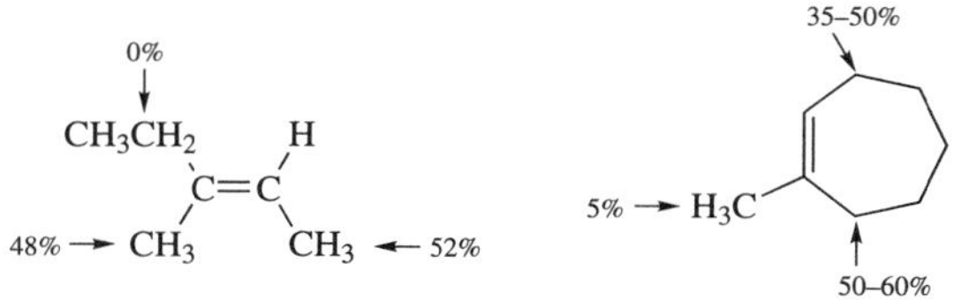

0%
CH_3CH_2 H
C=C
48% → CH_3 CH_3 ← 52%

35–50%
5% → H_3C
50–60%

Polar functional groups such as carbonyl, cyano, and sulfoxide, as well as silyl and stannyl groups, also exert a strong directing effect, favoring proton removal from a geminal methyl group.[126]

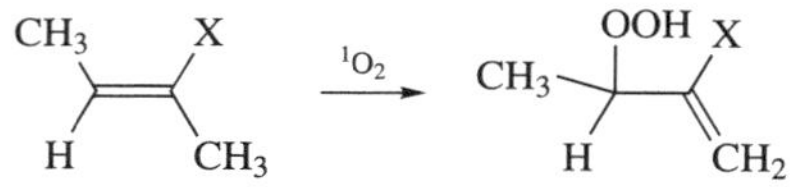

CH_3 X, H CH_3 $\xrightarrow{^1O_2}$ OOH X, CH_3, H CH_2

X = CO_2CH_3, CH=O, SOPh, $Si(CH_3)_3$, $Sn(CH_3)_3$

Hydroxyl groups favor *syn* stereoselectivity.[127] Amino groups have a similar directing effect.[128] This is similar to the substituent effects observed for peroxy acids and suggests that the substituents may stabilize the transition state by acting as electron donors.

The allylic hydroperoxides generated by singlet-oxygen oxidation are normally reduced to the corresponding allylic alcohol. The net synthetic transformation is then formation of an allylic alcohol with transposition of the double bond. Scheme 12.16 gives some examples of oxidations by singlet oxygen.

Certain compounds react with singlet oxygen in a different manner, giving dioxetanes as products.[129]

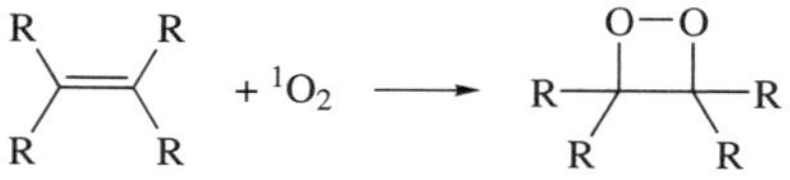

$R_2C=CR_2 + {}^1O_2 \longrightarrow$ O–O, R, R, R, R

This reaction is not usually a major factor with alkenes bearing only alkyl groups but is important for vinyl ethers and other alkenes with donor substituents. These reactions are believed to proceed via zwitterionic intermediates, which can be diverted by appropriate

125. M. Orfanopoulos, M. B. Grdina, and L. M. Stephenson, *J. Am. Chem. Soc.* **101**:275 (1979); K. H. Schulte-Elte, B. L. Muller, and V. Rautenstrauch, *Helv. Chim. Acta* **61**:2777 (1978), K. H. Schulte-Elte and V. Rautenstrauch, *J. Am. Chem. Soc.* **102**:1738 (1980).
126. E. L. Clennan, X. Chen, and J. J. Koola, *J. Am. Chem. Soc.* **112**:5193 (1990); M. Orfanopoulos, M. Stratakis, and Y. Elemes, *J. Am. Chem. Soc.* **112**:6417 (1990); W. Adam and M. J. Richter, *Tetrahedron Lett.* **34**:8423 (1993).
127. W. Adam and B. Nestler, *J. Am. Chem. Soc.* **114**:6549 (1992); W. Adam and M. Prein, *J. Am. Chem. Soc.* **115**:5041 (1993), W. Adam and B. Nestler, *J. Am. Chem. Soc.* **115**:5041 (1993); M. Stratakis, M. Orfanopoulos, and C. S. Foote, *Tetrahedron Lett.* **37**:7159 (1996).
128. H. G. Brünker and W. Adam, *J. Am. Chem. Soc.* **117**:3976 (1995).
129. W. Fenical, D. R. Kearns, and P. Radlick, *J. Am. Chem. Soc.* **91**:3396 (1969); S. Mazur and C. S. Foote, *J. Am. Chem. Soc.* **92**:3225 (1970); P. D. Bartlett and A. P. Schaap, *J. Am. Chem. Soc.* **92**:3223 (1970).

Scheme 12.16. Oxidation of Alkenes with Singlet Oxygen

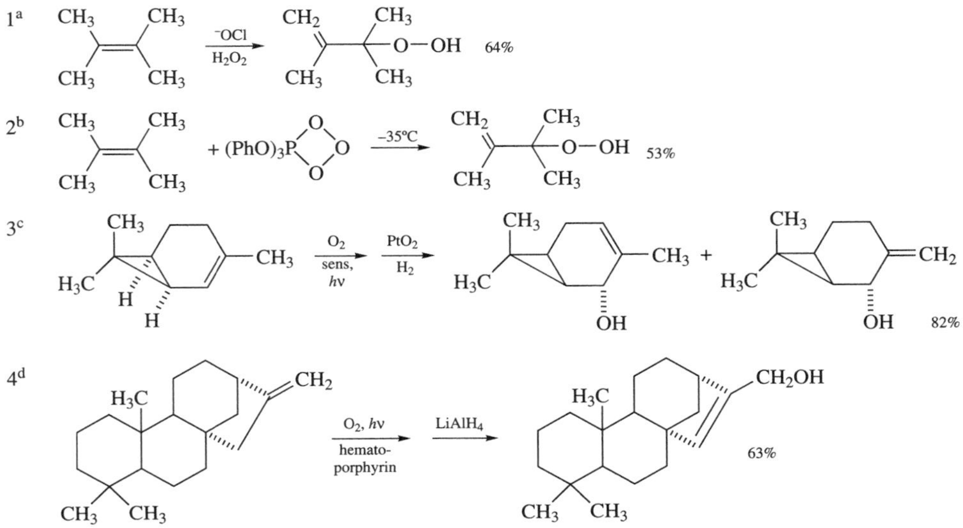

1[a] $(CH_3)_2C=C(CH_3)_2$ —(^-OCl, H_2O_2)→ $CH_2=C(CH_3)-C(CH_3)_2-O-OH$ 64%

2[b] $(CH_3)_2C=C(CH_3)_2$ + $(PhO)_3P$(ozonide) —(−35°C)→ $CH_2=C(CH_3)-C(CH_3)_2-O-OH$ 53%

3[c] —(O_2, sens, $h\nu$; PtO_2, H_2)→ 82%

4[d] —(O_2, $h\nu$, hematoporphyrin; $LiAlH_4$)→ CH_2OH 63%

a. C. S. Foote, S. Wexler, W. Ando, and R. Higgins, *J. Am. Chem. Soc.* **90:**975 (1968).
b. R. W. Murray and M. L. Kaplan, *J. Am. Chem. Soc.* **91:**5358 (1969).
c. K. Gollnick and G. Schade, *Tetrahedron Lett.* **1966:**2335.
d. R. A. Bell, R. E. Ireland, and L. N. Mander, *J. Org. Chem.* **31:**2536 (1966).

trapping reagents.[130]

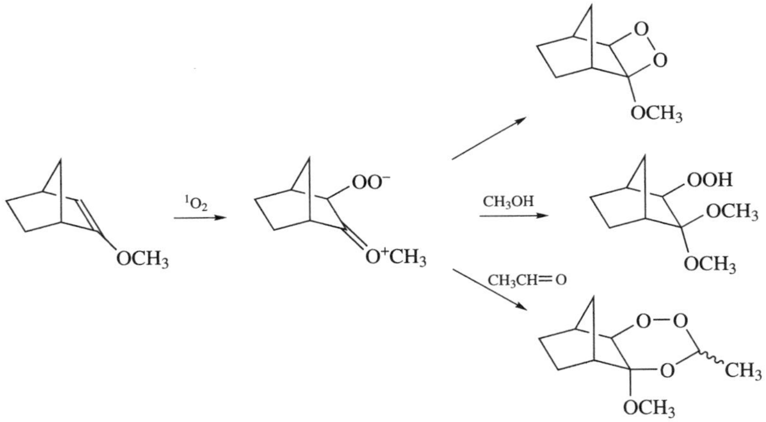

OCH_3 —($^1O_2$)→ OO^-, O^+CH_3 —(CH_3OH)→ OOH, OCH_3, OCH_3; —($CH_3CH=O$)→ O–O, O, CH_3, OCH_3; → O, O, OCH_3

Enamino ketones undergo a clean oxidative cleavage to α-diketones, presumably through a dioxetane intermediate.[131]

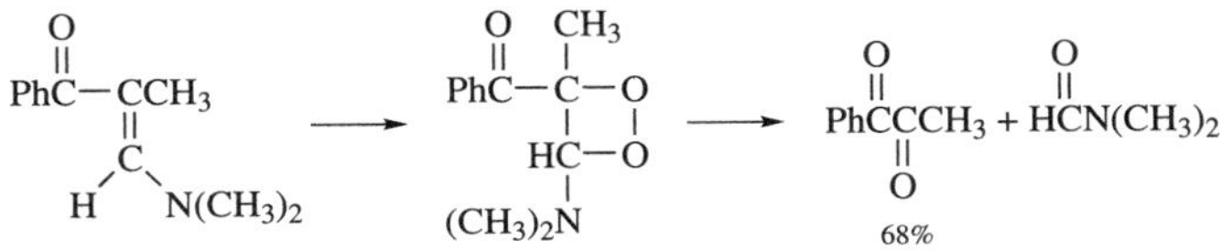

$PhC(=O)-C(CH_3)=CH-N(CH_3)_2$ → $PhC(=O)-C(CH_3)(-O-O-)CH-N(CH_3)_2$ → $PhC(=O)C(=O)CH_3 + HC(=O)N(CH_3)_2$ 68%

130. C. W. Jefford, S. Kohmoto, J. Boukouvalas, and U. Burger, *J. Am. Chem. Soc.* **105:**6498 (1983).
131. H. H. Wasserman and J. L. Ives, *J. Am. Chem. Soc.* **98:**7868 (1976).

Singlet oxygen undergoes [4 + 2] cycloaddition with dienes.

$$+ {}^1O_2 \longrightarrow$$ Ref. 132

$$+ {}^1O_2 \longrightarrow$$ Ref. 133

12.3. Cleavage of Carbon–Carbon Double Bonds

12.3.1. Transition-Metal Oxidants

The most selective methods for cleaving organic molecules at carbon–carbon double bonds involve glycols as intermediates. Oxidations of alkenes to glycols were discussed in Section 12.2.1. Cleavage of alkenes can be carried out in one operation under mild conditions by using a solution containing periodate ion and a catalytic amount of permanganate ion.[134] The permanganate ion effects the hydroxylation, and the glycol is then cleaved by reaction with periodate. A cyclic intermediate is believed to be involved in the periodate oxidation. Permanganate is regenerated by the oxidizing action of periodate.

$$RCH{=}CHR + KMnO_4 \longrightarrow RCH(OH){-}CH(OH)R \xrightarrow{IO_4^-} [\text{cyclic periodate ester}] \longrightarrow 2\,RCH{=}O + H_2O + IO_3^-$$

Osmium tetroxide used in combination with sodium periodate can also effect alkene cleavage.[135] Successful oxidative cleavage of double bonds using ruthenium tetroxide and sodium periodate has also been reported.[136] In these procedures, the osmium or ruthenium can be used in substoichiometric amounts because the periodate reoxidizes the metal to the tetroxide state. Entries 1–4 in Scheme 12.17 are examples of these procedures.

The strong oxidants Cr(VI) and MnO_4^- can also be used for oxidative cleavage of double bonds, provided there are no other sensitive groups in the molecule. The permanganate oxidation proceeds first to the diols and ketols, as described earlier (p. 757), and these are then oxidized to carboxylic acids or ketones. Good yields can be obtained provided care is taken to prevent subsequent oxidative degradation of the products. Entries 5 and 6 in Scheme 12.17 are illustrative.

132. C. S. Foote, S. Wexler, W. Ando, and R. Higgins, *J. Am. Chem. Soc.* **90:**975 (1968).
133. C. H. Foster and G. A. Berchtold, *J. Am. Chem. Soc.* **94:**7939 (1972).
134. R. U. Lemieux and E. von Rudloff, *Can. J. Chem.* **33:**1701, 1710 (1955); E. von Rudloff, *Can. J. Chem.* **33:**1714 (1955).
135. R. Pappo, D. S. Allen, Jr., R. U. Lemieux, and W. S. Johnson, *J. Org. Chem.* **21:**478 (1956); H. Vorbrueggen and C. Djerassi, *J. Am. Chem. Soc.* **84:**2990 (1962).
136. W. G. Dauben and L. E. Friedrich, *J. Org. Chem.* **37:**241 (1972); B. E. Rossiter, T. Katsuki, and K. B. Sharpless, *J. Am. Chem. Soc.* **103:**464 (1981); J. W. Patterson, Jr., and D. V. Krishna Murthy, *J. Org. Chem.* **48:**4413 (1983).

Scheme 12.17. Oxidative Cleavage of Carbon–Carbon Double Bonds with Transition-Metal Oxidants

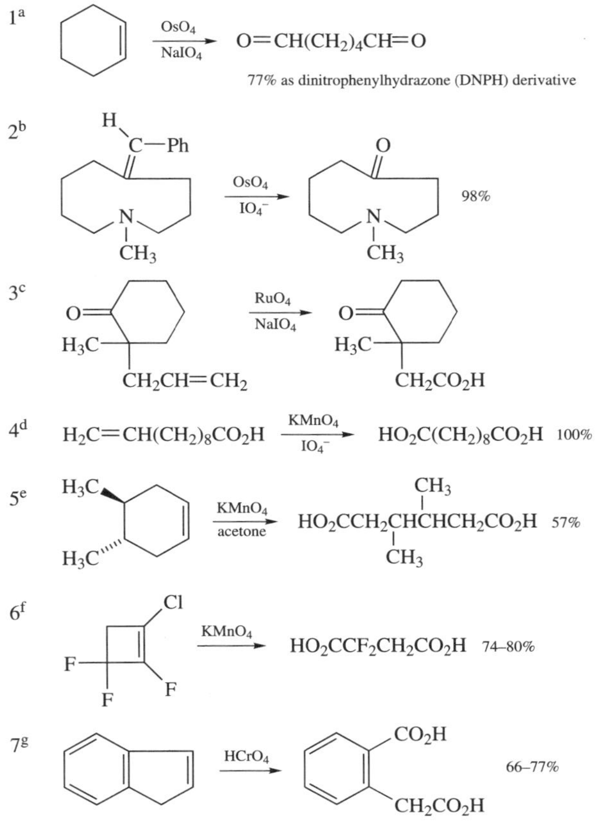

1[a] cyclohexene $\xrightarrow[NaIO_4]{OsO_4}$ O=CH(CH$_2$)$_4$CH=O
77% as dinitrophenylhydrazone (DNPH) derivative

2[b] $\xrightarrow[IO_4^-]{OsO_4}$ 98%
H, C—Ph, N, CH$_3$ / O, N, CH$_3$

3[c] $\xrightarrow[NaIO_4]{RuO_4}$
O=, H$_3$C—, CH$_2$CH=CH$_2$ / O=, H$_3$C—, CH$_2$CO$_2$H

4[d] H$_2$C=CH(CH$_2$)$_8$CO$_2$H $\xrightarrow[IO_4^-]{KMnO_4}$ HO$_2$C(CH$_2$)$_8$CO$_2$H 100%

5[e] H$_3$C, H$_3$C $\xrightarrow[acetone]{KMnO_4}$ HO$_2$CCH$_2$CH(CH$_3$)CH(CH$_3$)CH$_2$CO$_2$H 57%

6[f] Cl, F, F, F $\xrightarrow{KMnO_4}$ HO$_2$CCF$_2$CH$_2$CO$_2$H 74–80%

7[g] indene $\xrightarrow{HCrO_4}$ CO$_2$H, CH$_2$CO$_2$H 66–77%

a. R. U. Lemieux and E. von Rudloff, *Can. J. Chem.* **33**:1701 (1955).
b. M. G. Reinecke, L. R. Kray, and R. F. Francis, *J. Org. Chem.* **37**:3489 (1972).
c. A. A. Asselin, L. G. Humber, T. A. Dobson, J. Komlossy, and R. R. Martel, *J. Med. Chem.* **19**:787 (1976).
d. R. Pappo, D. S. Allen, Jr., R. U. Lemieux, and W. S. Johnson, *J. Org. Chem.* **21**:478 (1956).
e. W. C. M. C. Kokke and F. A. Varkvisser, *J. Org. Chem.* **39**:1535 (1974).
f. N. S. Raasch and J. E. Castle, *Org. Synth.* **42**:44 (1962).
g. O. Grummitt, R. Egan, and A. Buck, *Org. Synth.* **III**:449 (1955).

The oxidation of cyclic alkenes by Cr(VI) reagents can be a useful method for formation of dicarboxylic acids. The initial oxidation step appears to yield an epoxide, which then undergoes solvolytic ring opening to a glycol or glycol monoester, which is then oxidatively cleaved.[137] Two possible complications that can be encountered are competing allylic attack and skeletal rearrangement. Allylic attack can lead to eventual formation of a dicarboxylic acid that has lost one carbon atom. Pinacol-type rearrange-

137. J. Rocek and J. C. Drozd, *J. Am. Chem. Soc.* **92**:6668 (1970); A. K. Awasthy and J. Rocek, *J. Am. Chem. Soc.* **91**:991 (1969).

ments of the epoxide or glycol intermediates can give rise to rearranged products.

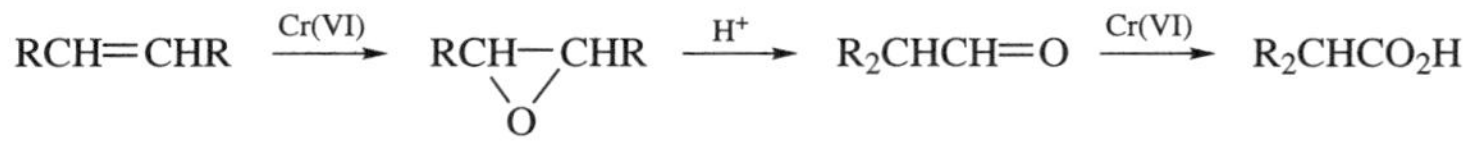

$$RCH{=}CHR \xrightarrow{Cr(VI)} RCH{-}CHR\ (\text{epoxide, O}) \xrightarrow{H^+} R_2CHCH{=}O \xrightarrow{Cr(VI)} R_2CHCO_2H$$

12.3.2. Ozonolysis

The reaction of alkenes with ozone constitutes an important method of cleaving carbon–carbon double bonds.[138] Application of low-temperature spectroscopic techniques has provided information about the rather unstable species that are intermediates in the ozonolysis process. These studies, along with isotope labeling results, have provided an understanding of the reaction mechanism.[139] The two key intermediates in ozonolysis are the 1,2,3-trioxolane, or initial ozonide, and the 1,2,4-trioxolane, or ozonide. The first step of the reaction is a cycloaddition to give the 1,2,3-trioxolane. This is followed by a fragmentation and recombination to give the isomeric 1,2,4-trioxolane. The first step is a 1,3-dipolar cycloaddition reaction. Ozone is expected to be a very electrophilic 1,3-dipole because of the accumulation of electronegative oxygen atoms in the ozone molecule. The cycloaddition, fragmentation, and recombination are all predicted to be exothermic on the basis of thermochemical considerations.[140]

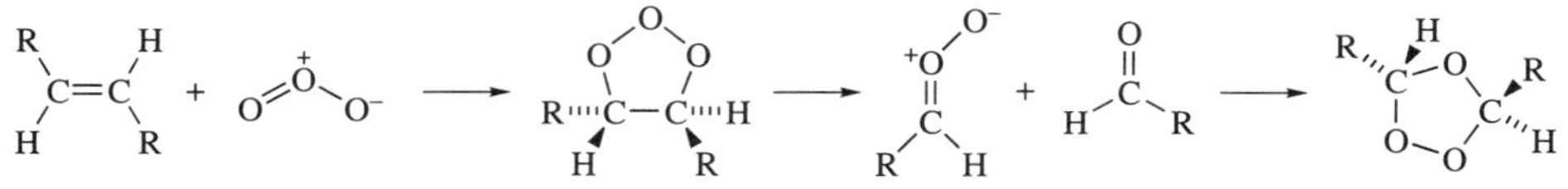

The actual products isolated after ozonolysis depend upon the conditions of workup. Simple hydrolysis leads to the carbonyl compounds and hydrogen peroxide, and these can react to give secondary oxidation products. It is usually preferable to include a mild reducing agent that is capable of reducing peroxidic bonds. The current practice is to use dimethyl sulfide, though numerous other reducing agents have been used, including zinc,[141] trivalent phosphorus compounds,[142] and sodium sulfite.[143] If the alcohols resulting from the reduction of the carbonyl cleavage products are desired, the reaction mixture can be reduced with $NaBH_4$.[144] Carboxylic acids are formed in good yields from aldehydes when the ozonolysis reaction mixture is worked up in the presence of excess hydrogen peroxide.[145]

138. P. S. Bailey, *Ozonization in Organic Chemistry*, Vol. 1, Academic Press, New York, 1978.
139. R. P. Lattimer, R. L. Kuckowski, and C. W. Gillies, *J. Am. Chem. Soc.* **96**:348 (1974); C. W. Gillies, R. P. Lattimer, and R. L. Kuczkowski, *J. Am. Chem. Soc.* **96**:1536 (1974); G. Klopman and C. M. Joiner, *J. Am. Chem. Soc.* **97**:5287 (1975), P. S. Bailey and T. M. Ferrell, *J. Am. Chem. Soc.* **100**:899 (1978), I. C. Histasune, K. Shinoda, and J. Heicklen, *J. Am. Chem. Soc.* **101**:2524 (1979); J.-I. Choe, M. Srinivasan, and R. L. Kuczkowski, *J. Am. Chem. Soc.* **105**:4703 (1983); R. L. Kuchzkowski, in *1,3-Dipolar Cycloaddition Chemistry*, A. Padwa, ed., Wiley-Interscience, New York, Vol. 2, Chapter 11, 1984; C. Geletneky and S. Berger, *Eur. J. Chem.* **1998**:1625.
140. P. S. Nangia and S. W. Benson, *J. Am. Chem. Soc.* **102**:3105 (1980).
141. S. M. Church, F. C. Whitmore, and R. V. McGrew, *J. Am. Chem. Soc.* **56**:176 (1934).
142. W. S. Knowles and Q. E. Thompson, *J. Org. Chem.* **25**:1031 (1960).
143. R. H. Callighan and M. H. Wilt, *J. Org. Chem.* **26**:4912 (1961).
144. F. L. Greenwood, *J. Org. Chem.* **20**:803 (1955).
145. A. L. Henne and P. Hill, *J. Am. Chem. Soc.* **65**:752 (1943).

When ozonolysis is done in alcoholic solvents, the carbonyl oxide fragmentation product can be trapped as an α-hydroperoxy ether.[146] Recombination to the ozonide is then prevented, and the carbonyl compound formed in the fragmentation step can also be isolated. If the reaction mixture is treated with dimethyl sulfide, the hydroperoxide is reduced and the second carbonyl compound is also formed in good yield.[147] This procedure prevents oxidation of the aldehyde by the peroxidic compounds present at the conclusion of ozonolysis.

$$R_2C{=}\overset{+}{O}{-}O^- + CH_3OH \longrightarrow R_2C(OCH_3)OOH$$

$$PhCH{=}CH_2 \xrightarrow[CH_3OH]{O_3} \underset{31\%}{PhCH(OCH_3)OOH} + \underset{23\%}{CH_2(OCH_3)OOH} + \underset{26\%}{PhCH{=}O} + \underset{27\%}{CH_2{=}O}$$

Especially reactive carbonyl compounds such as methyl pyruvate can be used to trap the carbonyl ylide component. For example, ozonolysis of cyclooctene in the presence of methyl pyruvate leads to **G**, which, when treated with triethylamine, is converted to **H**, in which the two carbons of the original double bond have been converted to different functionalities.[148]

$$\text{cyclooctene} \xrightarrow[CH_2Cl_2]{O_3,\ CH_3COCO_2CH_3} \underset{\mathbf{G}}{\text{ozonide of } CH_3O_2C(H_3C)C \text{ and } H{-}C{-}(CH_2)_6CH{=}O} \xrightarrow{(C_2H_5)_3N} \underset{\mathbf{H}}{HO_2C(CH_2)_6CH{=}O}$$

Ozonolysis in the presence of NaOH or $NaOCH_3$ in methanol with CH_2Cl_2 as a co-solvent leads to formation of esters. This transformation proceeds by trapping both the carbonyl oxide and aldehyde products of the fragmentation step.[149]

$$RCH{=}CHR' \xrightarrow{O_3} RCH{=}\overset{+}{O}{-}O^- \xrightarrow{CH_3O^-} RCH(OCH_3)OO^- \longrightarrow RCO_2CH_3$$

$$+\quad R'CH{=}O \xrightarrow{CH_3O^-} RCH(OCH_3)O^- \longrightarrow RCO_2CH_3$$

Cyclooctene gives dimethyl octanedioate under these conditions. Scheme 12.18 illustrates some cases in which ozonolysis reactions have been used in the course of synthesis.

146. W. P. Keaveney, M. G. Berger, and J. J. Pappas, *J. Org. Chem.* **32:**1537 (1967).
147. J. J. Pappas, W. P. Keaveney, E. Gancher, and M. Berger, *Tetrahedron Lett.* **1966:**4273.
148. Y.-S. Hon and J.-L. Yan, *Tetrahedron* **53:**5217 (1997).
149. J. A. Marshall and A. W. Garofalo, *J. Org. Chem.* **58:**3675 (1993).

Scheme 12.18. Ozonolysis Reactions

A. Reductive workup

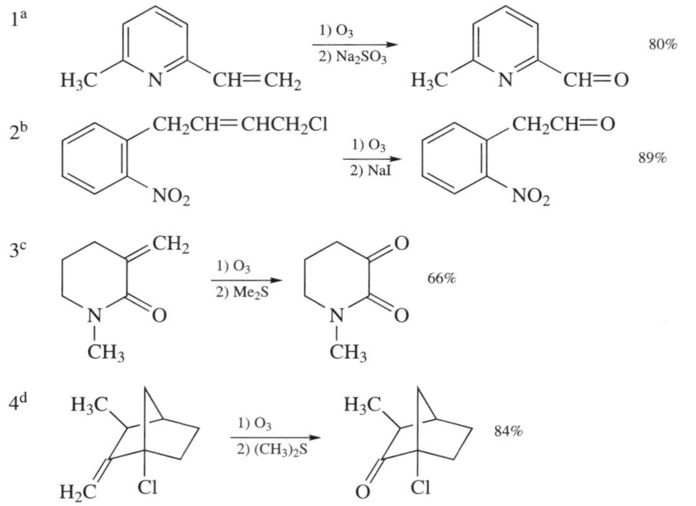

1[a] 2-methyl-6-vinylpyridine (H_3C, N, $CH{=}CH_2$) → 1) O_3; 2) Na_2SO_3 → 6-methylpyridine-2-carbaldehyde (H_3C, N, $CH{=}O$) 80%

2[b] o-nitro-$C_6H_4CH_2CH{=}CHCH_2Cl$ (NO_2) → 1) O_3; 2) NaI → o-nitro-$C_6H_4CH_2CH{=}O$ (NO_2) 89%

3[c] 3-methylene-1-methyl-2-piperidinone (CH_2, N, O, CH_3) → 1) O_3; 2) Me_2S → 1-methyl-2,3-piperidinedione (O, N, O, CH_3) 66%

4[d] (H_3C, H_2C, Cl) → 1) O_3; 2) $(CH_3)_2S$ → (H_3C, O, Cl) 84%

B. Oxidative workup

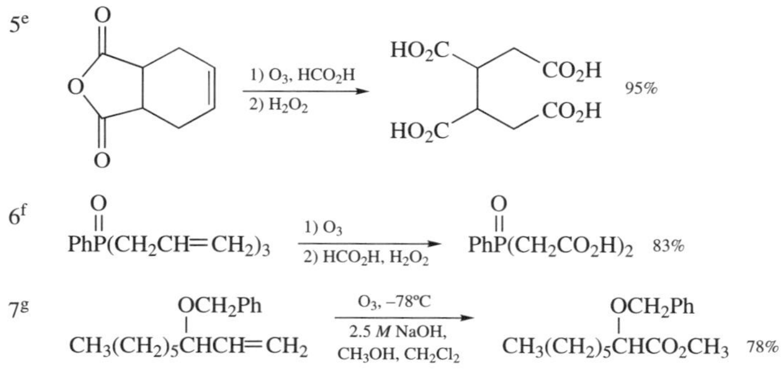

5[e] (O, O, O) → 1) O_3, HCO_2H; 2) H_2O_2 → (HO_2C, CO_2H, CO_2H, HO_2C) 95%

6[f] $PhP(O)(CH_2CH{=}CH_2)_3$ → 1) O_3; 2) HCO_2H, H_2O_2 → $PhP(O)(CH_2CO_2H)_2$ 83%

7[g] $CH_3(CH_2)_5CH(OCH_2Ph)CH{=}CH_2$ → O_3, –78°C; 2.5 *M* NaOH, CH_3OH, CH_2Cl_2 → $CH_3(CH_2)_5CH(OCH_2Ph)CO_2CH_3$ 78%

a. R. H. Calligham and M. H. Wilt, *J. Org. Chem.* **26:**4912 (1961).
b. W. E. Noland and J. H. Sellstedt, *J. Org. Chem.* **31:**345 (1966).
c. M. L. Rueppel and H. Rapoport, *J. Am. Chem. Soc.* **94:**3877 (1972).
d. J. V. Paukstelis and B. W. Macharia, *J. Org. Chem.* **38:**646 (1973).
e. J. E. Franz, W. S. Knowles, and C. Osuch, *J. Org. Chem.* **30:**4328 (1965).
f. J. L. Eichelberger and J. K. Stille, *J. Org. Chem.* **36:**1840 (1971).
g. J. A. Marshall and A. W. Garofalo, *J. Org. Chem.* **58:**3675 (1993).

12.4. Selective Oxidative Cleavages at Other Functional Groups

12.4.1. Cleavage of Glycols

As discussed in connection with cleavage of double bonds by permanganate–periodate or osmium tetroxide–periodate (see p. 757), the glycol unit is susceptible to oxidative cleavage under mild conditions. The most commonly used reagent for this oxidative cleavage is the periodate ion.[150] The fragmentation is believed to occur via a

150. C. A. Bunton, in *Oxidation in Organic Chemistry, Part A*, K. B. Wiberg, ed., Academic Press, New York, 1965, pp. 367–388; A. S. Perlin, in *Oxidation*, Vol. 1, R. L. Augustine, ed., Marcel Dekker, New York, 1969, pp. 189–204.

cyclic adduct of the glycol and the oxidant.

$$\mathrm{R{-}CH(OH){-}CH(OH){-}R} \xrightarrow{IO_4^-} \text{[cyclic periodate ester: R–CH–CH–R, O–I(=O)(O}^-\text{)(OH)(OH)–O]} \longrightarrow 2\,\mathrm{RCH{=}O} + H_2O + IO_3^-$$

Structural features that retard formation of the cyclic intermediate decrease the reaction rate. For example, *cis*-1,2-dihydroxycyclohexane is substantially more reactive than the *trans* isomer.[151] Glycols in which the geometry of the molecule precludes the possibility of a cyclic intermediate are essentially inert to periodate.

Certain other combinations of adjacent functional groups are also cleaved by periodate. Diketones are cleaved to carboxylic acids, and it has been proposed that a reactive cyclic intermediate is formed by nucleophilic attack on the diketone.[152]

$$H_3C{-}CO{-}CO{-}CH_3 + IO_4^- \xrightarrow{^-OH} H_3C{-}C(OH)(OIO_4^{2-}){-}C(=O)CH_3 \xrightarrow{H_2O} \text{[cyclic: } H_3C{-}C(OH){-}O{-}IO_4H_2^-{-}O{-}C(OH){-}CH_3] \longrightarrow 2\,CH_3CO_2H + IO_3^- + H_2O$$

α-Hydroxyketones and α-aminoalcohols are also subject to oxidative cleavage, presumably by a similar mechanism.

Lead tetraacetate is an alternative reagent to periodate for glycol cleavage. It is particularly useful for glycols that have low solubility in the aqueous media used for periodate reactions. A cyclic intermediate is suggested by the same kind of stereochemistry–reactivity relationships discussed for periodate.[153] With lead tetraacetate, unlike periodate, however, glycols that cannot form cyclic intermediates are eventually oxidized. For example, *trans*-9,10-dihydroxydecalin is oxidized, although the rate is 100 times less than for the *cis* isomer.[154] Thus, while a cyclic transition state appears to provide the lowest-energy pathway for this oxidative cleavage, it is not the only possible mechanism. Both the periodate cleavage and lead tetraacetate oxidation can be applied synthetically to the generation of medium-sized rings when the glycol is at the junction of two rings.

$$\text{[bicyclic diol (OH, OH; ring O)]} \xrightarrow{Pb(OAc)_4} \text{[ten-membered ring diketone (O, O; ring O)]}$$

Ref. 155

151. C. C. Price and M. Knell, *J. Am. Chem. Soc.* **64:**552 (1942).
152. C. A. Bunton and V. J. Shiner, *J. Chem. Soc.* **1960:**1593.
153. C. A. Bunton, in *Oxidation in Organic Chemistry*, K. Wiberg, ed., Academic Press, New York, 1965, pp. 398–405; W. S. Trahanovsky, J. R. Gilmore, and P. C. Heaton, *J. Org. Chem.* **38:**760 (1973).
154. R. Criegee, E. Höger, G. Huber, P. Kruck, F. Marktscheffel, and H. Schellenberger, *Justus Liebigs Ann. Chem.* **599:**81 (1956).
155. T. Wakamatsu, K. Akasaka, and Y. Ban, *Tetrahedron Lett.* **1977:**2751, 2755.

12.4.2. Oxidative Decarboxylation

Carboxylic acids are oxidized by lead tetraacetate. Decarboxylation occurs, and the product may be an alkene, alkane, or acetate ester or, under modified conditions, a halide. A free-radical mechanism operates, and the product composition depends on the fate of the radical intermediate.[156] The reaction is catalyzed by cupric salts, which function by oxidizing the intermediate radical to a carbocation (step 3 in the mechanism). Cu(II) is more reactive than $Pb(OAc)_4$ in this step.

$$Pb(OAc)_4 + RCO_2H \rightleftharpoons RCO_2Pb(OAc)_3 + CH_3CO_2H$$
$$RCO_2Pb(OAc)_3 \longrightarrow R\cdot + CO_2 + Pb(OAc)_3$$
$$R\cdot + Pb(OAc)_4 \longrightarrow R^+ + Pb(OAc)_3 + CH_3CO_2^-$$
and
$$R\cdot + Pb(OAc)_3 \longrightarrow R^+ + Pb(OAc)_2 + CH_3CO_2^-$$

Alkanes are formed when the intermediate radical abstracts hydrogen from solvent faster than it is oxidized to the carbocation. This reductive step is promoted by good hydrogen-donor solvents. It is also more prevalent for primary alkyl radicals because of the higher activation energy associated with formation of primary carbocations. The most favorable conditions for alkane formation involve photochemical decomposition of the carboxylic acid in chloroform, which is a relatively good hydrogen atom donor.

CO_2H (cyclobutane) $\xrightarrow[h\nu]{CHCl_3,\ Pb(OAc)_4}$ (cyclobutane) 65% Ref. 157

Normally, the dominant products are the alkene and acetate ester, which arise from the carbocation intermediate by, respectively, elimination of a proton and capture of an acetate ion. The presence of potassium acetate increases the alkene : ester ratio.[158]

CO_2H, CO_2H $\xrightarrow[KOAc]{Pb(OAc)_2,\ Cu(OAc)_2}$ O_2CCH_3, CH_3CO_2 93% Ref. 159

H_3C, H_3C, CH_3, CO_2H, O $\xrightarrow[Cu(OAc)_2]{Pb(OAc)_4}$ H_3C, H_3C, CH_3, O + H_3C, H_3C, CH_3, O Ref. 160

156. R. A. Sheldon and J. K. Kochi, *Org. React.* **19**:279 (1972).
157. J. K. Kochi and J. D. Bacha, *J. Org. Chem.* **33**:2746 (1968).
158. J. D. Bacha and J. K. Kochi, *Tetrahedron* **24**:2215 (1968).
159. P. Caluwe and T. Pepper, *J. Org. Chem.* **53**:1786 (1988).
160. D. D. Sternbach, J. W. Hughes, D. E. Bardi, and B. A. Banks, *J. Am. Chem. Soc.* **107**:2149 (1985).

In the presence of lithium chloride, the product is the corresponding chloride.[161]

$$\underset{\text{CH}_3\text{CHCH}_2\text{CO}_2\text{CH}_3}{\overset{\text{Cl}_2\text{CCO}_2\text{H}}{|}} \xrightarrow[\text{LiCl}]{\text{Pb(OAc)}_4} \underset{\text{CH}_3\text{CHCH}_2\text{CO}_2\text{CH}_3}{\overset{\text{Cl}_3\text{C}}{|}} \quad 77\%$$

Ref. 162

A related method for conversion of carboxylic acids to bromides with decarboxylation is the *Hunsdiecker reaction*.[163] The most convenient method for carrying out this transformation involves heating the carboxylic acid with mercuric oxide and bromine.

$$\text{c-C}_3\text{H}_5\text{–CO}_2\text{H} \xrightarrow[\text{Br}_2]{\text{HgO}} \text{c-C}_3\text{H}_5\text{–Br} \quad 41\text{–}46\%$$

Ref. 164

The overall transformation can also be accomplished by reaction of thallium(I) carboxylate with bromine.[165]

1,2-Dicarboxylic acids undergo bis-decarboxylation on reaction with lead tetraacetate to give alkenes. This reaction has been of occasional use for the synthesis of strained alkenes.

Pb(OAc)$_4$, pyridine, 80°C → 39%

Ref. 166

The reaction can occur by a concerted fragmentation process initiated by a two-electron oxidation.

$$\text{H–O–}\overset{\text{O}}{\overset{\|}{\text{C}}}\text{–}\underset{\text{R}}{\overset{\text{R}}{\text{C}}}\text{–}\underset{\text{R}}{\overset{\text{R}}{\text{C}}}\text{–}\overset{\text{O}}{\overset{\|}{\text{C}}}\text{–O–Pb(OAc)}_3 \longrightarrow 2\,\text{CO}_2 + \text{R}_2\text{C=CR}_2 + \text{Pb(OAc)}_2 + \text{CH}_3\text{CO}_2\text{H}$$

161. J. K. Kochi, *J. Org. Chem.* **30:**3265 (1965).
162. S. E. de Laszlo and P. G. Williard, *J. Am. Chem. Soc.* **107:**199 (1985).
163. C. V. Wilson, *Org. React.* **9:**332 (1957); R. A. Sheldon and J. Kochi, *Org. React.* **19:**279 (1972).
164. J. S. Meek and D. T. Osuga, *Org. Synth.* **V:**126 (1973).
165. A. McKillop, D. Bromley, and E. C. Taylor, *J. Org. Chem.* **34:**1172 (1969).
166. E. Grovenstein, Jr., D. V. Rao, and J. W. Taylor, *J. Am. Chem. Soc.* **83:**1705 (1961).

A concerted mechanism is also possible for α-hydroxycarboxylic acids, and these compounds readily undergo oxidative decarboxylation to ketones.[167]

$$R_2C(OH)C(=O)O{-}Pb(OAc)_3 \longrightarrow R_2C{=}O + CO_2 + Pb(OAc)_2 + CH_3CO_2H$$

γ-Keto carboxylic acids are oxidatively decarboxylated to enones.[168] This reaction is presumed to proceed through the usual oxidative decarboxylation, with the carbocation intermediate being efficiently deprotonated because of the developing conjugation.

HO$_2$C, CH$_3$ $\xrightarrow[Cu(OAc)_2]{Pb(OAc)_4}$ CH$_3$ 78% Ref. 168

12.5. Oxidation of Ketones and Aldehydes

12.5.1. Transition-Metal Oxidants

Ketones are oxidatively cleaved by Cr(VI) or Mn(VII) reagents. The reaction is sometimes of utility in the synthesis of difunctional molecules by ring cleavage. The mechanism for both reagents is believed to involve an enol intermediate.[169] A study involving both kinetic data and quantitative product studies has permitted a fairly complete description of the Cr(VI) oxidation of benzyl phenyl ketone.[170] The products include both oxidative-cleavage products and benzil, **3**, which results from oxidation α to the carbonyl. In addition, the dimeric product **4**, which is suggestive of a radical intermediate, is formed under some conditions.

$$PhCH_2C(=O)Ph \xrightarrow{Cr(VI)} \underset{\mathbf{3}}{PhC(=O)C(=O)Ph} + PhCH{=}O + PhCO_2H + \underset{\mathbf{4}}{PhCH(COPh){-}CH(COPh)Ph}$$

167. R. Criegee and E. Büchner, *Chem. Ber.* **73:**563 (1940).
168. J. E. McMurry and L. C. Blaszczak, *J. Org. Chem.* **39:**2217 (1974).
169. K. B. Wiberg and R. D. Geer, *J. Am. Chem. Soc.* **87:**5202 (1965); J. Rocek and A. Riehl, *J. Am. Chem. Soc.* **89:**6691 (1967).
170. K. B. Wiberg, O. Aniline, and A. Gatzke, *J. Org. Chem.* **37:**3229 (1972).

Both the diketone and the cleavage products were shown to arise from an α-hydroxy ketone intermediate (benzoin), **5**.

$$PhCH_2COPh \rightleftharpoons PhCH{=}C(OH)Ph \xrightarrow{H_2CrO_4} Ph{-}CH{=}CH(OCrO_3H){-}Ph \;(+H_2O) \longrightarrow PhCH(OH){-}C(=O)Ph + Cr(IV) \longrightarrow \text{products}$$

5

The coupling product is considered to arise from a radical intermediate formed by one-electron oxidation, probably effected by Cr(IV).

The oxidation of cyclohexanone involves 2-hydroxycyclohexanone and 1,2-cyclohexanedione as intermediates.[171]

$$\text{cyclohexanone} \xrightarrow{Cr(VI)} \text{2-hydroxycyclohexanone} \longrightarrow \text{1,2-cyclohexanedione} \longrightarrow HO_2C(CH_2)_4CO_2H$$

Because of the efficient oxidation of alcohols to ketones, alcohols can be used as the starting materials in oxidative cleavages. The conditions required are more vigorous than for the alcohol → ketone transformation (see Section 12.1.1).

Aldehydes can be oxidized to carboxylic acids by both Mn(VII) and Cr(VI). Fairly detailed mechanistic studies have been carried out for Cr(VI). A chromate ester of the aldehyde hydrate is believed to be formed, and this species decomposes in the rate-determining step by a mechanism similar to that which operates in alcohol oxidations:[172]

$$RCH{=}O + H_2CrO_4 \rightleftharpoons RC(OH)(H){-}OCrO_2H \longrightarrow RCO_2H + HCrO_3^- + H^+$$

Effective conditions for oxidation of aldehydes to carboxylic acids with $KMnO_4$ involve use of *t*-butanol and an aqueous NaH_2PO_4 buffer as the reaction medium.[173] Buffered sodium chlorite is also a convenient oxidant.[174] An older reagent for carrying out the aldehyde → carboxylic acid oxidation is silver oxide.

$$\text{4-HO-3-}CH_3O\text{-}C_6H_3CH{=}O \xrightarrow[2)\ HCl]{1)\ Ag_2O,\ NaOH} \text{4-HO-3-}CH_3O\text{-}C_6H_3CO_2H \quad 83\text{–}95\%$$ Ref. 175

The reaction of aldehydes with MnO_2 in the presence of cyanide ion in an alcoholic solvent is a convenient method of converting aldehydes directly to esters.[176] This reaction

171. J. Rocek and A. Riehl, *J. Org. Chem.* **32:**3569 (1967).
172. K. B. Wiberg, *Oxidation in Organic Chemistry, Part A*, Academic Press, New York, 1965, pp. 172–178.
173. A. Abiko, J. C. Roberts, T. Takemasa, and S. Masamune, *Tetrahedron Lett.* **27:**4537 (1986).
174. E. J. Corey and G. A. Reichard, *Tetrahedron Lett.* **34:**6973 (1993); P. M. Wovkulich, K. Shankaran, J. Kiegiel, and M. R. Uskokovic, *J. Org. Chem.* **58:**832 (1993).
175. I. A. Pearl, *Org. Synth.* **IV:**972 (1963).
176. E. J. Corey, N. W. Gilman, and B. E. Ganem, *J. Am. Chem. Soc.* **90:**5616 (1968).

involves the cyanohydrin as an intermediate. The initial oxidation product is an acyl cyanide, which is solvolyzed under the reaction conditions.

$$RCH{=}O + {}^{-}CN + H^+ \rightleftharpoons RCH(OH)CN$$

$$RCH(OH)CN + MnO_2 \longrightarrow RC(=O)CN \xrightarrow{R'OH} RC(=O)OR'$$

Lead tetraacetate can effect oxidation of carbonyl groups, leading to formation of α-acetoxyketones.[177] The yields are seldom high, however. Boron trifluoride can be used to catalyze these oxidations. It is presumed to function by catalyzing the formation of the enol, which is thought to be the reactive species.[178] With unsymmetrical ketones, products from oxidation at both α-methylene groups are found.[179]

$$R_2CHC(=O)R' \rightleftharpoons R_2C{=}C(OH)R' \xrightarrow{Pb(OAc)_4} [CH_3C(=O)O{-}Pb(OAc)_2{-}O{-}C(R'){=}CR_2] \longrightarrow R_2C(O_2CCH_3)C(=O)R'$$

With enol ethers, $Pb(O_2CCH_3)_4$ gives α-alkoxyketones.[180]

$$\text{1-methoxycyclohexene} \xrightarrow[BF_3]{Pb(O_2CCH_3)_4} \text{2-methoxycyclohexanone}$$

Introduction of oxygen α to a ketone function can also be carried out via the silyl enol ether. Lead tetraacetate gives the α-acetoxy ketone:[181]

$$\text{silyl enol ether } (OSi(CH_3)_3;\ H_3C, H_3C, CH_3) \xrightarrow{Pb(OAc)_4} \text{2-acetoxy ketone } (H_3C, H_3C, CH_3, O_2CCH_3)\quad 56\%$$

177. R. Criegee, in *Oxidation in Organic Chemistry, Part A*, K. B. Wiberg, ed., Academic Press, New York, 1965, pp. 305–312.
178. J. D. Cocker, H. B. Henbest, G. H. Philipps, G. P. Slater, and D. A. Thomas, *J. Chem. Soc.* **1965:**6.
179. S. Moon and H. Bohm, *J. Org. Chem.* **37:**4338 (1972).
180. V. S. Singh, C. Singh, and D. K. Dikshit, *Synth. Commun.* **28:**45 (1998).
181. G. M. Rubottom, J. M. Gruber, and K. Kincaid, *Synth. Commun.* **6:**59 (1976); G. M. Rubottom and J. M. Gruber, *J. Org. Chem.* **42:**1051 (1977); G. M. Rubottom and H. D. Juve, Jr., *J. Org. Chem.* **48:**422 (1983).

α-Hydroxy ketones can be obtained from silyl enol ethers by oxidation using a catalytic amount of OsO_4 with an amine oxide serving as the stoichiometric oxidant.[182]

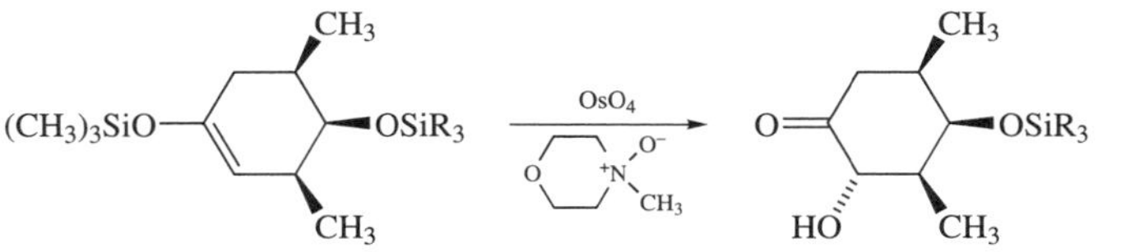

Ref. 183

The silyl enol ethers of ketones are also oxidized to α-hydroxy ketones by *m*-chloroperoxybenzoic acid. If the reaction workup includes acylation, α-acyloxy ketones are obtained.[184] These reactions proceed by initial epoxidation of the silyl enol ether, which then undergoes ring opening. Subsequent transfer of either the *O*-acyl or *O*-TMS substituent occurs, depending on the reaction conditions.

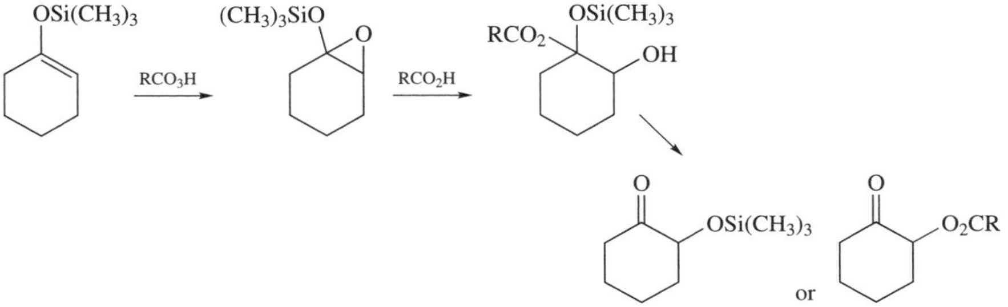

Another very useful series of reagents for oxidation of enolates to α-hydroxy ketones are *N*-sulfonyloxaziridines.[185] The best results are frequently achieved by using KHMDS to form the enolate. The hydroxylation occurs preferentially from the less hindered face of the enolate.

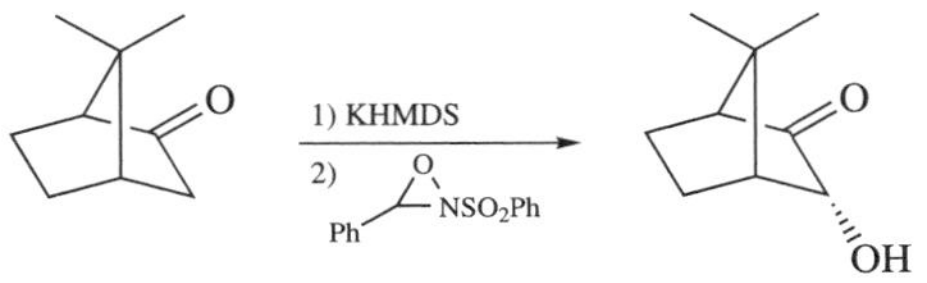

The mechanism of oxygen transfer is believed to involve nucleophilic opening of the oxaziridine, followed by collapse of the resulting *N*-sulfonylcarbinolamine.[186]

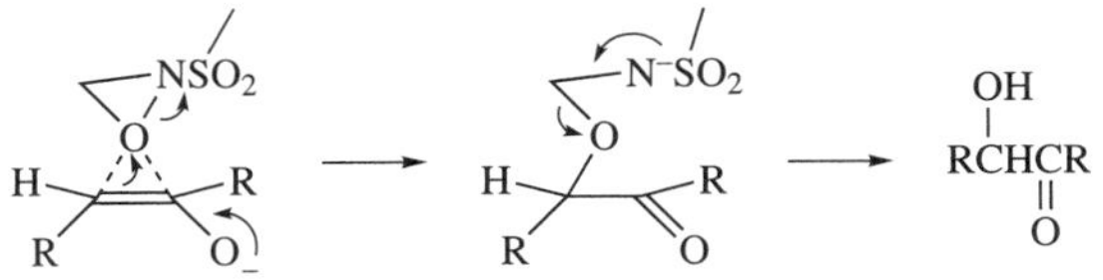

182. J. P. McCormick, W. Tomasik, and M. W. Johnson, *Tetrahedron Lett.* **22:**607 (1981).
183. R. K. Boeckman, Jr., J. E. Starrett, Jr., D. G. Nickell, and P.-E. Sun, *J. Am. Chem. Soc.* **108:**5549 (1986).
184. M. Rubottom, J. M. Gruber, R. K. Boeckman, Jr., M. Ramaiah, and J. B. Medwick, *Tetrahedron Lett.* **1978:**4603; G. M. Rubottom and J. M. Gruber, *J. Org. Chem.* **43:**1599 (1978), G. M. Rubottom, M. A. Vazquez, and D. R. Pelegrina, *Tetrahedron Lett.* **1974:**4319.
185. F. A. Davis, L. C. Vishwakarma, J. M. Billmers, and J. Finn, *J. Org. Chem.* **49:**3241 (1984); L. C. Vishwakarma, O. D. Stringer, and F. A. Davis, *Org. Synth.* **66:**203 (1988).
186. F. A. Davis, A. C. Sheppard, B.-C. Chen, and M. S. Haque, *J. Am. Chem. Soc.* **112:**6679 (1990).

These reagents exhibit good stereoselectivity toward chiral substrates, such as acyloxazolidines.[187] Chiral oxaziridine reagents, such as **A–C**, have been developed, and these can achieve enantioselective oxidation of enolates to α-hydroxy ketones.[188]

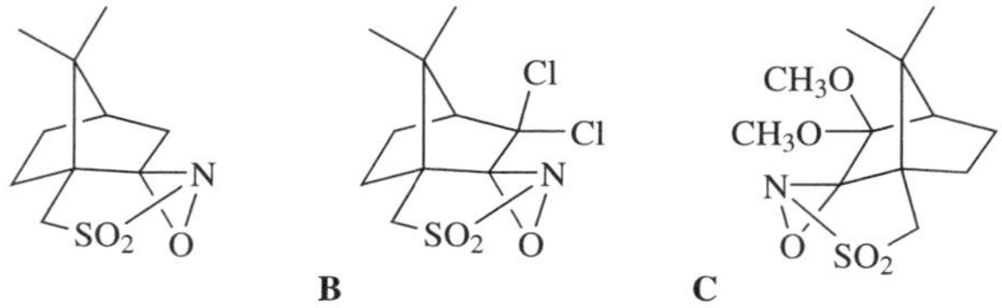

Scheme 12.19 gives some examples of enolate oxidation using *N*-sulfonyloxaziridines.

Other procedures for α oxidation of ketones are based on prior generation of the enolate. The most useful oxidant in these procedures is a molybdenum compound, MoO_5·pyridine·HMPA, which is prepared by dissolving MoO_3 in hydrogen peroxide, followed by addition of HMPA. This reagent oxidizes the enolates of aldehydes, ketones, esters, and lactones to the corresponding α-hydroxy compound.[189]

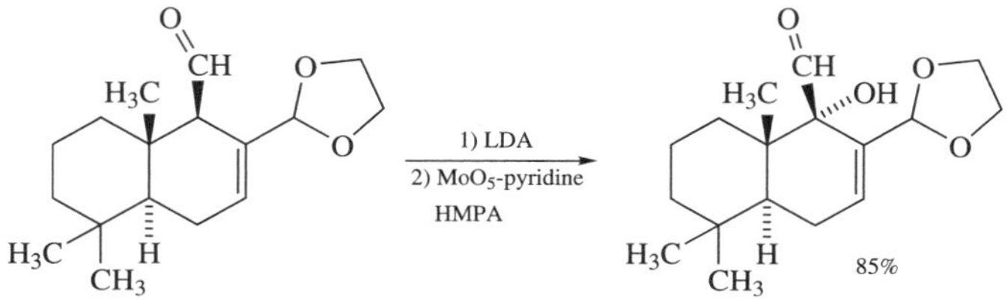

1) LDA
2) MoO_5-pyridine HMPA
85%
Ref. 190

12.5.2. Oxidation of Ketones and Aldehydes by Oxygen and Peroxidic Compounds

In the presence of acid catalysts, peroxy compounds are capable of oxidizing carbonyl compounds by insertion of an oxygen atom into one of the carbon–carbon bonds at the carbonyl group. This is known as the *Baeyer–Villiger oxidation*.[191] The insertion of oxygen is accomplished by a sequence of steps involving addition to the carbonyl group and migration to oxygen.

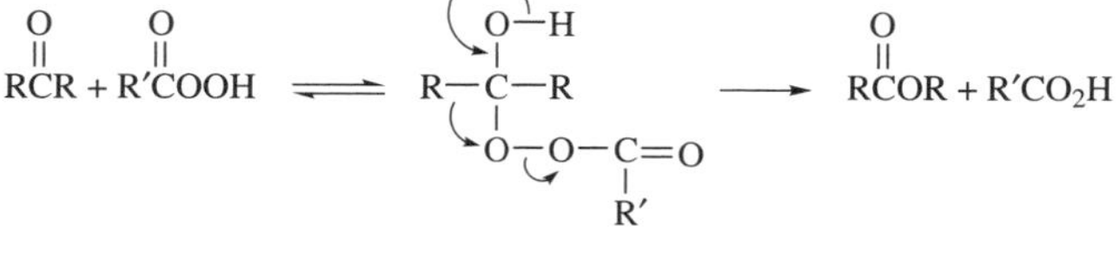

187. D. A. Evans, M. M. Morrissey, and R. L. Dorow, *J. Am. Chem. Soc.* **107:**4346 (1985).
188. F. A. Davis and B.-C. Chen, *Chem. Rev.* **92:**919 (1992).
189. E. Vedejs, *J. Am. Chem. Soc.* **96:**5945 (1974); E. Vedejs, D. A. Engler, and J. E. Telschow, *J. Org. Chem.* **43:**188 (1978); E. Vedejs and S. Larsen, *Org. Synth.* **64:**127 (1985).
190. S. P. Tanis and K. Nakanishi, *J. Am. Chem. Soc.* **101:**4398 (1979).
191. C. H. Hassall, *Org. React.* **9:**73 (1957); M. Renz and B. Meunier, *Eur. J. Org. Chem.* **1999:**737.

Scheme 12.19. Oxidation of Enolates by Oxaziridines

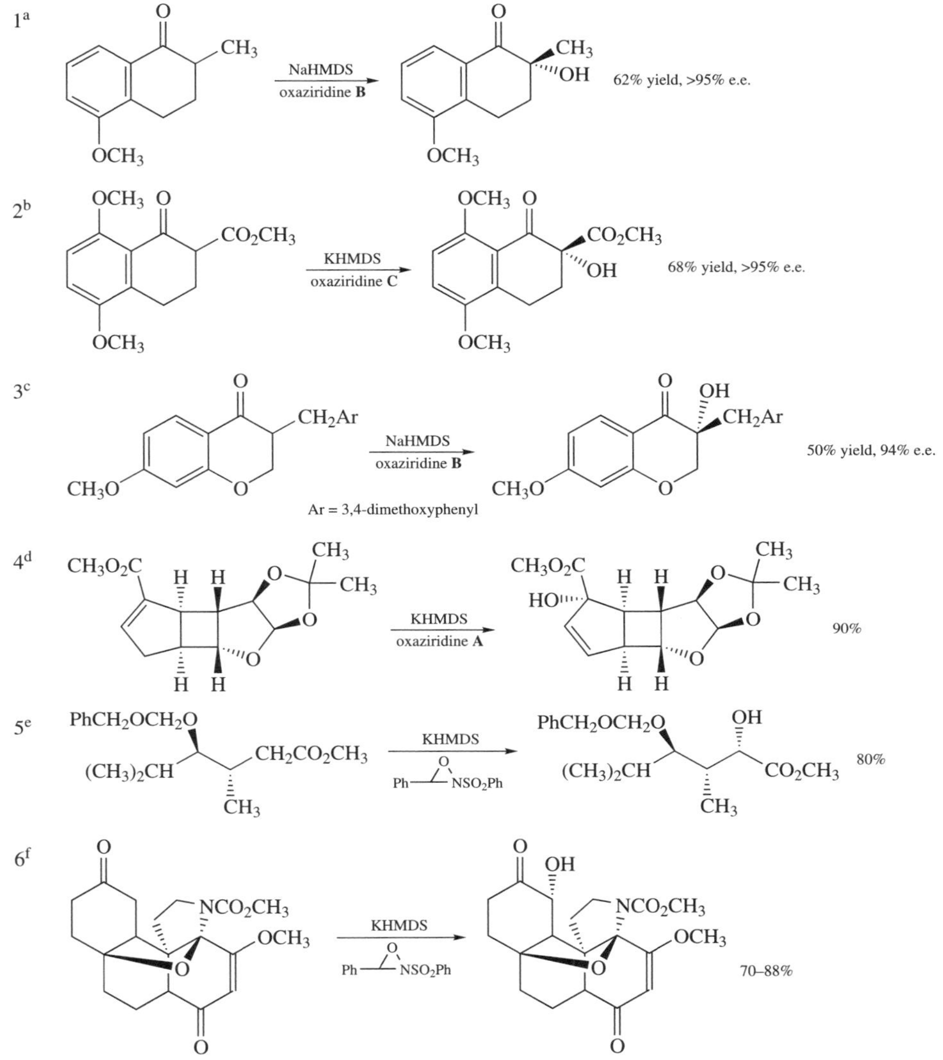

1[a] NaHMDS, oxaziridine **B** — 62% yield, >95% e.e.

2[b] KHMDS, oxaziridine **C** — 68% yield, >95% e.e.

3[c] NaHMDS, oxaziridine **B** — 50% yield, 94% e.e. Ar = 3,4-dimethoxyphenyl

4[d] KHMDS, oxaziridine **A** — 90%

5[e] KHMDS, Ph-oxaziridine-NSO_2Ph — 80%

6[f] KHMDS, Ph-oxaziridine-NSO_2Ph — 70–88%

a. F. A. Davis and M. C. Weismiller, *J. Org. Chem.* **55:**3715 (1990).
b. F. A. Davis, A. Kumar, and B.-C. Chen, *Tetrahedron Lett.* **32:**867 (1991).
c. F. A. Davis and B.-C. Chen, *J. Org. Chem.* **58:**1751 (1993).
d. A. B. Smith III, G. A. Sulikowski, M. M. Sulikowski, and K. Fujimoto, *J. Am. Chem. Soc.* **114:**2567 (1992).
e. S. Hanessian, Y. Gai, and W. Wang, *Tetrahedron Lett.* **37:**7473 (1996).
f. M. A. Tius and M. A. Kerr, *J. Am. Chem. Soc.* **114:**5959 (1992).

The concerted O–O heterolysis and migration is usually the rate-determining step.[192] The reaction is catalyzed by protic and Lewis acids,[193] including $Sc(O_3SCF_3)_3$.[194]

When the reaction involves an unsymmetrical ketone, the structure of the product depends on which group migrates. A number of studies have been directed at ascertaining

192. Y. Ogata and Y. Sawaki, *J. Org. Chem.* **37:**2953 (1972).
193. G. Strukul, *Angew. Chem. Int. Ed. Engl.* **37:**1199 (1998).
194. H. Kotsuki, K. Arimura, T. Araki, and T. Shinohara, *Synlett* **1999:**462.

the basis of migratory preference in the Baeyer–Villiger oxidation. From these studies, a general order of likelihood of migration has been established: *tert*-alkyl, *sec*-alkyl > benzyl, phenyl > *pri*-alkyl > cyclopropyl > methyl.[195] Thus, methyl ketones are uniformly found to give acetate esters resulting from migration of the larger group.[196] A major factor in determining which group migrates is the ability to accommodate partial positive charge. In *para*-substituted phenyl groups, electron-donor substituents favor migration.[197] Similarly, silyl substituents enhance migratory aptitude of alkyl groups.[198] Steric and conformational factors are also important, especially in cyclic systems.[199] As is generally true of migration to an electron-deficient center, the configuration of the migrating group is retained in Baeyer–Villiger oxidations. Some typical examples of Baeyer–Villiger oxidations are shown in Scheme 12.20.

Although ketones are essentially inert to molecular oxygen, enolates are susceptible to oxidation. The combination of oxygen and a strong base has found some utility in the introduction of an oxygen function at carbanionic sites.[200] Hydroperoxides are the initial products of such oxidations, but when DMSO or some other substance capable of reducing the hydroperoxide is present, the corresponding alcohol is isolated. A procedure that has met with considerable success involves oxidation in the presence of a trialkyl phosphite.[201] The intermediate hydroperoxide is efficiently reduced by the phospite ester.

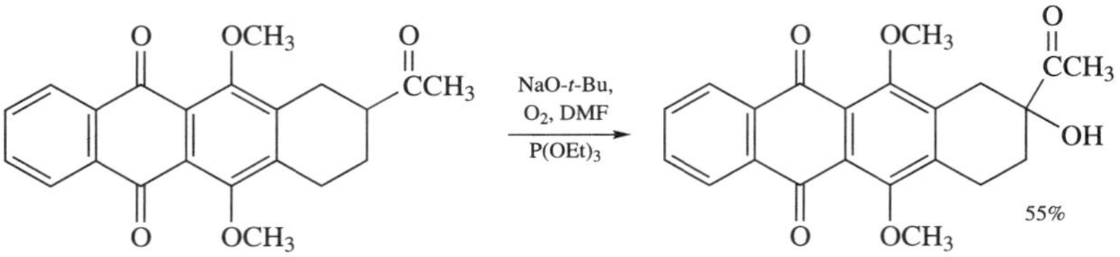

This oxidative process has been successful with ketones,[202] esters,[203] and lactones.[204] Hydrogen peroxide can also be used as the oxidant, in which case the alcohol is formed directly.[205] The mechanism for the oxidation of enolates by oxygen is a radical-chain autoxidation in which the propagation step involves electron transfer from the carbanion to

195. H. O. House, *Modern Synthetic Reactions*, 2nd ed., W. A. Benjamin, Menlo Park, California, 1972, p. 325.
196. P. A. S. Smith, in *Molecular Rearrangements*, P. de Mayo, ed., Interscience, New York, 1963, pp. 457–591.
197. W. E. Doering and L. Speers, *J. Am. Chem. Soc.* **72**:5515 (1950).
198. P. F. Hudrlik, A. M. Hudrlik, G. Nagendrappa, T. Yimenu, E. T. Zellers, and E. Chin, *J. Am. Chem. Soc.* **102**:6894 (1980).
199. M. F. Hawthorne, W. D. Emmons, and K. S. McCallum, *J. Am. Chem. Soc.* **80**:6393 (1958); J. Meinwald and E. Frauenglass, *J. Am. Chem. Soc.* **82**:5235 (1960); R. M. Goodman and Y. Kishi, *J. Am. Chem. Soc.* **120**:9392 (1998).
200. J. N. Gardner, T. L. Popper, F. E. Carlon, O. Gnoj, and H. L. Herzog, *J. Org. Chem.* **33**:3695 (1968).
201. J. N. Gardner, F. E. Carlon, and O. Gnoj, *J. Org. Chem.* **33**:3294 (1968).
202. F. A. J. Kerdesky. R. J. Ardecky, M. V. Lashmikanthan, and M. P. Cava, *J. Am. Chem. Soc.* **103**:1992 (1981).
203. E. J. Corey and H. E. Ensley, *J. Am. Chem. Soc.* **97**:6908 (1975).
204. J. J. Plattner, R. D. Gless, and H. Rapoport, *J. Am. Chem. Soc.* **94**:8613 (1972); R. Volkmann, S. Danishefsky, J. Eggler, and D. M. Solomon, *J. Am. Chem. Soc.* **93**:5576 (1971).
205. G. Büchi, K. E. Matsumoto, and H. Nishimura, *J. Am. Chem. Soc.* **93**:3299 (1971).

Scheme 12.20. Baeyer–Villiger Oxidations

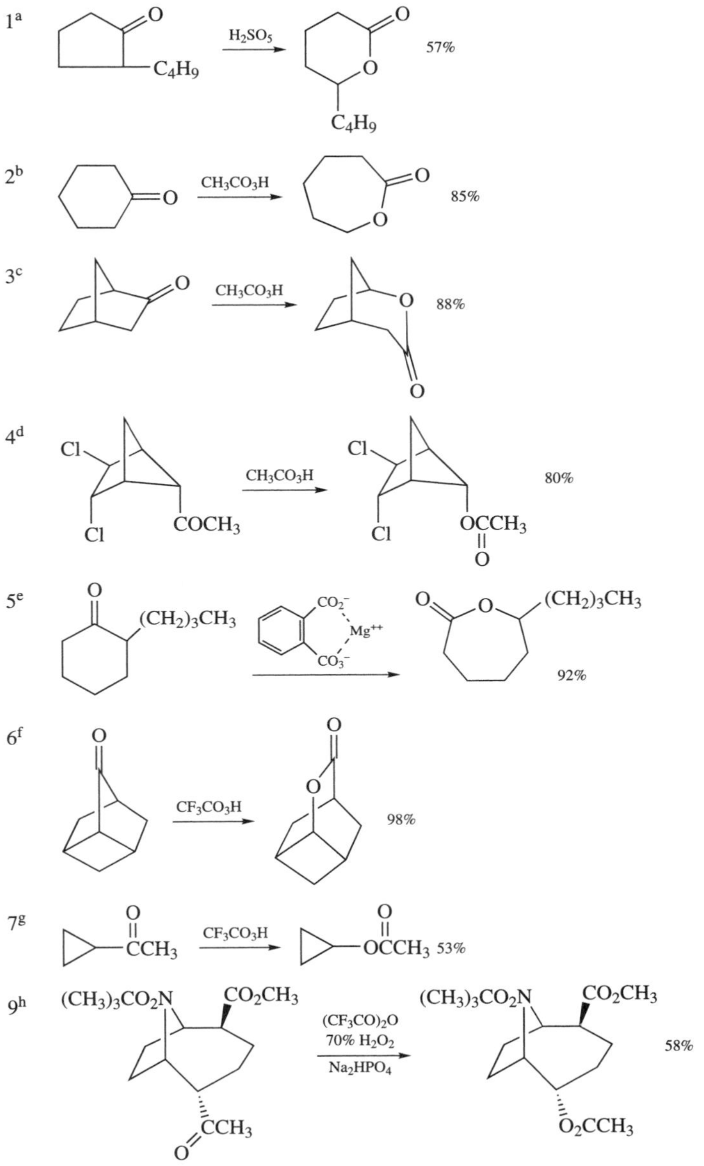
1[a]
H_2SO_5
57%
2[b]
CH_3CO_3H
85%
3[c]
CH_3CO_3H
88%
4[d]
CH_3CO_3H
80%
5[e]
92%
6[f]
CF_3CO_3H
98%
7[g]
CF_3CO_3H
53%
9[h]
$(CF_3CO)_2O$
70% H_2O_2
Na_2HPO_4
58%

a. T. H. Parliment, M. W. Parliment, and I. S. Fagerson, *Chem. Ind.* **1966:**1845.
b. P. S. Starcher and B. Phillips, *J. Am. Chem. Soc.* **80:**4079 (1958).
c. J. Meinwald and E. Frauenglass, *J. Am. Chem. Soc.* **82:**5235 (1960).
d. K. B. Wiberg and R. W. Ubersax, *J. Org. Chem.* **37:**3827 (1972).
e. M. Hirano, S. Yakabe, A. Satoh, J. H. Clark, and T. Morimoto, *Synth. Commun.* **26:**4591 (1996); T. Mino, S. Masuda, M. Nishio, and M. Yamashita, *J. Org. Chem.* **62:**2633 (1997).
f. S. A. Monti and S.-S. Yuan, *J. Org. Chem.* **36:**3350 (1971).
g. W. D. Emmons and G. B. Lucas, *J. Am. Chem. Soc.* **77:**2287 (1955).
h. F. J. Sardina, M. H. Howard, M. Morningstar, and H. Rapaport, *J. Org. Chem.* **55:**5025 (1990).

a hydroperoxy radical[206]:

$$RC(O^-)=CR_2 + O_2 \longrightarrow RC(=O)-\dot{C}R_2 + O_2^{\cdot -}$$

$$RC(=O)\dot{C}R_2 + O_2 \longrightarrow RC(=O)CR_2-O-O\cdot$$

$$RC(O^-)=CR_2 + RC(=O)CR_2-O-O\cdot \longrightarrow RC(=O)-\dot{C}R_2 + RC(=O)CR_2-O-O^-$$

Arguments for a non-chain reaction between the enolate and oxygen to give the hydroperoxide anion directly have also been advanced.[207]

12.5.3. Oxidation with Other Reagents

Selenium dioxide can be used to oxidize ketones and aldehydes to α-dicarbonyl compounds. The reaction often gives high yields of products when there is a single type of CH_2 group adjacent to the carbonyl group. In unsymmetrical ketones, oxidation usually occurs at the CH_2 group that is most readily enolized.[208]

Cyclohexanone $\xrightarrow{SeO_2}$ cyclohexane-1,2-dione 60% Ref. 209

$PhC(=O)CH_3 \xrightarrow{SeO_2} PhC(=O)CH=O$ 69–72% Ref. 210

The oxidation is regarded as taking place by an electrophilic attack of selenium dioxide (or selenous acid, H_2SeO_3, the hydrate) on the enol of the ketone or aldehyde. This is followed by hydrolytic elimination of the selenium.[211]

$$RC(OH)=CHR' \xrightarrow{SeO_2} RC(=O)-CHR'(Se(=O)OH) \xrightarrow{-H_2O} RC(=O)-CR'(=Se=O) \xrightarrow{H_2O} RC(=O)-CR'(OH)(Se(=O)H) \xrightarrow{-H_2SeO} RC(=O)-CR'(=O)$$

206. G. A. Russell and A. G. Bemix, *J. Am. Chem. Soc.* **88:**5491 (1966).
207. H. R. Gersmann and A. F. Bickel, *J. Chem. Soc. B* **1971:**2230.
208. E. N. Trachtenberg, in *Oxidation*, Vol. 1, R. L. Augustine, ed., Marcel Dekker, New York, 1969, Chapter 3.
209. C. C. Hach, C. V. Banks, and H. Diehl, Org. Synth. **IV:**229 (1963).
210. H. A. Riley and A. R. Gray, *Org. Synth.* **II:**509 (1943).
211. K. B. Sharpless and K. M. Gordon, *J. Am. Chem. Soc.* **98:**300 (1976).

Methyl ketones are degraded to the next lower carboxylic acid by reaction with hypochlorite or hypobromite ions. The initial step in these reactions involves base-catalyzed halogenation. The halo ketones are more reactive than their precursors, and rapid halogenation to the trihalo compound results. Trihalomethyl ketones are susceptible to alkaline cleavage because of the inductive stabilization provided by the halogen atoms.

$$RC(=O)CH_3 \xrightleftharpoons{slow} RC(O^-)=CH_2 \xrightarrow{^-OBr} RC(=O)CH_2Br \xrightarrow{^-OH} RC(O^-)=CHBr \xrightarrow{fast} RC(=O)CBr_3$$

$$RC(=O)CBr_3 + {}^-OH \rightleftharpoons RC(O^-)(OH)CBr_3 \longrightarrow RCO_2H + {}^-CBr_3 \rightleftharpoons RCO_2^- + HCBr_3$$

$$(CH_3)_3CC(=O)CH_3 \xrightarrow[Br_2]{NaOH} (CH_3)_3CCO_2H \quad 71\text{–}74\%$$ Ref. 212

$$(CH_3)_2C{=}CHC(=O)CH_3 \xrightarrow{KOCl} \xrightarrow{H^+} (CH_3)_2C{=}CHCO_2H \quad 49\text{–}53\%$$ Ref. 213

12.6. Allylic Oxidation

12.6.1. Transition Metal Oxidants

Carbon–carbon double bonds, besides being susceptible to addition of oxygen or cleavage, can also react at allylic positions. Synthetic utility requires that there be good regioselectivity. Among the transition-metal oxidants, the CrO_3–pyridine reagent in methylene chloride[214] and a related complex in which 3,5-dimethylpyrazole is used in place of pyridine[215] are the most satisfactory for allylic oxidation.

H3C CH3 → (CrO_3–3,5-dimethylpyrazole) → O= H3C CH3 Ref. 216

Several lines of mechanistic evidence implicate allylic radicals or cations as intermediates in these oxidations. Thus, ^{14}C in cyclohexene is located at both the 1,2- and 2,3-positions in the product cyclohexenone, indicating that a symmetrical allylic intermediate is involved at some stage:[217]

212. L. T. Sandborn and E. W. Bousquet, *Org. Synth.* **1:**512 (1932).
213. L. I. Smith, W. W. Prichard, and L. J. Spillane, *Org. Synth.* **III:**302 (1955).
214. W. G. Dauben, M. Lorber, and D. S. Fullerton, *J. Org. Chem.* **34:**3587 (1969).
215. W. G. Salmond, M. A. Barta, and J. L. Havens, *J. Org. Chem.* **43:**2057 (1978); R. H. Schlessinger, J. L. Wood, A. J. Poos, R. A. Nugent, and W. H. Parson, *J. Org. Chem.* **48:**1146 (1983).
216. A. B. Smith III and J. P. Konopelski, *J. Org. Chem.* **49:**4094 (1984).
217. K. B. Wiberg and S. D. Nielsen, *J. Org. Chem.* **29:**3353 (1964).

In many allylic oxidations, the double bond is found in a position indicating that an "allylic shift" occurs during the oxidation.

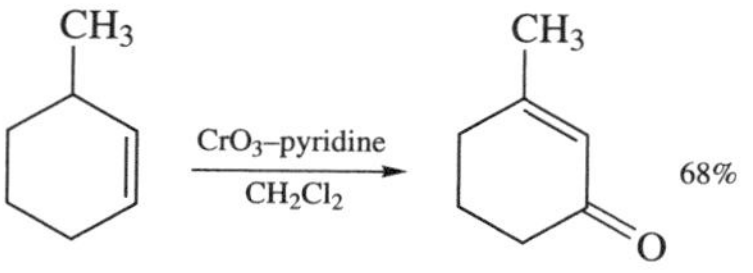

CH_3 $\xrightarrow[CH_2Cl_2]{CrO_3\text{–pyridine}}$ CH_3 O 68% Ref. 214

Detailed mechanistic understanding of the allylic oxidation has not been developed. One possibility is that an intermediate oxidation state of Cr, specifically Cr(IV), acts as the key reagent by abstracting hydrogen.[218]

Several catalytic systems based on copper can also achieve allylic oxidation. These reactions involve induced decomposition of peroxy esters. (See Part A, Section 12.8, for a discussion of the mechanism of this reaction.) When chiral copper ligands are used, enantioselectivity can be achieved. Table 12.1 shows some results for the oxidation of cyclohexene under these conditions.

Table 12.1. Enantioselective Copper-catalyzed Allylic Oxidation of Cyclohexene

Catalyst	Yield	e.e. (%)	Reference
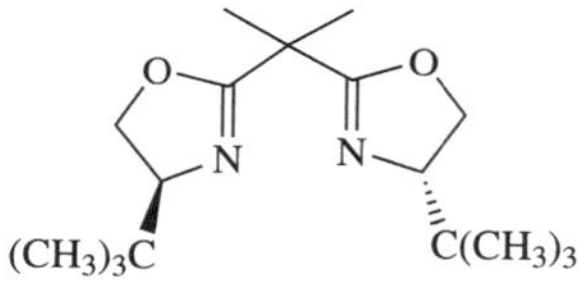 O N N O $(CH_3)_3C$ $C(CH_3)_3$	43	80	a
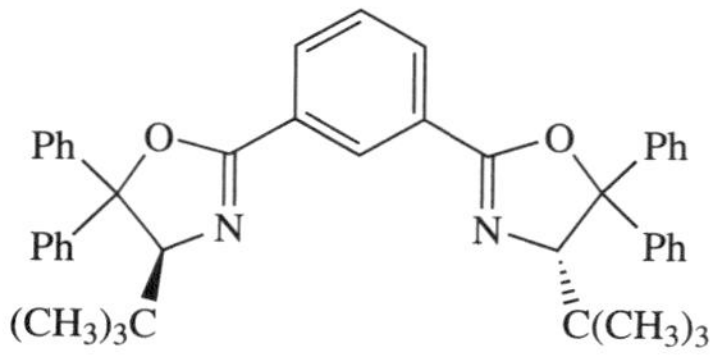 Ph O Ph N $(CH_3)_3C$ O Ph N Ph $C(CH_3)_3$	73	75	b
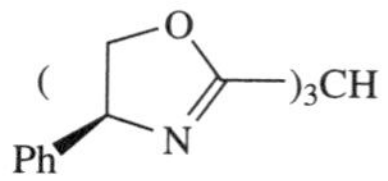 (O N Ph $)_3CH$	19	42	c
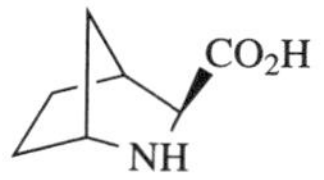 CO_2H NH	67	50	d

a. M. B. Andrus and X. Chen, *Tetrahedron* **53**:16229 (1997).
b. G. Sekar, A. Datta Gupta, and V. K. Singh, *J. Org. Chem.* **63**:2961 (1998).
c. K. Kawasaki and T. Katsuki, *Tetrahedron* **53**:6337 (1997).
d. M. J. Södergren and P. G. Andersson, *Tetrahedron Lett.* **37**:7577 (1996).

218. P. Müller and J. Rocek, *J. Am. Chem. Soc.* **96**:2836 (1974).

Selenium dioxide is a useful reagent for allylic oxidation of alkenes. The products can include enones, allylic alcohols, or allylic esters, depending on the reaction conditions. The basic mechanism consists of three essential steps: (a) an electrophilic "ene" reaction with SeO_2, (b) a sigmatropic rearrangement that restores the original location of the double bond, and (c) breakdown of the resulting selenium ester[219]:

$$RCH{=}CHCH_3 \xrightarrow{SeO_2} [\text{cyclic Se transition state}] \longrightarrow RCH{=}CHCH_2OSeOH \longrightarrow RCH{=}CHCH{=}O \text{ or } RCH{=}CHCH_2OH$$

The alcohols that are the initial oxidation products can be further oxidized to carbonyl groups by SeO_2, and the conjugated carbonyl compound is usually isolated. If the alcohol is the desired product, the oxidation can be run in acetic acid as solvent, in which case acetate esters are formed.

Although the traditional conditions for effecting SeO_2 oxidations involve use of a stoichiometric or excess amount of SeO_2, it is also possible to carry out the reaction with 1.5–2 mol% SeO_2, using *t*-butyl hydroperoxide as a stoichiometric oxidant. Under these conditions, the allylic alcohol is the principal reaction product. The use of a stoichiometric amount of SeO_2 and excess *t*-butyl hydroperoxide leads to good yields of allylic alcohols, even from alkenes that are poorly reactive under the traditional conditions.[220]

$$(CH_3)_2C{=}CHCH_2CH_2(CH_3)C{=}CHCH_2O_2CCH_3 \xrightarrow[t\text{-BuOOH}]{0.1\text{ mol }SeO_2} HOCH_2(CH_3)C{=}CHCH_2CH_2(H_3C)C{=}CHCH_2O_2CCH_3$$

50% + 5% aldehyde

Selenium dioxide exhibits a useful stereoselectivity in reactions with trisubstituted *gem*-dimethyl alkenes. The products are always predominantly the *E*-allylic alcohol or unsaturated aldehyde[221]:

$$(CH_3)_2C{=}CHCH_2CH_3 \xrightarrow{SeO_2} O{=}CH(CH_3)C{=}CHCH_2CH_3$$

45%

This stereoselectivity can be explained by a cyclic transition state for the sigmatropic rearrangement step. The observed stereochemistry results if the alkyl substituent adopts a

219. K. B. Sharpless and R. F. Lauer, *J. Am. Chem. Soc.* **94:**7154 (1972).
220. M. A. Umbreit and K. B. Sharpless, *J. Am. Chem. Soc.* **99:**5526 (1977).
221. U. T. Bhalerao and H. Rapoport, *J. Am. Chem. Soc.* **93:**4835 (1971); G. Büchi and H. W. Wüest, *Helv. Chim. Acta* **50:**2440 (1967).

pseudoequatorial conformation.

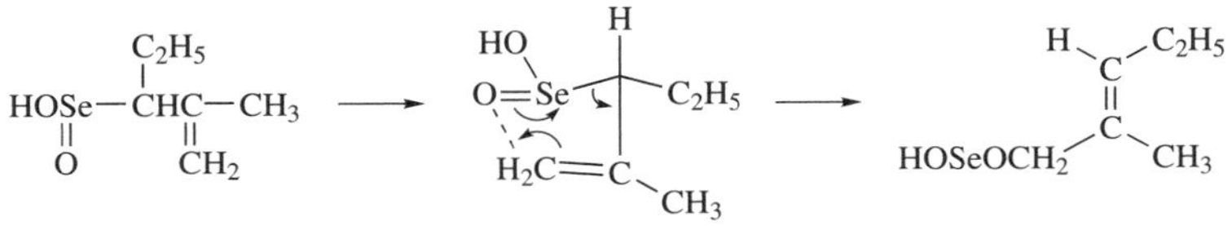

Trisubstituted alkenes are oxidized selectively at the more substituted end of the carbon–carbon double bond, indicating that the ene reaction step is electrophilic in character:

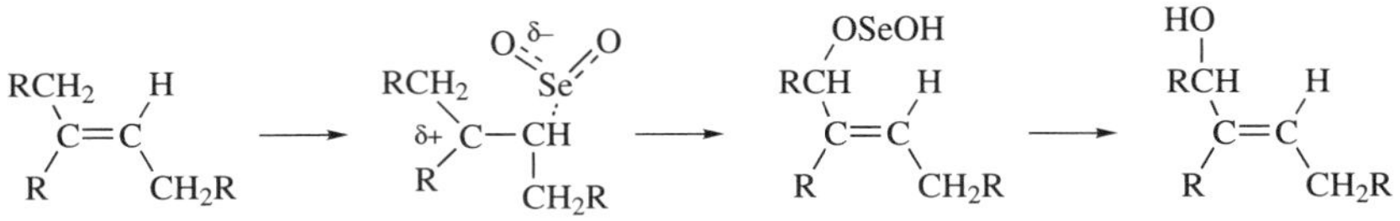

Thus, trisubstituted alkenes are preferentially oxidized at one of the allylic groups at the disubstituted carbon.

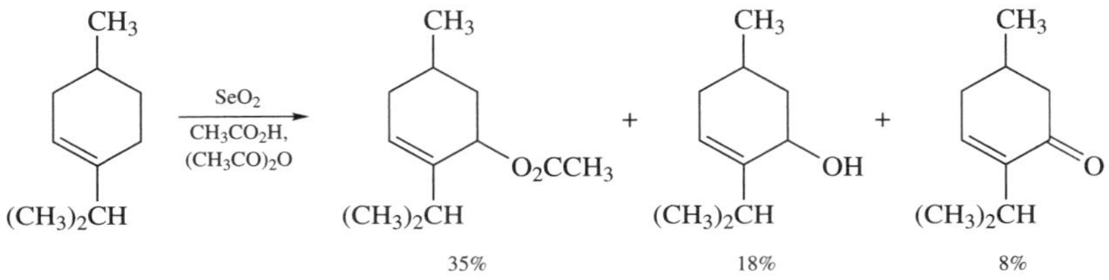

Ref. 222

The equivalent to allylic oxidation of alkenes, but with allylic transposition of the carbon–carbon double bond, can be carried out by an indirect oxidative process involving addition of an electrophilic arylselenenyl reagent, followed by oxidative elimination of selenium. In one procedure, addition of an arylselenenyl halide is followed by solvolysis and oxidation:

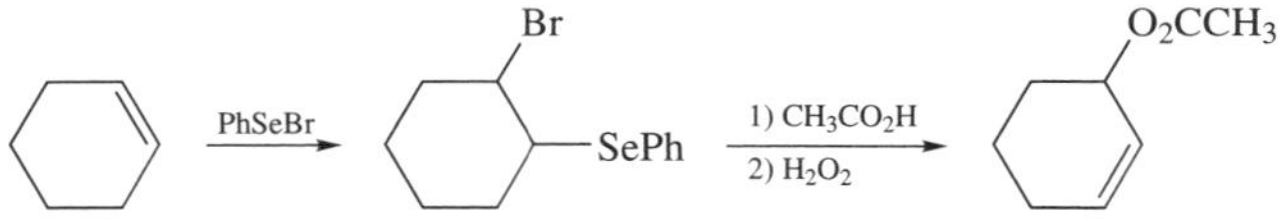

Ref. 223

This reaction depends upon the facile solvolysis of β-haloselenides and the oxidative elimination of selenium, which was discussed in Section 6.8.3. An alternative method, which is experimentally simpler, involves reaction of alkenes with a mixture of diphenyl diselenide and phenylseleninic acid.[224] The two selenium reagents generate an electro-

222. T. Suga, M. Sugimoto, and T. Matsuura, *Bull. Chem. Soc. Jpn.* **36:**1363 (1963).
223. K. B. Sharpless and R. F. Lauer, *J. Org. Chem.* **39:**429 (1974); D. L. J. Clive, *J. Chem. Soc., Chem. Commun.* **1974:**100.
224. T. Hori and K. B. Sharpless, *J. Org. Chem.* **43:**1689 (1978).

philic selenium species, phenylselenenic acid, PhSeOH.

$$\mathrm{RCH_2CH{=}CHR'} \xrightarrow{\mathrm{PhSeOH}} \mathrm{RCH_2\underset{\substack{|\\ \mathrm{PhSe}}}{C}H\overset{\substack{\mathrm{OH}\\ |}}{C}HR'} \xrightarrow{t\text{-BuOOH}} \mathrm{RCH{=}CH\underset{\substack{|\\ \mathrm{OH}}}{C}HR'}$$

The elimination is promoted by oxidation of the addition product to the selenoxide by *t*-butyl hydroperoxide. The regioselectivity in this reaction is such that the hydroxyl group becomes bound at the more substituted end of the carbon–carbon double bond. The origin of this regioselectivity is that the addition step follows Markownikoff's rule, with "$PhSe^+$" acting as the electrophile. The elimination step specifically proceeds away from the oxygen functionality.

12.7. Oxidations at Unfunctionalized Carbon

Attempts to achieve selective oxidations of hydrocarbons or other compounds when the desired site of attack is remote from an activating functional group are faced with several difficulties. With powerful transition-metal oxidants, the initial oxidation products are almost always more susceptible to oxidation than the starting material. Once a hydrocarbon is attacked, it is likely to be oxidized to a carboxylic acid, with chain cleavage by successive oxidation of alcohol and carbonyl intermediates. There are a few circumstances under which oxidations of hydrocarbons can be synthetically useful processes. One such set of circumstances involves catalytic industrial processes. Much work has been expended on the development of selective catalytic oxidation processes, and several have economic importance. We will focus on several reactions that are used on a laboratory scale.

The most general hydrocarbon oxidation is the oxidation of side chains on aromatic rings. Two factors contribute to making this a high-yield procedure, despite the use of strong oxidants. First, the benzylic position is particularly susceptible to hydrogen abstraction by the oxidants.[225] Second, the aromatic ring is resistant to attack by the Mn(VII) and Cr(VI) reagents which oxidize the side chain. Scheme 12.21 provides some examples of the familiar oxidation of aromatic alkyl substituents to carboxylic acid groups.

Selective oxidations are possible for certain bicyclic hydrocarbons.[226] Here, the bridgehead position is the preferred site of initial attack because of the order of reactivity of C–H bonds, which is $3° > 2° > 1°$. The tertiary alcohols which are the initial oxidation products are not easily further oxidized. The geometry of the bicyclic rings (Bredt's rule) prevents both dehydration of the tertiary bridgehead alcohols and further oxidation to ketones. Therefore, oxidation that begins at a bridgehead position stops at the alcohol stage. Chromic acid has been the most useful reagent for functionalizing unstrained bicyclic hydrocarbons. The reaction fails for strained bicyclic compounds such as norbornane because the reactivity of the bridgehead position is lowered by the unfavorable

225. K. A. Gardner, L. L. Kuehnert, and J. M. Mayer, *Inorg. Chem.* **36:**2069 (1997).
226. R. C. Bingham and P. v. R. Schleyer, *J. Org. Chem.* **36:**1198 (1971).

Scheme 12.21. Side-Chain Oxidation of Aromatic Compounds

1[a] 2-chlorotoluene → ($KMnO_4$) → 2-chlorobenzoic acid, 76–78%

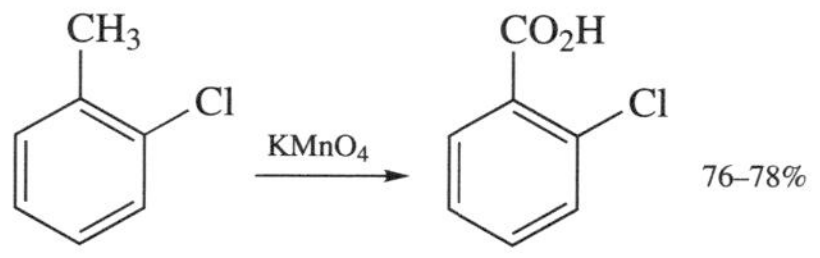

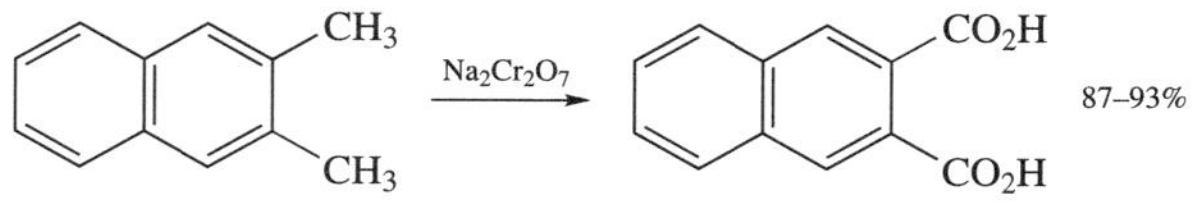

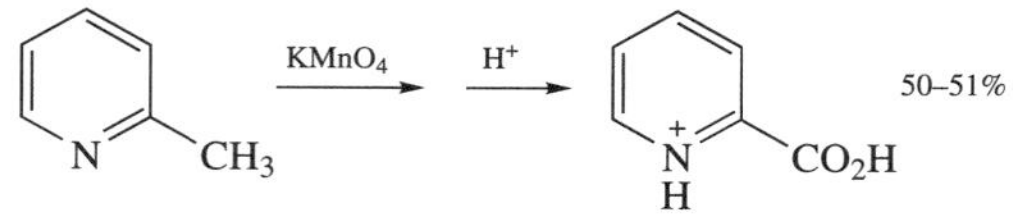

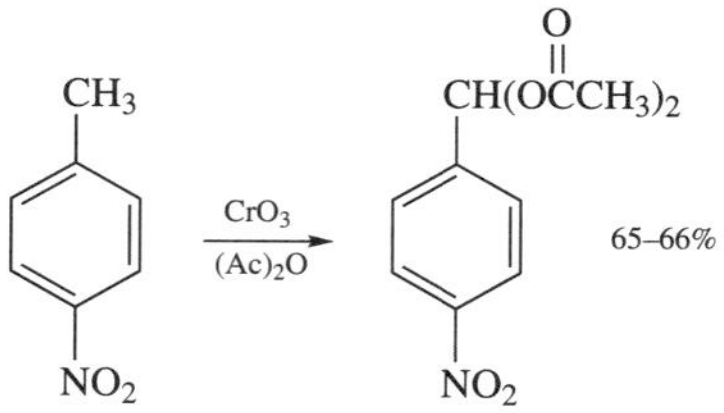

2[b] 2,3-dimethylnaphthalene → ($Na_2Cr_2O_7$) → naphthalene-2,3-dicarboxylic acid, 87–93%

3[c] 2-methylpyridine → ($KMnO_4$; H^+) → pyridinium-2-carboxylic acid, 50–51%

4[d] 4-nitrotoluene → (CrO_3, $(Ac)_2O$) → 4-nitrobenzylidene diacetate, $CH(OCCH_3)_2$, 65–66%

5[e] tetralin → (CrO_3, CH_3CO_2H, 20°C) → 1-tetralone, 55%

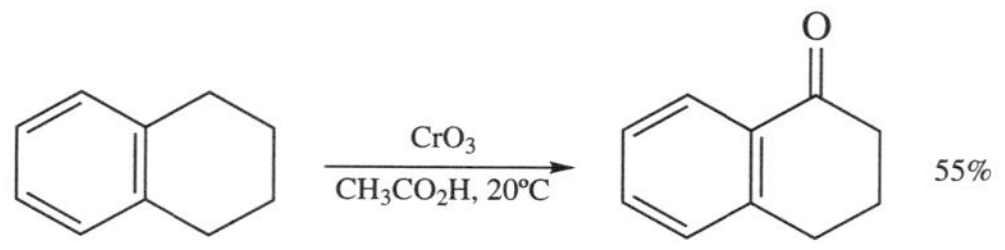

a. H. T. Clarke and E. R. Taylor, *Org. Synth.* **II:**135 (1943).
b. L. Friedman, *Org. Synth.* **43:**80 (1963); L. Friedman, D. L. Fishel, and H. Shechter, *J. Org. Chem.* **30:**1453 (1965).
c. A. W. Singer and S. M. McElvain, *Org. Synth.* **III:**740 (1955).
d. T. Nishimura, *Org. Synth.* **IV:**713 (1963).
e. J. W. Burnham, W. P. Duncan, E. J. Eisenbraun, G. W. Keen, and M. C. Hamming, *J. Org. Chem.* **39:**1416 (1974).

energy of radical or carbocation intermediates.

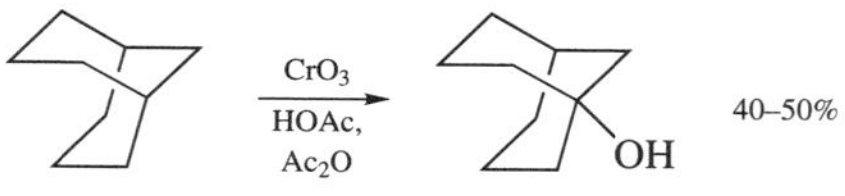

(CrO_3, HOAc, Ac_2O) → OH, 40–50%

Other successful selective oxidations of hydrocarbons by Cr(VI) have been reported—for example, the oxidation of *cis*-decalin to the corresponding alcohol—but careful attention to reaction conditions is required to obtain satisfactory yields.

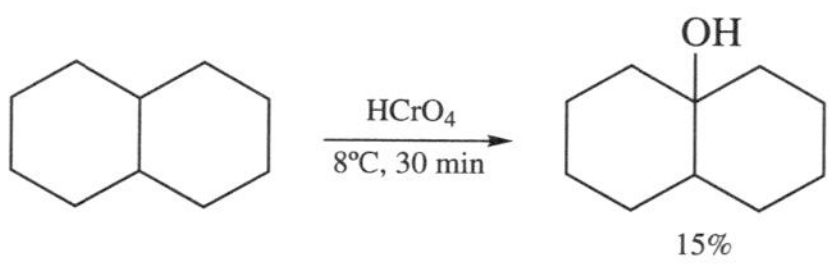

($HCrO_4$, 8°C, 30 min) → OH, 15% Ref. 227

227. K. B. Wiberg and G. Foster, *J. Am. Chem. Soc.* **83:**423 (1961).

Interesting hydrocarbon oxidations have been observed with Fe(II) catalysts with oxygen as the oxidant. These catalytic systems have become known as "Gif chemistry" after the location of their discovery in France.[228] An improved system involving Fe(III), picolinic acid, and H_2O_2, has been developed. The reactive species generated in these systems is believed to be at the Fe(V)=O oxidation level.[229]

$$Fe^{V}{=}O + RCH_2R \longrightarrow (HO)Fe{-}CHR_2 \xrightarrow{O_2} (HO)Fe{-}OOCHR_2$$

$$(HO)Fe{-}CHR_2 \xrightarrow{BrCCl_3} BrCHR_2 \qquad (HO)Fe{-}CHR_2 \xrightarrow{^{-}C{\equiv}O^{+},\ H_2O} R_2CHCO_2H$$

$$(HO)Fe{-}OOCHR_2 \longrightarrow HOCHR_2 \qquad (HO)Fe{-}OOCHR_2 \longrightarrow O{=}CR_2$$

The initial intermediates containing C–Fe bonds can be diverted by reagents such as $CBrCl_3$ or CO, among others.[230]

General References

W. Ando, ed., *Organic Peroxides*, John Wiley & Sons, New York, 1992.
R. L. Augustine, ed., *Oxidations*, Vol. 1, Marcel Dekker, New York, 1969.
R. L. Augustine and D. J. Trecker, eds., *Oxidations*, Vol. 2, Marcel Dekker, New York, 1971.
P. S. Bailey, *Ozonization in Organic Synthesis*, Vols. I and II, Academic Press, New York, 1978, 1982.
G. Cainelli and G. Cardillo, *Chromium Oxidations in Organic Chemistry*, Springer-Verlag, New York, 1984.
L. J. Chinn, *Selection of Oxidants in Synthesis*, Marcel Dekker, New York, 1971.
C. S. Foote, J. S. Valentine, A. Greenberg, and J. F. Liebman eds. *Active Oxygen in Chemistry*, Blackie Academic & Profesional, London, 1995.
A. H. Haines, *Methods for the Oxidation of Organic Compounds: Alkanes, Alkenes, Alkynes and Arenes*, Academic Press, Orlando, Florida, 1985.
M. Hudlicky, *Oxidations in Organic Chemistry*, American Chemical Society, Washington, D.C., 1990.
W. J. Mijs and C. R. H. de Jonge, *Organic Synthesis by Oxidation with Metal Compounds*, Plenum, New York, 1986.
W. Trahanovsky, ed., *Oxidations in Organic Chemistry*, Parts B–D, Academic Press, New York, 1973–1982.
H. H. Wasserman and R. W. Murray, eds., *Singlet Oxygen*, Academic Press, New York, 1979.
K. B. Wiberg, ed., *Oxidations in Organic Chemistry*, Part A, Academic Press, New York, 1965.

Problems

(*References for these problems will be found on page 941.*)

1. Indicate an appropriate oxidant for carrying out the following transformations.

(a) $(CH_3)_2C{=}CHCH_2CH_2CH(CH_3)CH_2CN \longrightarrow CH_2{=}C(CH_3)CH(OH)CH_2CH_2CH(CH_3)CH_2CN$

228. D. H. R. Barton and D. Doller, *Acc. Chem. Res.* **25:**504 (1992); D. H. R. Barton, *Chem. Soc. Rev.* **25:**237 (1996); D. H. R. Barton, *Tetrahedron* **54:**5805 (1998).
229. D. H. R. Barton, S. D. Beviere, W. Chavasiri, E. Csuhai, D. Doller, and W. G. Liu, *J. Am. Chem. Soc.* **114:**2147 (1992).
230. D. H. R. Barton, E. Csuhai, and D. Doller, *Tetrahedron Lett.* **33:**3413 (1992); D. H. R. Barton, E. Csuhai, and D. Doller, *Tetrahedron Lett.* **33:**4389 (1992).

(b) $PhCH_2OCH_2$ … CO_2R → $PhCH_2OCH_2$ … OH, CO_2R

(c) H, H_3C, $C{=}C(CH_3)_2$, H_3C, H, H, $CH_2O_2CCH_3$ → H, CH=O, H_3C, C=C, H_3C, CH_3, H, H, $CH_2O_2CCH_3$

(d) O, Ph → HO, O, Ph

(e) $CH_3CH_2CH_2$, H, C=C, H, $CH_2CH_2CH_3$ → HO, H, $CH_2CH_2CH_3$, $CH_3CH_2CH_2$, C—C, H, OH

(f) $PhCCH_2CH_3$ (O) → $PhCCHCH_3$ (O, OH)

(g) CH_2, CH_3, $CHCH_2CH_2OCPh_3$, CH_3 → O, CH_3, $CHCH_2CH_2OCPh_3$, CH_3

(h) → $OCCH_3$ (O)

(i) H, CH_3, H, C=C, $C{=}C(CH_3)_2$, $PhSO_2$, CH_2CH_2 → H, CH_3, H, CH=O, C=C, C=C, $PhSO_2$, CH_2CH_2, CH_3

(j) H_3C, CH_3, O → H_3C, CH_3, O, O

(k) CH_3, O, $OSiR_3$, CH_3 → CH_3, O, $OSiR_3$, HO, CH_3

(l)

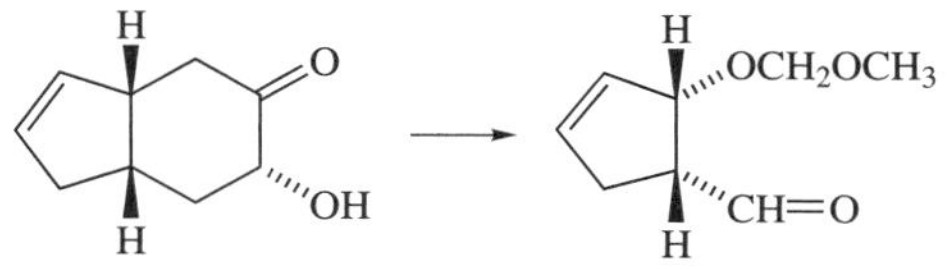

(m)

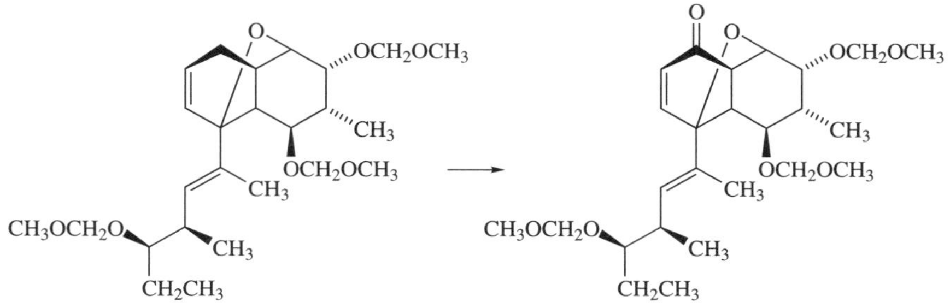

(n)

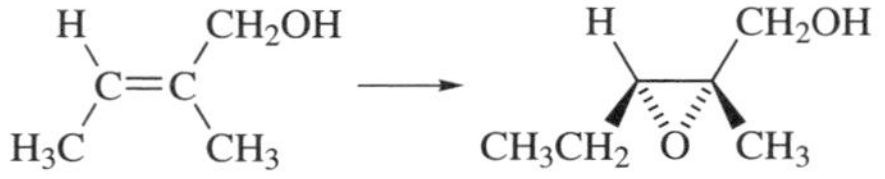

(o)

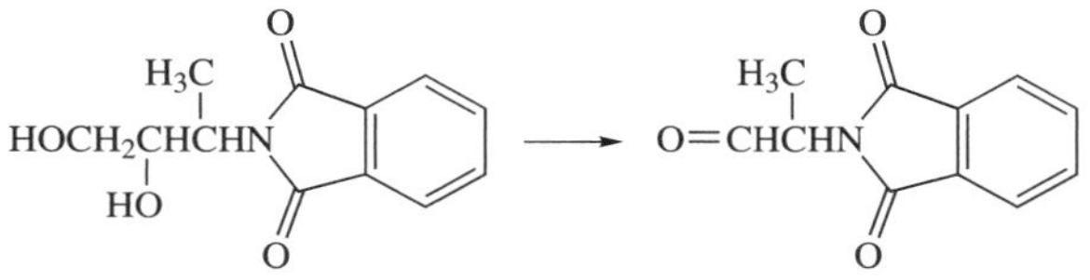

(p)

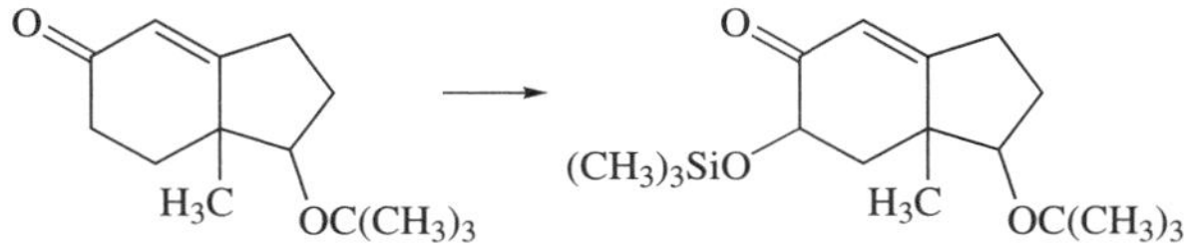

(q)

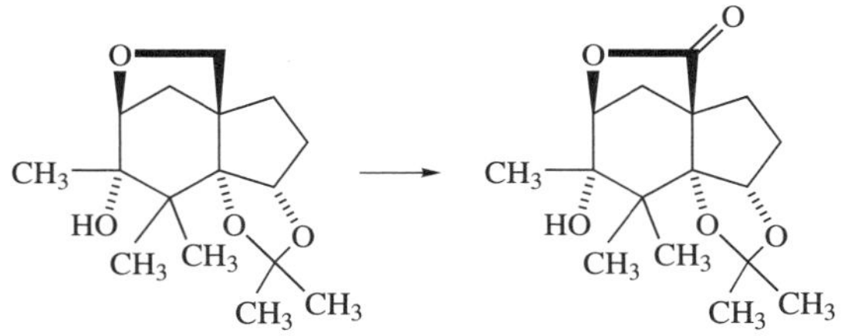

2. Predict the products of the following reactions. Be careful to consider all stereochemical aspects.

(a) OsO_4

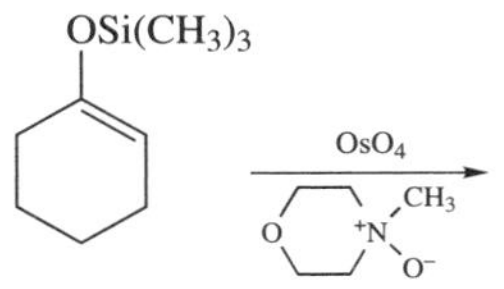

(b) *m*-chloroperoxybenzoic acid

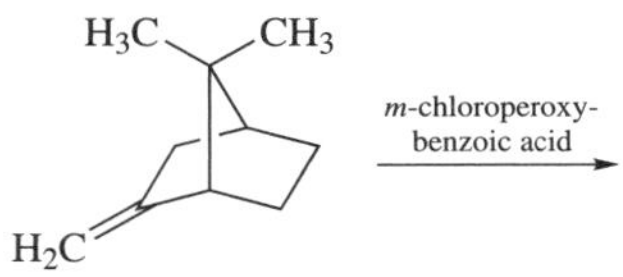

(c) H_3C, H_3C, HO, H, H, O — *m*-chloroperoxybenzoic acid →

(d) CH_3, O, CH_3 — $LiClO_4$ →

(e) H_3C, CO_2CH_3, O, CH_3, H, O — BF_3 →

(f) H_3C, H, O, $HOCH_2$, CH_3 — $Mo(CO)_6$, $(CH_3)_3COOH$ →

(g) H, CH_3, H, H, H, O, O — OsO_4 →

(h) $(H_3C)_2C{=}CHCH_2CH_2CH(CH_3)CH_2$–(1,3-dioxolan-2-yl) — SeO_2 →

(i) CH_2OCH_3, O, O — Collins reagent (excess) →

(j) CH_2CN, CH_3 — RuO_2, $NaIO_4$ →

(k) OH, CH_3 — *t*-BuOOH, $VO(acac)_2$ →

(l) 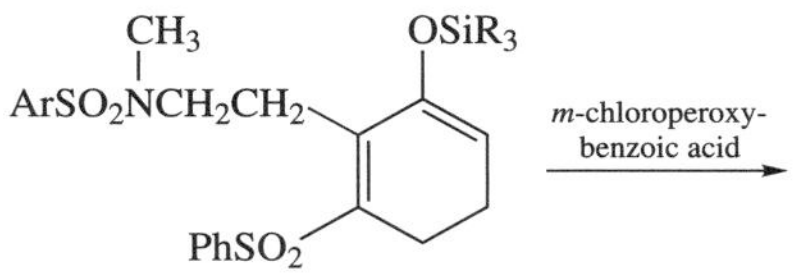 [structure: cyclohexene bearing $OSiR_3$, $ArSO_2N(CH_3)CH_2CH_2$–, and $PhSO_2$ substituents] $\xrightarrow{m\text{-chloroperoxybenzoic acid}}$

3. In chromic acid oxidation of isomeric cyclohexanols, it is usually found that axial hydroxyl groups react more rapidly than equatorial groups. For example, *trans*-4-*t*-butylcyclohexanol is less reactive (by a factor of 3.2) than the *cis* isomer. An even larger difference is noted with *cis*- and *trans*-3,3,5- trimethylcyclohexanol. The *trans* alcohol is more than 35 times more reactive than the *cis*. Are these data compatible with the mechanism given on p. 748 What additional detail do these data provide about the reaction mechanism? Explain.

4. Predict the products from opening of the two stereoisomeric epoxides derived from limonene shown below by reaction with (a) acetic acid, (b) dimethylamine, and (c) lithium aluminum hydride.

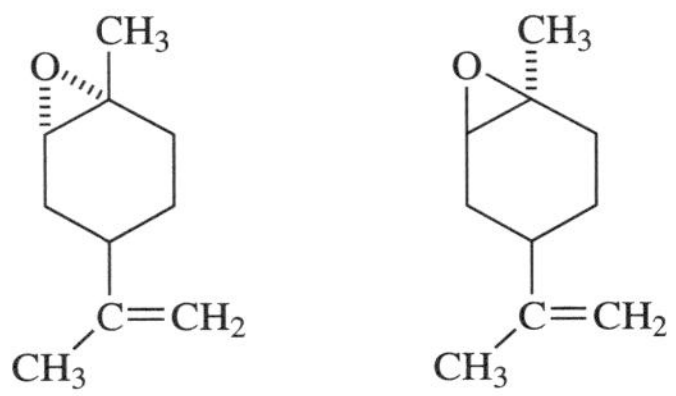

[structures: two stereoisomeric limonene 1,2-epoxides, each with CH_3 on the epoxide carbon and $C(CH_3)=CH_2$ at C-4]

5. The direct oxidative conversion of primary halides or tosylates to aldehydes can be carried out by reaction with dimethyl sulfoxide under alkaline conditions. Formulate a mechanism for this general reaction.

6. A method for synthesis of ozonides that involves no ozone has been reported. It consists of photosensitized oxidation of solutions of diazo compounds and aldehydes. Suggest a mechanism.

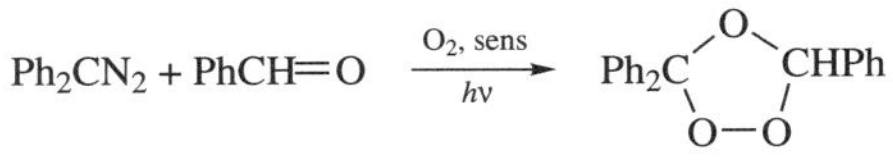

$$Ph_2CN_2 + PhCH{=}O \xrightarrow[h\nu]{O_2,\ \text{sens}} Ph_2C\overset{\;O\;}{\frown}CHPh \text{ (ring closed by } O{-}O)$$

7. Overoxidation of carbonyl products during ozonolysis can be prevented by addition of tetracyanoethylene to the reaction mixture. The stoichiometry of the reaction is then:

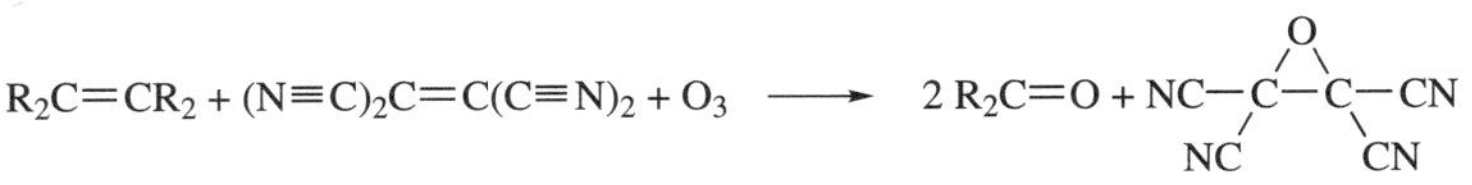

$$R_2C{=}CR_2 + (N{\equiv}C)_2C{=}C(C{\equiv}N)_2 + O_3 \longrightarrow 2\,R_2C{=}O + (NC)_2\overset{O}{C{-}C}(CN)_2$$

Propose a reasonable mechanism that would account for the effect of tetracyanoethylene. Does your mechanism suggest that tetracyanoethylene would be a particularly effective alkene for this purpose? Explain.

8. Suggest a mechanism by which the "abnormal" oxidations shown below might occur.

(a) [cyclohexene with CH_3, H_3C, CH_3 and $CH{=}CHCHCH_3$ / OH substituents] $\xrightarrow{^1O_2}$ [bicyclic lactone with CH_3, CH_3, CH_3, =O]

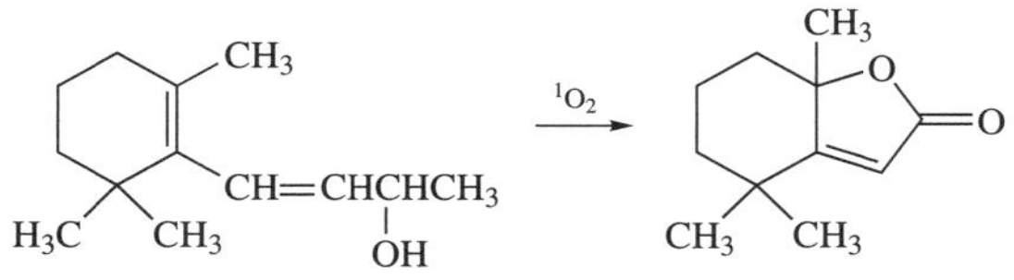

(b) $PhC(=O)C(CH_3)_2OH \xrightarrow[H_2O_2]{^-OH} PhCO_2H + (CH_3)_2C{=}O$

(c) [1-phenyl-2-indanone (Ph, H, =O)] $\xrightarrow[NaOCH_3]{O_2}$ [benzene with $C(=O)Ph$ and $CH_2CO_2CH_3$]

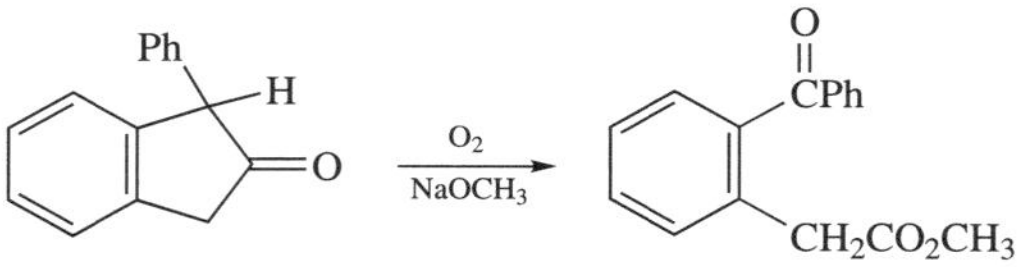

(d) [$THPO(CH_2)_3$, CH_3, H, CH_3 bicyclic ketone, O] $\xrightarrow[2)\ H^+]{1)\ H_2O_2,\ ^-OH}$ [$THPO(CH_2)_3$, CH_3, H, H, CH_3, =O, O, H bicyclic lactone]

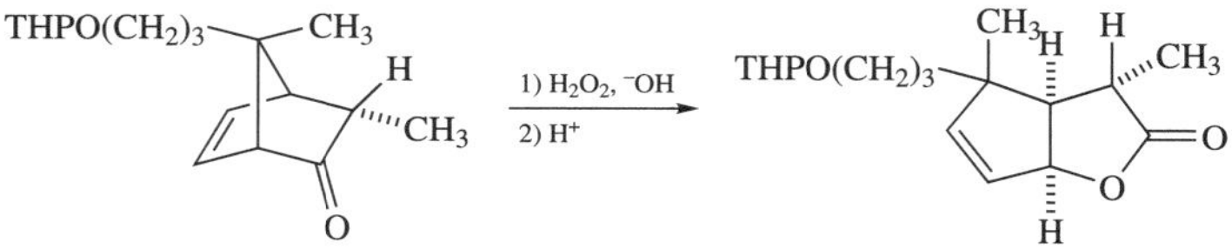

(e) [Ph–C(R)=C(H)–OCH_3] $\xrightarrow{O_2,\ h\nu,\ sens}$ [benzo-fused 1,2-dioxane with R, OCH_3, O–O]

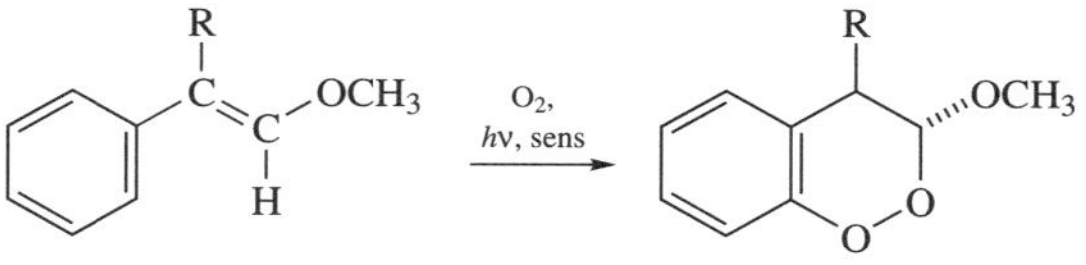

(f) [2-methylphenol (H_3C, OH)] + $CH_3S(=O)CH_3$ $\xrightarrow[H_3PO_4]{dicyclohexyl\text{-}carbodiimide}$ [phenol with H_3C, OH, CH_2SCH_3] (None of the *para* isomer is formed.)

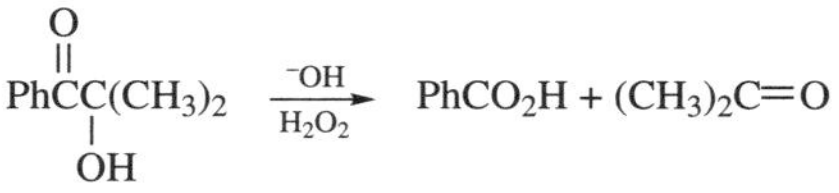

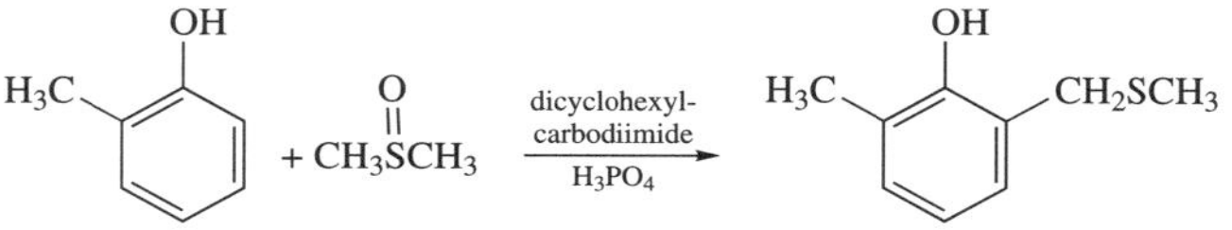

(g) [$CH_3C(=O)CH_2CH_2$, CH_3, $(CH_3)_3SiCH=$, HO_2C, CH_3 cyclohexane] $\xrightarrow[2)\ CF_3CO_2H]{1)\ O_3,\ MeOH,\ -78^\circ C}$ [tetracyclic peroxide lactone with H_3C, O, O, O, O, H, CH_3, H, CH_3, O]

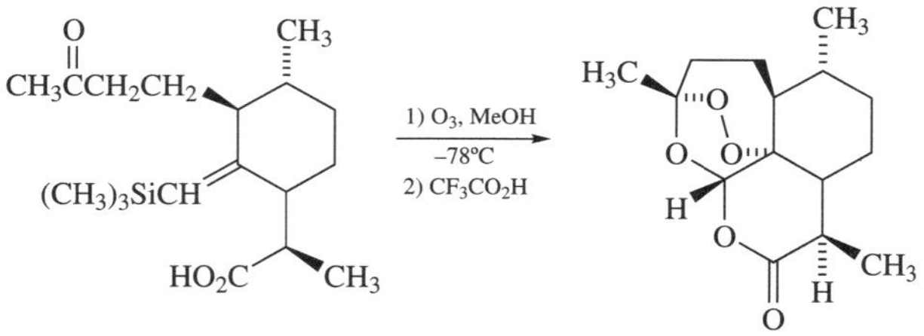

(h) [3,3-dimethylcyclohexanone with CH_3, CH_3, O, $Sn(CH_3)_3$] $\xrightarrow[CH_3CO_2^-Na^+]{H_2O_2,\ CH_3CO_2H}$ $CH_2{=}CH(CH_2)_2C(CH_3)_2CO_2H$

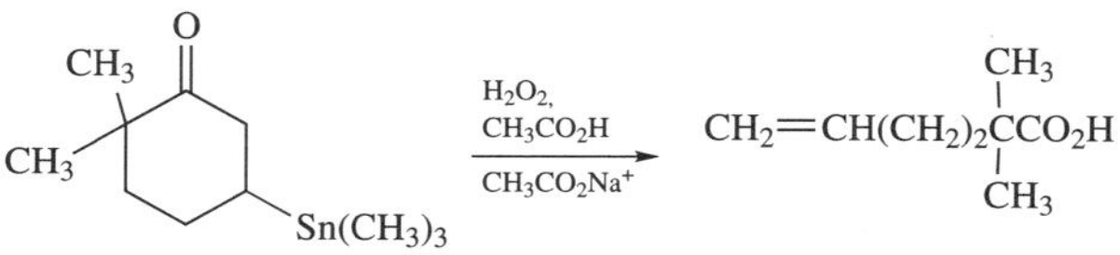

9. Indicate one or more satisfactory oxidants for effecting the following transformations. Each molecule poses problems of selectivity or the need to preserve a potentially

sensitive functional group. In most cases, a "single-pot" process is possible, and in no case are more than three steps required. Explain the basis for your choice of reagent.

(a)

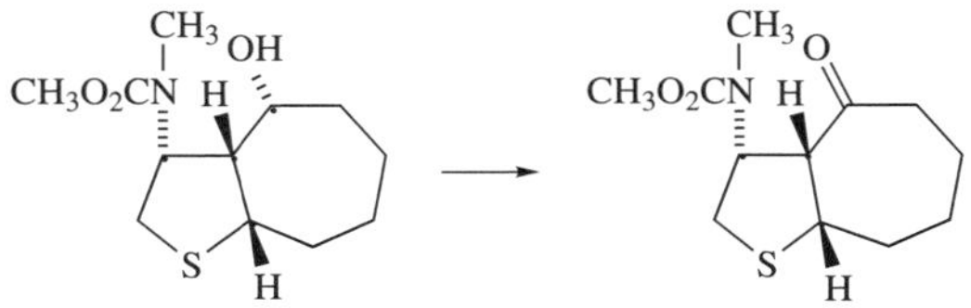

(b)

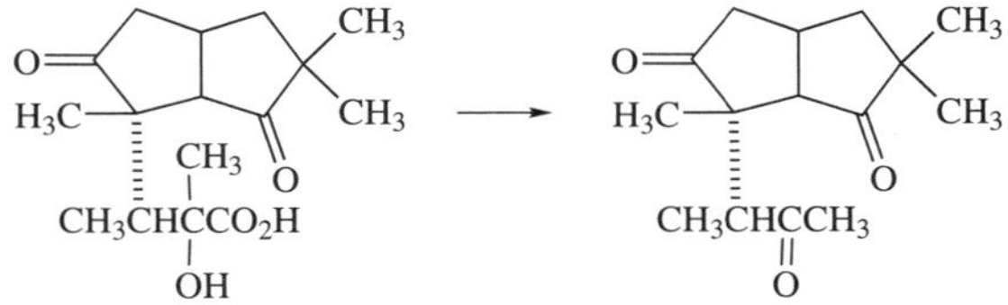

(c)

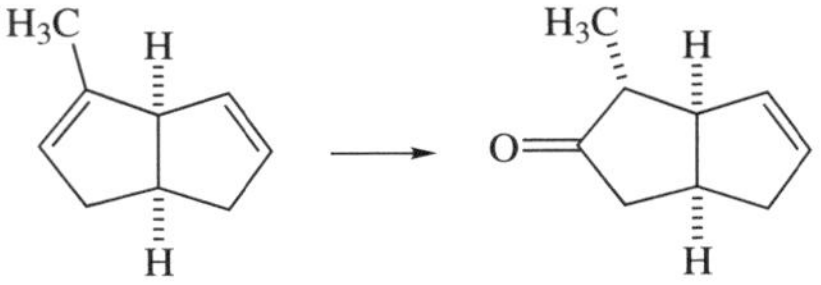

(d)

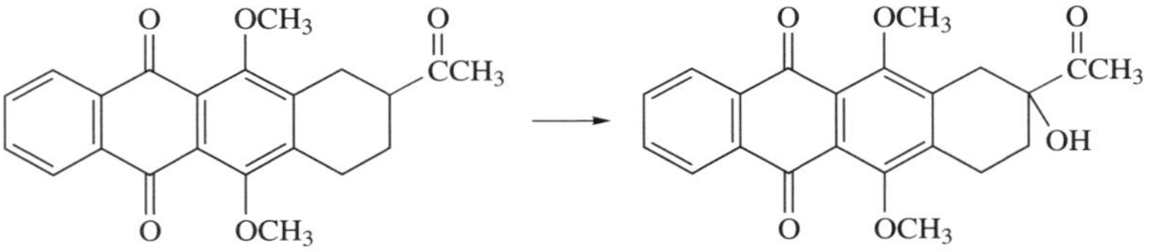

(e)

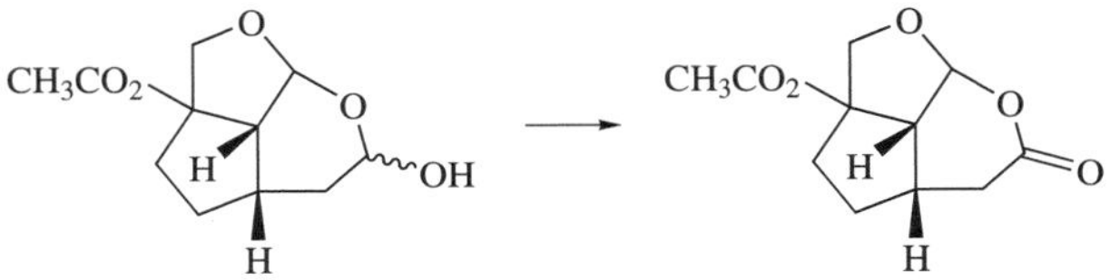

(f)

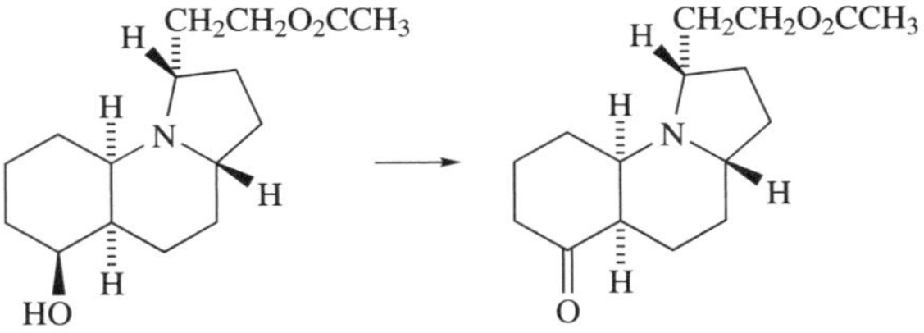

(g)

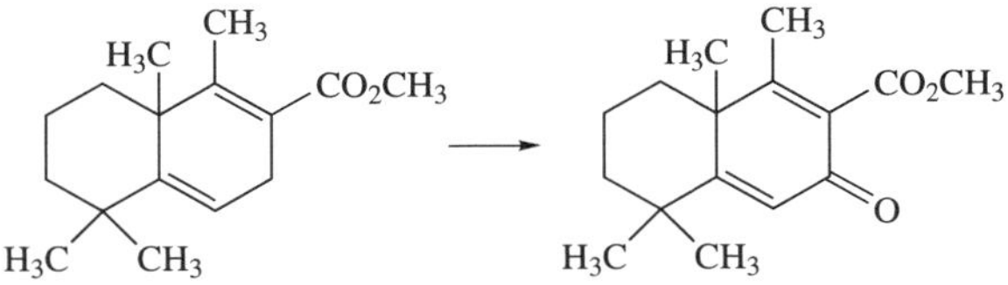

(h)

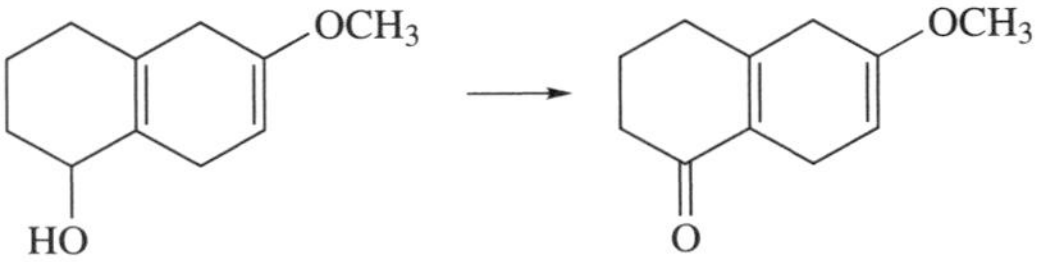

(i)

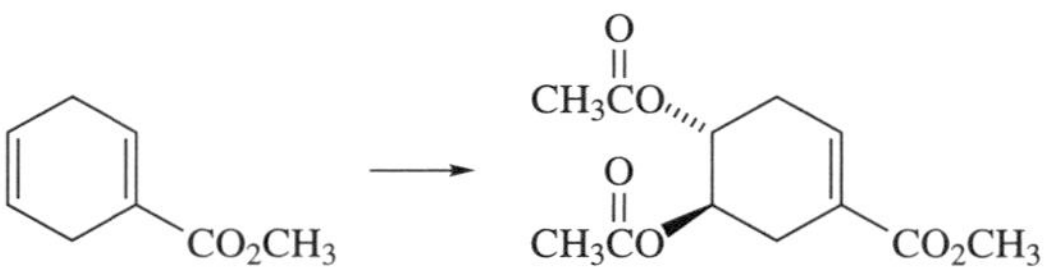

(j)

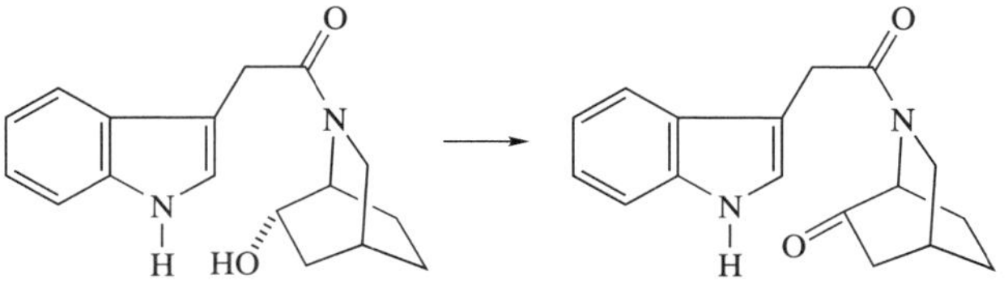

(k)

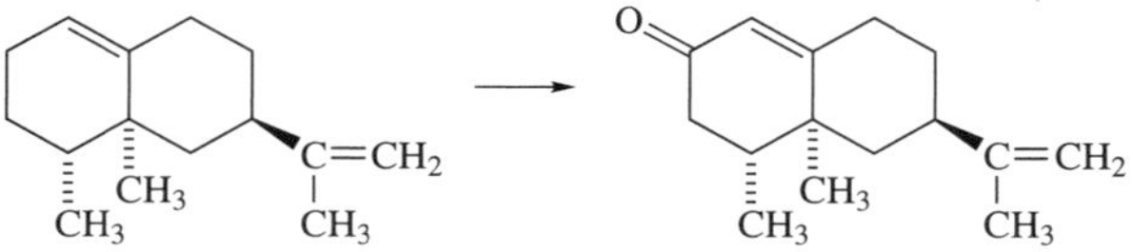

(l)

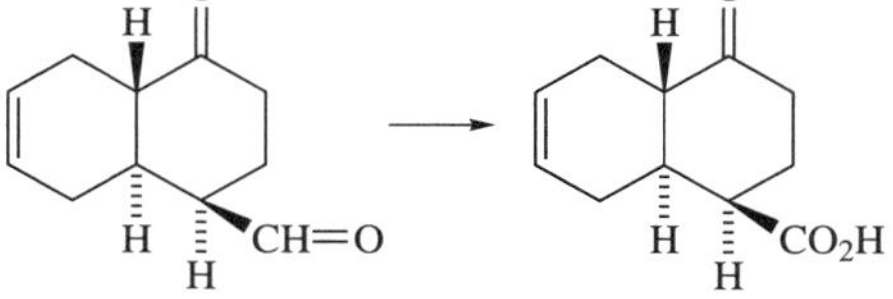

(m)

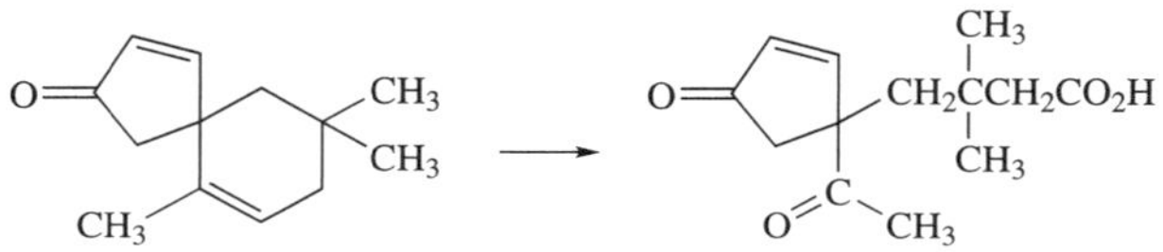

(n)

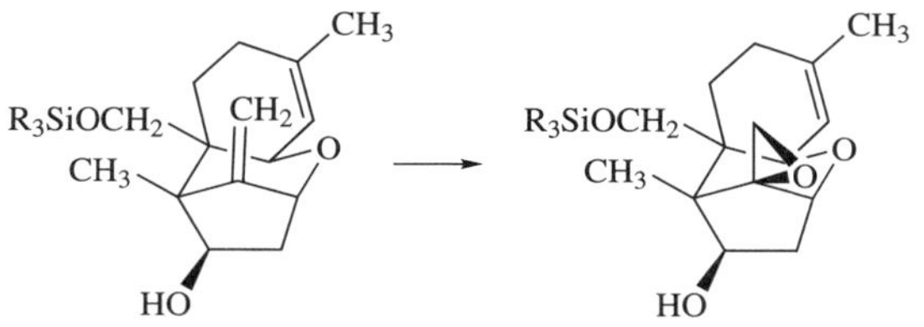

10. It has been noted that when unsymmetrical olefins are ozonized in methanol, there is often a large preference for one cleavage mode over the other. For example,

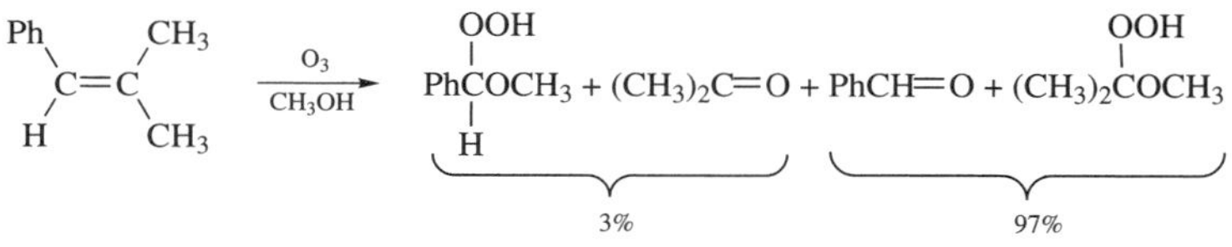

$$\text{PhCH}=\text{C(CH}_3)_2 \xrightarrow[\text{CH}_3\text{OH}]{\text{O}_3} \underbrace{\text{PhCH(OOH)OCH}_3 + (\text{CH}_3)_2\text{C}=\text{O}}_{3\%} + \underbrace{\text{PhCH}=\text{O} + (\text{CH}_3)_2\text{C(OOH)OCH}_3}_{97\%}$$

How would you explain this example of regioselective cleavage?

11. A method for oxidative cleavage of cyclic ketones involves a four-stage process. First, the ketone is converted to an α-phenylthio derivative (see Section 4.7). The ketone is then converted to an alcohol, either by reduction or addition of an organolithium reagent. This compound is then treated with lead tetraacetate to give an oxidation

product in which the hydroxyl group has been acetylated and an additional oxygen added to the β-thioalcohol. Aqueous hydrolysis of this intermediate in the presence of Hg^{2+} gives a dicarbonyl compound. Formulate a likely structure for the product of each reaction step and an overall mechanism for this process.

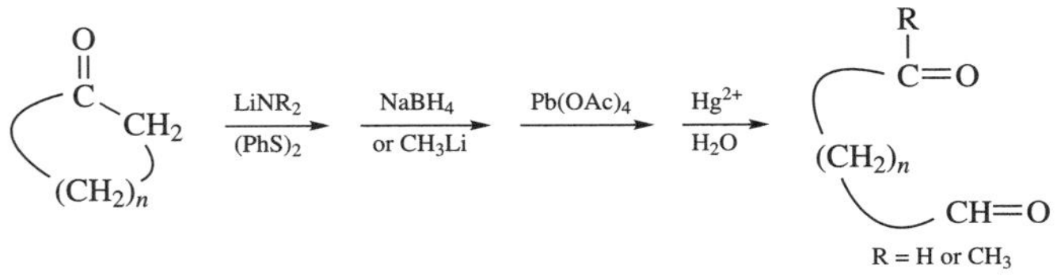

$\xrightarrow{LiNR_2,\ (PhS)_2}$ $\xrightarrow{NaBH_4\ \text{or}\ CH_3Li}$ $\xrightarrow{Pb(OAc)_4}$ $\xrightarrow{Hg^{2+},\ H_2O}$

R = H or CH_3

12. Certain thallium salts, particularly $Tl(NO_3)_3$, effect oxidation with an accompanying rearrangement. Especially good yields are found when the thallium salt is supported on inert material. Two examples are given. Formulate a mechanistic rationalization.

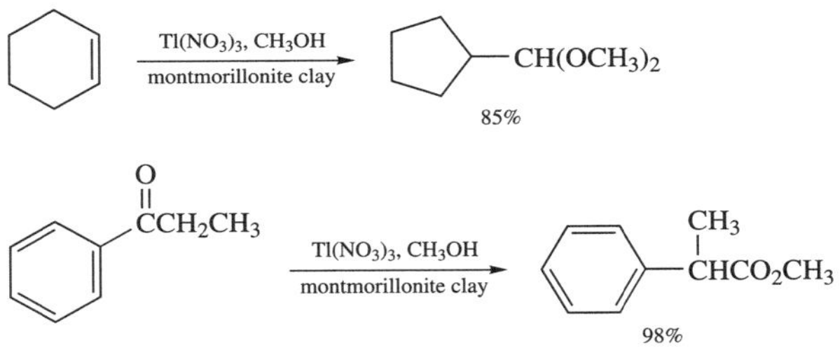

$\xrightarrow{Tl(NO_3)_3,\ CH_3OH,\ \text{montmorillonite clay}}$ $CH(OCH_3)_2$ 85%

CCH_2CH_3 $\xrightarrow{Tl(NO_3)_3,\ CH_3OH,\ \text{montmorillonite clay}}$ $CHCO_2CH_3$ (with CH_3) 98%

13. The two transformations shown below have been carried out by short reaction sequences involving several oxidative steps. Deduce a series of steps which could effect these transformations, and suggest reagents which might be suitable for each step.

(a) CH_3, $CH_3C{=}CH_2$ → CH_3

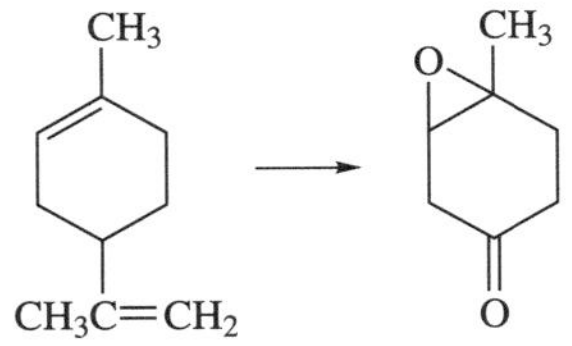

(b) → $HOCH_2$, CH_2CO_2H, H_3C, CH_3

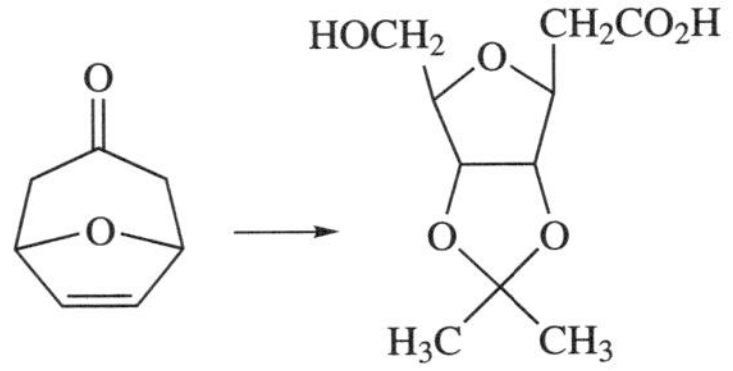

14. Provide mechanistic interpretations of the following reactions.

(a) Account for the products formed under the following conditions. Why does the inclusion of cupric acetate affect the course of the reaction?

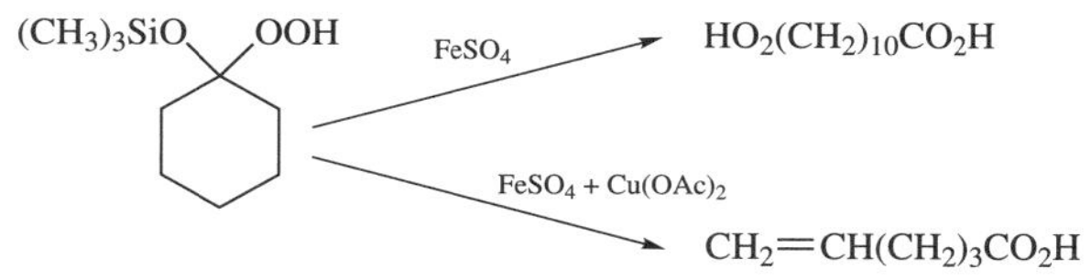

$(CH_3)_3SiO$, OOH (cyclohexane) — $FeSO_4$ → $HO_2(CH_2)_{10}CO_2H$

— $FeSO_4 + Cu(OAc)_2$ → $CH_2{=}CH(CH_2)_3CO_2H$

(b) CH_3N, CN, CH_3 — 1) LDA 2) O_2 3) $NaHSO_3$ → CH_3N, O, CH_3

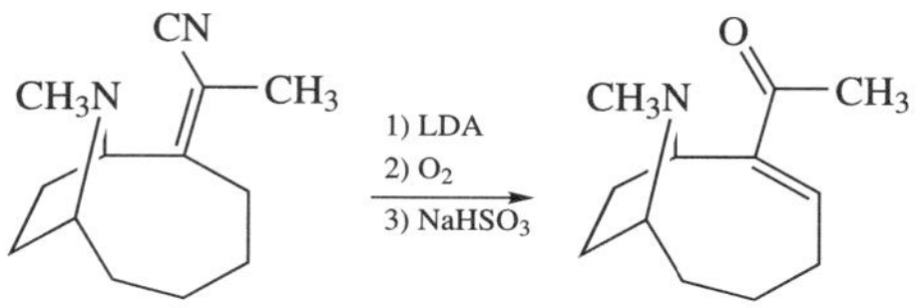

(c) $R_3MCH(R')CH(R'')CO_2H$ (M = Si, Sn) — $Pb(OAc)_4$ → $R'CH{=}CHR''$

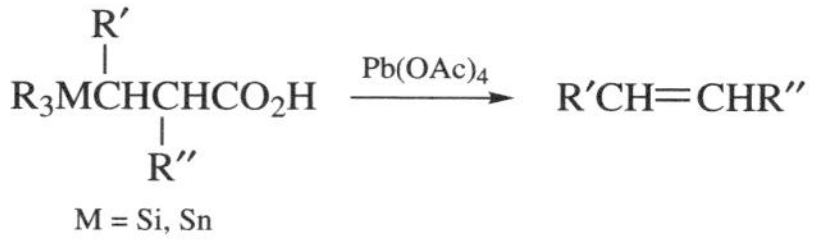

15. Devise a sequence of reactions which could accomplish the formation of the structure on the left from the potential precursor on the right. Pay close attention to stereochemical requirements.

(a) CH_3O, OCH_2Ph, CH_3, O, O, CH_3, H, $HOCH_2$, CH_3, H ⟹ CH_3O, OCH_2Ph, CH_3, OCH_3, O, H_3C, C, H, C=C, H, H, CH_2OH

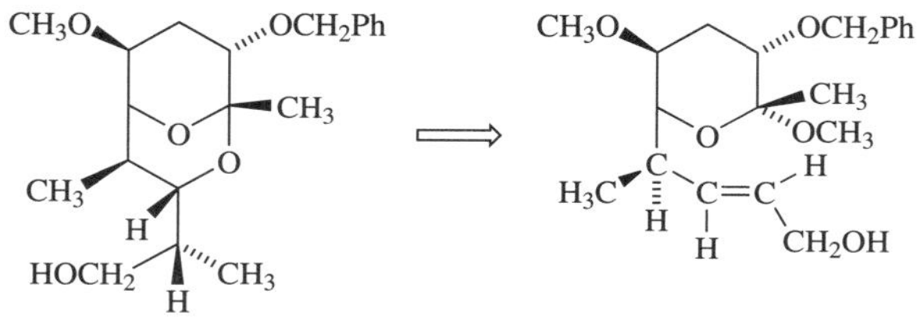

(b) HO, H, H_3C, O, H_3C, H ⟹ CH_3CH_2, H, O, O, H_3C, H, O

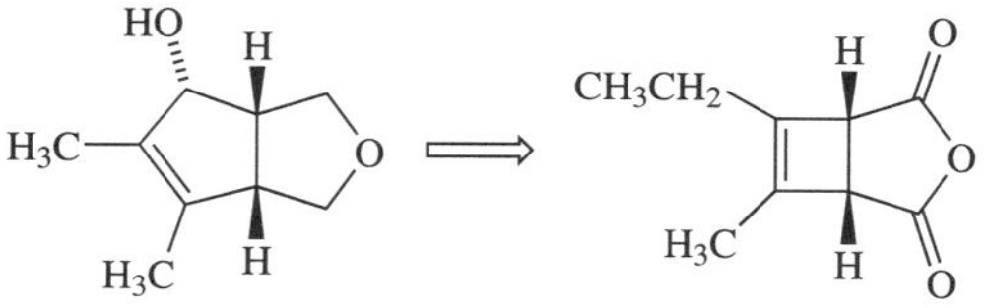

(c) $CH_3O_2C(CH_2)_4CH{=}CH_2$ ⟹ O, $CH_2Si(CH_3)_3$ (cyclohexanone)

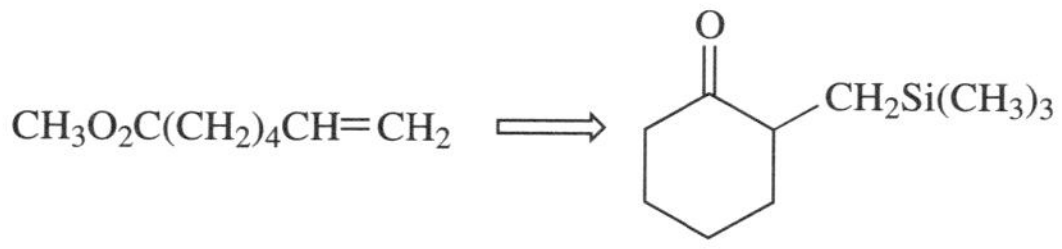

(d) O, O, CH_3, O ⟹ $CH_3O_2C(CH_2)_4CO_2CH_3$

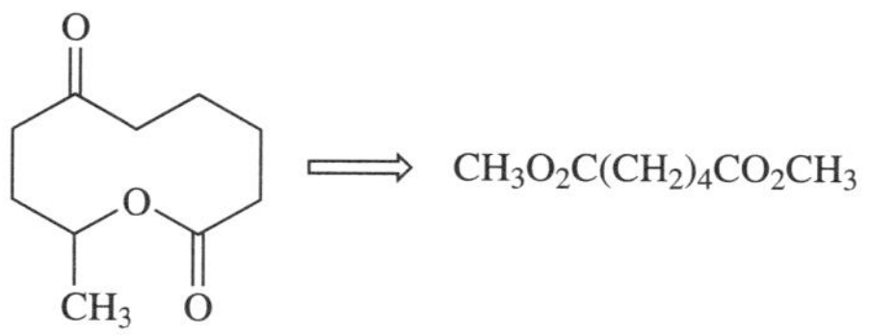

(e) $HOCH_2CH_2(H)C{=}C(CH_3)CH_2CH_2OH \Longrightarrow$ p-$CH_3C_6H_4OCH_3$

(f) $CH_3CH_2CH_2CH_2\overset{O}{\overset{\|}{C}}C{\equiv}CC_6H_5 \Longrightarrow C_6H_5C{\equiv}CH,\ CH_3CH_2CH_2CH_2CH_2OH$

(g) OH, CH_2OH $\Longrightarrow$ OH

(h) O $\Longrightarrow$

(i) $C_6H_5\underset{OH}{\underset{|}{C}}HCH_2{-}N\langle \Longrightarrow C_6H_5CH{=}CH_2$

(j) $CH_3CH_2CH{-}\overset{CH_3}{\overset{|}{C}}CH_2OH$ (O epoxide) $\Longrightarrow CH_3CH_2CH_2OH$

(k) CH_3, CH_3, O, O, $Si(CH_3)_3$, CH_3 $\Longrightarrow$ CH_3, CH_2, O, CH_3

(l) O, CH_3, O $\Longrightarrow$ $(CH_3)_3SiO$, $(CH_3)_3SiO$

(m) $FCH_2\overset{OH}{\overset{|}{C}}HCH_2OCH_2Ph \Longrightarrow CH_2{=}CHCH_2OCH_2Ph$

(n) $CH_3O_2CC{=}CHOCH_3$, H, H, $CH{=}CH_2$, O, O, H, O_2CCH_3 $\Longrightarrow$ $CH_3O_2CC{=}CHOCH_3$, O

(o) O, H, C_2H_5O, C_2H_5O, H, CH_3 $\Longrightarrow$ O, CO_2CH_3, CH_3

(p) $O{=}CH$, CH_3, CH_3, O, CH_3, $C(OCH_3)_2$, CH_3, R_3SiO, $OSiR_3$ $\Longrightarrow$ CH_3, $OSiR_3$, R_3SiO, CH_3, O, R_3SiO

16. Tomoxetine and fluoxetine are antidepressants. Both enantiomers of each compound can be prepared enantiospecifically, starting from cinnamyl alcohol. Give a reaction sequence which would accomplish this objective.

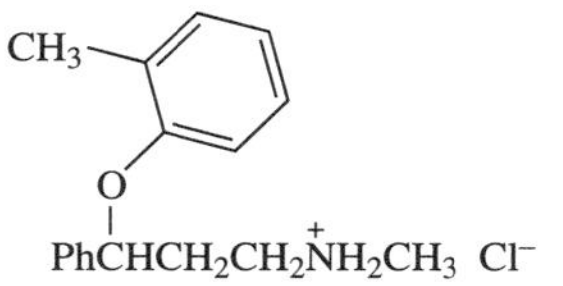

tomoxetine

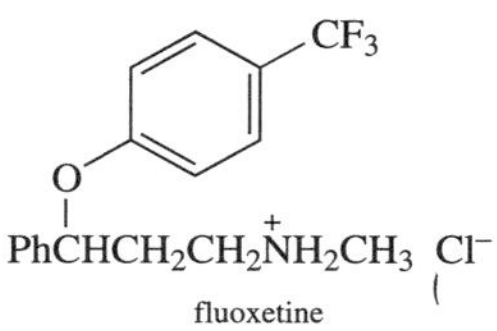

fluoxetine

17. The irradiation of **A** in the presence of rose bengal, oxygen, and acetaldehyde yields the mixture of products shown. Account for the formation of each product.

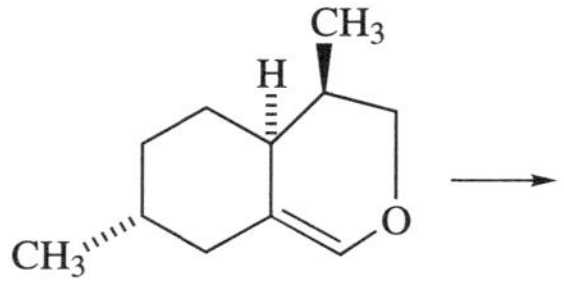

A

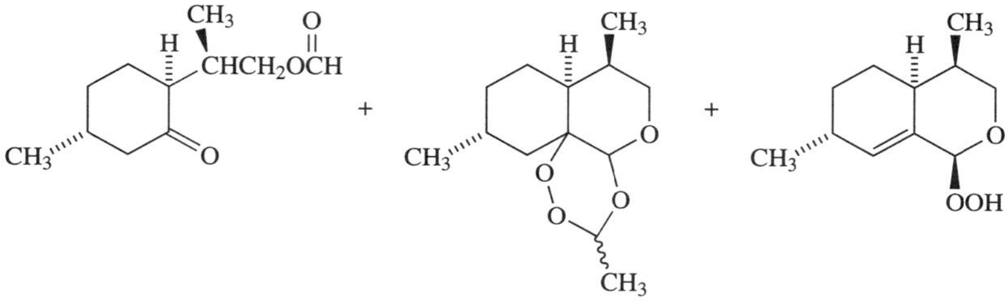

18. Analyze the following data on the product ratios obtained in the epoxidation of 3-substituted cyclohexenes. What are the principal factors which determine the stereoselectivity in each case?

Substituent	*trans* : *cis*[a]
OH	66 : 34[b]
OH	15 : 85
OCH_3	85 : 15
O_2CCH_3	62 : 38
CO_2CH_3	68 : 32
CO_2H	84 : 16
NHCOPh	3 : 97
CH_3	47 : 53
Ph	85 : 15

a. Solvent is 9 : 1 CCl_4–acetone mixture, except where otherwise noted.
b. Solvent is 9 : 1 MeOH–acetone mixture.

13

Planning and Execution of Multistep Syntheses

Introduction

The reactions that have been discussed in the preceding chapters provide the tools for synthesizing new and complex molecules. However, a strategy for using these tools is essential to successful synthesis of molecules by multistep synthetic sequences. The sequence of individual reactions must be planned so that they are mutually compatible. Certain functional groups can interfere with prospective reactions, and such problems must be avoided either by a modification of the sequence or by temporarily masking (protecting) the interfering group. *Protective groups* are used to temporarily modify functionality, which is then restored when the protecting group is removed. Another approach is to use *synthetic equivalent groups* in which a particular functionality is introduced as an alternative structure that can subsequently be converted to the desired group.

Protective groups and synthetic equivalent groups are tactical tools of multistep synthesis. They are the means, along with the individual synthetic methods, to reach the goal of a completed synthesis. These tactical steps must be incorporated into an overall synthetic plan. The synthetic plan is normally created on the basis of *retrosynthetic analysis*, which involves the identification of the particular bonds that need to be formed to obtain the desired molecule. Depending on the complexity of the synthetic target, the retrosynthetic analysis may be obvious or highly complicated. An eventual plan will identify potential starting materials and synthetic steps which could lead to the desired molecule. Most synthetic plans involve a combination of *linear sequences* and *convergent steps*. Linear sequences transform the starting material step-by-step by incremental transformations. Convergent steps bring together larger fragments of the molecule, which have been created by linear sequences. Because overall synthetic yields are the product of the yield for each of the individual steps in the synthesis, incorporation of convergent steps can improve overall yield by reducing the length of the individual linear sequences. After discussing protective groups, synthetic equivalent groups, and retrosynthetic analysis, this chapter will summarize several syntheses which demonstrate successful application of multi-step synthetic methods.

13.1. Protective Groups

The reactions that have been discussed to this point provide the tools for synthesis of organic compounds. When the synthetic target is a relatively complex molecule, a sequence of reactions that would lead to the desired product must be devised. At the present time, syntheses requiring 15–20 steps are common, and many that are even longer have been developed. In the planning and execution of such multistep syntheses, an important consideration is the compatibility of the functional groups that are already present with the reaction conditions required for subsequent steps. It is frequently necessary to modify a functional group in order to prevent interference with some reaction in the synthetic sequence. One way to do this is by use of a *protective group*. A protective group is a derivative that can be put in place, and then subsequently removed, in order to prevent an undesired reaction or other adverse influence. For example, alcohols are often protected as trisubstituted silyl ethers, and aldehydes as acetals. The use of the silyl group replaces the hydroxyl group with a less nucleophilic and aprotic silyl ether. The acetal group prevents both unwanted nucleophilic addition and enolate formation at an aldehyde site.

$$\mathrm{R{-}OH + R'_3SiX \longrightarrow R{-}O{-}SiR'_3}$$

$$\mathrm{RCH{=}O + R'OH \longrightarrow RCH(OR')_2}$$

Protective groups play a passive role in synthesis. Each operation of introduction and removal of a protective group adds steps to the synthetic sequence. It is thus desirable to minimize the number of such operations. Fortunately, the methods for protective group introduction and removal have been highly developed, and the yields are usually excellent.

Three considerations are important in choosing an appropriate protective group: (1) the nature of the group requiring protection; (2) the reaction conditions under which the protective group must be stable; and (3) the conditions that can be tolerated for removal of the protective group. No universal protective groups exist. The state of the art has been developed to a high level, however, and the many mutually complementary protective groups provide a great degree of flexibility in the design of syntheses for complex molecules.[1]

13.1.1. Hydroxyl-Protecting Groups

A common requirement in synthesis is that a hydroxyl group be masked as a derivative lacking a hydroxylic proton. An example of this requirement is in reactions involving Grignard or other organometallic reagents. The acidic hydrogen of a hydroxyl group will destroy one equivalent of a strongly basic organometallic reagent and possibly adversely affect the reaction in other ways. Conversion to an alkyl or silyl ether is the most common means of protecting hydroxyl groups. The choice of the most appropriate ether

1. The book *Protective Groups in Organic Synthesis*, which is listed in the general references, provides a thorough survey of protective groups and the conditions for introduction and removal.

group is largely dictated by the conditions that can be tolerated in subsequent removal of the protective group. An important method that is applicable when mildly acidic hydrolysis is an appropriate method for deprotection is to form a tetrahydropyranyl ether (THP group).[2]

ROH + (dihydropyran) $\xrightarrow{H^+}$ RO–THP

This protective group is introduced by an acid-catalyzed addition of the alcohol to the vinyl ether moiety in dihydropyran. *p*-Toluenesulfonic acid or its pyridinium salt is used most frequently as the catalyst,[3] although other catalysts are advantageous in special cases. The THP group can be removed by dilute aqueous acid. The chemistry involved in both the introduction and deprotection stages is the reversible acid-catalyzed formation and hydrolysis of an acetal (see Part A, Section 8.1).

introduction: ROH + (oxocarbenium ion) ⇌ $\mathrm{RO^+H}$–THP ⇌ RO–THP + H^+

removal: RO–THP $\xrightarrow{H^+}$ $\mathrm{RO^+H}$–THP $\xrightarrow{H_2O}$ ROH + HO–THP

The THP group, like other acetals and ketals, is inert to nucleophilic reagents and is unchanged under such conditions as hydride reduction, organometallic reactions, or base-catalyzed reactions in aqueous solution. It also protects the hydroxyl group against oxidation.

A disadvantage of the THP group is the fact that a stereogenic center is produced at C-2 of the tetrahydropyran ring. This presents no difficulties if the alcohol is achiral, since a racemic mixture results. However, if the alcohol is chiral, the reaction will give a mixture of diastereomeric ethers, which may complicate purification and characterization. One way of avoiding this problem is to use methyl 2-propenyl ether in place of dihydropyran. No new chiral center is introduced, and this acetal offers the further advantage of being hydrolyzed under somewhat milder conditions than those required for THP ethers.[4]

$$\mathrm{ROH + CH_2{=}C(CH_3){-}OCH_3 \xrightarrow{H^+} ROC(CH_3)_2OCH_3}$$

Ethyl vinyl ether is also useful for hydroxyl-group protection. The resulting derivative (1-ethoxyethyl ether) is abbreviated as the EE group.[5] As with the THP group, the EE group contains a stereogenic center.

2. W. E. Parham and E. L. Anderson, *J. Am. Chem. Soc.* **70:**4187 (1948).
3. J. H. van Boom, J. D. M. Herscheid, and C. B. Reese, *Synthesis* **1973:**169; M. Miyashita, A. Yoshikoshi, and P. A. Grieco, *J. Org. Chem.* **42:**3772 (1977).
4. A. F. Kluge, K. G. Untch, and J. H. Fried, *J. Am. Chem. Soc.* **94:**7827 (1972).
5. H. J. Sims, H. B. Parseghian, and P. L. DeBenneville, *J. Org. Chem.* **23:**724 (1958).

The methoxymethyl (MOM) and β-methoxyethoxymethyl (MEM) groups are used to protect alcohols and phenols as formaldehyde acetals. The groups are normally introduced by reaction of an alkali-metal salt of the alcohol with methoxymethyl chloride or β-methoxyethoxymethyl chloride.[6]

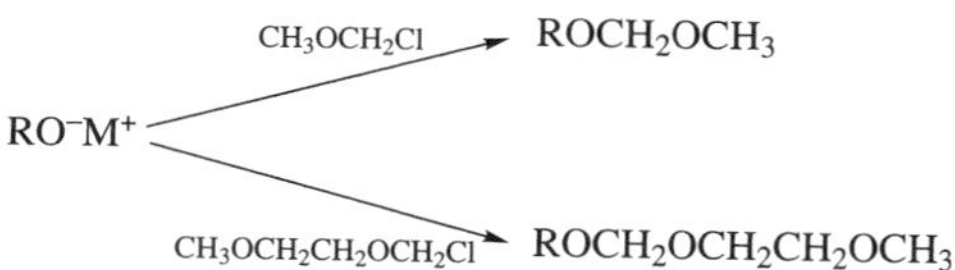

An attractive feature of the MEM group is the ease with which it can be removed under nonaqueous conditions. Lewis acids such as zinc bromide, magnesium bromide, titanium tetrachloride, dimethylboron bromide, and trimethylsilyl iodide permit its removal.[7] The MEM group is cleaved in preference to the MOM or THP groups under these conditions. Conversely, the MEM group is more stable to acidic aqueous hydrolysis than the THP group. These relative reactivity relationships allow the THP and MEM groups to be used in a complementary fashion when two hydroxyl groups must be deprotected at different points in a synthetic sequence.

$CH_2{=}CH$ … THPO … OMEM $\xrightarrow[35°C,\ 40\ h]{CH_3CO_2H,\ H_2O,\ THF}$ $CH_2{=}CH$ … HO … OMEM Ref. 8

The methylthiomethyl (MTM) group is a related alcohol-protecting group. There are several methods for introducing the MTM group. Alkylation of an alcoholate by methylthiomethyl chloride is efficient if catalyzed by iodide ion.[9] Alcohols are also converted to MTM ethers by reaction with dimethyl sulfoxide in the presence of acetic acid and acetic anhydride[10] or with benzoyl peroxide and dimethyl sulfide.[11] The latter two methods involve the generation of the methylthiomethylium ion by ionization of an acyloxysulfonium ion (*Pummerer reaction*).

$$RO^-M^+ + CH_3SCH_2Cl \xrightarrow{I^-} ROCH_2SCH_3$$

$$ROH + CH_3SOCH_3 \xrightarrow[(CH_3CO)_2O]{CH_3CO_2H} ROCH_2SCH_3$$

$$ROH + (CH_3)_2S + (PhCO_2)_2 \longrightarrow ROCH_2SCH_3$$

6. G. Stork and T. Takahashi, *J. Am. Chem. Soc.* **99:**1275 (1977); R. J. Linderman, M. Jaber, and B. D. Griedel, *J. Org. Chem.* **59:**6499 (1994); P. Kumar, S. V. N. Raju, R. S. Reddy, and B. Pandey, *Tetrahedron Lett.* **35:**1289 (1994).
7. E. J. Corey, J.-L. Gras, and P. Ulrich, *Tetrahedron Lett.* **1976:**809; Y. Quindon, H. E. Morton, and C. Yoakim, *Tetrahedron Lett.* **24:**3969 (1983); J. H. Rigby and J. Z. Wilson, *Tetrahedron Lett.* **25:**1429 (1984); S. Kim, Y. H. Park, and I. S. Lee, *Tetrahedron Lett.* **32:**3099 (1991).
8. E. J. Corey, R. L. Danheiser, S. Chandrasekaran, P. Siret, G. E. Keck, and J.-L. Gras, *J. Am. Chem. Soc.* **100:**8031 (1978).
9. E. J. Corey and M. G. Bock, *Tetrahedron Lett.* **1975:**3269.
10. P. M. Pojer and S. J. Angyal, *Tetrahedron Lett.* **1976:**3067.
11. J. C. Modina, M. Salomon, and K. S. Kyler, *Tetrahedron Lett.* **29:**3773 (1988).

The MTM group is selectively removed under nonacidic conditions in aqueous solutions containing Ag^+ or Hg^{2+} salts. The THP and MOM groups are stable under these conditions.[8] The MTM group can also be removed by reaction with methyl iodide, followed by hydrolysis of the resulting sulfonium salt in moist acetone.[9]

Two substituted alkoxymethoxy groups are designed for cleavage involving β elimination. The 2,2,2-trichloroethoxymethyl group can be cleaved by reducing agents, including zinc, samarium diiodide, and sodium amalgam.[12] The β elimination results in the formation of a formaldehyde hemiacetal, which decomposes easily.

$$Cl_3CCH_2OCH_2OR \xrightarrow{2e^-} Cl^- + Cl_2C{=}CH_2 + CH_2{=}O + {}^-OR$$

The 2-(trimethylsilyl)ethoxymethyl (SEM) group can be removed by various fluoride sources, including TBAF, pyridinium fluoride, and HF.[13]

$$(CH_3)_3SiCH_2CH_2OCH_2OR \xrightarrow{F^-} (CH_3)_3SiF + CH_2{=}CH_2 + CH_2{=}O + {}^-OR$$

The simple alkyl groups are generally not very useful for protection of alcohols as ethers. Although they can be introduced readily by alkylation, subsequent cleavage requires strongly electrophilic reagents such as boron tribromide (see Section 3.3). The *t*-butyl group is an exception and has found some use as a hydroxyl-protecting group. Because of the stability of the *t*-butyl cation, *t*-butyl ethers can be cleaved under moderately acidic conditions. Trifluoroacetic acid in an inert solvent is frequently used.[14] *t*-Butyl ethers can also be cleaved by acetic anhydride–$FeCl_3$ in ether.[15] The *t*-butyl group is normally introduced by reaction of the alcohol with isobutylene in the presence of an acid catalyst.[11,16] Acidic ion-exchange resins are effective catalysts.[17]

$$ROH + CH_2{=}C(CH_3)_2 \xrightarrow{H^+} ROC(CH_3)_3$$

The triphenylmethyl (trityl, abbreviated Tr) group is removed under even milder conditions than the *t*-butyl group and is an important hydroxyl-protecting group, especially in carbohydrate chemistry.[18] This group is introduced by reaction of the alcohol with triphenylmethyl chloride via an S_N1 substitution. Hot aqueous acetic acid suffices to remove the trityl group. The ease of removal can be increased by addition of electron-releasing substituents. The *p*-methoxy derivatives are used in this way.[19] Because of their steric bulk, triarylmethyl groups are usually introduced only at primary hydroxyl groups. Reactions at secondary hydroxyl groups can be achieved by using stronger organic bases such as DBU.[20]

The benzyl group can serve as a hydroxyl-protecting group when acidic conditions for ether cleavage cannot be tolerated. The benzyl C–O bond is cleaved by catalytic

12. R. M. Jacobson and J. W. Clader, *Synth. Commun.* **9**:57 (1979); D. A. Evans, S. W. Kaldor, T. K. Jones, J. Clardy, and T. J. Stout, *J. Am. Chem. Soc.* **112**:7001 (1990).
13. B. H. Lipshutz and J. J. Pegram, *Tetrahedron Lett.* **21**:3343 (1980); T. Kan, M. Hashimoto, M. Yanagiya, and H. Shirahama, *Tetrahedron Lett.* **29**:5417 (1988); J. D. White and M. Kawasaki, *J. Am. Chem. Soc.* **112**:4991 (1990); K. Sugita, K. Shigeno, C. F. Neville, H. Sasai, and M. Shibasaki, *Synlett:* **1994**:325.
14. H. C. Beyerman and G. J. Heiszwolf, *J. Chem. Soc.* **1963**:755.
15. B. Ganem and V. R. Small, Jr., *J. Org. Chem.* **39**:3728 (1974).
16. J. L. Holcombe and T. Livinghouse, *J. Org. Chem.* **51**:111 (1986).
17. A. Alexakis, M. Gardette, and S. Colin, *Tetrahedron Lett.* **29**:2951 (1988).
18. O. Hernandez, S. K. Chaudhary, R. H. Cox, and J. Porter, *Tetrahedron Lett.* **22**:1491 (1981); S. K. Chaudhary and O. Hernandez, *Tetrahedron Lett.* **20**:95 (1979).
19. M. Smith, D. H. Rammler, I. H. Goldberg, and H. G. Khorana, *J. Am. Chem. Soc.* **84**:430 (1962).
20. S. Colin-Messager, J.-P. Girard, and J.-C. Rossi, *Tetrahedron Lett.* **33**:2689 (1992).

hydrogenolysis[21] or by electron-transfer reduction using sodium in liquid ammonia or aromatic radical anions.[22] Benzyl ethers can also be cleaved using formic acid, cyclohexene, or cyclohexadiene as hydrogen sources in transfer hydrogenolysis catalyzed by platinum or palladium.[23]

Several nonreductive methods for cleavage of benzyl groups have also been developed. Treatment with *s*-butyllithium, followed by reaction with trimethyl borate and then hydrogen peroxide, liberates the alcohol.[24] The lithiated ether forms an alkyl boronate, which is oxidized as discussed in Section 4.9.2.

$$ROCH_2Ph \xrightarrow{s\text{-BuLi}} RO\underset{}{\overset{Li}{C}}HPh \xrightarrow{B(OCH_3)_2} (CH_3O)_2B\text{-}CH(OR)Ph \xrightarrow{H_2O_2} (CH_3O)_2BO\text{-}CH(OR)Ph \longrightarrow ROH + PhCH{=}O$$

Lewis acids such as $FeCl_3$ and $SnCl_4$ also cleave benzyl ethers.[25]

Benzyl groups having 4-methoxy or 3,5-dimethoxy substituents can be removed oxidatively by dichlorodicyanoquinone (DDQ).[26] These reactions presumably proceed through a benzyl cation, and the methoxy substituent is necessary to facilitate the oxidation.

$$CH_3O\text{-}C_6H_4\text{-}CH_2OR \xrightarrow[-H^+]{-2e^-} CH_3O\text{-}C_6H_4\text{-}\overset{+}{C}HOR \xrightarrow{H_2O} CH_3O\text{-}C_6H_4\text{-}CH(OH)OR \longrightarrow ROH$$

These reaction conditions do not affect most of the other common hydroxyl-protecting groups, and methoxybenzyl groups are therefore useful in synthetic sequences that require selective deprotection of different hydroxyl groups. 4-Methoxybenzyl ethers can also be selectively cleaved by dimethylboron bromide.[27]

Benzyl groups are usually introduced by the Williamson reaction (Section 3.2.4); they can also be prepared under nonbasic conditions if necessary. Benzyl alcohols are converted to trichloroacetimidates by reaction with trichloroacetonitrile. These then react with an

21. W. H. Hartung and R. Simonoff, *Org. React.* **7**:263 (1953).
22. E. J. Reist, V. J. Bartuska, and L. Goodman, *J. Org. Chem.* **29**:3725 (1964); R. E. Ireland, D. W. Norbeck, G. S. Mandel, and N. S. Mandel, *J. Am. Chem. Soc.* **107**:3285 (1985); R. E. Ireland and M. G. Smith, *J. Am. Chem. Soc.* **110**:854 (1988); H.-J. Liu, J. Yip, and K.-S. Shia, *Tetrahedron Lett.* **38**:2253 (1997).
23. B. ElAmin, G. M. Anatharamaiah, G. P. Royer, and G. E. Means, *J. Org. Chem.* **44**:3442 (1979); A. M. Felix, E. P. Heimer, T. J. Lambros, C. Tzougraki, and J. Meienhofer, *J. Org. Chem.* **43**:4194 (1978); A. E. Jackson and R. A. W. Johnstone, *Synthesis* **1976**:685; G. M. Anatharamaiah and K. M. Sivandaiah, *J. Chem. Soc., Perkins Trans. 1* **1977**:490.
24. D. A. Evans, C. E. Sacks, W. A. Kleschick, and T. R. Taber, *J. Am. Chem. Soc.* **101**:6789 (1979).
25. M. H. Park, R. Takeda, and K. Nakanishi, *Tetrahedron Lett.* **28**:3823 (1987).
26. Y. Oikawa, T. Yoshioka, and O. Yonemitsu, *Tetrahedron Lett.* **23**:885 (1982); Y. Oikawa, T. Tanaka, K. Horita, T. Yoshioka, and O. Yonemitsu, *Tetrahedron Lett.* **25**:5393 (1984); N. Nakajima, T. Hamada, T. Tanaka, Y. Oikawa, and O. Yonemitsu, *J. Am. Chem. Soc.* **108**:4645 (1986).
27. N. Hebert, A. Beck, R. B. Lennox, and G. Just, *J. Org. Chem.* **57**:1777 (1992).

alcohol to transfer the benzyl group.[28]

$$ArCH_2OH + Cl_3CCN \longrightarrow ArCH_2O\overset{NH}{\overset{\|}{C}}CCl_3 \xrightarrow{ROH} ROCH_2Ar + Cl_3C\overset{O}{\overset{\|}{C}}NH_2$$

Phenyldiazomethane can also be used to introduce benzyl groups.[29]

4-Methoxyphenyl (PMP) ethers find occasional use as hydroxyl-protecting groups. Unlike benzylic groups, they cannot be made directly from the alcohol. Instead, the phenoxy group must be introduced by a nucleophilic substitution.[30] Mitsunobu conditions are frequently used.[31] The PMP group can be cleaved by oxidation with ceric ammonium nitrate (CAN).

Allyl ethers can be cleaved by conversion to propenyl ethers, followed by acidic hydrolysis of the enol ether.

$$ROCH_2CH{=}CH_2 \longrightarrow ROCH{=}CHCH_3 \xrightarrow{H_3O^+} ROH + CH_3CH_2CH{=}O$$

The isomerization of an allyl ether to a propenyl ether can be achieved either by treatment with potassium *t*-butoxide in dimethyl sulfoxide[32] or by Wilkinson's catalyst, $(PPh_3)_3RhCl$[33] or $(PPh_3)_4RhH$.[34] Heating allyl ethers with Pd/C in acidic methanol can also effect cleavage.[35] This reaction, too, is believed to involve isomerization to the 1-propenyl ether. Other very mild conditions for allyl group cleavage include Wacker oxidation conditions[36] (see Section 8.2) and DiBAlH with catalytic $NiCl_2$(dppp).[37]

Silyl ethers play a very important role as hydroxyl-protecting groups.[38] Alcohols can be easily converted to trimethylsilyl (TMS) ethers by reaction with trimethylsilyl chloride in the presence of an amine or by heating with hexamethyldisilazane. *t*-Butyldimethylsilyl (TBDMS) ethers are also very useful. The increased steric bulk of the TBDMS group improves the stability of the group toward such reactions as hydride reduction and Cr(VI) oxidation. The TBDMS group is normally introduced using a tertiary amine as a catalyst for the reaction of the alcohol with *t*-butyldimethylsilyl chloride or triflate. Cleavage of the TBDMS group is slow under hydrolytic conditions, but anhydrous tetra-*n*-butylammonium

28. H.-P. Wessel, T. Iverson, and D. R. Bundle, *J. Chem. Soc., Perkin Trans. 1* **1985:**2247; N. Nakajima, K. Horita, R. Abe, and O. Yonemitsu, *Tetrahedron Lett.* **29:**4139 (1988); S. J. Danishefsky, S. DeNinno, and P. Lartey, *J. Am. Chem. Soc.* **109:**2082 (1987).
29. L. J. Liotta and B. Ganem, *Tetrahedron Lett.* **30:**4759 (1989).
30. Y. Masaki, K. Yoshizawa, and A. Itoh, *Tetrahedron Lett.* **37:**9321 (1996); S. Takano, M. Moriya, M. Suzuki, Y. Iwabuchi, T. Sugihara, and K. Ogaswawara, *Heterocycles* **31:**1555 (1990).
31. T. Fukuyama, A. A. Laird, and L. M. Hotchkiss, *Tetrahedron Lett.* **26:**6291 (1985); M. Petitou, P. Duchaussoy, and J. Choay, *Tetrahedron Lett.* **29:**1389 (1988).
32. R. Griggs and C. D. Warren, *J. Chem. Soc. C* **1968:**1903.
33. E. J. Corey and J. W. Suggs, *J. Org. Chem.* **38:**3224 (1973).
34. F. E. Ziegler, E. G. Brown, and S. B. Sobolov, *J. Org. Chem.* **55:**3691 (1990).
35. R. Boss and R. Scheffold, *Angew. Chem. Int. Ed. Engl.* **15:**558 (1976).
36. H. B. Mereyala and S. Guntha, *Tetrahedron Lett.* **34:**6929 (1993).
37. T. Taniguchi and K. Ogasawara, *Angew. Chem. Int. Ed. Engl.* **37:**1136 (1998).
38. J. F. Klebe, in *Advances in Organic Chemistry, Methods and Results*, Vol. 8, E. C. Taylor, ed., Wiley-Interscience, New York, 1972, pp. 97–178, A. E. Pierce, *Silylation of Organic Compounds*, Pierce Chemical Company, Rockford, Illinois, 1968.

Table 13.1. Common Hydroxyl-Protecting Groups

Structure	Name	Abbreviation
A. Ethers		
$C_6H_5CH_2OR$	Benzyl	Bn
$CH_3O-C_6H_4-CH_2OR$	*p*-Methoxybenzyl	PMB
$CH_2{=}CHCH_2OR$	Allyl	
Ph_3COR	Triphenylmethyl (trityl)	Tr
$CH_3O-C_6H_4-OR$	*p*-Methoxyphenyl	PMP
B. Acetals		
2-OR-tetrahydropyran (O, OR)	Tetrahydropyranyl	THP
CH_3OCH_2OR	Methoxymethyl	MOM
$CH_3CH_2OCH(CH_3)OR$	1-Ethoxyethyl	EE
$(CH_3)_2C(OCH_3)OR$	2-Methoxy-2-propyl	MOP
$Cl_3CCH_2OCH_2OR$	2,2,2-Trichloroethoxymethyl	
$CH_3OCH_2CH_2OCH_2OR$	2-Methoxyethoxymethyl	MEM
$(CH_3)_3SiCH_2CH_2OCH_2OR$	2-Trimethylsilylethoxymethyl	SEM
CH_3SCH_2OR	Methylthiomethyl	MTM
C. Silyl ethers		
$(CH_3)_3SiOR$	Trimethylsilyl	TMS
$(C_2H_5)_3SiOR$	Triethylsilyl	TES
$[(CH_3)_2CH]_3SiOR$	Triisopropylsilyl	TIPS
Ph_3SiOR	Triphenylsilyl	TPS
$(CH_3)_3CSi(CH_3)_2SiOR$	*t*-Butyldimethylsilyl	TBDMS
$(CH_3)_3CSi(Ph)_2SiOR$	*t*-Butyldiphenylsilyl	TBDPS
D. Esters		
CH_3CO_2R	Acetate	Ac
$PhCO_2R$	Benzoate	Bz
$CH_2{=}CHCH_2O_2COR$	Allyl carbonate	
$Cl_3CCH_2O_2COR$	2,2,2-Trichloroethyl carbonate	Troc
$(CH_3)_3SiCH_2CH_2O_2COR$	2-Trimethylsilylethyl carbonate	

fluoride,[39] methanolic NH_4F,[40] aqueous HF,[41] BF_3[42] or SiF_4[43] can be used for its removal. Other highly substituted silyl groups, such as dimethyl(1,2,2-trimethylpropyl)-silyl[44] and

39. E. J. Corey and A. Venkataswarlu, *J. Am. Chem. Soc.* **94:**6190 (1972).
40. W. Zhang and M. J. Robins, *Tetrahedron Lett.* **33:**1177 (1992).
41. R. F. Newton, D. P. Reynolds, M. A. W. Finch, D. R. Kelly, and S. M. Roberts, *Tetrahedron Lett.* **1979:**3981.
42. D. R. Kelly, S. M. Roberts, and R. F. Newton, *Synth. Commun.* **9:**295 (1979).
43. E. J. Corey and K. Y. Yi, *Tetrahedron Lett.* **33:**2289 (1992).

tri-isopropylsilyl[45] (TIPS), are even more sterically hindered than the TBDMS group and can be used when more stability is required. The triphenylsilyl (TPS) and *t*-butyldiphenylsilyl (TBDPS) groups are also used.[46] The hydrolytic stability of the various silyl protecting groups is TMS < TBDMS < TIPS < TBDPS.[47] All the groups are also susceptible to TBAF cleavage, but the TPS and TBDPS groups are cleaved more slowly than the trialkylsilyl groups.[48]

Table 13.1 gives the structures and common abbreviations of some of the most frequently used hydroxyl-protecting groups.

Diols represent a special case in terms of applicable protecting groups. Both 1,2- and 1,3-diols easily form cyclic acetals with aldehydes and ketones, unless cyclization is precluded by molecular geometry. The isopropylidene derivatives (also called acetonides) formed by reaction with acetone are a common example.

$$\mathrm{RCH(OH)CH(OH)R} + \mathrm{CH_3C(=O)CH_3} \xrightarrow{H^+} \text{RCH—CHR (ring: O–C(CH}_3)_2\text{–O)}$$

The isopropylidene group can also be introduced by acid-catalyzed exchange with 2,2-dimethoxypropane.[49]

$$\mathrm{RCH(OH)CH_2OH} + \mathrm{CH_3C(OCH_3)_2CH_3} \xrightarrow{H^+} \text{RCH—CH}_2\text{ (ring: O–C(CH}_3)_2\text{–O)} + 2\,\mathrm{CH_3OH}$$

This ketal protective group is resistant to basic and nucleophilic reagents but is readily removed by aqueous acid. Formaldehyde, acetaldehyde, and benzaldehyde can also used as the carbonyl component in the formation of cyclic acetals. They function in the same manner as acetone. A disadvantage in the case of acetaldehyde and benzaldehyde is the possibility of forming a mixture of diastereomers, because of the new stereogenic center at the acetal carbon.

Protection of an alcohol function by esterification sometimes offers advantages over use of acetal or ether groups. Generally, ester groups are stable under acidic conditions. Esters are especially useful in protection during oxidations. Acetates and benzoates are the most commonly used ester derivatives. They can be conveniently prepared by reaction of unhindered alcohols with acetic anhydride or benzoyl chloride, respectively, in the presence of pyridine or other tertiary amines. 4-Dimethylaminopyridine (DMAP) is often used as a catalyst. The use of *N*-acylimidazolides (see Section 3.4.1) allows the

44. H. Wetter and K. Oertle, *Tetrahedron Lett.* **26:**5515 (1985).
45. R. F. Cunico and L. Bedell, *J. Org. Chem.* **45:**4797 (1980).
46. S. Hanessian and P. Lavallee, *Can. J. Chem.* **53:**2975 (1975); S. A. Hardinger and N. Wijaya, *Tetrahedron Lett.* **34:**3821 (1993).
47. J. S. Davies, C. L. Higginbotham, E. J. Tremeer, C. Brown, and R. C. Treadgold, *J. Chem. Soc., Perkin Trans. 1* **1992:**3043.
48. J. W. Gillard, R. Fortin, H. E. Morton, C. Yoakim, C. A. Quesnelle, S. Daignault, and Y. Guindon, *J. Org. Chem.* **53:**2602 (1988).
49. M. Tanabe and B. Bigley, *J. Am Chem. Soc.* **83:**756 (1961).

acylation reaction to be carried out in the absence of added base.[50] Imidazolides are less reactive than the corresponding acyl chloride and can exhibit a higher degree of selectivity in reactions with a molecule possessing several hydroxyl groups:

Ph, O, O, O, HO, HO, OCH_3 + PhC(=O)–N (imidazole) $\xrightarrow[\Delta]{CHCl_3}$ Ph, O, O, O, HO, O, PhCO, OCH_3 78% Ref. 51

Hindered hydroxyl groups may require special acylation procedures. One approach is to increase the reactivity of the hydroxyl group by converting it to an alkoxide ion with strong base (e.g., *n*-BuLi or KH). When this conversion is not feasible, more reactive acylating reagents are used. Highly reactive acylating agents are generated *in situ* when carboxylic acids are mixed with trifluoroacetic anhydride. The mixed anhydride exhibits increased reactivity because of the high reactivity of the trifluoroacetate ion as a leaving group.[52] Dicyclohexylcarbodiimide is another reagent that serves to activate carboxy groups by forming the iminoanhydride **A** (see Section 3.4.1).

$$RC(=O)-O-C(=NC_6H_{11})-NHC_6H_{11} \quad \mathbf{A}$$

Ester groups can be removed readily by base-catalyzed hydrolysis. When basic hydrolysis is inappropriate, special acyl groups are required. Trichloroethyl carbonate esters, for example, can be reductively removed with zinc.[53]

$$ROC(=O)OCH_2CCl_3 \xrightarrow{Zn} ROH + CH_2{=}CCl_2 + CO_2$$

Allyl carbonate esters are also useful hydroxyl-protecting groups. They are introduced using allyl chloroformate. A number of Pd-based catalysts for allylic deprotection have been developed.[54] They are based on a catalytic cycle in which Pd(0) reacts by oxidative addition and activates the allylic bond to nucleophilic substitution. Various nucleophiles are effective, including dimedone,[55] pentane-2,4-dione,[56] and amines.[57]

$$ROC(=O)OCH_2CH{=}CH_2 \xrightarrow{Pd^0} ROC(=O)O^- \;(\eta^3\text{-allyl})Pd^{II} \xrightarrow{Nu:} ROH + CO_2 + Pd^0 + CH_2{=}CHCH_2Nu$$

50. H. A. Staab, *Angew. Chem.* **74:**407 (1962).
51. F. A. Carey and K. O. Hodgson, *Carbohyd. Res.* **12:**463 (1970).
52. R. C. Parish and L. M. Stock, *J. Org. Chem.* **30:**927 (1965); J. M. Tedder, *Chem. Rev.* **55:**787 (1955).
53. T. B. Windholz and D. B. R. Johnston, *Tetrahedron Lett.* **1967:**2555.
54. F. Guibe, *Tetrahedron* **53:**13509 (1997).
55. H. Kunz and H. Waldmann, *Angew. Chem. Int. Ed. Engl.* **23:**71 (1984).
56. A. De Mesmaeker, P. Hoffmann, and B. Ernst, *Tetrahedron Lett.* **30:**3773 (1989).
57. H. Kunz, H. Waldmann and H. Klinkhammer, *Helv. Chim. Acta* **71:**1868 (1988); S. Friedrich-Bochnitschek, H. Waldmann, and H. Kunz, *J. Org. Chem.* **54:**751 (1989); J. P. Genet, E. Blart, M. Savignac, S. Lemeune, and J.-M. Paris, *Tetrahedron Lett.* **34:**4189 (1993).

Cyclic carbonate esters are easily prepared from 1,2- and 1,3-diols. These are commonly prepared by reaction with N,N'-carbonyldiimidazole[58] or by transesterification with diethyl carbonate.

13.1.2. Amino-Protecting Groups

Primary and secondary amino groups are nucleophilic and easily oxidized. If either of these types of reactivity will cause a problem, the amino group must be protected. The most general way of masking nucleophilicity is by acylation. Carbamates are particularly useful. The most widely used group is the carbobenzyloxy (Cbz) group.[59] Because of the lability of the benzyl C–O bond toward hydrogenolysis, the amine can be regenerated from a Cbz derivative by hydrogenolysis, which is accompanied by spontaneous decarboxylation of the resulting carbamic acid.

$$\text{PhCH}_2\text{OC(=O)NR}_2 \xrightarrow[\text{cat}]{H_2} \left[\text{HOC(=O)NR}_2\right] + \text{toluene} \longrightarrow CO_2 + HNR_2$$

In addition to standard catalytic hydrogenolysis, methods for transfer hydrogenolysis using hydrogen donors such as ammonium formate or formic acid with Pd/C catalyst are available.[60] The Cbz group can also be removed by a combination of a Lewis acid and a nucleophile. Boron trifluoride can be used in conjunction with dimethyl sulfide or ethyl sulfide, for example.[61]

The *t*-butoxycarbonyl (*t*-Boc) group is another valuable amino-protecting group. The removal in this case is done with an acid such as trifluoroacetic acid or *p*-toluenesulfonic acid.[62] *t*-Butoxycarbonyl groups are introduced by reaction of amines with *t*-butoxypyrocarbonate or the mixed carbonate ester known as "BOC-ON".[63]

$$(CH_3)_3COC(=O)OC(=O)OC(CH_3)_3$$

t-butyl pyrocarbonate

$$(CH_3)_3COC(=O)ON{=}C(CN)Ph \quad \text{"BOC-ON"}$$

2-(*t*-butoxycarbonyloxyimino)-2-phenylacetonitrile

Allyl carbamates also can serve as amino-protecting groups. The allyloxy group is removed by Pd-catalyzed reduction or nucleophilic substitution. These reactions involve liberation of the carbamic acid by oxidative addition to the palladium. The allyl-palladium species is reductively cleaved by stannanes,[64] phenylsilane,[65] formic acid,[66] and $NaBH_4$.[67]

58. J. P. Kutney and A. H. Ratcliffe, *Synth. Commun.* **5**:47 (1975).
59. W. H. Hartung and R. Simonoff, *Org. React.* **7**:263 (1953).
60. S. Ram and L. D. Spicer, *Tetrahedron Lett.* **28**:515 (1987); B. El Amin, G. Anantharamaiah, G. Royer, and G. Means, *J. Org. Chem.* **44**:3442 (1979).
61. I. M. Sanchez, F. J. Lopez, J. J. Soria, M. I. Larraza, and H. J. Flores, *J. Am. Chem. Soc.* **105**:7640 (1983); D. S. Bose and D. E. Thurston, *Tetrahedron Lett.* **31**:6903 (1990).
62. E. Wünsch, *Methoden der Organischen Chemie*, 4th ed., Thieme, Stuttgart, 1975, Vol. 15.
63. O. Keller, W. Keller, G. van Look, and G. Wersin, *Org. Synth.* **63**:160 (1984); W. J. Paleveda, F. W. Holly, and D. F. Weber, *Org. Synth.* **63**:171 (1984).
64. O. Dangles, F. Guibe, G. Balavoine, S. Lavielle, and A. Marquet, *J. Org. Chem.* **52**:4984 (1987).
65. M. Dessolin, M.-G. Guillerez, N. T. Thieriet, F. Guibe, and A. Loffet, *Tetrahedron Lett.* **36**:5741 (1995).
66. I. Minami, Y. Ohashi, I. Shimizu, and J. Tsuji, *Tetrahedron Lett.* **26**:2449 (1985); Y. Hayakawa, S. Wakabayashi, H. Kato, and R. Noyori, *J. Am. Chem. Soc.* **112**:1691 (1990).
67. R. Beugelmans, L. Neuville, M. Bois-Choussy, J. Chastanet, and J. Zhu, *Tetrahedron Lett.* **36**:3129 (1995).

These reducing agents convert the allyl group to propene. Reagents used for nucleophilic cleavage include *N,N′*-dimethylbarbituric acid[68] and silylating agents, including $TMS{-}N_3/NH_4F$,[69] $TMSN(Me)_2$,[70] and $TMSN(CH_3)COCF_3$.[71] The silylated nucleophiles trap the deallylated product prior to hydrolytic workup.

$$RNHCO_2CH_2CH{=}CH_2 \xrightarrow{Pd^0} RNHCO_2^- \; (\pi\text{-allyl})Pd^{II} \;\begin{cases} \xrightarrow{H\text{-}D} RNH_2 + CO_2 + Pd^0 + CH_2{=}CHCH_2H \\ \xrightarrow{Nu\text{-}H} RNH_2 + CO_2 + Pd^0 + CH_2{=}CHCH_2Nu \end{cases}$$

2,2,2-Trichloroethyl carbamates can be reductively cleaved by zinc.[72]

Sometimes it is very useful to be able to remove a protecting group by photolysis. 2-Nitrobenzyl carbamates meet this requirement. The photoexcited nitro group abstracts a hydrogen from the benzylic position, which is then converted to an α-hydroxybenzyl carbamate that readily hydrolyzes.[73]

$$2\text{-}NO_2C_6H_4CH_2OCNR_2 \xrightarrow{h\nu} 2\text{-}NO_2C_6H_4CH(OH)OCNR_2 \longrightarrow 2\text{-}NO_2C_6H_4CH{=}O + CO_2 + H_2NR$$

Allyl groups attached directly to amine or amide nitrogen can be removed by isomerization and hydrolysis.[74] Catalysts that have been found to be effective include Wilkinson's catalyst,[75] other rhodium catalysts,[76] and iron pentacarbonyl.[76] Treatment of *N*-allyl amines with $Pd(PPh_3)_4$ and *N,N′*-dimethylbarbituric acid also cleaves the allyl group.[77]

N-Benzyl groups can be removed by reaction of amines with chloroformates. This can be a useful method for protecting-group manipulation if the resulting carbamate is also easily cleaved. A particularly effective reagent is α-chloroethyl chloroformate, which can be removed by subsequent solvolysis.[78]

Amide nitrogens can be protected by 4-methoxyphenyl or 2,4-dimethoxyphenyl groups. The protective group can be removed by oxidation with ceric ammonium nitrate.[79] 2,4-Dimethoxybenzyl groups can be removed using anhydrous trifluoroacetic acid.[80]

68. P. Braun, H. Waldmann, W. Vogt, and H. Kunz, *Synlett:* **1990:**105.
69. G. Shapiro and D. Buechler, *Tetrahedron Lett.* **35:**5421 (1994).
70. A. Merzouk, F. Guibe, and A. Loffet, *Tetrahedron Lett.* **33:**477 (1992).
71. M. Dessolin, M.-G. Guillerez, N. Thieriet, F. Guibe, and A. Loffet, *Tetrahedron Lett.* **36:**5741 (1995).
72. G. Just and K. Grozinger, *Synthesis:* **1976:**457.
73. J. F. Cameron and J. M. J. Frechet, *J. Am. Chem. Soc.* **113:**4303 (1991).
74. I. Minami, M. Yuhara, and J. Tsuji, *Tetrahedron Lett.* **28:**2737 (1987); M. Sakaitani, N. Kurokawa, and Y. Ohfune, *Tetrahedron Lett.* **27:**3753 (1986).
75. B. C. Laguzza and B. Ganem, *Tetrahedron Lett.* **22:**1483 (1981).
76. J. K. Stille and Y. Becker, *J. Org. Chem.* **45:**2139 (1980); R. J. Sundberg, G. S. Hamilton, and J. P. Laurino, *J. Org. Chem.* **53:**976 (1988).
77. F. Garro-Helion, A. Merzouk, and F. Guibe, *J. Org. Chem.* **58:**6109 (1993).
78. R. A. Olofson, J. T. Martz, J.-P. Senet, M. Piteau, and T. Malfroot, *J. Org. Chem.* **49:**2081 (1984).
79. M. Yamaura T. Suzuki, H. Hashimoto, J. Yoshimura, T. Okamoto, and C. Shin, *Bull. Chem. Soc. Jpn.* **58:**1413 (1985); R. M. Williams, R. W. Armstrong, and J.-S. Dung, *J. Med. Chem.* **28:**733 (1985).
80. R. H. Schlessinger, G. R. Bebernitz, P. Lin, and A. J. Poss, *J. Am. Chem. Soc.* **107:**1777 (1985); P. DeShong, S. Ramesh, V. Elango, and J. J. Perez, *J. Am. Chem. Soc.* **107:**5219 (1985).

Simple amides are satisfactory protective groups only if the rest of the molecule can resist the vigorous acidic or alkaline hydrolysis necessary for their removal. For this reason, only amides that can be removed under mild conditions have been found useful as amino-protecting groups. Phthalimides are used to protect primary amino groups. The phthalimides can be cleaved by treatment with hydrazine. This reaction proceeds by initial nucleophilic addition at an imide carbonyl, followed by an intramolecular acyl transfer.

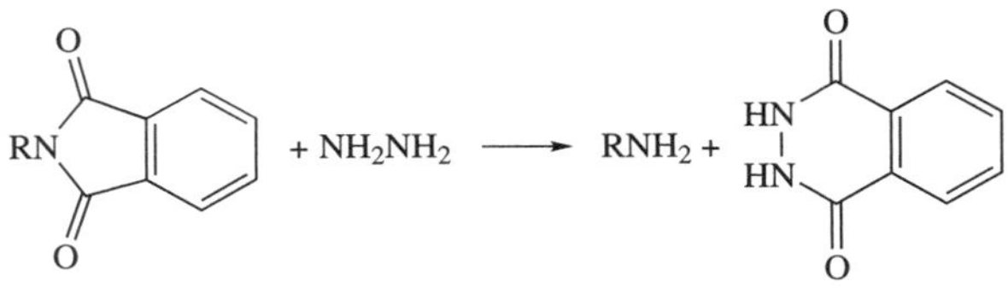

A similar sequence that takes place under milder conditions uses 4-nitrophthalimides as the protective group and *N*-methylhydrazine for deprotection.[81]

Reduction by $NaBH_4$ in aqueous ethanol is an alternative method for deprotection of phthalimides. This reaction involves formation of an *o*-hydroxymethylbenzamide in the reduction step. Intramolecular displacement of the amino group follows.[82]

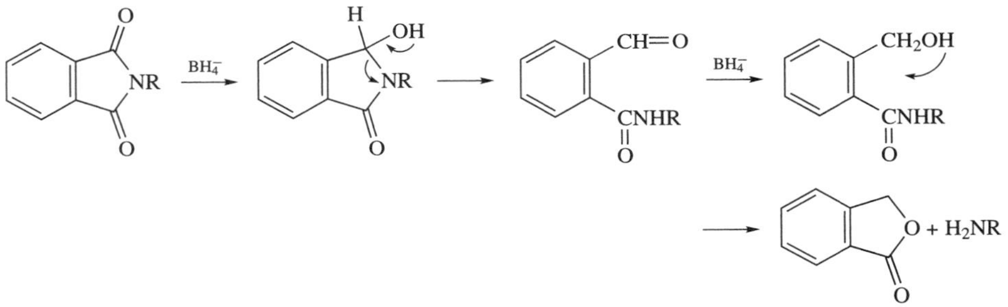

Because of the strong electron-withdrawing effect of the trifluoromethyl group, trifluoroacetamides are subject to hydrolysis under mild conditions. This has permitted trifluoroacetyl groups to be used as amino-protecting groups in some situations.

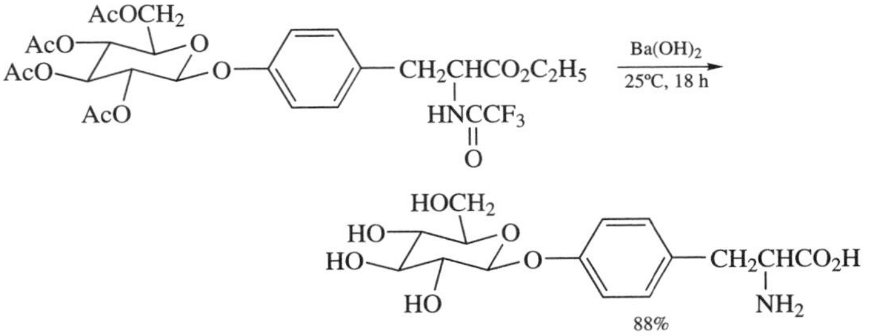

Ref. 83

81. H. Tsubouchi, K. Tsuji, and H. Ishikawa, *Synlett* **1994:**63.
82. J. O. Osborn, M. G. Martin, and B. Ganem, *Tetrahedron Lett.* **25:**2093 (1984).
83. A. Taurog, S. Abraham, and I. Chaikoff, *J. Am. Chem. Soc.* **75:**3473 (1953).

Amides can also be removed by partial reduction. If the reduction proceeds only to the carbinolamine stage, hydrolysis can liberate the deprotected amine. Trichloroacetamides are readily cleaved by sodium borohydride in alcohols by this mechanism.[84] Benzamides, and probably other simple amides, can be removed by careful partial reduction with diisobutylaluminum hydride.[85]

$$R_2NC(=O)Ph \xrightarrow{R_2AlH} R_2NCH(OAlR_2)Ph \xrightarrow[H_2O]{H^+} R_2NH + PhCH{=}O$$

The 4-pentenoyl group is easily removed from amides by I_2 and can be used as a protective group. The mechanism of cleavage involves iodocyclization and hydrolysis of the resulting iminolactone.[86]

$$RNHC(=O)CH_2CH_2CH{=}CH_2 \xrightarrow{I_2} RN{=}\text{(2-tetrahydrofuranylidene)}\text{-}CH_2I \xrightarrow{H_2O} RNH_2$$

Simple sulfonamides are very difficult to hydrolyze. However, a photoactivated reductive method for desulfonylation has been developed.[87] Sodium borohydride is used in conjunction with 1,2- or 1,4-dimethoxybenzene or 1,5-dimethoxynaphthalene. The photoexcited aromatic serves as an electron donor toward the sulfonyl group, which then fragments to give the deprotected amine. The $NaBH_4$ reduces the sulfonyl radical.

$$R_2NSO_2Ar + CH_3O\text{-}C_6H_4\text{-}OCH_3 \xrightarrow{h\nu} R_2N^- + [CH_3O\text{-}C_6H_4\text{-}OCH_3]^{+\cdot} + ArSO_2\cdot$$

Reagents which permit protection of primary amino groups as cyclic bis-silyl derivatives have been developed. Anilines, for example, can be converted to disilazolidines.[88]

$$ArNH_2 + (CH_3)_2SiHCH_2CH_2SiH(CH_3)_2 \xrightarrow[100^\circ C]{CsF,\ HMPA} \text{Ar–N}[Si(CH_3)_2CH_2CH_2Si(CH_3)_2]$$

$$ArNH_2 + 1,2\text{-}[(CH_3)_2SiH]_2C_6H_4 \xrightarrow{(PPh_3)_3RhCl} \text{Ar–N}[Si(CH_3)_2\text{-}C_6H_4\text{-}Si(CH_3)_2]$$

84. F. Weygand and E. Frauendorfer, *Chem. Ber.* **103:**2437 (1970).
85. J. Gutzwiller and M. Uskokovic, *J. Am. Chem. Soc.* **92:**204 (1970); K. Psotta and A. Wiechers, *Tetrahedron* **35:**255 (1979).
86. R. Madsen, C. Roberts, and B. Fraser-Reid, *J. Org. Chem.* **60:**7920 (1995).
87. T. Hamada, A. Nishida, and O. Yonemitsu, *Heterocycles* **12:**647 (1979); T. Hamada, A. Nishida, Y. Matsumoto, and O. Yonemitsu, *J. Am. Chem. Soc.* **102:**3978 (1980).
88. R. P. Bonar-Law, A. P. Davis, and B. J. Dorgan, *Tetrahedron Lett.* **31:**6721 (1990); R. P. Bonar-Law, A. P. Davis, B. J. Dorgan, M. T. Reetz, and A. Wehrsig, *Tetrahedron Lett.* **31:**6725 (1990); S. Djuric, J. Venit, and P. Magnus, *Tetrahedron Lett.* **22:**1787 (1981); T. L. Guggenheim, *Tetrahedron Lett.* **25:**1253 (1984); A. P. Davis and P. J. Gallagher, *Tetrahedron Lett.* **36:**3269 (1995).

These groups are stable to a number of reaction conditions, including generation and reaction of organometallic reagents.[89] They are readily removed by hydrolysis.

13.1.3. Carbonyl-Protecting Groups

Conversion to acetals or ketals is a very general method for protecting aldehydes and ketones against nucleophilic addition or reduction. Ethylene glycol, which gives a dioxolane derivative, is frequently employed for this purpose. The dioxolane derivative is usually prepared by heating the carbonyl compound with ethylene glycol in the presence of an acid catalyst, with provision for azeotropic removal of water.

$$RC(=O)R' + HOCH_2CH_2OH \xrightarrow{H^+} \text{(2-R-2-R'-1,3-dioxolane: } R(R')C(OCH_2)_2\text{)} + H_2O$$

Scandium triflate is also an effective catalyst for dioxolane formation.[90]

Dimethyl or diethyl acetals and ketals can be conveniently prepared by acid-catalyzed exchange with a ketal such as 2,2-dimethoxypropane or an ortho ester.[91]

$$RC(=O)R' + HC(OCH_3)_3 \xrightarrow{H^+} R{-}C(OCH_3)_2{-}R' + HCO_2CH_3$$

$$RC(=O)R' + (CH_3O)_2C(CH_3)_2 \xrightarrow{H^+} R{-}C(OCH_3)_2{-}R' + (CH_3)_2C{=}O$$

Acetals and ketals can be prepared under very mild conditions by reaction of the carbonyl compound with an alkoxytrimethylsilane, using trimethylsilyl trifluoromethylsulfonate as the catalyst.[92]

$$R_2C{=}O + 2\,R'OSi(CH_3)_3 \xrightarrow{Me_3SiO_3SCF_3} R_2C(OR')_2 + (CH_3)_3SiOSi(CH_3)_3$$

Dioxolanes and other acetals and ketals are generally inert to powerful nucleophiles, including organometallic reagents and hydride-transfer reagents. The carbonyl group can be deprotected by acid-catalyzed hydrolysis by the general mechanism for acetal hydrolysis (Part A, Section 8.1). Hydrolysis is also promoted by $LiBF_4$ in acetonitrile.[93]

If the carbonyl group must be regenerated under nonhydrolytic conditions, β-halo alcohols such as 3-bromo-1,2-dihydroxypropane or 2,2,2-trichloroethanol can be used for

89. R. P. Bonar-Law, A. P. Davis, and J. P. Dorgan, *Tetrahedron* **49:**9855 (1993); K. C. Grega, M. R. Barbachyn, S. J. Brickner, and S. A. Mizsak, *J. Org. Chem.* **60:**5255 (1995).
90. K. Ishihara, Y. Karumi, M. Kubota, and H. Yamamoto, *Synlett* **1996:**839.
91. C. A. MacKenzie and J. H. Stocker, *J. Org. Chem.* **20:**1695 (1955); E. C. Taylor and C. S. Chiang, *Synthesis* **1977:**467.
92. T. Tsunoda, M. Suzuki, and R. Noyori, *Tetrahedron Lett.* **21:**1357 (1980).
93. B. H. Lipshutz and D. F. Harvey, *Synth. Commun.* **12:**267 (1982).

acetal formation. These groups can be removed by reduction with zinc, which proceeds with β elimination.

$$\text{BrCH}_2\text{-(2,2-dialkyl-1,3-dioxolane)} \xrightarrow{\text{Zn}} R_2C{=}O + HOCH_2CH{=}CH_2 \qquad \text{Ref. 94}$$

$$RR'C(OCH_2CCl_3)_2 \xrightarrow[\text{THF}]{\text{Zn}} RC(=O)R' + CH_2{=}CCl_2 \qquad \text{Ref. 95}$$

Another carbonyl-protecting group is the 1,3-oxathiolane derivative, which can be prepared by reaction with mercaptoethanol in the presence of BF_3[96] or by heating with an acid catalyst with azeotropic removal of water.[97] The 1,3-oxathiolanes are useful when nonacidic conditions are required for deprotection. The 1,3-oxathiolane group can be removed by treatment with Raney nickel in alcohol, even under slightly alkaline conditions.[98] Deprotection can also be accomplished by treating with a mild halogenating agent, such as chloramine T.[99] This reagent oxidizes the sulfur to a chlorosulfonium salt and activates the ring to hydrolytic cleavage.

$$\text{R}_2\text{C(oxathiolane)} + H_3C\text{-}C_6H_4\text{-}SO_2N^-(Cl)\,Na^+ \longrightarrow \text{R}_2\text{C(oxathiolane-}S^+\text{-X)} \xrightarrow{H_2O} R_2C{=}O$$

chloramine T

X = Cl or NSO_2Ar

Dithioketals, especially the cyclic dithiolanes and dithianes, are also useful carbonyl-protecting groups. These can be formed from the corresponding dithiols by Lewis acid-catalyzed reactions. The catalysts that are used include BF_3, $Mg(O_3SCF_3)_2$, $Zn(O_3SCF_3)_2$, and $LaCl_3$.[100] *S*-Trimethylsilyl ethers of thiols and dithiols also react with ketones to form dithioketals.[101]

Di-*n*-butylstannyl dithiolates also serve as sources of dithiolanes and dithianes. These reactions are catalyzed by di-*n*-butylstannyl triflate.[102]

$$R_2C{=}O + (n\text{-Bu})_2Sn\langle S_2(CH_2)_n\rangle \xrightarrow{(n\text{-Bu})_2Sn(O_3SCF_3)_2} R_2C\langle S_2(CH_2)_n\rangle$$

94. E. J. Corey and R. A. Ruden, *J. Org. Chem.* **38:**834 (1973).
95. J. L. Isidor and R. M. Carlson, *J. Org. Chem.* **38:**544 (1973).
96. G. E. Wilson, Jr., M. G. Huang, and W. W. Scholman, Jr., *J. Org. Chem.* **33:**2133 (1968).
97. C. Djerassi and M. Gorman, *J. Am. Chem. Soc.* **75:**3704 (1953).
98. C. Djerassi, E. Batres, J. Romo, and G. Rosenkranz, *J. Am. Chem. Soc.* **74:**3634 (1952).
99. D. W. Emerson and H. Wynberg, *Tetrahedron Lett.* **1971:**3445.
100. L. F. Fieser, *J. Am. Chem. Soc.* **76:**1945 (1954); E. J. Corey and K. Shimoji, *Tetrahedron Lett.* **24:**169 (1983); L. Garlaschelli and G. Vidari, *Tetrahedron Lett.* **31:**5815 (1990).
101. D. A. Evans, L. K. Truesdale, K. G. Grimm, and S. L. Nesbitt, *J. Am. Chem. Soc.* **99:**5009 (1977).
102. T. Sato, J. Otero, and H. Nozaki, *J. Org. Chem.* **58:**4971 (1993).

Bis-trimethylsilyl sulfate in the presence of silica also promotes formation of dithiolanes.[103]

The regeneration of carbonyl compounds from dithioacetals and ketals is done best with reagents that oxidize or otherwise activate the sulfur as a leaving group and facilitate hydrolysis. Among the reagents that have been found effective are nitrous acid, *t*-butyl hypochlorite, $PhI(O_2CCF_3)_2$, DDQ, $SbCl_5$, and cupric salts.[104]

$$R_2C(SR')_2 + X^+ \longrightarrow R_2C(SR')(\overset{+}{S}R'X) \longrightarrow R_2C{=}\overset{+}{S}R' \xrightarrow{H_2O} R_2C(OH){-}SR' \longrightarrow R_2C{=}O$$

13.1.4. Carboxylic Acid-Protecting Groups

If only the O–H, as opposed to the carbonyl, of a carboxyl group needs to be masked, this can be readily accomplished by esterification. Alkaline hydrolysis is the usual way of regenerating the acid. *t*-Butyl esters, which are readily cleaved by acid, can be used if alkaline conditions must be avoided. 2,2,2-Trichloroethyl esters, which can be reductively cleaved with zinc, are another possibility.[105] Some esters can be cleaved by treatment with anhydrous TBAF. These reactions proceed best for esters of relatively acidic alcohols, such as 4-nitrobenzyl, 2,2,2-trichloroethyl, and cyanoethyl.[106]

The more difficult problem of protecting the carbonyl group can be accomplished by conversion to an oxazoline derivative. The most commonly used example is the 4,4-dimethyl derivative, which can be prepared from the acid by reaction with 2-amino-2-methylpropanol or with 2,2-dimethylaziridine.[107]

$$RCO_2H + HOCH_2C(CH_3)_2NH_2 \longrightarrow \text{R–(4,4-dimethyl-2-oxazolinyl)}$$

$$RCO_2H + HN\text{(2,2-dimethylaziridine)} \longrightarrow RC(=O){-}N\text{(2,2-dimethylaziridinyl)} \xrightarrow{H^+} \text{R–(4,4-dimethyl-2-oxazolinyl)}$$

The heterocyclic derivative successfully protects the acid from attack by Grignard reagents or hydride-transfer reagents. The carboxylic acid group can be regenerated by acidic

103. H. K. Patney, *Tetrahedron Lett.* **34:**7127 (1993).
104. M. T. M. El-Wassimy, K. A. Jorgensen, and S. O. Lawesson, *J. Chem. Soc., Perkin Trans. 1* **1983:**2201; J. Lucchetti and A. Krief, *Synth. Commun.* **13:**1153 (1983); G. Stork and K. Zhao, *Tetrahedron Lett.* **30:**287 (1989); L. Mathew and S. Sankararaman, *J. Org. Chem.* **58:**7576 (1993); M. Kamata, H. Otogawa, and E. Hasegawa, *Tetrahedron Lett.* **32:**7421 (1991).
105. R. B. Woodward, K. Heusler, J. Gostelli, P. Naegeli, W. Oppolzer, R. Ramage, S. Ranganathan, and H. Vorbrüggen, *J. Am. Chem. Soc.* **88:**852 (1966).
106. M. Namikoshi, B. Kundu, and K. L. Rinehart, *J. Org. Chem.* **56:**5464 (1991); Y. Kita, H. Maeda, F. Takahashi, S. Fukui, and T. Ogawa, *Chem. Pharm. Bull.* **42:**147 (1994).
107. A. I. Meyers, D. L. Temple, D. Haidukewych, and E. Mihelich, *J. Org. Chem.* **39:**2787 (1974).

hydrolysis or converted to an ester by acid-catalyzed reaction with the appropriate alcohol.

Carboxylic acids can also be protected as ortho esters. Ortho esters derived from simple alcohols are very easily hydrolyzed, and a more useful ortho ester protecting group is the 4-methyl-2,6,7-trioxabicyclo[2.2.2]octane structure. These bicyclic orthoesters can be prepared by exchange with other ortho esters, by reaction with iminoethers, or by rearrangement of the ester derived from 3-hydroxymethyl-3-methyloxetane.

$RC(OCH_3)_3$ —$(HOCH_2)_3CCH_3$→ Ref. 108

$RC(=NH)OR'$ —$(HOCH_2)_3CCH_3$→ R–(trioxabicyclooctane)–CH_3 Ref. 109

$RC(=O)OCH_2$–(3-methyloxetan-3-yl) —BF_3→ Ref. 110

Lactones can be protected as their dithioketals by using a method that is analogous to ketone protection. The required reagent is readily prepared from trimethylaluminum and ethanedithiol.

$$\text{lactone} + (CH_3)_2AlSCH_2CH_2SAl(CH_3)_2 \longrightarrow \text{spiro dithiolane}$$ Ref. 111

Acyclic esters react with this reagent to give ketenethioacetals.

$$R_2CHCO_2R' + (CH_3)_2AlSCH_2CH_2SAl(CH_2)_2 \longrightarrow R_2C{=}\text{(1,3-dithiolane)}$$

In general, the methods for protection and deprotection of carboxylic acids and esters are not as convenient as those for alcohols, aldehydes, and ketones. It is, therefore, common to carry potential carboxylic acids through synthetic schemes in the form of protected primary alcohols or aldehydes. The carboxylic acid can then be formed at a late stage in the synthesis by an appropriate oxidation. This strategy allows one to utilize the wider variety of alcohol and aldehyde protective groups indirectly for carboxylic acid protection.

108. M. P. Atkins, B. T. Golding, D. A. Howe and P. J. Sellars, *J. Chem. Soc., Chem. Commun.* **1980:**207.
109. E. J. Corey and K. Shimoji, *J. Am. Chem. Soc.* **105:**1662 (1983).
110. E. J. Corey and N. Raju, *Tetrahedron Lett.* **24:**5571 (1983).
111. E. J. Corey and D. J. Beames, *J. Am. Chem. Soc.* **95:**5829 (1973).

The protective groups discussed in the previous section play only a passive role during a synthetic sequence. The groups are introduced and removed at appropriate stages but do not directly contribute to construction of the target molecule. It is often advantageous to combine the need for masking of a functional group with a change in the reactivity of the functionality in question. As an example, suppose the transformation shown below was to be accomplished:

$$CH_3\overset{O}{\overset{\|}{C}}{:}^- \quad + \quad \text{cyclohexenone} \longrightarrow \text{3-acetylcyclohexanone} \; (CH_3\overset{\|}{\underset{O}{C}}-)$$

The electrophilic α,β-unsaturated ketone is reactive toward nucleophiles, but the nucleophile that is required, an acyl anion, is not normally an accessible entity. As will be discussed, however, there are several potential reagents that can introduce the desired acyl anion in a masked form. The masked functionality used in place of an inaccessible species is termed a *synthetic equivalent group*. Often, the concept of "umpolung" is involved in devising synthetic equivalent groups. The term *umpolung* refers to the reversal of the normal polarity of a functional group.[112] Acyl groups are normally electrophilic, but a synthetic operation may require the transfer of an acyl group as a nucleophile. The *acyl anion equivalent* would then be an umpolung of the acyl group.

Because of the great importance of carbonyl groups in synthesis, a substantial effort has been devoted to developing nucleophilic equivalents for introduction of acyl groups.[113] One successful method involves a three-step sequence in which an aldehyde is converted to an O-protected cyanohydrin. This α-alkoxynitrile is then deprotonated, generating a nucleophilic carbanion.[114] After carbon–carbon bond formation, the carbonyl group can be regenerated by hydrolysis of the cyanohydrin. This sequence has been used to solve the problem of introducing an acetyl group at the β position of cyclohexenone.[115]

$$CH_3\overset{\overset{\displaystyle OC_2H_5}{|}}{\overset{\displaystyle OCHCH_3}{|}}\!\!\!\!C^-(C{\equiv}N) \; + \; \text{cyclohexenone} \longrightarrow 3\text{-}[CH_3C(C{\equiv}N)(OCH(CH_3)OC_2H_5)]\text{cyclohexanone} \xrightarrow[H_2O]{H^+} 3\text{-}(CH_3CO)\text{cyclohexanone} \qquad \text{Ref. 114}$$

112. For a general discussion and many examples of the use of the umpolung concept, see D. Seebach, *Angew. Chem. Int. Ed. Engl.* **18:**239 (1979).
113. For a review of acyl anion synthons, see T. A. Hase and J. K. Koskimies, *Aldrichimica Acta* **15:**35 (1982).
114. G. Stork and L. Maldonado *J. Am. Chem. Soc.* **93:**5286 (1971); G. Stork and L. Maldonado, *J. Am. Chem. Soc.* **96:**5272 (1974).
115. For further discussion of synthetic applications of the carbanions of O-protected cyanohydrins, see J. D. Albright, *Tetrahedron* **39:**3207 (1983).

α-Lithio vinyl ethers provide another type of acyl anion equivalents.

$$CH_2{=}CHOCH_3 + t\text{-BuLi} \longrightarrow CH_2{=}C(Li)OCH_3 \qquad \text{Ref. 116}$$

$$CH_2{=}CHOC_2H_5 \xrightarrow[\text{2) CuI}]{\text{1) } t\text{-BuLi, }-65^\circ C} (CH_2{=}C(OC_2H_5))_2CuLi \qquad \text{Ref. 117}$$

These reagents are capable of adding the α-alkoxyvinyl group to electrophilic centers. Subsequent hydrolysis can generate the carbonyl group and complete the desired transformation.

Ref. 116: 88%, 86%

Ref. 117: 67%, 74%

Sulfur compounds have also proven to be useful as nucleophilic acyl equivalents. The first reagent of this type to find general use was 1,3-dithiane, which on lithiation provides a nucleophilic acyl equivalent. The lithio derivative is a reactive nucleophile toward alkyl halides, carbonyl compounds and enones.[118]

116. J. E. Baldwin, G. A. Höfle, and O. W. Lever, Jr., *J. Am. Chem. Soc.* **96:**7125 (1974).
117. R. K. Boeckman, Jr., and K. J. Bruza, *J. Org. Chem.* **44:**4781 (1979).
118. D. Seebach and E. J. Corey, *J. Org. Chem.* **40:**231 (1975); B. H. Lipshutz and E. Garcia, *Tetrahedron Lett.* **31:**7261 (1990).

Closely related procedures are based on α-alkylthiosulfoxides, with ethylthiomethyl ethyl sulfoxide being a particularly convenient example.[119]

$$CH_3CH_2\overset{O}{\overset{\|}{S}}-\underset{R}{\underset{|}{\bar{C}}}-SCH_2CH_3 \xrightarrow{R'X} CH_3CH_2\overset{O}{\overset{\|}{S}}-\overset{R'}{\overset{|}{\underset{R}{\underset{|}{C}}}}-SC_2H_5 \longrightarrow R\overset{O}{\overset{\|}{C}}R'$$

$$CH_3CH_2\overset{O}{\overset{\|}{S}}-\underset{R}{\underset{|}{\bar{C}}}-SCH_2CH_3 \xrightarrow{R'_2C=O} CH_3CH_2\overset{O}{\overset{\|}{S}}-\overset{R}{\overset{|}{\underset{SC_2H_5}{\underset{|}{C}}}}-\overset{OH}{\overset{|}{C}}R'_2 \longrightarrow R\overset{O}{\overset{\|}{C}}-\overset{OH}{\overset{|}{C}}R'_2$$

$$CH_3CH_2\overset{O}{\overset{\|}{S}}-\underset{R}{\underset{|}{\bar{C}}}-SCH_2CH_3 \xrightarrow{R'CH=CH\overset{O}{\overset{\|}{C}}R''} CH_3CH_2\overset{O}{\overset{\|}{S}}-\overset{R}{\overset{|}{\underset{SC_2H_5}{\underset{|}{C}}}}-\overset{R'}{\overset{|}{C}}HCH_2\overset{O}{\overset{\|}{C}}R'' \longrightarrow R\overset{O}{\overset{\|}{C}}\underset{R'}{\underset{|}{C}}HCH_2\overset{O}{\overset{\|}{C}}R''$$

The α-ethylthiosulfoxides can be converted to the corresponding carbonyl compounds by hydrolysis catalyzed by mercuric ion. In both the dithiane and alkylthiomethylsulfoxide systems, an umpolung is achieved on the basis of the carbanion-stabilizing ability of the sulfur substituents.

Scheme 13.1 summarizes some examples of synthetic sequences that employ acyl anion equivalents.

Another group of synthetic equivalents which have been developed correspond to the propanal "homoenolate", $^-CH_2CH_2CH{=}O$.[120] This structure is the umpolung equivalent of an important electrophilic reagent, the α,β-unsaturated aldehyde acrolein. Scheme 13.2 illustrates some of the propanal homoenolate equivalents that have been developed. In general, the reagents used for these transformations are reactive toward such electrophiles as alkyl halides and carbonyl compounds. Several general points can be made about the reagents in Scheme 13.2. First, it should be noted that all deliver the aldehyde functionality in a masked form, such as an acetal or enol ether. The aldehyde must be liberated in a final step from the protected precursor. Several of the reagents involve delocalized allylic anions. This gives rise to the possibility of electrophilic attack at either the α or γ position of the allylic group. In most cases, the γ attack that is necessary for the anion to function as a propanal homoenolate is dominant.

The concept of developing reagents that are the synthetic equivalent of inaccessible species can be taken another step by considering dipolar species. For example, structures **B** and **C** incorporate both electrophilic and nucleophilic centers. Such reagents might be incorporated into ring-forming schemes, because they have the ability, at least formally, of undergoing cycloaddition reactions.

$$H_5C_2O\underset{O}{\underset{\|}{C}}\bar{C}HCH_2\overset{+}{C}H_2 \qquad\qquad {}^-\overset{O}{\overset{\|}{C}}CH_2CH_2{}^+$$

B **C**

119. J. E. Richman, J. L. Herrmann, and R. H. Schlessinger, *Tetrahedron Lett.* **1973:**3267; J. L. Herrmann, J. E. Richman, and R. H. Schlessinger, *Tetrahedron Lett.* **1973:**3271; J. L. Herrmann, J. E. Richman, P. J. Wepplo, and R. H. Schlessinger, *Tetrahedron Lett.* **1973:**4707.
120. For reviews of homoenolate anions, see J. C. Stowell, *Chem. Rev.* **84:**409 (1984); N. H. Werstiuk, *Tetrahedron* **39:**205 (1983).

Scheme 13.1. Synthetic Sequences Used for Reaction of Acyl Anion Equivalents

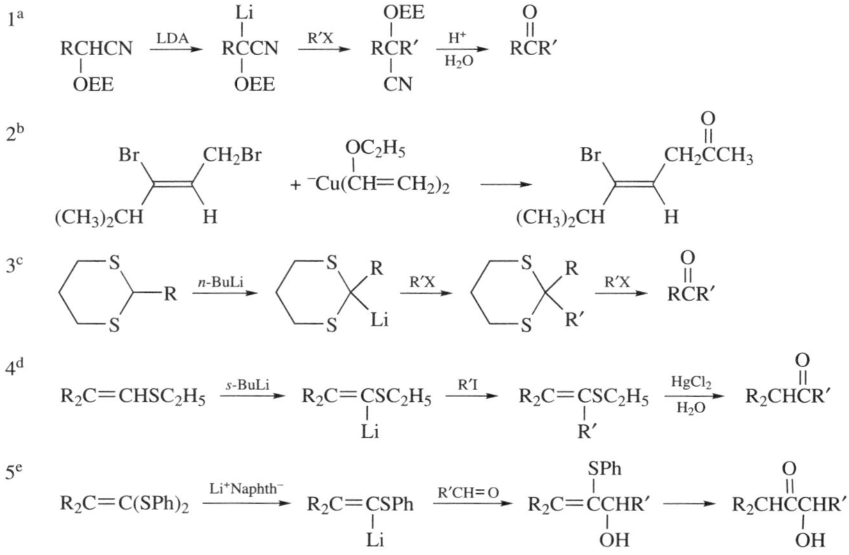

1[a] RCH(OEE)CN →(LDA) RC(Li)(OEE)CN →(R′X) RC(OEE)(CN)R′ →(H^+, H_2O) RC(=O)R′

2[b] (Z)-$(CH_3)_2CH$C(Br)=CHCH_2Br + $^-Cu(CH(OC_2H_5)=CH_2)_2$ → $(CH_3)_2CH$C(Br)=CH$CH_2C(=O)CH_3$

3[c] 2-R-1,3-dithiane →(n-BuLi) 2-R-2-Li-1,3-dithiane →(R′X) 2-R-2-R′-1,3-dithiane →(R′X) RC(=O)R′

4[d] $R_2C{=}CHSC_2H_5$ →(s-BuLi) $R_2C{=}C(Li)SC_2H_5$ →(R′I) $R_2C{=}C(R')SC_2H_5$ →($HgCl_2$, H_2O) $R_2CHC(=O)R'$

5[e] $R_2C{=}C(SPh)_2$ →($Li^+Naphth^-$) $R_2C{=}C(Li)SPh$ →($R'CH{=}O$) $R_2C{=}C(SPh)CH(OH)R'$ → $R_2CHC(=O)CH(OH)R'$

a. G. Stork and L. Maldonado, *J. Am. Chem. Soc.* **93:**5286 (1971).
b. P. Canonne, R. Boulanger, and P. Angers, *Tetrahedron Lett.* **32:**5861 (1991).
c. D. Seebach and E. J. Corey, *J. Org. Chem.* **40:**231 (1975).
d. K. Oshima, K. Shimoji, H. Takahashi, H. Yamamoto, and H. Nozaki, *J. Am. Chem. Soc.* **95:**2694 (1973).

Among the real chemical species that have been developed along these lines are the cyclopropylphosphonium salts **1** and **2**.

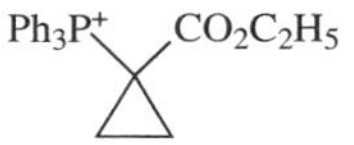

1: 1-(Ph_3P^+)-1-($CO_2C_2H_5$)cyclopropane **2**: 1-(Ph_3P^+)-1-(SPh)cyclopropane

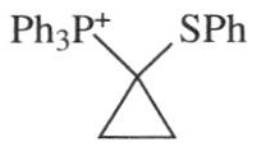

The phosphonium salt **1** reacts with β-keto esters and β-keto aldehydes to give excellent yields of cyclopentenecarboxylate esters.

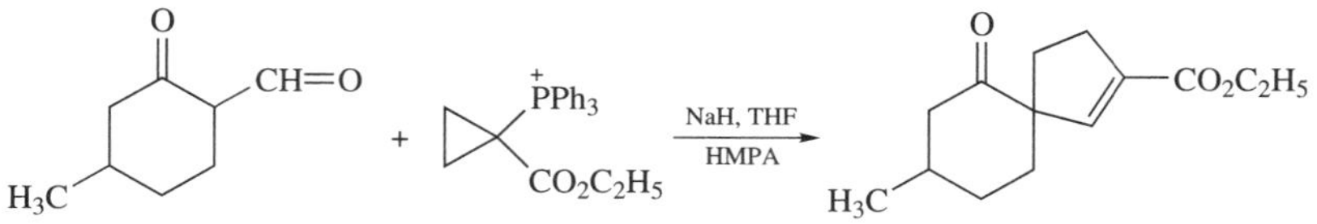

2-formyl-5-methylcyclohexanone (CH=O, H_3C) + 1-($^+PPh_3$)-1-($CO_2C_2H_5$)cyclopropane →(NaH, THF, HMPA) spiro cyclopentene ketone bearing $CO_2C_2H_5$ and H_3C — Ref. 121

1-($^+PPh_3$)-1-($CO_2C_2H_5$)cyclopropane + $CH_3CCH_2CO_2C_2H_5$ (C=O) →(NaH, THF, HMPA) $H_5C_2O_2C$–cyclopentene with H_3C and $CO_2C_2H_5$ — Ref. 122

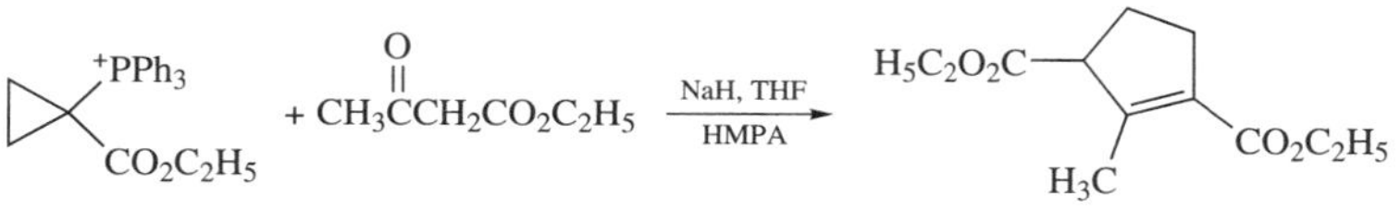

Several steps are involved. First, the enolate of the β-keto ester opens the cyclopropane ring. The polarity of this process corresponds to that in the formal synthon **B**. The product

121. W. G. Dauben and D. J. Hart, *J. Am. Chem. Soc.* **99:**7307 (1977).
122. P. L. Fuchs, *J. Am. Chem. Soc.* **96:**1607 (1974).

Scheme 13.2. Reaction Sequences Involving Propanal Homoenolate Anion Synthetic Equivalents

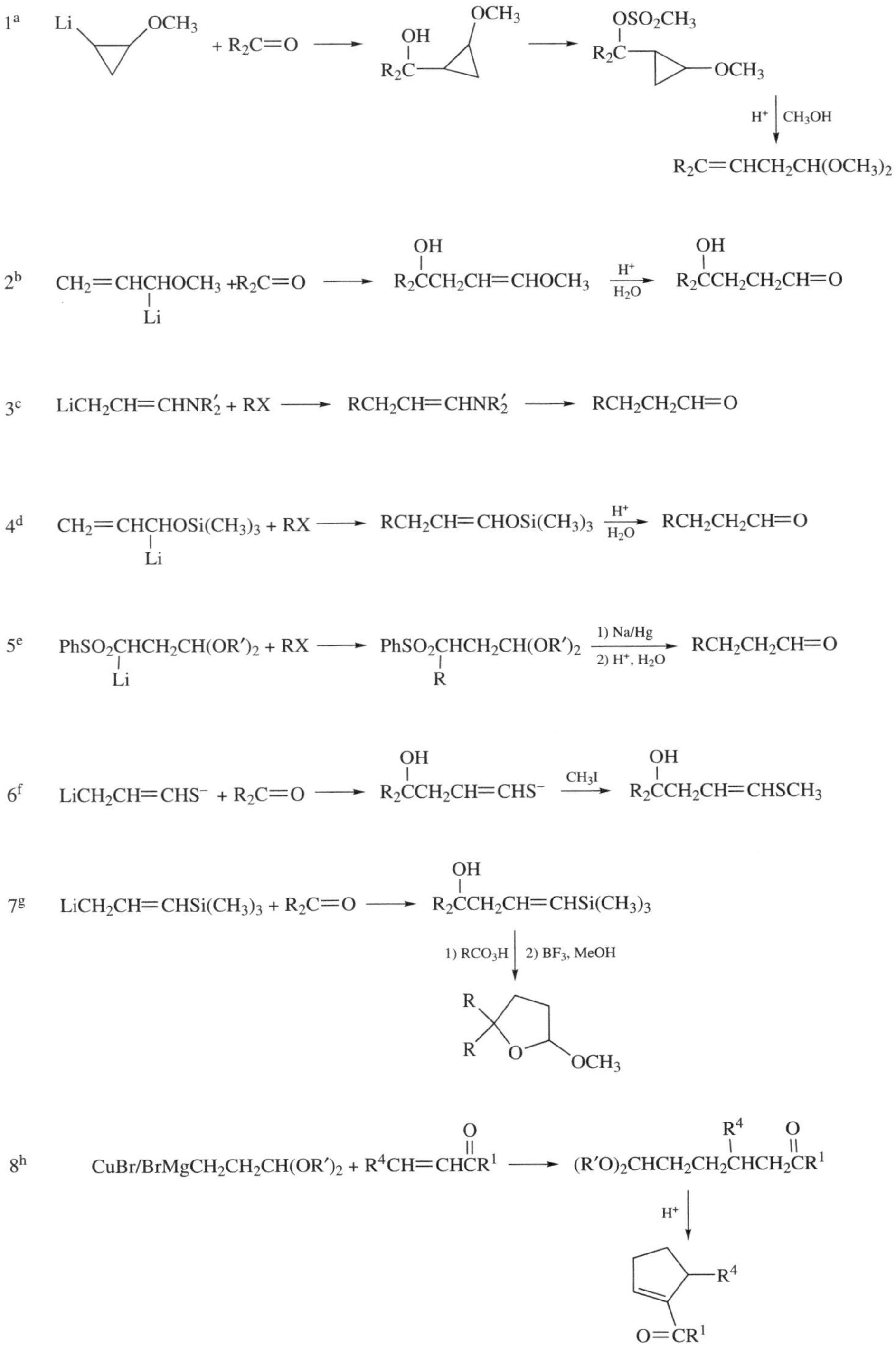

1[a] 2-Methoxycyclopropyllithium (Li and OCH_3 on the cyclopropane ring) + $R_2C{=}O$ → $R_2C(OH)$–(2-methoxycyclopropyl) → $R_2C(OSO_2CH_3)$–(2-methoxycyclopropyl) $\xrightarrow{H^+,\ CH_3OH}$ $R_2C{=}CHCH_2CH(OCH_3)_2$

2[b] $CH_2{=}CHCH(Li)OCH_3 + R_2C{=}O \longrightarrow R_2C(OH)CH_2CH{=}CHOCH_3 \xrightarrow[H_2O]{H^+} R_2C(OH)CH_2CH_2CH{=}O$

3[c] $LiCH_2CH{=}CHNR'_2 + RX \longrightarrow RCH_2CH{=}CHNR'_2 \longrightarrow RCH_2CH_2CH{=}O$

4[d] $CH_2{=}CHCH(Li)OSi(CH_3)_3 + RX \longrightarrow RCH_2CH{=}CHOSi(CH_3)_3 \xrightarrow[H_2O]{H^+} RCH_2CH_2CH{=}O$

5[e] $PhSO_2CH(Li)CH_2CH(OR')_2 + RX \longrightarrow PhSO_2CH(R)CH_2CH(OR')_2 \xrightarrow[2)\ H^+,\ H_2O]{1)\ Na/Hg} RCH_2CH_2CH{=}O$

6[f] $LiCH_2CH{=}CHS^- + R_2C{=}O \longrightarrow R_2C(OH)CH_2CH{=}CHS^- \xrightarrow{CH_3I} R_2C(OH)CH_2CH{=}CHSCH_3$

7[g] $LiCH_2CH{=}CHSi(CH_3)_3 + R_2C{=}O \longrightarrow R_2C(OH)CH_2CH{=}CHSi(CH_3)_3$ $\xrightarrow{1)\ RCO_3H\quad 2)\ BF_3,\ MeOH}$ 2,2-R,R-5-methoxytetrahydrofuran (ring O, OCH_3)

8[h] $CuBr/BrMgCH_2CH_2CH(OR')_2 + R^4CH{=}CHCR^1(=O) \longrightarrow (R'O)_2CHCH_2CH_2CH(R^4)CH_2CR^1(=O)$ $\xrightarrow{H^+}$ 1-acyl-5-R^4-cyclopentene ($O{=}CR^1$ on the ring double bond, R^4 adjacent)

Scheme 13.2. (*continued*)

9[i]

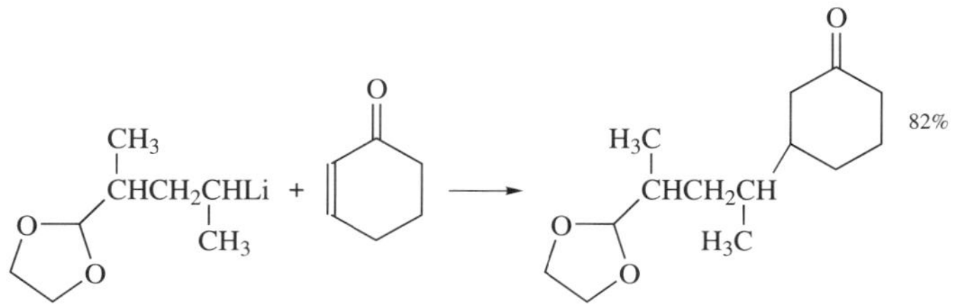

CH_3 | $CHCH_2CHLi$ (dioxolane) | CH_3 + cyclohexenone ⟶ H_3C | $CHCH_2CH$ (dioxolane) | H_3C (cyclohexanone) 82%

a. E. J. Corey and P. Ulrich, *Tetrahedron Lett.* **1975**:3685.
b. D. A. Evans, G. C. Andrews, and B. Buckwalter, *J. Am. Chem. Soc.* **96**:5560 (1974).
c. H. Ahlbrect and J. Eichler, *Synthesis* **1974**:672; S. F. Martin and M. T. DuPriest, *Tetrahedron Lett.* **1977**:3925; H. Ahlbrect, G. Bonnet, D. Enders, and G. Zimmerman, *Tetrahedron Lett.* **21**:3175 (1980).
d. W. C. Still and T. L. Macdonald, *J. Am. Chem. Soc.* **96**:5561 (1974).
e. M. Julia and B. Badet, *Bull. Soc. Chim. Fr.* **1975**:1363; K. Kondo and D. Tunemoto, *Tetrahedron Lett.* **1975**:1007.
f. K.-H. Geiss, B. Seuring, R. Pieter, and D. Seebach, *Angew. Chem. Int. Ed. Engl.* **13**:479 (1974); K.-H. Geiss, D. Seebach, and B. Seuring, *Chem. Ber.* **110**:1833 (1977).
g. E. Ehlinger and P. Magnus, *J. Am. Chem. Soc.* **102**:5004 (1980).
h. A. Marfat and P. Helquist, *Tetrahedron Lett.* **1978**:4217; A. Leone-Bay and L. A. Paquette, *J. Org. Chem.* **47**:4172 (1982).
i. J. P. Cherkaukas and T. Cohen, *J. Org. Chem.* **57**:6 (1992).

of the first step is a stabilized Wittig ylide, which goes on to react with the ketone carbonyl.

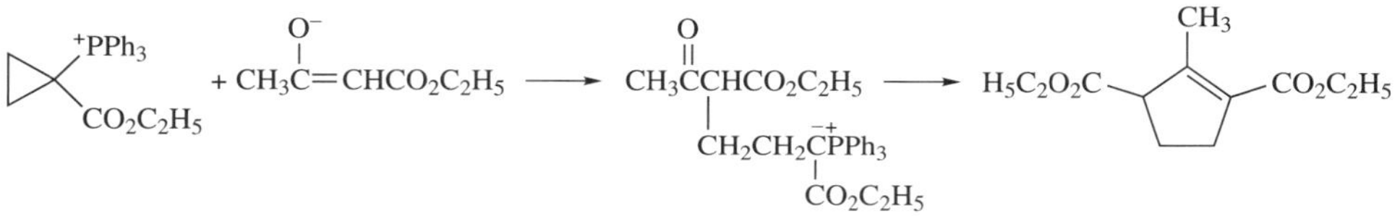

$^+PPh_3$, $CO_2C_2H_5$ (cyclopropane) + $CH_3\overset{O^-}{C}{=}CHCO_2C_2H_5$ ⟶ $CH_3\overset{O}{\overset{\|}{C}}CHCO_2C_2H_5$ | $CH_2CH_2\overset{-+}{C}PPh_3$ | $CO_2C_2H_5$ ⟶ $H_5C_2O_2C$, CH_3, $CO_2C_2H_5$ (cyclopentene)

The phosphonium salt **2** reacts similarly with enolates to give vinyl sulfides. The vinyl sulfide group can then be hydrolyzed to a ketone. The overall transformation corresponds to the reactivity of the dipolar synthon **C** (page 841).

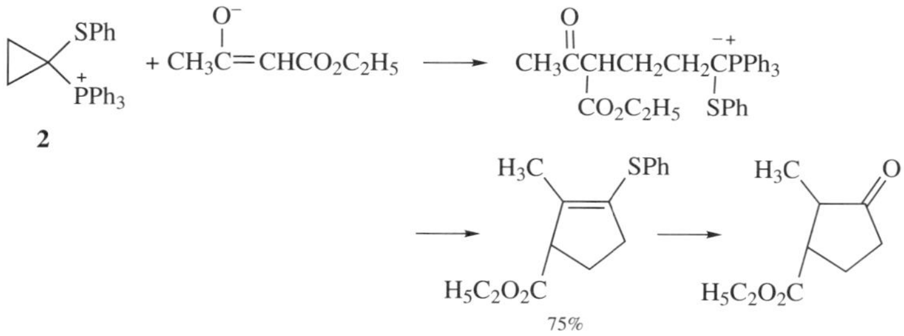

SPh, $\overset{+}{P}Ph_3$ (cyclopropane) **2** + $CH_3\overset{O^-}{C}{=}CHCO_2C_2H_5$ ⟶ $CH_3\overset{O}{\overset{\|}{C}}CHCH_2CH_2\overset{-+}{C}PPh_3$ | $CO_2C_2H_5$ SPh

⟶ H_3C, SPh, $H_5C_2O_2C$ (cyclopentene) 75% ⟶ H_3C, O, $H_5C_2O_2C$ (cyclopentanone) Ref. 123

Many other examples of synthetic equivalent groups have been developed. For example, in Chapter 6 the use of diene and dienophiles with masked functionality in the Diels–Alder reaction was discussed. It should be recognized that there is no absolute difference between what is termed a "reagent" and a "synthetic equivalent group." For

123. J. P. Marino and R. C. Landick, *Tetrahedron Lett.* **1975**:4531.

example, we think of potassium cyanide as a reagent, but the cyanide ion is a nucleophilic equivalent of a carboxyl group. This reactivity is evident in the classical preparation of carboxylic acids from alkyl halides via nitrile intermediates.

$$RX + KCN \longrightarrow RCN \xrightarrow[H^+]{H_2O} RCO_2H$$

The general point is that synthetic analysis and planning should not be restricted to the specific functionalities that must appear in the target molecules. These groups can be incorporated as masked equivalents by methods that would not be possible for the functional group itself.

13.3. Synthetic Analysis and Planning

The material covered to this point has been a description of the tools at the disposal of the synthetic chemist, consisting of the extensive catalog of reactions and the associated information on such issues as stereoselectivity and mutual reactivity. This knowledge permits a judgment of the applicability of a particular reaction in a synthetic sequence. Broad mechanistic insight is also crucial to synthetic analysis. The relative position of functional groups in a potential reactant may lead to special reactions. Mechanistic concepts are also the basis for developing new reactions that may be necessary in a particular situation. In this chapter, tactical tools of synthesis such as protective groups and synthetic equivalent groups have been introduced. The objective of synthetic analysis and planning is to develop a reaction sequence that will efficiently complete the desired synthesis within the constraints which apply.

The planning of a synthesis involves a critical comparative evaluation of alternative reaction sequences that could reasonably be expected to lead to the desired structure from appropriate starting materials. In general, the complexity of any synthetic plan will increase with the size of the molecule and with increasing numbers of functional groups and stereogenic centers. The goal of synthetic analysis is to recognize possible pathways to the target compound and to develop a suitable sequence of synthetic steps. In general, a large number of syntheses of any given compound are possible. The goal of synthetic planning is to identify a route which meets the specific criteria for the synthesis.

The restrictions that apply to a synthesis will depend on the reason the synthesis is being conducted. A synthesis of a material to be prepared in substantial quantity may impose a limitation on the cost of starting materials. Syntheses for commercial production must meet such criteria as economic feasibility, acceptability of by-products, and safety. Synthesis of complex structures with several stereogenic centers must deal with the problem of stereoselectivity. If an enantiomerically pure material is to be synthesized, the means of achieving enantioselectivity must be considered. The development of a satisfactory plan is the intellectual challenge to the chemist, and the task puts a premium on creativity and ingenuity. There is no single correct solution. Although there is no established routine by which a synthetic plan can be formulated, general principles which should guide synthetic analysis and planning have been described.[124]

The initial step in creating a synthetic plan should involve a *retrosynthetic analysis*. The structure of the molecule is dissected step-by-step along reasonable pathways to

124. E. J. Corey and X.-M. Cheng, *The Logic of Chemical Synthesis*, John Wiley & Sons, New York, 1989.

successively simpler compounds until molecules that are acceptable as starting materials are reached. Several factors enter into this process, and all are closely interrelated. The recognition of *bond disconnections* allows the molecule to be broken down into *key intermediates*. Such disconnections must be made in such a way that it is feasible to form the bonds by some synthetic process. The relative placement of potential functionality strongly influences which bond disconnections are preferred. To emphasize that these bond disconnections must correspond to transformations that can be conducted in the synthetic sense, they are sometimes called *antisynthetic transforms*, that is, the reverse of synthetic steps. An open arrow symbol, ⇒, is used to indicate an antisynthetic transform.

The overall synthetic plan will consist of a sequence of reactions designed to construct the total molecular framework from the key intermediates. The plan should take into account the advantages of a *convergent synthesis*. The purpose of making a synthesis more convergent is to decrease its overall length. In general, it is more desirable to construct the molecule from a few key fragments that can be combined late in the synthesis than to build the molecule step-by-step from a single starting material. The reason for this is that the overall yield is the multiplication product of the yields for all the individual steps. Overall yields decrease with the increasing number of steps to which the original starting material is subjected.[125]

Convergent synthesis: A + B ⟶ C $\xrightarrow{\text{D}}$ G; E + F ⟶ H; G + H ⟶ G—H $\xrightarrow{\text{I}}$ G—H—I

Linear synthesis: A + B ⟶ C $\xrightarrow{\text{D}}$ G $\xrightarrow{\text{E}}$ G—E $\xrightarrow{\text{F}}$ G—H $\xrightarrow{\text{I}}$ G—H—I

After a plan for assembly of the key intermediates into the molecular framework has been developed, the details of incorporation and transformation of functional groups must be considered. It is frequently necessary to interconvert functional groups. This may be done to develop a particular kind of reactivity at a center or to avoid interference with another reaction step. Protective groups and synthetic equivalent groups are important for planning of functional group transformations. Achieving the final array of functionality is often less difficult than establishing the overall molecular skeleton and stereochemistry, because of the large number of procedures for interconverting the common functional groups.

The care with which a synthesis is analyzed and planned will have a great impact on its likelihood of success. A single flaw can cause failure. The investment of material and effort which is made when the synthesis is begun may be lost if the plan is faulty. Even with the best of planning, however, unexpected problems are frequently encountered. This circumstance again tests the ingenuity of the chemist to devise a modified plan which can expeditiously overcome the unanticipated obstacle.

13.4. Control of Stereochemistry

The degree of control of stereochemistry that is necessary during synthesis depends on the nature of the molecule and the objective of the synthesis. The issue becomes of

125. A formal analysis of the concept of convergency has been presented by J. B. Hendrickson, *J. Am. Chem. Soc.* **99:**5439 (1977).

critical importance when the target molecule has several stereogenic centers, such as double bonds, ring junctures, and centers of chirality. The number of possible stereoisomers is 2^n, where n is the number of stereogenic centers. Failure to control stereochemistry of intermediates in the synthesis of a compound with several stereogenic centers will lead to a mixture of stereoisomers, which will, at best, lead to a reduced yield of the desired product and may generate inseparable mixtures.

We have considered stereochemical control for many of the synthetic methods that were discussed in the earlier chapters. In ring compounds, for example, stereoselectivity can frequently be predicted on the basis of conformational analysis of the reactant and consideration of the steric and stereoelectronic factors that will influence reagent approach. In the stereoselective synthesis of a chiral material in racemic form, it is necessary to control the *relative configuration* of all stereogenic centers. Thus, in planning a synthesis, the stereochemical outcome of all reactions that form new double bonds, ring junctions, and chiral centers must be incorporated into the synthetic plan. In a completely stereoselective synthesis, each successive center is introduced in the proper relationship to existing stereocenters. This ideal is often difficult to achieve. When a reaction is not completely stereoselective, the product will contain one or more diastereomers of the desired product. This requires either a purification or some manipulation to correct the stereochemistry. Fortunately, diastereomers are usually separable, but the overall efficiency of the synthesis is decreased with each such separation. Thus, high stereoselectivity is an important goal of synthetic planning.

If the compound is to be obtained in enantiomerically pure form, an enantioselective synthesis must be developed. (Review Section 2.1 of Part A for the basis of enantiomeric relationships.) There are four general approaches that are used to obtain enantiomerically pure material by synthesis. One is based on incorporating a *resolution* into the synthetic plan. This approach involves use of racemic or achiral starting materials and then resolving some intermediate in the synthesis. In a synthesis based on a resolution, the steps subsequent to the resolution step must meet two criteria: (1) they must not disturb the configuration at existing centers of stereochemistry, and (2) new stereogenic centers must be introduced with the correct configuration relative to the existing centers.

A second general approach is to use a starting material that is enantiomerically pure. There are a number of naturally occurring materials, or substances derived from them, that are available in enantiomerically pure form.[126] Again, a completely stereoselective synthesis must be capable of controlling the stereochemistry of all newly introduced stereogenic centers so that they have the proper relationship to the chiral centers existing in the starting material. When this is not achieved, the desired stereoisomer must be separated and purified.

A third method for enantioselective synthesis involves the use of a stoichiometric amount of a *chiral auxiliary*. This is an enantiomerically pure material that can control the stereochemistry of one or more reaction steps in such a way as to give product having the desired configuration. Once the chiral auxiliary has achieved its purpose, it can be eliminated from the molecule. As in syntheses involving resolution or enantiomerically pure starting materials, subsequent steps must be controlled to give the correct relative configuration of newly created stereogenic centers.

A fourth approach to enantioselective synthesis is to use a chiral catalyst in a reaction which creates one or more stereocenters. If the catalyst operates with complete efficiency,

126. For a discussion of this approach to enantioselective synthesis, see S. Hanessian, *Total Synthesis of Natural Products, the Chiron Approach*, Pergamon Press, New York, 1983.

an enantiomerically pure material will be obtained. Subsequent steps must control the relative configuration of newly introduced chiral centers.

In practice, any of these four approaches might be the most effective for a given synthesis. If they are judged on the basis of absolute efficiency in the use of chiral material, the ranking is resolution $<$ natural source $<$ chiral auxiliary $<$ enantioselective catalyst. A resolution process inherently employs only half of the original racemic material. A starting material from a natural source can, in principle, be used with 100% efficiency, but it is consumed and cannot be reused. A chiral auxiliary can, in principle, be recovered and reused, but it must be used in stoichiometric amount. A chiral catalyst can, in principle, produce an unlimited amount of an enantiomerically pure material.

The key issue for synthesis of pure stereoisomers, in either racemic or enantiomerically pure form, is that the configuration at newly created chiral centers must be controlled in some way. This may be accomplished in several ways. An existing functional group may control the approach of a reagent by coordination. An existing stereocenter may control reactant conformation, and thereby the direction of approach of a reagent. Whatever the detailed mechanism, the synthetic plan must include the means by which the required stereochemical control is to be achieved. When this cannot be done, the price to be paid is a separation of stereoisomers and the resulting reduction in overall yield.

13.5. Illustrative Syntheses

In this section, we will consider several syntheses of five illustrative compounds. We will examine the retrosynthetic plans and discuss crucial bond-forming steps and, where appropriate, the means of stereochemical control. In this discussion, we will have the benefit of hindsight in being able to look at successfully completed syntheses. This retrospective analysis can serve to illustrate the issues that arise in planning a synthesis and provide examples of solutions that have been developed. The individual syntheses also provide many examples of the synthetic transformations presented in the earlier chapters and of the use of protective groups in the synthesis of complex molecules.

13.5.1. Juvabione

Juvabione is a terpene-derived keto ester that has been isolated from various plant sources. There are two stereoisomers, both of which occur naturally with *R* configuration at C-4 of the cyclohexene ring, and which are referred to as *erythro*- and *threo*-juvabione. The 7*S* isomer is sometimes called epijuvabione. Juvabione exhibits "juvenile hormone" acitivity in insects; that is, it can modify the process of metamorphosis.[127]

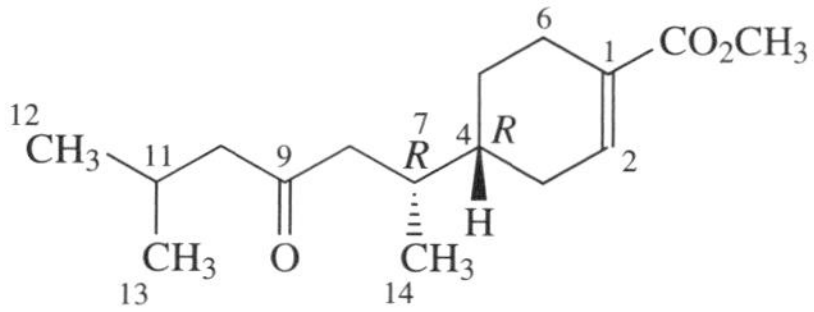

threo-juvabione

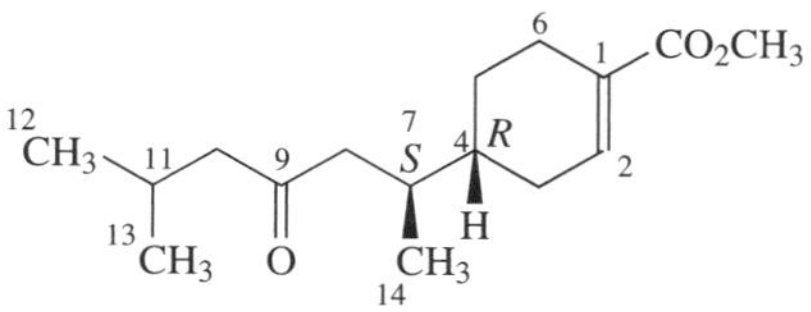

erythro-juvabione

127. For a review, see Z. Wimmer and M. Romanuk, *Collect. Czech. Chem. Commun.* **54:**2302 (1989).

In considering the retrosynthetic analysis of juvabione, two factors draw special attention to the bond between C-4 and C-7. First, this bond establishes the stereochemistry of the molecule. The C-4 and C-7 carbons are both chiral, and their relative configuration determines which diastereomeric structure will be obtained. In a stereocontrolled synthesis, it is necessary to establish the desired stereochemistry at C-4 and C-7. The C(4)–C(7) bond also connects the side chain to the cyclohexene ring. Because a cyclohexane derivative would make a logical candidate for one key intermediate, the C(4)–C(7) bond is a potential bond disconnection.

Other bonds which merit attention are those connecting C-7 through C-11. These could be formed by one of the many methods for synthesis of ketones. The only other point of functionality is the conjugated unsaturated ester. This functionality is remote from the stereochemical centers and the ketone functionality, and in most of the reported syntheses, it does not play a key role. Some of the existing syntheses use similar types of starting materials. Those in Schemes 13.4 and 13.5 lead back to a *para*-substituted aromatic ether. The syntheses in Schemes 13.7–13.9 begin with an accessible terpene intermediate. The syntheses in Schemes 13.11 and 13.12 start with cyclohexenone.

Scheme 13.3 presents a retrosynthetic analysis leading to the key intermediates used by the syntheses in Schemes 13.4 and 13.5. The first disconnection is that of the ester functionality. This corresponds to a decision that the ester group can be added late in the synthesis. Disconnection **2** identifies the C(9)–C(10) bond as one that can be readily formed by addition of some nucleophilic group corresponding to C(10)–C(13) to the carbonyl center at C-9. The third retrosynthetic transform recognizes that the cyclohexanone ring might be obtained by a Birch reduction of an appropriately substituted aromatic ether. The methoxy substituent would provide for correct placement of the cyclic carbonyl group. The final disconnection identifies a simple starting material, 4-methoxyacetophenone.

A synthesis corresponding to this pattern is shown in Scheme 13.4. It relies on well-known reaction types. The C(4)–C(7) bond is formed by a Reformatsky reaction, and this is followed by benzylic hydrogenolysis. Steps **B** and **C** introduce the C-10–C13 isobutyl group. The C(9)–C(10) bond connection is done in step **C** by a Grignard addition reaction. In this synthesis, the relative configuration at and C-4 and C-7 is established by the hydrogenation in step **E**. In principle, this reaction could be diastereoselective if the adjacent stereocenter at C-7 strongly influenced the direction of addition of hydrogen. In practice, the reduction is not very selective, and a mixture of isomers was obtained. Steps **F** and **G** introduce the C-1 ester group.

The synthesis in Scheme 13.5 also makes use of an aromatic starting material and follows a retrosynthetic plan corresponding to that in Scheme 13.3. This synthesis is

Scheme 13.3. Retrosynthetic Analysis of Juvabione with Disconnection to *p*-Methoxyacetophenone

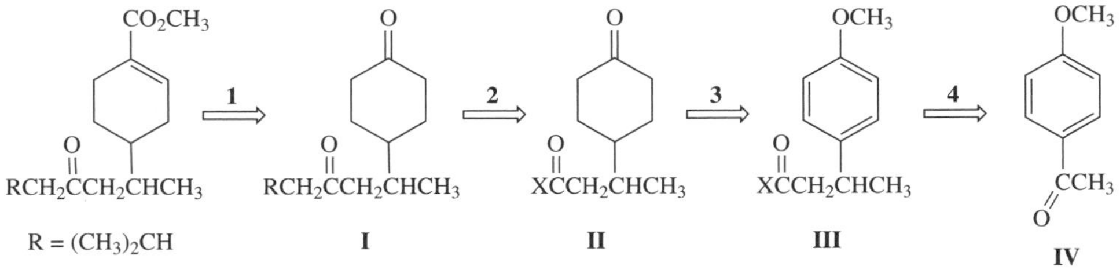

Scheme 13.4. Juvabione Synthesis: K. Mori and M. Matsui[a]

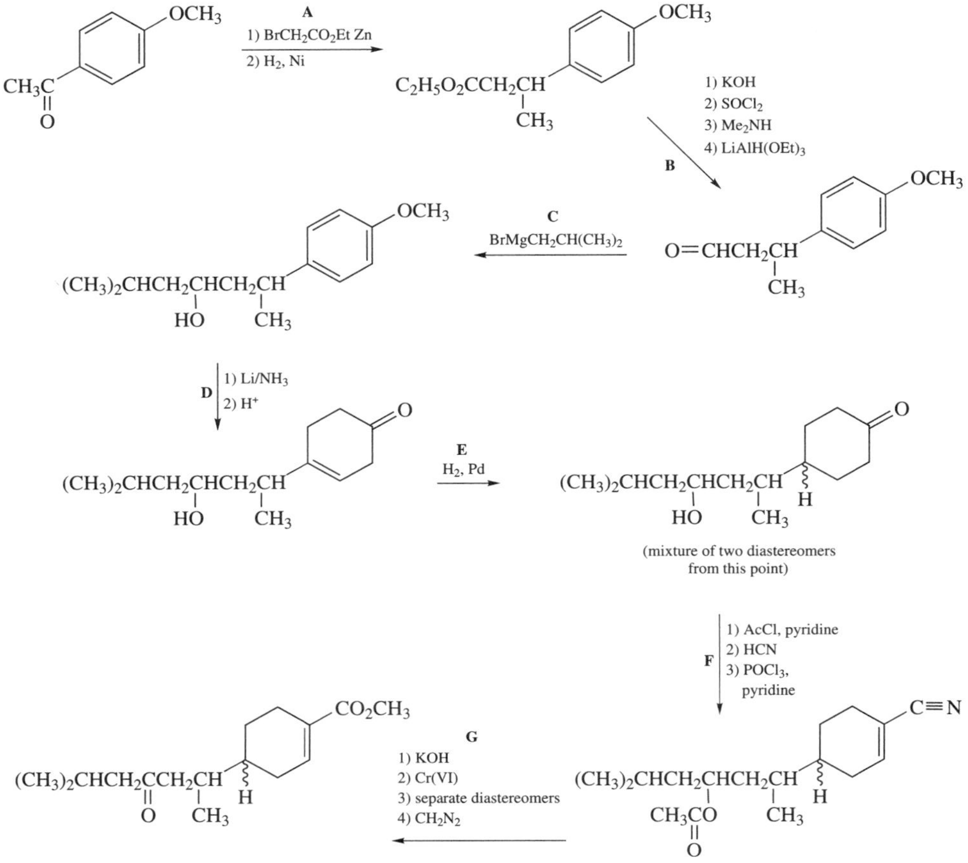

A: 1) $BrCH_2CO_2Et$ Zn; 2) H_2, Ni

$C_2H_5O_2CCH_2CH(CH_3)$–Ar (Ar = 4-$OCH_3C_6H_4$)

B: 1) KOH; 2) $SOCl_2$; 3) Me_2NH; 4) $LiAlH(OEt)_3$

$O{=}CHCH_2CH(CH_3)$–Ar

C: $BrMgCH_2CH(CH_3)_2$

$(CH_3)_2CHCH_2CH(OH)CH_2CH(CH_3)$–Ar

D: 1) Li/NH_3; 2) H^+

E: H_2, Pd

(mixture of two diastereomers from this point)

F: 1) AcCl, pyridine; 2) HCN; 3) $POCl_3$, pyridine

G: 1) KOH; 2) Cr(VI); 3) separate diastereomers; 4) CH_2N_2

a. K. Mori and M. Matsui, *Tetrahedron* **24:**3127 (1968).

Scheme 13.5. Juvabione Synthesis: K. S. Ayyar and G. S. K. Rao[a]

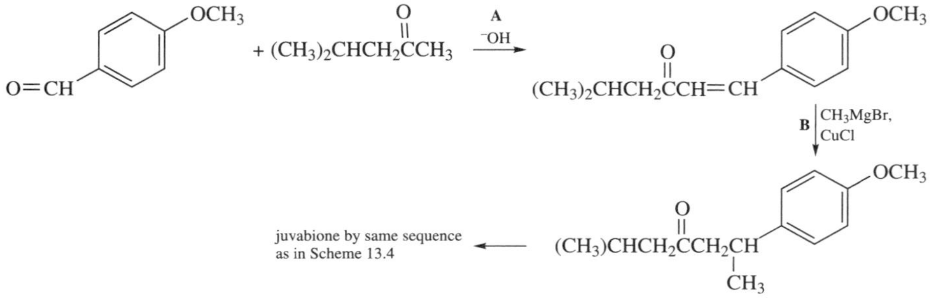

$O{=}CH$–Ar + $(CH_3)_2CHCH_2C(O)CH_3$

A: ^-OH

$(CH_3)_2CHCH_2C(O)CH{=}CH$–Ar

B: CH_3MgBr, CuCl

$(CH_3)CHCH_2C(O)CH_2CH(CH_3)$–Ar

juvabione by same sequence as in Scheme 13.4

a. K. S. Ayyar and G. S. K. Rao, *Can. J. Chem.* **46:**1467 (1968).

somewhat more convergent in that the entire side chain, except for C-14, is introduced as a single unit. The C-14 methyl is introduced by a copper-catalyzed conjugate addition in step **B**.

Scheme 13.6 is a retrosynthetic outline of the syntheses in Schemes 13.7 and 13.8. The common feature of these syntheses is the use of terpene-derived starting materials. The use of such a starting material is suggested by the terpenoid structure of juvabione, which can be divided into "isoprene units." Furthermore, the terpenoid precursors can establish the configuration at C-4.

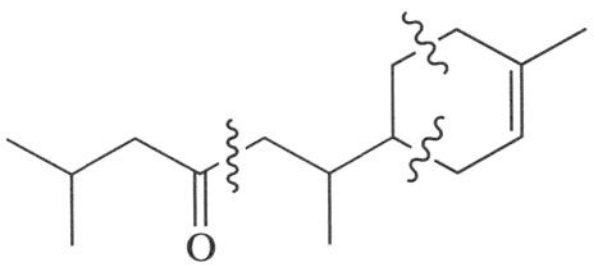
O

isoprene units in juvabione

The synthesis shown in Scheme 13.7 used limonene as the starting material ($R = CH_3$ in Scheme 13.6) whereas Scheme 13.8 uses the corresponding aldehyde

Scheme 13.6. Retrosynthetic Analysis of Juvabione with Disconnection to the Terpene Limonene

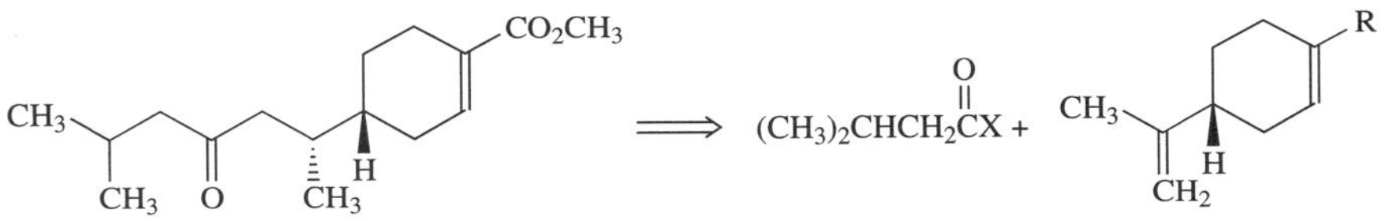
CO2CH3
CH3
CH3 O
H
CH3
O
(CH3)2CHCH2CX +
CH3
R
H
CH2

limonene (R = CH3)

Scheme 13.7. Juvabione Synthesis: B. A. Pawson, H.-C. Cheung, S. Gurbaxani, and G. Saucy[a]

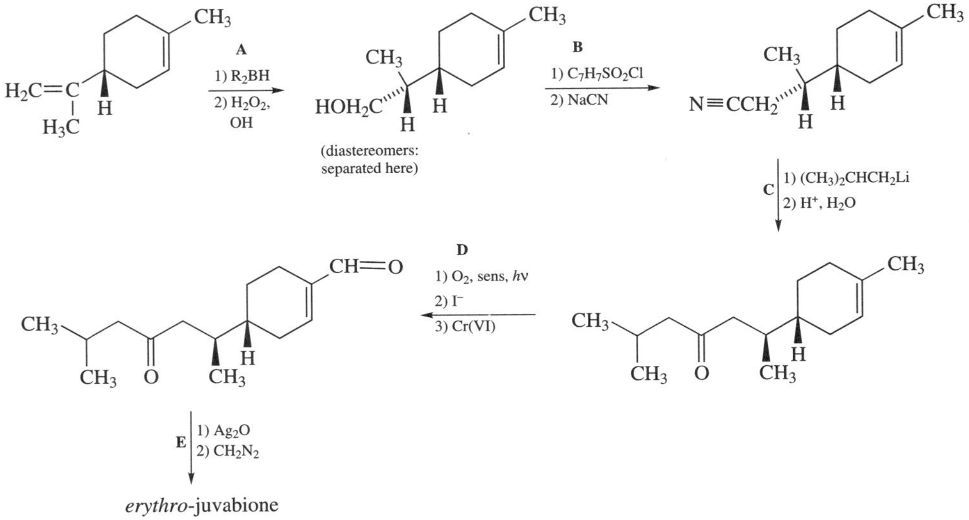
CH3
H2C=C
H
H3C
A
1) R2BH
2) H2O2,
OH
CH3
CH3
HOH2C
H
H
(diastereomers:
separated here)
B
1) C7H7SO2Cl
2) NaCN
CH3
CH3
N≡CCH2
H
H
C
1) (CH3)2CHCH2Li
2) H+, H2O
CH3
CH3
CH3 O
H
CH3
D
1) O2, sens, hv
2) I−
3) Cr(VI)
CH=O
CH3
CH3 O
H
CH3
E
1) Ag2O
2) CH2N2
erythro-juvabione

a. B. A. Pawson, H.-C. Cheung, S. Gurbaxani, and G. Saucy, *J. Am. Chem. Soc.* **92**:336 (1970).

Scheme 13.8. Juvabione Synthesis: E. Negishi, M. Sabanski, J. J. Katz, and H. C. Brown[a]

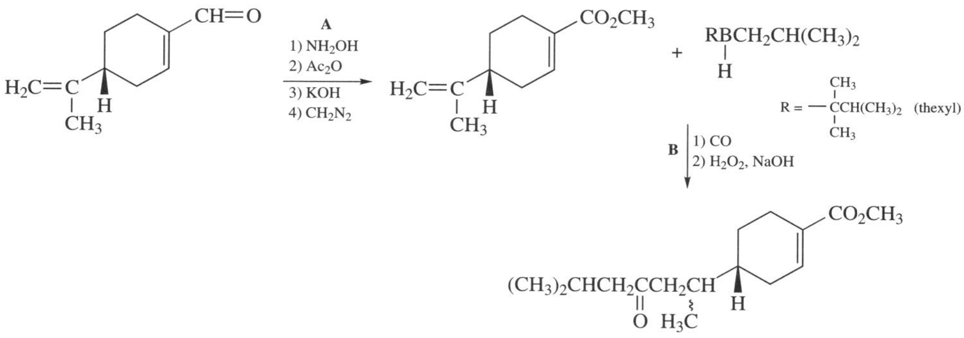

a. E. Negishi, M. Sabanski, J. J. Katz, and H. C. Brown, *Tetrahedron* **32**:925 (1976).

(R = CH=O) (perillaldehyde). The use of these starting materials focuses attention on the means of attaching the C-9–C-13 side chain. Furthermore, enantioselectivity controlled by the chiral center at C-4 of the starting material might be feasible. In the synthesis in Scheme 13.7, the C-4–C-7 stereochemistry is established in the hydroboration which is the first step of the synthesis. Unfortunately, this reaction shows only very modest stereoselectivity, and a 3 : 2 mixture of diastereomers was obtained and separated. The subsequent steps do not affect these chiral centers. The synthesis in Scheme 13.7 uses a three-step sequence to oxidize the C-15 methyl group at step **D**. The first reaction is oxidation by singlet oxygen to give a mixture of hydroperoxides, with oxygen bound mainly at C-2. The mixture is reduced to the corresponding alcohols, which are then oxidized to the acid via an aldehyde intermediate.

In Scheme 13.8, the side chain is added in one step by a borane carbonylation reaction. This synthesis is very short, and the first four steps are used to transform the aldehyde group in the starting material to a methyl ester. The stereochemistry at C-7 is established in the hydroboration in step **B**, where the C(7)–H bond is formed. A 1 : 1 mixture of diastereomers was formed, indicating that the configuration at C-4 did not influence the direction of approach of the borane reagent.

Another synthesis which starts with the same aldehyde has been completed more recently. This synthesis is shown in Scheme 13.9. In this synthesis, the C-7 stereochemistry is established by hydrogenation of a methylene group, but it also produces both stereoisomers.

The first stereocontrolled syntheses of juvabione are described in Schemes 13.11 and 13.12. Scheme 13.10 is a retrosynthetic analysis corresponding to these syntheses. These syntheses have certain similarities. Both start with cyclohexenone. There is a general similarity in the fragments that were utilized, but the order of construction differs. In the synthesis shown in Scheme 13.11, the crucial step for stereochemical control is step **B**. The first intermediate is constructed by a [2 + 2] cycloaddition between reagents of complementary polarity, the electron-rich enamine and the electron-poor enone. The cyclobutane ring is then opened in a process which corresponds to retrosynthetic step **IIa** ⇒ **IIIa** in

Scheme 13.9. Juvabione: A. A. Carveiro and I. G. P. Viera[a]

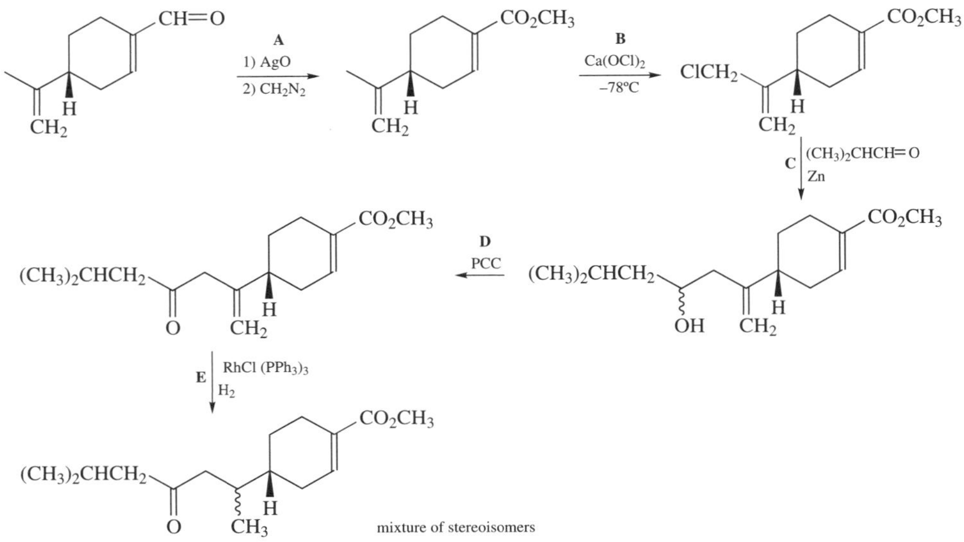

A: 1) AgO, 2) CH_2N_2; B: $Ca(OCl)_2$, –78°C; C: $(CH_3)_2CHCH{=}O$, Zn; D: PCC; E: $RhCl(PPh_3)_3$, H_2

mixture of stereoisomers

a. A. A. Carveiro and L. G. P. Viera, *J. Braz. Chem. Soc.* **3**:124 (1992).

Scheme 13.10. The stereoselectivity of this step results from preferential protonation of the enamine from the less hindered side of the bicyclic intermediate.

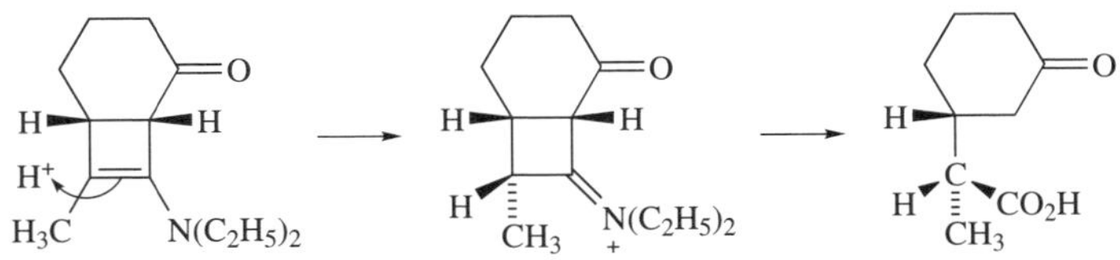

The cyclobutane ring is then cleaved by hydrolysis of the enamine and resulting β-diketone. The relative configuration of the chiral centers is unaffected by subsequent reaction steps, so the overall sequence is stereoselective. Another key step in this synthesis is step **D**, which corresponds to the transformation **IIa** $\Rightarrow$ **Ia** in the retrosynthesis. A protected cyanohydrin is used as a nucleophilic acyl anion equivalent in this step. The final steps of the synthesis in Scheme 13.11 employ the C-2 carbonyl group to introduce the carboxy group and the C-1–C-2 double bond.

The stereoselectivity achieved in the synthesis in Scheme 13.12 is the result of a preferred conformation for the base-catalyzed oxy-Cope rearrangement in step **C**. Although the intermediate used in step **C** is a mixture of stereoisomers, both give predominantly the desired relative stereochemistry at C-4 and C-7. The stereoselectivity

Scheme 13.10. Retrosynthetic Analysis of Juvabione with Alternate Disconnections to Cyclohexanone

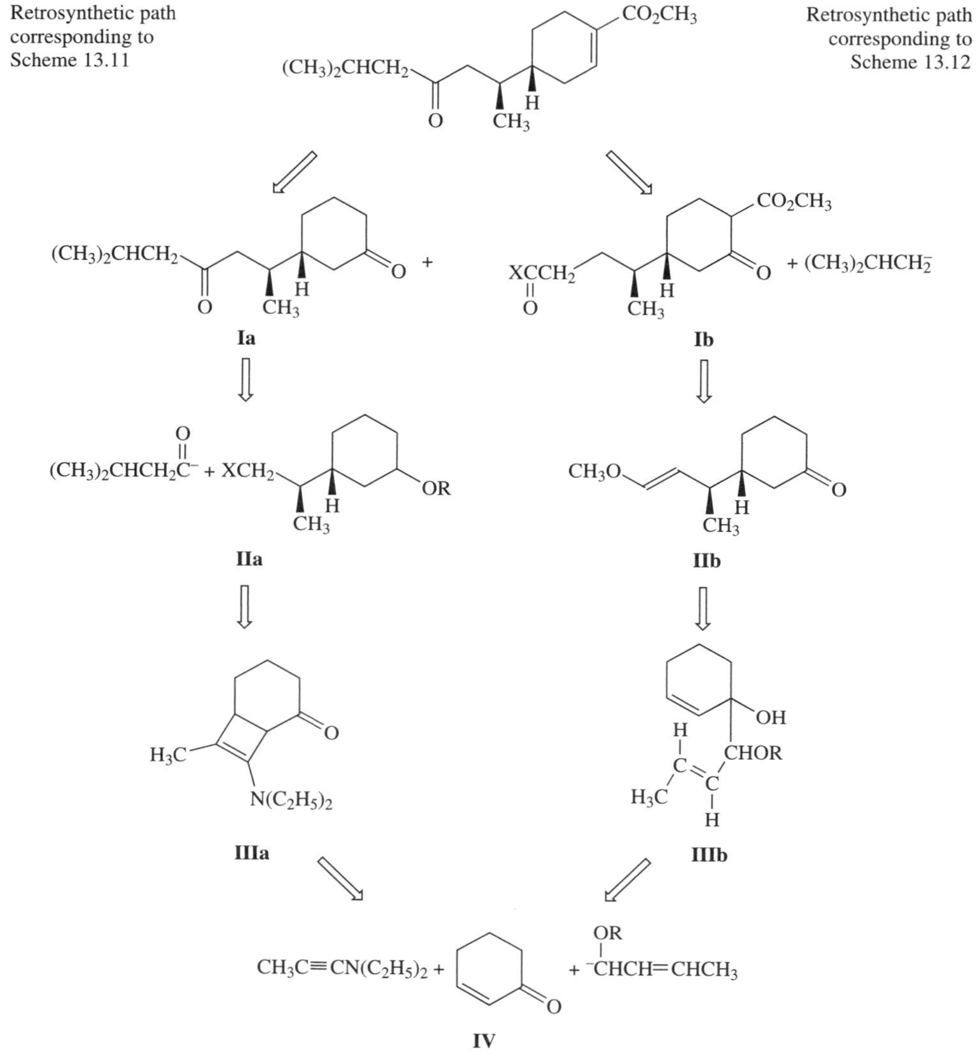

Retrosynthetic path corresponding to Scheme 13.11

Retrosynthetic path corresponding to Scheme 13.12

$(CH_3)_2CHCH_2$ CO_2CH_3 O H CH_3

$(CH_3)_2CHCH_2$ O H CH_3 O + **Ia**

$XCCH_2$ O H CH_3 CO_2CH_3 O $+ (CH_3)_2CHCH_2^-$ **Ib**

$(CH_3)_2CHCH_2C^-$ O $+ XCH_2$ H CH_3 OR **IIa**

CH_3O H CH_3 O **IIb**

H_3C O $N(C_2H_5)_2$ **IIIa**

OH H CHOR C C H_3C H **IIIb**

$CH_3C{\equiv}CN(C_2H_5)_2$ + O + $^-\overset{OR}{C}HCH{=}CHCH_3$ **IV**

is based on the preferred chair conformation for the transition state of the concerted rearrangement.

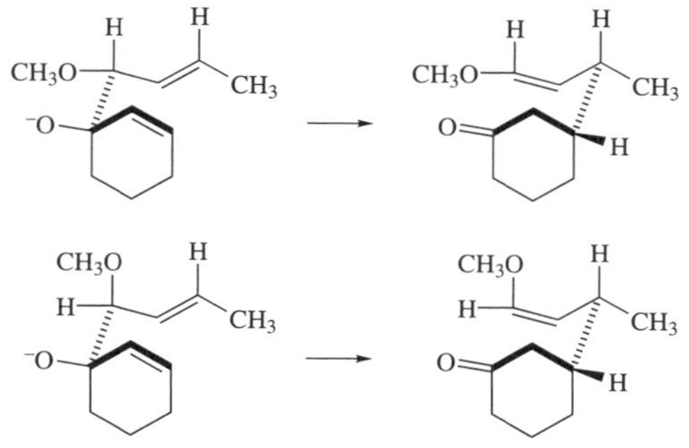

H H CH_3O CH_3 ^-O → H H CH_3O CH_3 O H

CH_3O H H CH_3 ^-O → CH_3O H H CH_3 O H

Scheme 13.11. Juvabione Synthesis: J. Ficini, J. D'Angelo, and J. Noire[a]

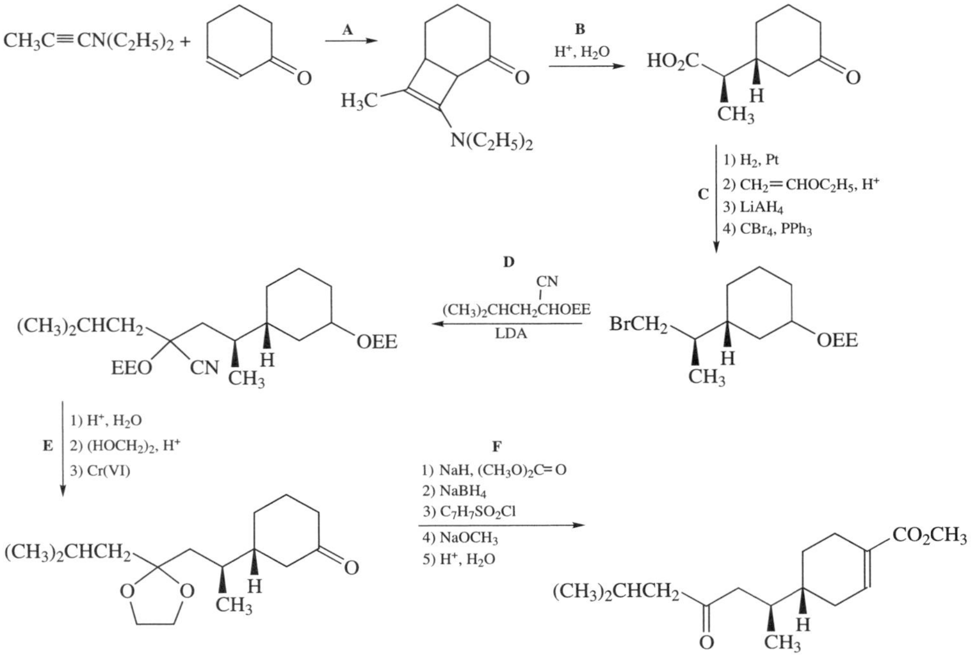

$CH_3C{\equiv}CN(C_2H_5)_2$ +

A

H_3C $N(C_2H_5)_2$ O

B
H^+, H_2O

HO_2C H CH_3 O

C
1) H_2, Pt
2) $CH_2{=}CHOC_2H_5$, H^+
3) $LiAlH_4$
4) CBr_4, PPh_3

$BrCH_2$ H CH_3 OEE

D
$(CH_3)_2CHCH_2CH(CN)OEE$
LDA

$(CH_3)_2CHCH_2$ EEO CN H CH_3 OEE

E
1) H^+, H_2O
2) $(HOCH_2)_2$, H^+
3) Cr(VI)

$(CH_3)_2CHCH_2$ O O H CH_3 O

F
1) NaH, $(CH_3O)_2C{=}O$
2) $NaBH_4$
3) $C_7H_7SO_2Cl$
4) $NaOCH_3$
5) H^+, H_2O

$(CH_3)_2CHCH_2$ O H CH_3 CO_2CH_3

a. J. Ficini, J. D'Angelo, and J. Noire, *J. Am. Chem. Soc.* **96:**1213 (1974).

Scheme 13.12. Juvabione Synthesis: D. A. Evans and J. V. Nelson[a]

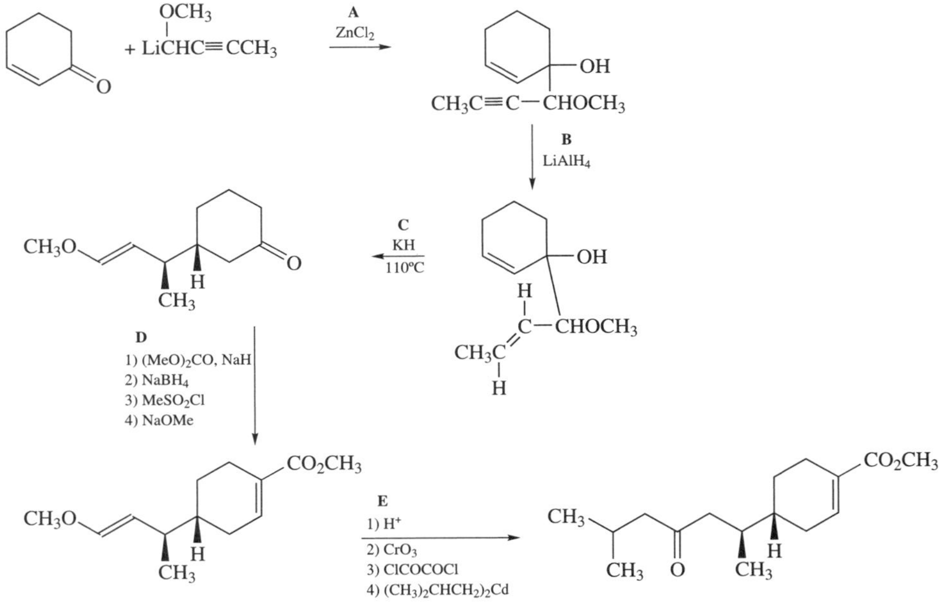

O + $LiCH(OCH_3)C{\equiv}CCH_3$

A
$ZnCl_2$

OH
$CH_3C{\equiv}C{-}CHOCH_3$

B
$LiAlH_4$

OH
H
$CH_3C{=}C{-}CHOCH_3$
H

C
KH
110°C

CH_3O H CH_3 O

D
1) $(MeO)_2CO$, NaH
2) $NaBH_4$
3) $MeSO_2Cl$
4) NaOMe

CH_3O H CH_3 CO_2CH_3

E
1) H^+
2) CrO_3
3) ClCOCOCl
4) $((CH_3)_2CHCH_2)_2Cd$

CH_3 CH_3 O H CH_3 CO_2CH_3

a. D. A. Evans and J. V. Nelson, *J. Am. Chem. Soc.* **102:**774 (1980).

The synthesis in Scheme 13.13 leads to the *erythro* stereoisomer. An intramolecular enolate alkylation in step **B** gives a bicyclic intermediate. The relative configuration of C-4 and C-7 is established by the hydrogenation in step **C**. The hydrogen is added from the less hindered *exo* face of the bicyclic enone. This synthesis is an example of the use of geometric constraints built into a ring system to control relative stereochemistry.

The *threo* stereoisomer is the major product obtained by the synthesis in Scheme 13.14. This stereochemistry is established by the conjugate addition in step **A**, where a significant (4–6 : 1) diastereoselectivity is observed. The C-4–C-7 stereochemical relationship is retained throughout the remainder of the synthesis. The other special features of this synthesis are in steps **B** and **C**. The mercuric acetate-mediated cyclopropane ring opening is facilitated by the alkoxy substituent.[128] The reduction by $NaBH_4$ accomplishes both demercuration and reduction of the aldehyde group.

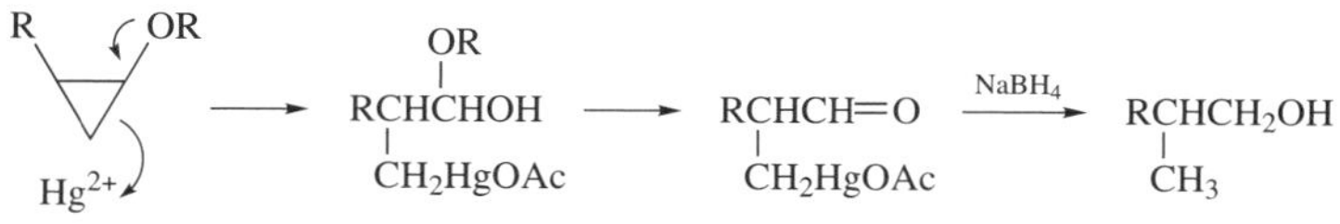

R, OR, Hg^{2+} → RCH(OR)CHOH(CH$_2$HgOAc) → RCHCH=O(CH$_2$HgOAc) —$NaBH_4$→ RCH(CH$_3$)CH$_2$OH

In step **C**, a dithiane anion is used as a nucleophilic acyl anion equivalent to introduce the C-10–C-13 isobutyl group.

In the synthesis shown in Scheme 13.15, racemates of both *erythro-* and *threo-*juvabione were synthesized by parallel routes. The isomeric intermediates are obtained in >10 : 1 selectivity by choice of the *E-* or *Z-*silanes used for conjugate addition to cyclohexenone. Further optimization of the stereoselectivity was obtained by choice of the silyl substituents. The purified intermediates were then converted to the juvabione stereoisomers.

Scheme 13.13. Juvabione Synthesis: A. G. Schultz and J. P. Dittami[a]

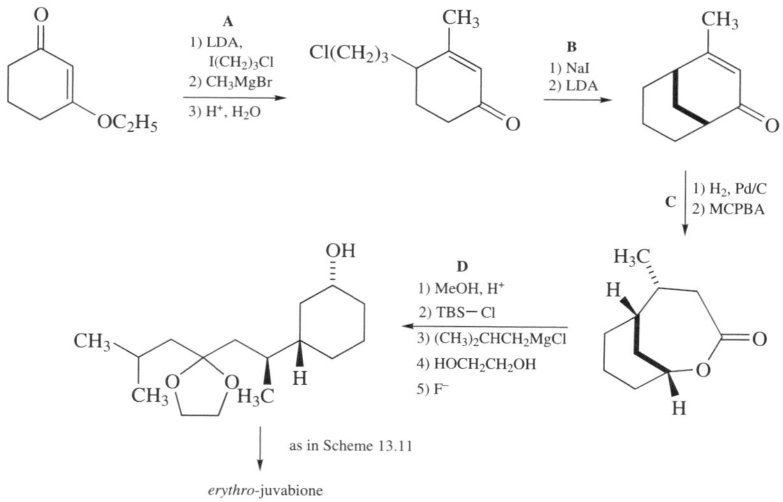

O, OC_2H_5 —**A** 1) LDA, $I(CH_2)_3Cl$ 2) CH_3MgBr 3) H^+, H_2O→ $Cl(CH_2)_3$, CH_3, O —**B** 1) NaI 2) LDA→ CH_3, O —**C** 1) H_2, Pd/C 2) MCPBA→ H_3C, H, O, O, H —**D** 1) MeOH, H^+ 2) TBS−Cl 3) $(CH_3)_2CHCH_2MgCl$ 4) $HOCH_2CH_2OH$ 5) F^-→ OH, CH_3, CH_3, O, O, H_3C, H —as in Scheme 13.11→ *erythro*-juvabione

a. A. G. Schultz and J. P. Dittami, *J. Org. Chem.* **49:**2615 (1984).

128. A. DeBoer and C. H. DePuy, *J. Am. Chem. Soc.* **92:**4008 (1970).

Scheme 13.14. Juvabione Synthesis: D. J. Morgans, Jr. and G. B. Feigelson[a]

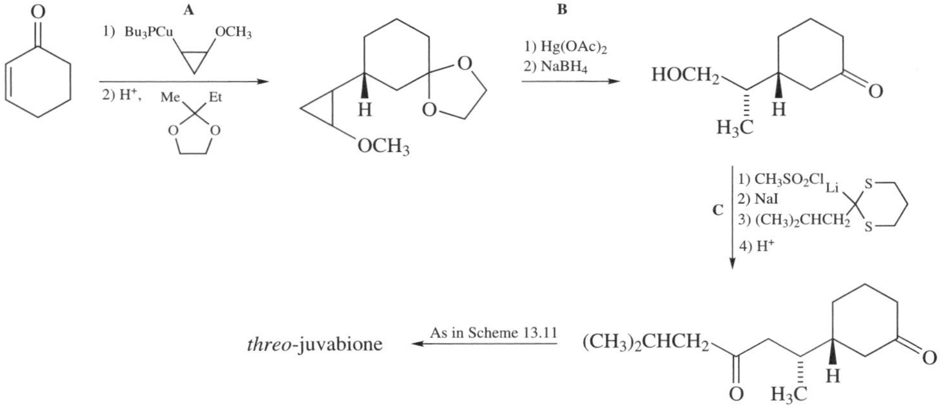

a. D. J. Morgan, Jr. and G. B. Feigelson, *J. Am. Chem. Soc.* **105:**5477 (1983).

Except for the syntheses using terpene-derived starting materials (Schemes 13.7–13.9), the previous juvabione syntheses all gave racemic materials. Two more recently developed juvabione syntheses are *enantiospecific*. The synthesis in Scheme 13.16 relies on a chiral sulfoxide which undergoes stereoselective addition to cyclohexenone to establish the correct relative and absolute configuration at C-4 and C-7. The origin of the stereoselectivity has not been established, but one transition state that would lead to the

Scheme 13.15. Juvabione Synthesis: T. Tokoroyama and L.-R. Pan[a]

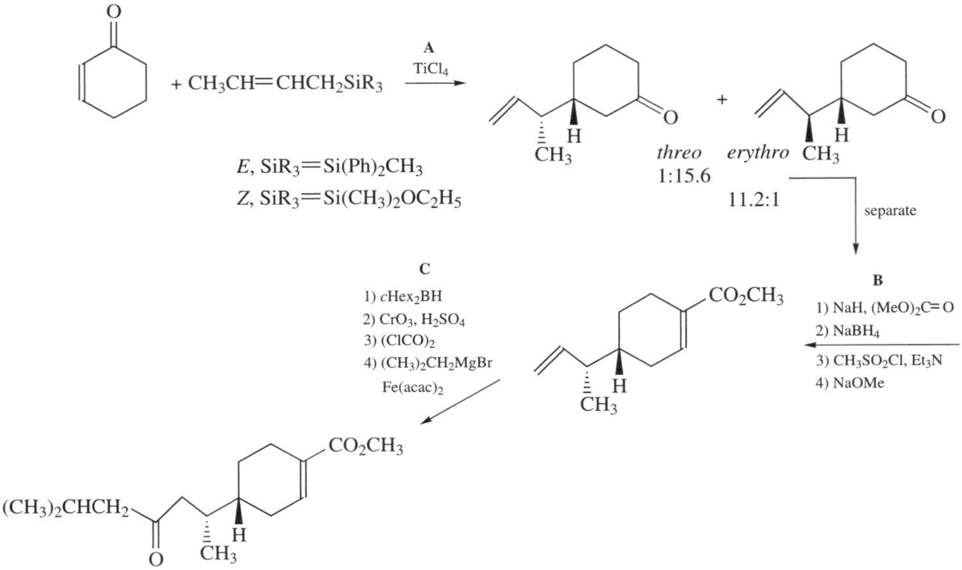

a. T. Tokoroyama and L.-R. Pan, *Tetrahedron Lett.* **30:**197 (1989).

Scheme 13.16. Juvabione Synthesis: H. Watanabe, H. Shimizu, and K. Mori[a]

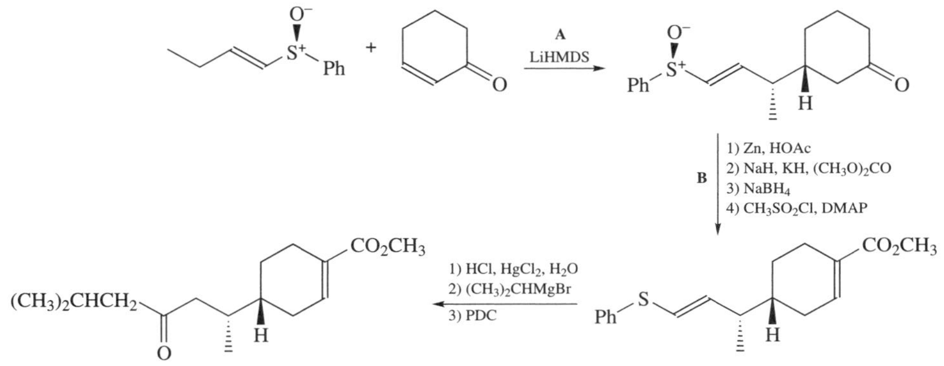
A
LiHMDS
B
1) Zn, HOAc
2) NaH, KH, (CH3O)2CO
3) NaBH4
4) CH3SO2Cl, DMAP
1) HCl, HgCl2, H2O
2) (CH3)2CHMgBr
3) PDC
(CH3)2CHCH2
CO2CH3
Ph
H

a. H. Watanabe, H. Shimizu, and K. Mori, *Synthesis* **1994:**1249.

observed product is shown below.

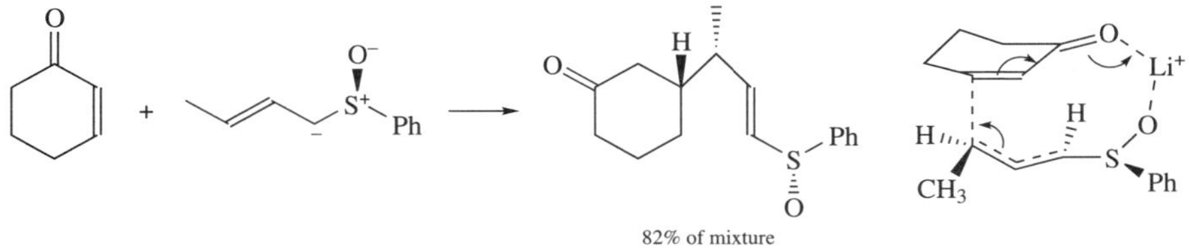
Ph
Li+
CH3
82% of mixture

Another enantioselective synthesis, shown in Scheme 13.17, is based on enantioselective reduction of bicyclo[2.2.2]octane-2,6-dione by baker's yeast.[129] The enantiomerically pure intermediate is then converted to the lactone intermediate by Baeyer–Villiger oxidation and an allylic rearrangement. The methyl group is then introduced stereoselectively from the *exo* face of the bicyclic lactone in step **C**-1. A final crucial step in this synthesis is a [2,3] sigmatropic rearrangement to complete sequence **D**.

Another enantioselective synthesis is shown in Scheme 13.18. This synthesis involves an early kinetic resolution of the the alcohol intermediate in step **B**-2 by lipase PS. The stereochemistry at the C-7 methyl group is then controlled by an *exo* addition in the conjugate addition (step **D**-1).

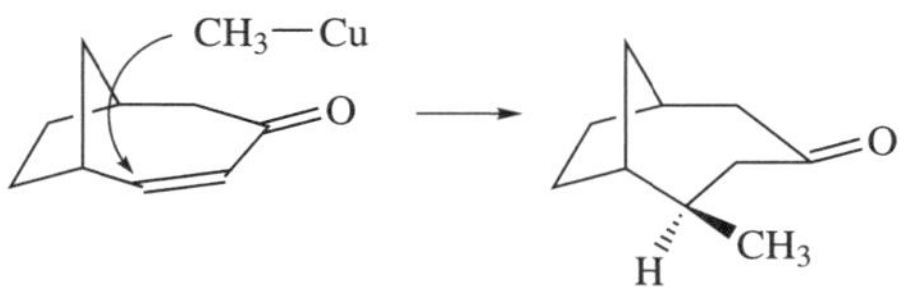
CH3—Cu
CH3

Another interesting feature of this synthesis is the ring expansion used in sequences **A** and **F**. Trimethylsilyl enol ethers are treated with Simmons–Smith reagent to form cyclopropyl silyl ethers. These rings undergo oxidative cleavage and ring expansion when treated with

129. K. Mori and F. Nagano, *Biocatalysis* **3:**25 (1990).

a. E. Nagano and K. Mori, *Biosci. Biotechnol. Biochem.* **56:**1589 (1992).

$FeCl_3$, and the α-chloroketones are then dehydrohalogenated by DBU.[130]

Several other syntheses of juvabione have also been completed.[131]

13.5.2. Longifolene

Longifolene is a tricyclic terpene. It is a typical terpene in terms of the structural complexity. Schemes 13.19 through 13.27 describe eight separate syntheses of longi-

130. V. Ito, S. Fujii, and T. Saegusa, *J. Org. Chem.* **41:**2073 (1976).
131. A. A. Drabkina and Y. S. Tsizin, *J. Gen. Chem. USSR (English transl.)* **43:**422, 691 (1973); R. J. Crawford, U.S. Patent, 3,676,506; *Chem. Abstr.* **77:**113889e (1972); A. J. Birch, P. L. Macdonald, and V. H. Powell, *J. Chem. Soc. C* **1970:**1469; M. Fujii, T. Aida, M. Yoshihara, and A. Ohno, *Bull. Chem. Soc. Jpn.* **63:**1255 (1990).

Scheme 13.18. Juvabione Synthesis: N. Nagata, T. Taniguchi, M. Kawamura, and K. Ogasawara[a]

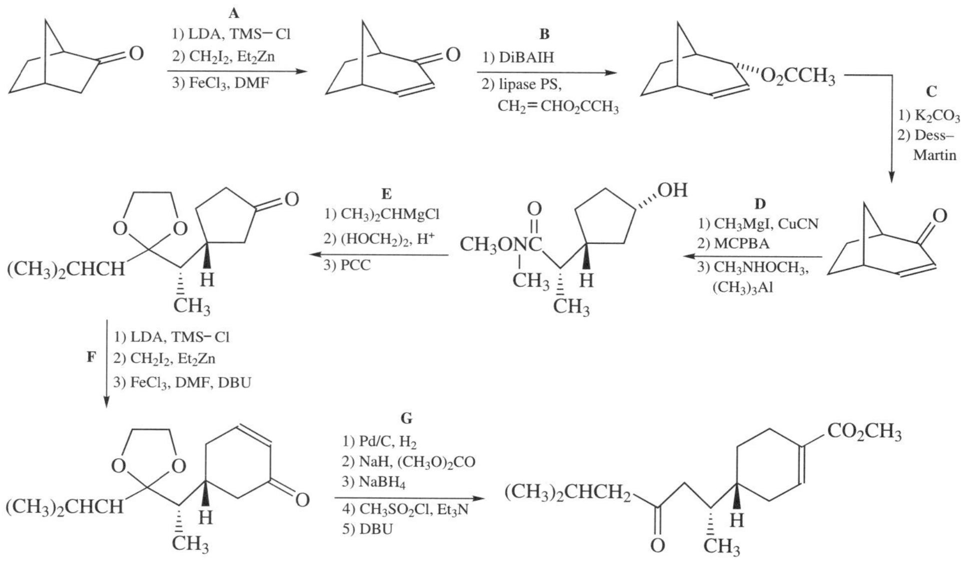

A: 1) LDA, TMS–Cl; 2) CH_2I_2, Et_2Zn; 3) $FeCl_3$, DMF

B: 1) DiBAlH; 2) lipase PS, $CH_2{=}CHO_2CCH_3$

O_2CCH_3

C: 1) K_2CO_3; 2) Dess–Martin

D: 1) CH_3MgI, CuCN; 2) MCPBA; 3) CH_3NHOCH_3, $(CH_3)_3Al$

OH; CH_3ONC; CH_3; H; CH_3

E: 1) $(CH_3)_2CHMgCl$; 2) $(HOCH_2)_2$, H^+; 3) PCC

$(CH_3)_2CHCH$; H; CH_3

F: 1) LDA, TMS–Cl; 2) CH_2I_2, Et_2Zn; 3) $FeCl_3$, DMF, DBU

G: 1) Pd/C, H_2; 2) NaH, $(CH_3O)_2CO$; 3) $NaBH_4$; 4) CH_3SO_2Cl, Et_3N; 5) DBU

$(CH_3)_2CHCH_2$; CO_2CH_3; O; H; CH_3

folene. We wish to particularly emphasize the methods for carbon–carbon bond formation used in these syntheses. There are four stereogenic centers in longifolene, but they are not independent of one another because the geometry of the ring system requires that they have a specific relative relationship. That does not mean that stereochemistry can be ignored, however, because the formation of the various rings will fail if the reactants do not have the proper stereochemistry.

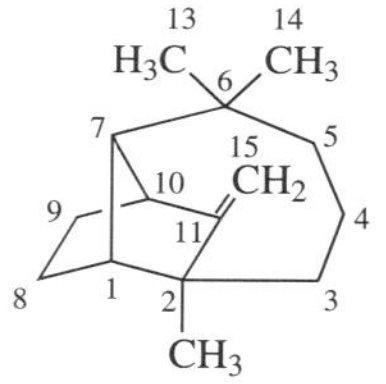

13 14 H_3C 6 CH_3 7 15 5 10 CH_2 9 11 4 8 1 2 3 CH_3

The first successful synthesis of longifolene was described in detail by E. J. Corey and co-workers in 1964. Scheme 13.19 presents a retrosynthetic analysis corresponding to this route. A key disconnection is made on going from **I** $\Rightarrow$ **II**. This transformation simplifies the tricyclic skeleton to a bicyclic one. For this disconnection to correspond to a reasonable synthetic step, the functionality in the intermediate to be cyclized must engender mutual reactivity between C-7 and C-10. This is achieved in diketone **II**, because an enolate generated by deprotonation at C-10 can undergo an intramolecular Michael addition to C-7. Retrosynthesic step **II** $\Rightarrow$ **III** is attractive, because it suggests a decalin derivative as a key intermediate. Methods for preparing structures of this type have been well developed, because they are useful intermediates in the synthesis of other terpenes and also steroids.

Scheme 13.19. Retrosynthesis of Longifolene Corresponding to the Synthesis in Scheme 13.20

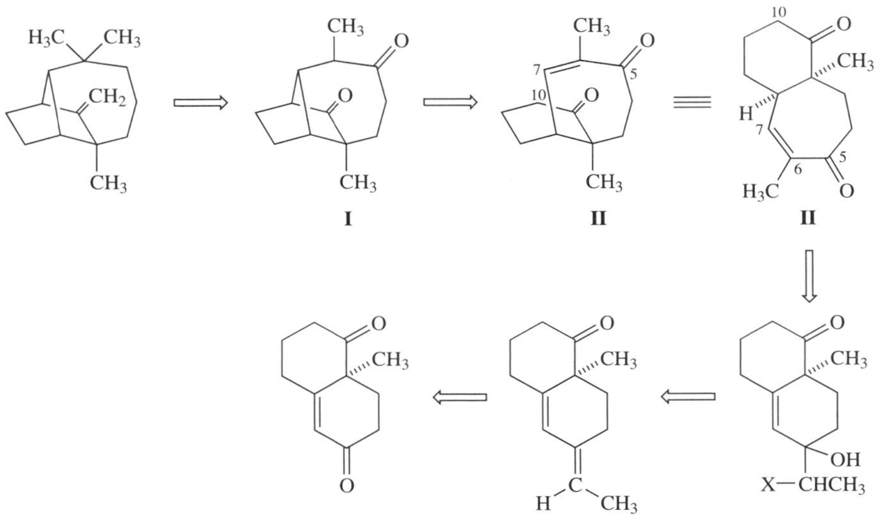

Can a chemical reaction be recognized which would permit **III** ⇒ **II** to proceed in the synthetic sense? The hydroxyl → carbonyl transformation with migration corresponds to the pinacol rearrangement (Section 10.1.2). The retrosynthetic transformation **II** ⇒ **III** would constitute a workable synthetic step if the group X in **III** is a leaving group that could promote the rearrangement. The other transformations in the retrosynthetic plan, **III** ⇒ **IV** ⇒ **V**, are straightforward in concept and lead to identification of **V** as a potential starting material.

The synthesis was carried out as is shown in Scheme 13.20. The key intramolecular Michael addition was accomplished using triethylamine under high-temperature conditions.

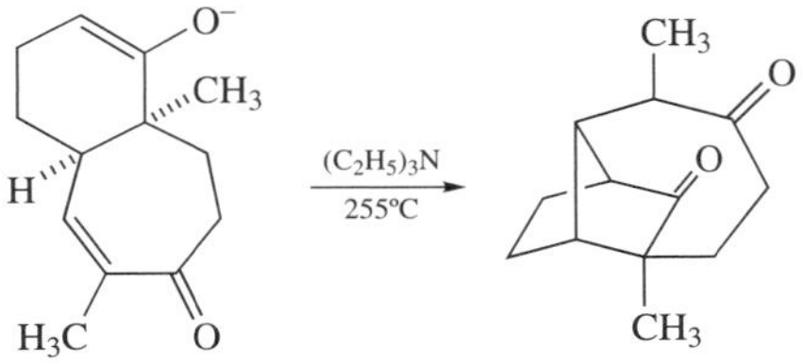

The cyclization requires that the intermediate have a *cis* ring fusion. The stereochemistry of the ring junction is established when the double bond is moved into conjugation in step **D**. This product was not stereochemically characterized, and need not be, because the stereochemically important site at C-1 can be epimerized under the basic cyclization conditions. Thus, the equilibration of the ring junction through a dienol can allow the cyclization to proceed to completion from either stereoisomer. Step **C** is the pinacol rearrangement corresponding to **II** ⇒ **III** in the retrosynthesis. A diol is formed and selectively tosylated at the secondary hydroxyl group (step **B**). Base then promotes the

Scheme 13.20. Longifolene Synthesis: E. J. Corey, R. B. Mitra, and P. A. Vatakencherry[a]

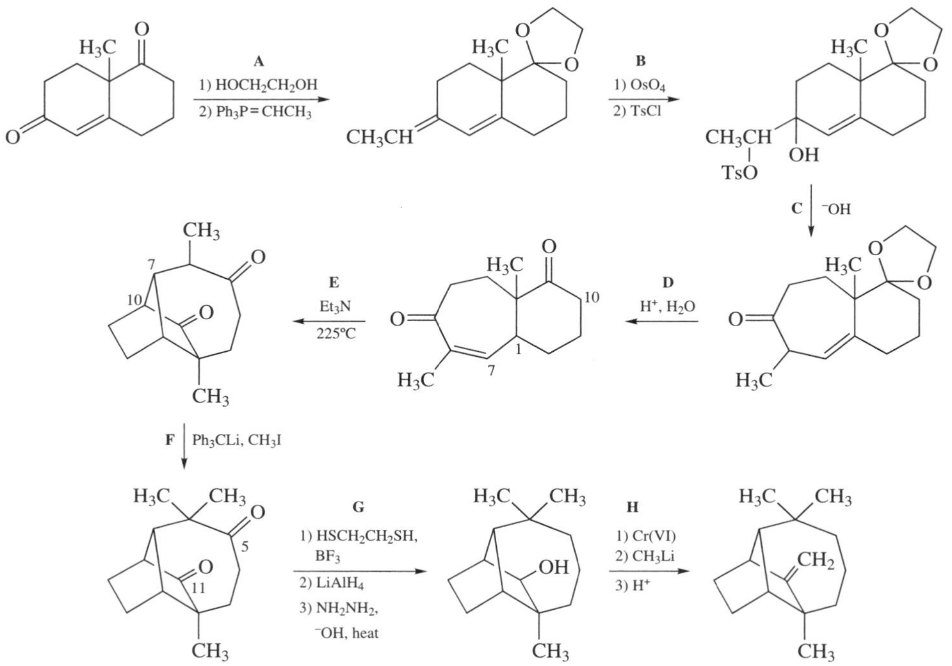

A: 1) $HOCH_2CH_2OH$ 2) $Ph_3P{=}CHCH_3$

B: 1) OsO_4 2) TsCl

C: ^-OH

D: H^+, H_2O

E: Et_3N, 225°C

F: Ph_3CLi, CH_3I

G: 1) $HSCH_2CH_2SH$, BF_3 2) $LiAlH_4$ 3) NH_2NH_2, ^-OH, heat

H: 1) Cr(VI) 2) CH_3Li 3) H^+

a. E. J. Corey, R. B. Mitra, and P. A. Vatakencherry, *J. Am. Chem. Soc.* **86**:478 (1964).

skeletal rearrangement in step **C**. The remaining transformations effect the addition of the remaining methyl and methylene groups by well-known methods. Step **G** accomplishes a selective reduction of one of the two carbonyl groups to a methylene by taking advantage of the difference in the steric environment of the two carbonyls. Selective protection of the less hindered C-5 carbonyl was done using a thioketal. The C-11 carbonyl was then reduced to give the alcohol, and finally C-5 was reduced to a methylene group under Wolff–Kishner conditions. The hydroxyl group at C-11 provides the reactive center necessary to introduce the C-15 methylene group in step **H**.

The key bond closure in Scheme 13.21 is somewhat similar to that used in Scheme 13.20 but is performed on a bicyclo[4.4.0]decane ring system. The ring juncture must be *cis* to permit the intramolecular epoxide ring opening. The required *cis* ring fusion is established during the catalytic hydrogenation in step **A**.

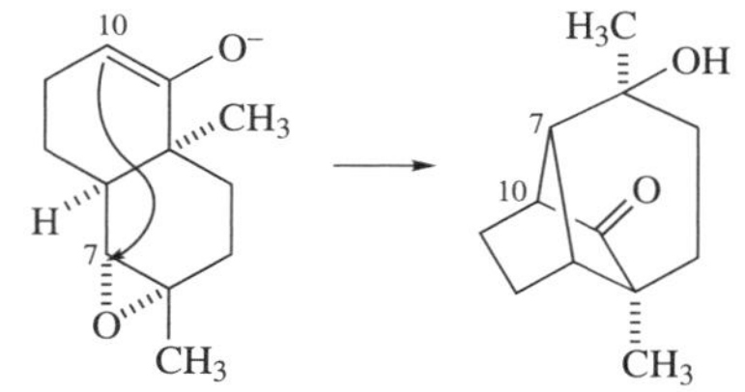

The cyclization is followed by a sequence of steps **F–H**, which effect a ring expansion via a carbene addition and cyclopropyl halide solvolysis. The products of steps **I** and **J** are interesting in that the tricyclic structures are largely converted to tetracyclic derivatives by intramolecular aldol reactions. The extraneous bond is broken in step **K**. First, a diol is

Scheme 13.21. Longifolene Synthesis: J. E. McMurry and S. J. Isser[a]

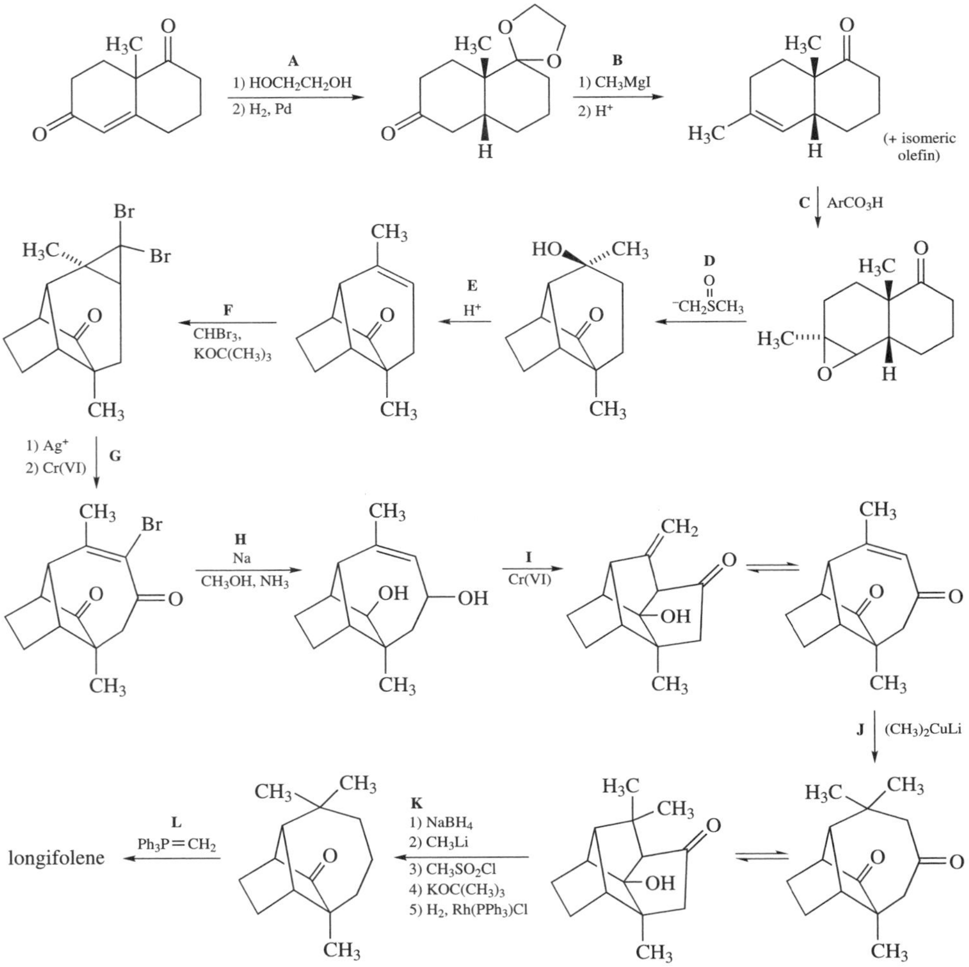

A: 1) $HOCH_2CH_2OH$ 2) H_2, Pd
B: 1) CH_3MgI 2) H^+ (+ isomeric olefin)
C: $ArCO_3H$
D: $^-CH_2SCH_3$ (sulfinyl, O)
E: H^+
F: $CHBr_3$, $KOC(CH_3)_3$
G: 1) Ag^+ 2) Cr(VI)
H: Na, CH_3OH, NH_3
I: Cr(VI)
J: $(CH_3)_2CuLi$
K: 1) $NaBH_4$ 2) CH_3Li 3) CH_3SO_2Cl 4) $KOC(CH_3)_3$ 5) H_2, $Rh(PPh_3)Cl$
L: $Ph_3P{=}CH_2$ → longifolene

a. J. E. McMurry and S. J. Isser, *J. Am. Chem. Soc.* **94**:7132 (1972).

formed by $NaBH_4$ reduction, and this is converted to a monomesylate. The resulting β-hydroxy mesylate is capable of a concerted fragmentation, which occurs on treatment with potassium *t*-butoxide.

A retrosynthetic analysis corresponding to the synthesis in Scheme 13.23 is given in Scheme 13.22. The striking feature of this synthesis is the structural simplicity of the key intermediate **IV**. A synthesis according to this scheme would generate the tricyclic skeleton in a single step from a monocyclic intermediate. The disconnection **III** ⇒ **IV** corresponds to a cationic cyclization of the highly symmetric cation **IVa**.

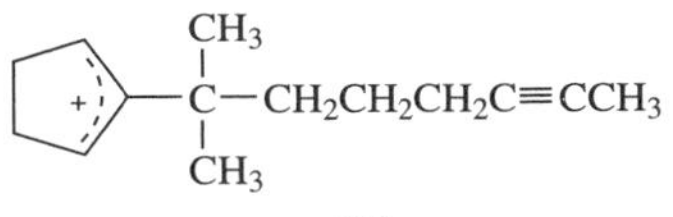

$$C_5H_4^{+}\text{-}C(CH_3)_2\text{-}CH_2CH_2CH_2C{\equiv}CCH_3$$

IVa

Scheme 13.22. Retrosynthetic Analysis Corresponding to Synthesis in Scheme 13.23

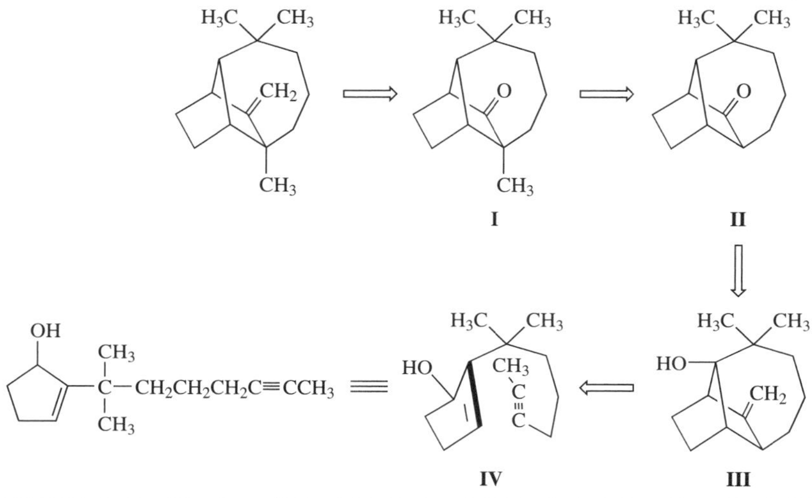

No issues of stereochemistry arise until the carbon skeleton has been formed, at which point all of the stereocenters would be in the proper relative relationship. The structures of the successive intermediates, assuming a stepwise mechanism for the cationic cyclization, are shown below.

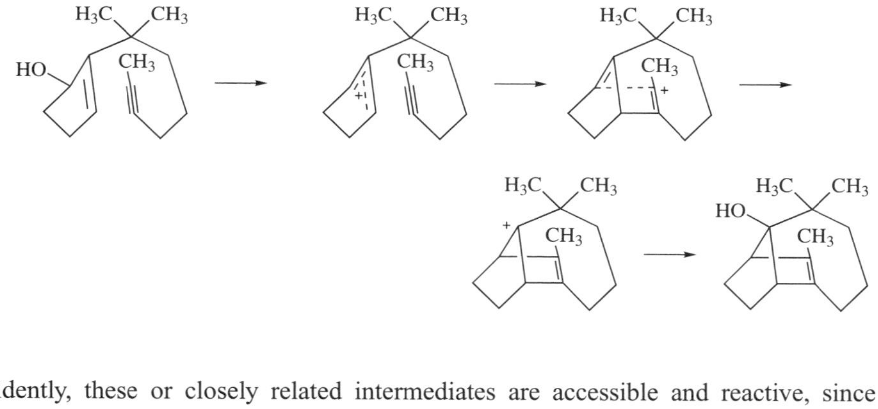

Evidently, these or closely related intermediates are accessible and reactive, since the synthesis was successfully achieved as outlined in Scheme 13.23. In addition to the key cationic cyclization in step **D**, interesting transformations are carried out in step **E**, where a bridgehead tertiary alcohol is reductively removed, and in step **F**, where a methylene group, which is eventually reintroduced, must be removed. The endocyclic double bond, which is strained because of its bridgehead location, is isomerized to the exocyclic position and then cleaved with RuO_4/IO_4^-. The enolate of the ketone is then used to introduce the C-12 methyl group.

Scheme 13.23. Longifolene Synthesis: R. A. Volkmann, G. C. Andrews, and W. S. Johnson[a]

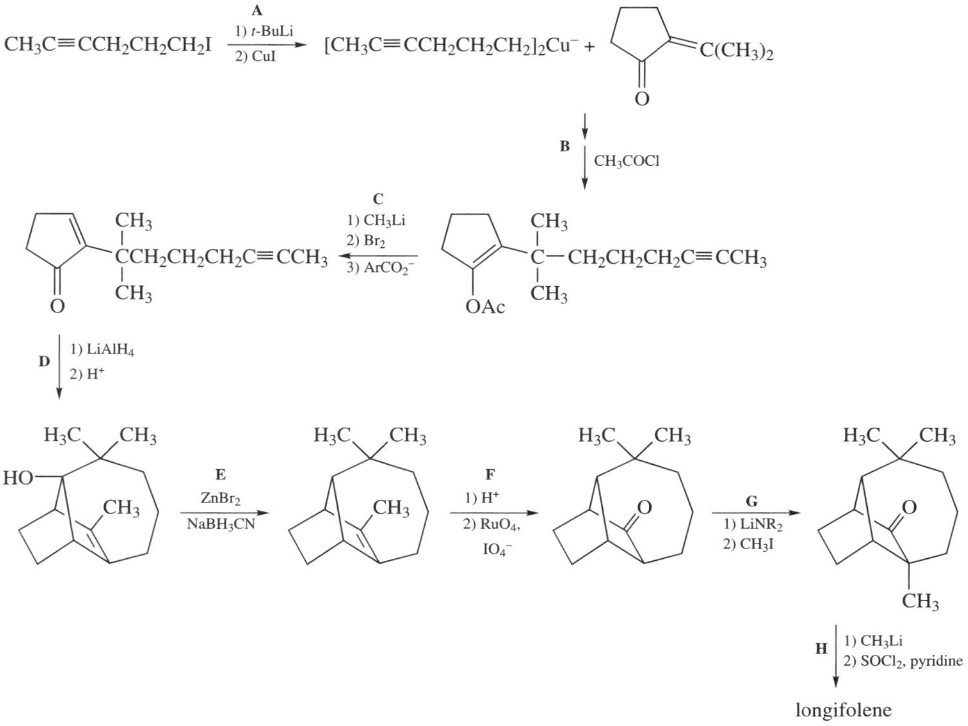

a. R. A. Volkmann, G. C. Andrews, and W. S. Johnson, *J. Am. Chem. Soc.* **97**:4777 (1975).

The synthesis in Scheme 13.24 also uses a remarkably simple starting material to achieve the construction of the tricyclic skeleton. A partial retrosynthesis is outlined below.

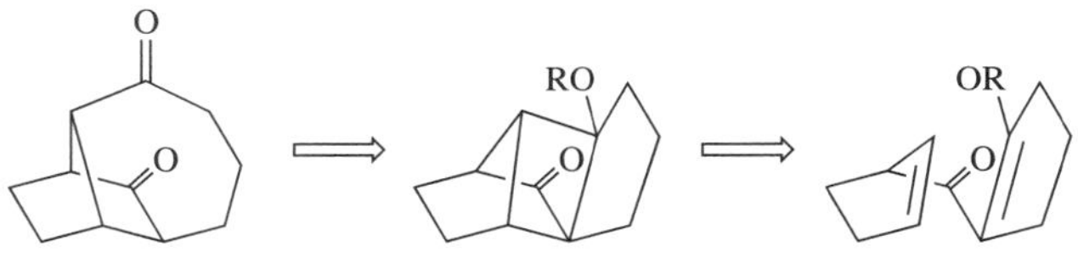

Intermediate **I** contains the tricyclic skeleton of longifolene, shorn of its substituent groups, but containing carbonyl groups suitably placed so that the methyl groups at C-2 and C-6 and the C-11 methylene can eventually be introduced. The retrosynthetic step **I** ⇒ **II** corresponds to an intramolecular aldol condensation. However, **II** clearly is strained relative to **I**, so **II** (with OR=OH) should open to **I**.

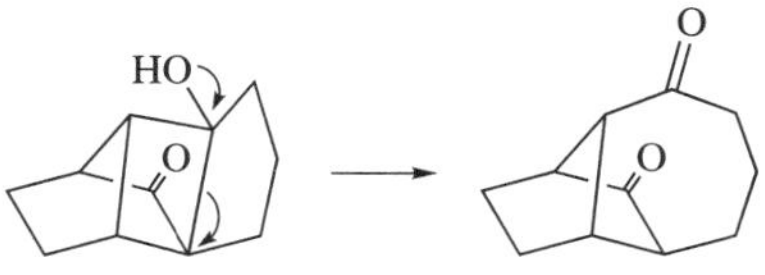

How might **II** be obtained? The four-membered ring suggests that a [2 + 2] (photochemical) cycloaddition might be useful, and this, in fact, was successful (step **B** in Scheme 13.24). After liberation of the hydroxyl group by hydrogenolysis in step **C**, the extra carbon–carbon bond between and C-2 and C-6 was broken by a spontaneous retro-aldol reaction. Step **D** in this synthesis is an interesting way of introducing the geminal dimethyl groups. It proceeds through a cyclopropane intermediate which is cleaved by hydrogenolysis.

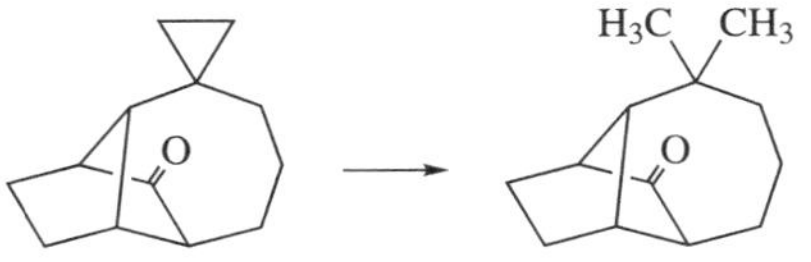

H_3C CH_3

The synthesis of longifolene in Scheme 13.25 commences with a Birch reduction and tandem alkylation of methyl 2-methoxybenzoate (see Section 5.5.1). Step **C** is an intramolecular cycloaddition of a diazoalkane which is generated from an aziridinoimine intermediate:

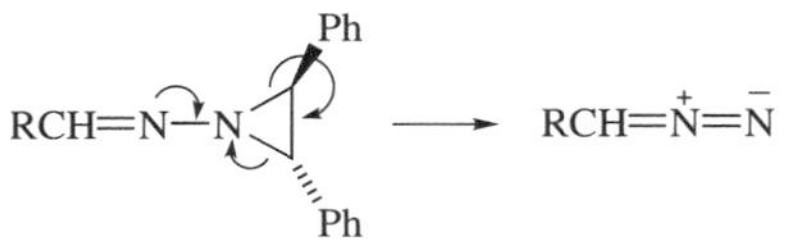

$RCH{=}N{-}N$ (Ph, Ph aziridine) $\longrightarrow$ $RCH{=}\overset{+}{N}{=}\overset{-}{N}$

The thermolysis in step **D** generates a diradical (or the corresponding dipolar intermediate), which then closes to generate the desired carbon skeleton.

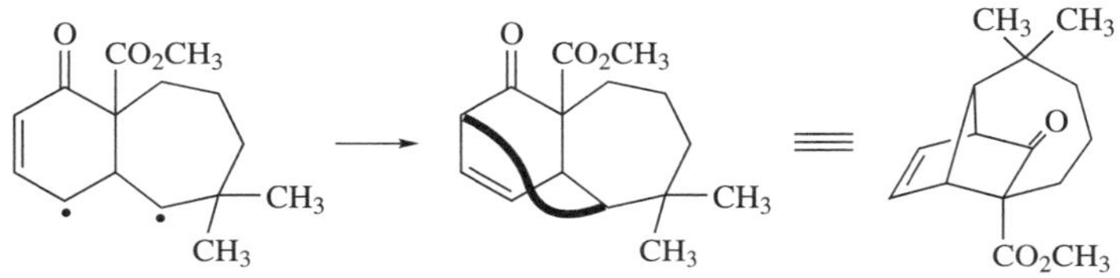

O CO_2CH_3 CH_3 CH_3 — O CO_2CH_3 CH_3 CH_3 ≡ CH_3 CH_3 O CO_2CH_3

Scheme 13.24. Longifolene Synthesis: W. Oppolzer and T. Godel

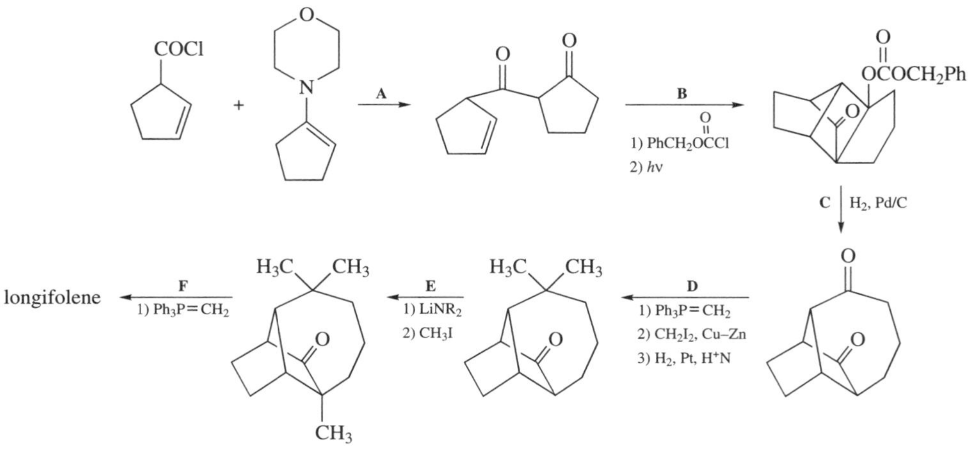

COCl + (morpholine enamine) —**A**→ (diketone) —**B**: 1) $PhCH_2OCCl$ (O); 2) $h\nu$ → $OCOCH_2Ph$ (O) —**C**: H_2, Pd/C → (ketone) —**D**: 1) $Ph_3P{=}CH_2$; 2) CH_2I_2, Cu–Zn; 3) H_2, Pt, H^+N → H_3C CH_3 O —**E**: 1) $LiNR_2$; 2) CH_3I → H_3C CH_3 O CH_3 —**F**: 1) $Ph_3P{=}CH_2$ → longifolene

a. W. Oppolzer and T. Godel, *J. Am. Chem. Soc.* **100**:2583 (1978).

Scheme 13.25. Longifolene Synthesis: A. G. Schultz and S. Puig[a]

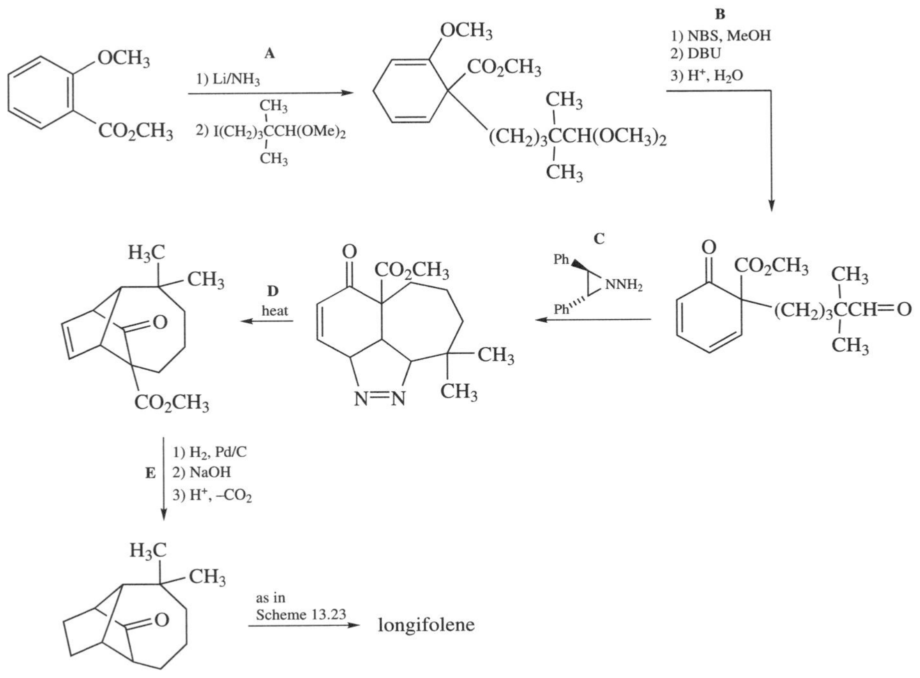

A: 1) Li/NH_3; 2) $I(CH_2)_3CCH(OMe)_2$ (CH_3, CH_3)
B: 1) NBS, MeOH; 2) DBU; 3) H^+, H_2O
C: (Ph, Ph aziridine NNH_2)
D: heat
E: 1) H_2, Pd/C; 2) NaOH; 3) H^+, $-CO_2$
as in Scheme 13.23 → longifolene

a. A. G. Schultz and S. Puig, *J. Org. Chem.* **50**:915 (1985).

The cyclization product is converted to an intermediate which was used in the longifolene synthesis described in Scheme 13.23.

The synthesis in Scheme 13.25 has also been done in such a way as to give enantiomerically pure longifolene. A starting material, whose chirality is derived from the amino acid L-proline, was enantioselectively converted to the product of step **A**.

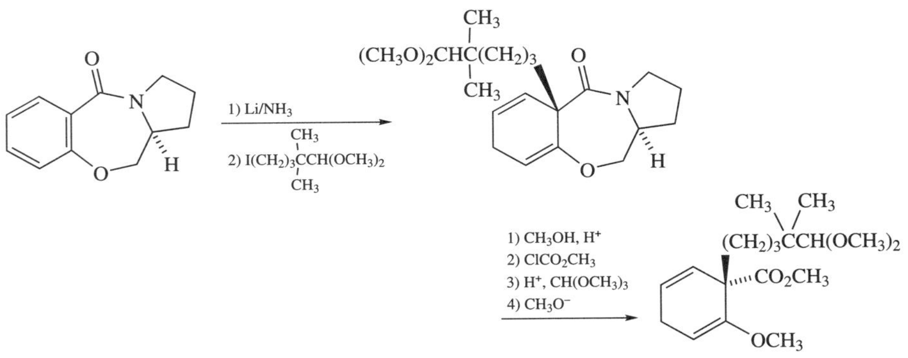

1) Li/NH_3; 2) $I(CH_2)_3CCH(OCH_3)_2$ (CH_3, CH_3)
1) CH_3OH, H^+; 2) $ClCO_2CH_3$; 3) H^+, $CH(OCH_3)_3$; 4) CH_3O^-

This enantiomerically pure intermediate, when carried through the reaction sequence in Scheme 13.25, generates the enantiomer of natural longifolene. Accordingly, D-proline would have to be used to generate the natural enantiomer.

Another enantiospecific synthesis of longifolene was done starting with camphor, a natural product available in enantiomerically pure form (Scheme 13.26). The tricyclic ring system is formed in step **C** by an intramolecular Mukaiyama reaction. The dimethyl substituents are formed in the first step of sequence **E** by hydrogenolysis of the cyclopropane ring. The final step of the synthesis involves a rearrangement of the tricyclic ring system that is induced by solvolysis of the mesylate intermediate.

MsO

H

Ms = CH_3SO_3

Another enantioselective synthesis of longifolene, shown in Scheme 13.27, uses an intramolecular Diels–Alder reaction as a key step. The alcohol intermediate is resolved in sequence **B** by formation and separation of a menthyl carbonate. After oxidation, the pyrone ring is introduced by γ addition of the ester enolate of methyl 3-methylbutenoate.

$LiCH_3$

O

O

CH_2OH

H

resolved as menthyl carbonate ester

1) MnO_2/C
2) LDA,
$(CH_3)_2C{=}CHCO_2CH_3$

O

O

H

The pyrone ring then acts as the dienophile in the intramolecular Diels–Alder cycloaddition, which was conducted in a microwave oven. The final step of this synthesis is a high-temperature ester pyrolysis to introduce the exocyclic double bond of longifolene.

These syntheses of longifolene provide good examples of the approaches that are available for construction of ring compounds of this type. In each case, a set of

Scheme 13.26. Longifolene Synthesis: D. L. Kuo and T. Money[a]

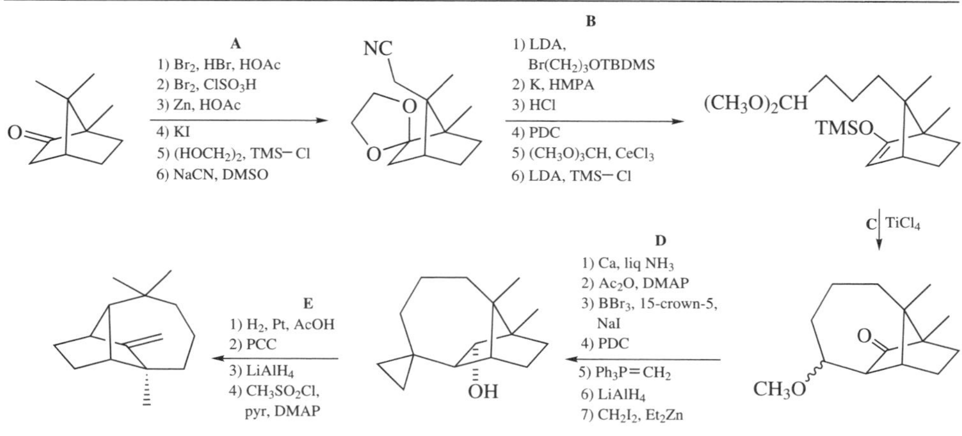

a. D. L. Kuo and T. Money, *Can. J. Chem.* **66:**1794 (1988).

Scheme 13.27. Longifolene Synthesis: B. Lei and A. G. Fallis[a]

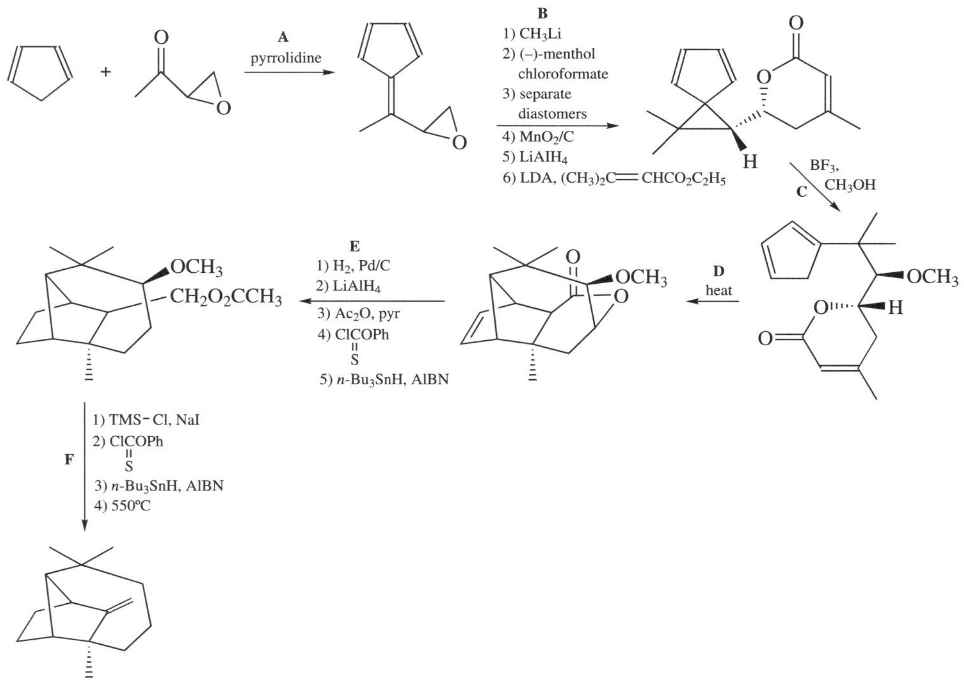

A: pyrrolidine

B: 1) CH_3Li; 2) (–)-menthol chloroformate; 3) separate diastomers; 4) MnO_2/C; 5) $LiAlH_4$; 6) LDA, $(CH_3)_2C{=}CHCO_2C_2H_5$

C: BF_3, CH_3OH

D: heat

E: 1) H_2, Pd/C; 2) $LiAlH_4$; 3) Ac_2O, pyr; 4) ClC(=S)OPh; 5) n-Bu_3SnH, AIBN

F: 1) TMS-Cl, NaI; 2) ClC(=S)OPh; 3) n-Bu_3SnH, AIBN; 4) 550°C

a. B. Lei and A. G. Fallis, *J. Am. Chem. Soc.* **112**:4609 (1990); *J. Org. Chem.* **58**:2186 (1993).

functionalities that have the potential for *intramolecular* reaction was assembled. After assembly of the carbon framework, the final functionality changes were effected. It is the necessity for the formation of the carbon skeleton which determines the functionalities that are present at the ring-closure stage. After the ring structure is established, necessary adjustments of the functionalities are made.

13.5.3. Prelog–Djerassi Lactone

The Prelog–Djerassi lactone (abbreviated as P-D-lactone) was originally isolated as a degradation product during structural investigation of antibiotics. Its open-chain precursor **1**, is typical of methyl-branched carbon chains that occur frequently in macrolide and polyether antibiotics.

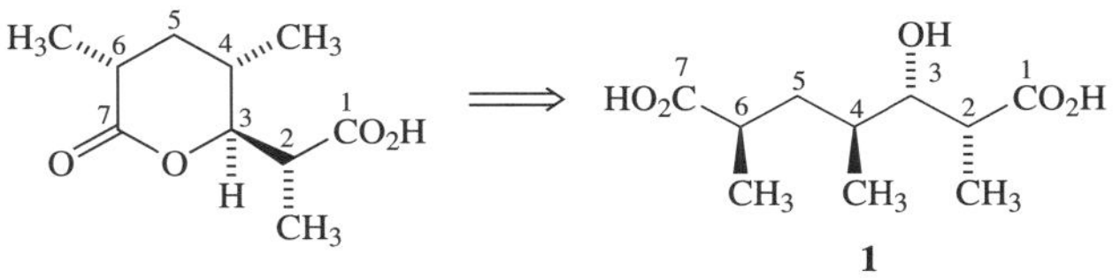

1

There have been more than 20 different syntheses of P-D-lactone.[132] We will focus here on some of those which provide enantiomerically pure product, since they illustrate several of the methods for enantioselective synthesis.[133]

The synthesis in Scheme 13.28 is based on a starting material that can be prepared in enantiomerically pure form. In the synthesis, C-7 of the norbornenone starting material becomes C-4 of P-D-lactone. The configuration of the C-3 hydroxyl and C-2 and C-6 methyl groups must then be established relative to the C-4 stereochemistry. The alkylation in step **A** establishes the configuration at C-2. The basis for the stereoselectivity is the preference for *exo* versus *endo* approach in the alkylation. The Baeyer–Villiger oxidation in step **B** is followed by a Lewis acid-mediated allylic rearrangement. This rearrangement is suprafacial. This stereochemistry is evidently dictated by the preference for maintaining a *cis* ring juncture at the five-membered rings.

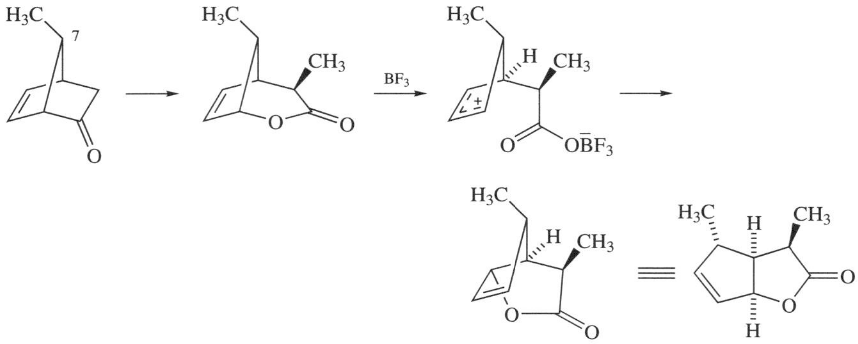

H_3C 7 CH_3 BF_3 H OBF_3 O

The stereochemistry of the C-3 hydroxyl is established in step **E**. The Baeyer–Villiger oxidation proceeds with retention of configuration of the migrating group (see Section 12.5.2), so the correct stereochemistry is established for the C–O bond. The final center for which configuration must be established is the methyl group at C-6. The methyl group is introduced by an enolate alkylation in step **F**, but this reaction is not highly stereoselective. However, because this center is adjacent to the lactone carbonyl, it can be epimerized through the enolate. The enolate is formed and quenched with acid. The kinetically preferred protonation from the axial direction provides the correct stereochemistry at C-6.

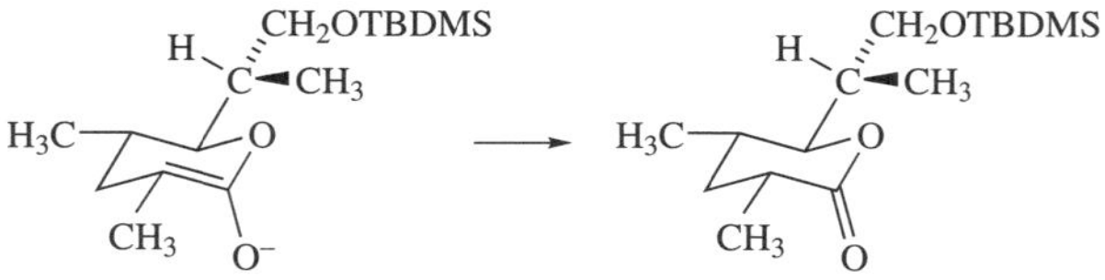

$CH_2OTBDMS$ H C CH_3 H_3C O CH_3 O^- → $CH_2OTBDMS$ H C CH_3 H_3C O CH_3 O

132. For references to these syntheses, see S. F. Martin and D. G. Guinn, *J. Org. Chem.* **52:**5588 (1987); H. F. Chow and I. Fleming, *Tetrahedron Lett.* **26:**397 (1985); S. F. Martin and D. E. Guinn, *Synthesis* **1991:**245.
133. For other syntheses of enantiomerically pure Prelog–Djerassi lactone, see F. E. Ziegler, A. Kneisley, J. K. Thottathil, and R. T. Wester, *J. Am. Chem. Soc.* **110:**5434 (1988); A. Nakano, S. Takimoto, J. Inanaga, T. Katsuki, S. Ouchida, K. Inoue, M. Aiga, N. Okukado, and M. Yamaguchi, *Chem. Lett.* **1979:**1019; K. Suzuki, K. Tomooko, T. Matsumoto, E. Katayama, and G. Tsuchihashi, *Tetrahedron Lett.* **26:**3711 (1985); M. Isobo, Y. Ichikawa, and T. Goto, *Tetrahedron Lett.* **22:**4287 (1981); M. Mori, T. Chuman, and K. Kato, *Carbohyd. Res.* **129:**73 (1984).

Scheme 13.28. Prelog–Djerassi Lactone Synthesis: P. A. Grieco, Y. Ohfune, Y. Yokoyama, and W. Owens[a]

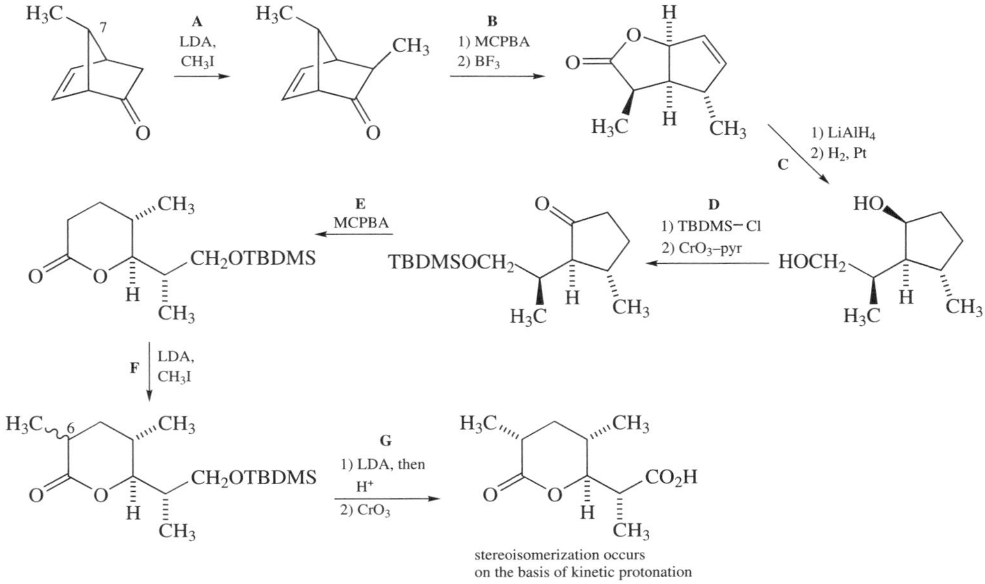

a. P. A. Grieco, Y. Ohfune, Y. Yokoyama, and W. Owens, *J. Am. Chem. Soc.* **101**:4749 (1979).

Another synthesis of P-D-lactone based on a resolved starting material is shown in Scheme 13.29. The stereocenter in the starting material is destined to become C-4 in the final product. Steps **A** and **B** serve to extend the chain to provide a seven-carbon diene. The configuration of two of the three remaining stereocenters is controlled by the hydroboration step. Hydroboration is a stereospecific *syn* addition (Section 4.9.1). In 1,5-dienes of this type, an intramolecular hydroboration occurs and establishes the configuration of the two newly formed C−B and C−H bonds.

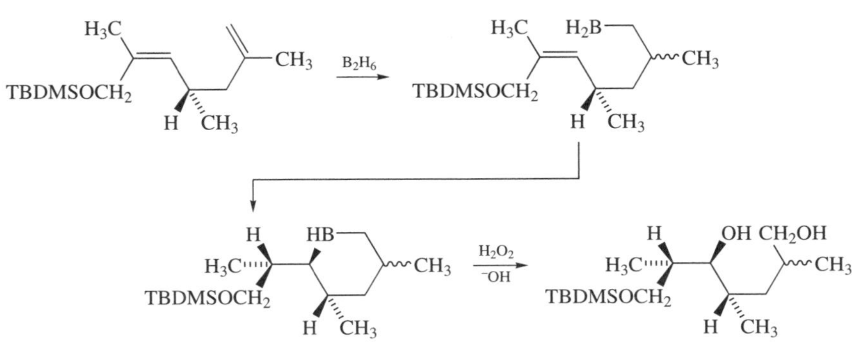

There is, however, no significant selectivity in the initial hydroboration of the terminal double bond. As a result, both configurations are formed at C-6. This problem was overcome using the epimerization process from Scheme 13.28.

Scheme 13.29. Prelog–Djerassi Lactone Synthesis: W. C. Still and K. R. Shaw[a]

a. W. C. Still and K. R. Shaw, *Tetrahedron Lett.* **22:**3725 (1981).
b. Epimerization carried out as in Scheme 13.28.

The syntheses in Schemes 13.30 and 13.31 are conceptually related. The starting material is prepared by reduction of the half-ester of *meso*-2,4-dimethylglutaric acid. The use of the *meso*-diacid ensures the correct relative configuration of the C-4 and C-6 methyl substituents. The half-acid is resolved, and the correct enantiomer is reduced to the aldehyde.

The synthesis in Scheme 13.30 uses stereoselective aldol condensation methodology. Both the lithium enolate and the boron enolate method were employed. The enol derivatives were used in enantiomerically pure form, so the condensations are examples of *double stereodifferentiation* (Section 2.1.3). The stereoselectivity observed in the reactions is that predicted for a cyclic transition state for the aldol condensations.

The synthesis in Scheme 13.31 also relies on *meso*-2,4-dimethylglutaric acid as the starting material. Both the resolved aldehyde employed in Scheme 13.30 and a resolved half-amide were successfully used as intermediates. The configuration at C-2 and C-3 was controlled by addition of a butenylborane to an aldehyde (see Section 9.1.2). The boronate ester was used in enantiomerically pure form so that stereoselectivity was enhanced by double stereodifferentiation. The allylic additions carried out by the butenylboronates do not appear to have been quite as highly stereoselective as the aldol condensations used in Scheme 13.30, because a minor diastereomer was formed in the boronate addition reactions.

The synthesis in Scheme 13.32 is based on an interesting kinetic differentiation in the reactivity of two centers that are structurally identical but are diastereomeric. A bis-amide of *meso*-2,4-dimethylglutaric acid and a chiral thiazoline is formed in step **A**. The thiazoline is derived from the amino acid cysteine. The two amide carbonyls in this bis-amide are nonequivalent by virtue of the diastereomeric relationship established by the

Scheme 13.30. Prelog–Djerassi Lactone Synthesis: S. Masamune, M. Hirama, S. Mori, S. A. Ali, and D. S. Garvey; S. Masamune, S. A. Ali, D. L. Snitman, and D. S. Garvey[a]

A: MeO_2C–CH(CH$_3$)–CH$_2$–CH(CH$_3$)–CH=O + lithium enolate (CH$_3$, OLi, OTMS, CH$_3$, C_6H_{11}) → MeO_2C ... (OH, O, OTMS, CH$_3$, cyclohexyl; CH$_3$ CH$_3$ CH$_3$)

B: 1) H$^+$ 2) Zn(BH$_4$)$_2$ → lactone (H$_3$C, CH$_3$, OH, OTMS, CH$_3$, O, O, H, CH$_3$)

A′: boron enolate (H$_3$C, OB, OTBDMS, H, C_6H_{11}) → MeO_2C ... (OH, O, OTBDMS, H, cyclohexyl; CH$_3$ CH$_3$ CH$_3$)

C: oxid.

D: 1) HF 2) IO_4^- → H$_3$C, CH$_3$, O, O, H, CO$_2$H, CH$_3$

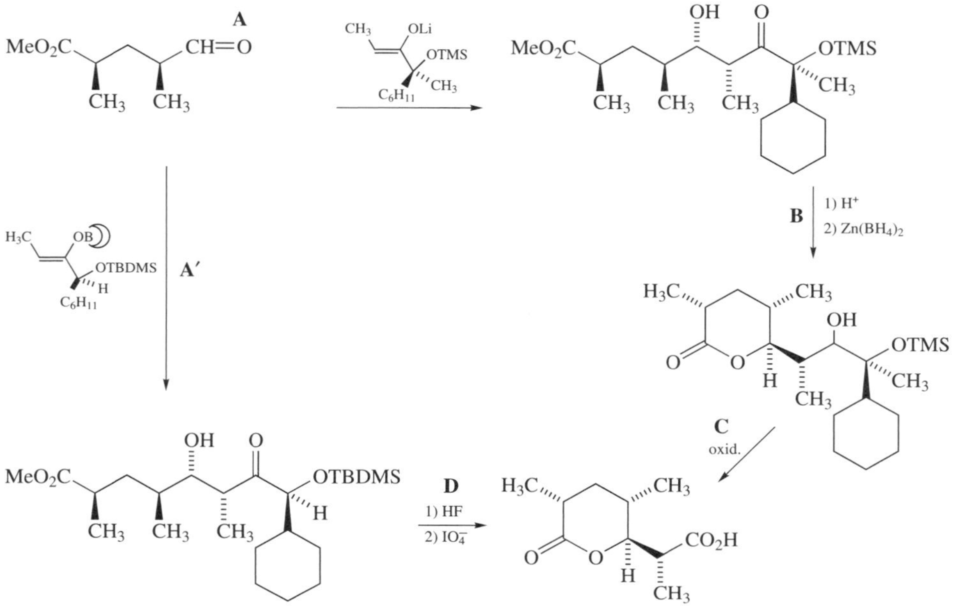

a. S. Masamune, S. A. Ali, D. L. Snitman, and D. S. Garvey, *Angew. Chem. Int. Ed. Engl.* **19:**557 (1980); S. Masamune, M. Hirama, S. Mori, S. A. Ali, and D. S. Garvey, *J. Am. Chem. Soc.* **103:**1568 (1981).

chiral centers at C-2 and C-4 in the glutaric acid portion of the structure. One of the centers reacts with a 97 : 3 preference with the achiral amine piperidine.

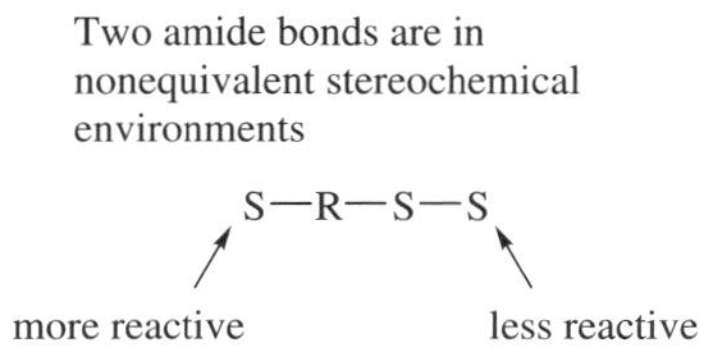

Two amide bonds are in nonequivalent stereochemical environments

S—R—S—S

more reactive — less reactive

In step **D**, a *chiral auxiliary*, also derived from cysteine, is used to achieve double stereodifferentiation in an aldol condensation. A tin enolate was used. The stereoselectivity of this reaction parallels that of aldol condensations carried out with lithium or zinc enolates. Once the configuration of all the centers has been established, the synthesis proceeds to P-D-lactone by functional group modifications.

There have been several syntheses of P-D lactone that are based on carbohydrate-derived starting materials. The starting material used in Scheme 13.33 had been prepared from a carbohydrate in earlier work.[134] The relative stereochemistry at C-4 and C-6 was established by the hydrogenation in step **B**. This *syn* hydrogenation is not completely stereoselective but provides a 4 : 1 mixture favoring the desired isomer. The stereoselec-

134. M. B. Yunker, D. E. Plaumann, and B. Fraser-Reid, *Can. J. Chem.* **55:**4002 (1977).

Scheme 13.31. Prelog–Djerassi Lactone Synthesis: R. W. Hoffmann, H.-J. Zeiss, W. Ladner, and S. Tabche[a]

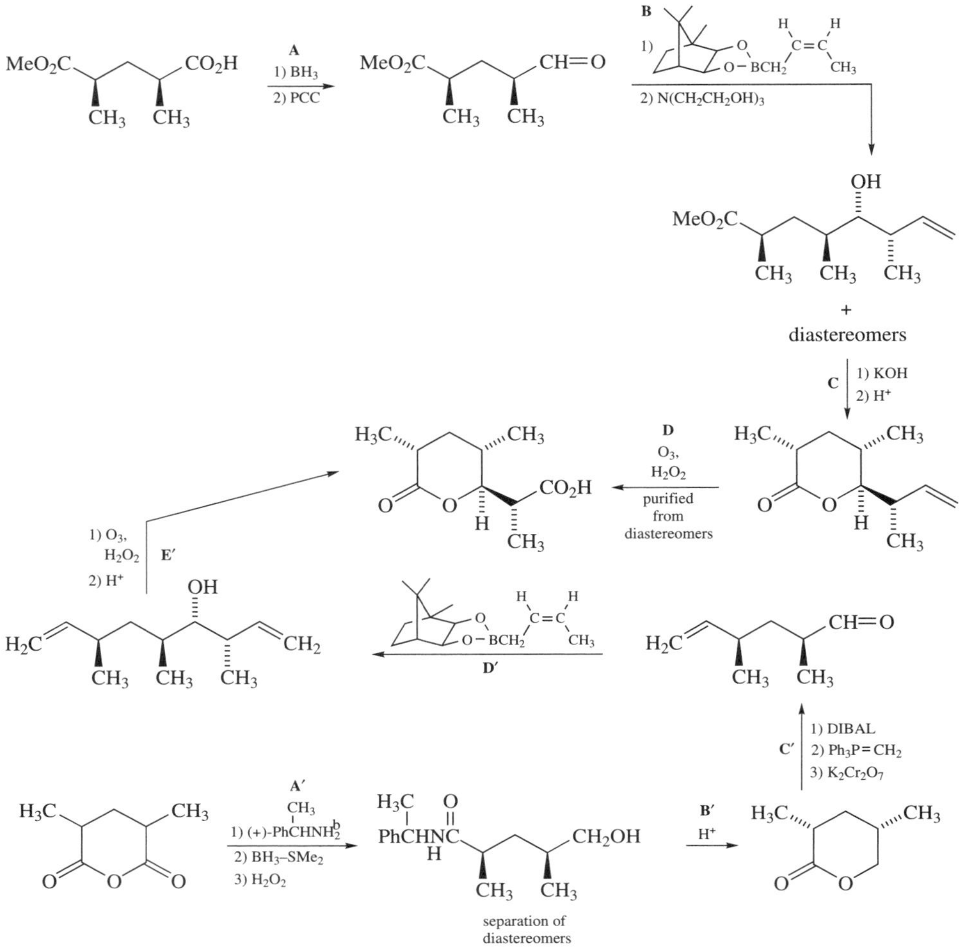

a. R. W. Hoffmann, H.-J. Zeiss, W. Ladner, and S. Tabche, *Chem. Ber.* **115:**2357 (1982).
b. Resolved via α-phenylethylamine salt; S. Masamune, S. A. Ali, D. L. Snitman, and D. S. Garvey, *Angew. Chem. Int. Ed. Engl.* **19:**557 (1980).

tivity is presumably the result of preferential absorption from the less hindered *β* face of the molecule. The configuraton of C-2 is established by protonation during the hydrolysis of the enol ether in step **D**. This step is not stereoselective, and so a separation of diastereomers after the oxidation in step **E** was required.

The synthesis in Scheme 13.34 also begins with carbohydrate-derived starting material and also uses catalytic hydrogenation to establish the stereochemical relationship between the C-4 and C-6 methyl groups. As was the case in Scheme 13.33, the configuration at C-2 is not controlled in this synthesis, and separation of the diastereomeric products was necessary.

The synthesis in Scheme 13.35 is also based on a carbohydrate-derived starting material. It controls the stereochemistry at C-2 by means of the stereoselectivity of the Ireland–Claisen rearrangement in step **A** (see Section 6.5). The ester enolate is formed

Scheme 13.32. Prelog–Djerassi Lactone Synthesis: Y. Nagao, T. Inoue, K. Hashimoto, Y. Hagiwara, M. Ochai, and E. Fujita[a]

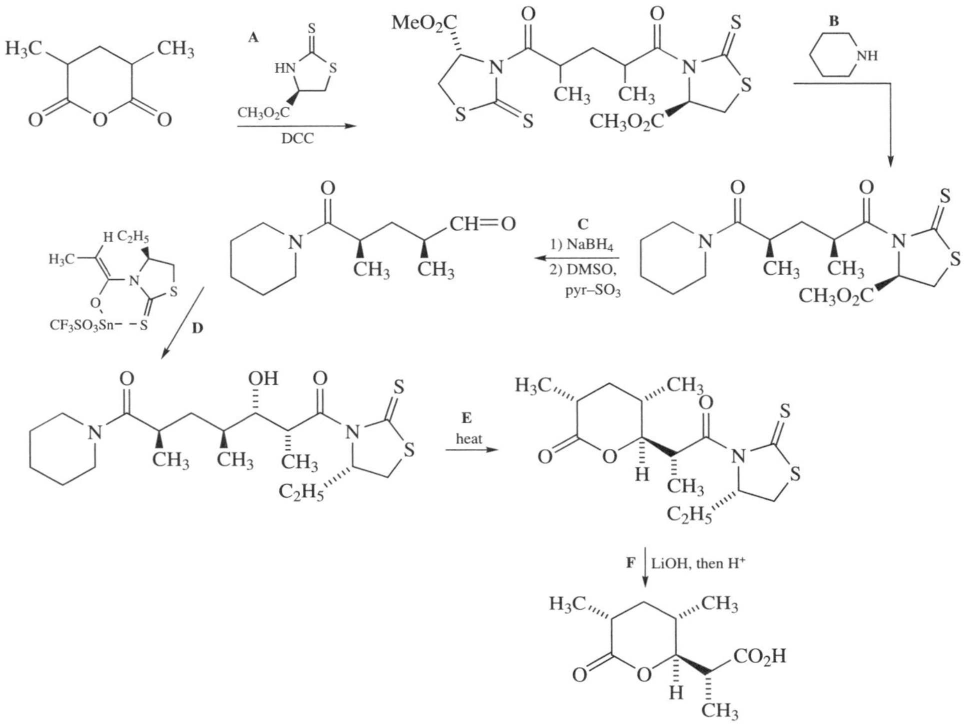

A: DCC

B

C: 1) $NaBH_4$ 2) DMSO, pyr–SO_3

D: CF_3SO_3Sn

E: heat

F: LiOH, then H^+

a. Y. Nagao, T. Inoue, K. Hashimoto, Y. Hagiwara, M. Ochai, and E. Fujita, *J. Chem. Soc., Chem. Commun.* **1985:**1419.

Scheme 13.33. Prelog–Djerassi Lactone Synthesis: S. Jarosz and B. Fraser-Reid[a]

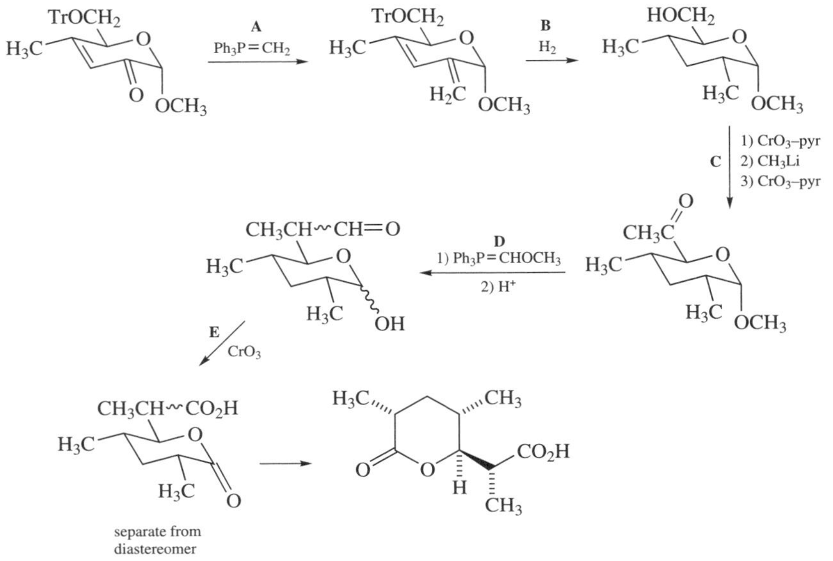

A: $Ph_3P{=}CH_2$

B: H_2

C: 1) CrO_3–pyr 2) CH_3Li 3) CrO_3–pyr

D: 1) $Ph_3P{=}CHOCH_3$ 2) H^+

E: CrO_3

separate from diastereomer

a. S. Jarosz and B. Fraser-Reid, *Tetrahedron Lett.* **22:**2533 (1981).

Scheme 13.34. Prelog–Djerassi Lactone Synthesis: N. Kawauchi and H. Hashimoto[a]

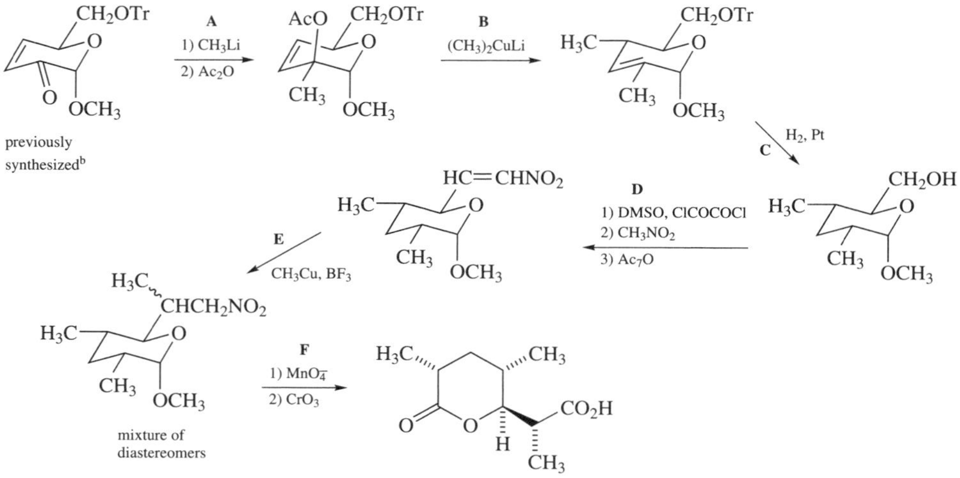

A: 1) CH_3Li 2) Ac_2O; B: $(CH_3)_2CuLi$; C: H_2, Pt; D: 1) DMSO, ClCOCOCl 2) CH_3NO_2 3) Ac_2O; E: CH_3Cu, BF_3; F: 1) MnO_4^- 2) CrO_3

previously synthesized[b]

mixture of diastereomers

a. N. Kawauchi and H. Hashimoto, *Bull. Chem. Soc. Jpn.* **60:**1441 (1987).
b. N. L. Holder and B. Fraser-Reid, *Can. J. Chem.* **51:**3357 (1973).

under conditions in which the *E*-enolate is expected to predominate. Heating the resulting silyl enol ether gave a 9 : 1 preference for the expected stereoisomer. The preferred transition state, which is boatlike, minimizes the steric interaction between the bulky silyl substituent and the ring structure.

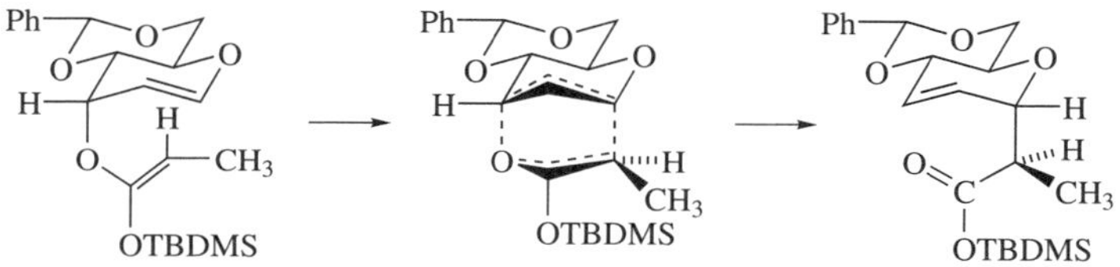

The stereochemistry at C-4 and C-6 is then established. The cuprate addition in step **C**, occurs *anti* to the substituent at C-2 of the pyran ring. After a Wittig methylenation, the catalytic hydrogenation in step **D**, establishes the stereochemistry at C-6.

The syntheses in Schemes 13.36 and 13.37 illustrate the use of chiral auxiliaries in enantioselective synthesis. Step **A** in Scheme 13.36 establishes the configuration at the carbon which becomes C-4 in the product. This is an enolate alkylation in which the steric effect of the oxazolinone substituents directs the approach of the alkylating group. Step **C** also uses the oxazolidinone structure. In this case, the enolborinate is formed and condensed with the aldehyde intermediate. This stereoselective aldol condensation establishes the configuration at C-2 and C-3. The configuration at the final stereocenter is established by the hydroboration in step **D**. The selectivity for the desired stereoisomer is 85 : 15. Stereoselectivity in the same sense has been observed for a number of other 2-methyl-alkenes in which the remainder of the alkene constitutes a relatively bulky

135. D. A. Evans, J. Bartroli, and T. Godel, *Tetrahedron Lett.* **23:**4577 (1982).

Scheme 13.35. Prelog–Djerassi Lactone Synthesis: R. E. Ireland and J. P. Daub[a]

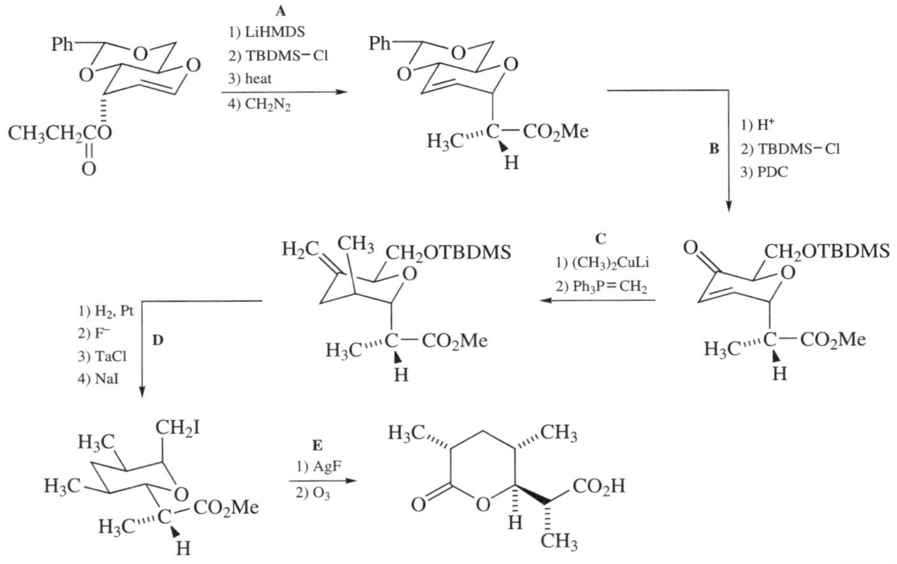

A: 1) LiHMDS 2) TBDMS−Cl 3) heat 4) CH_2N_2

B: 1) H^+ 2) TBDMS−Cl 3) PDC

C: 1) $(CH_3)_2CuLi$ 2) $Ph_3P{=}CH_2$

D: 1) H_2, Pt 2) F^- 3) TaCl 4) NaI

E: 1) AgF 2) O_3

a. R. E. Ireland and J. P. Daub, *J. Org. Chem.* **46:**479 (1981).

Scheme 13.36. Prelog–Djerassi Lactone Synthesis: D. A. Evans and J. Bartroli[a]

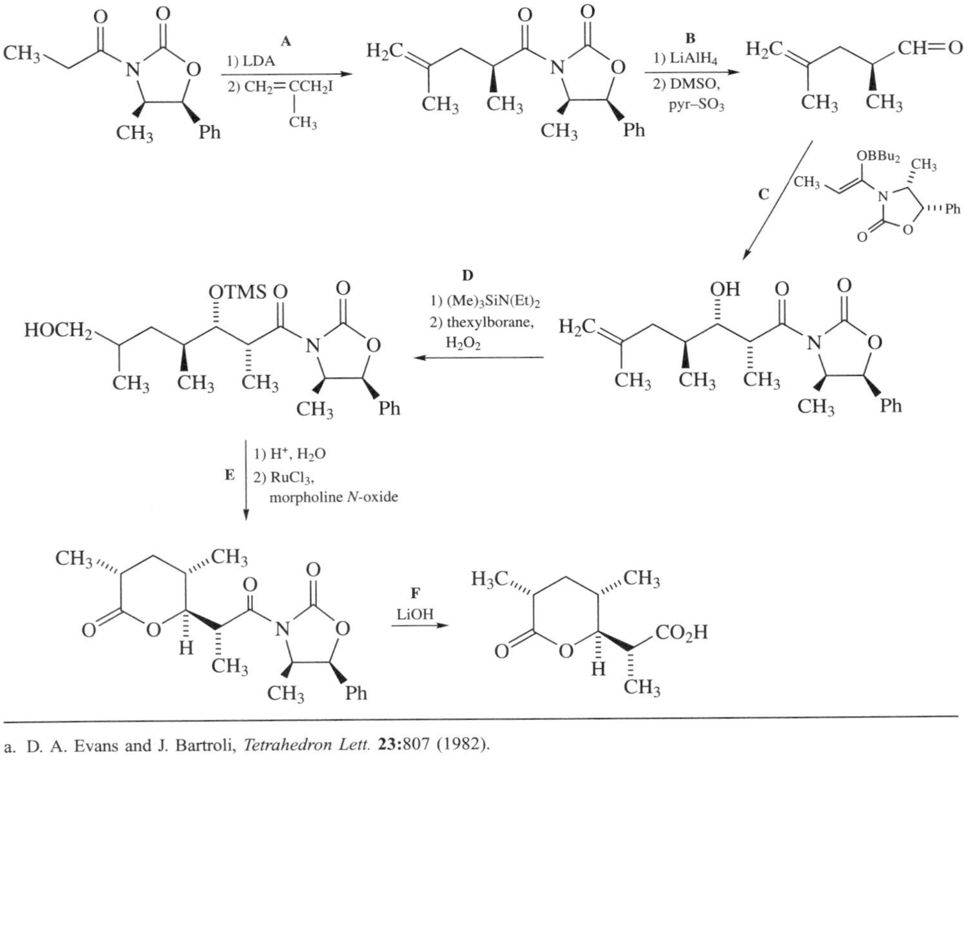

A: 1) LDA 2) $CH_2{=}C(CH_3)CH_2I$

B: 1) $LiAlH_4$ 2) DMSO, pyr–SO_3

C: (Z)-boron enolate ($OBBu_2$) of the propionyl oxazolidinone

D: 1) $(Me)_3SiN(Et)_2$ 2) thexylborane, H_2O_2

E: 1) H^+, H_2O 2) $RuCl_3$, morpholine *N*-oxide

F: LiOH

a. D. A. Evans and J. Bartroli, *Tetrahedron Lett.* **23:**807 (1982).

group.[135] Postulation of a transition state such as **A** can rationalize this result.

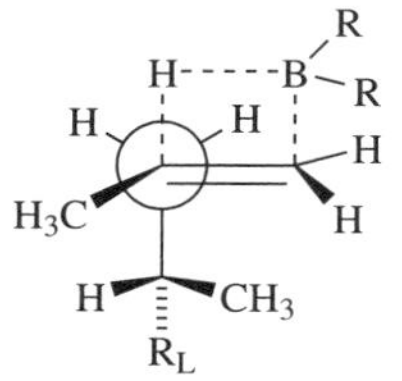

A

In the synthesis in Scheme 13.37, a stereoselective aldol condensation is used to establish the configuration at C-2 and C-3 in step **A**. The furan ring is then subjected to an electrophilic addition and solvolytic rearrangement in step **B**.

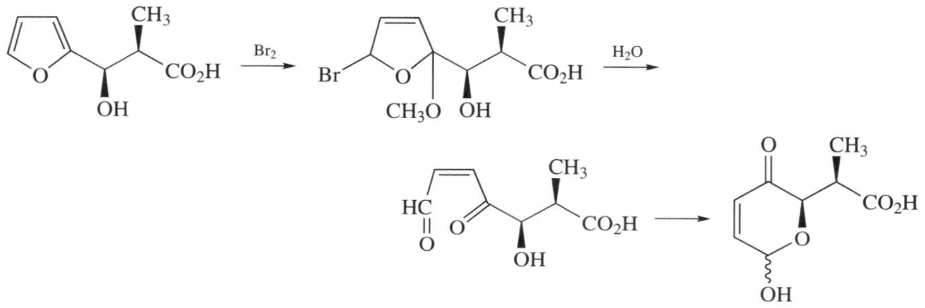

The protection of the hemiacetal hydroxyl in step **C** is followed by a purification of the dominant stereoisomer. The enone from step **E** is then subjected to a Wittig reaction. As in several of the other syntheses, the hydrogenation in step **E** is used to establish the configuration at C-4 and C-6.

The synthesis in Scheme 13.38 features a catalytic asymmetric epoxidation (see Section 12.2.1). By use of *meso*-2,4-dimethylglutaric anhydride as the starting material, the proper relative configuration at C-4 and C-6 is ensured. The epoxidation directed by the tartrate catalyst controls the configuration established at C-2 and C-3 by the epoxidation. Whereas the epoxidation is highly selective in establishing the configuration at C-2 and C-3, the configuration at C-4 and C-6 does not strongly influence the reaction, and a mixture of diastereomeric products is formed and must be separated at a later stage of the synthesis. The reductive ring opening in step **D** occurs with dominant inversion to establish the necessary configuration at C-2. The preference for 1,3-diol formation is characteristic of reductive ring opening by Red-Al of epoxides derived from allylic alcohols.[136] Presumably, initial coordination at the hydroxyl group and intramolecular

136. P. Ma, V. S. Martin, S. Masamune, K. B. Sharpless, and S. M. Viti, *J. Org. Chem.* **47:**1378 (1982); S. M. Viti, *Tetrahedron Lett.* **23:**4541 (1982); J. M. Finan and Y. Kishi, *Tetrahedron Lett.* **23:**2719 (1982).

Scheme 13.37. Prelog–Djerassi Lactone Synthesis: S. F. Martin and D. E. Guinn[a]

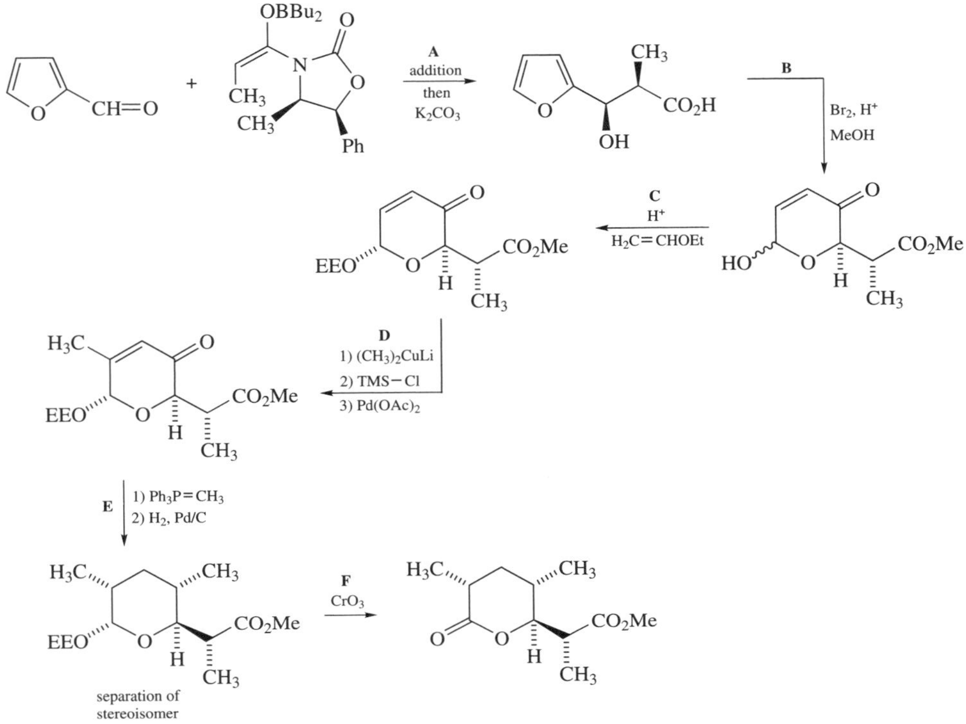

OBBu2 O
CH=O + N O CH3 CH3 Ph
A addition then K_2CO_3
CH3 CO2H OH
B Br_2, H^+ MeOH
O HO O H CO2Me CH3
C H^+ $H_2C{=}CHOEt$
O EEO O H CO2Me CH3
D 1) $(CH_3)_2CuLi$ 2) TMS−Cl 3) $Pd(OAc)_2$
H3C O EEO O H CO2Me CH3
E 1) $Ph_3P{=}CH_3$ 2) H_2, Pd/C
H3C CH3 EEO O H CO2Me CH3
separation of stereoisomer
F CrO_3
H3C CH3 O O H CO2Me CH3

a. S. F. Martin and D. E. Guinn, *J. Org. Chem.* **52:**5588 (1987).

Scheme 13.38. Prelog–Djerassi Lactone Synthesis: M. Honda, T. Katsuki, and M. Yamaguchi[a]

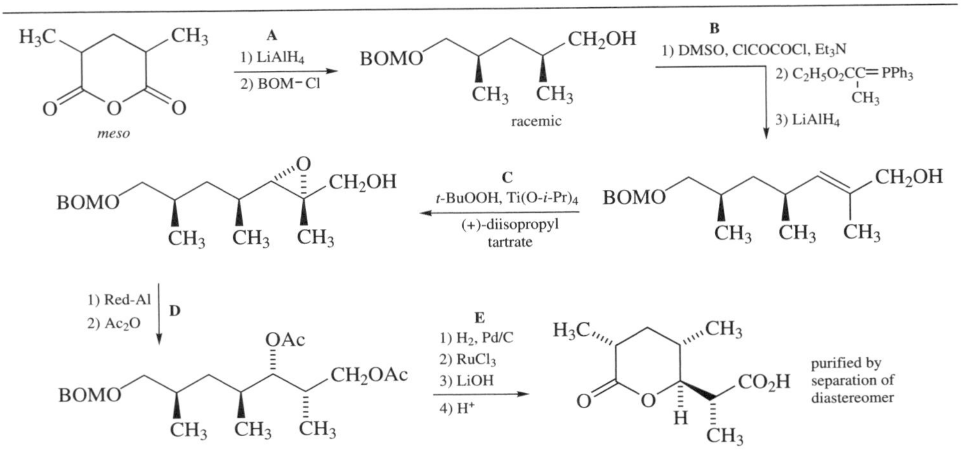

H3C CH3 O O O
meso
A 1) $LiAlH_4$ 2) BOM−Cl
BOMO CH2OH CH3 CH3
racemic
B 1) DMSO, ClCOCOCl, Et_3N 2) $C_2H_5O_2CC{=}PPh_3$ (CH3) 3) $LiAlH_4$
BOMO CH2OH CH3 CH3 CH3
C *t*-BuOOH, Ti(O-*i*-Pr)4 (+)-diisopropyl tartrate
BOMO O CH2OH CH3 CH3 CH3
D 1) Red-Al 2) Ac_2O
BOMO OAc CH2OAc CH3 CH3 CH3
E 1) H_2, Pd/C 2) $RuCl_3$ 3) LiOH 4) H^+
H3C CH3 O O H CO2H CH3
purified by separation of diastereomer

a. M. Honda, T. Katsuki, and M. Yamaguchi, *Tetrahedron Lett.* **25:**3857 (1984).

Scheme 13.39. Prelog–Djerassi Lactone Synthesis: A. J. Pearson and Y.-S. Lai[a]

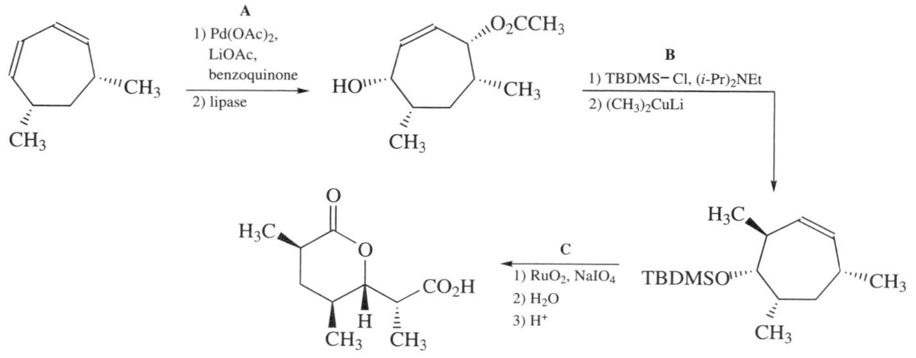

A: 1) $Pd(OAc)_2$, LiOAc, benzoquinone; 2) lipase

B: 1) TBDMS−Cl, $(i\text{-}Pr)_2NEt$; 2) $(CH_3)_2CuLi$

C: 1) RuO_2, $NaIO_4$; 2) H_2O; 3) H^+

a. A. J. Pearson and Y.-S. Lai, *J. Chem. Soc., Chem. Commun.* **1988:**442.

delivery of hydride is responsible for this stereoselectivity.

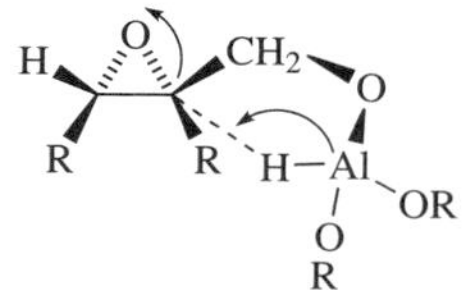

The synthesis in Scheme 13.39 is built on a rather different intermediate. The *cis*-dimethylcycloheptadiene intermediate is acetoxylated with $Pd(OAc)_2$, and the resulting diacetate is enantioselectively hydrolyzed with a lipase to give a monoacetate. An *anti* S_N2' displacement establishes the correct configuration of the C-2 methyl substituent. Oxidative ring cleavage and lactonization give the final product.

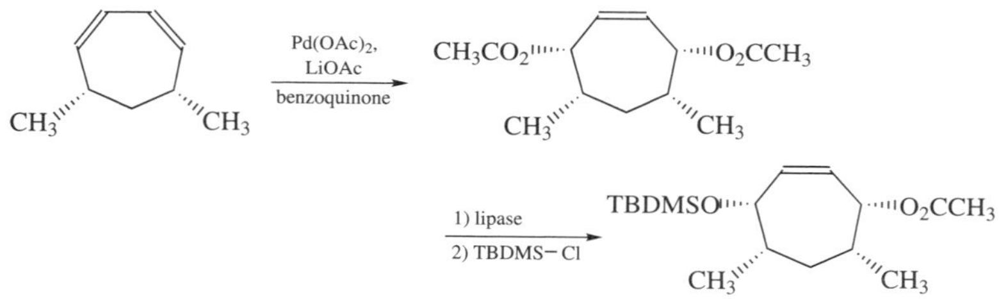

$Pd(OAc)_2$, LiOAc, benzoquinone → ; 1) lipase, 2) TBDMS−Cl →

The synthesis in Scheme 13.40 uses stereospecific ring opening of the epoxide to establish the stereochemistry of the C-4 methyl group. The starting material can be made by enantiospecific epoxidation of the corresponding allylic alcohol.[137]

The synthesis in Scheme 13.41 features use of an enantioselective allylboronate reagent derived from diisopropyl tartrate. The stereochemistry at the C-2 methyl group is influenced by the C-5 hydroxyl group, with 3 : 1 stereoselectivity for the desired stereoisomer.

137. H. Nagaoka and Y. Kishi, *Tetrahedron* **37:**3873 (1981).

Scheme 13.40. Prelog–Djerassi Lactone Synthesis: M. Miyashita, M. Hoshino, A. Yoshikoshi, K. Kawamine, K. Yoshihara, and H. Irie[a]

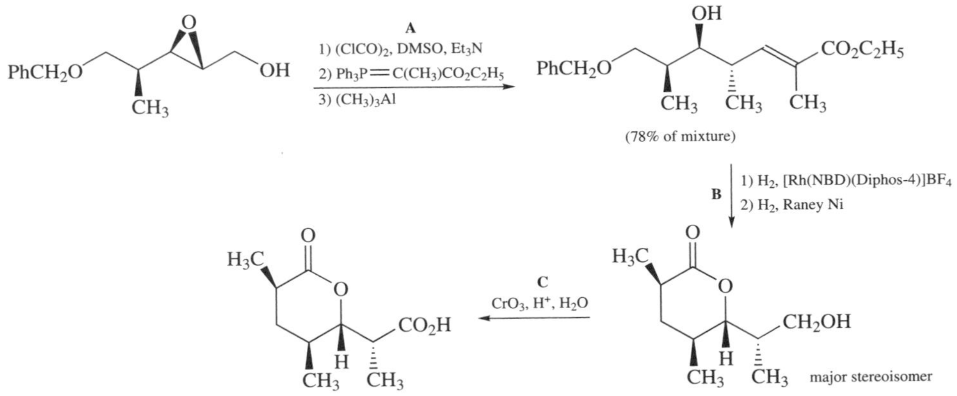
PhCH2O
CH3
OH
A
1) (ClCO)2, DMSO, Et3N
2) Ph3P=C(CH3)CO2C2H5
3) (CH3)3Al
CO2C2H5
(78% of mixture)
B
1) H2, [Rh(NBD)(Diphos-4)]BF4
2) H2, Raney Ni
H3C
CH2OH
major stereoisomer
C
CrO3, H+, H2O
CO2H

a. M. Miyashita, M. Hoshino, A. Yoshikoshi, K. Kawamine, K. Yoshihara, and H. Irie, *Chem. Lett.* **1992:**1101.

Scheme 13.41. Prelog–Djerassi Lactone Synthesis: J. Cossy, D. Bauer, and V. Bellosta[a]

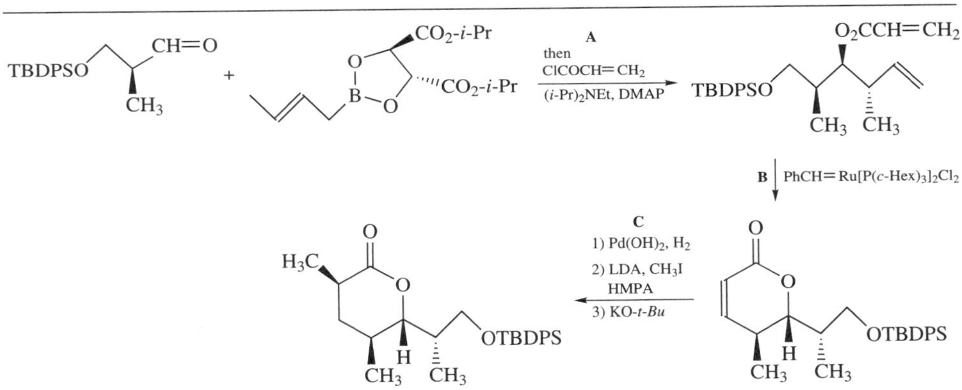
TBDPSO
CH=O
CH3
CO2-i-Pr
A
then
ClCOCH=CH2
(i-Pr)2NEt, DMAP
O2CCH=CH2
B
PhCH=Ru[P(c-Hex)3]2Cl2
C
1) Pd(OH)2, H2
2) LDA, CH3I
HMPA
3) KO-t-Bu
OTBDPS
H3C

a. J. Cossy, D. Bauer, and V. Bellosta, *Tetrahedron Lett.* **40:**4187 (1999).

13.5.4. Taxol

Taxol[138] was first discovered to have anticancer activity during screening of natural substances.[139] Several Taxol analogs differing in the side-chain substitution pattern also have good activity.[140] Production of Taxol directly from plant sources presented serious problems because the plants are slow-growing and the Taxol content is low. However, the tetracyclic ring system is found in a more available material, baccatin III, which can subsequently be converted to Taxol.[141] The combination of important biological activity,

138. Taxol is a registered trade name of Bristol-Myers Squibb. The generic name is paclitaxel.
139. M. C. Wani, H. L. Taylor, M. E. Wall, P. Coggon, and A. McPhail, *J. Am. Chem. Soc.* **93:**2325 (1971).
140. M. Suffness, ed., *Taxol: Science and Applications*, CRC Press, Boca Raton, Florida, 1995.
141. J.-N. Denis, A. E. Greene, D. Guenard, F. Gueritte-Vogelein, L. Mangatal, and P. Potier, *J. Am. Chem. Soc.* **110:**5917 (1988); R. A. Holton, Z. Zhang, P. A. Clarke, H. Nadizadeh, and D. J. Procter, *Tetrahedron Lett.* **39:**2883 (1998).

the limited natural sources, and the interesting structure made Taxol a target of great synthetic interest during the 1990s. Among the challenging aspects of the structure from a synthetic point of view are the eight-membered ring, the bridgehead double bond, and the large number of oxygen functional groups. Several synthesis of baccatin III and closely related Taxol precursors have been reported in the past few years.

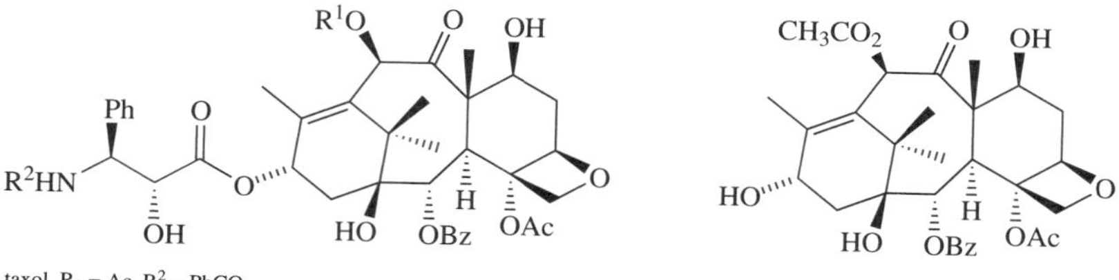

taxol $R_1 = Ac$, $R^2 = PhCO$

taxotere $R_1 = H$, $R^2 = (CH_3)_3CO_2C$

baccatin III

The first synthesis of Taxol was completed by Robert A. Holton and co-workers and is outlined in Scheme 13.42. One of the key steps occurs early in the synthesis in sequence **A** and effects fragmentation of **4** to **5**. The intermediate epoxide can be prepared from a terpene called "patchino."[142] The epoxide was then converted to **5** by a BF_3-mediated rearrangement.

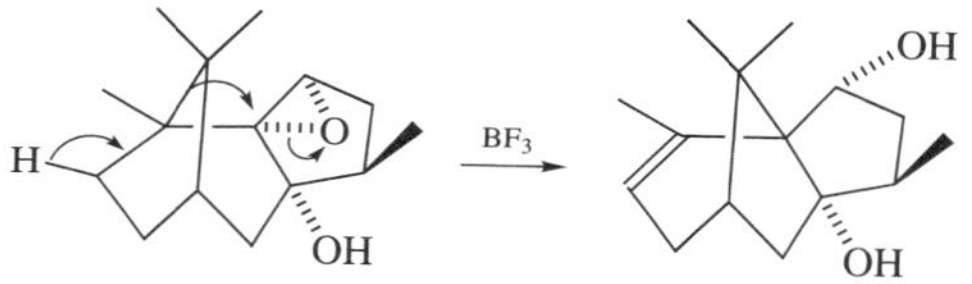

Another epoxidation followed by fragmentation, gives the bicyclic intermediate which contains the eight-membered ring and bridgehead double bond properly positioned for conversion to Taxol.

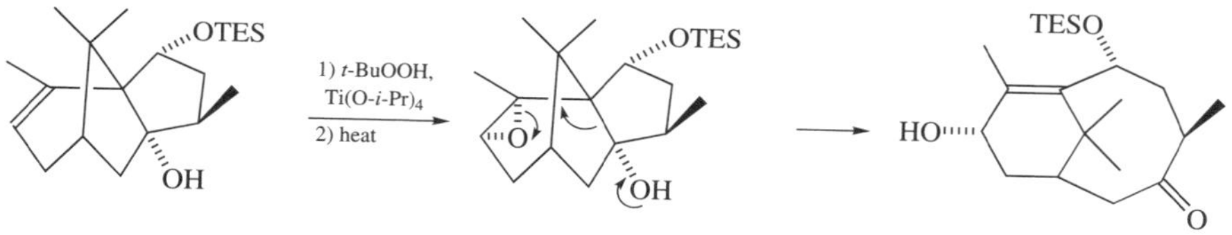

The next phase of the synthesis is construction of the C ring. An aldol addition was used to introduce a pentenyl group at C-8, and the product was trapped as a carbonate between the C-2 and C-7 oxygens. The Davis oxaziridine was then used to introduce an oxygen at C-2. After reduction of the C-3 oxygen, a carbonate is formed, and C-2 is converted to a

142. R. A. Holton, R. R. Juo, H. B. Kim, A. D. Williams, S. Harusawa, R. E. Lowenthal, and S. Yogai, *J. Am. Chem. Soc.* **110**:6558 (1988).

carbonyl group. In step **D**, this carbonate is rearranged to a lactone.

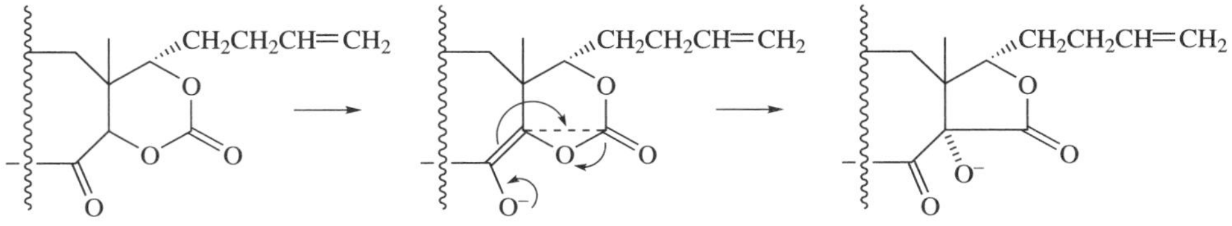

After oxidation of the vinyl group, the C ring is constructed by a Dieckmann cyclization. The resulting β-keto ester is subjected to nucleophilic decarboxylation by phenylthiolate [step **F** (4–6)].

In the later stages of the synthesis, the oxetane ring is constructed in step **H** (1–4). An exocyclic methylene group was introduced by a methyl Grignard addition followed by dehydration with Burgess reagent. The double bond was then hydroxylated with OsO_4, and a sequence of selective transformations of the triol provided the hydroxy tosylate, which undergoes intramolecular nucleophilic substitution to form the oxetane ring.

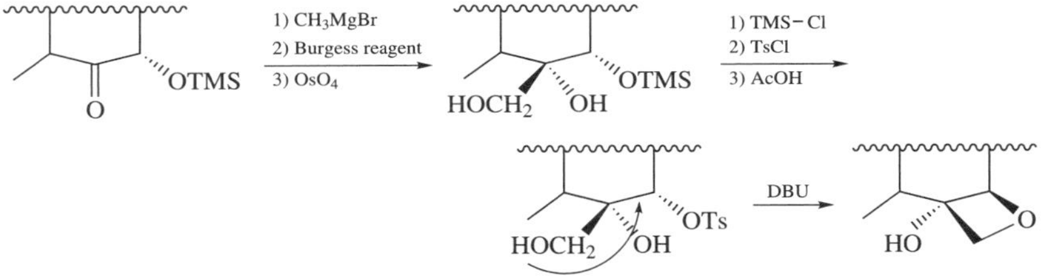

In step **H**-7, the addition of phenyllithium to the cyclic carbonate group neatly generates the C-2 benzoate group. A similar reaction was used in several other Taxol syntheses.

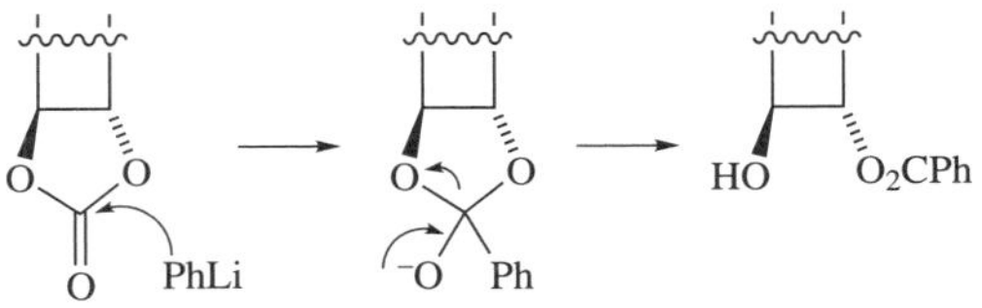

The final phase of the synthesis is introduction of the C-9 oxygen by phenylseleninic anhydride (step **H**-9).

The synthesis by K. C. Nicolaou and co-workers is summarized in Scheme 13.43. Diels–Alder reactions are prominent in forming the early intermediates. The formation of the A ring in steps **A** and **B** involves use of α-chloroacrylonitrile as a ketene synthon. In step **C**, the pyrone ring serves as diene. This reaction is facilitated by phenylboronic acid, which brings the diene and dienophile together as a boronate ester, permitting an intramolecular reaction.

Scheme 13.42. Taxol Synthesis: R. A. Holton and Co-workers[a]

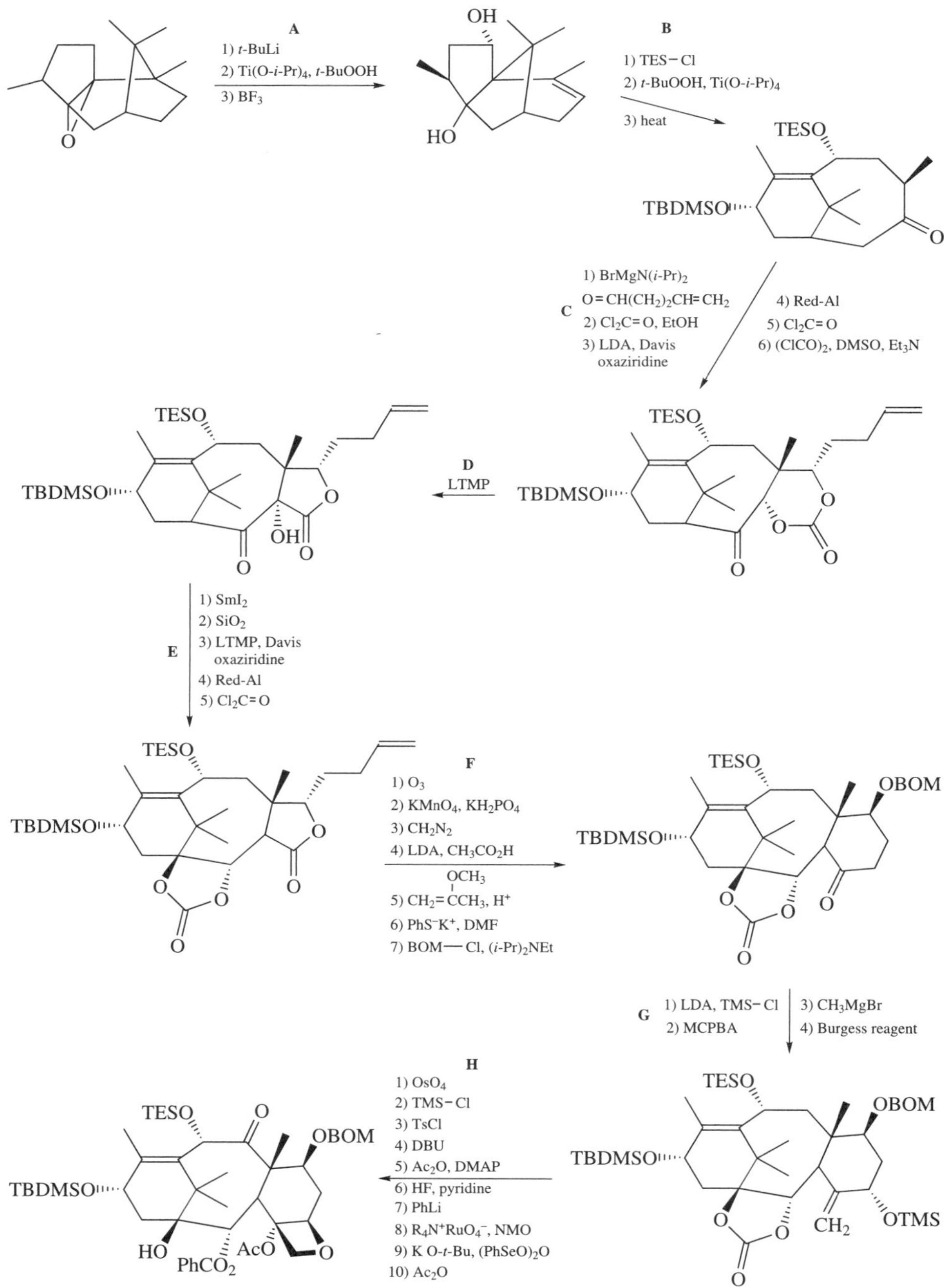

a. R. A. Holton, C. Somoza, H.-B. Kim, F. Liang, R. J. Biediger, P. D. Boatman, M. Shindo, C. C. Smith, S. Kim, H. Nadizadeh, Y. Suzuki, C. Tao, P. Vu, S. Tang, P. Zhang, K. K. Murthi, L. N. Gentile, and J. H. Lin, *J. Am. Chem. Soc.* **116:**1597 (1994); R. A. Holton, H.-B. Kim, C. Somoza, F. Liang, R. J. Biediger, P. D. Boatman, M. Shindo, C. C. Smith, S. Kim, H. Nadizadeh, Y. Suzuki, C. Tao, P. Vu, S. Tang, P. Zhang, K. K. Murthi, L. N. Gentile, and J. H. Liu, *J. Am. Chem. Soc.* **116:**1599 (1994).

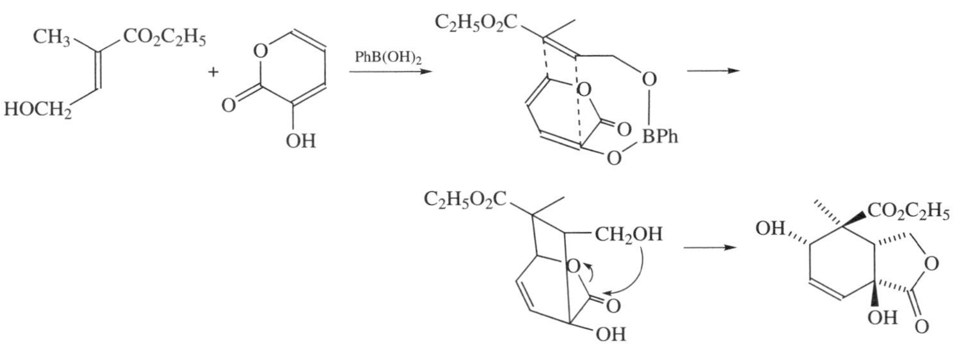

The A and C rings were brought together in step **G** by an organolithium addition to the aldehyde. The lithium reagent is generated by a Shapiro reaction. The eight-membered B ring was then closed by a titanium-mediated reductive coupling of a dialdehyde in step **I**-1. The C-13 oxygen is introduced very late in the synthesis by an allylic oxidation using PCC (step **L**-3).

The synthesis by Samuel Danishefsky's group is outlined in Scheme 13.44. The early stages of this synthesis construct an intermediate which contains the functionality of the C and D rings of baccatin III. Ring A is then introduced by the functionalized lithium reagent in step E. The closure of the B ring is done by an intramolecular Heck reaction involving a vinyl triflate at step **G**-4. The late functionalizations include the introduction of the C-10 and C-13 oxygens. These were done by phenylseleninic anhydride oxidation of the enolate in step **I**-5 and by allylic oxidation at C-13 in step **J**-1. These oxidative steps are presaged by similar transformations in the Holton and Nicolaou syntheses.

The synthesis of the baccatin III structure by Paul Wender and co-workers, shown in Scheme 13.45, begins with an oxidation product of the readily available terpene pinene. One of the key early steps is a photochemical rearrangement in step **B**.

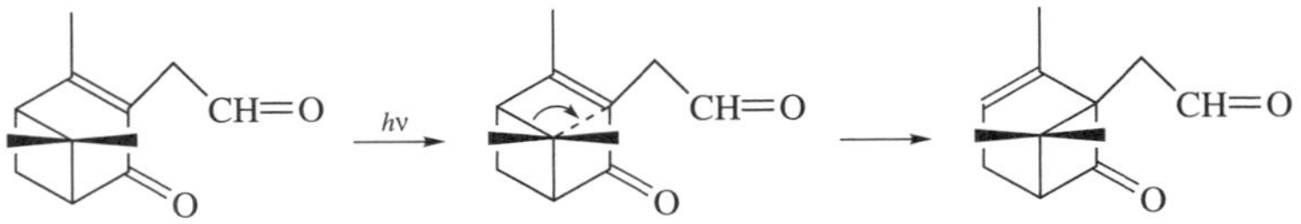

Another key step is the fragmentation induced by treating **6** first with MCPBA and then with 1,4-diazabicyclo[2.2.2]octane (DABCO) [step **E**-(1,2)]. The four-membered ring is fragmented, forming the eight-membered ring and providing the C-13 oxygen.

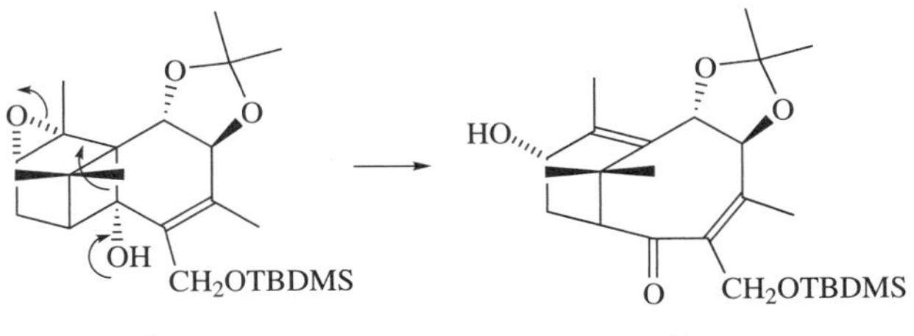

6 **7**

The C-1 oxygen was introduced at step **F**-1 by enolate oxidation. The C ring was constructed by building up a substituent at C-16 (steps **G** and **H**) and then performing an intramolecular aldol addition (step **I**).

Scheme 13.43. Taxol Synthesis: K. C. Nicolaou and Co-workers[a]

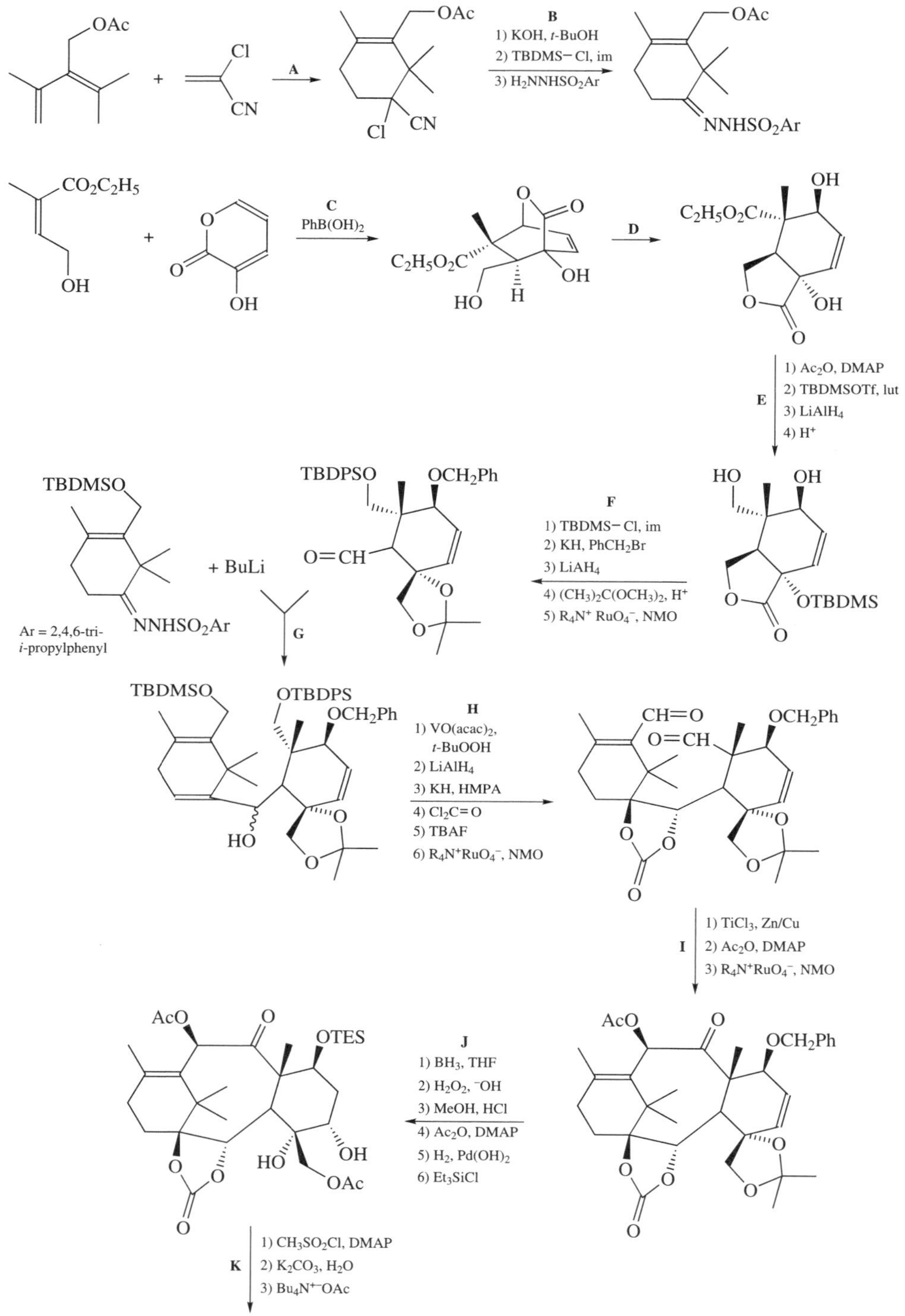

OAc + Cl, CN —A→ OAc, Cl, CN

B
1) KOH, *t*-BuOH
2) TBDMS−Cl, im
3) H_2NNHSO_2Ar

OAc, $NNHSO_2Ar$

$CO_2C_2H_5$, OH + O, O, OH

C
$PhB(OH)_2$

O, O, $C_2H_5O_2C$, HO, H, OH

D

OH, $C_2H_5O_2C$, O, O, OH

E
1) Ac_2O, DMAP
2) TBDMSOTf, lut
3) $LiAlH_4$
4) H^+

HO, OH, O, O, OTBDMS

F
1) TBDMS−Cl, im
2) KH, $PhCH_2Br$
3) $LiAH_4$
4) $(CH_3)_2C(OCH_3)_2$, H^+
5) R_4N^+ RuO_4^-, NMO

TBDPSO, OCH_2Ph, O=CH, O, O

TBDMSO, $NNHSO_2Ar$ + BuLi

Ar = 2,4,6-tri-*i*-propylphenyl

G

TBDMSO, OTBDPS, OCH_2Ph, HO, O, O

H
1) $VO(acac)_2$, *t*-BuOOH
2) $LiAlH_4$
3) KH, HMPA
4) $Cl_2C=O$
5) TBAF
6) $R_4N^+RuO_4^-$, NMO

CH=O, O=CH, OCH_2Ph, O, O, O, O, O

I
1) $TiCl_3$, Zn/Cu
2) Ac_2O, DMAP
3) $R_4N^+RuO_4^-$, NMO

AcO, O, OCH_2Ph, O, O, O, O, O

J
1) BH_3, THF
2) H_2O_2, ^-OH
3) MeOH, HCl
4) Ac_2O, DMAP
5) H_2, $Pd(OH)_2$
6) Et_3SiCl

AcO, O, OTES, O, O, O, HO, OH, OAc

K
1) CH_3SO_2Cl, DMAP
2) K_2CO_3, H_2O
3) $Bu_4N^{+-}OAc$

L
1) PhLi
2) Ac_2O, DMAP
3) PCC
4) $NaBH_4$

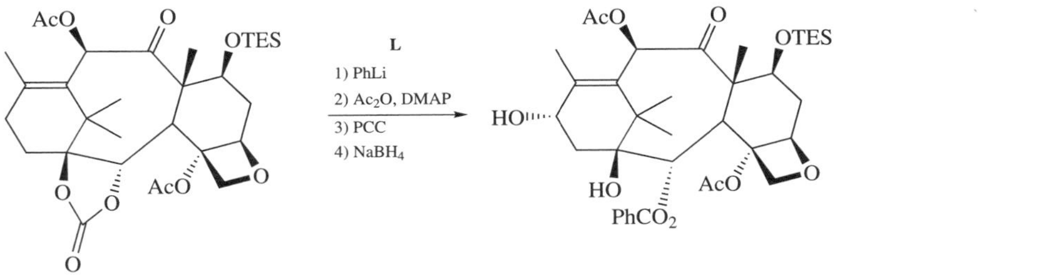

a. K. C. Nicolaou, P. G. Nantermet, H. Ueno, R. K. Guy, E. A. Couladouros, and E. J. Sorensen, *J. Am. Chem. Soc.* **117**:624 (1995); K. C. Nicolaou, J.-J. Liu, Z. Yang, H. Ueno, E. J. Sorensen, C. F. Claiborne, R. K. Guy, C.-K. Hwang, M. Nakada, and P. G. Nantermet, *J. Am. Chem. Soc.* **117**:634 (1995); K. C. Nicolaou, Z. Zhang, J.-J. Liu, P. G. Nantermet, C. F. Claiborne, J. Renaud, R. K. Guy, and K. S. Shibayama, *J. Am. Chem. Soc.* **117**:645 (1995); K. C. Nicolaou, H. Ueno, J.-J. Liu, P. G. Nantermet, Z. Yang, J. Renaud, K. Paulvannan, and R. Chadha, *J. Am. Chem. Soc.* **117**:653 (1995).

The Holton, Nicolaou, Danishefsky, and Wender syntheses of baccatin III structures employ various cyclic intermediates and take advantage of stereochemical features built into these rings to control subsequent reaction stereochemistry. These syntheses also provide numerous examples of the selective use of protective groups to differentiate between the several hydroxy groups that are present in the intermediates.

The synthesis of Taxol completed by a group led by the Japanese chemist Teruaki Mukaiyama and shown in Scheme 13.46 takes a rather different approach. Much of the stereochemistry is built into the B ring by a series of acyclic aldol condensations in steps **A–D**. The ring is closed by a samarium-mediated cyclization at step **E**-3.

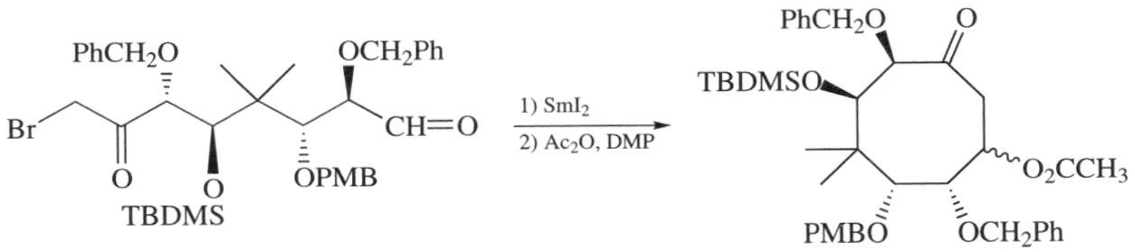

1) SmI_2
2) Ac_2O, DMP

The A ring is closed by a Ti-mediated reductive coupling (step **H**-5). The C(11)–C(12) double bond is introduced from the diol by deoxygenation of the thiocarbonate [step **I**-(1,2)]. The final sequence for conversion of the product from steps **I**-1–8 to baccatin III begins with a copper-mediated allylic oxidation and also involves an allylic rearrangement of the halide. The exocyclic double bond was then used to introduce the final oxygens needed to perform the oxetane ring closure.

Another Japanese group developed the Taxol synthesis shown in Scheme 13.47. The eight-membered B ring was closed early in the synthesis using a Lewis acid-induced Mukaiyama reaction (step **B**-1). Note that a trimethylsilyl dienol ether served as the nucleophile. The C-19 methyl group is introduced via a cyclopropanation in step **C**-5, followed by a reduction in step **D**-1.

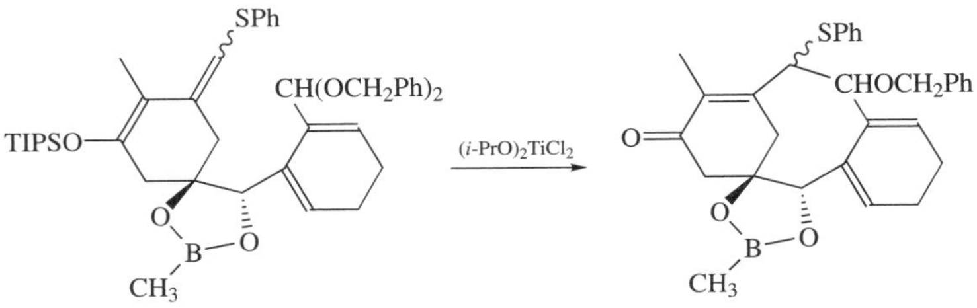

$(i\text{-PrO})_2TiCl_2$

Scheme 13.44. Taxol Synthesis: S. J. Danishefsky and Co-workers[a]

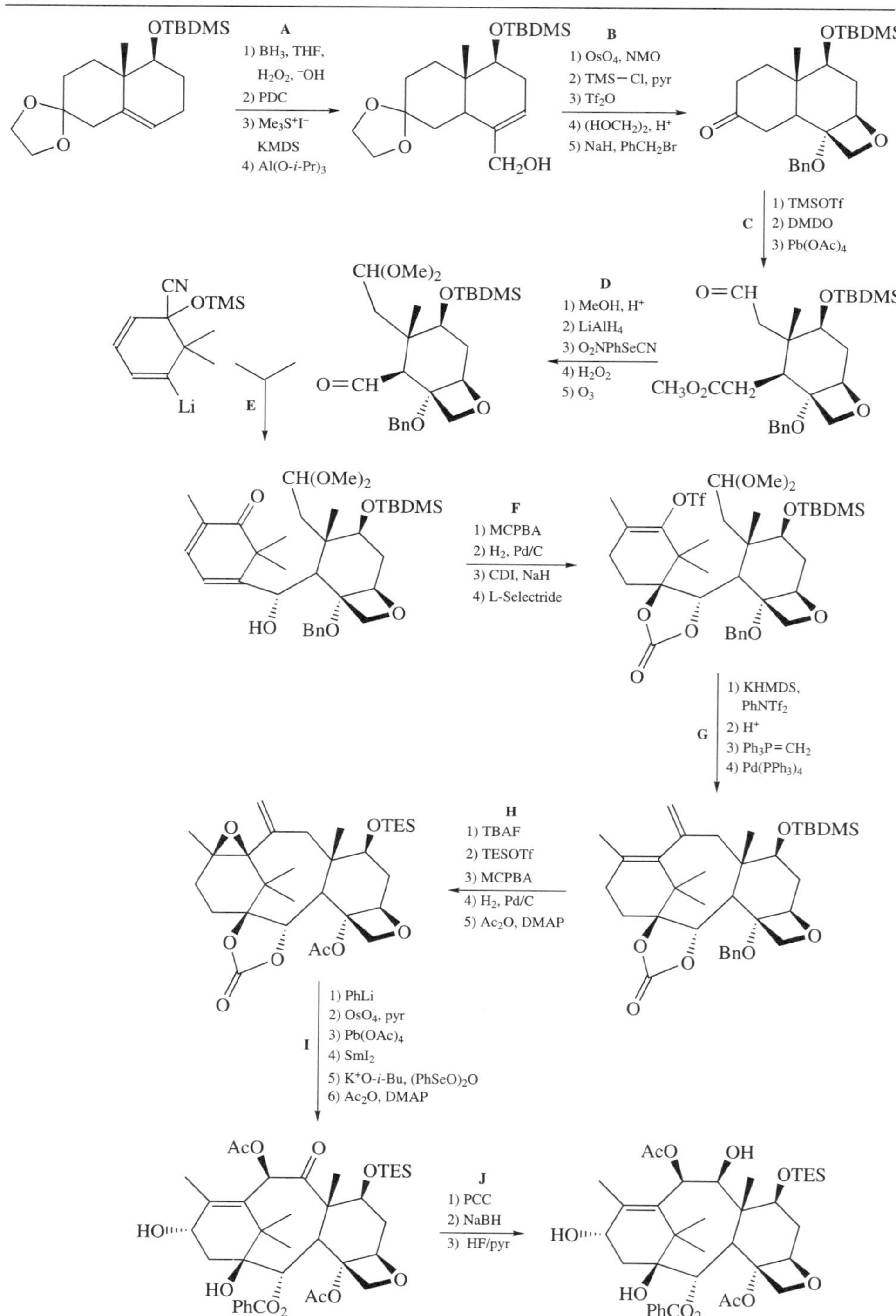

a. S. J. Danishefsky, J. J. Masters, W. B. Young, J. T. Link, L. B. Snyder, T. V. Magee, D. K. Jung, R. C. A. Isaacs, W. G. Bornmann, C. A. Alaimo, C. A. Coburn, and M. J. Di Grandi, *J. Am. Chem. Soc.* **118**:2843 (1996).

Scheme 13.45. Taxol Synthesis: P. A. Wender and Co-workers[a]

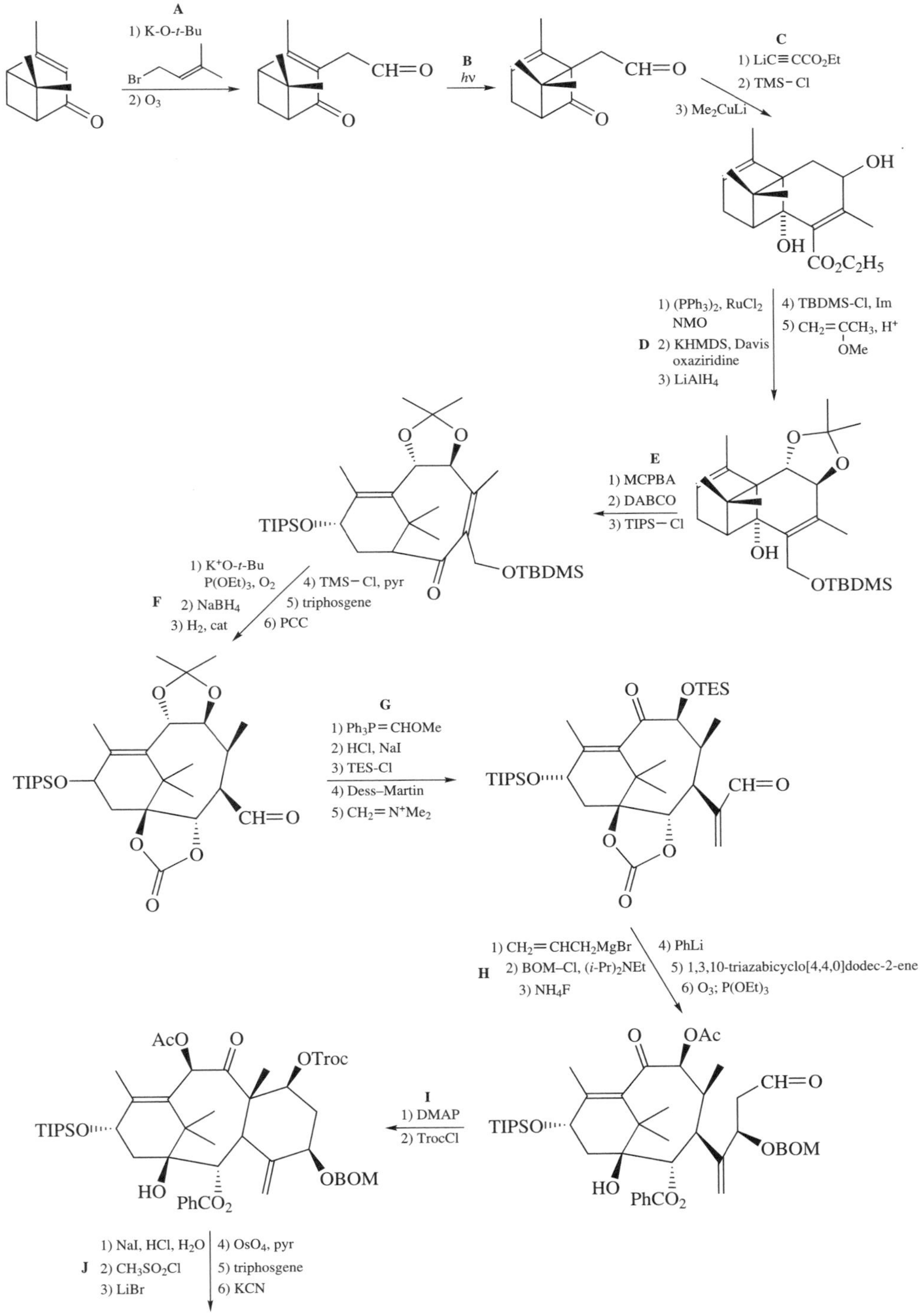

A
1) K-O-*t*-Bu, Br–$CH_2CH=C(CH_3)_2$
2) O_3

CH=O

B
$h\nu$

CH=O

C
1) LiC≡CCO_2Et
2) TMS−Cl
3) Me_2CuLi

OH, OH, $CO_2C_2H_5$

D
1) $(PPh_3)_2$, $RuCl_2$ NMO
2) KHMDS, Davis oxaziridine
3) $LiAlH_4$
4) TBDMS-Cl, Im
5) $CH_2=C(OMe)CH_3$, H^+

OH, OTBDMS

E
1) MCPBA
2) DABCO
3) TIPS−Cl

TIPSO, OTBDMS

F
1) K^+O-*t*-Bu, $P(OEt)_3$, O_2
2) $NaBH_4$
3) H_2, cat
4) TMS−Cl, pyr
5) triphosgene
6) PCC

TIPSO, CH=O

G
1) $Ph_3P=CHOMe$
2) HCl, NaI
3) TES-Cl
4) Dess–Martin
5) $CH_2=N^+Me_2$

TIPSO, OTES, CH=O

H
1) $CH_2=CHCH_2MgBr$
2) BOM–Cl, $(i\text{-Pr})_2NEt$
3) NH_4F
4) PhLi
5) 1,3,10-triazabicyclo[4,4,0]dodec-2-ene
6) O_3; $P(OEt)_3$

TIPSO, OAc, CH=O, OBOM, HO, $PhCO_2$

I
1) DMAP
2) TrocCl

AcO, OTroc, TIPSO, HO, $PhCO_2$, OBOM

J
1) NaI, HCl, H_2O
2) CH_3SO_2Cl
3) LiBr
4) OsO_4, pyr
5) triphosgene
6) KCN

Scheme 13.45. Taxol Synthesis: P. A. Wender and Co-workers[a]

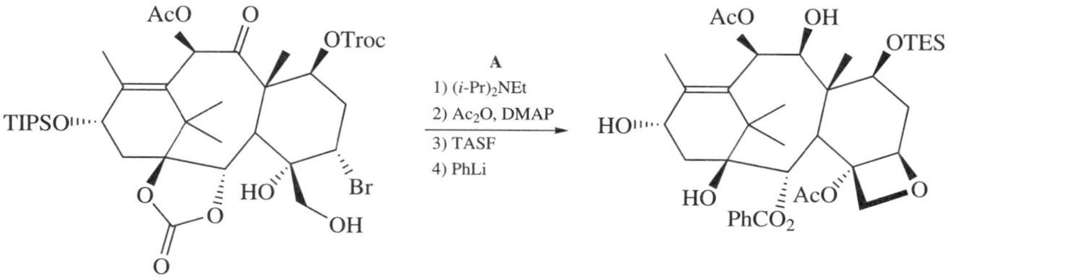

A
1) $(i\text{-Pr})_2NEt$
2) Ac_2O, DMAP
3) TASF
4) PhLi

a. P. A. Wender, N. F. Badham, S. P. Conway, P. E. Floreancig, T. E. Glass, C. Gränicher, J. B. Houze, J. Jänichen, D. Lee, D. G. Marquess, P. L. McGrane, W. Meng, T. P. Mucciaro, M. Muhlebach, M. G. Natchus, H. Paulsen, D. B. Rawlins, J. Satkofsky, A. J. Shuker, J. C. Sutton, R. E. Taylor, and K. Tomooka, *J. Am. Chem. Soc.* **119:**2755 (1997); P. A. Wender, N. F. Badham, S. P. Conway, P. E. Floreancig, T. E. Glass, J. B. Houze, N. E. Krauss, D. Lee, D. G. Marquess, P. L. McGrane, W. Meng, M. G. Natchus, A. J. Shuker, J. C. Sutton, and R. E. Taylor, *J. Am. Chem. Soc.* **119:**2757 (1997).

13.5.5. Epothilone A

The epothilones are 16-membered lactones that have been isolated from mycobacteria. Epothilones A–D differ in the presence of the C(12)–C(13) epoxide and in the C-12 methyl group. Although the epothilones are structurally very different from Taxol, the biochemical mechanism of anticancer action is related, and epothilone A and analogs are of substantial current interest as chemotherapeutic agents.[143] Schemes 13.48–13.51 summarize four syntheses of epothilone A. Syntheses of epothilone B have also been completed.[144]

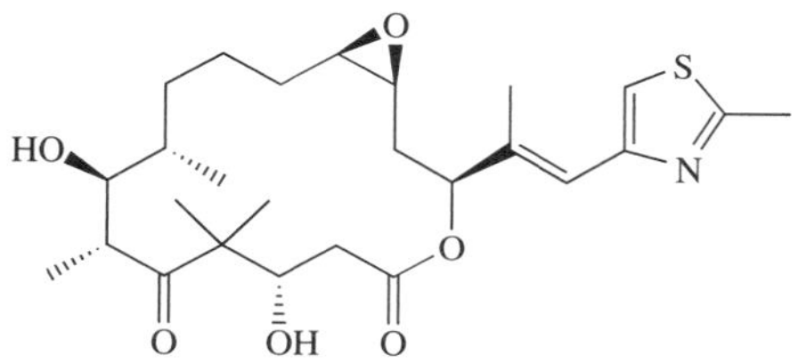

epothilone A

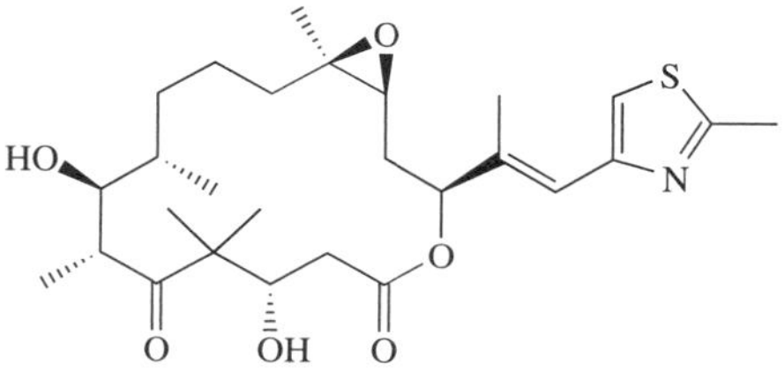

epothilone B

The group of K. C. Nicolaou at Scripps Research Institute has developed two synthetic routes to epothilone A. Because the 16-membered lactone ring is quite flexible, it is not likely to impose strong stereoselectivity. Instead, the stereoselective synthesis of epothilone A requires building the correct stereochemistry into acyclic precursors which are closed later in the synthesis. One of the Nicolaou syntheses involves closure of the lactone ring as a late step. This synthesis is shown in Scheme 13.48. Two enantiomerically

143. T.-C. Chou, X.-G. Zhang, C. R. Harris, S. D. Kuduk, A. Balog, K. A. Savin, J. R. Bertino, and S. J. Danishefsky, *Proc. Natl. Acad. Sci. U.S.A.* **95:**15798 (1998).

144. J. Mulzer, A. Mantoulidis, and E. Öhler, *Tetrahedron Lett.* **39:**8633 (1998); D. S. Sa, D. F. Meng, P. Bertinato, A. Balog, E. J. Sorensen, S. J. Danishefsky, Y.-H. Zheng, T.-C. Chou, L. He, and S. B. Horwitz, *Angew. Chem. Int. Ed. Engl.* **36:**757 (1997); A. Balog, C. Harris, K. Savin, S.-G. Zhang, T. C. Chou, and S. J. Danishefsky, *Angew. Chem. Int. Ed. Engl.* **37:**2675 (1998); D. Schinzer, A. Bauer, and J. Schieber, *Synlett* **1998:**861; S. A. May and P. A. Grieco, *J. Chem. Soc., Chem. Commun.* **1998:**1597; K. C. Nicolaou, S. Ninkovic, F. Sarabia, D. Vourloumis, Y. He, H. Vallberg, M. R. V. Finlay and Z. Yang, *J. Am. Chem. Soc.* **119:**7974 (1997); K. C. Nicolaou, D. Hepworth, M. R. V. Finlay, B. Wershkun, and A. Bigot, *J. Chem. Soc., Chem. Commun.* **1999:**519; D. Schinzer, A. Bauer, and J. Schieber, *Chem. Eur. J.* **5:**2492 (1999); J. D. White, R. G. Carter, and K. F. Sundermann, *J. Org. Chem.* **64:**684 (1999).

Scheme 13.46. Taxol Synthesis: T. Mukaiyama and Co-workers[a]

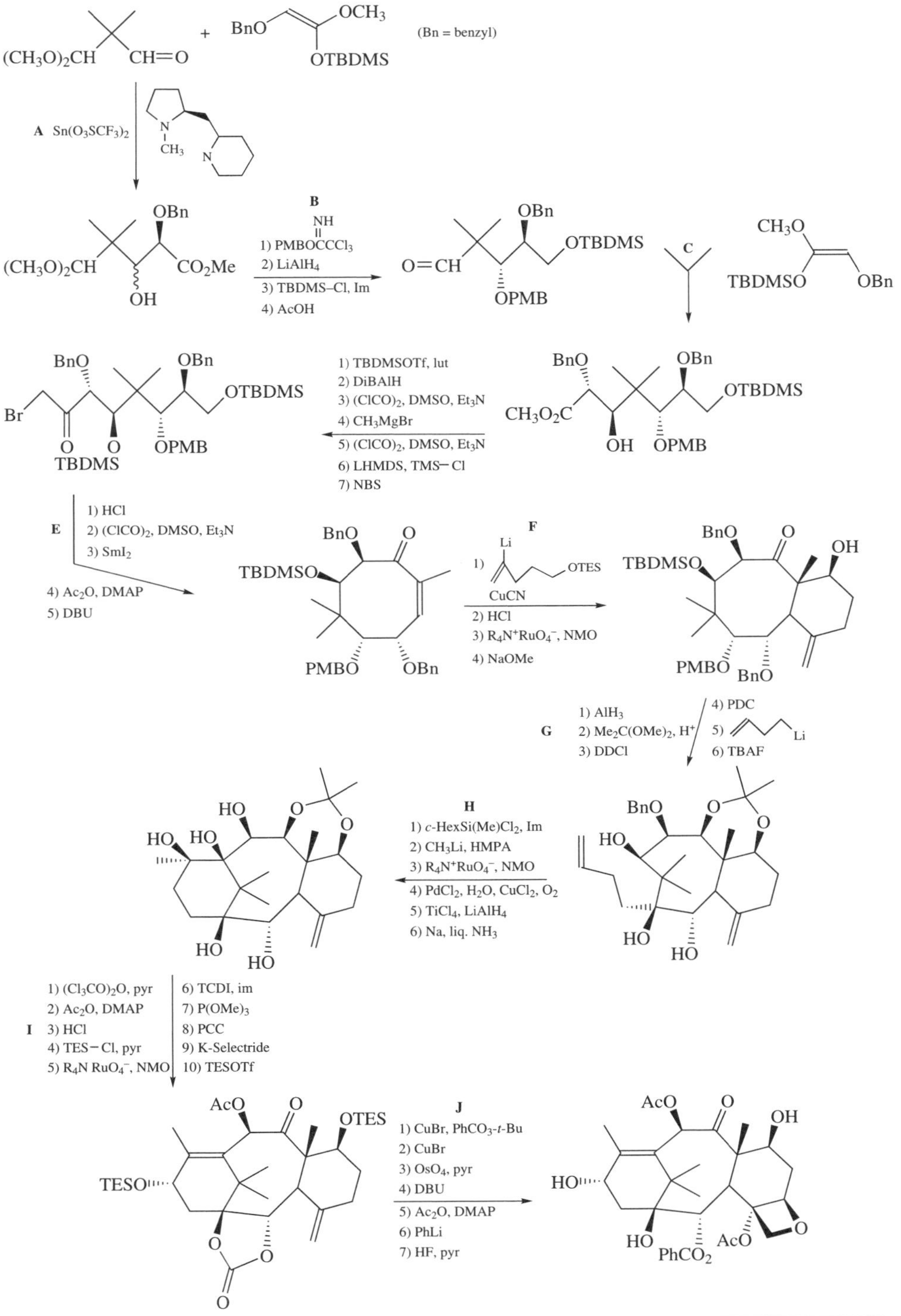

$(CH_3O)_2CH$ CH=O + BnO OCH$_3$ OTBDMS (Bn = benzyl)

A $Sn(O_3SCF_3)_2$ N CH$_3$ N

$(CH_3O)_2CH$ OBn CO$_2$Me OH

B 1) PMBOC(=NH)CCl$_3$ 2) LiAlH$_4$ 3) TBDMS–Cl, Im 4) AcOH

O=CH OBn OTBDMS OPMB

C CH$_3$O TBDMSO OBn

BnO OBn CH$_3$O$_2$C OTBDMS OH OPMB

1) TBDMSOTf, lut 2) DiBAlH 3) $(ClCO)_2$, DMSO, Et$_3$N 4) CH$_3$MgBr 5) $(ClCO)_2$, DMSO, Et$_3$N 6) LHMDS, TMS–Cl 7) NBS

BnO OBn Br O O OPMB OTBDMS TBDMS

E 1) HCl 2) $(ClCO)_2$, DMSO, Et$_3$N 3) SmI$_2$ 4) Ac$_2$O, DMAP 5) DBU

BnO O TBDMSO PMBO OBn

F 1) Li OTES CuCN 2) HCl 3) $R_4N^+RuO_4^-$, NMO 4) NaOMe

BnO O OH TBDMSO PMBO BnO

G 1) AlH$_3$ 2) Me$_2$C(OMe)$_2$, H$^+$ 3) DDCl 4) PDC 5) Li 6) TBAF

BnO O O HO HO HO

H 1) *c*-HexSi(Me)Cl$_2$, Im 2) CH$_3$Li, HMPA 3) $R_4N^+RuO_4^-$, NMO 4) PdCl$_2$, H$_2$O, CuCl$_2$, O$_2$ 5) TiCl$_4$, LiAlH$_4$ 6) Na, liq. NH$_3$

HO O O HO HO HO HO

I 1) $(Cl_3CO)_2O$, pyr 2) Ac$_2$O, DMAP 3) HCl 4) TES–Cl, pyr 5) $R_4N\ RuO_4^-$, NMO 6) TCDI, im 7) P(OMe)$_3$ 8) PCC 9) K-Selectride 10) TESOTf

AcO O OTES TESO O O O

J 1) CuBr, PhCO$_3$-*t*-Bu 2) CuBr 3) OsO$_4$, pyr 4) DBU 5) Ac$_2$O, DMAP 6) PhLi 7) HF, pyr

AcO O OH HO HO PhCO$_2$ AcO O

a. T. Mukaiyama, I. Shiina, H. Iwadare, M. Saitoh, T. Nishimura, N. Ohkawa, H. Sakoh, K. Nishimura, Y. Tani, M. Hasegawa, K. Yamada, and K. Saitoh, *Chem. Eur. J.* **5**:121 (1999).

Scheme 13.47. Taxol Synthesis: H. Kusama, I. Kuwajima, and Co-workers[a]

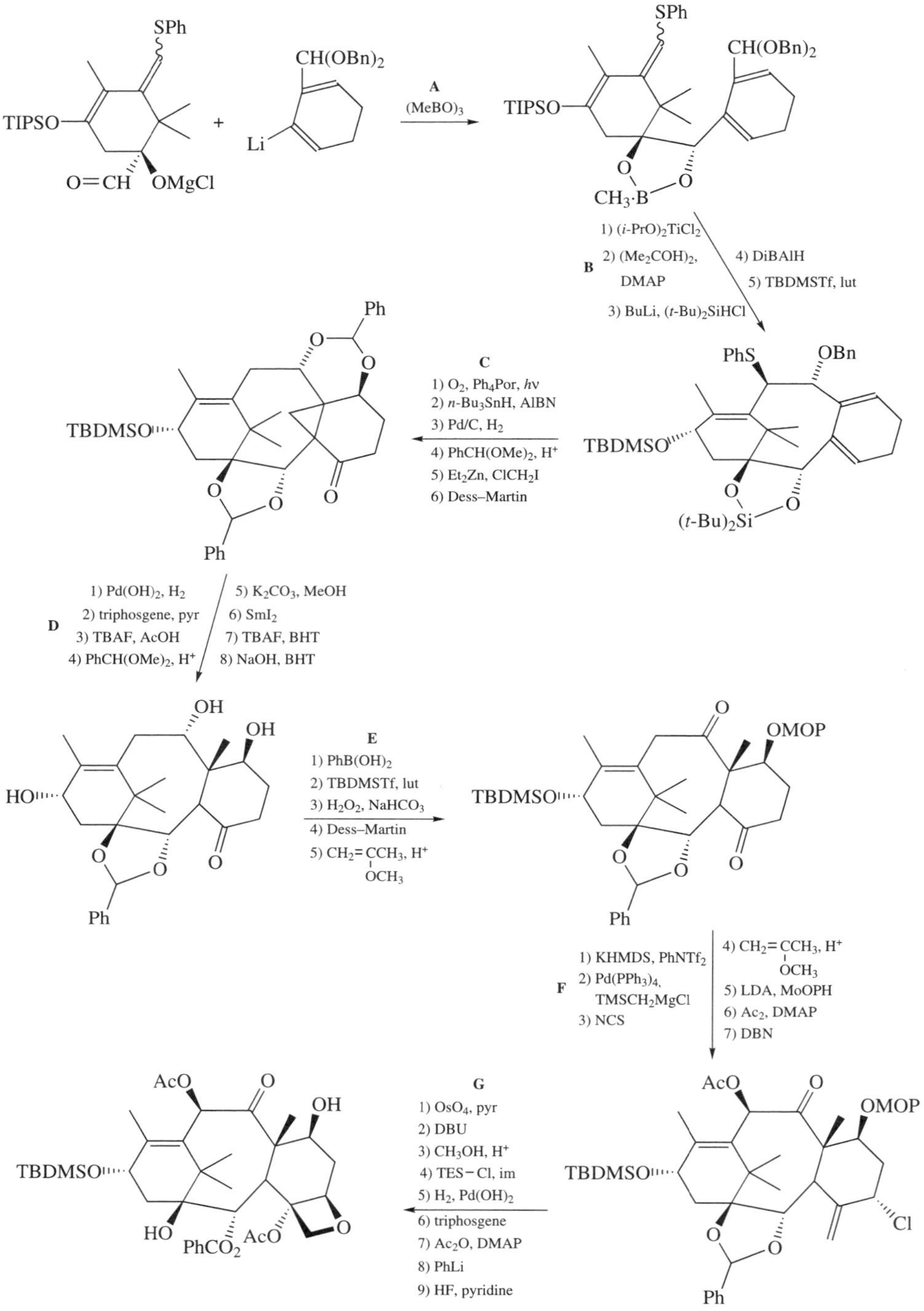

SPh
TIPSO
O=CH OMgCl
+
$CH(OBn)_2$
Li

A
$(MeBO)_3$

SPh
$CH(OBn)_2$
TIPSO
O O
CH_3·B

B
1) $(i\text{-PrO})_2TiCl_2$
2) $(Me_2COH)_2$, DMAP
3) BuLi, $(t\text{-Bu})_2SiHCl$
4) DiBAlH
5) TBDMSTf, lut

PhS OBn
TBDMSO
O O
$(t\text{-Bu})_2Si$

C
1) O_2, Ph_4Por, $h\nu$
2) $n\text{-}Bu_3SnH$, AIBN
3) Pd/C, H_2
4) $PhCH(OMe)_2$, H^+
5) Et_2Zn, $ClCH_2I$
6) Dess–Martin

Ph
O O
TBDMSO
O O O
Ph

D
1) $Pd(OH)_2$, H_2
2) triphosgene, pyr
3) TBAF, AcOH
4) $PhCH(OMe)_2$, H^+
5) K_2CO_3, MeOH
6) SmI_2
7) TBAF, BHT
8) NaOH, BHT

OH
OH
HO
O O O
Ph

E
1) $PhB(OH)_2$
2) TBDMSTf, lut
3) H_2O_2, $NaHCO_3$
4) Dess–Martin
5) $CH_2{=}C(OCH_3)CH_3$, H^+

O
OMOP
TBDMSO
O O O
Ph

F
1) KHMDS, $PhNTf_2$
2) $Pd(PPh_3)_4$, $TMSCH_2MgCl$
3) NCS
4) $CH_2{=}C(OCH_3)CH_3$, H^+
5) LDA, MoOPH
6) Ac_2, DMAP
7) DBN

AcO O
OMOP
TBDMSO
Cl
O O
Ph

G
1) OsO_4, pyr
2) DBU
3) CH_3OH, H^+
4) TES−Cl, im
5) H_2, $Pd(OH)_2$
6) triphosgene
7) Ac_2O, DMAP
8) PhLi
9) HF, pyridine

AcO O
OH
TBDMSO
HO $PhCO_2$ AcO O

a. K. Morihira, R. Hara, S. Kawahara, T. Nishimori, N. Nakamura, H. Kusama, and I. Kuwajima, *J. Am. Chem. Soc.* **120**:12980 (1998).

Scheme 13.48. Epothilone A, Macrolactonization: K. C. Nicolaou and Co-workers[a]

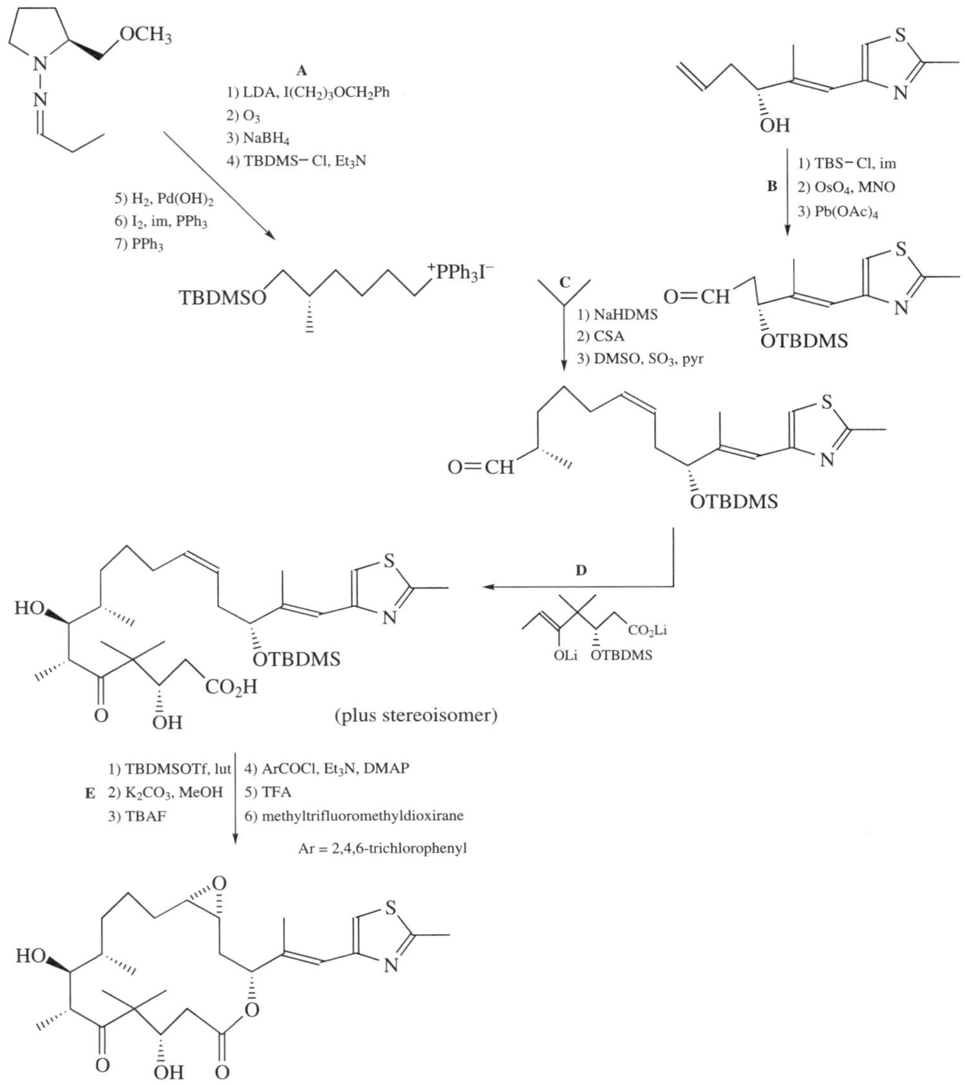

A
1) LDA, $I(CH_2)_3OCH_2Ph$
2) O_3
3) $NaBH_4$
4) TBDMS−Cl, Et_3N
5) H_2, $Pd(OH)_2$
6) I_2, im, PPh_3
7) PPh_3

B
1) TBS−Cl, im
2) OsO_4, MNO
3) $Pb(OAc)_4$

C
1) NaHDMS
2) CSA
3) DMSO, SO_3, pyr

D

(plus stereoisomer)

E
1) TBDMSOTf, lut
2) K_2CO_3, MeOH
3) TBAF
4) ArCOCl, Et_3N, DMAP
5) TFA
6) methyltrifluoromethyldioxirane

Ar = 2,4,6-trichlorophenyl

a. K. C. Nicolaou, F. Sarabia, S. Ninkovic, and Z. Yang, *Angew. Chem. Int. Ed. Engl.* **36:**525 (1997); K. C. Nicolaou, S. Ninkovic, F. Sarabia, D. Vourloumis, Y. He, H. Vallberg, M. R. V. Finlay, and Z. Yang, *J. Am. Chem. Soc.* **119:**7974 (1997).

pure starting materials are employed. These materials are synthesized in steps **A**-1–7 and **B**-1–3, respectively. These are brought together by a Wittig reaction in step **C**. The aldol addition in step **D**-3 is then stereoselective and creates the stereochemistry at C-6 and C-7. The lactonization (step **E**-4) is accomplished by a mixed anhydride (see Section 3.4.1). The final epoxidation is done using 3-methyl-3-trifluoromethyldioxirane.

The second Nicolaou synthesis is shown in Scheme 13.49. Stereoselective aldol additions are used to construct the fragments which are brought together by esterification at step **D**. The synthesis uses an *olefin metathesis* reaction to construct the 16-membered ring (step **E**).

Scheme 13.49. Epothilone A, Olefin Metathesis: K. C. Nicolaou and Co-workers[a]

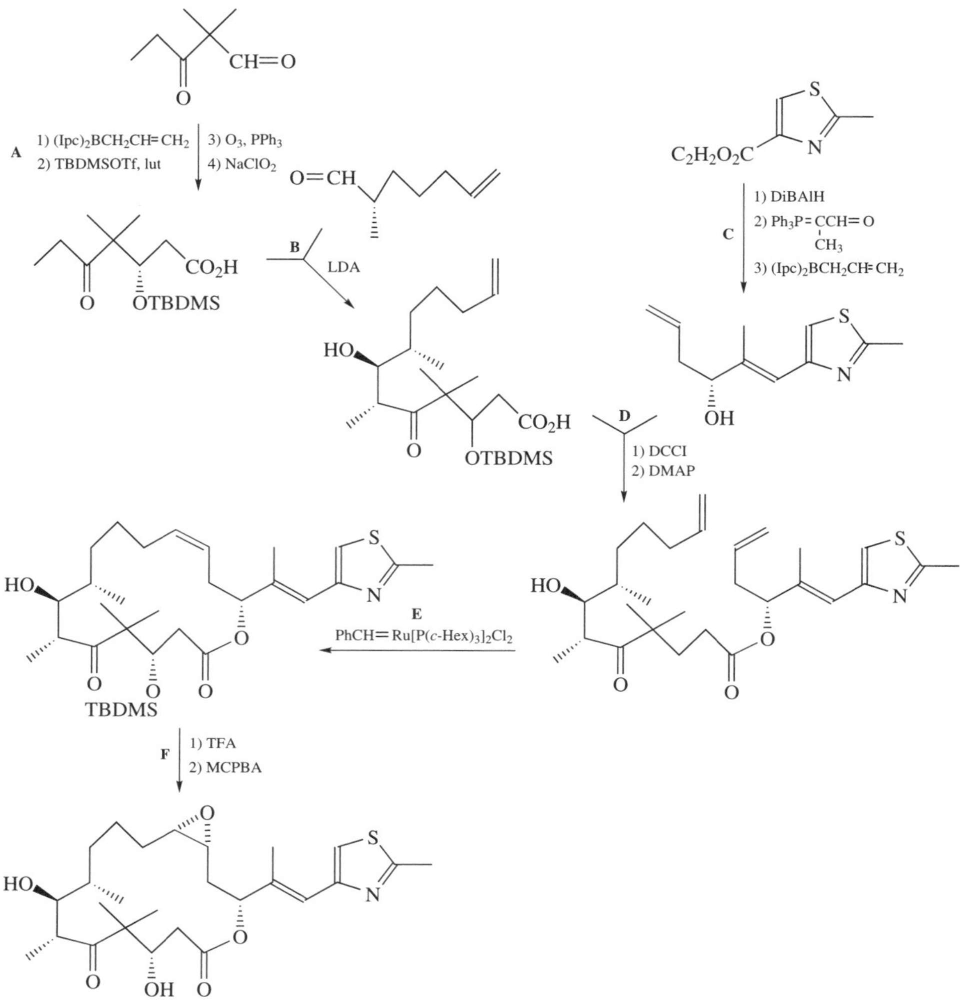

A: 1) $(Ipc)_2BCH_2CH{=}CH_2$ 2) TBDMSOTf, lut 3) O_3, PPh_3 4) $NaClO_2$

B: LDA

C: 1) DiBAlH 2) $Ph_3P{=}C(CH_3)CH{=}O$ 3) $(Ipc)_2BCH_2CH{=}CH_2$

D: 1) DCCI 2) DMAP

E: $PhCH{=}Ru[P(c\text{-Hex})_3]_2Cl_2$

F: 1) TFA 2) MCPBA

a. Z. Yang, Y. He, D. Vourloumis, H. Vallberg, and K. C. Nicolaou, *Angew. Chem. Int. Ed. Engl.* **36:**166 (1997).

The olefin metathesis reaction is also a key feature of the synthesis of epothilone A completed by a group at the Technical University in Braunschweig, Germany. This synthesis, shown in Scheme 13.50, employs a series of stereoselective additions to create the correct stereochemistry. Step **A**-1 uses a stereoselective aldol addition to bring together the two starting materials and also create the stereocenters at C-6 and C-7. This sequence establishes the correct configuration at C-3, C-6, C-7, and C-8. The thiazole ring, along with C(13)–C(15), is added by esterification at step **C**. The ring is closed by olefin metathesis, and the synthesis is completed by deprotection and epoxidation.

a. D. Schinzer, A. Limberg, A. Bauer, O. M. Bohm, and M. Cordes, *Angew. Chem. Int. Ed. Engl.* **36:**523 (1997); D. Schinzer, A. Bauer, O. M. Bohm, A. Limberg, and M. Cordes, *Chem. Eur. J.* **5:**2483 (1999).

The group of Samuel Danishefsky at the Sloan-Kettering Institute for Cancer Research in New York has also been active in the synthesis of the natural epothilones and biologically active analogs. One of these syntheses also uses the olefin metathesis reaction (not shown). The synthesis in Scheme 13.51 uses an alternative approach to create the macrocycle. One of the key steps is a Suzuki coupling carried out at step **H**-(1,2) between a vinylborane and vinyl iodide. The macrocyclization is an aldol addition reaction at step **H**-4. The enolate of the acetate adds to the aldehyde, creating the C(2)–C(3) bond of the macrolactone and also establishing the stereocenter at C-3.

Scheme 13.51. Epothilone, Macroaldol Cyclization: S. J. Danishefsky and Co-workers[a]

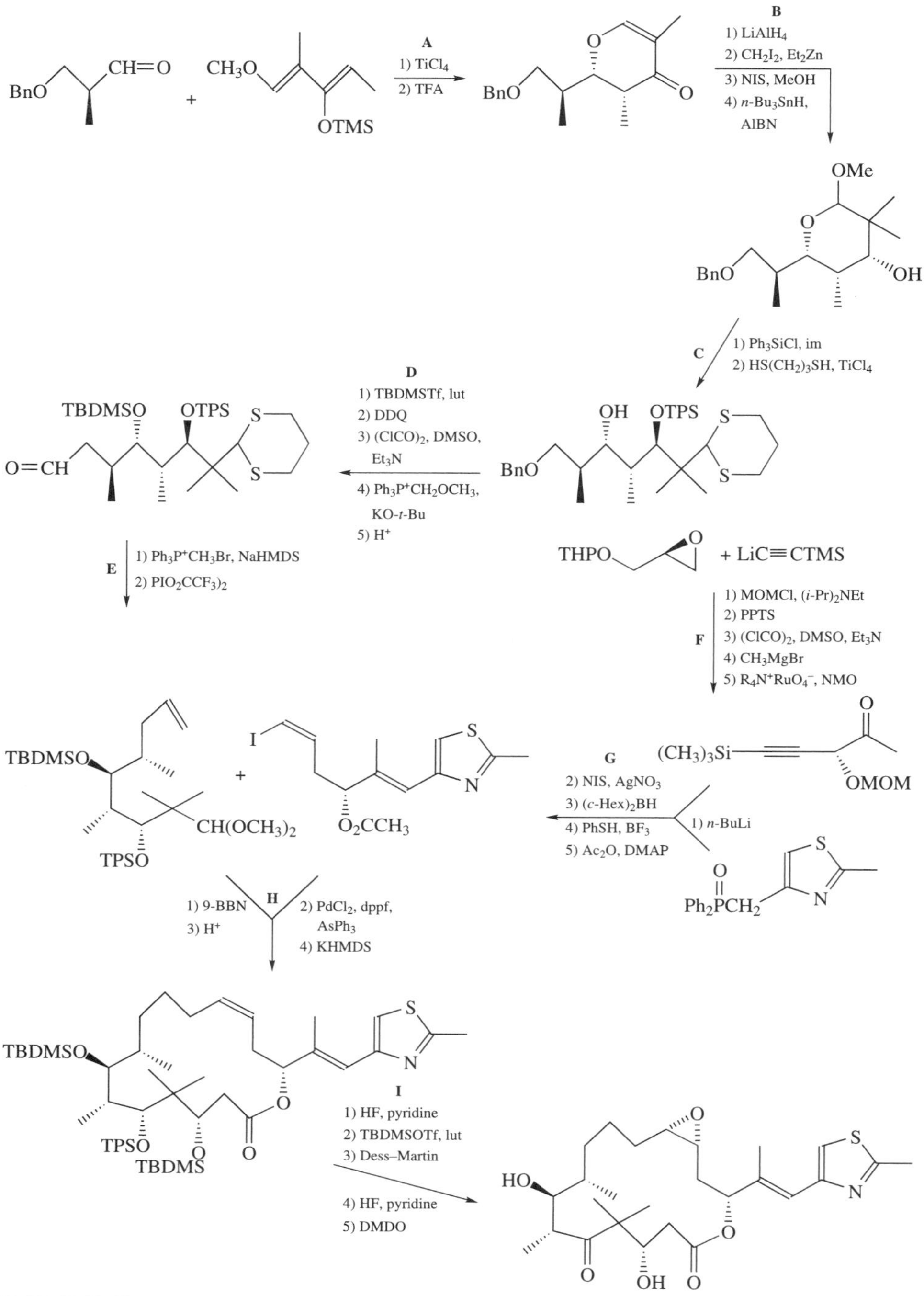

BnO CH=O + CH_3O OTMS

A
1) $TiCl_4$
2) TFA

BnO O O

B
1) $LiAlH_4$
2) CH_2I_2, Et_2Zn
3) NIS, MeOH
4) n-Bu_3SnH, AIBN

OMe O BnO OH

C
1) Ph_3SiCl, im
2) $HS(CH_2)_3SH$, $TiCl_4$

BnO OH OTPS S S

D
1) TBDMSTf, lut
2) DDQ
3) $(ClCO)_2$, DMSO, Et_3N
4) $Ph_3P^+CH_2OCH_3$, KO-t-Bu
5) H^+

O=CH TBDMSO OTPS S S

E
1) $Ph_3P^+CH_3Br$, NaHMDS
2) $PIO_2CCF_3)_2$

THPO O + LiC≡CTMS

F
1) MOMCl, $(i\text{-Pr})_2NEt$
2) PPTS
3) $(ClCO)_2$, DMSO, Et_3N
4) CH_3MgBr
5) $R_4N^+RuO_4^-$, NMO

$(CH_3)_3Si$ O OMOM

G
1) n-BuLi
2) NIS, $AgNO_3$
3) $(c\text{-Hex})_2BH$
4) PhSH, BF_3
5) Ac_2O, DMAP

Ph_2PCH_2 O S N

TBDMSO TPSO $CH(OCH_3)_2$ + I O_2CCH_3 S N

H
1) 9-BBN
2) $PdCl_2$, dppf, $AsPh_3$
3) H^+
4) KHMDS

TBDMSO TPSO O TBDMS O O S N

I
1) HF, pyridine
2) TBDMSOTf, lut
3) Dess–Martin
4) HF, pyridine
5) DMDO

HO O O OH O O S N

a. A. Balog, D. Meng, T. K. Kamenecka, P. Bertinato, D.-S. Su, E. J. Sorensen, and S. J. Danishefsky, *Angew. Chem. Int. Ed. Engl.* **35:**2801 (1996); D. Meng, P. Bertinato, A. Balog, D.-S. Su, T. Kamenecka, E. J. Sorensen, and S. J. Danishefsky, *J. Am. Chem. Soc.* **119:**10073 (1997).

13.6. Solid-Phase Synthesis

One of the most highly developed applications of the systematic use of protective groups is in the synthesis of polypeptides and oligonucleotides. Polypeptides and oligonucleotides consist of linear sequences of individual amino acids and nucleotides, respectively. The ability to synthesize polypeptides and oligonucleotides of known sequence is of great importance for a number of biological applications. Although these molecules can be synthesized by synthetic manipulations in solution, they are now usually synthesized by solid-phase methods, often by automated repetitive cycles of deprotection and coupling.

13.6.1. Solid-Phase Synthesis of Polypeptides

The techniques for automated solid-phase synthesis were first highly developed for polypeptides. Polypeptide synthesis requires the sequential coupling of the individual amino acids.

$$P'O_2C\overset{R^1}{\overset{|}{C}}HNH_2 + HO_2C\overset{R^2}{\overset{|}{C}}HNHP \xrightarrow[R^2]{\text{couple}} P'O_2C\overset{R^1}{\overset{|}{C}}HNH\underset{\underset{O}{\|}}{C}\overset{R^2}{\overset{|}{C}}HNHP \xrightarrow{\text{deprotect}} P'O_2C\overset{R^1}{\overset{|}{C}}HNH\underset{\underset{O}{\|}}{C}\overset{R^2}{\overset{|}{C}}HNH_2$$

(P, P′ = protective groups)

$$P'O_2C\overset{R^1}{\overset{|}{C}}HNH\underset{\underset{O}{\|}}{C}\overset{R^2}{\overset{|}{C}}HNH_2 \xrightarrow[R^3]{\text{couple}} P'O_2C\overset{R^1}{\overset{|}{C}}HNH\underset{\underset{O}{\|}}{C}\overset{R^2}{\overset{|}{C}}HNH\underset{\underset{O}{\|}}{C}\overset{R^3}{\overset{|}{C}}HNHP \xrightarrow[x\text{ times}]{\text{repeat}} P'O_2C\overset{R^1}{\overset{|}{C}}HNH(\underset{\underset{O}{\|}}{C}\overset{R^x}{\overset{|}{C}}HNH)_x\underset{\underset{O}{\|}}{C}\overset{R^n}{\overset{|}{C}}HNHP$$

$$P'O_2C\overset{R^1}{\overset{|}{C}}HNH(\underset{\underset{O}{\|}}{C}\overset{R^x}{\overset{|}{C}}HNH)_x\underset{\underset{O}{\|}}{C}\overset{R^n}{\overset{|}{C}}HNHP \xrightarrow[\text{deprotection}]{\text{terminal}} HO_2C\overset{R^1}{\overset{|}{C}}HNH(\underset{\underset{O}{\|}}{C}\overset{R^x}{\overset{|}{C}}HNH)_x\underset{\underset{O}{\|}}{C}\overset{R^n}{\overset{|}{C}}HNH_2$$

Excellent solution methods involving alternative cycles of deprotection and coupling are available for peptide synthesis.[145] The techniques have been adapted to solid-phase synthesis.[146] The N-protected carboxy terminal amino acid is linked to the solid support, which is usually polystyrene with divinylbenzene cross-linking. The amino group is then deprotected, and the second N-protected amino acid introduced and coupled. The sequence of deprotection and coupling is then continued until the synthesis is complete. Each deprotection and coupling step must go in very high yield. Because of the iterative nature of solid-phase synthesis, errors accumulate throughout the synthesis. For the polypeptide to be of high purity, the conversion must be very efficient at each step.

The first version of solid-phase peptide synthesis (SPPS) to be developed used the *t*-Boc group as the amino-protecting group. It can be cleaved with relatively mild acidic treatment, and TFA is usually used. The original coupling reagent was dicyclohexylcar-

145. M. Bodanszky and A. Bodanszky, *The Practice of Peptide Synthesis*, 2nd ed., Springer-Verlag, Berlin, 1994; V. J. Hruby and J.-P. Mayer, in *Bioorganic Chemistry: Peptides and Proteins*, S. Hecht, ed., Oxford University Press, Oxford, 1998, pp. 27–64.
146. R. B. Merrifield, *Methods. Enzymol.* **289:**3 (1997); K. B. Merrifield, in *Peptides: Synthesis, Structure, and Applications*, B. Gutte, ed., Academic Press, San Diego, 1995, p. 93; Atherton and R. C. Sheppard, *Solid Phase Peptide Synthesis*, IRL Press, Oxford, U.K., 1989; P. Lloyd-Williams, F. Albericio, and E. Giralt, *Chemical Synthesis of Peptides and Proteins*, CRC Press, Boca Raton, Florida, 1997.

Scheme 13.52. *t*-Boc Protocol for Solid-Phase Peptide Synthesis

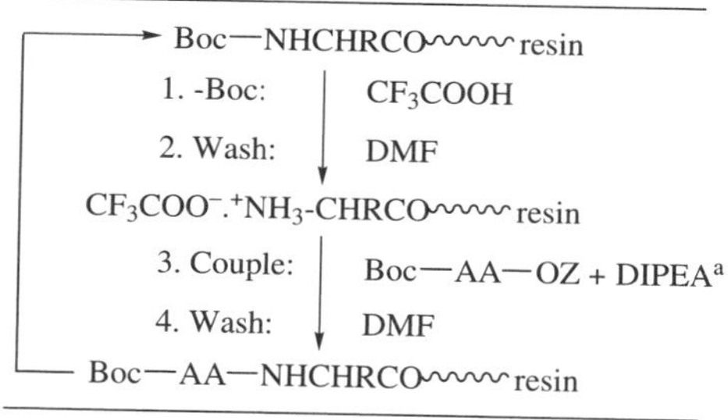

a. AA = amino acid; DIPEA = diisopropylethylamine.

bodiimide. The mechanism of peptide coupling by carbodiimides was discussed in Section 3.4.1. Currently optimized versions of the *t*-Boc protocol can provide polypeptides of 60–80 residues in high purity.[147] SPPS using *t*-Boc protection is outlined in Scheme 13.52.

A second method that has been developed more recently uses the 9-fluorenylmethylcarboxy (Fmoc) group.[148] The Fmoc group is stable to mild acid and to hydrogenation, but it is cleaved by basic reagents through deprotonation at the acidic 9-position of the fluorene ring.

B: H CH$_2$O$_2$CNHCHRCO$_2$P′ → (CH$_2$) + CO$_2$ + H$_2$NCHRCO$_2$P′

The Fmoc protocol for SPPS is outlined in Scheme 13.53.

In both the *t*-Boc and Fmoc versions of SPPS, the amino acids with functional groups in the side chain also require protective groups. These protective groups are designed to stay in place throughout the synthesis and then are removed when the synthesis is complete. The serine and threonine hydroxyl groups can be protected as benzyl ethers. The ϵ-amino group of lysine can be protected as the trifluoroacetyl derivative or as a sulfonamide derivative. The imidazole nitrogen of histidine can also be protected as a sulfonamide. The indole nitrogen of tryptophan is frequently protected as a formyl derivative. The exact choice of protective group depends upon the deprotection–coupling sequence being used.

The original coupling reagents used for SPPS were carbodiimides. In addition to dicyclohexylcarbodiimide (DCCI), *N,N′*-diisopropylcarbodiimide (DIPCDI) is frequently used. At the present time, the coupling is usually done via an activated ester (see Section 3.4.1). The coupling reagent and one of several *N*-hydroxy heterocycles are first allowed to react to form the activated ester, followed by coupling with the deprotected amino group.

147. M. Schnolzer, P. Alewood, A. Jones, D. Alewood, and S. B. H. Kent, *Int. J. Peptide Protein Res.* **40:**180 (1992); M. Schnolzer and S. B. H. Kent, *Science* **256:**221 (1992).

148. L. A. Carpino and G. Y. Han, *J. Org. Chem.* **37:**3404 (1972); G. B. Fields and R. L. Noble, *Int. J. Peptide Protein Res.* **35:**161 (1990); D. A. Wellings and E. Atherton, *Methods Enzymol.* **289:**44 (1997); W. C. Chan and P. D. White, ed., *Fmoc Solid Phase Peptide Synthesis: A Practical Approach*, Oxford University Press, Oxford, 2000.

Scheme 13.53. Fmoc Protocol for Solid-Phase Peptide Synthesis

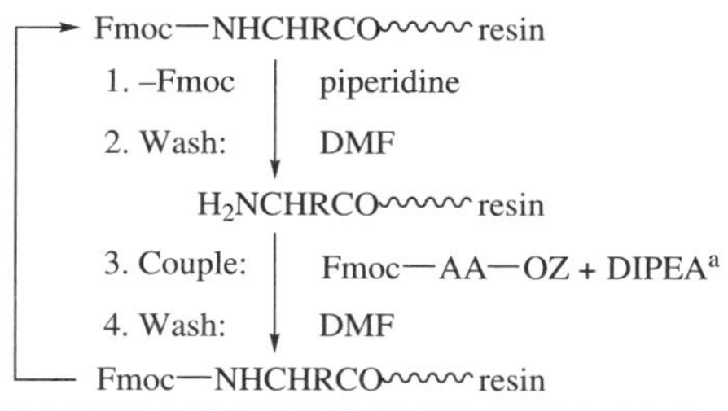

Fmoc—NHCHRCO~~~~resin

1. –Fmoc piperidine
2. Wash: DMF

$H_2NCHRCO$~~~~resin

3. Couple: Fmoc—AA—OZ + DIPEA[a]
4. Wash: DMF

Fmoc—NHCHRCO~~~~resin

a. AA = amino acid.

The most frequently used compounds are *N*-hydroxysuccinimide, 1-hydroxybenzotriazole (HOBt), and 1-hydroxy-7-azabenzotriazole (HOAt).[149]

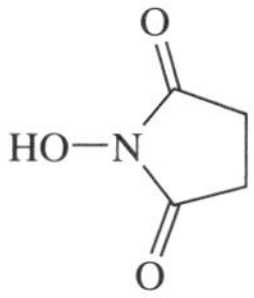

N-hydroxysuccinimide 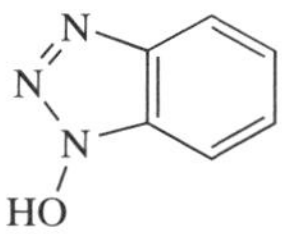 HOBt 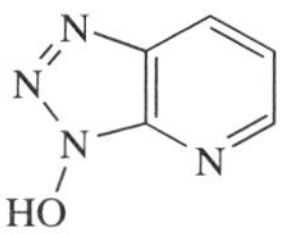 HOAt

Another family of coupling reagents is used frequently with the Fmoc method. These are related to benzotriazole and 7-azabenzotriazole but also incorporate amidinium or phosponium groups. The structures and abbreviations of these reagents are given in Scheme 13.54.

Whereas the original version of SPPS attached the carboxy terminal residue directly to the resin as a benzylic ester using chloromethyl groups attached to the polymer, it is now common to use "linking groups" which are added to the polymer. Two of the more common linking groups are the Wang[150] and the Rink linkers,[151] which are shown below. These groups have the advantage of permitting milder conditions for the final removal of the polypeptide from the solid support.

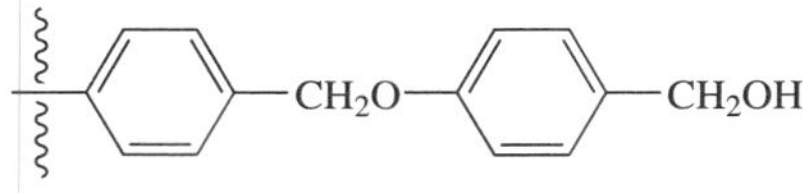

Wang linker 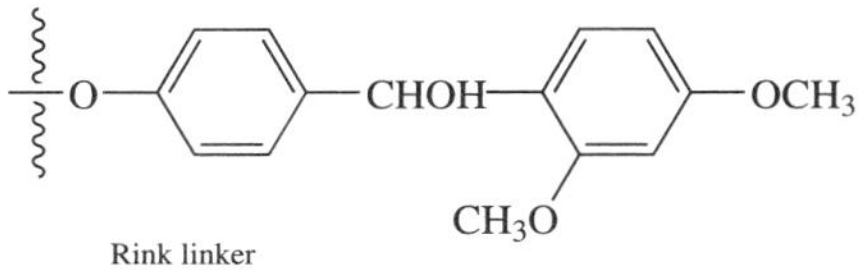 Rink linker

In the *t*-Boc protocol, the most common reagent for final removal of the peptide from the solid support is anhydrous hydrogen fluoride. Although this is a dangerous reagent, commercial systems designed for its safe handling are available. In the Fmoc protocol, milder acid reagents can be used for cleavage from the resin. The alkoxybenzyl ester group

149. F. Albericio and L. A. Carpino, *Methods Enzymol.* **289:**104 (1997).
150. S. Wang, *J. Am. Chem. Soc.* **95:**1328 (1993).
151. H. Rink, *Tetrahedron Lett.* **28:**3787 (1987); M. S. Bernatowicz, S. B. Daniels, and H. Koster, *Tetrahedron Lett.* **30:**4645 (1989); R. S. Garigipati, *Tetrahedron Lett.* **38:**6807 (1997).

Scheme 13.54. Coupling Reagents

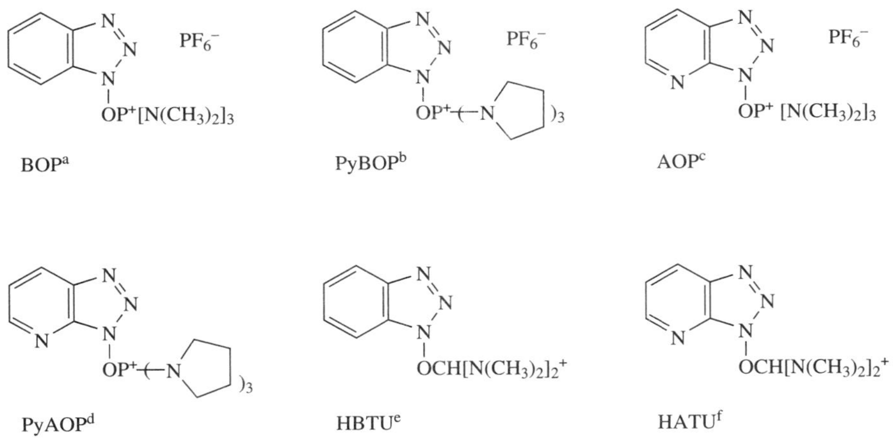

BOP[a] PyBOP[b] AOP[c]

PyAOP[d] HBTU[e] HATU[f]

a. B. Castro, J. R. Dormoy, G. Evin, and C. Selve, *Tetrahedron Lett.* **1975:**1219.
b. J. Coste, D. Le-Nguyen, and B. Castro, *Tetrahedron Lett.* **31:**205 (1990).
c. L. A. Carpino, A. El-Fahan, C. A. Minor, and F. Albericio, *J. Chem. Soc., Chem. Commun.* **1994:**201.
d. F. Albericio, M. Cases, J. Alsina, S. A. Triolo, L. A. Carpino, and S. A. Kates, *Tetrahedron Lett.* **38:**4853 (1997).
e. R. Knorr, A. Trzeciak, W. Barnwarth, and D. Gillessen, *Tetrahedron Lett.* **30:**1927 (1989).
f. L. A. Carpino, *J. Am. Chem. Soc.* **115:**4397 (1993).

at the linker can be cleaved by TFA. Often, a scavenger, such as thioanisole, is used to capture carbocations formed by cleavage of *t*-Boc protecting groups from side-chain substituents.

13.6.2. Solid-Phase Synthesis of Oligonucleotides

Synthetic oligonucleotides are very important tools in the study and manipulation of DNA, including such techniques as site-directed mutagenesis and DNA amplification by the polymerase chain reaction (PCR). The techniques for chemical synthesis of oligonucleotides have been highly developed. Very efficient and automated methodologies based on synthesis on a solid support are used widely in fields that depend on the availability of defined DNA sequences.[152]

The construction of oligonucleotides proceeds from the four nucleotides by formation of new phosphorus–oxygen bonds. The potentially interfering nucleophilic sites on the nucleotide bases are protected. The benzoyl group is usually used for the 6-amino group of adenine and the 4-amino group of cytosine, whereas the *i*-butyroyl group is used for the 2-amino group of guanine. These amides are cleaved by ammonia after the synthesis is completed. The nucleotides are protected at the 5′-hydroxyl group, usually as the 4,4′-dimethoxytrityl (DMT) group.

In the early solution-phase synthesis of oligonucleotides, coupling of phosphate diesters was used. A mixed 3′-ester with one aryl substituent, usually *o*-chlorophenyl, was coupled with a deprotected 5′-OH. The coupling reagents used were sulfonyl halides, particularly 2,4,6-triisopropylbenzenesulfonyl chloride.[153] The reactions proceed by

152. S. L. Beaucage and M. H. Caruthers, in *Bioorganic Chemistry: Nucleic Acids*, S. M. Hecht, ed., Oxford University Press, Oxford, 1996, pp. 36–74.
153. C. B. Reese, *Tetrahedron* **34:**3143 (1978).

formation of reactive sulfonate esters. Coupling conditions have subsequently been improved, and a particularly effective coupling reagent is 1-mesitylenesulfonyl-3-nitrotriazole (MSNT).[154]

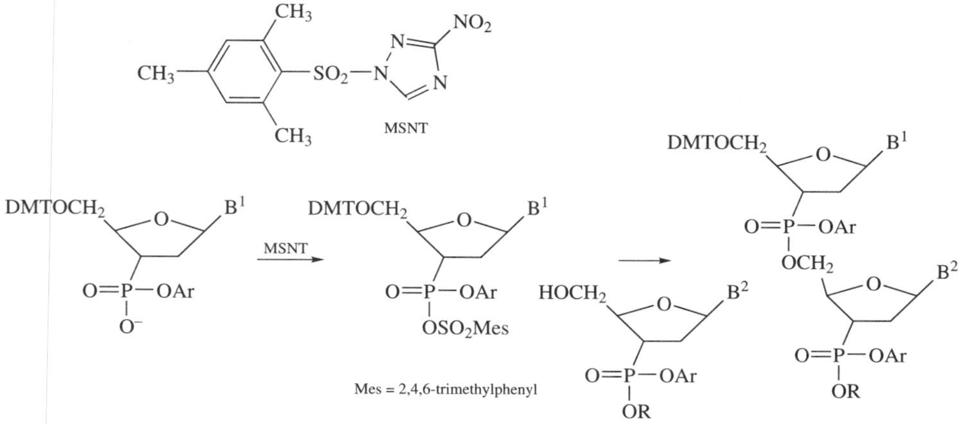

MSNT

Mes = 2,4,6-trimethylphenyl

Current solid-phase synthesis of oligonucleotides relies on coupling at the phosphite oxidation level. The individual nucleotides are introduced as phosphoramidites, and the method is called the phosphoramidite method.[155] The N,N-diisopropyl phosphoramidites are usually used. The third phosphorus substituent is methoxy or 2-cyanoethoxy. The cyanoethyl group is easily removed by mild base (β-elimination) after completion of the synthesis. The coupling is accomplished by tetrazole, which displaces the amine substituent to form a reactive phosphite that undergoes coupling. After coupling, the phosphorus is oxidized to the phosphoryl level by iodine or another oxidant. The most commonly used protective group for the 5′-OH is DMT, which is removed by mild acid. The typical cycle of deprotection, coupling, and oxidation is outlined in Scheme 13.55. One feature of oligonucleotide synthesis is the use of a capping step. This is an acetylation which follows coupling. The purpose of capping is to permanently block any 5′-OH groups that were not successfully coupled. This prevents the addition of a nucleotide at the site in the succeeding cycle and terminates the further growth of this particular oligonucleotide and prevents the synthesis of oligonucleotides with single base deletions. The capped oligomers can be more easily removed during purification.

Silica or porous glass is usually used as the solid phase in oligonucleotide synthesis. The support is functionalized through an amino group attached to the silica surface. There is a secondary linkage through a succinate ester to the terminal 3′-OH group.

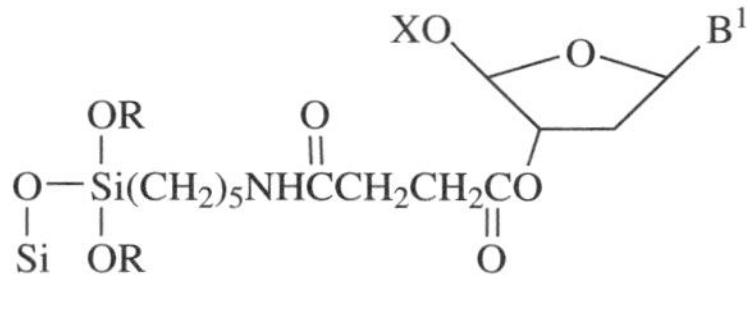

154. J. B. Chattopadhyara and C. B. Reese, *Tetrahedron Lett.* **20:**5059 (1979).

155. R. L. Letsinger and W. B. Lunsford, *J. Am. Chem. Soc.* **98:**3655 (1976); S. L. Beaucage and M. H. Caruthers, *Tetrahedron Lett.* **22:**1859 (1981); M. H. Caruthers, *J. Chem. Ed.* **66:**577 (1989); S. L. Beaucage and R. P. Iyer, *Tetrahedron* **48:**2223 (1992); S. L. Beaucage and M. Caruthers, in *Bioorganic Chemistry: Nucleic Acids*, S. M. Hecht, ed., Oxford University Press, New York, 1996, pp. 36–74.

Scheme 13.55. Automated Solid-Phase Synthesis at Oligonucleotides[a]

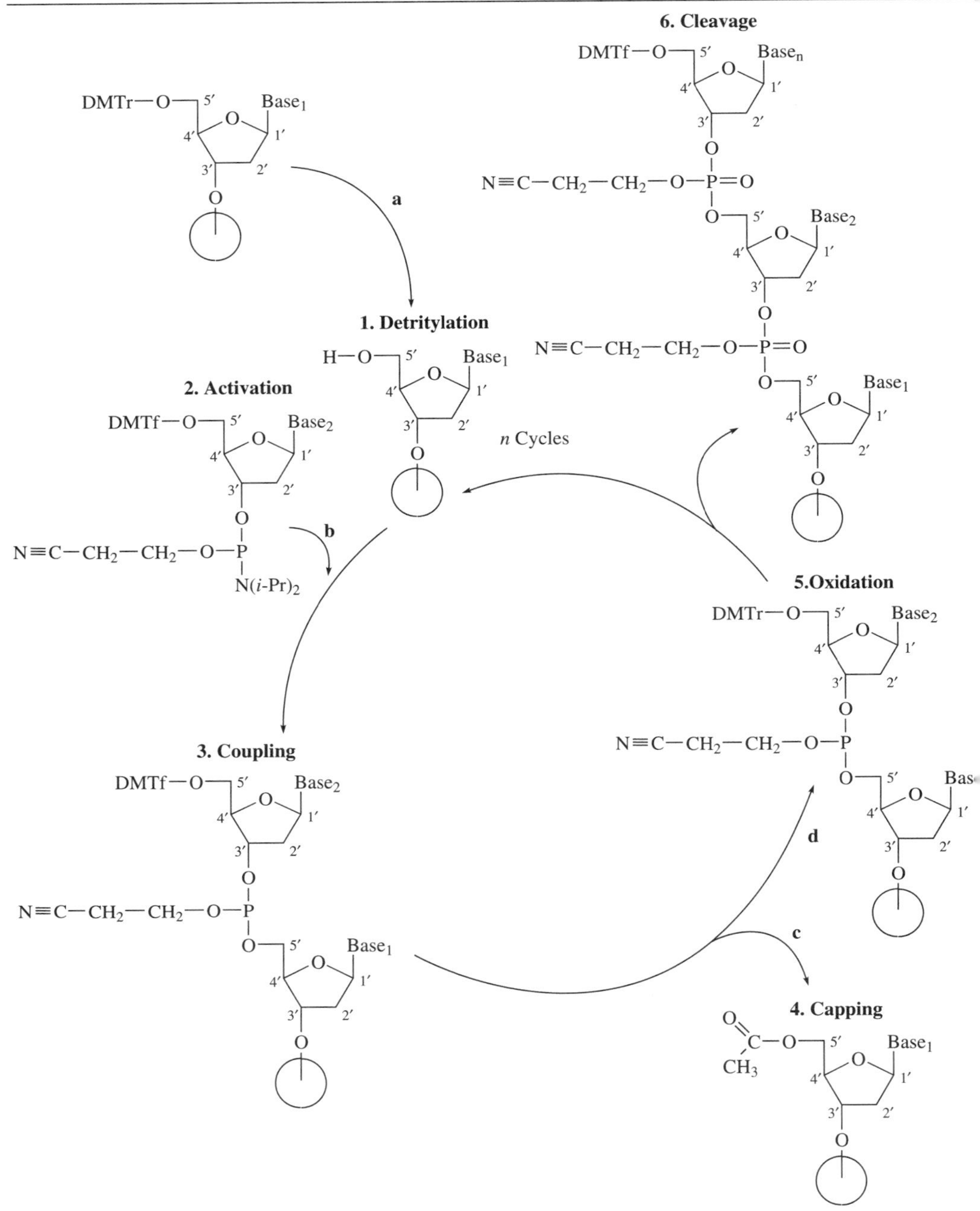

Reagents: **a**, 97% dichloromethane, 3% trichloroacetic acid; **b**, 97% acetonitrile, 3% tetrazole; **c**$_1$, 80% tetrahydrofuran, 10% acetic anhydri 10% 2,6-lutidine; **c**$_2$, 93% tetrahydrofuran, 7% 1-methylimidazole; **d**, 93% tetrahydrofuran, 3% iodine, 2% water, 2% pyridine.

a. G. A. Urbina, G. Grübler, A. Weiler, H. Echner, S. Stoeva, J. Schernthaler, W. Grass, and W. Voelter, *Z. Naturforsch. B* **B53:**1051 (199

While use of automated oligonucleotide synthesis is widespread, work continues on the optimization of protective groups, coupling conditions, and deprotection methods, as well as on the automated devices.[156]

13.7. Combinatorial Synthesis

Over the past decade, the techniques of *combinatorial synthesis* have received much attention. Solid-phase synthesis of polypeptides and oligonucleotides is particularly adaptable to combinatorial synthesis, but the method is not limited to these fields. The goal of combinatorial synthesis is to prepare a large number of related molecules by carrying out a synthetic sequence with several closely related starting materials. For example, if a three-step sequence is done with eight related reactants at each step, a total of 4096 different products are obtained.

$$\text{A (8) + B (8)} \xrightarrow{\text{Step 1}} \text{A–B (64)} \xrightarrow[\text{C (8)}]{\text{Step 2}} \text{A–B–C (512)} \xrightarrow[\text{D (8)}]{\text{Step 3}} \text{A–B–C–D (4096)}$$

Whereas the objective of traditional multistep synthesis is the preparation of a single pure compound, combinatorial synthesis is designed to make many closely related molecules.[157] The purpose is often to have a large collection (library) of related compounds for evaluation of biological activity. In the paragraphs which follow, we will consider examples of application of combinatorial methods to several kinds of compounds.

One approach to combinatorial synthesis is to carry out a series of conventional reactions in parallel with one another. For example, a matrix of 6 starting materials, each with 8 different reactants would generate 48 reaction products. By splitting each reaction mixture and using a different reactant for each portion, further expansion of the number of final compounds is achieved. However, relatively little savings in effort in isolation and purification of products is obtained by running the reactions in parallel, since each product must be separately isolated and purified. The reaction sequence below was used to create a 48-component library by reacting 6 amines with each of 8 epoxides. Several specific approaches were used to improve the purity of the product and maximize the efficiency of the process. First, the amines were monosilylated to minimize the potential for interference from dialkylation of the amine. The purification process was also chosen to improve efficiency. Because the desired products are basic, they are retained by acidic ion-exchange resins. The products were adsorbed on the resin, and nonbasic impurities were washed out, followed by elution of the products by methanolic ammonia.[158]

$$R_2C(R_1)(R_3)\text{–}NH_2 \xrightarrow{(CH_3)_3SiX} R_2C(R_1)(R_3)\text{–}NHSi(CH_3)_3 \xrightarrow{\text{epoxide-}R_4} \xrightarrow[\text{resin, followed by elution with methanolic ammonia}]{\text{purified by adsorption on sulfonic acid ion exchange}} R^2C(R^1)(R^3)\text{–}NHCH_2CH(OH)R^4$$

156. G. A. Urbina, G. Grubler, A. Weiber, H. Echner, S. Stoeva, J. Schernthaner, W. Gross, and W. Voelter, *Z. Naturforsch.* **B53:**1051 (1998); S. Rayner, S. Brignac, R. Bumeiester, Y. Belosludtsev, T. Ward, O. Grant, K. O'Brien, G. A. Evans, and H. R. Garner, *Genome Res.* **8:**741 (1998).
157. A. Furka, *Drug Dev. Res.* **36:**1 (1995).
158. A. J. Shuker, M. G. Siegel, D. P. Matthews, and L. O. Weigel, *Tetrahedron Lett.* **38:**6149 (1997).

A considerable improvement in efficiency can be achieved by solid-phase synthesis.[159] The first reactant is attached to a solid support through a linker group, as was described for polypeptide and oligonucleotide synthesis. The individual reaction steps are then conducted on the polymer-bound material. Use of solid-phase methodology has several advantages. Excess reagents can be used to drive individual steps to completion and obtain high yields. The purification after each step is also simplified, because reagents are simply rinsed from the solid support. The process can be automated, greatly reducing the manual effort required.

When the solid-phase method is combined with the sample-splitting method, there is a particularly useful outcome.[160] The solid support can be used in the form of small beads, and the starting point is a collection of beads, each with one initial starting material. After each reaction step, the beads are recombined and split again. As the collection of beads is split and recombined during the combinatorial synthesis, each bead acquires a particular compound, depending on its history of exposure to the reagents, *but all the beads in a particular split have the same compound, since their reaction history is identical*. Scheme 13.56 illustrates this approach for three steps, each using three different reactants. However, in the end, all of the beads are together, and there must be some way of establishing the identity of the compound attached to any particular bead. In some cases, it is possible to detect compounds with the desired property while they are still attached to the bead. This is true for some assays of biological or catalytic activity that can be performed under heterogeneous conditions.

Another method is to tag the beads with identifying markers that encode the sequence of reactants and thus the structure of the product attached to a particular bead.[161] One method of coding involves attachment of a chemically identifiable tag.[162] After each combinatorial step, a different chemical tag is applied to each of the splits before they are recombined. The tags used for this approach are a series of chlorinated aromatic ethers which can be detected and identified by mass spectrometry. The tags are attached to the polymer support by a Rh-catalyzed carbenoid insertion reaction. Detachment is done by oxidizing the methoxyphenyl linker with ceric ammonium nitrate. Any split which shows interesting biological activity can then be identified by analyzing the code provided by the chemical tags for that particular split.

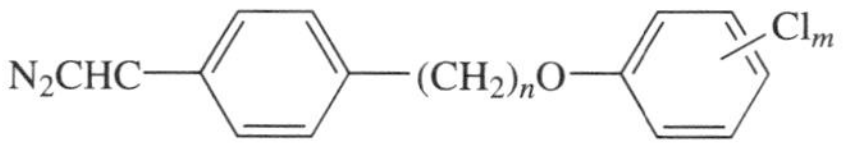

$N_2CHC-C_6H_4-(CH_2)_nO-C_6H_{5-m}Cl_m$

$n = 2–11$
$m = 2–5$

Scheme 13.57 illustrates the concept of the tagging method.

Combinatorial approaches can be applied to the synthesis of any kind of molecule that can be built up from a sequence of individual components, for example, in reactions forming heterocyclic rings.[163] The equations below represent an approach to preparing

159. A. R. Brown, P. H. H. Hermkens, H. C. J. Ottenheijm, and D. C. Rees, *Synlett:* **1998:**817.
160. A. Furka, F. Sebestyen, M. Asgedon, and G. Dibo, *Int. J. Peptide Protein Res.* **37:**487 (1991); K. S. Lam, M. Lebl, and V. Krchnak, *Chem. Rev.* **97:**411 (1997).
161. S. Brenner and R. A. Lerner, *Proc. Natl. Acad. Sci. U.S.A.* **89:**5381 (1993).
162. H. P. Nestler, P. A. Bartlett, and W. C. Still, *J. Org. Chem.* **59:**4723 (1994); W. C. Still, *Acc. Chem. Res.* **29:** 155 (1996). C. Barnes, R. H. Scott, and S. Balasubramanian, *Recent Res. Dev. Org. Chem.* **2:**367 (1998).
163. A. Netzi, J. M. Ostresh, and R. A. Houghten, *Chem. Rev.* **97:**449 (1997).

Scheme 13.56. Splitting Method for Combinatorial Synthesis on Solid Support[a]

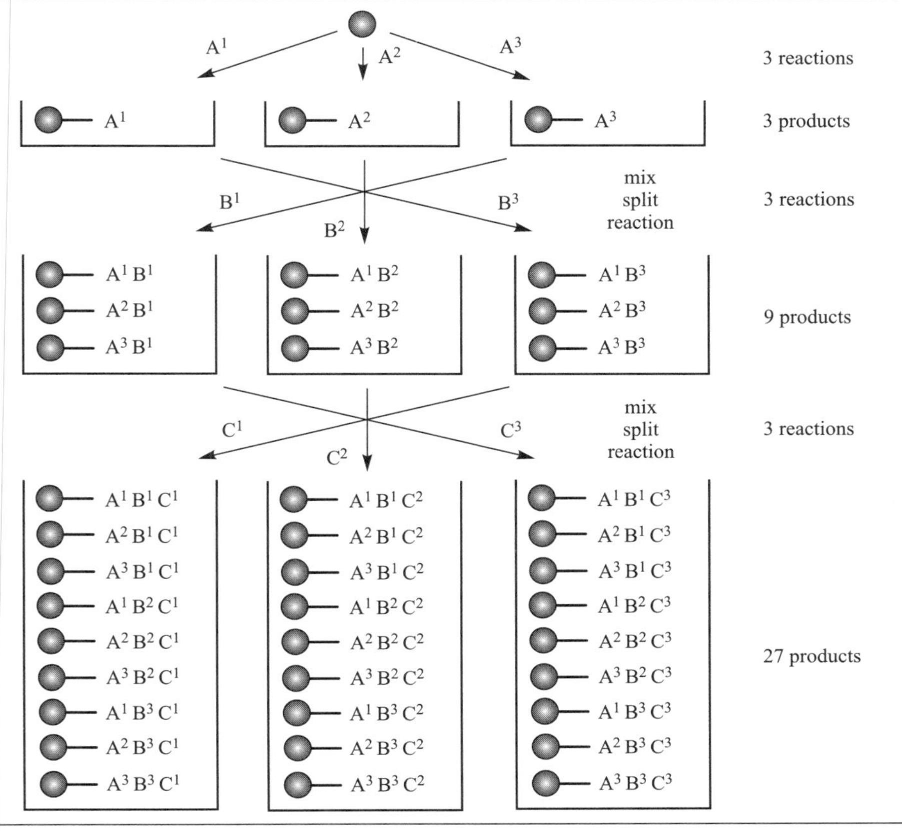

[a] Reproduced from F. Balkenhohl, C. von dem Bussche-Hünnefeld, A. Lansky and C. Zechel *Angew. Chem. Int. Ed. Engl*, **35:** 2288 (1996) with permission of VCH-Wiley Publishers.

differentially substituted indoles.

$$\text{Ⓟ}-NHCOCH{=}CHCH_2Br \xrightarrow[(i\text{-}Pr)_2NEt,\ DMF]{\text{X-2-iodoaniline}} \text{Ⓟ}-NHCOCH{=}CHCH_2NH(C_6H_3(I)X)$$

$$\xrightarrow[(i\text{-}Pr)_2NEt,\ DMF]{\text{Y-}C_6H_4CH_2Br} \text{Ⓟ}-NHCOCH{=}CHCH_2N(CH_2C_6H_4Y)(C_6H_3(I)X) \qquad \text{Ref. 164}$$

$$\xrightarrow[2)\ TFA,\ CH_2Cl_2]{1)\ Pd\ catalyst} \text{X-indole-3-}CH_2CONH_2\text{, N-}CH_2C_6H_4Y \qquad \text{Ref. 165}$$

164. H.-C. Zhang and B. E. Maryanoff, *J. Org. Chem.* **62:**1804 (1997).
165. H.-C. Zhang, K. K. Brumfield, and B. E. Maryanoff, *Tetrahedron Lett.* **38:**2439 (1997).

Scheme 13.57. Use of Chemical Tags to Encode Sequence in Combinatorial Synthesis on Solid Support[a]

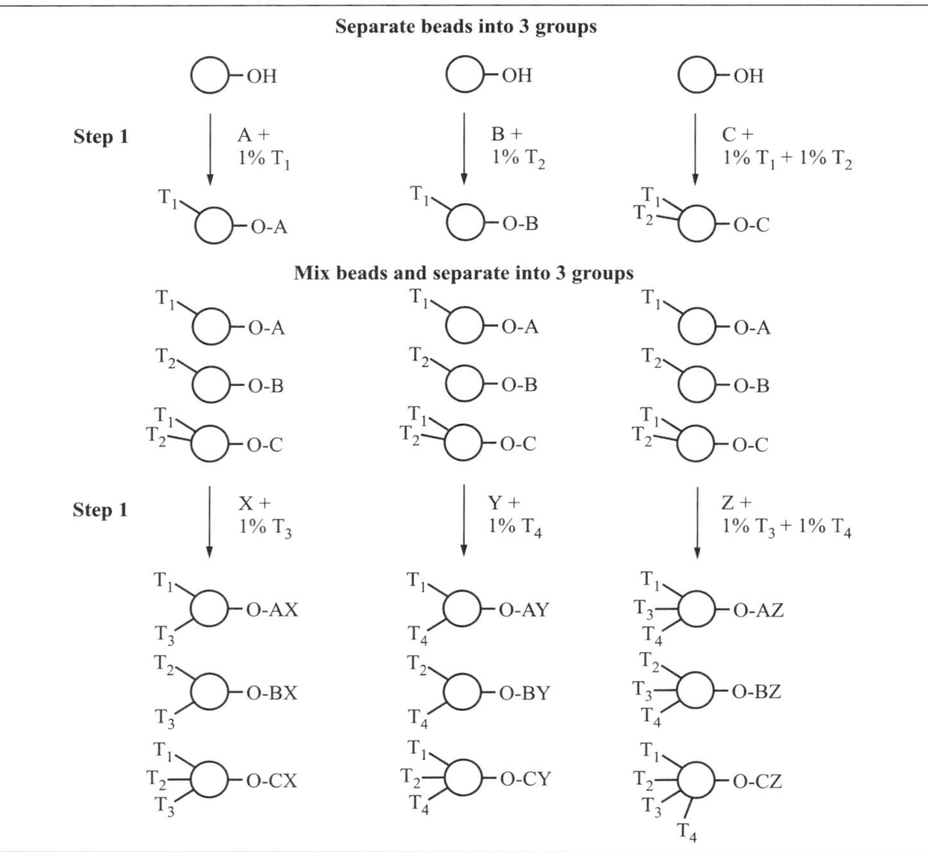

[a] Reproduced from W. C. Still, *Acc. Chem. Res.*, **29:** 155 (1996) by permision of the American Chemical Society.

There is nothing to prevent continuation to incorporate additional diversity by continuing to build on a side chain at one of the substituent sites.

Another kind of combinatorial synthesis can be applied to reactions that assemble the product from several components in a single step, a *multicomponent reaction*. A particularly interesting four-component reaction is the *Ugi reaction*, which generates dipeptides from an α-amino acid, an isocyanide, an aldehyde, an amine, and a carboxylic acid. Use of 10 different isocyanides and amines, along with 40 different aldehydes and carboxylic acids, has the potential to generate 160,000 different dipeptide products:

$$\underset{(10)}{R^1N{\equiv}C} + \underset{(40)}{R^2CH{=}O} + \underset{(10)}{R^3NH_2} + \underset{(40)}{R^4CO_2H} \longrightarrow \underset{(160.000)}{R^1NHC(=O)CH(R^2)N(R_3)C(=O)R^4}$$

In one study,[166] this system was explored by synthesizing arbitrarily chosen sets of 20 compounds in parallel. The biological assay data from these 20 combinations were then

166. L. Weber, S. Wallbaum, C. Broger, and K. Gubernator, *Angew. Chem. Int. Ed. Engl.* **34:**2280 (1995).

used to select the next 20 combinations for synthesis. The synthesis–assay–selection process was repeated 20 times. The data from the final biological assays yielded an average inhibitory concentration of $< 1\ \mu M$ for the final set of 20 dipeptide products, as compared to 1 mM for the 20 initial products.

The epothilone synthesis in Scheme 13.49 has been used as the basis for a combinatorial approach to epothilone analogs.[167] The acyclic precursors were synthesized and attached to a solid support resin by steps **A–E** in Scheme 13.58. The cyclization and disconnection from the resin were then done by the olefin metathesis reaction. The aldol condensation in step **D** is not highly stereoselective. Similarly, olefin metathesis gives a mixture of *E*- and *Z*-stereoisomers so that the product of each combinatorial sequence is a mixture of four isomers. These were separated by thin-layer chromatography prior to bioassay. In this project, reactants **A** (3 variations), **B** (3 variations), and **C** (5 variations) were used, generating 45 possible combinations. The stereoisomeric products increase this to 180 (45×4).

In this study, a nonchemical means of encoding the identity of each compound was used. The original polymer-bound reagent was placed in a porous microreactor equipped with a radio-frequency device that can be used for identification.[168] The porous micro-

Scheme 13.58. Combinatorial Synthesis of Epothilone Analogs Using Microreactors[a]

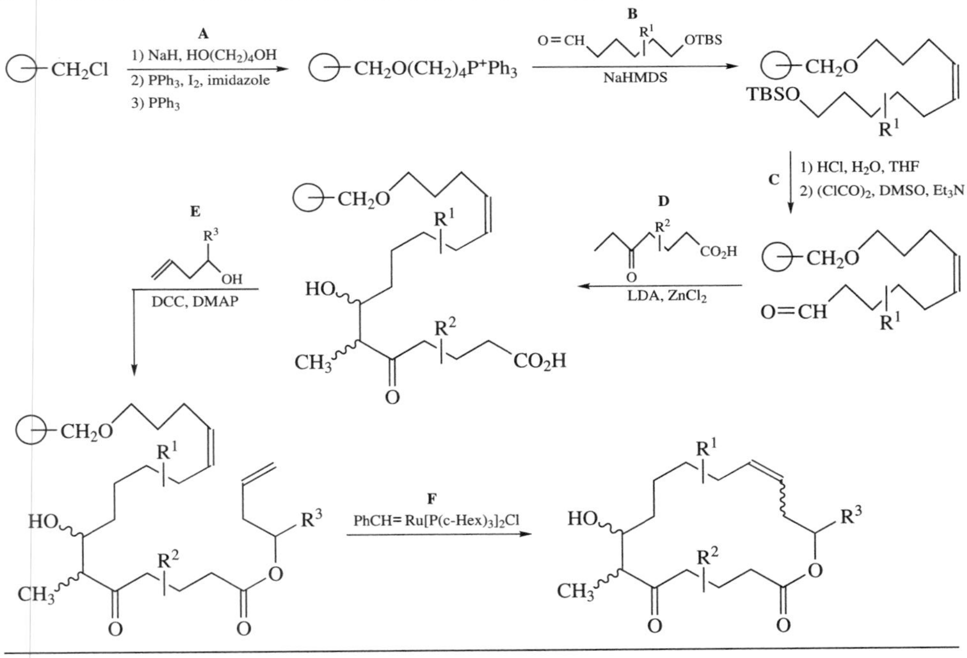

a. K. C. Nicolaou, D. Vourloumis, T. Li, J. Pastor, N. Winssinger, Y. He, S. Ninkovic, F. Sarabia, H. Vallberg, F. Roschangar, N. P. King, M. R. V. Finlay, P. Giannakakou, D. Verdier-Pinard, and E. Hamel, *Angew. Chem. Int. Ed. Engl.* **36**:2097 (1997).

167. K. C. Nicolaou, N. Winssinger, J. Pastor, S. Ninkovic, F. Sarabia, Y. He, D. Vourloumis, Z. Yang, T. Li, P. Giannakakou, and E. Hamel, *Nature* **387**:268 (1997); K. C. Nicolaou, D. Vourloumis, T. Li, J. Pastor, N. Winssinger, Y. He, S. Ninkovic, F. Sarabia, H. Vallberg, F. Roschangar, N. P. King, M. R. V. Finlay, P. Giannakakou, D. Verdier-Pinard, and E. Hamel, *Angew. Chem. Int. Ed. Engl.* **36**:2097 (1997).
168. K. C. Nicolaou, Y.-Y. Xiao, Z. Parandoosh, A. Senyei, and M. P. Nova, *Angew. Chem. Int. Ed. Engl.* **34**:2289 (1995); E. J. Moran, S. Sarshar, J. F. Cargill, M. M. Shahbaz, A. Lio, A. M. M. Mjalli, and R. W. Armstrong, *J. Am. Chem. Soc.* **117**:10787 (1995).

Scheme 13.59. Radio-Frequency Tagging of Microreactors for Combinatorial Synthesis on a Solid Support

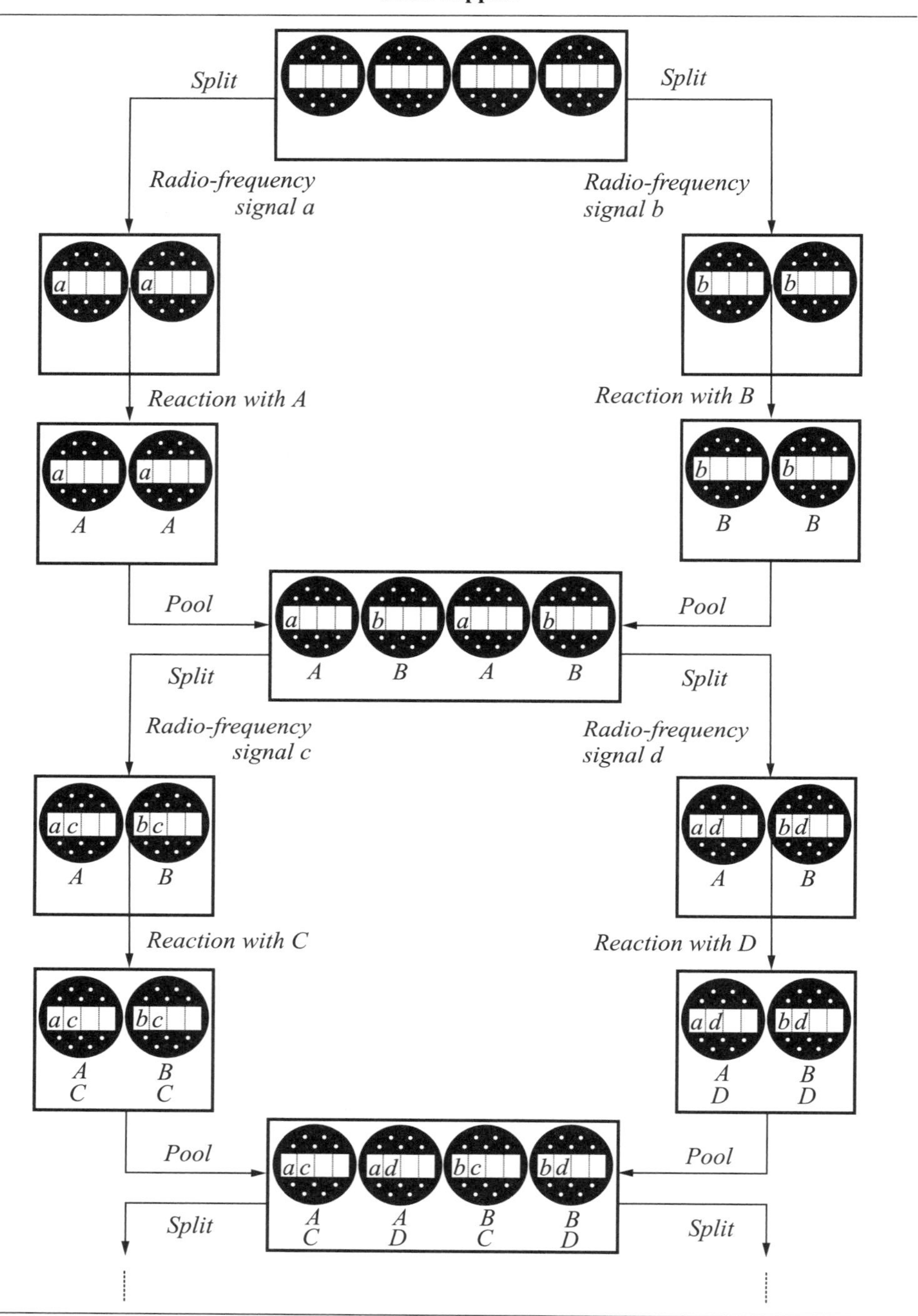

a. Reproduced from K. C. Nicolaou, X.-Y. Xiao, Z. Parandoosh, A. Senyei, and M. P. Nova, *Angew. Chem. Int. Ed. Engl.*, **34:** 2289 (1995) with permission of VCH-Wiley Publishers.

reactors permit reagents to diffuse into contact with the polymer-bound reactants, but the polymer cannot diffuse out. At each split, the individual microreactors are coded to identify the reagent that is used. When the synthesis is complete, the sequence of signals recorded in the radio-frequency device identifies the product that has been assembled in that particular reactor. Scheme 13.59 illustrates the principle of this coding method.

General References

Protective Groups

T. Greene and P. G. M. Wuts, *Protective Groups in Organic Synthesis*, 3rd ed., John Wiley & Sons, New York, 1999.
P. J. Kocienski, *Protecting Groups*, G. Thieme, Stuttgart, 1994.
J. F. W. McOmie, ed., *Protective Groups in Organic Synthesis*, Plenum Publishers, New York, 1973.

Synthetic Equivalents

T. A. Hase, ed., *Umpoled Synthons: A Survey of Sources and Uses in Synthesis*, John Wiley & Sons, New York, 1987.
A. Dondoni, ed., *Advances in the Use of Synthons in Organic Chemistry*, Vols. 1–3. JAI Press, Greenwich, Connecticut, 1993–1995.
D. Seebach, *Angew. Chem. Int. Ed. Engl.* **18:**239 (1979).

Synthetic Analysis and Planning

R. K. Bansal, *Synthetic Approaches to Organic Chemistry*, Jones and Bartlett, Sudbury, Massachusetts, 1998.
E. J. Corey and X.-M Chang, *The Logic of Chemical Synthesis*, John Wiley & Sons, New York, 1989.
J.-H. Furhop and G. Penzlin, *Organic Synthesis: Concepts, Methods, and Starting Materials*, Verlag Chemie, Weinheim, 1983.
T.-L. Ho, *Tactics of Organic Synthesis*, John Wiley & Sons, New York, 1994.
T.-L. Ho, *Tandem Organic Reactions*, John Wiley & Sons, New York, 1992.
T. Mukaiyama, *Challenges in Synthetic Organic Chemistry*, Clarendon Press, Oxford, 1990.
F. Serratosa and J. Xicart, *Organic Chemistry in Action: The Design of Organic Synthesis*, Elsevier, New York, 1996.
W. A. Smit, A. F. Bochkov, and R. Caple, *Organic Synthesis: The Science behind the Art*, Royal Society of Chemistry, Cambridge, U.K., 1998.
B. M. Trost, editor-in-chief, *Comprehensive Organic Synthesis: Selectivity, Strategy, and Efficiency in Modern Organic Chemistry*, Pergamon Press, New York, 1991.
S. Warren, *Organic Synthesis: The Disconnection Approach*, John Wiley & Sons, New York, 1982.

Stereoselective Synthesis

R. S. Atkinson, *Stereoselective Synthesis*, John Wiley & Sons, New York, 1995.
G. M. Coppola and H. F. Schuster, *Asymmetric Synthesis: Construction of Chiral Molecules Using Amino Acids*, Wiley-Interscience, New York, 1987.
S. Hanessian, *Total Synthesis of Natural Products, the Chiron Approach*, Pergamon Press, New York, 1983.
S. Nogradi, *Stereoselective Syntheses*, Verlag Chemie, Weinheim, 1987.
G. Procter, *Stereoselectivity in Organic Synthesis*, Oxford University Press, Oxford, 1998.

Descriptions of Total Syntheses

N. Anand, J. S. Bindra, and S. Ranganathan, *Art in Organic Synthesis*, 2nd ed., Wiley-Interscience, New York, 1988.

J. ApSimon, ed., *The Total Synthesis of Natural Products*, Vols. 1–9, Wiley-Interscience, New York, 1973–1992.

S. Danishefsky and S. E. Danishefsky, *Progress in Total Synthesis*, Meredith, New York, 1971.

I. Fleming, *Selected Organic Syntheses*, Wiley-Interscience, New York, 1973.

K. C. Nicolaou and E. J. Sorensen, *Classics in Total Synthesis: Targets, Strategies and Methods*, VCH Publishers, New York, 1996.

Solid-Phase Synthesis

K. Burgess, *Solid Phase Organic Synthesis*, John Wiley & Sons, New York, 2000.

Problems

(*References for these problems will be found on page 942.*)

1. Indicate conditions which would be appropriate for the following transformations involving introduction or removal of protecting groups.

(a) CH_3, CH_3, OH, $(CH_3)_2CH$ → CH_3, CH_3, $OCH_2OCH_2CH_2OCH_3$, $(CH_3)_2CH$

(b) O, CH_3, CH_2CH_2OH, CH_3O, O, OH, H_3C, O, OH → O, CH_3, $CH_2CH_2OTBDMS$, CH_3O, O, OH, H_3C, O, OH

(c) O, O, TBDMSO, CH_2OCH_2Ph → O, O, HO, CH_2OCH_2Ph

(d) O, O, H_3C, $CH_2CH{=}O$ → O, O, H_3C, CH_2, S, S

(e) [Reaction scheme: dioxolane with H_3C, CH_3; substituents $CH_3OC(CH_3)_2OCH_2$ and CH_2CH_3 → $HOCH_2$, CH_2CH_3]

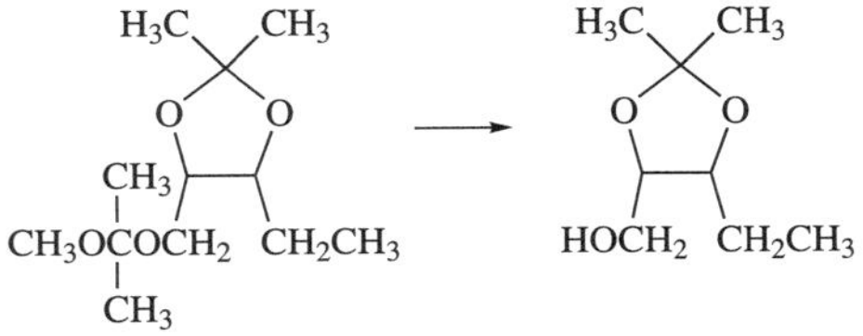

(f) [Reaction scheme: H_3C dioxolane with $CH_2CH_2CH_2$ linking to acetonide (H_3C, CH_3) bearing CH_2CH_3 → bicyclic ketal with CH_3, CH_2CH_3]

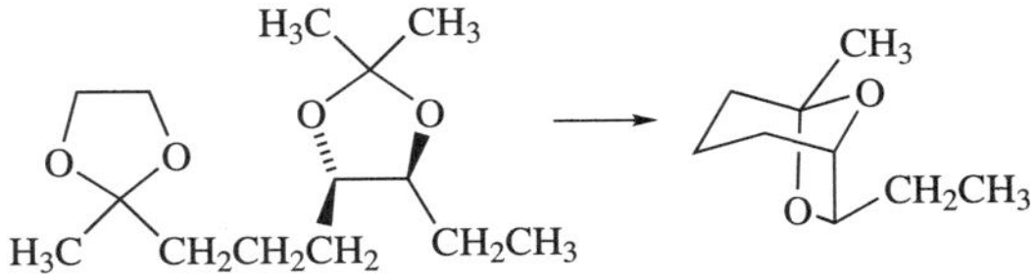

2. Indicate the product to be expected under the following reaction conditions.

(a) [Structure with O_2CCH_3, H_3C, OH, epoxide O, H_3C, CH_3] $\xrightarrow[\text{25°C, 3h}]{\text{dihydropyran, } CH_2Cl_2;\ \text{piperidinium } {}^+NH_2\,{}^-O_3SC_7H_7}$

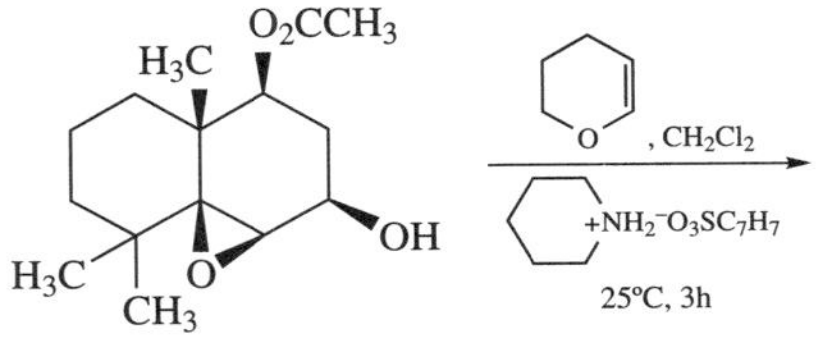

(b) [1,3-Dithiane–CH_2CH_2–CH(OTBDMS)–CH(CH_3)–CH_2OH] + $CH_2{=}C(OCH_3)CH_3$ $\xrightarrow[CH_2Cl_2]{POCl_3}$

(c) $PhCH_2OC(=O)NHCH_2C(=O)-N$ [pyrrolidine with HO_2C] $\xrightarrow[\text{ethanol}]{\text{Pd, cyclohexadiene}}$

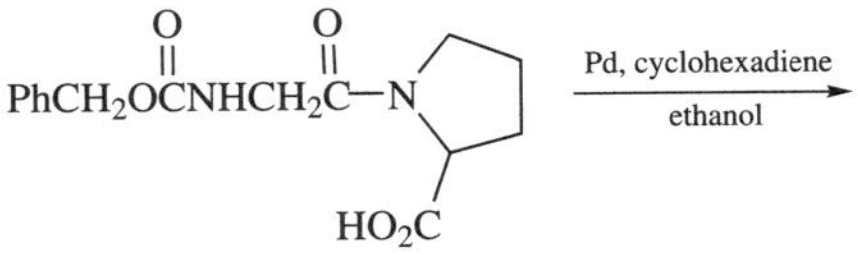

(d) H–C(NH_2)(CO_2H)–CH_2SH + PhCH=O ⟶

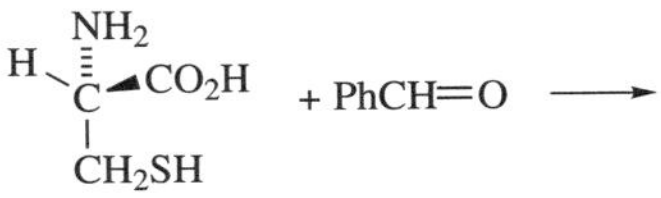

(e) $CH(SCH_2CH_3)_2$–CH(OH)–CH(OH)–CH(OH)–CH(OH)–CH_3 [Fischer projection: H–C–OH, H–C–OH, HO–C–H, HO–C–H] + $CH_3C(=O)CH_3$ $\xrightarrow{CuSO_4}$ formula is $C_{13}H_{26}O_4S_2$

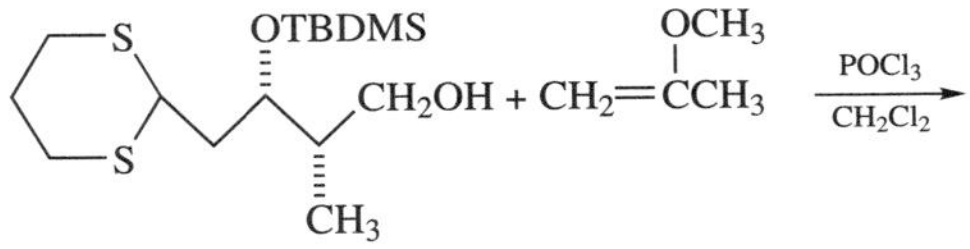

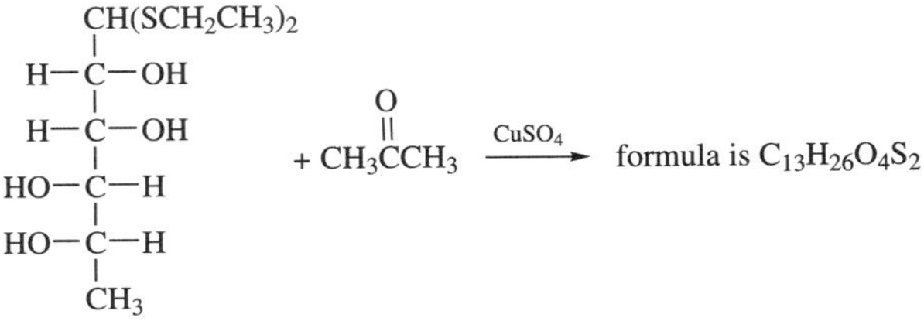

3. In each of the synthetic transformations shown, the reagents are appropriate, but the reactions will not be practical as they are written. What modification would be necessary to permit each transformation to be carried out to give the desired product?

(a) LiAlH$_4$

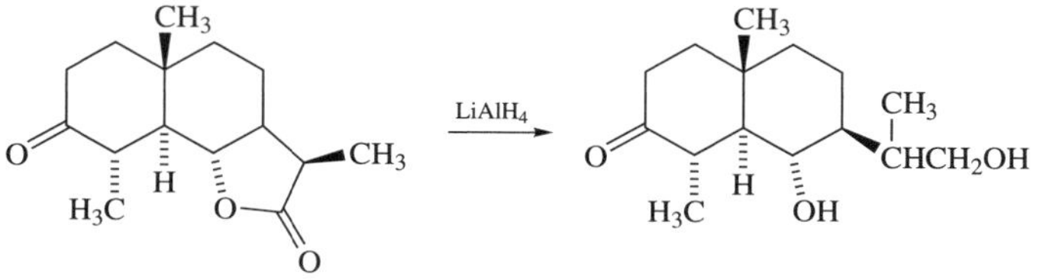

(b) 1) CrO–acetone 2) H_2NNH_2, KOH

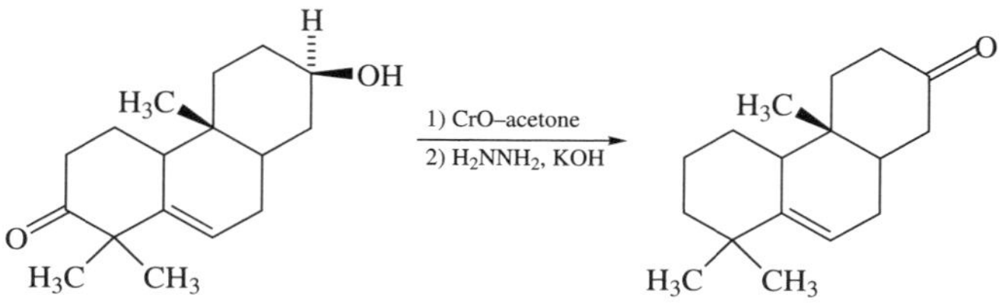

(c) $POCl_3$ pyridine

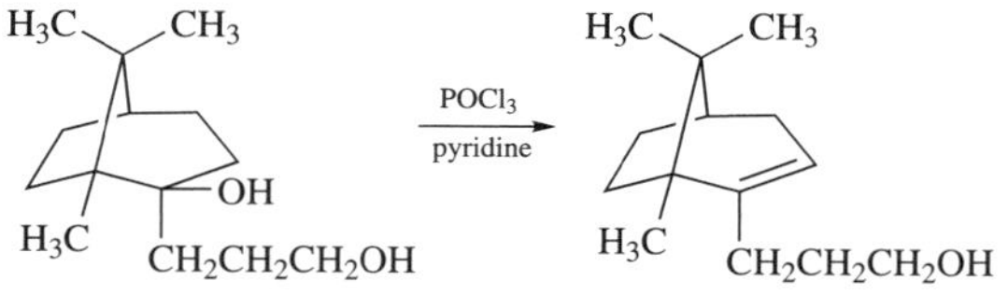

(d) CH_3I $NaNH_2$

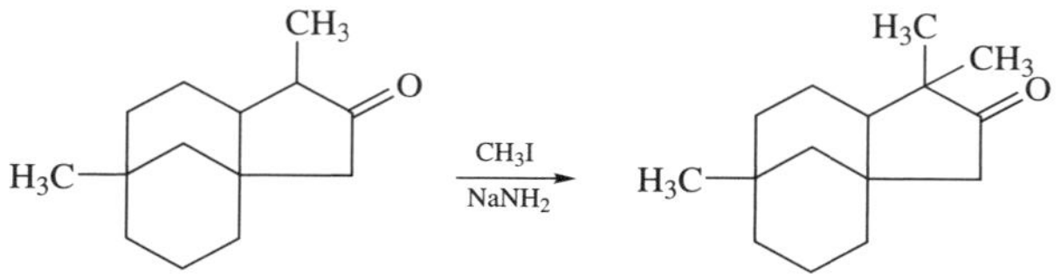

(e) $(CH_3)_2CHN{=}C{=}NCH(CH_3)_2$

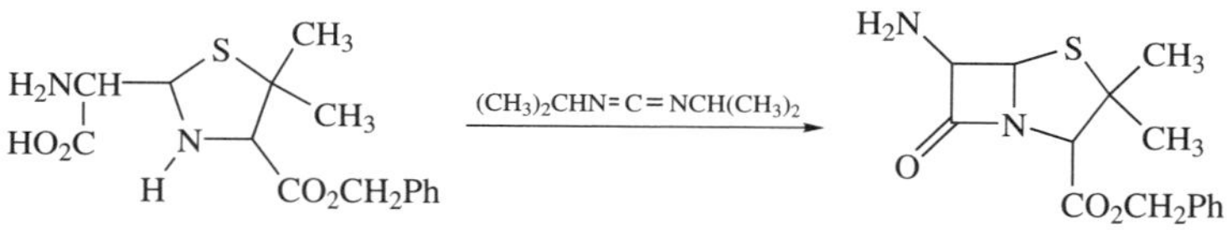

(f) $CH_3CH{=}PPh_3$

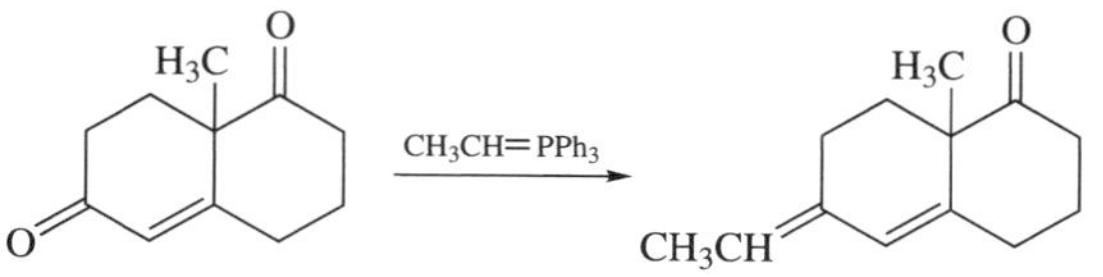

4. Under certain circumstances, each of the following groups can serve as a temporary protecting group for secondary amines by acting as a removable tertiary substituent. Suggest conditions which might be appropriate for subsequent removal of each group.

(a) $PhCH_2$— (b) $CH_2{=}CHCH_2$— (c) $CH_2{=}CH$— (d) $PhCH_2OCH_2$—

5. Show how synthetic equivalent groups might be used to efficiently carry out the following transformations.

(a)

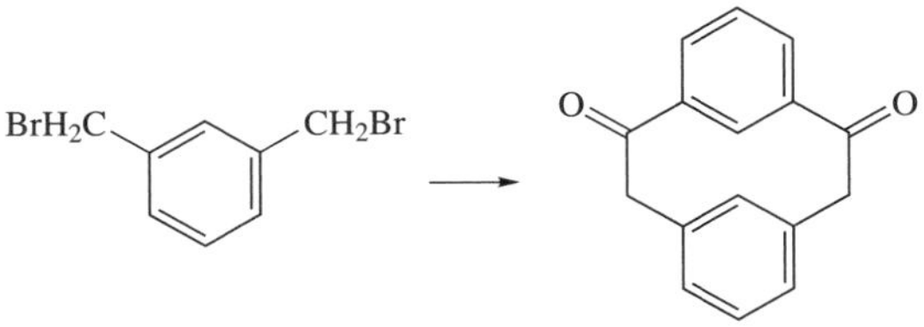

BrH$_2$C CH$_2$Br →

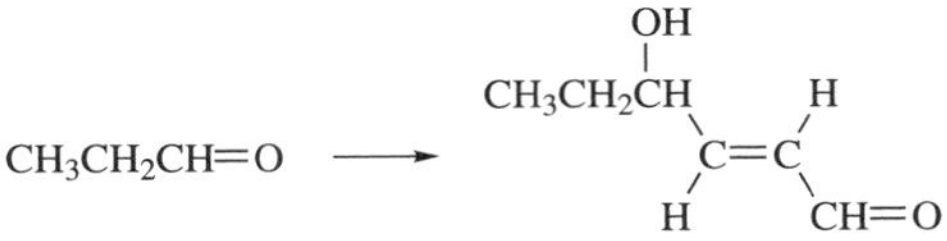

(b)

CH$_3$CH$_2$CH=O → CH$_3$CH$_2$CH(OH)CH=CHCH=O

(c)

→ CH$_2$OH OH

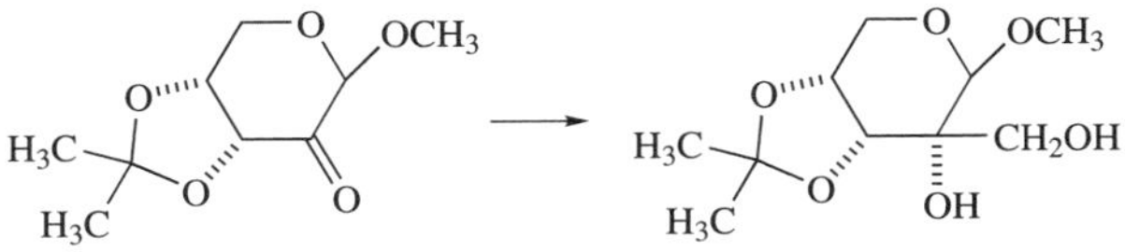

(d)

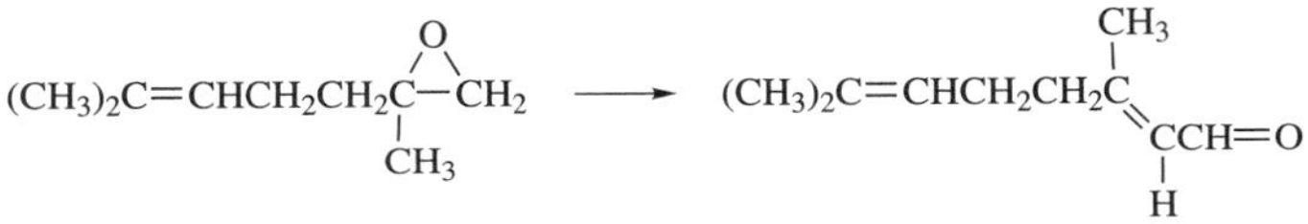

$(CH_3)_2C{=}CHCH_2CH_2C(CH_3)$—CH$_2$ (epoxide O) → $(CH_3)_2C{=}CHCH_2CH_2C(CH_3){=}C(H)CH{=}O$

(e)

CH$_3$C(=O)— OH

(f)

CH=O → CCH$_2$CH$_2$CN (C=O)

(g)

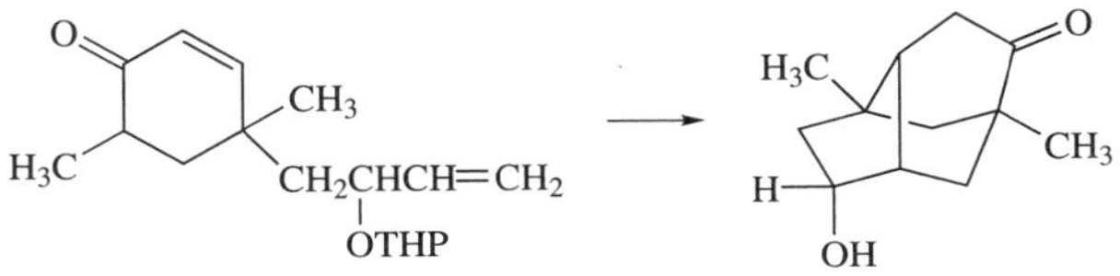

O CH$_3$ H$_3$C CH$_2$CHCH=CH$_2$ OTHP → H$_3$C O CH$_3$ H OH

(h)

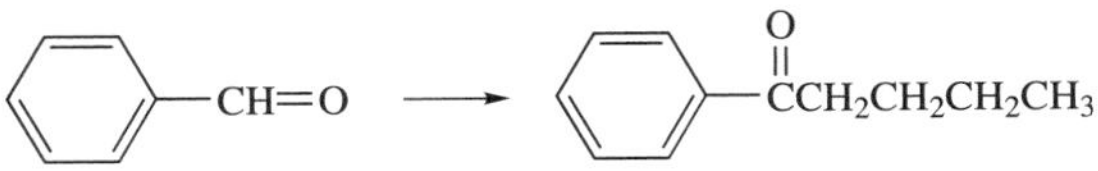

CH=O → CCH$_2$CH$_2$CH$_2$CH$_3$ (C=O)

(i)

CH$_3$ CH$_2$Cl CH=O CH$_3$ CH$_3$ CH$_3$ → CH$_3$ O CH$_3$ CH$_3$ CH$_3$

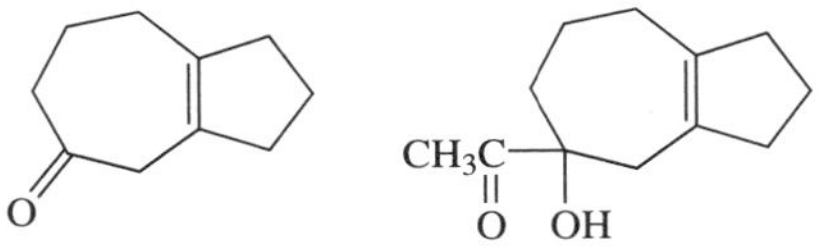

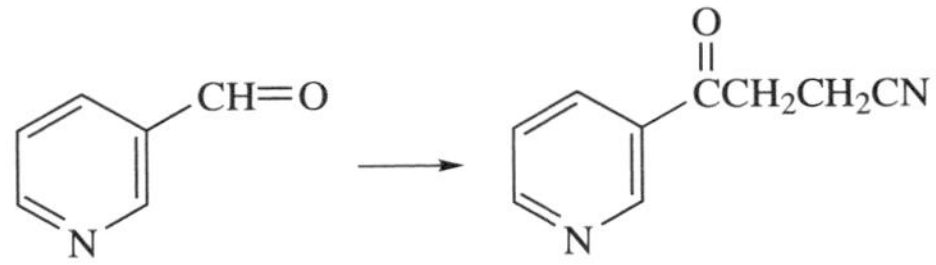

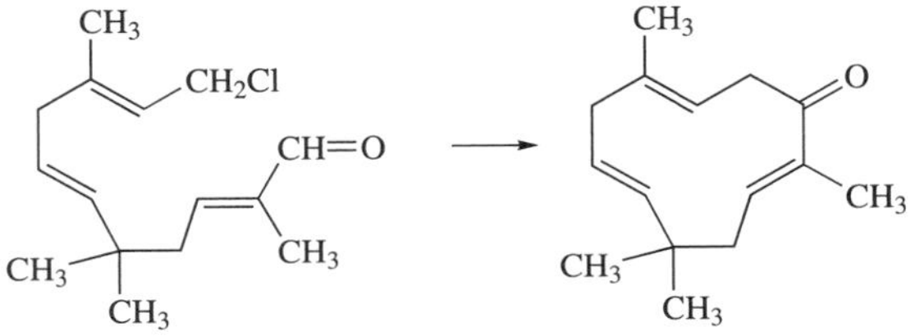

6. Indicate a reagent or short sequence that would accomplish each of the following synthetic transformations.

(a) $CH_3CCH{=}CH_2$ (C=O) ⟶ 4-acetylcyclohexanone (CCH_3, C=O)

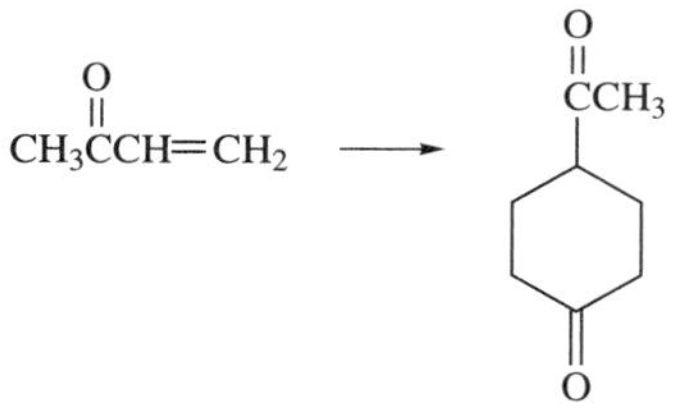

(b) $H_2C{=}C(CH_3)$–cyclohexenyl–$C(CH_3)_2OH$ ⟶ bicyclic enone with CH_3 and $C(CH_3)_2OH$

(c) cyclohexene oxide (O) ⟶ OH, $CH{=}CHCH{=}O$

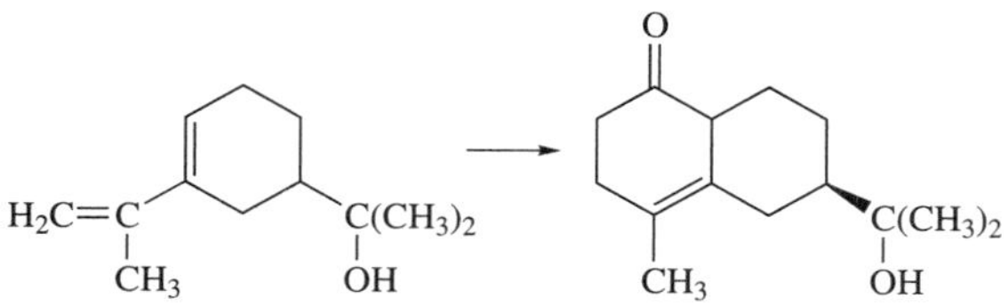

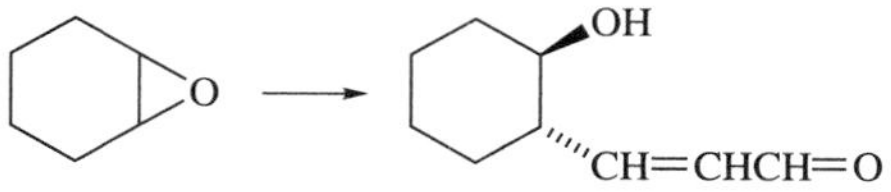

(d) 2-phenylcyclohexanone (O, Ph) ⟶ H_3C, O, Ph

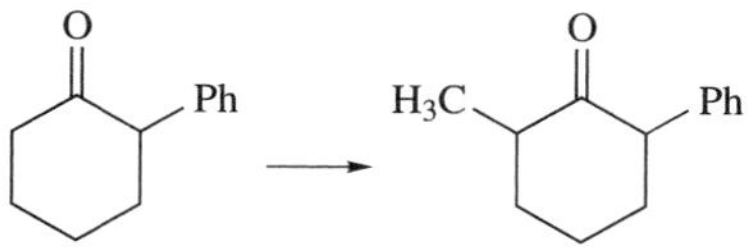

(e) CH_3CCH_2OH ($\|CH_2$) ⟶ $CH_3CCH_2CH_2CHC{=}CH_2$ (OH; $\|CH_2$; CH_3)

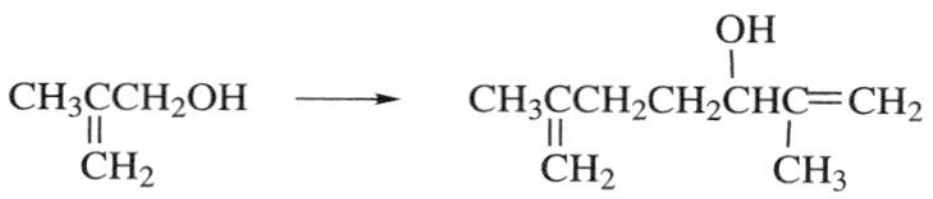

(f) CO_2CH_3 ⟶ O_2CCH_3

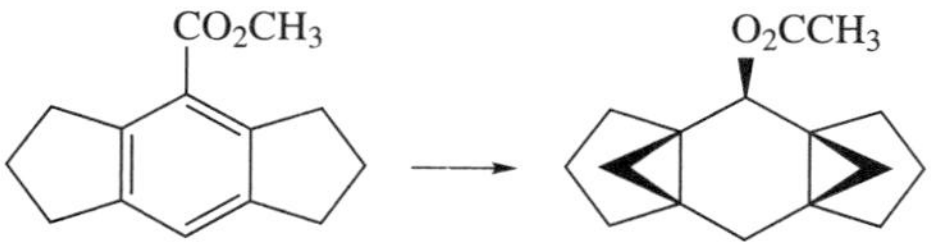

(g) CH=O, CH_2Cl ⟶ O

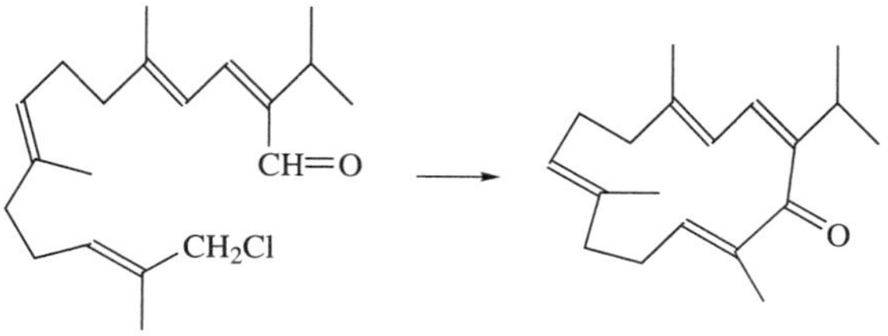

7. Indicate a reagent or short reaction sequence which could accomplish synthesis of the material shown on the left from the starting material on the right.

(a) CCH_3 (O), $CH_2CH_2CH(CH_3)_2$ ⟹ 2-cyclopentenone (O)

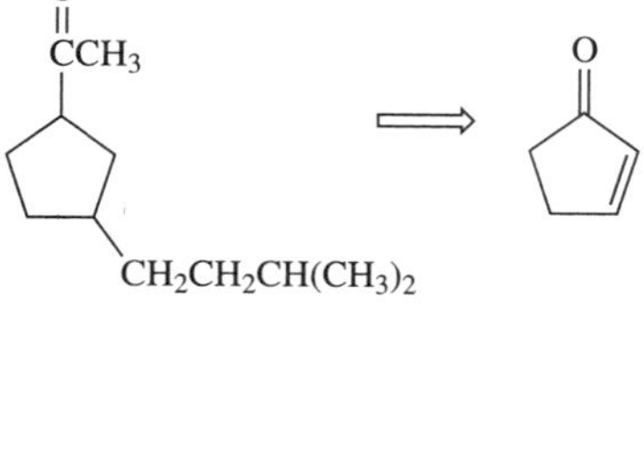

(b) 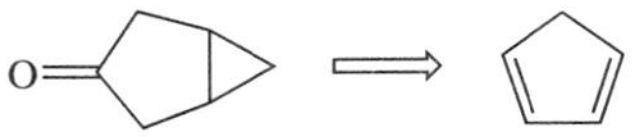 O= [bicyclo[3.1.0]hexan-3-one] ⟹ [cyclopentadiene]

(c)  [spiro[3.3]hepta-1,5-diene] ⟹ $C(CH_2Br)_4$

(d) 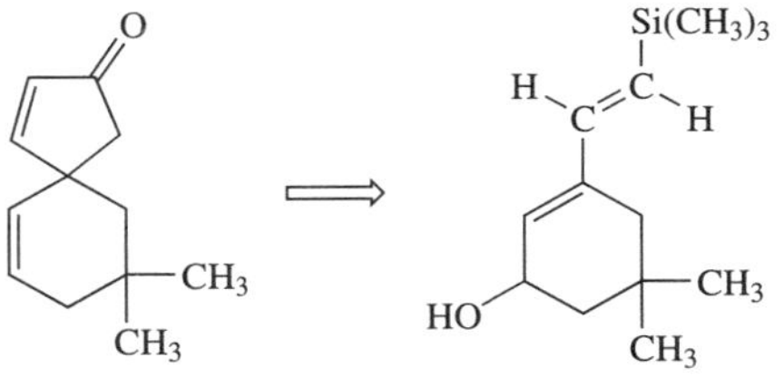 [spiro enone with CH_3, CH_3] ⟹ $Si(CH_3)_3$, H, C=C, H, HO, CH_3, CH_3

(e) 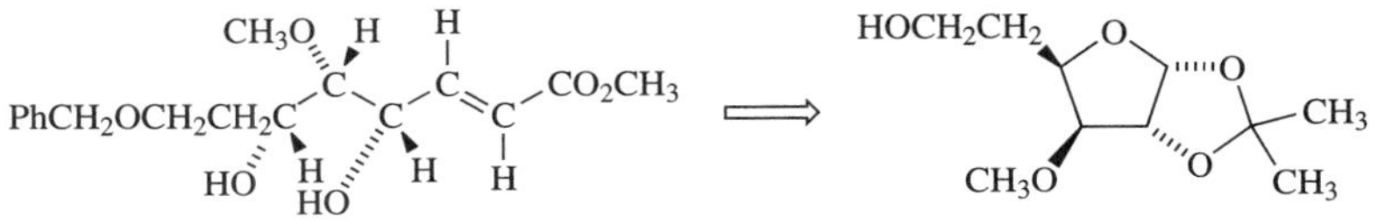 CH_3O H H, $PhCH_2OCH_2CH_2C$ C C C=C CO_2CH_3, HO H HO H H ⟹ $HOCH_2CH_2$ O O, CH_3O O CH_3 CH_3

(f) 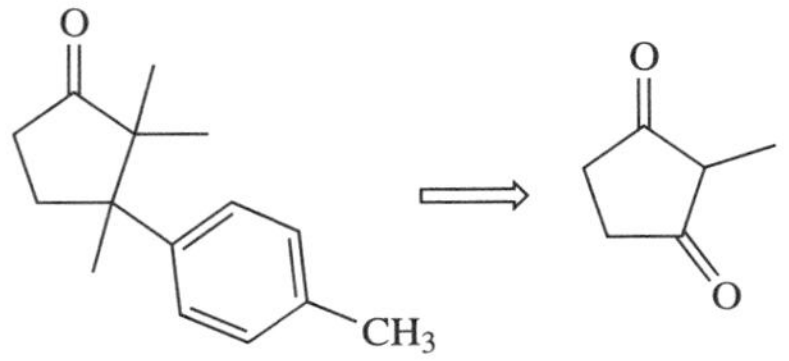 O, CH_3 ⟹ O, O

8. Because they are readily available from natural sources in enantiomerically pure form, carbohydrates are very useful starting materials for the synthesis of enantiomerically pure substances. However, the high number of similar functional groups present in carbohydrates requires versatile techniques for selective protection and selective reaction. Show how appropriate manipulation of protecting groups and/or selective reagents might be employed to effect the desired transformations.

(a) 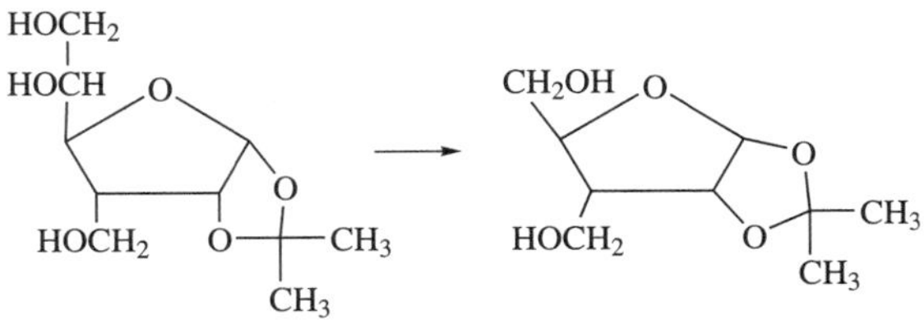 $HOCH_2$, HOCH, O, O, $HOCH_2$, O, CH_3, CH_3 ⟶ CH_2OH, O, O, $HOCH_2$, O, CH_3, CH_3

(b) 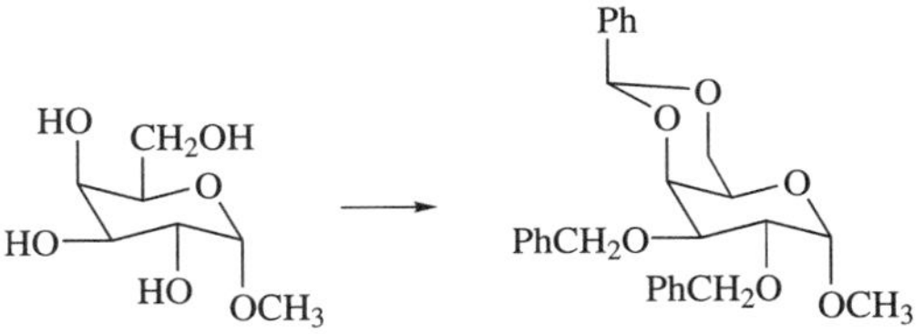 HO, CH_2OH, O, HO, HO, OCH_3 ⟶ Ph, O, O, O, $PhCH_2O$, $PhCH_2O$, OCH_3

(c) 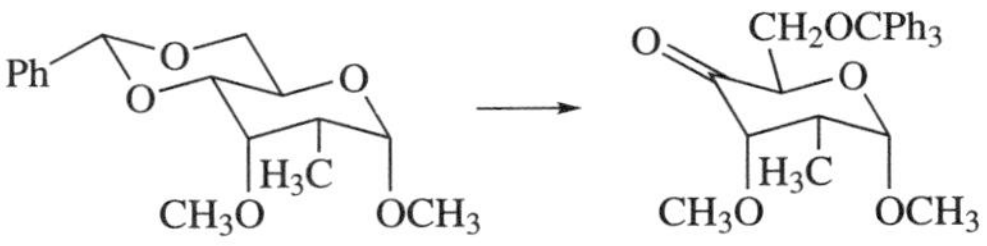 Ph, O, O, O, H_3C, CH_3O, OCH_3 ⟶ O, CH_2OCPh_3, O, H_3C, CH_3O, OCH_3

(d)

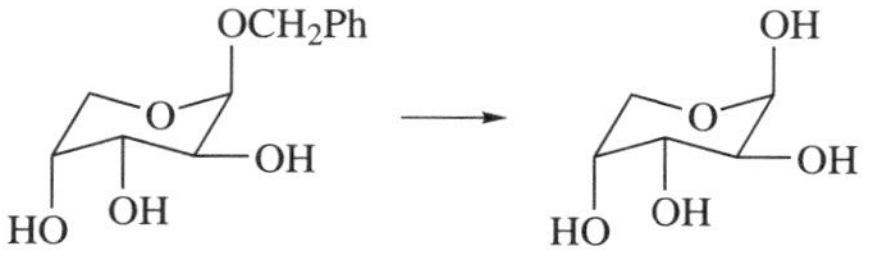

9. Synthetic transformations which are parts of total syntheses of natural products are outlined by a general retrosynthetic outline. For each retrosynthetic disconnection, suggest a reagent or short sequence of reactions which could accomplish the forward synthetic reaction. The proposed route should be diastereoselective but need not be enantioselective.

(a)

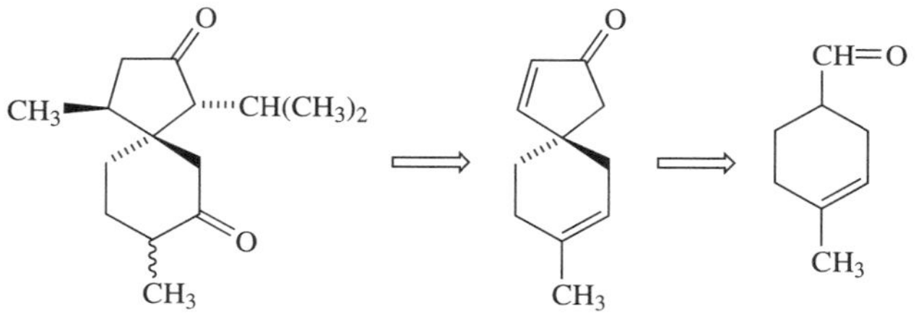

(b)

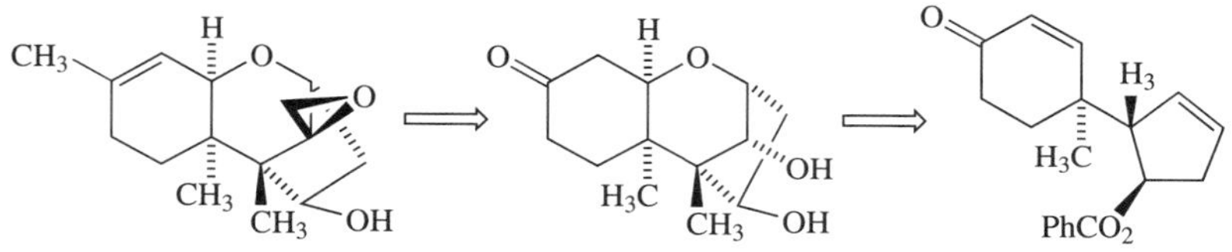

(c)

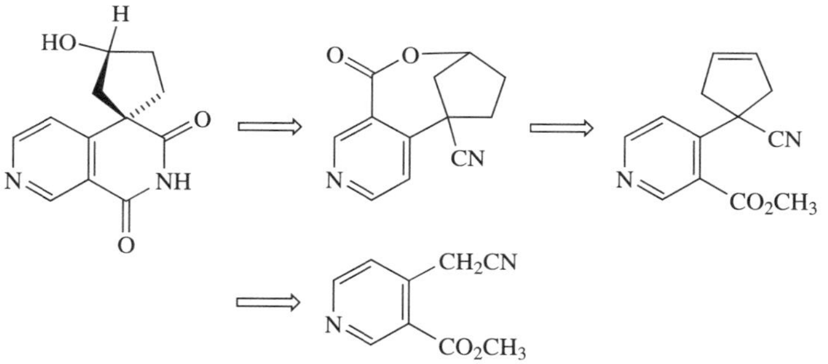

(d)

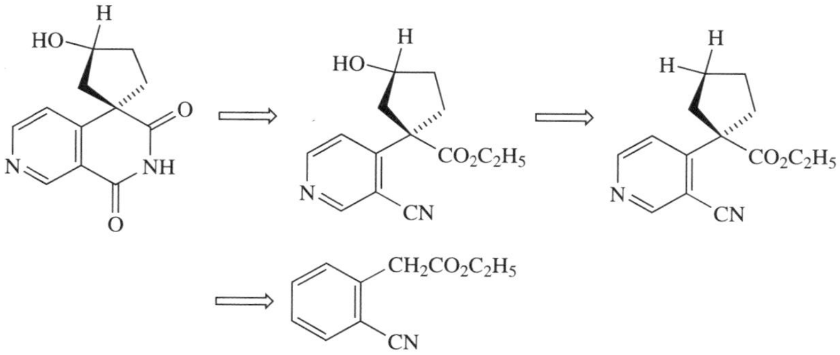

(e)

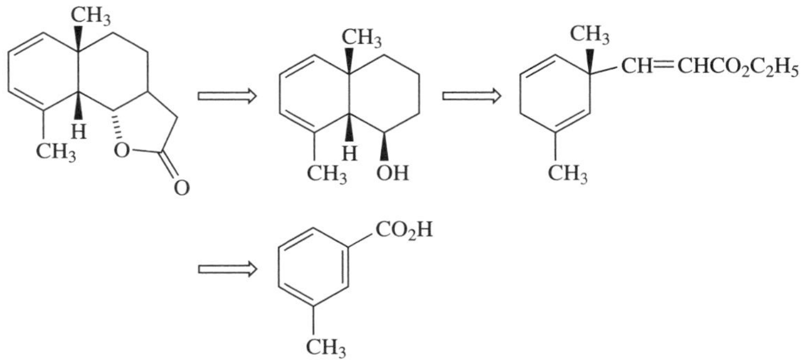

$\Longrightarrow$ $\Longrightarrow$ $\Longrightarrow$

10. Diels–Alder reactions are attractive for many synthetic applications, particularly because of their predictable stereochemistry. There are, however, significant limitations on the type of compound which can serve as a dienophile or diene. As a result, the idea of synthetic equivalency has been exploited in this area. For each of the reactive dienophiles and dienes given below, suggest one or more transformations which might be carried out on a Diels–Alder adduct derived from it that would lead to a product not directly attainable by a Diels–Alder reaction. Give the structure of the diene or dienophile "synthetic equivalent," and indicate why the direct Diels–Alder reaction would not be possible.

Dienophiles	Dienes
(a) $CH_2{=}CH\overset{+}{P}Ph_3$	(d) $CH_2{=}C(OSi(CH_3)_3){-}CH{=}CHOCH_3$
(b) $CH_2{=}CHS(=O)Ph$	(e) $CH_2{=}C(O_2CCH_3){-}C(SPh){=}CH_2$
(c) $CH_2{=}C(O_2CCH_3)CO_2C_2H_5$	

11. One approach to the synthesis of enantiomerically pure materials is to start with an available enantropure material and efffect the synthesis by a series of stereospecific reactions. Devise a sequence of reactions which would be appropriate for the following syntheses based on enantiomerically pure starting materials.

(a) from

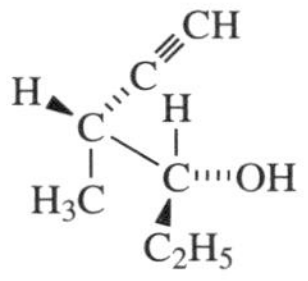

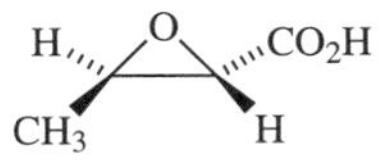

(b) 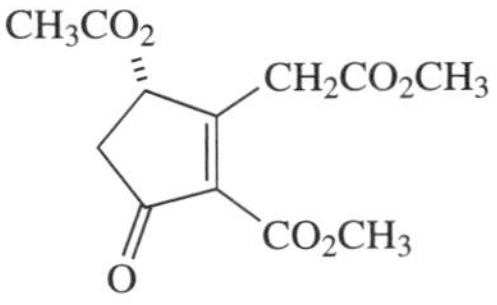 from

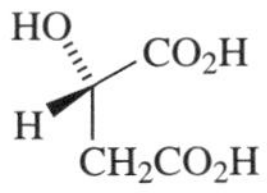

(c) CH_3O_2C, N, $CHCH{=}CH_2$, OH, Ph, S from H_2N–C–H, CO_2H, CH_2SH

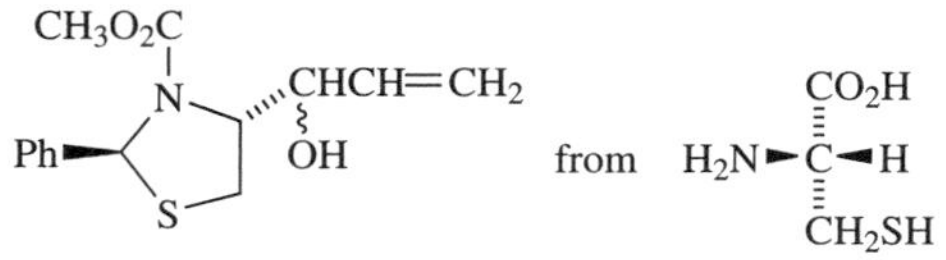

(d) CH=O, H–C–OH, CH_2OH from CH_2OH, HO–C–H, HO–C–H, H–C–OH, H–C–OH, CH_2OH

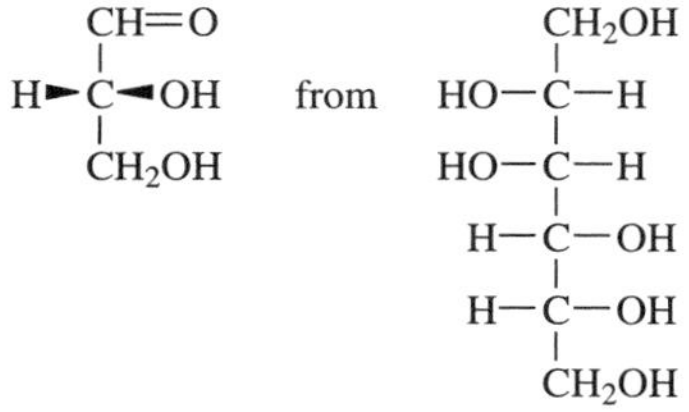

(e) $CH_3(CH_2)_{10}$, O, O from H, $CH_3(CH_2)_{10}$, O

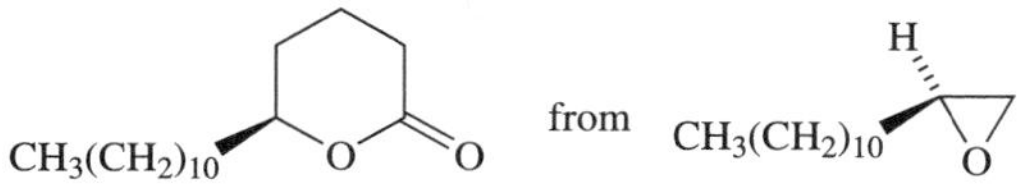

(f) CH_2OH, HO, OH, HO, O, CH_3 from H_3C, H_3C, O, O, OH, O, CH_3

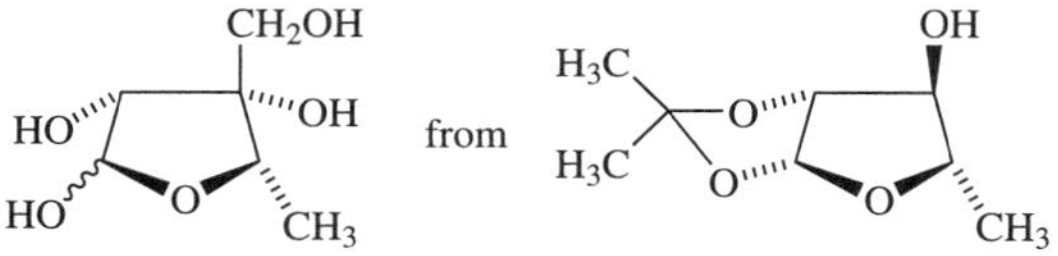

(g) O=CH, CH_3, $(CH_3)_2CH$, CH_3, CH=O from CH_3, O, $CH(CH_3)_2$

(h) O, H, O, CH_3 from CH_3, O, H, H, H

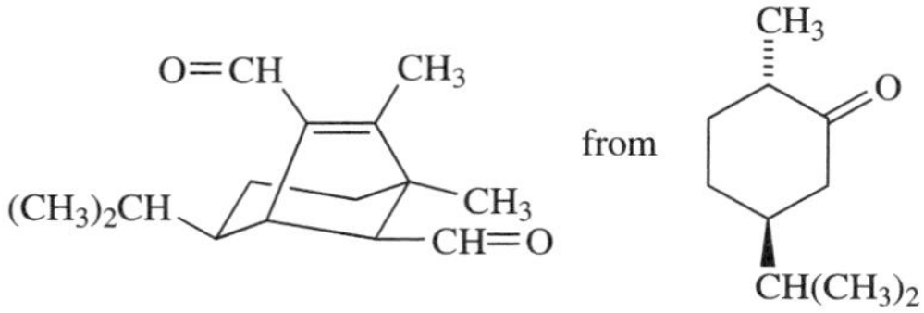

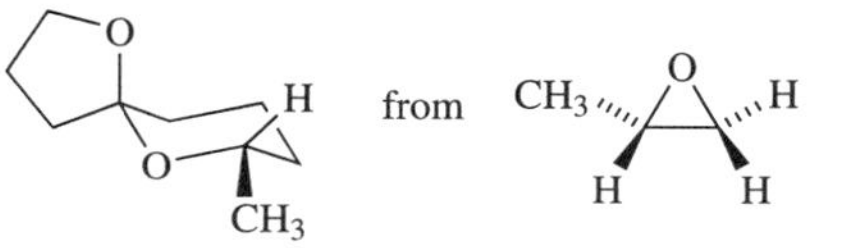

12. Several natural product syntheses are outlined in restrosynthetic form. Suggest a reaction or short reaction series which could accomplish each lettered transformation in the forward synthetic direction. The structures shown refer to racemic materials.

(a) Isotelekin

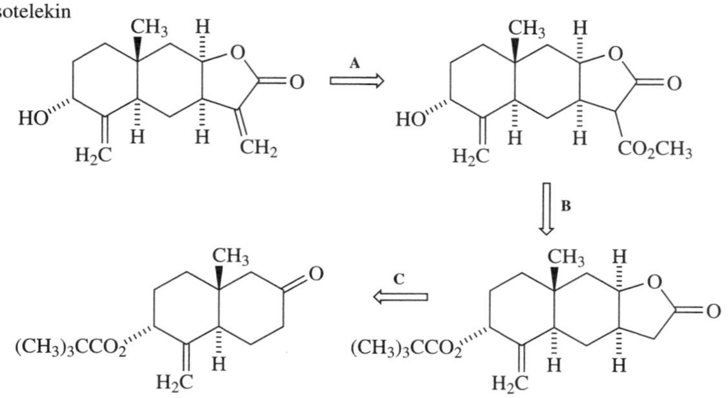

CH_3, H, O, O, HO, H, H, CH_2, H_2C ⟹ A CH_3, H, O, O, HO, H, H, CO_2CH_3, H_2C

⇓ B

CH_3, O, $(CH_3)_3CCO_2$, H, H_2C ⟸ C CH_3, H, O, O, $(CH_3)_3CCO_2$, H, H, H_2C

(continued)

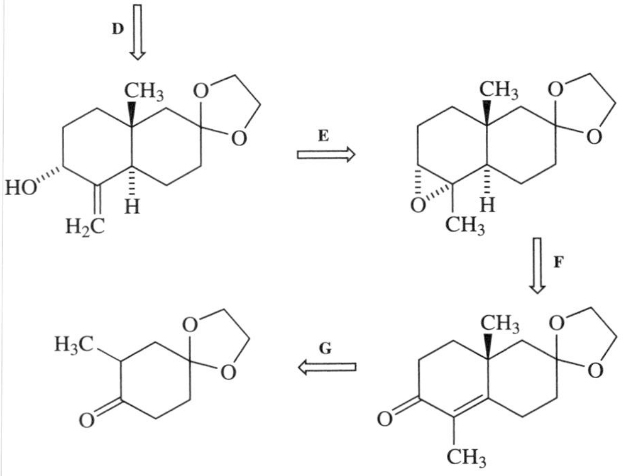

D

E

F

G

(b) Aromandrene

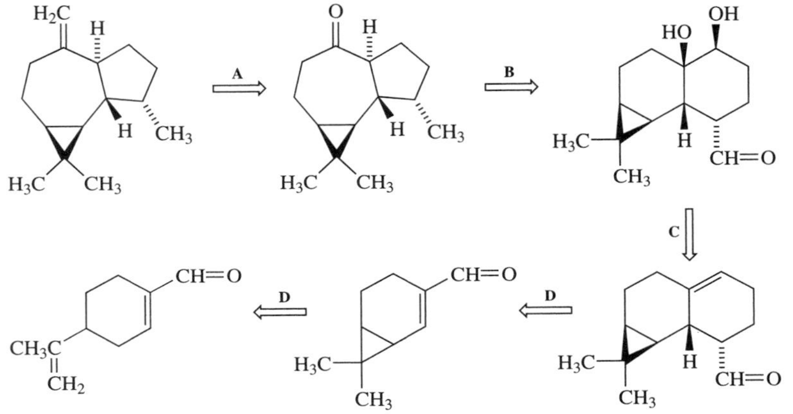

A B C D D

(c) α-Bourbonene

A B C D E

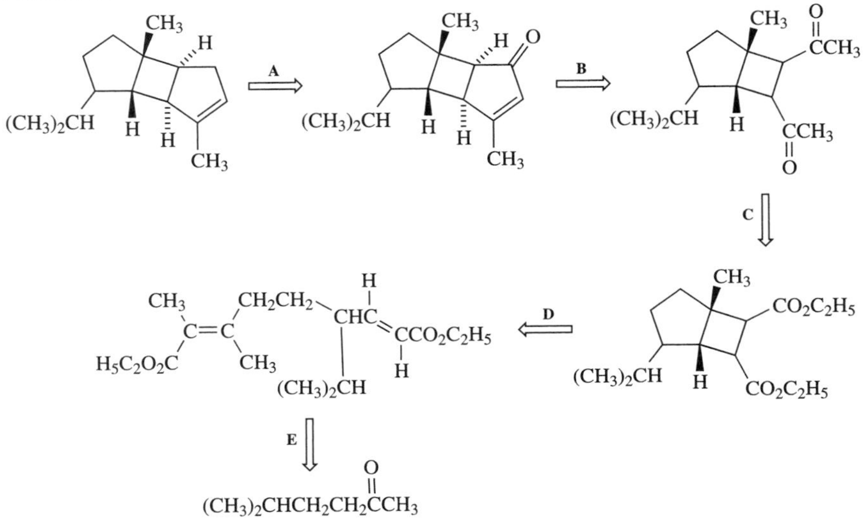

(d) Caryophyllene

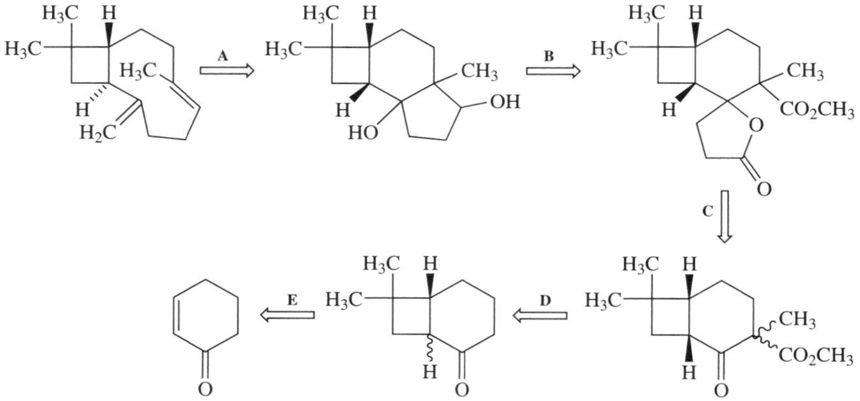

13. Perform a retrosynthetic analysis for each of the following molecules. Develop at least three outline schemes. Discuss the relative merits of the three schemes, and develop a fully elaborated synthetic plan for the one you consider to be most promising.

(a)

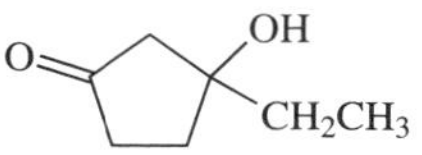

(b)

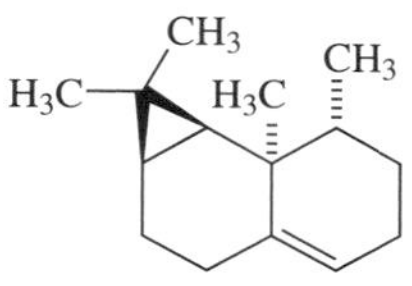

(c)

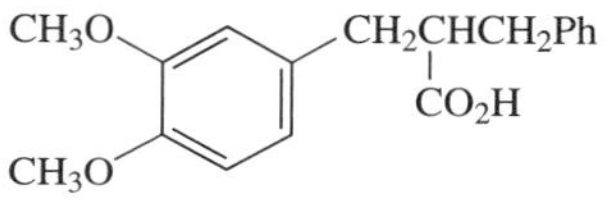

(d)

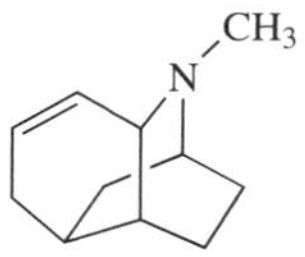

(e)

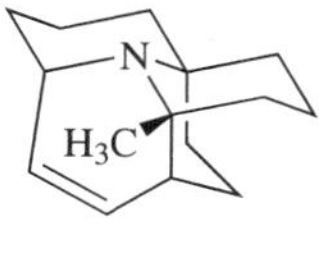

(f)

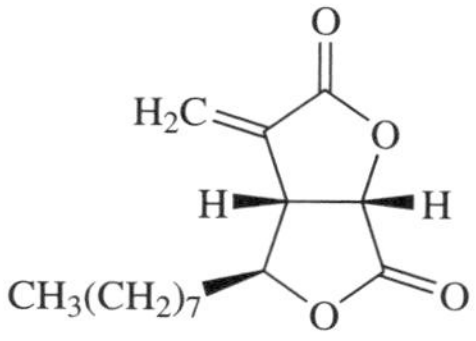

(g)

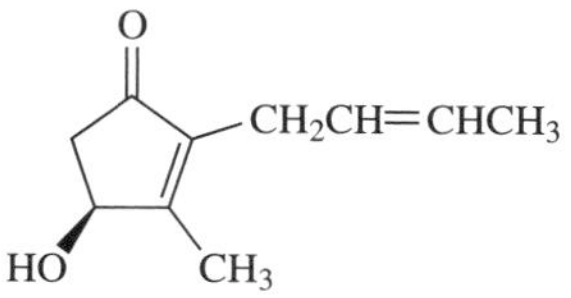

(h)

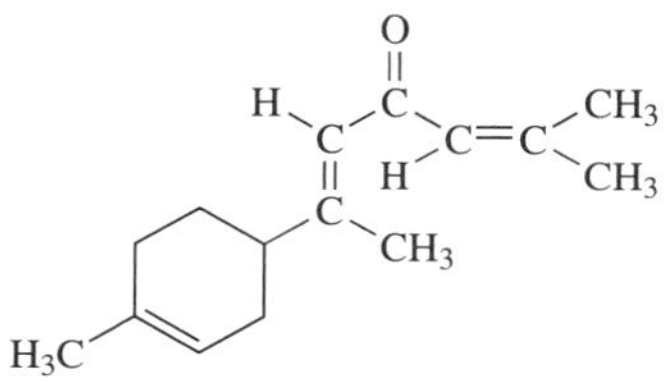

(i)

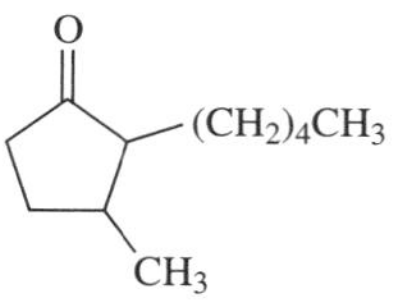

(j)

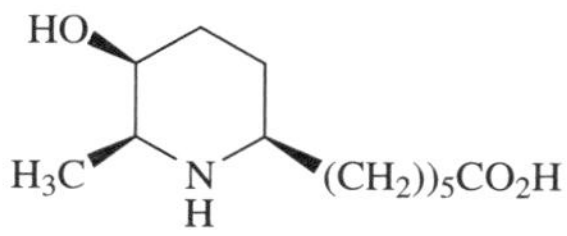

(k)

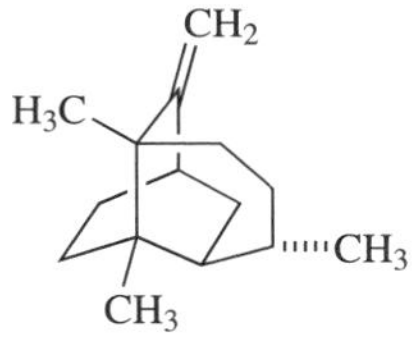

(l)

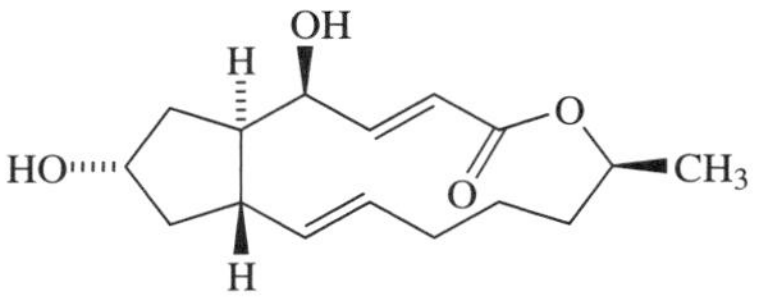

(m)

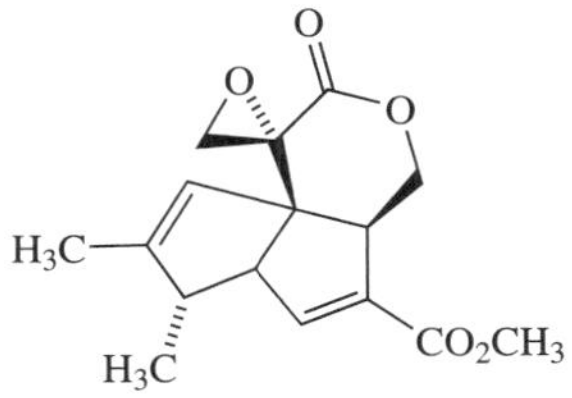

14. Suggest methods that would be expected to achieve a diastereoselective synthesis of the following compounds.

(a)

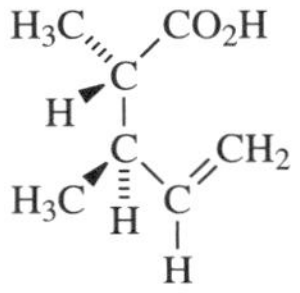

(b)

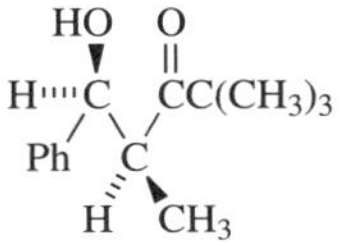

(c)

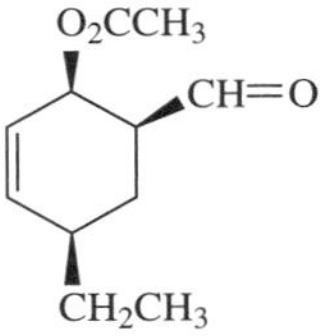

(d)

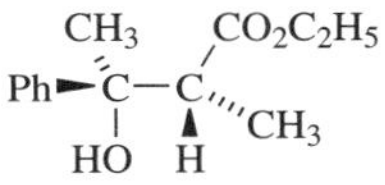

(e)

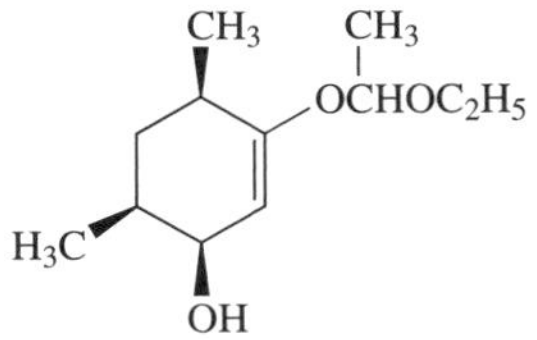

15. Devise a route for synthesis of the desired compound in high enantiomeric purity from the suggested starting material.

(a) from D-ribose

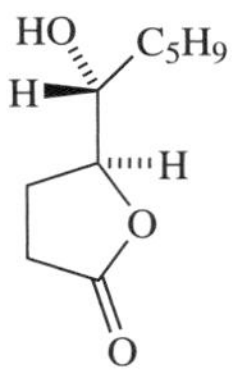

(b) from $ArSO_2NH\text{–}CH_2CH{=}CHCH_2OH$

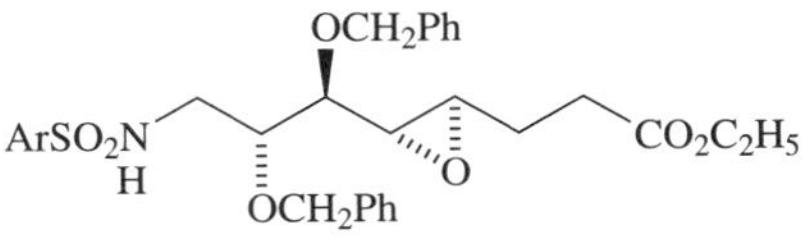

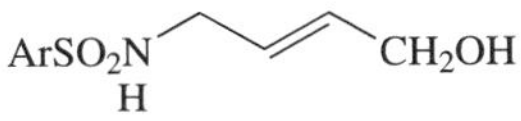

(c) from H_3C–cyclopentadiene

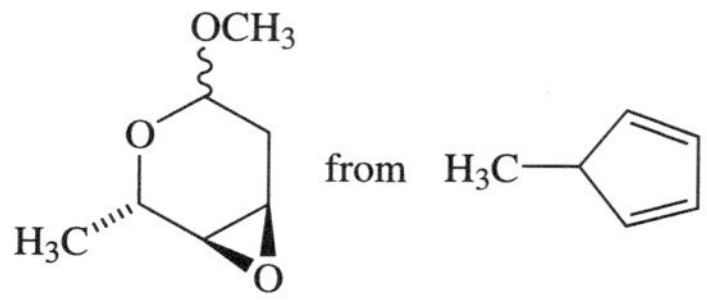

16. By careful consideration of transition-state geometry using molecular models, predict the *absolute configuration* of the major product for each reaction. Explain the basis of your prediction.

(a) 1) t-BuLi 2) CH_3I 3) HCl, H_2O → $PhCH(CH_3)CH_2CH{=}O$

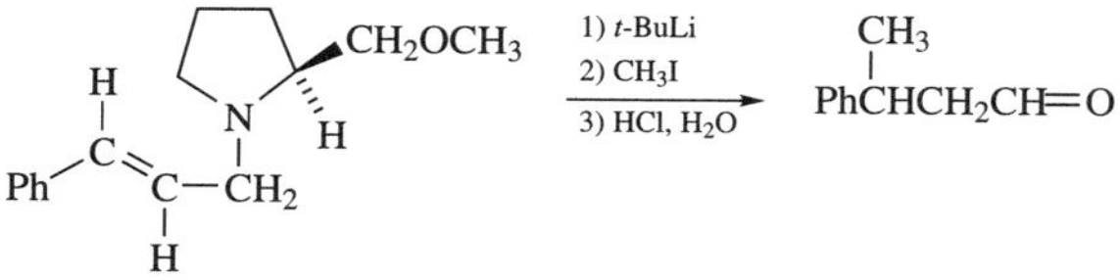

(b) $PhCH_2CH_2$–oxazoline (Ph, CH_2OCH_3) 1) $LiNR_2$ 2) $CH_3CH_2CH_2CH_2I$ 3) H^+, H_2O → $PhCH_2CH(CH_2CH_2CH_2CH_3)CO_2H$

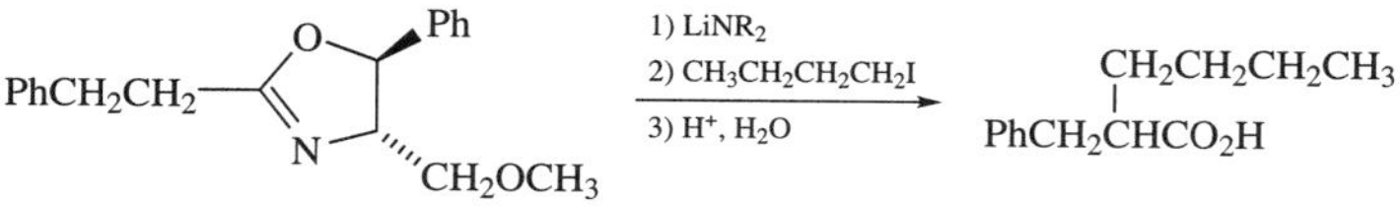

(c) $(CH_3)_3C$, H, $CO_2C(CH_3)_3$ 1) $LiNR_2$ 2) CH_3I → 2-methylcyclohexanone (CH_3)

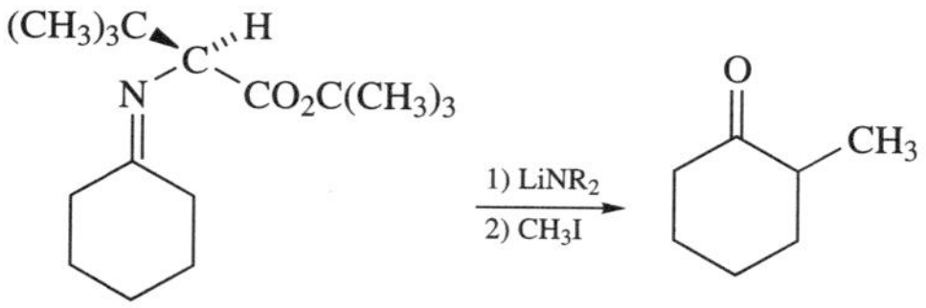

(d) PhC(=O)–oxazoline (Ph, CH_2OCH_3) 1) $(CH_3)_2CHCH_2MgBr$ 2) H^+, H_2O → $(CH_3)_2CHCH_2C(OH)(Ph)CO_2H$

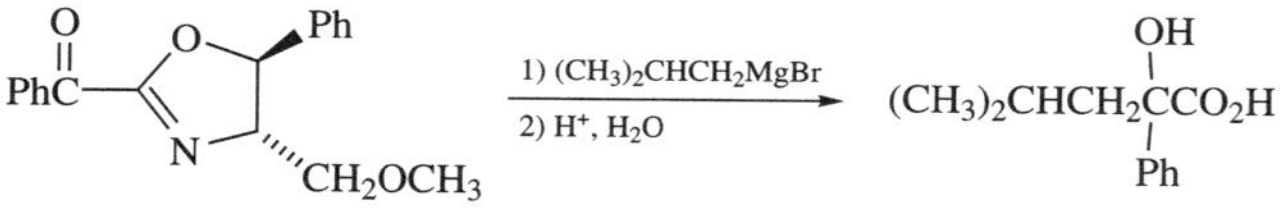

(e) Ph, H, CH_3, HO_2C 1) Br_2 2) $KOC(CH_3)_3$ → H_3C, Ph, H, CO_2H

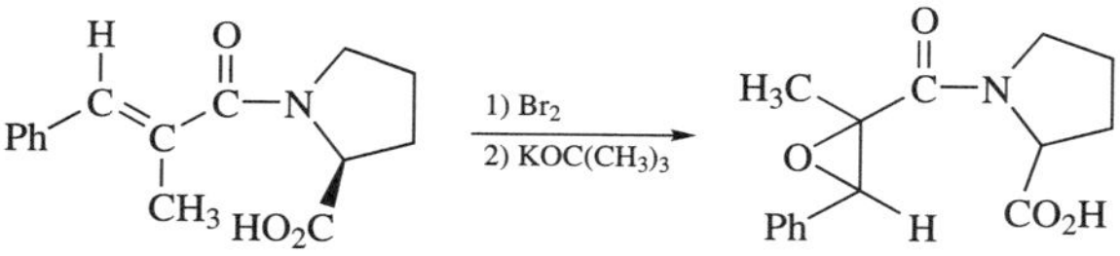

(f) $H_2C{=}CHCH{=}CHO_2C{-}C(Ph)(H)(OCH_3)$ + OH, O, O $B(O_2CCH_3)_3$ 0°C → OH, O, O, $OCHPh(OCH_3)$

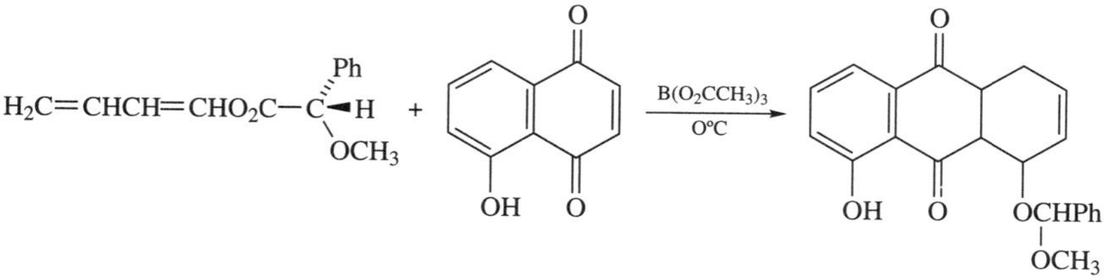

(g) $PhCH{=}CH$–oxazoline (Ph, CH_2OCH_3) 1) C_2H_5Li 2) H^+, H_2O → $PhCH(C_2H_5)CO_2H$

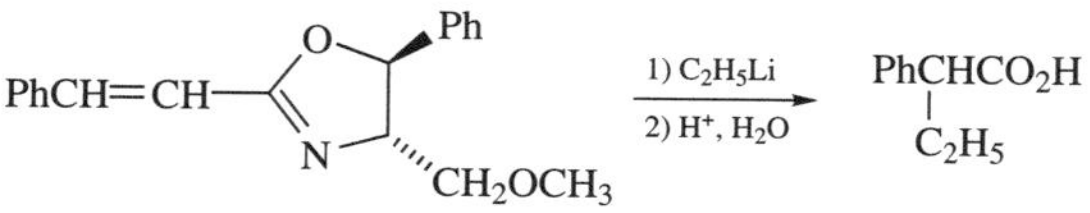

References for Problems

Chapter 1

1a. W. S. Matthews, J. E. Bares, J. E. Bartmess, F. G. Bordwell, F. J. Cornforth, G. E. Drucker, Z. Margolin, R. J. McCallum, G. J. McCollum, and N. E. Vanier, *J. Am. Chem. Soc.* **97:**7006 (1975).
b. H. D. Zook, W. L. Kelly, and I. Y. Posey, *J. Org. Chem.* **33:**3477 (1968).
2a. H. O. House and M. J. Umen, *J. Org. Chem.* **38:**1000 (1973).
b. W. C. Still and M.-Y. Tsai, *J. Am. Chem. Soc.* **102:**3654 (1980).
c. H. O. House and B. M. Trost, *J. Org. Chem.* **30**:1341 (1965).
d. D. Caine and T. L. Smith, Jr., *J. Am. Chem. Soc.* **102:**7568 (1980).
e. M. F. Semmelhack, S. Tomoda, and K. M. Hurst, *J. Am. Chem. Soc.* **102:**7567 (1980).
f. R. A. Lee, C. McAndrews, K. M. Patel, and W. Reusch, *Tetrahedron Lett.* **1973:**965.
g. R. H. Frazier, Jr., and R. L. Harlow, *J. Org. Chem.* **45:**5408 (1980).
3a. M. Gall and H. O. House, *Org. Synth.* **52:**39 (1972).
b. P. S. Wharton and C. E. Sundin, *J. Org. Chem.* **33:**4255 (1968).
c. B. W. Rockett and C. R. Hauser, *J. Org. Chem.* **29**:1394 (1964).
d. J. Meier, *Bull. Soc. Chim. Fr.* **1962:**290.
e. M. E. Jung and C. A. McCombs, *Org. Synth.* **58:**163 (1978).
f & g. H. O. House, T. S. B. Sayer, and C.-C. Yau, *J. Org. Chem.* **43:**2153 (1978).
4a. J. M. Harless and S. A. Monti, *J. Am. Chem. Soc.* **96:**4714 (1974).
b. A. Wissner and J. Meinwald, *J. Org. Chem.* **38:**1967 (1973).
c. W. J. Gensler and P. H. Solomon, *J. Org. Chem.* **38:**1726 (1973).
d. H. W. Whitlock, Jr., *J. Am. Chem. Soc.* **84:**3412 (1962).
e. C. H. Heathcock, R. A. Badger, and J. W. Patterson, Jr., *J. Am. Chem. Soc.* **89:**4133 (1967).
f. E. J. Corey and D. S. Watt, *J. Am. Chem. Soc.* **95:**2302 (1973).
5. W. G. Kofron and L. G. Wideman, *J. Org. Chem.* **37:**555 (1972).
6. C. R. Hauser, T. M. Harris, and T. G. Ledford, *J. Am. Chem. Soc.* **81:**4099 (1959).
7a. N. Campbell and E. Ciganek, *J. Chem. Soc.* **1956:**3834.
b. F. W. Sum and L. Weiler, *J. Am. Chem. Soc.*, **101:**4401 (1979).
c. K. W. Rosemund, H. Herzberg, and H. Schutt, *Chem. Ber.* **87:**1258 (1954).
d. T. Hudlicky, F. J. Koszyk, T. M. Kutchan, and J. P. Sheth, *J. Org. Chem.* **45:**5020 (1980).
e. C. R. Hauser and W. R. Dunnavant, *Org. Synth.* **40:**38 (1960).
f. G. Opitz, H. Milderberger, and H. Suhr, *Justus Liebigs Ann. Chem.* **649:**47 (1961).
g. K. Wiesner, K. K. Chan, and C. Demerson, *Tetrahedron Lett.* **1965:**2893.
h. K. Shimo, S. Wakamatsu, and T. Inoüe, *J. Org. Chem.* **26:**4868 (1961).
i. T. A. Spencer, K. K. Schmiegel, and K. L. Williamson, *J. Am. Chem. Soc.* **85:**3785 (1963).

j. G. R. Kieczkowski and R. H. Schlessinger, *J. Am. Chem. Soc.* **100:**1938 (1978).
8a–d. E. D. Bergmann, D. Ginsburg, and R. Pappo, *Org. React.* **10:**179 (1959).
e. L. Mandell, J. U. Piper, and K. P. Singh, *J. Org. Chem.* **28:**3440 (1963).
f. H. O. House, W. A. Kleschick, and E. J. Zaiko, *J. Org. Chem.* **43:**3653 (1978).
g. J. E. McMurry and J. Melton, *Org. Synth.* **56:**36 (1977).
h. D. F. Taber and B. P. Gunn, *J. Am. Chem. Soc.* **101:**3992 (1979).
i. H. Feuer, A. Hirschfield, and E. D. Bergmann, *Tetrahedron* **24:**1187 (1968).
j. A. Baradel, R. Longeray, J. Dreux, and J. Doris, *Bull. Soc. Chim. Fr.* **1970:**255.
k. H. H. Baer and K. S. Ong, *Can. J. Chem.* **46:**2511 (1968).
l. A. Wettstein, K. Heusler, H. Ueberwasser, and P. Wieland, *Helv. Chim. Acta* **40:**323 (1957).
9a. E. Wenkert and D. P. Strike, *J. Org. Chem.* **27:**1883 (1962).
b. S. J. Etheredge, *J. Org. Chem.* **31:**1990 (1966).
c. R. Deghenghi and R. Gaudry, *Tetrahedron Lett.* **1962:**489.
d. P. A. Grieco and C. C. Pogonowski, *J. Am. Chem. Soc.* **95:**3071 (1973).
e. E. M. Kaiser, W. G. Kenyon, and C. R. Hauser, *Org. Synth.* **V:**559 (1973).
f. J. Cason, *Org. Synth.* **IV:**630 (1963).
g. S. A. Glickman and A. C. Cope, *J. Am. Chem. Soc.* **67:**1012 (1945).
h. W. Steglich and L. Zechlin, *Chem. Ber.* **111:**3939 (1978).
i. S. F. Brady, M. A. Ilton, and W. S. Johnson, *J. Am. Chem. Soc.* **90:**2882 (1968).
j. R. P. Hatch, J. Shringarpure, and S. M. Weinreb, *J. Org. Chem.* **43:**4172 (1978).
10. S. Masamune, *J. Am. Chem. Soc.* **86:**288 (1964).
11. E. J. Corey, M. Ohno, R. B. Mitra, and P. A. Vatakencherry, *J. Am. Chem. Soc.* **86:**478 (1964).
12. J. Fried, in *Heterocyclic Compounds*, R. C. Elderfield, ed., Vol. 1, John Wiley & Sons, New York, 1950, p. 358.
13. R. Chapurlat, J. Huet, and J. Druex, *Bull. Soc. Chim. Fr.* **1967:**2446, 2450.
14a. F. Kuo and P. L. Fuchs, *J. Am. Chem. Soc.* **109:**1122 (1987).
b. L. A. Paquette, H.-S. Lin, D. T. Belmont, and J. P. Springer, *J. Org. Chem.* **54:**4807 (1986).
c. R. K. Boeckman, Jr., D. K. Heckenden, and R. L. Chinn, *Tetrahedron Lett.* **28:**3551 (1987).
d. D. Seebach, J. D. Aebi, M. Gander-Coquot, and R. Naef, *Helv. Chim. Acta* **70:**1194 (1987).
e. F. E. Ziegler, S. I. Klein, U. K. Pati, and T.-F. Wang, *J. Am. Chem. Soc.* **107:**2730 (1985).
f. M. E. Kuehne, *J. Org. Chem.* **35:**171 (1970).
g. D. A. Evans, S. L. Bender, and J. Morris, *J. Am. Chem. Soc.* **110:**2506 (1988).
h. K. Tomioka, Y.-S. Cho, F. Sato, and K. Koga, *J. Org. Chem.* **53:**4094 (1988).
i. K. Tomioka, H. Kawasaki, K. Yasuda, and K. Koga, *J. Am. Chem. Soc.* **110:**3597 (1988).
15a. T. Kametani, Y. Suzuki, H. Furuyama, and T. Honda, *J. Org. Chem.* **48:**31 (1983).
b. R. A. Kjonaas and D. D. Patel, *Tetrahedron Lett.* **25:**5467 (1984).
c. D. F. Taber and R. E. Ruckle, Jr., *J. Am. Chem. Soc.* **108:**7686 (1986).
d. M. Yamaguchi, M. Tsukamoto, and I. Hirao, *Tetrahedron Lett.* **26:**1723 (1985).
e. D. L. Snitman, M.-Y. Tsai, D. S. Watt, C. L. Edwards, and P. L. Stotter, *J. Org. Chem.* **44:**2838 (1979).
f. A. G. Schultz and J. P. Dittami, *J. Org. Chem.* **48:**2318 (1983).
16a. K. F. McClure and M. Z. Axt, *Biorg. Med. Chem. Lett.* **8:**143 (1998).
b. T. Honda, F. Ishikawa, K. Kanai, S. Sato, D. Kato, and H. Tominaga, *Heterocycles* **42:**109 (1996).
c. I. Vaulot, H.-J. Gais, N. Reuter, E. Schmitz, and R. K. L. Ossenkamp, *Eur. J. Org. Chem.* **1998:**805.
d. H. Pellissier, P.-Y. Michellys, and M. Santelli, *J. Org. Chem.* **62:**5588 (1997).
17. J. G. Henkel and L. A. Spurlock, *J. Am. Chem. Soc.* **95:**8339 (1973).
18. M. S. Newman, V. DeVries, and R. Darlak, *J. Org. Chem.* **31:**2171 (1966).
19. P. A. Manis and M. W. Rathke, *J. Org. Chem.* **45:**4952 (1980).
20. F. D. Lewis, T.-I. Ho, and R. J. DeVoe, *J. Org. Chem.* **45:**5283 (1980).
21. N. Langlois and H.-S. Wang, *Synth. Commun.* **27:**3133 (1997).

Chapter 2

1a. G. Ksander, J. E. McMurry, and N. Johnson, *J. Org. Chem.* **42:**1180 (1977).
b. J. Zabicky, *J. Chem. Soc.* **1961:**683.

c. G. Stork, G. A. Kraus, and G. A. Garcia, *J. Org. Chem.* **39:**3459 (1974).
d. H. Midorikawa, *Bull. Chem. Soc. Jpn.* **27:**210 (1954).
e. G. Stork and S. R. Dowd, *Org. Synth.* **55:**46 (1976).
f. E. C. Du Feu, F. J. McQuillin, and R. Robinson, *J. Chem. Soc.* **1937:**53.
g. E. Buchta, G. Wolfrum, and H. Ziener, *Chem. Ber.* **91:**1552 (1958).
h. L. H. Briggs and E. F. Orgias, *J. Chem. Soc. C* **1970:**1885.
i. J. A. Profitt and D. S. Watt, *Org. Synth.* **56:**1984 (1977).
j. U. Hengartner and V. Chu, *Org. Synth.* **58:**83 (1978).
k. E. Giacomini, M. A. Loreto, L. Pellacani, and P. A. Tardella, *J. Org. Chem.* **45:**519 (1980).
l. N. Narasimhan and R. Ammanamanchi, *J. Org. Chem.* **48:**3945 (1983).
m. M. P. Bosch, F. Camps, J. Coll, A. Guerro, T. Tatsouka, and J. Meinwald, *J. Org. Chem.* **51:**773 (1986).
2a. M. W. Rathke and D. F. Sullivan, *J. Am. Chem. Soc.* **95:**3050 (1973).
b. E. J. Corey, H. Yamamoto, D. K. Herron, and K. Achiwa, *J. Am. Chem. Soc.* **92:**6635 (1970).
c. E. J. Corey and D. E. Cane, *J. Org. Chem.* **36:**3070 (1971).
d. E. W. Yabkee and D. J. Cram, *J. Am. Chem. Soc.* **92:**6328 (1970).
e. W. G. Dauben, C. D. Poulter, and C. Suter, *J. Am. Chem. Soc.* **92:**7408 (1970).
f. P. A. Grieco and K. Hiroi, *J. Chem. Soc., Chem. Commun.* **1972:**1317.
g. T. Mukaiyama, M. Higo, and H. Takei, *Bull. Chem. Soc. Jpn.* **43:**2566 (1970).
h. I. Vlattas, I. T. Harrison, L. Tokes, J. H. Fried, and A. D. Cross, *J. Org. Chem.* **33:**4176 (1968).
i. A. T. Nielsen and W. R. Carpenter, *Org. Synth.* **V:**288 (1973).
j. M. L. Miles, T. M. Harris, and C. R. Hauser, *Org. Synth.* **V:**718 (1973).
k. A. P. Beracierta and D. A. Whiting, *J. Chem. Soc., Perkin Trans. 1* **1978:**1257.
l. T. Amatayakul, J. R. Cannon, P. Dampawan, T. Dechatiwongse, R. G. F. Giles, D. Huntrakul, K. Kusamran, M. Mokkhasamit, C. L. Raston, V. Reutrakul, and A. H. White, *Aust. J. Chem.* **32:**71 (1979).
m. R. M. Coates, S. K. Shah, and R. W. Mason, *J. Am. Chem. Soc.* **101:**6765 (1979).
n. K. A. Parker and T. H. Fedynyshyn, *Tetrahedron Lett.* **1979:**1657.
o. M. Miyashita and A. Yoshikishi, *J. Am. Chem. Soc.* **96:**1917 (1974).
p. W. R. Roush, *J. Am. Chem. Soc.* **102:**1390 (1980).
q. L. Fitjer and U. Quabeck, *Synth. Commun.* **15:**855 (1985).
r. A. Padwa, L. Brodsky, and S. Clough, *J. Am. Chem. Soc.* **94:**6767 (1972).
s. W. R. Roush, *J. Am. Chem. Soc.* **102:**1390 (1980).
t. C. R. Johnson, K. Mori, and A. Nakanishi, *J. Org. Chem.* **44:**2065 (1979).
u. T. Yanami, M. Miyashita, and A. Yoshikoshi, *J. Org. Chem.* **45:**607 (1980).
3a. K. D. Croft, E. L. Ghisalberti, P. R. Jefferies, and A. D. Stuart, *Aust. J. Chem.* **32:**2079 (1971).
b. L. H. Briggs and G. W. White, *J. Chem. Soc., C* **1971:**3077.
c. D. F. Taber and B. P. Gunn, *J. Am. Chem. Soc.* **101:**3992 (1979).
d. G. V. Kryshtal, V. V. Kulganek, V. F. Kucherov, and L. A. Yanovskaya, *Synthesis* **1979:**107.
e. S. F. Brady, M. A Ilton, and W. S. Johnson, *J. Am. Chem. Soc.* **90:**2882 (1968).
f. R. M. Coates and J. E. Shaw, *J. Am. Chem. Soc.* **92:**5657 (1970).
g. K. Mitsuhashi and S. Shiotoni, *Chem. Pharm. Bull.* **18:**75 (1970).
h. G. Wittig and H.-D. Frommeld, *Chem. Ber.* **97:**3548 (1964).
i. R. J. Sundberg, P. A. Bukowick, and F. O. Holcombe, *J. Org. Chem.* **32:**2938 (1967).
j. D. R. Howton, *J. Org. Chem.* **10:**277 (1945).
k. T. Yamane and K. Ogasawara, *Synlett* **1996:**925.
l. Y. Chan and W. W. Epstein, *Org. Synth.* **53:**48 (1973).
m. I. Fleming and M. Woolias, *J. Chem. Soc., Perkin Trans. 1* **1979:**827.
n. F. Johnson, K. G. Paul, D. Favara, R. Ciabatti, and U. Guzzi, *J. Am. Chem. Soc.* **104:**2190 (1982).
o. M. Ihara, M. Suzuki, K. Fukumoto, T. Kametani, and C. Kabuto, *J. Am. Chem. Soc.* **110:**1963 (1988).
p. D. J. Critcher, S. Connolly, and M. Wills, *J. Org. Chem.* **62:**6638 (1997).
q. S. P. Chavan and M. S. Venkatraman, *Tetrahedron Lett.* **39:**6745 (1998).
4a. W. A. Mosher and R. W. Soeder, *J. Org. Chem.* **36:**1561 (1971).
b. M. R. Roberts and R. H. Schlessinger, *J. Am. Chem. Soc.* **101:**7626 (1979).
c. J. E. McMurry and T. E. Glass, *Tetrahedron Lett.* **1971:**2575.
d. D. J. Cram, A. Langemann, and F. Hauck, *J. Am. Chem. Soc.* **81:**5750 (1959).
e. W. G. Dauben and J. Ipaktschi, *J. Am. Chem. Soc.* **95:**5088 (1973).
f. T. J. Curphey and H. L. Kim, *Tetrahedron Lett.* **1968:**1441.
g. K. P. Singh and L. Mandell, *Chem. Ber.* **96:**2485 (1963).

h. S. D. Lee, T. H. Chan, and K. S. Kwon, *Tetrahedron Lett.* **25:**3399 (1984).
i. J. F. Lavallee and P. Deslongchamps, *Tetrahedron Lett.* **29:**6033 (1988).
5. T. T. Howarth, G. P. Murphy, and T. M. Harris, *J. Am. Chem. Soc.* **91:**517 (1969).
6a. E. Vedejs, K. A. Snoble, and P. L. Fuchs, *J. Org. Chem.* **38:**1178 (1973).
b. P. B. Dervan and M. A. Shippey, *J. Am. Chem. Soc.* **98:**1265 (1976).
7a. E. E. Schweizer and G. J. O'Neil, *J. Org. Chem.* **30:**2082 (1965); E. E. Schweizer, *J. Am. Chem. Soc.* **86:**2744 (1984).
b. G. Büchi and H. Wüest, *Helv. Chim. Acta* **54:**1767 (1971).
c. G. H. Posner, S.-B. Lu, and E. Asirvathan, *Tetrahedron Lett.* **27:**659 (1986).
8. R. B. Woodward, F. Sondheimer, D. Taub, K. Heusler, and W. M. McLamore, *J. Am. Chem. Soc.* **74:**4223 (1952).
9. G. Stork, S. D. Darling, I. T. Harrison, and P. S. Wharton, *J. Am. Chem. Soc.* **84:**2018 (1962).
10. J. R. Pfister, *Tetrahedron Lett.* **1980:**1281.
11. R. M. Jacobson, G. P. Lahm, and J. W. Clader, *J. Org. Chem.* **45:**395 (1980).
12a. A. I. Meyers and N. Nazarenko, *J. Org. Chem.* **38:**175 (1973).
b. W. C. Still and F. L. Van Middlesworth, *J. Org. Chem.* **42:**1258 (1977).
13a. R. V. Stevens and A. W. M. Lee, *J. Am. Chem. Soc.* **101:**7032 (1979).
b. C. H. Heathcock, E. Kleinman, and E. S. Binkley, *J. Am. Chem. Soc.* **100:**8036 (1978).
14a. W. A. Kleschick and C. H. Heathcock, *J. Org. Chem.* **43:**1256 (1978).
b. S. D. Darling, F. N. Muralidharan, and V. B. Muralidharan, *Tetrahedron Lett.* **1979:**2761.
15a. M. Ertas and D. Seebach, *Helv. Chim. Acta* **68:**961 (1985).
b. S. Masamune, W. Choy, F. A. J. Kerdesky, and B. Imperiali, *J. Am. Chem. Soc.* **103:**1566 (1981).
c. C. H. Heathcock, C. T. Buse, W. A. Kleschick, M. C. Pirrung, J. E. Sohn, and J. Lampe, *J. Org. Chem.* **45:**1066 (1980).
d. R. Noyori, K. Yokoyama, J. Sakata, I. Kuwajima, E. Nakamura, and M. Shimizu, *J. Am. Chem. Soc.* **99:**1265 (1977).
e. D. A. Evans, E. Vogel, and J. V. Nelson, *J. Am. Chem. Soc.* **101:**6120 (1979); D. A. Evans, J. V. Nelson, E. Vogel, and T. R. Taber, *J. Am. Chem. Soc.* **103:**3099 (1981).
f. C. T. Buse and C. H. Heathcock, *J. Am. Chem. Soc.* **99:**8109 (1977).
g. R. Mahrwald, B. Costisella, and C. Gündogan, *Synthesis* **1998:**262.
h. S. F. Martin and D. E. Guinn, *J. Org. Chem.* **52:**5588 (1987).
16. A. Choudhury and E. R. Thornton, *Tetrahedron Lett.* **34:**2221 (1993).
17. D. Enders, O. F. Prokopenko, G. Raabe, and J. Runsink, *Synthesis* **1996:**1095.
18. H. Angert, R. Czerwonka, and H.-U. Reissig, *Liebigs Ann. Chem.* **1996:**259.
19. M. T. Reetz and A. Jung, *J. Am. Chem. Soc.* **105:**4833 (1983).

Chapter 3

1a. M. E. Kuehne and J. C. Bohnert, *J. Org. Chem.* **46:**3443 (1981).
b. B. C. Barot and H. W. Pinnick, *J. Org. Chem.* **46:**2981 (1981).
c. T. Mukaiyama, S. Shoda, and Y. Watanabe, *Chem. Lett.* **1977:**383.
d. H. Loibner and E. Zbiral, *Helv. Chim. Acta* **59:**2100 (1976).
e. E. J. Prisbe, J. Smejkal, J. P. H. Verheyden, and J. G. Moffatt, *J. Org. Chem.* **41:**1836 (1976).
f. B. D. MacKenzie, M. M. Angelo, and J. Wolinsky, *J. Org. Chem.* **44:**4042 (1979).
g. A. I. Meyers, R. K. Smith, and C. E. Whitten, *J. Org. Chem.* **44:**2250 (1979).
h. W. A. Bonner, *J. Org. Chem.* **32:**2496 (1967).
i. B. E. Smith and A. Burger, *J. Am. Chem. Soc.* **75:**5891 (1953).
j. W. D. Klobucar, L. A. Paquette, and J. F. Blount, *J. Org. Chem.* **46:**4021 (1981).
k. G. Grethe, V. Toome, H. L. Lee, M. Uskokovic, and A. Brossi, *J. Org. Chem.* **33:**504 (1968).
l. B. Neises and W. Steglich, *Org. Synth.* **63:**183 (1984).
2. A. W. Friederang and D. S. Tarbell, *J. Org. Chem.* **33:**3797 (1968).
3. H. R. Hudson and G. R. de Spinoza, *J. Chem. Soc., Perkin Trans. 1* **1976:**104.
4a. L. A. Paquette and M. K. Scott, *J. Am. Chem. Soc.* **94:**6760 (1972).
b. P. N. Confalone, G. Pizzolato, E. G. Baggiolini, D. Lollar, and M. R. Uskokovic, *J. Am. Chem. Soc.* **99:**7020 (1977).
c. E. L. Eliel, J. K. Koskimies, and B. Lohri, *J. Am. Chem. Soc.* **100:**1614 (1978).

d. H. Hagiwara, M. Numata, K. Konishi, and Y. Oka, *Chem. Pharm. Bull.* **13:**253 (1965).
e. A. S. Kende and T. P. Demuth, *Tetrehedron Lett.* **1980:**715.
f. P. A. Grieco, D. S. Clark, and G. P. Withers, *J. Org. Chem.* **44:**2945 (1979).
g. J. Yu, J. R. Falck, and C. Mioskowski, *J. Org. Chem.* **57:**3757 (1992).
h. J. Freedman, M. J. Vaal, and E. W. Huber, *J. Org. Chem.* **56:**670 (1991).
5a, b. D. Seebach, H.-O. Kalinowski, B. Bastani, G. Crass, H. Daum, H. Dorr, N. P. DuPreez, V. Ehring, W. Langer, C. Nussler, H.-A. Oei, and M. Schmidt, *Helv. Chim. Acta* **60:**301 (1977).
c. G. L. Baker, S. J. Fritschel, J. R. Stille, and J. K. Stille, *J. Org. Chem.* **46:**2954 (1981).
d. D. Seebach, H.-O. Kalinowski, B. Bastani, G. Crass, H. Daum, H. Dorr, N. P. DuPreez, V. Ehring, W. Langer, C. Nussler, H.-A. Oei, and M. Schmidt, *Helv. Chim. Acta* **60:**301 (1977).
e. M. D. Fryzuk and B. Bosnich, *J. Am. Chem. Soc.* **99:**6262 (1977).
f. S. Hanessian and R. Frenette, *Tetrahedron Lett.* **1979:**3391.
g. K. G. Paul, F. Johnson, and D. Favara, *J. Am. Chem. Soc.* **98:**1285 (1976).
h. S. D. Burke, J. Hong, J. R. Lennox, and A. P. Mongin, *J. Org. Chem.* **63:**6952 (1998).
i. G. Dujardin, S. Rossignol, and E. Brown, *Synthesis* **1998:**763.
6a. P. Henley-Smith, D. A. Whiting, and A. F. Wood, *J. Chem. Soc., Perkin Trans. 1* **1980:**614.
b. P. Beak and L. G. Carter, *J. Org. Chem.* **46:**2363 (1981).
c. M. E. Jung and T. J. Shaw, *J. Am. Chem. Soc.* **102:**6304 (1980).
d. P. N. Swepston, S.-T. Lin, A. Hawkins, S. Humphrey, S. Siegel, and A. W. Cordes, *J. Org. Chem.* **46:**3754 (1981).
e. P. J. Maurer and M. J. Miller, *J. Org. Chem.* **46:**2835 (1981).
f. N. A. Porter, J. D. Byers, A. E. Ali, and T. E. Eling, *J. Am. Chem. Soc.* **102:**1183 (1980).
g. G. A. Olah, B. G. B. Gupta, R. Malhotra, and S. C. Narang, *J. Org. Chem.* **45:**1638 (1980).
7a. A. K. Bose, B. Lal, W. Hoffman III, and M. S. Manhas, *Tetrahedron Lett.* **1973:**1619.
b. J. B. Hendrickson and S. M. Schwartzman, *Tetrahedron Lett.* **1975:**277.
c. J. F. King, S. M. Loosmore, J. D. Lock, and M. Aslam, *J. Am. Chem. Soc.* **100:**1637 (1978); C. N. Sukenik and R. G. Bergman, *J. Am. Chem. Soc.* **98:**6613 (1976).
d. R. S. Freedlander, T. A. Bryson, R. B. Dunlap, E. M. Schulman, and C. A. Lewis, Jr., *J. Org. Chem.* **46:**3519 (1981).
e. A. Trzeciak and W. Bannwarth, *Synthesis* **1996:**1433.
8a. J. Jacobus, M. Raban, and K. Mislow, *J. Org. Chem.* **33:**1142 (1968).
b. M. Schmid and R. Barner, *Helv. Chim. Acta* **62:**464 (1979).
c. V. Eswarakrishnan and L. Field, *J. Org. Chem.* **46:**4182 (1981).
d. R. F. Borch, A. J. Evans, and J. J. Wade, *J. Am. Chem. Soc.* **99:**1612 (1977).
e. H. S. Aaron and C. P. Ferguson, *J. Org. Chem.* **33:**684 (1968).
9. B. Koppenhoefer and V. Schuring, *Org. Synth.* **66:**151, 160 (1987).
10. B. E. Watkins and H. Rapoport, *J. Org. Chem.* **47:**4471 (1982).
11. A. Brändstrom, *Adv. Phys. Org. Chem.* **15:**267 (1977); D. Landini, A. Maia, and A. Pampoli, *J. Org. Chem.* **51:**5475 (1986).
12. R. M. Magid, O. S. Fruchey, W. L. Johnson, and T. G. Allen, *J. Org. Chem.* **44:**359 (1979).
13. R. B. Woodward, R. A. Olofson, and H. Mayer, *Tetrahedron Suppl.* **8:**321 (1966); R. B. Woodward and R. A. Olofson, *J. Am. Chem. Soc.* **83:**1007 (1961); B. Belleau and G. Malek, *J. Am. Chem. Soc.* **90:**1651 (1968).
14a. E. J. Corey, K. C. Nicolaou, and L. S. Melvin, Jr., *J. Am. Chem. Soc.* **97:**654 (1975).
b. L. M. Beacham III, *J. Org. Chem.* **44:**3100 (1979).
c. J. Huang and J. Meinwald, *J. Am. Chem. Soc.* **103:**861 (1981).
d. P. Beak and L. G. Carter, *J. Org. Chem.* **46:**2363 (1981).
15. T. Mukaiyama, S. Shoda, T. Nakatsuka, and K. Narasaka, *Chem. Lett.* **1978:**605.
16. R. U. Lemieux, K. B. Hendriks, R. V. Stick, and K. James, *J. Am. Chem. Soc.* **97:**4056 (1975).
17a. T. Mukaiyama, R. Matsueda, and M. Suzuki, *Tetrahedron Lett.* **1970:**1901.
b. E. J. Corey and D. A. Clark, *Tetrahedron Lett.* **1979:**2875.

Chapter 4

1a. N. Kharasch and C. M. Buess, *J. Am. Chem. Soc.* **71:**2724 (1949).
b. I. Heilbron, E. R. H. Jones, M. Julia, and B. C. L. Weedon, *J. Chem. Soc.* **1949:**1823.
c. A. J. Sisti, *J. Org. Chem.* **33:**3953 (1968).

d. H. C. Brown and G. Zweifel, *J. Am. Chem. Soc.* **83:**1241 (1961).
e. F. W. Fowler, A. Hassner, and L. A. Levy, *J. Am. Chem. Soc.* **89:**2077 (1967).
f. A. Hassner and F. W. Fowler, *J. Org. Chem.* **33:**2686 (1968).
g. A. Padwa, T. Blacklock, and A. Tremper, *Org. Synth.* **57:**83 (1977).
h. I. Ryu, S. Murai, I. Niwa, and N. Sonoda, *Synthesis* **1977:**874.
i. R. A. Amos and J. A. Katzenellenbogen, *J. Org. Chem.* **43:**560 (1978).
j. H. C. Brown and G. J. Lynch, *J. Org. Chem.* **46:**531 (1981).
k. R. C. Cambie, R. C. Hayward, P. S. Rutledge, T. Smith-Palmer, B. E. Swedlund, and P. D. Woodgate, *J. Chem. Soc., Perkin Trans. 1* **1979:**180.
l. A. B. Holmes, K. Russell, E. S. Stern, M. E. Stubbs, and N. K. Welland, *Tetrahedron Lett.* **25:**4183 (1984).
m. N. S. Zefirov, T. N. Velikokhat'ko, and N. K. Sadovaya, *Zh. Org. Khim. (Engl. trans.)* **19:**1407 (1983).
n. F. B. Gonzalez and P. A. Bartlett, *Org. Synth.* **64:**175 (1985).
2. D. J. Pasto and J. A. Gontarz, *J. Am. Chem. Soc.* **91:**2147 (1969).
3. D. J. Pasto and C. C. Cumbo, *J. Am. Chem. Soc.* **86:**4343 (1964).
4. D. J. Pasto and J. A. Gontarz, *J. Am. Chem. Soc.* **93:**6902 (1971).
5. R. Gleiter and G. Müller, *J. Org. Chem.* **53:**3912 (1988).
6. G. Stork and R. Borch, *J. Am. Chem. Soc.* **86:**935 (1964).
7. T. Hori and K. B. Sharpless, *J. Org. Chem.* **43:**1689 (1978).
8a. E. Kloster-Jensen, E. Kovats, A. Eschenmoser, and E. Heilbronner, *Helv. Chim. Acta* **39:**1051 (1956).
b. P. N. Rao, *J. Org. Chem.* **36:**2426 (1971).
c. R. A. Moss and E. Y. Chen, *J. Org.Chem.* **46:**1466 (1981).
d. J. M. Jerkunica and T. G. Traylor, *Org. Synth.* **53:**94 (1973).
e. G. Zweifel and C. C. Whitney, *J. Am. Chem. Soc.* **89:**2753 (1967).
f. R. E. Ireland and P. Bey, *Org. Synth.* **53:**63 (1973).
g. W. I. Fanta and W. F. Erman, *J. Org. Chem.* **33:**1656 (1968).
h. W. E. Billups, J. H. Cross, and C. V. Smith, *J. Am. Chem. Soc.* **95:**3438 (1973).
i. G. W. Kabalka and E. E. Gooch III, *J. Org. Chem.* **45:**3578 (1980).
j. E. J. Corey, G. Wess, Y. B. Xiang, and A. K. Singh, *J. Am. Chem. Soc.* **109:**4717 (1987).
k. I. Nakatsuka, N. L. Ferreira, W. C. Eckelman, B. E. Francis, W. J. Rzeszotarski, R. E. Gibson, E. M. Jagoda, and R. C. Reba, *J. Med. Chem.* **27:**1287 (1984).
l. W. Oppolzer, H. Hauth, P. Pfaffli, and R. Wenger, *Helv. Chim. Acta* **60:**1801 (1977).
m. G. H. Posner and P. W. Tang, *J. Org. Chem.* **43:**4131 (1978).
n. A. V. Bayquen and R. W. Read, *Tetrahedron* **52:**13467 (1996).
o. T. Fukuyama and G. Liu, *J. Am. Chem. Soc.* **118:**7426 (1996).
9a. E. J. Corey and H. Estreicher, *J. Am. Chem. Soc.* **100:**6294 (1978).
b. E. J. Corey and H. Estreicher, *Tetrahedron Lett.* **1980:**1113.
c. G. A. Olah and M. Nohima, *Synthesis* **1973:**785.
10. D. J. Pasto and F. M. Klein, *Tetrahedron Lett.* **1967:**963.
11. H. J. Reich, J. M. Renga, and I. L. Reich, *J. Am. Chem. Soc.* **97:**5434 (1975).
12. H. C. Brown, G. J. Lynch, W. J. Hammar, and L. C. Liu, *J. Org. Chem.* **44:**1910 (1979).
13. P. A. Bartlett and J. Myerson, *J. Am. Chem. Soc.* **100:**3950 (1978).
14a. R. Lavilla, O. Coll, M. Nicolas, and J. Bosch, *Tetrahedron Lett.* **39:**5089 (1998).
b. A. Garofalo, M. B. Hursthouse, K. M. A. Malik, H. F. Olivio, S. M. Roberts, and V. Sik, *J. Chem. Soc., Perkin Trans. 1* **1994:**1311.
c. H. Imagawa, T. Shigaraki, T. Suzuki, H. Takao, H. Yamada, T. Sugihara, and M. Nichizawa, *Chem. Pharm. Bull.* **46:**1341 (1998).
d. A. G. Schultz and S. J. Kirmich, *J. Org. Chem.* **61:**5626 (1996).
e. I. Fleming and N. J. Lawrence, *J. Chem. Soc., Perkin Trans. 1* **1992:**3309; **1998:**2679.
f. R. Bittman, H.-S. Byun, K. C. Reddy, P. Samadder, and G. Arthur, *J. Med. Chem.* **40:**1391 (1997).
15a. D. A. Evans, J. E. Ellman, and R. L. Dorow, *Tetrahedron Lett.* **28:**1123 (1987).
b. K. C. Nicolaou, R. L. Magolda, W. J. Sipio, W. E. Barnette, Z. Lysenko, and M. M. Joullie, *J. Am. Chem. Soc.* **102:**3784 (1980).
c. K. C. Nicolaou, W. E. Barnette, and R. L. Magolda, *J. Am. Chem. Soc.* **103:**3472 (1981).
d. S. Knapp, A. T. Levorse, and J. A. Potenza, *J. Org. Chem.* **53:**4773 (1988).
e. T. H. Jones and M. S. Blum, *Tetrahedron Lett.* **22:**4373 (1981).
16a. W. T. Smith and G. L. McLeod, *Org. Synth.* **IV:**345 (1963).
b. K. E. Harding, T. H. Marman, and D. Nam, *Tetrahedron Lett.* **29:**1627 (1988).

c. S. Terahima, M. Hayashi, and K. Koga, *Tetrahedron Lett.* **1980:**2733.
17. A. Toshimitsu, K. Terao, and S. Uemura, *J. Org. Chem.* **51:**1724 (1986).
18. W. Oppolzer and P. Dudfield, *Tetrahedron Lett.* **26:**5037 (1985); D. A. Evans, J. A. Ellman, and R. L. Dorow, *Tetrahedron Lett.* **28:**1123 (1987).
19. T. W. Bell, *J. Am. Chem. Soc.* **103:**1163 (1981).
20a. H. C. Brown and B. Singaran, *J. Am. Chem. Soc.* **106:**1794 (1984).
b. S. Masamune, B. M. Kim, J. S. Petersen, T. Sato, S. J. Veenstra, and T. Imai, *J. Am. Chem. Soc.* **107:**4549 (1985).
21. C. F. Palmer, K. D. Parry, S. M. Roberts, and V. Sik, *J. Chem. Soc., Perkin Trans. 1* **1992:**1021; C. F. Palmer and R. McCague, *J. Chem. Soc., Perkin Trans. 1* **1998:**2977; A. Toyota, A. Nishimura, and C. Kaneko, *Heterocycles* **45:**2105 (1997).
22a. M. Noguchi, H. Okada, M. Watanabe, K. Okuda, and O. Nakamura, *Tetrahedron* **52:**6581 (1996).
b. R. Madsen, C. Roberts, and B. Fraser-Reid, *J. Org. Chem.* **60:**7920 (1995).

Chapter 5

1a. W. R. Roush, *J. Am. Chem. Soc.* **102:**1390 (1980).
b. H. C. Brown, S. C. Kim, and S. Krishnamurthy, *J. Org. Chem.* **45:**1 (1980).
c. G. W. Kabalka, D. T. C. Yang, and J. D. Baker, Jr., *J. Org. Chem.* **41:**574 (1976).
d. J. K. Whitesell, R. S. Matthews, M. A. Minton, and A. M. Helbling, *J. Am. Chem. Soc.* **103:**3468 (1981).
e. K. S. Kim, M. W. Spatz, and F. Johnson, *Tetrahedron Lett.* **1979:**331.
f. M.-H. Rei, *J. Org. Chem.* **44:**2760 (1979).
g. R. O. Hutchins, D. Kandasamy, F. Dux III, C. A. Maryanoff, D. Rolstein, B. Goldsmith, W. Burgoyne, F. Cistone, J. Dalessandro, and J. Puglis, *J. Org. Chem.* **43:**2259 (1978).
h. H. Lindlar, *Helv. Chim. Acta* **35:**446 (1952).
i. E. Vedejs, R. A. Buchanan, R. Conrad, G. P. Meier, M. J. Mullins, and Y. Watanabe, *J. Am. Chem. Soc.* **109:**5878 (1987).
j. C. B. Jackson and G. Pattenden, *Tetrahedron Lett.* **26:**3393 (1985).
2. D. C. Wigfield and D. J. Phelps, *J. Am. Chem. Soc.* **96:**543 (1974).
3a. E. J. Corey, T. K. Schaaf, W. Huber, U. Koelliker, and N. M. Weinshenker, *J. Am. Chem. Soc.* **92:**397 (1970).
b. E. J. Corey and R. Noyori, *Tetrahedron Lett.* **1970:**311.
c. R. F. Borch, *Org. Synth.* **52:**124 (1972).
d. D. Seyferth and V. A. Mai, *J. Am. Chem. Soc.* **92:**7412 (1970).
e. R. V. Stevens and J. T. Lai, *J. Org. Chem.* **37:**2138 (1972).
f. M. J. Robins and J. S. Wilson, *J. Am. Chem. Soc.* **103:**932 (1981).
g. G. R. Pettit and J. R. Dias, *J. Org. Chem.* **36:**3207 (1971).
h. P. A. Grieco, T. Oguri, and S. Gilman, *J. Am. Chem. Soc.* **102:**5886 (1980).
i. M. F. Semmelhack, S. Tomoda, and K. M. Hurst, *J. Am. Chem. Soc.* **102:**7567 (1980).
j. H. C. Brown and P. Heim, *J. Org. Chem.* **38:**912 (1973).
k. R. O. Hutchins and N. R. Natale, *J. Org. Chem.* **43:**2299 (1978).
l. M. R. Detty and L. A. Paquette, *J. Am. Chem. Soc.* **99:**821 (1977).
m. C. A. Bunnell and P. L. Fuchs, *J. Am. Chem. Soc.* **99:**5184 (1977).
n. Y.-J. Wu and D. J. Burnell, *Tetrahedron Lett.* **29:**4369 (1988).
o. P. W. Collins, E. Z. Dajani, R. Pappo, A. F. Gasiecki, R. G. Bianchi, and E. M. Woods, *J. Med. Chem.* **26:**786 (1983).
4a. F. A. Carey, D. H. Ball, and L. Long, Jr., *Carbohydr. Res.* **3:**205 (1966).
b. D. J. Cram and R. A. Abd Elhafez, *J. Am. Chem. Soc.* **74:**5828 (1952).
c. R. N. Rej, C. Taylor, and G. Eadon, *J. Org. Chem.* **45:**126 (1980).
d. M. C. Dart and H. B. Henbest, *J. Chem. Soc.* **1960:**3563.
e. E. Piers, W. deWaal, and R. W. Britton, *J. Am. Chem. Soc.* **93:**5113 (1971).
f. A. L. J. Beckwith and C. Easton, *J. Am. Chem. Soc.* **100:**2913 (1978).
g. D. Horton and W. Weckerle, *Carbohydr. Res.* **44:**227 (1975).

h. R. A. Holton and R. M. Kennedy, *Tetrahedron Lett.* **28:**303 (1987).
i. H. Iida, N. Yamazaki, and C. Kibayashi, *J. Org. Chem.* **51:**1069, 3769 (1986).
j. D. A. Evans and M. M. Morrissey, *J. Am. Chem. Soc.* **106:**3866 (1984).
k. N. A. Porter, C. B. Ziegler, Jr., F. F. Khouri, and D. H. Roberts, *J. Org. Chem.* **50:**2252 (1985).
l. G. Stork and D. E. Kahne, *J. Am. Chem. Soc.* **105:**1072 (1983).
m. Y. Yamamoto, K. Matsuoka, and H. Nemoto, *J. Am. Chem. Soc.* **110:**4475 (1988).
n. G. Palmisano, B. Danieli, G. Lesma, D. Passarella, and L. Toma, *J. Org. Chem.* **56:**2380 (1991).
o. D. A. Evans, S. J. Miller, and M. D. Ennis, *J. Org. Chem.* **58:**471 (1993).
p. A. G. Schultz and N. J. Green, *J. Am. Chem. Soc.* **113:**4931 (1991).
q. Y. Yamamoto, K. Matsuoka, and H. Nemoto, *J. Am. Chem. Soc.* **110:**4475 (1988).
5a. D. Lenoir, *Synthesis* **1977:**553.
b. J. A. Marshall and A. E. Greene, *J. Org. Chem.* **36:**2035 (1971).
c. B. M. Trost, Y. Nishimura, and K. Yamamoto, *J. Am. Chem. Soc.* **101:**1328 (1979); J. E. McMurry, A. Andrus, G. M. Ksander, J. H. Musser, and M. A. Johnson, *J. Am. Chem. Soc.* **101:**1330 (1979).
d. R. E. Ireland and C. S. Wilcox, *J. Org. Chem.* **45:**197 (1980).
e. P. G. Gassman and T. J. Atkins, *J. Am. Chem. Soc.* **94:**7748 (1972).
f. A. Gopalan and P. Magnus, *J. Am. Chem. Soc.* **102:**1756 (1980).
g. R. M. Coates, S. K. Shah, and R. W. Mason, *J. Am. Chem. Soc.* **101:**6765 (1979); Y.-K. Han and L. A. Paquette, *J. Org. Chem.* **44:**3731 (1979).
h. L. P. Kuhn, *J. Am. Chem. Soc.* **80:**5950 (1958).
i. R. P. Hatch, J. Shringarpure, and S. M. Weinreb, *J. Org. Chem.* **43:**4172 (1978).
j. T. Shono, Y. Matsumura, S. Kashimura, and H. Kyutoko, *Tetrahedron Lett.* **1978:**1205.
6a, b. K. E. Wiegers and S. G. Smith, *J. Org. Chem.* **43:**1126 (1978).
c. D. C. Wiegfield and F. W. Gowland, *J. Org. Chem.* **45:**653 (1980).
7. D. Caine and T. L. Smith, Jr., *J. Org. Chem.* **43:**755 (1978).
8. R. E. A. Dear and F. L. M. Pattison, *J. Am. Chem. Soc.* **85:**622 (1963).
9a. S. Danishefsky, M. Hirama, K. Gombatz, T. Harayama, E. Berman, and P. F. Schuda, *J. Am. Chem. Soc.* **101:**7020 (1979).
b. A. S. Kende, M. L. King, and D. P. Curran, *J. Org. Chem.* **46:**2826 (1981).
c. A. P. Kozikowski and A. Ames, *J. Am. Chem. Soc.* **103:**3923 (1981).
d. E. J. Corey, S. G. Pyre, and W. Su, *Tetrahedron Lett.* **24:**4883 (1983).
e. T. Rosen and C. Heathcock, *J. Am. Chem. Soc.* **107:**3731 (1985).
f. T. Fujisawa and T. Sato, *Org. Synth.* **66:**121 (1987).
g. H. J. Liu and M. G. Kulkarni, *Tetrahedron Lett.* **26:**4847 (1985).
h. T. Fujisawa and T. Sato, *Org. Synth.* **66:**121 (1987).
i. D. A. Evans, S. J. Miller, and M. D. Ennis, *J. Org. Chem.* **58:**471 (1993).
j. D. L. J. Clive, K. S. K. Murthy, A. G. H. Wee, J. S. Prasad, G. V. J. da Silva, M. Majewski, P. C. Anderson, C. F. Evans, R. D. Haugen, L. D. Heerze, and J. R. Barrie, *J. Am. Chem. Soc.* **112:**3018 (1990).
10a. H. C. Brown and W. C. Dickason, *J. Am. Chem. Soc.* **92:**709 (1970).
b. D. Seyferth, H. Yamazaki, and D. L. Alleston, *J. Org. Chem.* **28:**703 (1963).
c. G. Stork and S. D. Darling, *J. Am. Chem. Soc.* **82:**1512 (1960).
11a. R. F. Borch, *Org. Synth.* **52:**124 (1972).
b. R. F. Borch, M. D. Bernstein, and H. D. Durst, *J. Am. Chem. Soc.* **93:**2897 (1971).
c. A. D. Harmon and C. R. Hutchinson, *Tetrahedron Lett.* **1973:**1293.
12. S.-K. Chung and F.-F. Chung, *Tetrahedron Lett.* **1979:**2473.
13. D. F. Taber, *J. Org. Chem.* **41:**2649 (1976).
14. D. R. Briggs and W. B. Whalley, *J. Chem. Soc., Perkin Trans. 1* **1976:**1382.
15. J. R. Flisak and S. S. Hall, *J. Am. Chem. Soc.* **112:**7299 (1990).
16. R. Yoneda, S. Harusawa, and T. Kurihara, *J. Org. Chem.* **56:**1827 (1991).
17. H.-C. Zhang, B. D. Harris, M. J. Costanzo. E, C. Lawson, C. A. Maryanoff, and B. E. Maryanoff, *J. Org. Chem.* **63:**7964 (1998).
18. C. M. Tice and C. H. Heathcock, *J. Org. Chem.* **46:**9 (1981).
19. S. Kaneko, N. Nakajima, M. Shikano. T. Katoh, and S. Terahima, *Tetrahedron* **54:**5485 (1998).
20a. N. J. Leonard and S. Gelfand, *J. Am. Chem. Soc.* **77:**3272 (1955).
b. P. S. Wharton and D. H. Bohlen, *J. Org. Chem.* **26:**3615 (1961); W. R. Benn and R. M. Dodson, *J. Org. Chem.* **29:**1142 (1964).
c. G. Lardelli and O. Jeger, *Helv. Chim. Acta* **32:**1817 (1949).

d. R. J. Petersen and P. S. Skell, *Org. Synth.* **V**:929 (1973).
21a. N. M. Yoon, C. S. Pak, H. C. Brown, S. Krishnamurthy, and T. P. Stocky, *J. Org. Chem.* **38**:2786 (1973).
b. D. J. Dawson and R. E. Ireland, *Tetrahedron Lett.* **1968**:1899.
c. R. O. Hutchins, C. A. Milewski, and B. A. Maryanoff, *J. Am. Chem. Soc.* **95**:3662 (1973).
d. M. J. Kornet, P. A. Thio, and S. I. Tan, *J. Org. Chem.* **33**:3637 (1968).
e. C. T. West, S. J. Donnelly, D. A. Kooistra, and M. P. Doyle, *J. Org. Chem.* **38**:2675 (1973).
f. M. R. Johnson and B. Rickborn, *J. Org. Chem.* **35**:1041 (1970).
g. N. Akubult and M. Balci, *J. Org. Chem.* **53**:3338 (1988).
22. H. Iida, N. Yamazaki, and C. Kibayashi, *J. Org. Chem.* **51**:3769 (1986).
23a. H. C. Brown, G. G. Pai, and P. K. Jadhav, *J. Am. Chem. Soc.* **106**:1531 (1984).
b. E. J. Corey, R. K. Bakshi, S. Shibata, C.-P. Chen, and V. K. Singh, *J. Am. Chem. Soc.* **109**:7925 (1987).
c. M. Srebnik, P. V. Ramachandran, and H. C. Brown, *J. Org. Chem.* **53**:2916 (1988).
24. L. A. Paquette, T. J. Nitz, R. J. Ross, and J. P. Springer, *J. Am. Chem. Soc.* **106**:1446 (1984).
25. J. A. Marshall, *Acc. Chem. Res.* **13**:213 (1980); J. A. Marshall and K. E. Flynn, *J. Am. Chem. Soc.* **106**:723 (1984); J. A. Marshall, J. C. Peterson, and L. Lebioda, *J. Am. Chem. Soc.* **106**:6006 (1984).
26a. J. P. Guidot, T. Le Gall, and C. Mioskowski, *Tetrahedron Lett.* **35**:6671 (1994).
b. M. Schwaebe and R. D. Little, *J. Org. Chem.* **61**:3240 (1996).
c. T. Kan, S. Hosokawa, S. Naja, M. Oikawa, S. Ito, F. Matsuda, and H. Shirahama, *J. Org. Chem.* **59**:5532 (1994).
d. E. J. Enholm, H. Satici, and A. Trivellas, *J. Org. Chem.* **54**:5841 (1989).
e. E. J. Enholm and A. Trivellas, *Tetrahedron Lett.* **35**:1627 (1994).

Chapter 6

1a. B. M. Trost, S. A. Godleski, and J. P. Genet, *J. Am. Chem.* **100**:3930 (1978).
b. M. E. Jung and C. A. McCombs, *J. Am. Chem. Soc.* **100**:5207 (1978).
c. L. E. Overman and P. J. Jessup, *J. Am. Chem. Soc.* **100**:5179 (1978).
d. C. Cupas, W. E. Watts, and P. von R. Schleyer, *Tetrahedron Lett.* **1964**:2503.
e. T. C. Jain, C. M. Banks, and J. E. McCloskey, *Tetrahedron Lett.* **1970**:841.
f. G. Büchi and J. E. Powell, Jr., *J. Am. Chem. Soc.* **89**:4559 (1967).
g. M. Raban, F. B. Jones, Jr., E. H. Carlson, E. Banucci, and N. A. LeBel, *J. Org. Chem.* **35**:1497 (1970).
h. H. Yamamoto and H. L. Sham, *J. Am. Chem. Soc.* **101**:1609 (1979).
i. H. O. House, T. S. B. Sayer, and C.-C. Yau, *J. Org. Chem.* **43**:2153 (1978).
j. M. C. Pirrung, *J. Am. Chem. Soc.* **103**:82 (1981).
k. M. Sevrin and A. Krief, *Tetrahedron Lett.* **1978**:187.
l. L. A. Paquette, G. D. Crouse, and A. K. Sharma, *J. Am. Chem. Soc.* **102**:3972 (1980).
m. N.-K. Chan and G. Saucy, *J. Org. Chem.* **42**:3838 (1977).
n, o. J. A. Marshall and J. Lebreton, *J. Org. Chem.* **53**:4108 (1988).
p. R. L. Funk, W. J. Daily, and M. Parvez, *J. Org. Chem.* **53**:4142 (1988).
q. J. D. Winkler, J. P. Hey, and P. G. Williard, *Tetrahedron Lett.* **29**:4691 (1988).
r. F. A. J. Kerdesky, R. J. Ardecky, M. V. Lakshmikathan, and M. P. Cava, *J. Am. Chem. Soc.* **103**:1992 (1981).
s. B. B. Snider and R. A. H. F. Hui, *J. Org. Chem.* **50**:5167 (1985).
t. L. Lambs, N. P. Singh, and J.-F. Biellmann, *J. Org. Chem.* **57**:6301 (1992).
u. M. T. Reetz and E. H. Lauterbach, *Tetrahedron Lett.* **32**:4481 (1991).
v. B. Coates, D. Montgomery, and P. J. Stevenson, *Tetrahedron Lett.* **32**:4199 (1991).
w. K. Honda, S. Inoue, and K. Sato, *J. Org. Chem.* **57**:428 (1992).
x. D. Kim, S. K. Ahn, H. Bae, W. J. Choi, and H. S. Kim, *Tetrahedron Lett.* **38**:4437 (1997).
y. K. Tanaka, T. Imase, and S. Iwata, *Bull. Chem. Soc. Jpn.* **69**:2243 (1996).
2a. W. Oppolzer and M. Petrzilka, *J. Am. Chem. Soc.* **98**:6722 (1976).
b. A. Padwa and N. Kamigata, *J. Am. Chem. Soc.* **99**:1871 (1977).
c. P. A. Jacobi, A. Brownstein, M. Martinelli, and K. Grozinger, *J. Am. Chem. Soc.* **103**:239 (1981).
d. H. W. Gschwend, A. O. Lee, and H.-P. Meier, *J. Org. Chem.* **38**:2169 (1973).

e. T. Kametani, M. Tsubuki, Y. Shiratori, H. Nemoto, M. Ihara, K. Fukumoto, F. Satoh, and H. Inoue, *J. Org. Chem.* **42**:2672 (1977).
f. J. L. Gras and M. Bertrand, *Tetrahedron Lett.* **1979**:4549.
g. H. Seto, M. Sakaguchi, Y. Fujimoto, T. Tatsuno, and H. Yoshioka, *Chem. Pharm. Bull.* **33**:412 (1985).
3a. K. C. Brannock, A. Bell, R. D. Burpitt, and C. A. Kelly, *J. Org. Chem.* **29**:801 (1964).
b. K. Ogura, S. Furukawa, and G. Tsuchihashi, *J. Am. Chem. Soc.* **102**:2125 (1980).
c. E. Vedejs, M. J. Arco, D. W. Powell, J. M. Renga, and S. P. Singer, *J. Org. Chem.* **43**:4831 (1978).
d. J. J. Tufariello and J. J. Tegeler, *Tetrahedron Lett.* **1976**:4037.
e. W. A. Thaler and B. Franzus, *J. Org. Chem.* **29**:2226 (1964).
f. B. B. Snider, *J. Org. Chem.* **41**:3061 (1976).
g. L. A. Paquette, *J. Org. Chem.* **29**:2851 (1964).
h. M. E. Monk and Y. K. Kim, *J. Am. Chem. Soc.* **86**:2213 (1964).
i. B. Cazes and S. Julia, *Bull. Soc. Chim. Fr.* **1977**:925.
j. D. L. Boger and D. D. Mullican, *Org. Synth.* **65**:98 (1987).
4a. P. E. Eaton and U. R. Chakraborty, *J. Am. Chem. Soc.* **100**:3634 (1978).
b. H. Hogeveen and B. J. Nusse, *J. Am. Chem. Soc.* **100**:3110 (1978).
c. T. Oida, S. Tanimoto, T. Sugimoto, and M. Okano, *Synthesis* **1980**:131.
d. J. N. Labovitz, C. A. Henrick, and V. L. Corbin, *Tetrahedron Lett.* **1975**:4209.
e. W. Steglich and L. Zechlin, *Chem. Ber.* **111**:3939 (1978).
f. F. D. Lewis and R. J. DeVoe, *J. Org. Chem.* **45**:948 (1980).
g. S. P. Tanis and K. Nakanishi, *J. Am. Chem. Soc.* **101**:4398 (1979).
h. S. Danishefsky, M. P. Prisbylla, and S. Hiner, *J. Am. Chem. Soc.* **100**:2918 (1978).
i. T. Hudlicky, F. J. Koszyk, T. M. Kutchan, and J. P. Sheth, *J. Org. Chem.* **45**:5020 (1980).
j. G. Li, Z. Li, and X. Fang, *Synth. Commun.* **26**:2569 (1996).
k. C. Chen and D. J. Hart, *J. Org. Chem.* **55**:6236 (1990).
l. S. Doye, T. Hotopp, and E. Winterfeldt, *J. Chem. Soc., Chem Commun.* **1997**:1491.
m. S. Chackalamannil, R. J. Davies, Y. Wang, T. Asberom, D. Doller, J. Wong, D. Leone, and A. T. McPhail, *J. Org. Chem.* **64**:1932 (1999).
5. H. E. Zimmerman, G. L. Grunewald, R. M. Paufler, and M. A. Sherwin, *J. Am. Chem. Soc.* **91**:2330 (1969).
6. C. J. Albisetti, N. G. Fisher, M. J. Hogsed, and R. M. Joyce, *J. Am. Chem. Soc.* **78**:2637 (1956).
7. N. Shimizu, M. Ishikawa, K. Ishikura, and S. Nishida, *J. Am. Chem. Soc.* **96**:6456 (1974).
8. R. Schug and R. Huisgen, *J. Chem. Soc., Chem. Commun.* **1975**:60.
9. W. L. Howard and N. B. Lorette, *Org. Synth.* **V**:25 (1973).
10. S. Danishefsky, M. Hirama, N. Fritsch, and J. Clardy, *J. Am. Chem. Soc.* **101**:7013 (1979).
11. D. A. Evans, C. A. Bryan and C. L. Sims, *J. Am. Chem. Soc.* **94**:2891 (1972).
12. J. Wolinsky and R. B. Login, *J. Org. Chem.* **35**:3205 (1970).
13. C. H. Heathcock and R. A. Badger, *J. Org. Chem.* **37**:234 (1972).
14. B. J. Arnold, S. M. Mellows, P. G. Sammes, and T. W. Wallace, *J. Chem. Soc., Perkin Trans. 1* **1974**:401; B. J. Arnold, P. G. Sammes, and T. W. Wallace, *J. Chem. Soc., Perkin Trans. 1* **1974**:409.
15. J. C. Gilbert and R. D. Selliah, *J. Org. Chem.* **58**:6255 (1993).
16. B. Bichan and M. Winnik, *Tetrahedron Lett.* **1974**:3857.
17a. R. A. Carboni and R. V. Lindsey, Jr., *J. Am. Chem. Soc.* **81**:4342 (1959).
b. L. A Carpino, *J. Am. Chem. Soc.* **84**:2196 (1962); **85**:2144 (1963).
18a. S. Hanessian, P. J. Roy, M. Petrini, P. J. Hodges, R. Di Fabio, and G. Carganico, *J. Org. Chem.* **55**:5766 (1990).
b. W. R. Roush and B. B. Brown, *J. Am. Chem. Soc.* **115**:2268 (1993).
c. J. W. Coe and W. R. Roush, *J. Org. Chem.* **54**:915 (1989).
19a. D. L. J. Clive, G. Chittattu, N. J. Curtis, and S. M. Menchen, *J. Chem. Soc., Chem. Commun.* **1978**:770.
b. B. W. Metcalf, P. Bey, C. Danzin, M. J. Jung, P. Casara, and J. P. Veveri, *J. Am. Chem. Soc.* **100**:2551 (1978).
c. T. Cohen, Z. Kosarych, K. Suzuki, and L.-C. Yu, *J. Org. Chem.* **50**:2965 (1985).
d. T. Cohen, M. Bhupathy, and J. R. Matz, *J. Am. Chem. Soc.* **105**:520 (1983).
e. R. G. Shea, J. N. Fitzner, J. E. Farkhauser, A. Spaltenstein, P. A. Carpino, R. M. Peevey, D. V. Pratt, B. J. Tenge, and P. B. Hopkins, *J. Org. Chem.* **51**:5243 (1986).
f. R. L. Funk, P. M. Novak, and M. M. Abelman, *Tetrahedron Lett.* **29**:1493 (1988).

g. R. M. Coates and C. H. Cummins, *J. Org. Chem.* **51:**1383 (1986).
h. K. Ogura, S. Furukawa, and G. Tsuchihashi, *J. Am. Chem. Soc.* **102:**2125 (1980).
i. H. F. Schmitthenner and S. M. Weinreb, *J. Org. Chem.* **45:**3372 (1980).
j. R. A. Gibbs and W. H. Okamura, *J. Am. Chem. Soc.* **110:**4062 (1988).
k. E. Vedejs, J. D. Rodgers, and S. J. Wittenberger, *J. Am. Chem. Soc.* **110:**4822 (1988).
l. J. Ahman, T. Jarevang, and P. I. Somfai, *J. Org. Chem.* **61:**8148 (1996).
m. E. Vedejs and M. Gingras, *J. Am .Chem. Soc.* **116:**579 (1994).
n. P. Beak, Z. Song, and J. E. Resek, *J. Org. Chem.* **57:**944 (1992).
o. T. A. Blumenkopf, G. C. Look, and L. E. Overman, *J. Am. Chem. Soc.* **112:**4399 (1990).
20a. N. Ono, A. Kanimura, and A. Kaji, *Tetrahedron Lett.* **27:**1595 (1986).
b. R. V. C. Carr, R. V. Williams, and L. A. Paquette, *J. Org. Chem.* **48:**4976 (1983).
c. C. H. DePuy and P. R. Story, *J. Am. Chem. Soc.* **82:**627 (1960).
d. S. Ranganathan, D. Ranganathan, and R. Iyengar, *Tetrahedron* **32:**961 (1976).
21a. D. J. Faulkner and M. R. Peterson, *J. Am. Chem. Soc.* **95:**553 (1973).
b. N. A. LeBel, N. D. Ojha, J. R. Menke, and R. J. Newland, *J. Org. Chem.* **37:**2896 (1972).
c. G. Büchi and H. Wüest, *J. Am. Chem. Soc.* **96:**7573 (1974).
d. C. A. Henrick, F. Schaub, and J. B. Siddall, *J. Am. Chem. Soc.* **94:**5374 (1972).
e. R. E. Ireland and R. H. Mueller, *J. Am. Chem. Soc.* **94:**5897 (1972).
f. E. J. Corey, R. B. Mitra, and H. Uda, *J. Am. Chem. Soc.* **86:**485 (1964).
g. J. E. McMurry and L. C. Blaszcak, *J. Org. Chem.* **39:**2217 (1974).
h. W. Sucrow, *Angew. Chem. Int. Ed. Engl.* **7:**629 (1968).
i. O. P. Vig, K. L. Matta, and I. Raj, *J. Indian Chem. Soc.* **41:**752 (1964).
j. W. Nagata, S. Hirai, T. Okumura, and K. Kawata, *J. Am. Chem. Soc.* **90:**1650 (1968).
k. H. O. House, J. Lubinkowski, and J. J. Good, *J. Org. Chem.* **40:**86 (1975).
l. L. A. Paquette, G. D. Grouse, and A. K. Sharma, *J. Am. Chem. Soc.* **102:**3972 (1980).
m. R. L. Funk and G. L. Bolton, *J. Org. Chem.* **49:**5021 (1984).
n. A. P. Kozikowski and C.-S. Li, *J. Org. Chem.* **52:**3541 (1987).
o. B. M. Trost, and A. C. Lavoie, *J. Am. Chem. Soc.* **105:**5075 (1983).
p. A. P. Marchand, S. C. Suri, A. D. Earlywine, D. R. Powell, and D. van der Helm, *J. Org. Chem.* **49:**670 (1984).
q. M. Kodoma, Y. Shiobara, H. Sumitomo, K. Fukuzumi, H. Minami, and Y. Miyamoto, *J. Org. Chem.* **53:**1437 (1988).
r. K. M. Werner, J. M. de los Santos, and S. M. Weinreb, *J. Org. Chem.* **64:**686 (1999).
s. D. Perez, G. Bures, E. Guitian, and L. Castedo, *J. Org. Chem.* **61:**1650 (1996).
22a. L. A. Paquette, S. K. Huber, and R. C. Thompson, *J. Org. Chem.* **58:**6874 (1993).
b. R. L. Funk, T. Olmstead, M. Parvez, and J. B. Stallman, *J. Org. Chem.* **58:**5873 (1993).
23a. J. J. Tufariello, A. S. Milowsky, M. Al-Nuri, and S. Goldstein, *Tetrahedron Lett.* **28:**263 (1987).
b. G. H. Posner, A. Haas, W. Harrison, and C. M. Kinter, *J. Org. Chem.* **52:**4836 (1987).
c. F. E. Ziegler, A. Nangia, and G. Schulte, *J. Am. Chem. Soc.* **109:**3987 (1987).
d. M. P. Edwards, S. V. Ley, S. G. Lister, B. D. Palmer and D. J. Williams, *J. Org. Chem.* **49:**3503 (1984).
e. R. E. Ireland and M. D. Varney, *J. Org. Chem.* **48:**1829 (1983).
f. D. J.-S. Tsai and M. M. Midland, *J. Am. Chem. Soc.* **107:**3915 (1985).
g. E. Vedejs, J. M. Dolphin, and H. Mastalerz, *J. Am. Chem. Soc.* **105:**127 (1983).
h. T. Zoller, D. Uguen, A. DeCian, J. Fischer, and S. Sable, *Tetrahedron Lett.* **38:**3409 (1997).
i. L. Grimaud, J.-P. Ferezou, J. Prunet, and J. Y. Lallemand, *Tetrahedron* **53:**9253 (1997).
j. P. M. Wovkulich, K. Shankaran, J. Kliegiel, and M. R. Uskokovic, *J. Org. Chem.* **58:**832 (1993).
24. P. A. Grieco and M. D. Kaufman, *Tetrahedron Lett.* **40:**1265 (1999); P. Grieco and Y. Dai, *J. Am. Chem. Soc.* **120:**5128 (1998).
25. S. Cossu, S. Battaggia, and O. DeLucchi, *J. Org. Chem.* **62:**4162 (1997).
26a. K. Nomura, K. Okazaki, H. Hori, and E. Yoshii, *Chem. Pharm. Bull.* **34:**3175 (1986).
b. Y. Tamura, M. Sasho, K. Nakagawa, T. Tsugoshi, and Y. Kita, *J. Org. Chem.* **49:**473 (1984).
27. S. D. Burke, D. M. Armistead, and K. Shankaran, *Tetrahedron Lett.* **27:**6295 (1986).
28a. C. Siegel and E. R. Thornton, *Tetrahedron Lett.* **29:**5225 (1988).
b. D. P. Curran, B. H. Kim, J. Daugherty, and T. A. Heffner, *Tetrahedron Lett.* **29:**3555 (1988).
c. H. Waldmann, *J. Org. Chem.* **53:**6133 (1988).

REFERENCES FOR PROBLEMS

1a. H. Neumann and D. Seebach, *Tetrahedron Lett.* **1976:**4839.
b. P. Canonne, G. Foscolos, and G. Lemay, *Tetrahedron Lett.* **1980:**155.
c. T. L. Shih, M. Wyvratt, and H. Mrozik, *J. Org. Chem.* **52:**2029 (1987).
d. R. K. Boeckman, Jr., and E. W. Thomas, *J. Am. Chem. Soc.* **101:**987 (1979).
e. G. M. Rubottom and C. Kim, *J. Org. Chem.* **48:**1550 (1983).
f. S. L. Buchwald, B. T. Watson, R. T. Lum, and W. A. Nugent, *J. Am. Chem. Soc.* **109:**7137 (1987).
g. T. Okazoe, K. Takai, K. Oshima, and K.Utimoto, *J. Org. Chem.* **52:**4410 (1987).
h. J. W. Frankenfeld and J. J. Werner, *J. Org. Chem.* **34:**3689 (1969).
i. E. R. Burkhardt and R. D. Rieke, *J. Org. Chem.* **50:**416 (1985).
j. G. Veeresa and A. Datta, *Tetrahedron* **54:**15673 (1998).
2. R. W. Herr and C. R. Johnson, *J. Am. Chem. Soc.* **92:**4979 (1970).
3a. J. S. Sawyer, A. Kucerovy, T. L. Macdonald, and G. J. McGarvey, *J. Am. Chem. Soc.* **110:**842 (1988).
b. T. Cohen and J. R. Matz, *J. Am. Chem. Soc.* **102:**6900 (1980).
c. C. R. Johnson and J. R. Medich, *J. Org. Chem.* **53:**4131 (1988).
d. B. M. Trost and T. N. Nanninga, *J. Am. Chem. Soc.* **107:**1293 (1985).
e. T. Morwick, *Tetrahedron Lett.* **21:**3227 (1980).
f. W. C. Still and C. Sreekumar, *J. Am. Chem. Soc.* **102:**1201 (1980).
g. R. F. Cunio and F. J. Clayton, *J. Org. Chem.* **41:**1480 (1976).
4a. J. J. Fitt and H. W. Gschwend, *J. Org. Chem.* **45:**4258 (1980).
b. S. Aikyama and J. Hooz, *Tetrahedron Lett.* **1973:**4115.
c. K. P. Klein and C. R. Hauser, *J. Org. Chem.* **32:**1479 (1967).
d. B. M. Graybill and D. A. Shirley, *J. Org. Chem.* **31:**1221 (1966).
e. M. M. Midland, A. Tramontano, and J. R. Cable, *J. Org. Chem.* **45:**28 (1980).
f. W. Fuhrer and H. W. Gschwend, *J. Org. Chem.* **44:**1133 (1979).
g. D. F. Taber and R. W. Korsmeyer, *J. Org. Chem.* **43:**4925 (1978).
h. R. R. Schmidt, J. Talbiersky, and P. Russegger, *Tetrahedron Lett.* **1979:**4273.
i. R. M. Carlson, *Tetrahedron Lett.* **1978:**111.
j. R. J. Sundberg, R. Broome, C. P. Walters, and D. Schnur, *J. Heterocycl. Chem.* **18:**807 (1981).
k. J. J. Eisch and J. N. Shah, *J. Org. Chem.* **56:**2955 (1991).
5a. M. P. Dreyfuss, *J. Org. Chem.* **28:**3269 (1963).
b. P. J. Pearce, D. H. Richards, and N. F. Scilly, *Org. Synth.* **52:**19 (1972).
c. U. Schöllkopf, H. Küppers, H.-J. Traencker, and W. Pitteroff, *Justus Liebigs Ann. Chem.* **704:**120 (1967).
d. J. V. Hay and T. M. Harris, *Org. Synth.* **53:**56 (1973).
e. F. Sato, M. Inoue, K. Oguro, and M. Sato, *Tetrahedron Lett.* **1979:**4303.
f. J. C. H. Hwa and H. Sims, *Org. Synth.* **V:**608 (1973).
6a. J. H. Rigby and C. Senanyake, *J. Am. Chem. Soc.* **109:**3147 (1987).
b. K. Takai, Y. Kataoka, T. Okazoe, and K. Utimoto, *Tetrahedron Lett.***29:**1065 (1988).
c. E. Nakamura, S. Aoki, K. Sekiya, H. Oshino, and I. Kuwajima, *J. Am. Chem. Soc.* **109:**8056 (1987).
d. H. A. Whaley, *J. Am. Chem. Soc.* **93:**3767 (1971).
7. J. Barluenga, F. J. Fananas, and M. Yus, *J. Org. Chem.* **44:**4798 (1979).
8a, b. W. C. Still and J. H. McDonald III, *Tetrahedron Lett.* **21:**1031 (1980).
c. E. Casadevall and Y. Povet, *Tetrahedron Lett.* **1976:**2841.
9. P. Beak, J. E. Hunter, Y. M. Jan, and A. P. Wallin, *J. Am. Chem. Soc.* **109:**5403 (1987).
10. C. J. Kowalski and M. S. Haque, *J. Org. Chem.* **50:**5140 (1985).
11. C. Fehr, J. Galindo, and R. Perret, *Helv. Chim. Acta* **70:**1745 (1987).
12. M. P. Cooke, Jr., and I. N. Houpis, *Tetrahedron Lett.* **26:**4987 (1985); E. Piers and P. C. Marais, *Tetrahedron Lett.* **29:**4053 (1988).
13a. C. Phillips, R. Jacobson, B. Abrahams, H. J. Williams, and C. R. Smith, *J. Org. Chem.* **45:**1920 (1980).
b. T. Cohen and J. R. Matz, *J. Am. Chem. Soc.* **102:**6900 (1980).
c. T. R. Govindachari, P. C. Parthasarathy, H. K. Desai, and K. S. Ramachandran, *Indian J. Chem.* **13:**537 (1975).
d. W. C. Still, *J. Am. Chem. Soc.* **100:**1481 (1978).
e. E. J. Corey and D. R. Williams, *Tetrahedron Lett.* **1977:**3847.

f. T. Okazoe, K. Takai, K. Oshiama, and K. Utimoto, *J. Org. Chem.* **52:**4410 (1987).
g. M. A. Adams, A. J. Duggan, J. Smolanoff, and J. Meinwald, *J. Am. Chem. Soc.* **101:**5364 (1979).
h. S. O. deSilva, M. Watanabe, and V. Snieckus, *J. Org. Chem.* **44:**4802 (1979).
14. M. Kitamura, S. Suga, K. Kawai, and R. Noyori, *J. Am. Chem. Soc.* **108:**6071 (1986); K. Soai, A. Ookawa, T. Kaba,and K. Ogawa, *J. Am. Chem. Soc.* **109:**7111 (1987); M. Kitamura, S. Okada, and R. Noyori, *J. Am. Chem. Soc.* **111:**4028 (1989).
15. C. A. Broka and T. Shen, *J. Am. Chem. Soc.* **111:**2981 (1989).
16. W.-L. Cheng, Y.-J. Shaw, S.-M. Yeh, P. P. Kanakamma, Y.-H. Chen, C. Chen, J.-C. Shieu, S.-J. Yiin, G.-H. Lee, Y. Wang, and T.-Y. Luh, *J. Org. Chem.* **64:**532 (1999).

Chapter 8

1a. C. Huynh, F. Derguini-Boumechal, and G. Linstrumelle, *Tetrahedron Lett.* **1979:**1503.
b. N. J. LaLima, Jr., and A. B. Levy, Jr., *J. Org. Chem.* **43:**1279 (1978).
c. A. Cowell and J. K. Stille, *J. Am. Chem. Soc.* **102:**4193 (1980).
d. T. Sato, M. Kawasima, and T. Fujisawa, *Tetrahedron Lett.* **1981:**2375.
e. H. P. Dang and G. Linstrumelle, *Tetrahedron Lett.* **1978:**191.
f. B. M. Trost and D. P. Curran, *J. Am. Chem. Soc.* **102:**5699 (1980).
g. D. J. Pasto, S.-K. Chou, E. Fritzen, R. H. Shults, A. Waterhouse, and G. F. Hennion, *J. Org. Chem.* **43:**1389 (1978).
h. B. H. Lipshutz, J. Kozlowski, and R. S. Wilhelm, *J. Am. Chem. Soc.* **104:**2305 (1982).
i. P. A. Grieco and C. V. Srinivasan, *J. Org. Chem.* **46:**2591 (1981).
j. C. Iwata, K. Suzuki, S. Aoki, K. Okamura, M. Yamashita, I. Takahashi, and T. Tanaka, *Chem. Pharm. Bull.* **34:**4939 (1988).
k. A. Alexakis, G. Cahiez, and J. F. Normant, *Org. Synth.* **62:**1 (1984).
l. J. Tsuji, Y. Kobayashi, H. Kataoka, and T. Takahashi, *Tetrahedron Lett.* **21:**1475 (1980).
m. W. A. Nugent and R. J. McKinney, *J. Org. Chem.* **50:**5370 (1985).
n. R. M. Wilson, K. A. Schnapp, R. K. Merwin, R. Ranganathan, D. L. Moats, and T. T. Conrad, *J. Org. Chem.* **51:**4028 (1986).
o. L. N. Pridgen, *J. Org. Chem.* **47:**4319 (1982).
p. R. Casas, C. Cave, and J. d'Anglelo, *Tetrahedron Lett.* **36:**1039 (1995).
q. N. Miyaura, K. Yamada, and A. Suzuki, *Tetrahedron Lett.* **1979:**3437.
r. F. K. Steffy, J. P. Godschalx, and J. K. Stille, *J. Am. Chem. Soc.* **106:**4833 (1984).
2a. B. H. Lipshutz, M. Koerner, D. A. Parker, *Tetrahedron Lett.* **28:**945 (1987).
b. B. H. Lipshutz, R. S. Wilheim, J. A. Kozlowski, and D. Parker, *J. Org. Chem.* **49:**3928 (1984).
c. J. P. Marino, R. Fernandez de la Pradilla, and E. Laborde, *J. Org. Chem.* **52:**4898 (1987).
d. C. R. Johnson and D. S. Dhanoa, *J. Org. Chem.* **52:**1887 (1987).
3a. H. Urata, A. Fujita, and T. Fuchikami, *Tetrahedron Lett.* **29:**4435 (1988).
b. Y. Itoh, H. Aoyama, T. Hirao, A. Mochizuki, and T. Saegusa, *J. Am. Chem. Soc.* **101:**494 (1979).
c. P. G. M. Wuts, M. L. Obrzut, and P. A. Thompson, *Tetrahedron Lett.* **25:**4051 (1984).
d. K. Kokubo, K. Matsumasa, M. Miura, and M. Nomura, *J. Org. Chem.* **61:**6941 (1996).
4a. R. J. Anderson , V. L. Corbin, G. Cotterrel, G. R. Cox, C. A. Henrick, F. Schaub, and J. B. Siddall, *J. Am. Chem. Soc.* **97:**1197 (1975).
b. P. deMayo, L. K. Sydnes, and G. Wenska, *J. Org. Chem.* **45:**1549 (1980).
c. Y. Yamamoto, H. Yatagai, and K. Maruyama, *J. Org. Chem.* **44:**1744 (1979).
d. H. Shostarez and L. A. Paquette, *J. Am. Chem. Soc.* **103:**722 (1981).
e. W. G. Dauben, G. H. Beasley, M. D. Broadhurst, B. Muller, D. J. Peppard, P. Pesnelle, and C. Suter, *J. Am. Chem. Soc.* **97:**4973 (1975).
f. L. Watts, J. D. Fitzpatrick, and R. Pettit, *J. Am. Chem. Soc.* **88:**623 (1966).
g. J. I. Kim, B. A. Patel, and R. F. Heck, *J. Org. Chem.* **46:**1067 (1981).
h. J. A. Marshall, W. F. Huffman, and J. A. Ruth, *J. Am. Chem. Soc.* **94:**4691 (1972).
i. H.-A. Hasseberg and H. Gerlach, *Helv. Chim. Acta* **71:**957 (1988).
j. R. Alvarez, M. Herrero, S. Lopez, and A. R. de Lera, *Tetrahedron* **54:**6793 (1998).
k. T. K. Chakraborty and D. Thippeswamy, *Synlett.* **1999:**150.

l. S. Jinno, T. Okita, and K. Inouye, *Biorg. Med. Chem. Lett.* **9:**1029 (1999).
m. J. Thibonnet, M. Abarbi, A. Duchene, and J.-L. Parrain, *Synlett* **1999:**141.
5a, b. B. O'Connor and G. Just, *J. Org. Chem.* **52:**1801 (1987); G. Just and B. O'Connor, *Tetrahedron Lett.* **26:**1799 (1985).
6. B. H. Lipshutz, R. S. Wilheim, J. A. Kozlowski, and D. Parker, *J. Org. Chem.* **49:**3928 (1984); E. C. Ashby, R. N. DePriest, A. Tuncay, and S. Srivasta, *Tetrahedron Lett.* **23:**5251 (1982).
7a. C. G. Chavdarian and C. H. Heathcock, *J. Am. Chem. Soc.* **97:**3822 (1975).
b. E. J. Corey and D. R. Williams, *Tetrahedron Lett.* **1977:**3847.
c. G. Mehta and K. S. Rao, *J. Am. Chem. Soc.* **108:**8015 (1986).
d. W. A. Nugent and F. W. Hobbs, Jr., *Org. Synth.* **66:**52 (1988).
e. G. F. Cooper, D. L. Wren, D. Y. Jackson, C. C. Beard, E. Galeazzi, A. R. Van Horn, and T. T. Li, *J. Org. Chem.* **58:**4280 (1993).
f. R. K. Dieter, J. W. Dieter, C. W. Alexander, and N. S. Bhinderwala, *J. Org. Chem.* **61:**2930 (1996).
g. C. R. Johnson and T. D. Penning, *J. Am. Chem. Soc.* **110:**4726 (1988).
h. T. Hudlicky and H. F. Olivo, *J. Am. Chem. Soc.* **114:**9694 (1992).
8a. C. M. Lentz and G. H. Posner, *Tetrahedron Lett.* **1978:**3769.
b. A. Marfat, P. R. McGuirk, R. Kramer, and P. Helquist, *J. Am. Chem. Soc.* **99:**253 (1977).
c. L. A. Paquette and Y.-K. Han, *J. Am. Chem. Soc.* **103:**1831 (1981).
d. A. Alexakis, J. Berlan, and Y. Besace, *Tetrahedron Lett.* **27:**1047 (1986).
e. M. Sletzinger, T. R. Verhoeven, R. P. Volante, J. M. McNamara, E. G. Corley, and T. M. H. Liu, *Tetrahedron Lett.* **26:**2951 (1985).
f. G. Giambastiani and G. Poli, *J. Org. Chem.* **63:**9608 (1998).
9a. E. J. Corey and E. Hamanaka, *J. Am. Chem. Soc.* **89:**2758 (1967).
b. Y. Kitagawa, A. Itoh, S. Hashimoto, H. Yamamoto, and H. Nozaki, *J. Am. Chem. Soc.* **99:**3864 (1977).
c. B. M. Trost and R. W. Warner, *J. Am. Chem. Soc.* **105:**5940 (1983).
d. S. Brandt, A. Marfat, and P. Helquist, *Tetrahedron Lett.* **1979:**2193.
e. A. Fürstner and H. Weintritt, *J. Am. Chem. Soc.* **120:**2817 (1998).
10. R. H. Grubbs and R. A. Grey, *J. Am. Chem. Soc.* **95:**5765 (1973).
11. H. L. Goering, E. P. Seitz, Jr., and C. C. Tseng, *J. Org. Chem.* **46:**5304 (1981).
12. A. Marfat, P. R. McGuirk, and P. Helquist, *J. Org. Chem.* **44:**1345 (1979).
13. N. Cohen, W. F. Eichel, R. J. Lopresti, C. Neukom, and G. Saucy, *J. Org. Chem.* **41:**3505 (1976).
14a. R. J. Linderman, A. Godfrey, and K. Horne, *Tetrahedron Lett.* **28:**3911 (1987).
b. H. Schostarez and L. A. Paquette, *J. Am. Chem. Soc.* **103:**722 (1981).
c. Y. Yamamoto, S. Yamamoto, H. Yatagai, Y. Ishihara, and K. Maruyama, *J. Org. Chem.* **47:**119 (1982).
d. T. Kawabata, P. A. Grieco, H.-L. Sham, H. Kim, J. Y. Law, and S. Tu, *J. Org. Chem.* **52:**3346 (1987).
15a. A. Minato, K. Suzuki, K. Tamao, and M. Kumada, *Tetrahedron Lett.* **25:**83 (1984).
b. E. R. Larson and R. A. Raphael, *Tetrahedron Lett.* **1979:**5401.
c. M. C. Pirrung and S. A. Thomson, *J. Org. Chem.* **53:**227 (1988).
d. J. Just and B. O'Connor, *Tetrahedron Lett.* **29:**753 (1988).
e. M. F. Semmelhack and A. Yamashita, *J. Am. Chem. Soc.* **102:**5924 (1980).
f. A. M. Echavarren and J. K. Stille, *J. Am. Chem. Soc.* **110:**4051 (1988).
16a. J. E. Bäckvall, S. E. Byström, and R. E. Nordberg, *J. Org. Chem.* **49:**4619 (1984).
b. M. F. Semmelhack and C. Bodurow, *J. Am. Chem. Soc.* **106:**1496 (1984).
c. D. Valentine, Jr., J. W. Tilley, and R. A. Le Mahieu, *J. Org. Chem.* **46:**4614 (1981).
d. A. S. Kende, B. Roth, P. J. SanFilippo, and T. J. Blacklock, *J. Am. Chem. Soc.* **104:**5808 (1982).
17. E. J. Corey, F. J. Hannon, and N. W. Boaz, *Tetrahedron* **45:**545 (1989).
18a. P. A. Bartlett, J. D. Meadows, and E. Ottow, *J. Am. Chem. Soc.* **106:**5304 (1984).
b. M. Larcheveque and Y. Petit, *Tetrahedron Lett.* **28:**1993 (1987).
c. B. M. Trost and J. D. Oslob, *J. Am. Chem. Soc.* **121:**3057 (1999).

Chapter 9

1a. P. Jacob III and H. C. Brown, *J. Am. Chem. Soc.* **98:**7832 (1976).
b. D. Milstein and J. K. Stille, *J. Org. Chem.* **44:**1613 (1979).

c. H. C. Brown and K. K. Wang, *J. Org. Chem.* **51:**4514 (1986).
d. H. Yatagai, Y. Yamamoto, and K. Maruyama, *J. Am. Chem. Soc.* **102:**4548 (1980).
e. R. Mohan and J. A. Katzenellenbogen, *J. Org. Chem.* **49:**1234 (1984).
f. B. M. Trost and A. Brandi, *J. Org. Chem.* **49:**4811 (1984).
g. H. C. Brown and T. Imai, *J. Am. Chem. Soc.* **105:**6285 (1983).
h. H. C. Brown, N. G. Bhat, and J. B. Campbell, Jr., *J. Org. Chem.* **51:**3398 (1986).
2a. H. C. Brown, M. M. Rogic, H. Nambu, and M. W. Rathke, *J. Am. Chem. Soc.* **91:**2147 (1969); H. C. Brown, H. Nambu, and M. M. Rogic, *J. Am. Chem. Soc.* **91:**6852 (1969).
b. H. C. Brown and R. A. Coleman, *J. Am. Chem. Soc.* **91:**4606 (1969).
c. H. C. Brown and G. W. Kabalka, *J. Am. Chem. Soc.* **92:**714 (1970).
d. G. Zweifel, R. P. Fisher, J. T. Snow, and C. C. Whitney, *J. Am. Chem. Soc.* **93:**6309 (1971).
e. H. C. Brown and M. W. Rathke, *J. Am. Chem. Soc.* **89:**2738 (1967).
f. H. C. Brown and M. M. Rogic, *J. Am. Chem. Soc.* **91:**2146 (1969).
3. See references in Scheme 9.1.
4a. H. C. Brown, H. D. Lee, and S. U. Kulkarni, *J. Org. Chem.* **51:**5282 (1986).
b. J. A. Sikorski, N. G. Bhat, T. E. Cole, K. K. Wang, and H. C. Brown, *J. Org. Chem.* **51:**4521 (1986).
c. S. U. Kulkarni, H. D. Lee, and H. C. Brown, *J. Org. Chem.* **45:**4542 (1980).
d. M. C. Welch and T. A. Bryson, *Tetrahedron Lett.* **29:**521 (1988).
5a. D. R. McKean, G. Parrinello. A. F. Renaldo, and J. K. Stille, *J. Org. Chem.* **52:**422 (1987).
b. L. Kuwajima and H. Urabe, *J. Am. Chem. Soc.* **104:**6830 (1982).
c. A. Pelter, K. J. Gould, and C. R. Harrison, *Tetrahedron Lett.* **1975:**3327.
d. A. Pelter and R. A. Drake, *Tetrahedron Lett.* **29:**4181 (1988).
e. L. E. Overman and M. J. Sharp, *J. Am. Chem. Soc.* **110:**612 (1988).
f. H. C. Brown and S. U. Kulkarni, *J. Org. Chem.* **44:**2422 (1979).
6a. W. R. Roush, M. A. Adam, and D. J. Harris, *J. Org. Chem.* **50:**2000 (1985).
b. S. J. Danishefsky, S. DeNinno, and P. Lartey, *J. Am. Chem. Soc.* **109:**2082 (1987).
c. J. Hooz and D. M. Gunn, *J. Am. Chem. Soc.* **91:**6195 (1969).
d. D. A. Heerding, C. Y. Hong, N. Kado, G. C. Look, and L. E. Overman, *J. Org. Chem.* **58:**6947 (1993).
7a, b. H. C. Brown and N. G. Bhat, *J. Org. Chem.* **53:**6009 (1988).
c, d. H. C. Brown, D. Basavaiah, S. U. Kulkarni, N. Bhat, and J. V. N. Vara Prasad, *J. Org. Chem.* **53:**239 (1988).
8a. J. A. Marshall, S. L. Crooks, and B. S. DeHoff, *J. Org. Chem.* **53:**1616 (1988).
b. B. M. Trost and T. Sato, *J. Am. Chem. Soc.* **107:**719 (1985).
9a. W. E. Fristad, D. S. Dime, T. R. Bailey, and L. A. Paquette, *Tetrahedron Lett.* **1979:**1999.
b. E. Piers and H. E. Morton, *J. Org. Chem.* **45:**4263 (1980).
c. J. C. Bottaro, R. N. Hanson, and D. E. Seitz, *J. Org. Chem.* **46:**5221 (1981).
d. Y. Yamamoto and A. Yanagi, *Heterocycles* **16:**1161 (1981).
e. M. B. Anderson and P. L. Fuchs, *Synth. Commun.* **17:**621 (1987); B. A. Narayanan and W. H. Bunelle, *Tetrahedron Lett.* **28:**6261 (1987).
f. A. Hosomi, M. Sato, and H. Sakurai, *Tetrahedron Lett.* **1979:**429.
10. P. A. Grieco and W. F. Fobare, *Tetrahedron Lett.* **27:**5067 (1986).
11. E. J. Corey and W. L. Seibel, *Tetrahedron Lett.* **27:**905 (1986).
12a. J. A. Marshall, S. L. Crooks, and B. S. DeHoff, *J. Org. Chem.* **53:**1616 (1988); J. A. Marshall and W. Y. Gung, *Tetrahedron Lett.* **29:**3899 (1988).
b. E. Moret and M. Schlosser, *Tetrahedron Lett.* **25:**4491 (1984).
c. L. K. Truesdale, D. Swanson, and R. C. Sun, *Tetrahedron Lett.* **26:**5009 (1985).
d. B. M. Trost and T. Sato, *J. Am. Chem. Soc.* **107:**719 (1985).
e. R. L. Funk and G. L. Bolton, *J. Org. Chem.* **49:**5021 (1984).
f. L. E. Overman, T. C. Malone, and G. P. Meier, *J. Am. Chem. Soc.* **105:**6993 (1983).
g. Y. Naruse, T. Esaki, and H. Yamamoto, *Tetrahedron Lett.* **29:**1417 (1988).
13a. H. C. Brown, T. Imai, M. C. Desai, and B. Singaran, *J. Am. Chem. Soc.* **107:**4980 (1985).
b, c. H. C. Brown, R. K. Bakshi, and B. Singaran, *J. Am. Chem. Soc.* **110:**1529 (1988).
d. H. C. Brown, M. Srebnik, R. R. Bakshi, and T. E. Cole, *J. Am. Chem. Soc.* **109:**5420 (1987).
14. K. K. Wang and K.-H. Chu, *J. Org. Chem.* **49:**5175 (1984).
15a. W. R. Roush, J. A. Straub, and M. S. VanNieuwenhze, *J. Org. Chem.* **56:**1636 (1991).
b. L. A. Paquette and G. D. Maynard, *J. Am. Chem. Soc.* **114:**5018 (1992).
c. Y. Nishigaichi, N. Ishida, M. Nishida, and A. Takuwa, *Tetrahedron Lett.* **37:**3701 (1996).
d. C. Y. Hong, N. Kado, and L. E. Overman, *J. Am. Chem. Soc.* **115:**11028 (1993).

e. P. V. Ramachandran, G.-M. Chen, and H. C. Brown, *Tetrahedron Lett.* **38:**2417 (1997).
f. A. B. Charette, C. Mellon, and M. Motamedi, *Tetrahedron Lett.* **36:**8561 (1995).
g. C. Masse, M. Yang, J. Solomon, and J. S. Panek, *J. Am. Chem. Soc.* **120:**4123 (1998).

Chapter 10

1a. S. Julia and A. Ginebreda, *Synthesis* **1977:**682.
b. R. Breslow and H. W. Chang, *J. Am. Chem. Soc.* **83:**2367 (1961).
c. D. J. Burton and J. L. Hahnfeld, *J. Org. Chem.* **42:**828 (1977).
d. D. Seyfreth and S. P. Hopper, *J. Org. Chem.* **37:**4070 (1972).
e. G. L. Closs, L. E. Closs, and W. A. Böll, *J. Am. Chem. Soc.* **85:**3796 (1963).
f. L. G. Mueller and R. G. Lawton, *J. Org. Chem.* **44:**4741 (1979).
g. F. G. Bordwell and M. W. Carlson, *J. Am. Chem. Soc.* **92:**3377 (1970).
h. A. Burger and G. H. Harnest, *J. Am. Chem. Soc.* **65:**2382 (1943).
i. E. Schmitz, D. Habish, and A. Stark, *Angew. Chem. Int. Ed. Engl.* **2:**548 (1963).
j. R. Zurflüh, E. N. Wall, J. B. Sidall, and J. A. Edwards, *J. Am. Chem. Soc.* **90:**6224 (1968).
k. M. Nishizawa, H. Takenaka, and Y. Hayashi, *J. Org. Chem.* **51:**806 (1986).
l. H. Nishiyama, K. Sakuta, and K. Itoh, *Tetrahedron Lett.* **25:**233 (1984).
m. H. Seto, M. Sakaguchi, and Y. Fujimoto, *Chem. Pharm. Bull.* **33:**412 (1985).
n. D. F. Taber and E. H. Petty, *J. Org. Chem.* **47:**4808 (1982).
o. B. Iddon, D. Price, H. Suschitzky, and D. J. C. Scopes, *Tetrahedron Lett.* **24:**413 (1983).
p. A. Chu and L. N. Mander, *Tetrahedron Lett.* **29:**2727 (1988).
q. G. E. Keck and D. F. Kachensky, *J. Org. Chem.* **51:**2487 (1986).
r. D. H. R. Barton, J. Guilhem, Y. Hervé, P. Potier, and J. Thierry, *Tetrahedron Lett.* **28:**1413 (1987).
s. A. M. Gomez, G. O. Danelon, S. Valverde, and J. C. Lopez, *J. Org. Chem.* **63:**9626 (1998).
t. S. D. Burke and D. N. Deaton, *Tetrahedron Lett.* **32:**4651 (1991).
u. J. A. Wendt and J. Aube, *Tetrahedron Lett.* **37:**1531 (1996).
2a. K. B. Wiberg, B. L. Furtek, and L. K. Olli, *J. Am. Chem. Soc.* **101:**7675 (1979).
b. A. E. Greene and J.-P. Depres, *J. Am. Chem. Soc.* **101:**4003 (1979).
c. R. A. Moss and E. Y. Chen, *J. Org. Chem.* **46:**1466 (1981).
d. B. M. Trost, R. M. Cory, P. H. Scudder, and H. B. Neubold, *J. Am. Chem. Soc.* **95:**7813 (1973).
e. T. J. Nitz, E. M. Holt, B. Rubin, and C. H. Stammer, *J. Org. Chem.* **46:**2667 (1981).
f. L. N. Mander, J. V. Turner, and B. G. Colmbe, *Aust. J. Chem.* **27:**1985 (1974).
g. P. J. Jessup, C. B. Petty, J. Roos, and L. E. Overman, *Org. Synth.* **59:**1 (1979).
h. H. Dürr, H. Nickels, L. A. Pacala, and M. Jones, Jr., *J. Org. Chem.* **45:**973 (1980).
i. G. A. Scheihser and J. D. White, *J. Org. Chem.* **45:**1864 (1980).
j. M. B. Groen and F. J. Zeelen, *J. Org. Chem.* **43:**1961 (1978).
k. R. C. Gadwood, R. M. Lett, and J. E. Wissinger, *J. Am. Chem. Soc.* **108:**6343 (1986).
l. V. B. Rao, C. F. George, S. Wolff, and W. C. Agosta, *J. Am. Chem. Soc.* **107:**5732 (1985).
m. Y. Araki, T. Endo, M. Tanji, J. Nagasawa, and Y. Ishido, *Tetrahedron Lett.* **28:**5853 (1987).
n. M. Newcomb and J. Kaplan, *Tetrahedron Lett.* **28:**1615 (1987).
o. G. Stork, P. M. Sher, and H.-L. Chen, *J. Am. Chem. Soc.* **108:**6384 (1986).
p. E. J. Corey and M. Kang, *J. Am. Chem. Soc.* **106:**5384 (1984).
q. G. E. Keck, D. F. Kachensky, and E. J. Enholm, *J. Org. Chem.* **50:**4317 (1985).
r. A. DeMesmaeker, P. Hoffmann, and B. Ernst, *Tetrahedron Lett.* **30:**57 (1989).
s. A. K. Singh, R. K. Bakshi, and E. J. Corey, *J. Am. Chem. Soc.* **109:**6187 (1987).
t. S. Danishefsky and J. S. Panek, *J. Am. Chem. Soc.* **109:**917 (1987).
3a. W. J. Hehre, J. A. Pople, W. A. Latham. L. Radom, E. Wasserman, and Z. R. Wasserman, *J. Am. Chem. Soc.* **98:**4378 (1976); N. C. Baird and K. F. Taylor, *J. Am. Chem. Soc.* **100:**1333 (1978); J. M. Bofill, J. Farrás, S. Olivella, A. Solé, and J. Vilarrasa, *J. Am. Chem. Soc.* **110:**1694 (1988).
b. P. H. Mueller, N. G. Rondan, K. N. Houk, J. F. Harrison, D. Hooper, B. H. Willen, and J. F. Liebman, *J. Am. Chem. Soc.* **103:**5049 (1981).
c, d. R. Gleiter and R. Hoffmann, *J. Am. Chem. Soc.* **90:**5457 (1968).
4. C. D. Poulter, E. C. Friedrich, and S. Winstein, *J. Am. Chem. Soc.* **91:**6892 (1969).
5. P. L. Barili, G. Berti, B. Macchia, and L. Monti, *J. Chem. Soc. C*, **1970:**1168.

6a. R. K. Hill and D. A. Cullison, *J. Am. Chem. Soc.* **95:**2923 (1973).
b. A. B. Smith III, B. H. Toder, S. J. Branca, and R. K. Dieter, *J. Am. Chem. Soc.* **103:**1996 (1981).
c. M. C. Pirrung and J. A. Werner, *J. Am. Chem. Soc.* **108:**6060 (1986).
7. E. W. Warnhoff, C. M. Wong, and W. T. Tai, *J. Am. Chem. Soc.* **90:**514 (1968).
8a. S. A. Godleski, P. v. R. Schleyer, E. Osawa, Y. Inamoto, and Y. Fujikura, *J. Org. Chem.* **41:**2596 (1976).
b. P. E. Eaton, Y. S. Or, and S. J. Branca, *J. Am. Chem. Soc.* **103:**2134 (1981).
c. G. H. Posner, K. A. Babiak, G. L. Loomis, W. J. Frazee, R. D. Mittal, and I. L. Karle, *J. Am. Chem. Soc.* **102:**7498 (1980).
d. T. Hudlicky, F. J. Koszyk, T. M. Kutchan, and J. P. Sheth, *J. Org. Chem.* **45:**5020 (1980).
e. L. A. Paquette and Y.-K. Han, *J. Am. Chem. Soc.* **103:**1835 (1981).
f. L. A. Paquette and R. W. Houser, *J. Am. Chem. Soc.* **91:**3870 (1969).
g. L. A. Paquette, S. Nakatani, T. M. Zydowski, S. D. Edmondson, L.-Q. Sun, and R. Skerlj, *J. Org. Chem.* **64:**3244 (1999).
9a. Y. Ito, S. Fujii, M. Nakatsuka, F. Kawamoto, and T. Saegusa, *Org. Synth.* **59:**113 (1979).
b. P. Nedenskov, H. Heide, and N. Clauson-Kass, *Acta Chem. Scand.* **16:**246 (1962).
c. L.-F. Tietze, *J. Am. Chem. Soc.* **96:**946 (1974).
d. E. G. Breitholle and A. G. Fallis, *J. Org. Chem.* **43:**1964 (1978).
e. E. Y. Chen, *J. Org. Chem.* **49:**3245 (1984).
f. G. Mehta and K. S. Rao, *J. Org. Chem.* **50:**5537 (1985).
g. T. V. Rajan Babu, *J. Org. Chem.* **53:**4522 (1988).
h. W. D. Klobucar, L. A. Paquette, and J. P. Blount, *J. Org. Chem.* **46:**4021 (1981).
i. F. E. Ziegler, S. I. Klein, U. K. Pati, and T.-F. Wang, *J. Am. Chem. Soc.* **107:**2730 (1985).
j. T. Hudlicky, F. J. Koszyk, D. M. Dochwat, and G. L. Cantrell, *J. Org. Chem.* **46:**2911 (1981).
k. R. E. Ireland, W. C. Dow, J. D. Godfrey, and S. Thaisrivongs, *J. Org. Chem.* **49:**1001 (1984).
l. C. P. Chuang and D. J. Hart, *J. Org. Chem.* **48:**1782 (1983).
10a. S. D. Larsen and S. A. Monti, *J. Am. Chem. Soc.* **99:**8015 (1977).
b. S. A. Monti and J. M. Harless, *J. Am. Chem. Soc.* **99:**2690 (1977).
c. F. T. Bond and C.-Y. Ho, *J. Org. Chem.* **41:**1421 (1976).
d. E. Wenkert, R. S. Greenberg, and H.-S. Kim, *Helv. Chim. Acta* **70:**2159 (1987).
e. B. B. Snider and M. A. Dombroski, *J. Org. Chem.* **52:**5487 (1987).
f. G. A. Kraus and K. Landgrebe, *Tetrahedron Lett.* **25:**3939 (1984).
g. S. Kim, S. Lee,and J. S. Koh, *J. Am. Chem. Soc.* **113:**5106 (1991).
h. S. Ando, K. P. Minor, and L. E. Overman, *J. Org. Chem.* **62:**6379 (1997).
i. A. Johns and J. A. Murphy, *Tetrahedron Lett.* **29:**837 (1988).
j. K. S. Feldman and A. K. K. Vong, *Tetrahedron Lett.* **31:**823 (1990).
k. D. L. J. Clive and S. Daigneault, *J. Org. Chem.* **56:**5285 (1991).
l. M. A. Brodney and A. Padwa, *J. Org. Chem.* **64:**556 (1999).
m. S.-H. Chen, S. Huang, and G. P. Roth, *Tetrahedron Lett.* **36:**8933 (1995).
n. J. B. Brogan, C. B. Bauer, R. D. Rogers, and C. K. Zerchner, *Tetrahedron Lett.* **37:**5053 (1996).
11. L. Blanco, N. Slougi, G. Rousseau, and J. M. Conia, *Tetrahedron Lett.* **1981:**645.
12. J. A. Marshall and J. A. Ruth, *J. Org. Chem.* **39:**1971 (1974).
13. S. D. Burke, M. E. Kort, S. M. S. Strickland, H. M. Organ, and L. A. Silks III, *Tetrahedron Lett.* **35:**1503 (1994).
14. C. A. Grob, H. R. Kiefer, H. J. Lutz, and H. J. Wilkens, *Helv. Chim. Acta* **50:**416 (1967).
15. M. P. Doyle, W. E. Buhro, and J. F. Dellaria, Jr., *Tetrahedron Lett.* **1979:**4429.
16. R. Tsang, J. K. Dickson, Jr., H. Pak, R. Walton, and B. Fraser-Reid, *J. Am. Chem. Soc.* **109:**3484 (1987).

Chapter 11

1a. L. Friedman and H. Shechter, *J. Org. Chem.* **26:**2522 (1961).
b. E. C. Taylor, F. Kienzle, R. L. Robey, and A. McKillop, *J. Am. Chem. Soc.* **92:**2175 (1970).
c. G. F. Hennion and S. F. de C. McLeese, *J. Am. Chem. Soc.* **64:**2421 (1942).
d. J. Koo, *J. Am. Chem. Soc.* **75:**1889 (1953).

e. E. C. Taylor, F. Kienzle, R. L. Robey, A. McKillop, and J. D. Hunt, *J. Am. Chem. Soc.* **93:**4845 (1971).
f. G. A. Ropp and E. C. Coyner, *Org. Synth.* **IV:**727 (1963).
g. M. Shiratsuchi, K. Kawamura, T. Akashi, M. Fujij, H. Ishihama, and Y. Uchida, *Chem. Pharm. Bull.* **35:**632 (1987).
h. D. C. Furlano and K. D. Kirk, *J. Org. Chem.* **51:**4073 (1986).
i. C. K. Bradsher, F. C. Brown, and H. K. Porter, *J. Am. Chem. Soc.* **76:**2357 (1954).
2a. E. C. Taylor, E. C. Bigham, and D. K. Johnson, *J. Org. Chem.* **42:**362 (1977).
b. P. Studt, *Justus Liebigs Ann. Chem.* **1978:**2105.
c. T. Jojima, H. Takeshiba, and T. Kinoto, *Bull. Chem. Soc. Jpn.* **52:**2441 (1979).
d. R. W. Bost and F. Nicholson, *J. Am. Chem. Soc.* **57:**2368 (1935).
e. H. Durr, H. Nickels, L. A. Pacala, and M. Jones, Jr., *J. Org. Chem.* **45:**973 (1980).
3a. C. L. Perrin and G. A. Skinner, *J. Am. Chem. Soc.* **93:**3389 (1971).
b. R. A. Rossi and J. F. Bunnett, *J. Am. Chem. Soc.* **94:**683 (1972).
c. M. Jones, Jr., and R. H. Levin, *J. Am. Chem. Soc.* **91:**6411 (1969).
d. Y. Naruta, Y. Nishigaichi, and K. Maruyama, *J. Org. Chem.* **53:**1192 (1988).
e. S. P. Khanapure, R. T. Reddy, and E. R. Biehl, *J. Org. Chem.* **52:**5685 (1987).
f. G. Büchi and J. C. Leung, *J. Org. Chem.* **51:**4813 (1986).
4a. M. P. Doyle, J. F. Dellaria, Jr., B. Siegfried, and S. W. Bishop, *J. Org. Chem.* **42:**3494 (1977).
b. T. Cohen, A. G. Dietz, Jr., and J. R. Miser, *J. Org. Chem.* **42:**2053 (1977).
c. M. P. Doyle, B. Siegfried, R. C. Elliot, and J. F. Dellaria, Jr., *J. Org. Chem.* **42:**2431 (1977).
d. G. D. Figuly and J. C. Martin, *J. Org. Chem.* **45:**3728 (1980).
e. E. McDonald and R. D. Wylie, *Tetrahedron* **35:**1415 (1979).
f. M. P. Doyle, B. Siegfried, and J. F. Dellaria, Jr., *J. Org. Chem.* **42:**2426 (1977).
g. A. P. Kozikowski, M. N. Greco, and J. P. Springer, *J. Am. Chem. Soc.* **104:**7622 (1982).
h. P. H. Gore and I. M. Khan, *J. Chem. Soc., Perkin Trans. 1* **1979:**2779.
i. A. A. Leon, G. Daub, and I. R. Silverman, *J. Org. Chem.* **49:**4544 (1984).
j. S. R. Wilson and L. A. Jacob, *J. Org. Chem.* **51:**4833 (1986).
k. S. A. Khan, M. A. Munawar, and M. Siddiq, *J. Org. Chem.* **53:**1799 (1988).
l. J. R. Beadle, S. H. Korzeniowski, D. E. Rosenberg, B. J. Garcia-Slanga, and G. W. Gokel, *J. Org. Chem.* **49:**1594 (1984).
5. B. L. Zenitz and W. H. Hartung, *J. Org. Chem.* **11:**444 (1946).
6. T. F. Buckley III and H. Rapoport, *J. Am. Chem. Soc.* **102:**3056 (1980).
7. G. A. Olah and J. A. Olah, *J. Am. Chem. Soc.* **98:**1839 (1976).
8. E. J. Corey, S. Barcza, and G. Klotmann, *J. Am. Chem. Soc.* **91:**4782 (1969).
9. E. C. Taylor, F. Kienzle, R. L. Robey, and A. McKillop, *J. Am. Chem. Soc.* **92:**2175 (1970).
10. M. Essiz, G. Guillaumet, J.-J. Brunet, and P. Caubere, *J. Org. Chem.* **45:**240 (1980).
11. S. P. Khanapure, L. Crenshaw, R. T. Reddy, and E. R. Biehl, *J. Org. Chem.* **53:**4915 (1988).
12a. J. H. Boyer and R. S. Burkis, *Org. Synth.* **V:**1067 (1973).
b. H. P. Schultz, *Org. Synth.* **IV:**364 (1963); F. D. Gunstone and S. H. Tucker, *Org. Synth.* **IV:**160 (1963).
c. D. H. Hey and M. J. Perkins, *Org. Synth.* **V:**51 (1973).
d. K. Rorig, J. D. Johnston, R. W. Hamilton, and T. J. Telinski, *Org. Synth.* **IV:**576 (1963).
e. K. G. Rutherford and W. Redmond, *Org. Synth.* **V:**133 (1973).
f. M. M. Robinson and B. L. Robinson, *Org. Synth.* **IV:**947 (1963).
g. R. Adams, W. Reifschneider, and A. Ferretti, *Org. Synth.* **V:**107 (1973).
h. G. H. Cleland, *Org. Synth.* **51:**1 (1971).
13a. R. E. Ireland, C. A. Lipinski, C. J. Kowalski, J. W. Tilley, and D. M. Walba, *J. Am. Chem. Soc.* **96:**3333 (1974).
b. J. J. Korst, J. D. Johnston, K. Butler, E. J. Bianco, L. H. Conover, and R. B. Woodward, *J. Am. Chem. Soc.* **90:**439 (1968).
c. K. A. Parker and J. Kallmerten, *J. Org. Chem.* **45:**2614, 2620 (1980).
d. F. A. Carey and R. M. Guiliano, *J. Org. Chem.* **46:**1366 (1981).
e. R. B. Woodward and T. R. Hoye, *J. Am. Chem. Soc.* **99:**8007 (1977).
f. E. C. Horning, J. Koo, M. S. Fish, and G. N. Walker, *Org. Synth.* **IV:**408 (1963); J. Koo, *Org. Synth.* **V:**550 (1973).
14. T. F. Buckley III and H. Rapoport, *J. Am. Chem. Soc.* **102:**3056 (1980).
15. H. C. Bell, J. R. Kalman, J. T. Pinhey, and S. Sternhell, *Tetrahedron Lett.* **1974:**3391.
16. B. Chauncy and E. Gellert, *Aust. J. Chem.* **22:**993 (1969); R. I. Duclos, Jr., J. S. Tung, and H. Rapoport, *J. Org. Chem.* **49:**5243 (1984).
17. W. G. Miller and C. U. Pittman, Jr., *J. Org. Chem.* **39:**1955 (1974).

18. W. Nagata, K. Okada, and T. Aoki, *Synthesis* **1979:**365.
19. T. J. Doyle, M. Hendrix, D. Van Derveer, S. Javanmard, and J. Haseltine, *Tetrahedron* **53:**11153 (1997).
20. T. P. Smyth and B. W. Corby, *Org. Process Res. Dev.* **1:**264 (1997).

Chapter 12

1a. Y. Butsugan, S. Yoshida, M. Muto, and T. Bito, *Tetrahedron Lett.* **1971:**1129.
b. E. J. Corey and H. E. Ensley, *J. Am. Chem. Soc.* **97:**6908 (1975).
c. R. G. Gaughan and C. D. Poulter, *J. Org. Chem.* **44:**2441 (1979).
d. E. Vedejs, D. A. Engler, and J. E. Telschow, *J. Org. Chem.* **43:**188 (1978).
e. K. Akashi, R. E. Palermo, and K. B. Sharpless, *J. Org. Chem.* **43:**2063 (1978).
f. A. Hassner, R. H. Reuss, and H. W. Pinnick, *J. Org. Chem.* **40:**3427 (1975).
g. R. N. Mirrington and K. J. Schmalzl, *J. Org. Chem.* **37:**2877 (1972).
h. K. B. Sharpless and R. F. Lauer, *J. Org. Chem.* **39:**429 (1974).
i. J. A. Marshall and R. C. Andrews, *J. Org. Chem.* **50:**1602 (1985).
j. R. H. Schlessinger, J. J. Wood, A. J. Poos, R. A. Nugent, and W. H. Parsons, *J. Org. Chem.* **48:**1146 (1983).
k. R. K. Boeckman, Jr., J. E. Starett, Jr., D.G. Nickell, and P.-E. Sum, *J. Am. Chem. Soc.* **108:**5549 (1986).
l. E. J. Corey and Y. B. Xiang, *Tetrahedron Lett.* **29:**995 (1988).
m. D. J. Plata and J. Kallmerten, *J. Am. Chem. Soc.* **110:**4041 (1988).
n. B. E. Rossiter, T. Katsuki, and K. B. Sharpless, *J. Am. Chem. Soc.* **103:**464 (1981).
o. J. Mulzer, A. Angermann, B. Schubert, and C. Seilz, *J. Org. Chem.* **51:**5294 (1986).
p. R. H. Schlessinger and R. A. Nugent, *J. Am. Chem. Soc.* **104:**1116 (1982).
q. H. Niwa, T. Mori, T. Hasegawa, and K. Yamada, *J. Org. Chem.* **51:**1015 (1986).
2a. J. P. McCormick, W. Tomasik, and M. W. Johnson, *Tetrahedron Lett.* **1981:**607.
b. H. C. Brown, J. H. Kawakami, and S. Ikegami, *J. Am. Chem. Soc.* **92:**6914 (1970).
c. R. M. Scarborough, Jr., B. H. Toder, and A. B. Smith III, *J. Am. Chem. Soc.* **102:**3904 (1980).
d. B. Rickborn and R. M. Gerkin, *J. Am. Chem. Soc.* **90:**4193 (1968).
e. J. A. Marshall and R. A. Ruden, *J. Org. Chem.* **36:**594 (1971).
f. G. A. Kraus and B. Roth, *J. Org. Chem.* **45:**4825 (1980).
g. T. Sakan and K. Abe, *Tetrahedron Lett.* **1968:**2471.
h. K. J. Clark, G. I. Fray, R. H. Jaeger, and R. Robinson, *Tetrahedron* **6:**217 (1959).
i. T. Kawabata, P. Grieco, H.-L. Sham, H. Kim, J. Y. Jaw, and S. Tu, *J. Org. Chem.* **52:**3346 (1987).
j. P. T. Lansbury, J. P. Galbo, and J. P. Springer, *Tetrahedron Lett.* **29:**147 (1988).
k. J. P. Marino, R. F. de la Pradilla, and E. Laborde, *J. Org. Chem.* **52:**4898 (1987).
l. J. E. Toth, P. R. Hamann, and P. L. Fuchs, *J. Org. Chem.* **53:**4694 (1988).
3. E. L. Eliel, S. H. Schroeter, T. J. Brett, F. J. Biros, and J.-C. Richer, *J. Am. Chem. Soc.* **88:**3327 (1966).
4. E. E. Royals and J. C. Leffingwell, *J. Org. Chem.* **31:**1927 (1966).
5. W. W. Epstein and F. W. Sweat, *Chem. Rev.* **67:**247 (1967).
6. D. P. Higley and R. W. Murray, *J. Am. Chem. Soc.* **96:**3330 (1974).
7. R. Criegee and P. Günther, *Chem. Ber.* **96:**1564 (1963).
8a. S. Isoe, S. Katsumura, S. B. Hyeon, and T. Sakan, *Tetrahedron Lett.* **1971:**1089.
b. Y. Ogata, Y. Sawaki, and M. Shiroyama, *J. Org. Chem.* **42:**4061 (1977).
c. F. G. Bordwell and A. C. Knipe, *J. Am. Chem. Soc.* **93:**3416 (1971).
d. B. M. Trost, P. R. Bernstein, and P. C. Funfschilling, *J. Am. Chem. Soc.* **101:**4378 (1979).
e. C. S. Foote, S. Mazur, P. A. Burns, and D. Lerdal, *J. Am. Chem. Soc.* **95:**586 (1973).
f. J. P. Marino, K. E. Pfitzner, and R. A. Olofson, *Tetrahedron* **27:**4181 (1971).
g. M. A. Avery, C. Jennings-White, and W. K. M. Chong, *Tetrahedron Lett.* **28:**4629 (1987).
h. S. Horvat, P. Karallas, and J. M. White, *J. Chem. Soc., Perkin Trans. 2* **1998:**2151.
9a. P. N. Confalone, C. Pizzolato, D. L. Confalone, and M. R. Uskokovic, *J. Am. Chem. Soc.* **102:**1954 (1980).

b. S. Danishefsky, R. Zamboni, M. Kahn, and S. J. Etheredge, *J. Am. Chem. Soc.* **103:**3460 (1981).
c. J. K. Whitesell, R. S. Matthews, M. A. Minton, and A. M. Helbling, *J. Am. Chem. Soc.* **103:**3468 (1981).
d. F. A. J. Kerdesky, R. J. Ardecky, M. V. Lakshmikanthan, and M. P. Cava, *J. Am. Chem. Soc.* **103:**1992 (1981).
e. J. K. Whitesell and R. S. Matthews, *J. Org. Chem.* **43:**1650 (1978).
f. R. Fujimoto, Y. Kishi, and J. F. Blount, *J. Am. Chem. Soc.* **102:**7154 (1980).
g. S. P. Tanis and K. Nakanishi, *J. Am. Chem. Soc.* **101:**4398 (1979).
h. R. B. Miller and R. D. Nash, *J. Org. Chem.* **38:**4424 (1973).
i. R. Grewe and I. Hinrichs, *Chem. Ber.* **97:**443 (1964).
j. W. Nagata, S. Hirai, K. Kawata, and T. Okumura, *J. Am. Chem. Soc.* **89:**5046 (1967).
k. W. G. Dauben, M. Lorber, and D. S. Fullerton, *J. Org. Chem.* **34:**3587 (1969).
l. E. E. van Tamelen, M. Shamma, A. W. Burgstahler, J. Wolinsky, R. Tamm, and P. E. Aldrich, *J. Am. Chem. Soc.* **80:**5006 (1958).
m. S. D. Burke, C. W. Murtishaw, J. O. Saunders, J. A. Oplinger, and M. S. Dike, *J. Am. Chem. Soc.* **106:**4558 (1984).
n. B. M. Trost, P. G. McDougal, and K. J. Haller, *J. Am. Chem. Soc.* **106:**383 (1984).
10. W. P. Keaveney, M. G. Berger, and J. J. Pappas, *J. Org. Chem.* **32:**1537 (1967).
11. B. M. Trost and K. Hiroi, *J. Am. Chem. Soc.* **97:**6911 (1975).
12. E. C. Taylor, C.-S. Chiang, A. McKillop, and J. F. White, *J. Am. Chem. Soc.* **98:**6750 (1976).
13a. F. Delay and G. Ohloff, *Helv. Chim. Acta* **62:**2168 (1979).
b. R. Noyori, T. Sato, and Y. Hayakawa, *J. Am. Chem. Soc.* **100:**2561 (1978).
14a. I. Saito, R. Nagata, K. Yubo, and Y. Matsuura, *Tetrahedron Lett.* **24:**4439 (1983).
b. J. R. Wiseman and S. Y. Lee, *J. Org. Chem.* **51:**2485 (1986).
c. H. Nishiyama, M. Matsumoto, H. Arai, H. Sakaguchi, and K. Itoh, *Tetrahedron Lett.* **27:**1599 (1986).
15a. R. E. Ireland, P. G. M. Wuts, and B. Ernst, *J. Am. Chem. Soc.* **103:**3205 (1981).
b. R. M. Scarborough, Jr., B. H. Tober, and A. B. Smith III, *J. Am. Chem. Soc.* **102:**3904 (1980).
c. P. F. Hudrlik, A. M. Hudrlik, G. Nagendrappa, T. Yimenù, E. T. Zellers, and E. Chin, *J. Am. Chem. Soc.* **102:**6894 (1980).
d. T. Wakamatsu, K. Akasaka, and Y. Ban, *J. Org. Chem.* **44:**2008 (1979).
e. D. A. Evans, C. E. Sacks, R. A. Whitney, and N. G. Mandel, *Tetrahedron Lett.* **1978:**727.
f. F. Bourelle-Wargnier, M. Vincent, and J. Chuche, *J. Org. Chem.* **45:**428 (1980).
g. J. A. Zalikowski, K. E. Gilbert, and W. T. Borden, *J. Org. Chem.* **45:**346 (1980).
h. E. Vogel, W. Klug, and A. Breuer, *Org. Synth.* **55:**86 (1976).
i. L. D. Spicer, M. W. Bullock, M. Garber, W. Groth, J. J. Hand, D. W. Long, J. L. Sawyer, and R. S. Wayne, *J. Org. Chem.* **33:**1350 (1968).
j. B. E. Rossiter, T. Katsuki, and K. B. Sharpless, *J. Am. Chem. Soc.* **103:**464 (1981).
k. L. A. Paquette and Y.-K. Han, *J. Am. Chem. Soc.* **103:**1831 (1981).
l. T. Wakamatsu, K. Akasaka, and Y. Ban, *Tetrahedron Lett.* **1977:**2755.
m. M. Muelbacher and C. D. Poulter, *J. Org. Chem.* **53:**1026 (1988).
n. P. T. W. Cheng and S. McLean, *Tetrahedron Lett.* **29:**3511 (1988).
o. A. B. Smith III and R. E. Richmond, *J. Am. Chem. Soc.* **105:**575 (1983).
p. R. K. Boeckman, Jr., J. E. Starrett, Jr., D. G. Nickell, and P.-E. Sum, *J. Am. Chem. Soc.* **108:**5549 (1986).
16. Y. Gao and K. B. Sharpless, *J. Org. Chem.* **53:**4081 (1988).
17. C. W. Jefford, Y. Wang, and G. Bernardinelli, *Helv. Chim. Acta* **71:**2042 (1988).
18. R. W. Murray, M. Singh, B. L. Williams, and H. M. Moncrieff, *J. Org. Chem.* **61:**1830 (1996).

Chapter 13

1a. E. J. Corey, J.-L. Gras, and P. Ulrich, *Tetrahedron Lett.* **1976:**809.
b. K. C. Nicolaou, S. P. Seitz, and M. R. Pavia, *J. Am. Chem. Soc.* **103:**1222 (1981).
c. E. J. Corey and A. Venkateswarlu, *J. Am. Chem. Soc.* **94:**6190 (1972).
d–f. H. H. Meyer, *Justus Liebigs Ann. Chem.* **1977:**732.

2a. M. Miyashita, A. Yoshikoshi, and P. A. Grieco, *J. Org. Chem.* **42:**3772 (1977).
b. E. J. Corey, L. O. Wiegel, D. Floyd, and M. G. Bock, *J. Am. Chem. Soc.* **100:**2916 (1978).
c. A. M. Felix, E. P. Heimer, T. J. Lambros, C. Tzougraki, and J. Meienhofer, *J. Org. Chem.* **43:**4194 (1978).
d. P. N. Confalone, G. Pizzolato, E. G. Baggiolini, D. Lollar, and M. R. Uskokovic, *J. Am. Chem. Soc.* **97:**5936 (1975).
e. A. B. Foster, J. Lehmann, and M. Stacey, *J. Chem. Soc.* **1961:**4649.
3a. D. M. Simonović, A. S. Rao, and S. C. Bhattacharyya, *Tetrahedron* **19:**1061 (1963).
b. R. E. Ireland and L. N. Mander, *J. Org. Chem.* **32:**689 (1967).
c. G. Büchi, W. D. MacLeod, Jr., and J. Padilla, *J. Am. Chem.. Soc.* **86:**4438 (1964).
d. P. Doyle, I. R. Maclean, W. Parker, and R. A. Raphael, *Proc. Chem. Soc.* **1963:**239.
e. J. C. Sheehan and K. R. Henry-Logan, *J. Am. Chem. Soc.* **84:**2983 (1962).
f. E. J. Corey, M. Ohno, R. B. Mitra, and P. A. Vatakencherry, *J. Am. Chem. Soc.* **86:**478 (1964).
4a. B. ElAmin, G. M. Anantharamaiah, G. P. Royer, and G. E. Means, *J. Org. Chem.* **44:**3442 (1979).
b. B. Moreay, S. Lavielle, and A. Marquet, *Tetrahedron Lett.* **1977:**2591; B. C. Laguzza and B. Ganem, *Tetrahedron Lett.* **1981:**1483.
c. J. I. Seeman, *Synthesis* **1977:**498; D. Spitzner, *Synthesis* **1977:**242.
d. H. J. Anderson and J. K. Groves, *Tetrahedron Lett.* **1971:**3165.
5a. T. Hylton and V. Boekelheide, *J. Am. Chem. Soc.* **90:**6987 (1968).
b. B. W. Erickson, *Org. Synth.* **53:**189 (1973).
c. H. Paulsen, V. Sinnwell, and P. Stadler, *Angew. Chem. Int. Ed. Engl.* **11:**149 (1972).
d. S. Torii, K. Uneyama, and M. Isihara, *J. Org. Chem.* **39:**3645 (1974).
e. J. A. Marshall and A. E. Greene, *J. Org. Chem.* **36:**2035 (1971).
f. E. Leete, M. R. Chedekel, and G. B. Bodem, *J. Org. Chem.* **37:**4465 (1972).
g. H. Yamamoto and H. L. Sham, *J. Am. Chem. Soc.* **101:**1609 (1979).
h. K. Deuchert, U. Hertenstein, S. Hünig, and G. Wehner, *Chem. Ber.* **112:**2045 (1979).
i. T. Takahashi, K. Kitamura, and J. Tsuji, *Tetrahedron Lett.* **24:**4695 (1983).
6a. S. Danishefsky and T. Kitahara, *J. Am. Chem. Soc.* **96:**7807 (1974).
b. P. S. Wharton, C. E. Sundin, D. W. Johnson, and H. C. Kluender, *J. Org. Chem.* **37:**34 (1972).
c. E. J. Corey, B. W. Erikson, and R. Noyori, *J. Am. Chem. Soc.* **93:**1724 (1971).
d. R. E. Ireland and J. A. Marshall, *J. Org. Chem.* **27:**1615 (1962).
e. W. S. Johnson, T. J. Brocksom, P. Loew, D. H. Rich, L. Werthemann, R. A. Arnold, T. Li, and D. J. Faulkner, *J. Am. Chem. Soc.* **92:**4463 (1970).
f. L. Birladeanu, T. Hanafusa, and S. Winstein, *J. Am. Chem. Soc.* **88:**2315 (1966); T. Hanafusa, L. Birladeanu, and S. Winstein, *J. Am. Chem. Soc.* **87:**3510 (1965).
g. H. Takayanagi, Y. Kitano, and Y. Morinaka, *J. Org. Chem.* **59:**2700 (1994).
7a. A. B. Smith III and W. C. Agosta, *J. Am. Chem. Soc.* **96:**3289 (1974).
b. R. S. Cooke and U. H. Andrews, *J. Am. Chem. Soc.* **96:**2974 (1974).
c. L. A. Hulshof and H. Wynberg, *J. Am. Chem. Soc.* **96:**2191 (1974).
d. S. D. Burke, C. W. Murtiashaw, M. S. Dike, S. M. S. Strickland, and J. O. Saunders, *J. Org. Chem.* **46:**2400 (1981).
e. K. C. Nicolaou, M. R. Pavia, and S. P. Seitz, *J. Am. Chem. Soc.* **103:**1224 (1981).
f. J. Cossy, B. Gille, S. BouzBouz, and V. Bellosta, *Tetrahedron Lett.* **38:**4069 (1997).
8a. E. M. Acton, R. N. Goerner, H. S. Uh, K. J. Ryan, D. W. Henry, C. E. Cass, and G. A. LePage, *J. Med. Chem.* **22:**518 (1979).
b. E. G. Gros, *Carbohydr. Res.* **2:**56 (1966).
c. S. Hanessian and G. Rancourt, *Can. J. Chem.* **55:**1111 (1977).
d. R.R. Schmidt and A. Gohl, *Chem. Ber.* **112:**1689 (1979).
9a. S. F. Martin and T. Chou, *J. Org. Chem.* **43:**1027 (1978).
b. W. C. Still and M.-Y.Tsai, *J. Am. Chem. Soc.* **102:**3654 (1980).
c. J. C. Bottaro and G. A. Berchtold, *J. Org. Chem.* **45:**1176 (1980).
d. A. S. Kende and T. P. Demuth, *Tetrahedron Lett.* **1980:**715.
e. J. A. Marshall and P. G. M. Wuts, *J. Org. Chem.* **43:**1086 (1978).
10a. R. Bonjouklian and R. A. Ruden, *J. Org. Chem.* **42:**4095 (1977).
b. L. A. Paquette, R. E. Moerck, B. Harirchian, and P. D. Magnus, *J. Am. Chem. Soc.* **100:**1597 (1978).
c. P. S. Wharton, C. E. Sundin, D. W. Johnson, and H. C. Kluender, *J. Org. Chem.* **37:**34 (1972).
d. S. Danishefsky, T. Kitahara, C. F. Yan, and J. Morris, *J. Am. Chem. Soc.* **101:**6996 (1979).
e. B. M. Trost, J. Ippen, and W. C. Vladuchick, *J. Am. Chem. Soc.* **99:**8116 (1977).

11a. E. J. Corey, E. J. Trybulski, L. S. Melvin, Jr., K. C. Nicolaou, J. A. Secrist, R. Lett, P. W. Sheldrake, J. R. Falck, D. J. Brunelle, M. F. Haslanger, S. Kim, and S. Yoo, *J. Am. Chem. Soc.* **100:**4618 (1978).

b. K. G. Paul, F. Johnson, and D. Favara, *J. Am. Chem. Soc.* **98:**1285 (1976).

c. P. N. Confalone, G. Pizzolato, E. G. Baggiolini, D. Lollar, and M. R. Uskokovic, *J. Am. Chem. Soc.* **97:**5936 (1975).

d. E. Baer, J. M. Grosheintz, and H. O. L. Fischer, *J. Am. Chem. Soc.* **61:**2607 (1939).

e. J. L. Coke and A. B. Richon, *J. Org. Chem.* **41:**3516 (1976).

f. J. R. Dyer, W. E. McGonigal, and K. C. Rice, *J. Am Chem. Soc.* **87:**654 (1965).

g. E. J. Corey and S. Nozoe, *J. Am. Chem. Soc.* **85:**3527 (1963).

h. R. Jacobson, R. J. Taylor, H. J. Williams, and L. R. Smith, *J. Org. Chem.* **47:**3140 (1982).

12a. R. B. Miller and E. S. Behare, *J. Am. Chem. Soc.* **96:**8102 (1974).

b. G. Büchi, W. Hofheinz, and J. V. Paukstelis, *J. Am. Chem. Soc.* **91:**6473 (1969).

c. M. Brown, *J. Org. Chem.* **33:**162 (1968).

d. E. J. Corey, R. B. Mitra, and H. Uda, *J. Am. Chem. Soc.* **86:**485 (1964).

13a. I. Fleming, *Selected Organic Syntheses*, John Wiley & Sons, London, 1973, pp. 3–6; J. E. McMurry and J. Melton, *J. Am. Chem. Soc.* **93:**5309 (1971).

b. R. M. Coates and J. E. Shaw, *J. Am. Chem. Soc.* **92:**5657 (1970).

c. T. F. Buckley III and H. Rapoport, *J. Am. Chem. Soc.* **102:**3056 (1980).

d. D. A. Evans, A. M. Golob, N. S. Mandel, and G. S. Mandel, *J. Am. Chem. Soc.* **100:**8170 (1978).

e. E. J. Corey and R. D. Balanson, *J. Am. Chem. Soc.* **96:**6516 (1974).

f. J. L. Herrmann, M. H. Berger, and R. H. Schlessinger, *J. Am. Chem. Soc.* **95:**7923 (1973).

g. R. F. Romanet and R. H. Schlessinger, *J. Am. Chem. Soc.* **96:**3701 (1974); R. A. LeMahieu, M. Carson, and R. W. Kierstead, *J. Org. Chem.* **33:**3660 (1968); G. Büchi, D. Minster, and J. C. F. Young, *J. Am. Chem. Soc.* **93:**4319 (1971).

h. J. H. Babler, D. O. Olsen, and W. H. Arnold, *J. Org. Chem.* **39:**1656 (1974); R. J. Crawford, W. F. Erman, and C. D. Broaddus, *J. Am. Chem. Soc.* **94:**4298 (1972).

i. C. S. Subramanian, P. J. Thomas, V. R. Mamdapur, and M. S. Chandra, *J. Chem. Soc., Perkin Trans. 1* **1979:**2346.

j. S. Hanessian and R. Frenette, *Tetrahedron Lett.* **1979:**3391.

k. E. Piers, R. W. Britton, and W. de Waal, *J. Am. Chem. Soc.* **93:**5113 (1971); K. J. Schmalzl and R. N. Mirrington, *Tetrahedron Lett.* **1970:**3219; N. Fukamiya, M. Kato, and A. Yoshikoshi, *J. Chem. Soc., Chem. Commun.* **1971:**1120; G. Frater, *Helv. Chim. Acta* **57:**172 (1974); K. Yamada, Y. Kyotani, S. Manabe, and M. Suzuki, *Tetrahedron* **35:**293 (1979); M. E. Jung, C. A. McCombs, Y. Takeda, and Y. G. Pan, *J. Am. Chem. Soc.* **103:**6677 (1981); S. C. Welch, J. M. Gruber, and P. A. Morrison, *J. Org. Chem.* **50:**2676 (1985); S. C. Welch, C. Chou, J. M. Gruber, and J. M. Assercq, *J. Org. Chem.* **50:**2668 (1985); H. Hagaiwara, A. Okano, and H. Uda, *J. Chem. Soc., Chem. Commun.* **1985:**1047; G. Stork and N. H. Baird, *Tetrahedron Lett.* **26:**5927 (1985).

l. E. J. Corey and R. H. Wollenberg, *Tetrahedron Lett.* **1976:**4705; R. Baudouy, P. Crabbe, A. E. Greene, C. LeDrain, and A. F. Orr, *Tetrahedron Lett.* **1977:**2973; A. E. Greene, C. LeDrian, and P. Crabbe, *J. Am. Chem. Soc.* **102:**7583 (1980); P. A. Bartlett and F. R. Green, *J. Am. Chem. Soc.* **100:**4858 (1978); T. Kitahara, K. Mori, and M. Matsui, *Tetrahedron Lett.* **1979:**3021; Y. Köksal, P. Raddatz, and E. Winterfeldt, *Angew. Chem. Int. Ed. Engl.* **19:**472 (1980); K. H. Marx, P. Raddatz, and E. Winterfeldt, *Justus Liebigs Ann. Chem.* **1984:**474; C. LeDrain and A. E. Green, *J. Am. Chem. Soc.* **104:**5473 (1982); T. Kitahara and K. Mori, *Tetrahedron* **40:**2935 (1984); K. Nakatani and S. Isoe, *Tetrahedron Lett.* **26:**2209 (1985); B. M. Trost and S. M. Mignani, *Tetrahedron Lett.* **27:**4137 (1986); B. M. Trost, J. Lunch, P. Renault, and D. H. Steinman, *J. Am. Chem. Soc.* **108:**284 (1986).

m. S. Danishefsky, M. Hirama, K. Gombatz, T. Harayam, E. Berman, and P. F. Schuda, *J. Am. Chem. Soc.* **101:**7020 (1979); W. H. Parsons, R. H. Schlessinger, and M. L. Quesada, *J. Am. Chem. Soc.* **102:**889 (1980); S. D. Burke, C. W. Murtiashaw, J. O. Saunders, and M. S. Dike, *J. Am. Chem. Soc.* **104:**872 (1982); L. A. Paquette, G. D. Amis, and H. Schostarez, *J. Am. Chem. Soc.* **104:**6646 (1982); M. C. Pirrung and S. A. Thompson, *J. Org. Chem.* **53:**227 (1988); T. Ohtsuka, H. Shirahama, and T. Matsumoto, *Tetrahedron Lett.* **24:**3851 (1983); D. E. Cane and P. J. Thomas, *J. Am. Chem. Soc.* **106:**5295 (1984); D. F. Taber and J. L. Schuchardt, *J. Am. Chem. Soc.* **107:**5289 (1985).

14a. R. E. Ireland, R. H. Mueller, and A. K. Willard, *J. Am. Chem. Soc.* **98:**2568 (1976).

b. W. A. Kleschick, C. T. Buse, and C. H. Heathcock, *J. Am. Chem. Soc.* **99:**247 (1977); P. Fellmann and J. E. Dubois, *Tetrahedron* **34:**1349 (1978).

c. B. M. Trost, S. A. Godleski, and J. P. Genêt, *J. Am. Chem. Soc.* **100:**3930 (1978).

d. M. Mousseron, M. Mousseron, J. Neyrolles, and Y. Beziat, *Bull. Chim. Soc. Fr.* **1963:**1483; Y. Beziat and M. Mousseron-Canet, *Bull. Chim. Soc. Fr.* **1968:**1187.
e. G. Stork and V. Nair, *J. Am. Chem. Soc.* **101:**1315 (1979).
15a. R. D. Cooper, V. B. Jigajimmi, and R. H. Wightman, *Tetrahedron Lett.* **25:**5215 (1984).
b. C. E. Adams, F. J. Walker, and K. B. Sharpless, *J. Org. Chem.* **50:**420 (1985).
c. G. Grethe, J. Sereno, T. H. Williams, and M. R. Uskokovic, *J. Org. Chem.* **48:**5315 (1983).
16a. H. Ahlbrecht, G. Bonnet, D. Enders, and G. Zimmermann, *Tetrahedron Lett.* **1980:**3175.
b. A. I. Meyers, G. Knaus, K. Kamata, and M. E. Ford, *J. Am. Chem. Soc.* **98:**567 (1976).
c. S. Hashimoto and K. Koga, *Tetrahedron Lett.* **1978:**573.
d. A. I. Meyers and J. Slade, *J. Org. Chem.* **45:**2785 (1980).
e. S. Terashima, M. Hayashi, and K. Koga, *Tetrahedron Lett.* **1980:**2733.
f. B. M. Trost, D. O'Krongly, and J. L. Balletire, *J. Am. Chem. Soc.* **102:**7595 (1980).
g. A. I. Meyers, R. K. Smith, and C. E. Whitten, *J. Org. Chem.* **44:**2250 (1979).

Index